# ADVANCED ENGINEERING MATHEMATICS

# ADVANCED ENGINEERING MATHEMATICS

[For the Students of M.E., B.E. and other Engineering Examinations]

**H.K. DASS**

M.Sc.

*Diploma in Specialist Studies (Maths.)*

*University of Hull*

*(England)*

**Secular India Award - 98** *for National Integration and Communal Harmony given by Prime Minister* **Shri Atal Behari Vajpayee on 12th June 1999.**

S. CHAND PUBLISHING
empowering minds

**S. CHAND & COMPANY PVT. LTD.**

**(AN ISO 9001 : 2008 COMPANY)**

**RAM NAGAR, NEW DELHI-110055**

# S. CHAND & COMPANY PVT. LTD.

S. CHAND PUBLISHING
empowering minds

**(An ISO 9001 : 2008 Company)**
*Head Office*: 7361, RAM NAGAR, NEW DELHI - 110 055
Phone: 23672080-81-82, 9899107446, 9911310888; Fax: 91-11-23677446
Shop at: **schandgroup.com;** e-mail: **info@schandgroup.com**

*BBranches* :

| | | |
|---|---|---|
| **AHMEDABAD** | : | 1st Floor, Heritage, Near Gujarat Vidhyapeeth, Ashram Road, **Ahmedabad** - 380 014, Ph: 27541965, 27542369, ahmedabad@schandgroup.com |
| **BENGALURU** | : | No. 6, Ahuja Chambers, 1st Cross, Kumara Krupa Road, **Bengaluru** - 560 001, Ph: 22268048, 22354008, bangalore@schandgroup.com |
| **BHOPAL** | : | Bajaj Tower, Plot No. 2&3, Lala Lajpat Rai Colony, Raisen Road, **Bhopal** - 462 011, Ph: 4274723, 4209587. bhopal@schandgroup.com |
| **CHANDIGARH** | : | S.C.O. 2419-20, First Floor, Sector - 22-C (Near Aroma Hotel), **Chandigarh** -160 022, Ph: 2725443, 2725446, chandigarh@schandgroup.com |
| **CHENNAI** | : | No.1, Whites Road, Opposite Express Avenue, Royapettah, **Chennai** - 600014 Ph. 28410027, 28410058, chennai@schandgroup.com |
| **COIMBATORE** | : | 1790, Trichy Road, LGB Colony, Ramanathapuram, **Coimbatore** -6410045, Ph: 2323620, 4217136 coimbatore@schandgroup.com **(Marketing Office)** |
| **CUTTACK** | : | 1st Floor, Bhartia Tower, Badambadi, **Cuttack** - 753 009, Ph: 2332580; 2332581, cuttack@schandgroup.com |
| **DEHRADUN** | : | 1st Floor, 20, New Road, Near Dwarka Store, **Dehradun** - 248 001, Ph: 2711101, 2710861, dehradun@schandgroup.com |
| **GUWAHATI** | : | Dilip Commercial (Ist floor), M.N. Road, Pan Bazar, **Guwahati** - 781 001, Ph: 2738811, 2735640 guwahati@schandgroup.com |
| **HALDWANI** | : | Bhatt Colony, Talli Bamori, Mukhani, **Haldwani** -263139 **(Marketing Office)** Mob. 09452294584 |
| **HYDERABAD** | : | Padma Plaza, H.No. 3-4-630, Opp. Ratna College, Narayanaguda, **Hyderabad** - 500 029, Ph: 27550194, 27550195, hyderabad@schandgroup.com |
| **JAIPUR** | : | 1st Floor, Nand Plaza, Hawa Sadak, Ajmer Road, **Jaipur** - 302 006, Ph: 2219175, 2219176, jaipur@schandgroup.com |
| **JALANDHAR** | : | Mai Hiran Gate, **Jalandhar** - 144 008, Ph: 2401630, 5000630, jalandhar@schandgroup.com |
| **KOCHI** | : | Kachapilly Square, Mullassery Canal Road, Ernakulam, **Kochi** - 682 011, Ph: 2378740, 2378207-08, cochin@schandgroup.com |
| **KOLKATA** | : | 285/J, Bipin Bihari Ganguli Street, **Kolkata** - 700 012, Ph: 22367459, 22373914, kolkata@schandgroup.com |
| **LUCKNOW** | : | Mahabeer Market, 25 Gwynne Road, Aminabad, **Lucknow** - 226 018, Ph: 4076971, 4026791, 4065646, 4027188, lucknow@schandgroup.com |
| **MUMBAI** | : | Blackie House, IInd Floor, 103/5, Walchand Hirachand Marg, Opp. G.P.O., **Mumbai** - 400 001, Ph: 22690881, 22610885, mumbai@schandgroup.com |
| **NAGPUR** | : | Karnal Bagh, Near Model Mill Chowk, **Nagpur** - 440 032, Ph: 2720523, 2777666 nagpur@schandgroup.com |
| **PATNA** | : | 104, Citicentre Ashok, Mahima Palace , Govind Mitra Road, **Patna** - 800 004, Ph: 2300489, 2302100, patna@schandgroup.com |
| **PUNE** | : | Sadguru Enclave, Ground floor, Survey No. 114/3, Plot no. 8 Alandi Road , Vishrantwadi **Pune** – 411015 Ph: 64017298 pune@schandgroup.com |
| **RAIPUR** | : | Kailash Residency, Plot No. 4B, Bottle House Road, Shankar Nagar, **Raipur** - 492 007, Ph: 2443142,Mb. : 09981200834, raipur@schandgroup.com **(Marketing Office)** |
| **RANCHI** | : | Shanti Deep Tower, Opp.Hotel Maharaja, Radium Road, **Ranchi**-834001 Mob. 09430246440 ranchi@schandgroup.com |
| **SILIGURI** | : | 122, Raja Ram Mohan Roy Road, East Vivekanandapally, P.O., Siliguri, **Siliguri**-734001, Dist., Jalpaiguri, (W.B.) Ph. 0353-2520750 **(Marketing Office)** siliguri@schandgroup.com |
| **VISAKHAPATNAM** | : | No. 49-54-15/53/8, Plot No. 7, 1st Floor, Opp. Radhakrishna Towers, Seethammadhara North Extn., **Visakhapatnam** - 530 013, Ph-2782609 (M) 09440100555, visakhapatnam@schandgroup.com **(Marketing Office)** |





*First Edition 1988*
*Subsequent Editions and Reprints 1990, 92, 93, 94, 96, 97, 98, 99, 2000 (Twice), 2001 (Twice), 2002, 2003 (Twice), 2004, 2005, 2006, 2007, 2008, 2009, 2010, 2011, 2012, 2013, 2014 (Twice)*
*Reprint with corrections 2015*

**ISBN** : 978-81-219-0345-5 **Code : 1010B 110**

PRINTED IN INDIA

*By Nirja Publishers & Printers Pvt. Ltd., 54/3/2, Jindal Paddy Compound, Kashipur Road, Rudrapur-263153, Uttarakhand and published by S. Chand & Company Pvt. Ltd., 7361, Ram Nagar, New Delhi -110 055.*

# PREFACE TO THE TWENTYFIRST REVISED EDITION

I am happy to be able to bring out this revised edition.

Four solved question papers of A.M.I.E.T.E. New Scheme and Old Scheme, December 2014 are included at the end of the book.

Misprints and errors which came to my notice have been corrected.

Suggestions and healthy criticism from students and teachers to improve the book shall be personally acknowledged and deeply appreciated to help me to make it an ideal book for all.

We are thankful to the Management Team and the Editorial Department of S. Chand & Company Pvt. Ltd. for all help and support in the publication of this book.

**H.K. DASS**

D-1/87, Janakpuri
New Delhi-110 058
Tel. 28525078, 28521776
Mob. 9350055078
hk_dass@yahoo.com

# PREFACE TO THE FIRST EDITION

It gives me great pleasure to present this textbook of Mathematics to the students pursuing I.E.T.E and various engineering courses.

This book has been written according to the new revised syllabus of Mathematics of I.E.T.E. and includes topics from the syllabi of the other engineering courses. There is not a single textbook which entirely covers the syllabus of I.E.T.E. and the students have all along been facing great difficulties. Endeavour has been made to cover the syllabus exhaustively and present the subject matter in a systematic and lucid style. More than 550 solved examples on various topics have been incorporated in the textbook for the better understanding of the students. Most of the examples have been taken from previous question papers of I.E.T.E. which should make the students familiar with the standard and trend of questions set in the examinations. Care has been taken to systematically grade these examples.

The author possesses very long and rich experience of teaching Mathematics to the students preparing for I.E.T.E. and other examinations of engineering and has first hand experience of the problems and difficulties that they generally face.

This book should satisfy both average and brilliant students. It would help the students to get through their examination and at the same time would arouse greater intellectual curiosity in them.

I am really thankful to my Publishers, Padamshree Lala Shyam Lal Gupta, Shri Ravindra Kumar Gupta for showing personal interest and his General Manager, Shri P.S. Bhatti and Km. Shashi Kanta for their co-operations. I am also thankful to the Production Manager, Shri Ravi Gupta for bringing out the book in a short period.

Suggestions for the improvement of the book will be gratefully acknowledged.

D-1/87, Janakpuri
New Delhi-110 058

**H.K. DASS**

# FOREWORD

On my recent visit to India, I happened to meet Prof. H.K. Dass, who has written quite a number of successful books on Mathematics for students at various levels.

During my meeting, Prof. H.K. Dass presented me with the book entitled "Advanced Engineering Mathematics" I am delighted to write this Foreword, as I am highly impressed on seeing the wide variety of its contents. The contents includes many key topics, for examples, advanced calculus, vector analysis, tensor analysis, fuzzy sets, various transforms and special functions, probability (curiously some tests of significance are given under that chapter), numerical methods; matrix algebra and transforms. In spite of this breadth , the development of the material is very lucid, simple and in plain English.

I know of quite a number of other textbooks on Engineering Mathematics but the material that has been included in this textbook is so comprehensive that the students of all the engineering streams will find this textbook useful. It contains problems, questions and their solutions which are useful both to the teachers and students, and I am not surprised that it has gone through various editions.The style reminds me of the popular books of Schaum's Series. I believe that this book will be also helpful to non-engineering students as a quick reference guide.

This book is a work of dedicated scholarship and vast learning of Mr. Dass, and I have no hesitation in recommending this book to the students for any Engineering degree world-wide.

Prof. K.V. Mardia
M.Sc. (Bombay), M.Sc.(Pune)
Ph.D. (Raj.), Ph.D. (N'cle),D.Sc.(N'cle)
Senior Research Professor
University of Leeds,
LEEDS (England)

# CONTENTS

# 1

# Partial Differentiation

## 1.1 INTRODUCTION

Area of a rectangle depends upon its length and breadth, hence we can say that area is the function of two variables *i.e.* its length and breadth.

$z$ is called a function of two variables $x$ and $y$ if $z$ has one definite value for every pair of values of $x$ and $y$. Symbolically, it is written as

$$z = f(x, y)$$

The variable $x$ and $y$ are called independent variables while $z$ is called the dependent variable.

Similarly, we can define $z$ as a function of more than two variables.

**Geometrically:** Let $z = f(x, y)$

where $x, y$ belong to an area $A$ of the $xy$-plane. For each point $(x, y)$ corresponds a value of $z$. These values of $(x, y, z)$ form a surface in space.

Hence, the function $z = f(x, y)$ represents a surface.

## 1.2 LIMIT

The function $f(x, y)$ is said to tend to the limit $l$ as $x \to a$ and $y \to b$ if and only if the limit $l$ is independent of the path followed by the point $(x, y)$ as $x \to a$ and $y \to b$. Then

$$\lim_{\substack{x \to a \\ y \to b}} f(x, y) = l$$

The function $f(x, y)$ in region $R$ is said to tend to the limit $l$ as $x \to a$ and $y \to b$ if and only if corresponding to a positive number $\in (a, b)$, there exists another positive number $\delta$ such that

$$|f(x, y) - l| < \in \quad \text{for} \quad 0 < (x - a)^2 + (y - b)^2 < \delta^2$$

for every point $(x, y)$ in $R$.

## 1.3 WORKING RULE TO FIND THE LIMIT

**Step 1.** Find the value of $f(x, y)$ along $x \to a$ and $y \to b$.

**Step 2.** Find the value of $f(x, y)$ along $y \to b$ and $x \to a$.

If the values of $f(x, y)$ in step 1 and step 2 remain the same, the limit exists otherwise not.

**Step 3.** If $a \to 0$ and $b \to 0$, find the limit along $y = mx$ or $y = mx^n$. If the value of the limit does not contain $m$ then limit exists. If it contains $m$, the limit does not exist.

**Note.** (*i*) Put $x = 0$ and then $y = 0$ in $f$. Find its value $f_1$.

(*ii*) Put $y = 0$ and then $x = 0$ in $f$. Find the value $f_2$.

If $f_1 \neq f_2$, limit does not exist.

If $f_1 = f_2$, then

(*iii*) Put $y = mx$ and find the limit $f_3$.
If $f_1 = f_2 \neq f_3$, then limit does not exist.
If $f_1 = f_2 = f_3$, then

(*iv*) Put $y = mx^2$ and find the limit $f_4$.
If $f_1 = f_2 = f_3 \neq f_4$, then limit does not exist.
If $f_1 = f_2 = f_3 = f_4$, then limit exists.

**Example 1.** *Evaluate* $\lim\limits_{\substack{x \to 0 \\ y \to 0}} \dfrac{x^2 y}{x^4 + y^2}$

**Solution.** (*i*) $$\lim_{\substack{x \to 0 \\ y \to 0}} \frac{x^2 y}{x^4 + y^2} = \lim_{y \to 0} \frac{0}{0 + y^2} = 0 = f_1 \quad \text{(say)}$$

(*ii*) $$\lim_{\substack{x \to 0 \\ y \to 0}} \frac{x^2 y}{x^4 + y^2} = \lim_{x \to 0} \frac{0}{x^4 + 0} = 0 = f_2 \quad \text{(say)}$$

Here, $f_1 = f_2$, therefore

(*iii*) Put $y = mx$

$$\lim_{\substack{x \to 0 \\ y \to 0}} \frac{x^2 mx}{x^4 + m^2 x^2} = \lim_{x \to 0} \frac{mx}{x^2 + m^2} = 0 = f_3 \quad \text{(say)}$$

Here, $f_1 = f_2 = f_3$, therefore

(*vi*) Put $y = mx^2$

$$\lim_{\substack{x \to 0 \\ y \to 0}} \frac{x^2 mx^2}{x^4 + m^2 x^4} = \frac{m}{1 + m^2} = f_4$$

Here, $f_1 = f_2 = f_3 \neq f_4$

Thus, limit does not exist. **Ans.**

**Example 2.** *Evaluate* $\lim\limits_{\substack{x \to 0 \\ y \to 0}} (x^3 + y^3)$.

**Solution.** (*i*) $$\lim_{\substack{x \to 0 \\ y \to 0}} (x^3 + y^3) = \lim_{y \to 0} (0 + y^3) = 0 = f_1 \quad \text{(say)}$$

(*ii*) $$\lim_{\substack{x \to 0 \\ y \to 0}} (x^3 + y^3) = \lim_{x \to 0} (x^3 + 0) = 0 = f_2 \quad \text{(say)}$$

Here, $f_1 = f_2$, therefore

(*iii*) Put $y = mx$

$$\lim_{\substack{x \to 0 \\ y \to 0}} (x^3 + y^3) = \lim_{x \to 0} \left[ \lim_{y \to mx} (x^3 + y^3) \right] = \lim_{x \to 0} (x^3 + m^3 x^3) = 0 = f_3 \quad \text{(say)}$$

Here, $f_1 = f_2 = f_3$, therefore

(*iv*) Put $y = mx^2$

$$\lim_{\substack{x \to 0 \\ y \to 0}} (x^3 + y^3) = \lim_{x \to 0} \left[ \lim_{y \to mx^2} (x^3 + y^3) \right] = \lim_{x \to 0} (x^3 + m^3 x^6)$$

$$= \lim_{x \to 0} x^3 (1 + m^3 x^3) = 0 = f_4 \quad \text{(say)}$$

Here, $f_1 = f_2 = f_3 = f_4$

Thus, limit exists with value 0. **Ans.**

**Example 3.** *Evaluate* $\lim\limits_{\substack{x \to 1 \\ y \to 2}} \dfrac{3x^2 y}{x^2 + y^2 + 5}$.

**Solution.** $\lim\limits_{\substack{x\to 1\\ y\to 2}} \dfrac{3x^2y}{x^2+y^2+5} = \lim\limits_{x\to 1}\left[\lim\limits_{y\to 2} \dfrac{3x^2y}{x^2+y^2+5}\right] = \lim\limits_{x\to 1} \dfrac{3x^2(2)}{x^2+(2)^2+5}$

$= \lim\limits_{x\to 1} \dfrac{6x^2}{x^2+9} = \dfrac{6}{1+9} = \dfrac{3}{5}$ **Ans.**

**Example 4.** *Evaluate* $\lim\limits_{\substack{x\to\infty\\ y\to 3}} \dfrac{2x-3}{x^3+4y^3}$.

**Solution.** (*i*) $\lim\limits_{\substack{x\to\infty\\ y\to 3}} \dfrac{2x-3}{x^3+4y^3} = \lim\limits_{y\to 3}\left[\lim\limits_{x\to\infty} \dfrac{2x-3}{x^3+4y^3}\right]$

$= \lim\limits_{y\to 3}\left[\lim\limits_{x\to\infty} \dfrac{\dfrac{2}{x^2}-\dfrac{3}{x^3}}{1+4\left(\dfrac{y}{x}\right)^3}\right] = \lim\limits_{y\to 3}\left(\dfrac{0-0}{1+4(0)}\right) = 0 = f_1$ **(say)**

(*ii*) $\lim\limits_{\substack{y\to 3\\ x\to\infty}} \dfrac{2x-3}{x^3+4y^3} = \lim\limits_{x\to\infty}\left[\lim\limits_{y\to 3} \dfrac{2x-3}{x^3+4y^3}\right]$

$= \lim\limits_{x\to\infty} \dfrac{2x-3}{x^3+108} = \lim\limits_{x\to\infty} \dfrac{\dfrac{2}{x^2}-\dfrac{3}{x^3}}{1+\dfrac{108}{x^3}} = \dfrac{0-0}{1+0} = 0 = f_2$ **(say)**

Here, $f_1 = f_2$.

Hence, the limit exists with value 0. **Ans.**

## EXERCISE 1.1

**Evaluate the following limits:**

1. $\lim\limits_{\substack{x\to 1\\ y\to 2}} \dfrac{2x^2+y^2}{2xy}$ **Ans.** $\dfrac{3}{4}$

2. $\lim\limits_{\substack{x\to 2\\ y\to 3}} \dfrac{x^3+y^2}{x^2-y}$, **Ans.** 17

3. $\lim\limits_{\substack{x\to\infty\\ y\to 3}} \dfrac{2xy-3}{x^3+4y^3}$, **Ans.** 0

4. $\lim\limits_{\substack{x\to 0\\ y\to 0}} \dfrac{xy}{y-x^2}$; $x \neq 0, y \neq 0$ **Ans.** Limit does not exist

5. $\lim\limits_{\substack{x\to 0\\ y\to 0}} \dfrac{x-y}{x^2+y^2}$; $x \neq 0, y \neq 0$ **Ans.** Limit does not exist

6. $\lim\limits_{\substack{x\to 1\\ y\to 1}} \dfrac{xy-2x}{xy-2y}$ **Ans.** 1

7. $\lim\limits_{\substack{x\to 0\\ y\to 0}} \dfrac{x^3+2y^3}{x^2+4y^2}$ $x \neq 0, y \neq 0$ **Ans.** 0

8. $\lim\limits_{\substack{x\to 0\\ y\to 0}} \dfrac{x^2y^3}{x^2+y^2}$ $x \neq 0, y \neq 0$ **Ans.** 0

9. $\lim\limits_{\substack{x\to 0\\ y\to 0}} \dfrac{xy+2}{x^2+y^2}$, $x \neq 0, y \neq 0$ **Ans.** 0

10. $\lim\limits_{\substack{x\to 1\\ y\to 1}} \dfrac{3x(y-2)}{2y(x-2)}$ **Ans.** $\dfrac{1}{2}$

## 1.4 CONTINUITY

A function $f(x, y)$ is said to be continuous at a point $(a, b)$ if

$$\lim_{(x, y)\to(a, b)} f(x, y) = f(a\ b)$$

A function is said to be continuous in a domain if it is continuous at every point of the domain.

## 1.5 WORKING RULE FOR CONTINUITY AT A POINT (*a*, *b*)

**Step 1.** $f(a, b)$ should be well defined

**Step 2.** $\lim_{(x,y)\to(a,b)} f(x, y)$ should exist.

**Step 3.** $\lim_{(x,y)\to(a,b)} f(x, y) = f(a, b)$

**Example 5.** *Test the function* $f(x, y) = \begin{cases} \dfrac{x^3 - y^3}{x^2 + y^2} & \text{when } x \neq 0, \quad y \neq 0 \\ 0 & \text{when } x = 0, \quad y = 0 \end{cases}$

*for continuity.*

**Solution. Step 1.** The function is well defined at (0, 0).

**Step 2.**
$$\lim_{(x,y)\to(0,0)} f(x, y) = \lim_{(x,y)\to(0,0)} \frac{x^3 - y^3}{x^2 + y^2} = \lim_{x\to 0}\left[\lim_{y\to mx} \frac{x^3 - y^3}{x^2 + y^2}\right]$$
$$= \lim_{x\to 0} \frac{x^3 - m^3x^3}{x^2 + m^2x^2} = \lim_{x\to 0} \frac{x(1 - m^3)}{1 + m^2} = 0$$

Thus, limit exists at (0, 0).

**Step 3.** limit of $f(x)$ at origin = value of the function at origin.
$$\lim_{(x,y)\to(0,0)} \frac{x^3 - y^3}{x^2 + y^2} = f(0, 0) = 0$$

Hence, the function $f$ is continuous at the origin. **Ans.**

**Example 6.** *Discuss the continuity of* $f(x, y) = \begin{cases} \dfrac{x}{\sqrt{x^2 + y^2}}, & x \neq 0, \quad y \neq 0 \\ 2, & x = 0, \quad y = 0 \end{cases}$

*at the origin.*

**Solution.** Here, we $f(x, y) = \begin{cases} \dfrac{x}{\sqrt{x^2 + y^2}}, & x \neq 0, \quad y \neq 0 \\ 2, & x = 0, \quad y = 0 \end{cases}$

**Step 1.** The function $f(x, y)$ at (0, 0) is well defined.

**Step 2.**
$$\lim_{(x,y)\to(0,0)} \frac{x}{\sqrt{x^2 + y^2}} = \lim_{x\to 0}\left[\lim_{y\to mx} \frac{x}{\sqrt{x^2 + y^2}}\right] = \lim_{x\to 0} \frac{x}{\sqrt{x^2 + m^2x^2}} = \lim_{x\to 0} \frac{1}{\sqrt{1 + m^2}}$$

For different values of $m$ the limit $f$ is not unique.

so the $\lim_{(x,y)\to(0,0)} \dfrac{x}{\sqrt{x^2 + y^2}}$ does not exist.

Hence $f(x, y)$ is not continuous at origin. **Ans.**

## 1.6 TYPES OF DISCONTINUITY

(*Gujarat Univ. I sem. Jan. 2009*)

**1. First Kind.** $f(x)$ is said to have discontinuity of first kind at the point $x = x_1$ if Right limit $f(x_1 + 0)$ and left limit $f(x_1 - 0)$ exist but are not equal.

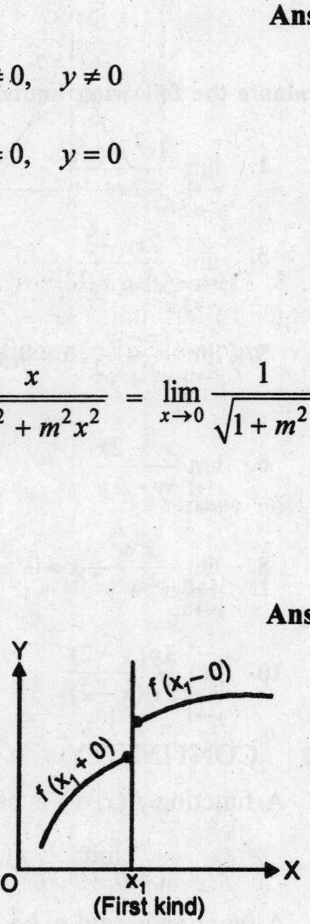

(First kind)

**2. Second Kind.** $f(x)$ is said to have discontinuity of the second kind at $x = x_1$ if neither right limit $f(x_1 + 0)$ exists nor left limit $f(x_1 - 0)$ exists.

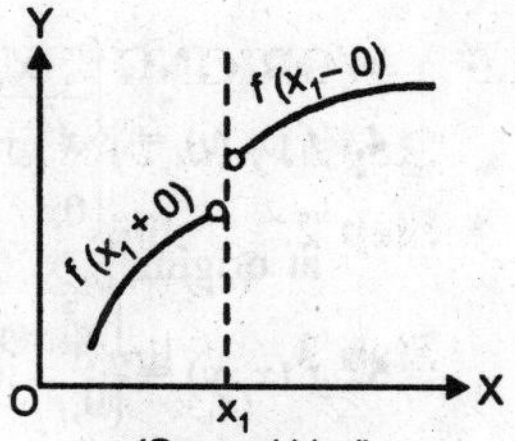

(Second kind)

**3. Third Kind (Mixed discontinuity).** $f(x)$ is said to have mixed discontinuity at the point $x = x_1$ if only one of the two limits right limit $f(x_1 + 0)$ and left limit $f(x_1 - 0)$ exists and not the other.

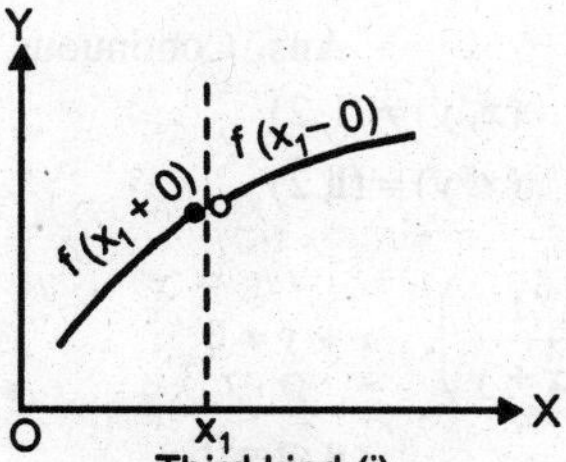

Third kind (i)

OR

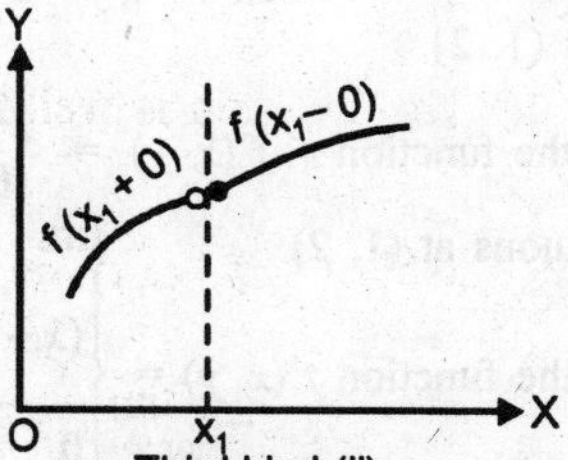

Third kind (ii)

**4. Fourth Kind (Infinite discontinuity).** $f(x)$ is said to have infinite discontinuity at the point $x = x_1$ if either one or both limits right limit and left limit $f(x_1 - 0)$ is infinite.

If both limits do not exist and if $f(x_1 \pm h)$ oscillates between limits one of which is infinite as $\pm h \to 0$. It is also a point of infinite discontinuity.

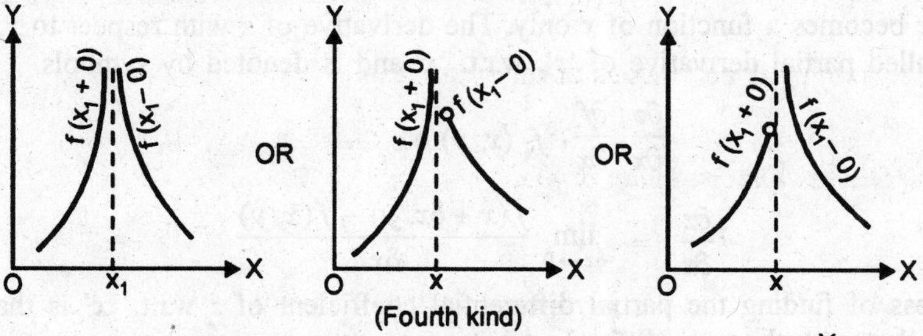

(Fourth kind)

**5. Fifth Kind (Removable discontinuity).** If right limit $f(x_1 + 0)$ is equal to left limit $f(x_1 - 0)$ is not equal to $f(x_1)$, then $f(x)$ is said to have removable discontinuity.

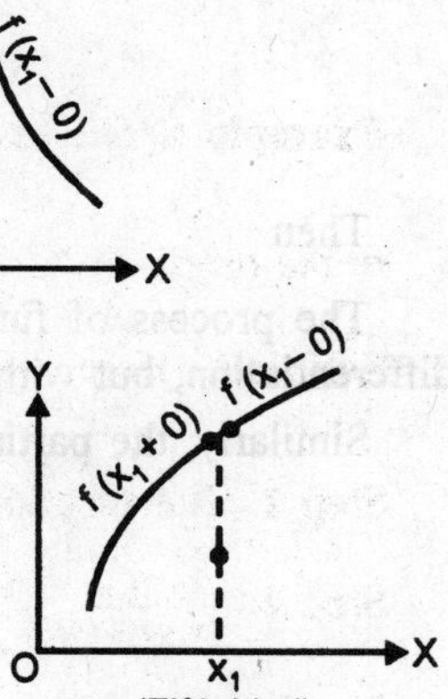

(Fifth kind)

## EXERCISE 1.2

**Test for continuity:**

**1.** $f(x, y) = \begin{cases} \dfrac{xy(x^2 - y^2)}{x^2 + y^2}, & \text{when } x \neq 0, y \neq 0 \\ 0, & \text{when } x = 0, y = 0 \end{cases}$

at origin. **Ans.** Continuous at origin.

**2.** $f(x, y) = \begin{cases} \dfrac{x^2 - y^2}{x^2 + y^2}, & \text{when } x \neq 0, y \neq 0 \\ 0, & \text{when } x = 0, y = 0 \end{cases}$

at origin. **Ans.** Not continuous at origin.

**3.** $f(x, y) = \begin{cases} \dfrac{x^3 y^3}{x^3 + y^3}, & \text{when } x \neq 0, y \neq 0 \\ 0, & \text{when } x = 0, y = 0 \end{cases}$ at origin. **Ans.** Not continuous at origin.

**4.** $f(x, y) = \begin{cases} \dfrac{xy}{x^2+y^2}, & \text{when } x \neq 0, y \neq 0 \\ 0, & \text{when } x = 0, y = 0 \end{cases}$

at origin. **Ans.** Not continuous at origin.

**5.** $f(x, y) = \begin{cases} x^3 + y^3, & \text{when } x \neq 0, y \neq 0 \\ 0, & \text{when } x = 0, y = 0 \end{cases}$

at origin. **Ans.** Continuous at origin.

**6.** $f(x, y) = \begin{cases} \dfrac{x^2+2y}{x+y^2}, & \\ 1 & \text{when } x = 1, y = 2 \end{cases}$

at the point (1, 2). **Ans.** Continuous at (1, 2).

**7.** Show that the function $f(x, y) = \begin{cases} 2x^2 + y, & (x, y) \neq (1, 2) \\ 0, & (x, y) = (1, 2) \end{cases}$

is discontinuous at (1, 2).

**8.** Show that the function $f(x, y) = \begin{cases} (x+y)\sin\left(\dfrac{1}{x+y}\right), & x + y \neq 0 \\ 0, & x + y = 0 \end{cases}$

is continuous at (0, 0) but its partial derivatives of first order do not exist at (0, 0).

*(A.M.I.E.T.E., Dec. 2007)*

## 1.7 PARTIAL DERIVATIVES

Let $z = f(x, y)$ be function of two independent variables $x$ and $y$. If we keep $y$ constant and $x$ varies then $z$ becomes a function of $x$ only. The derivative of $z$ with respect to $x$, keeping $y$ as constant is called partial derivative of '$z$', w.r.t. '$x$' and is denoted by symbols.

$$\frac{\partial z}{\partial x}, \frac{\partial f}{\partial x}, f_x(x, y) \text{ etc.}$$

Then
$$\frac{\partial z}{\partial x} = \lim_{\delta x \to 0} \frac{f(x+\delta x, y) - f(x, y)}{\delta x}$$

The process of finding the partial differential coefficient of $z$ w.r.t. '$x$' is that of ordinary differentiation, but with the only difference that we treat $y$ as constant.

Similarly, the partial derivative of '$z$' w.r.t. '$y$' keeping $x$ as constant is denoted by

$$\frac{\partial z}{\partial y}, \frac{\partial f}{\partial y}, f_y(x, y) \text{ etc.}$$

$$\frac{\partial z}{\partial y} = \lim_{\delta y \to 0} \frac{f(x, y+\delta y) - f(x, y)}{\delta y}$$

**Notation.** $\dfrac{\partial z}{\partial x} = p, \quad \dfrac{\partial z}{\partial y} = q, \quad \dfrac{\partial^2 z}{\partial x^2} = r, \quad \dfrac{\partial^2 z}{\partial x\, \partial y} = s, \quad \dfrac{\partial^2 z}{\partial y^2} = t$

**Example 7.** *If* $u = \sin^{-1}\left(\dfrac{x}{y}\right) + \tan^{-1}\left(\dfrac{y}{x}\right)$, *then find the value of* $x\dfrac{\partial u}{\partial x} + y\dfrac{\partial u}{\partial y}$.

**Solution.** $u = \sin^{-1}\left(\dfrac{x}{y}\right) + \tan^{-1}\left(\dfrac{y}{x}\right)$

$$\frac{\partial u}{\partial x} = \frac{1}{\sqrt{1-\left(\frac{x}{y}\right)^2}} \cdot \frac{1}{y} + \frac{1}{1+\left(\frac{y}{x}\right)^2} \cdot \left(-\frac{y}{x^2}\right) = \frac{1}{\sqrt{y^2-x^2}} - \frac{y}{x^2+y^2}$$

$$x\frac{\partial u}{\partial x} = \frac{x}{\sqrt{y^2-x^2}} - \frac{xy}{x^2+y^2} \quad ...(1)$$

$$\frac{\partial u}{\partial y} = \frac{1}{\sqrt{1-\left(\frac{x}{y}\right)^2}}\left(-\frac{x}{y^2}\right) + \frac{1}{1+\left(\frac{y}{x}\right)^2}\frac{1}{x} = -\frac{x}{y\sqrt{y^2-x^2}} + \frac{x}{x^2+y^2}$$

$$y.\frac{\partial u}{\partial y} = -\frac{x}{\sqrt{y^2-x^2}} + \frac{xy}{x^2+y^2} \quad ...(2)$$

On adding (1) and (2), we have $x.\frac{\partial u}{\partial x} + y.\frac{\partial u}{\partial y} = 0$ **Ans.**

**Example 8.** *Find* $\frac{\partial u}{\partial r}$ *and* $\frac{\partial u}{\partial \theta}$ *if* $u = e^{r\cos\theta}.\cos(r\sin\theta)$

**Solution.** $u = e^{r\cos\theta}.\cos(r\sin\theta)$

$$\frac{\partial u}{\partial r} = e^{r\cos\theta}.[-\sin(r\sin\theta).\sin\theta] + [\cos\theta.e^{r\cos\theta}].\cos(r\sin\theta)$$

(keeping θ as constant)

$$= e^{r\cos\theta}.[-\sin(r\sin\theta).\sin\theta + \cos(r\sin\theta).\cos\theta]$$

$$= e^{r\cos\theta}.\cos(r\sin\theta+\theta)$$ **Ans.**

$$\frac{\partial u}{\partial \theta} = e^{r\cos\theta}.[-\sin(r\sin\theta).r\cos\theta] + [-r\sin\theta.e^{r\cos\theta}].\cos(r\sin\theta)$$

(keeping $r$ as constant)

$$= -r\,e^{r\cos\theta}.[\sin(r\sin\theta).\cos\theta + \sin\theta\cos(r\sin\theta)]$$

$$= -r\,e^{r\cos\theta}.\sin(r\sin\theta+\theta)$$ **Ans.**

**Example 9.** *If* $u = (1-2xy+y^2)^{-1/2}$ *prove that,* $x\frac{\partial u}{\partial x} - y\frac{\partial u}{\partial y} = y^2u^3$.

**Solution.** $u = (1-2xy+y^2)^{-1/2}$ ...(1)

Differentiating (1) partially w.r.t. '$x$', we get

$$\frac{\partial u}{\partial x} = -\frac{1}{2}(1-2xy+y^2)^{-3/2}(-2y)$$

$$x\frac{\partial u}{\partial x} = xy(1-2xy+y^2)^{-3/2} \quad ...(2)$$

Differentiating (1) partially w.r.t. '$y$', we get

$$\frac{\partial u}{\partial y} = -\frac{1}{2}(1-2xy+y^2)^{-3/2}(-2x+2y)$$

$$y\frac{\partial u}{\partial y} = (xy-y^2)(1-2xy+y^2)^{-3/2} \quad ...(3)$$

Subtracting (3) from (2), we get

$$x\frac{\partial u}{\partial x} - y\frac{\partial u}{\partial y} = xy(1-2xy+y^2)^{-3/2} - (xy-y^2)(1-2xy+y^2)^{-3/2}$$

$$= y^2(1-2xy+y^2)^{-3/2} = y^2u^3.$$ **Proved.**

**Example 10.** *If* $z = e^{ax+by}.f(ax-by)$, *prove that* $b\frac{\partial z}{\partial x} + a\frac{\partial z}{\partial y} = 2abz$.

(*A.M.I.E.T.E., Summer* 2004)

**Solution.** $z = e^{ax+by} . f(ax - by)$ ...(1)

Differentiating (1) w.r.t. '$x$', we get

$$\frac{\partial z}{\partial x} = a\, e^{ax+by} . f(ax - by) + e^{ax+by} . a\, f'(ax - by)$$

$$b\frac{\partial z}{\partial x} = a\, b\, e^{ax+by} . f(ax - by) + a\, b\, e^{ax+by} . f'(ax - by) \quad ...(2)$$

Differentiating (1) w.r.t. '$y$', we get

$$\frac{\partial z}{\partial y} = b\, e^{ax+by} . f(ax - by) + e^{ax+by} . (-b)\, f'(ax - by)$$

$$a\frac{\partial z}{\partial y} = a\, b\, e^{ax+by} . f(ax - by) - a\, b\, e^{ax+by} . f'(ax - by) \quad ...(3)$$

On adding (2) and (3), we get

$$b\frac{\partial z}{\partial x} + a\frac{\partial z}{\partial y} = 2ab\, e^{ax+by}\, f(ax - by)$$

$$\Rightarrow \quad b\frac{\partial z}{\partial x} + a\frac{\partial z}{\partial y} = 2a\, b\, z$$ **Proved.**

## 1.8 PARTIAL DERIVATIVES OF HIGHER ORDERS

Let $z = f(x, y)$ then $\frac{\partial z}{\partial x}$ and $\frac{\partial z}{\partial y}$ being the functions of $x$ and $y$ can further be differentiated partially with respect to $x$ and $y$.

Symbolically

$$\frac{\partial}{\partial x}\left(\frac{\partial z}{\partial x}\right) = \frac{\partial^2 z}{\partial x^2} \quad \text{or} \quad \frac{\partial^2 f}{\partial x^2} \quad \text{or} \quad f_{xx}$$

$$\frac{\partial}{\partial y}\left(\frac{\partial z}{\partial x}\right) = \frac{\partial^2 z}{\partial y\, \partial x} \quad \text{or} \quad \frac{\partial^2 f}{\partial y\, \partial x} \quad \text{or} \quad f_{yx}$$

$$\frac{\partial}{\partial x}\left(\frac{\partial z}{\partial y}\right) = \frac{\partial^2 z}{\partial x\, \partial y} \quad \text{or} \quad \frac{\partial^2 f}{\partial x\, \partial y} \quad \text{or} \quad f_{xy}$$

**Note.** $$\frac{\partial^2 z}{\partial y \partial x} = \frac{\partial^2 z}{\partial x \partial y}$$

**Example 11.** *Prove that $y = f(x + at) + g(x - at)$ satisfies*

$$\frac{\partial^2 y}{\partial t^2} = a^2\left(\frac{\partial^2 y}{\partial x^2}\right)$$

*where $f$ and $g$ are assumed to be at least twice differentiable and $a$ is any constant.*

*(U.P., I Sem; Jan 2011, A.M.I.E., Summer* 2000*)*

**Solution.** $y = f(x + at) + g(x - at)$ ...(1)

Differentiating (1) w.r.t. '$x$' partially, we get

$$\frac{\partial y}{\partial x} = f'(x + at) + g'(x - at)$$

$$\frac{\partial^2 y}{\partial x^2} = f''(x + at) + g''(x - at)$$

Differentiating (1) w.r.t. '$t$' partially, we get

$$\frac{\partial y}{\partial t} = f'(x + at).a + g'(x - at)\,(-a)$$

$$\frac{\partial^2 y}{\partial t^2} = a^2 f''(x + at) + g''(x - at)\, a^2$$

$$= a^2 [f''(x + at) + g''(x - at)] = a^2 \frac{\partial^2 y}{\partial x^2}$$

Hence $$\frac{\partial^2 y}{\partial t^2} = a^2 \frac{\partial^2 y}{dx^2}$$ **Proved.**

**Example 12.** *If* $z = x^2 \tan^{-1}\left(\frac{y}{x}\right) - y^2 \tan^{-1}\left(\frac{x}{y}\right)$, *prove that* $\frac{\partial^2 z}{\partial y\, \partial x} = \frac{x^2 - y^2}{x^2 + y^2}$.

**Solution.** $$z = x^2 \tan^{-1}\left(\frac{y}{x}\right) - y^2 \tan^{-1}\left(\frac{x}{y}\right)$$ *(U.P., I Semester Comp. 2002)*

$$\frac{\partial z}{\partial x} = 2x \tan^{-1}\left(\frac{y}{x}\right) + x^2 \frac{1}{1+\frac{y^2}{x^2}}\left(-\frac{y}{x^2}\right) - y^2 \frac{1}{1+\frac{x^2}{y^2}}\left(\frac{1}{y}\right)$$

$$= 2x \tan^{-1}\left(\frac{y}{x}\right) - \frac{x^2 y}{x^2 + y^2} - \frac{y^3}{x^2 + y^2}$$

$$= 2x \tan^{-1}\left(\frac{y}{x}\right) - y\frac{(x^2 + y^2)}{x^2 + y^2} = 2x \tan^{-1}\left(\frac{y}{x}\right) - y$$

$$\frac{\partial^2 z}{\partial y\, \partial x} = 2x.\frac{1}{1+\frac{y^2}{x^2}}.\left(\frac{1}{x}\right) - 1 = 2\frac{x^2}{x^2 + y^2} - 1 = \frac{x^2 - y^2}{x^2 + y^2}$$ **Proved.**

**Example 13.** *If* $u = e^{xyz}$, *find the value of* $\frac{\partial^3 u}{\partial x\, \partial y\, \partial z}$. *(A.M.I.E. Winter 2000)*

**Solution.** $$u = e^{xyz}$$

$$\frac{\partial u}{\partial z} = e^{xyz}\,(x\, y)$$

$$\Rightarrow \quad \frac{\partial^2 u}{\partial y \partial z} = e^{xyz}\,(x) + e^{xyz}\,(x\, z)\,(x\, y) = e^{xyz}\,(x + x^2 y\, z)$$

$$\frac{\partial^3 u}{\partial x \partial y \partial z} = e^{xyz}\,(1 + 2x\, y\, z) + e^{xyz}\,(y\, z).(x + x^2 y\, z)$$

$$= e^{xyz}\,[1 + 2\, x\, y\, z + x\, y\, z + x^2 y^2 z^2]$$

$$= e^{xyz}\,[1 + 3\, x\, y\, z + x^2 y^2 z^2]$$ **Ans.**

**Example 14.** *If* $v = (x^2 + y^2 + z^2)^{\frac{m}{2}}$, *then find the value of m* $(m \neq 0)$ *which will make*

$$\frac{\partial^2 v}{\partial x^2} + \frac{\partial^2 v}{\partial y^2} + \frac{\partial^2 v}{\partial z^2} = 0.$$

**Solution.** We have, $$v = (x^2 + y^2 + z^2)^{\frac{m}{2}}$$

$$\frac{\partial v}{\partial x} = \frac{m}{2}(x^2 + y^2 + z^2)^{\frac{m}{2}-1}(2x) = mx\,(x^2 + y^2 + z^2)^{\frac{m}{2}-1}$$

$$\frac{\partial^2 v}{\partial x^2} = m\left(\frac{m}{2} - 1\right)x(x^2 + y^2 + z^2)^{\frac{m}{2}-2}(2x) + m(x^2 + y^2 + z^2)^{\frac{m}{2}-1}$$

$$= m(m-2)x^2(x^2+y^2+z^2)^{\frac{m}{2}-2} + m(x^2+y^2+z^2)^{\frac{m}{2}-1}$$

$$= m(x^2+y^2+z^2)^{\frac{m}{2}-2} [(m-2)x^2 + x^2 + y^2 + z^2] \quad ...(1)$$

Similarly, $$\frac{\partial^2 v}{\partial y^2} = m(x^2+y^2+z^2)^{\frac{m}{2}-2} .[(m-2)y^2 + x^2 + y^2 + z^2] \quad ...(2)$$

$$\frac{\partial^2 v}{\partial z^2} = m(x^2+y^2+z^2)^{\frac{m}{2}-2} .[(m-2)z^2 + x^2 + y^2 + z^2] \quad ...(3)$$

On adding (1), (2) and (3), we get

$$\frac{\partial^2 v}{\partial x^2} + \frac{\partial^2 v}{\partial y^2} + \frac{\partial^2 v}{\partial z^2} = m(x^2+y^2+z^2)^{\frac{m}{2}-2} [(m-2)(x^2+y^2+z^2) + 3(x^2+y^2+z^2)]$$

$$0 = m(x^2+y^2+z^2)^{\frac{m}{2}-1} [m-2+3] \qquad \left[\frac{\partial^2 v}{\partial x^2} + \frac{\partial^2 v}{\partial y^2} + \frac{\partial^2 v}{\partial z^2} = 0\right]$$

$$0 = m(m+1)(x^2+y^2+z^2)^{\frac{m}{2}-1}$$

$$0 = m(m+1) \Rightarrow m = 0, -1 \qquad (m \neq 0)$$

Hence, $m = -1$ **Ans.**

**Example 15.** *If* $u = (1-2xy+y^2)^{-\frac{1}{2}}$, *prove that* $\frac{\partial}{\partial x}\left((1-x^2)\frac{\partial u}{\partial x}\right) + \frac{\partial}{\partial y}\left(y^2 \frac{\partial u}{\partial y}\right) = 0.$

**Solution.** We have, $u = (1-2xy+y^2)^{-\frac{1}{2}}$ ...(1)

$$\frac{\partial u}{\partial x} = -\frac{1}{2}(1-2xy+y^2)^{-\frac{3}{2}}(-2y) = \frac{y}{(1-2xy+y^2)^{\frac{3}{2}}}$$

$$(1-x^2)\frac{\partial u}{\partial x} = \frac{(1-x^2)y}{(1-2xy+y^2)^{\frac{3}{2}}}$$

$$\frac{\partial}{\partial x}\left((1-x^2)\frac{\partial u}{\partial x}\right) = \frac{(1-2xy+y^2)^{\frac{3}{2}}(-2xy) - (1-x^2)\,y\left(\frac{3}{2}(1-2xy+y^2)^{\frac{1}{2}}(-2y)\right)}{(1-2xy+y^2)^3}$$

Cancelling $(1-2xy+y^2)^{\frac{1}{2}}$ from numerator and denominator, we have

$$= \frac{(1-2xy+y^2)(-2xy) + 3(1-x^2)y^2}{(1-2xy+y^2)^{\frac{5}{2}}} = \frac{-2xy + 4x^2y^2 - 2xy^3 + 3y^2 - 3x^2y^2}{(1-2xy+y^2)^{\frac{5}{2}}}$$

$$= \frac{x^2y^2 - 2xy^3 - 2xy + 3y^2}{(1-2xy+y^2)^{\frac{5}{2}}} \quad ...(2)$$

Differentiating (1) partially w.r.t. $y$, we get

$$\frac{\partial u}{\partial y} = -\frac{1}{2}(1-2xy+y^2)^{-\frac{3}{2}}(-2x+2y) = \frac{x-y}{(1-2xy+y^2)^{\frac{3}{2}}}$$

$$y^2\frac{\partial u}{\partial y}=\frac{xy^2-y^3}{(1-2xy+y^2)^{\frac{3}{2}}}$$

$$\frac{\partial}{\partial y}\left(y^2\frac{\partial u}{\partial y}\right)=\frac{(1-2xy+y^2)^{\frac{3}{2}}(2xy-3y^2)-(xy^2-y^3)\left(\frac{3}{2}(1-2xy+y^2)^{\frac{1}{2}}(-2x+2y)\right)}{(1-2xy+y^2)^3}$$

Dividing numerator and denominator by $(1-2xy+y^2)^{\frac{1}{2}}$, we get

$$\frac{\partial}{\partial y}\left(y^2\frac{\partial u}{\partial y}\right)=\frac{(1-2xy+y^2)(2xy-3y^2)+(xy^2-y^3)\,3\,(x-y)}{(1-2xy+y^2)^{\frac{5}{2}}}$$

$$=\frac{(2xy-4x^2y^2+2xy^3-3y^2+6xy^3-3y^4)+3x^2y^2-3xy^3-3xy^3+3y^4}{(1-2xy+y^2)^{\frac{5}{2}}}$$

$$=\frac{-x^2y^2+2xy^3+2xy-3y^2}{(1-2xy+y^2)^{\frac{5}{2}}} \qquad \ldots(3)$$

On adding (2) and (3), we get

$$\frac{\partial}{\partial x}(1-x^2)\frac{\partial u}{\partial x}+\frac{\partial}{\partial y}\left(y^2\frac{\partial u}{\partial y}\right)=\frac{x^2y^2-2xy^3-2xy+3y^2}{(1-2xy+y^2)^{\frac{5}{2}}}+\frac{-x^2y^2+2xy^3+2xy-3y^2}{(1-2xy+y^2)^{\frac{5}{2}}}=0$$

**Proved.**

**Example 16.** *Prove that if* $f(x, y)=\frac{1}{\sqrt{y}}e^{-\frac{(x-a)^2}{4y}}$ *then*

$$f_{xy}(x, y)=f_{yx}(x, y).$$

**Solution.** $$f(x, y)=\frac{1}{\sqrt{y}}e^{-\frac{(x-a)^2}{4y}} \qquad \ldots(1)$$

Differentiating $f(x, y)$ partially w.r.t. $x$, we get

$$f_x(x, y)=\frac{1}{\sqrt{y}}\cdot\frac{[-2(x-a)]}{4y}e^{-\frac{(x-a)^2}{4y}}=\frac{-(x-a)}{2y^{3/2}}e^{-\frac{(x-a)^2}{4y}}$$

Differentiating again partially w.r.t. '$y$' by product rule, we have

$$f_{yx}(x, y)=\frac{3(x-a)}{4y^{5/2}}.e^{-\frac{(x-a)^2}{4y}}-\frac{(x-a)^3}{8y^{7/2}}.e^{-\frac{(x-a)^2}{4y}}$$

$$=\frac{(x-a)}{8y^{7/2}}.e^{-\frac{(x-a)^2}{4y}}.[6y-(x-a)^2] \qquad \ldots(2)$$

Differentiating (1) partially w.r.t. '$y$', we have

$$f_y(x, y)=-\frac{1}{2y^{3/2}}.e^{-\frac{(x-a)^2}{4y}}+\frac{(x-a)^2}{4y^{5/2}}e^{-\frac{(x-a)^2}{4y}}$$

Differentiating again partially w.r.t. '$x$', we have

$$f_{xy}(x, y) = \frac{(x-a)}{4y^{5/2}} e^{-\frac{(x-a)^2}{4y}} + \frac{(x-a)}{2y^{5/2}} e^{-\frac{(x-a)^2}{4y}} - \frac{(x-a)^3}{8y^{7/2}} e^{-\frac{(x-a)^2}{4y}}$$

$$= \frac{(x-a)}{8y^{7/2}} e^{-\frac{(x-a)^2}{4y}} [2y + 4y - (x-a)^2]$$

$$= \frac{(x-a)}{8y^{7/2}} e^{-\frac{(x-a)^2}{4y}} [6y - (x-a)^2] \qquad ...(3)$$

From (2) and (3), we have $f_{xy}(x, y) = f_{yx}(x, y)$ **Proved.**

**Example 17.** *If* $u = x^y$, *show that* $\frac{\partial^3 u}{\partial x^2 \partial y} = \frac{\partial^3 u}{\partial x \partial y \partial x}$.

**Solution.** $u = x^y$

$\Rightarrow \quad \log u = \log x^y = y \log x$

Differentiating partially, we have

$$\frac{1}{u}.\frac{\partial u}{\partial x} = \frac{y}{x}, \quad \text{and} \quad \frac{1}{u}.\frac{\partial u}{\partial y} = \log x$$

$$\Rightarrow \quad \frac{\partial u}{\partial x} = u\frac{y}{x}, \quad \Rightarrow \quad \frac{\partial u}{\partial y} = u \log x$$

$$\frac{\partial^2 u}{\partial y \partial x} = \frac{1}{x}.\left[u + y.\frac{\partial u}{\partial y}\right] = \frac{u}{x} + \frac{y}{x}.\frac{\partial u}{\partial y} = \frac{u}{x} + \frac{uy.\log x}{x} \qquad \left(\text{As } \frac{\partial u}{\partial y} = u \log x\right)$$

$$\frac{\partial^3 u}{\partial x \partial y \partial x} = -\frac{u}{x^2} + \frac{1}{x}.\frac{\partial u}{\partial x} + y.\left\{\frac{x.\left(\frac{\partial u}{\partial x}.\log x + \frac{u}{x}\right) - u \log x}{x^2}\right\}$$

$$= -\frac{u}{x^2} + \frac{1}{x}\frac{\partial u}{\partial x} + \frac{y \log x}{x}\frac{\partial u}{\partial x} + \frac{uy}{x^2} - \frac{uy \log x}{x^2}$$

$$= -\frac{u}{x^2} + \frac{uy}{x^2} + \frac{uy^2 \log x}{x^2} + \frac{uy}{x^2} - \frac{uy \log x}{x^2}$$

$$= -\frac{u}{x^2} + \frac{2uy}{x^2} + \frac{uy^2 \log x}{x^2} - \frac{uy \log x}{x^2} \qquad ...(1)$$

$$\frac{\partial u}{\partial y} = u \log x$$

$$\frac{\partial^2 u}{\partial x \partial y} = \frac{u}{x} + \log x.\frac{\partial u}{\partial x} = \frac{u}{x} + \log x.\frac{uy}{x} \qquad \left[\frac{\partial u}{\partial x} = \frac{uy}{x}\right]$$

$$\frac{\partial^3 u}{\partial x^2 \partial y} = -\frac{u}{x^2} + \frac{1}{x}.\frac{\partial u}{\partial x} + y.\frac{x.\left(\frac{u}{x} + \log x.\frac{\partial u}{\partial x}\right) - u \log x \text{ (1)}}{x^2}$$

$$= -\frac{u}{x^2} + \frac{uy}{x^2} + \frac{uy}{x^2} + \frac{y \log x}{x}\frac{\partial u}{\partial x} - \frac{uy \log x}{x^2}$$

$$= -\frac{u}{x^2} + \frac{2uy}{x^2} + \frac{y \log x}{x}\frac{uy}{x} - \frac{uy \log x}{x^2}$$

$$= -\frac{u}{x^2} + \frac{2uy}{x^2} + \frac{uy^2 \log x}{x^2} - \frac{uy \log x}{x^2} \qquad ...(2)$$

From (1) and (2), we get $\dfrac{\partial^3 u}{\partial x^2 \partial y} = \dfrac{\partial^3 u}{\partial x \partial y \partial x}$ **Proved.**

**Example 18.** *If* $u = \log(x^3 + y^3 + z^3 - 3xyz)$, *show that*

$$\left(\frac{\partial}{\partial x} + \frac{\partial}{\partial y} + \frac{\partial}{\partial z}\right)^2 u = -\frac{9}{(x+y+z)^2}$$

(*U.P. I Semester, winter 2003*)

**Solution.** $u = \log(x^3 + y^3 + z^3 - 3xyz)$ ...(1)

Differentiating (1) partially w.r.t. '$x$', we get

$$\frac{\partial u}{\partial x} = \frac{3x^2 - 3yz}{x^3 + y^3 + z^3 - 3xyz} \qquad ...(2)$$

Similarly,

$$\frac{\partial u}{\partial y} = \frac{3y^2 - 3zx}{x^3 + y^3 + z^3 - 3xyz} \qquad ...(3)$$

$$\frac{\partial u}{\partial z} = \frac{3z^2 - 3xy}{x^3 + y^3 + z^3 - 3xyz} \qquad ...(4)$$

On adding (2), (3) and (4), we get

$$\frac{\partial u}{\partial x} + \frac{\partial u}{\partial y} + \frac{\partial u}{\partial z} = \frac{3(x^2 + y^2 + z^2 - xy - yz - zx)}{x^3 + y^3 + z^3 - 3xyz}$$

$$= \frac{3(x^2 + y^2 + z^2 - xy - yz - zx)}{(x+y+z)(x^2 + y^2 + z^2 - xy - yz - zx)} = \frac{3}{x+y+z}$$

$$\Rightarrow \quad \left(\frac{\partial}{\partial x} + \frac{\partial}{\partial y} + \frac{\partial}{\partial z}\right) u = \frac{3}{x+y+z}$$

$$\left(\frac{\partial}{\partial x} + \frac{\partial}{\partial y} + \frac{\partial}{\partial z}\right)^2 u = \left(\frac{\partial}{\partial x} + \frac{\partial}{\partial y} + \frac{\partial}{\partial z}\right) \frac{3}{x+y+z}$$

$$= \frac{\partial}{\partial x} \frac{3}{x+y+z} + \frac{\partial}{\partial y} \frac{3}{x+y+z} + \frac{\partial}{\partial z} \frac{3}{x+y+z}$$

$$= -3(x+y+z)^{-2} - 3(x+y+z)^{-2} - 3(x+y+z)^{-2}$$

$$= -\frac{9}{(x+y+z)^2}$$ **Proved.**

## 1.9 WHICH VARIABLE IS TO BE TREATED AS CONSTANT

Let $x = r\cos\theta, \qquad y = r\sin\theta$

To find $\dfrac{\partial r}{\partial x}$, we need a relation between $r$ and $x$.

$$r = x \sec\theta \qquad ...(1)$$

Differentiating (1) w.r.t. '$x$' keeping $\theta$ as constant

$$\frac{\partial r}{\partial x} = \sec\theta \qquad ...(2)$$

Here, we have $r^2 = x^2 + y^2$ ...(3)

Differentiating (3) w.r.t. '$x$' keeping $y$ as constant.

$$2r\frac{\partial r}{\partial x} = 2x \quad \text{or} \quad \frac{\partial r}{\partial x} = \frac{x}{r} = \cos\theta \qquad ...(4)$$

From (2), $\frac{\partial r}{\partial x} = \sec\theta$ and from (4), $\frac{\partial r}{\partial x} = \cos\theta$. These two values of $\frac{\partial r}{\partial x}$ make confusion. To avoid the confusion we use the following notations.

**Notation.** (*i*) $\left(\frac{\partial r}{\partial x}\right)_\theta$ means the partial derivative of $r$ with respect to $x$, keeping $\theta$ as constant.

From (3), $\left(\frac{\partial r}{\partial x}\right)_\theta = \sec\theta$

(*ii*) $\left(\frac{\partial r}{\partial x}\right)_y$ means the partial derivative of $r$ with respect to $x$ keeping $y$ as constant.

From (4), $\left(\frac{\partial r}{\partial x}\right)_y = \cos\theta$

(*iii*) When no indication is given regarding the variables to be treated as constant

$$\frac{\partial}{\partial x} \text{ means } \left(\frac{\partial}{\partial x}\right)_y, \quad \frac{\partial}{\partial y} \text{ means } \left(\frac{\partial}{\partial y}\right)_x.$$

$$\frac{\partial}{\partial r} \text{ means } \left(\frac{\partial}{\partial r}\right)_\theta, \quad \frac{\partial}{\partial \theta} \text{ means } \left(\frac{\partial}{\partial \theta}\right)_r.$$

**Example 19.** *If* $x = r\cos\theta$, $y = r\sin\theta$, *find*

(*i*) $\left(\frac{\partial x}{\partial r}\right)_\theta$ (*ii*) $\left(\frac{\partial y}{\partial \theta}\right)_r$ (*iii*) $\left(\frac{\partial r}{\partial x}\right)_y$ (*iv*) $\left(\frac{\partial \theta}{\partial y}\right)_x$

**Solution.** (*i*) $\left(\frac{\partial x}{\partial r}\right)_\theta$ means the partial derivative of $x$ with respect to $r$, keeping $\theta$ as constant.

$$x = r\cos\theta \qquad \left(\frac{\partial x}{\partial r}\right)_\theta = \cos\theta$$

(*ii*) $\left(\frac{\partial y}{\partial \theta}\right)_r$ means the partial derivative of $y$ with respect to $\theta$, treating $r$ as constant.

$$y = r\sin\theta \qquad \left(\frac{\partial y}{\partial \theta}\right)_r = r\cos\theta$$

(*iii*) $\left(\frac{\partial r}{\partial x}\right)_y$ means the partial derivative of $r$ with respect to $x$, treating $y$ as constant. We have to express $r$ as a function of $x$ and $y$.

$$r = \sqrt{x^2+y^2} \qquad \text{(From the given equations)}$$

$$\left(\frac{\partial r}{\partial x}\right)_y = \frac{1}{2}\frac{1}{\sqrt{x^2+y^2}}.2x = \frac{x}{\sqrt{x^2+y^2}}$$

(*iv*) Before finding $\left(\frac{\partial \theta}{\partial y}\right)_x$ we have to express $\theta$ in terms of $x$ and $y$.

$$\theta = \tan^{-1}\frac{y}{x} \qquad \text{(From the given equations)}$$

$$\left(\frac{\partial \theta}{\partial y}\right)_x = \frac{1}{1+\frac{y^2}{x^2}}.\frac{1}{x} = \frac{x}{x^2+y^2} \qquad \textbf{Ans.}$$

## EXERCISE 1.3

**1.** If $z^3 - 3yz - 3x = 0$, show that $z\dfrac{\partial z}{\partial x} = \dfrac{\partial z}{\partial y}$; $z\left[\dfrac{\partial^2 z}{\partial x \partial y} + \left(\dfrac{\partial z}{\partial x}\right)^2\right] = \dfrac{\partial^2 z}{\partial y^2}$

**2.** If $z(z^2 + 3x) + 3y = 0$, prove that $\dfrac{\partial^2 z}{\partial x^2} + \dfrac{\partial^2 z}{\partial y^2} = \dfrac{2z(x-1)}{(z^2+x)^3}$.

**3.** If $z = \log(e^x + e^y)$, show that $rt - s^2 = 0$.

**4.** If $f(x, y) = x^3y - xy^3$, find $\left[\dfrac{1}{\frac{\partial f}{\partial x}} + \dfrac{1}{\frac{\partial f}{\partial y}}\right]_{\substack{x=1\\y=2}}$ **Ans.** $-\dfrac{13}{22}$

**5.** If $\theta = t^n e^{-\frac{r^2}{4t}}$, find what value of $n$ will make $\dfrac{1}{r^2}\dfrac{\partial}{\partial r}\left(r^2\dfrac{\partial \theta}{\partial r}\right) = \dfrac{\partial \theta}{\partial t}$. **Ans.** $n = -\dfrac{3}{2}$

**6.** Show that the function $u = \arctan(y/x)$ satisfies the Laplace equation $\dfrac{\partial^2 u}{\partial x^2} + \dfrac{\partial^2 u}{\partial y^2} = 0$.

**7.** If $z = y\,f(x^2 - y^2)$ show that $y\dfrac{\partial z}{\partial x} + x\dfrac{\partial z}{\partial y} = \dfrac{xz}{y}$.

**8.** Show that $\dfrac{\partial^2 z}{\partial x^2} - 2\dfrac{\partial^2 z}{\partial x \partial y} + \dfrac{\partial^2 z}{\partial y^2} = 0$, where $z = x\,.f(x+y) + y\,.\,g(x+y)$.

**9.** If $u = \log(x^2 + y^2) + \tan^{-1}\left(\dfrac{y}{x}\right)$. Show that $\dfrac{\partial^2 z}{\partial x^2} + \dfrac{\partial^2 u}{\partial y^2} = 0$.

**10.** If $u(x, y, z) = \dfrac{1}{x^2 + y^2 + z^2}$, find the value of $\dfrac{\partial^2 u}{\partial x^2} + \dfrac{\partial^2 u}{\partial y^2} + \dfrac{\partial^2 u}{\partial z^2}$. **Ans.** $\dfrac{2}{(x^2+y^2+z^2)^2}$

**11.** If $x = e^{r\cos\theta}\cos(r\sin\theta)$ and $y = e^{r\cos\theta}\sin(r\sin\theta)$

Prove that $\dfrac{\partial x}{\partial r} = \dfrac{1}{r}\dfrac{\partial y}{\partial \theta}$, $\dfrac{\partial y}{\partial r} = -\dfrac{1}{r}\dfrac{\partial x}{\partial \theta}$

Hence deduce that $\dfrac{\partial^2 x}{\partial r^2} + \dfrac{1}{r}\dfrac{\partial x}{\partial r} + \dfrac{1}{r}\dfrac{\partial^2 x}{\partial \theta^2} = 0$

**12.** If $x = r\cos\theta$, $y = r\sin\theta$, prove that

(*i*) $\dfrac{\partial r}{\partial x} = \dfrac{\partial x}{\partial r}$, $r.\dfrac{\partial \theta}{\partial x} = \dfrac{1}{r}.\dfrac{\partial x}{\partial \theta}$ (*ii*) $\dfrac{\partial^2 \theta}{\partial x^2} + \dfrac{\partial^2 \theta}{\partial y^2} = 0$ (*c*) $\dfrac{\partial^2 r}{\partial x^2} + \dfrac{\partial^2 r}{\partial y^2} = \dfrac{1}{r}\left[\left(\dfrac{\partial r}{\partial x}\right)^2 + \left(\dfrac{\partial r}{\partial y}\right)^2\right]$

**13.** If $v = (x^2 - y^2).\,f(xy)$, prove that $\dfrac{\partial^2 v}{\partial x^2} + \dfrac{\partial^2 v}{\partial y^2} = (x^4 - y^4)\,f''(xy)$

**14.** If $ux + vy = 0$ and $\dfrac{u}{x} + \dfrac{v}{y} = 1$, show that $\dfrac{\partial u}{\partial x} - \dfrac{\partial v}{\partial y} = \dfrac{x^2 + y^2}{y^2 - x^2}$

**15.** If $z = x^y + y^x$, verify that $\dfrac{\partial^2 z}{\partial x \partial y} = \dfrac{\partial^2 z}{\partial y \partial x}$

**16.** If $u = f(ax^2 + 2h\,x\,y + by^2)$ and $v = \varphi(ax^2 + 2hx\,y + by^2)$ show that $\dfrac{\partial}{\partial y}\left(u\dfrac{\partial v}{\partial x}\right) = \dfrac{\partial}{\partial x}\left(u\dfrac{\partial v}{\partial y}\right)$.

**17.** If $u = r^m$, where $r^2 = x^2 + y^2 + z^2$, find the value of $\dfrac{\partial^2 u}{\partial x^2} + \dfrac{\partial^2 u}{\partial y^2} + \dfrac{\partial^2 u}{\partial z^2}$. **Ans.** $m(m+1)r^{m-2}$

**18.** If $x = \dfrac{r}{2}(e^\theta + e^{-\theta})$, $y = \dfrac{r}{2}(e^\theta - e^{-\theta})$ prove that $\dfrac{\partial x}{\partial r} = \dfrac{\partial r}{\partial x}$

19. If $u(x, t) = a\,e^{-gx} \sin(nt - gx)$, satisfies the equation $\dfrac{\partial u}{\partial t} = a^2 \dfrac{\partial^2 u}{\partial x^2}$, show that $ag = \sqrt{\dfrac{n}{2}}$.

20. If $u = \log(\tan x + \tan y)$, prove that, $\sin 2x \dfrac{\partial u}{\partial x} + \sin 2y \dfrac{\partial u}{\partial y} = 2$.

21. If $u = \dfrac{1}{x}[(x - y) + \psi(x + y)]$, then show that $\dfrac{\partial}{\partial x}\left(\dfrac{x^2 \partial u}{\partial x}\right) = x^2 \dfrac{\partial^2 u}{\partial y^2}$

22. If $u = e^{xyz} f\left(\dfrac{xz}{y}\right)$, prove that

(i) $x\dfrac{\partial u}{\partial x} + y\dfrac{\partial u}{\partial y} = 2\,x\,y\,z\,u$, (ii) $y\dfrac{\partial u}{\partial y} + z\dfrac{\partial u}{\partial z} = 2\,x\,y\,z\,u$

Also deduce that $x\dfrac{\partial^2 u}{\partial z \partial x} = y\dfrac{\partial^2 u}{\partial z \partial y}$ *(A.M.I.E., Summer 2001)*

23. If $u = f(x, y)$, $x = r\cos\theta$, $y = r\sin\theta$, then show that

$$\left(\frac{\partial u}{\partial x}\right)^2 + \left(\frac{\partial u}{\partial y}\right)^2 = \left(\frac{\partial u}{\partial r}\right)^2 + \frac{1}{r^2}\left(\frac{\partial u}{\partial \theta}\right)^2$$

## 1.10 HOMOGENEOUS FUNCTION

A function $f(x, y)$ is said to be homogeneous function in which the power of each term is the same.

A function $f(x, y)$ is a homogeneous function of order $n$, if the degree of each of its terms in $x$ and $y$ is equal to $n$. Thus

$$a_0 x^n + a_1 x^{n-1} y + a_2 x^{n-2} y^2 + \dots + a_{n-1} x y^{n-1} + a_n y^n \qquad \text{...(1)}$$

is a homogeneous function of order $n$.

The polynomial function (1) which can be written as

$$x^n\left[a_0 + a_1\left(\frac{y}{x}\right) + a_2\left(\frac{y}{x}\right)^2 + \dots + a_{n-1}\left(\frac{y}{x}\right)^{n-1} + a_n\left(\frac{y}{x}\right)^n\right] = x^n \varphi\left(\frac{y}{x}\right) \qquad \text{...(2)}$$

(i) The function $x^3\left[1 + \dfrac{y}{x} + 3\left(\dfrac{y}{x}\right)^2 + 5\left(\dfrac{y}{x}\right)^3\right]$ is a homogeneous function of order 3.

(ii) $\dfrac{\sqrt{x} + \sqrt{y}}{x^2 + y^2} = \dfrac{\sqrt{x}\left[1 + \sqrt{\dfrac{y}{x}}\right]}{x^2\left[1 + \left(\dfrac{y}{x}\right)^2\right]} = x^{-3/2} \cdot \dfrac{1 + \sqrt{\dfrac{y}{x}}}{1 + \left(\dfrac{y}{x}\right)^2}$ is a homogeneous function of order – 3/2

(iii) $\sin^{-1}\dfrac{\sqrt{x} + \sqrt{y}}{x^2 + y^2}$ is not a homogeneous function as it cannot be written in the form of $x^n f\left(\dfrac{y}{x}\right)$ so that its degree may be pronounced. It is a function of homogeneous expression.

## 1.11 EULER'S THEOREM ON HOMOGENEOUS FUNCTION

***(U.P. I Semester, Dec. 2006)***

**Statement.** If $z$ is a homogeneous function of $x$, $y$ of order $n$, then

$$x.\frac{\partial z}{\partial x} + y.\frac{\partial z}{\partial y} = n\,z$$

**Proof.** Since $z$ is a homogeneous function of $x$, $y$ of order $n$.

$\therefore$ $z$ can be written in the form

$$z = x^n . f\left(\frac{y}{x}\right) \quad ...(1)$$

Differentiating (1) partially w.r.t. '$x$', we have

$$\frac{\partial z}{\partial x} = nx^{n-1}.f\left(\frac{y}{x}\right) + x^n.f'\left(\frac{y}{x}\right)\left(-\frac{y}{x^2}\right)$$

$$\Rightarrow \quad \frac{\partial z}{\partial x} = nx^{n-1}.f\left(\frac{y}{x}\right) - x^{n-2}y.f'\left(\frac{y}{x}\right)$$

Multiplying both sides by $x$, we have

$$x\frac{\partial z}{\partial x} = n\,x^n.f\left(\frac{y}{x}\right) - x^{n-1}y.f'\left(\frac{y}{x}\right) \quad ...(2)$$

Differentiating (1) partially w.r.t. '$y$', we have

$$\frac{\partial z}{\partial y} = x^n f'\left(\frac{y}{x}\right).\frac{1}{x}$$

Multiplying both sides by $y$, we get

$$y.\frac{\partial z}{\partial y} = x^{n-1}y.f'\left(\frac{y}{x}\right) \quad ...(3)$$

Adding (2) and (3), we have

$$x.\frac{\partial z}{\partial x} + y.\frac{\partial z}{\partial y} = nx^n f\left(\frac{y}{x}\right)$$

$$\Rightarrow \quad x.\frac{\partial z}{\partial x} + y.\frac{\partial z}{\partial y} = n\,z$$

**Proved.**

**Note.** If $u$ is a homogeneous function of $x$, $y$, $z$ of degree $n$, then

$$x\frac{\partial u}{\partial x} + y\frac{\partial u}{\partial y} + z\frac{\partial u}{\partial z} = nu$$

**I. Deduction from Euler's theorem**

If $z$ is a homogeneous function of $x$, $y$ of degree $n$ and $z = f(u)$, then

$$x\frac{\partial u}{\partial x} + y\frac{\partial u}{\partial y} = n\frac{f(u)}{f'(u)}$$

*(Nagpur University, Winter 2003)*

**Proof.** Since $z$ is a homogeneous function of $x$, $y$ of degree $n$, we have, by Euler's theorem,

$$x\frac{\partial z}{\partial x} + y\frac{\partial z}{\partial y} = nz \quad ...(1)$$

Now $z = f(u)$, given

$$\therefore \quad \frac{\partial z}{\partial x} = f'(u)\frac{\partial u}{\partial x} \quad \text{and} \quad \frac{\partial z}{\partial y} = f'(u)\frac{\partial u}{\partial y}$$

Substituting in (1), we get

$$x\,f'(u)\frac{\partial u}{\partial x} + y\,f'(u)\frac{\partial u}{\partial y} = nf(u)$$

$$\Rightarrow \quad x\frac{\partial u}{\partial x} + y\frac{\partial u}{\partial y} = n\frac{f(u)}{f'(u)}$$

**Note.** If $v = f(u)$ where $v$ is a homogeneous function in $x$, $y$, $z$ of degree $n$, then

$$x\frac{\partial u}{\partial x} + y\frac{\partial u}{\partial y} + z\frac{\partial u}{\partial z} = \frac{nf(u)}{f'(u)}$$

**Example 20.** *Verify Euler's theorem for* $z = \dfrac{x^{1/3} + y^{1/3}}{x^{1/2} + y^{1/2}}$. *(U.P. Ist Semester, Dec. 2009)*

**Solution.** Here, we have

$$z = \frac{x^{1/3} + y^{1/3}}{x^{1/2} + y^{1/2}} = \frac{x^{1/3}\left[1 + \left(\dfrac{y}{x}\right)^{1/3}\right]}{x^{1/2}\left[1 + \left(\dfrac{y}{x}\right)^{1/2}\right]} = x^{-\frac{1}{6}}\,\phi\left(\frac{y}{x}\right) \qquad \text{...(1)}$$

Thus $z$ is homogeneous function of degree $-\dfrac{1}{6}$.

By Euler's theorem $x\dfrac{\partial z}{\partial x} + y\dfrac{\partial z}{\partial y} = -\dfrac{1}{6}z.$ ...(2)

Differentiating (1) w.r.t. '$x$', we get

$$\frac{\partial z}{\partial x} = \frac{\left(x^{\frac{1}{2}} + y^{\frac{1}{2}}\right)\left(\frac{1}{3}x^{-\frac{2}{3}}\right) - \left(x^{\frac{1}{3}} + y^{\frac{1}{3}}\right)\left(\frac{1}{2}x^{-\frac{1}{2}}\right)}{\left(x^{\frac{1}{2}} + y^{\frac{1}{2}}\right)^2} = \frac{\frac{1}{3}x^{\frac{-1}{6}} + \frac{1}{3}x^{-\frac{2}{3}}y^{\frac{1}{2}} - \frac{1}{2}x^{-\frac{1}{6}} - \frac{1}{2}x^{-\frac{1}{2}}y^{\frac{1}{3}}}{\left(x^{\frac{1}{2}} + y^{\frac{1}{2}}\right)^2}$$

$$x\frac{\partial z}{\partial x} = \frac{\frac{1}{3}x^{\frac{5}{6}} + \frac{1}{3}x^{\frac{1}{3}}y^{\frac{1}{2}} - \frac{1}{2}x^{\frac{5}{6}} - \frac{1}{2}x^{\frac{1}{2}}y^{\frac{1}{3}}}{\left(x^{\frac{1}{2}} + y^{\frac{1}{2}}\right)^2} \qquad \text{...(3)}$$

$$\frac{\partial z}{\partial y} = \frac{\left(x^{\frac{1}{2}} + y^{\frac{1}{2}}\right)\left(\frac{1}{3}y^{\frac{-2}{3}}\right) - \left(x^{\frac{1}{3}} + y^{\frac{1}{3}}\right)\left(\frac{1}{2}y^{\frac{-1}{2}}\right)}{\left(x^{\frac{1}{2}} + y^{\frac{1}{2}}\right)^2} = \frac{\frac{1}{3}x^{\frac{1}{2}}y^{\frac{-2}{3}} + \frac{1}{3}y^{-\frac{1}{6}} - \frac{1}{2}x^{\frac{1}{3}}y^{-\frac{1}{2}} - \frac{1}{2}y^{\frac{-1}{6}}}{\left(x^{\frac{1}{2}} + y^{\frac{1}{2}}\right)^2}$$

$$\Rightarrow \quad y\frac{\partial z}{\partial y} = \frac{\frac{1}{3}x^{\frac{1}{2}}y^{\frac{1}{3}} + \frac{1}{3}y^{\frac{5}{6}} - \frac{1}{2}x^{\frac{1}{3}}y^{\frac{1}{2}} - \frac{1}{2}y^{\frac{5}{6}}}{\left(x^{\frac{1}{2}} + y^{\frac{1}{2}}\right)^2} \qquad \text{...(4)}$$

Adding (3) and (4), we get

$$x\frac{\partial z}{\partial x} + y\frac{\partial z}{\partial y} = \frac{\frac{1}{3}x^{\frac{5}{6}} + \frac{1}{3}x^{\frac{1}{3}}y^{\frac{1}{2}} - \frac{1}{2}x^{\frac{5}{6}} - \frac{1}{2}x^{\frac{1}{2}}y^{\frac{1}{3}} + \frac{1}{3}x^{\frac{1}{2}}y^{\frac{1}{3}} + \frac{1}{3}y^{\frac{5}{6}} - \frac{1}{2}x^{\frac{1}{3}}y^{\frac{1}{2}} - \frac{1}{2}y^{\frac{5}{6}}}{\left(x^{\frac{1}{2}} + y^{\frac{1}{2}}\right)^2}$$

$$= \frac{-\frac{1}{6}\left[x^{\frac{5}{6}} + y^{\frac{5}{6}} + x^{\frac{1}{3}}y^{\frac{1}{2}} + x^{\frac{1}{2}}y^{\frac{1}{3}}\right]}{\left(x^{\frac{1}{2}} + y^{\frac{1}{2}}\right)^2} = \frac{-\frac{1}{6}\left[x^{\frac{1}{2}}\left(x^{\frac{1}{3}} + y^{\frac{1}{3}}\right) + y^{\frac{1}{2}}\left(x^{\frac{1}{3}} + y^{\frac{1}{3}}\right)\right]}{\left(x^{\frac{1}{2}} + y^{\frac{1}{2}}\right)^2}$$

$$= \frac{-\frac{1}{6}\left(x^{\frac{1}{2}}+y^{\frac{1}{2}}\right)\left(x^{\frac{1}{3}}+y^{\frac{1}{3}}\right)}{\left(x^{\frac{1}{2}}+y^{\frac{1}{2}}\right)^2} = -\frac{1}{6}\frac{x^{\frac{1}{3}}+y^{\frac{1}{3}}}{x^{\frac{1}{2}}+y^{\frac{1}{2}}}$$

$$x\frac{\partial z}{\partial x} + y\frac{\partial z}{\partial y} = -\frac{1}{6}z \qquad ...(5)$$

From (2) and (5), Euler's theorem is verified. **Verified.**

**Example 21.** *If* $u = \cos^{-1}\left(\frac{x+y}{\sqrt{x}+\sqrt{y}}\right)$, *show that*

$$x\frac{\partial u}{\partial x} + y\frac{\partial u}{\partial y} + \frac{1}{2}\cot u = 0.$$

*(U.P. Ist Semester, Dec. 2009)*

**Solution.** Here, we have, $u = \cos^{-1}\left(\frac{x+y}{\sqrt{x}+\sqrt{y}}\right)$

$u$ is not a homogeneous function but if $z = \cos u$, then

$$u = \cos^{-1} z = \frac{x+y}{\sqrt{x}+\sqrt{y}} = \frac{x\left(1+\frac{y}{x}\right)}{\sqrt{x}\left(1+\sqrt{\frac{y}{x}}\right)} = x^{\frac{1}{2}}\frac{\left(1+\frac{y}{x}\right)}{\left(1+\sqrt{\frac{y}{x}}\right)} = x^{\frac{1}{2}}\phi\left(\frac{y}{x}\right).$$

$z$ is a homogeneous function in $x$, $y$ of degree $\frac{1}{2}$.

By Euler's theorem, we have $x\frac{\partial z}{\partial x} + y\frac{\partial z}{\partial y} = \frac{1}{2}z$

$$x\frac{\partial z}{\partial u}\frac{\partial u}{\partial x} + y\frac{\partial z}{\partial u}\frac{\partial u}{\partial y} = \frac{1}{2}z$$

$$x\frac{\partial u}{\partial x}(-\sin u) + y\frac{\partial u}{\partial y}(-\sin u) = \frac{1}{2}\cos u$$

$$x\frac{\partial u}{\partial x} + y\frac{\partial u}{\partial y} = -\frac{1}{2}\cot u. \quad \Rightarrow \quad x\frac{\partial u}{\partial x} + y\frac{\partial u}{\partial y} + \frac{1}{2}\cot u = 0$$

**Proved.**

**Example 22.** *If* $u = \sin^{-1}\left[\frac{x+2y+3z}{\sqrt{x^8+y^8+z^8}}\right]$, *show that*

$$x\frac{\partial u}{\partial x} + y\frac{\partial u}{\partial y} + z\frac{\partial u}{\partial z} + 3\tan u = 0.$$

*(U.P. I Sem., Winter 2003)*

**Solution.** We have, $u = \sin^{-1}\left[\frac{x+2y+3z}{\sqrt{x^8+y^8+z^8}}\right]$

Here, $u$ is not a homogeneous function but if $v = \sin u = \frac{x+2y+3z}{\sqrt{x^8+y^8+z^8}}$ then

$v$ is a homogeneous function in $x$, $y$, $z$ of degree $-3$.

By Euler's Theorem $\qquad x\frac{\partial v}{\partial x} + y\frac{\partial v}{\partial y} + z\frac{\partial v}{\partial z} = n\,v$

$$x\frac{\partial v}{\partial u}\frac{\partial u}{\partial x} + y\frac{\partial v}{\partial u}\frac{\partial u}{\partial y} + z\frac{\partial v}{\partial u}\frac{\partial u}{\partial z} = -3\,v \qquad ...(1)$$

Putting the value of $\frac{\partial v}{\partial u}$ in (1), we get

$$x \cos u \frac{\partial u}{\partial x} + y \cos u \frac{\partial u}{\partial y} + z \cos u \frac{\partial u}{\partial z} = -3 \sin u$$

$$x \frac{\partial u}{\partial x} + y \frac{\partial u}{\partial y} + z \frac{\partial u}{\partial z} = -3 \frac{\sin u}{\cos u} = -3 \tan u$$

$$x \frac{\partial u}{\partial x} + y \frac{\partial u}{\partial y} + z \frac{\partial u}{\partial z} + 3 \tan u = 0$$

**Proved.**

**Example 23.** *If* $u = \log_e \left( \frac{x^4 + y^4}{x + y} \right)$, *show that* $x \frac{\partial u}{\partial x} + y \frac{\partial u}{\partial y} = 3.$

*(Nagpur University, Summer 2008, Uttarakhand, I Semester 2008)*

**Solution.** We have, $u = \log_e \left( \frac{x^4 + y^4}{x + y} \right)$

Here, $u$ is not a homogeneous function but if

$$z = e^u = \frac{x^4 + y^4}{x + y} = \frac{x^4 \left[ 1 + \left( \frac{y}{x} \right)^4 \right]}{x \left[ 1 + \left( \frac{y}{x} \right) \right]} = x^3 \varphi \left( \frac{y}{x} \right)$$

Then $z$ is a homogeneous function of degree 3.
By Euler's Deduction formula I

$$x \frac{\partial u}{\partial x} + y \frac{\partial u}{\partial y} = n \frac{f(u)}{f'(u)} = 3 \frac{e^u}{e^u} = 3$$

**Proved.**

**Example 24.** *If* $f(x, y) = \frac{1}{x^2} + \frac{1}{xy} + \frac{\log x - \log y}{x^2 + y^2}$, *prove that*

$$x \frac{\partial f}{\partial x} + y \frac{\partial f}{\partial y} + 2f = 0.$$

*(A.M.I.E. Summer 2004)*

**Solution.**

$$f(x, y) = \frac{1}{x^2} + \frac{1}{xy} + \frac{\log x - \log y}{x^2 + y^2}$$

$$= \frac{1}{x^2} \left( \frac{y}{x} \right)^0 + \frac{1}{x^2} \frac{1}{\left( \frac{y}{x} \right)} - \frac{1}{x^2} \frac{\log \frac{y}{x}}{\left[ 1 + \left( \frac{y}{x} \right)^2 \right]}$$

$f(x, y)$ is a homogeneous function of degree $-2$.
By Euler's Theorem

$$x \frac{\partial f}{\partial x} + y \frac{\partial f}{\partial y} = -2f \Rightarrow x \frac{\partial f}{\partial x} + y \frac{\partial f}{\partial y} + 2f = 0$$

**Proved.**

**Example 25.** *If z be a homogeneous function of degree n, show that*

(*i*) $x . \frac{\partial^2 z}{\partial x^2} + y . \frac{\partial^2 z}{\partial x \partial y} = (n - 1) \frac{\partial z}{\partial x}$ (*ii*) $x . \frac{\partial^2 z}{\partial x \partial y} + y . \frac{\partial^2 z}{\partial y^2} = (n - 1) \frac{\partial z}{\partial y}$

(*iii*) $x^2 . \frac{\partial^2 z}{\partial x^2} + 2xy . \frac{\partial^2 z}{\partial x \partial y} + y^2 . \frac{\partial^2 z}{\partial y^2} = n(n - 1) z.$

*(Uttarakhand Ist Semester, Dec. 2006)*

**Solution.** By Euler's Theorem $x.\dfrac{\partial z}{\partial x}+y\dfrac{\partial z}{\partial y} = n\,z$ ...(1)

Differentiating (1), partially w.r.t. '$x$', we get

$$\frac{\partial z}{\partial x}+x.\frac{\partial^2 z}{\partial x^2}+y\frac{\partial^2 z}{\partial x\partial y} = n\frac{\partial z}{\partial x}$$

$$\Rightarrow \quad x.\frac{\partial^2 z}{\partial x^2}+y\frac{\partial^2 z}{\partial x\partial y} = (n-1)\frac{\partial z}{\partial x}$$ **Proved** (*i*) ...(2)

Differentiating (1), partially w.r.t. '$y$', we have

$$x.\frac{\partial^2 z}{\partial y\partial x}+\frac{\partial z}{\partial y}+y\frac{\partial^2 z}{\partial y^2} = n\frac{\partial z}{\partial y}$$

$$\Rightarrow \quad \frac{x\partial^2 z}{\partial x\partial y}+\frac{y\partial^2 z}{\partial y^2} = (n-1)\frac{\partial z}{\partial y}$$ **Proved** (*ii*) ...(3)

Multiplying (2) by $x$, we have

$$x^2.\frac{\partial^2 z}{\partial x^2}+xy.\frac{\partial^2 z}{\partial x\partial y} = (n-1)\,x\frac{\partial z}{\partial x}$$ ...(4)

Multiplying (3) by $y$, we have

$$xy.\frac{\partial^2 z}{\partial y\partial x}+y^2.\frac{\partial^2 z}{\partial y^2} = (n-1)\,y\frac{\partial z}{\partial y}$$ ...(5)

Adding (4) and (5), we get

$$x^2.\frac{\partial^2 z}{\partial x^2}+2xy.\frac{\partial^2 z}{\partial x\partial y}+y^2.\frac{\partial^2 z}{\partial y^2} = (n-1)\left(x\frac{\partial z}{\partial x}+y\frac{\partial z}{\partial y}\right)$$

$$= (n-1)\,n\,z$$ [From (1)]

$$= n(n-1)\,z$$ **Proved** (*iii*)

**Example 26.** *If $f(x, y)$ and $\phi(x, y)$ are homogeneous functions of $x$, $y$ of degree $p$ and $q$ respectively and $u = f(x, y) + \phi(x, y)$, show that*

$$f(x, y) = \frac{1}{P(P-q)}\left[x^2\frac{\partial^2 u}{\partial x^2}+2xy\frac{\partial^2 u}{\partial x\partial y}+y^2\frac{\partial^2 u}{\partial y^2}\right]-\frac{q-1}{P(P-q)}\left[x\frac{\partial u}{\partial x}+y\frac{\partial u}{\partial y}\right]$$

(*A.M.I.E.T.E. Winter 2000*)

**Solution.** Since $f$ and $\phi$ are homogeneous functions of degree $p$ and $q$ respectively, we have

$$x\frac{\partial f}{\partial x}+y\frac{\partial f}{\partial y} = P.f$$ ...(1)

$$x\frac{\partial \phi}{\partial x}+y\frac{\partial \phi}{\partial y} = q.\phi$$ ...(2)

On adding (1) and (2), we get

$$x\left[\frac{\partial f}{\partial x}+\frac{\partial \phi}{\partial x}\right]+y\left[\frac{\partial f}{\partial y}+\frac{\partial \phi}{\partial y}\right] = P\,f+q\,\phi$$

*i.e.,* $$x\frac{\partial u}{\partial x}+y\frac{\partial u}{\partial y} = P\,f+q\,\phi$$ ...(3)

$$\left|\begin{array}{l} u = f(x, y)+\phi(x, y) \\ \dfrac{\partial u}{\partial x} = \dfrac{\partial f}{\partial x}+\dfrac{\partial \phi}{\partial x} \\ \dfrac{\partial u}{\partial y} = \dfrac{\partial f}{\partial y}+\dfrac{\partial \phi}{\partial y}\end{array}\right.$$

Also $$x^2\frac{\partial^2 f}{\partial x^2}+2xy\frac{\partial^2 f}{\partial x\partial y}+y^2\frac{\partial^2 f}{\partial y^2} = P\,(P-1)\,f$$ ...(4)

And $$x^2\frac{\partial^2\phi}{\partial x^2}+2xy\frac{\partial^2\phi}{\partial x\partial y}+y^2\frac{\partial^2\phi}{\partial y^2} = q(q-1)\phi \quad ...(5)$$

On adding (4) and (5), we obtain

$$x^2\left(\frac{\partial^2 f}{\partial x^2}+\frac{\partial^2\phi}{\partial x^2}\right)+2xy\left(\frac{\partial^2 f}{\partial x\partial y}+\frac{\partial^2\phi}{\partial x\partial y}\right)+y^2\left(\frac{\partial^2 f}{\partial y^2}+\frac{\partial^2\phi}{\partial y^2}\right) = P(P-1)f+q(q-1)\phi$$

$$x^2\frac{\partial^2 u}{\partial x^2}+2xy\frac{\partial^2 u}{\partial x\partial y}+y^2\frac{\partial^2 u}{\partial y^2} = P(p-1)f+q(q-1)\phi$$

$$\therefore u = f+\phi$$

$$\frac{\partial^2 u}{\partial x^2}=\frac{\partial^2 f}{\partial x^2}+\frac{\partial^2\phi}{\partial x^2}$$

$$\frac{\partial^2 u}{\partial y^2}=\frac{\partial^2 f}{\partial y^2}+\frac{\partial^2\phi}{\partial y^2}$$

Dividing by $P(P-q)$, we get

$$\Rightarrow \frac{1}{P(P-q)}\left[x^2\frac{\partial^2 u}{\partial x^2}+2xy\frac{\partial^2 u}{\partial x\partial y}+y^2\frac{\partial^2 u}{\partial y^2}\right] = \frac{1}{P(P-q)}[P(P-1)f+q(q-1)\phi]$$

Subtracting $\frac{q-1}{P(P-q)}\left(x\frac{\partial u}{\partial x}+y\frac{\partial u}{\partial y}\right)$ from both sides, we get

$$\Rightarrow \frac{1}{P(P-q)}\left[x^2\frac{\partial^2 u}{\partial x^2}+2xy\frac{\partial^2 u}{\partial x\partial y}+y^2\frac{\partial^2 u}{\partial y^2}\right]-\frac{(q-1)}{P(P-q)}\left[x\frac{\partial u}{\partial x}+y\frac{\partial u}{\partial y}\right]$$

$$= \frac{1}{P(P-q)}[P(P-1)f+q(q-1)\phi]-\frac{(q-1)}{P(P-q)}\left[x\frac{\partial u}{\partial x}+y\frac{\partial u}{\partial y}\right]$$

$$= \frac{1}{P(P-q)}[P(P-1)f+q(q-1)\phi-(q-1)[Pf+q\phi]] \quad \text{[From (3)]}$$

$$= \frac{1}{P(P-q)}[P^2-P-Pq+P)f+(q^2-q-q^2+q)\phi]$$

$$= \frac{1}{P(P-q)}[(P^2-Pq)f] = \frac{P(P-q)}{P(P-q)}f = f(x,y)$$ **Proved.**

**Example 27.** *If* $z = x^n f\left(\frac{y}{x}\right)+y^{-n}\phi\left(\frac{x}{y}\right)$ *then prove that*

$$x^2\frac{\partial^2 z}{\partial x^2}+2xy\frac{\partial^2 z}{\partial x\partial y}+y^2\frac{\partial^2 z}{\partial y^2}+x\frac{\partial z}{\partial x}+y\frac{\partial z}{\partial y} = n^2 z.$$ *(Nagpur University, Summer 2003)*

**Solution.** $$z = x^n f\left(\frac{y}{x}\right)+y^{-n}\phi\left(\frac{x}{y}\right)$$

$$z = u+v \quad ...(1)$$

where, $u = x^n f\left(\frac{y}{x}\right)$ and $v = y^{-n}\phi\left(\frac{x}{y}\right)$

Since $u$ is a homogeneous function of $x$, $y$ of degree $n$.

$$x\frac{\partial u}{\partial x}+y\frac{\partial u}{\partial y} = nu \quad ...(2)$$

and $$x^2\frac{\partial^2 u}{\partial x^2}+2xy\frac{\partial^2 u}{\partial x\partial y}+y^2\frac{\partial^2 u}{\partial y^2} = n(n-1)u \quad ...(3)$$

As $v$ is a homogeneous function of $x$, $y$ of degree $-n$.

$$x\frac{\partial v}{\partial x}+y\frac{\partial v}{\partial y} = -nv \quad ...(4)$$

and $$x^2\frac{\partial^2 v}{\partial x^2}+2xy\frac{\partial^2 v}{\partial x\partial y}+y^2\frac{\partial^2 v}{\partial y^2} = -n(-n-1)v = n(n+1)v \quad ...(5)$$

On adding (2) and (4), we get

$$x\left(\frac{\partial u}{\partial x}+\frac{\partial v}{\partial x}\right)+y\left(\frac{\partial u}{\partial y}+\frac{\partial v}{\partial y}\right) = nu - nv$$

$$\left| \begin{array}{l} z = u + v \\ \frac{\partial z}{\partial x} = \frac{\partial u}{\partial x} + \frac{\partial v}{\partial x} \\ \frac{\partial z}{\partial y} = \frac{\partial u}{\partial y} + \frac{\partial v}{\partial y} \end{array} \right.$$

$$x\frac{\partial z}{\partial x}+y\frac{\partial z}{\partial y} = nu - nv \quad ...(6) \quad \text{[From (1)]}$$

On adding (3) and (5), we get

$$x^2\left(\frac{\partial^2 u}{\partial x^2}+\frac{\partial^2 v}{\partial x^2}\right)+2xy\left(\frac{\partial^2 u}{\partial x\partial y}+\frac{\partial^2 v}{\partial x\partial y}\right)+y^2\left(\frac{\partial^2 u}{\partial y^2}+\frac{\partial^2 v}{\partial y^2}\right) = n(n-1)u + n(n+1)v$$

$$\left| \frac{\partial^2 z}{\partial x^2} = \frac{\partial^2 u}{\partial x^2} + \frac{\partial^2 v}{\partial x^2} \right.$$

$$x^2\frac{\partial^2 z}{\partial x^2}+2xy\frac{\partial^2 z}{\partial x\partial y}+y^2\frac{\partial^2 z}{\partial y^2} = n(n-1)u + n(n+1)v \quad ...(7) \quad \text{[From (1)]}$$

On adding (6) and (7), we have

$$x^2\frac{\partial^2 z}{\partial x^2}+2xy\frac{\partial^2 z}{\partial x\partial y}+y^2\frac{\partial^2 z}{\partial y^2}+x\frac{\partial z}{\partial x}+y\frac{\partial z}{\partial y} = n(n-1)u + n(n+1)v + nu - nv$$

$$= nu(n-1+1) + nv(n+1-1)$$

$$= n^2u + n^2v = n^2(u+v) = n^2 z$$

**Proved.**

**II. Deduction:** Prove that

$$x^2\frac{\partial^2 u}{\partial x^2}+2xy\frac{\partial^2 u}{\partial x\partial y}+y^2\frac{\partial^2 u}{\partial y^2} = g(u)\,[g'(u)-1]$$ (*Nagpur University, Winter 2003*)

where, $$g(u) = n\frac{f(u)}{f'(u)}$$

**Proof.** By Euler's deduction formula I

$$x\frac{\partial u}{\partial x}+y\frac{\partial u}{\partial y} = n.\frac{f(u)}{f'(u)} \qquad \left(\text{Given } n\frac{f(u)}{f'(u)} = g(u)\right)$$

$$x\frac{\partial u}{\partial x}+y\frac{\partial u}{\partial y} = g(u) \quad ...(1)$$

Differentiating (1) partially w.r.t. '$x$', we have

$$\left(x\frac{\partial^2 u}{\partial x^2}+\frac{\partial u}{\partial x}.1\right)+y\frac{\partial^2 u}{\partial x\partial y} = g'(u)\frac{\partial u}{\partial x}$$

$$\Rightarrow \qquad x\frac{\partial^2 u}{\partial x^2}+y.\frac{\partial^2 u}{\partial x\partial y} = [g'(u)-1]\frac{\partial u}{\partial x} \quad ...(2)$$

Similarly, on differentiating (1) partially w.r.t. '$y$', we have

$$y\frac{\partial^2 u}{\partial y^2}+x.\frac{\partial^2 u}{\partial y\partial x} = [g'(u)-1]\frac{\partial u}{\partial y} \quad ...(3)$$

Multiplying (2) by $x$, (3) by $y$ and adding, we get

$$x^2\frac{\partial^2 u}{\partial x^2}+2xy\frac{\partial^2 u}{\partial x\partial y}+y^2\frac{\partial^2 u}{\partial y^2} = [g'(u)-1]\left[x\frac{\partial u}{\partial x}+y\frac{\partial u}{\partial y}\right]$$

$$= [g'(u)-1]g(u) \quad \text{[From (1)]}$$

$$= g(u)\,[g'(u)-1]$$

**Proved.**

**Example 28.** *If* $u = \tan^{-1}\left(\frac{x^3+y^3}{x-y}\right)$, *prove that*

(*i*) $x.\frac{\partial u}{\partial x} + y.\frac{\partial u}{\partial y} = \sin 2u$ (*A.M.I.E., Winter 2001*)

(*ii*) $x^2.\frac{\partial^2 u}{\partial x^2} + 2xy\frac{\partial^2 u}{\partial x\,\partial y} + y^2\frac{\partial^2 u}{\partial y^2} = 2\cos 3u \sin u.$ (*M.U. 2009; Nagpur University, 2002*)

**Solution.** Here $u$ is not a homogeneous function. We however write

$$z = \tan u = \frac{x^3+y^3}{x-y} = \frac{x^3\left[1+\left(\frac{y}{x}\right)^3\right]}{x\left[1-\left(\frac{y}{x}\right)\right]} = x^2.\frac{1+\left(\frac{y}{x}\right)^3}{1-\left(\frac{y}{x}\right)} = x^2\varphi\left(\frac{y}{x}\right)$$

so that $z$ is a homogeneous function of $x$, $y$ of order 2.

**(*i*) By Euler's Theorem** [Here $f(u) = \tan u$]

$$\therefore \quad x\frac{\partial u}{\partial x} + y.\frac{\partial u}{\partial y} = \frac{n\,f(u)}{f'(u)} \quad \text{...(1)}$$

$$= \frac{2\tan u}{\sec^2 u} = \frac{2\sin u\cos^2 u}{\cos u} = 2\sin u\cos u = \sin 2u$$

**(*ii*) By deduction II**

$$x^2.\frac{\partial^2 u}{\partial x^2} + 2xy\frac{\partial^2 u}{\partial x\partial y} + y^2\frac{\partial^2 u}{\partial y^2} = g(u)[g'(u)-1]$$

Here $\sin 2u = g(u)$

$$\therefore \quad x^2.\frac{\partial^2 u}{\partial x^2} + 2xy\frac{\partial^2 u}{\partial x\partial y} + y^2\frac{\partial^2 u}{\partial y^2} = \sin 2u\,(2\cos 2u - 1) = 2\sin 2u\cos 2u - \sin 2u$$

$$= \sin 4u - \sin 2u = 2\cos 3u\sin u$$ **Proved.**

**Example 29.** *If* $u = \sin^{-1}\left[\frac{x+y}{\sqrt{x}+\sqrt{y}}\right]$

*Prove that* $x^2\frac{\partial^2 u}{\partial x^2} + 2xy\frac{\partial^2 u}{\partial x\partial y} + y^2\frac{\partial^2 u}{\partial y^2} = \frac{-\sin u\cos 2u}{4\cos^3 u}$.

**Solution.** We have, $u = \sin^{-1}\frac{x+y}{\sqrt{x}+\sqrt{y}}$

Let $z = \sin u = \frac{x+y}{\sqrt{x}+\sqrt{y}} = \frac{x\left[1+\frac{y}{x}\right]}{\sqrt{x}\left[1+\sqrt{\frac{y}{x}}\right]} = x^{1/2}\,\varphi(x)$

$z = f(u) = \sin u$

$z$ is a homogeneous function of degree $\frac{1}{2}$.

By Euler's deduction I

$$x\frac{\partial u}{\partial x} + y\frac{\partial u}{\partial y} = n\frac{f(u)}{f'(u)} \Rightarrow x\frac{\partial u}{\partial x} + y\frac{\partial u}{\partial y} = \frac{1}{2}\frac{\sin u}{\cos u}$$

$$x\frac{\partial u}{\partial x} + y\frac{\partial u}{\partial y} = \frac{1}{2}\tan u$$

Let $g(u) = \frac{1}{2}\tan u$

By Euler's deduction II

$$x^2\frac{\partial^2 u}{\partial x^2}+2xy\frac{\partial^2 u}{\partial x\partial y}+y^2\frac{\partial^2 u}{\partial y^2} = g(u)\,[g'(u)-1] = \frac{1}{2}\tan u\left(\frac{1}{2}\sec^2 u-1\right)$$

$$= \frac{1}{4}\frac{\sin u}{\cos u}\left(\frac{1}{\cos^2 u}-2\right) = \frac{1}{4}\frac{\sin u}{\cos^3 u}(1-2\cos^2 u) = \frac{-\sin u\cos 2u}{4\cos^3 u}$$ **Proved.**

## EXERCISE 1.4

1. Verify Euler's theorem in case
   (i) $f(x, y) = ax^2 + 2hxy + by^2$ (ii) $u = (\sqrt{x}+\sqrt{y})(x^n+y^n)$
2. If $v = \frac{x^3y^3}{x^3+y^3}$, show that $x.\frac{\partial v}{\partial x}+y.\frac{\partial v}{\partial y}=3v$.
3. If $u = \log\frac{x^3+y^3}{x^2+y^2}$, prove that $x.\frac{\partial u}{\partial x}+y.\frac{\partial u}{\partial y}=1$.
4. If $z = (x^2+y^2)/\sqrt{(x+y)}$, prove that $x\frac{\partial z}{\partial x}+y.\frac{\partial z}{\partial x}=\frac{3}{2}z$.
5. If $f(x, y) = x^4y^2\sin^{-1}\frac{y}{x}$, then find the value of $x\frac{\partial f}{\partial x}+y\frac{\partial f}{\partial y}$.
   (*A.M.I.E.T.E., Winter 2001*) **Ans.** $6f(x, y)$
6. If $u = \sin^{-1}\frac{\sqrt{x}-\sqrt{y}}{\sqrt{x}+\sqrt{y}}$, show that $\frac{\partial u}{\partial x}=-\frac{y}{x}\frac{\partial u}{\partial y}$.
7. If $u = \sec^{-1}\left(\frac{x^3-y^3}{x+y}\right)$, show that $x.\frac{\partial u}{\partial x}+y.\frac{\partial u}{\partial y}=2\cot u$, then evaluate
   $x^2\frac{\partial^2 u}{\partial x^2}+2xy\frac{\partial^2 u}{\partial x\,\partial y}+y^2\frac{\partial^2 u}{\partial y^2}$ (*A.M.I.E.T.E., Winter 2001*) **Ans.** $-2\cot u\,(2\,\mathrm{cosec}^2 u+1)$.
8. If $x = e^u\tan v$, $y = e^u\sec v$, find the value of
   $\left(x.\frac{\partial u}{\partial x}+y.\frac{\partial u}{\partial y}\right).\left(x.\frac{\partial v}{\partial x}+y.\frac{\partial v}{\partial y}\right)$. (*A.M.I.E., Summer 2001*) **Ans.** 0
   [**Hint:** Eliminate $u$ and apply formula I. Again eliminate $v$ and apply the formula]
9. If $u = \sin^{-1}\left[\frac{x^{1/4}+y^{1/4}}{x^{1/6}+y^{1/6}}\right]$, prove that
   $$x^2\frac{\partial^2 u}{\partial x^2}+2xy\frac{\partial^2 u}{\partial x\,\partial y}+y^2\frac{\partial^2 u}{\partial y^2} = \frac{1}{144}\tan u\,[\tan^2 u-11].$$
10. Find the value of $x^2\frac{\partial^2 u}{\partial x^2}+2xy\frac{\partial^2 u}{\partial x\,\partial y}+y^2\frac{\partial^2 u}{\partial y^2}$ if $u = \sin^{-1}(x^3+y^3)^{2/5}$.
    **Ans.** $\frac{5}{6}\tan u\left(\frac{6}{5}\sec^2 u-1\right)$
11. If $u = \tan^{-1}\left(\frac{y^2}{x}\right)$, find
    (i) $x.\frac{\partial u}{\partial x}+y.\frac{\partial u}{\partial y}$, and (ii) $x^2.\frac{\partial^2 u}{\partial x^2}+2xy.\frac{\partial^2 u}{\partial x\partial y}+y^2.\frac{\partial^2 u}{\partial y^2}$ **Ans.** (i) $\frac{xy^2}{x^2+y^4}$ (ii) $\frac{-2xy^6}{(x^2+y^4)^2}$
12. If $u = \tan^{-1}\frac{\sqrt{x^3+y^3}}{\sqrt{x}+\sqrt{y}}$, find the value of $x^2\frac{\partial^2 u}{\partial x^2}+2xy\frac{\partial^2 u}{\partial x\,\partial y}+y^2\frac{\partial^2 u}{\partial y^2}$ **Ans.** $-2\sin^3 u\cos u$

13. If $u = f\left(\frac{y}{x}\right) + \sqrt{x^2 + y^2}$, find the value of $x^2 \frac{\partial^2 u}{\partial x^2} + 2xy \frac{\partial^2 u}{\partial x \, \partial y} + y^2 \frac{\partial^2 u}{\partial y^2}$. **Ans.** 0

14. If $z = xy/(x + y)$, find the value of $x^2 \frac{\partial^2 z}{\partial x^2} + 2xy \frac{\partial^2 z}{\partial x \, \partial y} + y^2 \frac{\partial^2 z}{\partial y^2}$. **Ans.** 0

15. Verify Euler's theorem on homogeneous function when $f(x, y, z) = 3x^2yz + 5xy^2z + 4z^4$.

16. If $u = x\phi\left(\frac{Y}{X}\right) + \psi\left(\frac{Y}{X}\right)$, prove by Euler's theorem on homogeneous function that

$$X^2 \frac{\partial^2 u}{\partial x^2} + 2XY \frac{\partial^2 u}{\partial x \, \partial y} + Y^2 \frac{\partial^2 u}{\partial y^2} = 0.$$

17. Given $F(u) = V(x, y, z)$ where $V$ is a homogeneous function of $x, y, z$ of degree $n$, prove that

$$x \frac{\partial u}{\partial x} + y \frac{\partial u}{\partial y} + z \frac{\partial u}{\partial z} = n \frac{F(u)}{F'(u)}$$

18. State and prove Euler's theorem, and verify for $u = \frac{x}{y} + \frac{y}{z} + \frac{z}{x}$ *(A.M.I.E., Summer 2000)*

19. If $u = \frac{x^2y^2z^2}{x^2 + y^2 + z^2} + \cos \frac{xy + yz}{x^2 + y^2 + z^2}$, show that $x \frac{\partial u}{\partial x} + y \frac{\partial u}{\partial y} + z \frac{\partial u}{\partial z} = \frac{4x^2y^2z^2}{x^2 + y^2 + z^2}$

## 1.12 TOTAL DIFFERENTIAL

In partial differentiation of a function of two or more variables, only one variable varies. But in total differentiation, increments are given in all the variables.

## 1.13 TOTAL DIFFERENTIAL CO-EFFICIENT

Let $z = f(x, y)$ ...(1)

If $\delta x$, $\delta y$ be the increments in $x$ and $y$ respectively, let $\delta z$ be the corresponding increment in $z$.

Then $z + \delta z = f(x + \delta x, y + \delta y)$ ...(2)

Subtracting (1) from (2), we have

$$\delta z = f(x + \delta x, y + \delta y) - f(x, y) \quad ...(3)$$

Adding and subtracting $f(x, y + \delta y)$ on R.H.S. of (3), we have

$$\delta z = f(x + \delta x, y + \delta y) - f(x, y + \delta y) + f(x, y + \delta y) - f(x, y)$$

$$\delta z = \frac{f(x + \delta x, y + \delta y) - f(x, y + \delta y)}{\delta x} \delta x + \frac{f(x, y + \delta y) - f(x, y)}{\delta y} \delta y$$

On taking limit when $\delta x \to 0$ and $\delta y \to 0$

$$dz = \frac{\partial f}{\partial x} dx + \frac{\partial f}{\partial y} dy \quad ...(4)$$ **[Remember]**

$dz$ is called as the total differential of $z$.

## 1.14 CHANGE OF TWO INDEPENDENT VARIABLES $x$ AND $y$ BY ANY OTHER VARIABLE $t$.

Differentiation of composite function

If $z = f(x, y)$

Where $x = \varphi(t)$

$y = \psi(t)$

Here $z$ is composite function of $t$.

Dividing (4) by $dt$, we have

$$\frac{dz}{dt} = \frac{\partial z}{\partial x}\frac{dx}{dt} + \frac{\partial z}{\partial y}\frac{dy}{dt} \qquad ...(5) \quad \textbf{[Remember]}$$

Then $\frac{dz}{dt}$ is called the total differential co-efficient of $z$.

## 1.15 CHANGE IN THE INDEPENDENT VARIABLES *x* AND *y* BY OTHER TWO VARIABLES *u* AND *v*.

Let $z = f(x, y)$

where $x = \varphi(u, v)$

$y = \psi(u, v)$

Then from (5), we obtain

$$\frac{\partial z}{\partial u} = \frac{\partial f}{\partial x}\cdot\frac{\partial x}{\partial u} + \frac{\partial f}{\partial y}\cdot\frac{\partial y}{\partial u} \qquad ...(6)$$

and
$$\frac{\partial z}{\partial v} = \frac{\partial f}{\partial x}\cdot\frac{\partial x}{\partial v} + \frac{\partial f}{\partial y}\cdot\frac{\partial y}{\partial v} \qquad ...(7)$$

**Example 30.** *If $u = x^3 + y^3$ where, $x = a \cos t$, $y = b \sin t$, find $\frac{du}{dt}$ and verify the result.*

**Solution.** We have, $u = x^3 + y^3$

$x = a \cos t$

$y = b \sin t$

$$\frac{dz}{dt} = \frac{\partial u}{\partial x}\frac{dx}{dt} + \frac{\partial u}{\partial y}\frac{dy}{dt}$$

$$= (3x^2)(-a \sin t) + (3y^2)(b \cos t)$$

$$= -3a^3 \cos^2 t \sin t + 3b^3 \sin^2 t \cos t \qquad ...(1)$$

**Verification.** $u = x^3 + y^3$

$$= a^3 \cos^3 t + b^3 \sin^3 t$$

$$\frac{du}{dt} = -3a^3 \cos^2 t \sin t + 3b^3 \sin^2 t \cos t \qquad ...(2)$$

Results (1) and (2) are the same. **Verified.**

**Example 31.** *If $z = f(x, y)$ where $x = e^u \cos v$ and $y = e^u \sin v$, show that*

(*i*) $y\frac{\partial z}{\partial u} + x\frac{\partial z}{\partial v} = e^{2u}\frac{\partial z}{\partial y}$. *(M.U. 2009; Nagpur Univesity 2002)*

(*ii*) $\left(\frac{\partial z}{\partial x}\right)^2 + \left(\frac{\partial z}{\partial y}\right)^2 = e^{-2u}\left[\left(\frac{\partial z}{\partial u}\right)^2 + \left(\frac{\partial z}{\partial v}\right)^2\right]$ *(M.U. 2009)*

**Solution.** (*i*) We have,

$$x = e^u \cos v, \qquad y = e^u \sin v$$

$$\frac{\partial z}{\partial u} = \frac{\partial z}{\partial x}\frac{\partial x}{\partial u} + \frac{\partial z}{\partial y}\frac{\partial y}{\partial u} = \frac{\partial z}{\partial x}e^u \cos v + \frac{\partial z}{\partial y}e^u \sin v = x\frac{\partial z}{\partial x} + y\frac{\partial z}{\partial y}$$

$$y\frac{\partial z}{\partial u} = xy\frac{\partial z}{\partial x} + y^2\frac{\partial z}{\partial y} \qquad ...(1)$$

And
$$\frac{\partial z}{\partial v} = \frac{\partial z}{\partial x}\frac{\partial x}{\partial v} + \frac{\partial z}{\partial y}\frac{\partial y}{\partial v}$$

$$= \frac{\partial z}{\partial x}(-e^u \sin v) + \frac{\partial z}{\partial y}(e^u \cos v) = -y\frac{\partial z}{\partial x} + x\frac{\partial z}{\partial y}$$

$$x\frac{\partial z}{\partial v} = -xy\frac{\partial z}{\partial x} + x^2\frac{\partial z}{\partial y} \qquad ...(2)$$

On adding (1) and (2), we get

$$y\frac{\partial z}{\partial u} + x\frac{\partial z}{\partial v} = (x^2+y^2)\frac{\partial z}{\partial y} = (e^{2u}\cos^2 v + e^{2u}\sin^2 v)\frac{\partial z}{\partial y}$$

$$= e^{2u}(\cos^2 v + \sin^2 v)\frac{\partial z}{\partial y} = e^{2u}\frac{\partial z}{\partial y}$$ **Proved.**

(*ii*)
$$\frac{\partial z}{\partial u} = \frac{\partial z}{\partial x}\frac{\partial x}{\partial u} + \frac{\partial z}{\partial y}\frac{\partial y}{\partial u}$$

$$= \frac{\partial z}{\partial x}(e^u \cos v) + \frac{\partial z}{\partial y}e^u \sin v$$

$$e^{-u}\frac{\partial z}{\partial u} = \frac{\partial z}{\partial x}\cos v + \frac{\partial z}{\partial y}\sin v$$

$x = e^u \cos v$

$\Rightarrow \frac{\partial x}{\partial u} = e^u \cos v$ and $\frac{\partial x}{\partial v} = -e^u \sin v$

$y = e^u \sin v$

$\Rightarrow \frac{\partial y}{\partial u} = e^u \sin v$ and $\frac{\partial y}{\partial v} = e^u \cos v$

On squaring, we get

$$e^{-2u}\left(\frac{\partial z}{\partial u}\right)^2 = \left(\frac{\partial z}{\partial x}\right)^2\cos^2 v + \left(\frac{\partial z}{\partial y}\right)^2\sin^2 v + 2\left(\frac{\partial z}{\partial x}\right)\left(\frac{\partial z}{\partial y}\right)\sin v\cos v \qquad ...(3)$$

Again
$$\frac{\partial z}{\partial v} = \frac{\partial z}{\partial x}\frac{\partial x}{\partial v} + \frac{\partial z}{\partial y}\frac{\partial y}{\partial v}$$

$$= \frac{\partial z}{\partial x}(-e^u \sin v) + \frac{\partial z}{\partial y}(e^u \cos v)$$

$$e^{-u}\left(\frac{\partial z}{\partial v}\right) = -\frac{\partial z}{\partial x}\sin v + \frac{\partial z}{\partial y}\cos v$$

On squaring, we get

$$e^{-2u}\left(\frac{\partial z}{\partial v}\right)^2 = \left(\frac{\partial z}{\partial x}\right)^2\sin^2 v + \left(\frac{\partial z}{\partial y}\right)^2\cos^2 v - 2\left(\frac{\partial z}{\partial x}\right)\left(\frac{\partial z}{\partial y}\right)\sin v\cos v \qquad ...(4)$$

On adding (3) and (4), we get

$$e^{-2u}\left[\left(\frac{\partial z}{\partial u}\right)^2 + \left(\frac{\partial z}{\partial v}\right)^2\right] = \left(\frac{\partial z}{\partial x}\right)^2(\sin^2 v + \cos^2 v) + \left(\frac{\partial z}{\partial y}\right)^2(\sin^2 v + \cos^2 v)$$

$$= \left(\frac{\partial z}{\partial x}\right)^2 + \left(\frac{\partial z}{\partial y}\right)^2$$ **Proved.**

**Example 32.** *If* $u = u\,(y - z,\ \ z - x,\ \ x - y)$, *prove that* $\frac{\partial u}{\partial x} + \frac{\partial u}{\partial y} + \frac{\partial u}{\partial z} = 0$

*(Nagpur University, Winter 2002, U.P., I Sem., Winter 2002, A.M.I.E winter 2001)*

**Solution.** Let $r = y - z, \quad s = z - x, \quad t = x - y$

so that $u = u\,(r, s, t)$

$$\frac{\partial u}{\partial x} = \frac{\partial u}{\partial r}\frac{\partial r}{\partial x} + \frac{\partial u}{\partial s}\frac{\partial s}{\partial x} + \frac{\partial u}{\partial t}\frac{\partial t}{\partial x}$$

$$= \frac{\partial u}{\partial r}(0) + \frac{\partial u}{\partial s}(-1) + \frac{\partial u}{\partial t}(1) = -\frac{\partial u}{\partial s} + \frac{\partial u}{\partial t} \quad \text{...(1)}$$

$$\frac{\partial u}{\partial y} = \frac{\partial u}{\partial r}\frac{\partial r}{\partial y} + \frac{\partial u}{\partial s}\frac{\partial s}{\partial y} + \frac{\partial u}{\partial t}\frac{\partial t}{\partial y}$$

$$= \frac{\partial u}{\partial r}(1) + \frac{\partial u}{\partial s}(0) + \frac{\partial u}{\partial t}(-1) = \frac{\partial u}{\partial r} - \frac{\partial u}{\partial t} \quad \text{...(2)}$$

$$\frac{\partial u}{\partial z} = \frac{\partial u}{\partial r}\frac{\partial r}{\partial z} + \frac{\partial u}{\partial s}\frac{\partial s}{\partial z} + \frac{\partial u}{\partial t}\frac{\partial t}{\partial z}$$

$$= \frac{\partial u}{\partial r}(-1) + \frac{\partial u}{\partial s}(1) + \frac{\partial u}{\partial t}(0) = -\frac{\partial u}{\partial r} + \frac{\partial u}{\partial s} \quad \text{...(3)}$$

Adding (1), (2) and (3), we get $\frac{\partial u}{\partial x} + \frac{\partial u}{\partial y} + \frac{\partial u}{\partial z} = 0$ **Proved.**

**Example 33.** *If* $u = u\left(\frac{y-x}{xy}, \frac{z-x}{xz}\right)$*, show that* $x^2\frac{\partial u}{\partial x} + y^2\frac{\partial u}{\partial y} + z^2\frac{\partial u}{\partial z} = 0$

*(U. P. I. Sem., Dec. 2004, Nagpur University, Summer 2000)*

**Solution.** Here, we have $u = u\left(\frac{y-x}{xy}, \frac{z-x}{zx}\right) = u\,(r, s)$

where $r = \frac{y-x}{xy}$, and $s = \frac{z-x}{zx}$

$r = \frac{1}{x} - \frac{1}{y}$ and $s = \frac{1}{x} - \frac{1}{z}$

$\frac{\partial r}{\partial x} = -\frac{1}{x^2}$ and $\frac{\partial s}{\partial x} = -\frac{1}{x^2}$

$\frac{\partial r}{\partial y} = \frac{1}{y^2}$ and $\frac{\partial s}{\partial z} = \frac{1}{z^2}$

$\frac{\partial r}{\partial z} = 0$ and $\frac{\partial s}{\partial y} = 0$

We know that,

$$\frac{\partial u}{\partial x} = \frac{\partial u}{\partial r}\frac{\partial r}{\partial x} + \frac{\partial u}{\partial s}\frac{\partial s}{\partial x}$$

$$= \frac{\partial u}{\partial r}\left(-\frac{1}{x^2}\right) + \frac{\partial u}{\partial s}\left(-\frac{1}{x^2}\right) = -\frac{1}{x^2}\frac{\partial u}{\partial r} - \frac{1}{x^2}\frac{\partial u}{\partial s}$$

$$\Rightarrow \quad x^2\frac{\partial u}{\partial x} = -\frac{\partial u}{\partial r} - \frac{\partial u}{\partial s} \quad \text{...(1)}$$

$$\frac{\partial u}{\partial y} = \frac{\partial u}{\partial r}\frac{\partial r}{\partial y} + \frac{\partial u}{\partial s}\frac{\partial s}{\partial y} = \frac{\partial u}{\partial r}\frac{1}{y^2} + \frac{\partial u}{\partial s} \times 0 = \frac{1}{y^2}\frac{\partial u}{\partial r}$$

$$\Rightarrow \quad y^2\frac{\partial u}{\partial y} = \frac{\partial u}{\partial r} \quad \text{...(2)}$$

$$\frac{\partial u}{\partial z} = \frac{\partial u}{\partial r}\frac{\partial r}{\partial z} + \frac{\partial u}{\partial s}\frac{\partial s}{\partial z} = \frac{\partial u}{\partial r} \times 0 + \frac{\partial u}{\partial s} \times \frac{1}{z^2} = \frac{1}{z^2}\frac{\partial u}{\partial s}$$

$$\Rightarrow \qquad z^2 \frac{\partial u}{\partial z} = \frac{\partial u}{\partial s} \qquad ...(3)$$

On adding (1), (2) and (3), we get

$$x^2 \frac{\partial u}{\partial x} + y^2 \frac{\partial u}{\partial y} + z^2 \frac{\partial u}{\partial z} = 0$$

**Proved.**

**Example 34.** *If* φ *(cx – az, cy – bz) = 0 show that ap + bq = c :*

$$\text{where } p = \frac{\partial z}{\partial x} \text{ and } q \equiv \frac{\partial z}{\partial y}$$

**Solution.** Here, we have [x and y are independent but z is dependent on x and y]

$$\varphi\,(cx - az,\ cy - bz) = 0$$

$$\varphi\,(r, s) = 0$$

where $r = cx - az, \qquad s = cy - bz$

$$\frac{\partial r}{\partial x} = c - a\frac{\partial z}{\partial x}, \qquad \frac{\partial r}{\partial y} = -a\frac{\partial z}{\partial y}$$

$$\frac{\partial s}{\partial x} = -b\frac{\partial z}{\partial x}, \qquad \frac{\partial s}{\partial y} = c - b\frac{\partial z}{\partial y}$$

We know that,

$$\frac{\partial \varphi}{\partial r} = \frac{\partial \varphi}{\partial r}\frac{\partial r}{\partial x} + \frac{\partial \varphi}{\partial s}\frac{\partial s}{\partial x}$$

$$0 = \frac{\partial \varphi}{\partial r}\left(c - a\frac{\partial z}{\partial x}\right) + \frac{\partial \varphi}{\partial s}\left(-b\frac{\partial z}{\partial x}\right)$$

$$\Rightarrow \qquad 0 = c\frac{\partial \varphi}{\partial r} + \frac{\partial z}{\partial x}\left(-a\frac{\partial \varphi}{\partial r} - b\frac{\partial \varphi}{\partial s}\right)$$

$$c\frac{\partial \phi}{\partial r} = \frac{\partial z}{\partial x}\left(a\frac{\partial \phi}{\partial r} + b\frac{\partial \phi}{\partial s}\right) \Rightarrow a\frac{\partial z}{\partial x} = \frac{ac\dfrac{\partial \varphi}{\partial r}}{a\dfrac{\partial \varphi}{\partial r} + b\dfrac{\partial \varphi}{\partial s}} \qquad ...(1)$$

Again

$$\frac{\partial \varphi}{\partial y} = \frac{\partial \varphi}{\partial r}\frac{\partial r}{\partial y} + \frac{\partial \varphi}{\partial s}\frac{\partial s}{\partial y}$$

$$0 = \frac{\partial \varphi}{\partial r}\left(-a\frac{\partial z}{\partial y}\right) + \frac{\partial \varphi}{\partial s}\left(c - b\frac{\partial z}{\partial y}\right)$$

$$0 = c\frac{\partial \varphi}{\partial s} - \frac{\partial z}{\partial y}\left(a\frac{\partial \varphi}{\partial r} + b\frac{\partial \varphi}{\partial s}\right) \Rightarrow c\frac{\partial \phi}{\partial s} = \frac{\partial z}{\partial y}\left(a\frac{\partial \phi}{\partial r} + b\frac{\partial \phi}{\partial s}\right)$$

$$\Rightarrow \qquad b\frac{\partial z}{\partial y} = \frac{bc\dfrac{\partial \varphi}{\partial s}}{a\dfrac{\partial \varphi}{\partial r} + b\dfrac{\partial \varphi}{\partial s}} \qquad ...(2)$$

Adding (1) and (2), we get

$$a\frac{\partial z}{\partial x} + b\frac{\partial z}{\partial y} = \frac{ac\dfrac{\partial \varphi}{\partial r} + bc\dfrac{\partial \phi}{\partial s}}{a\dfrac{\partial \varphi}{\partial r} + b\dfrac{\partial \varphi}{\partial s}}$$

$$\Rightarrow \qquad a\frac{\partial z}{\partial x} + b\frac{\partial z}{\partial y} = c \qquad \Rightarrow \quad ap + bq = c$$

**Proved.**

## 1.16 CHANGE IN BOTH THE INDEPENDENT AND DEPENDENT VARIABLES, (POLAR COORDINATES)

**Example 35.** *If $w = f(x,y)$, $x = r\cos\theta$, $y = r\sin\theta$, show that*

$$\left(\frac{\partial w}{\partial r}\right)^2 + \frac{1}{r^2}\left(\frac{\partial w}{\partial \theta}\right)^2 = \left(\frac{\partial f}{\partial x}\right)^2 + \left(\frac{\partial f}{\partial y}\right)^2$$

**Solution.** Here, $x = r\cos\theta$, $y = r\sin\theta$

$$\frac{\partial x}{\partial r} = \cos\theta \qquad \frac{\partial y}{\partial r} = \sin\theta$$

$$\frac{\partial x}{\partial \theta} = -r\sin\theta \qquad \frac{\partial y}{\partial \theta} = r\cos\theta$$

Now,

$$\frac{\partial w}{\partial r} = \frac{\partial f}{\partial x}\cdot\frac{\partial x}{\partial r} + \frac{\partial f}{\partial y}\cdot\frac{\partial y}{\partial r}$$

$$\frac{\partial w}{\partial r} = \frac{\partial f}{\partial x}\cdot(\cos\theta) + \frac{\partial f}{\partial y}\cdot(\sin\theta) \qquad \text{...(1)}$$

$$\frac{\partial w}{\partial \theta} = \frac{\partial f}{\partial x}\cdot\frac{\partial x}{\partial \theta} + \frac{\partial f}{\partial y}\cdot\frac{\partial y}{\partial \theta} = \frac{\partial f}{\partial x}\cdot(-r\sin\theta) + \frac{\partial f}{\partial y}\cdot(r\cos\theta)$$

$$\Rightarrow \quad \frac{1}{r}\frac{\partial w}{\partial \theta} = -\frac{\partial f}{\partial x}\sin\theta + \frac{\partial f}{\partial y}\cos\theta \qquad \text{...(2)}$$

Squaring (1) and (2) and adding, we obtain

$$\left(\frac{\partial w}{\partial r}\right)^2 + \frac{1}{r^2}\left(\frac{\partial w}{\partial \theta}\right)^2 = \left(\frac{\partial f}{\partial x}\right)^2 + \left(\frac{\partial f}{\partial y}\right)^2$$

**Proved.**

**Example 36.** *Transform the equation* $\frac{\partial^2 u}{\partial x^2} + \frac{\partial^2 u}{\partial y^2} = 0$ *into polar co-ordinates.*

**Solution.** We have, $x = r\cos\theta$, $y = r\sin\theta$

$$r^2 = x^2 + y^2, \qquad \theta = \tan^{-1}\frac{y}{x}$$

$$\frac{\partial u}{\partial x} = \frac{\partial u}{\partial r}\frac{\partial r}{\partial x} + \frac{\partial u}{\partial \theta}\frac{\partial \theta}{\partial x} = \frac{\partial u}{\partial r}\frac{x}{r} + \frac{\partial u}{\partial \theta}\frac{-y}{x^2+y^2} = \frac{\partial u}{\partial r}\cos\theta - \frac{\partial u}{\partial \theta}\frac{\sin\theta}{r}$$

$$\frac{\partial^2 u}{\partial x^2} = \frac{\partial}{\partial x}\left(\frac{\partial u}{\partial x}\right)$$

$$= \left(\cos\theta\frac{\partial}{\partial r} - \frac{\sin\theta}{r}\frac{\partial}{\partial \theta}\right)\left(\cos\theta\frac{\partial u}{\partial r} - \frac{\sin\theta}{r}\frac{\partial u}{\partial \theta}\right)$$

$$= \cos\theta\frac{\partial}{\partial r}\left(\cos\theta\frac{\partial u}{\partial r} - \frac{\sin\theta}{r}\frac{\partial u}{\partial \theta}\right) - \frac{\sin\theta}{r}\frac{\partial}{\partial \theta}\left(\cos\theta\frac{\partial u}{\partial r} - \frac{\sin\theta}{r}\frac{\partial u}{\partial \theta}\right)$$

$$= \cos\theta\left(\cos\theta\frac{\partial^2 u}{\partial r^2} + \frac{\sin\theta}{r^2}\frac{\partial u}{\partial \theta} - \frac{\sin\theta}{r}\frac{\partial^2 u}{\partial r\partial \theta}\right)$$

$$-\frac{\sin\theta}{r}\left(-\sin\theta\frac{\partial u}{\partial r} + \cos\theta\frac{\partial^2 u}{\partial \theta\partial r} - \frac{\cos\theta}{r}\frac{\partial u}{\partial \theta} - \frac{\sin\theta}{r}\frac{\partial^2 u}{\partial \theta^2}\right)$$

$$= \cos^2\theta\frac{\partial^2 u}{\partial r^2} + \frac{\sin\theta\cos\theta}{r^2}\frac{\partial u}{\partial \theta} - \frac{\sin\theta\cos\theta}{r}\frac{\partial^2 u}{\partial r\partial \theta}$$

$$+\frac{\sin^2\theta}{r}\frac{\partial u}{\partial r} - \frac{\sin\theta\cos\theta}{r}\frac{\partial^2 u}{\partial \theta\partial r} + \frac{\sin\theta\cos\theta}{r^2}\frac{\partial u}{\partial \theta} + \frac{\sin^2\theta}{r^2}\frac{\partial^2 u}{\partial \theta^2}$$

$$= \cos^2\theta \frac{\partial^2 u}{\partial r^2} + \frac{2\sin\theta\cos\theta}{r^2}\frac{\partial u}{\partial \theta} - \frac{2\sin\theta\cos\theta}{r}\frac{\partial^2 u}{\partial r\,\partial\theta} + \frac{\sin^2\theta}{r}\frac{\partial u}{\partial r} + \frac{\sin^2\theta}{r^2}\frac{\partial^2 u}{\partial\theta^2} \quad ..(1)$$

$$\frac{\partial u}{\partial y} = \frac{\partial u}{\partial r}\frac{\partial r}{\partial y} + \frac{\partial u}{\partial\theta}\frac{\partial\theta}{\partial y} = \frac{\partial u}{\partial r}\frac{y}{r} + \frac{\partial u}{\partial\theta}\frac{x}{x^2+y^2} = \frac{\partial u}{\partial r}\sin\theta + \frac{\partial u}{\partial\theta}\frac{\cos\theta}{r}$$

$$\frac{\partial^2 u}{\partial y^2} = \frac{\partial}{\partial y}\left(\frac{\partial u}{\partial y}\right) = \left(\sin\theta\frac{\partial}{\partial r} + \frac{\cos\theta}{r}\frac{\partial}{\partial\theta}\right)\left(\sin\theta\frac{\partial u}{\partial r} + \frac{\cos\theta}{r}\frac{\partial u}{\partial\theta}\right)$$

$$= \sin\theta\frac{\partial}{\partial r}\left(\sin\theta\frac{\partial u}{\partial r} + \frac{\cos\theta}{r}\frac{\partial u}{\partial\theta}\right) + \frac{\cos\theta}{r}\frac{\partial}{\partial\theta}\left(\sin\theta\frac{\partial u}{\partial r} + \frac{\cos\theta}{r}\frac{\partial u}{\partial\theta}\right)$$

$$= \sin\theta\left[\sin\theta\frac{\partial^2 u}{\partial r^2} - \frac{\cos\theta}{r^2}\frac{\partial u}{\partial\theta} + \frac{\cos\theta}{r}\frac{\partial^2 u}{\partial r\,\partial\theta}\right]$$

$$+ \frac{\cos\theta}{r}\left[\cos\theta\frac{\partial u}{\partial r} + \sin\theta\frac{\partial^2 u}{\partial\theta\,\partial r} - \frac{\sin\theta}{r}\frac{\partial u}{\partial\theta} + \frac{\cos\theta}{r}\frac{\partial^2 u}{\partial\theta^2}\right]$$

$$= \sin^2\theta\frac{\partial^2 u}{\partial r^2} - \frac{\sin\theta\cos\theta}{r^2}\frac{\partial u}{\partial\theta} + \frac{\sin\theta\cos\theta}{r}\frac{\partial^2 u}{\partial r\,\partial\theta}$$

$$+ \frac{\cos^2\theta}{r}\frac{\partial u}{\partial r} + \frac{\sin\theta\cos\theta}{r}\frac{\partial^2 u}{\partial\theta\,\partial r} - \frac{\sin\theta\cos\theta}{r^2}\frac{\partial u}{\partial\theta} + \frac{\cos^2\theta}{r^2}\frac{\partial^2 u}{\partial\theta^2}$$

$$= \sin^2\theta\frac{\partial^2 u}{\partial r^2} - \frac{2\sin\theta\cos\theta}{r^2}\frac{\partial u}{\partial\theta} + \frac{2\sin\theta\cos\theta}{r}\frac{\partial^2 u}{\partial r\,\partial\theta}$$

$$+ \frac{\cos^2\theta}{r}\frac{\partial u}{\partial r} + \frac{\cos^2\theta}{r^2}\frac{\partial^2 u}{\partial\theta^2} \quad ...(2)$$

Adding (1) and (2), we get

$$\frac{\partial^2 u}{\partial x^2} + \frac{\partial^2 u}{\partial y^2} = (\sin^2\theta + \cos^2\theta)\frac{\partial^2 u}{\partial r^2} + (\sin^2\theta + \cos^2\theta)\frac{1}{r}\frac{\partial u}{\partial r} + (\sin^2\theta + \cos^2\theta)\frac{1}{r^2}\frac{\partial^2 u}{\partial\theta^2}$$

$$= \frac{\partial^2 u}{\partial r^2} + \frac{1}{r}\frac{\partial u}{\partial r} + \frac{1}{r^2}\frac{\partial^2 u}{\partial\theta^2}$$

**Ans.**

**Example 37.** *If* $u = f(r)$ *and* $x = r\cos\theta$, $y = r\sin\theta$, *prove that*

$$\frac{\partial^2 u}{\partial x^2} + \frac{\partial^2 u}{\partial y^2} = f''(r) + \frac{1}{r}f'(r).$$

(*Nagpur University, Winter 2004*)

(*A.M.I.E.T.E., Winter 2003, U.P. I Semester, Winter 2005, 2000*)

**Solution.** Here, we have

$x = r\cos\theta$

$y = r\sin\theta$

$r^2 = x^2 + y^2$ so that $\frac{\partial r}{\partial x} = \frac{x}{r}$

$u = f(r)$

$$\frac{\partial u}{\partial x} = \frac{df}{dr}\cdot\frac{\partial r}{\partial x} \Rightarrow \frac{\partial u}{\partial x} = \frac{df}{dr}\cdot\frac{x}{r}$$

Differentiating again *w.r.t. x*, we get

$$\frac{\partial^2 u}{\partial x^2} = \left(\frac{d^2 f}{dr^2}\cdot\frac{\partial r}{\partial x}\right)\cdot\frac{x}{r} + \frac{df}{dr}\cdot\left[\frac{r.1 - x\frac{\partial r}{\partial x}}{r^2}\right] = \left(\frac{d^2 f}{dr^2}\frac{x}{r}\right)\cdot\frac{x}{r} + \frac{df}{dr}\cdot\left[\frac{r.1 - x.\frac{x}{r}}{r^2}\right]$$

$$= \frac{d^2 f}{dr^2}\frac{x^2}{r^2} + \frac{df}{dr}\cdot\frac{r^2 - x^2}{r^3} = \frac{d^2 f}{dr^2}\frac{x^2}{r^2} + \frac{df}{dr}\frac{y^2}{r^3} \qquad ...(1)$$

Similarly,

$$\frac{\partial^2 u}{\partial y^2} = \frac{d^2 f}{dr^2}\frac{y^2}{r^2} + \frac{df}{dr}\frac{x^2}{r^3} \qquad ...(2)$$

On adding (1) and (2), we get

$$\frac{\partial^2 u}{\partial x^2} + \frac{\partial^2 u}{\partial y^2} = \frac{d^2 f}{dr^2}\frac{x^2+y^2}{r^2} + \frac{df}{dr}\frac{x^2+y^2}{r^3}$$

$$= \frac{d^2 f}{dr^2} + \frac{df}{dr}\frac{1}{r} = f''(r) + \frac{1}{r}f'(r)$$ **Proved.**

**Example 38.** *A function f (x, y) is rewritten in terms of new variables*

$$u = e^x \cos y, \qquad v = e^x \sin y$$

*Show that (i)* $\dfrac{\partial f}{\partial x} = u\dfrac{\partial f}{\partial u} + v\dfrac{\partial f}{\partial v}$ *and (ii)* $\dfrac{\partial f}{\partial y} = -v\dfrac{\partial f}{\partial u} + u\dfrac{\partial f}{\partial v}$

*and hence deduce that*

$$(iii)\quad \frac{\partial^2 f}{\partial x^2} + \frac{\partial^2 f}{\partial y^2} = (u^2+v^2)\left(\frac{\partial^2 f}{\partial u^2} + \frac{\partial^2 f}{\partial v^2}\right)$$

**Solution.** $u = e^x \cos y,\ \dfrac{\partial u}{\partial x} = e^x \cos y = u, \Rightarrow \dfrac{\partial u}{\partial y} = -e^x \sin y = -v$

$$v = e^x \sin y,\ \frac{\partial v}{\partial x} = e^x \sin y = v,\ \frac{\partial v}{\partial y} = e^x \cos y = u$$

(*i*) We know that $\dfrac{\partial f}{\partial x} = \dfrac{\partial f}{\partial u}\dfrac{\partial u}{\partial x} + \dfrac{\partial f}{\partial v}\dfrac{\partial v}{\partial x}$

$$\frac{\partial f}{\partial x} = \frac{\partial f}{\partial u}u + \frac{\partial f}{\partial v}v$$ ... (1) **Proved.**

(*ii*) $\dfrac{\partial f}{\partial y} = \dfrac{\partial f}{\partial u}\dfrac{\partial u}{\partial y} + \dfrac{\partial f}{\partial v}\dfrac{\partial v}{\partial y} = \dfrac{\partial f}{\partial u}.(-v) + \dfrac{\partial f}{\partial v}u = -v.\dfrac{\partial f}{\partial u} + u.\dfrac{\partial f}{\partial v}$ ... (2) **Proved.**

(*iii*) $$\frac{\partial^2 f}{\partial x^2} = \frac{\partial}{\partial x}\left(\frac{\partial f}{\partial x}\right)$$

$$= \left(u.\frac{\partial}{\partial u} + v.\frac{\partial}{\partial v}\right)\left(u.\frac{\partial f}{\partial u} + v.\frac{\partial f}{\partial v}\right) \qquad \text{[From (1)]}$$

$$= u\frac{\partial}{\partial u}\left(u\frac{\partial f}{\partial u} + v\frac{\partial f}{\partial v}\right) + v\frac{\partial}{\partial v}\left(u\frac{\partial f}{\partial u} + v\frac{\partial f}{\partial v}\right)$$

$$= u\frac{\partial}{\partial u}\left(u\frac{\partial f}{\partial u}\right) + u\frac{\partial}{\partial u}\left(v\frac{\partial f}{\partial v}\right) + v\frac{\partial}{\partial v}\left(u\frac{\partial f}{\partial u}\right) + v\frac{\partial}{\partial v}\left(v\frac{\partial f}{\partial v}\right)$$

$$= u\left(u\frac{\partial^2 f}{\partial u^2} + \frac{\partial f}{\partial u}\right) + u\left(v\frac{\partial^2 f}{\partial u \partial v}\right) + v\left(u\frac{\partial^2 f}{\partial v \partial u}\right) + v\left(v\frac{\partial^2 f}{\partial v^2} + \frac{\partial f}{\partial v}\right)$$

$$= u^2\frac{\partial^2 f}{\partial u^2} + u\frac{\partial f}{\partial u} + uv\frac{\partial^2 f}{\partial v \partial u} + uv\frac{\partial^2 f}{\partial u \partial v} + v^2\frac{\partial^2 f}{\partial v^2} + v\frac{\partial f}{\partial v} \qquad ...(3)$$

$$\frac{\partial^2 f}{\partial y^2} = \frac{\partial}{\partial y}\left(\frac{\partial f}{\partial y}\right)=\left(-v\frac{\partial}{\partial u}+u\frac{\partial}{\partial v}\right)\left(-v\frac{\partial f}{\partial u}+u\frac{\partial f}{\partial v}\right) \quad \text{[From (2)]}$$

$$= -v\frac{\partial}{\partial u}\left(-v\frac{\partial f}{\partial u}+u\frac{\partial f}{\partial v}\right)+u\frac{\partial}{\partial v}\left(-v\frac{\partial f}{\partial u}+u\frac{\partial f}{\partial v}\right)$$

$$= -v\frac{\partial}{\partial u}\left(-v\frac{\partial f}{\partial u}\right)-v\frac{\partial}{\partial u}\left(u\frac{\partial f}{\partial v}\right)+u\frac{\partial}{\partial v}\left(-v\frac{\partial f}{\partial u}\right)+u\frac{\partial}{\partial v}\left(u\frac{\partial f}{\partial v}\right)$$

$$= -v\left(-v\frac{\partial^2 f}{\partial u^2}\right)-v\left(u\frac{\partial^2 f}{\partial u\partial v}+\frac{\partial f}{\partial v}\right)+u\left(-v\frac{\partial^2 f}{\partial v\partial u}-\frac{\partial f}{\partial u}\right)+u\left(u\frac{\partial^2 f}{\partial v^2}\right)$$

$$= v^2\frac{\partial^2 f}{\partial u^2}-uv\frac{\partial^2 f}{\partial u\partial v}-v\frac{\partial f}{\partial v}-uv\frac{\partial^2 f}{\partial u\partial v}-u\frac{\partial f}{\partial u}+u^2\frac{\partial^2 f}{\partial v^2} \quad ...(4)$$

On adding (3) and (4), we obtain

$$\frac{\partial^2 f}{\partial x^2}+\frac{\partial^2 f}{\partial y^2} = u^2\frac{\partial^2 f}{\partial u^2}+v^2\frac{\partial^2 f}{\partial v^2}+v^2\frac{\partial^2 f}{\partial u^2}+u^2\frac{\partial^2 f}{\partial v^2}$$

$$= (u^2+v^2)\frac{\partial^2 f}{\partial u^2}+(u^2+v^2)\frac{\partial^2 f}{\partial v^2}=(u^2+v^2)\left(\frac{\partial^2 f}{\partial u^2}+\frac{\partial^2 f}{\partial v^2}\right)$$ **Proved.**

**Example 39.** *If $x + y = 2e^{\theta}\cos\phi$ and $x - y = 2ie^{\theta}\sin\phi$*

*Show that :* $\dfrac{\partial^2 V}{\partial\theta^2}+\dfrac{\partial^2 V}{\partial\phi^2}=4xy\dfrac{\partial^2 V}{\partial x\partial y}$ *(A.M.I.E.T.E., Winter 2007)*

*(Nagpur University, Summer 2001, Winter 2000, U.P., I Semester, Winter 2001)*

**Solution.** We have, $x + y = 2e^{\theta}\cos\phi$ ...(1)

$x - y = 2ie^{\theta}\sin\phi$ ...(2)

By adding and subtracting equations (1) and (2), we have

$2x = 2e^{\theta}(\cos\phi + i\sin\phi) \Rightarrow x = e^{\theta+i\phi}$

and $2y = 2e^{\theta}(\cos\phi - i\sin\phi) \Rightarrow y = e^{\theta-i\phi}$ ...(3)

It is clear that $V = f(x, y)$ and $x, y$ are functions of $\theta$ and $\phi$. Hence $V$ is a composite function of $\theta$ and $\phi$.

We want to convert $V, \theta, \phi$ in $V, x, y$ respectively.

From equation (3), we have

$$\frac{\partial x}{\partial\theta} = e^{\theta+i\phi}=x, \qquad \frac{\partial x}{\partial\phi}=ie^{\theta+i\phi}=ix$$

$$\frac{\partial y}{\partial\theta}=e^{\theta-i\phi}=y, \qquad \text{and} \qquad \frac{\partial y}{\partial\phi} = -ie^{\theta-i\phi}=-iy$$

Now, $$\frac{\partial V}{\partial\theta} = \frac{\partial V}{\partial x}\cdot\frac{\partial x}{\partial\theta}+\frac{\partial V}{\partial y}\cdot\frac{\partial y}{\partial\theta}=x\frac{\partial V}{\partial x}+y\frac{\partial V}{\partial y}$$

and $$\frac{\partial^2 V}{\partial\theta^2} = \frac{\partial}{\partial\theta}\left(\frac{\partial V}{\partial\theta}\right)=\left(x\frac{\partial}{\partial x}+y\frac{\partial}{\partial y}\right)\left(x\frac{\partial V}{\partial x}+y\frac{\partial V}{\partial y}\right)$$

$$= x\frac{\partial}{\partial x}\left[x\frac{\partial V}{\partial x}+y\frac{\partial V}{\partial y}\right]+y\frac{\partial}{\partial y}\left[x\frac{\partial V}{\partial x}+y\frac{\partial V}{\partial y}\right]$$

$$= x\left[x\frac{\partial^2 V}{\partial x^2}+\frac{\partial V}{\partial x}+y\frac{\partial^2 V}{\partial x\partial y}\right]+y\left[x\frac{\partial^2 V}{\partial x\partial y}+y\frac{\partial^2 V}{\partial y^2}+\frac{\partial V}{\partial y}\right]$$

$$\Rightarrow \quad \frac{\partial^2 V}{\partial \theta^2} = x^2 \frac{\partial^2 V}{\partial x^2} + y^2 \frac{\partial^2 V}{\partial y^2} + 2xy \frac{\partial^2 V}{\partial x \partial y} + \left( x\frac{\partial V}{\partial x} + y\frac{\partial V}{\partial y} \right) \qquad ...(4)$$

Again $$\frac{\partial V}{\partial \phi} = \frac{\partial V}{\partial x}\frac{\partial x}{\partial \phi} + \frac{\partial V}{\partial y}\frac{\partial y}{\partial \phi} = i\left[ x\frac{\partial V}{\partial x} - y\frac{\partial V}{\partial y} \right]$$

$$\frac{\partial}{\partial \phi} = i\left[ x\frac{\partial}{\partial x} - y\frac{\partial}{\partial y} \right]$$

$$\frac{\partial^2 V}{\partial \phi^2} = \frac{\partial}{\partial \phi}\left( \frac{\partial V}{\partial \phi} \right) = i\left( x\frac{\partial}{\partial x} - y\frac{\partial}{\partial y} \right) i\left( x\frac{\partial V}{\partial x} - y\frac{\partial V}{\partial y} \right)$$

$$= ix\frac{\partial}{\partial x}.i\left[ x\frac{\partial V}{\partial x} - y\frac{\partial V}{\partial y} \right] - iy\frac{\partial}{\partial y}.i\left[ x\frac{\partial V}{\partial x} - y\frac{\partial V}{\partial y} \right]$$

$$= -x\frac{\partial}{\partial x}\left[ x\frac{\partial V}{\partial x} - y\frac{\partial V}{\partial y} \right] + y\frac{\partial}{\partial y}\left[ x\frac{\partial V}{\partial x} - y\frac{\partial V}{\partial y} \right]$$

$$= -x\left[ \frac{\partial V}{\partial x} + x\frac{\partial^2 V}{\partial x^2} - y\frac{\partial^2 V}{\partial x \partial y} \right] + y\left[ x\frac{\partial^2 V}{\partial x \partial y} - y\frac{\partial^2 V}{\partial y^2} - \frac{\partial V}{\partial y} \right]$$

$$\Rightarrow \quad \frac{\partial^2 V}{\partial \phi^2} = -\left[ x\frac{\partial V}{\partial x} + y\frac{\partial V}{\partial y} \right] + 2xy\frac{\partial^2 V}{\partial x \partial y} - \left[ x^2\frac{\partial^2 V}{\partial x^2} + y^2\frac{\partial^2 V}{\partial y^2} \right] \qquad ...(5)$$

Adding (4) and (5), we get $\dfrac{\partial^2 V}{\partial \theta^2} + \dfrac{\partial^2 V}{\partial \phi^2} = 4xy\dfrac{\partial^2 V}{\partial x \partial y}$ **Proved.**

## EXERCISE 1.5

**1.** If $z = u^2 + v^2$ and $u = at^2$, $v = 2at$, find $\dfrac{dz}{dt}$. **Ans.** $4a^2t(t^2 + 2)$

**2.** If $z = \sin^{-1}(x - y)$, $x = 3t$, $y = 4t^3$; show that $\dfrac{dz}{dt} = \dfrac{3}{\sqrt{1-t^2}}$.

**3.** If $w = f(u, v)$, where $u = x + y$ and $v = x - y$, show that $\dfrac{dw}{dx} + \dfrac{dw}{dy} = 2\dfrac{dw}{du}$

**4.** If $u = xe^yz$, where $y = \sqrt{a^2 - x^2}$, $z = \sin^3 x$. Find $\dfrac{du}{dx}$ **Ans.** $e^yz\left(1 - \dfrac{x^2}{y} + 3x\cot x\right)$

**5.** If $u = x^2 + y^2 + z^2 - 2xyz = 1$, show that $\dfrac{dx}{\sqrt{1-x^2}} + \dfrac{dy}{\sqrt{1-y^2}} + \dfrac{dz}{\sqrt{1-z^2}} = 0$

[**Hint.** $du = \dfrac{\partial u}{\partial x}dx + \dfrac{\partial u}{\partial y}dy + \dfrac{\partial u}{\partial z}dz = 0$

$= 2(x - yz)\,dx + 2(y - zx)\,dy + 2(z - xy)\,dz = 0$

But $x^2 + y^2 + z^2 - 2xyz = 1$, $\Rightarrow y^2 - 2xyz = 1 - x^2 - z^2$

$y^2 - 2xyz + x^2z^2 = 1 + x^2z^2 - x^2 - z^2$ $\Rightarrow (y - xz)^2 = (1 - x^2)(1 - z^2)$]

**6.** If $z = z(u, v)$, $u = x^2 - 2xy - y^2$ and $v = y$. Show that

$$(x + y)\frac{\partial z}{\partial x} + (x - y)\frac{\partial z}{\partial y} = 0 \text{ is equivalent to } \frac{\partial z}{\partial v} = 0$$

**7.** If $u = f(x^2 + 2yz,\ y^2 + 2zx)$, prove that

$$(y^2 - zx)\frac{\partial u}{\partial x} + (x^2 - yz)\frac{du}{dy} + (z^2 - xy)\frac{du}{dz} = 0$$

**8.** By changing the independent variables $x$ and $t$ to $u$ and $v$ by means of the relationships
$u = x - at, \qquad v = x + at$

Show that $a^2 \dfrac{\partial^2 y}{\partial x^2} - \dfrac{\partial^2 y}{\partial t^2} = 4a^2 \dfrac{\partial^2 y}{\partial u \, \partial v}$

**9.** If $x^2 = au + bv$, $y = au - bv$, prove that $\left(\dfrac{du}{dx}\right)_y \cdot \left(\dfrac{dx}{du}\right)_v = \dfrac{1}{2} = \left(\dfrac{\partial v}{\partial y}\right)_x \cdot \left(\dfrac{\partial y}{\partial v}\right)_u$

**10.** If $z = f(x, y)$ where $x = uv$, $y = \dfrac{u+v}{u-v}$, show that $2x \dfrac{\partial z}{dx} = u \dfrac{\partial z}{du} + v \dfrac{\partial z}{dv}$.

**11.** If $u = x \cos \dfrac{y}{z}$, $x = 3r^2 + 2s$, $y = 4r - 2s^3$, $z = 2r^2 - 3s^2$ find $\dfrac{\partial u}{\partial r}, \dfrac{\partial u}{\partial s}$.

**Ans.** $\dfrac{\partial u}{\partial r} = 6r \cos \dfrac{y}{z} - \dfrac{4x}{z} \sin \dfrac{y}{z} + \dfrac{4xyr}{z^2} \sin \dfrac{y}{z}, \dfrac{\partial u}{\partial s} = 2 \cos \dfrac{y}{z} + \dfrac{6xs^2}{z} \sin \dfrac{y}{z} - \dfrac{6xys}{z^2} \sin \dfrac{y}{z}$

**12.** If $z = f(x, y)$ where $x = e^u \cos v$, $y = e^u \sin v$. Prove that

$$\left(\frac{\partial f}{\partial x}\right)^2 + \left(\frac{\partial f}{\partial y}\right)^2 = e^{-2u}\left[\left(\frac{\partial f}{\partial u}\right)^2 + \left(\frac{\partial f}{\partial v}\right)^2\right]$$

**13.** If $x = \dfrac{\cos\theta}{u}, y = \dfrac{\sin\theta}{u}$ and $z = f(x, y)$, then show that

$$\frac{\partial^2 z}{\partial x^2} + \frac{\partial^2 z}{\partial y^2} = u^4 \frac{\partial^2 z}{\partial u^2} + u^3 \frac{\partial z}{\partial u} + u^4 \frac{\partial^2 z}{\partial \theta^2}$$

**14.** If $x = z \sin^{-1} \dfrac{y}{x}$ where $x = 3r^2 + 2s$, $y = 4r - 2s^3$, $z = 2r^2 - 3s^2$, find $\dfrac{\partial u}{\partial r}, \dfrac{\partial u}{\partial s}$

**Ans.** $\dfrac{\partial u}{\partial r} = -\dfrac{6ryz}{x\sqrt{x^2 - y^2}} + \dfrac{4z}{\sqrt{x^2 - y^2}} + 4r \sin \dfrac{y}{x}, \quad \dfrac{\partial u}{\partial s} = \dfrac{2yz}{x\sqrt{x^2 - y^2}} - \dfrac{6s^2 z}{\sqrt{x^2 - y^2}} - 6s \sin \dfrac{y}{x}$

**15.** If $z = f(u, v)$ where $u = x \cos\theta - y \sin\theta$, $v = x \sin\theta + y \cos\theta$, show that

$$x\frac{\partial z}{\partial x} + y\frac{\partial z}{\partial y} = u\frac{\partial x}{\partial u} + v\frac{\partial z}{\partial v}, \theta \text{ being constant.}$$

**16.** Given the transformation $x = \cosh\xi \cos\eta$, $y = \sinh\xi \sin\eta$
establish the following equation for the function $u$ (function of $x$, $y$ and also of $\xi$, $\eta$):

$$\frac{\partial^2 u}{\partial \xi^2} + \frac{\partial^2 u}{\partial \eta^2} = (\sinh^2\xi + \sin^2\eta)\left(\frac{\partial^2 u}{\partial x^2} + \frac{\partial^2 u}{\partial y^2}\right)$$

**17.** If $z = f(x, y)$, where $x = r\cos\theta$, $y = r\sin\theta$, prove that

$$\left(\frac{\partial z}{\partial x}\right)^2 + \left(\frac{\partial z}{\partial y}\right)^2 = \left(\frac{\partial z}{\partial r}\right)^2 + \frac{1}{r^2}\left(\frac{\partial z}{\partial \theta}\right)^2 \text{ and } \frac{\partial^2 z}{\partial x^2} + \frac{\partial^2 z}{\partial y^2} = \frac{\partial^2 z}{\partial r^2} + \frac{1}{r^2}\left(\frac{\partial^2 z}{\partial \theta^2}\right) + \frac{1}{r}\left(\frac{\partial z}{\partial r}\right)$$

**18.** If by substituting $u = x^2 - y^2$, $v = 2xy$, $f(x, y) = \varphi(u, v)$. Show that

$$\frac{\partial^2 f}{\partial x^2} + \frac{\partial^2 f}{\partial y^2} = 4(u^2 + v^2)\left(\frac{\partial^2 \varphi}{\partial u^2} + \frac{\partial^2 \varphi}{\partial v^2}\right)$$

**19.** If $x = r \sin\theta \cos\phi$, $y = r \sin\theta \sin\phi$, $z = r\cos\theta$, $v = v(x, y, z)$, prove that

$$\left(\frac{\partial v}{\partial x}\right)^2 + \left(\frac{\partial v}{\partial y}\right)^2 + \left(\frac{\partial v}{\partial z}\right)^2 = \left(\frac{\partial v}{\partial r}\right)^2 + \left(\frac{1}{r}\frac{\partial v}{\partial \theta}\right)^2 + \left(\frac{1}{r\sin\theta}\frac{\partial v}{\partial \varphi}\right)^2$$

**20.** If $v$ be a potential function such that $v = v(r)$ and $r^2 = x^2 + y^2 + z^2$, show that

$$\frac{\partial^2 v}{\partial x^2}+\frac{\partial^2 v}{\partial y^2}+\frac{\partial^2 v}{\partial z^2}=\frac{d^2 v}{d r^2}+\frac{2}{r}\frac{d v}{d r}$$

**21.** Given that $w = x + 2y + z^2$, $x = r/s$, $y = r^2 + e^s$, and $z = 2r$, show that $r\frac{\partial w}{\partial r}+s\frac{\partial w}{\partial s}=12 r^2+2 s e^s$.

**22.** Find $\frac{\partial w}{\partial v}$ when $u = 0,\ v = 0$

If $w = (x^2 + y - 2)^4 + (x - y + 2)^3$, $x = u - 2v + 1$,

and $y = 2u + v - 2$ **Ans.** 99

**23.** If $x = u + v + w$, $y = vw + wu + uv$, $z = uvw$ and $F$ is a function of $x, y, z$, then show that

$$u\frac{\partial F}{\partial u}+v\frac{\partial F}{\partial v}+w\frac{\partial F}{\partial w}=x\frac{\partial F}{\partial x}+2 y\frac{\partial F}{\partial y}+3 z\frac{\partial F}{\partial z}$$

**24.** If $u = x + ay$ and $v = x + by$, transform the equation

$2\frac{\partial^2 z}{\partial x^2}-5\frac{\partial^2 z}{\partial x\,\partial y}+3\frac{\partial^2 z}{\partial y^2}=0$ into the equation $\frac{\partial^2 z}{\partial u\,\partial v}=0$, find the values of $a$ and $b$.

*(A.M.I.E.T.E., Summer 2000)* **Ans.** $\left(a=1, b=\frac{2}{3}\right),\left(a=\frac{2}{3}, b=1\right)$

## 1.17 IMPORTANT DEDUCTIONS

Let $z = f(x, y)$, then

$$dz = \frac{\partial f}{\partial x}dx+\frac{\partial f}{\partial y}dy$$

If $z = 0,\ dz = 0$

$$0 = \frac{\partial f}{\partial x}dx+\frac{\partial f}{\partial y}dy \quad\Rightarrow\quad \frac{\partial f}{\partial y}\cdot\frac{dy}{dx}=-\frac{\partial f}{\partial x}$$

$$\Rightarrow \quad \boxed{\frac{dy}{dx} = -\frac{\frac{\partial f}{\partial x}}{\frac{\partial f}{\partial y}}} \qquad \textbf{[Remember]} \quad ...(1)$$

We can find $\frac{d^2 y}{d x^2}$ by differentiating (1).

Let $\frac{\partial f}{\partial x} = p,\ \frac{\partial f}{\partial y}=q,\ \frac{\partial^2 f}{\partial x^2}=r,\ \frac{\partial^2 f}{\partial x\partial y}=s,\ \frac{\partial^2 f}{\partial y^2}=t$

From (1) $\frac{\partial y}{\partial x} = -\frac{p}{q}$.

On differentiating again, we obtain $\frac{d^2 y}{d x^2}=-\frac{q\frac{dp}{dx}-p\frac{dq}{dx}}{q^2}$ ...(2)

$$\text{But}\quad \frac{dp}{dx} = \frac{\partial p}{\partial x}+\frac{\partial p}{\partial y}\frac{dy}{dx}$$

$$\Rightarrow \quad \frac{dp}{dx} = \frac{\partial}{\partial x}\left(\frac{\partial f}{\partial x}\right)+\frac{\partial}{\partial y}\left(\frac{\partial f}{\partial x}\right)\cdot\frac{dy}{dx}$$

$$\frac{dp}{dx} = \frac{\partial^2 f}{\partial x^2}+\frac{\partial^2 f}{\partial y\,\partial x}\frac{dy}{dx}=\frac{\partial^2 f}{\partial x^2}+\frac{\partial^2 f}{\partial y\,\partial x}\left(-\frac{\frac{\partial f}{\partial x}}{\frac{\partial f}{\partial y}}\right)=r-s\frac{p}{q}=\frac{qr-ps}{q}$$

$$\frac{dq}{dx} = \frac{\partial q}{\partial x}+\frac{\partial q}{\partial y}\frac{dy}{dx}=\frac{\partial}{\partial x}\left(\frac{\partial f}{\partial y}\right)+\frac{\partial}{\partial y}\left(\frac{\partial f}{\partial y}\right)\left(\frac{-p}{q}\right)$$

$$= \frac{\partial^2 f}{\partial x\,\partial y}-\frac{\partial^2 f}{\partial y^2}\frac{p}{q}=s-\frac{tp}{q}=\frac{qs-tp}{q}$$

Making substitutions in (2), we obtain

$$\frac{d^2 y}{dx^2} = -\frac{q\dfrac{qr-ps}{q}-p\dfrac{qs-tp}{q}}{q^2} = -\frac{q^2 r - pqs - pqs + p^2 t}{q^3}$$

$$\Rightarrow \boxed{\frac{d^2 y}{dx^2} = -\frac{q^2 r - 2pqs + p^2 t}{q^3}}$$

**Example 40.** *If* $x^3 + 3x^2y + 6xy^2 + y^3 = 1$, *find* $\dfrac{dy}{dx}$.

**Solution.** Let $f(x, y) = x^3 + 3x^2y + 6xy^2 + y^3 - 1 = 0$

$$\frac{\partial f}{\partial x} = 3x^2 + 6xy + 6y^2$$

$$\frac{\partial f}{\partial y} = 3x^2 + 12xy + 3y^2$$

$$\frac{dy}{dx} = -\frac{\dfrac{\partial f}{\partial x}}{\dfrac{\partial f}{\partial y}} = -\frac{3x^2+6xy+6y^2}{3x^2+12xy+3y^2} = -\frac{x^2+2xy+2y^2}{x^2+4xy+y^2}$$ **Ans.**

**Example 41.** *If* $y^3 - 3ax^2 + x^3 = 0$, *then prove that*

$$\frac{d^2 y}{dx^2} = -\frac{2a^2x^2}{y^5}$$

**Solution.** Let $f(x, y) = y^3 - 3ax^2 + x^3$ ...(1)

$$p = \frac{\partial f}{\partial x} = -6ax + 3x^2, \quad q = \frac{\partial f}{\partial y} = 3y^2$$

$$r = \frac{\partial^2 f}{\partial x^2} = -6a + 6x, \quad s = \frac{\partial^2 f}{\partial x\,\partial y} = 0, \quad t = \frac{\partial^2 f}{\partial y^2} = 6y$$

$$\frac{\partial^2 y}{\partial x^2} = -\frac{q^2 r - 2pqs + p^2 t}{q^3}$$ ...(2) [Art 1.17]

Putting the values of $p$, $q$, $r$, $s$ and $t$ in (2), we get

$$\frac{\partial^2 y}{\partial x^2} = -\frac{(3y^2)^2(-6a+6x) - 2(-6ax+3x^2)(3y^2)(0) + (-6ax+3x^2)^2(6y)}{(3y^2)^3}$$

$$= -\frac{54y^4(-a+x) + 54(-2ax+x^2)^2\, y}{27y^6}$$

$$\frac{\partial^2 y}{\partial x^2} = -\frac{2y^3(-a+x) + 2(4a^2x^2 + x^4 - 4ax^3)}{y^5}$$ ...(3)

Putting the value of $y^3 = 3ax^2 - x^3$ from (1) in (3), we get

$$\frac{\partial^2 y}{\partial x^2} = -\frac{2(3ax^2 - x^3)(-a+x) + 2(4a^2x^2 + x^4 - 4ax^3)}{y^5}$$

$$= -\frac{-6a^2x^2+6ax^3+2ax^3-2x^4+8a^2x^2+2x^4-8ax^3}{y^5}$$

$$\frac{\partial^2 y}{\partial x^2} = -\frac{2a^2x^2}{y^5}$$ **Proved.**

**Example 42.** *If $x^3 + y^3 - 3axy = 0$, find $\frac{dy}{dx}$ and $\frac{d^2y}{dx^2}$.*

**Solution.** Let $f(x, y) = x^3 + y^3 - 3axy = 0$

$$p = \frac{\partial f}{\partial x} = 3x^2 - 3ay, \quad q = \frac{\partial f}{\partial y} = 3y^2 - 3ax$$

$$r = \frac{\partial^2 f}{\partial x^2} = 6x, \quad s = \frac{\partial^2 f}{\partial x \partial y} = -3a, \quad t = \frac{\partial^2 f}{\partial y^2} = 6y$$

$$\frac{dy}{dx} = -\frac{\frac{\partial f}{\partial x}}{\frac{\partial f}{\partial y}} = -\frac{3x^2-3ay}{3y^2-3ax} = \frac{ay-x^2}{y^2-ax}$$

$$\frac{d^2y}{dx^2} = -\frac{q^2r - 2pqs + p^2t}{q^3} \qquad \text{...(1) [Art. 1.17]}$$

Putting the values of $p$, $q$, $r$, $s$ and $t$ in (1), we get

$$\frac{d^2y}{dx^2} = -\frac{(3y^2-3ax)^2\,6x - 2(3x^2-3ay)(3y^2-3ax)(-3a) + (3x^2-3ay)^2(6y)}{(3y^2-3ax)^3}$$

$$= -\frac{2x(y^2-ax)^2 + 2a(x^2-ay)(y^2-ax) + 2y(x^2-ay)^2}{(y^2-ax)^3}$$

$$\frac{d^2y}{dx^2} = \frac{2a^3xy}{(ax-y^2)^3}$$ **Ans.**

**Example 43.** *Find $\frac{dy}{dx}$ when $(\cos x)^y = (\sin y)^x$*

**Solution.** Given equation can be written as : $(\cos x)^y - (\sin y)^x = 0$

Here $f(x, y) = (\cos x)^y - (\sin y)^x = 0$

$$\frac{\partial f}{\partial x} = y(\cos x)^{y-1}(-\sin x) - (\sin y)^x \log \sin y$$

$$= -[y \sin x (\cos x)^{y-1} + (\sin y)^x \log \sin y]$$

$$\frac{\partial f}{\partial y} = (\cos x)^y \log \cos x - x(\sin y)^{x-1} \cos y$$

$$\frac{dy}{dx} = -\frac{\frac{df}{dx}}{\frac{df}{dy}} \qquad \text{[Art. 1.17]}$$

$$\frac{dy}{dx} = \frac{y \sin x (\cos x)^{y-1} + (\sin y)^x \log \sin y}{(\cos x)^y \log \cos x - x(\sin y)^{x-1} \cos y} \qquad \text{...(1)}$$

In (1), put $(\cos x)^y$ for $(\sin y)^x$

$$\frac{dy}{dx} = \frac{y \sin x (\cos x)^{y-1} + (\cos x)^y \log \sin y}{(\cos x)^y \log \cos x - \frac{x(\cos x)^y}{\sin y} . \cos y}$$

$$= \frac{(\cos x)^y\,[y\tan x + \log\sin y]}{(\cos x)^y\,[\log\cos x - x\cot y]} = \frac{y\tan x + \log\sin y}{\log\cos x - x\cot y}$$ **Ans.**

**Example 44.** *If $f(x, y) = 0$ and $\phi(y, z) = 0$, show that* $\frac{\partial f}{\partial y}\cdot\frac{\partial \varphi}{\partial z}\cdot\frac{dz}{dx} = \frac{\partial f}{\partial x}\cdot\frac{\partial \varphi}{\partial y}$.

**Solution.** $f(x, y) = 0$ ...(1)

$\varphi(y, z) = 0$ ...(2)

Differentiating (1) w.r.t. $x$, we get

$$0 = \frac{\partial f}{\partial x} + \frac{\partial f}{\partial y}\cdot\frac{dy}{dx} \Rightarrow \frac{dy}{dx} = -\frac{\frac{\partial f}{\partial x}}{\frac{\partial f}{\partial y}} \quad ...(3)$$

Differentiating (2) w.r.t. '$y$', we get

$$0 = \frac{\partial \phi}{\partial y} + \frac{\partial \phi}{\partial z}\cdot\frac{dz}{dy} \Rightarrow \frac{dz}{dy} = -\frac{\frac{\partial \varphi}{\partial y}}{\frac{\partial \varphi}{\partial z}} \quad ...(4)$$

Multiplying (3) and (4), we get

$$\frac{dy}{dx}\times\frac{dz}{dy} = \left(-\frac{\frac{\partial f}{\partial x}}{\frac{\partial f}{\partial y}}\right)\left(-\frac{\frac{\partial \varphi}{\partial y}}{\frac{\partial \varphi}{\partial z}}\right) \Rightarrow \frac{dz}{dx} = \frac{\frac{\partial f}{\partial x}\times\frac{\partial \varphi}{\partial y}}{\frac{\partial f}{\partial y}\times\frac{\partial \varphi}{\partial z}}$$

$$\Rightarrow \frac{\partial f}{\partial y}\cdot\frac{\partial \varphi}{\partial z}\cdot\frac{dz}{dx} = \frac{\partial f}{\partial x}\cdot\frac{\partial \varphi}{\partial y}$$ **Proved.**

**Example 45.** *If $u = x\log xy$ where $x^3 + y^3 + 3xy = 1$. Find* $\frac{du}{dx}$

*(U.P. I Sem., Dec. 2005, Com. 2002)*

**Solution.** We have, $u = x\log xy$

$$\frac{\partial u}{\partial x} = x\left(\frac{1}{xy}.y\right) + 1\log xy$$

$$\Rightarrow \frac{\partial u}{\partial x} = 1 + \log xy \quad ...(1)$$

$$\frac{\partial u}{\partial y} = x\frac{1}{xy}.x = \frac{x}{y} \quad ...(2)$$

$$x^3 + y^3 + 3xy = 1$$

On differentiating, we get

$$3x^2 + 3y^2\frac{dy}{dx} + 3x\frac{dy}{dx} + 3y = 0$$

$$\Rightarrow \frac{dy}{dx} = -\frac{x^2+y}{x+y^2} \quad ...(3)$$

We know that $$\frac{du}{dx} = \frac{\partial u}{\partial x}\frac{dx}{dx} + \frac{\partial u}{\partial y}\frac{dy}{dx}$$

$$= (1+\log xy).1 + \frac{x}{y}\left(-\frac{x^2+y}{x+y^2}\right)$$ [From (1), (2), (3)]

$$= 1 + \log xy - \frac{x}{y}\cdot\frac{x^2+y}{x+y^2}$$ **Ans.**

## EXERCISE 1.6

Find $\frac{dy}{dx}$ in the following cases :

1. $x \sin (x - y) - (x + y) = 0$ **Ans.** $[y + x^2 \cos (x - y)] / [x + x^2 \cos (x - y)]$
2. $x^y = y^x$ **Ans.** $y (y - x \log y)/ x (x - y \log x)$
3. If $ax^2 + 2\,hxy + by^2 = 1$, find $\frac{d^2 y}{dx^2}$ **Ans.** $\frac{h^2 - ab}{(hx + by)^3}$
4. If $u = x^2y + y^2z + z^2x$ and if $z$ is defined implicitly as a function of $x$ and $y$ by the equation $x^2 + yz + z^3 = 0$

   find $\frac{\partial u}{\partial x}$, where $u$ is considered as a function of $x$ and $y$ alone.

   **Ans.** $\frac{\partial u}{\partial x} = 2xy + z^2 - (y^2 + 2zx)\left(\frac{2x}{y + 3z^2}\right)$
5. Find $\frac{dy}{dx}$, if $\tan^{-1} \frac{x}{y} + y^3 + 1 = 0$ **Ans.** $\frac{y}{x - 3x^2y^2 - 3y^4}$
6. If $f(x, y, z) = 0$, prove that

   $$\left(\frac{dx}{dy}\right)_z \cdot \left(\frac{dy}{dz}\right)_x \cdot \left(\frac{dz}{dx}\right)_y = -1$$

   $\left[\textbf{Hint.} \left(\frac{dx}{dy}\right)_z = -\frac{\partial f}{\partial y} \div \frac{\partial f}{\partial x}\right]$

## 1.18 TYPICAL CASES

**Example 46.** *If* $x = f(u, v)$, $y = \phi (u, v)$, *find* $\frac{\partial u}{\partial x}, \frac{\partial u}{\partial y}, \frac{\partial v}{\partial x}, \frac{\partial v}{\partial y}$.

**Solution.**
$$x = f(u, v) \qquad ...(1)$$
$$y = \phi (u, v) \qquad ...(2)$$

Differentiating (1), (2) w.r.t. $x$ (treating $y$ as constant), we obtain

$$1 = \frac{\partial f}{\partial u} . \frac{\partial u}{\partial x} + \frac{\partial f}{\partial v} . \frac{\partial v}{\partial x} \qquad ...(3)$$

$$0 = \frac{\partial \varphi}{\partial u} . \frac{\partial u}{\partial x} + \frac{\partial \varphi}{\partial v} . \frac{\partial v}{\partial x} \qquad ...(4)$$

Solving the equations (3) and (4) for $\frac{\partial u}{\partial x}$ and $\frac{\partial v}{\partial x}$, we obtain

$$\frac{\partial u}{\partial x} = \frac{\frac{\partial \varphi}{\partial v}}{\frac{\partial f}{\partial u} . \frac{\partial \varphi}{\partial v} - \frac{\partial f}{\partial v} . \frac{\partial \varphi}{\partial u}} \qquad \textbf{Ans.}$$

$$\frac{\partial v}{\partial x} = \frac{-\frac{\partial \varphi}{\partial u}}{\frac{\partial f}{\partial u} . \frac{\partial \varphi}{\partial v} - \frac{\partial f}{\partial v} . \frac{\partial \varphi}{\partial u}} \qquad \textbf{Ans.}$$

Similarly, differentiating (1) and (2) w.r.t. y, we get

$$0 = \frac{\partial f}{\partial u} . \frac{\partial u}{\partial y} + \frac{\partial f}{\partial v} . \frac{\partial v}{\partial y} \qquad ...(5)$$

$$1 = \frac{\partial \varphi}{\partial u} . \frac{\partial u}{\partial y} + \frac{\partial \varphi}{\partial v} . \frac{\partial v}{\partial y} \qquad ...(6)$$

Solving the equations (5) and (6) for $\frac{\partial u}{\partial y}$ and $\frac{\partial v}{\partial y}$, we obtain

$$\frac{\partial u}{\partial y} = \frac{-\dfrac{\partial f}{\partial v}}{\dfrac{\partial f}{\partial u}\cdot\dfrac{\partial \varphi}{\partial v} - \dfrac{\partial f}{\partial v}\cdot\dfrac{\partial \varphi}{\partial u}}$$ **Ans.**

$$\frac{\partial v}{\partial y} = \frac{\dfrac{\partial f}{\partial u}}{\dfrac{\partial f}{\partial u}\cdot\dfrac{\partial \varphi}{\partial v} - \dfrac{\partial f}{\partial v}\cdot\dfrac{\partial \varphi}{\partial u}}$$ **Ans.**

**Example 47.** *If* $x = u^2 - v^2$ *and* $y = uv$, *find* $\dfrac{\partial u}{\partial x}, \dfrac{\partial u}{\partial y}, \dfrac{\partial v}{\partial x}$ *and* $\dfrac{\partial v}{\partial y}$ (*Nagpur University, Winter 2003*)

**Solution.** Here, we have $x = u^2 - v^2$ ...(1)

$y = uv$ ...(2)

$$\frac{\partial x}{\partial u} = 2u; \qquad \frac{\partial y}{\partial u} = v$$

$$\frac{\partial x}{\partial v} = -2v, \qquad \frac{\partial y}{\partial v} = u$$

Differentiating (1) w.r.t. $x$, we get

$$\Rightarrow \quad 1 = 2u\frac{\partial u}{\partial x} - 2v\frac{\partial v}{\partial x} \quad ...(3)$$

Similarly differentiating (2) w.r.t. $x$, we get

$$0 = v\frac{\partial u}{\partial x} + u\frac{\partial v}{\partial x} \quad ...(4)$$

On solving (3) and (4), we get

$$\frac{\partial u}{\partial x} = \frac{u}{2u^2 + 2v^2}, \qquad \frac{\partial v}{\partial x} = -\frac{v}{2u^2 + 2v^2}$$ **Ans.**

On differentiating (1) and (2) w.r.t. $y$, we get

$$0 = 2u\frac{\partial u}{\partial y} - 2v\frac{\partial v}{\partial y} \quad ...(5)$$

$$1 = v\frac{\partial u}{\partial y} + u\frac{\partial v}{\partial y} \quad ...(6)$$

On solving (5) and (6), we get

$$\frac{\partial v}{\partial y} = \frac{u}{u^2 + v^2}, \quad \text{and} \quad \frac{\partial u}{\partial y} = \frac{v}{u^2 + v^2}$$ **Ans.**

**Example 48.** *Find p and q, if* $x = \sqrt{a}\,(\sin u + \cos v)$

$y = \sqrt{a}\,(\cos u - \sin v)$

$z = 1 + \sin(u - v)$

*where p and q mean* $\dfrac{\partial z}{\partial x}$ *and* $\dfrac{\partial z}{\partial y}$ *respectively.*

**Solution.** We have, $x = \sqrt{a}\,(\sin u + \cos v)$ ...(1)

$y = \sqrt{a}\,(\cos u - \sin v)$ ...(2)

$z = 1 + \sin(u - v)$ ...(3)

Differentiating (3) w.r.t. $x$, we get

$$\frac{\partial z}{\partial x} = \cos(u - v)\left[\frac{\partial u}{\partial x} - \frac{\partial v}{\partial x}\right] \quad ...(4)$$

Differentiating (1) partially w.r.t. $x$, we get

$$1 = \sqrt{a}\left(\cos u\frac{\partial u}{\partial x}-\sin v\frac{\partial v}{\partial x}\right) \qquad ...(5)$$

Differentiating (2) partially w.r.t. $x$, we get

$$0 = \sqrt{a}\left(-\sin u\frac{\partial u}{\partial x}-\cos v\frac{\partial v}{\partial x}\right) \qquad ...(6)$$

Solving (5) and (6) for $\frac{\partial u}{\partial x}$ and $\frac{\partial v}{\partial x}$, we get

$$\frac{\partial u}{\partial x} = \frac{1}{\sqrt{a}}\frac{\cos v}{\cos(u-v)},\ \frac{\partial v}{\partial x}=-\frac{\sin u}{\sqrt{a}\cos(u-v)}$$

Putting the values of $\frac{\partial u}{\partial x}$ and $\frac{\partial v}{\partial x}$ in (4), we get

$$p=\frac{\partial z}{\partial x} = \cos(u-v)\left[\frac{1}{\sqrt{a}}\frac{\cos v}{\cos(u-v)}+\frac{\sin u}{\sqrt{a}\cos(u-v)}\right]$$

$$p=\frac{\partial z}{\partial x} = \frac{1}{\sqrt{a}}(\sin u+\cos v)$$ **Ans.**

Differentiating (3) w.r.t. $y$, we get

$$\frac{\partial z}{\partial y} = \cos(u-v)\left[\frac{\partial u}{\partial y}-\frac{\partial v}{\partial y}\right] \qquad ...(7)$$

Differentiating (1) and (2) partially w.r.t. $y$, we get

$$0 = \sqrt{a}\left(\cos u\frac{\partial u}{\partial y}-\sin v\frac{\partial v}{\partial y}\right) \qquad ...(8)$$

$$1 = \sqrt{a}\left(-\sin u\frac{\partial u}{\partial y}-\cos v\frac{\partial v}{\partial y}\right) \qquad ...(9)$$

Solving (8) and (9) for $\frac{\partial u}{\partial y}$ and $\frac{\partial v}{\partial y}$, we get

$$\frac{\partial u}{\partial y} = -\frac{\sin v}{\sqrt{a}\cos(u-v)},\ \text{and}\ \frac{\partial v}{\partial y}=-\frac{\cos u}{\sqrt{a}\cos(u-v)}$$

Putting the values of $\frac{\partial u}{\partial y}$, and $\frac{\partial v}{\partial y}$ in (7), we have

$$q=\frac{\partial z}{\partial y} = \cos(u-v)\left[-\frac{\sin v}{\sqrt{a}\cos(u-v)}+\frac{\cos u}{\sqrt{a}\cos(u-v)}\right]$$

$$q = \frac{1}{\sqrt{a}}(-\sin v+\cos u)$$ **Ans.**

## EXERCISE 1.7

**1. Fill in the blanks**

*(i)* If $f(x, y, z) = 0$, then $\frac{\partial x}{\partial y}\cdot\frac{\partial y}{\partial z}\cdot\frac{\partial z}{\partial x}$ is equal to......

*(ii)* If $z = f(x, y)$, where $x = \varphi(t)$, $y = \psi(t)$, then $\frac{dz}{dt}$ =......

(*iii*) If $f(x, y) = 0$, then $\frac{dy}{dx} = \ldots\ldots$

(*iv*) If $u = x^2 + y^2,\ x = s + 3t,\ y = 2s - t$, then $\frac{du}{ds} =$

(*v*) If $f(x, y) = 0$ and $\varphi(y,z) = 0$, then $\frac{\partial f}{\partial y} \cdot \frac{\partial \varphi}{\partial z} \cdot \frac{\partial z}{\partial x} = \ldots\ldots$

(*vi*) If $f(x, y) = 0$, then $\frac{d^2 y}{dx^2} = \ldots\ldots$ **Ans.** (*i*) $-1$ (*ii*) $\frac{\partial z}{\partial x}\frac{dx}{dt} + \frac{\partial z}{\partial y}\frac{dy}{dt}$ (*iii*) $-\frac{\frac{\partial f}{\partial x}}{\frac{\partial f}{\partial y}}$

(*iv*) $2x + 4y$ (*v*) $\frac{\partial f}{\partial x} \cdot \frac{\partial \varphi}{\partial y}$ (*vi*) $-\frac{q^2 r - 2pqs + p^2 t}{q^3}$

## 1.19 GEOMETRICAL INTERPRETATION OF $\frac{\partial z}{\partial x}$ AND $\frac{\partial z}{\partial y}$

(*Gujarat, I Semester, Jan. 2009*)

Let $z = f(x, y)$ be a surface S.

Let $y = k$ be a plane parallel to $XZ$ – plane, passing through $P(x, k, z)$ cutting the surface $z = f(x, y)$ along the curve APB.

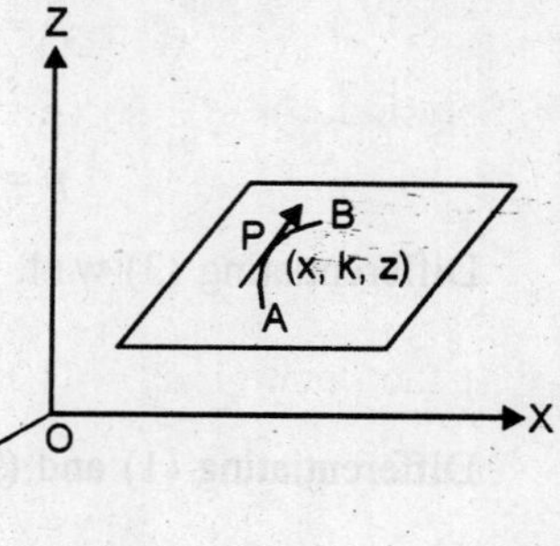


This section $APB$ is a plane curve whose equations are

$$z = f(x, y)$$
$$y = k$$

The slope of the tangent to this curve is given by $\frac{\partial z}{\partial x}$.

**Similarly,** $\frac{\partial z}{\partial y}$ is the slope of the tangent to the curve of intersection of the surface $z = f(x, y)$ with a plane parallel to $YZ$-plane.

## 1.20 TANGENT PLANE TO A SURFACE

Let $f(x, y, z) = 0$ be the equation of a surface $S$. Now we wish to find out the equation of a tangent plane to $S$ at the point $P(x_1, y_1, z_1)$.

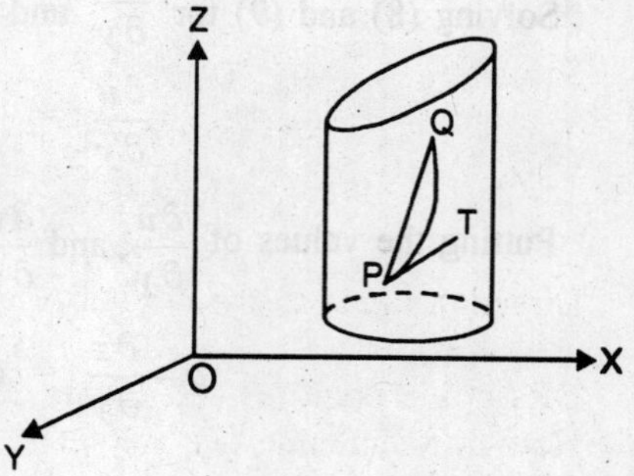


Let $Q(x_1 + \delta x_1,\ y_1 + \delta y_1,\ z_1 + \delta z_1)$ be a neighbouring point to $P$. Let the arc $PQ$ be $\delta s$ and the chord $PQ$ be $\delta c$.

The direction cosines of $PQ$ are

$$\frac{\delta x}{\delta c},\ \frac{\delta y}{\delta c},\ \frac{\delta z}{\delta c}$$

$$\Rightarrow \quad \frac{\delta x}{\delta s}\cdot\frac{\delta s}{\delta c},\ \frac{\delta y}{\delta s}\cdot\frac{\delta s}{\delta c},\ \frac{\delta z}{\delta s}\cdot\frac{\delta s}{\delta c}$$

As $\delta s \to 0$, $Q \to P$ and $PQ$ tends to a tangent line $PT$. The direction cosines of $PT$ are

$$\frac{dx}{ds},\ \frac{dy}{ds},\ \frac{dz}{ds} \qquad \ldots(1)$$

Differentiating $F(x, y, z) = 0$ w.r.t. '$s$', we get

$$\frac{\partial F}{\partial x}\frac{dx}{ds} + \frac{\partial F}{\partial y}\frac{dy}{ds} + \frac{\partial F}{\partial z}\frac{dz}{ds} = 0 \qquad \ldots(2)$$

From (1) and (2) it is clear that the tangent whose direction cosines are $\frac{dx}{ds}, \frac{dy}{ds}, \frac{dz}{ds}$ is perpendicular to a line having direction ratios

$$\frac{\partial F}{\partial x}, \quad \frac{\partial F}{\partial y}, \quad \frac{\partial F}{\partial z} \qquad ...(3)$$

There are a number of tangent lines at P to the curves joining P and Q. All these tangents will be perpendicular to the line having direction ratios as given by (3).

Hence all these tangent lines will lie in a plane known as tangent plane.

**Equation of tangent plane**

$$(x-x_1)\frac{\partial F}{\partial x}+(y-y_1)\frac{\partial F}{\partial y}+(z-z_1)\frac{\partial F}{\partial z} = 0$$

Equation of the normal to the plane.

$$\frac{x-x_1}{\frac{\partial F}{\partial x}} = \frac{y-y_1}{\frac{\partial F}{\partial y}} = \frac{z-z_1}{\frac{\partial F}{\partial z}}$$

**Example 49.** *Find the equation of the tangent plane and normal line to the surface* $x^2 + 2y^2 + 3z^2 = 12$ *at* $(1, 2, -1)$.

**Solution.** $F(x, y, z) = x^2 + 2y^2 + 3z^2 - 12$

$$\frac{\partial F}{\partial x} = 2x, \quad \frac{\partial F}{\partial y} = 4y, \quad \frac{\partial F}{\partial z} = 6z$$

At the point $(1, 2, -1)$ $\quad \frac{\partial F}{\partial x} = 2, \quad \frac{\partial F}{\partial y} = 8, \quad \frac{\partial F}{\partial z} = -6$

Hence the equation of the tangent plane at $(1, 2, -1)$ is

$$2(x-1) + 8(y-2) - 6(z+1) = 0$$

$$\Rightarrow \quad 2x + 8y - 6z = 24 \quad \Rightarrow \quad x + 4y - 3z = 12$$

Equation of normal is $\frac{x-1}{2} = \frac{y-2}{8} = \frac{z+1}{-6} \Rightarrow \frac{x-1}{1} = \frac{y-2}{4} = \frac{z+1}{-3}$ **Ans.**

**Example 50.** *Show that the surface* $x^2 - 2yz + y^3 = 4$ *is perpendicular to any number of the family of surfaces* $x^2 + 1 = (2-4a)y^2 + az^2$ *at the point of intersection* $(1, -1, 2)$.

**Solution.**

$$f(x, y, z) = x^2 - 2yz + y^3 - 4 = 0 \qquad ...(1)$$

$$F(x, y, z) = x^2 + 1 - (2-4a)y^2 - az^2 = 0 \qquad ...(2)$$

$$\frac{\partial f}{\partial x} = 2x, \quad \frac{\partial f}{\partial y} = -2z+3y^2, \quad \frac{\partial f}{\partial z} = -2y$$

Direction ratios to the normal of the tangent plane to (1) are

$$2x, -2z+3y^2, -2y$$

DRs at the point $(1, -1, 2)$ are $2, -1, 2$.

Now differentiating (2), we get

$$\frac{\partial F}{\partial x} = 2x, \quad \frac{\partial F}{\partial y} = -2(2-4a)y, \quad \frac{\partial F}{\partial z} = -2az.$$

Direction ratios to the normal of the tangent plane to (2) are

$$2x, (-4+8a)y, -2az.$$

DRs at the point $(1, -1, 2)$ are $2, 4-8a, -4a$

Now

$$l_1 l_2 + m_1 m_2 + n_1 n_2 = (2)(2) + (-1)(4-8a) + 2(-4a)$$
$$= 4 - 4 + 8a - 8a = 0.$$

Hence, the given surfaces are perpendicular at $(1, -1, 2)$. **Ans.**

### EXERCISE 1.8

1. Find the equation of tangent plane and the normal line to the surface

$x\,y\,z = 6$ at (1, 2, 3). **Ans.** $6x + 3y + 2z = 18,\ \frac{x-1}{6} = \frac{y-2}{3} = \frac{z-3}{2}$

2. Find the equations of the tangent plane and the normal to the surface $z^2 = 4(1 + x^2 + y^2)$ at (2, 2, 6). **Ans.** $4x + 4y - 3z + 2 = 0,\ \frac{x-2}{4} = \frac{y-2}{4} = \frac{z-6}{-3}$

3. Find the equations of the tangent plane and the normal to the surface

$\frac{x^2}{2} - \frac{y^2}{3} = z$ at (2, 3, – 1) **Ans.** $2x - 2y - z + 1 = 0,\ \frac{x-2}{2} = \frac{y-3}{-2} = \frac{z+6}{-1}$

4. Show that the plane $3x + 12y - 6z - 17 = 0$, touches the conicoid $3x^2 - 6y^2 + 9z^2 + 17 = 0$. Find also the point of contact. **Ans.** $\left(1, 2, \frac{2}{3}\right)$

5. Show that the plane $ax + by + cz + d = 0$ touches the surface $px^2 + qy^2 + 2z = 0$, if $\frac{a^2}{p} + \frac{b^2}{q} + 2\,c\,d = 0$.

# Applications of differential Calculus
## (Error, Jacobians, Taylor's Series, Maxima and Minima)

### 1.21 ERROR DETERMINATION

We know that $$\lim_{\partial x \to 0} \frac{\partial y}{\partial x} = \frac{dy}{dx}$$

$$\frac{\partial y}{\partial x} = \frac{dy}{dx} \text{ approximately} \Rightarrow \delta y = \left(\frac{dy}{dx}\right).\delta x \text{ approximately}$$

**Definitions:**

(*i*) $\delta x$ is known as *absolute error* in *x*. (*ii*) $\frac{\delta x}{x}$ is known as *relative error* in *x*.

(*iii*) $\left(\frac{\delta x}{x}\right) \times 100$ is known as *percentage error* in *x*.

**Example 51.** *The power dissipated in a resistor is given by* $P = \frac{E^2}{R}$. *Find by using Calculus the approximate percentage change in P when E is increased by 3% and R is decreased by 2%.* (*A.M.I.E., Summer 2001*)

**Solution.** Here, we have $P = \frac{E^2}{R} \Rightarrow \log P = 2 \log E - \log R$

On differentiating, we get

$$\frac{\delta P}{P} = \frac{2}{E}\delta E - \frac{\delta R}{R} \Rightarrow 100\frac{\delta P}{P} = 2 \times \frac{100\,\delta E}{E} - \frac{100\,\delta R}{R}$$

$$100\frac{\delta P}{P} = 2(3) - (-2) = 8 \qquad \left[\text{Given, } \frac{100\,\delta E}{E} = 2\%, \frac{100\,\delta R}{R} = -2\%\right]$$

Percentage change in $P = 8\%$ **Ans.**

**Example 52.** *The diameter and altitude of a can in the shape of a right circular cylinder are measured as 40 and 64 cm respectively. The possible error in each measurement is ± 5%. Find approximately the maximum possible error in the computed value for the volume and the lateral surface. Find the corresponding percentage error.*

**Solution.** Here we have, Diameter of the can ($D$) = 40 cm

$$\frac{100\,\delta D}{D} = \frac{100\,\delta h}{h} = \pm 5\%$$

$$V = \pi r^2 h = \frac{\pi (D^2) h}{4} = \frac{\pi}{4} D^2 h$$

$$\log V = \log \frac{\pi}{4} + 2 \log D + \log h$$

$$\frac{\delta V}{V} = 0 + \frac{2\,\delta D}{D} + \frac{\delta h}{h}$$

$$\frac{\delta V}{V} 100 = \frac{2\delta D}{D} 100 + \frac{\delta h}{h} 100 = 2(\pm 5) + (\pm 5) = \pm 15$$ **Ans.**

Again $$S = 2\pi r l = \pi D h$$

$$\log S = \log \pi + \log D + \log h$$

$$\frac{\delta S}{S} = 0 + \frac{\delta D}{D} + \frac{\delta h}{h}$$

$$\frac{\delta S}{S} 100 = \frac{\delta D}{D} 100 + \frac{\delta h}{h} 100 = (\pm 5) + (\pm 5) = \pm 10$$ **Ans.**

**Example 53.** *The period T of a simple pendulum is*

$$T = 2\pi \sqrt{\frac{l}{8}}.$$

*Find the maximum error in T due to possible errors upto 1% in l and 2% in g.*
(*U.P. I semester winter 2003*)

**Solution.** We have, $T = 2\pi \sqrt{\dfrac{l}{g}}.$

$\Rightarrow$ $$\log T = \log 2\pi + \frac{1}{2} \log l - \frac{1}{2} \log g$$

Differentiating, we get

$$\frac{\partial T}{T} = 0 + \frac{1}{2}\frac{\partial l}{l} - \frac{1}{2}\frac{\partial g}{g}$$

$\Rightarrow$ $$\left(\frac{\partial T}{T}\right) \times 100 = \frac{1}{2}\left[\left(\frac{\partial l}{l}\right) \times 100 - \left(\frac{\partial g}{g}\right) \times 100\right]$$

But $$\frac{\partial l}{l} \times 100 = 1, \frac{\partial g}{g} \times 100 = 2$$

$\therefore$ $$\left(\frac{\partial T}{T}\right) \times 100 = \frac{1}{2}[1 \pm 2] = \frac{3}{2}$$

Maximum error in $T = 1.5\%$ **Ans.**

**Example 54.** *A balloon is in the form of right circular cylinder of radius 1.5 m and length 4 m and is surmounted by hemispherical ends. If the radius is increased by 0.01 m and the length by 0.05 m, find the percentage change in the volume of the balloon.*
(*U.P. I Sem., Dec., 2005, Comp 2002*)

**Solution.** Radius of the cylinder $(r)$ = 1.5 m
Length of the cylinder $(h)$ = 4 m

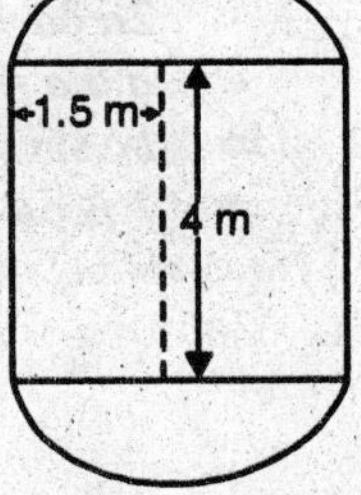


Volume of the balloon = Volume of cylinder + Volume of two hemispheres

$$\text{Volume } (V) = \pi r^2 h + \frac{2}{3}\pi r^3 + \frac{2}{3}\pi r^3 = \pi r^2 h + \frac{4}{3}\pi r^3$$

$$\delta V = \pi\, 2r\, \delta r . h + \pi r^2 . \delta h + \frac{4}{3}\pi\, 3r^2 . \delta r$$

$$\frac{\delta V}{V} = \frac{\pi r[2\,\delta r.h + r.\delta h + 4r\,\delta r]}{\pi r^2 h + \frac{4}{3}\pi r^3} = \frac{2.\delta r.h + r.\delta h + 4r.\delta r}{r h + \frac{4}{3} r^2}$$

$$= \frac{2\times 0.01\times 4 + 1.5\times 0.05 + 4\times 1.5\times 0.01}{1.5\times 4 + \frac{4}{3}(1.5)^2}$$

$$= \frac{0.08 + 0.075 + 0.06}{6+3} = \frac{0.215}{9}$$

$$100\frac{\delta V}{V} = \frac{100\times 0.215}{9} = \frac{21.5}{9} = 2.389\%$$ **Ans.**

**Example 55.** *In estimating the number of bricks in a pile which is measured to be (5m × 10m × 5m), count of bricks is taken as 100 bricks per m³. Find the error in the cost when the tape is stretched 2% beyond its standard length. The cost of bricks is ₹ 2,000 per thousand bricks.* *(U.P., I Semester, Winter 2000)*

**Solution.** Volume $V = x\,y\,z$

$$\log V = \log x + \log y + \log z$$

Differentiating, we get

$$\frac{\delta V}{V} = \frac{\delta x}{x} + \frac{\delta y}{y} + \frac{\delta z}{z}$$

$$100\frac{\delta V}{V} = \frac{100\,\delta x}{x} + \frac{100\,\delta y}{y} + \frac{100\,\delta z}{z} = 2 + 2 + 2$$

$$\frac{100\,\delta V}{V} = 6$$

$$\delta V = \frac{6V}{100} = \frac{6\,(5\times 10\times 5)}{100} = 15 \text{ cubicmetre.}$$

Number of bricks in $\delta V = 15 \times 100 = 1500$

$$\text{Error in cost} = \frac{1500\times 2000}{1000} = 3000$$

Thus error in cost, a loss to the seller of bricks = ₹ 3000. **Ans.**

**Example 56.** *The angles of a triangle are calculated from the sides a, b, c. If small changes δ a, δ b, δ c are made in the sides, show that approximately*

$$\delta A = \frac{a}{2\,\Delta}\left[\delta a - \delta b.\cos C - \delta c.\cos B\right]$$

*where Δ is the area of the triangle and A, B, C are the angles opposite to a, b, c respectively. Verify that δ A + δ B + δ C = 0* *(U.P., I Sem., Winter 2001, A.M.I.E.T.E., 2001)*

**Solution.** We know that

$$\cos A = \frac{b^2 + c^2 - a^2}{2\,b\,c}$$

$\Rightarrow$ $$a^2 = b^2 + c^2 - 2\,b\,c\cos A \qquad \text{...(1)}$$

Differentiating both sides of (1), we get

$$2a\,\delta a = 2b\,\delta b + 2\,c\,\delta c - 2b\,\delta c\cos A - 2\,\delta b\,c\cos A + 2b\,c\sin A\,\delta A \qquad \text{(approx.)}$$

$$a\,\delta a = b\,\delta b + c\,\delta c - b\,\delta c\cos A - \delta b\,c\cos A + b\,c\sin A\,\delta A$$

$\Rightarrow$ $$bc\sin A\,\delta A = a\,\delta a - (b - c\cos A)\,\delta b - (c - b\cos A)\,\delta c$$

$\Rightarrow$ $$2\,\Delta\,\delta A = a\,\delta a - (a\cos C + c\cos A - c\cos A)\,\delta b - (a\cos B + b\cos A - b\cos A)\,\delta c$$

$$\left[\begin{array}{l} \because \Delta = \frac{1}{2}bc\sin A \\ b\cos C + c\cos B = a \end{array}\right]$$

$$2\,\Delta\,\delta A = a\,\delta a - a\,\delta b\cos C - a\,\delta c\cos B$$
$$= a\,(\delta a - \delta b\cos C - \delta c\cos B)$$

$$\Rightarrow \qquad \delta A = \frac{a}{2\Delta}[\delta a - \delta b.\cos C - \delta c.\cos B] \qquad ...(2)$$

Similarly,

$$\delta B = \frac{b}{2\Delta}[\delta b - \delta c.\cos A - \delta a.\cos C] \qquad ...(3)$$

$$\delta C = \frac{c}{2\Delta}[\delta c - \delta a.\cos B - \delta b.\cos A] \qquad ...(4)$$

On adding (2), (3) and (4), we get

$$[\delta A + \delta B + \delta C] = \frac{1}{2\Delta}[(a - b\cos C - c\cos B)\,\delta a + (b - a\cos C - c\cos A)\,\delta b$$
$$+ (c - a\cos B - b\cos A)\,\delta c]$$
$$= \frac{1}{2\Delta}[(a-a)\,\delta a + (b-b)\,\delta b + (c-c)\,\delta c]$$
$$= 0 \qquad [\because\ b\cos C + c\cos B = a]$$ **Verified.**

**Example 57.** *The height h and semi-vertical angle α of a cone are measured, and from there A, the total area of the cone, including the base, is calculated. If h and α are in error by small quantities δh and δα respectively, find the corresponding error in the area. Show further that, if $\alpha = \frac{\pi}{6}$, an error of + 1 per cent in h will be approximately compensated by an error of – 19.8′ in α.* (*A.M.I.E.T.E., Summer 2003*)

**Solution.** Let $l$ be the slant height of the cone and $r$ its radius

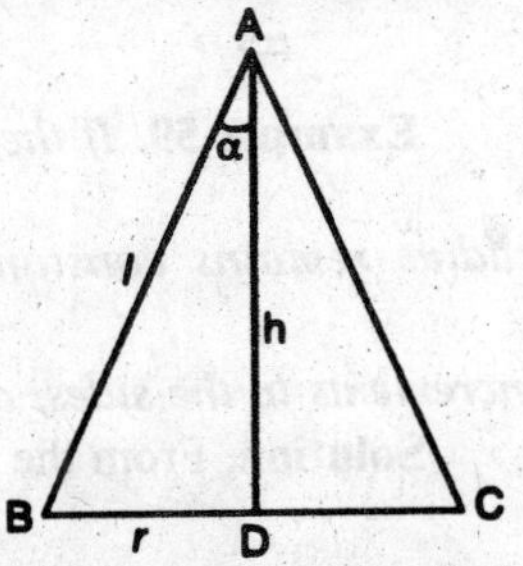


$$l = h\sec\alpha$$
$$r = h\tan\alpha$$
$$A = \pi r^2 + \pi r l$$
$$= \pi h^2\tan^2\alpha + \pi\,(h\tan\alpha)\,(h\sec\alpha)$$
$$= \pi h^2[\tan^2\alpha + \tan\alpha\sec\alpha]$$
$$\delta A = 2\pi h\,\delta h\,[\tan^2\alpha + \tan\alpha\sec\alpha]$$
$$+ \pi h^2[2\tan\alpha\sec^2\alpha\,\delta\alpha + \sec^2\alpha.\,\delta\alpha.\sec\alpha + \tan\alpha\sec\alpha\tan\alpha\,\delta\alpha]$$
$$\delta A = 2\pi h\,[\tan\alpha + \sec\alpha]\tan\alpha.\,\delta h + \pi h^2\,[2\tan\alpha\sec\alpha + \sec^2\alpha + \tan^2\alpha].\sec\alpha.\,\delta\alpha$$
$$\delta A = 2\pi h\,[\tan\alpha + \sec\alpha]\tan\alpha.\,\delta h + \pi h^2\,[\tan\alpha + \sec\alpha]^2\sec\alpha.\,\delta\alpha$$
$$\delta A = \pi h^2\,[\tan\alpha + \sec\alpha]\left[2\tan\alpha\,\frac{\delta h}{h} + (\tan\alpha + \sec\alpha)\sec\alpha.\,\delta\alpha\right]$$

On putting $\delta A = 0,\ \Rightarrow\ \alpha = \frac{\pi}{6},\ \frac{\delta h}{h}\times 100 = 1$, we get

$$\Rightarrow \qquad 0 = \pi h^2\left[\tan\frac{\pi}{6} + \sec\frac{\pi}{6}\right]\left[\left(2\tan\frac{\pi}{6}\right)\frac{1}{100} + \left(\tan\frac{\pi}{6} + \sec\frac{\pi}{6}\right)\sec\frac{\pi}{6}\,\delta\alpha\right]$$

$$\Rightarrow \qquad 0 = \left[\left(2\tan\frac{\pi}{6}\right)\frac{1}{100} + \left(\tan\frac{\pi}{6} + \sec\frac{\pi}{6}\right)\sec\frac{\pi}{6}\,\delta\alpha\right]$$

$$\Rightarrow \qquad 0 = \frac{2}{\sqrt{3}}\,\frac{1}{100} + \left(\frac{1}{\sqrt{3}} + \frac{2}{\sqrt{3}}\right)\frac{2}{\sqrt{3}}\,\delta\alpha \Rightarrow \frac{2}{\sqrt{3}}\,\frac{1}{100} - \left(\frac{1}{\sqrt{3}} + \frac{2}{\sqrt{3}}\right)\frac{2}{\sqrt{3}}\,\delta\alpha$$

$$\Rightarrow \quad \frac{1}{100} = -\left(\frac{1}{\sqrt{3}} + \frac{2}{\sqrt{3}}\right)\delta\alpha \Rightarrow \frac{1}{100} = \frac{-3}{\sqrt{3}}\delta\alpha \Rightarrow \delta\alpha = -\frac{1}{100\sqrt{3}}$$

$$\Rightarrow \quad \delta\alpha = -\frac{1}{100\sqrt{3}}\frac{180}{\pi}\text{ degree} = \left(-\frac{9}{5\sqrt{3}}\right)\frac{60}{\pi}\text{ minutes} = -19.8 \text{ minutes} \quad \textbf{Ans.}$$

**Example 58.** *Find the possible percentage error in computing the parallel resistance r of three resistances $r_1, r_2, r_3$ from the formula $\frac{1}{r} = \frac{1}{r_1} + \frac{1}{r_2} + \frac{1}{r_3}$ if $r_1, r_2, r_3$ are each in error by plus 1.2%.*

**Solution.** Here, $$\frac{1}{r} = \frac{1}{r_1} + \frac{1}{r_2} + \frac{1}{r_3} \quad \ldots(1)$$

Differentiating, we get

$$-\frac{1}{r^2}dr = -\frac{1}{r_1^2}dr_1 - \frac{1}{r_2^2}dr_2 - \frac{1}{r_3^2}dr_3$$

$$\Rightarrow \quad \frac{1}{r}\left(\frac{100\,dr}{r}\right) = \frac{1}{r_1}\left(\frac{100\,dr_1}{r_1}\right) + \frac{1}{r_2}\left(\frac{100\,dr_2}{r_2}\right) + \frac{1}{r_3}\left(\frac{100\,dr_3}{r_3}\right)$$

$$= \frac{1}{r_1}(1.2) + \frac{1}{r_2}(1.2) + \frac{1}{r_3}(1.2) = (1.2)\left[\frac{1}{r_1} + \frac{1}{r_2} + \frac{1}{r_3}\right]$$

$$= 1.2\left(\frac{1}{r}\right) \quad \text{[From (1).]}$$

$$\Rightarrow \quad \frac{100\,dr}{r} = 1.2\% \quad \textbf{Ans.}$$

**Example 59.** *If the sides and angles of a plane triangle vary in such a way that its circum radius remains constant, prove that $\frac{da}{\cos A} + \frac{db}{\cos B} + \frac{dc}{\cos C} = 0$, where da, db, dc are small increments in the sides, a, b, c respectively.*

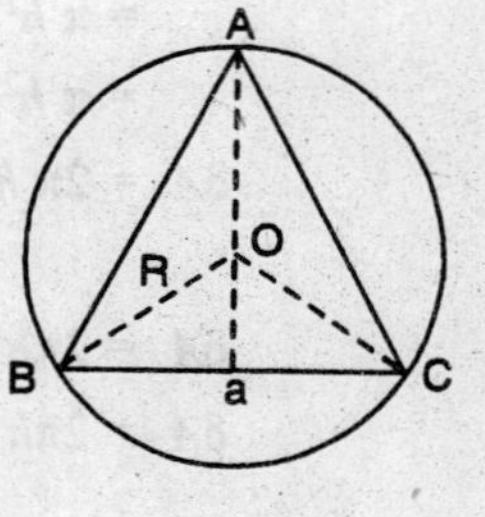


**Solution.** From the sine rule,

$$\frac{a}{\sin A} = \frac{b}{\sin B} = \frac{c}{\sin C}$$

We know that $$R = \frac{a}{2\sin A}, \quad \ldots(1)$$

Differentiating, we get

$$\frac{\partial R}{\partial A} = -\frac{a\cos A}{2\sin^2 A}$$

$$\frac{\partial R}{\partial a} = \frac{1}{2\sin A}$$

By total differentiation $$dR = \frac{\partial R}{\partial A}dA + \frac{\partial R}{\partial a}da$$

$$0 = -\frac{a\cos A}{2\sin^2 A}.\,dA + \frac{1}{2\sin A}.\,da, \quad R \text{ being constant}$$

$$\Rightarrow \quad \frac{\cos A}{\sin A}dA = \frac{1}{\sin A}da$$

$$\Rightarrow \quad \frac{da}{\cos A} = \frac{a}{\sin A}.\, dA = 2R.\, dA \qquad \text{[Using (1)]}$$

$$\Rightarrow \quad \frac{da}{\cos A} = 2\,R\,d\,A \qquad ...(1)$$

Similarly,
$$\frac{db}{\cos B} = 2\,R\,d\,B \qquad ...(2)$$

and
$$\frac{dc}{\cos C} = 2\,R\,d\,C \qquad ...(3)$$

Adding (1), (2) and (3), we have

$$\frac{da}{\cos A} + \frac{db}{\cos B} + \frac{dc}{\cos C} = 2R\,[dA + dB + dC] \qquad ...(4)$$

But in any triangle ABC, $A + B + C = \pi$

Hence, $dA + dB + dC = 0$

Putting value of $dA + dB + dC = 0$ in (4), we get

$$\frac{da}{\cos A} + \frac{db}{\cos B} + \frac{dc}{\cos C} = 2R\,(0) = 0 \Rightarrow \frac{da}{\cos A} + \frac{db}{\cos B} + \frac{dc}{\cos C} = 0 \qquad \textbf{Proved.}$$

**Example 60.** *Compute an approximate value of* $(1.04)^{3.01}$

**Solution.** Let $f(x, y) = x^y$

We have $\frac{\partial f}{\partial x} = y\,x^{y-1}, \quad \frac{\partial f}{\partial y} = x^y \log x$

Here, let
$$x = 1,\ \delta x = 0.04,$$
$$y = 3,\ \delta x = 0.01 \qquad ...(1)$$

Now
$$d\,f = \frac{\partial f}{\partial x}\,dx + \frac{\partial f}{\partial y}\,dy \qquad ...(2)$$
$$= y\,x^{y-1} + x^y \log x$$

Substituting the values from (1) in (2), we get

$$d\,f = (3)\,(1)^{3-1}\,(0.04) + (1)^3 \log(1)\,(0.01) = 0.12$$
$$(1.04)^{3.01} = f(1, 3) + d\,f = 1 + 0.12 = 1.12 \qquad \textbf{Ans.}$$

**Example 61.** *Find* $[(3.82)^2 + 2(2.1)^3]^{\frac{1}{5}}$

**Solution.** Let $f(x, y) = (x^2 + 2y^3)^{\frac{1}{5}}$

Taking
$$x = 4,\ \delta x = 3.82 - 4 = -0.18$$
$$y = 2,\ \delta x = 2.1 - 2 = 0.1$$

$$\frac{\partial f}{\partial x} = \frac{1}{5}[x^2 + 2y^3]^{-\frac{4}{5}}\,(2x) = \frac{2}{5}\,(4)\,[16 + 2(2)^3]^{-\frac{4}{5}} = \frac{8}{5}\left(\frac{1}{16}\right) = \frac{1}{10}$$

$$\frac{\partial f}{\partial y} = \frac{1}{5}[x^2 + 2y^3]^{-\frac{4}{5}}\,(6y^2) = \frac{6}{5}\,(2)^2\,[16 + 2(2)^3]^{-\frac{4}{5}} = \frac{24}{5} \times \frac{1}{16} = \frac{3}{10}$$

By total differentiation, we get

$$df = \frac{\partial f}{\partial x}\partial x + \frac{\partial f}{\partial y}\partial y = \frac{1}{10}(-0.18) + \frac{3}{10}(0.1) = -0.018 + 0.03 = 0.012$$

$$[(3.82)^2 + 2(2.1)^3]^{\frac{1}{5}} = f(4, 2) + df$$

$$= [(4)^2 + 2(2)^3]^{\frac{1}{5}} + 0.012 = 2 + 0.012 = 2.012 \qquad \textbf{Ans.}$$

## EXERCISE 1.9

**1.** If the density $\rho$ of a body be inferred from its weights $W$, $\omega$ in air and water respectively, show that the relative error in $\rho$ due to errors $\delta W$, $\delta \omega$ in $W$, $\omega$ is $\frac{\delta\rho}{\rho} = \frac{-\omega}{W-\omega} \cdot \frac{\delta W}{W} + \frac{\delta\omega}{W-\omega}$.

**2.** The period of oscillation of a pendulum is computed by the formula

$T = 2\pi\sqrt{\frac{l}{g}}$.

Show that the percentage error in $T = \frac{1}{2}$ [% error in $l$ – % error in $g$]

If $l$ = ·6 cm and relative error in $g$ is equal to $\frac{1}{160}$, find the error in the determination of $T$.

(Given $g = 981$ cm/sec$^2$) **Ans.** – 0.00153

**3.** The indicated horse power I of an engine is calculated from the formula.

$I = PLAN/33000$

where $A = \frac{\pi}{4} d^2$. Assuming that errors of $r$ percent may have been made in measuring $P$. $L$, $N$ and $d$. Find the greatest possible error in $I$. **Ans.** 5 $r$ %

**4.** The dimensions of a cone are radius 4 cm, height 6 cm. What is the error in its volume if the scale used in taking the measurement is short by 0.01 cm per cm. **Ans.** 0.96 $\pi$ cm$^3$.

**5.** The work that must be done to propel a ship of displacement $D$ for a distance $s$ in time $t$ is proportional to $s^2 D^{2/3} t^2$.

Find approximately the percentage increase of work necessary when the displacement is increased by 1%, the time is diminished by 1% and the distance is increased by 3%. **Ans.** $\frac{14}{3}$%

**6.** The power $P$ required to propel a ship of length $l$ moving with a velocity $V$ is given by $P = kV^3 l^2$. Find the percentage increase in power if increase in velocity is 3% and increase in length is 4%.

**Ans.** 17%

**7.** In estimating the cost of a pile of bricks measured as 2m × 15 m × 1.2 m, the tape is stretched 1% beyond the standard length if the count is 450 bricks to 1 m$^3$ and bricks cost ₹ 1300 per 1000, find the approximate error in the cost. **Ans.** ₹ 631.80

**8.** In estimating the cost of a pile of bricks measured as 6′ × 50′ × 4′, the tape is stretched 1% beyond the standard length. If the count is 12 bricks to $ft^3$, and bricks cost ₹ 100 per 1000, find the approximate error in the cost. *(U.P. I Sem., Dec. 2004)* **Ans.** 720 bricks, ₹ 25.20

**9.** The sides of a triangle are measured as 12 cm and 15 cm and the angle included between them as 60°. If the lengths can be measured within 1% accuracy while the angle can be measured within 2% accuracy. Find the percentage error in determining (*i*) area of the triangle (*ii*) length of opposite side of the triangle. *(A.M.I.E.T.E., Winter 2002)*

**10.** The voltage $V$ across a resistor is measured with error $h$, and the resistance $R$ is measured with an error $k$. Show that the error in calculating the power $W(V, R) = \frac{V^2}{R}$ generated in the resistor is $\frac{V}{R^2}(2Rh - Vk)$. If $V$ can be measured to an accuracy of 0.5 p.c. and $R$ to an accuracy of 1 p.c., what is the approximate possible percentage error in $W$? **Ans.** Zero percent

**11.** Find the possible percentage error in computing the parallel resistance $r$ of two resistance $r_1$ and $r_2$ from the formula $\frac{1}{r}=\frac{1}{r_1}+\frac{1}{r_2}$, where $r_1$ and $r_2$ are both is error by + 2% each. **Ans.** 2%

**12.** In the manufacture of closed cylindrical boxes with specified sides $a, b, c$ $(a \neq b \neq c)$, small changes of $A\%$, $B\%$, $C\%$ occurred in $a, b, c$, respectively from box to box from the specified dimension. However, the volume and surface area of all boxes were according to specification, show that:

$$\frac{A}{a(b-c)}=\frac{B}{b(c-a)}=\frac{C}{c(a-b)}$$

**13.** Find the percentage error in calculating the area of ellipse $x^2/a^2 + y^2/b^2 = 1$, when error of + 1% is made in measuring the major and minor axes. **Ans.** 2%

*(U.P., I Sem, Jan 2011)*

**14.** If $f = x^2\, y^2\, z^{\frac{1}{10}}$, find the approximate value of $f$, when $x = 1.99$, $y = 3.01$ and $z = 0.98$.

**Ans.** 107.784

**15.** A diameter and altitude of a can in the form of right circular cylinder are measured as 4 cm and 6 cm respectively. The possible error in each measurement is 0.1 cm. Find approximately the maximum possible error in the value computed for the volume and lateral surface.

*(A.M.I.E., Summer 2001)* **Ans.** 5.0336 cm$^3$, 3.146 cm$^2$

**16.** Prove that the relative error of a quotient does not exceed the sum of the relative errors of the dividend and the divisor. *(A.M.I.E., Winter 2001)*

## 1.22 JACOBIANS

If $u$ and $v$ are functions of the two independent variables $x$ and $y$, then the determinant

$$\begin{vmatrix} \frac{\partial u}{\partial x} & \frac{\partial u}{\partial y} \\ \frac{\partial v}{\partial x} & \frac{\partial v}{\partial y} \end{vmatrix}$$

is called the jacobian of $u$, $v$ with respect to $x$, $y$ and is written as

$$\frac{\partial (u, v)}{\partial (x, y)} \text{ or } J\left(\frac{u, v}{x, y}\right)$$

Similarly, the jacobian of $u$, $v$, $w$ with respect to $x$, $y$, $z$ is

$$\frac{\partial (u, v, w)}{\partial (x, y, z)} = \begin{vmatrix} \frac{\partial u}{\partial x} & \frac{\partial u}{\partial y} & \frac{\partial u}{\partial z} \\ \frac{\partial v}{\partial x} & \frac{\partial v}{\partial y} & \frac{\partial v}{\partial z} \\ \frac{\partial w}{\partial x} & \frac{\partial w}{\partial y} & \frac{\partial w}{\partial z} \end{vmatrix}$$

**Example 62.** *If* $x = r\cos\theta$, $y = r\sin\theta$ ; *evaluate* $\frac{\partial (x, y)}{\partial (r, \theta)}$, *and* $\frac{\partial (r, \theta)}{\partial (x, y)}$

**Solution.** We have,

$$x = r\cos\theta, \qquad y = r\sin\theta$$

$$\frac{\partial x}{\partial r} = \cos\theta, \qquad \frac{\partial y}{\partial r} = \sin\theta$$

$$\frac{\partial x}{\partial \theta} = -r\sin\theta, \qquad \frac{\partial y}{\partial \theta} = r\cos\theta$$

$$\frac{\partial (x, y)}{\partial (r, \theta)} = \begin{vmatrix} \frac{\partial x}{\partial r} & \frac{\partial x}{\partial \theta} \\ \frac{\partial y}{\partial r} & \frac{\partial y}{\partial \theta} \end{vmatrix} = \begin{vmatrix} \cos\theta & -r\sin\theta \\ \sin\theta & r\cos\theta \end{vmatrix}$$

$$= r\cos^2\theta + r\sin^2\theta = r(\cos^2\theta + \sin^2\theta) = r$$

Now, $r^2 = x^2 + y^2, \qquad \theta = \tan^{-1}\frac{y}{x}$

$$\frac{\partial r}{\partial x} = \frac{x}{r}, \quad \frac{\partial \theta}{\partial x} = \frac{-y}{x^2+y^2} = -\frac{y}{r^2}$$

$$\frac{\partial r}{\partial y} = \frac{y}{r}, \quad \frac{\partial \theta}{\partial y} = \frac{x}{x^2+y^2} = \frac{x}{r^2}$$

$$\frac{\partial(r,\theta)}{\partial(x,y)} = \begin{vmatrix} \frac{\partial r}{\partial x} & \frac{\partial r}{\partial y} \\ \frac{\partial \theta}{\partial x} & \frac{\partial \theta}{\partial y} \end{vmatrix} = \begin{vmatrix} \frac{x}{r} & \frac{y}{r} \\ \frac{-y}{r^2} & \frac{x}{r^2} \end{vmatrix} = \frac{x^2}{r^3} + \frac{y^2}{r^3} = \frac{x^2+y^2}{r^3} = \frac{r^2}{r^3} = \frac{1}{r} \quad \textbf{Ans.}$$

**Note :** $\frac{\partial(x,y)}{\partial(r,\theta)} \times \frac{\partial(r,\theta)}{\partial(x,y)} = r \cdot \frac{1}{r} = 1$

**Example 63.** *If $x = a\cosh\xi\cos\eta$, $y = a\sinh\xi\sin\eta$, show that*

$$\frac{\partial(x,y)}{\partial(\xi,\eta)} = \frac{a^2}{2}(\cosh 2\xi - \cos 2\eta)$$

**Solution.** Here, we have, $x = a\cosh\xi\cos\eta$

$y = a\sinh\xi\sin\eta$

$$\frac{\partial(x,y)}{\partial(\xi,\eta)} = \begin{vmatrix} \frac{\partial x}{\partial \xi} & \frac{\partial x}{\partial \eta} \\ \frac{\partial y}{\partial \xi} & \frac{\partial y}{\partial \eta} \end{vmatrix} = \begin{vmatrix} a\sinh\xi\cos\eta & -a\cosh\xi\sin\eta \\ a\cosh\xi\sin\eta & a\sinh\xi\cos\eta \end{vmatrix}$$

$$= a^2\begin{vmatrix} \sinh\xi\cos\eta & -\cosh\xi\sin\eta \\ \cosh\xi\sin\eta & \sinh\xi\cos\eta \end{vmatrix} = a^2[\sinh^2\xi\cos^2\eta + \cosh^2\xi\sin^2\eta]$$

$$= a^2[\sinh^2\xi(1-\sin^2\eta) + (1+\sinh^2\xi)\sin^2\eta]$$

$$= a^2[\sinh^2\xi - \sinh^2\xi\ \sin^2\eta + \sin^2\eta + \sinh^2\xi\ \sin 2\eta]$$

$$= a^2[\sinh^2\xi + \sin^2\eta] = \frac{a^2}{2}[\cosh 2\xi - 1 + 1 - \cos 2\eta] = \frac{a^2}{2}[\cosh 2\xi - \cos 2\eta] \quad \textbf{Proved.}$$

**Example 64.** *If $y_1 = \frac{x_2x_3}{x_1}$, $y_2 = \frac{x_3x_1}{x_2}$, $y_3 = \frac{x_1x_2}{x_3}$.*

*Show that the Jacobian of $y_1, y_2, y_3$ with respect to $x_1, x_2, x_3$ is 4.*

(*U.P. I Sem. Jan 2011;* 2004, *Comp.* 2002, *A.M.I.E., Summer* 2002, 2000, *Winter* 2001)

**Solution.** Here, we have $y_1 = \frac{x_2\,x_3}{x_1}, y_2 = \frac{x_3\,x_1}{x_2}, y_3 = \frac{x_1\,x_2}{x_3}$

$$\frac{\partial(y_1,y_2,y_3)}{\partial(x_1,x_2,x_3)} = \begin{vmatrix} \frac{\partial y_1}{\partial x_1} & \frac{\partial y_1}{\partial x_2} & \frac{\partial y_1}{\partial x_3} \\ \frac{\partial y_2}{\partial x_1} & \frac{\partial y_2}{\partial x_2} & \frac{\partial y_2}{\partial x_3} \\ \frac{\partial y_3}{\partial x_1} & \frac{\partial y_3}{\partial x_2} & \frac{\partial y_3}{\partial x_3} \end{vmatrix} = \begin{vmatrix} -\frac{x_2\,x_3}{x_1^2} & \frac{x_3}{x_1} & \frac{x_2}{x_1} \\ \frac{x_3}{x_2} & -\frac{x_3\,x_1}{x_2^2} & \frac{x_1}{x_2} \\ \frac{x_2}{x_3} & \frac{x_1}{x_3} & -\frac{x_1\,x_2}{x_3^2} \end{vmatrix}$$

$$= \frac{1}{x_1^2\,x_2^2\,x_3^2}\begin{vmatrix} -x_2\,x_3 & x_3\,x_1 & x_1\,x_2 \\ x_2\,x_3 & -x_3\,x_1 & x_1\,x_2 \\ x_2\,x_3 & x_3\,x_1 & -x_1\,x_2 \end{vmatrix} = \frac{x_1^2\,x_2^2\,x_3^2}{x_1^2\,x_2^2\,x_3^2}\begin{vmatrix} -1 & 1 & 1 \\ 1 & -1 & 1 \\ 1 & 1 & -1 \end{vmatrix}$$

$$= -1(1-1) - 1(-1-1) + 1(1+1) = 0 + 2 + 2 = 4 \quad \textbf{Proved.}$$

**Example 65.** *If* $x = r \sin\theta \cos\varphi$,
$y = r \sin\theta \sin\varphi$,
$z = r \cos\theta$,

*Show that* $\dfrac{\partial(x, y, z)}{\partial(r, \theta, \varphi)} = r^2 \sin\theta$. (*U.P., I Semester, Winter* 2000)

**Solution.** We have, $x = r \sin\theta \cos\phi$, $y = r \sin\theta \sin\phi$, $z = r\cos\theta$

$$\frac{\partial x}{\partial r} = \sin\theta\cos\phi, \quad \frac{\partial y}{\partial r} = \sin\theta\sin\phi, \quad \frac{\partial z}{\partial r} = \cos\theta$$

$$\frac{\partial x}{\partial \theta} = r\cos\theta\cos\phi, \quad \frac{\partial y}{\partial \theta} = r\cos\theta\sin\phi, \quad \frac{\partial z}{\partial \theta} = -r\sin\theta$$

$$\frac{\partial x}{\partial \varphi} = -r\sin\theta\sin\phi, \quad \frac{\partial y}{\partial \varphi} = r\sin\theta\cos\phi, \quad \frac{\partial z}{\partial \varphi} = 0$$

$$\frac{\partial(x, y, z)}{\partial(r, \theta, \varphi)} = \begin{vmatrix} \frac{\partial x}{\partial r} & \frac{\partial x}{\partial \theta} & \frac{\partial x}{\partial \varphi} \\ \frac{\partial y}{\partial r} & \frac{\partial y}{\partial \theta} & \frac{\partial y}{\partial \varphi} \\ \frac{\partial z}{\partial r} & \frac{\partial z}{\partial \theta} & \frac{\partial z}{\partial \varphi} \end{vmatrix} = \begin{vmatrix} \sin\theta\cos\varphi & r\cos\theta\cos\varphi & -r\sin\theta\sin\varphi \\ \sin\theta\sin\varphi & r\cos\theta\sin\varphi & r\sin\theta\cos\varphi \\ \cos\theta & -r\sin\theta & 0 \end{vmatrix}$$

$$= r^2\sin\theta \begin{vmatrix} \sin\theta\cos\phi & \cos\theta\cos\phi & -\sin\phi \\ \sin\theta\sin\phi & \cos\theta\sin\phi & \cos\phi \\ \cos\theta & -\sin\theta & 0 \end{vmatrix}$$

$= r^2 \sin\theta\, [\sin\theta\cos\phi\,(0 + \sin\theta\cos\phi) - \cos\theta\cos\phi\,(0 - \cos\phi\cos\theta)$
$- \sin\phi\,(-\sin^2\theta\sin\phi - \cos^2\theta\sin\phi)]$
$= r^2\sin\theta\,[\sin^2\theta\cos^2\phi + \cos^2\theta\cos^2\phi + \sin^2\theta\sin^2\phi + \cos^2\theta\sin^2\phi]$
$= r^2\sin\theta\,[(\sin^2\theta + \cos^2\theta)\cos^2\phi + (\sin^2\theta + \cos^2\theta)\sin^2\phi]$
$= r^2\sin\theta\,[\cos^2\phi + \sin^2\phi] = r^2\sin\theta$ **Ans.**

## EXERCISE 1.10

1. If $u = x^2$, $v = y^2$, find $\dfrac{\partial(u, v)}{\partial(x, y)}$ **Ans.** $4xy$

2. If $u = \dfrac{y - x}{1 + xy}$ and $v = \tan^{-1} y - \tan^{-1} x$, find $\dfrac{\partial(u, v)}{\partial(x, y)}$ **Ans.** 0

3. If $u = xyz$, $v = xy + yz + zx$, $w = x + y + z$, compute $\dfrac{\partial(u, v, w)}{\partial(x, y, z)}$ **Ans.** $(x - y)(y - z)(z - x)$

4. If $x = r\cos\theta$, $y = r\sin\theta$, $z = z$ find $\dfrac{\partial(x, y, z)}{\partial(r, \theta, z)}$.

5. If $u_1 = \dfrac{x_1}{x_n}$, $u_2 = \dfrac{x_2}{x_n}$, ...... $u_{n-1} = \dfrac{x_{n-1}}{x_n}$ and $x_1^2 + x_2^2 + x_3^2 + \ldots\ldots x_n^2 = 1$ find $\dfrac{\partial(u_1, u_2, \ldots\ldots u_{n-1})}{\partial(x_1, x_2, \ldots\ldots x_{n-1})}$
**Ans.** $\dfrac{1}{x_n^{n-1}}$

6. Find the value of $\dfrac{\partial(y_1, y_2, y_3)}{\partial(x_1, x_2, x_3)}$, $y_1 = (1 - x_1)$, $y_2 = x_1(1 - x_2)$, $y_3 = x_1x_2(1 - x_3)$ **Ans.** $-x_1^2 x_2$

7. If $u = \dfrac{x}{y - z}$, $v = \dfrac{y}{z - x}$, $w = \dfrac{z}{x - y}$ then show that $\dfrac{\partial(u, v, w)}{\partial(x, y, z)} = 0$

8. Fill in the blanks

(*i*) If $x = r\cos\theta$, $y = r\sin\theta$, then the value of Jacobian $\dfrac{\partial(x, y)}{\partial(r, \theta)}$ is ........ **Ans.** $r$

(*ii*) If $u = x(1-y)$, $v = xy$, then the value of the Jacobian $\dfrac{\partial(u,v)}{\partial(x,y)} = \ldots\ldots\ldots$ **Ans.** $x$

## 1.23 PROPERTIES OF JACOBIANS

### (1) First Property

**If $u$ and $v$ are the functions of $x$ and $y$, then**

$$\frac{\partial(u,v)}{\partial(x,y)} \times \frac{\partial(x,y)}{\partial(u,v)} = 1$$

(*U. P. I Semester Dec. 2005*)

**Proof.** Let $u = f(x, y)$ ...(1)

$v = \varphi(x, y)$ ...(2)

$$\frac{\partial(u,v)}{\partial(x,y)} \times \frac{\partial(x,y)}{\partial(u,v)} = \begin{vmatrix} \frac{\partial u}{\partial x} & \frac{\partial u}{\partial y} \\ \frac{\partial v}{\partial x} & \frac{\partial v}{\partial y} \end{vmatrix} \times \begin{vmatrix} \frac{\partial x}{\partial u} & \frac{\partial x}{\partial v} \\ \frac{\partial y}{\partial u} & \frac{\partial y}{\partial v} \end{vmatrix}$$

On interchanging the rows and columns of second determinant

$$= \begin{vmatrix} \frac{\partial u}{\partial x} & \frac{\partial u}{\partial y} \\ \frac{\partial v}{\partial x} & \frac{\partial v}{\partial y} \end{vmatrix} \times \begin{vmatrix} \frac{\partial x}{\partial u} & \frac{\partial y}{\partial u} \\ \frac{\partial x}{\partial v} & \frac{\partial y}{\partial v} \end{vmatrix} = \begin{vmatrix} \frac{\partial u}{\partial x}\cdot\frac{\partial x}{\partial u} + \frac{\partial u}{\partial y}\cdot\frac{\partial y}{\partial u} & \frac{\partial u}{\partial x}\cdot\frac{\partial x}{\partial v} + \frac{\partial u}{\partial y}\cdot\frac{\partial y}{\partial v} \\ \frac{\partial v}{\partial x}\cdot\frac{\partial x}{\partial u} + \frac{\partial v}{\partial y}\cdot\frac{\partial y}{\partial u} & \frac{\partial v}{\partial x}\cdot\frac{\partial x}{\partial v} + \frac{\partial v}{\partial y}\cdot\frac{\partial y}{\partial v} \end{vmatrix} \quad \ldots(3)$$

On differentiating (1) and (2) w. r. t. $u$ and $v$, we get

$$\left.\begin{aligned} \frac{\partial u}{\partial u} &= 1 = \frac{\partial u}{\partial x}\cdot\frac{\partial x}{\partial u} + \frac{\partial u}{\partial y}\cdot\frac{\partial y}{\partial u} \\ \frac{\partial u}{\partial v} &= 0 = \frac{\partial u}{\partial x}\cdot\frac{\partial x}{\partial v} + \frac{\partial u}{\partial y}\cdot\frac{\partial y}{\partial v} \\ \frac{\partial v}{\partial v} &= 1 = \frac{\partial v}{\partial x}\cdot\frac{\partial x}{\partial v} + \frac{\partial v}{\partial y}\cdot\frac{\partial y}{\partial v} \\ \frac{\partial v}{\partial u} &= 0 = \frac{\partial v}{\partial x}\cdot\frac{\partial x}{\partial u} + \frac{\partial v}{\partial y}\cdot\frac{\partial y}{\partial u} \end{aligned}\right\} \quad \ldots(4)$$

On making substitutions from (4) in (3), we get

$$\frac{\partial(u,v)}{\partial(x,y)} \times \frac{\partial(x,y)}{\partial(u,v)} = \begin{vmatrix} 1 & 0 \\ 0 & 1 \end{vmatrix} = 1$$

**Proved.**

**Example 66.** *If* $x = uv$, $y = \dfrac{u+v}{u-v}$, *find* $\dfrac{\partial(u,v)}{\partial(x,y)}$.

**Solution.** Here it is easy to find $\dfrac{\partial x}{\partial u}, \dfrac{\partial x}{\partial v}, \dfrac{\partial y}{\partial u}, \dfrac{\partial y}{\partial v}$. But to find $\dfrac{\partial u}{\partial x}, \dfrac{\partial u}{\partial y}, \dfrac{\partial v}{\partial x}, \dfrac{\partial v}{\partial y}$ is comparatively difficult. So we first find $\dfrac{\partial(x,y)}{\partial(u,v)}$

$$\frac{\partial(x,y)}{\partial(u,v)} = \begin{vmatrix} \frac{\partial x}{\partial u} & \frac{\partial x}{\partial v} \\ \frac{\partial y}{\partial u} & \frac{\partial y}{\partial v} \end{vmatrix} = \begin{vmatrix} v & u \\ \frac{-2v}{(u-v)^2} & \frac{2u}{(u-v)^2} \end{vmatrix} = \frac{uv}{(u-v)^2}\begin{vmatrix} 1 & 1 \\ -2 & 2 \end{vmatrix} = \frac{uv}{(u-v)^2}(2+2) = \frac{4uv}{(u-v)^2}$$

But $$\frac{\partial(u,v)}{\partial(x,y)} \times \frac{\partial(x,y)}{\partial(u,v)} = 1 \Rightarrow \frac{\partial(u,v)}{\partial(x,y)} \times \frac{4uv}{(u-v)^2} = 1 \Rightarrow \frac{\partial(u,v)}{\partial(x,y)} = \frac{(u-v)^2}{4uv}$$ **Ans.**

**Example 67.** *If* $u = xyz$, $v = x^2 + y^2 + z^2$, $w = x + y + z$, *find* $J = \dfrac{\partial(x,y,z)}{\partial(u,v,w)}$.

**Solution.** Since $u$, $v$, $w$ are explicitly given, so first we evaluate (*U.P. I Sem., Winter 2002*)

$$J' = \frac{\partial(u,v,w)}{\partial(x,y,z)}$$

$$J' = \begin{vmatrix} \frac{\partial u}{\partial x} & \frac{\partial u}{\partial y} & \frac{\partial u}{\partial z} \\ \frac{\partial v}{\partial x} & \frac{\partial v}{\partial y} & \frac{\partial v}{\partial z} \\ \frac{\partial w}{\partial x} & \frac{\partial w}{\partial y} & \frac{\partial w}{\partial z} \end{vmatrix} = \begin{vmatrix} yz & zx & xy \\ 2x & 2y & 2z \\ 1 & 1 & 1 \end{vmatrix}$$

$$= yz(2y - 2z) - zx(2x - 2z) + xy(2x - 2y) = 2[yz(y - z) - zx(x - z) + xy(x - y)]$$
$$= 2[x^2y - x^2z - xy^2 + xz^2 + y^2z - yz^2] = 2[x^2(y - z) - x(y^2 - z^2) + yz(y - z)]$$
$$= 2(y - z)[x^2 - x(y + z) + yz] = 2(y - z)[y(z - x) - x(z - x)]$$
$$= 2(y - z)(z - x)(y - x) = -2(x - y)(y - z)(z - x)$$

Hence, by $JJ' = 1$, we have

$$J = \frac{\partial(x,y,z)}{\partial(u,v,w)} = \frac{-1}{2(x-y)(y-z)(z-x)}$$ **Ans.**

## EXERCISE 1.11

1. Given $u = x^2 - y^2$, $v = 2xy$, calculate $\dfrac{\partial(x,y)}{\partial(u,v)}$ **Ans.** $\dfrac{1}{4(x^2+y^2)}$
2. If $x = uv$, $y = \dfrac{u+v}{u-v}$, find $\dfrac{\partial(u,v)}{\partial(x,y)}$ **Ans.** $\dfrac{(u-v)^2}{4uv}$
3. If $x = r\sin\theta\cos\phi$, $y = r\sin\theta\sin\phi$, $z = r\cos\theta$, find $\dfrac{\partial(r,\theta,\varphi)}{\partial(x,y,z)}$ **Ans.** $\dfrac{1}{r^2\sin\theta}$
4. Verify $JJ' = 1$, if $x = uv$, $y = \dfrac{u}{v}$
5. Verify $JJ' = 1$, if $x = e^v \sec u$, $y = e^v \tan u$.
6. Verify $JJ' = 1$, if $x = \sin\theta\cos\phi$, $y = \sin\theta\sin\phi$

**(2) Second Property (Chain Rule)**

**If $u$, $v$ are the functions of $r$, $s$ where $r$, $s$ are functions of $x$, $y$, then**

$$\frac{\partial(u,v)}{\partial(x,y)} = \frac{\partial(u,v)}{\partial(r,s)} \times \frac{\partial(r,s)}{\partial(x,y)}$$ (*U.P. I Sem. Jan 2011*)

**Proof.**
$$\frac{\partial(u,v)}{\partial(r,s)} \times \frac{\partial(r,s)}{\partial(x,y)} = \begin{vmatrix} \frac{\partial u}{\partial r} & \frac{\partial u}{\partial s} \\ \frac{\partial v}{\partial r} & \frac{\partial v}{\partial s} \end{vmatrix} \times \begin{vmatrix} \frac{\partial r}{\partial x} & \frac{\partial r}{\partial y} \\ \frac{\partial s}{\partial x} & \frac{\partial s}{\partial y} \end{vmatrix}$$

**On interchanging the columns and rows in second determinant**

$$= \begin{vmatrix} \frac{\partial u}{\partial r} & \frac{\partial u}{\partial s} \\ \frac{\partial v}{\partial r} & \frac{\partial v}{\partial s} \end{vmatrix} \times \begin{vmatrix} \frac{\partial r}{\partial x} & \frac{\partial s}{\partial x} \\ \frac{\partial r}{\partial y} & \frac{\partial s}{\partial y} \end{vmatrix} = \begin{vmatrix} \frac{\partial u}{\partial r}\cdot\frac{\partial r}{\partial x} + \frac{\partial u}{\partial s}\cdot\frac{\partial s}{\partial x} & \frac{\partial u}{\partial r}\cdot\frac{\partial r}{\partial y} + \frac{\partial u}{\partial s}\cdot\frac{\partial s}{\partial y} \\ \frac{\partial v}{\partial r}\cdot\frac{\partial r}{\partial x} + \frac{\partial v}{\partial s}\cdot\frac{\partial s}{\partial x} & \frac{\partial v}{\partial r}\cdot\frac{\partial r}{\partial y} + \frac{\partial v}{\partial s}\cdot\frac{\partial s}{\partial y} \end{vmatrix}$$

$$= \begin{vmatrix} \frac{\partial u}{\partial x} & \frac{\partial u}{\partial y} \\ \frac{\partial v}{\partial x} & \frac{\partial v}{\partial y} \end{vmatrix} = \frac{\partial (u, v)}{\partial (x, y)}$$ **Proved.**

Similarly, $$\frac{\partial (u, v, w)}{\partial (x, y, z)} = \frac{\partial (u, v, w)}{\partial (r, s, t)} \times \frac{\partial (r, s, t)}{\partial (x, y, z)}$$

**Example 68.** *Find the value of the Jacobian* $\frac{\partial (u, v)}{\partial (r, \theta)}$, *where* $u = x^2 - y^2$, $v = 2x\,y$ *and* $x = r \cos \theta$, $y = r \sin \theta$.

**Solution.** $u = x^2 - y^2$, $v = 2\,x\,y$

$$\frac{\partial (u, v)}{\partial (x, y)} = \begin{vmatrix} \frac{\partial u}{\partial x} & \frac{\partial u}{\partial y} \\ \frac{\partial v}{\partial x} & \frac{\partial v}{\partial y} \end{vmatrix} = \begin{vmatrix} 2x & -2y \\ 2y & 2x \end{vmatrix} = 4\,(x^2 + y^2) = 4\,r^2$$

$$\frac{\partial (x, y)}{\partial (r, \theta)} = \begin{vmatrix} \frac{\partial x}{\partial r} & \frac{\partial x}{\partial \theta} \\ \frac{\partial y}{\partial r} & \frac{\partial y}{\partial \theta} \end{vmatrix} = \begin{vmatrix} \cos \theta & -r \sin \theta \\ \sin \theta & r \cos \theta \end{vmatrix} = r \cos^2 \theta + r \sin^2 \theta = r$$

$$\frac{\partial (u, v)}{\partial (r, \theta)} = \frac{\partial (u, v)}{\partial (x, y)} \times \frac{\partial (x, y)}{\partial (r, \theta)} = 4r^2 \,.\, r = 4r^3$$ **Ans.**

## EXERCISE 1.12

1. If $u = e^x \cos y$, $v = e^x \sin y$, *where* $x = lr + sm$, $y = mr - sl$, verify $\frac{\partial (u, v)}{\partial (x, y)} \times \frac{\partial (x, y)}{\partial (r, s)} = \frac{\partial (u, v)}{\partial (r, s)}$

2. If $u = x\,(1 - r^2)^{-\frac{1}{2}}$, $v = y\,(1 - r^2)^{-\frac{1}{2}}$

   $w = z\,(1 - r^2)^{-\frac{1}{2}}$ where $r^2 = x^2 + y^2 + z^2$

   Show that $\frac{\partial (u, v, w)}{\partial (x, y, z)} = (1 - r^2)^{-\frac{5}{2}}$ *(Q. Bank, U. P. 2001)*

3. If $u = x + y + z$, $u^2\,v = y + z$, $u^3\,w = z$, show that $\frac{\partial (u, v, w)}{\partial (x, y, z)} = u^{-5}$

   **Hint.** Put $r = x + y + z$, $s = y + z$, $t = z$

   $u = r \quad u^2 v = s \quad u^3 w = t$

   $$\frac{\partial (u, v, w)}{\partial (x, y, z)} = \frac{\partial (u, v, w)}{\partial (r, s, t)} \times \frac{\partial (r, s, t)}{\partial (x, y, z)} = \frac{1}{\left(\frac{\partial (r, s, t)}{\partial (u, v, w)}\right)} \times \left(\frac{\partial (r, s, t)}{\partial (x, y, z)}\right)$$

4. If $u = x + y + z$, $uv = y + z$, $uvw = z$. Evaluate $\frac{\partial (x, y, z)}{\partial (u, v, w)}$ *( U. P. I Sem. Winter 2003 )* **Ans.** $u^2 v$

5. If $u^3 + v^3 = x + y$, $u^2 + v^2 = x^3 + y^3$, show that $\frac{\partial (u, v)}{\partial (x, y)} = \frac{1}{2} \frac{(y^2 - x^2)}{uv(u - v)}$

**(3) Third Property**

If functions $u$, $v$, $w$ of three independent variables $x$, $y$, $z$ are not independent, then

$$\frac{\partial (u, v, w)}{\partial (x, y, z)} = 0$$

**Proof.** As $u, v, w$ are not indepentent, then $f(u, v, w) = 0$ ...(1)

Differentiating (1) w.r.t $x, y, z$, we get

$$\frac{\partial f}{\partial u}\cdot\frac{\partial u}{\partial x}+\frac{\partial f}{\partial v}\cdot\frac{\partial v}{\partial x}+\frac{\partial f}{\partial w}\cdot\frac{\partial w}{\partial x}=0 \quad ...(2)$$

$$\frac{\partial f}{\partial u}\cdot\frac{\partial u}{\partial y}+\frac{\partial f}{\partial v}\cdot\frac{\partial v}{\partial y}+\frac{\partial f}{\partial w}\cdot\frac{\partial w}{\partial y}=0 \quad ...(3)$$

$$\frac{\partial f}{\partial u}\cdot\frac{\partial u}{\partial z}+\frac{\partial f}{\partial v}\cdot\frac{\partial v}{\partial z}+\frac{\partial f}{\partial w}\cdot\frac{\partial w}{\partial z}=0 \quad ...(4)$$

Eliminating $\frac{\partial f}{\partial u},\frac{\partial f}{\partial v},\frac{\partial f}{\partial w}$ from (2), (3) and (4), we have

$$\begin{vmatrix}\frac{\partial u}{\partial x} & \frac{\partial v}{\partial x} & \frac{\partial w}{\partial x}\\ \frac{\partial u}{\partial y} & \frac{\partial v}{\partial y} & \frac{\partial w}{\partial y}\\ \frac{\partial u}{\partial z} & \frac{\partial v}{\partial z} & \frac{\partial w}{\partial z}\end{vmatrix}=0$$

On interchanging rows and columns, we get

$$\Rightarrow \quad \begin{vmatrix}\frac{\partial u}{\partial x} & \frac{\partial u}{\partial y} & \frac{\partial u}{\partial z}\\ \frac{\partial v}{\partial x} & \frac{\partial v}{\partial y} & \frac{\partial v}{\partial z}\\ \frac{\partial w}{\partial x} & \frac{\partial w}{\partial y} & \frac{\partial w}{\partial z}\end{vmatrix}=0$$

$$\Rightarrow \quad \frac{\partial(u,v,w)}{\partial(x,y,z)}=0$$ **Proved.**

**Converse (The sufficient condition)**

If it is given that $\frac{\partial(u,v,w)}{\partial(x,y,z)}=0$ and $u, v, w$ are not independent of one another then they are connected by a relation $f(u, v, w) = 0$.

**Example 69.** *If $u = xy + yz + zx$, $v = x^2 + y^2 + z^2$ and $w = x + y + z$, determine whether there is a functional relationship between u, v, w and if so, find it.*

**Solution.** We have, $u = xy + yz + zx$, $v = x^2 + y^2 + z^2$, $w = x + y + z$

$$\frac{\partial(u,v,w)}{\partial(x,y,z)}=\begin{vmatrix}\frac{\partial u}{\partial x} & \frac{\partial u}{\partial y} & \frac{\partial u}{\partial z}\\ \frac{\partial v}{\partial x} & \frac{\partial v}{\partial y} & \frac{\partial v}{\partial z}\\ \frac{\partial w}{\partial x} & \frac{\partial w}{\partial y} & \frac{\partial w}{\partial z}\end{vmatrix}=\begin{vmatrix}y+z & z+x & x+y\\ 2x & 2y & 2z\\ 1 & 1 & 1\end{vmatrix}$$

$$=2\begin{vmatrix}y+z & z+x & x+y\\ x & y & z\\ 1 & 1 & 1\end{vmatrix}=2\begin{vmatrix}x+y+z & x+y+z & x+y+z\\ x & y & z\\ 1 & 1 & 1\end{vmatrix} \quad R_1 \to R_1 + R_2$$

$$= 2(x+y+z)\begin{vmatrix} 1 & 1 & 1 \\ x & y & z \\ 1 & 1 & 1 \end{vmatrix} = 0 \quad (R_1 = R_3)$$

Hence, the functional relationship exists between $u$, $v$ and $w$.

Now, $w^2 = (x+y+z)^2 = x^2 + y^2 + z^2 + 2(xy + yz + zx)$

$w^2 = v + 2u$

$w^2 - v - 2u = 0$ which is the required relationship. **Ans.**

**Example 70.** *Verify whether the following functions are functionally dependent, and if so, find the relation between them.*

$$u = \frac{x+y}{1-xy}, \qquad v = \tan^{-1} x + \tan^{-1} y$$

**Solution.** $$\frac{\partial(u, v)}{\partial(x, y)} = \begin{vmatrix} \frac{\partial u}{\partial x} & \frac{\partial u}{\partial y} \\ \frac{\partial v}{\partial x} & \frac{\partial v}{\partial y} \end{vmatrix} = \begin{vmatrix} \frac{1+y^2}{(1-xy)^2} & \frac{1+x^2}{(1-xy)^2} \\ \frac{1}{1+x^2} & \frac{1}{1+y^2} \end{vmatrix} = \frac{1}{(1-xy)^2} - \frac{1}{(1-xy)^2} = 0$$

Hence $u$, $v$ are functionally related.

$$\tan^{-1} x + \tan^{-1} y = \tan^{-1} \frac{x+y}{1-xy}$$

$$v = \tan^{-1} u$$

$\Rightarrow$ $u = \tan v$. **Ans.**

## EXERCISE 1.13

1. Verify whether $u = \dfrac{x-y}{x+y}$, $v = \dfrac{x+y}{x}$ are functionally dependent, and if so, find the relation between them. **Ans.** $u = \dfrac{2-v}{v}$

2. Determine functional dependence and find relation between $u = \dfrac{x-y}{x+y}$, $v = \dfrac{xy}{(x+y)^2}$ **Ans.** $4v = 1 - u^2$

3. Are $x+y-z$, $x-y+z$, $x^2+y^2+z^2-2yz$ functionally dependent ? If so, find a relation between them. **Ans.** $u^2 + v^2 = 2w$

4. If $u = x+y+z$, $v = x^2+y^2+z^2$, $w = x^3+y^3+z^3-3xyz$, prove that $u$, $v$, $w$ are not independent and find the relation between them. **Ans.** $2w = u(3v - u^2)$

5. Are the following two functions of $x$, $y$, $z$ functionally dependent ? If so find the relation between them.
$u = \dfrac{x-y}{x+z}$, $v = \dfrac{x+z}{y+z}$ **Ans.** $v = \dfrac{1}{1-u}$

6. If $u = \dfrac{x+y}{z}$, $v = \dfrac{y+z}{x}$, $w = \dfrac{y(x+y+z)}{xz}$, show that $u$, $v$, $w$ are not independent and find the relation between them. *(U.P., Ist Semester, 2009)* **Ans.** $uv - w = 1$

## 1.24 JACOBIAN OF IMPLICIT FUNCTIONS

The variables $x$, $y$, $u$, $v$ are connected by implicit functions

$$f_1(x, y, u, v) = 0 \qquad \text{... (1)}$$

$$f_2(x, y, u, v) = 0 \qquad \text{... (2)}$$

where $u$, $v$ are implicit functions of $x$, $y$.

Differentiating (1) and (2) w.r.t. $x$ and $y$, we get

$$\frac{\partial f_1}{\partial x}+\frac{\partial f_1}{\partial u}\frac{\partial u}{\partial x}+\frac{\partial f_1}{\partial v}\frac{\partial v}{\partial x}=0 \qquad \dots (3)$$

$$\frac{\partial f_1}{\partial y}+\frac{\partial f_1}{\partial u}\frac{\partial u}{\partial y}+\frac{\partial f_1}{\partial v}\frac{\partial v}{\partial y}=0 \qquad \dots (4)$$

$$\frac{\partial f_2}{\partial x}+\frac{\partial f_2}{\partial u}\frac{\partial u}{\partial x}+\frac{\partial f_2}{\partial v}\frac{\partial v}{\partial x}=0 \qquad \dots (5)$$

$$\frac{\partial f_2}{\partial y}+\frac{\partial f_2}{\partial u}\frac{\partial u}{\partial y}+\frac{\partial f_2}{\partial v}\frac{\partial v}{\partial y}=0 \qquad \dots (6)$$

Now, we have

$$\frac{\partial(f_1,f_2)}{\partial(u,v)}\times\frac{\partial(u,v)}{\partial(x,y)}=\begin{vmatrix}\frac{\partial f_1}{\partial u} & \frac{\partial f_1}{\partial v}\\ \frac{\partial f_2}{\partial u} & \frac{\partial f_2}{\partial v}\end{vmatrix}\times\begin{vmatrix}\frac{\partial u}{\partial x} & \frac{\partial u}{\partial y}\\ \frac{\partial v}{\partial x} & \frac{\partial v}{\partial y}\end{vmatrix}$$

$$=\begin{vmatrix}\frac{\partial f_1}{\partial u}\frac{\partial u}{\partial x}+\frac{\partial f_1}{\partial v}\frac{\partial v}{\partial x} & \frac{\partial f_1}{\partial u}\frac{\partial u}{\partial y}+\frac{\partial f_1}{\partial v}\frac{\partial v}{\partial y}\\ \frac{\partial f_2}{\partial u}\frac{\partial u}{\partial x}+\frac{\partial f_2}{\partial v}\frac{\partial v}{\partial x} & \frac{\partial f_2}{\partial u}\frac{\partial u}{\partial y}+\frac{\partial f_2}{\partial v}\frac{\partial v}{\partial y}\end{vmatrix}=\begin{vmatrix}-\frac{\partial f_1}{\partial x} & -\frac{\partial f_1}{\partial y}\\ -\frac{\partial f_2}{\partial x} & -\frac{\partial f_2}{\partial y}\end{vmatrix}$$

[From (3), (4), (5), (6)]

$$=(-1)^2\frac{\partial(f_1,f_2)}{\partial(x,y)}$$

$$\frac{\partial(u,v)}{\partial(x,y)}=(-1)^2\frac{\partial(f_1,f_2)/\partial(x,y)}{\partial(f_1,f_2)/\partial(u,v)}$$

In general, the variables $x_1, x_2, \dots x_n$ are connected with $u_1, u_2, \dots u_n$ implicitly as $f_1(x_1, x_2, \dots x_n, u_1, u_2, \dots u_n)=0$, $f_2(x_1, x_2, \dots x_n, u_1, u_2, \dots u_n)=0$ ... ... ... ... $f_n(x_1, x_2, \dots x_n, u_1, u_2, \dots u_n)=0$

Then we have

$$\frac{\partial(u_1,u_2,\dots u_n)}{\partial(x_1,x_2,\dots x_n)}=(-1)^n\frac{\partial(f_1,f_2,\dots f_n)/\partial(x_1,x_2,\dots x_n)}{\partial(f_1,f_2,\dots f_n)/\partial(u_1,u_2,\dots u_n)}$$

**Example 71.** *If* $x^2+y^2+u^2-v^2=0$ *and* $uv+xy=0$, *prove that* $\frac{\partial(u,v)}{\partial(x,y)}=\frac{x^2-y^2}{u^2+v^2}$

**Solution.** Let $\quad f_1=x^2+y^2+u^2-v^2, \quad f_2=uv+xy$

$$\frac{\partial(f_1,f_2)}{\partial(x,y)}=\begin{vmatrix}\frac{\partial f_1}{\partial x} & \frac{\partial f_1}{\partial y}\\ \frac{\partial f_2}{\partial x} & \frac{\partial f_2}{\partial y}\end{vmatrix}=\begin{vmatrix}2x & 2y\\ y & x\end{vmatrix}=2(x^2-y^2)$$

$$\frac{\partial(f_1,f_2)}{\partial(u,v)}=\begin{vmatrix}\frac{\partial f_1}{\partial u} & \frac{\partial f_1}{\partial v}\\ \frac{\partial f_2}{\partial u} & \frac{\partial f_2}{\partial v}\end{vmatrix}=\begin{vmatrix}2u & -2v\\ v & u\end{vmatrix}=2(u^2+v^2)$$

But $$\frac{\partial(u,v)}{\partial(x,y)}=(-1)^2\frac{\dfrac{\partial(f_1,f_2)}{\partial(x,y)}}{\dfrac{\partial(f_1,f_2)}{\partial(u,v)}}=\frac{2(x^2-y^2)}{2(u^2+v^2)}=\frac{x^2-y^2}{u^2+v^2}$$ **Proved.**

**Example 72.** *If* $u^3 + v + w = x + y^2 + z^2,\ u + v^3 + w = x^2 + y + z^2,\ u + v + w^3 = x^2 + y^2 + z$, *prove that* $\dfrac{\partial (u, v, w)}{\partial (x, y, z)} = \dfrac{1 - 4(xy + yz + zx) + 16\,xyz}{2 - 3(u^2 + v^2 + w^2) + 27\,u^2 v^2 w^2}$

**Solution.** Let

$$f_1 = u^3 + v + w - x - y^2 - z^2$$
$$f_2 = u + v^3 + w - x^2 - y - z^2$$
$$f_3 = u + v + w^3 - x^2 - y^2 - z$$

Now,
$$\frac{\partial (f_1, f_2, f_3)}{\partial (x, y, z)} = \begin{vmatrix} -1 & -2y & -2z \\ -2x & -1 & -2z \\ -2x & -2y & -1 \end{vmatrix} = -1 + 4(yz + zx + xy) - 16\,xyz$$

and
$$\frac{\partial (f_1, f_2, f_3)}{\partial (u, v, w)} = \begin{vmatrix} 3u^2 & 1 & 1 \\ 1 & 3v^2 & 1 \\ 1 & 1 & 3w^2 \end{vmatrix} = 2 - 3(u^2 + v^2 + w^2) + 27\,u^2 v^2 w^2$$

$$\frac{\partial (u, v, w)}{\partial (x, y, z)} = (-1)^3 \frac{\partial (f_1, f_2, f_3)/\partial (x, y, z)}{\partial (f_1, f_2, f_3)/\partial (u, v, w)} = \frac{1 - 4(yz + zx + xy) + 16\,xyz}{2 - 3(u^2 + v^2 + w^2) + 27\,u^2 v^2 w^2}$$ **Proved.**

**Example 73.** If $x + y + z = u,\ y + z = uv,\ z = uvw$, *show that* $\dfrac{\partial (x, y, z)}{\partial (u, v, w)} = u^2 v$

**Solution.** Let

$$f_1 = x + y + z - u$$
$$f_2 = y + z - uv$$
$$f_3 = z - uvw$$

$$\frac{\partial (f_1, f_2, f_3)}{\partial (x, y, z)} = \begin{vmatrix} \frac{\partial f_1}{\partial x} & \frac{\partial f_1}{\partial y} & \frac{\partial f_1}{\partial z} \\ \frac{\partial f_2}{\partial x} & \frac{\partial f_2}{\partial y} & \frac{\partial f_2}{\partial z} \\ \frac{\partial f_3}{\partial x} & \frac{\partial f_3}{\partial y} & \frac{\partial f_3}{\partial z} \end{vmatrix} = \begin{vmatrix} 1 & 1 & 1 \\ 0 & 1 & 1 \\ 0 & 0 & 1 \end{vmatrix} = 1$$

$$\frac{\partial (f_1, f_2, f_3)}{\partial (u, v, w)} = \begin{vmatrix} \frac{\partial f_1}{\partial u} & \frac{\partial f_1}{\partial v} & \frac{\partial f_1}{\partial w} \\ \frac{\partial f_2}{\partial u} & \frac{\partial f_2}{\partial v} & \frac{\partial f_2}{\partial w} \\ \frac{\partial f_3}{\partial u} & \frac{\partial f_3}{\partial v} & \frac{\partial f_3}{\partial w} \end{vmatrix} = \begin{vmatrix} -1 & 0 & 0 \\ -v & -u & 0 \\ -vw & -uw & -uv \end{vmatrix} = -u^2 v$$

But
$$\frac{\partial (x, y, z)}{\partial (u, v, w)} = (-1)^3 \frac{\dfrac{\partial (f_1, f_2, f_3)}{\partial (u, v, w)}}{\dfrac{\partial (f_1, f_2, f_3)}{\partial (x, y, z)}} = -\frac{-u^2 v}{1} = u^2 v$$ **Proved.**

**Example 74.** *If u, v, w are the roots of the equation* $(\lambda - x)^3 + (\lambda - y)^3 + (\lambda - z)^3 = 0$ *in* $\lambda$ *find* $\dfrac{\partial (u, v, w)}{\partial (x, y, z)}$.

*(U.P. I Sem. Jan, 2011; Winter* 2001)

**Solution.** $(\lambda - x)^3 + (\lambda - y)^3 + (\lambda - z)^3 = 0$

$$\Rightarrow 3\lambda^3 - 3(x+y+z)\lambda^2 + 3(x^2+y^2+z^2)\lambda - (x^3+y^3+z^3) = 0$$

Sum of the roots $= u + v + w = x + y + z$ ... (1)

Product of the roots $= uv + vw + wu = x^2 + y^2 + z^2$ ... (2)

$$uvw = \frac{1}{3}(x^3 + y^3 + z^3) \qquad \text{... (3)}$$

Equations (1), (2) and (3) can be rewritten as

$$f_1 = u + v + w - x - y - z$$
$$f_2 = uv + vw + wu - x^2 - y^2 - z^2$$
$$f_3 = uvw - \frac{1}{3}(x^3 + y^3 + z^3)$$

$$\frac{\partial(f_1, f_2, f_3)}{\partial(x, y, z)} = \begin{vmatrix} \frac{\partial f_1}{\partial x} & \frac{\partial f_1}{\partial y} & \frac{\partial f_1}{\partial z} \\ \frac{\partial f_2}{\partial x} & \frac{\partial f_2}{\partial y} & \frac{\partial f_2}{\partial z} \\ \frac{\partial f_3}{\partial x} & \frac{\partial f_3}{\partial y} & \frac{\partial f_3}{\partial z} \end{vmatrix} = \begin{vmatrix} -1 & -1 & -1 \\ -2x & -2y & -2z \\ -x^2 & -y^2 & -z^2 \end{vmatrix}$$

$$= (-1)(-2)(-1)\begin{vmatrix} 1 & 1 & 1 \\ x & y & z \\ x^2 & y^2 & z^2 \end{vmatrix} = -2\begin{vmatrix} 0 & 0 & 1 \\ x-y & y-z & z \\ x^2-y^2 & y^2-z^2 & z^2 \end{vmatrix} \quad \begin{matrix} C_1 \to C_1 - C_2 \\ C_2 \to C_2 - C_3 \end{matrix}$$

$$= -2(x-y)(y-z)\begin{vmatrix} 0 & 0 & 1 \\ 1 & 1 & z \\ x+y & y+z & z^2 \end{vmatrix} \begin{aligned} &= -2(x-y)(y-z)(y+z-x-y) \\ &= -2(x-y)(y-z)(z-x) \end{aligned}$$

$$\frac{\partial(f_1, f_2, f_3)}{\partial(u, v, w)} = \begin{vmatrix} \frac{\partial f_1}{\partial u} & \frac{\partial f_1}{\partial v} & \frac{\partial f_1}{\partial w} \\ \frac{\partial f_2}{\partial u} & \frac{\partial f_2}{\partial v} & \frac{\partial f_2}{\partial w} \\ \frac{\partial f_3}{\partial u} & \frac{\partial f_3}{\partial v} & \frac{\partial f_3}{\partial w} \end{vmatrix} = \begin{vmatrix} 1 & 1 & 1 \\ v+w & u+w & u+v \\ vw & wu & uv \end{vmatrix}$$

$$= \begin{vmatrix} 0 & 0 & 1 \\ v-u & w-v & u+v \\ w(v-u) & u(w-v) & uv \end{vmatrix} \quad \begin{matrix} C_1 \to C_1 - C_2 \\ C_2 \to C_2 - C_3 \end{matrix}$$

$$= (v-u)(w-v)\begin{vmatrix} 0 & 0 & 1 \\ 1 & 1 & u+v \\ w & u & uv \end{vmatrix} \begin{aligned} &= (v-u)(w-v)(u-w) \\ &= -(u-v)(v-w)(w-u) \end{aligned}$$

$$\frac{\partial(u, v, w)}{\partial(x, y, z)} = -\frac{\dfrac{\partial(f_1, f_2, f_3)}{\partial(x, y, z)}}{\dfrac{\partial(f_1, f_2, f_3)}{\partial(u, v, w)}} = -\frac{-2(x-y)(y-z)(z-x)}{-(u-v)(v-w)(w-u)} = -\frac{2(x-y)(y-z)(z-x)}{(u-v)(v-w)(w-u)}$$

**Ans.**

## EXERCISE 1.14

1. If $u^3 + v^3 = x + y$, $u^2 + v^2 = x^3 + y^3$, then prove that $\dfrac{\partial (u, v)}{\partial (x, y)} = \dfrac{1}{2}\,\dfrac{y^2 - x^2}{2\,uv(u - v)}$

2. If $u^3 + v^3 + w^3 = x + y + z$, $\quad u^2 + v^2 + w^2 = x^3 + y^3 + z^3$, $\quad u + v + w = x^2 + y^2 + z^2$,
show that $\dfrac{\partial (u, v, w)}{\partial (x, y, z)} = \dfrac{(x - y)(y - z)(z - x)}{(u - v)(v - w)(w - u)}$

3. If $u = \dfrac{x}{\sqrt{1 - r^2}}$, $v = \dfrac{y}{\sqrt{1 - r^2}}$, $w = \dfrac{z}{\sqrt{1 - r^2}}$, show that
$\dfrac{\partial (u, v, w)}{\partial (x, y, z)} = \dfrac{1}{(1 - r^2)^{5/2}}$ where $r^2 = x^2 + y^2 + z^2$

4. If $u_1 = x_1 + x_2 + x_3 + x_4$, $\quad u_1u_2 = x_2 + x_3 + x_4$, $\quad u_1u_2u_3 = x_3 + x_4$, $\quad u_1u_2u_3u_4 = x_4$
show that $\dfrac{\partial (x_1, x_2, x_3, x_4)}{\partial (u_1, u_2, u_3, u_4)} = u_1^3 \cdot u_2^2 \cdot u_3$

5. If $u, v, w$ are the roots of the equation in $\lambda$ and $\dfrac{x}{a + \lambda} + \dfrac{y}{b + \lambda} + \dfrac{z}{c + \lambda} = 1$, then find $\dfrac{\partial (x, y, z)}{\partial (u, v, w)}$.

**Ans.** $\dfrac{(u - v)(v - w)(w - u)}{(a - b)(b - c)(c - a)}$

## 1.25 PARTIAL DERIVATIVES OF IMPLICIT FUNCTIONS BY JACOBIAN

Given $f_1 (x, y, u, v) = 0, f_2 (x, y, u, v) = 0$

$$\frac{\partial f_1}{\partial u}\frac{\partial u}{\partial x} + \frac{\partial f_1}{\partial v}\frac{\partial v}{\partial x} + \frac{\partial f_1}{\partial x} \cdot 1 = 0 \qquad \text{... (1)}$$

$$\frac{\partial f_2}{\partial u}\frac{\partial u}{\partial x} + \frac{\partial f_2}{\partial v}\frac{\partial v}{\partial x} + \frac{\partial f_2}{\partial x} \cdot 1 = 0 \qquad \text{... (2)}$$

Solving (1) and (2) , we get

$$\frac{\dfrac{\partial u}{\partial x}}{\dfrac{\partial f_1}{\partial v}\cdot\dfrac{\partial f_2}{\partial x} - \dfrac{\partial f_1}{\partial x}\cdot\dfrac{\partial f_2}{\partial v}} = \frac{\dfrac{\partial v}{\partial x}}{\dfrac{\partial f_1}{\partial x}\cdot\dfrac{\partial f_2}{\partial u} - \dfrac{\partial f_1}{\partial u}\cdot\dfrac{\partial f_2}{\partial x}} = \frac{1}{\dfrac{\partial f_1}{\partial u}\cdot\dfrac{\partial f_2}{\partial v} - \dfrac{\partial f_1}{\partial v}\cdot\dfrac{\partial f_2}{\partial u}}$$

$$\frac{\partial u}{\partial x} = \frac{\dfrac{\partial f_1}{\partial v}\cdot\dfrac{\partial f_2}{\partial x} - \dfrac{\partial f_1}{\partial x}\cdot\dfrac{\partial f_2}{\partial v}}{\dfrac{\partial f_1}{\partial u}\cdot\dfrac{\partial f_2}{\partial v} - \dfrac{\partial f_1}{\partial v}\cdot\dfrac{\partial f_2}{\partial u}} = -\frac{\dfrac{\partial (f_1, f_2)}{\partial (x, v)}}{\dfrac{\partial (f_1, f_2)}{\partial (u, v)}}$$

$$\frac{\partial v}{\partial x} = \frac{\dfrac{\partial f_1}{\partial x}\cdot\dfrac{\partial f_2}{\partial u} - \dfrac{\partial f_1}{\partial u}\cdot\dfrac{\partial f_2}{\partial x}}{\dfrac{\partial f_1}{\partial u}\cdot\dfrac{\partial f_2}{\partial v} - \dfrac{\partial f_1}{\partial v}\cdot\dfrac{\partial f_2}{\partial u}} = \frac{\dfrac{\partial (f_1, f_2)}{\partial (x, u)}}{\dfrac{\partial (f_1, f_2)}{\partial (v, u)}}$$

and if, $\quad f_1 (x, y, z, u, v, w) = 0,\ f_2 = (x, y, z, u, v, w) = 0$
$f_3 (x, y, z, u, v, w) = 0$

$$\frac{\partial x}{\partial u} = -\frac{\partial (f_1, f_2, f_3)/\partial (u, y, z)}{\partial (f_1, f_2, f_3)/\partial (x, y, z)}$$

and so on.

**Note.** First we write the Jacobian in the denominator and then we write the Jacobian in the numerator by replacing $x$ by $u$.

**Example 75.** *Use Jacobians to find* $\left(\frac{\partial u}{\partial x}\right)_v$ *if* :

$u^2 + xv^2 - xy = 0$ and $u^2 + xyv + v^2 = 0$

**Solution.** Let $f_1 = u^2 + xv^2 - xy,\ f_2 = u^2 + xyv + v^2$

$$\frac{\partial(f_1, f_2)}{\partial(u, v)} = \begin{vmatrix} \frac{\partial f_1}{\partial u} & \frac{\partial f_1}{\partial v} \\ \frac{\partial f_2}{\partial u} & \frac{\partial f_2}{\partial v} \end{vmatrix} = \begin{vmatrix} 2u & 2xv \\ 2u & xy + 2v \end{vmatrix} = 2uxy + 4uv - 4uxv$$

$$\frac{\partial(f_1, f_2)}{\partial(x, v)} = \begin{vmatrix} \frac{\partial f_1}{\partial x} & \frac{\partial f_1}{\partial v} \\ \frac{\partial f_2}{\partial x} & \frac{\partial f_2}{\partial v} \end{vmatrix} = \begin{vmatrix} v^2 - y & 2xv \\ yv & xy + 2v \end{vmatrix}$$

$$= x\,y\,v^2 + 2v^3 - xy^2 - 2yv - 2xyv^2 = -xyv^2 + 2v^3 - xy^2 - 2yv$$

$$\frac{\partial u}{\partial x} = -\frac{\partial(f_1, f_2)/\partial(x, v)}{\partial(f_1, f_2)/\partial(u, v)} = \frac{xyv^2 - 2v^3 + xy^2 + 2yv}{2uxy + 4uv - 2xuv}$$ **Proved.**

**Example 76.** *If* $u = x + y^2,\ v = y + z^2,\ w = z + x^2$, *prove that*

(*i*) $\frac{\partial x}{\partial u} = \frac{1}{1 + 8\,xyz}$ (*ii*) *Also find* $\frac{\partial^2 x}{\partial u^2}$.

**Solution.** (*i*) Here $f_1 \equiv u - x - y^2,\ f_2 = v - y - z^2$

$f_3 \equiv w - z - x^2.$

Now
$$\frac{\partial(f_1, f_2, f_3)}{\partial(u, y, z)} = \begin{vmatrix} 1 & -2y & 0 \\ 0 & -1 & -2z \\ 0 & 0 & -1 \end{vmatrix} = 1\ ;$$

$$\frac{\partial(f_1, f_2, f_3)}{\partial(x, y, z)} = \begin{vmatrix} -1 & -2y & 0 \\ 0 & -1 & -2z \\ -2x & 0 & -1 \end{vmatrix} = -1\,(1 + 0) + 2y\,(0 - 4zx) = -1 - 8\,xyz$$

$$\frac{\partial x}{\partial u} = -\frac{\partial(f_1, f_2, f_3)/\partial(u, y, z)}{\partial(f_1, f_2, f_3)/\partial(x, y, z)}$$

$$\frac{\partial x}{\partial u} = -\left(\frac{1}{-1 - 8\,xyz}\right) = \frac{1}{1 + 8\,xyz}$$ **Proved.**

(*ii*)
$$\frac{\partial^2 x}{\partial u^2} = \frac{\partial}{\partial u}\left(\frac{\partial x}{\partial u}\right) = \frac{\partial}{\partial u}\left(\frac{1}{1 + 8\,xyz}\right) = -\frac{1}{(1 + 8\,xyz)^2}\frac{\partial}{\partial u}(1 + 8\,xyz)$$

$$= -\frac{1}{(1 + 8\,xyz)^2}\left[0 + 8\left(\frac{\partial x}{\partial u}\,y\,z + \frac{\partial y}{\partial u}\,z\,x + \frac{\partial z}{\partial u}\,x\,y\right)\right]$$

$$= \frac{-8}{(1 + 8\,xyz)^2}\left[\frac{\partial x}{\partial u}\,y\,z + \frac{\partial y}{\partial u}\,z\,x + \frac{\partial z}{\partial u}\,x\,y\right] \quad \text{... (1)}$$

We have,
$$\frac{\partial(f_1, f_2, f_3)}{\partial(x, u, z)} = \begin{vmatrix} -1 & 1 & 0 \\ 0 & 0 & -2z \\ -2x & 0 & -1 \end{vmatrix} = 4\,zx\ ;$$

$$\frac{\partial(f_1, f_2, f_3)}{\partial(x, y, u)} = \begin{vmatrix} -1 & -2y & 1 \\ 0 & -1 & 0 \\ -2x & 0 & 0 \end{vmatrix} = -2x.$$

Now, $$\frac{\partial y}{\partial u} = -\frac{\partial(f_1, f_2, f_3)/\partial(x, u, z)}{\partial(f_1, f_2, f_3)/\partial(x, y, z)}$$

$$\therefore \quad \frac{\partial y}{\partial u} = -\frac{4zx}{-1-8\,xyz} = \frac{4zx}{1+8\,xyz}$$

$$\frac{\partial z}{\partial u} = -\frac{\partial(f_1, f_2, f_3)/\partial(x, y, u)}{\partial(f_1, f_2, f_3)/\partial(x, y, z)}$$

$$\therefore \quad \frac{\partial z}{\partial u} = -\frac{-2x}{-1-8\,xyz} = \frac{-2x}{1+8\,xyz}$$

Substituting in (1), we have

$$\frac{\partial^2 x}{\partial u^2} = \frac{-8}{(1+8\,xyz)^2}\left[\frac{yz}{1+8\,xyz} + \frac{4\,z^2x^2}{1+8\,xyz} + \frac{-2\,x^2y}{1+8\,xyz}\right]$$

$$= \frac{-8\,(yz + 4\,z^2x^2 - 2\,x^2y)}{(1+8\,xyz)^3}$$ **Ans.**

**Example 77.** *Given,* $x = u + v + w$, $y = u^2 + v^2 + w^2$, $z = u^3 + v^3 + w^3$

*show that* $\dfrac{\partial u}{\partial x} = \dfrac{vw}{(u-v)(u-w)}$

**Solution.** Let
$$f_1 = u + v + w - x = 0$$
$$f_2 = u^2 + v^2 + w^2 - y = 0$$
$$f_3 = u^3 + v^3 + w^3 - z = 0$$

$$\frac{\partial u}{\partial x} = -\frac{\partial(f_1, f_2, f_3)/\partial(x, v, w)}{\partial(f_1, f_2, f_3)/\partial(u, v, w)} \quad \ldots (1)$$

$$\frac{\partial(f_1, f_2, f_3)}{\partial(x, v, w)} = \begin{vmatrix} -1 & 1 & 1 \\ 0 & 2v & 2w \\ 0 & 3v^2 & 3w^2 \end{vmatrix} = -6\,vw\begin{vmatrix} 1 & 1 \\ v & w \end{vmatrix} = 6\,vw\,(v-w) \quad \ldots (2)$$

$$\frac{\partial(f_1, f_2, f_3)}{\partial(u, v, w)} = \begin{vmatrix} 1 & 1 & 1 \\ 2u & 2v & 2w \\ 3u^2 & 3v^2 & 3w^2 \end{vmatrix} = 6\,(v-u)\,(w-u)\,(w-v) \quad \ldots (3)$$

Thus from (1), (2) and (3), we get

$$\frac{\partial u}{\partial x} = -\frac{6\,vw\,(v-w)}{6\,(v-u)\,(w-u)\,(w-v)} = \frac{vw}{(u-v)\,(u-w)}$$ **Proved.**

## EXERCISE 1.15

1. If $u^2 + xv^2 - uxy = 0$, $v^2 - xy^2 + 2uv + u^2 = 0$ find $\dfrac{\partial u}{\partial x}$. **Ans.** $-\dfrac{(v^2 - uy)(u+v) + xvy^2}{(u+v)(2u - xy - 2xv)}$

2. If $x = u + e^{-v}\sin u$, $y = v + e^{-v}\cos u$, find $\dfrac{\partial u}{\partial y}, \dfrac{\partial v}{\partial x}$. **Ans.** $\dfrac{\partial u}{\partial y} = \dfrac{\partial v}{\partial x} = \dfrac{e^{-v}\sin u}{1 - e^{-3v}}$

3. If $x = u^2 - v^2$, $y = 2uv$, find $\dfrac{\partial u}{\partial x}; \dfrac{\partial u}{\partial y}; \dfrac{\partial v}{\partial x}; \dfrac{\partial v}{\partial y}$ and $\dfrac{\partial(u, v)}{\partial(x, y)}$

**Ans.** $\dfrac{u}{2(u^2+v^3)}; \dfrac{v}{2(u^2+v^3)}; \dfrac{-v}{2(u^2+v^2)}; \dfrac{u}{2(u^2+v^2)}; \dfrac{1}{4(u^2+v^2)}$

**4.** If $u^3 + xv^2 - uy = 0,\ u^2 + xyv + v^2 = 0$, find $\dfrac{\partial u}{\partial x}$ **Ans.** $\dfrac{xyv^2 - 2v^3}{2xyu^2 - xy^2 + 6u^2 v - 2vy - 4xuv}$

**5.** If $u^2 + xv^2 = x + y,\ v^2 + yu^2 = x - y$, find $\dfrac{\partial u}{\partial x}, \dfrac{\partial v}{\partial y}$ **Ans.** $\dfrac{1 - x - v^2}{2u(1 - xy)}, \dfrac{1 + y + u^2}{-2v(1 - xy)}$

**6.** If $u = xyz,\ v = x^2 + y^2 + z^2,\ w = x + y + z$ find $\dfrac{\partial x}{\partial u}$ **Ans.** $\dfrac{1}{(x - y)(x - z)}$

**7.** If $u = x^2 + y^2 + z^2,\ v = xyz$, find $\dfrac{\partial x}{\partial u}$ **Ans.** $\dfrac{x}{2(2x^2 - y^2)}$

## 1.26 TAYLOR'S SERIES OF TWO VARIABLES

If $f(x, y)$ and all its partial derivatives upto the $n$th order are finite and continuous for all points $(x, y)$, where

$$a \le x \le a + h,\ b \le y \le b + k$$

Then $$f(a+h, b+k) = f(a,b) + \left(h\frac{\partial}{\partial x} + k\frac{\partial}{\partial y}\right) f + \frac{1}{2!}\left(h\frac{\partial}{\partial x} + k\frac{\partial}{\partial y}\right)^2 f + \frac{1}{3!}\left(h\frac{\partial}{\partial x} + k\frac{\partial}{\partial y}\right)^3 f + \ldots$$

**Proof.** Suppose that $f(x + h, y + k)$ is a function of one variable only, say $x$ where $y$ is assumed as constant. Expanding by Taylor's Theorem for one variable, we have

$$f(x + \delta x, y + \delta y) = f(x, y+\delta y) + \delta x \frac{\partial f(x, y+\delta y)}{\partial x} + \frac{(\delta x)^2}{2!}\frac{\partial^2 f(x, y+\delta y)}{\partial x^2} + \ldots\ldots$$

Now expanding for $y$, we get

$$= \left[f(x,y) + \delta y \frac{\partial f(x,y)}{\partial y} + \frac{(\delta y)^2}{2!}\frac{\partial^2 f(x,y)}{\partial y^2} + \ldots\ldots\right] + \delta x \cdot \frac{\partial}{\partial x}\left[f(x,y) + \delta y \frac{\partial f(x,y)}{\partial y} + \ldots\ldots\right]$$

$$+ \frac{(\delta x)^2}{2!}\frac{\partial^2}{\partial x^2}\left[f(x, y) + \delta y \frac{\partial}{\partial y} f(x, y) + \ldots\ldots\right] + \ldots\ldots$$

$$= \left[f(x, y) + \delta y \frac{\partial f(x, y)}{\partial y} + \frac{(\delta y)^2}{2!}\cdot\frac{\partial^2 f(x, y)}{\partial y^2} + \ldots\ldots\right] +$$

$$+ \delta x\left[\frac{\partial f(x, y)}{\partial x} + \delta y \cdot \frac{\partial^2 f(x, y)}{\partial x \partial y}\right] + \frac{(\delta x)^2}{2!}\left[\frac{\partial^2 f(x, y)}{\partial x^2} + \ldots\right] + \ldots$$

$$= f(x, y) + \left[\delta x \frac{\partial f(x, y)}{\partial x} + \delta y \cdot \frac{\partial f(x, y)}{\partial y}\right] + \frac{1}{2!}\left[(\delta x)^2 \frac{\partial^2 f(x, y)}{\partial x^2} + 2\,\delta x . \delta y . \frac{\partial^2 f(x, y)}{\partial x \partial y} + (\delta y)^2 . \frac{\partial^2 f(x, y)}{\partial y^2}\right] + \ldots$$

$$\Rightarrow f(a + h, b + k) = f(a, b) + \left[h\frac{\partial f}{\partial x} + k\frac{\partial f}{\partial y}\right] + \frac{1}{2!}\left[h^2 \frac{\partial^2 f}{\partial x^2} + 2hk\frac{\partial^2 f}{\partial x \partial y} + k^2 \frac{\partial^2 f}{\partial y^2}\right] + \ldots\ldots$$

$$\Rightarrow\ f(a + h, b + k) = f(a, b) + \left[h\frac{\partial}{\partial x} + k\frac{\partial}{\partial y}\right] f + \frac{1}{2!}\left[h\frac{\partial}{\partial x} + k\frac{\partial}{\partial y}\right]^2 f + \ldots\ldots$$

On putting $a = 0,\ b = 0,\ h = x,\ k = y$, we get

$$f(x, y) = f(0, 0) + \left(x\frac{\partial f}{\partial x} + y\frac{\partial f}{\partial y}\right) + \frac{1}{2!}\left(x^2 \frac{\partial^2 f}{\partial x^2} + 2xy\frac{\partial^2 f}{\partial x \partial y} + y^2 \frac{\partial^2 f}{\partial y^2}\right) + \ldots\ldots$$

**Example 78.** *Expand $e^x . \sin y$ in powers of x and y, x = 0, y = 0 as far as terms of third degree.*

**Solution.**

| | | $x = 0, y = 0$ |
|---|---|---|
| $f(x, y)$ | $e^x \sin y$, | 0 |
| $f_x(x, y)$ | $e^x \sin y$, | 0 |
| $f_y(x, y)$ | $e^x \cos y$, | 1 |
| $f_{xx}(x, y)$ | $e^x \sin y$, | 0 |
| $f_{xy}(x, y)$ | $e^x \cos y$, | 1 |
| $f_{yy}(x, y)$ | $-e^x \sin y$, | 0 |
| $f_{xxx}(x, y)$ | $e^x \sin y$, | 0 |
| $f_{xxy}(x, y)$ | $e^x \cos y$, | 1 |
| $f_{xyy}(x, y)$ | $-e^x \sin y$, | 0 |
| $f_{yyy}(x, y)$ | $-e^x \cos y$, | $-1$ |

By Taylor's theorem

$$f(x, y) = f(0,0) + \left(x\frac{\partial}{\partial x} + y\frac{\partial}{\partial y}\right) f(0,0) + \frac{1}{2!}\left(x\frac{\partial}{\partial x} + y\frac{\partial}{\partial y}\right)^2 f(0,0)$$

$$+ \frac{1}{3!}\left(x\frac{\partial}{\partial x} + y\frac{\partial}{\partial y}\right)^3 f(0,0) + \ldots$$

$$= f(0,0) + x f_x(0,0) + y f_y(0,0) + \frac{x^2}{2!} f_{xx}(0,0) + \frac{2xy}{2!} f_{xy}(0,0) + \frac{y^2}{2!} f_{yy}(0,0)$$

$$+ \frac{1}{3!}x^3 f_{xxx}(0,0) + \frac{3x^2 y}{3!} f_{xxy}(0,0) + \frac{3}{3!} x y^2 f_{xyy}(0,0) + \frac{1}{3!} y^3 f_{yyy}(0,0) + \ldots$$

$$e^x \sin y = 0 + x(0) + y(1) + \frac{x^2}{2}(0) + xy(1) + \frac{y^2}{2}(0) + \frac{x^3}{6}(0) + \frac{3x^2 y}{6}(1) + \frac{3xy^2}{6}(0) + \frac{y^3}{6}(-1) + \ldots$$

$$= y + xy + \frac{x^2 y}{2} - \frac{y^3}{6} + \ldots\ldots$$

**Ans.**

**Example 79.** *Find the expansion for cos x cos y in powers of x, y upto fourth order terms.*

**Solution.**

By Taylor's Series

$$f(x, y) = f(0,0) + x f_x(0,0) + y f_y(0,0) + \frac{1}{2!}\left[x^2 f_x^2(0,0) + 2xy f_{xy}(0,0) + y^2 f_{yy}(0,0)\right]$$

$$+ \frac{1}{3!}\left[x^3 f_x^3(0,0) + 3x^2 y f_{xy}^2(0,0) + 3xy^2 f_{xy}^{\ 2}(0,0) + y^3 f_y^3(0,0)\right]$$

$$+ \frac{1}{4!}\left[x^4 f_x^4(0,0) + 4x^3 y f_{xy}^3(0,0) + 6x^2 y^2 f_{xy}^{22}(0,0) + 4xy^3 f_{xy}^{\ 3}(0,0) + y^4 f_y^4(0,0)\right] + \ldots$$

$$\cos x \cos y = 1 + 0 + 0 + \frac{1}{2}(-x^2 + 0 - y^2) + \frac{1}{6}(0 + 0 + 0 + 0) + \frac{1}{24}(x^4 + 0 + 6x^2 y^2 + 0 + y^4)$$

$$= 1 - \frac{x^2}{2} - \frac{y^2}{2} + \frac{x^4}{24} + \frac{x^2 y^2}{4} + \frac{y^2}{24} + \ldots$$

**Ans.**

| | | $x = 0, y = 0$ |
|---|---|---|
| $f(x, y)$ | $\cos x \cos y,$ | 1 |
| $f_x$ | $-\sin x \cos y,$ | 0 |
| $f_y$ | $-\cos x \sin y,$ | 0 |
| $f_{xx}$ | $-\cos x \cos y,$ | $-1$ |
| $f_{xy}$ | $\sin x \sin y,$ | 0 |
| $f_{yy}$ | $-\cos x \cos y,$ | $-1$ |
| $f_{xxx}$ | $\sin x \cos y,$ | 0 |
| $f_{xxy}$ | $\cos x \sin y,$ | 0 |
| $f_{xyy}$ | $\sin x \cos y,$ | 0 |
| $f_{yyy}$ | $\cos x \sin y,$ | 0 |
| $f_{xxxx}$ | $\cos x \cos y,$ | 1 |
| $f_{xxxy}$ | $-\sin x \sin y,$ | 0 |
| $f_{xxyy}$ | $\cos x \cos y,$ | 1 |
| $f_{xyyy}$ | $-\sin x \sin y,$ | 0 |
| $f_{yyyy}$ | $\cos x \cos y,$ | 1 |

**Example 80.** *Find the first six terms of the expansion of the function* $e^x \log (I + y)$ *in a Taylor's series in the neighbourhood of the point* (0, 0).

**Solution.**

Taylor's series is

$$f(x, y) = f(0, 0) + \left(x\frac{\partial f}{\partial x} + y\frac{\partial f}{\partial y}\right)$$

$$+\frac{1}{2!}\left(x^2\frac{\partial^2 f}{\partial x^2} + 2xy\frac{\partial^2 f}{\partial x \partial y} + y^2\frac{\partial^2 f}{\partial y^2}\right) + \ldots$$

$$\Rightarrow \quad e^x \log (1 + y) = 0 + (x \times 0 + y \times 1)$$

$$+\frac{1}{2!}[x^2 \times (0) + 2xy \times 1 + y^2 \times (-1)] + ..$$

$$\Rightarrow \quad e^x \log (1 + y) = y + xy - \frac{y^2}{2} \qquad \textbf{Ans.}$$

| | | $x = 0, y = 0$ |
|---|---|---|
| $f(x, y)$ | $e^x \log (1 + y)$ | 0 |
| $\frac{\partial f}{\partial x}$ | $e^x \log (1 + y)$ | 0 |
| $\frac{\partial f}{\partial y}$ | $\frac{e^x}{1+y}$ | 1 |
| $\frac{\partial^2 f}{\partial x^2}$ | $e^x \log (1 + y)$ | 0 |
| $\frac{\partial^2 f}{\partial y^2}$ | $-\frac{e^x}{(1+y)^2}$ | $-1$ |
| $\frac{\partial^2 f}{\partial x \partial y}$ | $\frac{e^x}{(1+y)}$ | 1 |

## EXERCISE 1.16

1. Expand $e^x \cos y$ at (0, 0) upto three terms. **Ans.** $1 + x + \frac{1}{2}(x^2 - y^2) + \ldots\ldots$
2. Expand $z = e^{2x} \cos 3y$ in power series of $x$ and $y$ upto quadratic terms. *(AMIE Summer 2004)* **Ans.** $1 + 2x + 2x^2 - \frac{9}{2}y^2 + \ldots$
3. Show that $e^y \log (1 + x) = x + xy - \frac{x^2}{2}$ approximately.
4. Verify $\sin (x + y) = x + y - \frac{(x + y)^3}{3} + \ldots\ldots$

**Example 81.** *Expand* sin (*xy*) *in powers of* $(x-1)$ *and* $\left(y-\frac{\pi}{2}\right)$ *as far as the terms of second degree.*

*(Nagpur University, Summer 2003)*

**Solution.** We have, $f(x,y) = \sin(xy)$

Here $\left[\begin{array}{l} a+h = x \text{ and } h = x-1 \\ \Rightarrow a+(x-1) = x \Rightarrow a = 1 \end{array}\right.$

$\left[\begin{array}{l} b+k = y \text{ and } k = y-\frac{\pi}{2} \\ \Rightarrow b+y-\frac{\pi}{2} = y \Rightarrow b = \frac{\pi}{2} \end{array}\right.$

| | | $x=1, y=\frac{\pi}{2}$ |
|---|---|---|
| $f(x,y)$ | $\sin(xy)$ | 1 |
| $f_x(x,y)$ | $y\cos(xy)$, | 0 |
| $f_y(x,y)$ | $x\cos(xy)$, | 0 |
| $f_{xx}(x,y)$ | $-y^2\sin(xy)$; | $-\frac{\pi^2}{4}$ |
| $f_{xy}(x,y)$ | $\cos(xy) - xy\sin(xy)$, | $-\frac{\pi}{2}$ |
| $f_{yy}(x,y)$ | $-x^2\sin(xy)$, | $-1$ |

By Taylor's theorem for a function of two variables, we have

$$f(a+h, b+k) = f(a,b) + hf_x(a,b) + kf_y(a,b) + \frac{1}{2!}\left\{h^2 f_{xx}(a,b) + 2hkf_{xy}(a,b) + k^2 f_{yy}(a,b)\right\}$$

$$\Rightarrow f(x,y) = f\left(1,\frac{\pi}{2}\right) + (x-1)\, f_x\left(1,\frac{\pi}{2}\right) + \left(y-\frac{\pi}{2}\right) f_y\left(1,\frac{\pi}{2}\right)$$

$$+\frac{1}{2!}\left\{(x-1)^2 f_{xx}\left(1,\frac{\pi}{2}\right) + 2(x-1)\left(y-\frac{\pi}{2}\right) f_{xy}\left(1,\frac{\pi}{2}\right) + \left(y-\frac{\pi}{2}\right)^2 f_{yy}\left(1,\frac{\pi}{2}\right)\right\}$$

$$\Rightarrow \sin(xy) = 1 + (x-1).0 + \left(y-\frac{\pi}{2}\right).0 + \frac{1}{2!}\left\{(x-1)^2\left(-\frac{\pi^2}{4}\right) + 2(x-1)\left(y-\frac{\pi}{2}\right)\left(-\frac{\pi}{2}\right) + \left(y-\frac{\pi}{2}\right)^2(-1)\right\} + \ldots$$

$$\Rightarrow \sin(xy) = 1 - \frac{\pi^2}{8}(x-1)^2 - \frac{\pi}{2}(x-1)\left(y-\frac{\pi}{2}\right) - \frac{1}{2}\left(y-\frac{\pi}{2}\right)^2 + \ldots$$

**Ans.**

**Example 82.** *Expand* $e^x \cos y$ *near the point* $\left(1,\frac{\pi}{4}\right)$ *by Taylor's Theorem.*

*(U.P., I Semester Dec. 2007)*

**Solution.** $f(x+h, y+k) = f(x,y) + \left[h\frac{\partial}{\partial x} + k\frac{\partial}{\partial y}\right] f$

$$+\frac{1}{2!}\left(h\frac{\partial}{\partial x} + k\frac{\partial}{\partial y}\right)^2 f + \frac{1}{3!}\left(h\frac{\partial}{\partial x} + k\frac{\partial}{\partial y}\right)^3 f + \ldots$$

$$e^x\cos y = f(x,y) = f\left[1+(x-1), \frac{\pi}{4} + \left(y-\frac{\pi}{4}\right)\right]$$

where $h = x-1$, $k = y-\frac{\pi}{4}$ $= f\left(1+h, \frac{\pi}{4}+k\right)$

| | | $x=1, y=\frac{\pi}{4}$ |
|---|---|---|
| $f(x,y)$ | $e^x\cos y$ | $\frac{e}{\sqrt{2}}$ |
| $\frac{\partial f}{\partial x}$ | $e^x\cos y$, | $\frac{e}{\sqrt{2}}$ |
| $\frac{\partial f}{\partial y}$ | $-e^x\sin y$, | $\frac{-e}{\sqrt{2}}$ |
| $\frac{\partial^2 f}{\partial x^2}$ | $e^x\cos y$, | $\frac{e}{\sqrt{2}}$ |
| $\frac{\partial^2 f}{\partial y^2}$ | $-e^x\cos y$, | $\frac{-e}{\sqrt{2}}$ |
| $\frac{\partial^2 f}{\partial x\,\partial y}$ | $-1\, e^x\sin y$, | $\frac{-e}{\sqrt{2}}$ |

Putting these values in Taylor's Theorem, we get

$$e^x\cos y = \frac{e}{\sqrt{2}} + \left[(x-1)\frac{e}{\sqrt{2}} + \left(y-\frac{\pi}{4}\right)\left(\frac{-e}{\sqrt{2}}\right)\right]$$

$$+\frac{1}{2!}\left[(x-1)^2\frac{e}{\sqrt{2}}+2(x-1)\left(y-\frac{\pi}{4}\right)\left(\frac{-e}{\sqrt{2}}\right)+\left(y-\frac{\pi}{4}\right)^2\left(\frac{-e}{\sqrt{2}}\right)\right]+\ldots.$$

$$=\frac{e}{\sqrt{2}}\left[1+(x-1)-\left(y-\frac{\pi}{4}\right)+\frac{(x-1)^2}{2}-(x-1)\left(y-\frac{\pi}{4}\right)-\left(y-\frac{\pi}{4}\right)^2+\ldots.\right] \quad \textbf{Ans.}$$

**Example 83.** *If $f(x, y) = \tan^{-1}(x y)$, compute an approximate value of $f(0.9, -1.2)$.*

**Solution.** We have,

$$f(x, y) = \tan^{-1}(x y)$$

Let us expand $f(x, y)$ near the point $(1, -1)$

$$f(0.9, -1.2) = f(1-0.1, -1-0.2)$$

$$= f(1,-1)+\left[(-0.1)\frac{\partial f}{\partial x}+(-0.2)\frac{\partial f}{\partial y}\right]+\frac{1}{2!}\left[(-0.1)^2\frac{\partial^2 f}{\partial x^2}\right.$$

$$\left.+2(-0.1)(-0.2)\frac{\partial^2 f}{\partial x\,\partial y}+(-0.2)^2\frac{\partial^2 f}{\partial y^2}\right]+\ldots \quad \ldots(1)$$

| | | $x=1,\ y=-1$ |
|---|---|---|
| $f(x, y)$ | $\tan^{-1}(xy)$ | $-\frac{\pi}{4}$ |
| $\frac{\partial f}{\partial x}$ | $\frac{y}{1+x^2y^2}$, | $-\frac{1}{2}$ |
| $\frac{\partial f}{\partial y}$ | $\frac{x}{1+x^2y^2}$, | $\frac{1}{2}$ |
| $\frac{\partial^2 f}{\partial x^2}$ | $-\frac{(2x)y}{(1+x^2y^2)^2}$, | $\frac{1}{2}$ |
| $\frac{\partial^2 f}{\partial y\,\partial x}$ | $\frac{1+x^2y^2-x(2xy^2)}{(1+x^2y^2)^2}=\frac{1-x^2y^2}{(1+x^2y^2)^2}$ | 0 |
| $\frac{\partial^2 f}{\partial y^2}$ | $\frac{-x(2x^2y)}{(1+x^2y^2)^2}$, | $\frac{1}{2}$ |

Substituting the values of $\frac{\partial f}{\partial x}, \frac{\partial f}{\partial y}$ etc. in (1), we get

$$f(0.9,-1.2) = -\frac{\pi}{4}+(-0.1)\left(-\frac{1}{2}\right)+(-0.2)\left(\frac{1}{2}\right)+\frac{1}{2}\left[(-0.1)^2\left(\frac{1}{2}\right)\right.$$

$$\left.+2(-0.1)(-0.2)\,0+(-0.2)^2\left(\frac{1}{2}\right)\right]+\ldots$$

$$=-\frac{22}{28}+0.05-0.1+\frac{1}{2}(0.005+0.02)$$

$$=-0.786+0.05-0.1+0.0125=-0.8235 \quad \textbf{Ans.}$$

**Example 84.** *Obtain Taylor's expansion of* $\tan^{-1}\dfrac{y}{x}$ *about (1, 1) upto and including the second degree terms. Hence compute f (1. 1, 0.9).* *(U.P., I Sem. Winter 2005, 2002)*

**Solution.**

| | | $x = 1, y = 1$ |
|---|---|---|
| $f(x, y)$ | $\tan^{-1}\dfrac{y}{x}$ | $\dfrac{\pi}{4}$ |
| $\dfrac{\partial f}{\partial x}$ | $\dfrac{1}{1+\dfrac{y^2}{x^2}}\left(-\dfrac{y}{x^2}\right)=\dfrac{-y}{x^2+y^2},$ | $-\dfrac{1}{2}$ |
| $\dfrac{\partial f}{\partial y}$ | $\dfrac{1}{1+\dfrac{y^2}{x^2}}\left(\dfrac{1}{x}\right)=\dfrac{x}{x^2+y^2},$ | $\dfrac{1}{2}$ |
| $\dfrac{\partial^2 f}{\partial x^2}$ | $\dfrac{y(2x)}{(x^2+y^2)^2}=\dfrac{2xy}{(x^2+y^2)^2},$ | $\dfrac{1}{2}$ |
| $\dfrac{\partial^2 f}{\partial y^2}$ | $\dfrac{-x(2y)}{(x^2+y^2)^2}=\dfrac{-2xy}{(x^2+y^2)^2},$ | $-\dfrac{1}{2}$ |
| $\dfrac{\partial^2 f}{\partial y\partial x}$ | $\dfrac{(x^2+y^2)-(x)(2x)}{(x^2+y^2)^2}=\dfrac{y^2-x^2}{(x^2+y^2)^2},$ | 0 |

By Taylor's Theorem

$$f(x, y) = f(a, b)+\left[(x-a)\frac{\partial f}{\partial x}+(y-b)\frac{\partial f}{\partial y}\right]+\frac{1}{2!}\left[(x-a)^2\frac{\partial^2 f}{\partial x^2}+2(x-a)(y-b)\frac{\partial^2 f}{\partial x\partial y}+(y-b)^2\frac{\partial^2 f}{\partial y^2}\right]+\ldots$$

Here, $a = 1, b = 1$

$$\tan^{-1}\frac{y}{x} = \frac{\pi}{4}+(x-1)\left(-\frac{1}{2}\right)+(y-1)\frac{1}{2}+\frac{1}{2!}\left[(x-1)^2\left(\frac{1}{2}\right)+2(x-1)(y-1)(0)+(y-1)^2\left(-\frac{1}{2}\right)\right]+\ldots$$

$$\tan^{-1}\frac{y}{x} = \frac{\pi}{4}-\frac{1}{2}(x-1)+\frac{1}{2}(y-1)+\frac{1}{4}(x-1)^2-\frac{1}{4}(y-1)^2+\ldots \qquad \ldots(1)$$

Putting $(x-1) = 1.1-1 = 0.1$, $(y-1) = 0.9-1 = -0.1$ in (1), we get

$$f(1.1, 0.9) = \frac{\pi}{4}-\frac{1}{2}(0.1)-\frac{1}{2}(-0.1)+\frac{1}{4}(0.1)^2-\frac{1}{4}(-0.1)^2$$

$$= 0.786 - 0.05 + 0.05 + 0.0025 - 0.0025 = 0.786$$ **Ans.**

**Example 85.** *Expand* $\dfrac{(x+h)(y+k)}{x+h+y+k}$ *in powers of h, k upto and inclusive of the second degree terms.* *(A.M.I.E., Summer 2001)*

**Solution.**
$$f(x+h, y+k) = \frac{(x+h)(y+k)}{x+h+y+k}$$

$$f(x, y) = \frac{xy}{x+y}$$

$$\frac{\partial f}{\partial x} = \frac{(x+y)\,y - x\,y}{(x+y)^2} = \frac{y^2}{(x+y)^2}$$

$$\frac{\partial f}{\partial y} = \frac{(x+y)\,x - x\,y}{(x+y)^2} = \frac{x^2}{(x+y)^2}$$

$$\frac{\partial^2 f}{\partial x^2} = \frac{-2y^2}{(x+y)^3}$$

$$\frac{\partial^2 f}{\partial y\,\partial x} = \frac{(x+y)^2\,2x - 2\,(x+y)\,x^2}{(x+y)^4} = \frac{(x+y)\,2x - 2x^2}{(x+y)^3} = \frac{2x\,y}{(x+y)^3}$$

$$\frac{\partial^2 f}{\partial y^2} = \frac{-2x^2}{(x+y)^3}$$

$$f(x+h, y+k) = f(x,y) + \left(h\frac{\partial}{\partial x} + k\frac{\partial}{\partial y}\right) f(x,y) + \frac{1}{2!}\left(h\frac{\partial}{\partial x} + k\frac{\partial}{\partial y}\right)^2 f(x,y) + \ldots$$

$$\frac{(x+h)(y+k)}{x+h+y+k} = \frac{x\,y}{x+y} + h\frac{y^2}{(x+y)^2} + k\frac{x^2}{(x+y)^2}$$

$$+ \frac{h^2}{2!}\frac{(-2y^2)}{(x+y)^3} + \frac{1}{2!}2hk\frac{2x\,y}{(x+y)^3} + \frac{1}{2!}k^2\frac{(-2x^2)}{(x+y)^3} + \ldots$$

$$= \frac{x\,y}{x+y} + \frac{h\,y^2}{(x+y)^2} + \frac{kx^2}{(x+y)^2} - \frac{h^2 y^2}{(x+y)^3} + \frac{2hkxy}{(x+y)^3} - \frac{k^2 x^2}{(x+y)^3} + \ldots \quad \textbf{Ans.}$$

**Example 86.** *Expand $x^2y + 3y - 2$ in powers of $x - 1$ and $y + 2$ using Taylor's Theorem.*

*(A.M.I.E.T.E., Winter 2003, A.M.I.E., Summer 2004, 2003)*

**Solution.** $f(x, y) = x^2y + 3y - 2$

Here $a + h = x$ and $h = x - 1$, so $a = 1$

$b + k = y$ and $k = y + 2$ so $b = -2$

| | | $x = 1, y = -2$ |
|---|---|---|
| $f(x, y)$ | $x^2y + 3y - 2$, | $-10$ |
| $f_x(x, y)$ | $2xy$, | $-4$ |
| $f_y(x, y)$ | $x^2 + 3$, | $4$ |
| $f_{xx}(x, y)$ | $2y$, | $-4$ |
| $f_{xy}(x, y)$ | $2x$, | $2$ |
| $f_{yy}(x, y)$ | $0$, | $0$ |
| $f_{xxx}(x, y)$ | $0$, | $0$ |
| $f_{xxy}(x, y)$ | $2$, | $2$ |
| $f_{xyy}(x, y)$ | $0$, | $0$ |
| $f_{yyy}(x, y)$ | $0$, | $0$ |

Now Taylor's Theorem is

$$f(a+h, b+k) = f(a,b) + \left(h\frac{\partial f}{\partial x} + k\frac{\partial f}{\partial y}\right)_{(a,b)} + \frac{1}{2!}\left[h^2\frac{\partial^2 f}{\partial x^2} + 2hk\frac{\partial^2 f}{\partial x \partial y} + k^2\frac{\partial^2 f}{\partial y^2}\right]_{(a,b)}$$

$$+\frac{1}{3!}\left(h^3\frac{\partial^3 f}{\partial x^3} + 3h^2k\frac{\partial^3 f}{\partial x^2\,\partial y} + 3hk^2\frac{\partial^2 f}{\partial x \partial y^2} + k^3\frac{\partial^3 f}{\partial y^3}\right) + \ldots$$

Putting the values of $f(a, b)$ etc. in Taylor's Theorem, we get

$$x^2y + 3y - 2 = -10 + [(x-1)(-4) + (y+2)(4)]$$

$$+\frac{1}{2!}[(x-1)^2(-4) + 2(x-1)(y+2)(2) + (y+2)^2(0)]$$

$$+\frac{1}{3!}[(x-1)^3(0) + 3(x-1)^2(y+2)(2) + 3(x-1)(y+2)^2(0) + (y+2)^3(0)]$$

$$x^2y + 3y - 2 = -10 - 4(x-1) + 4(y+2) - 2(x-1)^2 + 2(x-1)(y+2) + (x-1)^2(y+2)$$ **Ans.**

## EXERCISE 1.17

1. Expand $e^{xy}$ at (1, 1) upto three terms.

**Ans.** $e\left[1 + (x-1) + (y-1) + \frac{1}{2!}[(x-1)^2 + 4(x-1)(y-1) + (y-1)^2]\right]$

2. Expand $y^x$ at (1, 1) upto second term **Ans.** $1 + (y-1) + (x-1)(y-1) + \ldots\ldots$

3. Expand $e^{ax}\sin by$ in powers of $x$ and $y$ as far as the terms of third degree. *(U.P. I sem. Jan 2011)*

**Ans.** $by + abxy + \frac{1}{3!}\left(3a^2bx^2y - b^3y^3\right) + \ldots$

4. Expand $(x^2y + \sin y + e^x)$ in powers of $(x-1)$ and $(x-\pi)$.

**Ans.** $\pi + e + (x-1)(2\pi + e) + \frac{1}{2}(x-1)^2(2\pi + e) + 2(x-1)(y-\pi)$.

5. Expand $(1 + x + y^2)^{1/2}$ at (1, 0). **Ans.** $\sqrt{2}\left[1 + \frac{x-1}{4} - \frac{(x-1)^2}{32} + \frac{y^2}{4} + \ldots\right]$

6. Obtain the linearised form $T(x, y)$ of the function $f(x, y) = x^2 - xy + \frac{1}{2}y^2 + 3$ at the point (3, 2), using the Taylor's series expansion. Find the maximum error in magnitude in the approximation $f(x, y) \approx T(x, y)$ over the rectangle R: $|x-3| < 0.1, |y-2| < 0.1$.

**Ans.** $8 + 4(x-3) - (y-2)$., Error 0.04.

7. Expand $\sin(x+h)(y+k)$ by Taylor's Theorem.

**Ans.** $\sin xy + h(x+y)\cos xy + hk\cos xy - \frac{1}{2}h^2(x+y)^2\sin xy + \ldots$

8. Fill in the blank:

$f(x, y) = f(2, 3) + \ldots\ldots\ldots\ldots$ **Ans.** $\left[(x-2)\frac{\partial}{\partial x} + (y-3)\frac{\partial}{\partial y}\right]f + \frac{1}{2!}\left[(x-2)\frac{\partial}{\partial x} + (y-3)\frac{\partial}{\partial y}\right]^2 f + \ldots$

9. If $f(x) = f(0) + kf_1(0) + \frac{k^2}{2!}f_2(\theta k), 0 < \theta < 1$ then the value of $\theta$ when $k = 1$ and $f(x) = (1-x)^{3/2}$ is given as ......... *(U.P. Ist Semester, Dec 2008)*

## 1.27 MAXIMUM VALUE

A function $f(x, y)$ is said to have a maximum value at $x = a$, $y = b$, if there exists a small neighbourhood of $(a, b)$ such that,

$$f(a, b) > f(a+h, b+k)$$

**Minimum Value.** A function $f(x, y)$ is said to have a minimum value for $x = a$, $y = b$, if there exists a small neighbourhood of $(a, b)$ such that

$$f(a, b) < f(a+h, b+k)$$

The maximum and minimum values of a function are also called extreme or extremum values of the function.

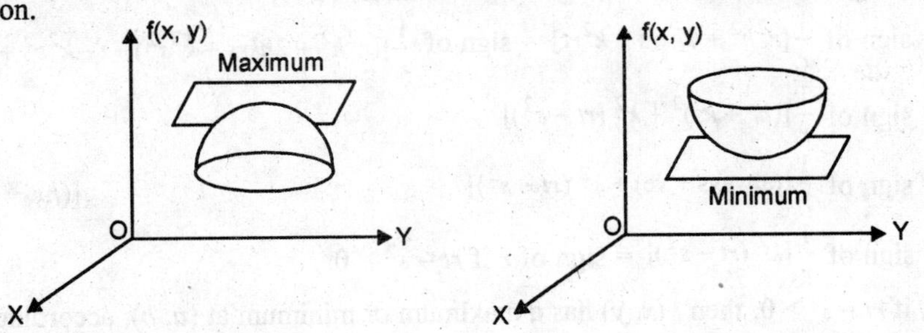


Maximum value of f(x, y) at (a, b)

Minimum value of f(x, y) at (a, b)

**Saddle point or Minimax.** It is a point where a function is neither maximum nor minimum.

**Geometrical Interpretation.** Such a surface (looks like the leather seat on the back of a horse) forms a ridge rising in one direction and falling in another direction.

## 1.28 CONDITIONS FOR EXTREMUM VALUES

If $f(a+h, b+k) - f(a, b)$ remains of the same sign for all values (positive or negative) of $h$, $k$ then $f(a, b)$ is said to be extremum value of $f(x, y)$ at $(a, b)$

(*i*) If $f(a+h, b+k) - f(a, b) < 0$, then $f(a, b)$ is maximum.

(*ii*) If $f(a+h, b+k) - f(a, b) > 0$, then $f(a, b)$ is minimum.

By Taylor's Theorem

$$f(a+h, b+k) = f(a, b) + \left(h\frac{\partial f}{\partial x} + k\frac{\partial f}{\partial y}\right)_{(a,b)} + \frac{1}{2!}\left[h^2\frac{\partial^2 f}{\partial x^2} + 2hk\frac{\partial^2 f}{\partial x \partial y} + k^2\frac{\partial^2 f}{\partial y^2}\right] + \ldots$$

$$\Rightarrow \quad f(a+h, b+k) - f(a, b) = \left(h\frac{\partial f}{\partial x} + k\frac{\partial f}{\partial y}\right)_{(a,b)} + \frac{1}{2!}\left[h^2\frac{\partial^2 f}{\partial x^2} + 2hk\frac{\partial^2 f}{\partial x \partial y} + k^2\frac{\partial^2 f}{\partial y^2}\right] \quad \ldots(1)$$

$$\Rightarrow \quad f(a+h, b+k) - f(a, b) = \left(h\frac{\partial f}{\partial x} + k\frac{\partial f}{\partial y}\right)_{(a,b)} \quad \ldots(2)$$

For small values of $h$, $k$, the second and higher order terms are still smaller and hence may be neglected.

The sign of L.H.S. of (2) is governed by $h\frac{\partial f}{\partial x} + k\frac{\partial f}{\partial y}$ which may be positive or negative depending on $h$, $k$.

Hence, the necessary condition for $f(a, b)$ to be a maximum or minimum is that

$$\left(h\frac{\partial f}{\partial x} + k\frac{\partial f}{\partial y}\right) = 0 \Rightarrow \frac{\partial f}{\partial x} = 0, \quad \frac{\partial f}{\partial y} = 0$$

By solving the equations, we get, point $x = a$, $y = b$ which may be maximum or minimum value.

Then from (1)

$$f(a+h, b+k) - f(a, b) = \frac{1}{2!}\left[h^2\frac{\partial^2 f}{\partial x^2} + 2hk\frac{\partial^2 f}{\partial x \partial y} + k^2\frac{\partial^2 f}{\partial y^2}\right]$$

$$= \frac{1}{2!}[h^2 r + 2hks + k^2 t] \quad ...(3)$$

where $r = \dfrac{\partial^2 f}{\partial x^2}, s = \dfrac{\partial^2 f}{\partial x \partial y}, t = \dfrac{\partial^2 f}{\partial y^2}$ at $(a, b)$

Now the sign of L.H.S. of (3) is sign of $[rh^2 + 2hks + k^2 t]$

$$= \text{sign of } \frac{1}{r}[r^2h^2 + 2hkrs + k^2rt] = \text{sign of } \frac{1}{r}[(r^2h^2 + 2hkrs + k^2s^2) + (-k^2s^2 + k^2rt)]$$

$$= \text{sign of } \frac{1}{r}[(hr + ks)^2 + k^2(rt - s^2)]$$

$$= \text{sign of } \frac{1}{r}[(\text{always} + \text{ve}) + k^2(rt - s^2)] \qquad [(hr + ks)^2 = +\text{ ve}]$$

$$= \text{sign of } \frac{1}{r}[k^2(rt - s^2)] = \text{sign of } r \text{ if } rt - s^2 > 0$$

Hence, if $rt - s^2 > 0$, then $f(x, y)$ has a maximum or minimum at $(a, b)$, according as $r < 0$ or $r > 0$.

**Note:** (*i*) If $rt - s^2 < 0$, then L.H.S. will change with $h$ and $k$ hence there is no maximum or minimum at $(a, b)$, *i.e.*, it is a saddle point.

(*ii*) If $rt - s^2 = 0$, then $rh^2 + 2shk + tk^2 = \frac{1}{r}[(rh + sk)^2 + k^2(rt - s^2)]$

$$= \frac{1}{r}(rh + sk)^2 \text{ which is zero for values of } h, k, \text{ such that}$$

$$\Rightarrow \quad \frac{h}{k} = -\frac{s}{r}$$

This is, therefore, a doubtful case, further investigation is required.

## 1.29 WORKING RULE TO FIND EXTREMUM VALUES

(*i*) Differentiate $f(x, y)$ and find out $\dfrac{\partial f}{\partial x}, \dfrac{\partial f}{\partial y}, \dfrac{\partial^2 f}{\partial x^2}, \dfrac{\partial^2 f}{\partial x \partial y}, \dfrac{\partial^2 f}{\partial y^2}$

(*ii*) Put $\dfrac{\partial f}{\partial x} = 0$ and $\dfrac{\partial f}{\partial y} = 0$ and solve these equations for $x$ and $y$. Let $(a, b)$ be the values of $(x, y)$.

(*iii*) Evaluate $r = \dfrac{\partial^2 f}{\partial x^2}, s = \dfrac{\partial^2 f}{\partial x \partial y}, t = \dfrac{\partial^2 f}{\partial y^2}$ for these values $(a, b)$.

(*iv*) If $rt - s^2 > 0$ and

(*a*) $r < 0$, then $f(x, y)$ has a maximum value.

(*b*) $r > 0$, then $f(x, y)$ has a minimum value.

(*v*) If $rt - s^2 < 0$, then $f(x, y)$ has no extremum value at the point $(a, b)$.

(*vi*) If $rt - s^2 = 0$, then the case is doubtful and needs further investigation.

**Note:** The point $(a, b)$ which are the roots of $\dfrac{\partial f}{\partial x} = 0, \dfrac{\partial f}{\partial y} = 0$, are called stationary points.

**Example 87.** *Discuss the maximum and minimum of* $x^2 + y^2 + 6x + 12$.

**Solution.** We have, $f(x, y) = x^2 + y^2 + 6x + 12$

$$\frac{\partial f}{\partial x} = 2x + 6, \frac{\partial f}{\partial y} = 2y, \frac{\partial^2 f}{\partial x^2} = 2, \frac{\partial^2 f}{\partial y^2} = 2, \frac{\partial^2 f}{\partial x \partial y} = 0$$

For maxima and minima, $\dfrac{\partial f}{\partial x} = 0$ and $\dfrac{\partial f}{\partial y} = 0$

$\Rightarrow \qquad 2x + 6 = 0, \text{ and } 2y = 0$

$\Rightarrow \qquad x = -3, \text{ and } y = 0$

At $(-3, 0)$ $\qquad rt - s^2 = 2 \times 2 - 0 = 4 > 0$

$$r = 2 > 0$$

Hence $f(x, y)$ is minimum when $x = -3$ and $y = 0$

Minimum value $= f(-3, 0) = 9 + 0 - 18 + 12 = 3$ **Ans.**

**Example 88.** *Find the absolute maximum and minimum values of*

$$f(x, y) = 2 + 2x + 2y - x^2 - y^2$$

*on triangular plate in the first quadrant, bounded by the lines $x = 0$, $y = 0$ and $y = 9 - x$.*

*(Gujarat, I semester, Jan. 2009)*

**Solution.** We have, $\quad f(x, y) = 2 + 2x + 2y - x^2 - y^2$

$$\frac{\partial f}{\partial x} = 2 - 2x, \quad \frac{\partial f}{\partial y} = 2 - 2y$$

$$\frac{\partial^2 f}{\partial x^2} = -2, \quad \frac{\partial^2 f}{\partial x \partial y} = 0, \quad \frac{\partial^2 f}{\partial y^2} = -2$$

For maxima and minima,

$$\frac{\partial f}{\partial x} = 0 \Rightarrow 2 - 2x = 0 \Rightarrow x = 1$$

$$\frac{\partial f}{\partial y} = 0 \Rightarrow 2 - 2y = 0 \Rightarrow y = 1$$

At $(1, 1)$ $\qquad rt - s^2 = (-2)(-2) - 0 = 4 > 0$

Hence $f(x, y)$ is maximum at $(1, 1)$.

Maximum value of $f(x, y) = 2 + 2 + 2 - 1 - 1 = 4$ **Ans.**

**Example 89.** *Examine $f(x, y) = x^3 + y^3 - 3axy$ for maximum and minimum values.*

*(U.P. I Sem., Dec. 2004), (M.U. 2004, 2003)*

**Solution.** We have, $\quad f(x, y) = x^3 + y^3 - 3axy$

$$p = \frac{\partial f}{\partial x} = 3x^2 - 3ay, \qquad q = \frac{\partial f}{\partial y} = 3y^2 - 3ax$$

$$r = \frac{\partial^2 f}{\partial x^2} = 6x, \qquad s = \frac{\partial^2 f}{\partial x \partial y} = -3a, \quad t = \frac{\partial^2 f}{\partial y^2} = 6y$$

For maxima and minima

$$\frac{\partial f}{\partial x} = 0 \qquad \text{and} \qquad \frac{\partial f}{\partial y} = 0$$

$$3x^2 - 3ay = 0. \qquad\qquad 3y^2 - 3ax = 0$$

$\Rightarrow \qquad x^2 = ay \Rightarrow y = \dfrac{x^2}{a} \quad ...(1) \qquad\qquad \Rightarrow y^2 = ax \quad ...(2)$

Putting the value of $y$ from (1) in (2), we get

$$x^4 = a^3 x \Rightarrow x(x^3 - a^3) = 0$$

$\Rightarrow \qquad x(x - a)(x^2 + ax + a^2) = 0$

$\Rightarrow \qquad x = 0, a$

Putting $x = 0$ in (1), we get $y = 0$,

Putting $x = a$ in (1), we get $y = a$,

| *Stationary pairs* | (0, 0) | (a, a) |
|---|---|---|
| $r$ | 0 | $6a$ |
| $s$ | $-3a$ | $-3a$ |
| $t$ | 0 | $6a$ |
| $rt - s^2$ | $-9a^2 < 0$ | $27a^2 > 0$ |

At (0, 0) there is no extremum value, since $rt - s^2 < 0$.

At $(a, a)$, $rt - s^2 > 0$, $r > 0$

Therefore $(a, a)$ is a point of minimum value.

The minimum value of $f(a, a) = a^3 + a^3 - 3a^3 = -a^3$ **Ans.**

**Example 90.** *Show that the function*

$$f(x,y) = x^3 + y^3 - 63(x + y) + 12xy$$

*is maximum at (– 7, – 7) and minimum at (3, 3).*

**Solution.** We have, $f(x,y) = x^3 + y^3 - 63(x + y) + 12xy$ ...(1)

$$\frac{\partial f}{\partial x} = 3x^2 - 63 + 12y, \qquad \frac{\partial f}{\partial y} = 3y^2 - 63 + 12x$$

$$\frac{\partial^2 f}{\partial x^2} = 6x, \quad \frac{\partial^2 f}{\partial x \partial y} = 12, \qquad \frac{\partial^2 f}{\partial y^2} = 6y$$

For extremum, we have

$$p = \frac{\partial f}{\partial x} = 3x^2 - 63 + 12y = 0 \qquad \Rightarrow x^2 + 4y - 21 = 0 \quad ...(2)$$

$$q = \frac{\partial f}{\partial y} = 3y^2 - 63 + 12x = 0 \qquad \Rightarrow y^2 + 4x - 21 = 0 \quad ...(3)$$

$$r = \frac{\partial^2 f}{\partial x^2} = 6x \quad \Rightarrow \quad s = \frac{\partial^2 f}{\partial x \partial y} = 12 \quad \Rightarrow \quad t = \frac{\partial^2 f}{\partial y^2} = 6y$$

We have to solve (2) and (3) for $x$ and $y$.

On subtracting (3) from (2), we have

$$x^2 - y^2 - 4(x - y) = 0 \qquad \Rightarrow \qquad (x - y)(x + y - 4) = 0$$

$$x = y \text{ and } x + y = 4 \qquad ...(4)$$

If $x = y$ then (2) becomes, $x^2 + 4x - 21 = 0$, $(x + 7)(x - 3) = 0$

$$x = -7, \text{ and } x = 3$$

$$y = -7, \text{ and } y = 3$$

Two stationary points are (–7, –7) and (3, 3).

On solving (2) and (4), we get

$$x^2 + 4(4 - x) - 21 = 0, \Rightarrow x^2 - 4x - 5 = 0$$

$$\Rightarrow \qquad (x - 5)(x + 1) = 0$$

$$x = -1, x = 5$$

$$y = 5, y = -1$$

Two more stationary points are (–1, 5) and (5, –1).

Hence four possible extremum points of $f(x, y)$ are (– 7, – 7), (3, 3), (–1, 5) and (5, – 1) may be.

| *Stationary pairs* | | $(-7, -7)$ | $(3, 3)$ | $(-1, 5)\,(5, -1)$ |
|---|---|---|---|---|
| $r = 6x$ | $-42$ | $+18$ | $-6$ | $30$ |
| $s = 12$ | $12$ | $12$ | $12$ | $12$ |
| $t = 6y$ | $-42$ | $18$ | $30$ | $-6$ |
| $rt - s^2$ | $+1620$ | $+180$ | $-324$ | $-324$ |

At $(-7, -7)$

$$r = -\text{ ve, and } rt - s^2 = +\text{ ve}$$

Hence, $f(x, y)$ is maximum at $(-7, -7)$,

At $(3, 3)$ $\qquad r = +\text{ ve, and } rt - s^2 = +\text{ ve}$

Hence, $f(x, y)$ is minimum at $(3, 3)$. **Proved.**

**Example 91.** *Find the extreme values of* $u = x^2 y^2 - 5x^2 - 8xy - 5y^2$.

**Solution.** We have,

$$u = x^2 y^2 - 5x^2 - 8xy - 5y^2 \quad \Rightarrow \quad p = \frac{\partial u}{\partial x} = 2xy^2 - 10x - 8y$$

$$q = \frac{\partial u}{\partial y} = 2x^2 y - 8x - 10y \quad \Rightarrow \quad r = \frac{\partial^2 u}{\partial x^2} = 2y^2 - 10$$

$$s = \frac{\partial^2 u}{\partial x\, \partial y} = 4xy - 8 \quad \Rightarrow \quad t = \frac{\partial^2 u}{\partial y^2} = 2x^2 - 10$$

For extreme values of $u$, $\quad \dfrac{\partial u}{\partial x} = 0, \quad \dfrac{\partial u}{\partial y} = 0$

$$2xy^2 - 10x - 8y = 0 \Rightarrow x = \frac{8y}{2y^2 - 10}$$

$$2x^2 y - 8x - 10y = 0$$

$$\Rightarrow \quad 2\left(\frac{8y}{2y^2 - 10}\right)^2 y - 8\left(\frac{8y}{2y^2 - 10}\right) - 10y = 0$$

$$\Rightarrow \quad \frac{128\, y^2}{(2y^2 - 10)^2} - \frac{64}{2y^2 - 10} - 10 = 0, \quad y = 0 \text{ then } x = 0 \Rightarrow \frac{16\, y^2}{(y^2 - 5)^2} - \frac{16}{y^2 - 5} - 5 = 0$$

$$\Rightarrow \quad 16\, y^2 - 16\,(y^2 - 5) - 5\,(y^2 - 5)^2 = 0$$

$$\Rightarrow \quad 16\, y^2 - 16\, y^2 + 80 - 5\,(y^2 - 5)^2 = 0$$

$$\Rightarrow \quad (y^2 - 5)^2 = 16 \Rightarrow y^2 - 5 = \pm 4 \Rightarrow y^2 = 9 \text{ and } 1 \text{ and } \Rightarrow y = \pm 3, \pm 1$$

Now, $\qquad x = \dfrac{4y}{y^2 - 5}$

If $y = 1$ then $x = -1$; $\qquad$ If $y = -1$ then $x = 1$

If $y = 3$ then $x = 3$; $\qquad$ If $y = -3$ then $x = -3$

| *Stationary pairs* | $(0, 0)$ | $(1, -1)$ | $(-1, 1)$ | $(3, 3)$ | $(-3, -3)$ |
|---|---|---|---|---|---|
| $r = 2y^2 - 10$ | $-10$ | $-8$ | $-8$ | $8$ | $8$ |
| $s = 4xy - 8$ | $-8$ | $-12$ | $-12$ | $28$ | $28$ |
| $t = 2x^2 - 10$ | $-10$ | $-8$ | $-8$ | $8$ | $8$ |
| $rt - s^2$ | $+36$ | $-80$ | $-80$ | $-720$ | $-720$ |

At (0, 0), $r$ is – ve.

Origin (0, 0) is the only point at which $r\,t - s^2 > 0$.

Hence, the function $u$ is maximum at origin. **Ans.**

**Example 92.** *A rectangular box, open at the top, is to have a volume of 32 c.c. Find the dimensions of the box requiring least material for its construction.*

*(M.U. 2009; U.P. I semester Jan. 2011; Dec. 2005, A.M.I.E Summer 2001)*

**Solution.** Let $l$, $b$ and $h$ be the length, breadth, and height of the box respectively and $S$ its surface area and $V$ the volume.

$$V = 32 \text{ c.c.}$$

$$\Rightarrow \qquad l\,b\,h = 32 \Rightarrow b = \frac{32}{lh}$$

$$S = 2\,(l + b)\,h + l\,b \qquad \text{...(1)}$$

Putting the value of $b$ in (1), we get

$$S = 2\left(l + \frac{32}{l\,h}\right)h + l\left(\frac{32}{l\,h}\right)$$

$$S = 2\,l\,h + \frac{64}{l} + \frac{32}{h} \qquad \text{...(2)}$$

Differentiating (2) partially w.r.t. $l$, we get

$$\frac{\partial S}{\partial l} = 2\,h - \frac{64}{l^2} \qquad \text{...(3)}$$

Differentiating (2) partially w.r.t. $h$, we get

$$\frac{\partial S}{\partial h} = 2\,l - \frac{32}{h^2} \qquad \text{...(4)}$$

For maximum and minimum $S$, we get

$$\frac{\partial S}{\partial l} = 0 \Rightarrow 2\,h - \frac{64}{l^2} = 0 \Rightarrow h = \frac{32}{l^2} \qquad \text{...(5)}$$

$$\frac{\partial S}{\partial h} = 0 \Rightarrow 2\,l - \frac{32}{h^2} = 0 \Rightarrow l = \frac{16}{h^2} \qquad \text{...(6)}$$

From (5) and (6), $l = 4$, $h = 2$ and $b = 4$

$$\frac{\partial^2 S}{\partial l^2} = \frac{128}{l^3} = \frac{128}{64} = 2$$

$$\frac{\partial^2 S}{\partial l\,\partial h} = 2$$

$$\frac{\partial^2 S}{\partial h^2} = \frac{64}{h^3} = \frac{64}{8} = 8$$

$$\frac{\partial^2 S}{\partial l^2}\cdot\frac{\partial^2 S}{\partial h^2} - \left(\frac{\partial^2 S}{\partial l\,\partial h}\right)^2 = (2)\,(8) - (2)^2 = +\,12$$

$\dfrac{\partial^2 S}{\partial l^2} = +\,2$, so $S$ is minimum for $l = 4$, $b = 4$, $h = 2$ **Ans.**

## EXERCISE 1.18

Find the stationary points of the following functions

1. $f(x, y) = y^2 + 4\,xy + 3\,x^2 + x^3$ **Ans.** $\left(\frac{2}{3}, -\frac{4}{3}\right)$, Minimum

2. $f(x, y) = x^3 y^2 (1 - x - y)$ [*A.M.I.E., Summer 2004*] **Ans.** $\left(\frac{1}{2}, \frac{1}{3}\right)$, Maximum

3. $f(x, y) = x^3 + 3xy^2 - 15x^2 - 15y^3 + 72x$. (*M.U. 2007, 2005, 2004*) **Ans.** (6, 0), (4, 0)

4. $f(x, y) = x^2 + 2xy + 2y^2 + 2x + 3y$ such that $x^2 - y = 1$. **Ans.** $\left(-\frac{3}{4}, -\frac{7}{16}\right), -\frac{155}{128}$.

5. $xy\, e^{-(2x + 3y)}$ (*A.M.I.E., Winter 2000*)

6. Find the extreme value of the function $f(x, y) = x^2 + y^2 + xy + x - 4y + 5$. State whether this value is a relative maximum or a relative minimum.
**Ans.** Minimum value of $f(x, y)$ at $(-2, 3) = -2$.

7. Find the values of $x$ and $y$ for which $x^2 + y^2 + 6x = 12$ has a minimum value and find this minimum value. **Ans.** (– 3, 0), 3.

8. Find a point within a triangle such that the sum of the square of its distances from the three angular points is a minimum.

9. A tapering log has a square cross-section whose side varies uniformly and is equal to $a$ at the top and $b\left(b > \frac{3a}{2}\right)$ at the bottom. Show that the volume of the greatest conical frustum that can be obtained from the log is $\frac{\pi b^3 l}{27(b-a)}$, where $l$ is the length of the log.

10. A tree trunk of length $l$ metres has the shape of a frustum of a circular cone with radii of its ends $a$ and $b$ metres where $a > b$. Find the length of a beam of uniform square cross section which can be cut from the tree trunk so that the beam has the greatest volume. **Ans.** $\frac{8a^3 l}{27(a-b)}$

## 1.30 LAGRANGE METHOD OF UNDETERMINED MULTIPLIERS

Let $f(x, y, z)$ be a function of three variables $x, y, z$ and the variables be connected by the relation.

$$\phi(x, y, z) = 0 \qquad ...(1)$$

$\Rightarrow$ $f(x, y, z)$ to have stationary values,

$$\frac{\partial f}{\partial x} = 0, \quad \frac{\partial f}{\partial y} = 0, \quad \frac{\partial f}{\partial z} = 0 \quad \Rightarrow \quad \frac{\partial f}{\partial x}dx + \frac{\partial f}{\partial y}dy + \frac{\partial f}{\partial z}dz = 0 \qquad ...(2)$$

By total differentiation of (1), we get $\frac{\partial \phi}{\partial x}dx + \frac{\partial \phi}{\partial y}dy + \frac{\partial \phi}{\partial z}dz = 0$ ...(3)

Multiplying (3) by $\lambda$ and adding to (2), we get

$$\left(\frac{\partial f}{\partial x}dx + \lambda\frac{\partial \phi}{\partial x}dx\right) + \left(\frac{\partial f}{\partial y}dy + \lambda\frac{\partial \phi}{\partial y}dy\right) + \left(\frac{\partial f}{\partial z}dz + \lambda\frac{\partial \phi}{\partial z}dz\right) = 0$$

$$\left(\frac{\partial f}{\partial x} + \lambda\frac{\partial \phi}{\partial x}\right)dx + \left(\frac{\partial f}{\partial y} + \lambda\frac{\partial \phi}{\partial y}\right)dy + \left(\frac{\partial f}{\partial z} + \lambda\frac{\partial \phi}{\partial z}\right)dz = 0$$

This equation will hold good if

$$\frac{\partial f}{\partial x} + \lambda\frac{\partial \phi}{\partial x} = 0 \qquad ...(4)$$

$$\frac{\partial f}{\partial y} + \lambda\frac{\partial \phi}{\partial y} = 0 \qquad ...(5)$$

$$\frac{\partial f}{\partial z} + \lambda\frac{\partial \phi}{\partial z} = 0 \qquad ...(6)$$

On solving (1), (4), (5), (6), we can find the values of $x$, $y$, $z$ and $\lambda$ for which $f(x, y, z)$ has stationary value.

**Draw Back** in Lagrange method is that the nature of stationary point cannot be determined.

**Example 93.** *Find the point upon the plane $ax + by + cz = p$ at which the function*

$$f = x^2 + y^2 + z^2$$ (*Nagpur University, Winter* 2000)

*has a minimum value and find this minimum f.*

**Solution.** We have, $$f = x^2 + y^2 + z^2 \quad \text{...(1)}$$

$$ax + by + cz = p \Rightarrow \phi = ax + by + cz - p \quad \text{...(2)}$$

$$\frac{\partial f}{\partial x} + \lambda \frac{\partial \varphi}{\partial x} = 0 \Rightarrow 2x + \lambda a = 0 \Rightarrow x = \frac{-\lambda a}{2}$$

$$\frac{\partial f}{\partial y} + \lambda \frac{\partial \varphi}{\partial y} = 0 \Rightarrow 2y + \lambda b = 0 \Rightarrow y = \frac{-\lambda b}{2}$$

$$\frac{\partial f}{\partial z} + \lambda \frac{\partial \varphi}{\partial z} = 0 \Rightarrow 2z + \lambda c = 0 \Rightarrow z = \frac{-\lambda c}{2}$$

Substituting the values of $x$, $y$, $z$ in (2), we get

$$a\left(\frac{-\lambda a}{2}\right) + b\left(\frac{-\lambda b}{2}\right) + c\left(\frac{-\lambda c}{2}\right) = p$$

$$\lambda(a^2 + b^2 + c^2) = -2p \Rightarrow \lambda = \frac{-2p}{a^2 + b^2 + c^2}$$

$$\therefore \quad x = \frac{ap}{a^2 + b^2 + c^2}, \quad y = \frac{bp}{a^2 + b^2 + c^2}, \quad z = \frac{cp}{a^2 + b^2 + c^2}$$

The minimum value of $$f = \frac{a^2 p^2}{(a^2 + b^2 + c^2)^2} + \frac{b^2 p^2}{(a^2 + b^2 + c^2)^2} + \frac{c^2 p^2}{(a^2 + b^2 + c^2)^2}$$

$$= \frac{p^2(a^2 + b^2 + c^2)}{(a^2 + b^2 + c^2)^2} = \frac{p^2}{a^2 + b^2 + c^2}$$ **Ans.**

**Example 94.** *Find the maximum value of $u = x^p y^q z^r$ when the variables x, y, z are subject to the condition $ax + by + cz = p + q + r$.*

**Solution.** Here, we have $$u = x^p y^q z^r \quad \text{...(1)}$$

If $$\log u = p \log x + q \log y + r \log z \quad \text{...(2)}$$

$$\frac{1}{u}\frac{\partial u}{\partial x} = \frac{p}{x} \Rightarrow \frac{\partial u}{\partial x} = \frac{pu}{x}$$

$$\frac{1}{u}\frac{\partial u}{\partial y} = \frac{q}{y} \Rightarrow \frac{\partial u}{\partial y} = \frac{qu}{y}$$

$$\frac{1}{u}\frac{\partial u}{\partial z} = \frac{r}{z} \Rightarrow \frac{\partial u}{\partial z} = \frac{ru}{z}$$

$$ax + by + cz = p + q + r$$

$$\phi(x, y, z) = ax + by + cz - p - q - r$$

$$\frac{\partial \phi}{\partial x} = a, \quad \frac{\partial \phi}{\partial y} = b, \quad \frac{\partial \phi}{\partial z} = c$$

Lagranges equations are

$$\frac{\partial u}{\partial x}+\lambda\frac{\partial \phi}{\partial x} = 0 \Rightarrow \frac{pu}{x}+\lambda a = 0 \Rightarrow x = -\frac{pu}{\lambda a}$$

$$\frac{\partial u}{\partial y}+\lambda\frac{\partial \phi}{\partial y} = 0 \Rightarrow \frac{qu}{y}+\lambda b = 0 \Rightarrow y = -\frac{qu}{\lambda b}$$

$$\frac{\partial u}{\partial z}+\lambda\frac{\partial \phi}{\partial z} = 0 \Rightarrow \frac{ru}{z}+\lambda c = 0 \Rightarrow z = -\frac{ru}{\lambda c}$$

Putting in (2), we have

$$-\frac{pu}{\lambda}-\frac{qu}{\lambda}-\frac{ru}{\lambda} = p+q+r$$

$$-\frac{u}{\lambda}(p+q+r) = p+q+r \Rightarrow -\frac{u}{\lambda} = 1 \Rightarrow \lambda = -u$$

$$x = -\frac{pu}{\lambda a} = \frac{-pu}{-ua} = \frac{p}{a}$$

$$y = -\frac{qu}{\lambda b} = \frac{-qu}{-ub} = \frac{q}{b}$$

$$z = -\frac{ru}{\lambda c} = \frac{-ru}{-uc} = \frac{r}{c}$$

Putting in (1), we have

Maximum value of $u = \left(\frac{p}{a}\right)^p\left(\frac{q}{b}\right)^q\left(\frac{r}{c}\right)^r$ **Ans.**

**Example 95.** *Show that the rectangular solid of maximum volume that can be inscribed in a sphere is a cube.*

**Solution.** Let $2x$, $2y$, $2z$ be the length, breadth and height of the rectangular solid.

Let R be the radius of the sphere.

Volume of solid $V = 8\,x\,.\,y\,.\,z$ ...(1)

$$x^2 + y^2 + z^2 = R^2 \quad ....(2)$$

$$\Rightarrow \quad \phi(x, y, z) = x^2 + y^2 + z^2 - R^2 = 0$$

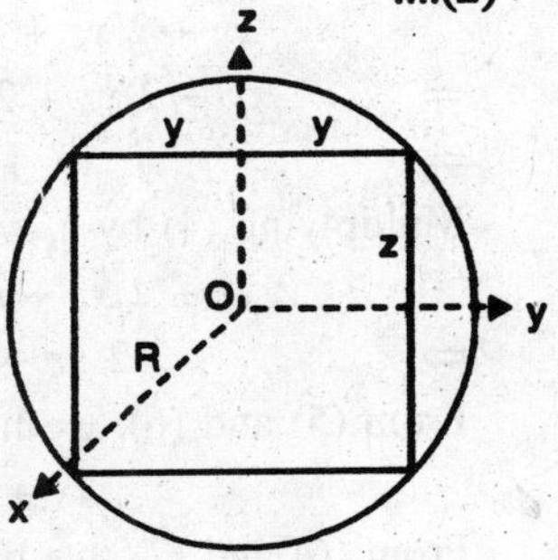


$$\frac{\partial V}{\partial x} + \lambda\frac{\partial \phi}{\partial x} = 0 \Rightarrow 8yz + \lambda(2x) = 0 \quad ...(3)$$

$$\frac{\partial V}{\partial y} + \lambda\frac{\partial \phi}{\partial y} = 0 \Rightarrow 8xz + \lambda(2y) = 0 \quad ...(4)$$

$$\frac{\partial V}{\partial z} + \lambda\frac{\partial \phi}{\partial z} = 0 \Rightarrow 8xy + \lambda(2z) = 0 \quad ...(5)$$

From (3) $\quad 2\lambda x = -8yz \quad \Rightarrow \quad 2\lambda x^2 = -8xyz$

From (4) $\quad 2\lambda y = -8xz \quad \Rightarrow \quad 2\lambda y^2 = -8xyz$

From (5) $\quad 2\lambda z = -8xy \quad \Rightarrow \quad 2\lambda z^2 = -8xyz$

$$2\lambda x^2 = 2\lambda y^2 = 2\lambda z^2$$

$$\Rightarrow \quad x^2 = y^2 = z^2$$

$$\Rightarrow \quad x = y = z$$

Hence, rectangular solid is a cube. **Proved.**

**Example 96.** *A rectangular box, which is open at the top, has a capacity of 256 cubic feet. Determine the dimensions of the box such that the least material is required for the construction of the box. Use Lagrange's method of multipliers to obtain the solution.*

**Solution.** Let $x$, $y$, $z$ be the length, breadth and height of the box.

$$\Rightarrow \quad \text{Volume} = xyz = 256 \Rightarrow xyz - 256 = 0 \quad \text{...(1)}$$

$$\Rightarrow \quad \phi(x, y, z) = xyz - 256$$

Let S be the material surface of the box.

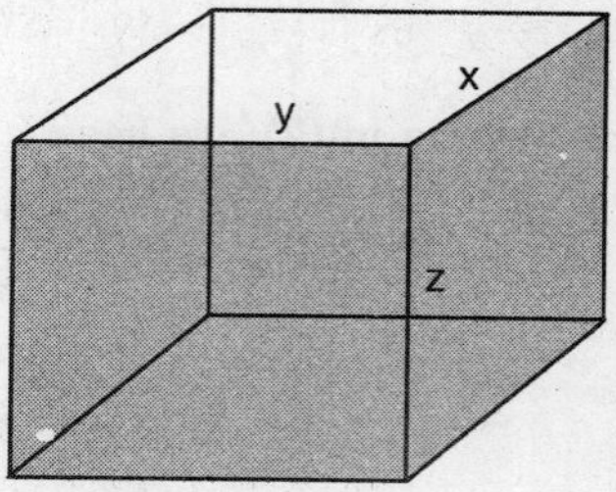


$$S = xy + 2yz + 2zx$$

$$\frac{\partial S}{\partial x} = y + 2z \quad \text{and} \quad \frac{\partial \phi}{\partial x} = yz$$

$$\frac{\partial S}{\partial y} = x + 2z \quad \text{and} \quad \frac{\partial \phi}{\partial y} = xz$$

$$\frac{\partial S}{\partial z} = 2y + 2x \quad \text{and} \quad \frac{\partial \phi}{\partial z} = xy$$

By Lagrange's method of multiplier, we have

$$\frac{\partial S}{\partial x} + \lambda \frac{\partial \phi}{\partial x} = 0 \Rightarrow y + 2z + \lambda yz = 0 \quad \text{...(2)}$$

$$\frac{\partial S}{\partial y} + \lambda \frac{\partial \phi}{\partial y} = 0 \Rightarrow x + 2z + \lambda xz = 0 \quad \text{...(3)}$$

$$\frac{\partial S}{\partial z} + \lambda \frac{\partial \phi}{\partial z} = 0 \Rightarrow 2y + 2x + \lambda xy = 0 \quad \text{...(4)}$$

Multiplying (2) by $x$, we get

$$xy + 2xz + \lambda xyz = 0$$

$$\Rightarrow \quad xy + 2xz + 256\lambda = 0 \quad (xyz = 256)$$

$$\Rightarrow \quad xy + 2xz = -256\lambda \quad \text{...(5)}$$

Multiplying (3) by $y$, we get

$$xy + 2yz + \lambda xyz = 0$$

$$\Rightarrow \quad xy + 2yz + 256\lambda = 0$$

$$\Rightarrow \quad xy + 2yz = -256\lambda \quad \text{...(6)}$$

Multiplying (4) by $z$, we get

$$2yz + 2xz + \lambda xyz = 0 \Rightarrow 2yz + 2xz + 256\lambda = 0$$

$$\Rightarrow \quad 2yz + 2zx = -256\lambda \quad \text{...(7)}$$

From (5) and (6), we have

$$xy + 2xz = xy + 2yz \Rightarrow 2xz = 2yz \Rightarrow x = y$$

From (6) and (7), we have

$$xy + 2yz = 2yz + 2xz \Rightarrow xy = 2xz \Rightarrow y = 2z$$

From (1) $\quad xyz = 256$

$$\Rightarrow \quad (y)(y)\left(\frac{y}{2}\right) = 256 \Rightarrow y^3 = 512 \Rightarrow y = 8$$

$$x = 8, \quad y = 8, \quad z = 4$$

Hence, length = breadth = 8′, height = 4′. **Ans.**

**Example 97.** *Use the method of the Lagrange's multipliers to find the volume of the largest rectangular parallelopiped that can be inscribed in the ellipsoid* $\frac{x^2}{a^2}+\frac{y^2}{b^2}+\frac{z^2}{c^2}=1$.

*(Nagpur Univesity, Summer 2008, Winter 2003)*

*(A.M.I.E.T.E., Summer 2004, U.P., I Semester, Winter 2002, 2000)*

**Solution.** Here, we have $\frac{x^2}{a^2}+\frac{y^2}{b^2}+\frac{z^2}{c^2} = 1$

$$\Rightarrow \qquad \phi(x, y, z) = \frac{x^2}{a^2}+\frac{y^2}{b^2}+\frac{z^2}{c^2}-1=0 \qquad ...(1)$$

Let $2x$, $2y$, $2z$ be the length, breadth and height of the rectangular parallelopiped inscribed in the ellipsoid.

$$V = (2x)(2y)(2z) = 8\,xyz$$

$$\frac{\partial V}{\partial x} = 8\,yz; \quad \frac{\partial V}{\partial y} = 8xz, \quad \frac{\partial V}{\partial z} = 8xy$$

$$\frac{\partial \phi}{\partial x} = \frac{2x}{a^2}, \quad \frac{\partial \phi}{\partial y} = \frac{2y}{b^2}, \quad \frac{\partial \phi}{\partial z} = \frac{2z}{c^2}$$

Lagrange's equations are

$$\frac{\partial V}{\partial x}+\lambda\frac{\partial \phi}{\partial x} = 0 \quad \Rightarrow \quad 8\,yz+\lambda\frac{2x}{a^2} = 0 \qquad ...(1)$$

$$\frac{\partial V}{\partial y}+\lambda\frac{\partial \phi}{\partial y} = 0 \quad \Rightarrow \quad 8\,xz+\lambda\frac{2y}{b^2} = 0 \qquad ...(2)$$

$$\frac{\partial V}{\partial z}+\lambda\frac{\partial \phi}{\partial z} = 0 \quad \Rightarrow \quad 8\,xy+\lambda\frac{2z}{c^2} = 0 \qquad ...(3)$$

Multiplying (1), (2) and (3) by $x$, $y$, $z$ respectively and adding, we get

$$24xyz+2\lambda\left[\frac{x^2}{a^2}+\frac{y^2}{b^2}+\frac{z^2}{c^2}\right] = 0 \qquad \left[\frac{x^2}{a^2}+\frac{y^2}{b^2}+\frac{z^2}{c^2}=1\right]$$

$$\Rightarrow \quad 24\,xyz+2\,\lambda\,(1) = 0 \quad \Rightarrow \quad \lambda = -\,12\,xyz$$

Putting the value of $\lambda$ in (1), we get

$$8\,yz+(-\,12\,xyz)\frac{2x}{a^2} = 0 \quad \Rightarrow \quad 1-\frac{3x^2}{a^2}=0 \quad \Rightarrow \quad x=\frac{a}{\sqrt{3}}$$

Similarly from (2) and (3), we have

$$y = \frac{b}{\sqrt{3}}, \quad z=\frac{c}{\sqrt{3}}$$

Volume of the largest rectangular parallelopiped $= 8\,xyz$

$$= 8\left(\frac{a}{\sqrt{3}}\right)\left(\frac{b}{\sqrt{3}}\right)\left(\frac{c}{\sqrt{3}}\right)=\frac{8\,abc}{3\sqrt{3}}$$ **Ans.**

**Example 98.** *The shape of a hole pored by a drill is a cone surmounted by cylinder. If the cylinder be of height h and radius r and the semi-vertical angle of the cone be* $\alpha$, *where* $\tan\alpha=\frac{h}{r}$ *show that for a total height H of the hole, the volume removed is maximum if* $h = H(\sqrt{7}+1)/6$.

*(R.G.P.V., Bhopal I sem. 2003)*

**Solution.** Let *ABCD* be the given cylinder of height '*h*' and radius '*r*' and *DPCO be the cone of course*, of radius *r*.

Now, since $\alpha$ is the semi-vertical angle of the cone.

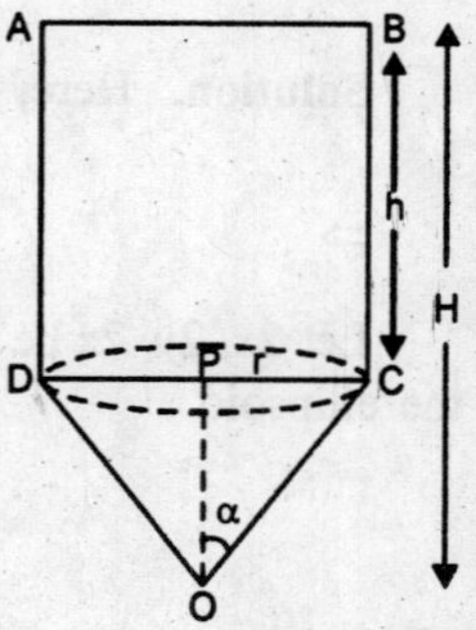


$$\therefore \quad \tan\alpha = \frac{PC}{OP} = \frac{r}{OP} \qquad \text{...(1)}$$

but, given that $\quad \tan\alpha = \dfrac{h}{r} \qquad$ ...(2)

From (1) and (2), we have $\dfrac{h}{r} = \dfrac{r}{OP} \Rightarrow OP = \dfrac{r^2}{h} \qquad$ ...(3)

Total height of the hole = H

$$\Rightarrow \quad H = h + OP \Rightarrow OP = H - h \qquad \text{...(4)}$$

From (3) and (4), we have

$$\frac{r^2}{h} = H - h \qquad \text{...(5)}$$

Again, let $\quad \phi = H - h - \dfrac{r^2}{h} \qquad$ ...(6)

In drilling a hole, the volume of the removed portion

$$\frac{\partial\phi}{\partial r} = -\frac{2r}{h}, \quad \frac{\partial\phi}{\partial h} = -1 + \frac{r^2}{h^2}$$

$V$ = Volume of the cylinder + Volume of the cone.

$$= \pi r^2 h + \frac{1}{3}\pi r^2 (OP) = \pi r^2 h + \frac{1}{3}\pi r^2 . \frac{r^2}{h} \qquad \text{[From (3)]}$$

$$V = \pi r^2 h + \frac{\pi r^4}{3h},$$

$$\frac{\partial V}{\partial r} = 2\pi r h + \frac{4\pi r^3}{3h} \qquad \text{...(7)}$$

By Lagrange's Method

$$\frac{\partial V}{\partial r} + \lambda\frac{\partial\phi}{\partial r} = 0 \quad \Rightarrow \quad 2\pi r h + \frac{4\pi r^3}{3h} + \lambda\left(\frac{-2r}{h}\right) = 0 \qquad \text{...(8)}$$

$$\frac{\partial V}{\partial h} + \lambda\frac{\partial\phi}{\partial h} = 0 \quad \Rightarrow \quad \pi r^2 - \frac{\pi r^4}{3h^2} + \lambda\left(-1 + \frac{r^2}{h^2}\right) = 0 \qquad \text{...(9)}$$

Multiplying (8) by *r* and (9) by 2*h*, we get

$$2\pi r^2 h + \frac{4\pi r^4}{3h} + \lambda\left(\frac{-2r^2}{h}\right) = 0$$

$$2\pi r^2 h - \frac{2\pi r^4}{3h} + 2\lambda\left(-h + \frac{r^2}{h}\right) = 0$$

On subtracting, we get

$$\frac{6\pi r^4}{3h} + \lambda\left(\frac{-2r^2}{h} + 2h - \frac{2r^2}{h}\right) = 0 \quad \Rightarrow \quad \frac{6\pi r^4}{3h} + \lambda\left(\frac{-4r^2}{h} + 2h\right) = 0$$

$$2\pi r^4 + \lambda(-4r^2 + 2h^2) = 0$$

$$\Rightarrow \quad \pi r^4 + \lambda(-2r^2 + h^2) = 0 \quad \Rightarrow \quad \lambda = \frac{\pi r^4}{-h^2 + 2r^2}$$

Putting the value of $\lambda$ in (8), we get

$$2\pi rh+\frac{4\pi r^3}{3h}+\left(\frac{\pi r^4}{-h^2+2r^2}\right)\left(-\frac{2r}{h}\right)=0 \Rightarrow h+\frac{2r^2}{3h}-\frac{r^4}{h(h^2-2r^2)}=0 \quad \left[\frac{r^2}{h}=H-h\right]$$

$$h+\frac{2}{3}(H-h)+\frac{h^2(H-h)^2}{h[h^2-2h(H-h)]}=0 \quad \Rightarrow \quad h+\frac{2}{3}(H-h)+\frac{(H^2+h^2-2hH)}{h-2H+2h}=0$$

$$h+\frac{2}{3}(H-h)+\frac{H^2+h^2-2hH}{3h-2H}=0$$

$$3h^2-2Hh+\frac{2}{3}(H-h)(3h-2H)+H^2+h^2-2hH=0$$

$$9h^2-6Hh+6Hh-4H^2-6h^2+4Hh+3H^2+3h^2-6hH=0$$

$$6h^2-2hH-H^2=0$$

$$h=\frac{2H\pm\sqrt{4H^2+24H^2}}{12}$$

$$h=\frac{H\pm H\sqrt{7}}{6}=H\frac{[\sqrt{7}+1]}{6} \text{ (–ve is not possible)}$$ **Proved.**

**Example 99.** *A tent of a given volume has a square base of side 2a, has its four-side vertical of length b and is surmounted by a regular pyramid of height h. Find the values of a and b in terms of h such that the canvas required for its construction is minimum.*

**Solution.** Let $V$ be the volume and $S$ be the surface of the tent.

$$V=4a^2b+\frac{1}{3}(4a^2)h \qquad \left[\text{Volume of pyramid}=\frac{1}{3}\text{ Area of the base}\times\text{height}\right]$$

$$S=8ab+4a\sqrt{a^2+h^2} \qquad \left[\text{Surface Area of pyramid}=\frac{1}{2}\text{ perimeter}\times\text{slant height}\right]$$

$$\frac{\partial S}{\partial a}+\lambda\frac{\partial V}{\partial a}=0$$

$$\Rightarrow \quad 8b+4\sqrt{a^2+h^2}+\frac{4a^2}{\sqrt{a^2+h^2}}+\lambda\left[8ab+\frac{8ah}{3}\right]=0 \qquad \text{...(1)}$$

$$\frac{\partial S}{\partial b}+\lambda\frac{\partial V}{\partial b}=0 \quad \Rightarrow \quad 8a+4\lambda a^2=0 \qquad \text{...(2)}$$

$$\frac{\partial S}{\partial h}+\lambda\frac{\partial V}{\partial h}=0 \quad \Rightarrow \quad \frac{4ah}{\sqrt{a^2+h^2}}+\frac{4}{3}\lambda a^2=0 \qquad \text{...(3)}$$

From (2) $\quad \lambda a+2=0 \quad \Rightarrow \quad \lambda a=-2 \qquad$ ...(4)

From (3) $\quad 12ah+4\lambda a^2\sqrt{a^2+h^2}=0$

$$\Rightarrow \quad 3h+\lambda a\sqrt{a^2+h^2}=0 \qquad \text{...(5)}$$

Substituting the value of $\lambda a$ from (4) in (5), we get

$$3h-2\sqrt{a^2+h^2}=0 \Rightarrow 9h^2=4a^2+4h^2 \Rightarrow 4a^2=5h^2$$

$$a=\frac{\sqrt{5}}{2}h$$

Substituting $\lambda\, a = -2$ and $a = \frac{\sqrt{5}}{2}h$ in (1) and simplifying, we get

$$8b + 4\sqrt{\frac{5h^2}{4} + h^2} + \frac{5h^2}{\sqrt{\frac{5h^2}{4} + h^2}} - 2\left[8b + \frac{8h}{3}\right] = 0$$

$$\Rightarrow\ 8b + 6h + \frac{10h}{3} - 16b - \frac{16h}{3} = 0 \quad \Rightarrow\ -8b + 4h = 0 \quad \Rightarrow\ b = \frac{h}{2}.$$

Thus, when $a = \frac{\sqrt{5}}{2}h$ and $b = \frac{h}{2}$ we get the stationary value of $S$. **Ans.**

**Example 100.** *Find the maximum and minimum distances of the point* (3, 4, 12) *from the sphere* $x^2 + y^2 + z^2 = 1$.

**Solution.** Let the co-ordinates of the given point be $(x, y, z)$, then its distance ($D$) from (3, 4, 12).

$$D = \sqrt{(x-3)^2 + (y-4)^2 + (z-12)^2}$$

$$\Rightarrow\quad F(x, y, z) = (x-3)^2 + (y-4)^2 + (z-12)^2$$

$$x^2 + y^2 + z^2 = 1$$

$$\phi(x, y, z) = x^2 + y^2 + z^2 - 1$$

$$\frac{\partial F}{\partial x} + \lambda \frac{\partial \phi}{\partial x} = 2(x-3) + 2\lambda x = 0 \qquad ...(1)$$

$$\frac{\partial F}{\partial y} + \lambda \frac{\partial \phi}{\partial y} = 2(y-4) + 2\lambda y = 0 \qquad ...(2)$$

$$\frac{\partial F}{\partial z} + \lambda \frac{\partial \phi}{\partial z} = 2(z-12) + 2\lambda z = 0 \qquad ...(3)$$

Multiplying (1) by $x$, (2) by $y$ and (3) by $z$ and adding, we get

$$(x^2 + y^2 + z^2) - 3x - 4y - 12z + \lambda.(x^2 + y^2 + z^2) = 0$$

$$1 - 3x - 4y - 12z + \lambda = 0 \qquad ...(4)$$

From (1) $\quad x = \frac{3}{1+\lambda}$ ...(5)

From (2) $\quad y = \frac{4}{1+\lambda}$ ...(6)

From (3) $\quad z = \frac{12}{1+\lambda}$ ...(7)

Putting these values of $x, y, z$ in (4), we have

$$1 + \lambda - \frac{9}{1+\lambda} - \frac{16}{1+\lambda} - \frac{144}{1+\lambda} = 0 \Rightarrow (1+\lambda)^2 = 169 \quad \Rightarrow \quad 1 + \lambda = \pm 13$$

Putting the value of $1 + \lambda$ in (5), (6) and (7) we have the points

$$\left(\frac{3}{13}, \frac{4}{13}, \frac{12}{13}\right) \text{ and } \left(\frac{-3}{13}, \frac{-4}{13}, \frac{-12}{13}\right).$$

The minimum distance $= \sqrt{\left(3 - \frac{3}{13}\right)^2 + \left(4 - \frac{4}{13}\right)^2 + \left(12 - \frac{12}{13}\right)^2} = 12$

The maximum distance $= \sqrt{\left(3 + \frac{3}{13}\right)^2 + \left(4 + \frac{4}{13}\right)^2 + \left(12 + \frac{12}{13}\right)^2} = 14$ **Ans.**

**Example 101.** *If $u = ax^2 + by^2 + cz^2$ where $x^2 + y^2 + z^2 = 1$ and $lx + my + nz = 0$ prove that stationary values of 'u' satisfy the equation*

$$\frac{l^2}{a-u}+\frac{m^2}{b-u}+\frac{n^2}{c-u} = 0$$

**Solution.** We have, $\quad u = ax^2 + by^2 + cz^2 \qquad ...(1)$

Let $\quad \phi = x^2 + y^2 + z^2 - 1 \qquad ...(2)$

$\quad \psi = lx + my + nz \qquad ...(3)$

$$\frac{\partial u}{\partial x} = 2\,a\,x, \quad \frac{\partial u}{\partial y} = 2\,b\,y, \quad \frac{\partial u}{\partial z} = 2\,c\,z$$

$$\frac{\partial \phi}{\partial x} = 2\,x, \quad \frac{\partial \phi}{\partial y} = 2\,y, \quad \frac{\partial \phi}{\partial z} = 2\,z$$

$$\frac{\partial \psi}{\partial x} = l, \quad \frac{\partial \psi}{\partial y} = m, \quad \frac{\partial \psi}{\partial z} = n$$

By Lagrange's method

$$\frac{\partial u}{\partial x}+\lambda_1\frac{\partial \phi}{\partial x}+\lambda_2\frac{\partial \psi}{\partial x} = 0, \qquad 2\,a\,x + 2\,x\,\lambda_1 + \lambda_2\,l = 0 \qquad ...(4)$$

$$\frac{\partial u}{\partial y}+\lambda_1\frac{\partial \phi}{\partial y}+\lambda_2\frac{\partial \psi}{\partial y} = 0, \qquad 2\,b\,y + 2\,y\,\lambda_1 + \lambda_2\,m = 0 \qquad ...(5)$$

$$\frac{\partial u}{\partial z}+\lambda_1\frac{\partial \phi}{\partial z}+\lambda_2\frac{\partial \psi}{\partial z} = 0, \qquad 2\,c\,z + 2\,z\,\lambda_1 + \lambda_2\,n = 0 \qquad ...(6)$$

Multiplying (4), (5) and (6) by $x$, $y$ and $z$ respectively and adding, we get

$$(2\,ax^2 + 2\,by^2 + 2\,cz^2) + (2\,x^2 + 2\,y^2 + 2\,z^2)\,\lambda_1 + (lx + my + nz)\,\lambda_2 = 0$$

$$2u + 2\lambda_1 = 0, \qquad \lambda_1 = -\,u$$

Putting the value of $\lambda_1$ in (4), (5) and (6), we get

$$2\,a\,x - 2\,x\,u + \lambda_2\,l = 0, \qquad x = \frac{-\lambda_2 l}{2(a-u)}$$

$$2\,b\,y - 2\,y\,u + \lambda_2\,m = 0, \qquad y = \frac{-\lambda_2\,m}{2(b-u)}$$

$$2\,c\,z - 2\,z\,u + \lambda_2\,n = 0, \qquad z = \frac{-\lambda_2\,n}{2(c-u)}$$

Putting the values of $x$, $y$, $z$ in (3), we get

$$\frac{-\lambda_2\,l^2}{2(a-u)}+\frac{-\lambda_2\,m^2}{2(b-u)}+\frac{-\lambda_2 n^2}{2(c-u)} = 0 \;\Rightarrow\; \frac{l^2}{a-u}+\frac{m^2}{b-u}+\frac{n^2}{c-u} = 0 \qquad \textbf{Proved.}$$

## EXERCISE 1.19

1. Show that the greatest value of $x^m\, y^n$ where $x$ and $y$ are positive and $x + y = a$ is $\dfrac{m^m.n^n.a^{m+n}}{(m+n)^{m+n}}$, where $a$ is constant.
2. Using Lagrange's method (of multipliers), find the critical (stationary values) of the function $f(x, y, z) = x^2 + y^2 + z^2$, given that $z^2 = xy + 1$. **Ans.** (0, 0, – 1), (0, 0, 1).

3. Decompose a positive number '$a$' into three parts so that their product is maximum.

**Ans.** $\frac{a}{3}, \frac{a}{3}, \frac{a}{3}$

4. The sum, of three numbers is constant. Prove that their product is a maximum when they are equal.
5. Using the method of Lagrange's multipliers, find the largest product of the numbers $x$, $y$ and $z$ when $x + y + z^2 = 16$. **Ans.** $\frac{4096}{25\sqrt{5}}$
6. Using the method of Lagrange's multipliers, find the largest product of the numbers $x$, $y$ and $z$ when $x^2 + y^2 + z^2 = 9$. **Ans.** $3\sqrt{3}$
7. Find a point in the plane $x + 2y + 3z = 13$ nearest to the point (1, 1, 1) using the method of Lagrange's multipliers. **Ans.** $\left(\frac{3}{2}, 2, \frac{5}{2}\right)$
8. Using the Lagrange's method (of multipliers), find the shortest distance from the point (1, 2, 2) to the sphere $x^2 + y^2 = 36$. **Ans.** 3
9. Find the shortest and the longest distances from the point (1, 2, – 1) to the $x^2 + y^2 + z^2 = 24$. **Ans.** $\sqrt{6}, 3\sqrt{6}$
10. The sum of the surfaces of a sphere and a cube is given. Show that when the sum of the volumes is least, the diameter of the sphere is equal to the edge of the cube.
11. The electric time constant of a cylindrical coil of wire can be expressed approximately by

$$K = \frac{mxyz}{ax + by + cz}$$

where $z$ is the axial length of the coil, $y$ is the difference between the external and internal radii and $x$ is the mean radius ; $a$, $b$, $m$ and $c$ represent positive constants. If the volume of the coil is fixed, find the values of $x$ and $y$ which make the time constant $K$ as large as possible.

12. If $u = \frac{a^3}{x^2} + \frac{b^3}{y^2} + \frac{c^3}{z^2}$, where $x + y + z = 1$, prove that the stationary value of $u$ is given by

$$x = \frac{a}{a+b+c}, \quad y = \frac{b}{a+b+c}, \quad z = \frac{c}{a+b+c}$$

13. Find maximum value of the expression $\sum_{i=1}^{n} a_i x_i$ with $\sum_{i=1}^{n} x_i^2 = 1$, where $a_1, a_2, a_3 \ldots\ldots a_n$ are positive constants. **Ans.** $(a_1^2 + a_2^2 + \ldots\ldots + a_n^2)^{\frac{1}{2}}$
14. If $r$ is the distance of a point on conic $ax^2 + by^2 + cz^2 = 1$, $lx + my + nz = 0$ from origin, then the stationary values of $r$ are given by the equation.

$$\frac{l^2}{1-ar^2} + \frac{m^2}{1-br^2} + \frac{n^2}{1-cr^2} = 0$$

*(A.M.I.E.T.E., Winter 2002)*

15. If $x$ and $y$ satisfy the relation $ax^2 + by^2 = ab$, prove that the extreme values of function $u = x^2 + xy + y^2$ are given by the roots of the equation $4(u - a)(u - b) = ab$

*(A.M.I.E.T.E., Winter 2000)*

16. Use the Lagranges method of undetermined multipliers to find the minimum value of $x^2 + y^2 + z^2$ subject to the conditions $x + y + z = 1$, $xyz + 1 = 0$.
17. Test the function $f(x, y) = (x^2 + y^2)e^{-(e^2+y^2)}$ for maxima and minima for points not on the circle $x^2 + y^2 = 1$.
18. Find the absolute maximum and minimum values of the function $f(x, y) = cx^2 + y^2 - x$ over the region to $2x^2 + y^2 \leq 1$ *(AMIETE, Dec. 2008)*

# 2

# Multiple Integral

## 2.1 DOUBLE INTEGRATION

We know that

$$\int_a^b f(x)\,dx = \lim_{\substack{n\to\infty \\ \delta x\to 0}} [f(x_1)\,\delta x_1 + f(x_2)\,\delta x_2 + f(x_3)\,\delta x_3 + \ldots + f(x_n)\,\delta x_n]$$

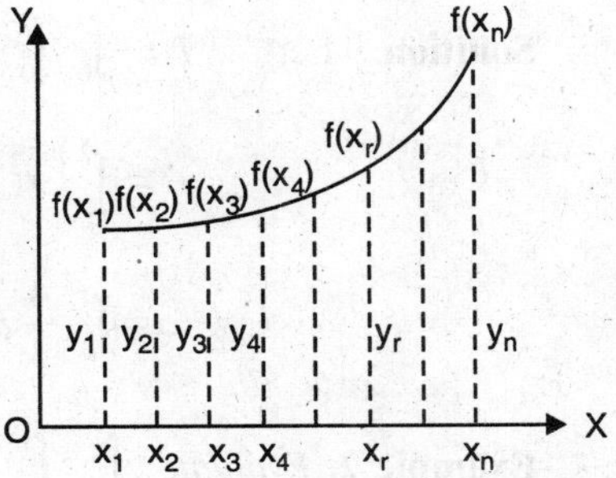


Let us consider a function $f(x, y)$ of two variable $x$ and $y$ defined in the finite region $A$ of $xy$-plane. Divide the region $A$ into elementary areas.

$$\delta A_1,\ \delta A_2,\ \delta A_3,\ \ldots\ldots\ \delta A_n$$

Then $\iint_A f(x, y)\,dA$

$$= \lim_{\substack{n\to\infty \\ \delta A\to 0}} [f(x_1, x_1)\,\delta A_1 + f(x_2, y_2)\,\delta A_2 + \ldots\ldots + f(x_n, y_n)\,\delta A_n]$$

## 2.2 EVALUATION OF DOUBLE INTEGRAL

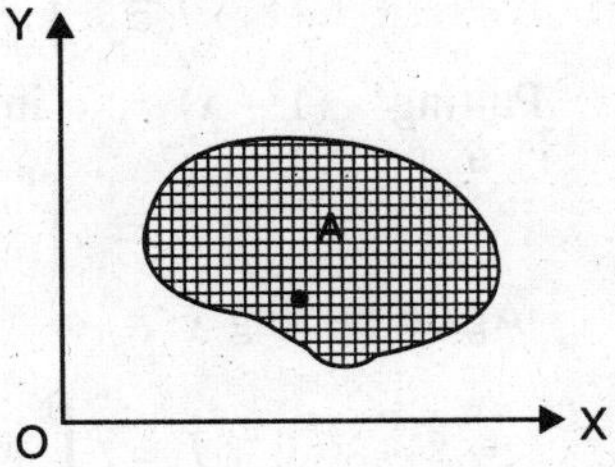


Double integral over region $A$ may be evaluated by two successive integrations.

If $A$ is described as $f_1(x) \le y \le f_2(x)$ $[y_1 \le y \le y_2]$

and $a \le x \le b$,

Then $$\iint_A f(x, y)\,dA = \int_a^b \int_{y_1}^{y_2} f(x, y)\,dx\,dy$$

**(1) First Method**

$$\iint_A f(x, y)\,dA = \int_a^b \boxed{\int_{y_1}^{y_2} f(x, y)\,dy}\,dx$$

$f(x, y)$ is first integratred with respect to $y$ treating $x$ as constant between the limits $a$ and $b$.

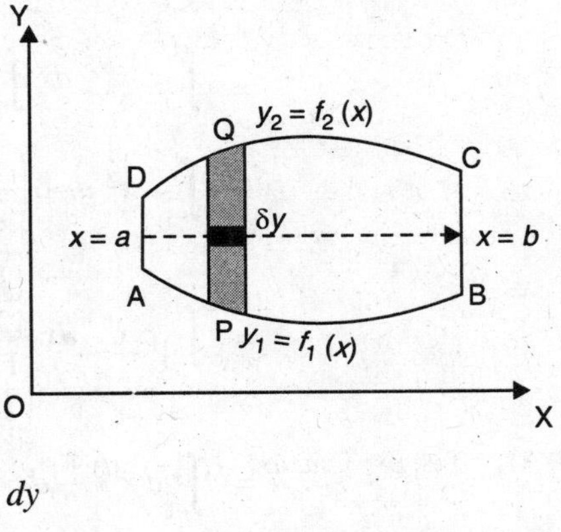


In the region we take an elementary area $\delta x \delta y$. Then integration w.r.t $y$ ($x$ keeping constant). converts small rectangle $\delta x \delta y$ into a strip $PQ$ ($y\,\delta x$). While the integration of the result w.r.t. $x$ corresponding to the sliding to the strip $PQ$, from $AD$ to $BC$ covering the while region $ABCD$.

**Second method**

$$\iint_A f(x, y)\,dxdy = \int_c^d \boxed{\int_{x_1}^{x_2} f(x, y)\,dx}\,dy$$

Here $f(x,y)$ is first integrated w.r.t $x$ keeping $y$ constant between the limits $x_1$ and $x_2$ and then the resulting expression is integrated with respect to $y$ between the limits $c$ and $d$

Take a small area $\delta x \delta y$. The integration w.r.t. $x$ between the limits $x_1$, $x_2$ keeping $y$ fixed indicates that integration is done, along $PQ$. Then the integration of result w.r.t $y$ corresponds to sliding the strips $PQ$ from $BC$ to $AD$ covering the whole region $ABCD$.

Note. For constant limits, it does not matter whether we first integrate w.r.t $x$ and then w.r.t $y$ or vice versa.

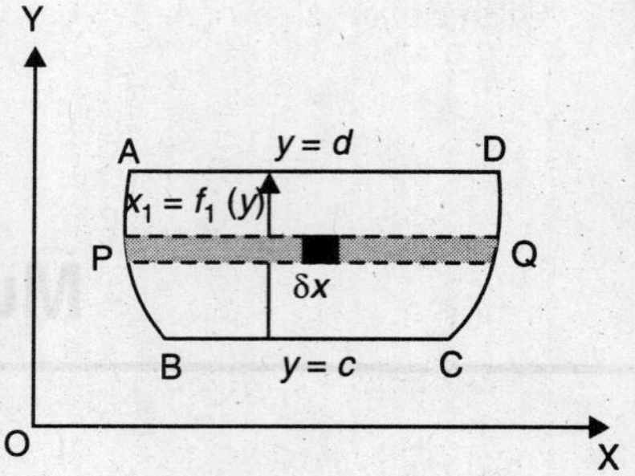


**Example 1.** *Evaluate* $\int_0^1 \int_0^x (x^2+y^2)\, dA$, *where dA indicates small area in xy-plane.*

*(Gujarat, I Semester, Jan. 2009)*

**Solution.** Let $$I = \int_0^1 \int_0^x (x^2+y^2)\, dy\, dx = \int_0^1 \left[x^2 y + \frac{y^3}{3}\right]_0^x dx$$

$$= \int_0^1 \left[x^2(x-0) + \frac{1}{3}(x^3-0)\right] dx = \int_0^1 \left[x^3 + \frac{x^3}{3}\right] dx$$

$$= \int_0^1 \frac{4}{3} x^3\, dx = \frac{4}{3}\left[\frac{x^4}{4}\right]_0^1 = \frac{1}{3}[1-0] = \frac{1}{3} \text{ sq. units.}$$ **Ans.**

**Example 2.** *Evaluate* $\int_{-1}^1 \int_0^{1-x} x^{1/3}\, y^{-1/2}\, (1-x-y)^{1/2}\, dy\, dx$. *(M.U., II Semester 2002)*

**Solution.** Here, we have

$$I = \int_{-1}^1 \int_0^{1-x} x^{1/3}\, y^{-1/2}\, (1-x-y)^{1/2}\, dy\, dx \qquad ...(1)$$

Putting $(1-x) = c$ in (1), we get

$$I = \int_{-1}^1 x^{1/3}\, dx \int_0^c y^{-1/2}\,(c-y)^{1/2}\, dy \qquad ...(2)$$

Again putting $y = ct \Rightarrow dy = c\, dt$ in (2), we get

$$I = \int_{-1}^1 x^{\frac{1}{3}}\, dx \int_0^1 c^{-\frac{1}{2}}\, t^{-\frac{1}{2}}\, (c-ct)^{\frac{1}{2}}\, c\, dt$$

$$= \int_{-1}^1 x^{1/3}\, dx \int_0^1 c^{-1/2}\, t^{-1/2}\, c^{1/2}\, (1-t)^{1/2}\, c\, dt$$

$$= \int_{-1}^1 c\, x^{1/3}\, dx \int_0^c t^{-1/2}(1-t)^{1/2}\, dt = \int_{-1}^1 c\, x^{1/3}\, dx \int_0^1 t^{1/2-1}\,(1-t)^{3/2-1}\, dt$$

$$= \int_{-1}^1 c\, x^{\frac{1}{3}}\, dx\, \beta\left(\frac{1}{2}, \frac{3}{2}\right) \qquad \left[\int_0^1 x^{l-1}(1-x)^{m-1}\, dx = \beta(l, m)\right]$$

$$= \int_{-1}^1 c\, x^{1/3}\, dx \frac{\Gamma\left(\frac{1}{2}\right)\Gamma\left(\frac{3}{2}\right)}{\Gamma\left(\frac{1}{2}+\frac{3}{2}\right)} = \int_{-1}^1 c\, x^{1/3}\, dx \frac{\Gamma\left(\frac{1}{2}\right)\cdot\frac{1}{2}\Gamma\left(\frac{1}{2}\right)}{\Gamma(2)} = \int_{-1}^1 c\, x^{1/3}\, dx \frac{\sqrt{\pi}\,\frac{1}{2}\,\sqrt{\pi}}{1}$$

$$= \int_{-1}^1 c\, x^{1/3}\, \frac{\pi}{2}\, dx = \frac{\pi}{2} \int_{-1}^1 x^{1/3} . c\, dx$$

Putting the value of $c$, we get

$$I = \frac{\pi}{2}\int_{-1}^{1} x^{1/3}(1-x)\,dx = \frac{\pi}{2}\int_{-1}^{1}(x^{1/3}-x^{4/3})\,dx = \frac{\pi}{2}\left[\frac{x^{4/3}}{\frac{4}{3}} - \frac{x^{7/3}}{\frac{7}{3}}\right]_{-1}^{1}$$

$$= \frac{\pi}{2}\left[\frac{3}{4}(1) - \frac{3}{7}(1) - \frac{3}{4}(-1) + \frac{3}{7}(-1)\right] = \frac{\pi}{2}\left[\frac{9}{14}\right] = \frac{9\pi}{28}$$ **Ans.**

**Example 3.** *Evaluate* $\iint_R (x+y)\,dy\,dx$, *R is the region bounded by* $x = 0$, $x = 2$, $y = x$, $y = x + 2$. *(Gujarat, I Semester, Jan. 2009)*

**Solution.** Let $I = \iint_R (x+y)\,dy\,dx$

The limits are $x = 0$, $x = 2$, $y = x$ and $y = x + 2$

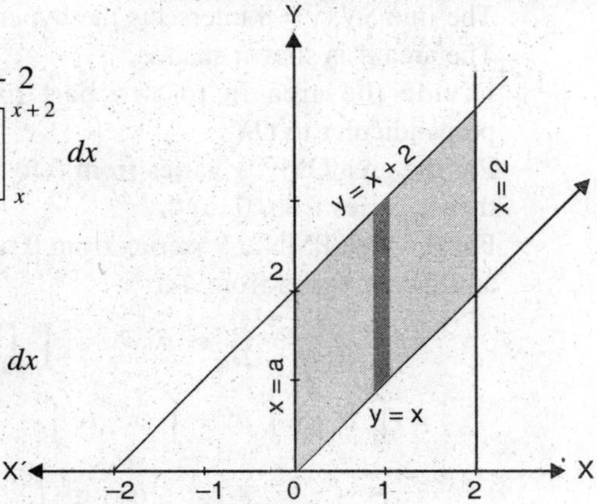


$$I = \int_0^2 dx \int_x^{x+2}(x+y)\,dy = \int_0^2 \left[xy + \frac{y^2}{2}\right]_x^{x+2} dx$$

$$= \int_0^2 \left[x(x+2) + \frac{1}{2}(x+2)^2 - x^2 - \frac{x^2}{2}\right] dx$$

$$= \int_0^2 \left[x^2 + 2x + \frac{1}{2}(x^2+4x+4) - x^2 - \frac{x^2}{2}\right] dx$$

$$= \int_0^2 [2x + 2x + 2]\,dx$$

$$= 2\int_0^2 (2x+1)\,dx = 2[x^2 + x]_0^2 = 2\,[4 + 2] = 12$$ **Ans.**

**Example 4.** *Evaluate* $\iint_R xy\,dx\,dy$

*where R is the quadrant of the circle* $x^2 + y^2 = a^2$ *where* $x \geq 0$ *and* $y \geq 0$. *(A.M.I.E.T.E, Summer 2004, 1999)*

**Solution.** Let the region of integration be the first quadrant of the circle $OAB$.

$$\iint_D xy\,dx\,dy \quad (x^2 + y^2 = a^2 \Rightarrow y = \sqrt{a^2 - x^2})$$

First we integrate w.r.t. $y$ and then w.r.t. $x$.

The limits for $y$ are 0 and $\sqrt{a^2-x^2}$ and for $x$, 0 to $a$.

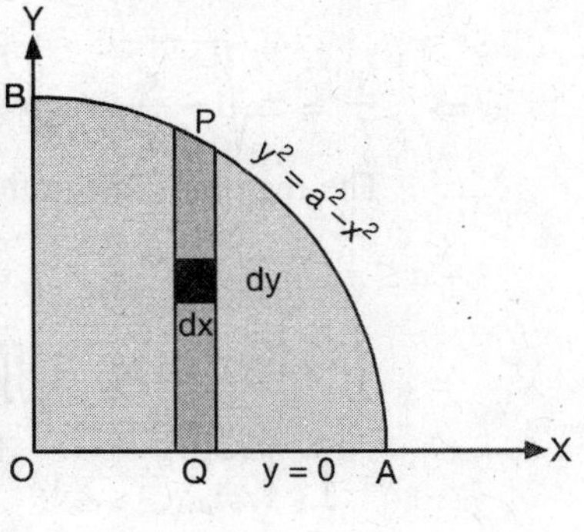


$$= \int_0^a x\,dx \int_0^{\sqrt{a^2-x^2}} y\,dy = \int_0^a x\,dx \left[\frac{y^2}{2}\right]_0^{\sqrt{a^2-x^2}}$$

$$= \frac{1}{2}\int_0^a x(a^2-x^2)\,dx = \frac{1}{2}\left[\frac{a^2x^2}{2} - \frac{x^4}{4}\right]_0^a = \frac{a^4}{8}$$ **Ans.**

**Example 5.** *Evaluate* $\iint_S \sqrt{xy - y^2}\,dy\,dx$,

*where S is a triangle with vertices (0, 0), (10, 1) and (1, 1).*

**Solution.** Let the verties of a triangle $OBA$ be (0, 0) (10, 1) and (1, 1).

Equation of $OA$ is $x = y$.

Equation of $OB$ is $x = 10\,y$.

The region of $\Delta OBA$, given by the limits

$y \leq x \leq 10\,y$ and $0 \leq y \leq 1$.

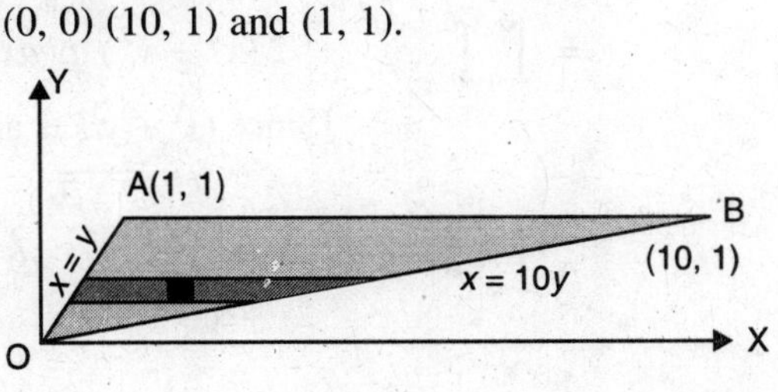


$$\iint_S \sqrt{xy - y^2}\,dy\,dx = \int_0^1 dy \int_y^{10y} (xy - y^2)^{1/2}\,dx$$

$$= \int_0^1 dy\left[\frac{2}{3}\frac{1}{y}(xy-y^2)^{3/2}\right]_y^{10y} = \int_0^1 \frac{2}{3}\frac{1}{y}(9y^2)^{3/2}\,dy = 18\int_0^1 y^2\,dy$$

$$= 18\left[\frac{y^3}{3}\right]_0^1 = \frac{18}{3} = 6$$ **Ans.**

**Example 6.** *Evaluate* $\iint_A x^2\,dx\,dy$, *where A is the region in the first quadrant bounded by the hyperbola xy = 16 and the lines y = x, y = 0 and x = 8.* *(A.M.I.E., Summer 2001)*

**Solution.** The line *OP*, $y = x$ and the curve *PS*, $xy = 16$ intersect at (4, 4).

The line *SN*, $x = 8$ intersects the hyperbola at *S* (8, 2). $y = 0$ is *x*-axis.

The area *A* is shown shaded.

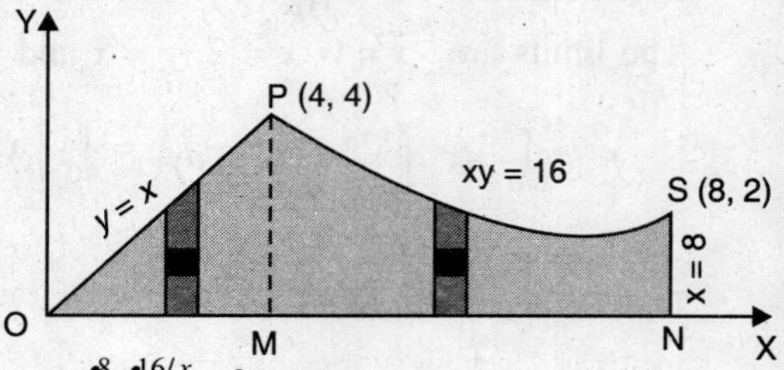


Divide the area in to two part by *PM* perpendicular to *OX*.

For the area OMP, *y* varies from 0 to *x*, and then *x* varies from 0 to 4.

For the area PMNS, *y* -series from 0 to 16/*x* and then *x* varies from 4 to 8.

$$\therefore \quad \iint_A x^2\,dx\,dy = \int_0^4\int_0^x x^2\,dx\,dy + \int_4^8\int_0^{16/x} x^2\,dx\,dy$$

$$= \int_0^4 x^2\,dx\int_0^x dy + \int_4^8 x^2\,dx\int_0^{16/x} dy = \int_0^4 x^2\,[y]_0^x\,dx + \int_4^8 x^2\,[y]_0^{16/x}\,dx$$

$$= \int_0^4 x^3\,dx + \int_4^8 16\,x\,dx = \left[\frac{x^4}{4}\right]_0^4 + 16\left[\frac{x^2}{2}\right]_4^8 = 64 + 8\,(8^2 - 4^2) = 64 + 384 = 448.$$ **Ans.**

**Example 7.** *Evaluate* $\iint (x+y)^2\,dx\,dy$ *over the area bounded by the ellipse* $\frac{x^2}{a^2}+\frac{y^2}{b^2}=1$

*(U.P. Ist Semester Compartment 2004)*

**Solution.** For the ellipse $\frac{x^2}{a^2}+\frac{y^2}{b^2}=1$

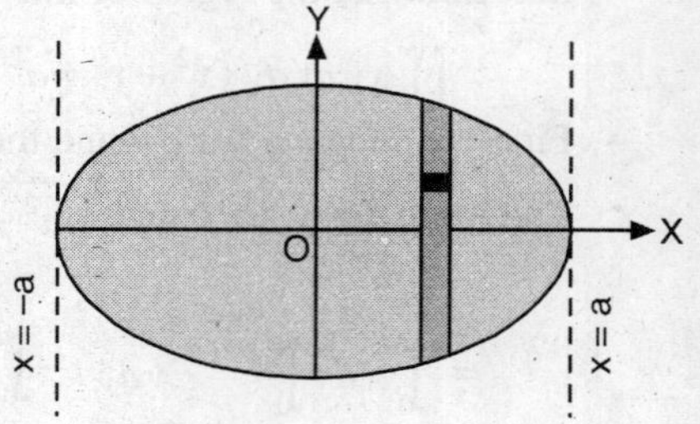


$$\Rightarrow \quad \frac{y}{b} = \pm\sqrt{1-\frac{x^2}{a^2}} \quad \Rightarrow \quad y = \pm\frac{b}{a}\sqrt{a^2-x^2}$$

$\therefore$ The region of integration can be expressed as

$$-a \le x \le a \text{ and } -\frac{b}{a}\sqrt{a^2-x^2} \le y \le \frac{b}{a}\sqrt{a^2-x^2}$$

$$\therefore \quad \iint (x+y)^2\,dx\,dy = \iint (x^2+y^2+2xy)\,dx\,dy$$

$$= \int_{-a}^{a}\int_{(-b/a)\sqrt{a^2-x^2}}^{b/a\sqrt{a^2-x^2}} (x^2+y^2+2xy)\,dx\,dy$$

$$= \int_{-a}^{a}\int_{(-b/a)\sqrt{a^2-x^2}}^{b/a\sqrt{a^2-x^2}} (x^2+y^2)\,dx\,dy \quad \int_{-a}^{a}\int_{(-b/a)\sqrt{a^2-x^2}}^{b/a\sqrt{a^2-x^2}} 2xy\,dy\,dx$$

$$= \int_{-a}^{a}\int_{0}^{b/a\sqrt{a^2-x^2}} 2\,(x^2+y^2)\,dy\,dx + 0$$

[Since $(x^2 + y^2)$ is an even function of *y* and 2*xy* is an odd function of *y*]

$$= \int_{-a}^{a}\left[2\left(x^2y+\frac{y^3}{3}\right)\right]_0^{\left(\frac{b}{a}\right)\sqrt{a^2-x^2}} dx$$

$$= 2\int_{-a}^{a}\left[x^2 \times \frac{b}{a}\sqrt{a^2 - x^2} + \frac{1}{3}\frac{b^3}{a^3}(a^2 - x^2)^{3/2}\right]dx$$

$$= 4\int_{0}^{a}\left[\frac{b}{a}x^2\sqrt{a^2 - x^2} + \frac{b^3}{3a^3}(a^2 - x^2)^{3/2}\right]dx$$

[On putting $x = a \sin\theta$ and $dx = a\cos\theta\, d\theta$]

$$= 4\int_{0}^{\frac{\pi}{2}}\left(\frac{b}{a}.a^2\sin^2\theta.a\cos\theta + \frac{b^3}{3a^3}a^3\cos^3\theta\right)\times a\cos\theta\, d\theta$$

$$= 4\int_{0}^{\frac{\pi}{2}}\left(a^3 b\sin^2\theta\cos^2\theta + \frac{ab^3}{3}\cos^4\theta\right)d\theta = 4\left[a^3 b.\frac{1}{4}.\frac{1}{2}.\frac{\pi}{2} + \frac{ab^3}{3}.\frac{3}{4}.\frac{1}{2}.\frac{\pi}{2}\right]$$

$$= \frac{\pi}{4}(a^3 b + ab^3) = \frac{\pi}{4}ab(a^2 + b^2)$$ **Ans.**

**Example 8.** *Evaluate* $\iint (x^2 + y^2)\, dx\, dy$ *throughout the area enclosed by n the curves* $y = 4x$, $x + y = 3$, $y = 0$ *and* $y = 2$.

**Solution.** Let $OC$ represent $y = 4x$; $BD$, $x + y = 3$; $OB$, $y = 0$, and $CD$, $y = 2$. The given integral is to be evaluated over the area A of the trapezium $OCDB$.

Area $OCDB$ consists of area $OCE$, area $ECDF$ and area $FDB$.

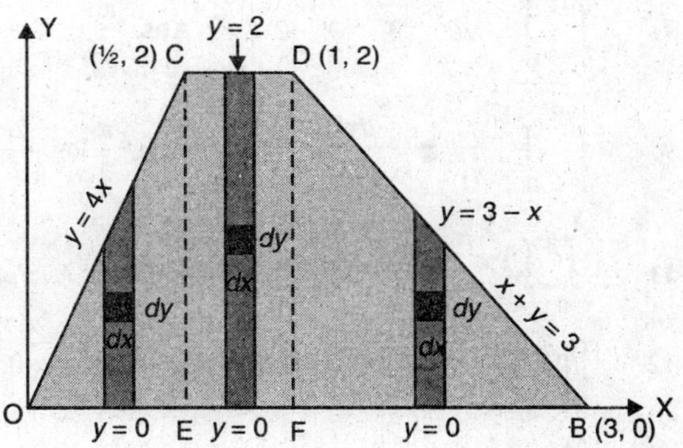


The co-ordinates of C, D and B are $\left(\frac{1}{2}, 2\right)$ (1, 2) and (3, 0) respectively.

$$\therefore \quad \iint_A (x^2 + y^2)\, dy\, dx$$

$$= \iint_{OCE}(x^2 + y^2)\, dy\, dx + \iint_{ECDE}(x^2 + y^2)\, dy\, dx + \iint_{FDB}(x^2 + y^2)\, dy\, dx$$

$$= \underbrace{\int_0^{1/2} dx \int_0^{4x}(x^2 + y^2)\, dy}_{I_1} + \underbrace{\int_{1/2}^{1} dx \int_0^{2}(x^2 + y^2)\, dy}_{I_2} + \underbrace{\int_1^{3} dx \int_0^{3-x}(x^2 + y^2)\, dy}_{I_3}$$

Now, $$I_1 = \int_0^{1/2} dx \int_0^{4x}(x^2 + y^2)\, dy = \int_0^{1/2}\left[x^2 y + \frac{y^3}{3}\right]_0^{4x} dx = \int_0^{1/2}\frac{76}{3}x^3\, dx$$

$$= \frac{76}{3}\int_0^{1/2} x^3\, dx = \frac{76}{3}\left[\frac{x^4}{4}\right]_0^{1/2} = \frac{76}{3}\left[\frac{1}{4}.\frac{1}{16}\right] = \frac{19}{48}$$

$$I_2 = \int_{1/2}^{1} dx \int_{1/2}^{1}(x^2 + y^2)\, dy = \int_{1/2}^{1}\left[x^2 y + \frac{y^3}{3}\right]_0^{2} dx = \int_{1/2}^{1}\left(2x^2 + \frac{8}{3}\right)dx$$

$$= \left[\frac{2x^3}{3} + \frac{8}{3}x\right]_{1/2}^{1} = \left[\left(\frac{2}{3} + \frac{8}{3}\right) - \left(\frac{2}{3}.\frac{1}{8} + \frac{8}{3}.\frac{1}{2}\right)\right] = \frac{23}{12}$$

$$I_3 = \int_1^{3} dx \int_0^{3-x}(x^2 + y^2)\, dy = \int_1^{3}\left[x^2 y + \frac{y^3}{3}\right]_0^{3-x} dx = \int_0^{3}\left[x^2(3 - x) + \frac{(3 - x)^3}{3}\right]dx$$

$$= \int_1^{3}\left[3x^2 - x^3 + \frac{(3 - x)^3}{3}\right]dx = \left[x^3 - \frac{x^4}{4} - \frac{(3 - x)^4}{3}\right]_1^{3}$$

$$= \left[27 - \frac{81}{4} - 0 - 1 + \frac{1}{4} + \frac{16}{12}\right] = \frac{22}{3}$$

$$\therefore \quad \int_A\int (x^2 + y^2)\, dy\, dx = I_1 + I_2 + I_3 = \frac{19}{48} + \frac{23}{12} + \frac{22}{3} = \frac{463}{48} = 9\frac{31}{48}. \qquad \textbf{Ans.}$$

## EXERCISE 2.1

**Evaluate**

1. $\int_0^2\int_0^{x^2} e^{\frac{y}{x}}\, dy\, dx$ **Ans.** $e^2 - 1$

2. $\int_0^a\int_0^{\sqrt{ay}} xy\, dx\, dy$ **Ans.** $\frac{a^4}{6}$

3. $\int_0^a\int_0^{\sqrt{a^2-y^2}} dx\, dy$ **Ans.** $\frac{\pi a^2}{4}$

4. $\int_0^1\int_{y^2}^{y} (1 + xy^2)\, dx\, dy$ **Ans.** $\frac{41}{210}$

5. $\int_0^{2a}\int_0^{\sqrt{2ax-x}} xy\, dy\, dx$ **Ans.** $\frac{2a^4}{3}$

6. $\int_0^{2a}\int_0^{\sqrt{2ax-x^2}} x^2\, dy\, dx$ **Ans.** $\frac{5\pi a^4}{8}$

7. $\int_0^a\int_0^{\sqrt{a^2-x^2}} \sqrt{a^2 - x^2 - y^2}\, dy\, dx$ **Ans.** $\frac{\pi a^3}{4}$

8. $\int_0^1\int_0^{\sqrt{\frac{1}{2}(1-y^2)}} \frac{dx\, dy}{\sqrt{1 - x^2 - y^2}}$ **Ans.** $\frac{\pi}{4}$

9. $\int_0^a\int_0^{\sqrt{a^2-x^2}} \frac{dx\, dy}{(1 + e^y)\sqrt{a^2 - x^2 - y^2}}$ **Ans.** $\frac{\pi}{2}\log\frac{2e^a}{1 + e^a}$

10. $\int_0^a\int_y^a \frac{x\, dx\, dy}{\sqrt{x^2 + y^2}}$ **Ans.** $\frac{a^2}{2}\log(\sqrt{2} + 1)$

11. $\int_{x=0}^{1}\int_{y=0}^{2} (x^2 + 3xy^2)\, dx\, dy$ *(A.M.I.E.T.E., June 2009)* **Ans.** $\frac{14}{3}$

12. $\iint_A (5 - 2x - y)\, dx\, dy$, where $A$ is given by $y = 0$, $x + 2y = 3$, $x = y^2$. **Ans.** $\frac{217}{60}$

13. $\iint_A xy\, dx\, dy$, where $A$ is given by $x^2 + y^2 - 2x = 0$, $y^2 = 2x$, $y = x$. **Ans.** $\frac{7}{12}$

14. $\iint_A \sqrt{4x^2 - y^2}\, dx\, dy$, where $A$ is the triangle given by $y = 0$, $y = x$ and $x = 1$. **Ans.** $\frac{1}{3}\left(\frac{\pi}{3} + \frac{\sqrt{3}}{2}\right)$

15. $\iint_R x^2\, dx\, dy$, where $R$ is the two-dimensional region bounded by the curves $y = x$ and $y = x^2$. **Ans.** $\frac{1}{20}$

16. $\iint_A \sqrt{xy}\,(1 + x - y)\, dx\, dy$ where $A$ is the area bounded by $x = 0$, $y = 0$ and $x + y = 1$. **Ans.** $\frac{2\pi}{105}$

## 2.3 EVALUATION OF DOUBLE INTEGRALS IN POLAR CO-ORDINATES

We have to evalaute $\int_{\theta_1}^{\theta_2}\int_{r_1(\theta)}^{r_2(\theta)} f(r, \theta)\, dr\, d\theta$ over the region bounded by the staight lines

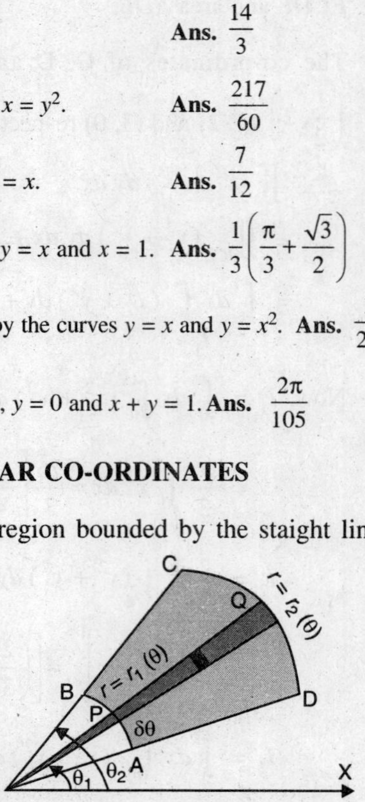


$\theta = \theta_1$ and $\theta = \theta_2$ and the curves $r = r_1$ (q) and $r = r_2$ ($\theta$). We first integrate with respect to $r$ between the limits $r = r_1(\theta)$ and $r = r_2(\theta)$ and taking $\theta$ as constant. Then the resulting expression is integrated with respect to $\theta$ between the limits $\theta = \theta_1$ and $\theta = \theta_2$.

The area of integration is $ABCD$. On integrating first with respect to $r$, the strip extends from $P$ to $Q$ and the integration with respect to $\theta$ means the rotation ot this strip $PQ$ from $AD$ to $BC$.

**Example 9.** *Transform the integral to cartesian form and hence evaluate*

$\int_0^{\pi}\int_0^{a} r^3 \sin\theta\cos\theta\, dr\, d\theta.$ (*M.U., II Semester 2000*)

**Solution.** Here, we have

$$\int_0^{\pi}\int_0^{a} r^3 \sin\theta\cos\theta\, dr\, d\theta \quad ...(1)$$

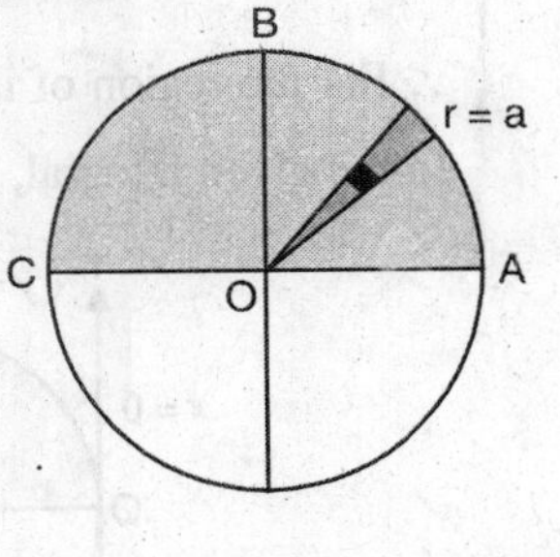


Here the region *i.e.*, semicircle *ABC* of integration is bounded by $r = 0$, *i.e.*, *x*-axis.

$r = a$ *i.e.*, circle, $\theta = 0$ and $\theta = \pi$ *i.e.*, *x*-axis in the second quadrant.

$$\iint (r\sin\theta)(r\cos\theta)(r\, d\theta\, dr)$$

Putting $x = r\cos\theta$, $y = r\sin\theta$, $dx\, dy = r\, d\theta\, dr$ in (1), we get

$$\int_{-a}^{a}\int_0^{\sqrt{a^2-x^2}} xy\, dy\, dx = \int_{-a}^{a} x\, dx \int_0^{\sqrt{a^2-x^2}} y\, dy$$

$$= \int_{-a}^{a} x\, dx\left[\frac{y^2}{2}\right]_0^{\sqrt{a^2-x^2}} = \int_{-a}^{a} x\, dx\, \frac{(a^2-x^2)}{2}$$

$$= \frac{1}{2}\int_{-a}^{a}(a^2 x - x^3)\, dx = 0 \text{ **Ans.**} \quad \left[\begin{array}{l}\text{Since } f(x) \text{ is odd function} \\ \int_{-a}^{a} f(x)\, dx = 0\end{array}\right]$$

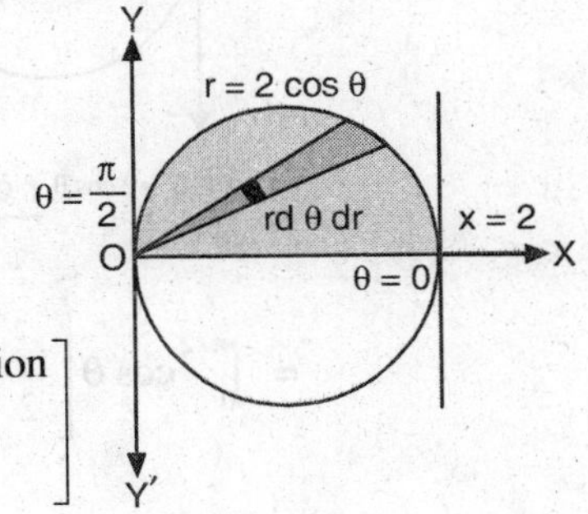


**Example 10.** *Evaluate* $\int_0^2\int_0^{\sqrt{2x-x^2}} (x^2+y^2)\, dy\, dx$

**Solution.** $\int_0^2\int_0^{\sqrt{2x-x^2}} (x^2+y^2)\, dy\, dx$

Limits of $\quad y = \sqrt{2x-x^2} \Rightarrow y^2 = 2x - x^2 \Rightarrow x^2 + y^2 - 2x = 0 \quad ...(1)$

(1) represents a circle whose centre is (1, 0) and radius = 1.

Lower limit of *y* is 0 *i.e.*, *x*-axis.

Region of integration is upper half circle.

Let us convert (1) into polar co-ordinate by putting

$$x = r\cos\theta,\ y = r\sin\theta$$

$$r^2 - 2r\cos\theta = 0 \Rightarrow r = 2\cos\theta$$

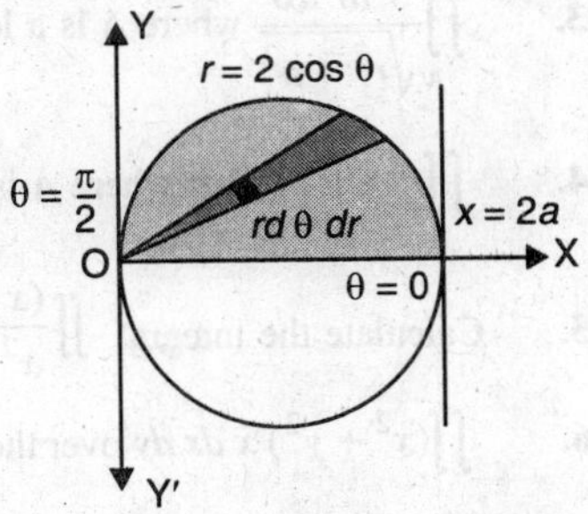


Limits of *r* are 0 to $2\cos\theta$

Limits of $\theta$ are 0 to $\frac{\pi}{2}$

$$\int_0^2\int_0^{\sqrt{2x-x^2}} (x^2+y^2)\, dy\, dx = \int_0^{\frac{\pi}{2}}\int_0^{2\cos\theta} r^2\,(r\, d\theta\, dr) = \int_0^{\frac{\pi}{2}} d\theta \int_0^{2\cos\theta} r^3\, dr = \int_0^{\frac{\pi}{2}} d\theta\left[\frac{r^4}{4}\right]_0^{2\cos\theta}$$

$$= 4\int_0^{\frac{\pi}{2}} \cos^4\theta\, d\theta = 4\times\frac{3\times 1\times\pi}{4\times 2\times 2} = \frac{3\pi}{4} \quad \textbf{Ans.}$$

**Example 11.** *Evaluate* $\int_0^2\int_0^{\sqrt{2x-x^2}} \frac{x\, dy\, dx}{\sqrt{x^2+y^2}}$ *by changing to polar coordinates.*

**Solution.** In the given integral, *y* varies from 0 to $\sqrt{2x-x^2}$ and *x* varies from 0 to 2.

$$y = \sqrt{2x-x^2}$$

$\Rightarrow \qquad y^2 = 2x - x^2$

$\Rightarrow \qquad x^2 + y^2 = 2x$

In polar co-ordinates, we have $r^2 = 2r \cos\theta \quad \Rightarrow \quad r = 2\cos\theta$.

$\therefore$ For the region of integarion, $r$ varies from 0 to $2\cos\theta$ and $\theta$ varies from 0 to $\frac{\pi}{2}$.

In the given integral, replacing $x$ by $r\cos\theta$, $y$ by $r\sin\theta$, $dy\,dx$ by $r\,dr\,d\theta$, we have

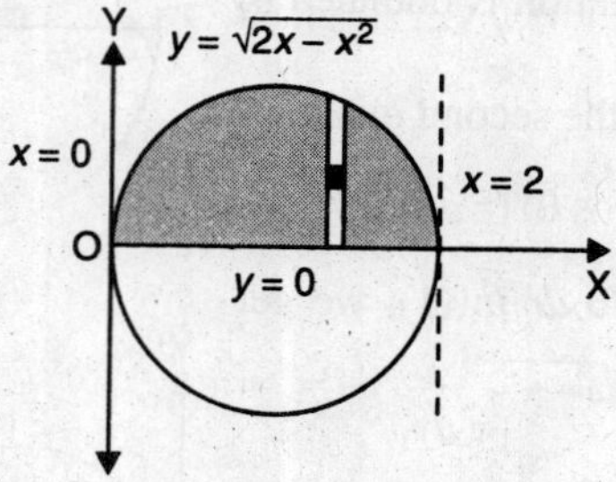


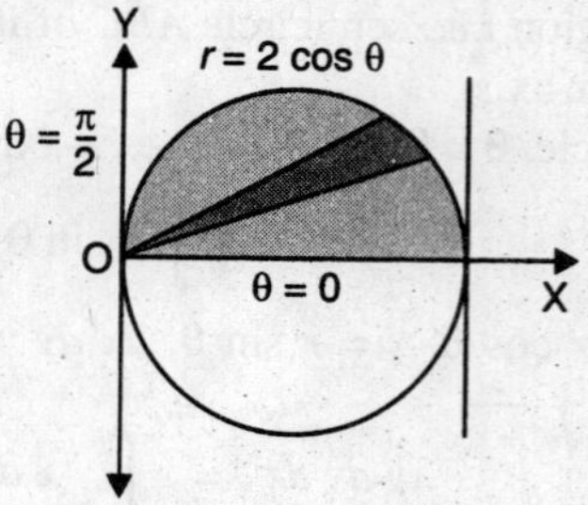


$$I = \int_0^{\pi/2}\int_0^{2\cos\theta} \frac{r\cos\theta.\, r\,dr\,d\theta}{r} = \int_0^{\pi/2}\int_0^{2\cos\theta} r\cos\theta\,dr\,d\theta$$

$$= \int_0^{\pi/2} \cos\theta \left[\frac{r^2}{2}\right]_0^{2\cos\theta} d\theta = \int_0^{\pi/2} 2\cos^3\theta\,d\theta = 2.\frac{2}{3} = \frac{4}{3}. \qquad \textbf{Ans.}$$

## EXERCISE 2.2

**Evaluate the folloing:**

**1.** $\int_0^{\pi}\int_0^{a(1-\cos\theta)} 2\pi r^2 \sin\theta\, d\theta\, dr$ **Ans.** $\frac{8}{3}\pi a^3$

**2.** $\int_0^{\pi}\int_0^{a(1+\cos\theta)} r^2 \cos\theta\, dr\, d\theta$ **Ans.** $\frac{5}{8}\pi a^3$

**3.** $\iint_A \frac{r\,dr\,d\theta}{\sqrt{r^2+a^2}}$ where $A$ is a loop of $r^2 = a^2\cos 2\theta$ **Ans.** $2a - \frac{\pi a}{2}$

**4.** $\iint_A r^2 \sin\theta\, d\theta\, dr$ where $A$ is $r = 2a\cos\theta$ above initial line. *(A.M.I.E. Winter 2001)* **Ans.** $\frac{2a^3}{3}$

**5.** Calculate the integral $\iint \frac{(x-y)^2}{x^2+y^2}\,dx\,dy$ over the circle $x^2 + y^2 \le 1$. **Ans.** $\pi - 2$

**6.** $\iint (x^2 + y^2)\, x\, dx\, dy$ over the positive quadrant of the circle $x^2 + y^2 = a^2$ by changing to polar coordinates. **Ans.** $\frac{a^2}{5}$

**7.** $\iint_R \sqrt{x^2+y^2}\,dx\,dy$ by changing to polar coordinates, R is the region in the $xy$-plane bounded by the circles $x^2 + y^2 = 4$ *(AMIETE, Dec. 2009)* **Ans.** $\frac{38\pi}{3}$

**8.** Convert into polar coordinates

$$\int_0^{2a}\int_0^{2ax-x^2} dx\,dy \qquad \textbf{Ans.} \int_0^{\pi/2}\int_0^{2a\cos\theta} r\,d\theta\,dr$$

**9.** $\iint r^3\,dr\,d\theta$, over the area bounded between the circles $r = 2b\cos\theta$ and $r = 2b\cos\theta$. **Ans.** $\frac{3\pi}{2}(a^4 - b^4)$

**10.** $\iint r\sin\theta\,dr\,d\theta$ over the area of the cardiod $r = a(1 + \cos\theta)$ above the initial line. **Ans.** $\frac{5}{8}\pi a^3$

**11.** $\iint_A x^2\, dr\, d\theta$, where $A$ is the area between the circles $r = a\cos\theta$ and $r = 2a\cos\theta$. **Ans.** $\frac{28a^3}{9}$

**12.** Transform the integral $\int_0^1 \int_0^x f(x, y)\, dy\, dx$ to the integral in polar co-ordinates.

**Ans.** $\int_0^{\pi/4} \int_0^{\sec\theta} f(r, \theta)\, r\, d\theta\, dr$

## 2.4 CHANGE OF ORDER OF INTEGRATION

On changing the order of integration, the limits of integration change. To find the new limits, we draw the rough sketch of the region of integration.

Some of the problems connected with double integrals, which seem to be complicated, can be made easy to handle by a change in the order of integration.

**Example 12.** *Evaluate* $\int_0^a \int_y^a \frac{x}{x^2 + y^2}\, dx\, dy$ *by changing the order of integration.*

*(AMIETE, June 2010, Nagpur University, Summer 2008)*

**Solution.** Here we have

$$I = \int_0^a \int_y^a \frac{x}{x^2 + y^2}\, dx\, dy$$

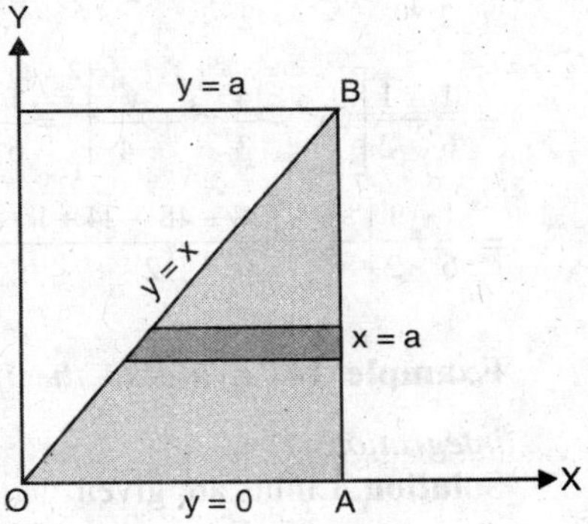


Here $x = a, x = y, y = 0$ and $y = a$

The area of integration is $OAB$.

On changing the order of integration Lower limit of $y = 0$ and

upper limit is $y = x$.

Lower limit of $x = 0$ and upper limit is $x = a$.

$$I = \int_0^a x\,dx \int_0^{y=x} \frac{1}{x^2 + y^2}\, dy$$

$$= \int_0^a x\,dx \left[\frac{1}{x}\tan^{-1}\frac{y}{x}\right]_0^{y=x}$$

$$= \int_0^a \frac{x}{x}\, dx \left(\tan^{-1}\frac{x}{x} - \tan^{-1} 0\right)$$

$$= \int_0^a dx \left(\frac{\pi}{4}\right) = \frac{\pi}{4}[x]_0^a = \frac{a\pi}{4} \quad \textbf{Ans.}$$

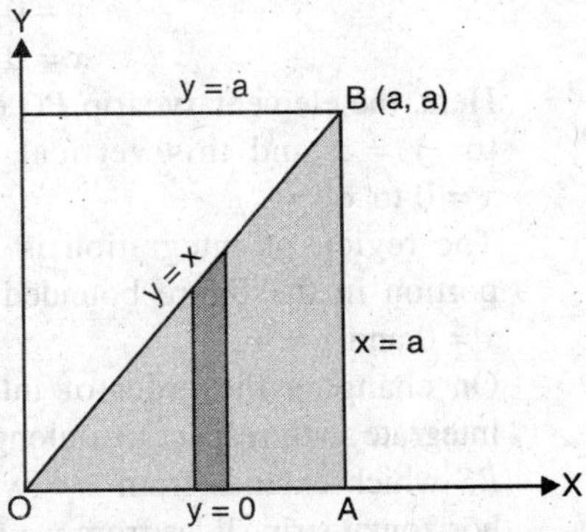


**Example 13.** *Change the order of integration in*

$I = \int_0^1 \int_{x^2}^{2-x} xy\, dx\, dy$ *and hence evaluate the same.*

*(A.M.I.E.T.E., June 2010, 2009, U.P. I Sem., Dec., 2004)*

**Solution.** $I = \int_0^1 \int_{x^2}^{2-x} xy\, dx\, dy$

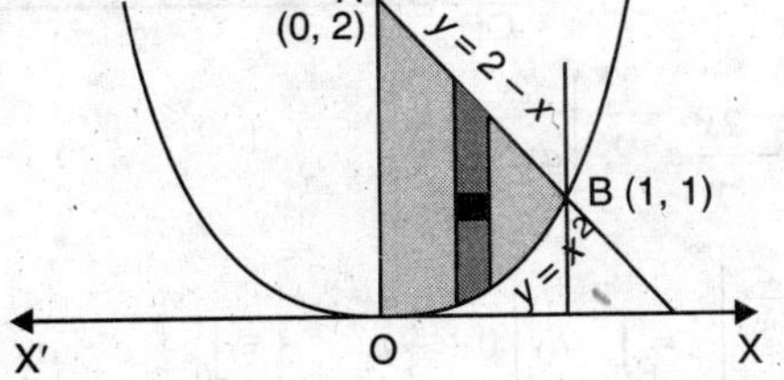


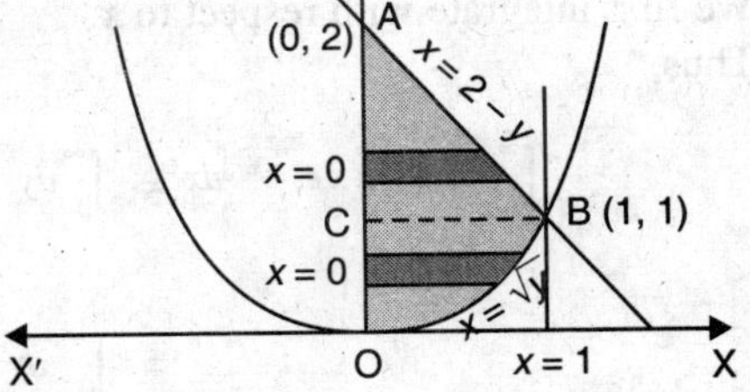

The region of integration is shown by shaded portion in the figure bounded by parabola $y = x^2$ and the line $y = 2 - x$.

The point of intersection of the parabola $y = x^2$ and the line $y = 2 - x$ is B (1, 1).

In the figure below (left) we have taken a strip parallel to $y$-axis and the order of integration is

$$\int_0^1 x\,dx \int_{x^2}^{2-x} y\,dy$$

In the second figure above we have taken a strip parallel to $x$-axis in the area $OBC$ and second strip in the area $ABC$. The limits of $x$ in the area $OBC$ are 0 and $\sqrt{y}$ and the limits of $x$ in the area $ABC$ are 0 and $2 - y$.

$$= \int_0^1 y\,dy \int_0^{\sqrt{y}} x\,dx + \int_1^2 y\,dx \int_0^{2-y} x\,dx = \int_0^1 y\,dy\left[\frac{x^2}{2}\right]_0^{\sqrt{y}} + \int_0^{\sqrt{y}} y\,dy\left[\frac{x^2}{2}\right]_0^{2-y}$$

$$= \frac{1}{2}\int_0^1 y^2\,dy + \frac{1}{2}\int_1^2 y\,(2-y)^2\,dy = \frac{1}{2}\left[\frac{y^3}{3}\right]_0^1 + \frac{1}{2}\int_1^2 (4y - 4y^2 + y^3)\,dy$$

$$= \frac{1}{6} + \frac{1}{2}\left[2y^2 - \frac{4}{3}y^3 + \frac{y^4}{4}\right]_1^2 = \frac{1}{6} + \frac{1}{2}\left[8 - \frac{32}{3} + 4 - 2 + \frac{4}{3} - \frac{1}{4}\right]$$

$$= \frac{1}{6} + \frac{1}{2}\left[\frac{96 - 128 + 48 - 24 + 16 - 3}{12}\right] = \frac{1}{6} + \frac{5}{24} = \frac{9}{24} = \frac{3}{8}$$ **Ans.**

**Example 14.** *Evaluate the integral* $\int_0^\infty \int_0^x x \exp\left(-\frac{x^2}{y}\right) dx\,dy$ *by changing the order of integration* *(U.P. I Semester Dec., 2005)*

**Solution.** Limits are given

$$y = 0 \text{ and } y = x$$
$$x = 0 \text{ and } x = \infty$$

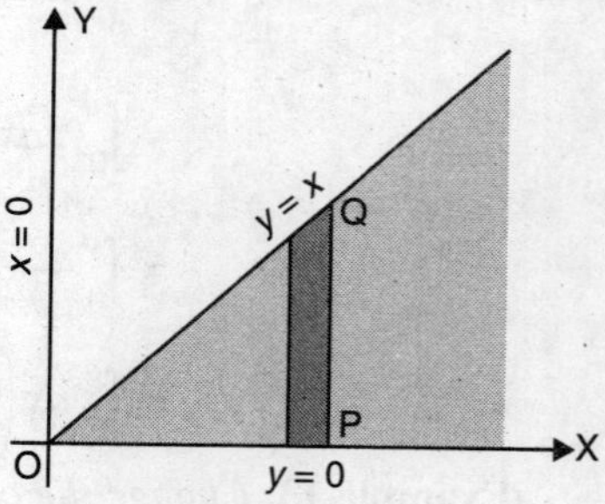


Here, the elementary strip $PQ$ extends from $y = 0$ to $y = x$ and this vertical strip slides from $x = 0$ to $x = \infty$.

The region of integration is shown by shaded portion in the figure bounded by $y = 0$, $y = x$, $x = 0$ and $x = \infty$.

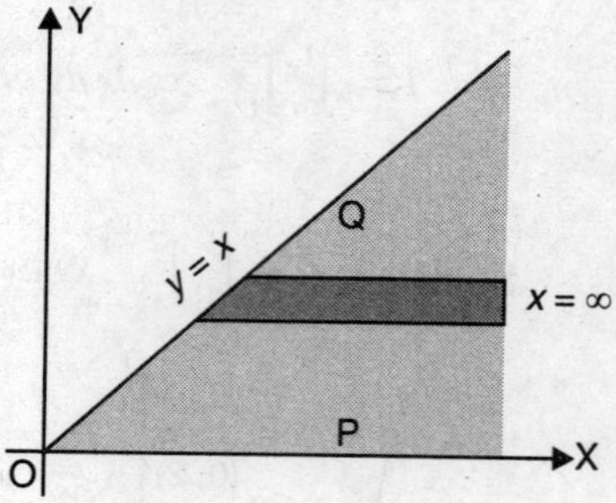


On changing the order of integration, we first integrate with respect to $x$ along a horizontal strip $RS$ which extends from $x = y$ to $x = \infty$ and this horizontal strip slides from $y = 0$ to $y = \infty$ to cover the given region of integration.

New limits :

$$x = y \quad \text{and} \quad x = \infty$$
$$y = 0 \quad \text{and} \quad y = \infty$$

We first integrate with respect to $x$.

Thus,

$$\int_0^\infty dy \int_y^\infty x e^{-\frac{x^2}{y}}\,dx = \int_0^\infty dy \int_y^\infty -\frac{y}{2}\left(-\frac{2x}{y}e^{-\frac{x^2}{y}}\right) dx$$

$$= \int_0^\infty dy\left[-\frac{y}{2}e^{-\frac{x^2}{y}}\right]_y^\infty = \int_0^\infty dy\left[0 + \frac{y}{2}e^{-\frac{y^2}{2}}\right] = \int_0^\infty \frac{y}{2}e^{-y}\,dy$$

$$= \left[\frac{y}{2}(-e^{-y}) - \left(\frac{1}{2}\right)(e^{-y})\right]_0^{\infty} \quad \text{(Integrating by parts)}$$

$$= \left[(0-0) - \left(0 - \frac{1}{2}\right)\right] = \frac{1}{2} \quad \textbf{Ans.}$$

**Example 15.** *Change the order of the integration*

$$\int_0^{\infty} \int_0^{x} e^{-xy}\, y\, dy\, dx$$

*(B.P.U.T.; I Semester 2008)*

**Solution.** Here, we have

$$\int_0^{\infty} \int_0^{x} e^{-xy}\, y\, dy\, dx$$

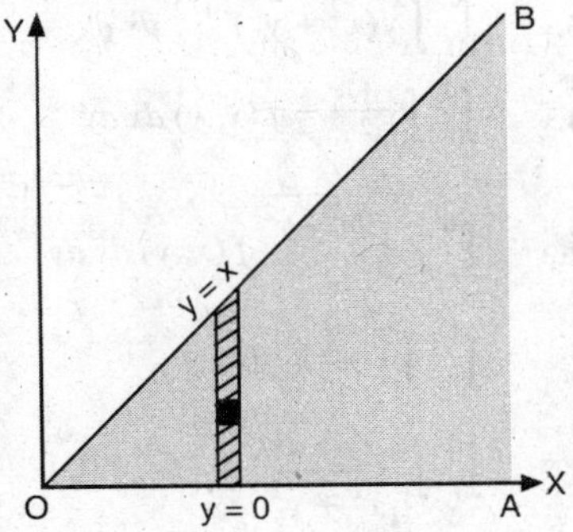


Here the region $OAB$ of integration is bounded by $y = 0$ ($x$-axis), $y = x$ (a straight line), $x = 0$, *i.e.*, $y$ axis. A strip is drawn parallel to $y$-axis, $y$ varies 0 to $x$ and $x$ varies 0 to $\infty$.

On changing the order of integration, first we integrate w.r.t. $x$ and then w.r.t. $y$.

A strip is drawn parallel to $x$-axis. On this strip $x$ varies from $y$ to $\infty$ and $y$ varies from 0 to $\infty$.

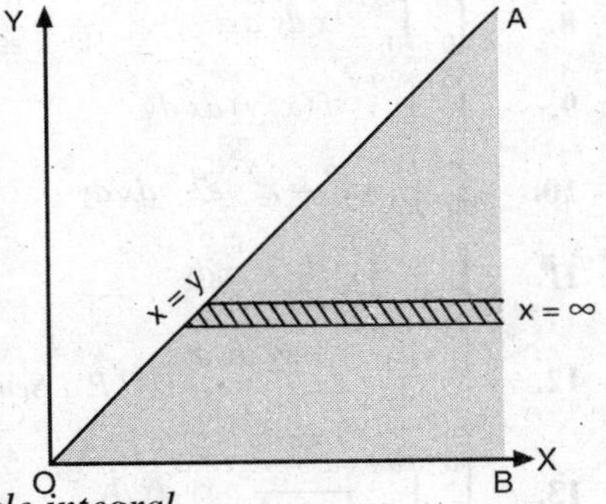


Hence $\displaystyle\int_0^{\infty} \int_0^{x} e^{-xy}\, y\, dy\, dx = \int_0^{\infty} y\, dy \int_y^{\infty} e^{-xy} dx$

$$= \int_0^{\infty} y\, dy \left(\frac{e^{-xy}}{-y}\right)_y^{\infty}$$

$$= \int_0^{\infty} \frac{y\, dy}{-y}[0 - e^{y^2}]$$

$$= \int_0^{\infty} e^{-y^2}\, dy = \frac{1}{2}\sqrt{\pi} \quad \textbf{Ans.}$$

**Example 16.** *Change the order of integration in the double integral*

$$\int_0^{2a} \int_{\sqrt{2ax - x^2}}^{\sqrt{2ax}} V\, dx\, dy$$

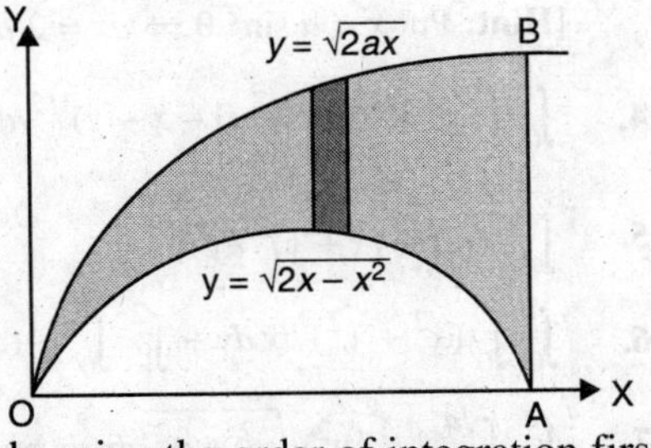


**Solution.** Limits are given as

$$x = 0,\ x = 2a$$

$$y = \sqrt{2ax}$$

and $\quad y = \sqrt{2ax - x^2} \quad \Rightarrow \quad y^2 = 2ax$

and $\quad (x-a)^2 + y^2 = a^2$

The area of integration is the shaded portion $OAB$. On changing the order of integration first we have to integrate w.r.t. $x$, The area of integration has three portions $BCE$, $ODE$ and $ACD$.

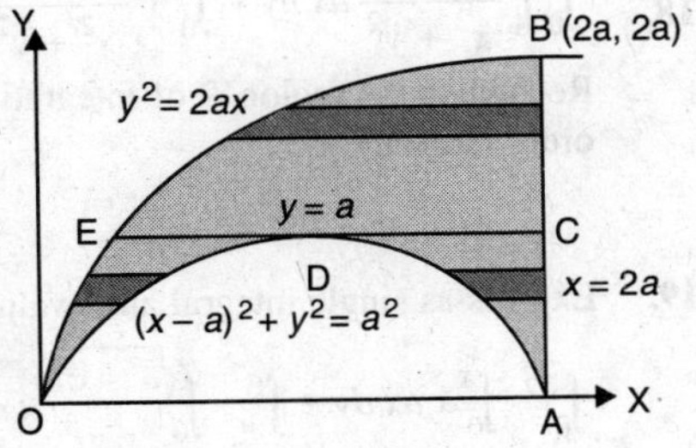


$$\int_0^{2a} dx \int_{\sqrt{2ax - x^2}}^{\sqrt{2ax}} V\, dy$$

$$= \int_0^{2a} dy \int_{y^2/2a}^{2a} V\, dx + \int_0^{a} dy \int_{y^2/2a}^{a+\sqrt{a^2+y^2}} V\, dx$$

$$+ \int_0^{a} dy \int_{a+\sqrt{a^2-y^2}}^{2a} V\, dx \quad \textbf{Ans.}$$

## EXERCISE 2.3

**Change the order of integration and hence evaluate the following:**

**1.** $\int_0^a \int_0^x \frac{\cos y\, dy}{\sqrt{(a-x)(a-y)}}\, dx$ **Ans.** (a) $\int_0^a dy \int_y^a \frac{\cos y\, dx}{\sqrt{(a-x)(a-y)}}$ (b) $2 \sin a$.

**2.** $\int_0^{2a} \int_{\frac{x^2}{4a}}^{3a-x} (x^2+y^2)\, dy\, dx$ **Ans.** (a) $\int_0^a dy \int_0^{2\sqrt{ay}} (x^2+y^2)\, dx + \int_a^{3a} dy \int_0^{3a-y} (x^2+y^2)\, dx$ (b) $\frac{314\, a^4}{35}$.

**3.** $\int_0^1 \int_{x^2}^{x} (x^2+y^2)^{-1/2}\, dy\, dx$ **Ans.** $\int_0^1 dy \int_y^{\sqrt{y}} (x^2+y^2)^{-1/2}\, dx$.

**4.** $\int_0^a \int_{\sqrt{a^2-y^2}}^{y+a} f(x, y)\, dx\, dy$ **Ans.** $\int_0^a dx \int_{\sqrt{a^2-x^2}}^{a} f(x, y)\, dy + \int_a^{2a} dx \int_{x-a}^{a} f(x, y)\, dy$

**5.** $\int_{-a}^{a} \int_0^{\sqrt{a^2-y^2}} f(x, y)\, dx\, dy$ **Ans.** $\int_0^a dx \int_{-\sqrt{a^2-x^2}}^{\sqrt{a^2-x^2}} f(x, y)\, dy$

**6.** $\int_0^1 \int_x^{2-x} \frac{x}{y}\, dy\, dx$ **Ans.** $\int_0^a \frac{dy}{y} \int_0^y x dx + \int_1^2 \frac{dy}{y} \int_0^{2-y} x dx$; $\log \frac{4}{e}$

**7.** $\int_0^b \int_y^a \frac{x\, dy\, dx}{x^2+y^2}$ *(M.P. 2003)*

**8.** $\int_0^a \int_0^{bx/a} x\, dy\, dx$ **Ans.** (a) $\int_0^b dy \int_{ay/b}^{a} x\, dx$ (b) $\frac{1}{3} a^2 b$

**9.** $\int_0^5 \int_{2-x}^{2+x} f(x, y)\, dx\, dy$ **Ans.** $\int_0^2 dy \int_{2-y}^{5} f(x, y)\, dx + \int_2^7 dy \int_{y-2}^{5} f(x, y)\, dx$

**10.** $\int_0^{\infty} \int_{-y}^{y} (y^2-x^2)\, e^{-y}\, dx\, dy$ **Ans.** $\int_{-\infty}^{\infty} dx \int_{-x}^{x} (y^2-x^2)\, e^{-y}\, dy$ *(A.M.I.E., Summer 2000)*

**11.** $\int_{y=0}^{1} \int_{x=\sqrt{y}}^{2-y} \times yx\, dy$ *(A.M.I.E.T.E., June 2009)*

**12.** $\int_0^a \int_{\frac{x^2}{a}}^{2a-x} xy\, dx\, dy$ *(U.P. I Semester, Dec., 2007)* **Ans.** $\int_0^a \int_0^{\sqrt{ay}} xy\, dx\, dy + \int_0^{2a-y} xy\, dx\, dy$, $\frac{3a^2}{8}$

**13.** $\int_0^a \int_{a-\sqrt{a^2-y^2}}^{a+\sqrt{a^2-y^2}} xy\, dx\, dy$ **Ans.** $\int_0^{2a} x\, dx \int_0^{\sqrt{a^2-(x-a)^2}} y\, dy$, $\frac{2}{3} a^4$

[**Hint:** Put $x = a \sin^2 \theta \Rightarrow dx = 2\, a \sin \theta \cos \theta\, d\, \theta$]

**14.** $\int_0^1 \int_{-1}^{1-y} x^{1/3}\, y^{-1/2}\, (1-x-y)^{1/2}\, dx\, dy$ **Ans.** $\int_{-1}^{1} x^{\frac{1}{3}}\, dx \int_0^{1-x} y^{-\frac{1}{2}}\, (1-x-y)^{\frac{1}{2}}\, dy, -\frac{3\pi}{7}$

**15.** $\int_0^{2a} dx \int_0^{\frac{x^2}{4a}} (x+y)^3\, dy$ **Ans.** $\int_0^a dy \int_{\sqrt{4a\, y}}^{2a} (x+y)^3\, dx$

**16.** $\int_0^1 \int_0^y (x^2+y^2)\, dx\, dy + \int_1^2 \int_0^{2-y} (x^2+y^2)\, dx\, dy$ **Ans.** $\int_0^1 dx \int_x^{2-x} (x^2+y^2)\, dy, \frac{5}{3}$

**17.** $\int_0^a \int_0^{\sqrt{a^2-x^2}} y^2 \sqrt{x^2+y^2}\, dx\, dy$ by changing into polar coordinates. **Ans.** $\frac{\pi\, a^5}{20}$

*(U.P., I Semester, Dec. 2007, A.M.I.E., Summer 2001)*

**18.** $\int_0^1 \int_1^2 \frac{1}{x^2+y^2}\, dx\, dy + \int_0^2 \int_y^2 \frac{1}{x^2+y^2}\, dx\, dy = \int_R \int \frac{1}{x^2+y^2}\, dy\, dx$

Recognise the region $R$ of integration on the R.H.S. and then evaluate the integral on the right in the order indicated. *(AMIETE, Dec. 2004)*

**Ans.** Region $R$ is $x = 0$, $x = y$, $y = 1$ and $y = 2$, $\frac{\pi}{4} \log 2$.

**19.** Express as single integral and evaluate :

$\int_0^{\frac{a}{\sqrt{2}}} \int_0^x x\, dx\, dy + \int_{\frac{a}{\sqrt{2}}}^{a} \int_0^{\sqrt{a^2-x^2}} x\, dx\, dy$ **Ans.** $\int_0^{\frac{a}{\sqrt{2}}} dy \int_y^{\sqrt{a^2-y^2}} x\, dx, \frac{5a^3}{6\sqrt{2}}$

**20.** Express as single integral and evaluate :

$\int_0^1 \int_0^y (x^2 + y^2)\, dx\, dy + \int_1^2 \int_0^{2-y} (x^2 + y^2)\, dx\, dy$ **Ans.** $\int_0^1 dx \int_x^{2-x} (x^2 + y^2)\, dy, \frac{5}{3}$

**21.** If $f(x, y)\, dx\, dy$, where $R$ is the circle $x^2 + y^2 = a^2$, is R equivalent to the repeated integral.

(*AMIE winter 2001*) [**Ans.** $\int_0^{2\pi} \int_0^1 (r, \theta)\, r\, dr\, d\theta.$]

## 2.5 CHANGE OF VARIABLES

Sometimes the problems of double integration can be solved easily by change of independent variables. Let the double integral as be $\iint_R f(x, y)\, dx\, dy$. It is to be changed by the new variables $u, v$.

The relation of $x, y$ with $u, v$ are given as $x = f(u, v)$, $y = \Psi(u, v)$. Then the double integration is converted into.

$\iint_{R'} f\{\phi(u, v), \Psi(u, v)\} |J|\, du\, dv$, where

$$dx\, dy = |J|\, du\, dv = \frac{\partial(x, y)}{\partial(u, v)}\, du\, dv = \begin{vmatrix} \frac{\partial x}{\partial u} & \frac{\partial x}{\partial v} \\ \frac{\partial y}{\partial u} & \frac{\partial y}{\partial v} \end{vmatrix} du\, dv$$

**Example 17.** *Evaluate* $\iint_R (x + y)^2\, dx\, dy$, *where R is the parallelogram in the xy-plane with vertices (1, 0), (3, 1), (2, 2), (0, 1), using the transformation* $u = x + y$ *and* $v = x - 2y$.

(*U.P., I Semester, 2003*)

**Solution.** The region of integration is a parallelogram *ABCD*, where *A* (1, 0), *B* (3, 1), *C* (2, 2) and *D* (0, 1) in *xy*-plane.

The new region of integration is a rectangle *A′B′C′D′* in *uv*-plane.

| | | | | |
|---|---|---|---|---|
| *xy*-plane | $A \equiv (x, y)$<br>$A \equiv (1, 0)$ | $B \equiv (x, y)$<br>$B \equiv (3, 1)$ | $C \equiv (x, y)$<br>$C \equiv (2, 2)$ | $D \equiv (x, y)$<br>$D \equiv (0, 1)$ |
| *uv*-plane | $A' \equiv (u, v)$<br>$A' \equiv (x + y, x - 2y)$<br>$A' \equiv (1 + 0, 1 - 2 \times 0)$<br>$A' \equiv (1, 1)$ | $B' \equiv (u, v)$<br>$B' \equiv (x + y, x - 2y)$<br>$B' \equiv (3 + 1, 3 - 2 \times 1)$<br>$B' \equiv (4, 1)$ | $C' \equiv (u, v)$<br>$C' \equiv (u, v)$<br>$C' \equiv (2 + 2, 2 - 2 \times 2)$<br>$C' \equiv (4, -2)$ | $D' \equiv (u, v)$<br>$D' \equiv (0 + 1, 0 - 2 \times 1)$<br>$D' \equiv (1, -2)$ |

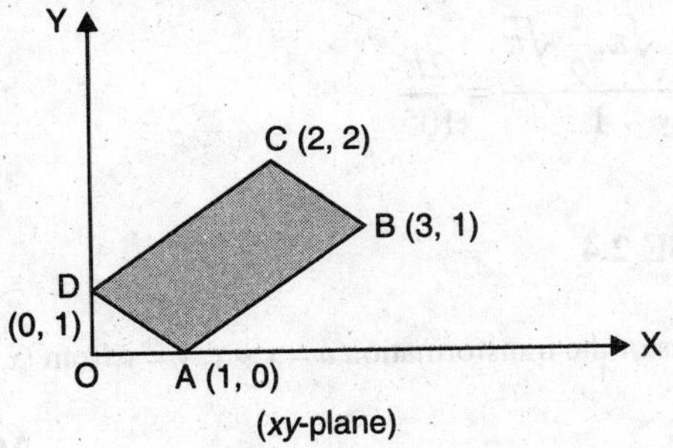

(*xy*-plane)

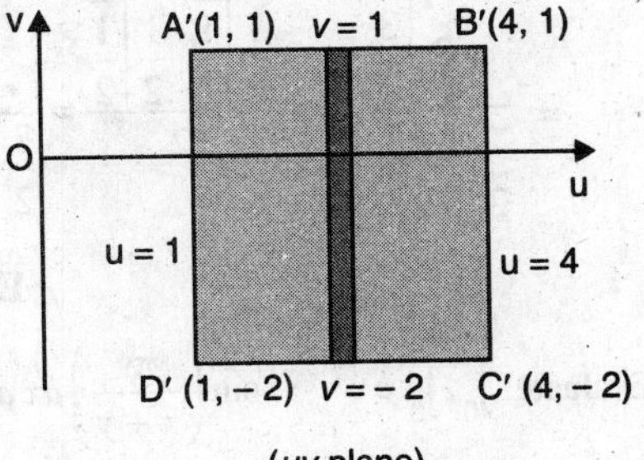

(*uv*-plane)

and $\begin{Bmatrix} u = x + y \\ v = x - 2y \end{Bmatrix} \Rightarrow$ $x = \frac{1}{3}(2u + v)$ and $y = \frac{1}{3}(u - v)$

$$J = \frac{\partial(x, y)}{\partial(u, v)} = \begin{vmatrix} \frac{\partial x}{\partial u} & \frac{\partial x}{\partial v} \\ \frac{\partial y}{\partial u} & \frac{\partial y}{\partial v} \end{vmatrix} = \begin{vmatrix} \frac{2}{3} & \frac{1}{3} \\ \frac{1}{3} & -\frac{1}{3} \end{vmatrix} = -\frac{1}{3}$$

$$dx\,dy = |J|\,du\,dv = \frac{1}{3}\,du\,dv$$

$$\iint_R (x+y)^2\,dx\,dy = \int_{-2}^{1}\int_{1}^{4} u^2 \cdot \frac{1}{3}\,du\,dv = \int_{-2}^{1}\frac{1}{3}\left[\frac{u^3}{3}\right]_1^4 dv = \int_{-2}^{1} 7\,dv = 7\,[v]_{-2}^{1} = 7\times 3 = 21 \text{ **Ans.**}$$

**Example 18.** *Using the transformation* $x + y = u$, $y = uv$, *show that*

$$\iint [xy\,(1-x-y)]^{1/2}\,dx\,dy = \frac{2\pi}{105},$$ *integration being taken over the area of the tringle bounded by the lines* $x = 0$, $y = 0$, $x + y = 1$.

**Solution.** $\iint [xy\,(1-x-y)]^{1/2}\,dx\,dy$

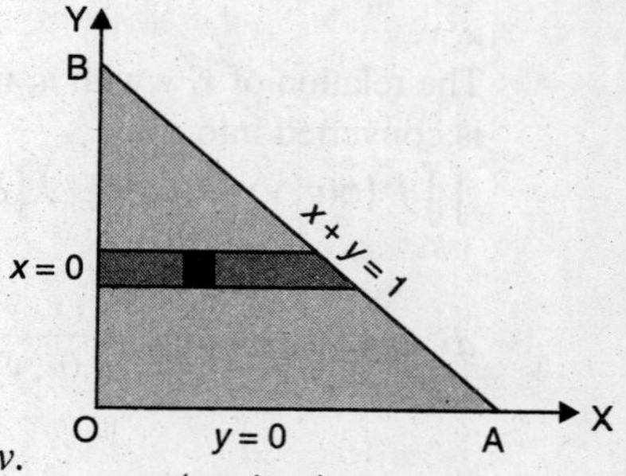

(xy-plane)

$$x + y = u \text{ or } x = u - y = u - uv,$$

$$dx\,dy = \frac{\partial(x, y)}{\partial(u, v)}\,du\,dv = \begin{vmatrix} \frac{\partial x}{\partial u} & \frac{\partial x}{\partial v} \\ \frac{\partial y}{\partial u} & \frac{\partial y}{\partial v} \end{vmatrix} du\,dv$$

$$dx\,dy = \begin{vmatrix} 1-v & -u \\ v & u \end{vmatrix} du\,dv = u\,du\,dv.$$

$$x = 0 \Rightarrow u(1-v) = 0$$
$$\Rightarrow u = 0,\ v = 1$$
$$y = 0 \Rightarrow uv = 0$$
$$\Rightarrow u = 0,\ v = 0$$
$$x + y = 1 \Rightarrow u = 1$$

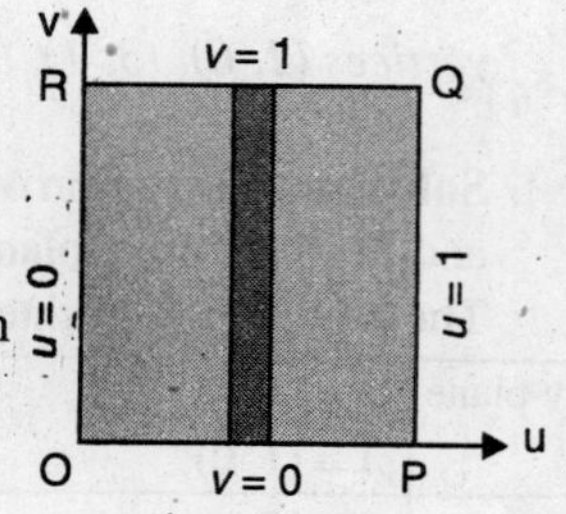

(uv-plane)

Hence, the limits of $u$ are from 0 to 1 and the limits of $v$ are from
The area of integration is a square $OPQR$ in $uv$-plane.
On putting $x = u - uv$, $y = uv$, $dx\,dy = u\,du\,dv$ in (1), we get

$$\iint (u - uv)^{1/2}\,(uv)^{1/2}\,(1-v)^{1/2}\,u\,du\,dv$$

$$= \int_0^1 u^2\,(1-u)^{1/2}\,du \int_0^1 v^{1/2}\,(1-v)^{1/2}\,dv = \frac{\lceil 3 \,\lceil \frac{3}{2}}{\lceil \frac{9}{2}} \times \frac{\lceil \frac{3}{2}\,\lceil \frac{3}{2}}{\lceil 5}$$

$$= \frac{2.\lceil \frac{3}{2}}{\frac{7}{2}.\frac{5}{2}.\frac{3}{2}.\lceil \frac{3}{2}} \times \frac{\frac{1}{2}\lceil \frac{1}{2}.\frac{1}{2}\lceil \frac{1}{2}}{2} = \frac{1}{\frac{7}{2}.\frac{5}{2}.\frac{3}{2}} \times \frac{\frac{1}{2}\sqrt{\pi}.\frac{1}{2}\sqrt{\pi}}{1} = \frac{2\pi}{105}$$ **Ans.**

## EXERCISE 2.4

1. Evaluate $\int_0^{\infty}\int_0^{\infty} e^{-(x+y)} \sin\left(\frac{\pi y}{x+y}\right) dx\,dy$ by means of the transformation $u = x + y$, $v = y$ from $(x, y)$ to $(u, v)$ **Ans.** $\frac{1}{\pi}$

2. Using the transformation $x + y = u$, $y = uv$, show that $\int_0^1 \int_0^{1-x} \frac{y}{e^{x+y}}\,dy\,dx = \frac{1}{2}(e-1)$ *(A.M.I.E. Winter 2001)*

3. Using the transformation $u = x - y$, $v = x + y$, prove that $\iint_R \cos\frac{x-y}{x+y}\,dx\,dy = \frac{1}{2}\sin 1$ where $R$ is bounded by $x = 0$, $y = 0$, $x + y = 1$

   $\left[\textbf{Hint: } x = \frac{1}{2}(u+v),\ y = \frac{1}{2}(v-u) \text{ so that } |J| = \frac{1}{2}\right]$

## 2.6 AREA IN CARTESIAN CO-ORDINATES

Let the curves $AB$ and $CD$ be $y_1 = f_1(x)$ and $y_2 = f_2(x)$.

Let the ordinates $AD$ and $BC$ be $x = a$ and $x = b$.

So the area enclosed by the two curves $y_1 = f_1(x)$ and $y_2 = f_2(x)$ and $x = a$ and $x = b$ is $ABCD$.

Let $P(x, y)$ and $Q(x + \delta x, y + \delta y)$ be two neighbouring points, thent the area of the small rectangle $PQ = \delta x . \delta y$.

Area of the vertical strip $= \lim\limits_{\partial y \to 0} \sum\limits_{y_1}^{y_2} \delta x \, \delta y = \delta x \int_{y_1}^{y_2} dy \, \delta x$ the width of the strip is constant throughout.

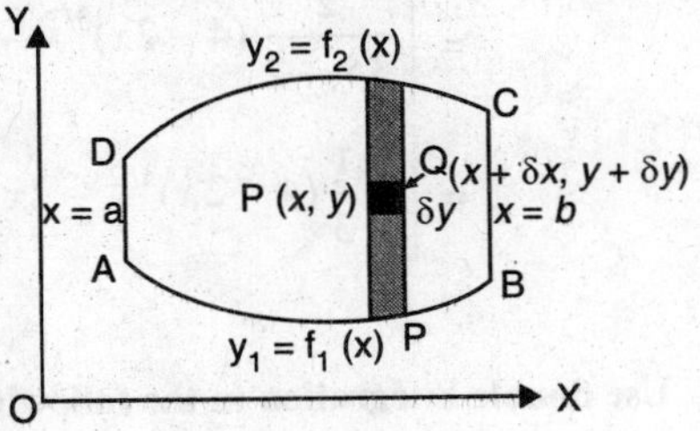


If we add all the strips from $x = a$ to $x = b$, we get

$$\text{The area } ABCD = \lim_{\partial x \to 0} \sum_{a}^{b} \delta x \int_{y_1}^{y_2} dy = \int_a^b dx \int_{y_1}^{y_2} dy$$

$$\boxed{\textbf{Area} = \int_a^b \int_{y_1}^{y_2} dx\, dy}$$

**Example 19.** *Find the area bounded by the parabola $y^2 = 4ax$ and its latus rectum.*

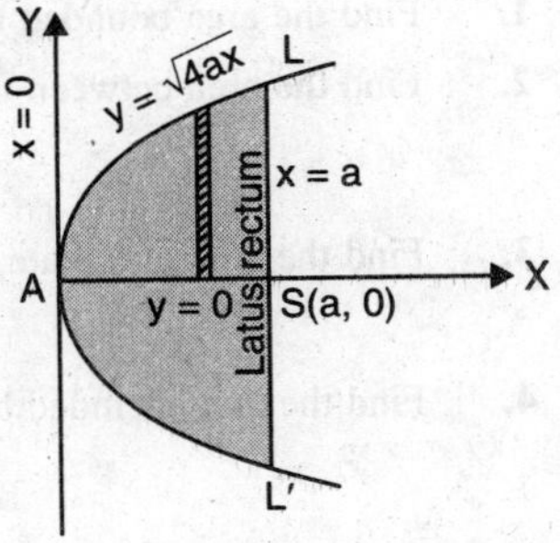


**Solution.** Required area = 2 (area (ASL)

$$= 2 \int_0^a \int_0^{2\sqrt{ax}} dy\, dx$$

$$= 2 \int_0^a 2\sqrt{ax}\, dx$$

$$= 4\sqrt{a} \left( \frac{x^{3/2}}{3/2} \right)_0^a = \frac{8a^2}{3}$$

**Example 20.** *Find the area between the parabolas $y^2 = 4\, ax$ and $x^2 = 4\, ay$.*

**Solution.**

$$y^2 = 4ax \qquad \text{...(1)}$$

$$x^2 = 4ay \qquad \text{...(2)}$$

On solving the equations (1) and (2) we get the point of intersection $(4a, 4a)$.

Divide the area into horizontal strips of width $\delta y$, $x$ varies from $P, \dfrac{y^2}{4a}$ to $Q, \sqrt{4\, ay}$ and then $y$ varies from $O(y = 0)$ to $A\ (y = 4a)$.

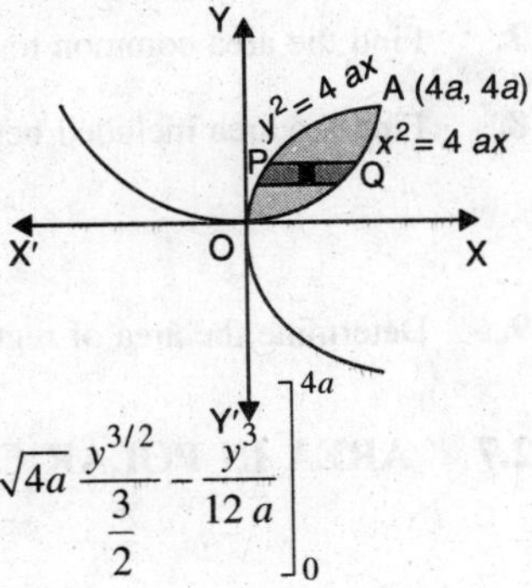


$$\therefore \quad \text{The required area} = \int_0^{4a} dy \int_{y^2/4a}^{\sqrt{4ay}} dx$$

$$= \int_0^{4a} dy\, [x]_{y^2/4a}^{\sqrt{4ay}} = \int_0^{4a} dy \left[ \sqrt{4ay} - \frac{y^2}{4a} \right] = \left[ \sqrt{4a}\, \frac{y^{3/2}}{\frac{3}{2}} - \frac{y^3}{12\, a} \right]_0^{4a}$$

$$= \left[ \frac{4\sqrt{a}}{3} (4a)^{3/2} - \frac{(4a)^3}{12a} \right] = \left[ \frac{32}{3} a^2 - \frac{16}{3} a^2 \right] = \frac{16}{3} a^2 \qquad \textbf{Ans.}$$

**Example 21.** *Find by double integration the area enclosed by the pair of curves*

$$y = 2 - x \text{ and } y^2 = 2(2 - x)$$

**Solution.**

$$y = 2 - x$$

$$y^2 = 2\,(2 - x)$$

On solving the equations (1) and (2), we get the points of intersection (2, 0) and (0, 2).

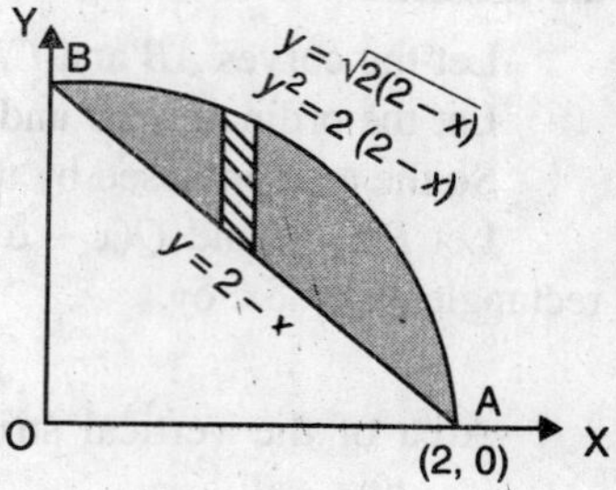


$$A = \iint dx\, dy$$

$$\text{The required area} = \int_0^2 dx \int_{2-x}^{\sqrt{2(2-x)}} dy$$

$$= \int_0^2 dx\, [y]_{2-x}^{\sqrt{2(2-x)}} = \int_0^2 dx\, [\sqrt{4-2x} - 2 + x]$$

$$= \left[\frac{2}{3 \times -2}(4-2x)^{3/2} - 2x + \frac{x^2}{2}\right]_0^2$$

$$= \left[-\frac{1}{3}(4-2x)^{3/2} - 2x + \frac{x^2}{2}\right]_0^2 = \left(-4 + \frac{4}{2}\right) + \frac{8}{3} = \frac{2}{3}$$ **Ans.**

## EXERCISE 2.5

**Use double integration in the following questions:**

**1.** Find the area bounded by $y = x - 2$ and $y^2 = 2x + 4$. **Ans.** 18.

**2.** Find the area between the circle $x^2 + y^2 = a^2$ and the line $x + y = a$ in the first quadrant. **Ans.** $(\pi - 2)a^2/4$

**3.** Find the area of a plate in the form of quadrant of the ellipse $\frac{x^2}{a^2} + \frac{y^2}{b^2} = 1$. **Ans.** $\frac{\pi ab}{4}$

**4.** Find the area included between the curves $y^2 = 4a(x + a)$ and $y^2 = 4b(b - x)$. **Ans.** $\frac{8\sqrt{ab}}{3}$

(*A.M.I.E.T.E., Summer 2001*)

**5.** Find the area bounded by (*a*) $y^2 = 4 - x$ and $y^2 = x$. **Ans.** $\frac{16\sqrt{2}}{3}$

(*b*) $x - 2y + 4 = 0,\ x + y - 5 = 0,\ y = 0$ (*A.M.I.E., Winter 2001*) **Ans.** $\frac{27}{2}$

**6.** Find the area enclosed by the leminscate $r^2 = a^2 \cos 2\theta$. **Ans.** $a^2$

**7.** Find the area common to the circles $x^2 + y^2 = a^2$ and $x^2 + y^2 = 2ax$. **Ans.** $\left[\frac{\pi}{3} - \frac{\sqrt{3}}{4}\right]a^2$

**8.** Find the area included between the curves $y = x^2 - 6x + 3$ and $y = 2x + 9$.

(*A.M.I.E., Summer 2001*) **Ans.** $\frac{88\sqrt{22}}{3}$

**9.** Determine the area of region bounded by the curves $xy = 2$, $4y = x^2$, $y = 4$. **Ans.** $\frac{28}{3} - 4\log 2$

(*U.P. I Semester 2003*)

## 2.7 AREA IN POLAR CO-ORDINATES

$$\text{Area} = \iint r\, d\theta\, dr$$

Let us consider the area enclosed by the curve $r = f(\theta)$.

Let $P(r, \theta)$, $Q(r + \delta r, \theta + \delta\theta)$ be two neighbouring points.

Draw ares $PL$ and $QM$, radii $r$ and $r + \delta r$.

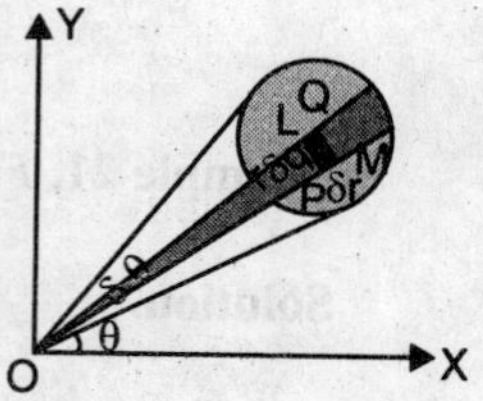


$$PL = r\delta\theta,\ PM = \delta r$$

Area of rectangle $PLQM = PL \times PM$

$$= (r\delta\theta)(d\delta r) = r\,\delta\theta\,\delta r.$$

The whole area $A$ is composed of such small rectangles.

Hence,

$$A = \lim_{\substack{\partial r \to 0 \\ \delta\theta \to 0}} \sum \sum r\,\delta\theta.\,\delta r = \iint r\,d\theta\,dr$$

**Example 22.** *Find by double integration, the area lying inside the cardioid r = a (1 + cos θ) and outside the circle r = a.* *(Nagpur University, Winter 2000)*

**Solution.**
$$r = a\,(1 + \cos\theta) \quad \ldots(1)$$
$$r = a \quad \ldots(2)$$

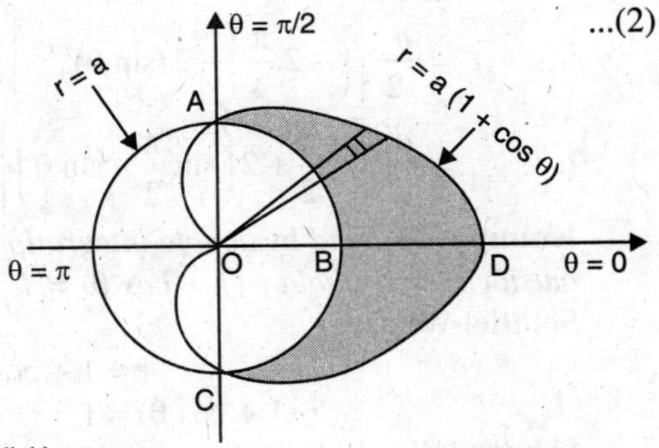


Solving (1) and (2), by eliminating $r$, we get

$$a(1 + \cos\theta) = a \quad \Rightarrow \quad 1 + \cos\theta = 1$$

$$\cos\theta = 0 \quad \Rightarrow \quad \theta = -\frac{\pi}{2} \text{ or } \frac{\pi}{2}$$

limits of $r$ are $a$ and $a(1 + \cos\theta)$

limits of $\theta$ are $-\frac{\pi}{2}$ to $\frac{\pi}{2}$

Required area = Area ABCDA

$$= \int_{-\pi/2}^{\pi/2} \int_{r \text{ for circle}}^{\text{for cardioid}} r\,d\theta\,dr$$

$$= \int_{-\pi/2}^{\pi/2} \int_{a}^{a(1+\cos\theta)} r\,d\theta\,dr \qquad = \int_{-\pi/2}^{\pi/2} \left(\frac{r^2}{2}\right)_a^{a(1+\cos\theta)} d\theta$$

$$= \frac{a^2}{2} \int_{-\pi/2}^{\pi/2} [(1+\cos\theta)^2 - 1]\,d\theta \qquad = \frac{a^2}{2} \int_{-\pi/2}^{\pi/2} (\cos^2\theta + 2\cos\theta)\,d\theta$$

$$= a^2 \int_0^{\pi/2} (\cos^2\theta + 2\cos\theta)\,d\theta \qquad = a^2 \left[\int_0^{\pi/2} \cos^2\theta\,d\theta + 2\int_0^{\pi/2} \cos\theta\,d\theta\right]$$

$$= a^2 \left[\frac{\pi}{4} + 2\,(\sin\theta)_0^{\pi/2}\right] = a^2 \left[\frac{\pi}{4} + 2\right] = \frac{a^2}{4}(\pi + 8) \qquad \textbf{Ans.}$$

**Example 23.** *Find by double integration, the area lying inside the circle r = a sin θ and outside the cardioid r = a (1 – cos θ).*

**Solution.** We have,

$$r = a\sin\theta \quad \ldots(1)$$
$$r = a\,(1 - \cos\theta) \quad \ldots(2)$$

Solving (1) and (2) by eliminating $r$, we have

$$\sin\theta = 1 - \cos\theta \Rightarrow \sin\theta + \cos\theta = 1$$

Squaring above, we get

$$\sin^2\theta + \cos^2\theta + 2\sin\theta\cos\theta = 1$$

$$\Rightarrow \qquad 1 + \sin 2\theta = 1 \Rightarrow \sin 2\theta = 0 \Rightarrow 2\theta = 0 \text{ or } \pi = \theta = 0 \text{ or } \frac{\pi}{2}$$

The required area is shaded portion in the fig.

Limits of $r$ are $a(1 - \cos\theta)$ and $a\sin\theta$, limits of $\theta$ are 0 and $\frac{\pi}{2}$.

$$\text{Required area} = \int_0^{\frac{\pi}{2}} \int_{a(1-\cos\theta)}^{a\sin\theta} r\,dr\,d\theta$$

$$= \int_0^{\frac{\pi}{2}} \left[\frac{r^2}{2}\right]_{a(1-\cos\theta)}^{a\sin\theta} d\theta = \frac{1}{2}\int_0^{\pi/2} a^2\,[\sin^2\theta - (1-\cos\theta)^2]\,d\theta$$

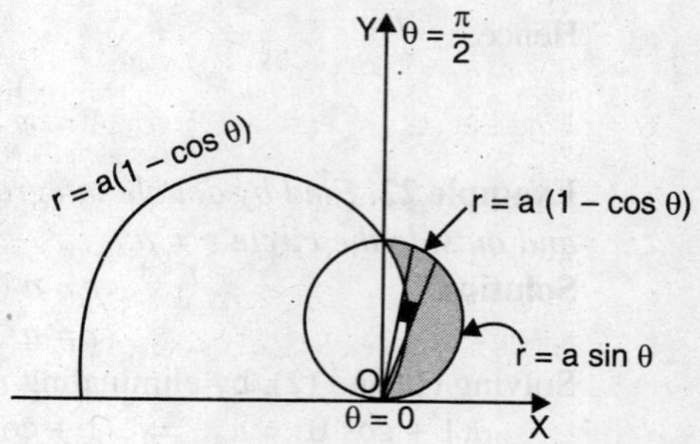


$$= \frac{a^2}{2}\int_0^{\pi/2} (\sin^2\theta - 1 - \cos^2\theta + 2\cos\theta]\, d\theta$$

$$= \frac{a^2}{2}\int_0^{\pi/2} (-2\cos^2\theta + 2\cos\theta)\, d\theta$$

$$= \frac{a^2}{2}\left[\int_0^{\pi/2} -2\cos^2\theta\, d\theta + \int_0^{\pi/2} 2\cos\theta\, d\theta\right]$$

$$= \frac{a^2}{2}\left[\left(-2.\frac{\pi}{4}\right) + 2\,(\sin\theta)_0^{\pi/2}\right]$$

$$= \frac{a^2}{2}\left[-\frac{\pi}{2} + 2\left(\sin\frac{\pi}{2} - \sin 0\right)\right] = \frac{a^2}{2}\left[-\frac{\pi}{2} + 2\right] = a^2\left(1 - \frac{\pi}{4}\right)$$ **Ans.**

**Example 24.** *Find by double integration, the area lying inside a cardioid r = 1 + cos θ and outside the parabola r (1 + cos θ) = 1.*

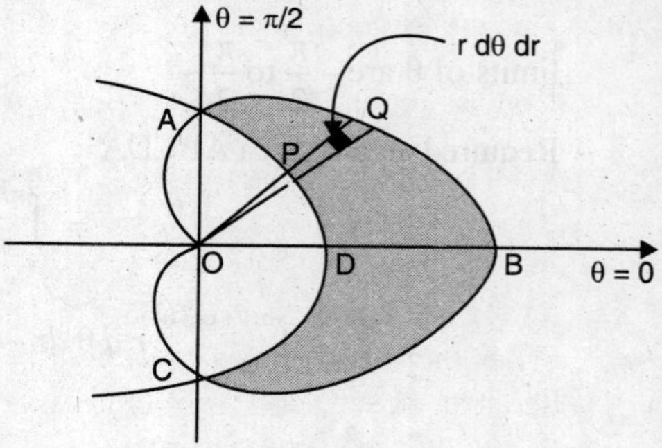


**Solutio.** We have,

$$r = 1 + \cos\theta \quad ...(1)$$

$$r\,(1 + \cos\theta) = 1 \quad ...(2)$$

Solving (1) and (2), we get

$$(1 + \cos\theta)\,(1 + \cos\theta) = 1$$

$$(1 + \cos\theta)^2 = 1$$

$$1 + \cos\theta = 1$$

$$\cos\theta = 0 \Rightarrow \theta = \pm\frac{\pi}{2}$$

limits of $r$ are $1 + \cos\theta$ and $\frac{1}{1+\cos\theta}$ limits of θ are $-\frac{\pi}{2}$ to $\frac{\pi}{2}$.

Required area = Area *ADCBA* (Shaded portion)

$$= \int_{-\pi/2}^{\pi/2}\int_{\frac{1}{1+\cos\theta}}^{1+\cos\theta} r\, d\theta\, dr = \int_{\frac{-\pi}{2}}^{\frac{\pi}{2}}\left(\frac{r^2}{2}\right)_{\frac{1}{1+\cos\theta}}^{1+\cos\theta} d\theta = \frac{1}{2}\int_{-\pi/2}^{\pi/2}\left[(1+\cos\theta)^2 - \frac{1}{(1+\cos\theta)^2}\right] d\theta$$

$$= \frac{1}{2}\int_{-\pi/2}^{\pi/2}\left[(1+\cos^2\theta + 2\cos\theta) - \frac{1}{\left(2\cos^2\frac{\theta}{2}\right)^2}\right] d\theta$$

$$= 2\times\frac{1}{2}\int_0^{\pi/2}\left[(1+\cos^2\theta + 2\cos\theta) - \frac{1}{4}\sec^4\frac{\pi}{2}\right] d\theta$$

$$= \int_0^{\pi/2}\left[(1+\cos^2\theta + 2\cos\theta) - \frac{1}{4}\left(1+\tan^2\frac{\theta}{2}\right)\sec^2\frac{\theta}{2}\right] d\theta$$

$$= \int_0^{\pi/2}\left[\left(1 + \frac{1+\cos 2\theta}{2} + 2\cos\theta\right) - \frac{1}{4}\left(1+\tan^2\frac{\pi}{2}\right)\sec^2\frac{\pi}{2}\right] d\theta$$

$$= \int_0^{\pi/2}\left[1 + \frac{1}{2} + \frac{\cos 2\theta}{2} + 2\cos\theta - \frac{1}{4}\left(\sec^2\frac{\theta}{2} + \tan^2\frac{\theta}{2}\times\sec^2\frac{\theta}{2}\right)\right] d\theta$$

$$= \left[\theta + \frac{\theta}{2} + \frac{\sin 2\theta}{4} + 2\sin\theta - \frac{1}{4}\left(2\tan\frac{\theta}{2} + \frac{2}{3}\tan^3\frac{\theta}{2}\right)\right]_0^{\frac{\pi}{2}}$$

$$= \left[\frac{\pi}{2} + \frac{\pi}{4} + 0 + 2\sin\frac{\pi}{2} - \frac{1}{2}\tan\frac{\pi}{4} - \frac{1}{6}\tan^3\frac{\pi}{4}\right] = \left[\frac{3\pi}{4} + 2 - \frac{1}{2} - \frac{1}{6}\right] = \left[\frac{3\pi}{4} + \frac{4}{3}\right]$$ **Ans.**

## EXERCISE 2.6

1. Find the area of cardioid $r = a(1 + \cos\theta)$. **Ans.** $\dfrac{3\pi a^2}{2}$
2. Find the area of the curve $r^2 = a^2 \cos 2\theta$. **Ans.** $a^2$
3. Find the area enclosed by the curve $r = 2a\cos\theta$ **Ans.** $\pi a^2$
4. Find the area enclosed by the curve $r = 3 + 2\cos\theta$. **Ans.** $11\pi$
5. Find the area enclosed by the curve

   $r^3 = a^2 \cos^2\theta + b^2 \sin^2\theta$. **Ans.** $\dfrac{\pi}{2}(a^2 + b^2)$
6. Show that the area of the region included between the cardioides $r = a(1 + \cos\theta)$ and $r = a(1 - \cos\theta)$ is $\dfrac{a^2}{2}(3\pi - 8)$.
7. Find the area outside the circle $r = 2$ and inside the cardioid $r = 2(1 + \cos\theta)$. **Ans.** $(\pi + 8)$
8. Find the area inside the circle $r = 2a\cos\theta$ and outside the circle $r = a$. **Ans.** $2a^2\left(\dfrac{\pi}{3} + \dfrac{\sqrt{3}}{4}\right)$
9. Find the area inside the circle $r = 4\sin\theta$ and outside the lemniscate $r^2 = 8\cos 2\theta$.

   **Ans.** $\left(\dfrac{8}{3}\pi + 4\sqrt{3} - 4\right)$

## 2.8 VOLUME OF SOLID BY ROTATION OF AN AREA (DOUBLE INTEGRAL)

When the area enclosed by a curve $y = f(x)$ is revolved about an axis, a solid is generated, we have to find out the volume of solid generated.

Volume of the solid generated about $x$-axis

$$= \int_a^b \int_{y_1(x)}^{y_2(x)} 2\pi\, PQ\, dx\, dy$$

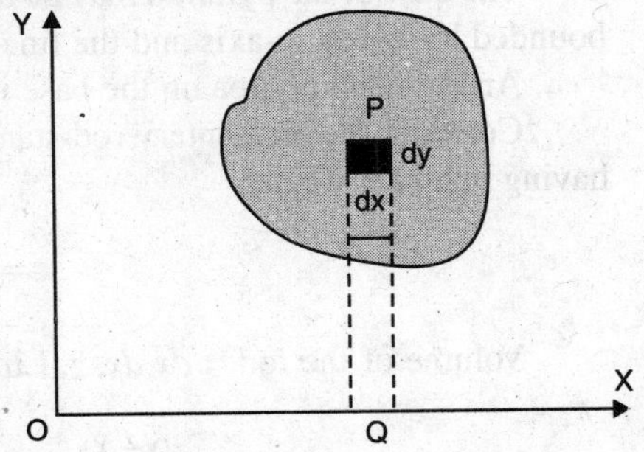


**Example 25.** *Find the volume of the torus generated by revolving the circle $x^2 + y^2 = 4$ about the line $x = 3$.*

**Solution.** $x^2 + y^2 = 4$

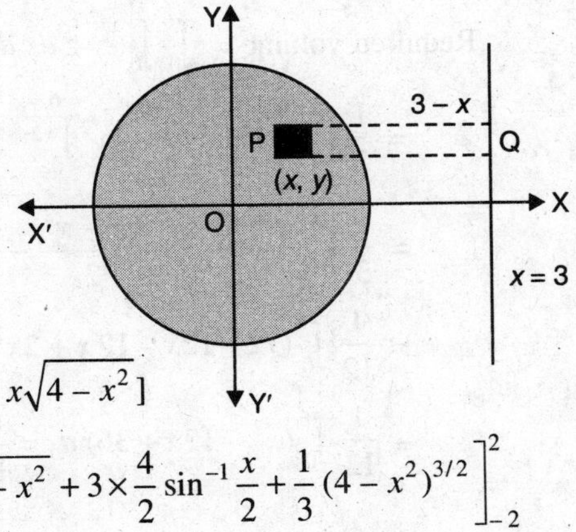


$$V = \int\int (2\pi\, PQ)\, dx\, dy = 2\pi \int\int (3 - x)\, dx\, dy$$

$$= 2\pi \int_{-2}^{+2} dx \int_{-\sqrt{4-x^2}}^{+\sqrt{4-x^2}} (3 - x)\, dy$$

$$= 2\pi \int_{-2}^{+2} dx\, (3y - xy)_{-\sqrt{4-x^2}}^{+\sqrt{4-x^2}}$$

$$= 2\pi \int_{-2}^{+2} dx\, [3\sqrt{4-x^2} - x\sqrt{4-x^2} + 3\sqrt{4-x^2} - x\sqrt{4-x^2}]$$

$$= 4\pi [3\sqrt{4-x^2} - x\sqrt{4-x^2}]\, dx = 4\pi \left[3\frac{x}{2}\sqrt{4-x^2} + 3\times\frac{4}{2}\sin^{-1}\frac{x}{2} + \frac{1}{3}(4 - x^2)^{3/2}\right]_{-2}^{2}$$

$$= 4\pi\left[6\times\frac{\pi}{2} + 6\times\frac{\pi}{2}\right] = 24\pi^2$$ **Ans.**

**Example 26.** *Calculate by double integration the volume generated by the revolution of the cardioid $r = a(1 - \cos\theta)$ about its axis.* (AMIETE, June 2010)

**Solution.** $r = a(1 - \cos\theta)$

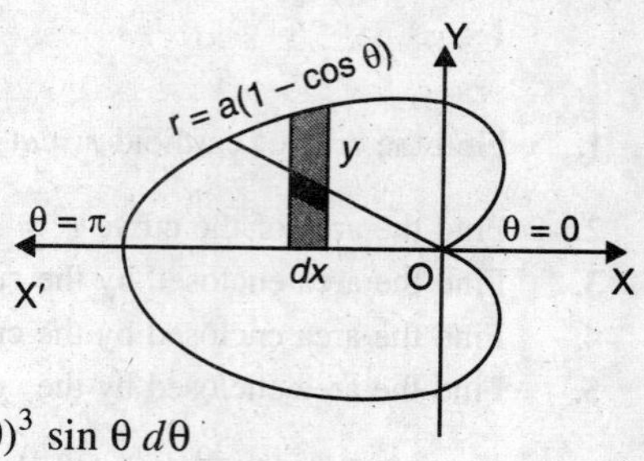


$$V = 2\pi \int\int y\,dx\,dy \Rightarrow V = 2\pi \int\int (r\,d\theta\,dr)\,y$$

$$= 2\pi \int d\theta \int r\,dr\,(r\sin\theta)$$

$$= 2\pi \int_0^{\pi} \sin\theta\,d\theta \int_0^{a(1-\cos\theta)} r^2\,dr$$

$$= 2\pi \int_0^{\pi} \sin\theta\,d\theta \left[\frac{r^3}{3}\right]_0^{a(1-\cos\theta)} = \frac{2\pi}{3}\int_0^{\pi} a^3 (1-\cos\theta)^3 \sin\theta\,d\theta$$

$$= \frac{2\pi a^3}{3}\left[\frac{(1-\cos\theta)^4}{4}\right]_0^{\pi} = \frac{2\pi a^3}{12}[16] = \frac{8}{3}\pi a^3$$ **Ans.**

**Example 27.** *A pyramid is bounded by the three co-ordinate planes and the plane $x + 2y + 3z = 6$. Compute this volume by double integration.*

**Solution.** $x + 2y + 3z = 6$ ...(1)

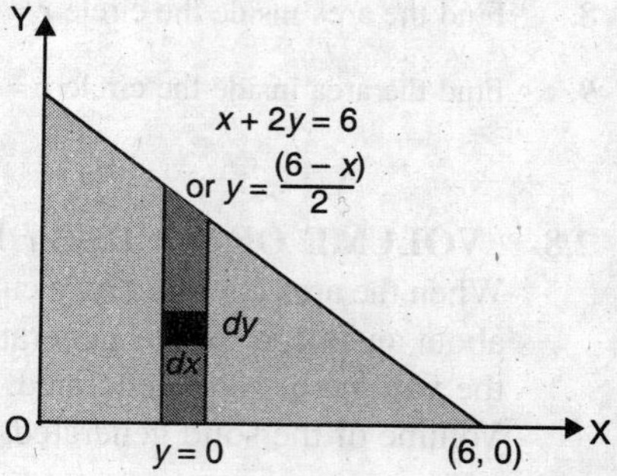


$x = 0, y = 0, z = 0$ are co-ordinate planes.

The line of intersection of plane (1) and $xy$ plane ($z = 0$) is

$$x + 2y = 6 \qquad ...(2)$$

The base of the pyramid may be taken to be the triangle bounded by $x$-axis, $y$-axis and the line (2).

An elementary area on the base is $dx\,dy$.

Consider the elementary rod standing on this area and having height $z$, where

$$3z = 6 - x - 2y \text{ or } z = \frac{6 - x - 2y}{3}$$

Volume of the rod $= dx\,dy,\, z$, Limits for $z$ are 0 and $\frac{6 - x - 2y}{3}$.

Limits of $y$ are 0 and $\frac{6-x}{2}$ and limits of $x$ are 0 and 6.

$$\text{Required volume} = \int_0^6 \int_0^{\frac{6-x}{2}} z\,dx\,dy = \int_0^6 dx \int_0^{\frac{6-x}{2}} \frac{6 - x - 2y}{3}\,dy$$

$$= \frac{1}{3}\int_0^6 dx\left(6x - xy - y^2\right)_0^{\frac{6-x}{2}} = \frac{1}{3}\int_0^6 \left(\frac{6(6-x)}{2} - \frac{x(6-x)}{2} - \left(\frac{6-x}{2}\right)^2\right) dx$$

$$= \frac{1}{3}\int_0^6 \left(\frac{36-6x}{2} - \frac{6x - x^2}{2} - \frac{36 + x^2 - 12x}{4}\right) dx$$

$$= \frac{1}{12}\int_0^6 (72 - 12x - 12x + 2x^2 - 36 - x^2 + 12x)\,dx$$

$$= \frac{1}{12}\int_0^6 (x^2 - 12x + 36)\,dx = \frac{1}{12}\left[\frac{x^3}{3} - \frac{12x^2}{2} + 36x\right]_0^6 = \frac{1}{12}[72 - 216 + 216] = 6$$ **Ans.**

## EXERCISE 2.7

1. Find the volume of the sphere $x^2 + y^2 + z^2 = a^2$ by revolving area of the circle $x^2 + y^2 = a^2$. **Ans.** $\frac{4}{3}\pi a^3$

## 2.9 CENTRE OF GRAVITY

$$\bar{x} = \frac{\int\int \rho\,x\,dx\,dy}{\int\int \rho\,dx\,dy}, \quad \bar{y} = \frac{\int\int \rho\,y\,dx\,dy}{\int\int \rho\,dx\,dy}$$

**Example 28.** *Find the position of the C.G. of a semi-circular lamina of radius a if its density varies as the square of the distance from the diameter.* (*AMIETE, Dec. 2010*)

**Solution.** Let the bounding diameter be as the $x$-axis and a line perpendicular to the diameter and passing through the centre is $y$-axis. Equation of the circle is $x^2 + y^2 = a^2$. By symmetry $\overline{x} = 0$.

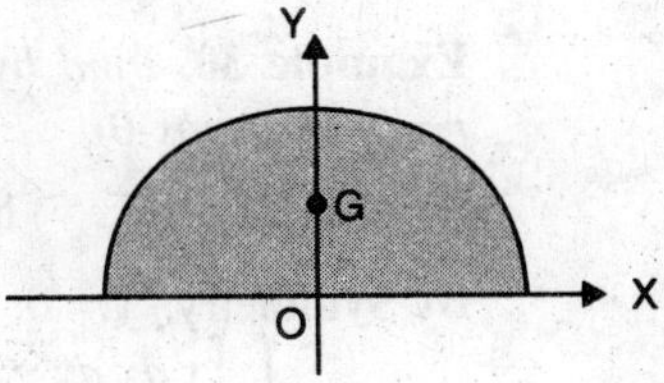


$$\overline{y} = \frac{\int\int y\,\rho\,dx\,dy}{\int\int \rho\,dx\,dy} = \frac{\int\int (\lambda y^2)\,y\,dx\,dy}{\int\int (\lambda y^2)\,dx\,dy} = \frac{\int_{-a}^{a} dx \int_0^{\sqrt{a^2-x^2}} y^3\,dy}{\int_{-a}^{a} dx \int_0^{\sqrt{a^2-x^2}} y^2\,dy}$$

$$= \frac{\int_{-a}^{a} dx \left[\frac{y^4}{4}\right]_0^{\sqrt{a^2-x^2}}}{\int_{-a}^{a} dx \left(\frac{y^3}{3}\right)_0^{\sqrt{a^2-x^2}}} = \frac{3\int_{-a}^{a} (a^2-x^2)^2\,dx}{4\int_{-a}^{a} (a^2-x^2)^{3/2}\,dx}$$

Put $x = a\sin\theta$

$$= \frac{3\int_{-\frac{\pi}{2}}^{\frac{\pi}{2}} (a^2 - a^2\sin^2\theta)^2\,a\cos\theta\,d\theta}{4\int_{-\frac{\pi}{2}}^{\frac{\pi}{2}} (a^2 - a^2\sin^2\theta)^{3/2}\,a\cos\theta\,d\theta} = \frac{3\int_{-\frac{\pi}{2}}^{\frac{\pi}{2}} a^5\cos^5\theta\,d\theta}{4\int_{-\frac{\pi}{2}}^{\frac{\pi}{4}} a^4\cos^4\theta\,d\theta}$$

$$= \frac{3a}{4}\,\frac{\frac{4\times 2}{5\times 3}}{\frac{3\times 1}{4\times 2}\,\frac{\pi}{2}} = \left(\frac{3a}{4}\right)\left(\frac{8}{15}\right)\left(\frac{16}{3\pi}\right) = \frac{32a}{15\pi}$$

Hence C.G. is $\left(0, \frac{32a}{15\pi}\right)$ **Ans.**

**Example 29.** *Find C.G. of the area in the positive quadrant of the curve* $x^{2/3} + y^{2/3} = a^{2/3}$.

**Solution.** For C.G. of area; $\overline{x} = \frac{\int\int x\,dx\,dy}{\int\int dx\,dy}$, $\overline{y} = \frac{\int\int y\,dx\,dy}{\int\int dx\,dy}$

$$\overline{x} = \frac{\int_0^a x\,dx \int_0^{(a^{2/3}-x^{2/3})^{3/2}} dy}{\int_0^a dx \int_0^{(a^{2/3}-x^{2/3})^{3/2}} dy} = \frac{\int_0^a x\,dx\,[y]_0^{(a^{2/3}-x^{2/3})^{3/2}}}{\int_0^a dx\,[y]_0^{(a^{2.3}-x^{2/3})^{3/2}}}$$

[Put $x = a\cos^3\theta$]

$$= \frac{\int_0^a x\,dx\,(a^{2/3} - x^{2/3})^{3/2}}{\int_0^a dx\,(a^{2/3} - x^{2/3})^{3/2}} = \frac{\int_{\frac{\pi}{2}}^{0} a\cos^3\theta\,(a^{2/3} - a^{2/3}\cos^2\theta)^{3/2}\,(-3a\cos^2\theta\sin\theta\,d\theta)}{\int_{\frac{\pi}{2}}^{0} (a^{2/3} - a^{2/3}\cos^2\theta)^{3/2}\,(-3a\cos^2\theta\sin\theta\,d\theta)}$$

$$= \frac{\int_0^{\frac{\pi}{2}} 3a^3\cos^3\theta\sin^3\theta\cos^2\theta\sin\theta\,d\theta}{\int_0^{\frac{\pi}{2}} 3a^2\sin^3\theta\cos^2\theta\sin\theta\,d\theta} = \frac{a\int_0^{\frac{\pi}{2}} \sin^4\theta\cos^5\theta\,d\theta}{\int_0^{\frac{\pi}{2}} \sin^4\theta\cos^2\theta\,d\theta} = \frac{\frac{\Gamma\frac{5}{2}\,\Gamma\frac{6}{2}}{2\,\Gamma\frac{4}{2}}\,a}{\frac{\Gamma\frac{5}{2}\,\Gamma\frac{3}{2}}{2\,\Gamma 4}}$$

$$= \frac{\overline{|3}\;\overline{|4}\; a}{\overline{\left|\frac{3}{2}\right.}\;\overline{\left|\frac{11}{2}\right.}} = \frac{(2)\,(6)\,a}{\frac{1}{2}\cdot\frac{9}{2}\cdot\frac{7}{2}\cdot\frac{5}{2}\cdot\frac{3}{2}\cdot\frac{1}{2}\,\pi} = \frac{256\,a}{315\,\pi}, \quad \text{Similarly, } \bar{y} = \frac{256\,a}{315\,\pi}$$

Hence, C.G. of the area is $\left(\dfrac{256\,a}{315\,\pi}, \dfrac{256\,a}{315\,\pi}\right)$.

**Example 30.** *Find by double integration, the centre of gravity of the area of the cardioid $r = a\,(1 + \cos\theta)$.*

**Solution.** Let $(\bar{x}, \bar{y})$ be the C.G. the cardioid

By Symmetry, $\bar{y} = 0$.

$$\bar{x} = \frac{\int\int_A x\,dx\,dy}{\int\int_A dx\,dy} = \frac{\int\int_A x\,dx\,dy}{\int\int_A dx\,dy}$$

$$= \frac{\int_{-\pi}^{\pi}\int_0^{a(1+\cos\theta)} (r\cos\theta)\,(r\,d\theta\,dr)}{\int_{-\pi}^{\pi}\int_0^{a(1+\cos\theta)} r\,d\theta\,dr} = \frac{\int_{-\pi}^{\pi}\cos\theta\,d\theta\int_0^{a\,(1+\cos\theta)} r^2\,dr}{\int_{-\pi}^{\pi} d\theta\int_0^{a(1+\cos\theta)} r\,dr}$$

$$= \frac{\int_{-\pi}^{\pi}\cos\theta\,d\theta\left[\frac{r^3}{3}\right]_0^{a\,(a+\cos\theta)}}{\int_{-\pi}^{\pi} d\theta\left(\frac{r^2}{2}\right)_0^{a\,(1+\cos\theta)}} = \frac{\int_{-\pi}^{\pi}\cos\theta\,d\theta.\,\frac{a^3}{3}\,(1+\cos\theta)^3}{\int_{-\pi}^{\pi} d\theta\,\frac{a^2}{2}\,(1+\cos\theta)^2}$$

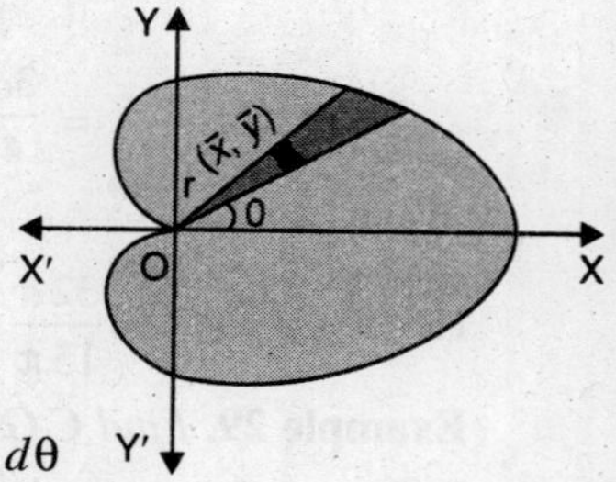


$$= \frac{\frac{a^3}{3}\int_{-\pi}^{\pi}\left(2\cos^2\frac{\theta}{2}-1\right)\left(1+2\cos^2\frac{\theta}{2}-1\right)^3 d\theta}{\frac{a^2}{2}\int_{-\pi}^{\pi}\left(1+2\cos^2\frac{\theta}{2}-1\right)d\theta}$$

$$= \frac{a^3}{3}\int_{-\pi}^{\pi}\left(2\cos^2\frac{\theta}{2}-1\right)\left(8\cos^6\frac{\theta}{2}\right)d\theta \div \frac{a^2}{2}\int_{-\pi}^{\pi} 4\cos^4\frac{\theta}{2}\,d\theta$$

$$= \frac{8a^3}{3}\int_{-\pi}^{\pi}\left(2\cos^8\frac{\theta}{2} - \cos^6\frac{\theta}{2}\right)d\theta \div 2a^2\int_{-\pi}^{\pi}\cos^4\frac{\theta}{2}\,d\theta$$

$$= \frac{2\times 8a^3}{3}\int_0^{\pi}\left(2\cos^8\frac{\theta}{2} - \cos^6\frac{\theta}{2}\right)d\theta \div 4a^2\int_0^{\pi}\cos^4\frac{\theta}{2}\,d\theta$$

$$= \frac{16a^3}{3}\int_0^{\pi/2}(2\cos^8 t - \cos^6 t)\,(2\,dt) \div 4a^2\int_0^{\pi/2}\cos^4 t\,(2\,dt)$$

$$= \frac{32\,a^3}{3}\left[\frac{2\times 7\times 5\times 3\times 1}{8\times 6\times 4\times 2}\,\frac{\pi}{2} - \frac{5\times 3\times 1}{6\times 4\times 2}\,\frac{\pi}{2}\right] \div 8a^2\left(\frac{3\times 1}{4\times 2}\,\frac{\pi}{2}\right)$$

$$= \frac{32a^3}{3}\left(\frac{35\pi}{128} - \frac{5\pi}{32}\right) \div 8a^2\left(\frac{3\pi}{16}\right) = \frac{8a^3}{3}\times\frac{15\pi}{128}\times\frac{16}{8a^2\times 3\pi} = \frac{5a}{24}$$ **Ans.**

## 2.10 CENTRE OF GRAVITY OF AN ARC

**Example 31.** *Find the C.G. of the arc of the curve*

*$x = a\,(\theta + \sin\theta)$, $y = a(1 - \cos\theta)$ in the positive quadrant.*

**Solution.** We know that, $\bar{x} = \dfrac{\int x\,ds}{\int ds}$, $\bar{y} = \dfrac{\int y\,ds}{\int ds}$

Now, $ds = \sqrt{\left(\frac{dx}{d\theta}\right)^2 + \left(\frac{dy}{d\theta}\right)^2}\, d\theta$

$= \sqrt{\{a^2 (1+\cos\theta)^2 + a^2 \sin^2\theta\}}\, d\theta = a\sqrt{1+2\cos\theta+\cos^2\theta+\sin^2\theta}\, d\theta$

$= a\sqrt{1+2\cos\theta+1}\, d\theta = a\sqrt{2(1+\cos\theta)}\, d\theta = a\sqrt{4\cos^2\frac{\theta}{2}}\, d\theta = 2a\cos\frac{\theta}{2}\, d\theta$

$$\bar{x} = \frac{\int x\,dx}{\int ds} = \frac{\int_0^\pi a(\theta+\sin\theta)\, 2a\cos\frac{\theta}{2}\, d\theta}{\int_0^\pi 2a\cos\frac{\theta}{2}\, d\theta} = \frac{a\int_0^\pi\left(\theta + 2\sin\frac{\theta}{2}\cos\frac{\theta}{2}\, d\theta\right)}{\left[2\sin\frac{\pi}{2}\right]_0^\pi}$$

$$= \frac{a}{2}\int_0^\pi\left[\theta\cos\frac{\theta}{2} + 2\sin\frac{\theta}{2}\cos^2\frac{\theta}{2}\right] d\theta = \frac{a}{2}\int_0^{\frac{\pi}{2}}(2t\cos t + 2\sin t\cos^2 t)\, 2\, dt$$

$$= 2a\left[t\sin t + \cos t - \frac{\cos^3 t}{3}\right]_0^{\frac{\pi}{2}} = 2a\left[\frac{\pi}{2} - 1 + \frac{1}{3}\right] = a\left[\pi - \frac{4}{3}\right]$$

$$\bar{y} = \frac{\int y\,ds}{\int ds} = \frac{\int_0^\pi a(1-\cos\theta)\, 2a\cos\frac{\theta}{2}\, d\theta}{\int_0^\pi 2a\cos\frac{\theta}{2}\, d\theta} = \frac{a\int_0^\pi 2\sin^2\frac{\theta}{2}\cos\frac{\theta}{2}\, d\theta}{\int_0^\pi \cos\frac{\theta}{2}\, d\theta}$$

$$= \frac{a\, r\left[\sin^3\frac{\theta}{2}\right]_0^\pi}{3\left[2\sin\frac{\theta}{2}\right]_0^\pi} = \frac{4a}{3\times 2} = \frac{2a}{3}$$ Hence, C.G. of the arc is $\left[a\left(\pi - \frac{4}{3}\right), \frac{2a}{3}\right]$ **Ans.**

## EXERCISE 2.8

**1.** Find the centre of gravity of the area bounded by the parabola $y^2 = x$ and the line $x + y = 2$.

**Ans.** $\left(\frac{8}{5}, -\frac{1}{2}\right)$

**2.** Find the centroid of the tetrahedron bounded by the coordinate planes and the plane $x + y + z = 1$, the density at any point varying as its distance from the face $z = 0$. **Ans.** $\left(\frac{1}{5}, \frac{1}{5}, \frac{2}{5}\right)$

**3.** Find the centroid of the area enclosed by the parabola $y^2 = 4ax$, the axis of $x$ and latus rectum.

**Ans.** $\left(\frac{3a}{20}, \frac{3a}{16}\right)$

**4.** Find the centroid of the loop of curve $r^2 = a^2 \cos 2\theta$. **Ans.** $\left(\frac{\pi a\sqrt{2}}{8}, 0\right)$

**5.** Find the centroid of solid formed by revolving about the $x$-axis that part of the area of the ellipse $\frac{x^2}{a^2} + \frac{y^2}{b^2} = 1$ which lies in the first quadrant. **Ans.** $\left(\frac{3a}{8}, 0\right)$

**6.** Find the average density of the sphere of radius $a$ whose density at a distance $r$ from the centre of the sphere is $\rho = \rho_0\left[1 + k\frac{r^3}{a^3}\right]$. $\rho_0\left(1 + \frac{k}{2}\right)$

**7.** The density at a point on a circular lamina varies as the distance from a point O on the circumference. Show that the C.G. divides the diameter through O in the ratio 3 : 2.

## 1.11 TRIPLE INTEGRATION

Let a function $f(x, y, z)$ be a continuous at every point of a finite region $S$ of three dimensional space. Consider $n$ sub-spaces $\delta S_1, \delta S_2, \delta S_3, \dots \delta S_n$ of the space $S$.

If $(x_r, y_r, z_r)$ be a point in the $r$th subspace.

The limit of the sum $\sum_{r=1}^{n} f(x_r, y_r, z_r)\,\delta S_r$, as $n \to \infty$, $\delta S_r \to 0$ is known as the triple integral of $f(x, y, z)$ over the space $S$.

Symbolically, it is denoted by

$$\iiint_S f(x, y, z)\, dS$$

It can be calculated as $\int_{x_1}^{x_2}\int_{y_1}^{y_2}\int_{z_1}^{z_2} f(x, y, z)\, dx\, dy\, dz$. First we integrate with respect to $z$ treating $x$, $y$ as constant between the limits $z_1$ and $z_2$. The resulting expression (function of $x$, $y$) is integrated with respect to $y$ keeping $x$ as constant between the limits $y_1$ and $y_2$. At the end we integrate the resulting expression (function of $x$ only) within the limits $x_1$ and $x_2$.

$$\int_{x_1=a}^{x_2=b} \Psi(x)\, dx \left[ \int_{y_1=\phi_1(x)}^{y_2=\phi_2(x)} \phi(x, y)\, dy \left[ \int_{z_1=f_1(x,y)}^{z_2=f_2(x,y)} f(x, y, z)\, dz \right]\right]$$

First we integrate from inner most integral w.r.t. $z$, then we integrate with respect to $y$ and finally the outer most with respect to $x$.

But the above order of integration is immaterial provided the limits change accordingly.

**Example 32.** *Evaluate* $\iiint_R (x+y+z)\, dx\, dy\, dz$, *where* $R : 0 \le x \le 1,\ 1 \le y \le 2,\ 2 \le z \le 3$.

**Solution.** $\int_0^1 dx \int_1^2 dy \int_2^3 (x+y+z)\, dz = \int_0^1 dx \int_1^2 dy \left[\frac{(x+y+z)^2}{2}\right]_2^3$

$$= \frac{1}{2}\int_0^1 dx \int_1^2 dy\,[(x+y+3)^2 - (x+y+2)^2] = \frac{1}{2}\int_0^1 dx \int_1^2 (2x+2y+5)\,.\,1.dy$$

$$= \frac{1}{2}\int_0^1 dx \left[\frac{(2x+2y+5)^2}{4}\right]_1^2 = \frac{1}{8}\int_0^1 dx\,[(2x+4+5)^2 - (2x+2+5)^2]$$

$$= \frac{1}{8}\int_0^1 (4x+16)\,.\,2\,dx = \int_0^1 (x+4)\,dx = \left[\frac{x^2}{2}+4x\right]_0^1 = \frac{1}{2}+4 = \frac{9}{2}$$ **Ans.**

**Example 33.** *Evaluate the integral :* $\int_0^{\log 2}\int_0^{x}\int_0^{x+\log y} e^{x+y+z}\, dz\, dy\, dx$.

**Solution.** $\int_0^{\log 2}\int_0^{x}\int_0^{x+\log y} e^{x+y+z}\, dz\, dy\, dx$.

$$= \int_0^{\log 2} e^x dx \int_0^x e^y dy \int_0^{x+\log y} e^z dz = \int_0^{\log 2} e^x\, dx \int_0^x e^y dy (e^z)_0^{x+\log y}$$

$$= \int_0^{\log 2} e^x dx \int_0^x e^y dy\,(e^{x+\log y} - 1) = \int_0^{\log 2} e^x\, dx \int_0^x e^y dy\,(e^{\log y}\,.\,e^x - 1)$$

$$= \int_0^{\log 2} e^x dx \int_0^x e^y\,(y e^x - 1)\, dy = \int_0^{\log 2} e^x dx \left[(ye^x - 1)\,e^y - \int e^x\,.\,e^y\, dy\right]_0^x$$

$$= \int_0^{\log 2} e^x dx \left[(ye^x - 1)\,e^y - e^{x+y}\right]_0^x = \int_0^{\log 2} e^x dx\,[(xe^x - 1)\,e^x - e^{2x} + 1 + e^x]$$

$$= \int_0^{\log 2} e^x dx\,[xe^{2x} - e^x - e^{2x} + 1 + e^x] = \int_0^{\log 2} (xe^{3x} - e^{3x} + e^x)\, dx$$

$$=\left[x\frac{e^{3x}}{3}-\int 1.\frac{e^{3x}}{3}dx-\frac{e^{3x}}{3}+e^x\right]_0^{\log 2}=\left[\frac{x}{3}e^{3x}-\frac{e^{3x}}{9}-\frac{e^{3x}}{3}+e^x\right]_0^{\log 2}$$

$$=\frac{\log 2}{3}e^{3\log 2}-\frac{e^{3\log 2}}{9}-\frac{e^{3\log 2}}{3}+e^{\log 2}+\frac{1}{9}+\frac{1}{3}-1$$

$$=\frac{\log 2}{3}e^{\log 2^3}-\frac{e^{\log 2^3}}{9}-\frac{e^{\log 2^3}}{3}+e^{\log 2}+\frac{1}{9}+\frac{1}{3}-1$$

$$=\frac{8}{3}\log 2-\frac{8}{9}-\frac{8}{3}+2+\frac{1}{9}+\frac{1}{3}-1=\frac{8}{3}\log 2-\frac{19}{9}$$ **Ans.**

**Example 34.** *Evaluate* $\int_0^{\log 2}\int_0^x\int_0^{x+y} e^{x+y+z}\,dx\,dy\,dz.$

(*M.U. II Semester, 2005, 2003, 2002*)

**Solution.** $I=\int_0^{\log 2}\int_0^x e^{x+y}\left[e^z\right]_0^{x+y}dx\,dy$

$$=\int_0^{\log 2}\int_0^x e^{x+y}(e^{x+y}-1)\,dx\,dy=\int_0^{\log 2}\int_0^x\left[e^{2(x+y)}-e^{(x+y)}\right]dx\,dy$$

$$=\int_0^{\log 2}\left[e^{2x}.\frac{e^{2y}}{2}-e^x.e^y\right]_0^x dx=\int_0^{\log 2}\left(\frac{e^{4x}}{2}-e^{2x}-\frac{e^{2x}}{2}+e^x\right)dx$$

$$=\left[\frac{e^{4x}}{8}-\frac{e^{2x}}{2}-\frac{e^{2x}}{4}+e^x\right]_0^{\log 2}=\left[\frac{e^{4\log 2}}{8}-\frac{e^{2\log 2}}{2}-\frac{e^{2\log 2}}{4}+e^{\log 2}\right]-\left(\frac{1}{8}-\frac{1}{2}-\frac{1}{4}+1\right)$$

$$=\left(\frac{e^{\log 16}}{8}-\frac{e^{\log 4}}{2}-\frac{e^{\log 4}}{4}+e^{\log 2}\right)-\left(\frac{1}{8}-\frac{1}{2}-\frac{1}{4}+1\right)$$

$$=\left(\frac{16}{8}-\frac{4}{2}-\frac{4}{4}+2\right)-\left(\frac{1}{8}-\frac{1}{2}-\frac{1}{4}+1\right)=\frac{5}{8}$$ **Ans.**

**Example 35.** *Evaluate* $\iiint_R (x^2+y^2+z^2)\,dx\,dy\,dz$

*where R denotes the region bounded by x = 0, y = 0, z = 0 and x + y + z = a, (a > 0)*

**Solution.** $\iiint_R (x^2+y^2+z^2)\,dx\,dy\,dz$

$x+y+z=a$ or $z=a-x-y$

Upper limit of $z=a-x-y$

On x-y plane, $x+y+z=a$ becomes $x+y=a$ as shown in the figure.

Upper limit of $y=a-x$

Upper limit of $x=a$

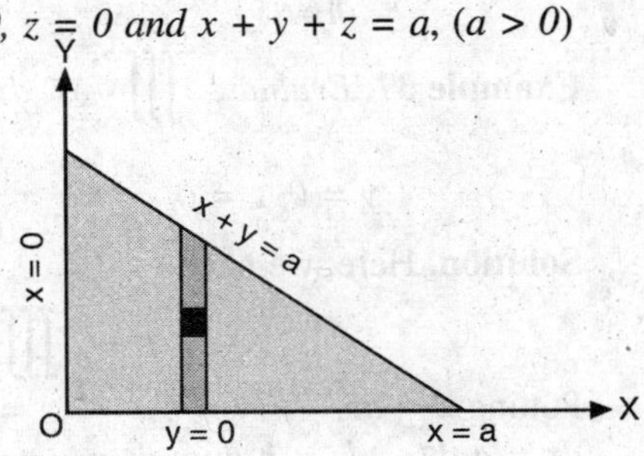


$$=\int_{x=0}^a dx\int_{y=0}^{a-x}dy\int_{z=0}^{a-x-y}(x^2+y^2+z^2)\,dz=\int_0^a dx\int_0^{a-x}dy\left(x^2z+y^2z+\frac{z^3}{3}\right)_0^{a-x-y}$$

$$=\int_0^a dx\int_0^{a-x}dy\left[x^2(a-x-y)+y^2(a-x-y)+\frac{(a-x-y)^3}{3}\right]$$

$$=\int_0^a dx\int_0^{a-x}\left[x^2(a-x)-x^2y+(a-x)\,y^2-y^3+\frac{(a-x-y)^3}{3}\right]dy$$

$$=\int_0^a dx\left[x^2(a-x)\,y-\frac{x^2y^2}{2}+(a-x)\frac{y^3}{3}-\frac{y^4}{4}-\frac{(a-x-y)^4}{12}\right]_0^{a-x}$$

$$= \int_0^a dx\left[x^2(a-x)^2 - \frac{x^2}{2}(a-x)^2 + (a-x)\frac{(a-x)^3}{3} - \frac{(a-x)^4}{4} + \frac{(a-x)^4}{12}\right]$$

$$= \int_0^a \left[\frac{x^2}{2}(a-x)^2 + \frac{(a-x)^4}{6}\right] dx = \int_0^a \left[\frac{1}{2}(a^2x^2 - 2ax^3 + x^4) + \frac{(a-x)^4}{6}\right] dx$$

$$= \left[\frac{1}{2}a^2\frac{x^3}{3} - \frac{ax^4}{4} + \frac{x^5}{10} - \frac{(a-x)^5}{30}\right]_0^a = \frac{a^5}{6} - \frac{a^5}{4} + \frac{a^5}{10} + \frac{a^5}{30} = \frac{a^5}{20}$$ **Ans.**

**Example 36.** *Compute* $\iiint \frac{dx\,dy\,dz}{(x+y+z+1)^3}$ *if the region of integration is bounded by the coordinate planes and the plane* $x + y + z = 1$. *(M.U., II Semester 2007, 2006)*

**Solution.** Let the given region be $R$, then $R$ is expressed as

$$0 \le z \le 1 - x - y, \quad 0 \le y \le 1 - x, \quad 0 \le x \le 1.$$

$$\iiint_R \frac{dx\,dy\,dz}{(x+y+z+1)^3} = \int_0^1 dx \int_0^{1-x} dy \int_0^{1-x-y} \frac{dz}{(x+y+z+1)^3}$$

$$= \int_0^1 dx \int_0^{1-x} dy \left[\frac{1}{-2(x+y+z+1)^2}\right]_0^{1-x-y}$$

$$= -\frac{1}{2}\int_0^1 dx \int_0^{1-x} dy \left[\frac{1}{(x+y+1-x-y+1)^2} - \frac{1}{(x+y+1)^2}\right]$$

$$= -\frac{1}{2}\int_0^1 dx \int_0^{1-x} \left[\frac{1}{4} - \frac{1}{(x+y+1)^2}\right] dy = -\frac{1}{2}\int_0^1 dx \left[\frac{y}{4} + \frac{1}{x+y+1}\right]_0^{1-x}$$

$$= -\frac{1}{2}\int_0^1 dx \left[\frac{1-x}{4} + \frac{1}{x+1+1-x} - \frac{1}{x+1}\right] = -\frac{1}{2}\int_0^1 \left[\frac{1-x}{4} + \frac{1}{2} - \frac{1}{x+1}\right] dx$$

$$= -\frac{1}{2}\left[-\frac{(1-x)^2}{8} + \frac{x}{2} - \log(x+1)\right]_0^1 = -\frac{1}{2}\left[\frac{1}{2} - \log 2 + \frac{1}{8}\right] = -\frac{1}{2}\left[\frac{5}{8} - \log 2\right]$$

$$= \frac{1}{2}\log 2 - \frac{5}{16}$$ **Ans.**

**Example 37.** *Evaluate* $\iiint x^2yz\ dx\,dy\,dz$ *throughout the volume bounded by the planes* $x = 0$, $y = 0$, $z = 0$, $\frac{x}{a} + \frac{y}{b} + \frac{z}{c} = 1$. *(M.U. II Semester 2003, 2002, 2001)*

**Solution.** Here, we have

$$I = \iiint x^2yz\ dx\,dy\,dz \quad ...(1)$$

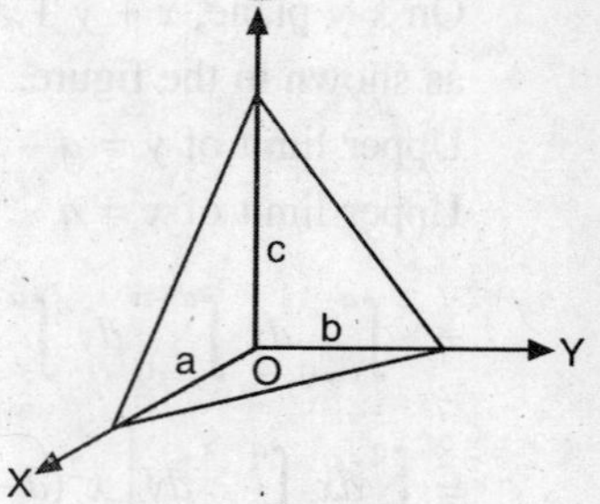


Putting $x = au$, $\quad y = bv$, $\quad z = cw$

$dx = a\,du$, $\quad dy = b\,dv$, $\quad dz = c\,dw$ in (1), we get

$$I = \iiint a^2bc\,u^2vw\,a\,bc\,du\,dv\,dw$$

Limits are for $u = 0, 1$ for $v = 0, 1 - u$ and for $w = 0, 1 - u - v$

$$u + v + w = 1$$

$$I = \int_{u=o}^1 \int_{v=0}^{1-u} \int_{w=0}^{1-u-v} a^3b^2c^2\,u^2vw\,du\,dv\,dw = \int_0^1 \int_0^{1-u} a^3b^2c^2\,u^2v\left[\frac{w^2}{2}\right]_0^{1-u-v} du\,dv$$

$$= \frac{a^3b^2c^2}{2}\int_0^1 \int_0^{1-u} u^2v(1-u-v)^2\,du\,dv$$

$$= \frac{a^3b^2c^2}{2}\int_0^1 \int_0^{1-u} u^2v\left[(1-u)^2 - 2(1-u)v + v^2\right] du\,dv$$

$$= \frac{a^3b^2c^2}{2}\int_0^1\int_0^{1-u} u^2[(1-u)^2 v - 2(1-u)v^2 + v^3]\, du\, dv$$

$$= \frac{a^3b^2c^2}{2}\int_0^1 u^2\left[(1-u)^2\frac{v^2}{2} - 2(1-u)\frac{v^3}{3} + \frac{v^4}{4}\right]_0^{1-u} du$$

$$= \frac{a^3b^2c^2}{2}\int_0^1 u^2\left[\frac{(1-u)^4}{2} - \frac{2(1-u)^4}{3} + \frac{(1-u)^4}{4}\right] du$$

$$= \frac{a^3b^2c^2}{2}\int_0^1 \frac{u^2(1-u)^4}{12}\, du = \frac{a^3b^2c^2}{24}\int_0^1 u^{3-1}(1-u)^{5-1}\, du$$

$$= \frac{a^3b^2c^2}{24}\beta(3,5) = \frac{a^3b^2c^2}{24}\cdot\frac{\overline{|3}\,\overline{|5}}{\overline{|8}} = \frac{a^3b^2c^2}{24}\cdot\left(\frac{2!4!}{7!}\right) = \frac{a^3b^2c^2}{2520}.$$ **Ans.**

## 2.12 INTEGRATION BY CHANGE OF CARTESIAN COORDINATES INTO SPHERICAL COORDINATES

Sometime it becomes easy to integrate by changing the cartesian coordinates into spherical coordinates.

The relations between the cartesian and spherical polar co-ordinates of a point are given by the relations

$$\begin{aligned} x &= r\sin\theta\cos\phi \\ y &= r\sin\theta\sin\phi \\ z &= r\cos\theta \\ dx\,dy\,dz &= |J|\,dr\,d\theta\,d\phi \\ &= r^2\sin\theta\,dr\,d\theta\,d\phi \end{aligned}$$

**Note. 1.** Spherical coordinates are very useful if the expression $x^2 + y^2 + z^2$ is involved in the problem.

**2.** In a sphere $x^2 + y^2 + z^2 = a^2$ the limits of $r$ are 0 and $a$ and limits of $\theta$ are 0, $\pi$ and that of $\phi$ are 0 and $2\pi$.

**Example 38.** *Evaluate the integral* $\iiint (x^2 + y^2 + z^2)\, dx\, dy\, dz$ *taken over the volume enclosed by the sphere* $x^2 + y^2 + z^2 = 1$.

**Solution.** Let us convert the given integral into spherical polar co-ordinates. By putting

$$x = r\sin\theta\cos\phi\,;\quad y = r\sin\theta\sin\phi\,;\quad z = r\cos\theta$$

$$\iiint (x^2+y^2+z^2)\, dx\, dy\, dz = \int_0^{2\pi}\int_0^{\pi}\int_0^1 r^2(r^2\sin\theta\, d\theta\, d\phi\, dr)$$

$$= \int_0^{2\pi} d\phi\int_0^{\pi}\sin\theta\, d\theta\int_0^1 r^4 dr = \int_0^{2\pi} d\phi\int_0^{\pi}\sin\theta\, d\theta\left(\frac{r^5}{5}\right)_0^1 = \frac{1}{5}\int_0^{2\pi} d\phi\,[-\cos\theta]_0^{\pi} = \frac{2}{5}\int_0^{2\pi} d\phi$$

$$= \frac{2}{5}(\phi)_0^{2\pi} = \frac{4\pi}{5}$$ **Ans.**

**Example 39.** *Evaluate* $\iiint (x^2 + y^2 + z^2)\, dx\, dy\, dz$ *over the first octant of the sphere* $x^2 + y^2 + z^2 = a^2$. *(M.U. II Semester 2007)*

**Solution.** Here, we have

$$I = \iiint (x^2 + y^2 + z^2)\, dx\, dy\, dz \qquad ...(1)$$

Putting $x = r\sin\theta\cos\phi$, $y = r\sin\theta\sin\phi$, $z = r\cos\theta$ and $dx\,dy\,dz = r^2\sin\theta\, dr\, d\theta\, d\phi$ in (1), we get

Limits of $r$ are 0, $a$ for $\theta$ are 0, $\frac{\pi}{2}$ for $\phi$ are 0, $\frac{\pi}{2}$.

$$I = \int_0^{\frac{\pi}{2}} \int_0^{\frac{\pi}{2}} \int_0^a r^2 . r^2 \sin\theta \, dr \, d\theta \, d\phi = \int_0^{\frac{\pi}{2}} d\phi \int_0^{\frac{\pi}{2}} \sin\theta \, d\theta \int_0^a r^4 \, dr$$

$$\left( \begin{array}{c} x^2 + y^2 + z^2 = r^2 \sin^2\theta \cos^2\phi + r^2 \sin^2\theta \sin^2\phi + r^2 \cos^2\theta \\ = r^2 \sin^2\theta + r^2 \cos^2\theta = r^2 \end{array} \right)$$

$$= [\phi]_0^{\pi/2} [-\cos\theta]_0^{\pi/2} \left[\frac{r^5}{5}\right]_0^a = \frac{\pi}{2} . (1) . \frac{a^5}{5} = \pi . \frac{a^5}{10}.$$ **Ans.**

**Example 40.** *Evaluate* $\iiint \frac{dx\,dy\,dz}{x^2 + y^2 + z^2}$ *throughout the volume of the sphere* $x^2 + y^2 + z^2 = a^2$.

(*M.U. II Semester 2002, 2001*)

**Solution.** Here, we have

$$I = \iiint \frac{dx\,dy\,dz}{x^2 + y^2 + z^2} \qquad \ldots(1)$$

Putting $x = r\sin\theta\cos\phi$, $y = r\sin\theta\sin\phi$, $z = r\cos\theta$ and $dx\,dy\,dz = r^2\sin\theta\,dr\,d\theta\,d\phi$ in (1), we get

The limits of $r$ are 0 and $a$, for $\theta$ are 0 and $\frac{\pi}{2}$ for $\phi$ are 0 and $\frac{\pi}{2}$ in first octant.

$$I = 8 \int_0^{\frac{\pi}{2}} \int_0^{\frac{\pi}{2}} \int_0^a \frac{r^2 \sin\theta \, dr \, d\theta \, d\phi}{r^2}$$ [Sphere $x^2 + y^2 + z^2$ lies in 8 quadrants]

$$I = 8 \int_0^{\frac{\pi}{2}} d\phi \int_0^{\frac{\pi}{2}} \sin\theta \, d\theta \int_0^a dr = 8 [\phi]_0^{\pi/2} [-\cos\theta]_0^{\pi/2} [r]_0^a = 8\left(\frac{\pi}{2} - 0\right)(0 + 1)(a + 0)$$

$$= 8\frac{\pi}{2} . 1 . a = 4\pi a$$ **Ans.**

## EXERCISE 2.9

**Evaluate the following :**

1. $\int_{-1}^{1} \int_{-2}^{2} \int_{-3}^{3} dx\,dy\,dz$ (*M.U., II Semester 2002*) **Ans.** 48
2. $\int_0^4 \int_0^x \int_0^{x+y} z\,dz\,dy\,dx$ (*R.G.P.V. Bhopal I Sem. 2003*) **Ans.** 70
3. $\int_1^2 \int_0^1 \int_{-1}^1 (x^2 + y^2 + z^2)\,dx\,dy\,dz$ **Ans.** 6
4. $\int_0^1 \int_0^1 \int_0^1 (x^2 + y^2 + z^2)\,dz\,dy\,dx$ (*AMIETE, June 2006*) **Ans.** 1
5. $\int_{-1}^1 \int_0^z \int_{x-z}^{x+z} (x - y + z)\,dx\,dy\,dz$ (*AMIETE, Summer 2004*) **Ans.** 0
6. $\iiint_R (x - y - z)\,dx\,dy\,dz$, where $R : 1 \le x \le 2;\ 2 \le y \le 3;\ 1 \le z \le 3$ **Ans.** 2
7. $\int_0^2 \int_1^3 \int_1^2 xy^2z\,dx\,dy\,dz$ (*AMIETE, Dec. 2007*) **Ans.** 26
8. $\int_0^1 dx \int_0^2 dy \int_1^2 x^2 yz\,dz$ **Ans.** 1
9. $\iiint x^2yz\ dx\,dy\,dz$ throughout the volume bounded by $x = 0$, $y = 0$, $z = 0$, $x + y + z = 1$.

   (*M.U. II Semester, 2003*) **Ans.** $\frac{1}{2520}$
10. $\int_0^1 \int_0^{1-x} \int_0^{1-x^2-y^2} dz\,dy\,dx$ **Ans.** $\frac{1}{3}$
11. $\int_1^e \int_1^{\log y} \int_1^{e^x} \log z\,dz\,dx\,dy$ **Ans.** $\frac{1}{2}(e^2 - 8e + 13)$

12. $\iiint_T y\,dx\,dy\,dz,$ where $T$ is the region bounded by the surfaces $x = y^2$, $x = y + 2$, $4z = x^2 + y^2$ and $z = y + 3$. *(AMIETE Dec. 2008)*

13. $\int_0^2 \int_0^x \int_0^{2x+2y} e^{x+y+z}\,dz\,dy\,dx$ **Ans.** $\frac{1}{3}\left[\frac{e^{12}}{6}-\frac{e^6}{3}-\frac{1}{6}+\frac{1}{3}\right]-\frac{1}{2}[e^4-1]+[e^2-1]$ *(M.U. II Sem., 2003)*

14. $\iiint (x + y + z)\;dx\;dy\;dz$ over the tetrahedron bounded by the planes $x = 0$, $y = 0$, $z = 0$ and $x + y + z = 1$. **Ans.** $\frac{1}{8}$

15. $\int_0^a \int_0^{a-x} \int_0^{a-x-y} x^2\,dx\,dy\,dz$ **Ans.** $\frac{a^5}{60}$ 16. $\int_{-2}^{2} \int_{-\sqrt{(4-x^2)/2}}^{\sqrt{(4-x^2)/2}} \int_{x^2+3y^2}^{8-x^2-y^2} dz\,dy\,dx$ **Ans.** $8\sqrt{2}\pi$

17. $\int_{-1}^{1} \int_0^z \int_{x-z}^{x+z} (x + y + z)\,dz\,dx\,dy$ *(M.U. II Semester, 2000, 02)* **Ans.** 0

18. $\int_0^2 \int_0^y \int_{x-y}^{x+y} (x + y + z)\,dx\,dy\,dz$ *(M.U. II Semester 2004)* **Ans.** 16

19. $\iiint \sqrt{1-\frac{x^2}{a^2}-\frac{y^2}{b^2}-\frac{z^2}{c^2}}\;dx\;dy\;dz$ throughout the volume of the ellipsoid $\frac{x^2}{a^2}+\frac{y^2}{b^2}+\frac{z^2}{c^2}=1.$

**Ans.** $\frac{\pi^2}{4}abc$

20. $\iiint \sqrt{\frac{x^2}{a^2}+\frac{y^2}{b^2}+\frac{z^2}{c^2}}\;dx\;dy\;dz$ over the volume of the ellipsoid $\frac{x^2}{a^2}+\frac{y^2}{b^2}+\frac{z^2}{c^2}=1.$ **Ans.** $\frac{4\pi}{3}abc$

21. $\iiint x^{l-1} y^{m-1} z^{n-1}\;dx\;dy\;dz$ throughout the volume of the tetrahedron

$x \ge 0,\ y \ge 0,\ z \ge 0,\ x + y + z \le 1.$

**Ans.** $\frac{1}{(l+m+n)}\cdot\frac{\overline{|l}\,\overline{|m}\,\overline{|n}}{\overline{|l+m+n}}$

22. $\iiint \frac{dx\,dy\,dz}{\sqrt{1-x^2-y^2-z^2}}$ taken throughout the volume of the sphere $x^2 + y^2 + z^2 = 1$, lying in the first octant. **Ans.** $\frac{\pi^2}{8}$

23. $\int_0^{\pi} 2d\theta \int_0^{a(1+\cos\theta)} r\,dr \int_0^h \left[1-\frac{r}{a(1+\cos\theta)}\right] dz$ **Ans.** $\frac{\pi a^2}{2}h$

24. $\int_0^{\pi/2} \int_0^{a\sin\theta} \int_0^{(a^2-r^2)/a} r\,d\theta\,dr\,dz$ **Ans.** $\frac{5a^3}{64}$

25. $\iiint z^2\,dx\,dy\,dz$ over the volume common to the sphere $x^2 + y^2 + z^2 = a^2$ and the cylinder $x^2 + y^2 + z^2 = ax$. **Ans.** $\frac{2a^5\pi}{15}$

26. $\iiint_V \frac{dx\,dy\,dz}{(1+x^2+y^2+z^2)^2}$ where $V$ is the volume in the first octant. **Ans.** $\frac{\pi^2}{8}$

27. $\iiint \frac{dx\,dy\,dz}{(x^2+y^2+z^2)^{3/2}}$ over the volume bounded by the spheres $x^2 + y^2 + z^2 = 16$ and $x^2 + y^2 + z^2 = 25$. *(M.U. II Semester, 2001, 03)* **Ans.** $4\pi \log(5/4)$

28. $\iiint_T z^2\,dx\,dy\,dz$ over the volume bounded by the cylinder $x^2 + y^2 = a^2$ and the paraboloid $x^2 + y^2 = z$ and the plane $z = 0$. **Ans.** $\frac{\pi a^8}{12}$

## 2.13 VOLUME = $\iiint dx\, dy\, dz.$

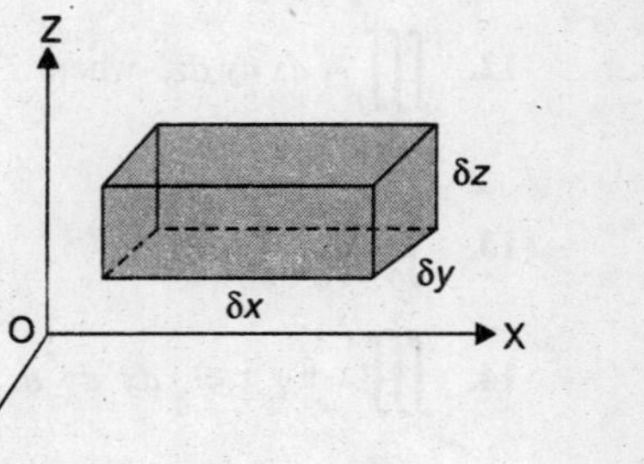


The elementary volume $\delta v$ is $\delta x \,.\, \delta y \,.\, \delta z$ and therefore the volume of the whole solid is obtained by evaluating the triple integral.

$$\delta V = \delta x\ \delta y\ \delta z$$

$$\boxed{V = \iiint dx\ dy\ dz.}$$

**Note :** (*i*) Mass = volume × density = $\iiint \rho\ dx\ dy\ dz$ if $\rho$ is the density.

(*ii*) In cylindrical co-ordinates, we have $\boxed{V = \iiint_V r\ dr\ d\varphi\ dz}$

(*iii*) In spherical polar co-ordinates, we have $\boxed{V = \iiint_V r^2 \sin\theta dr\ d\theta\ d\varphi}$

**Example 41.** *Find the volume of the tetrahedron bounded by the planes $x = 0$, $y = 0$, $z = 0$ and $x + y + z = a$.* (*M.U. II Semester, 2005, 2000*)

**Solution.** Here, we have a solid which is bounded by $x = 0$, $y = 0$, $z = 0$ and $x + y + z = a$ planes.

The limits of $z$ are 0 and $a - x - y$, the limits of $y$ are 0 and $1 - x$, the limits of $x$ are 0 and $a$.

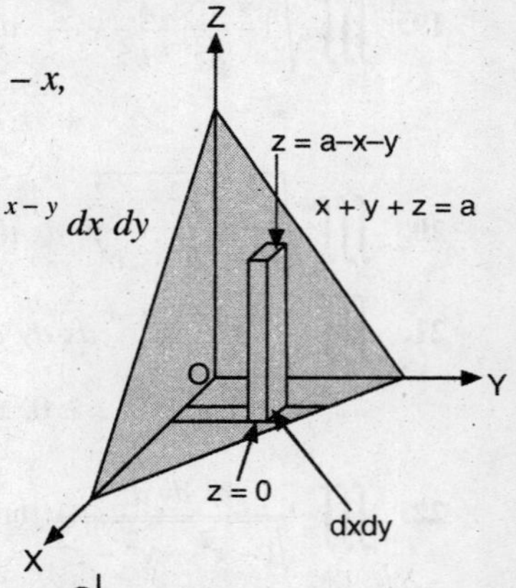


$$V = \int_{x=0}^{a} \int_{y=0}^{a-x} \int_{z=0}^{a-x-y} dx\ dy\ dz = \int_{x=0}^{a} \int_{y=0}^{a-x} [z]_0^{a-x-y}\ dx\ dy$$

$$= \int_{x=0}^{a} \int_{y=0}^{a-x} (a - x - y)\ dx\ dy$$

$$= \int_{x=0}^{a} \left[ay - xy - \frac{y^2}{2}\right]_0^{a-x} dx$$

$$= \int_0^a \left[a(a-x) - x(a-x) - \frac{(a-x)^2}{2}\right] dx$$

$$= \int_0^a \left[a^2 - ax - ax + x^2 - \frac{a^2}{2} + ax - \frac{x^2}{2}\right] dx$$

$$= \int_0^a \left(\frac{a^2}{2} - ax + \frac{x^2}{2}\right) dx$$

$$= \left[\frac{a^2}{2}.x - \frac{ax^2}{2} + \frac{x^3}{6}\right]_0^a = a^3\left(\frac{1}{2} - \frac{1}{2} + \frac{1}{6}\right) = \frac{a^3}{6}. \quad \textbf{Ans.}$$

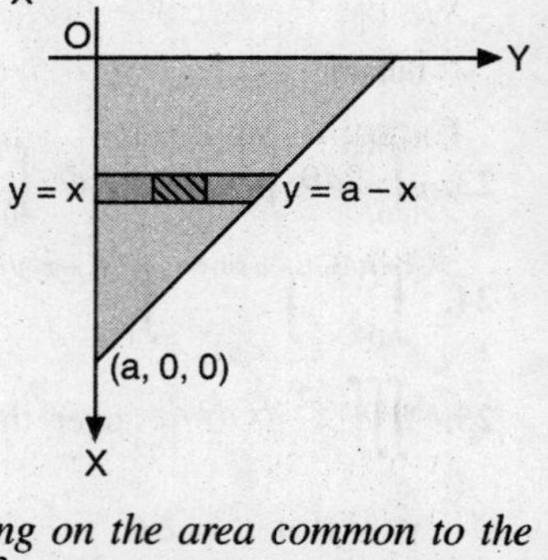


**Example 42.** *Find the volume of the cylindrical column standing on the area common to the parabolas $y^2 = x$, $x^2 = y$ and cut off by the surface $z = 12 + y - x^2$.* (*U.P., II Sem., Summer 2001*)

**Solution.** We have,

$$y^2 = x$$
$$x^2 = y$$
$$z = 12 + y - x^2$$

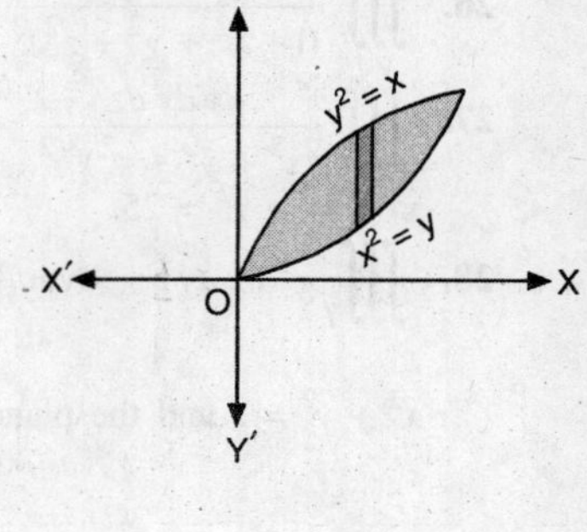


$$V = \int_0^1 dx \int_{x^2}^{\sqrt{x}} dy \int_0^{12+y-x^2} dz = \int_0^1 dx \int_{x^2}^{\sqrt{x}} (12 + y - x^2)\ dy$$

$$= \int_0^1 dx \left(12y + \frac{y^2}{2} - x^2 y\right)_{x^2}^{\sqrt{x}}$$

$$= \int_0^1 \left(12\sqrt{x} + \frac{x}{2} - x^{5/2} - 12x^2 - \frac{x^4}{2} + x^4\right) dx$$

$$= \left[\frac{2}{3}\times 12x^{3/2} + \frac{x^2}{4} - \frac{2}{7}x^{7/2} - 4x^3 - \frac{x^5}{10} + \frac{x^5}{5}\right]_0^1$$

$$= 8 + \frac{1}{4} - \frac{2}{7} - 4 - \frac{1}{10} + \frac{1}{5} = 4 + \frac{1}{4} - \frac{2}{7} - \frac{1}{10} + \frac{1}{5} = \frac{560 + 35 - 40 - 14 + 28}{140} = \frac{569}{140}$$ **Ans.**

**Example 43.** *A triangular prism is formed by planes whose equations are ay = bx, y = 0 and x = a. Find the volume of the prism between the planes z = 0 and surface z = c + xy.*

*(M.U. II Semester 2000; U.P., Ist Semester, 2009 (C.O) 2003)*

**Solution.** Required volume $= \int_0^a \int_0^{\frac{bx}{a}} \int_0^{c+xy} dz\, dy\, dx$

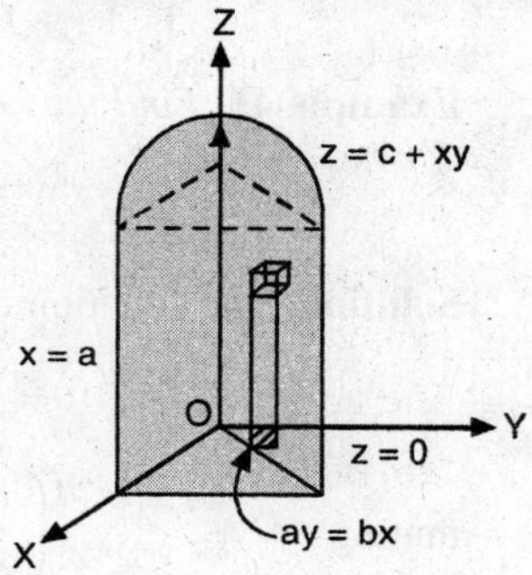


$$= \int_0^a \int_0^{\frac{bx}{a}} (c + xy)\, dy\, dx$$

$$= \int_0^a \left(cy + \frac{xy^2}{2}\right)_0^{\frac{bx}{a}} dx$$

$$= \int_0^a \left(\frac{cbx}{a} + \frac{b^2}{2a^2}x^3\right) dx = \frac{bc}{a}\left(\frac{x^2}{2}\right)_0^a + \frac{b^2}{2a^2}\left(\frac{x^4}{4}\right)_0^a$$

$$= \frac{abc}{2} + \frac{b^2a^2}{8} = \frac{ab}{8}(4c + ab)$$ **Ans.**

## 2.14 VOLUME OF SOLID BOUNDED BY SPHERE OR BY CYLINDER

We use spherical coordinates $(r, \theta, \phi)$ and the cylindrical coordinates are $(\rho, \phi, z)$ and the relations are $x = \rho \cos \phi$, $y = \rho \sin \phi$.

**Example 44.** *Find the volume of a solid bounded by the spherical surface $x^2 + y^2 + z^2 = 4a^2$ and the cylinder $x^2 + y^2 - 2ay = 0$.*

**Solution.** $x^2 + y^2 + z^2 = 4a^2$ ...(1)

$x^2 + y^2 - 2ay = 0$ ...(2)

Considering the section in the positive quadrant of the *xy*-plane and taking *z* to be positive (that is volume above the *xy*-plane) and changing to polar co-ordinates, (1) becomes

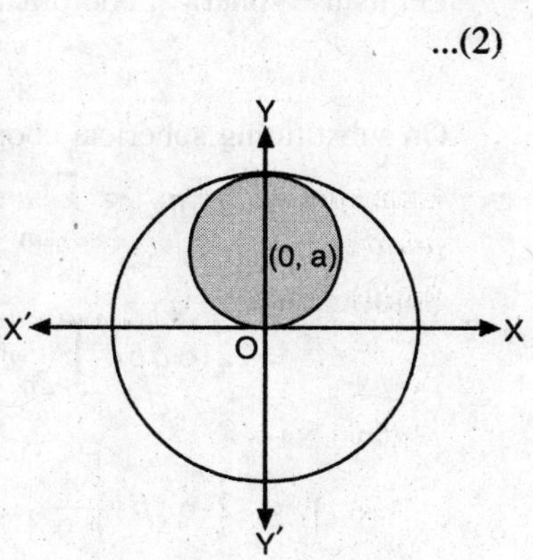


$$r^2 + z^2 = 4a^2 \quad \Rightarrow \quad z^2 = 4a^2 - r^2$$

$$\therefore \quad z = \sqrt{4a^2 - r^2}$$

(2) becomes $r^2 - 2ar\sin\theta = 0 \Rightarrow r = 2a\sin\theta$

$$\text{Volume} = \iiint dx\, dy\, dz$$

$$= 4\int_0^{\pi/2} d\theta \int_0^{2a\sin\theta} r\, dr \int_0^{\sqrt{4a^2 - r^2}} dz$$ (Cylindrical coordinates)

$$= 4\int_0^{\pi/2} d\theta \int_0^{2a\sin\theta} r\,dr\,[z]_0^{\sqrt{4a^2-r^2}} = 4\int_0^{\pi/2} d\theta \int_0^{2a\sin\theta} r\,dr\,.\sqrt{4a^2-r^2}$$

$$= 4\int_0^{\pi/2} d\theta \left[-\frac{1}{3}(4a^2-r^2)^{3/2}\right]_0^{2a\sin\theta} = \frac{4}{3}\int_0^{\pi/2}\left[-(4a^2-4a^2\sin^2\theta)^{3/2}+8a^3\right] d\theta$$

$$= \frac{4}{3}\int_0^{\pi/2}(-8a^3\cos^3\theta + 8a^3)\,d\theta = \frac{8\times 4a^3}{3}\int_0^{\pi/2}(1-\cos^3\theta)\,d\theta$$

$$= \frac{32a^3}{3}\int_0^{\pi/2}\left(1-\frac{1}{4}\cos 3\theta - \frac{3}{4}\cos\theta\right) d\theta$$

$$= \frac{32a^3}{3}\left[\theta - \frac{1}{12}\sin 3\theta - \frac{3}{4}\sin\theta\right]_0^{\pi/2} = \frac{32a^3}{3}\left(\frac{\pi}{2}+\frac{1}{12}-\frac{3}{4}\right) = \frac{32a^3}{3}\left[\frac{\pi}{2}-\frac{2}{3}\right]$$ **Ans.**

**Example 45.** *Find the volume enclosed by the solid*

$$\left(\frac{x}{a}\right)^{2/3} + \left(\frac{y}{b}\right)^{2/3} + \left(\frac{z}{c}\right)^{2/3} = 1$$

**Solution.** The equation of the solid is

$$\left(\frac{x}{a}\right)^{2/3} + \left(\frac{y}{b}\right)^{2/3} + \left(\frac{z}{c}\right)^{2/3} = 1$$

Putting
$$\left(\frac{x}{a}\right)^{1/3} = u \quad\Rightarrow\quad x = a\,u^3 \quad\Rightarrow\quad dx = 3\,au^2\,du$$

$$\left(\frac{y}{b}\right)^{1/3} = v \quad\Rightarrow\quad y = b\,v^3 \quad\Rightarrow\quad dy = 3\,bv^2\,dv$$

$$\left(\frac{z}{c}\right)^{1/3} = w \quad\Rightarrow\quad z = c\,w^3 \quad\Rightarrow\quad dz = 3\,c\,w^2\,dw$$

The equation of the solid becomes

$$u^2 + v^2 + w^2 = 1 \qquad \text{...(1)}$$

$$V = \iiint dx\,dy\,dz \qquad \text{...(2)}$$

On putting the values of $dx$, $dy$ and $dz$ in (2), we get

$$V = \iiint 27abc\,u^2v^2w^2 du\,dv\,dw \qquad \text{...(3)}$$

(1) represents a sphere.
Let us use spherical coordinates.

$$u = r\sin\theta\cos\phi, \qquad v = r\sin\theta\sin\phi,$$

$$w = r\cos\theta, \qquad du\,dv\,dw = r^2\sin\theta\,dr\,d\theta\,d\phi$$

On substituting spherical coordinates in (3), we have

$$V = 27abc\,.\,8\int_{r=0}^{1}\int_{\phi=0}^{\pi/2}\int_{\theta=0}^{\pi/2} r^2\sin^2\theta\cos^2\phi\,.\,r^2\sin^2\theta\sin^2\phi\,.\,r^2\cos^2\theta\,.\,r^2\sin\theta\,dr\,d\theta\,d\phi$$

$$= 216\,abc\int_{r=0}^{1} r^8\,dr\int_{\phi=0}^{\pi/2}\sin^2\phi\cos^2\phi\,d\phi\int_{\theta=0}^{\pi/2}\sin^5\theta\cos^2\theta\,d\theta$$

$$= 216abc\left[\frac{r^9}{9}\right]_0^1 .\left(\frac{\overline{|\tfrac{3}{2}}\,\overline{|\tfrac{3}{2}}}{2\,\overline{|3}}\right)\left(\frac{\overline{|3}\,\overline{|\tfrac{3}{2}}}{2\,\overline{|\tfrac{9}{2}}}\right) = 24\,abc\,.\frac{1}{2}.\frac{\overline{|\tfrac{3}{2}}\,\overline{|\tfrac{3}{2}}}{\overline{|3}}.\frac{1}{2}.\frac{\overline{|3}\,\overline{|\tfrac{3}{2}}}{\overline{|\tfrac{9}{2}}}$$

$$= 6abc.\frac{\left[\left(\frac{1}{2}\right)\sqrt{\frac{1}{2}}\right]^2}{2!}.\frac{2!\sqrt{\frac{3}{2}}}{\left(\frac{7}{2}\right)\left(\frac{5}{2}\right)\frac{3}{2}\sqrt{\frac{3}{2}}} = 6abc.\frac{1}{4}.\pi\frac{1}{\left(\frac{7}{2}\right)\left(\frac{5}{2}\right)\left(\frac{3}{2}\right)} = \frac{4}{35}abc\,\pi$$

**Ans.**

**Example 46.** *Find the volume bounded above by the sphere $x^2 + y^2 + z^2 = a^2$ and below by the cone $x^2 + y^2 = z^2$.* (*U.P. II Semester 2002*)

**Solution.** The equation of the sphere is $x^2 + y^2 + z^2 = a^2$ ...(1)

and that of the cone is $x^2 + y^2 = z^2$ ...(2)

In polar coordinates $x = r\sin\theta\cos\phi,\ y = r\sin\theta\sin\phi,\ z = r\cos\theta$

The equation (1) in polar co-ordinates is

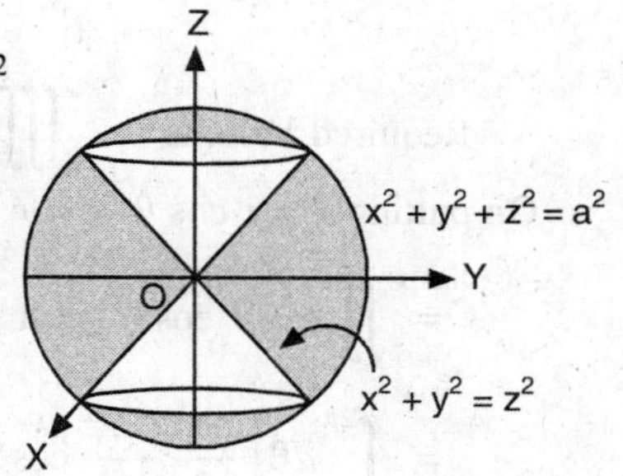


$$(r\sin\theta\cos\phi)^2 + (r\sin\theta\sin\phi)^2 + (r\cos\theta)^2 = a^2$$

$$\Rightarrow \quad r^2\sin^2\theta\cos^2\phi + r^2\sin^2\theta\sin^2\phi + r^2\cos^2\theta = a^2$$

$$\Rightarrow \quad r^2\sin^2\theta(\cos^2\phi + \sin^2\phi) + r^2\cos^2\theta = a^2$$

$$\Rightarrow \quad r^2\sin^2\theta + r^2\cos^2\theta = a^2$$

$$\Rightarrow \quad r^2(\sin^2\theta + \cos^2\theta) = a^2$$

$$\Rightarrow \quad r^2 = a^2 \Rightarrow r = a$$

The equation (2) in polar co-ordinates is

$$(r\sin\theta\cos\phi)^2 + (r\sin\theta\sin\phi)^2 = (r\cos\theta)^2$$

$$\Rightarrow \quad r^2\sin^2\theta(\cos^2\phi + \sin^2\phi) = r^2\cos^2\theta \Rightarrow r^2\sin^2\theta = r^2\cos^2\theta$$

$$\Rightarrow \quad \tan^2\theta = 1 \Rightarrow \tan\theta = 1 \Rightarrow \quad \theta = \pm\frac{\pi}{4}$$

Thus equations (1) and (2) in polar coordinates are respectively,

$$r = a \quad \text{and} \quad \theta = \pm\frac{\pi}{4}$$

The volume in the first octant is one fourth only.

Limits in the first octant : $r$ varies 0 to $a$, $\theta$ from 0 to $\frac{\pi}{4}$ and $\phi$ from 0 to $\frac{\pi}{2}$.

The required volume lies between $x^2 + y^2 + z^2 = a^2$ and $x^2 + y^2 = z^2$.

$$V = 4\int_0^{\frac{\pi}{2}}\int_0^{\frac{\pi}{4}}\int_0^a r^2\sin\theta\,dr\,d\theta\,d\phi = 4\int_0^{\frac{\pi}{2}}d\phi\int_0^{\frac{\pi}{4}}\sin\theta\,d\theta\left[\frac{r^3}{3}\right]_0^a$$

$$= 4\int_0^{\frac{\pi}{2}}d\phi\int_0^{\frac{\pi}{4}}\sin\theta\,d\theta.\frac{a^3}{3} = \frac{4a^3}{3}\int_0^{\frac{\pi}{2}}d\phi\,[-\cos\theta]_0^{\frac{\pi}{4}} = \frac{4a^3}{3}(\phi)_0^{\frac{\pi}{2}}\left[-\frac{1}{\sqrt{2}}+1\right]$$

$$= \frac{2}{3}\pi a^3\left(1-\frac{1}{\sqrt{2}}\right)$$

**Ans.**

## 2.15 VOLUME OF SOLID BOUNDED BY CYLINDER OR CONE

We use cylindrical coordinates $(r, \theta, z)$.

**Example 47.** *Find the volume of the solid bounded by the parabolic $y^2 + z^2 = 4x$ and the plane $x = 5$.*

**Solution.** $y^2 + z^2 = 4x,\ x = 5$

$$V = \int_0^5 dx\int_{-2\sqrt{x}}^{2\sqrt{x}} dy\int_{-\sqrt{4x-y^2}}^{\sqrt{4x-y^2}} dz = 4\int_0^5 dx\int_0^{2\sqrt{x}} dy\int_0^{\sqrt{4x-y^2}} dz$$

$$= 4\int_0^5 dx \int_0^{2\sqrt{x}} dy\,[z]_0^{\sqrt{4x-y^2}} = 4\int_0^5 dx \int_0^{2\sqrt{x}} dy\,\sqrt{4x-y^2}$$

$$= 4\int_0^5 dx\left[\frac{y}{2}\sqrt{4x-y^2}+\frac{4x}{2}\sin^{-1}\frac{y}{2\sqrt{x}}\right]_0^{2\sqrt{x}} = 4\int_0^5\left[0+2x\left(\frac{\pi}{2}\right)\right]dx = 4\pi\int_0^5 x\,dx$$

$$= 4\pi\left[\frac{x^2}{2}\right]_0^5 = 50\pi \qquad \textbf{Ans.}$$

**Example 48.** *Calculate the volume of the solid bounded by the following surfaces :*

$z = 0,\quad x^2 + y^2 = 1,\quad x + y + z = 3$

**Solution.**
$$x^2 + y^2 = 1 \qquad ...(1)$$
$$x + y + z = 3 \qquad ...(2)$$
$$z = 0 \qquad ...(3)$$

$$\text{Required Volume} = \iiint dx\,dy\,dz = \iint dx\,dy\,[z]_0^{3-x-y} = \iint (3-x-y)\,dx\,dy$$

On putting $x = r\cos\theta$, $y = r\sin\theta$, $dx\,dy = r\,d\theta\,dr$, we get

$$= \iint (3 - r\cos\theta - r\sin\theta)\,r\,d\theta\,dr = \int_0^{2\pi} d\theta \int_0^1 (3r - r^2\cos\theta - r^2\sin\theta)\,dr$$

$$= \int_0^{2\pi} d\theta\left(\frac{3r^2}{2}-\frac{r^3}{3}\cos\theta-\frac{r^3}{3}\sin\theta\right)_0^1 = \int_0^{2\pi}\left(\frac{3}{2}-\frac{1}{3}\cos\theta-\frac{1}{3}\sin\theta\right)d\theta$$

$$= \left[\frac{3}{2}\theta-\frac{1}{3}\sin\theta+\frac{1}{3}\cos\theta\right]_0^{2\pi} = 3\pi-\frac{1}{3}\sin 2\pi+\frac{1}{3}\cos 2\pi-\frac{1}{3} = 3\pi \qquad \textbf{Ans.}$$

**Example 49.** *Find the volume bounded by the cylinder $x^2 + y^2 = 4$ and the planes $y + z = 4$ and $z = 0$.*

**Solution.** $x^2 + y^2 = 4 \Rightarrow y = \pm\sqrt{4-x^2}$

$y + z = 4 \Rightarrow z = 4 - y$ and $z = 0$

$x$ varies from $-2$ to $+2$.

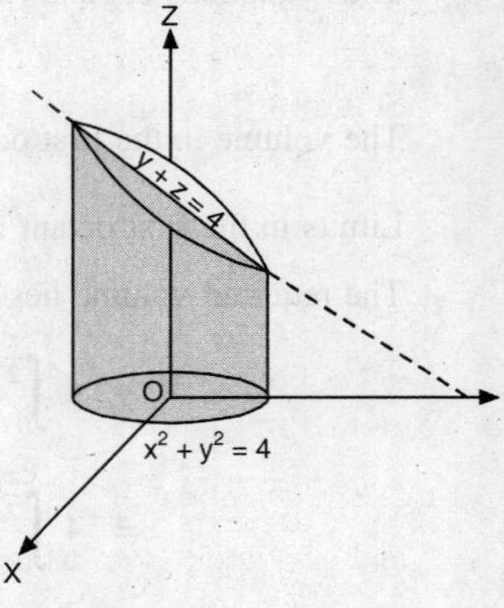


$$V = \iiint dx\,dy\,dz = \int_{-2}^{2} dx \int_{-\sqrt{4-x^2}}^{\sqrt{4-x^2}} dy \int_0^{4-y} dz$$

$$= \int_{-2}^{2} dx \int_{-\sqrt{4-x^2}}^{\sqrt{4-x^2}} dy\,[z]_0^{4-y}$$

$$= \int_{-2}^{2} dx \int_{-\sqrt{4-x^2}}^{\sqrt{4-x^2}} dy\,(4-y) = \int_{-2}^{2} dx\left[4y-\frac{y^2}{2}\right]_{-\sqrt{4-x^2}}^{\sqrt{4-x^2}}$$

$$= \int_{-2}^{2} dx\left[4\sqrt{4-x^2}-\frac{1}{2}(4-x^2)+4\sqrt{4-x^2}+\frac{1}{2}(4-x^2)\right]$$

$$= 8\int_{-2}^{2}\sqrt{4-x^2}\,dx = 8\left[\frac{x}{2}\sqrt{4-x^2}+\frac{4}{2}\sin^{-1}\frac{x}{2}\right]_{-2}^{2} = 16\pi \qquad \textbf{Ans.}$$

**Example 50.** *Find the volume in the first octant bounded by the cylinder $x^2 + y^2 = 2$ and the planes $z = x + y$, $y = x$, $z = 0$ and $x = 0$.* (*M.U. II Semester 2005*)

**Solution.** Here, we have the solid bounded by

$$x^2 + y^2 = 2 \text{ (cylinder)}$$
$$(\text{or } r^2 = 2)$$
$$z = x + y \quad \Rightarrow \quad z = r(\cos\theta + \sin\theta) \quad \text{(plane)}$$
$$y = x \quad \Rightarrow r\sin\theta = r\cos\theta \quad \text{(plane)}$$
$$\Rightarrow \quad \tan\theta = 1 \quad \Rightarrow \quad \theta = \frac{\pi}{4}$$

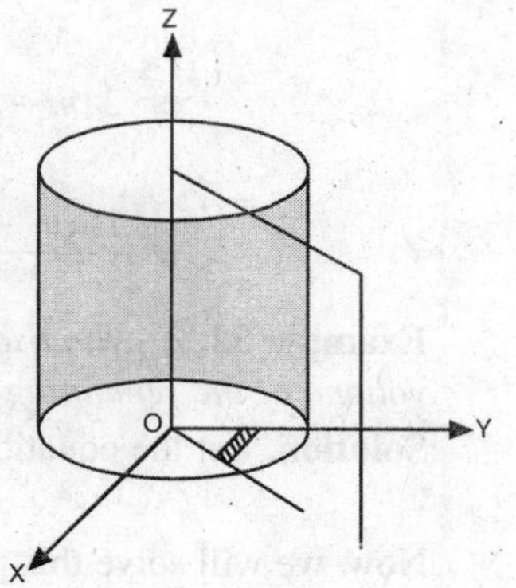


$$x = 0 \Rightarrow r\cos\theta = 0 \Rightarrow \cos\theta = 0 \Rightarrow \theta = \frac{\pi}{2}$$

$z$ varies from 0 to $r(\cos\theta + \sin\theta)$

$r$ varies from 0 to $\sqrt{2}$

$\theta$ varies from $\frac{\pi}{4}$ to $\frac{\pi}{2}$

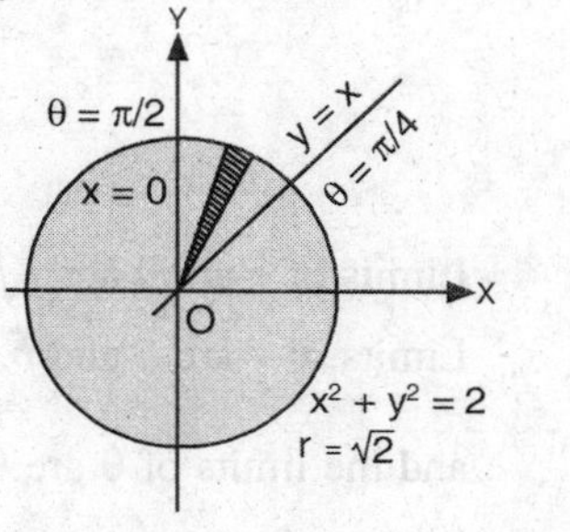


$$\therefore \quad V = \int_{\theta=\pi/4}^{\pi/2} \int_{r=0}^{\sqrt{2}} \int_{z=0}^{r(\cos\theta+\sin\theta)} r\, dr\, d\theta\, dz$$

$$= \int_{\theta=\pi/4}^{\pi/2} \int_{r=0}^{\sqrt{2}} r[z]_0^{r(\cos\theta+\sin\theta)}\, dr\, d\theta$$

$$= \int_{\theta=\pi/4}^{\pi/2} \int_{r=0}^{\sqrt{2}} r^2(\cos\theta + \sin\theta)\, dr\, d\theta$$

$$= \int_{\theta=\pi/4}^{\pi/2} (\cos\theta + \sin\theta)\left[\frac{r^3}{3}\right]_0^{\sqrt{2}} d\theta = \frac{2\sqrt{2}}{3}\int_{\theta=\pi/4}^{\pi/2} (\cos\theta + \sin\theta)\, d\theta$$

$$= \frac{2\sqrt{2}}{3}[\sin\theta - \cos\theta]_{\pi/4}^{\pi/2} = \frac{2\sqrt{2}}{3}\left[(1-0) - \left(\frac{1}{\sqrt{2}} - \frac{1}{\sqrt{2}}\right)\right] = \frac{2\sqrt{2}}{3} \qquad \textbf{Ans.}$$

**Example 51.** *Show that the volume of the wedge intercepted between the cylinder $x^2 + y^2 = 2ax$ and planes $z = mx$, $z = nx$ is $\pi(m - n)\, a^3$.* *(M.U. II Semester, 2000)*

**Solution.** The equation of the cylinder is $x^2 + y^2 = 2\,a\,x$

we convert the cartesian coordinates into cylindrical coordinates.

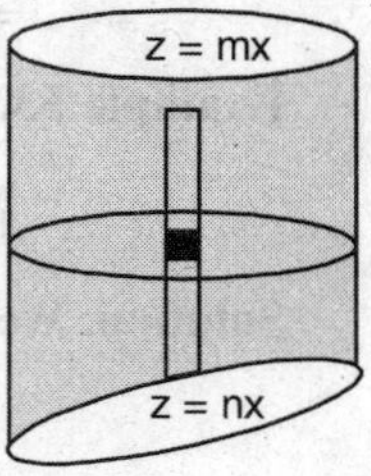


$$x = r\cos\theta$$
$$y = r\sin\theta$$
$$x^2 + y^2 = 2ax \quad \Rightarrow \quad r^2 = 2ar\cos\theta$$
$$\Rightarrow \quad r = 2a\cos\theta$$

$r$ varies from 0 to $2a\cos\theta$

$\theta$ varies from $-\frac{\pi}{2}$ to $\frac{\pi}{2}$

and $z$ varies from $z = nx$ ($z = nr\cos\theta$) to $z = m\,x$ ($z = m\,r\cos\theta$)

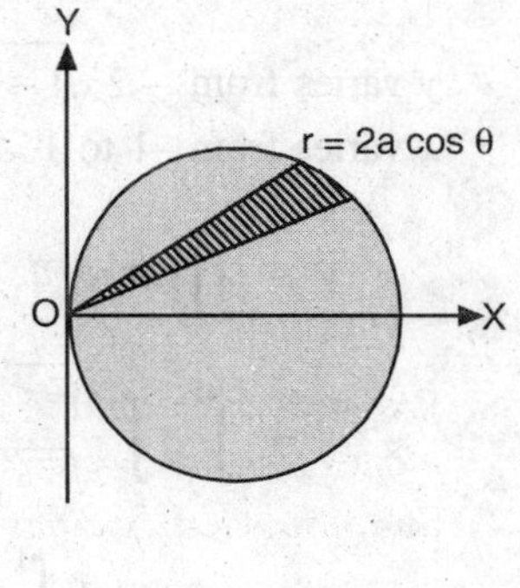


$$V = 2\int_{\theta=0}^{\pi/2} \int_{r=0}^{2a\cos\theta} \int_{z=nr\cos\theta}^{mr\cos\theta} r\, dr\, d\theta\, dz$$

$$= 2\int_{\theta=0}^{\pi/2} \int_{r=0}^{2a\cos\theta} r[z]_{nr\cos\theta}^{mr\cos\theta}\, dr\, d\theta$$

$$= 2\int_{\theta=0}^{\pi/2} \int_{r=0}^{2a\cos\theta} r.(m-n)\, r\cos\theta\, dr\, d\theta$$

$$= 2(m-n)\int_{\theta=0}^{\pi/2} \int_{r=0}^{2a\cos\theta} r^2\cos\theta\, dr\, d\theta$$

$$= 2(m-n)\int_{\theta=0}^{\pi/2}\left[\frac{r^3}{3}\right]_0^{2a\cos\theta}\cos\theta\,d\theta = 2(m-n)\int_{\theta=0}^{\pi/2}\frac{8a^3}{3}\cos^3\theta\cos\theta\,d\theta$$

$$= \frac{16(m-n)}{3}a^3\int_{\theta=0}^{\pi/2}\cos^4\theta\,d\theta = \frac{16(m-n)}{3}.a^3.\frac{3}{4}.\frac{1}{2}.\frac{\pi}{2} = (m-n)\pi a^3 \quad \textbf{Ans.}$$

**Example 52.** *A cylindrical hole of radius b is bored through a sphere of radius a. Find the volume of the remaining solid.* (*M.U. II Semester 2004*)

**Solution.** Let the equation of the sphere be

$$x^2 + y^2 + z^2 = a^2$$

Now, we will solve this problem using cylindrical coordinates

$$x = r\cos\theta$$
$$y = r\sin\theta$$
$$z = z$$

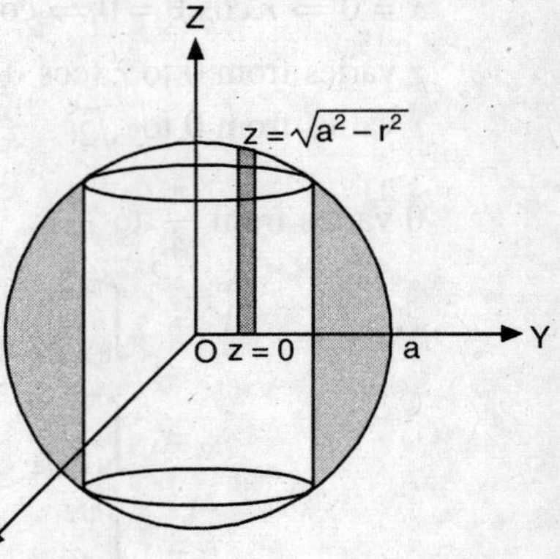


Limits of $z$ are 0 and $\sqrt{a^2-(x^2+y^2)}$ *i.e.*, $\sqrt{a^2-r^2}$

Limits of $r$ are $a$ and $b$.

and the limits of $\theta$ are 0 and $\frac{\pi}{2}$

$$V = 8\int_{\theta=0}^{\pi/2}\int_{r=b}^{a}\int_{z=0}^{\sqrt{a^2-r^2}} r\,dr\,d\theta\,dz = 8\int_{\theta=0}^{\pi/2}\int_{r=b}^{a}[z]_0^{\sqrt{a^2-r^2}}\,r\,dr\,d\theta$$

$$= 8\int_{\theta=0}^{\pi/2}\int_{r=b}^{a}(a^2-r^2)^{1/2}.r\,dr\,d\theta$$

$$= 8\int_{\theta=0}^{\pi/2}\left[\frac{(a^2-r^2)^{3/2}}{3/2}.\left(-\frac{1}{2}\right)\right]_b^a d\theta = -\frac{8}{3}\int_0^{\pi/2}-(a^2-b^2)^{\frac{3}{2}}\,d\theta$$

$$= \frac{8}{3}(a^2-b^2)^{\frac{3}{2}}[\theta]_0^{\pi/2} = \frac{4\pi}{3}(a^2-b^2)^{\frac{3}{2}} \quad \textbf{Ans.}$$

**Example 53.** *Find the volume cut off from the paraboloid*

$x^2 + \frac{y^2}{4} + z = 1$ *by the plane* $z = 0$. (*M.U. II Semester 2005*)

**Solution.** We have

$$x^2 + \frac{y^2}{4} + z = 1 \quad \text{(Paraboloid)} \quad ...(1)$$

$$z = 0 \quad (x\text{-}y \text{ plane}) \quad ...(2)$$

$z$ varies from 0 to $1 - x^2 - \frac{y^2}{4}$

$y$ varies from $-2\sqrt{1-x^2}$ to $2\sqrt{1-x^2}$

$x$ varies from $-1$ to 1.

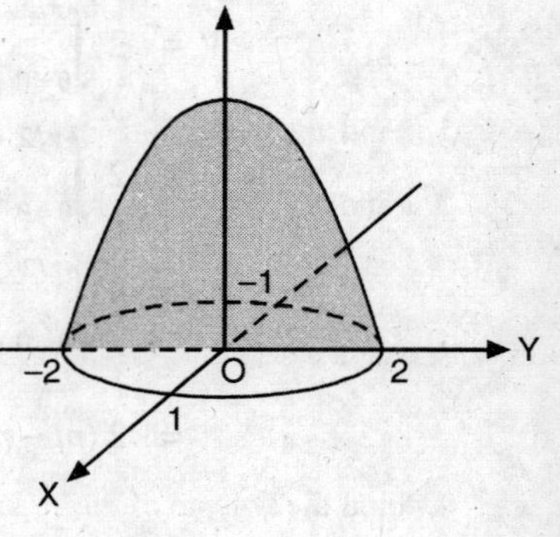


$$V = \iiint dx\,dy\,dz = \int_{-1}^{1}dx\int_{-2\sqrt{1-x^2}}^{2\sqrt{1-x^2}}dy\int_0^{\left(1-x^2-\frac{y^2}{4}\right)}dz$$

$$= \int_{-1}^{1}\int_{-2\sqrt{1-x^2}}^{2\sqrt{1-x^2}}\left(1-x^2-\frac{y^2}{4}\right)dx\,dy$$

$$= 4\int_0^1\int_0^{2\sqrt{1-x^2}}\left(1-x^2-\frac{y^2}{4}\right)dx\,dy$$

$$= 4\int_0^1 \left[(1-x^2)\,y - \frac{y^3}{12}\right]_0^{2\sqrt{1-x^2}} .\,dx$$

$$= 4\int_0^1 \left[(1-x^2)\,.\,2\sqrt{1-x^2} - \frac{8}{12}(1-x^2)^{3/2}\right] dx$$

$$= 4\int_0^1 \left[2\,(1-x^2)^{3/2} - \frac{2}{3}(1-x^2)^{3/2}\right] d\,x$$

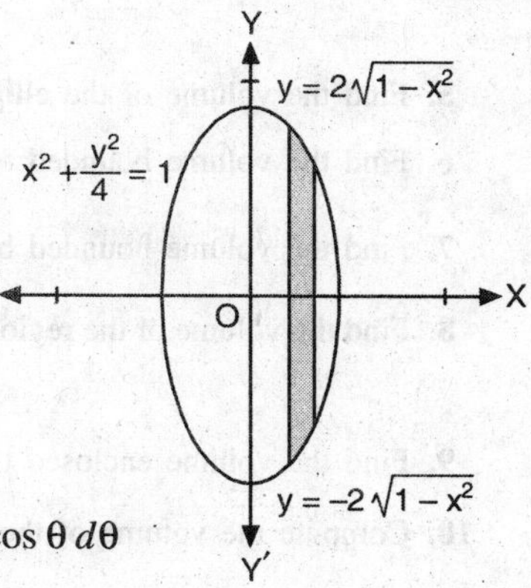


On putting $x = \sin\theta$, we get

$$V = 4\int_0^1 \frac{4}{3}(1-x^2)^{3/2}\,dx = \frac{16}{3}\int_0^{\pi/2} (-\sin^2\theta)^{3/2}\cos\theta\,d\theta$$

$$= \frac{16}{3}\int_0^{\pi/2} \cos^4\theta\,d\theta = \frac{16}{3}.\frac{3}{4}.\frac{1}{2}.\frac{\pi}{2} = \pi \qquad \textbf{Ans.}$$

**Example 54.** *Find the volume enclosed between the cylinders $x^2 + y^2 = a\,x$, and $z^2 = a\,x$.*

**Solution.** Here, we have $x^2 + y^2 = ax$ ...(1)

$z^2 = ax$ ...(2)

$$V = \iiint dx\,dy\,dz$$

$$= \int_0^a dx \int_{-\sqrt{ax-x^2}}^{\sqrt{ax-x^2}} dy \int_{-\sqrt{ax}}^{\sqrt{ax}} dz = 2\int_0^a dx \int_{-\sqrt{a^2-x^2}}^{\sqrt{ax-x^2}} dy \int_0^{\sqrt{ax}} dz$$

$$= 2\int_0^a dx \int_{-\sqrt{ax-x^2}}^{\sqrt{ax-x^2}} dy\,(z)_0^{\sqrt{ax}} = 2\int_0^a dx \int_{-\sqrt{ax-x^2}}^{\sqrt{ax-x^2}} dy\,\sqrt{ax} = 2\int_0^a \sqrt{ax}\,dx\,[y]_{-\sqrt{ax-x^2}}^{\sqrt{ax-x^2}}$$

$$= 2\int_0^a \sqrt{ax}\,dx\left(2\sqrt{ax-x^2}\right) = 4\sqrt{a}\int_0^a x\sqrt{a-x}\,dx$$

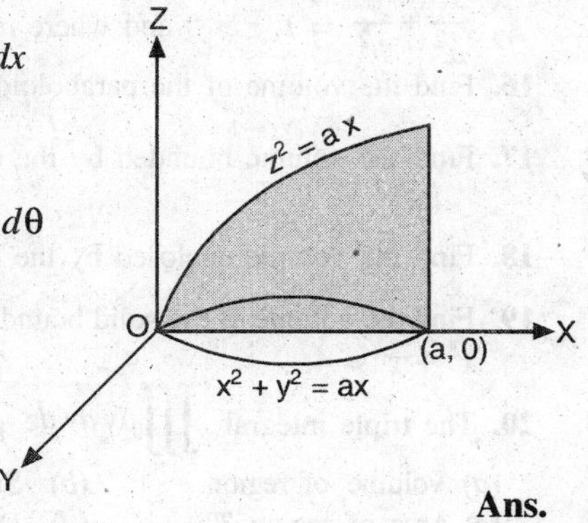


Putting $x = a\sin^2\theta$ so that $dx = 2a\sin\theta\cos\theta\,d\theta$, we get

$$V = 4\sqrt{a}\int_0^{\pi/2} a\sin^2\theta\,\sqrt{a - a\sin^2\theta}\,.\,2a\sin\theta\cos\theta\,d\theta$$

$$= 8a^3\int_0^{\pi/2} \sin^3\theta\cos^2\theta\,d\theta$$

$$= 8a^3\,\frac{\Gamma 2\,\Gamma\frac{3}{2}}{2\,\Gamma\frac{7}{2}} = 4a^3\,\frac{\Gamma\frac{3}{2}}{\frac{5}{2}.\frac{3}{2}\Gamma\frac{3}{2}} = \frac{16a^3}{15} \qquad \textbf{Ans.}$$

## EXERCISE 2.10

**1.** Find the volume bounded by the coordinate planes and the plane. $\frac{x}{a} + \frac{y}{b} + \frac{z}{c} = 1$ **Ans.** $\frac{abc}{6}$

**2.** Find the volume bounded by the cylinders $y^2 = x$ and $x^2 = y$ between the planes $z = 0$ and $x + y + z = 2$. **Ans.** $\frac{11}{30}$

**3.** Find the volume bounded by the co-ordinate planes and the plane.
$l\,x + m\,y + n\,z = 1$ *(A.M.I.E.T.E. Winter 2001)* **Ans.** $\frac{1}{6\,l\,m\,n}$

**4.** Find the volume of the sphere $x^2 + y^2 + z^2 = a^2$ by triple integration. *(AMIETE, June 2009)* **Ans.** $\frac{4}{3}\pi a^3$

**5.** Find the volume of the ellipsoid $\frac{x^2}{a^2}+\frac{y^2}{b^2}+\frac{z^2}{c^2}=1$ **Ans.** $\frac{4\pi abc}{3}$

**6.** Find the volume bounded by the cylinder $x^2+y^2=a^2$ and the planes $y+z=2a$ and $z=0$. *(M.U. II Semester 2000, 02, 06)* **Ans.** $2\pi a^3$

**7.** Find the volume bounded by the cylinder $x^2+y^2=a^2$ and the planes $z=0$ and $y+z=b$. **Ans.** $\pi a^2 b$

**8.** Find the volume of the region bounded by $z=x^2+y^2$, $z=0$, $x=-a$, $x=a$ and $y=-a$, $y=a$. **Ans.** $\frac{8}{3}a^4$

**9.** Find the volume enclosed by the cylinder $x^2+y^2=9$ and the planes $x+z=5$ and $z=0$. **Ans.** $45\pi-36$

**10.** Compute the volume of the solid bounded by $x^2+y^2=z$, $z=2x$.*(A.M.I.E., Summer 2000)* **Ans.** $2\pi$

**11.** Find the volume cut from the paraboloid $4z=x^2+y^2$ by plane $z=4$. *(U.P. I Semester, Dec. 2005)* **Ans.** $32\pi$

**12.** By using triple integration find the volume cut off from the sphere $x^2+y^2+z^2=16$ by the plane $z=0$ and the cylinder $x^2+y^2=4x$. **Ans.** $\frac{64}{9}(3\pi-4)$

**13.** The sphere $x^2+y^2+z^2=a^2$ is pierced by the cylinder $x^2+y^2=a^2(x^2-y^2)$.

Prove that the volume of the sphere that lies inside the cylinder is $\frac{8}{3}\left[\frac{\pi}{4}+\frac{5}{3}-\frac{4\sqrt{2}}{3}\right]a^3$.

**14.** Find the volume of the solid bounded by the surfaces $z=0$, $3z=x^2+y^2$ and $x^2+y^2=9$. *(A.M.I.E.T.E., Summer 2005)* **Ans.** $\frac{27\pi}{2}$

**15.** Obtain the volume bounded by the surface $z=c\left(1-\frac{x}{a}\right)\left(1-\frac{y}{b}\right)$ and a quadrant of the elliptic cylinder $\frac{x^2}{a^2}+\frac{y^2}{b^2}=1$, $z>0$ and where $a, b>0$. **Ans.** $\pi abc$ *(A.M.I.E.T.E., Dec. 2005)*

**16.** Find the volume of the paraboloid $x^2+y^2=4z$ cut off by the plane $z=4$. **Ans.** $32\pi$

**17.** Find the volume bounded by the cone $z^2=x^2+y^2$ and the paraboloid $z=x^2+y^2$. **Ans.** $\frac{\pi}{6}$

**18.** Find the volume enclosed by the cylinders $x^2+y^2=2ax$ and $z^2=2ax$. **Ans.** $\frac{128a^3}{15}$

**19.** Find the volume of the solid bounded by the plane $z=0$, the paraboloid $z=x^2+y^2+2$ and the cylinder $x^2+y^2=4$. **Ans.** $16\pi$

**20.** The triple integral $\iiint dx\,dy\,dz$ gives

(*a*) Volume of region (*b*) Surface area of region T
(*c*) Area of region *T* (*d*) Density of region *T*. *(A.M.I.E.T.E., Dec. 2006, 2002)* **Ans.** (*a*)

## 2.16 SURFACE AREA

Let $z=f(x,y)$ be the surface *S*. Let its projection on the *x-y* plane be the region *A*. Consider an element $8x$. $\delta y$ in the region *A*. Erect a cylinder on the element $\delta x$. $\delta y$ having its generator parallel to *OZ* and meeting the surface *S* in an element of area $\delta s$.

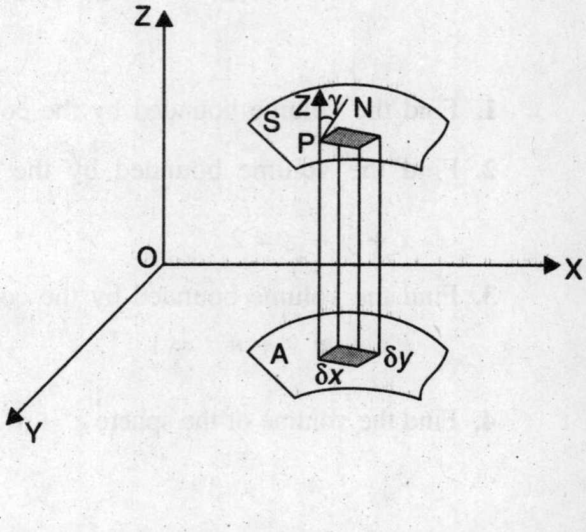


$$\therefore \quad \delta x\,\delta y=\delta s\cos\gamma,$$

Where $\gamma$ is the angle between the *xy*-plane and the tangent plane to *S* at *P*, *i.e.*, it is the angle between the *Z*-axis and the normal to *S* at *P*.

The direction cosines of the normal to the surface $F(x, y, z) = 0$ are proportional to

$$\frac{\partial F}{\partial x}, \frac{\partial F}{\partial y}, \frac{\partial F}{\partial z}$$

$\therefore$ The direction of the normal to $S$ $[F = f(x, y) - z]$ are proportional to $-\frac{\partial z}{\partial x}, -\frac{\partial z}{\partial y}, 1$ and those of the $Z$-axis are 0, 0 , 1.

$$\text{Direction cosines} = \frac{-\frac{\partial z}{\partial x}}{\sqrt{\left(\frac{\partial z}{\partial x}\right)^2 + \left(\frac{\partial z}{\partial y}\right)^2 + 1}}, \frac{-\frac{\partial z}{\partial y}}{\sqrt{\left(\frac{\partial z}{\partial x}\right)^2 + \left(\frac{\partial z}{\partial y}\right)^2 + 1}}, \frac{1}{\sqrt{\left(\frac{\partial z}{\partial x}\right)^2 + \left(\frac{\partial z}{\partial y}\right)^2 + 1}},$$

Hence
$$\cos\gamma = \frac{1}{\sqrt{\left[\left(\frac{\partial z}{\partial x}\right)^2 + \left(\frac{\partial z}{\partial y}\right)^2 + 1\right]}} \qquad (\cos\theta = l_1 l_2 + m_1 m_2 + n_1 n_2)$$

$$\delta S = \frac{\delta x\,\delta y}{\cos\gamma} = \sqrt{\left[\left(\frac{\partial z}{\partial x}\right)^2 + \left(\frac{\partial z}{\partial y}\right)^2 + 1\right]}\,\delta x\,\delta y\,; \quad S = \iint_A \sqrt{\left[\left(\frac{\partial z}{\partial x}\right)^2 + \left(\frac{\partial z}{\partial y}\right)^2 + 1\right]}\,dx\,dy$$

**Example 55.** *Find the surface area of the cylinder $x^2 + z^2 = 4$ inside the cylinder $x^2 + y^2 = 4$.*

**Solution.** $x^2 + y^2 = 4$

$$2x + 2z\frac{\partial z}{\partial x} = 0 \text{ or } \frac{\partial z}{\partial x} = -\frac{x}{z}, \quad \frac{\partial z}{\partial y} = 0$$

$$\left(\frac{\partial z}{\partial x}\right)^2 + \left(\frac{\partial z}{\partial y}\right)^2 + 1 = \frac{x^2}{z^2} + 1 = \frac{x^2 + z^2}{z^2} = \frac{4}{4 - x^2}$$

Hence, the required surface area $= 8\int_0^2\int_0^{\sqrt{4-x^2}} \sqrt{\left[\left(\frac{\partial z}{\partial x}\right)^2 + \left(\frac{\partial z}{\partial y}\right)^2 + 1\right]}\,dx\,dy$

$$= 8\int_0^2\int_0^{\sqrt{4-x^2}} \frac{2}{\sqrt{4-x^2}}\,dx\,dy = 16\int_0^2 \frac{1}{\sqrt{4-x^2}}[y]_0^{\sqrt{4-x^2}}\,dx = 16\int_0^2 \frac{1}{\sqrt{4-x^2}}[\sqrt{4-x^2}\,]\,dx$$

$$= 16\int_0^2 dx = 16\,(x)_0^2 = 32$$

**Ans.**

**Example 56.** *Find the surface area of the sphere $x^2 + y^2 + z^2 = 9$ lying inside the cylinder $x^2 + y^2 = 3y$.*

**Solution.**
$$x^2 + y^2 + z^2 = 9$$
$$2x + 2z\frac{\partial z}{\partial x} = 0, \quad \frac{\partial z}{\partial x} = -\frac{x}{z}$$
$$2x + 2z\frac{\partial z}{\partial y} = 0, \quad \frac{\partial z}{\partial y} = -\frac{y}{z}$$

$$\left[\left(\frac{\partial z}{\partial x}\right)^2 + \left(\frac{\partial z}{\partial y}\right)^2 + 1\right] = \frac{x^2}{z^2} + \frac{y^2}{z^2} + 1 = \frac{x^2 + y^2 + z^2}{z^2} = \frac{9}{9 - x^2 - y^2} = \frac{9}{9 - r^2} \quad \begin{bmatrix} x = r\cos\theta \\ y = r\sin\theta \end{bmatrix}$$

$$x^2 + y^2 = 3y \text{ or } r^2 = 3r\sin\theta \text{ or } r = 3\sin\theta.$$

Hence, the required surface area

$$= \iint \sqrt{\left[\left(\frac{\partial z}{\partial x}\right)^2 + \left(\frac{\partial z}{\partial y}\right)^2 + 1\right]}\,dx\,dy = 4\int_0^{\pi/2}\int_0^{3\sin\theta} \frac{3}{\sqrt{9-r^2}}\,r\,d\theta\,dr = 12\int_0^{\pi/2} d\theta \int_0^{3\sin\theta} \frac{r\,dr}{\sqrt{9-r^2}}$$

$$= 12\int_0^{\pi/2} d\theta\,[-\sqrt{9-r^2}\,]_0^{3\sin\theta} = 12\int_0^{\pi/2} [-\sqrt{9 - 9\sin^2\theta} + 3]\,d\theta$$

$$= 36\int_0^{\pi/2}(-\cos\theta+1)\,d\theta = 36(-\sin\theta+\theta)_0^{\pi/2} = 36\left(-1+\frac{\pi}{2}\right) = 18(\pi-2)$$ **Ans.**

**Example 57.** *Find the surface area of the section of the cylinder $x^2 + y^2 = a^2$ made by the plane $x + y + z = a$.*

**Solution.**
$$x^2 + y^2 = a^2 \quad \ldots (1)$$
$$x + y + z = a \quad \ldots (2)$$

The projection of the surface area on *xy*-plane is a circle
$x^2 + y^2 = a^2$

$$1+\frac{\partial z}{\partial x} = 0 \quad \text{or} \quad \frac{\partial z}{\partial x} = -1$$
$$1+\frac{\partial z}{\partial y} = 0 \quad \text{or} \quad \frac{\partial z}{\partial y} = -1$$
$$\sqrt{\left(\frac{\partial z}{\partial x}\right)^2+\left(\frac{\partial z}{\partial y}\right)^2+1} = \sqrt{(-1)^2+(-1)^2+1} = \sqrt{3}$$

Hence the required surface area

$$= 4\int_0^a\int_0^{\sqrt{a^2-x^2}}\sqrt{\left(\frac{\partial z}{\partial x}\right)^2+\left(\frac{\partial z}{\partial y}\right)^2+1}\;dx\,dy = 4\int_0^a\int_0^{\sqrt{a^2-x^2}}\sqrt{3}\,dx\cdot dy$$

$$= 4\sqrt{3}\int_0^a[y]_0^{\sqrt{a^2-x^2}}\,dx = 4\sqrt{3}\int_0^a\sqrt{a^2-x^2}\,dx$$

$$= 4\sqrt{3}\left[\frac{x}{2}\sqrt{a^2-x^2}+\frac{a^2}{2}\sin^{-1}\frac{x}{a}\right]_0^a = 4\sqrt{3}\left[0+\frac{a^2}{2}\frac{\pi}{2}\right] = 4\sqrt{3}\left(\frac{a^2\pi}{4}\right) = \sqrt{3}\,\pi a^2$$ **Ans.**

**Example 58.** *Find the area of that part of the surface of the paraboloid of the paraboloid $y^2 + z^2 = 2\,ax$, which lies between the cylinder, $y^2 = ax$ and the plane $x = a$.*

**Solution.**
$$y^2 + z^2 = 2\,ax \quad \ldots (1)$$
$$y^2 = ax \quad \ldots (2)$$
$$x = a \quad \ldots (3)$$

Differentiating (1), we get

$$2z\frac{\partial z}{\partial x} = 2a, \quad \frac{\partial z}{\partial x} = \frac{a}{z}$$
$$2y+2z\frac{\partial z}{\partial y} = 0, \quad \frac{\partial z}{\partial y} = -\frac{y}{z}$$

$$\left(\frac{\partial z}{\partial x}\right)^2+\left(\frac{\partial z}{\partial y}\right)^2+1 = \frac{a^2}{z^2}+\frac{y^2}{z^2}+1 = \frac{a^2+y^2}{z^2}+1 \qquad \left[\begin{array}{l} y^2+z^2 = 2\,ax \\ z^2 = 2\,ax-y^2 \end{array}\right]$$

$$= \frac{a^2+y^2}{2\,a\,x-y^2}+1 = \frac{a^2+y^2+2\,a\,x-y^2}{2\,a\,x-y^2} = \frac{a^2+2\,a\,x}{2\,a\,x-y^2}$$

$$S = \int_0^a\int_{-\sqrt{ax}}^{\sqrt{ax}}\sqrt{\left(\frac{\partial z}{\partial x}\right)^2+\left(\frac{\partial z}{\partial y}\right)^2+1}\;dx\,dy = \int_0^a\int_{-\sqrt{ax}}^{\sqrt{ax}}\sqrt{\frac{a^2+2\,ax}{2\,ax-y^2}}\,dx\,dy \qquad \left[\begin{array}{l} y^2 = ax \\ y = \pm\sqrt{ax} \end{array}\right]$$

$$= \sqrt{a}\int_0^a\int_{-\sqrt{ax}}^{\sqrt{ax}}\sqrt{\frac{a+2x}{2\,ax-y^2}}\,dx\,dy = \sqrt{a}\int_0^a\sqrt{a+2x}\,dx\int_{-\sqrt{ax}}^{\sqrt{ax}}\frac{1}{\sqrt{2\,ax+y^2}}\,dy$$

$$= \sqrt{a}\int_0^a \sqrt{a+2x}\,dx\left[\sin^{-1}\frac{y}{\sqrt{2ax}}\right]_{-\sqrt{ax}}^{\sqrt{ax}}$$

$$= \sqrt{a}\int_0^a \sqrt{a+2x}\,dx\left[\sin^{-1}\frac{1}{\sqrt{2}} - \sin^{-1}\left(-\frac{1}{\sqrt{2}}\right)\right] = \sqrt{a}\int_0^a \sqrt{a+2x}\,dx\left[\frac{\pi}{4}-\left(\frac{\pi}{4}\right)\right]$$

$$= \sqrt{a}\,\frac{\pi}{2}\int_0^a \sqrt{a+2x}\,dx = \frac{\pi}{2}\cdot\frac{\sqrt{a}}{2}\cdot\frac{2}{3}[(a+2x)^{3/2}]_0^a$$

$$= \frac{\pi\sqrt{a}}{6}[(3a)^{3/2} - a^{3/2}] = \frac{\pi a^2}{6}[3\sqrt{3}-1]$$ **Ans.**

## EXERCISE 2.11

1. Find the surface area of sphere $x^2 + y^2 + z^2 = 16$. **Ans.** $64\pi$
2. Find the surface area of the portion of the cylinder $x^2 + y^2 = 4y$ lying inside the sphere $x^2 + y^2 + z^2 = 16$. **Ans.** 64.
3. Show that the area of surfaces cz = xy intercepted by the cylinder $x^2 + y^2 = b^2$ is $\iint_A \frac{\sqrt{c^2+x^2+y^2}}{c}\,dx\,dy$, where A is the area of the circle $x^2 + y^2 = b^2$, $z = 0$

   **Ans.** $\frac{2\pi}{3c}\left[(c^2+b^2)^{\frac{1}{2}} - c^2\right]$
4. Find the area of the portion of the sphere $x^2 + y^2 + z^2 = a^2$ lying inside the cylinder $x^2 + y^2 = ax$. **Ans.** $2(\pi - 2)a^2$
5. Find the area of the surface of the cone $z^2 = 3(x^2 + y^2)$ cut out by the paraboloid $z = x^2 + y^2$ using surface integral. **Ans.** $6\pi$

## 2.17 CALCULATION OF MASS

We have,

Volume = $\iint_V dx\,dy\,dz$ Density = $\rho = f(x, y, z)$

[Density = Mass per unit volume] Mass = Volume × Density

Mass = $\iint_V dx\,dy\,dz$ $\boxed{Mass = \iint_V f(x,y,z)\,dx\,dy\,dz}$

**Example 59.** Find the mass of a plate which is formed by the co-ordinate planes and the plane $\frac{x}{a}+\frac{y}{b}+\frac{z}{c}=1$, the density is given by $\rho = kxyz$. *(U.P., I Semester, Dec., 2003)*

**Solution.** The plate is bounded by the planes $x = 0$, $y = 0$, $z = 0$ and $\frac{x}{a}+\frac{y}{b}+\frac{z}{c}=1$.

$$\text{Mass} = \iiint dx\,dy\,dz\,\rho = \int_0^c\int_0^{b\left(1-\frac{z}{c}\right)}\int_0^{a\left(1-\frac{y}{b}-\frac{z}{c}\right)} dx\,dy\,dz\,(k\,xyz)$$

$$= k\int_0^c z\,dz\int_0^{b\left(1-\frac{z}{c}\right)} y\,dy\int_0^{a\left(1-\frac{y}{b}-\frac{z}{c}\right)} x\,dx = k\int_0^c z\,dz\int_0^{b\left(1-\frac{z}{c}\right)} y\,dy\left(\frac{x^2}{2}\right)_0^{a\left(1-\frac{y}{b}-\frac{z}{c}\right)}$$

$$= k\int_0^c z\,dz\int_0^{b\left(1-\frac{z}{c}\right)} y\,dy\,\frac{a^2}{2}\left(1-\frac{y}{b}-\frac{z}{c}\right)^2 = \frac{ka^2}{2}\int_0^c z\,dz\int_0^{b\left(1-\frac{z}{c}\right)} y\left[\left(1-\frac{z}{c}\right)-\frac{y}{b}\right]^2 dy$$

$$= \frac{ka^2}{2}\int_0^c z\,dz\int_0^{b\left(1-\frac{z}{c}\right)}\left[y\left(1-\frac{z}{c}\right)^2 + \frac{y^3}{b^2} - \frac{2y^2}{b}\left(1-\frac{z}{c}\right)\right] dy$$

$$= \frac{k\,a^2}{2}\int_0^c z\,dz\left[\frac{y^2}{2}\left(1-\frac{z}{c}\right)^2+\frac{y^4}{4\,b^2}-\frac{2\,y^3}{3\,b}\left(1-\frac{z}{c}\right)\right]_0^{b\left(1-\frac{z}{c}\right)}$$

$$= \frac{k\,a^2}{2}\int_0^c z\,dz\left[\frac{b^2}{2}\left(1-\frac{z}{c}\right)^4+\frac{b^4}{4\,b^2}\left(1-\frac{z}{c}\right)^4-\frac{2}{3}\cdot\frac{b^3}{b}\left(1-\frac{z}{c}\right)^4\right]$$

$$= \frac{k\,a^2}{2}\int_0^c z\left[\frac{b^2}{2}+\frac{b^2}{4}-\frac{2b^2}{3}\right]\left(1-\frac{z}{c}\right)^4 dz = \frac{k\,a^2}{2}\frac{b^2}{12}\int_0^c\left(1-\frac{z}{c}\right)^4 dz \qquad [\text{Put } z = c\sin^2\theta]$$

$$= \frac{k\,a^2\,b^2\,c^2}{12}\int_0^{\frac{\pi}{2}} c\sin^2\theta\,(1-\sin^2\theta)^4\,(2\,c\sin\theta\cos\theta\,d\theta)$$

$$= \frac{k^2\,a^2\,b^2\,c^2}{12}\int_0^{\pi/2}\sin^2\theta\,(\cos^8\theta)\sin\theta\cos\theta\,d\theta = \frac{k^2\,a^2\,b^2\,c^2}{12}\int_0^{\pi/2}\sin^3\theta\cos^9\theta\,d\theta$$

$$= \frac{k^2\,a^2\,b^2\,c^2}{12}\;\frac{\left|\overline{\frac{3+1}{2}}\right.\left|\overline{\frac{9+1}{2}}\right.}{2\left|\overline{\frac{3+9+2}{2}}\right.} = \frac{k\,a^2\,b^2\,c^2}{12}\cdot\frac{|\overline{2}\,|\overline{5}}{2\,|\overline{7}} = \frac{k\,a^2\,b^2\,c^2}{12}\;\frac{(1)\,(|\overline{5})}{2\times 6\times 5\,|\overline{5}} = \frac{k\,a^2\,b^2\,c^2}{720}$$ **Ans.**

## 2.18 CENTRE OF GRAVITY

$$\bar{x} = \frac{\iiint x\,\rho\,dx\,dy\,dz}{\iiint \rho\,dx\,dy\,dz},\quad \bar{y} = \frac{\iiint y\,\rho\,dx\,dy\,dz}{\iiint \rho\,dx\,dy\,dz},\quad \bar{z} = \frac{\iiint z\,\rho\,dx\,dy\,dz}{\iiint \rho\,dx\,dy\,dz}$$

**Example 60.** *Find the co-ordinates of the centre of gravity of the positive octant of the sphere $x^2 + y^2 + z^2 = a^2$, density being given $= k\,xyz$.*

**Solution.** $$\bar{x} = \frac{\iiint_V x\,\rho\,dx\,dy\,dz}{\iiint_V \rho\,dx\,dy\,dz} = \frac{\iiint z\,\rho\,dx\,dy\,dz}{\iiint \rho\,dx\,dy\,dz} = \frac{\iiint_V x^2\,yz\,dx\,dy\,dz}{\iiint_V xyz\,dx\,dy\,az}$$

Converting into polar co-ordinates, $x = r\sin\theta\cos\phi$, $y = r\sin\theta\sin\phi$, $z = r\cos\theta$,
$dx\,dy\,dz = r^2\sin\theta\,dr\,d\theta\,d\phi$

$$\bar{x} = \frac{\int_0^{\pi/2}\int_0^{\pi/2}\int_0^a (r\sin\theta\cos\phi)^2\,(r\sin\theta\sin\phi)\,(r\cos\theta)\,(r^2\sin\theta\,dr\,d\theta\,d\phi)}{\int_0^{\pi/2}\int_0^{\pi/2}\int_0^a (r\sin\theta\cos\phi)\,(r\sin\theta\sin\phi)\,(r\cos\theta)\,(r^2\sin\theta\,dr\,d\theta\,d\phi)}$$

$$= \frac{\int_0^{\pi/2}\int_0^{\pi/2}\int_0^a r^6\sin^4\theta\cos\theta\sin\phi\cos^2\phi\,dr\,d\theta\,d\phi}{\int_0^{\pi/2}\int_0^{\pi/2}\int_0^a r^5\sin^3\theta\cos\theta\sin\phi\cos\phi\,dr\,d\theta\,d\phi}$$

$$= \frac{\int_0^{\pi/2}\sin\phi\cos^2\phi\,d\phi\int_0^{\pi/2}\sin^4\theta\cos\theta\,d\theta\int_0^a r^6\,dr}{\int_0^{\pi/2}\sin\phi\cos\phi\,d\phi\int_0^{\pi/2}\sin^3\theta\cos\theta\,d\theta\int_0^a r^5\,dr}$$

$$= \frac{\left[-\frac{\cos^3\phi}{3}\right]_0^{\pi/2}\left[\frac{\sin^5\theta}{5}\right]_0^{\pi/2}\left[\frac{r^7}{7}\right]_0^a}{-\left[\frac{\cos^2\phi}{2}\right]_0^{\pi/2}\left[\frac{\sin^4\theta}{4}\right]_0^{\pi/2}\left[\frac{r^6}{6}\right]_0^a} = \frac{\left(\frac{1}{3}\right)\left(\frac{1}{5}\right)\left(\frac{a^7}{7}\right)}{\left(\frac{1}{2}\right)\left(\frac{1}{4}\right)\left(\frac{a^6}{6}\right)} = \frac{16\,a}{35}$$

Similarly, $\bar{y} = \bar{z} = \frac{16\,a}{35}$; Hence, C.G. is $\left(\frac{16\,a}{35}, \frac{16\,a}{35}, \frac{16\,a}{35}\right)$ **Ans.**

## 2.19 MOMENT OF INERTIA OF A SOLID

Let the mass of an element of a solid of volume $V$ be $\rho\ \delta x\ \delta y\ \delta z$.

Perpendicular distance of this element from the $x$-axis $= \sqrt{y^2 + z^2}$

$M.I.$ of this element about the $x$-axis $= \rho\ \delta x\ \delta y\ \delta z\ \sqrt{y^2 + z^2}$

$M.I.$ of the solid about $x$-axis $= \iiint_V \rho\,(y^2 + z^2)\,dx\,dy\,dz$

$M.I.$ of the solid about $y$-axis $= \iiint_V \rho\,(x^2 + z^2)\,dx\,dy\,dz$

$M.I.$ of the solid about $z$-axis $= \iiint_V \rho\,(x^2 + y^2)\,dx\,dy\,dz$

**The Perpendicular Axes Theorem**

If $I_{ox}$ and $I_{oy}$ be the moments of inertia of a lamina about $x$-axis and $y$-axis respectively and $I_{oz}$ be the moment of inertia of the lamina about an axis perpendicular to the lamina and passing through the point of intersection of the axes $OX$ and $OY$.

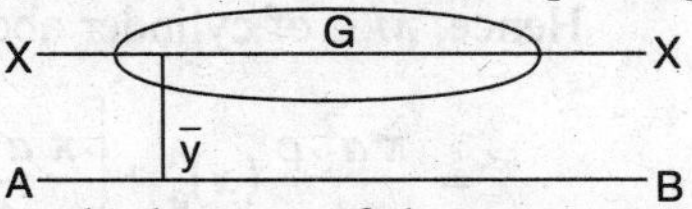


$$I_{OZ} = I_{OX} + I_{OY}$$

**The Parallel Axes Theorem**

M.I. of a lamina about an axis in the plane of the lamina equals the sum of the moment of inertia about a parallel centroidal axis in the plane of lamina together with the product of the mass of the lamin $a$ and square of the distance between the two axes.

$$I_{AB} = I_{XX} + My^{-2}$$

**Example 61.** *Find M.I. of a sphere about diameter.*

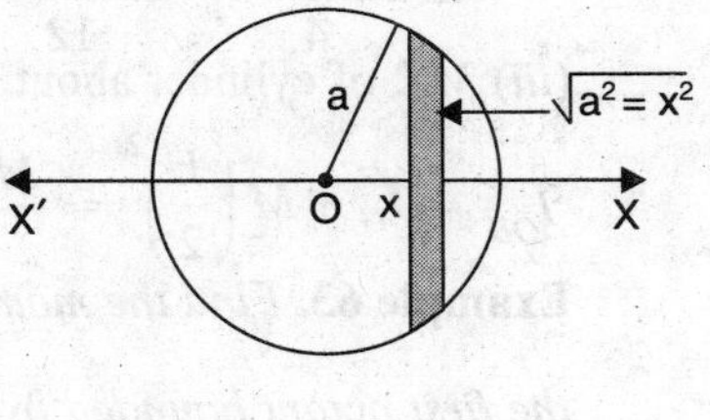


**Solution.** Let a circular disc of $\delta\ x$ thickness be perpendicular to the given diameter $XX'$ at a distance $x$ from it.

The radius of the disc $= \sqrt{a^2 - x^2}$

Mass of the disc $= \rho\,\pi\,(a^2 - x^2)$

Moment of inertia of the disc about a diameter perpendicular on it

$$= \frac{1}{2}MR^2 = \frac{1}{2}[\rho\,\pi\,(a^2 - x^2)]\,(a^2 - x^2) = \frac{1}{2}\rho\,\pi\,(a^2 - x^2)^2$$

$$\text{M.I. of the sphere} = \int_{-a}^{a} \frac{1}{2}\rho\,\pi\,(a^2 - x^2)^2\,dx = 2\left(\frac{1}{2}\rho\,\pi\right)\int_0^a [a^4 - 2\,a^2\,x^2 + x^4]\,dx$$

$$= \rho\,\pi\left[a^4 x - \frac{2\,a^2\,x^3}{3} + \frac{x^5}{5}\right]_0^a = \rho\,\pi\left[a^5 - \frac{2\,a^5}{3} + \frac{a^5}{5}\right]$$

$$= \frac{8}{15}\pi\,\rho\,a^5 = \frac{2}{5}\left(\frac{4\,\pi}{3}a^3\,\rho\right)a^2 = \frac{2}{5}M\,a^2 \qquad \textbf{Ans.}$$

**Example 62.** *The mass of a solid right circular cylinder of radius a and height h* is *M. Find the moment of inertia of the cylinder about (i) its axis (ii) a line through its centre of gravity perpendicular to its axis (iii) any diameter through its base.*

**Solution.** *To find M.I. about OX. Consider a disc at a distance x from O at the base.*

$$\text{M.I. } \textit{of the about OX,} = \frac{(\pi\,a^2\,\rho\,dx)\,a^2}{2} = \frac{\pi\,\rho\,a^4\,dx}{2}$$

(*i*) M.I. of the cylinder about OX

$$\int_0^h \frac{\pi\,\rho\,a^4 dx}{2} = \frac{\pi\,\rho\,a^4}{2}(x)_0^h = \frac{\pi\,\rho\,a^4\,h}{2} = (\pi\,a^2\,h)\,\rho\cdot\frac{a^2}{2} = \frac{M\,a^2}{2}$$

(*ii*) *M.I.* of the disc about a line through *C.G.* and perpendicular to *OX*.

$$I_{OX} + I_{OY} = I_{OZ}$$
$$I_{OX} + I_{OX} = I_{OZ}$$
$$I_{OX} = \frac{1}{2} I_{OZ}$$

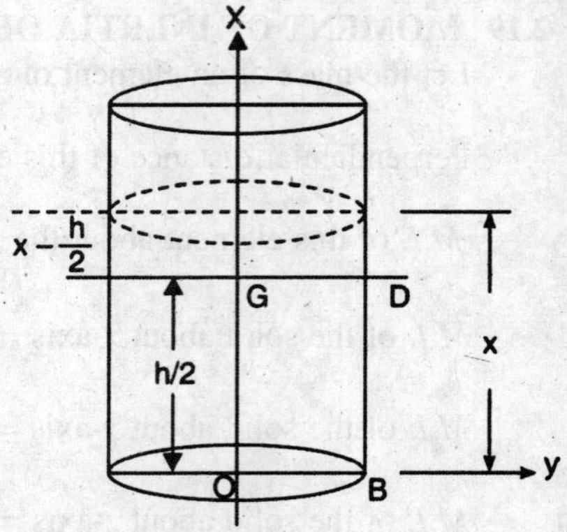


M.I. of the disc about a line through

$$C.G. = \frac{1}{2}\left(\frac{M\,a^2}{2}\right) = \frac{M\,a^2}{4}$$

M.I. of the disc about the diameter $= \left(\frac{\pi\, a^2\, \rho\, dx}{4}\right) a^2$

M.I. of the disc about line $GD = \frac{\pi\, a^2\, \rho\, dx}{4} + (\pi\, a^2\, \rho\, dx)\left(x - \frac{h}{2}\right)^2$

Hence, *M.I.* of cylinder about $GD = \int_0^h \frac{\pi\, a^2\, \rho}{4}\, dx + \int_0^h (\pi\, a^2\, \rho\, dx)\left(x - \frac{h}{2}\right)^2$

$$= \frac{\pi\, a^2\, \rho}{4}(x)_0^h + \left[\frac{\pi\, a^2\, \rho}{4}\left(x - \frac{h}{2}\right)^3\right]_0^h = \frac{\pi\, a^2\, \rho\, h}{4} + \left[\frac{\pi\, a^2\, \rho}{3}\left(\frac{h}{2}\right)^3 + \frac{\pi\, a^2\, \rho}{3}\left(\frac{h}{2}\right)^3\right]$$

$$= \frac{\pi\, a^2\, \rho\, h}{4} + \frac{\pi\, a^2\, \rho\, h^3}{12} = \frac{M\, a^2}{4} + \frac{M\, h^2}{12}$$

(*iii*) *M.I.* of cylinder about line *OB* (through) base

$$I_{OB} = I_G + M\left(\frac{h}{2}\right)^2 = \frac{M\, a^2}{4} + \frac{M\, h^2}{12} + \frac{M\, h^2}{4} = \frac{M\, a^2}{4} + \frac{M\, h^2}{3}$$ **Ans.**

**Example 63.** *Find the moment of inertia and radius of gyration about z-axis of the region in the first octant bounded by* $\frac{x}{a} + \frac{y}{b} + \frac{z}{c} = 1$.

**Solution.** Let *r* be the density. M.I. of tetrahedron about *z*-axis

$$= \iiint (\rho\, dx\, dy\, dz)\,(x^2 + y^2)$$

$$= \rho \int_0^a dx \int_0^{b\left(1-\frac{x}{a}\right)} (x^2 + y^2)\, dy \int_0^{c\left(1-\frac{x}{a}-\frac{y}{b}\right)} dz = \rho \int_0^a dx \int_0^{b\left(1-\frac{x}{a}\right)} (x^2 + y^2)\, dy\, (z)_0^{c\left(1-\frac{x}{a}-\frac{y}{b}\right)}$$

$$= \rho \int_0^a dx \int_0^{b\left(1-\frac{x}{a}\right)} (x^2 + y^2)\, dy\, c\, c\left(1 - \frac{x}{a} - \frac{y}{b}\right)$$

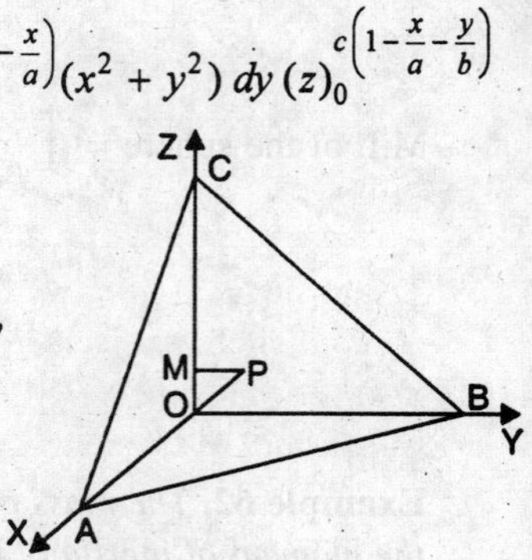


$$= \rho c \int_0^a dx \int_0^{b\left(1-\frac{x}{a}\right)} \left[x^2\left(1 - \frac{x}{a}\right) - \frac{x^2 y}{b} + y^2\left(1 - \frac{x}{a}\right) - \frac{y^3}{b}\right] dy$$

$$= \rho c \int_0^a dx \left[x^2\left(1 - \frac{x}{a}\right) y - \frac{x^2 y^2}{2\,b} + \frac{y^3}{3}\left(1 - \frac{x}{a}\right) - \frac{y^4}{4\,b}\right]_0^{b\left(1-\frac{x}{a}\right)}$$

$$= \rho c \int_0^a dx \left[x^2\left(1 - \frac{x}{a}\right) b\left(1 - \frac{x}{a}\right) - \frac{x^2}{2\,b}\, b^2\left(1 - \frac{x}{a}\right)^2 + \frac{b^3}{3}\left(1 - \frac{x}{a}\right)^3\left(1 - \frac{x}{a}\right) - \frac{b^4}{4\,b}\left(1 - \frac{x}{a}\right)^4\right]$$

$$= b\rho c \int_0^a \left[x^2\left(1 - \frac{x}{a}\right)^2 - \frac{x^2}{2}\left(1 - \frac{x}{a}\right)^2 - \frac{b^2}{3}\left(1 - \frac{x}{a}\right)^4 - \frac{b^2}{4}\left(1 - \frac{x}{a}\right)^4\right] dx$$

$$= \rho bc \int_0^a \left[\frac{x^2}{2}\left(1-\frac{x}{a}\right)^2 + \frac{b^2}{12}\left(1-\frac{x}{a}\right)^4\right] dx$$

$$= \rho bc \int_0^a \left[\frac{1}{2}\left(x^2 - \frac{2x^3}{a} + \frac{x^4}{a^2}\right) + \frac{b^2}{12}\left(1 - \frac{4x}{a} + \frac{6x^2}{a^2} - \frac{4x^3}{a^3} + \frac{x^4}{a^4}\right)\right] dx$$

$$= \rho bc \int_0^a \left[\frac{1}{2}\left(\frac{x^3}{3} - \frac{x^4}{2a} + \frac{x^5}{5a^2}\right) + \frac{b^2}{12}\left(x - \frac{2x^2}{a} + \frac{6x^2}{a^2} - \frac{4x^3}{a^3} + \frac{x^4}{a^4}\right)\right]_0^a dx$$

$$= \rho bc \int_0^a \left[\frac{1}{2}\left(\frac{a^3}{3} - \frac{a^3}{2} + \frac{a^3}{5}\right) + \frac{b^2}{12}\left(a - 2a + 2a - a + \frac{a}{5}\right)\right]$$

$$= \rho bc \left[\frac{a^3}{60} + \frac{ab^2}{60}\right] = \rho \frac{abc}{60}(a^2 + b^2)$$

$$\text{Radius of gyration} = \sqrt{\frac{M.I.}{Mass}} = \frac{\sqrt{\frac{\rho abc}{60}(a^2+b^2)}}{\frac{\rho abc}{6}} = \sqrt{\frac{1}{10}(a^2+b^2)}$$ **Ans.**

## 2.20 CENTRE OF PRESSURE

The centre of pressure of a plane area immersed in a fluid is the point at which the resultant force acts on the area.

Consider a plane area $A$ immersed vertically in a homogeneous liquid. Let $x$-axis be the line of intersection of the plane with the free surface. Any line in this plane and perpendicular to $x$-axis is the *y-axis.*

Let $P$ be the pressure at the point $(x, y)$. Then the pressure on elementary area $\delta x\, \delta y$ is $P\, \delta x\, \delta y$.

Let $(\bar{x}, \bar{y})$ be the centre of pressure. Taking moment about $y$-axis.

$$\bar{x} \cdot \iint_A P\, dx\, dy = \iint_A Px\, dx\, dy$$

$$\bar{x} = \frac{\iint_A Px\, dx\, dy}{\iint_A P\, dx\, dy}$$

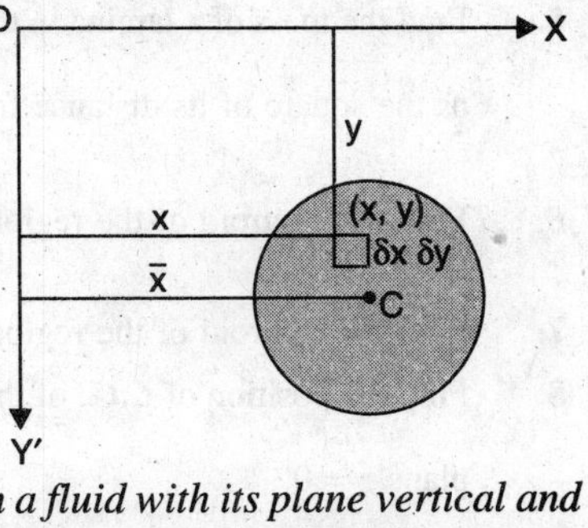


Similarly, $$\bar{y} = \frac{\iint_A Py\, dx\, dy}{\iint_A P\, dx\, dy}$$

**Example 64.** *A uniform semi-circular lamina is immersed in a fluid with its plane vertical and its bounding diameter on the free surface. If the density at any point of the fluid varies as the depth of the point below the free surface, find the position of the centre of pressure of the lamina.*

**Solution.** Let the semi-circular lamina be

$$x^2 + y^2 = a^2$$

By symmetry its centre of pressure lies on $OY$. Let $ky$ be the density of the fluid.

$$\bar{y} = \frac{\iint_A Py\, dx\, dy}{\iint_A P\, dx\, dy} = \frac{\iint_A (\rho y)\, y\, dx\, dy}{\iint_A (\rho y)\, dx\, dy} \qquad (\because \rho = ky)$$

$$= \frac{\iint_A (ky \,.\, y)\, y\, dx\, dy}{\iint_A (ky \,.\, y)\, dx\, dy} = \frac{\iint_A y^3\, dx\, dy}{\iint_A y^2\, dx\, dy} = \frac{\int_{-a}^{a} dx \int_0^{\sqrt{a^2-x^2}} y^3\, dy}{\int_{-a}^{a} dx \int_0^{\sqrt{a^2-x^2}} y^2\, dy}$$

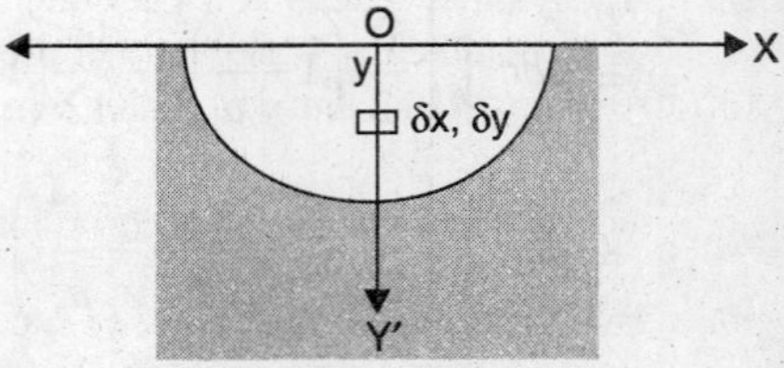


$$= \frac{\int_{-a}^{a} dx \left[\frac{y^4}{4}\right]_0^{\sqrt{a^2-x^2}}}{\int_{-a}^{a} dx \left[\frac{y^3}{3}\right]_0^{\sqrt{a^2-x^2}}} = \frac{3}{4}\,\frac{\int_{-a}^{a} dx\,(a^2-x^2)^2}{\int_{-a}^{a} dx\,(a^2-x^2)^{3/2}}$$

$$= \frac{3}{4}\,\frac{\int_{-\pi/2}^{\pi/2} (a\cos\theta\, d\theta)\,(a^2 - a^2\sin^2\theta)^2}{\int_{-\pi/2}^{\pi/2} (a\cos\theta\, d\theta)\,(a^2 - a^2\sin^2\theta)^{3/2}} \qquad \text{(Put } x = a\sin\theta)$$

$$= \frac{3a}{4}\,\frac{\int_{-\pi/2}^{\pi/2} \cos^5\theta\, d\theta}{\int_{-\pi/2}^{\pi/2} \cos^4\theta\, d\theta} = \frac{3a}{4}\,\frac{2\int_0^{\pi/2} \cos^5\theta\, d\theta}{2\int_0^{\pi/2} \cos^4\theta\, d\theta} = \frac{3a}{4}\,\frac{\frac{4\times 2}{5\times 3}}{\frac{3\times 1}{4\times 2}\frac{\pi}{2}} = \frac{32a}{15\pi} \qquad \textbf{Ans.}$$

## EXERCISE 2.12

1. Find the mass of the solid bounded by the ellipsoid $\frac{x^2}{a^2}+\frac{y^2}{b^2}+\frac{z^2}{c^2}=1$ and the co-ordinate planes, where the density at any point $P(x, y, z)$ is $k\,xyz$. **Ans.** $P$
2. If the density at a point varies as the square of the distance of the point from $XOY$ plane, find the mass of the volume common to the sphere $x^2 + y^2 + z^2 = a^2$ and cylinder $x^2 + y^2 = ax$. **Ans.** $\frac{4k}{15}a^5\left(\frac{\pi}{2}-\frac{8}{15}\right)$
3. Find the mass of the plate in the form of one loop of leminscate $r^2 = a^2 \sin 2\theta$, where $\rho = k\,r^2$. **Ans.** $\frac{k\pi a^4}{16}$
4. Find the mass of the plate which is inside the circle $r = 2a\cos\theta$ and outside the circle $r = a$, if the density varies as the distance from the pole.
5. Find the mass of a lamina in the form of the cardioid $r = a(1 + \cos\theta)$ whose density at any point varies as the square of its distance from the initial line. **Ans.** $\frac{21\pi k a^4}{32}$
6. Find the centroid of the region in the first octant bounded by $\frac{x}{a}+\frac{y}{b}+\frac{z}{c}=1$. **Ans.** $\left(\frac{a}{4},\frac{b}{4},\frac{c}{4}\right)$
7. Find the centroid of the region bounded by $z = 4 - x^2 - y^2$ and xy-plane. **Ans.** $\left(0, 0, \frac{4}{3}\right)$
8. Find the position of $C.G.$ of the volume intercepted between the parallelepiped $x^2+y^2 = a(a - z)$ and the plane $z = 0$. **Ans.** $\left(0, 0, \frac{a}{3}\right)$
9. A solid is cut off the cylinder $x^2 + y^2 = a^2$ by the plane $z = 0$ and that part of the olane $z = mx$ for which $z$ is positive. The density of the solid cut off at any point varies as the height of the point above plane $z = 0$. Find $C.G.$ of the solid. **Ans.** $\bar{z} = \frac{64\,ma}{45\,\pi}$
10. If an area is bounded by two concentric semi-circles with their common bounding diameter in a free surface, prove that the depth of the centre of pressure is $\frac{3\pi}{16}\frac{(a+b)(a^2+b^2)}{a^2+ab+b^2}$
11. An ellipse $\frac{x^2}{a^2}+\frac{y^2}{b^2}=1$ is immersed vertically in a fluid with its major axis horizontal. If its centre be at depth $h$, find the depth of its centre of pressure. **Ans.** $h+\frac{b^2}{4h}$

12. A horizontal boiler has a flat bottom and its ends are plane and semi-circular. If it is just full of water, show that the depth of centre of pressure of either end is 0.7 × total depth approximately.

13. A quadrant of a circle of radius $a$ is just immersed vertically in a homogeneous liquid with one edge in the surface. Determine the co-ordinates of the centre of pressure. **Ans.** $\left(\frac{3a}{8}, \frac{3\pi a}{16}\right)$

14. Find the product of inertia of an equilateral triangle about two perpendicular axes in its plane at a vertex, one of the axes being along a side.

15. Find the *M.I.* of a right circular cylinder of radius $a$ and height $h$ about axis if density varies as distance from the axis. **Ans.** $\frac{2}{5} k\pi a^5 h$

16. Compute the moment of inertia of a right circular cone whose altitude is $h$ and base radius $r$, about (*i*) the axis of symmetry (*ii*) the diameter of the base. **Ans.** (*i*) $\frac{\pi h r^4}{10}$ (*ii*) $\frac{\pi h r^2}{60}(2h^2 + 3r^2)$

17. Find the moment of inertia for the area of the cardioid $r = a(1 - \cos\theta)$ relative to the pole. **Ans.** $\frac{35\pi a^4}{16}$

18. Find the M.I. about the line $\theta = \frac{\pi}{2}$ of the area enclosed by $r = a(1 + \cos\theta)$.

19. Find the moment of inertia of the uniform solid in the form of octant of the ellipsoid $\frac{x^2}{a^2} + \frac{y^2}{b^2} + \frac{z^2}{c^2} = 1$ about $OX$ **Ans.** $\frac{M}{5}(b^2 + c^2)$

20. Prove that the moment of inertia of the area included between the curves $y^2 = 4ax$ and $x^2 = 4ay$ about the $x$-axis is $\frac{144}{35} M a^2$, where $M$ is the mass of area included between the curves.

21. A solid body of density $p$ is the shape of solid formed by revolution of the cardioid $r = a(1 + \cos\theta)$ about the initial line. Show that its moment of inertia about a straight line through the pole perpendicular to the initial line is $\left(\frac{352}{105}\right)\pi l a^5$. (*U. P. II Semester, Summer 2001*)

22. Find the product of inertia of a disc in the form of a quadrant of a circle of radius '$a$' about bounding radii. (*U. P. II Semester, Summer 2002*) **Ans.** $\rho\frac{a^4}{4}$

23. Show that the principal axes at the origin of the triangle enclosed by $x = 0, y = 0, \frac{x}{a} + \frac{y}{b} = 1$ are inclined at angles $\alpha$ and $\alpha + \frac{\pi}{2}$ to the $x$-axis, where $a = \frac{1}{2}\tan^{-1}\left(\frac{ab}{a^2 - b^2}\right)$ (*U.P. II Semester Summer* 2001)

**Choose the correct answer:**

24. The triple integral $\iiint_T dx\, dy\, dz$ gives

(*i*) Volume of region $T$ (*ii*) Surface area of region $T$
(*ii*) Area of region $T$ (*iv*) Density of region $T$. (*A.M.I.E.T.E. 2002*)

**Ans.** (*i*)

25. The volume of the solid under the surface $az = x^2 + y^2$ and whose base $R$ is the circle $x^2 + y^2 = a^2$ is given as

(*i*) $\frac{\pi}{2a}$ (*ii*) $\frac{\pi a^3}{2}$ **Ans.** (*ii*)
(*iii*) $\frac{4}{3}\pi a^3$ (*iv*) None of the above. [*U.P., I. Sem. Dec. 2008*]

# 3

# Differential Equations

## 3.1 DEFINITION

An equation which involves differential co-efficient is called a differential equation.

*For example,*

1. $\frac{dy}{dx} = \frac{1+x^2}{1-y^2}$ 2. $\frac{d^2y}{dx^2} = 2\frac{dy}{dx} - 8y = 0$ 3. $\left[1+\left(\frac{dy}{dx}\right)^2\right]^{3/2} = k\frac{d^2y}{dx^2}$

4. $x\frac{\partial u}{\partial x} + y\frac{\partial u}{\partial y} = nu,$ 5. $\frac{\partial^2 z}{\partial x \partial y} = \frac{\partial z}{\partial y}$

There are two types of differential equations :

**(1) Ordinary Differential Equation**

A differential equation involving derivatives with respect to a single independent variable is called an ordinary differential equation.

**(2) Partial Differential Equation**

A differential equation involving partial derivatives with respect to more than one independent variable is called a partial differential equation.

## 3.2 ORDER AND DEGREE OF A DIFFERENTIAL EQUATION

The *order* of a differential equation is the order of the highest differential co-efficient present in the equation. Consider

1. $L\frac{d^2q}{dt^2} + R\frac{dq}{dt} + \frac{q}{c} = E \sin wt.$ 2. $\cos x \frac{d^2y}{dx^2} + \sin x\left(\frac{dy}{dx}\right)^2 + 8y = \tan x$

3. $\left[1+\left(\frac{dy}{dx}\right)^2\right]^3 = \left(\frac{d^2y}{dx^2}\right)^2$

The order of the above equations is 2.

The degree of a differential equation is the degree of the highest derivative after removing the radical sign and fraction.

The *degree* of the equation (1) and (2) is 1. The degree of the equation (3) is 2.

## 3.3 FORMATION OF DIFFERENTIAL EQUATIONS

The differential equations can be formed by differentiating the ordinary equation and eliminating the arbitrary constants.

**Example 1.** *Form the differential equation by eliminating arbitrary constants, in the following cases and also write down the order of the differential equations obtained.*

(*a*) $y = Ax + A^2$ (*b*) $y = A \cos x + B \sin x$ (*c*) $y^2 = Ax^2 + Bx + C.$

*(R.G.P.V. Bhopal, June 2008)*

**Solution.** (*a*) $y = Ax + A^2$ ... (1)

On differentiation $\frac{dy}{dx} = A$

Putting the value of $A$ in (1), we get $y = x\frac{dy}{dx} + \left(\frac{dy}{dx}\right)^2$ **Ans.**

On eliminating one constant $A$ we get the differential equation of order 1.

(*b*) $y = A\cos x + B\sin x$

On differentiation $\frac{dy}{dx} = -A\sin x + B\cos x$

Again differentiating

$$\frac{d^2y}{dx^2} = -A\cos x - B\sin x \Rightarrow \frac{d^2y}{dx^2} = -(A\cos x + B\sin x)$$

$$\Rightarrow \quad \frac{d^2y}{dx^2} = -y \Rightarrow \frac{d^2y}{dx^2} + y = 0 \quad \textbf{Ans.}$$

This is differential equation of order 2 obtained by eliminating two constants $A$ and $B$.

(*c*) $y^2 = Ax^2 + Bx + C$

On differentiation $2y\frac{dy}{dx} = 2Ax + B$

Again differentiating $2y\frac{d^2y}{dx^2} + 2\left(\frac{dy}{dx}\right)^2 = 2A$

On differentiating again $y\frac{d^3y}{dx^3} + \frac{dy}{dx}\frac{d^2y}{dx^2} + 2\frac{dy}{dx}\frac{d^2y}{dx^2} = 0 \Rightarrow y\frac{d^3y}{dx^3} + 3\frac{dy}{dx}\frac{d^2y}{dx^2} = 0$ **Ans.**

This is the differential equation of order 3, obtained by eliminating three constants *A*, *B*, *C*.

## EXERCISE 3.1

**1.** Write the order and the degree of the following differential equations.

(*i*) $\frac{d^2y}{dx^2} + a^2x = 0$; (*ii*) $\left[1 + \left(\frac{dy}{dx}\right)^2\right]^{\frac{3}{2}} = \frac{d^2y}{dx^2}$; (*iii*) $x^2\left(\frac{d^2y}{dx^2}\right)^3 + y\left(\frac{dy}{dx}\right)^4 + y^4 = 0.$

**Ans.** (*i*) 2,1 (*ii*) 2,2 (*iii*) 2,3

**2.** Give an example of each of the following type of differential equations.

(*i*) A linear-differential equation of second order and first degree **Ans.** *Q*, 1 (*i*)

(*ii*) A non-linear differential equation of second order and second degree **Ans.** *Q*, 1 (*ii*)

(*iii*) Second order and third degree. **Ans.** *Q* 1 (*iii*)

**3.** Obtain the differential equation of which $y^2 = 4a(x + a)$ is a solution.

**Ans.** $y^2\left(\frac{dy}{dx}\right)^2 + 2xy\frac{dy}{dx} - y^2 = 0$

**4.** Obtain the differential equation associated with the primitive $Ax^2 + By^2 = 1$.

**Ans.** $xy\frac{d^2y}{dx^2} + x\left(\frac{dy}{dx}\right)^2 - y\frac{dy}{dx} = 0$

**5.** Find the differential equation corresponding to

$y = a\,e^{3x} + b\,e^x$. **Ans.** $\frac{d^2y}{dx^2} - 4\frac{dy}{dx} + 3y = 0$

6. By the elimination of constants $A$ and $B$, find the differential equation of which

$y = e^x (A \cos x + B \sin x)$ is a solution. **Ans.** $\frac{d^2y}{dx^2} - 2\frac{dy}{dx} + 2y = 0$

7. Find the differential equation whose solution is $y = a \cos (x = 3)$. (*A.M.I.E., Summer 2000*)

**Ans.** $\frac{dy}{dx} = -\tan(x+3)$

8. Show that set of function $\left\{x, \frac{1}{x}\right\}$ forms a basis of the differential equation $x^2y'' + xy' - y = 0$.

Obtain a particular solution when $y(1) = 1$, $y(1) = 2$. **Ans.** $y = \frac{3x}{2} - \frac{1}{2x}$

## 3.4 SOLUTION OF A DIFFERENTIAL EQUATION

In the example 1(*b*), $y = A \cos x + B \sin x$, on eliminating $A$ and $B$ we get the differential equation

$$\frac{d^2y}{dx^2} + y = 0$$

$y = A \cos x + B \sin x$ is called the solution of the differential equation $\frac{d^2y}{dx^2} + y = 0$.

The order of the differential equation $\frac{d^2y}{dx^2} + y = 0$ is two and the solution $y = A \cos x + B \sin x$ contains two arbitrary constants. The number of arbitrary constants in the solution is equal to the order of the differential equation.

An equation containing dependent variable ($y$) and independent variable ($x$) and free from derivative, which satisfies the differential equation, is called the solution (primative) of the differential equation.

## 3.5 DIFFERENTIAL EQUATIONS OF THE FIRST ORDER AND FIRST DEGREE

We will discuss the standard methods of solving the differential equations of the following types:

(*i*) Equations solvable by separation of the variables. (*ii*) Homogeneous equations.
(*iii*) Linear equations of the first order. (*iv*) Exact differential equations.

## 3.6 VARIABLES SEPARABLE

If a differential equation can be written in the form

$$f(y)\,dy = \phi(x)\,dx$$

We say that variables are separable, $y$ on left hand side and $x$ on right hand side.
We get the solution by integrating both sides.

**Working Rule:**

**Step 1.** Separate the variables as $f(y)\,dy = \phi(x)\,dx$

**Step 2.** Integrate both sides as $\int f(y)\,dy = \int \phi(x)\,dx$

**Step 3.** Add an arbitrary constant $C$ on R.H.S.

**Example 2.** *Solve* : $\frac{dy}{dx} = \frac{x(2\log x + 1)}{\sin y + y\cos y}$ (*UP, II 2008, U.P.B. Pharm (C.O.) 2005*)

**Solution.** We have, $\frac{dy}{dx} = \frac{x(2\log x + 1)}{\sin y + y\cos y}$

Separating the variables, we get

$$(\sin y + y\cos y)\,dy = \{x\,(2\log x + 1)\}\,dx$$

Integrating both the sides, we get $\int(\sin y + y\cos y)dy = \int\{x(2\log x+1)\}\,dx + C$

$$-\cos y + y\sin y - \int(1)\cdot\sin y\,dy = 2\int\log x\cdot x\,dx + \int x\,dx + C$$

$$\Rightarrow \quad -\cos y + y\sin y + \cos y = 2\left[\log x\cdot\frac{x^2}{2} - \int\frac{1}{x}\cdot\frac{x^2}{2}dx\right] + \frac{x^2}{2} + C$$

$$\Rightarrow \quad y\sin y = 2\log x\cdot\frac{x^2}{2} - \int x\,dx + \frac{x^2}{2} + C$$

$$\Rightarrow \quad y\sin y = 2\log x\cdot\frac{x^2}{2} - \frac{x^2}{2} + \frac{x^2}{2} + C$$

$$\Rightarrow \quad y\sin y = x^2\log x + C$$ **Ans.**

**Example 3.** *Solve the differential equation.*

$$x^4\frac{dy}{dx} + x^3y = -\sec(xy).$$ (*A.M.I.E.T.E., Winter 2003*)

**Solution.** $x^4\frac{dy}{dx} + x^3y = -\sec(xy) \quad \Rightarrow \quad x^3\left(x\frac{dy}{dx} + y\right) = -\sec xy$

Put $v = xy, \ \frac{dv}{dx} = x\frac{dy}{dx} + y \quad \Rightarrow \quad x^3\frac{dv}{dx} = -\sec v$

$\Rightarrow \quad \frac{dv}{\sec v} = -\frac{dx}{x^3} \quad \Rightarrow \quad \int\cos v\,dv = -\int\frac{dx}{x^3} + c$

$\Rightarrow \ \sin v = \frac{1}{2x^2} + c \quad \Rightarrow \ \sin xy = \frac{1}{2x^2} + c$ **Ans.**

**Example 4.** *Solve* : $\cos(x + y)dy = dx$

**Solution.** $\cos(x + y)\,dy = dx \quad \Rightarrow \quad \frac{dy}{dx} = \sec(x + y)$

On putting $x + y = z$

So that $1 + \frac{dy}{dx} = \frac{dz}{dx} \quad \Rightarrow \quad \frac{dy}{dx} = \frac{dz}{dx} - 1$

$$\frac{dz}{dx} - 1 = \sec z \quad \Rightarrow \quad \frac{dz}{dx} = 1 + \sec z$$

Separating the variables, we get

$$\frac{dz}{1+\sec z} = dx$$

On integrating,

$$\int\frac{\cos z}{\cos z + 1}dz = \int dx \quad \Rightarrow \quad \int\left[1 - \frac{1}{\cos z + 1}\right]dz = x + C$$

$$\int\left(1 - \frac{1}{2\cos^2\frac{z}{2} - 1 + 1}\right)dz = x + C$$

$$\int\left(1 - \frac{1}{2}\sec^2\frac{z}{2}\right)dz = x + C \quad \Rightarrow \quad z - \tan\frac{z}{2} = x + C$$

$$x + y - \tan\frac{x+y}{2} = x + C$$

$$y - \tan\frac{x+y}{2} = C$$ **Ans.**

**Example 5.** *Solve the equation.*

$(2x^2 + 3y^2 - 7)\, x\, dx - (3x^2 + 2y^2 - 8)\, y\, dy = 0$ (*U.P. II Semester, Summer 2005*)

**Solution.** We have

$$(2x^2 + 3y^2 - 7)\, x\, dx - (3x^2 + 2y^2 - 8)\, y\, dy = 0$$

Re-arranging (1), we get $\dfrac{x\,dx}{y\,dy} = \dfrac{3x^2 + 2y^2 - 8}{2x^2 + 3y^2 - 7}$

Applying componendo and dividendo rule, we get

$$\frac{x\,dx + y\,dy}{x\,dx - y\,dy} = \frac{5x^2 + 5y^2 - 15}{x^2 - y^2 - 1} \quad \Rightarrow \quad \frac{x\,dx + y\,dy}{x^2 + y^2 - 3} = 5\left(\frac{x\,dx - y\,dy}{x^2 - y^2 - 1}\right)$$

Multiplying by 2 both the sides, we get

$$\Rightarrow \quad \left(\frac{2x\,dx + 2y\,dy}{x^2 + y^2 - 3}\right) = 5\left(\frac{2x\,dx - 2y\,dy}{x^2 - y^2 - 1}\right)$$

Integrating both sides, we get

$$\log(x^2 + y^2 - 3) = 5\log(x^2 - y^2 - 1) + \log C$$

$$\Rightarrow \quad x^2 + y^2 - 3 = C(x^2 - y^2 - 1)^5$$ **Ans.**

where $C$ is arbitrary constant of integration.

## EXERCISE 3.2

**Solve the following differential equations :**

1. $\dfrac{dx}{x} = \tan y \cdot dy$ **Ans.** $x \cos y = C$
2. $\dfrac{dy}{dx} = \dfrac{\sqrt{1-y^2}}{\sqrt{1-x^2}}$ **Ans.** $\sin^{-1} y = \sin^{-1} x + C$
3. $y(1+x^2)^{1/2}\, dy + x\sqrt{1+y^2}\, dx = 0$ **Ans.** $\sqrt{1+y^2} + \sqrt{1+x^2} = C$
4. $\sec^2 x \tan y\, dx + \sec^2 y \tan x\, dy = 0$ **Ans.** $\tan x \tan y = C$
5. $(1 + x^2)\, dy - x\, y\, dx = 0$ **Ans.** $y^2 = C(1 + x^2)$
6. $(e^y + 1)\cos x\, dx + e^y \sin x\, dy = 0$ **Ans.** $(e^y + 1)\sin x = C$
7. $3e^x \tan y\, dx + (1 - e^x)\sec^2 y\, dy = 0$ **Ans.** $(1 - e^x)^3 = C \tan y$
8. $(e^y + 2)\sin x\, dx - e^y \cos x\, dy = 0$ **Ans.** $(e^y + 2)\cos x = C$
9. $\dfrac{dy}{dx} = e^{x-y} + x^2 e^{-y}$ **Ans.** $e^y = e^x + \dfrac{x^3}{3} + C$
10. $\dfrac{dy}{dx} = 1 + \tan(y - x)$ [Put $y - x = z$] **Ans.** $\sin(y - x) = e^{x+c}$
11. $(4x + y)^2 \dfrac{dx}{dy} = 1$ **Ans.** $\tan^{-1}\dfrac{4x+y}{2} = 2x + C$
12. $\dfrac{dy}{dx} = (4x + y + 1)^2$ [**Hint.** Put $4x + y + 1 = z$] **Ans.** $\tan^{-1}\dfrac{4x+y+1}{2} = 2x + C$

## 3.7 HOMOGENEOUS DIFFERENTIAL EQUATIONS

A differential equation of the form $\dfrac{dy}{dx} = \dfrac{f(x, y)}{\phi(x, y)}$

is called a homogeneous equation if each term of $f(x, y)$ and $\phi(x, y)$ is of the same degree *i.e.*,

$$\frac{dy}{dx}=\frac{3xy+y^2}{3x^2+xy}$$

In such case we put $\boxed{y = vx \text{ and } \frac{dy}{dx}=v+x\frac{dv}{dx}}$

The reduced equation involves $v$ and $x$ only. This new differential equation can be solved by *variables separable* method.

**Working Rule**

**Step 1.** Put $y = vx$ so that $\frac{dy}{dx}=v+x\frac{dv}{dx}$

**Step 2.** Separate the variables.

**Step 3.** Integrate both the sides.

**Step 4.** Put $v=\frac{y}{x}$ and simplify.

**Example 6.** *Solve the following differential equation*

$$(2xy + x^2)\, y = 3y^2 + 2xy \qquad \text{(A.M.I.E.T.E. Dec. 2006)}$$

**Solution.** We have, $(2xy+x^2)\frac{dy}{dx}=3y^2+2xy \Rightarrow \frac{dy}{dx}=\frac{3y^2+2xy}{2xy+x^2}$

Put $y = vx$ so that $\frac{dy}{dx}=v+x\frac{dv}{dx}$

On substituting, the given equation becomes $v+x\frac{dv}{dx}=\frac{3v^2x^2+2vx^2}{2vx^2+x^2}=\frac{3v^2+2v}{2v+1}$

$$\Rightarrow \quad x\frac{dv}{dx}=\frac{3v^2+2v-2v^2-v}{2v+1} \qquad \Rightarrow \quad x\frac{dv}{dx}=\frac{v^2+v}{2v+1} \quad \Rightarrow \quad \left(\frac{2v+1}{v^2+v}\right)dv=\frac{dx}{x}$$

$$\Rightarrow \quad \int\left(\frac{2v+1}{v^2+v}\right)dv=\int\frac{dx}{x} \qquad \Rightarrow \quad \log(v^2+v)\log x+\log c$$

$$\Rightarrow \quad v^2+v=cx \qquad \Rightarrow \quad \frac{y^2}{x^2}+\frac{y}{x}=cx$$

$$\Rightarrow \quad y^2+xy=cx^3$$

**Example 7.** *Solve the equation* :

$$\frac{dy}{dx}=\frac{y}{x}+x\sin\frac{y}{x}$$

**Solution.**

$$\frac{dy}{dx}=\frac{y}{x}+x\sin\frac{y}{x} \qquad \text{... (1)}$$

Put $y = vx$ in (1) so that $\frac{dy}{dx}=v+x\frac{dv}{dx}$

$$v+x\frac{dv}{dx}=v+x\sin v$$

$$\Rightarrow \quad x\frac{dv}{dx}=x\sin v \qquad \Rightarrow \quad \frac{dv}{dx}=\sin v$$

Separating the variable, we get

$$\Rightarrow \quad \frac{dv}{\sin v}=dx \qquad \Rightarrow \quad \int \operatorname{cosec} v\, dv=\int dx+C$$

$$\log\tan\frac{v}{2}=x+C \qquad \Rightarrow \quad \log\tan\frac{y}{2x}=x+C \qquad \textbf{Ans.}$$

## EXERCISE 3.3

*Solve the following differential equations:*

1. $(y^2 - xy)\,dx + x^2 dy = 0$ **Ans.** $\dfrac{x}{y} = \log x + C$
2. $(x^2 - y^2)\,dx + 2xy\,dy = 0$ *(AMIETE, June 2009)* **Ans.** $x^2 + y^2 = ax$
3. $x(y-x)\dfrac{dy}{dx} = y(y+x)$. **Ans.** $\dfrac{y}{x} - \log xy = a$
4. $x(x-y)\,dy + y^2\,dx = 0$ *(U.P. B. Pharm (C.O.) 2005)* **Ans.** $y = x \log C y$
5. $\dfrac{dy}{dx} + \dfrac{x-2y}{2x-y} = 0$ **Ans.** $y - x = C(x+y)^3$
6. $\dfrac{dy}{dx} = \tan\dfrac{y}{x} + \dfrac{y}{x}$ **Ans.** $\sin\dfrac{y}{x} = Cx$
7. $\dfrac{dy}{dx} = \dfrac{3xy + y^2}{3x^2}$ **Ans.** $3x + y\log x + Cy = 0$
8. $\dfrac{dy}{dx} = \dfrac{x^2 - 2y^2}{2xy}$ **Ans.** $4y^2 - x^2 = \dfrac{C}{x^2}$
9. $(x^2 + y^2)\,dy = xy\,dx$ **Ans.** $-\dfrac{x^2}{2y^2} + \log y = C$
10. $x^2 y\,dx - (x^3 + y^3)\,dy = 0$ **Ans.** $\dfrac{-x^3}{3y^3} + \log y = C$
11. $(y^2 + 2xy)\,dx + (2x^2 + 3xy)\,dy = 0$ *(AMIETE, Summer 2004)* **Ans.** $xy^2(x+y) = C$
12. $(2xy^2 - x^3)\,dy + (y^3 - 2yx^2)\,dx = 0$ **Ans.** $y^2(y^2 - x^2) = Cx^{-2}$
13. $(x^3 - 3xy^2)\,dx + (y^3 - 3x^2y)\,dy = 0$, $y(0) = 1$ **Ans.** $x^4 - 6x^2y^2 + y^4 = 1$
14. $2xy^2\,dy - (x^3 + 2y^3)\,dx = 0$ **Ans.** $2y^3 = 3x^3 \log x + 3x^3 + C$
15. $x\sin\dfrac{y}{x}\,dy = \left(y\sin\dfrac{y}{x} - x\right)dx$ **Ans.** $\cos\dfrac{y}{x} = \log x + C$
16. $\left\{x\cos\dfrac{y}{x} + y\sin\dfrac{y}{x}\right\}y - \left\{y\sin\dfrac{y}{x} - x\cos\dfrac{y}{x}\right\}x\dfrac{dy}{dx} = 0$ **Ans.** $xy\cos\dfrac{y}{x} = a$
17. $\dfrac{dy}{dx} = \dfrac{y + \sqrt{x^2+y^2}}{x}$ **Ans.** $y + \sqrt{x^2+y^2} = Cx^2$
18. $x\dfrac{dy}{dx} = y(\log y - \log x + 1)$ *(AMIETE, Summer 2004)* **Ans.** $\log\dfrac{y}{x} = Cx$
19. $xy\log\dfrac{x}{y}\,dx + \left(y^2 - x^2\log\dfrac{x}{y}\right)dy = 0$ given that $y(1) = 0$

    **Ans.** $\dfrac{x^2}{2y^2}\log\dfrac{x}{y} - \dfrac{x^2}{4y^2} + \log y = 1 - \dfrac{3}{4e^2}$
20. $\left(1 + e^{\frac{x}{y}}\right)dx + e^{\frac{x}{y}}\left(1 - \dfrac{x}{y}\right)dy = 0$ *(AMIETE, June 2009)* **Ans.** $e^{\frac{x}{y}} + \dfrac{x}{y} = e^{-y} + C$

## 3.8 EQUATIONS REDUCIBLE TO HOMOGENEOUS FORM

**Case I.** $\dfrac{a}{A} + \dfrac{b}{B}$

The equations of the form

$$\boxed{\frac{dy}{dx} = \frac{ax + by + c}{Ax + By + C}}$$

can be reduced to the homogeneous form by the substitution if $\dfrac{a}{A} + \dfrac{b}{B}$

$$x = X + h,\ \ y = Y + k \qquad (h, k \text{ being constants})$$

$$\therefore \quad \frac{dy}{dx} = \frac{dY}{dX}$$

The given differential equation reduces to

$$\frac{dY}{dX} = \frac{a(X+h)+b(Y+k)+c}{A(X+h)+B(Y+k)+C} = \frac{aX+bY+ah+bk+c}{AX+BY+Ah+Bk+C}$$

Choose $h, k$ so that $\quad ah + bk + c = 0$

$$Ah + Kk + C = 0$$

Then the given equation becomes homogeneous $\dfrac{dY}{dX} = \dfrac{aX+bY}{AX+BY}$

**Case II.** If $\dfrac{a}{A} = \dfrac{b}{B}$ then the value of $h, k$ will not be finite.

$$\frac{a}{A} = \frac{b}{B} = \frac{1}{m} \quad \text{(say)}$$

$$A = am,\ B = bm$$

The given equation becomes $\dfrac{dy}{dx} = \dfrac{ax+by+c}{m(ax+by)+c}$

Now put $ax + by = z$ and apply the method of variables separable.

**Example 8.** *Solve* : $\dfrac{dy}{dx} = \dfrac{x+2y-3}{2x+y-3}$

**Solution.** Put $\quad x = X + h, \qquad y = Y + k.$

The given equation reduces to

$$\therefore \quad \frac{dY}{dX} = \frac{(X+h)+2(Y+k)-3}{2(X+h)+(Y+k)-3} \qquad \left(\frac{1}{2} \neq \frac{2}{1}\right)$$

$$= \frac{X+2Y+(h+2k-3)}{2X+Y+(2h+k-3)} \qquad \text{.... (1)}$$

Now choose $h$ and $k$ so that $h + 2k - 3 = 0,\ 2h + k - 3 = 0$

Solving these equations we get $h = k = 1$

$$\therefore \quad \frac{dY}{dX} = \frac{X+2Y}{2X+Y} \qquad \text{... (2)}$$

Put $Y = vX$, so that $\dfrac{dY}{dX} = v + X\dfrac{dv}{dX}$

The equation (2) is transformed as

$$v + X\frac{dv}{dX} = \frac{X+2vX}{2X+vX} = \frac{1+2v}{2+v}$$

$$X\frac{dv}{dX} = \frac{1+2v}{2+v} - v = \frac{1-v^2}{2+v} \quad \Rightarrow \quad \left(\frac{2+v}{1-v^2}\right)dv = \frac{dX}{X}$$

$$\Rightarrow \quad \frac{1}{2}\frac{1}{(1+v)}dv + \frac{3}{2}\frac{1}{1-v}dv = \frac{dX}{X} \qquad \text{(Partial fractions)}$$

On integrating, we have

$$\frac{1}{2}\log(1+v) - \frac{3}{2}\log(1-v) = \log X + \log C$$

$$\Rightarrow \quad \log\frac{1+v}{(1-v)^3}=\log C^2X^2 \qquad \Rightarrow \quad \frac{1+v}{(1-v)^3}=C^2X^2$$

$$\frac{1+\frac{Y}{X}}{\left(1-\frac{Y}{X}\right)^3}=C^2X^2 \qquad \Rightarrow \quad \frac{X+Y}{(X-Y)^3}=C^2 \quad \text{or} \quad X+Y=C^2(X-Y)^3$$

Put $X = x - 1$ and $Y = y - 1$ $\Rightarrow$ $x + y - 2 = a(x - y)^3$ **Ans.**

**Example 9.** *Solve* : $(x + 2y)(dx - dy) = dx + dy$

**Solution.** $(x + 2y)(dx - dy) = dx + dy \Rightarrow (x + 2y - 1)\,dx - (x + 2y + 1)\,dy = 0$

$$\Rightarrow \quad \frac{dy}{dx}=\frac{x+2y-1}{x+2y+1} \qquad ...(1)$$

Hence $\frac{a}{A}=\frac{b}{B}$ *i.e.*, $\left(\frac{1}{1}=\frac{2}{2}\right)$ (Case of failure)

Now put $x + 2y = z$ so that $1+2\frac{dy}{dx}=\frac{dz}{dx}$

Equation (1) becomes

$$\frac{1}{2}\frac{dz}{dx}-\frac{1}{2}=\frac{z-1}{z+1} \qquad \Rightarrow \quad \frac{dz}{dx}=2\frac{(z-1)}{z+1}+1=\frac{3z-1}{z+1}$$

$$\Rightarrow \quad \frac{z+1}{3z-1}dz=dx \qquad \Rightarrow \quad \left(\frac{1}{3}+\frac{4}{3}\frac{1}{3z-1}\right)dz=dx$$

On integrating, $\frac{z}{3}+\frac{4}{9}\log(3z-1)=x+C$

$$3z + 4\log(3z - 1) = 9x + 9C$$

$$\Rightarrow \quad 3(x + 2y) + 4\log(3x + 6y - 1) = 9x + 9C$$

$$3x - 3y + a = 2\log(3x + 6y - 1)$$ **Ans.**

## EXERCISE 3.4

Solve the following differential equations :

1. $\frac{dy}{dx}=\frac{2x+9y-20}{6x+2y-10}$ **Ans.** $(2x - y)^2 = C(x + 2y - 5)$
2. $\frac{dy}{dx}=\frac{y-x+1}{y+x+5}$ **Ans.** $\log[(y+3)^2+(x+2)^2]+2\tan^{-1}\frac{y+3}{x+2}=a$
3. $\frac{dy}{dx}=\frac{x-y-2}{x+y+6}$ **Ans.** $(y + 4)^2 + 2(x + 2)(y + 4) - (x + 2)^2 = a^2$
4. $\frac{dy}{dx}=\frac{y+x-2}{y-x-4}$ *(AMIETE, Dec. 2009)* **Ans.** $-(y - 3)^2 + 2(x + 1)(y - 3) + (x + 1)^2 = a$
5. $\frac{dy}{dx}=\frac{2x-5y+3}{2x+4y-6}$ **Ans.** $(x - 4y + 3)(2x + y - 3) = a$
6. $(2x + y + 1)\,dx + (4x + 2y - 1)\,dy = 0$ **Ans.** $2(2x + y) + \log(2x + y - 1) = 3x + C$
7. $(x - y - 2)\,dx - (2x - 2y - 3)\,dy = 0$ **Ans.** $\log(x - y - 1) = x - 2y + C$ *(U.P. B. Pharm (C.O.) 2005)*
8. $(6x - 4y + 1)\,dy - (3x - 2y + 1)\,dx = 0$ *(A.M.I.E.T. E., Dec. 2006)* **Ans.** $4x - 8y - \log(12x - xy + 1) = c$

9. $\frac{dy}{dx} = -\frac{3y-2x+7}{7y-3x+3}$ (*A.M.I.E.T.E., Summer 2004*) **Ans.** $(x+y-1)^5 (x-y-1)^2 = 1$

10. $\frac{dy}{dx} = \frac{2y-x-4}{y-3x+3}$ (*AMIETE, Dec. 2010*)

**Ans.** $X^2 - 5XY + Y^2 = c\left[\frac{2Y+(-5+\sqrt{21})\,X}{2Y-(5+\sqrt{21})\,X}\right]^{\frac{1}{\sqrt{21}}}$, $X = x-2$, $Y = y-3$

## 3.9 LINEAR DIFFERENTIAL EQUATIONS

A differential equation of the form

$$\boxed{\frac{dy}{dx} + Py = Q} \quad \text{... (1)}$$

is called a linear differential equation, where $P$ and $Q$, are functions of $x$ (but not of $y$) or constants.

In such case, multiply both sides of (1) by $e^{\int P dx}$

$$e^{\int P dx}\left(\frac{dy}{dx} + Py\right) = Q e^{\int P dx} \quad \text{... (2)}$$

The left hand side of (2) is

$$\frac{d}{dx}\left[y.e^{\int P dx}\right]$$

(2) becomes

$$\frac{d}{dx}\left[y.e^{\int P dx}\right] = Q.e^{\int P dx}$$

Integrating both sides, we get

$$y.e^{\int P dx} = \int Q.e^{\int P dx}\,dx + C$$

This is the required solution.

**Note.** $e^{\int P dx}$ is called the integrating factor.

Solution is $\boxed{y \times [I.F.] = \int Q\,[I.F.]\,dx + C}$

**Working Rule**

**Step 1.** Convert the given equation to the standard form of linear differential equation

*i.e.* $\frac{dy}{dx} + Py = Q$

**Step 2.** Find the integrating factor i.e. I.F. = $e^{\int P dx}$

**Step 3.** Then the solution is $y(I.F.) = \int Q\,(I.F.)dx + C$

**Example 10.** *Solve*: $(x+1)\frac{dy}{dx} - y = e^x (x+1)^2$ (*A.M.I.E.T.E., Summer 2002*)

**Solution.** $\frac{dy}{dx} - \frac{y}{x+1} = e^x (x+1)$

Integrating factor $= e^{-\int \frac{dx}{x+1}} = e^{-\log(x+1)} = e^{\log(x+1)^{-1}} = \frac{1}{x+1}$

The solution is $y.\frac{1}{x+1} = \int e^x.(x+1).\frac{1}{x+1}\,dx = \int e^x\,dx$

$$\frac{y}{x+1} = e^x + C$$ **Ans.**

**Example 11.** *Solve differential equation*

$$(x^3-x)\frac{dy}{dx}-(3x^2-1)y=x^5-2x^3+x. \qquad \textit{(Nagpur University, Summer 2008)}$$

**Solution.** We have $(x^3-x)\dfrac{dy}{dx}-(3x^2-1)y=x^5-2x^3+x$

$$\Rightarrow \quad \frac{dy}{dx}-\frac{3x^2-1}{x^3-x}y = \frac{x^5-2x^3+x}{x^3-x} \quad \Rightarrow \quad \frac{dy}{dx}-\frac{3x^2-1}{x^3-x}y = x^2-1$$

$$\text{I.F.} = e^{\int -\frac{3x^2-1}{x^3-x}dx} = e^{-\log(x^3-x)} = e^{\log(x^3-x)^{-1}} = \frac{1}{x^3-x}$$

Its solution is

$$y(\text{I.F.}) = \int Q(I.F.)\,dx+C \quad \Rightarrow \quad y\left(\frac{1}{x^3-x}\right) = \int \frac{x^2-1}{x^3-x}dx+C$$

$$\Rightarrow \quad \frac{y}{x^3-x} = \int \frac{x^2-1}{x(x^2-1)}dx+C \quad \Rightarrow \quad \frac{y}{x^3-x} = \int \frac{1}{x}dx+C$$

$$\Rightarrow \quad \frac{y}{x^3-x} = \log x + C \quad \Rightarrow \quad y = (x^3-x)\log x + (x^3-x)C \qquad \textbf{Ans.}$$

**Example 12.** *Solve* $\sin x\dfrac{dy}{dx}+2y=\tan^3\left(\dfrac{x}{2}\right)$ *(Nagpur University, Summer 2004)*

**Solution.** Given equation : $\sin x\dfrac{dy}{dx}+2y=\tan^3\dfrac{x}{2} \Rightarrow \dfrac{dy}{dx}+\dfrac{2}{\sin x}y=\dfrac{\tan^3\frac{x}{2}}{\sin x}$

This is linear form of $\dfrac{dy}{dx}+Py=Q$

$$\therefore \quad P=\frac{2}{\sin x} \quad \text{and} \quad Q=\frac{\tan^3\frac{x}{2}}{\sin x}$$

$$\therefore \quad \text{I.F.} = e^{\int Pdx} = e^{\int \frac{2}{\sin x}dx} = e^{2\int \operatorname{cosec} x\,dx} = e^{2\log\tan\frac{x}{2}} = \tan^2\frac{x}{2}$$

$\therefore$ Solution is $\quad y.(\text{I.F.}) = \int \text{I.F.}(Q\,dx)+C$

$$y\tan^2\frac{x}{2} = \int \tan^2\frac{x}{2}\cdot\frac{\tan^3\frac{x}{2}}{2\sin\frac{x}{2}\cdot\cos\frac{x}{2}}+C = \frac{1}{2}\int\frac{\tan^4\frac{x}{2}}{\cos^2\frac{x}{2}}dx+C$$

$$= \frac{1}{2}\int \tan^4\frac{x}{2}.\sec^2\frac{x}{2}\,dx+C \qquad \dots (1)$$

Putting $\tan\dfrac{x}{2}=t$ so that $\dfrac{1}{2}\sec^2\dfrac{x}{2}dx=dt$ on R.H.S. (1), we get

$$y.\tan^2\frac{x}{2} = \frac{1}{2}\int t^4(2dt)+C \quad \Rightarrow \quad y\tan^2\frac{x}{2}=\frac{t^5}{5}+C$$

$$y\tan^2\frac{x}{2} = \frac{\tan^5\frac{x}{2}}{5}+C \qquad \textbf{Ans.}$$

## EXERCISE 3.5

Solve the following differential equations:

1. $\dfrac{dy}{dx}+\dfrac{1}{x}y=x^3-3$ **Ans.** $xy=\dfrac{x^5}{5}-\dfrac{3x^2}{2}+C$

2. $(2y - 3x)\, dx + x\, dy = 0$ **Ans.** $y x^2 = x^3 + C$

3. $\dfrac{dy}{dx} + y \cot x = \cos x$ **Ans.** $y \sin x = \dfrac{\sin^2 x}{2} + C$

4. $\dfrac{dy}{dx} + y \sec x = \tan x$ **Ans.** $y = \dfrac{C - x}{\sec x + \tan x} + 1$

5. $\cos^2 x \dfrac{dy}{dx} + y = \tan x$ **Ans.** $y = \tan x - 1 + Ce^{-\tan x}$

6. $(x + a)\dfrac{dy}{dx} - 3y = (x + a)^5$ **Ans.** $2y = (x + a)^5 + 2C\,(x + a)^3$

7. $x \cos x \dfrac{dy}{dx} + y(x \sin x + \cos x) = 1$ **Ans.** $x\, y = \sin x + C \cos x$

8. $x \log x \dfrac{dy}{dx} + y = 2 \log x$ **Ans.** $y \log x = (\log x)^2 + C$

9. $x\dfrac{dy}{dx} + 2y = x^2 \log x$ **Ans.** $yx^2 = \dfrac{x^4}{4} \log x - \dfrac{x^4}{16} + C$

10. $dr + (2r \cot \theta + \sin 2\theta)\, d\theta = 0$ **Ans.** $r \sin^2 \theta = \dfrac{-\sin^4 \theta}{2} + C$

11. $\dfrac{dy}{dx} + y \cos x = \dfrac{1}{2} \sin 2x$ **Ans.** $y = \sin x - 1 + Ce^{-\sin x}$

12. $(1 - x^2)\dfrac{dy}{dx} + 2xy = x(1 - x^2)^{1/2}$ **Ans.** $y = \sqrt{1 - x^2} + C(1 - x^2)$

13. $\sec x \dfrac{dy}{dx} = y + \sin x$ *(A.M.I.E.T.E., Dec 2005)* **Ans.** $y = -\sin x - 1 + ce^{\sin x}$

14. $y' + y \tan x = \cos x, y(0) = 0$ *(A.M.I.E.T.E., June 2006)* **Ans.** $y = x \cos x$

15. Solve $(1 + y^2)\, dx = (\tan^{-1} y - x)\, dy$ *(AMIETE, Dec. 2009)* **Ans.** $x = -\tan^{-1} y - 1 + ce^{\tan^{-1} y}$

16. Find the value of α so that $e^2$ is an integrating factor of differential equation $x(1 - y)\, dx - dy = 0$. *(A.M.I.E.T.E., Summer 2005)* **Ans.** $\alpha = \dfrac{1}{2}$

17. Slove the differential equation $\cot 3x \dfrac{dy}{dx} - 3y = \cos 3x + \sin 3x,\ 0 < x < \dfrac{\pi}{2}$.

*(AMIETE, Dec. 2009)* **Ans.** $y \cos 3x = \dfrac{1}{12}[6x - \sin 6x - \cos 6x]$

18. The value of α so that $e^{\alpha y^2}$ is an integrating factor of the differential equation $(e^{\frac{-y^2}{2}} - xy)\, dy - dx = 0$ is *(A.M.I.E.T.E. Dec., 2005)*

(*a*) $-1$ (*b*) $1$ (*c*) $\dfrac{1}{2}$ (*d*) $-\dfrac{1}{2}$ **Ans.** (*c*)

19. The solution of the differential equation $(y + x)^2 \dfrac{dy}{dx} = a^2$ is given by

(*a*) $y + x = a \tan\left(\dfrac{y - c}{a}\right)$ (*b*) $y - x = \tan\left(\dfrac{y - c}{a}\right)$

(*c*) $y - x = a \tan (y - c)$ (*d*) $a\,(y - x) = \tan\left(y - \dfrac{c}{a}\right)$ **Ans.** (*a*)

*(AMIETE, June 2010)*

## 3.10 EQUATIONS REDUCIBLE TO THE LINEAR FORM (BERNOULLI EQUATION)

The equation of the form

$$\frac{dy}{dx}+Py=Qy^n \qquad ...(1)$$

where **P** and **Q** are constants or functions of $x$ can be reduced to the linear form on dividing by $y^n$ and substituting $\frac{1}{y^{n-1}}=z$

On dividing bothsides of (1) by $y^n$, we get

$$\frac{1}{y^n}\frac{dy}{dx}+\frac{1}{y^{n-1}}P=Q \qquad ...(2)$$

Put $\frac{1}{y^{n-1}}=z$, so that $\frac{(1-n)}{y^n}\frac{dy}{dx}=\frac{dz}{dx} \Rightarrow \frac{1}{y^n}\frac{dy}{dx}=\frac{1}{1-n}\frac{dz}{dx}$

$\therefore$ (2) becomes $\frac{1}{1-n}\frac{dz}{dx}+Pz=Q$ or $\frac{dz}{dx}+P(1-n)z=Q(1-n)$

which is a linear equation and can be solved easily by the previous method discussed in article 3.8 on page 144.

**Example 13.** *Solve* $x^2dy + y(x + y)\, dx = 0$ (*U.P. II Semester Summer 2006*)

**Solution.** We have, $x^2\, dy + y\,(x + y)\, dx = 0$

$$\Rightarrow \quad \frac{dy}{dx}+\frac{y}{x}=-\frac{y^2}{x^2} \quad \Rightarrow \quad \frac{1}{y^2}\frac{dy}{dx}+\frac{1}{xy}=-\frac{1}{x^2}$$

Put $-\frac{1}{y}=z$ so that $\frac{1}{y^2}\frac{dy}{dx}=\frac{dz}{dx}$

The given equation reduces to a linear differential equation in $z$.

$$\frac{dz}{dx}-\frac{z}{x}=-\frac{1}{x^2}$$

$$\text{I.F.}=e^{-\int\frac{1}{x}dx}=e^{-\log x}=e^{\log 1/x}=\frac{1}{x}.$$

Hence the solution is

$$z.\frac{1}{x}=\int -\frac{1}{x^2}.\frac{1}{x}\,dx+C \quad \Rightarrow \quad \frac{z}{x}=\int -x^{-3}\,dx+C$$

$$\Rightarrow \quad -\frac{1}{xy}=-\frac{x^{-2}}{-2}+C \quad \Rightarrow \quad \frac{1}{xy}=-\frac{1}{2x^2}-C \qquad \textbf{Ans.}$$

**Example 14.** *Solve*: $x\frac{dy}{dx}+y\log y=xy\,e^x$ (*A.M.I.E., Summer 2000*)

**Solution.** $x\frac{dy}{dx}+y\log y=xye^x$

Dividing by $xy$, we get

$$\frac{1}{y}\frac{dy}{dx}+\frac{1}{x}\log y=e^x \qquad ...(1)$$

Put $\log y = z$, so that $\frac{1}{y}\frac{dy}{dx}=\frac{dz}{dx}$

Equation (1) becomes, $\frac{dz}{dx}+\frac{z}{x}=e^x$

$$\text{I.F.} = e^{\int \frac{1}{x} dx} = e^{\log x} = x$$

Solution is $\quad zx = \int x e^x \, dx + C$

$$zx = xe^x - e^x + C$$

$\Rightarrow \quad x \log y = xe^x - e^x + C$ **Ans.**

**Example 15.** *Solve*: $\dfrac{dy}{dx} - \dfrac{\tan y}{1+x} = (1+x)e^x \sec y.$ *(Nagpur University, Summer 2000)*

**Solution.** $\quad \dfrac{dy}{dx} - \dfrac{\tan y}{1+x} = (1+x)e^x \sec y$

$\Rightarrow \quad \cos y \dfrac{dy}{dx} - \dfrac{\sin y}{1+x} = (1+x)e^x$ ...(1)

Put $\quad \sin y = z,$ so that $\cos y \dfrac{dy}{dx} = \dfrac{dz}{dx}$

(1) becomes $\quad \dfrac{dz}{dx} - \dfrac{z}{1+x} = (1+x)e^x$

$$I.F. = e^{-\int \frac{1}{1+x} dx} = e^{-\log(1+x)} = e^{\log 1/_{1+x}} = \frac{1}{1+x}$$

Solution is $\quad z.\dfrac{1}{1+x} = \int (1+x)e^x . \dfrac{1}{1+x} dx + C = \int e^x dx + C$

$$\frac{\sin y}{1+x} = e^x + C$$ **Ans.**

**Example 16.** *Solve*: $\tan y \dfrac{dy}{dx} + \tan x = \cos y \cos^2 x$ *(Nagpur University, Summer 2000)*

**Solution.** $\quad \tan y \dfrac{dy}{dx} + \tan x = \cos y \cos^2 x$

$$\sec y \tan y \frac{dy}{dx} + \sec y \tan x = \cos^2 x$$

Writing $\quad z = \sec y,$ so that $\dfrac{dz}{dx} = \sec y \tan y \dfrac{dy}{dx}$

The equation becomes $\dfrac{dz}{dx} + z \tan x = \cos^2 x$

$$\text{I.F.} = e^{\int \tan x \, dx} = e^{\log \sec x} = \sec x$$

$\therefore \quad$ The solution of the equation is

$$z \sec x = \int \cos^2 x \sec x \, dx + C$$

$$\sec y \sec x = \int \cos x \, dx + C = \sin x + C$$

$$\sec y = (\sin x + C) \cos x$$ **Ans.**

**Example 17.** $\quad x\left[\dfrac{dx}{dy} + y\right] = 1 - y$ *(Nagpur University, Summer 2004)*

**Solution.** $\quad x\left(\dfrac{dy}{dx} + y\right) = (1 - y)$

$\Rightarrow \quad \dfrac{dy}{dx} + y = \dfrac{1}{x} - \dfrac{y}{x} \quad \Rightarrow \quad \dfrac{dy}{dx} + \left(1 + \dfrac{1}{x}\right) y = \dfrac{1}{x}$

which is in linear form of $\frac{dy}{dx}+Py=Q.$

$$\therefore \qquad P=\left(1+\frac{1}{x}\right), \qquad Q=\frac{1}{x}$$

$$\text{I.F.}=e^{\int P dx}=e^{\int\left(1+\frac{1}{x}\right)dx}=e^{x+\log x}=e^{x}.e^{\log x}=e^{x}.x=xe^{x}$$

Its solution is

$$y(\text{I.F.})=\int \text{I.F.}(Q\,dx)+C$$

$$y(x.e^{x})=\int (x.e^{x})\times\frac{1}{x}dx+C \Rightarrow y(x.e^{x})=\int e^{x}dx+C$$

$$y\,(x\,.\,e^{x})=e^{x}+C$$

$$\therefore \qquad y=\frac{1}{x}+\frac{C}{x}e^{-x}$$ **Ans.**

**Example 18.** *Solve the differential equation.*

$$y\log y\,dx+(x-\log y)\,dy=0$$ *(Uttarakhand II Semester, June 2007)*

**Solution.** We have,

$$y\log y\,dx+(x-\log y)\,dy=0$$

$$\Rightarrow \quad \frac{dx}{dy}=\frac{-x+\log y}{y\log y} \quad \Rightarrow \quad \frac{dx}{dy}=\frac{-x}{y\log y}+\frac{\log y}{y\log y}$$

$$\Rightarrow \quad \frac{dx}{dy}+\frac{x}{y\log y}=\frac{1}{y}$$

$$\text{I.F.}=e^{\int\frac{1}{y\log y}dy}=e^{\log(\log y)}=\log y$$

Its solution is $\quad x.\log y=\int\frac{1}{y}(\log y)\,dy$

$$x.\log y=\frac{(\log y)^{2}}{2}+C$$ **Ans.**

**Example 19.** *Solve*: $(1+y^2)\,dx=(\tan^{-1}y-x)\,dy.$

*(AMIETE, June 2010, 2004, R.G.P.V., Bhopal, April 2010, June 2008, U.P. (B. Pharm) 2005)*

**Solution.** $(1+y^2)\,dx=(\tan^{-1}y-x)\,dy$

$$\frac{dx}{dy}=\frac{\tan^{-1}y-x}{1+y^{2}} \quad \Rightarrow \quad \frac{dx}{dy}+\frac{x}{1+y^{2}}=\frac{\tan^{-1}y}{1+y^{2}}$$

This is a linear differential equation.

$$\text{I.F.}=e^{\int\frac{1}{1+y^{2}}dy}=e^{\tan^{-1}y}$$

Its solution is $\quad x.e^{\tan^{-1}y}=\int e^{\tan^{-1}y}\frac{\tan^{-1}y}{1+y^{2}}dy+C$

Put $\quad \tan^{-1}y=t$ on R.H.S., so that $\frac{1}{1+y^{2}}dy=dt$

$$x.e^{\tan^{-1}y}=\int e^{t}.t\,dt+C=t.e^{t}-e^{t}+C=e^{\tan^{-1}y}\left(\tan^{-1}y-1\right)+C$$

$$x=\left(\tan^{-1}y-1\right)+Ce^{-\tan^{-1}y}$$ **Ans.**

**Example 20.** *Solve* : $r\sin\theta - \frac{dr}{d\theta}\cos\theta = r^2$ *(Nagpur University, Summer 2005)*

**Solution.** The given equation can be written as $-\frac{dr}{d\theta}\cos\theta + r\sin\theta = r^2$ ... (1)

Dividing (1) by $r^2\cos\theta$, we get $-r^{-2}\frac{dr}{d\theta} + r^{-1}\tan\theta = \sec\theta$ ... (2)

Putting $r^{-1} = v$ so that $-r^{-2}\frac{dr}{d\theta} = \frac{dv}{d\theta}$ in (2), we get

$$\frac{dv}{d\theta} + v\tan\theta = \sec\theta$$

$$\text{I.F.} = e^{\int \tan\theta\, d\theta} = e^{\log\sec\theta} = \sec\theta$$

Solution is $v\sec\theta = \int \sec\theta,\ \sec\theta + C \quad\Rightarrow\quad v\sec\theta = \int \sec^2\theta\, d\theta + C$

$$\frac{\sec\theta}{r} = \tan\theta + C \quad\Rightarrow\quad r^{-1} = (\sin\theta + C\cos\theta)$$

$$\therefore \quad r = \frac{1}{\sin\theta + C\cos\theta}$$ **Ans.**

## EXERCISE 3.6

**Solve the following differential equations:**

1. $\frac{1}{y^2}\frac{dy}{dx} - \frac{1}{y} = 2xe^{-x}$ **Ans.** $e^x + x^2y + Cy = 0$
2. $3\frac{dy}{dx} + 3\frac{y}{x} = 2x^4y^4$ **Ans.** $\frac{1}{y^3} = x^5 + Cx^3$
3. $\frac{dy}{dx} = y\tan x - y^2\sec x$ **Ans.** $\sec x = (\tan x + C)\, y$
4. $\frac{dy}{dx} = 2y\tan x + y^2\tan^2 x$, if $y = 1$ at $x = 0$ **Ans.** $\frac{1}{y}\sec^2 x = -\frac{\tan^3 x}{3} + 1$
5. $\frac{dy}{dx} + \tan x\tan y = \cos x\sec y$ **Ans.** $\sin y\sec x = x + C$
6. $dy + y\tan x\,.\,dx = y^2\sec x\,.\,dx$ **Ans.** $y(x + C) + \cos x = 0$
7. $(x^2y^2 + xy)\,y\,dx + (x^2y^2 - 1)\,x\,dy = 0$ **Ans.** $xy = \log Cy$
8. $(x^2 + y^2 + x)\,dx + xy\,dy = 0$ **Ans.** $x^2y^2 = -\frac{x^4}{2} - \frac{2x^3}{3} + C$
9. $\frac{dy}{dx} + y = 3e^xy^3$ **Ans.** $\frac{1}{y^2} = 6e^x + Ce^{2x}$
10. $(x - y^2)\,dx + 2xy\,dy = 0$ **Ans.** $\frac{y^2}{x} + \log x = C$
11. $e^y\left(\frac{dy}{dx} + 1\right) = e^x$ **Ans.** $e^{x+y} = \frac{e^{2x}}{2} + C$
12. $x^2y - x^3\frac{dy}{dx} = y^4\cos x$ **Ans.** $x^3 = y^3(3\sin x - C)$
13. $3\frac{dy}{dx} + \frac{2}{x+1}.y = \frac{x^2}{y^2}$ **Ans.** $y^3(x+1)^2 = \frac{x^5}{5} + \frac{x^4}{2} + \frac{x^3}{3} + C$
14. $\cos x\frac{dy}{dx} + 4y\sin x = 4\sqrt{y}\sec x$ **Ans.** $\sqrt{y}\sec^2 x = 2\left[\tan x + \frac{\tan^3 x}{3}\right] + C$

15. $\frac{dy}{dx}+x\sin 2y=x^3\cos^2 y$ **Ans.** $\tan y=\frac{1}{2}(x^2-1)+Ce^{-x^2}$

16. $\frac{1}{1+y^2}\frac{dy}{dx}+2x\tan^{-1}y=x^3$ **Ans.** $\tan^{-1}y=\frac{1}{2}(x^2-1)+Ce^{-x^2}$

17. $e^{-y}\sec^2 y\,dy = dx + x\,dy$ **Ans.** $x\,e^y = \tan y + C$

18. $(x+y+1)\frac{dy}{dx}=1$ **Ans.** $x + y + 2 = C\,e^y$

19. $\frac{dy}{dx}=\frac{y^3}{e^{2x}+y^2}$ **Ans.** $e^{-2x}y^2 + 2\log y + C = 0$

20. $dx - xy(1 + xy^2)\,dy = 0$ **Ans.** $-\frac{1}{x}=y^2-2+Ce^{-y^2/2}$

21. $\frac{dy}{dx}+\frac{y}{x}\log y=\frac{y}{x^2}(\log y)^2$ *(A.M.I.E.T.E., Summer 2004, 2003, Winter 2003, 2001)*

**Ans.** $\frac{1}{x\log y}=\frac{1}{2x^2}+C$

22. $3\frac{dy}{dx}+xy=xy^{-2}$ *(A.M.I.E.T.E., June 2009)* **Ans.** $y^3=1+Ce^{-x^2/2}$

27. $x\frac{dy}{dx}+y=x^3y^6$ *(AMIETE, June 2010)* **Ans.** $\frac{1}{y^5x^5}=\frac{5}{2x^2}+C$

23. General solution of linear differential equation of first order $\frac{dx}{dy}+Px=Q$ (where $P$ and $Q$ are constants or functions of $y$) is

(*a*) $ye^{\int P.dx}=\int Qe^{\int P.dx}\,dx+c$ (*b*) $xe^{\int P.dy}=\int Qe^{\int P.dy}\,dy+c$

(*c*) $y=\int Qe^{\int P.dx}\,dx+c$ (*d*) $x=\int Qe^{\int P.dy}\,dy+c$ *(AMIETE, June, 2010)* **Ans.** (*b*)

## 3.11 EXACT DIFFERENTIAL EQUATION

An exact differential equation is formed by directly differentiating its primitive (solution) without any other process

$$Mdx + Ndy = 0$$

is said to be an exact differential equation if it satisfies the following condition

$$\boxed{\frac{\partial M}{\partial y}=\frac{\partial N}{\partial x}}$$

where $\frac{\partial M}{\partial y}$ denotes the differential co-efficient of $M$ with respect to $y$ keeping $x$ constant and $\frac{\partial N}{\partial x}$, the differential co-efficient of $N$ with respect to $x$, keeping $y$ constant.

**Method for Solving Exact Differential Equations**

**Step I.** Integrate $M$ w.r.t. $x$ keeping $y$ constant

**Step II.** Integrate w.r.t. $y$, only those terms of $N$ which do not contain $x$.

**Step III.** Result of I + Result of II = Constant.

**Example 21.** *Solve* :

$$(5x^4 + 3x^2y^2 - 2xy^3)\,dx + (2x^3y - 3x^2y^2 - 5y^4)\,dy = 0$$

**Solution.** Here, $M = 5x^4 + 3x^2y^2 - 2xy^3$, $N = 2x^3y - 3x^2y^2 - 5y^4$

$$\frac{\partial M}{\partial y} = 6x^2y - 6xy^2, \quad \frac{\partial N}{\partial x} = 6x^2y - 6xy^2$$

Since, $\dfrac{\partial M}{\partial y} = \dfrac{\partial N}{\partial x}$, the given equation is exact.

Now $\int M\,dx + \int(\text{terms of } N \text{ is not containing } x)\,dy = C$ ($y$ constant)

$$\int\left(5x^4 + 3x^2y^2 - 2xy^3\right)dx + \int -5y^4\,dy = C$$

$$\Rightarrow \quad x^5 + x^3y^2 - x^2y^3 - y^5 = C$$ **Ans.**

**Example 22.** *Solve*: $\left\{2xy\cos x^2 - 2xy + 1\right\}dx + \left\{\sin x^2 - x^2 + 3\right\}dy = 0$

*(Nagpur University, Summer 2000)*

**Solution.** Here we have

$$\left\{2xy\cos x^2 - 2xy + 1\right\}dx + \left\{\sin x^2 - x^2 + 3\right\}dy = 0 \quad \text{... (1)}$$

$$M\,dx + N\,dy = 0 \quad \text{... (2)}$$

Comparing (1) and (2), we get

$$M = 2xy\cos x^2 - 2xy + 1 \quad \Rightarrow \quad \frac{\partial M}{\partial y} = 2x\cos x^2 - 2x$$

$$N = \sin x^2 - x^2 + 3 \quad \Rightarrow \quad \frac{\partial N}{\partial x} = 2x\cos x^2 - 2x$$

Here, $\therefore \quad \dfrac{\partial M}{\partial y} = \dfrac{\partial N}{\partial x}$

So the given differential equation is exact differential equation.

Hence solution is $\int_{y \text{ as const}} M\,dx + \int (\text{terms of } N \text{ not containing } x)\,dy = C$

$$\int (2xy\cos x^2 - 2xy + 1)\,dx + \int 3\,dy = C$$

$$\Rightarrow \quad \int[y(2x\cos x^2) - y(2x) + 1]\,dx + 3\int dy = C$$

$$\Rightarrow \quad y\int 2x\cos x^2\,dx - y\int 2x\,dx + \int 1\,dx + 3\int y\,dy = C$$

Put $x^2 = t$ so that $2x\,dx = dt$

$$y\int \cos t\,dt - 2y\frac{x^2}{2} + x + 3y = C$$

$$\Rightarrow \quad y\sin t - x^2y + x + 3y = C$$

$$y\sin x^2 - yx^2 + x + 3y = C$$ **Ans.**

**Example 23.** *Solve* :

$$(1 + e^{x/y}) + e^{x/y}\left(1 - \frac{x}{y}\right)\frac{dy}{dx} = 0$$

*(Nagpur University, Summer 2008, A.M.I.E.T.E. June, 2009)*

**Solution.** We have,

$$\left(1 + e^{\frac{x}{y}}\right) + e^{\frac{x}{y}}\left(1 - \frac{x}{y}\right)\frac{dy}{dx} = 0 \quad \Rightarrow \quad \left(1 + e^{\frac{x}{y}}\right)dx + \left(e^{\frac{x}{y}} - e^{\frac{x}{y}}\frac{x}{y}\right)dy = 0$$

$$M = 1 + e^{\frac{x}{y}} \quad \Rightarrow \quad \frac{\partial M}{\partial y} = -\frac{x}{y^2}e^{\frac{x}{y}}$$

$$N = e^{\frac{x}{y}} - e^{\frac{x}{y}}\frac{x}{y} \quad \Rightarrow \quad \frac{\partial N}{\partial x} = \frac{1}{y}e^{\frac{x}{y}} - \frac{1}{y}e^{\frac{x}{y}} - \frac{x}{y^2}e^{\frac{x}{y}} = -\frac{x}{y^2}e^{\frac{x}{y}}$$

$$\Rightarrow \quad \frac{\partial M}{\partial y} = \frac{\partial N}{\partial x}$$

$\therefore$ Given equation is exact.

Its solution is $\int\left(1+e^{\frac{x}{y}}\right)dx + \int(\text{terms of } N \text{ not containing } x)\,dy = C$

$$\Rightarrow \quad \int\left(1+e^{\frac{x}{y}}\right)dx + \int 0\,dy = C \quad \Rightarrow \quad x + ye^{\frac{x}{y}} = C$$ **Ans.**

**Example 24.** *Solve*: $[1+\log(x\,y)]\,dx + \left[1+\frac{x}{y}\right]dy = 0$ *(Nagpur University, Winter 2003)*

**Solution.** $[1+\log x\,y]\,dx + \left[1+\frac{x}{y}\right]dy = 0$

$$\therefore \quad [1+\log x+\log y]\,dx + \left[1+\frac{x}{y}\right]dy = 0$$

which is in the form $M\,dx + N\,dy = 0$

$$M = [1 + \log x + \log y] \quad \text{and} \quad N = 1+\frac{x}{y}$$

$$\Rightarrow \quad \frac{\partial M}{\partial y} = \frac{1}{y} \quad \text{and} \quad \Rightarrow \quad \frac{\partial N}{\partial x} = \frac{1}{y} \quad \Rightarrow \quad \frac{\partial M}{\partial y} = \frac{\partial N}{\partial x}$$

Hence the given differential equation is exact.

$\therefore$ Solution is $\int_{y\text{ constant}} M\,dx + \int N\,(\text{terms not containing } x)\,dy = C$

$y$ constant

$$\therefore \quad \int(1+\log x+\log y)\,dx + \int dy = C$$

$$\Rightarrow \quad x + \int \log x\,dx + \int \log y\,dx + y = C \quad \ldots (1)$$

Now, $\int \log x\,dx = \int \log x.(1)\,dx = (\log x)x - \int\left[\frac{d}{dx}(\log x)x\right]dx = x\log x - \int \frac{1}{x}.x\,dx$

$$= x\log x - \int dx = x\log x - x = x[\log x - 1]$$

$\therefore$ Equation (1) becomes $\Rightarrow$ $x + x\log x - x + x\log y + y = C$

$$x\,[\log x + \log y] + y = C \Rightarrow x\log xy + y = C$$ **Ans.**

## EXERCISE 3.7

*Solve the following differential equations (1 – 12).*

1. $(x + y - 10)\,dx + (x - y - 2)\,dy = 0$ **Ans.** $\frac{x^2}{2} + xy - 10x - \frac{y^2}{2} - 2y = C$

2. $(y^2 - x^2)\,dx + 2x\,y\,dy = 0$ **Ans.** $\frac{x^3}{3} = x\,y^2 + C$

3. $\left(1+3e^{x/y}\right)dx + 3e^{x/y}\left(1-\frac{x}{y}\right)dy = 0$ *(R.G.P.V. Bhopal, Winter 2010)* **Ans.** $x + 3y\,e^{x/y} = C$

4. $(2x - y)\,dx = (x - y)\,dy$ **Ans.** $xy = x^2 + \dfrac{y^2}{2} + C$

5. $(y \sec^2 x + \sec x \tan x)\,dx + (\tan x + 2y)\,dy = 0$ **Ans.** $y \tan x + \sec x + y^2 = C$

6. $(ax + hy + g)\,dx + (hx + by + f)\,dy = 0$ **Ans.** $ax^2 + 2h\,xy + by^2 + 2gx + 2fy + C = 0$

7. $(x^4 - 2xy^2 + y^4)\,dx - (2x^2y - 4xy^3 + \sin y)\,dy = 0$ **Ans.** $\dfrac{x^5}{5} - x^2y^2 + xy^4 + \cos y = C$

8. $(2xy + e^y)\,dx + (x^2 + xe^y)\,dy = 0$ **Ans.** $x^2y + xe^y = C$

9. $(x^2 + 2ye^{2x})\,dy + (2xy + 2y^2e^{2x})\,dx = 0$ **Ans.** $x^2y + y^2 e^{2x} = C$

10. $\left[y\left(1+\dfrac{1}{x}\right)+\cos y\right]dx+(x+\log x-x\sin y)\,dy=0$ *(M.D.U., 2010)*

**Ans.** $y\,(x + \log x) + x \cos y = C$

11. $(x^3 - 3xy^2)\,dx + (y^3 - 3x^3y)\,dy = 0,\ y(0) = 1$ **Ans.** $x^4 - 6x^2y^2 + y^4 = 1$

12. The differential equation $M(x, y)\,dx + N(x, y)\,dy = 0$ is an exact differential equation if

(*a*) $\dfrac{\partial M}{\partial y}+\dfrac{\partial N}{\partial x}=0$ (*b*) $\dfrac{\partial M}{\partial y}-\dfrac{\partial N}{\partial x}=0$ (*c*) $\dfrac{\partial M}{\partial y}\times\dfrac{\partial N}{\partial x}=1$ (*d*) None of the above

*(A.M.I.E.T.E. Dec. 2010, Dec 2006)* **Ans.** (*b*)

## 3.12 EQUATIONS REDUCIBLE TO THE EXACT EQUATIONS

Sometimes a differential equation which is not exact may become so, on multiplication by a suitable function known as the integrating factor.

**Rule 1.** If $\dfrac{\dfrac{\partial M}{\partial y}-\dfrac{\partial N}{\partial x}}{N}$ is a function of $x$ alone, say $f(x)$, then $\text{I.F.} = e^{\int f(x)dx}$

**Example 25.** *Solve* $(2x \log x - xy)\,dy + 2y\,dx = 0$ ... (1)

**Solution.** $M = 2y,\qquad N = 2x \log x - xy$

$$\frac{\partial M}{\partial y} = 2,\qquad \frac{\partial N}{\partial x} = 2(1+\log x) - y$$

Here, $$\frac{\dfrac{\partial M}{\partial y}-\dfrac{\partial N}{\partial x}}{N}=\frac{2-2-2\log x+y}{2x\log x-xy}=\frac{-(2\log x-y)}{x\,(2\log x-y)}=-\frac{1}{x}=f(x)$$

$$\text{I.F.} = e^{\int f(x)dx} = e^{\int -\frac{1}{x}dx} = e^{-\log x} = e^{\log x^{-1}} = x^{-1} = \frac{1}{x}$$

On multiplying the given differential equation (1) by $\dfrac{1}{x}$, we get

$$\frac{2y}{x}dx + (2\log x - y)dy = 0 \quad\Rightarrow\quad \int\frac{2y}{x}dx + \int -y\,dy = c$$

$$\Rightarrow\qquad 2y\log x - \frac{1}{2}y^2 = c$$ **Ans.**

### EXERCISE 3.8

*Solve the following differential equations:*

1. $(y \log y)\,dx + (x - \log y)\,dy = 0$ **Ans.** $2x \log y = c + (\log y)^2$

2. $\left(y+\dfrac{1}{3}y^3+\dfrac{1}{2}x^2\right)dx+\dfrac{1}{4}\left(1+y^2\right)x\,dy=0$ **Ans.** $\dfrac{yx^4}{4}+\dfrac{y^3x^4}{12}+\dfrac{x^6}{12}=c$

3. $(y - 2x^3)\,dx - x\,(1 - xy)\,dy = 0$ **Ans.** $-\dfrac{y}{x} - x^2 + \dfrac{y^2}{2} = c$

4. $(x \sec^2 y - x^2 \cos y)\,dy = (\tan y - 3x^4)\,dx$ **Ans.** $-\dfrac{1}{x}\tan y - x^3 + \sin y = c$

5. $(x - y^2)\,dx + 2xy\,dy = 0$ **Ans.** $y^2 = cx - x \log x$

**Rule II.** If $\dfrac{\dfrac{\partial N}{\partial x} - \dfrac{\partial M}{\partial y}}{M}$ is a function of $y$ alone, say $f(y)$, then

$$\text{I.F.} = e^{\int f(y)dy}$$

**Example 26.** *Solve* $(y^4 + 2y)\,dx + (xy^3 + 2y^4 - 4x)\,dy = 0$

**Solution.** Here $M = y^4 + 2y$; $\qquad N = xy^3 + 2y^4 - 4x$ ...(1)

$$\therefore \quad \frac{\partial M}{\partial y} = 4y^3 + 2; \qquad \frac{\partial N}{\partial x} = y^3 - 4$$

$$\therefore \quad \frac{\dfrac{\partial N}{\partial x} - \dfrac{\partial M}{\partial y}}{M} = \frac{(y^3 - 4) - (4y^3 + 2)}{y^4 + 2y} = \frac{-3(y^3 + 2)}{y(y^3 + 2)} = -\frac{3}{y} = f(y)$$

$$\text{I.F.} = e^{\int f(y)dy} = e^{\int -\frac{3}{y}dy} = e^{-3\log y} = e^{\log y^{-3}} = y^{-3} = \frac{1}{y^3}$$

On multiplying the given equation (1) by $\dfrac{1}{y^3}$ we get the exact differential equation.

$$\left(y + \frac{2}{y^2}\right)dx + \left(x + 2y - \frac{4x}{y^3}\right)dy = 0$$

$$\int\left(y + \frac{2}{y^2}\right)dx + \int 2y\,dy = c \qquad \Rightarrow \qquad x\left(y + \frac{2}{y^2}\right) + y^2 = c \qquad \textbf{Ans.}$$

## EXERCISE 3.9

*Solve the following differential equations:*

1. $(3x^2y^4 + 2xy)\,dx + (2x^3y^3 - x^2)\,dy = 0$ **Ans.** $x^3y^2 + \dfrac{x^2}{y} = c$

2. $(xy^3 + y)\,dx + 2(x^2y^2 + x + y^4)\,dy = 0$ **Ans.** $\dfrac{x^2y^4}{2} + xy^2 + \dfrac{y^6}{3} = c$

3. $y(x^2y + e^x)dx - e^x dy = 0$ **Ans.** $\dfrac{x^3}{3} + \dfrac{e^x}{y} = c$

4. $(2x^4y^4e^y + 2xy^3 + y)\,dx + (x^2y^4e^y - x^2y^2 - 3x)\,dy = 0$ **Ans.** $x^2e^y + \dfrac{x^2}{y} + \dfrac{x}{y^3} = c$

**Rule III.** If $M$ is of the form $M = y\,f_1(xy)$ and N is of the form $N = x\,f_2(xy)$

Then $\qquad \text{I.F.} = \dfrac{1}{M.x - N.y}$

**Example 27.** *Solve* $y\,(xy + 2x^2y^2)\,dx + x\,(xy - x^2y^2)\,dy = 0$

**Solution.** $\qquad y\,(xy + 2x^2y^2)\,dx + x\,(xy - x^2y^2)\,dy = 0$ ... (1)

Dividing (1) by $xy$, we get

$$y(1+2xy)\,dx + x(1-xy)\,dy = 0 \quad \text{... (2)}$$

$$M = y f_1(xy), \quad N = x f_2(xy)$$

$$\text{I.F.} = \frac{1}{Mx - Ny} = \frac{1}{xy(1+2xy) - xy(1-xy)} = \frac{1}{3x^2y^2}$$

On multiplying (2) by $\frac{1}{3x^2y^2}$, we have an exact differential equation

$$\left(\frac{1}{3x^2y} + \frac{2}{3x}\right)dx + \left(\frac{1}{3xy^2} - \frac{1}{3y}\right)dy = 0 \quad \Rightarrow \quad \int\left(\frac{1}{3x^2y} + \frac{2}{3x}\right)dx + \int -\frac{1}{3y}\,dy = c$$

$$\Rightarrow \quad -\frac{1}{3xy} + \frac{2}{3}\log x - \frac{1}{3}\log y = c \quad \Rightarrow \quad -\frac{1}{xy} + 2\log x - \log y = b \qquad \textbf{Ans.}$$

## EXERCISE 3.10

*Solve the following differential equations*

1. $(y - xy^2)\,dx - (x + x^2y)\,dy = 0$ **Ans.** $\log\left(\frac{x}{y}\right) - xy = A$

2. $y(1 + xy)\,dx + x(1 - xy)\,dy = 0$ **Ans.** $xy\log\left(\frac{y}{x}\right) = cxy - 1$

2. $y(1 + xy)\,dx + x(1 + xy + x^2y^2)\,dy = 0$ **Ans.** $\frac{1}{2x^2y^2} + \frac{1}{xy} - \log y = c$

4. $(xy\sin xy + \cos xy)\,y\,dx + (xy\sin xy - \cos xy)\,x\,dy = 0$ **Ans.** $y\cos xy = cx$

**Rule IV.** For of this type of $x^m y^n(ay\,dx + bx\,dy) + x^{m'}y^{n'}(a'y\,dx + b'x\,dy) = 0$, the integrating factor is $x^h y^k$.

where $\frac{m+h+1}{a} = \frac{n+k+1}{b}$, and $\frac{m'+h+1}{a'} = \frac{n'+k+1}{b'}$

**Example 28.** *Solve* $(y^3 - 2x^2y)\,dx + (2xy^2 - x^3)\,dy = 0$

**Solution.** $(y^3 - 2x^2y)\,dx + (2xy^2 - x^3)\,dy = 0$

$$y^2(y\,dx + 2x\,dy) + x^2(-2y\,dx - x\,dy) = 0$$

Here $m = 0,\ h = 2,\ a = 1,\ b = 2, \quad m' = 2,\ n' = 0,\ a' = -2,\ b' = -1$

$$\frac{0+h+1}{1} = \frac{2+k+1}{2} \quad \text{and} \quad \frac{2+h+1}{-2} = \frac{0+k+1}{-1}$$

$$\Rightarrow \quad 2h + 2 = 2 + k + 1 \text{ and } h + 3 = 2k + 2$$

$$\Rightarrow \quad 2h - k = 1 \text{ and } h - 2k = -1$$

On solving $h = k = 1$. Integrating Factor $= xy$

Multiplying the given equation by $xy$, we get

$$(xy^4 - 2x^3y^2)\,dx + (2x^2y^3 - x^4y)\,dy = 0$$

which is an exact differential equation.

$$\int(xy^4 - 2x^3y^2)\,dx = C \quad \Rightarrow \quad \frac{x^2y^4}{2} - \frac{2x^4y^2}{4} = C$$

$$\Rightarrow \qquad x^2y^4 - x^4y^2 = C' \qquad \Rightarrow \quad x^2y^2(y^2 - x^2) = C' \qquad \textbf{Ans.}$$

**Example 29.** Solve $(3y - 2xy^3)\,dx + (4x - 3x^2y^2)\,dy = 0$. *(U.P., II Semester, June 2007)*

**Solution.** $(3y - 2xy^3)\,dx + (4x - 3x^2y^2)\,dy = 0$

$$\Rightarrow \quad (3y\,dx + 4x\,dy) + xy^2(-2y\,dx - 3x\,dy) = 0 \qquad ...(1)$$

Comparing the coefficients of (1) with

$$x^m y^n (a\,y\,dx + b\,x\,dy) + x^{m'} y^{n'} (a'\,y\,dx + b'\,x\,dy) = 0, \text{ we get}$$

$$m = 0,\ n = 0,\ a = 3,\ b = 4$$

$$m' = 1,\ n' = 2,\ a' = -2,\ b' = -3$$

**To find the integrating factor $x^h y^k$**

$$\frac{m+h+1}{a} = \frac{n+k+1}{b} \text{ and } \frac{m'+h+1}{a'} = \frac{n'+k+1}{b'}$$

$$\frac{0+h+1}{3} = \frac{0+k+1}{4} \text{ and } \frac{1+h+1}{-2} = \frac{2+k+1}{-3}$$

$$\Rightarrow \quad \frac{h+1}{3} = \frac{k+1}{4} \text{ and } \frac{h+2}{2} = \frac{k+3}{3} \Rightarrow 4h - 3k + 1 = 0 \qquad ...(2)$$

and $$3h - 2k = 0 \quad \Rightarrow \quad h = \frac{2k}{3} \qquad ...(3)$$

Putting the value of $h$ from (3) in (2), we get

$$\frac{8k}{3} - 3k + 1 = 0 \quad \Rightarrow \quad -\frac{k}{3} + 1 = 0 \quad \Rightarrow \quad k = 3$$

Putting $k = 3$ in (2), we get $h = \frac{2k}{3} = \frac{2 \times 3}{3} = 2$

$$\text{I.F.} = x^h y^k = x^2y^3$$

On multiplying the given differential equation by $x^2y^3$, we get

$$x^2y^3\,(3y - 2xy^3)dx + x^2y^3(4x - 3x^2y^2)\,dy = 0$$

$$(3x^2y^4 - 2x^3y^6)\,dx + (4x^3y^3 - 3x^4y^5)\,dy = 0$$

This is the exact differential equation.

Its solution is $\int (3x^2y^4 - 2x^3y^6)dx = 0 \quad \Rightarrow \quad x^3y^4 - \frac{x^4}{2}y^6 = C$ **Ans.**

## EXERCISE 3.11

*Solve the following differential equations.*

1. $(2y\,dx + 3x\,dy) + 2xy\,(3y\,dx + 4x\,dy) = 0$ **Ans.** $x^2y^3(1 + 2xy) = c$

2. $(y^2 + 2yx^2)\,dx + (2x^3 - xy)\,dy = 0$ **Ans.** $4(xy)^{1/2} - \frac{2}{3}\left(\frac{y}{x}\right)^{3/2} = c$

3. $(3x + 2y^2)y\,dx + 2x\,(2x + 3y^2)\,dy = 0$ **Ans.** $x^2y^4(x + y^2) = c$

4. $(2x^2y^2 + y)\,dx - (x^3y - 3x)\,dy = 0$ **Ans.** $\frac{7}{5}x^{10/7}y^{-5/7} - \frac{7}{4}x^{-4/7}y^{-12/7} = c$

5. $x\,(3y\,dx + 2x\,dy) + 8y^4\,(y\,dx + 3x\,dy) = 0$ **Ans.** $x^3y^2 + 4x^2y^6 = c$

**Rule V.**

If the given equation $M\,dx + N\,dy = 0$ is homogeneous equation and $Mx + Ny \neq 0$, then $\frac{1}{Mx+Ny}$ is an integrating factor.

**Example 30.** *Solve* $\frac{dy}{dx} = \frac{x^3+y^3}{xy^2}$

**Solution.** $(x^3 + y^3)\,dx - (xy^2)\,dy = 0$ ... (1)

Here $M = x^3 + y^3$, $N = -xy^2$

$$\text{I.F.} = \frac{1}{Mx+Ny} = \frac{1}{x(x^3+y^3)-xy^2(y)} = \frac{1}{x^4}$$

Multiplying (1) by $\frac{1}{x^4}$ we get $\frac{1}{x^4}(x^3+y^3)dx + \frac{1}{x^4}(-xy^2)dy = 0$

$$\Rightarrow \quad \left(\frac{1}{x}+\frac{y^3}{x^4}\right)dx - \frac{y^2}{x^3}dy = 0, \text{ which is an exact differential equation.}$$

$$\int\left(\frac{1}{x}+\frac{y^3}{x^4}\right)dx = c \quad \Rightarrow \quad \log x - \frac{y^3}{3x^3} = c \quad \textbf{Ans.}$$

## EXERCISE 3.12

*Solve the following differential equations:*

1. $x^2y\,dx - (x^3 + y^3)\,dy = 0$ **Ans.** $-\frac{x^3}{3y^3} + \log y = c$
2. $(y^3 - 3xy^2)\,dx + (2x^2y - xy^2)\,dy = 0$ **Ans.** $\frac{y}{x} + 3\log x - 2\log y = c$
3. $(x^2y - 2xy^2)\,dx - (x^3 - 3x^2y)\,dy = 0$ **Ans.** $\frac{x}{y} - 2\log x + 3\log y = c$
4. $(y^3 - 2yx^2)\,dx + (2xy^2 - x^3)\,dy = 0$ **Ans.** $x^2y^4 - x^4y^2 = c$

## 3.13 EQUATIONS OF FIRST ORDER AND HIGHER DEGREE

The differential equations will involve $\frac{dy}{dx}$ in higher degree and $\frac{dy}{dx}$ will be denoted by $p$. The differential equation will be of the form $f(x, y, p) = 0$.

**Case 1. Equations solvable for *p*.**

**Example 31.** *Solve* : $x^2 = 1 + p^2$

**Solution.** $x^2 = 1 + p^2 \quad \Rightarrow \quad p^2 = x^2 - 1$

$$\Rightarrow \quad p = \pm\sqrt{x^2-1} \quad \Rightarrow \quad \frac{dy}{dx} = \pm\sqrt{x^2-1} \quad \Rightarrow \quad dy = \pm\sqrt{x^2-1}\,dx$$

which gives on integration $y = \pm\frac{x}{2}\sqrt{x^2-1} \mp \frac{1}{2}\log\left(x+\sqrt{x^2-1}\right) + c$ **Ans.**

**Case II. Equations solvable for *y*.**

(*i*) Differentiate the given equation w.r.t. "*x*".

(*ii*) Eliminate *p* from the given equation and the equation obtained as above.

(*iii*) The eliminant is the required solution.

**Example 32.** *Solve*: $y = (x - a)\,p - p^2$.

**Solution.** $y = (x - a)\,p - p^2$ ... (1)

Differentiating (1) w.r.t. "$x$" we obtain

$$\frac{dy}{dx} = p + (x-a)\frac{dp}{dx} - 2p\frac{dp}{dx}$$

$$p = p + (x-a)\frac{dp}{dx} - 2p\frac{dp}{dx}$$

$$\Rightarrow \quad 0 = (x-a)\frac{dp}{dx} - 2p\frac{dp}{dx}$$

$$\Rightarrow \quad 0 = \frac{dp}{dx}[x-a-2p] \quad \Rightarrow \quad \frac{dp}{dx} = 0$$

On integration, we get $p = c$.

Putting the value of $p$ in (1), we get

$$y = (x-a)\,c - c^2$$ **Ans.**

**Case III. Equations solvable for $x$**

(*i*) Differentiate the given equation w.r.t. "$y$".

(*ii*) Solve the equation obtained as in (1) for $p$.

(*iii*) Eliminate $p$, by putting the value of $p$ in the given equation.

(*iv*) The eliminant is the required solution.

**Example 33.** *Solve:* $y = 2px + yp^2$

**Solution.** $y = 2px + yp^2$ ... (1)

$$\Rightarrow \quad 2px = y - yp^2 \quad \Rightarrow \quad 2x = \frac{y}{p} - yp \qquad \text{... (2)}$$

Differentiating (2) w.r.t. "$y$" we get

$$2\frac{dx}{dy} = \frac{1}{p} - \frac{y}{p^2}\frac{dp}{dy} - p - y\frac{dp}{dy}$$

$$\Rightarrow \quad \frac{2}{p} = \frac{1}{p} - \frac{y}{p^2}\frac{dp}{dy} - p - y\frac{dp}{dy} \quad \Rightarrow \quad \frac{1}{p} + p = -\frac{y}{p^2}\frac{dp}{dy} - y\frac{dp}{dy}$$

$$\Rightarrow \quad \frac{1}{p} + p = -y\left(\frac{1}{p^2} + 1\right)\frac{dp}{dy} \quad \Rightarrow \quad \frac{1+p^2}{p} = -y\frac{1+p^2}{p^2}\frac{dp}{dy}$$

$$\Rightarrow \quad 1 = -\frac{y}{p}\frac{dp}{dy} \quad \Rightarrow \quad -\frac{dy}{y} = \frac{dp}{p} \quad \Rightarrow \quad -\log y = \log p + \log c'$$

$$\Rightarrow \quad \log p\,y = \log c \Rightarrow \quad p\,y = c \quad \Rightarrow \quad p = \frac{c}{y}$$

Putting the value of $p$ in (1), we get

$$y = 2\left(\frac{c}{y}\right)x + y\left(\frac{c^2}{y^2}\right) \quad \Rightarrow \quad y^2 = 2\,cx + c^2$$

$$\Rightarrow \quad y^2 = c(2x + c)$$ **Ans.**

**Class IV. Clairaut's Equation.**

The equation $y = px + f(p)$ is known as Clairaut's equation. ... (1)

Differentiating (1) w.r.t. "$x$", we get

$$\frac{dy}{dx} = p + x\frac{dp}{dx} + f'(p)\frac{dp}{dx}$$

$$\Rightarrow \quad p = p + x\frac{dp}{dx} + f'(p)\frac{dp}{dx} \quad \Rightarrow \quad 0 = x\frac{dp}{dx} + f'(p)\frac{dp}{dx}$$

$$\Rightarrow \quad [x + f'(p)]\frac{dp}{dx} = 0 \quad \Rightarrow \quad \frac{dp}{dx} = 0 \quad \Rightarrow \quad p = a \text{ (constant)}$$

Putting the value of $p$ in (1), we have

$$y = ax + f(a)$$

which is the required solution.

**Method.** In the Clairaut's equation, on replacing $p$ by a (constant), we get the solution of the equation.

**Example 34.** *Solve* : $p = \log(p x - y)$

**Solution.** $p = \log(p x - y)$ or $e^p = p x - y$ or $y = p x - e^p$

Which is Clairaut's equation.

Hence its solution is $y = a x - e^a$ **Ans.**

## EXERCISE 3.13

*Solve the following differential equations.*

1. $xp^2 + x = 2yp$ **Ans.** $2cy = c^2x^2 + 1$
2. $x(1 + p^2) = 1$ **Ans.** $y - c = \sqrt{(x - x^2)} - \tan^{-1}\sqrt{\frac{1-x}{x}}$
3. $x^2p^2 + xyp - 6y^2 = 0$ **Ans.** $y = \frac{c}{x^3}, y = c_1x^2$
4. $\frac{dy}{dx} - \frac{dx}{dy} = \frac{x}{y} - \frac{y}{x}$ **Ans.** $xy = c,\ x^2 - y^2 = c$
5. $y = px + p^3$ **Ans.** $y = ax + a^3$
6. $x^2(y - px) = yp^2$ **Ans.** $y^2 = cx^2 + c^2$

## 3.14 ORTHOGONAL TRAJECTORIES

Two families of curves are such that every curve of either family cuts each curve of the other family at right angles. They are called orthogonal trajectories of each other.

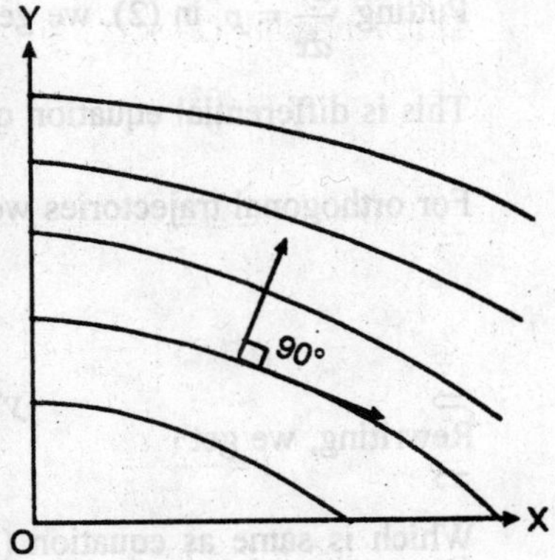


Orthogonal trajectories are very useful in engineering problems.

**For example:**

(*i*) The path of an electric field is perpendicular to equi-potential curves.

(*ii*) In fluid flow, the stream lines and equipotential lines are orthogonal trajectories.

(*iii*) The lines of heat flow is perpendicular to isothermal curves.

**Working rule to find orthogonal trajectories of curves**

**Step 1.** By differenciating the equaton of curves find the differential equations in the form $f\left(x, y, \frac{dy}{dx}\right) = 0$

**Step 2.** Replace $\frac{dy}{dx}$ by $-\frac{dx}{dy}$ $(M_1 . M_2 = -1)$

**Step 3.** Solve the differential equation of the orthogonal trajectories *i.e.,* $f\left(x, y - \frac{dx}{dy}\right) = 0$

**Self-orthogonal.** A given family of curves is said to be 'self-orthogonal' if the family of orthogonal trajectory is the same as the given family of curves.

**Example 35.** Find the orthogonal trajectories of the family of curves $xy = c$.

**Solution.** Here, we have

$$xy = c \quad \text{... (1)}$$

Differentiating (1), w.r.t., "x", we get

$$y+x\frac{dy}{dx}=0 \quad \Rightarrow \quad \frac{dy}{dx}=-\frac{y}{x}$$

On replacing $\frac{dy}{dx}$ by $-\frac{dx}{dy}$, we get

$$\Rightarrow \quad -\frac{dx}{dy}=-\frac{y}{x} \quad \Rightarrow \quad \frac{dy}{dx}=\frac{x}{y}$$

$$y\,dy = x\,dx \quad \text{... (2)}$$

Integrating (2), we get $\quad \frac{y^2}{2}=\frac{x^2}{2}+c$

$$\Rightarrow \quad y^2 - x^2 = 2c$$ **Ans.**

**Example 36.** *Show that the family of porabolas $y^2 = 2cx + c^2$ is "self-orthogonal."*

**Solution.** Here we have

$$y^2 = 2cx + c^2 \quad \text{... (1)}$$

Differentiating (1), we get $2y\frac{dy}{dx}=2c \quad \Rightarrow \quad c=y\frac{dy}{dx}$

Putting the value of $c$ in (1), we have $y^2=2\left(y\frac{dy}{dx}\right)x+\left(y\frac{dy}{dx}\right)^2$ ... (2)

Putting $\frac{dy}{dx}=p$ in (2), we get

$$y^2 = 2ypx + y^2p^2 \quad \text{... (3)}$$

This is differential equation of give n family of parabolas.

For orthogonal trajectories we put $-\frac{1}{p}$ for $p$ in (3)

$$y^2=2y\left(-\frac{1}{p}\right)x+y^2\left(-\frac{1}{p}\right)^2 \quad \Rightarrow \quad y^2=-\frac{2yx}{p}+\frac{y^2}{p^2}$$

$$\Rightarrow \quad y^2p^2 = -2pyx + y^2$$

Rewriting, we get

$$\Rightarrow \quad y^2 = 2ypx + y^2p^2$$

Which is same as equation (3). Thus (2) is D.E. for the given family and its orthogonal trajectories.

Hence, the given family is self-orthogonal. **Proved.**

## EXERCISE 3.14

*Find the orthogonal trajectories of the following family of curves:*

1. $y^2 = cx^3$ **Ans.** $(x+1)^2+y^2=a^2$
2. $x^2-y^2=cx$ **Ans.** $y(y^2+3x^2)=c$
3. $x^2-y^2=c$ **Ans.** $xy=c$
4. $(a+x)y^2=x^2(3a-x)$ **Ans.** $(x^2+y^2)^5=cy^3(5x^2+y^2)$
5. $y=ce^{-2x}+3x$, passing through the point (0, 3) **Ans.** $9x-3y+5=-4e^{6(3-y)}$
6. $16x^2+y^2=c$ **Ans.** $y^{16}=kx$
7. $y=\tan x+c$ **Ans.** $2x+4y+\sin 2x=k$
8. $y=ax^2$ **Ans.** $x^2+2y^2=c$

**9.** $x^2 + (y - c)^2 = c^2$ **Ans.** $x^2 + y^2 = cx$

**10.** $x^2 + y^2 + 2gx + 2fy + c = 0$ **Ans.** $x^2 + y^2 + 2fy - c = 0$

## 3.15 POLAR EQUATION OF THE FAMILY OF CURVES

Let the polar equation of the family of curves be $f(r,\theta,c) = 0$ ... (1)

**Working Rule**

**Step 1.** On differentiating and eliminating the arbitrary constant $c$ between (1) and $f'(r,\theta,c) = 0$ we get the differential equation of (1) *i.e.,*

$$F\left(r, \theta, \frac{dr}{d\theta}\right) = 0 \qquad \text{... (2)}$$

**Step 2.** Replace $\frac{dr}{d\theta}$ by $-r^2\frac{d\theta}{dr}$ in (2). Here we will get the differential equation of orthogonal trajectory *i.e.,*

$$F\left(r, \theta - r^2\frac{d\theta}{dr}\right) = 0 \qquad \text{... (3)}$$

**Step 3.** Integrating (3) to get the equation of the orthogonal trajectory.

**Example 37.** *Find the orthogonal trajectory of the cardioids $r = a\,(1 - \cos\theta)$.*

**Solution.** We have, $r = a(1 - \cos\theta)$ ... (1)

Differentiating (1) w.r.t. $\theta$, we get $\frac{dr}{d\theta} = a\sin\theta$ ... (2)

Dividing (2) by (1) to eliminate $a$, we get

$$\frac{1}{r}\frac{dr}{d\theta} = \frac{\sin\theta}{1-\cos\theta} = \frac{2\sin\frac{\theta}{2}\cos\frac{\theta}{2}}{1-1+2\sin^2\frac{\theta}{2}} = \cot\frac{\theta}{2} \qquad \text{... (3)}$$

which is the differential equation of (1).

Replacing $\frac{dr}{d\theta}$ by $-r^2\frac{d\theta}{dr}$ in (3), we get $\frac{1}{r}\left(-r^2\frac{d\theta}{dr}\right) = \cot\frac{\theta}{2}$

$$r\frac{d\theta}{dr} = -\cot\frac{\theta}{2}$$

Separating the variables we get $\frac{dr}{r} = -\tan\frac{\theta}{2}\,d\theta$ ... (4)

Integrating (4), we get $\log r = 2\log\cos\frac{\theta}{2} + \log c = \log c\cos^2\frac{\theta}{2}$

$\Rightarrow$ $r = c\cos^2\frac{\theta}{2}$ $\Rightarrow$ $r = \frac{c}{2}(1+\cos\theta)$

Which is the required trajectory. **Ans.**

**Example 38.** *Find the orthogonal trajectory the family of curves*

$$r^2 = c\sin 2\theta$$

**Solution.** We have

$$r^2 = c\sin 2\theta \qquad \text{... (1)}$$

Differentiating (1), we get $2r\frac{dr}{d\theta} = 2c\cos 2\theta$ ... (2)

Dividing (2) by (1), to eliminate '$c$' we get $\frac{2}{r}\frac{dr}{d\theta} = 2\cot 2\theta$ ... (3)

Replacing $\frac{dr}{d\theta}$ by $-r^2\frac{d\theta}{dr}$ in (3), we have $\frac{2}{r}\left(-r^2\frac{d\theta}{dr}\right) = 2\cot 2\theta$

$$-2r\frac{d\theta}{dr} = 2\cot 2\theta \quad \text{... (4)}$$

Separating the variables of (4), we obtain $\frac{dr}{r} = -\tan 2\theta\, d\theta$ ... (5)

Integrating (5), we get $\log r = \frac{1}{2}\log\cos 2\theta + \log c$

$$2\log r = \log c \cos\theta$$

$$r^2 = c\cos 2\theta$$

which is the required trajectory **Ans.**

## EXERCISE 3.15

*Find the orthogonal trajectory of the following families of the curves:*

1. $r = ce^{\theta}$ **Ans.** $r = ke^{-\theta}$
2. $r = c\theta^2$ **Ans.** $r = ke^{-\frac{\theta^2}{4}}$
3. $r = a(1+\cos\theta)$ **Ans.** $r = c(1-\cos\theta)$
4. $r^n \sin n\theta = a^n$ **Ans.** $r^n \cos n\theta = c^n$
5. $r = a\cos^2\theta$ **Ans.** $r^2 = c\sin\theta$
6. $r = 2a(\sin\theta+\cos\theta)$ **Ans.** $r = 2c(\sin\theta - \cos\theta)$
7. $r = c(1+\sin^2\theta)$ **Ans.** $r^2 = k\cos\theta.\cot\theta$
8. $r = \frac{a}{1+2\cos\theta}$ **Ans.** $r^2\sin^3\theta = (1+\cos\theta)$

## 3.16 ELECTRICAL CIRCUIT

We will consider circuits made up of

(*i*) Voltage source which may be a battery or a generator.

(*ii*) Resistance, inductance and capacitance.

The formation of differential equation for an electric circuit depends upon the following laws.

(*i*) $i = \frac{dq}{dt}$, (*ii*) Voltage drop across resistance $R = Ri$

(*iii*) Voltage drop across inductance $L = L.\frac{di}{dt}$

(*iv*) Voltage drop across capacitance $C = \frac{q}{C}$

**Kirchhoff's laws**

**I. Voltage law.** The algebraic sum of the voltage drop around any closed circuit is equal to the resultant electromotive force in the circuit.

**II. Current law.** At a junction or node, current coming is equal to current going.

**(*i*) L - R series circuit.** Let *i* be the current flowing in the circuit containing resistance *R* and inductance L in series, with voltage source E, at any time *t*.

By voltage law $Ri + L\frac{di}{dt} = E \Rightarrow \frac{di}{dt} + \frac{R}{L}i = \frac{E}{L}$ ...(1) (*M.U. II Semester, 2009*)

This is the linear differential equation

$$\text{I.F.} = e^{\int \frac{R}{L}dt} = e^{\frac{R}{L}t}$$

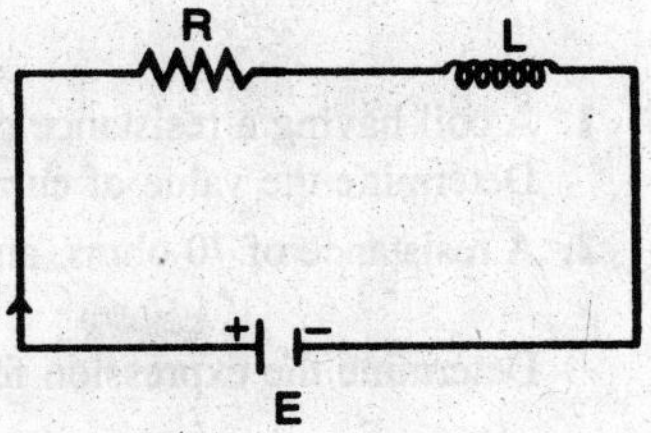


Its solution is $\quad i.e^{\frac{R}{L}t} = \int \frac{E}{L} e^{\frac{R}{L}t}\, dt + A$

$\Rightarrow \quad i.e^{\frac{R}{L}t} = \frac{E}{L} \times \frac{L}{R} e^{\frac{R}{L}t} + A$

$\Rightarrow \quad i = \frac{E}{R} + A e^{-\frac{Rt}{L}} \quad \text{...(2)}$

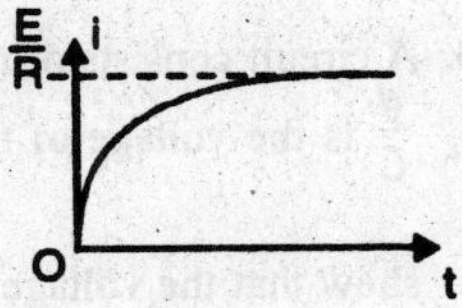


At $t = 0$, $\quad i = 0 \Rightarrow A = -\frac{E}{R}$

Thus, (2) becomes $i = \frac{E}{R}\left[1 - e^{-\frac{Rt}{L}}\right]$

***(ii)* C-R series circuit.** Let $i$ be current in the circuit containing resistance $R$, $L$, and capacitance $C$ in series with voltage source $E$, at any time $t$.

By voltage law

$$Ri + \frac{q}{C} = E \qquad \left[i = \frac{dq}{dt}\right]$$

$$\Rightarrow \quad R\frac{dq}{dt} + \frac{q}{C} = E$$

**Example 39.** *Solve the equation* $L\frac{di}{dt} + Ri = E_0 \sin wt$

*where* $L$, $R$ *and* $E_0$ *are constants and discuss the case when* $t$ *increases indefinitely.*

**Solution.** $L\frac{di}{dt} + Ri = E_0 \sin wt$

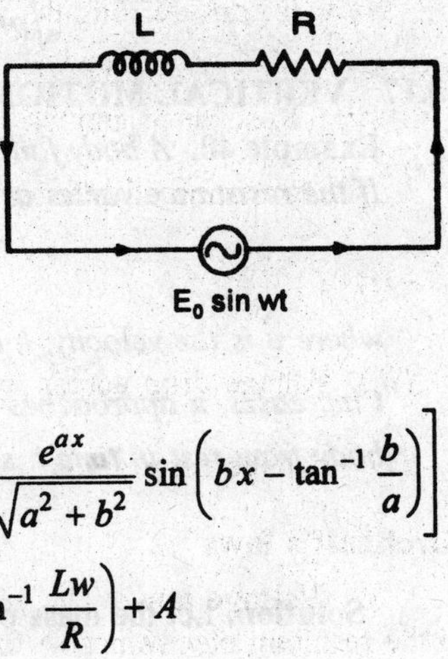


$$\Rightarrow \quad \frac{di}{dt} + \frac{R}{L} i = \frac{E_0}{L} \sin wt$$

$$\text{I.F.} = e^{\int \frac{R}{L} dt} = e^{\frac{R}{L}t}$$

Solution is $\quad i.e^{\frac{R}{L}t} = \frac{E_0}{L} \int e^{\frac{R}{L}t} \sin wt\, dt + A$

$$\left[\int e^{ax} \sin bx\, dx = \frac{e^{ax}}{\sqrt{a^2 + b^2}} \sin\left(bx - \tan^{-1}\frac{b}{a}\right)\right]$$

$$\Rightarrow \quad i.e^{\frac{R}{L}t} = \frac{E_0}{L} \frac{e^{\frac{R}{L}t}}{\sqrt{\frac{R^2}{L^2} + w^2}} \sin\left(wt - \tan^{-1}\frac{Lw}{R}\right) + A$$

$$i = \frac{E_0}{\sqrt{R^2 + L^2 w^2}} \sin\left(wt - \tan^{-1}\frac{Lw}{R}\right) + Ae^{-\frac{R}{L}t}$$

As $t$ increases indefinitely, then $Ae^{-\frac{Rt}{L}}$ tends to zero.

so $\quad i = \frac{E_0}{\sqrt{R^2 + L^2 w^2}} \sin\left(wt - \tan^{-1}\frac{Lw}{R}\right)$ **Ans.**

## EXERCISE 3.16

1. A coil having a resistance of 15 ohms and an inductance of 10 henries is connected to 90 volts supply. Determine the value of current after 2 seconds. ($e^{-3} = 0.05$) **Ans.** 5.985 amp.

2. A resistance of 70 ohms, an inductance of 0.80 henry are connected in series with a battery of 10 volts. Determine the expression for current as a function of time after $t = 0$. **Ans.** $i = \frac{1}{7}\left(1 - e^{-\frac{175}{2}t}\right)$

3. A circuit consists of resistance $R$ ohms and a condenser of $C$ farads connected to a constant e.m.f. $E$; if $\frac{q}{C}$ is the voltage of the condenser at time $t$ after closing the circuit Show that $\frac{q}{C} = E - Ri$ and hence show that the voltage at time $t$ is $E\left(1 - e^{-\frac{t}{CR}}\right)$.

4. Show that the current $i = \frac{q}{CR} e^{-\frac{t}{R.C.}}$ during the discharge of a condenser of charge $Q$ coulomb through a resistance $R$ ohms.

5. A condenser of capacity $C$ farads with voltage $v_0$ is discharged through a resistance $R$ ohms. Show that if $q$ coulomb is the charge on the condenser, $i$ ampere the current and $v$ the voltage at time $t$.

   $q = Cv$, $v = Ri$ and $i = -\frac{dq}{dt}$, hence show that $v = v_0\, e^{-\frac{1}{Rc}}$.

6. Solve $L\frac{di}{dt} + Ri = E\cos wt$ **Ans.** $i = \frac{E}{L^2w^2 + R^2}\left(R\cos wt + Lw\sin wt - \text{Re}^{-\frac{Rt}{L}}\right)$

7. A circuit consists of a resistance $R$ ohms and an inductance of $L$ henry connected to a generator of $E\cos(wt + \alpha)$ voltage. Find the current in the circuit. ($i = 0$, when $t = 0$).

   **Ans.** $i = \frac{E}{\sqrt{R^2 + L^2w^2}}\cos\left[wt + \alpha - \tan^{-1}\frac{Lw}{R}\right] - \frac{E}{\sqrt{R^2 + L^2w^2}}.e^{-\frac{R}{L}t}\cos\left[\alpha - \tan^{-1}\frac{Lw}{R}\right]$

## 3.17 VERTICAL MOTION

**Example 40.** *A body falling vertically under gravity encounters resistance of the atmosphere. If the resistance varies as the velocity, show that the equation of motion is given by*

$$\frac{du}{dt} = g - ku$$

*where u is the velocity, k is a constant and g is the acceleration due to gravity. Show that as t increases, u approaches the value g/k. Also, if* $u = \frac{dx}{dt}$ *where x is the distance fallen by the body from rest in time t, show that*

$$x = \frac{gt}{k} - \frac{g}{k^2}(1 - e^{-kt})$$

**Solution.** Let the mass of the falling body be unity.

$$\text{Acceleration} = \frac{du}{dt}$$

$$\text{Force acting downward} = 1.\frac{du}{dt} = \frac{du}{dt}$$

$$\text{Force of resistance} = ku$$

$$\text{Net force acting downward} = g - ku \Rightarrow \frac{du}{dt} = g - ku \qquad ...(1)\textbf{Proved.}$$

$$\Rightarrow \frac{du}{g - ku} = dt$$

Integrating, we get $\int \frac{du}{g-ku} = \int dt$

$$\Rightarrow \quad t = -\frac{1}{k}\log(g-ku) + \log A = \log(g-ku)^{-1/k} A$$

$$A(g-ku)^{-1/k} = e^t \quad \Rightarrow \quad (g-ku) = A^k e^{-kt}$$

$$\Rightarrow \quad u = \frac{g}{k} - \frac{A^k}{k} e^{-kt}$$

If $t$ increases very large then $\frac{A^k}{k} e^{-kt} = 0$

$$\Rightarrow \quad u = \frac{g}{k} \quad \text{when } t \to \infty$$ **Proved.**

Given $\quad u = \frac{dx}{dt} \quad \Rightarrow \quad \frac{du}{dt} = \frac{d^2x}{dt^2}$

Putting the values of $\frac{du}{dt}$ and $u$ in (1), we get

$$\frac{d^2x}{dt^2} + k\frac{dx}{dt} = g \quad \Rightarrow \quad (D^2 + kD)\,x = g$$

A.E. is $\quad m(m+k) = 0 \quad \Rightarrow \quad m = 0,\ m = -k$

$$\text{C.F.} = A_1 + A_2 e^{-kt}$$

$$\text{P.I.} = \frac{1}{D^2 + kD} g = t\frac{1}{2D+k} g$$

$$= \frac{t}{k}\frac{1}{\left(1+\frac{2D}{k}\right)} g = \frac{t}{k}\left(1+\frac{2D}{k}\right)^{-1} g = \frac{t}{k}\left(1-\frac{2D}{k}\right) g = \frac{t\,g}{k}$$

$$x = A_1 + A_2 e^{-kt} + \frac{gt}{k} \qquad \text{...(2)}$$

Putting the values of $t = 0$ and $x = 0$ in (2), we get

$$0 = A_1 + A_2 \quad \Rightarrow \quad A_2 = -A_1$$

(2) becomes $\quad x = A_1 - A_1 e^{-kt} + \frac{gt}{k} \qquad \text{...(3)}$

On differentiating (3), we get $\frac{dx}{dt} = A_1 k e^{-kt} + \frac{g}{k} \qquad \text{...(4)}$

On putting $\frac{dx}{dt} = 0$, when $t = 0$ in (4), we get $0 = A_1 k + \frac{g}{k} \quad \Rightarrow \quad A_1 = -\frac{g}{k^2}$

Putting the value of $A_1$ in (3), we get

$$x = -\frac{g}{k^2} + \frac{g}{k^2} e^{-kt} + \frac{gt}{k} \quad \Rightarrow \quad x = \frac{gt}{k} - \frac{g}{k^2}(1 - e^{-kt})$$ **Proved.**

**Example 41.** *The acceleration and velocity of a body falling in the air approximately satisfy the equation :*
*Acceleration* $= g - kv^2$, *where v is the velocity of the body at any time t, and g, k are constants. Find the distance traversed as a function of the time, if the body falls from rest.*

*Show that value of v will never exceed* $\sqrt{\frac{g}{k}}$.

**Solution** Acceleration $= g - k v^2 \Rightarrow \dfrac{dv}{dt} = g - kv^2 \Rightarrow \dfrac{dv}{g - kv^2} = dt.$

$$\Rightarrow \quad \frac{1}{2\sqrt{g}}\left[\frac{1}{\sqrt{g}+\sqrt{k}.v} + \frac{1}{\sqrt{g}-\sqrt{k}.v}\right] dv = dt$$

On integrating, we get

$$\frac{1}{2\sqrt{g}}\frac{1}{\sqrt{k}}\log(\sqrt{g}+\sqrt{k}.v) - \frac{1}{2\sqrt{gk}}\log(\sqrt{g}-\sqrt{k}.v) = t + A$$

$$\Rightarrow \quad \frac{1}{2\sqrt{gk}}\log\frac{\sqrt{g}+\sqrt{k}.v}{\sqrt{g}-\sqrt{k}.v} = t + A \qquad ...(1)$$

On putting $t = 0$, $v = 0$ in (1), we get $\dfrac{1}{2\sqrt{gk}}\log 1 = 0 + A \Rightarrow A = 0$

Equation (1) becomes $\dfrac{1}{2\sqrt{gk}}\log\dfrac{\sqrt{g}+\sqrt{k}.v}{\sqrt{g}-\sqrt{k}.v} = t \Rightarrow \log\dfrac{\sqrt{g}+\sqrt{k}.v}{\sqrt{g}-\sqrt{k}.v} = 2\sqrt{gk}\,t$

$$\Rightarrow \quad \frac{\sqrt{g}+\sqrt{k}.v}{\sqrt{g}-\sqrt{k}.v} = e^{2\sqrt{gk}\,t}$$

By componendo and dividendo, we have

$$\frac{\sqrt{k}.v}{\sqrt{g}} = \frac{e^{2\sqrt{gkt}}-1}{e^{2\sqrt{gkt}}+1} = \frac{e^{\sqrt{gk}\,t}-e^{-\sqrt{gk}\,t}}{e^{\sqrt{gk}\,t}+e^{-\sqrt{gk}\,t}} = \tan h\sqrt{gk}\,t$$

$$\Rightarrow \quad v = \sqrt{\frac{g}{k}}\tan h\sqrt{gk}\,t$$

Whatever the value of $t$ may be $\tanh\sqrt{gk}\,t \le 1.$

Hence the value of $v$ will never exceed $\sqrt{\dfrac{g}{k}}.$ **Proved.**

$$\frac{dx}{dt} = \sqrt{\frac{g}{k}}\tanh\sqrt{gk}\,t$$

Integrating again, we get $x = \sqrt{\dfrac{g}{k}}\displaystyle\int\tanh\sqrt{gk}\,t\,dt = \dfrac{1}{k}\log\cosh\sqrt{gk}\,t + B$

when $t = 0$, $x = 0$ then $B = 0$

$$\therefore \quad x = \frac{1}{k}\log\cosh\sqrt{gk}\,t \qquad \textbf{Ans.}$$

## EXERCISE 3.17

1. A moving body is opposed by a force proportional to the displacement and by a resistance proportional to the square of velocity. Prove that the velocity is given by

$V^2 = ae - \dfrac{cx}{b} + \dfrac{c}{ab^2}$ **Hint.** Equation of motion is $m\dfrac{VdV}{dx} = -K_1x - K_2V^2$

2. A particle of mass $m$ is projected vertically upward with an initial velocity $v_0$. The resisting force at any time is $K$ times the velocity. Formulate the differential equation of motion and show that the distance $s$ covered by the particle at any time $t$ is given by

$$s = \left(\frac{g}{K^2} + \frac{v_0}{K}\right)(1 - e^{-Kt}) - \frac{g}{K}t$$

**3.** A particle falls in a vertical line under gravity (supposed constant) and the force of air resistance to its motion is proportional to its velocity. Show that its velocity cannot exceed a particular limit.

**Ans.** $V = \frac{g}{K}$

**4.** A body falling from rest is subjected to a force of gravity and an air resistance of $\frac{n^2}{g}$ times the square of velocity. Show that the distance travelled by the body in $t$ seconds in $\frac{g}{n^2}$ log cosh $nt$.

**5.** A body of mass $m$, falling from rest is subject to the force of gravity and an air resistance proportional to the square of the velocity $Kv^2$. If it falls through a distance $x$ and possesses a velocity $v$, at the instant, prove that

$$\frac{2kx}{m} = \log\left(\frac{a^2}{a^2 - v^2}\right) \quad \text{where} \quad \frac{mg}{k} = a^2$$

*(A.M.I.E.T.E., June 2009)*

## HEAT CONDUCTION

**Example 42.** *The rate at which a body cools is proportional to the difference between the temperature of the body and that of the surrounding air. If a body in air at 25°C will cool from 100° to 75° in one minute, find its temperature at the end of three minutes.*

**Solution.** Let temperature of the body be T°C.

$$\frac{dT}{dt} = k\,(T - 25) \quad \text{or} \quad \frac{dT}{T-25} = k\,dt$$

$$\log\,(T - 25) = kt + \log A \quad \text{or} \quad \log\frac{T-25}{A} = kt$$

$$T - 25 = A\,e^{kt} \qquad \text{...(1)}$$

When $t = 0$, then $T = 100$, from (1) A = 75

When $t = 1$, then $T = 75$ and A = 75, From (1) $\frac{2}{3} = e^k$

$\therefore$ (1) becomes $\qquad T = 25 + 75\,e^{kt}$

When $t = 3$, $\quad$ then $T = 25 + 75\,e^{3k} = 25 + 75 \times 8/27 = 47.22$ **Ans.**

**Example 43.** *The rate at which the ice melts is proportional to the amount of ice at the instant. Find the amount of ice left after 2 hours if half the quantity melts in 30 minutes.*

**Solution.** Let $m$ be the amount of ice at any time $t$.

$$\therefore \quad \frac{dm}{dt} = km \quad \Rightarrow \quad \frac{dm}{m} = k\,dt$$

$$\int\frac{dm}{m} = k\int dt + C \quad \Rightarrow \quad \log m = kt + C \qquad \text{...(1)}$$

At $t = 0$, $m = M$

$$\log M = 0 + C \quad \Rightarrow \quad C = \log M$$

On putting the value of C, (1) becomes,

$$\log m = kt + \log M \qquad \text{...(2)}$$

$m = M/2$ $\quad$ when $t = 1/2$ hour

$$\log\frac{M}{2} = \frac{k}{2} + \log M \quad \Rightarrow \quad \log\frac{M}{2M} = \frac{k}{2}$$

$$\Rightarrow \quad \log\frac{1}{2} = \frac{k}{2} \quad \text{or} \quad k = 2\log\frac{1}{2}$$

On putting the value of $k$ in (2), we have $\log m = \left(2\log\frac{1}{2}\right)t + \log M$ ...(3)

On putting $t = 2$ hours in (3), we have $\log m = 4\log\frac{1}{2} + \log M$

$$\Rightarrow \quad \log\frac{m}{M} = \log\left(\frac{1}{2}\right)^4 \quad \text{or} \quad \frac{m}{M} = \frac{1}{16} \quad \text{or} \quad m = \frac{M}{16}$$

After 2 hours, amount of ice left = $\frac{1}{16}$ of the amount of ice at the beginning. **Ans.**

**CHEMICAL ACTION:**

**Example 44.** *Under certain conditions, cane sugar is converted into dextrose at a rate, which is proportional to the amount unconverted at any time. If out of 75 grams of sugar at t = 0, 8 grams are converted during the first 3 minutes, find the amount converted in $1\frac{1}{2}$ hours.*

**Solution.** Let $M$ be the amount of cane sugar initially, $m$ be the amount of cane sugar converted into dextrose.

Then according to problem,

$$\frac{dm}{dt} = K(M - m) \quad \text{or} \quad \frac{dm}{dt} + Km = KM$$

which is a linear differential equation.

$$\text{I.F.} = e^{\int K\,dt} = e^{k.t}$$

Solution is $\quad m\,e^{k.t} = \int KMe^{kt}\,dt = Me^{kt} + C$

$$\Rightarrow \quad m = M + Ce^{-k.t} \quad \text{...(1)}$$

(*i*) At $t = 0$, $m = 0$, $M = 75$

(1) becomes $\quad m = 75 - 75\,e^{-k.t}$ ...(2)

$0 = 75 + C \Rightarrow C = -75$

(*ii*) At $t = 30$, $m = 8$

$$8 = 75 - 75\,e^{-30k} \Rightarrow 67 = 75e^{-30k}$$

$$\Rightarrow \quad e^{-30k} = \frac{67}{75} \quad \text{...(3)}$$

(*iii*) At $t = 90$, (2) becomes

$$m = 75 - 75\,e^{-90k} = 75 - 75\left(\frac{67}{75}\right)^3 \text{ from(3)} \quad \text{...(4)}$$

$$= 75 - \frac{(67)^3}{(75)^2} = 75 - \frac{300763}{5625} = 75 - 53.45 = 21.55$$ **Ans.**

**Example 45.** *Uranium disintegrates at a rate proportional to the amount present at any instant. If $m_1$ and $m_2$ grams of uranium are present at time $t_1$ and $t_2$ respectively, show that half life of uranium is*

$$\frac{(t_1 - t_2)\log 2}{\log\frac{m_1}{m_2}}$$

**Solution.** Let $m$ be the amount of uranium at any time $t$.

$$\frac{dm}{dt} = -km$$

$$\therefore \quad \int_{m_1}^{m_2}\frac{dm}{m} = -k\int_{t_1}^{t_2} dt \Rightarrow \log\frac{m_1}{m_2} = k(t_2 - t_1) \quad \text{...(1)}$$

Let the mass $m$ reduce to $\frac{m}{2}$ in time $t$. Also $\int_m^{\frac{m}{2}} \frac{dm}{m} = -k \int_0^{t_2} dt$

$$\therefore \quad \log \frac{m}{2} - \log m = -kt \Rightarrow kt = \log 2 = \log 2 \Rightarrow k = \frac{\log 2}{t}$$

Substituting the value of $k$ in (1), we get

$$\log \frac{m_1}{m_2} = \frac{\log 2}{t}(t_2 - t_1), \Rightarrow t = \frac{(t_2 - t_1)\log 2}{\log \frac{m_1}{m_2}}$$ **Proved.**

## EXERCISE 3.18

1. Radium decomposes at a rate proportional to the amount present. If 5% of the original amount disappears in 50 years, how much will remain after 100 years? **Ans.** 90.25%
2. If a thermometer is taken outdoors where the temperature is 0°C from a room in which the temperature is 21°C and the reading drops to 10°C is 1 minute, how long after its removal will the reading be 5°C ? **Ans.** 2 minutes, 13 seconds.
3. In one dimensional steady state heat conduction for a hollow cylinder with constant thermal conductivity $k$ in the region $a \le r \le b$, the temperature $T_r$ at a distance $r$ $(a \le r \le b)$ is given by

$$\frac{d}{dr}\left[r\frac{dT_r}{dr}\right] = 0,$$

with $T_r = T_1$ where $r = a$ and $T_r = T_2$ where $r = b$. Use this to determine steady state temperature distribution $T_r$ in the cylinder in terms of $r$. **Ans.** $T_r = \frac{T_1 - T_2}{\log(r_1/r_2)}\log r + \frac{T_2 \log r_1 - T_1 \log r_2}{\log(r_1/r_2)}$

## MISCELLANEOUS QUESTION

**Example 46.** *If the population of a country doubles in 50 years, in how many years will it treble, assuming that the rate of increase is proportional to the number of inhabitants?*

**Solution.** Let $t$ = time in years,

$y$ = population after $t$ years

$P$ = original population (when $t = 0$).

The rate of increase of population is proportional to the population, so that

$$\frac{dy}{dt} = ky, \text{ where } k \text{ is a constant. or } k\,dt = dy/y$$

Integrating, $kt = c + \log y$ ... (1)

When $t = 0, y = P$

When $t = 50, y = 2P$

Substituting in (1), $0 = c + \log P$, and $50k = c + \log 2P$

Solving, $c = -\log P$

$$50k = -\log P + \log 2P = \log 2 \text{ or } k = \frac{1}{50}\log 2$$

The value of $t$ when population has trebled is obtained by putting $y = 3P$ in (1). We get

$$kt = c + \log 3P = -\log P + \log 3P = \log 3$$

$$t = \frac{1}{k}\log 3 = \frac{50}{\log 2}\cdot \log 3 \text{ years.}$$ **Ans.**

## 3.18 LINEAR DIFFERENTIAL EQUATIONS OF SECOND ORDER WITH CONSTANT COEFFICIENTS

The general form of the linear differential equation of second order is

$$\frac{d^2y}{dx^2}+P\frac{dy}{dx}+Qy=R$$

where $P$ and $Q$ are constants and $R$ is a function of $x$ or constant.

**Differential operator.** Symbol $D$ stands for the operation of differential *i.e.*,

$$Dy=\frac{dy}{dx},\quad D^2y=\frac{d^2y}{dx^2}$$

$\frac{1}{D}$ stands for the operation of integration.

$\frac{1}{D^2}$ stands for the operation of integration twice.

$\frac{d^2y}{dx^2}+P\frac{dy}{dx}+Qy=R$ can be written in the operator form.

$$D^2y+P\,Dy+Qy=R \quad\Rightarrow\quad (D^2+PD+Q)\,y=R$$

## 3.19 COMPLETE SOLUTION = COMPLEMENTARY FUNCTION + PARTICULAR INTEGRAL

Let us consider a linear differential equation of the first order

$$\frac{dy}{dx}+Py=Q \qquad ...(1)$$

Its solution is $\quad ye^{\int Pdx}=\int(Q\,e^{\int Pdx})\,dx+C$

$\Rightarrow \quad y=Ce^{-\int Pdx}+e^{-\int Pdx}\int(Qe^{\int Pdx})\,dx$

$\Rightarrow \quad y=cu+v\,(\text{say}) \qquad ...(2)$

where $u=e^{-\int Pdx}$ and $v=e^{-\int Pdx}\int Q\,e^{\int Pdx}dx$

(*i*) Now differentiating $u=e^{-\int Pdx}$ w.r.t. $x$, we get $\frac{du}{dx}=-Pe^{-\int Pdx}=-Pu$

$\Rightarrow \quad \frac{du}{dx}+Pu=0 \quad\Rightarrow\quad \frac{d(cu)}{dx}+P(cu)=0$

which shows that $y=c.u$ is the solution of $\frac{dy}{dx}+Py=0$

(*ii*) Differentiating $v=e^{-\int Pdx}\int(Qe^{\int Pdx})\,dx$ with respect to $x$, we get

$$\frac{dv}{dx}=-Pe^{\int Pdx}\int(Qe^{\int Pdx})\,dx+e^{-\int Pdx}\,Qe^{\int Pdx} \quad\Rightarrow\quad \frac{dv}{dx}=-Pv+Q$$

$\Rightarrow \quad \frac{dv}{dx}+Pv=Q$ which shows that $y=v$ is the solution of $\boxed{\frac{dy}{dx}+Py=Q}$

Solution of the differential equation (1) is (2) consisting of two parts *i.e.* $cu$ and $v$. $cu$ is the solution of the differential equation whose R.H.S. is zero. $cu$ is known as *complementary function*. Second part of (2) is $v$ free from any arbitrary constant and is known as *particular integral*.

Complete Solution = Complementary Function + Particular Integral.

$\Rightarrow$ $$\boxed{y = C.F. + P.I.}$$

## 3.20 METHOD FOR FINDING THE COMPLEMENTARY FUNCTION

(1) In finding the complementary function, R.H.S. of the given equation is replaced by zero.

(2) Let $y = C_1 e^{mx}$ be the C.F. of

$$\frac{d^2y}{dx^2} + P\frac{dy}{dx} + Qy = 0 \qquad \text{...(1)}$$

Putting the values of $y$, $\frac{dy}{dx}$ and $\frac{d^2y}{dx^2}$ in (1) then $C_1 e^{mx}(m^2 + Pm + Q) = 0$

$\Rightarrow$ $m^2 + Pm + Q = 0$. It is called **Auxiliary equation.**

(3) Solve the auxiliary equation :

**Case I : Roots, Real and Different.** If $m_1$ and $m_2$ are the roots, then the C.F. is

$$y = C_1 e^{m_1 x} + C_2 e^{m_2 x}$$

**Case II : Roots, Real and Equal.** If both the roots are $m_1$, $m_1$ then the C.F. is

$$y = (C_1 + C_2 x)\, e^{m_1 x}$$

Equation (1) can be written as

$$(D - m_1)(D - m_1)y = 0 \qquad \text{... (2)}$$

Replacing $(D - m_1)y = v$ in (2), we get

$$(D - m_1)v = 0 \qquad \text{... (3)}$$

$$\frac{dv}{dx} - m_1 v = 0 \quad \Rightarrow \quad \frac{dv}{v} = m_1 dx \quad \Rightarrow \quad \log v = m_1 x + \log c_2 \quad \Rightarrow \quad v = c_2 e^{m_1 x}$$

$$v = c_2 e^{m_1 x}$$

From (3) $\quad (D - 1)y = c_2 e^{m_1 x}$

This is the linear differential equation.

$$\text{I.F.} = e^{-m_1 \int dx} = e^{-m_1 x}$$

Solution is

$$y.e^{-m_1 x} = \int (c_2 e^{m_1 x})(e^{-m_1 x})\, dx + c_1 = \int c_2\, dx + c_1 = c_2 x + c_1$$

$$y = (c_2 x + c_1) e^{m_1 x}$$

$$\text{C.F.} = (c_1 + c_2 x)\, e^{m_1 x}$$

**Example 47.** *Solve:* $\frac{d^2y}{dx^2} - 8\frac{dy}{dx} + 15y = 0.$

**Solution.** Given equation can be written as

$$(D^2 - 8D + 15)\,y = 0$$

Here auxiliary equation is $m^2 - 8m + 15 = 0$

$\Rightarrow$ $(m - 3)(m - 5) = 0$ $\qquad \therefore\ m = 3, 5$

Hence, the required solution is

$$y = C_1 e^{3x} + C_2 e^{5x}$$ **Ans.**

**Example 48.** *Solve:* $\frac{d^2y}{dx^2} - 6\frac{dy}{dx} + 9y = 0$

**Solution.** Given equation can be written as

$$(D^2 - 6D + 9)\,y = 0$$

A.E. is $m^2 - 6m + 9 = 0 \quad \Rightarrow \quad (m-3)^2 = 0 \quad \Rightarrow \quad m = 3, 3$

Hence, the required solution is

$$y = (C_1 + C_2 x)\, e^{3x}$$ **Ans.**

**Example 49.** *Solve:* $\dfrac{d^2y}{dx^2} + 4\dfrac{dy}{dx} + 5y = 0,$

$$y = 2 \text{ and } \frac{dy}{dx} = \frac{d^2y}{dx^2} \text{ when } x = 0.$$

**Solution.** Here the auxiliary equation is

$$m^2 + 4m + 5 = 0$$

Its root are $-2 \pm i$

The complementary function is

$$y = e^{-2x}(A\cos x + B\sin x) \quad ...(1)$$

On putting $y = 2$ and $x = 0$ in (1), we get

$$2 = A$$

On putting $A = 2$ in (1), we have

$$y = e^{-2x}[2\cos x + B\sin x] \quad ...(2)$$

On differentiating (2), we get

$$\frac{dy}{dx} = e^{-2x}[-2\sin x + B\cos x] - 2e^{-2x}[2\cos x + B\sin x]$$

$$= e^{-2x}[(-2B-2)\sin x + (B-4)\cos x]$$

$$\frac{d^2y}{dx^2} = e^{-2x}[(-2B-2)\cos x - (B-4)\sin x]$$

$$- 2e^{-2x}[(-2B-2)\sin x + (B-4)\cos x]$$

$$= e^{-2x}[(-4B+6)\cos x + (3B+8)\sin x]$$

But $\qquad \dfrac{dy}{dx} = \dfrac{d^2y}{dx^2}$

$$e^{-2x}[(-2B-2)\sin x + (B-4)\cos x] = e^{-2x}[(-4B+6)\cos x + (3B+8)\sin x]$$

On putting $x = 0$, we get

$$B - 4 = -4B + 6 \quad \Rightarrow \quad B = 2$$

(2) becomes, $\qquad y = e^{-2x}[2\cos x + 2\sin x]$

$$y = 2e^{-2x}[\sin x + \cos x]$$ **Ans.**

### Exercise 3.19

*Solve the following equations :*

1. $\dfrac{d^2y}{dx^2} - 3\dfrac{dy}{dx} + 2y = 0$ **Ans.** $y = C_1 e^x + C_2 e^{2x}$

2. $\dfrac{d^2y}{dx^2} + \dfrac{dy}{dx} - 30y = 0$ **Ans.** $y = C_1 e^{5x} + C_2 e^{-6x}$

3. $\dfrac{d^2y}{dx^2} - 8\dfrac{dy}{dx} + 16y = 0$ **Ans.** $y = (C_1 + C_2 x)\, e^{4x}$

4. $\dfrac{d^2y}{dx^2} + \mu^2 y = 0$ **Ans.** $y = C_1\cos\mu x + C_2\sin\mu x$

5. $(D^2 + 2D + 2)y = 0,\ y(0) = 0,\ y'(0) = 1$ *(A.M.I.E.T.E., June 2006)* **Ans.** $y = e^{-x}\sin x$

6. $\dfrac{d^3y}{dx^3} - 2\dfrac{d^2y}{dx^2} + 4\dfrac{dy}{dx} - 8y = 0$ **Ans.** $y = C_1 e^{2x} + C_2\cos 2x + C_3\sin 2x$

7. $\dfrac{d^4y}{dx^4} - 32\dfrac{d^2y}{dx^2} + 256 = 0$ *(AMIETE., Dec. 2004)* **Ans.** $y = (C_1 + C_2 x)\cos 4x + (C_3 + C_4 x)\sin 4x$

8. $\frac{d^4y}{dx^4}-4\frac{d^3y}{dx^3}+8\frac{d^2y}{dx^2}-8\frac{dy}{dx}+4y=0$ **Ans.** $y=e^x[(C_1+C_2x)\cos x+(C_3+C_4x)\sin x]$

9. $\frac{d^4y}{dx^4}+\frac{d^2y}{dx^2}=0,\ y(0)=y'(0)=y''(0)=0,\ y'''(0)=1$ **Ans.** $y=x-\sin x$

10. The equation for the bending of a strut is EI $\frac{d^2y}{dx^2}+Py=0$

If $y=0$ when $x=0$, and $y=a$ when $x=\frac{1}{2}$, find $y$. **Ans.** $y=\dfrac{a\sin\sqrt{\frac{P}{EI}}x}{\sin\sqrt{\frac{P}{EI}}\frac{1}{2}}$

11. $\frac{d^3y}{dx^3}+6\frac{d^2y}{dx^2}+12\frac{dy}{dx}+8y=0,\ y(0)=0$, and $y'(0)=0$ and $y''(0)=2$

*(A.M.I.E.T.E. Dec. 2008)* **Ans.** $y=x^2e^{-2x}$

12. $\frac{d^3y}{dx^3}+\frac{d^2y}{dx^2}+\frac{4dy}{dx}+4y=0,\ y(0)=0,\ y'(0)=0,\ y''(0)=-5,$ **Ans.** $y=-e^x+\cos 2x-\frac{1}{2}\sin 2x$

13. $(D^8+6D^6-32D^2)y=0$ *(A.M.I.E.T.E., Summer 2005)*

**Ans.** $y=C_1+C_2x+C_3e^{\sqrt{2}x}+C_4e^{-\sqrt{2}x}+C_5\cos 2x+C_6\sin 2x$

14. Show that non-trivial solutions of the boundary value problem $y^{(iv)}-w^4y=0,\ y(0)=0=y''(0),$ $y(L)=0,\ y''(L)=0$ are $y(x)=\sum_{n=1}^{\infty}D_n\sin\left(\frac{n\pi x}{L}\right)$ where $D_n$ are constants. *(AMIETE, Dec. 2005)*

15. Solve the initial value problem $y'''+6y''+11y'+6y=0,\ y(0)=0,\ y'(0)=1,\ y''(0)=-1.$

*(A.M.I.E.T.E., Dec. 2006)* **Ans.** $y=2e^{-x}-3e^{-2x}+e^{-3x}$.

16. Let $y_1, y_2$ be two linearly independent solutions of the differential equation $yy''-(y')^2=0$. Then, $c_1y_1+c_2y_2$, where $c_1, c_2$ are constants is a solution of this differential equation for
(*a*) $c_1=c_2=0$ only. (*b*) $c_1=0$ or $c_2=0$ (*c*) no value of $c_1, c_2$. (*d*) all real $c_1, c_2$
*(A.M.I.E.T.E., Dec. 2004)*

## 3.21 RULES TO FIND PARTICULAR INTEGRAL

(*i*) $\frac{1}{f(D)}e^{ax}=\frac{1}{f(a)}e^{ax}$ If $f(a)=0$ then $\frac{1}{f(D)}\cdot e^{ax}=x\cdot\frac{1}{f'(a)}\cdot e^{ax}$

If $f'(a)=0$ then $\frac{1}{f(D)}\cdot e^{ax}=x^2\frac{1}{f''(a)}\cdot e^{ax}$

(*ii*) $\frac{1}{f(D)}x^n=[f(D)]^{-1}x^n$ Expand $[f(D)]^{-1}$ and then operate.

(*iii*) $\frac{1}{f(D^2)}\sin ax=\frac{1}{f(-a^2)}\sin ax$ and $\frac{1}{f(D^2)}\cos ax=\frac{1}{f(-a^2)}\cos ax$

If $f(-a^2)=0$ then $\frac{1}{f(D^2)}\sin ax=x\cdot\frac{1}{f'(-a^2)}\cdot\sin ax$

(*iv*) $\frac{1}{f(D)}e^{ax}\cdot\phi(x)=e^{ax}\cdot\frac{1}{f(D+a)}\phi(x)$

(*v*) $\frac{1}{D+a}\phi(x)=e^{-ax}\int e^{ax}\cdot\phi(x)\,dx$

**3.22** $\boxed{\dfrac{1}{f(D)} e^{ax} = \dfrac{1}{f(a)} e^{ax}.}$

We know that, $D.e^{ax} = a.e^{ax}$, $D^2 e^{ax} = a^2.e^{ax}, \ldots\ldots, D^n e^{ax} = a^n e^{ax}$

Let $f(D)\, e^{ax} = (D^n + K_1 D^{n-1} + \ldots + K_n)\, e^{ax} = (a^n + K_1 a^{n-1} + \ldots + K_n) e^{ax} = f(a)\, e^{ax}$.

Operating both sides by $\dfrac{1}{f(D)}$

$$\frac{1}{f(D)} \cdot f(D) e^{ax} = \frac{1}{f(D)} \cdot f(a) e^{ax}$$

$$\Rightarrow \qquad e^{ax} = f(a) \frac{1}{f(D)} \cdot e^{ax} \quad \Rightarrow \quad \frac{1}{f(D)} e^{ax} = \frac{1}{f(a)} e^{ax}$$

If $f(a) = 0$, then the above rule fails.

Then $\dfrac{1}{f(D)} e^{ax} = x \cdot \dfrac{1}{f'(D)} e^{ax} = x \dfrac{1}{f'(a)} e^{ax} \Rightarrow \boxed{\dfrac{1}{f(D)} e^{ax} = x \,.\, \dfrac{1}{f'(a)} e^{ax}}$

If $f'(a) = 0$ then $\boxed{\dfrac{1}{f(D)} e^{ax} = x^2 \dfrac{1}{f''(a)} e^{ax}}$

**Example 50.** *Solve the differential equation*

$$\frac{d^2 x}{dt^2} + \frac{g}{t} x = \frac{g}{l} L$$

*where g, l, L are constants subject to the conditions,*

$$x = a, \quad \frac{dx}{dt} = 0 \quad at\ t = 0.$$

**Solution.** We have, $\dfrac{d^2 x}{dt^2} + \dfrac{g}{t} x = \dfrac{g}{l} L \quad \Rightarrow \quad \left(D^2 + \dfrac{g}{l}\right) x = \dfrac{g}{l} L$

A.E. is $\quad m^2 + \dfrac{g}{l} = 0 \quad \Rightarrow \quad m = \pm\, i \sqrt{\dfrac{g}{l}}$

$$\text{C.F.} = C_1 \cos \sqrt{\frac{g}{l}}\, t + C_2 \sin \sqrt{\frac{g}{l}}\, t$$

$$\text{P.I.} = \frac{1}{D^2 + \frac{g}{l}} \cdot \frac{g}{l} L = \frac{g}{l} L \frac{1}{D^2 + \frac{g}{l}} e^{0t} = \frac{g}{l} L \frac{1}{0 + \frac{g}{l}} = L \qquad [D = 0]$$

$\therefore$ General solution is = C.F. + P.I.

$$x = C_1 \cos\left(\sqrt{\frac{g}{l}}\right) t + C_2 \sin\left(\sqrt{\frac{g}{l}}\right) t + L \qquad \ldots(1)$$

$$\frac{dx}{dt} = -C_1 \sqrt{\frac{g}{l}} \sin\left(\sqrt{\frac{g}{l}}\right) t + C_2 \sqrt{\frac{g}{l}} \cos\left(\sqrt{\frac{g}{l}}\right) t$$

Put $t = 0$ and $\dfrac{dx}{dt} = 0$

$$0 = C_2 \sqrt{\frac{g}{l}} \qquad \therefore \quad C_2 = 0$$

(1) becomes $x = C_1 \cos\sqrt{\frac{g}{l}}\,t + L$ ...(2)

Put $x = a$ and $t = 0$ in (2), we get

$a = C_1 + L$ or $C_1 = a - L$

On putting the value of $C_1$ in (2), we get $x = (a-L)\cos\left(\sqrt{\frac{g}{l}}\right)t + L$ **Ans.**

**Example 51.** *Solve* : $\frac{d^2y}{dx^2} + 6\frac{dy}{dx} + 9y = 5e^{3x}$

**Solution.** $(D^2 + 6D + 9)y = 5e^{3x}$

Auxiliary equation is $m^2 + 6m + 9 = 0 \quad \Rightarrow \quad (m+3)^2 = 0 \quad \Rightarrow \quad m = -3, -3,$

$$\text{C.F.} = (C_1 + C_2x)\,e^{-3x}$$

$$\text{P.I.} = \frac{1}{D^2+6D+9}.5.e^{3x} = 5\frac{e^{3x}}{(3)^2+6(3)+9} = \frac{5e^{3x}}{36}$$

The complete solution is $y = (C_1 + C_2x)e^{-3x} + \frac{5e^{3x}}{36}$ **Ans.**

**Example 52.** *Solve* : $\frac{d^2y}{dx^2} - 6\frac{dy}{dx} + 9y = 6e^{3x} + 7e^{-2x} - \log 2$

**Solution.** $(D^2 - 6D + 9)y = 6e^{3x} + 7e^{-2x} - \log 2$

A.E. is $(m^2 - 6m + 9) = 0 \quad \Rightarrow \quad (m-3)^2 = 0, \quad \Rightarrow \quad m = 3, 3$

$$\text{C.F.} = (C_1 + C_2x)e^{3x}$$

$$\text{P.I.} = \frac{1}{D^2-6D+9}6e^{3x} + \frac{1}{D^2-6D+9}7e^{-2x} + \frac{1}{D^2-6D+9}(-\log 2)$$

$$= x\frac{1}{2D-6}6e^{3x} + \frac{1}{4+12+9}7e^{-2x} - \log 2\frac{1}{D^2-6D+9}e^{0x}$$

$$= x^2\frac{1}{2}\cdot 6\cdot e^{3x} + \frac{7}{25}e^{-2x} - \log 2\left(\frac{1}{9}\right) = 3x^2e^{3x} + \frac{7}{25}e^{-2x} - \frac{1}{9}\log 2$$

Complete solution is $y = (C_1 + C_2\,x)e^{3x} + 3x^2e^{3x} + \frac{7}{25}e^{-2x} - \frac{1}{9}\log 2$ **Ans.**

## EXERCISE 3.20

*Solve the following differential equations:*

1. $[D^2 + 5D + 6]\,[y] = e^x$ **Ans.** $C_2\,e^{-2x} + C_2e^{-3x} + \frac{e^x}{12}$

2. $\frac{d^2y}{dx^2} - 3\frac{dy}{dx} + 2y = e^{3x}$ **Ans.** $C_1e^x + C_2e^{2x} + \frac{e^{3x}}{2}$ *(A.M.I.E.T.E. June 2010, 2007)*

3. $(D^3 + 2D^2 - D - 2)\,y = e^x$ **Ans.** $C_1e^x + C_2e^{-x} + C_3e^{-2x} + \frac{x}{6}e^x$

4. $\frac{d^2y}{dx^2} + 2\frac{dy}{dx} + 2y = \sinh x$ **Ans.** $e^{-x}[C_1\cos x + C_2\sin x] + \frac{e^x}{10} - \frac{e^{-x}}{2}$

5. $\frac{d^2y}{dx^2} + 4\frac{dy}{dx} + 5y = -2\cosh x$ **Ans.** $e^{-2x}(C_1\cos x + C_2\sin x) - \frac{1}{10}e^x - \frac{e^{-x}}{2}$

6. $(D^3 - 2D^2 - 5D + 6)\, y = e^{3x}$ **Ans.** $C_1 e^x + C_2 e^{-2x} + C_3 e^{3x} + \frac{x.e^{3x}}{10}$

7. $\frac{d^3y}{dx^3} - \frac{d^2y}{dx^2} + 4\frac{dy}{dx} - 4y = e^x$ **Ans.** $C_1 e^x + C_2 \cos 2x + C_3 \sin 2x + \frac{x e^x}{5}$

8. $\frac{d^2y}{dx^2} - 6\frac{dy}{dx} + 9y = e^{3x}$ **Ans.** $(C_1 + C_2 x) e^{3x} + \frac{x^2}{2} e^{3x}$

9. $\frac{d^3y}{dx^3} + 3\frac{d^2y}{dx^2} + 3\frac{dy}{dx} + y = e^{-x}$ **Ans.** $(C_1 + C_2 x + C_3 x^2) e^{-x} + \frac{x^3}{6} e^{-x}$

10. $\frac{d^2y}{dx^2} - \frac{dy}{dx} - 6y = e^x \cosh 2x$ **Ans.** $C_1 e^{3x} + C_2 e^{-2x} + \frac{1}{10} x e^{3x} - \frac{1}{8} e^{-x}$

11. $(D - 2)(D + 1)^2\, y = e^{2x} + e^x$ **Ans.** $C_1 e^{2x} + (C_2 + C_3 x) e^{-x} + \frac{x}{9} e^{2x} - \frac{e^x}{4}$

12. $(D - 1)^3\, y = 16\, e^{3x}$ **Ans.** $(C_1 + C_2 x + C_3 x^2)\, e^x + 2e^{3x}$

## 3.23 $\frac{1}{f(D)} x^n = [f(D)]^{-1} x^n.$

Expand $[f(D)]^{-1}$ by the Binomial theorem in ascending powers of $D$ as far as the result of operation on $x^n$ is zero.

**Example 53.** *Solve the differential equation* $\frac{d^2y}{dx^2} + a^2 y = \frac{a^2 R}{p}(l - x)$

*where a, R, p and l are constants subject to the conditions* $y = 0$, $\frac{dy}{dx} = 0$ *at* $x = 0$.

**Solution.** $\frac{d^2y}{dx^2} + a^2 y = \frac{a^2}{p} R(l - x) \quad \Rightarrow \quad (D^2 + a^2) y = \frac{a^2}{p} R(l - x)$

A.E. is $m^2 + a^2 = 0 \Rightarrow m = \pm ia$

$$\text{C.F.} = C_1 \cos ax + C_2 \sin ax$$

$$\text{P.I.} = \frac{1}{D^2 + a^2} \frac{a^2}{p} R(l - x) = \frac{a^2 R}{p} \frac{1}{a^2} \left[\frac{1}{1 + \frac{D^2}{a^2}}\right](l - x) = \frac{R}{p}\left[1 + \frac{D^2}{a^2}\right]^{-1} (l - x)$$

$$= \frac{R}{p}\left[1 - \frac{D^2}{a^2}\right](l - x) = \frac{R}{p}(l - x)$$

$$y = C_1 \cos ax + C_2 \sin ax + \frac{R}{p}(l - x) \qquad \text{...(1)}$$

On putting $y = 0$, and $x = 0$ in (1), we get $0 = C_1 + \frac{R}{p} l \Rightarrow \quad C_1 = -\frac{Rl}{p}$

On differentiating (1), we get $\frac{dy}{dx} = -a\, C_1 \sin ax + a\, C_2 \cos ax - \frac{R}{p}$ ...(2)

On putting $\frac{dy}{dx} = 0$ and $x = 0$ in (2), we have

$$0 = a\, C_2 - \frac{R}{p} \quad \Rightarrow \quad C_2 = \frac{R}{a.p}$$

On putting the values of $C_1$ and $C_2$ in (1), we get

$$y = -\frac{R}{p} l \cos a.x + \frac{R}{a.p} \sin a.x + \frac{R}{p}(l - x) \Rightarrow y = \frac{R}{p}\left[\frac{\sin ax}{a} - l \cos a.x + l - x\right]$$ **Ans.**

## EXERCISE 3.21

*Solve the following equations :*

1. $(D^2 + 5D + 4)\, y = 3 - 2x$ **Ans.** $C_1 e^{-x} + C_2 e^{-4x} + \frac{1}{8}(11 - 4x)$

2. $\frac{d^2y}{dx^2} + 2\frac{dy}{dx} + y = x$ **Ans.** $(C_1 + C_2 x)\, e^{-x} + x - 2$

3. $(2D^2 + 3D + 4)\, y = x^2 - 2x$ **Ans.** $e^{-\frac{3}{4}x}\left[A\cos\frac{\sqrt{23}}{4}x + B\sin\frac{\sqrt{23}}{4}x\right] + \frac{1}{32}[8x^2 - 28x + 13]$

4. $(D^2 - 4D + 3)\, y = x^3$ **Ans.** $C_1 e^{x} + C_2 e^{3x} + \frac{1}{27}(9x^3 + 36x^2 + 78x + 80).$

5. $5\frac{d^3y}{dx^3} - \frac{d^2y}{dx^2} - 6\frac{dy}{dx} = 1 + x^2.$ **Ans.** $A + Be^{-2x} + Ce^{3x} - \frac{1}{36}\left(2x^3 - x^2 + \frac{25}{3}x\right)$

6. $\frac{d^4y}{dx^4} + 4y = x^4$ **Ans.** $e^{x}(C_1 \cos x + C_2 \sin x) + e^{-x}(C_3 \cos x + C_4 \sin x) + \frac{1}{4}(x^4 - 6)$

7. $\frac{d^2y}{dx^2} + 2p\frac{dy}{dx} + (p^2 + q^2)y = e^{cx} + p.q\, x^2$

**Ans.** $e^{-px}[C_1 \cos qx + C_2 \sin qx] + \frac{e^{Cx}}{(p+C)^2 + q^2} + \frac{pq}{p^2 + q^2}\left[x^2 - \frac{4px}{p^2 + q^2} + \frac{6p^2 - 2q^2}{(p^2 + q^2)^2}\right]$

8. $D^2(D^2 + 4)\, y = 96\, x^2$ **Ans.** $C_1 + C_2 x + C_3 \cos 2x + C_4 \sin 2x + 2x^2(x^2 - 3)$

**3.24** $\boxed{\frac{1}{f(D^2)}\sin ax = \frac{\sin ax}{f(-a^2)}}$ $\boxed{\frac{1}{f(D^2)}\cdot\cos ax = \frac{\cos ax}{f(-a^2)}}$

$D(\sin ax) = a.\cos ax,\ D^2(\sin ax) = D(a\cos ax) = -a^2.\sin ax$

$D^4(\sin ax) = D^2.D^2(\sin ax) = D^2(-a^2 \sin ax) = (-a^2)^2 \sin ax$

$(D^2)^n \sin ax = (-a^2)^n \sin ax$

Hence, $f(D^2)\sin ax = f(-a^2)\sin ax$

$$\frac{1}{f(D^2)}\cdot f(D^2)\sin ax = \frac{1}{f(D^2)}.f(-a^2).\sin ax$$

$$\sin ax = f(-a^2)\frac{1}{f(D^2)}\sin ax \quad\Rightarrow\quad \frac{1}{f(D^2)}\cdot\sin ax = \frac{\sin ax}{f(-a^2)}$$

Similarly, $$\frac{1}{f(D^2)}\cos ax = \frac{\cos ax}{f(-a^2)}$$

If $f(-a^2) = 0$ then above rule fails.

$$\frac{1}{f(D^2)}\sin ax = x\frac{\sin ax}{f'(-a^2)}$$

If $f'(-a^2) = 0$ then, $$\frac{1}{f(D^2)}\sin ax = x^2\frac{\sin ax}{f''(-a^2)}$$

**Example 54.** *Solve* : $(D^2 + 4)\, y = \cos 2x$

*(R.G.P.V., Bhopal June, 2008, A.M.I.E.T.E. Dec 2008)*

**Solution.** $(D^2 + 4)\, y = \cos 2x$

Auxiliary equation is $m^2 + 4 = 0$

$m = \pm 2i,$ C.F. $= A\cos 2x + B\sin 2x$

$$\text{P.I.} = \frac{1}{D^2+4}\cos 2x = x.\frac{1}{2D}\cos 2x = \frac{x}{2}\left(\frac{1}{2}\sin 2x\right) = \frac{x}{4}\sin 2x$$

Complete solution is $y = A\cos 2x + B\sin 2x + \frac{x}{4}\sin 2x$ **Ans.**

**Example 55.** *Solve* : $\frac{d^3y}{dx^3} - 3\frac{d^2y}{dx^3} + 4\frac{dy}{dx} - 2y = e^x + \cos x$ *(U.P., II Semester, Summer* 2006, 2001)

**Solution.** Given $(D^3 - 3D^2 + 4D - 2)\, y = e^x + \cos x$

A.E. is $m^3 - 3m^2 + 4m - 2 = 0$

$\Rightarrow$ $(m-1)(m^2 - 2m + 2) = 0$, *i.e.*, $m = 1,\ 1 \pm i$

$\therefore$ C.F. $= C_1 e^x + e^x (C_2 \cos x + C_3 \sin x)$

$$\text{P.I.} = \frac{1}{(D-1)(D^2-2D+2)}e^x + \frac{1}{D^3-3D^2+4D-2}\cos x$$

$$= \frac{1}{(D-1)(1-2+2)}e^x + \frac{1}{(-1)D-3(-1)+4D-2}\cos x$$

$$= \frac{1}{(D-1)}e^x + \frac{1}{3D+1}\cos x = x\frac{1}{1}e^x + \frac{3D-1}{9D^2-1}\cos x$$

$$= e^x.x + \frac{(-3\sin x - \cos x)}{-9-1} = e^x.x + \frac{1}{10}(3\sin x + \cos x)$$

Hence, complete solution is

$$y = C_1e^x + e^x(C_2\cos x + C_3 \sin x) + x\,e^x + \frac{1}{10}(3\sin x + \cos x)$$ **Ans.**

**Example 56.** *Solve* : $(D^3+1)y = \cos^2\left(\frac{x}{2}\right) + e^{-x}$ *(Nagpur University, Summer* 2004)

**Solution.** $(D^3+1)y = \cos^2\left(\frac{x}{2}\right) + e^{-x}$

A.E. is $m^3 + 1 = 0$

$$(m+1)(m^2 - m + 1) = 0 \quad \Rightarrow \quad m = -1$$

or $$m = \frac{-(-1) \pm \sqrt{1-4}}{2} = \frac{1 \pm i\sqrt{3}}{2} \quad \Rightarrow \quad m = \frac{1}{2} \pm i\frac{\sqrt{3}}{2}$$

$$\therefore \quad \text{C.F.} = C_1e^{-x} + e^{\frac{x}{2}}\left[C_2\cos\frac{\sqrt{3}}{2}x + C_3\sin\frac{\sqrt{3}}{2}x\right]$$

$$\text{P.I.} = \frac{1}{D^3+1}\left[\cos^2\left(\frac{x}{2}\right) + e^{-x}\right] = \frac{1}{D^3+1}\cos^2\left(\frac{x}{2}\right) + \frac{1}{D^3+1}e^{-x} \qquad [\text{Put } D = -1]$$

$$= \frac{1}{D^3+1}\left(\frac{1+\cos x}{2}\right) + x\frac{1}{3D^2}e^{-x}$$

$$= \frac{1}{2}\frac{1}{D^3+1}e^{0x} + \frac{1}{2}\frac{1}{D^3+1}\cos x + x\frac{1}{3(-1)^2}e^{-x} = \frac{1}{2} + \frac{1}{2}\frac{1}{-D+1}\cos x + \frac{x}{3}e^{-x}$$

$$= \frac{1}{2} - \frac{1}{2}\frac{(D+1)\cos x}{(D-1)(D+1)} + \frac{x}{3}e^{-x} = \frac{1}{2} - \frac{1}{2}\frac{(-\sin x + \cos x)}{(D^2-1)} + \frac{x}{3}e^{-x}$$

$$= \frac{1}{2} + \frac{1}{2}\frac{\sin x}{(D^2-1)} - \frac{1}{2}\frac{1}{(D^2-1)}\cos x + \frac{x}{3}e^{-x}$$

$$= \frac{1}{2} + \frac{1}{2}\frac{\sin x}{(-1-1)} - \frac{1}{2}\frac{1}{(-1-1)}\cos x + \frac{x}{3}e^{-x} = \frac{1}{2} - \frac{\sin x}{4} + \frac{\cos x}{4} + \frac{x}{3}e^{-x}$$

P.I. =

$$\frac{1}{2} + \frac{1}{4}(\cos x - \sin x) + \frac{x}{3}e^{-x}$$

Hence, the complete solution is

$$y = C_1 e^{-x} + e^{\frac{x}{2}}\left[C_2 \cos\frac{\sqrt{3}}{2}x + C_3 \sin\frac{\sqrt{3}}{2}x\right] + \frac{1}{2} + \frac{1}{4}(\cos x - \sin x) + \frac{x}{3}e^{-x}$$ **Ans.**

## EXERCISE 3.22

*Solve the following differential equations :*

**1.** $\frac{d^2y}{dx^2} + 6y = \sin 4x$ **Ans.** $C_1 \cos\sqrt{6}x + C_2 \sin\sqrt{6}x - \frac{1}{10}\sin 4x$

**2.** $\frac{d^2x}{dt^2} + 2\frac{dx}{dt} + 3x = \sin t$ **Ans.** $e^{-t}[A\cos\sqrt{2}t + B\sin\sqrt{2}t] - \frac{1}{4}(\cos t - \sin t)$

**3.** $\frac{d^2x}{dt^2} + 2\frac{dx}{dt} + 5x = \sin 2t$, given that when $t = 0$, $x = 3$ and $\frac{dx}{dt} = 0$

**Ans.** $e^{-t}\left[\frac{55}{17}\cos 2t + \frac{53}{34}\sin 2t\right] - \frac{1}{17}(4\cos 2t - \sin 2t)$

**4.** $\frac{d^2y}{dx^2} - 7\frac{dy}{dx} + 6y = 2\sin 3x$, given that $y = 1$, $\frac{dy}{dx} = 0$ when $x = 0$.

**Ans.** $-\frac{13}{75}e^{6x} + \frac{27}{25}e^{x} + \frac{1}{75}(7\cos 3x - \sin 3x)$

**5.** $(D^3 + 1)\,y = 2\cos^2 x$

**Ans.** $C_1 e^{-x} + e^{\frac{1}{2}x}\left(C_2 \cos\frac{\sqrt{3}}{2}x + C_3 \sin\frac{\sqrt{3}}{2}x\right) + 1 + \frac{1}{65}(-8\sin 2x + \cos 2x)$

**6.** $(D^2 + a^2)\,y = \sin ax$ *(A.M.I.E.T.E., June 2009)* **(Ans.** $C_1 \cos ax + C_2 \sin ax - \frac{x}{2a}\cos ax$

**7.** $(D^4 + 2a^2D^2 + a^4)\,y = 8\cos ax$ **Ans.** $(C_1 + C_2x + C_3 \cos ax + C_4 \sin ax) - \frac{x^2}{a^2}\cos ax$

**8.** $\frac{d^2y}{dx^2} + 3\frac{dy}{dx} + 2y = \sin 2x$ *(A.M.I.E.T.E., Summer 2002)*

**Ans.** $C_1 e^{-x} + C_2 e^{-2x} - \frac{1}{20}(3\cos 2x + \sin 2x)$

**9.** $\frac{d^2y}{dx^2} + y = \sin 3x \cos 2x$ **Ans.** $C_1 \cos x + C_2 \sin x + \frac{1}{48}[-\sin 5x - 12x\cos x]$

**10.** $\frac{d^2y}{dx^2} - 2\frac{dy}{dx} - 3y = 2e^{2x} + 10\sin 3x$ given that $y(0) = 2$ and $y'(0) = 4$

**Ans.** $\frac{29}{12}e^{3x} - \frac{1}{12}e^{-x} - \frac{2}{3}e^{2x} + \frac{1}{3}[\cos 3x - 2\sin 3x]$

**11.** $\frac{d^2y}{dx^2} + 3\frac{dy}{dx} + 2y = 4\cos^2 x$ *(R.G.P.V., Bhopal, I Semester, June 2007)*

**Ans.** $C_1 e^{-x} + C_2 e^{-2x} - e^{2x} + \frac{1}{10}(3\sin 2x - \cos 2x) + 1$

**12.** $\frac{d^2y}{dx^2} - 2\frac{dy}{dx} + 3y = \cos x + x^2$

**Ans.** $e^{x}[C_1 \cos\sqrt{2}\,x + C_2 \sin\sqrt{2}\,x] + \frac{1}{4}(\cos x - \sin x) + \frac{1}{3}(x^2 + \frac{4}{3}x + \frac{2}{9})$

13. $(D^3-3D^2+4D-2)y=e^x+\cos x$ **Ans.** $(C_1+C_2\cos x+C_3\sin x)e^x+\frac{1}{10}(3\sin x+\cos x)$

14. $(D^3-4D^2+13D)y=1+\cos 2x$

**Ans.** $C_1+e^{2x}(C_2\cos 3x+C_3\sin 3x)+\frac{1}{290}(9\sin 2x+8\cos 2x)+\frac{x}{13}$

15. $(D^2-4D+4)y=e^{2x}+x^3+\cos 2x$

**Ans.** $(C_1+C_2x)e^{2x}+\frac{1}{2}x^2e^{2x}+\frac{1}{8}(2x^3+6x^2+9x+6)-\frac{1}{8}\sin 2x$

16. $\frac{d^2y}{dx^2}+n^2y=h\sin px$ $(P\neq n)$

where $h, p$ and $n$ are constants satisfying the conditions

$y=a,\ \frac{dy}{dx}=b$ for $x=0$ **Ans.** $a\cos nx+\left(\frac{b}{n}-\frac{ph}{n(n^2-p^2)}\right)\sin nx+\frac{h\sin px}{(n^2-p^2)}$

17. $y''+y'-2y=-6\sin 2x-18\cos 2x,\ y(0)=2,\ y'(0)=2$ **Ans.** $-e^{-2x}+3\cos 2x$

**3.25** $\boxed{\frac{1}{f(D)}.e^{ax}\cdot\phi(x)=e^{ax}.\frac{1}{f(D+a)}.\phi(x)}$

$$D[e^{ax}\phi(x)]=e^{ax}D\phi(x)+ae^{ax}\phi(x)=e^{ax}(D+a)\phi(x)$$

$$D^2[e^{ax}\phi(x)]=D[e^{ax}(D+a)\phi(x)]=e^{ax}(D^2+aD)\phi(x)+ae^{ax}(D+a)\phi(x)$$

$$=e^{ax}(D^2+2aD+a^2)\phi(x)=e^{ax}(D+a)^2\phi(x)$$

Similarly, $D^n[e^{ax}\phi(x)]=e^{ax}(D+a)^n\phi(x)$

$$f(D)[e^{ax}\phi(x)]=e^{ax}f(D+a)\phi(x)$$

$$e^{ax}\phi(x)=\frac{1}{f(D)}.[e^{ax}f(D+a)\phi(x)] \qquad ...(1)$$

Put $f(D+a)\phi(x)=X$, so that $\phi(x)=\frac{1}{f(D+a)}.X$

Substituting these values in (1), we get

$$e^{ax}\frac{1}{f(D+a)}X=\frac{1}{f(D)}[e^{ax}.X] \Rightarrow \frac{1}{f(D)}[e^{ax}.\phi(x)]=e^{ax}\frac{1}{f(D+a)}\phi(x)$$

**Example 57.** *Solve* : $(D^2-4D+4)y=x^3e^{2x}$

**Solution.** $(D^2-4D+4)y=x^3e^{2x}$

A.E. is $m^2-4m+4=0 \Rightarrow (m-2)^2=0 \Rightarrow m=2, 2$

C.F. $=(C_1+C_2x)e^{2x}$

$$\text{P.I.}=\frac{1}{D^2-4D+4}x^3\cdot e^{2x}=e^{2x}\frac{1}{(D+2)^2-4(D+2)+4}.x^3$$

$$=e^{2x}\frac{1}{D^2}x^3=e^{2x}.\frac{1}{D}\left(\frac{x^4}{4}\right)=e^{2x}\cdot\frac{x^5}{20}$$

The complete solution is $y=(C_1+C_2x)e^{2x}+e^{2x}.\frac{x^5}{20}$ **Ans.**

**Example 58.** *Solve the differential equation :*

$\frac{d^3y}{dx^3}-7\frac{d^2y}{dx^2}+10\frac{dy}{dx}=e^{2x}\sin x$ *(AMIETE, June 2010, Nagpur University, Summer* 2005)

**Solution.** $\dfrac{d^3y}{dx^3}-7\dfrac{d^2y}{dx^2}+10\dfrac{dy}{dx}=e^{2x}\sin x \quad \Rightarrow D^3y-7D^2y+10\,Dy=e^{2x}\sin x$

A.E. is

$$m^3-7m^2+10\,m=0 \quad \Rightarrow \quad (m-2)(m^2-5m)=0$$

$$\Rightarrow \quad m(m-2)(m-5)=0 \quad \Rightarrow \quad m=0,2,5$$

$$\text{C.F}=C_1e^{0x}+C_2e^{2x}+C_3e^{5x}$$

$$\text{P.I.}=\frac{1}{D^3-7D^2+10D}e^{2x}\sin x=e^{2x}\frac{1}{(D+2)^3-7(D+2)^2+10(D+2)}.\sin x$$

$$=e^{2x}\frac{1}{D^3+6D^2+12D+8-7D^2-28D-28+10D+20}.\sin x$$

$$=e^{2x}\frac{1}{D^3-D^2-6D}\sin x=e^{2x}\frac{1}{(-1^2)\,D-(-1^2)-6D}\sin x$$

$$=e^{2x}\frac{1}{-D+1-6D}\sin x=e^{2x}\frac{1}{1-7D}\sin x=e^{2x}\frac{1+7D}{1-49D^2}\sin x=e^{2x}\frac{1+7D}{1-49(-1^2)}\sin x$$

$$=e^{2x}\frac{1+7D}{50}\sin x=\frac{e^{2x}}{50}(\sin x+7\cos x)$$

Complete solution is

$$y=\text{C.F.}+\text{P.I.}$$

$$\Rightarrow \quad y=C_1+C_2\,e^{2x}+C_3\,e^{5x}+\frac{e^{2x}}{50}(\sin x+7\cos x) \qquad \textbf{Ans.}$$

**Example 59.** *Solve* $(D^2+6D+9)y=\dfrac{e^{-3x}}{x^3}$.

*(Nagpur University, Summer 2002, A.M.I.E.T.E., June 2009)*

**Solution** A.E. is $m^2+6m+9=0$

$(m+3)^2=0 \quad \therefore \quad m=-3,-3$

C.F. $=(C_1+C_2x)\,e^{-3x}$

$$\text{P.I.}=\frac{1}{D^2+6D+9}\frac{e^{-3x}}{x^3}=e^{-3x}\frac{1}{(D-3)^2+6(D-3)+9}\frac{1}{x^3}$$

$$=e^{-3x}\frac{1}{D^2-6D+9+6D-18+9}\frac{1}{x^3}=e^{-3x}\frac{1}{D^2}(x^{-3})$$

$$=e^{-3x}\frac{1}{D}\left(\frac{x^{-2}}{-2}\right)=e^{-3x}\frac{x^{-1}}{(-2)(-1)}=\frac{e^{-3x}x^{-1}}{2}=\frac{e^{-3x}}{2x}$$

Hence, the solution is $y=(C_1+C_2x)e^{-3x}+\dfrac{e^{-3x}}{2x}$ **Ans.**

**Example 60.** *Solve* $(D^2+5D+6)\,y=e^{-2x}\sec^2x\,(1+2\tan x)$ *(A.M.I.E.T.E., Summer 2003)*

**Solution.** $(D^2+5D+6)\,y=e^{-2x}\sec^2x\,(1+2\tan x)$

Auxiliary Equation is $m^2+5m+6=0$

$$\Rightarrow \quad (m+2)(m+3)=0 \quad \Rightarrow \quad m=-2,\text{ and } m=-3$$

Hence, complementry function (C.F.) $=C_1e^{-2x}+C_2e^{-3x}$

$$\text{P.I.}=\frac{1}{D^2+5D+6}e-2x\sec 2x(1+2\tan x)=e^{-2x}\frac{1}{(D-2)^2+5(D-2)+6}\sec^2x(1+2\tan x)$$

$$= e^{-2x} \frac{1}{D^2 - 4D + 4 + 5D - 10 + 6} \sec^2 x(1 + 2\tan x)$$

$$= e^{-2x} \frac{1}{D^2 + D} \sec^2 x(1 + 2\tan x) = e^{-2x}\left[\frac{\sec^2 x}{D^2 + D} + \frac{2\tan x \sec^2 x}{D^2 + D}\right]$$

$$= e^{-2x} \frac{1}{D(D+1)} \sec^2 + \frac{1}{D(D+1)} 2\tan x \sec^2 x$$

$$= e^{-2x}\left[\left(\frac{1}{D} - \frac{1}{D+1}\right)\sec^2 x + \right]\left(\frac{1}{D} - \frac{1}{D+1}\right) 2\tan x \sec^2 x$$

$$= e^{-2x}\left[\frac{1}{D}\sec^2 x - \frac{1}{D+1}\sec^2 x + \frac{1}{D} 2\tan x \sec^2 x - \frac{1}{D+1} 2\tan x \sec^2 x\right]$$

$$= e^{-2x}\left[\tan x - e^{-x}\int ex.\sec 2x\,dx + \tan^2 x - e^{-x}\int 2e^x \tan x \sec 2x\,dx\right]$$

Now, $= e^{-2x}\int e^x \sec^2 x\,dx = e^x \sec^2 x - \int e^x.2\sec x \sec x \tan x.dx$

$= e^x \sec^2 x - 2\int ex \sec^2 x.\tan x\,dx$

$\therefore$ P.I. $= e^{-2x}\left[\tan x^{-e} - x.e^x \sec^2 x + 2e^{-x}\int e^x \sec x \tan x\,dx + \tan^2 x - 2e^{-x}\int e^x \sec^2 x \tan x\,dx\right]$

$= e^{-2x}[\tan x - \sec^2 x + \tan^2 x] = e^{-2x}[\tan x - (\sec^2 x - \tan 2x] = e^{-2x}(\tan x - 1)$

$\therefore$ Complete solution is

$$\Rightarrow y = C.F. + P.I. = C_1 e^{-2x} + C_2 e^{-3x} + e^{-2x}(\tan x - 1)$$

**Example 61.** *Solve the differential equation* $(D^2 - 4D + 4)\,y = 8x^2 e^{2x} \sin 2x$

*(U.P. II Semester, Summer 2008, Uttrakhand 2007, 2005, 2004; Nagpur University June 2008)*

**Solution.** $(D^2 - 4D + 4)\,y = 8x^2 e^{2x} \sin 2x$

A.E. is $(m^2 - 4m + 4) = 0 \quad \Rightarrow \quad (m-2)^2 = 0 \quad \Rightarrow \quad m = 2, 2$

C.F. $= (C_1 + C_2 x)\,e^{2x}$

$$\text{P.I.} = \frac{1}{D^2 - 4D + 4} 8x^2 e^{2x} \sin 2x = 8\frac{1}{(D-2)^2} x^2 e^{2x} \sin 2x$$

$$= 8e^{2x} \frac{1}{(D-2+2)^2} x^2 \sin 2x = 8e^{2x} \frac{1}{D^2} x^2 \sin 2x$$

$$= 8e^{2x} \frac{1}{D}\left[x^2 \frac{(-\cos 2x)}{2} - 2x\left(-\frac{\sin 2x}{4}\right) + 2\frac{\cos 2x}{8}\right] = 8e^{2x} \frac{1}{D}\left[-\frac{x^2}{2}\cos 2x + \frac{x \sin 2x}{2} + \frac{\cos 2x}{4}\right]$$

$$= 8e^{2x}\left[-\frac{x^2}{2}\left(\frac{\sin 2x}{2}\right) - \left(\frac{-2x}{2}\right)\left(-\frac{\cos 2x}{4}\right) + (-1)\left(-\frac{\sin 2x}{8}\right) + \frac{x}{2}\left(-\frac{\cos 2x}{2}\right) - \left(\frac{1}{2}\right)\left(-\frac{\sin 2x}{4}\right) + \frac{\sin 2x}{8}\right]$$

$= e^{2x}[-2x^2 \sin 2x - 2x \cos 2x + \sin 2x - 2x \cos 2x + \sin 2x + \sin 2x]$

$= e^{2x}[-2x^2 \sin 2x - 4x \cos 2x + 3\sin 2x] = -e^{2x}[4x \cos 2x + (2x^2 - 3)\sin 2x]$

Complete solution is, $y$ = C.F. + P.I.

$y = (C_1 + C_2 x)\,e^{2x} - e^{2x}[4x \cos 2x + (2x^2 - 3)\sin 2x]$ **Ans.**

## EXERCISE 3.23

*Solve the following equations :*

1. $(D^2 - 5D + 6)\,y = e^x \sin x$ **Ans.** $y = C_1 e^{2x} + C_2 e^{3x} + \frac{e^x}{10}(3\cos x + \sin x)$

2. $\frac{d^2y}{dx^2}-7\frac{dy}{dx}+10y=e^{2x}\sin x$ **Ans.** $y=C_1e^{2x}+C_2e^{5x}+\frac{e^{2x}}{10}(3\cos x-\sin x)$

3. $\frac{d^3y}{dx^3}-2\frac{dy}{dx}+4y=e^x\cos x$ **Ans.** $y=C_1e^{-2x}+e^x(C_2\cos x+C_3\sin x)+\frac{xe^x}{20}(3\sin x-\cos x)$

4. $(D^2-4D+3)\,y=2xe^{3x}+3e^{3x}\cos 2x$

**Ans.** $y=C_1e^x+C_2e^{3x}+\frac{1}{2}e^{3x}(x^2-x)+\frac{3}{8}e^{3x}(\sin 2x-\cos 2x)$

5. $\frac{d^2y}{dx^2}+2\frac{dy}{dx}+y=\frac{e^{-x}}{x^2}$ **Ans.** $y=(C_1+C_2x)\,e^{-x}-e^{-x}\log x$

6. $(D^2-4)\,y=x^2e^{3x}$ **Ans.** $y=C_1e^{2x}+C_2e^{-2x}+\frac{e^{3x}}{5}\left[x^2-\frac{12x}{5}+\frac{62}{25}\right]$

7. $(D^2-3D+2)\,y=2x^2e^{4x}+5e^{3x}$ **Ans.** $y=C_1e^x+C_2e^{2x}+\frac{e^{4x}}{54}[18x^2-30x+19]+\frac{5}{2}e^{3x}$

8. $\frac{d^2y}{dx^2}-4y=x\sinh x$ **Ans.** $y=C_1e^{2x}+C_2e^{-2x}-\frac{x}{3}\sinh x-\frac{2}{9}\cosh x$

9. $\frac{d^2y}{dt^2}+2h\frac{dy}{dt}+(h^2+p^2)y=ke^{-ht}\cos pt$ **Ans.** $y=e^{-ht}[A\cos pt+B\sin pt]+\frac{k}{2p}te^{-ht}\sin pt$

## 3.26 TO FIND THE VALUE OF $\frac{1}{f(D)}x^n\sin ax$.

Now $\frac{1}{f(D)}x^n(\cos ax+i\sin ax)=\frac{1}{f(D)}x^ne^{iax}=e^{iax}\frac{1}{f(D+ia)}x^n$

$$\frac{1}{f(D)}\cdot x^n\sin ax=\text{Imaginary part of } e^{iax}\frac{1}{f(D+ia)}\cdot x^n$$

$$\frac{1}{f(D)}\cdot x^n\cos ax=\text{Real part of } e^{iax}\frac{1}{f(D+ia)}\cdot x^n$$

**Example 62.** *Solve* $\frac{d^2y}{dx^2}-2\frac{dy}{dx}+y=x\sin x$

**Solution.** Auxiliary equation is $m^2-2m+1=0$ or $m=1, 1$

C.F. $=(C_1+C_2x)\,e^x$

P.I. $=\frac{1}{D^2-2D+1}x\cdot\sin x$ $(e^{ix}=\cos x+i\sin x)$

$=$ Imaginary part of $\frac{1}{D^2-2D+1}x(\cos x+i\sin x)$ $=$ Imaginary part of $\frac{1}{D^2-2D+1}x\cdot e^{ix}$

$=$ Imaginary part of $e^{ix}\frac{1}{(D+i)^2-2(D+i)+1}\cdot x$ $=$ Imaginary part of $e^{ix}\frac{1}{D^2-2(1-i)D-2i}\cdot x$

$=$ Imaginary part of $e^{ix}\frac{1}{-2i}\left[1-(1+i)D-\frac{1}{2i}D^2\right]^{-1}\cdot x$

$=$ Imaginary part of $(\cos x+i\sin x)\left(\frac{i}{2}\right)[1+(1+i)D]x=$ Imaginary part of $\frac{1}{2}(i\cos x-\sin x)[x+1+i]$

P.I. $=\frac{1}{2}x\cos x+\frac{1}{2}\cos x-\frac{1}{2}\sin x$

Complete solution is $y=(C_1+C_2x)e^x+\frac{1}{2}(x\cos x+\cos x-\sin x)$ **Ans.**

## EXERCISE 3.24

*Solve the following differential equations :*

1. $(D^2+4)y = 3x\sin x$ **Ans.** $C_1\cos 2x + C_2\sin 2x + x\sin x - \frac{2}{3}\cos x$

2. $\frac{d^2y}{dx^2} - y = x\sin 3x + \cos x$ **Ans.** $C_1e^x + C_2e^{-x} - \frac{1}{10}\left[\frac{3}{5}\cos 3x + x\sin 3x + 5\cos x\right]$

3. $\frac{d^2y}{dx^2} - y = x\sin x + e^x + x^2e^x$ **Ans.** $C_1e^x + C_2e^{-x} - \frac{1}{2}[x\sin x + \cos x] + \frac{x}{12}e^x(2x^2 - 3x + 9)$

4. $(D^4 + 2D^2 + 1)\,y = x^2\cos x$

**Ans.** $(C_1 + C_2x)\cos x + (C_3 + C_4x)\sin x + \frac{1}{12}x^3\sin x - \frac{1}{48}(x^4 - 9x^2)\cos x$

## 3.27 GENERAL METHOD OF FINDING THE PARTICULAR INTEGRAL OF ANY FUNCTION $\phi(x)$

$$\text{P.I.} = \frac{1}{D-a}\phi(x) = y \qquad \text{...(1)}$$

or $$(D-a)\frac{1}{D-a}\cdot\phi(x) = (D-a)\cdot y$$

$$\phi(x) = (D-a)y \quad \text{or} \quad \phi(x) = Dy - ay$$

$$\frac{dy}{dx} - ay = \phi(x) \quad \text{which is the linear differential equation.}$$

Its solution is $$ye^{-\int a\,dx} = \int e^{-\int a\,dx}\cdot\phi(x)\,dx \quad \text{or} \quad ye^{-ax} = \int e^{-ax}\cdot\phi(x)\,dx$$

$$y = e^{ax}\int e^{-ax}\cdot\phi(x)\,dx \qquad \boxed{\frac{1}{D-a}\cdot\phi(x) = e^{ax}\int e^{-ax}\cdot\phi(x)\,dx}$$

**Example 63.** *Solve* $\frac{d^2y}{dx^2} + 9y = \sec 3x.$

**Solution.** Auxiliary equation is $m^2 + 9 = 0$ or $m = \pm 3i$,

$$\text{C.F.} = C_1\cos 3x + C_2\sin 3x$$

$$\text{P.I.} = \frac{1}{D^2+9}\cdot\sec 3x = \frac{1}{(D+3i)(D-3i)}\cdot\sec 3x = \frac{1}{6i}\left[\frac{1}{D-3i} - \frac{1}{D+3i}\right]\cdot\sec 3x$$

$$= \frac{1}{6i}\cdot\frac{1}{D-3i}\cdot\sec 3x - \frac{1}{6i}\frac{1}{D+3i}\cdot\sec 3x \qquad \text{...(1)}$$

Now, $$\frac{1}{D-3i}\sec 3x = e^{3ix}\int e^{-3ix}\sec 3x\,dx \qquad \left[\frac{1}{D-a}\phi(x) = e^{ax}\int e^{-ax}\phi(x)\,dx\right]$$

$$= e^{3ix}\int\frac{\cos 3x - i\sin 3x}{\cos 3x}\,dx = e^{3ix}\int(1 - i\tan 3x)\,dx = e^{3ix}\left(x + \frac{i}{3}\log\cos 3x\right)$$

Changing $i$ to $-i$, we have $\frac{1}{D+3i}\sec 3x = e^{-3ix}\left(x - \frac{i}{3}\log\cos 3x\right)$

Putting these values in (1), we get

$$\text{P.I.} = \frac{1}{6i}\left[e^{3ix}\left(x + \frac{i}{3}\log\cos 3x\right) - e^{-3ix}\left(x - \frac{i}{3}\log\cos 3x\right)\right]$$

$$= \frac{x}{6i}e^{3ix} + \frac{e^{3ix}\log\cos 3x}{18} - \frac{xe^{-3ix}}{6i} + \frac{e^{-3ix}}{18}\log\cos 3x$$

$$= \frac{x}{3}\frac{e^{3ix} - e^{-3ix}}{2i} + \frac{1}{9}\cdot\frac{e^{3ix} + e^{-3ix}}{2}\log\cos 3x = \frac{x}{3}\sin 3x + \frac{1}{9}\cos 3x\cdot\log\cos 3x$$

Hence, complete solution is $y = C_1\cos 3x + C_2\sin 3x + \frac{x}{3}\sin 3x + \frac{1}{9}\cos 3x\cdot\log\cos 3x$ **Ans.**

## EXERCISE 3.25

*Solve the following differential equations :*

1. $\frac{d^2y}{dx^2}+a^2y=\sec ax$ (*R.G.P.V., Bhopal April, 2010*)

   **Ans.** $C_1\cos ax+C_2\sin ax+\frac{x}{a}\sin ax+\frac{1}{a^2}\cos ax\cdot\log\cos ax$

2. $\frac{d^2y}{dx^2}+y=\operatorname{cosec} x$ **Ans.** $C_1\cos x+C_2\sin x-x\cos x+\sin x\log\sin x$

3. $(D^2+4)\,y=\tan 2x$ **Ans.** $C_1\cos 2x+C_2\sin 2x-\frac{1}{4}\cos 2x\log(\sec 2x+\tan 2x)$

4. $\frac{d^2y}{dx^2}+y=(x-\cot x)$ (*A.M.I.E. Winter* 2002)

   **Ans.** $C_1\cos x+C_2\sin x-x\cos^2 x-\sin x\log(\operatorname{cosec} x-\cot x)$

## 3.28 CAUCHY EULER HOMOGENEOUS LINEAR EQUATIONS

$$a_n x_n\frac{d^ny}{dx^n}+a_{n-1}x^{n-1}\frac{d^{n-1}y}{dx^{n-1}}+\ldots.+a_0\,y=\phi(x) \qquad \ldots(1)$$

where $a_0, a_1, a_2, \ldots$ are constants, is called a homogeneous equation.

Put $x=e^z,\quad z=\log_e x,\quad \frac{d}{dz}\equiv D$

$$\frac{dy}{dx}=\frac{dy}{dz}\cdot\frac{dz}{dx}=\frac{1}{x}\frac{dy}{dz}\Rightarrow x\frac{dy}{dx}=\frac{dy}{dz}\Rightarrow x\frac{dy}{dx}=Dy$$

Again, $$\frac{d^2y}{dx^2}=\frac{d}{dx}\left(\frac{dy}{dx}\right)=\frac{d}{dx}\left(\frac{1}{x}\frac{dy}{dz}\right)=-\frac{1}{x^2}\frac{dy}{dz}+\frac{1}{x}\frac{d^2y}{dz^2}\frac{dz}{dx}$$

$$=-\frac{1}{x^2}\frac{dy}{dz}+\frac{1}{x}\frac{d^2y}{dz^2}\frac{1}{x}=\frac{1}{x^2}\left(\frac{d^2y}{dz^2}-\frac{dy}{dz}\right)=\frac{1}{x^2}(D^2-D)\,y;\ x^2\frac{d^2y}{dx^2}=(D^2-D)\,y$$

or $\boxed{x^2\frac{d^2y}{dx^2}=D(D-1)\,y}$ Similarly, $\boxed{x^3\frac{d^3y}{dx^3}=D(D-1)(D-2)\,y}$

The substitution of these values in (1) reduces the given homogeneous equation to a differential equation with constant coefficients.

**Example 64.** *Solve*: $x^2\frac{d^2y}{dx^2}-2x\frac{dy}{dx}-4y=x^4$ (*A.M.I.E. Summer* 2000)

**Solution.** We have, $$x^2\frac{d^2y}{dx^2}-2x\frac{dy}{dx}-4y=x^4 \qquad \ldots(1)$$

Putting $x=e^z,\quad D\equiv\frac{d}{dz},\quad x\frac{dy}{dx}=Dy,\quad x^2\frac{d^2y}{dx^2}=D(D-1)\,y$ in (1), we get

$$D(D-1)\,y-2\,Dy-4y=e^{4z}\quad\text{or}\quad(D^2-3D-4)\,y=e^{4z}$$

A.E. is $m^2-3m-4=0 \Rightarrow (m-4)(m+1)=0 \Rightarrow m=-1, 4$

$$\text{C.F.}=C_1e^{-z}+C_2e^{4z}\quad \text{P.I.}=\frac{1}{D^2-3D-4}e^{4z}\quad \textbf{[Rule Fails]}$$

$$=z\frac{1}{2D-3}e^{4z}=z\frac{1}{2(4)-3}e^{4z}=\frac{ze^{4z}}{5}$$

Thus, the complete solution is given by

$$y = C_1 e^{-z} + C_2 e^{4z} + \frac{z e^{4z}}{5} \quad \Rightarrow \quad y = \frac{C_1}{x} + C_2 x^4 + \frac{1}{5} x^4 \log x$$ **Ans.**

**Example 65.** *Solve* $x^2 \frac{d^2 y}{dx^2} + x \frac{dy}{dx} + y = \sin(\log x^2)$ (*Nagpur University, Summer* 2005)

**Solution.** We have, $x^2 \frac{d^2 y}{dx^2} + x \frac{dy}{dx} + y = \sin(\log x^2)$ ... (1)

Let $x = e^z$, so that $z = \log x$, $D \equiv \frac{d}{dz}$

(1) becomes

$$D(D-1)y + Dy + y = \sin(2z) \quad \Rightarrow \quad (D^2 + 1)y = \sin 2z$$

A.E. is $m^2 + 1 = 0$ or $m = \pm i$

$$C.F. = C_1 \cos z + C_2 \sin z$$

$$\text{P.I} = \frac{1}{D^2 + 1} \sin 2z = \frac{1}{-4+1} \sin 2z = -\frac{1}{3} \sin 2z$$

$$y = \text{C.F.} + \text{P.I.} = C_1 \cos z + C_2 \sin z - \frac{1}{3} \sin 2z$$

$$= C_1 \cos(\log x) + C_2 \sin(\log x) - \frac{1}{3} \sin(\log x^2)$$ **Ans.**

**Example 66.** *Solve:* $x^2 \frac{d^3 y}{dx^3} + 3x \frac{d^2 y}{dx^2} + \frac{dy}{dx} = x^2 \log x$ (*Nagpur University, Summer* 2003)

**Solution.** We have, $x^3 \frac{d^3 y}{dx^3} + 3x^2 \frac{d^2 y}{dx^2} + x \frac{dy}{dx} = x^3 \log x$

Let $x = e^z$ so that $z = \log x$, $D \equiv \frac{d}{dz}$

The equation becomes after substitution

$$[D(D-1)(D-2) + 3D(D-1) + D]\, y = z e^{3z} \quad \Rightarrow \quad D^3 y = z e^{3z}$$

Auxiliary equation is $m^3 = 0 \Rightarrow m = 0, 0, 0.$

C.F. $= C_1 + C_2 z + C_3 z^2 = C_1 + C_2 \log x + C_3 (\log x)^2$

$$\text{P.I.} = \frac{1}{D^3} \cdot z e^{3z} = e^{3z} \cdot \frac{1}{(D+3)^3} \cdot z$$

$$= e^{3z} \frac{1}{27} \left(1 + \frac{D}{3}\right)^{-3} z = \frac{e^{3z}}{27} (1 - D) z = \frac{e^{3z}}{27} (z - 1) = \frac{x^3}{27} (\log x - 1)$$

Complete solution is $y = C_1 + C_2 \log x + C_3 (\log x)^2 + \frac{x^3}{27} (\log x - 1)$ **Ans.**

## 3.29 LEGENDRE'S HOMOGENEOUS DIFFERENTIAL EQUATIONS

A linear differential equation of the form

$$(a + bx)^n \frac{d^n y}{dx^n} + a_1 (a + bx)^{n-1} \frac{d^{n-1} y}{dx^{n-1}} + \ldots + a_n y = X \qquad \ldots (1)$$

where $a, b, a_1, a_2, \ldots a_n$ are constants and $X$ is a function of $x$, is called Legendre's linear equation.

Equation (1) can be reduced to linear differential equation with constant coefficients by the substitution.

$$a + bx = e^z \qquad \Rightarrow \qquad z = \log (a + bx)$$

so that $$\frac{dy}{dx} = \frac{dy}{dz} \cdot \frac{dz}{dx} = \frac{b}{a+bx} \cdot \frac{dy}{dz}$$

$$\Rightarrow \qquad (a+bx)\frac{dy}{dx} = b\frac{dy}{dz} = b\,Dy, \quad D \equiv \frac{d}{dz} \quad \Rightarrow \quad (a+bx)\frac{dy}{dx} = b\,Dy$$

where

Again $$\frac{d^2y}{dx^2} = \frac{d}{dx}\left(\frac{dy}{dx}\right) = \frac{d}{dx}\left(\frac{b}{a+bx} \cdot \frac{dy}{dz}\right)$$

$$= -\frac{b^2}{(a+bx)^2}\frac{dy}{dz} + \frac{b}{(a+bx)} \cdot \frac{d^2y}{dz^2} \cdot \frac{dz}{dx}$$

$$= -\frac{b^2}{(a+bx)^2}\frac{dy}{dz} + \frac{b}{(a+bx)} \cdot \frac{d^2y}{dz^2} \cdot \frac{b}{(a+bx)}$$

$$\Rightarrow \qquad (a+bx)^2\frac{d^2y}{dx^2} = -b^2\frac{dy}{dz} + b^2\frac{d^2y}{dz^2}$$

$$= b^2\left(\frac{d^2y}{dz^2} - \frac{dy}{dz}\right) = b^2(D^2y - Dy) = b^2 D(D-1)y$$

$$\Rightarrow \qquad (a+bx)^2\frac{d^2y}{dx^2} = b^2 D(D-1)$$

Similarly, $(a+bx)^3\dfrac{d^3y}{dx^3} = b^3 D(D-1)(D-2)y$

..........................................................................

$$(a+bx)^n\frac{d^ny}{dx^n} = b^n D(D-1)(D-2)\ldots\ldots(D-n+1)y$$

Substituting these values in equation (1), we get a linear differential equation with constant coefficients, which can be solved by the method given in the previous section.

**Example 67.** *Solve* $(1+x)^2\dfrac{d^2y}{dx^2} + (1+x)\dfrac{dy}{dx} + y = \sin 2\{\log(1+x)\}$

**Solution.** We have, $(1+x)^2\dfrac{d^2y}{dx^2} + (1+x)\dfrac{dy}{dx} + y = \sin 2\{\log(1+x)\}$

Put $1 + x = e^z$ or $\log(1+x) = z$

$(1+x)\dfrac{dy}{dx} = Dy$ and $(1+x)^2\dfrac{d^2y}{dx^2} = D(D-1)y$, where $D \equiv \dfrac{d}{dz}$

Putting these values in the given differential equation, we get

$$D(D-1)y + Dy + y = \sin 2z \qquad \text{or} \qquad (D^2 - D + D + 1)y = \sin 2z$$

$(D^2 + 1)y = \sin 2z$

A.E. is $m^2 + 1 = 0 \quad \Rightarrow \quad m = \pm i$

$$C.F. = A\cos z + B\sin z$$

$$\text{P.I.} = \frac{1}{D^2+1}\sin 2z = \frac{1}{-4+1}\sin 2z = -\frac{1}{3}\sin 2z$$

Now, complete solution is $y$ = C.F. + P.I.

$$\Rightarrow \quad y = A\cos z + B\sin z - \frac{1}{3}\sin 2z$$

$$\Rightarrow \quad y = A\cos\{\log(1+x)\} + B\sin\{\log(1+x)\} - \frac{1}{3}\sin 2\{\log(1+x)\} \quad \textbf{Ans.}$$

## EXERCISE 3.26

**Solve the following differential equations:**

1. $x^2\frac{d^2y}{dx^2} - 4x\frac{dy}{dx} + 6y = \frac{42}{x^4}$ **Ans.** $C_1x^2 + C_2x^3 + \frac{1}{x^4}$

2. $(x^2D^2 - 3xD + 4)\,y = 2x^2$ **Ans.** $(C_1 + C_2\log x)\,x^2 + x^2(\log x)^2$

3. $x^2\frac{d^2y}{dx^2} - x\frac{dy}{dx} + y = \log x$ *(AMIETE, June 2010)* **Ans.** $(C_1 + C_2\log x)\,x + \log x + 2$

4. $\frac{d^2y}{dx^2} + \frac{1}{x}\frac{dy}{dx} = \frac{12\log x}{x^2}$ **Ans.** $C_1 + C_2\log x + 2(\log x)^3$

5. $(x^2D^2 - xD - 3)\,y = x^2\log x$ **Ans.** $\frac{C_1}{x} + C_2x^3 - \frac{x^2}{3}\left(\log x + \frac{2}{3}\right)$
(*A.M.I.E. Winter* 2001, *Summer* 2001)

6. $x^2\frac{d^2y}{dx^2} - 2x\frac{dy}{dx} + 2y = x^2 + \sin(5\log x)$

**Ans.** $c_1x + c_2x^2 + x^2\log x + \frac{1}{754}[15\cos(5\log x) - 23\sin(5\log x)]$

7. $x^2\frac{d^2y}{dx^2} - 3x\frac{dy}{dx} + y = \log x\frac{\sin(\log x) + 1}{x}$ *(AMIETE, Dec. 2009)*

**Ans.** $y + C_1x^{2+\sqrt{3}} + C_2x^{2-\sqrt{3}} + \frac{1}{x}\left[\frac{382}{61}\cos\log x + \frac{54}{61}\sin(\log x) + 6\log x\cos(\log x) + 5\log x\sin(\log x)\right] + \frac{1}{6x}$

8. $(1+x)^2\frac{d^2y}{dx^2} + (1+x)\frac{dy}{dx} + y = 2\sin\log(1+x)$

**Ans.** $y = C_1\cos\log(1+x) + C_2\sin\log(1+x) - \log(1+x)\cos\log(1+x)$

9. Which of the basis of solutions are for the differential equation $x^2\frac{d^2y}{dx^2} - x\frac{dy}{dx} + y = 0$

(*a*) $x, x\,I_n\,x$, (*b*) $I_n\,x, e^x$ (*c*) $\frac{1}{x}, \frac{1}{x^2}$, (*d*) $\frac{1}{x^2}e^x, x\,I_n\,x$

(*A.M.I.E., Winter* 2001) **Ans.** (*a*)

10. The general solution of $x^2\frac{d^2y}{dx^2} - 5x\frac{dy}{dx} + 9y = 0$ is

(*a*) $(C_1 + C_2x)\,e^{3x}$ (*b*) $(C_1 + C_{2n}\,x)\,x^3$ (*c*) $(C_1 + C_2x)\,x^3$ (*d*) $(C_1 + C_2 l_n\,x)\,e^{x^3}$

*(AMIETE, Dec. 2009)* **Ans.** (*b*)

11. To transform $x\frac{d^2y}{dx^2} + \frac{dy}{dx} = \frac{1}{x}$ into a linear differential equation with constant coefficients, the required substitution is

(*a*) $x = \sin t$ (b) $x = t^2 + 1$ (*c*) $x = \log t$ (*d*) $x = e^t$

*(AMIETE, June 2010)* **Ans.** (*d*)

## 3.30 METHOD OF VARIATION OF PARAMETERS

To find particular integral of $\frac{d^2y}{dx^2} + b\frac{dy}{dx} + cy = X$ ... (1)

Let complementary function $= Ay_1 + By_2$, so that $y_1$ and $y_2$ satisfy

$$\frac{d^2y}{dx^2} + b\frac{dy}{dx} + cy = 0 \qquad \text{... (2)}$$

Let us assume particular integral $y = uy_1 + vy_2$, ... (3)

where $u$ and $v$ are unknown function of $x$.

Differentiation (3) *w.r.t.* $x$, we have $y' = uy_1' + vy_2' + u'y_1 + v'y_2$ assuming that $u, v$ satisfy the equation

$$u'y_1 + v'y_2 = 0 \qquad \text{... (4)}$$

then $$y' = uy_1' + vy_2' \qquad \text{... (5)}$$

Differentiating (5) *w.r.t.x*, we have $y'' = uy_1'' + u'y_1' + vy_2'' + v'y_2'$

Substituting the values of $y$, $y'$ and $y''$ in (1), we get

$$(uy_1'' + u'y_1' + vy_2'' + v'y_2') + b(uy_1' + vy_2') + c(uy_1 + vy_2) = X$$

$$\Rightarrow \quad u(y_1'' + by_1' + cy_1) + v(ay_2'' + by_2' + cy_2) + (u'y_1' + v'y_2') = X \qquad \text{... (6)}$$

$y_1$ and $y_2$ will satisfy equation (1)

$$\therefore \quad y_1'' + by_1' + cy_1 = 0 \qquad \text{... (7)}$$

and $$y_2'' + by_2' + cy_2 = 0 \qquad \text{... (8)}$$

Putting the values of expressions from (7) and (8) in (6), we get

$$\Rightarrow \quad u'y_1' + v'y_2' = X \qquad \text{... (9)}$$

Solving (4) and (9), we get

$$u' = \begin{vmatrix} 0 & y_2 \\ X & y_2' \end{vmatrix} \div \begin{vmatrix} y_1 & y_2 \\ y_1' & y_2' \end{vmatrix} = \frac{-y_2 X}{y_1 y_2' - y_1' y_2}$$

$$v' = \begin{vmatrix} y_1 & 0 \\ y_1' & X \end{vmatrix} \div \begin{vmatrix} y_1 & y_2 \\ y_1' & y_2' \end{vmatrix} = \frac{y_1 X}{y_1 y_2' - y_1' y_2}$$

$$\boxed{u = \int \frac{-y_2 X}{y_1 y_2' - y_1' y_2}\,dx} \qquad \boxed{v = \int \frac{y_1 X}{y_1 y_2' - y_1' y_2}\,dx}$$

General solution = complementary function + particular integral.

**Working Rule**

**Step 1.** Find out the *C.F. i.e.*, $A\,y_1 + B\,y_2$ **Step 2.** Particular integral $= u\,y_1 + v\,y_2$

**Step 3.** Find $u$ and $v$ by the formulae

$$u = \int \frac{-y_2 X}{y_1 y_2' - y_1' y_2}\,dx, \quad v = \int \frac{y_1 X}{y_1 y_2' - y_1' y_2}\,dx$$

**Example 68.** *Solve* $\frac{d^2y}{dx^2} + y = \operatorname{cosec} x.$ (*Nagpur University, Summer* 2005)

**Solution.** $(D^2 + 1)\,y = \operatorname{cosec} x$

A.E. is $m^2 + 1 = 0 \quad \Rightarrow \quad m = \pm i$

C.F. $= A\cos x + B\sin x$

Here $$y_1 = \cos x, \qquad y_2 = \sin x$$

$$\text{P.I.} = y_1\, u + y_2\, v$$

where $$u = \int \frac{-y_2 \,.\, \operatorname{cosec} x\, dx}{y_1 \cdot y'_2 - y'_1 \cdot y_2} = \int \frac{-\sin x \,.\, \operatorname{cosec} x\, dx}{\cos x\,(\cos x) - (-\sin x)(\sin x)}$$

$$= \int \frac{-\sin x \,.\, \dfrac{1}{\sin x}\, dx}{\cos^2 x + \sin^2 x} = -\int dx = -x$$

$$v = \int \frac{y_1 \,.\, X\, dx}{y_1 \cdot y'_2 - y'_1\, y_2} = \int \frac{\cos x \,.\, \operatorname{cosec} x\, dx}{\cos x\,(\cos x) - (-\sin x)(\sin x)}$$

$$= \int \frac{\cos x \,.\, \dfrac{1}{\sin x}}{\cos^2 x + \sin^2 x}\, dx = \int \frac{\cot x\, dx}{1} = \log \sin x$$

$$P.I. = uy_1 + vy_2 = -x \cos x + \sin x\,(\log \sin x)$$

General solution = C.F. + P.I.

$$y = A \cos x + B \sin x - x \cos x + \sin x\,.(\log \sin x)$$ **Ans.**

**Example 69.** *Apply the mehtod of variation of parameters to solve*

$$\frac{d^2y}{dx^2} + y = \tan x$$ *(A. M. I. E. T. E., Dec. 2010, Winter* 2001, *Summer* 2000)

**Solution.** We have, $$\frac{d^2y}{dx^2} + y = \tan x$$

$$(D^2 + 1)y = \tan x$$

A.E. is $m^2 = -1$ or $m = \pm i$

C. F. $$y = A \cos x + B \sin x$$

Here, $$y_1 = \cos x, \qquad y_2 = \sin x$$

$$y_1 \cdot y'_2 - y'_1 \cdot y_2 = \cos x\,(\cos x) - (-\sin x) \sin x = \cos^2 x + \sin^2 x = 1$$

P. I. $= u.\, y_1 + v.\, y_2$ where

$$u = \int \frac{-y_2 \tan x}{y_1 \cdot y'_2 - y'_1 \cdot y_2}\, dx = -\int \frac{\sin x \tan x}{1}\, dx = -\int \frac{\sin^2 x}{\cos x}\, dx = -\int \frac{1 - \cos^2 x}{\cos x}\, dx$$

$$= \int (\cos x - \sec x)\, dx = \sin x - \log(\sec x + \tan x)$$

$$v = \int \frac{y_1 \tan x}{y_1 \cdot y'_2 - y'_1 \cdot y_2}\, dx = \int \frac{\cos x \cdot \tan x}{1}\, dx = \int \sin x\, dx = -\cos x$$

P. I. $= u.\, y_1 + v.\, y_2$

$$= [\sin x - \log(\sec x + \tan x)] \cos x - \cos x \sin x = -\cos x \log(\sec x + \tan x)$$

Complete solution is

$$y = A \cos x + B \sin x - \cos x \log(\sec x + \tan x)$$ **Ans.**

**Example 70.** *Solve by method of variation of parameters:*

$$\frac{d^2y}{dx^2} - y = \frac{2}{1 + e^x}$$ *(Uttarakhand, II Semester, June* 2007, *A.M.I.E.T.E., Summer* 2001)

*(Nagpur University, Summer* 2001)

**Solution.** $\dfrac{d^2y}{dx^2} - y = \dfrac{2}{1+e^x}$

A. E. is $(m^2 - 1) = 0$

$$m^2 = 1, \quad m = \pm 1$$

$$C.F. = C_1 e^x + C_2 e^{-x}$$

$\therefore$ $P.I. = uy_1 + vy_2$

Here, $y_1 = e^x, \quad y_2 = e^{-x}$

and $y_1 . y'_2 - y'_1 . y_2 = -e^x . e^{-x} - e^x . e^{-x} = -2$

$$u = \int \frac{-y_2 X}{y_1.y'_2 - y'_1.y_2} dx = -\int \frac{e^{-x}}{-2} \times \frac{2}{1+e^x} dx$$

$$= \int \frac{e^{-x}}{1+e^x} dx = \int \frac{dx}{e^x(1+e^x)} = \int \left(\frac{1}{e^x} - \frac{1}{1+e^x}\right) dx$$

$$= \int e^{-x} dx - \int \frac{e^{-x}}{e^{-x}+1} dx = -e^{-x} + \log(e^{-x}+1)$$

$$v = \int \frac{y_1 X}{y_1.y'_2 - y'_1.y_2} dx = \int \frac{e^x}{-2} \frac{2}{1+e^x} dx = -\int \frac{e^x}{1+e^x} dx = -\log(1+e^x)$$

$$\text{P.I.} = u.y_1 + v.y_2 = [-e^{-x} + \log(e^{-x}+1)]e^x - e^{-x}\log(1+e^x)$$

$$= -1 + e^x \log(e^{-x}+1) - e^{-x}\log(e^x+1)$$

Complete solution $= y = C_1 e^x + C_2 e^{-x} - 1 + e^x \log(e^{-x}+1) - e^{-x}\log(e^x+1)$ **Ans.**

## EXERCISE 3.27

Solve the following equations by variation of parameters method.

1. $\dfrac{d^2y}{dx^2} - 4y = e^{2x}$ **Ans.** $y = C_1 e^{2x} + C_2 e^{-2x} + \dfrac{x}{4} e^{2x} - \dfrac{e^{2x}}{16}$

2. $\dfrac{d^2y}{dx^2} + y = \sin x$ **Ans.** $y = C_1 \cos x + C_2 \sin x - \dfrac{x}{2} \cos x + \dfrac{1}{4} \sin x$

3. $\dfrac{d^2y}{dx^2} - 3\dfrac{dy}{dx} + 2y = \sin x$ **Ans.** $y = C_1 e^x + C_2 e^{2x} + \dfrac{1}{10}(3\cos x + \sin x)$

4. $\dfrac{d^2y}{dx^2} + y = \sec x \tan x$ **Ans.** $y = C_1 \cos x + C_2 \sin x + x\cos x + \sin x \log \sec x - \sin x$

5. $y'' - 6y' + 9y = \dfrac{e^{3x}}{x^2}$ *(AMIETE, June 2010, 2009)* **Ans.** $y = (C_1 + x\,C_2)\,e^{3x} - e^{3x}\log x$

## 3.31 SIMULTANEOUS DIFFERENTIAL EQUATIONS

If two or more dependent variables are functions of a single independent variable, the equations involving their deriavatives are called simultaneous equations, *e.g.*

$$\frac{dx}{dt} + 4y = t \quad \Rightarrow \quad \frac{dy}{dt} + 2x = e^t$$

The method of solving these equations is based on the process of elimination, as we solve algebraic simultaneous equations.

**Example 71.** *The equations of motions of a particle are given by*

$$\frac{dx}{dt} + \omega y = 0 \quad \Rightarrow \quad \frac{dy}{dt} - \omega x = 0$$

*Find the path of the particle and show that it is a circle.*

(*R.G.P.V. Bhopal, Feb. 2006, U.P. II Semester summer* 2009)

**Solution.** On putting $\frac{d}{dt} \equiv D$ in the equations, we have

$$Dx + \omega y = 0 \quad \text{...(1)}$$
$$-\omega x + Dy = 0 \quad \text{...(2)}$$

On multiplying (1) by w and (2) by *D*, we get

$$\omega Dx + \omega^2 y = 0 \quad \text{...(3)}$$
$$-\omega Dx + D^2 y = 0 \quad \text{...(4)}$$

On adding (3) and (4), we obtain

$$\omega^2 y + D^2 y = 0 \quad \Rightarrow \quad (D^2 + \omega^2)\, y = 0 \quad \text{...(5)}$$

Now, we have to solve (5) to get the value of *y*.

A.E. is $m^2 + \omega^2 = 0 \quad \Rightarrow \quad m^2 = -\omega^2 \quad \Rightarrow \quad m = \pm i\omega$

$$\therefore \quad y = A \cos wt + B \sin wt \quad \text{...(6)}$$

$\Rightarrow Dy = -A\omega \sin \omega t + B\omega \cos wt$

On putting the value of *Dy* in (2), we get $-\omega x - A\omega \sin \omega t + B\omega \cos \omega t = 0$

$$\Rightarrow \quad \omega x = -A\omega \sin \omega t + B\omega \cos \omega t$$
$$\Rightarrow \quad x = -A \sin \omega t + B \cos \omega t \quad \text{...(7)}$$

On squaring (6) and (7) and adding , we get

$$x^2 + y^2 = A^2(\cos^2 \omega t + \sin^2 \omega t) + B^2 (\cos^2 \omega t + \sin^2 \omega t)$$
$$\Rightarrow \quad x^2 + y^2 = A^2 + B^2$$

This is the equation of circle. **Proved.**

**Example 72.** *Solve the following differential equation*

$$\frac{dx}{dt} = y + 1, \quad \frac{dy}{dt} = x + 1$$

(*U.P. II Semester, 2009*)

**Solution.** Here, we have

$$Dx - y = 1 \quad \text{... (1)}$$
$$-x + Dy = 1 \quad \text{...(2)}$$

Multiplying (1) by D, we get

$$D^2 x - Dy = D.1 \quad \text{...(3)}$$

Adding (2) and (3), we get

$$(D^2 - 1)\, x = 1 + D.1$$
$$\Rightarrow \quad (D^2 - 1)\, x = 1 \text{ or } (D^2 - 1)x = e^0 \quad [D.\,(1) = 0]$$

A.E. is $m^2 - 1 = 0 \quad \Rightarrow \quad m = \pm 1$

$$\therefore \quad \text{C.F.} = c_1 e^t + c_2 e^{-t}$$

$$\text{P.I.} = \frac{1}{D^2 - 1} . e^0 = \frac{1}{0 - 1} e^0 = -1$$

$$\therefore \quad x = \text{C.F.} + \text{P.I.} = c_1 e^t + c_2 e^{-t} - 1$$

From (1), $y = \frac{dx}{dt} - 1 \quad \Rightarrow \quad y = \frac{d}{dt}(c_1 e^t + c_2 e^{-t} - 1) - 1$

$$\Rightarrow \quad \left.\begin{aligned} y &= c_1 e^t - c_2 e^{-t} - 1 \\ \text{and} \quad x &= c_1 e^t + c_2 e^{-t} - 1 \end{aligned}\right\}$$

**Ans.**

**Example 73.** *Solve*:

$$\frac{dx}{dt} + y = \sin t$$

$$\frac{dy}{dt} + x = \cos t \qquad \text{where } y(0) = 0, \quad x(0) = 2$$

*(R.G.P.V., Bhopal, I Semester, April, 2010 June 2007)*

**Solution.** We have,

$$\frac{dx}{dt} + y = \sin t \quad \Rightarrow \quad Dx + y = \sin t \qquad \text{... (1)}$$

$$\frac{dy}{dt} + x = \cos t \quad \Rightarrow \quad Dy + x = \cos t \qquad \text{... (2)}$$

Multiplying (2) by D, we get

$$D^2 y + Dx = D \cos t$$

$$D^2 y + Dx = -\sin t \qquad \text{... (3)}$$

Subtracting (1) from (3), we have

$$D^2 y - y = -2 \sin t$$

$$\Rightarrow \quad (D^2 - 1) y = -2 \sin t$$

A.E. is $m^2 - 1 = 0 \quad \Rightarrow \quad m^2 = 1 \quad \Rightarrow \quad m = \pm 1$

$$\text{C.F.} = C_1 e^t + C_2 e^{-t}$$

$$\text{P.I.} = \frac{1}{D^2 - 1}(-2 \sin t)$$

$$\Rightarrow \quad \text{P.I.} = \frac{1}{-1-1}(-2 \sin t) = \sin t$$

Complete solution = C.F. + P.I.

$$y = C_1 e^t + C_2 e^{-t} + \sin t \qquad \text{... (4)}$$

Putting $y = 0$ and $t = 0$ in (4), we get

$$0 = C_1 + C_2 \quad \text{or} \quad C_2 = -C_1$$

On putting $C_2 = -C_1$ in (4), we get

$$y = C_1 e^t - C_1 e^{-t} + \sin t$$

On putting the value of $y$ in (2), we get

$$D(C_1 e^t - C_1 e^{-t} + \sin t) + x = \cos t$$

$$C_1 e^t + C_1 e^{-t} + \cos t + x = \cos t$$

$$x = -C_1 e^t - C_1 e^{-t} \qquad \text{...(5)}$$

On putting $x = 2$, $t = 0$ in (5), we get

$$2 = -C_1 - C_1 \quad \Rightarrow \quad C_1 = -1$$

Putting the value of $C_1$ in (5) and (4), we have

$$x = e^t + e^{-t}$$

$$y = -e^t + e^{-t} + \sin t$$

Which is the required solution. **Ans.**

**Example 74.** *Solve:* $\frac{dx}{dt} + 4x + 3y = t$

$$\frac{dy}{dt} + 2x + 5y = e^t$$ [*U.P. II Semester, 2006*]

**Solution.** Here, we have

$$(D + 4)x + 3y = t \qquad ...(1)$$

$$2x + (D + 5)y = e^t \qquad ...(2) \left(D \equiv \frac{d}{dt}\right)$$

To eliminate *y*, operating (1) by (*D* + 5) and multiplying (2) by 3 then subtracting, we get

$$(D + 5)(D + 4)x + 3(D + 5)y - 3(2x) - 3(D + 5)y = (D + 5)t - 3e^t$$

$$[(D + 4)(D + 5) - 6]x = (D + 5)t - 3e^t$$

$$(D^2 + 9D + 14)x = 1 + 5t - 3e^t$$

Auxiliary equation is

$$m^2 + 9m + 14 = 0 \Rightarrow m = -2, -7$$

$$\therefore \qquad \text{C.F.} = c_1e^{-2t} + c_2e^{-7t}$$

$$\text{P.I.} = \frac{1}{D^2 + 9D + 14}(1 + 5t - 3e^t)$$

$$= \frac{1}{D^2 + 9D + 14}e^{0t} + 5\frac{1}{D^2 + 9D + 14}t - 3\frac{1}{D^2 + 9D + 14}e^t$$

$$= \frac{1}{0^2 + 9(0) + 14}e^{0t} + 5.\frac{1}{14\left(1 + \frac{9D}{14} + \frac{D^2}{14}\right)}t - 3\frac{1}{1^2 + 9(1) + 14}e^t$$

$$= \frac{1}{14} + \frac{5}{14}\left[1 + \left(\frac{9D}{14} + \frac{D^2}{14}\right)\right]^{-1} t - \frac{1}{8}e^t = \frac{1}{14} + \frac{5}{14}\left[1 - \left(\frac{9D}{14} + \frac{D^2}{14}\right) + \cdots\right]t - \frac{1}{8}e^t$$

$$= \frac{1}{14} + \frac{5}{14}\left(t - \frac{9}{14}\right) - \frac{1}{8}e^t = \frac{5}{14}t - \frac{31}{196} - \frac{1}{8}e^t$$

$$x = c_1e^{-2t} + c_2e^{-7t} + \frac{5}{14}t - \frac{31}{196} - \frac{1}{8}e^t$$

$$3y = t - \frac{dx}{dt} - 4x \qquad \text{[From (1)]}$$

$$= t - \frac{d}{dt}\left[c_1e^{-2t} + c_2e^{-7t} + \frac{5}{14}t - \frac{31}{196} - \frac{1}{8}e^t\right] - 4\left[c_1e^{-2t} + c_2e^{-7t} + \frac{5}{14}t - \frac{31}{196} - \frac{e^t}{8}\right]$$

$$3y = t + 2c_1e^{-2t} + 7c_2e^{-7t} - \frac{5}{14} + \frac{1}{8}e^t - 4c_1e^{-2t} - 4c_2e^{-7t} - \frac{10}{7}t + \frac{31}{49} + \frac{1}{2}e^t$$

$$\therefore y = \frac{1}{3}\left[-2c_1e^{-2t} + 3c_2e^{-7t} - \frac{3}{7}t + \frac{27}{98} + \frac{5}{8}e^t\right]$$

Hence,

$$x = c_1e^{-2t} + c_2e^{-7t} + \frac{5}{14}t - \frac{31}{196} - \frac{1}{8}e^t$$

$$y = -\frac{2}{3}c_1e^{-2t} + c_2e^{-7t} - \frac{1}{7}t + \frac{9}{98} + \frac{5}{24}e^t$$ **Ans.**

**Example 75.** *Solve* $\frac{dx}{dt}=2y$, $\frac{dy}{dt}=2z$, $\frac{dz}{dt}=2x$ *(Uttarakhand, II Semester, June 2007)*

**Solution.** Here, we have

$$\frac{dx}{dt}=2y \Rightarrow Dx=2y \quad ...(1)$$

$$\frac{dy}{dt}=2z \Rightarrow Dy=2z \quad ...(2)$$

$$\frac{dz}{dt}=2x \Rightarrow Dz=2x \quad ...(3)$$

From (1), we have $\frac{dx}{dt}=2y$

$$\Rightarrow \frac{d^2x}{dt^2}=\frac{2\,dy}{dt}=2\,(2z)=4z \qquad \left[\text{Using (2), } \frac{dy}{dt}=2z\right]$$

$$\frac{d^3x}{dt^3}=4\frac{dz}{dt}=4\,(2x)=8x \qquad \left[\text{Using (3), } \frac{dz}{dt}=2x\right]$$

$$\Rightarrow \frac{d^3x}{dt^3}-8x=0 \Rightarrow (D^3-8)\,x=0$$

A.E. is $m^3-8=0 \Rightarrow (m-2)\,(m^2+2m+4)=0$

$\Rightarrow m-2=0 \Rightarrow m=2$

or $m^2+2m+4=0 \Rightarrow m=\frac{-2\pm\sqrt{4-16}}{2}=\frac{-2\pm i\sqrt{12}}{2}=-1\pm i\sqrt{3}$

So the C.F. of $x$ is

$$x=C_1e^{2t}+e^{-t}\,(A\cos\sqrt{3}t+B\sin\sqrt{3}t) \quad ...(4)$$

$$[A=C_2\cos\alpha,\ B=C_2\sin\alpha] \qquad \left[\tan\alpha=\frac{B}{A},\ \alpha=\tan^{-1}\left(\frac{B}{A}\right)\right]$$

$$x=C_1e^{2t}+e^{-t}\,[C_2\cos\alpha\cos\sqrt{3}t+C_2\sin\alpha\sin\sqrt{3}t]$$

$$x=C_1e^{2t}+e^{-t}\,C_2\cos(\sqrt{3}t-\alpha)=C_1e^{2t}+C_2e^{-t}\cos(\sqrt{3}t-\alpha)$$

From (3), we have $\frac{dz}{dt}=2x$

$$\Rightarrow \frac{dz}{dt}=2C_1e^{2t}+2C_2e^{-t}\cos(\sqrt{3}t-\alpha) \qquad \text{[On putting the value of } x]$$

$$z=C_1e^{2t}+2C_2\frac{e^{-t}}{\sqrt{1+3}}\cos(\sqrt{3}t-\alpha-\beta) \qquad \left[\int e^{ax}\cos bx\,dx=\frac{e^{ax}}{\sqrt{a^2+b^2}}\cos(bx-\beta)\right]$$

$$\Rightarrow z=C_1e^{2t}+2C_2\frac{e^{-t}}{\sqrt{1+3}}\cos\left[\sqrt{3}t-\alpha-\frac{2\pi}{3}\right] \qquad \left[\beta=\tan^{-1}\frac{\sqrt{3}}{-1}=\frac{2\pi}{3}\right]$$

$$\Rightarrow z=C_1e^{2t}+C_2e^{-t}\cos\left(\sqrt{3}t-\alpha+\frac{4\pi}{3}\right) \quad ...(5) \qquad \left[-\frac{2\pi}{3}=\frac{4\pi}{3}\right]$$

From (2), we have $\frac{dy}{dx} = 2z$

$$\Rightarrow \quad \frac{dy}{dt} = 2C_1 e^{2t} + 2C_2 e^{-t} \cos\left(\sqrt{3}t - \alpha + \frac{4\pi}{3}\right) \quad \text{[On putting the value of } z\text{]}$$

$$\Rightarrow \quad y = \int 2C_1 e^{2t}\, dt + 2C_2 \int e^{-t} \cos\left(\sqrt{3}t - \alpha + \frac{4\pi}{3}\right) dt \quad \left(\gamma = \tan^{-1}\frac{\sqrt{3}}{-1} = \frac{2\pi}{3}\right)$$

$$y = C_1 e^{2t} + 2C_2 \frac{e^{-x}}{\sqrt{1+3}} \cos\left(\sqrt{3}t - \alpha + \frac{4\pi}{3} - \gamma\right)$$

$$\Rightarrow \quad y = C_1 e^{2t} + 2C_2 \frac{e^{-t}}{\sqrt{1+3}} \cos\left(\sqrt{3}t - \alpha + \frac{4\pi}{3} - \frac{2\pi}{3}\right)$$

$$y = C_1 e^{2t} + C_2 e^{-t} \cos\left(\sqrt{3}t - \alpha + \frac{2\pi}{3}\right) \quad \ldots(6)$$

Relations (4), (5) and (6) are the required solutions. **Ans.**

**Example 76.** *Solve the following simultaneous equations :*

$$\frac{d^2x}{dt^2} - 3x - 4y = 0, \quad \frac{d^2y}{dt^2} + x + y = 0$$ (*U.P. II Semester, Summer* 2005)

**Solution.** We have, $\frac{d^2x}{dt^2} - 3x - 4y = 0$

$$\frac{d^2y}{dt^2} + x + y = 0$$

$$(D^2 - 3)x - 4y = 0 \quad \ldots(1)$$

$$x + (D^2 + 1)\, y = 0 \quad \ldots(2)$$

Operating equation (2) by $(D^2 - 3)$, we get

$$(D^2 - 3)x + (D^2 - 3)(D^2 + 1)\, y = 0 \quad \ldots(3)$$

Subtracting (3) from (1), we get

$$-4y - (D^2 - 3)(D^2 + 1)\, y = 0 = \Rightarrow -4y - (D^4 - 2D^2 - 3)\, y = 0$$

$$\Rightarrow \quad (D^4 - 2D^2 - 3 + 4)\, y = 0 \quad \Rightarrow (D^4 - 2D^2 + 1)\, y = 0$$

$$\Rightarrow \quad (D^2 - 1)^2\, y = 0$$

A.E. is $(m^2 - 1)^2 = 0 \quad \Rightarrow \quad (m^2 - 1) = 0 \Rightarrow \quad m = \pm 1$

$$y = (c_1 + c_2 t)e^t + (c_3 + c_4 t)e^{-t} \quad \ldots(4)$$

From (2), we have

$$x = -(D^2 + 1)y \quad = -D^2 y - y$$

$$= -D^2 [(c_1 + c_2 t)\, e^t + (c_3 + c_4 t)e^{-t}] - [(c_1 + c_2 t)e^t + (c_3 + c_4 t)e^{-t}]$$

$$= -D\, [\{(c_1 + c_2 t)e^t + c_2 e^t\} + \{(c_3 + c_4 t)(-e^{-t}) + c_4 e^{-t}\}] - [(c_1 + c_2 t)e^t + (c_3 + c_4 t)e^{-t}]$$

$$= -[(c_1 + c_2 t)\, e^t + c_2 e^t + c_2 e^t + (c_3 + c_4 t)(e^{-t}) - c_4 e^{-t} - c_4 e^{-t}] - [(c_1 + c_2 t)\, e^t + (c_3 + c_4 t)e^{-t}]$$

$$= -[(c_1 + c_2 t + 2c_2 + c_1 + c_2 t)\, e^t + (c_3 + c_4 t - 2c_4 + c_3 + c_4 t)\, e^{-t}]$$

$$= -[(2c_1 + 2c_2 + 2c_2 t)e^t + (2c_3 - 2c_4 + 2c_4 t)e^{-t}] \quad \ldots(5)$$

Relations (4) and (5) are the required solutions. **Ans.**

## EXERCISE 3.28

Solve the following simultaneous equations:

1. $\frac{dx}{dt}+2x-3y=0$, $\frac{dy}{dt}-3x+2y=0$ **Ans.** $x=c_1e^t-c_2e^{-5t}$, $y=c_1e^t+c_2e^{-5t}$

2. $\frac{d^2y}{dt^2}=x$, $\frac{d^2x}{dt^2}=y$ **Ans.** $x=c_1e^t+c_2e^{-t}+(c_3\cos t+c_4\sin t)$

$y=c_1e^t+c_2e^{-t}-(c_3\cos t+c_4\sin t)$

3. $\frac{dx}{dt}+5x-2y=t$, $\frac{dy}{dt}+2x+y=0$ **Ans.** $x=-\frac{1}{27}(1+6t)e^{-3t}+\frac{1}{27}(1+3t)$

So that $x=y=0$ when $t=0$ *(AMIETE, June 2009, U.P., II Semester, June 2008)*

**Ans.** $y=-\frac{2}{27}(2+3t)e^{-3t}+\frac{2}{27}(2-3t)$

4. $\frac{dx}{dt}-y=t$, $\frac{dy}{dt}=t^2-x$ **Ans.** $x=c_1\cos t+c_2\sin t+t^2-1$ ; $y=-c_1\sin t+c_2\cos t+t$

5. $\frac{dx}{dt}+2y+\sin t=0$

$\frac{dy}{dt}-2x-\cos t=0$ **Ans.** $x=c_1\cos 2t+c_2\sin 2t-\cos t$ ; $y=c_1\sin 2t-c_2\cos 2t-\sin t$

6. $4\frac{dx}{dt}-\frac{dy}{dt}+3x=\sin t$ ; $\frac{dx}{dt}+y=\cos t$ **Ans.** $x=c_1e^{-t}+c_2e^{-3t}$, $y=c_1e^{-t}+3c_2e^{-3t}+\cos t$

7. $\frac{dy}{dx}=x$ and $\frac{dx}{dt}=y+e^{2t}$ **Ans.** $x=C_1e^t+C_2e^{-t}+\frac{2}{3}e^{2t}$, $y=C_1e^t-C_2e^{-t}+\frac{1}{3}e^{2t}$

8. $\frac{dx}{dt}=y+t$, $\frac{dy}{dx}=-2x+3y+1$ **Ans.** $x=c_1e^t+\frac{1}{2}c_2e^{2t}-\frac{3}{2}t-\frac{5}{4}$, $y=c_1e^t+c_2e^{2t}-t-\frac{3}{2}$

9. $t\frac{dx}{dt}+y=0$, $t\frac{dy}{dx}+x=0$ **Ans.** $x=c_1t+c_2t^{-1}$, $y=c_2t^{-1}-c_1t$

given $x(1)=1$ and $y(-1)=0$

10. $\frac{dx}{dt}+y=\sin t$, $\frac{dy}{dt}+x=\cos t$, given that $x=2$, $y=0$ when $t=0$ *(U.P., II Semester, 2004)*

**Ans.** $x=e^t+e^{-t}$, $y=\sin t-e^t+e^{-t}$

11. $(D-1)x+Dy=2t+1$; $(2D+1)x+2Dy=t$ **Ans.** $x=-t-\frac{2}{3}$, $y=\frac{t^2}{2}+\frac{4}{3}t+C$

12. $\frac{dx}{dt}+\frac{2}{t}(x-y)=1$, *(U.P., II Semester, Summer (C.O.) 2005)*

$\frac{dy}{dt}+\frac{1}{t}(x+5y)=t$ **Ans.** $x=At^{-4}+Bt^{-3}+\frac{t^2}{15}+\frac{3y}{10}$, $y=-At^{-4}-\frac{1}{2}Bt^{-3}+\frac{2t^2}{15}-\frac{t}{20}$

13. $(D^2-1)x+8Dy=16e^t$ and $Dx+3(D^2+1)y=0$ *(Q. Bank U.P.T.U. 2001)*

**Ans.** $y=c_1\cos\frac{t}{\sqrt{3}}+c_2\sin\frac{t}{\sqrt{3}}+c_3\cosh\sqrt{3}\,t+c_4\sin h\sqrt{3}\,t+2e^t$

$x=\sqrt{3}\,c_1\sin\frac{t}{\sqrt{3}}-\sqrt{3}\,c_2\cos\frac{t}{\sqrt{3}}-3\sqrt{3}\,c_3\sinh\sqrt{3}\,t-3\sqrt{3}\,c_4\cosh\sqrt{3}\,t-6e^t-3t.$

14. $\frac{dx}{dt}+\frac{2}{t}(x-y)=1$, *(U.P. II Semester, 2005)*

$\frac{dy}{dt}+\frac{1}{t}(x+5y)=t.$ **Ans.** $x=At^{-4}+Bt^{-3}+\frac{t^2}{15}+\frac{3t}{10}$, $y=-At^{-4}-\frac{1}{2}Bt^{-3}+\frac{2t^2}{15}-\frac{t}{20}$

### 3.32 EQUATION OF THE TYPE $\boxed{\dfrac{d^n y}{dx^n} = f(x)}$

This type of exact differential equations are solved by successive integration.

**Example 77.** *Solve* $\dfrac{d^2 y}{dx^2} = x^2 \sin x.$

**Solution.** We have $\dfrac{d^2 y}{dx^2} = x^2 \sin x$ ...(1)

Integrating the differential equation (1), we get

$$\frac{dy}{dx} = x^2(-\cos x) - (2x)(-\sin x) + (2)(\cos x) + c_1$$

$$\frac{dy}{dx} = x^2 \cos x + 2x \sin x + 2\cos x + c_1$$

Integrating again, we have $y = [(-x^2)(\sin x) - (-2x)(-\cos x) + (-2)(-\sin x)]$

$$+ [(2x)(-\cos x) - 2(-\sin x)] + 2\sin x + c_1 x + c_2$$

$$= -x^2 \sin x - 4x \cos x + 6 \sin x + c_1 x + c_2$$ **Ans.**

**Example 78.** *Solve* $\dfrac{d^3 y}{dx^3} = x + \log x.$

**Solution.** We have, $\dfrac{d^3 y}{dx^3} = x + \log x$ ...(1)

Integrating the differential equation (1), we get $\dfrac{d^2 y}{dx^2} = \dfrac{x^2}{2} + (\log x)(x) - \int \dfrac{1}{x} \cdot x \, dx + c_1$

$$\frac{d^2 y}{dx^2} = \frac{x^2}{2} + x \log x - x + c_1 \quad ...(2)$$

Again integrating (2), we have,

$$\frac{dy}{dx} = \frac{x^3}{6} + (\log x)\left(\frac{x^2}{2}\right) - \int \frac{1}{x} \cdot \frac{x^2}{2} dx - \frac{x^2}{2} + c_1 x + c_2$$

$$\frac{dy}{dx} = \frac{x^3}{6} + \frac{x^2}{2} \log x - \frac{x^2}{4} - \frac{x^2}{2} + c_1 x + c_2 \quad ...(3)$$

Again integrating (3), we obtain

$$y = \frac{x^4}{24} + (\log x)\frac{x^3}{6} - \int \frac{1}{x} \frac{x^3}{6} dx - \frac{x^3}{12} - \frac{x^3}{6} + c_1 \frac{x^2}{2} + c_2 x + c_3$$

$$\Rightarrow \quad y = \frac{x^4}{24} + \frac{x^3}{6} \log x - \frac{x^3}{18} - \frac{x^3}{12} - \frac{x^3}{6} + \frac{c_1 x^2}{2} + c_2 x + c_3$$

$$\Rightarrow \quad y = \frac{x^4}{24} + \frac{x^3}{6} \log x - \frac{11x^3}{36} + c_1 \frac{x^2}{2} + c_2 x + c_3$$ **Ans.**

## EXERCISE 3.29

**Solve the following differential equations:**

1. $\dfrac{d^5 y}{dx^5} = x$ **Ans.** $y = \dfrac{x^6}{720} + \dfrac{c_1 x^4}{24} + \dfrac{c_2 x^3}{6} + \dfrac{c_3 x^2}{2} + c_4 x + c_5$

2. $\dfrac{d^2 y}{dx^2} = x e^x$ **Ans.** $y = (x - 2) e^x + c_1 x + c_2$

3. $\dfrac{d^4 y}{dx^4} = x + e^{-x} - \cos x$ **Ans.** $y = \dfrac{x^5}{120} + e^{-x} - \cos x + c_1 \dfrac{x^3}{6}$

4. $x^2\dfrac{d^2y}{dx^2}=\log x$ **Ans.** $y=-\dfrac{1}{2}(\log x)^2+\log x-c_1x+c_2$

5. $\dfrac{d^3y}{dx^3}=\log x$ **Ans.** $y=\dfrac{1}{36}[6x^3\log x-11x^3+c_1x^2+c_2x+c_3]$

6. $\dfrac{d^3y}{dx^3}=\sin^2 x$ **Ans.** $y=\dfrac{x^3}{12}+\dfrac{\sin 2x}{16}+\dfrac{c_1x^2}{2}+c_2x+c_3$

## 3.33 EQUATION OF THE TYPE $\boxed{\dfrac{d^ny}{dx^n}=f(y)}$

Multiplying by $2\dfrac{dy}{dx}$, we get $2\dfrac{dy}{dx}\dfrac{d^2y}{dx^2}=2f(y)\dfrac{dy}{dx}$ ...(1)

Integrating (1), we have $\left(\dfrac{dy}{dx}\right)^2=2\int f(y)\,dy+c=\phi(y)$ (say)

$$\frac{dy}{dx}=\sqrt{\phi(y)}\Rightarrow\frac{dy}{\sqrt{\phi(y)}}=dx\quad\Rightarrow\int\frac{dy}{\sqrt{\phi(y)}}=x+c$$

**Example 79.** *Solve* $\dfrac{d^2y}{dx^2}=\sqrt{y}$, *under the condition* $y=1.\dfrac{dy}{dx}=\dfrac{2}{\sqrt{3}}$ *at* $x=0$

**Solution.** We have $\dfrac{d^2y}{dx^2}=\sqrt{y}$ ...(1)

Multiplying (1) by $2\dfrac{dy}{dx}$, we get $2\dfrac{dy}{dx}\dfrac{d^2y}{dx^2}=2\sqrt{y}\dfrac{dy}{dx}$ ...(2)

Integrating (2), we get $\left(\dfrac{dy}{dx}\right)^2=\dfrac{4}{3}y^{3/2}+c_1$ ...(3)

On putting $y=1$ and $\dfrac{dy}{dx}=\dfrac{2}{\sqrt{3}}$, we have $c_1=0$

Equation (3) becomes $\left(\dfrac{dy}{dx}\right)^2=\dfrac{4}{3}y^{3/2}$ or $\dfrac{dy}{dx}=\dfrac{2}{\sqrt{3}}y^{3/4}$ or $y^{-3/4}\,dy=\dfrac{2}{\sqrt{3}}dx$

Again integrating $\dfrac{y^{1/4}}{\frac{1}{4}}=\dfrac{2}{\sqrt{3}}x+c_2\quad\Rightarrow 4y^{\frac{1}{4}}=\dfrac{2}{\sqrt{3}}x+c_2$ ...(4)

On putting $x=0$, $y=1$, we get $c_2=4$

(4) becomes $4y^{1/4}=\dfrac{2}{\sqrt{3}}x+4$ **Ans.**

**Example 80.** *Solve* $\dfrac{d^2y}{dx^2}=\sec^2 y\tan y$ *under the condition* $y=0$ *and* $\dfrac{dy}{dx}=1$ *when* $x=0.$

**Solution.** $\dfrac{d^2y}{dx^2}=\sec^2 y\tan y\quad\Rightarrow\quad 2\dfrac{dy}{dx}\dfrac{d^2y}{dx^2}=2\sec^2 y\tan y\dfrac{dy}{dx}$

$$\int 2\frac{dy}{dx}\frac{d^2y}{dx^2}=\int 2\sec^2 y\tan y\frac{dy}{dx}$$

$$\left(\frac{dy}{dx}\right)^2=\tan^2 y+c_1\quad\text{or}\quad\frac{dy}{dx}=\sqrt{\tan^2 y+c_1}$$

On putting $y=0$, and $\dfrac{dy}{dx}=1$, we get $c_1=1$

Now, $$\frac{dy}{dx} = \sqrt{\tan^2 y + 1} = \sec y$$

$\Rightarrow$ $\cos y \, dy = dx$

On integrating we get $\sin y = x + c$

On putting $y = 0$, $x = 0$, we have $c = 0$

$$\sin y = x \Rightarrow y = \sin^{-1} x$$ **Ans.**

**Example 81.** *Solve* $\frac{d^2y}{dx^2} = 2(y^3 + y)$, *under the condition* $y = 0, \frac{dy}{dx} = 1$ *when* $x = 0$.

(*U.P., II Semester, Summer 2003*)

**Solution.** $$\frac{d^2y}{dx^2} = 2(y^3 + y) \text{ or } 2\frac{dy}{dx}\frac{d^2y}{dx^2} = 4(y^3 + y)\frac{dy}{dx}$$

Integrating, we get

$$\left(\frac{dy}{dx}\right)^2 = 4\left(\frac{y^4}{4} + \frac{y^2}{2}\right) + c_1 = y^4 + 2y^2 + c_1 \quad \text{...(1)}$$

On putting $y = 0$ and $\frac{dy}{dx} = 1$ in (1), we get $1 = c_1$

Equation (1) becomes $$\left(\frac{dy}{dx}\right)^2 = y^4 + 2y^2 + 1 = (y^2 + 1)^2$$

$$\frac{dy}{dx} = y^2 + 1 \text{ or } \frac{dy}{1 + y^2} = dx$$

Again integraiting, we get $\tan^{-1} y = x + c_2$ ...(2)

On putting $y = 0$ and $x = 0$ in (2), we have $0 = c_2$

Equation (2) is reduced to $\tan^{-1} y = x \Rightarrow y = \tan x$ **Ans.**

**Example 82.** *A motion is governed by* $\frac{d^2x}{dt^2} = 36x^{-2}$, *given that at* $t = 0$, $x = 8$ *and* $\frac{dx}{dt} = 0$, *find the displacement at any time t.*

**Solution.** We have $$\frac{d^2x}{dt^2} = 36x^{-2} \Rightarrow 2\frac{d^2x}{dt^2}\frac{dx}{dt} = 2 \times 36x^{-2}\frac{dx}{dt} \quad \text{...(1)}$$

Integrating (1), we hae $$\left(\frac{dx}{dt}\right)^2 = -72x^{-1} + c_1 \quad \text{...(2)}$$

Putting $x = 8$ and $\frac{dx}{dt} = 0$ in (2), we get $0 = -\frac{72}{8} + c_1$ or $c_1 = 9$

(2) becomes $$\left(\frac{dx}{dt}\right)^2 = -\frac{72}{x} + 9 \text{ or } \left(\frac{dx}{dt}\right)^2 = \frac{-72 + 9x}{x} \Rightarrow \frac{dx}{dt} = 3\sqrt{\frac{(x-8)}{x}}$$

$$\Rightarrow \int \frac{\sqrt{x}\,dx}{\sqrt{x-8}} = 3\int dt + c_2 \Rightarrow \int \frac{x\,dx}{\sqrt{x^2 - 8x}} = 3t + c_2$$

$$\frac{1}{2}\int \frac{2x - 8 + 8}{\sqrt{x^2 - 8x}}\,dx = 3t + c_2$$

$$\frac{1}{2}\int \frac{2x - 8}{\sqrt{x^2 - 8x}}\,dx + 4\int \frac{1}{\sqrt{(x-4)^2 - (4)^2}}\,dx = 3t + c_2$$

$$\sqrt{x^2 - 8x} + 4\cos h^{-1}\frac{x-4}{4} = 3t + c_2 \quad \text{...(3)}$$

On putting $x = 8$ and $t = 0$ in (3), we get $c_2 = 0$

(3) becomes $\sqrt{x^2 - 8x + 4} \cosh^{-1}\frac{x-4}{4} = 3t$ **Ans.**

**EXERCISE 3.30**

**1.** $y^3 \frac{d^2y}{dx^2} = a$ **Ans.** $c_1 y^2 = (c_1 x + c_2)^2$

**2.** $e^{2y}\frac{d^2y}{dx^2} = 1$ **Ans.** $c_1 e^y = \cosh(c_1 x + c_2)$

**3.** $\sin^3 y \frac{d^2y}{dx^2} = \cos y$ **Ans.** $\sin[(x+c_2)\sqrt{(1+c_1)}] + \sqrt{\left(\frac{1+c_1}{c_1}\right)}\cos y = 0$

**4.** A particle is acted upon by a force $\mu\left(x + \frac{a^4}{x^3}\right)$ per unit mass towards the origin where $x$ is the distance from the origin at time $t$. If it starts that it will arrive at the origin in time $\frac{\pi}{4\sqrt{\mu}}$.

**5.** In the case of a stretched elastic string which has one end fixed and a particle of mass $m$ attached to the other end, the equation of motion is $\frac{d^2s}{dt^2} = -\frac{mg}{e}(s-l)$

where $l$ is the natural length of the string and $e$ its elongwith due to a weight $mg$. Find $s$ and $v$ determining the constants, so that $s = s_0$ at the time $t = 0$ and $v = 0$ when $t = 0$.

**Ans.** $v = -\sqrt{\left(\frac{g}{e}\right)}[(s_0 - l)^2 - (s-l)^2]^{1/2}, \; s - l = (s_0 - l)\cos\left[\sqrt{\left(\frac{g}{e}\right)}.t\right]$

## 3.34 EQUATIONS WHICH DO NOT CONTAIN '$y$' DIRECTLY

The equation which do not contain $y$ directly, can be written

$$f\left(\frac{d^n y}{dx^n}, \frac{d^{n-1}y}{dx^{n-1}}, \ldots\ldots \frac{dy}{dx}, x\right) = 0 \qquad \ldots(1)$$

On substituting $\frac{dy}{dx} = P$ *i.e.*, $\frac{d^2y}{dx^2} = \frac{dP}{dx}, \frac{d^3y}{dx^3} = \frac{d^2P}{dx^2}$ etc. in (1), we get $f\left(\frac{d^{n-1}P}{dx^{n-1}}, \ldots\ldots P, x\right) = 0$

**Example 83.** *Solve* $\frac{d^2y}{dx^2} + \frac{dy}{dx} + \left(\frac{dy}{dx}\right)^3 = 0$

**Solution.** On putting $\frac{dy}{dx} = P$ and $\frac{d^2y}{dx^2} = \frac{dP}{dx}$, equation (1) becomes

$$\frac{dP}{dx} + P + P^3 = 0 \text{ or } \frac{dP}{dx} + P(1+P^2) = 0$$

$$\frac{dP}{dx} = -P(1+P^2) \text{ or } \frac{dP}{P(1+P^2)} = -dx \Rightarrow \left(\frac{1}{P} - \frac{P}{1+P^2}\right)dP = -dx$$

On integrating, we have $\log P - \frac{1}{2}\log(1+P^2) = -x + c_1$ or $\log\frac{P}{\sqrt{1+P^2}} = -x + c_1$

$$\frac{P}{\sqrt{1+P^2}} = e^{-x+c_1} \text{ or } \frac{P^2}{1+P^2} = a^2 e^{-2x} \Rightarrow P^2 = (1+P^2)a^2 e^{-2x}$$

$$\Rightarrow P^2(1 - a^2 e^{-2x}) = a^2 e^{-2x} \Rightarrow P = \frac{ae^{-x}}{\sqrt{1 - a^2 e^{-2x}}} \Rightarrow \frac{dy}{dx} = \frac{ae^{-x}}{\sqrt{1-a^2 e^{-2x}}}$$

$$dy = \frac{ae^{-x}}{\sqrt{1-a^2e^{-2x}}}dx$$

On integration, we get $y = -\sin^{-1}(a e^{-x}) + b$ **Ans.**

**Example 84.** *Solve* $\frac{d^2y}{dx^2} = \left[1 - \left(\frac{dy}{dx}\right)^2\right]^{1/2}$ *(U.P. Second Sem., 2002)*

**Solution.** We have, $$\frac{d^2y}{dx^2} = \left[1 - \left(\frac{dy}{dx}\right)^2\right]^{1/2} \quad ...(1)$$

Putting $P = \frac{dy}{dx} \Rightarrow \frac{dP}{dx} = \frac{d^2y}{dx^2}$ in (1), we get $\frac{dP}{dx} = \sqrt{1-P^2} \Rightarrow \frac{dP}{\sqrt{1-P^2}} = dx$

On integrating, we have
$$\sin^{-1} P = x + c \quad \Rightarrow \quad P = \sin(x + c)$$
$$\frac{dy}{dx} = \sin(x + c)$$
On integrating, we have $\quad y = -\cos(x + c) + c_1$ **Ans.**

**Example 85.** *Solve* $x\frac{d^2y}{dx^2} + x\left(\frac{dy}{dx}\right)^2 - \frac{dy}{dx} = 0$ *(U.P. II Semester, 2010)*

**Solution.** On putting $\frac{dy}{dx} = P$ and $\frac{d^2y}{dx^2} = \frac{dP}{dx}$ in the given equation, we get
$$x\frac{dP}{dx} + xP^2 - P = 0 \Rightarrow \frac{1}{P^2}\frac{dP}{dx} - \frac{1}{P}\frac{1}{x} = -1 \quad ...(1)$$
Again putting $\frac{1}{P} = z$ so that $-\frac{1}{P^2}\frac{dP}{dx} = \frac{dz}{dx}$

Equation (1) becomes $\frac{-dz}{dx} - \frac{z}{x} = -1 \Rightarrow \frac{dz}{dx} + \frac{z}{x} = 1$
$$\text{I.F.} = e^{\int \frac{1}{x} dx} = e^{\log x} = x$$
Hence, solution is $\quad zx = \int x\, dx + c_1$ or $zx = \frac{x^2}{2} + c_1$ or $\frac{1}{P}x = \frac{x^2}{2} + c_1$
$$\Rightarrow \quad \frac{x}{P} = \frac{x^2 + 2c_1}{2} \Rightarrow P = \frac{2x}{x^2 + 2c_1} \Rightarrow \frac{dy}{dx} = \frac{2x}{x^2 + 2c_1} \Rightarrow dy = \frac{2x}{x^2 + 2c_1}dx$$
On integrating, we have $\quad y = \log(x^2 + 2c_1) + c_2$ **Ans.**

## EXERCISE 3.31

**Solving the following differential equations:**

1. $(1+x^2)\frac{d^2y}{dx^2} + x\frac{dy}{dx} + ax = 0$ **Ans.** $y = c_2 - ax + c_1 \log[x + \sqrt{(1+x^2)}]$
2. $(1+x^2)\frac{d^2y}{dx^2} + 1 + \left(\frac{dy}{dx}\right)^2 = 0$ **Ans.** $y = -\frac{x}{k} + \frac{1+k^2}{k^2}\log(1+kx) + a$
3. $\frac{d^4y}{dx^4} - \cot x \frac{d^3y}{dx^3} = 0$ **Ans.** $y = c_1 \cos x + c_2x^2 + c_3x + c_4$
4. $2x\frac{d^3y}{dx^3}\cdot\frac{d^2y}{dx^2} = \left[\frac{d^2y}{dx^2}\right]^2 - a^2$ **Ans.** $15c_1^2 y = 4(c_1x + a^2)^{5/2} + c_2x + c_3$
5. $e^{x^2/2}\left[x\frac{d^2y}{dx^2} - \frac{dy}{dx}\right] = x^3$ **Ans.** $y = e^{-x^2/2} + c_1\frac{x^2}{2} + c_2$

## 3.35 EQUATIONS THAT DO NOT CONTAIN 'x' DIRECTLY

The equations that do not contain $x$ directly are of the form

$$f\left(\frac{d^n y}{dx^n}, \frac{d^{n-1}y}{dx^{n-1}}, \ldots\ldots, \frac{dy}{dx}, y\right) = 0 \qquad \ldots(1)$$

On substituting $\frac{dy}{dx} = P, \frac{d^2y}{dx^2} = \frac{dP}{dx} = \frac{dP}{dx}\cdot\frac{dy}{dx} = \frac{dP}{dy}P$ in the equation (1), we get

$$\left[\frac{dP^{n-1}}{dy^{n-1}}, \ldots\ldots P, y\right] = 0 \qquad \ldots(2)$$

Equation (2) is solved for $P$. Let

$$P = f_1(y) \Rightarrow \frac{dy}{dx} = f_1(y) \text{ or } \frac{dy}{f_1(y)} = dx \Rightarrow \int \frac{dy}{f_1(y)} = x + c$$

**Example 86.** *Solve* $y\frac{d^2y}{dx^2} + \left(\frac{dy}{dx}\right)^2 = \frac{dy}{dx}$ ...(1)

**Solution.** Put $\frac{dy}{dx} = P, \frac{d^2y}{dx^2} = \frac{dP}{dx} = \frac{dP}{dy}\cdot\frac{dy}{dx} = P\frac{dP}{dy}$ in equation (1)

$$yP\frac{dP}{dy} + P^2 = P \Rightarrow y\frac{dP}{dy} = 1 - P$$

$$\Rightarrow \quad \frac{dp}{1-P} = \frac{dy}{y} \Rightarrow -\log(1-P) = \log y + \log c_1$$

$$\Rightarrow \quad \frac{1}{1-P} = c_1 y \Rightarrow P = 1 - \frac{1}{c_1 y} \text{ or } \frac{dy}{dx} = \frac{c_1 y - 1}{c_1 y}$$

$$\Rightarrow \quad \frac{c_1 y}{c_1 y - 1}\, dy = dx \Rightarrow \left(1 + \frac{1}{c_1 y - 1}\right) dy = dx$$

$$y + \frac{1}{c_1}\log(c_1 y - 1) = x + c_1$$ **Ans.**

**Example 87.** *Solve* $y\frac{d^2y}{dx^2} + \left(\frac{dy}{dx}\right)^2 = y^2$ ...(1)

**Solution.** Put $\frac{dy}{dx} = P, \frac{d^2y}{dx^2} = \frac{dP}{dx} = \frac{dP}{dy}\frac{dy}{dx} = P\frac{dP}{dy}$ in (1)

$$yP\frac{dP}{dy} + P^2 = y^2 \text{ or } P\frac{dP}{dy} + \frac{P^2}{y} = y \qquad \ldots(2)$$

Put $P^2 = z$ or $2P\frac{dP}{dy} = \frac{dz}{dy}$ in (2), $\frac{1}{2}\frac{dz}{dy} + \frac{z}{y} = y$ or $\frac{dz}{dy} + \frac{2z}{y} = 2y$

$$\text{I.F.} = e^{\int \frac{2}{y}dy} = e^{2\log y} = e^{\log y^2} = y^2$$

Hence, the solution is $\quad z\,y^2 = \int 2y.(y^2)\, dy + c$

$$\Rightarrow \quad P^2 y^2 = \frac{y^4}{2} + c$$

$$\Rightarrow \quad 2P^2 y^2 = y^4 + k \text{ or } \sqrt{2}\, y\, P = \sqrt{y^4 + k}$$ [Put $2c = k$]

$$\Rightarrow \quad \sqrt{2}\, y\frac{dy}{dx} = \sqrt{y^4+k} \text{ or } \sqrt{2}\frac{y\,dy}{\sqrt{y^4+k}} = dx$$

$$\Rightarrow \quad \frac{1}{\sqrt{2}}\frac{dt}{\sqrt{t^2+k}} = dx \quad [\text{Put } y^2 = t,\ 2y\,dy = dt] \quad \Rightarrow \frac{1}{\sqrt{2}}\sin h^{-1}\frac{t}{\sqrt{k}} = x + c$$

$$\Rightarrow \quad \sin h^{-1}\frac{y^2}{\sqrt{k}} = \sqrt{2}\,x + c \text{ or } y^2 = \sqrt{k}\sin h\,(\sqrt{2}\,x + c)$$ **Ans.**

## EXERCISE 3.32

**Solve the following differential equations:**

1. $\frac{d^2y}{dx^2} + \left(\frac{dy}{dx}\right)^2 = 1$ **Ans.** $y^2 = x^2 + ax + b$ 2. $y\frac{d^2y}{dx^2} - \left(\frac{dy}{dx}\right)^2 + 2\frac{dy}{dx} = 0$ **Ans.** $c^y + 2 = d\,e^{cx^a}$

3. $2y\frac{d^2y}{dx^2} - 3\left(\frac{dy}{dx}\right)^2 - 4y^2 = 0$ **Ans.** $y = a\sec^2(x+b)$ 4. $y\frac{d^2y}{dx^2} + \frac{dy}{dx} + \left(\frac{dy}{dx}\right)^3 = 0$ **Ans.** $y = a - \sin^{-1}(b\,e^{-x})$

5. $y\frac{d^2y}{dx^2} = \left\{\frac{dy}{dx}\right\}^2\left[1 - \frac{dy}{dx}\cos y + y\frac{dy}{dx}\sin y\right]$ **Ans.** $x = c_1 + c_2\log y + \sin y$

6. $y\frac{d^2y}{dx^2} - \left(\frac{dy}{dx}\right)^2 = y^2\log y$ **Ans.** $\log y = b.\,e^x + a\,e^{-x}$

## 3.36 EQUATION WHOSE ONE SOLUTION IS KNOWN

**If $y = u$ is given solution belonging to the complementary function of the differential equation. Let the other solution be $y = v$. Then $y = u.\,v$ is complete solution of the differential equation.**

Let $\frac{d^2y}{dx^2} + P\frac{dy}{dx} + Qy = R$ ....(1), be the differential equation and $u$ is the solution included in the complementary function of (1)

$$\therefore \quad \frac{d^2u}{dx^2} + P\frac{du}{dx} + Qu = 0 \qquad ...(2)$$

$$y = u.\,v$$

$$\frac{dy}{dx} = v\frac{du}{dx} + u\frac{dv}{dx}$$

$$\frac{d^2y}{dx^2} = v\frac{d^2u}{dx^2} + 2\frac{dv}{dx}\frac{du}{dx} + u\frac{d^2v}{dx^2}$$

Substituting the values of $y, \frac{dy}{dx}, \frac{d^2y}{dx^2}$ in (1), we get

$$v\frac{d^2u}{dx^2} + 2\frac{dv}{dx}\frac{du}{dx} + u\frac{d^2v}{dx^2} + P\left(v\frac{du}{dx} + u\frac{dv}{dx}\right) + Qu.v = R$$

On arranging

$$\Rightarrow \quad v\left[\frac{d^2u}{dx^2} + P\frac{du}{dx} + Qu\right] + u\left[\frac{d^2v}{dx^2} + P\frac{dv}{dx}\right] + 2\frac{du}{dx}\cdot\frac{dv}{dx} = R$$

The first bracket is zero by virtue of relation (2), and the remaining is divided by $u$.

$$\frac{d^2v}{dx^2} + P\frac{dv}{dx} + \frac{2}{u}\frac{du}{dx}\frac{dv}{dx} = \frac{R}{u}$$

$$\Rightarrow \quad \frac{d^2v}{dx^2} + \left[P + \frac{2}{u}\frac{du}{dx}\right]\frac{dv}{dx} = \frac{R}{u} \qquad ...(3)$$

Let $\frac{dv}{dx} = z$, so that $\frac{d^2v}{dx^2} = \frac{dz}{dx}$

Equation (3) becomes $\boxed{\frac{dz}{dx} + \left[P + \frac{2}{u}\frac{du}{dx}\right] z = \frac{R}{u}}$

This is the linear differential equation of first order and can be solved ($z$ can be found), which will contain one constant.

On integration $z = \frac{dv}{dx}$, we can get $v$.

Having found $v$, the solution is $y = uv$.

**Note:** Rule to find out the integral belonging to the complementary function

| *Rule* | *Condition* | *u* |
|---|---|---|
| 1 | $1 + P + Q = 0$ | $e^x$ |
| 2 | $1 - P + Q = 0$ | $e^{-x}$ |
| 3 | $1 + \frac{P}{a} + \frac{Q}{a^2} = 0$ | $e^{ax}$ |
| 4 | $P + Qx = 0$ | $x$ |
| 5 | $2 + 2Px + Qx^2 = 0$ | $x^2$ |
| 6 | $n(n-1) + Pnx + Qx^2 = 0$ | $x^n$ |

**Example 88.** *Solve $y'' - 4xy' + (4x^2 - 2)y = 0$ given that $y = e^{x^2}$ is an integral included in the complementary function.* *(U.P., II Semester, 2004)*

**Solution.** $y'' - 4xy' + (4x^2 - 2)y = 0$ ...(1)

On putting $y = v.e^{x^2}$ in (1), the reduced equation as in the article 3.36

$$\frac{d^2v}{dx^2} + \left[P + \frac{2}{u}\frac{du}{dx}\right]\frac{dv}{dx} = 0 \qquad [P = -4x,\ Q = 4x^2 - 2,\ R = 0]$$

$$\Rightarrow \frac{d^2v}{dx^2} + \left[-4x + \frac{2}{e^{x^2}}(2x\, e^{x^2})\right]\frac{dv}{dx} = 0$$

$$\Rightarrow \frac{d^2v}{dx^2} + [-4x + 4x]\frac{dv}{dx} = 0 \qquad \Rightarrow \qquad \frac{d^2v}{dx^2} = 0 \Rightarrow \frac{dv}{dx} = c, \Rightarrow v = c_1x + c_2$$

$$\therefore \qquad y = uv \qquad [u = e^{x^2}]$$

$$y = e^{x^2}(c_1x + c_2)$$

**Ans.**

**Example 89.** *Solve* $x\frac{d^2y}{dx^2} - (2x-1)\frac{dy}{dx} + (x-1)y = 0$

*given that $y = e^x$ is an integral included in the complementary function.*

**Solution.** $x\frac{d^2y}{dx^2} - (2x-1)\frac{dy}{dx} + (x-1)\,y = 0$

$$\Rightarrow \frac{d^2y}{dx^2} - \frac{2x-1}{x}\frac{dy}{dx} + \frac{x-1}{x}y = 0 \qquad [1 + P + Q = 0] \qquad ...(1)$$

By putting $y = ve^x$ in (1),we get the reduced equation as in the article 3.36.

$$\frac{d^2v}{dx^2} + \left[P + \frac{2}{u}\frac{du}{dx}\right]\frac{dv}{dx} = 0 \qquad ...(2)$$

Putting $u = e^x$ and $\frac{dv}{dx} = z$ in (2), we get $\frac{dz}{dx} + \left[-\frac{2x-1}{x} + \frac{2}{e^x}e^x\right]z = 0$

$$\Rightarrow \quad \frac{dz}{dx} + \frac{-2x+1+2x}{x}z = 0 \quad \Rightarrow \quad \frac{dz}{dx} + \frac{z}{x} = 0$$

$$\Rightarrow \quad \frac{dz}{z} = -\frac{dx}{x} \Rightarrow \log z = -\log x + \log c_1$$

$$\Rightarrow \quad z = \frac{c_1}{x} \text{ or } \frac{dv}{dx} = \frac{c_1}{x} \text{ or } dv = c_1\frac{dx}{x} \Rightarrow v = c_1 \log x + c_2$$

$$y = u.\ v = e^x(c_1 \log x + c_2)$$ **Ans.**

**Example 90.** *Solve* $x^2\frac{d^2y}{dx^2} - 2x[1+x]\frac{dy}{dx} + 2(1+x)y = x^3$

**Solution.** $x^2\frac{d^2y}{dx^2} - 2x(1+x)\frac{dy}{dx} + 2(1+x)y = x^3$

$$\Rightarrow \quad \frac{d^2y}{dx^2} - \frac{2x(1+x)}{x^2}\frac{dy}{dx} + \frac{2(1+x)y}{x^2} = x \quad \text{...(1)}$$

Here $$P + Qx = -\frac{2x(1+x)}{x^2} + \frac{2(1+x)}{x^2}x = 0$$

Hence $y = x$ is a solution of the C.F. and the other solution is $v$.
Putting $y = vx$ in (1), we get the reduced equation as in article 3.36

$$\frac{d^2v}{dx^2} + \left\{P + \frac{2}{u}\frac{du}{dx}\right\}\frac{du}{dx} = \frac{x}{u}$$

$$\frac{d^2v}{dx^2} + \left[\frac{-2x(1+x)}{x^2} + \frac{2}{x}(1)\right]\frac{dv}{dx} = \frac{x}{x}$$

$$\Rightarrow \quad \frac{d^2v}{dx^2} - 2\frac{dv}{dx} = 1 \Rightarrow \frac{dz}{dx} - 2z = 1 \qquad \left[\frac{dv}{dx} = z\right]$$

which is a linear differential equation of first order and $I.F. = e^{\int -2\,dx = e^{-2x}}$
Its solution is $z\,e^{-2x} = \int e^{-2x}\,dx + c_1$

$$z\,e^{-2x} = \frac{e^{-2x}}{-2} + c_1 \text{ or } z = \frac{-1}{2} + c_1 e^{2x}$$

$$\Rightarrow \quad \frac{dv}{dx} = -\frac{1}{2} + c_1 e^{2x} \text{ or } dv = \left(-\frac{1}{2} + c_1 e^{2x}\right)dx \Rightarrow v = \frac{-x}{2} + \frac{c_1}{2}e^{2x} + c_2$$

$$y = uv = x\left(\frac{-x}{2} + \frac{c_1}{2}e^{2x} + c_2\right)$$ **Ans.**

**Example 91.** *Verify that* $y = e^2 x$ *is a solution of* $(2x+1)y'' - 4(x+1)y' + 4y = 0$. *Hence find the general solution.*

**Solution.** We have $(2x+1)\frac{d^2y}{dx^2} - 4(x+1)\frac{dy}{dx} + 4y = 0$ ... (1)

$$y = e^{2x}, \quad y' = 2e^{2x}, \quad y'' = 4e^{2x}$$

Substiting the values of $y$, $y'$ and $y''$ in (1), we get

$$(2x+1)4e^{2x} - 4(x+1)2e^{2x} + 4e^{2x} = 0$$

or $$[8x + 4 - 8x - 8 + 4]\,e^{2x} = 0 \quad \Rightarrow \quad 0 = 0.$$

Thus $y_1 = e^{2x}$ is a solution

Equation (1) in the standard form is $y'' - \dfrac{4(x+1)}{(2x+1)} y' + \dfrac{4}{(2x+1)} y = 0$

So $P(x) = -\dfrac{4(x+1)}{(2x+1)}$. Then $\omega(x) = \dfrac{1}{y_1^2} e^{-\int P\,dx}$

Now $\int -P\,dx = -\int -\dfrac{4(x+1)}{(2x+1)}\,dx = \int\left(\dfrac{4x+2}{2x+1} + \dfrac{2}{2x+1}\right)dx$

$= 2x + \ln(2x+1)$

Then $\omega = \dfrac{1}{(e^{2x})^2} e^{2x + \ln(2x+1)} = \dfrac{e^{2x}}{(e^{2x})^2} \cdot (2x+1)$

$\omega(x) = \dfrac{2x+1}{e^{2x}}$ Now $v(x) = \int \omega(x)\,dx = \int \dfrac{2x+1}{e^{2x}}\,dx$

Integrating by parts $v(x) = (2x+1)\dfrac{e^{-2x}}{-2} - 2 \cdot \dfrac{e^{-2x}}{4}$

The required second solution

$$y_2(x) = y_1(x)\,v(x) = e^{2x}\left[-\frac{2x+1}{2} \cdot \frac{1}{e^{2x}} - \frac{1}{2}\frac{1}{e^{2x}}\right] = -x - 1 = -(x+1)$$

Then the general solution is

$y(x) = c_1 y_1(x) + c_2 y_2(x) = c_1 e^{2x} - c_2(x+1)$ **Ans.**

**Example 92.** *Solve* $x^2 y'' - (x^2 + 2x)y' + (x+2)y = x^3 e^x$ *given that* $y = x$ *is a solution.*

**Solution.** $x^2 y'' - (x^2 + 2x)y' + (x+2)y = x^3 e^x$

$$\Rightarrow \quad y'' - \frac{x^2 + 2x}{x^2} y' + \frac{x+2}{x^2} y = x e^x \qquad ...(1)$$

On putting $y = vx$ in (1), the reduced equation as in the article 3.36.

$$\frac{d^2v}{dx^2} + \left\{P + \frac{2}{u}\frac{du}{dx}\right\}\frac{dv}{dx} = \frac{R}{u} \Rightarrow \frac{d^2v}{dx^2} + \left[-\frac{x^2+2x}{x^2} + \frac{2}{x}(1)\right]\frac{dv}{dx} = \frac{xe^x}{x}$$

$$\Rightarrow \quad \frac{d^2v}{dx^2} - \frac{dv}{dx} = e^x \Rightarrow \frac{dz}{dx} - z = e^x \qquad \left(z = \frac{dv}{dx}\right)$$

which is a linear differential equation

$I.F. = e^{-\int dx} = e^{-x} \Rightarrow z e^{-x} = \int e^x . e^{-x}\,dx + c$

$z e^{-x} = x + c$ or $z = e^x . x + c\,e^x \Rightarrow \dfrac{dv}{dx} = e^x . x + c\,e^x$

$v = x.e^x - e^x + c\,e^x + c_1 \Rightarrow v = (x-1)e^x + c\,e^x + c_1$

$y = vx = (x^2 - x + cx)e^x + c_1 x$ **Ans.**

**Example 93.** *Solve* $(x+2)\dfrac{d^2y}{dx^2} - (2x+5)\dfrac{dy}{dx} + 2y = (x+1)e^x$

Here, $P = \dfrac{2x+5}{x+2}$, $Q = \dfrac{2}{x+2}$, $R = \dfrac{(x+1)}{x+2} e^x$

**Solution.** $\dfrac{d^2y}{dx^2} - \dfrac{2x+5}{x+2}\dfrac{dy}{dx} + \dfrac{2y}{x+2} = \dfrac{(x+1)e^x}{x+2}$ ...(1)

Here $1 + \dfrac{P}{a} + \dfrac{Q}{a^2} = 0$, Choosing $a = 2$

$$1-\frac{2x+5}{2x+4}+\frac{2}{4x+8}=0$$

Hence $y = e^{2x}$ is a part of C.F.

Putting $y = e^{2x}v$ in (1), the reduced equation as in the article 3.36.

$$\frac{d^2v}{dx^2}+\left[P+\frac{2}{u}\frac{du}{dx}\right]\frac{dv}{dx}=\frac{(x+1)\,e^x}{u\,(x+2)}$$

$$\Rightarrow \quad \frac{d^2v}{dx^2}+\left[-\frac{2x+5}{x+2}+\frac{2}{e^{2x}}2e^{2x}\right]\frac{dv}{dx}=\frac{(x+1)\,e^x}{e^{2x}(x+2)}$$

$$\Rightarrow \quad \frac{d^2v}{dx^2}+\left[-\frac{2x+5}{x+2}+4\right]\frac{dv}{dx}=\frac{x+1}{x+2}e^{-x}$$

$$\Rightarrow \quad \frac{dz}{dx}+\frac{2x+3}{x+2}\,z=\frac{x+1}{x+2}e^{-x} \qquad \left(\frac{dv}{dx}=z\right)$$

which is a linear differential equation,

$$I.F. = e^{\int\frac{2x+3}{x+2}dx}=e^{\int\left(2-\frac{1}{x+2}\right)dx}=e^{2x-\log(x+2)}=\frac{e^{2x}}{x+2}$$

Its solution is

$$z.\frac{e^{2x}}{x+2}=\int\frac{e^{2x}}{x+2}\,\frac{x+1}{x+2}e^{-x}\,dx+c$$

$$=\int\frac{e^x\,(x+1)}{(x+2)^2}dx+c=\int e^x\left[\frac{1}{x+2}-\frac{1}{(x+2)^2}\right]dx+c=\int\frac{e^x\,dx}{x+2}-\int\frac{e^x\,dx}{(x+2)^2}+c$$

$$=\frac{e^x}{x+2}+\int\frac{e^x\,dx}{(x+2)^2}-\int\frac{e^x\,dx}{(x+2)^2}+c=\frac{e^x}{x+2}+c$$

$$\Rightarrow \quad z=e^{-x}+c\,(x+2)e^{-2x} \Rightarrow \frac{dv}{dx}=e^{-x}+c(x+2)e^{-2x}$$

$$v=\int e^{-x}\,dx+c\int(x+2)e^{-2x}\,dx+c_1=-e^{-x}+c\left[\frac{(x+2)\,e^{-2x}}{-2}-\frac{e^{-2x}}{4}\right]+c_1$$

$$=-e^{-x}-\frac{ce^{-2x}}{4}[2x+5]+c_1$$

$$y=u.v$$

$$y=e^{2x}\left[-e^{-x}-\frac{ce^{-2x}}{4}(2x+5)+c_1\right] \Rightarrow y=-e^x+\frac{c}{4}(2x+5)+c_1\,e^{2x} \qquad \textbf{Ans.}$$

## EXERCISE 3.33

**Solve the following differential equations:**

**1.** $(3-x)\dfrac{d^2y}{dx^2}-(9-4x)\dfrac{dy}{dx}+(6-3x)\,y=0$, given $y = e^x$ is a solution.

**Ans.** $y=\dfrac{c_1}{8}e^{3x}\,(4x^3-42x^2+150x-183)+c_2e^x$

**2.** $x\dfrac{d^2y}{dx^2}-\dfrac{dy}{dx}+(1-x)y=x^2\,e^{-x}$ given $y = e^x$ is an integral included in C.F.

**Ans.** $y=c_2\,e^x+c_1\,(2x+1)\,e^{-x}-\dfrac{1}{4}(2x^2+2x+1)e^{-x}$

**3.** $(1-x^2)\dfrac{d^2y}{dx^2}+x\dfrac{dy}{dx}-y=x(1-x^2)^{3/2}$, given $y = x$ is part of C.F.

**Ans.** $y=-\dfrac{x}{9}(1-x^2)^{3/2}-c_1\,[\sqrt{(1-x^2)}+x\sin^{-1}x]+c_2x.$

4. $\sin^2 x \dfrac{d^2y}{dx^2} = 2y$, given that $y = \cot x$ is a solution **Ans.** $cy = 1 + (c_1 - x)\cot x$

5. $\dfrac{d^2y}{dx^2} - x^2\dfrac{dy}{dx} + xy = x$, given $y = x$ is a part of C.F. **Ans.** $y = 1 + c_1 x \int \dfrac{1}{x^2} e^{\frac{x^3}{3}} dx + c_2 x$

6. $(x\sin x + \cos x)\dfrac{d^2y}{dx^2} - x\cos x\dfrac{dy}{dx} + y\cos x = 0$ given $y = x$ is solution.

**Ans.** $y = c_2 x - c_1 \cos x$

7. $x^2\dfrac{d^2y}{dx^2} + x\dfrac{dy}{dx} - y = 0$, given that $y = x + \dfrac{1}{x}$ is one integral. **Ans.** $y = c_2\left(x + \dfrac{1}{x}\right) + \dfrac{c_1}{x}$

8. $x^2\dfrac{d^2y}{dx^2} + 2x\dfrac{dy}{dx} - 12y = x^3\log x$ *(U.P., II Semester 2004)*

[**Hint.** $(n(n-1) + pnx + Qx^2 = 0)$, n = 3, satisfies this equation. Put $y = vx^3, \dfrac{dv}{dx} = z$]

**Ans.** $y = \left(c_1 x^3 + \dfrac{c_2}{x^4}\right) + \dfrac{x^3}{98}\log x\,(7\log x - 2)$

## 3.37 NORMAL FORM (REMOVAL OF FIRST DERIVATIVE)

Consider the differential equation $\dfrac{d^2y}{dx^2} + P\dfrac{dy}{dx} + Qy = R$ ...(1)

Put $y = uv$ where $v$ is not an integral solution of C.F.

$$\frac{dy}{dx} = v\frac{du}{dx} + u\frac{du}{dx}$$

$$\frac{d^2y}{dx^2} = u\frac{d^2v}{dx^2} + 2\frac{du}{dx}\frac{dv}{dx} + v\frac{d^2u}{dx^2}$$

On putting the values of $y, \dfrac{dy}{dx}, \dfrac{d^2y}{dx^2}$ in (1) we get

$$\left(u\frac{d^2v}{dx^2} + 2\frac{dv}{dx}\frac{du}{dx} + v\frac{d^2u}{dx^2}\right) + P\left(u\frac{du}{dx} + v\frac{du}{dx}\right) + Q.uv = R$$

$$\Rightarrow \quad v\frac{d^2u}{dx^2} + \frac{du}{dx}\left(Pv + 2\frac{dv}{dx}\right) + u\left(\frac{d^2v}{dx^2} + P\frac{dv}{dx} + Q.v\right) = R$$

$$\Rightarrow \quad \frac{d^2u}{dx^2} + \frac{du}{dx}\left(P + \frac{2}{u}\frac{dv}{dx}\right) + \frac{u}{v}\left(\frac{d^2v}{dx^2} + P\frac{dv}{dx} + Q.v\right) = \frac{R}{v} \qquad ...(2)$$

Here in the last bracket on L.H.S. is not zero $y =$ $v$ is not a part of C.F.

Here we shall remove the first derivative.

$$P + \frac{2}{v}\frac{dv}{dx} = 0 \text{ or } \frac{dv}{v} = -\frac{1}{2}P\,dx \text{ or } \log v = \frac{-1}{2}\int P\,dx$$

$$v = e^{-\frac{1}{2}\int P\,dx}$$

In (2) we have to find out the value of the last bracket *i.e.*, $\dfrac{d^2v}{dx^2} + P\dfrac{dv}{dx} + Qv$

$$\frac{dv}{dx} = -\frac{P}{2}e^{-\frac{1}{2}\int P\,dx} = -\frac{1}{2}Pv \qquad \left[\therefore v = e^{-1/2\int p\,dx}\right]$$

$$\frac{d^2v}{dx^2} = -\frac{1}{2}\frac{dP}{dx}v - \frac{P}{2}\frac{dv}{dx} = -\frac{1}{2}\frac{dP}{dx}v - \frac{P}{2}\left(-\frac{1}{2}Pv\right) = -\frac{1}{2}\frac{dP}{dx}v + \frac{1}{4}P^2v$$

$$\therefore \frac{d^2v}{dx^2}+P\frac{dv}{dx}+Qv = -\frac{1}{2}\frac{dP}{dx}v+\frac{1}{4}P^2v+P\left(-\frac{1}{2}Pv\right)+Qv = v\left[Q-\frac{1}{2}\frac{dP}{dx}-\frac{1}{4}P^2\right]$$

Equation (1) is transformed as

$$\frac{d^2u}{dx^2}+\frac{u}{v}v\left\{Q-\frac{1}{2}\frac{dP}{dx}-\frac{P^2}{4}\right\}=\frac{R}{v}$$

$$\Rightarrow \quad \frac{d^2u}{dx^2}+u\left\{Q-\frac{1}{2}\frac{dP}{dx}-\frac{P^2}{4}\right\}=Re^{\frac{1}{2}\int P\,dx}$$

$$\frac{d^2u}{dx^2}+Q_1u=R_1 \qquad \text{where } Q_1=\left[Q-\frac{1}{2}\frac{dP}{dx}-\frac{P^2}{4}\right],\ R_1=Re^{\frac{1}{2}\int P\,dx} \text{ or } \frac{R}{v}$$

$$y=uv \quad \text{and} \quad v=e^{-\frac{1}{2}\int P\,dx}$$ **Ans.**

**Example 94.** *Solve* $\frac{d}{dx}\left[\cos^2 x\frac{dy}{dx}\right]+\cos^2 x.\,y=0$

**Solution.** We have, $\frac{d}{dx}\left(\cos^2 x\frac{dy}{dx}\right)+\cos^2 x.\,y=0$

$$\Rightarrow \quad \frac{d^2y}{dx^2}\cos^2 x-2\cos x\sin x\frac{dy}{dx}+(\cos^2 x)y=0 \Rightarrow \frac{d^2y}{dx^2}-2\tan x.\frac{dy}{dx}+y=0$$

Here, $\quad P=-2\tan x,\ Q=1,\ R=0$

$$Q_1=Q-\frac{1}{2}\frac{dP}{dx}-\frac{P^2}{4}=1-\frac{1}{2}(-2\sec^2 x)-\frac{4\tan^2 x}{4}$$

$$=1+\sec^2x-\tan^2x=1+1=2$$

$$R_1=Re^{\frac{1}{2}\int P\,dx}=0$$

$$v=e^{-\frac{1}{2}\int P\,dx}=e^{-\frac{1}{2}\int(-2\tan x)dx}=e^{\int\tan x\,dx}=e^{\log\sec x}=\sec x$$

Normal equation is

$$\frac{d^2u}{dx^2}+Q_1u=R_1$$

$$\frac{d^2u}{dx^2}+2u=0 \quad \text{or} \quad (D_2+2)\,u=0 \quad \Rightarrow \quad D=\pm i\sqrt{2}$$

$$u=c_1\cos\sqrt{2}\,x+c_2\sin\sqrt{2}\,x$$

$$y=u.v$$

$$=[c_1\cos\sqrt{2}\,x+c_2\sin\sqrt{2}x]\sec x$$ **Ans.**

**Example 95.** *Solve* $x^2\frac{d^2y}{dx^2}-2(x^2+x)\frac{dy}{dx}+(x^2+2x+2)y=0$

**Solution.** We have, $\frac{d^2y}{dx^2}-\frac{2(x^2+x)}{x^2}\frac{dy}{dx}+\left(\frac{x^2+2x+2}{x^2}\right)y=0$ ...(1)

Here $p=-2\left(1+\frac{1}{x}\right),\ Q=\frac{x^2+2x+2}{x^2},\ \ R=0$

In order to remove the first derivative, we put $y=u.v$ in (1) to get the normal equation

$$\frac{d^2v}{dx^2}+Q_1v=R_1$$ ...(2)

where $v = e^{-\frac{1}{2}\int p\,dx} = e^{-\frac{1}{2}\int -2\left(1+\frac{1}{x}\right)dx} = e^{\int\left(1+\frac{1}{x}\right)dx} = e^x \cdot e^{\log x} = x\,e^x$

$$Q_1 = Q - \frac{1}{2}\frac{dp}{dx} - \frac{p^2}{4} = \frac{x^2+2x+2}{x^2} - \frac{1}{2}\left(\frac{2}{x^2}\right) - \frac{4}{4}\left(1+\frac{1}{x}\right)^2 = 1 + \frac{2}{x} + \frac{2}{x^2} - \frac{1}{x^2} - 1 - \frac{1}{x^2} - \frac{2}{x}$$

$$R_1 = R e^{\frac{1}{2}\int p\,dx} = 0$$

On putting the values of $Q_1$ and $R_1$ in (2), we get $\frac{d^2u}{dx^2} + 0(u) = 0 \Rightarrow \frac{d^2u}{dx^2} = 0$

$$\frac{du}{dx} = c_1 \Rightarrow u = c_1 x + c_2$$

$\therefore$ $$y = u.v = (c_1 x + c_2)\, x\, e^x$$ **Ans.**

**Example 96.** *Solve* $\frac{d^2y}{dx^2} - 4x\frac{dy}{dx} + (4x^2-1)y = -3e^{x^2}\sin 2x$ (*U.P. II Semester, (C.O.) 2004*)

**Solution.** We have, $\frac{d^2y}{dx^2} = 4x\frac{dy}{dx} + (4x^2-1)\,y = -3e^{x^2}\sin 2x$ ...(1)

Here $p = -4x$, $Q = 4x^2 - 1$, $R = -3e^{x^2}\sin 2x$

In order to remove the first derivative, $v = e^{-\frac{1}{2}\int p\,dx} = e^{-\frac{1}{2}\int -4x\,dx} = e^{2\int x\,dx} = e^{x^2}$

On putting $y = uv$, the normal equation is $\frac{d^2u}{dx^2} + Q_1 u = R_1$ ...(2)

where $Q_1 = Q - \frac{1}{2}\frac{dp}{dx} - \frac{p^2}{4} = (4x^2-1) - \frac{1}{2}(-4) - \frac{16x^2}{4} = 4x^2 - 1 + 2 - 4x^2 = 1$

$$R_1 = \frac{R}{v} = \frac{-3e^{x^2}\sin 2x}{e^{x^2}} = -3\sin 2x$$

Equation (2) becomes $\frac{d^2u}{dx^2} + u = -3\sin 2x$

$$(D^2+1)u = -3\sin 2x$$

A.E. is $D^2 + 1 = 0 \Rightarrow D = \pm i \Rightarrow C.F. = c_1\cos x + c_2\sin x$

$$P.I. = \frac{1}{D_2+1}(-3\sin 2x) = \frac{-3\sin 2x}{-4+1} = \sin 2x$$

$$u = c_1\cos x + c_2\sin x + \sin 2x$$

$$y = u.\,v = (c_1\cos x + c_2\sin x + \sin 2x)e^{x^2}$$ **Ans.**

## EXERCISE 3.34

**Solve the following differential equations:**

**1.** $\frac{d^2y}{dx^2} - 2\tan x.\,y - 5y = 0$ **Ans.** $y = (a\,e^{2x} + e^{-3x})\sec x$

**2.** $\frac{d^2y}{dx^2} - 4x\frac{dy}{dx} + (4x^2-3)\,y = e^{x^2}$ **Ans.** $y = (c_1 e^x + c_2 e^{-x} - 1)$

**3.** $\frac{d^2y}{dx^2} - 2x\frac{dy}{dx} + (x^2+2)y = e^{\frac{1}{2}(x^2+2x)}$ **Ans.** $y = (c_1\cos\sqrt{3}x + c_2\sin\sqrt{3}\,x)\,e^{\frac{x^2}{2}} + \frac{1}{4}e^x.e^{\frac{x^2}{2}}$

**4.** $\frac{d^2y}{dx^2} - 2x\frac{dy}{dx} + \left(n^2 + \frac{2}{x^2}\right)y = 0$ **Ans.** $y = (c_1\cos nx + c_2\sin nx)x$

**5.** $\frac{d^2y}{dx^2} + \frac{2}{x}\frac{dy}{dx} - n^2y = 0$ **Ans.** $y = (c_1 e^{nx} + c_2 + e^{-nx})\frac{1}{x}$

6. $\dfrac{d^2y}{dx^2}+\dfrac{1}{x^{\frac{1}{3}}}\dfrac{dy}{dx}+\left(\dfrac{1}{4x^{2/3}}-\dfrac{1}{x^{4/3}}-\dfrac{6}{x^2}\right)y=0$ **Ans.** $y=(c_1x^3+c_2\,x^{-2})e^{-\frac{3}{4}x^{\frac{2}{3}}}$

7. $\dfrac{d^2y}{dx^2}-\dfrac{1}{\sqrt{x}}\dfrac{dy}{dx}+\dfrac{y}{4x^2}(-8+\sqrt{x}+x)=0$ **Ans.** $y=(c_1x^2+c_2x^{-1})\,e^{\sqrt{x}}$

## 3.38 METHOD OF SOLVING LINEAR DIFFERENTIAL EQUATIONS BY CHANGING THE INDEPENDENT VARIABLE

Consider, $$\frac{d^2y}{dx^2}+P\frac{dy}{dx}+Qy=R \qquad ...(1)$$

Let us change the independent variable $x$ to $z$ and $z=f(x)$

$$\frac{dy}{dx}=\frac{dy}{dz}\frac{dz}{dx} \Rightarrow \frac{d^2y}{dx^2}=\frac{d^2y}{dz^2}\left(\frac{dz}{dx}\right)^2+\frac{dy}{dz}\frac{d^2z}{dx^2}$$

Putting the values of $\dfrac{dy}{dx}$ and $\dfrac{d^2y}{dx^2}$ in (1), we get

$$\left(\frac{d^2y}{dz^2}\left(\frac{dz}{dx}\right)^2+\frac{dy}{dz}\frac{d^2z}{dx^2}\right)+P\left(\frac{dy}{dz}\frac{d^2z}{dx^2}\right)+Qy=R$$

$$\Rightarrow \frac{d^2y}{dz^2}\left(\frac{dz}{dx}\right)^2+\left(P\frac{dz}{dx}+\frac{d^2z}{dx^2}\right)\frac{dy}{dz}+Qy=R$$

$$\Rightarrow \frac{d^2y}{dz^2}+\frac{P\left(\frac{dz}{dx}+\frac{d^2z}{dx^2}\right)}{\left(\frac{dz}{dx}\right)^2}\frac{dy}{dz}+\frac{Qy}{\left(\frac{dz}{dx}\right)^2}=\frac{R}{\left(\frac{dz}{dx}\right)^2} \Rightarrow \frac{d^2y}{dz^2}+P_1\frac{dy}{dz}+Q_1y=R_1 \qquad ...(2)$$

where $$P_1=\frac{P\left(\frac{dz}{dx}+\frac{d^2z}{dx^2}\right)}{\left(\frac{dz}{dx}\right)^2},\quad Q_1=\frac{Q}{\left(\frac{dz}{dx}\right)^2},\quad R_1=\frac{R}{\left(\frac{dz}{dx}\right)^2}$$

Equation (2) is solved either by taking $P_1=0$ or $Q_1=a$ constant.

Equation (2) can be solved by by two methods, by taking

**First Method,** $P_1=0$

**Second Method,** $Q=$ constant

**Working Rule**

**Step 1.** Coefficient of $\dfrac{d^2y}{dx^2}$ should be made as 1 if it is not so.

**Step 2.** To get $P$, $Q$ and $R$, compare the given differential equation with the standard form $y''+Py'+Qy=R$.

**Step 3.** Find $P_1$, $Q_1$ and $R_1$ by the following formulae.

$$P_1=\frac{\frac{d^2y}{dx^2}+p\frac{dz}{dx}}{\left(\frac{dz}{dx}\right)^2},\quad R_1=\frac{R}{\left(\frac{dz}{dx}\right)^2}$$

**Step 4.** Find out the value of $z$ by taking

**First Method,** $P_1=0$ **Second Method.** $Q_1=$ constant

**Step 5.** We get a reduced equation $\frac{d^2y}{dz^2} + P_1 \frac{dy}{dz} + Q_1 y = R_1$

On solving this equation we can find out the value of $y$ in terms of $z$.
Then write down the solution in terms of $x$ by replacing the value of $z$.

**Example 97.** *Solve* $\frac{d^2y}{dx^2} + \cot x \frac{dy}{dx} + 4\,y \operatorname{cosec}^2 x = 0$

**Solution.** We have, $\frac{d^2y}{dx^2} + \cot x \frac{dy}{dx} + 4\,y \operatorname{cosec}^2 x = 0$ ...(1)

Here, $P = \cot x$, $Q = 4 \operatorname{cosec}^2 x$ and $R = 0$

Changing the independent variable from $x$ to $z$, the equation becomes

$$\frac{d^2y}{dx^2} + P_1 \frac{dy}{dz} + Q_1 y = 0 \qquad \text{...(2)}$$

where $$P_1 = \frac{P\frac{dz}{dx} + \frac{d^2z}{dx^2}}{\left(\frac{dz}{dx}\right)^2}, \quad Q_1 = \frac{Q}{\left(\frac{dz}{dx}\right)^2}$$

**Case I.** Let us take $P_1 = 0$

$$\frac{p\frac{dz}{dx} + \frac{d^2z}{dx^2}}{\left(\frac{dz}{dx}\right)^2} = 0 \text{ or } P\frac{dz}{dx} + \frac{d^2z}{dx^2} = 0 \Rightarrow \frac{d^2z}{dx^2} + \cot x \frac{dz}{dx} = 0 \qquad \text{...(3)}$$

Put $$\frac{dz}{dx} = v, \quad \frac{d^2z}{dx^2} = \frac{dv}{dx}$$

(3) becomes $$\frac{dv}{dx} + (\cot x)\,v = 0 \Rightarrow \frac{dv}{v} = -\cot x.\,dx$$

$$\Rightarrow \quad \log v = -\log \sin x + \log c = \log c \log c \operatorname{cosec} x \Rightarrow v = c \operatorname{cosec} x$$

$$\frac{dz}{dx} = c \operatorname{cosec} x \Rightarrow dz = (c \operatorname{cosec} x)\,dx \Rightarrow z = c \log \tan \frac{x}{2}$$

$$Q_1 = \frac{Q}{\left(\frac{dz}{dx}\right)^2} = \frac{4 \operatorname{cosec}^2 x}{c^2 \operatorname{cosec}^2 x} = \frac{4}{c^2} \text{ which is constant}$$

Hence the equation (2) reduces to

$$\frac{d^2y}{dz^2} + 0\frac{dy}{dz} + \frac{4}{c^2}y = 0 \text{ or } \frac{d^2y}{dz^2} + \frac{4}{c^2}y = 0 \qquad \left[\because P_1 = 0, Q_1 = \frac{4}{c^2}\right]$$

$$\Rightarrow \quad \left(D^2 + \frac{4}{c^2}\right)y = 0, \qquad \text{A.E. is } m^2 + \frac{4}{c^2} = 0 \Rightarrow m = \pm i\frac{2}{c}$$

$$\text{C.F.} = c_1 \cos\frac{2z}{c} + c_2 \sin\frac{2z}{c} \qquad \left(z = c \log \tan\frac{x}{2}\right)$$

$$\Rightarrow \quad y = c_1 \cos\left(2 \log \tan\frac{x}{2}\right) + c_2 \sin\left(2 \log \tan\frac{x}{2}\right)$$ **Ans.**

**Example 98.** *Solve* $x^6 \dfrac{d^2y}{dx^2} + 3x^5 \dfrac{dy}{dx} + a^2 y = \dfrac{1}{x^2}$

**Solution.** We have, $\dfrac{d^2y}{dx^2} + \dfrac{3}{x}\dfrac{dy}{dx} + a^2 \dfrac{y}{x^6} = \dfrac{1}{x^8}$ ...(1)

Here $P = \dfrac{3}{x}$ and $Q = \dfrac{a^2}{x^6}$

On changing the independent variable $x$ to $z$, the equation (1) is reduced to

$$\frac{d^2y}{dz^2} + P_1 \frac{dy}{dz} + Q_1 y = R_1 \quad ...(2)$$

**Using Second Method**

Let $Q_1 = a_2$ (constant) $\quad Q_1 = \dfrac{Q}{\left(\dfrac{dz}{dx}\right)^2} = \dfrac{a^2}{x^6\left(\dfrac{dz}{dx}\right)^2}$ = constant = $a^2$ (say)

$$\therefore \quad x^6\left(\frac{dz}{dx}\right)^2 = 1 \Rightarrow x^3\left(\frac{dz}{dx}\right) = 1 \Rightarrow \frac{dz}{dx} = \frac{1}{x^3} \Rightarrow dz = \frac{dx}{x^3} \Rightarrow z = \frac{x^{-2}}{-2} + c$$

On differentiating twice, we have $\dfrac{d^2z}{dx^2} = \dfrac{-3}{x^4}$

$$P_1 = \frac{P\dfrac{dz}{dx} + \dfrac{d^2z}{dx^2}}{\left(\dfrac{dz}{dx}\right)^2} = \frac{\dfrac{3}{x}\cdot\dfrac{1}{x^3} + \left(\dfrac{-3}{x^4}\right)}{\left(\dfrac{1}{x^3}\right)^2} = 0 \Rightarrow R_1 = \frac{R}{\left(\dfrac{dz}{dx}\right)^2} = \frac{\dfrac{1}{x^8}}{\dfrac{1}{x^6}} = \frac{1}{x^2} = -2z$$

On putting the values of $P_1$, $Q_1$ and $R_1$ in (2), we get

$$\frac{d^2y}{dz^2} + a^2 y = -2z \quad \Rightarrow \quad (D^2 + a^2)\,y = -2z$$

A.E. is $m^2 + a^2 = 0$, $m = \pm i a$, $\Rightarrow$ C.F. $= c_1 \cos az + c_2 \sin az$

$$\text{P.I.} = \frac{1}{D^2 + a^2}(-2z) = \frac{1}{a^2}\left[1 + \frac{D^2}{a^2}\right]^{-1}(-2z) = \frac{1}{a^2}\left[1 - \frac{D^2}{a^2}\right](-2z) = \frac{-2z}{a^2} = \frac{1}{a^2x^2}$$

$$y = \text{C.F.} + \text{P.I.}$$

$$y = c_1 \cos\frac{a}{2x^2} - c_2 \sin\frac{a}{2x^2} + \frac{1}{a^2x^2}$$ **Ans.**

**Example 99.** *Solve* $\dfrac{d^2y}{dx^2} - \dfrac{1}{x}\dfrac{dy}{dx} + 4x^2 y = x^4$ *(U.P., I Semester Summer 2003, 2002)*

**Solution.** We have, $\dfrac{d^2y}{dx^2} - \dfrac{1}{x}\dfrac{dy}{dx} + 4x^2 y = x^4$ ...(1)

Hence $P = -\dfrac{1}{x}$, $Q = 4x^2$, $R = x^4$

On changing the independent variable $x$ to $z$, the equation (1) is transformed as

$$\frac{d^2y}{dz^2} + P_1 \frac{dy}{dz} + Q_1 y = R_1 \quad ...(2)$$

Let $Q_1 = 1$ (constant)

but $$Q_1 = \frac{Q}{\left(\frac{dz}{dx}\right)^2} = \frac{4x^2}{\left(\frac{dz}{dx}\right)^2} = \text{constant} = 4 \quad \text{(say)}$$

$$\left(\frac{dz}{dx}\right)^2 = x^2 \Rightarrow \frac{dz}{dx} = x$$

$$\Rightarrow \quad dz = x\,dx \Rightarrow z = \frac{x^2}{2} + c \Rightarrow x^2 = 2z - c \qquad ...(3)$$

$$P_1 = \frac{F\left(\frac{dz}{dx}\right) + \frac{d^2z}{dx^2}}{\left(\frac{dz}{dx}\right)^2} = \frac{-\frac{1}{x}\frac{dz}{dx} + \frac{d^2z}{dx^2}}{\left(\frac{dz}{dx}\right)^2} = \frac{-\frac{1}{x}(x) + 1}{(x)^2} = 0$$

$$R_1 = \frac{R}{\left(\frac{dz}{dx}\right)^2} = \frac{x^4}{x^2} = x^2 = 2(z - c) \qquad \text{[Using (3)]}$$

On putting the value of $P_1$, $Q_1$ and $R_1$ in (2), we get

$$\frac{d^2y}{dz^2} + (0)\frac{dy}{dz} + (4)y = 2(z-c) \Rightarrow \frac{d^2y}{dz^2} + 4y = 2(z-c) \Rightarrow (D^2 + 4)y = 2(z-c)$$

A.E. is $m^2 + 4 = 0 \quad \Rightarrow \quad m = \pm 2i$

C.F. $= c_1 \cos 2z + c_2 \sin 2z$

$$\text{P.I.} = \frac{1}{D^2 + 4} 2(z-c) = \frac{1}{4}\left(1 + \frac{D^2}{4}\right)^{-1} 2(z-c) = \frac{1}{2}\left(1 + \frac{D^2}{4}\right)(z-c) = \frac{z-c}{2}$$

Now complete solution = C.F. + P.I.

$$\Rightarrow \quad y = c_1 \cos 2z + c_2 \sin 2z + \frac{z-c}{2} \Rightarrow y = c_1 \cos x^2 + c_2 \sin x^2 + \frac{x^2}{4} \qquad \textbf{Ans.}$$

## EXERCISE 3.35

**Solve the following differential equations:**

**1.** $x^4 \frac{d^2y}{dx^2} + 2x^3 \frac{dy}{dx} + a^2 y = 0$ **Ans.** $y = c_1 \cos \frac{a}{x} + c_2 \sin \frac{a}{x}$

**2.** $\cos x \frac{d^2y}{dx^2} + \frac{dy}{dx} \sin x - 2y\cos^3 x = 2\cos^5 x$ **Ans.** $y = c_1 e^{\sqrt{2}\sin x} + c_2 e^{-\sqrt{2}\sin x} + \sin^2 x$

**3.** $\frac{d^2y}{dx^2} + \tan x \frac{dy}{dx} + y\cos^2 x = 0$ **Ans.** $y = c_1 \cos(\sin x) + c_2 \sin(\sin x)$

**4.** $x\frac{d^2y}{dx^2} - \frac{dy}{dx} - 4x^3 y = 8x^2 \sin x^2$ **Ans.** $y = c_1 e^{x^2} + c_2 e^{\cos x} + \frac{1}{6}e^{-\cos x}$

**5.** $\frac{d^2y}{dx^2} + (3\sin x - \cot x)\frac{dy}{dx} + 2y\sin^2 x = e^{-\cos x}\sin^2 x$ **Ans.** $y = c_1 e^{2\cos x} + c_2 e^{\cos x} + \frac{1}{6}e^{-\cos x}$

**6.** $\frac{d^2y}{dx^2} + (\tan x - 1)^2 \frac{dy}{dx} - n(n-1)y\sec^4 x = 0$ **Ans.** $y = C_1 e^{-n\tan x} + C_2 e^{(n-1)\tan x}$

**7.** $\frac{d^2y}{dx^2} - \cot x \frac{dy}{dx} - y\sin^2 x = \cos x - \cos^3 x$ **Ans.** $y = C_1 e^{-\cos x} + C_2 e^{\cos x} - \cos x$

## 3.39 Applications of Differential Equations of Second Order

**Example 100.** *The differential equation satisfied by a beam uniformly loaded (W kg/metre), with one end fixed and the second end subjected to tensile force P, is given by*

$$E.I.\frac{d^2y}{dx^2} = Py - \frac{1}{2}Wx^2$$

*Show that the elastic curve for the beam with conditions*

$$y = 0 = \frac{dy}{dx} \text{ at } x = 0, \textit{ is given by}$$

$$y = \frac{W}{Pn^2}(1 - \cosh nx) + \frac{Wx^2}{2P} \quad \textit{where } n^2 = \frac{P}{EI}$$

**Solution.** We have, $E.I.\dfrac{d^2y}{dx^2} = Py - \dfrac{1}{2}W.x^2$ ...(1)

$$\Rightarrow \quad \frac{d^2y}{dx^2} - \frac{P}{E.I.}y = -\frac{W}{2\,E.I.}x^2 \quad \Rightarrow \quad \left(D^2 - \frac{P}{E.I.}\right)y = -\frac{W}{2\,E.I.}x^2$$

A.E. is $m^2 - \dfrac{P}{E.I.} = 0 \Rightarrow m^2 = \dfrac{P}{E.I.} = n^2 \Rightarrow m = \pm n$ $\left(n^2 = \dfrac{P}{EI}\right)$

C.F. $= c_1 e^{nx} + c_2 e^{-nx}$

$$\text{P.I.} = \frac{1}{D^2 - \dfrac{P}{E.I.}}\left(-\frac{W}{2\,E.I.}\right)x^2 = -\frac{W}{2\,E.I.}\frac{1}{D^2 - n^2}.x^2$$

$$= \frac{W}{2n^2.\,E.I.}\left(1 - \frac{D^2}{n^2}\right)^{-1}.x^2 = \frac{W}{2n^2\,E.I.}\left(1 + \frac{D^2}{n^2}\right).x^2 = \frac{W}{2n^2\,E.I.}\left(x^2 + \frac{2}{n^2}\right)$$

$$\therefore \quad y = c_1 e^{nx} + c_2 e^{-nx} + \frac{W}{2n^2\,E.I.}\left(x^2 + \frac{2}{n^2}\right) \quad ...(2)$$

Differentiating (2) w.r.t. $x$, we get

$$\frac{dy}{dx} = n c_1 e^{nx} - n c_2 e^{-nx} + \frac{W}{2n^2\,E.I.}(2x) \quad ...(3)$$

Putting $x = 0$, $\dfrac{dy}{dx} = 0$ in (3), we get

$$0 = n c_1 - n c_2 \quad \Rightarrow \quad c_1 = c_2$$

Putting $x = 0$, $y = 0$ in (2), we get

$$0 = c_1 + c_2 + \frac{W}{2n^2\,E.I.}\frac{2}{n^2} \quad \Rightarrow \quad 0 = c_1 + c_2 + \frac{W}{n^4\,E.I.} \quad ...(4)$$

Putting $c_1 = c_2$ in (4), we get $0 = 2c_1 + \dfrac{W}{n^4\,E.I.} \Rightarrow c_1 = -\dfrac{W}{2n^4\,E.I.}$,

Now, $$n^2 = \frac{P}{E.I.} \quad \Rightarrow \quad n^2\,E.I. = P$$

$$\Rightarrow \quad c_1 = c_2 = -\frac{W}{2n^2 P}$$

Putting the values of $c_1$ and $c_2$ in (2), we get

$$y = \frac{-W}{2n^2 P}(e^{nx} + e^{-nx}) + \frac{W}{2P}\left(x^2 + \frac{2}{n^2}\right)$$

$$y = \frac{-W}{n^2 P}\cosh nx + \frac{W}{2P}x^2 + \frac{W}{Pn^2} \Rightarrow y = \frac{W}{P.n^2}(1-\cosh nx) + \frac{W x^2}{2P}$$ **Ans.**

## EXERCISE 3.36

1. A beam of length $l$ and of uniform cross-section has the differential equation of its elastic curve as

$$E.I.\frac{d^2y}{dx^2} = \frac{w}{2}\left(\frac{l^2}{4} - x^2\right)$$

where $E$ is the modulus of elasticity, $I$ is the moment of inertia of the cross-section, $w$ is weight per unit length and $x$ is measured from the centre of span.

If at $x = 0, \frac{dy}{dx} = 0$. Prove that the equation of the elastic curve is

$$y = \frac{1}{2}\cdot\frac{2}{E.I.}\left(\frac{l^3.x^2}{8} - \frac{x^4}{12}\right) + \frac{5w.l^4}{384\,E.I.}$$

2. A horizontal tie rod of length $l$ is freely pinned at each end. It carries a uniform load $w$ kg per unit length and has a horizontal pull $P$. Find the central deflection and the maximum bending moment, taking the origin at one of its ends. **Ans.** $\frac{w}{a}\left(\sec h\frac{al}{2} - 1\right)$ where $a^2 = \frac{P}{EI}$

3. A light horizontal strut $AB$ is freely pinned at $A$ and $B$. It is under the action of equal and opposite compressive forces $P$ at its ends and it carries a load $W$ at its centre. Then for $0 < x < \frac{l}{2}$,

$$EI\frac{d^2y}{dx^2} + Py + \frac{1}{2}Wx = 0$$

Also $y = 0$ at $x = 0$ and $\frac{dy}{dx} = 0$ at $x = \frac{l}{2}$. Prove that $y = \frac{W}{2P}\left(\frac{\sin ax}{a\cos\frac{al}{2}} - x\right)$, where $a^2 = \frac{P}{EI}$

4. A horizontal tie-rod of length $2l$ with concentrated load $W$ at its centre and ends freely hinged satisfies the differential equation $EI\frac{d^2y}{dx^2} = Py - \frac{W}{2}x$. With conditions $x = 0$, $y = 0$ and $x = l, \frac{dy}{dx} = 0$. Prove that the deflection $\delta$ and bending moment $M$ at the centre ($x = l$) are given by $\delta = \frac{W}{2Pn}(nl - \tan nl)$ and $M = -\frac{W}{2n}\tan h\,nl$, where $n^2\,EI = P$.

**Example 101.** *The voltage V and the current i at a distance x from the sending end of the transmission line satisfy the equations.*

$$-\frac{dV}{dx} = Ri, \qquad -\frac{di}{dx} = GV$$

*where R and G are constants. If $V = V_0$ at the sending end (x = 0) and V = 0 at receiving end (x = l). Show that*

$$V = V_0\left\{\frac{\sinh n(l-x)}{\sinh nl}\right\}$$

When $n^2 = RG$

**Solution.** $$-\frac{dV}{dx} = Ri \quad \ldots(1)$$

$$-\frac{di}{dx} = GV \quad \ldots(2)$$

When $x = 0$, $V = V_0$
When $x = l$ $V = 0$

Putting the value of $i$ from (1) into (2), we get

$$-\frac{d}{dx}\left(-\frac{dV}{dx}\frac{1}{R}\right)=GV \quad\Rightarrow\quad \frac{d^2V}{dx^2}=RGV$$

$$\Rightarrow\quad \frac{d^2V}{dx^2}-(RG)V=0 \quad\Rightarrow\quad (D^2-RG)V=0 \qquad (RG=n^2)$$

A. E. is $m^2 - n^2 = 0$, $m = \pm n$

$\therefore$ $V = Ae^{nx} + Be^{-nx}$ ... (3)

Now, we have to find out the value of $A$ and $B$ with the help of given conditions.

On putting $x = 0$ and $V = V_0$ in (3), we get $V_0 = A + B$ ... (4)

On putting $x = 1$ and $V = 0$ in (3), we get $0 = Ae^{nl} + Be^{-nl}$

On solving (4) and (5), we have $A=\frac{V_0}{1-e^{2nl}}$, $B=\frac{-V_0e^{2nl}}{1-e^{2nl}}$

Substituting the values of $A$ and $B$ in (3), we have

$$V=\frac{V_0e^{nx}}{1-e^{2nl}}-\frac{V_0e^{2nl}e^{-nx}}{1-e^{2nl}}=\frac{V_0[e^{nx}-e^{2nl-nx}]}{1-e^{2nl}}=\frac{V_0[e^{(nl-nx)}-e^{-(nl-nx)}]}{e^{nl}-e^{-nl}}=V_0\left\{\frac{\sinh n(l-x)}{\sinh nl}\right\}$$ **Proved.**

## EXERCISE 3.37

1. An *e.m.f.* $E \sin pt$ is applied at $t = 0$ to a circuit containing a condenser C and inductance $L$ in series. The current $x$ satisfies the equation

$$L\frac{dx}{dt}+\frac{1}{C}\int x\,dt=E\sin pt$$

If $p^2=\frac{1}{LC}$ and initially the current $x$ and the charge $q$ are zero, show that the current in the circuit at time $t$ is given by $\frac{E}{2l}\,t\sin pt$, where $x=-\frac{dq}{dt}$.

**Fill in the blanks:**

2. The integrating factor of $\cos^2 x\frac{dy}{dx}+y=\tan x$ is ..............
3. The integrating factor of $x(x-1)\frac{dy}{dx}-(x-2)y=x^2(2x-1)$ is ............
4. Solution of $(x+y+1)\frac{dy}{dx}=1$ is ...........
5. $Mdx + Ndy = 0$ is an exact differential equation if .......
6. P. I. of $(D^2 + 4)y = \sin 3x$ is ..........
7. P. I. of $(D^2 - 2D + 1)y = e^x$ is .........
8. On putting $x = e^z$, the transformed differential equation of $x^2\frac{d^2y}{dx^2}+\frac{xdy}{dx}+y=x$ is ............
9. If the C.F. of $ay' + by' + cy + X$ is $Ay_1 + By_2$, then P. I. $= uy_1 + vy_2$ where $u =$ ....... and $v=$ .......

**Ans. 2.** $e^{\tan x}$ **3.** $\frac{x-1}{x^2}$ **4.** $x + y + 2 = ce^y$ **5.** $\frac{\partial M}{dy}=\frac{\partial N}{\partial x}$

# 4

# Determinants and Matrices

## 4.1 INTRODUCTION

In Engineering Mathematics, solution of simultaneous equations is very important. In this chapter we shall study the system of linear equations with emphasis on their solution by means of determinants.

## 4.2 DETERMINANT

The notation of determinants arises from the process of elimination of the unknowns of simultaneous linear equations.

Consider the two linear equations in $x$,

$$a_1 x + b_1 = 0 \qquad \text{... (1)}$$

$$a_2 x + b_2 = 0 \qquad \text{... (2)}$$

From (1) $\qquad x = -\dfrac{b_1}{a_1}$

Substituting the value of $x$ in (2); we get the eliminant

$$a_2\left(-\frac{b_1}{a_1}\right) + b_2 = 0$$

or $\qquad a_1 b_2 - a_2 b_1 = 0 \qquad$ ... (3)

From (1) and (2) by suppressing $x$, the eliminant is written as

$$\begin{vmatrix} a_1 & b_1 \\ a_2 & b_2 \end{vmatrix} = 0 \qquad \text{... (4)}$$

when the two **rows** of $a_1$, $b_1$ and $a_2$, $b_2$ are enclosed by two vertical bars then it is called a **determinant of second order.**

$$\begin{vmatrix} a_1 \\ \\ a_2 \end{vmatrix} \text{ and } \begin{vmatrix} b_1 \\ \\ b_2 \end{vmatrix}$$

Column 1      Column 2

Row 1 → $a_1 \;\; b_1$

Row 2 → $a_2 \;\; b_2$

Each quantity $a_1$, $b_1$, $a_2$, $b_2$ is called an **element** or a constituent of the determinant.

From (3) and (4), we know that both expressions are eliminant, so we equate them.

$$\begin{vmatrix} a_1 & b_1 \\ a_2 & b_2 \end{vmatrix} = a_1 b_2 - a_2 b_1 \qquad \text{or} \qquad \begin{vmatrix} a_1 & b_1 \\ a_2 & b_2 \end{vmatrix} = a_1 b_2 - a_2 b_1$$

$a_1b_2 - a_2b_1$ is called the expansion of the determinant of $\begin{vmatrix} a_1 & b_1 \\ a_2 & b_2 \end{vmatrix}$.

**Example 1.** *Expand the determinant* $\begin{vmatrix} 3 & 2 \\ 6 & 7 \end{vmatrix}$.

**Solution.** $\begin{vmatrix} \overset{+}{3} & \overset{-}{2} \\ 6 & 7 \end{vmatrix} = (3) \times (7) - (2) \times (6) = 21 - 12 = 9.$ **Ans.**

### EXERCISE 4.1

**Expand the following determinants :**

**1.** $\begin{vmatrix} 4 & 6 \\ 2 & 5 \end{vmatrix}$ **Ans.** 8 **2.** $\begin{vmatrix} -3 & 7 \\ 2 & 4 \end{vmatrix}$ **Ans.** $-26$

**3.** $\begin{vmatrix} 8 & 5 \\ 3 & 1 \end{vmatrix}$ **Ans.** $-7$ **4.** $\begin{vmatrix} 5 & -2 \\ 4 & 3 \end{vmatrix}$ **Ans.** 23

## 4.3. DETERMINANT AS ELIMINANT

Consider the following three equations having three unknowns, $x$, $y$ and $z$.

$$a_1 x + b_1 y + c_1 z = 0 \quad \text{...(1)}$$
$$a_2 x + b_2 y + c_2 z = 0 \quad \text{...(2)}$$
$$a_3 x + b_3 y + c_3 z = 0 \quad \text{...(3)}$$

From (2) and (3) by cross-multiplication, we get

$$\frac{x}{b_2 c_3 - b_3 c_2} = \frac{y}{a_3 c_2 - a_2 c_3} = \frac{z}{a_2 b_3 - a_3 b_2} = k \text{ (say)}$$

$$x = (b_2 c_3 - b_3 c_2)\, k$$
$$y = (a_3 c_2 - a_2 c_3)\, k$$

and $$z = (a_2 b_3 - a_3 b_2)\, k$$

Substituting the values of $x$, $y$ and $z$ in (1), we get the eliminant

$$a_1 (b_2c_3 - b_3c_2)\, k + b_1 (a_3c_2 - a_2c_3)\, k + c_1 (a_2b_3 - a_3b_2)\, k = 0$$

or $$a_1 (b_2c_3 - b_3c_2) - b_1 (a_2c_3 - a_3c_2) + c_1 (a_2b_3 - a_3b_2) = 0 \quad \text{...(4)}$$

From (1), (2) and (3) by suppressing $x, y, z$ the remaining can be written in the determinant as

$$\begin{vmatrix} a_1 & b_1 & c_1 \\ a_2 & b_2 & c_2 \\ a_3 & b_3 & c_3 \end{vmatrix} = 0 \quad \text{...(5)}$$

This is determinant of third order.

As (4) and (5) both are the eliminant of the same equations.

$$\therefore \quad \begin{vmatrix} a_1 & b_1 & c_1 \\ a_2 & b_2 & c_2 \\ a_3 & b_3 & c_3 \end{vmatrix} = a_1 (b_2c_3 - b_3c_2) - b_1 (a_2c_3 - a_3c_2) + c_1 (a_2b_3 - a_3b_2) = 0$$

or
$$\begin{vmatrix} a_1 & b_1 & c_1 \\ a_2 & b_2 & c_2 \\ a_3 & b_3 & c_3 \end{vmatrix} = a_1\begin{vmatrix} b_2 & c_2 \\ b_3 & c_3 \end{vmatrix} - b_1\begin{vmatrix} a_2 & c_2 \\ a_3 & c_3 \end{vmatrix} + c_1\begin{vmatrix} a_2 & b_2 \\ a_3 & b_3 \end{vmatrix}$$

## 4.4. MINOR

**The minor of an element is defined as a determinant obtained by deleting the row and column containing the element.**

**Thus the minors $a_1$, $b_1$ and $c_1$ are respectively.**

$$\begin{vmatrix} b_2 & c_2 \\ b_3 & c_3 \end{vmatrix}, \quad \begin{vmatrix} a_2 & c_2 \\ a_3 & c_3 \end{vmatrix} \text{ and } \begin{vmatrix} a_2 & b_2 \\ a_3 & b_3 \end{vmatrix}$$

Thus

$$\begin{vmatrix} a_1 & b_1 & c_1 \\ a_2 & b_2 & c_2 \\ a_3 & b_3 & c_3 \end{vmatrix} = a_1 \text{ (minor of } a_1) - b_1 \text{ (minor of } b_1) + c_1 \text{ (minor of } c_1).$$

## 4.5. COFACTOR

Cofactor = $(-1)^{r+c}$ Minor

where $r$ is the number of rows of the element and $c$ is the number of columns of the element.

The cofactor of any element of $j$th row and $i$th column is

$$(-1)^{i+j} \text{ minor}$$

Thus the cofactor of $a_1$ $= (-1)^{1+1}(b_2c_3 - b_3c_2) = +(b_2c_3 - b_3c_2)$

The cofactor of $b_1$ $= (-1)^{1+2}(a_2c_3 - a_3c_2) = -(a_2c_3 - a_3c_2)$

The cofactor of $c_1$ $= (-1)^{1+3}(a_2b_3 - a_3b_2) = +(a_2b_3 - a_3b_2)$

The determinant $= a_1$ (cofactor of $a_1$) + $a_2$ (cofactor of $a_2$) + $a_3$ (cofactor of $a_3$).

**Example 2.** *Write down the minors and cofactors of each element and also evaluate the determinant.*

$$\begin{vmatrix} 1 & 3 & -2 \\ 4 & -5 & 6 \\ 3 & 5 & 2 \end{vmatrix}$$

**Solution.** $M_{11}$ = Minor of element (1) $= \begin{vmatrix} 1 & 3 & -2 \\ 4 & -5 & 6 \\ 3 & 5 & 2 \end{vmatrix}$

$$= \begin{vmatrix} -5 & 6 \\ 5 & 2 \end{vmatrix} = (-5) \times 2 - 6 \times 5 = -10 - 30 = -40$$

Cofactor of element (1) $= A_{11} = (-1)^{1+1} M_{11} = (-1)^2(-40) = -40$

$M_{12}$ = Minor of element (3)

$$= \begin{vmatrix} 1 & 3 & -2 \\ 4 & -5 & 6 \\ 3 & 5 & 2 \end{vmatrix} = \begin{vmatrix} 4 & 6 \\ 3 & 2 \end{vmatrix} = 4 \times 2 - 3 \times 6 = 8 - 18 = -10$$

$\Rightarrow$ Cofactor of element (–2) $= A_{12} = (-1)^{1+2}(-10) = 10$

$M_{13}$ = Minor of element (– 2)

$$= \begin{vmatrix} 1 & 3 & -2 \\ 4 & -5 & 6 \\ 3 & 5 & 2 \end{vmatrix} = \begin{vmatrix} 4 & -5 \\ 3 & 5 \end{vmatrix} = 4\times5-(-5)\times3 = 20+15 = 35$$

$\Rightarrow$ Cofactor of element $(-2) = A_{13} = (-1)^{1+3} M_{13} = (-1)^4\, 35 = 35$

$M_{21}$ = Minor of element (4)

$$= \begin{vmatrix} 1 & 3 & -2 \\ 4 & -5 & 6 \\ 3 & 5 & 2 \end{vmatrix} = \begin{vmatrix} 3 & -2 \\ 5 & 2 \end{vmatrix} = 3\times2-(-2)\times5 = 6+10 = 16$$

$\Rightarrow$ Cofactor of element $(4) = A_{21} = (-1)^{2+1} M_{21} = (-1)^{2+1}(16) = -16$

$M_{22}$ = Minor of element $(-5)$

$$= \begin{vmatrix} 1 & 3 & -2 \\ 4 & -5 & 6 \\ 3 & 5 & 2 \end{vmatrix} = \begin{vmatrix} 1 & -2 \\ 3 & 2 \end{vmatrix} = 1\times2-(-2)\times3 = 2+6 = 8$$

$\Rightarrow$ Cofactor of element $(-5) = A_{22} = (-1)^{2+2} M_{22} = (-1)^{2+2}(8) = 8$

$M_{23}$ = Minor of element (6)

$$= \begin{vmatrix} 1 & 3 & -2 \\ 4 & -5 & 6 \\ 3 & 5 & 2 \end{vmatrix} = \begin{vmatrix} 1 & 3 \\ 3 & 5 \end{vmatrix} = 1\times5-3\times3 = 5-9 = -4$$

$\Rightarrow$ Cofactor of element $(6) = A_{23} = (-1)^{2+3} M_{23} = (-1)^{2+3}(-4) = 4$

$M_{31}$ = Minor of element (3)

$$= \begin{vmatrix} 1 & 3 & -2 \\ 4 & -5 & 6 \\ 3 & 5 & 2 \end{vmatrix} = \begin{vmatrix} 3 & -2 \\ -5 & 6 \end{vmatrix} = 3\times6-(-2)\times(-5) = 18-10 = 8$$

$\Rightarrow$ Cofactor of element $(3) = A_{31} = (-1)^{3+1} M_{31} = (-1)^{3+1}\, 8 = 8$

$M_{32}$ = Minor of element (5)

$$= \begin{vmatrix} 1 & 3 & -2 \\ 4 & -5 & 6 \\ 3 & 5 & 2 \end{vmatrix} = \begin{vmatrix} 1 & -2 \\ 4 & 6 \end{vmatrix} = 1\times6-(-2)\times4 = 6+8 = 14$$

$\Rightarrow$ Cofactor of element $(5) = A_{32} = (-1)^{3+2} M_{32} = (-1)^{3+2}\, 14 = -14$

$M_{33}$ = Minor of element (2)

$$= \begin{vmatrix} 1 & 3 & -2 \\ 4 & -5 & 6 \\ 3 & 5 & 2 \end{vmatrix} = \begin{vmatrix} 1 & 3 \\ 4 & -5 \end{vmatrix} = 1\times(-5)-4\times3 = -5-12 = -17$$

Cofactor of element $(2) = A_{33} = (-1)^{3+3} M_{33} = (-1)^{3+3}(-17) = -17.$

$$\begin{vmatrix} 1 & 3 & -2 \\ 4 & -5 & 6 \\ 3 & 5 & 2 \end{vmatrix} = 1\times(\text{cofactor of } 1) + 3\times(\text{cofactor of } 3) + (-2)\times[\text{cofactor of } (-2)].$$

$$= 1\times(-40) + 3\times(10) + (-2)\times(35)$$

$$= -40 + 30 - 70$$

$$= -80$$

**Ans.**

**Example 3.** *Find :*
(*i*) *Minors* (*ii*) *Cofactors of the elements of the first row of the determinant*

$$\begin{vmatrix} 2 & 3 & 5 \\ 4 & 1 & 0 \\ 6 & 2 & 7 \end{vmatrix}$$

**Solution.**
(*i*) The minor of the element (2) is

$$\begin{vmatrix} 2 & 3 & 5 \\ 4 & 1 & 0 \\ 6 & 2 & 7 \end{vmatrix} = \begin{vmatrix} 1 & 0 \\ 2 & 7 \end{vmatrix} = (1)\times(7)-(0)\times(2) = 7-0 = 7$$

The minor of the element (3) is

$$\begin{vmatrix} 2 & 3 & 5 \\ 4 & 1 & 0 \\ 6 & 2 & 7 \end{vmatrix} = \begin{vmatrix} 4 & 0 \\ 6 & 7 \end{vmatrix} = (4)\times(7)-(0)\times(6) = 28-0 = 28$$

The minor of the element (5) is

$$\begin{vmatrix} 2 & 3 & 5 \\ 4 & 1 & 0 \\ 6 & 2 & 7 \end{vmatrix} = \begin{vmatrix} 4 & 1 \\ 6 & 2 \end{vmatrix} = (4)\times(2)-(1)\times(6) = 8-6 = 2$$

(*ii*) The cofactor of (2) $= (-1)^{1+1}\,(7) = +7$
The cofactor of (3) $= (-1)^{1+2}\,(28) = -28$ **Ans.**
The cofactor of (5) $= (-1)^{1+3}\,(2) = +2$.

**Example 4.** *Expand the determinant* $\begin{vmatrix} 6 & 2 & 3 \\ 2 & 3 & 5 \\ 4 & 2 & 1 \end{vmatrix}$

**Solution.** $\begin{vmatrix} 6 & 2 & 3 \\ 2 & 3 & 5 \\ 4 & 2 & 1 \end{vmatrix} = 6\,(\text{cofactor of } 6) + 2\,(\text{cofactor of } 2) + 3\,(\text{cofactor of } 3).$

$$= 6\,(3\times1-5\times2)-2\,(2\times1-4\times5)+3\,(2\times2-3\times4)$$
$$= 6\,(3-10)-2\,(2-20)+3\,(4-12)$$
$$= 6\,(-7)-2\,(-18)+3\,(-8)$$
$$= -42+36-24$$
$$= -30.$$

**Ans.**

**Example 5.** *Evaluate the determinant* $\begin{vmatrix} 1 & 0 & 4 \\ 3 & 5 & -1 \\ 0 & 1 & 2 \end{vmatrix}$.

(*i*) *With the help of second row,* (*ii*) *with the help of third column.*

**Solution.**

(*i*) $\begin{vmatrix} 1 & 0 & 4 \\ 3 & 5 & -1 \\ 0 & 1 & 2 \end{vmatrix} = 3\times(\text{cofactor of } 3) + 5\times(\text{cofactor of } 5) + (-1)\,(\text{cofactor of } -1).$

$$= 3 \times (-1)^{2+1} \begin{vmatrix} 0 & 4 \\ 1 & 2 \end{vmatrix} + 5 \times (-1)^{2+2} \begin{vmatrix} 1 & 4 \\ 0 & 2 \end{vmatrix} + (-1) \times (-1)^{2+3} \begin{vmatrix} 1 & 0 \\ 0 & 1 \end{vmatrix}$$

$$= -3 \times (0-4) + 5\,(2-0) + (1-0)$$

$$= 12 + 10 + 1 = 23$$ **Ans.**

(*ii*) $\begin{vmatrix} 1 & 0 & 4 \\ 3 & 5 & -1 \\ 0 & 1 & 2 \end{vmatrix} = 4 \times (\text{cofactor of } 4) + (-1)\,(\text{cofactor of } (-1)) + 2 \times (\text{cofactor of } 2)$

$$= 4 \times (-1)^{1+3} \begin{vmatrix} 3 & 5 \\ 0 & 1 \end{vmatrix} + (-1)\,(-1)^{2+3} \begin{vmatrix} 1 & 0 \\ 0 & 1 \end{vmatrix} + 2 \times (-1)^{3+3} \begin{vmatrix} 1 & 0 \\ 3 & 5 \end{vmatrix}$$

$$= 4 \times (3-0) + (1-0) + 2\,(5-0)$$

$$= 12 + 1 + 10 = 23$$ **Ans.**

**Example 6.** *Expand the fourth order determinant* $\begin{vmatrix} 0 & 1 & 2 & 3 \\ 1 & 0 & 2 & 0 \\ 2 & 0 & 1 & 3 \\ 1 & 2 & 1 & 0 \end{vmatrix}$

**Solution.** Given determinant $= (0)\,(-1)^{1+1} \begin{vmatrix} 0 & 2 & 0 \\ 0 & 1 & 3 \\ 2 & 1 & 0 \end{vmatrix} + 1\,(-1)^{1+2} \begin{vmatrix} 1 & 2 & 0 \\ 2 & 1 & 3 \\ 1 & 1 & 0 \end{vmatrix}$

$$+ 2\,(-1)^{1+3} \begin{vmatrix} 1 & 0 & 0 \\ 2 & 0 & 3 \\ 1 & 2 & 0 \end{vmatrix} + 3\,(-1)^{1+4} \begin{vmatrix} 1 & 0 & 2 \\ 2 & 0 & 1 \\ 1 & 2 & 1 \end{vmatrix}$$

$$= 0 - \begin{vmatrix} 1 & 2 & 0 \\ 2 & 1 & 3 \\ 1 & 1 & 0 \end{vmatrix} + 2 \begin{vmatrix} 1 & 0 & 0 \\ 2 & 0 & 3 \\ 1 & 2 & 0 \end{vmatrix} - 3 \begin{vmatrix} 1 & 0 & 2 \\ 2 & 0 & 1 \\ 1 & 2 & 1 \end{vmatrix}$$

Now $\begin{vmatrix} 1 & 2 & 0 \\ 2 & 1 & 3 \\ 1 & 1 & 0 \end{vmatrix} = 1(1 \times 0 - 3 \times 1) - 2\,(2 \times 0 - 3 \times 1) + 0(2 \times 1 - 1 \times 1)$
$= -3 + 6 + 0 = 3$

$\begin{vmatrix} 1 & 0 & 0 \\ 2 & 0 & 3 \\ 1 & 2 & 0 \end{vmatrix} = 1\,(0 \times 0 - 3 \times 2) - 0\,(2 \times 0 - 3 \times 1) + 0(2 \times 2 - 0 \times 1)$
$= -6$

$\begin{vmatrix} 1 & 0 & 2 \\ 2 & 0 & 1 \\ 1 & 2 & 1 \end{vmatrix} = 1\,(0 \times 1 - 1 \times 2) - 0(2 \times 1 - 1 \times 1) + 2(2 \times 2 - 0 \times 1)$
$= -2 - 0 + 8 = 6$

Now $\begin{vmatrix} 0 & 1 & 2 & 3 \\ 1 & 0 & 2 & 0 \\ 2 & 0 & 1 & 3 \\ 1 & 2 & 1 & 0 \end{vmatrix} = -3 + 2(-6) - 3(6)$
$= -3 - 12 - 18 = -33$ **Ans.**

## EXERCISE 4.2

***Write the minors and co factors of each element of the following determinants and also evalutate the determinant in each case :***

**1.** $\begin{vmatrix} -2 & 3 \\ 4 & -9 \end{vmatrix}$ $M_{11} = -9,\ M_{12} = 4,\ M_{21} = 3,\ M_{22} = -2$
$A_{11} = -9,\ A_{12} = -4,\ A_{21} = -3,\ A_{22} = -2 \quad |A| = 6$ **Ans.**

**2.** $\begin{vmatrix} \cos\theta & -\sin\theta \\ \sin\theta & \cos\theta \end{vmatrix}$ $M_{11} = \cos\theta,\ M_{12} = \sin\theta,\ M_{21} = -\sin\theta,\ M_{22} = \cos\theta$
$A_{11} = \cos\theta, A_{12} = -\sin\theta, A_{21} = \sin\theta,\ A_{22} = \cos\theta, \quad |A| = 1$ **Ans.**

**3.** $\begin{vmatrix} 42 & 1 & 6 \\ 28 & 7 & 4 \\ 14 & 3 & 2 \end{vmatrix}$ $M_{11} = 2,\ M_{12} = 0,\ M_{13} = -14,\ M_{21} = -16,\ M_{22} = 0$
$M_{23} = 112,\ M_{31} = -38, M_{32} = 0,\ M_{33} = 266$
$A_{11} = 2,\ A_{12} = 0,\ A_{13} = -14,\ A_{21} = 16,\ A_{22} = 0$.
$A_{23} = -112,\ A_{31} = -38,\ A_{32} = 0,\ A_{33} = 266, \quad |A| = 0$ **Ans.**

**4.** $\begin{vmatrix} 1 & a & bc \\ 1 & b & ca \\ 1 & c & ab \end{vmatrix}$
$M_{11} = (ab^2 - ac^2),\ M_{12} = (ab - ac),\ M_{13} = (c - b),\ M_{21} = a^2b - bc^2$
$M_{22} = (ab - bc),\ M_{23} = (c - a),\ M_{31} = (ca^2 - cb^2),\ M_{32} = ca - bc, M_{33} = (b - a)$,
$A_{11} = (ab^2 - ac^2),\ A_{12} = (ac - ab), A_{13} = (c - b), A_{21} = bc^2 - a^2b$
$A_{22} = (ab - bc),\ A_{23} = (a - c), A_{31} = (ca^2 - cb^2),\ A_{32} = (bc - ca), A_{33} = (b - a)$
$|A| = (a - b)(b - c)(c - a)$. **Ans.**

***Expand the following determinants :***

**5.** $\begin{vmatrix} 2 & -3 & 4 \\ 5 & 1 & -6 \\ -7 & 8 & -9 \end{vmatrix}$ **Ans.** $|A| = 5$

**6.** $\begin{vmatrix} 5 & 0 & 7 \\ 8 & -6 & -4 \\ 2 & 3 & 9 \end{vmatrix}$ **Ans.** $|A| = 42$

**7.** $\begin{vmatrix} a & h & g \\ h & b & f \\ g & f & c \end{vmatrix}$ **Ans.** $|A| = abc + 2fgh - af^2 - bg^2 - ch^2$

***Expand the following determinants by two methods :***

(*i*) along the-third row.

(*ii*) along the-third column.

**8.** $\begin{vmatrix} 1 & -3 & 2 \\ 4 & -1 & 2 \\ 3 & 5 & 2 \end{vmatrix}$ **Ans.** $|A| = 40$

**9.** $\begin{vmatrix} 3 & -2 & 4 \\ 1 & 2 & 1 \\ 0 & 1 & -1 \end{vmatrix}$ **Ans.** $|A| = -7$

**10.** $\begin{vmatrix} 2 & 3 & -2 \\ 1 & 2 & 3 \\ -2 & 1 & -3 \end{vmatrix}$ **Ans.** $|A| = -37$

**11.** $\begin{vmatrix} \log_3 512 & \log_4 3 \\ \log_3 8 & \log_4 9 \end{vmatrix}$ **Ans.** $|A| = \dfrac{15}{2}$

**12.** If $a, b, c$ are all positive and are the $p$th, $q$th, $r$th terms of a *G.P.* respectively; then prove that

$$\begin{vmatrix} \log a & p & 1 \\ \log b & q & 1 \\ \log c & r & 1 \end{vmatrix} = 0$$

**13.** $\begin{vmatrix} 3 & 2 & 5 & 7 \\ -1 & -4 & -3 & 0 \\ 6 & 4 & 2 & -1 \\ 2 & -1 & 0 & 3 \end{vmatrix}$ **Ans.** 96

## 4.6 RULES OF SARRUS (For third order determinants only).

After writing the determinant, repeat the first two columns as below

$$\begin{vmatrix} a_1 & b_1 & c_1 \\ a_2 & b_2 & c_2 \\ a_3 & b_3 & c_3 \end{vmatrix} = \begin{array}{|ccc|cc} + & + & + & - & - \quad - \\ a_1 & b_1 & c_1 & a_1 & b_1 \\ a_2 & b_2 & c_2 & a_2 & b_2 \\ a_3 & b_3 & c_3 & a_3 & b_3 \end{array}$$

$$= (a_1b_2c_3 + b_1c_2a_3 + c_1a_2b_3) + (-\,c_1b_2a_3 - a_1c_2b_3 - b_1a_2c_3)$$

**Example 7.** *Expand the determinant*

$$\Delta = \begin{vmatrix} 2 & 3 & 4 \\ 1 & 5 & 3 \\ 3 & 0 & 5 \end{vmatrix} \text{ by Rule of Sarrus.}$$

**Solution.**

$$\Delta = \begin{array}{|ccc|cc} + & + & + & - & - \quad - \\ 2 & 3 & 4 & 2 & 3 \\ 1 & 5 & 3 & 1 & 5 \\ 3 & 0 & 5 & 3 & 0 \end{array}$$

$=(2) \times (5) \times (5) + (3) \times (3) \times (3) + (4) \times (1) \times (0) - (4) \times (5) \times (3) - (2) \times (3) \times (0) - (3) \times (1) \times (5)$

$= 50 + 27 + 0 - 60 - 0 - 15 = 2$ **Ans.**

### EXERCISE 4.3

***Expand the following determinants by Rule of Sarrus.***

**1.** $\begin{vmatrix} 3 & 2 & -4 \\ 5 & 1 & -1 \\ -2 & 6 & 7 \end{vmatrix}$ **Ans.** $-155$

**2.** $\begin{vmatrix} 1 & 4 & 2 \\ 2 & 5 & 3 \\ 3 & 6 & 4 \end{vmatrix}$ **Ans.** 0

**3.** $\begin{vmatrix} 6 & 3 & 7 \\ 32 & 13 & 37 \\ 10 & 4 & 11 \end{vmatrix}$ **Ans.** 10

**4.** $\begin{vmatrix} 9 & 25 & 6 \\ 7 & 13 & 5 \\ 9 & 23 & 6 \end{vmatrix}$ **Ans.** 6

**5.** If $a + b + c = 0$, solve the equation $\begin{vmatrix} a-x & c & b \\ c & b-x & a \\ b & a & c-x \end{vmatrix} = 0$

**Ans.** $x = \pm\sqrt{(a^2 + b^2 + c^2 - ab - bc - ca)}\,,\ x = 0$

## 4.7 PROPERTIES OF DETERMINANTS

**Property (*i*)** The value of a determinant remains unaltered, if the rows are interchanged into columns (or the columns into rows).

Consider the determinant.

$$\Delta = \begin{vmatrix} a_1 & b_1 & c_1 \\ a_2 & b_2 & c_2 \\ a_3 & b_3 & c_3 \end{vmatrix}$$

$$= a_1 (b_2c_3 - b_3c_2) - b_1 (a_2c_3 - a_3c_2) + c_1 (a_2b_3 - a_3b_2)$$

$$= a_1b_2c_3 - a_1b_3c_2 - a_2b_1c_3 + a_3b_1c_2 + a_2b_3c_1 - a_3b_2c_1$$

$$= (a_1b_2c_3 - a_1b_3c_2) - (a_2b_1c_3 - a_2b_3c_1) + (a_3b_1c_2 - a_3b_2c_1)$$

$$= a_1 (b_2c_3 - b_3c_2) - a_2 (b_1c_3 - b_3c_1) + a_3 (b_1c_2 - b_2c_1)$$

$$= \begin{vmatrix} a_1 & a_2 & a_3 \\ b_1 & b_2 & b_3 \\ c_1 & c_2 & c_3 \end{vmatrix}$$ **Proved.**

**Property (*ii*)** If two rows (or two columns) of a determinant are interchanged, the sign of the value of the determinant changes.

Interchanging the first two rows of Δ, we get

$$\Delta' = \begin{vmatrix} a_2 & b_2 & c_2 \\ a_1 & b_1 & c_1 \\ a_3 & b_3 & c_3 \end{vmatrix}$$

$$= a_2 (b_1c_3 - b_3c_1) - b_2 (a_1c_3 - a_3c_1) + c_2 (a_1b_3 - a_3b_1)$$

$$= a_2b_1c_3 - a_2b_3c_1 - a_1b_2c_3 + a_3b_2c_1 + a_1b_3c_2 - a_3b_1c_2$$

$$= -[(a_1b_2c_3 - a_1b_3c_2) - (a_2b_1c_3 - a_3b_1c_2) + (a_2b_3c_1 - a_3b_2c_1)]$$

$$= -[(a_1 (b_2c_3 - b_3c_2) - b_1 (a_2c_3 - a_3c_2) + c_1 (a_2b_3 - a_3b_2)]$$

$$= - \begin{vmatrix} a_1 & b_1 & c_1 \\ a_2 & b_2 & c_2 \\ a_3 & b_3 & c_3 \end{vmatrix} = -\Delta$$ **Proved.**

**Property (*iii*)** If two rows (or columns) of a determinant are identical, the value of the determinant is zero.

Let $\Delta = \begin{vmatrix} a_1 & b_1 & c_1 \\ a_1 & b_1 & c_1 \\ a_3 & b_3 & c_3 \end{vmatrix}$, so that the first two rows are identical.

By interchanging the first two rows, we get the same determinant Δ.

By property (*ii*), on interchanging the rows, the sign of the determinant changes.

or $\Delta = -\Delta$ or $2\Delta = 0$ or $\Delta = 0$ **Proved.**

**Property (*iv*)** If the elements of any row (or column) of a determinant be each multiplied by the same number, the determinant is multiplied by that number.

$$\Delta' = \begin{vmatrix} ka_1 & kb_1 & kc_1 \\ a_2 & b_2 & c_2 \\ a_3 & b_3 & c_3 \end{vmatrix}$$

$$= ka_1\,(b_2c_3 - b_3c_2) - kb_1\,(a_2c_3 - a_3c_2) + kc_1\,(a_2b_3 - a_3b_2)$$
$$= k\,[a_1\,(b_2c_3 - b_3c_2) - b_1\,(a_2c_3 - a_3c_2) + c_1\,(a_2b_3 - a_3b_2)]$$
$$= k\begin{vmatrix} a_1 & b_1 & c_1 \\ a_2 & b_2 & c_2 \\ a_3 & b_3 & c_3 \end{vmatrix} = k\,\Delta.$$

**Example 8.** *Prove that*

$$\begin{vmatrix} a^2 & a & bc \\ b^2 & b & ca \\ c^2 & c & ab \end{vmatrix} = -\begin{vmatrix} 1 & 1 & 1 \\ a^2 & b^2 & c^2 \\ a^3 & b^3 & c^3 \end{vmatrix}$$

**Solution.** $\begin{vmatrix} a^2 & a & bc \\ b^2 & b & ca \\ c^2 & c & ab \end{vmatrix}$

By multiplying $R_1$, $R_2$, $R_3$ by $a$, $b$ and $c$ respectively we get

$$= \frac{1}{abc}\begin{vmatrix} a^3 & a^2 & abc \\ b^3 & b^2 & abc \\ c^3 & c^2 & abc \end{vmatrix} = \frac{abc}{abc}\begin{vmatrix} a^3 & a^2 & 1 \\ b^3 & b^2 & 1 \\ c^3 & c^2 & 1 \end{vmatrix}$$

$$= \begin{vmatrix} a^3 & a^2 & 1 \\ b^3 & b^2 & 1 \\ c^3 & c^2 & 1 \end{vmatrix} = -\begin{vmatrix} 1 & a^2 & a^3 \\ 1 & b^2 & b^3 \\ 1 & c^2 & c^3 \end{vmatrix}$$

$$= -\begin{vmatrix} 1 & 1 & 1 \\ a^2 & b^2 & c^2 \\ a^3 & b^3 & c^3 \end{vmatrix}$$ By changing rows into columns

**Proved**

**Example 9.** *Without expanding and or evaluating, show that*

$$\begin{vmatrix} a^2 & a & 1 & bcd \\ b^2 & b & 1 & cda \\ c^2 & c & 1 & dab \\ d^2 & d & 1 & abc \end{vmatrix} = \begin{vmatrix} a^3 & a^2 & a & 1 \\ b^3 & b^2 & b & 1 \\ c^3 & c^2 & c & 1 \\ d^3 & d^2 & d & 1 \end{vmatrix}$$

**Solution**

$$\begin{vmatrix} a^2 & a & 1 & bcd \\ b^2 & b & 1 & cda \\ c^2 & c & 1 & dab \\ d^2 & d & 1 & abc \end{vmatrix} = \frac{1}{abcd}\begin{vmatrix} a^3 & a^2 & a & abcd \\ b^3 & b^2 & b & abcd \\ c^3 & c^2 & c & abcd \\ d^3 & d^2 & d & abcd \end{vmatrix} \quad \begin{matrix} R_1 \to aR_1 \\ R_2 \to bR_2 \\ R_3 \to cR_3 \\ R_4 \to dR_4 \end{matrix}$$

$$= \frac{abcd}{abcd}\begin{vmatrix} a^3 & a^2 & a & 1 \\ b^3 & b^2 & b & 1 \\ c^3 & c^2 & c & 1 \\ d^3 & d^2 & d & 1 \end{vmatrix} C_4 \to \frac{1}{abcd} C_4 = \begin{vmatrix} a^3 & a^2 & a & 1 \\ b^3 & b^2 & b & 1 \\ c^3 & c^2 & c & 1 \\ d^3 & d^2 & d & 1 \end{vmatrix}$$ **Proved**

**Example 10.** *Prove that* $\begin{vmatrix} 1 & a & a^2 \\ 1 & b & b^2 \\ 1 & c & c^2 \end{vmatrix} = \begin{vmatrix} 1 & a & bc \\ 1 & b & ca \\ 1 & c & ab \end{vmatrix}$ *(Try yourself)*

**Property ($v$)** The value of the determinant remains unaltered if to the elements of one row (or column) be added any constant multiple of the corresponding elements of any other row (or column) respectively.

Let $$\Delta = \begin{vmatrix} a_1 & b_1 & c_1 \\ a_2 & b_2 & c_2 \\ a_3 & b_3 & c_3 \end{vmatrix}$$

On multiplying the second column by $l$ and the third column by $m$ and adding to the first column we get

$$\Delta' = \begin{vmatrix} a_1 + lb_1 + mc_1 & b_1 & c_1 \\ a_2 + lb_2 + mc_2 & b_2 & c_2 \\ a_3 + lb_3 + mc_3 & b_3 & c_3 \end{vmatrix}$$

$$= \begin{vmatrix} a_1 & b_1 & c_1 \\ a_2 & b_2 & c_2 \\ a_3 & b_3 & c_3 \end{vmatrix} + l\begin{vmatrix} b_1 & b_1 & c_1 \\ b_2 & b_2 & c_2 \\ b_3 & b_3 & c_3 \end{vmatrix} + m\begin{vmatrix} c_1 & b_1 & c_1 \\ c_2 & b_2 & c_2 \\ c_3 & b_3 & c_3 \end{vmatrix}$$

$= \Delta + 0 + 0$ (Since columns are identical)

$= \Delta$ **Proved**

**Example 11.** *Evaluate, using the properties of determinant* $\begin{vmatrix} 9 & 9 & 12 \\ 1 & 3 & -4 \\ 1 & 9 & 12 \end{vmatrix}$

**Solution.** Let $$\Delta = \begin{vmatrix} 9 & 9 & 12 \\ 1 & 3 & -4 \\ 1 & 9 & 12 \end{vmatrix}$$

Applying : $R_1 \to R_1 + 3R_2$ and $R_3 \to R_3 + 3R_2$, we get

$$\Delta = \begin{vmatrix} 12 & 18 & 0 \\ 1 & 3 & -4 \\ 4 & 18 & 0 \end{vmatrix} = 6 \times 2 \begin{vmatrix} 2 & 3 & 0 \\ 1 & 3 & -4 \\ 2 & 9 & 0 \end{vmatrix}$$

Expand by $C_3$ $$\Delta = 6 \times 2 \times 4 \begin{vmatrix} 2 & 3 \\ 2 & 9 \end{vmatrix}$$

$= 48\,(2 \times 9 - 2 \times 3) = 48 \times 12 = 576.$ **Ans.**

**Example 12.** *Without expanding evaluate the determinant* $\Delta = \begin{vmatrix} 265 & 240 & 219 \\ 240 & 225 & 198 \\ 219 & 198 & 181 \end{vmatrix}$

**Solution.** Applying $C_1 \to C_1 - C_3$ and $C_2 \to C_2 - C_3$, we get

$$\Delta = \begin{vmatrix} 46 & 21 & 219 \\ 42 & 27 & 198 \\ 38 & 17 & 181 \end{vmatrix}$$

Applying $C_1 \to C_1 - 2C_2$ and $C_3 \to C_3 - 10\,C_2$, we get

$$\Delta = \begin{vmatrix} 4 & 21 & 9 \\ -12 & 27 & -72 \\ 4 & 17 & 11 \end{vmatrix}$$

Applying $R_1 \to R_1 - R_3$ and $R_2 \to R_2 + 3R_3$

$$\Delta = \begin{vmatrix} 0 & 4 & -2 \\ 0 & 78 & -39 \\ 4 & 17 & 11 \end{vmatrix} = 2(39)\begin{vmatrix} 0 & 2 & -1 \\ 0 & 2 & -1 \\ 4 & 17 & 11 \end{vmatrix}$$ [Taking 2 common from $R_1$ and 39 common from $R_2$]

$= 78 \times 0 = 0$ (Since $R_1$ and $R_2$ are identical) **Ans.**

**Example 13.** *Show that* $\Delta = \begin{vmatrix} b-c & c-a & a-b \\ c-a & a-b & b-c \\ a-b & b-c & c-a \end{vmatrix} = 0$

**Solution.** Let $\Delta = \begin{vmatrix} b-c & c-a & a-b \\ c-a & a-b & b-c \\ a-b & b-c & c-a \end{vmatrix}$

Applying $C_1 \to C_1 + C_2 + C_3$, we get

$$\Delta = \begin{vmatrix} 0 & c-a & a-b \\ 0 & a-b & b-c \\ 0 & b-c & c-a \end{vmatrix} = 0$$ [$\because$ $C_1$ consists of all zeros.]

**Example 14.** *Without expanding, evaluate the determinant* $\begin{vmatrix} \sin\alpha & \cos\alpha & \sin(\alpha+\delta) \\ \sin\beta & \cos\beta & \sin(\beta+\delta) \\ \sin\gamma & \cos\gamma & \sin(\gamma+\delta) \end{vmatrix}$.

**Solution.** Let $\Delta = \begin{vmatrix} \sin\alpha & \cos\alpha & \sin(\alpha+\delta) \\ \sin\beta & \cos\beta & \sin(\beta+\delta) \\ \sin\gamma & \cos\gamma & \sin(\gamma+\delta) \end{vmatrix}$

$$\Rightarrow \quad \Delta = \begin{vmatrix} \sin\alpha & \cos\alpha & \sin\alpha\cos\delta + \cos\alpha\sin\delta \\ \sin\beta & \cos\beta & \sin\beta\cos\delta + \cos\beta\sin\delta \\ \sin\gamma & \cos\gamma & \sin\gamma\cos\delta + \cos\gamma\sin\delta \end{vmatrix}$$

[$\because$ $\sin(A+B) = \sin A\cos B + \cos A\sin B$]

$$\Rightarrow \quad \Delta = \begin{vmatrix} \sin\alpha & \cos\alpha & 0 \\ \sin\beta & \cos\beta & 0 \\ \sin\gamma & \cos\gamma & 0 \end{vmatrix} \quad [\text{Applying } C_3 \to C_3 - \cos\delta.C_1 - \sin\delta.C_2]$$

$$\Rightarrow \quad \Delta = 0 \qquad [\because C_3 \text{ consists of all zeros}]$$ **Ans.**

**Example 15.** *Solve the determinantal equation* $\begin{vmatrix} 2x-1 & x+7 & x+4 \\ x & 6 & 2 \\ x-1 & x+1 & 3 \end{vmatrix} = 0$

**Solution.** Given equation $\begin{vmatrix} 2x-1 & x+7 & x+4 \\ x & 6 & 2 \\ x-1 & x+1 & 3 \end{vmatrix} = 0$

By applying $R_1 \to R_1 - (R_2 + R_3)$, we get $\begin{vmatrix} 0 & 0 & x-1 \\ x & 6 & 2 \\ x-1 & x+1 & 3 \end{vmatrix} = 0$

On expanding by first row, we get

$(x-1)(x^2 + x - 6x + 6) = 0 \Rightarrow (x-1)(x-2)(x-3) = 0 \Rightarrow x = 1, 2, 3$ **Ans.**

**Example 16.** *Using the properties of determinants, show that*

$$\begin{vmatrix} x+y & x & x \\ 5x+4y & 4x & 2x \\ 10x+8y & 8x & 3x \end{vmatrix} = x^3.$$

**Solution.** Let $\Delta = \begin{vmatrix} x+y & x & x \\ 5x+4y & 4x & 2x \\ 10x+8y & 8x & 3x \end{vmatrix}$

Operate : $R_2 \to R_2 - 2R_1$ ; $R_3 \to R_3 - 3R_1$

$$\Delta = \begin{vmatrix} x+y & x & x \\ 3x+2y & 2x & 0 \\ 7x+5y & 5x & 0 \end{vmatrix} \quad \text{Expand by } C_3 \quad \Delta = x\begin{vmatrix} 3x+2y & 2x \\ 7x+5y & 5x \end{vmatrix}$$

$= x\,[5x(3x + 2y) - 2x(7x + 5y)]$

$= x\,[15x^2 + 10\,xy - (14x^2 + 10\,xy)] = x^3.$ **Proved.**

**Example 17.** *Using the properties of determinants, evaluate the following :*

$$\begin{vmatrix} 0 & ab^2 & ac^2 \\ a^2b & 0 & bc^2 \\ a^2c & cb^2 & 0 \end{vmatrix}$$

**Solution.** Let $\Delta = \begin{vmatrix} 0 & ab^2 & ac^2 \\ a^2b & 0 & bc^2 \\ a^2c & cb^2 & 0 \end{vmatrix}$

Take $a^2$, $b^2$ and $c^2$ common from $C_1$, $C_2$ and $C_3$ respectively, $\Delta = a^2b^2c^2\begin{vmatrix} 0 & a & a \\ b & 0 & b \\ c & c & 0 \end{vmatrix}$

Operate : $C_2 \to C_2 - C_3$, $\Delta = a^2 b^2 c^2 \begin{vmatrix} 0 & 0 & a \\ b & -b & b \\ c & c & 0 \end{vmatrix}$

Expand by $R_1$, $\Delta = a^2 b^2 c^2 . a \begin{vmatrix} b & -b \\ c & c \end{vmatrix} = a^3 b^2 c^2 (bc + bc) = 2a^3 b^3 c^3$. **Ans.**

**Example 18.** *Using properties of determinants, prove that*

$$\begin{vmatrix} x & y & z \\ x^2 & y^2 & z^2 \\ x^3 & y^3 & z^3 \end{vmatrix} = x y z (x - y)(y - z)(z - x).$$

**Solution.** Let $\Delta = \begin{vmatrix} x & y & z \\ x^2 & y^2 & z^2 \\ x^3 & y^3 & z^3 \end{vmatrix} = xyz \begin{vmatrix} 1 & 1 & 1 \\ x & y & z \\ x^2 & y^2 & z^2 \end{vmatrix}$

Operate : $C_1 \to C_1 - C_2$ ; $C_2 \to C_2 - C_3$, $\Delta = xyz \begin{vmatrix} 0 & 0 & 1 \\ x-y & y-z & z \\ x^2 - y^2 & y^2 - z^2 & z^2 \end{vmatrix}$

On expanding by $R_1$, $\Delta = xyz \begin{vmatrix} x - y & y - z \\ x^2 - y^2 & y^2 - z^2 \end{vmatrix} = xyz (x - y)(y - z) \begin{vmatrix} 1 & 1 \\ x + y & y + z \end{vmatrix}$

$= xyz (x - y)(y - z)(z - x)$. **Proved.**

**Example 19.** *Using the properties of determinants, show that*

$$\begin{vmatrix} a + x & y & z \\ x & a + y & z \\ x & y & a + z \end{vmatrix} = a^2 (a + x + y + z).$$

**Solution.** Let $\Delta = \begin{vmatrix} a + x & y & z \\ x & a + y & z \\ x & y & a + z \end{vmatrix}$

Operate : $R_1 \to R_1 - R_2$, $\Delta = \begin{vmatrix} a & -a & 0 \\ x & a + y & z \\ x & y & a + z \end{vmatrix}$

Operate : $C_2 \to C_2 + C_1$, $\Delta = \begin{vmatrix} a & 0 & 0 \\ x & a + y + x & z \\ x & y + x & a + z \end{vmatrix}$

On expanding by $R_1$ $\Delta = a \begin{vmatrix} a + y + x & z \\ y + x & a + z \end{vmatrix} = a [(a + y + x)(a + z) - (y + x) z]$

$= a [a^2 + az + (y + x) a + (y + x) z - (y + x) z]$

$= a^2 (a + x + y + z)$. **Proved.**

**Example 20.** *If $\omega$ is the one of the imaginary cube roots of unity, find the value of the determinant*

$$\begin{vmatrix} 1 & \omega & \omega^2 \\ \omega & \omega^2 & 1 \\ \omega^2 & 1 & \omega \end{vmatrix}$$

**Solution.** The given determinant $= \begin{vmatrix} 1 & \omega & \omega^2 \\ \omega & \omega^2 & 1 \\ \omega^2 & 1 & \omega \end{vmatrix}$

By $R_1 \to R_1 + R_2 + R_3$, we get

$$= \begin{vmatrix} 1+\omega+\omega^2 & 1+\omega+\omega^2 & 1+\omega+\omega^2 \\ \omega & \omega^2 & 1 \\ \omega^2 & 1 & \omega \end{vmatrix} = \begin{vmatrix} 0 & 0 & 0 \\ \omega & \omega^2 & 1 \\ \omega^2 & 1 & \omega \end{vmatrix} \quad [\because 1+\omega+\omega^2=0]$$

$= 0$ (Since each entry in $R_1$ is zero) **Ans.**

**Example 21.** *Without expanding the determinant, show that $(a+b+c)$ is a factor of* $\begin{vmatrix} a & b & c \\ b & c & a \\ c & a & b \end{vmatrix}$.

**Solution.** Let
$$\Delta = \begin{vmatrix} a & b & c \\ b & c & a \\ c & a & b \end{vmatrix}$$

Operate : $C_1 \to C_1 + C_2 + C_3$, $\Delta = \begin{vmatrix} a+b+c & b & c \\ a+b+c & c & a \\ a+b+c & a & b \end{vmatrix} = (a+b+c)\begin{vmatrix} 1 & b & c \\ 1 & c & a \\ 1 & a & b \end{vmatrix}$

$\Rightarrow$ $(a+b+c)$ is a factor of $\Delta$. **Proved.**

**Example 22.** *Using properties of determinants, prove that* :

$$\begin{vmatrix} x+4 & x & x \\ x & x+4 & x \\ x & x & x+4 \end{vmatrix} = 16(3x+4)$$

**Solution.** Let
$$\Delta = \begin{vmatrix} x+4 & x & x \\ x & x+4 & x \\ x & x & x+4 \end{vmatrix}$$

Operate : $C_1 \to C_1 + C_2 + C_3$, $\Delta = \begin{vmatrix} 3x+4 & x & x \\ 3x+4 & x+4 & x \\ 3x+4 & x & x+4 \end{vmatrix}$

$$= (3x+4)\begin{vmatrix} 1 & x & x \\ 1 & x+4 & x \\ 1 & x & x+4 \end{vmatrix} = (3x+4)\begin{vmatrix} 1 & x & x \\ 0 & 4 & 0 \\ 0 & 0 & 4 \end{vmatrix}\begin{matrix} \\ R_2-R_1 \\ R_3-R_1 \end{matrix} = (3x+4)\begin{vmatrix} 4 & 0 \\ 0 & 4 \end{vmatrix}$$

$= 16(3x+4)$ **Proved.**

**Example 23.** *Without expanding the determinant, prove that* $\begin{vmatrix} 1 & a & b+c \\ 1 & b & c+a \\ 1 & c & a+b \end{vmatrix} = 0.$

**Solution.** Let
$$\Delta = \begin{vmatrix} 1 & a & b+c \\ 1 & b & c+a \\ 1 & c & a+b \end{vmatrix}$$

Operate : $C_3 \to C_3 + C_2$,
$$\Delta = \begin{vmatrix} 1 & a & a+b+c \\ 1 & b & a+b+c \\ 1 & c & a+b+c \end{vmatrix} = (a+b+c)\begin{vmatrix} 1 & a & 1 \\ 1 & b & 1 \\ 1 & c & 1 \end{vmatrix}$$

$$= 0 \qquad (\because\ C_1 \text{ and } C_3 \text{ are identical}).$$ **Proved.**

**Example 24.** *Without expanding the determinant, prove that* $\begin{vmatrix} \frac{1}{a} & a^2 & bc \\ \frac{1}{b} & b^2 & ca \\ \frac{1}{c} & c^2 & ab \end{vmatrix} = 0$

**Solution.** Let
$$\Delta = \begin{vmatrix} \frac{1}{a} & a^2 & bc \\ \frac{1}{b} & b^2 & ca \\ \frac{1}{c} & c^2 & ab \end{vmatrix}$$

Multiply $R_1$ by $a$, $R_2$ by $b$ and $R_3$ by $c$.
$$\Delta = \frac{1}{abc}\begin{vmatrix} 1 & a^3 & abc \\ 1 & b^3 & abc \\ 1 & c^3 & abc \end{vmatrix} = \frac{1}{abc}.abc\begin{vmatrix} 1 & a^3 & 1 \\ 1 & b^3 & 1 \\ 1 & c^3 & 1 \end{vmatrix} = 1 \times 0 = 0.$$

(Since $C_1$ and $C_3$ are identical) **Proved.**

**Example 25.** *Evaluate* $\begin{vmatrix} 1 & a & a^2 \\ 1 & b & b^2 \\ 1 & c & c^2 \end{vmatrix}$

**Solution.** Let $\Delta$ be the given determinant. Applying $R_2 \to R_2 - R_1$ and $R_3 \to R_3 - R_1$, we get,

$$\Delta = \begin{vmatrix} 1 & a & a^2 \\ 0 & b-a & b^2-a^2 \\ 0 & c-a & c^2-a^2 \end{vmatrix} = (b-a)(c-a)\begin{vmatrix} 1 & a & a^2 \\ 0 & 1 & b+a \\ 0 & 1 & c+a \end{vmatrix}$$ [Taking out $(b-a)$ common from $R_2$ and $(c-a)$ from $R_3$]

$$= (b-a)(c-a)\begin{vmatrix} 1 & a & a^2 \\ 0 & 1 & b+a \\ 0 & 0 & c-b \end{vmatrix}$$ [Applying $R_3 \to R_3 - R_2$]

$$= (b-a)(c-a)\begin{vmatrix} 1 & b+a \\ 0 & c-b \end{vmatrix}$$ [Expanding along $C_1$]

$$= (b-a)(c-a)(c-b).$$ **Ans.**

**Example 26.** *Using properties of determinants, prove that :*

$$\begin{vmatrix} 1 & a & a^3 \\ 1 & b & b^3 \\ 1 & c & c^3 \end{vmatrix} = (a-b)(b-c)(c-a)(a+b+c)$$

**Solution.** Let $\Delta = \begin{vmatrix} 1 & a & a^3 \\ 1 & b & b^3 \\ 1 & c & c^3 \end{vmatrix}$

Operate : $R_1 \to R_1 - R_2$ ; $R_2 \to R_2 - R_3$, $\Delta = \begin{vmatrix} 0 & a-b & a^3-b^3 \\ 0 & b-c & b^3-c^3 \\ 1 & c & c^3 \end{vmatrix} = 1.\begin{vmatrix} a-b & a^3-b^3 \\ b-c & b^3-c^3 \end{vmatrix}$

(Expanding by $C_1$)

$$= (a-b)(b-c)\begin{vmatrix} 1 & a^2+ab+b^2 \\ 1 & b^2+bc+c^2 \end{vmatrix}$$

Operate : $R_1 \to R_1 - R_2$, $\Delta = (a-b)(b-c)\begin{vmatrix} 0 & (a^2-c^2)+(ab-bc) \\ 1 & b^2+bc+c^2 \end{vmatrix}$

$$= (a-b).(b-c).(-1)[(a^2-c^2)+b(a-c)]$$

$$= (a-b).(b-c)(c-a)(a+b+c).$$ **Proved.**

**Example 27.** *Evaluate* $\begin{vmatrix} a-b-c & 2a & 2a \\ 2b & b-c-a & 2b \\ 2c & 2c & c-a-b \end{vmatrix}$

**Solution.** By $R_1 \to R_1 + R_2 + R_3$, we get $\begin{vmatrix} a+b+c & a+b+c & a+b+c \\ 2b & b-c-a & 2b \\ 2c & 2c & c-a-b \end{vmatrix}$

$$= (a+b+c)\begin{vmatrix} 1 & 1 & 1 \\ 2b & b-c-a & 2b \\ 2c & 2c & c-a-b \end{vmatrix}$$

$$= (a+b+c)\begin{vmatrix} 1 & 0 & 0 \\ 2b & -(a+b+c) & 0 \\ 2c & 0 & -(a+b+c) \end{vmatrix} \begin{matrix} \\ C_2 - C_1 \\ C_3 - C_1 \end{matrix}$$

On expanding by first row $= (a+b+c)(a+b+c)^2 = (a+b+c)^3$. **Ans.**

**Example 28.** *Show, without expanding* $\begin{vmatrix} 1 & 1 & 1 \\ x & y & z \\ x^2 & y^2 & z^2 \end{vmatrix} = (x-y)(y-z)(z-x)$

**Solution.** By $C_1 - C_2, C_2 - C_3$, we get $= \begin{vmatrix} 0 & 0 & 1 \\ x-y & y-z & z \\ x^2-y^2 & y^2-z^2 & z^2 \end{vmatrix} = \begin{vmatrix} x-y & y-z \\ x^2-y^2 & y^2-z^2 \end{vmatrix}$

On expanding by first row, we get

$$= (x-y)(y-z)\begin{vmatrix} 1 & 1 \\ x+y & y+z \end{vmatrix} = (x-y)(y-z)(y+z-x-y) = (x-y)(y-z)(z-x).$$

**Proved.**

**Example 29.** *Prove that* $\begin{vmatrix} \alpha & \beta & \gamma \\ \alpha^2 & \beta^2 & \gamma^2 \\ \beta+\gamma & \gamma+\alpha & \alpha+\beta \end{vmatrix} = (\alpha-\beta)(\beta-\gamma)(\gamma-\alpha)(\alpha+\beta+\gamma)$

**Solution.** Let $\Delta = \begin{vmatrix} \alpha & \beta & \gamma \\ \alpha^2 & \beta^2 & \gamma^2 \\ \beta+\gamma & \gamma+\alpha & \alpha+\beta \end{vmatrix}.$

$$\Delta = \begin{vmatrix} \alpha & \beta & \gamma \\ \alpha^2 & \beta^2 & \gamma^2 \\ \alpha+\beta+\gamma & \alpha+\beta+\gamma & \alpha+\beta+\gamma \end{vmatrix} \quad \text{Applying } R_3 \to R_1 + R_3$$

$$= (\alpha+\beta+\gamma)\begin{vmatrix} \alpha & \beta & \gamma \\ \alpha^2 & \beta^2 & \gamma^2 \\ 1 & 1 & 1 \end{vmatrix} \quad [\text{Taking out } (\alpha+\beta+\gamma) \text{ common from } R_3]$$

$$= (\alpha+\beta+\gamma)\begin{vmatrix} \alpha & \beta-\alpha & \gamma-\alpha \\ \alpha^2 & \beta^2-\alpha^2 & \gamma^2-\alpha^2 \\ 1 & 0 & 0 \end{vmatrix} \quad \text{Applying } C_2 \to C_2 - C_1, \; C_3 \to C_3 - C_1$$

$$= (\alpha+\beta+\gamma)(\beta-\alpha)(\gamma-\alpha)\begin{vmatrix} \alpha & 1 & 1 \\ \alpha^2 & \beta+\alpha & \gamma+\alpha \\ 1 & 0 & 0 \end{vmatrix}$$

$$= (\alpha+\beta+\gamma)(\beta-\alpha)(\gamma-\alpha).1\begin{vmatrix} 1 & 1 \\ \beta+\alpha & \gamma+\alpha \end{vmatrix} \quad [\text{Expanding along } R_3]$$

$$= (\alpha+\beta+\gamma)(\beta-\alpha)(\gamma-\alpha)(\gamma+\alpha-\beta-\alpha)$$

$$= (\alpha+\beta+\gamma)(\beta-\gamma)(\gamma-\alpha)(\alpha-\beta)$$

**Proved.**

**Example 30.** *Prove that* $\begin{vmatrix} -a^2 & ab & ac \\ ba & -b^2 & bc \\ ac & bc & -c^2 \end{vmatrix} = 4\,a^2b^2c^2$

**Solution.** Let $\Delta = \begin{vmatrix} -a^2 & ab & ac \\ ba & -b^2 & bc \\ ac & bc & -c^2 \end{vmatrix}$

Taking $a, b, c$ common from $R_1, R_2$ and $R_3$ respectively, we get $\Delta = abc\begin{vmatrix} -a & a & a \\ b & -b & b \\ c & c & -c \end{vmatrix}$

$$= a^2b^2c^2\begin{vmatrix} -1 & 1 & 1 \\ 1 & -1 & 1 \\ 1 & 1 & -1 \end{vmatrix}$$ [Taking $a, b, c$ common from $C_1, C_2$ and $C_3$ respectively]

$$= a^2b^2c^2\begin{vmatrix} -1 & 0 & 0 \\ 1 & 0 & 2 \\ 1 & 2 & 0 \end{vmatrix}$$ [Applying $C_2 \to C_2 + C_1,\ C_3 \to C_3 + C_1$]

$$= a^2b^2c^2(-1)\begin{vmatrix} 0 & 2 \\ 2 & 0 \end{vmatrix}$$ [Expanding along $R_1$]

$$= a^2b^2c^2(-1)(0-4) = 4a^2b^2c^2$$ **Proved.**

**Example 31.** *Show that* $\begin{vmatrix} 3a & -a+b & -a+c \\ -b+a & 3b & -b+c \\ -c+a & -c+b & 3c \end{vmatrix} = 3(a+b+c)(ab+bc+ca)$

**Solution.** Let $$\Delta = \begin{vmatrix} 3a & -a+b & -a+c \\ -b+a & 3b & -b+c \\ -c+a & -c+b & 3c \end{vmatrix}$$

Applying $C_1 \to C_1 + C_2 + C_3$, we get $$\Delta = \begin{vmatrix} a+b+c & -a+b & -a+c \\ a+b+c & 3b & -b+c \\ a+b+c & -c+b & 3c \end{vmatrix}$$

$$= (a+b+c)\begin{vmatrix} 1 & -a+b & -a+c \\ 1 & 3b & -b+c \\ 1 & -c+b & 3c \end{vmatrix}$$ [Taking $(a+b+c)$ common from $C_1$]

$$= (a+b+c)\begin{vmatrix} 1 & -a+b & -a+c \\ 0 & 2b+a & -b+a \\ 0 & -c+a & 2c+a \end{vmatrix}$$ [Applying $R_2 \to R_2 - R_1,\ R_3 \to R_3 - R_1$]

$$= (a+b+c)\begin{vmatrix} 2b+a & -b+a \\ -c+a & 2c+a \end{vmatrix}$$ [Expanding along $C_1$]

$$= (a+b+c)[(2b+a)(2c+a) - (-b+a)(-c+a)]$$
$$= (a+b+c)\{(4bc + 2ab + 2ca + a^2 - (bc - ab - ac + a^2)\}$$
$$= (a+b+c)(3bc + 3ab + 3ca)$$
$$= 3(a+b+c)(ab+bc+ca)$$ **Proved.**

**Property (*vi*)** If each element of a row (or column) of a determinant consists of the algebraic sum of $n$ terms, the determinant can be expressed as the sum of $n$ determinants,

Let $$\Delta = \begin{vmatrix} a_1 + p_1 + q_1 & b_1 & c_1 \\ a_2 + p_2 + q_2 & b_2 & c_2 \\ a_3 + p_3 + q_3 & b_3 & c_3 \end{vmatrix}.$$

$$= (a_1 + p_1 + q_1)(b_2c_3 - b_3c_2) - (a_2 + p_2 + q_2)(b_1c_3 - b_3c_1) + (a_3 + p_3 + q_3)(b_1c_2 - b_2c_1)$$

$$= a_1(b_2c_3 - b_3c_2) - a_2(b_1c_3 - b_3c_1) + a_3(b_1c_2 - b_2c_1)$$
$$+ p_1(b_2c_3 - b_3c_2) - p_2(b_1c_3 - b_3c_1) + p_3(b_1c_2 - b_2c_1)$$
$$+ q_1(b_2c_3 - b_3c_2) - q_2(b_1c_3 - b_3c_1) + q_3(b_1c_2 - b_2c_1)$$

$$= \begin{vmatrix} a_1 & b_1 & c_1 \\ a_2 & b_2 & c_2 \\ a_3 & b_3 & c_3 \end{vmatrix} + \begin{vmatrix} p_1 & b_1 & c_1 \\ p_2 & b_2 & c_2 \\ p_3 & b_3 & c_3 \end{vmatrix} + \begin{vmatrix} q_1 & b_1 & c_1 \\ q_2 & b_2 & c_2 \\ q_3 & b_3 & c_3 \end{vmatrix}$$ **Proved.**

**Example 32.** *If* $\begin{vmatrix} a & a^2 & a^3-1 \\ b & b^2 & b^3-1 \\ c & c^2 & c^3-1 \end{vmatrix} = 0$*, prove that* $abc = 1$.

**Solution.** $\begin{vmatrix} a & a^2 & a^3-1 \\ b & b^2 & b^3-1 \\ c & c^2 & c^3-1 \end{vmatrix} = 0 \Rightarrow \begin{vmatrix} a & a^2 & a^3 \\ b & b^2 & b^3 \\ c & c^2 & c^3 \end{vmatrix} + \begin{vmatrix} a & a^2 & -1 \\ b & b^2 & -1 \\ c & c^2 & -1 \end{vmatrix} = 0$

$$\Rightarrow \quad abc\begin{vmatrix} 1 & a & a^2 \\ 1 & b & b^2 \\ 1 & c & c^2 \end{vmatrix} - \begin{vmatrix} a & a^2 & 1 \\ b & b^2 & 1 \\ c & c^2 & 1 \end{vmatrix} = 0$$

(Taking out common $a, b, c$ from $R_1$, $R_2$ and $R_3$ from 1st determinant)

$$\Rightarrow \quad abc\begin{vmatrix} 1 & a & a^2 \\ 1 & b & b^2 \\ 1 & c & c^2 \end{vmatrix} + \begin{vmatrix} a & 1 & a^2 \\ b & 1 & b^2 \\ c & 1 & c^2 \end{vmatrix} = 0$$ (Interchanging $C_2$ and $C_3$)

$$\Rightarrow \quad abc\begin{vmatrix} 1 & a & a^2 \\ 1 & b & b^2 \\ 1 & c & c^2 \end{vmatrix} - \begin{vmatrix} 1 & a & a^2 \\ 1 & b & b^2 \\ 1 & c & c^2 \end{vmatrix} = 0$$ (Interchanging $C_1$ and $C_2$)

$$\Rightarrow \quad (abc - 1)\begin{vmatrix} 1 & a & a^2 \\ 1 & b & b^2 \\ 1 & c & c^2 \end{vmatrix} = 0$$

$$\Rightarrow \quad abc - 1 = 0$$

$$\Rightarrow \quad abc = 1$$ **Proved.**

**Example 33.** *Show that* $\begin{vmatrix} b+c & c+a & a+b \\ q+r & r+p & p+q \\ y+z & z+x & x+y \end{vmatrix} = 2\begin{vmatrix} a & b & c \\ p & q & r \\ x & y & z \end{vmatrix}$

**Solution.** The above determinant can be expressed as the sum of 8 determinants as given below:

$$\begin{vmatrix} b+c & c+a & a+b \\ q+r & r+p & p+q \\ y+z & z+x & x+y \end{vmatrix} = \begin{vmatrix} b & c & a \\ q & r & p \\ y & z & x \end{vmatrix} + \begin{vmatrix} b & a & a \\ q & p & p \\ y & x & x \end{vmatrix} + \begin{vmatrix} b & c & b \\ q & r & q \\ y & z & y \end{vmatrix} + \begin{vmatrix} b & a & b \\ q & p & q \\ y & x & y \end{vmatrix}$$

$$+\begin{vmatrix} c & c & a \\ r & r & p \\ z & z & x \end{vmatrix}+\begin{vmatrix} c & a & a \\ r & p & p \\ z & x & x \end{vmatrix}+\begin{vmatrix} c & c & b \\ r & r & q \\ z & z & y \end{vmatrix}+\begin{vmatrix} c & a & b \\ r & p & q \\ z & x & y \end{vmatrix}$$

$$=\begin{vmatrix} b & c & a \\ q & r & p \\ y & z & x \end{vmatrix}+0+0+0+0+0+0+\begin{vmatrix} c & a & b \\ r & p & q \\ z & x & y \end{vmatrix}$$

$$=(-1)^2\begin{vmatrix} a & b & c \\ p & q & r \\ x & y & z \end{vmatrix}+(-1)^2\begin{vmatrix} a & b & c \\ p & q & r \\ x & y & z \end{vmatrix}=2\begin{vmatrix} a & b & c \\ p & q & r \\ x & y & z \end{vmatrix}$$

**Example 34.** *Prove that* $\begin{vmatrix} 2\alpha & \alpha+\beta & \alpha+\gamma \\ \beta+\alpha & 2\beta & \beta+\gamma \\ \gamma+\alpha & \gamma+\beta & 2\gamma \end{vmatrix}=0$ **Proved.**

**Solution.** Given determinant $=\begin{vmatrix} \alpha+\alpha & \alpha+\beta & \alpha+\gamma \\ \beta+\alpha & \beta+\beta & \beta+\gamma \\ \gamma+\alpha & \gamma+\beta & \gamma+\gamma \end{vmatrix}$

The above determinant can be expressed as the sum of 8 determinants.

Each of the 8 determinants has either two identical columns or identical rows.

∴ Each of the resulting determinant is zero. Hence the result. **Proved.**

**Example 35.** *Prove that* $\begin{vmatrix} x & l & m & 1 \\ \alpha & x & n & 1 \\ \alpha & \beta & x & 1 \\ \alpha & \beta & \gamma & 1 \end{vmatrix}=(x-\alpha)(x-\beta)(x-\gamma)$

**Solution.** $\begin{vmatrix} x & l & m & 1 \\ \alpha & x & n & 1 \\ \alpha & \beta & x & 1 \\ \alpha & \beta & \gamma & 1 \end{vmatrix}=\begin{vmatrix} x-\alpha & l & m & 1 \\ 0 & x & n & 1 \\ 0 & \beta & x & 1 \\ 0 & \beta & \gamma & 1 \end{vmatrix}$ $(C_1 \to C_1-\alpha C_4)$

[On expanding by first column we get]

$$=(x-\alpha)\begin{vmatrix} x & n & 1 \\ \beta & x & 1 \\ \beta & \gamma & 1 \end{vmatrix}=(x-\alpha)\begin{vmatrix} x-\beta & n & 1 \\ 0 & x & 1 \\ 0 & \gamma & 1 \end{vmatrix} \quad (C_1 \to C_1-\beta C_3)$$

$=(x-\alpha)(x-\beta)(x-\gamma)$ [On expanding by first column] **Proved.**

**Example 36.** *Show that* $x=-(a+b+c)$ *is one root of the equation:*

$$\begin{vmatrix} x+a & b & c \\ b & x+c & a \\ c & a & x+b \end{vmatrix}=0 \text{ and solve the equation completely.}$$

**Solution.** By $C_1 \to C_1+C_2+C_3$, we get $\begin{vmatrix} x+a+b+c & b & c \\ x+a+b+c & x+c & a \\ x+a+b+c & a & x+b \end{vmatrix}=0$

$$\Rightarrow \quad (x+a+b+c)\begin{vmatrix} 1 & b & c \\ 1 & x+c & a \\ 1 & a & x+b \end{vmatrix} = 0$$

$$\Rightarrow \quad (x+a+b+c)\begin{vmatrix} 1 & b & c \\ 0 & x-b+c & a-c \\ 0 & a-b & x+b-c \end{vmatrix} = 0, \; R_2 \to R_2 - R_1; \; R_3 \to R_3 - R_1$$

On expanding by first column, we get

$$(x+a+b+c)\,[(x-b+c)(x+b-c)-(a-b)(a-c)] = 0$$

$$\Rightarrow \quad (x+a+b+c)\,[x^2-(b-c)^2-(a^2-ac-ab+bc)] = 0$$

$$\Rightarrow \quad (x+a+b+c)\,(x^2-b^2-c^2+2bc-a^2+ac+ab-bc] = 0$$

$$\Rightarrow \quad (x+a+b+c)\,(x^2-a^2-b^2-c^2+ab+bc+ca) = 0$$

Either $x+a+b+c=0 \Rightarrow \quad x=-(a+b+c)$

or $\quad x^2-a^2-b^2-c^2+ab+bc+ca=0$

$$\Rightarrow \quad x = \pm\sqrt{a^2+b^2+c^2-ab-bc-ca}$$

Hence, $x=-(a+b+c)$ is one root of the given equation. **Proved.**

**Example 37.** *Find the value of* $\begin{vmatrix} (b+c)^2 & a^2 & a^2 \\ b^2 & (c+a)^2 & b^2 \\ c^2 & c^2 & (a+b)^2 \end{vmatrix}$

**Solution.** By $C_1 - C_3$, $C_2 - C_3$, we get $\begin{vmatrix} (b+c)^2-a^2 & a^2-a^2 & a^2 \\ b^2-b^2 & (c+a)^2-b^2 & b^2 \\ c^2-(a+b)^2 & c^2-(a+b)^2 & (a+b)^2 \end{vmatrix}$

$$= \begin{vmatrix} (a+b+c)(b+c-a) & 0 & a^2 \\ 0 & (a+b+c)(c+a-b) & b^2 \\ (a+b+c)(c-a-b) & (a+b+c)(c-a-b) & (a+b)^2 \end{vmatrix}$$

On taking out $(a+b+c)$ as common from 1st and 2nd column, we get

$$= (a+b+c)^2\begin{vmatrix} b+c-a & 0 & a^2 \\ 0 & c+a-b & b^2 \\ c-a-b & c-a-b & (a+b)^2 \end{vmatrix}$$

$$= (a+b+c)^2\begin{vmatrix} -a+b+c & 0 & a^2 \\ 0 & a-b+c & b^2 \\ -2b & -2a & 2ab \end{vmatrix} \; R_3 \to R_3 - (R_1 + R_2)$$

$$= -2(a+b+c)^2 \begin{vmatrix} -a+b+c & 0 & a^2 \\ 0 & a-b+c & b^2 \\ b & a & -ab \end{vmatrix}$$

On expanding by first row, we get

$$= -2(a+b+c)^2 [(-a+b+c)\{-ab(a-b+c)-ab^2\} + a^2\{0-b(a-b+c)\}]$$
$$= -2(a+b+c)^2 [(-a+b+c)(-a^2b-abc)-a^2b(a-b+c)]$$
$$= -2ab(a+b+c)^2 [(-a+b+c)(-a-c)-a(a-b+c)]$$
$$= -2ab(a+b+c)^2 (a^2+ac-ab-bc-ac-c^2-a^2+ab-ac]$$
$$= -2ab(a+b+c)^2(-bc-ac-c^2)$$
$$= 2abc(a+b+c)^2(b+a+c)$$
$$= 2abc(a+b+c)^3.$$

**Ans.**

**Example 38.** *Using properties of determinants, solve for x :* $\begin{vmatrix} a+x & a-x & a-x \\ a-x & a+x & a-x \\ a-x & a-x & a+x \end{vmatrix} = 0$

**Solution.** Given that $\begin{vmatrix} a+x & a-x & a-x \\ a-x & a+x & a-x \\ a-x & a-x & a+x \end{vmatrix} = 0$

Applying $C_1 \to C_1 + C_2 + C_3$ $\begin{vmatrix} 3a-x & a-x & a-x \\ 3a-x & a+x & a-x \\ 3a-x & a-x & a+x \end{vmatrix} = 0$

$$\Rightarrow (3a-x)\begin{vmatrix} 1 & a-x & a-x \\ 1 & a+x & a-x \\ 1 & a-x & a+x \end{vmatrix} = 0$$

Now, $R_2 \to R_2 - R_1$ and $R_3 \to R_3 - R_1$, $\Rightarrow (3a-x)\begin{vmatrix} 1 & a-x & a-x \\ 0 & 2x & 0 \\ 0 & 0 & 2x \end{vmatrix} = 0$

Expanding by $C_1$, we get $(3a-x)(4x^2-0) = 0$

$\Rightarrow 4x^2(3a-x) = 0$ $\Rightarrow$ If $4x^2 = 0$, then $x = 0$

$\Rightarrow$ If $3a - x = 0$, then $x = 3a$

Hence, $x = 0$ or $3a$ **Ans.**

**Example 39.** *Using properties of determinants, prove the following*

$$\begin{vmatrix} 1+a & 1 & 1 \\ 1 & 1+b & 1 \\ 1 & 1 & 1+c \end{vmatrix} = abc\left(\frac{1}{a}+\frac{1}{b}+\frac{1}{c}+1\right)$$

**Solution.** Let $\Delta = \begin{vmatrix} 1+a & 1 & 1 \\ 1 & 1+b & 1 \\ 1 & 1 & 1+c \end{vmatrix}$

$$\Delta = abc\begin{vmatrix} \frac{1+a}{a} & \frac{1}{a} & \frac{1}{a} \\ \frac{1}{b} & \frac{1+b}{b} & \frac{1}{b} \\ \frac{1}{c} & \frac{1}{c} & \frac{1+c}{c} \end{vmatrix} \Rightarrow \Delta = abc\begin{vmatrix} 1+\frac{1}{a} & \frac{1}{a} & \frac{1}{a} \\ \frac{1}{b} & 1+\frac{1}{b} & \frac{1}{b} \\ \frac{1}{c} & \frac{1}{c} & 1+\frac{1}{c} \end{vmatrix}$$

Operate : $R_1 \to R_1 + R_2 + R_3$, $\Delta = abc\begin{vmatrix} 1+\frac{1}{a}+\frac{1}{b}+\frac{1}{c} & 1+\frac{1}{a}+\frac{1}{b}+\frac{1}{c} & 1+\frac{1}{a}+\frac{1}{b}+\frac{1}{c} \\ \frac{1}{b} & 1+\frac{1}{b} & \frac{1}{b} \\ \frac{1}{c} & \frac{1}{c} & 1+\frac{1}{c} \end{vmatrix}$

Taking $\left(1+\frac{1}{a}+\frac{1}{b}+\frac{1}{c}\right)$ common from $R_1$, we get

$$\Delta = abc\left(1+\frac{1}{a}+\frac{1}{b}+\frac{1}{c}\right)\begin{vmatrix} 1 & 1 & 1 \\ \frac{1}{b} & 1+\frac{1}{b} & \frac{1}{b} \\ \frac{1}{c} & \frac{1}{c} & 1+\frac{1}{c} \end{vmatrix}$$

Operate : $C_2 \to C_2 - C_1$ ; $C_3 \to C_3 - C_1$, $\Delta = abc\left(\frac{1}{a}+\frac{1}{b}+\frac{1}{c}+1\right)\begin{vmatrix} 1 & 0 & 0 \\ \frac{1}{b} & 1 & 0 \\ \frac{1}{c} & 0 & 1 \end{vmatrix}$

$= abc\left(\frac{1}{a}+\frac{1}{b}+\frac{1}{c}+1\right)$ (On expanding by $R_1$) **Proved.**

**Example 40.** *Prove that :* $\begin{vmatrix} a & a^2 & bc \\ b & b^2 & ac \\ c & c^2 & ab \end{vmatrix} = (a-b)(b-c)(c-a)(ab+bc+ac).$

**Solution.** Let $\Delta = \begin{vmatrix} a & a^2 & bc \\ b & b^2 & ac \\ c & c^2 & ab \end{vmatrix} = \frac{1}{abc}\begin{vmatrix} a^2 & a^3 & abc \\ b^2 & b^3 & abc \\ c^2 & c^3 & abc \end{vmatrix} = \frac{1}{abc}.abc\begin{vmatrix} a^2 & a^3 & 1 \\ b^2 & b^3 & 1 \\ c^2 & c^3 & 1 \end{vmatrix}$

Operate : $R_1 \to R_1 - R_2$ ; $R_2 \to R_2 - R_3$, $\Delta = \begin{vmatrix} a^2-b^2 & a^3-b^3 & 0 \\ b^2-c^2 & b^3-c^3 & 0 \\ c^2 & c^3 & 1 \end{vmatrix}$

$$= (a-b)(b-c)\begin{vmatrix} a+b & a^2+ab+b^2 & 0 \\ b+c & b^2+bc+c^2 & 0 \\ c^2 & c^3 & 1 \end{vmatrix}$$

Expand by $C_3$ $\Delta = (a-b)(b-c).1\begin{vmatrix} a+b & a^2+ab+b^2 \\ b+c & b^2+bc+c^2 \end{vmatrix}$

Operate : $R_2 \to R_2 - R_1$ $\Delta = (a-b)(b-c)\begin{vmatrix} a+b & a^2+ab+b^2 \\ c-a & b(c-a)+(c^2-a^2) \end{vmatrix}$

$$= (a-b)(b-c)(c-a)\begin{vmatrix} a+b & a^2+ab+b^2 \\ 1 & b+c+a \end{vmatrix}$$

$$= (a-b)(b-c)(c-a)\,[(a+b)(a+b+c) - 1\,.\,(a^2+ab+b^2)]$$

$$= (a-b)(b-c)(c-a)(ab+bc+ac).$$ **Proved.**

## EXERCISE 4.4

***Expand the following determinants, using properties of the determinants :***

**1.** $\begin{vmatrix} 1 & 3 & 7 \\ 4 & 9 & 1 \\ 2 & 7 & 6 \end{vmatrix}$ **Ans.** 51.

**2.** Prove that $\begin{vmatrix} x & a & a \\ a & x & a \\ a & a & x \end{vmatrix} = (x+2a)(x-a)^2.$

**3.** Solve the equation $\begin{vmatrix} x^3-a^3 & x^2 & x \\ b^3-a^3 & b^2 & b \\ c^3-a^3 & c^2 & c \end{vmatrix} = 0, b \neq c, bc \neq 0$ **Ans.** $x = \dfrac{a^3}{bc}, x = b, x = c$

**4.** Show that zero is one of the roots of the equation $\begin{vmatrix} 0 & x-a & x-b \\ x+a & 0 & x-c \\ x+b & x-c & 0 \end{vmatrix} = 0$

**5.** Without expanding the determinant, prove that $\begin{vmatrix} \frac{1}{a} & a & bc \\ \frac{1}{b} & b & ca \\ \frac{1}{c} & c & ab \end{vmatrix} = 0$

**6.** Without expanding the determinant, prove that : $\begin{vmatrix} x+y & y+z & z+x \\ z & x & y \\ 1 & 1 & 1 \end{vmatrix} = 0.$

**7.** Using properties of determinant prove that : $\begin{vmatrix} x+4 & 2x & 2x \\ 2x & x+4 & 2x \\ 2x & 2x & x+4 \end{vmatrix} = (5x+4)(4-x)^2$

**8.** $\begin{vmatrix} a+b+2c & a & b \\ c & b+c+2a & b \\ c & a & c+a+2b \end{vmatrix} = 2(a+b+c)^3$

**9.** $\begin{vmatrix} 1 & x+y & x^2+y^2 \\ 1 & y+z & y^2+z^2 \\ 1 & z+x & z^2+x^2 \end{vmatrix} = (x-y)(y-z)(z-x).$

10. $\begin{vmatrix} b^2c^2 & bc & b+c \\ c^2a^2 & ca & c+a \\ a^2b^2 & ab & a+b \end{vmatrix} = 0$

11. $\begin{vmatrix} 1 & a & a^2-bc \\ 1 & b & b^2-ca \\ 1 & c & c^2-ab \end{vmatrix} = 0.$

12. $\begin{vmatrix} a & a+b & a+b+c \\ 2a & 3a+2b & 4a+3b+2c \\ 3a & 6a+3b & 10a+6b+3c \end{vmatrix} = a^3.$

13. $\begin{vmatrix} 1 & 1 & 1 \\ \alpha & \beta & \gamma \\ \beta\gamma & \gamma\alpha & \alpha\beta \end{vmatrix} = (\beta-\gamma)(\gamma-\alpha)(\alpha-\beta).$

14. $\begin{vmatrix} 1 & a & a^2 \\ 1 & b & b^2 \\ 1 & c & c^2 \end{vmatrix} = \begin{vmatrix} 1 & a & bc \\ 1 & b & ca \\ 1 & c & ab \end{vmatrix}$

15. $\begin{vmatrix} a^2 & bc & ac+c^2 \\ a^2+ab & b^2 & ac \\ ab & b^2+bc & c^2 \end{vmatrix} = 4a^2b^2c^2$

16. $\begin{vmatrix} a+b+c & -c & -b \\ -c & a+b+c & -a \\ -b & -a & a+b+c \end{vmatrix} = 2(a+b)(b+c)(c+a).$

## 4.8 FACTOR THEOREM

**If the elements of a determinant are polynomials in a variable $x$ and if the substitution $x = a$ makes two rows (or columns) identical, then $(x - a)$ is a factor of the determinant.**

**When two rows are identical, the value of the determinant is zero. The expansion of a determinant being polynomial in $x$ vanishes on putting $x = a$, then $x - a$ is its factor by the Remainder theorem.**

**Example 41.** *Show that* $\begin{vmatrix} 1 & 1 & 1 \\ x & y & z \\ x^2 & y^2 & z^2 \end{vmatrix} = (x-y)(y-z)(z-x)$

**Solution.** If we put $x = y$, $y = z$, $z = x$ then in each case two columns become identical and the determinant vanishes.

$\therefore$ $(x - y)$, $(y - z)$, $(z - x)$ are the factors.

Since the determinant is of third degree, the other factor can be numerical only $k$ (say).

$$\begin{vmatrix} 1 & 1 & 1 \\ x & y & z \\ x^2 & y^2 & z^2 \end{vmatrix} = k(x-y)(y-z)(z-x) \quad \dots (1)$$

This leading term (product of the elements of the diagonal elements) in the given determinant is $yz^2$ and in the expansion

$$k(x-y)(y-z)(z-x) \text{ we get } kyz^2$$

Equating the coefficient of $yz^2$ on both sides of (1), we have

$$k = 1$$

Hence the expansion $= (x - y)(y - z)(z - x)$. **Proved.**

**Example 42.** *Factorize* $\Delta = \begin{vmatrix} 1 & 1 & 1 \\ a^2 & b^2 & c^2 \\ a^3 & b^3 & c^3 \end{vmatrix}$

**Solution.** Putting $a = b$, $C_1 = C_2$ and hence $\Delta = 0$.

$\therefore$ $a - b$ is a factor of $\Delta$. Similarly $b - c$, $c - a$ are also factors of $\Delta$.

$\therefore$ $(a - b)(b - c)(c - a)$ is a third degree factor of $\Delta$ which itself is of the fifth degree as is judged from the leading term $b^2c^3$.

$\therefore$ The remaining factor must be of the second degree. As $\Delta$ is symmetrical in a, b, c the remaining factor must, therefore, be of the form $k(a^2 + b^2 + c^2) + l(ab + bc + ca)$

$\therefore$ $\Delta = (a - b)(b - c)(c - a)\{k(a^2 + b^2 + c^2) + l(ab + bc + ca)\}$

If $k \neq 0$, we shall get terms like $a^4b$, $b^4c$ etc. which do not occur in $\Delta$. Hence, $k$ must be zero.

$\therefore$ $\Delta = (a - b)(b - c)(c - a)\{0 + l(ab + bc + ca)\}$

or $\Delta = l(a - b)(b - c)(c - a)(ab + bc + ca)$

The leading term in $\Delta = b^2c^3$. The corresponding term on R.H.S $= l\, b^2c^3$

$\therefore$ $l = 1$

Hence, $\Delta = (a - b)(b - c)(c - a)(ab + bc + ca)$. **Ans.**

**Example 43.** Show that $\begin{vmatrix} x & x^2 & x^3 \\ y & y^2 & y^3 \\ z & z^2 & z^3 \end{vmatrix} = xyz(x-y)(y-z)(z-x)$.

**Solution.** $\begin{vmatrix} x & x^2 & x^3 \\ y & y^2 & y^3 \\ z & z^2 & z^3 \end{vmatrix} = xyz \begin{vmatrix} 1 & x & x^2 \\ 1 & y & y^2 \\ 1 & z & z^2 \end{vmatrix} = xyz\,(x - y)\,(y - z)\,(z - x)$ (see example 42).

**Proved.**

**Example 44.** Show that $\begin{vmatrix} x^3 & x^2 & x & 1 \\ \alpha^3 & \alpha^2 & \alpha & 1 \\ \beta^3 & \beta^2 & \beta & 1 \\ \gamma^3 & \gamma^2 & \gamma & 1 \end{vmatrix} = (x-\alpha)(x-\beta)(x-\gamma)(\alpha-\beta)(\beta-\gamma)(\alpha-\gamma)$

**Solution.** If we put $x = \alpha$; $x = \beta$; $x = \gamma$; $\alpha = \beta$, $\beta = \gamma$; $\gamma = \alpha$ then two rows become identical and the determinant vanishes.

$\therefore$ $(x - \alpha)$; $(x - \beta)$; $(x - \gamma)$; $(\alpha - \beta)$; $(\beta - \gamma)$; $(\alpha - \gamma)$ are the factors.

Since the determinant is of six degree the other factor can be numerical only say $k$.

$$\begin{vmatrix} x^3 & x^2 & x & 1 \\ \alpha^3 & \alpha^2 & \alpha & 1 \\ \beta^3 & \beta^2 & \beta & 1 \\ \gamma^3 & \gamma^2 & \gamma & 1 \end{vmatrix} = k(x-\alpha)(x-\beta)(x-\gamma)(\alpha-\beta)(\beta-\gamma)(\gamma-\alpha)$$

The leading term is $x^3\alpha^2\beta$. And in the expansion it is $kx^3(-\alpha^2)\beta$.

$\therefore \quad k = -1$ Hence the expansion $= -(x-\alpha)\,(x-\beta)(x-\gamma)(\alpha-\beta)(\beta-\gamma)\,(\gamma-\alpha)$

**Proved.**

## EXERCISE 4.5

**1.** Evaluate, without expanding $\begin{vmatrix} a & a^2 & 1+a^3 \\ b & b^2 & 1+b^3 \\ c & c^2 & 1+c^3 \end{vmatrix}$ **Ans.** $(a-b)(b-c)(c-a)(1+abc)$

**2.** Without expanding, show that

$$\Delta = \begin{vmatrix} (a-x)^2 & (a-y)^2 & (a-z)^2 \\ (b-x)^2 & (b-y)^2 & (b-z)^2 \\ (c-x)^2 & (c-y)^2 & (c-z)^2 \end{vmatrix} = 2(a-b)(b-c)(c-a)(x-y)(y-z)(z-x).$$

**3.** Show (without expanding) that

$$\begin{vmatrix} bc & a^2 & a^2 \\ b^2 & ca & b^2 \\ c^2 & c^2 & ab \end{vmatrix} = \begin{vmatrix} bc & ab & ca \\ ab & ca & bc \\ ca & bc & ab \end{vmatrix} = -\frac{1}{2}(ab+bc+ca)[(ab-bc)^2+(bc-ca)^2+(ca-ab)^2]$$

## 4.9 PIVOTAL CONDENSATION METHOD

The condensation process of reducing $n^{\text{th}}$ order determinant to $(n-1)^{\text{th}}$ order determinant is shown below :

$$\text{Consider } n^{\text{th}} \text{ order determinant } D = \begin{vmatrix} a_1 & b_1 & c_1 & d_1 & \dots & \dots \\ a_2 & b_2 & c_2 & d_2 & \dots & \dots \\ a_3 & b_3 & c_3 & d_3 & \dots & \dots \\ a_4 & b_4 & c_4 & d_4 & \dots & \dots \\ \dots & \dots & \dots & \dots & \dots & \dots \\ \dots & \dots & \dots & \dots & \dots & \dots \\ a_n & b_n & c_n & d_n & \dots & \dots \end{vmatrix}$$

Add such a multiple of first column in the other columns so that at the places of $b_1, c_1, d_1$ ......., we get zero. Hence subtracting $\frac{b_1}{a_1}, \frac{c_1}{a_1}, \frac{d_1}{a_1}, \dots,$ times the first column from the 2nd, 3rd, 4th... columns respectively, we get

$$D = \begin{vmatrix} a_1 & 0 & 0 & 0 & \dots & \dots \\ a_2 & b_2 - \frac{b_1}{a_1}.a_2 & c_2 - \frac{c_1}{a_1}a_2 & d_2 - \frac{d_1}{a_1}.a_2 & \dots & \dots \\ a_3 & b_3 - \frac{b_1}{a_1}.a_3 & c_3 - \frac{c_1}{a_1}a_3 & d_3 - \frac{d_1}{a_1}.a_3 & \dots & \dots \\ a_4 & b_4 - \frac{b_1}{a_1}.a_4 & c_4 - \frac{c_1}{a_1}a_4 & d_4 - \frac{d_1}{a_1}.a_4 & \dots & \dots \\ & \dots & \dots & \dots & & \\ & \dots & \dots & \dots & & \\ a_n & b_n - \frac{b_1}{a_1}.a_n & c_n - \frac{c_1}{a_1}.a_n & d_n - \frac{d_1}{a_1}.a_n & \dots & \dots \end{vmatrix}$$

$$D = a_1 \begin{vmatrix} b_2 - \frac{b_1}{a_1}.a_2 & c_2 - \frac{c_1}{a_1}.a_2 & d_2 - \frac{d_1}{a_1}.a_2 & \cdots & \cdots \\ b_3 - \frac{b_1}{a_1}.a_3 & c_3 - \frac{c_1}{a_1}.a_3 & d_3 - \frac{d_1}{a_1}.a_3 & & \\ b_4 - \frac{b_1}{a_1}.a_4 & c_4 - \frac{c_1}{a_1}.a_4 & d_4 - \frac{d_1}{a_1}.a_4 & \cdots & \cdots \\ \cdots & \cdots & \cdots & & \\ \cdots & \cdots & \cdots & & \\ b_n - \frac{b_1}{a_1}.a_n & c_n - \frac{c_1}{a_1}.a_n & d_n - \frac{d_1}{a_1}.a_n \cdots & & \cdots \end{vmatrix} \quad \left[\begin{array}{l}\text{On expanding along} \\ \text{the first row}\end{array}\right]$$

Which is a determinant of $(n-1)$th order. Now,

$$D = a_1 \begin{vmatrix} \frac{a_1b_2 - b_1a_2}{a_1} & \frac{a_1c_2 - c_1a_2}{a_1} & \frac{a_1d_2 - d_1a_2}{a_1} & \cdots & \cdots \\ \frac{a_1b_3 - b_1a_3}{a_1} & \frac{a_1c_3 - c_1a_3}{a_1} & \frac{a_1d_3 - d_1a_3}{a_1} & \cdots & \cdots \\ \frac{a_1b_4 - b_1a_4}{a_1} & \frac{a_1c_4 - c_1a_4}{a_1} & \frac{a_1d_4 - d_1a_4}{a_1} & \cdots & \cdots \\ \cdots & \cdots & \cdots & & \\ \cdots & & \cdots & & \\ \frac{a_1b_n - b_1a_n}{a_1} & \frac{a_1c_n - c_1a_n}{a_1} & \frac{a_1d_n - d_1a_n}{a_1} & \cdots & \end{vmatrix}$$

$$D = a_1 . \frac{1}{(a_1)^{n-1}} \begin{vmatrix} a_1b_2 - b_1a_2 & a_1c_2 - c_1a_2 & a_1d_2 - d_1a_2 & \cdots & \cdots \\ a_1b_3 - b_1a_3 & a_1c_3 - c_1a_3 & a_1d_3 - d_1a_3 & \cdots & \cdots \\ a_1b_4 - b_1a_4 & a_1c_4 - c_1a_4 & a_1d_4 - d_1a_4 & \cdots & \cdots \\ \cdots & \cdots & \cdots & \cdots & \cdots \\ \cdots & \cdots & \cdots & \cdots & \cdots \\ a_1b_n - b_1a_n & a_1c_n - c_1a_n & a_1d_n - d_1a_n & \cdots & \cdots \end{vmatrix}$$

as the determinant is of $(n-1)^{\text{th}}$ order and $\frac{1}{a_1}$ is common in every row (or column)

$$= \frac{1}{(a_1)^{n-2}} \begin{vmatrix} \begin{vmatrix} a_1 & b_1 \\ a_2 & b_2 \end{vmatrix} & \begin{vmatrix} a_1 & c_1 \\ a_2 & c_2 \end{vmatrix} & \begin{vmatrix} a_1 & d_1 \\ a_2 & d_2 \end{vmatrix} & & \\ \begin{vmatrix} a_1 & b_1 \\ a_3 & b_3 \end{vmatrix} & \begin{vmatrix} a_1 & c_1 \\ a_3 & c_3 \end{vmatrix} & \begin{vmatrix} a_1 & d_1 \\ a_3 & d_3 \end{vmatrix} & \cdots & \cdots \\ \begin{vmatrix} a_1 & b_1 \\ a_4 & b_4 \end{vmatrix} & \begin{vmatrix} a_1 & c_1 \\ a_4 & c_4 \end{vmatrix} & \begin{vmatrix} a_1 & d_1 \\ a_4 & d_4 \end{vmatrix} & \cdots & \cdots \\ \cdots & \cdots & \cdots & \cdots & \cdots \\ \cdots & \cdots & \cdots & \cdots & \cdots \\ \begin{vmatrix} a_1 & b_1 \\ a_n & b_n \end{vmatrix} & \begin{vmatrix} a_1 & c_1 \\ a_n & c_n \end{vmatrix} & \begin{vmatrix} a_1 & d_1 \\ a_n & d_n \end{vmatrix} & \cdots & \cdots \end{vmatrix}$$

Thus, the $n^{th}$ order determinant is condensed to $(n-1)$th order determinant. Repeated application of this method ultimately results in a determinant of 2nd order which can be evaluated.

It is obvious that the leading element $a_1$ behaves like a pivot in the condensation process (*i.e.*, reduction from $n$ to $(n-1)$) and hence the method is pivotal condensation.

If the leading element is zero, it can be made non-zero by interchanging the columns.

**Example 45.** Condense the following determinants to second order and hence evaluate them:

(*i*) $\begin{vmatrix} 10 & 2 & -3 \\ 5 & 12 & 15 \\ 7 & -6 & 4 \end{vmatrix}$ (*ii*) $D = \begin{vmatrix} 2 & 1 & 3 & 5 \\ 4 & -2 & 7 & 6 \\ -8 & 3 & 1 & 0 \\ 5 & 7 & 2 & -6 \end{vmatrix}$

**Solution.** (*i*) Using the leading element as pivot, we get

$$D = \frac{1}{(10)^{3-2}} \begin{vmatrix} 120-10 & 150+15 \\ -60-14 & 40+21 \end{vmatrix} \quad [\because \text{ order } = 3]$$

$$\Rightarrow D = \frac{1}{10} \begin{vmatrix} 110 & 165 \\ -74 & 61 \end{vmatrix} = \frac{55}{10} \begin{vmatrix} 2 & 3 \\ -74 & 61 \end{vmatrix} = \frac{11}{2}(122+222) = \frac{11}{2} \times 344 = 11 \times 172 = \mathbf{1892}.$$ **Ans.**

(*ii*) $\dfrac{1}{(2)^{4-2}} \begin{vmatrix} -4-4 & 14-12 & 12-20 \\ 6+8 & 2+24 & 0+40 \\ 14-5 & 4-15 & -12-25 \end{vmatrix}$ as the order is 4.

$$= \frac{1}{4} \begin{vmatrix} -8 & 2 & -8 \\ 14 & 26 & 40 \\ 9 & -11 & -37 \end{vmatrix} = \frac{2 \times 2}{4} \begin{vmatrix} -4 & 1 & -4 \\ 7 & 13 & 20 \\ 9 & -11 & -37 \end{vmatrix} = \frac{1}{(-4)^{3-2}} \begin{vmatrix} -52-7 & -80+28 \\ 44-9 & 148+36 \end{vmatrix}$$

$$= -\frac{1}{4} \begin{vmatrix} -59 & -52 \\ 35 & 184 \end{vmatrix} = \frac{4}{4} \begin{vmatrix} 59 & 52 \\ 35 & 46 \end{vmatrix} = 59 \times 46 - 13 \times 35 = \mathbf{2259}.$$ **Ans.**

**Example 46.** Condense and hence evaluate the determinant, $\begin{vmatrix} 0 & 4 & 1 & 2 \\ 5 & 3 & 7 & 8 \\ 4 & 1 & 2 & 3 \\ 1 & 2 & 5 & 5 \end{vmatrix}$

**Solution.** As the leading element is zero, hence interchanging the 1st and second columns, we get

$$\begin{vmatrix} 0 & 4 & 1 & 2 \\ 5 & 3 & 7 & 8 \\ 4 & 1 & 2 & 3 \\ 1 & 2 & 5 & 5 \end{vmatrix} = - \begin{vmatrix} 4 & 0 & 1 & 2 \\ 3 & 5 & 7 & 8 \\ 1 & 4 & 2 & 3 \\ 2 & 1 & 5 & 5 \end{vmatrix} = -\frac{1}{4^2} \begin{vmatrix} 20-0 & 28-3 & 32-6 \\ 16-0 & 8-1 & 12-2 \\ 4-0 & 20-2 & 20-4 \end{vmatrix} = -\frac{1}{16} \begin{vmatrix} 20 & 25 & 26 \\ 16 & 7 & 10 \\ 4 & 18 & 16 \end{vmatrix}$$

$$= -\frac{4 \times 2}{16} \begin{vmatrix} 5 & 25 & 13 \\ 4 & 7 & 5 \\ 1 & 18 & 8 \end{vmatrix} = -\frac{1}{2} \cdot \frac{1}{5} \begin{vmatrix} 35-100 & 25-52 \\ 90-25 & 40-13 \end{vmatrix} = -\frac{1}{10} \begin{vmatrix} -65 & -27 \\ 65 & 27 \end{vmatrix} = \frac{1}{10} \begin{vmatrix} 65 & 27 \\ 65 & 27 \end{vmatrix} = 0.$$

**Example 47.** By condensing the given determinant evaluate $x$, $\begin{vmatrix} x-1 & 7 & 9 & 3 \\ 1 & 0 & 2 & 5 \\ 2x+2 & 6 & 8 & 3 \\ -2 & 1 & 1 & 0 \end{vmatrix} = 0.$

**Solution.** $D = \begin{vmatrix} x-1 & 7 & 9 & 3 \\ 1 & 0 & 2 & 5 \\ 2x+2 & 6 & 8 & 3 \\ -2 & 1 & 1 & 0 \end{vmatrix} = -\begin{vmatrix} 7 & x-1 & 9 & 3 \\ 0 & 1 & 2 & 5 \\ 6 & 2x+2 & 8 & 3 \\ 1 & -2 & 1 & 0 \end{vmatrix}$

$$= -\frac{1}{7^2}\begin{vmatrix} 7 & 14 & 35 \\ 14x+14-6x+6 & 56-54 & 21-18 \\ -14-x+1 & 7-9 & 0-3 \end{vmatrix} = -\frac{1}{7^2}\times 7\begin{vmatrix} 1 & 2 & 5 \\ 8x+20 & 2 & 3 \\ -x-13 & -2 & -3 \end{vmatrix}$$

$$= +\frac{2}{7}\begin{vmatrix} 1 & 1 & 5 \\ 8x+20 & 1 & 3 \\ x+13 & 1 & 3 \end{vmatrix} = \frac{2}{7}\begin{vmatrix} 1-8x-20 & 3-40x-100 \\ 1-x-13 & 3-5x-65 \end{vmatrix}$$

$$= \frac{2}{7}\begin{vmatrix} -8x-19 & -40x-97 \\ -x-12 & -5x-62 \end{vmatrix} = \frac{2}{7}\left[(8x+19)(5x+62)-(40x+97)(x+12)\right]$$

$$= \frac{2}{7}\left[40x^2+95x+496x+1178-40x^2-97x-480x-1164\right] = \frac{2}{7}[14x+14] = 4x+4$$

Thus $4x+4=0 \quad \Rightarrow \quad x+1=0 \quad \Rightarrow x=-1$ **Ans.**

## EXERCISE 4.6

Using the leading element as pivots, condense the following determinants to second order and hence evaluate them.

**1.** $\begin{vmatrix} 1 & 3 & 7 \\ 4 & 9 & 1 \\ 2 & 7 & 6 \end{vmatrix}$ **Ans.** 51

**2.** $\begin{vmatrix} 2 & 0 & 2 \\ 3 & 7 & 4 \\ -2 & -5 & 1 \end{vmatrix}$ **Ans.** 52

**3.** $\begin{vmatrix} 5 & 2 & 7 \\ 9 & 1 & 10 \\ -2 & 3 & 4 \end{vmatrix}$ **Ans.** $-39$

**4.** $\begin{vmatrix} 1 & 2 & 1 & 3 \\ 3 & 4 & 2 & 5 \\ 6 & 1 & 7 & 1 \\ 4 & 3 & 9 & 2 \end{vmatrix}$ **Ans.** 75

**5.** $\begin{vmatrix} 4 & -2 & 3 & 0 \\ 1 & 0 & 2 & 7 \\ -5 & 1 & 6 & 1 \\ 2 & 3 & 5 & -4 \end{vmatrix}$ **Ans.** $-1334$

**6.** $\begin{vmatrix} 1 & 2 & 3 & 4 & 5 \\ 3 & 7 & 2 & 1 & 1 \\ 9 & 4 & 1 & 2 & 3 \\ 8 & 1 & 3 & 7 & 2 \\ 4 & 2 & 0 & 3 & 1 \end{vmatrix}$ **Ans.** $-2276$

**7.** Condense the following determinant and hence evaluate $x$,

$\begin{vmatrix} 3 & 2 & 1 & 5 \\ 4 & 7 & 6 & 2 \\ 2 & 1 & x=1 & -4 \\ 5 & 3 & 4 & 1 \end{vmatrix} = 0.$ **Ans.** $x=4$

## 4.10 CONJUGATE ELEMENTS

Two equidistant elements lying on a line perpendicular to the leading diagonal are said to be conjugate.

In the determinant $\begin{vmatrix} a_1 & b_1 & c_1 \\ a_2 & b_2 & c_2 \\ a_3 & b_3 & c_3 \end{vmatrix}$, $a_2, b_1$; $a_3, c_1$; $b_3, c_2$; are pairs of conjugate elements.

## 4.11 SPECIAL TYPES OF DETERMINANTS

***(i)* Orthosymmetric Determinant.** If every element of the leading diagonal is the same and the conjugate elements are equal, then the determinant is said to be orthosymmetric determinant.

$$\begin{vmatrix} a & h & g \\ h & a & f \\ g & f & a \end{vmatrix}$$

***(ii)* Skew-Symmetric Determinant.** If the elements of the leading diagonal are all zero and every other element is equal to its conjugate with sign changed, the determinant is said to be Skew-symmetric.

$$\begin{vmatrix} 0 & -a & -b \\ a & 0 & -c \\ b & c & 0 \end{vmatrix}$$

**Property 1.** A Skew-symmetric determinant of odd order vanishes.

**Example 48.** *Prove that* $\Delta = \begin{vmatrix} 0 & -a & -b \\ a & 0 & -c \\ b & c & 0 \end{vmatrix} = 0$

**Solution.** Taking out (– 1) common from each of the three columns

$$\Delta = (-1)^3 \begin{vmatrix} 0 & a & b \\ -a & 0 & c \\ -b & -c & 0 \end{vmatrix}$$

Changing rows into columns $\Delta = (-1)^3 \begin{vmatrix} 0 & -a & -b \\ a & 0 & -c \\ b & c & 0 \end{vmatrix} = (-1)^3\, \Delta = -\Delta$

or $2\Delta = 0$ or $\Delta = 0$ **Proved.**

**Property 2.** A skew-symmetric determinant of even order is a perfect square.

**Example 49.** Prove that $\begin{vmatrix} 0 & x & y & z \\ -x & 0 & c & b \\ -y & -c & 0 & a \\ -z & -b & -a & 0 \end{vmatrix} = (ax - by + cz)^2$

**Solution.** Multiplying column 2 by $a$ the given determinant is $= \dfrac{1}{a}\begin{vmatrix} 0 & ax & y & z \\ -x & 0 & c & b \\ -y & -ac & 0 & a \\ -z & -ab & -a & 0 \end{vmatrix}$

On expanding by column 2, we get

$$= \frac{-(ax - by + cz)}{a}\begin{vmatrix} -x & c & b \\ -y & 0 & a \\ -z & -a & 0 \end{vmatrix} = \frac{(ax - by + cz)}{a}\begin{vmatrix} x & c & b \\ y & 0 & a \\ z & -a & 0 \end{vmatrix}$$

$$= \frac{(ax-by+cz)}{a\times a}\begin{vmatrix} ax & ac & ab \\ y & 0 & a \\ z & -a & 0 \end{vmatrix} = \frac{(ax-by+cz)}{a^2}\begin{vmatrix} ax-by+cz & ac-ac & ab-ab \\ y & 0 & a \\ z & -a & 0 \end{vmatrix}$$

$[\because R_1 \to R_1 - bR_2 + cR_3]$

$$= \frac{(ax-by+cz)}{a^2}\begin{vmatrix} ax-by+cz & 0 & 0 \\ y & 0 & a \\ z & -a & 0 \end{vmatrix} = \frac{(ax-by+cz)}{a^2}(ax-by+cz)(a^2)$$

$= (ax - by + cz)^2$ **Proved.**

## 4.12 LAPLACE METHOD FOR THE EXPANSION OF A DETERMINANT IN TERMS OF FIRST TWO ROWS

(*i*) Make all possible determinants from first two rows by taking any two columns.

(*ii*) Multiply each of them by corresponding determinant which is left by suppressing the rows and columns intersecting at them.

(*iii*) Add them with proper signs.

**Here we count the number of movements of columns of the determinant by shifting to the place of the first determinant. If the number of movement is odd then negative sign, if even then positive sign.**

**Example 50.** Expand the determinant $\begin{vmatrix} a_1 & b_1 & c_1 & d_1 \\ a_2 & b_2 & c_2 & d_2 \\ a_3 & b_3 & c_3 & d_3 \\ a_4 & b_4 & c_4 & d_4 \end{vmatrix}$ by Laplace method.

**Solution.** $\Delta = \begin{vmatrix} a_1 & b_1 \\ a_2 & b_2 \end{vmatrix}\begin{vmatrix} c_3 & d_3 \\ c_4 & d_4 \end{vmatrix} - \begin{vmatrix} a_1 & c_1 \\ a_2 & c_2 \end{vmatrix}\begin{vmatrix} b_3 & d_3 \\ b_4 & d_4 \end{vmatrix} + \begin{vmatrix} a_1 & d_1 \\ a_2 & d_2 \end{vmatrix}\begin{vmatrix} b_3 & c_3 \\ b_4 & c_4 \end{vmatrix} + \begin{vmatrix} b_1 & c_1 \\ b_2 & c_2 \end{vmatrix}\begin{vmatrix} a_3 & d_3 \\ a_4 & d_4 \end{vmatrix}$

$- \begin{vmatrix} b_1 & d_1 \\ b_2 & d_2 \end{vmatrix}\begin{vmatrix} a_3 & c_3 \\ a_4 & c_4 \end{vmatrix} + \begin{vmatrix} c_1 & d_1 \\ c_2 & d_2 \end{vmatrix}\begin{vmatrix} a_3 & b_3 \\ a_4 & b_4 \end{vmatrix}$ Explanation : $\begin{vmatrix} a_1 & c_1 \\ a_2 & c_2 \end{vmatrix}$

Now the c column being 3rd can be made 2nd by one movement of column; "*a*" column is in the position of first column so that the total number of movements is one i.e. odd; hence the sign will be –ve.

**Ans.**

**Example 51.** Expand the following determinant by Laplace method : $\begin{vmatrix} a_1 & b_1 & 0 & 0 \\ a_2 & b_2 & 0 & 0 \\ a_3 & b_3 & c_3 & d_3 \\ a_4 & b_4 & c_4 & d_4 \end{vmatrix}$

**Solution.** $\Delta = \begin{vmatrix} a_1 & b_1 \\ a_2 & b_2 \end{vmatrix}\begin{vmatrix} c_3 & d_3 \\ c_4 & d_4 \end{vmatrix} - \begin{vmatrix} a_1 & 0 \\ a_2 & 0 \end{vmatrix}\begin{vmatrix} b_3 & d_3 \\ b_4 & d_4 \end{vmatrix}$

$+ \begin{vmatrix} a_1 & 0 \\ a_2 & 0 \end{vmatrix}\begin{vmatrix} b_3 & c_3 \\ b_4 & c_4 \end{vmatrix} + \begin{vmatrix} b_1 & 0 \\ b_2 & 0 \end{vmatrix}\begin{vmatrix} a_3 & d_3 \\ a_4 & d_4 \end{vmatrix} - \begin{vmatrix} b_1 & 0 \\ b_2 & 0 \end{vmatrix}\begin{vmatrix} a_3 & c_3 \\ a_4 & c_4 \end{vmatrix} + \begin{vmatrix} 0 & 0 \\ 0 & 0 \end{vmatrix}\begin{vmatrix} a_3 & b_3 \\ a_4 & b_4 \end{vmatrix}$

$= \begin{vmatrix} a_1 & b_1 \\ a_2 & b_2 \end{vmatrix}\begin{vmatrix} c_3 & d_3 \\ c_4 & d_4 \end{vmatrix} = (a_1b_2 - \alpha_2 b_1)(c_3d_4 - c_4d_3)$ **Ans.**

## 4.13 APPLICATION OF DETERMINANTS

**Area of triangle.** We know that the area of a triangle, whose vertices are $(x_1, y_1)$, $(x_2, y_2)$ and $(x_3, y_3)$ is given by

$$\Delta = \frac{1}{2}\left[x_1(y_2 - y_3) - x_2(y_1 - y_3) + x_3(y_1 - y_2)\right]$$

$$= \frac{1}{2}\left[x_1\begin{vmatrix} y_2 & 1 \\ y_3 & 1 \end{vmatrix} - x_2\begin{vmatrix} y_1 & 1 \\ y_3 & 1 \end{vmatrix} + x_3\begin{vmatrix} y_1 & 1 \\ y_2 & 1 \end{vmatrix}\right] = \frac{1}{2}\begin{vmatrix} x_1 & y_1 & 1 \\ x_2 & y_2 & 1 \\ x_3 & y_3 & 1 \end{vmatrix}$$

**Note.** Since area is always a positive quantity, therefore we always take the absolute value of the determinant for the area.

**Condition of collinearity of three points.** Let $A(x_1, y_1)$, $B(x_2, y_2)$ and $C(x_3, y_3)$ be three points. Then,

$A, B, C$ are collinear $\Leftrightarrow$ area of triangle $ABC = 0$

$$\Leftrightarrow \frac{1}{2}\begin{vmatrix} x_1 & y_1 & 1 \\ x_2 & y_2 & 1 \\ x_3 & y_3 & 1 \end{vmatrix} = 0 \quad \Leftrightarrow \quad \begin{vmatrix} x_1 & y_1 & 1 \\ x_2 & y_2 & 1 \\ x_3 & y_3 & 1 \end{vmatrix} = 0$$

**Proved.**

**Example 52.** *Using determinants, find the area of the triangle with vertices (– 3, 5), (3, – 6) and (7, 2).*

**Solution.** The area of the given triangle $= \frac{1}{2}\begin{vmatrix} -3 & 5 & 1 \\ 3 & -6 & 1 \\ 7 & 2 & 1 \end{vmatrix}$

Operate : $R_1 \to R_1 - R_2$ ; $R_2 \to R_2 - R_3$ $= \frac{1}{2}\begin{vmatrix} -6 & 11 & 0 \\ -4 & -8 & 0 \\ 7 & 2 & 1 \end{vmatrix}$

Expand by $C_3$ $= \frac{1}{2}.1.\begin{vmatrix} -6 & 11 \\ -4 & -8 \end{vmatrix} = \frac{1}{2}(48 + 44) = 46$ sq. units **Ans.**

**Example 53.** *Using determinants, show that the points (11, 7), (5, 5) and (– 1, 3) are collinear.*

**Solution.** The area of the triangle formed by the given points $= \frac{1}{2}\begin{vmatrix} 11 & 7 & 1 \\ 5 & 5 & 1 \\ -1 & 3 & 1 \end{vmatrix}$

Operate : $R_1 \to R_1 - R_2$ ; $R_2 \to R_2 - R_3$

$$= \frac{1}{2}\begin{vmatrix} 6 & 2 & 0 \\ 6 & 2 & 0 \\ -1 & 3 & 1 \end{vmatrix} = \frac{1}{2}.0 = 0. \qquad (\because R_1 \text{ and } R_2 \text{ are identical})$$

$\Rightarrow$ The three given points are collinear. **Proved.**

**Example 54.** *Using determinants, find the area of the triangle whose vertices are (1, 4) (2, 3) and (– 5, – 3). Are the given points collinear?*

**Solution.** Area of the required triangle $= \frac{1}{2}\begin{vmatrix} 1 & 4 & 1 \\ 2 & 3 & 1 \\ -5 & -3 & 1 \end{vmatrix}$

$$= \frac{1}{2}[1(3+3) - 4(2+5) + 1(-6+15)] = \frac{1}{2}(6 - 28 + 9) = \frac{13}{2} \neq 0$$

Hence, the given points are not collinear. **Ans.**

## EXERCISE 4.7

**Using determinants, find the area of the triangle with vertices:**

**1.** (2,– 7), (1, 3), (10, 8). **Ans.** Area $= \frac{95}{2}$ **2.** (– 2, 4), (2, – 6) and (5, 4). **Ans.** Area = 35

**3.** (– 1, – 3), (2, 4) and (3, – 1). **Ans.** Area = 11 **4.** (1, – 1), (2, 4) and (– 3, 5). **Ans.** Area = 13

**5.** Using determinants, show that the points (3, 8), (– 4, 2) and (10, 14) are collinear.

**6.** Find the value of α, so that the points (1, – 5), (– 4, 5) and (α, 7) are collinear. **Ans.** α = –5

**7.** Find the value of *x*, if the area of Δ is 35 square cms with vertices (*x*, 4), (2, – 6), (5, 4). **Ans.** *x* = – 2, 12

**8.** Using determinants find the value of *k*, so that the points (*k*, 2 – 2*k*), (– *k* + 1, 2*k*) and (– 4 – *k*, 6 – 2 *k*) may be collinear. **Ans. k = –1, $\frac{1}{2}$**

**9.** If the points (*x*, – 2), (5, 2) and (8, 8) are collinear, find *x* using determinants. **Ans.** *x* = 3

**10.** If the points (3, – 2), (*x*, 2) and (8, 8) are collinear, find *x* using determinants. **Ans.** *x* = 1

## 4.14. SOLUTION OF SIMULTANEOUS LINEAR EQUATIONS BY DETERMINANTS (CRAMER'S RULE)

Let us solve the following equations.

$$a_1x + b_1y + c_1z = d_1$$
$$a_2x + b_2y + c_2z = d_2$$
$$a_3x + b_3y + c_3z = d_3$$

Let $D = \begin{vmatrix} a_1 & b_1 & c_1 \\ a_2 & b_2 & c_2 \\ a_3 & b_3 & c_3 \end{vmatrix}$ or $x\,D = \begin{vmatrix} a_1x & b_1 & c_1 \\ a_2x & b_2 & c_2 \\ a_3x & b_3 & c_3 \end{vmatrix}$

Multiplying the 2nd column by *y* and 3rd column by *z* and adding to the 1st column, we get

$$x\,D = \begin{vmatrix} a_1x + b_1y + c_1z & b_1 & c_1 \\ a_2x + b_2y + c_2z & b_2 & c_2 \\ a_3x + b_3y + c_3z & b_3 & c_3 \end{vmatrix} \Rightarrow x\,D = \begin{vmatrix} d_1 & b_1 & c_1 \\ d_2 & b_2 & c_2 \\ d_3 & b_3 & c_3 \end{vmatrix}$$

$$\Rightarrow \quad x = \frac{\begin{vmatrix} d_1 & b_1 & c_1 \\ d_2 & b_2 & c_2 \\ d_3 & b_1 & c_3 \end{vmatrix}}{\begin{vmatrix} a_1 & b_1 & c_1 \\ a_2 & b_2 & c_2 \\ a_3 & b_3 & c_3 \end{vmatrix}} = \frac{D_1}{D} \quad \text{Similarly,} \quad y = \frac{D_2}{D} = \frac{\begin{vmatrix} a_1 & d_1 & c_1 \\ a_2 & d_2 & c_2 \\ a_3 & d_3 & c_3 \end{vmatrix}}{\begin{vmatrix} a_1 & b_1 & c_1 \\ a_2 & b_2 & c_2 \\ a_3 & b_3 & c_3 \end{vmatrix}}$$

$$z = \frac{D_3}{D} = \frac{\begin{vmatrix} a_1 & b_1 & d_1 \\ a_2 & b_2 & d_2 \\ a_3 & b_3 & d_3 \end{vmatrix}}{\begin{vmatrix} a_1 & b_1 & c_1 \\ a_2 & b_2 & c_2 \\ a_3 & b_3 & c_3 \end{vmatrix}}$$

$$x = \frac{D_1}{D}, \qquad y = \frac{D_2}{D}, \qquad z = \frac{D_3}{D}$$ **Ans.**

**Example 55.** *Solve the following system of equations using Cramer's rule :*

$$5x - 7y + z = 11$$
$$6x - 8y - z = 15$$
$$3x + 2y - 6z = 7$$

**Solution.** The given equations are

$$5x - 7y + z = 11$$
$$6x - 8y - z = 15$$
$$3x + 2y - 6z = 7$$

Here, $$D = \begin{vmatrix} 5 & -7 & 1 \\ 6 & -8 & -1 \\ 3 & 2 & -6 \end{vmatrix} = 5\,(48 + 2) + 7\,(-36 + 3) + 1\,(12 + 24) = 55\ (\neq 0)$$

$$D_1 = \begin{vmatrix} 11 & -7 & 1 \\ 15 & -8 & -1 \\ 7 & 2 & -6 \end{vmatrix} = 11\,(48 + 2) + 7\,(-90 + 7) + 1\,(30 + 56) = 55$$

$$D_2 = \begin{vmatrix} 5 & 11 & 1 \\ 6 & 15 & -1 \\ 3 & 7 & -6 \end{vmatrix} = 5\,(-90 + 7) - 11\,(-36 + 3) + 1\,(42 - 45) = -55$$

$$D_3 = \begin{vmatrix} 5 & -7 & 11 \\ 6 & -8 & 15 \\ 3 & 2 & 7 \end{vmatrix} = 5\,(-56 - 30) + 7\,(42 - 45) + 11\,(12 + 24) = -55$$

By Cramer's Rule $x = \frac{D_1}{D} = \frac{55}{55} = 1$, $y = \frac{D_2}{D} = \frac{-55}{55} = -1$, $z = \frac{D_3}{D} = \frac{-55}{55} = -1$

Hence, $x = 1$, $y = -1$, $z = -1$ **Ans.**

**Example 56.** *Solve, by determinants, the following set of simultaneous equations :*

$$5x - 6y + 4z = 15$$
$$7x + 4y - 3z = 19$$
$$2x + y + 6z = 46$$

**Solution.** $$D = \begin{vmatrix} 5 & -6 & 4 \\ 7 & 4 & -3 \\ 2 & 1 & 6 \end{vmatrix} = 419$$

$$D_1 = \begin{vmatrix} 15 & -6 & 4 \\ 19 & 4 & -3 \\ 46 & 1 & 6 \end{vmatrix} = 1257, \quad D_2 = \begin{vmatrix} 5 & 15 & 4 \\ 7 & 19 & -3 \\ 2 & 46 & 6 \end{vmatrix} = 1676, \quad D_3 = \begin{vmatrix} 5 & -6 & 15 \\ 7 & 4 & 19 \\ 2 & 1 & 46 \end{vmatrix} = 2514$$

By Cramer's Rule: $x = \frac{D_1}{D} = \frac{1257}{419} = 3, \quad y = \frac{D_2}{D} = \frac{1676}{419} = 4, \quad z = \frac{D_3}{D} = \frac{2514}{419} = 6.$

Hence, $x = 3, \quad y = 4, \quad z = 6$ **Ans.**

**Example 57.** *Solve the following system of equations using Cramer's Rule :*

$$2x - 3y + 4z = -9$$
$$-3x + 4y + 2z = -12$$
$$4x - 2y - 3z = -3$$

**Solution.** The given equations are

$$2x - 3y + 4z = -9$$
$$-3x + 4y + 2z = -12$$
$$4x - 2y - 3z = -3$$

Here $D = \begin{vmatrix} 2 & -3 & 4 \\ -3 & 4 & 2 \\ 4 & -2 & -3 \end{vmatrix} = 2(-12+4) + 3(9-8) + 4(6-16) = -53$

$$D_1 = \begin{vmatrix} -9 & -3 & 4 \\ -12 & 4 & 2 \\ -3 & -2 & -3 \end{vmatrix} = -9(-12+4) + 3(36+6) + 4(24+12) = -342$$

$$D_2 = \begin{vmatrix} 2 & -9 & 4 \\ -3 & -12 & 2 \\ 4 & -3 & -3 \end{vmatrix} = 2(36+6) + 9(9-8) + 4(9+48) = -321$$

$$D_3 = \begin{vmatrix} 2 & -3 & -9 \\ -3 & 4 & -12 \\ 4 & -2 & -3 \end{vmatrix} = 2(-12-24) + 3(9+48) - 9(6-16) = -189$$

By Cramer's Rule,

$$x = \frac{D_1}{D} = \frac{-342}{-53} = \frac{342}{53}, \quad y = \frac{D_2}{D} = \frac{-321}{-53} = \frac{321}{53}, \quad z = \frac{D_3}{D} = \frac{-189}{-53} = \frac{189}{53}$$

Hence, $x = \frac{342}{53}, \quad y = \frac{321}{53}, \quad z = \frac{189}{53}$ **Ans.**

**Example 58.** *Solve the following system of equations by using determinants :*

$$x + y + z = 1$$
$$ax + by + cz = k$$
$$a^2x + b^2y + c^2z = k^2$$

**Solution.** We have $D = \begin{vmatrix} 1 & 1 & 1 \\ a & b & c \\ a^2 & b^2 & c^2 \end{vmatrix}$

$$= \begin{vmatrix} 1 & 0 & 0 \\ a & b-a & c-a \\ a^2 & b^2-a^2 & c^2-a^2 \end{vmatrix} \text{ [Applying } C_2 \to C_2 - C_1 \text{ and } C_3 \to C_3 - C_1]$$

$$= (b-a)(c-a) \begin{vmatrix} 1 & 0 & 0 \\ a & 1 & 1 \\ a^2 & b+a & c+a \end{vmatrix}$$

$$= (b-a)(c-a).1.\begin{vmatrix} 1 & 1 \\ b+a & c+a \end{vmatrix} \qquad \text{[Expanding along } R_1]$$

$$= (b-a)\,(c-a)\,(c+a-b-a)$$

$$= (b-c)\,(c-a)\,(a-b) \qquad ...(1)$$

$$D_1 = \begin{vmatrix} 1 & 1 & 1 \\ k & b & c \\ k^2 & b^2 & c^2 \end{vmatrix} = (b-c)(c-k)(k-b) \qquad \text{[Replacing } a \text{ by } k \text{ in (1)]}$$

$$D_2 = \begin{vmatrix} 1 & 1 & 1 \\ a & k & c \\ a^2 & k^2 & c^2 \end{vmatrix} = (k-c)(c-a)(a-k) \qquad \text{[Replacing } b \text{ by } k \text{ in (1)]}$$

and $$D_3 = \begin{vmatrix} 1 & 1 & 1 \\ a & b & k \\ a^2 & b^2 & k^2 \end{vmatrix} = (a-b)(b-k)(k-a) \qquad \text{[Replacing } c \text{ by } k \text{ in (1)]}$$

$\therefore$ $$x = \frac{D_1}{D} = \frac{(b-c)(c-k)(k-b)}{(b-c)(c-a)(a-b)}, \quad y = \frac{D_2}{D} = \frac{(k-c)(c-a)(a-k)}{(b-c)(c-a)(a-b)}$$

and $$z = \frac{D_3}{D} = \frac{(a-b)(b-k)(k-a)}{(a-b)(b-c)(c-a)}$$

Hence, $$x = \frac{(c-k)(k-b)}{(c-a)(a-b)}, \quad y = \frac{(k-c)(a-k)}{(b-c)(a-b)} \quad \text{and} \quad z = \frac{(b-k)(k-a)}{(b-c)(c-a)}$$ **Ans.**

**Example 59.** *The sum of three numbers is 6. If we multiply the third number by 2 and add the first number to the result, we get 7. By adding second and third numbers to three times the first number we get 12. Use determinants to find the numbers.*

**Solution.** Let the three numbers be $x$, $y$ and $z$. Then, from the given conditions, we have

$$\left.\begin{aligned} x+y+z&=6 \\ x+2z&=7 \\ 3x+y+z&=12 \end{aligned}\right\} \quad \text{or} \quad \left\{\begin{aligned} x+y+z&=6 \\ x+0.y+2z&=7 \\ 3x+y+z&=12 \end{aligned}\right.$$

Here, $$D = \begin{vmatrix} 1 & 1 & 1 \\ 1 & 0 & 2 \\ 3 & 1 & 1 \end{vmatrix} = 1(0-2) - 1(1-6) + 1(1-0) = -2+5+1 = 4$$

$$D_1 = \begin{vmatrix} 6 & 1 & 1 \\ 7 & 0 & 2 \\ 12 & 1 & 1 \end{vmatrix} = 6(0-2) - 1(7-24) + (7-0) = -12+17+7 = 12$$

$$D_2 = \begin{vmatrix} 1 & 6 & 1 \\ 1 & 7 & 2 \\ 3 & 12 & 1 \end{vmatrix} = 1(7-24) - 6(1-6) + 1(12-21) = -17 + 30 - 9 = 4$$

and $$D_3 = \begin{vmatrix} 1 & 1 & 6 \\ 1 & 0 & 7 \\ 3 & 1 & 12 \end{vmatrix} = 1(0-7) - 1(12-21) + 6(1-0) = -7 + 9 + 6 = 8$$

$$\therefore \quad x = \frac{D_1}{D} = \frac{12}{4} = 3, \quad y = \frac{D_2}{D} = \frac{4}{4} = 1, \text{ and } z = \frac{D_3}{D} = \frac{8}{4} = 2$$

Thus, the three numbers are 3, 1 and 2. **Ans.**

## EXERCISE 4.8

**Using Cramer's Rule, solve the following system of equations :**

**1.** $2x - 3y = 7$
$7x - 3y = 10$
**Ans.** $x = \frac{3}{5}, y = -\frac{29}{15}$

**2.** $2x + y = 1$
$x - 2y = 8$
**Ans.** $x = 2, y = -3$

**3.** $2x + 3y = 10$
$x + 6y = 4.$
**Ans.** $x = \frac{16}{3}, y = -\frac{2}{9}$

**4.** $5x + 2y = 3$
$3x + 2y = 5.$
**Ans.** $x = -1, y = 4$

**5.** $7x - 2y = -7$
$2x - y = 1.$
**Ans.** $x = -3, y = -7$

**6.** $x - 2y = 4$
$-3x + 5y = -7$
**Ans.** $x = -6, y = -5$

**7.** $x - 4y - z = 11$
$2x - 5y + 2z = 39$
$-3x + 2y + z = 1.$
**Ans.** $x = -1, y = -5, z = 8$

**8.** $x + 3y - 2z = 5$
$2x + y + 4z = 8$
$6x + y - 3z = 5.$
**Ans.** $x = 1, y = 2, z = 1$

**9.** $x + 2y + 5z = 23$
$3x + y + 4z = 26$
$6x + y + 7z = 47$
**Ans.** $x = 4, y = 2, z = 3$

**10.** $x + y + z = 1$
$3x + 5y + 6z = 4$
$9x + 2y - 36x = 17$
**Ans.** $x = \frac{1}{3}, y = 1, z = -\frac{1}{3}$

**11.** $2y - z = 0$
$x + 3y = -4$
$3x + 4y = 3$
**Ans.** $x = 5, y = -3, z = -6$

**12.** $x + y + z = -1$
$x + 2y + 3z = -4$
$x + 3y + 4z = -6$
**Ans.** $x = 1, y = -1, z = -1$

**13.** $x + y + z = 1$
$x + 2y + 3z = k$
$1^2x + 2^2y + 3^2z = k^2$ **Ans.** $x = \frac{(2-k)(3-k)}{2}, \; y = \frac{(1-k)(3-k)}{-1}, \; z = \frac{(1-k)(2-k)}{2}$

**14.** Show that there are three real values of $\lambda$ for which the equations:
$(a - \lambda)x + by + cz = 0$
$bx + (c - \lambda)y + az = 0$
$cx + ay + (b - \lambda)z = 0$
are simultaneously true, and that the product of these values of $\lambda$ is $\begin{vmatrix} a & b & c \\ b & c & a \\ c & a & b \end{vmatrix}$

**15.** Solve the following system of equations by using the Cramer's Rule

$x_1 + x_2 = 1;\; x_2 + x_3 = 0;\; x_3 + x_4 = 0;\; x_4 + x_5 = 0;\; x_5 + x_1 = 0$ (*A.M.I.E.T.E., Summer 2005*)

**Ans.** $x_1 = \frac{1}{2}, x_2 = \frac{1}{2}, x_3 = -\frac{1}{2}, x_4 = \frac{1}{2}, x_5 = -\frac{1}{2}$

## 4.15 RULE FOR MULTIPLICATION OF TWO DETERMINANTS

Multiply the elements of the first row of $\Delta_1$ with the corresponding elements of the first, the second and the third row of $\Delta_2$ respectively.

Their respective sums form the elements of the first row of $\Delta_1\Delta_2$. Similarly multiply the elements of the second row of $\Delta_1$ with the corresponding elements of first, second and third row of $\Delta_2$ to form the second row of $\Delta_1\Delta_2$ and so on.

**Example 60.** *Find the product* $\begin{vmatrix} a_1 & b_1 & c_1 \\ a_2 & b_2 & c_2 \\ a_3 & b_3 & c_3 \end{vmatrix} \times \begin{vmatrix} \alpha_1 & \beta_1 & \gamma_1 \\ \alpha_2 & \beta_2 & \gamma_2 \\ \alpha_3 & \beta_3 & \gamma_3 \end{vmatrix}$

**Solution.** Product of the given determinants

$$= \begin{vmatrix} a_1\alpha_1 + b_1\beta_1 + c_1\gamma_1 & a_1\alpha_2 + b_1\beta_2 + c_1\gamma_2 & a_1\alpha_3 + b_1\beta_3 + c_1\gamma_3 \\ a_2\alpha_1 + b_2\beta_1 + c_2\gamma_1 & a_2\alpha_2 + b_2\beta_2 + c_2\gamma_2 & a_2\alpha_3 + b_2\beta_3 + c_2\gamma_3 \\ a_3\alpha_1 + b_3\beta_1 + c_3\gamma_1 & a_3\alpha_2 + b_3\beta_2 + c_3\gamma_2 & a_3\alpha_3 + b_3\beta_3 + c_3\gamma_3 \end{vmatrix}$$ **Ans.**

**Example 61.** Show that $\begin{vmatrix} a & b & c \\ b & c & a \\ c & a & b \end{vmatrix} \times \begin{vmatrix} -a & c & b \\ -b & a & c \\ -c & b & a \end{vmatrix}$

$$= \begin{vmatrix} 2bc - a^2 & c^2 & b^2 \\ c^2 & 2ca - b^2 & a^2 \\ b^2 & a^2 & 2ab - c^2 \end{vmatrix} = (a^3 + b^3 + c^3 - 3abc)^2$$

**Solution.** Product of the given determinants

$$\begin{vmatrix} a & b & c \\ b & c & a \\ c & a & b \end{vmatrix} \times \begin{vmatrix} -a & c & b \\ -b & a & c \\ -c & b & a \end{vmatrix} = \begin{vmatrix} -a^2 + bc + bc & -ab + ab + c^2 & -ac + b^2 + ac \\ -ab + c^2 + ab & -b^2 + ac + ac & -bc + bc + a^2 \\ -ca + ca + b^2 & -bc + a^2 + bc & -c^2 + ab + ab \end{vmatrix}$$

$$= \begin{vmatrix} 2bc - a^2 & c^2 & b^2 \\ c^2 & 2ca - b^2 & a^2 \\ b^2 & a^2 & 2ab - c^2 \end{vmatrix} \qquad \text{... (1)}$$

Now, $\begin{vmatrix} -a & c & b \\ -b & a & c \\ -c & b & a \end{vmatrix} = (-1)^2 \begin{vmatrix} a & b & c \\ b & c & a \\ c & a & b \end{vmatrix}$

$$= a(bc - a^2) - b(b^2 - ac) + c(ab - c^2) = -(a^3 + b^3 + c^3 - 3abc)$$

$$\Rightarrow \begin{vmatrix} a & b & c \\ b & c & a \\ c & a & b \end{vmatrix} \begin{vmatrix} -a & c & b \\ -b & a & c \\ -c & b & a \end{vmatrix} = (a^3 + b^3 + c^3 - 3abc)^2 \qquad \text{... (2)}$$

From (1) and (2), we get the required result. **Proved.**

**Example 62.** *Prove that the determinant*

$$\begin{vmatrix} 2b_1+c_1 & c_1+3a_1 & 2a_1+3b_1 \\ 2b_2+c_2 & c_2+3a_2 & 2a_2+3b_2 \\ 2b_3+c_3 & c_3+3a_3 & 2a_3+3b_3 \end{vmatrix}$$

*is a multiple of the determinant*

$$\begin{vmatrix} a_1 & b_1 & c_1 \\ a_2 & b_2 & c_2 \\ a_3 & b_3 & c_3 \end{vmatrix} \text{ and find the other factor.}$$

**Solution.** $$\begin{vmatrix} 2b_1+c_1 & c_1+3a_1 & 2a_1+3b_1 \\ 2b_2+c_2 & c_2+3a_2 & 2a_2+3b_2 \\ 2b_3+c_3 & c_3+3a_3 & 2a_3+3b_3 \end{vmatrix}$$

$$= \begin{vmatrix} a_1 & b_1 & c_1 \\ a_2 & b_2 & c_2 \\ a_3 & b_3 & c_3 \end{vmatrix} \times \begin{vmatrix} 0 & 2 & 1 \\ 3 & 0 & 1 \\ 2 & 3 & 0 \end{vmatrix}$$ **Ans.**

**Example 63.** *Prove that* $$\begin{vmatrix} 1 & \cos(\beta-\alpha) & \cos(\gamma-\alpha) \\ \cos(\alpha-\beta) & 1 & \cos(\gamma-\beta) \\ \cos(\alpha-\gamma) & \cos(\beta-\gamma) & 1 \end{vmatrix} = 0$$

**Solution.** $$\begin{vmatrix} \cos\alpha & \sin\alpha & 0 \\ \cos\beta & \sin\beta & 0 \\ \cos\gamma & \sin\gamma & 0 \end{vmatrix} \times \begin{vmatrix} \cos\alpha & \sin\alpha & 0 \\ \cos\beta & \sin\beta & 0 \\ \cos\gamma & \sin\gamma & 0 \end{vmatrix} = 0$$

or $$\begin{vmatrix} \cos^2\alpha+\sin^2\alpha & \cos\alpha\cos\beta+\sin\alpha\sin\beta & \cos\alpha\cos\gamma+\sin\alpha\sin\gamma \\ \cos\beta\cos\alpha+\sin\beta\sin\alpha & \cos^2\beta+\sin^2\beta & \cos\beta\cos\gamma+\sin\beta\sin\gamma \\ \cos\gamma\cos\alpha+\sin\gamma\sin\alpha & \cos\gamma\cos\beta+\sin\gamma\sin\beta & \cos^2\gamma+\sin^2\gamma \end{vmatrix} = 0$$

or $$\begin{vmatrix} 1 & \cos(\beta-\alpha) & \cos(\gamma-\alpha) \\ \cos(\alpha-\beta) & 1 & \cos(\gamma-\beta) \\ \cos(\alpha-\gamma) & \cos(\beta-\gamma) & 1 \end{vmatrix} = 0$$ **Proved.**

**Example 64.** *If $A_1, A_2, A_3, B_1, B_2, B_3, C_1, C_2, C_3$ are cofactors of the elements $a_1, a_2, a_3, b_1, b_2,$ $b_3, c_1, c_2, c_3$ respectively of the determinant* $\begin{vmatrix} a_1 & b_1 & c_1 \\ a_2 & b_2 & c_2 \\ a_3 & b_3 & c_3 \end{vmatrix}$, *show that*

$$\begin{vmatrix} A_1 & B_1 & C_1 \\ A_2 & B_2 & C_2 \\ A_3 & B_3 & C_3 \end{vmatrix} = \begin{vmatrix} a_1 & b_1 & c_1 \\ a_2 & b_2 & c_2 \\ a_3 & b_3 & c_3 \end{vmatrix}^2.$$ **(Try Yourself)**

## 4.16 CONDITION FOR CONSISTENCY OF A SYSTEM OF SIMULTANEOUS HOMOGENEOUS EQUATIONS

**Case I** For a system of homogeneous equations.

$$a_{11}x + a_{12}y + a_{13}z = 0$$
$$a_{21}x + a_{22}y + a_{23}z = 0$$
$$a_{31}x + a_{32}y + a_{33}z = 0\ ,\qquad D = \begin{vmatrix} a_{11} & a_{12} & a_{13} \\ a_{21} & a_{22} & a_{23} \\ a_{31} & a_{32} & a_{33} \end{vmatrix}$$

1. If $D = 0$, then the system of equations are consistent with infinite solutions.
2. If $D \neq 0$, then the system of equations is consistent with trivial solution.

Homogeneous Equations

| $D = 0$ | $D \neq 0$ |
|---|---|
| Consistent with infinitely many solutions | Consistent with unique solution $x = 0,\ y = 0,\ z = 0$ (Trivial solution) |

**Example 65.** *Find values of λ, for which the following system of equations is consistent and has nontrivial solutions:*

$$(\lambda - 1)x + (3\lambda + 1)y + 2\lambda z = 0$$
$$(\lambda - 1)x + (4\lambda - 2)y + (\lambda + 3)z = 0$$
$$2x + (3\lambda + 1)y + 3(\lambda - 1)z = 0$$

**Solution.**

$$(\lambda - 1)x + (3\lambda + 1)y + 2\lambda z = 0$$
$$(\lambda - 1)x + (4\lambda - 2)y + (\lambda + 3)z = 0$$
$$2x + (3\lambda + 1)y + 3(\lambda - 1)z = 0$$

This is a system of homogeneous equations.

For infinite solutions,

$$D = \begin{vmatrix} \lambda - 1 & 3\lambda + 1 & 2\lambda \\ \lambda - 1 & 4\lambda - 2 & \lambda + 3 \\ 2 & 3\lambda + 1 & 3\lambda - 3 \end{vmatrix} = 0$$

$$\Rightarrow \begin{vmatrix} 0 & -\lambda + 3 & \lambda - 3 \\ \lambda - 1 & 4\lambda - 2 & \lambda + 3 \\ 2 & 3\lambda + 1 & 3\lambda - 3 \end{vmatrix} = 0 \qquad [R_1 \to R_1 - R_2]$$

$$\begin{vmatrix} 0 & 0 & \lambda - 3 \\ \lambda - 1 & 5\lambda + 1 & \lambda + 3 \\ 2 & 6\lambda - 2 & 3\lambda - 3 \end{vmatrix} = 0 \qquad [C_2 \to C_2 + C_3]$$

$$\Rightarrow (\lambda - 3)\,[(\lambda - 1)(6\lambda - 2) - 2(5\lambda + 1)] = 0 \Rightarrow [6\lambda^2 - 8\lambda + 2 - 10\lambda - 2] = 0$$

$$\Rightarrow 6\lambda^2 - 18\lambda = 0 \Rightarrow 6\lambda(\lambda - 3) = 0 \Rightarrow \lambda = 3,\ 0$$ **Ans.**

## 4.17 FOR A SYSTEM OF THREE SIMULTANEOUS LINEAR EQUATIONS WITH THREE UNKNOWNS

(*i*) If $D \neq 0$, then the given system of equations is consistent and has a unique solution given by $x = \dfrac{D_1}{D}$, $y = \dfrac{D_2}{D}$, and $z = \dfrac{D_3}{D}$.

(*ii*) If $D = 0$ and $D_1 = D_2 = D_3 = 0$, then the given system of equations is consistent, and it has infinitely many solutions.

(*iii*) If $D = 0$ and at least one of the determinants $D_1, D_2, D_3$ is non zero, then the given system of equations is inconsistent.

**Equations with three unknowns**

- $D \neq 0$: **Consistent with unique solution**
- $D = 0$
  - $D_1 = D_2 = D_3 = 0$: **Consistent with infinitely many solutions**
  - $D_1$ or $D_2$ or $D_3 \neq 0$: **Inconsistent**

**Example 66.** *Test the consistency of the following equations and solve them if possible:*

$$3x + 3y + 2z = 1, \quad x + 2y = 4, \quad 10y + 3z = -2$$

**Solution.** The system of equations is

$$3x + 3y + 2z = 1$$
$$x + 2y + 0z = 4$$
$$0x + 10y + 3z = -2$$

Therefore
$$D = \begin{vmatrix} 3 & 3 & 2 \\ 1 & 2 & 0 \\ 0 & 10 & 3 \end{vmatrix}$$
$$= 3(6-0) - 3(3-0) + 2(10-0)$$
$$= 18 - 9 + 20 = 29 \neq 0$$

Since $D \neq 0$, so the system of simultaneous equations is consistent with unique solution. Now let us solve the system of equation.

$$D_1 = \begin{vmatrix} 1 & 3 & 2 \\ 4 & 2 & 0 \\ -2 & 10 & 3 \end{vmatrix} = 1(6-0) - 3(12-0) + 2(40+4) = 6 - 36 + 88 = 58$$

$$D_2 = \begin{vmatrix} 3 & 1 & 2 \\ 1 & 4 & 0 \\ 0 & -2 & 3 \end{vmatrix} = 3(12-0) - 1(3-0) + 2(-2-0) = 36 - 3 - 4 = 29$$

$$D_3 = \begin{vmatrix} 3 & 3 & 1 \\ 1 & 2 & 4 \\ 0 & 10 & -2 \end{vmatrix} = 3(-4-40) - 3(-2-0) + 1(10-0) = -132 + 6 + 10 = -116$$

By Cramer's Rule

$$x = \frac{D_1}{D} = \frac{58}{29} = 2, \quad y = \frac{D_2}{D} = \frac{29}{29} = 1, \quad z = \frac{D_3}{D} = \frac{-116}{29} = -4$$

Hence $x = 2$, $y = 1$, $z = -4$. **Ans.**

**Example 67.** *Show that the system of equations*

$$2x + 6y = -11, \quad 6x + 20y - 6z = -3, \quad 6y - 18z = -1 \text{ is not consistent.}$$

**Solution.**
$$2x + 6y + 0z = -11$$
$$6x + 20y - 6z = -3$$
$$0x + 6y - 18z = -1$$

$$D = \begin{vmatrix} 2 & 6 & 0 \\ 6 & 20 & -6 \\ 0 & 6 & -18 \end{vmatrix} = 2(-360+36) - 6(-108) = -648 + 648 = 0$$

$$D_1 = \begin{vmatrix} -11 & 6 & 0 \\ -3 & 20 & -6 \\ -1 & 6 & -18 \end{vmatrix} = -11(-360+36) - 6(54-6) = 3564 - 288 = 3276$$

Here $D = 0$ and $D_1 \neq 0$

Hence the system of equations is not consistent. **Proved**

## EXERCISE 4.9

***Find, whether the following system of equations is consistent or inconsistent. If consistent solve them.***

1. Find the value of $k$, for which the following system of equations
$3x_1 - 2x_2 + 2x_3 = 3,\ x_1 + kx_2 - 3x_3 = 0,\ 4x_1 + x_2 + 2x_3 = 7$ is consistent. **Ans.** $k = \frac{1}{4}$,

2. Find the value of $\lambda$, for which the system of equations
$x + y + 4z = 1,\ x + 2y - 2z = 1,\ \lambda x + y + z = 1$ will have a unique solution. **Ans.** $\lambda \neq \frac{7}{10}$

3. For what values of $\lambda$ and $\mu$, the following system of equations
$2x + 3y + 5z = 9,\ 7x + 3y - 2z = 8,\ 2x + 3y + \lambda z = \mu$ will have
(*i*) unique solution; (*ii*) no solution. **Ans.** (*i*) $\lambda \neq 5$ (*ii*) $\lambda = 5,\ \mu \neq 9$

4. Determine the values of $a$ and $b$ for which the system $\begin{bmatrix} 3 & -2 & 1 \\ 5 & -8 & 9 \\ 2 & 1 & a \end{bmatrix}\begin{bmatrix} x \\ y \\ z \end{bmatrix} = \begin{bmatrix} b \\ 3 \\ -1 \end{bmatrix}$
(*i*) has a unique solution, (*ii*) has no solution and (*iii*) has infinitely many solutions.
**Ans.** (*i*) $a \neq -3$ (*ii*) $a = -3,\ b \neq \frac{1}{3}$ (*iii*) $a = -3,\ b = \frac{1}{3}$

5. Find the condition on $\lambda$ for which the system of equations
$3x - y + 4z = 3,\ x + 2y - 3z = -2,\ 6x + 5y + \lambda z = -3$
has a unique solution. Find the solution for $\lambda = -5$.
**Ans.** $\lambda \neq -5,\ x = -\frac{5k}{7} + \frac{4}{7},\ y = \frac{13k}{7} - \frac{9}{7},\ z = k$

## EXERCISE OF OBJECTIVE QUESTIONS

**Choose the Correct Answers :**

1. The value of $\begin{vmatrix} 5^2 & 5^3 & 5^4 \\ 5^3 & 5^4 & 5^5 \\ 5^4 & 5^6 & 5^7 \end{vmatrix}$ is

(*a*) $5^2$ (*b*) 0 (*c*) $5^{13}$ (*d*) $5^9$ **Ans.** (*b*)

**2.** If $\begin{vmatrix} a_1 & b_1 & c_1 \\ a_2 & b_2 & c_2 \\ a_3 & b_3 & c_3 \end{vmatrix} = 5$ then the value of $\begin{vmatrix} b_2c_3 - b_3c_2 & a_3c_2 - a_2c_3 & a_2b_3 - a_3b_2 \\ b_3c_1 - b_1c_3 & a_1c_3 - a_3c_1 & a_1b_1 - a_1b_3 \\ b_1c_2 - b_2c_1 & a_2c_1 - a_1c_2 & a_1b_2 - a_2b_1 \end{vmatrix}$ is

(*a*) 5 (*b*) 25 (*c*) 125 (*d*) 0 **Ans.** (*b*)

**3.** If $\begin{vmatrix} 1+ax & 1+bx & 1+cx \\ 1+a_1x & 1+b_1x & 1+c_1x \\ 1+a_2x & 1+b_2x & 1+c_2x \end{vmatrix} = A_0 + A_1x + A_2x_2 + A_3x_3$, then $A_1$ is equal to

(*a*) *abc* (*b*) 0 (*c*) 1 (*d*) none of these **Ans.** (*b*)

**4.** If 1, $\omega, \omega^2$ are the cube roots of unity, then $\Delta = \begin{vmatrix} 1 & \omega^n & \omega^{2n} \\ \omega^n & \omega^{2n} & 1 \\ \omega^{2n} & 1 & \omega^n \end{vmatrix}$ is equal to

(*a*) 0 (*b*) 1 (*c*) $\omega$ (*d*) $\omega^2$ **Ans.** (*a*)

**5.** The determinant $\begin{vmatrix} xp+y & x & y \\ yp+z & y & z \\ 0 & xp+y & yp+z \end{vmatrix} = 0$ if

(*a*) *x*, *y*, *z* are in A.P. (*b*) *x*, *y*, *z* are in G.P.
(*c*) *x*, *y*, *z* are in H.P. (*d*) *xy*, *yz*, *zx* are in A.P. **Ans.** (*b*)

**6.** If the determinant $\begin{vmatrix} a & b & 2a\alpha+3b \\ b & c & 2b\alpha+3c \\ 2a\alpha+3b & 2b\alpha+3c & 0 \end{vmatrix} = 0$, then

(*a*) *a*, *b*, *c* are in H.P. (*b*) $\alpha$ is root of $4ax^2 + 12bx + 9c = 0$ or *a*, *b*, *c* are in G.P.
(*c*) *a*, *b*, *c* are in G.P. only (*d*) *a*, *b*, *c* are in A.P. **Ans.** (*a*)

**7.** If *l*, *m*, *n* are the $p^{th}$, $q^{th}$ and $r^{th}$ term of a G.P. all positive, then $\begin{vmatrix} \log l & p & 1 \\ \log m & q & 1 \\ \log n & r & 1 \end{vmatrix}$ equals

(*a*) 3 (*b*) 2 (*c*) 1 (*d*) zero **Ans.** (*d*)

**8.** If the system of linear equations

$$x + 2ay + az = 0$$
$$x + 3by + bz = 0$$
$$x + 4cy + cz = 0$$

has a non-zero solution, then *a*, *b*, *c*

(*a*) are in A.P. (*b*) are in G.P. (*c*) are in H.P. (*d*) satisfy $a + 2b + 3c = 0$
**Ans.** (*c*)

**9.** If $\alpha, \beta$ and $\gamma$ are the roots of the equation $x^3 + px + q = 0$, then the value of the determinant $\begin{vmatrix} \alpha & \beta & \gamma \\ \beta & \gamma & \alpha \\ \gamma & \alpha & \beta \end{vmatrix}$ is

(*a*) $q$ (*b*) 0 (*c*) $p$ (*d*) $p^2-2q$. **Ans.** (*a*)

**10.** The number of values of $k$ for which the system of equations

$$(k+1)x+8y=4k\ ,\qquad kx+(k+3)y=3k-1$$

has infinitely many solutions is

(*a*) 0 (*b*) 1 (*c*) 2 (*d*) infinite

**Ans.** (*b*) [**Hint :** Here $\Delta=0$ for $k=3,\ 1,\ \Delta_x=0$ for $k=2,\ 1,\ \Delta_y=0$ for $k=1$. Hence $k=1$. Alternatively, for infinitely many solutions the two equations become identical

$$\Rightarrow \frac{k+1}{k}=\frac{8}{k+3}=\frac{4k}{3k-1}\quad\Rightarrow\quad k=1]$$

**11.** The system $x+\alpha y=0$, $y+\alpha z=0$, $z+\alpha x=0$ has infinitely many solutions when

(*a*) $\alpha=1$ (*b*) $\alpha=-1$ (*c*) $\alpha=0$ (*d*) no real value of $\alpha$

**Ans.** (*b*) [**Hint :** $\begin{vmatrix} 1 & \alpha & 0 \\ 0 & 1 & \alpha \\ \alpha & 0 & 1 \end{vmatrix}=0$. Solve for $\alpha$.]

**12.** If the system of equations $x-ky-z=0$, $kx-y-z=0$, $x+y-z=0$ has a non-zero solution, then the possible values of $k$ are

(*a*) – 1, 2 (*b*) 1, 2 (*c*) 0, 1 (*d*) – 1, 1

**Ans.** (*d*)

**13.** The value of $\lambda$ for which the system of equations $2x-y-2z=2$, $x-2y+z=-4$, $x+y+\lambda z=4$ has no solution is

(*a*) 3 (*b*) – 3 (*c*) 2 (*d*) – 2

**Ans.** (*b*) [**Hint :** The Coefficient determinant = $\begin{vmatrix} 2 & -1 & -2 \\ 1 & -2 & 1 \\ 1 & 1 & \lambda \end{vmatrix}=-3\lambda-9$.

For no solution the necessary condition is $-3\lambda-9=0$

$\Rightarrow \lambda=-3$

For $\lambda=-3$, there is no solution for the given system of equations].

**14.** If the system of equations $x+2y-3z=1$, $(\lambda+3)z=3$, $(2\lambda+1)x+z=0$ is inconsistent, then the value of $\lambda$ is equal to

(*a*) $-\frac{1}{2}$ (*b*) – 3 (*c*) 2 (*d*) 0 **Ans.** (*b*)

**15.** $A=\begin{vmatrix} a & 0 & 1 \\ 1 & c & b \\ 1 & d & b \end{vmatrix}$, $B=\begin{vmatrix} a & 1 & 1 \\ 0 & c & d \\ f & g & h \end{vmatrix}$, $U=\begin{vmatrix} f \\ g \\ h \end{vmatrix}$, $V=\begin{vmatrix} a^2 \\ 0 \\ 0 \end{vmatrix}$. If there is a vector matrix $X$ such that $AX=U$ has infinitely many solutions, then prove that $BX=V$ cannot have a unique solution. If $afd\neq 0$, prove that $BX=V$ has no solution.

## 4.18 MATRICES

Let us consider a set of simultaneous equations,

$$x + 2y + 3z + 5t = 0$$
$$4x + 2y + 5z + 7t = 0$$
$$3x + 4y + 2z + 6t = 0.$$

Now we write down the coefficients of $x, y, z, t$ of the above equations and enclose them within brackets and then we get

$$A = \begin{bmatrix} 1 & 2 & 3 & 5 \\ 4 & 2 & 5 & 7 \\ 3 & 4 & 2 & 6 \end{bmatrix}$$

The above system of numbers, arranged in a rectangular array in rows and columns and bounded by the brackets, is called a matrix.

It has got 3 rows and 4 columns and in all $3 \times 4 = 12$ elements. It is termed as $3 \times 4$ matrix, to be read as [3 by 4 matrix]. In the double subscripts of an element, the first subscript determines the row and the second subscript determines the column in which the element lies, $a_{ij}$ lies in the $i$th row and $j$th column.

## 4.19 VARIOUS TYPES OF MATRICES

(*a*) **Row Matrix.** If a matrix has only one row and any number of columns, it is called a *Row matrix, e.g.,*

$$[2\ 7\ 3\ 9]$$

(*b*) **Column Matrix.** A matrix, having one column and any number of rows, is called a *Column matrix, e.g.,* $\begin{bmatrix} 1 \\ 2 \\ 3 \end{bmatrix}$

(*c*) **Null Matrix or Zero Matrix.** Any matrix, in which all the elements are zeros, is called a *Zero matrix* or *Null matrix e.g.,*

$$\begin{bmatrix} 0 & 0 & 0 & 0 \\ 0 & 0 & 0 & 0 \end{bmatrix}$$

(*d*) **Square Matrix.** A matrix, in which the number of rows is equal to the number of columns, is called a square matrix *e.g.,*

$$\begin{bmatrix} 2 & 5 \\ 1 & 4 \end{bmatrix}$$

(*e*) **Diagonal Matrix.** A square matrix is called a diagonal matrix, if all its non-diagonal elements are zero *e.g.,*

$$\begin{bmatrix} 1 & 0 & 0 \\ 0 & 3 & 0 \\ 0 & 0 & 4 \end{bmatrix}$$

(*f*) **Scalar matrix.** A diagonal matrix in which all the diagonal elements are equal to a scalar, say ($k$) is called a scalar matrix.

*For example;*

$$\begin{bmatrix} 2 & 0 & 0 \\ 0 & 2 & 0 \\ 0 & 0 & 2 \end{bmatrix}, \begin{bmatrix} -6 & 0 & 0 & 0 \\ 0 & -6 & 0 & 0 \\ 0 & 0 & -6 & 0 \\ 0 & 0 & 0 & -6 \end{bmatrix}$$

*i.e.*, $A = [a_{ij}]_{n \times n}$ is a scalar matrix if $a_{ij} = \begin{cases} 0, & \text{when } i \neq j \\ k, & \text{when } i = j \end{cases}$

(*g*) **Unit or Identity Matrix.** A square matrix is called a unit matrix if all the diagonal elements are unity and non-diagonal elements are zero *e.g.*,

$$\begin{bmatrix} 1 & 0 & 0 \\ 0 & 1 & 0 \\ 0 & 0 & 1 \end{bmatrix}, \begin{bmatrix} 1 & 0 \\ 0 & 1 \end{bmatrix}$$

(*h*) **Symmetric Matrix.** A square matrix will be called symmetric, if for all values of *i* and *j*, $a_{ij} = a_{ji}$ *i.e.*, $A' = A$

$$e.g., \begin{bmatrix} a & h & g \\ h & b & f \\ g & f & c \end{bmatrix}$$

(*i*) **Skew Symmetric Matrix.** A square matrix is called skew symmetric matrix, if (1) $a_{ij} = -a_{ji}$ for all values of *i* and *j*, or $A' = -A$
(2) All diagonal elements are zero, *e.g.*,

$$\begin{bmatrix} 0 & -h & -g \\ h & 0 & -f \\ g & f & 0 \end{bmatrix}$$

(*j*) **Triangular Matrix.** (Echelon form) A square matrix, all of whose elements below the leading diagonal are zero, is called an *upper triangular matrix.* A square matrix, all of whose elements above the leading diagonal are zero, is called *a lower triangular matrix e.g.,*

$$\begin{bmatrix} 1 & 3 & 2 \\ 0 & 4 & 1 \\ 0 & 0 & 6 \end{bmatrix} \qquad \begin{bmatrix} 2 & 0 & 0 \\ 4 & 1 & 0 \\ 5 & 6 & 7 \end{bmatrix}$$

Upper triangular matrix Lower triangular matrix

(*k*) **Transpose of a Matrix.** If in a given matrix *A,* we interchange the rows and the corresponding columns, the new matrix obtained is called the transpose of the matrix *A* and is denoted by $A'$ or $A^T$ *e.g.,*

$$A = \begin{bmatrix} 2 & 3 & 4 \\ 1 & 0 & 5 \\ 6 & 7 & 8 \end{bmatrix}, \; A' = \begin{bmatrix} 2 & 1 & 6 \\ 3 & 0 & 7 \\ 4 & 5 & 8 \end{bmatrix}$$

(*l*) **Orthogonal Matrix.** A square matrix *A* is called an orthogonal matrix if the product of the matrix *A* and the transpose matrix *A'* is an identity matrix *e.g.,*

$$A. A' = I$$

if $|A| = 1$, matrix *A* is proper.

(*m*) **Conjugate of a Matrix**

Let $$A = \begin{bmatrix} 1+i & 2-3i & 4 \\ 7+2i & -i & 3-2i \end{bmatrix}$$

Conjugate of matrix *A* is $\overline{A}$

$$\overline{A} = \begin{bmatrix} 1-i & 2+3i & 4 \\ 7-2i & i & 3+2i \end{bmatrix}$$

(*n*) **Matrix $A^\theta$.** Transpose of the conjugate of a matrix *A* is denoted by $A^\theta$.

Let $$A = \begin{bmatrix} 1+i & 2-3i & 4 \\ 7+2i & -i & 3-2i \end{bmatrix}$$

$$\bar{A} = \begin{bmatrix} 1-i & 2+3i & 4 \\ 7-2i & +i & 3+2i \end{bmatrix}$$

$$(\bar{A})' = \begin{bmatrix} 1-i & 7-2i \\ 2+3i & i \\ 4 & 3+2i \end{bmatrix}$$

$$A^{\theta} = \begin{bmatrix} 1-i & 7-2i \\ 2+3i & i \\ 4 & 3+2i \end{bmatrix}$$

(*o*) **Unitary Matrix.** A square matrix $A$ is said to be unitary if

$$A^{\theta} A = I$$

*e.g.* $$A = \begin{bmatrix} \frac{1+i}{2} & \frac{-1+i}{2} \\ \frac{1+i}{2} & \frac{1-i}{2} \end{bmatrix}, \quad A^{\theta} = \begin{bmatrix} \frac{1-i}{2} & \frac{1-i}{2} \\ \frac{-1-i}{2} & \frac{1+i}{2} \end{bmatrix}, \quad A \cdot A^{\theta} = I$$

(*p*) **Hermitian Matrix.** A square matrix $A = (a_{ij})$ is called Hermitian matrix, if every *i-jth* element of $A$ is equal to conjugate complex *j-ith* element of $A$.

In other words, $a_{ij} = \bar{a}_{ji}$

*e.g.* $$\begin{bmatrix} 1 & 2+3i & 3+i \\ 2-3i & 2 & 1-2i \\ 3-i & 1+2i & 5 \end{bmatrix}$$

Necessary and sufficient condition for a matrix $A$ to be Hermitian is that $A = A^{\theta}$ *i.e.* conjugate transpose of $A$

$$\Rightarrow \qquad A = (\bar{A})'.$$

(*q*) **Skew Hermitian Matrix.** A square matrix $A = (a_{ij})$ will be called a Skew Hermitian matrix if every *i-jth* element of $A$ is equal to negative conjugate complex of *j-ith* element of $A$.

In other words, $a_{ij} = -\bar{a}_{ji}$

All the elements in the principal diagonal will be of the form

$$a_{ii} = -\bar{a}_{ii} \quad \text{or} \quad a_{ii} + \bar{a}_{ii} = 0$$

If $a_{ii} = a + ib$ then $\bar{a}_{ii} = a - ib$

$$(a + ib) + (a - ib) = 0 \quad \Rightarrow \quad 2a = 0 \Rightarrow a = 0$$

So, $a_{ii}$ is pure imaginary $\Rightarrow a_{ii} = 0$.

Hence, all the diagonal elements of a Skew Hermitian Matrix are either zeros or pure imaginary.

*e.g.* $$\begin{bmatrix} i & 2-3i & 4+5i \\ -(2+3i) & 0 & 2i \\ -(4-5i) & 2i & -3i \end{bmatrix}$$

The necessary and sufficient condition for a matrix $A$ to be Skew Hermitian is that

$$A^{\theta} = -A$$

$$(\bar{A})' = -A$$

(*r*) **Idempotent Matrix.** A matrix, such that $A^2 = A$ is called Idempotent Matrix.

$$e.g.\ A = \begin{bmatrix} 2 & -2 & -4 \\ -1 & 3 & 4 \\ 1 & -2 & -3 \end{bmatrix},\ A^2 = \begin{bmatrix} 2 & -2 & -4 \\ -1 & 3 & 4 \\ 1 & -2 & -3 \end{bmatrix}\begin{bmatrix} 2 & -2 & -4 \\ -1 & 3 & 4 \\ 1 & -2 & -3 \end{bmatrix} = \begin{bmatrix} 2 & -2 & -4 \\ -1 & 3 & 4 \\ 1 & -2 & -3 \end{bmatrix} = A$$

(*s*) **Periodic Matrix.** A matrix $A$ will be called a Periodic Matrix, if

$$A^{k+1} = A$$

where $k$ is a +ve integer. If $k$ is the least + ve integer, for which $A^{k+1} = A$, then $k$ is said to be the period of $A$. If we choose $k = 1$, we get $A^2 = A$ and we call it to be idempotent matrix.

(*t*) **Nilpotent Matrix.** A matrix will be called a Nilpotent matrix, if $A^k = 0$ (null matrix) where $k$ is a +ve integer ; if however $k$ is the least +ve integer for which $A^k = 0$, then $k$ is the *index* of the nilpotent matrix.

$$e.g.,\ A = \begin{bmatrix} ab & b^2 \\ -a^2 & -ab \end{bmatrix},\ A^2 = \begin{bmatrix} ab & b^2 \\ -a^2 & -ab \end{bmatrix}\begin{bmatrix} ab & b^2 \\ -a^2 & -ab \end{bmatrix} = \begin{bmatrix} 0 & 0 \\ 0 & 0 \end{bmatrix} = 0$$

$A$ is nilpotent matrix whose index is 2.

(*u*) **Involuntary Matrix.** A matrix $A$ will be called an Involuntary matrix, if $A^2 = I$ (unit matrix). Since $I^2 = I$ always $\therefore$ Unit matrix is involuntary.

(*v*) **Equal Matrices.** Two matrices are said to be equal if

(*i*) They are of the same order.

(*ii*) The elements in the corresponding positions are equal.

Thus if $$A = \begin{bmatrix} 2 & 3 \\ 1 & 4 \end{bmatrix},\ B = \begin{bmatrix} 2 & 3 \\ 1 & 4 \end{bmatrix}$$

Here $A = B$

(*w*) **Singular Matrix.** If the determinant of the matrix is zero, then the matrix is known as singular matrix e.g. $A = \begin{bmatrix} 1 & 2 \\ 3 & 6 \end{bmatrix}$ is singular matrix, because $|A| = 6 - 6 = 0$.

**Example 1.** *Find the values of x, y, z and 'a' which satisfy the matrix equation.*

$$\begin{bmatrix} x+3 & 2y+x \\ z-1 & 4a-6 \end{bmatrix} = \begin{bmatrix} 0 & -7 \\ 3 & 2a \end{bmatrix}$$

**Solution.** As the given matrices are equal, so their corresponding elements are equal.

$$x + 3 = 0 \quad \Rightarrow \quad x = -3 \qquad ...(1)$$

$$2y + x = -7 \qquad ...(2)$$

$$z - 1 = 3 \quad \Rightarrow \quad z = 4 \qquad ...(3)$$

$$4a - 6 = 2a \quad \Rightarrow \quad a = 3 \qquad ...(4)$$

Putting the value of $x = -3$ from (1) into (2), we have

$$2y - 3 = -7 \quad \Rightarrow \quad y = -2$$

Hence. $x = -3,\ y = -2,\ z = 4,\ a = 3$ **Ans.**

## 4.20 ADDITION OF MATRICES

If $A$ and $B$ be two matrices of the same order, then their sum, $A + B$ is defined as the matrix, each element of which is the sum of the corresponding elements of $A$ and $B$.

Thus if $$A = \begin{bmatrix} 4 & 2 & 5 \\ 1 & 3 & -6 \end{bmatrix},\ B = \begin{bmatrix} 1 & 0 & 2 \\ 3 & 1 & 4 \end{bmatrix}$$

then $A + B = \begin{bmatrix} 4+1 & 2+0 & 5+2 \\ 1+3 & 3+1 & -6+4 \end{bmatrix} = \begin{bmatrix} 5 & 2 & 7 \\ 4 & 4 & -2 \end{bmatrix}$

If $A = [a_{ij}],\ B = [b_{ij}]$ then $A + B = [a_{ij} + b_{ij}]$

**Example 2.** *Show that any square matrix can be expressed as the sum of two matrices, one symmetric and the other anti-symmetric.*

**Solution.** Let $A$ be a given square matrix.

Then $A = \frac{1}{2}(A + A') + \frac{1}{2}(A - A')$

Now, $(A + A')' = A' + A = A + A'$.

$\therefore$ $A + A'$ is a symmetric matrix.

Also, $(A - A')' = A' - A = -(A - A')$

$\therefore$ $A - A'$ or $\frac{1}{2}(A - A')$ is an anti-symmetric matrix.

$$\boxed{A = \frac{1}{2}(A + A') + \frac{1}{2}(A - A')}$$

Square matrix = Symmetric matrix + Anti-symmetric matrix **Proved.**

**Example 3.** *Write matrix A given below as the sum of a symmetric and a skew symmetric matrix.*

$$A = \begin{pmatrix} 1 & 2 & 4 \\ -2 & 5 & 3 \\ -1 & 6 & 3 \end{pmatrix}$$

**Solution.** $A = \begin{bmatrix} 1 & 2 & 4 \\ -2 & 5 & 3 \\ -1 & 6 & 3 \end{bmatrix}$ On transposing, we get $A' = \begin{bmatrix} 1 & -2 & -1 \\ 2 & 5 & 6 \\ 4 & 3 & 3 \end{bmatrix}$

On adding A and A′, we have

$$A + A' = \begin{bmatrix} 1 & 2 & 4 \\ -2 & 5 & 3 \\ -1 & 6 & 3 \end{bmatrix} + \begin{bmatrix} 1 & -2 & -1 \\ 2 & 5 & 6 \\ 4 & 3 & 3 \end{bmatrix} = \begin{bmatrix} 2 & 0 & 3 \\ 0 & 10 & 9 \\ 3 & 9 & 6 \end{bmatrix} \quad ...(1)$$

On subtracting $A'$ from $A$, we get

$$A - A' = \begin{bmatrix} 1 & 2 & 4 \\ -2 & 5 & 3 \\ -1 & 6 & 3 \end{bmatrix} - \begin{bmatrix} 1 & -2 & -1 \\ 2 & 5 & 6 \\ 4 & 3 & 3 \end{bmatrix} = \begin{bmatrix} 0 & 4 & 5 \\ -4 & 0 & -3 \\ -5 & 3 & 0 \end{bmatrix} \quad ...(2)$$

On adding (1) and (2), we have

$$2A = \begin{bmatrix} 2 & 0 & 3 \\ 0 & 10 & 9 \\ 3 & 9 & 6 \end{bmatrix} + \begin{bmatrix} 0 & 4 & 5 \\ -4 & 0 & -3 \\ -5 & 3 & 0 \end{bmatrix}$$

$$A = \begin{bmatrix} 1 & 0 & \frac{3}{2} \\ 0 & 5 & \frac{9}{2} \\ \frac{3}{2} & \frac{9}{2} & 3 \end{bmatrix} + \begin{bmatrix} 0 & 2 & \frac{5}{2} \\ -2 & 0 & -\frac{3}{2} \\ \frac{-5}{2} & \frac{3}{2} & 0 \end{bmatrix}$$

A = [Symmetric matrix] + [Skew symmetric matrix.] **Ans.**

## 4.21 PROPERTIES OF MATRIX ADDITION

Only matrices of the same order can be added or subtracted.

(*i*) **Commutative Law.** $A + B = B + A$.

(*ii*) **Associative law.** $A + (B + C) = (A + B) + C$.

## 4.22 SUBTRACTION OF MATRICES

The difference of two matrices is a matrix, each element of which is obtained by subtracting the elements of the second matrix from the corresponding element of the first.

$$A - B = [a_{ij} - b_{ij}]$$

Thus $\begin{bmatrix} 8 & 6 & 4 \\ 1 & 2 & 0 \end{bmatrix} - \begin{bmatrix} 3 & 5 & 1 \\ 7 & 6 & 2 \end{bmatrix} = \begin{bmatrix} 8-3 & 6-5 & 4-1 \\ 1-7 & 2-6 & 0-2 \end{bmatrix} = \begin{bmatrix} 5 & 1 & 3 \\ -6 & -4 & -2 \end{bmatrix}$ **Ans.**

## 4.23 SCALAR MULTIPLE OF A MATRIX

If a matrix is multiplied by a scalar quantity *k*, then each element is multiplied by *k*, *i.e.*

$$A = \begin{bmatrix} 2 & 3 & 4 \\ 4 & 5 & 6 \\ 6 & 7 & 9 \end{bmatrix}$$

$$3A = 3\begin{bmatrix} 2 & 3 & 4 \\ 4 & 5 & 6 \\ 6 & 7 & 9 \end{bmatrix} = \begin{bmatrix} 3\times 2 & 3\times 3 & 3\times 4 \\ 3\times 4 & 3\times 5 & 3\times 6 \\ 3\times 6 & 3\times 7 & 3\times 9 \end{bmatrix} = \begin{bmatrix} 6 & 9 & 12 \\ 12 & 15 & 18 \\ 18 & 21 & 27 \end{bmatrix}$$

## EXERCISE 4.10

**1.** (*i*) If $A = \begin{bmatrix} -1 & 7 & 1 \\ 2 & 3 & 4 \\ 5 & 0 & 5 \end{bmatrix}$, represent it as A = B + C where B is a symmetric and C is a skew-symmetric matrix.

(*b*) Express $\begin{bmatrix} 1 & 2 & 0 \\ 3 & 7 & 1 \\ 5 & 9 & 3 \end{bmatrix}$ as a sum of symmetric and skew-symmetric matrix.

**Ans.** (*i*) $A = \begin{bmatrix} -1 & \frac{9}{2} & 3 \\ \frac{9}{2} & 3 & 2 \\ 3 & 2 & 5 \end{bmatrix} + \begin{bmatrix} 0 & \frac{5}{2} & -2 \\ \frac{-5}{2} & 0 & 2 \\ 2 & -2 & 0 \end{bmatrix}$ (*b*) $A = \begin{bmatrix} 1 & \frac{5}{2} & \frac{5}{2} \\ \frac{5}{2} & 7 & 5 \\ \frac{5}{2} & 5 & 3 \end{bmatrix} + \begin{bmatrix} 0 & -\frac{1}{2} & -\frac{5}{2} \\ \frac{1}{2} & 0 & -4 \\ \frac{5}{2} & 4 & 0 \end{bmatrix}$

**2.** Matrices *A* and *B* are such that

$$3A - 2B = \begin{bmatrix} 2 & 1 \\ -2 & -1 \end{bmatrix} \text{ and } -4A + B = \begin{bmatrix} -1 & 2 \\ -4 & 3 \end{bmatrix}$$

Find A and B. **Ans.** $A = \begin{bmatrix} 0 & -1 \\ 2 & -1 \end{bmatrix}$, $B = \begin{bmatrix} -1 & -2 \\ 4 & -1 \end{bmatrix}$

**3.** Given $3\begin{bmatrix} x & y \\ z & w \end{bmatrix} = \begin{bmatrix} x & 6 \\ -1 & 2w \end{bmatrix} + \begin{bmatrix} 4 & x+y \\ z+w & 3 \end{bmatrix}$

Find *x*, *y*, *z* and *w*. **Ans.** $x = 2,\ y = 4,\ z = 1,\ w = 3$

**4.** If $A = \begin{bmatrix} 0 & 2 & 0 \\ 1 & 0 & 3 \\ 1 & 1 & 2 \end{bmatrix}$, $B = \begin{bmatrix} 1 & 2 & 1 \\ 2 & 1 & 0 \\ 0 & 0 & 3 \end{bmatrix}$ **Ans.** (*i*) = $\begin{bmatrix} 3 & 10 & 3 \\ 8 & 3 & 6 \\ 2 & 2 & 13 \end{bmatrix}$, (*ii*) = $\begin{bmatrix} -4 & -2 & -4 \\ -5 & -4 & 9 \\ 3 & 3 & -6 \end{bmatrix}$

Find (*i*) $2A + 3B$ (*ii*) $3A - 4B$.

## 4.24 MULTIPLICATION

The product of two matrices $A$ and B is only possible if the number of columns in $A$ is equal to the number of rows in $B$.

Let $A = [a_{ij}]$ be an $m \times n$ matrix and $B = [b_{ij}]$ be an $n \times p$ matrix. Then the product $AB$ of these matrices is an $m \times p$ matrix $C = [c_{ij}]$ where

$$c_{ij} = a_{i1}\, b_{1j} + a_{i2}\, b_{2j} + a_{i3}\, b_{3j} + \dots + a_{in}\, b_{nj}$$

## 4.25 $(AB)' = B'A'$

*If A and B are two matrices conformal for product AB, then show that $(AB)' = B'A'$, where dash represents transpose of a matrix.*

**Solution.** Let $A = (a_{ij})$ be an $m \times n$ matrix and $B = (b_{ij})$ be $n \times p$ matrix.

Since $AB$ is $m \times p$ matrix, $(AB)'$ is a $p \times m$ matrix.

Further $B'$ is $p \times n$ matrix and $A'$ an $n \times m$ matrix and therefore $B'\,A'$ is a $p \times m$ matrix.

Then $(AB)'$ and $B'\,A'$ are matrices of the same order.

Now the $(j, i)$th element of $(AB)' = (i, j)$th element of $(AB) = \sum_{k=1}^{n} a_{ik}b_{kj}$ ...(1)

Also the $j$th row of $B'$ is $b_{1j}, b_{2j} \dots b_{nj}$ and $i$th column of $A'$ is $a_{i1}, a_{i2}, a_{i3} \dots a_{in}$.

$\therefore$ $(j, i)$th element of $B'A' = \sum_{k=1}^{n} b_{kj}a_{ik}$ ...(2)

From (1) and (2), we have $(j, i)$th element of $(AB)' = (j, i)$ th element of $B'A'$.

As the matrices $(AB)'$ and $B'A'$ are of the same order and their corresponding elements are equal, we have $(AB)' = B'A'$. **Proved.**

## 4.26 PROPERTIES OF MATRIX MULTIPLICATION

1. Multiplication of matrices is not commutative. $AB \neq BA$
2. Matrix multiplication is associative, if conformability is assured. $A\,(BC) = (AB)\,C$
3. Matrix multiplication is distributive with respect to addition. $A\,(B + C) = AB + AC$
4. Multiplication of matrix $A$ by unit matrix. $AI = IA = A$
5. Multiplicative inverse of a matrix exists if $|A| \neq 0$. $A\,.\,A^{-1} = A^{-1}\,.\,A = I$
6. If A is a square then $A \times A = A^2$, $A \times A \times A = A^3$.
7. $A^0 = I$
8. $I^n = I$, where $n$ is positive integer.

**Example 4.** *If* $A = \begin{bmatrix} 0 & 1 & 2 \\ 1 & 2 & 3 \\ 2 & 3 & 4 \end{bmatrix}$ *and* $B = \begin{bmatrix} 1 & -2 \\ -1 & 0 \\ 2 & -1 \end{bmatrix}$

*obtain the product AB and explain why BA is not defined.*

**Solution.** The number of columns in $A$ is 3 and the number of rows in $B$ is also 3, therefore the product $AB$ is defined.

$$AB = \begin{bmatrix} 0 & 1 & 2 \\ 1 & 2 & 3 \\ 2 & 3 & 4 \end{bmatrix} \begin{matrix} R_1 \\ R_2 \\ R_3 \end{matrix} \times \begin{matrix} C_1 \quad C_2 \\ \begin{bmatrix} 1 & -2 \\ -1 & 0 \\ 2 & -1 \end{bmatrix} \end{matrix} = \begin{bmatrix} R_1 C_1 & R_1 C_2 \\ R_2 C_1 & R_2 C_2 \\ R_3 C_1 & R_3 C_2 \end{bmatrix}$$

$R_1, R_2, R_3$ are rows of $A$ and $C_1, C_2$ are columns of $B$.

$$= \begin{bmatrix} \boxed{0\ 1\ 2}\begin{bmatrix}1\\-1\\2\end{bmatrix} & \boxed{0\ 1\ 2}\begin{bmatrix}-2\\0\\-1\end{bmatrix} \\ \boxed{1\ 2\ 3}\begin{bmatrix}1\\-1\\2\end{bmatrix} & \boxed{1\ 2\ 3}\begin{bmatrix}-2\\0\\-1\end{bmatrix} \\ \boxed{2\ 3\ 4}\begin{bmatrix}1\\-1\\2\end{bmatrix} & \boxed{2\ 3\ 4}\begin{bmatrix}-2\\0\\-1\end{bmatrix} \end{bmatrix}$$

For convenience of multiplication, we write the columns in horizontal rectangles.

$$= \begin{bmatrix} \boxed{0\ \ 1\ \ 2} & \boxed{0\ \ 1\ \ 2} \\ \boxed{1\ -1\ \ 2} & \boxed{-2\ \ 0\ -1} \\ \boxed{1\ \ 1\ \ 3} & \boxed{1\ \ 2\ \ 3} \\ \boxed{1\ -1\ \ 2} & \boxed{-2\ \ 0\ -1} \\ \boxed{2\ \ 3\ \ 4} & \boxed{2\ \ 3\ \ 4} \\ \boxed{1\ -1\ \ 2} & \boxed{-2\ \ 0\ -1} \end{bmatrix} = \begin{bmatrix} 0\times 1+1\times(-1)+2\times 2 & 0\times(-2)+1\times 0+2\times(-1) \\ 1\times 1+2\times(-1)+3\times 2 & 1\times(-2)+2\times 0+3\times(-1) \\ 2\times 1+3\times(-1)+4\times 2 & 2\times(-2)+3\times 0+4\times(-1) \end{bmatrix}$$

$$= \begin{bmatrix} 0-1+4 & 0+0-2 \\ 1-2+6 & -2+0-3 \\ 2-3+8 & -4+0-4 \end{bmatrix} = \begin{bmatrix} 3 & -2 \\ 5 & -5 \\ 7 & -8 \end{bmatrix}$$ **Ans.**

Since, the number of columns of $B$ is (2) $\neq$ the number of rows of $A$ is 3, $BA$ is not defined.

**Example 5.** *If* $A = \begin{bmatrix} 1 & -2 & 3 \\ 2 & 3 & -1 \\ -3 & 1 & 2 \end{bmatrix}$ *and* $B = \begin{bmatrix} 1 & 0 & 2 \\ 0 & 1 & 2 \\ 1 & 2 & 0 \end{bmatrix}$

*from the products AB and BA, and show that* $AB \neq BA$.

**Solution.** Here,

$$AB = \begin{bmatrix} 1 & -2 & 3 \\ 2 & 3 & -1 \\ -3 & 1 & 2 \end{bmatrix} \begin{bmatrix} 1 & 0 & 2 \\ 0 & 1 & 2 \\ 1 & 2 & 0 \end{bmatrix} = \begin{bmatrix} 1-0+3 & 0-2+6 & 2-4+0 \\ 2+0-1 & 0+3-2 & 4+6-0 \\ -3+0+2 & 0+1+4 & -6+2+0 \end{bmatrix} = \begin{bmatrix} 4 & 4 & -2 \\ 1 & 1 & 10 \\ -1 & 5 & -4 \end{bmatrix}$$

$$BA = \begin{bmatrix} 1 & 0 & 2 \\ 0 & 1 & 2 \\ 1 & 2 & 0 \end{bmatrix} \begin{bmatrix} 1 & -2 & 3 \\ 2 & 3 & -1 \\ -3 & 1 & 2 \end{bmatrix} = \begin{bmatrix} 1+0-6 & -2+0+2 & 3-0+4 \\ 0+2-6 & 0+3+2 & 0-1+4 \\ 1+4+0 & -2+6+0 & 3-2+0 \end{bmatrix} = \begin{bmatrix} -5 & 0 & 7 \\ -4 & 5 & 3 \\ 5 & 4 & 1 \end{bmatrix}$$

$AB \neq BA$ **Proved.**

**Example 6.** *If* $A = \begin{bmatrix} 1 & 2 \\ -2 & 3 \end{bmatrix}$, $B = \begin{bmatrix} 2 & 1 \\ 2 & 3 \end{bmatrix}$ *and* $C = \begin{bmatrix} -3 & 1 \\ 2 & 0 \end{bmatrix}$

*Verify that* $(AB)\,C = A\,(BC)$ *and* $A\,(B + C) = AB + AC$.

**Solution.** We have,

$$AB = \begin{bmatrix} 1 & 2 \\ -2 & 3 \end{bmatrix} \times \begin{bmatrix} 2 & 1 \\ 2 & 3 \end{bmatrix} = \begin{bmatrix} (1)(2)+(2)(2) & (1)(1)+(2)(3) \\ (-2)(2)+(3)(2) & (-2)(1)+(3)(3) \end{bmatrix} = \begin{bmatrix} 6 & 7 \\ 2 & 7 \end{bmatrix}$$

$$BC = \begin{bmatrix} 2 & 1 \\ 2 & 3 \end{bmatrix} \times \begin{bmatrix} -3 & 1 \\ 2 & 0 \end{bmatrix} = \begin{bmatrix} -6+2 & 2+0 \\ -6+6 & 2+0 \end{bmatrix} = \begin{bmatrix} -4 & 2 \\ 0 & 2 \end{bmatrix}$$

$$AC = \begin{bmatrix} 1 & 2 \\ -2 & 3 \end{bmatrix} \times \begin{bmatrix} -3 & 1 \\ 2 & 0 \end{bmatrix} = \begin{bmatrix} -3+4 & 1+0 \\ 6+6 & -2+0 \end{bmatrix} = \begin{bmatrix} 1 & 1 \\ 12 & -2 \end{bmatrix}$$

$$B + C = \begin{bmatrix} 2+(-3) & 1+1 \\ 2+2 & 3+0 \end{bmatrix} = \begin{bmatrix} -1 & 2 \\ 4 & 3 \end{bmatrix}$$

(*i*) $$(AB)\,C = \begin{bmatrix} 6 & 7 \\ 2 & 7 \end{bmatrix} \times \begin{bmatrix} -3 & 1 \\ 2 & 0 \end{bmatrix} = \begin{bmatrix} -18+14 & 6+0 \\ -6+14 & 2+0 \end{bmatrix} = \begin{bmatrix} -4 & 6 \\ 8 & 2 \end{bmatrix} \quad ...(1)$$

and $$A\,(BC) = \begin{bmatrix} 1 & 2 \\ -2 & 3 \end{bmatrix} \begin{bmatrix} -4 & 2 \\ 0 & 2 \end{bmatrix} = \begin{bmatrix} -4+0 & 2+4 \\ 8+0 & -4+6 \end{bmatrix} = \begin{bmatrix} -4 & 6 \\ 8 & 2 \end{bmatrix} \quad ...(2)$$

Thus from (1) and (2), we get

$$(AB)\,C = A\,(BC)$$

(*ii*) $$A\,(B+C) = \begin{bmatrix} 1 & 2 \\ -2 & 3 \end{bmatrix} \begin{bmatrix} -1 & 2 \\ 4 & 3 \end{bmatrix}$$

$$= \begin{bmatrix} -1+8 & 2+6 \\ 2+12 & -4+9 \end{bmatrix} = \begin{bmatrix} 7 & 8 \\ 14 & 5 \end{bmatrix} \quad ...(3)$$

$$AB + AC = \begin{bmatrix} 6+1 & 7+1 \\ 2+12 & 7-2 \end{bmatrix} = \begin{bmatrix} 7 & 8 \\ 14 & 5 \end{bmatrix} \quad ...(4)$$

Thus from (3) and (4), we get

$$A\,(B+C) = AB + AC$$ **Verified.**

**Example 7.** *If* $A = \begin{pmatrix} 1 & 2 & 2 \\ 2 & 1 & 2 \\ 2 & 2 & 1 \end{pmatrix}$ *show that* $A^2 - 4A - 5I = 0$ *where I, 0 are the unit matrix and the null matrix of order 3 respectively. Use this result to find* $A^{-1}$. *(A.M.I.E., Summer 2004)*

**Solution.** Here, we have $$A = \begin{bmatrix} 1 & 2 & 2 \\ 2 & 1 & 2 \\ 2 & 2 & 1 \end{bmatrix}$$

$$A^2 = \begin{bmatrix} 1 & 2 & 2 \\ 2 & 1 & 2 \\ 2 & 2 & 1 \end{bmatrix} \begin{bmatrix} 1 & 2 & 2 \\ 2 & 1 & 2 \\ 2 & 2 & 1 \end{bmatrix} = \begin{bmatrix} 9 & 8 & 8 \\ 8 & 9 & 8 \\ 8 & 8 & 9 \end{bmatrix}$$

$$A^2 - 4A - 5I = \begin{bmatrix} 9 & 8 & 8 \\ 8 & 9 & 8 \\ 8 & 8 & 9 \end{bmatrix} - 4 \begin{bmatrix} 1 & 2 & 2 \\ 2 & 1 & 2 \\ 2 & 2 & 1 \end{bmatrix} - 5 \begin{bmatrix} 1 & 0 & 0 \\ 0 & 1 & 0 \\ 0 & 0 & 1 \end{bmatrix}$$

$$A^2 - 4A - 5I = \begin{bmatrix} 9-4-5 & 8-8-0 & 8-8-0 \\ 8-8-0 & 9-4-5 & 8-8-0 \\ 8-8-0 & 8-8-0 & 9-4-5 \end{bmatrix} = \begin{bmatrix} 0 & 0 & 0 \\ 0 & 0 & 0 \\ 0 & 0 & 0 \end{bmatrix}$$

$$A^2 - 4A - 5I = 0 \quad \Rightarrow \quad 5I = A^2 - 4A$$

Multiplying by $A^{-1}$, we get

$$5A^{-1} = A - 4I$$

$$= \begin{bmatrix} 1 & 2 & 2 \\ 2 & 1 & 2 \\ 2 & 2 & 1 \end{bmatrix} - 4\begin{bmatrix} 1 & 0 & 0 \\ 0 & 1 & 0 \\ 0 & 0 & 1 \end{bmatrix} = \begin{bmatrix} -3 & 2 & 2 \\ 2 & -3 & 2 \\ 2 & 2 & -3 \end{bmatrix}$$

$$A^{-1} = \frac{1}{5}\begin{bmatrix} -3 & 2 & 2 \\ 2 & -3 & 2 \\ 2 & 2 & -3 \end{bmatrix}$$ **Ans.**

**Example 8.** *Show by means of an example that in matrices AB = 0 does not necessarily mean that either A = 0 or B = 0, where 0 stands for the null matrix.*

**Solution.** Let $A = \begin{bmatrix} 1 & -1 & 1 \\ -3 & 2 & -1 \\ -2 & 1 & 0 \end{bmatrix}$, $B = \begin{bmatrix} 1 & 2 & 3 \\ 2 & 4 & 6 \\ 1 & 2 & 3 \end{bmatrix}$

$$AB = \begin{bmatrix} 1-2+1 & 2-4+2 & 3-6+3 \\ -3+4-1 & -6+8-2 & -9+12-3 \\ -2+2+0 & -4+4+0 & -6+6+0 \end{bmatrix} = \begin{bmatrix} 0 & 0 & 0 \\ 0 & 0 & 0 \\ 0 & 0 & 0 \end{bmatrix}$$

$$AB = 0.$$

But here neither $A = 0$ nor $B = 0$. **Proved.**

**Example 9.** *If AB = AC, it is not necessarily true that B = C i.e. like ordinary algebra, the equal matrices in the identity cannot be cancelled.*

**Solution.** Let $AB = \begin{bmatrix} 1 & -3 & 2 \\ 2 & 1 & -3 \\ 4 & -3 & -1 \end{bmatrix}\begin{bmatrix} 1 & 4 & 1 & 0 \\ 2 & 1 & 1 & 1 \\ 1 & -2 & 1 & 2 \end{bmatrix} = \begin{bmatrix} -3 & -3 & 0 & 1 \\ 1 & 15 & 0 & -5 \\ -3 & 15 & 0 & -5 \end{bmatrix}$

$$AC = \begin{bmatrix} 1 & -3 & 2 \\ 2 & 1 & -3 \\ 4 & -3 & -1 \end{bmatrix}\begin{bmatrix} 2 & 1 & -1 & -2 \\ 3 & -2 & -1 & -1 \\ 2 & -5 & -1 & 0 \end{bmatrix} = \begin{bmatrix} -3 & -3 & 0 & 1 \\ 1 & 15 & 0 & -5 \\ -3 & 15 & 0 & -5 \end{bmatrix}$$ **Proved.**

Here, $AB = AC$. But $B \neq C$.

**Example 10.** *Represent each of the transformations*

$x_1 = 3y_1 + 2y_2$, $y_1 = z_1 + 2z_2$ *and* $x_2 = -y_1 + 4y_2$, $y_2 = 3z_1$

*by the use of matrices and find the composite transformation which expresses* $x_1, x_2$ *in terms* of $z_1, z_2$.

**Solution.** The equations in the matrix form are

$$\begin{bmatrix} x_1 \\ x_2 \end{bmatrix} = \begin{bmatrix} 3 & 2 \\ -1 & 4 \end{bmatrix}\begin{bmatrix} y_1 \\ y_2 \end{bmatrix} \qquad ...(1)$$

$$\begin{bmatrix} y_1 \\ y_2 \end{bmatrix} = \begin{bmatrix} 1 & 2 \\ 3 & 0 \end{bmatrix}\begin{bmatrix} z_1 \\ z_2 \end{bmatrix} \qquad ...(2)$$

Substituting the values of $y_1, y_2$ in (1), we get

$$\begin{bmatrix} x_1 \\ x_2 \end{bmatrix} = \begin{bmatrix} 3 & 2 \\ -1 & 4 \end{bmatrix}\begin{bmatrix} 1 & 2 \\ 3 & 0 \end{bmatrix}\begin{bmatrix} z_1 \\ z_2 \end{bmatrix} = \begin{bmatrix} 9 & 6 \\ 11 & -2 \end{bmatrix}\begin{bmatrix} z_1 \\ z_2 \end{bmatrix} = \begin{bmatrix} 9z_1 + 6z_2 \\ 11z_1 - 2z_2 \end{bmatrix}$$

$$x_1 = 9z_1 + 6z_2, \quad x_2 = 11z_1 - 2z_2$$ **Ans.**

**Example 11.** *Prove that the product of two matrices*

$$\begin{bmatrix} \cos^2\theta & \cos\theta\sin\theta \\ \cos\theta\sin\theta & \sin^2\theta \end{bmatrix} \text{ and } \begin{bmatrix} \cos^2\phi & \cos\phi\sin\phi \\ \cos\phi\sin\phi & \sin^2\phi \end{bmatrix}$$

*is zero when $\theta$ and $\phi$ differ by an odd multiple of $\frac{\pi}{2}$.*

**Solution.** $= \begin{bmatrix} \cos^2\theta & \cos\theta\sin\theta \\ \cos\theta\sin\theta & \sin^2\theta \end{bmatrix} \times \begin{bmatrix} \cos^2\phi & \cos\phi\sin\phi \\ \cos\phi\sin\phi & \sin^2\phi \end{bmatrix}$

$$= \begin{bmatrix} \cos^2\theta\cos^2\phi + \cos\theta\sin\theta\cos\phi\sin\phi & \cos^2\theta\cos\phi\sin\phi + \cos\theta\sin\theta\sin^2\phi \\ \cos\theta\sin\theta\cos^2\phi + \sin^2\theta\cos\phi\sin\phi & \cos\theta\sin\theta\cos\phi\sin\phi + \sin^2\theta\sin^2\phi \end{bmatrix}$$

$$= \begin{bmatrix} \cos\theta\cos\phi(\cos\theta\cos\phi + \sin\theta\sin\phi) & \cos\theta\sin\phi(\cos\theta\cos\phi + \sin\theta\sin\phi) \\ \sin\theta\cos\phi(\cos\theta\cos\phi + \sin\theta\sin\phi) & \sin\theta\sin\phi(\cos\theta\cos\phi + \sin\theta\sin\phi) \end{bmatrix}$$

$$= \begin{bmatrix} \cos\theta\cos\phi\cos(\theta-\phi) & \cos\theta\sin\phi\cos(\theta-\phi) \\ \sin\theta\cos\phi\cos(\theta-\phi) & \sin\theta\sin\phi\cos(\theta-\phi) \end{bmatrix}$$

Given $\qquad \theta - \phi = (2n+1)\frac{\pi}{2}$

$$\cos(\theta-\phi) = \cos(2n+1)\frac{\pi}{2} = 0$$

$\therefore$ The product $= \begin{bmatrix} 0 & 0 \\ 0 & 0 \end{bmatrix} = 0.$ **Proved.**

**Example 12.** *Verify that*

$$A = \frac{1}{3}\begin{bmatrix} 1 & 2 & 2 \\ 2 & 1 & -2 \\ -2 & 2 & -1 \end{bmatrix} \text{ is orthogonal.}$$

**Solution.** $A = \frac{1}{3}\begin{bmatrix} 1 & 2 & 2 \\ 2 & 1 & -2 \\ -2 & 2 & -1 \end{bmatrix} \therefore A' = \frac{1}{3}\begin{bmatrix} 1 & 2 & -2 \\ 2 & 1 & 2 \\ 2 & -2 & -1 \end{bmatrix}$

$$AA' = \frac{1}{9}\begin{bmatrix} 1 & 2 & 2 \\ 2 & 1 & -2 \\ -2 & 2 & -1 \end{bmatrix}\begin{bmatrix} 1 & 2 & -2 \\ 2 & 1 & 2 \\ 2 & -2 & -1 \end{bmatrix} = \frac{1}{9}\begin{bmatrix} 9 & 0 & 0 \\ 0 & 9 & 0 \\ 0 & 0 & 9 \end{bmatrix}\begin{bmatrix} 1 & 0 & 0 \\ 0 & 1 & 0 \\ 0 & 0 & 1 \end{bmatrix} = I$$

Hence, $A$ is an orthogonal matrix. **Verified.**

**Example 13.** *Determine the values of $\alpha$, $\beta$, $\gamma$ when*

$$\begin{bmatrix} 0 & 2\beta & \gamma \\ \alpha & \beta & -\gamma \\ \alpha & -\beta & \gamma \end{bmatrix} \text{ is orthogonal.}$$

**Solution.** Let $A = \begin{bmatrix} 0 & 2\beta & \gamma \\ \alpha & \beta & -\gamma \\ \alpha & -\beta & \gamma \end{bmatrix}$

On transposing $A$, we have

$$A' = \begin{bmatrix} 0 & \alpha & \alpha \\ 2\beta & \beta & -\beta \\ \gamma & -\gamma & \gamma \end{bmatrix}$$

If $A$ is orthogonal, then $AA' = I$

$$\begin{bmatrix} 0 & 2\beta & \gamma \\ \alpha & \beta & -\gamma \\ \alpha & -\beta & \gamma \end{bmatrix} \begin{bmatrix} 0 & \alpha & \alpha \\ 2\beta & \beta & -\beta \\ \gamma & -\gamma & \gamma \end{bmatrix} = \begin{bmatrix} 1 & 0 & 0 \\ 0 & 1 & 0 \\ 0 & 0 & 1 \end{bmatrix}$$

$$\Rightarrow \begin{bmatrix} 4\beta^2+\gamma^2 & 2\beta^2-\gamma^2 & -2\beta^2+\gamma^2 \\ 2\beta^2-\gamma^2 & \alpha^2+\beta^2+\gamma^2 & \alpha^2-\beta^2-\gamma^2 \\ -2\beta^2+\gamma^2 & \alpha^2-\beta^2-\gamma^2 & \alpha^2+\beta^2+\gamma^2 \end{bmatrix} = \begin{bmatrix} 1 & 0 & 0 \\ 0 & 1 & 0 \\ 0 & 0 & 1 \end{bmatrix}$$

Equating the corresponding elements, we have

$$\left.\begin{array}{l} 4\beta^2+\gamma^2 = 1 \\ 2\beta^2-\gamma^2 = 0 \end{array}\right\} \Rightarrow \beta = \pm\frac{1}{\sqrt{6}}, \gamma = \pm\frac{1}{\sqrt{3}}$$

But $\alpha^2+\beta^2+\gamma^2 = 1$ as $\beta = \pm\frac{1}{\sqrt{6}}, \gamma = \pm\frac{1}{\sqrt{3}}, \alpha = \pm\frac{1}{\sqrt{2}}$ **Ans.**

**Example 14.** *Prove that*

$$(AB)^n = A^n \cdot B^n, \text{ if } A \cdot B = B \cdot A$$

**Solution.**

$$(AB)^1 = AB = (A) \cdot (B)$$
$$(AB)^2 = (AB) \cdot (AB) = (ABA) \cdot B = \{ A (AB) \} \cdot B$$
$$= (A^2 B) \cdot B = A^2 (B \cdot B) = A^2 \cdot B^2$$

Suppose that

$$(AB)^n = A^n \cdot B^n$$
$$(AB)^{n+1} = (AB)^n \cdot (AB) = (A^n \cdot B^n) \cdot (AB) = A^n \cdot (B^n A) \cdot B$$
$$= A^n \cdot (B^{n-1} \cdot BA) \cdot B = A^n \cdot (B^{n-1} \cdot AB) \cdot B$$
$$= A^n \cdot (B^{n-2} \cdot B \cdot AB) \cdot B = A^n \cdot (B^{n-2} \cdot AB \cdot B) \cdot B$$
$$= A^n \cdot (B^{n-2} \cdot AB^2) \cdot B, \text{ continuing the process } n \text{ times.}$$
$$= A^n \cdot (A \cdot B^n) \cdot B = A^n \cdot (A \cdot B^{n+1}) = A^{n+1} \cdot B^{n+1}$$

Hence, taking the above to be true for $n = n$, we have shown that it is true for $n = n + 1$ and also it was true for $n = 1, 2, ....$ so it is universally true. **Proved.**

## EXERCISE 4.11

1. Compute $AB$, if

$$A = \begin{bmatrix} 1 & 2 & 3 \\ 4 & 5 & 6 \end{bmatrix} \text{ and } B = \begin{bmatrix} 2 & 5 & 3 \\ 3 & 6 & 4 \\ 4 & 7 & 5 \end{bmatrix}$$

**Ans.** $\begin{bmatrix} 20 & 38 & 26 \\ 47 & 92 & 62 \end{bmatrix}$

2. If $A = \begin{bmatrix} 1 & 3 & 0 \\ -1 & 2 & 1 \\ 0 & 0 & 2 \end{bmatrix}$, $B = \begin{bmatrix} 2 & 3 & 4 \\ 1 & 2 & 3 \\ -1 & 1 & 2 \end{bmatrix}$. From the product $AB$ and $BA$. Show that $AB \neq BA$

3. If $A = \begin{bmatrix} 0 & 1 & 0 \\ 0 & 0 & 1 \\ 0 & 0 & 0 \end{bmatrix}$, $B = \begin{bmatrix} 0 & 0 & 0 \\ 1 & 0 & 0 \\ 0 & 1 & 0 \end{bmatrix}$

(*i*) Calculate $AB$ and $BA$. Hence evaluate $A^2 B + B^2 A$

(*ii*) Show that for any number $k$, $(A + kB^2)^3 = KI$, where $I$ is the unit matrix.

**4.** If $A = \begin{bmatrix} 0 & 1 \\ -1 & 0 \end{bmatrix}$ choose α and β so that $(\alpha I + \beta A)^2 = A$ **Ans.** $\alpha = \beta = \pm\frac{1}{\sqrt{2}}$

**5.** If $A = \begin{bmatrix} 2 \\ 0 \end{bmatrix}, B = [-1 \quad 1 \quad 0]$ and $C = \begin{bmatrix} 3 & 0 \\ -2 & 1 \\ 0 & 1 \end{bmatrix}$

verify that $(ABC)^T = C^T B^T A^T$, where T denotes the transpose.

**6.** Write the following transformation in matrix form :

$$x_1 = \frac{\sqrt{3}}{2} y_1 + \frac{1}{2} y_2 \; ; x_2 = -\frac{1}{2} y_1 + \frac{\sqrt{3}}{2} y_2$$

Hence, find the transformation in matrix form which expresses $y_1, y_2$ in terms of $x_1, x_2$.

**Ans.** $y_1 = \frac{\sqrt{3}}{2} x_1 - \frac{1}{2} x_2, \; y_2 = \frac{1}{2} x_1 + \frac{\sqrt{3}}{2} x_2$

**7.** The linear transformation $\begin{pmatrix} y_1 \\ y_2 \end{pmatrix} = \begin{pmatrix} 0 & 0 \\ 0 & 1 \end{pmatrix} \begin{pmatrix} x_1 \\ x_2 \end{pmatrix}$, represents

(*a*) reflection about $x_1$ -axis (*b*) reflection about $x_2$ -axis (*c*) clockwise rotation through angle $\frac{\pi}{2}$
(*d*) orthogonal projection on to $x_2$ axis. (*A.M.IE.T.E., Summer 2005*) **Ans.** (*d*)

**8.** If $A = \begin{bmatrix} 0 & -\tan\frac{\alpha}{2} \\ \tan\frac{\alpha}{2} & 0 \end{bmatrix}$ and $I$ is a unit matrix, show that $I + A = (I - A)\begin{bmatrix} \cos\alpha & -\sin\alpha \\ \sin\alpha & \cos\alpha \end{bmatrix}$

**9.** If $f(x) = x^3 - 20x + 8$, find $f(A)$ where $A = \begin{bmatrix} 1 & 1 & 3 \\ 1 & 3 & -3 \\ -2 & -4 & -4 \end{bmatrix}$ **Ans.** 0

**10.** Show that $\begin{bmatrix} \cos\theta & -\sin\theta \\ \sin\theta & \cos\theta \end{bmatrix} = \begin{bmatrix} 1 & -\tan\frac{\theta}{2} \\ \tan\frac{\theta}{2} & 1 \end{bmatrix} \begin{bmatrix} 1 & \tan\frac{\theta}{2} \\ -\tan\frac{\theta}{2} & 1 \end{bmatrix}^{-1}$

**11.** If $A = \begin{bmatrix} 3 & -3 & 4 \\ 2 & -3 & 4 \\ 0 & -1 & 1 \end{bmatrix}$ then show that $A^3 = A^{-1}$.

**12.** Verify whether the matrix $A = \frac{1}{3}\begin{bmatrix} 2 & 2 & 1 \\ 2 & -1 & 2 \\ -1 & 2 & 2 \end{bmatrix}$ is orthogonal.

**13.** Verify that $\frac{1}{3}\begin{bmatrix} 1 & -2 & 2 \\ -2 & 1 & 2 \\ -2 & -2 & -1 \end{bmatrix}$ is an orthogonal matrix.

**14.** Show that $\begin{bmatrix} \cos\phi & 0 & \sin\phi \\ \sin\theta\sin\phi & \cos\theta & -\sin\theta\cos\phi \\ -\cos\theta\sin\phi & \sin\theta & \cos\theta\cos\phi \end{bmatrix}$ is an orthogonal matrix.

**15.** Show that $A = \begin{bmatrix} \cos\theta & 0 & \sin\theta \\ 0 & 1 & 0 \\ -\sin\theta & 0 & \cos\theta \end{bmatrix}$ is an orthogonal matrix. (*A.M.I.E., Summer 2004*)

**16.** If $A$ and $B$ are square matrices of the same order, explain in general

(*i*) $(A + B)^2 \neq A^2 + 2AB + B^2$ (*ii*) $(A - B)^2 \neq A^2 - 2AB + B^2$ (*iii*) $(A + B)(A - B) \neq A^2 - B^2$

**17.** Let A and B be any two matrices such that AB = 0 and A is non- singular.
then (a) B = 0; (b) B is also non-singular; (c) B = A; (d) B is singular.

**Ans.** (*d*)

**18.** If $A^2 = A$ then matrix A is called
(a) Idempotent Matrix (b) Null Matrix (c) Transpose Matrix (d) Identity Matrix
(A.M,I.E.T.E.,Dec.,2006) **Ans.** (*a*)

## 4.27 MATHEMATICAL INDUCTION

By mathematical induction we can prove results for all positive integers. If the result to be proved for the positive integer *n* then we apply the following method.

**Working Rule:**

**Step 1.** Verify the result for $n = 1$

**Step 2.** Assume the result to be true for $n = k$ and then prove that it is true for $n = k + 1$.

**Explanation.** By step 1, the result is true for $n = k = 1$

By step 2, the result is true for $n = k + 1 = 1 + 1 = 2$ $(k = 1)$

Again, the result is also true for $n = k + 1 = 2 + 1 = 3$ $(k = 2)$

Similarly, the result is also true for $n = k + 1 = 3 + 1 = 4$ $(k = 3)$

Hence, in this way the result is true for all positive integer *n*.

**Example 15.** *By mathematical induction,*

$$\text{if } A = \begin{bmatrix} \cos\alpha & \sin\alpha \\ -\sin\alpha & \cos\alpha \end{bmatrix}, \text{ show that } A^n = \begin{bmatrix} \cos n\alpha & \sin n\alpha \\ -\sin n\alpha & \cos n\alpha \end{bmatrix}$$

Where *n* is a positive integer.

**Solution.** We prove the result by mathematical induction :

$$A^n = \begin{bmatrix} \cos n\alpha & \sin n\alpha \\ -\sin n\alpha & \cos n\alpha \end{bmatrix}$$

Let us verify the result for $n = 1$.

$$A^1 = \begin{bmatrix} \cos 1\alpha & \sin 1\alpha \\ -\sin 1\alpha & \cos 1\alpha \end{bmatrix} = \begin{bmatrix} \cos\alpha & \sin\alpha \\ -\sin\alpha & \cos\alpha \end{bmatrix} = A \qquad \text{[Given]}$$

The result is true when $n = 1$.

Let us assume that the result is true for any positive integer *k*.

$$A^k = \begin{bmatrix} \cos k\alpha & \sin k\alpha \\ -\sin k\alpha & \cos k\alpha \end{bmatrix}$$

Now,
$$A^{k+1} = A^k . A = \begin{bmatrix} \cos k\alpha & \sin k\alpha \\ -\sin k\alpha & \cos k\alpha \end{bmatrix}\begin{bmatrix} \cos\alpha & \sin\alpha \\ -\sin\alpha & \cos\alpha \end{bmatrix}$$

$$= \begin{bmatrix} \cos k\alpha \cos\alpha - \sin k\alpha \sin\alpha & \cos k\alpha \sin\alpha + \sin k\alpha \cos\alpha \\ -\sin k\alpha \cos\alpha - \cos k\alpha \sin\alpha & -\sin k\alpha \sin\alpha + \cos k\alpha \cos\alpha \end{bmatrix}$$

$$= \begin{bmatrix} \cos(k\alpha + \alpha) & \sin(k\alpha + \alpha) \\ -\sin(k\alpha + \alpha) & \cos(k\alpha + \alpha) \end{bmatrix} = \begin{bmatrix} \cos(k+1)\alpha & \sin(k+1)\alpha \\ -\sin(k+1)\alpha & \cos(k+1)\alpha \end{bmatrix}$$

The result is true for $n = k + 1$.

Hence, by mathematical induction the result is true for all positive integer *n*. **Proved.**

## 4.28 ADJOINT OF A SQUARE MATRIX

Let the determinant of the square matrix $A$ be $|A|$.

If $A = \begin{bmatrix} a_1 & a_2 & a_3 \\ b_1 & b_2 & b_3 \\ c_1 & c_2 & c_3 \end{bmatrix}$, Than $|A| = \begin{bmatrix} a_1 & a_2 & a_3 \\ b_1 & b_2 & b_3 \\ c_1 & c_2 & c_3 \end{bmatrix}$

The matrix formed by the co-factors of the elements in

$$|A| \text{ is } \begin{bmatrix} A_1 & A_2 & A_3 \\ B_1 & B_2 & B_3 \\ C_1 & C_2 & C_3 \end{bmatrix}.$$

where $A_1 = \begin{vmatrix} b_2 & b_3 \\ c_2 & c_3 \end{vmatrix} = b_2c_3 - b_3c_2$, $\quad A_2 = -\begin{vmatrix} b_1 & b_3 \\ c_1 & c_3 \end{vmatrix} = -b_1c_3 + b_3c_1$

$A_3 = \begin{vmatrix} b_1 & b_2 \\ c_1 & c_2 \end{vmatrix} = b_1c_2 - b_2c_1$, $\quad B_1 = -\begin{vmatrix} a_2 & a_3 \\ c_2 & c_3 \end{vmatrix} = -a_2c_3 + a_3c_2$

$B_2 = \begin{vmatrix} a_1 & a_3 \\ c_1 & c_3 \end{vmatrix} = a_1c_3 - a_3c_1$, $\quad B_3 = -\begin{vmatrix} a_1 & a_2 \\ c_1 & c_2 \end{vmatrix} = -a_1c_2 + a_2c_1$

$C_1 = \begin{vmatrix} a_2 & a_3 \\ b_2 & b_3 \end{vmatrix} = a_2b_3 - a_3b_2$, $\quad C_2 = -\begin{vmatrix} a_1 & a_3 \\ b_1 & b_3 \end{vmatrix} = -a_1b_3 + a_3b_1$

$$C_3 = \begin{vmatrix} a_1 & a_2 \\ b_1 & b_2 \end{vmatrix} = a_1b_2 - a_2b_1$$

Then the transpose of the matrix of co-factors

$$\begin{bmatrix} A_1 & B_1 & C_1 \\ A_2 & B_2 & C_2 \\ A_3 & B_3 & C_3 \end{bmatrix}$$

is called the adjoint of the matrix $A$ and is written as adj $A$.

## 4.29 PROPERTY OF ADJOINT MATRIX

***The product of a matrix A and its adjoint is equal to unit matrix multiplied by the determinant A.***

**Proof.** If $A$ be a square matrix, then (Adjoint $A$) $\cdot A = A \cdot$ (Adjoint $A$) $= |A| \cdot I$

Let $A = \begin{bmatrix} a_1 & a_2 & a_3 \\ b_1 & b_2 & b_3 \\ c_1 & c_2 & c_3 \end{bmatrix}$ and adj . $A = \begin{bmatrix} A_1 & B_1 & C_1 \\ A_2 & B_2 & C_2 \\ A_3 & B_3 & C_3 \end{bmatrix}$

$$A \,.\, (\text{adj. } A) = \begin{bmatrix} a_1 & a_2 & a_3 \\ b_1 & b_2 & b_3 \\ c_1 & c_2 & c_3 \end{bmatrix} \times \begin{bmatrix} A_1 & B_1 & C_1 \\ A_2 & B_2 & C_2 \\ A_3 & B_3 & C_3 \end{bmatrix}$$

$$= \begin{bmatrix} a_1A_1 + a_2A_2 + a_3A_3 & a_1B_1 + a_2B_2 + a_3B_3 & a_1C_1 + a_2C_2 + a_3C_3 \\ b_1A_1 + b_2A_2 + b_3A_3 & b_1B_1 + b_2B_2 + b_3B_3 & b_1C_1 + b_2C_2 + b_3C_3 \\ c_1A_1 + c_2A_2 + c_3A_3 & c_1B_1 + c_2B_2 + c_3B_3 & c_1C_1 + c_2C_2 + c_3C_3 \end{bmatrix}$$

$$= \begin{bmatrix} |A| & 0 & 0 \\ 0 & |A| & 0 \\ 0 & 0 & |A| \end{bmatrix} = |A| \begin{bmatrix} 1 & 0 & 0 \\ 0 & 1 & 0 \\ 0 & 0 & 1 \end{bmatrix} = |A|\, I$$

*(A.M.I.E., Summer 2004)*

## 4.30 INVERSE OF A MATRIX

If $A$ and B are two square matrices of the same order, such that

$$AB = BA = I \qquad (I = \text{unit matrix})$$

then $B$ is called the inverse of $A$ *i.e.* $B = A^{-1}$ and $A$ is the inverse of $B$.

*Condition for a square matrix A to possess an inverse is that matrix A is non-singular, i.e.,* $|A| \neq 0$

If $A$ is a square matrix and $B$ be its inverse, then $AB = I$

Taking determinant of both sides, we get $|AB| = |I|$ or $|A||B| = I$

From this relation it is clear that $|A| \neq 0$

*i.e.* the matrix $A$ is non-singular.

**To find the inverse matrix with the help of adjoint matrix**

We know that $\quad A \cdot (Adj.\ A) = |A|\ I$

$$\Rightarrow \qquad A \cdot \frac{1}{|A|}(A\,dj.\ A) = I \qquad [\text{Provided } |A| \neq 0] \qquad \ldots(1)$$

and $$A \cdot A^{-1} = I \qquad \ldots(2)$$

From (1) and (2), we have

$$\therefore \qquad \boxed{A^{-1} = \frac{1}{|A|}(Adj.\ A)}$$

**Example 16.** *If* $A = \begin{bmatrix} 3 & -3 & 4 \\ 2 & -3 & 4 \\ 0 & -1 & 1 \end{bmatrix}$, *find* $A^{-1}$. *(A.M.I.E. Summer 2004)*

**Solution.** $A = \begin{bmatrix} 3 & -3 & 4 \\ 2 & -3 & 4 \\ 0 & -1 & 1 \end{bmatrix}$

$$|A| = 3(-3+4) + 3(2-0) + 4(-2-0) = 3 + 6 - 8 = 1$$

The co-factors of elements of various rows of $|A|$ are

$$\begin{bmatrix} (-3+4) & (-2-0) & (-2-0) \\ (3-4) & (3-0) & (3-0) \\ (-12+12) & (-12+8) & (-9+6) \end{bmatrix}$$

Therefore, the matrix formed by the co-factors of $|A|$ is

$$\begin{bmatrix} 1 & -2 & -2 \\ -1 & 3 & 3 \\ 0 & -4 & -3 \end{bmatrix}, \quad Adj.\ A = \begin{bmatrix} 1 & -1 & 0 \\ -2 & 3 & -4 \\ -2 & 3 & -3 \end{bmatrix}$$

$$\therefore \qquad A^{-1} = \frac{1}{|A|} Adj.\ A = \frac{1}{1}\begin{bmatrix} 1 & -1 & 0 \\ -2 & 3 & -4 \\ -2 & 3 & -3 \end{bmatrix} = \begin{bmatrix} 1 & -1 & 0 \\ -2 & 3 & -4 \\ -2 & 3 & -3 \end{bmatrix} \qquad \textbf{Ans.}$$

**Example 17.** *If* $A = \frac{1}{9}\begin{bmatrix} -8 & 1 & 4 \\ 4 & 4 & 7 \\ 1 & -8 & 4 \end{bmatrix}$, *prove that* $A^{-1} = A'$, *A' being the transpose of A.*

*(A.M.I.E., Winter 2000)*

**Solution.** We have, $A = \frac{1}{9}\begin{bmatrix} -8 & 1 & 4 \\ 4 & 4 & 7 \\ 1 & -8 & 4 \end{bmatrix}$, $A' = \frac{1}{9}\begin{bmatrix} -8 & 4 & 1 \\ 1 & 4 & -8 \\ 4 & 7 & 4 \end{bmatrix}$

$$AA' = \frac{1}{9}\begin{bmatrix} -8 & 1 & 4 \\ 4 & 4 & 7 \\ 1 & -8 & 4 \end{bmatrix} \cdot \frac{1}{9}\begin{bmatrix} -8 & 4 & 1 \\ 1 & 4 & -8 \\ 4 & 7 & 4 \end{bmatrix}$$

$$= \frac{1}{81}\begin{bmatrix} 64+1+16 & -32+4+28 & -8-8+16 \\ -32+4+28 & 16+16+49 & 4-32+28 \\ -8-8+16 & 4-32+28 & 1+64+16 \end{bmatrix}$$

$$= \frac{1}{81}\begin{bmatrix} 81 & 0 & 0 \\ 0 & 81 & 0 \\ 0 & 0 & 81 \end{bmatrix} = \begin{bmatrix} 1 & 0 & 0 \\ 0 & 1 & 0 \\ 0 & 0 & 1 \end{bmatrix} \text{ or } AA' = I$$

$A' = A^{-1}$ **Proved.**

**Example 18.** *If a matrix A satisfies a relation $A^2 + A - I = 0$ proved that $A^{-1}$ exists and that $A^{-1} = I + A$, I being an identity matrix.* *(A M I E Winter 2003)*

**Solution.** Here $A^2 + A - I = 0$ or $A^2 + AI = I$ or $A(A+I) = I$

$$|A||A+I| = |I|$$

$\therefore$ $|A| \neq 0$ and so $A^{-1}$exists.

Again $A^2 + A - I = 0$ or $A^2 + A = I$ ...(1)

Multiplying (1) by $A^{-1}$,we get

$$A^{-1}(A^2 + A) = A^{-1}I \quad \text{or} \quad A + I = A^{-1}$$

$\Rightarrow$ $A^{-1} = I + A$ **Proved.**

**Example 19.** *If A and B are non-singular matrices of the same order then,*

$$(AB)^{-1} = B^{-1} \cdot A^{-1}$$

*Hence prove that $(A^{-1})^m = (A^m)^{-1}$ for any positive integer m.*

**Solution.** We know that,

$$(AB) \cdot (B^{-1} A^{-1}) = [(AB) B^{-1}] \cdot A^{-1} = [A (BB^{-1}] \cdot A^{-1}$$
$$= [AI] A^{-1} = A \cdot A^{-1} = I$$

Also, $B^{-1} A^{-1} \cdot (AB) = B^{-1}[A^{-1} \cdot (AB)] = B^{-1} [(A^{-1} A) \cdot B]$

$$= B^{-1} [I . B] = B^{-1} \cdot B = I$$

By definition of the inverse of a matrix, $B^{-1} A^{-1}$ is inverse of $AB$.

$\Rightarrow$ $B^{-1} A^{-1} = (AB)^{-1}$ **Proved.**

$$(A^m)^{-1} = [A \cdot A^{m-1}]^{-1} = (A^{m-1})^{-1} A^{-1}$$
$$= (A \cdot A^{m-2})^{-1} \cdot A^{-1} = [(A^{m-2})^{-1} \cdot A^{-1}] \cdot A^{-1} = (A^{m-2})^{-1} (A^{-1})^2$$
$$= (A \cdot A^{m-3})^{-1} \cdot (A^{-1})^2 = [(A^{m-3})^{-1} \cdot A^{-1}] (A^{-1})^2 = (A^{m-3})^{-1} (A^{-1})^3$$
$$= A^{-1} (A^{-1})^{m-1} = (A^{-1})^m$$

**Proved.**

**Example 20.** *Find A satisfying the Matrix equation.*

$$\begin{bmatrix} 2 & 1 \\ 3 & 2 \end{bmatrix} A \begin{bmatrix} -3 & 2 \\ 5 & -3 \end{bmatrix} = \begin{bmatrix} -2 & 4 \\ 3 & -1 \end{bmatrix}$$

**Solution.** $\begin{bmatrix} 2 & 1 \\ 3 & 2 \end{bmatrix} A \begin{bmatrix} -3 & 2 \\ 5 & -3 \end{bmatrix} = \begin{bmatrix} -2 & 4 \\ 3 & -1 \end{bmatrix}$

Both sides of the equation are pre-multiplied by the inverse of $\begin{bmatrix} 2 & 1 \\ 3 & 2 \end{bmatrix}$ *i.e.*, $\begin{bmatrix} 2 & -1 \\ -3 & 2 \end{bmatrix}$

$$\begin{bmatrix} 2 & -1 \\ -3 & 2 \end{bmatrix}\begin{bmatrix} 2 & 1 \\ 3 & 2 \end{bmatrix} A \begin{bmatrix} -3 & 2 \\ 5 & -3 \end{bmatrix} = \begin{bmatrix} 2 & -1 \\ -3 & 2 \end{bmatrix}\begin{bmatrix} -2 & 4 \\ 3 & -1 \end{bmatrix}$$

$$\begin{bmatrix} 1 & 0 \\ 0 & 1 \end{bmatrix} A \begin{bmatrix} -3 & 2 \\ 5 & -3 \end{bmatrix} = \begin{bmatrix} -7 & 9 \\ 12 & -14 \end{bmatrix}$$

$$A \begin{bmatrix} -3 & 2 \\ 5 & -3 \end{bmatrix} = \begin{bmatrix} -7 & 9 \\ 12 & -14 \end{bmatrix}$$

Again both sides are post-multiplied by the inverse of $\begin{bmatrix} -3 & 2 \\ 5 & -3 \end{bmatrix}$ *i.e.* $\begin{bmatrix} 3 & 2 \\ 5 & 3 \end{bmatrix}$

$$A \begin{bmatrix} -3 & 2 \\ 5 & -3 \end{bmatrix}\begin{bmatrix} 3 & 2 \\ 5 & 3 \end{bmatrix} = \begin{bmatrix} -7 & 9 \\ 12 & -14 \end{bmatrix}\begin{bmatrix} 3 & 2 \\ 5 & 3 \end{bmatrix}$$

$$A \begin{bmatrix} 1 & 0 \\ 0 & 1 \end{bmatrix} = \begin{bmatrix} 24 & 13 \\ -34 & -18 \end{bmatrix} \Rightarrow A = \begin{bmatrix} 24 & 13 \\ -34 & -18 \end{bmatrix}$$ **Ans.**

## EXERCISE 4.12

Find the adjoint and inverse of the following matrices: (1 - 3)

**1.** $\begin{bmatrix} 2 & 5 & 3 \\ 3 & 1 & 2 \\ 1 & 2 & 1 \end{bmatrix}$ **Ans.** $\frac{1}{4}\begin{bmatrix} -3 & 1 & 7 \\ -1 & -1 & 5 \\ 5 & 1 & -13 \end{bmatrix}$ **2.** $\begin{bmatrix} 1 & 1 & 2 \\ 1 & 9 & 3 \\ 1 & 4 & 2 \end{bmatrix}$ **Ans.** $-\frac{1}{3}\begin{bmatrix} 6 & 6 & -15 \\ 1 & 0 & -1 \\ -5 & -3 & 8 \end{bmatrix}$

**3.** $\begin{bmatrix} 1 & 0 & -1 \\ 3 & 4 & 5 \\ 0 & -6 & -7 \end{bmatrix}$ **Ans.** $\frac{1}{20}\begin{bmatrix} 2 & 6 & 4 \\ 21 & -7 & -8 \\ -18 & 6 & 4 \end{bmatrix}$

**4.** If $A = \begin{bmatrix} 3 & -4 \\ 1 & -1 \end{bmatrix}$, then show that $A^n = \begin{bmatrix} 1+2n & -4n \\ n & 1-2n \end{bmatrix}$

**5.** If $A = \begin{bmatrix} 1 & 1 & -2 \\ -1 & 2 & 1 \\ 0 & 1 & -1 \end{bmatrix}$, $P = \begin{bmatrix} 1 & 1 & 3 \\ 0 & 3 & 2 \\ 1 & 1 & 1 \end{bmatrix}$, show that $P^{-1}AP = \begin{bmatrix} -1 & 0 & 0 \\ 0 & 2 & 0 \\ 0 & 0 & 1 \end{bmatrix}$

**6.** If $A = \begin{bmatrix} 1 & 1 & 1 \\ 1 & 2 & 3 \\ 1 & 4 & 9 \end{bmatrix}$, $B = \begin{bmatrix} 2 & 5 & 3 \\ 3 & 1 & 2 \\ 1 & 2 & 1 \end{bmatrix}$, show that $(AB)^{-1} = B^{-1}A^{-1}$.

**7.** Given the matrix $A = \begin{bmatrix} 3 & 2 & 2 \\ 1 & 3 & 1 \\ 5 & 3 & 4 \end{bmatrix}$ compute det $(A)$, $A^{-1}$ and the matrix $B$ such that $AB = \begin{bmatrix} 3 & 4 & 2 \\ 1 & 6 & 1 \\ 5 & 6 & 4 \end{bmatrix}$

Also compute $BA$. Is $AB = BA$ ?

**Ans.** $5, \frac{1}{5}\begin{bmatrix} 9 & -2 & -4 \\ 1 & 2 & -1 \\ -12 & 1 & 7 \end{bmatrix}$. $B = \begin{bmatrix} 1 & 0 & 0 \\ 0 & 2 & 0 \\ 0 & 0 & 1 \end{bmatrix}$, $AB \neq BA$

**8.** Find the condition of $k$ such that the matrix

$A = \begin{bmatrix} 1 & 3 & 4 \\ 3 & k & 6 \\ -1 & 5 & 1 \end{bmatrix}$ has an inverse. Obtain $A^{-1}$ for $k = 1$. **Ans.** $k \neq -\frac{3}{5}$, $A^{-1} = \frac{1}{8}\begin{bmatrix} -29 & 17 & 14 \\ -9 & 5 & 6 \\ 16 & -8 & -8 \end{bmatrix}$

**9.** Prove that $(A^{-1})^T = (A^T)^{-1}$.

**10.** If $A\begin{bmatrix}0 & 1\\ 2 & -1\end{bmatrix} = \begin{bmatrix}2 & 1\\ -1 & 0\end{bmatrix}$ where $A = \begin{bmatrix}a & b\\ c & d\end{bmatrix}$, then $A$ is

(*a*) $\begin{bmatrix}2 & 1\\ 0 & 0\end{bmatrix}$ (*b*) $\begin{bmatrix}0 & 1\\ 2 & -1\end{bmatrix}$ (*c*) $\begin{bmatrix}2 & 1\\ -1 & 0\end{bmatrix}$ (*d*) $\begin{bmatrix}2 & 1\\ -\frac{1}{2} & -\frac{1}{2}\end{bmatrix}$ (*AMIETE, June 2010*) **Ans.** (*d*)

**11.** For what values of $x$, the matrix $\begin{bmatrix}3-x & 2 & 2\\ 2 & 4-x & 1\\ -2 & -4 & -1-x\end{bmatrix}$ is singular ? (*A.M.I.E.T.E. Summer 2004*) **Ans.** 0,3

**12.** Prove that $(A^{-1})^T = (A^T)^{-1}$

**13.** Let I be the unit matrix of order n and adj. $(2I) = 2^k I$. Then k equals

(*a*) 1 (*b*) 2 (*c*) $n-1$ (*d*) $n$. **Ans** (*c*)

**14.** Let $T$ be a linear transformation defined by

$$T\left[\begin{pmatrix}1 & 1\\ 1 & 1\end{pmatrix}\right] = \begin{pmatrix}1\\ 2\\ 3\end{pmatrix}, T\left[\begin{pmatrix}0 & 0\\ 1 & 1\end{pmatrix}\right] = \begin{pmatrix}1\\ -2\\ -3\end{pmatrix}, T\left[\begin{pmatrix}0 & 1\\ 1 & 1\end{pmatrix}\right] = \begin{pmatrix}1\\ -2\\ 3\end{pmatrix}, T\left[\begin{pmatrix}0 & 0\\ 0 & 1\end{pmatrix}\right] = \begin{pmatrix}-1\\ 2\\ 3\end{pmatrix}, \text{ Find } T\left[\begin{pmatrix}4 & 5\\ 3 & 8\end{pmatrix}\right].$$

(*AMIETE Dec. 2005*)

## 4.31 ELEMENTARY TRANSFORMATIONS

Any one of the following operations on a matrix is called an elementary transformation.

1. Interchanging any two rows (or columns). This transformation is indicated by $R_{ij}$, if the *i*th and *j*th rows are interchanged.
2. Multiplication of the elements of any row $R_i$ (or column) by a non-zero scalar quantity $k$ is denoted by $(k.R_i)$.
3. Addition of constant multiplication of the elements of any row $R_j$ to the corresponding elements of any other row $R_i$ is denoted by $(R_i + kR_j)$.

If a matrix $B$ is obtained from a matrix $A$ by one or more E-operations, then $B$ is said to be equivalent to $A$. The symbol ~ is used for equivalence.

*i.e.*, $A \sim B$.

**Example 21.** *Reduce the following matrix to upper triangular form* (*Echelon form*) :

$$\begin{bmatrix}1 & 2 & 3\\ 2 & 5 & 7\\ 3 & 1 & 2\end{bmatrix}$$

**Solution.** *Upper triangular matrix.* If in a square matrix, all the elements below the principal diagonal are zero, the matrix is called an upper triangular matrix.

$$\begin{bmatrix}1 & 2 & 3\\ 2 & 5 & 7\\ 3 & 1 & 2\end{bmatrix} \sim \begin{bmatrix}1 & 2 & 3\\ 0 & 1 & 1\\ 0 & -5 & -7\end{bmatrix} \begin{matrix}R_2 \to R_2 - 2R_1\\ R_3 \to R_3 - 3R_1\end{matrix} \sim \begin{bmatrix}1 & 2 & 3\\ 0 & 1 & 1\\ 0 & 0 & -2\end{bmatrix} R_3 \to R_3 + 5R_2 \quad \textbf{Ans.}$$

**Example 22.** *Transform* $\begin{bmatrix}1 & 3 & 3\\ 2 & 4 & 10\\ 3 & 8 & 4\end{bmatrix}$ *into a unit matrix.* (*Q. Bank U.P., 2001*)

**Solution.** $\begin{bmatrix} 1 & 3 & 3 \\ 2 & 4 & 10 \\ 3 & 8 & 4 \end{bmatrix} \sim \begin{bmatrix} 1 & 3 & 3 \\ 0 & -2 & 4 \\ 0 & -1 & -5 \end{bmatrix} \begin{matrix} \\ R_2 \to R_2 - 2R_1 \\ R_3 \to R_3 - 3R_1 \end{matrix}$

$$\sim \begin{bmatrix} 1 & 3 & 3 \\ 0 & 1 & -2 \\ 0 & -1 & -5 \end{bmatrix} R_2 \to -\frac{1}{2} R_2 \sim \begin{bmatrix} 1 & 0 & 9 \\ 0 & 1 & -2 \\ 0 & 0 & -7 \end{bmatrix} \begin{matrix} R_1 \to R_1 - 3R_2 \\ \\ R_3 \to R_3 + R_2 \end{matrix}$$

$$\sim \begin{bmatrix} 1 & 0 & 9 \\ 0 & 1 & -2 \\ 0 & 0 & 1 \end{bmatrix} R_3 \to -\frac{1}{7} R_3 \sim \begin{bmatrix} 1 & 0 & 0 \\ 0 & 1 & 0 \\ 0 & 0 & 1 \end{bmatrix} \begin{matrix} R_1 \to R_1 - 9R_3 \\ R_2 \to R_2 + 2R_3 \\ \\ \end{matrix}$$

## 4.32 ELEMENTARY MATRICES

A matrix obtained from a unit matrix by a single elementary transformation is called elementary matrix.

$$I = \begin{bmatrix} 1 & 0 & 0 \\ 0 & 1 & 0 \\ 0 & 0 & 1 \end{bmatrix}$$

Consider the matrix obtained by $R_2 + 3R_1$

$$\begin{bmatrix} 1 & 0 & 0 \\ 3 & 1 & 0 \\ 0 & 0 & 1 \end{bmatrix} \text{ is called the elementary matrix.}$$

## 4.33 THEOREM

Every elementary row transformation of a matrix can be affected by pre-multiplication with the corresponding elementary matrix.

Consider the matrix $A = \begin{bmatrix} 2 & 3 & 4 \\ 5 & 6 & 7 \\ 3 & 5 & 9 \end{bmatrix}$

Let us apply row transformation $R_3 + 4R_1$ and we get a matrix $B$.

$$B = \begin{bmatrix} 2 & 3 & 4 \\ 5 & 6 & 7 \\ 11 & 17 & 25 \end{bmatrix}$$

Now we shall show that pre-multiplication of $A$ by corresponding elementary matrix $R_3 + 4R_1$ will give us $B$.

Now, if $I = \begin{bmatrix} 1 & 0 & 0 \\ 0 & 1 & 0 \\ 0 & 0 & 1 \end{bmatrix}$ then, Elementary matrix $= \begin{bmatrix} 1 & 0 & 0 \\ 0 & 1 & 0 \\ 4 & 0 & 1 \end{bmatrix}_{(R_3 + 4R_1)}$

$$\therefore \quad \text{Elementary matrix} \times A = \begin{bmatrix} 1 & 0 & 0 \\ 0 & 1 & 0 \\ 4 & 0 & 1 \end{bmatrix} \times \begin{bmatrix} 2 & 3 & 4 \\ 5 & 6 & 7 \\ 3 & 5 & 9 \end{bmatrix} = \begin{bmatrix} 2 & 3 & 4 \\ 5 & 6 & 7 \\ 11 & 17 & 25 \end{bmatrix} = B$$

Similarly, we can show that every elementary column transformation of a matrix can be affected by post-multiplication with the corresponding elementary matrix.

## 4.34 TO COMPUTE THE INVERSE OF A MATRIX FROM ELEMENTARY MATRICES (Gauss-jordan Method)

If $A$ is reduced to $I$ by elementary transformation then

$$PA = I \quad \text{where} \quad P = P_n P_{n-1} \dots P_2 P_1$$

$$\therefore \quad P = A^{-1} \quad = \text{Elementary matrix.}$$

**Working rule.** Write $A = IA$. Perform elementary row transformation on $A$ of the left side and on $I$ of the right hand side so that $A$ is reduced to $I$ and $I$ of right hand side is reduced to $P$ getting $I = PA$.

Then $P$ is the inverse of $A$.

## 4.35 THE INVERSE OF A SYMMETRIC MATRIX

The elementary transformations are to be transformed so that the property of being symmetric is preserved. This requires that the transformations occur in pairs, a row transformation must be followed immediately by the same column transformation.

**Example 24.** *Find the inverse of the following matrix employing elementary transformations*:

$$\begin{bmatrix} 3 & -3 & 4 \\ 2 & -3 & 4 \\ 0 & -1 & 1 \end{bmatrix}$$

*(U.P., I Semester, Compartment 2002)*

**Solution.** The given matrix is $A = \begin{bmatrix} 3 & -3 & 4 \\ 2 & -3 & 4 \\ 0 & -1 & 1 \end{bmatrix}$

$$\begin{bmatrix} 3 & -3 & 4 \\ 2 & -3 & 4 \\ 0 & -1 & 1 \end{bmatrix} = \begin{bmatrix} 1 & 0 & 0 \\ 0 & 1 & 0 \\ 0 & 0 & 1 \end{bmatrix} A \quad \Rightarrow \begin{bmatrix} 1 & -1 & \frac{4}{3} \\ 2 & -3 & 4 \\ 0 & -1 & 1 \end{bmatrix} = \begin{bmatrix} \frac{1}{3} & 0 & 0 \\ 0 & 1 & 0 \\ 0 & 0 & 1 \end{bmatrix} A \quad R_1 \to \frac{R_1}{3}$$

$$\Rightarrow \begin{bmatrix} 1 & -1 & \frac{4}{3} \\ 0 & -1 & \frac{4}{3} \\ 0 & -1 & 1 \end{bmatrix} = \begin{bmatrix} \frac{1}{3} & 0 & 0 \\ -\frac{2}{3} & 1 & 0 \\ 0 & 0 & 1 \end{bmatrix} A \quad R_2 \to R_2 - 2R_1 \quad \Rightarrow \begin{bmatrix} 1 & -1 & \frac{4}{3} \\ 0 & 1 & -\frac{4}{3} \\ 0 & -1 & 1 \end{bmatrix} = \begin{bmatrix} \frac{1}{3} & 0 & 0 \\ \frac{2}{3} & -1 & 0 \\ 0 & 0 & 1 \end{bmatrix} A \quad R_2 \to -R_2$$

$$\Rightarrow \begin{bmatrix} 1 & -1 & \frac{4}{3} \\ 0 & 1 & -\frac{4}{3} \\ 0 & 0 & -\frac{1}{3} \end{bmatrix} = \begin{bmatrix} \frac{1}{3} & 0 & 0 \\ \frac{2}{3} & -1 & 0 \\ \frac{2}{3} & -1 & 1 \end{bmatrix} A \quad R_3 \to R_3 + R_2 \quad \Rightarrow \begin{bmatrix} 1 & -1 & \frac{4}{3} \\ 0 & 1 & -\frac{4}{3} \\ 0 & 0 & 1 \end{bmatrix} = \begin{bmatrix} \frac{1}{3} & 0 & 0 \\ \frac{2}{3} & -1 & 0 \\ -2 & 3 & -3 \end{bmatrix} A \quad R_3 \to -3R_3$$

$$\Rightarrow \begin{bmatrix} 1 & -1 & 0 \\ 0 & 1 & 0 \\ 0 & 0 & 1 \end{bmatrix} = \begin{bmatrix} 3 & -4 & 4 \\ -2 & 3 & -4 \\ -2 & 3 & -3 \end{bmatrix} A \quad \begin{matrix} R_1 \to R_1 - \frac{4}{3} R_3 \\ R_2 \to R_2 + \frac{4}{3} R_3 \end{matrix}$$

$$\Rightarrow \begin{bmatrix} 1 & 0 & 0 \\ 0 & 1 & 0 \\ 0 & 0 & 1 \end{bmatrix} = \begin{bmatrix} 1 & -1 & 0 \\ -2 & 3 & -4 \\ -2 & 3 & -3 \end{bmatrix} A \quad R_1 \to R_1 + R_2 \qquad \text{Hence, } A^{-1} = \begin{bmatrix} 1 & -1 & 0 \\ -2 & 3 & -4 \\ -2 & 3 & -3 \end{bmatrix} \quad \textbf{Ans.}$$

**Example 23.** *Find the inverse of the matrix M by applying elementary transformations*

$$\begin{bmatrix} 0 & 2 & 1 & 3 \\ 1 & 1 & -1 & -2 \\ 1 & 2 & 0 & 1 \\ -1 & 1 & 2 & 6 \end{bmatrix}.$$

*[U.P.T.U.(C.O.) 2003]*

**Solution.** Here, we have $A = \begin{bmatrix} 0 & 2 & 1 & 3 \\ 1 & 1 & -1 & -2 \\ 1 & 2 & 0 & 1 \\ -1 & 1 & 2 & 6 \end{bmatrix}$

Let $\begin{bmatrix} 0 & 2 & 1 & 3 \\ 1 & 1 & -1 & -2 \\ 1 & 2 & 0 & 1 \\ -1 & 1 & 2 & 6 \end{bmatrix} \sim \begin{bmatrix} 1 & 0 & 0 & 0 \\ 0 & 1 & 0 & 0 \\ 0 & 0 & 1 & 0 \\ 0 & 0 & 0 & 1 \end{bmatrix} A$

$$\begin{bmatrix} 1 & 1 & -1 & -2 \\ 0 & 2 & 1 & 3 \\ 1 & 2 & 0 & 1 \\ -1 & 1 & 2 & 6 \end{bmatrix} \sim \begin{bmatrix} 0 & 1 & 0 & 0 \\ 1 & 0 & 0 & 0 \\ 0 & 0 & 1 & 0 \\ 0 & 0 & 0 & 1 \end{bmatrix} A \quad R_1 \leftrightarrow R_2$$

$$\begin{bmatrix} 1 & 1 & -1 & -2 \\ 0 & 2 & 1 & 3 \\ 0 & 1 & 1 & 3 \\ 0 & 2 & 1 & 4 \end{bmatrix} \sim \begin{bmatrix} 0 & 1 & 0 & 0 \\ 1 & 0 & 0 & 0 \\ 0 & -1 & 1 & 0 \\ 0 & 1 & 0 & 1 \end{bmatrix} A \quad \begin{matrix} R_3 \to R_3 - R_1 \\ R_4 \to R_4 + R_1 \end{matrix}$$

$$\begin{bmatrix} 1 & 1 & -1 & -2 \\ 0 & 1 & 1 & 3 \\ 0 & 2 & 1 & 3 \\ 0 & 2 & 1 & 4 \end{bmatrix} = \begin{bmatrix} 0 & 1 & 0 & 0 \\ 0 & -1 & 1 & 0 \\ 1 & 0 & 0 & 0 \\ 0 & 1 & 0 & 1 \end{bmatrix} A \quad R_3 \leftrightarrow R_2$$

$$\begin{bmatrix} 1 & 1 & -1 & -2 \\ 0 & 1 & 1 & 3 \\ 0 & 0 & -1 & -3 \\ 0 & 0 & -1 & -2 \end{bmatrix} = \begin{bmatrix} 0 & 1 & 0 & 0 \\ 0 & -1 & 1 & 0 \\ 1 & 2 & -2 & 0 \\ 0 & 3 & -2 & 1 \end{bmatrix} A \quad \begin{matrix} R_3 \to R_3 - 2R_2 \\ R_4 \to R_4 - 2R_2 \end{matrix}$$

$$\begin{bmatrix} 1 & 1 & -1 & -2 \\ 0 & 1 & 1 & 3 \\ 0 & 0 & -1 & -3 \\ 0 & 0 & 0 & 1 \end{bmatrix} = \begin{bmatrix} 0 & 1 & 0 & 0 \\ 0 & -1 & 1 & 0 \\ 1 & 2 & -2 & 0 \\ -1 & 1 & 0 & 1 \end{bmatrix} A \quad R_4 \to R_4 - R_3$$

$$\begin{bmatrix} 1 & 1 & -1 & -2 \\ 0 & 1 & 1 & 3 \\ 0 & 0 & 1 & 3 \\ 0 & 0 & 0 & 1 \end{bmatrix} = \begin{bmatrix} 0 & 1 & 0 & 0 \\ 0 & -1 & 1 & 0 \\ -1 & -2 & 2 & 0 \\ -1 & 1 & 0 & 1 \end{bmatrix} A \quad R_3 \to -R_3$$

$$\begin{bmatrix} 1 & 1 & -1 & 0 \\ 0 & 1 & 1 & 0 \\ 0 & 0 & 1 & 0 \\ 0 & 0 & 0 & 1 \end{bmatrix} = \begin{bmatrix} -2 & 3 & 0 & 2 \\ 3 & -4 & 1 & -3 \\ 2 & -5 & 2 & -3 \\ -1 & 1 & 0 & 1 \end{bmatrix} A \quad \begin{matrix} R_1 \to R_1 + 2R_4 \\ R_2 \to R_2 - 3R_4 \\ R_3 \to R_3 - 3R_4 \end{matrix}$$

$$\begin{bmatrix} 1 & 1 & 0 & 0 \\ 0 & 1 & 0 & 0 \\ 0 & 0 & 1 & 0 \\ 0 & 0 & 0 & 1 \end{bmatrix} = \begin{bmatrix} 0 & -2 & 2 & -1 \\ 1 & 1 & -1 & 0 \\ 2 & -5 & 2 & -3 \\ -1 & 1 & 0 & 1 \end{bmatrix} A \quad \begin{matrix} R_1 \to R_1 + R_3 \\ R_2 \to R_2 - R_3 \end{matrix}$$

$$\begin{bmatrix} 1 & 0 & 0 & 0 \\ 0 & 1 & 0 & 0 \\ 0 & 0 & 1 & 0 \\ 0 & 0 & 0 & 1 \end{bmatrix} = \begin{bmatrix} -1 & -3 & 3 & -1 \\ 1 & 1 & -1 & 0 \\ 2 & -5 & 2 & -3 \\ -1 & 1 & 0 & 1 \end{bmatrix} A \quad R_1 \to R_1 - R_2$$

$$I = A^{-1} A$$

Hence,
$$A^{-1} = \begin{bmatrix} -1 & -3 & 3 & -1 \\ 1 & 1 & -1 & 0 \\ 2 & -5 & 2 & -3 \\ -1 & 1 & 0 & 1 \end{bmatrix}$$
**Ans.**

## EXERCISE 4.13

**Reduce the matrices to triangular form:**

**1.** $A = \begin{bmatrix} 1 & 2 & 3 \\ 2 & 5 & 7 \\ 3 & 1 & 2 \end{bmatrix}$ **Ans.** $\begin{bmatrix} 1 & 2 & 3 \\ 0 & 1 & 1 \\ 0 & 0 & -2 \end{bmatrix}$

**2.** $\begin{bmatrix} 3 & 1 & 4 \\ 1 & 2 & -5 \\ 0 & 1 & 5 \end{bmatrix}$ **Ans.** $\begin{bmatrix} 0 & 1 & 4 \\ 0 & 5 & -19 \\ 0 & 0 & 22 \end{bmatrix}$

**Find the inverse of the following matrices:**

**3.** $\begin{bmatrix} 1 & 3 & 3 \\ 1 & 4 & 3 \\ 1 & 3 & 4 \end{bmatrix}$ **Ans.** $\begin{bmatrix} 7 & -3 & -3 \\ -1 & 1 & 0 \\ -1 & 0 & 1 \end{bmatrix}$

**4.** $\begin{bmatrix} 1 & -1 & 1 \\ 4 & 1 & 0 \\ 8 & 1 & 1 \end{bmatrix}$ **Ans.** $\begin{bmatrix} 1 & 2 & -1 \\ -4 & -7 & 4 \\ -4 & -9 & 5 \end{bmatrix}$

**5.** Use elementary row operations to find inverse of $A = \begin{bmatrix} 1 & 1 & 3 \\ 1 & 3 & -3 \\ -2 & -4 & -4 \end{bmatrix}$ **Ans.** $\frac{1}{4}\begin{bmatrix} 12 & 4 & 6 \\ -5 & -1 & -3 \\ -1 & -1 & -1 \end{bmatrix}$

*(AMIETE, June 2010)*

**6.** $\begin{bmatrix} 2 & 1 & -1 & 2 \\ 1 & 3 & 2 & -3 \\ -1 & 2 & 1 & -1 \\ 2 & -3 & -1 & 4 \end{bmatrix}$ **Ans.** $\frac{1}{18}\begin{bmatrix} 2 & 5 & -7 & 1 \\ 5 & -1 & 5 & -2 \\ -7 & 5 & 11 & 10 \\ 1 & -2 & 10 & 5 \end{bmatrix}$

**7.** $\begin{bmatrix} 1 & 2 & 3 & 1 \\ 1 & 3 & 3 & 2 \\ 2 & 4 & 3 & 3 \\ 1 & 1 & 1 & 1 \end{bmatrix}$ (*Q. Bank U.P. II Semester 2001*) **Ans.** $\begin{bmatrix} 1 & -2 & 1 & 0 \\ 1 & -2 & 2 & -3 \\ 0 & 1 & -1 & 1 \\ -2 & 3 & -2 & 3 \end{bmatrix}$

**8.** $\begin{bmatrix} 2 & 1 & -1 & 2 \\ 1 & 3 & 2 & -3 \\ -1 & 2 & 1 & -1 \\ 2 & -3 & -1 & 4 \end{bmatrix}$ **Ans.** $\dfrac{1}{18}\begin{bmatrix} 2 & 5 & -7 & 1 \\ 5 & -1 & 5 & -2 \\ -7 & 5 & 11 & 10 \\ 1 & -2 & 10 & 5 \end{bmatrix}$ **9.** $\begin{bmatrix} 2 & -6 & -2 & -3 \\ 5 & -13 & -4 & -7 \\ -1 & 4 & 1 & 2 \\ 0 & 1 & 0 & 1 \end{bmatrix}$ **Ans.** $\begin{bmatrix} -2 & 1 & 0 & 1 \\ 1 & 0 & 2 & -1 \\ -4 & 1 & -3 & 1 \\ -1 & 0 & -2 & 2 \end{bmatrix}$

**10.** $\begin{bmatrix} 1 & 3 & 3 & 2 & 1 \\ 1 & 4 & 3 & 3 & -1 \\ 1 & 3 & 4 & 1 & 1 \\ 1 & 1 & 1 & 1 & -1 \\ 1 & -2 & -1 & 2 & 2 \end{bmatrix}$ **Ans.** $\dfrac{1}{15}\begin{bmatrix} 30 & -20 & -15 & 25 & -5 \\ 30 & -11 & -18 & 7 & -8 \\ -30 & 12 & 21 & -9 & 6 \\ -15 & 12 & 6 & -9 & 6 \\ 15 & -7 & -6 & -1 & -1 \end{bmatrix}$

**11.** If $X$, $Y$ are non-singular matrices and $B = \begin{bmatrix} X & O \\ O & Y \end{bmatrix}$, show that $B^{-1} = \begin{bmatrix} X^{-1} & O \\ O & Y^{-1} \end{bmatrix}$ where $O$ is a null matrix.

## 4.36 RANK OF A MATRIX

The rank of a matrix is said to be $r$ if

(*a*) It has at least one non-zero minor of order $r$.

(*b*) Every minor of $A$ of order higher than $r$ is zero.

**Note:** (*i*) Non-zero row is that row in which all the elements are not zero.

(*ii*) The rank of the product matrix $AB$ of two matrices $A$ and $B$ is less than the rank of either of the matrices $A$ and $B$.

(*iii*) Corresponding to every matrix $A$ of rank $r$, there exist non-singular matrices $P$ and $Q$ such that $\quad PAQ = \begin{bmatrix} I_r & 0 \\ 0 & 0 \end{bmatrix}$

## 4.37 NORMAL FORM (CANONICAL FORM)

By performing elementary transformation, any non-zero matrix $A$ can be reduced to one of the following four forms, called the Normal form of $A$ :

(*i*) $I_r$ (*ii*) $[I_r \;\; 0]$ (*iii*) $\begin{bmatrix} I_r \\ 0 \end{bmatrix}$ (*iv*) $\begin{bmatrix} I_r & 0 \\ 0 & 0 \end{bmatrix}$.

The number $r$ so obtained is called the rank of $A$ and we write $\rho(A) = r$. The form $\begin{bmatrix} I_r & 0 \\ 0 & 0 \end{bmatrix}$ is called first canonical form of $A$. Since both row and column transformations may be used here, the element 1 of the first row obtained can be moved in the first column. Then both the first row and first column can be cleared of other non-zero elements. Similarly, the element 1 of the second row can be brought into the second column, and so on.

**Example 24.** *Reduce to normal form the following matrix*

$$A = \begin{bmatrix} 1 & 2 & 3 & 4 \\ 2 & 1 & 4 & 3 \\ 3 & 0 & 5 & -10 \end{bmatrix}$$

$R_3 - 2R_2,\; R_3 - 3R_1$

**Solution.** $A=\begin{bmatrix}1&2&3&4\\2&1&4&3\\3&0&5&-10\end{bmatrix}\sim\begin{bmatrix}1&2&3&4\\0&-3&-2&-5\\0&-6&-4&-22\end{bmatrix}$

$C_2-2C_1, C_3-3C_1, C_4-4C_1, -\frac{1}{3}R_2, -\frac{1}{6}R_3 \qquad R_3-R_2$

$$\sim\begin{bmatrix}1&0&0&0\\0&-3&-2&-5\\0&-6&-4&-22\end{bmatrix}\sim\begin{bmatrix}1&0&0&0\\0&1&\frac{2}{3}&\frac{5}{3}\\0&1&\frac{2}{3}&\frac{11}{3}\end{bmatrix}\sim\begin{bmatrix}1&0&0&0\\0&1&\frac{2}{3}&\frac{5}{3}\\0&0&0&2\end{bmatrix}$$

$C_3-\frac{2}{3}C_2, C_4-\frac{5}{3}C_2 C_3\leftrightarrow C_4 \qquad \frac{1}{2}C_3$

$$\sim\begin{bmatrix}1&0&0&0\\0&1&0&0\\0&0&0&2\end{bmatrix}\sim\begin{bmatrix}1&0&0&0\\0&1&0&0\\0&0&2&0\end{bmatrix}\sim\begin{bmatrix}1&0&0&0\\0&1&0&0\\0&0&1&0\end{bmatrix}$$

$= [I_3\ 0]$ is the normal form of $A$. **Ans.**

**Example 25.** *Find the rank of the following matrix by reducing it to normal form –*

$$A=\begin{bmatrix}1&2&-1&3\\4&1&2&1\\3&-1&1&2\\1&2&0&1\end{bmatrix}$$ *(U.P. I Sem., Com. 2002, Winter 2001)*

**Solution.** $\begin{bmatrix}1&2&-1&3\\4&1&2&1\\3&-1&1&2\\1&2&0&1\end{bmatrix}\sim\begin{bmatrix}1&2&-1&3\\0&-7&6&-11\\0&-7&4&-7\\0&0&1&-2\end{bmatrix}$ $\begin{matrix}R_2\to R_2-4R_1\\R_3\to R_3-3R_1\\R_4\to R_4-R_1\end{matrix}$

$$\sim\begin{bmatrix}1&0&0&0\\0&-7&6&-11\\0&-7&4&-7\\0&0&1&-2\end{bmatrix}\sim\begin{bmatrix}1&0&0&0\\0&-7&6&-11\\0&0&-2&4\\0&0&1&-2\end{bmatrix}\ R_3\to R_3-R_2$$

$C_2\to C_2-2C_1, C_3\to C_3+C_1, C_4\to C_4-3C_1$

$$\sim\begin{bmatrix}1&0&0&0\\0&-7&0&0\\0&0&2&4\\0&0&1&-2\end{bmatrix}\sim\begin{bmatrix}1&0&0&0\\0&-7&0&0\\0&0&-2&4\\0&0&0&0\end{bmatrix}\ R_4\to R_4+\frac{1}{2}R_3$$

$C_3\to C_3+\frac{6}{7}C_2, C_4\to C_4-\frac{11}{7}C_2,$

$$C_4\to C_4+2C_3\sim\begin{bmatrix}1&0&0&0\\0&-7&0&0\\0&0&-2&0\\0&0&0&0\end{bmatrix}\sim\begin{bmatrix}1&0&0&0\\0&1&0&0\\0&0&1&0\\0&0&0&0\end{bmatrix}\ \begin{matrix}R_2\to-1/7\,R_2\\R_3\to-1/2\,R_3\end{matrix}$$

Rank of $A=3$ **Ans.**

**Example 26.** *Reduce the matrix A to its normal form, when*

$$A = \begin{bmatrix} 1 & 2 & -1 & 4 \\ 2 & 4 & 3 & 4 \\ 1 & 2 & 3 & 4 \\ -1 & -2 & 6 & -7 \end{bmatrix}$$

*Hence, find the rank of A.* (*U.P., I Semester, Dec. 2004, Winter 2001*)

**Solution.** The given matrix is $A = \begin{bmatrix} 1 & 2 & -1 & 4 \\ 2 & 4 & 3 & 4 \\ 1 & 2 & 3 & 4 \\ -1 & -2 & 6 & -7 \end{bmatrix}$

$$\sim \begin{bmatrix} 1 & 2 & -1 & 4 \\ 0 & 0 & 5 & -4 \\ 0 & 0 & 4 & 0 \\ 0 & 0 & 5 & -3 \end{bmatrix} \begin{matrix} R_2 \to R_2 - 2R_1 \\ R_3 \to R_3 - R_1 \\ R_4 \to R_4 + R_1 \end{matrix} \sim \begin{bmatrix} 1 & 0 & 0 & 0 \\ 0 & 0 & 5 & -4 \\ 0 & 0 & 4 & 0 \\ 0 & 0 & 5 & -3 \end{bmatrix} \begin{matrix} C_2 \to C_2 - 2C_1 \\ C_3 \to C_3 + C_1 \\ C_4 \to C_4 - 4C_1 \end{matrix}$$

$$\sim \begin{bmatrix} 1 & 0 & 0 & 0 \\ 0 & 5 & 0 & -4 \\ 0 & 4 & 0 & 0 \\ 0 & 5 & 0 & -3 \end{bmatrix} C_3 \leftrightarrow C_2 \sim \begin{bmatrix} 1 & 0 & 0 & 0 \\ 0 & 5 & 0 & -4 \\ 0 & 0 & 0 & \frac{16}{5} \\ 0 & 0 & 0 & 1 \end{bmatrix} \begin{matrix} R_3 \to R_3 - \frac{4}{5} R_2 \\ R_4 \to R_4 - R_2 \end{matrix}$$

$$\sim \begin{bmatrix} 1 & 0 & 0 & 0 \\ 0 & 5 & -4 & 0 \\ 0 & 0 & \frac{16}{5} & 0 \\ 0 & 0 & 1 & 0 \end{bmatrix} C_4 \leftrightarrow C_3 \sim \begin{bmatrix} 1 & 0 & 0 & 0 \\ 0 & 5 & 0 & 0 \\ 0 & 0 & \frac{16}{5} & 0 \\ 0 & 0 & 0 & 0 \end{bmatrix} \begin{matrix} R_2 \to R_2 + \frac{5}{4} R_3 \\ \\ R_4 \to R_4 - \frac{5}{16} R_3 \end{matrix}$$

$$\sim \begin{bmatrix} 1 & 0 & 0 & 0 \\ 0 & 1 & 0 & 0 \\ 0 & 0 & 1 & 0 \\ 0 & 0 & 0 & 0 \end{bmatrix} \begin{matrix} R_2 \to 1/5\, R_2 \\ R_3 \to 5/16\, R_3 \end{matrix} \sim \begin{bmatrix} I_3 & 0 \\ 0 & 0 \end{bmatrix}$$

Which is the required normal form.

And since, the non-zero rows are 3 hence, the rank of the given matrix is 3. **Ans.**

**Example 27.** *Find non-singular matrices P, Q so that PAQ is a normal form where*

$$A = \begin{bmatrix} 2 & 1 & -3 & -6 \\ 3 & -3 & 1 & 2 \\ 1 & 1 & 1 & 2 \end{bmatrix}$$ (*R.G.P.V., Bhopal, April, 2010, U.P., I Sem. Winter 2002*)

*and hence find its rank.*

**Solution.** Order of $A$ is $3 \times 4$

Total number of rows in $A = 3$; $\therefore$ Consider unit matrix $I_3$.

Total number of columns in $A = 4$

Hence, consider unit matrix $I_4$,

$$\therefore \qquad A_{3 \times 4} = I_3 A I_4$$

$$\begin{bmatrix} 2 & 1 & -3 & -6 \\ 3 & -3 & 1 & 2 \\ 1 & 1 & 1 & 2 \end{bmatrix} = \begin{bmatrix} 1 & 0 & 0 \\ 0 & 1 & 0 \\ 0 & 0 & 1 \end{bmatrix} A \begin{bmatrix} 1 & 0 & 0 & 0 \\ 0 & 1 & 0 & 0 \\ 0 & 0 & 1 & 0 \\ 0 & 0 & 0 & 1 \end{bmatrix}$$

$$\begin{bmatrix} 1 & 1 & 1 & 2 \\ 3 & -3 & 1 & 2 \\ 2 & 1 & -3 & -6 \end{bmatrix} = \begin{bmatrix} 0 & 0 & 1 \\ 0 & 1 & 0 \\ 1 & 0 & 0 \end{bmatrix} A \begin{bmatrix} 1 & 0 & 0 & 0 \\ 0 & 1 & 0 & 0 \\ 0 & 0 & 1 & 0 \\ 0 & 0 & 0 & 1 \end{bmatrix} \quad R_1 \leftrightarrow R_3$$

$$\begin{bmatrix} 1 & 1 & 1 & 2 \\ 0 & -6 & -2 & -4 \\ 0 & -1 & -5 & -10 \end{bmatrix} = \begin{bmatrix} 0 & 0 & 1 \\ 0 & 1 & -3 \\ 1 & 0 & -2 \end{bmatrix} A \begin{bmatrix} 1 & 0 & 0 & 0 \\ 0 & 1 & 0 & 0 \\ 0 & 0 & 1 & 0 \\ 0 & 0 & 0 & 1 \end{bmatrix} \quad \begin{matrix} R_2 \to R_2 - 3R_1 \\ R_3 \to R_3 - 2R_1 \end{matrix}$$

$$C_2 \to C_2 - C_1,\ C_3 \to C_3 - C_1,\ C_4 \to C_4 - 2C_1$$

$$\begin{bmatrix} 1 & 0 & 0 & 0 \\ 0 & -6 & -2 & -4 \\ 0 & -1 & -5 & -10 \end{bmatrix} = \begin{bmatrix} 0 & 0 & 1 \\ 0 & 1 & -3 \\ 1 & 0 & -2 \end{bmatrix} A \begin{bmatrix} 1 & -1 & -1 & -2 \\ 0 & 1 & 0 & 0 \\ 0 & 0 & 1 & 0 \\ 0 & 0 & 0 & 1 \end{bmatrix}$$

$$\begin{bmatrix} 1 & 0 & 0 & 0 \\ 0 & 6 & 2 & 4 \\ 0 & 1 & 5 & 10 \end{bmatrix} = \begin{bmatrix} 0 & 0 & 1 \\ 0 & -1 & 3 \\ -1 & 0 & 2 \end{bmatrix} A \begin{bmatrix} 1 & -1 & -1 & -2 \\ 0 & 1 & 0 & 0 \\ 0 & 0 & 1 & 0 \\ 0 & 0 & 0 & 1 \end{bmatrix} \quad \begin{matrix} R_2 \to (-1)R_2 \\ R_3 \to (-1)R_3 \end{matrix}$$

$$\begin{bmatrix} 1 & 0 & 0 & 0 \\ 0 & 1 & 5 & 10 \\ 0 & 6 & 2 & 4 \end{bmatrix} = \begin{bmatrix} 0 & 0 & 1 \\ -1 & 0 & 2 \\ 0 & -1 & 3 \end{bmatrix} A \begin{bmatrix} 1 & -1 & -1 & -2 \\ 0 & 1 & 0 & 0 \\ 0 & 0 & 1 & 0 \\ 0 & 0 & 0 & 1 \end{bmatrix} \quad R_2 \leftrightarrow R_3$$

$$\begin{bmatrix} 1 & 0 & 0 & 0 \\ 0 & 1 & 5 & 0 \\ 0 & 0 & -28 & -56 \end{bmatrix} = \begin{bmatrix} 0 & 0 & 1 \\ -1 & 0 & 2 \\ 6 & -1 & -9 \end{bmatrix} A \begin{bmatrix} 1 & -1 & -1 & -2 \\ 0 & 1 & 0 & 0 \\ 0 & 0 & 1 & 0 \\ 0 & 0 & 0 & 1 \end{bmatrix} \quad R_3 \to R_3 - 6R_2$$

$$C_3 \to C' - 5C_2, \qquad C_4 \to C_4 - 10C_2$$

$$\begin{bmatrix} 1 & 0 & 0 & 0 \\ 0 & 1 & 0 & 0 \\ 0 & 0 & -28 & -56 \end{bmatrix} = \begin{bmatrix} 0 & 0 & 1 \\ -1 & 0 & 2 \\ 6 & -1 & -9 \end{bmatrix} A \begin{bmatrix} 1 & -1 & 4 & 8 \\ 0 & 1 & -5 & -10 \\ 0 & 0 & 1 & 0 \\ 0 & 0 & 0 & 1 \end{bmatrix}$$

$$\begin{bmatrix} 1 & 0 & 0 & 0 \\ 0 & 1 & 0 & 0 \\ 0 & 0 & 1 & 2 \end{bmatrix} = \begin{bmatrix} 0 & 0 & 1 \\ -1 & 0 & 2 \\ -\frac{6}{28} & \frac{1}{28} & \frac{9}{28} \end{bmatrix} A \begin{bmatrix} 1 & -1 & 4 & 8 \\ 0 & 1 & -5 & -10 \\ 0 & 0 & 1 & 0 \\ 0 & 0 & 0 & 1 \end{bmatrix} \quad R_3 \to -\frac{1}{28} R_3$$

$$\begin{bmatrix} 1 & 0 & 0 & 0 \\ 0 & 1 & 0 & 0 \\ 0 & 0 & 1 & 0 \end{bmatrix} = \begin{bmatrix} 0 & 0 & 1 \\ -1 & 0 & 2 \\ -\frac{6}{28} & \frac{1}{28} & \frac{9}{28} \end{bmatrix} A \begin{bmatrix} 1 & -1 & 4 & 0 \\ 0 & 1 & -5 & 0 \\ 0 & 0 & 1 & -2 \\ 0 & 0 & 0 & 1 \end{bmatrix} C_4 \to C_4 - 2C_3$$

$$N = PAQ$$

$$P = \begin{bmatrix} 0 & 0 & 1 \\ -1 & 0 & 2 \\ -\frac{3}{14} & \frac{1}{28} & \frac{9}{28} \end{bmatrix}, Q = \begin{bmatrix} 1 & -1 & 4 & 0 \\ 0 & 1 & -5 & 0 \\ 0 & 0 & 1 & -2 \\ 0 & 0 & 0 & 1 \end{bmatrix}$$ **Ans.**

**Note.** $P$ and $Q$ are not unique.

Normal form of the given matrix is $\begin{bmatrix} 1 & 0 & 0 & 0 \\ 0 & 1 & 0 & 0 \\ 0 & 0 & 1 & 0 \end{bmatrix}$

The number of non zero rows in the normal matrix = 3
Hence Rank = 3 **Ans.**

**Example 28.** *If* $A = \begin{bmatrix} 3 & -3 & 4 \\ 2 & -3 & 4 \\ 0 & -1 & 1 \end{bmatrix}$, *Find two non singular matrices P and Q such that* $PAQ = I$. *Hence find* $A^{-1}$.

**Solution.** $A_{3\times 3} = I_3 A I_3$

$$\begin{bmatrix} 3 & -3 & 4 \\ 2 & -3 & 4 \\ 0 & -1 & 1 \end{bmatrix} = \begin{bmatrix} 1 & 0 & 0 \\ 0 & 1 & 0 \\ 0 & 0 & 1 \end{bmatrix} A \begin{bmatrix} 1 & 0 & 0 \\ 0 & 1 & 0 \\ 0 & 0 & 1 \end{bmatrix}$$

$$\begin{bmatrix} 1 & 0 & 0 \\ 2 & -3 & 4 \\ 0 & -1 & 1 \end{bmatrix} = \begin{bmatrix} 1 & -1 & 0 \\ 0 & 1 & 0 \\ 0 & 0 & 1 \end{bmatrix} A \begin{bmatrix} 1 & 0 & 0 \\ 0 & 1 & 0 \\ 0 & 0 & 1 \end{bmatrix} R_1 \to R_1 - R_2$$

$$\begin{bmatrix} 1 & 0 & 0 \\ 0 & -3 & 4 \\ 0 & -1 & 1 \end{bmatrix} = \begin{bmatrix} 1 & -1 & 0 \\ -2 & 3 & 0 \\ 0 & 0 & 1 \end{bmatrix} A \begin{bmatrix} 1 & 0 & 0 \\ 0 & 1 & 0 \\ 0 & 0 & 1 \end{bmatrix} R_3 \to R_2 - 2R_1$$

$$\begin{bmatrix} 1 & 0 & 0 \\ 0 & 3 & 4 \\ 0 & 1 & 1 \end{bmatrix} = \begin{bmatrix} 1 & -1 & 0 \\ -2 & 3 & 0 \\ 0 & 0 & 1 \end{bmatrix} A \begin{bmatrix} 1 & 0 & 0 \\ 0 & -1 & 0 \\ 0 & 0 & 1 \end{bmatrix} C_2 \to -C_2$$

$$\begin{bmatrix} 1 & 0 & 0 \\ 0 & 1 & 1 \\ 0 & 3 & 4 \end{bmatrix} = \begin{bmatrix} 1 & -1 & 0 \\ 0 & 0 & 1 \\ -2 & 3 & 0 \end{bmatrix} A \begin{bmatrix} 1 & 0 & 0 \\ 0 & -1 & 0 \\ 0 & 0 & 1 \end{bmatrix} R_2 \leftrightarrow R_3$$

$$\begin{bmatrix} 1 & 0 & 0 \\ 0 & 1 & 1 \\ 0 & 0 & 1 \end{bmatrix} = \begin{bmatrix} 1 & -1 & 0 \\ 0 & 0 & 1 \\ -2 & 3 & -3 \end{bmatrix} A \begin{bmatrix} 1 & 0 & 0 \\ 0 & -1 & 0 \\ 0 & 0 & 1 \end{bmatrix} R_3 \to R_3 - 3R_2$$

$$\begin{bmatrix}1&0&0\\0&1&0\\0&0&1\end{bmatrix}=\begin{bmatrix}1&-1&0\\0&0&1\\-2&3&-3\end{bmatrix}A\begin{bmatrix}1&0&0\\0&-1&1\\0&0&1\end{bmatrix}\quad C_3\to C_3-C_2$$

$$I_3=PAQ$$

$$A^{-1}=QP,\qquad A^{-1}=\begin{bmatrix}1&0&0\\0&-1&1\\0&0&1\end{bmatrix}\begin{bmatrix}1&-1&0\\0&0&1\\-2&3&-3\end{bmatrix}\qquad \left[\begin{array}{l}I=PAQ\\P^{-1}=AQ\\P^{-1}Q^{-1}=A\\(P^{-1}Q^{-1})^{-1}=A^{-1}\\QP=A^{-1}\end{array}\right.$$

$$\Rightarrow\qquad A^{-1}=\begin{bmatrix}1&-1&0\\-2&3&-4\\-2&3&-3\end{bmatrix}$$ **Ans.**

**Exercise 4.14**

Find non singular matrices P and Q such that PAQ is normal form

1. $\begin{bmatrix}1&2&3\\3&1&2\end{bmatrix}$ **Ans.** $p=\begin{bmatrix}1&0\\-3&1\end{bmatrix}$, $Q=\begin{bmatrix}1&\frac{2}{5}&\frac{-1}{5}\\0&\frac{-1}{5}&\frac{-7}{5}\\0&0&1\end{bmatrix}$

2. $\begin{bmatrix}1&1&2\\1&2&3\\0&-1&-1\end{bmatrix}$ **Ans.** $p=\begin{bmatrix}1&0&0\\-1&1&0\\-1&1&1\end{bmatrix}$, $Q=\begin{bmatrix}1&-1&-1\\0&1&-1\\0&0&1\end{bmatrix}$

3. $\begin{bmatrix}1&2&3&-2\\2&-2&1&3\\3&0&4&1\end{bmatrix}$ **Ans.** $P=\begin{bmatrix}1&0&0\\-2&1&0\\-1&-1&1\end{bmatrix}$, $Q=\begin{bmatrix}1&\frac{1}{3}&\frac{4}{15}&\frac{-1}{21}\\0&-\frac{1}{6}&\frac{1}{6}&\frac{1}{6}\\0&0&-\frac{1}{5}&0\\0&0&0&\frac{1}{7}\end{bmatrix}$

## 4.38 RANK OF MATRIX BY TRIANGULAR FORM

Rank = Number of non-zero row in upper triangular matrix.

**Note.** Non-zero row is that row which does not contain all the elements as zero.

**Example 29.** *Find the rank of the matrix*

$$\begin{bmatrix}1&2&3&2\\2&3&5&1\\1&3&4&5\end{bmatrix}$$

*(U.P., I Semester, Winter 2003, 2000)*

**Solution.**
$$\begin{bmatrix}1 & 2 & 3 & 2\\ 2 & 3 & 5 & 1\\ 1 & 3 & 4 & 5\end{bmatrix} \sim \begin{bmatrix}1 & 2 & 3 & 2\\ 0 & -1 & -1 & -3\\ 0 & 1 & 1 & 3\end{bmatrix} \begin{matrix} \\ R_2 \to R_2 - 2R_1 \\ R_3 \to R_3 - R_1\end{matrix}$$

$$\sim \begin{bmatrix}1 & 2 & 3 & 2\\ 0 & -1 & -1 & -3\\ 0 & 0 & 0 & 0\end{bmatrix} R_3 \to R_3 + R_2$$

Rank = Number of non zero rows = 2. **Ans.**

**Example 30.** *Find the rank of the matrix* $\begin{bmatrix}-1 & 2 & 3 & -2\\ 2 & -5 & 1 & 2\\ 3 & -8 & 5 & 2\\ 5 & -12 & -1 & 6\end{bmatrix}$

**Solution.**
$$\begin{bmatrix}-1 & 2 & 3 & -2\\ 2 & -5 & 1 & 2\\ 3 & -8 & 5 & 2\\ 5 & -12 & -1 & 6\end{bmatrix} \sim \begin{bmatrix}-1 & 2 & 3 & -2\\ 0 & -1 & 7 & -2\\ 0 & -2 & 14 & -4\\ 0 & -2 & 14 & -4\end{bmatrix} \begin{matrix} \\ R_2 \to R_2 + 2R_1 \\ R_3 \to R_3 + 3R_1 \\ R_4 \to R_4 + 5R_1\end{matrix}$$

$$\sim \begin{bmatrix}-1 & 2 & 3 & -2\\ 0 & -1 & 7 & -2\\ 0 & 0 & 0 & 0\\ 0 & 0 & 0 & 0\end{bmatrix} \begin{matrix} R_3 \to R_3 - 2R_2 \\ R_4 \to R_4 - 2R_2\end{matrix}$$

Here the 4th order and 3rd order minors are zero. But a minor of second order

$$\begin{vmatrix}3 & -2\\ 7 & -2\end{vmatrix} = -6 + 14 = 8 \neq 0$$

Rank = Number of non-zero rows = 2. **Ans.**

**Example 31.** *Find the rank of matrix*

$$\begin{bmatrix}2 & 3 & -2 & 4\\ 3 & -2 & 1 & 2\\ 3 & 2 & 3 & 4\\ -2 & 4 & 0 & 5\end{bmatrix}$$

(*U.P., I Semester, Dec., 2006*)

**Solution.** Multiplying $R_1$ by $\frac{1}{2}$, we get 1 as pivotal element

$$\sim \begin{bmatrix}1 & \frac{3}{2} & -1 & 2\\ 3 & -2 & 1 & 2\\ 3 & 2 & 3 & 4\\ -2 & 4 & 0 & 5\end{bmatrix}$$

$$\sim \begin{bmatrix} 1 & \frac{3}{2} & -1 & 2 \\ 0 & -\frac{13}{2} & 4 & -4 \\ 0 & -\frac{5}{2} & 6 & -2 \\ 0 & 7 & -2 & 9 \end{bmatrix} \begin{matrix} \\ R_2 \to R_2 - 3R_1 \\ R_3 \to R_3 - 3R_1 \\ R_4 \to R_4 + 2R_1 \end{matrix} \sim \begin{bmatrix} 1 & 0 & 0 & 0 \\ 0 & 1 & -\frac{8}{13} & \frac{8}{13} \\ 0 & -\frac{5}{2} & 6 & -2 \\ 0 & 7 & -2 & 9 \end{bmatrix} \begin{matrix} R_2 \to -\frac{2}{13} R_2 \\ C_2 \to C_2 - \frac{3}{2} C_1 \\ C_3 \to C_3 + C_1 \\ C_4 \to C_4 - 2C_1 \end{matrix}$$

$$\sim \begin{bmatrix} 1 & 0 & 0 & 0 \\ 0 & 1 & -\frac{8}{13} & \frac{8}{13} \\ 0 & 0 & \frac{58}{13} & -\frac{6}{13} \\ 0 & 0 & \frac{30}{13} & \frac{61}{13} \end{bmatrix} \begin{matrix} \\ \\ R_3 \to R_3 + \frac{5}{2} R_2 \\ R_4 \to R_4 - 7 R_2 \end{matrix} \sim \begin{bmatrix} 1 & 0 & 0 & 0 \\ 0 & 1 & 0 & 0 \\ 0 & 0 & \frac{58}{13} & -\frac{6}{13} \\ 0 & 0 & \frac{30}{13} & \frac{61}{13} \end{bmatrix} \begin{matrix} \\ \\ C_3 \to C_3 + \frac{8}{13} C_2 \\ C_4 \to C_4 - \frac{8}{13} C_2 \end{matrix}$$

$$\sim \begin{bmatrix} 1 & 0 & 0 & 0 \\ 0 & 1 & 0 & 0 \\ 0 & 0 & 1 & -\frac{3}{29} \\ 0 & 0 & \frac{30}{13} & \frac{61}{13} \end{bmatrix} R_3 \to \frac{13}{58} R_3 \sim \begin{bmatrix} 1 & 0 & 0 & 0 \\ 0 & 1 & 0 & 0 \\ 0 & 0 & 1 & -\frac{3}{29} \\ 0 & 0 & 0 & \frac{143}{29} \end{bmatrix} R_4 \to R_4 - \frac{30}{13} R_3$$

$$\sim \begin{bmatrix} 1 & 0 & 0 & 0 \\ 0 & 1 & 0 & 0 \\ 0 & 0 & 1 & 0 \\ 0 & 0 & 0 & \frac{143}{29} \end{bmatrix} C_4 \to C_4 + \frac{3}{29} C_3 \sim \begin{bmatrix} 1 & 0 & 0 & 0 \\ 0 & 1 & 0 & 0 \\ 0 & 0 & 1 & 0 \\ 0 & 0 & 0 & 1 \end{bmatrix} R_4 \to \frac{29}{143} R_4$$

$$= I_4$$

Hence, the rank of the given matrix = 4 **Ans.**

**Example 32.** *Use elementary transformation to reduce the following matrix A to triangular from and hence find the rank of A.*

$$A = \begin{bmatrix} 2 & 3 & -1 & -1 \\ 1 & -1 & -2 & -4 \\ 3 & 1 & 3 & -2 \\ 6 & 3 & 0 & -7 \end{bmatrix}$$

*(R.G.P.V., Bhopal, June 2007, Winter 2003, U.P., I Semester, Dec. 2005)*

**Solution.** We have,

$$A = \begin{bmatrix} 2 & 3 & -1 & -1 \\ 1 & -1 & -2 & -4 \\ 3 & 1 & 3 & -2 \\ 6 & 3 & 0 & -7 \end{bmatrix} \approx \begin{bmatrix} 1 & -1 & -2 & -4 \\ 2 & 3 & -1 & -1 \\ 3 & 1 & 3 & -2 \\ 6 & 3 & 0 & -7 \end{bmatrix} R_1 \longleftrightarrow R_2$$

$$\approx \begin{bmatrix} 1 & -1 & -2 & -4 \\ 0 & 5 & 3 & 7 \\ 0 & 4 & 9 & 10 \\ 0 & 9 & 12 & 17 \end{bmatrix} \begin{matrix} R_2 \to R_2 - 2R_1 \\ R_3 \to R_3 - 3R_1 \\ R_4 \to R_4 - 6R_1 \end{matrix} \approx \begin{bmatrix} 1 & -1 & -2 & -4 \\ 0 & 5 & 3 & 7 \\ 0 & 0 & 33/5 & 22/5 \\ 0 & 0 & 33/5 & 22/5 \end{bmatrix} \begin{matrix} R_3 \to R_3 - 4/5\, R_2 \\ R_4 \to R_4 - 9/5\, R_2 \end{matrix}$$

$$\approx \begin{bmatrix} 1 & -1 & -2 & -4 \\ 0 & 5 & 3 & 7 \\ 0 & 0 & 33/5 & 22/5 \\ 0 & 0 & 0 & 0 \end{bmatrix} R_4 \to R_4 - R_3$$

$R(A)$ = Number of non-zero rows.

$\Rightarrow \quad R(A) = 3$ **Ans.**

## EXERCISE 4.15

**Find the rank of the following matrices:**

**1.** $\begin{bmatrix} 1 & 2 & 3 \\ 2 & 4 & 7 \\ 3 & 6 & 10 \end{bmatrix}$ **Ans.** 2

**2.** $\begin{bmatrix} 1 & 2 & 1 \\ -1 & 0 & 2 \\ 2 & 1 & -3 \end{bmatrix}$ **Ans.** 3

**3.** $\begin{bmatrix} 0 & 1 & 2 & -2 \\ 4 & 0 & 2 & 6 \\ 2 & 1 & 3 & 1 \end{bmatrix}$ **Ans.** 2

**4.** $\begin{bmatrix} 2 & 4 & 3 & -2 \\ -3 & -2 & -1 & 4 \\ 6 & -1 & 7 & 2 \end{bmatrix}$ **Ans.** 3

**5.** $\begin{bmatrix} 3 & 4 & 1 & 1 \\ 2 & 4 & 3 & 6 \\ -1 & -2 & 6 & 4 \\ 1 & -1 & 2 & -3 \end{bmatrix}$ **Ans.** 4

**6.** $\begin{bmatrix} 1 & 4 & 3 & -2 & 1 \\ -2 & -3 & -1 & 4 & 3 \\ -1 & 6 & 7 & 2 & 9 \\ -3 & 3 & 6 & 6 & 12 \end{bmatrix}$ **Ans.** 2

**Reduce the following matrices to Echelon form and find out the rank:**

**7.** $\begin{bmatrix} 1 & 1 & 2 \\ 1 & 2 & 2 \\ 2 & 2 & 3 \end{bmatrix}$ **Ans.** $\begin{bmatrix} 1 & 0 & 0 \\ 0 & 1 & 0 \\ 0 & 0 & 1 \end{bmatrix}$, Rank = 3

**8.** $\begin{bmatrix} 1 & 2 & 3 & 0 \\ 2 & 4 & 3 & 2 \\ 3 & 2 & 1 & 3 \\ 6 & 8 & 7 & 5 \end{bmatrix}$ **Ans.** $\begin{bmatrix} I_3 & 0 \\ 0 & 0 \end{bmatrix}$, Rank = 3

**9.** $\begin{bmatrix} 3 & 2 & 5 & 7 & 12 \\ 1 & 1 & 2 & 3 & 5 \\ 3 & 3 & 6 & 9 & 15 \end{bmatrix}$ **Ans.** $\begin{bmatrix} I_2 & 0 \\ 0 & 0 \end{bmatrix}$, Rank = 2

**10.** $\begin{bmatrix} 2 & -4 & 3 & 1 & 0 \\ 1 & -2 & 1 & -4 & 2 \\ 0 & 1 & -1 & 3 & 1 \\ 4 & -7 & 4 & -4 & 5 \end{bmatrix}$ **Ans.** $\begin{bmatrix} I_3 & 0 \\ 0 & 0 \end{bmatrix}$, Rank = 3

**Using elementary transformations, reduce the following matrices to the canonical form (or row-reduced Echelon form):**

**11.** $A = \begin{bmatrix} 0 & 0 & 0 & 0 & 0 \\ 0 & 1 & 2 & 3 & 4 \\ 0 & 2 & 3 & 4 & 1 \\ 0 & 3 & 4 & 1 & 2 \end{bmatrix}$ **Ans.** $\begin{bmatrix} 1 & 0 & 0 & 0 \\ 0 & 1 & 0 & 0 \\ 0 & 0 & 1 & 0 \end{bmatrix}$

**12.** $A = \begin{bmatrix} 0 & 4 & -12 & 8 & 9 \\ 0 & 2 & -6 & 2 & 5 \\ 0 & 1 & -3 & 6 & 4 \\ 0 & -8 & 24 & 3 & 1 \end{bmatrix}$ **Ans.** $\begin{bmatrix} 1 & 0 & 0 & 0 \\ 0 & 1 & 0 & 0 \\ 0 & 0 & 1 & 0 \end{bmatrix}$

**Using elementary transformations, reduce the following matrices to the normal form:**

**13.** $A = \begin{bmatrix} 1 & 2 & 0 & -1 \\ 3 & 4 & 1 & 2 \\ -2 & 3 & 2 & 5 \end{bmatrix}$ **Ans.** $\begin{bmatrix} 1 & 0 & 0 & 0 \\ 0 & 1 & 0 & 0 \\ 0 & 0 & 1 & 0 \end{bmatrix}$

**14.** $A = \begin{bmatrix} 1 & 2 & 3 & 4 \\ 3 & 4 & 1 & 2 \\ 4 & 3 & 1 & 2 \end{bmatrix}$ **Ans.** $\begin{bmatrix} 1 & 0 & 0 & 0 \\ 0 & 1 & 0 & 0 \\ 0 & 0 & 1 & 0 \end{bmatrix}$

**Obtain a matrix N in the normal form equivalent to**

**15.** $A = \begin{bmatrix} 0 & 0 & 0 & 0 & 0 \\ 0 & 4 & 5 & 0 & 0 \\ 0 & 9 & 1 & -1 & 2 \\ 0 & 10 & 0 & 1 & 11 \end{bmatrix}$

Hence find non-singular matrices $P$ and $Q$ such that $PAQ = N$.

**16.** $\begin{bmatrix} 1 & -3 & 1 & 2 \\ 0 & 1 & 2 & 3 \\ 3 & 4 & 1 & -2 \end{bmatrix}$

**Find the rank of the following matrix by reducing it into normal form:**

**17.** $A = \begin{bmatrix} 1 & 3 & 2 & 5 & 1 \\ 2 & 2 & -1 & 6 & 3 \\ 1 & 1 & 2 & 3 & -1 \\ 0 & 2 & 5 & 2 & -3 \end{bmatrix}$ **Ans. 3**

**18.** $A = \begin{bmatrix} 1 & 2 & 3 & 1 \\ 1 & 3 & 3 & 2 \\ 2 & 4 & 3 & 3 \\ 1 & 1 & 1 & 1 \end{bmatrix}$ **Ans. 4**

**19.** Rank of matrix $A = \begin{bmatrix} 1 & 2 & 3 \\ 1 & 4 & 2 \\ 2 & 6 & 5 \end{bmatrix}$ is

(*a*) 0 (*b*) 1 (*c*) 3 (*d*) 2 (*AMIETE, June 2009*) **Ans. (*d*)**

**20.** For which value of 'b' the rank of the matrix $A = \begin{bmatrix} 1 & 5 & 4 \\ 0 & 3 & 2 \\ b & 13 & 10 \end{bmatrix}$ is

(*a*) 1 (*b*) 2 (*c*) 3 (*d*) 0 (*AMIETE, Dec. 2009*) **Ans. (*b*)**

## 4.39 SOLUTION OF SIMULTANEOUS EQUATIONS

The matrix of the coefficients of $x, y, z$ is reduced into Echelon form by elementary row transformations. At the end of the row transformation the value of $z$ is calculated from the last equation and value of $y$ and the value of $x$ are calculated by the backward substitution.

**Example 33.** *Solve the following equations*

$$x - y + 2z = 3, \quad x + 2y + 3z = 5, \quad 3x - 4y - 5z = -13$$

**Solution.** In the matrix form, the equations are written in the following form.

$$\begin{bmatrix} 1 & -1 & 2 \\ 1 & 2 & 3 \\ 3 & -4 & -5 \end{bmatrix} \begin{bmatrix} x \\ y \\ z \end{bmatrix} = \begin{bmatrix} 3 \\ 5 \\ -13 \end{bmatrix} \text{ or } \begin{bmatrix} 1 & -1 & 2 \\ 0 & 3 & 1 \\ 0 & -1 & -11 \end{bmatrix} \begin{bmatrix} x \\ y \\ z \end{bmatrix} = \begin{bmatrix} 3 \\ 2 \\ -22 \end{bmatrix} \begin{matrix} \\ R_2 \to R_2 - R_1 \\ R_3 \to R_3 - 3R_1 \end{matrix}$$

$$\begin{bmatrix} 1 & -1 & 2 \\ 0 & 3 & 1 \\ 0 & 0 & -\frac{32}{3} \end{bmatrix} \begin{bmatrix} x \\ y \\ z \end{bmatrix} = \begin{bmatrix} 3 \\ 2 \\ -\frac{64}{3} \end{bmatrix} \quad R_3 \to R_3 + \frac{1}{3} R_2$$

$$x - y + 2z = 3 \quad \ldots(1)$$

$$3y + z = 2 \quad \ldots(2)$$

$$\frac{-32}{3} z = \frac{-64}{3} \Rightarrow z = 2$$

Putting the value of $z$ in (2), we get $3y + 2 = 2 \Rightarrow y = 0$

Putting the value of $y$, $z$ in (1), we get $x - 0 + 4 = 3 \Rightarrow x = -1$

$$x = -1, y = 0, z = 2$$ **Ans.**

**Example 34.** *Find all the solutions of the system of equations*

$$x_1 + 2x_2 - x_3 = 1,\ 3x_1 - 2x_2 + 2x_3 = 2,\ 7x_1 - 2x_2 + 3x_3 = 5$$

**Solution.** $$\begin{bmatrix} 1 & 2 & -1 \\ 3 & -2 & 2 \\ 7 & -2 & 3 \end{bmatrix}\begin{bmatrix} x_1 \\ x_2 \\ x_3 \end{bmatrix} = \begin{bmatrix} 1 \\ 2 \\ 5 \end{bmatrix}$$

$$R_2 \to R_2 - 3R_1, R_3 \to R_3 - 7R_1$$

$$\begin{bmatrix} 1 & 2 & -1 \\ 0 & -8 & 5 \\ 0 & -16 & 10 \end{bmatrix}\begin{bmatrix} x_1 \\ x_2 \\ x_3 \end{bmatrix} = \begin{bmatrix} 1 \\ -1 \\ -2 \end{bmatrix}, \begin{bmatrix} 1 & 2 & -1 \\ 0 & -8 & 5 \\ 0 & 0 & 0 \end{bmatrix}\begin{bmatrix} x_1 \\ x_2 \\ x_3 \end{bmatrix} = \begin{bmatrix} 1 \\ -1 \\ 0 \end{bmatrix} R_3 \to R_3 - 2R_2$$

$$x_1 + 2x_2 - x_3 = 1 \quad ...(1)$$

$$-8x_2 + 5x_3 = -1 \quad ...(2)$$

Let $$x_3 = k$$

Putting $x_3 = k$ in (2), we get $-8x_2 + 5k = -1 \Rightarrow x_2 = \frac{1}{8}(5k+1)$

Substituting the values of $x_3$, $x_1$ in (1), we get $x_1 + \frac{1}{4}(5k+1) - k = 1$

$$\therefore \quad x_1 = 1 + k - \frac{5k}{4} - \frac{1}{4} = -\frac{k}{4} + \frac{3}{4}$$

$$\therefore \quad x_1 = -\frac{k}{4} + \frac{3}{4}, x_2 = \frac{5k}{8} + \frac{1}{8}, x_3 = k$$

The equations have infinite solution. **Ans.**

## 4.40 Gauss - Jordan Method

*(R.G.P.V., Bhopal, III Semester, Dec. 2007)*

This is modification of the Gaussian elimination method.

By this method we eliminate unknowns not only from the equations below but also from the equations above. In this way the system is reduced to a diagonal matrix.

Finally each equation consists of only one unknown and thus, we get the solution. Here, the labour of backward substitution for finding the unknowns is saved

Gauss-Jordan method is modification of Gaussian elimination method.

**Example 35.** *Express the following system of equations in matrix form and solve them by the elimination method ( Gauss Jordan Method)*

$$2x_1 + x_2 + 2x_3 + x_4 = 6$$
$$6x_1 - 6x_2 + 6x_3 + 12x_4 = 36$$
$$4x_1 + 3x_2 + 3x_3 - 3x_4 = -1$$
$$2x_1 + 2x_2 - x_3 + x_4 = 10$$

**Solution.** The equations are expressed in matrix form as

$$\begin{bmatrix} 2 & 1 & 2 & 1 \\ 6 & -6 & 6 & 12 \\ 4 & 3 & 3 & -3 \\ 2 & 2 & -1 & 1 \end{bmatrix}\begin{bmatrix} x_1 \\ x_2 \\ x_3 \\ x_4 \end{bmatrix} = \begin{bmatrix} 6 \\ 36 \\ -1 \\ 10 \end{bmatrix}$$

$$\Rightarrow \begin{bmatrix} 2 & 1 & 2 & 1 \\ 0 & -9 & 0 & 9 \\ 0 & 1 & -1 & -5 \\ 0 & 1 & -3 & 0 \end{bmatrix} \begin{bmatrix} x_1 \\ x_2 \\ x_3 \\ x_4 \end{bmatrix} = \begin{bmatrix} 6 \\ 18 \\ -13 \\ 4 \end{bmatrix} \begin{matrix} R_2 \to R_2 - 3R_1 \\ R_3 \to R_3 - 2R_1 \\ R_4 \to R_4 - R_1 \end{matrix}$$

$$\begin{bmatrix} 2 & 1 & 2 & 1 \\ 0 & 1 & 0 & -1 \\ 0 & 1 & -1 & -5 \\ 0 & 1 & -3 & 0 \end{bmatrix} \begin{bmatrix} x_1 \\ x_2 \\ x_3 \\ x_4 \end{bmatrix} = \begin{bmatrix} 6 \\ -2 \\ -13 \\ 4 \end{bmatrix} \quad R_2 \to \frac{R_2}{-9}$$

$$\begin{bmatrix} 2 & 1 & 2 & 1 \\ 0 & 1 & 0 & -1 \\ 0 & 0 & -1 & -4 \\ 0 & 0 & -3 & 1 \end{bmatrix} \begin{bmatrix} x_1 \\ x_2 \\ x_3 \\ x_4 \end{bmatrix} = \begin{bmatrix} 6 \\ -2 \\ -11 \\ 6 \end{bmatrix} \begin{matrix} R_3 \to R_3 - R_2 \\ \\ R_4 \to R_4 - R_2 \end{matrix}$$

$$\begin{bmatrix} 2 & 1 & 2 & 1 \\ 0 & 1 & 0 & -1 \\ 0 & 0 & -1 & -4 \\ 0 & 0 & 0 & 13 \end{bmatrix} \begin{bmatrix} x_1 \\ x_2 \\ x_3 \\ x_4 \end{bmatrix} = \begin{bmatrix} 6 \\ -2 \\ -11 \\ 39 \end{bmatrix} \quad R_4 \to R_4 - 3R_3$$

$$2x_1 + x_2 + 2x_3 + x_4 = 6 \qquad \text{...(1)}$$
$$x_2 - x_4 = -2 \qquad \text{...(2)}$$
$$-x_3 - 4x_4 = -11 \qquad \text{...(3)}$$
$$13x_4 = 39 \Rightarrow x_4 = 3$$

Putting the value of $x_4$ in (3), we get

$$-x_3 - 12 = -11 \Rightarrow x_3 = -1$$

Putting the value of $x_4$ in (2), we get

$$x_2 - 3 = -2 \Rightarrow x_2 = 1$$

Substituting the values of $x_4$, $x_3$ and $x_2$ in (1), we get

$$2x_1 + 1 - 2 + 3 = 6 \text{ or } 2x_1 = 4 \Rightarrow x_1 = 2$$

$\therefore$ $$x_1 = 2, x_2 = 1, x_3 = -1, x_4 = 3$$ **Ans.**

**Example 36.** *Find the general solution of the system of equations:*

$$3x_1 + 2x_3 + 2x_4 = 0$$
$$-x_1 + 7x_2 + 4x_3 + 9x_4 = 0$$
$$7x_1 - 7x_2 - 5x_4 = 0$$

**Solution.** The system of equations in the matrix form is expressed as

$$\begin{bmatrix} 3 & 0 & 2 & 2 \\ -1 & 7 & 4 & 9 \\ 7 & -7 & 0 & -5 \end{bmatrix} \begin{bmatrix} x_1 \\ x_2 \\ x_3 \\ x_4 \end{bmatrix} = \begin{bmatrix} 0 \\ 0 \\ 0 \end{bmatrix}$$

$$\begin{bmatrix} -1 & 7 & 4 & 9 \\ 3 & 0 & 2 & 2 \\ 7 & -7 & 0 & -5 \end{bmatrix} \begin{bmatrix} x_1 \\ x_2 \\ x_3 \\ x_4 \end{bmatrix} = \begin{bmatrix} 0 \\ 0 \\ 0 \end{bmatrix} \quad R_1 \leftrightarrow R_2$$

$$\begin{bmatrix} -1 & 7 & 4 & 9 \\ 0 & 21 & 14 & 29 \\ 0 & 42 & 28 & 58 \end{bmatrix} \begin{bmatrix} x_1 \\ x_2 \\ x_3 \\ x_4 \end{bmatrix} = \begin{bmatrix} 0 \\ 0 \\ 0 \end{bmatrix} \quad \begin{matrix} R_2 \to R_2 + 3R_1 \\ R_3 \to R_3 + 7R_1 \end{matrix}$$

$$\begin{bmatrix} -1 & 7 & 4 & 9 \\ 0 & 21 & 14 & 29 \\ 0 & 0 & 0 & 0 \end{bmatrix} \begin{bmatrix} x_1 \\ x_2 \\ x_3 \\ x_4 \end{bmatrix} = \begin{bmatrix} 0 \\ 0 \\ 0 \end{bmatrix} \quad R_3 \to R_3 - 2R_2$$

$$-x_1 + 7x_2 + 4x_3 + 9x_4 = 0 \qquad \ldots(1)$$

$$21x_2 + 14x_3 + 29x_4 = 0 \qquad \ldots(2)$$

Let $x_4 = a,\ x_3 = b$

From (2), $21x_2 + 14b + 29a = 0$ or $x_2 = -\frac{2b}{3} - \frac{29a}{21}$

From (1), $-x_1 + 7\left(-\frac{2b}{3} - \frac{29a}{21}\right) + 4b + 9a = 0$

$$x_1 = -\frac{2a}{3} - \frac{2b}{3}$$

$$x_1 = -\frac{2}{3}(a+b),\ x_2 = -\frac{1}{21}(29a + 14b)$$

$$x_3 = b,\ x_4 = a$$ **Ans.**

## 4.41 TYPES OF LINEAR EQUATIONS

**(1) Consistent.** A system of equations is said to be *consistent,* if they have one or more solution *i.e.*

| $x + 2y = 4$ | $x + 2y = 4$ |
|---|---|
| $3x + 2y = 2$ | $3x + 6y = 12$ |
| Unique solution | Infinite solution |

**(2) Inconsistent.** If a system of equation has no solution, it is said to be inconsistent *i.e.*

$$x + 2y = 4$$
$$3x + 6y = 5$$

## 4.42 CONSISTENCY OF A SYSTEM OF LINEAR EQUATIONS

$$a_{11}x_1 + a_{12}x_2 + \ldots a_{1n}x_n = b_1$$
$$a_{21}x_1 + a_{22}x_2 + \ldots a_{2n}x_n = b_2$$
$$\ldots\ldots\ldots\ldots\ldots\ldots\ldots\ldots\ldots\ldots$$
$$a_{m1}x_1 + a_{m2}x_2 + \ldots a_{mn}x_n = b_m$$

$$\Rightarrow \begin{bmatrix} a_{11} & a_{12} \ldots\ldots & a_{1n} \\ a_{21} & a_{22} \ldots\ldots & a_{2n} \\ \ldots & \ldots & \ldots \\ a_{m1} & a_{m2} \ldots\ldots & a_{mn} \end{bmatrix} \begin{bmatrix} x_1 \\ x_2 \\ \ldots \\ x_m \end{bmatrix} = \begin{bmatrix} b_1 \\ b_2 \\ \ldots \\ b_m \end{bmatrix} \quad \text{and} \quad C = [A, B] = \begin{bmatrix} a_{11} & a_{12} \ldots\ldots & a_{1n} & b_1 \\ a_{21} & a_{22} \ldots\ldots & a_{2n} & b_2 \\ \ldots & \ldots & \ldots & \ldots \\ a_{m1} & a_{m2} \ldots\ldots & a_{mn} & b_m \end{bmatrix}$$

$$\Rightarrow \qquad AX = B$$

is called the **augmented** matrix.

$$\boxed{[A:B] = C}$$

(*a*) **Consistent equations.** If Rank $A$ = Rank $C$

(*i*) *Unique solution:* Rank $A$ = Rank $C = n$ where $n$ = number of unknown.

(*ii*) *Infinite solution:* Rank $A$ = Rank $C = r$, $r < n$

(*b*) **Inconsistent equations.** If Rank $A \neq$ Rank $C$.

**In Brief :**

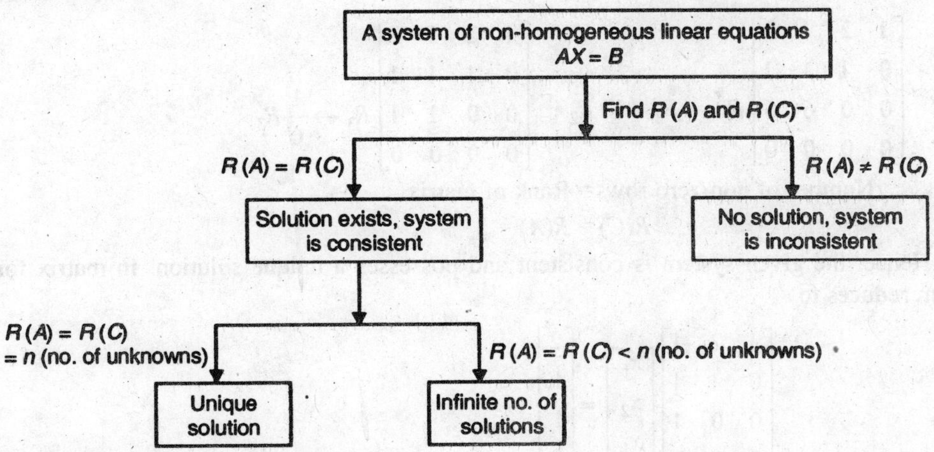


**Example 37.** *Show that the equations*

$$2x + 6y = -11,\ 6x + 20y - 6z = -3,\ 6y - 18z = -1$$

*are not consistent.*

**Solution.** Augmented matrix $C = [A, B]$

$$= \begin{bmatrix} 2 & 6 & 0 & : & -11 \\ 6 & 20 & -6 & : & -3 \\ 0 & 6 & -18 & : & -1 \end{bmatrix} \sim \begin{bmatrix} 2 & 6 & 0 & : & -11 \\ 0 & 2 & -6 & : & 30 \\ 0 & 6 & -18 & : & -1 \end{bmatrix} R_2 \to R_2 - 3R_1$$

$$\sim \begin{bmatrix} 2 & 6 & 0 & : & -11 \\ 0 & 2 & -6 & : & 30 \\ 0 & 0 & 0 & : & -91 \end{bmatrix} R_3 \to R_3 - 3R_2$$

The rank of $C$ is 3 and the rank of $A$ is 2.

Rank of $A \neq$ Rank of $C$. The equations are not consistent. **Ans.**

**Example 38.** *Test the consistency and hence solve the following set of equation.*

$$x_1 + 2x_2 + x_3 = 2$$
$$3x_1 + x_2 - 2x_3 = 1$$
$$4x_1 - 3x_2 - x_3 = 3$$
$$2x_1 + 4x_2 + 2x_3 = 4$$

*(U.P., I Semester, Compartment 2002)*

**Solution.** The given set of equations is written in the matrix form:

$$\begin{bmatrix} 1 & 2 & 1 \\ 3 & 1 & -2 \\ 4 & -3 & -1 \\ 2 & 4 & 2 \end{bmatrix} \begin{bmatrix} x_1 \\ x_2 \\ x_3 \end{bmatrix} = \begin{bmatrix} 2 \\ 1 \\ 3 \\ 4 \end{bmatrix}$$

$$AX = B$$

Here, we have augmented matrix $C = [A, B] \sim \begin{bmatrix} 1 & 2 & 1 & 2 \\ 3 & 1 & -2 & 1 \\ 4 & -3 & -1 & 3 \\ 2 & 4 & 2 & 4 \end{bmatrix}$

$$\sim\begin{bmatrix}1 & 2 & 1 & 2\\ 0 & -5 & -5 & -5\\ 0 & -11 & -5 & -5\\ 0 & 0 & 0 & 0\end{bmatrix}\begin{matrix}R_2 \to R_2 - 3R_1\\ R_3 \to R_3 - 4R_1\\ R_4 \to R_4 - 2R_1\end{matrix} \sim\begin{bmatrix}1 & 2 & 1 & 2\\ 0 & 1 & 1 & 1\\ 0 & -11 & -5 & -5\\ 0 & 0 & 0 & 0\end{bmatrix} R_2 \to -\frac{1}{5}R_2$$

$$\sim\begin{bmatrix}1 & 2 & 1 & 2\\ 0 & 1 & 1 & 1\\ 0 & 0 & 6 & 6\\ 0 & 0 & 0 & 0\end{bmatrix} R_3 \to R_3 + 11R_2 \sim\begin{bmatrix}1 & 2 & 1 & 2\\ 0 & 1 & 1 & 1\\ 0 & 0 & 1 & 1\\ 0 & 0 & 0 & 0\end{bmatrix} R_3 \to \frac{1}{6}R_3$$

Number of non-zero rows = Rank of matrix.

$$\Rightarrow \qquad R(C) = R(A) = 3$$

Hence, the given system is consistent and possesses a unique solution. In matrix form the system reduces to

$$\begin{bmatrix}1 & 2 & 1\\ 0 & 1 & 1\\ 0 & 0 & 1\\ 0 & 0 & 0\end{bmatrix}\begin{bmatrix}x_1\\ x_2\\ x_3\end{bmatrix} = \begin{bmatrix}2\\ 1\\ 1\\ 0\end{bmatrix}$$

$$x_1 + 2x_2 + x_3 = 2 \qquad \text{...(1)}$$

$$x_2 + x_3 = 1 \qquad \text{...(2)}$$

$$x_3 = 1$$

From (2), $\quad x_2 + 1 = 1 \Rightarrow x_2 = 0$

From (1), $\quad x_1 + 0 + 1 = 2 \Rightarrow x_1 = 1$

Hence, $\quad x_1 = 1, x_2 = 0$ and $x_3 = 1$ **Ans.**

**Example 39.** *Test for consistency and solve :*

$$5x + 3y + 7z = 4,\ 3x + 26y + 2z = 9,\ 7x + 2y + 10z = 5$$

**Solution.** The augmented matrix $C = [A, B]$ *(R.G. P.V. Bhopal I. Sem. April 2009-08-03)*

$$\begin{bmatrix}5 & 3 & 7 & : & 4\\ 3 & 26 & 2 & : & 9\\ 7 & 2 & 10 & : & 5\end{bmatrix} \sim \begin{bmatrix}1 & \frac{3}{5} & \frac{7}{5} & : & \frac{4}{5}\\ 3 & 26 & 2 & : & 9\\ 7 & 2 & 10 & : & 5\end{bmatrix} R_1 \to \frac{1}{5}R_1$$

$$\sim\begin{bmatrix}1 & \frac{3}{5} & \frac{7}{5} & : & \frac{4}{5}\\ 0 & \frac{121}{5} & -\frac{11}{5} & : & \frac{33}{5}\\ 0 & -\frac{11}{5} & \frac{1}{5} & : & -\frac{3}{5}\end{bmatrix}\begin{matrix}R_2 \to R_2 - 3R_1\\ R_3 \to R_3 - 7R_1\end{matrix} \sim\begin{bmatrix}1 & \frac{3}{5} & \frac{7}{5} & : & \frac{4}{5}\\ 0 & \frac{121}{5} & -\frac{11}{5} & : & \frac{33}{5}\\ 0 & 0 & 0 & : & 0\end{bmatrix} R_3 \to R_3 + \frac{1}{11}R_2$$

Rank of $A = 2 =$ Rank of $C$

Hence, the equations are consistent. But the rank is less than 3 *i.e.* number of unknows. So its solutions are infinite.

$$\begin{bmatrix} 1 & \frac{3}{5} & \frac{7}{5} \\ 0 & \frac{121}{5} & -\frac{11}{5} \\ 0 & 0 & 0 \end{bmatrix} \begin{bmatrix} x \\ y \\ z \end{bmatrix} = \begin{bmatrix} \frac{4}{5} \\ \frac{33}{5} \\ 0 \end{bmatrix}$$

$$x + \frac{3}{5}y + \frac{7}{5}z = \frac{4}{5}$$

$$\frac{121}{5}y - \frac{11z}{5} = \frac{33}{5} \quad \text{or } 11y - z = 3$$

Let $z = k$ then $\quad 11y - k = 3$ or $y = \frac{3}{11} + \frac{k}{11}$

$$x + \frac{3}{5}\left[\frac{3}{11} + \frac{k}{11}\right] + \frac{7}{5}k = \frac{4}{5} \quad \text{or } x = -\frac{16}{11}k + \frac{7}{11}$$ **Ans.**

**Example 40.** *Discuss the consistency of the following system of equations*

*$2x + 3y + 4z = 11$, $x + 5y + 7z = 15$, $3x + 11y + 13z = 25$.*

*If found consistent, solve it.* *(A.M.I.E.T.E., Winter 2001)*

**Solution.** The augmented matrix $C = [A, B]$

$$\begin{bmatrix} 2 & 3 & 4 & 11 \\ 1 & 5 & 7 & 15 \\ 3 & 11 & 13 & 25 \end{bmatrix} \sim \begin{bmatrix} 1 & 5 & 7 & 15 \\ 2 & 3 & 4 & 11 \\ 3 & 11 & 13 & 25 \end{bmatrix} R_1 \leftrightarrow R_2$$

$R_2 \to R_2 - 2R_1,\ R_3 \to R_3 - 3R_1, \quad R_2 \to -\frac{1}{7}R_2,\ R_3 \to -\frac{1}{4}R_3, \quad R_3 \to R_3 - R_2$

$$\begin{bmatrix} 1 & 5 & 7 & 15 \\ 0 & -7 & -10 & -19 \\ 0 & -4 & -8 & -20 \end{bmatrix} \sim \begin{bmatrix} 1 & 5 & 7 & 15 \\ 0 & 1 & \frac{10}{7} & \frac{19}{7} \\ 0 & 1 & 2 & 5 \end{bmatrix} \sim \begin{bmatrix} 1 & 5 & 7 & 15 \\ 0 & 1 & \frac{10}{7} & \frac{19}{7} \\ 0 & 0 & \frac{4}{7} & \frac{16}{7} \end{bmatrix}$$

Rank of $C = 3 =$ Rank of $A$

Hence, the system of equations is consistent with unique solution.

Now, $$\begin{bmatrix} 1 & 5 & 7 \\ 0 & 1 & \frac{10}{7} \\ 0 & 0 & \frac{4}{7} \end{bmatrix} \begin{bmatrix} x \\ y \\ z \end{bmatrix} = \begin{bmatrix} 15 \\ \frac{19}{7} \\ \frac{16}{7} \end{bmatrix}$$

$\Rightarrow$ $$x + 5y + 7z = 15 \qquad \text{...(1)}$$

$$y + \frac{10z}{7} = \frac{19}{7} \qquad \text{...(2)}$$

$$\frac{4z}{7} = \frac{16}{7} \Rightarrow z = 4$$

From (2), $\quad y + \frac{10}{7} \times 4 = \frac{19}{7} \Rightarrow y = -3$

From (1), $\quad x + 5(-3) + 7(4) = 15 \Rightarrow x = 2$

$$x = 2,\ y = -3,\ z = 4$$ **Ans.**

**Example 41.** *Find for what values of $\lambda$ and $\mu$ the system of linear equations:*

$$x + y + z = 6$$
$$x + 2y + 5z = 10$$
$$2x + 3y + \lambda z = \mu$$

*has (i) a unique solution (ii) no solution*

*(iii) infinite solutions. Also find the solution for $\lambda = 2$ and $\mu = 8$.*

(*Uttarakhand, 1st semester, Dec. 2006*)

**Solution.** $\begin{bmatrix} 1 & 1 & 1 \\ 1 & 2 & 5 \\ 2 & 3 & \lambda \end{bmatrix}\begin{bmatrix} x \\ y \\ z \end{bmatrix} = \begin{bmatrix} 6 \\ 10 \\ \mu \end{bmatrix}$

$$AX = B$$

$$C = (A, B) = \begin{bmatrix} 1 & 1 & 1 & : & 6 \\ 1 & 2 & 5 & : & 10 \\ 2 & 3 & \lambda & : & \mu \end{bmatrix} \sim \begin{bmatrix} 1 & 1 & 1 & : & 6 \\ 0 & 1 & 4 & : & 4 \\ 0 & 1 & \lambda - 2 & : & \mu - 12 \end{bmatrix} \begin{matrix} \\ R_2 \to R_2 - R_1 \\ R_3 \to R_3 - 2R_1 \end{matrix}$$

$$\sim \begin{bmatrix} 1 & 1 & 1 & : & 6 \\ 0 & 1 & 4 & : & 4 \\ 0 & 0 & \lambda - 6 & : & \mu - 16 \end{bmatrix} R_3 \to R_3 - R_2 \qquad \text{...(1)}$$

***(i)*** A unique solution

If $R(A) = R(C) = 3$

then $\lambda - 6 \neq 0 \Rightarrow \lambda \neq 6$ and $\mu - 16 \neq 0 \Rightarrow \mu \neq 16$

***(ii)*** No solutions

If $R(A) \neq R(C)$, then $R(A) = 2$ and $R(C) = 3$

$\lambda - 6 = 0 \Rightarrow \lambda = 6$ and $\mu - 16 \neq 0 \Rightarrow \mu \neq 16$

***(iii)*** Infinite solutions

If $R(A) = R(C) = 2$

then $\lambda - 6 = 0$ and $\mu - 16 = 0$

$\Rightarrow \quad \lambda = 6$ and $\mu = 16$

***(iv)*** Putting $\lambda = 2$ and $\mu = 8$ in (1), we get

$$\begin{bmatrix} 1 & 1 & 1 & : & 6 \\ 0 & 1 & 4 & : & 4 \\ 0 & 0 & -4 & : & -8 \end{bmatrix} \Rightarrow \begin{bmatrix} 1 & 1 & 1 \\ 0 & 1 & 4 \\ 0 & 0 & -4 \end{bmatrix}\begin{bmatrix} x \\ y \\ z \end{bmatrix} = \begin{bmatrix} 6 \\ 4 \\ -8 \end{bmatrix}$$

$$x + y + z = 6$$
$$y + 4z = 4$$
$$-4z = -8 \quad \Rightarrow \quad z = 2$$

Putting $z = 2$ in (3), we get

$$y + 8 = 4 \quad \Rightarrow \quad y = -4$$

Putting $y = -4$, $z = 2$ in (2), we get

$$x - 4 + 2 = 6 \quad \Rightarrow \quad x = 8$$

Hence, $\quad x = 8, \quad y = -4, \quad z = 2$ **Ans.**

**Example 42.** *Find for what values of k the set of equations*

$$2x - 3y + 6z - 5t = 3,\ \ y - 4z + t = 1,\ \ 4x - 5y + 8z - 9t = k$$

*has (i) no solution (ii) infinite number of solutions.* (*A.M.I.E.T.E., Summer 2004*)

**Solution.** The augmented matrix $C = [A, B]$

$$R_3 \to R_3 - 2R_1$$

$$\begin{bmatrix} 2 & -3 & 6 & -5 & \cdot & 3 \\ 0 & 1 & -4 & 1 & \cdot & 1 \\ 4 & -5 & 8 & -9 & \cdot & k \end{bmatrix} \sim \begin{bmatrix} 2 & -3 & 6 & -5 & \cdot & 3 \\ 0 & 1 & -4 & 1 & \cdot & 1 \\ 0 & 1 & -4 & 1 & \cdot & k-6 \end{bmatrix}$$

$$\sim \begin{bmatrix} 2 & -3 & 6 & -5 & \cdot & 3 \\ 0 & 1 & -4 & 1 & \cdot & 1 \\ 0 & 0 & 0 & 0 & \cdot & k-7 \end{bmatrix} \quad R_3 \to R_3 - R_2$$

(*i*) There is no solution if $R(A) \neq R(C)$

$$k - 7 \neq 0 \text{ or } k \neq 7, R(A) = 2 \text{ and } R(C) = 3.$$

(*ii*) There are infinite solutions if $R(A) = R(C) = 2$

$$k - 7 = 0 \Rightarrow k = 7$$ **Ans.**

$$\begin{bmatrix} 2 & -3 & 6 & -5 \\ 0 & 1 & -4 & 1 \end{bmatrix} \begin{bmatrix} x \\ y \\ z \\ t \end{bmatrix} = \begin{bmatrix} 3 \\ 1 \end{bmatrix}$$

$$2x - 3y + 6z - 5t = 3 \quad ...(1)$$

$$y - 4z + t = 1 \quad ...(2)$$

Let $t = k_1$ and $z = k_2$.

From (2), $y - 4k_2 + k_1 = 1$ or $y = 1 + 4k_2 - k_1$

From (l), $2x - 3 - 12k_2 + 3k_1 + 6k_2 - 5k_1 = 3$

$\Rightarrow$ $$2x = 6 + 6k_2 + 2k_1 \Rightarrow x = 3 + 3k_2 + k_1$$

$$y = 1 + 4k_2 - k_1 \Rightarrow z = k_2, \; t = k_1$$ **Ans.**

## 4.43. HOMOGENEOUS EQUATIONS

For a system of homogeneous linear equations $AX = O$

(*i*) $X = O$ is always a solution. This solution in which each unknown has the value zero is called the **Null Solution** or the **Trivial solution.** Thus a homogeneous system is always consistent.

A system of homogeneous linear equations has either the trivial solution or an infinite number of solutions.

(*ii*) If $R(A)$ = number of unknowns, the system has only the trivial solution.

(*iii*) If $R(A) <$ number of unknowns, the system has an infinite number of non-trivial solutions.

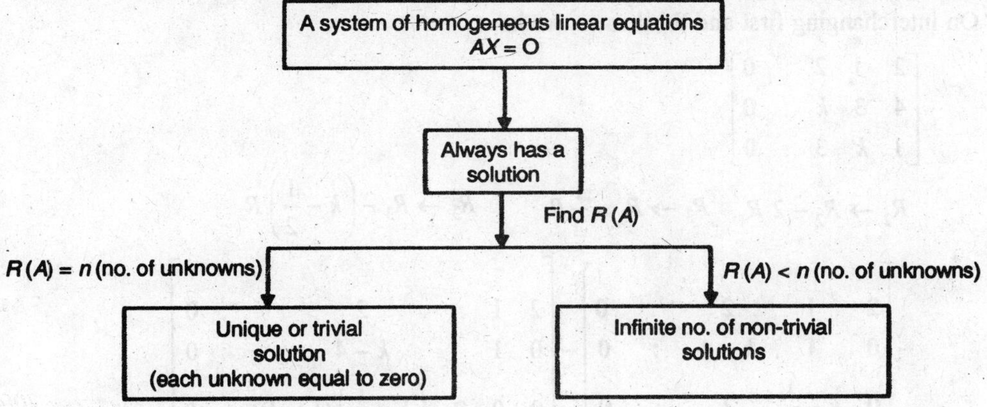

**Example 43.** *Determine 'b' such that the system of homogeneous equations*

$$2x + y + 2z = 0\ ;$$
$$x + y + 3z = 0\ ;$$
$$4x + 3y + bz = 0$$

*has (i) Trivial solution*

*(ii) Non-Trivial solution . Find the Non-Trivial solution using matrix method.*

(*U.P., I Sem Dec 2008*)

**Solution.** Here, we have

$$2x + y + 2z = 0$$
$$x + y + 3z = 0$$
$$4x + 3y + bz = 0$$

(*i*) **For trivial solution:** We know that $x = 0$, $y = 0$ and $z = 0$. So, $b$ can have any value.

(*ii*) **For non-trivial solution:** The given equations are written in the matrix form as :

$$\begin{bmatrix} 2 & 1 & 2 \\ 1 & 1 & 3 \\ 4 & 3 & b \end{bmatrix}\begin{bmatrix} x \\ y \\ z \end{bmatrix} = \begin{bmatrix} 0 \\ 0 \\ 0 \end{bmatrix}$$

$$AX = B$$

$$R_1 \leftrightarrow R_2, \qquad R_2 \to R_2 - 2R_1,\ R_3 \to R_3 - 4R_1, \qquad R_3 \to R_3 - R_2$$

$$C = \begin{bmatrix} 2 & 1 & 2 & : 0 \\ 1 & 1 & 3 & : 0 \\ 4 & 3 & b & : 0 \end{bmatrix} \sim \begin{bmatrix} 1 & 1 & 3 & : 0 \\ 2 & 1 & 2 & : 0 \\ 4 & 3 & b & : 0 \end{bmatrix} \sim \begin{bmatrix} 1 & 1 & 3 & : 0 \\ 0 & -1 & -4 & : 0 \\ 0 & -1 & b-12 & : 0 \end{bmatrix} \sim \begin{bmatrix} 1 & 1 & 3 & : 0 \\ 0 & -1 & -4 & : 0 \\ 0 & 0 & b-8 & : 0 \end{bmatrix}$$

For non trivial solution or infinite solutions $R(C) = R(A) = 2 <$ Number of unknowns

$$b - 8 = 0, \quad b = 8$$ **Ans.**

**Example 44.** *Find the values of k such that the system of equations*

$$x + ky + 3z = 0, \quad 4x + 3y + kz = 0, \quad 2x + y + 2z = 0$$

*has non-trivial solution.*

**Solution.** The set of equations is written in the form of matrices

$$\begin{bmatrix} 1 & k & 3 \\ 4 & 3 & k \\ 2 & 1 & 2 \end{bmatrix}\begin{bmatrix} x \\ y \\ z \end{bmatrix} = \begin{bmatrix} 0 \\ 0 \\ 0 \end{bmatrix}, \quad AX = B, \quad C = [A : B] = \begin{bmatrix} 1 & k & 3 & : & 0 \\ 4 & 3 & k & : & 0 \\ 2 & 1 & 2 & : & 0 \end{bmatrix}$$

On interchanging first and third rows, we have

$$\begin{bmatrix} 2 & 1 & 2 & : & 0 \\ 4 & 3 & k & : & 0 \\ 1 & k & 3 & : & 0 \end{bmatrix}$$

$$R_2 \to R_2 - 2R_1, \quad R_3 \to R_3 - \frac{1}{2}R_1 \qquad R_3 \to R_3 - \left(k - \frac{1}{2}\right)R_2$$

$$\sim \begin{bmatrix} 2 & 1 & 2 & : & 0 \\ 0 & 1 & k-4 & : & 0 \\ 0 & k-\frac{1}{2} & 2 & : & 0 \end{bmatrix} \sim \begin{bmatrix} 2 & 1 & 2 & : & 0 \\ 0 & 1 & k-4 & : & 0 \\ 0 & 0 & 2-\left(k-\frac{1}{2}\right)(k-4) & : & 0 \end{bmatrix}$$

For a non-trivial solution or for infinite solution, $R(A) = R(C) = 2$

so $$2-\left(k-\frac{1}{2}\right)(k-4) = 0 \;\Rightarrow\; 2-k^2+4k+\frac{k}{2}-2 = 0$$

$$\Rightarrow\; -k^2+\frac{9}{2}k = 0 \;\Rightarrow\; k\left(-k+\frac{9}{2}\right) = 0 \;\Rightarrow\; k = \frac{9}{2},\; k = 0$$ **Ans.**

## 4.44 CRAMER'S RULE

**Example 45.** *Find values of λ for which the following system of equations is consistent and has non-trivial solutions. Solve equations for all such values of λ.*

$$(\lambda-1)x+(3\lambda+1)y+2\lambda z = 0$$
$$(\lambda-1)x+(4\lambda-2)y+(\lambda+3)z = 0$$
$$2x+(3\lambda+1)y+3(\lambda-1)z = 0$$

*(A.M.I.E.T.E., Summer 2010, 2001)*

**Solution.** $$\begin{bmatrix}(\lambda-1) & (3\lambda+1) & 2\lambda\\ (\lambda-1) & (4\lambda-2) & (\lambda+3)\\ 2 & (3\lambda+1) & (3\lambda-3)\end{bmatrix}\begin{bmatrix}x\\y\\z\end{bmatrix} = \begin{bmatrix}0\\0\\0\end{bmatrix} \quad ...(1)$$

$$AX = 0$$

For infinite solutions, $|A| = 0$

$$\begin{vmatrix}\lambda-1 & 3\lambda+1 & 2\lambda\\ \lambda-1 & 4\lambda-2 & \lambda+3\\ 2 & 3\lambda+1 & 3\lambda-3\end{vmatrix} = 0, \quad \begin{vmatrix}0 & -\lambda+3 & \lambda-3\\ \lambda-1 & 4\lambda-2 & \lambda+3\\ 2 & 3\lambda+1 & 3\lambda-3\end{vmatrix} = 0,$$

$$\begin{vmatrix}0 & 0 & \lambda-3\\ \lambda-1 & 5\lambda+1 & \lambda+3\\ 2 & 6\lambda-2 & 3\lambda-3\end{vmatrix} = 0,$$

$$(\lambda-3)\,[(\lambda-1)(6\lambda-2)-2(5\lambda+1)] = 0$$

$$[6\lambda^2-8\lambda+2-10\lambda-2] = 0 \text{ or } 6\lambda^2-18\lambda = 0 \text{ or } 6\lambda(\lambda-3) = 0,\; \lambda = 3$$

On putting $\lambda = 3$ in (1), we get

$$\begin{bmatrix}2 & 10 & 6\\ 2 & 10 & 6\\ 2 & 10 & 6\end{bmatrix}\begin{bmatrix}x\\y\\z\end{bmatrix} = \begin{bmatrix}0\\0\\0\end{bmatrix} \Rightarrow \begin{bmatrix}2 & 10 & 6\\ 0 & 0 & 0\\ 0 & 0 & 0\end{bmatrix}\begin{bmatrix}x\\y\\z\end{bmatrix} = \begin{bmatrix}0\\0\\0\end{bmatrix}$$

$$2x+10y+6z = 0 \Rightarrow x+5y+3z = 0$$

Let $$x = k_1,\; y = k_2,\; 3z = -k_1-5k_2 \Rightarrow z = \frac{-k_1}{3}-\frac{5k_2}{3}$$ **Ans.**

## EXERCISE 4.16

Test the consistency of the following equations and solve them if possible.

**1.** $3x+3y+2z = 1,\quad x+2y = 4,\quad 10y+3z = -2,\quad 2x-3y-z = 5$

**Ans.** Consistent, $x = 2,\; y = 1,\; z = -4$ *(R.G.P.V. Bhopal 1st Sem 2001)*

**2.** $x_1-x_2+x_3-x_4+x_5 = 1,\quad 2x_1-x_2+3x_3+4x_5 = 2,$

$3x_1-2x_2+2x_3+x_4+x_5 = 1,\quad x_1+x_3+2x_4+x_5 = 0$ *(A.M.I.E.T.E., Winter 2003)*

**Ans.** $x_1 = -3k_1+k_2-1,\; x_2 = -3k_1-1,\; x_3 = k_1-2k_2+1,\; x_4 = k_1,\; x_4 = k_1,\; x_5 = k_2 1$

3. Find the value of $k$ for which the following system of equations is consistent.

$3x_1 - 2x_2 + 2x_3 = 3,\ x_1 + kx_2 - 3x_3 = 0,\ 4x_1 + x_2 + 2x_3 = 7$ **Ans.** $k = \frac{1}{4}$

4. Find the value of $\lambda$ for which the system of equations

$x + y + 4z = 1, x + 2y - 2z = 1,\ \lambda x + y + z = 1$

will have a unique solution. *(A.M.I.E., Winter 2000)* **Ans.** $\lambda \neq \frac{7}{10}$

5. Determine the values of $a$ and $b$ for which the system $\begin{bmatrix} 3 & -2 & 1 \\ 5 & -8 & 9 \\ 2 & 1 & a \end{bmatrix}\begin{bmatrix} x \\ y \\ z \end{bmatrix} = \begin{bmatrix} b \\ 3 \\ -1 \end{bmatrix}$

(*i*) has a unique solution, (*ii*) has no solution and, (*iii*) has infinitely many solutions.

**Ans.** (*i*) $a \neq -3$, (*ii*) $a = -3,\ b \neq \frac{1}{3}$, (*iii*) $a = -3,\ b = \frac{1}{3}$

6. Choose $\lambda$ that makes the following system of linear equations consistent and find the general solution of the system for that $\lambda$.

$x + y - z + t = 2,\ 2y + 4z + 2t = 3,\ x + 2y + z + 2t = \lambda$

**Ans.** $\lambda = \frac{7}{2}$, $x = \frac{1}{2} + 3k_2$, $y = \frac{3}{2} - 2k_2 - k_1$, $z = k_2$, $t = k_1$

7. Show that the equations

$3x + 4y + 5z = a,\ 4x + 5y + 6z = b,\ 5x + 6y + 7z = c$

don't have a solution unless $a + c = 2b$.

Solve the equations when $a = b = c = -1$ **Ans.** $x = k + 1,\ y = -2k - 1, z = k$

8. Find the values of $k$, such that the system of equations

$4x_1 + 9x_2 + x_3 = 0,\ kx_1 + 3x_2 + kx_3 = 0,\ x_1 + 4x_2 + 2x_3 = 0$

has non-trivial solution. Hence, find the solution of the system.

**Ans.** $k = 1,\ x_1 = 2\lambda,\ x_2 = -\lambda,\ x_3 = \lambda$

9. Find values of $\lambda$ for which the following system of equations has a non-trivial solution.

$3x_1 + x_2 - \lambda x_3 = 0,\ 2x_1 + 4x_2 + \lambda x_3 = 0,\ 8x_1 - 4x_2 - 6x_3 = 0$ **Ans.** $\lambda = 1$

10. Find value of $\lambda$ so that the following system of homogeneous equations have exactly two linearly independent solutions

$\lambda x_1 - x_2 - x_3 = 0,\ -x_1 + \lambda x_2 - x_3 = 0,\ -x_1 - x_2 + \lambda x_3 = 0,$ **Ans.** $\lambda = -1$

11. Find the values of $k$ for which the following system of equations has a non-trivial solution.

$(3k - 8)x + 3y + 3z = 0,\ 3x + (3k - 8)y + 3z = 0,\ 3x + 3y + (3k - 8)z = 0$ *(AMIETE, June 2010)*

**Ans.** $k = \frac{2}{3}, \frac{11}{3}$

12. Solve the homogeneous system of equations :

$4x + 3y - z = 0,\ 3x + 4y + z = 0,\ x - y - 2z = 0,\ 5x + y - 4z = 0$ **Ans.** $x = k,\ y = -k,\ z = k$

13. If $A = \begin{bmatrix} -1 & 2 & 1 \\ 3 & -1 & 2 \\ 0 & 1 & \lambda \end{bmatrix}$ **Ans.** (*i*) $\lambda \neq 1$, (*ii*) $\lambda - 1$

find the values of $\lambda$ for which equation $AX = 0$ has (*i*) a unique solution, (*ii*) more than one solution.

14. Show that the following system of equations:

$x + 2y - 2u = 0,\ 2x - y - u = 0,\ x + 2z - u = 0,\ 4x - y + 3z - u = 0$

do not have a non-trivial solution.

15. Determine the values of $\lambda$ and $\mu$ such that the following system has (*i*) no solution (*ii*) a unique solution (*iii*) infinite number of solutions:

$2x - 5y + 2z = 8,\quad 2x + 4y + 6z = 5,\quad x + 2y + \lambda z = \mu$

**Ans.** (*i*) $\lambda = 3,\ \mu \neq \frac{5}{2}$ (*ii*) $\lambda \neq 3$, (*iii*) $\lambda = 3,\ \mu = \frac{5}{2}$

**16.** Test the following system of equations for consistency. If possible, solve for non-trivial solutions.

$3x + 4y - z - 6t = 0,\ 2x + 3y + 2z - 3t = 0,\ 2x + y - 14z - 9t = 0,\ x + 3y + 13z + 3t = 0$

(*A.M.I.E.T.E., Winter 2000*) **Ans.** $x = 11k_1 + 6k_2,\ y = -8k_1 - 3k_2,\ z = k_1\ \ t = k_2$

**17.** Given the following system of equations

$2x - 2y + 5z + 3w = 0,\ 4x - y + z + w = 0,\ 3x - 2y + 3z + 4w = 0,\ x - 3y + 7z + 6w = 0$

Reduce the coefficient matrix A into Echelon form and find the rank utilising the property of rank, test the given system of equation for consistency and if possible find the solution of the given system.

(*A.M.I.E.T.E., Summer 2001*) **Ans.** $x = 5k,\ y = 36k,\ z = 7k,\ w = 9k$

**18.** Find the values of $\lambda$ for which the equations

$(2 - \lambda)\, x + 2y + 3 = 0,\quad 2x + (4 - \lambda)\, y + 7 = 0,\quad 2x + 5y + (6 - \lambda) = 0$

are consistent and find the values of $x$ and $y$ corresponding to each of these values of $\lambda$.

(*R.G.P.V, Bhopal I sem. 2003, 2001*) **Ans.** $\lambda = 1, -1, 12.$

## 4.45 LINEAR DEPENDENCE AND INDEPENDENCE OF VECTORS

Vectors (matrices) $X_1, X_2, \ldots, X_n$ are said to be dependent if

(1) all the vectors (row or column matrices) are of the same order.

(2) $n$ scalars $\lambda_1, \lambda_2, \ldots \lambda_n$ (not all zero) exist such that

$$\lambda_1 X_1 + \lambda_2 X_2 + \lambda_3 X_3 + \ldots + \lambda_n X_n = 0$$

Otherwise they are linearly independent.

**Remember:** If in a set of vectors, any vector of the set is the combination of the remaining vectors, then the vectors are called dependent vectors.

**Example 46.** *Examine the following vectors for linear dependence and find the relation if it exists.*

$X_1 = (1, 2, 4),\ X_2 = (2, -1, 3),\ X_3 = (0, 1, 2),\ X_4 = (-3, 7, 2)$ (*U.P., I Sem. Winter* 2002)

**Solution.** Consider the matrix equation

$$\lambda_1 X_1 + \lambda_2 X_2 + \lambda_3 X_3 + \lambda_4 X_4 = 0$$

$$\Rightarrow \lambda_1 (1, 2, 4) + \lambda_2 (2, -1, 3) + \lambda_3 (0, 1, 2) + \lambda_4 (-3, 7, 2) = 0$$

$$\lambda_1 + 2\lambda_2 + 0\lambda_3 - 3\lambda_4 = 0$$

$$2\lambda_1 - \lambda_2 + \lambda_3 + 7\lambda_4 = 0$$

$$4\lambda_1 + 3\lambda_2 + 2\lambda_3 + 2\lambda_4 = 0$$

This is the homogeneous system

$$\begin{bmatrix} 1 & 2 & 0 & -3 \\ 2 & -1 & 1 & 7 \\ 4 & 3 & 2 & 2 \end{bmatrix} \begin{bmatrix} \lambda_1 \\ \lambda_2 \\ \lambda_3 \\ \lambda_4 \end{bmatrix} = \begin{bmatrix} 0 \\ 0 \\ 0 \end{bmatrix} \quad \text{or } A\,\lambda = 0$$

$$\begin{bmatrix} 1 & 2 & 0 & -3 \\ 0 & -5 & 1 & 13 \\ 0 & -5 & 2 & 14 \end{bmatrix} \begin{bmatrix} \lambda_1 \\ \lambda_2 \\ \lambda_3 \\ \lambda_4 \end{bmatrix} = \begin{bmatrix} 0 \\ 0 \\ 0 \end{bmatrix} \qquad \begin{matrix} R_2 \to R_2 - 2\,R_1 \\ R_3 \to R_3 - 4\,R_1 \end{matrix}$$

$$\begin{bmatrix} 1 & 2 & 0 & -3 \\ 0 & -5 & 1 & 13 \\ 0 & 0 & 1 & 1 \end{bmatrix} \begin{bmatrix} \lambda_1 \\ \lambda_2 \\ \lambda_3 \\ \lambda_4 \end{bmatrix} = \begin{bmatrix} 0 \\ 0 \\ 0 \end{bmatrix} \qquad R_3 \to R_3 - R_2$$

$$\lambda_1 + 2\lambda_2 - 3\lambda_4 = 0$$
$$-5\lambda_2 + \lambda_3 + 13\lambda_4 = 0$$
$$\lambda_3 + \lambda_4 = 0$$

Let $\lambda_4 = t,\ \lambda_3 + t = 0,\ \lambda_3 = -t$

$$-5\lambda_2 - t + 13t = 0,\ \lambda_2 = \frac{12t}{5}$$

$$\lambda_1 + \frac{24t}{5} - 3t = 0 \text{ or } \lambda_1 = \frac{-9t}{5}$$

Hence, the given vectors are linearly dependent.

Substituting the values of $\lambda$ in (1), we get

$$-\frac{9tX_1}{5} + \frac{12t}{5}X_2 - tX_3 + tX_4 = 0 \Rightarrow -\frac{9X_1}{5} + \frac{12X_2}{5} - X_3 + X_4 = 0$$

$$\Rightarrow \quad 9X_1 - 12X_2 + 5X_3 - 5X_4 = 0$$ **Ans.**

**Example 47.** *Define linear dependence and independence of vectors.*

*Examine for linear dependence [1, 0, 2, 1], [3, 1, 2, 1], [4, 6, 2, –4], [–6, 0, –3, –4] and find the relation between them, if possible.*

**Solution.** Consider the matrix equation

$$\lambda_1 X_1 + \lambda_2 X_2 + \lambda_3 X_3 + \lambda_4 X_4 = 0 \qquad ...(1)$$

$$\lambda_1(1, 0, 2, 1) + \lambda_2(3, 1, 2, 1) + \lambda_3(4, 6, 2, -4) + \lambda_4(-6, 0, -3, -4) = 0$$

$$\lambda_1 + 3\lambda_2 + 4\lambda_3 - 6\lambda_4 = 0$$
$$0\lambda_1 + \lambda_2 + 6\lambda_3 + 0\lambda_4 = 0$$
$$2\lambda_1 + 2\lambda_2 + 2\lambda_3 - 3\lambda_4 = 0$$
$$\lambda_1 + \lambda_2 - 4\lambda_3 - 4\lambda_4 = 0$$

$$\begin{bmatrix} 1 & 3 & 4 & -6 \\ 0 & 1 & 6 & 0 \\ 2 & 2 & 2 & -3 \\ 1 & 1 & -4 & -4 \end{bmatrix} \begin{bmatrix} \lambda_1 \\ \lambda_2 \\ \lambda_3 \\ \lambda_4 \end{bmatrix} = \begin{bmatrix} 0 \\ 0 \\ 0 \\ 0 \end{bmatrix}$$

$$\begin{bmatrix} 1 & 3 & 4 & -6 \\ 0 & 1 & 6 & 0 \\ 0 & -4 & -6 & 9 \\ 0 & -2 & -8 & 2 \end{bmatrix} \begin{bmatrix} \lambda_1 \\ \lambda_2 \\ \lambda_3 \\ \lambda_4 \end{bmatrix} = \begin{bmatrix} 0 \\ 0 \\ 0 \\ 0 \end{bmatrix} \quad \begin{matrix} R_3 \to R_3 - 2R_1 \\ R_4 \to R_4 - R_1 \end{matrix}$$

$$\begin{bmatrix} 1 & 3 & 4 & -6 \\ 0 & 1 & 6 & 0 \\ 0 & 0 & 18 & 9 \\ 0 & 0 & 4 & 2 \end{bmatrix} \begin{bmatrix} \lambda_1 \\ \lambda_2 \\ \lambda_3 \\ \lambda_4 \end{bmatrix} = \begin{bmatrix} 0 \\ 0 \\ 0 \\ 0 \end{bmatrix} \quad \begin{matrix} R_3 \to R_3 + 4R_2 \\ R_4 \to R_4 + 2R_2 \end{matrix}$$

$$\begin{bmatrix} 1 & 3 & 4 & -6 \\ 0 & 1 & 6 & 0 \\ 0 & 0 & 18 & 9 \\ 0 & 0 & 0 & 0 \end{bmatrix} \begin{bmatrix} \lambda_1 \\ \lambda_2 \\ \lambda_3 \\ \lambda_4 \end{bmatrix} = \begin{bmatrix} 0 \\ 0 \\ 0 \\ 0 \end{bmatrix} \quad R_4 \to R_4 - \frac{2}{9}R_3$$

$$\lambda_1 + 3\lambda_2 + 4\lambda_3 - 6\lambda_4 = 0$$
$$\lambda_2 + 6\lambda_3 = 0$$
$$18\lambda_3 + 9\lambda_4 = 0$$

Let $\lambda_4 = t$, $\quad 18\lambda_3 + 9t = 0$ or $\lambda_3 = \frac{-t}{2}$

$$\lambda_2 - 3t = 0 \text{ or } \lambda_2 = 3t$$
$$\lambda_1 + 9t - 2t - 6t = 0$$
$$\lambda_1 = -t$$

Substituting the values of $\lambda_1$, $\lambda_2$, $\lambda_3$ and $\lambda_4$ in (1), we get

$$-tX_1 + 3tX_2 - \frac{t}{2}X_3 + tX_4 = 0 \text{ or } 2X_1 - 6X_2 + X_3 - 2X_4 = 0$$ **Ans.**

## 4.46 LINEARLY DEPENDENCE AND INDEPENDENCE OF VECTORS BY RANK METHOD

1. If the rank of the matrix of the given vectors is equal to number of vectors, then the vectors are linearly independent.
2. If the rank of the matrix of the given vectors is less than the number of vectors, then the vectors are linearly dependent.

**Example 48.** *Is the system of vector*

$$X_1 = (2,2,1)^T, X_2 = (1,3,1)^T, X_3 = (1,2,2)^T$$

*linearly dependent.*

**Solution.** Here $X_1 = \begin{bmatrix} 2 \\ 2 \\ 1 \end{bmatrix}, X_2 = \begin{bmatrix} 1 \\ 3 \\ 1 \end{bmatrix} X_3 = \begin{bmatrix} 1 \\ 2 \\ 2 \end{bmatrix}$ ($T$ stands for transposition)

Consider the matrix equation

$$\lambda_1 X_1 + \lambda_2 X_2 + \lambda_3 X_3 = 0 \qquad ...(1)$$

$$\lambda_1 \begin{bmatrix} 2 \\ 2 \\ 1 \end{bmatrix} + \lambda_2 \begin{bmatrix} 1 \\ 3 \\ 1 \end{bmatrix} + \lambda_3 \begin{bmatrix} 1 \\ 2 \\ 2 \end{bmatrix} = \begin{bmatrix} 0 \\ 0 \\ 0 \end{bmatrix}$$

$$2\lambda_1 + \lambda_2 + \lambda_3 = 0$$
$$2\lambda_1 + 3\lambda_2 + 2\lambda_3 = 0$$
$$\lambda_1 + \lambda_2 + 2\lambda_3 = 0$$

which is the homogeneous equation.

$$R_1 \leftrightarrow R_3$$

$$\begin{bmatrix} 2 & 1 & 1 \\ 2 & 3 & 2 \\ 1 & 1 & 2 \end{bmatrix} \begin{bmatrix} \lambda_1 \\ \lambda_2 \\ \lambda_3 \end{bmatrix} = \begin{bmatrix} 0 \\ 0 \\ 0 \end{bmatrix} \text{ or } \begin{bmatrix} 1 & 1 & 2 \\ 2 & 3 & 2 \\ 2 & 1 & 1 \end{bmatrix} \begin{bmatrix} \lambda_1 \\ \lambda_2 \\ \lambda_3 \end{bmatrix} = \begin{bmatrix} 0 \\ 0 \\ 0 \end{bmatrix}$$

$$R_2 - 2R_1, R_3 - 2R_1 \qquad R_3 + R_2$$

$$\begin{bmatrix} 1 & 1 & 2 \\ 0 & 1 & -2 \\ 0 & -1 & -3 \end{bmatrix} \begin{bmatrix} \lambda_1 \\ \lambda_2 \\ \lambda_3 \end{bmatrix} = \begin{bmatrix} 0 \\ 0 \\ 0 \end{bmatrix} \text{ or } \begin{bmatrix} 1 & 1 & 2 \\ 0 & 1 & -2 \\ 0 & 0 & -5 \end{bmatrix} \begin{bmatrix} \lambda_1 \\ \lambda_2 \\ \lambda_3 \end{bmatrix} = \begin{bmatrix} 0 \\ 0 \\ 0 \end{bmatrix}$$

$$\lambda_1 + \lambda_2 + 2\lambda_3 = 0$$
$$\lambda_2 - 2\lambda_3 = 0$$
$$-5\lambda_3 = 0 \Rightarrow \lambda_3 = 0$$
$$\therefore \quad \lambda_2 = 0 \text{ and } \lambda_1 = 0$$

Thus non-zero values of $\lambda_1, \lambda_2, \lambda_3$ do not exist which can satisfy (1). Hence by definition, the given system of vectors is not linearly dependent. **Ans.**

**Example 49.** *Show using a matrix that the set of vectors*

*$X = [1, 2, -3, 4]$, $Y = [3, -1, 2, 1]$, $Z = [1, -5, 8, -7]$ is linearly dependent.*

**Solution.** Here, we have

$X = [1, 2, -3, 4]$, $\quad Y = [3, -1, 2, 1]$, $\quad Z = [1, -5, 8, -7]$

Let us form a matrix of the above vectors

$$\begin{bmatrix} 1 & 2 & -3 & 4 \\ 3 & -1 & 2 & 1 \\ 1 & -5 & 8 & -7 \end{bmatrix} \sim \begin{bmatrix} 1 & 2 & -3 & 4 \\ 0 & -7 & 11 & -11 \\ 0 & -7 & 11 & -11 \end{bmatrix} \begin{matrix} \\ R_2 \to R_2 - 3R_1 \\ R_3 \to R_3 - R_1 \end{matrix} \sim \begin{bmatrix} 1 & 2 & -3 & 4 \\ 0 & -7 & 11 & -11 \\ 0 & 0 & 0 & 0 \end{bmatrix} R_3 \to R_3 - R_2$$

Here the rank of the matrix = 2 < Number of vectors

Hence, vectors are linearly dependent. **Proved.**

**Example 50.** *Show using a matrix that the set of vectors : [2, 5, 2, –3], [3, 6, 5, 2], [4, 5, 14, 14], [5, 10, 8, 4] is linearly independent.*

**Solution.** Here, the given vectors are

[2, 5, 2, –3], [3, 6, 5, 2], [4, 5, 14, 14], [5, 10, 8, 4]

Let us form a matrix of the above vectors :

$$\begin{bmatrix} 2 & 5 & 2 & -3 \\ 3 & 6 & 5 & 2 \\ 4 & 5 & 14 & 14 \\ 5 & 10 & 8 & 4 \end{bmatrix} \sim \begin{bmatrix} 2 & 5 & 2 & -3 \\ 1 & 1 & 3 & 5 \\ 1 & -1 & 9 & 12 \\ 1 & 5 & -6 & -10 \end{bmatrix} \begin{matrix} \\ R_2 \to R_2 - R_1 \\ R_3 \to R_3 - R_2 \\ R_4 \to R_4 - R_3 \end{matrix}$$

$$\sim \begin{bmatrix} 1 & 1 & 3 & 5 \\ 2 & 5 & 2 & -3 \\ 1 & -1 & 9 & 12 \\ 1 & 5 & -6 & -10 \end{bmatrix} \begin{matrix} R_1 \leftrightarrow R_2 \\ R_2 \leftrightarrow R_1 \\ \\ \\ \end{matrix} \sim \begin{bmatrix} 1 & 1 & 3 & 5 \\ 0 & 3 & -4 & -13 \\ 0 & -2 & 6 & 7 \\ 0 & 4 & -9 & -15 \end{bmatrix} \begin{matrix} \\ R_2 \to R_2 - 2R_1 \\ R_3 \to R_3 - R_1 \\ R_4 \to R_4 - R_1 \end{matrix}$$

$$\sim \begin{bmatrix} 1 & 1 & 3 & 5 \\ 0 & 3 & -4 & -13 \\ 0 & 0 & \frac{10}{3} & \frac{-5}{3} \\ 0 & 0 & \frac{-11}{3} & \frac{7}{3} \end{bmatrix} \begin{matrix} \\ \\ R_3 \to R_3 + \frac{2}{3}R_2 \\ R_4 \to R_4 - \frac{4}{3}R_2 \end{matrix} \sim \begin{bmatrix} 1 & 1 & 3 & 5 \\ 0 & 3 & -4 & -13 \\ 0 & 0 & \frac{10}{3} & \frac{-5}{3} \\ 0 & 0 & 0 & \frac{1}{2} \end{bmatrix} R_4 \to R_4 + \frac{11}{10}R_3$$

Here, the rank of the matrix = 4 = Number of vectors

Hence, the vectors are linearly independent. **Proved.**

## EXERCISE 4.17

Examine the following system of vectors for linear dependence. If dependent, find the relation between them.

**1.** $X_1 = (1, -1, 1)$, $X_2 = (2, 1, 1)$, $X_3 = (3, 0, 2)$. **Ans.** Dependent, $X_1 + X_2 - X_3 = 0$

**2.** $X_1 = (1, 2, 3)$, $X_2 = (2, -2, 6)$. **Ans.** Independent

3. $X_1 = (3, 1, -4), X_2 = (2, 2, -3), X_3 = (0, -4, 1)$. **Ans.** Dependent, $2X_1 - 3X_2 - X_3 = 0$

4. $X_1 = (1, 1, 1, 3), X_2 = (1, 2, 3, 4), X_3 = (2, 3, 4, 7)$. **Ans.** Dependent, $X_1 + X_2 - X_3 = 0$

5. $X_1 = (1, 1, -1, 1), X_2 = (1, -1, 2, -1), X_3 = (3, 1, 0, 1)$. **Ans.** Dependent, $2X_1 + X_2 - X_3 = 0$

6. $X_1 = (1, -1, 2, 0), X_2 = (2, 1, 1, 1), X_3 = (3, -1, 2, -1), X_4 = (3, 0, 3, 1)$.
**Ans.** Dependent, $X_1 + X_2 - X_4 = 0$

7. Show that the column vectors of following matrix A are linearly independent:

$$A = \begin{bmatrix} 1 & 0 & 0 \\ 6 & 2 & 1 \\ 4 & 3 & 2 \end{bmatrix}$$

8. Show that the vectors $x_1 = (2, 3, 1, -1), x_2 = (2, 3, 1, -2), x_3 = (4, 6, 2, 1)$ are linearly dependent. Express one of the vectors as linear combination of the others.

9. Find whether or not the following set of vectors are linearly dependent or independent:

(*i*) (1, –2), (2, 1), (3, 2) (*ii*) (1, 1, 1, 1), (0, 1, 1, 1), (0, 0, 1, 1), (0, 0, 0, 1).

**Ans.** (*i*) Dependent (*ii*) Independent

10. Show that the vectors $x_1 = (a_1, b_1), x_2 = (a_2, b_2)$ are linearly dependent if $a_1 b_2 - a_2 b_1 = 0$.

## 4.47 ANOTHER METHOD (ADJOINT METHOD) TO SOLVE LINEAR EQUATIONS

Let the equations be

$$a_1 x + a_2 y + a_3 z = d_1$$
$$b_1 x + b_2 y + b_3 z = d_2$$
$$c_1 x + c_2 y + c_3 z = d_3$$

We write the above equations in the matrix form

$$\begin{bmatrix} a_1 x + a_2 y + a_3 z \\ b_1 x + b_2 y + b_3 z \\ c_1 x + c_2 y + c_3 z \end{bmatrix} = \begin{bmatrix} d_1 \\ d_2 \\ d_3 \end{bmatrix} \text{ or } \begin{bmatrix} a_1 & a_2 & a_3 \\ b_1 & b_2 & b_3 \\ c_1 & c_2 & c_3 \end{bmatrix} \begin{bmatrix} x \\ y \\ z \end{bmatrix} = \begin{bmatrix} d_1 \\ d_2 \\ d_3 \end{bmatrix}$$

$$AX = B \qquad \ldots(1)$$

where $A = \begin{bmatrix} a_1 & a_2 & a_3 \\ b_1 & b_2 & b_3 \\ c_1 & c_2 & c_3 \end{bmatrix}, X = \begin{bmatrix} x \\ y \\ z \end{bmatrix}$ and $B = \begin{bmatrix} d_1 \\ d_2 \\ d_3 \end{bmatrix}$

Multiplying (1) by $A^{-1}$.

$A^{-1} AX = A^{-1} B$ or $IX = A^{-1} B$ or $X = A^{-1} B$.

**Example 51.** *Solve, with the help of matrices, the simultaneous equations*

$x + y + z = 3, \; x + 2y + 3z = 4, \; x + 4y + 9z = 6$ (*A.M.I.E., Summer* 2004, 2003)

**Solution.** The given equations in the matrix form are written as below:

$$\begin{bmatrix} 1 & 1 & 1 \\ 1 & 2 & 3 \\ 1 & 4 & 9 \end{bmatrix} \begin{bmatrix} x \\ y \\ z \end{bmatrix} = \begin{bmatrix} 3 \\ 4 \\ 6 \end{bmatrix}$$

$$AX = B$$

where $A = \begin{bmatrix} 1 & 1 & 1 \\ 1 & 2 & 3 \\ 1 & 4 & 9 \end{bmatrix}, X = \begin{bmatrix} x \\ y \\ z \end{bmatrix}, B = \begin{bmatrix} 3 \\ 4 \\ 6 \end{bmatrix}$

Now we have to find out the $A^{-1}$.

$$|A| = 1 \times 6 + 1 \times (-6) + 1 \times 2 = 6 - 6 + 2 = 2$$

$$\text{Matrix of co-factors} = \begin{bmatrix} 6 & -6 & 2 \\ -5 & 8 & -3 \\ 1 & -2 & 1 \end{bmatrix}, \; Adjoint\, A = \begin{bmatrix} 6 & -5 & 1 \\ -6 & 8 & -2 \\ 2 & -3 & 1 \end{bmatrix}$$

$$A^{-1} = \frac{1}{|A|}\, Adjoint\, A = \frac{1}{2}\begin{bmatrix} 6 & -5 & 1 \\ -6 & 8 & -2 \\ 2 & -3 & 1 \end{bmatrix}$$

$$X = A^{-1}B = \frac{1}{2}\begin{bmatrix} 6 & -5 & 1 \\ -6 & 8 & -2 \\ 2 & -3 & 1 \end{bmatrix}\begin{bmatrix} 3 \\ 4 \\ 6 \end{bmatrix}$$

$$\Rightarrow \begin{bmatrix} x \\ y \\ z \end{bmatrix} = \frac{1}{2}\begin{bmatrix} 18-20+6 \\ -18+32-12 \\ 6-12+6 \end{bmatrix} = \frac{1}{2}\begin{bmatrix} 4 \\ 2 \\ 0 \end{bmatrix} = \begin{bmatrix} 2 \\ 1 \\ 0 \end{bmatrix}$$

$x = 2,\ y = 1,\ z = 0$ **Ans.**

**Example 52.** *Given the matrices*

$$A \equiv \begin{bmatrix} 1 & 2 & 3 \\ 3 & -1 & 1 \\ 4 & 2 & 1 \end{bmatrix}, X \equiv \begin{bmatrix} x \\ y \\ z \end{bmatrix} \text{ and } C \equiv \begin{bmatrix} 1 \\ 2 \\ 3 \end{bmatrix}$$

*Write down the linear equations given by AX = C and solve for x, y, z by the matrix method.*

**Solution.** $AX = C$

$$\begin{bmatrix} 1 & 2 & 3 \\ 3 & -1 & 1 \\ 4 & 2 & 1 \end{bmatrix}\begin{bmatrix} x \\ y \\ z \end{bmatrix} = \begin{bmatrix} 1 \\ 2 \\ 3 \end{bmatrix}$$

$$X = A^{-1}.C$$

$$\begin{bmatrix} x \\ y \\ z \end{bmatrix} = \begin{bmatrix} 1 & 2 & 3 \\ 3 & -1 & 1 \\ 4 & 2 & 1 \end{bmatrix}^{-1}\begin{bmatrix} 1 \\ 2 \\ 3 \end{bmatrix}$$

$$\text{Matrix of co-factors of } A = \begin{bmatrix} -3 & 1 & 10 \\ 4 & -11 & 6 \\ 5 & 8 & -7 \end{bmatrix}$$

$$|A| = 1(-3) + 2(1) + 3(10) = -3 + 2 + 30 = 29$$

$$\text{Adj. } A = \begin{bmatrix} -3 & 4 & 5 \\ 1 & -11 & 8 \\ 10 & 6 & -7 \end{bmatrix}$$

$$A^{-1} = \frac{1}{|A|}\, Adj.\, A = \frac{1}{29}\begin{bmatrix} -3 & 4 & 5 \\ 1 & -11 & 8 \\ 10 & 6 & -7 \end{bmatrix}$$

$$X = A^{-1}C$$

$$\begin{bmatrix} x \\ y \\ z \end{bmatrix} = \frac{1}{29}\begin{bmatrix} -3 & 4 & 5 \\ 1 & -11 & 8 \\ 10 & 6 & -7 \end{bmatrix}\begin{bmatrix} 1 \\ 2 \\ 3 \end{bmatrix}$$

$$\begin{bmatrix} x \\ y \\ z \end{bmatrix} = \frac{1}{29}\begin{bmatrix} -3 & +8 & +15 \\ 1 & -22 & +24 \\ 10 & +12 & -21 \end{bmatrix} = \frac{1}{29}\begin{bmatrix} 20 \\ 3 \\ 1 \end{bmatrix} = \begin{bmatrix} \frac{20}{29} \\ \frac{3}{29} \\ \frac{1}{29} \end{bmatrix}$$

Hence, $x = \frac{20}{29}, y = \frac{3}{29}, z = \frac{1}{29}$ **Ans.**

**Example 53.** Let $y_1 = 5x_1 + 3x_2 + 3x_3,\quad y_2 = 3x_1 + 2x_2 - 2x_3,\quad y_3 = 2x_1 - x_2 + 2x_3$ be a linear transformation from $(x_1, x_2, x_3)$ to $(y_1, y_2, y_3)$ and

$$z_1 = 4x_1 + 2x_3,\quad z_2 = x_2 + 4x_3,\quad z_3 = 5x_3$$

be a linear transformation from $(x_1, x_2, x_3)$ to $(z_1, z_2, z_3)$. Find the linear transformation from $(z_1, z_2, z_3)$ to $(y_1, y_2, y_3)$ by inverting appropriate matrix and matrix multiplication.

(*A.M.I.E.T.E., Dec. 2004*)

**Solution:** Hence $\begin{bmatrix} y_1 \\ y_2 \\ y_3 \end{bmatrix} = \begin{bmatrix} 5 & 3 & 3 \\ 3 & 2 & -2 \\ 2 & -1 & 2 \end{bmatrix}\begin{bmatrix} x_1 \\ x_2 \\ x_3 \end{bmatrix}$ ...(1)

and $\begin{bmatrix} z_1 \\ z_2 \\ z_3 \end{bmatrix} = \begin{bmatrix} 4 & 0 & 2 \\ 0 & 1 & 4 \\ 0 & 0 & 5 \end{bmatrix}\begin{bmatrix} x_1 \\ x_2 \\ x_3 \end{bmatrix}$

$$\Rightarrow \begin{bmatrix} x_1 \\ x_2 \\ x_3 \end{bmatrix} = \begin{bmatrix} 4 & 0 & 2 \\ 0 & 1 & 4 \\ 0 & 0 & 5 \end{bmatrix}^{-1}\begin{bmatrix} z_1 \\ z_2 \\ z_3 \end{bmatrix} = \frac{1}{20}\begin{bmatrix} 5 & 0 & -2 \\ 0 & 20 & -16 \\ 0 & 0 & 4 \end{bmatrix}\begin{bmatrix} z_1 \\ z_2 \\ z_3 \end{bmatrix} \quad ...(2)$$

Putting the value of $\begin{bmatrix} x_1 \\ x_2 \\ x_3 \end{bmatrix}$ from (2) in (1), we get

$$\begin{bmatrix} y_1 \\ y_2 \\ y_3 \end{bmatrix} = \begin{bmatrix} 5 & 3 & 3 \\ 3 & 2 & -2 \\ 2 & -1 & 2 \end{bmatrix} \times \frac{1}{20}\begin{bmatrix} 5 & 0 & -2 \\ 0 & 20 & -16 \\ 0 & 0 & 4 \end{bmatrix}\begin{bmatrix} z_1 \\ z_2 \\ z_3 \end{bmatrix} = \frac{1}{20}\begin{bmatrix} 25 & 60 & -46 \\ 15 & 40 & -46 \\ 10 & -20 & 20 \end{bmatrix}\begin{bmatrix} z_1 \\ z_2 \\ z_3 \end{bmatrix}$$ **Ans.**

## EXERCISE 4.18

**Solve the following equations**

1. $3x + y + 2z = 3,\ 2x - 3y - z = -3,\ x + 2y + z = 4$ (*A.M.I.E. Winter* 2001)
   **Ans.** $x = 1,\ y = 2,\ z = -1$

2. $x + 2y + 3z = 1,\ 2x + 3y + 8z = 2,\ x + y + z = 3$
   **Ans.** $x = \frac{9}{2}, y = -1, z = -\frac{1}{2}$

3. $4x + 2y - z = 9,\ x - y + 3z = -4,\ 2x + z = 1$ **Ans.** $x = 1, y = 2, z = -1$

4. $5x + 3y + 3z = 48,\ 2x + 6y - 3z = 18,\ 8x - 3y + 2z = 21$ **Ans.** $x = 3, y = 5, z = 6$

5. $x + y + z = 6,\ x - y + 2z = 5,\ 3x + y + z = 8$ **Ans.** $x = 1, y = 2, z = 3$

6. $x + 2y + 3z = 1,\ 3x - 2y + z = 2,\ 4x + 2y + z = 3$ **Ans.** $x = \frac{7}{10}, y = \frac{3}{40}, z = \frac{1}{20}$

7. $9x + 4y + 3z = -1,\ 5x + y + 2z = 1,\ 7x + 3y + 4z = 1$ **Ans.** $x = 0, y = -1, z = 1$

8. $x + y + z = 8,\ x - y + 2z = 6,\ 9x + 5y - 7z = 14$ **Ans.** $x = 5, y = \frac{5}{3}, z = \frac{4}{3}$

9. $3x + 2y + 4z = 7,\ 2x + y + z = 4,\ x + 3y + 5z = 2$ **Ans.** $x = \frac{9}{4}, y = -\frac{9}{8}, z = \frac{5}{8}$

10. Represent each of the transformations

$$x_1 = 3y_1 + 2y_2,\ x_2 = -y_1 + 4y_2 \quad \text{and} \quad y_1 = z_1 + 2z_2,\ y_2 = -3z_1$$

by the use of matrices, find the composite transformation which expresses $x_1$, $x_2$ in terms of $z_1$, $z_2$. **Ans.** $x_1 = -3z_1 + 6z_2,\ x_2 = -13z_1 - 2z_2$

## 4.48 PARTITIONING OF MATRICES

**Sub matrix.** A matrix obtained by deleting some of the rows and columns of a matrix $A$ is said to be sub matrix.

*For example,* $A = \begin{bmatrix} 4 & 1 & 0 \\ 5 & 2 & 1 \\ 6 & 3 & 4 \end{bmatrix}$, then $\begin{bmatrix} 4 & 1 \\ 5 & 2 \end{bmatrix}, \begin{bmatrix} 5 & 2 \\ 6 & 3 \end{bmatrix}, \begin{bmatrix} 1 & 0 \\ 2 & 1 \end{bmatrix}$ are the sub matrices.

**Partitioning:** A matrix may be subdivided into sub matrices by drawing lines parallel to its rows and columns. These sub matrices may be considered as the elements of the original matrix.

*For example,*
$$A = \begin{bmatrix} 2 & 1 & : & 0 & 4 & 1 \\ 1 & 0 & : & 2 & 3 & 4 \\ \cdots & \cdots & : & \cdots & \cdots & \cdots \\ 4 & 5 & : & 1 & 6 & 5 \end{bmatrix}$$

$$A_{11} = \begin{bmatrix} 2 & 1 \\ 1 & 0 \end{bmatrix}, \qquad A_{12} = \begin{bmatrix} 0 & 4 & 1 \\ 2 & 3 & 4 \end{bmatrix}$$

$$A_{21} = [4\ \ 5], \qquad A_{22} = [1\ \ 6\ \ 5]$$

Then we may write
$$A = \begin{bmatrix} A_{11} & A_{12} \\ A_{21} & A_{22} \end{bmatrix}$$

So, the matrix is partitioned. The dotted lines divide the matrix into sub-matrices. $A_{11}, A_{12}, A_{21}, A_{22}$ are the sub-matrices but behave like elements of the original matrix $A$. The matrix $A$ can be partitioned in several ways.

**Addition by submatrices:** Let $A$ and $B$ be two matrices of the same order and are partitioned identically.

*For example;*

$$A = \begin{bmatrix} 2 & 3 & 4 & : & 5 \\ 0 & 1 & 2 & : & 3 \\ \cdots & \cdots & \cdots & : & \cdots \\ 3 & 4 & 5 & : & 6 \\ \cdots & \cdots & \cdots & : & \cdots \\ 4 & 5 & 0 & : & 1 \end{bmatrix} \qquad B = \begin{bmatrix} 3 & 1 & 4 & : & 6 \\ 2 & 1 & 0 & : & 4 \\ \cdots & \cdots & \cdots & : & \cdots \\ 4 & 5 & 1 & : & 2 \\ \cdots & \cdots & \cdots & : & \cdots \\ 1 & 3 & 4 & : & 5 \end{bmatrix}$$

$$A = \begin{bmatrix} A_{11} & A_{12} \\ A_{21} & A_{22} \\ A_{31} & A_{32} \end{bmatrix}, \quad B = \begin{bmatrix} B_{11} & B_{12} \\ B_{21} & B_{22} \\ B_{31} & B_{32} \end{bmatrix}$$

$$A + B = \begin{bmatrix} A_{11} + B_{11} & A_{12} + B_{12} \\ A_{21} + B_{21} & A_{22} + B_{22} \\ A_{31} + B_{31} & A_{32} + B_{32} \end{bmatrix}$$

## 4.49 MULTIPLICATION BY SUB-MATRICES

Two matrices $A$ and $B$, which are conformable to the product $AB$ are partitioned in such a way that the columns of $A$ partitioned in the same way as the rows of $B$ are partitioned. But the rows of $A$ and columns of $B$ can be partitioned in any way..

*For example,* Here $A$ is a $3 \times 4$ matrix and $B$ is $4 \times 3$ matrix.

$$A = \begin{bmatrix} 1 & 2 & 3 & \vdots & 4 \\ 0 & 1 & 2 & \vdots & 3 \\ 1 & 4 & 1 & \vdots & 2 \end{bmatrix} \text{ and } B = \begin{bmatrix} 4 & 5 & 6 \\ 3 & 2 & 1 \\ 1 & 0 & 4 \\ \cdots & \cdots & \cdots \\ 2 & 5 & 3 \end{bmatrix}$$

The partitioning of the columns of $A$ is the same as the partitioning of the rows of $B$. Here, $A$ is partitioned after third column, $B$ has been partitioned after third row.

**Example 54.** *If C and D are two non-singular matrices, show that if*

$$A = \begin{bmatrix} C & 0 \\ 0 & D \end{bmatrix}, \quad \textit{then } A^{-1} = \begin{bmatrix} C^{-1} & 0 \\ 0 & D^{-1} \end{bmatrix}$$

**Solution.** Let $$A^{-1} = \begin{bmatrix} E & F \\ G & H \end{bmatrix} \qquad ...(1)$$

Then $$AA^{-1} = \begin{bmatrix} C & 0 \\ 0 & D \end{bmatrix} \begin{bmatrix} E & F \\ G & H \end{bmatrix} = \begin{bmatrix} CE + 0G & CF + 0H \\ 0E + DG & 0F + DH \end{bmatrix}$$

So that $$\begin{bmatrix} CE + 0G & CF + 0H \\ 0E + DG & 0F + DH \end{bmatrix} = \begin{bmatrix} I & 0 \\ 0 & I \end{bmatrix}$$

$$CE + 0G = I \Rightarrow CE = I$$
$$CF + 0H = 0 \Rightarrow CF = 0$$
$$0E + DG = 0 \Rightarrow DG = 0$$
$$OF + DH = I \Rightarrow DH = I$$

Since, C is non singular and $CF = 0$, $\therefore F = 0$

$$CE = I \Rightarrow E = C^{-1}$$

Similarly, $D$ is non singular and $DG = 0 \Rightarrow G = 0$ and $DH = I \Rightarrow H = D^{-1}$

Putting these values in (1), we get

$$A^{-1} = \begin{bmatrix} C^{-1} & 0 \\ 0 & D^{-1} \end{bmatrix}$$

**Proved.**

**4.50 Inverse By Partitioning:** Let the matrix $B$ be the inverse of the matrix $A$. Matrices $A$ and $B$ are partitioned as

$$A = \begin{bmatrix} A_{11} & A_{12} \\ A_{21} & A_{22} \end{bmatrix}, \quad B = \begin{bmatrix} B_{11} & B_{12} \\ B_{21} & B_{22} \end{bmatrix}$$

Since, $$AB = BA = I$$

$$\begin{bmatrix} A_{11} & A_{12} \\ A_{21} & A_{22} \end{bmatrix}\begin{bmatrix} B_{11} & B_{12} \\ B_{21} & B_{22} \end{bmatrix} = \begin{bmatrix} B_{11} & B_{12} \\ B_{21} & B_{22} \end{bmatrix}\begin{bmatrix} A_{11} & A_{12} \\ A_{21} & A_{22} \end{bmatrix} = \begin{bmatrix} I & 0 \\ 0 & I \end{bmatrix}$$

$$\begin{bmatrix} A_{11} B_{11} + A_{12} B_{21} & A_{11} B_{12} + A_{12} B_{22} \\ A_{21} B_{11} + A_{22} B_{21} & A_{21} B_{12} + A_{22} B_{22} \end{bmatrix} = \begin{bmatrix} B_{11} A_{11} + B_{12} A_{21} & B_{11} A_{12} + B_{12} A_{22} \\ B_{21} A_{11} + B_{22} A_{21} & B_{21} A_{12} + B_{22} A_{22} \end{bmatrix} = \begin{bmatrix} I & 0 \\ 0 & I \end{bmatrix}$$

Let us solve the equations for $B_{11}$, $B_{12}$, $B_{21}$ and $B_{22}$.

Let, $\quad B_{22} = M^{-1}$

From (2), $\quad B_{12} = -A_{11}^{-1}(A_{22} B_{22}) = -(A_{11}^{-1} A_{22}) M^{-1}$

From (3), $\quad B_{21} = -(B_{22} A_{21}) A_{11}^{-1} = -M^{-1}(A_{21} A_{11}^{-1})$

From (1), $\quad B_{11} = A_{11}^{-1} - A_{11}^{-1}(A_{12} B_{21}) = A_{11}^{-1} - (A_{11}^{-1} A_{12}) B_{21}$

$$= A_{11}^{-1} + (A_{11}^{-1} A_{12}) M^{-1} (A_{21} A_{11}^{-1})$$

Here $\quad M = A_{22} - A_{21}(A_{11}^{-1} A_{22})$

**Note:** A is usually taken of order $n - 1$.

**Example 55.** *Find the inverse of the following matrix by partitioning*

$$\begin{bmatrix} 1 & 3 & 3 \\ 1 & 4 & 3 \\ 1 & 3 & 4 \end{bmatrix}$$

**Solution.** Let the matrix be partitioned into four submatrices as follows:

Let $\quad A = \begin{bmatrix} 1 & 3 & 3 \\ 1 & 4 & 3 \\ 1 & 3 & 4 \end{bmatrix}$

$$A_{11} = \begin{bmatrix} 1 & 3 \\ 1 & 4 \end{bmatrix}; \quad A_{12} = \begin{bmatrix} 3 \\ 3 \end{bmatrix}$$

$$A_{21} = [1 \quad 3]; \quad A_{22} = [4]$$

We have to find $A^{-1} = \begin{bmatrix} B_{11} & B_{12} \\ B_{21} & B_{22} \end{bmatrix}$ where

$$B_{11} = A_{11}^{-1} + (A_{11}^{-1} A_{12})(M^{-1})(A_{21} A_{11}^{-1})$$

$$B_{21} = -M^{-1}(A_{21} A_{11}^{-1})$$

$$B_{12} = -A_{11}^{-1} A_{12} M^{-1}; \quad B_{22} = M^{-1}$$

and $\quad M = A_{22} - A_{21}(A_{11}^{-1} A_{12})$

Now

$$A_{11}^{-1} = \begin{bmatrix} 4 & -3 \\ -1 & 1 \end{bmatrix}; \quad A_{11}^{-1} A_{12} = \begin{bmatrix} 4 & -3 \\ -1 & 1 \end{bmatrix}\begin{bmatrix} 3 \\ 3 \end{bmatrix} = \begin{bmatrix} 3 \\ 0 \end{bmatrix}$$

$$A_{21} A_{11}^{-1} = [1 \quad 3]\begin{bmatrix} 4 & -3 \\ -1 & 1 \end{bmatrix} = [1 \quad 0]$$

$$M = [4]-[1\ -3]\begin{bmatrix}3\\0\end{bmatrix} = [4]-[3] = [1]$$

$$M^{-1} = [3]$$

$$\therefore \quad B_{11} = \begin{bmatrix}4 & -3\\-1 & 1\end{bmatrix}+\begin{bmatrix}3\\0\end{bmatrix}[1\ \ 0] = \begin{bmatrix}4 & -3\\-1 & 1\end{bmatrix}+\begin{bmatrix}3 & 0\\0 & 0\end{bmatrix} \Rightarrow B_{11} = \begin{bmatrix}7 & -3\\-1 & 1\end{bmatrix}$$

$$B_{21} = -[1][1\ \ 0] = -[1\ \ 0]$$

$$B_{12} = -\begin{bmatrix}3\\0\end{bmatrix}$$

$$B_{22} = [1]$$

$$A^{-1} = \begin{bmatrix}B_{11} & B_{12}\\B_{21} & B_{22}\end{bmatrix} = \begin{bmatrix}7 & -3 & -3\\-1 & 1 & 0\\-1 & 0 & 1\end{bmatrix}$$ **Ans.**

**Example 56.** *Find the inverse of* $A = \begin{bmatrix}1 & 2 & 3 & 1\\1 & 3 & 3 & 2\\2 & 4 & 3 & 3\\1 & 1 & 1 & 1\end{bmatrix}$ *by partitioning.*

**Solution.** (*a*) Take $G_3 = \begin{bmatrix}1 & 2 & 3\\1 & 3 & 3\\2 & 4 & 3\end{bmatrix}$ and partition so that

$$A_{11} = \begin{bmatrix}1 & 2\\1 & 3\end{bmatrix},\ A_{12} = \begin{bmatrix}3\\3\end{bmatrix},\ A_{21} = [2\ \ 4],\ \text{and } A_{22} = [3]$$

Now,
$$A_{11}^{-1} = \begin{bmatrix}3 & -2\\-1 & 1\end{bmatrix},\ A_{11}^{-1}A_{12} = \begin{bmatrix}3 & -2\\-1 & 1\end{bmatrix}\begin{bmatrix}3\\3\end{bmatrix} = \begin{bmatrix}3\\0\end{bmatrix},$$

$$A_{21}A_{11}^{-1} = [2\ \ 4]\begin{bmatrix}3 & -2\\-1 & 1\end{bmatrix} = [2\ \ 0]$$

$$M = A_{22}-A_{21}(A_{11}^{-1}A_{12}) = [3]-[2\ \ 4]\begin{bmatrix}3\\0\end{bmatrix} = [-3],\ \text{And } M^{-1} = [-1/3]$$

Then

$$B_{11} = A_{11}^{-1}+(A_{11}^{-1}A_{12})M^{-1}(A_{21}A_{11}^{-1}) = \begin{bmatrix}3 & -2\\-1 & 1\end{bmatrix}+\begin{bmatrix}3\\0\end{bmatrix}\left[-\frac{1}{3}\right][2\ \ 0] = \begin{bmatrix}3 & -2\\-1 & 1\end{bmatrix}-\begin{bmatrix}2 & 0\\0 & 0\end{bmatrix}$$

$$= \frac{1}{3}\begin{bmatrix}3 & -6\\-3 & 3\end{bmatrix}$$

$$B_{12} = -(A_{11}^{-1}A_{12})M^{-1} = \frac{1}{3}\begin{bmatrix}3\\0\end{bmatrix}$$

$$B_{21} = -M^{-1}(A_{21}A_{11}^{-1}) = \frac{1}{3}[2\ \ 0]$$

$$B_{22} = M^{-1} = \left[-\frac{1}{3}\right]$$

and
$$G_3^{-1} = \begin{bmatrix} B_{11} & B_{12} \\ B_{21} & B_{22} \end{bmatrix} = \frac{1}{3}\begin{bmatrix} 3 & -6 & 3 \\ -3 & 3 & 0 \\ 2 & 0 & -1 \end{bmatrix}$$

(*b*) Partition $A$ so that $A_{11} = \begin{bmatrix} 1 & 2 & 3 \\ 1 & 3 & 3 \\ 2 & 4 & 3 \end{bmatrix}$, $A_{12} = \begin{bmatrix} 1 \\ 2 \\ 3 \end{bmatrix}$, $A_{21} = [1 \;\; 1 \;\; 1]$, and $A_{22} = [1]$.

Now, $A_{11}^{-1} = \frac{1}{3}\begin{bmatrix} 3 & -6 & 3 \\ -3 & 3 & 0 \\ 2 & 0 & -1 \end{bmatrix}$, $A_{11}^{-1} A_{12} = \frac{1}{3}\begin{bmatrix} 0 \\ 3 \\ -1 \end{bmatrix}$, $A_{21}\, A_{11}^{-1} = \frac{1}{3}[2 \;\; -3 \;\; 2]$

$$M = [1] - [1 \;\; 1 \;\; 1]\left(\frac{1}{3}\right)\begin{bmatrix} 0 \\ 3 \\ -1 \end{bmatrix} = \left[\frac{1}{3}\right]. \text{ and } M^{-1} = [3]$$

Then
$$B_{11} = \frac{1}{3}\begin{bmatrix} 3 & -6 & 3 \\ -3 & 3 & 0 \\ 2 & 0 & -1 \end{bmatrix} + \frac{1}{3}\begin{bmatrix} 0 \\ 3 \\ -1 \end{bmatrix}[3]\frac{1}{3}[2 \;\; -3 \;\; 2]$$

$$= \frac{1}{3}\begin{bmatrix} 3 & -6 & 3 \\ -3 & 3 & 0 \\ 2 & 0 & -1 \end{bmatrix} + \frac{1}{3}\begin{bmatrix} 0 & 0 & 0 \\ 6 & -9 & 6 \\ -2 & 3 & -2 \end{bmatrix} = \begin{bmatrix} 1 & -2 & 1 \\ 1 & -2 & 2 \\ 0 & 1 & -1 \end{bmatrix}$$

$$B_{12} = \begin{bmatrix} 0 \\ -3 \\ 1 \end{bmatrix}, \quad B_{21} = [-2 \;\; 3 \;\; -2], \quad B_{22} = [3]$$

$$A^{-1} = \begin{bmatrix} B_{11} & B_{12} \\ B_{21} & B_{22} \end{bmatrix} = \begin{bmatrix} 1 & -2 & 1 & 0 \\ 1 & -2 & 2 & -3 \\ 0 & 1 & -1 & 1 \\ -2 & 3 & -2 & 3 \end{bmatrix}$$ **Ans.**

## EXERCISE 4.19

**1.** Compute A + B using partitioning

$$A = \begin{bmatrix} 4 & 1 & 0 & 5 \\ 6 & 7 & 8 & 1 \\ 0 & 2 & 1 & 1 \\ 1 & 2 & 0 & 1 \end{bmatrix}, \quad B = \begin{bmatrix} 3 & 2 & 1 & 1 \\ 1 & 0 & 1 & 1 \\ 2 & 1 & 2 & 1 \\ 0 & 1 & 2 & 3 \end{bmatrix}$$

**2.** Compute AB using partitioning

$$A = \begin{bmatrix} 1 & 2 & 0 & 1 \\ 4 & 1 & 3 & 2 \\ 2 & 1 & 3 & 0 \end{bmatrix}, \quad B = \begin{bmatrix} 2 & 3 & 1 \\ 0 & 1 & 4 \\ 4 & 1 & 2 \\ 2 & 1 & 2 \end{bmatrix}$$ **Ans.** $\begin{bmatrix} 4 & 6 & 11 \\ 24 & 18 & 18 \\ 16 & 10 & 12 \end{bmatrix}$

**3.** Find the inverse of $\begin{bmatrix} A & B \\ C & 0 \end{bmatrix}$

where $B$, $C$ are non-singular. **Ans.** $\begin{bmatrix} 0 & C^{-1} \\ B^{-1} & -B^{-1}AC^{-1} \end{bmatrix}$

**Find the inverse of the following metrices by partitioning:**

**4.** $\begin{bmatrix} 2 & 1 & -1 \\ 1 & 3 & 2 \\ -1 & 2 & 1 \end{bmatrix}$ **Ans.** $\frac{1}{10}\begin{bmatrix} 1 & 3 & -5 \\ 3 & -1 & 5 \\ -5 & 5 & -5 \end{bmatrix}$ **5.** $\begin{bmatrix} 1 & 2 & -1 \\ -1 & 1 & 2 \\ 2 & -1 & 1 \end{bmatrix}$ **Ans.** $\frac{1}{14}\begin{bmatrix} 3 & -1 & 5 \\ 5 & 3 & -1 \\ -1 & 5 & 3 \end{bmatrix}$

**6.** $\begin{bmatrix} 2 & 3 & 4 \\ 4 & 3 & 1 \\ 1 & 2 & 4 \end{bmatrix}$ **Ans.** $\frac{1}{5}\begin{bmatrix} -10 & 4 & 9 \\ 15 & -4 & -14 \\ -5 & 1 & 6 \end{bmatrix}$ **7.** $\begin{bmatrix} 1 & 2 & 3 \\ 2 & 4 & 5 \\ 3 & 5 & 6 \end{bmatrix}$ **Ans.** $\begin{bmatrix} 1 & -3 & 2 \\ -3 & 3 & -1 \\ 2 & -1 & 0 \end{bmatrix}$

**8.** $\begin{bmatrix} 3 & 4 & 2 & 7 \\ 2 & 3 & 3 & 2 \\ 52 & 7 & 3 & 9 \\ 2 & 3 & 2 & 3 \end{bmatrix}$ **Ans.** $\frac{1}{2}\begin{bmatrix} -1 & 11 & 7 & -26 \\ -1 & -7 & -3 & 16 \\ 1 & 1 & -1 & 0 \\ 1 & -1 & -1 & 2 \end{bmatrix}$

**Choose the correct answer:**

**9.** If $3x + 2y + z = 0$, $x + 4y + z = 0$, $2x + y + 4z = 0$, be a system of equations then

(*i*) System is inconsistent (*ii*) it has only trivial solution

(*iii*) it can be reduced to a single equation thus solution does not exist

(*iv*) Determinant of the coefficient matrix is zero. (*AMIETE, June 2010*) **Ans.** (*ii*)

## 4.51 EIGEN VALUES

Let $\begin{bmatrix} a_{11} & a_{12} & a_{13} & \cdots & a_{1n} \\ a_{21} & a_{22} & a_{23} & \cdots & a_{2n} \\ a_{31} & a_{32} & a_{33} & \cdots & a_{3n} \\ \cdots & \cdots & \cdots & \cdots & \cdots \\ a_{n1} & a_{n2} & a_{n3} & \cdots & a_{nn} \end{bmatrix}\begin{bmatrix} x_1 \\ x_2 \\ x_3 \\ \vdots \\ x_n \end{bmatrix} = \begin{bmatrix} y_1 \\ y_2 \\ y_3 \\ \vdots \\ y_n \end{bmatrix}$

$$AX = Y \qquad ...(1)$$

Where $A$ is the matrix , $X$ is the column vector and $Y$ is also column vector.

Here column vector $X$ is transformed into the column vector $Y$ by means of the square matrix $A$.

Let $X$ be a such vector which transforms into $\lambda X$ by means of the transformation (1). Suppose the linear transformation $Y = AX$ transforms $X$ into a scalar multiple of itself i.e. $\lambda X$.

$$AX = Y = \lambda X$$
$$AX - \lambda IX = 0$$
$$(A - \lambda I)X = 0 \qquad ...(2)$$

Thus the unknown scalar $\lambda$ is known as an eigen value of the matrix $A$ and the corresponding non zero vector $X$ as **eigen vector.**

The eigen values are also called characteristic values or proper values or latent values.

Let $A = \begin{bmatrix} 2 & 2 & 1 \\ 1 & 3 & 1 \\ 1 & 2 & 2 \end{bmatrix}$

$$A - \lambda I = \begin{bmatrix} 2 & 2 & 1 \\ 1 & 3 & 1 \\ 1 & 2 & 2 \end{bmatrix} - \lambda\begin{bmatrix} 1 & 0 & 0 \\ 0 & 1 & 0 \\ 0 & 0 & 1 \end{bmatrix} = \begin{bmatrix} 2-\lambda & 2 & 1 \\ 1 & 3-\lambda & 1 \\ 1 & 2 & 2-\lambda \end{bmatrix}$$ characteristic matrix

(*b*) **Characteristic Polynomial:** The determinant $| A - \lambda I |$ when expanded will give a polynomial, which we call as characteristic polynomial of matrix *A*.

*For example*; $\begin{vmatrix} 2-\lambda & 2 & 1 \\ 1 & 3-\lambda & 1 \\ 1 & 2 & 2-\lambda \end{vmatrix}$

$$= (2-\lambda)(6-5\lambda+\lambda^2-2) - 2(2-\lambda-1) + 1(2-3+\lambda)$$
$$= -\lambda^3 + 7\lambda^2 - 11\lambda + 5$$

(*c*) **Characteristic Equation:** The equation $| A - \lambda I | = 0$ is called the characteristic equation of the matrix *A* *e.g.*

$$\lambda^3 - 7\lambda^2 + 11\lambda - 5 = 0$$

(*d*) **Characteristic Roots or Eigen Values:** The roots of characteristic equation $| A - \lambda I | = 0$ are called characteristic roots of matrix *A*. *e.g.*

$$\lambda^3 - 7\lambda^2 + 11\lambda - 5 = 0$$

$\Rightarrow \quad (\lambda - 1)(\lambda - 1)(\lambda - 5) = 0 \qquad \therefore \lambda = 1, 1, 5$

Characteristic roots are 1, 1, 5.

**Some Important Properties of Eigen Values** *(AMIETE, Dec. 2009)*

(1) Any square matrix *A* and its transpose *A′* have the same eigen values.

**Note.** The sum of the elements on the principal diagonal of a matrix is called the **trace** of the matrix.

(2) The sum of the eigen values of a matrix is equal to the **trace** of the matrix.

(3) The product of the eigen values of a matrix *A* is equal to the **determinant** of *A*.

(4) If $\lambda_1, \lambda_2, \ldots \lambda_n$ are the eigen values of *A*, then the eigen values of

(*i*) $kA$ are $k\lambda_1, \; k\lambda_2, \; \ldots, \; k\lambda_n$ (*ii*) $A^m$ are $\lambda_1^m, \lambda_2^m, \ldots, \lambda_n^m$

(*iii*) $A^{-1}$ are $\frac{1}{\lambda_1}, \frac{1}{\lambda_2}, \ldots, \frac{1}{\lambda_n}$.

**Example 57.** *Find the characteristic roots of the matrix* $\begin{bmatrix} 6 & -2 & 2 \\ -2 & 3 & -1 \\ 2 & -1 & 3 \end{bmatrix}$

**Solution.** The characteristic equation of the given matrix is

$$\begin{vmatrix} 6-\lambda & -2 & 2 \\ -2 & 3-\lambda & -1 \\ 2 & -1 & 3-\lambda \end{vmatrix} = 0$$

$\Rightarrow \quad (6-\lambda)(9-6\lambda+\lambda^2-1) + 2(-6+2\lambda+2) + 2(2-6+2\lambda) = 0$

$\Rightarrow \quad -\lambda^3 + 12\lambda^2 - 36\lambda + 32 = 0$

By trial, $\lambda = 2$ is a root of this equation.

$\Rightarrow \quad (\lambda-2)(\lambda^2 - 10\lambda + 16) = 0 \Rightarrow (\lambda-2)(\lambda-2)(\lambda-8) = 0$

$\Rightarrow \quad \lambda = 2, 2, 8$ are the characteristic roots or Eigen values. **Ans.**

**Example 58.** *The matrix A is defined as* $A = \begin{bmatrix} 1 & 2 & -3 \\ 0 & 3 & 2 \\ 0 & 0 & -2 \end{bmatrix}$

*Find the eigen values of* $3A^3 + 5A^2 - 6A + 2I$.

**Solution.** $| A - \lambda I | = 0$

$$\begin{vmatrix} 1-\lambda & 2 & -3 \\ 0 & 3-\lambda & 2 \\ 0 & 0 & -2-\lambda \end{vmatrix} = 0$$

$\Rightarrow \quad (1 - \lambda)(3 - \lambda)(-2 - \lambda) = 0$ or $\lambda = 1, 3, -2$

Eigen values of $A^3 = 1, 27, -8$; Eigen values of $A^2 = 1, 9, 4$

Eigen values of $A = 1, 3, -2$; Eigen values of $I = 1, 1, 1$

$\therefore$ Eigen values of $3A^3 + 5A^2 - 6A + 2I$

First eigen value $= 3(1)^3 + 5(1)^2 - 6(1) + 2(1) = 4$

Second eigen value $= 3(27) + 5(9) - 6(3) + 2(1) = 110$

Third eigen value $= 3(-8) + 5(4) - 6(-2) + 2(1) = 10$

Required eigen values are 4, 110, 10 **Ans.**

**Example 59.** *If $\lambda_1, \lambda_2, \ldots \lambda_n$ are the eigen values of A, find the eigen values of the martrix $(A - \lambda I)^2$.*

**Solution.** $(A - \lambda I)^2 = A^2 - 2\lambda AI + \lambda^2 I^2 = A^2 - 2\lambda A + \lambda^2 I$

Eigen values of $A^2$ are $\lambda_1^2, \lambda_2^2, \lambda_3^2 \ldots \lambda_n^2$

Eigen values of $2\lambda A$ are $2\lambda\lambda_1, \quad 2\lambda\lambda_2, \quad 2\lambda\lambda_3 \ldots 2\lambda\lambda_n$.

Eigen values of $\lambda^2 I$ are $\lambda^2$.

$\therefore$ Eigen values of $A^2 - 2\lambda A + \lambda^2 I$

$$\lambda_1^2 - 2\lambda\lambda_1 + \lambda^2, \quad \lambda_2^2 - 2\lambda\lambda_2 + \lambda^2, \quad \lambda_3^2 - 2\lambda\lambda_3 + \lambda^2 \ldots\ldots$$

$$\Rightarrow \quad (\lambda_1 - \lambda)^2, \quad (\lambda_2 - \lambda)^2, \quad (\lambda_3 - \lambda)^2, \ldots (\lambda_n - \lambda)^2$$ **Ans.**

**Example 60.** *Prove that a matrix A and its transpose A′ have the same characteristic roots.*

**Solution.** Characteristic equation of matrix $A$ is

$$|A - \lambda I| = 0 \quad \ldots(1)$$

Characteristic equation of matrix $A'$ is

$$|A' - \lambda I| = 0 \quad \ldots(2)$$

Clearly both (1) and (2) are same, as we know that

$$|A| = |A'|$$

*i.e.*, a determinant remains unchanged when rows be changed into columns and columns into rows. **Proved.**

**Example 61.** *If A and P be square matrices of the same type and if P be invertible, show that the matrices A and $P^{-1}AP$ have the same characteristic roots.*

**Solution.** Let us put $B = P^{-1}AP$ and we will show that characteristic equations for both $A$ and $B$ are the same and hence they have the same characteristic roots.

$$B - \lambda I = P^{-1}AP - \lambda I = P^{-1}AP - P^{-1}\lambda IP = P^{-1}(A - \lambda I)P$$

$$\therefore \quad |B - \lambda I| = |P^{-1}(A - \lambda I)P| = |P^{-1}||A - \lambda I||P|$$

$$= |A - \lambda I||P^{-1}||P| = |A - \lambda I||P^{-1}P|$$

$$= |A - \lambda I||I| = |A - \lambda I| \text{ as } |I| = 1$$

Thus the matrices $A$ and $B$ have the same characteristic equations and hence the same characteristic roots. **Proved.**

**Example 62.** *If A and B be two square invertible matrices, then prove that AB and BA have the same characteristic roots.*

**Solution.** Now $\quad AB = IAB = B^{-1}B(AB) = B^{-1}(BA)B \quad \ldots(1)$

But by Ex. 8, matrices $BA$ and $B^{-1}(BA)B$ have same characteristic roots or matrices $BA$ and $AB$ by (1) have same characteristic roots. **Proved.**

**Example 63.** *If $A$ and $B$ be n rowed square matrices and if $A$ be invertible, show that the matrices $A^{-1}B$ and $BA^{-1}$ have the same characteristics roots.*

**Solution.** $A^{-1}B = A^{-1}BI = A^{-1}B(A^{-1}A) = A^{-1}(BA^{-1})A.$ ...(1)

But by Ex. 8, matrices $BA^{-1}$ and $A^{-1}(BA^{-1})A$ have same characteristic roots or matrices $BA^{-1}$ and $A^{-1}B$ by (1) have same characteristic roots. **Proved.**

**Example 64.** *Show that 0 is a characteristic root of a matrix, if and only if, the matrix is singular.*

**Solution.** Characteristic equation of matrix $A$ is given by

$$|A - \lambda I| = 0$$

If $\lambda = 0$, then from above it follows that $|A| = 0$ *i.e.* Matrix $A$ is singular.

Again if Matrix $A$ is singular *i.e.*, $|A| = 0$ then

$$|A - \lambda I| = 0 \Rightarrow |A| - \lambda|I| = 0, 0 - \lambda \cdot 1 = 0 \Rightarrow \lambda = 0.$$ **Proved.**

**Example 65.** *Show that characteristic roots of a triangular matrix are just the diagonal elements of the matrix.*

**Solution.** Let us consider the triangular matrix.

$$A = \begin{bmatrix} a_{11} & 0 & 0 & 0 \\ a_{21} & a_{22} & 0 & 0 \\ a_{31} & a_{32} & a_{33} & 0 \\ a_{41} & a_{42} & a_{43} & a_{44} \end{bmatrix}$$

Characteristic equation is $|A - \lambda I| = 0$

or $$\begin{vmatrix} a_{11}-\lambda & 0 & 0 & 0 \\ a_{21} & a_{22}-\lambda & 0 & 0 \\ a_{31} & a_{32} & a_{33}-\lambda & 0 \\ a_{41} & a_{42} & a_{43} & a_{44}-\lambda \end{vmatrix} = 0$$

On expansion it gives $(a_{11} - \lambda)(a_{22} - \lambda)(a_{33} - \lambda)(a_{44} - \lambda) = 0$

$\therefore$ $$\lambda = a_{11}, a_{22}, a_{33}, a_{44}$$

which are diagonal elements of matrix $A$. **Proved.**

**Example 66.** *If $\lambda$ is an eigen value of an orthogonal matrix, then $\frac{1}{\lambda}$ is also eigen value.*

[**Hint:** $AA' = I$ if $\lambda$ is the eigen value of $A$, then $\lambda^2 = 1$, $\lambda = \frac{1}{\lambda}$]

**Example 67.** *Find the eigen values of the orthogonal matrix.*

$$B = \frac{1}{3}\begin{bmatrix} 1 & 2 & 2 \\ 2 & 1 & -2 \\ 2 & -2 & 1 \end{bmatrix}$$

**Solution.** The characteristic equation of

$$A = \begin{bmatrix} 1 & 2 & 2 \\ 2 & 1 & -2 \\ 2 & -2 & 1 \end{bmatrix} \text{ is } \begin{vmatrix} 1-\lambda & 2 & 2 \\ 2 & 1-\lambda & -2 \\ 2 & -2 & 1-\lambda \end{vmatrix} = 0$$

$$\Rightarrow \quad (1-\lambda)[(1-\lambda)(1-\lambda)-4] - 2[2(1-\lambda)+4] + 2[-4-2(1-\lambda)] = 0$$

$$\Rightarrow \quad (1-\lambda)(1 - 2\lambda + \lambda^2 - 4) - 2(2 - 2\lambda + 4) + 2(-4 - 2 + 2\lambda) = 0$$

$$\Rightarrow \quad \lambda^3 - 3\lambda^2 - 9\lambda + 27 = 0$$

$$\Rightarrow \quad (\lambda - 3)^2 (\lambda + 3) = 0$$

The eigen values of $A$ are 3, 3, –3, so the eigen values of $B = \frac{1}{3} A$ are 1, 1, –1.

**Note.** If $\lambda = 1$ is an eigen value of $B$ then its reciprocal $\frac{1}{\lambda} = \frac{1}{1} = 1$ is also an eigen value of $B$. **Ans.**

## EXERCISE 4.20

Show that, for any square matrix $A$.

**1.** If $\lambda$ be an eigen value of a non singular matrix A, show that $\frac{|A|}{\lambda}$ is an eigen value of the matrix adj $A$.

**2.** There are infinitely many eigen vectors corresponding to a single eigen value.

**3.** Find the product of the eigen values of the matrix $\begin{bmatrix} 3 & -3 & 3 \\ 2 & 1 & 1 \\ 1 & 5 & 6 \end{bmatrix}$ **Ans.** 18

**4.** Find the sum of the eigen values of the matrix $\begin{bmatrix} 3 & 2 & 1 \\ 1 & 3 & 2 \\ 4 & 1 & 5 \end{bmatrix}$ **Ans.** 11

**5.** Find the eigen value of the inverse of the matrix $\begin{bmatrix} 4 & 6 & 6 \\ 1 & 3 & 2 \\ -1 & -4 & -3 \end{bmatrix}$ **Ans.** $-1, 1, \frac{1}{4}$

**6.** Find the eigen values of the square of the matrix $\begin{bmatrix} 1 & 0 & -1 \\ 1 & 2 & 1 \\ 2 & 2 & 3 \end{bmatrix}$ **Ans.** 1, 4, 9

**7.** Find the eigen values of the matrix $\begin{bmatrix} 3 & 1 & 4 \\ 0 & 2 & 6 \\ 0 & 0 & 5 \end{bmatrix}^3$ **Ans.** 8, 27, 125

**8.** The sum and product of the eigen values of the matrix $A = \begin{bmatrix} 2 & 2 & 1 \\ 1 & 3 & 1 \\ 1 & 2 & 2 \end{bmatrix}$ are respectively

(*a*) 7 and 7 (*b*) 7 and 5 (*c*) 7 and 6 (*d*) 7 and 8 (*AMIETE, June 2010*) **Ans.** (*b*)

## 4.52 CAYLEY-HAMILTON THEOREM

**Satement.** Every square matrix satisfies its own characteristic equation.

If $|A - \lambda I| = (-1)^n (\lambda^n + a_1\lambda^{n-1} + a_2\lambda^{n-2} + \cdots + a_n)$ be the characteristic polynomial of $n \times n$ matrix $A = (a_{ij})$, then the matrix equation

$$X^n + a_1 X^{n-1} + a_2 X^{n-2} + \ldots + a_n I = 0 \text{ is satisfied by } X = A \text{ i.e.,}$$

$$A^n + a_1 A^{n-1} + a_2 A^{n-2} + \ldots + a_n I = 0$$

**Proof.** Since the elements of the matrix $A - \lambda I$ are at most of the first degree in $\lambda$, the elements of adj. $(A - \lambda I)$ are at most degree $(n-1)$ in $\lambda$. Thus, adj. $(A - \lambda I)$ may be written as a matrix polynomial in $\lambda$, given by

$$Adj(A-\lambda I)=B_0\lambda^{n-1}+B_1\lambda^{n-2}+\ldots+B_{n-1}$$

where $B_0, B_1, \ldots, B_{n-1}$ are $n \times n$ matrices, their elements being polynomial in $\lambda$.

We know that

$$(A-\lambda I)Adj(A-\lambda I)=|A-\lambda I|\,I$$

$$(A-\lambda I)\left(B_0\lambda^{n-1}+B_1\lambda^{n-2}+\ldots.+B_{n-1}\right)=(-1)^n\left(\lambda^n+a_1\lambda^{n-1}+\ldots+a_n\right)I$$

Equating coefficient of like power of $\lambda$ on both sides, we get

$$-IB_0=(-1)^n I$$

$$AB_0-IB_1=(-1)^n a_1 I$$

$$AB_1-IB_2=(-1)^n a_2 I$$

..............................

$$AB_{n-1}=(-1)^n a_n I$$

On multiplying the equation by $A^n, A^{n-1}, \ldots, I$ respectively and adding, we obtain

$$0=(-1)^n\left[A^n+a_1A^{n-1}+\ldots+a_nI\right]$$

Thus $\qquad A^n+a_1A^{n-1}+\ldots+a_nI=0$

for example, Let $A$ be square matrix and if

$$\lambda^3-2\lambda^2+3\lambda-4=0 \qquad \ldots(1)$$

be its characteristic equation, then according to Cayley Hamilton Theorem (1) is satisfied by $A$.

$$A^3-2A^2+3A-4I=0 \qquad \ldots(2)$$

We can find out $A^{-1}$ from (2). On premultiplying (2) by $A^{-1}$, we get

$$A^2-2A+3I-4A^{-1}=0$$

$$A^{-1}=\frac{1}{4}\left[A^2-2A+3I\right]$$

**Example 68.** *Find the characteristic equation of the symmetric matrix*

$$A=\begin{bmatrix}2 & -1 & 1\\ -1 & 2 & -1\\ 1 & -1 & 2\end{bmatrix}$$

*and verify that it is satisfied by A and hence obtain $A^{-1}$.*

Express $A^6-6A^5+9A^4-2A^3-12A^2+23A-9I$ in linear polynomial in $A$.

*(A.M.I.E.T.E., Summer 2000)*

**Solution.** Characteristic equation is $|A-\lambda I|=0$

$$\begin{bmatrix}2-\lambda & -1 & 1\\ -1 & 2-\lambda & -1\\ 1 & -1 & 2-\lambda\end{bmatrix}=0$$

$$(2-\lambda)\,[(2-\lambda)^2-1]+1\,[-2+\lambda+1]+1\,[1-2+\lambda]=0$$

or $\qquad (2-\lambda)^3-(2-\lambda)+\lambda-1+\lambda-1=0$

or $\qquad (2-\lambda)^3-2+\lambda+\lambda-1+\lambda-1=0$ or $(2-\lambda)^3+3\lambda-4=0$

or $\quad 8-\lambda^3-12\lambda+6\lambda^2+3\lambda-4=0$

or $\quad -\lambda^3+6\lambda^2-9\lambda+4=0$ or $\lambda^3-6\lambda^2+9\lambda-4=0$

By Cayley-Hamilton Theorem $A^3-6A^2+9A-4I=0$ ... (1)

*Verification:*

$$A^2=\begin{bmatrix}2&-1&1\\-1&2&-1\\1&-1&2\end{bmatrix}\begin{bmatrix}2&-1&1\\-1&2&-1\\1&-1&2\end{bmatrix}$$

$$=\begin{pmatrix}4+1+1&-2-2-1&2+1+2\\-2-2-1&1+4+1&-1-2-2\\2+1+2&-1-2-2&1+1+4\end{pmatrix}=\begin{pmatrix}6&-5&5\\-5&6&-5\\5&-5&6\end{pmatrix}$$

$$A^3=\begin{pmatrix}6&-5&5\\-5&6&-5\\5&-5&6\end{pmatrix}\begin{pmatrix}2&-1&1\\-1&2&-1\\1&-1&2\end{pmatrix}$$

$$=\begin{pmatrix}12+5+5&-6-10+5&6+5+10\\-10-6-5&5+12+5&-5-6-10\\10+5+6&-5-10-6&5+5+12\end{pmatrix}=\begin{pmatrix}22&-21&21\\-21&22&-21\\21&-21&22\end{pmatrix}$$

$A^3-6A^2+9A-4I$

$$=\begin{pmatrix}22&-21&21\\-21&22&-21\\21&-21&22\end{pmatrix}-6\begin{pmatrix}6&-5&5\\-5&6&-5\\5&-5&6\end{pmatrix}+9\begin{pmatrix}2&-1&1\\-1&2&-1\\1&-1&2\end{pmatrix}-4\begin{pmatrix}1&0&0\\0&1&0\\0&0&1\end{pmatrix}$$

$$=\begin{pmatrix}22-36+18-4&-21+30-9-0&21-30+9-0\\-21+30-9-0&22-36+18-4&-21+30-9-0\\21-30+9-0&-21+30-9-0&22-36+18-4\end{pmatrix}=\begin{pmatrix}0&0&0\\0&0&0\\0&0&0\end{pmatrix}=0$$

So it is verified that the characteristic equation (1) is satisfied by $A$.

**Inverse of Matrix A,**

$$A^3-6A^2+9A-4I=0$$

On multiplying by $A^{-1}$, we get

$$A^2-6A+9I-4A^{-1}=0 \quad \text{or} \quad 4A^{-1}=A^2-6A+9I$$

or $$4A^{-1}=\begin{pmatrix}6&-5&5\\-5&6&-5\\5&-5&6\end{pmatrix}-6\begin{pmatrix}2&-1&1\\-1&2&-1\\1&-1&2\end{pmatrix}+9\begin{pmatrix}1&0&0\\0&1&0\\0&0&1\end{pmatrix}$$

$$=\begin{pmatrix}6-12+9&-5+6+0&5-6+0\\-5+6+0&6-12+9&-5+6+0\\5-6+0&-5+6+0&6-12+9\end{pmatrix},\quad A^{-1}=\frac{1}{4}\begin{pmatrix}3&1&-1\\1&3&1\\-1&1&3\end{pmatrix}$$ **Ans.**

$A^6-6A^5+9A^4-2A^3-12A^2+23A-9I$

$=A^3(A^3-6A^2+9A-4I)+2(A^3-6A^2+9A-4I)+5A-I$

$=5A-I$ **Ans.**

**Example 69.** *Find the characteristic equation of the matrix* $A=\begin{bmatrix}2 & 1 & 1\\ 0 & 1 & 0\\ 1 & 1 & 2\end{bmatrix}$

*Verify Cayley Hamilton Theorem and hence prove that :*

$$A^8-5A^7+7A^6-3A^5+A^4-5A^3+8A^2-2A+I=\begin{bmatrix}8 & 5 & 5\\ 0 & 3 & 0\\ 5 & 5 & 8\end{bmatrix}$$

(*Gujarat, II Semester, June 2009*)

**Solution.** Characteristic equation of the matrix $A$ is

$$\begin{vmatrix}2-\lambda & 1 & 1\\ 0 & 1-\lambda & 0\\ 1 & 1 & 2-\lambda\end{vmatrix}=0$$

$$\Rightarrow \quad (2-\lambda)[(1-\lambda)(2-\lambda)]-1(0)+1(0-1+\lambda)=0 \quad \Rightarrow \quad \lambda^3-5\lambda^2+7\lambda-3=0$$

According to Cayley-Hamilton Theorem

$$A^3-5A^2+7A-3I=0 \qquad ...(1)$$

We have to verify the equation (1).

$$A^2=\begin{bmatrix}2 & 1 & 1\\ 0 & 1 & 0\\ 1 & 1 & 2\end{bmatrix}\begin{bmatrix}2 & 1 & 1\\ 0 & 1 & 0\\ 1 & 1 & 2\end{bmatrix}=\begin{bmatrix}5 & 4 & 4\\ 0 & 1 & 0\\ 4 & 4 & 5\end{bmatrix}$$

$$A^3=A^2.A=\begin{bmatrix}2 & 1 & 1\\ 0 & 1 & 0\\ 1 & 1 & 2\end{bmatrix}\begin{bmatrix}5 & 4 & 4\\ 0 & 1 & 0\\ 4 & 4 & 5\end{bmatrix}=\begin{bmatrix}14 & 13 & 13\\ 0 & 1 & 0\\ 13 & 13 & 14\end{bmatrix}$$

$$A^3-5A^2+7A-3I=\begin{bmatrix}14 & 13 & 13\\ 0 & 1 & 0\\ 13 & 13 & 14\end{bmatrix}-5\begin{bmatrix}5 & 4 & 4\\ 0 & 1 & 0\\ 4 & 4 & 5\end{bmatrix}+7\begin{bmatrix}2 & 1 & 1\\ 0 & 1 & 0\\ 1 & 1 & 2\end{bmatrix}-3\begin{bmatrix}1 & 0 & 0\\ 0 & 1 & 0\\ 0 & 0 & 1\end{bmatrix}$$

$$=\begin{bmatrix}14-25+14-3 & 13-20+7+0 & 13-20+7+0\\ 0+0+0+0 & 1-5+7-3 & 0-0+0-0\\ 13-20+7+0 & 13-20+7-0 & 14-25+14-3\end{bmatrix}=\begin{bmatrix}0 & 0 & 0\\ 0 & 0 & 0\\ 0 & 0 & 0\end{bmatrix}=0$$

Hence Cayley Hamilton Theorem is verified.

Now, $A^8-5A^7+7A^6-3A^5+A^4-5A^3+8A^2-2A+I$

$$=A^5\left(A^3-5A^2+7A-3I\right)+A\left(A^3-5A^2+7A-3I\right)+A^2+A+I$$

$$=A^5\times O+A\times O+A^2+A+I \;=A^2+A+I$$

$$=\begin{bmatrix}5 & 4 & 4\\ 0 & 1 & 0\\ 4 & 4 & 5\end{bmatrix}+\begin{bmatrix}2 & 1 & 1\\ 0 & 1 & 0\\ 1 & 1 & 2\end{bmatrix}+\begin{bmatrix}1 & 0 & 0\\ 0 & 1 & 0\\ 0 & 0 & 1\end{bmatrix}$$

$$=\begin{bmatrix}5+2+1 & 4+1+0 & 4+1+0\\ 0+0+0 & 1+1+1 & 0+0+0\\ 4+1+0 & 4+1+0 & 5+2+1\end{bmatrix}=\begin{bmatrix}8 & 5 & 5\\ 0 & 3 & 0\\ 5 & 5 & 8\end{bmatrix}$$

**Proved.**

## 4.53 POWER OF MATRIX (by Cayley Hamilton Theorem)

**Any positive integral power $A^m$ of matrix $A$ is linearly expressible in terms of those of lower degree, where $m$ is a positive integer and $n$ is the degree of characteristic equation such that $m > n$.**

**Example 70.** *Find $A^4$ with the help of Cayley Hamilton Theorem, if*

$$A = \begin{bmatrix} 1 & 0 & -1 \\ 1 & 2 & 1 \\ 2 & 2 & 3 \end{bmatrix}$$

**Solution.** Here, we have
$$A = \begin{bmatrix} 1 & 0 & -1 \\ 1 & 2 & 1 \\ 2 & 2 & 3 \end{bmatrix}$$

Characteristic equation of the matrix $A$ is

$$\begin{vmatrix} 1-\lambda & 0 & -1 \\ 1 & 2-\lambda & 1 \\ 2 & 2 & 3-\lambda \end{vmatrix} = 0 \quad \Rightarrow \quad \lambda^3 - 6\lambda^2 - 11\lambda - 6 = 0 \quad \Rightarrow \quad (\lambda-1)(\lambda-2)(\lambda-3) = 0$$

Eigen values of $A$ are 1, 2, 3.

Let $\lambda^4 = \left(\lambda^3 - 6\lambda^2 - 11\lambda - 6\right)Q(\lambda) + \left(a\lambda^2 + b\lambda + c\right) = 0$ ...(1)

(where $Q(\lambda)$ is quotient)

Put $\lambda = 1$ in (1), $(1)^4 = a + b + c$ $\Rightarrow$ $a + b + c = 1$ ...(2)

Put $\lambda = 2$ in (1), $(2)^4 = 4a + 2b + c$ $\Rightarrow$ $4a + 2b + c = 16$ ... (3)

Put $\lambda = 3$ in (1), $(3)^4 = 9a + 3b + c$ $\Rightarrow$ $9a + 3b + c = 81$ ... (4)

Solving (2), (3) and (4), we get

$a = 25, \quad b = -60, \quad c = 36$

Replacing $\lambda$ by matrix A in (1), we get

$$A^4 = \left(A^3 - 6A^2 + 11A - 6\right)Q(A) + \left(aA^2 + bA + c\right)$$
$$= O + aA^2 + bA + cI$$

$$= 25\begin{bmatrix} 1 & 0 & -1 \\ 1 & 2 & 1 \\ 2 & 2 & 3 \end{bmatrix}\begin{bmatrix} 1 & 0 & -1 \\ 1 & 2 & 1 \\ 2 & 2 & 3 \end{bmatrix} + (-60)\begin{bmatrix} 1 & 0 & -1 \\ 1 & 2 & 1 \\ 2 & 2 & 3 \end{bmatrix} + 36\begin{bmatrix} 1 & 0 & 0 \\ 0 & 1 & 0 \\ 0 & 0 & 1 \end{bmatrix}$$

$$= \begin{bmatrix} -25 & -50 & -100 \\ 125 & 150 & 100 \\ 250 & 250 & 225 \end{bmatrix} + \begin{bmatrix} -60 & 0 & 60 \\ -60 & -120 & -60 \\ -120 & -120 & -180 \end{bmatrix} + \begin{bmatrix} 36 & 0 & 0 \\ 0 & 36 & 0 \\ 0 & 0 & 36 \end{bmatrix}$$

$$= \begin{bmatrix} -25-60+36 & -50+0+0 & -100+60+0 \\ 125-60+0 & 150-120+36 & 100-60+0 \\ 250-120+0 & 250-120+0 & 225-180+36 \end{bmatrix} = \begin{bmatrix} -49 & -50 & -40 \\ 65 & 66 & 40 \\ 130 & 130 & 81 \end{bmatrix}$$

**(It is also solved by diagonalization method on page 496 Example 38.)**

## EXERCISE 4.21

1. Find the characteristic polynomial of the matrix

$$A = \begin{bmatrix} 3 & 1 & 1 \\ -1 & 5 & -1 \\ 1 & -1 & 3 \end{bmatrix}$$

Verify Cayley-Hamilton Theorem for this matrix. Hence find $A^{-1}$. **Ans.** $A^{-1} = \frac{1}{20}\begin{bmatrix} 7 & -2 & -3 \\ 1 & 4 & 1 \\ -2 & 2 & 8 \end{bmatrix}$

2. Use Cayley-Hamilton Theorem to find the inverse of the matrix

$$\begin{bmatrix} \cos\theta & \sin\theta \\ -\sin\theta & \cos\theta \end{bmatrix}.$$

**Ans.** $\begin{bmatrix} \cos\theta & -\sin\theta \\ \sin\theta & \cos\theta \end{bmatrix}$

3. Using Cayley-Hamilton Theorem, find $A^{-1}$, given that

$$A = \begin{bmatrix} 2 & -1 & 3 \\ 1 & 0 & 2 \\ 4 & -2 & 1 \end{bmatrix}$$

**Ans.** $-\frac{1}{5}\begin{bmatrix} 4 & -5 & -2 \\ 7 & -10 & -1 \\ -2 & 0 & 1 \end{bmatrix}$

4. Using Cayley-Hamilton Theorem, find the inverse of the matrix

$$\begin{bmatrix} 5 & -1 & 5 \\ 0 & 2 & 0 \\ -5 & 3 & -15 \end{bmatrix}$$

**Ans.** $\frac{1}{10}\begin{bmatrix} 3 & 0 & 1 \\ 0 & 5 & 0 \\ -1 & 1 & -1 \end{bmatrix}$

5. Find the characteristic equation of the matrix

$$A = \begin{bmatrix} 1 & 3 & 7 \\ 4 & 2 & 3 \\ 1 & 2 & 1 \end{bmatrix}$$

*(R.G.P.V., Bhopal, Summer 2004)*

and show that the equation is also satisfied by $A$. **Ans.** $\lambda^3 - 4\lambda^2 - 20\lambda - 35 = 0$

6. Find the eigenvalues of the matrix

$$\begin{bmatrix} 2 & -3 & 1 \\ 3 & 1 & 3 \\ -5 & 2 & -4 \end{bmatrix}$$

**Ans.** Eigenvalues are 0, +1, –2

7. Using, Cayley-Hamilton Theorem obtain the inverse of the matrix

$$\begin{bmatrix} 1 & 1 & 3 \\ 1 & 3 & -3 \\ -2 & -4 & -4 \end{bmatrix}$$

*(R.G.P.V. Bhopal, I Sem., 2003)*

**Ans.** $\frac{1}{8}\begin{bmatrix} 24 & 8 & 12 \\ -10 & -2 & -6 \\ -2 & -2 & -2 \end{bmatrix}$

8. Show that the matrix $A = \begin{bmatrix} 1 & -2 & 2 \\ 1 & 2 & 3 \\ 0 & -1 & 2 \end{bmatrix}$ **Ans.** $\frac{1}{9}\begin{bmatrix} 7 & 2 & -10 \\ -2 & 2 & -1 \\ -1 & 1 & 4 \end{bmatrix}$

satisfies its characteristic equation. Hence find $A^{-1}$.

9. Use Cayley Hamilton Theorem to find the inverse of

$$A = \begin{bmatrix} 1 & 2 & 4 \\ -1 & 0 & 3 \\ 3 & 1 & -2 \end{bmatrix}$$

**Ans.** $A^{-1} = \frac{1}{7}\begin{bmatrix} -3 & 8 & 6 \\ 7 & -14 & -7 \\ -1 & 5 & 2 \end{bmatrix}$

**10.** Verify Cayley-Hamilton Theorem for the matrix

$$A=\begin{bmatrix}1 & 1 & 2\\ 3 & 1 & 1\\ 2 & 3 & 1\end{bmatrix}$$ Hence evaluate $A^{-1}$. **Ans.** $\frac{1}{11}\begin{bmatrix}-2 & 5 & -1\\ -1 & -3 & 5\\ 7 & -1 & -2\end{bmatrix}$

**11.** If $A=\begin{bmatrix}1 & 4\\ 2 & 3\end{bmatrix}$, then express $A^5-4A^4-7A^3+11A^2-A-10I$ in terms of A.

*(A.M.I.E.T.E., Winter 2001)* **Ans.** $A+5I$

**12.** If $\lambda_1$, $\lambda_2$ and $\lambda_3$ are the eigenvalues of the matrix

$$\begin{bmatrix}-2 & -9 & 5\\ -5 & -10 & 7\\ -9 & -21 & 14\end{bmatrix}$$ then $\lambda_1+\lambda_2+\lambda_3$ is equal to

(*i*) –16 (*ii*) 2 (*iii*) –6 (*iv*) –14 **Ans.** (*ii*)

**13.** The matrix $A=\begin{bmatrix}1 & 0\\ 2 & 4\end{bmatrix}$ is given. The eivenvalues of $4A^{-1}+3A+2I$ are

(A) 6, 15; (B) 9, 12 (C) 9, 15; (D) 7, 15 **Ans.** (C)

**14.** A(3 × 3) real matrix has an eigenvalue *i*, then its other two eigenvalues can be

(A) 0, 1 (B) –1, *i* (C) 2*i*, –2*i* (D) 0, –*i* *(A.M.I.E.T.E, Dec. 2004)*

**15.** Verify Cayley-Hamilton theorem for the matrix

$$A=\begin{bmatrix}1 & -2 & 3\\ 2 & 4 & -2\\ -1 & 1 & 2\end{bmatrix}$$

**16.** Find adj. *A* by using Cayley-Hamilton thmeorem where *A* is given by

$$A=\begin{bmatrix}1 & 2 & 1\\ 0 & 1 & -1\\ 3 & -1 & 1\end{bmatrix}$$ *(R.G.P.V., Bhopal, April 2010)* **Ans.** $\begin{bmatrix}0 & -3 & -3\\ -3 & -2 & 1\\ -3 & 7 & 1\end{bmatrix}$

**17.** If a matrix $A=\begin{bmatrix}1 & 0 & 0\\ 0 & -1 & 0\\ 1 & 0 & 1\end{bmatrix}$, find the matrix $A^{32}$, using Cayley Hamilton Theorem. **Ans.** $\begin{bmatrix}1 & 0 & 0\\ 0 & 1 & 0\\ 32 & 0 & 1\end{bmatrix}$

## 4.54 CHARACTERISTIC VECTORS OR EIGEN VECTORS

As we have discussed in Art 21.2,

A column vector *X* is transformed into column vector *Y* by means of a square matrix A.

Now we want to multiply the column vector *X* by a scalar quantity $\lambda$ so that we can find the same transformed column vector *Y*.

*i.e.*, $AX=\lambda X$

*X* is known as eigen vector.

**Example 71.** *Show that the vector (1, 1, 2) is an eigen vector of the matrix*

$$A=\begin{bmatrix}3 & 1 & -1\\ 2 & 2 & -1\\ 2 & 2 & 0\end{bmatrix}$$ *corresponding to the eigen value 2.*

**Solution.** Let $X = (1, 1, 2)$.

Now, $$AX = \begin{bmatrix} 3 & 1 & -1 \\ 2 & 2 & -1 \\ 2 & 2 & 0 \end{bmatrix}\begin{bmatrix} 1 \\ 1 \\ 2 \end{bmatrix} = \begin{bmatrix} 3+1-2 \\ 2+2-2 \\ 2+2+0 \end{bmatrix} = \begin{bmatrix} 2 \\ 2 \\ 4 \end{bmatrix} = 2\begin{bmatrix} 1 \\ 1 \\ 2 \end{bmatrix} = 2X$$

Corresponding to each characteristic root $\lambda$, we have a corresponding non-zero vector $X$ which satisfies the equation $[A-\lambda I]X = 0$. The non-zero vector $X$ is called characteristic vector or Eigen vector.

## 4.55 PROPERTIES OF EIGEN VECTORS

1. The eigen vector $X$ of a matrix $A$ is not unique.
2. If $\lambda_1, \lambda_2, \ldots, \lambda_n$ be distinct eigen values of an $n \times n$ matrix then corresponding eigen vectors $X_1, X_2, \ldots, X_n$ form a linearly independent set.
3. If two or more eigen values are equal it may or may not be possible to get linearly independent eigen vectors corresponding to the equal roots.
4. Two eigen vectors $X_1$ and $X_2$ are called orthogonal vectors if $X_1' X_2 = 0$.
5. Eigen vectors of a symmetric matrix corresponding to different eigen values are orthogonal.

**Normalised form of vectors.** To find normalised form of $\begin{bmatrix} a \\ b \\ c \end{bmatrix}$, we divide each element by $\sqrt{a^2+b^2+c^2}$.

*For example,* normalised form of $\begin{bmatrix} 1 \\ 2 \\ 2 \end{bmatrix}$ is $\begin{bmatrix} 1/3 \\ 2/3 \\ 2/3 \end{bmatrix}$ $\left[\sqrt{1^2+2^2+2^2} = 3\right]$

## 4.56 NON-SYMMETRIC MATRICES WITH NON-REPEATED EIGEN VALUES

**Example 72.** *Find the eigen values and eigen vectors of matrix* $A = \begin{bmatrix} 3 & 1 & 4 \\ 0 & 2 & 6 \\ 0 & 0 & 5 \end{bmatrix}$

**Solution.** $$|A-\lambda I| = \begin{vmatrix} 3-\lambda & 1 & 4 \\ 0 & 2-\lambda & 6 \\ 0 & 0 & 5-\lambda \end{vmatrix} = (3-\lambda)(2-\lambda)(5-\lambda)$$

Hence the characteristic equation of matrix $A$ is given by

$$|A-\lambda I| = 0 \quad \Rightarrow \quad (3-\lambda)(2-\lambda)(5-\lambda) = 0$$

$\therefore \quad \lambda = 2, 3, 5.$

Thus the eigen values of matrix $A$ are 2, 3, 5.

The eigen vectors of the matrix $A$ corresponding to the eigen value $\lambda$ is given by the non-zero solution of the equation $(A-\lambda I)X = 0$

or $$\begin{bmatrix} 3-\lambda & 1 & 4 \\ 0 & 2-\lambda & 6 \\ 0 & 0 & 5-\lambda \end{bmatrix}\begin{bmatrix} x_1 \\ x_2 \\ x_3 \end{bmatrix} = \begin{bmatrix} 0 \\ 0 \\ 0 \end{bmatrix} \qquad \ldots (1)$$

**When $\lambda = 2$,** the corresponding eigen vector is given by

$$\begin{bmatrix} 3-2 & 1 & 4 \\ 0 & 2-2 & 6 \\ 0 & 0 & 5-2 \end{bmatrix}\begin{bmatrix} x_1 \\ x_2 \\ x_3 \end{bmatrix} = \begin{bmatrix} 0 \\ 0 \\ 0 \end{bmatrix}$$

$$\Rightarrow \begin{bmatrix} 1 & 1 & 4 \\ 0 & 0 & 6 \\ 0 & 0 & 3 \end{bmatrix}\begin{bmatrix} x_1 \\ x_2 \\ x_3 \end{bmatrix} = \begin{bmatrix} 0 \\ 0 \\ 0 \end{bmatrix}$$

$$\Rightarrow \begin{aligned} x_1 + x_2 + 4x_3 &= 0 \\ 0x_1 + 0x_2 + 6x_3 &= 0 \end{aligned}$$

$$\frac{x_1}{6-0} = \frac{x_2}{0-6} = \frac{x_3}{0-0} = k \quad \Rightarrow \quad \frac{x_1}{1} = \frac{x_2}{-1} = \frac{x_3}{0} = k \quad \Rightarrow x_1 = k,\ x_2 = -k,\ x_3 = 0$$

Hence $X_1 = \begin{bmatrix} k \\ -k \\ 0 \end{bmatrix} = k\begin{bmatrix} 1 \\ -1 \\ 0 \end{bmatrix}$ can be taken as an eigen vector of A corresponding to the eigen value $\lambda = 2$

**When** $\lambda = 3$, substituting in (1), the corresponding eigen vector is given by

$$\begin{bmatrix} 3-3 & 1 & 4 \\ 0 & 2-3 & 6 \\ 0 & 0 & 5-3 \end{bmatrix}\begin{bmatrix} x_1 \\ x_2 \\ x_3 \end{bmatrix} = \begin{bmatrix} 0 \\ 0 \\ 0 \end{bmatrix} \quad \Rightarrow \quad \begin{bmatrix} 0 & 1 & 4 \\ 0 & -1 & 6 \\ 0 & 0 & 2 \end{bmatrix}\begin{bmatrix} x_1 \\ x_2 \\ x_3 \end{bmatrix} = \begin{bmatrix} 0 \\ 0 \\ 0 \end{bmatrix}$$

$$\begin{aligned} 0x_1 + x_2 + 4x_3 &= 0 \\ 0x_1 - x_2 + 6x_3 &= 0 \end{aligned}$$

$$\frac{x_1}{6+4} = \frac{x_2}{0-0} = \frac{x_3}{0-0} \quad \Rightarrow \quad \frac{x_1}{10} = \frac{x_2}{0} = \frac{x_3}{0} = \frac{k}{10}$$

$x_1 = k,\ x_2 = 0,\ x_3 = 0$

Hence, $X_2 = \begin{bmatrix} k \\ 0 \\ 0 \end{bmatrix} = k\begin{bmatrix} 1 \\ 0 \\ 0 \end{bmatrix}$ can be taken as an eigen vector of A corresponding to the eigen value $\lambda = 3$.

**When** $\lambda = 5$.

Again, when $\lambda = 5$, substituting in (1), the corresponding eigen vector is given by

$$\begin{bmatrix} 3-5 & 1 & 4 \\ 0 & 2-5 & 6 \\ 0 & 0 & 5-5 \end{bmatrix}\begin{bmatrix} x_1 \\ x_2 \\ x_3 \end{bmatrix} = \begin{bmatrix} 0 \\ 0 \\ 0 \end{bmatrix} \quad \Rightarrow \quad \begin{bmatrix} -2 & 1 & 4 \\ 0 & -3 & 6 \\ 0 & 0 & 0 \end{bmatrix}\begin{bmatrix} x_1 \\ x_2 \\ x_3 \end{bmatrix} = \begin{bmatrix} 0 \\ 0 \\ 0 \end{bmatrix}$$

$$\begin{aligned} -2x_1 + x_2 + 4x_3 &= 0 \\ -3x_2 + 6x_3 &= 0 \end{aligned}$$

By cross-multiplication method, we have

$$\frac{x_1}{6+12} = \frac{x_2}{0+12} = \frac{x_3}{6-0} \quad \Rightarrow \quad \frac{x_1}{18} = \frac{x_2}{12} = \frac{x_3}{6} \quad \Rightarrow \quad \frac{x_1}{3} = \frac{x_2}{2} = \frac{x_3}{1} = k$$

$x_1 = 3k,\quad x_2 = 2k,\quad x_3 = k$

Hence, $X_3 = \begin{bmatrix} 3k \\ 2k \\ k \end{bmatrix} = k\begin{bmatrix} 3 \\ 2 \\ 1 \end{bmatrix}$ can be taken as an eigen vector of A corresponding to the eigen value $\lambda = 5$. **Ans.**

## EXERCISE 4.22

Non-symmetric matrix with different eigen values:

Find the eigen values and the corresponding eigen vectors for the following matrices:

**1.** $\begin{bmatrix} 1 & 1 & -2 \\ -1 & 2 & 1 \\ 0 & 1 & -1 \end{bmatrix}$ *(A.M.I.E.T., June 2006)* **Ans.** $-1, 1, 2, \begin{bmatrix} 1 \\ 0 \\ 1 \end{bmatrix}, \begin{bmatrix} 3 \\ 2 \\ 1 \end{bmatrix}, \begin{bmatrix} 1 \\ 3 \\ 1 \end{bmatrix}$

**2.** $\begin{bmatrix} 4 & 2 & -2 \\ -5 & 3 & 2 \\ -2 & 4 & 1 \end{bmatrix}$ **Ans.** $1, 2, 5; \begin{bmatrix} 2 \\ 1 \\ 4 \end{bmatrix}, \begin{bmatrix} 1 \\ 1 \\ 2 \end{bmatrix}, \begin{bmatrix} 0 \\ 1 \\ 0 \end{bmatrix}$

**2.** $\begin{bmatrix} 2 & -2 & 3 \\ 1 & 1 & 1 \\ 1 & 3 & -1 \end{bmatrix}$ **Ans.** $-2, 1, 3; \begin{bmatrix} 11 \\ 1 \\ 14 \end{bmatrix}, \begin{bmatrix} -1 \\ 1 \\ 1 \end{bmatrix}, \begin{bmatrix} 1 \\ 1 \\ 1 \end{bmatrix}$

**3.** $\begin{bmatrix} -9 & 2 & 6 \\ 5 & 0 & -3 \\ -16 & 4 & 11 \end{bmatrix}$ **Ans.** $-1, 1, 2; \begin{bmatrix} 2 \\ -1 \\ 3 \end{bmatrix}, \begin{bmatrix} 1 \\ -1 \\ 2 \end{bmatrix}, \begin{bmatrix} 2 \\ -1 \\ 4 \end{bmatrix}$

**4.** $\begin{bmatrix} 4 & 6 & 6 \\ 1 & 3 & 2 \\ -1 & -4 & -3 \end{bmatrix}$ **Ans.** $-1, 1, 4; \begin{bmatrix} -6 \\ -2 \\ 7 \end{bmatrix}, \begin{bmatrix} 0 \\ 1 \\ -1 \end{bmatrix}, \begin{bmatrix} 3 \\ 1 \\ -1 \end{bmatrix}$

**4.** $\begin{bmatrix} 1 & 1 & 1 \\ 1 & 2 & 1 \\ 3 & 2 & 3 \end{bmatrix}$ **Ans.** $0, 1, 5; \begin{bmatrix} 0 \\ -1 \\ 1 \end{bmatrix}, \begin{bmatrix} -1 \\ 0 \\ 1 \end{bmatrix}, \begin{bmatrix} 4 \\ 5 \\ 11 \end{bmatrix}$

**5.** $\begin{bmatrix} -2 & 1 & 1 \\ -11 & 4 & 5 \\ -1 & 1 & 0 \end{bmatrix}$ **Ans.** $-1, 1, 2; \begin{bmatrix} 0 \\ -1 \\ 1 \end{bmatrix}, \begin{bmatrix} 1 \\ 2 \\ 1 \end{bmatrix}, \begin{bmatrix} 2 \\ 3 \\ 1 \end{bmatrix}$

**8.** Show that the matrices A and $A^T$ have the same eigenvalues. Further if l, m are two distinct eigenvalues, then show that the eigenvector corresponding to l for A is orthogonal to eigenvector corresponding to m for $A^T$.

## 4.57 NON-SYMMETRIC MATRIX WITH REPEATED EIGEN VALUES

**Example 73.** *Find all the Eigen values and Eigen vectors of the matrix*

$$A = \begin{bmatrix} -2 & 2 & -3 \\ 2 & 1 & -6 \\ -1 & -2 & 0 \end{bmatrix}$$ *(AMIETE, Dec. 2009)*

**Solution.** Characteristic equation of $A$ is

$$\begin{vmatrix} -2-\lambda & 2 & -3 \\ 2 & 1-\lambda & -6 \\ -1 & -2 & 0-\lambda \end{vmatrix} = 0$$

$$\Rightarrow \quad (-2-\lambda)[-\lambda+\lambda^2-12]-2(-2\lambda-6)-3(-4+1-\lambda)=0$$

$$\lambda^3+\lambda^2-21\lambda-45=0 \qquad \text{.... (1)}$$

By trial: If $\lambda = -3$, then $-27+9+63-45=0$, so $(\lambda+3)$ is one factor of (1).

The remaining factors are obtained on dividing (1) by $\lambda + 3$.

| | | | | |
|---|---|---|---|---|
| −3 | 1 | 1 | −21 | −45 |
| | | −3 | 6 | 45 |
| | 1 | −2 | −15 | 0 |

$$\lambda^2-2\lambda-15=0 \qquad \Rightarrow (\lambda-5)(\lambda+3)=0$$

$$\Rightarrow (\lambda+3)(\lambda+3)(\lambda-5)=0 \Rightarrow \lambda = 5, -3, -3$$

To find the eigen vectors for corresponding eigen values, we will consider the matrix equation

$$(A-\lambda I)X=0 \quad i.e., \quad \begin{bmatrix} -2-\lambda & 2 & -3 \\ 2 & 1-\lambda & -6 \\ -1 & -2 & 0-\lambda \end{bmatrix}\begin{bmatrix} x \\ y \\ z \end{bmatrix}=\begin{bmatrix} 0 \\ 0 \\ 0 \end{bmatrix} \quad \ldots (2)$$

On putting $\lambda=5$ in eq. (2), it becomes $\begin{bmatrix} -7 & 2 & -3 \\ 2 & -4 & -6 \\ -1 & -2 & -5 \end{bmatrix}\begin{bmatrix} x \\ y \\ z \end{bmatrix}=\begin{bmatrix} 0 \\ 0 \\ 0 \end{bmatrix}$

We have $-7x+2y-3z=0,$
$2x-4y-6z=0$

$$\frac{x}{-12-12}=\frac{y}{-6-42}=\frac{z}{28-4} \quad \text{or} \quad \frac{x}{-24}=\frac{y}{-48}=\frac{z}{24} \quad \text{or} \quad \frac{x}{1}=\frac{y}{2}=\frac{z}{-1}=k$$

$$x=k, \quad y=2k, \quad z=-k$$

Hence, the eigen vector $X_1=\begin{bmatrix} k \\ 2k \\ -k \end{bmatrix}=k\begin{bmatrix} 1 \\ 2 \\ -1 \end{bmatrix}$

Put $\lambda=-3$ in eq. (2), it becomes $\begin{bmatrix} 1 & 2 & -3 \\ 2 & 4 & -6 \\ -1 & -2 & 3 \end{bmatrix}\begin{bmatrix} x \\ y \\ z \end{bmatrix}=\begin{bmatrix} 0 \\ 0 \\ 0 \end{bmatrix}$

We have $x+2y-3z=0,$
$2x+4y-6z=0,$
$-x-2y+3z=0$

Here first, second and third equations are the same.

Let $x=k_1, y=k_2$ then $z=\frac{1}{3}(k_1+2k_2)$

Hence, the eigen vector is $\begin{bmatrix} k_1 \\ k_2 \\ \frac{1}{3}(k_1+2k_2) \end{bmatrix}$

Let $k_1=0, k_2=3$, Hence $X_2=\begin{bmatrix} 0 \\ 3 \\ 2 \end{bmatrix}$

Since the matrix is non-symmetric, the corresponding eigen vectors $X_2$ and $X_3$ must be linearly independent. This can be done by choosing

$k_1=3,\ k_2=0$, and Hence $X_3=\begin{bmatrix} 3 \\ 0 \\ 1 \end{bmatrix}$

Hence, $X_1=\begin{bmatrix} 1 \\ 2 \\ -1 \end{bmatrix}, \quad X_2=\begin{bmatrix} 0 \\ 3 \\ 2 \end{bmatrix}, \quad X_3=\begin{bmatrix} 3 \\ 0 \\ 1 \end{bmatrix}.$ **Ans.**

## EXERCISE 4.23

**Non-symmetric matrices with repeated eigen values**

**Find the eigen values and eigen vectors of the following matrices:**

**1.** $\begin{bmatrix} 2 & -2 & 2 \\ 1 & 1 & 1 \\ 1 & 3 & -1 \end{bmatrix}$ **Ans.** $-2, 2, 2;$ $\begin{bmatrix} -4 \\ -1 \\ 7 \end{bmatrix}, \begin{bmatrix} 0 \\ 1 \\ 1 \end{bmatrix}$

**2.** $\begin{bmatrix} 2 & 2 & 1 \\ 1 & 3 & 1 \\ 1 & 2 & 2 \end{bmatrix}$ **Ans.** $1, 1, 5;$ $\begin{bmatrix} 1 \\ 2 \\ -5 \end{bmatrix}, \begin{bmatrix} 1 \\ 1 \\ 1 \end{bmatrix}$

**3.** $\begin{bmatrix} 2 & 1 & 1 \\ 2 & 3 & 2 \\ 3 & 3 & 4 \end{bmatrix}$ **Ans.** $1, 1, 7;$ $\begin{bmatrix} 0 \\ 1 \\ -1 \end{bmatrix}, \begin{bmatrix} 1 \\ 0 \\ -1 \end{bmatrix}, \begin{bmatrix} 1 \\ 2 \\ 3 \end{bmatrix}$

**4.** $\begin{bmatrix} -9 & 4 & 4 \\ -8 & 3 & 4 \\ -16 & 8 & 7 \end{bmatrix}$ **Ans.** $-1, -1, 3;$ $\begin{bmatrix} 0 \\ 1 \\ -1 \end{bmatrix}, \begin{bmatrix} 1 \\ 1 \\ 1 \end{bmatrix}, \begin{bmatrix} 1 \\ 1 \\ 2 \end{bmatrix}$

**5.** $\begin{bmatrix} 1 & 1 & 0 \\ 0 & 1 & 0 \\ 0 & 0 & 1 \end{bmatrix}$ *(AMIETE, Dec. 2010)* **Ans.** $1, 1, 1,$ $\begin{bmatrix} 1 \\ 0 \\ 1 \end{bmatrix}$

## 4.58 SYMMETRIC MATRICES WITH NON REPEATED EIGEN VALUES

**Example 74.** *Find the eigen values and the corresponding eigen vectors of the matrix*

$$\begin{bmatrix} -2 & 5 & 4 \\ 5 & 7 & 5 \\ 4 & 5 & -2 \end{bmatrix}$$

**Solution.** $|A - \lambda I| = 0$

$$\begin{vmatrix} -2-\lambda & 5 & 4 \\ 5 & 7-\lambda & 5 \\ 4 & 5 & -2-\lambda \end{vmatrix} = 0 \quad \Rightarrow \quad \lambda^3 - 3\lambda^2 - 90\lambda - 216 = 0$$

**By trial:** Take $\lambda = -3$, then $-27 - 27 + 270 - 216 = 0$

By synthetic division

| | | | | |
|---|---|---|---|---|
| −3 | 1 | −3 | −90 | −216 |
| | | −3 | 18 | 216 |
| | 1 | −6 | −72 | 0 |

$$\lambda^2 - 6\lambda - 72 = 0 \Rightarrow (\lambda - 12)(\lambda + 6) = 0 \Rightarrow \lambda = -3, -6, 12$$

Matrix equation for eigen vectors $[A - \lambda I]X = 0$

$$\begin{bmatrix} -2-\lambda & 5 & 4 \\ 5 & 7-\lambda & 5 \\ 4 & 5 & -2-\lambda \end{bmatrix}\begin{bmatrix} x \\ y \\ z \end{bmatrix} = \begin{bmatrix} 0 \\ 0 \\ 0 \end{bmatrix} \quad ...(1)$$

**Eigen Vector**

On putting $\lambda = -3$ in (1), it will become

$$\begin{bmatrix} 1 & 5 & 4 \\ 5 & 10 & 5 \\ 4 & 5 & 1 \end{bmatrix}\begin{bmatrix} x \\ y \\ z \end{bmatrix} = \begin{bmatrix} 0 \\ 0 \\ 0 \end{bmatrix} \Rightarrow \begin{cases} x + 5y + 4z = 0 \\ 5x + 10y + 5z = 0 \end{cases}$$

$$\frac{x}{25-40}=\frac{y}{20-5}=\frac{z}{10-25} \quad \text{or} \quad \frac{x}{1}=\frac{y}{-1}=\frac{z}{1}$$

Eigen vector $X_1=\begin{bmatrix}1\\-1\\1\end{bmatrix}$.

Eigen vector corresponding to eigen value $\lambda = -6$.

Equation (1) becomes

$$\begin{bmatrix}4 & 5 & 4\\5 & 13 & 5\\4 & 5 & 4\end{bmatrix}\begin{bmatrix}x\\y\\z\end{bmatrix}=\begin{bmatrix}0\\0\\0\end{bmatrix} \quad \text{or} \quad \begin{cases}4x+5y+4z=0\\5x+13y+5z=0\end{cases}$$

$$\frac{x}{25+52}=\frac{y}{20+20}=\frac{z}{52-25} \quad \text{or} \quad \frac{x}{1}=\frac{y}{0}=\frac{z}{-1}$$

eigen vector $X_2=\begin{bmatrix}1\\0\\-1\end{bmatrix}$

Eigen vector corresponding to eigen value $\lambda = 12$.

Equation (1) becomes

$$\begin{bmatrix}-14 & 5 & 4\\5 & -5 & 5\\4 & 5 & -14\end{bmatrix}\begin{bmatrix}x\\y\\z\end{bmatrix}=\begin{bmatrix}0\\0\\0\end{bmatrix} \quad \text{or} \quad \begin{cases}-14x+5y+4z=0\\5x-5y+5z=0\end{cases}$$

$$\frac{x}{25+20}=\frac{y}{20+70}=\frac{z}{70-25} \quad \text{or} \quad \frac{x}{1}=\frac{y}{2}=\frac{z}{1}$$

Eigen vector $X_3=\begin{bmatrix}1\\2\\1\end{bmatrix}$ **Ans.**

## EXERCISE 4.24

**Symmetric matrices with non-repeated eigen values**

**Find the eigen values and eigen vectors of the following matrices:**

1. $\begin{bmatrix}5 & 0 & 1\\0 & -2 & 0\\1 & 0 & 5\end{bmatrix}$ **Ans.** $-2, 4, 6;\ \begin{bmatrix}0\\1\\0\end{bmatrix}, \begin{bmatrix}1\\0\\-1\end{bmatrix}, \begin{bmatrix}1\\0\\1\end{bmatrix}$

2. $\begin{bmatrix}3 & -1 & 1\\-1 & 5 & -1\\1 & -1 & 3\end{bmatrix}$ **Ans.** $2, 3, 6;\ \begin{bmatrix}-1\\0\\1\end{bmatrix}, \begin{bmatrix}1\\1\\1\end{bmatrix}, \begin{bmatrix}1\\-2\\1\end{bmatrix}$

3. $\begin{bmatrix}8 & -6 & 2\\-6 & 7 & -4\\2 & -4 & 3\end{bmatrix}$ (*U.P., I Semester, Jan 2011*) **Ans.** $0, 3, 15;\ \begin{bmatrix}1\\2\\2\end{bmatrix}, \begin{bmatrix}2\\1\\-2\end{bmatrix}, \begin{bmatrix}2\\-2\\1\end{bmatrix}$

4. $\begin{bmatrix}2 & 4 & -6\\4 & 2 & -6\\-6 & -6 & -15\end{bmatrix}$ **Ans.** $-2, 9, -18;\ \begin{bmatrix}1\\-1\\0\end{bmatrix}, \begin{bmatrix}2\\2\\-1\end{bmatrix}, \begin{bmatrix}1\\1\\4\end{bmatrix}$

5. $\begin{bmatrix}1 & 1 & 3\\1 & 5 & 1\\3 & 1 & 1\end{bmatrix}$ **Ans.** $-2, 3, 6;\ \begin{bmatrix}-1\\0\\1\end{bmatrix}, \begin{bmatrix}1\\-1\\1\end{bmatrix}, \begin{bmatrix}1\\2\\1\end{bmatrix}$

## 4.59 SYMMETRIC MATRICES WITH REPEATED EIGEN VALUES

**Example 75.** *Find all the eigen values and eigen vectors of the matrix*

$$\begin{bmatrix} 2 & -1 & 1 \\ -1 & 2 & -1 \\ 1 & -1 & 2 \end{bmatrix}$$

**Solution.** The characteristic equation is $\begin{vmatrix} 2-\lambda & -1 & 1 \\ -1 & 2-\lambda & -1 \\ 1 & -1 & 2-\lambda \end{vmatrix} = 0$

$$\Rightarrow \quad (2-\lambda)[(2-\lambda)^2-1]+1[-2+\lambda+1]+1[1-2+\lambda]=0$$

$$\Rightarrow \quad (2-\lambda)(4-4\lambda+\lambda^2-1)+(\lambda-1)+\lambda-1=0$$

$$\Rightarrow \quad 8-8\lambda+2\lambda^2-2-4\lambda+4\lambda^2-\lambda^3+\lambda+2\lambda-2=0$$

$$\Rightarrow \quad -\lambda^3+6\lambda^2-9\lambda+4=0$$

$$\Rightarrow \quad \lambda^3-6\lambda^2+9\lambda-4=0 \qquad \text{... (1)}$$

On putting $\lambda = 1$ in (1), the equation (1) is satisfied. So $\lambda - 1$ is one factor of the equation (1).

The other factor $(\lambda^2 - 5\lambda + 4)$ is got on dividing (1) by $\lambda - 1$.

$$\Rightarrow \quad (\lambda-1)(\lambda^2-5\lambda+4)=0 \text{ or } (\lambda-1)(\lambda-1)(\lambda-4)=0 \quad \Rightarrow \lambda = 1, 1, 4$$

The eigen values are 1, 1, 4.

When $\lambda = 4$ $\begin{pmatrix} 2-4 & -1 & 1 \\ -1 & 2-4 & -1 \\ 1 & -1 & 2-4 \end{pmatrix}\begin{pmatrix} x_1 \\ x_2 \\ x_3 \end{pmatrix} = \begin{bmatrix} 0 \\ 0 \\ 0 \end{bmatrix} \Rightarrow \begin{pmatrix} -2 & -1 & 1 \\ -1 & -2 & -1 \\ 1 & -1 & -2 \end{pmatrix}\begin{pmatrix} x_1 \\ x_2 \\ x_3 \end{pmatrix} = \begin{bmatrix} 0 \\ 0 \\ 0 \end{bmatrix}$

$$-2x_1 - x_2 + x_3 = 0$$

$$x_1 - x_2 - 2x_3 = 0$$

$$\Rightarrow \quad \frac{x_1}{2+1} = \frac{x_2}{1-4} = \frac{x_3}{2+1} \quad \Rightarrow \quad \frac{x_1}{1} = \frac{x_2}{-1} = \frac{x_3}{1} = k$$

$$x_1 = k, \quad x_2 = -k, \quad x_3 = k$$

$$X_1 = \begin{bmatrix} k \\ -k \\ k \end{bmatrix} = k\begin{bmatrix} 1 \\ -1 \\ 1 \end{bmatrix} \quad \text{or} \quad X_1 = \begin{bmatrix} 1 \\ -1 \\ 1 \end{bmatrix}$$

When $\lambda = 1$ $\begin{pmatrix} 2-1 & -1 & 1 \\ -1 & 2-1 & -1 \\ 1 & -1 & 2-1 \end{pmatrix}\begin{pmatrix} x_1 \\ x_2 \\ x_3 \end{pmatrix} = 0$

$$\Rightarrow \quad \begin{pmatrix} 1 & -1 & 1 \\ -1 & 1 & -1 \\ 1 & -1 & 1 \end{pmatrix}\begin{pmatrix} x_1 \\ x_2 \\ x_3 \end{pmatrix} = 0 \Rightarrow \begin{pmatrix} 1 & -1 & 1 \\ 0 & 0 & 0 \\ 0 & 0 & 0 \end{pmatrix}\begin{pmatrix} x_1 \\ x_2 \\ x_3 \end{pmatrix} = 0, \ R_2 \to R_2 + R_1, \ R_3 \to R_3 - R_1$$

$$x_1 - x_2 + x_3 = 0$$

Let $x_1 = k_1$ and $x_2 = k_2$

$$k_1 - k_2 + x_3 = 0 \quad \text{or} \quad x_3 = k_2 - k_1$$

$$X_2 = \begin{bmatrix} k_1 \\ k_2 \\ k_2 - k_1 \end{bmatrix} \Rightarrow X_2 = \begin{bmatrix} 1 \\ 1 \\ 0 \end{bmatrix} \quad \begin{bmatrix} k_1 = 1 \\ k_2 = 1 \end{bmatrix}$$

Let $X_3 = \begin{bmatrix} l \\ m \\ n \end{bmatrix}$

As $X_3$ is orthogonal to $X_1$ since the given matrix is symmetric

$$[1, -1, 1]\begin{bmatrix} l \\ m \\ n \end{bmatrix} = 0 \quad \text{or} \quad l - m + n = 0 \qquad \text{... (2)}$$

As $X_3$ is orthogonal to $X_2$ since the given matrix is symmetric

$$[1, 1, 0]\begin{bmatrix} l \\ m \\ n \end{bmatrix} = 0 \quad \text{or} \quad l + m + 0 = 0 \qquad \text{... (3)}$$

Solving (2) and (3), we get $\dfrac{l}{0-1} = \dfrac{m}{1-0} = \dfrac{n}{1+1} \Rightarrow \dfrac{l}{-1} = \dfrac{m}{1} = \dfrac{n}{2}$

$$X_3 = \begin{bmatrix} -1 \\ 1 \\ 2 \end{bmatrix}$$ **Ans.**

## EXERCISE 4.25

**Symmetric matrices with repeated eigen values**

**Find the eigen values and the corresponding eigen vectors of the following matrices:**

**1.** $\begin{bmatrix} 1 & 2 & 3 \\ 2 & 4 & 6 \\ 3 & 6 & 9 \end{bmatrix}$ **Ans.** 0, 0, 14; $\begin{bmatrix} -2 \\ 1 \\ 0 \end{bmatrix}, \begin{bmatrix} 3 \\ 6 \\ -5 \end{bmatrix}, \begin{bmatrix} 1 \\ 2 \\ 3 \end{bmatrix}$ **2.** $\begin{bmatrix} 2 & 0 & 1 \\ 0 & 3 & 0 \\ 1 & 0 & 2 \end{bmatrix}$ **Ans.** 1, 3, 3; $\begin{bmatrix} 1 \\ 0 \\ -1 \end{bmatrix}, \begin{bmatrix} 1 \\ 1 \\ 1 \end{bmatrix}, \begin{bmatrix} 1 \\ -2 \\ 1 \end{bmatrix}$

**3.** $\begin{bmatrix} 6 & -2 & 2 \\ -2 & 3 & -1 \\ 2 & -1 & 3 \end{bmatrix}$ **Ans.** 8, 2, 2; $\begin{bmatrix} 2 \\ -1 \\ 1 \end{bmatrix}, \begin{bmatrix} 1 \\ 0 \\ -2 \end{bmatrix}, \begin{bmatrix} 1 \\ 2 \\ 0 \end{bmatrix}$ **4.** $\begin{bmatrix} 6 & -3 & 3 \\ -3 & 6 & -3 \\ 3 & -3 & 6 \end{bmatrix}$ **Ans.** 3, 3, 12

**4.** Choose the correct or the best of the answers given in the following Parts;

(*i*) Two of the eigenvalues of a $3 \times 3$ matrix, whose determinant equals, 4, are $-1$ and $+2$ the third eigen value of the matrix is equal to

(*a*) $-2$ (*b*) $-1$ (*c*) 1 (*d*) 2

(*ii*) If a square matrix $A$ has an eigenvalue $\lambda$, then an eigenvalue of the matrix $(kA)^T$ where, $k \neq 0$, is a scalar is

(*a*) $\lambda / k$ (*b*) $k / \lambda$ (*c*) $k\,\lambda$ (*d*) None of these

(*iii*) An eigenvalue of a square matrix $A$ is $\lambda = 0$. Then

(*a*) $|A| \neq 0$; (*b*) $A$ is symmetric (*c*) $A$ is singular;

(*d*) $A$ is skew-symmetric; (*e*) $A$ is an even order matrix; (*f*) $A$ is an odd order matrix.

(*iv*) The matrix $A$ is defined as $A = \begin{bmatrix} -1 & 0 & 0 \\ 2 & -3 & 0 \\ 1 & 4 & 2 \end{bmatrix}$. The eigenvalues of $A^2$ are

(*a*) –1, –9, –4, (*b*) 1, 9, 4 (*c*) –1, –3, 2, (*d*) 1, 3, –2.

(*v*) If the matrix is $A = \begin{bmatrix} -1 & 2 & 3 \\ 0 & 3 & 5 \\ 0 & 0 & -2 \end{bmatrix}$ then the eigenvalues of $A^3 + 5A + 8I$, are

(*a*) –1, 27, –8; (*b*) –1, 3, –2; (*c*) 2, 50, –10, (*d*) 2, 50, 10.

(*vi*) The matrix $A$ has eigen values $\lambda_i \neq 0$. Then $A^{-1} - 2I + A$ has eigenvalues

(*a*) $1 + 2\lambda_i + \lambda_i^2$ (*b*) $\frac{1}{\lambda_i} - 2 + \lambda_i$ (*c*) $1 - 2\lambda_i + \lambda_i^2$ (*d*) $1 - \frac{2}{\lambda_i} + \frac{1}{\lambda_i^2}$

(*viii*) The eigen values of a matrix $A$ are 1,–2, 3. The eigen of $3I - 2A + A^2$ are

(*a*) 2, 11, 6 (*b*) 3, 11, 18 (*c*) 2, 3, 6 (*d*) 6, 3, 11

**Ans.** (*i*)(*b*), (*ii*)(*c*), (*iii*)(*c*), (*iv*)(*b*), (*v*)(*c*), (*vi*)(*b*), (*vii*)(*a*)

## 4.60 DIAGONALISATION OF A MATRIX

Diagonalisation of a matrix $A$ is the process of reduction of $A$ to a diagonal form '$D$'. If $A$ is related to $D$ by a similarity transformation such that $D = P^{-1}AP$ then $A$ is reduced to the diagonal matrix $D$ through *modal matrix P*. $D$ is also called *spectral matrix* of $A$.

## 4.61 THEOREM ON DIAGONALIZATION OF A MATRIX

**Theorem.** *If a square matrix A of order n has n linearly independent eigen vectors, then a matrix P can be found such that $P^{-1}AP$ is a diagonal matrix.*

**Proof.** We shall prove the theorem for a matrix of order 3. The proof can be easily extended to matrices of higher order.

Let
$$A = \begin{bmatrix} a_1 & b_1 & c_1 \\ a_2 & b_2 & c_2 \\ a_3 & b_3 & c_3 \end{bmatrix}$$

and let $\lambda_1, \lambda_2, \lambda_3$ be its eigen values and $X_1, X_2, X_3$ the corresponding eigen vectors, where

$$X_1 = \begin{bmatrix} x_1 \\ y_1 \\ z_1 \end{bmatrix}, \quad X_2 = \begin{bmatrix} x_2 \\ y_2 \\ z_2 \end{bmatrix}, \quad X_3 = \begin{bmatrix} x_3 \\ y_3 \\ z_3 \end{bmatrix}$$

For the eigen value $\lambda_1$, the eigen vector is given by

$$\left.\begin{aligned} (a_1 - \lambda_1)x_1 + b_1 y_1 + c_1 z_1 &= 0 \\ a_2 x_1 + (b_2 - \lambda_1)y_1 + c_2 z_1 &= 0 \\ a_3 x_1 + b_3 y_1 + (c_3 - \lambda_1)z_1 &= 0 \end{aligned}\right\} \quad ...(1)$$

Similarly for $\lambda_2$ and $\lambda_3$ we have

$$\left.\begin{aligned} a_1x_2 + b_1y_2 + c_1z_2 &= \lambda_2 x_2 \\ a_2x_2 + b_2y_2 + c_2z_2 &= \lambda_2 y_2 \\ a_3x_2 + b_3y_2 + c_3z_2 &= \lambda_2 z_2 \end{aligned}\right\} \quad ...(3)$$

and

$$\left.\begin{aligned} a_1x_3 + b_1y_3 + c_1z_3 &= \lambda_3 x_3 \\ a_2x_3 + b_2y_3 + c_2z_3 &= \lambda_3 y_3 \\ a_3x_3 + b_3y_3 + c_3z_3 &= \lambda_3 z_3 \end{aligned}\right\} \quad ...(4)$$

We consider the matrix $P = \begin{bmatrix} x_1 & x_2 & x_3 \\ y_1 & y_2 & y_3 \\ z_1 & z_2 & z_3 \end{bmatrix}$

Whose columns are the eigenvectors of A.

Then $A\,P = \begin{bmatrix} a_1 & b_1 & c_1 \\ a_2 & b_2 & c_2 \\ a_3 & b_3 & c_3 \end{bmatrix}\begin{bmatrix} x_1 & x_2 & x_3 \\ y_1 & y_2 & y_3 \\ z_1 & z_2 & z_3 \end{bmatrix}$

$$= \begin{pmatrix} a_1x_1 + b_1y_1 + c_1z_1 & a_1x_2 + b_1y_2 + c_1z_2 & a_1x_3 + b_1y_3 + c_1z_3 \\ a_2x_1 + b_2y_1 + c_2z_1 & a_2x_2 + b_2y_2 + c_2z_2 & a_2x_3 + b_2y_3 + c_2z_3 \\ a_3x_1 + b_3y_1 + c_3z_1 & a_3x_2 + b_3y_2 + c_3z_2 & a_3x_3 + b_3y_3 + c_3z_3 \end{pmatrix}$$

$$= \begin{pmatrix} \lambda_1x_1 & \lambda_2x_2 & \lambda_3x_3 \\ \lambda_1y_1 & \lambda_2y_2 & \lambda_3y_3 \\ \lambda_1z_1 & \lambda_2z_2 & \lambda_3z_3 \end{pmatrix} \qquad \text{[Using results (2), (3) and (4)]}$$

$$= \begin{pmatrix} x_1 & x_2 & x_3 \\ y_1 & y_2 & y_3 \\ z_1 & z_2 & z_3 \end{pmatrix}\begin{pmatrix} \lambda_1 & 0 & 0 \\ 0 & \lambda_2 & 0 \\ 0 & 0 & \lambda_3 \end{pmatrix} = PD$$

where $D$ is the Diagonal matrix $\begin{pmatrix} \lambda_1 & 0 & 0 \\ 0 & \lambda_2 & 0 \\ 0 & 0 & \lambda_3 \end{pmatrix}$.

$\therefore \quad AP = PD$

$\Rightarrow \quad P^{-1}AP = P^{-1}PD = D$

**Notes 1.** The square matrix $P$, which diagonalises $A$, is found by grouping the eigen vectors of $A$ into square-matrix and the resulting diagonal matrix has the eigen values of $A$ as its diagonal elements.

**2.** The transformation of a matrix $A$ to $P^{-1}AP$ is known as a *similarity transformation.*

**3.** The reduction of $A$ to a diagonal matrix is, obviously, a particular case of similarity transformation.

**4.** The matrix $P$ which diagonalises $A$ is called the *modal matrix* of $A$ and the resulting diagonal matrix $D$ is known as the *spectra matrix* of $A$.

**Example 76.** *Let* $A = \begin{bmatrix} 6 & -2 & 2 \\ -2 & 3 & -1 \\ 2 & -1 & 3 \end{bmatrix}$ *Find matrix P such that* $P^{-1}AP$ *is diagonal matrix.*

**Solution.** The characteristic equation of the matrix $A$ is

$$\begin{vmatrix} 6-\lambda & -2 & 2 \\ -2 & 3-\lambda & -1 \\ 2 & -1 & 3-\lambda \end{vmatrix} = 0$$

$$\Rightarrow \quad (6-\lambda)[9+\lambda^2-6\lambda-1]+2[-6+2\lambda+2]+2[2-6+2\lambda]=0$$

$$\Rightarrow \quad (6-\lambda)(\lambda^2-6\lambda+8)-8+4\lambda-8+4\lambda=0$$

$$\Rightarrow \quad 6\lambda^2-36\lambda+48-\lambda^3+6\lambda^2-8\lambda-16+8\lambda=0$$

$$\Rightarrow \quad -\lambda^3+12\lambda^2-36\lambda+32=0 \quad \Rightarrow \quad \lambda^3-12\lambda^2+36\lambda-32=0$$

$$\Rightarrow \quad (\lambda-2)^2(\lambda-8)=0 \quad \Rightarrow \quad \lambda = 2, 2, 8$$

Eigen vector for $\lambda = 2$

$$\begin{bmatrix} 4 & -2 & 2 \\ -2 & 1 & -1 \\ 2 & -1 & 1 \end{bmatrix}\begin{bmatrix} x_1 \\ x_2 \\ x_3 \end{bmatrix} = \begin{bmatrix} 0 \\ 0 \\ 0 \end{bmatrix} \text{ or } \begin{bmatrix} 2 & -1 & 1 \\ -2 & 1 & -1 \\ 2 & -1 & 1 \end{bmatrix}\begin{bmatrix} x_1 \\ x_2 \\ x_3 \end{bmatrix} = \begin{bmatrix} 0 \\ 0 \\ 0 \end{bmatrix} \quad \begin{matrix} R_2 \to R_1 + R_2 \\ R_3 \to R_2 + R_3 \end{matrix}$$

$$\begin{bmatrix} 2 & -1 & 1 \\ 0 & 0 & 0 \\ 0 & 0 & 0 \end{bmatrix}\begin{bmatrix} x_1 \\ x_2 \\ x_3 \end{bmatrix} = \begin{bmatrix} 0 \\ 0 \\ 0 \end{bmatrix} \text{ or } 2x_1 - x_2 + x_3 = 0$$

This equation is satisfied by $x_1 = 0, x_2 = 1, x_3 = 1$

$$X_1 = \begin{bmatrix} 0 \\ 1 \\ 1 \end{bmatrix}$$

and again $x_1 = 1, x_2 = 3, x_3 = 1.$

$$X_2 = \begin{bmatrix} 1 \\ 3 \\ 1 \end{bmatrix}$$

Eigen vector for $\lambda = 8$

$$\begin{bmatrix} -2 & -2 & 2 \\ -2 & -5 & -1 \\ 2 & -1 & -5 \end{bmatrix}\begin{bmatrix} x_1 \\ x_2 \\ x_3 \end{bmatrix} = \begin{bmatrix} 0 \\ 0 \\ 0 \end{bmatrix}$$

$$-2x_1 - 2x_2 + 2x_3 = 0$$

$$-2x_1 - 5x_2 - x_3 = 0$$

$$\frac{x_1}{2+10} = \frac{x_2}{-4-2} = \frac{x_3}{10-4} \Rightarrow \frac{x_1}{12} = \frac{x_2}{-6} = \frac{x_3}{6} \Rightarrow \frac{x_1}{2} = \frac{x_2}{-1} = \frac{x_3}{1}$$

$$X_3 = \begin{bmatrix} 2 \\ -1 \\ 1 \end{bmatrix}$$

$$\therefore \quad P = \begin{bmatrix} 0 & 1 & 2 \\ 1 & 3 & -1 \\ 1 & 1 & 1 \end{bmatrix}, \quad P^{-1} = -\frac{1}{6}\begin{bmatrix} 4 & 1 & -7 \\ -2 & -2 & 2 \\ -2 & 1 & -1 \end{bmatrix}$$

Now $P^{-1}AP = -\frac{1}{6}\begin{bmatrix} 4 & 1 & -7 \\ -2 & -2 & 2 \\ -2 & 1 & -1 \end{bmatrix}\begin{bmatrix} 6 & -2 & 2 \\ -2 & 3 & -1 \\ 2 & -1 & 3 \end{bmatrix}\begin{bmatrix} 0 & 1 & 2 \\ 1 & 3 & -1 \\ 1 & 1 & 1 \end{bmatrix} = \begin{bmatrix} 2 & 0 & 0 \\ 0 & 2 & 0 \\ 0 & 0 & 8 \end{bmatrix}$ **Ans.**

**Example 77.** *The matrix* $A = \begin{bmatrix} a & h \\ h & b \end{bmatrix}$ *is transformed to the diagonal form* $D = T^{-1}AT$, *where* $T = \begin{bmatrix} \cos\theta & \sin\theta \\ -\sin\theta & \cos\theta \end{bmatrix}$. *Find the value of* $\theta$ *which gives this diagonal transformation.*

**Solution.** $T = \begin{bmatrix} \cos\theta & \sin\theta \\ -\sin\theta & \cos\theta \end{bmatrix}$ $\therefore\ T^{-1} = \begin{bmatrix} \cos\theta & -\sin\theta \\ \sin\theta & \cos\theta \end{bmatrix}$

Now $T^{-1}AT = \begin{bmatrix} \cos\theta & -\sin\theta \\ \sin\theta & \cos\theta \end{bmatrix}\begin{bmatrix} a & h \\ h & b \end{bmatrix}\begin{bmatrix} \cos\theta & \sin\theta \\ -\sin\theta & \cos\theta \end{bmatrix}$

$$= \begin{bmatrix} a\cos\theta - h\sin\theta & h\cos\theta - b\sin\theta \\ a\sin\theta + h\cos\theta & h\sin\theta + b\cos\theta \end{bmatrix}\begin{bmatrix} \cos\theta & \sin\theta \\ -\sin\theta & \cos\theta \end{bmatrix}$$

$$= \begin{bmatrix} a\cos^2\theta - 2h\sin\theta\cos\theta + b\sin^2\theta & (a-b)\sin\theta\cos\theta - h\sin^2\theta + h\cos^2\theta \\ (a-b)\sin\theta\cos\theta + h\cos^2\theta - h\sin^2\theta & a\sin^2\theta + 2h\sin\theta\cos\theta + b\cos^2\theta \end{bmatrix}$$

$$= \begin{bmatrix} a\cos^2\theta - h\sin 2\theta + b\sin^2\theta & (a-b)\sin\theta\cos\theta + h\cos 2\theta \\ (a-b)\sin\theta\cos\theta + h\cos 2\theta & a\sin^2\theta + h\sin 2\theta + b\cos^2\theta \end{bmatrix}$$

$$= \begin{bmatrix} d_1 & 0 \\ 0 & d_2 \end{bmatrix} \text{ being diagonal matrix}$$

$\therefore\quad (a-b)\sin\theta\cos\theta + h\cos 2\theta = 0$

$\Rightarrow\quad \frac{a-b}{2}\sin 2\theta + h\cos 2\theta = 0 \quad \Rightarrow\quad \frac{a-b}{2}\sin 2\theta = -h\cos 2\theta$

$\Rightarrow\quad \tan 2\theta = \frac{2h}{b-a} \quad \Rightarrow\quad \theta = \frac{1}{2}\tan^{-1}\frac{2h}{b-a}$ **Ans.**

## EXERCISE 4.26

**1.** Find the matrix $B$ which transforms the matrix

$A = \begin{bmatrix} 8 & -8 & -2 \\ 4 & -3 & -2 \\ 3 & -4 & 1 \end{bmatrix}$ to a diagonal matrix. **Ans.** $B = \begin{bmatrix} 4 & 3 & 2 \\ 3 & 2 & 1 \\ 2 & 1 & 1 \end{bmatrix}$

**2.** For the matrix $A = \begin{bmatrix} 4 & 1 & 0 \\ 1 & 4 & 1 \\ 0 & 1 & 4 \end{bmatrix}$, determine a matrix P such that $P^{-1}AP$ is diagonal matrix.

**Ans.** $P = \begin{bmatrix} -1 & 1 & 1 \\ 0 & -\sqrt{2} & \sqrt{2} \\ 1 & 1 & 1 \end{bmatrix}$

**3.** Determine the eigen values and the corresponding eigen vectors of the matrix $A = \begin{bmatrix} 5 & 7 & -5 \\ 0 & 4 & -1 \\ 2 & 8 & -3 \end{bmatrix}$

Hence find the matrix $P$ such that $P^{-1}AP$ is diagonal matrix. **Ans.** $P = \begin{bmatrix} 2 & 1 & -1 \\ 1 & 1 & 1 \\ 3 & 2 & 1 \end{bmatrix}$

**4.** Reduce the following matrix $A$ into a diagonal matrix

$$A=\begin{bmatrix} 8 & -6 & 2 \\ -6 & 7 & -4 \\ 2 & -4 & 3 \end{bmatrix}$$ **Ans.** $\begin{bmatrix} 0 & 0 & 0 \\ 0 & 3 & 0 \\ 0 & 0 & 15 \end{bmatrix}$

**5.** Prove that similar matrices have the same eigenvalues. Also give the relationship between the eigenvectors of two similar matrices. (*A.M.I.E.T.E, June 2005*)

**6.** Let a 4 × 4 matrix A have eigenvalues 1, –1, 2, –2 and matrix $B = 2A + A^{-1} - I$ Find (*i*) determinant of matrix B. (*ii*) trace of matrix $B$. (*A.M.I.E.T.E, June 2005*)

## 4.62 POWERS OF A MATRIX (By diagonalisation)

We can obtain powers of a matrix by using diagonalisation.

We know that $D = P^{-1} AP$

Where $A$ is the square matrix and $P$ is a non-singular matrix.

$$D^2 = (P^{-1} AP)(P^{-1} AP) = P^{-1} A (P P^{-1}) AP = P^{-1} A^2 P$$

Similarly $D^3 = P^{-1} A^3 P$

In general $D^n = P^{-1} A^n P$ ...(1)

Pre-multiply (1) by $P$ and post-multiply by $P^{-1}$

$$P D^n P^{-1} = P (P^{-1} A^n P) P^{-1}$$
$$= (P P^{-1}) A^n (P P^{-1})$$
$$= A^n$$

**Procedure:** (1) Find eigen values for a square matrix $A$.

(2) Find eigen vectors to get the modal matrix $P$.

(3) Find the diagonal matrix $D$, by the formula $D = P^{-1}$ AP

(4) Obtain $A^n$ by the formula $A^n = P D^n P^{-1}$.

**Example 78.** *Find a matrix P which transform the matrix* $A=\begin{bmatrix} 1 & 0 & -1 \\ 1 & 2 & 1 \\ 2 & 2 & 3 \end{bmatrix}$ *to diagonal form. Hence $A^4$.*

**Solution.** Characteristic equation of the matrix $A$ is

$$\begin{vmatrix} 1-\lambda & 0 & -1 \\ 1 & 2-\lambda & 1 \\ 2 & 2 & 3-\lambda \end{vmatrix} = 0$$

or $\lambda^3 - 6\lambda^2 + 11\lambda - 6 = 0$

or $(\lambda-1)(\lambda-2)(\lambda-3) = 0$

$\Rightarrow \lambda = 1, 2, 3$

For $\lambda = 1$, eigen vector is given by

$$\begin{bmatrix} 1-1 & 0 & -1 \\ 1 & 2-1 & 1 \\ 2 & 2 & 3-1 \end{bmatrix}\begin{bmatrix} x_1 \\ x_2 \\ x_3 \end{bmatrix} = \begin{bmatrix} 0 \\ 0 \\ 0 \end{bmatrix} \Rightarrow \begin{bmatrix} 0 & 0 & -1 \\ 1 & 1 & 1 \\ 2 & 2 & 2 \end{bmatrix}\begin{bmatrix} x_1 \\ x_2 \\ x_3 \end{bmatrix} = \begin{bmatrix} 0 \\ 0 \\ 0 \end{bmatrix}$$

$$\left.\begin{array}{l} 0x_1 + 0x_2 - x_3 = 0 \\ x_1 + x_2 + x_3 = 0 \end{array}\right] \Rightarrow \frac{x_1}{0+1} = \frac{x_2}{-1+0} = \frac{x_3}{0} \text{ or } x_1 = 1, x_2 = -1, x_3 = 0$$

Eigen vector is [1, –1, 0].

For $\lambda = 2$, eigen vector is given by

$$\begin{bmatrix} 1-2 & 0 & -1 \\ 1 & 2-2 & 1 \\ 2 & 2 & 3-2 \end{bmatrix}\begin{bmatrix} x_1 \\ x_2 \\ x_3 \end{bmatrix} = \begin{bmatrix} 0 \\ 0 \\ 0 \end{bmatrix} \Rightarrow \begin{bmatrix} -1 & 0 & -1 \\ 1 & 0 & 1 \\ 2 & 2 & 1 \end{bmatrix}\begin{bmatrix} x_1 \\ x_2 \\ x_3 \end{bmatrix} = \begin{bmatrix} 0 \\ 0 \\ 0 \end{bmatrix}$$

$$\Rightarrow \begin{bmatrix} 0 & 0 & 0 \\ 1 & 0 & 1 \\ 2 & 2 & 1 \end{bmatrix}\begin{bmatrix} x_1 \\ x_2 \\ x_3 \end{bmatrix} = \begin{bmatrix} 0 \\ 0 \\ 0 \end{bmatrix} \quad R_1 \to R_1 + R_2$$

$$\left.\begin{array}{l} x_1 + 0x_2 + x_3 = 0 \\ 2x_1 + 2x_2 + x_3 = 0 \end{array}\right]$$

$$\Rightarrow \frac{x_1}{0-2} = \frac{x_2}{2-1} = \frac{x_3}{2-0} \quad \Rightarrow \quad x_1 = -2, \quad x_2 = 1, \quad x_3 = 2$$

Eigen vector is [–2, 1, 2].

For $\lambda = 3$, eigen vector is given by

$$\begin{bmatrix} 1-3 & 0 & -1 \\ 1 & 2-3 & 1 \\ 2 & 2 & 3-3 \end{bmatrix}\begin{bmatrix} x_1 \\ x_2 \\ x_3 \end{bmatrix} = \begin{bmatrix} 0 \\ 0 \\ 0 \end{bmatrix} \Rightarrow \begin{bmatrix} -2 & 0 & -1 \\ 1 & -1 & 1 \\ 2 & 2 & 0 \end{bmatrix}\begin{bmatrix} x_1 \\ x_2 \\ x_3 \end{bmatrix} = \begin{bmatrix} 0 \\ 0 \\ 0 \end{bmatrix}$$

$$\left.\begin{array}{l} -2x_1 + 0x_2 - x_3 = 0 \\ x_1 - x_2 + x_3 = 0 \end{array}\right]$$

$$\Rightarrow \frac{x_1}{0-1} = \frac{x_2}{-1+2} = \frac{x_3}{2-0} \quad \Rightarrow \quad x_1 = -1, \quad x_2 = 1, \quad x_3 = 2$$

Eigen vector is [–1, 1, 2].

$$\text{Modal matrix } P = \begin{bmatrix} 1 & -2 & -1 \\ -1 & 1 & 1 \\ 0 & 2 & 2 \end{bmatrix} \text{ and } P^{-1} = -\frac{1}{2}\begin{bmatrix} 0 & 2 & -1 \\ 2 & 2 & 0 \\ -2 & -2 & -1 \end{bmatrix}$$

$$\text{Now} \quad P^{-1}AP = \begin{bmatrix} 0 & -1 & \frac{1}{2} \\ -1 & -1 & 0 \\ 1 & 1 & \frac{1}{2} \end{bmatrix}\begin{bmatrix} 1 & 0 & -1 \\ 1 & 2 & 1 \\ 2 & 2 & 3 \end{bmatrix}\begin{bmatrix} 1 & -2 & -1 \\ -1 & 1 & 1 \\ 0 & 2 & 2 \end{bmatrix} = \begin{bmatrix} 1 & 0 & 0 \\ 0 & 2 & 0 \\ 0 & 0 & 3 \end{bmatrix} = D$$

$$A^4 = PD^4P^{-1} = \begin{bmatrix} 1 & -2 & -1 \\ -1 & 1 & 1 \\ 0 & 2 & 2 \end{bmatrix}\begin{bmatrix} 1 & 0 & 0 \\ 0 & 16 & 0 \\ 0 & 0 & 81 \end{bmatrix}\begin{bmatrix} 0 & -1 & \frac{1}{2} \\ -1 & -1 & 0 \\ 1 & 1 & \frac{1}{2} \end{bmatrix} = \begin{bmatrix} -49 & -50 & -40 \\ 65 & 66 & 40 \\ 130 & 130 & 81 \end{bmatrix} \quad \textbf{Ans.}$$

## EXERCISE 4.27

Find a matrix $P$ which transforms the following matrices to diagonal form. Hence calculate the power matrix.

**1.** If $A = \begin{bmatrix} 1 & 1 & 3 \\ 1 & 5 & 1 \\ 3 & 1 & 1 \end{bmatrix}$, calculate $A^4$. **Ans.** $\begin{bmatrix} 251 & 405 & 235 \\ 405 & 891 & 405 \\ 235 & 405 & 251 \end{bmatrix}$

**2.** If $A = \begin{bmatrix} 3 & -1 & 1 \\ -1 & 5 & -1 \\ 1 & -1 & 3 \end{bmatrix}$, calculate $A^4$. **Ans.** $\begin{bmatrix} 251 & -405 & 235 \\ -405 & 891 & -405 \\ 235 & -405 & 251 \end{bmatrix}$

**3.** If $A = \begin{bmatrix} 2 & -1 & 1 \\ -1 & 2 & -1 \\ 1 & -1 & 2 \end{bmatrix}$, calculate $A^6$. **Ans.** $\begin{bmatrix} 1366 & -1365 & 1365 \\ -1365 & 1366 & -1365 \\ 1365 & -1365 & 1366 \end{bmatrix}$

4. If $A=\begin{bmatrix}1 & 1 & 1\\ 0 & 2 & 1\\ -4 & 4 & 3\end{bmatrix}$, calculate $A^8$. **Ans.** $\begin{bmatrix}-12099 & 12355 & 6305\\ -12100 & 12356 & 6305\\ -13120 & 13120 & 6561\end{bmatrix}$

5. Show that the matrix $A$ is diagonalisable $A=\begin{bmatrix}3 & 1 & -1\\ -2 & 1 & 2\\ 0 & 1 & 2\end{bmatrix}$. If so obtain the matrix $P$ such that $P^{-1}AP$ is a diagonal matrix. *(AMIETE, June 2010)*

## 4.63 SYLVESTER THEOREM

Let $\quad P(A) = C_0 A^n + C_1 A^{n-1} + C_2 A^{n-2} + \ldots + C_{n-1} A + C_n I$

and $\quad |\lambda I - A| = f(\lambda)$ and Adjoint matrix of $[\lambda I - A] = [f(\lambda)]$

$$z(\lambda) = \frac{[f(\lambda)]}{f'(\lambda)} = \frac{\text{Adjoint matrix of } [\lambda I - A]}{f'(\lambda)}$$

Then according to Sylvester's theorem

$$P(A) = P(\lambda_1).Z(\lambda_1) + P(\lambda_2).Z(\lambda_2) + P(\lambda_3).Z(\lambda_3) + \ldots$$

$$= \sum_{r=1}^{n} P(\lambda_r).Z(\lambda_r)$$

**Example 79.** *If* $A=\begin{bmatrix}2 & 0\\ 0 & 1\end{bmatrix}$, *find* $A^{100}$.

**Solution.** $f(\lambda) = |\lambda I - A| = \begin{vmatrix}\lambda & 0\\ 0 & \lambda\end{vmatrix} - \begin{vmatrix}2 & 0\\ 0 & 1\end{vmatrix} = \begin{vmatrix}\lambda-2 & 0\\ 0 & \lambda-1\end{vmatrix} = 0$

$\Rightarrow \quad f(\lambda) = (\lambda-2)(\lambda-1) = 0$ or $\lambda_1 = 1,\ \lambda_2 = 2$

$$f(\lambda) = \lambda^2 - 3\lambda + 2,\quad f'(\lambda) = 2\lambda - 3$$

$$f'(2) = 4 - 3 = 1,\ f'(1) = 2 - 3 = -1$$

$$[f(\lambda)] = \text{Adjoint matrix of the matrix } [\lambda I - A] = \begin{bmatrix}\lambda-1 & 0\\ 0 & \lambda-2\end{bmatrix}$$

$$Z(\lambda_1) = Z(1) = \frac{[f(1)]}{f'(1)} = \frac{1}{-1}\begin{bmatrix}0 & 0\\ 0 & -1\end{bmatrix} = \begin{bmatrix}0 & 0\\ 0 & 1\end{bmatrix}$$

$$Z(\lambda_2) = Z(2) = \frac{[f(2)]}{f'(2)} = \frac{1}{1}\begin{bmatrix}1 & 0\\ 0 & 0\end{bmatrix}$$

By Sylvester theorem $P(A) = P(\lambda_1).Z(\lambda_1) + P(\lambda_2).Z(\lambda_2)$

$$A^{100} = P(\lambda_1)Z(\lambda_1) + P(\lambda_2)Z(\lambda_2)$$

$$= \lambda_1^{100}\begin{bmatrix}0 & 0\\ 0 & 1\end{bmatrix} + \lambda_2^{100}\begin{bmatrix}1 & 0\\ 0 & 0\end{bmatrix} = 1^{100}\begin{bmatrix}0 & 0\\ 0 & 1\end{bmatrix} + 2^{100}\begin{bmatrix}1 & 0\\ 0 & 0\end{bmatrix}$$

$$= \begin{bmatrix}0 & 0\\ 0 & 1\end{bmatrix} + \begin{bmatrix}2^{100} & 0\\ 0 & 0\end{bmatrix} = \begin{bmatrix}2^{100} & 0\\ 0 & 1\end{bmatrix}$$ **Ans.**

## EXERCISE 4.28

1. Verify Sylvesters theorem for $A^3$, where $A=\begin{bmatrix}1 & 4\\ 3 & 2\end{bmatrix}$.

**Use sylvesters theorem in solving the following:**

2. Given $A=\begin{bmatrix}1 & 0\\ 0 & 3\end{bmatrix}$, find $A^{256}$. **Ans.** $\begin{bmatrix}1 & 0\\ 0 & 3^{256}\end{bmatrix}$

3. Given $A=\begin{bmatrix}\lambda_1 & 0\\ 0 & \lambda_2\end{bmatrix}$, show that $e^A=\begin{bmatrix}e^{\lambda_1} & 0\\ 0 & e^{\lambda_2}\end{bmatrix}$.

4. Given $A=\begin{bmatrix}-1 & 3\\ 1 & 1\end{bmatrix}$, show that 2 sin $A$ = | sin 2 | $A$.

5. Prove that 3 tan $A = A$ tan (3) where $A=\begin{bmatrix}-1 & 4\\ 2 & 1\end{bmatrix}$

6. Prove that $\sin^2 A+\cos^2 A=1$, where $A=\begin{bmatrix}1 & 2\\ -1 & 4\end{bmatrix}$

7. Given $A=\begin{bmatrix}1 & -2 & -3\\ 0 & 2 & 0\\ 0 & 0 & 3\end{bmatrix}$, find $A^{-1}$. **Ans.** $\begin{bmatrix}1 & -1 & 1\\ 0 & -\frac{1}{2} & 0\\ 0 & 0 & \frac{1}{3}\end{bmatrix}$

8. Given $A=\begin{bmatrix}1 & 20 & 0\\ -1 & 7 & 1\\ 3 & 0 & -2\end{bmatrix}$, find tan $A$.

**Ans.** $\frac{\tan 1}{2}\begin{bmatrix}-18 & 60 & 20\\ 0 & 0 & 0\\ -18 & 60 & 20\end{bmatrix}+\frac{\tan 2}{-1}\begin{bmatrix}-20 & 80 & 20\\ -1 & 4 & 1\\ -15 & 60 & 15\end{bmatrix}+\frac{\tan 3}{2}\begin{bmatrix}-20 & 100 & 20\\ -2 & 10 & 2\\ -12 & 60 & 12\end{bmatrix}$

## 4.64 QUADRATIC FORMS

The quadratic forms are defined as a homogeneous polynomial of second degree in any number of variables.

For example

1. **Two variables** $ax^2+2hxy+by^2=Q(x, y)$
2. **Three variables** $ax^2+2hxy+by^2+cz^2+2hxy+2gyz+2fzx=Q(x, y, z)$
3. **Four variables**
$ax^2+by^2+cz^2+dw^2+2hxy+2gyz+2fzx+2lxw+2myw+2nzw=Q(x, y, z, w)$
4. ***n* variables** $=Q(x_1, x_2, \ldots\ldots x_n)$

## 4.65 QUADRATIC FORM EXPRESSED IN MATRICES

Quadratic form can be expressed as a product of matrices.

Quadratic form $=Q(x)=X'AX$

where $X=\begin{bmatrix}x_1\\ x_2\\ x_3\end{bmatrix}$ and $A=\begin{bmatrix}a_{11} & a_{12} & a_{13}\\ a_{21} & a_{22} & a_{23}\\ a_{31} & a_{32} & a_{33}\end{bmatrix}$

$X'$ is the transpose of $X$.

$$X'AX=\begin{bmatrix}x_1 & x_2 & x_3\end{bmatrix}\begin{bmatrix}a_{11} & a_{12} & a_{13}\\ a_{21} & a_{22} & a_{23}\\ a_{31} & a_{32} & a_{33}\end{bmatrix}\begin{bmatrix}x_1\\ x_2\\ x_3\end{bmatrix}$$

$$= \begin{bmatrix} a_{11}x_1 + a_{21}x_2 + a_{31}x_3 & a_{12}x_1 + a_{22}x_2 + a_{32}x_3 & a_{13}x_1 + a_{23}x_2 + a_{33}x_3 \end{bmatrix} \begin{bmatrix} x_1 \\ x_2 \\ x_3 \end{bmatrix}$$

$$= a_{11}x_1^2 + a_{21}x_1x_2 + a_{31}x_1x_3 + a_{12}x_1x_2 + a_{22}x_2^2 + a_{32}x_2x_3 + a_{13}x_1x_3 + a_{23}x_2x_3 + a_{33}x_3^2$$

$$= a_{11}x_1^2 + a_{22}x_2^2 + a_{33}x_3^2 + (a_{12} + a_{21})x_1x_2 + (a_{23} + a_{32})x_2x_3 + (a_{31} + a_{13})x_1x_3$$

$a_{12}$ and $a_{21}$ are the coefficients of $x_1x_2$ it means $(a_{12} + a_{21})$ are the coefficient of $x_1x_2$. In general $a_{ij}$ and $a_{ji}$ are the both coefficients of $x_ix_j$ $(i \neq j)$.

So $(a_{ij} + a_{ji})$ are the coefficient of $x_i\,x_j$

Let us have new coefficients of $x_i\,x_j$.

$$c_{ij} = c_{ji} = \frac{1}{2}(a_{ij} + a_{ji})$$

We know $\frac{1}{2}(A + A') =$ symmetric matrix $C$

Thus, the coefficient matrix in quadratic form is always symmetric matrix without loss of generality.

Then $X'AX = c_{11}x_1^2 + c_{22}x_2^2 + c_{33}x^2 + 2c_{12}x_1x_2 + 2c_{23}x_2x_3 + 2c_{31}x_1x_3$

Matrix $A$ is known as the coefficient matrix or matrix of quadratic form and $(R)$ is the discriminant of the quadratic form.

**Example 80.** *Write down the quadratic form corresponding to the matrix*

$$A = \begin{bmatrix} 1 & 2 & 5 \\ 2 & 0 & 3 \\ 5 & 3 & 4 \end{bmatrix}$$

**Solution.** Quadratic form $= X'AX$

$$= [x_1\ x_2\ x_3] \begin{bmatrix} 1 & 2 & 5 \\ 2 & 0 & 3 \\ 5 & 3 & 4 \end{bmatrix} \begin{bmatrix} x_1 \\ x_2 \\ x_3 \end{bmatrix} = [x_1 + 2x_2 + 5x_3,\ \ 2x_1 + 3x_3, 5x_1 + 3x_2 + 4x_3] \begin{bmatrix} x_1 \\ x_2 \\ x_3 \end{bmatrix}$$

$$= x_1^2 + 2x_1x_2 + 5x_3x_1 + 2x_1x_2 + 3x_2x_3 + 5x_1x_3 + 3x_2x_3 + 4x_3^2$$

$$= x_1^2 + 4x_3^2 + 4x_1x_2 + 10x_1x_3 + 6x_2x_3$$ **Ans.**

**Example 81.** *Find a real symmetric matrix C of the quadratic form*:

$$Q(x_1, x_2, x_3) = x_1^2 + 4x_2^2 + 6x_3^2 + 2x_1x_2 + x_2x_3 + 3x_1x_3$$

**Solution.** On Comparing the coefficients in the given quadratic form, with the standard quadratic form, we get

Here $a_{11} = 1,\ a_{22} = 4,\ a_{33} = 6,\ a_{12} = 2,\ a_{21} = 0,\ a_{23} = 1,\ a_{32} = 0,\ a_{13} = 3,\ a_{31} = 0$

$$Q(x_1, x_2, x_3) = [x_1\ \ x_2\ \ x_3] \begin{bmatrix} 1 & 2 & 3 \\ 0 & 4 & 1 \\ 0 & 0 & 6 \end{bmatrix} \begin{bmatrix} x_1 \\ x_2 \\ x_3 \end{bmatrix} = [x_1\ \ 2x_1 + 4x_2\ \ 3x_1 + x_2 + 6x_3] \begin{bmatrix} x_1 \\ x_2 \\ x_3 \end{bmatrix}$$

$$= x_1^2 + 2x_1x_2 + 4x_2^2 + 3x_1x_3 + x_2x_3 + 6x_3^2$$

$$C = \frac{1}{2}(A + A') = \frac{1}{2}\begin{bmatrix} 1 & 2 & 3 \\ 0 & 4 & 1 \\ 0 & 0 & 6 \end{bmatrix} + \begin{bmatrix} 1 & 0 & 0 \\ 2 & 4 & 0 \\ 3 & 1 & 6 \end{bmatrix} = \frac{1}{2}\begin{bmatrix} 2 & 2 & 3 \\ 2 & 8 & 1 \\ 3 & 1 & 12 \end{bmatrix} = \begin{bmatrix} 1 & 1 & \frac{3}{2} \\ 1 & 4 & \frac{1}{2} \\ \frac{3}{2} & \frac{1}{2} & 6 \end{bmatrix}$$ **Ans.**

## 4.66 LINEAR TRANSFORMATION OF QUADRATIC FORM

(Diagonalisation of the matrix)

Let the given quadratic form be $X'AX$ where $A$ is a symmetric matrix.

Consider the linear transformation $X = PY$

Then $\quad X' = (PY)' = Y' P'$

$\therefore \quad X'AX = (Y'P')\, A\, (PY) = Y'(P'AP)\, Y = Y'BY$

where $\quad B = P'AP \qquad$ (Transformed quadratic form)

Now $\quad B' = (P'AP)' = P'AP = B$

Rank $(B)$ = Rank $(A)$

Therefore, $A$ and $B$ are *congruent matrices* and the transformation $X = PY$ is known as *congruent transformation.*

## 4.67 CANONICAL FORM OF SUM OF THE SQUARES FORM USING LINEAR TRANSFORMATION

When a quadratic form is linearly transformed then the transformed quadratic of new variable is called canonical form of the given quadratic form.

When $X'AX$ is linearly transformed then the transformed quadratic $Y'BY$ is called the canonical form of the given quadratic $X'AX$.

If $B = P'AP = \text{Diag}\ (\lambda_1, \lambda_2, \lambda_3 \ldots\ldots \lambda_n)$ than $X'AX = Y'BY = \sum_{i=1}^{n} \lambda_i Y_i^2$

**Remarks.** (1) $\lambda_i$ (eigen values) can be positive or negative or zero.

(2) If Rank $(A) = r$, then the quadratic form $X'AX$ will contain only $r$ terms.

## 4.68 CANONICAL FORM OF SUM OF THE SQUARES FORM USING ORTHOGONAL TRANSFORMATION

Real symmetric matrix $A$ can be reduced to a diagonal form $M'AM = D$ ...(1)

where $M$ is the normalised orthogonal *modal matrix* of $A$ and $D$ is its *spectoral matrix.*

Let the orthogonal transformation be

$$X = MY$$

$$Q = X'AX = (MY)'\, A\, (MY) = (Y'M')\, A\, (MY) = Y'\, (M'AM)\, Y$$

$$= Y'DY \qquad [\because M'AM = D]$$

$$= Y'\ \text{Diag.}\ (\lambda_1\ \lambda_2 \ldots \lambda_n)\, Y$$

$$= [y_1\ \ y_2 \ldots\ldots y_n] \begin{bmatrix} \lambda_1 & 0 \ldots\ldots 0 \\ 0 & \lambda_2 \ldots\ldots 0 \\ \vdots & \vdots \quad\ \ \vdots \\ 0 & 0 \ldots\ldots \lambda_n \end{bmatrix} \begin{bmatrix} y_1 \\ y_2 \\ \vdots \\ y_n \end{bmatrix} = [\lambda_1 y_1\ \ \lambda_2 y_2 \ldots\ldots \lambda_n y_n] \begin{bmatrix} y_1 \\ y_2 \\ \vdots \\ y_n \end{bmatrix}$$

$$= \lambda_1 y_1^2 + \lambda_2 y_2^2 + \ldots\ldots + \lambda_n y_n^2, \text{ which is called canonical form.}$$

Now, we have seen that quadratic form $X'AX$ can be reduced to the sum of the squares by the transformation $X = PI$ where $P$ is the normalised modal matrix of $A$.

**Canonical form.** $B$ is a diagonal matrix, then the transformed quadratic is a sum of square terms, known as canonical form.

**Index.** The number of positive terms in canonical form of a quadratic form is known as index $(s)$ of the form.

**Rank of form.** Rank $(r)$ of matrix $B$ (or $A$) is called the rank of the form.

**Signature of quadratic form.** The difference of positive terms ($s$) and negative terms ($r-s$) is known as the signature of quadratic form.

$$\text{Signature} = s-(r-s) = s-r+s = 2s-r$$

## 4.69 CLASSIFICATION OF DEFINITENESS OF A QUADRATIC FORM $A$

Let $Q$ be $X'AX$ and variables $(x_1, x_2, x_3 \ldots x_n)$,

$$\text{Rank }(A) = r,$$
$$\text{Index} = s$$

**1. Positive definite**

If rank and index are equal *i.e.*, $r = n$, $s = n$ or if all the eigen values of $A$ are positive.

**2. Negative definite**

If index = 0, *i.e.*, $r = n$, $s = 0$ or if all the eigen values of $A$ are negative.

**3. Positive semi-definite**

If rank and index are equal but less than $n$, *i.e.*, $s = r < n$ $[\,|A| = 0]$

or all eigen values of A are positive at least one eigen value is zero.

**4. Negative semi-definite**

If index is zero, *i.e.*, $s = 0$, $r < n$ $[\,|A| = 0]$

or all eigen values of A are negative and at least one eigen value is zero.

**5. Indefinite**

If some eigen values are positive and some eigen values are negative.

**6. Notes :**

(1) If $Q$ is negative definite (semi-definite) then $-Q$ is positive definite (semi-definite).

(2) The classification of the definiteness of a quadratic form depends upon the location of eigen values of $A$.

**Example 82.** *Reduce to diagonal form the following symmetric matrix by congruent transformation and interpret the result in terms of quadratic form*

$$A = \begin{bmatrix} 3 & 2 & -1 \\ 2 & 2 & 3 \\ -1 & 3 & 1 \end{bmatrix}$$

**Solution.** $A = \begin{bmatrix} 3 & 2 & -1 \\ 2 & 2 & 3 \\ -1 & 3 & 1 \end{bmatrix}$

Let us reduce $A$ into diagonal matrix.

$$I\,A\,I = A$$

$$\underset{R}{\begin{bmatrix} 1 & 0 & 0 \\ 0 & 1 & 0 \\ 0 & 0 & 1 \end{bmatrix}} A \underset{S}{\begin{bmatrix} 1 & 0 & 0 \\ 0 & 1 & 0 \\ 0 & 0 & 1 \end{bmatrix}} = \begin{bmatrix} 3 & 2 & -1 \\ 2 & 2 & 3 \\ -1 & 3 & 1 \end{bmatrix} \qquad |A| \neq 0$$

Row transformation carried out on R.H.S. will be applied on $R$ prefactor matrix.
Column transformation applied on R.H.S. will be applied on $S$ post factor matrix.

$$\begin{bmatrix} 1 & 0 & 0 \\ -\frac{2}{3} & 1 & 0 \\ \frac{1}{3} & 0 & 1 \end{bmatrix} A \begin{bmatrix} 1 & 0 & 0 \\ 0 & 1 & 0 \\ 0 & 0 & 1 \end{bmatrix} = \begin{bmatrix} 3 & 2 & -1 \\ 0 & \frac{2}{3} & \frac{11}{3} \\ 0 & \frac{11}{3} & \frac{2}{3} \end{bmatrix} \qquad R_2 - \frac{2}{3}R_1,\; R_3 + \frac{R_1}{3}$$

$$\begin{bmatrix} 1 & 0 & 0 \\ -\frac{2}{3} & 1 & 0 \\ \frac{1}{3} & 0 & 1 \end{bmatrix} A \begin{bmatrix} 1 & -\frac{2}{3} & \frac{1}{3} \\ 0 & 1 & 0 \\ 0 & 0 & 1 \end{bmatrix} = \begin{bmatrix} 3 & 0 & 0 \\ 0 & \frac{2}{3} & \frac{11}{3} \\ 0 & \frac{11}{3} & \frac{2}{3} \end{bmatrix} \qquad C_2 - \frac{2}{3}C_1,\ C_3 + \frac{1}{3}C_1$$

$$\begin{bmatrix} 1 & 0 & 0 \\ -\frac{2}{3} & 1 & 0 \\ 4 & -\frac{11}{2} & 1 \end{bmatrix} A \begin{bmatrix} 1 & -\frac{2}{3} & \frac{1}{3} \\ 0 & 1 & 0 \\ 0 & 0 & 1 \end{bmatrix} = \begin{bmatrix} 3 & 0 & 0 \\ 0 & \frac{2}{3} & \frac{11}{3} \\ 0 & 0 & -\frac{39}{2} \end{bmatrix} \qquad R_3 - \frac{11}{2}R_2$$

$$\begin{bmatrix} 1 & 0 & 0 \\ -\frac{2}{3} & 1 & 0 \\ 4 & -\frac{11}{2} & 1 \end{bmatrix} A \begin{bmatrix} 1 & -\frac{2}{3} & 4 \\ 0 & 1 & -\frac{11}{2} \\ 0 & 0 & 1 \end{bmatrix} = \begin{bmatrix} 3 & 0 & 0 \\ 0 & \frac{2}{3} & 0 \\ 0 & 0 & -\frac{39}{2} \end{bmatrix} \qquad C_3 - \frac{11}{2}C_2$$

Thus the matrix $A$ is reduced to the diagonal form $B$.

$$P'AP = \begin{bmatrix} 3 & 0 & 0 \\ 0 & \frac{2}{3} & 0 \\ 0 & 0 & -\frac{39}{2} \end{bmatrix} \quad \text{where } P = \begin{bmatrix} 1 & -\frac{2}{3} & 4 \\ 0 & 1 & -\frac{11}{2} \\ 0 & 0 & 1 \end{bmatrix}$$

The canonical form (sum of the squares) is

$$Q = Y'BY = [y_1\ y_2\ y_3] \begin{bmatrix} 3 & 0 & 0 \\ 0 & \frac{2}{3} & 0 \\ 0 & 0 & -\frac{39}{2} \end{bmatrix} \begin{bmatrix} y_1 \\ y_2 \\ y_3 \end{bmatrix} = 3y_1^2 + \frac{2}{3}y_2^2 - \frac{39}{2}y_3^2$$

$$X = PY \qquad i.e. \quad \begin{bmatrix} x_1 \\ x_2 \\ x_3 \end{bmatrix} = \begin{bmatrix} 1 & -\frac{2}{3} & 4 \\ 0 & 1 & -\frac{11}{2} \\ 0 & 0 & 1 \end{bmatrix} \begin{bmatrix} y_1 \\ y_2 \\ y_3 \end{bmatrix}$$

$$x_1 = y_1 - \frac{2}{3}y_2 + 4y_3, \qquad x_2 = y_2 - \frac{11}{2}y_3, \qquad x_3 = y_3$$

The rank of $A$ $(r) = 3$

The index of quadratic form $(s) = 2$

The signature of quadratic form $[r - (r - P)] = 2 - (3 - 2) = 1$ **Ans.**

**Example 83.** *Reduce the quadratic form* $6x_1^2+3x_2^2+3x_3^2-4x_1x_2-2x_2x_3+4x_3x_1$ *to the sum of square, by Lagrange Reduction Method*

**Solution.** $Q=6x_1^2+3x_2^2-4x_1x_2+4x_1x_3-2x_2x_3$

$$=6\left[x_1^2-\frac{2}{3}x_1(x_2-x_3)\right]+3x_2^2+3x_3^2-2x_2x_3$$

$$=6\left[x_1-\frac{1}{3}(x_2-x_3)\right]^2+3x_2^2+3x_3^2-2x_2x_3-\frac{2}{3}(x_2-x_3)^2$$

$$=6\left(x_1-\frac{1}{3}x_2+\frac{1}{3}x_3\right)^2+\frac{7}{3}x_2^2-\frac{2}{3}x_2x_3+\frac{7}{3}x_3^2$$

$$=6\left(x_1-\frac{1}{3}x_2+\frac{1}{3}x_3\right)^2+\frac{7}{3}\left(x_2^2-\frac{2}{7}x_2x_3\right)+\frac{7}{3}x_3^2$$

$$=6\left(x_1-\frac{1}{3}x_2+\frac{1}{3}x_3\right)^2+\frac{7}{3}\left(x_2-\frac{1}{7}x_3\right)^2+\frac{7}{3}x_3^2-\frac{7}{3}\times\frac{1}{49}x_3^2$$

$$=6\left(x_1-\frac{1}{3}x_2+\frac{1}{3}x_3\right)^2+\frac{7}{3}\left(x_2-\frac{1}{7}x_3\right)^2+\frac{16}{7}x_3^2=6y_1^2+\frac{7}{3}y_2^2+\frac{16}{7}y_3^2$$

where
$$\left.\begin{aligned}y_1&=x_1-\frac{1}{3}x_2+\frac{1}{3}x_3\\ y_2&=x_2-\frac{1}{7}x_3\\ y_3&=x_3\end{aligned}\right] \Rightarrow \begin{aligned}x_3&=y_3\\ x_2&=y_2+\frac{1}{7}y_3\\ x_1&=y_1+\frac{1}{3}y_2-\frac{2}{7}y_3\end{aligned}$$

## EXERCISE 4.29

**1.** Express the quadratic form $x_1^2+2x_2^2+2x_3^2-2x_1x_2-2x_2x_3$ as product of matrices. **Ans.** $\begin{bmatrix}x_1 & x_2 & x_3\end{bmatrix}\begin{bmatrix}1 & -1 & 0\\ -1 & 2 & -1\\ 0 & -1 & 2\end{bmatrix}\begin{bmatrix}x_1\\ x_2\\ x_3\end{bmatrix}$

**2.** Write down the matrix of the quadratic form

$x_1^2+2x_2^2-7x_3^2+x_4^2-4x_1x_2+8x_1x_3-6x_3x_4$ **Ans.** $\begin{bmatrix}1 & -2 & 4 & 0\\ -2 & 2 & 0 & 0\\ 4 & 0 & -7 & -3\\ 0 & 0 & -3 & 1\end{bmatrix}$

**3.** Find the transformation that will transform $10x^2+2y^2+5z^2+6yz-10zx-4xy$ into a sum of square and find its reduced form.

**Ans.** $Q=10y_1^2+\frac{8}{5}y_2^2,\ P=\begin{bmatrix}1 & \frac{1}{5} & \frac{1}{4}\\ 0 & 1 & -\frac{5}{4}\\ 0 & 0 & 1\end{bmatrix}$

4. Find the transformation which will transform the following form into a sum of squares and find the reduced form :

$Q = 4x^2 + 3y^2 + z^2 - 8\,xy - 6\,yz + 4\,xz$ **Ans.** $Q = 4y_1^2 - y_2^2 + y_3^2,\ P = \begin{bmatrix} 1 & 1 & \frac{-3}{2} \\ 0 & 1 & -1 \\ 0 & 0 & 1 \end{bmatrix}$

5. Reduce to sum to squares

$Q = x_1^2 + 2x_2^2 - 7x_3^2 - 4x_1x_2 + 8x_1x_3$ **Ans.** $Q = y_1^2 - 2y_2^2 + 9y_3^2,\ P = \begin{bmatrix} 1 & 2 & 4 \\ 0 & 1 & 4 \\ 0 & 0 & 1 \end{bmatrix}$

6. Express the following quadratic form as "sum of squares" by congruent transformation and write down the corresponding linear transformation

$$Q = 10x_1^2 + x_2^2 + x_3^2 - 6x_1x_2 - 2x_2x_3 + x_3x_1$$

**Ans.** $10y_1^2 + \frac{1}{10}y_2^2,\ x_1 = y_1 + \frac{3}{10}y_2,\ x_2 = y_2 + y_3,\ x_3 = y_3$

7. Reduce to the diagonal matrix by rational congruent transformation and interpret the result in terms of quadratic form.

$A = \begin{bmatrix} 1 & 2 & -1 \\ 2 & 1 & 3 \\ -1 & 3 & 1 \end{bmatrix}$ **Ans.** $Q = x_1^2 + x_2^2 + x_3^2 + 4x_1x_2 + 6x_2x_3 - 2x_3x_1;\ Q = y_1^2 - 4y_2^2 + \frac{25}{4}y_3^2$

Determine the definiteness of the quadratic forms.

8. $Q = x_1^2 + 2x_2^2 + 3x_3^2 + 2x_2x_3 - 2x_3x_1 + 2x_1x_2$ **Ans.** Indefinite.

9. $Q = 4x_1^2 + x_2^2 + 15x_3^2 - 4x_1x_2$. **Ans.** Positive semi-definite

10. $Q = 5x_1^2 + 26x_2^2 + 10x_3^2 + 4x_2x_3 + 14x_3x_1 + 6x_1x_2$ **Ans.** Positive semi-definite

11. $Q = x_1^2 + 5x_2^2 + x_3^2 + 2x_1x_2 + 2x_2x_3 + 6x_3x_1$. **Ans.** Indefinite

12. $Q = 8x_1^2 + 7x_2^2 + 3x_3^2 - 12x_1x_2 - 8x_2x_3 + 4x_3x_2$ **Ans.** Positive semi-definite

13. $Q = -4x_1^2 - 2x_2^2 - 13x_3^2 - 4x_1x_2 - 8x_2x_3 - 4x_3x_1$. **Ans.** Negative definite

14. Find the eigenvalues and corresponding eigen vector of $A = \begin{bmatrix} 3 & -2 & 4 \\ -2 & -2 & 6 \\ 4 & 6 & -1 \end{bmatrix}$

Verify that the eigen vectors are orthogonal and write down an orthogonal matrix $M$ such that $M'AM = D$, where $D$ is diagonal matrix.

**Ans.** – 9, 6, 3, [12 – 2]′, [212]′, [–2 2 1]′, $M = \frac{P}{3}$, where $P$ is modal matrix

## 4.70 DIFFERENTIATION AND INTEGRATION OF MATRICES

If the elements of a matrix $A$ are function of scalar variable $t$, the matrix is called a matrix function of $t$.

$$A = A(t) = [a_{ij}(t)]$$

The differential coefficient of $A$ w.r.t. "$t$" is defined as

$$\frac{d}{dt}(A)=\left[\frac{d}{dt}(a_{ij})\right]$$

Hence the elements of the differentiated matrix $\frac{dA}{dt}$ are the derivatives of the corresponding elements of $A$.

$$\frac{d}{dt}A(t)=\begin{bmatrix}\frac{da_{11}}{dt} & \frac{da_{12}}{dt} & \cdots & \frac{da_{1n}}{dt}\\ \frac{da_{21}}{dt} & \frac{da_{22}}{dt} & \cdots & \frac{da_{2n}}{dt}\\ \frac{da_{n1}}{dt} & \frac{da_{n2}}{dt} & \cdots & \frac{da_{nn}}{dt}\end{bmatrix}$$

It is easy to prove that

$$\frac{d}{dt}(A)(B)=A.\frac{dB}{dt}+B\frac{da}{dt}$$

**The integral of the matrix $A$ is defined as**

$$\int A dt=\left[\int(a_{ij})\,dt\right]$$

Thus the integral of A is obtained by integrating each element of $A$.

**Power series.** Let $A$ be a square matrix with all eigenvalues less than 1 in absolute value, then $a_0I + a_aA + a_2A^2 + \ldots$ is convergent.

The following series are also convergent

$$e^A=I+\frac{A}{\lfloor 1}+\frac{1}{\lfloor 2}A^2+\ldots.$$

$$\cos A=1-\frac{1}{\lfloor 2}A^2+\frac{1}{\lfloor 4}A^4-\ldots.$$

$$\sin A=A-\frac{1}{\lfloor 3}A^3+\frac{1}{\lfloor 5}A^5-\ldots.$$

$$(1-A)^{-1}=1+A+A^2+\ldots.$$

**Example 84.** *Prove that* $\frac{d}{dt}(e^{tA})=Ae^{tA}$ *if* $e^{tA}=I+\frac{tA}{\lfloor 1}+\frac{1}{\lfloor 2}(tA)^2+\frac{1}{\lfloor 3}(tA)^3+\ldots.$

**Solution.**

$$\frac{d}{dt}(e^{tA})=\frac{d}{dt}\left[1+\frac{tA}{\lfloor 1}+\frac{1}{\lfloor 2}t^2A^2+\frac{1}{\lfloor 3}t^3A^3+\ldots.\right]$$

$$=\frac{d}{dt}(1)+\frac{d}{dt}(tA)+\frac{d}{dt}\left(\frac{1}{\lfloor 2}t^2A^2\right)+\left(\frac{1}{\lfloor 3}t^3A^3\right)+\ldots..$$

$$=0+A+\frac{1}{\lfloor 1}tA^2+\frac{1}{\lfloor 2}t^2A^3+\ldots.$$

$$=A\left[1+\frac{1}{\lfloor 1}tA+\frac{1}{\lfloor 2}t^2A^2+\ldots.\right]=Ae^{tA}$$ **Ans.**

**Example 85.** *Solve* $\frac{d^2x}{dt^2}+4\frac{dx}{dt}-12x=0.$

$x(0)=0; x'(0)=8$ *by Matrix Method.* ...(1)

**Solution.** Let $$x = x_1 \text{ and } \frac{dx_1}{dt} = x_2 \quad ...(2)$$

(1) becomes

$$\therefore \frac{d}{dt}\left(\frac{dx_1}{dt}\right) = -4\frac{dx_1}{dt} + 12x_1 \quad ...(3)$$

or $$\frac{dx_2}{dt} = 12x_1 - 4x_2$$

(2) and (3) are written in matrix form.

$$\frac{d}{dt}\begin{bmatrix} x_1 \\ x_2 \end{bmatrix} = \begin{bmatrix} 0 & 1 \\ 12 & -4 \end{bmatrix}\begin{bmatrix} x_1 \\ x_2 \end{bmatrix}$$

From R.H.S. we have to find eigenvector.

Characteristic equation is $\begin{bmatrix} 0-\lambda & 1 \\ 12 & -4-\lambda \end{bmatrix} = 0$

or $$-\lambda(-4-\lambda) - 12 = 0 \text{ or } \lambda^2 + 4\lambda - 12 = 0$$

or $$(\lambda - 2)(\lambda + 6) = 0 \Rightarrow \lambda = 2, -6$$

For $\lambda = 2$ and $\lambda = -6$, eigenvectors are $\begin{bmatrix} 1 \\ 2 \end{bmatrix}$ and $\begin{bmatrix} 1 \\ -6 \end{bmatrix}$

Matrix of eigen vectors $= P = \begin{bmatrix} 1 & 1 \\ 2 & -6 \end{bmatrix}, P^{-1} = \frac{1}{8}\begin{bmatrix} 6 & 1 \\ 2 & -1 \end{bmatrix}$

Now $$Pe^{\lambda t}P^{-1} = \begin{bmatrix} 1 & 1 \\ 2 & -6 \end{bmatrix}\begin{bmatrix} e^{2t} & 0 \\ 0 & e^{-6t} \end{bmatrix}\frac{1}{8}\begin{bmatrix} 6 & 1 \\ 2 & -1 \end{bmatrix}.$$

$$= \frac{1}{8}\begin{bmatrix} e^{2t} & e^{-6t} \\ 2e^{2t} & -6e^{-6t} \end{bmatrix}\begin{bmatrix} 6 & 1 \\ 2 & -1 \end{bmatrix} = \frac{1}{8}\begin{bmatrix} 6e^{2t} + 2e^{-6t} & e^{2t} - e^{-6t} \\ 12e^{2t} - 12e^{-6t} & 2e^{2t} + 6e^{-6t} \end{bmatrix}$$

By initial conditions $x(0) = 0$, $x'(0) = 8$

$$\begin{bmatrix} x_1 \\ x_2 \end{bmatrix} = \frac{1}{8}\begin{bmatrix} 6e^{2t} + 2e^{-6t} & e^{2t} - e^{-6t} \\ 12e^{2t} - 12e^{-6t} & 2e^{2t} + 6e^{-6t} \end{bmatrix}\begin{bmatrix} 0 \\ 8 \end{bmatrix} = \begin{bmatrix} e^{2t} - e^{-6t} \\ 2e^{2t} + 6e^{-6t} \end{bmatrix}.$$

$$x_1 = x = e^{2t} - e^{-6t}, x_2 = \frac{dx}{dt} = 2e^{2t} + 6e^{-6t}$$ **Ans.**

**Example 86.** *Solve by matrix method.*

$$\frac{d^2x}{dt^2} - 5\frac{dx}{dt} + 6x = 0, x(0) = 1. x'(0) = 2.$$

**Solution.** $$\frac{d^2x}{dt^2} - 5\frac{dx}{dt} + 6x = 0 \quad ...(1)$$

Let $x = x_1$ and $\frac{dx_1}{dt} = x_2$ ...(2)

On substitution (1) becomes

$$\frac{dx_2}{dt} = 5x_2 - 6x_1 \text{ or } \frac{dx_2}{dt} = -6x_1 + 5x_2 \quad ...(3)$$

Equations (2) and (3) are written in a single matrix equation

$$\frac{d}{dt}\begin{bmatrix} x_1 \\ x_2 \end{bmatrix} = \begin{bmatrix} 0 & 1 \\ -6 & 5 \end{bmatrix}\begin{bmatrix} x_1 \\ x_2 \end{bmatrix}$$

From R.H.S we have to find eigen vector.

Characteristic equation is $\begin{bmatrix} 0-\lambda & 1 \\ -6 & 5-\lambda \end{bmatrix} = 0$

$$-\lambda(5-\lambda)+6=0 \quad \text{or} \quad \lambda^2 - 5\lambda + 6 = 0 \Rightarrow \lambda = 2, 3$$

Eigen vectors for $\lambda = 2$ and $\lambda = 3$ are $\begin{bmatrix} 1 \\ 2 \end{bmatrix}$ and $\begin{bmatrix} 1 \\ 3 \end{bmatrix}$

Matrix of eigen vectors $= P = \begin{bmatrix} 1 & 1 \\ 2 & 3 \end{bmatrix}$, $\quad P^{-1} = \begin{bmatrix} 3 & -1 \\ -2 & 1 \end{bmatrix}$

$$Pe^{\lambda t}P^{-1} = \begin{bmatrix} 1 & 1 \\ 2 & 3 \end{bmatrix}\begin{bmatrix} e^{2t} & 0 \\ 0 & e^{3t} \end{bmatrix}\begin{bmatrix} 3 & -1 \\ -2 & 1 \end{bmatrix}$$

$$= \begin{bmatrix} e^{2t} & e^{3t} \\ 2e^{2t} & 3e^{3t} \end{bmatrix}\begin{bmatrix} 3 & -1 \\ -2 & 1 \end{bmatrix} = \begin{bmatrix} 3e^{2t}-2e^{3t} & -e^{2t}+e^{3t} \\ 6e^{2t}-6e^{3t} & -2e^{2t}+3e^{3t} \end{bmatrix}$$

By initial conditions $x(0) = 1$ and $x'(0) = 2$

$$\begin{bmatrix} x_1 \\ x_2 \end{bmatrix} = \begin{bmatrix} 3e^{2t}-2e^{3t} & -e^{2t}+e^{3t} \\ 6e^{2t}-6e^{3t} & -2e^{2t}+3e^{3t} \end{bmatrix}\begin{bmatrix} 1 \\ 2 \end{bmatrix}$$

$$= \begin{bmatrix} 3e^{2t}-2e^{3t} & -2e^{2t}+2e^{3t} \\ 6e^{2t}-6e^{3t} & -4e^{2t}+6e^{3t} \end{bmatrix}\begin{bmatrix} e^{2t} \\ 2e^{2t} \end{bmatrix}$$

$$x_1 = x = e^{2t}$$

$$x_2 = \frac{dx}{dt} = 2e^{2t}$$

**Ans.**

**Example 87.** *Use matrices to solve the differential equation*

$$\frac{d^2y}{dx^2} + 4y = 0, \quad y(0) = 1, \quad y'(0) = 0$$

**Solution.** Let $\qquad y = y_1, \quad \frac{dy_1}{dx} = y_2 \qquad$ ...(1)

$$\frac{d^2y}{dx^2} + 4y = 0, \text{ or } \frac{d}{dx}\left(\frac{dy_1}{dx}\right) + 4y_1 = 0 \text{ or } \frac{dy_2}{dx} = -4y_1 \qquad \text{...(2)}$$

Differential equations (1) and (2) are written in matrix form

$$\frac{d}{dx}\begin{bmatrix} y_1 \\ y_2 \end{bmatrix} = \begin{bmatrix} 0 & 1 \\ -4 & 0 \end{bmatrix}\begin{bmatrix} y_1 \\ y_2 \end{bmatrix}$$

The characteristic equation of $\begin{bmatrix} 0 & 1 \\ -4 & 0 \end{bmatrix}$ is

$$\begin{bmatrix} 0-\lambda & 1 \\ -4 & 0-\lambda \end{bmatrix} = 0 \Rightarrow \lambda^2 + 4 = 0 \Rightarrow \lambda = \pm i2$$

Eigenvector for $\lambda = -i2$

$$\begin{bmatrix} 0-i2 & 1 \\ -4 & 0-i2 \end{bmatrix}\begin{bmatrix} x_1 \\ x_2 \end{bmatrix} = \begin{bmatrix} 0 \\ 0 \end{bmatrix} \text{ or } \begin{bmatrix} -i2 & 1 \\ 0 & 0 \end{bmatrix}\begin{bmatrix} x_1 \\ x_2 \end{bmatrix} = \begin{bmatrix} 0 \\ 0 \end{bmatrix} \qquad R_2 \to R_2 + 2iR_1$$

or $$-i2x_1 + x_2 = 0 \text{ or } \begin{bmatrix} x_1 \\ x_2 \end{bmatrix} = \begin{bmatrix} 1 \\ i2 \end{bmatrix}$$

Eigenvector for $\lambda = -i2$

$$\begin{bmatrix} 0+i2 & 1 \\ -4 & 0+i2 \end{bmatrix}\begin{bmatrix} x_1 \\ x_2 \end{bmatrix} = \begin{bmatrix} 0 \\ 0 \end{bmatrix} \text{ or } \begin{bmatrix} i2 & 1 \\ -4 & i2 \end{bmatrix}\begin{bmatrix} x_1 \\ x_2 \end{bmatrix} = \begin{bmatrix} 0 \\ 0 \end{bmatrix}$$

$$\begin{bmatrix} i2 & 1 \\ 0 & 0 \end{bmatrix}\begin{bmatrix} x_1 \\ x_2 \end{bmatrix} = \begin{bmatrix} 0 \\ 0 \end{bmatrix} \text{ or } i2x_1 + x_2 = 0 \text{ or } \begin{bmatrix} x_1 \\ x_2 \end{bmatrix} = \begin{bmatrix} 1 \\ -i2 \end{bmatrix}$$

Let $P = \begin{bmatrix} 1 & 1 \\ 2i & -2i \end{bmatrix}$, $P^{-1} = \begin{bmatrix} \frac{1}{2} & \frac{1}{4i} \\ \frac{1}{2} & -\frac{1}{4i} \end{bmatrix}$

$$Pe^{\lambda x}P^{-1} = \begin{bmatrix} 1 & 1 \\ 2i & -2i \end{bmatrix}\begin{bmatrix} e^{i2x} & 0 \\ 0 & e^{-i2x} \end{bmatrix}\begin{bmatrix} \frac{1}{2} & \frac{1}{4i} \\ \frac{1}{2} & -\frac{1}{4i} \end{bmatrix} = \begin{bmatrix} e^{i2x} & e^{-i2x} \\ 2ie^{i2x} & -2ie^{-i2x} \end{bmatrix}\begin{bmatrix} \frac{1}{2} & \frac{1}{4i} \\ \frac{1}{2} & -\frac{1}{4i} \end{bmatrix}$$

$$= \begin{bmatrix} \frac{1}{2}e^{i2x} + \frac{1}{2}e^{-i2x} & \frac{1}{4i}e^{i2x} - \frac{1}{4i}e^{-i2x} \\ ie^{i2x} - ie^{-i2x} & \frac{1}{2}e^{i2x} + \frac{1}{2}e^{-i2x} \end{bmatrix} = \begin{bmatrix} \cos 2x & \frac{1}{2}\sin sx \\ -2\sin 2x & \cos 2x \end{bmatrix}$$

**Applying the initial conditions, we get**

$$\begin{bmatrix} y_1 \\ y_2 \end{bmatrix} = \begin{bmatrix} \cos 2x & \frac{1}{2}\sin 2x \\ -2\sin 2x & \cos 2x \end{bmatrix}\begin{bmatrix} 1 \\ 0 \end{bmatrix} = \begin{bmatrix} \cos 2x \\ -2\sin 2x \end{bmatrix} \text{ or } y_1 = \cos 2x \text{ and } y_2 = -2\sin 2x \qquad \textbf{Ans.}$$

## EXERCISE 4.30

**Solve the following differential equations by matrix method:**

**1.** $\frac{d^2y}{dx^2} - \frac{4dy}{dx} + 3y = 0,\ y(0) = 2,\ y'(0) = 1$ **Ans.** $y = \frac{5}{2}e^x - \frac{e^{3x}}{2}$

**2.** $\frac{d^2y}{dx^2} - \frac{3dy}{dx} + 2y = 0,\ y(0) = 5,\ y'(0) = 8$ **Ans.** $y = 2e^x + 3e^{2x}$

**3.** $\frac{d^2y}{dx^2} + \frac{5dy}{dx} - 14y = 0,\ y(0) = 2,\ y'(0) = -5$ **Ans.** $y = e^{2x} + e^{-7x}$

**4.** $\frac{d^2y}{dx^2} + \mu^2 y = 0,\ y(0) = 1,\ y'(0) = \mu$ **Ans.** $y = \cos x + \sin x$

**5.** $\frac{d^2y}{dx^2} + 9y = 0,\ y(0) = 1,\ y'(0) = 3$ **Ans.** $y = \cos x + \sin x$

## 4.71 COMPLEX MATRICES

### Conjugate of a Complex Number

$z = x + iy$ is called a complex number where $\sqrt{-1} = i$, $x, y$ are real numbers. $\bar{z} = x - iy$ is called the conjugate of the complex number $z$, e.g.,

| Complex number | Conjugate number |
|---|---|
| $2 + 3i$ | $2 - 3i$ |
| $-4 - 5i$ | $-4 + 5i$ |
| $6i$ | $-6i$ |
| $2$ | $2$ |

**Conjugate of a matrix.** The matrix formed by replacing the elements of a matrix by their respective conjugate numbers is called the conjugate of $A$ and is denoted by $\bar{A}$.

$$A = (a_{ij})_{m\times n}, \text{ then } \bar{A} = (\bar{a}_{ij})_{m\times n}$$

Example

If $A = \begin{bmatrix} 3+4i & 2-i & 4 \\ i & 2 & -3i \end{bmatrix}$ then $\bar{A} = \begin{bmatrix} 3-4i & 2+i & 4 \\ -i & 2 & 3i \end{bmatrix}$

## 4.72 THEOREM

If $A$ and $B$ be two matrices and their conjugate matrices are $\bar{A}$ and $\bar{B}$ respectively, then

(*i*) $\overline{(\bar{A})} = A$ (*ii*) $\overline{(A+B)} = \bar{A} + \bar{B}$ (*iii*) $\overline{(kA)} = \bar{k}\,\bar{A}$ (*iv*) $\overline{(AB)} = \bar{A}\bar{B}$

**Proof.** Let $A = [a_{ij}]_{m\times n}$, then

$\bar{A} = [\bar{a}_{ij}]_{m\times n}$ where $\bar{a}_{ij}$ is the conjugate complex of $a_{ij}$.

The $(i, j)$ th element of $\overline{(\bar{A})}$ = the conjugate complex of the $(i, j)$th element of $\bar{A}$

= the conjugate complex of $\bar{a}_{ij}$

= $a_{ij}$ = the $(i, j)$th element of $A$.

Hence $\overline{(\bar{A})} = A$. **Proved.**

(*ii*) Let $A = [a_{ij}]_{m\times n}$ and $B = [b_{ij}]_{m\times n}$

$\bar{A} = [\bar{a}_{ij}]_{m\times n}$ and $\bar{B} = [\bar{b}_{ij}]_{m\times n}$

$(i, j)$ th element of $\overline{(A+B)}$ = conjugate complex of $(i, j)$ th element of $(A + B)$

= conjugate complex of $(a_{ij} + b_{ij}) = \overline{(a_{ij} + b_{ij})} = \bar{a}_{ij} + \bar{b}_{ij}$

= $(i, j)$th element of $\bar{A}$ + $(i, j)$th element of $\bar{B}$

= $(i, j)$th element of $(\bar{A} + \bar{B})$

Hence, $\overline{(A+B)} = \bar{A} + \bar{B}$ **Proved.**

(*iii*) Let $A = [a_{ij}]_{m\times n}$, let $k$ be any complex number.

The $(i, j)$th element of $\overline{(kA)}$ = conjugate complex of the $(i, j)$th element of $kA$

= conjugate complex of $ka_{ij}$

$= \overline{ka_{ij}} = \bar{k}\cdot\bar{a}_{ij} = \bar{k}\cdot(i, j)\text{th element of } \bar{A} = (i, j)\text{th element of } \bar{k}.\ \bar{A}$

Hence, $\overline{kA} = \bar{k}\cdot\bar{A}$ **Proved.**

(*iv*) Let $A = [a_{ij}]_{m\times n}$, $B = [b_{ij}]_{n\times p}$

Then $\overline{A} = [\overline{a_{ij}}]_{m\times n}$, $\overline{B} = [\overline{b_{ij}}]_{n\times p}$

The $(i, j)$th element of $(\overline{AB})$ = conjugate complex of $(i, j)$th element of $AB$

$$= \text{conjugate complex of } \sum_{j=1}^{n} a_{ij}\, b_{jk} = \left(\sum_{j=1}^{n} \overline{a_{ij} b_{jk}}\right) = \sum_{j=1}^{n} \overline{a_{ij}} \cdot \overline{b_{jk}}$$

$= (i, j)$th element of $\overline{A} \cdot \overline{B}$

Hence, $(\overline{AB}) = \overline{A} \cdot \overline{B}$ **Proved.**

## 4.73 TRANSPOSE OF CONJUGATE OF A MATRIX

The transpose of a conjugate of a matrix $A$ is denoted by $A^{\theta}$ or $A^{*}$.

$$(\overline{A})' = A^{\theta}$$

The $(i, j)$th element of $A^{\theta}$ = $(j, i)$th element of $\overline{A}$
= conjugate complex of $(j, i)$th element of $A$.

**Example 88.** *If* $A = \begin{bmatrix} 2+3i & 1-2i & 2+4i \\ 3-4i & 4+3i & 2-6i \\ 5 & 5+6i & 3 \end{bmatrix}$, *find* $A^{\theta}$

**Solution.** We have, $A = \begin{bmatrix} 2+3i & 1-2i & 2+4i \\ 3-4i & 4+3i & 2-6i \\ 5 & 5+6i & 3 \end{bmatrix} \Rightarrow \overline{A} = \begin{bmatrix} 2-3i & 1+2i & 2-4i \\ 3+4i & 4-3i & 2+6i \\ 5 & 5-6i & 3 \end{bmatrix}$

$$A^{\theta} = (\overline{A})' = \begin{bmatrix} 2-3i & 3+4i & 5 \\ 1+2i & 4-3i & 5-6i \\ 2-4i & 2+6i & 3 \end{bmatrix}$$ **Ans.**

## EXERCISE 4.31

**1.** If the matrix $A = \begin{bmatrix} 1+i & 3-5i \\ 2i & 5 \end{bmatrix}$, find *(i)* $\overline{A}$ *(ii)* $(\overline{A})'$ *(iii)* $A^{\theta}$ *(iv)* $(A^{\theta})^{\theta}$

**Ans.** *(i)* $\overline{A} = \begin{bmatrix} 1-i & 3+5i \\ -2i & 5 \end{bmatrix}$ *(ii)* $(\overline{A})' = \begin{bmatrix} 1-i & -2i \\ 3+5i & 5 \end{bmatrix}$

*(iii)* $A^{\theta} = \begin{bmatrix} 1-i & -2i \\ 3+5i & 5 \end{bmatrix}$ *(iv)* $(A^{\theta})^{\theta} = \begin{bmatrix} 1+i & 3-5i \\ 2i & 5 \end{bmatrix}$

## 4.74 HERMITIAN MATRIX

**Definition.** A square matrix $A = [a_{ij}]$ is said to be Hermitian if the $(i, j)$th element of A, *i.e.*,

$$a_{ij} = \overline{a}_{ji} \text{ for all } i \text{ and } j.$$

*For example,* $\begin{bmatrix} 2 & 3+4i \\ 3-4i & 1 \end{bmatrix}$, $\begin{bmatrix} a & b-id \\ b+id & c \end{bmatrix}$

Hence all the elements of the principal diagonal are real.

A necessary and sufficient condition for a matrix $A$ to be Hermitian is that $A = A^{\theta}$.

**Example 89.** *The characteristic roots of a Hermitian matrix are all real.*

*(A.M.I.E.T.E., June 2006)*

**Solution.** We know that matrix $A$ is Hermitian if

$$A^{\theta} = A \text{ i.e., where } A^{\theta} = (\overline{A}') \text{ or } (\overline{A})'$$

Also $(\lambda A)^\theta = \bar{\lambda} A^\theta$ and $(AB)^\theta = B^\theta A^\theta$

If $\lambda$ is a characteristic root of matrix $A$ then $AX = \lambda X$. ... (1)

$\therefore$ $(AX)^\theta = (\lambda X)^\theta$ or $X^\theta A^\theta = \lambda X^\theta$.

But $A$ is Hermitian $\therefore$ $A^\theta = A$.

$\therefore$ $X^\theta A = \bar{\lambda} X^\theta \therefore X^\theta AX = \bar{\lambda} X^\theta X$ ... (2)

Again from (1) $I X^\theta AX = X^\theta \lambda X = \lambda X^\theta X$. ... (3)

Hence from (2) and (3) we conclude that $\bar{\lambda} = \lambda$ showing that $\lambda$ is real.

**Deduction 1.** *From above we conclude that characteristic roots of real symmetric matrix are all real, as in this case, real symmetric matrix will be Hermitian.*

For symmetric, we know that $A' = A$. $(\bar{A}') = \bar{A}$.

or $A^\theta = A$ $\therefore \bar{A} = A$ as $A$ is real. Rest as above.

**Example 90.** *Prove that the following*

$$(i)\ (A^\theta)^\theta = A \quad (ii)\ (A+B)^\theta = A^\theta + B^\theta \quad (iii)\ (kA)^\theta = \bar{k} A^\theta \quad (iv)\ (AB)^\theta = B^\theta \cdot A^\theta$$

*where $A^\theta$ and $B^\theta$ be the transposed conjugates of A and B respectively, A and B being conformable to multiplication.*

**Solution.**

(i) $(A^\theta)^\theta = [\overline{\{(\bar{A})'\}}]' = [\bar{\bar{A}}] = A$ as $\overline{\{(\bar{A})'\}}' = \bar{A}$

(ii) $(A+B)^\theta = (\overline{A+B})' = (\bar{A}+\bar{B})' = (\bar{A})' + (\bar{B})' = A^\theta + B^\theta$

(iii) $(kA)^\theta = (\overline{kA})' = (\bar{k}\bar{A})' = \bar{k}(\bar{A})' = \bar{k} A^\theta$

(iv) $(AB)^\theta = (\overline{AB})' = (\bar{A}\cdot\bar{B})' = (\bar{B})'\cdot(\bar{A})' = B^\theta \cdot A^\theta$ **Proved.**

**Example 91.** *Prove that matrix* $A = \begin{bmatrix} 1 & 1-i & 2 \\ 1+i & 3 & i \\ 2 & -i & 0 \end{bmatrix}$ *is Hermitian.*

**Solution.** $\bar{A} = \begin{bmatrix} 1 & 1+i & 2 \\ 1-i & 3 & -i \\ 2 & i & 0 \end{bmatrix} \Rightarrow (\bar{A})' = \begin{bmatrix} 1 & 1-i & 2 \\ 1+i & 3 & i \\ 2 & -i & 0 \end{bmatrix}$

$\Rightarrow$ $A^\theta = A$ $\Rightarrow$ $A$ is Hermitian matrix. **Proved.**

**Example 92.** *Show that* $A = \begin{bmatrix} -i & 3+2i & -2-i \\ -3+2i & 0 & 3-4i \\ 2-i & -3-4i & -2i \end{bmatrix}$ *is Skew-Hermitian matrix.*

**Solution.** $\bar{A} = \begin{bmatrix} i & 3-2i & -2+i \\ -3-2i & 0 & 3+4i \\ 2+i & -3+4i & 2i \end{bmatrix}$

$$(\bar{A})' = \begin{bmatrix} i & -3-2i & 2+i \\ 3-2i & 0 & -3+4i \\ -2+i & 3+4i & 2i \end{bmatrix}$$

$$\Rightarrow \quad A^\theta = \begin{bmatrix} i & -3-2i & 2+i \\ 3-2i & .0 & -3+4i \\ -2+i & 3+4i & 2i \end{bmatrix} \qquad [\because\ A^\theta = (\bar{A})']$$

$$= -\begin{bmatrix} -i & 3+2i & -2-i \\ -3+2i & 0 & 3-4i \\ 2-i & -3-4i & -2i \end{bmatrix} = -A$$

$A^\theta = -A \Rightarrow A$ is Skew-Hermitian matrix. **Proved.**

**Example 93.** *Show that the matrix $B^\theta AB$ is Hermitian or Skew-Hermitian according as A is Hermitian or Skew-Hermitian.*

**Solution.** (*i*) Let $A$ be Hermitian $\Rightarrow\ A^\theta = A$

Now
$$(B^\theta AB)^\theta = (AB)^\theta\,(B^\theta)^\theta$$
$$= B^\theta \cdot A^\theta \cdot B$$
$$= B^\theta \cdot A \cdot B \qquad (A^\theta = A)$$

Hence, $A^\theta AB$ is Hermitian.

(*ii*) Let A be Skew-Hermitian $\Rightarrow\ A^\theta = -A$

Now,
$$(B^\theta AB)^\theta = (AB)^\theta \cdot (B^\theta)^\theta$$
$$= B^\theta \cdot A^\theta \cdot B$$
$$= -B^\theta A \cdot B \qquad (A^\theta = -A)$$

Hence, $B^\theta\ AB$ is Skew-Hermitian. **Proved.**

## 4.75 SKEW-HERMITIAN MATRIX

**Definition.** A square matrix $A = (a_{ij})$ is said to be Skew-Hermitian matrix if the $(i, j)$th element of $A$ is equal to the negative of the conjugate complex of the $(j, i)$th element of $A$, *i.e.*,

$a_{ij} = -\bar{a}_{ji}$ for all $i$ and $j$.

If $A$ is a Skew-Hermitian matrix, then

$$a_{ii} = -\bar{a}_{ii}$$

$$a_{ii} + \bar{a}_{ii} = 0$$

Obviously, $a_{ii}$ is either a pure imaginary number or must be zero.

For example, $\begin{bmatrix} 0 & -3+4i \\ 3+4i & 0 \end{bmatrix}$ and $\begin{bmatrix} 0 & a-ib \\ -a-ib & 0 \end{bmatrix}$ are Skew-Hermitian matrixes.

A necessary and sufficient condition for a matrix $A$ to be Skew-Hermitian is that $A^\theta = -A$.

**Deduction 2.** *Characteristic roots of a skew Hermitian matrix is either zero or a pure imaginary numbers.*

If $A$ is skew Hermitian, then $iA$ is Hermitian.

Also $\lambda$ be a characteristic root of $A$ then $AX = \lambda X$.

$$\therefore \qquad (i\,.A)\,X = (i\lambda)\,X.$$

Above shows that $i\lambda$ is characteristic root of matrix $iA$, which is Hermitian and hence $i\lambda$ should be real, which will be possible if $\lambda$ is either pure imaginary or zero.

**Example 94.** *Show that every square matrix can be expressed as R + iS uniquely where R and S are Hermitian matrices.*

**Solution.** Let $A$ be any square matrix. It can be rewritten as

$$A=\left\{\frac{1}{2}(A+A^{\theta})\right\}+i\left\{\frac{1}{2i}(A-A^{\theta})\right\}=R+iS$$

where $R=\frac{1}{2}(A+A^{\theta})$, $S=\frac{1}{2i}(A-A^{\theta})$

Now we have to show that $R$ and $S$ are Hermitian matrices.

$$R^{\theta}=\frac{1}{2}(A+A^{\theta})^{\theta}=\frac{1}{2}[A^{\theta}+(A^{\theta})^{\theta}]=\frac{1}{2}(A^{\theta}+A)=\frac{1}{2}(A+A^{\theta})=R$$

Thus $R$ is Hermitian matrix.

Now, $$S^{\theta}=\left[\frac{1}{2i}(A-A^{\theta})\right]^{\theta}=-\frac{1}{2i}(A-A^{\theta})^{\theta}$$

$$=-\frac{1}{2i}[A^{\theta}-(A^{\theta})^{\theta}]=\frac{-1}{2i}(A^{\theta}-A)=\frac{1}{2i}(A-A^{\theta})=S$$

Thus $S$ is a Hermitian matrix.

Hence $A = R + iS$, where $R$ and $S$ are Hermitian matrices.

Now, we have to show its **uniqueness**.

Let $A = P + iQ$ be another expression, where $P$ and $Q$ are Hermitian matrices, *i.e.*,

$$P^{\theta}=P,\ Q^{\theta}=Q$$

Then $$A^{\theta}=(P+iQ)^{\theta}=P^{\theta}+(iQ)^{\theta}=P^{\theta}-iQ^{\theta}=P-iQ$$

$$A=P+iQ \text{ and } A^{\theta}=P-iQ$$

$\Rightarrow$ $$P=\frac{1}{2}(A+A^{\theta})=R \text{ and } Q=\frac{1}{2i}(A-A^{\theta})=S$$

Hence $A = R + iS$ is the unique expression, where $R$ and $S$ are Hermitian matrices. **Proved.**

**Example 95.** *Express the matrix* $A=\begin{bmatrix}1+i & 2 & 5-5i\\ 2i & 2+i & 4+2i\\ -1+i & -4 & 7\end{bmatrix}$ *as the sum of Hermitian matrix and Skew-Hermitian matrix.*

**Solution.** $$A=\begin{bmatrix}1+i & 2 & 5-5i\\ 2i & 2+i & 4+2i\\ -1+i & -4 & 7\end{bmatrix} \Rightarrow \overline{A}=\begin{bmatrix}1-i & 2 & 5+5i\\ -2i & 2-i & 4-2i\\ -1-i & -4 & 7\end{bmatrix} \quad ...(1)$$

$$(\overline{A})'=\begin{bmatrix}1-i & -2i & -1-i\\ 2 & 2-i & -4\\ 5+5i & 4-2i & 7\end{bmatrix} \Rightarrow A^{\theta}=\begin{bmatrix}1-i & -2i & -1-i\\ 2 & 2-i & -4\\ 5+5i & 4-2i & 7\end{bmatrix} \quad ...(2)$$

On adding (1) and (2), we get

$$A+A^{\theta}=\begin{bmatrix}2 & 2-2i & 4-6i\\ 2+2i & 4 & 2i\\ 4+6i & -2i & 14\end{bmatrix}$$

Let $$R=\frac{1}{2}(A+A^{\theta})=\begin{bmatrix}1 & 1-i & 2-3i\\ 1+i & 2 & i\\ 2+3i & -i & 7\end{bmatrix} \quad ...(3)$$

On subtracting (2) from (1), we get

$$A - A^{\theta} = \begin{bmatrix} 2i & 2+2i & 6-4i \\ -2+2i & 2i & 8+2i \\ -6-4i & -8+2i & 0 \end{bmatrix}$$

Let $$S = \frac{1}{2}(A - A^{\theta}) = \begin{bmatrix} i & 1+i & 3-2i \\ -1+i & i & 4+i \\ -3-2i & -4+i & 0 \end{bmatrix} \quad \text{...(4)}$$

From (3) and (4), we have

$$A = \underbrace{\begin{bmatrix} 1 & 1-i & 2-3i \\ 1+i & 2 & i \\ 2+3i & -i & 7 \end{bmatrix}}_{\text{Hermitian matrix}} + \underbrace{\begin{bmatrix} i & 1+i & 3-2i \\ -1+i & i & 4+i \\ -3-2i & -4+i & 0 \end{bmatrix}}_{\text{Skew-Hermitian matrix}} \quad \textbf{Ans.}$$

**Example 96.** *For any square matrix, if $AA^{\theta} = I$ show that $A^{\theta}A = I$.*

**Solution.** $AA^{\theta} = I$ (given)

So $A$ is invertible.

Let $B$ be another matrix such that

$$AB = BA = I \quad \text{...(1)}$$

Now $B = BI = B(AA^{\theta})$ $(AA^{\theta} = I)$

$= (BA)\,A^{\theta}$

$= IA^{\theta} = A^{\theta}$ [Using (1)]

We know that $BA = I$ [From (1)]

Putting the value of B from (2) in (1), we get

$\Rightarrow$ $A^{\theta}A = I$ **Proved.**

## 4.76 PERIODIC MATRIX

A square matrix is said to be periodic, if $A^{k+1} = A$, where $k$ is a positive integer. If $k$ is the least positive integer for which $A^{k+1} = A$, then $A$ is said to be of period $k$.

## 4.77 IDEMPOTENT MATRIX

A square matrix is said to be idempotent provided $A^2 = A$.

**Example 97.** *Determine all the idempotent diagonal matrices of order n.*

**Solution.** Let $A = \text{diag.}\,[d_1, d_2, d_3, \ldots d_n]$ be an idempotent matrix of order $n$.

Here, for the matrix '$A$' to be idempotent $A^2 = A$

$$\begin{bmatrix} d_1 & 0 & 0\ldots\ldots 0 \\ 0 & d_2 & 0\ldots\ldots 0 \\ 0 & 0 & d_3\ldots\ldots 0 \\ \hline 0 & 0 & 0\ldots\ldots d_n \end{bmatrix} \begin{bmatrix} d_1 & 0 & 0\ldots\ldots 0 \\ 0 & d_2 & 0\ldots\ldots 0 \\ 0 & 0 & d_3\ldots\ldots 0 \\ \hline 0 & 0 & 0\ldots\ldots d_n \end{bmatrix} = \begin{bmatrix} d_1 & 0 & 0\ldots\ldots 0 \\ 0 & d_2 & 0\ldots\ldots 0 \\ 0 & 0 & d_3\ldots\ldots 0 \\ \hline 0 & 0 & 0\ldots\ldots d_n \end{bmatrix}$$

$$\Rightarrow \quad \begin{bmatrix} d_1^2 & 0 & 0\ldots\ldots 0 \\ 0 & d_2^2 & 0\ldots\ldots 0 \\ 0 & 0 & d_3^2\ldots\ldots 0 \\ \hline 0 & 0 & 0\ldots\ldots d_n^2 \end{bmatrix} = \begin{bmatrix} d_1 & 0 & 0\ldots\ldots 0 \\ 0 & d_2 & 0\ldots\ldots 0 \\ 0 & 0 & d_3\ldots\ldots 0 \\ \hline 0 & 0 & 0\ldots\ldots d_n \end{bmatrix}$$

$\therefore \qquad d_1^2 = d_1;\; d_2^2 = d_2 \ldots\ldots d_n^2 = d_n$

*i.e.*, $\qquad d_1 = 0, 1;\; d_2 = 0, 1;\; d_3 = 0, 1 \;\ldots\ldots\ldots\; d_n = 0, 1.$

Hence diag. $[d_1, d_2, d_3 \ldots d_n]$, is the required idempotent matrix where

$$d_1 = d_2 = d_3 = \ldots d_n = 0 \text{ or } 1.$$ **Ans.**

## EXERCISE 4.32

**1.** Which of the following matrices are Hermitian:

(*a*) $\begin{bmatrix} 1 & 2+i & 3-i \\ 2+i & 2 & 4-i \\ 3+i & 4+i & 3 \end{bmatrix}$ (*b*) $\begin{bmatrix} 2i & 3 & 1 \\ 4 & -1 & 6 \\ 3 & 7 & 2i \end{bmatrix}$ (*c*) $\begin{bmatrix} 4 & 2-i & 5+2i \\ 2+i & 1 & 2-5i \\ 5-2i & 2+5i & 2 \end{bmatrix}$ (*d*) $\begin{bmatrix} 0 & i & 3 \\ -7 & 0 & 5i \\ 3i & 1 & 0 \end{bmatrix}$ **Ans.** (*c*)

**2.** Which of the following matrices are Skew-Hermitian:

(*a*) $\begin{bmatrix} 2i & -3 & 4 \\ 3 & 3i & -5 \\ -4 & 5 & 4i \end{bmatrix}$ (*b*) $\begin{bmatrix} 3i & -1 & 2 \\ 1 & 2i & -6 \\ 4 & 6 & -3i \end{bmatrix}$ (*c*) $\begin{bmatrix} 0 & 1-i & 2+3i \\ -1-i & 0 & 6i \\ -2+3i & 6i & 4i \end{bmatrix}$ (*d*) $\begin{bmatrix} 1 & 3 & 7+i \\ 3i & -i & 6 \\ 7-i & 8 & 0 \end{bmatrix}$

**Ans.** (*a*), (*c*)

**3.** Give an example of a matrix which is Skew-symmetric but not Skew-Hermitian.

**Ans.** $\begin{bmatrix} 0 & 2+3i \\ -2-3i & 0 \end{bmatrix}$

**4.** If $A$ be a Hermitian matrix, show that $iA$ is Skew-Hermitian. Also show that if $B$ be a Skew-Hermitian matrix, then $iB$ must be Hermitian.

**5.** If $A$ and $B$ are Hermitian matrices, then show that $AB + BA$ is Hermitian and $AB - BA$ is Skew-Hermitian.

**6.** If $A$ is any square matrix, show that $A + A^{\theta}$ is Hermitian.

**7.** If $H = \begin{bmatrix} 3 & 5+2i & -3 \\ 5-2i & 7 & 4i \\ -3 & -4i & 5 \end{bmatrix}$, show that $H$ is a Hermitian matrix.

Verify that $iH$ is a Skew-Hermitian matrix.

**8.** Show that for any complex square matrix $A$,

(*i*) $(A + A^*)$ is a Hermitian matrix, where $A^* = \overline{A}^T$ (*ii*) $(A - A^*)$ is Skew-Hermitian matrix.

(*iii*) $AA^*$ and $A^*A$ are Hermitian matrices.

**9.** Show that any complex square matrix can be uniquely expressed as the sum of a Hermitian matrix and a Skew-Hermitian matrix.

**10.** Express $A = \begin{bmatrix} i & 2-3i & 4+5i \\ 6+i & 0 & 4-5i \\ -i & 2-i & 2+i \end{bmatrix}$ as the sum of Hermitian and Skew-Hermitian matrices.

**11.** Prove that the latent roots of a Hermitian matrix are all real.

**12.** If A = $\begin{bmatrix} 2+i & 3 & -1+3i \\ -5 & i & 4-2i \end{bmatrix}$ show that $AA^*$ is a Hermitian matrix; where $A^*$ is the conjugate transpose of $A$. (*AMIETE, June 2010*)

## 4.78 UNITARY MATRIX

A square matrix $A$ is said to be unitary matrix if $A \cdot A^{\theta} = A^{\theta} A = I$

**Example 98.** *If $A$ is a unitary matrix, show that $A^T$ is also unitary.*

**Solution.** $A \cdot A^\theta = A^\theta A = I$, since $A$ is a unitary matrix.

$$(AA^\theta)^\theta = (A^\theta A)^\theta = I^\theta \qquad (I^\theta = I)$$

$$(AA^\theta)^\theta = (A^\theta A)^\theta = I$$

$$(A^\theta)^\theta A^\theta = A^\theta (A^\theta)^\theta = I$$

$$AA^\theta = A^\theta A = I \qquad [\text{since } (A^\theta)^\theta = A]$$

$$(AA^\theta)^T = (A^\theta A)^T = (I)^T$$

$$(A^\theta)^T A^T = A^T (A^\theta)^T = I$$

$$(A^T)^\theta \cdot A^T = A^T (A^T)^\theta = I$$

Hence, $A^T$ is a unitary matrix. **Proved.**

**Example 99.** *If $A$ is a unitary matrix, show that $A^{-1}$ is also unitary.*

**Solution.** $AA^\theta = A^\theta A = I$, since $A$ is a unitary matrix.

$$(AA^\theta)^{-1} = (A^\theta \cdot A)^{-1} = (I)^{-1} \quad \text{taking inverse}$$

$$(A^\theta)^{-1} \cdot A^{-1} = A^{-1} (A^\theta)^{-1} = I$$

$$(A^{-1})^\theta \cdot A^{-1} = A^{-1} (A^{-1})^\theta = I$$

Hence, $A^{-1}$ is a unitary matrix. **Proved.**

**Example 100.** *If $A$ and $B$ are two unitary matrices, show that $AB$ is a unitary matrix.*

**Solution.** $A \cdot A^\theta = A^\theta A = I$ since $A$ is a unitary matrix. ...(1)

Similarly, $\quad B \cdot B^\theta = B^\theta B = I$ ...(2)

Now, $\quad (AB)(AB)^\theta = (AB)(B^\theta \cdot A^\theta) = A(BB^\theta) \cdot A^\theta$

$= A I A^\theta$ [From (2)]

$= AA^\theta = I$ [From (1)]

Again, $\quad (AB)^\theta \cdot (AB) = (B^\theta \cdot A^\theta)(AB)$

$= B^\theta (A^\theta A) B$ [From (1)]

$= B^\theta I B = B^\theta B$

$= I$ [From (2)]

Hence, $AB$ is a unitary matrix. **Proved.**

**Example 101.** *Prove that the matrix* $\frac{1}{\sqrt{3}}\begin{bmatrix} 1 & 1+i \\ 1-i & -1 \end{bmatrix}$ *is unitary.*

**Solution.** Let $A = \frac{1}{\sqrt{3}}\begin{bmatrix} 1 & 1+i \\ 1-i & -1 \end{bmatrix}$

$$A^\theta = \frac{1}{\sqrt{3}}\begin{bmatrix} 1 & 1+i \\ 1-i & -1 \end{bmatrix}$$

$$A^\theta \cdot A = \frac{1}{\sqrt{3}}\begin{bmatrix} 1 & 1+i \\ 1-i & -1 \end{bmatrix} \times \frac{1}{\sqrt{3}}\begin{bmatrix} 1 & 1+i \\ 1-i & -1 \end{bmatrix}$$

$$= \frac{1}{3}\begin{bmatrix} 1+(1+1) & (1+i)-(1+i) \\ (1-i)-1(1-i) & (1+1)+1 \end{bmatrix} = \frac{1}{3}\begin{bmatrix} 3 & 0 \\ 0 & 3 \end{bmatrix} = \begin{bmatrix} 1 & 0 \\ 0 & 1 \end{bmatrix} = I$$

Hence, $A$ is a unitary matrix. **Proved.**

**Example 102.** *Define a unitary matrix. If* $N = \begin{bmatrix} 0 & 1+2i \\ -1+2i & 0 \end{bmatrix}$ *is a matrix, then show that*

$(I - N)(I + N)^{-1}$ *is a unitary matrix, where I is an identity matrix.*

(*U.P., I Semester, Winter* 2000)

**Solution. Unitary matrix:** A square matrix '$A$' is said to be unitary if $A^\theta A = I$, where $A^\theta = (\bar{A})^T$ and $I$ is an identity matrix.

we have $N = \begin{bmatrix} 0 & 1+2i \\ -1+2i & 0 \end{bmatrix}$

$$I - N = \begin{bmatrix} 1 & 0 \\ 0 & 1 \end{bmatrix} - \begin{bmatrix} 0 & 1+2i \\ -1+2i & 0 \end{bmatrix} = \begin{bmatrix} 1 & -1+2i \\ 1-2i & 1 \end{bmatrix} \quad ...(1)$$

Now we have to find $(I + N)^{-1}$

$$I + N = \begin{bmatrix} 1 & 0 \\ 0 & 1 \end{bmatrix} + \begin{bmatrix} 0 & 1+2i \\ -1+2i & 0 \end{bmatrix} = \begin{bmatrix} 1 & 1+2i \\ -1+2i & 1 \end{bmatrix}$$

$$|I + N| = 1 - (-1 - 4) = 6$$

$$\text{Adj. } (I + N) = \begin{bmatrix} 1 & -1-2i \\ 1-2i & 1 \end{bmatrix}$$

$$(I + N)^{-1} = \frac{Adj(I + N)}{|I + N|} = \frac{1}{6}\begin{bmatrix} 1 & -1-2i \\ 1-2i & 1 \end{bmatrix} \quad ...(2)$$

For unitary matrix, $A^\theta A = I$

From (1) and (2), we get

$$\therefore \quad (I - N)(I + N)^{-1} = \frac{1}{6}\begin{bmatrix} 1 & -1-2i \\ 1-2i & 1 \end{bmatrix}\begin{bmatrix} 1 & -1-2i \\ 1-2i & 1 \end{bmatrix} = \frac{1}{6}\begin{bmatrix} -4 & -2-4i \\ 2-4i & -4 \end{bmatrix} = B \text{ (say)}$$

Now $$(\bar{B})^T = \frac{1}{6}\begin{bmatrix} -4 & 2+4i \\ -2+4i & -4 \end{bmatrix}$$

$$\therefore \quad (\bar{B})^T B = \frac{1}{36}\begin{bmatrix} -4 & 2+4i \\ -2+4i & -4 \end{bmatrix}\begin{bmatrix} -4 & -2-4i \\ 2-4i & -4 \end{bmatrix} = \frac{1}{36}\begin{bmatrix} 36 & 0 \\ 0 & 36 \end{bmatrix} = I.$$

Hence the result. **Proved.**

## 4.79 THE MODULUS OF EACH CHARACTERISTIC ROOT OF A UNITARY MATRIX IS UNITY.

(*U.P., I Semester, Compartment* 2002)

**Solution.** Suppose $A$ is a unitary matrix. Then

$$A^\theta A = I.$$

Let $\lambda$ be a characteristic root of $A$. Then

$$AX = \lambda X \quad ...(1)$$

Taking conjugate transpose of both sides of (1), we get

$$(AX)^\theta = \bar{\lambda} X^\theta \quad ...(2)$$

$$\Rightarrow \quad X^\theta A^\theta = \bar{\lambda} X^\theta$$

From (1) and (2), we have

$$(X^{\theta}A^{\theta})(AX) = \bar{\lambda}\lambda X^{\theta}X$$

$$\Rightarrow \quad X^{\theta}(A^{\theta}A)X = \bar{\lambda}\lambda X^{\theta}X$$

$$\Rightarrow \quad X^{\theta}IX = \bar{\lambda}\lambda X^{\theta}X \qquad (\because A^{\theta}.A = I)$$

$$\Rightarrow \quad X^{\theta}X = \bar{\lambda}\lambda X^{\theta}X$$

$$\Rightarrow \quad X^{\theta}X(\bar{\lambda}\lambda - 1) = 0 \qquad \text{...(3)}$$

Since, $X^{\theta}\ X \neq 0$ therefore (3) gives

$$\lambda\bar{\lambda} - 1 = 0. \text{ or } \lambda\bar{\lambda} = 1 \quad \text{or} \quad |\lambda|^2 = 1 \quad \Rightarrow \quad |\lambda| = 1$$ **Proved.**

## EXERCISE 4.33

1. Show that the matrix $A = \frac{1}{\sqrt{2}}\begin{bmatrix} 1 & i \\ -i & -1 \end{bmatrix}$ is unitary.

2. Prove that a real matrix is unitary if it is orthogonal.

3. Prove that the following matrix is unitary:

$$\begin{bmatrix} \frac{1}{2}(1+i) & \frac{1}{2}(-1+i) \\ \frac{1}{2}(1+i) & \frac{1}{2}(1-i) \end{bmatrix}$$

4. Show that $U = \frac{1}{\sqrt{3}}\begin{bmatrix} 1 & 1 & 1 \\ 1 & \omega & \omega^2 \\ 1 & \omega^2 & \omega \end{bmatrix}$ is a unitary matrix, where $\omega$ is the complex cube root of unity.

5. Prove that the latent roots of a unitary matrix have unit modulus.

6. Verify that the matrix

$$A = \frac{1}{2}\begin{bmatrix} 1+i & 1-i \\ 1-i & 1+i \end{bmatrix}$$

has eigen values with unit modulus.

# 5

# Vectors

## 5.1 VECTORS

A vector is a quantity having both magnitude and direction such as force, velocity acceleration, displacement etc.

## 5.2 ADDITION OF VECTORS

Let $\vec{a}$ and $\vec{b}$ be two given vectors

$\vec{OA} = \vec{a}$ and $\vec{AB} = \vec{b}$ then vector $\vec{OB}$ is called the sum of $\vec{a}$ and $\vec{b}$.

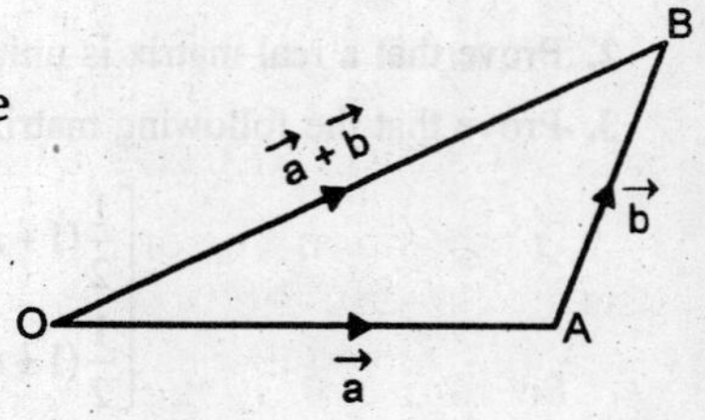


Symbolically

$$\overrightarrow{OA} + \overrightarrow{AB} = \overrightarrow{OB}$$

$$\vec{a} + \vec{b} = \overrightarrow{OB}$$

## 5.3 RECTANGULAR RESOLUTION OF A VECTOR

Let $OX, OY, OZ$ be the three rectangular axes. Let $\hat{i}, \hat{j}, \hat{k}$ be three unit vectors and parallel to three axes.

If $\vec{OP} = \hat{n}$ and the co-ordinates of $P$ be $(x, y, z)$

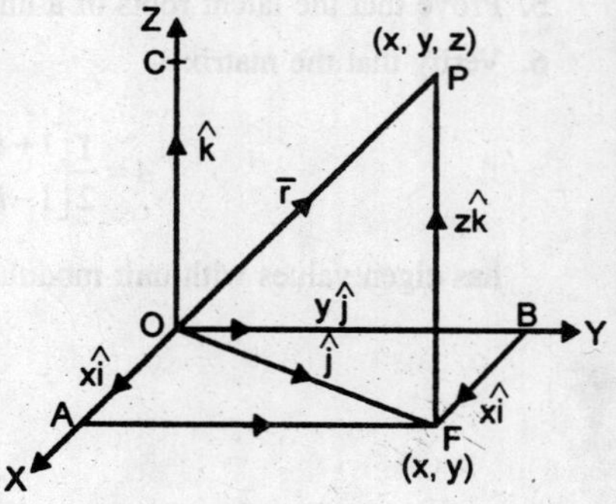


$$\vec{OA} = x\hat{i}, \quad \vec{OB} = y\hat{j} \quad \text{and} \quad \vec{OC} = z\hat{k}$$

$$\vec{OP} = \vec{OF} + \vec{FP}$$

$$\Rightarrow \quad \vec{OP} = (\vec{OA} + \vec{AF}) + \vec{FP}$$

$$\Rightarrow \quad \vec{OP} = \vec{OA} + \vec{OB} + \vec{OC}$$

$$\vec{r} = x\hat{i} + y\hat{j} + z\hat{k}$$

$$\Rightarrow \quad OP^2 = OF^2 + FP^2$$

$$= (OA^2 + AF^2) + FP^2 = OA^2 + OB^2 + OC^2 = x^2 + y^2 + z^2$$

$$OP = \sqrt{x^2 + y^2 + z^2}$$

$$|\vec{r}| = \sqrt{x^2 + y^2 + z^2}$$

## 5.4 UNIT VECTOR

Let a vector be $x\hat{i} + y\hat{j} + z\hat{k}$.

$$\text{Unit vector} = \frac{x\hat{i} + y\hat{j} + z\hat{k}}{\sqrt{x^2 + y^2 + z^2}}$$

**Example 1.** *If $\vec{a}$ and $\vec{b}$ be two unit vectors and α be the angle between them, then find the value of α such that $\vec{a} + \vec{b}$ is a unit vector.* (*Nagpur, University, Winter 2001*)

**Solution.** Let $\vec{OA} = \vec{a}$ be a unit vector and $\vec{AB} = \vec{b}$ is another unit vector and α be the angle between $\vec{a}$ and $\vec{b}$.

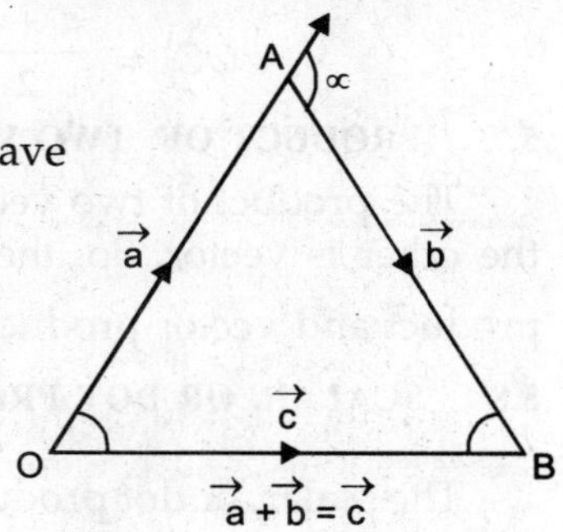


If $\vec{OB} = \vec{c} = \vec{a} + \vec{b}$ is also a unit vector then, we have

$$|\vec{OA}| = 1$$

$$|\vec{AB}| = 1$$

$$|\vec{OB}| = 1$$

*OAB* is an equilateral triangle.

So, each angle of Δ *OAB* is $\frac{\pi}{3}$ **Ans.**

Hence $\alpha = \pi - \frac{\pi}{3} = \frac{2\pi}{3}$

## 5.5 POSITION VECTOR OF A POINT

The position vector of a point *A* with respect to origin *O* is the vector $\vec{OA}$ which is used to specify the position of *A* w.r.t. *O*.

**To find $\overrightarrow{AB}$ if the position vectors of the point *A* and point *B* are given.**

If the position vectors of *A* and *B* are $\vec{a}$ and $\vec{b}$. Let the origin be *O*.

Then $\vec{OA} = \vec{a}, \quad \vec{OB} = \vec{b}$

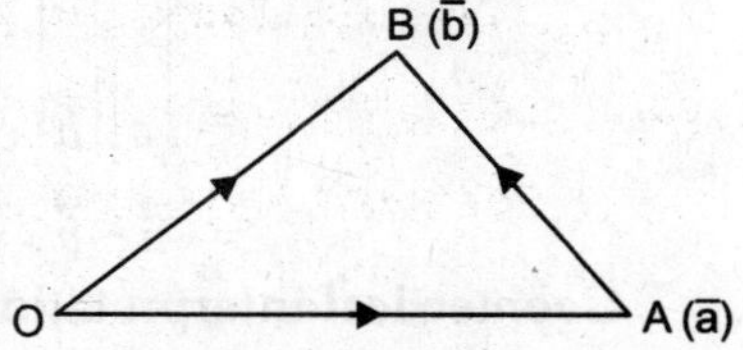


$$\vec{OA} + \vec{AB} = \vec{OB}$$

$$\vec{AB} = \vec{OB} - \vec{OA}$$

$$\Rightarrow \quad \vec{AB} = \vec{b} - \vec{a}$$

$\vec{AB}$ = **Position vector of *B* – Position vector of *A***

**Example 2.** *If A and* B *are* (3, 4, 5) *and* (6, 8, 9), *find* $\vec{AB}$.

**Solution.** $\vec{AB}$ = Position vector of *B* – Position vector of *A*

$$= (6\hat{i} + 8\hat{j} + 9\hat{k}) - (3\hat{i} + 4\hat{j} + 5\hat{k}) = 3\hat{i} + 4\hat{j} + 4\hat{k}$$ **Ans.**

## 5.6 RATIO FORMULA

To find the position vector of the point which divides the line joining two given points.

Let *A* and *B* be two points and a point *C* divides *AB* in the ratio of *m* : *n*.

Let *O* be the origin, then

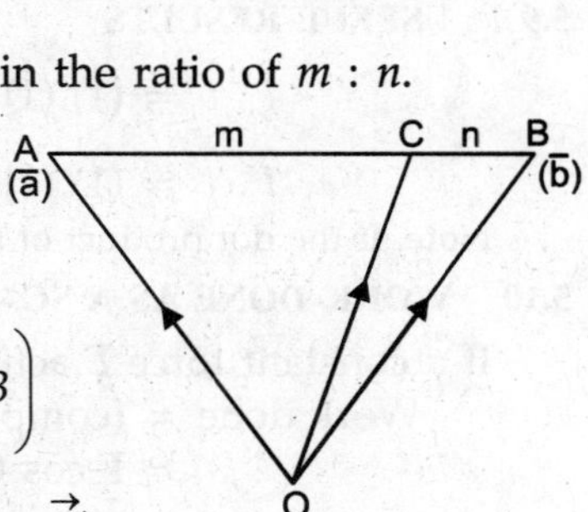


$$\vec{OA} = \vec{a}, \quad \text{and} \quad \vec{OB} = \vec{b}, \quad \vec{OC} = ?$$

$$\vec{OC} = \vec{OA} + \vec{AC}$$

$$= \vec{OA} + \frac{m}{m+n} AB \quad \left( \therefore AC = \frac{m}{m+n} AB \right)$$

$$= \vec{a} + \frac{m}{m+n} . (\vec{b} - \vec{a}) \quad (\therefore \vec{AB} = \vec{b} - \vec{a})$$

$$\vec{OC} = \frac{m\vec{b} + n\vec{a}}{m+n}$$

**Cor.** If $m = n = 1$, then $C$ will be the mid-point, and

$$\vec{OC} = \frac{\vec{a} + \vec{b}}{2}$$

## 5.7 PRODUCT OF TWO VECTORS

The product of two vectors results in two different ways, the one is a number and the other is vector. So, there are two types of product of two vectors, namely scalar product and vector product. They are written as $\vec{a} \cdot \vec{b}$ and $\vec{a} \times \vec{b}$.

## 5.8 SCALAR, OR DOT PRODUCT

The scalar, or dot product of two vectors $\vec{a}$ and $\vec{b}$ is defined to be $|\vec{a}||\vec{b}| \cos\theta$ *i.e.*, scalar where $\theta$ is the angle between $\vec{a}$ and $\vec{b}$.

Symbolically, $\vec{\mathbf{a}} \cdot \vec{\mathbf{b}} = |\vec{\mathbf{a}}||\vec{\mathbf{b}}| \cos\theta$

Due to a dot between $\vec{a}$ and $\vec{b}$ this product is also called dot product.
The scalar product is commutative

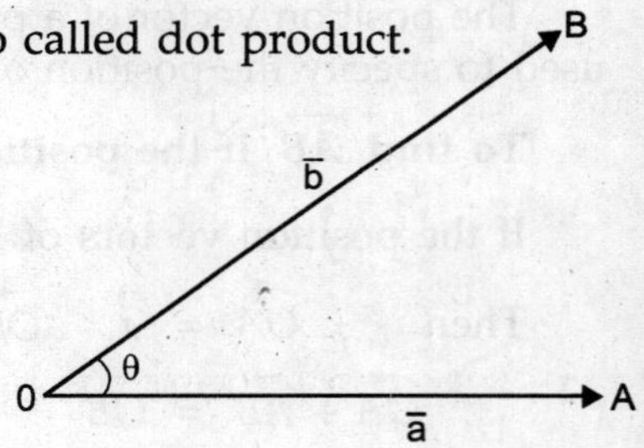


**To Prove.** $\vec{a} \cdot \vec{b} = \vec{b} \cdot \vec{a}$

**Proof.** $\vec{b} \cdot \vec{a} = |\vec{b}||\vec{a}| \cos(-\theta)$

$= |\vec{a}||\vec{b}| \cos\theta$

$= \vec{a} \cdot \vec{b}$ **Proved.**

**Geometrical interpretation.** The scalar product of two vectors is the product of one vector and the length of the projection of the other in the direction of the first.

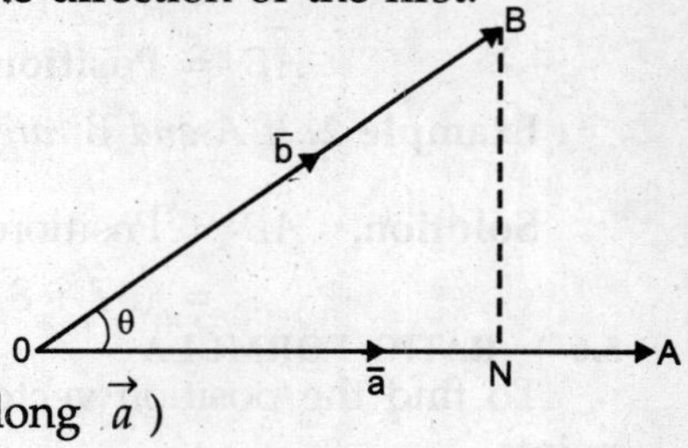


Let $\vec{OA} = \vec{a}$ and $\vec{OB} = \vec{b}$

then $\vec{a} \cdot \vec{b} = (OA) \cdot (OB) \cos\theta$

$= OA \cdot OB \cdot \dfrac{ON}{OB}$

$= OA \cdot ON$

$=$ (Length of $\vec{a}$) (projection of $\vec{b}$ along $\vec{a}$)

## 5.9 USEFUL RESULTS

$\hat{i} \cdot \hat{i} = (1)(1) \cos 0° = 1$ Similarly, $\hat{j} \cdot \hat{j} = 1$, $\hat{k} \cdot \hat{k} = 1$

$\hat{i} \cdot \hat{j} = (1)(1) \cos 90° = 0$ Similarly, $\hat{j} \cdot \hat{k} = 0$, $\hat{k} \cdot \hat{i} = 0$

**Note.** If the dot product of two vectors is zero then vectors are prependicular to each other.

## 5.10 WORK DONE AS A SCALAR PRODUCT

If a constant force $F$ acting on a particle displaces it from $A$ to $B$ then,

Work done = (component of $F$ along $AB$). Displacement

$= F \cos\theta \cdot AB$

$= \vec{F} \cdot \vec{AB}$

**Work done = Force . Displacement**

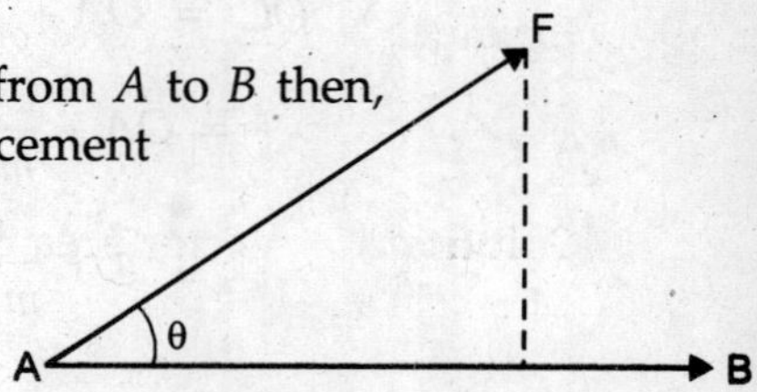

## 5.11 VECTOR PRODUCT OR CROSS PRODUCT

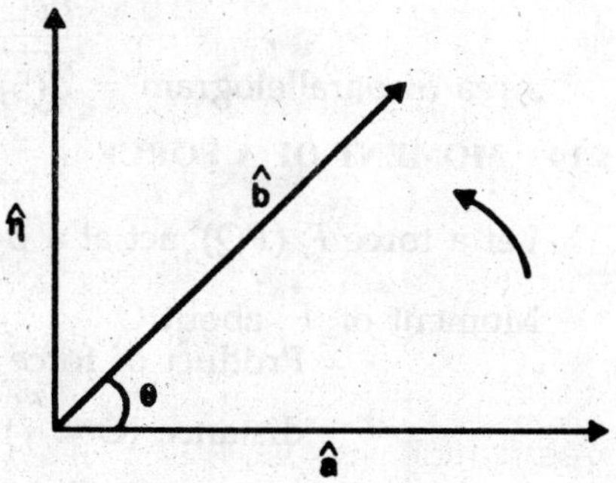


**1. The vector, or cross product of two vectors $\vec{a}$ and $\vec{b}$ is defined to be a vector such that**

(*i*) Its magnitude is $|\vec{a}||\vec{b}|\sin\theta$, where $\theta$ is the angle between $\vec{a}$ and $\vec{b}$.

(*ii*) Its direction is perpendicular to both vectors $\vec{a}$ and $\vec{b}$.

(*iii*) It forms with a right handed system.

Let $\hat{\eta}$ be a unit vector perpendicular to both the vectors $\vec{a}$ and $\vec{b}$.

$$\vec{a} \times \vec{b} = |\vec{a}||\vec{b}|\sin\theta.\hat{\eta}$$

**2. Useful results**

Since $\hat{i}, \hat{j}, \hat{k}$ are three mutually perpendicular unit vectors, then

$$\hat{i}\times\hat{i} = \hat{j}\times\hat{j} = \hat{k}\times\hat{k} = 0$$

$$\hat{i}\times\hat{j} = -\hat{j}\times\hat{i} = \hat{k} \qquad \hat{j}\times\hat{i} = -\hat{i}\times\hat{j}$$

$$\hat{j}\times\hat{k} = -\hat{k}\times\hat{j} = \hat{i} \quad \text{and} \quad \hat{k}\times\hat{j} = -\hat{j}\times\hat{k}$$

$$\hat{k}\times\hat{i} = -\hat{i}\times\hat{k} = \hat{j} \qquad \hat{i}\times\hat{k} = -\hat{k}\times\hat{i}$$

## 5.12 VECTOR PRODUCT EXPRESSED AS A DETERMINANT

If
$$\vec{a} = a_1\hat{i} + a_2\hat{j} + a_3\hat{k}$$
$$\vec{b} = b_1\hat{i} + b_2\hat{j} + b_3\hat{k}$$

$$\vec{a}\times\vec{b} = (a_1\hat{i} + a_2\hat{j} + a_3\hat{k})\times(b_1\hat{i} + b_2\hat{j} + b_3\hat{k})$$

$$= a_1b_1(\hat{i}\times\hat{i}) + a_1b_2(\hat{i}\times\hat{j}) + a_1b_3(\hat{i}\times\hat{k}) + a_2b_1(\hat{j}\times\hat{i}) + a_2b_2(\hat{j}\times\hat{j})$$
$$+ a_2b_3(\hat{j}\times\hat{k}) + a_3b_1(\hat{k}\times\hat{i}) + a_3b_2(\hat{k}\times\hat{j}) + a_3b_3(\hat{k}\times\hat{k})$$

$$= a_1b_2\hat{k} - a_1b_3\hat{j} - a_2b_1\hat{k} + a_2b_3\hat{i} + a_3b_1\hat{j} - a_3b_2\hat{i}$$

$$= (a_2b_3 - a_3b_2)\hat{i} - (a_1b_3 - a_3b_1)\hat{j} + (a_1b_2 - a_2b_1)\hat{k}$$

$$= \begin{vmatrix} \hat{i} & \hat{j} & \hat{k} \\ a_1 & a_2 & a_3 \\ b_1 & b_2 & b_3 \end{vmatrix}$$

## 5.13 AREA OF PARALLELOGRAM

**Example 3.** *Find the area of a parallelogram whose adjacent sides are i – 2j + 3 k and 2i + j – 4k.*

**Solution.** Vector area of || gm $= \begin{vmatrix} \hat{i} & \hat{j} & \hat{k} \\ 1 & -2 & 3 \\ 2 & 1 & -4 \end{vmatrix}$

$$= (8-3)\hat{i} - (-4-6)\hat{j} + (1+4)\hat{k} = 5\hat{i} + 10\hat{j} + 5\hat{k}$$

Area of parallelogram $= \sqrt{(5)^2 + (10)^2 + (5)^2} = 5\sqrt{6}$ **Ans.**

## 5.14 MOMENT OF A FORCE

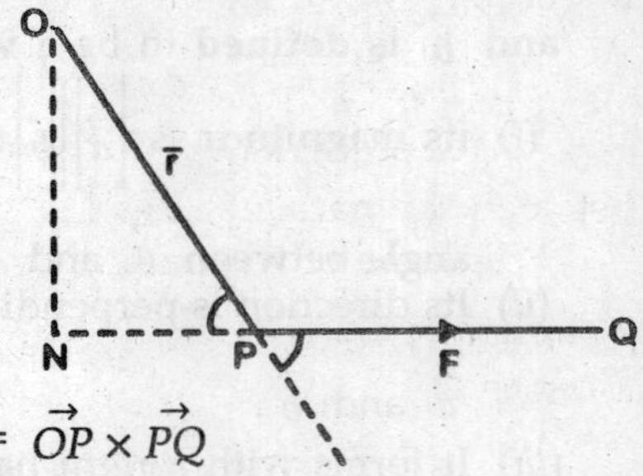


Let a force $F\ (\overrightarrow{PQ})$ act at a point $P$.

Moment of $\vec{F}$ about $O$

= Product of force $F$ and perpendicular distance $(ON.\hat{\eta})$

$$= (PQ)\ (ON)(\hat{\eta}) = (PQ)\ (OP)\sin\theta\ (\hat{\eta}) = \overrightarrow{OP} \times \overrightarrow{PQ}$$

$$\Rightarrow \quad \vec{M} = \vec{r} \times \vec{F}$$

## 5.15 ANGULAR VELOCITY

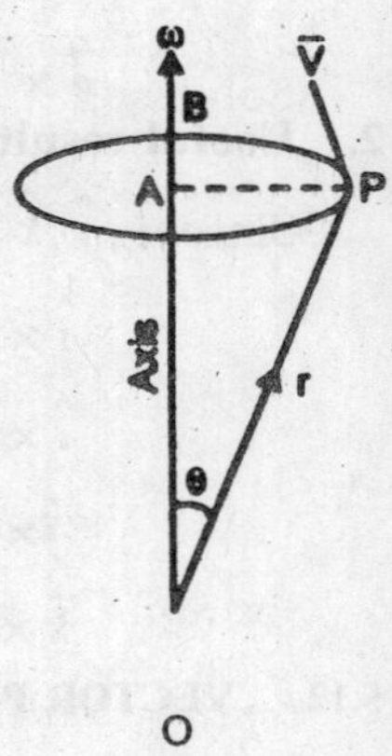


Let a rigid body be rotating about the axis $OA$ with the angular velocity $\omega$ which is a vector and its magnitude is $\omega$ radians per second and its direction is parallel to the axis of rotation $OA$.

Let $P$ be any point on the body such that $\overline{OP} = \vec{r}$ and $\angle AOP = \theta$ and $AP \perp OA$. Let the velocity of $P$ be $V$.

Let $\hat{\eta}$ be a unit vector perpendicular to $\vec{\omega}$ and $\vec{r}$.

$$\vec{\omega} \times \vec{r} = (\omega\, r \sin\theta)\ \hat{\eta} = (\omega\, AP)\ \hat{\eta} = (\text{Speed of } P)\ \hat{\eta}$$

$$= \text{Velocity of } P \perp \text{ to } \vec{\omega} \text{ and } r$$

Hence $\boxed{\vec{V} = \vec{\omega} \times \vec{r}}$

## 5.16 SCALAR TRIPLE PRODUCT

Let $\vec{a}, \vec{b}, \vec{c}$ be three vectors then their dot product is written as $\vec{a} \cdot (\vec{b} \times \vec{c})$ or $[\vec{a}\ \vec{b}\ \vec{c}]$.

If $\vec{a} = a_1\hat{i} + a_2\hat{j} + a_3\hat{k}$, $\vec{b} = b_1\hat{i} + b_2\hat{j} + b_3\hat{k}$, and $\vec{c} = c_1\hat{i} + c_2\hat{j} + c_3\hat{k}$

$$\vec{a} \cdot (\vec{b} \times \vec{c}) = (a_1\hat{i} + a_2\hat{j} + a_3\hat{k}) \cdot [(b_1\hat{i} + b_2\hat{j} + b_3\hat{k}) \times (c_1\hat{i} + c_2\hat{j} + c_3\hat{k})]$$

$$= (a_1\hat{i} + a_2\hat{j} + a_3\hat{k}) \cdot [(b_2c_3 - b_3c_2)\hat{i} + (b_3c_1 - b_1c_3)\hat{j} + (b_1c_2 - b_2c_1)\hat{k}]$$

$$= a_1(b_2c_3 - b_3c_2) + a_2(b_3c_1 - b_1c_3) + a_3(b_1c_2 - b_2c_1)$$

$$= \begin{vmatrix} a_1 & a_2 & a_3 \\ b_1 & b_2 & b_3 \\ c_1 & c_2 & c_3 \end{vmatrix}$$

Similarly, $\vec{b} \cdot (\vec{c} \times \vec{a})$ and $\vec{c} \cdot (\vec{a} \times \vec{b})$ have the same value.

$$\therefore \quad \vec{a} \cdot (\vec{b} \times \vec{c}) = \vec{b} \cdot (\vec{c} \times \vec{a}) = \vec{c} \cdot (\vec{a} \times \vec{b})$$

The value of the product depends upon the cyclic order of the vector, but is independent of the position of the dot and cross. These may be interchanged.

The value of the product changes if the order is non-cyclic.

**Note.** $\vec{a} \times (\vec{b} \cdot \vec{c})$ and $(\vec{a} \cdot \vec{b}) \times \vec{c}$ are meaningless.

## 5.17 GEOMETRICAL INTERPRETATION

The scalar triple product $\vec{a}\cdot(\vec{b}\times\vec{c})$ represents the volume of the parallelopiped having $\vec{a}, \vec{b}, \vec{c}$ as its co-terminous edges.

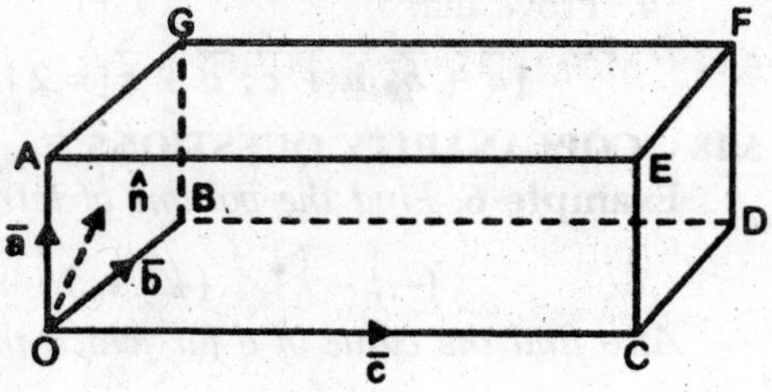


$\vec{a}\cdot(\vec{b}\times\vec{c}) = \vec{a}$ . Area of || gm $OBDC\ \hat{\eta}$

= Area of || gm $OBDC$ × perpendicular distance between the parallel faces $OBDC$ and $AEFG$.

= Volume of the parallelopiped

**Note.** (1) If $\vec{a}\cdot(\vec{b}\times\vec{c}) = 0$, then $\vec{a}, \vec{b}, \vec{c}$ are coplanar.

(2) Volume of tetrahedron $\frac{1}{6}(\vec{a}\ \vec{b}\ \vec{c})$.

**Example 4.** *Find the volume of parallelopiped if*

$$\vec{a} = -3\hat{i} + 7\hat{j} + 5\hat{k}, \quad \vec{b} = -3\hat{i} + 7\hat{j} - 3\hat{k}, \text{ and } \vec{c} = 7\hat{i} - 5\hat{j} - 3\hat{k}$$

*are the three co-terminous edges of the parallelopiped.*

**Solution.**

$$\text{Volume} = \vec{a}\cdot(\vec{b}\times\vec{c})$$

$$= \begin{vmatrix} -3 & 7 & 5 \\ -3 & 7 & -3 \\ 7 & -5 & -3 \end{vmatrix} = -3(-21-15) - 7(9+21) + 5(15-49)$$

$$= 108 - 210 - 170 = -272$$

Volume = 272 cube units. **Ans.**

**Example 5.** *Show that the volume of the tetrahedron having $\vec{A}+\vec{B}, \vec{B}+\vec{C}, \vec{C}+\vec{A}$ as concurrent edges is twice the volume of the tetrahendron having $\vec{A}, \vec{B}, \vec{C}$ as concurrent edges.*

**Solution.** Volume of tetrahendron $= \frac{1}{6}(\vec{A}+\vec{B})\cdot[(\vec{B}+\vec{C})\times(\vec{C}+\vec{A})]$

$$= \frac{1}{6}(\vec{A}+\vec{B})\cdot[\vec{B}\times\vec{C} + \vec{B}\times\vec{A} + \vec{C}\times\vec{C} + \vec{C}\times\vec{A}] \qquad [\vec{C}\times\vec{C} = 0]$$

$$= \frac{1}{6}(\vec{A}+\vec{B})\cdot(\vec{B}\times\vec{C} + \vec{B}\times\vec{A} + \vec{C}\times\vec{A})$$

$$= \frac{1}{6}[\vec{A}\cdot(\vec{B}\times\vec{C}) + \vec{A}\cdot(\vec{B}\times\vec{A}) + \vec{A}\cdot(\vec{C}\times\vec{A}) + \vec{B}\cdot(\vec{B}\times\vec{C}) + \vec{B}\cdot(\vec{B}\times\vec{A}) + \vec{B}\cdot(\vec{C}\times\vec{A})]$$

$$= \frac{1}{6}[\vec{A}\cdot(\vec{B}\times\vec{C}) + \vec{B}\cdot(\vec{C}\times\vec{A})] = \frac{1}{3}\vec{A}\cdot(\vec{B}\times\vec{C})$$

$$= 2\times\frac{1}{6}[\vec{A}\ \vec{B}\ \vec{C}]$$

= 2 Volume of tetrahedron having $\vec{A}, \vec{B}, \vec{C}$, as concurrent edges. **Proved.**

## EXERCISE 5.1

1. Find the volume of the parallelopiped with adjacent sides.

$$\overline{OA} = 3\hat{i} - \hat{j}, \quad \overline{OB} = \hat{j} + 2\hat{k}, \quad \text{and} \quad \overline{OC} = \hat{i} + 5\hat{j} + 4\hat{k}$$

extending from the origin of co-ordinates $O$. **Ans.** 20

2. Find the volume of the tetrahedron whose vertices are the points $A$ (2, –1, –3), $B$ (4, 1, 3) $C$ (3, 2, –1) and $D$ (1, 4, 2). **Ans.** $7\frac{1}{3}$

3. Choose $y$ in order that the vectors $\vec{a} = 7\hat{i} + y\hat{j} + \hat{k}$, $\vec{b} = 3\hat{i} + 2\hat{j} + \hat{k}$, $\vec{c} = 5\hat{i} + 3\hat{j} + \hat{k}$ are linearly dependent. **Ans.** $y = 4$

4. Prove that

$$[\vec{a}+\vec{b},\ \vec{b}+\vec{c},\ \vec{c}+\vec{a}] = 2[\vec{a}\ \vec{b}\ \vec{c}]$$

## 5.18 COPLANARITY QUESTIONS

**Example 6.** *Find the volume of tetrahedron having vertices*

$$(-\hat{j}-\hat{k}),\quad (4\hat{i}+5\hat{j}+q\hat{k}),\quad (3\hat{i}+9\hat{j}+4\hat{k}) \text{ and } 4(-\hat{i}+\hat{j}+\hat{k}).$$

*Also find the value of q for which these four points are coplanar.*

*(Nagpur University, Summer 2004, 2003, 2002)*

**Solution.** Let $\vec{A} = -\hat{j}-\hat{k},\ \bar{B} = 4\hat{i}+5\hat{j}+q\hat{k},\ \bar{C} = 3\hat{i}+9\hat{j}+4\hat{k},\ \bar{D} = 4(-\hat{i}+\hat{j}+\hat{k})$

$$\overline{AB} = \vec{B}-\vec{A} = 4\hat{i}+5\hat{j}+q\hat{k}-(-\hat{j}-\hat{k}) = 4\hat{i}+6\hat{j}+(q+1)\hat{k}$$

$$\overline{AC} = \vec{C}-\vec{A} = (3\hat{i}+9\hat{j}+4\hat{k})-(-\hat{j}-\hat{k}) = 3\hat{i}+10\hat{j}+5\hat{k}$$

$$\overline{AD} = \vec{D}-\vec{A} = 4(-\hat{i}+\hat{j}+\hat{k})-(-\hat{j}-\hat{k}) = -4\hat{i}+5\hat{j}+5\hat{k}$$

$$\text{Volume of the tetrahedron} = \frac{1}{6}[\overline{AB}\ \overline{AC}\ \overline{AD}]$$

$$= \frac{1}{6}\begin{vmatrix} 4 & 6 & q+1 \\ 3 & 10 & 5 \\ -4 & 5 & 5 \end{vmatrix} = \frac{1}{6}\{4(50-25)-6(15+20)+(q+1)(15+40)\}$$

$$= \frac{1}{6}\{100-210+55(q+1)\} = \frac{1}{6}(-110+55+55q)$$

$$= \frac{1}{6}(-55+55q) = \frac{55}{6}(q-1)$$

If four points $A$, $B$, $C$ and $D$ are coplanar, then $(\overline{AB}\ \overline{AC}\ \overline{AD}) = 0$

*i.e.,* Volume of the tetrahedron = 0

$$\Rightarrow \quad \frac{55}{6}(q-1) = 0 \quad \Rightarrow \quad q = 1$$ **Ans.**

**Example 7.** *If four points whose position vectors are $\bar{a}, \bar{b}, \bar{c}, \bar{d}$ are coplanar, show that*

$$[\vec{a}\ \vec{b}\ \vec{c}] = [\vec{a}\ \vec{d}\ \vec{b}] + [\vec{a}\ \vec{d}\ \vec{c}] + [\vec{d}\ \vec{b}\ \vec{c}]$$ *(Nagpur University, Summer 2005)*

**Solution.** Let $A, B, C, D$ be four points whose position vectors are $\vec{a}, \vec{b}, \vec{c}, \vec{d}$.

$$\vec{AD} = \vec{d}-\vec{a},\quad \vec{BD} = \vec{d}-\vec{b} \quad \text{and} \quad \vec{CD} = \vec{d}-\vec{c}$$

If $\vec{AD}, \vec{BD}, \vec{CD}$ are coplanar, then

$$\vec{AD}\cdot(\vec{BD}\times\vec{CD}) = 0$$

$$\Rightarrow \quad (\vec{d}-\vec{a})\cdot[(\vec{d}-\vec{b})\times(\vec{d}-\vec{c})] = 0$$

$$\Rightarrow (\vec{d}-\vec{a})\cdot[\vec{d}\times\vec{d}-\vec{d}\times\vec{c}-\vec{b}\times\vec{d}+\vec{b}\times\vec{c}] = 0$$

$$\Rightarrow \quad (\vec{d}-\vec{a})\cdot[-\vec{d}\times\vec{c}-\vec{b}\times\vec{d}+\vec{b}\times\vec{c}] = 0$$

$$\Rightarrow -\vec{d}\cdot(\vec{d}\times\vec{c})-\vec{d}\cdot(\vec{b}\times\vec{d})+\vec{d}\cdot(\vec{b}\times\vec{c})+\vec{a}\cdot(\vec{d}\times\vec{c})+\vec{a}\cdot(\vec{b}\times\vec{a})-\vec{a}\cdot(\vec{b}\times\vec{c}) = 0$$

$$\Rightarrow -0+0+[\vec{d}\ \vec{b}\ \vec{c}]+[\vec{a}\ \vec{d}\ \vec{c}]+[\vec{d}\ \vec{b}\ \vec{d}]-[\vec{a}\ \vec{b}\ \vec{c}] = 0$$

$$\Rightarrow \quad [\vec{a}\ \vec{b}\ \vec{c}] = [\vec{a}\ \vec{b}\ \vec{d}]+[\vec{a}\ \vec{d}\ \vec{c}]+[\vec{d}\ \vec{b}\ \vec{c}]$$ **Proved.**

## EXERCISE 5.2

1. Determine $\lambda$ such that

$\bar{a} = \hat{i} + \hat{j} + \hat{k}$, $\bar{b} = 2\hat{i} - 4\hat{k}$, and $\bar{c} = \hat{i} + \lambda\hat{j} + 3\hat{k}$ are coplanar. **Ans.** $\lambda = 5/3$

2. Show that the four points

$-6\hat{i} + 3\hat{j} + 2\hat{k}$, $3\hat{i} - 2\hat{j} + 4\hat{k}$, $5\hat{i} + 7\hat{j} + 3\hat{k}$ and $-13\hat{i} + 17\hat{j} - \hat{k}$ are coplanar.

3. Find the constant $a$ such that the vectors

$2\hat{i} - \hat{j} + \hat{k}$, $\hat{i} + 2\hat{j} - 3\hat{k}$, and $3\hat{i} + a\hat{j} + 5\hat{k}$ are coplanar. **Ans.** $-4$

4. Prove that four points

$4\hat{i} + 5\hat{j} + \hat{k}$, $-(\hat{j} + \hat{k})$, $3\hat{i} + 9\hat{j} + 4\hat{k}$, $4(-\hat{i} + \hat{j} + \hat{k})$ are coplanar.

5. If the vectors $\vec{a}$, $\vec{b}$ and $\vec{c}$ are coplanar, show that

$$\begin{vmatrix} \vec{a} & \vec{b} & \vec{c} \\ \vec{a}\cdot\vec{a} & \vec{a}\cdot\vec{b} & \vec{a}\cdot\vec{c} \\ \vec{b}\cdot\vec{a} & \vec{b}\cdot\vec{b} & \vec{b}\cdot\vec{c} \end{vmatrix} = 0$$

## 5.19 VECTOR PRODUCT OF THREE VECTORS

*(A.M.I.E.T.E., Summer, 2004, 2000)*

Let $\vec{a}$, $\vec{b}$ and $\vec{c}$ be three vectors then their vector product is written as $\vec{a} \times (\vec{b} \times \vec{c})$.

Let
$$\vec{a} = a_1\hat{i} + a_2\hat{j} + a_3\hat{k},$$
$$\vec{b} = b_1\hat{i} + b_2\hat{j} + b_3\hat{k},$$
$$\vec{c} = c_1\hat{i} + c_2\hat{j} + c_3\hat{k}$$

$$\vec{a} \times (\vec{b} \times \vec{c}) = (a_1\hat{i} + a_2\hat{j} + a_3\hat{k}) \times (b_1\hat{i} + b_2\hat{j} + b_3\hat{k}) \times (c_1\hat{i} + c_2\hat{j} + c_3\hat{k})$$
$$= (a_1\hat{i} + a_2\hat{j} + a_3\hat{k}) \times [(b_2c_3 - b_3c_2)\hat{i} + (b_3c_1 - b_1c_3)\hat{j} + (b_1c_2 - b_2c_1)\hat{k}]$$
$$= [a_2(b_1c_2 - b_2c_1) - a_3(b_3c_1 - b_1c_3)]\hat{i} + [a_3(b_2c_3 - b_3c_2) - a_1(b_1c_2 - b_2c_1)]\hat{j}$$
$$+ [a_1(b_3c_1 - b_1c_3) - a_2(b_2c_3 - b_3c_2)\hat{k}]$$
$$= (a_1c_1 + a_2c_2 + a_3c_3)(b_1\hat{i} + b_2\hat{j} + b_3\hat{k}) - (a_1b_1 + a_2b_2 + a_3b_3)(c_1\hat{i} + c_2\hat{j} + c_3\hat{k})$$
$$= (\vec{a}\cdot\vec{c})\,\vec{b} - (\vec{a}\cdot\vec{b})\,\vec{c}.$$
**Ans.**

**Example 8.** *Prove that :*

$$\vec{a} \times (\vec{b} \times \vec{c}) + \vec{b} \times (\vec{c} \times \vec{a}) + \vec{c} \times (\vec{a} \times \vec{b}) = 0$$ *(Nagpur University, Winter 2008)*

**Solution.** Here, we have

$$\vec{a} \times (\vec{b} \times \vec{c}) + \vec{b} \times (\vec{c} \times \vec{a}) + \vec{c} \times (\vec{a} \times \vec{b})$$
$$= [(\vec{a}\cdot\vec{c})\vec{b} - (\vec{a}\cdot\vec{b})\vec{c}] + [(\vec{b}\cdot\vec{a})\vec{c} - (\vec{b}\cdot\bar{c})\vec{a}] + [(\vec{c}\cdot\bar{b})\vec{b} - (\vec{c}\cdot\vec{a})\vec{b}]$$
$$= [(\vec{b}\cdot\vec{a})\vec{c} - (\vec{a}\cdot\vec{b})\vec{c}] + [(\vec{c}\cdot\vec{b})\vec{a} - (\vec{b}\cdot\vec{c})\vec{a}] + [(\vec{a}\cdot\vec{c})\vec{b} - (\vec{c}\cdot\vec{a})\vec{b}]$$
$$= [(\vec{a}\cdot\vec{b})\vec{c} - (\vec{a}\cdot\vec{b})\vec{c}] + [(\vec{b}\cdot\vec{c})\vec{a} - (\vec{b}\cdot\vec{c})\vec{a}] + [(\vec{c}\cdot\vec{a})\vec{b} - (\vec{c}\cdot\vec{a})\vec{b}]$$
$$= 0 + 0 + 0 = 0$$
**Proved.**

**Example 9.** *Prove that :*

$$\hat{i} \times (\vec{a} \times \hat{i}) + \hat{j} \times (\vec{a} \times \hat{j}) + \hat{k} \times (\vec{a} \times \hat{k}) = 2\,\vec{a}$$ *(Nagpur University, Winter 2003)*

**Solution.** Let $\vec{a} = a_1\hat{i} + a_2\hat{j} + a_3\hat{k}$

Now, L.H.S. $= \hat{i} \times (\vec{a} \times \hat{i}) + \hat{j} \times (\vec{a} \times \hat{j}) + \hat{k} \times (\vec{a} \times \hat{k})$

$$= \hat{i} \times \left[(a_1 \hat{i} + a_2 \hat{j} + a_3 \hat{k}) \times \hat{i}\right] + \hat{j} \times \left[(a_1 \hat{i} + a_2 \hat{j} + a_3 \hat{k}) \times \hat{j}\right] + \hat{k} \times \left[(a_1 \hat{i} + a_2 \hat{j} + a_3 \hat{k}) \times \hat{k}\right]$$

$$= \hat{i} \times \left[a_1(\hat{i} \times \hat{i}) + a_2(\hat{j} \times \hat{i}) + a_3(\hat{k} \times \hat{i})\right] + \hat{j} \times \left[a_1(\hat{i} \times \hat{j}) + a_2(\hat{j} \times \hat{j}) + a_3(\hat{k} \times \hat{j})\right] + \hat{k} \times \left[a_1(\hat{i} \times \hat{k}) + a_2(\hat{j} \times \hat{k}) + a_3(\hat{k} \times \hat{k})\right]$$

$$= \hat{i} \times \left[0 - a_2 \hat{k} + a_3 \hat{j}\right] + \hat{j} \times \left[a_1 \hat{k} + 0 - a_3 \hat{i}\right] + \hat{k} \times \left[-a_1 \hat{j} + a_2 \hat{i} + 0\right]$$

$$= -a_2(\hat{i} \times \hat{k}) + a_3(\hat{i} \times \hat{j}) + a_1(\hat{j} \times \hat{k}) - a_3(\hat{j} \times \hat{i}) - a_1(\hat{k} \times \hat{j}) + a_2(\hat{k} \times \hat{i})$$

$$= a_2 \hat{j} + a_3 \hat{k} + a_1 \hat{i} + a_3 \hat{k} + a_1 \hat{i} + a_2 \hat{j} = 2a_1 \hat{i} + 2a_2 \hat{j} + 2a_3 \hat{k}$$

$$= 2\,(a_1 \hat{i} + a_2 \hat{j} + a_3 \hat{k}) = 2\,\vec{a}$$ **Proved.**

**Example 10.** *Show that for any scalar λ, the vectors* $\vec{x}, \vec{y}$ *given by*

$$\vec{x} = \lambda\,\vec{a} + \frac{q\,(\vec{a} \times \vec{b})}{a^2},\ \vec{y} = \frac{(1 - p\,\lambda)}{q}\,\vec{a} - \frac{p\,(\vec{a} \times \vec{b})}{a^2}$$ *satisfy the equations*

$$p\,\vec{x} + q\,\vec{y} = \vec{a} \text{ and } \vec{x} \times \vec{y} = \vec{b}.$$ *(Nagpur University, Winter 2004)*

**Solution.** The given equations are

$$p\vec{x} + q\vec{y} = \vec{a} \quad ...(1)$$

$$\vec{x} \times \vec{y} = \vec{b} \quad ...(2)$$

Multiplying equation (1) vectorially by $\vec{x}$, we get

$$\vec{x} \times (p\vec{x} + q\vec{y}) = \vec{x} \times \vec{a}$$

$$p\,(\vec{x} \times \vec{x}) + q\,(\vec{x} \times \vec{y}) = \vec{x} \times \vec{a}$$

$$q \times (\vec{x} \times \vec{y}) = \vec{x} \times \vec{a}, \qquad \text{as } \vec{x} \times \vec{x} = 0$$

$$\vec{x} \times \vec{a} = q\vec{b}, \qquad [\text{From (2) } \vec{x} \times \vec{y} = \vec{b}] \quad ...(3)$$

Multiplying (3) vectorially by $\vec{a}$, we have

$$\vec{a} \times (\vec{x} \times \vec{a}) = \vec{a} \times q\,\vec{b}$$

$$(\vec{a}\,.\,\vec{a})\,\vec{x} - (\vec{a}\,.\,\vec{x})\,\vec{a} = q\,(\vec{a} \times \vec{b})$$

$$a^2\,\vec{x} - (\vec{a}\,.\,\vec{x})\,\vec{a} = q\,(\vec{a} \times \vec{b}) \quad \Rightarrow a^2\,\vec{x} = (\vec{a}\,.\,\vec{x})\,\vec{a} + q\,(\vec{a} \times \vec{b})$$

$$\vec{x} = \frac{(\vec{a}\,.\,\vec{x})\,\vec{a}}{a^2} + \frac{q\,(\vec{a} \times \vec{b})}{a^2}$$

$$\vec{x} = \lambda\, a + \frac{q\,(\vec{a} \times \vec{b})}{a^2} \qquad \text{where } \lambda = \frac{\vec{a}\cdot\vec{x}}{a^2}$$

Substituting the value of $\vec{x}$ in (1), we get $p\left\{\lambda\,\vec{a} + \frac{q\,(\vec{a}\times\vec{b})}{a^2}\right\} + q\,\vec{y} = \vec{a}$

$$q\,\vec{y} = \vec{a} - p\left\{\lambda\,\vec{a} + \frac{q\,(\vec{a}\times\vec{b})}{a^2}\right\}$$

$$\vec{y} = \frac{(1-p\lambda)\,\vec{a}}{q} - \frac{p\,(\vec{a}\times\vec{b})}{a^2}$$ **Ans.**

## EXERCISE 5.3

1. Show that $\vec{a}\times(\vec{b}\times\vec{a}) = (\vec{a}\times\vec{b})\times\vec{a}$
2. Write the correct answer

   (*a*) $(\vec{A}\times\vec{B})\times\vec{C}$ lies in the plane of

   (*i*) $\vec{A}$ and $\vec{B}$ (*ii*) $\vec{B}$ and $\vec{C}$ (*iii*) $\vec{C}$ and $\vec{A}$ **Ans.** (*ii*)

   (*b*) The value of $\vec{a}\cdot(\vec{b}+\vec{c})\times(\vec{a}+\vec{b}+\vec{c})$ is

   (*i*) Zero (*ii*) $[\vec{a},\vec{b},\vec{c}]+[\vec{b},\vec{c},\vec{a}]$ (*iii*) $[\vec{a},\vec{b},\vec{c}]$ (*iv*) None of these **Ans.** (*ii*)

## 5.20 SCALAR PRODUCT OF FOUR VECTORS

Prove the identity

$$(\vec{a}\times\vec{b})\cdot(\vec{c}\times\vec{d}) = (\vec{a}\cdot\vec{c})(\vec{b}\cdot\vec{d}) - (\vec{a}\cdot\vec{d})(\vec{b}\cdot\vec{c})$$

**Proof.** $(\vec{a}\times\vec{b})\cdot(\vec{c}\times\vec{d}) = (\vec{a}\times\vec{b})\cdot\vec{r}$

$= \vec{a}\cdot(\vec{b}\times\vec{r})$ dot and cross can be interchanged. Put $\vec{c}\times\vec{d} = \vec{r}$

$= \vec{a}\cdot[\vec{b}\times(\vec{c}\times\vec{d})] = \vec{a}\cdot[(\vec{b}\cdot\vec{d})\,\vec{c} - (\vec{b}\cdot\vec{c})\,\vec{d}]$

$= (\vec{a}\cdot\vec{c})(\vec{b}\cdot\vec{d}) - (\vec{a}\cdot\vec{d})(\vec{b}\cdot\vec{c})$

$= \begin{vmatrix} \vec{a}\cdot\vec{c} & \vec{a}\cdot\vec{d} \\ \vec{b}\cdot\vec{c} & \vec{b}\cdot\vec{d} \end{vmatrix}$ **Proved.**

## EXERCISE 5.4

1. If $\vec{a} = 2i+3j-k$, $\vec{b} = -i+2j-4k$, $\vec{c} = i+j+k$, find $(\vec{a}\times\vec{b})\cdot(\vec{a}\times\vec{c})$. **Ans.** –74
2. Prove that $(\vec{a}\times\vec{b})\cdot(\vec{a}\times\vec{c}) = a^2(\vec{b}\cdot\vec{c}) - (\vec{a}\cdot b)(\vec{a}\cdot\vec{c})$.

## 5.21 VECTOR PRODUCT OF FOUR VECTORS

Let $\vec{a}, \vec{b}, \vec{c}$ and $\vec{d}$ be four vectors then their vector product is written as

$$(\vec{a}\times\vec{b})\times(\vec{c}\times\vec{d})$$

Now, $(\vec{a}\times\vec{b})\times(\vec{c}\times\vec{d}) = \vec{r}\times(\vec{c}\times\vec{d})$ [Put $\vec{a}\times\vec{b} = \vec{r}$ ]

$= (\vec{r}\cdot\vec{d})\,\vec{c} - (\vec{r}\cdot\vec{c})\,\vec{d}$

$$= [(\vec{a} \times \vec{b}) \cdot \vec{d}]\, \vec{c} - [(\vec{a} \times \vec{b}) \cdot \vec{c}]\, \vec{d}$$

$$= [\vec{a}\ \vec{b}\ \vec{d}]\, \vec{c} - [\vec{a}\ \vec{b}\ \vec{c}]\, \vec{d}$$

$\therefore$ $(\vec{a} \times \vec{b}) \times (\vec{c} \times \vec{d})$ lies in the plane of $\vec{c}$ and $\vec{d}$. ...(1)

Again, $(\vec{a} \times \vec{b}) \times (\vec{c} \times \vec{d}) = (\vec{a} \times \vec{b}) \times \vec{s}$ [Put $= \vec{c} \times \vec{d} = \vec{s}$]

$$= -\vec{s} \times (\vec{a} \times \vec{b}) = -(\vec{s} \cdot \vec{b})\, \vec{a} + (\vec{s} \cdot \vec{a})\, \vec{b}$$

$$= -[(\vec{c} \times \vec{d}) \cdot \vec{b}]\, \vec{a} + [(\vec{c} \times \vec{d}) \cdot \vec{a}]\, \vec{b} = -[(\vec{b}\ \vec{c}\ \vec{d}]\, \vec{a} + [\vec{a}\ \vec{c}\ \vec{d}]\, \vec{b}$$

$\therefore$ $(\vec{a} \times \vec{b}) \times (\vec{c} \times \vec{d})$ lies in the plane of $\vec{a}$ and $\vec{b}$. ...(2)

**Geometrical interpretation :** From (1) and (2) we conclude that $(\vec{a} \times \vec{b}) \times (\vec{c} \times \vec{d})$ is a vector parallel to the line of intersection of the plane containing $\vec{a}$, $\vec{b}$ and plane containing $\vec{c}$, $\vec{d}$.

**Example 11.** *Show that*

$$(\vec{B} \times \vec{C}) \times (\vec{A} \times \vec{D}) + (\vec{C} \times \vec{A}) \times (\vec{B} \times \vec{D}) + (\vec{A} \times \vec{B}) \times (\vec{C} \times \vec{D}) = -2\, (\vec{A}\ \vec{B}\ \vec{C})\, \vec{D}$$

**Solution.** L.H.S. $= (\vec{B} \times \vec{C}) \times (\vec{A} \times \vec{D}) + (\vec{C} \times \vec{A}) \times (\vec{B} \times \vec{D}) + (\vec{A} \times \vec{B}) \times (\vec{C} \times \vec{D})$

$$= [(\vec{B}\ \vec{C}\ \vec{D})\, \vec{A} - (\vec{B}\ \vec{C}\ \vec{A})\, \vec{D}] + [(\vec{C}\ \vec{A}\ \vec{D})\, \vec{B} - (\vec{C}\ \vec{A}\ \vec{B})\, \vec{D}] + [(-\vec{B}\ \vec{C}\ \vec{D})\, \vec{A} + (\vec{A}\ \vec{C}\ \vec{D})\, \vec{B}]$$

$$= (\vec{B}\ \vec{C}\ \vec{D})\, \vec{A} - (\vec{B}\ \vec{C}\ \vec{D})\, \vec{A} + (\vec{C}\ \vec{A}\ \vec{D})\, \vec{B} + (\vec{A}\ \vec{C}\ \vec{D})\, \vec{B} - (\vec{B}\ \vec{C}\ \vec{A})\, \vec{D} - (\vec{C}\ \vec{A}\ \vec{B})\, \vec{D}$$

$$= -(\vec{A}\ \vec{C}\ \vec{D})\, \vec{B} + (\vec{A}\ \vec{C}\ \vec{D})\, \vec{B} - (\vec{A}\ \vec{B}\ \vec{C})\, \vec{D} - (\vec{A}\ \vec{B}\ \vec{C})\, \vec{D}$$

$= -2\, (\vec{A}\ \vec{B}\ \vec{C})\, \vec{D}$ = R.H.S. **Proved.**

## EXERCISE 5.5

Show that:

1. $(\vec{b} \times \vec{c}) \times (\vec{c} \times \vec{a}) = \vec{c}\, (\vec{a}\ \vec{b}\ \vec{c})$ when $(\vec{a}\ \vec{b}\ \vec{c})$ stands for scalar triple product.
2. $[\vec{b} \times \vec{c},\quad \vec{c} \times \vec{a},\quad \vec{a} \times \vec{b})] = [\vec{a}\ \vec{v}\ \vec{c}]^2$
3. $\vec{d}\, [\vec{a} \times \{\vec{b} \times (\vec{c} \times \vec{d})\}] = [(\vec{b} \cdot \vec{d})[\vec{a} \cdot (\vec{c} \times \vec{d})]$
4. $\vec{a}\, [\vec{a} \times [\vec{a} \times (\vec{a} \times \vec{b})] = a^2\, (\vec{b} \times \vec{a})$
5. $[(\vec{a} \times \vec{b}) \times (\vec{a} \times \vec{c})] \cdot \vec{d} = (\vec{a} \cdot \vec{d})\, [\vec{a}\ \vec{b}\ \vec{c}]$
6. $2a^2 = \left|\vec{a} \times \hat{i}\right|^2 + \left|\vec{a} \times \hat{j}\right|^2 + \left|\vec{a} \times \hat{k}\right|^2$
7. $\vec{a} \times \vec{b} = [(\hat{i} \times \vec{a}) \cdot \vec{b}]\, \hat{i} + [(\hat{j} \times \vec{a}) \cdot \vec{b}]\, \hat{j} + [(\hat{k} \times \vec{a}) \cdot \vec{b}]\, \hat{k}$
8. $\vec{p} \times [(\vec{a} \times \vec{q}) \times (\vec{b} \times \vec{r})] + \vec{q} \times [(\vec{a} \times \vec{r}) \times (\vec{b} \times \vec{p})] + \vec{r} \times [(\vec{a} \times \vec{p}) \times (\vec{b} \times \vec{q})] = 0$

## 5.22 VECTOR FUNCTION

If vector $r$ is a function of a scalar variable $t$, then we write

$$\vec{r} = \vec{r}(t)$$

If a particle is moving along a curved path then the position vector $\vec{r}$ of the particle is a function of $t$. If the component of $f(t)$ along $x$-axis, $y$-axis, $z$-axis are $f_1(t), f_2(t), f_3(t)$ respectively. Then,

$$\overrightarrow{f(t)} = f_1(t)\,\hat{i} + f_2(t)\,\hat{j} + f_3(t)\,\hat{k}$$

## 5.23 DIFFERENTIATION OF VECTORS

Let $O$ be the origin and $P$ be the position of a moving particle at time $t$.

Let $\overrightarrow{OP} = \vec{r}$

Let $Q$ be the position of the particle at the time $t + \delta t$ and

the position vector of $Q$ is $\overrightarrow{OQ} = \vec{r} + \delta\vec{r}$

$$\overrightarrow{PQ} = \overrightarrow{OQ} - \overrightarrow{OP}$$

$$= (\vec{r} + \delta\vec{r}) - \vec{r} = \delta\vec{r}$$

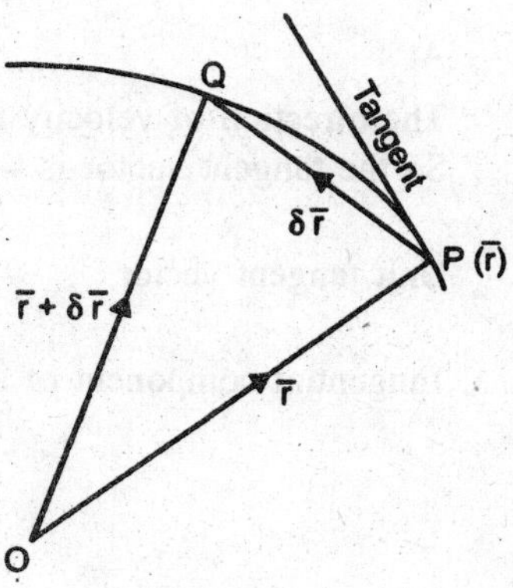


$\dfrac{\delta\vec{r}}{\delta t}$ is a vector. As $\delta t \to 0$, $Q$ tends to $P$ and the chord becomes the tangent at $P$.

We define $\quad \dfrac{d\vec{r}}{dt} = \lim\limits_{\delta t \to 0} \dfrac{\delta\vec{r}}{\delta t}$, then

$\dfrac{d\vec{r}}{dt}$ is a vector in the direction of the *tangent* at $P$.

$\dfrac{d\vec{r}}{dt}$ is also called the differential coefficient of $\vec{r}$ with respect to '$t$'.

Similarly, $\dfrac{d^2\vec{r}}{dt^2}$ is the second order derivative of $\vec{r}$.

$\dfrac{d\vec{r}}{dt}$ gives the velocity of the particle at $P$, which is along the tangent to its path. Also $\dfrac{d^2\vec{r}}{dt^2}$ gives the *acceleration* of the particle at $P$.

## 5.24 FORMULAE OF DIFFERENTIATION

(*i*) $\dfrac{d}{dt}(\vec{F} + \vec{G}) = \dfrac{d\vec{F}}{dt} + \dfrac{d\vec{G}}{dt}$ (*ii*) $\dfrac{d}{dt}(\vec{F}\phi) = \dfrac{d\vec{F}}{dt}\phi + \vec{F}\dfrac{d\phi}{dt}$ (*U.P. I semester, Dec. 2005*)

(*iii*) $\dfrac{d}{dt}(\vec{F}.\vec{G}) = \vec{F}.\dfrac{d\vec{G}}{dt} + \dfrac{d\vec{F}}{dt}.\vec{G}$ (*iv*) $\dfrac{d}{dt}(\vec{F} \times \vec{G}) = \vec{F} \times \dfrac{d\vec{G}}{dt} + \dfrac{d\vec{F}}{dt} \times \vec{G}$

(*v*) $\dfrac{d}{dt}[\vec{a}\ \vec{b}\ \vec{c}] = \left[\dfrac{d\vec{a}}{dt}\ \vec{b}\ \vec{c}\right] + \left[\vec{a}\ \dfrac{d\vec{b}}{dt}\ \vec{c}\right] + \left[\vec{a}\ \vec{b}\ \dfrac{d\vec{c}}{dt}\right]$

(*vi*) $\dfrac{d}{dt}[\vec{a} \times (\vec{b} \times \vec{c})] = \dfrac{d\vec{a}}{dt} \times (\vec{b} \times \vec{c}) + \vec{a} \times \left(\dfrac{d\vec{b}}{dt} \times \vec{c}\right) + \vec{a} \times \left(\vec{b} \times \dfrac{d\vec{c}}{dt}\right)$

The order of the functions $\vec{F}, \vec{G}$ is not to be changed.

**Example 12.** *A particle moves along the curve* $\overline{r} = (t^3 - 4t)\hat{i} + (t^2 + 4t)\hat{j} + (8t^2 - 3t^3)\hat{k}$, *where t is the time. Find the magnitude of the tangential components of its acceleration at t = 2.*

(*Nagpur University, Summer 2005*)

**Solution.** We have, $\vec{r} = (t^3 - 4t)\hat{i} + (t^2 + 4t)\hat{j} + (8t^2 - 3t^3)\hat{k}$

$$\text{Velocity} = \frac{d\vec{r}}{dt} = (3t^2 - 4)\hat{i} + (2t + 4)\hat{j} + (16t - 9t^2)\hat{k}$$

At $t = 2$, Velocity $= 8\hat{i} + 8\hat{j} - 4\hat{k}$

$$\text{Acceleration} = \vec{a} = \frac{d^2\vec{r}}{dt^2} = 6t\,\hat{i} + 2\hat{j} + (16 - 18t)\hat{k}$$

At $t = 2$ $\vec{a} = 12\hat{i} + 2\hat{j} - 20\hat{k}$

**The direction of velocity is along tangent.**
**So the tangent vector is velocity.**

$$\text{Unit tangent vector, } \hat{T} = \frac{\vec{v}}{|v|} = \frac{8\hat{i} + 8\hat{j} - 4\hat{k}}{\sqrt{64 + 64 + 16}} = \frac{8\hat{i} + 8\hat{j} - 4\hat{k}}{12} = \frac{2\hat{i} + 2\hat{j} - \hat{k}}{3}$$

**Tangential component of acceleration,** $a_t = \overline{a}.\hat{T}$

$$= (12\hat{i} + 2\hat{j} - 20\hat{k}).\frac{2\hat{i} + 2\hat{j} - \hat{k}}{3} = \frac{24 + 4 + 20}{3} = \frac{48}{3} = 16 \textbf{ Ans.}$$

**Example 13.** *If* $\frac{d\vec{a}}{dt} = \vec{u} \times \vec{a}$ *and* $\frac{d\vec{b}}{dt} = \vec{u} \times \vec{b}$ *then prove that* $\frac{d}{dt}[\vec{a} \times \vec{b}] = \vec{u} \times (\vec{a} \times \vec{b})$

(*M.U. 2009*)

**Solution.** We have,

$$\frac{d}{dt}[\vec{a} \times \vec{b}] = \vec{a} \times \frac{d\vec{b}}{dt} + \frac{d\vec{a}}{dt} \times \vec{b} = \vec{a} \times (\vec{u} \times \vec{b}) + (\vec{u} \times \vec{a}) \times \vec{b}$$

$$= \vec{a} \times (\vec{u} \times \vec{b}) - \vec{b} \times (\vec{u} \times \vec{a})$$

$$= (\vec{a}.\vec{b})\,\vec{u} - (\vec{a}.\vec{u})\,\vec{b} - [(\vec{b}.\vec{a})\,\vec{u} - (\vec{b}.\vec{u})\,\vec{a}]$$

(Vector triple product)

$$= (\vec{a}.\vec{b})\,\vec{u} - (\vec{u}.\vec{a})\,\vec{b} - (\vec{a}.\vec{b})\,\vec{u} + (\vec{u}.\vec{b})\,\vec{a}$$

$$= (\vec{u}.\vec{b})\,\vec{a} - (\vec{u}.\vec{a})\,\vec{b}$$

$$= \vec{u} \times (\vec{a} \times \vec{b})$$

**Proved.**

**Example 14.** *Find the angle between the surface* $x^2 + y^2 + z^2 = 9$ and $z = x^2 + y^2 - 3$ at (2, –1, 2).

(*M.D.U. Dec. 2009*)

**Solution.** Here, we have

$$x^2 + y^2 + z^2 = 9 \qquad \ldots(1)$$

$$z = x^2 + y^2 - 3 \qquad \ldots(2)$$

Normal to (1) $\eta_1 = \nabla(x^2 + y^2 + z^2 - 9)$

$$= \left(\hat{i}\frac{\partial}{\partial x} + \hat{j}\frac{\partial}{\partial y} + \hat{k}\frac{\partial}{\partial z}\right)(x^2 + y^2 + z^2 - 9) = 2x\hat{i} + 2y\hat{j} + 2z\hat{k}$$

Normal to (1) at (2, – 1, 2), $\eta_1 = 4\hat{i} - 2\hat{j} + 4\hat{k}$ ...(3)

Normal to (2), $\eta_2 = \nabla(z - x^2 - y^2 + 3)$

$$= \left(\hat{i}\frac{\partial}{\partial x} + \hat{j}\frac{\partial}{\partial y} + \hat{k}\frac{\partial}{\partial z}\right)(z - x^2 - y^2 + 3) = -2x\hat{i} - 2y\hat{j} + \hat{k}$$

Normal to (2) at (2, – 1, 2), $\eta_2 = -4\hat{i} + 2\hat{j} + \hat{k}$ ...(4)

$$\eta_1 . \eta_2 = |\eta_1||\eta_2|\cos\theta$$

$$\cos\theta = \frac{\eta_1.\eta_2}{|\eta_1||\eta_2|} = \frac{(4\hat{i} - 2\hat{j} + 4\hat{k}).(-4\hat{i} + 2\hat{j} + \hat{k})}{|4\hat{i} - 2\hat{j} + 4\hat{k}||-4\hat{i} + 2\hat{j} + \hat{k}|} = \frac{-16 - 4 + 4}{\sqrt{16+4+16}\sqrt{16+4+1}}$$

$$= \frac{-16}{6\sqrt{21}} = \frac{-8}{3\sqrt{21}}$$

$$\theta = \cos^{-1}\left(\frac{-8}{3\sqrt{21}}\right)$$

Hence the angle between (1) and (2) $\cos^{-1}\left(\frac{-8}{3\sqrt{21}}\right)$ **Ans**

## EXERCISE 5.6

1. The coordinates of a moving particle are given by $x = 4t - \frac{t^2}{2}$ and $y = 3 + 6t - \frac{t^3}{6}$. Find the velocity and acceleration of the particle when $t = 2$ secs. **Ans.** 4.47, 2.24
2. A particle moves along the curve

$$x = 2t^2, \quad y = t^2 - 4t \quad \text{and} \quad z = 3t - 5$$

where $t$ is the time. Find the components of its velocity and acceleration at time $t = 1$, in the direction $\hat{i} - 3\hat{j} + 2\hat{k}$. (*Nagpur, Summer 2001*) **Ans.** $\frac{8\sqrt{14}}{7}, -\frac{\sqrt{14}}{7}$

3. Find the unit tangent and unit normal vector at $t = 2$ on the curve $x = t^2 - 1$, $y = 4t - 3$, $z = 2t^2 - 6t$ where $t$ is any variable. **Ans.** $\frac{1}{3}(2\hat{i} + 2\hat{j} + \hat{k})$, $\frac{1}{3\sqrt{5}}(2\hat{i} + 2\hat{k})$
4. Prove that $\frac{d}{dt}(\vec{F} \times \vec{G}) = \vec{F} \times \frac{d\vec{G}}{dt} + \frac{d\vec{F}}{dt} \times \vec{G}$
5. Find the angle between the tangents to the curve $\bar{r} = t^2\hat{i} + 2t\hat{j} - t^3\hat{k}$, at the points $t = \pm 1$. **Ans.** $\cos^{-1}\left(\frac{9}{17}\right)$
6. If the surface $5x^2 - 2byz = 9x$ be orthogonal to the surface $4x^2y + z^3 = 4$ at the point (1, –1, 2) then $b$ is equal to

(*a*) 0 (*b*) 1 (*c*) 2 (*d*) 3 (*AMIETE, Dec. 2009*) **Ans.** (*b*)

## 5.25 SCALAR AND VECTOR POINT FUNCTIONS

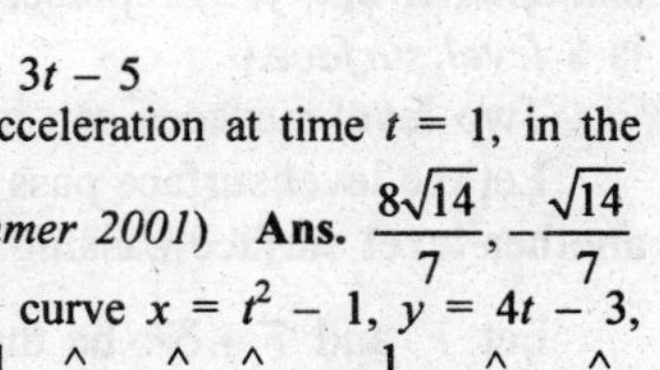


**Point function.** A variable quantity whose value at any point in a region of space depends upon the position of the point, is called a *point function*. There are two types of point functions.

(*i*) **Scalar point function.** If to each point $P$ ($x$, $y$, $z$) of a region $R$ in space there corresponds a unique scalar $f$ ($P$), then $f$ is called a scalar point function. *For example*, the temperature distribution in a heated body, density of a body and potential due to gravity are the examples of a scalar point function.

(*ii*) **Vector point function.** If to each point $P$ ($x$, $y$, $z$) of a region $R$ in space there corresponds a unique vector $f$ ($P$), then $f$ is called a *vector point function*. The velocity of a moving fluid, gravitational force are the examples of vector point function.

(*U.P., I Semester, Winter 2000*)

**Vector Differential Operator Del *i.e.* ∇**

The vector differential operator Del is denoted by ∇. It is defined as

$$\nabla = \hat{i}\frac{\partial}{\partial x}+\hat{j}\frac{\partial}{\partial y}+\hat{k}\frac{\partial}{\partial z}$$

## 5.26 GRADIENT OF A SCALAR FUNCTION

If $\phi(x, y, z)$ be a scalar function then $\hat{i}\frac{\partial\phi}{\partial x}+\hat{j}\frac{\partial\phi}{\partial y}+\hat{k}\frac{\partial\phi}{\partial z}$ is called the gradient of the scalar function $\phi$.

And is denoted by grad $\phi$.

Thus,
$$\text{grad } \phi = \hat{i}\frac{\partial\phi}{\partial x}+\hat{j}\frac{\partial\phi}{\partial y}+\hat{k}\frac{\partial\phi}{\partial z}$$

$$\text{gard } \phi = \left(\hat{i}\frac{\partial}{\partial x}+\hat{j}\frac{\partial}{\partial y}+\hat{k}\frac{\partial}{\partial z}\right)\phi(x, y, z)$$

$$\text{gard } \phi = \nabla \phi \qquad (\nabla \text{ is read del or nebla})$$

## 5.27 GEOMETRICAL MEANING OF GRADIENT, NORMAL

*(U.P. Ist Semester, Dec 2006)*

If a surface $\phi(x, y, z) = c$ passes through a point $P$. The value of the function at each point on the surface is the same as at $P$. Then such a surface is called a *level surface* through $P$. *For example,* If $\phi(x, y, z)$ represents potential at the point $P$, then *equipotential surface* $\phi(x, y, z) = c$ is a *level surface.*

Two level surfaces can not intersect.

Let the level surface pass through the point $P$ at which the value of the function is $\phi$. Consider another level surface passing through $Q$, where the value of the function is $\phi + d\phi$.

Let $\bar{r}$ and $\bar{r}+\delta\bar{r}$ be the position vector of $P$ and $Q$ then $\overrightarrow{PQ} = \delta\bar{r}$

$$\nabla\phi.d\bar{r} = \left(\hat{i}\frac{\partial\phi}{\partial x}+\hat{j}\frac{\partial\phi}{\partial y}+\hat{k}\frac{\partial\phi}{\partial z}\right).(\hat{i}\,dx+\hat{j}\,dy+\hat{k}\,dz)$$

$$= \frac{\partial\phi}{\partial x}dx+\frac{\partial\phi}{\partial y}dy+\frac{\partial\phi}{\partial z}dz = d\phi \qquad ...(1)$$

If $Q$ lies on the level surface of $P$, then $d\phi = 0$

Equation (1) becomes $\nabla\phi \,.\, dr = 0$. Then $\overline{\nabla\phi}$ is ⊥ to $d\bar{r}$ (tangent).

Hence, $\nabla\phi$ is **normal** to the surface $\phi(x, y, z) = c$

Let $\nabla\phi = |\nabla\phi|\widehat{N}$, where $\widehat{N}$ is a unit normal vector. Let $\delta n$ be the perpendicular distance between two level surfaces through $P$ and $R$. Then the rate of change of $\phi$ in the direction of the normal to the surface through $P$ is $\frac{\partial\phi}{\partial n}$.

$$\frac{d\phi}{dn} = \lim_{\delta n\to 0}\frac{\delta\phi}{\delta n} = \lim_{\delta n\to 0}\frac{\nabla\phi.d\vec{r}}{\delta n}$$

$$= \lim_{\delta n\to 0}\frac{|\nabla\phi|\,\hat{N}.d\vec{r}}{\delta n} \qquad \left\{\begin{aligned}\widehat{N}.\delta\vec{r} &= |\vec{N}||\delta\vec{r}|\cos\theta\\ &= |\delta\vec{r}|\cos\theta = \delta n\end{aligned}\right\}$$

$$= \lim_{\delta n\to 0}\frac{|\nabla\phi|\,\delta n}{\delta n} = |\nabla\phi|$$

$$\therefore \qquad |\nabla\phi| = \frac{\partial\phi}{\partial n}$$

Hence, gradient $\phi$ is a vector normal to the surface $\phi = c$ and has a magnitude equal to the rate of change of $\phi$ along this normal.

## 5.28 NORMAL AND DIRECTIONAL DERIVATIVE

(*i*) **Normal.** If $\phi(x, y, z) = c$ represents a family of surfaces for different values of the constant $c$. On differentiating $\phi$, we get $d\phi = 0$

But $\quad d\phi = \nabla\phi \,.\, d\vec{r} \quad$ so $\quad \nabla\phi \,.\, dr = 0$

The scalar product of two vectors $\nabla\phi$ and $d\vec{r}$ being zero, $\nabla\phi$ and $d\vec{r}$ are perpendicular to each other. $d\vec{r}$ is in the direction of tangent to the given surface.

Thus $\nabla\phi$ is a vector *normal* to the surface $\phi(x, y, z) = c$.

(*ii*) **Directional derivative.** The component of $\nabla\phi$ in the direction of a vector $\vec{d}$ is equal to $\nabla\phi.\hat{d}$ and is called the directional derivative of $\phi$ in the direction of $\vec{d}$.

$$\frac{\partial\phi}{\partial r} = \lim_{\delta r \to 0} \frac{\delta\phi}{\delta r} \qquad \text{where, } \delta r = PQ$$

$\dfrac{\partial\phi}{\partial r}$ is called the *directional derivative* of $\phi$ at $P$ in the direction of $PQ$.

Let a unit vector along $PQ$ be $\hat{N}'$.

$$\frac{\delta n}{\delta r} = \cos\theta \Rightarrow \delta r = \frac{\delta n}{\cos\theta} = \frac{\delta n}{\hat{N}\cdot\hat{N}'} \qquad \text{...(1)}$$

Now
$$\frac{\partial\phi}{\partial r} = \lim_{\delta r \to 0}\left[\frac{\delta\phi}{\dfrac{\delta n}{\hat{N}.\hat{N}'}}\right] = \hat{N}.\hat{N}'\frac{\partial\phi}{\partial n} \qquad \left[\text{From (1), } \delta r = \frac{\delta n}{\hat{N}.\hat{N}'}\right]$$

$$= \hat{N}'.\hat{N}\,|\nabla\phi| = \hat{N}'.\nabla\phi \qquad (\because \hat{N}\,|\nabla\phi| = \nabla\phi)$$

Hence, $\dfrac{\partial\phi}{\partial r}$, directional derivative is the component of $\nabla\phi$ in the direction $\hat{N}'$

$$\frac{\partial\phi}{\partial r} = \hat{N}'.\nabla\phi = |\nabla\phi|\cos\theta \le |\nabla\phi|$$

Hence, $\nabla\phi$ is the maximum rate of change of $\phi$.

**Example 15.** *For the vector field (i)* $\vec{A} = m\hat{i}$ *and (ii)* $\vec{A} = m\,\vec{r}$. *Find* $\nabla.\vec{A}$ *and* $\nabla\times\vec{A}$. *Draw the sketch in each case.* (*Gujarat, I Semester, Jan. 2009*)

**Solution.** (*i*) Vector $\vec{A} = m\,\hat{i}$ is represented in the figure (*i*).

(*ii*) $\quad \vec{A} = m\,\vec{r}$ is represented in the figure (*ii*).

(*iii*) $$\nabla.\vec{A} = \left(\hat{i}\frac{\partial}{\partial x} + \hat{j}\frac{\partial}{\partial y} + \hat{k}\frac{\partial}{\partial z}\right).(x\hat{i} + y\hat{j} + z\hat{k}) = 1 + 1 + 1 = 3$$

$\nabla.\vec{A} = 3$ is represented on the number line at 3.

(*iv*) $$\nabla\times\vec{A} = \left(\hat{i}\frac{\partial}{\partial x} + \hat{j}\frac{\partial}{\partial y} + \hat{k}\frac{\partial}{\partial z}\right)\times(x\hat{i} + y\hat{j} + z\hat{k})$$

$$= \begin{vmatrix} \hat{i} & \hat{j} & \hat{k} \\ \dfrac{\partial}{\partial x} & \dfrac{\partial}{\partial y} & \dfrac{\partial}{\partial z} \\ x & y & z \end{vmatrix} = 0$$

are represented in the adjoining figure.

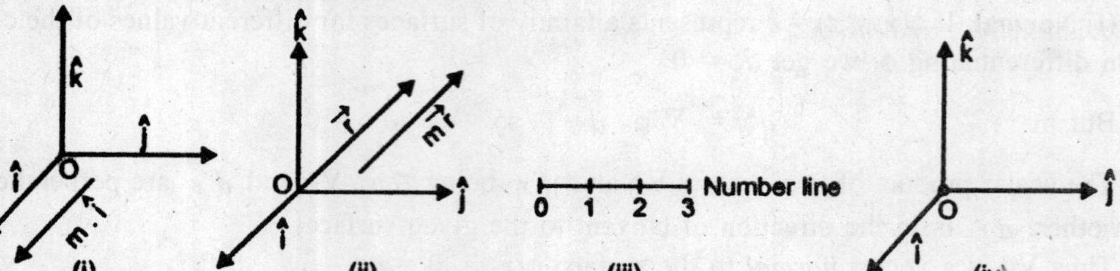

(i) (ii) (iii) (iv)

**Example 16.** *If $\phi = 3x^2y - y^3z^2$; find grad $\phi$ at the point (1, –2, –1).*

*(AMIETE, June 2009, U.P., I Semester, Dec. 2006)*

**Solution.**

$$\text{grad } \phi = \nabla\phi$$

$$= \left(\hat{i}\frac{\partial}{\partial x} + \hat{j}\frac{\partial}{\partial y} + \hat{k}\frac{\partial}{\partial z}\right)(3x^2y - y^3z^2)$$

$$= \hat{i}\frac{\partial}{\partial x}(3x^2y - y^3z^2) + \hat{j}\frac{\partial}{\partial y}(3x^2y - y^3z^2) + \hat{k}\frac{\partial}{\partial z}(3x^2y - y^3z^2)$$

$$= \hat{i}\,(6xy) + \hat{j}\,(3x^2 - 3y^2z^2) + \hat{k}\,(-2y^3z)$$

$$\text{grad } \phi \text{ at } (1, -2, -1) = \hat{i}\,(6)(1)(-2) + \hat{j}\,[(3)(1) - 3(4)(1)] + \hat{k}\,(-2)(-8)(-1)$$

$$= -12\hat{i} - 9\hat{j} - 16\hat{k}$$

**Ans.**

**Example 17.** *If $u = x + y + z$, $v = x^2 + y^2 + z^2$, $w = yz + zx + xy$ prove that grad u, grad v and grad w are coplanar vectors.* [*U.P., I Semester, 2001*]

**Solution.** We have,

$$\text{grad } u = \left(\hat{i}\frac{\partial}{\partial x} + \hat{j}\frac{\partial}{\partial y} + \hat{k}\frac{\partial}{\partial z}\right)(x + y + z) = \hat{i} + \hat{j} + \hat{k}$$

$$\text{grad } v = \left(\hat{i}\frac{\partial}{\partial x} + \hat{j}\frac{\partial}{\partial y} + \hat{k}\frac{\partial}{\partial z}\right)(x^2 + y^2 + z^2) = 2x\,\hat{i} + 2y\,\hat{j} + 2z\,\hat{k}$$

$$\text{grad } w = \left(\hat{i}\frac{\partial}{\partial x} + \hat{j}\frac{\partial}{\partial y} + \hat{k}\frac{\partial}{\partial z}\right)(yz + zx + xy) = \hat{i}(z + y) + \hat{j}(z + x) + \hat{k}(y + x)$$

[For vectors to be coplanar, their scalar triple product is 0]

$$\text{Now, grad } u.(\text{grad } v \times \text{grad } w) = \begin{vmatrix} 1 & 1 & 1 \\ 2x & 2y & 2z \\ z+y & z+x & y+x \end{vmatrix} = 2\begin{vmatrix} 1 & 1 & 1 \\ x & y & z \\ z+y & z+x & y+x \end{vmatrix}$$

$$= 2\begin{vmatrix} 1 & 1 & 1 \\ x+y+z & x+y+z & x+y+z \\ z+y & z+x & y+x \end{vmatrix} \qquad [\text{Applying } R_2 \to R_2 + R_3]$$

$$= 2(x + y + z)\begin{vmatrix} 1 & 1 & 1 \\ 1 & 1 & 1 \\ y+z & z+x & x+y \end{vmatrix} = 0$$

Since the scalar product of grad $u$, grad $v$ and grad $w$ are zero, hence these vectors are coplanar vectors. **Proved.**

**Example 18.** *Find the directional derivative of $x^2y^2z^2$ at the point (1, 1, –1) in the direction of the tangent to the curve $x = e^t$, $y = \sin 2t + 1$, $z = 1 - \cos t$ at $t = 0$.*

*(Nagpur University, Summer 2005)*

**Solution.** Let $\phi = x^2 y^2 z^2$

Directional Derivative of $\phi$

$$= \nabla\phi = \left(\hat{i}\frac{\partial}{\partial x} + \hat{j}\frac{\partial}{\partial y} + \hat{k}\frac{\partial}{\partial z}\right)(x^2y^2z^2)$$

$$\nabla\phi = 2xy^2z^2\,\hat{i} + 2yx^2z^2\,\hat{j} + 2zx^2y^2\,\hat{k}$$

Directional Derivative of $\phi$ at (1, 1, –1)

$$= 2(1)(1)^2(-1)^2\,\hat{i} + 2(1)(1)^2(-1)^2\,\hat{j} + 2(-1)(1)^2(1)^2\,\hat{k}$$

$$= 2\,\hat{i} + 2\,\hat{j} - 2\,\hat{k} \qquad ...(1)$$

$$\vec{r} = x\,\hat{i} + y\,\hat{j} + z\,\hat{k} = e^t\,\hat{i} + (\sin 2t + 1)\,\hat{j} + (1 - \cos t)\,\hat{k}$$

Tangent vector, $\vec{T} = \dfrac{d\vec{r}}{dt} = e^t\,\hat{i} + 2\cos 2t\,\hat{j} + \sin t\,\hat{k}$

$$\text{Tangent(at } t = 0) = e^0\,\hat{i} + 2(\cos 0)\,\hat{j} + (\sin 0)\,\hat{k} = \hat{i} + 2\,\hat{j} \qquad ...(2)$$

Required directional derivative along tangent $= (2\hat{i} + 2\hat{j} - 2\hat{k})\dfrac{(\hat{i} + 2\hat{j})}{\sqrt{1+4}}$

[From (1), (2)]

$$= \frac{2+4+0}{\sqrt{5}} = \frac{6}{\sqrt{5}} \qquad \textbf{Ans.}$$

**Example 19.** *Find the unit normal to the surface $xy^3z^2 = 4$ at (–1, –1, 2).* *(M.U. 2008)*

**Solution.** Let $\phi(x, y, z) = xy^3z^2 = 4$

We know that $\nabla\phi$ is the vector normal to the surface $\phi(x, y, z) = c$.

$$\text{Normal vector} = \nabla\phi = \hat{i}\frac{\partial\phi}{\partial x} + \hat{j}\frac{\partial\phi}{\partial y} + \hat{k}\frac{\partial\phi}{\partial z}$$

Now
$$= \hat{i}\frac{\partial}{\partial x}(xy^3z^2) + \hat{j}\frac{\partial}{\partial y}(xy^3z^2) + \hat{k}\frac{\partial}{\partial z}(xy^3z^2)$$

$$\Rightarrow \quad \text{Normal vector} = y^3z^2\,\hat{i} + 3xy^2z^2\,\hat{j} + 2xy^3z\,\hat{k}$$

Normal vector at (–1, –1, 2) $= -4\hat{i} - 12\hat{j} + 4\hat{k}$

Unit vector normal to the surface at (–1, –1, 2).

$$= \frac{\nabla\phi}{|\nabla\phi|} = \frac{-4\hat{i} - 12\hat{j} + 4\hat{k}}{\sqrt{16+144+16}} = -\frac{1}{\sqrt{11}}(\hat{i} + 3\hat{j} - \hat{k}) \qquad \textbf{Ans.}$$

**Example 20.** *Find the rate of change of $\phi = xyz$ in the direction normal to the surface $x^2y + y^2x + yz^2 = 3$ at the point (1, 1, 1).* *(Nagpur University, Summer 2001)*

**Solution.** Rate of change of $\phi = \Delta\,\phi$

$$= \left(\hat{i}\frac{\partial}{\partial x} + \hat{j}\frac{\partial}{\partial y} + \hat{k}\frac{\partial}{\partial z}\right)(x\,y\,z) = \hat{i}\,yz + \hat{j}\,xz + \hat{k}\,xy$$

Rate of change of $\phi$ at (1, 1, 1) = $(\hat{i}+\hat{j}+\hat{k})$

Normal to the surface $\Psi = x^2y + y^2x + yz^2 - 3$ is given as -

$$\nabla\psi = \left(\hat{i}\frac{\partial}{\partial x}+\hat{j}\frac{\partial}{\partial y}+\hat{k}\frac{\partial}{\partial z}\right)(x^2y+y^2x+yz^2-3)$$

$$= \hat{i}(2xy+y^2)+\hat{j}(x^2+2xy+z^2)+\hat{k}\,2yz$$

$$(\nabla\psi)_{(1,\,1,\,1)} = 3\hat{i}+4\hat{j}+2\hat{k}$$

$$\text{Unit normal} = \frac{3\hat{i}+4\hat{j}+2\hat{k}}{\sqrt{9+16+4}}$$

Required rate of change of $\phi = (\hat{i}+\hat{j}+\hat{k}).\dfrac{(3\hat{i}+4\hat{j}+2\hat{k})}{\sqrt{9+16+4}} = \dfrac{3+4+2}{\sqrt{29}} = \dfrac{9}{\sqrt{29}}$ **Ans.**

**Example 21.** *Find the constants m and n such that the surface $m\,x^2 - 2nyz = (m+4)x$ will be orthogonal to the surface $4x^2y + z^3 = 4$ at the point (1, –1, 2).*

*(M.D.U. Dec. 2009, Nagpur University, Summer 2002)*

**Solution.** The point *P* (1, –1, 2) lies on both surfaces. As this point lies in

$$mx^2 - 2nyz = (m+4)x, \text{ so we have}$$

$$m - 2n\,(-2) = (m+4)$$

$$\Rightarrow \quad m + 4n = m + 4 \quad \Rightarrow \quad n = 1$$

$$\therefore \quad \text{Let } \phi_1 = mx^2 - 2yz - (m+4)x \text{ and } \phi_2 = 4x^2y + z^3 - 4$$

$$\text{Normal to } \phi_1 = \nabla\phi_1$$

$$= \left(\hat{i}\frac{\partial}{\partial x}+\hat{j}\frac{\partial}{\partial y}+\hat{k}\frac{\partial}{\partial z}\right)[mx^2-2yz-(m+4)x]$$

$$= \hat{i}(2mx-m-4)-2z\,\hat{j}-2y\,\hat{k}$$

$$\text{Normal to } \phi_1 \text{ at } (1,-1,2) = \hat{i}(2m-m-4)-4\hat{j}+2\hat{k} = (m-4)\hat{i}-4\hat{j}+2\hat{k}$$

$$\text{Normal to } \phi_2 = \nabla\phi_2$$

$$= \left(\hat{i}\frac{\partial}{\partial x}+\hat{j}\frac{\partial}{\partial y}+\hat{k}\frac{\partial}{\partial z}\right)(4x^2y+z^3-4) = \hat{i}\,8xy+4x^2\,\hat{j}+3z^2\,\hat{k}$$

Normal to $\phi_2$ at (1, –1, 2) = $-8\hat{i}+4\hat{j}+12\hat{k}$

Sinec $\phi_1$ and $\phi_2$ are orthogonal, then normals are perpendicular to each other.

$$\nabla\phi_1\,.\,\nabla\phi_2 = 0$$

$$\Rightarrow \quad [(m-4)\hat{i}-4\hat{j}+2\hat{k}].[-8\hat{i}+4\hat{j}+12\hat{k}] = 0$$

$$\Rightarrow \quad -8\,(m-4) - 16 + 24 = 0$$

$$\Rightarrow \quad m - 4 = -2 + 3 \quad \Rightarrow \quad m = 5$$

Hence $m = 5$, $n = 1$ **Ans.**

**Example 22.** *Find the values of constants $\lambda$ and $\mu$ so that the surfaces $\lambda x^2 - \mu yz = (\lambda+2)\,x$, $4x^2y + z^3 = 4$ intersect orthogonally at the point (1, – 1, 2).*

*(AMIETE, II Sem., Dec. 2010, June 2009)*

**Solution.** Here, we have

$$\lambda x^2 - \mu yz = (\lambda+2)\,x \quad ...(1)$$

$$4x^2\,y + z^3 = 4 \quad ...(2)$$

Normal to the surface (1), $= \nabla\left[\lambda x^2 - \mu yz - (\lambda+2)x\right]$

$$= \left[\hat{i}\frac{\partial}{\partial x} + \hat{j}\frac{\partial}{\partial y} + \hat{k}\frac{\partial}{\partial z}\right]\left[\lambda x^2 - \mu yz - (\lambda+2)x\right]$$

$$= \hat{i}\,(2\lambda x - \lambda - 2) + \hat{j}\,(-\mu z) + \hat{k}(-\mu y)$$

Normal at (1, –1, 2) $= \hat{i}\ \ (2\lambda - \lambda - 2) - \hat{j}\ (-2\mu) + \hat{k}\,\mu$ ...(3)

$$= \hat{i}\ \ (\lambda - 2) + \hat{j}\,z\ (2\mu) + \hat{k}\,\mu$$

Normal at the surface (2)

$$= \left(\hat{i}\frac{\partial}{\partial x} + \hat{j}\frac{\partial}{\partial y} + \hat{k}\frac{\partial}{\partial z}\right)(4x^2y + z^3 - 4)$$

$$= \hat{i}\ \ (8 \times y) + \hat{j}\ \ (4x^2) + \hat{k}\ \ (3z^2)$$

Normal at the point (1, –1, 2) $= -\,8\,\hat{i}\ + 4\,\hat{j}\ + 12\,\hat{k}$ ...(4)

Since (3) and (4) are orthogonal so

$$\left[\hat{i}\,(\lambda-2) + \hat{j}\,(2\mu) + \hat{k}\mu\right].\left[-8\hat{i} + 4\hat{j} + 12\hat{k}\right] = 0$$

$$-8\,(\lambda-2) + 4(2\mu) + 12\mu = 0 \quad \Rightarrow \quad -8\lambda + 16 + 8\mu + 12\mu = 0$$

$$-8\lambda - 20\mu + 16 = 0 \qquad \Rightarrow 4(-2\lambda + 5\mu + 4) = 0$$

$$-2\lambda + 5\mu + 4 = 0 \qquad \Rightarrow \quad 2\lambda - 5\mu = 4 \qquad \text{...(5)}$$

Point (1, – 1, 2) will satisfy (1)

$\therefore \qquad \lambda(1)^2 - \mu(-1)(2) = (\lambda+2)\ (1) \Rightarrow \quad \lambda + 2\mu = \lambda + 2 \ \Rightarrow \ \mu = 1$

Putting $\mu = 1$ in (5), we get

$$2\lambda - 5 = 4 \ \Rightarrow \ \lambda = \frac{9}{2}$$

Hence $\quad \lambda = \dfrac{9}{2}$ and $\mu = 1$ **Ans.**

**Example 23.** *Find the angle between the surfaces $x^2 + y^2 + z^2 = 9$ and $z = x^2 + y^2 - 3$ at the point (2, –1, 2).* (*Nagpur University, Summer 2002*)

**Solution.** Normal on the surface $(x^2 + y^2 + z^2 - 9 = 0)$

$$\nabla\phi = \left(\hat{i}\frac{\partial}{\partial x} + \hat{j}\frac{\partial}{\partial y} + \hat{k}\frac{\partial}{\partial z}\right)(x^2 + y^2 + z^2 - 9) = (2x\,\hat{i} + 2y\,\hat{j} + 2z\,\hat{k})$$

Normal at the point (2, –1, 2) $= 4\hat{i} - 2\hat{j} + 4\hat{k}$ ...(1)

Normal on the surface $(z = x^2 + y^2 - 3) = \left(\hat{i}\dfrac{\partial}{\partial x} + \hat{j}\dfrac{\partial}{\partial y} + \hat{k}\dfrac{\partial}{\partial z}\right)(x^2 + y^2 - z - 3)$

$$= 2x\,\hat{i} + 2y\,\hat{j} - \hat{k}$$

Normal at the point (2, –1, 2) $= 4\hat{i} - 2\hat{j} - \hat{k}$ ...(2)

Let $\theta$ be the angle between normals (1) and (2).

$$(4\hat{i} - 2\hat{j} + 4\hat{k}).(4\hat{i} - 2\hat{j} - \hat{k}) = \sqrt{16+4+16}\,\sqrt{16+4+1}\,\cos\theta$$

$$16 + 4 - 4 = 6\sqrt{21}\cos\theta \qquad \Rightarrow \qquad 16 = 6\sqrt{21}\cos\theta$$

$$\Rightarrow \quad \cos\theta = \frac{8}{3\sqrt{21}} \quad \Rightarrow \quad \theta = \cos^{-1}\frac{8}{3\sqrt{21}}$$ **Ans.**

**Example 24.** *Find the directional derivative of $\frac{1}{r}$ in the direction $\overline{r}$ where $\overline{r} = x\hat{i} + y\hat{j} + z\hat{k}$.*

*(Nagpur University, Summer 2004, U.P., I Semester, Winter 2005, 2002)*

**Solution.** Here, $\phi(x, y, z) = \frac{1}{r} = \frac{1}{\sqrt{x^2+y^2+z^2}} = (x^2+y^2+z^2)^{-\frac{1}{2}}$

Now $\nabla\left(\frac{1}{r}\right) = \left(\hat{i}\frac{\partial}{\partial x} + \hat{j}\frac{\partial}{\partial y} + \hat{k}\frac{\partial}{\partial z}\right)(x^2+y^2+z^2)^{-\frac{1}{2}}$

$$= \frac{\partial}{\partial x}(x^2+y^2+z^2)^{-\frac{1}{2}}\hat{i} + \frac{\partial}{\partial y}(x^2+y^2+z^2)^{-\frac{1}{2}}\hat{j} + \frac{\partial}{\partial z}(x^2+y^2+z^2)^{-\frac{1}{2}}\hat{k}$$

$$= \left\{-\frac{1}{2}(x^2+y^2+z^2)^{-\frac{3}{2}}2x\right\}\hat{i} + \left(-\frac{1}{2}(x^2+y^2+z^2)^{-\frac{3}{2}}2y\right)\hat{j} + \left\{-\frac{1}{2}(x^2+y^2+z^2)^{-\frac{3}{2}}2z\right\}\hat{k}$$

$$= \frac{-(x\hat{i}+y\hat{j}+z\hat{k})}{(x^2+y^2+z^2)^{3/2}} \quad \text{...(1)}$$

and $\hat{r}$ = unit vector in the direction of $x\hat{i}+y\hat{j}+z\hat{k}$

$$= \frac{x\hat{i}+y\hat{j}+z\hat{k}}{\sqrt{x^2+y^2+z^2}} \quad \text{...(2)}$$

So, the required directional derivative

$$= \nabla\phi.\hat{r} = -\frac{x\hat{i}+y\hat{j}+z\hat{k}}{(x^2+y^2+z^2)^{3/2}}.\frac{x\hat{i}+y\hat{j}+z\hat{k}}{(x^2+y^2+z^2)^{1/2}} = \frac{x^2+y^2+z^2}{(x^2+y^2+z^2)^2} \quad \text{[From (1), (2)]}$$

$$= \frac{1}{x^2+y^2+z^2} = \frac{1}{r^2}$$ **Ans.**

**Example 25.** *Find the direction in which the directional derivative of $\phi(x, y) = \frac{x^2+y^2}{xy}$ at (1, 1) is zero and hence find out component of velocity of the vector $\overline{r} = (t^3+1)\hat{i} + t^2\hat{j}$ in the same direction at t = 1.* *(Nagpur University, Winter 2000)*

**Solution.** Directional derivative $= \nabla\phi = \left(\hat{i}\frac{\partial}{\partial x} + \hat{j}\frac{\partial}{\partial y} + \hat{k}\frac{\partial}{\partial z}\right)\left(\frac{x^2+y^2}{xy}\right)$

$$= \hat{i}\left[\frac{xy.2x-(x^2+y^2)y}{x^2y^2}\right] + \hat{j}\left[\frac{xy.2y-x(y^2+x^2)}{x^2y^2}\right]$$

$$= \hat{i}\left[\frac{x^2y-y^3}{x^2y^2}\right] + \hat{j}\left[\frac{xy^2-x^3}{x^2y^2}\right]$$

Directional Derivative at (1, 1) $= \hat{i}\,0 + \hat{j}\,0 = 0$

Since $(\nabla\phi)_{(1,1)} = 0$, the directional derivative of $\phi$ at (1, 1) is zero in any direction.

Again $\quad \overline{r} = (t^3+1)\hat{i} + t^2\hat{j}$

Velocity, $\bar{v} = \dfrac{d\bar{r}}{dt} = 3t^2\,\hat{i} + 2t\,\hat{j}$

Velocity at $t = 1$ is $= 3\hat{i} + 2\hat{j}$

The component of velocity in the same direction of velocity

$$= (3\hat{i} + 2\hat{j}).\left(\frac{3\hat{i} + 2\hat{j}}{\sqrt{9+4}}\right) = \frac{9+4}{\sqrt{13}} = \sqrt{13}$$ **Ans.**

**Example 26.** *Find the directional derivative of $\phi(x, y, z) = x^2 y z + 4 x z^2$ at $(1, -2, 1)$ in the direction of $2\hat{i} - \hat{j} - 2\hat{k}$. Find the greatest rate of increase of $\phi$.*

*(Uttarakhand, I Semester, Dec. 2006)*

**Solution.** Here, $\phi(x, y, z) = x^2 y z + 4xz^2$

Now, $\nabla\phi = \left(\hat{i}\dfrac{\partial}{\partial x} + \hat{j}\dfrac{\partial}{\partial y} + \hat{k}\dfrac{\partial}{\partial z}\right)(x^2yz + 4xz^2)$

$$= (2xyz + 4z^2)\hat{i} + (x^2 z)\hat{j} + (x^2 y + 8xz)\hat{k}$$

$\nabla\phi$ at $(1, -2, 1) = \{2(1)(-2)(1) + 4(1)^2\}\hat{i} + (1\times 1)\hat{j} + \{1(-2) + 8(1)(1)\}\hat{k}$

$$= (-4+4)\hat{i} + \hat{j} + (-2+8)\hat{k} = \hat{j} + 6\hat{k}$$

Let $\hat{a} = \text{unit vector} = \dfrac{2\hat{i} - \hat{j} - 2\hat{k}}{\sqrt{4+1+4}} = \dfrac{1}{3}(2\hat{i} - \hat{j} - 2\hat{k})$

So, the required directional derivative at (1, –2, 1)

$$= \nabla\phi.\hat{a} = (\hat{j} + 6\hat{k}).\frac{1}{3}(2\hat{i} - \hat{j} - 2\hat{k}) = \frac{1}{3}(-1-12) = \frac{-13}{3}$$

Greatest rate of increase of $\phi = \left|\hat{j} + 6\hat{k}\right| = \sqrt{1+36}$

$$= \sqrt{37}$$ **Ans.**

**Example 27.** *Find the directional derivative of the function $\phi = x^2 - y^2 + 2z^2$ at the point P (1, 2, 3) in the direction of the line PQ where Q is the point (5, 0, 4).*

*(AMIETE, Dec. 20010, Nagpur University, Summer 2008, U.P., I Sem., Winter 2000)*

**Solution.** Directional derivative $= \bar{\nabla}\phi$

$$= \left(\hat{i}\frac{\partial}{\partial x} + \hat{j}\frac{\partial}{\partial y} + \hat{k}\frac{\partial}{\partial z}\right)(x^2 - y^2 + 2z^2) = 2x\hat{i} - 2y\hat{j} + 4z\hat{k}$$

Directional Derivative at the point $P$ (1, 2, 3) $= 2\hat{i} - 4\hat{j} + 12\hat{k}$ ...(1)

$\overline{PQ} = \bar{Q} - \bar{P} = (5, 0, 4) - (1, 2, 3) = (4, -2, 1)$ ...(2)

Directional Derivative along $PQ = (2\hat{i} - 4\hat{j} + 12\hat{k}).\dfrac{(4\hat{i} - 2\hat{j} + \hat{k})}{\sqrt{16+4+1}}$ [From (1) and (2)]

$$= \frac{8+8+12}{\sqrt{21}} = \frac{28}{\sqrt{21}}$$ **Ans.**

**Example 28.** *For the function $\phi(x, y) = \dfrac{x}{x^2 + y^2}$, find the magnitude of the directional derivative along a line making an angle 30° with the positive x-axis at (0, 2).*

*(A.M.I.E.T.E., Winter 2002)*

**Solution.** Directional derivative $= \vec{\nabla}\phi$

$$= \left(\hat{i}\frac{\partial}{\partial x}+\hat{j}\frac{\partial}{\partial y}+\hat{k}\frac{\partial}{\partial z}\right)\frac{x}{x^2+y^2} = \hat{i}\left(\frac{1}{x^2+y^2}-\frac{x(2x)}{(x^2+y^2)^2}\right)-\hat{j}\frac{x(2y)}{(x^2+y^2)^2}$$

$$= \hat{i}\frac{y^2-x^2}{(x^2+y^2)^2}-\hat{j}\frac{2xy}{(x^2+y^2)^2}$$

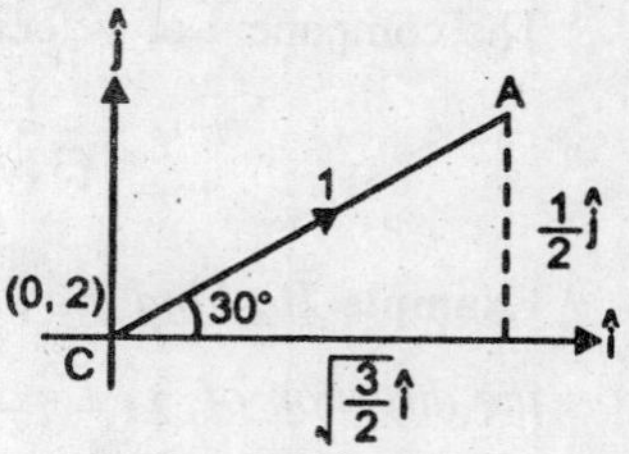


Directional derivative at the point (0, 2)

$$= \hat{i}\frac{4-0}{(0+4)^2}-\hat{j}\frac{2(0)(2)}{(0+4)^2}=\frac{\hat{i}}{4}$$

Directional derivative at the point (0, 2) in the direction $\overrightarrow{CA}$ *i.e.* $\left(\frac{\sqrt{3}}{2}\hat{i}+\frac{1}{2}\hat{j}\right)$.

$$= \frac{\hat{i}}{4}\cdot\left(\frac{\sqrt{3}}{2}\hat{i}+\frac{1}{2}\hat{j}\right) \qquad \left\{\begin{aligned}\overrightarrow{CA}=\overrightarrow{OB}+\overrightarrow{BA}&=\hat{i}\cos 30^\circ+\hat{j}\sin 30^\circ\\ &=\left(\frac{\sqrt{3}}{2}\hat{i}+\frac{1}{2}\hat{j}\right)\end{aligned}\right\}$$

$$= \frac{\sqrt{3}}{8}$$ **Ans.**

**Example 29.** *Find the directional derivative of $\vec{V}^2$, where $\vec{V}=xy^2\hat{i}+zy^2\hat{j}+xz^2\hat{k}$, at the point (2, 0, 3) in the direction of the outward normal to the sphere $x^2+y^2+z^2=14$ at the point (3, 2, 1).* (*A.M.I.E.T.E., Dec. 2007*)

**Solution.** $V^2 = \vec{V}.\vec{V}$

$$= (xy^2\hat{i}+zy^2\hat{j}+xz^2\hat{k}).(xy^2\hat{i}+zy^2\hat{j}+xz^2\hat{k}) = x^2y^4+z^2y^4+x^2z^4$$

Directional derivative $= \vec{\nabla}V^2$

$$= \left(\hat{i}\frac{\partial}{\partial x}+\hat{j}\frac{\partial}{\partial y}+\hat{k}\frac{\partial}{\partial z}\right)(x^2y^4+z^2y^4+x^2z^4)$$

$$= (2xy^4+2xz^4)\hat{i}+(4x^2y^3+4y^3z^2)\hat{j}+(2y^4z+4x^2z^3)\hat{k}$$

Directional derivative at (2, 0, 3) $= (0+2\times 2\times 81)\hat{i}+(0+0)\hat{j}+(0+4\times 4\times 27)\hat{k}$

$$= 324\hat{i}+432\hat{k}=108(3\hat{i}+4\hat{k}) \qquad ...(1)$$

Normal to $x^2+y^2+z^2-14 = \nabla\phi$

$$= \left(\hat{i}\frac{\partial}{\partial x}+\hat{j}\frac{\partial}{\partial y}+\hat{k}\frac{\partial}{\partial z}\right)(x^2+y^2+z^2-14)$$

$$= (2x\hat{i}+2y\hat{j}+2z\hat{k})$$

Normal vector at (3, 2, 1) $= 6\hat{i}+4\hat{j}+2\hat{k}$ ...(2)

Unit normal vector $= \dfrac{6\hat{i}+4\hat{j}+2\hat{k}}{\sqrt{36+16+4}}=\dfrac{2(3\hat{i}+2\hat{j}+\hat{k})}{2\sqrt{14}}=\dfrac{3\hat{i}+2\hat{j}+\hat{k}}{\sqrt{14}}$ [From (1), (2)]

Directional derivative along the normal $= 108(3\hat{i}+4\hat{k}).\dfrac{3\hat{i}+2\hat{j}+\hat{k}}{\sqrt{14}}$

$$= \frac{108\times(9+4)}{\sqrt{14}}=\frac{1404}{\sqrt{14}}$$ **Ans.**

**Example 30.** *Find the directional derivative of $\nabla(\nabla f)$ at the point (1, – 2, 1) in the direction of the normal to the surface $xy^2z = 3x + z^2$, where $f = 2x^3y^2z^4$. (U.P., I Semester, Dec 2008)*

**Solution.** Here, we have

$$f = 2x^3y^2z^4$$

$$\nabla f = \left(\hat{i}\frac{\partial}{\partial x} + \hat{j}\frac{\partial}{\partial y} + \hat{k}\frac{\partial}{\partial z}\right)(2x^3y^2z^4) = 6x^2y^2z^4\hat{i} + 4x^3yz^4\hat{j} + 8x^3y^2z^3\hat{k}$$

$$\nabla(\nabla f) = \left(\hat{i}\frac{\partial}{\partial x} + \hat{j}\frac{\partial}{\partial y} + \hat{k}\frac{\partial}{\partial z}\right)(6x^2y^2z^4\hat{i} + 4x^3yz^4\hat{j} + 8x^3y^2z^3\hat{k})$$

$$= 12xy^2z^4 + 4x^3z^4 + 24x^3y^2z^2$$

Directional derivative of $\nabla(\nabla f)$

$$= \left(\hat{i}\frac{\partial}{\partial x} + \hat{j}\frac{\partial}{\partial y} + \hat{k}\frac{\partial}{\partial z}\right)(12xy^2z^4 + 4x^3z^4 + 24x^3y^2z^2)$$

$$= (12y^2z^4 + 12x^2z^4 + 72x^2y^2z^2)\hat{i} + (24xyz4 + 48x3yz2)\hat{j}$$
$$+ (48xy^2z^3 + 16x^3z^3 + 48x^3y^2z)\hat{k}$$

Directional derivative at (1, – 2, 1) = $(48 + 12 + 288)\hat{i} + (-48 - 96)\hat{j} + (192 + 16 + 192)\hat{k}$

$$= 348\hat{i} - 144\hat{j} + 400\hat{k}$$

Normal to$(xy^2z - 3x - z^2) = \nabla(xy^2z - 3x - z^2)$

$$= \left(\hat{i}\frac{\partial}{\partial x} + \hat{j}\frac{\partial}{\partial y} + \hat{k}\frac{\partial}{\partial z}\right)(xy^2z - 3x - z^2)$$

$$= (y^2z - 3)\hat{i} + (2xyz)\hat{j} + (xy^2 - 2z)\hat{k}$$

Normal at(1, – 2, 1) = $\hat{i} - 4\hat{j} + 2\hat{k}$

Unit Normal Vector = $\dfrac{\hat{i} - 4\hat{j} + 2\hat{k}}{\sqrt{1+16+4}} = \dfrac{1}{\sqrt{21}}(\hat{i} - 4\hat{j} + 2\hat{k})$

Directional derivative in the direction of normal

$$= (348\hat{i} - 144\hat{j} + 400\hat{k})\frac{1}{\sqrt{21}}(\hat{i} - 4\hat{j} + 2\hat{k})$$

$$= \frac{1}{\sqrt{21}}(348 + 576 + 800) = \frac{1724}{\sqrt{21}}$$ **Ans.**

**Example 31.** *If the directional derivative of $\phi = a\,x^2y + b\,y^2z + c\,z^2x$ at the point (1, 1, 1) has maximum magnitude 15 in the direction parallel to the line $\dfrac{x-1}{2} = \dfrac{y-3}{-2} = \dfrac{z}{1}$, find the values of a, b and c.* *(U.P. I Semester, June 2007, Winter 2001)*

**Solution.** Given $\phi = a\,x^2y + b\,y^2z + c\,z^2x$

$$\therefore \quad \bar{\nabla}\phi = \left(\hat{i}\frac{\partial}{\partial x} + \hat{j}\frac{\partial}{\partial y} + \hat{k}\frac{\partial}{\partial z}\right)(a\,x^2y + b\,y^2z + c\,z^2x)$$

$$= \hat{i}(2axy + cz^2) + \hat{j}(ax^2 + 2byz) + \hat{k}(by^2 + 2czx)$$

$$\bar{\nabla}\phi \text{ at the point } (1, 1, 1) = \hat{i}(2a + c) + \hat{j}(a + 2b) + \hat{k}(b + 2c) \quad \ldots(1)$$

We know that the maximum value of the directional derivative is in the direction of $\bar{\nabla}\phi$.

*i.e.* $|\nabla\phi| = 15 \Rightarrow (2a + c)^2 + (2b + a)^2 + (2c + b)^2 = (15)^2$

But, the directional derivative is given to be maximum parallel to the line

$$\frac{x-1}{2} = \frac{y-3}{-2} = \frac{z}{1} \text{ i.e., parallel to the vector } 2\hat{i} - 2\hat{j} + \hat{k}. \quad ...(2)$$

On comparing the coefficients of (1) and (2)

$$\Rightarrow \quad \frac{2a+c}{2} = \frac{2b+a}{-2} = \frac{2c+b}{1}$$

$$\Rightarrow \quad 2a + c = -2b - a \Rightarrow 3a + 2b + c = 0 \quad ...(3)$$

and $\quad 2b + a = -2(2c + b)$

$$\Rightarrow \quad 2b + a = -4c - 2b \Rightarrow a + 4b + 4c = 0 \quad ...(4)$$

Rewriting (3) and (4), we have

$$\left.\begin{aligned} 3a + 2b + c &= 0 \\ a + 4b + 4c &= 0 \end{aligned}\right\} \Rightarrow \frac{a}{4} = \frac{b}{-11} = \frac{c}{10} = k \text{ (say)}$$

$$\Rightarrow \quad a = 4k, \quad b = -11k \quad \text{and} \quad c = 10k.$$

Now, we have

$$(2a + c)^2 + (2b + a)^2 + (2c + b)^2 = (15)^2$$

$$\Rightarrow \quad (8k + 10k)^2 + (-22k + 4k)^2 + (20k - 11k)^2 = (15)^2$$

$$\Rightarrow \quad k = \pm\frac{5}{9}$$

$$\Rightarrow \quad a = \pm\frac{20}{9}, \quad b = \pm\frac{55}{9} \quad \text{and} \quad c = \pm\frac{50}{9} \qquad \textbf{Ans.}$$

**Example 32.** *If* $\overline{r} = x\hat{i} + y\hat{j} + z\hat{k}$, *show that :*

*(i) grad* $r = \frac{\vec{r}}{r}$ *(ii) grad* $\left(\frac{1}{r}\right) = -\frac{\vec{r}}{r^3}$. (*Nagpur University, Summer 2002*)

**Solution.** (*i*) $\overline{r} = x\hat{i} + y\hat{j} + z\hat{k} \Rightarrow r = \sqrt{x^2 + y^2 + z^2} \Rightarrow r^2 = x^2 + y^2 + z^2$

$$\therefore \quad 2r\frac{\partial r}{\partial x} = 2x \Rightarrow \frac{\partial r}{\partial x} = \frac{x}{r}$$

Similarly, $\frac{\partial r}{\partial y} = \frac{y}{r}$ and $\frac{\partial r}{\partial z} = \frac{z}{r}$

$$\text{grad } r = \nabla r = \left(\hat{i}\frac{\partial}{\partial x} + \hat{j}\frac{\partial}{\partial y} + \hat{k}\frac{\partial}{\partial z}\right) r = \hat{i}\frac{\partial r}{\partial x} + \hat{j}\frac{\partial r}{\partial y} + \hat{k}\frac{\partial r}{\partial z}$$

$$= \hat{i}\frac{x}{r} + \hat{j}\frac{y}{r} + \hat{k}\frac{z}{r} = \frac{x\hat{i} + y\hat{j} + z\hat{k}}{r} = \frac{\overline{r}}{r} \qquad \textbf{Proved.}$$

$$(ii) \text{ grad}\left(\frac{1}{r}\right) = \nabla\left(\frac{1}{r}\right) = \left(\hat{i}\frac{\partial}{\partial x} + \hat{j}\frac{\partial}{\partial y} + \hat{k}\frac{\partial}{\partial z}\right)\left(\frac{1}{r}\right) = \hat{i}\frac{\partial}{\partial x}\left(\frac{1}{r}\right) + \hat{j}\frac{\partial}{\partial y}\left(\frac{1}{r}\right) + \hat{k}\frac{\partial}{\partial z}\left(\frac{1}{r}\right)$$

$$= \hat{i}\left(-\frac{1}{r^2}\frac{\partial r}{\partial x}\right) + \hat{j}\left(\frac{-1}{r^2}\frac{\partial r}{\partial y}\right) + \hat{k}\left(-\frac{1}{r^2}\frac{\partial r}{\partial z}\right)$$

$$= \hat{i}\left(-\frac{1}{r^2}\frac{x}{r}\right) + \hat{j}\left(-\frac{1}{r^2}\frac{y}{r}\right) + \hat{k}\left(-\frac{1}{r^2}\frac{z}{r}\right) = -\frac{x\hat{i} + y\hat{j} + z\hat{k}}{r^3} = -\frac{\overline{r}}{r^3} \qquad \textbf{Proved.}$$

**Example 33.** *Prove that* $\nabla^2 f(r) = f''(r) + \frac{2}{r} f'(r)$. (*K. University, Dec. 2008*)

**Solution.**

$$\nabla f(r) = \left(i\frac{\partial}{\partial x} + j\frac{\partial}{\partial y} + k\frac{\partial}{\partial z}\right) f(r)$$

$$\left[r^2 = x^2 + y^2 + z^2 \Rightarrow 2r\frac{\partial r}{\partial x} = 2x \Rightarrow \frac{\partial r}{\partial x} = \frac{x}{r},\ \frac{\partial r}{\partial y} = \frac{y}{r} \text{ and } \frac{\partial r}{\partial z} = \frac{z}{r}\right]$$

$$= i\, f'(r)\frac{\partial r}{\partial x} + j\, f'(r)\frac{\partial r}{\partial y} + k\, f'(r)\frac{\partial r}{\partial z} = f'(r)\left[i\frac{x}{r} + j\frac{y}{r} + k\frac{z}{r}\right]$$

$$= f'(r)\frac{x\,i + y\,j + z\,k}{r}$$

$$\nabla^2 f(r) = \nabla[\nabla f(r)] = \left(i\frac{\partial}{\partial x} + j\frac{\partial}{\partial y} + k\frac{\partial}{\partial z}\right)\left[f'(r)\frac{xi + yj + zk}{r}\right]$$

$$= \frac{\partial}{\partial x}\left[f'(r)\frac{x}{r}\right] + \frac{\partial}{\partial y}\left[f'(r)\frac{y}{r}\right] + \frac{\partial}{\partial z}\left[f'(r)\frac{z}{r}\right]$$

$$= \left(f''(r)\frac{\partial r}{\partial x}\right)\left(\frac{x}{r}\right) + f'(r)\frac{r1 - x\dfrac{\partial r}{\partial x}}{r^2} + \left(f''(r)\frac{\partial r}{\partial y}\right)\left(\frac{y}{r}\right) + f'(r)\frac{r.1 - y\dfrac{\partial r}{\partial y}}{r^2} +$$

$$\left(f''(r)\frac{\partial r}{\partial z}\right)\left(\frac{z}{r}\right) + f'(r)\frac{r.1 - z\dfrac{\partial z}{\partial r}}{r^2}$$

$$= \left(f''(r)\frac{x}{r}\right)\left(\frac{x}{r}\right) + f'(r)\frac{r - \dfrac{x^2}{r}}{r^2} + \left(f''(r)\frac{y}{r}\right)\left(\frac{y}{r}\right) + f'(r)\frac{r - \dfrac{y^2}{r}}{r^2} + \left(f''(r)\frac{z}{r}\right)\left(\frac{z}{r}\right) + f'(r)\frac{r - \dfrac{z^2}{r}}{r^2}$$

$$= \left(f''(r)\frac{x}{r}\right)\left(\frac{x}{r}\right) + f'(r)\frac{r^2 - x^2}{r^3} + \left(f''(r)\frac{y}{r}\right) + f'(r)\frac{r^2 - y^2}{r^3} + \left(f''(r)\frac{z}{r}\right)\left(\frac{z}{r}\right) + f'(r)\frac{r^2 - z^2}{r^3}$$

$$= f''(r)\frac{x^2}{r^2} + f'(r)\frac{y^2 + z^2}{r^3} + f''(r)\frac{y^2}{r^2} + f'(r)\frac{x^2 + z^2}{r^3} + f''(r)\frac{z^2}{r^2} + f'(r)\frac{x^2 + y^2}{r^3}$$

$$= f''(r)\left[\frac{x^2}{r^2} + \frac{y^2}{r^2} + \frac{z^2}{r^2}\right] + f'(r)\left[\frac{y^2 + z^2}{r^3} + \frac{z^2 + x^2}{r^3} + \frac{x^2 + y^2}{r^3}\right]$$

$$= f''(r)\frac{x^2 + y^2 + x^2}{r^2} + f'(r)\frac{2(x^2 + y^2 + z^2)}{r^3} = f''(r)\frac{r^2}{r^2} + f'(r)\frac{2r^2}{r^3}$$

$$= f''(r) + f'(r)\frac{2}{r}$$ **Ans.**

## EXERCISE 5.7

1. Evaluate grad $\phi$ if $\phi = \log(x^2 + y^2 + z^2)$ **Ans.** $\dfrac{2(x\hat{i} + y\hat{j} + z\hat{k})}{x^2 + y^2 + z^2}$
2. Find a unit normal vector to the surface $x^2 + y^2 + z^2 = 5$ at the point (0, 1, 2). **Ans.** $\dfrac{1}{\sqrt{5}}(\hat{j} + 2\hat{k})$ *(AMIETE, June 2010)*
3. Calculate the directional derivative of the function $\phi(x, y, z) = xy^2 + yz^3$ at the point (1, –1, 1) in the direction of (3, 1, –1) *(A.M.I.E.T.E. Winter 2009, 2000)* **Ans.** $\dfrac{5}{\sqrt{11}}$
4. Find the direction in which the directional derivative of $f(x, y) = (x^2 - y^2)/xy$ at (1, 1) is zero. *(Nagpur Winter 2000)* **Ans.** $\dfrac{\hat{i} + \hat{j}}{\sqrt{2}}$

5. Find the directional derivative of the scalar function of $(x, y, z) = xyz$ in the direction of the outer normal to the surface $z = xy$ at the point (3, 1, 3). **Ans.** $\frac{27}{\sqrt{11}}$

6. The temperature of the points in space is given by $T(x, y, z) = x^2 + y^2 - z$. A mosquito located at (1, 1, 2) desires to fly in such a direction that it will get warm as soon as possible. In what direction should it move? **Ans.** $\frac{1}{3}(2\hat{i} + 2\hat{j} - \hat{k})$

7. If $\phi(x, y, z) = 3xz^2y - y^3z^2$, find grad $\phi$ at the point (1, –2, –1) **Ans.** $-(16\hat{i} + 9\hat{j} + 4\hat{k})$

8. Find a unit vector normal to the surface $x^2y + 2xz = 4$ at the point (2, –2, 3).
**Ans.** $\frac{1}{3}(-\hat{i} + 2\hat{j} + 2\hat{k})$

9. What is the greatest rate of increase of the function $u = xyz^2$ at the point (1, 0, 3)? **Ans.** 9

10. If $\theta$ is the acute angle between the surfaces $xyz^2 = 3x + z^2$ and $3x^2 - y^2 + 2z = 1$ at the point (1, –2, 1) show that $\cos\theta = 3/7\sqrt{6}$.

11. Find the values of constants $a, b, c$ so that the maximum value of the directional directive of $\phi = axy^2 + byz + cz^2x^3$ at (1, 2, –1) has a maximum magnitude 64 in the direction parallel to the axis of $z$. **Ans.** $a = b, b = 24, c = -8$

12. Find the values of $\lambda$ and $\mu$ so that surfaces $\lambda x^2 - \mu yz = (\lambda + 2)x$ and $4x^2y + z^3 = 4$ intersect orthogonally at the point (1, –1, 2). **Ans.** $\lambda = \frac{9}{2}, \mu = 1$

13. The position vector of a particle at time $t$ is $R = \cos(t-1)\, i + \sinh(t-1)\, j + at^2k$. If at $t = 1$, the acceleration of the particle be perpendicular to its position vector, then a is equal to

(*a*) 0 (*b*) 1 (*c*) $\frac{1}{2}$ (*d*) $\frac{1}{\sqrt{2}}$ (*AMIETE, Dec. 2009*) **Ans.** (*d*)

## 5.29 DIVERGENCE OF A VECTOR FUNCTION

The divergence of a vector point function $\vec{F}$ is denoted by *div F* and is defined as below.

Let $\vec{F} = F_1\hat{i} + F_2\hat{j} + F_3\hat{k}$

$$\text{div}\,\vec{F} = \vec{\nabla}.\vec{F} = \left(\hat{i}\frac{\partial}{\partial x} + \hat{j}\frac{\partial}{\partial y} + \hat{k}\frac{\partial}{\partial z}\right)(\hat{i}F_1 + \hat{j}F_2 + \hat{k}F_3) = \frac{\partial F_1}{\partial x} + \frac{\partial F_2}{\partial y} + \frac{\partial F_3}{\partial z}$$

It is evident that div $F$ is scalar function.

## 5.30 PHYSICAL INTERPRETATION OF DIVERGENCE

Let us consider the case of a fluid flow. Consider a small rectangular parallelopiped of dimensions $dx$, $dy$, $dz$ parallel to $x$,$y$ and $z$ axes respectively.

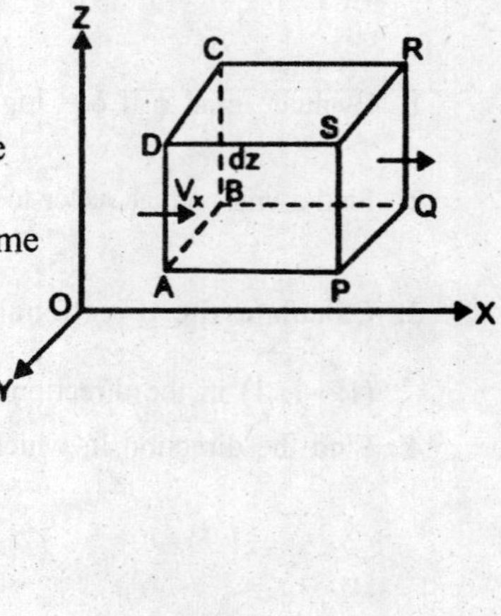


Let $\vec{V} = V_x\hat{i} + V_y\hat{j} + V_z\hat{k}$ be the velocity of the fluid at $P(x, y, z)$.

∴ Mass of fluid flowing in through the face $ABCD$ in unit time

= Velocity × Area of the face $= V_x\,(dy\,dz)$

Mass of fluid flowing out across the face $PQRS$ per unit time

$$= V_x(x + dx)\,(dy\,dz)$$

$$= \left(V_x + \frac{\partial V_x}{\partial x}dx\right)(dy\,dz)$$

Net decrease in mass of fluid in the parallelopiped corresponding to the flow along $x$-axis per unit time

$$= V_x\, dy\, dz - \left(V_x + \frac{\partial V_x}{\partial x} dx\right) dy\, dz$$

$$= -\frac{\partial V_x}{\partial x} dx\, dy\, dz \qquad \text{(Minus sign shows decrease)}$$

Similarly, the decrease in mass of fluid to the flow along $y$-axis $= \frac{\partial V_y}{\partial y} dx\, dy\, dz$

and the decrease in mass of fluid to the flow along $z$-axis $= \frac{\partial V_z}{\partial z} dx\, dy\, dz$

Total decrease of the amount of fluid per unit time $= \left(\frac{\partial V_x}{\partial x} + \frac{\partial V_y}{\partial y} + \frac{\partial V_z}{\partial z}\right) dx\, dy\, dz$

Thus the rate of loss of fluid per unit volume $= \frac{\partial V_x}{\partial x} + \frac{\partial V_y}{\partial y} + \frac{\partial V_z}{\partial z}$

$$= \left(\hat{i}\frac{\partial}{\partial x} + \hat{j}\frac{\partial}{\partial y} + \hat{k}\frac{\partial}{\partial z}\right).(\hat{i}V_x + \hat{j}V_y + \hat{k}V_z) = \overline{\nabla}.\overline{V} = \text{div}\, \overline{V}$$

If the fluid is compressible, there can be no gain or loss in the volume element. Hence

$$\text{div}\, \vec{V} = 0 \qquad ...(1)$$

and $V$ is called a *Solenoidal* vector function.
Equation (1) is also called the *equation of continuity or conservation of mass.*

**Example 34.** *If* $\vec{v} = \frac{x\hat{i} + y\hat{j} + z\hat{k}}{\sqrt{x^2 + y^2 + z^2}}$, *find the value of div* $\overline{v}$.

*(U.P., I Semester, Winter 2000)*

**Solution.** We have, $\vec{v} = \frac{x\hat{i} + y\hat{j} + z\hat{k}}{\sqrt{x^2 + y^2 + z^2}}$

$$\text{div}\, \vec{v} = \vec{\nabla}.\vec{v} = \left(\hat{i}\frac{\partial}{\partial x} + \hat{j}\frac{\partial}{\partial y} + \hat{k}\frac{\partial}{\partial z}\right).\left(\frac{x\hat{i} + y\hat{j} + z\hat{k}}{(x^2 + y^2 + z^2)^{1/2}}\right)$$

$$= \frac{\partial}{\partial x}\frac{x}{(x^2 + y^2 + z^2)^{1/2}} + \frac{\partial}{\partial y}\frac{y}{(x^2 + y^2 + z^2)^{1/2}} + \frac{\partial}{\partial z}\frac{z}{(x^2 + y^2 + z^2)^{1/2}}$$

$$= \frac{\left[(x^2 + y^2 + z^2)^{1/2} - x.\frac{1}{2}(x^2 + y^2 + z^2)^{-\frac{1}{2}}.2x\right]}{(x^2 + y^2 + z^2)}$$

$$+ \frac{\left[(x^2 + y^2 + z^2)^{\frac{1}{2}} - y.\frac{1}{2}(x^2 + y^2 + z^2)^{-\frac{1}{2}} \times 2y\right]}{(x^2 + y^2 + z^2)} + \frac{\left[(x^2 + y^2 + z^2)^{1/2} - z.\frac{1}{2}(x^2 + y^2 + z^2)^{-1/2}.2z\right]}{(x^2 + y^2 + z^2)}$$

$$= \frac{(x^2 + y^2 + z^2) - x^2}{(x^2 + y^2 + z^2)^{3/2}} + \frac{(x^2 + y^2 + z^2) - y^2}{(x^2 + y^2 + z^2)^{3/2}} + \frac{(x^2 + y^2 + z^2) - z^2}{(x^2 + y^2 + z^2)^{3/2}}$$

$$= \frac{y^2 + z^2 + x^2 + z^2 + x^2 + y^2}{(x^2 + y^2 + z^2)^{3/2}} = \frac{2(x^2 + y^2 + z^2)}{(x^2 + y^2 + z^2)^{3/2}} = \frac{2}{\sqrt{(x^2 + y^2 + z^2)}} \qquad \textbf{Ans.}$$

**Example 35.** *If* $u = x^2 + y^2 + z^2$, *and* $\overline{r} = x\hat{i} + y\hat{j} + z\hat{k}$, *then find div* $(u\overline{r})$ *in terms of u.*

*(A.M.I.E.T.E., Summer 2004)*

**Solution.** $\text{div}\ (u\vec{r}) = \left(\hat{i}\frac{\partial}{\partial x} + \hat{j}\frac{\partial}{\partial y} + \hat{k}\frac{\partial}{\partial z}\right).[(x^2+y^2+z^2)(x\hat{i}+y\hat{j}+z\hat{k})]$

$$= \left(\hat{i}\frac{\partial}{\partial x} + \hat{j}\frac{\partial}{\partial y} + \hat{k}\frac{\partial}{\partial z}\right).[(x^2+y^2+z^2)x\hat{i}+(x^2+y^2+z^2)y\hat{j}+(x^2+y^2+z^2)z\hat{k}]$$

$$= \frac{\partial}{\partial x}(x^3+xy^2+xz^2) + \frac{\partial}{\partial y}(x^2y+y^3+yz^2) + \frac{\partial}{\partial z}(x^2z+y^2z+z^3)$$

$$= (3x^2+y^2+z^2) + (x^2+3y^2+z^2) + (x^2+y^2+3z^2) = 5\,(x^2+y^2+z^2) = 5\,u \quad \textbf{Ans.}$$

**Example 36.** *Find the value of n for which the vector* $r^n\vec{r}$ *is solenoidal, where* $\bar{r} = x\hat{i}+y\hat{j}+z\hat{k}$.

**Solution.** Divergence $\vec{F} = \vec{\nabla}.\vec{F} = \vec{\nabla}.r^n\vec{r} = \nabla.(x^2+y^2+z^2)^{n/2}(x\hat{i}+y\hat{j}+z\hat{k})$

$$= \left[\hat{i}\frac{\partial}{\partial x} + \hat{j}\frac{\partial}{\partial y} + \hat{k}\frac{\partial}{\partial z}\right].[(x^2+y^2+z^2)^{n/2}x\hat{i}+(x^2+y^2+z^2)^{n/2}y\hat{j}+(x^2+y^2+z^2)^{n/2}z\hat{k}]$$

$$= \frac{n}{2}(x^2+y^2+z^2)^{n/2-1}(2x^2) + (x^2+y^2+z^2)^{n/2} + \frac{n}{2}(x^2+y^2+z^2)^{n/2-1}(2y^2)$$

$$+ (x^2+y^2+z^2)^{n/2} + \frac{n}{2}(x^2+y^2+z^2)^{n/2-1}(2z^2) + (x^2+y^2+z^2)^{n/2}$$

$$= n(x^2+y^2+z^2)^{n/2-1}(x^2+y^2+z^2) + 3\,(x^2+y^2+z^2)^{n/2}$$

$$= n(x^2+y^2+z^2)^{n/2} + 3(x^2+y^2+z^2)^{n/2} = (n+3)\,(x^2+y^2+z^2)^{n/2}$$

If $r^n\vec{r}$ is solenoidal, then $(n+3)\,(x^2+y^2+z^2)^{n/2} = 0$ or $n+3=0$ or $n=-3$. **Ans.**

**Example 37.** *Show that* $\nabla\left[\frac{(\vec{a}.\vec{r})}{r^n}\right] = \frac{\vec{a}}{r^n} - \frac{n(\vec{a}.\vec{r})\,\vec{r}}{r^{n+2}}$. *(M.U. 2005)*

**Solution.** We have, $\frac{\vec{a}.\vec{r}}{r^n} = \frac{(a_1\hat{i}+a_2\hat{j}+a_3\hat{k}).(x\hat{i}+y\hat{j}+z\hat{k})}{r^n} = \frac{a_1x+a_2y+a_3z}{r^n}$

Let $\phi = \frac{\vec{a}.\vec{r}}{r^n} = \frac{a_1x+a_2y+a_3z}{r^n}$

$$\therefore \quad \frac{\partial\phi}{\partial x} = \frac{r^n.a_1 - (a_1x+a_2y+a_3z)\,n\,r^{n-1}(\partial r/\partial x)}{r^{2n}}$$

But $r^2 = x^2+y^2+z^2 \Rightarrow 2r\frac{\partial r}{\partial x} = 2x \Rightarrow \frac{\partial r}{\partial x} = \frac{x}{r}$

$$\therefore \quad \frac{\partial\phi}{\partial x} = \frac{a_1r^n - (a_1x+a_2y+a_3z).n\,r^{n-2}.x}{r^{2n}} = \frac{a_1}{r^n} - \frac{n(a_1x+a_2y+a_3z)x}{r^{n+2}}$$

$$\therefore \quad \nabla\phi = \frac{\partial\phi}{\partial x}\hat{i} + \frac{\partial\phi}{\partial y}\hat{j} + \frac{\partial\phi}{\partial z}\hat{k}$$

$$= \frac{1}{r^n}(a_1\hat{i}+a_2\hat{j}+a_3\hat{k}) - \frac{n}{r^{n+2}}[(a_1x+a_2y+a_3z)(x\hat{i}+y\hat{j}+z\hat{k})]$$

$$= \frac{\bar{a}}{r^n} - \frac{n}{r^{n+2}}(\bar{a}.\bar{r})\bar{r}$$

**Example 38.** *Let $\vec{r} = x\hat{i} + y\hat{j} + z\hat{k}, r = |\vec{r}|$ and $\vec{a}$ is a constant vector. Find the value of*

$$div\left(\frac{\vec{a}\times\vec{r}}{r^n}\right)$$

**Solution.** Let $\vec{a} = a_1\hat{i} + a_2\hat{j} + a_3\hat{k}$

$$\vec{a}\times\vec{r} = (a_1\hat{i} + a_2\hat{j} + a_3\hat{k})\times(x\hat{i} + y\hat{j} + z\hat{k})$$

$$= \begin{vmatrix} \hat{i} & \hat{j} & \hat{k} \\ a_1 & a_2 & a_3 \\ x & y & z \end{vmatrix} = (a_2z - a_3y)\hat{i} - (a_1z - a_3x)\hat{j} + (a_1y - a_2x)\hat{k}$$

$$\frac{\vec{a}\times\vec{r}}{|\vec{r}|^n} = \frac{(a_2z - a_3y)\hat{i} - (a_1z - a_3x)\hat{j} + (a_1y - a_2x)\hat{k}}{(x^2 + y^2 + z^2)^{n/2}}$$

$$\text{div}\left(\frac{\vec{a}\times\vec{r}}{|\vec{r}|^n}\right) = \vec{\nabla}.\frac{\vec{a}\times\vec{r}}{|\vec{r}|^n}$$

$$= \left(\hat{i}\frac{\partial}{\partial x} + \hat{j}\frac{\partial}{\partial y} + \hat{k}\frac{\partial}{\partial z}\right).\frac{(a_2z - a_3y)\hat{i} - (a_1z - a_3x)\hat{j} + (a_1y - a_2x)\hat{k}}{(x^2 + y^2 + z^2)^{n/2}}$$

$$= \frac{\partial}{\partial x}\frac{a_2z - a_3y}{(x^2 + y^2 + z^2)^{n/2}} - \frac{\partial}{\partial y}\frac{a_1z - a_3x}{(x^2 + y^2 + z^2)^{n/2}} + \frac{\partial}{\partial z}\frac{(a_1y - a_2x)}{(x^2 + y^2 + z^2)^{n/2}}$$

$$= -\frac{n}{2}\frac{(a_2z - a_3y)\,2x}{(x^2 + y^2 + z^2)^{\frac{n+2}{2}}} + \frac{n}{2}\frac{(a_1z - a_3x)\,2y}{(x^2 + y^2 + z^2)^{\frac{n+2}{2}}} - \frac{n}{2}\frac{(a_1y - a_2x)\,2z}{(x^2 + y^2 + z^2)^{\frac{n+2}{2}}}$$

$$= -\frac{n}{(x^2 + y^2 + z^2)^{\frac{n+2}{2}}}[(a_2z - a_3y)x - (a_1z - a_3x)y + (a_1y - a_2x)z]$$

$$= -\frac{n}{(x^2 + y^2 + z^2)^{\frac{n+2}{2}}}[a_2zx - a_3xy - a_1yz + a_3xy + a_1yz - a_2zx] = 0$$ **Ans.**

**Example 39.** *Find the directional derivative of div $(\vec{u})$ at the point (1, 2, 2) in the direction of the outer normal of the sphere $x^2 + y^2 + z^2 = 9$ for $\vec{u} = x^4\hat{i} + y^4\hat{j} + z^4\hat{k}$.*

**Solution.** $\text{div}\,(\vec{u}) = \nabla.\vec{u}$

$$= \left(\hat{i}\frac{\partial}{\partial x} + \hat{j}\frac{\partial}{\partial y} + \hat{k}\frac{\partial}{\partial z}\right).(x^4\hat{i} + y^4\hat{j} + z^4\hat{k}) = 4x^3 + 4y^3 + 4z^3$$

Outer normal of the sphere $= \nabla(x^2 + y^2 + z^2 - 9)$

$$= \left(\hat{i}\frac{\partial}{\partial x} + \hat{j}\frac{\partial}{\partial y} + \hat{k}\frac{\partial}{\partial z}\right)(x^2 + y^2 + z^2 - 9) = 2x\hat{i} + 2y\hat{j} + 2z\hat{k}$$

Outer normal of the sphere at (1, 2, 2) $= 2\hat{i} + 4\hat{j} + 4\hat{k}$ ...(1)

Directional derivative $= \vec{\nabla}(4x^3 + 4y^3 + 4z^3)$

$$= \left(\hat{i}\frac{\partial}{\partial x} + \hat{j}\frac{\partial}{\partial y} + \hat{k}\frac{\partial}{\partial z}\right)(4x^3 + 4y^3 + 4z^3) = 12x^2\hat{i} + 12y^2\hat{j} + 12z^2\hat{k}$$

Directional derivative at (1, 2, 2) $= 12\hat{i} + 48\hat{j} + 48\hat{k}$ ...(2)

Directional derivative along the outer normal $= (12\hat{i} + 48\hat{j} + 48\hat{k}).\dfrac{2\hat{i} + 4\hat{j} + 4\hat{k}}{\sqrt{4+16+16}}$ [From (1), (2)]

$$= \frac{24 + 192 + 192}{6} = 68$$ **Ans.**

**Example 40.** *Show that div (grad $r^n$) = n (n + 1)$r^{n-2}$, where*

$$r = \sqrt{x^2 + y^2 + z^2}$$

*Hence, show that* $\Delta^2\left(\dfrac{1}{r}\right) = 0.$ (*U.P. I Semester, Dec. 2004, Winter 2002*)

**Solution.** $\text{grad}(r^n) = \hat{i}\dfrac{\partial}{\partial x}r^n + \hat{j}\dfrac{\partial}{\partial y}r^n + \hat{k}\dfrac{\partial}{\partial z}r^n$ by definition

$$= \hat{i}\, n\, r^{n-1}\frac{\partial r}{\partial x} + \hat{j}\, n\, r^{n-1}\frac{\partial r}{\partial y} + \hat{k}.n\, r^{n-1}\frac{\partial r}{\partial z} = n\, r^{n-1}\left[\hat{i}\frac{\partial r}{\partial x} + \hat{j}\frac{\partial r}{\partial y} + \hat{k}\frac{\partial r}{\partial z}\right]$$

$$= n\, r^{n-1}\left[\hat{i}\left(\frac{x}{r}\right) + \hat{j}\left(\frac{y}{r}\right) + \hat{k}\left(\frac{z}{r}\right)\right] = nr^{n-2}(x\hat{i} + y\hat{j} + z\hat{k}) = nr^{n-2}\bar{r}.$$

$$\left[\because r^2 = x^2 + y^2 + z^2 \Rightarrow 2r\frac{\partial r}{\partial x} = 2x \Rightarrow \frac{\partial r}{\partial x} = \frac{x}{r} \text{ etc.}\right]$$

Thus, $\text{grad}(r^n) = n\, r^{n-2}\, x\hat{i} + n\, r^{n-2}\, y\hat{j} + n\, r^{n-2}\, z\hat{k}$ ...(1)

$\therefore$ $\text{div grad } r^n = \text{div}\,[n\, r^{n-2}\, x\hat{i} + n\, r^{n-2}\, y\hat{j} + n\, r^{n-2}\, z\hat{k}]$

$$= \left(\hat{i}\frac{\partial}{\partial x} + \hat{j}\frac{\partial}{\partial y} + \hat{k}\frac{\partial}{\partial z}\right).(nr^{n-2}\, x\hat{i} + nr^{n-2}\, y\hat{j} + nr^{n-2}\, z\hat{k})$$ [From (1)]

$$= \frac{\partial}{\partial x}(n\, r^{n-2}x) + \frac{\partial}{\partial y}(n\, r^{n-2}y) + \frac{\partial}{\partial z}(n\, r^{n-2}z)$$ (By definition)

$$= \left(n\, r^{n-2} + nx\,(n-2)\, r^{n-3}\frac{\partial r}{\partial x}\right) + \left(n\, r^{n-2} + ny(n-2)\, r^{n-3}\frac{\partial r}{\partial y}\right) + \left(n\, r^{n-2} + nz\,(n-2)\, r^{n-3}\frac{\partial r}{\partial z}\right)$$

$$= 3n\, r^{n-2} + n(n-2)r^{n-3}\left[x\frac{\partial r}{\partial x} + y\frac{\partial r}{\partial y} + z\frac{\partial r}{\partial z}\right]$$

$$= 3n\, r^{n-2} + n(n-2)r^{n-3}\left[x\left(\frac{x}{r}\right) + y\left(\frac{y}{r}\right) + z\left(\frac{z}{r}\right)\right]$$

$$\left[\because r^2 = x^2 + y^2 + z^2 \Rightarrow 2r\frac{\partial r}{\partial x} = 2x \Rightarrow \frac{\partial r}{\partial x} = \frac{x}{r} \text{ etc.}\right]$$

$$= 3nr^{n-2} + n\,(n-2)r^{n-4}\,[x^2 + y^2 + z^2]$$

$$= 3nr^{n-2} + n\,(n-2)\, r^{n-4}.r^2$$ $(\because r^2 = x^2 + y^2 + z^2)$

$$= r^{n-2}\,[3n + n^2 - 2n] = r^{n-2}\,(n^2 + n) = n(n+1)\, r^{n-2}$$

If we put $n = -1$

div grad $(r^{-1}) = -1\,(-1 + 1)\, r^{-1-2}$

$\Rightarrow$ $\nabla^2\left(\dfrac{1}{r}\right) = 0$

**Ques.** If $\vec{r} = x\hat{i} + y\hat{j} + z\hat{k}$, and $r = |r|$ find div $\left(\dfrac{r}{r^2}\right)$. (*U.P. I Sem., Dec. 2006*) **Ans.** $\dfrac{1}{r^2}$

## EXERCISE 5.8

1. If $r = x\hat{i} + y\hat{j} + z\hat{k}$ and $r = |\vec{r}|$, show that (*i*) div $\left(\dfrac{\vec{r}}{|\vec{r}|^3}\right) = 0$,

   (*ii*) div (grad $r^n$) = $n(n+1)\,r^{n-2}$ (*AMIETE, June 2010*) (*iii*) div $(r\,\phi) = 3\phi + r$ grad $\phi$.
2. Show that the vector $V = (x+3y)\hat{i} + (y-3z)\hat{j} + (x-2z)\hat{k}$ is solenoidal. (*R.G.P.V., Bhopal, Dec. 2003*)
3. Show that $\nabla.(\phi\, A) = \nabla\phi.A + \phi(\nabla.A)$
4. If $\rho$, $\phi$, $z$ are cylindrical coordinates, show that grad (log $\rho$) and grad $\phi$ are solenoidal vectors.
5. Obtain the expression for $\nabla^2 f$ in spherical coordinates from their corresponding expression in orthogonal curvilinear coordinates.

**Prove the following:**

6. $\vec{\nabla}.(\phi\vec{F}) = (\vec{\nabla}\phi).\vec{F} + \phi(\vec{\nabla}.\vec{F})$
7. (*a*) $\nabla.(\nabla\phi) = \nabla^2\phi$ (*b*) $\vec{\nabla}\times\dfrac{(\vec{A}\times\vec{R})}{r^n} = \dfrac{(2-n)\vec{A}}{r^n} + \dfrac{n(\vec{A}.\vec{R})\vec{R}}{r^{n+2}}, r = |\vec{R}|$
8. div $(f\,\nabla g)$ − div $(g\,\nabla f) = f\,\nabla^2 g - g\,\nabla^2 f$

## 5.31 CURL

(*U.P., I semester, Dec. 2006*)

The curl of a vector point function $F$ is defined as below

$$\text{curl } \vec{F} = \vec{\nabla}\times\vec{F} \qquad (\vec{F} = F_1\hat{i} + F_2\hat{j} + F_3\hat{k})$$

$$= \left(\hat{i}\frac{\partial}{\partial x} + \hat{j}\frac{\partial}{\partial y} + \hat{k}\frac{\partial}{\partial z}\right)\times(F_1\hat{i} + F_2\hat{j} + F_3\hat{k})$$

$$= \begin{vmatrix} \hat{i} & \hat{j} & \hat{k} \\ \dfrac{\partial}{\partial x} & \dfrac{\partial}{\partial y} & \dfrac{\partial}{\partial z} \\ F_1 & F_2 & F_3 \end{vmatrix} = \hat{i}\left(\frac{\partial F_3}{\partial y} - \frac{\partial F_2}{\partial z}\right) - \hat{j}\left(\frac{\partial F_3}{\partial x} - \frac{\partial F_1}{\partial z}\right) + \hat{k}\left(\frac{\partial F_2}{\partial x} - \frac{\partial F_1}{\partial y}\right)$$

Curl $\vec{F}$ is a vector quantity.

## 5.32 PHYSICAL MEANING OF CURL

(*M.D.U., Dec. 2009, U.P. I Semester, Winter 2009, 2000*)

We know that $\vec{V} = \vec{\omega}\times\vec{r}$, where $\omega$ is the angular velocity, $\vec{V}$ *is the linear velocity and* $\vec{r}$ *is the position vector of a point on the rotating body.*

$$\text{Curl } \vec{V} = \vec{\nabla}\times\vec{V} \qquad \left[\begin{aligned} \vec{\omega} &= \omega_1\hat{i} + \omega_2\hat{j} + \omega_3\hat{k} \\ \vec{r} &= x\hat{i} + y\hat{j} + z\hat{k} \end{aligned}\right]$$

$$= \vec{\nabla}\times(\vec{\omega}\times\vec{r}) = \vec{\nabla}\times[(\omega_1\hat{i} + \omega_2\hat{j} + \omega_3\hat{k})\times(x\hat{i} + y\hat{j} + z\hat{k})]$$

$$= \vec{\nabla}\times\begin{vmatrix} \hat{i} & \hat{j} & \hat{k} \\ \omega_1 & \omega_2 & \omega_3 \\ x & y & z \end{vmatrix} = \vec{\nabla}\times[(\omega_2 z - \omega_3 y)\hat{i} - (\omega_1 z - \omega_3 x)\hat{j} + (\omega_1 y - \omega_2 x)\hat{k}]$$

$$= \left(\hat{i}\frac{\partial}{\partial x} + \hat{j}\frac{\partial}{\partial y} + \hat{k}\frac{\partial}{\partial z}\right)\times[(\omega_2 z - \omega_3 y)\hat{i} - (\omega_1 z - \omega_3 x)\hat{j} + (\omega_1 y - \omega_2 x)\hat{k}]$$

$$= \begin{vmatrix} \hat{i} & \hat{j} & \hat{k} \\ \frac{\partial}{\partial x} & \frac{\partial}{\partial y} & \frac{\partial}{\partial z} \\ \omega_2 z - \omega_3 y & \omega_3 x - \omega_1 z & \omega_1 y - \omega_2 x \end{vmatrix}$$

$$= (\omega_1 + \omega_2)\hat{i} - (-\omega_2 - \omega_2)\hat{j} + (\omega_3 + \omega_3)\hat{k} \;=\; 2(\omega_1 \hat{i} + \omega_2 \hat{j} + \omega_3 \hat{k}) = 2\omega$$

Curl $\vec{V} = 2\omega$ which shows that curl of a vector field is connected with rotational properties of the vector field and justifies the name *rotation* used for curl.

If Curl $\bar{F} = 0$, the field $F$ is termed as *irrotational.*

**Example 41.** *Find the divergence and curl of* $\vec{v} = (xyz)\hat{i} + (3x^2 y)\hat{j} + (xz^2 - y^2 z)\hat{k}$ *at* $(2, -1, 1)$ *(Nagpur University, Summer 2003)*

**Solution.** Here, we have

$$\vec{v} = (x\,y\,z)\hat{i} + (3x^2 y)\hat{j} + (xz^2 - y^2 z)\hat{k}$$

$$\text{Div. } \vec{v} = \nabla\phi$$

$$\text{Div } \vec{v} = \frac{\partial}{\partial x}(x\,y\,z) + \frac{\partial}{\partial y}(3x^2 y) + \frac{\partial}{\partial z}(xz^2 - y^2 z)$$

$$= yz + 3x^2 + 2x\,z - y^2 \qquad = -1 + 12 + 4 - 1 = 14 \text{ at } (2, -1, 1)$$

$$\text{Curl } \bar{v} = \begin{vmatrix} \hat{i} & \hat{j} & \hat{k} \\ \frac{\partial}{\partial x} & \frac{\partial}{\partial y} & \frac{\partial}{\partial z} \\ xyz & 3x^2 y & xz^2 - y^2 z \end{vmatrix} \qquad = -2yz\,\hat{i} - (z^2 - xy)\hat{j} + (6xy - xz)\hat{k}$$

$$= -2yz\,\hat{i} + (xy - z^2)\hat{j} + (6xy - xz)\hat{k}$$

Curl at (2, −1, 1)

$$= -2(-1)(1)\hat{i} + \{(2)(-1) - 1\}\hat{j} + \{6(2)(-1) - 2(1)\}\hat{k}$$

$$= 2\hat{i} - 3\hat{j} - 14\hat{k}$$

**Ans.**

**Example 42.** *If* $\vec{V} = \dfrac{x\hat{i} + y\hat{j} + z\hat{k}}{\sqrt{x^2 + y^2 + z^2}}$, *find the value of curl* $\vec{V}$.

*(U.P., I Semester, Winter 2000)*

**Solution.** Curl $\vec{V} = \vec{\nabla} \times \vec{V}$

$$= \left(\hat{i}\frac{\partial}{\partial x} + \hat{j}\frac{\partial}{\partial y} + \hat{k}\frac{\partial}{\partial z}\right) \times \left(\frac{x\hat{i} + y\hat{j} + z\hat{k}}{(x^2 + y^2 + z^2)^{1/2}}\right)$$

$$= \begin{vmatrix} \hat{i} & \hat{j} & \hat{k} \\ \frac{\partial}{\partial x} & \frac{\partial}{\partial y} & \frac{\partial}{\partial z} \\ \frac{x}{(x^2 + y^2 + z^2)^{1/2}} & \frac{y}{(x^2 + y^2 + z^2)^{1/2}} & \frac{z}{(x^2 + y^2 + z^2)^{1/2}} \end{vmatrix}$$

$$= \hat{i}\left[\frac{\partial}{\partial y}\left(\frac{z}{(x^2+y^2+z^2)^{1/2}}\right)-\frac{\partial}{\partial z}\left(\frac{y}{(x^2+y^2+z^2)^{1/2}}\right)\right]-\hat{j}\left[\frac{\partial}{\partial x}\left(\frac{z}{(x^2+y^2+z^2)^{1/2}}\right)\right.$$

$$\left.-\frac{\partial}{\partial z}\left(\frac{x}{(x^2+y^2+z^2)^{1/2}}\right)\right]+\hat{k}\left[\frac{\partial}{\partial x}\left(\frac{y}{(x^2+y^2+z^2)^{1/2}}\right)-\frac{\partial}{\partial y}\left(\frac{x}{(x^2+y^2+z^2)^{1/2}}\right)\right]$$

$$= \hat{i}\left[\frac{-yz}{(x^2+y^2+z^2)^{3/2}}+\frac{y.z}{(x^2+y^2+z^2)^{3/2}}\right]-\hat{j}\left[\frac{-zx}{(x^2+y^2+z^2)^{3/2}}+\frac{zx}{(x^2+y^2+z^2)^{3/2}}\right]$$

$$+\hat{k}\left[\frac{-xy}{(x^2+y^2+z^2)^{3/2}}+\frac{xy}{(x^2+y^2+z^2)^{3/2}}\right]=0$$ **Ans.**

**Example 43.** *Prove that* $(y^2-z^2+3yz-2x)\hat{i}+(3xz+2xy)\hat{j}+(3xy-2xz+2z)\hat{k}$ *is both solenoidal and irrotational.* (*U.P., I Sem, Dec. 2008*)

**Solution.** Let $\vec{F}=(y^2-z^2+3yz-2x)\hat{i}+(3xz+2xy)\hat{j}+(3xy-2xz+2z)\hat{k}$

For solenoidal, we have to prove $\vec{\nabla}.\vec{F}=0.$

Now, $\vec{\nabla}.\vec{F}=\left[\hat{i}\frac{\partial}{\partial x}+\hat{j}\frac{\partial}{\partial y}+\hat{k}\frac{\partial}{\partial z}\right]\cdot\left[(y^2-z^2+3yz-2x)\hat{i}+(3xz+2xy)\hat{j}+(3xy-2xz+2z)\hat{k}\right]$

$$=-2+2x-2x+2=0$$

Thus, $\vec{F}$ is solenoidal. For irrotational, we have to prove Curl $\overline{F}=0.$

Now, 
$$\text{Curl }\vec{F}=\begin{vmatrix}\hat{i} & \hat{j} & \hat{k}\\ \frac{\partial}{\partial x} & \frac{\partial}{\partial y} & \frac{\partial}{\partial z}\\ y^2-z^2+3yz-2x & 3xz+2xy & 3xy-2xz+2z\end{vmatrix}$$

$$=(3z+2y-2y+3z)\hat{i}-(-2z+3y-3y+2z)\hat{j}+(3z+2y-2y-3z)\hat{k}$$

$$=0\hat{i}+0\hat{j}+0\hat{k}=0$$

Thus, $\vec{F}$ is irrotational.

Hence, $\vec{F}$ is both solenoidal and irrotational. **Proved.**

**Example 44.** *Determine the constants a and b such that the curl of vector*

$$\overline{A}=(2xy+3yz)\hat{i}+(x^2+axz-4z^2)\hat{j}-(3xy+byz)\hat{k}\ \textit{is zero.}$$

(*U.P. I Semester, Dec 2008*)

**Solution.** 
$$\text{Curl } A=\left(\hat{i}\frac{\partial}{\partial x}+\hat{j}\frac{\partial}{\partial y}+\hat{k}\frac{\partial}{\partial z}\right)\times[(2xy+3yz)\hat{i}+(x^2+axz-4z^2)\hat{j}-(3xy+byz)\hat{k}]$$

$$=\begin{vmatrix}\hat{i} & \hat{j} & \hat{k}\\ \frac{\partial}{\partial x} & \frac{\partial}{\partial y} & \frac{\partial}{\partial z}\\ 2xy+3yz & x^2+axz-4z^2 & -3xy-byz\end{vmatrix}$$

$$= [-3x - bz - ax + 8z]\hat{i} - [-3y - 3y]\hat{j} + [2x + az - 2x - 3z]\hat{k}$$

$$= [-x(3+a) + z(8-b)]\hat{i} + 6y\,\hat{j} + z(-3+a)\hat{k}$$

$$= 0 \quad \text{(given)}$$

*i.e.*, $3 + a = 0$ and $8 - b = 0$, $-3 + a = 0$

$a = -3, 3$ $\quad b = 8$ $\quad \Rightarrow \quad a = 3$ **Ans.**

**Example 45.** *If a vector field is given by* $\vec{F} = (x^2 - y^2 + x)\hat{i} - (2xy + y)\hat{j}$. *Is this field irrotational ? If so, find its scalar potential.*

(*U.P. I Semester, Dec 2009*)

**Solution.** Here, we have

$$\vec{F} = (x^2 - y^2 + x)\hat{i} - (2xy + y)\hat{j}$$

$$\text{Curl } F = \nabla \times \vec{F}$$

$$= \left(\hat{i}\frac{\partial}{\partial x} + \hat{j}\frac{\partial}{\partial y} + \hat{k}\frac{\partial}{\partial z}\right) \times (x^2 - y^2 + x)\hat{i} - (2xy + y)\hat{j}$$

$$= \begin{vmatrix} \hat{i} & \hat{j} & \hat{k} \\ \frac{\partial}{\partial x} & \frac{\partial}{\partial y} & \frac{\partial}{\partial z} \\ x^2 - y^2 + x & -2xy - y & 0 \end{vmatrix} = \hat{i}(0-0) - \hat{j}(0-0) + \hat{k}(-2y + 2y) = 0$$

Hence, vector field $\vec{F}$ is irrotational.

To find the scalar potential function $\phi$

$$\vec{F} = \nabla\phi$$

$$d\phi = \frac{\partial\phi}{\partial x}dx + \frac{\partial\phi}{\partial y}dy + \frac{\partial\phi}{\partial z}dz = \left|\hat{i}\frac{\partial\phi}{\partial x} + \hat{j}\frac{\partial\phi}{\partial y} + \hat{k}\frac{\partial\phi}{\partial z}\right| \cdot (\hat{i}\,dx + \hat{j}\,dy + \hat{k}\,dz)$$

$$= \left(\hat{i}\frac{\partial}{\partial x} + \hat{j}\frac{\partial}{\partial y} + \hat{k}\frac{\partial}{\partial z}\right)\phi \cdot (\vec{d}\,\vec{r}) = \nabla\phi \cdot \vec{d}\,\vec{r} = \vec{F} \cdot \vec{d}\,\vec{r}$$

$$= [(x^2 - y^2 + x)\hat{i} - (2xy + y)\hat{j}] \cdot (\hat{i}\,dx + \hat{j}\,dy + \hat{k}\,dz)$$

$$= (x^2 - y^2 + x)dx - (2xy + y)dy.$$

$$\phi = \int [(x^2 - y^2 + x)dx - (2xy + y)dy] + c$$

$$= \int \left[x^2\,dx + x\,dx - y\,dy - y^2\,dx - 2\,xy\,dy\right] + c = \frac{x^3}{3} + \frac{x^2}{2} - \frac{y^2}{2} - xy^2 + c$$

Hence, the scalar potential is $\frac{x^3}{3} + \frac{x^2}{2} - \frac{y^2}{2} - xy^2 + c$ **Ans.**

**Example 46.** *Find the scalar potential function f for* $\vec{A} = y^2\,\hat{i} + 2xy\,\hat{j} - z^2\,\hat{k}$.

(*Gujarat, I Semester, Jan. 2009*)

**Solution.** We have,

$$\vec{A} = y^2\,\hat{i} + 2xy\,\hat{j} - z^2\,\hat{k}$$

$$\text{Curl } \vec{A} = \nabla \times \vec{A} = \left(\hat{i}\frac{\partial}{\partial x} + \hat{j}\frac{\partial}{\partial y} + \hat{k}\frac{\partial}{\partial z}\right) \times (y^2\,\hat{i} + 2xy\,\hat{j} - z^2\,\hat{k})$$

$$= \begin{vmatrix} \hat{i} & \hat{j} & \hat{k} \\ \frac{\partial}{\partial x} & \frac{\partial}{\partial y} & \frac{\partial}{\partial z} \\ y^2 & 2xy & -z^2 \end{vmatrix} = \hat{i}(0) - \hat{j}(0) + \hat{k}(2y - 2y) = 0$$

Hence, $\vec{A}$ is irrotational. To find the scalar potential function $f$.

$$\vec{A} = \nabla f$$

$$df = \frac{\partial f}{\partial x} dx + \frac{\partial f}{\partial y} dy + \frac{\partial f}{\partial z} dz \quad = \left(\hat{i}\frac{\partial f}{\partial x} + \hat{j}\frac{\partial f}{\partial y} + \hat{k}\frac{\partial f}{\partial z}\right).(\hat{i}\,dx + \hat{j}\,dy + \hat{k}\,dz)$$

$$= \left(\hat{i}\frac{\partial}{\partial x} + \hat{j}\frac{\partial}{\partial y} + \hat{k}\frac{\partial}{\partial z}\right) f.dr = \nabla f.d\vec{r}$$

$$= \vec{A}.dr \qquad (A = \nabla f)$$

$$= (y^2\,\hat{i} + 2xy\,\hat{j} - z^2\,\hat{k}).(\hat{i}\,dx + \hat{j}\,dy + \hat{k}\,dz)$$

$$= y^2\,dx + 2xy\,dy - z^2\,dz = d\,(xy^2) - z^2\,dz$$

$$f = \int d\,(xy^2) - \int z^2\,dz = xy^2 - \frac{z^3}{3} + C$$ **Ans.**

**Example 47.** *A vector field is given by* $\vec{A} = (x^2 + xy^2)\,\hat{i} + (y^2 + x^2y)\,\hat{j}$. *Show that the field is irrotational and find the scalar potential.(Nagpur Univeristy, Summer 2003, Winter 2002)*

**Solution.** $\vec{A}$ is irrotational if curl $\vec{A} = 0$

$$\text{Curl } \vec{A} = \nabla \times \vec{A} = \begin{vmatrix} \hat{i} & \hat{j} & \hat{k} \\ \frac{\partial}{\partial x} & \frac{\partial}{\partial y} & \frac{\partial}{\partial z} \\ x^2 + xy^2 & y^2 + x^2y & 0 \end{vmatrix} = \hat{i}(0-0) - \hat{j}(0-0) + \hat{k}(2xy - 2xy) = 0$$

Hence, $\vec{A}$ is irrotational. If $\phi$ is the scalar potential, then

$$\vec{A} = \text{grad}\,\phi$$

$$d\,\phi = \frac{\partial \phi}{\partial x} dx + \frac{\partial \phi}{\partial y} dy + \frac{\partial \phi}{\partial z} dz \qquad \text{[Total differential coefficient]}$$

$$= \left(\hat{i}\frac{\partial \phi}{\partial x} + \hat{j}\frac{\partial \phi}{\partial y} + \hat{k}\frac{\partial \phi}{\partial z}\right).(\hat{i}\,dx + \hat{j}\,dy + \hat{k}\,dz) = \text{grad}\,\phi\,.\,dr$$

$$= \vec{A}.dr = [(x^2 + xy^2)\,\hat{i} + (y^2 + x^2y)\,\hat{j}].(\hat{i}\,dx + \hat{j}\,dy + \hat{k}\,dz)$$

$$= (x^2 + xy^2)\,dx + (y^2 + x^2y)\,dy = x^2\,dx + y^2\,dy + (x\,dx)y^2 + (x^2)\,(y\,dy)$$

$$\phi = \int x^2 dx + \int y^2 dy + \int [(x\,dx)\,y^2 + (x^2)\,(y\,dy)] = \frac{x^3}{3} + \frac{y^3}{3} + \frac{x^2y^2}{2} + c$$ **Ans.**

**Example 48.** *Show that* $\vec{V}(x, y, z) = 2x\,y\,z\,\hat{i} + (x^2z + 2y)\,\hat{j} + x^2\,y\,\hat{k}$ *is irrotational and find a scalar function u(x, y, z) such that* $\vec{V}$ *= grad (u).*

**Solution.** $\vec{V}\,(x, y, z) = 2x\,y\,z\,\hat{i} + (x^2z + 2y)\,\hat{j} + x^2\,y\,\hat{k}$

$$\text{Curl } \vec{V} = \left(\hat{i}\frac{\partial}{\partial x} + \hat{j}\frac{\partial}{\partial y} + \hat{k}\frac{\partial}{\partial z}\right) \times [2xyz\,\hat{i} + (x^2 z + 2y)\,\hat{j} + x^2 y\,\hat{k}]$$

$$= \begin{vmatrix} \hat{i} & \hat{j} & \hat{k} \\ \dfrac{\partial}{\partial x} & \dfrac{\partial}{\partial y} & \dfrac{\partial}{\partial z} \\ 2xyz & x^2 z + 2y & x^2 y \end{vmatrix}$$

$$= (x^2 - x^2)\,\hat{i} - (2xy - 2xy)\,\hat{j} + (2xz - 2xz)\,\hat{k} = 0$$

Hence, $\vec{V}\,(x, y, z)$ is irrotational.

To find corresponding scalar function $u$, consider the following relations given

$$\vec{V} = \text{grad}\,(u)$$

or $$\vec{V} = \vec{\nabla}(u) \qquad \text{...(1)}$$

$$du = \frac{\partial u}{\partial x}dx + \frac{\partial u}{\partial y}dy + \frac{\partial u}{\partial z}dz \qquad \text{(Total differential coefficient)}$$

$$= \left(\hat{i}\frac{\partial u}{\partial x} + \hat{j}\frac{\partial u}{\partial y} + \hat{k}\frac{\partial u}{\partial z}\right).(\hat{i}\,dx + \hat{j}\,dy + \hat{k}\,dz)$$

$$= \vec{\nabla} u.d\vec{r} = \vec{V}.d\vec{r} \qquad \text{[From (1)]}$$

$$= [2xyz\,\hat{i} + (x^2 z + 2y)\,\hat{j} + x^2 y\,\hat{k}].(\hat{i}\,dx + \hat{j}\,dy + \hat{k}\,dz)$$

$$= 2\,x\,y\,z\,dx + (x^2 z + 2y)\,dy + x^2 y\,dz$$

$$= y(2x\,z\,dx + x^2\,dz) + (x^2 z)\,dy + 2y\,dy$$

$$= [yd\,(x^2 z) + (x^2 z)\,dy] + 2y\,dy = d(x^2 yz) + 2y\,dy$$

Integrating, we get $$u = x^2 yz + y^2$$ **Ans.**

**Example 49.** *A fluid motion is given by* $\vec{v} = (y+z)\,\hat{i} + (z+x)\,\hat{j} + (x+y)\,\hat{k}$. *Show that the motion is irrotational and hence find the velocity potential.*

*(Uttarakhand, I Semester 2006; U.P., I Semester, Winter 2003)*

**Solution.** Curl $\vec{v} = \nabla \times \vec{v}$

$$= \left(\hat{i}\frac{\partial}{\partial x} + \hat{j}\frac{\partial}{\partial y} + \hat{k}\frac{\partial}{\partial z}\right) \times [(y+z)\,\hat{i} + (z+x)\,\hat{j} + (x+y)\,\hat{k}]$$

$$= \begin{vmatrix} \hat{i} & \hat{j} & \hat{k} \\ \dfrac{\partial}{\partial x} & \dfrac{\partial}{\partial y} & \dfrac{\partial}{\partial z} \\ y+z & z+x & x+y \end{vmatrix} = (1-1)\,\hat{i} - (1-1)\,\hat{j} + (1-1)\,\hat{k} = 0$$

Hence, $\vec{v}$ is irrotational.

To find the corresponding velocity potential $\phi$, consider the following relation.

$$\bar{v} = \nabla\phi$$

$$d\phi = \frac{\partial\phi}{\partial x}dx + \frac{\partial\phi}{\partial y}dy + \frac{\partial\phi}{\partial z}dz \qquad \text{[Total Differential coefficient]}$$

$$= \left(\hat{i}\frac{\partial\phi}{\partial x}+\hat{j}\frac{\partial\phi}{\partial y}+\hat{k}\frac{\partial\phi}{\partial z}\right).(\hat{i}\,dx+\hat{j}\,dy+\hat{k}\,dz) = \left(\hat{i}\frac{\partial}{\partial x}+\hat{j}\frac{\partial}{\partial y}+\hat{k}\frac{\partial}{\partial z}\right)\phi.d\vec{r} = \nabla\phi.d\vec{r} = \vec{v}.d\vec{r}$$

$$= [(y+z)\hat{i}+(z+x)\hat{j}+(x+y)\hat{k}].(\hat{i}\,dx+\hat{j}\,dy+\hat{k}\,dz)$$

$$= (y+z)\,dx+(z+x)\,dy+(x+y)\,dz$$

$$= y\,dx+z\,dx+z\,dy+x\,dy+x\,dz+y\,dz$$

$$\phi = \int(y\,dx+x\,dy)+\int(z\,dy+y\,dz)+\int(z\,dx+x\,dz)$$

$$\phi = xy+yz+zx+c$$

Velocity potential $= xy + yz + zx + c$ **Ans.**

**Example 50.** *A fluid motion is given by*

$$\vec{v} = (y\sin z-\sin x)\hat{i}+(x\sin z+2yz)\hat{j}+(xy\cos z+y^2)\hat{k}$$

*is the motion irrotational? If so, find the velocity potential.*

**Solution.** Curl $\vec{v} = \vec{\nabla}\times\vec{v}$

$$= \left(\hat{i}\frac{\partial}{\partial x}+\hat{j}\frac{\partial}{\partial y}+\hat{k}\frac{\partial}{\partial z}\right)\times(y\sin z-\sin x)\hat{i}+(x\sin z+2yz)\hat{j}+(xy\cos z+y^2)\hat{k}$$

$$= \begin{vmatrix}\hat{i} & \hat{j} & \hat{k}\\ \frac{\partial}{\partial x} & \frac{\partial}{\partial y} & \frac{\partial}{\partial z}\\ y\sin z-\sin x & x\sin z+2yz & xy\cos z+y^2\end{vmatrix}$$

$$= (x\cos z+2y-x\cos z-2y)\hat{i}-[y\cos z-y\cos z]\hat{j}+(\sin z-\sin z)\hat{k} = 0$$

Hence, the motion is irrotational.

So, $\bar{v} = \nabla\phi$ where $\phi$ is called velocity potential.

$$d\phi = \frac{\partial\phi}{\partial x}dx+\frac{\partial\phi}{\partial y}dy+\frac{\partial\phi}{\partial z}dz \qquad \text{[Total differential coefficient]}$$

$$= \left(\hat{i}\frac{\partial\phi}{\partial x}+\hat{j}\frac{\partial\phi}{\partial y}+\hat{k}\frac{\partial\phi}{\partial z}\right).(\hat{i}\,dx+\hat{j}\,dy+\hat{k}\,dz) = \vec{\nabla}\phi.d\vec{r} = \vec{v}.d\vec{r}$$

$$= [(y\sin z-\sin x)\hat{i}+(x\sin z+2yz)\hat{j}+(xy\cos z+y^2)\hat{k}].[\hat{i}\,dx+\hat{j}\,dy+\hat{k}\,dz]$$

$$= (y\sin z-\sin x)\,dx+(x\sin z+2yz)\,dy+(xy\cos z+y^2)\,dz$$

$$= (y\sin z\,dx+x\,dy\sin z+xy\cos z\,dz)-\sin x\,dx+(2yz\,dy+y^2\,dz)$$

$$= d(xy\sin z)+d(\cos x)+d(y^2z)$$

$$\phi = \int d(xy\sin z)+\int d(\cos x)+\int d(y^2z)$$

$$\phi = xy\sin z+\cos x+y^2z+c$$

Hence, Velocity potential $= xy\sin z+\cos x+y^2z+c.$ **Ans.**

**Example 51.** *Prove that* $\vec{F} = r^2\vec{r}$ *is conservative and find the scalar potential* $\phi$ *such that* $\vec{F} = \nabla\phi$. *(Nagpur University, Summer 2004)*

**Solution.** Given $\vec{F} = r^2\vec{r} = r^2(x\hat{i}+y\hat{j}+z\hat{k}) = r^2x\hat{i}+r^2y\hat{j}+r^2z\hat{k}$

Consider $\vec{\nabla}\times\vec{F} = \begin{vmatrix}\hat{i} & \hat{j} & \hat{k}\\ \frac{\partial}{\partial x} & \frac{\partial}{\partial y} & \frac{\partial}{\partial z}\\ r^2x & r^2y & r^2z\end{vmatrix}$

$$= \hat{i}\left[\frac{\partial}{\partial y}r^2 z - \frac{\partial}{\partial z}r^2 y\right] - \hat{j}\left[\frac{\partial}{\partial x}r^2 z - \frac{\partial}{\partial z}r^2 x\right] + \hat{k}\left[\frac{\partial}{\partial x}r^2 y - \frac{\partial}{\partial y}r^2 x\right]$$

$$= \hat{i}\left[2rz\frac{\partial r}{\partial y} - 2ry\frac{\partial r}{\partial z}\right] - \hat{j}\left[2rz\frac{\partial r}{\partial x} - 2rx\frac{\partial r}{\partial z}\right] + \hat{k}\left[2ry\frac{\partial r}{\partial x} - 2rx\frac{\partial r}{\partial y}\right]$$

$$\left[\text{But } r^2 = x^2 + y^2 + z^2, \frac{\partial r}{\partial x} = \frac{x}{r}, \quad \frac{\partial r}{\partial y} = \frac{y}{r}, \quad \frac{\partial r}{\partial z} = \frac{z}{r}\right]$$

$$= \hat{i}\left[2rz\frac{y}{r} - 2ry\frac{z}{r}\right] - \hat{j}\left[2rz\frac{x}{r} - 2rx\frac{z}{r}\right] + \hat{k}\left[2ry\frac{x}{r} - 2rx\frac{y}{r}\right]$$

$$= \hat{i}(2yz - 2yz) - \hat{j}(2zx - 2zx) + \hat{k}(2xy - 2xy) = 0\hat{i} - 0\hat{j} + 0\hat{k} = 0$$

$\therefore \quad \nabla \times \vec{F} = 0$

$\therefore \quad \vec{F}$ is irrotational $\quad \therefore \quad F$ is conservative.

Consider scalar potential $\phi$ such that $\vec{F} = \nabla\phi$.

$$d\phi = \frac{\partial\phi}{\partial x}dx + \frac{\partial\phi}{\partial y}dy + \frac{\partial\phi}{\partial z}dz \qquad \text{[Total differential coefficient]}$$

$$= \left(\hat{i}\frac{\partial\phi}{\partial x} + \hat{j}\frac{\partial\phi}{\partial y} + \hat{k}\frac{\partial\phi}{\partial z}\right).(\hat{i}\,dx + \hat{j}\,dy + \hat{k}\,dz)$$

$$= \left(\hat{i}\frac{\partial}{\partial x} + \hat{j}\frac{\partial}{\partial y} + \hat{k}\frac{\partial}{\partial z}\right)\phi.(\hat{i}\,dx + \hat{j}\,dy + \hat{k}\,dz) = \nabla\phi.(\hat{i}\,dx + \hat{j}\,dy + \hat{k}\,dz)$$

$$= \vec{F}.(\hat{i}\,dx + \hat{j}\,dy + \hat{k}\,dz) = r^2\vec{r}.(\hat{i}\,dx + \hat{j}\,dy + \hat{k}\,dz) \qquad (\nabla\phi = \vec{F})$$

$$= (x^2 + y^2 + z^2)(\hat{i}\,x + \hat{j}\,y + \hat{k}\,z).(\hat{i}\,dx + \hat{j}\,dy + \hat{k}\,dz)$$

$$= (x^2 + y^2 + z^2)(x\,dx + y\,dy + z\,dz)$$

$$= x^3\,dx + y^3\,dy + z^3\,dz + (x\,dx)\,y^2 + (x^2)\,(y\,dy) + (x\,dx)z^2 + z^2\,(y\,dy) + x^2\,(z\,dz) + y^2\,(z\,dz)$$

$$\phi = \int x^3\,dx + \int y^3\,dy + \int z^3\,dz + \int[(x\,dx)y^2 + (y\,dy)x^2] + \int[(x\,dx)z^2 + (z\,dz)x^2] + \int[(y\,dy)z^2 + (z\,dz)y^2]$$

$$= \frac{x^4}{4} + \frac{y^4}{4} + \frac{z^4}{4} + \frac{1}{2}x^2y^2 + \frac{1}{2}x^2z^2 + \frac{1}{2}y^2z^2 + c$$

$$= \frac{1}{4}(x^4 + y^4 + z^4 + 2x^2y^2 + 2x^2z^2 + 2y^2z^2) + c \qquad \textbf{Ans.}$$

**Example 52.** *Show that the vector field* $\vec{F} = \dfrac{\vec{r}}{|\vec{r}|^3}$ *is irrotational as well as solenoidal. Find the scalar potential.*

*(Nagpur University, Summer 2008, 2001, U.P. I Semester Dec. 2005, 2001)*

**Solution.** $\quad F = \dfrac{\vec{r}}{|\vec{r}|^3} = \dfrac{x\hat{i} + y\hat{j} + z\hat{k}}{(x^2 + y^2 + z^2)^{3/2}}$

$$\text{Curl } \vec{F} = \vec{\nabla} \times \vec{F} = \left(\hat{i}\frac{\partial}{\partial x} + \hat{j}\frac{\partial}{\partial y} + \hat{k}\frac{\partial}{\partial z}\right) \times \left(\frac{x\hat{i} + y\hat{j} + z\hat{k}}{(x^2 + y^2 + z^2)^{3/2}}\right)$$

$$= \begin{vmatrix} \hat{i} & \hat{j} & \hat{k} \\ \frac{\partial}{\partial x} & \frac{\partial}{\partial y} & \frac{\partial}{\partial z} \\ \frac{x}{(x^2+y^2+z^2)^{3/2}} & \frac{y}{(x^2+y^2+z^2)^{3/2}} & \frac{z}{(x^2+y^2+z^2)^{3/2}} \end{vmatrix}$$

$$= \hat{i}\left[\frac{-3}{2}\frac{2yz}{(x^2+y^2+z^2)^{5/2}} + \frac{3}{2}\frac{2yz}{(x^2+y^2+z^2)^{5/2}}\right]$$

$$-\hat{j}\left[\frac{-3}{2}\frac{2xz}{(x^2+y^2+z^2)^{5/2}} - \left(-\frac{3}{2}\right)\frac{2xz}{(x^2+y^2+z^2)^{5/2}}\right]$$

$$+\hat{k}\left[-\frac{3}{2}\frac{2xy}{(x^2+y^2+z^2)^{5/2}} - \left(-\frac{3}{2}\right)\frac{2xy}{(x^2+y^2+z^2)^{5/2}}\right]$$

$$= 0$$

Hence, $\vec{F}$ is irrotational.

$\Rightarrow \quad \vec{F} = \vec{\nabla}\phi$, where $\phi$ is called scalar potential

$$d\phi = \frac{\partial\phi}{\partial x}dx + \frac{\partial\phi}{\partial y}dy + \frac{\partial\phi}{\partial z}dz \qquad \text{[Total differential coefficient]}$$

$$= \left(\hat{i}\frac{\partial\phi}{\partial x} + \hat{j}\frac{\partial\phi}{\partial y} + \hat{k}\frac{\partial\phi}{\partial z}\right).(\hat{i}\,dx + \hat{j}\,dy + \hat{k}\,dz) = \vec{\nabla}\phi.d\vec{r} = \vec{F}.d\vec{r}$$

$$= \frac{x\hat{i} + y\hat{j} + z\hat{k}}{(x^2+y^2+z^2)^{3/2}}.(\hat{i}\,dx + \hat{j}\,dy + \hat{k}\,dz) = \frac{x\,dx + y\,dy + z\,dz}{(x^2+y^2+z^2)^{3/2}}$$

$$\phi = \frac{1}{2}\int\frac{2x\,dx + 2y\,dy + 2z\,dz}{(x^2+y^2+z^2)^{3/2}}$$

$$= \frac{1}{2}\left(-\frac{2}{1}\right)(x^2+y^2+z^2)^{-\frac{1}{2}} = \frac{1}{(x^2+y^2+z^2)^{\frac{1}{2}}} = -\frac{1}{|\vec{r}|} \qquad \textbf{Ans.}$$

Now, $\quad \text{Div}\,\vec{F} = \vec{\nabla}.\vec{F}$

$$= \left(\hat{i}\frac{\partial}{\partial x} + \hat{j}\frac{\partial}{\partial y} + \hat{k}\frac{\partial}{\partial z}\right).\frac{x\hat{i} + y\hat{j} + z\hat{k}}{(x^2+y^2+z^2)^{3/2}}$$

$$= \frac{\partial}{\partial x}\frac{x}{(x^2+y^2+z^2)^{3/2}} + \frac{\partial}{\partial y}\frac{y}{(x^2+y^2+z^2)^{3/2}} + \frac{\partial}{\partial z}\frac{z}{(x^2+y^2+z^2)^{3/2}}$$

$$= \frac{(x^2+y^2+z^2)^{3/2}\,(1) - x\left(\frac{3}{2}\right)(x^2+y^2+z^2)^{1/2}\,(2x)}{(x^2+y^2+z^2)^3}$$

$$+ \frac{(x^2+y^2+z^2)^{3/2}\,(1) - y\left(\frac{3}{2}\right)(x^2+y^2+z^2)^{1/2}\,(2y)}{(x^2+y^2+z^2)^3}$$

$$+ \frac{(x^2+y^2+z^2)^{3/2}\,(1) - z\left(\frac{3}{2}\right)(x^2+y^2+z^2)^{1/2}\,(2z)}{(x^2+y^2+z^2)^3}$$

$$= \frac{(x^2+y^2+z^2)^{1/2}}{(x^2+y^2+z^2)^3}\,[x^2+y^2+z^2-3x^2+x^2+y^2+z^2-3y^2+x^2+y^2+z^2-3z^2]$$

$$= 0$$

Hence, $\vec{F}$ is solenoidal. **Proved.**

**Example 53.** *Given the vector field* $\vec{V} = (x^2 - y^2 + 2xz)\hat{i} + (xz - xy + yz)\hat{j} + (z^2 + x^2)\hat{k}$ *find curl V. Show that the vectors given by curl V at* $P_0$ *(1, 2, –3) and* $P_1$ *(2, 3, 12) are orthogonal.*

**Solution.** $\overline{\text{Curl}\,\vec{V}} = \vec{\nabla} \times \vec{V}$

$$= \left(\hat{i}\frac{\partial}{\partial x} + \hat{j}\frac{\partial}{\partial y} + \hat{k}\frac{\partial}{\partial z}\right) \times [(x^2 - y^2 + 2xz)\hat{i} + (xz - xy + yz)\hat{j} + (z^2 + x^2)\hat{k}]$$

$$\text{curl}\,\vec{V} = \begin{vmatrix} \hat{i} & \hat{j} & \hat{k} \\ \dfrac{\partial}{\partial x} & \dfrac{\partial}{\partial y} & \dfrac{\partial}{\partial z} \\ x^2 - y^2 + 2xz & xz - xy + yz & z^2 + x^2 \end{vmatrix}$$

$$= -(x+y)\hat{i} - (2x - 2x)\hat{j} + (z - y + 2y)\hat{k} = -(x+y)\hat{i} + (y+z)\hat{k}$$

$$\text{curl}\,\vec{V} \text{ at } P_0\,(1, 2, -3) = -(1+2)\hat{i} + (2-3)\hat{k} = -3\hat{i} - \hat{k}$$

$$\text{curl}\,\vec{V} \text{ at } P_1\,(2, 3, 12) = -(2+3)\hat{i} + (3+12)\hat{k} = -5\hat{i} + 15\hat{k}$$

The curl $\vec{V}$ at (1, 2, –3) and (2, 3, 12) are perpendicular since

$$(-3\hat{i} - \hat{k}).(-5\hat{i} + 15\hat{k}) = +15 - 15 = 0$$ **Proved.**

**Example 54.** *Find the constants a, b, c, so that*

$$\vec{F} = (x + 2y + az)\hat{i} + (bx - 3y - z)\hat{j} + (4x + cy + 2z)\hat{k} \quad ...(1)$$

*is irrotational and hence find function* $\phi$ *such that* $\vec{F} = \nabla\phi$.

(*Nagpur University, Summer 2005, Winter 2000; R.G.P.V., Bhopal 2009*)

**Solution.** We have,

$$\therefore \quad \nabla \times \vec{F} = \begin{vmatrix} \hat{i} & \hat{j} & \hat{k} \\ \dfrac{\partial}{\partial x} & \dfrac{\partial}{\partial y} & \dfrac{\partial}{\partial z} \\ (x + 2y + az) & (bx - 3y - z) & (4x + cy + 2z) \end{vmatrix}$$

$$= (c+1)\hat{i} - (4-a)\hat{j} + (b-2)\hat{k}$$

As $\vec{F}$ is irrotational, $\nabla \times \vec{F} = \vec{0}$

*i.e.,* $(c+1)\hat{i} - (4-a)\hat{j} + (b-2)\hat{k} = 0\hat{i} + 0\hat{j} + 0\hat{k}$

$\therefore$ $c + 1 = 0$, $4 - a = 0$ and $b - 2 = 0$

*i.e.,* $a = 4$, $b = 2$, $c = -1$

Putting the values of *a, b, c* in (1), we get

$$\vec{F} = (x + 2y + 4z)\hat{i} + (2x - 3y - z)\hat{j} + (4x - y + 2z)\hat{k}$$

Now we have to find $\phi$ such that $\vec{F} = \nabla\phi$

We know that

$$d\phi = \frac{\partial\phi}{\partial x}dx + \frac{\partial\phi}{\partial y}dy + \frac{\partial\phi}{\partial z}dz \quad \text{[Total differential coefficient]}$$

$$= \left(\hat{i}\frac{\partial\phi}{\partial x} + \hat{j}\frac{\partial\phi}{\partial y} + \hat{k}\frac{\partial\phi}{\partial z}\right).(\hat{i}\,dx + \hat{j}\,dy + \hat{k}\,dz)$$

$$= \left(\hat{i}\frac{\partial}{\partial x} + \hat{j}\frac{\partial}{\partial y} + \hat{k}\frac{\partial}{\partial z}\right)\phi.(\hat{i}\,dx + \hat{j}\,dy + \hat{k}\,dz) = \nabla\phi.(\hat{i}\,dx + \hat{j}\,dy + \hat{k}\,dz)$$

$$= \bar{F}.(\hat{i}\,dx + \hat{j}\,dy + \hat{k}\,dz)$$

$$= [(x + 2y + 4z)\hat{i} + (2x - 3y - z)\hat{j} + (4x - y + 2z)\hat{k}].(\hat{i}\,dx + \hat{j}\,dy + \hat{k}\,dz)$$

$$= (x + 2y + 4z)\,dx + (2x - 3y - z)\,dy + (4x - y + 2z)\,dz$$

$$= x\,dx - 3y\,dy + 2z\,dz + (2y\,dx + 2x\,dy) + (4z\,dx + 4x\,dz) + (-z\,dy - y\,dz)$$

$$\phi = \int x\,dx - 3\int y\,dy + 2\int z\,dz + \int(2y\,dx + 2x\,dy) + \int(4z\,dx + 4x\,dz) - \int(z\,dy + y\,dz)$$

$$= \frac{x^2}{2} - \frac{3y^2}{2} + z^2 + 2xy + 4zx - yz + c$$ **Ans.**

**Example 55.** *Let $\vec{V}(x, y, z)$ be a differentiable vector function and $\phi(x, y, z)$ be a scalar function. Derive an expression for div $(\phi\vec{V})$ in terms of $\phi$ . $\vec{V}$, div $\vec{V}$ and $\nabla\phi$.*

*(U.P. I Semester, Winter 2003)*

**Solution.** Let $\vec{V} = V_1\hat{i} + V_2\hat{j} + V_3\hat{k}$

$$\text{div}\,(\phi\vec{V}) = \vec{\nabla}.(\phi\vec{F})$$

$$= \left(\hat{i}\frac{\partial}{\partial x} + \hat{j}\frac{\partial}{\partial y} + \hat{k}\frac{\partial}{\partial z}\right).[\phi V_1\hat{i} + \phi V_2\hat{j} + \phi V_3\hat{k}] = \frac{\partial}{\partial x}(\phi V_1) + \frac{\partial}{\partial y}(\phi V_2) + \frac{\partial}{\partial z}(\phi V_3)$$

$$= \left(\phi\frac{\partial V_1}{\partial x} + \frac{\partial\phi}{\partial x}V_1\right) + \left(\phi\frac{\partial V_2}{\partial y} + \frac{\partial\phi}{\partial y}V_2\right) + \left(\phi\frac{\partial V_3}{\partial z} + \frac{\partial\phi}{\partial z}V_3\right)$$

$$= \phi\left(\frac{\partial V_1}{\partial x} + \frac{\partial V_2}{\partial y} + \frac{\partial V_3}{\partial z}\right) + \left(\frac{\partial\phi}{\partial x}V_1 + \frac{\partial\phi}{\partial y}V_2 + \frac{\partial\phi}{\partial z}V_3\right)$$

$$= \phi\left(\hat{i}\frac{\partial}{\partial x} + \hat{j}\frac{\partial}{\partial y} + \hat{k}\frac{\partial}{\partial z}\right).(V_1\hat{i} + V_2\hat{j} + V_3\hat{k}) + \left(\hat{i}\frac{\partial\phi}{\partial x} + \hat{j}\frac{\partial\phi}{\partial y} + \hat{k}\frac{\partial\phi}{\partial z}\right).(V_1\hat{i} + V_2\hat{j} + V_3\hat{k})$$

$$= \phi(\vec{\nabla}.\vec{V}) + (\vec{\nabla}\phi).\vec{V} = \phi\,(\text{div}\,\vec{V}) + (\text{grad}\,\phi).\vec{V}$$ **Ans.**

**Example 56.** *If $\vec{A}$ is a constant vector and $\vec{R} = x\hat{i} + y\hat{j} + z\hat{k}$, then prove that*

$$Curl\left[\left(\vec{A}.\vec{R}\right)\vec{A}\right] = \vec{A}\times\vec{R}$$

*(K. University, Dec. 2009)*

**Solution.** Let $\vec{A} = A_1\hat{i} + A_2\hat{j} + A_3\hat{k}$, $\quad \vec{R} = x\hat{i} + y\hat{j} + z\hat{k}$

$$\vec{A}.\vec{R} = (A_1\hat{i} + A_2\hat{j} + A_3\hat{k}).(x\hat{i} + y\hat{j} + z\hat{k}) = A_1x + A_2y + A_3z$$

$$[\vec{A}.\vec{R}]\vec{R} = (A_1x + A_2y + A_3z)(x\hat{i} + y\hat{j} + z\hat{k})$$

$$= (A_1x^2 + A_2xy + A_3zx)\,\hat{i} + (A_1xy + A_2y^2 + A_3yz)\,\hat{j} + (A_1xz + A_2yz + A_3z^2)\,\hat{k}$$

$$\text{Curl}\left[(\vec{A}.\vec{R})\,\vec{R}\right] = \begin{vmatrix} \hat{i} & \hat{j} & \hat{k} \\ \frac{\partial}{\partial x} & \frac{\partial}{\partial y} & \frac{\partial}{\partial z} \\ A_1x^2 + A_2xy + A_3zx & A_2xy + A_2y^2 + A_3yz & A_1xz + A_2yz + A_3z^2 \end{vmatrix}$$

$$= (A_2 z - A_3 y)\ \hat{i}\ - [A_1 z - A_3 x)\ \hat{j}\ [A_1 y - A_2 x]\ \hat{k} \qquad \ldots (1)$$

$$\text{L.H.S.} = \vec{A} \times \vec{R}$$

$$= (A_1 \hat{i}\ + A_2 \hat{j}\ + A_3 k) \times (x \hat{i}\ + y \hat{j}\ + z\ \hat{k})$$

$$= \begin{vmatrix} \hat{i} & \hat{j} & k \\ A_1 & A_2 & A_3 \\ x & y & z \end{vmatrix}$$

$$= (A_2 z - A_3 y)\ \hat{i}\ - (A_1 z - A_3 x)\ \hat{j}\ + (A_1 y - A_2 x)\ \hat{k}$$

= R.H.S. [From (1)]

**Example 57.** *Suppose that $\vec{U}, \vec{V}$ and f are continuously differentiable fields then Prove that, div $(\vec{U} \times \vec{V}) = \vec{V}.curl\,\vec{U} - \vec{U}.curl\,\vec{V}$.* *(M.U. 2003, 2005)*

**Solution.** Let $\vec{U} = u_1 \hat{i} + u_2 \hat{j} + u_3 \hat{k},\ \vec{V} = v_1 \hat{i} + v_2 \hat{j} + v_3 \hat{k}$

$$\vec{U} \times \vec{V} = \begin{vmatrix} \hat{i} & \hat{j} & \hat{k} \\ u_1 & u_2 & u_3 \\ v_1 & v_2 & v_3 \end{vmatrix}$$

$$= (u_2v_3 - u_3v_2)\hat{i} - (u_1v_3 - u_3v_1)\hat{j} + (u_1v_2 - u_2v_1)\hat{k}$$

$$\text{div}\ (\vec{U} \times \vec{V}) = \left(\hat{i}\frac{\partial}{\partial x} + \hat{j}\frac{\partial}{\partial y} + \hat{k}\frac{\partial}{\partial z}\right).[(u_2v_3 - u_3v_2)\hat{i} - (u_1v_3 - u_3v_1)\hat{j} + (u_1v_2 - u_2v_1)\hat{k}]$$

$$= \frac{\partial}{\partial x}(u_2v_3 - u_3v_2) + \frac{\partial}{\partial y}(-u_1v_3 + u_3v_1) + \frac{\partial}{\partial z}(u_1v_2 - u_2v_1)$$

$$= \left[u_2\frac{\partial v_3}{\partial x} + v_3\frac{\partial u_2}{\partial x} - u_3\frac{\partial v_2}{\partial x} - v_2\frac{\partial u_3}{\partial x}\right] + \left[-u_1\frac{\partial v_3}{\partial y} - v_3\frac{\partial u_1}{\partial y} + u_3\frac{\partial v_1}{\partial y} + v_1\frac{\partial u_3}{\partial y}\right]$$
$$+ \left[u_1\frac{\partial v_2}{\partial z} + v_2\frac{\partial u_1}{\partial z} - u_2\frac{\partial v_1}{\partial z} - v_1\frac{\partial u_2}{\partial z}\right]$$

$$= v_1\left(\frac{\partial u_3}{\partial y} - \frac{\partial u_2}{\partial z}\right) + v_2\left(-\frac{\partial u_3}{\partial x} + \frac{\partial u_1}{\partial z}\right) + v_3\left(\frac{\partial u_2}{\partial x} - \frac{\partial u_1}{\partial y}\right)$$
$$+ u_1\left(-\frac{\partial v_3}{\partial y} + \frac{\partial v_2}{\partial z}\right) + u_2\left(\frac{\partial v_3}{\partial x} - \frac{\partial v_1}{\partial z}\right) + u_3\left(\frac{\partial v_1}{\partial y} - \frac{\partial v_2}{\partial x}\right)$$

$$= (v_1\hat{i} + v_2\hat{j} + v_3\hat{k}).\left[\hat{i}\left(\frac{\partial u_3}{\partial y} - \frac{\partial u_2}{\partial z}\right) + \hat{j}\left(\frac{\partial u_1}{\partial z} - \frac{\partial u_3}{\partial x}\right) + \hat{k}\left(\frac{\partial u_2}{\partial x} - \frac{\partial u_1}{\partial y}\right)\right]$$
$$-(u_1\hat{i} + u_2\hat{j} + u_3\hat{k}).\left[\hat{i}\left(-\frac{\partial v_2}{\partial z} + \frac{\partial v_3}{\partial y}\right) + \hat{j}\left(-\frac{\partial v_3}{\partial x} + \frac{\partial v_1}{\partial z}\right) + \hat{k}\left(-\frac{\partial v_1}{\partial y} + \frac{\partial v_2}{\partial x}\right)\right]$$

$$= V.(\vec{\nabla} \times \vec{U}) - \vec{U}.(\vec{\nabla} \times V) = \vec{V}.\text{curl}\,\vec{U} - \vec{U}.\text{curl}\,\vec{V}$$

**Proved.**

**Example 58.** *Prove that*

$$\vec{\nabla}\times(\vec{F}\times\vec{G}) = \vec{F}(\vec{\nabla}.\vec{G})-\vec{G}(\vec{\nabla}.\vec{F})+(\vec{G}.\vec{\nabla})\vec{F}-(\vec{F}.\vec{\nabla})\vec{G} \qquad \text{(M.U. 2004, 2005)}$$

**Solution.**

$$\vec{\nabla}\times(\vec{F}\times\vec{G}) = \Sigma\hat{i}\times\frac{\partial}{\partial x}(\vec{F}\times\vec{G})$$

$$= \Sigma\hat{i}\times\left(\frac{\partial F}{\partial x}\times\vec{G}+\vec{F}\times\frac{\partial G}{\partial x}\right)=\Sigma\hat{i}\times\left(\frac{\partial F}{\partial x}\times\vec{G}\right)+\Sigma\hat{i}\times\left(\vec{F}\times\frac{\partial G}{\partial x}\right)$$

$$= \Sigma\left[(\hat{i}.\vec{G})\frac{\partial F}{\partial x}-\left(\hat{i}\frac{\partial F}{\partial x}\right)\vec{G}\right]+\Sigma\left[\left(\hat{i}\frac{\partial G}{\partial x}\right)\vec{F}-(\hat{i}.\vec{F})\frac{\partial G}{\partial x}\right]$$

$$= \Sigma(\vec{G}.\hat{i})\frac{\partial F}{\partial x}-\vec{G}\Sigma\left(\hat{i}.\frac{\partial F}{\partial x}\right)+\vec{F}\Sigma\left(\hat{i}.\frac{\partial G}{\partial x}\right)-\Sigma(\vec{F}.\hat{i})\frac{\partial G}{\partial x}$$

$$= \vec{F}\left(\Sigma\hat{i}\frac{\partial G}{\partial x}\right)-\vec{G}\Sigma\left(\hat{i}.\frac{\partial F}{\partial x}\right)+\Sigma(\vec{G}.\hat{i})\frac{\partial F}{\partial x}-\Sigma(\vec{F}.\hat{i})\frac{\partial G}{\partial x}$$

$$= \vec{F}(\vec{\nabla}\vec{G})-\vec{G}(\vec{\nabla}.\vec{F})+(\vec{G}.\vec{\nabla})\vec{F}-(\vec{F}.\vec{\nabla})\vec{G}$$

**Proved.**

**Questions for practice:**

Prove that

$$\vec{\nabla}(\vec{F}.\vec{G}) = (\vec{G}.\vec{\nabla})\vec{F}+(\vec{F}.\vec{\nabla})\vec{G}+\vec{G}\times(\vec{\nabla}\times\vec{F})+\vec{F}\times(\vec{\nabla}\times\vec{G})$$

**Example 59.** *Prove that, for every field $\vec{V}$; div curl $\vec{V}$ = 0.*

*(Nagpur University, Summer 2004; AMIETE, Sem II, June 2010)*

**Solution.** Let

$$V = V_1\hat{i}+V_2\hat{j}+V_3\hat{k}$$

$$\text{div}\ (\overline{\text{curl}\ \vec{V}}) = \vec{\nabla}.(\vec{\nabla}\times\vec{V})$$

$$= \vec{\nabla}.\begin{vmatrix}\hat{i} & \hat{j} & \hat{k}\\ \frac{\partial}{\partial x} & \frac{\partial}{\partial y} & \frac{\partial}{\partial z}\\ V_1 & V_2 & V_3\end{vmatrix}$$

$$= \left(\hat{i}\frac{\partial}{\partial x}+\hat{j}\frac{\partial}{\partial y}+\hat{k}\frac{\partial}{\partial z}\right).\left[\hat{i}\left(\frac{\partial V_3}{\partial y}-\frac{\partial V_2}{\partial z}\right)-\hat{j}\left(\frac{\partial V_3}{\partial x}-\frac{\partial V_1}{\partial z}\right)+\hat{k}\left(\frac{\partial V_2}{\partial x}-\frac{\partial V_1}{\partial y}\right)\right]$$

$$= \frac{\partial}{\partial x}\left(\frac{\partial V_3}{\partial y}-\frac{\partial V_2}{\partial z}\right)-\frac{\partial}{\partial y}\left(\frac{\partial V_3}{\partial x}-\frac{\partial V_1}{\partial z}\right)+\frac{\partial}{\partial z}\left(\frac{\partial V_2}{\partial x}-\frac{\partial V_1}{\partial y}\right)$$

$$= \frac{\partial^2 V_3}{\partial x\partial y}-\frac{\partial^2 V_2}{\partial x\partial z}-\frac{\partial^2 V_3}{\partial y\partial x}+\frac{\partial^2 V_1}{\partial y\partial z}+\frac{\partial^2 V_2}{\partial z\partial x}-\frac{\partial^2 V_1}{\partial z\partial y}$$

$$= \left(\frac{\partial^2 V_1}{\partial y\partial z}-\frac{\partial^2 V_1}{\partial z\partial y}\right)+\left(\frac{\partial^2 V_2}{\partial z\partial x}-\frac{\partial^2 V_2}{\partial x\partial z}\right)+\left(\frac{\partial^2 V_3}{\partial x\partial y}-\frac{\partial^2 V_3}{\partial y\partial x}\right)$$

$$= 0$$

**Ans.**

**Example 60.** *If $\vec{a}$ is a constant vector, show that*

$$\vec{a}\times(\vec{\nabla}\times\vec{r}) = \vec{\nabla}(\vec{a}.\vec{r})-(\vec{a}.\vec{\nabla})\vec{r}. \qquad \text{(U.P., Ist Semester, Dec. 2007)}$$

**Solution.**

$$\vec{a} = a_1\hat{i}+a_2\hat{j}+a_3\hat{k},\quad \vec{r}=r_1\hat{i}+r_2\hat{j}+r_3\hat{k}$$

$$\vec{\nabla}\times\vec{r} = \begin{vmatrix} \hat{i} & \hat{j} & \hat{k} \\ \frac{\partial}{\partial x} & \frac{\partial}{\partial y} & \frac{\partial}{\partial z} \\ r_1 & r_2 & r_3 \end{vmatrix} = \left(\frac{\partial r_3}{\partial y}-\frac{\partial r_2}{\partial z}\right)\hat{i}-\left(\frac{\partial r_3}{\partial x}-\frac{\partial r_1}{\partial z}\right)\hat{j}+\left(\frac{\partial r_2}{\partial x}-\frac{\partial r_1}{\partial y}\right)\hat{k}$$

$$\vec{a}\times(\vec{\nabla}\times\vec{r}) = \begin{vmatrix} \hat{i} & \hat{j} & \hat{k} \\ a_1 & a_2 & a_3 \\ \frac{\partial r_3}{\partial y}-\frac{\partial r_2}{\partial z} & -\frac{\partial r_3}{\partial x}+\frac{\partial r_1}{\partial z} & \frac{\partial r_2}{\partial x}-\frac{\partial r_1}{\partial y} \end{vmatrix}$$

$$= \left[\left(a_2\frac{\partial r_2}{\partial x}-a_2\frac{\partial r_1}{\partial y}\right)-\left(-a_3\frac{\partial r_3}{\partial x}+a_3\frac{\partial r_1}{\partial z}\right)\right]\hat{i}-\left[a_1\frac{\partial r_2}{\partial x}-a_1\frac{\partial r_1}{\partial y}-a_3\frac{\partial r_3}{\partial y}+a_3\frac{\partial r_2}{\partial z}\right]\hat{j}$$
$$+\left[-a_1\frac{\partial r_3}{\partial x}+a_1\frac{\partial r_1}{\partial z}-a_2\frac{\partial r_3}{\partial y}+a_2\frac{\partial r_2}{\partial z}\right]\hat{k}$$

$$= \left[\left(a_1\hat{i}\frac{\partial r_1}{\partial x}+a_2\hat{i}\frac{\partial r_2}{\partial x}+a_3\hat{i}\frac{\partial r_3}{\partial x}\right)+\left(a_1\hat{j}\frac{\partial r_1}{\partial y}+a_2\hat{j}\frac{\partial r_2}{\partial y}+a_3\hat{j}\frac{\partial r_3}{\partial y}\right)\right.$$
$$\left.+\left(a_1\hat{k}\frac{\partial r_1}{\partial z}+a_2\hat{k}\frac{\partial r_2}{\partial z}+a_3\hat{k}\frac{\partial r_3}{\partial z}\right)\right]-\left[\left(a_1\hat{i}\frac{\partial r_1}{\partial x}+a_1\hat{j}\frac{\partial r_2}{\partial x}+a_1\hat{k}\frac{\partial r_3}{\partial x}\right)\right.$$
$$\left.+\left(a_2\hat{i}\frac{\partial r_1}{\partial y}+a_2\hat{j}\frac{\partial r_2}{\partial y}+a_2\hat{k}\frac{\partial r_3}{\partial y}\right)+\left(a_3\hat{i}\frac{\partial r_1}{\partial z}+a_3\hat{j}\frac{\partial r_2}{\partial z}+a_3\hat{k}\frac{\partial r_3}{\partial z}\right)\right]$$

$$= \left(\hat{i}\frac{\partial}{\partial x}+\hat{j}\frac{\partial}{\partial y}+\hat{k}\frac{\partial}{\partial z}\right)(a_1r_1+a_2r_2+a_3r_3)-\left[a_1\frac{\partial}{\partial x}+a_2\frac{\partial}{\partial y}+a_3\frac{\partial}{\partial z}\right](r_1\hat{i}+r_2\hat{j}+r_3\hat{k})$$

$$= \nabla(\vec{a}.\vec{r})-(\vec{a}.\vec{\nabla})\vec{r}$$ **Proved.**

**Example 61.** *If r is the distance of a point (x, y, z) from the origin, prove that* $Curl\left(k\times grad\frac{1}{r}\right)+grad\left(k.grad\frac{1}{r}\right)=0$, *where k is the unit vector in the direction OZ.*

*(U.P., I Semester, Winter 2000)*

**Solution.** $r^2 = (x-0)^2+(y-0)^2+(z-0)^2 = x^2+y^2+z^2$

$$\Rightarrow \quad \frac{1}{r} = (x^2+y^2+z^2)^{-1/2}$$

$$\text{grad}\,\frac{1}{r} = \vec{\nabla}\frac{1}{r} = \left(\hat{i}\frac{\partial}{\partial x}+\hat{j}\frac{\partial}{\partial y}+\hat{k}\frac{\partial}{\partial z}\right)(x^2+y^2+z^2)^{-1/2}$$

$$= -\frac{1}{2}(x^2+y^2+z^2)^{-3/2}(2x\hat{i}+2y\hat{j}+2z\hat{k})$$

$$= -(x^2+y^2+z^2)^{-3/2}(x\hat{i}+y\hat{j}+z\hat{k})$$

$$k\times\text{grad}\,\frac{1}{r} = k\times[-(x^2+y^2+z^2)^{-3/2}(x\hat{i}+y\hat{j}+z\hat{k})]$$

$$= -(x^2+y^2+z^2)^{-3/2}(x\hat{j}-y\hat{i})$$

$$\text{curl}\left(k\times\text{grad}\frac{1}{r}\right) = \vec{\nabla}\times\left(k\times\text{grad}\frac{1}{r}\right)$$

$$= \left(\hat{i}\frac{\partial}{\partial x}+\hat{j}\frac{\partial}{\partial y}+\hat{k}\frac{\partial}{\partial z}\right)\times[-(x^2+y^2+z^2)^{-3/2}\ (x\hat{j}-y\hat{i})]$$

$$= \begin{vmatrix} \hat{i} & \hat{j} & \hat{k} \\ \frac{\partial}{\partial x} & \frac{\partial}{\partial y} & \frac{\partial}{\partial z} \\ \frac{y}{(x^2+y^2+z^2)^{3/2}} & \frac{-x}{(x^2+y^2+z^2)^{3/2}} & 0 \end{vmatrix}$$

$$= -\left(-\frac{3}{2}\right)\frac{(-x)(2z)}{(x^2+y^2+z^2)^{5/2}}\hat{i}+-\frac{3}{2}\frac{y(2z)}{(x^2+y^2+z^2)^{5/2}}\hat{j}+\left[-\frac{3}{2}\frac{(-x)(2x)}{(x^2+y^2+z^2)^{5/2}}\right.$$
$$\left.-\frac{1}{(x^2+y^2+z^2)^{3/2}}-\frac{(-3/2)(y)(2y)}{(x^2+y^2+z^2)^{5/2}}-\frac{1}{(x^2+y^2+z^2)^{3/2}}\right]\hat{k}$$

$$= \frac{-3xz}{(x^2+y^2+z^2)^{5/2}}\hat{i}-\frac{3yz}{(x^2+y^2+z^2)^{5/2}}\hat{j}+\frac{(3x^2-x^2-y^2-z^2+3y^2-x^2-y^2-z^2)}{(x^2+y^2+z^2)^{5/2}}\hat{k}$$

$$= \frac{-3xz\,\hat{i}-3yz\,\hat{j}+(x^2+y^2-2z^2)\hat{k}}{(x^2+y^2+z^2)^{5/2}} \qquad \ldots(1)$$

$$k\,.\,\text{grad}\ \frac{1}{r} = k.[-(x^2+y^2+z^2)^{-3/2}\ (x\hat{i}+y\hat{j}+z\hat{k})] = \frac{-z}{(x^2+y^2+z^2)^{3/2}}$$

$$\text{grad}\left(k.\text{grad}\frac{1}{r}\right) = \left(\hat{i}\frac{\partial}{\partial x}+\hat{j}\frac{\partial}{\partial y}+\hat{k}\frac{\partial}{\partial z}\right)\frac{-z}{(x^2+y^2+z^2)^{3/2}}$$

$$= -\frac{3}{2}\frac{\hat{i}(-z)(2x)}{(x^2+y^2+z^2)^{5/2}}+-\frac{3}{2}\frac{\hat{j}(-z)(2y)}{(x^2+y^2+z^2)^{5/2}}$$
$$+\left[-\frac{3}{2}\frac{(-z)(2z)}{(x^2+y^2+z^2)^{5/2}}-\frac{1}{(x^2+y^2+z^2)^{3/2}}\right]\hat{k}$$

$$= \frac{3xz\,\hat{i}+3yz\,\hat{j}+(3z^2-x^2-y^2-z^2)\hat{k}}{(x^2+y^2+z^2)^{5/2}} = \frac{3xz\,\hat{i}+3yz\,\hat{j}-(x^2+y^2-2z^2)\hat{k}}{(x^2+y^2+z^2)^{5/2}} \qquad \ldots(2)$$

Adding (1) and (2), we get

$$\text{Curl}\left(k\times\text{grad}\frac{1}{r}\right)+\text{grad}\left(k.\text{grad}\frac{1}{r}\right) = 0$$ **Proved.**

**Example 62.** *Prove that* $\nabla\times\left(\frac{\vec{a}\times\vec{r}}{r^n}\right)=\frac{(2-n)\vec{a}}{r^n}+\frac{n(\vec{a}.\vec{r})\vec{r}}{r^{n+2}}.$

*(M.U. 2009, 2005, 2003, 2002; AMIETE, II Sem. June 2010)*

**Solution.** We have,

$$\frac{\vec{a}\times\vec{r}}{r^n} = \frac{1}{r^n}\begin{vmatrix} \hat{i} & \hat{j} & \hat{k} \\ a_1 & a_2 & a_3 \\ x & y & z \end{vmatrix}$$

$$= \frac{1}{r^n}(a_2z-a_3y)\hat{i}+\frac{1}{r^n}(a_3x-a_1z)\hat{j}+\frac{1}{r^n}(a_1y-a_2x)\hat{k}$$

$$\nabla \times \frac{(\vec{a} \times \vec{r})}{r^n} = \begin{vmatrix} \hat{i} & \hat{j} & \hat{k} \\ \frac{\partial}{\partial x} & \frac{\partial}{\partial y} & \frac{\partial}{\partial z} \\ \frac{a_2 z - a_3 y}{r^n} & \frac{a_3 x - a_1 z}{r^n} & \frac{a_1 y - a_2 x}{r^n} \end{vmatrix}$$

$$= \hat{i}\left[\frac{\partial}{\partial y}\left(\frac{a_1 y - a_2 x}{r^n}\right) - \frac{\partial}{\partial z}\left(\frac{a_3 x - a_1 z}{r^n}\right)\right] - \hat{j}\left[\frac{\partial}{\partial x}\left(\frac{a_1 y - a_2 x}{r^n}\right) - \frac{\partial}{\partial z}\left(\frac{a_2 z - a_3 y}{r^n}\right)\right]$$
$$+ \hat{k}\left[\frac{\partial}{\partial x}\left(\frac{a_3 x - a_1 z}{r^n}\right) - \frac{\partial}{\partial y}\left(\frac{a_2 z - a_3 y}{r^n}\right)\right]$$

Now, $r^2 = x^2 + y^2 + z^2 \Rightarrow 2r\frac{\partial r}{\partial x} = 2x \Rightarrow \frac{\partial r}{\partial x} = \frac{x}{r}$

Similarly, $\frac{\partial r}{\partial y} = \frac{y}{r}$, $\frac{\partial r}{\partial z} = \frac{z}{r}$

$$\therefore \nabla \times \left(\frac{\vec{a} \times \vec{r}}{r^n}\right) = \hat{i}\left[\left\{-nr^{-n-1}\left(\frac{y}{r}\right)(a_1 y - a_2 x) + \frac{1}{r^n} a_1\right\}\right.$$
$$\left.- \left\{-nr^{-n-1}\left(\frac{z}{r}\right)(a_3 x - a_1 z) + \frac{1}{r^n}(-a_1)\right\}\right] + \text{two similar terms}$$

$$= \hat{i}\left[-\frac{n}{r^{n+2}}(a_1 y^2 - a_2 xy) + \frac{a_1}{r^n} + \frac{n}{r^{n+2}}(a_3 xz - a_1 z^2) + \frac{a_1}{r^n}\right] + \text{two similar terms}$$

$$= \hat{i}\left[\frac{2a_1}{r^n} - \frac{n}{r^{n+2}} a_1 (y^2 + z^2) + \frac{n}{r^{n+2}}(a_2 xy + a_3 xz)\right] + \text{two similar terms}$$

Adding and subtracting $\frac{n}{r^{n+2}} a_1 x^2$ to third and from second term, we get

$$\vec{\nabla} \times \left(\frac{\vec{a} \times \vec{r}}{r^n}\right) = \hat{i}\left[\frac{2a_1}{r^n} - \frac{na_1}{r^{n+2}}(x^2 + y^2 + z^2) + \frac{n}{r^{n+2}}(a_1 x^2 + a_2 xy + a_3 xz)\right] + \text{two similar terms}$$

$$= \hat{i}\left[\frac{2a_1}{r^n} - \frac{na_1}{r^{n+2}} r^2 + \frac{n}{r^{n+2}} x(a_1 x + a_2 y + a_3 z)\right] + \text{two similar terms}$$

$$= \hat{i}\left[\frac{2a_1}{r^n} - \frac{na_1}{r^n} + \frac{n}{r^{n+2}} x(a_1 x + a_2 y + a_3 z)\right] + \hat{j}\left[\frac{2a_2}{r^n} - \frac{na_2}{r^n} + \frac{n}{r^{n+2}} y(a_2 y + a_3 z + a_1 x)\right]$$
$$+ \hat{k}\left[\frac{2a_3}{r^n} - \frac{na_3}{r^n} + \frac{n}{r^{n+2}} z(a_3 z + a_1 x + a_2 y)\right]$$

$$= \frac{2}{r^n}(a_1 \hat{i} + a_2 \hat{j} + a_3 \hat{k}) - \frac{n}{r^n}(a_1 \hat{i} + a_2 \hat{j} + a_3 \hat{k}) + \frac{n}{r^{n+2}}(a_1 x + a_2 y + a_3 z)(x\hat{i} + y\hat{j} + z\hat{k})$$

$$= \frac{2-n}{r^n}(a_1 \hat{i} + a_2 \hat{j} + a_3 \hat{k}) + \frac{n}{r^{n+2}}(a_1 x + a_2 y + a_3 z)(x\hat{i} + y\hat{j} + z\hat{k})$$

$$= \frac{2-n}{r^n}\vec{a} + \frac{n}{r^{n+2}}(\vec{a}\cdot\vec{r})\vec{r}$$

**Proved.**

**Example 63.** *If f and g are two scalar point functions, prove that*

*div* $(f \nabla g) = f \nabla^2 g + \nabla f \nabla g$. *(U.P., I Semester, compartment, Winter 2001)*

**Solution.** We have, $\nabla g = \frac{\partial g}{\partial x}\hat{i} + \frac{\partial g}{\partial y}\hat{j} + \frac{\partial g}{\partial z}\hat{k}$

$$\Rightarrow \quad f\nabla g = f\frac{\partial g}{\partial x}\hat{i} + f\frac{\partial g}{\partial y}\hat{j} + f\frac{\partial g}{\partial z}\hat{k}$$

$$\Rightarrow \quad \text{div}(f\nabla g) = \frac{\partial}{\partial x}\left(f\frac{\partial g}{\partial x}\right) + \frac{\partial}{\partial y}\left(f\frac{\partial g}{\partial y}\right) + \frac{\partial}{\partial z}\left(f\frac{\partial g}{\partial z}\right)$$

$$= f\left(\frac{\partial^2 g}{\partial x^2} + \frac{\partial^2 g}{\partial y^2} + \frac{\partial^2 g}{\partial z^2}\right) + \left(\frac{\partial f}{\partial x}\frac{\partial g}{\partial x} + \frac{\partial f}{\partial y}\frac{\partial g}{\partial y} + \frac{\partial f}{\partial z}\frac{\partial g}{\partial z}\right)$$

$$= f\left(\frac{\partial^2}{\partial x^2} + \frac{\partial^2}{\partial y^2} + \frac{\partial^2}{\partial z^2}\right)g + \left(\frac{\partial f}{\partial x}\hat{i} + \frac{\partial f}{\partial y}\hat{j} + \frac{\partial f}{\partial z}\hat{k}\right).\left(\frac{\partial g}{\partial x}\hat{i} + \frac{\partial g}{\partial y}\hat{j} + \frac{\partial g}{\partial z}\hat{k}\right)$$

$$= f\nabla^2 g + \nabla f.\nabla g$$ **Proved.**

**Example 64.** *For a solenoidol vector $\vec{F}$, show that curl curl curl curl $\vec{F} = \nabla^4 \vec{F}$.*

*(M.D.U., Dec. 2009)*

**Solution.** Since vector $\vec{F}$ is solenoidal, so div $\vec{F} = 0$ ... (1)

We know that curl curl $\vec{F}$ = grad div $(\vec{F} - \nabla^2 \vec{F})$ ... (2)

Using (1) in (2), grad div $\vec{F}$ = grad (0) = 0 ... (3)

On putting the value of grad div F in (2), we get

curl curl $\vec{F} = -\nabla^2 \vec{F}$ ... (4)

Now, curl curl curl curl $\vec{F}$ = curl curl $(-\nabla^2 \vec{F})$ [Using (4)]

$= -$ curl curl $(\nabla^2 \vec{F}) = -$ [grad div $(\nabla^2 \vec{F}) - \nabla^2 (\nabla^2 \vec{F})$] [Using (2)]

$= -$ grad $(\nabla . \nabla^2 \vec{F}) + \nabla^2 (\nabla^2 \vec{F}) = -$ grad $(\nabla^2 \nabla . \vec{F}) + \nabla^4 \vec{F}$ $[\nabla . \vec{F} = 0]$

$= 0 + \nabla^4 \vec{F} = \nabla^4 \vec{F}$ [Using (1)] **Proved.**

## EXERCISE 5.9

**1.** Find the divergence and curl of the vector field $V = (x^2 - y^2)\hat{i} + 2xy\hat{j} + (y^2 - xy)\hat{k}$.

**Ans.** Divergence $= 4x$, Curl $= (2y - x)\hat{i} + y\hat{j} + 4y\hat{k}$

**2.** If $a$ is constant vector and $r$ is the radius vector, prove that

(*i*) $\nabla(\vec{a}.\vec{r}) = \vec{a}$ (*ii*) div $(\vec{r} \times \vec{a}) = 0$ (*iii*) curl $(\vec{r} \times \vec{a}) = -2\vec{a}$

where $\vec{r} = x\hat{i} + y\hat{j} + z\hat{k}$ and $\vec{a} = a_1\hat{i} + a_2\hat{j} + a_3\hat{k}$.

**3.** Prove that:

(*i*) $\nabla.(\phi A) = \nabla\phi.A + \phi(\nabla.A)$

(*ii*) $\nabla(A.B) = (A.\nabla)B + (B.\nabla)A + A \times (\nabla \times B) + B \times (\nabla \times A)$ *(R.G.P.V. Bhopal, June 2004)*

(*iii*) $\nabla \times (A \times B) = (B.\nabla)A - B(\nabla.A) - (A.\nabla)B + A(\nabla.B)$

**4.** If $F = (x + y + 1)\hat{i} + \hat{j} - (x + y)\hat{k}$, show that $F$.curl $F = 0$.

*(R.G.P.V. Bhopal, Feb. 2006, June 2004)*

**Prove that**

**5.** $\vec{\nabla} \times (\phi\vec{F}) = (\vec{\nabla}\phi) \times \vec{F} + \phi(\vec{\nabla} \times \vec{F})$ **6.** $\nabla.(\vec{F} \times \vec{G}) = \vec{G}.(\nabla \times \vec{F}) - \vec{F}.(\nabla \times \vec{G})$

**7.** Evaluate div $(\vec{A} \times \vec{r})$ if curl $\vec{A} = 0$. **8.** Prove that curl $(\vec{a} \times \vec{r}) = 2a$

9. Find div $\vec{F}$ and curl $F$ where $F$ = grad $(x^3 + y^3 + z^3 - 3xyz)$. *(R.G.P.V. Bhopal Dec. 2003)*

**Ans.** div $\vec{F} = 6(x + y + z)$, curl $\vec{F} = 0$

10. Find out values of *a, b, c* for which $\vec{v} = (x + y + az)\hat{i} + (bx + 3y - z)\hat{j} + (3x + cy + z)\hat{k}$ is irrotational.

**Ans.** $a = 3,\ b = 1,\ c = -1$

11. Determine the constants *a, b, c,* so that $\vec{F} = (x + 2y + az)\hat{i} + (bx - 3y - z)\hat{j} + (4x + cy + 2z)\hat{k}$ is irrotational. Hence find the scalar potential $\phi$ such that $\vec{F}$ = grad $\phi$.

*(R.G.P.V. Bhopal, Feb. 2005)* **Ans.** $a = 4,\ b = 2,\ c = 1$

Potential $\phi = \left(\frac{x^2}{2} - \frac{3y^2}{2} + z^2 + 2xy - yz + 4zx\right)$

**Choose the correct alternative:**

12. The magnitude of the vector drawn in a direction perpendicular to the surface $x^2 + 2y^2 + z^2 = 7$ at the point $(1, -1, 2)$ is

(*i*) $\frac{2}{3}$ (*ii*) $\frac{3}{2}$ (*iii*) 3 (*iv*) 6 *(A.M.I.E.T.E., Summer 2000)* **Ans.** (*iv*)

13. If $u = x^2 - y^2 + z^2$ and $\vec{V} = x\hat{i} + y\hat{j} + z\hat{k}$ then $\nabla\cdot(u\vec{V})$ is equal to

(*i*) $5u$ (*ii*) $5|\vec{V}|$ (*iii*) $5(u - |\vec{V}|)$ (*iv*) $5(u - |\vec{V}|)$ *(A.M.I.E.T.E., June 2007)*

14. A unit normal to $x^2 + y^2 + z^2 = 5$ at $(0, 1, 2)$ is equal to

(*i*) $\frac{1}{\sqrt{5}}(\hat{i} + \hat{j} + \hat{k})$ (*ii*) $\frac{1}{\sqrt{5}}(\hat{i} + \hat{j} - \hat{k})$ (*iii*) $\frac{1}{\sqrt{5}}(\hat{j} + 2\hat{k})$ (*iv*) $\frac{1}{\sqrt{5}}(\hat{i} - \hat{j} + \hat{k})$

*(A.M.I.E.T.E., Dec. 2008)*

15. The directional derivative of $\phi = x\,y\,z$ at the point $(1, 1, 1)$ in the direction $\hat{i}$ is:

(*i*) $-1$ (*ii*) $-\frac{1}{3}$ (*iii*) 1 (*iv*) $\frac{1}{3}$ **Ans.** (*iii*)

*(R.G.P.V. Bhopal, II Sem., June 2007)*

16. If $\vec{r} = x\hat{i} + y\hat{j} + z\hat{k}$ and $r = |\vec{r}|$ then $\nabla\phi(r)$ is:

(*i*) $\phi'(r)\vec{r}$ (*ii*) $\frac{\phi(r)\vec{r}}{r}$ (*iii*) $\frac{\phi'(r)\vec{r}}{r}$ (*iv*) None of these **Ans.** (*iii*)

*(R.G.P.V. Bhopal, II Semester, Feb. 2006)*

17. If $\vec{r} = x\hat{i} + y\hat{j} + z\hat{k}$ is position vector, then value of $\nabla(\log r)$ is *(U.P., I Sem, Dec 2008)*

(*i*) $\frac{\vec{r}}{r}$ (*ii*) $\frac{\vec{r}}{r^2}$ (*iii*) $-\frac{\vec{r}}{r^3}$ (*iv*) none of the above. **Ans.** (*ii*)

18. If $\vec{r} = x\hat{i} + y\hat{j} + z\hat{k}$ and $|\vec{r}| = r$, then div $\vec{r}$ is:

(*i*) 2 (*ii*) 3 (*iii*) $-3$ (*iv*) $-2$ **Ans.** (*ii*)

*(R.G.P.V. Bhopal, II Semester, Feb. 2006)*

19. If $\vec{V} = xy^2\hat{i} + 2yx^2z\hat{j} - 3yz^2\hat{k}$ then curl $\vec{V}$ at point $(1, -1, 1)$ is

(*i*) $-(\hat{j} + 2\hat{k})$ (*ii*) $(\hat{i} + 3\hat{k})$ (*iii*) $-(\hat{i} + 2\hat{k})$ (*iv*) $(\hat{i} + 2\hat{j} + \hat{k})$

*(R.G.P.V. Bhopal, II Semester, Feb 2006)*
**Ans.** (*iii*)

20. If $\vec{A}$ is such that $\nabla \times \vec{A} = 0$ then $\vec{A}$ is called

(*i*) Irrotational (*ii*) Solenoidal (*iii*) Rotational (*iv*) None of these

*(A.M.I.E.T.E., Dec. 2008)*

21. If $\vec{F}$ is a conservative force field, then the value of curl $\vec{F}$ is

(*i*) 0 (*ii*) 1 (*iii*) $\overline{\nabla F}$ (*iv*) $-1$ *(A.M.I.E.T.E., June 2007)*

**22.** If $\nabla^2$ $[(1-x).(1-2x)]$ is equal to
(*i*) 2 (*ii*) 3 (*iii*) 4 (*iv*) 6 (*A.M.I.E.T.E., Dec. 2009*) **Ans.** (*iii*)

**23.** If $\vec{R} = xi + yj + zk$ and $\vec{A}$ is a constant vector, curl $(\vec{A} \times \vec{R})$ is equal to

(*i*) $\vec{R}$ (*ii*) $2\vec{R}$ (*iii*) $\vec{A}$ (*iv*) $2\vec{A}$ (*A.M.I.E.T.E., Dec. 2009*) **Ans.** (*iv*)

**24.** If $r$ is the distance of a point (x, y, z) from the origin, the value of the expression $\hat{j} \times \text{grad}\,\frac{1}{2}$ equals

(*i*) $(x^2+y^2+z^2)^{-\frac{3}{2}}(\hat{j}z - \hat{k}x)$ (*ii*) $(x^2+y^2+z^2)^{-\frac{3}{2}}(\hat{j}z - \hat{i}z)$

(*iii*) zero (*iv*) $(x^2+y^2+z^2)^{\frac{3}{2}}(\hat{j}y - \hat{k}x)$

(*AMIETE, Dec. 2010*) **Ans.** (*ii*)

## 5.33 LINE INTEGRAL

Let $\vec{F}(x, y, z)$ be a vector function and a curve $AB$.

Line integral of a vector function $\vec{F}$ along the curve $AB$ is defined as integral of the component of $\vec{F}$ along the tangent to the curve $AB$.

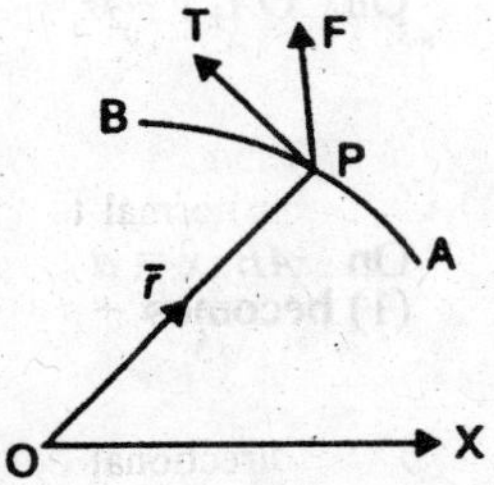


Component of $\vec{F}$ along a tangent $PT$ at $P$

$$= \text{Dot product of } \vec{F} \text{ and unit vector along PT}$$

$$= \vec{F} \cdot \frac{\vec{dr}}{ds} \quad \left(\frac{\vec{dr}}{ds} \text{ is a unit vector along tangent PT}\right)$$

$$\text{Line integral} = \sum \vec{F} \cdot \frac{\vec{dr}}{ds} \text{ from A to B along the curve}$$

$$\therefore \text{Line integral} = \int_c \left(\vec{F} \cdot \frac{\vec{dr}}{ds}\right) ds = \int_c \vec{F} \cdot \vec{dr}$$

**Note (1) Work.** If $\overline{F}$ represents the variable force acting on a particle along arc AB, then the total work done $= \int_A^B \vec{F} \cdot \vec{dr}$

**(2) Circulation.** If $\vec{v}$ represents the velocity of a liquid then $\oint_c \vec{V} \cdot \vec{dr}$ is called the circulation of $V$ round the closed curve $c$.

If the circulation of V round every closed curve is zero then V is said to be irrotational there.

**(3)** When the path of integration is a closed curve then notation of integration is $\oint$ in place of $\int$.

**Example 65.** *If a force* $\vec{F} = 2x^2y\hat{i} + 3xy\hat{j}$ *displaces a particle in the xy-plane from* (0, 0) *to* (1, 4) *along a curve* $y = 4x^2$. *Find the work done.*

**Solution.** Work done $= \int_c \vec{F} . \vec{dr}$

$$\left[\begin{array}{l} \vec{r} = x\hat{i} + y\hat{j} \\ \vec{dr} = dx\hat{i} + dy\,\hat{j} \end{array}\right]$$

$$= \int_c (2x^2y\hat{i} + 3xy\,\hat{j}).(dx\,\hat{i} + dy\,\hat{j})$$

$$= \int_c (2x^2y\,dx + 3xy\,dy)$$

Putting the values of $y$ and $dy$, we get $\left(\begin{array}{l} y = 4x^2 \\ dy = 8x\,dx \end{array}\right)$

$$= \int_0^1 \cdot [2x^2 (4x^2)\,dx + 3x(4x^2)\,8x\,dx]$$

$$= 104 \int_0^1 x^4\,dx = 104 \left(\frac{x^5}{5}\right)_0^1 = \frac{104}{5}$$ **Ans.**

**Example 66.** *Evaluate $\int_C \vec{F}.\vec{dr}$ where $\vec{F} = x^2\hat{i} + xy\hat{j}$ and C is the boundary of the square in the plane z = 0 and bounded by the lines x = 0, y = 0, x = a and y = a.*

(*Nagpur University, Summer 2001*)

**Solution.** $\int_C \vec{F}.\vec{dr} = \int_{OA} \vec{F}.\vec{dr} + \int_{AB} \vec{F}.\vec{dr} + \int_{BC} \vec{F}.\vec{dr} + \int_{CO} \vec{F}.\vec{dr}$

Here $\vec{r} = x\hat{i} + y\hat{j}, \quad \vec{dr} = dx\hat{i} + dy\hat{j}, \quad \vec{F} = x^2\hat{i} + xy\,\hat{j}$

$$\vec{F}.\vec{dr} = x^2dx + xydy \qquad ...(1)$$

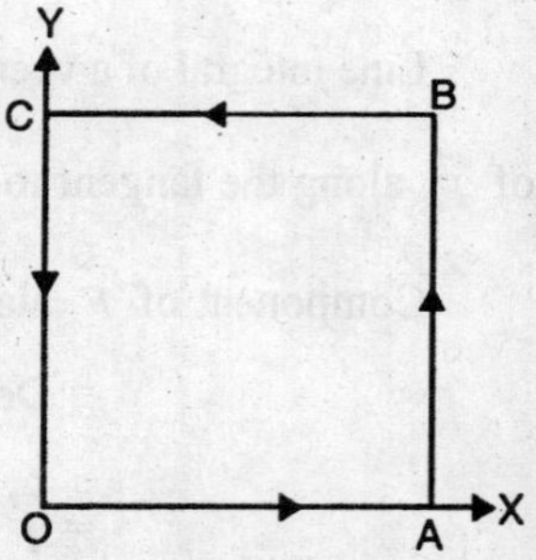


On $OA$, $y = 0$ $\quad \therefore \vec{F}.\vec{dr} = x^2dx$

$$\int_{OA} \vec{F}.\vec{dr} = \int_0^a x^2dx = \left[\frac{x^3}{3}\right]_0^a = \frac{a^3}{3} \qquad ...(2)$$

On $AB$, $x = a$ $\quad \therefore dx = 0$

(1) becomes

$$\therefore \vec{F}.\vec{dr} = aydy$$

$$\int_{Ab} \vec{F}.\vec{dr} = \int_0^a aydy = a\left[\frac{y^2}{2}\right]_0^a = \frac{a^3}{2} \qquad ...(3)$$

On $BC$, $y = a$ $\quad \therefore dy = 0$

$\Rightarrow$ (1) becomes $\quad \vec{F}.\vec{dr} = x^2dx$

$$\int_{BC} \vec{F}.\vec{dr} = \int_a^0 x^2dx = \left[\frac{x^3}{3}\right]_a^0 = \frac{-a^3}{3} \qquad ...(4)$$

On $CO$, $x = 0$, $\quad \therefore \vec{F}.\vec{dr} = 0$

(1) becomes

$$\int_{CO} \vec{F}.\vec{dr} = 0 \qquad ...(5)$$

On adding (2), (3), (4) and (5), we get $\int_C \vec{F}.\vec{dr} = \frac{a^3}{3} + \frac{a^3}{2} - \frac{a^3}{3} + 0 = \frac{a^3}{2}$ **Ans.**

**Example 67.** *A vector field is given by*

*$\vec{F} = (2y+3)\,\hat{i} + xz\hat{j} + (yz - x)\,\hat{k}$. Evaluate $\int_C \vec{F}.\vec{dr}$ along the path c is x = 2t, y = t, z = $t^3$ from t = 0 to t = 1.*

(*Nagpur University, Winter 2003*)

**Solution.** $\int_C \vec{F}.\vec{dr} = \int_C (2y+3)\,dx + (xz)\,dy + (yz - x)\,dz$

$$\left[\begin{array}{lll} \text{Since } x = 2t & y = t & z = t^3 \\ \therefore \frac{dx}{dt} = 2 & \frac{dy}{dt} = 1 & \frac{dz}{dt} = 3t^2 \end{array}\right]$$

$$= \int_0^1 (2t+3)(2\,dt) + (2t)(t^3)\,dt + (t^4 - 2t)(3t^2\,dt) = \int_0^1 (4t + 6 + 2t^4 + 3t^6 - 6t^3)\,dt$$

$$= \left[4\frac{t^2}{2} + 6t + \frac{2}{5}t^5 + \frac{3}{7}t^7 - \frac{6}{4}t^4\right]_0^1 = \left[2t^2 + 6t + \frac{2}{5}t^5 + \frac{3}{7}t^7 - \frac{3}{2}t^4\right]_0^1$$

$$= 2 + 6 + \frac{2}{5} + \frac{3}{7} - \frac{3}{2} = 7.32857.$$ **Ans.**

**Example 68.** *The acceleration of a particle at time t is given by*

$$\vec{a} = 18\cos 3t\,\hat{i} - 8\sin 2t\,\hat{j} + 6t\,\hat{k}.$$

*If the velocity* $\vec{v}$ *and displacement* $\vec{r}$ *be zero at t = 0, find* $\vec{v}$ *and* $\vec{r}$ *at any point t.*

**Solution.** Here, $\vec{a} = \frac{d^2\vec{r}}{dt^2} = 18\cos 3t\,\hat{i} - 8\sin 2t\,\hat{j} + 6t\,\hat{k}.$

On integrating, we have

$$\vec{v} = \frac{d\vec{r}}{dt} = \hat{i}\int 18\cos 3t\,dt + \hat{j}\int -8\sin 2t\,dt + \hat{k}\int 6t\,dt$$

$$\Rightarrow \quad \vec{v} = 6\sin 3t\,\hat{i} + 4\cos 2t\,\hat{j} + 3t^2\,\hat{k} + \vec{c} \quad \text{...(1)}$$

At $t = 0,\ \vec{v} = \vec{0}$

Putting $t = 0$ and $\vec{v} = 0$ in (1), we get

$$\vec{0} = 4\hat{j} + \vec{c} \Rightarrow \vec{c} = -4\hat{j}$$

$$\therefore \quad \vec{v} = \frac{d\vec{r}}{dt} = 6\sin 3t\,\hat{i} + 4(\cos 2t - 1)\,\hat{j} + 3t^2\hat{k}$$

Again integrating, we have

$$\vec{r} = \hat{i}\int 6\sin 3t\,dt + \hat{j}\int 4(\cos 2t - 1)\,dt + \hat{k}\int 3t^2\,dt$$

$$\Rightarrow \quad \vec{r} = -2\cos 3t\,\hat{i} + (2\sin 2t - 4t)\hat{j} + t^3\,\hat{k} + \vec{c_1} \quad \text{...(2)}$$

At, $t = 0,\ \vec{r} = 0$

Putting $t = 0$ and $\vec{r} = 0$ in (2), we get

$$\therefore \quad \vec{0} = -2\hat{i} + \vec{C_1} \Rightarrow \vec{C_1} = 2\hat{i}$$

Hence, $$\vec{r} = 2(1 - \cos 3t)\,\hat{i} + 2(\sin 2t - 2t)\hat{j} + t^3\hat{k}$$ **Ans.**

**Example 69.** *If* $\vec{A} = (3x^2 + 6y)\,\hat{i} - 14yz\hat{j} + 20xz^2\,\hat{k}$, *evaluate the line integral* $\oint \vec{A}.d\vec{r}$ *from* (0, 0, 0) to (1, 1, 1) *along the curve C.*

$x = t,\ y = t^2,\ z = t^3.$ *(Uttarakhand, I Semester, Dec. 2006)*

**Solution.** We have,

$$\int_C \vec{A}.d\vec{r} = \int_C [(3x^2 + 6y)\,\hat{i} - 14yz\hat{j} + 20xz^2\hat{k}].[\hat{i}\,dx + \hat{j}\,dy + \hat{k}\,dz]$$

$$= \int_C [(3x^2 + 6y)\,dx - 14yz dy + 20xz^2 dz]$$

If $x = t,\ y = t^2,\ z = t^3$, then points (0, 0, 0) and (1, 1, 1) correspond to $t = 0$ and $t = 1$ respectively.

Now, $$\int_C \vec{A}.d\vec{r} = \int_{t=0}^{t=1} [(3t^2 + 6t^2)\,d(t) - 14t^2\,t^3\,d(t^2) + 20t\,(t^3)^2\,d(t^3)]$$

$$= \int_{t=0}^{t=1} [9t^2\,dt - 14t^5 .2t\,dt + 20t^7 .3t^2\,dt] = \int_0^1 (9t^2 - 28t^6 + 60t^9)\,dt$$

$$= \left[9\left(\frac{t^3}{3}\right) - 28\left(\frac{t^7}{7}\right) + 60\left(\frac{t^{10}}{10}\right)\right]_0^1 \qquad = 3 - 4 + 6 = 5 \qquad \textbf{Ans.}$$

**Example 70.** *Evaluate* $\iint_S \vec{A}.\hat{n}\, ds$ *where* $\vec{A} = (x+y^2)\,\hat{i} - 2x\hat{j} + 2yz\hat{k}$ *and S is the surface of the plane* $2x + y + 2z = 6$ *in the first octant.* *(Nagpur University, Summer 2000)*

**Solution.** A vector normal to the surface "S" is given by

$$\nabla(2x+y+2z) = \left(\hat{i}\frac{\partial}{\partial x} + \hat{j}\frac{\partial}{\partial y} + \hat{k}\frac{\partial}{\partial z}\right)(2x+y+2z) = 2\hat{i} + \hat{j} + 2\hat{k}$$

And $\hat{n}$ = a unit vector normal to surface $S$

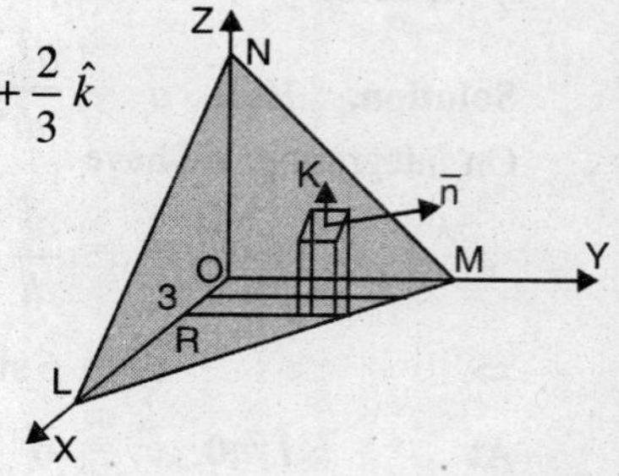


$$= \frac{2\hat{i} + \hat{j} + 2\hat{k}}{\sqrt{4+1+4}} = \frac{2}{3}\hat{i} + \frac{1}{3}\hat{j} + \frac{2}{3}\hat{k}$$

$$\hat{k}.\hat{n} = \hat{k}.\left(\frac{2}{3}\hat{i} + \frac{1}{3}\hat{j} + \frac{2}{3}\hat{k}\right) = \frac{2}{3}$$

$$\therefore \qquad \iint_S \bar{A}.\hat{n}\, ds = \iint_R \bar{A}.\hat{n}\frac{dx\,dy}{\hat{k}.\bar{n}}$$

Where $R$ is the projection of $S$.

Now, $\vec{A}.\hat{n} = [(x+y^2)\,\hat{i} - 2x\hat{j} + 2yz\hat{k}].\left(\frac{2}{3}\hat{i} + \frac{1}{3}\hat{j} + \frac{2}{3}\hat{k}\right)$

$$= \frac{2}{3}(x+y^2) - \frac{2}{3}x + \frac{4}{3}yz = \frac{2}{3}y^2 + \frac{4}{3}yz \qquad \text{...(1)}$$

Putting the value of $z$ in (1), we get

$$\vec{A}.\hat{n} = \frac{2}{3}y^2 + \frac{4}{3}y\left(\frac{6-2x-y}{2}\right) \qquad \left(\because \text{ on the plane } 2x+y+2z=6,\; z = \frac{(6-2x-y)}{2}\right)$$

$$\vec{A}.\hat{n} = \frac{2}{3}y(y+6-2x-y) = \frac{4}{3}y(3-x) \qquad \text{...(2)}$$

Hence, $$\iint_S \vec{A}.\hat{n}\,ds = \iint_R \bar{A}.\bar{n}\frac{dx\,dy}{|\hat{k}.\bar{n}|} \qquad \text{...(3)}$$

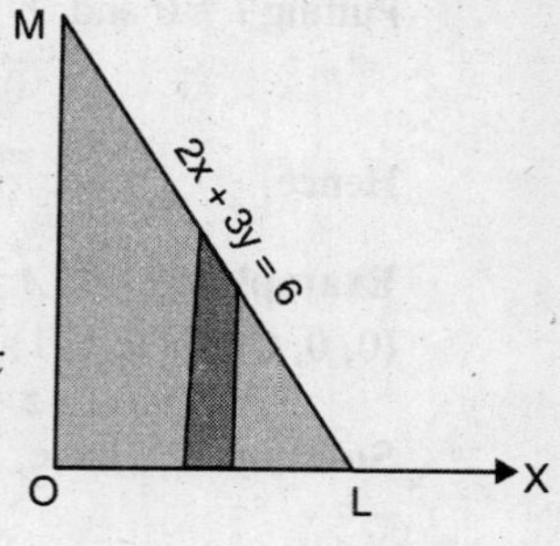


Putting the value of $\vec{A}.\hat{n}$ from (2) in (3), we get

$$\iint_S \vec{A}.\hat{n}\,ds = \iint_R \frac{4}{3}y(3-x).\frac{3}{2}\,dx\,dy = \int_0^3\int_0^{6-2x} 2y(3-x)\,dy\,dx$$

$$= \int_0^3 2(3-x)\left[\frac{y^2}{2}\right]_0^{6-2x} dx$$

$$= \int_0^3 (3-x)(6-2x)^2\,dx = 4\int_0^3 (3-x)^3\,dx$$

$$= 4.\left[\frac{(3-x)^4}{4(-1)}\right]_0^3 = -(0-81) = 81 \qquad \textbf{Ans.}$$

**Example 71.** *Compute* $\int_c \vec{F}.\vec{dr}$, *where* $\vec{F} = \dfrac{\hat{i}y - \hat{j}x}{x^2+y^2}$ *and c is the circle* $x^2 + y^2 = 1$ *traversed counter clockwise.*

**Solution.** $\vec{r} = \hat{i}\,x + \hat{j}\,y + \hat{k}\,z,\ d\overline{r} = \hat{i}\,dx + \hat{j}\,dy + \hat{k}\,dz$

$$\int_c \vec{F}\cdot d\vec{r} = \int_c \frac{\hat{i}y - \hat{j}x}{x^2+y^2}\cdot(\hat{i}dx + \hat{j}dy + \hat{k}dz)$$

$$= \int_c \frac{ydx - xdy}{x^2+y^2} = \int_c (ydx - xdy) \qquad \text{...(1) } [\because x^2+y^2=1]$$

Parametric equation of the circle are $x = \cos\theta$, $y = \sin\theta$.

Putting $x = \cos\theta$, $y = \sin\theta$, $dx = -\sin\theta\, d\theta$, $dy = \cos\theta\, d\theta$ in (1), we get

$$\int_C \vec{F}\, d\vec{r} = \int_0^{2\pi} \sin\theta\,(-\sin\theta\, d\theta) - \cos\theta\,(\cos\theta\, d\theta)$$

$$= -\int_0^{2\pi} (\sin^2\theta + \cos^2\theta)\, d\theta = -\int_0^{2\pi} d\theta = -(\theta)_0^{2\pi} = -2\pi \qquad \textbf{Ans.}$$

**Example 72.** *Show that the vector field* $\vec{F} = 2x(y^2+z^3)\hat{i} + 2x^2y\hat{j} + 3x^2z^2\hat{k}$ *is conservative. Find its scalar potential and the work done in moving a particle from* (–1, 2, 1) to (2, 3, 4).

*(A.M.I.E.T.E. June 2010, 2009)*

**Solution.** Here, we have

$$\vec{F} = 2x(y^2+z^3)\,\hat{i} + 2x^2y\,\hat{j} + 3x^2z^2\hat{k}$$

$$\text{Curl } \vec{F} = \nabla\times\vec{F}$$

$$= \begin{vmatrix} \hat{i} & \hat{j} & \hat{k} \\ \dfrac{\partial}{\partial x} & \dfrac{\partial}{\partial y} & \dfrac{\partial}{\partial z} \\ 2x(y^2+z^3) & 2x^2y & 3x^2z^2 \end{vmatrix} = (0-0)i - (6xz^2 - 6xz^2)\hat{j} + (4xy - 4xy)\hat{k} = 0$$

Hence, vector field $\vec{F}$ is irrotational.

To find the scalar potential function $\phi$

$$\vec{F} = \vec{\nabla}\phi$$

$$d\phi = \frac{\partial\phi}{\partial x}dx + \frac{\partial\phi}{\partial y}dy + \frac{\partial\phi}{\partial z}dz = \left(\hat{i}\frac{\partial\phi}{\partial x} + \hat{j}\frac{\partial\phi}{\partial y} + \hat{k}\frac{\partial\phi}{\partial z}\right).\left(\hat{i}dx + \hat{j}dy + \hat{k}dz\right)$$

$$= \left(\hat{i}\frac{\partial}{\partial x} + \hat{j}\frac{\partial}{\partial y} + \hat{k}\frac{\partial}{\partial z}\right)\phi.\left(d\vec{r}\right) = \nabla\phi.d\vec{r} = \vec{F}.d\vec{r}$$

$$= \left[2x(y^2+z^3)\hat{i} + 2x^2y\hat{j} + 3x^2z^2\hat{k}\right](\hat{i}dx + \hat{j}dy + \hat{k}dz)$$

$$= 2x(y^2+z^3)\,dx + 2x^2y\,dy + 3x^2z^2\,dz$$

$$\phi = \int\left[2x(y^2+z^3)dx + 2x^2ydy + 3x^2z^2dz\right] + C$$

$$\int(2xy^2dx + 2x^2ydy) + (2xz^3dx + 3x^2z^2dz) + C = x^2y^2 + x^2z^3 + C$$

Hence, the scalar potential is $x^2y^2 + x^2z^3 + C$

Now, for conservative field

$$\text{Work done} = \int_{(-1,2,1)}^{(2,3,4)} \vec{F}.d\vec{r} = \int_{(-1,2,1)}^{(2,3,4)} d\phi = [\phi]_{(-1,2,1)}^{(2,3,4)} = \left[x^2y^2 + x^2z^3 + c\right]_{(-1,2,1)}^{(2,3,4)}$$

$$= (36 + 256) - (2 - 1) = 291 \qquad \textbf{Ans.}$$

**Example 73.** *A vector field is given by $\vec{F} = (\sin y)\,\hat{i} + x\,(1 + \cos y)\,\hat{j}$. Evaluate the line integral over a circular path $x^2 + y^2 = a^2$, $z = 0$.* (Nagpur University, Winter 2001)

**Solution.** We have,

$$\text{Work done} = \int_C \vec{F} \cdot \vec{dr}$$

$$= \int_C [(\sin y)\,\hat{i} + x\,(1 + \cos y)\,\hat{j}] \cdot [dx\hat{i} + dy\hat{j}] \quad (\because z = 0 \text{ hence } dz = 0)$$

$$\Rightarrow \int_C \vec{F} \cdot \vec{dr} = \int_C \sin y\, dx + x\,(1 + \cos y)\, dy = \int_C (\sin y\, dx + x \cos y\, dy + x\, dy)$$

$$= \int_C d\,(x \sin y) + \int_C x\, dy$$

(where $d$ is differential operator).

The parametric equations of given path

$x^2 + y^2 = a^2$ are $x = a\cos\theta$, $y = a\sin\theta$,

Where $\theta$ varies form 0 to $2\pi$

$$\therefore \int_C \vec{F} \cdot \vec{dr} = \int_0^{2\pi} d\,[\,a\cos\theta \sin(a\sin\theta)\,] + \int_0^{2\pi} a\cos\theta \,.\, a\cos\theta\, d\theta$$

$$= \int_0^{2\pi} d\,[a\cos\theta \sin(a\sin\theta)\,] + \int_0^{2\pi} a^2 \cos^2\theta \,.\, d\theta$$

$$= [a\cos\theta \sin(a\sin\theta)\,]_0^{2\pi} + \int_0^{2\pi} a^2 \cos^2\theta\, d\theta$$

$$= 0 + a^2 \int_0^{2\pi} \left(\frac{1 + \cos 2\theta}{2}\right) d\theta = \frac{a^2}{2}\left[\theta + \frac{\sin 2\theta}{2}\right]_0^{2\pi}$$

$$= \frac{a^2}{2} \,.\, 2\pi = \pi a^2$$

**Ans.**

**Example 74.** *Determine whether the line integral* $\int (2xyz^2)\, dx + (x^2z^2 + z\cos yz)\, dy + (2x^2yz + y\cos yz)\, dz$ *is independent of the path of integration ? If so, then evaluate it from (1, 0, 1) to* $\left(0, \frac{\pi}{2}, 1\right)$.

**Solution.** $\int_c (2xyz^2)\, dx + (x^2z^2 + z\cos yz)\, dy + (2x^2yz + y\cos yz)\, dz$

$$= \int_c [(2xyz^2\hat{i}) + (x^2z^2 + z\cos yz)\,\hat{j} + (2x^2yz + y\cos yz)\,\hat{k}] \cdot (\hat{i}dx + \hat{j}dy + \hat{k}dz)$$

$$= \int_c \vec{F} \cdot \vec{dr}$$

This integral is independent of path of integration if

$$\vec{F} = \nabla\phi \Rightarrow \nabla \times \vec{F} = 0$$

$$\nabla \times \overline{F} = \begin{vmatrix} \hat{i} & \hat{j} & \hat{k} \\ \dfrac{\partial}{\partial x} & \dfrac{\partial}{\partial y} & \dfrac{\partial}{\partial z} \\ 2xyz^2 & x^2z^2 + z\cos yz & 2x^2yz + y\cos yz \end{vmatrix}$$

$= (2x^2z + \cos yz - yz\sin yz - 2x^2z - \cos yz + yz\sin yz) = \hat{i} - (4xyz - 4x\,yz)\,\hat{j} + (2xz^2 - 2xz^2)\,\hat{k}$
$= 0$

Hence, the line integral is independent of path.

$$d\phi = \frac{\partial\phi}{\partial x} dx + \frac{\partial\phi}{\partial y} dy + \frac{\partial\phi}{\partial z} dz \qquad \text{(Total differentiation)}$$

$$= \left(\hat{i}\frac{\partial \phi}{\partial x} + \hat{j}\frac{\partial \phi}{\partial y} + \hat{k}\frac{\partial \phi}{\partial z}\right)\cdot(\hat{i}dx + \hat{j}dy + \hat{k}dz) \quad = \nabla\phi\cdot dr = \vec{F}\cdot\vec{d}\,r$$

$$= [(2xyz^2)\,\hat{i} + (x^2z^2 + z\cos y\,z)\,\hat{j} + (2x^2yz + y\cos yz)\,\hat{k}].\,(\hat{i}dx + \hat{j}dy + \hat{k}dz)$$

$$= 2xyz^2\,dx + (x^2z^2 + z\cos y\,z)\,dy + (2x^2yz + y\cos yz)\,dz$$

$$= [(2x\,dx)\,yz^2 + x^2\,(dy)\,z^2 + x^2y\,(2z\,dz)] + [(\cos yz\,dy)\,z + (\cos yz\,dz)\,y]$$

$$= d\,(x^2yz^2) + d\,(\sin yz)$$

$$\phi = \int d\,(x^2yz^2) + \int d\,(\sin yz) = x^2yz^2 + \sin yz$$

$$[\phi]_A^B = \phi\,(B) - \phi\,(A)$$

$$= [x^2yz^2 + \sin yz]_{(0,\frac{\pi}{2},1)} - [x^2yz^2 + \sin yz]_{(1,0,1)} = \left[0 + \sin\left(\frac{\pi}{2}\times 1\right)\right] - [0+0]$$

$$= 1$$ **Ans.**

**Example 75.** *Evaluate* $\iint_S \vec{A}.\hat{n}\,d\,S$, *where* $\vec{A} = 18z\hat{i} - 12\hat{j} + 3y\,\hat{k}$ *and S is the part of the plane* $2x + 3y + 6z = 12$ *included in the first octant.* *(Uttarakhand, I semester, Dec. 2006)*

**Solution.** Here, $\vec{A} = 18z\hat{i} - 12\hat{j} + 3y\hat{k}$

Given surface $f\,(x, y, z) = 2x + 3y + 6z - 12$

$$\text{Normal vector} = \nabla f = \left(\hat{i}\frac{\partial}{\partial x} + \hat{j}\frac{\partial}{\partial y} + \hat{k}\frac{\partial}{\partial z}\right)(2x + 3y + 6z - 12) = 2\hat{i} + 3\hat{j} + 6\hat{k}$$

$\hat{n}$ = unit normal vector at any point $(x, y, z)$ of $2x + 3y + 6z = 12$

$$= \frac{2\hat{i} + 3\hat{j} + 6\hat{k}}{\sqrt{4+9+36}} = \frac{1}{7}(2\hat{i} + 3\hat{j} + 6\hat{k})$$

$$dS = \frac{dx\,dy}{\hat{n}.\hat{k}} = \frac{dx\,dy}{\frac{1}{7}(2\hat{i} + 3\hat{j} + 6\hat{k}).\hat{k}} = \frac{dxdy}{\frac{6}{7}} = \frac{7}{6}\,dx\,dy$$

$$\text{Now, } \iint \vec{A}.\hat{n}\,dS = \iint (18z\hat{i} - 12\hat{j} + 3y\hat{k}).\frac{1}{7}(2\hat{i} + 3\hat{j} + 6\hat{k})\frac{7}{6}\,dx\,dy$$

$$= \iint (36z - 36 + 18y)\frac{dx\,dy}{6} \quad = \iint (6z - 6 + 3y)\,dx\,dy.$$

Putting the value of $6z = 12 - 2x - 3y$, we get

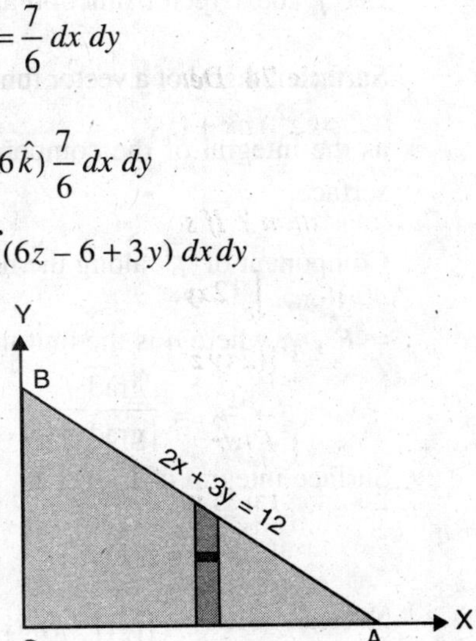


$$= \int_0^6 \int_0^{\frac{1}{3}(12-2x)} (12 - 2x - 3y - 6 + 3y)\,dx\,dy$$

$$= \int_0^6 \int_0^{\frac{1}{3}(12-2x)} (6 - 2x)\,dx\,dy$$

$$= \int_0^6 (6 - 2x)\,dx \int_0^{\frac{1}{3}(12-2x)} dy$$

$$= \int_0^6 (6 - 2x)\,dx\,(y)_0^{\frac{1}{3}(12-2x)}$$

$$= \int_0^6 (6 - 2x)\,\frac{1}{3}(12 - 2x)\,dx = \frac{1}{3}\int_0^6 (4x^2 - 36x + 72)\,dx$$

$$= \frac{1}{3}\left[\frac{4x^3}{3} - 18x^2 + 72x\right]_0^6 = \frac{1}{3}[4\times 36\times 2 - 18\times 36 + 72\times 6] = \frac{72}{3}[4 - 9 + 6] = 24$$ **Ans.**

## EXERCISE 5.10

1. Find the work done by a force $y\hat{i} + x\hat{j}$ which displaces a particle from origin to a point $(\hat{i} + \hat{j})$. **Ans.** 1
2. Find the work done when a force $\overline{F} = (x^2 - y^2 + x)\,\hat{i} - (2xy + y)\,\hat{j}$ moves a particle from origin to (1, 1) along a parabola $y^2 = x$. **Ans.** $\frac{2}{3}$
3. Show that $\vec{V} = (2xy + z^3)\,\hat{i} + x^2\,\hat{j} + 3xz^2\,\hat{k}$ is a conservative field. Find its scalar potential $\phi$ such that $\vec{V}$ = grad $\phi$. Find the work done by the force $\vec{V}$ in moving a particle from (1, – 2, 1) to (3, 1, 4). **Ans.** $x^2y + xz^3$, 202
4. Show that the line integral $\int_c (2xy + 3)\,dx + (x^2 - 4z)\,dy - 4y\,dz$ where $c$ is any path joining (0, 0, 0) to (1, – 1, 3) does not depend on the path $c$ and evaluate the line integral. **Ans.** 14
5. Find the work done in moving a particle once round the ellipse $\frac{x^2}{25} + \frac{y^2}{16} = 1$, $z = 0$, under the field of force given by $F = (2x - y + z)\,\hat{i} + (x + y - z^2)\,\hat{j} + (3x - 2y + 4z)\,\hat{k}$. Is the field of force conservative? (*A.M.I.E.T.E., Winter 2000*) **Ans.** $40\pi$
6. If $\vec{\nabla}\phi = (y^2 - 2xyz^3)\,\hat{i} + (3 + 2xy - x^2z^3)\,\hat{j} + (z^3 - 3x^2yz^2)\,\hat{k}$, find $\phi$. **Ans.** $3y + \frac{z^4}{4} + xy^2 - x^2yz^3$
7. $\int_C \vec{R}\,.\,d\vec{R}$ is independent of the path joining any two point if it is. (*A.M.I.E.T.E., June 2010*)
(*i*) irrotational field (*ii*) solenoidal field (*iii*) rotational field (*iv*) vector field. **Ans.** (*i*)

### 5.34 SURFACE INTEGRAL

A surface $r = f(u, v)$ is called smooth if $f\ (u, v)$ posses continous first order partial derivative.

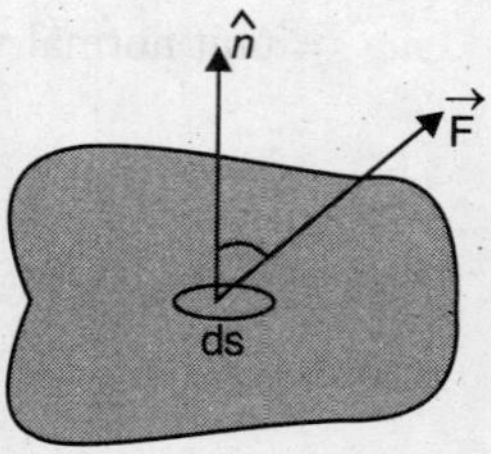


Let $\vec{F}$ be a vector function and $S$ be the given surface.

Surface integral of a vector function $\vec{F}$ over the surface $S$ is defined as the integral of the components of $\vec{F}$ along the normal to the surface.

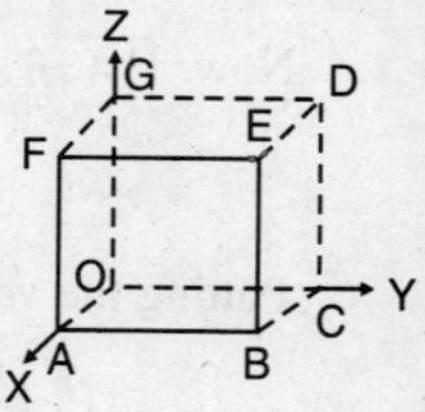


Component of $\vec{F}$ along the normal

$= \vec{F}\,.\,\hat{n}$, where $n$ is the unit normal vector to an element $ds$ and

$$\hat{n} = \frac{\text{grad } f}{|\text{grad } f|} \qquad ds = \frac{dx\,dy}{(\hat{n}\cdot\hat{k})}$$

Surface integral of F over S

$$= \Sigma\,\vec{F}\cdot\hat{n} \qquad = \iint_S (\vec{F}\cdot\hat{n})\,ds$$

**Note.** (1) Flux = $\iint_S (\vec{F}\cdot\hat{n})\,ds$ where, $\overline{F}$ represents the velocity of a liquid.

If $\iint_S (\vec{F}\cdot\hat{n})\,ds = 0$, then $\vec{F}$ is said to be a *solenoidal* vector point function.

**Example 76.** *Evaluate* $\iint_S (yz\hat{i} + zx\hat{j} + xy\hat{k})\cdot\vec{ds}$ *where S is the surface of the sphere* $x^2 + y^2 + z^2 = a^2$ *in the first octant.* (*U.P., I Semester, Dec. 2004*)

**Solution.** Here, $\phi = x^2 + y^2 + z^2 - a^2$

$$\text{Vector normal to the surface} = \nabla\phi = \hat{i}\frac{\partial\phi}{\partial x} + \hat{j}\frac{\partial\phi}{\partial y} + \hat{k}\frac{\partial\phi}{\partial z}$$

$$= \left(\hat{i}\frac{\partial}{\partial x} + \hat{j}\frac{\partial}{\partial y} + \hat{k}\frac{\partial}{\partial z}\right)(x^2 + y^2 + z^2 - a^2) = 2x\hat{i} + 2y\hat{j} + 2z\hat{k}$$

$$\hat{n} = \frac{\nabla\phi}{|\nabla\phi|} = \frac{2x\hat{i} + 2y\hat{j} + 2z\hat{k}}{\sqrt{4x^2 + 4y^2 + 4z^2}} = \frac{x\hat{i} + y\hat{j} + z\hat{k}}{\sqrt{x^2 + y^2 + z^2}}$$

$$= \frac{x\hat{i} + y\hat{j} + z\hat{k}}{a} \qquad [\because x^2 + y^2 + z^2 = a^2]$$

Here, $\vec{F} = yz\hat{i} + zx\hat{j} + xy\hat{k}$

$$\vec{F}\cdot\hat{n} = (yz\hat{i} + zx\hat{j} + xy\hat{k})\cdot\left(\frac{x\hat{i} + y\hat{j} + z\hat{k}}{a}\right) = \frac{3xyz}{a}$$

Now, $$\iint_S F\cdot\hat{n}\, ds = \iint_S (\vec{F}\cdot\hat{n})\frac{dx\,dy}{|\hat{k}\cdot\hat{n}|} = \int_0^a \int_0^{\sqrt{a^2 - x^2}} \frac{3xyz\, dx\, dy}{a\left(\frac{z}{a}\right)}$$

$$= 3\int_0^a \int_0^{\sqrt{a^2 - x^2}} xy\, dy\, dx = 3\int_0^a x\left(\frac{y^2}{2}\right)_0^{\sqrt{a^2 - x^2}} dx$$

$$= \frac{3}{2}\int_0^a x(a^2 - x^2)\, dx = \frac{3}{2}\left(\frac{a^2x^2}{2} - \frac{x^4}{4}\right)_0^a = \frac{3}{2}\left(\frac{a^4}{2} - \frac{a^4}{4}\right) = \frac{3a^4}{8}. \quad \textbf{Ans.}$$

**Example 77.** *Show that* $\iint_S \vec{F}\cdot\hat{n}\, ds = \frac{3}{2}$, *where* $\vec{F} = 4xz\hat{i} - y^2\hat{j} + yz\hat{k}$ *and S is the surface of the cube bounded by the planes,* $x = 0, x = 1, y = 0, y = 1, z = 0, z = 1.$

| S.No. | Surface | Outward normal | ds | |
|---|---|---|---|---|
| 1 | OABC | $-k$ | $dx\,dy$ | $z = 0$ |
| 2 | DEFG | $k$ | $dx\,dy$ | $z = 1$ |
| 3 | OAGF | $-j$ | $dx\,dz$ | $y = 0$ |
| 4 | BCED | $j$ | $dx\,dz$ | $y = 1$ |
| 5 | ABDG | $i$ | $dy\,dz$ | $x = 1$ |
| 6 | OCEF | $-i$ | $dy\,dz$ | $x = 0$ |

**Solution.** $$\iint_S \vec{F}\cdot\hat{n}\, ds = \iint_{OABC} \vec{F}\cdot\hat{n}\, ds$$

$$+ \iint_{DEFG} \vec{F}\cdot\hat{n}\, ds + \iint_{OAGF} \vec{F}\cdot\hat{n}\, ds$$

$$+ \iint_{BCED} \vec{F}\cdot\hat{n}\, ds + \iint_{ABDG} \vec{F}\cdot\hat{n}\, ds$$

$$+ \iint_{OCEF} \vec{F}\cdot\hat{n}\, ds \qquad \ldots(1)$$

Now, $$\iint_{OABC} \vec{F}\cdot n\, ds = \iint_{OABC} (4xz\hat{i} - y^2\hat{j} + yz\hat{k})(-k)\, dx\, dy = \int_0^1\int_0^1 -yz\, dx\, dy = 0 \text{ (as } z = 0)$$

$$\iint_{DEFG} (4xz\hat{i} - y^2\hat{j} + yz\hat{k})\cdot\hat{k}\, dx\, dy$$

$$= \iint_{DEFG} yz\, dx\, dy = \int_0^1\int_0^1 y(1)\, dx\, dy$$

$$= \int_0^1 dx\left[\frac{y^2}{2}\right]_0^1 = [x]_0^1\frac{1}{2} = \frac{1}{2}$$

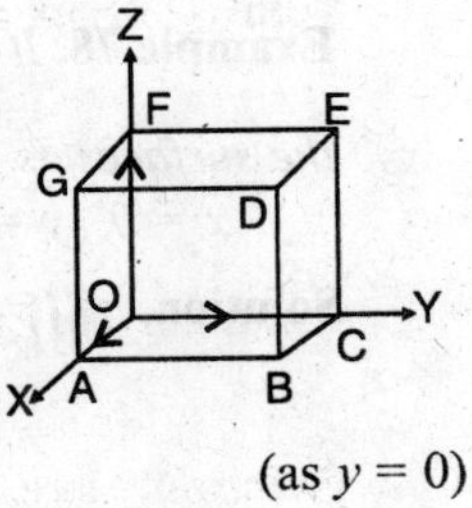


$$\iint_{OAGF} (4xz\hat{i} - y^2\hat{j} + yz\hat{k})\cdot(-j)\, dx\, dz = \iint_{OAGF} y^2\, dx\, dz = 0 \qquad (\text{as } y = 0)$$

$$\iint_{BCED}(4xz\,\hat{i} - y^2\,\hat{j} + yz\,\hat{k})\cdot\hat{j}\,dx\,dz = \iint_{BCED}(-y^2)\,dx\,dz$$

$$= -\int_0^1 dx\int_0^1 dz = -(x)_0^1\,(z)_0^1 = -1 \qquad \text{(as } y = 1\text{)}$$

$$\iint_{ABDG}(4xz\hat{i} - y^2\hat{j} + yz\hat{k})\cdot\hat{i}\,dy\,dz = \iint 4xz\,dy\,dz = \int_0^1\int_0^1 4\,(1)\,z\,dy\,dz$$

$$= 4\,(y)_0^1\left(\frac{z^2}{2}\right)_0^1 = 4\,(1)\left(\frac{1}{2}\right) = 2$$

$$\iint_{OCEF}(4xz\,\hat{i} - y^2\,\hat{j} + yz\,\hat{k})\,(-\hat{i})\,dy\,dz = \int_0^1\int_0^1 -4xz\,dy\,dz = 0 \qquad \text{(as } x = 0\text{)}$$

On putting these values in (1), we get

$$\iint_S \bar{F}\cdot\hat{n}\,ds = 0 + \frac{1}{2} + 0 - 1 + 2 + 0 = \frac{3}{2} \qquad \textbf{Proved.}$$

## EXERCISE 5.11

1. Evaluate $\iint_S \vec{A}.\hat{n}\,ds$, where $\vec{A} = (x + y^2)\,\hat{i} - 2x\hat{j} + 2yz\hat{k}$ and $S$ is the surface of the plane $2x + y + 2z = 6$ in the first octant. **Ans.** 81

2. Evaluate $\iint_S \vec{A}.\hat{n}\,ds$, where $\vec{A} = z\hat{i} + x\hat{j} - 3y^2z\hat{k}$ and $S$ is the surface of the cylinder $x^2 + y^2 = 16$ included in the first octant between $z = 0$ and $z = 5$. **Ans.** 90

3. If $\vec{r} = t\hat{i} - t^2\hat{j} + (t-1)\,\hat{k}$ and $\vec{S} = 2t^2\hat{i} + 6t\hat{k}$, evaluate $\int_0^2 \vec{r}\cdot\vec{S}\,dt$. **Ans.** 12

4. Evaluate $\iint_S \vec{F}\cdot\hat{n}\,dS$, where, $\vec{F} = 18z\,\hat{i} - 12\,\hat{j} + 3y\hat{k}$ and $S$ is the surface of the plane $2x + 3y + 6z = 12$ in the first octant. **Ans.** 24

5. Evaluate $\iint_S \vec{F}\cdot\hat{n}\,ds$, where, $F = 2yx\hat{i} - yz\hat{j} + x^2\hat{k}$ over the surface $S$ of the cube bounded by the coordinate planes and planes $x = a$, $y = a$ and $z = a$. **Ans.** $\frac{1}{2}a^4$

6. If $\vec{F} = 2y\hat{i} - 3\hat{j} + x^2\hat{k}$ and $S$ is the surface of the parabolic cylinder $y^2 = 8x$ in the first octant bounded by the planes $y = 4$, and $z = 6$, then evaluate $\iint_S \vec{F}\cdot\hat{n}\,dS$. **Ans.** 132

## 5.35 VOLUME INTEGRAL

Let $\vec{F}$ be a vector point function and volume $V$ enclosed by a closed surface.

The volume integral $= \iiint_V \vec{F}\,dv$

**Example 78.** *If $\vec{F} = 2z\,\hat{i} - x\,\hat{j} + y\,\hat{k}$, evaluate $\iiint_V \vec{F}\,dv$ where, v is the region bounded by the surfaces*

*$x = 0,\ y = 0,\ x = 2,\ y = 4,\ z = x^2,\ z = 2.$*

**Solution.** $\iiint_V \vec{F}\,dv = \iiint(2z\,\hat{i} - x\hat{j} + y\hat{k})\,dx\,dy\,dz$

$$= \int_0^2 dx\int_0^4 dy\int_{x^2}^2 (2z\hat{i} - x\hat{j} + y\hat{k})\,dz = \int_0^2 dx\int_0^4 dy\,[z^2\,\hat{i} - xz\hat{j} + yz\hat{k}]_{x^2}^2$$

$$= \int_0^2 dx\int_0^4 dy\,[4\hat{i} - 2x\hat{j} + 2y\hat{k} - x^4\hat{i} + x^3\,\hat{j} - x^2y\hat{k}]$$

$$= \int_0^2 dx \left[ 4y\hat{i} - 2xy\hat{j} + y^2\hat{k} - x^4 y\hat{i} + x^3 y\hat{j} - \frac{x^2y^2}{2}\hat{k} \right]_0^4$$

$$= \int_0^2 (16\hat{i} - 8x\hat{j} + 16\hat{k} - 4x^4\hat{i} + 4x^3\hat{j} - 8x^2\hat{k})\, dx$$

$$= \left[ 16x\hat{i} - 4x^2\hat{j} + 16x\hat{k} - \frac{4x^5}{5}\hat{i} + x^4\hat{j} - \frac{8x^3}{3}\hat{k} \right]_0^2$$

$$= 32\hat{i} - 16\hat{j} + 32\hat{k} - \frac{128}{5}\hat{i} + 16\hat{j} - \frac{64}{3}\hat{k} = \frac{32\hat{i}}{5} + \frac{32\hat{k}}{3} = \frac{32}{15}(3\hat{i} + 5\hat{k})$$ **Ans.**

## EXERCISE 5.12

1. If $\vec{F} = (2x^2 - 3z)\hat{i} - 2xy\hat{j} - 4x\hat{k}$, then evaluate $\iiint_V \nabla \vec{F}\, dV$, where $V$ is bounded by the plane $x = 0$, $y = 0$, $z = 0$ and $2x + 2y + z = 4$. **Ans.** $\frac{8}{3}$

2. Evaluate $\iiint_V \phi\, dV$, where $\phi = 45\, x^2 y$ and V is the closed region bounded by the planes $4x + 2y + z = 8$, $x = 0$, $y = 0$, $z = 0$ **Ans.** 128

3. If $\vec{F} = (2x^2 - 3z)\,\hat{i} - 2xy\,\hat{j} - 4x\hat{k}$, then evaluate $\iiint_V \nabla \times \vec{F}\, dV$, where V is the closed region bounded by the planes $x = 0$, $y = 0$, $z = 0$ and $2x + 2y + z = 4$. **Ans.** $\frac{8}{3}(\hat{j} - \hat{k})$

4. Evaluate $\iiint_V (2x + y)\, dV$, where $V$ is closed region bounded by the cylinder $z = 4 - x^2$ and the planes $x = 0$, $y = 0$, $y = 2$ and $z = 0$. **Ans.** $\frac{80}{3}$

5. If $\vec{F} = 2xz\hat{i} - x\hat{j} + y^2\hat{k}$, evaluate $\iiint \vec{F}\, dV$ over the region bounded by the surfaces $x = 0$, $y = 0$, $y = 6$ and $z = x^2$, $z = 4$. **Ans.** $(16\hat{i} - 3\hat{j} + 48\hat{k})$

## 5.36 GREEN'S THEOREM (For a plane)

**Statement.** *If $\phi(x, y)$, $\psi(x, y)$, $\frac{\partial \phi}{\partial y}$ and $\frac{\partial \psi}{\partial x}$ be continuous functions over a region R bounded by simple closed curve C in x – y plane, then*

$$\oint_C (\phi\, dx + \psi\, dy) = \iint_R \left( \frac{\partial \psi}{\partial x} - \frac{\partial \phi}{\partial y} \right) dx\, dy$$ *(AMIETE, June 2010, U.P., I Semester, Dec. 2007)*

**Proof.** Let the curve $C$ be divided into two curves $C_1$ $(ABC)$ and $C_1$ $(CDA)$.
Let the equation of the curve $C_1$ $(ABC)$ be $y = y_1(x)$ and equation of the curve $C_2$ $(CDA)$ be $y = y_2(x)$.
Let us see the value of

$$\iint_R \frac{\partial \phi}{\partial y}\, dx\, dy = \int_{x=a}^{x=c} \left[ \int_{y=y_1(x)}^{y=y_2(x)} \frac{\partial \phi}{\partial y}\, dy \right] dx = \int_a^c [\phi(x, y)]_{y=y_1(x)}^{y=y_2(x)}\, dx$$

$$= \int_a^c [\phi(x, y_2) - \phi(x, y_1)]\, dx = -\int_c^a \phi(x, y_2)\, dx - \int_a^c \phi(x, y_1)\, dx$$

$$= -\left[ \int_c^a \phi(x, y_2)\, dx + \int_a^c \phi(x, y_1)\, dx \right]$$

$$= -\left[ \int_{C2} \phi(x, y)\, dx + \int_{C1} \phi(x, y)\, dx \right] = -\oint_c \phi(x, y)\, dx$$

Thus, $$\oint_c \phi\, dx = -\iint_R \frac{\partial \phi}{\partial y}\, dx\, dy \qquad ...(1)$$

Similarly, it can be shown that

$$\oint_c \psi\, dy = \iint \frac{\partial \psi}{\partial x}\, dx\, dy \qquad ...(2)$$

On adding (1) and (2), we get

$$\oint (\phi\, dx + \psi\, dy) = \iint_R \left(\frac{\partial \psi}{\partial x} - \frac{\partial \phi}{\partial y}\right) dx\, dy$$ **Proved.**

**Note.** Green's Theorem in vector form

$$\int_c \vec{F}\cdot d\vec{r} = \iint_R (\nabla \times \vec{F})\cdot \hat{k}\, dR$$

where, $\vec{F} = \phi\,\hat{i} + \psi\,\hat{j}$, $\vec{r} = x\hat{i} + y\hat{j}$, $\hat{k}$ is a unit vector along $z$-axis and $dR = dx\, dy$.

**Example 79.** *A vector field $\vec{F}$ is given by $\vec{F} = \sin y\hat{i} + x(1+\cos y)\hat{j}$.*

*Evaluate the line integral $\int_C \vec{F}\cdot d\vec{r}$ where C is the circular path given by $x^2 + y^2 = a^2$.*

**Solution.** $\vec{F} = \sin y\hat{i} + x(1+\cos y)\hat{j}$

$$\int_C \vec{F}\cdot d\vec{r} = \int_C [\sin y\hat{i} + x(1+\cos y)\hat{j}]\cdot(\hat{i}dx + \hat{j}dy) = \int_C \sin y\, dx + x(1+\cos y)\, dy$$

On applying Green's Theorem, we have

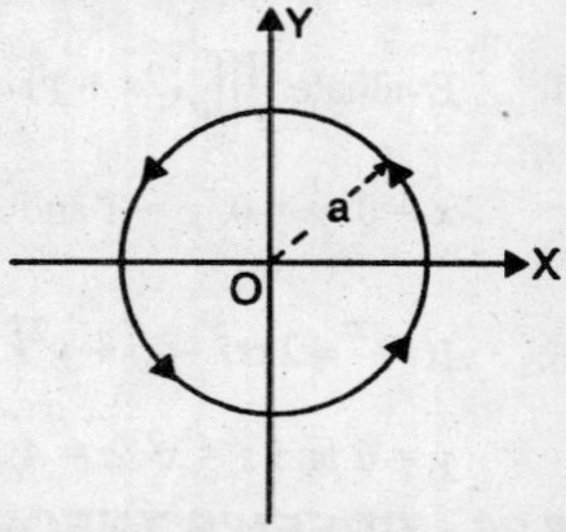


$$\oint_c (\phi\, dx + \psi\, dy) = \iint_s \left(\frac{\partial \psi}{\partial x} - \frac{\partial \phi}{\partial y}\right) dx\, dy$$

$$= \iint_s [(1+\cos y) - \cos y]\, dx\, dy$$

*where s is the circular plane surface of radius a.*

$$= \iint_s dx\, dy = \text{Area of circle} = \pi a^2.$$ **Ans.**

**Example 80.** *Using Green's Theorem, evaluate $\int_c (x^2 y dx + x^2 dy)$, where c is the boundary described counter clockwise of the triangle with vertices (0, 0), (1, 0), (1, 1).*

*(U.P., I Semester, Winter 2003)*

**Solution.** By Green's Theorem, we have

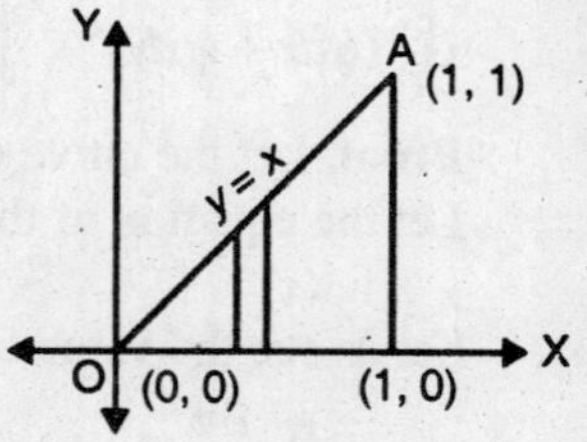


$$\int_c (\phi\, dx + \psi\, dy) = \iint_R \left(\frac{\partial \psi}{\partial x} - \frac{\partial \phi}{\partial y}\right) dx\, dy$$

$$\int_c (x^2 y\, dx + x^2\, dy) = \iint_R (2x - x^2)\, dx\, dy$$

$$= \int_0^1 (2x - x^2)\, dx \int_0^x dy = \int_0^1 (2x - x^2)\, dx\, [y]_0^x$$

$$= \int_0^1 (2x - x^2)(x)\, dx = \int_0^1 (2x^2 - x^3)\, dx = \left(\frac{2x^3}{3} - \frac{x^4}{4}\right)_0^1$$

$$= \left(\frac{2}{3} - \frac{1}{4}\right) = \frac{5}{12}$$ **Ans.**

**Example 81.** *State and verify Green's Theorem in the plane for $\oint (3x^2 - 8y^2)\, dx + (4y - 6xy)\, dy$ where C is the boundary of the region bounded by $x \geq 0$, $y \leq 0$ and $2x - 3y = 6$.*

*(Uttarakhand, I Semester, Dec. 2006)*

**Solution. Statement:** See Article 24.4 on page 576.

Here the closed curve $C$ consists of straight lines $OB$, $BA$ and $AO$, where coordinates of $A$ and $B$ are (3, 0) and (0, – 2) respectively. Let $R$ be the region bounded by $C$.

Then by Green's Theorem in plane, we have

$$\oint[(3x^2 - 8y^2)\,dx + (4y - 6xy)\,dy]$$

$$= \iint_R\left[\frac{\partial}{\partial x}(4y - 6xy) - \frac{\partial}{\partial y}(3x^2 - 8y^2)\right]dx\,dy \qquad ...(1)$$

$$= \iint_R(-6y + 16y)\,dx\,dy = \iint_R 10y\,dx\,dy$$

$$= 10\int_0^3 dx\int_{\frac{1}{3}(2x-6)}^0 y\,dy = 10\int_0^3 dx\left[\frac{y^2}{2}\right]_{\frac{1}{3}(2x-6)}^0 = -\frac{5}{9}\int_0^3 dx\,(2x-6)^2$$

$$= -\frac{5}{9}\left[\frac{(2x-6)^3}{3\times 2}\right]_0^3 = -\frac{5}{54}(0+6)^3 = -\frac{5}{54}(216) = -20 \qquad ...(2)$$

Now we evaluate L.H.S. of (1) along $OB$, $BA$ and $AO$.

Along $OB$, $x = 0$, $dx = 0$ and $y$ varies form 0 to –2.

Along $BA$, $x = \frac{1}{2}(6 + 3y)$, $dx = \frac{3}{2}\,dy$ and $y$ varies from –2 to 0.

and along $AO$, $y = 0$, $dy = 0$ and $x$ varies from 3 to 0.

L.H.S. of (1) $= \oint[(3x^2 - 8y^2)\,dx + (4y - 6xy)\,dy]$

$$= \int_{OB}[(3x^2 - 8y^2)\,dx + (4y - 6xy)\,dy] + \int_{BA}[(3x^2 - 8y^2)\,dx + (4x - 6xy)\,dy]$$
$$+ \int_{AO}[(3x^2 - 8y^2)\,dx + (4y - 6xy)\,dy]$$

$$= \int_0^{-2} 4y\,dy + \int_{-2}^0\left[\frac{3}{4}(6+3y)^2 - 8y^2\right]\left(\frac{3}{2}dy\right) + [4y - 3(6+3y)\,y]\,dy + \int_3^0 3x^2\,dx$$

$$= [2y^2]_0^{-2} + \int_{-2}^0\left[\frac{9}{8}(6+3y)^2 - 12y^2 + 4y - 18y - 9y^2\right]dy + (x^3)_3^0$$

$$= 2[4] + \int_{-2}^0\left[\frac{9}{8}(6+3y)^2 - 21y^2 - 14y\right]dy + (0 - 27)$$

$$= 8 + \left[\frac{9}{8}\,\frac{(6+3y)^3}{3\times 3} - 7y^3 - 7y^2\right]_{-2}^0 - 27 = -19 + \left[\frac{216}{8} + 7(-2)^3 + 7(-2)^2\right]$$

$$= -19 + 27 - 56 + 28 = -20 \qquad ...(3)$$

With the help of (2) and (3), we find that (1) is true and so Green's Theorem is verified.

**Example 82.** *Apply Green's Theorem to evaluate* $\int_C[(2x^2 - y^2)\,dx + (x^2 + y^2)\,dy]$, *where C is the boundary of the area enclosed by the x-axis and the upper half of circle* $x^2 + y^2 = a^2$.

*(M.D.U. Dec. 2009, U.P., I Sem., Dec. 2004)*

**Solution.** $\int_C[(2x^2 - y^2)\,dx + (x^2 + y^2)\,dy]$

By Green's Theorem, we've $\int_C(\phi\,dx + \psi\,dy) = \iint_S\left(\frac{\partial\psi}{\partial x} - \frac{\partial\phi}{\partial y}\right)dx\,dy$

$$= \int_{-a}^{a}\int_0^{\sqrt{a^2-x^2}}\left[\frac{\partial}{\partial x}(x^2+y^2)-\frac{\partial}{\partial y}(2x^2-y^2)\right]dx\,dy$$

$$= \int_{-a}^{a}\int_0^{\sqrt{a^2-x^2}}(2x+2y)\,dx\,dy = 2\int_{-a}^{a}dx\int_0^{\sqrt{a^2-x^2}}(x+y)\,dy$$

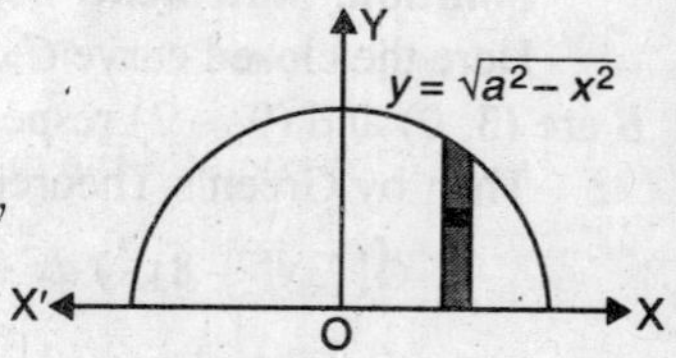


$$=2\int_{-a}^{a}dx\left(xy+\frac{y^2}{2}\right)_0^{\sqrt{a^2-x^2}} = 2\int_{-a}^{a}\left(x\sqrt{a^2-x^2}+\frac{a^2-x^2}{2}\right)dx$$

$$= 2\int_{-a}^{a}x\sqrt{a^2-x^2}\,dx + \int_{-a}^{a}(a^2-x^2)\,dx \qquad \left[\begin{array}{ll}\int_{-a}^{a}f(x)\,dx = 2\int_0^a f(x)\,dx, & f \text{ is even}\\ = 0, & f \text{ is odd}\end{array}\right]$$

$$= 0+2\int_0^a(a^2-x^2)\,dx = 2\left(a^2x-\frac{x^3}{3}\right)_0^a = 2\left(a^3-\frac{a^3}{3}\right) = \frac{4a^3}{3}$$ **Ans.**

**Example 83.** *Evaluate* $\oint_C -\frac{y}{x^2+y^2}dx+\frac{x}{x^2+y^2}dy$, *where* $C = C_1 \cup C_2$ *with* $C_1 : x^2+y^2 = 1$ *and* $C_2 : x = \pm 2,\ y = \pm 2$. *(Gujarat, I Semester, Jan 2009)*

**Solution.** $\oint_C -\frac{y}{x^2+y^2}\,dx+\frac{x}{x^2+y^2}\,dy$

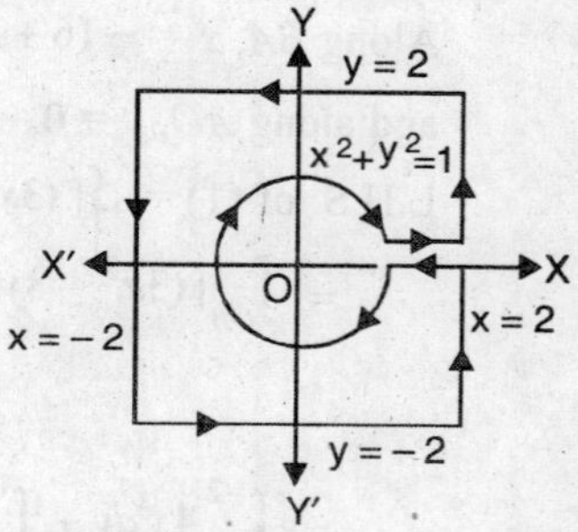


$$= \iint\left(\frac{\partial}{\partial x}\frac{x}{x^2+y^2}+\frac{\partial}{\partial y}\frac{y}{x^2+y^2}\right)dx\,dy$$

$$= \iint\left[\frac{(x^2+y^2)1-2x(x)}{(x^2+y^2)^2}+\frac{(x^2+y^2)1-2y(y)}{(x^2+y^2)^2}\right]dx\,dy$$

$$= \iint\left[\frac{x^2+y^2-2x^2}{(x^2+y^2)^2}+\frac{x^2+y^2-2y^2}{(x^2+y^2)^2}\right]dx\,dy$$

$$= \iint\left[\frac{y^2-x^2}{(x^2+y^2)^2}+\frac{x^2-y^2}{(x^2+y^2)^2}\right]dx\,dy = \iint\frac{0}{(x^2+y^2)^2}\,dx\,dy = 0$$ **Ans.**

## 5.37 AREA OF THE PLANE REGION BY GREEN'S THEOREM

**Proof.** We know that

$$\int_C M\,dx+N\,dy = \iint_A\left(\frac{\partial N}{\partial x}-\frac{\partial M}{\partial y}\right)dx\,dy \qquad ...(1)$$

On putting $N = x\ \left(\frac{\partial N}{\partial x}=1\right)$ and $M = -y\ \left(\frac{\partial M}{\partial y}=1\right)$ in (1), we get

$$\int_C -y\,dx+x\,dy = \iint_A[1-(-1)]\,dx\,dy = 2\iint dx\,dy = 2A$$

$$\text{Area} = \frac{1}{2}\int_C(x\,dy-y\,dx)$$

**Example 84.** *Using Green's theorem, find the area of the region in the first quadrant bounded by the curves*

$$y = x,\ y = \frac{1}{x},\ y = \frac{x}{4}$$ *(U.P. I, Semester, Dec. 2008)*

**Solution.** By Green's Theorem Area A of the region bounded by a closed curve $C$ is given by

$$A = \frac{1}{2}\oint_C (xdy - ydx)$$

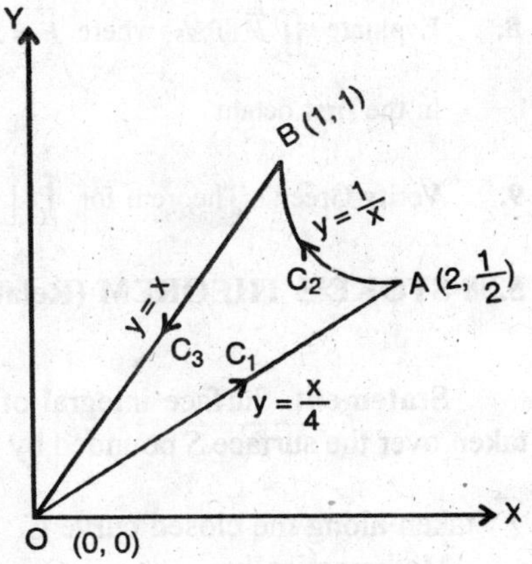


Here, $C$ consists of the curves $C_1 : y = \frac{x}{4}$, $C_2 : y = \frac{1}{x}$

and $C_3 : y = x$ So

$$\left[A = \frac{1}{2}\oint_C = \frac{1}{2}\left[\int_{C_1} + \int_{C_2} + \int_{C_3}\right] = \frac{1}{2}(I_1 + I_2 + I_3)\right]$$

Along $\quad C_1 : y = \frac{x}{4}, dy = \frac{1}{4}dx, x : 0 \text{ to } 2$

$$I_1 = \int_{C_1} (xdy - ydx) = \int_{C_1}\left(x\frac{1}{4}dx - \frac{x}{4}dx\right) = 0$$

Along $\quad C_2 : y = \frac{1}{x}, dy = -\frac{1}{x^2}dx, x : 2 \text{ to } 1$

$$I_2 = \int_{C_2} (xdy - ydx) = \int_2^1\left[x\left(-\frac{1}{x^2}\right)dx - \frac{1}{2}dx\right] = [-2\log x]_2^1 = 2\log 2$$

Along $\quad C_3 : y = x, dy = dx \,;\, x : 1 \text{ to } 0\,;$

$$I_3 = \int_{C_3} (xdy - ydx) = \int (xdx - xdx) = 0$$

$$A = \frac{1}{2}(I_1 + I_2 + I_3) = \frac{1}{2}(0 + 2\log 2 + 0) = \log 2$$ **Ans.**

## EXERCISE 5.13

1. Evaluate $\int_c [(3x^2 - 6yz)\,dx + (2y + 3xz)\,dy + (1 - 4xyz^2)\,dz]$ from (0, 0, 0) to (1, 1, 1) along the path $c$ given by the straight line from (0, 0, 0) to (0, 0, 1) then to (0, 1, 1) and then to (1, 1, 1).

2. Verify Green's Theorem in plane for $\int_C (x^2 + 2xy)\,dx + (y^2 + x^3 y)\,dy$, where $c$ is a square with the vertices $P$ (0, 0), $Q$ (1, 0), $R$ (1, 1) and $S$ (0, 1). **Ans.** $-\frac{1}{2}$

3. Verify Green's Theorem for $\int_c (x^2 - 2xy)\,dx + (x^2 y + 3)\,dy$ around the boundary $c$ of the region $y^2 = 8x$ and $x = 2$.

4. Use Green's Theorem in a plane to evaluate the integral $\int_c [(2x^2 - y^2)\,dx + (x^2 + y^2)\,dy]$, where $c$ is the boundary in the $xy$-plane of the area enclosed by the $x$-axis and the semi-circle $x^2 + y^2 = 1$ in the upper half $xy$-plane. **Ans.** $\frac{4}{3}$

5. Apply Green's Theoem to evaluate $\int_c [(y - \sin x)\,dy + \cos x\,dy]$, where $c$ is the plane triangle enclosed by the lines $y = 0$, $x = \frac{\pi}{2}$ and $y = \frac{2x}{\pi}$. **Ans.** $-\frac{\pi^2 + 8}{4\pi}$

6. Either directly or by Green's Therorem, evaluate the line integral $\int_c e^{-x}(\cos y\,dx - \sin y\,dy)$, where $c$ is the rectangle with vertices (0, 0), ($\pi$, 0,), $\left(\pi, \frac{\pi}{2}\right)$ and $\left(0, \frac{\pi}{2}\right)$. **Ans.** $2(1 - e^{-\pi})$

*(AMIETE II Sem June 2010)*

7. Verify the Green's Theorem to evaluate the line integral $\int_c (2y^2\,dx + 3x\,dy)$, where $c$ is the boundary of the closed region bounded by $y = x$ and $y = x^2$.

*(U.P., I Semester, Dec. 20005, AMIETE Summer 2004, Winter 2001)* **Ans.** $\frac{27}{4}$

8. Evaluate $\iint_S \overline{F}\cdot\hat{n}\,ds$, where $\vec{F} = xy\,\hat{i} - x^2\hat{j} + (x+z)\,\hat{k}$ and $s$ is the region of the plane $2x + 2y + z = 6$ in the first octant. (*A.M.I.E.T.E., Summer 2004, Winter 2001*) **Ans.** $\frac{27}{4}$

9. Verify Green's Theorem for $\int_C \left[(xy+y^2)\,dx + x^2 dy\right]$ where $C$ is the boundary by $y = x$ and $y = x^2$. (*AMIETE, June 2010*)

## 5.38 STOKE'S THEOREM (Relation between Line Integral and Surface Integral)

(*Uttarakhand, I Sem. 2008, U.P., Ist Semester, Dec. 2006*)

**Statement.** Surface integral of the component of curl $\overline{F}$ along the normal to the surface $S$, taken over the surface $S$ bounded by curve $C$ is equal to the line integral of the vector point function $\vec{F}$ taken along the closed curve $C$.

Mathematically

$$\oint \vec{F}\cdot d\,\overline{r} = \iint_S \text{curl}\,\vec{F}\cdot\hat{n}\,ds$$

where $\hat{n} = \cos\alpha\,\hat{i} + \cos\beta\,\hat{j} + \cos\gamma\,\hat{k}$ is a unit external normal to any surface $ds$,

**Proof.** Let

$$\vec{r} = x\hat{i} + y\hat{j} + z\hat{k}$$

$$d\vec{r} = \hat{i}\,dx + \hat{j}\,dy + \hat{k}\,dz$$

$$F = F_1\,\hat{i} + F_2\,\hat{j} + F_3\,\hat{k}$$

On putting the values of $\vec{F}, d\,\vec{r}$ in the statement of the theorem

$$\oint_c (F_1\,\hat{i} + F_2\,\hat{j} + F_3\,\hat{k})\cdot(\hat{i}\,dx + \hat{j}\,dy + \hat{k}\,dz)$$

$$= \iint_S \left(i\frac{\partial}{\partial x} + j\frac{\partial}{\partial y} + k\frac{\partial}{\partial z}\right) \times (F_1\,\hat{i} + F_2\,\hat{j} + F3\,\hat{k}).\,(\cos\alpha\,\hat{i} + \cos\beta\,\hat{j} + \cos\gamma\,\hat{k})\,ds$$

$$\oint (F_1\,dx + F_2\,dy + F_3\,dz) = \iint_S \left[\left(\frac{\partial F_3}{\partial y} - \frac{\partial F_2}{\partial z}\right)\hat{i} + \left(\frac{\partial F_1}{\partial z} - \frac{\partial F_3}{\partial x}\right)\hat{j} + \left(\frac{\partial F_2}{\partial x} - \frac{\partial F_1}{\partial y}\right)\hat{k}\right].$$

$$(\hat{i}\cos\alpha + \hat{j}\cos\beta + \hat{k}\cos\gamma)\,ds$$

$$= \iint_S \left[\left(\frac{\partial F_3}{\partial y} - \frac{\partial F_2}{\partial z}\right)\cos\alpha + \left(\frac{\partial F_1}{\partial z} - \frac{\partial F_3}{\partial x}\right)\cos\beta + \left(\frac{\partial F_2}{\partial x} - \frac{\partial F_1}{\partial y}\right)\cos\gamma\right] ds \quad \text{...(1)}$$

Let us first prove

$$\oint_c F_1\,dx = \iint_S \left[\left(\frac{\partial F_1}{\partial z}\cos\beta - \frac{\partial F_1}{\partial y}\cos\gamma\right)\right] ds \quad \text{...(2)}$$

Let the equation of the surface $S$ be $z = g(x, y)$. The projection of the surface on $x - y$ plane is region $R$.

$$\oint_c F_1(x, y, z)\,dx = \oint_c F_1[x, y, g(x, y)]\,dx$$

$$= -\iint_R \frac{\partial}{\partial y} F_1(x, y, g)\,dx\,dy \qquad \text{[By Green's Theorem]}$$

$$= -\iint_R \left(\frac{\partial F_1}{\partial y} + \frac{\partial F_1}{\partial z}\frac{\partial g}{\partial y}\right) dx\,dy \quad \text{...(3)}$$

The direction consines of the normal to the surface $z = g(x, y)$ are given by

$$\frac{\cos\alpha}{-\frac{\partial g}{\partial x}} = \frac{\cos\beta}{-\frac{\partial g}{\partial y}} = \frac{\cos\gamma}{1}$$

And $dx\,dy$ = projection of $ds$ on the $xy$-plane = $ds \cos\gamma$

Putting the values of $ds$ in R.H.S. of (2)

$$\iint_S \left(\frac{\partial F_1}{\partial z}\cos\beta - \frac{\partial F_1}{\partial y}\cos\gamma\right) ds = \iint_R \left(\frac{\partial F_1}{\partial z}\cos\beta - \frac{\partial F_1}{\partial y}\cos\gamma\right)\frac{dx\,dy}{\cos\gamma}$$

$$= \iint_R \left(\frac{\partial F_1}{\partial z}\frac{\cos\beta}{\cos\gamma} - \frac{\partial F_1}{\partial y}\right) dx\,dy = \iint_R \left(\frac{\partial F_1}{\partial z}\left(-\frac{\partial g}{\partial y}\right) - \frac{\partial F_1}{\partial y}\right) dx\,dy$$

$$= -\iint_R \left(\frac{\partial F_1}{\partial y} + \frac{\partial F_1}{\partial z}\frac{\partial g}{\partial y}\right) dx\,dy \qquad \text{...(4)}$$

From (3) and (4), we get

$$\oint_c F_1\,dx = \iint_S \left(\frac{\partial F_1}{\partial z}\cos\beta - \frac{\partial F_1}{\partial y}\cos\gamma\right) ds \qquad \text{...(5)}$$

Similarly,
$$\oint_c F_2\,dy = \iint_S \left(\frac{\partial F_2}{\partial x}\cos\gamma - \frac{\partial F_2}{\partial z}\cos\alpha\right) ds \qquad \text{...(6)}$$

and
$$\oint_c F_3\,dz = \iint_S \left(\frac{\partial F_3}{\partial y}\cos\alpha - \frac{\partial F_3}{\partial x}\cos\beta\right) ds \qquad \text{...(7)}$$

On adding (5), (6) and (7), we get

$$\oint_c (F_1\,dx + F_2\,dy + F_3\,dz) = \iint_S \left(\frac{\partial F_1}{\partial z}\cos\beta - \frac{\partial F_1}{\partial y}\cos\gamma + \frac{\partial F_2}{\partial x}\cos\gamma - \frac{\partial F_2}{\partial z}\cos\alpha + \frac{\partial F_3}{\partial y}\cos\alpha - \frac{\partial F_3}{\partial x}\cos\beta\right) ds$$

**Proved.**

## 5.39 ANOTHER METHOD OF PROVING STOKE'S THEOREM

*The circulation of vector F around a closed curve C is equal to the flux of the curve of the vector through the surface S bounded by the curve C.*

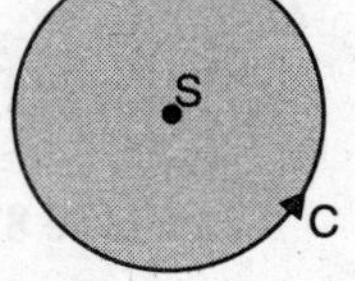


$$\oint_c \overline{F}\cdot d\overline{r} = \iint_S curl\ \vec{F}\cdot\hat{n}\,d\vec{s} = \iint_S curl\ \vec{F}\cdot d\vec{S}$$

**Proof :** The projection of any curved surface over $xy$-plane can be treated as kernal of the surface integral over actual surface

Now,
$$\iint_S (\nabla\times\vec{F})\cdot\hat{k}\,d\vec{S} = \iint_S (\nabla\times\vec{F})\cdot(\vec{i}\times\vec{j})\,dx\,dy \qquad [\hat{k} = \hat{i}\times\hat{j}]$$

$$= \iint_S [(\nabla\cdot\hat{i})(\vec{F}\cdot\hat{j}) - (\nabla\cdot\hat{j})(\vec{F}\cdot\hat{i})]\,dx\,dy = \iint_S \left[\frac{\partial}{\partial x}(F_y) - \frac{\partial}{\partial y}(F_x)\right] dx\,dy$$

$$= \iint_S [F_x\,dx + F_y\,dy] \text{ [By Green's theorem]}$$

$$= \iint_S [\hat{i}\,F_x + \hat{j}\,F_y]\cdot(\hat{i}\,dx + \hat{j}\,dy) = \oint_c \vec{F}\cdot\vec{dr}$$

$$\iint_S curl\ \vec{F}\cdot\hat{n}\,dS = \oint_c \vec{F}\cdot\vec{d}\,r.$$

where, $\vec{F} = F_x\,\hat{i} + F_y\,\hat{j} + F_z\,\hat{k}$ and $\vec{dr} = dx\,\hat{i} + dy\,\hat{j} + dz\,\hat{k}$

**Example 85.** *Evaluate by Strokes theorem* $\oint_C (yz\,dx + zx\,dy + xy\,dz)$ *where C is the curve* $x^2 + y^2 = 1,\ z = y^2.$ *(M.D.U., Dec 2009)*

**Solution.** Here we have $\oint yz\,dx + zx\,dy + xy\,dz$

$$= \int (yz\hat{i} + zx\hat{j} + xy\hat{k}).(\hat{i}dx + \hat{j}dy + kdz)$$

$$= \oint F.dx \qquad \text{Curl } \overline{F} = \begin{vmatrix} \hat{i} & \hat{j} & \hat{k} \\ \frac{\partial}{\partial x} & \frac{\partial}{\partial y} & \frac{\partial}{\partial z} \\ yz & zx & xy \end{vmatrix}$$

$$= \int curl\ F.\eta\ ds \qquad = (x - x)\ \hat{i} + (y - y)\ \hat{j} + (z - z)\ \hat{k}$$

$$= 0 \qquad = 0$$

**Ans.**

**Example 86.** *Using Stoke's theorem or otherwise, evaluate*

$$\int_c [(2x - y)\,dx - yz^2 dy - y^2 z\,dz]$$

*where c is the circle $x^2 + y^2 = 1$, corresponding to the surface of sphere of unit radius.* *(U.P., I Semester, Winter 2001)*

**Solution.** $\int_c [(2x - y)\,dx - yz^2 dy - y^2 z\,dz]$

$$= \int_c [(2x - y)\,\hat{i} - yz^2\,\hat{j} - y^2 z\,\hat{k}] \cdot (\hat{i}\,dx + \hat{j}\,dy + \hat{k}\,dz)$$

By Stoke's theorem $\oint \vec{F} \cdot d\vec{r} = \iint_S \text{Curl } \vec{F} \cdot \vec{n}\ ds$ ...(1)

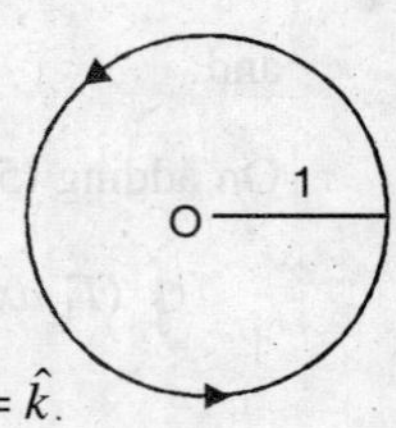


$$\text{Curl } \vec{F} = \nabla \times \vec{F} = \begin{vmatrix} \hat{i} & \hat{j} & \hat{k} \\ \frac{\partial}{\partial x} & \frac{\partial}{\partial y} & \frac{\partial}{\partial z} \\ 2x - y & -yz^2 & -y^2 z \end{vmatrix}$$

$$= (-2yz + 2yz)\,\hat{i} - (0 - 0)\,\hat{j} + (0 + 1)\,\hat{k} = \hat{k}.$$

Putting the value of curl $\vec{F}$ in (1), we get

$$= \iint \hat{k} \cdot \hat{n}\ ds = \iint \hat{k} \cdot \hat{n} \frac{dx\,dy}{\hat{n} \cdot \hat{k}} = \iint dx\,dy = \text{Area of the circle} = \pi \qquad \left[\because ds = \frac{dx\,dy}{(\hat{n} \cdot \hat{k})}\right]$$

**Example 87.** *Evaluate* $\int_C \vec{F} . d\vec{r}$, *where* $F(x, y, z) = -y^2\hat{i} + x\hat{j} + z^2\hat{k}$ *and C is the curve of intersection of the plane* $y + z = 2$ *and the cylinder* $x^2 + y^2 = 1$. *(Gujarat, I sem. Jan. 2009)*

**Solution.** $\oint_C \vec{F} . d\vec{r} = \iint_S \text{curl } \vec{F} . \hat{n}\,ds = \iint_S \text{curl } (-y^2\,\hat{i} + x\,\hat{j} + z^2\,\hat{k})\,\hat{n}\,ds$ ...(1)

$$F(x, y, z) = -y^2\,\hat{i} + x\,\hat{j} + z^2\,\hat{k}$$

*(By Stoke's Theorem)*

$$\text{Curl } \vec{F} = \begin{vmatrix} \hat{i} & \hat{j} & \hat{k} \\ \frac{\partial}{\partial x} & \frac{\partial}{\partial y} & \frac{\partial}{\partial z} \\ -y^2 & x & z^2 \end{vmatrix}$$

$$= \hat{i}\,(0 - 0) - \hat{j}\,(0 - 0) + \hat{k}\,(1 + 2y) = (1 + 2y)\,\hat{k}$$

Normal vector $= \nabla \vec{F}$

$$= \left(\hat{i}\frac{\partial}{\partial x} + \hat{j}\frac{\partial}{\partial y} + \hat{k}\frac{\partial}{\partial z}\right)(y + z - 2) = \hat{j} + \hat{k}$$

Unit normal vector $\hat{n} = \dfrac{\hat{j} + \hat{k}}{\sqrt{2}}$

$$ds = \frac{dx\,dy}{\hat{\eta} . \hat{k}}$$

$x^2 + y^2 = 1$

On putting the values of curl $\vec{F}, \hat{n}$ and $ds$ in (1), we get

$$\int_C \vec{F}\,.\,d\vec{r} = \iint_S (1+2y)\,\hat{k}\,.\,\frac{\hat{j}+\hat{k}}{\sqrt{2}}\,\frac{dx\,dy}{\left(\frac{\hat{j}+\hat{k}}{\sqrt{2}}\right).\hat{k}}$$

$$= \iint \frac{1+2y}{\sqrt{2}}\,\frac{dx\,dy}{\frac{1}{\sqrt{2}}} = \iint (1+2y)\,dx\,dy = \int_0^{2\pi}\int_0^1 (1+2r\sin\theta)\,r\,d\,\theta\,d\,r$$

$$= \int_0^{2\pi}\int_0^1 (r+2r^2\sin\theta)\,d\,\theta\,d\,r$$

$$= \int_0^{2\pi} d\,\theta\left[\frac{r^2}{2}+\frac{2r^3}{3}\sin\theta\right]_0^1 = \int_0^{2\pi}\left[\frac{1}{2}+\frac{2}{3}\sin\theta\right]d\,\theta$$

$$= \left[\frac{\theta}{2}-\frac{2}{3}\cos\theta\right]_0^{2\pi} = \left(\pi-\frac{2}{3}-0+\frac{2}{3}\right) = \pi \qquad \textbf{Ans.}$$

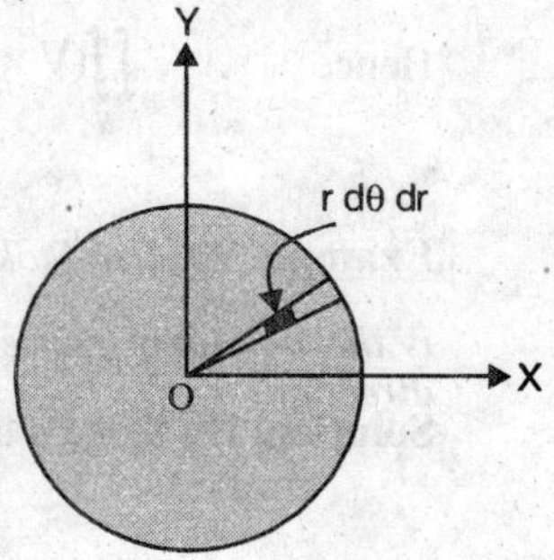


**Example 88.** *Apply Stoke's Theorem to find the value of*

$$\int_c (y\,dx + z\,dy + x\,dz)$$

*where c is the curve of intersection of $x^2 + y^2 + z^2 = a^2$ and $x + z = a$. (Nagpur, Summer 2001)*

**Solution.** $\int_c (y\,dx + z\,dy + x\,dz)$

$$= \int_c (y\hat{i} + z\hat{j} + x\hat{k})\cdot(\hat{i}\,dx + \hat{j}\,dy + \hat{k}\,dz) = \int_C (y\hat{i} + z\hat{j} + x\hat{k})\cdot d\vec{r}$$

$$= \iint_S \text{curl}\,(y\hat{i} + z\hat{j} + x\hat{k})\cdot\hat{n}\,ds \qquad \text{(By Stoke's Theorem)}$$

$$= \iint_S \left(\hat{i}\frac{\partial}{\partial x}+\hat{j}\frac{\partial}{\partial y}+\hat{k}\frac{\partial}{\partial z}\right)\times(y\hat{i}+z\hat{j}+x\hat{k})\cdot\hat{n}\,ds = \iint_S -(\hat{i}+\hat{j}+\hat{k})\cdot\hat{n}\,ds \quad ...(1)$$

*where S is the circle formed by the intersection of $x^2 + y^2 + z^2 = a^2$* and $x + z = a$.

$$\hat{n} = \frac{\nabla\phi}{|\nabla\phi|} = \frac{\left(\hat{i}\frac{\partial}{\partial x}+\hat{j}\frac{\partial}{\partial y}+\hat{k}\frac{\partial}{\partial z}\right)(x+z-a)}{|\nabla\phi|} = \frac{\hat{i}+\hat{k}}{\sqrt{1+1}}$$

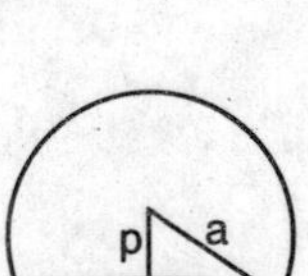


$$\therefore \qquad \hat{n} = \frac{\hat{i}}{\sqrt{2}}+\frac{\hat{k}}{\sqrt{2}}$$

Putting the vlaue of $\hat{n}$ in (1), we have

$$= \iint_S -(\hat{i}+\hat{j}+\hat{k})\cdot\left(\frac{\hat{i}}{\sqrt{2}}+\frac{\hat{k}}{\sqrt{2}}\right)ds$$

$$= \iint_S -\left(\frac{1}{\sqrt{2}}+\frac{1}{\sqrt{2}}\right)ds \qquad \left[\text{Use } r^2 = R^2 - p^2 = a^2 - \frac{a^2}{2} = \frac{a^2}{2}\right]$$

$$= \frac{-2}{\sqrt{2}}\iint_S ds = \frac{-2}{\sqrt{2}}\pi\left(\frac{a}{\sqrt{2}}\right)^2 = -\frac{\pi a^2}{\sqrt{2}} \qquad \textbf{Ans.}$$

**Example 89.** *Directly or by Stoke's Theorem, evaluate $\iint_s curl\ \vec{v}\cdot\hat{n}\,ds,\ \vec{v} = \hat{i}y + \hat{j}z + \hat{k}x$, s is the surface of the paraboloid $z = 1 - x^2 - y^2,\ z^3 \ge 0$ and $\hat{n}$ is the unit vector normal to s.*

**Solution.**
$$\vec{\nabla}\times\vec{v} = \begin{vmatrix} \hat{i} & \hat{j} & \hat{k} \\ \frac{\partial}{\partial x} & \frac{\partial}{\partial y} & \frac{\partial}{\partial z} \\ y & z & x \end{vmatrix} = -\hat{i}-\hat{j}-\hat{k}$$

Obviously $\quad \hat{n} = \hat{k}.$

Therefore $\quad (\nabla\times\vec{v})\cdot\hat{n} = (-\hat{i}-\hat{j}-\hat{k}).\,\hat{k} = -1$

Hence
$$\iint_S (\nabla\times\vec{v})\cdot\hat{n}\,ds = \iint_S (-1)\,dx\,dy = -\iint_S dx\,dy$$
$$= -\pi\,(1)^2 = -\pi. \qquad \text{(Area of circle} = \pi r^2\text{)}$$ **Ans.**

**Example 90.** *Use Stoke's Theorem to evaluate* $\int_c \vec{v}\cdot d\vec{r}$, *where* $\vec{v} = y^2\hat{i} + xy\hat{j} + xz\hat{k}$, *and c is the bounding curve of the hemisphere* $x^2 + y^2 + z^2 = 9$, $z > 0$, *oriented in the positive direction.*

**Solution.** By Stoke's theorem

$$\int_c \vec{v}\cdot d\vec{r} = \iint_S (\text{curl } \vec{v})\cdot\hat{n}\,ds = \iint_S (\nabla\times\vec{v})\cdot\hat{n}\,ds$$

$$\nabla\times\vec{v} = \begin{vmatrix} \hat{i} & \hat{j} & \hat{k} \\ \frac{\partial}{\partial x} & \frac{\partial}{\partial y} & \frac{\partial}{\partial z} \\ y^2 & xy & xz \end{vmatrix} = (0-0)\,\hat{i} - (z-0)\,\hat{j} + (y-2y)\,\hat{k} = -z\hat{j} - y\hat{k}$$

$$\hat{n} = \frac{\nabla\phi}{|\nabla\phi|} = \frac{\left(i\frac{\partial}{\partial x} + j\frac{\partial}{\partial y} + k\frac{\partial}{\partial z}\right)(x^2+y^2+z^2-9)}{|\nabla\phi|}$$

$$= \frac{2x\hat{i} + 2y\hat{j} + 2z\hat{k}}{\sqrt{4x^2+4y^2+4z^2}} = \frac{x\hat{i}+y\hat{j}+z\hat{k}}{\sqrt{x^2+y^2+z^2}} = \frac{x\hat{i}+y\hat{j}+z\hat{k}}{3}$$

$$(\nabla\times\vec{v})\cdot\hat{n} = (-z\hat{j} - y\hat{k})\cdot\frac{x\hat{i}+y\hat{j}+z\hat{k}}{3} = \frac{-yz-yz}{3} = \frac{-2yz}{3}$$

$$\hat{n}\cdot\hat{k}\,ds = dx\,dy \Rightarrow \frac{x\hat{i}+y\hat{j}+z\hat{k}}{3}.\hat{k}\;dx = dx\,dy \Rightarrow \frac{z}{3}\,ds = dx\,dy$$

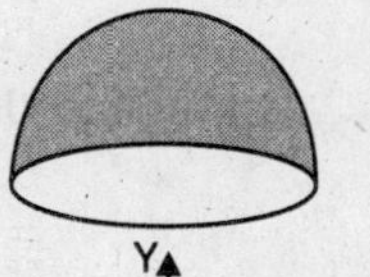

$\therefore$
$$ds = \frac{3}{z}\,dx\,dy$$

$$\iint_S (\nabla\times\vec{v})\cdot\hat{n}\,ds = \iint\left(\frac{-2yz}{3}\right)\left(\frac{3}{z}\,dx\,dy\right) = -\iint 2y\,dx\,dy$$

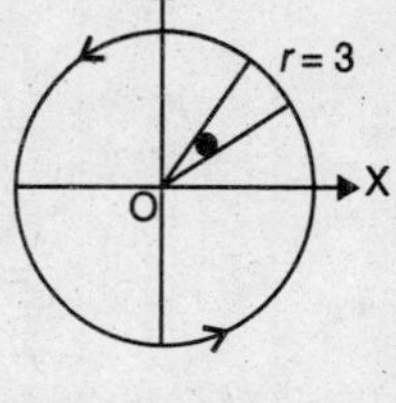


$$= -\iint 2r\sin\theta\,r\,d\theta\,dr = -2\int_0^{2\pi}\sin\theta\,d\theta\int_0^3 r^2\,dr$$

$$= -2\,(-\cos\theta)_0^{2\pi}\cdot\left[\frac{r^3}{3}\right]_0^3 = -2\,(-1+1)\,9 = 0$$ **Ans.**

**Example 91.** *Evaluate the surface integral* $\iint_S \text{curl } \vec{F}.\hat{n}\,dS$ *by transforming it into a line integral, S being that part of the surface of the paraboloid* $z = 1 - x^2 - y^2$ *for which* $z \geq 0$ and $\vec{F} = y\hat{i} + z\hat{j} + x\hat{k}\cdot$ *(K. University, Dec. 2008)*

**Solution.** $\vec{\nabla}\times\vec{F} = \begin{vmatrix} \hat{i} & \hat{j} & \hat{k} \\ \dfrac{\partial}{\partial x} & \dfrac{\partial}{\partial y} & \dfrac{\partial}{\partial z} \\ y & z & x \end{vmatrix} = -\hat{i}-\hat{j}-\hat{k}$

Obviously $\hat{n} = \hat{k}.$

Therefore $(\nabla\times\vec{F})\cdot\hat{n} = (-\hat{i}-\hat{j}-\hat{k}).\,\hat{k} = -1$

Hence $\iint_S (\nabla\times\vec{F})\cdot\hat{n}\,ds = \iint_S (-1)\,dx\,dy = -\iint_S dx\,dy$

$= -\pi\,(1)^2 = -\pi.$ (Area of circle $= \pi\,r^2$) **Ans.**

**Example 92.** *Evaluate $\oint_C \vec{F}\cdot\vec{dr}$ by Stoke's Theorem, where $\vec{F} = y^2\,\hat{i} + x^2\,\hat{j} - (x+z)\,\hat{k}$ and C is the boundary of triangle with vertices at (0, 0, 0), (1, 0, 0) and (1, 1, 0).*
*(U.P., I Semester, Winter 2000)*

**Solution.** We have, curl $\vec{F} = \nabla\times\vec{F}$

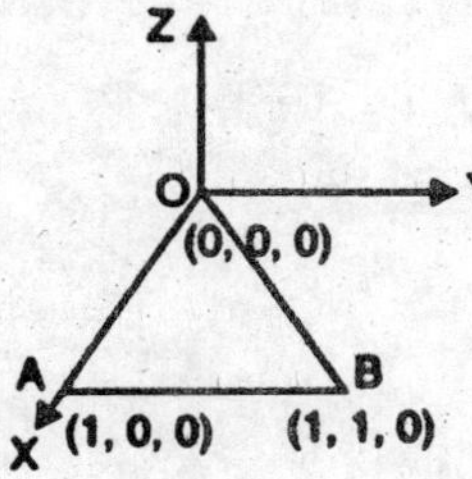


$= \begin{vmatrix} \hat{i} & \hat{j} & \hat{k} \\ \dfrac{\partial}{\partial x} & \dfrac{\partial}{\partial y} & \dfrac{\partial}{\partial z} \\ y^2 & x^2 & -(x+z) \end{vmatrix} = 0.\,\hat{i} + \hat{j}\,2\,(x-y)\,\hat{k}.$

We observe that $z$ co-ordinate of each vertex of the triangle is zero.
Therefore, the triangle lies in the $xy$-plane.

$\therefore \quad \hat{n} = \hat{k}$

$\therefore \quad \text{curl } \vec{F}\cdot\hat{n} = [\hat{j} + 2\,(x-y)\,\hat{k}]\,.\,\hat{k} = 2\,(x-y).$

In the figure, only $xy$-plane is considered.
The equation of the line OB is $y = x$
By Stoke's theorem, we have

$$\oint_C \vec{F}\cdot\vec{dr} = \iint_S (\text{curl } \vec{F}\cdot\hat{n})\,ds$$

$$= \int_{x=0}^{1}\int_{y=0}^{x} 2(x-y)\,dx\,dy = 2\int_0^1 \left[xy - \frac{y^2}{2}\right]_0^x dx$$

$$= 2\int_0^1 \left[x^2 - \frac{x^2}{2}\right] dx = 2\int_0^1 \frac{x^2}{2}\,dx = \int_0^1 x^2\,dx = \left[\frac{x^3}{3}\right]_0^1 = \frac{1}{3}.$$ **Ans.**

**Example 93.** *Evaluate $\oint_C \vec{F}\cdot\vec{dr}$ by Stoke's Theorem, where $\vec{F} = (x^2+y^2)\,\hat{i} - 2\,xy\,\hat{j}$ and C is the boundary of the rectangle $x = \pm\,a$, $y = 0$ and $y = b$.* *(U.P., I Semester, Winter 2002)*

**Solution.** Since the $z$ co-ordinate of each vertex of the given rectangle is zero, hence the given rectangle must lie in the $xy$-plane.

Here, the co-ordinates of $A$, $B$, $C$ and $D$ are $(a, 0)$, $(a, b)$, $(-a, b)$ and $(-a, 0)$ respectively.

$$\therefore \quad \text{Curl } \overline{F} = \begin{vmatrix} \hat{i} & \hat{j} & \hat{k} \\ \dfrac{\partial}{\partial x} & \dfrac{\partial}{\partial y} & \dfrac{\partial}{\partial z} \\ x^2+y^2 & -2xy & 0 \end{vmatrix} = -4\,y\,k$$

Here, $\hat{n} = \hat{k}$, so by Stoke's theorem, we've

$$\oint_C \vec{F} \cdot \vec{dr} = \iint_S \text{curl}\, \overline{F} \cdot \hat{n}\, ds$$

$$= \iint_S (-4y\hat{k}) \cdot (\hat{k})\, dx\, dy = -4 \int_{x=-a}^{a} \int_{y=0}^{b} y\, dx\, dy$$

$$= -4 \int_{-a}^{a} \left[ \frac{y^2}{2} \right]_0^b dx = -2b^2 \int_{-a}^{a} dx = -4ab^2$$ **Ans.**

**Example 94.** *Apply Stoke's Theorem to calculate* $\int_c 4y\, dx + 2z\, dy + 6y\, dz$ *where c is the curve of intersection of* $x^2 + y^2 + z^2 = 6z$ *and* $z = x + 3$.

**Solution.** $\int_c \vec{F} \cdot \vec{dr} = \int_c 4y\, dx + 2z\, dy + 6y\, dz$

$$= \int_c (4y\hat{i} + 2z\hat{j} + 6y\hat{k}) \cdot (\hat{i}dx + \hat{j}dy + \hat{k}dz)$$

$$\vec{F} = 4y\hat{i} + 2z\hat{j} + 6y\hat{k}$$

$$\nabla \times \vec{F} = \begin{vmatrix} \hat{i} & \hat{j} & \hat{k} \\ \frac{\partial}{\partial x} & \frac{\partial}{\partial y} & \frac{\partial}{\partial z} \\ 4y & 2z & 6y \end{vmatrix} = (6-2)\hat{i} - (0-0)\hat{j} + (0-4)\hat{k} = 4\hat{i} - 4\hat{k}$$

S is the surface of the circle $x^2 + y^2 + z^2 = 6z$, $z = x + 3$, $\hat{n}$ is normal to the plane $x - z + 3 = 0$

$$\hat{n} = \frac{\nabla\phi}{|\nabla\phi|} = \frac{\left(\hat{i}\frac{\partial}{\partial x} + \hat{j}\frac{\partial}{\partial y} + \hat{k}\frac{\partial}{\partial z}\right)(x - z + 3)}{|\nabla\phi|} = \frac{\hat{i} - \hat{k}}{\sqrt{1+1}} = \frac{\hat{i} - \hat{k}}{\sqrt{2}}$$

$$(\nabla \times F) \cdot \hat{n} = (4\hat{i} - 4\hat{k}) \cdot \frac{\hat{i} - \hat{k}}{\sqrt{2}} = \frac{4+4}{\sqrt{2}} = 4\sqrt{2}$$

$$\int_c \vec{F} \cdot \vec{dr} = \iint_S (\text{curl}\, F) \cdot \hat{n}\, ds = \iint_S 4\sqrt{2}\, (dx\, dz) = 4\sqrt{2}\ (\text{area of circle})$$

Centre of the sphere $x^2 + y^2 + (z-3)^2 = 9$, $(0, 0, 3)$ lies on the plane $z = x + 3$. It means that the given circle is a great circle of sphere, where radius of the circle is equal to the radius of the sphere.

Radius of circle = 3, Area = $\pi (3)^2 = 9\pi$

$$\iint_S (\nabla \times F) \cdot \hat{n}\, ds = 4\sqrt{2}(9\pi) = 36\sqrt{2}\,\pi$$ **Ans.**

**Example 95.** *Verify Stoke's Theorem for the function* $\overline{F} = z\hat{i} + x\hat{j} + y\hat{k}$, *where C is the unit circle in xy-plane bounding the hemisphere* $z = \sqrt{(1 - x^2 - y^2)}$. *(U.P., I Semester Comp. 2002)*

**Solution.** Here $\overline{F} = z\hat{i} + x\hat{j} + y\hat{k}$. ...(1)

Also, $\overline{r} = x\hat{i} + y\hat{j} + z\hat{k} \Rightarrow \overline{dr} = dx\hat{i} + dy\,\hat{j} + dz\,\hat{k}$.

$\therefore$ $\overline{F} \cdot \overline{dr} = z\, dx + x\, dy + y\, dz$.

$\therefore$ $\oint_C \overline{F} \cdot \overline{dr} = \oint_C (z\, dx + x\, dy + y\, dz)$. ...(2)

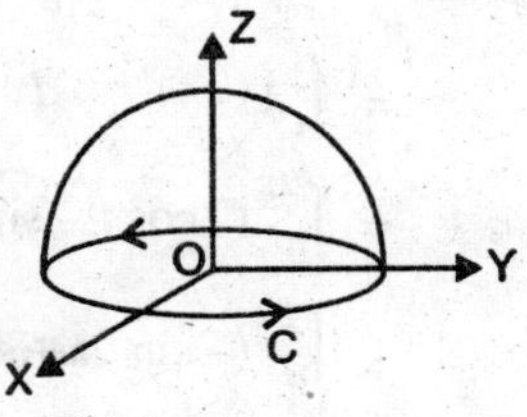


On the circle $C$, $x^2 + y^2 = 1$, $z = 0$ on the $xy$-plane. Hence on $C$, we have $z = 0$ so that $dz = 0$. Hence (2) reduces to

$$\oint_C \overline{F} \cdot \overline{dr} = \oint_C x\, dy. \qquad ...(3)$$

Now the parametric equations of $C$, *i.e.*, $x^2 + y^2 = 1$ are

$$x = \cos\phi,\ y = \sin\phi. \qquad ...(4)$$

Using (4), (3) reduces to $\oint_C \vec{F} \cdot \vec{dr} = \int_{\phi=0}^{2\pi} \cos\phi \cos\phi\, d\phi = \int_0^{2\pi} \frac{1+\cos 2\phi}{2}\, d\phi$

$$= \frac{1}{2}\left[\phi + \frac{\sin 2\phi}{2}\right]_0^{2\pi} = \pi \qquad ...(5)$$

Let $P(x, y, z)$ be any point on the surface of the hemisphere $x^2 + y^2 + z^2 = 1$, O origin is the centre of the sphere.

Radius $= \text{OP} = x\hat{i} + y\hat{j} + z\hat{k}$ Normal $= x\hat{i} + y\hat{j} + z\hat{k}$

$$\hat{n} = \frac{x\hat{i} + y\hat{j} + z\hat{k}}{\sqrt{x^2 + y^2 + z^2}} = x\hat{i} + y\hat{j} + z\hat{k}$$

(Radius is $\perp$ to tangent *i.e.* Radius is normal)

$$x = \sin\theta\cos\phi,\quad y = \sin\theta\sin\phi,\quad z = \cos\theta \qquad ...(6)$$

$$\hat{n} = \sin\theta\cos\phi\ \hat{i} + \sin\theta\sin\phi\ \hat{j} + \cos\theta\ \hat{k}$$

Also,

$$\text{Curl}\ \vec{F} = \begin{vmatrix} \hat{i} & \hat{j} & \hat{k} \\ \partial/\partial x & \partial/\partial y & \partial/\partial z \\ z & x & y \end{vmatrix} = \hat{i} + \hat{j} + \hat{k} \qquad ...(7)$$

$$\text{Curl}\ \vec{F}\cdot\hat{n} = (\hat{i} + \hat{j} + \hat{k}).(\sin\theta\cos\phi\,\hat{i} + \sin\theta\sin\phi\,\hat{j} + \sin\theta\,\hat{k})$$
$$= \sin\theta\cos\phi + \sin\theta\sin\phi + \cos\theta$$

$\therefore$

$$\iint_S \text{Curl}\ \vec{F}\cdot\hat{n}\, dS = \int_{\theta=0}^{\pi/2}\int_{\phi=0}^{2\pi} (\hat{i} + \hat{j} + \hat{k})$$
$$.(\sin\theta\cos\phi\ \hat{i} + \sin\theta\sin\phi\ \hat{j} + \cos\theta\ \hat{k})\sin\theta\, d\theta\, d\phi$$
$$= \int_{\theta=0}^{\pi/2} \sin\theta\, d\theta \int_{\phi=0}^{2\pi} (\sin\theta\cos\phi + \sin\theta\sin\phi + \cos\theta)\, d\phi$$

[$\because$ $dS$ = Elementary area on hemisphere $= \sin\theta\, d\theta\, d\phi$]

$$= \int_0^{\pi/2} \sin\theta\, d\theta\, [\sin\theta\sin\phi + \sin\theta(-\cos\phi) + \phi\cos\theta]_0^{2\pi} = \int_0^{\pi/2} \sin\theta\, d\theta$$

$$= \int_0^{\pi/2} (0 + 0 + 2\pi\sin\theta\cos\theta)\, d\theta = \pi\int_0^{\pi/2} \sin 2\theta\, d\theta = \pi\left[-\frac{\cos 2\theta}{2}\right]_0^{\pi/2}$$

$$= -(\pi/2)\,[-1 - 1] = \pi.$$

From (5) and (8), $\oint_C \vec{F}\cdot\vec{dr} = \iint_S \text{curl}\,\overline{F}\cdot\hat{n}\, dS$, which verifies Stokes's theorem.

**Example 96.** *Verify Stoke's theorem for the vector field* $\overline{F} = (2x - y)\,\hat{i} - yz^2\,\hat{j} - y^2 z\,\hat{k}$ *over the upper half of the surface* $x^2 + y^2 + z^2 = 1$ *bounded by its projection on xy- plane.*

(*Nagpur University, Summer 2001*)

**Solution.** Let $S$ be the upper half surface of the sphere $x^2 + y^2 + z^2 = 1$. The boundary $C$ or $S$ is a circle in the $xy$ plane of radius unity and centre $O$. The equation of C are $x^2 + y^2 = 1$, $z = 0$ whose parametric form is

$$x = \cos t,\ y = \sin t,\ z = 0,\ 0 < t < 2\pi$$

$$\int_C \vec{F}\cdot\vec{dr} = \int_C [(2x - y)\,\hat{i} - yz^2\,\hat{j} - y^2 z\,\hat{k}].[\hat{i}\, dx + \hat{j}\, dy + \hat{k}\, dz]$$

$$= \int_C [(2x - y)\,dx - yz^2\,dy - y^2 z\,dz] = \int_C (2x - y)\,dx, \quad \text{since on } C,\ z = 0 \text{ and } 2z = 0$$

$$= \int_0^{2\pi} (2\cos t - \sin t)\frac{dx}{dt}\,dt = \int_0^{2\pi} (2\cos t - \sin t)(-\sin t)\,dt$$

$$= \int_0^{2\pi} (-\sin 2t + \sin^2 t)\,dt = \int_0^{2\pi}\left(-\sin 2t + \frac{1 - \cos 2t}{2}\right)dt$$

$$= \left[\frac{\cos 2t}{2} + \frac{t}{2} - \frac{\sin 2t}{4}\right]_0^{2\pi} = \frac{1}{2} + \pi - \frac{1}{2} = \pi \qquad \text{...(1)}$$

$$\text{Curl } \vec{F} = \begin{vmatrix} \hat{i} & \hat{j} & \hat{k} \\ \frac{\partial}{\partial x} & \frac{\partial}{\partial y} & \frac{\partial}{\partial z} \\ 2x - y & -yz^2 & -y^2 z \end{vmatrix} = (-2yz + 2yz)\,\hat{i} + (0 - 0)\,\hat{j} + (0 + 1)\,\hat{k} = \hat{k}$$

$$\text{Curl } \vec{F}.\hat{n} = \hat{k}.\hat{n} = \hat{n}.\hat{k}$$

$$\iint_S \text{Curl } \vec{F}.\hat{n}\,ds = \iint_S \hat{n}.\hat{k}\,ds = \iint_R \hat{n}.\hat{k}.\frac{dx\ dy}{\hat{n}.\hat{k}}$$

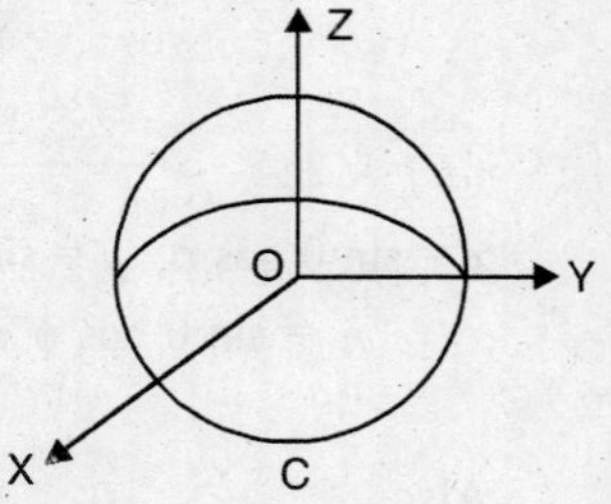


Where $R$ is the projection of $S$ on $xy$-plane.

$$= \int_{-1}^{1}\int_{-\sqrt{1-x^2}}^{\sqrt{1-x^2}} dx\,dy = \int_{-1}^{1} 2\sqrt{1 - x^2}\,dx = 4\int_0^1 \sqrt{1 - x^2}\,dx$$

$$= 4\left[\frac{x}{2}\sqrt{1 - x^2} + \frac{1}{2}\sin^{-1} x\right]_0^1 = 4\left[\frac{1}{2}.\frac{\pi}{2}\right] = \pi \qquad \text{...(2)}$$

From (1) and (2), we have

$$\therefore \quad \int_C \vec{F}.\vec{dr} = \iint \text{Curl } \bar{F}.\hat{n}\,ds \text{ which is the Stoke's theorem.}$$

**Ans.**

**Example 97.** *Verify Stoke's Theorem for* $\vec{F} = (x^2 + y - 4)\,\hat{i} + 3\,xy\hat{j} + (2\,xz + z^2)\,\hat{k}$ *over the surface of hemisphere* $x^2 + y^2 + z^2 = 16$ *above the xy-plane.*

**Solution.** $\int_c \vec{F}\cdot\vec{dr}$, where $c$ is the boundary of the circle $x^2 + y^2 + z^2 = 16$ (bounding the hemispherical surface)

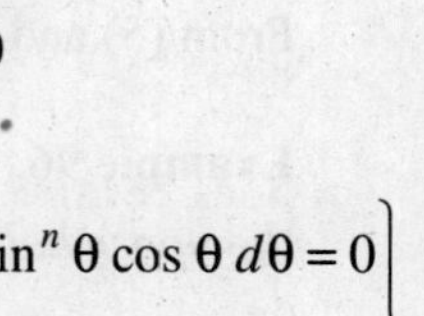


$$= \int_c [(x^2 + y - 4)\hat{i} + 3xy\hat{j} + (2xz + z^2)\hat{k}]\cdot(\hat{i}dx + \hat{j}dy)$$

$$= \int_c [(x^2 + y - 4)\,dx + 3\,xy\,dy)]$$

Putting $x = 4\cos\theta,\ y = 4\sin\theta,\ dx = -4\sin\theta\,d\theta,\ dy = 4\cos\theta\,d\theta$

$$= \int_0^{2\pi} [(16\cos^2\theta + 4\sin\theta - 4)(-4\sin\theta\,d\theta) + (192\sin\theta\cos^2\theta\,d\theta)]$$

$$= 16\int_0^{2\pi} [-4\cos^2\theta\sin\theta - \sin^2\theta + \sin\theta + 12\sin\theta\cos^2\theta]\,d\theta$$

$$= 16\int_0^{2\pi} (8\sin\theta\cos^2\theta - \sin^2\theta + \sin\theta)\,d\theta$$

$$= -16\int_0^{2\pi} \sin^2\theta\,d\theta \qquad \left\{\begin{array}{l} \int_0^{2\pi} \sin^n\theta\cos\theta\,d\theta = 0 \\ \int_0^{2\pi} \cos^n\theta\sin\theta\,d\theta = 0 \end{array}\right\}$$

$$= -16 \times 4\int_0^{\frac{\pi}{2}} \sin^2\theta\,d\theta = -64\left(\frac{1}{2}\frac{\pi}{2}\right) = -16\,\pi.$$

To evaluate surface integral
$$\nabla \times \vec{F} = \begin{vmatrix} \hat{i} & \hat{j} & \hat{k} \\ \frac{\partial}{\partial x} & \frac{\partial}{\partial y} & \frac{\partial}{\partial z} \\ x^2 + y - 4 & 3xy & 2xz + z^2 \end{vmatrix}$$

$$= (0-0)\,\hat{i} - (2z-0)\,\hat{j} + (3y-1)\,\hat{k} = -2z\hat{j} + (3y-1)\,k$$

$$\hat{n} = \frac{\nabla\phi}{|\nabla\phi|} = \frac{\left(\hat{i}\frac{\partial}{\partial x} + \hat{j}\frac{\partial}{\partial y} + \hat{k}\frac{\partial}{\partial z}\right)(x^2+y^2+z^2-16)}{|\nabla\phi|}$$

$$= \frac{2x\hat{i} + 2y\hat{j} + 2z\hat{k}}{\sqrt{4x^2+4y^2+4z^2}} = \frac{x\hat{i}+y\hat{j}+z\hat{k}}{\sqrt{x^2+y^2+z^2}} = \frac{x\hat{i}+y\hat{j}+z\hat{k}}{4}$$

$$(\nabla\times\vec{F})\cdot\hat{n} = [-2z\hat{j} + (3y-1)\,\hat{k}]\cdot\frac{x\hat{i}+y\hat{j}+z\hat{k}}{4} = \frac{-2yz+(3y-1)z}{4}$$

$$\hat{k}\cdot\hat{n}\cdot ds = dx\,dy \Rightarrow \frac{x\hat{i}+y\hat{j}+z\hat{k}}{4}.\,k\,ds = dx\,dy \Rightarrow \frac{z}{4}ds = dx\,dy$$

$$\therefore \qquad ds = \frac{4}{z}\,dx\,dy$$

$$\iint(\nabla\times F)\cdot\hat{n}\,ds = \iint\frac{-2yz+(3y-1)z}{4}\left(\frac{4}{z}\,dx\,dy\right) = \iint[-2y+(3y-1)]\,dx\,dy = \iint(y-1)\,dx\,dy$$

On putting $x = r\cos\theta$, $y = r\sin\theta$, $dx\,dy = r\,d\theta\,dr$, we get

$$= \iint(r\sin\theta - 1)\,r\,d\theta\,dr = \int d\theta\int(r^2\sin\theta - r)\,dr$$

$$= \int_0^{2\pi} d\theta\left(\frac{r^3}{3}\sin\theta - \frac{r^2}{2}\right)_0^4 = \int_0^{2\pi} d\theta\left(\frac{64}{3}\sin\theta - 8\right)$$

$$= \left(-\frac{64}{3}\cos\theta - 8\theta\right)_0^{2\pi} = \frac{-64}{3} - 16\pi + \frac{64}{3} = -16\pi$$

The line integral is equal to the surface integral, hence Stoke's Theorem is verified. **Proved.**

**Example 98.** *Verify Stoke's theorem for a vector field defined by* $\vec{F} = (x^2-y^2)\,\hat{i} + 2xy\,\hat{j}$ *in the rectangular in xy-plane bounded by lines x = 0, x = a, y = 0, y = b.*

*(Nagpur University, Summer 2000)*

**Solution.** Here we have to verify Stoke's theorem $\int_C \vec{F}.\,\vec{dr} = \iint_S(\nabla\times\vec{F})\,.\,\hat{n}\,ds$

Where 'C' be the boundary of rectangle (*ABCD*) and *S* be the surface enclosed by curve *C*.

$$\vec{F} = (x^2-y^2)\,\hat{i} + (2xy)\,\hat{j}$$

$$\vec{F}.\vec{dr} = [(x^2-y^2)\,\hat{i} + 2xy\,\hat{j}].[\hat{i}\,dx + \hat{j}\,dy]$$

$$\Rightarrow \qquad \vec{F}.\vec{dr} = (x^2+y^2)\,dx + 2xy\,dy \qquad ...(1)$$

Now,
$$\int_C \vec{F}.\vec{dr} = \int_{OA}\vec{F}.\vec{dr} + \int_{AB}\vec{F}.\vec{dr} + \int_{BC}\vec{F}.\vec{dr} + \int_{CO}\vec{F}.\vec{dr} \qquad ...(2)$$

Along *OA*, put $y = 0$ so that $k\,dy = 0$ in (1) and $\vec{F}.\vec{dr} = x^2\,dx$,
Where *x* is from 0 to *a*.

$$\therefore \qquad \int_{OA}\vec{F}.\vec{dr} = \int_0^a x^2\,dx = \left[\frac{x^3}{3}\right]_0^a = \frac{a^3}{3} \qquad ...(3)$$

Along *AB*, put $x = a$ so that $dx = 0$ in (1), we get $\vec{F}.\vec{dr} = 2ay\,dy$
Where *y* is from 0 to *b*.

$$\therefore \qquad \int_{AB}\vec{F}.\vec{dr} = \int_0^b 2ay\,dy = [ay^2]_0^b = ab^2 \qquad ...(4)$$

Along $BC$, put $y = b$ and $dy = 0$ in (1) we get $\bar{F} . \bar{d}r = (x^2 - b^2)\, dx$, where $x$ is from $a$ to 0.

$$\therefore \quad \int_{BC} \vec{F} . \vec{dr} = \int_a^0 (x^2 - b^2)\, dx = \left[\frac{x^3}{3} - b^2 x\right]_a^0 = \frac{-a^3}{3} + b^2 a \quad ...(5)$$

Along CO, put $x = 0$ and $dx = 0$ in (1), we get $\vec{F} . \vec{dr} = 0$

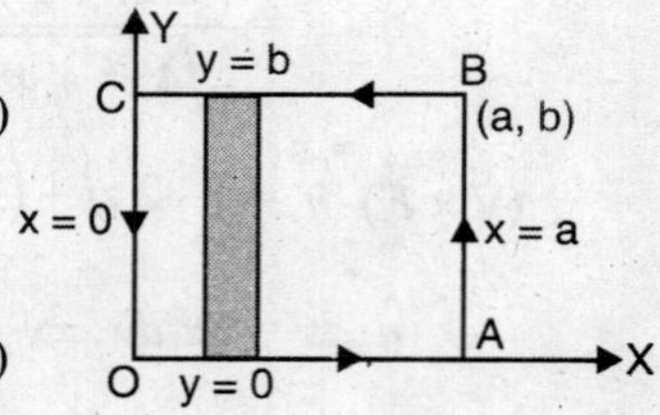


$$\therefore \quad \int_{CO} \vec{F} . \vec{dr} = 0 \quad ...(6)$$

Putting the values of integrals (3), (4), (5) and (6) in (2), we get

$$\int_C \vec{F} . \vec{dr} = \frac{a^3}{3} + ab^2 - \frac{a^3}{3} + ab^2 + 0 = 2ab^2 \quad ...(7)$$

Now we have to evaluate R.H.S. of Stoke's Theorem *i.e.* $\iint_S (\nabla \times \vec{F}) . \hat{n}\, ds$

We have,

$$\nabla \times \vec{F} = \begin{vmatrix} \hat{i} & \hat{j} & \hat{k} \\ \frac{\partial}{\partial x} & \frac{\partial}{\partial y} & \frac{\partial}{\partial z} \\ x^2 - y^2 & 2xy & 0 \end{vmatrix} = (2y + 2y)\, \hat{k} = 4y\, \hat{k}$$

Also the unit vector normal to the surface $S$ in outward direction is $\hat{n} = k$

($\because$ $z$-axis is normal to surface $S$)

Also in $xy$-plane $ds = dx\, dy$

$$\therefore \quad \iint_S (\nabla \times \vec{F}) . \hat{n} . ds = \iint_R 4y\hat{k} . \hat{k}\, dx\, dy = \iint_R 4y\, dx\, dy.$$

Where $R$ be the region of the surface $S$.

Consider a strip parallel to $y$-axis. This strip starts on line y = 0 (*i.e.* $x$-axis) and end on the line $y = b$, We move this strip from $x = 0$ ($y$-axis) to $x = a$ to cover complete region $R$.

$$\therefore \quad \iint_S (\nabla \times \vec{F}) . \hat{n} . ds = \int_0^a \left[\int_0^b 4y\, dy\right] dx = \int_0^a [2y^2]_0^b\, dx$$

$$= \int_0^a 2b^2\, dx = 2b^2\, [x]_0^a = 2ab^2 \quad ...(8)$$

$\therefore$ From (7) and (8), we get

$\int_C \vec{F} . \vec{dr} = \iint_S (\nabla \times \bar{F}) . \hat{n}\, ds$ and hence the Stoke's theorem is verified.

**Example 99.** *Verify Stoke's Theorem for the function*

$$\vec{F} = x^2\hat{i} - xy\hat{j}$$

*integrated round the square in the plane $z = 0$ and bounded by the lines*

*$x = 0$, $y = 0$, $x = a$, $y = a$.*

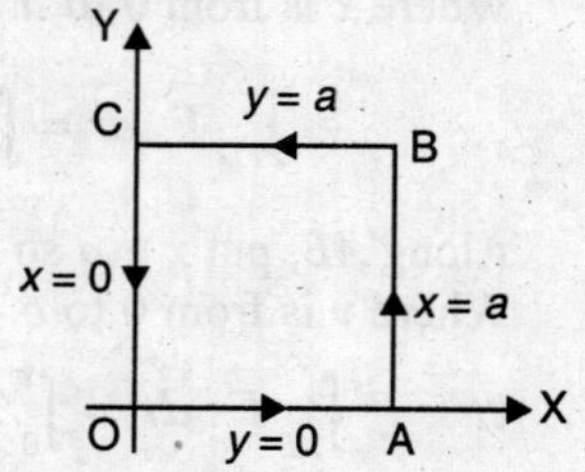


**Solution.** We have, $\vec{F} = x^2\hat{i} - xy\hat{j}$

$$\nabla \times \vec{F} = \begin{vmatrix} \hat{i} & \hat{j} & \hat{k} \\ \frac{\partial}{\partial x} & \frac{\partial}{\partial y} & \frac{\partial}{\partial z} \\ x^2 & -xy & 0 \end{vmatrix}$$

$$= (0 - 0)\, \hat{i} - (0 - 0)\, \hat{j} + (-y - 0)\, \hat{k} = -y\hat{k} \qquad (\hat{n} \perp \text{ to } xy \text{ plane } i.e.\ \hat{k})$$

$$\iint_S (\nabla \times \vec{F}) \cdot \hat{n}\, ds = \iint_S (-yk) \cdot k\, dx\, dy$$

$$= \int_0^a dx \int_0^a -y dy = \int_0^a dx \left[-\frac{y^2}{2}\right]_0^a = -\frac{a^2}{2}(x)_0^a = -\frac{a^3}{2} \qquad ...(1)$$

To obtain line integral

$$\int_C \vec{F} \cdot \vec{dr} = \int_C (x^2 \hat{i} - xy\hat{j}) \cdot (\hat{i}\, dx + \hat{j}\, dy) = \int_C (x^2\, dx - xy\, dy)$$

where $c$ is the path $OABCO$ as shown in the figure.

$$\text{Also, } \int_C \vec{F} \cdot \vec{dr} = \int_{OABCO} \vec{F} \cdot dr = \int_{OA} \vec{F} \cdot dr + \int_{AB} \vec{F} \cdot dr + \int_{BC} \vec{F} \cdot dr + \int_{CO} \vec{F} \cdot dr \qquad ...(2)$$

| line | Eq. of line | | Lower limit | Upper limit |
|---|---|---|---|---|
| $OA$ | $y = 0$ | $dy = 0$ | $x = 0$ | $x = a$ |
| $AB$ | $x = a$ | $dx = 0$ | $y = 0$ | $y = a$ |
| $BC$ | $y = a$ | $dy = 0$ | $x = a$ | $x = 0$ |
| $CO$ | $x = 0$ | $dx = 0$ | $y = a$ | $y = 0$ |

Along $OA$, $y = 0$, $dy = 0$

$$\int_{OA} \vec{F} \cdot \vec{dr} = \int_{OA} (x^2 dx - xydy)$$

$$= \int_0^a x^2 dx = \left[\frac{x^3}{3}\right]_0^a = \frac{a^3}{3}$$

Along $AB$, $x = a$, $dx = 0$

$$\int_{AB} \vec{F} \cdot \vec{dr} = \int_{AB} (x^2\, dx - x y\, d y)$$

$$= \int_0^a -a y\, d y = -a\left[\frac{y^2}{2}\right]_0^a = -\frac{a^3}{2}$$

Along $BC$, $y = a$, $dy = 0$

$$\int_{BC} \vec{F} \cdot \vec{dr} = \int_{BC} (x^2 dx - xy\, dy) = \int_a^0 x^2 dx = \left[\frac{x^3}{3}\right]_a^0 = -\frac{a^3}{3}$$

Along $CO$, $x = 0$, $dx = 0$

$$\int_{CO} \vec{F} \cdot \vec{dr} = \int_{CO} (x^2 dx - xy\, dy) = 0$$

Putting the values of these integrals in (2), we have

$$\int_C \vec{F} \cdot \vec{dr} = \frac{a^3}{3} - \frac{a^3}{2} - \frac{a^3}{3} + 0 = -\frac{a^3}{2} \qquad ...(3)$$

From (1) and (3), $\iint_S (\vec{\nabla} \times \vec{F}) \cdot \hat{n}\, ds = \int_C \vec{F} \cdot \vec{dr}$

Hence, Stoke's Theorem is verified. **Ans.**

**Example 100.** *Verify Stoke's Theorem for* $\vec{F} = (x + y)\, \hat{i} + (2x - z)\, \hat{j} + (y + z)\, \hat{k}$ *for the surface of a triangular lamina with vertices (2, 0, 0), (0, 3, 0) and (0, 0, 6).*

*(Nagpur University 2004, K. U. Dec. 2009, 2008, A.M.I.E.T.E., Summer 2000)*

**Solution.** Here the path of integration $c$ consists of the straight lines AB, BC, CA where the co-ordinates of A, B, C and (2, 0, 0), (0, 3, 0) and (0, 0, 6) respectively. Let $S$ be the plane surface of triangle ABC bounded by $C$. Let $\hat{n}$ be unit normal vector to surface $S$. Then by Stoke's Theorem, we must have

$$\oint_c \vec{F} \cdot \vec{dr} = \iint_s \text{curl}\, \vec{F} \cdot \hat{n}\, ds \qquad ...(1)$$

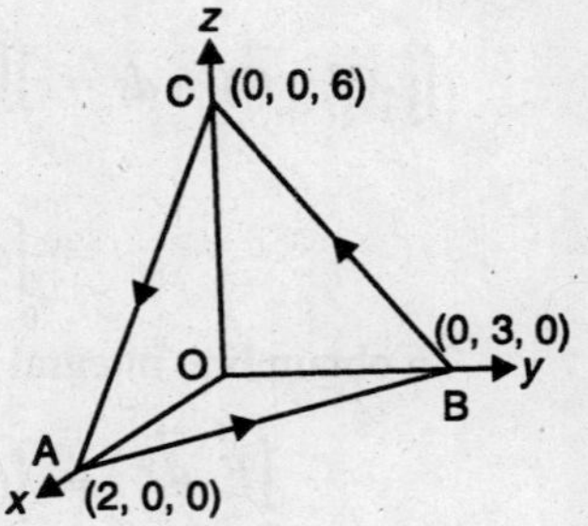


L.H.S. of (1)= $\int_{ABC}^{c} \vec{F}\cdot\vec{dr} = \int_{AB} \vec{F}\cdot\vec{dr} + \int_{BC} \vec{F}\cdot\vec{dr} + \int_{CA} \vec{F}\cdot\vec{dr}$

Along line $AB$, $z = 0$, equation of $AB$ is $\frac{x}{2} + \frac{y}{3} = 1$

$$\Rightarrow \qquad y = \frac{3}{2}(2 - x),\ dy = -\frac{3}{2}dx$$

At $A$, $x = 2$, At $B$, $x = 0$, $\bar{r} = x\hat{i} + y\hat{j}$

$$\int_{AB} \vec{F}\cdot\vec{dr} = \int_{AB} [(x + y)\hat{i} + 2x\hat{j} + y\hat{k}]\cdot(\hat{i}dx + \hat{j}dy)$$

$$= \int_{AB} (x + y)\,dx + 2x\,dy$$

$$= \int_{AB} \left(x + 3 - \frac{3x}{2}\right) dx + 2x\left(-\frac{3}{2}dx\right)$$

$$= \int_{2}^{0} \left(-\frac{7x}{2} + 3\right) dx = \left(-\frac{7x^2}{4} + 3x\right)_{2}^{0}$$

$$= (7 - 6) = +1$$

| line | Eq. of line | | Lower limit | Upper limit |
|---|---|---|---|---|
| $AB$ | $\frac{x}{2} + \frac{y}{3} = 1$, $z = 0$ | $dy = -\frac{3}{2}dx$ | At $A$ $x = 2$ | At $B$ $x = 0$ |
| $BC$ | $\frac{y}{3} + \frac{z}{6} = 1$, $x = 0$ | $dz = -2dy$ | At $B$ $y = 3$ | At $C$ $y = 0$ |
| $CA$ | $\frac{x}{2} + \frac{z}{6} = 1$, $y = 0$ | $dz = -3dx$ | At $C$ $x = 0$ | At $A$ $x = 2$ |

Along line $BC$, $x = 0$, Equation of $BC$ is $\frac{y}{3} + \frac{z}{6} = 1$ or $z = 6 - 2y$, $dz = -2dy$

At $B$, $y = 3$, At $C$, $y = 0$, $\bar{r} = y\hat{j} + z\hat{k}$

$$\int_{BC} \vec{F}\cdot\vec{dr} = \int_{BC} [yi + zj + (y + z)k]\cdot(jdy + kdz) = \int_{BC} -zdy + (y + z)\,dz$$

$$= \int_{3}^{0} (-6 + 2y)\,dy + (y + 6 - 2y)(-2dy)$$

$$= \int_{3}^{0} (4y - 18)\,dy = (2y^2 - 18y)_{3}^{0} = 36$$

Along line CA, $y = 0$, Eq. of CA, $\frac{x}{2} + \frac{z}{6} = 1$ or $z = 6 - 3x$, $dz = -3dx$

At C, $x = 0$, at A, $x = 2$, $\vec{r} = x\hat{i} + z\hat{k}$

$$\int_{CA} \vec{F}\cdot\vec{dr} = \int_{CA} [x\hat{i} + (2x - z)\hat{j} + z\hat{k}]\cdot[dx\hat{i} + dz\hat{k}] = \int_{CA} (xdx + zdz)$$

$$= \int_{0}^{2} xdx + (6 - 3x)(-3dx) = \int_{0}^{2} (10x - 18)\,dx = [5x^2 - 18x]_{0}^{2} = -16$$

$$\text{L.H.S. of (1)} = \int_{ABC} \vec{F}\cdot \vec{dr} = \int_{AB} \vec{F}\cdot \vec{dr} + \int_{BC} \vec{F}\cdot \vec{dr} + \int_{CA} \vec{F}\cdot \vec{dr} = 1 + 36 - 16 = 21 \quad ...(2)$$

$$\text{Curl } \overline{F} = \nabla \times \overline{F} = \left(\hat{i}\frac{\partial}{\partial x} + \hat{j}\frac{\partial}{\partial y} + \hat{k}\frac{\partial}{\partial z}\right) \times [(x+y)\,\hat{i} + (2x - z)\,\hat{j} + (y+z)\,k]$$

$$= \begin{vmatrix} \hat{i} & \hat{j} & \hat{k} \\ \frac{\partial}{\partial x} & \frac{\partial}{\partial y} & \frac{\partial}{\partial z} \\ x+y & 2x-z & y+z \end{vmatrix} = (1+1)\,\hat{i} - (0-0)\,\hat{j} + (2-1)\,\hat{k} = 2\hat{i} + \hat{k}$$

Equation of the plane of ABC is $\frac{x}{2} + \frac{y}{3} + \frac{z}{6} = 1$

Normal to the plane ABC is

$$\nabla \phi = \left(\hat{i}\frac{\partial}{\partial x} + \hat{j}\frac{\partial}{\partial y} + \hat{k}\frac{\partial}{\partial z}\right)\left(\frac{x}{2} + \frac{y}{3} + \frac{z}{6} - 1\right) = \frac{\hat{i}}{2} + \frac{\hat{j}}{3} + \frac{\hat{k}}{6}$$

$$\text{Unit Normal Vector} = \frac{\frac{\hat{i}}{2} + \frac{\hat{j}}{3} + \frac{\hat{k}}{6}}{\sqrt{\frac{1}{4} + \frac{1}{9} + \frac{1}{36}}}$$

$$\hat{n} = \frac{1}{\sqrt{14}}(3\hat{i} + 2\hat{j} + \hat{k})$$

$$\text{R.H.S. of (1)} = \iint_s \text{curl } \vec{F}\cdot \hat{n}\, ds = \iint_s (2\hat{i} + \hat{k}) \cdot \frac{1}{\sqrt{4}}(3\hat{i} + 2\hat{j} + \hat{k}) \frac{dx\, dy}{\frac{1}{\sqrt{14}}(3\hat{i} + 2\hat{j} + \hat{k}).\hat{k}}$$

$$= \iint_s \frac{(6+1)}{\sqrt{14}} \frac{dx\, dy}{\frac{1}{\sqrt{14}}} = 7 \iint dx\, dy = 7 \text{ Area of } \Delta \text{ OAB}$$

$$= 7\left(\frac{1}{2} \times 2 \times 3\right) = 21 \quad ...(3)$$

with the help of (2) and (3) we find (1) is true and so Stoke's *Theorem* is verified.

**Example 101.** *Verify Stoke's Theorem for*

$$\vec{F} = (y - z + 2)\,\hat{i} + (yz + 4)\,\hat{j} - (xz)\,\hat{k}$$

*over the surface of a cube x = 0, y = 0, z = 0, x = 2, y = 2, z = 2 above the XOY plane (open the bottom).*

**Solution.** Consider the surface of the cube as shown in the figure. Bounding path is *OABCO* shown by arrows.

$$\int_c \vec{F}\cdot \vec{d}\, r = \int [(y - z + 2)\,\hat{i} + (yz + 4)\,\hat{j} - (xz)\,\hat{k}] \cdot (\hat{i}dx + \hat{j}dy + \hat{k}dz)$$

$$= \int_c (y - z + 2)\, dx + (yz + 4)\, dy - xzdz$$

$$\int_c \vec{F}\cdot \vec{dr} = \int_{OA} \vec{F}\cdot \vec{dr} + \int_{AB} \vec{F}\cdot \vec{dr} + \int_{BC} \vec{F}\cdot \vec{dr} + \int_{CO} \vec{F}\cdot \vec{dr} \quad ...(1)$$

(1) Along *OA*, y = 0, dy = 0, z = 0, dz = 0

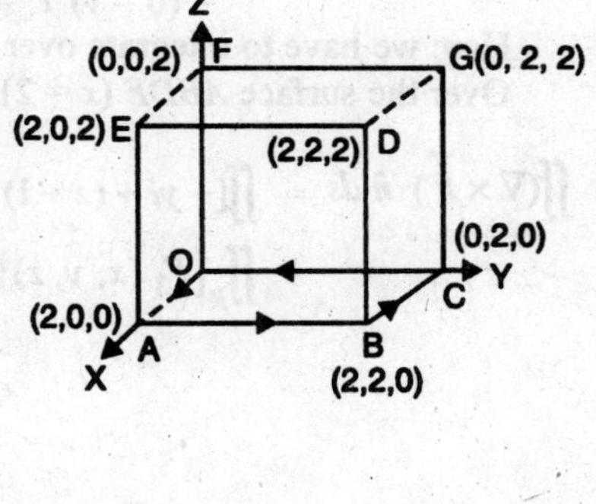

| | Line | Equ. of line | | Lower limit | Upper limit | $\vec{F}.\vec{dr}$ |
|---|---|---|---|---|---|---|
| 1 | $OA$ | $y=0$, $z=0$ | $dy=0$, $dz=0$ | $x=0$ | $x=2$ | $2\,dx$ |
| 2 | $AB$ | $x=2$, $z=0$ | $dx=0$, $dz=0$ | $y=0$ | $y=2$ | $4\,dy$ |
| 3 | $BC$ | $y=2$, $z=0$ | $dy=0$, $dz=0$ | $x=2$ | $x=0$ | $4\,dx$ |
| 4 | $CO$ | $x=0$, $z=0$ | $dx=0$, $dz=0$ | $y=2$ | $y=0$ | $4\,dy$ |

$$\int_{OA} \vec{F}\cdot\vec{dr} = \int_0^2 2\,dx = [2x]_0^2 = 4$$

(2) Along $AB$, $x=2$, $dx=0$, $z=0$, $dz=0$

$$\int_{AB} \vec{F}\cdot\vec{dr} = \int_0^2 4\,dy = 4\,(y)_0^2 = 8$$

(3) Along $BC$, $y=2$, $dy=0$, $z=0$, $dz=0$

$$\int_{BC} \vec{F}\cdot\vec{dr} = \int_0^2 (2-0+2)\,dx = (4x)_2^0 = -8$$

(4) Along $CO$, $x=0$, $dx=0$, $z=0$, $dz=0$

$$\int_{CO} \vec{F}\cdot\vec{dr} = \int (y-0+2)\times 0 + (0+4)\,dy - 0$$

$$= 4\int dy = 4\,(y)_2^0 = -8$$

On putting the values of these integrals in (1), we get

$$\int_c \vec{F}\cdot\vec{dr} = 4+8-8-8 = -4$$

To obtain surface integral

$$\nabla\times\vec{F} = \begin{vmatrix} \hat{i} & \hat{j} & \hat{k} \\ \dfrac{\partial}{\partial x} & \dfrac{\partial}{\partial y} & \dfrac{\partial}{\partial z} \\ y-z+2 & yz+4 & -xz \end{vmatrix}$$

$$= (0-y)\,\hat{i} - (-z+1)\,\hat{j} + (0-1)\,\hat{k} = -y\hat{i} + (z-1)\,\hat{j} - \hat{k}$$

Here we have to integrate over the five surfaces, $ABDE$, $OCGF$, $BCGD$, $OAEF$, $DEFG$.
Over the surface $ABDE$ $(x=2)$, $\hat{n} = i$, $ds = dy\,dz$

$$\iint (\nabla\times\vec{F})\cdot\hat{n}\,ds = \iint [-yi + (z-1)\,j - k]\cdot i\,dx\,dz = \iint -y\,dy\,dz$$

$$= \iint_R [F_3(x,y,z)]_{z=f_1(x,y)}^{z=f_2(x,y)}\,dx\,dy$$

| | Surface | Outward normal | ds | |
|---|---|---|---|---|
| 1 | *ABDE* | $i$ | $dy\,dz$ | $x = 2$ |
| 2 | *OCGF* | $-i$ | $dy\,dz$ | $x = 0$ |
| 3 | *BCGD* | $j$ | $dx\,dz$ | $y = 2$ |
| 4 | *OAEF* | $-j$ | $dx\,dz$ | $y = 0$ |
| 5 | *DEFG* | $k$ | $dx\,dy$ | $z = 2$ |

$$= -\int_0^2 y\,dy \int_0^2 dz = -\left[\frac{y^2}{2}\right]_0^2 [z]_0^2 = -4$$

Over the surface OCGF $(x = 0)$, $\hat{n} = -i$, $ds = dy\,dz$

$$\iint(\nabla \times \vec{F})\cdot\hat{n}\,ds = \iint[-y\hat{i} + (z-1)\,\hat{j} - \hat{k}]\cdot(-\hat{i})\,dy\,dz$$

$$= \iint y\,dy\,dz = \int_0^2 y\,dy \int_0^2 dz = 2\left[\frac{y^2}{2}\right]_0^2 = 4$$

(3) Over the surface *BCGD*, $(y = 2)$, $\hat{n} = j$, $ds = dx\,dz$

$$\iint(\nabla \times \vec{F})\cdot\hat{n}\,ds = \iint[-y\hat{i} + (z-1)\,\hat{j} - \hat{k}]\cdot\hat{j}\,dx\,dz$$

$$= -\iint(z-1)\,dx\,dz = \int_0^2 dx \int_0^2 (z-1)\,dz = -(x)_0^2\left(\frac{z^2}{2} - z\right)_0^2 = 0$$

(4) Over the surface *OAEF*, $(y = 0)$, $\hat{n} = -\hat{j}$, $ds = dx\,dz$

$$\iint(\nabla \times \vec{F})\cdot\hat{n}\,ds = \iint[-y\hat{i} + (z-1)\hat{j} - \hat{k}]\cdot(-\hat{j})\,dx\,dz$$

$$= -\iint(z-1)\,dx\,dz = -\int_0^2 dx \int_0^2 (z-1)\,dz = -(x)_0^2\left(\frac{z^2}{2} - z\right)_0^2 = 0$$

(5) Over the surface *DEFG*, $(z = 2)$, $\hat{n} = k$, $ds = dx\,dy$

$$\iint(\nabla \times \vec{F})\cdot\hat{n}\,ds = \iint[-y\hat{i} + (z-1)\hat{j} - \hat{k}]\cdot\hat{k}\,dx\,dy = -\iint dx\,dy$$

$$= -\int_0^2 dx \int_0^2 dy = -[x]_0^2\,[y]_0^2 = -4$$

Total surface integral $= -4 + 4 + 0 + 0 - 4 = -4$

Thus $\iint_S \text{curl}\,\vec{F}\cdot\hat{n}\,ds = \int_c \vec{F}\cdot d\vec{r} = -4$

which verifies Stoke's Theorem. **Ans.**

## EXERCISE 5.14

1. Use the Stoke's Theorem to evaluate $\int_C y^2 dx + xy\, dy + xz\, dz$, where $C$ is the bounding curve of the hemisphere $x^2 + y^2 + z^2 = 1$, $z \geq 0$, oriented in the positive direction. **Ans.** 0

2. Evaluate $\int_s (\text{curl } F) \cdot \hat{n}\, dA$, using the Stoke's Theorem, where $\vec{F} = y\hat{i} + z\hat{j} + x\hat{k}$ and $s$ is the paraboloid $z = f(x, y) = 1 - x^2 - y^2$, $z \geq 0$. **Ans.** $\pi$

3. Evaluate the integral for $\int_C y^2 dx + z^2 dy + x^2 dz$, where C is the triangular closed path joining the points (0, 0, 0), (0, $a$, 0) and (0, 0, $a$) by transforming the integral to surface integral using Stoke's Theorem. **Ans.** $\frac{a^3}{3}$.

4. Verify Stoke's Theorem for $\vec{A} = 3y\hat{i} - xz\hat{j} + yz^2\hat{k}$, where $S$ is the surface of the paraboloid $2z = x^2 + y^2$ bounded by $z = 2$ and $c$ is its boundary traversed in the clockwise direction. **Ans.** $-20\pi$

5. Evaluate $\int_C \vec{F} \cdot \vec{d}R$ where $\vec{F} = y\hat{i} + xz^3\hat{j} - zy^3\hat{k}$, $C$ is the circl $x^2 + y^2 = 4$, $z = 1.5$ **Ans.** $\frac{19}{2}\pi$

6. If $S$ is the surface of the sphere $x^2 + y^2 + z^2 = 9$. Prove that $\int_S \text{curl } \vec{F} \cdot ds = 0$.

7. Verify Stoke's Theorem for the vector field
$$\vec{F} = (2y + z)\,\hat{i} + (x - z)\,\hat{j} + (y - x)\,\hat{k}$$
over the portion of the plane $x + y + z = 1$ cut off by the co-ordinate planes.

8. Evaluate $\int_c \vec{F} \cdot dr$ by Stoke's Theorem for $\vec{F} = yz\,\hat{i} + zx\,\hat{j} + xy\,k$ and $C$ is the curve of intersection of $x^2 + y^2 = 1$ and $y = z^2$. **Ans.** 0

9. If $\vec{F} = (x - z)\,\hat{i} + (x^3 + yz)\,\hat{j} + 3xy^2\,\hat{k}$ and $S$ is the surface of the cone $z = a - \sqrt{(x^2 + y^2)}$ above the $xy$-plane, show that $\iint_s \text{curl } \vec{F} \cdot dS = 3\pi a^4 / 4$.

10. If $\vec{F} = 3y\hat{i} - xy\hat{j} + yz^2\hat{k}$ and $S$ is the surface of the paraboloid $2z = x^2 + y^2$ bounded by $z = 2$, show by using Stoke's Theorem that $\iint_s (\nabla \times \vec{F}) \cdot dS = 20\pi$.

11. If $\vec{F} = (y^2 + z^2 - x^2)\,\hat{i} + (z^2 + x^2 - y^2)\,\hat{j} + (x^2 + y^2 - z^2)\,\hat{k}$, evaluate $\int \text{curl } \vec{F} \cdot \hat{n}\, ds$ integrated over the portion of the surface $x^2 + y^2 - 2ax + az = 0$ above the plane $z = 0$ and verify Stoke's Theorem; where $\hat{n}$ is unit vector normal to the surface. *(A.M.I.E.T.E., Winter 20002)* **Ans.** $2\pi a^3$

12. Evaluate by using Stoke's Theorem $\int_C [\sin z\, dx - \cos x\, dy + \sin y\, dz]$ where C is the boundary of rectangle $0 \leq x \leq \pi$, $0 \leq y \leq 1$, $z = 3$. *(AMIETE, June 2010)*

## 5.40 GAUSS'S THEOREM OF DIVERGENCE

**(Relation between surface integral and volume integral)**

*(U.P., Ist Semester, Jan., 2011, Dec, 2006)*

**Statement.** The surface integral of the normal component of a vector function $F$ taken around a closed surface $S$ is equal to the integral of the divergence of $F$ taken over the volume $V$ enclosed by the surface $S$.

Mathematically

$$\iint_S \vec{F} . \hat{n}\, ds = \iiint_V div\, \vec{F} dw$$

**Proof.** Let $\vec{F} = F_1\hat{i} + F_2\hat{j} + F_3\hat{k}$.

Putting the values of $\vec{F}, \hat{n}$ in the statement of the divergence theorem, we have

$$\iint_S F_1\,\hat{i} + F_2\,\hat{j} + F_3\,\hat{k}\cdot\hat{n}\,ds = \iiint_V \left(\hat{i}\frac{\partial}{\partial x} + \hat{j}\frac{\partial}{\partial y} + \hat{k}\frac{\partial}{\partial z}\right)(F_1\,\hat{i} + F_2\,\hat{j} + F_3\,\hat{k})\,dx\,dy\,dz.$$

$$= \iiint_V \left(\frac{\partial F_1}{\partial x} + \frac{\partial F_2}{\partial y} + \frac{\partial F_3}{\partial z}\right) dx\,dy\,dz \qquad \text{...(1)}$$

We require to prove (1).

Let us first evaluate $\iiint_V \frac{\partial F_3}{\partial z}\,dx\,dy\,dz$.

$$\iiint_V \frac{\partial F_3}{\partial z}\,dx\,dy\,dz = \iint_R \left[\int_{z=f_1(x,y)}^{z=f_2(x,y)} \frac{\partial F_3}{\partial z}\,dz\right] dx\,dy$$

$$= \iint_R [F_3\,(x, y, f_2) - F_3\,(x, y, f_1)]\,dx\,dy \qquad \text{...(2)}$$

For the upper part of the surface *i.e.* $S_2$, we have

$$dx\,dy = ds_2 \cos r_2 = \hat{n}_2.\hat{k}\;ds_2$$

Again for the lower part of the surface *i.e.* $S_1$, we have,

$$dx\,dy = -\cos r_1,\, ds_1 = \hat{n}_1.\;\hat{k}\,ds_1$$

$$\iint_R F_3\,(x, y, f_2)\,dx\,dy = \iint_{S_2} F_3\,\hat{n}_2\cdot\hat{k}\,ds_2$$

and $$\iint_R F_3\,(x, y, f_1)\,dx\,dy = -\iint_{S_1} F_3\,\hat{n}_1\cdot\hat{k}\,ds_1$$

Putting these values in (2), we have

$$\iiint_V \frac{\partial F_3}{\partial z}\,dv = \iint_{S_2} F_3\,\hat{n}_2\cdot\hat{k}\,ds_2 + \iint_{S_1} F_3\,\hat{n}_1\cdot\hat{k}\,ds_1 = \iint_S F_3\,\hat{n}\cdot\hat{k}\,ds \qquad \text{...(3)}$$

Similarly, it can be shown that

$$\iiint_V \frac{\partial F_2}{\partial y}\,dv = \iint_S F_2\,\hat{n}\cdot\hat{j}\,ds \qquad \text{...(4)}$$

$$\iiint_V \frac{\partial F_1}{\partial x}\,dv = \iint_S F_1\,\hat{n}\cdot\hat{i}\,ds \qquad \text{...(5)}$$

Adding (3), (4) & (5), we have

$$\iiint_V \left(\frac{\partial F_1}{\partial x} + \frac{\partial F_2}{\partial y} + \frac{\partial F_3}{\partial z}\right) dv$$

$$= \iint_S (F_1\,\hat{i} + F_2\,\hat{j} + F_3\,\hat{k})\cdot\hat{n}\cdot ds$$

$$\Rightarrow \iiint_V (\nabla\cdot\overline{F})\,dv = \iint_S \overline{F}\cdot\hat{n}\cdot ds \quad \textbf{Proved.}$$

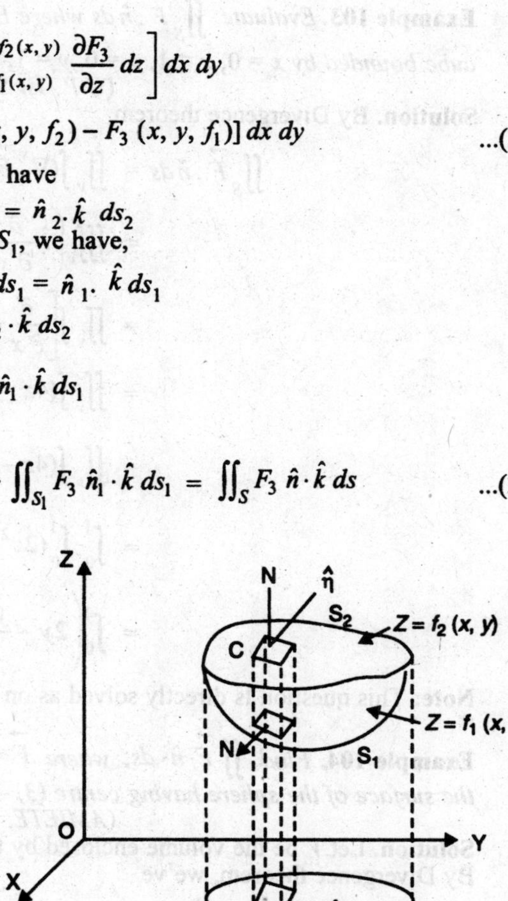


**Example 102.** *State Gauss's Divergence theorem* $\iint_S \vec{F}.\hat{n}\,ds = \iiint Div\,\vec{F}\,dv$ *where S is the surface of the sphere* $x^2 + y^2 + z^2 = 16$ *and* $\vec{F} = 3x\hat{i} + 4y\,\hat{j} + 5z\,\hat{k}$.

(*Nagpur University, Winter 2004*)

**Solution.** Statement of Gauss's Divergence theorem is given in Art 24.8 on page 597.
Thus by Gauss's divergence theorem,

$$\iint_S \vec{F}.\hat{n}\,ds = \iint_v \int \nabla.\vec{F}\,dv \qquad \text{Here } \vec{F} = 3x\hat{i} + 4y\,\hat{j} + 5z\,\hat{k}$$

$$\nabla . \vec{F} = \left( \hat{i} \frac{\partial}{\partial x} + \hat{j} \frac{\partial}{\partial y} + \hat{k} \frac{\partial}{\partial z} \right) . (3x\hat{i} + 4y\hat{j} + 5z\hat{k})$$

$$\nabla . \vec{F} = 3 + 4 + 5 = 12$$

Putting the value of $\nabla$ . F, we get

$$\iint_S \vec{F} . \hat{n}\, ds = \iint_v \int 12 . dv \qquad \text{where } v \text{ is volume of a sphere}$$

$$= 12\, v$$

$$= 12 \frac{4}{3} \pi (4)^3 = \frac{1024\pi}{3} \qquad \textbf{Ans.}$$

**Example 103.** *Evaluate* $\iint_S \vec{F} . \hat{n}\, ds$ *where* $\vec{F} = 4xz\,\hat{i} - y^2\,\hat{j} + yz\,\hat{k}$ *and S is the surface of the cube bounded by x = 0, x = 1, y = 0, y = 1, z = 0, z = 1.*

*(U.P., Ist Semester, 2009, Nagpur University, Winter 2003)*

**Solution.** By Divergence theorem,

$$\iint_S \vec{F} . \hat{n}\, ds = \iint_v \int (\nabla . \vec{F})\, dv$$

$$= \iiint_v \left( \hat{i} \frac{\partial}{\partial x} + \hat{j} \frac{\partial}{\partial y} + \hat{k} \frac{\partial}{\partial z} \right) . (4xz\,\hat{i} - y^2\,\hat{j} + yz\,\hat{k})\, dv$$

$$= \iint_v \int \left[ \frac{\partial}{\partial x}(4\,xz) + \frac{\partial}{\partial y}(-y^2) + \frac{\partial}{\partial z}(yz) \right] dx\, dy\, dz$$

$$= \iint_v \int (4z - 2y + y)\, dx\, dy\, dz$$

$$= \iint_v \int (4z - y)\, dx\, dy\, dz = \int_0^1 \int_0^1 \left( \frac{4z^2}{2} - yz \right)_0^1 dx\, dy$$

$$= \int_0^1 \int_0^1 (2z^2 - yz)_0^1\, dx\, dy = \int_0^1 \int_0^1 (2 - y)\, dx\, dy$$

$$= \int_0^1 \left( 2y - \frac{y^2}{2} \right)_0^1 dx = \frac{3}{2} \int_0^1 dx = \frac{3}{2} [x]_0^1 = \frac{3}{2}(1) = \frac{3}{2} \qquad \textbf{Ans.}$$

**Note:** This question is directly solved as on example 14 on Page 574.

**Example 104.** *Find* $\iint \vec{F} \cdot \hat{n} \cdot ds$, *where* $\vec{F} = (2x + 3z)\,\hat{i} - (xz + y)\,\hat{j} + (y^2 + 2z)\,\hat{k}$ *and S is the surface of the sphere having centre (3, – 1, 2) and radius 3.*

*(AMIETE, Dec. 2010, U.P., I Semester, Winter 2005, 2000)*

**Solution.** Let $V$ be the volume enclosed by the surface $S$.

By Divergence theorem, we've

$$\iint_S \vec{F} \cdot \hat{n} \cdot ds = \iiint_V div\, \vec{F}\, dv.$$

Now, $div\, \vec{F} = \frac{\partial}{\partial x}(2x + 3z) + \frac{\partial}{\partial y}[-(xz + y)] + \frac{\partial}{\partial z}(y^2 + 2z) = 2 - 1 + 2 = 3$

$$\therefore \quad \iint_S \vec{F} \cdot \hat{n} \cdot ds = \iiint_V 3\, dv = 3 \iiint_V dv = 3V.$$

Again V is the volume of a sphere of radius 3. Therefore

$$V = \frac{4}{3} \pi r^3 = \frac{4}{3} \pi (3)^3 = 36\, \pi.$$

$$\therefore \quad \iint_S \vec{F} \cdot \hat{n} \cdot ds = 3V = 3 \times 36\, \pi = 108\, \pi \qquad \textbf{Ans.}$$

**Example 105.** *Use Divergence Theorem to evaluate* $\iint_S \vec{A}\cdot\vec{ds}$,

*where* $\vec{A} = x^3\hat{i} + y^3\hat{j} + z^3\hat{k}$ *and S is the surface of the sphere* $x^2 + y^2 + z^2 = a^2$.

(*AMIETE, Dec. 2009*)

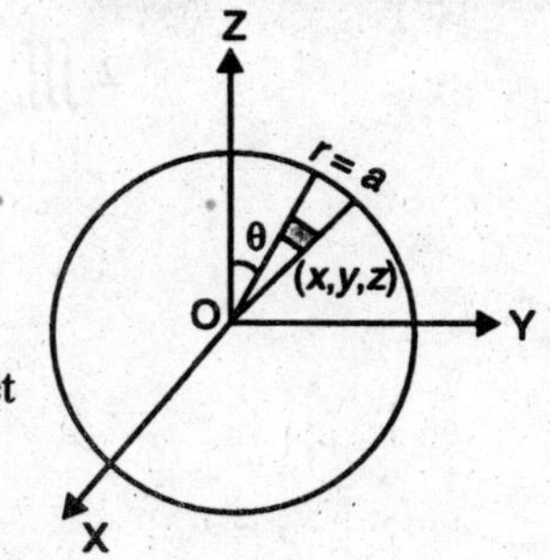


**Solution.** $\iint_S \vec{A}\cdot\vec{ds} = \iiint_V \text{div } \vec{A}\, dV$

$$= \iiint_V \left(\hat{i}\frac{\partial}{\partial x} + \hat{j}\frac{\partial}{\partial y} + \hat{k}\frac{\partial}{\partial z}\right)\cdot(x^3\hat{i} + y^3\hat{j} + z^3\hat{k})\,dV$$

$$= \iiint_V (3x^2 + 3y^2 + 3z^2)\,dV = 3\iiint_V (x^2 + y^2 + z^2)\,dV$$

On putting $x = r\sin\theta\cos\phi$, $y = r\sin\theta\sin\phi$, $z = r\cos\theta$, we get

$$= 3\iiint_V r^2\,(r^2\sin\theta\,dr\,d\theta\,d\phi) = 3\times 8\int_0^{\frac{\pi}{2}} d\phi \int_0^{\frac{\pi}{2}} \sin\theta\,d\theta \int_0^a r^4\,dr$$

$$= 24\,(\phi)_0^{\frac{\pi}{2}}\,(-\cos\theta)_0^{\frac{\pi}{2}}\left(\frac{r^5}{5}\right)_0^a = 24\left(\frac{\pi}{2}\right)(-0+1)\left(\frac{a^5}{5}\right) = \frac{12\pi a^5}{5}$$ **Ans.**

**Example 106.** *Use divergence Theorem to show that*

$$\iint_S \nabla(x^2 + y^2 + z^2)\,d\vec{s} = 6V$$

*where S is any closed surface enclosing volume V.* (*U.P., I Semester, Winter 2002*)

**Solution.** Here $\nabla\ (x^2 + y^2 + z^2) = \left(\hat{i}\frac{\partial}{\partial x} + \hat{j}\frac{\partial}{\partial y} + \hat{k}\frac{\partial}{\partial z}\right)\cdot(x^2 + y^2 + z^2)$

$$= 2x\hat{i} + 2y\hat{j} + 2z\hat{k} = 2(x\hat{i} + y\hat{j} + z\hat{k})$$

$$\therefore\quad \iint_S \nabla(x^2 + y^2 + z^2)\cdot ds = \iint_S \nabla(x^2 + y^2 + z^2)\cdot\hat{n}\,ds$$

$\hat{n}$ **being outward drawn unit normal vector to** $S$

$$= \iint_S 2(x\hat{i} + y\hat{j} + z\hat{k})\cdot\hat{n}\,ds$$

$$= 2\iiint_V div\,(x\hat{i} + y\hat{j} + z\hat{k})\,dv \quad ...(1)$$

**(By Divergence Theorem)**
**(V being volume enclosed by** $S$**)**

Now, div. $(x\hat{i} + y\hat{j} + z\hat{k}) = \left(\hat{i}\frac{\partial}{\partial x} + \hat{j}\frac{\partial}{\partial y} + \hat{k}\frac{\partial}{\partial z}\right)\cdot(x\hat{i} + y\hat{j} + z\hat{k})$

$$= \frac{\partial x}{\partial x} + \frac{\partial y}{\partial y} + \frac{\partial z}{\partial z} = 3 \quad ...(2)$$

From (1) & (2), we have

$$\iint \nabla(x^2 + y^2 + z^2)\cdot dS = 2\iiint_V 3\,dv = 6\iiint_V dv = 6V$$ **Proved.**

**Example 107.** *Evaluate* $\iint_S (y^2z^2\hat{i} + z^2x^2\hat{j} + z^2y^2\hat{k})\,\hat{n}\,dS$, *where S is the part of the sphere* $x^2 + y^2 + z^2 = 1$ *above the xy-plane and bounded by this plane.*

**Solution.** Let $V$ be the volume enclosed by the surface $S$. Then by divergence Theorem, we have

$$\iint_S (y^2z^2\hat{i} + z^2x^2\hat{j} + z^2y^2\hat{k})\,\hat{n}\,dS = \iiint_V div\,(y^2z^2\hat{i} + z^2x^2\hat{j} + z^2y^2\hat{k})\,dV$$

$$= \iiint_V \left[\frac{\partial}{\partial x}(y^2z^2) + \frac{\partial}{\partial y}(z^2x^2) + \frac{\partial}{\partial z}(z^2y^2)\right]dV = \iint_V 2z\,y^2\,dV = 2\iint_V zy^2\,dV$$

Changing to spherical polar coordinates by putting

$$x = r \sin\theta \cos\phi, \quad y = r \sin\theta \sin\phi, \quad z = r\cos\theta, \; dV = r^2 \sin\theta \, dr\, d\theta\, d\phi$$

To cover $V$, the limits of $r$ will be 0 to 1, those of $\theta$ will be 0 to $\frac{\pi}{2}$ and those of $\phi$ will be 0 to $2\pi$.

$$\therefore \quad 2\iiint_V zy^2 \, dV = 2\int_0^{2\pi}\int_0^{\pi/2}\int_0^1 (r\cos\theta)(r^2\sin^2\theta\sin^2\phi)\, r^2\sin\theta \cdot dr\, d\theta\, d\phi$$

$$= 2\int_0^{2\pi}\int_0^{\pi/2}\int_0^1 r^5 \sin^3\theta\cos\theta\sin^2\phi\, dr\, d\theta\, d\phi$$

$$= 2\int_0^{2\pi}\int_0^{\frac{\pi}{2}} \sin^3\theta\cos\theta\sin^2\phi \left[\frac{r^6}{6}\right]_0^1 d\theta\, d\phi$$

$$= \frac{2}{6}\int_0^{2\pi}\sin^2\phi \cdot \frac{2}{4.2}\, d\phi = \frac{1}{12}\int_0^{2\pi}\sin^2\phi\, d\phi = \frac{\pi}{12} \qquad \textbf{Ans.}$$

**Example 108.** *Use Divergence Theorem to evaluate* $\iint_S \vec{F}\cdot d\vec{S}$ *where* $\vec{F} = 4x\hat{i} - 2y^2\hat{j} + z^2\hat{k}$ *and S is the surface bounding the region* $x^2 + y^2 = 4$, $z = 0$ *and* $z = 3$.

*(A.M.I.E.T.E , Summer 2003, 2001)*

**Solution.** By Divergence Theorem,

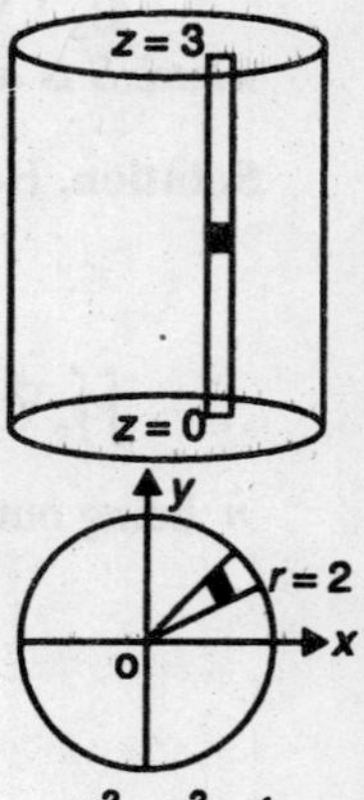

$x^2 + y^2 = 4$

$$\iint_S \vec{F}\cdot dS = \iiint_V div\,\vec{F}\, dV$$

$$= \iiint_V \left(\hat{i}\frac{\partial}{\partial x} + \hat{j}\frac{\partial}{\partial y} + \hat{k}\frac{\partial}{\partial z}\right)\cdot(4x\hat{i} - 2y^2\hat{j} + z^2\hat{k})\, dV$$

$$= \iiint_V (4 - 4y + 2z)\, dx\, dy\, dz$$

$$= \iint dx\, dy \int_0^3 (4 - 4y + 2z)dz = \iint dx\, dy\, [4z - 4yz + z^2]_0^3$$

$$= \iint (12 - 12y + 9)\, dx\, dy = \iint (21 - 12y)\, dx\, dy$$

Let us put $x = r\cos\theta$, $y = r\sin\theta$

$$= \iint (21 - 12r\sin\theta)\, r\, d\theta\, dr = \int_0^{2\pi} d\theta \int_0^2 (21r - 12r^2\sin\theta)\, dr$$

$$= \int_0^{2\pi} d\theta \left[\frac{21r^2}{2} - 4r^3\sin\theta\right]_0^2 = \int_0^{2\pi} d\theta\,(42 - 32\sin\theta) = (42\,\theta + 32\cos\theta)_0^{2\pi}$$

$$= 84\pi + 32 - 32 = 84\pi \qquad \textbf{Ans.}$$

**Example 109.** *Apply the Divergence Theorem to compute* $\iint \vec{u}\cdot\hat{n}\, ds$, *where s is the surface of the cylinder* $x^2 + y^2 = a^2$ *bounded by the planes* $z = 0$, $z = b$ *and where* $u = \hat{i}x - \hat{j}y + \hat{k}z$.

**Solution.** By Gauss's Divergence Theorem

$x^2 + y^2 = a^2$

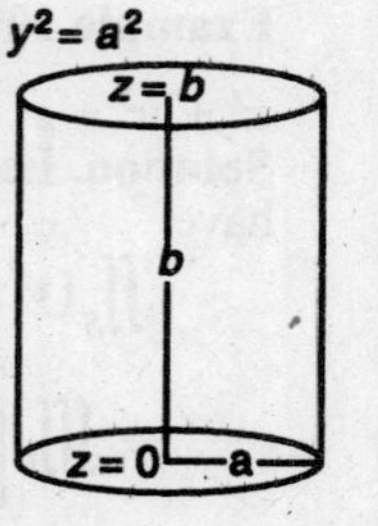


$$\iint \vec{u}\cdot\hat{n}ds = \iiint_V (\nabla\cdot\vec{u})\, dv$$

$$= \iiint_V \left(\hat{i}\frac{\partial}{\partial x} + \hat{j}\frac{\partial}{\partial y} + \hat{k}\frac{\partial}{\partial z}\right)\cdot(\hat{i}x - \hat{j}y + \hat{k}z)dv$$

$$= \iiint_V \left(\frac{\partial x}{\partial x} - \frac{\partial y}{\partial y} + \frac{\partial z}{\partial z}\right) dv = \iiint_V (1 - 1 + 1)\, dv$$

$$= \iiint_V dv = \iiint_V dx\, dy\, dz = \text{Volume of the cylinder} = \pi a^2 b \qquad \textbf{Ans.}$$

**Example 110.** *Apply Divergence Theorem to evaluate* $\iiint_V \vec{F}\,.\,\hat{n}\,ds$, *where* $\vec{F} = 4x^3\hat{i} - x^2 y\,\hat{j} + x^2 z\hat{k}$ *and S is the surface of the cylinder* $x^2 + y^2 = a^2$ *bounded by the planes* $z = 0$ *and* $z = b$. *(U.P. Ist Semester, Dec. 2006)*

**Solution.** We have,

$$\vec{F} = 4x^3\hat{i} - x^2 y\,\hat{j} + x^2 z\hat{k}$$

$$\therefore \quad div\,\vec{F} = \left(\hat{i}\frac{\partial}{\partial x} + \hat{j}\frac{\partial}{\partial y} + \hat{k}\frac{\partial}{\partial z}\right)\cdot(4x^3\hat{i} - x^2 y\hat{j} + x^2 z\hat{k})$$

$$= \frac{\partial}{\partial x}(4x^3) + \frac{\partial}{\partial y}(-x^2 y) + \frac{\partial}{\partial z}(x^2 z) = 12x^2 - x^2 + x^2 = 12\,x^2$$

Now, $$\iiint_V div\,\vec{F}\,dV = 12\int_{x=-a}^{a}\int_{y=-\sqrt{a^2-x^2}}^{\sqrt{a^2-x^2}}\int_{z=0}^{b} x^2\,dz\,dy\,dx$$

$$= 12\int_{x=-a}^{a}\int_{y=-\sqrt{a^2-x^2}}^{\sqrt{a^2-x^2}} x^2\,(z)_0^b\,dy\,dx = 12\,b\int_{-a}^{a} x^2\,(y)_{-\sqrt{a^2-x^2}}^{\sqrt{a^2-x^2}}\,dx$$

$$= 12\,b\int_{-a}^{a} x^2 . 2\sqrt{a^2-x^2}\,dx = 24\,b\int_{-a}^{a} x^2\sqrt{a^2-x^2}\,dx$$

$$= 48\,b\int_0^a x^2\sqrt{a^2-x^2}\,dx \qquad [\text{Put } x = a\sin\theta,\ dx = a\cos\theta\,d\theta]$$

$$= 48\,b\int_0^{\pi/2} a^2\sin^2\theta\;a\cos\theta\;a\cos\theta\;d\theta$$

$$= 48\,ba^4\int_0^{\pi/2}\sin^2\theta\cdot\cos^2\theta\,d\theta = 48\,ba^4\,\frac{\left|\frac{3}{2}\right.\left|\frac{3}{2}\right.}{2\left|3\right.}$$

$$= 48\,ba^4\,\frac{\frac{1}{2}\sqrt{\pi}\cdot\frac{1}{2}\sqrt{\pi}}{2\cdot 2} = 3\,b\,a^4\pi$$ **Ans.**

**Example 111.** *Evaluate surface integral* $\iint \vec{F}\cdot\hat{n}\,ds$, *where* $\vec{F} = (x^2 + y^2 + z^2)\,(\hat{i} + \hat{j} + \hat{k})$, *S is the surface of the tetrahedron* $x = 0$, $y = 0$, $z = 0$, $x + y + z = 2$ *and n is the unit normal in the outward direction to the closed surface S.*

**Solution.** By Divergence theorem

$$\iint_S \vec{F}\cdot\hat{n}\,ds = \iiint_V \text{div}\,\vec{F}\cdot dv$$

where $S$ is the surface of tetrahedron $x = 0$, $y = 0$, $z = 0$, $x + y + z = 2$

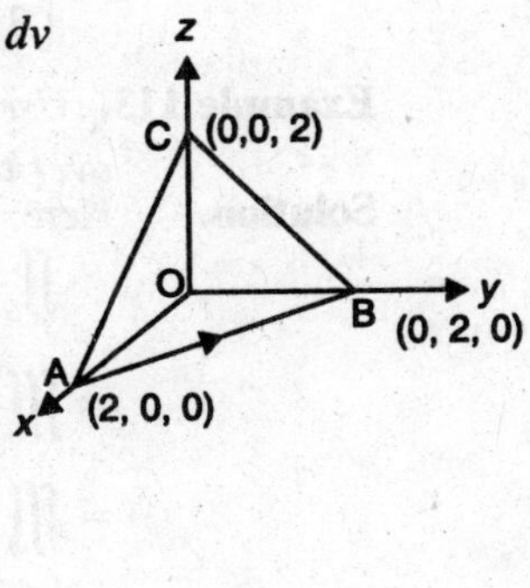


$$= \iiint_V\left(\hat{i}\frac{\partial}{\partial x} + \hat{j}\frac{\partial}{\partial y} + \hat{k}\frac{\partial}{\partial z}\right)\cdot(x^2 + y^2 + z^2)(\hat{i} + \hat{j} + \hat{k})\,dv$$

$$= \iiint_V (2x + 2y + 2z)\,dv$$

$$= 2\iiint_V (x + y + z)\,dx\,dy\,dz$$

$$= 2\int_0^2 dx\int_0^{2-x} dy\int_0^{2-x-y}(x + y + z)\,dz$$

$$= 2\int_0^2 dx\int_0^{2-x} dy\left(xz + yz + \frac{z^2}{2}\right)_0^{2-x-y}$$

$$= 2\int_0^2 dx \int_0^{2-x} dy \left(2x - x^2 - xy + 2y - xy - y^2 + \frac{(2-x-y)^2}{2}\right]$$

$$= 2\int_0^2 dx \left[2xy - x^2y - xy^2 + y^2 - \frac{y^3}{3} - \frac{(2-x-y)^3}{6}\right]_0^{2-x}$$

$$= 2\int_0^2 dx \left[2x(2-x) - x^2(2-x) - x(2-x)^2 + (2-x)^2 - \frac{(2-x)^3}{3} + \frac{(2-x)^3}{6}\right]$$

$$= 2\int_0^2 \left(4x - 2x^2 - 2x^2 + x^3 - 4x + 4x^2 - x^3 + (2-x)^2 - \frac{(2-x)^3}{3} + \frac{(2-x)^3}{6}\right]$$

$$= 2\left[2x^2 - \frac{4x^3}{3} + \frac{x^4}{4} - 2x^2 + \frac{4x^3}{3} - \frac{x^4}{4} - \frac{(2-x)^3}{3} + \frac{(2-x)^4}{12} - \frac{(2-x)^4}{24}\right]_0^2$$

$$= 2\left[-\frac{(2-x)^3}{3} + \frac{(2-x)^4}{12} - \frac{(2-x)^4}{24}\right]_0^2 = 2\left[\frac{8}{3} - \frac{16}{12} + \frac{16}{24}\right] = 4 \qquad \textbf{Ans.}$$

**Example 112.** *Use the Divergence Theorem to evaluate*

$$\iint_S (x\,dy\,dz + y\,dz\,dx + z\,dx\,dy)$$

*where S is the portion of the plane $x + 2y + 3z = 6$ which lies in the first Octant.*

*(U.P., I Semester, Winter 2003)*

**Solution.** $\iint_S (f_1\,dy\,dz + f_2\,dx\,dz + f_3\,dx\,dy)$

$$= \iiint_V \left(\frac{\partial f_1}{\partial x} + \frac{\partial f_2}{\partial y} + \frac{\partial f_3}{\partial z}\right) dx\,dy\,dz$$

where $S$ is a closed surface bounding a volume V.

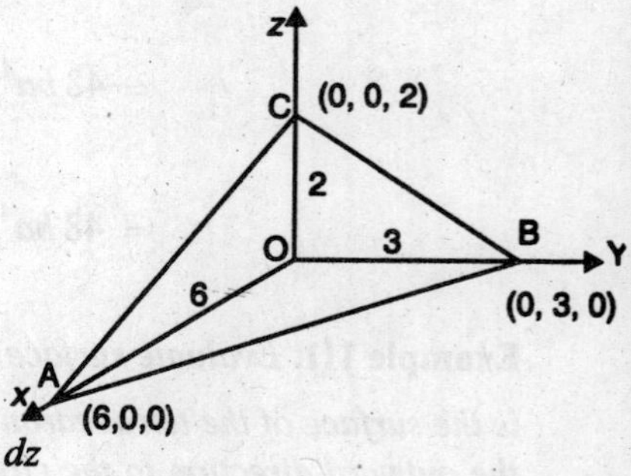


$$\therefore \qquad \iint_S (x\,dy\,dz + y\,dz\,dx + z\,dx\,dy)$$

$$= \iiint_V \left[\frac{\partial x}{\partial x} + \frac{\partial y}{\partial y} + \frac{\partial z}{\partial z}\right] dx\,dy\,dz$$

$$= \iiint_V (1+1+1)\,dx\,dy\,dz = 3\iiint_V dx\,dy\,dz$$

= 3 (Volume of tetrahedron $OABC$)

$$= 3\left[\left(\frac{1}{3}\text{ Area of the base } \Delta\, OAB\right) \times \text{height } OC\right]$$

$$= 3\left[\frac{1}{3}\left(\frac{1}{2} \times 6 \times 3\right) \times 2\right] = 18 \qquad \textbf{Ans.}$$

**Example 113.** *Use Divergence Theorem to evaluate :* $\iint (x\,dy\,dz + y\,dz\,dx + z\,dx\,dy)$ *over the surface of a sphere radius a.* *(K. University, Dec. 2009)*

**Solution.** Here, we have

$$\iint_S [x\,dy\,dz + y\,dx\,dz + z\,dx\,dy]$$

$$= \iiint_V \left(\frac{\partial f_1}{\partial x} + \frac{\partial f_2}{\partial y} + \frac{\partial f_3}{\partial z}\right) dx\,dy\,dz \quad = \iiint_V \left(\frac{\partial x}{\partial x} + \frac{\partial y}{\partial y} + \frac{\partial z}{\partial z}\right) dx\,dy\,dz$$

$$= \iiint_V (1+1+1)\,dx\,dy\,dz = 3 \text{ (volume of the sphere)}$$

$$= 3\left(\frac{4}{3}\pi a^3\right) = 4\pi a^3 \qquad \textbf{Ans.}$$

**Example 114.** *Using the divergence theorem, evaluate the surface integral* $\iint_S (yz\,dy\,dz + zx\,dz\,dx + xy\,dy\,dx)$ *where* $S : x^2 + y^2 + z^2 = 4.$

(AMIETE, Dec. 2010, UP, I Sem., Dec 2008)

**Solution.** $\iint_S (f_1\, dy\,dz + f_2\, dx\,dz + f_3\, dx\,dy)$

$$= \iiint_V \left(\frac{\partial f_1}{\partial x} + \frac{\partial f_2}{\partial y} + \frac{\partial f_3}{\partial z}\right) dx\,dy\,dz$$

where $S$ is closed surface bounding a volume $V$.

$\therefore \quad \iint_S (yz\,dy\,dz + zx\,dx\,dz + xy\,dx\,dy)$

$$= \iiint_V \left(\frac{\partial (yz)}{\partial x} + \frac{\partial (zx)}{\partial y} + \frac{\partial (xy)}{\partial z}\right) dx\,dy\,dz = \iiint_V (0+0+0)\,dx\,dy\,dz$$

$= 0$ **Ans.**

**Example 115.** *Evaluate* $\iint_S xz^2\, dy\,dz + (x^2 y - z^3)\,dz\,dx + (2xy + y^2 z)\,dx\,dy$ *where S is the surface of hemispherical region bounded by*

$$z = \sqrt{a^2 - x^2 - y^2} \text{ and } z = 0.$$

**Solution.** $\iint_S (f_1\, dy\,dz + f_2\, dz\,dx + f_3\, dx\,dy) = \iiint_V \left(\frac{\partial f_1}{\partial x} + \frac{\partial f_2}{\partial y} + \frac{\partial f_3}{\partial z}\right) dx\,dy\,dz$

where S is a closed surface bounding a volume V.

$\therefore \quad \iint_S xz^2\, dy\,dz + (x^2 y - z^3)\,dz\,dx + (2xy + y^2 z)\,dx\,dy$

$$= \iiint_V \left[\frac{\partial}{\partial x}(xz^2) + \frac{\partial}{\partial y}(x^2 y - z^3) + \frac{\partial}{\partial z}(2xy + y^2 z)\right] dx\,dy\,dz$$

(Here $V$ is the volume of hemisphere)

$$= \iiint_V (z^2 + x^2 + y^2)\,dx\,dy\,dz$$

Let $x = r\sin\theta\cos\phi,\ y = r\sin\theta\sin\phi,\ z = r\cos\theta$

$$= \iiint r^2 (r^2 \sin\theta\, dr\, d\theta\, d\phi) = \int_0^{2\pi} d\phi \int_0^{\frac{\pi}{2}} \sin\theta\, d\theta \int_0^a r^4\, dr$$

$$= (\phi)_0^{2\pi}\,(-\cos\theta)_0^{\pi/2}\left(\frac{r^5}{5}\right)_0^a = 2\pi(-0+1)\frac{a^5}{5} = \frac{2\pi a^5}{5}$$ **Ans.**

**Example 116.** *Evaluate* $\iint_S \vec{F}.\hat{n}\,ds$ *over the entire surface of the region above the xy-plane bounded by the cone* $z^2 = x^2 + y^2$ *and the plane* $z = 4$, *if* $F = 4xz\,\hat{i} + xyz^2\,\hat{j} + 3z\,\hat{k}$.

**Solution.** If V is the volume enclosed by S, then V is bounded by the surfaces $z = 0.\ z = 4,\ z^2 = x^2 + y^2$.

By divergence theorem, we have

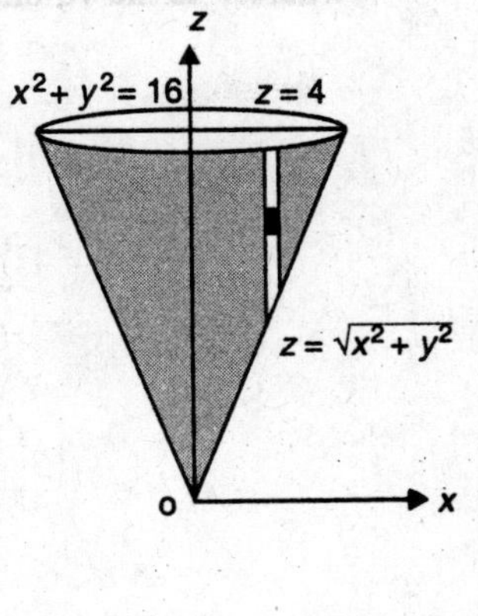


$$\iint_S \overline{F}.\hat{n}\,ds = \iiint_V div\,\vec{F}\,dx\,dy\,dz$$

$$= \iiint_V \left[\frac{\partial}{\partial x}(4xz) + \frac{\partial}{\partial y}(xyz^2) + \frac{\partial}{\partial z}(3z)\right] dx\,dy\,dz$$

$$= \iiint_V (4z + xz^2 + 3)\,dx\,dy\,dz$$

Limits of $z$ are $\sqrt{x^2 + y^2}$ and 4.

$$\iiint_{\sqrt{x^2+y^2}}^{4} (4z + xz^2 + 3)\, dz\, dy\, dx = \iint \left[ 2z^2 + \frac{xz^3}{3} + 3z \right]_{\sqrt{x^2+y^2}}^{4} dy\, dx$$

$$= \iint \left[ \left( 32 + \frac{64x}{3} + 12 \right) - \{2(x^2+y^2) + x(x^2+y^2)^{3/2} + 3\sqrt{x^2+y^2}\} \right] dy\, dx$$

$$= \iint \left( 44 + \frac{64x}{3} - 2(x^2+y^2) - x(x^2+y^2)^{3/2} - 3\sqrt{x^2+y^2} \right) dy\, dx$$

Putting $x = r\cos\theta$ and $y = r\sin\theta$, we have

$$= \iint \left( 44 + \frac{64 r\cos\theta}{3} - 2r^2 - r\cos\theta\, r^3 - 3r \right) r\, d\theta\, dr$$

Limits of $r$ are 0 to 4.
and limits of $\theta$ are 0 to $2\pi$.

$$= \int_0^{2\pi} \int_0^4 \left( 44r + \frac{64 r^2 \cos\theta}{3} - 2r^3 - r^5\cos\theta - 3r^2 \right) d\theta\, dr$$

$$= \int_0^{2\pi} \left[ 22r^2 + \frac{64 \times r^3 \cos\theta}{9} - \frac{r^4}{2} - \frac{r^6}{6}\cos\theta - r^3 \right]_0^4 d\theta$$

$$= \int_0^{2\pi} \left[ 22(4)^2 + \frac{64 \times (4)^3 \cos\theta}{9} - \frac{(4)^4}{2} - \frac{(4)^6}{6}\cos\theta - (4)^3 \right] d\theta$$

$$= \int_0^{2\pi} \left[ 352 + \frac{64 \times 64}{9}\cos\theta - 128 - \frac{(4)^6}{6}\cos\theta - 64 \right] d\theta$$

$$= \int_0^{2\pi} \left[ 160 + \left( \frac{64 \times 64}{9} - \frac{(4)^6}{6} \right) \cos\theta \right] d\theta$$

$$= \left[ 160\,\theta + \left( \frac{64 \times 64}{9} - \frac{(4)^6}{6} \right) \sin\theta \right]_0^{2\pi} = 160\,(2\pi) + \left( \frac{64 \times 64}{9} - \frac{(4)^6}{6} \right) \sin 2\pi$$

$= 320\,\pi$ **Ans.**

**Example 117.** *The vector field* $\vec{F} = x^2\hat{i} + z\hat{j} + yz\hat{k}$ *is defined over the volume of the cuboid given by* $0 \le x \le a$, $0 \le y \le b$, $0 \le z \le c$, *enclosing the surface S. Evaluate the surface integral* $\iint_S \vec{F} \cdot \vec{ds}$ (*U.P., I Semester, Winter 2001*)

**Solution.** By Divergence Theorem, we have

$$\iint_S (x^2\hat{i} + z\hat{j} + yz\hat{k}) \cdot ds = \iiint_V div\,(x^2\hat{i} + z\hat{j} + yz\hat{k})\, dv,$$

where $V$ is the volume of the cuboid enclosing the surface $S$.

$$= \iiint_V \left( \hat{i}\frac{\partial}{\partial x} + \hat{j}\frac{\partial}{\partial y} + \hat{k}\frac{\partial}{\partial z} \right) \cdot (x^2\hat{i} + z\hat{j} + yz\hat{k})\, dv$$

$$= \iiint_V \left\{ \frac{\partial}{\partial x}(x^2) + \frac{\partial}{\partial y}(z) + \frac{\partial}{\partial z}(yz) \right\} dx\, dy\, dz$$

$$= \int_{x=0}^{a} \int_{y=0}^{b} \int_{z=0}^{c} (2x + y)\, dx\, dy\, dz = \int_0^a dx \int_0^b dy \int_0^c (2x+y)\, dz$$

$$= \int_0^a dx \int_0^b [2xz + yz]_0^c\, dy = \int_0^a dx \int_0^b (2xc + yc)\, dy$$

$$= c\int_0^a dx\int_0^b (2x+y)\,dy = c\int_0^a \left[2xy+\frac{y^2}{2}\right]_0^b dx = c\int_0^a\left(2bx+\frac{b^2}{2}\right)dx$$

$$= c\left[\frac{2bx^2}{2}+\frac{b^2x}{2}\right]_0^a = c\left[a^2b+\frac{ab^2}{2}\right] = abc\left(a+\frac{b}{2}\right)$$ **Ans.**

**Example 118.** *Verify the divergence Theorem for the function* $\overline{F} = 2x^2yi - y^2j + 4xz^2k$ *taken over the region in the first octant bounded by* $y^2 + z^2 = 9$ *and* $x = 2$.

**Solution.** $\iiint_V \nabla\cdot\vec{F}\,dV = \iiint\left(\hat{i}\frac{\partial}{\partial x}+\hat{j}\frac{\partial}{\partial y}+\hat{k}\frac{\partial}{\partial z}\right)\cdot(2x^2y\hat{i}-y^2\hat{j}+4xz^2\hat{k})\,dV$

$$= \iiint(4xy-2y+8xz)\,dx\,dy\,dz = \int_0^2 dx\int_0^3 dy\int_0^{\sqrt{9-y^2}}(4xy-2y+8xz)\,dz$$

$$= \int_0^2 dx\int_0^3 dy\,(4xyz-2yz+4xz^2)_0^{\sqrt{9-y^2}}$$

$$= \int_0^2 dx\int_0^3 [4xy\sqrt{9-y^2}-2y\sqrt{9-y^2}+4x(9-y^2)]\,dy$$

$$= \int_0^2 dx\left[-\frac{4x}{2}\frac{2}{3}(9-y^2)^{3/2}+\frac{2}{3}(9-y^2)^{3/2}+36xy-\frac{4xy^3}{3}\right]_0^3$$

$$= \int_0^2(0+0+108x-36x+36x-18)\,dx = \int_0^2(108x-18)\,dx = \left[108\frac{x^2}{2}-18x\right]_0^2$$

$= 216 - 36 = 180$ ...(1)

Here $\iint_S \vec{F}\cdot\hat{n}\,ds = \iint_{OABC}\vec{F}\cdot\hat{n}\,ds + \iint_{OCE}\vec{F}\cdot\hat{n}\,ds + \iint_{OADE}\vec{F}\cdot\hat{n}\,ds + \iint_{ABD}\vec{F}\cdot\hat{n}\,ds + \iint_{BDEC}\vec{F}\cdot\hat{n}\,ds$

$$\iint_{BDEC}\overline{F}\cdot\hat{n}\,ds = \iint_{BDEC}(2x^2y\hat{i}-y^2\hat{j}+4xz^2\hat{k})\cdot\hat{n}\,ds$$

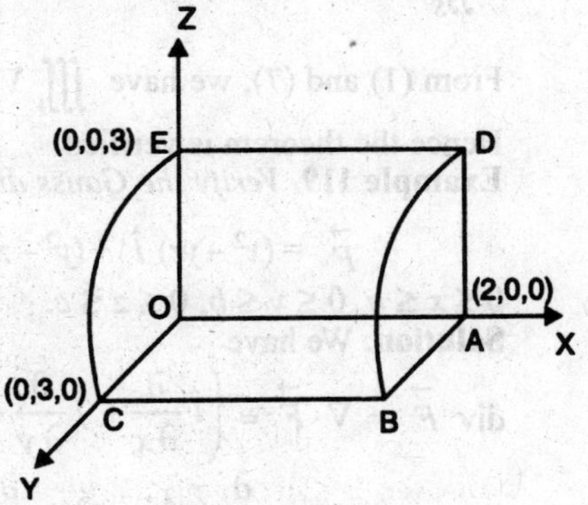


Normal vector

$$= \nabla\phi = \left(\hat{i}\frac{\partial}{\partial x}+\hat{j}\frac{\partial}{\partial y}+\hat{k}\frac{\partial}{\partial z}\right)(y^2+z^2-9)$$

$$= 2y\hat{j}+2z\hat{k}$$

Unit normal vector $= \hat{n} = \dfrac{2y\hat{j}+2z\hat{k}}{\sqrt{4y^2+4z^2}} = \dfrac{y\hat{j}+z\hat{k}}{\sqrt{y^2+z^2}}$

$$= \frac{y\hat{j}+z\hat{k}}{\sqrt{9}} = \frac{y\hat{j}+z\hat{k}}{3}$$

$$\iint_{BDEC}(2x^2y\hat{i}-y^2\hat{j}+4xz^2\hat{k})\cdot\frac{y\hat{j}+zk}{3}\,ds = \frac{1}{3}\iint_{BDEC}(-y^3+4xz^3)\,ds$$

$$\left[dx\,dy = ds\,(\hat{n}\cdot k) = ds\left(\frac{y\hat{j}+z\hat{k}}{3}\cdot\hat{k}\right) = ds\,\frac{z}{3}\text{ or } ds = \frac{dx\,dy}{\frac{z}{3}}\right]$$

$$= \frac{1}{3}\iint_{BDEC}(-y^3+4xz^3)\frac{dxdy}{\frac{z}{3}} = \int_0^2 dx\int_0^3\left(-\frac{y^3}{z}+4xz^2\right)dy \qquad \begin{pmatrix} y = 3\sin\theta, \\ z = 3\cos\theta \end{pmatrix}$$

$$= \int_0^2 dx\int_0^{\frac{\pi}{2}}\left[\frac{-27\sin^3\theta}{3\cos\theta}+4x(9\cos^2\theta)\right]$$

$$= \int_0^2 dx\left(-27\times\frac{2}{3}+108\,x\times\frac{2}{3}\right) = \int_0^2(-18+72\,x)\,dx$$

$$= \left[-18x+36x^2\right]_0^2 = 108 \qquad ...(2)$$

$$\iint_{OABC} \vec{F}\cdot\hat{n}\,ds = \iint_{OABC}(2x^2y\hat{i}-y^2\hat{j}+4xz^2\hat{k})\cdot(-\hat{k})\,ds$$

$$= \iint_{OABC} 4xz^2\,ds = 0$$ ...(3) because in $OABC$ $xy$-plane, $z = 0$

$$\iint_{OADE} \vec{F}\cdot\hat{n}\,ds = \iint_{OADE}(2x^2y\hat{i}-y^2\hat{j}+4xz^2\hat{k})\cdot(-\hat{j})\,ds = \iint_{OADE} y^2\,ds = 0 \qquad ...(4)$$

because in $OADE$ $xz$-plane, $y = 0$

$$\iint_{OCE} \vec{F}\cdot\hat{n}\,ds = \iint_{OCE}(2x^2y\hat{i}-y^2\hat{j}+4xz^2\hat{k})\cdot(-\hat{i})\,ds = \iint_{OCE} -2x^2y\,ds = 0 \qquad ...(5)$$

because in $OCE$ $yz$-plane, $x = 0$

$$\iint_{ABD} \vec{F}\cdot\hat{n}\,ds = \iint_{ABD}(2x^2y\hat{i}-y^2\hat{j}+4xz^2\hat{k})\cdot(\hat{i})\,ds = \iint_{ABD} 2x^2y\,ds$$

$$= \iint 2x^2y\,dy\,dz = \int_0^3 dz\int_0^{\sqrt{9-z^2}} 2\,(2)^2\,y\,dy$$ because in $ABD$ plane, $x = 2$

$$= 8\int_0^3 dz\left[\frac{y^2}{2}\right]_0^{\sqrt{9-z^2}} = 4\int_0^3 dz\,(9-z^2) = 4\left[9z-\frac{z^3}{3}\right]_0^3 = 4\,[27-9] = 72 \qquad ...(6)$$

On adding (2), (3), (4), (5) and (6), we get

$$\iint_S \vec{F}\cdot\hat{n}\,ds = 108+0+0+0+72 = 180 \qquad ...(7)$$

From (1) and (7), we have $\iiint_V \nabla\cdot\vec{F}\,dV = \iint_S \vec{F}\cdot\hat{n}\,ds$

Hence the theorem is verified.

**Example 119.** *Verify the Gauss divergence Theorem for*

$\vec{F} = (x^2-yz)\,\hat{i} + (y^2-zx)\,\hat{j} + (z^2-xy)\,\hat{k}$ *taken over the rectangular parallelopiped* $0 \le x \le a,\ 0 \le y \le b,\ 0 \le z \le c.$ *(U.P., I Semester, Compartment 2002)*

**Solution.** We have

$$\text{div}\ \vec{F} = \nabla\cdot\vec{F} = \left(\hat{i}\frac{\partial}{\partial x}+\hat{j}\frac{\partial}{\partial y}+\hat{k}\frac{\partial}{\partial z}\right)\cdot[(x^2-yz)\,\hat{i}+(y^2-zx)\,\hat{j}+(z^2-xy)\,\hat{k}]$$

$$= \frac{\partial}{\partial x}(x^2-yz)+\frac{\partial}{\partial y}(y^2-zx)+\frac{\partial}{\partial z}(z^2-xy) = 2x+2y+2z$$

$$\therefore \text{Volume integral} = \iiint_V \nabla\cdot\vec{F}\,dV = \iiint_V 2\,(x+y+z)\,dV$$

$$= 2\int_{x=0}^{a}\int_{y=0}^{b}\int_{z=0}^{c}(x+y+z)\,dx\,dy\,dz = 2\int_0^a dx\int_0^b dy\int_0^c (x+y+z)\,dz$$

$$= 2\int_0^a dx\int_0^b dy\left(xz+yz+\frac{z^2}{2}\right)_0^c = 2\int_0^a dx\int_0^b dy\left(cx+cy+\frac{c^2}{2}\right)$$

$$= 2\int_0^a dx\left(cxy+c\frac{y^2}{2}+\frac{c^2y}{2}\right)_0^b = 2\int_0^a dx\left(bcx+\frac{b^2c}{2}+\frac{bc^2}{2}\right)$$

$$= 2\left[\frac{bcx^2}{2}+\frac{b^2cx}{2}+\frac{bc^2x}{2}\right]_0^a = [a^2bc + ab^2c + abc^2]$$

$$= abc\,(a+b+c) \qquad \text{...(A)}$$

To evaluate $\iint_S \vec{F}\cdot\hat{n}\,ds$, where $S$ consists of six plane surfaces.

$$\iint_S \vec{F}\cdot\hat{n}\,ds = \iint_{OABC}\vec{F}\cdot\hat{n}\,ds + \iint_{DEFG}\vec{F}\cdot\hat{n}\,ds + \iint_{OAFG}\vec{F}\cdot\hat{n}\,ds$$

$$+ \iint_{BCDE}\vec{F}\cdot\hat{n}\,ds + \iint_{ABEF}\vec{F}\cdot\hat{n}\,ds + \iint_{OCDG}\vec{F}\cdot\hat{n}\,ds$$

$$\iint_{OABC}\vec{F}\cdot\hat{n}\,ds = \iint_{OABC}\{(x^2-yz)\hat{i}+(y^2-xz)\hat{j}+(z^2-xy)\hat{k}\}(-\hat{k})\,dx\,dy$$

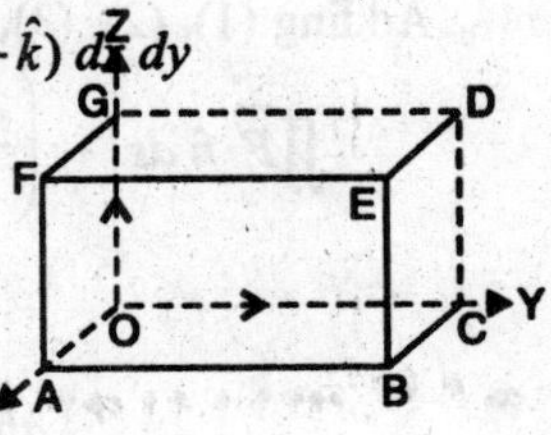


$$= -\int_0^a\int_0^b (z^2-xy)\,dx\,dy$$

$$= -\int_0^a\int_0^b (0-xy)\,dx\,dy = \frac{a^2b^2}{4} \qquad \text{...(1)}$$

$$\iint_{DEFG}\vec{F}\cdot\hat{n}\,ds = \iint_{DEFG}\{(x^2-yz)\hat{i}+(y^2-xz)\hat{j}+(z^2-xy)\hat{k}\}(\hat{k})\,dx\,dy$$

$$= \int_0^a\int_0^b (z^2-xy)\,dx\,dy = \int_0^a\int_0^b (c^2-xy)\,dx\,dy$$

$$= \int_0^a\left[c^2y-\frac{xy^2}{2}\right]_0^b dx = \int_0^a\left(c^2b-\frac{xb^2}{2}\right)dx$$

$$= \left[c^2bx-\frac{x^2b^2}{4}\right]_0^a = abc^2-\frac{a^2b^2}{4} \qquad \text{...(2)}$$

| S.No. | Surface | Outward normal | $ds$ | |
|---|---|---|---|---|
| 1 | OABC | $-k$ | $dx\,dy$ | $z=0$ |
| 2 | DEFG | $k$ | $dx\,dy$ | $z=c$ |
| 3 | OAFG | $-j$ | $dx\,dz$ | $y=0$ |
| 4 | BCDE | $j$ | $dx\,dz$ | $y=b$ |
| 5 | ABEF | $i$ | $dy\,dz$ | $x=a$ |
| 6 | OCDG | $-i$ | $dy\,dz$ | $x=0$ |

$$\iint_{OAFG}\vec{F}\cdot\hat{n}\,ds = \iint_{OAFG}\{(x^2-yz)\hat{i}+(y^2-zx)\hat{j}+(z^2-xy)\hat{k}\}(-\hat{j})\,dx\,dz$$

$$= -\iint_{OAFG}(y^2-zx)\,dx\,dz$$

$$= -\int_0^a dx\int_0^c (0-zx)\,dz = \int_0^a dx\left(\frac{xz^2}{2}\right)_0^c = \int_0^a \frac{xc^2}{2}\,dx = \left[\frac{x^2c^2}{4}\right]_0^a = \frac{a^2c^2}{4} \qquad \text{...(3)}$$

$$\iint_{BCDE}\vec{F}\cdot\hat{n}\,ds = \iint\{(x^2-yz)\hat{i}+(y^2-zx)\hat{j}+(z^2-xy)\hat{k}\}\cdot\hat{j}\,dx\,dz = \iint_{BCDE}(y^2-xz)\,dx\,dz$$

$$= -\int_0^a dx\int_0^c (b^2-xz)\,dz = \int_0^a\left(b^2z-\frac{xz^2}{2}\right)_0^c dx = \int_0^a\left(b^2c-\frac{xc^2}{2}\right)dx$$

$$= \left[b^2cx-\frac{x^2c^2}{4}\right]_0^a = ab^2c-\frac{a^2c^2}{4} \qquad \text{...(4)}$$

$$\iint_{ABEF}\vec{F}\cdot\hat{n}\,ds = \iint_{ABEF}\{(x^2-yz)\hat{i}+(y^2-xz)\hat{j}+(z^2-xy)\hat{k}\}\cdot\hat{i}\,dy\,dz$$

$$= \iint_{ABEF}(x^2-yz)\,dy\,dz = \int_0^b dy\int_0^c (a^2-yz)\,dz = \int_0^b dy\left(a^2z-\frac{yz^2}{2}\right)_0^c$$

$$= \int_0^b \left( a^2 c - \frac{y c^2}{2} \right) dy = \left[ a^2 c y - \frac{y^2 c^2}{4} \right]_0^b = a^2 b c - \frac{b^2 c^2}{4} \qquad \text{...(5)}$$

$$\iint_{OCDG} \vec{F} \cdot \hat{n}\, ds = \iint_{OCDG} \{(x^2 - yz)\hat{i} + (y^2 - zx)\hat{j} + (z^2 - xy)\hat{k}\} \cdot (-\hat{i})\, dy\, dz$$

$$= \int_0^b \int_0^c (x^2 - yz)\, dy\, dz = -\int_0^b dy \int_0^c (-yz)\, dz = -\int_0^b dy \left[ \frac{-y z^2}{2} \right]_0^c$$

$$= \int_0^b \frac{y c^2}{2} dy = \left[ \frac{y^2 c^2}{4} \right]_0^b = \frac{b^2 c^2}{4} \qquad \text{...(6)}$$

Adding (1), (2), (3), (4), (5) and (6), we get

$$\iint \vec{F} \cdot \hat{n}\, ds = \left( \frac{a^2 b^2}{4} \right) + \left( abc^2 - \frac{a^2 b^2}{4} \right) + \left( \frac{a^2 c^2}{4} \right) + \left( a b^2 c - \frac{a^2 c^2}{4} \right) + \left( \frac{b^2 c^2}{4} \right) + \left( a^2 bc - \frac{b^2 c^2}{4} \right)$$

$$= abc^2 + ab^2c + a^2bc$$

$$= abc\,(a + b + c) \qquad \text{...(B)}$$

From (A) and (B), Gauss divergence Theorem is verified. **Verified.**

**Example 120.** *Verify Divergence Theorem, given that* $\vec{F} = 4xz\hat{i} - y^2\hat{j} + yz\hat{k}$ *and S is the surface of the cube bounded by the planes* $x = 0, x = 1, y = 0, y = 1, z = 0, z = 1$.

**Solution.** $\nabla \cdot \vec{F} = \left( \hat{i} \frac{\partial}{\partial x} + \hat{j} \frac{\partial}{\partial y} + \hat{k} \frac{\partial}{\partial z} \right) \cdot (4zx\hat{i} - y^2\hat{j} + yz\hat{k})$

$$= 4z - 2y + y$$

$$= 4z - y$$

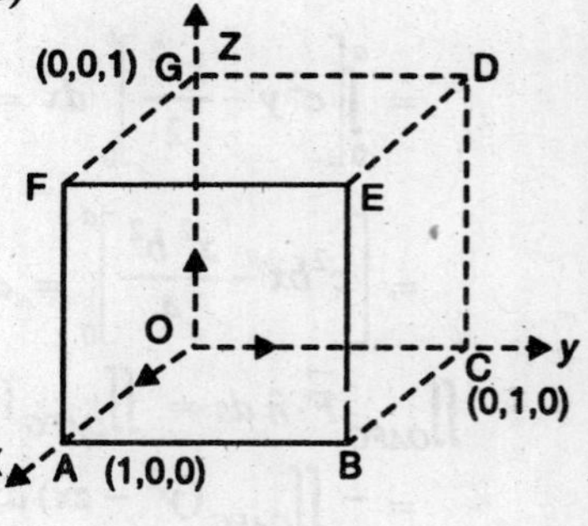


Volume Integral $= \iiint \nabla \cdot \vec{F}\, dv$

$$= \iiint (4z - y)\, dx\, dy\, dz$$

$$= \int_0^1 dx \int_0^1 dy \int_0^1 (4z - y)\, dz$$

$$= \int_0^1 dx \int_0^1 dy\, (2z^2 - yz)_0^1 = \int_0^1 dx \int_0^1 dy\, (2 - y)$$

$$= \int_0^1 dx \left( 2y - \frac{y^2}{2} \right)_0^1 = \int_0^1 dx \left( 2 - \frac{1}{2} \right) = \frac{3}{2} \int_0^1 dx = \frac{3}{2} (x)_0^1 = \frac{3}{2} \qquad \text{...(1)}$$

To evaluate $\iint_S \vec{F} \cdot \hat{n}\, ds$, where $S$ consists of six plane surfaces.

Over the face $OABC$, $z = 0$, $dz = 0$, $\hat{n} = -\hat{k}$, $ds = dx\, dy$

$$\iint \vec{F} \cdot \hat{n}\, ds = \int_0^1 \int_0^1 (-y^2\hat{j}) \cdot (-\hat{k})\, dx\, dy = 0$$

Over the face $BCDE$, $y = 1$, $dy = 0$

$$\iint \vec{F}\cdot\hat{n}\, ds = \int_0^1\int_0^1 (4\,xz\hat{i} - \hat{j} + z\hat{k})\cdot(\hat{j})\, dx\, dz$$

$$\hat{n} = \hat{j},\ ds = dx\, dz \quad = \int_0^1\int_0^1 -\, dx dz$$

$$= -\int_0^1 dx \int_0^1 dz = -(x)_0^1\,(z)_0^1 = -(1)(1) = -1$$

Over the face *DEFG*, $z = 1$, $dz = 0$, $\hat{n} = \hat{k}$, $ds = dx\, dy$

$$\iint \vec{F}\cdot\hat{n}\, ds = \int_0^1\int_0^1 [4\,x(1) - y^2\hat{j} + y(1)\,\hat{k}]\cdot(\hat{k})\, dx\, dy$$

$$= \int_0^1\int_0^1 y\, dx\, dy = \int_0^1 dx \int_0^1 y\, dy = (x)_0^1 \left(\frac{y^2}{2}\right)_0^1 = \frac{1}{2}$$

Over the face *OCDG*, $x = 0$, $dx = 0$, $\hat{n} = -\hat{i}$, $ds = dy\, dz$

$$\iint \vec{F}\cdot\hat{n}\, ds = \int_0^1\int_0^1 (0\hat{i} - y^2\hat{j} + yz\hat{k})\cdot(-\hat{i})\, dy\, dz = 0$$

Over the face *AOGF*, $y = 0$, $dy = 0$, $\hat{n} = -\hat{j}$, $ds = dx\, dz$

$$\iint \vec{F}\cdot\hat{n}\, ds = \int_0^1\int_0^1 (4\,xz\hat{i})\cdot(-\hat{j})\, dx\, dz = 0$$

Over the face *ABEF*, $x = 1$, $dx = 0$, $\hat{n} = \hat{i}$, $ds = dy\, dz$

$$\iint \vec{F}\cdot\hat{n}\, ds = \int_0^1\int_0^1 [(4\,z\hat{i} - y^2\hat{j} + yz\hat{k})\cdot(\hat{i})]\, dy\, dz = \int_0^1\int_0^1 4\,z\, dy\, dz$$

$$= \int_0^1 dy \int_0^1 4\,z\, dz = \int_0^1 dy\,(2\,z^2)_0^1 = 2\int_0^1 dy = 2\,(y)_0^1 = 2$$

On adding we see that over the whole surface

$$\iint \vec{F}\cdot\hat{n}\, ds = \left(0 - 1 + \frac{1}{2} + 0 + 0 + 2\right) = \frac{3}{2} \qquad \text{...(2)}$$

From (1) and (2), we have $\iiint_V \nabla\cdot\vec{F}\, dv = \iint_S \vec{F}\cdot\hat{n}\, ds$ **Verified.**

## EXERCISE 5.15

1. Use Divergence Theorem to evaluate $\iint_S (y^2z^2\hat{i} + z^2x^2\hat{j} + x^2y^2\hat{k}).\overrightarrow{ds}$, where $S$ is the upper part of the sphere $x^2 + y^2 + z^2 = 9$ above $xy$- plane. **Ans.** $\frac{243\pi}{8}$

2. Evaluate $\iint_S (\nabla\times\vec{F}).\, ds$, where $S$ is the surface of the paraboloid $x^2 + y^2 + z = 4$ above the $xy$-plane and $\vec{F} = (x^2 + y - 4)\,\hat{i} + 3\,xy\hat{j} + (2\,xz + z^2)\,\hat{k}$. **Ans.** $-4\pi$

3. Evaluate $\iint_S [xz^2\, dy\, dz + (x^2y - z^3)\, dz\, dx + (2xy + y^2z)\, dx\, dy]$, where $S$ is the surface enclosing a region bounded by hemisphere $x^2 + y^2 + z^2 = 4$ above $XY$-plane.

4. Verify Divergence Theorem for $\vec{F} = x^2\hat{i} + z\hat{j} + yz\hat{k}$, taken over the cube bounded by $x = 0$, $x = 1$, $y = 0$, $y = 1$, $z = 0$, $z = 1$.

5. Evaluate $\iint_S (2xy\hat{i} + yz^2\hat{j} + xz\,\hat{k})\cdot\overrightarrow{ds}$ over the surface of the region bounded by $x = 0$, $y = 0$, $y = 3$, $z = 0$ and $x + 2\,z = 6$ **Ans.** $\frac{351}{2}$

6. Verify Divergence Theorem for $\vec{F} = (x + y^2)\,\hat{i} - 2\,x\hat{j} + 2\,yz\hat{k}$ and the volume of a tetrahedron bounded by co-ordinate planes and the plane $2\,x + y + 2\,z = 6$.

*(Nagpur, Winter 2000, A.M.I.E.T.E.. Winter 2000)*

7. Verify Divergence Theorem for the function $\vec{F} = y\hat{i} + x\hat{j} + z^2\,\hat{k}$ over the region bounded by $x^2 + y^2 = 9$, $z = 0$ and $z = 2$.

8. Use the Divergence Theorem to evaluate $\iint_S x^3\,dy\,dz + x^2 y\,dz\,dx + x^2 z\,dx\,dy$,

where $S$ is the surface of the region bounded by the closed cylinder

$x^2 + y^2 = a^2$, $(0 \le z \le b)$ and $z = 0$, $z = b$. **Ans.** $\dfrac{5\pi a^4 b}{4}$

9. Evaluate the integral $\iint_S (z^2 - x)\,dy\,dz - xy\,dx\,dz + 3z\,dx\,dy$, where S is the surface of closed region bounded by $z = 4 - y^2$ and planes $x = 0$, $x = 3$, $z = 0$ by transforming it with the help of Divergence Theorem to a triple integral. **Ans.** 16

10. Evaluate $\iint_S \dfrac{ds}{\sqrt{a^2x^2 + b^2y^2 + c^2z^2}}$ over the closed surface of the ellipsoid $ax^2 + by^2 + cz^2 = 1$ by applying Divergence Theorem. **Ans.** $\dfrac{4\pi}{\sqrt{(a\,b\,c)}}$

11. Apply Divergence Theorem to evaluate $\iint (l\,x^2 + m\,y^2 + n\,z^2)\,ds$ taken over the sphere $(x - a)^2 + (y - b)^2 + (z - c)^2 = r^2$, $l, m, n$ being the direction cosines of the external normal to the sphere. *(AMIETE June 2010, 2009)* **Ans.** $\dfrac{8\pi}{3}(a + b + c)\,r^3$

12. Show that $\iiint_V (u\,\nabla \cdot \vec{V} + \nabla u \cdot \vec{V})\,dv = \iint_S u\vec{V} \cdot ds.$

13. If $E = \text{grad}\ \phi$ and $\nabla^2 \phi = 4\,\pi\,\rho$, prove that $\iint_S \vec{E} \cdot \vec{n}\,ds = -4\,\pi \iiint_V \rho\,dv$

where $\vec{n}$ is the outward unit normal vector, while $dS$ and $dV$ are respectively surface and volume elements.

**Pick up the correct option from the following:**

14. If $\vec{F}$ is the velocity of a fluid particle then $\int_C \vec{F} . \vec{dr}$ represents.

(*a*) Work done (*b*) Circulation (*c*) Flux (*d*) Conservative field.

*(U.P. Ist Semester, Dec 2009)* **Ans. (*b*)**

15. If $\vec{f} = ax\,\vec{i} + by\,\vec{j} + cz\,\vec{k}$, $a, b, c$, constants, then $\iint f . dS$ where $S$ is the surface of a unit sphere is

(*a*) $\dfrac{\pi}{3}(a + b + c)$ (*b*) $\dfrac{4}{3}\pi\,(a + b + c)$ (*c*) $2\,\pi\,(a + b + c)$ (*d*) $\pi\,(a + b + c)$

*(U.P., Ist Semester, 2009)* **Ans. (*b*)**

16. A force field $\vec{F}$ is said to be conservative if

(*a*) Curl $\vec{F} = 0$ (*b*) grad $\vec{F} = 0$ (*c*) Div $\vec{F} = 0$ (*d*) Curl (grad $\vec{F}$) = 0

*(AMIETE, Dec. 2006)* **Ans. (*a*)**

17. The line integral $\int_c x^2\,dx + y^2\,dy$, where $C$ is the boundary of the region $x^2 + y^2 < a^2$ equals

(*a*) 0, (*b*) $a$ (*c*) $\pi\,a^2$ (*d*) $\dfrac{1}{2}\pi\,a^2$

*(AMIETE, Dec. 2006)* **Ans. (*b*)**

# 6

# Complex Numbers

## 6.1 INTRODUCTION

We have learnt the complex numbers in the previous class. Here we will review the complex number. In this chapter we will learn how to add, subtract, multiply and divide complex numbers.

## 6.2 COMPLEX NUMBERS

A number of the form $a + i\,b$ is called a complex number when $a$ and $b$ are real numbers and $i = \sqrt{-1}$. We call '$a$' the real part and '$b$' the imaginary part of the complex number $a + ib$. If $a = 0$ the number $i\,b$ is said to be purely imaginary, if $b = 0$ the number $a$ is real.

A complex number $x + iy$ is denoted by $z$.

## 6.3 GEOMETRICAL REPRESENTATION OF IMAGINARY NUMBERS

Let $OA$ be positive numbers which is represented by $x$ and $OA'$ by $-x$.

And $$-x = (i)^2\,x = i\,(ix) \text{ is on } OX'.$$

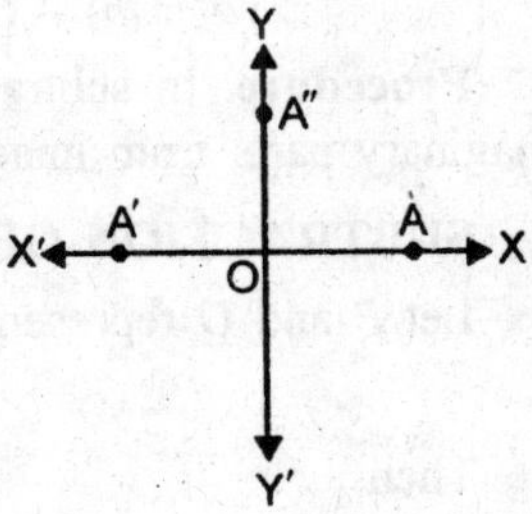


It means that the multiplication of the real number $x$ by $i$ twice amounts to the rotation of $OA$ through two right angles to reach $OA'$.

Thus, it means that multiplication of $x$ by $i$ is equivalent to the rotation of $x$ through one right angle to reach $OA''$.

Hence, $y$-axis is known as imaginary axis.

Multiplication by $i$ rotates its direction through a right angle.

## 6.4 ARGAND DIAGRAM

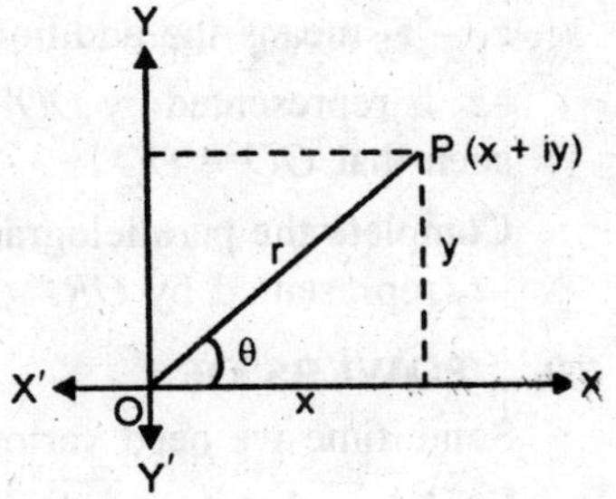


Mathematician Argand represented a complex number in a diagram known as Argand diagram. A complex number $x + iy$ can be represented by a point $P$ whose co-ordinate are $(x, y)$. The axis of $x$ is called the real axis and the axis of $y$ the imaginary axis. The distance $OP$ is the **modulus** and the angle, $OP$ makes with the $x$-axis, is the **argument** of $x + iy$.

## 6.5 EQUAL COMPLEX NUMBERS

If two complex numbers $a + ib$ and $c + id$ are equal, prove that

$$a = c \quad \text{and} \quad b = d$$

**Solution.** We have, $$a + ib = c + id \;\Rightarrow\; a - c = i(d - b)$$

$$(a - c)^2 = -(d - b)^2 \;\Rightarrow\; (a - c)^2 + (d - b)^2 = 0$$

Here sum of two positive numbers is zero. This is only possible if each number is zero.

*i.e.*, $(a - c)^2 = 0 \;\Rightarrow\; a = c$ and $(d - b)^2 = 0 \;\Rightarrow\; b = d$ **Ans.**

## 6.6 ADDITION OF COMPLEX NUMBERS

Let $a + ib$ and $c + id$ be two complex numbers, then

$$(a + ib) + (c + id) = (a + c) + i (b + d)$$

**Procedure.** In addition of complex numbers we add real parts with real parts and imaginary parts with imaginary parts.

## 6.7 ADDITION OF COMPLEX NUMBERS BY GEOMETRY

Let $z_1 = x_1 + i\,y_1$ and $z_2 = x_2 + i\,y_2$ be two complex numbers represented by the points $P$ and $Q$ on the Argand diagram.

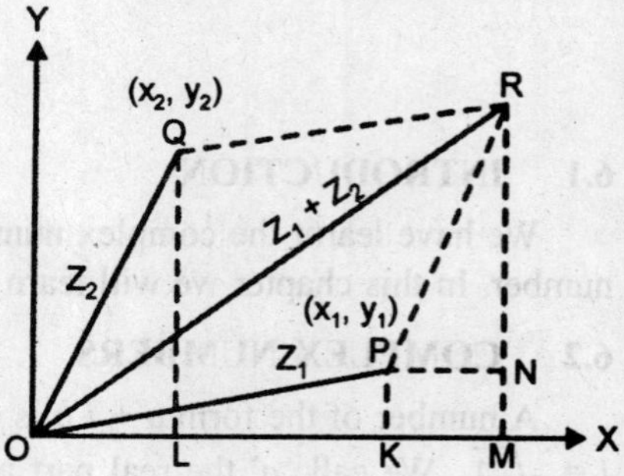


Complete the parallelogram $OPRQ$.

Draw $PK$, $RM$, $QL$, perpendiculars on $OX$.

Also draw $PN \perp$ to $RM$.

$$OM = OK + KM = OK + OL = x_1 + x_2$$

and $$RM = MN + NR = KP + LQ = y_1 + y_2$$

$\therefore$ The co-ordinates of $R$ are $(x_1 + x_2, y_1 + y_2)$ and it represents the complex number.

$$(x_1 + x_2) + i\,(y_1 + y_2) = (x_1 + iy_1) + (x_2 + iy_2)$$

Thus the sum of two complex numbers is represented by the extremity of the diagonal of the parallelogram formed by $OP$ $(z_1)$ and $OQ$ $(z_2)$ as adjacent sides.

$$|z_1 + z_2| = OR \quad \text{and} \quad \text{amp } (z_1 + z_2) = \angle ROM.$$

## 6.8 SUBTRACTION

$$(a + ib) - (c + id) = (a - c) + i(b - d)$$

**Procedure.** In subtraction of complex numbers we subtract real parts from real parts and imaginary parts from imaginary parts.

### SUBTRACTION OF COMPLEX NUMBERS BY GEOMETRY.

Let $P$ and $Q$ represent two complex numbers

$$z_1 = x_1 + iy_1 \text{ and } z_2 = x_2 + iy_2.$$

Then $$z_1 - z_2 = z_1 + (-z_2)$$

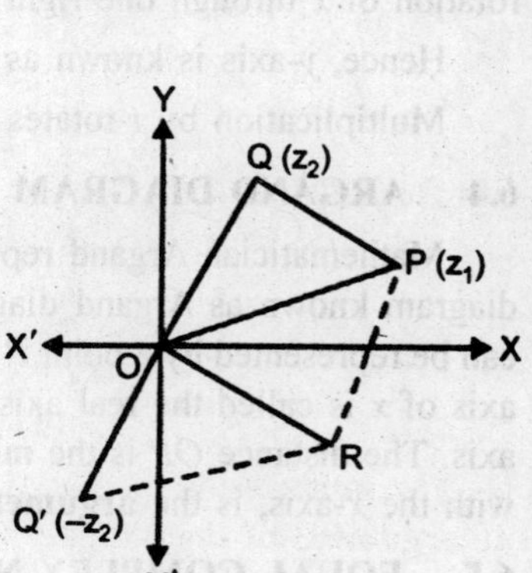


$z_1 - z_2$ means the addition of $z_1$ and $-z_2$.

$-z_2$ is represented by $OQ'$ formed by producing $OQ$ to $OQ'$ such that $OQ = OQ'$.

Complete the parallelogram $OPRQ'$, then the sum of $z_1$ and $-z_2$ represented by $OR$.

## 6.9 POWERS OF $i$

Some time we need various powers of $i$.

We know that $i = \sqrt{-1}$.

On squaring both sides, we get

$$i^2 = -1$$

Multiplying by $i$ both sides, we get

$$i^3 = -i$$

Again, $$i^4 = (i^3)\,(i) = (-i)\,(i) = -(i^2) = -(-1) = 1$$

$$i^5 = (i^4)\,(i) = (1)\,(i) = i$$

$$i^6 = (i^4)(i^2) = (1)(-1) = -1$$
$$i^7 = (i^4)(i^3) = 1(-i) = -i$$
$$i^8 = (i^4)(i^4) = (1)(1) = 1.$$

**Example 1.** *Simplify the following:* (*a*) $i^{49}$, (*b*) $i^{103}$.

**Solution.** (*a*) We divide 49 by 4 and we get

$$49 = 4 \times 12 + 1$$
$$i^{49} = i^{4 \times 12 + 1} = (i^4)^{12}(i^1) = (1)^{12}(i) = i$$

(*b*) we divide 103 by 4, we get

$$103 = 4 \times 25 + 3$$
$$i^{103} = i^{4 \times 25 + 3} = (i^4)^{25}(i^3) = (1)^{25}(-i) = -i$$ **Ans.**

## 6.10 MULTIPLICATION

$$(a + ib) \times (c + id) = ac - bd + i(ad + bc)$$

**Proof.** $(a + ib) \times (c + id) = ac + iad + ibc + i^2bd$

$$= ac + i(ad + bc) + (-1)bd \quad [\because i^2 = -1]$$
$$= (ac - bd) + (ad + bc)i$$

**Example 2.** *Multiply* $3 + 4i$ *by* $7 - 3i$.

**Solution.** Let $z_1 = 3 + 4i$ and $z_2 = 7 - 3i$

$$z_1.z_2 = (3 + 4i).(7 - 3i)$$
$$= 21 - 9i + 28i - 12i^2$$
$$= 21 - 9i + 28i - 12(-1) \quad [\because i^2 = -1]$$
$$= 21 - 9i + 28i + 12$$
$$= 33 + 19i$$ **Ans.**

**Multiplication of complex numbers (Polar form) :**

**Let** $z_1 = r_1(\cos\theta_1 + i\sin\theta_1)$ **and** $z_2 = r_2(\cos\theta_2 + i\sin\theta_2)$

$$x_1 = r_1\cos\theta_1, \quad y_1 = r_1\sin\theta_1$$
$$x_2 = r_2\cos\theta_2, \quad y_2 = r_2\sin\theta_2$$
$$z_1 = x_1 + iy_1 = r_1(\cos\theta_1 + i\sin\theta_1) \qquad |z_1| = r_1$$
$$z_2 = x_2 + iy_2 = r_2(\cos\theta_2 + i\sin\theta_2) \qquad |z_2| = r_2$$
$$z_1 . z_2 = r_1r_2(\cos\theta_1 + i\sin\theta_1)(\cos\theta_2 + i\sin\theta_2)$$
$$= r_1r_2[\cos\theta_1\cos\theta_2 - \sin\theta_1\sin\theta_2 + i(\sin\theta_1\cos\theta_2 + \cos\theta_1\sin\theta_2)]$$
$$= r_1r_1[\cos(\theta_1 + \theta_2) + i\sin(\theta_1 + \theta_2)], \qquad |z_1z_2| = r_1r_2$$

**The modulus of the product of two complex numbers is the product of their moduli and the argument of the product is the sum of their arguments.**

**Graphical method**

Let $P$, $Q$ represent the complex numbers.

$$z_1 = x_1 + iy_1$$
$$= r_1(\cos\theta_1 + i\sin\theta_1)$$
$$z_2 = x_2 + iy_2$$
$$= r_2(\cos\theta_2 + i\sin\theta_2)$$

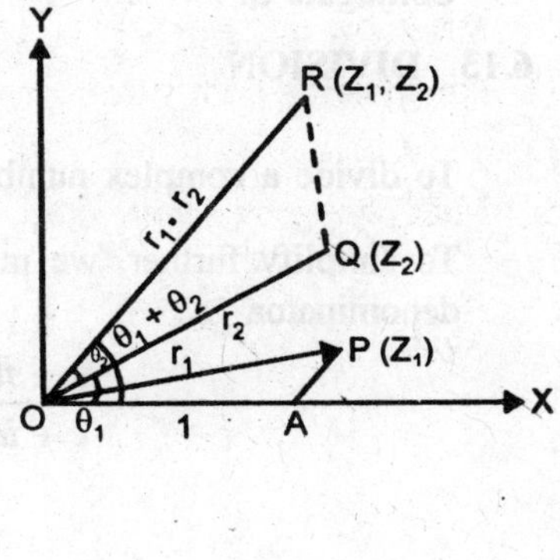


Cut off $OA = 1$ along $x$-axis. Construct $\Delta\, ORQ$ on $OQ$ similar to $\Delta\, OAP$.

So that $\frac{OR}{OP} = \frac{OQ}{OA} \Rightarrow \frac{OR}{OP} = \frac{OQ}{1} \Rightarrow OR = OP \,.\, OQ = r_1 r_2$

$$\angle XOR = \angle AOQ + \angle QOR = \theta_2 + \theta_1$$

Hence the product of two complex numbers $z_1$, $z_2$ is represented by the point $R$, such that

(*i*) $|z_1 \,.\, z_2| = |z_1|.|z_2|$ (*ii*) $\text{Arg}\,(z_1 \,.\, z_2) = \text{Arg}\,(z_1) + \text{Arg}\,(z_2)$

## 6.11 *i* (IOTA) AS AN OPERATOR

Multipliction of a complex number by *i*.

Let $z = x + iy = r(\cos\theta + i\sin\theta)$

$$i = 0 + i \,.\, 1 = \left[\cos\frac{\pi}{2} + i\sin\frac{\pi}{2}\right]$$

$$i \,.\, z = r(\cos\theta + i\sin\theta) \,.\, \left[\cos\frac{\pi}{2} + i\sin\frac{\pi}{2}\right]$$

$$= r\left[\cos\left(\theta + \frac{\pi}{2}\right) + i\sin\left(\theta + \frac{\pi}{2}\right)\right]$$

Hence a complex number multiplied by *i* results :

The rotation of the complex number by $\frac{\pi}{2}$ in anticlockwise direction without change in magnitude.

## 6.12 CONJUGATE OF A COMPLEX NUMBER

Two complex numbers which differ only in the sign of imaginary parts are called conjugate of each other.

A pair of complex number $a + ib$ and $a - ib$ are said to be conjugate of each other.

**Theorem.** Show that the sum and product of a complex number and its conjugate complex are both real.

**Proof.** Let $x + iy$ be a complex number and $x - iy$ its conjugate complex.

$$\text{Sum} = (x + iy) + (x - iy) = 2x \qquad \text{(Real)}$$

$$\text{Product} = (x + iy).(x - iy) = x^2 + y^2. \qquad \text{(Real)}$$

**Proved.**

**Note.** Let a complex number be $z$. Then the conjugate complex number is denoted by $\bar{z}$.

**Example 3.** *Find out the conjugate of a complex number 7 + 6i.*

**Solution.** Let $z = 7 + 6i$

To find conjugate complex number of $7 + 6i$ we change the sign of imaginary number.

Conjugate of $z = \bar{z} = 7 - 6i$ **Ans.**

## 6.13 DIVISION

To divide a complex number $a + ib$ by $c + id$, we write it as $\frac{a + ib}{c + id}$.

To simplify further, we multiply the numerator and denominator by the conjugate of the denominator.

$$\frac{a + ib}{c + id} = \frac{(a + ib)}{(c + id)} \times \frac{(c - id)}{(c - id)} = \frac{ac - iad + ibc - i^2 bd}{(c)^2 - (id)^2}$$

$$= \frac{ac - i(ad - bc) + bd}{c^2 - d^2 i^2} \qquad [\because i^2 = -1]$$

$$= \frac{ac + bd}{c^2 + d^2} + \frac{bc - ad}{c^2 + d^2} i$$

**Example 4.** *Divide $1 + i$ by $3 + 4i$.*

**Solution.** $\frac{1+i}{3+4i} = \frac{1+i}{3+4i} \times \frac{3-4i}{3-4i}$

$$= \frac{3 - 4i + 3i - 4i^2}{9 - 16i^2}$$

$$= \frac{3 - i + 4}{9 + 16} = \frac{7}{25} - \frac{1}{25} i$$ **Ans.**

**DIVISION (By Algebra)**

Let $z_1 = r_1(\cos\theta_1 + i\sin\theta_1)$ and $z_2 = r_2(\cos\theta_2 + i\sin\theta_2)$

$$\frac{z_1}{z_2} = \frac{r_1(\cos\theta_1 + i\sin\theta_1)}{r_2(\cos\theta_2 + i\sin\theta_2)} = \frac{r_1(\cos\theta_1 + i\sin\theta_1)(\cos\theta_2 - i\sin\theta_2)}{r_2(\cos\theta_2 + i\sin\theta_2)(\cos\theta_2 - i\sin\theta_2)}$$

$$= \frac{r_1[(\cos\theta_1\cos\theta_2 + \sin\theta_1\sin\theta_2) + i(\sin\theta_1\cos\theta_2 - \sin\theta_2\cos\theta_1)]}{r_2(\cos^2\theta_2 + \sin^2\theta_2)}$$

$$= \frac{r_1}{r_2}[\cos(\theta_1 - \theta_2) + i\sin(\theta_1 - \theta_2)]$$

The modulus of the quotient of two complex numbers is the quotient of their moduli, and the argument of the quotient is the difference of their arguments.

## 6.14 DIVISION OF COMPLEX NUMBERS BY GEOMETRY

Let $P$ and $Q$ represent the complex numbers.

$$z_1 = x_1 + i y_1 = r_1(\cos\theta_1 + i\sin\theta_1)$$
$$z_2 = x_2 + i y_2 = r_2(\cos\theta_2 + i\sin\theta_2)$$

Cut off $OA = 1$, construct $\Delta\, OAR$ on $OA$ similar to $\Delta\, OQP$.

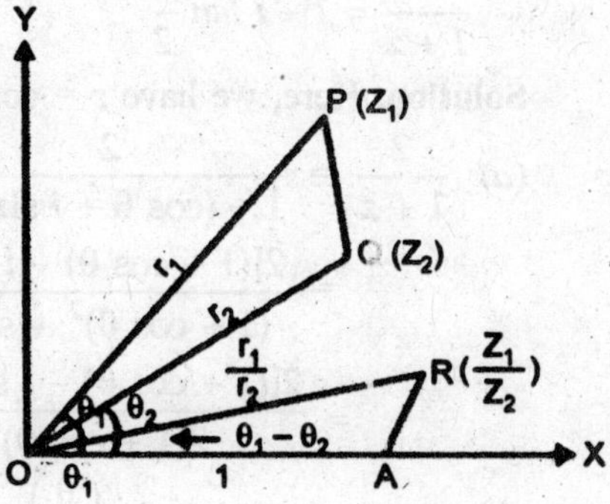


So that $\frac{OR}{OA} = \frac{OP}{OQ} \Rightarrow \frac{OR}{1} = \frac{OP}{OQ}$

$$OR = \frac{OP}{OQ} = \frac{r_1}{r_2}$$

$$\angle AOR = \angle QOP = \angle AOP - \angle AOQ = \theta_1 - \theta_2$$

$\therefore$ $R$ represents the number $\frac{r_1}{r_2}[\cos(\theta_1 - \theta_2) + i\sin(\theta_1 - \theta_2)]$

Hence the complex number $\frac{z_1}{z_2}$ is represented by the point $R$.

(*i*) $\left|\frac{z_1}{z_2}\right| = \frac{|z_1|}{|z_2|}$ (*ii*) Arg. $\left(\frac{z_1}{z_2}\right)$ = Arg. $(z_1)$ – Arg. $(z_2)$.

**Example 5.** *If $a = \cos\theta + i\sin\theta$, prove that $1 + a + a^2 = (1 + 2\cos\theta)(\cos\theta + i\sin\theta)$.*

**Solution.** Here we have $a = \cos\theta + i\sin\theta$

$$\begin{aligned}1 + a + a^2 &= 1 + (\cos\theta + i\sin\theta) + (\cos\theta + i\sin\theta)^2\\ &= 1 + \cos\theta + i\sin\theta + \cos^2\theta + 2i\sin\theta\cos\theta - \sin^2\theta\\ &= (\cos\theta + i\sin\theta) + (1 - \sin^2\theta) + \cos^2\theta + 2i\sin\theta\cos\theta\\ &= (\cos\theta + i\sin\theta) + \cos^2\theta + \cos^2\theta + 2i\sin\theta\cos\theta\\ &= (\cos\theta + i\sin\theta) + 2\cos^2\theta + 2i\sin\theta\cos\theta\\ &= (\cos\theta + i\sin\theta) + 2\cos\theta(\cos\theta + i\sin\theta)\\ &= (\cos\theta + i\sin\theta)(1 + 2\cos\theta)\end{aligned}$$

**Proved.**

**Example 6.** *If $a^2 + b^2 + c^2 = 1$ and $b + ic = (1 + a)z$, prove that $\dfrac{a + ib}{1 + c} = \dfrac{1 + iz}{1 - iz}$.*

**Solution.** Here, we have $b + ic = (1 + a)z \Rightarrow z = \dfrac{b + ic}{1 + a}$

$$\frac{1 + iz}{1 - iz} = \frac{1 + i\dfrac{b + ic}{1 + a}}{1 - i\dfrac{b + ic}{1 + a}} = \frac{1 + a + ib - c}{1 + a - ib + c}$$

$$= \frac{[(1 + a + ib) - c]}{(1 + a + c - ib)} \times \frac{(1 + a + ib + c)}{(1 + a + c + ib)} = \frac{(1 + a + ib)^2 - c^2}{(1 + a + c)^2 + b^2}$$

$$= \frac{1 + a^2 - b^2 + 2a + 2ib + 2iab - c^2}{1 + a^2 + c^2 + 2a + 2c + 2ac + b^2} = \frac{1 + a^2 - b^2 - c^2 + 2a + 2ib + 2iab}{1 + (a^2 + b^2 + c^2) + 2a + 2c + 2ac}$$

Putting the value of $a^2 + b^2 + c^2 = 1$ in the above, we get

$$= \frac{1 + a^2 - (1 - a^2) + 2a + 2ib + 2iab}{1 + 1 + 2a + 2c + 2ac} = \frac{2(a^2 + a + ib + iab)}{2(1 + a + c + ac)} = \frac{2(1 + a)(a + ib)}{2(1 + a)(1 + c)} = \frac{a + ib}{1 + c}$$

**Proved.**

**Example 7.** *If $z = \cos\theta + i\sin\theta$, prove that*

$$\frac{2}{1 + z} = 1 - i\tan\frac{\theta}{2}$$

**Solution.** Here, we have $z = \cos\theta + i\sin\theta$

(*a*) $$\frac{2}{1 + z} = \frac{2}{1 + (\cos\theta + i\sin\theta)} = \frac{2}{(1 + \cos\theta) + i\sin\theta} \times \frac{(1 + \cos\theta) - i\sin\theta}{(1 + \cos\theta) - i\sin\theta}$$

$$= \frac{2[(1 + \cos\theta) - i\sin\theta]}{(1 + \cos\theta)^2 + \sin^2\theta}$$

$$= \frac{2[(1 + \cos\theta) - i\sin\theta]}{2(1 + \cos\theta)} = 1 - \frac{i\sin\theta}{1 + \cos\theta}$$

$$= 1 - i\frac{2\sin\left(\dfrac{\theta}{2}\right)\cos\left(\dfrac{\theta}{2}\right)}{2\cos^2\left(\dfrac{\theta}{2}\right)} = 1 - i\tan\left(\frac{\theta}{2}\right)$$

$$\begin{aligned}&(1 + \cos\theta)^2 + \sin^2\theta\\ &= 1 + \cos^2\theta + 2\cos\theta + \sin^2\theta\\ &= 1 + (\sin^2\theta + \cos^2\theta) + 2\cos\theta\\ &= 1 + 1 + 2\cos\theta\\ &= 2 + 2\cos\theta\\ &= 2(1 + \cos\theta)\end{aligned}$$

**Proved.**

**Example 8.** *If $x = \cos\theta + i\sin\theta$, $y = \cos\phi + i\sin\phi$, prove that*

$$\frac{x - y}{x + y} = i\tan\left(\frac{\theta - \phi}{2}\right)$$

(*M.U. 2008*)

**Solution.** We have,

$$\frac{x - y}{x + y} = \frac{(\cos\theta + i\sin\theta) - (\cos\phi + i\sin\phi)}{(\cos\theta + i\sin\theta) + (\cos\phi + i\sin\phi)}$$

$$= \frac{(\cos\theta - \cos\phi) + i(\sin\theta - \sin\phi)}{(\cos\theta + \cos\phi) + i(\sin\theta + \sin\phi)}$$

$$= \frac{\left[-2\sin\left(\frac{\theta+\phi}{2}\right)\sin\left(\frac{\theta-\phi}{2}\right) + 2i\cos\left(\frac{\theta+\phi}{2}\right)\sin\left(\frac{\theta-\phi}{2}\right)\right]}{\left[2\cos\left(\frac{\theta+\phi}{2}\right)\cos\left(\frac{\theta-\phi}{2}\right) + 2i\sin\left(\frac{\theta+\phi}{2}\right)\cos\left(\frac{\theta-\phi}{2}\right)\right]}$$

$$= \frac{2i\sin\left(\frac{\theta-\phi}{2}\right)\left[\cos\left(\frac{\theta+\phi}{2}\right) + i\sin\left(\frac{\theta+\phi}{2}\right)\right]}{2\cos\left(\frac{\theta-\phi}{2}\right)\left[\cos\left(\frac{\theta+\phi}{2}\right) + i\sin\left(\frac{\theta+\phi}{2}\right)\right]} = i\tan\left(\frac{\theta-\phi}{2}\right)$$ **Proved.**

## EXERCISE 6.1

**1.** If $z = 1 + i$, find (*i*) $z^2$ (*ii*) $\frac{1}{z}$ and plot them on the Argand diagram. **And.** (*i*) $2i$, (*ii*) $\frac{1}{2} - \frac{i}{2}$

**Express the following in the form $a + ib$, where $a$ and $b$ are real (2 – 4):**

**2.** $\frac{2-3i}{4-i}$ **Ans.** $\frac{11}{17} - \frac{10}{17}i$ **3.** $\frac{(3+4i)(2+i)}{1+i}$ **Ans.** $\frac{13}{2} + \frac{9}{2}i$

**4.** $\frac{(1+2i)^3}{(1+i)(2-i)}$ **Ans.** $-\frac{7}{2} + \frac{1}{2}i$

**5.** The points $A, B, C$ represent the complex numbers $z_1, z_2, z_3$ respectively, and $G$ is the centroid of the triangle $ABC$, if $4z_1 + z_2 + z_3 = 0$, show that the origin is the mid-point of AG.

**6.** $ABCD$ is a parallelogram on the Argand plane. The affixes of $A, B, C$ are $8 + 5i, -7 - 5i, -5 + 5i$, respectively. Find the affix of $D$. **Ans.** $10 + 15i$

**7.** If $z_1, z_2, z_3$ are three complex numbers and

$$a_1 = z_1 + z_2 + z_3$$
$$b_1 = z_1 + \omega z_2 + \omega^2 z_3$$
$$c_1 = z_1 + \omega^2 z_2 + \omega z_3$$

show that $|a_1|^2 + |b_1|^2 + |c_1|^2 = 3\{|z_1|^2 + |z_2|^2 + |z_3|^2\}$

where $\omega, \omega^2$ are cube roots of unity.

**8.** Find the complex conjugate of $\frac{2+3i}{1-i}$. **Ans.** $-\frac{1}{2} - \frac{5}{2}i$

**9.** If $x + iy = \frac{1}{a+ib}$, prove that $(x^2 + y^2)(a^2 + b^2) = 1$

**10.** Find the value of $x^2 - 6x + 13$, when $x = 3 + 2i$. **Ans.** 0

**11.** If $\alpha - i\beta = \frac{1}{a - ib}$, prove that $(\alpha^2 + \beta^2)(a^2 + b^2) = 1$. (*M.U. 2008*)

**12.** If $\frac{1}{\alpha + i\beta} + \frac{1}{a + ib} = 1$, where $\alpha, \beta, a, b$ are real, express $b$ in terms of $\alpha, \beta$.

**Ans.** $\frac{-\beta}{\alpha^2 + \beta^2 - 2\alpha + 1}$

**13.** If $(x + iy)^{1/3} = a + ib$, then show that $4(a^2 - b^2) = \frac{x}{a} + \frac{y}{b}$.

**14.** If $(x + iy)^3 = u + iv$, then show that $\frac{u}{x} + \frac{v}{y} = 4(x^2 - y^2)$.

**15.** Find the values of $x$ and $y$, if $\frac{(1+i)x - 2i}{3+i} + \frac{(2-3i)y + i}{3-i} = i$. **Ans.** $x = 3$ and $y = -1$

**16.** If $a + ib = \frac{(x+i)^2}{2x^2+1}$, prove that $a^2 + b^2 = \frac{(x^2+1)^2}{(2x^2+1)^2}$.

## 6.15 MODULUS AND ARGUMENT

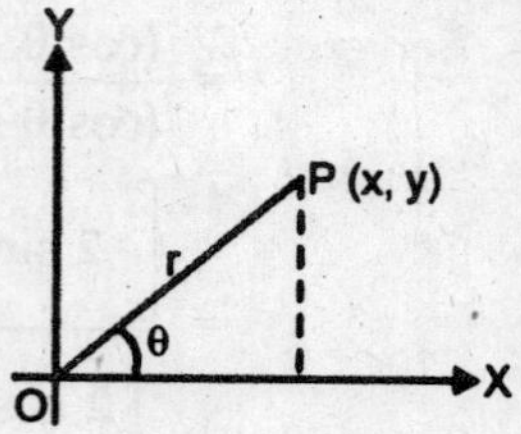


Let $x + iy$ be a complex number.

Putting $x = r\cos\theta$ and $y = r\sin\theta$ so that $r = \sqrt{x^2 + y^2}$

$$\cos\theta = \frac{x}{\sqrt{x^2 + y^2}} \quad \text{and} \quad \sin\theta = \frac{y}{\sqrt{x^2 + y^2}}$$

the positive value of the root being taken.

Then $r$ called the *modulus* or absolute value of the complex number $x + iy$ and is denoted by $|x + iy|$.

The angle $\theta$ is called the *argument* or *amplitude* of the complex number $x + iy$ and is denoted by arg. $(x + iy)$.

It is clear that $\theta$ will have infinite number of values differing by multiples of $2\pi$. The values of $\theta$ lying in the range $-\pi < \theta \le \pi$ $[(0 < \theta < \pi)$ or $(-\pi < \theta < 0)]$ is called the *principal value* of the argument.

The principal value of $\theta$ is written either between 0 and $\pi$ or between 0 and $-\pi$.

A complex number $x + iy$ is denoted by a single letter $z$. The number $x - iy$ (conjugate) is denoted by $\bar{z}$. The complex number in polar form is $r(\cos\theta + i\sin\theta)$.

Modulus of $z$ is denoted by $|z|$ and $|z|^2 = x^2 + y^2$.

| Angle θ | Principal value of θ |
|---|---|
| 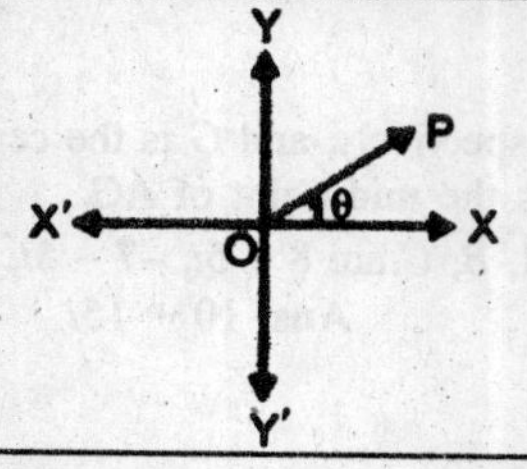 | 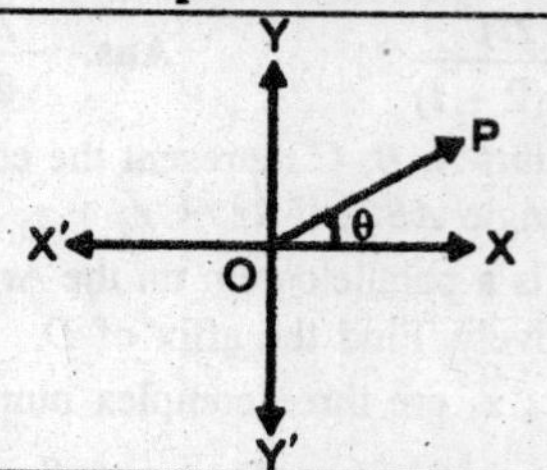 |
| 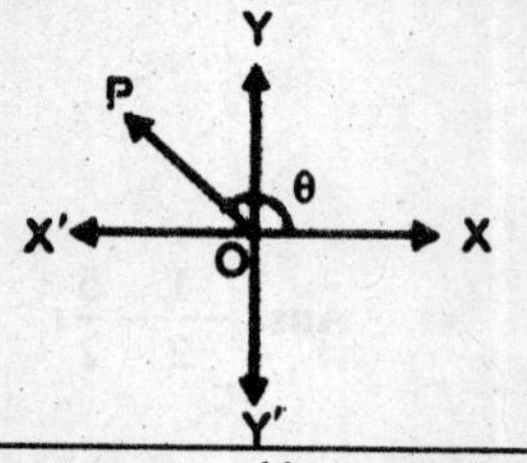 | 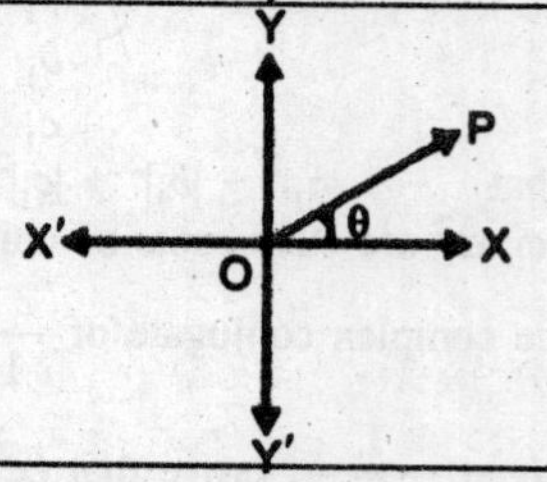 |
| 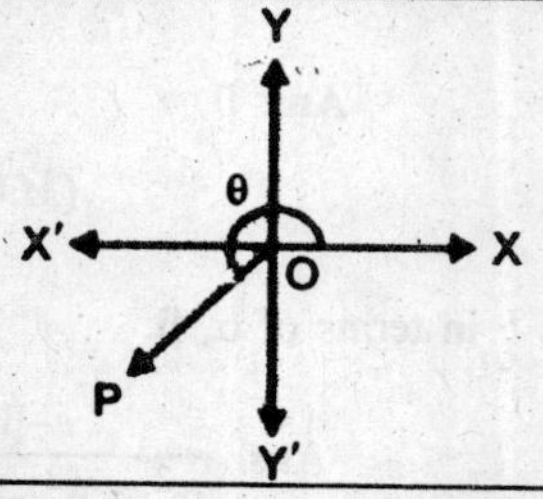 | 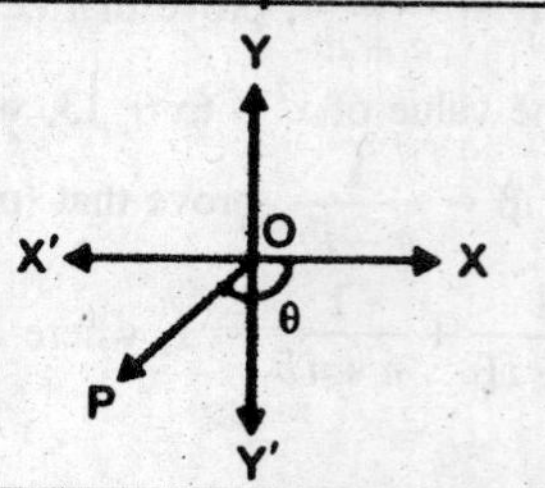 |
| 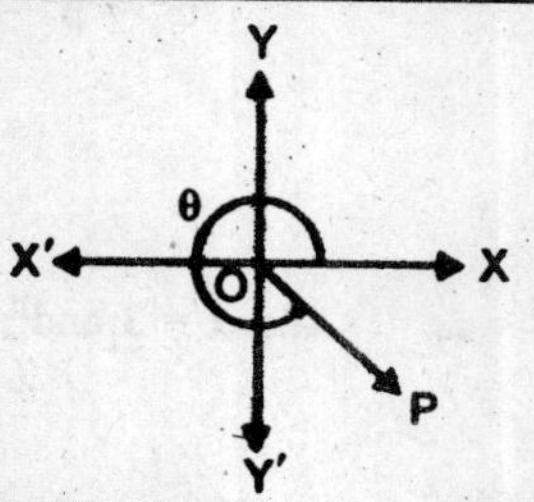 | 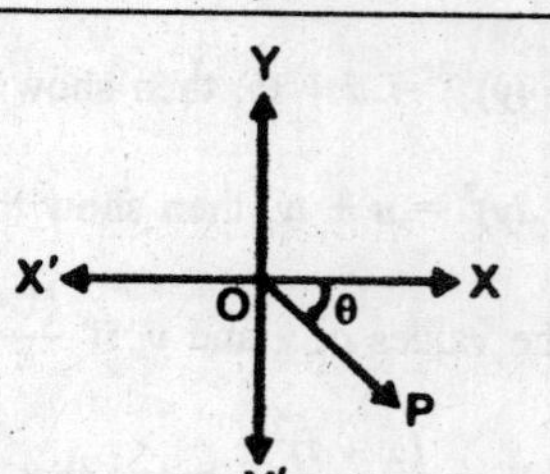 |

*For example* (*i*) the principal value of 240° is –120°.

(*ii*) the principal value of 330° is –30°.

**Example 9.** *Find the modulus and principal argument of the complex number*

$$\frac{1+2i}{1-(1-i)^2}.$$

**Solution.** $\dfrac{1+2i}{1-(1-i)^2} = \dfrac{1+2i}{1-(1-1-2i)} = \dfrac{1+2i}{1+2i} = 1 = 1 + 0i$

$\therefore \quad \left|\dfrac{1+2i}{1-(1-i)^2}\right| = |1+0i| = \sqrt{1^2} = 1$ **Ans.**

Principal argument of $\dfrac{1+2i}{1-(1-i)^2}$ = Principal argument of $1 + 0i$

$$= \tan^{-1}\frac{0}{1} = \tan^{-1} 0 = 0°.$$

Hence modulus = 1 and principal argument = 0°. **Ans.**

**Example 10.** *Find the modulus and principal argument of the complex number :*

$$1 + \cos\alpha + i\sin\alpha. \qquad \left(0 < \alpha < \frac{\pi}{2}\right)$$

**Solution.** Let $(1 + \cos\alpha) + i\sin\alpha = r(\cos\theta + i\sin\theta)$

Equating real and imaginary parts, we get

$$1 + \cos\alpha = r\cos\theta \qquad ...(1)$$

And $$\sin\alpha = r\sin\theta \qquad ...(2)$$

Squaring and adding (1) and (2), we get

$$r^2(\cos^2\theta + \sin^2\theta) = (1+\cos\alpha)^2 + (\sin\alpha)^2$$

$$\Rightarrow \quad r^2 = 1 + \cos^2\alpha + 2\cos\alpha + \sin^2\alpha = 1 + 2\cos\alpha + 1$$

$$= 2(1+\cos\alpha) = 2\left(1 + 2\cos^2\frac{\alpha}{2} - 1\right) = 4\cos^2\frac{\alpha}{2}$$

$$\Rightarrow \quad r = 2\cos\frac{\alpha}{2}$$

From (1), we have, $$\cos\theta = \frac{1+\cos\alpha}{r} = \frac{1+2\cos^2\frac{\alpha}{2}-1}{2\cos\frac{\alpha}{2}} = \cos\frac{\alpha}{2} \qquad ...(3)$$

From (2), we have, $$\sin\theta = \frac{\sin\alpha}{r} = \frac{\sin\alpha}{2\cos\frac{\alpha}{2}} = \frac{2\sin\frac{\alpha}{2}\cos\frac{\alpha}{2}}{2\cos\frac{\alpha}{2}} = \sin\frac{\alpha}{2} \qquad ...(4)$$

$$\text{Argument} = \tan^{-1}\frac{\sin\alpha}{1+\cos\alpha} = \tan^{-1}\frac{2\sin\frac{\alpha}{2}\cos\frac{\alpha}{2}}{1+2\cos^2\frac{\alpha}{2}-1} = \tan^{-1}\tan\frac{\alpha}{2} = \frac{\alpha}{2}$$

General value of argument $= 2\pi k + \dfrac{\alpha}{2}$

$\theta = \dfrac{\alpha}{2}$ satisfied both equations, (1) and (2),

Arg $(1 + \cos\alpha + i\sin\alpha) = \dfrac{\alpha}{2}$ and modulus of $(1 + \cos\alpha + i\sin\alpha) = r = 2\cos\dfrac{\alpha}{2}$ **Ans.**

## EXERCISE 6.2

**Find the modulus and principal argument of the following complex numbers:**

**1.** $-\sqrt{3}-i$ **Ans.** $2, -\dfrac{5\pi}{6}$ **2.** $\dfrac{(1+i)^2}{1-i}$ **Ans.** $\sqrt{2}, \dfrac{3\pi}{4}$

**3.** $\sqrt{\left(\dfrac{1+i}{1-i}\right)}$ **Ans.** $1, \dfrac{\pi}{4}$ **4.** $\tan\alpha - i$ **Ans.** $\sec\alpha, -\left(\dfrac{\pi}{2}-\alpha\right)$

**5.** $1-\cos\alpha + i\sin\alpha$ **Ans.** $2\sin\dfrac{\alpha}{2}, \dfrac{\pi-\alpha}{2}$

**6.** $(4+2i)(-3+\sqrt{2}i)$ **Ans.** $2\sqrt{55}, \tan^{-1}\left(\dfrac{3-2\sqrt{2}}{6+\sqrt{2}}\right)$

**Find the modulus of the following complex numbers :**

**7.** $\overline{(7+i^2)} + (6-i) - (4-3i^3)$ **Ans.** $4\sqrt{5}$

**8.** $\overline{(5-6i)} - (5+6i) + (8-i)$ **Ans.** $\sqrt{185}$

**9.** $(8-i^3) - (7i^2+5) + \overline{(9-i)}$ **Ans.** $\sqrt{365}$

**10.** $(5+6i^{11}) + (8i^3+i^5) + (i^2-i^4)$ **Ans.** $\sqrt{178}$

**11.** If arg. $(z+2i) = \dfrac{\pi}{4}$ and arg. $(z-2i) = \dfrac{3\pi}{4}$, find $z$. **Ans.** $z = 2$

**Example 11.** *If $z_1$ and $z_2$ are any two complex numbers, prove that*

$$|z_1+z_2|^2 + |z_1-z_2|^2 = 2[|z_1|^2 + |z_2|^2]$$

**Solution.** Let
$$z_1 = x_1 + iy_1$$
$$z_2 = x_2 + iy_2$$

$$|z_1+z_2|^2 = |(x_1+iy_1)+(x_2+iy_2)|^2$$
$$= |(x_1+x_2)+i(y_1+y_2)|^2$$
$$= (x_1+x_2)^2 + (y_1+y_2)^2 \quad \text{...(1)}$$

Similarly $|z_1-z_2|^2 = (x_1-x_2)^2+(y_1-y_2)^2$ ...(2)

and $|z_1|^2 = x_1^2 + y_1^2$ ...(3)

$|z_2|^2 = x_2^2 + y_2^2$ ...(4)

L.H.S. $= |z_1+z_2|^2 + |z_1-z_2|^2 = (x_1+x_2)^2 + (y_1+y_2)^2 + (x_1-x_2)^2 + (y_1-y_2)^2$ [Using (1) and (2)]

$$= x_1^2 + 2x_1x_2 + x_2^2 + y_1^2 + 2y_1y_2 + y_2^2$$
$$+ x_1^2 - 2x_1x_2 + x_2^2 + y_1^2 - 2y_1y_2 + y_2^2$$
$$= 2[x_1^2 + x_2^2 + y_1^2 + y_2^2] = 2[(x_1^2+y_1^2)+(x_2^2+y_2^2)] \quad \text{...(5)}$$
$$= 2[|z_1|^2 + |z_2|^2] = \text{R.H.S.}$$
**Proved.**

**Example 12.** *If $z_1$ and $z_2$ are two complex numbers such that $|z_1+z_2| = |z_1-z_2|$, prove that*

$$\arg. z_1 - \arg. z_2 = \frac{\pi}{2}$$
*(M.U. 2002, 2007)*

**Solution.** Let
$$z_1 = x_1 + iy_1$$
$$z_2 = x_2 + iy_2$$

Given that $|z_1+z_2| = |z_1-z_2|$

$\Rightarrow$ $|(x_1+iy_1)+(x_2+iy_2)| = |(x_1+iy_1)-(x_2+iy_2)|$

$\Rightarrow$ $|(x_1+x_2)+i(y_1+y_2)| = |(x_1-x_2)+(y_1-y_2)i|$

$\Rightarrow$ $(x_1+x_2)^2 + (y_1+y_2)^2 = (x_1-x_2)^2 + (y_1-y_2)^2$

$\Rightarrow x_1^2 + x_2^2 + 2x_1x_2 + y_1^2 + y_2^2 + 2y_1y_2 = x_1^2 + x_2^2 - 2x_1x_2 + y_1^2 + y_2^2 - 2y_1y_2$

$$\Rightarrow \quad 2x_1x_2 + 2y_1y_2 = -2x_1x_2 - 2y_1y_2$$

$$\Rightarrow \quad 4x_1x_2 + 4y_1y_2 = 0$$

$$\Rightarrow \quad x_1x_2 + y_1y_2 = 0 \qquad ...(1)$$

Now, $\quad \arg. z_1 - \arg. z_2 = \tan^{-1}\left(\frac{y_1}{x_1}\right) - \tan^{-1}\left(\frac{y_2}{x_2}\right)$

$$= \tan^{-1}\left[\frac{\frac{y_1}{x_1} - \frac{y_2}{x_2}}{1 + \left(\frac{y_1}{x_1}\right)\left(\frac{y_2}{x_2}\right)}\right] = \tan^{-1}\left(\frac{x_2y_1 - x_1y_2}{x_1x_2 + y_1y_2}\right)$$

$$= \tan^{-1}\left(\frac{x_2y_1 - x_1y_2}{0}\right) = \tan^{-1}\infty = \frac{\pi}{2} \qquad \text{[Using (1)]}$$

$$\arg. z_1 - \arg. z_2 = \frac{\pi}{2}$$ **Proved.**

**Example 13.** *Find the complex number z if arg (z + 1) =* $\frac{\pi}{6}$ *and arg (z – 1) =* $\frac{2\pi}{3}$.

**(*M.U.* 2009, 2000, 01, 02, 03)**

**Solution.** Let $\quad z = x + iy \qquad ...(1)$

$\therefore \quad z + 1 = (x + 1) + iy$

**We also given that**

$$\text{Arg}(z + 1) = \tan^{-1}\left(\frac{y}{x+1}\right) = \frac{\pi}{6}$$

$$\therefore \quad \frac{y}{x+1} = \tan 30° = \frac{1}{\sqrt{3}}$$

$$\therefore \quad \sqrt{3}y = x + 1 \qquad ...(2)$$

**Now** $\quad z - 1 = (x - 1) + iy \qquad$ **[From (1)]**

**and** $\quad \tan^{-1}\left(\frac{y}{x-1}\right) = \frac{2\pi}{3} \Rightarrow \frac{y}{x-1} = \tan 120°$

$$\Rightarrow \quad \frac{y}{x-1} = -\cot 30° = -\sqrt{3}$$

$$\therefore \quad -y = \sqrt{3}x - \sqrt{3}$$

$$\Rightarrow \quad -\sqrt{3}y = 3x - 3 \qquad ...(3)$$

**Adding (2) and (3), we get**

$$0 = 4x - 2 \Rightarrow 4x = 2 \Rightarrow x = \frac{1}{2}$$

**Putting** $x = \frac{1}{2}$ **in (2), we get**

$$\sqrt{3}y = \frac{1}{2} + 1 \Rightarrow \sqrt{3}y = \frac{3}{2} \Rightarrow y = \frac{\sqrt{3}}{2}$$

**Putting the values of *x* and *y* in (1), we get**

$$z = \frac{1}{2} + \frac{\sqrt{3}}{2}i$$ **Ans.**

**Example 14.** *Prove that*

(*i*) $|z_1 + z_2| \le |z_1| + |z_2|$ (*ii*) $|z_1 - z_2| \ge |z_1| - |z_2|$

**Solution.** (*a*) (By Geometry) Let $z_1 = x_1 + iy_1$ and

$z_2 = x_2 + iy_2$ be the two complex numbers shown in the figure

$$|z_1| = OP, \quad |z_2| = OQ$$

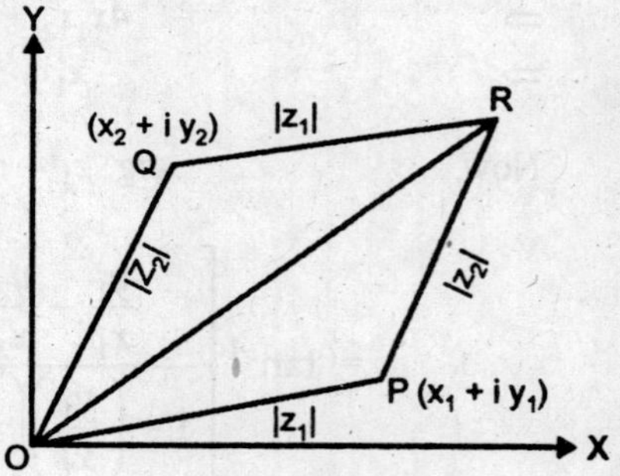


(*i*) Since in a traingle any side is less than the sum of the other two.

In Δ *OPR*, $OR < OP + PR$, $OR < OP + OQ$

$$\Rightarrow \quad |z_1 + z_2| < |z_1| + |z_2|$$

$OR = OP + PR$ if $O, P, R$ are collinear.

or $$|z_1 + z_2| = |z_1| + |z_2|$$

(*ii*) Again, any side of a triangle is greater than the difference between the other two, we have

In Δ *OPR*

$OR > OP - PR, \quad \Rightarrow \quad OR > OP - OQ$

$$|z_1 - z_2| > |z_1| - |z_2|$$

**Proved.**

(*b*) By Algebra. $\quad z_1 + z_2 = x_1 + iy_1 + x_2 + iy_2 = (x_1 + x_2) + i(y_1 + y_2)$

(*i*) $$|z_1 + z_2|^2 = (x_1 + x_2)^2 + (y_1 + y_2)^2$$

$$= x_1^2 + x_2^2 + 2x_1x_2 + y_1^2 + y_2^2 + 2y_1y_2 = (x_1^2 + y_1^2) + (x_2^2 + y_2^2) + 2(x_1x_2 + y_1y_2)$$

$$= (x_1^2 + y_1^2) + (x_2^2 + y_2^2) + 2\sqrt{(x_1x_2 + y_1y_2)^2}$$

$$= |z_1|^2 + |z_2|^2 + 2\sqrt{x_1^2x_2^2 + y_1^2y_2^2 + 2x_1x_2y_1y_2}$$

$$[\because (x_1y_2 - x_2y_1)^2 \ge 0 \text{ or } x_1^2y_2^2 + x_2^2y_1^2 \ge 2x_1x_2y_1y_2]$$

$$|z_1 + z_2|^2 \le |z_1|^2 + |z_2|^2 + 2\sqrt{x_1^2x_2^2 + y_1^2y_2^2 + x_1^2y_2^2 + x_2^2y_1^2}$$

$$\le |z_1|^2 + |z_2|^2 + 2\sqrt{(x_1^2 + y_1^2)(x_2^2 + y_2^2)}$$

$$\le |z_1|^2 + |z_2|^2 + 2|z_1||z_2|$$

$$\le \{|z_1| + |z_2|\}^2$$

$$|z_1 + z_2| \le |z_1| + |z_2|$$

(*ii*) $$|z_1| = |(z_1 - z_2) + z_2| \le |z_1 - z_2| + |z_2|$$

$$|z_1| - |z_2| \le |z_1 - z_2|$$

$$|z_1 - z_2| \ge |z_1| - |z_2|$$

**Proved.**

## EXERCISE 6.3

**1.** If $z = x + iy$, prove that $\left(\dfrac{z}{\bar{z}} + \dfrac{\bar{z}}{z}\right) = 2\left(\dfrac{x^2 - y^2}{x^2 + y^2}\right)$.

**2.** If $z = a \cos\theta + ia \sin\theta$, prove that $\left(\dfrac{z}{\bar{z}} + \dfrac{\bar{z}}{z}\right) = 2\cos 2\theta$.

**3.** Prove that $\left|\dfrac{z - 1}{\bar{z} - 1}\right| = 1$.

**4.** Let $z_1 = 2 - i$, $z_2 = -2 + i$, find

(*i*) $\text{Re}\left[\dfrac{z_1\, z_2}{\bar{z}_1}\right]$ (*ii*) $\text{Im}\left[\dfrac{1}{z_1\, \bar{z}_2}\right]$ **Ans.** (*i*) $-\dfrac{2}{5}$, (*ii*) 0

**5.** If $|z| = 1$, prove that $\dfrac{z - 1}{z + 1}$ $(z \ne 1)$ is a pure imaginary number, what will you conclude, if

**Ans.** If $z = 1$, $\dfrac{z - 1}{z + 1} = 0$, which is purely real.

## 6.16 POLAR FORM

Polar form of a complex number as we have discussed above

$$x = r\cos\theta \quad \text{and} \quad y = r\sin\theta$$

$$\Rightarrow \quad x + iy = r(\cos\theta + i\sin\theta)$$

$$= r\,e^{i\theta} \quad \text{(Exponential form)} \qquad (e^{i\theta} = \cos\theta + i\sin\theta)$$

**Procedure.** To convert $x + iy$ into polar.

We write $\quad x = r\cos\theta$

$$y = r\sin\theta$$

On solving these equations, we get the value of $\theta$ which satisfy both the equations and $r = \sqrt{x^2 + y^2}$.

## 6.17 TYPES OF COMPLEX NUMBERS

1. Cartesian form : $x + iy$
2. Polar form : $r(\cos\theta + i\sin\theta)$
3. Exponential form : $re^{i\theta}$

**Example 15.** *Express in polar form* : $1 - \sqrt{2} + i$

**Solution.** Let $(1 - \sqrt{2}) + i = r(\cos\theta + i\sin\theta)$

$$\therefore \quad 1 - \sqrt{2} = r\cos\theta \qquad \text{...(1)}$$

$$1 = r\sin\theta \qquad \text{...(2)}$$

Squaring and adding (1) and (2), we get

$$r^2(\cos^2\theta + \sin^2\theta) = (1 - \sqrt{2})^2 + 1^2$$

$$\Rightarrow \quad r^2 = 1 - 2\sqrt{2} + 2 + 1 \quad \Rightarrow r = \sqrt{4 - 2\sqrt{2}}$$

Putting the value of $r$ in (1) and (2), we get

$$\cos\theta = \frac{1 - \sqrt{2}}{\sqrt{4 - 2\sqrt{2}}} \quad \text{and} \quad \sin\theta = \frac{1}{\sqrt{4 - 2\sqrt{2}}}$$

Hence, the polar form is $\sqrt{4 - 2\sqrt{2}}\left\{\frac{1 - \sqrt{2}}{\sqrt{4 - 2\sqrt{2}}} + i\frac{1}{\sqrt{4 - 2\sqrt{2}}}\right\}$ **Ans.**

**Example 16.** *Find the smallest positive integer n for which*

$$\left(\frac{1+i}{1-i}\right)^n = 1. \qquad \textit{(Nagpur University, Winter 2004)}$$

**Solution.** $\left[\frac{1+i}{1-i}\right]^n = 1$

$$\left[\frac{1+i}{1-i} \times \frac{1+i}{1+i}\right]^n = 1 \quad \Rightarrow \quad \left(\frac{1 - 1 + 2i}{1+1}\right)^n = 1$$

$$(i)^n = 1 = (i)^4 \Rightarrow n = 4 \qquad \textbf{Ans.}$$

### EXERCISE 6.4

Express the following complex numbers into polar form :

1. $\frac{1+i}{1-i}$ **Ans.** $\cos\frac{\pi}{2} + i\sin\frac{\pi}{2}$

2. $\frac{-35 + 5i}{4\sqrt{2} + 3\sqrt{2}i}$ **Ans.** $5\left(\cos\frac{3\pi}{4} + i\sin\frac{3\pi}{4}\right)$

3. $\frac{3(-4 - \sqrt{3} + 4\sqrt{3}i - i)}{8 + 2i}$ **Ans.** $3\left(\cos\frac{2\pi}{3} + i\sin\frac{2\pi}{3}\right)$

4. $\frac{2 + 6\sqrt{3}i}{5 + \sqrt{3}i}$ **Ans.** $2\left(\cos\frac{\pi}{3} + i\sin\frac{\pi}{3}\right)$

5. $\dfrac{2+3i}{3-7i}$ **Ans.** $r = \sqrt{754}$, $\theta = \tan^{-1}\left(-\dfrac{23}{15}\right)$ 6. $\left(\dfrac{4-5i}{2+3i}\right)\cdot\left(\dfrac{3+2i}{7+i}\right)$ **Ans.** 0.905, $\theta = \tan^{-1}(-7.2)$

7. $\dfrac{(2+5i)(-3+i)}{(1-2i)^2}$ **Ans.** $\dfrac{\sqrt{290}}{5}$, $\tan^{-1}\left(-\dfrac{1}{17}\right)$ 8. $\dfrac{1+7i}{(2-i)^2}$ **Ans.** $\sqrt{2}\left(\cos\dfrac{3\pi}{4} + i\sin\dfrac{3\pi}{4}\right)$

9. $\dfrac{1+3i}{1-2i}$ **Ans.** $\sqrt{2}\left(\cos\dfrac{3\pi}{4} + i\sin\dfrac{3\pi}{4}\right)$ 10. $\dfrac{i-1}{\cos\dfrac{\pi}{3} + i\sin\dfrac{\pi}{3}}$ **Ans.** $\sqrt{2}\left(\cos\dfrac{5\pi}{12} + i\sin\dfrac{5\pi}{12}\right)$

## 6.18 SQUARE ROOT OF A COMPLEX NUMBER

Let $a + ib$ be a complex number and its square root is $x + iy$.

*i.e.*, $$\sqrt{a+ib} = x + iy \qquad ...(1)$$

where $x$ and $y \in R$.

Squaring both sides of (1), we get

$$a + ib = (x + iy)^2$$

$$\Rightarrow \quad a + ib = x^2 + i^2y^2 + i\,2xy$$

$$\Rightarrow \quad a + ib = (x^2 - y^2) + i\,2xy \qquad [\because i^2 = -1] \qquad ...(2)$$

Equating real and imaginary parts of (2), we get

$$x^2 - y^2 = a \qquad ...(3)$$

and $$2xy = b \qquad ...(4)$$

Also, we know that

$$(x^2 + y^2)^2 = (x^2 - y^2)^2 + 4x^2y^2$$

$$\Rightarrow \quad (x^2 + y^2)^2 = a^2 + b^2 \qquad \text{[Using (3) and (4)]}$$

$$\Rightarrow \quad x^2 + y^2 = \sqrt{a^2 + b^2} \qquad ...(5)$$

Adding (3) and (5), we get

$$2x^2 = a + \sqrt{a^2 + b^2} \quad \Rightarrow \quad x = \sqrt{\frac{a + \sqrt{a^2 + b^2}}{2}}$$

**Example 17.** *Find the square root of the complex number 5 + 12i.*

**Solution.** Let $\sqrt{5 + 12i} = x + iy$ ...(1)

Squaring both sides of (1), we get $5 + 12i = (x + iy)^2 = (x^2 - y^2) + i\,2xy$ ...(2)

Equating real and imaginary parts of (2), we get

$$x^2 - y^2 = 5 \qquad ...(3)$$

and $$2xy = 12 \qquad ...(4)$$

Now, $$x^2 + y^2 = \sqrt{(x^2 - y^2)^2 + 4x^2y^2} = \sqrt{(5)^2 + (12)^2}$$

$$= \sqrt{25 + 144} = \sqrt{169} = 13$$

$$\Rightarrow \quad x^2 + y^2 = 13 \qquad ...(5)$$

Adding (3) and (5), we get $2x^2 = 5 + 13 = 18 \Rightarrow x = \sqrt{\dfrac{18}{2}} = \sqrt{9} = \pm 3$

Subtracting (3) from (5), we get $2y^2 = 13 - 5 = 8 \Rightarrow y = \sqrt{\dfrac{8}{2}} = \sqrt{4} = \pm 2$

Since, $xy$ is positive, so $x$ and $y$ are of same sign. Hence, $x = \pm 3$, $y = \pm 2$

$\therefore \quad \sqrt{5 + 12i} = \pm 3 \pm 2i$ *i.e.* $(3 + 2i)$ or $-(3 + 2i)$ **Ans.**

**Example 18.** *Prove that if the sum and product of two complex numbers are real then the two numbers must be either real or conjugate.* *(M.U. 2008)*

**Solution.** Let $z_1$ and $z_2$ be the two complex numbers.

We are given that $z_1 + z_2 = a$ (real)

and $\quad z_1 . z_2 = b$ (real)

If sum and product of the roots of a quadratic equation are given. Then the equation becomes

$$x^2 - (\text{sum of the roots})\, x + \text{product of the roots} = 0$$

$$x^2 - ax + b = 0$$

$$\text{Root} = x = \frac{a \pm \sqrt{a^2 - 4b}}{2}$$

**Case I.** If $a^2 > 4b$ Then both the roots are real

**Case II.** If $a^2 < 4b$

Then

$$\text{one root} = \frac{a}{2} + i\frac{\sqrt{4b - a^2}}{2}$$

$$\text{Second root} = \frac{a}{2} - i\frac{\sqrt{4b - a^2}}{2}$$

These roots are conjugate to each other. **Proved.**

## EXERCISE 6.5

**Find the square root of the following :**

**1.** $1 + i$ **Ans.** $\left\{\pm\sqrt{\frac{\sqrt{2}+1}{2}} \pm \sqrt{\frac{\sqrt{2}-1}{2}}\, i\right\}$ **2.** $1 - i$ **Ans.** $\left\{\pm\sqrt{\frac{\sqrt{2}+1}{2}} \mp \sqrt{\frac{\sqrt{2}-1}{2}}\, i\right\}$

**3.** $i$ **Ans.** $\left\{\pm\frac{1}{\sqrt{2}} \pm \frac{1}{\sqrt{2}}\, i\right\}$ **4.** $15 - 8i$ **Ans.** $1 - 4i, -1 + 4i$

**5.** $-2 + 2\sqrt{3}i$ **Ans.** $\pm(1 + \sqrt{3}i)$ **6.** $3 + 4\sqrt{7}i$ **Ans.** $\pm(\sqrt{7} + 2i)$

**7.** $\frac{2+3i}{5-4i} + \frac{2-3i}{5+4i}$ **Ans.** $\pm\frac{2}{\sqrt{41}}\, i$ **8.** $x^2 - 1 + i\, 2x$ **Ans.** $\pm(x + i)$

**9.** $3 - 4i$ **Ans.** $\pm(2 - i)$

## 6.19 EXPONENTIAL AND CIRCULAR FUNCTIONS OF COMPLEX VARIABLES

**Proof.** $\boxed{\cos\theta + i\sin\theta = e^{i\theta}}$

$$e^z = 1 + z + \frac{z^2}{2!} + \frac{z^3}{3!} + \frac{z^4}{4!} + \ldots \qquad \ldots(1)$$

$$\sin z = z - \frac{z^3}{3!} + \frac{z^5}{5!} - \frac{z^7}{7!} + \ldots \qquad \ldots(2)$$

$$\cos z = 1 - \frac{z^2}{2!} + \frac{z^4}{4!} - \frac{z^6}{6!} + \ldots \qquad \ldots(3)$$

From (2) and (3), we have

$$\cos z + i\sin z = \left(1 - \frac{z^2}{2!} + \frac{z^4}{4!} - \frac{z^6}{6!} + \ldots\right) + i\left(z - \frac{z^3}{3!} + \frac{z^5}{5!} - \ldots\right)$$

$$= 1 + \frac{(iz)^1}{1!} + \frac{(iz)^2}{2!} + \frac{(iz)^3}{3!} + \ldots = e^{iz}$$

Therefore, $\quad \cos z + i\sin z = e^{iz} \qquad \ldots(4)$

Similarly, $\quad \cos z - i\sin z = e^{-iz} \qquad \ldots(5)$

From (4) and (5), we have

$$\cos z = \frac{e^{iz} + e^{-iz}}{2} \qquad \ldots(6)$$

$$\sin z = \frac{e^{iz} - e^{-iz}}{2i} \qquad \ldots(7)$$

## 6.20 DE MOIVRE'S THEOREM (By Exponential Function)

$$(\cos\theta + i\sin\theta)^n = \cos n\theta + i\sin n\theta$$

**Proof.** We know that

$$e^{i\theta} = \cos\theta + i\sin\theta$$
$$(e^{i\theta})^n = (\cos\theta + i\sin\theta)^n$$
$$e^{in\theta} = (\cos\theta + i\sin\theta)^n$$
$$(\cos n\theta + i\sin n\theta) = (\cos\theta + i\sin\theta)^n$$

**Proved.**

If $n$ is a fraction, then $\cos n\theta + i\sin n\theta$ is one of the values of $(\cos\theta + i\sin\theta)^n$

## 6.21 DE MOIVRE'S THEOREM (BY INDUCTION)

**Statement:** For any rational number $n$ the value or one of the values of

$$(\cos\theta + i\sin\theta)^n = \cos n\theta + i\sin n\theta$$

**Proof. Case I.** Let $n$ be a non-negative integer. By actual multiplication,

$$(\cos\theta_1 + i\sin\theta_1)(\cos\theta_2 + i\sin\theta_2) = (\cos\theta_1\cos\theta_2 - \sin\theta_1\sin\theta_2) + i(\cos\theta_1\sin\theta_2 + \sin\theta_1\cos\theta_2)$$
$$= \cos(\theta_1 + \theta_2) + i\sin(\theta_1 + \theta_2) \quad ...(1)$$

Similarly we can prove that

$$(\cos\theta_1 + i\sin\theta_1)(\cos\theta_2 + i\sin\theta_2)(\cos\theta_3 + i\sin\theta_3)$$
$$= \cos(\theta_1 + \theta_2 + \theta_3) + i\sin(\theta_1 + \theta_2 + \theta_3)$$

Continuing in this way, we can prove that

$$(\cos\theta_1 + i\sin\theta_1)(\cos\theta_2 + i\sin\theta_2)\ldots(\cos\theta_n + i\sin\theta_n)$$
$$= \cos(\theta_1 + \theta_2 \ldots + \theta_n) + i\sin(\theta_1 + \theta_2 + \ldots + \theta_n)$$

Putting $\theta_1 = \theta_2 = \theta_3 = \ldots \theta_n = \theta$, we get

$$(\cos\theta + i\sin\theta)^n = (\cos n\theta + i\sin n\theta)$$

**Case II.** Let $n$ be a negative integer, say $n = -m$ where $m$ is a positive integer. Then.

$$(\cos\theta + i\sin\theta)^n = (\cos\theta + i\sin\theta)^{-m}$$

$$= \frac{1}{(\cos\theta + i\sin\theta)^m} = \frac{1}{(\cos m\theta + i\sin m\theta)} \quad \text{[By case I]}$$

$$= \frac{1}{(\cos m\theta + i\sin m\theta)} \cdot \frac{(\cos m\theta - i\sin m\theta)}{(\cos m\theta - i\sin m\theta)} = \frac{\cos m\theta - i\sin m\theta}{\cos^2 m\theta + \sin^2 m\theta}$$

$$= \cos m\theta - i\sin m\theta \quad [\because \cos^2 m\theta + \sin^2 m\theta = 1]$$

$$= \cos(-m\theta) + i\sin(-m\theta) = \cos n\theta + i\sin n\theta$$

Hence, the theorem is true for negative integers also.

**Case III.** Let $n$ be a proper fraction $\frac{p}{q}$ where $p$ and $q$ are integers. Without loss of generality we can select $q$ to be positive integer, $p$ may be a positive or negative integer.

Since $q$ is a positive integer

$$\text{Now,} \quad \left(\cos\frac{\theta}{q} + i\sin\frac{\theta}{q}\right)^q = \cos q\cdot\frac{\theta}{q} + i\sin q\cdot\frac{\theta}{q} \quad \text{[By case I]}$$
$$= \cos\theta + i\sin\theta$$

Taking the $q$ th root of both sides, we get

$$(\cos\theta + i\sin\theta)^{\frac{1}{q}} = \cos\frac{\theta}{q} + i\sin\frac{\theta}{q}$$

Raising both sides to the power $p$,

$$(\cos\theta + i\sin\theta)^{\frac{p}{q}} = \left(\cos\frac{\theta}{q} + i\sin\frac{\theta}{q}\right)^p = \cos p.\frac{\theta}{q} + i\sin p.\frac{\theta}{q} \quad \text{[By case I and II]}$$

Hence, one of the values of $(\cos\theta + i\sin\theta)^n$ is $\cos n\theta + i\sin n\theta$ when n is a proper fraction.

Thus, the theorem is true for all rational values of $n$.

**Example 19.** *Express* $\dfrac{(\cos\theta + i\sin\theta)^8}{(\sin\theta + i\cos\theta)^4}$ *in the form* $(x + iy)$.

**Solution.** $$\frac{(\cos\theta + i\sin\theta)^8}{(\sin\theta + i\cos\theta)^4} = \frac{(\cos\theta + i\sin\theta)^8}{(i)^4\left(\cos\theta + \frac{1}{i}\sin\theta\right)^4}$$

$$= \frac{(\cos\theta + i\sin\theta)^8}{(\cos\theta - i\sin\theta)^4} = \frac{(\cos\theta + i\sin\theta)^8}{[\cos(-\theta) + i\sin(-\theta)]^4}$$

$$= \frac{(\cos\theta + i\sin\theta)^8}{[(\cos\theta + i\sin\theta)^{-1}]^4} = \frac{(\cos\theta + i\sin\theta)^8}{(\cos\theta + i\sin\theta)^{-4}} = (\cos\theta + i\sin\theta)^{12}$$

$$= \cos 12\theta + i\sin 12\theta$$ **Ans.**

**Example 20.** *Prove that* $(1 + \cos\theta + i\sin\theta)^n + (1 + \cos\theta - i\sin\theta)^n = 2^{n+1}\cos^n\frac{\theta}{2}\cos\frac{n\theta}{2}$ *where n is an integer.*

**Solution.** L.H.S. $= (1 + \cos\theta + i\sin\theta)^n + (1 + \cos\theta - i\sin\theta)^n$

$$= \left[1 + 2\cos^2\frac{\theta}{2} - 1 + 2i\sin\frac{\theta}{2}\cos\frac{\theta}{2}\right]^n + \left[1 + 2\cos^2\frac{\theta}{2} - 1 - i2\sin\frac{\theta}{2}\cos\frac{\theta}{2}\right]^n$$

$$= \left[2\cos^2\frac{\theta}{2} + 2i\sin\frac{\theta}{2}\cos\frac{\theta}{2}\right]^n + \left[2\cos^2\frac{\theta}{2} - 2i\sin\frac{\theta}{2}\cos\frac{\theta}{2}\right]^n$$

$$= \left(2\cos\frac{\theta}{2}\right)^n\left[\cos\frac{\theta}{2} + i\sin\frac{\theta}{2}\right]^n + \left(2\cos\frac{\theta}{2}\right)^n\left[\cos\frac{\theta}{2} - i\sin\frac{\theta}{2}\right]^n$$

$$= 2^n\cos^n\frac{\theta}{2}\left[\cos\frac{n\theta}{2} + i\sin\frac{n\theta}{2}\right] + 2^n\cos^n\frac{\theta}{2}\left[\cos\frac{n\theta}{2} - i\sin\frac{n\theta}{2}\right]$$

$$= 2^n\cos^n\frac{\theta}{2}\left[\cos\frac{n\theta}{2} + i\sin\frac{n\theta}{2} + \cos\frac{n\theta}{2} - i\sin\frac{n\theta}{2}\right]$$

$$= 2^n\cos^n\frac{\theta}{2}\left(2\cos\frac{n\theta}{2}\right) = 2^{n+1}\cos^n\frac{\theta}{2}\cos\frac{n\theta}{2} = \text{R.H.S.}$$ **Proved.**

**Example 21.** *Evaluate* $\left(\dfrac{1 + \sin\alpha + i\cos\alpha}{1 + \sin\alpha - i\cos\alpha}\right)^n$ *(M.U. 2001, 2004, 2005)*

**Solution.** We know that,

$$1 = \sin^2\alpha + \cos^2\alpha$$

$$\Rightarrow \quad 1 = \sin^2\alpha - i^2\cos^2\alpha$$

$$\Rightarrow \quad 1 = (\sin\alpha + i\cos\alpha)(\sin\alpha - i\cos\alpha) \quad \ldots(1)$$

Adding $\sin\alpha + i\cos\alpha$ both sides of (1), we get

$$1 + \sin\alpha + i\cos\alpha = (\sin\alpha + i\cos\alpha)(\sin\alpha - i\cos\alpha) + (\sin\alpha + i\cos\alpha)$$

$$= (\sin\alpha + i\cos\alpha)(\sin\alpha - i\cos\alpha + 1)$$

$$\therefore \quad \frac{1 + \sin\alpha + i\cos\alpha}{1 + \sin\alpha - i\cos\alpha} = \sin\alpha + i\cos\alpha$$

$$= \cos\left(\frac{\pi}{2} - \alpha\right) + i\sin\left(\frac{\pi}{2} - \alpha\right) \quad \ldots(2)$$

$$\Rightarrow \quad \left(\frac{1 + \sin\alpha + i\cos\alpha}{1 + \sin\alpha - i\cos\alpha}\right)^n = \left\{\cos\left(\frac{\pi}{2} - \alpha\right) + i\sin\left(\frac{\pi}{2} - \alpha\right)\right\}^n$$

$$= \cos n\left(\frac{\pi}{2} - \alpha\right) + i\sin n\left(\frac{\pi}{2} - \alpha\right)$$ **Ans.**

**Example 22.** *Prove that the general value of θ which satisfies the equation*

$(\cos\theta + i\sin\theta).(\cos 2\theta + i\sin 2\theta) \ldots (\cos n\theta + i\sin n\theta) = 1$ is $\dfrac{4m\pi}{n(n+1)}$, *where m is any integer.*

**Solution.**

$$(\cos\theta + i\sin\theta)(\cos 2\theta + i\sin 2\theta)\ldots(\cos n\theta + i\sin n\theta) = 1$$

$$(\cos\theta + i\sin\theta)(\cos\theta + i\sin\theta)^2\ldots(\cos\theta + i\sin\theta)^n = 1$$

$$(\cos\theta + i\sin\theta)^{1+2\ldots+n} = 1$$

$$(\cos\theta + i\sin\theta)^{\frac{n(n+1)}{2}} = (\cos 2m\pi + i\sin 2m\pi)$$

$$\cos\frac{n(n+1)}{2}\theta + i\sin\frac{n(n+1)}{2}\theta = \cos 2m\pi + i\sin 2m\pi$$

$$\frac{n(n+1)}{2}\theta = 2m\pi \Rightarrow \theta = \frac{4m\pi}{n(n+1)}$$ **Proved.**

**Example 23.** *If $(a_1 + ib_1).(a_2 + ib_2)\ldots(a_n + ib_n) = A + iB$, then prove that*

(i) $\tan^{-1}\dfrac{b_1}{a_1} + \tan^{-1}\dfrac{b_2}{a_2} + \ldots + \tan^{-1}\dfrac{b_n}{a_n} = \tan^{-1}\dfrac{B}{A}$

(ii) $(a_1^2 + b_1^2)(a_2^2 + b_2^2) \ldots (a_n^2 + b_n^2) = A^2 + B^2$

**Solution.** Let

$$a_1 = r_1\cos\alpha_1, \quad b_1 = r_1\sin\alpha_1$$
$$a_2 = r_2\cos\alpha_2, \quad b_2 = r_2\sin\alpha_2$$
$$\ldots\ldots\ldots\ldots\ldots\ldots\ldots\ldots\ldots$$
$$a_n = r_n\cos\alpha_n, \quad b_n = r_n\sin\alpha_n$$
$$A = R\cos\theta, \quad B = R\sin\theta,$$

$$(a_1 + ib_1).(a_2 + ib_2)\ldots(a_n + ib_n) = A + iB \qquad \text{(Given)}$$

$$r_1(\cos\alpha_1 + i\sin\alpha_1)\, r_2(\cos\alpha_2 + i\sin\alpha_2)\ldots r_n(\cos\alpha_n + i\sin\alpha_n) = R(\cos\theta + i\sin\theta)$$

$$r_1 r_2 \ldots r_n[\cos(\alpha_1 + \alpha_2 + \ldots \alpha_n) + i\sin(\alpha_1 + \alpha_2 + \ldots \alpha_n)] = R(\cos\theta + i\sin\theta)$$

$\therefore$ $$r_1 r_2 \ldots\ldots\ldots\ldots r_n = R$$

$\Rightarrow$ $$(a_1^2 + b_1^2)(a_2^2 + b_2^2)\ldots(a_n^2 + b_n^2) = A^2 + B^2$$

And $$\alpha_1 + \alpha_2 + \ldots + \alpha_n = \theta$$

$\Rightarrow$ $$\tan^{-1}\frac{b_1}{a_1} + \tan^{-1}\frac{b_2}{a_2} + \ldots + \tan^{-1}\frac{b_n}{a_n} = \tan^{-1}\frac{B}{A}$$ **Proved.**

**Example 24.** *If $\cos\alpha + \cos\beta + \cos\gamma = 0 = \sin\alpha + \sin\beta + \sin\gamma$, then prove that*

(i) $\sin^2\alpha + \sin^2\beta + \sin^2\gamma = \cos^2\alpha + \cos^2\beta + \cos^2\gamma = \dfrac{3}{2}$

(ii) $\cos 2\alpha + \cos 2\beta + \cos 2\gamma = 0$

(iii) $\cos(\alpha + \beta) + \cos(\beta + \gamma) + \cos(\gamma + \alpha) = 0$

(iv) $\sin(\alpha + \beta) + \sin(\beta + \gamma) + \sin(\gamma + \alpha) = 0$ *(M.U. 2009)*

**Solution.** Here, we have

$$(\cos\alpha + \cos\beta + \cos\gamma) + i(\sin\alpha + \sin\beta + \sin\gamma) = 0$$

$$(\cos\alpha + i\sin\alpha) + (\cos\beta + i\sin\beta) + (\cos\gamma + i\sin\gamma) = 0$$

$\therefore$ $$a + b + c = 0 \text{ say} \qquad \ldots(1)$$

where, $a = \cos\alpha + i\sin\alpha$, $b = \cos\beta + i\sin\beta$ and $c = \cos\gamma + i\sin\gamma$

Also we can write

$$(\cos\alpha - i\sin\alpha) + (\cos\beta - i\sin\beta) + (\cos\gamma - i\sin\gamma) = 0$$

$\Rightarrow \quad (\cos\alpha + i\sin\alpha)^{-1} + (\cos\beta + i\sin\beta)^{-1} + (\cos\gamma + i\sin\gamma)^{-1} = 0$

$\Rightarrow \quad \frac{1}{a} + \frac{1}{b} + \frac{1}{c} = 0$

$\Rightarrow \quad \frac{bc + ca + ab}{abc} = 0 \Rightarrow ab + bc + ca = 0$ ...(2)

But $\quad (a + b + c)^2 = (a^2 + b^2 + c^2) + 2(ab + bc + ca)$

$0 = (a^2 + b^2 + c^2) + 0$ [From (1) and (2)]

$\Rightarrow \quad a^2 + b^2 + c^2 = 0$

$\Rightarrow \quad (\cos\alpha + i\sin\alpha)^2 + (\cos\beta + i\sin\beta)^2 + (\cos\gamma + i\sin\gamma)^2 = 0$

$\Rightarrow \quad (\cos 2\alpha + i\sin 2\alpha) + (\cos 2\beta + i\sin 2\beta) + (\cos 2\gamma + i\sin 2\gamma) = 0$

$\Rightarrow \quad (\cos 2\alpha + \cos 2\beta + \cos 2\gamma) + i(\sin 2\alpha + \sin 2\beta + \sin 2\gamma) = 0$

$\Rightarrow \quad \cos 2\alpha + \cos 2\beta + \cos 2\gamma = 0.$ ...(3)

$\Rightarrow \quad 2\cos^2\alpha - 1 + 2\cos^2\beta - 1 + 2\cos^2\gamma - 1 = 0$

$\Rightarrow \quad \cos^2\alpha + \cos^2\beta + \cos^2\gamma = \frac{3}{2}$ ...(4)

Further $\quad 1 - \sin^2\alpha + 1 - \sin^2\beta + 1 - \sin^2\gamma = \frac{3}{2}$

$\Rightarrow \quad \sin^2\alpha + \sin^2\beta + \sin^2\gamma = \frac{3}{2}$ ...(5)

Again consider $\quad ab + bc + ca = 0$ [From (2)]

$\Rightarrow \quad (\cos\alpha + i\sin\alpha)(\cos\beta + i\sin\beta) + (\cos\beta + i\sin\beta)(\cos\gamma + i\sin\gamma)$
$+ (\cos\gamma + i\sin\gamma)(\cos\alpha + i\sin\alpha) = 0$

$\Rightarrow \quad [\cos(\alpha + \beta) + i\sin(\alpha + \beta)] + [\cos(\beta + \gamma) + i\sin(\beta + \gamma)]$
$+ [\cos(\gamma + \alpha) + i\sin(\gamma + \alpha)] = 0$

Equating real and imaginary parts, we get

$\cos(\alpha + \beta) + \cos(\beta + \gamma) + \cos(\gamma + \alpha) = 0$

$\sin(\alpha + \beta) + \sin(\beta + \gamma) + \sin(\gamma + \alpha) = 0$ **Proved.**

## EXERCISE 6.6

1. If $n$ is a positive integer show that $(a + ib)^n + (a - ib)^n = 2r^n \cos n\theta$ where $r^2 = a^2 + b^2$ and $\theta = \tan^{-1}\left(\frac{b}{a}\right)$. Hence deduce that $\left(1 + i\sqrt{3}\right)^8 + \left(1 - i\sqrt{3}\right)^8 = -2^8$.

2. If $n$ be a positive integer, prove that $(1 + i)^n + (1 - i)^n = 2^{\frac{n+2}{2}} \cos\frac{n\pi}{4}$

3. Show that $(a + ib)^{m/n} + (a - ib)^{m/n} = 2(a^2 + b^2)^{\frac{m}{2n}} \cos\left(\frac{m}{n}\tan^{-1}\frac{b}{a}\right)$.

4. If $P = \cos\theta + i\sin\theta$, $q = \cos\phi + i\sin\phi$, show that

   (*i*) $\frac{P - q}{P + q} = i\tan\frac{\theta - \phi}{2}$ (*ii*) $\frac{(P + q)(Pq - 1)}{(P - q)(Pq + 1)} = \frac{\sin\theta + \sin\phi}{\sin\theta - \sin\phi}$

5. If $x = \cos\theta + i\sin\theta$, show that (*i*) $x^m + \frac{1}{x^m} = 2\cos m\theta$, (*ii*) $x^m - \frac{1}{x^m} = 2i\sin m\theta$.

6. Prove that $\tanh(\log\sqrt{3}) = \frac{1}{2}$

7. Prove that $[\sin(\alpha + \theta) - e^{i\alpha}\sin\theta]^n = \sin^n\alpha\, e^{-in\theta}$

8. If $x + \frac{1}{x} = 2\cos\theta$, $y + \frac{1}{y} = 2\cos\phi$, $z + \frac{1}{z} = 2\cos\psi$, show that

$$xyz + \frac{1}{xyz} = 2\cos(\theta + \phi + \psi)$$

## 6.22 ROOTS OF A COMPLEX NUMBER

We know that $\cos\theta + i\sin\theta = \cos(2m\pi + \theta) + i\sin(2m\pi + \theta),\ m \in I$

$$[\cos\theta + i\sin\theta]^{1/n} = [\cos(2m\pi + \theta) + i\sin(2m\pi + \theta)]^{1/n}$$

$$= \cos\frac{(2m\pi + \theta)}{n} + i\sin\frac{(2m\pi + \theta)}{n}$$

Giving $m$ the values 0, 1, 2, 3, .... $n - 1$ successively, we get the following $n$ values of $(\cos\theta + i\sin\theta)^{1/n}$.

when $m = 0$, $\quad \cos\frac{\theta}{n} + i\sin\frac{\theta}{n}$

When $m = 1$, $\quad \cos\left(\frac{2\pi + \theta}{n}\right) + i\sin\left(\frac{2\pi + \theta}{n}\right)$

When $m = 2$, $\quad \cos\left(\frac{4\pi + \theta}{n}\right) + i\sin\left(\frac{4\pi + \theta}{n}\right)$

When $m = n - 1$, $\quad \cos\left(\frac{2(n-1)\pi + \theta}{n}\right) + i\sin\left(\frac{2(n-1)\pi + \theta}{n}\right)$

When $m = n$, $\quad \cos\frac{2n\pi + \theta}{n} + i\sin\frac{2n\pi + \theta}{n} = \cos\left(2\pi + \frac{\theta}{n}\right) + i\sin\left(2\pi + \frac{\theta}{n}\right)$

$$= \cos\frac{\theta}{n} + i\sin\frac{\theta}{n}$$

which is the same as the value for $m = 0$. Thus, the values of $(\cos\theta + i\sin\theta)^{1/n}$ for $m = n$, $n + 1$, $n + 2$ etc., are the mere repetition of the first $n$ values as obtained above.

**Example 25.** *Solve $x^4 + i = 0$.* *(M.U. 2008)*

**Solution.** Here, we have

$$x^4 = -i = \cos\frac{\pi}{2} - i\sin\frac{\pi}{2}$$

$$x^4 = \cos\left(2n\pi + \frac{\pi}{2}\right) - i\sin\left(2n\pi + \frac{\pi}{2}\right)$$

$$\Rightarrow \quad x = \left[\cos\left(2n\pi + \frac{\pi}{2}\right) - i\sin\left(2n\pi + \frac{\pi}{2}\right)\right]^{\frac{1}{4}}$$

$$= \cos(4n+1)\frac{\pi}{8} - i\sin(4n+1)\frac{\pi}{8}$$

Putting $n = 0, 1, 2, 3$ we get the roots as

$$x_1 = \cos\frac{\pi}{8} - i\sin\frac{\pi}{8}, \quad x_2 = \cos\frac{5\pi}{8} - i\sin\frac{5\pi}{8}$$

$$x_3 = \cos\frac{9\pi}{8} - i\sin\frac{9\pi}{8}, \quad x_4 = \cos\frac{13\pi}{8} - i\sin\frac{13\pi}{8}$$ **Ans.**

**Example 26.** *Solve $x^5 = 1 + i$ and find the continued product of the roots.* *(M.U. 2005, 2004)*

**Solution.** $\quad x^5 = 1 + i$

$$= \sqrt{2}\left(\cos\frac{\pi}{4} + i\sin\frac{\pi}{4}\right) \Rightarrow x = 2^{\frac{1}{10}}\left(\cos\frac{\pi}{4} + i\sin\frac{\pi}{4}\right)^{\frac{1}{5}}$$

$$\Rightarrow \quad x = 2^{\frac{1}{10}}\left[\cos\left(2k\pi + \frac{\pi}{4}\right).\frac{1}{5} + i\sin\left(2k\pi + \frac{\pi}{4}\right).\frac{1}{5}\right]$$

$$= 2^{\frac{1}{10}}\left[\cos(8k+1)\frac{\pi}{20} + i\sin(8k+1)\frac{\pi}{20}\right]$$

The roots are obtained by putting $k = 0, 1, 2, 3, 4...$

$$x_1 = 2^{\frac{1}{10}}\left[\cos\frac{\pi}{20} + i\sin\frac{\pi}{20}\right], \qquad x_2 = 2^{\frac{1}{10}}\left[\cos\frac{9\pi}{20} + i\sin\frac{9\pi}{20}\right]$$

$$x_3 = 2^{\frac{1}{10}}\left[\cos\frac{17\pi}{20} + i\sin\frac{17\pi}{20}\right], \qquad x_4 = 2^{\frac{1}{10}}\left[\cos\frac{5\pi}{4} + i\sin\frac{5\pi}{4}\right]$$

$$x_5 = 2^{\frac{1}{10}}\left[\cos\frac{33\pi}{20} + i\sin\frac{33\pi}{20}\right]$$

$$x_1 . x_2 . x_3 . x_4 . x_5 = \left(2^{\frac{1}{10}}\right)^5\left(\cos\frac{\pi}{20} + i\sin\frac{\pi}{20}\right)\left(\cos\frac{9\pi}{20} + i\sin\frac{9\pi}{20}\right)\left(\cos\frac{17\pi}{20} + i\sin\frac{17\pi}{20}\right)$$
$$\left(\cos\frac{5\pi}{4} + i\sin\frac{5\pi}{4}\right)\left(\cos\frac{33\pi}{20} + i\sin\frac{33\pi}{20}\right)$$

$$= 2^{\frac{1}{2}}\left[\cos\left(\frac{\pi}{20} + \frac{9\pi}{20} + \frac{17\pi}{20} + \frac{5\pi}{4} + \frac{33\pi}{20}\right) + i\sin\left(\frac{\pi}{20} + \frac{9\pi}{20} + \frac{17\pi}{20} + \frac{5\pi}{4} + \frac{33\pi}{20}\right)\right]$$

$$= \sqrt{2}\left[\cos\frac{17\pi}{4} + i\sin\frac{17\pi}{4}\right] = \sqrt{2}\left[\cos\left(4\pi + \frac{\pi}{4}\right) + i\sin\left(4\pi + \frac{\pi}{4}\right)\right]$$

$$= \sqrt{2}\left[\cos\frac{\pi}{4} + i\sin\frac{\pi}{4}\right] = \sqrt{2}\left[\frac{1}{\sqrt{2}} + i\frac{1}{\sqrt{2}}\right] = 1 + i$$ **Ans.**

**Example 27.** *If $\alpha, \alpha^2, \alpha^3, \alpha^4$, are the roots of $x^5 - 1 = 0$ find them and show that $(1-\alpha)(1-\alpha^2)(1-\alpha^3)(1-\alpha^4) = 5$.* *(M.U. 2007)*

**Solution.** Here, we have

$$x^5 - 1 = 0$$

$$\Rightarrow \quad x^5 = 1 = \cos 0 + i\sin 0$$

$$\Rightarrow \quad x^5 = \cos(2k\pi) + i\sin(2k\pi)$$

$$\Rightarrow \quad x = (\cos 2k\pi + i\sin 2k\pi)^{1/5} = \cos\frac{2k\pi}{5} + i\sin\frac{2k\pi}{5}$$

Putting $k = 0, 1, 2, 3, 4$, we get the five roots as below

$$x_0 = \cos 0 + i\sin 0, \quad x_1 = \cos\frac{2\pi}{5} + i\sin\frac{2\pi}{5}$$

$$x_2 = \cos\frac{4\pi}{5} + i\sin\frac{4\pi}{5}, \quad x_3 = \cos\frac{6\pi}{5} + i\sin\frac{6\pi}{5}$$

$$x_4 = \cos\frac{8\pi}{5} + i\sin\frac{8\pi}{5}$$

Putting $x_1 = \cos\frac{2\pi}{5} + i\sin\frac{2\pi}{5} = \alpha$, we see that

$$x_2 = \left(\cos\frac{4\pi}{3} + i\sin\frac{4\pi}{3}\right) = \left(\cos\frac{2\pi}{3} + i\sin\frac{2\pi}{3}\right)^2 = \alpha^2$$

Similarly, $x_3 = \alpha^3$ and $x_4 = \alpha^4$

$\therefore$ The roots are $1, \alpha, \alpha^2, \alpha^3, \alpha^4$

Hence
$$x^5 - 1 = (x - 1)(x - \alpha)(x - \alpha^2)(x - \alpha^3)(x - \alpha^4)$$

$$\Rightarrow \quad \frac{x^5 - 1}{x - 1} = (x - \alpha)(x - \alpha^2)(x - \alpha^3)(x - \alpha^4)$$

On dividing $x^5 - 1$ by $x - 1$, we get

$$x^4 + x^3 + x^2 + x + 1 = (x - \alpha)(x - \alpha^2)(x - \alpha^3)(x - \alpha^4)$$

$$\therefore (x - \alpha)(x - \alpha^2)(x - \alpha^3)(x - \alpha^4) = x^4 + x^3 + x^2 + x + 1$$

Putting $x = 1$, we get

$(1 - \alpha)(1 - \alpha^2)(1 - \alpha^3)(1 - \alpha^4) = 1 + 1 + 1 + 1 + 1 = 5.$ **Proved.**

**Example 28.** *If $\omega$ is a cube root of unity, prove that*

$$(1 - \omega)^6 = -27$$ *(M.U. 2003)*

**Solution.** Let $x^3 = 1$

$$\Rightarrow \quad x = (1)^{1/3} = (\cos 0 + i \sin 0)^{1/3} = (\cos 2n\pi + i \sin 2n\pi)^{1/3}$$

$$= \cos\left(\frac{2n\pi}{3}\right) + i \sin\left(\frac{2n\pi}{3}\right)$$

Putting $n = 0, 1, 2$ the roots of unity are

$$x_0 = 1$$

$$x_1 = \cos\frac{2\pi}{3} + i \sin\frac{2\pi}{3} = \omega \text{ (say)}$$

$$x_2 = \cos\frac{4\pi}{3} + i \sin\frac{4\pi}{3} = \left[\cos\frac{2\pi}{3} + i \sin\frac{2\pi}{3}\right]^2 = \omega^2$$

Now,
$$1 + \omega + \omega^2 = 1 + \cos\frac{2\pi}{3} + i \sin\frac{2\pi}{3} + \cos\frac{4\pi}{3} + i \sin\frac{4\pi}{3}$$

$$= 1 + \cos\left(\pi - \frac{\pi}{3}\right) + i \sin\left(\pi - \frac{\pi}{3}\right) + \cos\left(\pi + \frac{\pi}{3}\right) + i \sin\left(\pi + \frac{\pi}{3}\right)$$

$$= 1 - \cos\frac{\pi}{3} + i \sin\frac{\pi}{3} - \cos\frac{\pi}{3} - i \sin\frac{\pi}{3}$$

$$= 1 - 2\cos\frac{\pi}{3} = 1 - 2\left(\frac{1}{2}\right) = 0$$

$$\Rightarrow \quad 1 + \omega + \omega^2 = 0$$

$$\Rightarrow \quad 1 + \omega^2 = -\omega \quad \text{...(1)}$$

Now,
$$(1 - \omega)^6 = [(1 - \omega)^2]^3 = [1 - 2\omega + \omega^2]^3 = [-\omega - 2\omega]^3$$

$$= (-3\omega)^3 = -27\omega^3 = -27$$ [Using (1)] **Proved.**

**Example 29.** *Use De Moivre's theorem to solve the equation $x^4 - x^3 + x^2 - x + 1 = 0$.*

**Solution.** $x^4 - x^3 + x^2 - x + 1 = 0$

$(x + 1)(x^4 - x^3 + x^2 - x + 1) = 0$

$$x^5 + 1 = 0$$

$$x^5 = -1 = (\cos \pi + i \sin \pi) = \cos(2n\pi + \pi) + i \sin(2n\pi + \pi)$$

$$x = [\cos(2n + 1)\pi + i \sin(2n + 1)\pi]^{1/5}$$

$$= \cos\frac{(2n + 1)\pi}{5} + i \sin\frac{(2n + 1)\pi}{5}$$

When $n = 0, 1, 2, 3, 4$, the values are

$$\cos\frac{\pi}{5} + i\sin\frac{\pi}{5},\ \cos\frac{3\pi}{5} + i\sin\frac{3\pi}{5},\ \cos\pi + i\sin\pi,\ \cos\frac{7\pi}{5} + i\sin\frac{7\pi}{5},$$
$$\cos\frac{9\pi}{5} + i\sin\frac{9\pi}{5}.$$

$\cos\pi + i\sin\pi = -1$, which is rejected as it is corresponding to $x + 1 = 0$.
Hence, the required roots are

$$\cos\frac{\pi}{5} + i\sin\frac{\pi}{5},\ \cos\frac{3\pi}{5} + i\sin\frac{3\pi}{5},\ \cos\frac{7\pi}{5} + i\sin\frac{7\pi}{5},\ \cos\frac{9\pi}{5} + i\sin\frac{9\pi}{5}.$$ **Ans.**

## EXERCISE 6.7

Find the values of:

1. $(1 + i)^{1/5}$. **Ans.** $2^{1/10}\left[\cos\frac{1}{5}\left(2n\pi + \frac{\pi}{4}\right) + i\sin\frac{1}{5}\left(2n\pi + \frac{\pi}{4}\right)\right]$, where $n = 0, 1, 2, 3, 4$

2. $(1 + \sqrt{-3})^{3/4}$ **Ans.** $(2)^{3/4}\left[\cos\frac{3}{4}\left(2n\pi + \frac{\pi}{3}\right) + \sin\frac{3}{4}\left(2n\pi + \frac{\pi}{3}\right)\right]$, where $n = 0, 1, 2, 3$.

3. $(-i)^{1/6}$ **Ans.** $\cos(4n + 1)\frac{\pi}{12} - i\sin(4n + 1)\frac{\pi}{12}$, where $n = 0, 1, 2, 3, 4, 5$.

4. $(1 + i)^{2/3}$ **Ans.** $2^{1/3}\left[\cos\left(\frac{4n\pi}{3} + \frac{\pi}{6}\right) + i\sin\left(\frac{4n\pi}{3} + \frac{\pi}{6}\right)\right]$, where $n = 0, 1, 2$

5. Solve the equation with the help of De Moivre's theore $x^7 - 1 = 0$

   **Ans.** $\cos\frac{2n\pi}{7} + i\sin\frac{2n\pi}{7}$ where $n = 0, 1, 2, 3, 4, 5, 6$.

6. Find the roots of the equation $x^3 + 8 = 0$.

   **Ans.** $2\left[\cos\left(\frac{2n\pi + \pi}{3}\right) + i\sin\left(\frac{2n\pi + \pi}{3}\right)\right]$, where $n = 0, 1, 2$.

7. Use De-Moivre's theorem to solve $x^9 - x^5 + x^4 - 1 = 0$

   **Ans.** $\left[\cos(2n + 1)\frac{\pi}{5} + i\sin(2n + 1)\frac{\pi}{5}\right]$, wehre $n = 0, 1, 2, 3, 4$,

   and $\left[\cos\frac{n\pi}{2} + i\sin\frac{n\pi}{2}\right]$, where $n = 0, 1, 2, 3$.

8. Show that the roots of $(x + 1)^6 + (x - 1)^6 = 0$ are given by

   $i\cot\frac{(2n + 1)}{12}$, $n = 0, 1, 2, 3, 4, 5$. Deduce $\tan^2\frac{\pi}{12} + \tan^2\frac{3\pi}{12} + \tan^2\frac{5\pi}{12} = 15$.

9. Show that all the roots of $(x + 1)^7 = (x - 1)^7$ are given by $\pm i\cot\left(\frac{n\pi}{7}\right)$, where $n = 1, 2, 3$, why $r \neq 0$.

## 6.23 CIRCULAR FUNCTIONS OF COMPLEX NUMBERS

We have already discussed circular functions in terms of exponential functions i.e, Euler's exponential form of circular functions:

$$\boxed{\cos\theta = \frac{e^{i\theta} + e^{-i\theta}}{2}},\qquad \boxed{\sin\theta = \frac{e^{i\theta} - e^{-i\theta}}{2i}}$$

If $\theta = z$, then $\boxed{\cos z = \frac{e^{iz} + e^{-iz}}{2}}$ and $\boxed{\sin z = \frac{e^{iz} - e^{-iz}}{2i}}$

## 6.24 HYPERBOLIC FUNCTIONS

(*i*) $\sinh x = \frac{e^x - e^{-x}}{2}$ (*ii*) $\cosh x = \frac{e^x + e^{-x}}{2}$ (*iii*) $\tanh x = \frac{e^x - e^{-x}}{e^x + e^{-x}}$

(iv) coth $x = \dfrac{e^x + e^{-x}}{e^x - e^{-x}}$ (v) sech $x = \dfrac{2}{e^x + e^{-x}}$ (*vi*) cosech $x = \dfrac{2}{e^x - e^{-x}}$

(*vii*) $\cosh x + \sinh x = \dfrac{e^x + e^{-x}}{2} + \dfrac{e^x - e^{-x}}{2} = e^x$

(*viii*) $\cosh x - \sinh x = \dfrac{e^x + e^{-x}}{2} - \dfrac{e^x - e^{-x}}{2} = e^{-x}$

(*ix*) $(\cosh x + \sinh x)^n = \cosh nx + \sinh nx.$

## 6.25 RELATION BETWEEN CIRCULAR AND HYPERBOLIC FUNCTIONS

$\sin ix = i \sinh x$ $\sinh ix = i \sin x$

$\cos ix = \cosh x$ $\cosh ix = \cos x$

$\tan ix = i \tanh x$ $\tanh ix = i \tan x$

## 6.26 FORMULAE OF HYPERBOLIC FUNCTIONS

**A.** (1) $\cosh^2 x - \sinh^2 x = 1,$ (2) $\text{sech}^2 x = 1 - \tanh^2 x,$

(3) $\text{cosech}^2 x = \coth^2 x - 1$

**B.** (1) $\sinh (x \pm y) = \sinh x \cosh y \pm \cosh x \sinh y$

(2) $\cosh (x \pm y) = \cosh x \cosh y \pm \sinh x \sinh y$

(3) $\tanh (x \pm y) = \dfrac{\tanh x \pm \tanh y}{1 \pm \tanh x \tanh y}$

**C.** (1) $\sinh 2x = 2 \sinh x \cosh x$ (2) $\cosh 2x = \cosh^2 x + \sinh^2 x$

(3) $\cosh 2x = 2 \cosh^2 x - 1$ (4) $\cosh 2x = 1 + 2 \sinh^2 x$

(5) $\tanh 2x = \dfrac{2 \tanh x}{1 + \tanh^2 x}$

**D.** (1) $\sinh x + \sinh y = 2 \sinh \dfrac{x+y}{2} \cosh \dfrac{x-y}{2}$

(2) $\sinh x - \sinh y = 2 \cosh \dfrac{x+y}{2} \sinh \dfrac{x-y}{2}$

(3) $\cosh x + \cosh y = 2 \cosh \dfrac{x+y}{2} \cosh \dfrac{x-y}{2}$

(4) $\cosh x - \cosh y = 2 \sinh \dfrac{x+y}{2} \sinh \dfrac{x-y}{2}$

**Note:** For proof, put $\sinh x = \dfrac{e^x - e^{-x}}{2}$ and $\cosh x = \dfrac{e^x + e^{-x}}{2}$

**Example 30.** *Prove that*

$$(\cosh x - \sinh x)^n = \cosh nx - \sinh nx.$$ (*M.U.* 2001, 2002)

**Solution.** L.H.S. $= (\cosh x - \sinh x)^n$

$$= \left[\frac{e^x + e^{-x}}{2} - \frac{e^x - e^{-x}}{2}\right]^n = \left[\frac{2e^{-x}}{2}\right]^n = (e^{-x})^n = e^{-nx} \quad ...(1)$$

R.H.S. $= \cosh nx - \sinh nx$

$$= \left(\frac{e^{nx} + e^{-nx}}{2} - \frac{e^{nx} + e^{-nx}}{2}\right) = \frac{2e^{-nx}}{2} = e^{-nx} \quad ...(2)$$

From (1) and (2), we have

L.H.S. = R.H.S. **Proved.**

**Example 31.** *If tan (θ + iφ) = cos α + i sin α, prove that:*

$$\theta = \frac{n\pi}{2} + \frac{\pi}{4} \text{ and } \phi = \frac{1}{2} \log \tan\left(\frac{\pi}{4} + \frac{\alpha}{2}\right)$$ *(Nagpur University, Summer 2002, Winter 2001)*

**Solution.** We have, $\tan(\theta + i\phi) = \cos\alpha + i \sin\alpha$

$\therefore \quad \tan(\theta - i\phi) = \cos\alpha - i\sin\alpha$

But $\quad \tan 2\theta = \tan[(\theta + i\phi) + (\theta - i\phi)]$

$$= \frac{\tan(\theta + i\phi) + \tan(\theta - i\phi)}{1 - \tan(\theta + i\phi)\tan(\theta - i\phi)} = \frac{\cos\alpha + i\sin\alpha + \cos\alpha - i\sin\alpha}{1 - (\cos\alpha + i\sin\alpha)(\cos\alpha - i\sin\alpha)}$$

$$= \frac{2\cos\alpha}{1 - (\cos^2\alpha + \sin^2\alpha)} = \frac{2\cos\alpha}{1 - 1} = \infty = \tan\frac{\pi}{2}$$

$\therefore \quad 2\theta = \frac{\pi}{2}$ or for general values,

$$2\theta = n\pi + \frac{\pi}{2} \Rightarrow \theta = \frac{n\pi}{2} + \frac{\pi}{4}$$

Again, $\tan(2i\phi) = \tan[(\theta + i\phi) - (\theta - i\phi)] = \dfrac{\tan(\theta + i\phi) - \tan(\theta - i\phi)}{1 + \tan(\theta + i\phi)\tan(\theta - i\phi)}$

$$= \frac{\cos\alpha + i\sin\alpha - (\cos\alpha - i\sin\alpha)}{1 + (\cos\alpha + i\sin\alpha)(\cos\alpha - i\sin\alpha)}$$

$$= \frac{2i\sin\alpha}{1 + \cos^2\alpha + \sin^2\alpha} = \frac{2i\sin\alpha}{1+1} = \frac{2i\sin\alpha}{2} = i\sin\alpha$$

$\Rightarrow \quad i\tanh 2\phi = i\sin\alpha \qquad (\because \tan ix = i\tanh x)$

$\Rightarrow \quad \tanh 2\phi = \sin\alpha$

*i.e.,* $$\frac{e^{2\phi} - e^{-2\phi}}{e^{2\phi} + e^{-2\phi}} = \frac{\sin\alpha}{1}$$

$$\therefore \frac{e^{2\phi} - e^{-2\phi} + e^{2\phi} + e^{-2\phi}}{(e^{2\phi} + e^{-2\phi}) - (e^{2\phi} - e^{-2\phi})} = \frac{1 + \sin\alpha}{1 - \sin\alpha}$$ (Componendo and dividendo)

*i.e.* $$\frac{2e^{2\phi}}{2e^{-2\phi}} = \frac{1 - \cos\left(\frac{\pi}{2} + \alpha\right)}{1 + \cos\left(\frac{\pi}{2} + \alpha\right)} \qquad \because \cos\left(\frac{\pi}{2} + \theta\right) = -\sin\theta$$

$$\Rightarrow \quad e^{4\phi} = \frac{2\sin^2\left(\frac{\pi}{4} + \frac{\alpha}{2}\right)}{2\cos^2\left(\frac{\pi}{4} + \frac{\alpha}{2}\right)}$$

$$\Rightarrow \quad e^{4\phi} = \tan^2\left(\frac{\pi}{4} + \frac{\alpha}{2}\right) \Rightarrow e^{2\phi} = \tan\left(\frac{\pi}{4} + \frac{\alpha}{2}\right)$$

Hence, $\quad 2\phi = \log_e \tan\left(\frac{\pi}{4} + \frac{\alpha}{2}\right) \Rightarrow \phi = \frac{1}{2}\log_e \tan\left(\frac{\pi}{4} + \frac{\alpha}{2}\right)$ **Proved.**

**Example 32.** *If cosh x = sec θ, prove that:*

*(i)* $$\theta = \frac{\pi}{2} - 2\tan^{-1}(e^{-x})$$

*(ii)* $$\tanh\frac{\pi}{2} = \tan\frac{\theta}{2}$$ *(M.U. 2003, 2005)*

**Solution.** (*i*) Let $\tan^{-1} e^{-x} = \alpha$

$\Rightarrow \quad e^{-x} = \tan\alpha$ and $\alpha = \tan^{-1}(e^{-x})$ ...(1)

$\Rightarrow \quad e^{x} = \cot\alpha$ ...(2)

Now, $\sec\theta = \cosh x = \dfrac{e^x + e^{-x}}{2}$ ...(3) (Given)

Putting the values of $e^{-x}$ and $e^x$ from (1) and (2) in (3), we get

$$\sec\theta = \frac{\cot\alpha + \tan\alpha}{2}$$

$$\therefore \quad 2\sec\theta = \cot\alpha + \tan\alpha = \frac{\cos\alpha}{\sin\alpha} + \frac{\sin\alpha}{\cos\alpha} = \frac{\cos^2\alpha + \sin^2\alpha}{\sin\alpha\cos\alpha}$$

$$= \frac{2}{2\sin\alpha\cos\alpha} \qquad [\because \cos^2\alpha + \sin^2\alpha = 1]$$

$$= \frac{2}{\sin 2\alpha}$$

$$\therefore \quad \cos\theta = \sin 2\alpha$$

$$\Rightarrow \quad \cos\theta = \cos\left(\frac{\pi}{2} - 2\alpha\right)$$

$$\therefore \quad \theta = \frac{\pi}{2} - 2\alpha = \frac{\pi}{2} - 2\tan^{-1}(e^{-x}) \qquad \text{[From (1)]}$$ **Proved.**

(*ii*) We have,

$$\cosh x = \sec\theta \qquad \text{(Given)}$$

$$\Rightarrow \quad \frac{e^x + e^{-x}}{2} = \sec\theta \qquad \left[\because \cosh x = \frac{e^x + e^{-x}}{2}\right]$$

$$\therefore \quad e^x - 2\sec\theta + e^{-x} = 0$$

$$\therefore \quad (e^x)^2 - 2e^x\sec\theta + 1 = 0$$

Solving the quadratic equation in $e^x$.

$$e^x = \frac{2\sec\theta \pm \sqrt{4\sec^2\theta - 4}}{2}$$

$$\Rightarrow \quad e^x = \sec\theta \pm \sqrt{\sec^2\theta - 1}$$

$$\Rightarrow \quad e^x = \sec\theta \pm \tan\theta \qquad \text{...(4)}$$

Now, $$\tanh\frac{x}{2} = \frac{e^{\frac{x}{2}} - e^{-\frac{x}{2}}}{e^{\frac{x}{2}} + e^{-\frac{x}{2}}} = \frac{e^x - 1}{e^x + 1} \qquad \text{...(5)}$$

Putting the value of $e^x$ from (4) in (5), we get

$$\tanh\frac{x}{2} = \frac{\sec\theta + \tan\theta - 1}{\sec\theta + \tan\theta + 1} \qquad \text{[Using (1)]}$$

$$= \frac{1 + \sin\theta - \cos\theta}{1 + \sin\theta + \cos\theta} = \frac{(1 - \cos\theta) + \sin\theta}{(1 + \cos\theta) + \sin\theta}$$

$$= \frac{2\sin^2\frac{\theta}{2} + 2\sin\frac{\theta}{2}\cos\frac{\theta}{2}}{2\cos^2\frac{\theta}{2} + 2\sin\frac{\theta}{2}\cos\frac{\theta}{2}} = \frac{\sin\frac{\theta}{2}}{\cos\frac{\theta}{2}} = \tan\frac{\theta}{2}$$ **Proved.**

## EXERCISE 6.8

1. If $\tan\left(\frac{\pi}{8} + i\alpha\right) = x + iy$, prove that $x^2 + y^2 + 2x = 1$.

2. If $\cot\left(\frac{\pi}{8} + i\alpha\right) = x + iy$, prove that $x^2 + y^2 - 2x = 1$.

3. Prove that if $(1 + i \tan\alpha)^{1 + i \tan\beta}$ can have real values, one of them is $(\sec\alpha)^{\sec^2\beta}$.

4. If $\dfrac{(1 + i)^{x+iy}}{(1 - i)^{x-iy}} = \alpha + i\beta$, prove that the value of $\tan^{-1}\dfrac{\beta}{\alpha}$ *is* $\dfrac{\pi x}{2} + y \log 2$.

5. If $\tanh x = \frac{1}{2}$, find the value of $\sinh 2x$. **Ans.** $\frac{4}{3}$

6. If $\sin\alpha \cosh\beta = \frac{x}{2}$, $\cos\alpha \sinh\beta = \frac{y}{2}$, show that

   (*i*) $\operatorname{cosec}(\alpha - i\beta) + \operatorname{cosec}(\alpha + i\beta) = \dfrac{4x}{x^2 + y^2}$

   (*ii*) $\operatorname{cosec}(\alpha - i\beta) - \operatorname{cosec}(\alpha + i\beta) = \dfrac{4iy}{x^2 + y^2}$

7. Show that $\tan\left(\dfrac{u + iv}{2}\right) = \dfrac{\sin u + i \sinh v}{\cos u + \cosh v}$

8. If $\cot(\alpha + i\beta) = x + iy$, prove that

   (*i*) $x^2 + y^2 - 2x \cot 2\alpha = 1$

   (*ii*) $x^2 + y^2 + 2y \coth 2\beta + 1 = 0$

10. Solve the following equation for real values of $x$.

$$17 \cosh x + 18 \sinh x = 1$$ **Ans.** $-\log 5$

## 6.27 SEPARATION OF REAL AND IMAGINARY PARTS OF CIRCULAR FUNCTIONS

**Example 33.** *Separate the following into real and imaginary parts:*

(*i*) $\sin(x + iy)$ (*ii*) $\cos(x + iy)$ (*iii*) $\tan(x + iy)$

**Solution.** (*i*) $\sin(x + iy) = \sin x \cos iy + \cos x \sin(iy) = \sin x \cosh y + i \cos x \sinh y$.

(*ii*) $\cos(x + iy) = \cos x \cos(iy) - \sin x \sin(iy) = \cos x \cosh y - i \sin x \sinh y$.

(*iii*) $$\tan(x + iy) = \frac{\sin(x + iy)}{\cos(x + iy)} = \frac{2\sin(x + iy)\cos(x - iy)}{2\cos(x + iy)\cos(x - iy)}$$

$$= \frac{\sin 2x + \sin(2iy)}{\cos 2x + \cos 2iy} = \frac{\sin 2x + i \sinh 2y}{\cos 2x + \cosh 2y}$$

$$\left\{\begin{matrix}\because\ 2\sin A.\cos B = \sin(A + B) + \sin(A - B) \\ \text{and } 2\cos A.\cos B = \cos(A + B) + \cos(A - B)\end{matrix}\right\}$$

**Example 34.** *If* $\tan(A + iB) = x + iy$, *prove that*

$\tan 2A = \dfrac{2x}{1 - x^2 - y^2}$ *and* $\tanh 2B = \dfrac{2x}{1 + x^2 + y^2}$ *(Nagpur University, Summer 2000)*

**Solution.** $\tan(A + iB) = x + iy$ ; $\tan(A - iB) = x - iy$

$$\tan 2A = \tan(A + iB + A - iB)$$

$$= \frac{\tan(A+iB) + \tan(A-iB)}{1 - \tan(A+iB)\tan(A-iB)}$$

$$\tan 2A = \frac{(x+iy)+(x-iy)}{1-(x+iy)(x-iy)} = \frac{2x}{1-(x^2+y^2)} = \frac{2x}{1-x^2-y^2}$$

Again $$\tan 2iB = \tan(A + iB - A + iB) = \frac{\tan(A+iB) - \tan(A-iB)}{1 + \tan(A+iB)\tan(A-iB)}$$

$$\tan 2iB = \frac{(x+iy)-(x-iy)}{1+(x+iy)(x-iy)} = \frac{(2y)i}{1+x^2+y^2}$$

$$\tanh 2B = \frac{2y}{1+x^2+y^2} \qquad \tan ix = i \tanh x \qquad \textbf{Proved.}$$

**Example 35.** *If sin (α + iβ) = x + iy, prove that*

$$(a)\ \frac{x^2}{\cosh^2\beta}+\frac{y^2}{\sinh^2\beta}=1 \qquad (b)\ \frac{x^2}{\sin^2\alpha}-\frac{y^2}{\cos^2\alpha}=1$$

**Solution.** (*a*) $x + iy = \sin(\alpha + i\beta) = \sin\alpha\cosh\beta + i\cos\alpha\sinh\beta$.

Equating real and imaginary parts, we get

$$x = \sin\alpha\cosh\beta,\ y = \cos\alpha\sinh\beta$$

$$\sin\alpha = \frac{x}{\cosh\beta} \text{ and } \cos\alpha = \frac{y}{\sinh\beta}$$

Squaring and adding, $\sin^2\alpha + \cos^2\alpha = \dfrac{x^2}{\cosh^2\beta}+\dfrac{y^2}{\sinh^2\beta}$

$$\Rightarrow \qquad 1 = \frac{x^2}{\cosh^2\beta}+\frac{y^2}{\sinh^2\beta} \qquad \textbf{Proved.}$$

(*b*) Again $\qquad \cosh\beta = \dfrac{x}{\sin\alpha}$ and $\sinh\beta = \dfrac{y}{\cos\alpha}$

$$\cosh^2\beta - \sinh^2\beta = \frac{x^2}{\sin^2\alpha}-\frac{y^2}{\cos^2\alpha}$$

$$1 = \frac{x^2}{\sin^2\alpha}-\frac{y^2}{\cos^2\alpha} \qquad \textbf{Proved.}$$

## 6.28 SEPARATION OF REAL AND IMAGINARY PARTS OF HYPERBOLIC FUNCTIONS

**Example 36.** *Separate the following into real and imaginary parts of hyperbolic functions.*

*(a) sinh (x + iy)* *(b) cosh (x + iy)* *(c) tanh (x + iy)*

**Solution.** (*a*) $\sinh(x+iy) = \sinh x\cosh(iy) + \cosh x\sinh(iy)$

$$= \sinh x\cos y + i\sin y\cosh x. \qquad \textbf{Ans.}$$

(*b*) $\cosh(x+iy) = \cosh x\cosh(iy) - \sinh x\sinh iy = \cosh x\cos y - i\sinh x\sin y.$ **Ans.**

(*c*) $\tanh(x+iy) = \dfrac{\sinh(x+iy)}{\cosh(x+iy)} = \dfrac{-i\sin i(x+iy)}{\cos i(x+iy)}$

$$= \frac{-i\sin(ix-y)}{\cos(ix-y)} = \frac{-i\,2\sin(ix-y)\cos(ix+y)}{2\cos(ix-y)\cos(ix+y)} \qquad \text{(Note this step)}$$

$$= -i\frac{\sin 2ix - \sin 2y}{\cos 2ix + \cos 2y} = -i\frac{i\sinh 2x - \sin 2y}{\cosh 2x + \cos 2y} = \frac{\sinh 2x + i\sin 2y}{\cosh 2x + \cos 2y}$$

$$= \frac{\sinh 2x}{\cosh 2x + \cos 2y} + i\frac{\sin 2y}{\cosh 2x + \cos 2y} \qquad \textbf{Ans.}$$

**Example 37.** *If tan (x + iy) = sin (u + iv), prove that*

$$\frac{\sin 2x}{\sinh 2y} = \frac{\tan u}{\tanh v}$$

**Solution.** Now $\tan(x + iy) = \sin(u + iv)$ separating the real and imaginary parts of both sides, we have

$$\frac{\sin 2x}{\cos 2x + \cosh 2y} + \frac{i \sinh 2y}{\cos 2x + \cosh 2y} = \sin u \cosh v + i \cos u \sinh v$$

Equating real and imaginary parts, we get

$$\frac{\sin 2x}{\cos 2x + \cosh 2y} = \sin u \cosh v \qquad ...(1)$$

and $$\frac{\sinh 2y}{\cos 2x + \cosh 2y} = \cos u \sinh v \qquad ...(2)$$

Dividing (1) by (2), we obtain

$$\frac{\sin 2x}{\sinh 2y} = \frac{\sin u \cosh v}{\cos u \sinh v}$$

$$\Rightarrow \quad \frac{\sin 2x}{\sinh 2y} = \frac{\tan u}{\tanh v}$$ **Proved.**

**Example 38.** *If $\sin(\theta + i\phi) = \tan\alpha + i\sec\alpha$, show that $\cos 2\theta \cosh 2\phi = 3$*

**Solution.** $\sin(\theta + i\phi) = \tan\alpha + i\sec\alpha$

$\sin\theta\cosh\phi + i\cos\theta\sinh\phi = \tan\alpha + i\sec\alpha$

Equating real and imaginary parts, we get

$$\sin\theta\cosh\phi = \tan\alpha \qquad ...(1)$$

$$\cos\theta\sinh\phi = \sec\alpha \qquad ...(2)$$

We know that

$$\sec^2\alpha - \tan^2\alpha = 1$$

$\cos^2\theta\sinh^2\phi - \sin^2\theta\cosh^2\phi = 1$ [From (1) and (2)]

$$\left(\frac{1 + \cos 2\theta}{2}\right)\left(\frac{\cosh 2\phi - 1}{2}\right) - \left(\frac{1 - \cos 2\theta}{2}\right)\left(\frac{\cosh 2\phi + 1}{2}\right) = 1$$

$[-1 + \cosh 2\phi - \cos 2\theta + \cos 2\theta\cosh 2\phi] - [\cosh 2\phi + 1 - \cos 2\theta\cosh 2\phi - \cos 2\theta] = 4$

$\Rightarrow \quad -2 + 2\cos 2\theta\cosh 2\phi = 4$

$\Rightarrow \quad 2\cos 2\theta\cosh 2\phi = 6 \Rightarrow \cos 2\theta\cosh 2\phi = 3$ **Proved.**

**Example 39.** *If $e^z = \sin(u + iv)$ and $z = x + iy$, prove that*

$$2e^{2x} = \cosh 2v - \cos 2u$$ (*M.U.* 2006)

**Solution.** We have, $e^z = \sin(u + iv)$

$\Rightarrow \quad e^{x+iy} = \sin(u + iv)$

$\Rightarrow \quad e^x . e^{iy} = \sin u \cos iv + \cos u \sin iv$

$\Rightarrow \quad e^x(\cos y + i\sin y) = \sin u \cosh v + i\cos u \sinh v$

Equating real and imaginary parts, we get

$$e^x \cos y = \sin u \cosh v$$

and $$e^x \sin y = \cos u \sinh v$$

Squaring and adding, we get

$$e^{2x}(\cos^2 y + \sin^2 y) = \sin^2 u \cosh^2 v + \cos^2 u \sinh^2 v$$

$\Rightarrow \quad e^{2x} = (1 - \cos^2 u)\cosh^2 v + \cos^2 u(\cosh^2 v - 1)$

$\Rightarrow \quad e^{2x} = \cosh^2 v - \cos^2 u$

$\Rightarrow \quad e^{2x} = \frac{1}{2}(1 + \cosh 2v) - \frac{1}{2}(1 + \cos 2u)$

$$\Rightarrow \qquad e^{2x} = \frac{1}{2}(\cosh 2v - \cos 2u)$$

$$\Rightarrow \qquad 2e^{2x} = \cosh 2v - \cos 2u$$ **Proved.**

**Example 40.** *If* $\sin(\theta + i\phi) = \cos\alpha + i\sin\alpha$, *prove that*

$$\cos^4\theta = \sin^2\alpha = \sinh^4\phi.$$ (*M.U.* 2003, 2004)

**Solution.** Here, we have

$$\sin(\theta + i\phi) = \cos\alpha + i\sin\alpha$$

$$\Rightarrow \quad \sin\theta\cosh\phi + i\cos\theta\sinh\phi = \cos\alpha + i\sin\alpha$$

Equating real and imaginary parts, we get

$$\sin\theta\cosh\phi = \cos\alpha \quad \Rightarrow \quad \cosh\phi = \frac{\cos\alpha}{\sin\theta} \qquad ...(1)$$

and
$$\cos\theta\sinh\phi = \sin\alpha \quad \Rightarrow \quad \sinh\phi = \frac{\sin\alpha}{\cos\theta} \qquad ...(2)$$

But
$$\cosh^2\phi - \sinh^2\phi = 1$$

$$\Rightarrow \quad \frac{\cos^2\alpha}{\sin^2\theta} - \frac{\sin^2\alpha}{\cos^2\theta} = 1 \qquad \text{[Using (1) and (2)]}$$

$$\Rightarrow \quad \cos^2\alpha.\cos^2\theta - \sin^2\alpha.\sin^2\theta = \sin^2\theta\cos^2\theta$$

$$\Rightarrow \quad (1 - \sin^2\alpha)\cos^2\theta - \sin^2\alpha.\sin^2\theta = (1 - \cos^2\theta)\cos^2\theta$$

$$\Rightarrow \quad \cos^2\theta - \sin^2\alpha(\cos^2\theta + \sin^2\theta) = \cos^2\theta - \cos^4\theta$$

$$\Rightarrow \quad \sin^2\alpha = \cos^4\theta \qquad ...(3)$$

Again
$$\sin^2\theta + \cos^2\theta = 1$$

$$\therefore \quad \frac{\cos^2\alpha}{\cosh^2\phi} + \frac{\sin^2\alpha}{\sinh^2\phi} = 1 \qquad \text{[Using (1) and (2)]}$$

$$\Rightarrow \quad \cos^2\alpha.\sinh^2\phi + \sin^2\alpha\cosh^2\phi = \sinh^2\phi\cosh^2\phi$$

$$\Rightarrow \quad (1 - \sin^2\alpha)\sinh^2\phi + \sin^2\alpha(1 + \sinh^2\phi) = \sinh^2\phi(1 + \sinh^2\phi)$$

$$\Rightarrow \quad \sinh^2\phi - \sin^2\alpha\sinh^2\phi + \sin^2\alpha + \sin^2\alpha\sinh^2\phi = \sinh^2\phi + \sinh^4\phi$$

$$\Rightarrow \quad \sin^2\alpha = \sinh^4\phi. \qquad ...(4)$$

From (3) and (4), we have $\cos^4\theta = \sin^2\alpha = \sinh^4\phi$ **Proved.**

**Example 41.** *If* $\operatorname{cosec}\left(\frac{\pi}{4} + ix\right) = u + iv$, *prove that*

$$(u^2 + v^2)^2 = 2(u^2 - v^2)$$ (*M.U.* 2009)

**Solution.** Here, we have

$$u + iv = \operatorname{cosec}\left(\frac{\pi}{4} + ix\right)$$

$$= \frac{1}{\sin\left(\frac{\pi}{4} + ix\right)} \Rightarrow = \frac{1}{\sin\frac{\pi}{4}\cos ix + \cos\frac{\pi}{4}\sin ix}$$

$$= \frac{1}{\frac{1}{\sqrt{2}}\cosh x + \frac{1}{\sqrt{2}} i\sinh x} = \frac{\sqrt{2}}{\cosh x + i\sinh x}$$

$$= \frac{\sqrt{2}\,(\cosh x - i \sinh x)}{\cosh^2 x + \sinh^2 x} = \frac{\sqrt{2}\,(\cosh x - i \sinh x)}{\cosh 2x}$$

Equating real and imaginary parts, we get $u = \dfrac{\sqrt{2}\cosh x}{\cos h\, 2x}$, $v = -\dfrac{\sqrt{2}\sinh x}{\cosh 2x}$

Squaring and adding, we get

$$u^2 + v^2 = \frac{2\,(\cosh^2 x + \sinh^2 x)}{\cosh^2 2x} = \frac{2\cosh 2x}{\cosh^2 2x}$$

$$\Rightarrow \quad (u^2 + v^2)^2 = \left(\frac{2}{\cosh 2x}\right)^2 = \frac{4}{\cosh^2 2x} \qquad \ldots(1)$$

Also,

$$u^2 - v^2 = \frac{2}{\cosh^2 2x}(\cosh^2 x - \sinh^2 x) = \frac{2}{\cosh^2 2x} \qquad \ldots(2)$$

From (1) and (2), we have

$$(u^2 + v^2)^2 = 2\,(u^2 - v^2)$$

**Proved.**

**Example 42.** *Separate into real and imaginary parts* $\sqrt{i}^{\sqrt{i}}$ *(M.U. 2008)*

**Solution.** We have,

$$\sqrt{i} = i^{\frac{1}{2}} = \left(\cos\frac{\pi}{2} + i\sin\frac{\pi}{2}\right)^{\frac{1}{2}} = \cos\frac{\pi}{4} + i\sin\frac{\pi}{4} = \frac{1}{\sqrt{2}} + i\frac{1}{\sqrt{2}}$$

Also,

$$\sqrt{i} = \left(\cos\frac{\pi}{2} + i\sin\frac{\pi}{2}\right)^{\frac{1}{2}} = \left(e^{i\frac{\pi}{2}}\right)^{\frac{1}{2}} = e^{i\frac{\pi}{4}}$$

$$\therefore \quad \left(\sqrt{i}\right)^{\sqrt{i}} = \left(e^{i\frac{\pi}{4}}\right)^{\left(\frac{1}{\sqrt{2}} + i.\frac{1}{\sqrt{2}}\right)} = e^{i\frac{\pi}{4\sqrt{2}} - \frac{\pi}{4\sqrt{2}}}$$

$$= e^{-\frac{\pi}{4\sqrt{2}}}.e^{i\frac{\pi}{4\sqrt{2}}} = e^{-\frac{\pi}{4\sqrt{2}}}\left(\cos\frac{\pi}{4\sqrt{2}} + i\sin\frac{\pi}{4\sqrt{2}}\right)$$

$$\therefore \quad \text{Real part} = e^{-\frac{\pi}{4\sqrt{2}}}\cos\left(\frac{\pi}{4\sqrt{2}}\right)$$

$$\text{Imaginary part} = e^{-\frac{\pi}{4\sqrt{2}}}\sin\left(\frac{\pi}{4\sqrt{2}}\right)$$

**Ans.**

## EXERCISE 6.9

Separate into real and imaginary parts.

**1.** sech $(x + iy)$ **Ans.** $\dfrac{2\cosh x\cos y - 2i\sinh x\sin y}{\cosh 2x + \cos 2y}$

**2.** coth $i\,(x + iy)$ **Ans.** $\dfrac{-\sinh 2y - i\sin 2x}{\cosh 2x - \cos 2y}$

**3.** coth $(x + iy)$ **Ans.** $\dfrac{\sinh 2x - i\sin 2y}{\cosh 2x - \cos 2y}$

**4.** If $\sin(\theta + i\phi) = p(\cos\alpha + i\sin\alpha)$, prove that

$$p^2 = \frac{1}{2}\left[\cosh 2\phi - \cos 2\theta\right], \ \tan\alpha = \tanh\phi\cot\theta$$

**5.** If $\sin(\alpha + i\beta) = x + iy$, prove that $x^2\,\text{sech}^2\beta + y^2\,\text{cosech}^2\beta = 1$ and $x^2\,\text{cosec}^2\alpha - y^2\sec^2\alpha = 1$

6. If $\cos(\theta + i\phi) = r(\cos\alpha + i\sin\alpha)$, prove that $\theta = \frac{1}{2}\log\left[\frac{\sin(\theta - \alpha)}{\sin(\theta + \alpha)}\right]$

7. If $\tan\left(\frac{\pi}{6} + i\alpha\right) = x + iy$, prove that $x^2 + y^2 + \frac{2x}{\sqrt{3}} = 1$

8. If $\tan(A + B) = \alpha + i\beta$, show that $\frac{1-(\alpha^2+\beta^2)}{1+(\alpha^2+\beta^2)} = \frac{\cos 2A}{\cosh 2B}$

9. If $\frac{x + iy - c}{x + iy + c} = e^{u + iv}$, prove that

$$x = -\frac{c\sinh u}{\cosh u - \cos v}, \qquad y = \frac{c\sinh v}{\cosh u - \cos v}$$

Further, if $v = (2n + 1)\frac{\pi}{2}$, prove that $x^2 + y^2 = c^2$ where $n$ is an integer.

10. If $\frac{u-1}{u+1} = \sin(x + iy)$, where $u = \alpha + i\beta$ show that the argument of $u$ is $\theta + \phi$ where

$$\tan\theta = \frac{\cos x\sinh y}{1+\sin x\cosh y} \quad \text{and}, \quad \tan\phi = \frac{\cos x\sinh y}{1-\sin x\sinh y}$$

11. If $A + iB = C\tan(x + iy)$, prove that $\tan 2x = \frac{2CA}{C^2 - A^2 - B^2}$

12. If $\cosh(\alpha + i\beta) = x + iy$, prove that

(a) $\frac{x^2}{\cosh^2\alpha} + \frac{y^2}{\sinh^2\alpha} = 1$ (b) $\frac{x^2}{\cos^2\beta} - \frac{y^2}{\sin^2\beta} = 1$

13. If $\cos(\theta + i\phi) = R(\cos\alpha + i\sin\alpha)$, prove that $\phi = \frac{1}{2}\log_e\left[\frac{\sin(\theta-\alpha)}{\sin(\theta+\alpha)}\right]$

14. If $\cos(\alpha + i\beta)\cos(\gamma + i\delta) = 1$, prove that $\tanh^2\delta\cosh^2\beta = \sin^2\alpha$

15. If $\frac{u-1}{u+1} = \sin(x + iy)$, find $u$. **Ans.** $\tan^{-1}\frac{2\cos x\sinh y}{\cos^2 x - \sinh^2 y}$

## 6.29 LOGARITHMIC FUNCTION OF A COMPLEX VARIABLE

**Example 43.** *Define logarithm of a complex number.*

**Solution.** If $z$ and $w$ are two complex numbers and $z = e^w$ then $w = \log z$, and if $w = \log z$; then $z = e^w$

Here $\log z$ is a many valued function. General value of $\log z$ is defined by Log $z$, where Log $z = \log z + 2n\pi i$.

**Example 44.** *Separate log $(x + iy)$ into its real and imaginary parts.*

**Solution.** Let $x = r\cos\theta$ ...(1)

and $y = r\sin\theta$ ...(2)

Squaring and adding (1) and (2) we have $x^2 + y^2 = r^2$

$$\therefore \quad r = \sqrt{x^2+y^2},$$

We have, $\tan\theta = \frac{y}{x} \Rightarrow \theta = \tan^{-1}\left(\frac{y}{x}\right)$ [Dividing (2) by (1)]

$$\therefore \quad \log(x + iy) = \log[r(\cos\theta + i\sin\theta)]$$

$$= [\log r + \log(\cos\theta + i\sin\theta)]$$

$$\log(x + iy) = \log r + \log[\cos(2n\pi + \theta) + i\sin(2n\pi + \theta)]$$

$$= \log r + \log e^{i(2n\pi+\theta)} = \log r + i(2n\pi + \theta)$$

$$\text{Log}(x+iy) = \log\sqrt{x^2+y^2} + i\left(2n\pi + \tan^{-1}\frac{y}{x}\right)$$

and $$\log(x+iy) = \log\sqrt{x^2+y^2} + i\tan^{-1}\frac{y}{x}$$ **Ans.**

**Example 45.** *Show that* $\log\dfrac{x+iy}{x-iy} = 2i\tan^{-1}\dfrac{y}{x}$. *(Nagpur University, Winter 2003)*

**Solution.** Let $\log(x+iy) = \log(r\cos\theta + ir\sin\theta) = \log re^{i\theta}$

$$= \log r + i\theta \qquad \begin{bmatrix} x = r\cos\theta \\ y = r\sin\theta \end{bmatrix}$$

Similarly, $$\log(x-iy) = \log r - i\theta$$

$$\log\frac{x+iy}{x-iy} = \log(x+iy) - \log(x-iy) = (\log r + i\theta) - (\log r - i\theta) = 2i\theta$$

$$= 2i\tan^{-1}\frac{y}{x}.$$ **Proved.**

**Example 46.** *Show that for real values of a and b*

$$e^{2ai\cot^{-1}b}\left[\frac{bi-1}{bi+1}\right]^{-a} = 1$$ *(M.U. 2008)*

**Solution.** Consider $\dfrac{bi-1}{bi+1} = \dfrac{bi+i^2}{bi-i^2} = \dfrac{b+i}{b-i}$

$$\Rightarrow \left(\frac{bi-1}{bi+1}\right)^{-a} = \left(\frac{b+i}{b-i}\right)^{-a}$$

$$\log\left[\frac{bi-1}{bi+1}\right]^{-a} = \log\left(\frac{b+i}{b-i}\right)^{-a} = -a[\log(b+i) - \log(b-i)]$$

$$= -a\left[\log\sqrt{b^2+1} + i\tan^{-1}\frac{1}{b} - \log\sqrt{b^2+1} + i\tan^{-1}\frac{1}{b}\right]$$

$$= -2ai\tan^{-1}\frac{1}{b}$$

$$\left(\frac{bi-1}{bi+1}\right)^{-a} = e^{-2ai\tan^{-1}\left(\frac{1}{b}\right)} \qquad \begin{bmatrix} \text{If } \cot\theta = b,\ \tan\theta = \dfrac{1}{b} \\ \text{Since } \cot^{-1}b = \tan^{-1}\left(\dfrac{1}{b}\right) \end{bmatrix}$$

$$e^{2ai\cot^{-1}b}\left(\frac{bi-1}{bi+1}\right)^{-a} = \left[e^{2ai\tan^{-1}\left(\frac{1}{b}\right)}\right]\cdot\left[e^{-2ai\tan^{-1}\left(\frac{1}{b}\right)}\right] = 1$$ **Proved.**

## EXERCISE 6.10

1. Find the general value of Log $i$. **Ans.** $(4n+1)\dfrac{\pi i}{2}$
2. Express Log $(-5)$ in terms of $a + ib$. **Ans.** $\log 5 + i(2n+1)\pi$
3. Find the value of z if

(a) $\cos z = 2$. **Ans.** $z = 2n\pi \pm i\log\left(2+\sqrt{3}\right)$

(*b*) cosh $z = -1$. **Ans.** $z = (2n + 1)\pi i$

**4.** Find the general and principal values of $i^i$ **Ans.** $e^{-\left(2n\pi + \frac{\pi}{2}\right)},\ e^{-\frac{\pi}{2}}$

**5.** If $i^{(\alpha + i\beta)} = x + iy$, prove that $x^2 + y^2 = e^{-(4m+1)\pi\theta}$.

**6.** Prove that $\log \dfrac{1}{l - e^{i\theta}} = \log\left(\dfrac{1}{2}\operatorname{cosec}\theta\right) + i\left(\dfrac{\pi}{2} - \dfrac{\theta}{2}\right)$

**7.** Show that $\log \sin(x + iy) = \dfrac{1}{2}\log\dfrac{\cosh 2y - \cos 2x}{2} + i\tan^{-1}(\cot x \tanh y)$.

**8.** Prove that $\tan\left[i\log\dfrac{a - ib}{a + ib}\right] = \dfrac{2ab}{a^2 - b^2}$.

**9.** $\log \dfrac{\cos(x - iy)}{\cos(x + iy)} = 2i\tan^{-1}(\tan x \tanh y)$.

**10.** Separate $i^{(1+i)}$ into real and imaginary parts. **Ans.** $ie^{-\frac{\pi}{2}}$

## 6.30 INVERSE FUNCTIONS

If $\sin\theta = \dfrac{1}{2}$ then $\theta = \sin^{-1}\left(\dfrac{1}{2}\right)$, so here $\theta$ is called inverse sine of $\dfrac{1}{2}$.

Similarly, we can define inverse hyperbolic function sinh, cosh, tanh, etc. If $\cosh\theta = z$ then $\theta = \cosh^{-1} z$.

## 6.31 INVERSE HYPERBOLIC FUNCTIONS

**Example 47.** *Prove that* $\sinh^{-1} x = \log\left(x + \sqrt{x^2 + 1}\right)$ *(M.U. 2009)*

**Solution.** Let $\sinh^{-1} x = y \quad \Rightarrow \quad x = \sinh y$

$$x = \frac{e^y - e^{-y}}{2}$$

$$\Rightarrow \quad e^y - e^{-y} = 2x$$

$$\Rightarrow \quad e^{2y} - 2x\,e^y - 1 = 0$$

This is quadratic in $e^y$.

$$e^y = \frac{2x \pm \sqrt{4x^2 + 4}}{2} = x \pm \sqrt{x^2 + 1}$$

$$y = \log\left(x + \sqrt{x^2 + 1}\right) \quad \text{(Taking positive sign only)}$$

$$\sinh^{-1} x = \log\left(x + \sqrt{x^2 + 1}\right)$$ **Proved.**

**Example 48.** *Prove that* $\cosh^{-1} x = \log\left(x + \sqrt{x^2 - 1}\right)$ *(M.U. 2009)*

**Solution.** Let $y = \cosh^{-1} x \quad \Rightarrow \quad x = \cosh y$

$$x = \frac{e^y + e^{-y}}{2} \quad \Rightarrow \quad 2x = e^y + e^{-y}$$

$$\Rightarrow \quad e^{2y} - 2x\,e^y + 1 = 0 \quad \text{(This is quadratic in } e^y\text{)}$$

$$\Rightarrow \quad e^y = \frac{2x \pm \sqrt{4x^2 - 4}}{2} = x \pm \sqrt{x^2 - 1}$$

$$\Rightarrow \quad y = \log\left(x + \sqrt{x^2 - 1}\right) \quad \text{(Taking positive sign only)}$$

$$\Rightarrow \quad \cosh^{-1} x = \log\left(x + \sqrt{x^2 - 1}\right)$$ **Proved.**

**Example 49.** *Prove that* $\operatorname{sech}^{-1} x = \log\dfrac{1 + \sqrt{1 - x^2}}{x}$

**Solution.** Let $y = \operatorname{sech}^{-1} x \quad \Rightarrow \quad x = \operatorname{sech} y$

$$x = \frac{2}{e^y + e^{-y}} \quad \Rightarrow \quad x = \frac{2e^y}{e^{2y}+1}$$

$$\Rightarrow \quad xe^{2y} - 2e^y + x = 0 \quad \Rightarrow \quad e^y = \frac{2 \pm \sqrt{4-4x^2}}{2x} = \frac{1 \pm \sqrt{1-x^2}}{x}$$

We take only positive sign

$$e^y = \frac{1+\sqrt{1-x^2}}{x} \quad \Rightarrow y = \log \frac{1+\sqrt{1-x^2}}{x}$$

$$\text{sech}^{-1}\, x = \log \frac{1+\sqrt{1-x^2}}{x}$$

Similarly, $\quad \text{cosech}^{-1}\, x = \log \dfrac{1+\sqrt{1+x^2}}{x}$ **Proved.**

**Example 50.** *If $x + iy = \cos(\alpha + i\beta)$ or if $\cos^{-1}(x + iy) = \alpha + i\beta$ express x and y in terms of $\alpha$ and $\beta$. Hence show that $\cos^2 \alpha$ and $\cosh^2 \beta$ are the roots of the equation $\lambda^2 - (x^2 + y^2 + 1)\lambda + x^2 = 0$.* *(M.U. 2002, 2004)*

**Solution.** Here, we have

$$\cos(\alpha + i\beta) = x + iy$$

$$\Rightarrow \cos\alpha \cos i\beta - \sin\alpha \sin i\beta = x + iy$$

$$\Rightarrow \cos\alpha \cosh\beta - i \sin\alpha \sinh\beta = x + iy$$

Equating real and imaginary parts, we get

$$\cos\alpha \cosh\beta = x \text{ and } \sin\alpha \sinh\beta = -y$$

We want to find the equation whose roots are $\cos^2\alpha$ and $\cosh^2\beta$.

Now, $\quad x^2 + y^2 + 1 = \cos^2\alpha \cos h^2\beta + \sin^2\alpha \sinh^2\beta + 1$

$$= \cos^2\alpha \cosh^2\beta + (1 - \cos^2\alpha)(\cosh^2\beta - 1) + 1$$

$$= \cos^2\alpha \cosh^2\beta + \cosh^2\beta - 1 - \cos^2\alpha \cosh^2\beta + \cos^2\alpha + 1$$

$$= \cos^2\alpha + \cosh^2\beta$$

Sum of the roots $= \cos^2\alpha + \cosh^2\beta$

$$= x^2 + y^2 + 1$$

And product of the roots $= \cos^2\alpha \cosh^2\beta$

$$= x^2$$

Hence, the equation whose roots are $\cos^2\alpha$, $\cosh^2\beta$ is

$\lambda^2 - (x^2 + y^2 + 1)\lambda + x^2 = 0$ **Proved.**

**Example 51.** *Separate into real and imaginary part $\cos^{-1}\left(\dfrac{3i}{4}\right)$* *(M.U. 2003)*

**Solution.** Let $\cos^{-1}\left(\dfrac{3i}{4}\right) = x + iy$

$$\Rightarrow \quad \frac{3i}{4} = \cos(x + iy) \quad \Rightarrow \quad \frac{3i}{4} = \cos x \cosh y - i \sin x \sinh y$$

Equating real and imaginary parts, we get

$$\therefore \quad \cos x \cosh y = 0 \quad \Rightarrow \quad \cos x = 0 \quad \Rightarrow \quad x = \frac{\pi}{2}$$

and $\qquad -\sin x \sinh y = \frac{3}{4}$

$$-1 \sinh y = \frac{3}{4} \qquad\qquad \sin x = \sin\left(\frac{\pi}{2}\right) = 1$$

$\therefore \qquad \sinh y = -\frac{3}{4}$

$$\Rightarrow \; y = \log\left(\frac{-3}{4} + \sqrt{1+\frac{9}{16}}\right) \Rightarrow \; y = \log\left(\frac{-3}{4}+\frac{5}{4}\right) = -\log 2 = \log\left(\frac{1}{2}\right)$$

$\therefore$ Real part $= \frac{\pi}{2}$ and imaginary Part $= -\log 2$ **Proved.**

## 6.32 SOME OTHER INVERSE FUNCTIONS

**Example 52.** *Separate* $\tan^{-1}(\cos\theta + i\sin\theta)$ *into real and imaginary parts.* (*M.U. 2009*)

**Solution.** Let $\tan^{-1}(\cos\theta + i\sin\theta) = x + iy$

$$\Rightarrow \quad \cos\theta + i\sin\theta = \tan(x+iy)$$

Similarly, $\quad \cos\theta - i\sin\theta = \tan(x-iy)$

$$\tan 2x = \tan[(x+iy)+(x-iy)] = \frac{\tan(x+iy)+\tan(x-iy)}{1-\tan(x+iy)\tan(x-iy)}$$

$$= \frac{(\cos\theta+i\sin\theta)+(\cos\theta-i\sin\theta)}{1-(\cos\theta+i\sin\theta)(\cos\theta-i\sin\theta)} = \frac{2\cos\theta}{1-(\cos^2\theta+\sin^2\theta)}$$

$$= \frac{2\cos\theta}{1-1} = \frac{2\cos\theta}{0} = \infty = \tan\frac{\pi}{2}$$

$$\tan 2x = \tan\left(n\pi + \frac{\pi}{2}\right) \Rightarrow \; 2x = n\pi + \frac{\pi}{2} \Rightarrow \; x = \frac{n\pi}{2} + \frac{\pi}{4}$$

Now, $\tan 2iy = \tan[(x+iy)-(x-iy)] = \dfrac{\tan(x+iy)-\tan(x-iy)}{1+\tan(x+iy)\tan(x-iy)}$

$$= \frac{(\cos\theta+i\sin\theta)-(\cos\theta-i\sin\theta)}{1+(\cos\theta+i\sin\theta)(\cos\theta-i\sin\theta)} = \frac{2i\sin\theta}{1+(\cos^2\theta+\sin^2\theta)} = \frac{2i\sin\theta}{1+1} = i\sin\theta$$

$$i\tanh 2y = i\sin\theta \Rightarrow \frac{e^{2y}-e^{-2y}}{e^{2y}+e^{-2y}} = \frac{\sin\theta}{1}$$

By componendo and dividendo, we have

$$\frac{2e^{2y}}{2e^{-2y}} = \frac{1+\sin\theta}{1-\sin\theta} \Rightarrow e^{4y} = \frac{1+\cos\left(\frac{\pi}{2}-\theta\right)}{1-\cos\left(\frac{\pi}{2}-\theta\right)} = \frac{1+2\cos^2\left(\frac{\pi}{4}-\frac{\theta}{2}\right)-1}{1-\left[1-2\sin^2\left(\frac{\pi}{4}-\frac{\theta}{2}\right)\right]}$$

$$= \frac{\cos^2\left(\frac{\pi}{4}-\frac{\theta}{2}\right)}{\sin^2\left(\frac{\pi}{4}-\frac{\theta}{2}\right)} = \cot^2\left(\frac{\pi}{4}-\frac{\theta}{2}\right) \Rightarrow e^{2y} = \cot\left(\frac{\pi}{4}-\frac{\theta}{2}\right)$$

$$\Rightarrow \quad 2y = \log\cot\left(\frac{\pi}{4}-\frac{\theta}{2}\right) \Rightarrow y = \frac{1}{2}\log\cot\left(\frac{\pi}{4}-\frac{\theta}{2}\right)$$

$$\text{Imaginary part} = \frac{1}{2}\log\cot\left(\frac{\pi}{4}-\frac{\theta}{2}\right)$$

$$\text{Real part} = \frac{n\pi}{2}+\frac{\pi}{4}$$

$$\tan^{-1}(\cos\theta + i\sin\theta) = \frac{n\pi}{2}+\frac{\pi}{4}+\frac{i}{2}\log\cot\left(\frac{\pi}{4}-\frac{\theta}{2}\right) \quad \textbf{Ans.}$$

**Example 53.** *Separate $\tan^{-1}(a + i\,b)$ into real and imaginary parts.*

*(Nagpur University, Summer 2008, 2004)*

**Solution.** Let $\tan^{-1}(a + i\,b) = x + i\,y$ ...(1)

$\therefore \quad \tan(x + i\,y) = a + i\,b$

On both sides for $i$ write $-i$ we get,

$\therefore \quad \tan(x - i\,y) = a - i\,b$

Now, $\tan 2x = \tan[(x + i\,y) + (x - i\,y)]$

$$= \frac{\tan(x+i\,y)+\tan(x-i\,y)}{1-\tan(x+i\,y)\tan(x-i\,y)} = \frac{a+i\,b+a-i\,b}{1-(a+i\,b)(a-i\,b)} = \frac{2a}{1-a^2-b^2}$$

$$2x = \tan^{-1}\left[\frac{2a}{1-a^2-b^2}\right] \Rightarrow x = \frac{1}{2}\tan^{-1}\left[\frac{2a}{1-a^2-b^2}\right] \quad ...(2)$$

and $\tan(2\,y\,i) = \tan[(x + i\,y) - (x - i\,y)]$

$$= \frac{\tan(x+i\,y)-\tan(x-i\,y)}{1+\tan(x+i\,y)\tan(x-i\,y)} = \frac{a+b\,i-a+b\,i}{1+(a+b\,i)(a-b\,i)}$$

$$i\tanh 2y = \frac{2\,b\,i}{1+a^2+b^2} \quad \text{so, } \tanh 2y = \frac{2\,b}{1+a^2+b^2}$$

$$2y = \tanh^{-1}\left[\frac{2\,b}{1+a^2+b^2}\right]$$

$$\text{so } y = \frac{1}{2}\tanh^{-1}\left[\frac{2\,b}{1+a^2+b^2}\right] \quad ...(3)$$

From (1), (2) and (3), we have

$$\tan^{-1}(a+ib) = \frac{1}{2}\tan^{-1}\left[\frac{2\,a}{1-a^2-b^2}\right]+\frac{i}{2}\tanh^{-1}\left[\frac{2\,b}{1+a^2+b^2}\right] \quad \textbf{Ans.}$$

**Example 54.** *Show that* $\tan^{-1} i\left(\frac{x-a}{x+a}\right) = \frac{i}{2}\log\left(\frac{x}{a}\right)$. *(M.U. 2006, 2002)*

**Solution.** Let $\tan^{-1} i\left(\frac{x-a}{x+a}\right) = u + i\,v$ ...(1)

$$\Rightarrow \quad \tan(u+iv) = i\left(\frac{x-a}{x+a}\right) \text{ and } \tan(u-iv) = -i\left(\frac{x-a}{x+a}\right)$$

$$\tan 2u = \tan[(u+iv)+(u-iv)] = \frac{\tan(u+iv)+\tan(u-iv)}{1-\tan(u+iv)\tan(u-iv)} = \frac{ix-ia-ix+ia}{x+a} = 0$$

$\therefore \quad \tan 2u = 0 \Rightarrow 2u = 0 \Rightarrow u = 0$

**Putting the value of $u$ in (1), we get**

$\therefore \quad \tan^{-1} i\left(\dfrac{x-a}{x+a}\right) = i\,v \qquad \therefore \quad i\left(\dfrac{x-a}{x+a}\right) = \tan i\,v = i \tanh v$

$\therefore \quad \dfrac{x-a}{x+a} = \tanh v = \dfrac{e^v - e^{-v}}{e^v + e^{-v}}$

**By Componendo and dividendo, we get**

$$\frac{2x}{2a} = \frac{2e^v}{2e^{-v}} \Rightarrow \frac{x}{a} = e^{2v} \Rightarrow v = \frac{1}{2}\log\left(\frac{x}{a}\right)$$

$\therefore \tan^{-1} i\left(\dfrac{x-a}{x+a}\right) = u + iv = 0 + \dfrac{i}{2}\log\dfrac{x}{a} = \dfrac{i}{2}\log\left(\dfrac{x}{a}\right)$ **Proved.**

**Example 55.** *Prove that*

(*i*) $\cosh^{-1}\sqrt{1+x^2} = \sinh^{-1} x$ (*M.U. 2007*)

(*ii*) $\cosh^{-1}\sqrt{1+x^2} = \tanh^{-1}\left(\dfrac{x}{\sqrt{1+x^2}}\right)$ (*M.U. 2002*)

**Solution.** (*i*) Let $\cosh^{-1}\sqrt{1+x^2} = y$ ...(1)

$\Rightarrow \quad \sqrt{1+x^2} = \cosh y$ ...(2)

On squaring both sides, we get

$$1 + x^2 = \cosh^2 y$$

$\therefore \quad x^2 = \cosh^2 y - 1 \Rightarrow x^2 = \sinh^2 y$

$\Rightarrow \quad x = \sinh y$ ...(3)

$\Rightarrow \quad y = \sinh^{-1} x$

$\Rightarrow \quad \cosh^{-1}\sqrt{1+x^2} = \sinh^{-1} x$ [Using (1)] **Proved.**

(*ii*) Dividing (3) by (2), we get

$$\frac{\sinh y}{\cosh y} = \frac{x}{\sqrt{1+x^2}}$$

$\Rightarrow \quad \tanh y = \dfrac{x}{\sqrt{1+x^2}} \Rightarrow y = \tanh^{-1}\left(\dfrac{x}{\sqrt{1+x^2}}\right)$

$\Rightarrow \quad \cosh^{-1}\sqrt{1+x^2} = \tanh^{-1}\left(\dfrac{x}{\sqrt{1+x^2}}\right)$ [Using (1)] **Proved.**

## EXERCISE 6.11

1. Prove that $\sin^{-1}(\operatorname{cosec}\theta) = \dfrac{\pi}{2} + i\log\cot\dfrac{\theta}{2}$.

2. If $\tan(\alpha + i\beta) = x + iy$, prove that

(*a*) $x^2 + y^2 + 2x\cot 2\alpha = 1$ (*b*) $x^2 + y^2 - 2y\coth 2\beta = -1$.

**3.** If $\tan(\theta + i\,\phi) = \sin(x + iy)$, then prove that $\coth y \sinh 2\,\phi = \cot x \sin 2\,\theta$.

**4.** If $\sin^{-1}(\cos\theta + i\sin\theta) = x + iy$, show that.

(*a*) $x = \cos^{-1}\sqrt{\sin\theta}$ (*b*) $y = \log\left[\sqrt{\sin\theta} + \sqrt{1 + \sin\theta}\right]$.

**5.** Prove that $\tan^{-1} i\dfrac{x-a}{x+a} = -\dfrac{i}{2}\log\dfrac{a}{x}$

**6.** Separate into real and imaginary parts of $\cos^{-1}\dfrac{3i}{4}$.

**7.** Separate into real and imaginary parts $\sin^{-1}(e^{i\theta})$ **Ans.** $\cos^{-1}\sqrt{\sin\theta} + i\log\left[\sqrt{\sin\theta} + \sqrt{1+\sin\theta}\right]$

**8.** Prove that

$$\tan^{-1}\left(\frac{\tan 2\theta + \tan 2\phi}{\tan 2\theta - \tan 2\phi}\right) + \tan^{-1}\left(\frac{\tan\theta - \tan\phi}{\tan\theta + \tan\phi}\right) = \tan^{-1}(\cot\theta\coth\phi)$$

**9.** Prove that $\tanh^{-1} x = \sinh^{-1}\dfrac{x}{\sqrt{1-x^2}}$.

**10.** Prove that $\tanh^{-1}(\sin\theta) = \cosh^{-1}(\sec\theta)$

**11.** Prove that

$$\cosh^{-1}\left(\frac{b + a\cos x}{a + b\cos x}\right) = \log\left[\frac{\sqrt{b+a} + \sqrt{b-a}\tan\frac{x}{2}}{\sqrt{b+a} - \sqrt{b-a}\tan\frac{x}{a}}\right]$$

**12.** Prove that $\tan^{-1}(e^{i\theta}) = \dfrac{n\pi}{2} + \dfrac{\pi}{4} = \log\tan\left(\dfrac{\pi}{4} - \dfrac{\theta}{2}\right)$

**13.** If $\cosh^{-1}(x + iy) + \cosh^{-1}(x - iy) = \cosh^{-1} a$, prove that $2(a - 1)x^2 + 2(a + 1)y^2 = a^2 - 1$.

**14.** Prove that : $\tanh^{-1}\cos\theta = \cosh^{-1}\operatorname{cosec}\theta$

**15.** Prove that : $\sinh^{-1}\tan\theta = \log(\sec\theta + \tan\theta)$

**16.** Prove that : $\sinh^{-1}\tan\theta = \log\tan\left(\dfrac{\theta}{2} + \dfrac{\pi}{4}\right)$

**Separate into real and imaginary parts**

**17.** $\cos^{-1} e^{i\theta}$ or $\cos^{-1}(\cos\theta + i\sin\theta)$ **Ans.** $\sin^{-1}\sqrt{\sin\theta} + i\log\left(\sqrt{1+\sin\theta} - \sqrt{\sin\theta}\right)$

**18.** If $\sinh^{-1}(x + iy) + \sinh^{-1}(x - iy) = \sinh^{-1} a$, prove that $2(x^2 + y^2)\sqrt{a^2 + 1} = a^2 - 2x^2 - 2y^2$.

# 7

# Functions of a Complex Variable

## 7.1 INTRODUCTION

The theory of functions of a complex variable is of atmost importance in solving a large number of problems in the field of engineering and science. Many complicated integrals of real functions are solved with the help of functions of a complex variable.

## 7.2 COMPLEX VARIABLE

$x + iy$ is a complex variable and it is denoted by $z$.

(1) $z = x + iy$ where $i = \sqrt{-1}$ (Cartesian form)

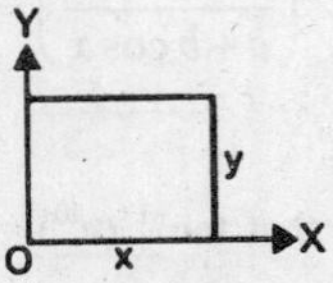


(2) $z = r(\cos\theta + i\sin\theta)$ (Polar form)

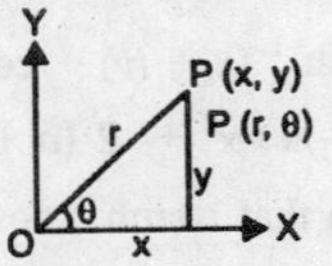


(3) $z = re^{i\theta}$ (Exponential form)

## 7.3 FUNCTIONS OF A COMPLEX VARIABLE

$f(z)$ is a function of a complex variable $z$ and is denoted by $w$.

$$w = f(z)$$

$$w = u + iv$$

where $u$ and $v$ are the real and imaginary parts of $f(z)$.

## 7.4 NEIGHBOURHOOD OF $Z_0$

Let $z_0$ is a point in the complex plane and let $z$ be any positive number, then the set of points $z$ such that

$$|z - z_0| < \epsilon$$

is called $\epsilon$-neighbourhood of $z_0$.

**Closed set**

A set $S$ is said to be closed if it contains all of its limits point.

**Interior Point**

A point $z_0$ is called a interior point of a point set $S$ if there exists a neighbourhood of $z_0$ lying wholly in $S$.

## 7.5 LIMIT OF A FUNCTION OF A COMPLEX VARIABLE

Let $f(z)$ be a single valued function defined at all points in some neighbourhood of point $z_0$. Then $f(z)$ is said to have the limit $l$ as $z$ approaches $z_0$ along any path if given an arbitrary real number $\in > 0$, however small there exists a real number $\delta > 0$, such that

$|f(z) - l| < \in$ whenever $0 < |z - z_0| < \delta$

i.e. for every $z \neq z_0$ in $\delta$-disc (dotted) of $z$-plane, $f(z)$ has a value lying in the $\in$-disc of $w$-plane

In symbolic form, $\lim\limits_{z \to z_0} f(z) = l$

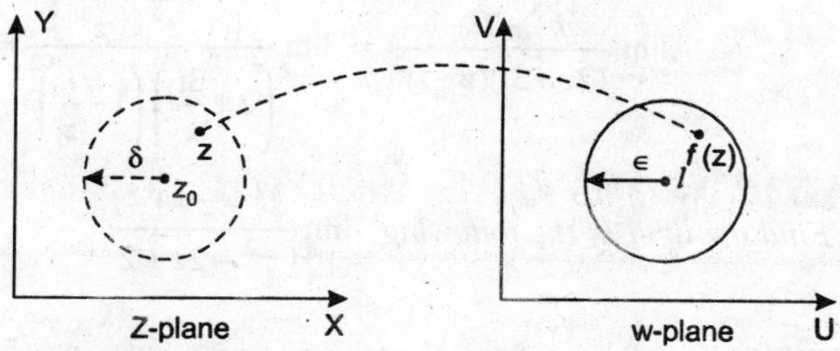


**Note:** (I) $\delta$ usually depends upon $\in$.

(II) $z \to z_0$ implies that $z$ approaches $z_0$ along any path. The limits must be independent of the manner in which z approaches $z_0$

If we get two different limits as $z \to z_0$ along two different paths then limits does not exist.

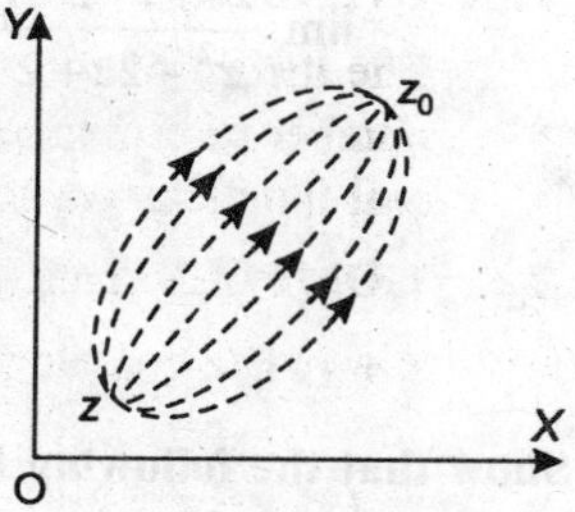


**Example 1.** *Prove that* $\lim\limits_{z \to 1-i} \dfrac{(z^2+4z+3)}{z+1} = 4 - i$

**Solution.** $\lim\limits_{z \to 1-i} \dfrac{z^2+4z+3}{z+1} = \lim\limits_{z \to 1-i} \dfrac{(z+1)(z+3)}{(z+1)} = \lim\limits_{z \to 1-i} (z+3) = (1-i)+3 = 4-i$ **Proved.**

**Example 2.** *Show that* $\lim\limits_{z \to 0} \dfrac{z}{|z|}$ *does not exist.*

**Solution.** $\lim\limits_{z \to 0} \dfrac{z}{|z|} = \lim\limits_{\substack{x \to 0 \\ y \to 0}} \dfrac{x+iy}{\sqrt{x^2+y^2}}$

Let $y = mx$,

$$= \lim_{x \to 0} \frac{x+imx}{\sqrt{x^2+m^2x}} = \lim_{x \to 0} \frac{1+im}{\sqrt{1+m^2}} = \frac{1+mi}{\sqrt{1+m^2}}$$

The value of $\dfrac{1+mi}{\sqrt{1+m^2}}$ are different for different values of $m$.

Hence, limit of the function does not exist. **Proved.**

**Example 3.** *Prove that* $\lim\limits_{z \to 0} \dfrac{z}{\bar{z}}$ *does not exist.*

**Solution. Case I.** $\lim\limits_{z \to 0} \dfrac{z}{\bar{z}} = \lim\limits_{\substack{x \to 0 \\ y \to 0}} \dfrac{x+iy}{x-iy} = \lim\limits_{x \to 0} \left[ \lim\limits_{y \to 0} \dfrac{x+iy}{x-iy} \right]$

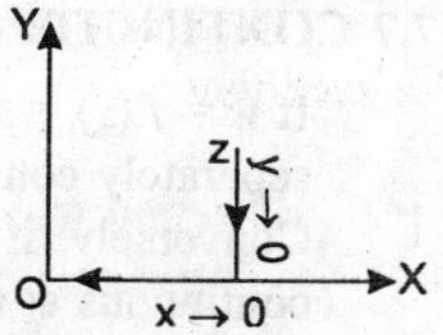


$$= \lim_{x \to 0} \frac{x}{x} = 1$$

Here the path is $y \to 0$ and then $x \to 0$

**Case II.** Again $\lim\limits_{z \to 0} \dfrac{z}{\bar{z}} = \lim\limits_{y \to 0} \left[ \lim\limits_{x \to 0} \dfrac{x+iy}{x-iy} \right] = \lim\limits_{y \to 0} \dfrac{iy}{-iy} = -1$

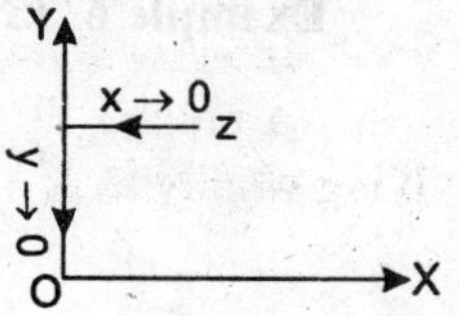

In this case, we have a different path first $x \to 0$, then $y \to 0$
As $z \to 0$ along two different paths we get different limits.
Hence the limit does not exist. **Proved.**

**Example 4.** *Find the limit of the following* $\lim_{z\to\infty} \frac{iz^3 + iz - 1}{(2z+3i)(z-i)^2}$

**Solution.** On dividing numerator and denominator by $z^3$, we get

$$\lim_{z\to\infty} \frac{iz^3 + iz - 1}{(2z+3i)(z-i)^2} = \lim_{z\to\infty} \frac{\left(i + \frac{i}{z^2} - \frac{1}{z^3}\right)}{\left(2 + \frac{3i}{z}\right)\left(1 - \frac{i}{z}\right)^2} = \frac{i}{2}$$ **Ans.**

**Example 5.** *Find the limit of the following* $\lim_{z\to 1+i} \frac{z^2 - z + 1 - i}{z^2 - 2z + 2}$

**Solution.**

$$\lim_{z\to 1+i} \frac{z^2 + 1 - z - i}{z^2 - 2z + 2} = \lim_{z\to 1+i} \frac{(z+i)(z-i) - 1(z+i)}{(z-1-i)(z+1+i)} = \lim_{z\to 1+i} \frac{(z+i)(z-i-1)}{(z-1-i)(z-1+i)} = \lim_{z\to 1+i} \frac{z+i}{z-1+i}$$

$$= \frac{1+i+i}{1+i-1+i} = \frac{1+2i}{2i} = \frac{(1+2i)(-i)}{2(i)(-i)} = \frac{-i+2}{2(1)} = \frac{2-i}{2} = 1 - \frac{i}{2}$$ **Ans.**

## EXERCISE 7.1

**Show that the following limits do not exist:**

**1.** $\lim_{z\to 0} \frac{\text{Im}(z)^3}{\text{Re}(z)^3}$ **2.** $\lim_{z\to -i} \frac{z^2}{z+i}$ **3.** $\lim_{z\to 0} \frac{\text{Re}(z)^2}{\text{Im}\, z^2}$ **4.** $\lim_{z\to 0} \frac{z}{(\bar{z})^2}$

**Find the Limits of the following:**

**5.** $\lim_{z\to 0} \frac{\text{Re}(z)^2}{|z|}$ **Ans.** 0 **6.** $\lim_{z\to 1+i} \frac{2z^3}{\text{Im}(z)^2}$ **Ans.** $2(-1+i)$ **7.** $\lim_{z\to 0} \frac{z^2 + 6z + 3}{z^2 + 2z + 2}$ **Ans.** $\frac{3}{2}$

## 7.6 CONTINUITY

The function $f(z)$ of a complex variable $z$ is said to be continuous at the point $z_0$ if for any given positive number $\in$, we can find a number $\delta$ such that $|f(z) - f(z_0)| < \in$ for all points $z$ of the domain satisfying

$$|z - z_0| < \delta$$

$f(z)$ is said to be continuous at $z = z_0$ if

$$\lim_{z\to z_0} f(z) = f(z_0)$$

## 7.7 CONTINUITY IN TERMS OF REAL AND IMAGINARY PARTS

If $w = f(z) = u(x, y) + iv(x, y)$ is continuous function at $z = z_0$ then $u(x, y)$ and $v(x, y)$ are separately continuous functions of $x, y$ at $(x_0, y_0)$ where $z_0 = x_0 + i\,y_0$.
Conversely, if $u(x, y)$ and $v(x, y)$ are continuous functions of $x, y$ at $(x_0, y_0)$ then $f(z)$ is continuous at $z = z_0$.

**Example 6.** *Examine the continuity of the following*

$$f(z) = \begin{cases} \dfrac{z^3 - iz^2 + z - i}{z - i}, & z \neq i \\ 0, & z = i \end{cases}$$ **at $z = i$**

**Solution.** $\lim\limits_{z\to i}\dfrac{z^3-iz^2+z-i}{z-i}=\lim\limits_{z\to i}\dfrac{z^2(z-i)+1(z-i)}{z-i}$

$$=\lim_{z\to i}\frac{(z-i)(z^2+1)}{z-i}=\lim_{z\to i}(z^2+1)=-1+1=0$$

$f(i)=0$

$$\boxed{\lim_{z\to i}f(z)=f(i)}$$

Hence $f(z)$ is continuous at $z=i$ **Ans.**

**Example 7.** *Show that the function f(z) defined by*

$$f(z)=\begin{cases}\dfrac{\text{Re}(z)}{z}, & z\neq 0\\ 0, & z=0\end{cases}$$

*is not continuous at z = 0*

**Solution.** Here $f(z)=\dfrac{\text{Re}(z)}{z}$ when $z\neq 0$

$$\lim_{z\to 0}\frac{\text{Re}(z)}{z}=\lim_{\substack{x\to 0\\ y\to 0}}\frac{x}{x+iy}=\lim_{x\to 0}\left[\lim_{y\to 0}\frac{x}{x+iy}\right]=\lim_{x\to 0}\frac{x}{x}=1$$

Again $\lim\limits_{z\to 0}\dfrac{\text{Re}(z)}{z}=\lim\limits_{y\to 0}\left[\lim\limits_{x\to 0}\dfrac{x}{x+iy}\right]=0$

As $z\to 0$, for two different paths limit have two different values. So, limit does not exist. Hence $f(z)$ is not continuous at $z=0$ **Proved.**

## EXERCISE 7.2

**Examine the continuity of the following functions.**

1. $f(z)=\begin{cases}\dfrac{\text{Im}(z)}{|z|}, & z\neq 0\\ 0 & z=0\end{cases}$ at $z=0$ **Ans. Not Continuous**

2. $f(z)=\dfrac{z^2+3z+4}{z^2+i}$ at $z=1-i$ **Ans. Continuous**

3. Show that the following functions are continuous for $z$

(*i*) $\cos z$ (*ii*) $e^{2z}$

## 7.8 DIFFERENTIABILITY

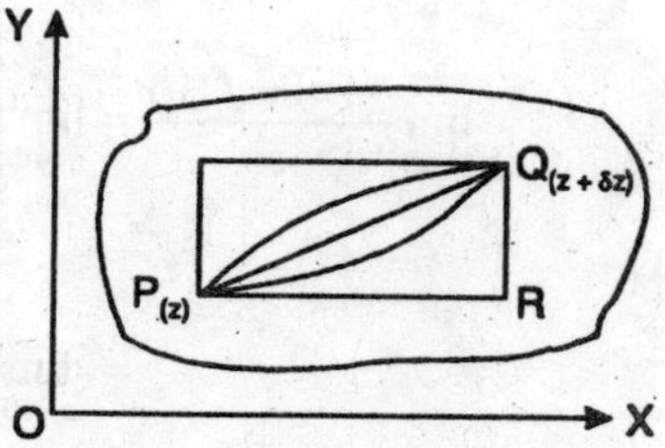


Let $f(z)$ be a single valued function of the variable $z$, then

$$f'(z)=\lim_{\delta z\to 0}\frac{f(z+\delta z)-f(z)}{\delta z}$$

provided that the limit exists and is independent of the path along which $\delta z\to 0$.

Let $P$ be a fixed point and $Q$ be a neighbouring point. The point $Q$ may approach $P$ along any straight line or curved path.

**Example 8.** *Consider the function*

$$f(z)=4x+y+i(-x+4y)$$

*and discuss* $\dfrac{df}{dz}$

**Solution.** Here, $f(z) = 4x + y + i(-x + 4y) = u + iv$

so $u = 4x + y$ and $v = -x + 4y$

$$f(z+\delta z) = 4(x+\delta x)+(y+\delta y)-i(x+\delta x)+4i(y+\delta y)$$

$$f(z+\delta z)-f(z) = 4(x+\delta x)+(y+\delta y)-i(x+\delta x)+4i(y+\delta y)-4x-y+ix-4iy$$

$$= 4\delta x+\delta y-i\,\delta x+4i\delta y$$

$$\frac{f(z+\delta z)-f(z)}{\delta z} = \frac{4\delta x+\delta y-i\,\delta x+4i\delta y}{\delta x+i\delta y}$$

$$\Rightarrow \qquad \frac{\delta f}{\delta z} = \frac{4\delta x+\delta y-i\delta x+4i\delta y}{\delta x+i\delta y} \qquad \dots (1)$$

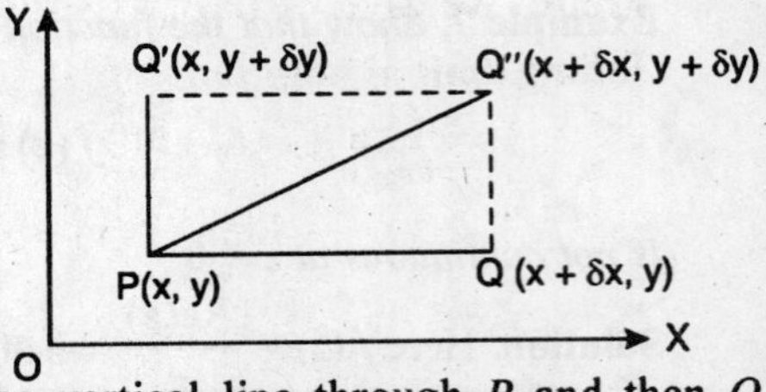


(*a*) **Along real axis:** If $Q$ is taken on the horizontal line through $P\ (x, y)$ and $Q$ then approaches $P$ along this line, we shall have $\delta y = 0$ and $\delta z = \delta x$.

$$\frac{\delta f}{\delta z} = \frac{4\delta x - i\delta x}{\delta x} = 4 - i$$

(*b*) **Along imaginary axis:** If $Q$ is taken on the vertical line through $P$ and then $Q$ approaches $P$ along this line, we have

$$z = x + iy = 0 + iy,\ \delta z = i\delta y,\ \delta x = 0.$$

Putting these values in (1), we have

$$\frac{\delta f}{\delta z} = \frac{\delta y + 4i\delta y}{i\delta y} = \frac{1}{i}(1+4i) = 4 - i$$

(*c*) **Along a line $y = x$ :** If $Q$ is taken on a line $y = x$.

$$z = x + iy = x + ix = (1 + i)x$$

$$\delta z = (1 + i)\delta x \quad \text{and} \quad \delta y = \delta x$$

On putting these values in (1), we have

$$\frac{\delta f}{\delta z} = \frac{4\delta x+\delta x-i\delta x+4i\delta x}{\delta x+i\delta x} = \frac{4+1-i+4i}{1+i} = \frac{5+3i}{1+i} = \frac{(5+3i)(1-i)}{(1+i)(1-i)} = 4-i$$

In all the three different paths approaching $Q$ from $P$, we get the same values of $\frac{\delta f}{\delta z} = 4 - i.$

In such a case, the function is said to be differentiable at the point $z$ in the given region.

**Example 9.** *If* $f(z) = \begin{cases} \dfrac{x^3 y(y-ix)}{x^6+y^2}, & z \neq 0, \\ 0, & z = 0 \end{cases}$ *then discuss* $\dfrac{df}{dz}$ *at* $z = 0$.

**Solution.** If $z \to 0$ along radius vector $y = mx$

$$\lim_{z\to 0}\frac{f(z)-f(0)}{z} = \lim_{z\to 0}\left[\frac{\dfrac{x^3y(y-ix)}{x^6+y^2}-0}{x+iy}\right] = \lim_{z\to 0}\left[\frac{-ix^3y(x+iy)}{(x^6+y^2)(x+iy)}\right]$$

$$= \lim_{z\to 0}\left[\frac{-ix^3y}{x^6+y^2}\right] = \lim_{x\to 0}\left[\frac{-ix^3(mx)}{x^6+m^2x^2}\right] \qquad [\because y = mx]$$

$$= \lim_{x\to 0}\left[\frac{-imx^2}{x^4+m^2}\right] = 0$$

But along $y = x^3$

$$\lim_{z\to 0}\frac{f(z)-f(0)}{z}=\lim_{z\to 0}\left[\frac{-ix^3y}{x^6+y^2}\right]=\lim_{x\to 0}\frac{-ix^3(x^3)}{x^6+(x^3)^2}=-\frac{i}{2}$$

**In different paths we get different values of $\frac{df}{dz}$ *i.e.* 0 and $\frac{-i}{2}$. In such a case, the function is not differentiable at $z = 0$.**

**Theorem: Continuity is a necessary condition but not sufficient condition for the existence of a finite derivative.**

**Proof.** We have, $f(z_0+\delta z)-f(z_0)=\delta z\left\{\frac{f(z_0+\delta z)-f(z_0)}{\delta z}\right\}$ ... (1)

**Taking limit of both sides of (1), as $\delta z\to 0$, we get**

$$\lim_{\delta z\to 0}[f(z_0+\delta z)-f(z_0)]=0.f'(z_0)\Rightarrow \lim_{\delta z\to 0}[f(z_0+\delta z)-f(z_0)]=0$$

$$\Rightarrow \lim_{z\to z_0}[f(z)-f(z_0)]=0 \quad\Rightarrow\quad \lim_{z\to z_0}f(z)=f(z_0)$$

$\Rightarrow$ $f(z)$ is continuous at $z = z_0$. **Proved.**

**The converse of the above theorem is not true.**

**This can be shown by the following example.**

**Example 10.** ***Prove that the function $f(z) = |z|^2$ is continuous everywhere but no where differentiable except at the origin.***

**Solution.** Here, $f(z) = |z|^2$.

$\therefore$ But $|z| = \sqrt{(x^2+y^2)}$ $\Rightarrow$ $|z|^2 = x^2+y^2$

**Since $x^2$ and $y^2$ are polynomial so $x^2 + y^2$ is continuous everywhere, therefore, $|z|^2$ is continuous everywhere.**

Now, we have $f'(z)=\lim_{\delta z\to 0}\frac{f(z+\delta z)-f(z)}{\delta z}$

$$f'(z)=\lim_{\delta z\to 0}\frac{|z+\delta z|^2-|z|^2}{\delta z} \qquad (z\bar{z}=|z|^2)$$

$$=\lim_{\delta z\to 0}\frac{(z+\delta z)(\bar{z}+\delta\bar{z})-z\bar{z}}{\delta z}=\lim_{\delta z\to 0}\frac{z\bar{z}+z\delta\bar{z}+\delta z.\bar{z}+\delta z.\delta\bar{z}-z\bar{z}}{\delta z}$$

$$=\lim_{\delta z\to 0}\frac{z.\delta\bar{z}+\delta z.\bar{z}+\delta z\,\delta\bar{z}}{\delta z}=\lim_{\delta z\to 0}\left\{\bar{z}+\delta\bar{z}+z\frac{\delta\bar{z}}{\delta z}\right\}=\lim_{\delta z\to 0}\left\{\bar{z}+z\frac{\delta\bar{z}}{\delta z}\right\} \quad ...(1)$$

[Since, $\delta z\to 0$ so $\delta\bar{z}\to 0$]

Let $\delta z = r(\cos\theta + i\sin\theta)$ and $\delta\bar{z} = r(\cos\theta - i\sin\theta)$

$$\Rightarrow \frac{\delta\bar{z}}{\delta z}=\frac{\cos\theta-i\sin\theta}{\cos\theta+i\sin\theta} \quad\Rightarrow\quad \frac{\delta\bar{z}}{\delta z}=(\cos\theta-i\sin\theta)(\cos\theta+i\sin\theta)^{-1}$$

$$\Rightarrow \frac{\delta\bar{z}}{\delta z}=(\cos\theta-i\sin\theta)(\cos\theta-i\sin\theta) \quad\Rightarrow\quad \frac{\delta\bar{z}}{\delta z}=(\cos\theta-i\sin\theta)^2$$

$$\Rightarrow \frac{\delta\bar{z}}{\delta z}=\cos 2\theta-i\sin 2\theta$$

Since $\frac{\delta\bar{z}}{\delta z}$ depends on $\theta$. It means for different values of $\theta$, $\frac{\delta\bar{z}}{\delta z}$ has different values.

It means $\frac{\delta\bar{z}}{\delta z}$ has different values for different $z$.

Therefore $\lim_{\delta z \to 0} \dfrac{\delta \bar{z}}{\delta z}$ does not tend to a unique limit when $z \neq 0$.

Thus, from (1), it follows that $f'(z)$ is not unique and hence $f(z)$ is not differentiable when $z \neq 0$.

But when $z = 0$ then $f'(z) = 0$ *i.e.*, $f'(0) = 0$ and is unique.

Hence, the function is differentiable at $z = 0$. **Proved.**

## 7.9 ANALYTIC FUNCTION

A function $f(z)$ is said to be **analytic** at a point $z_0$, if $f$ is differentiable not only at $z_0$ but at every point of some neighbourhood of $z_0$.

A function $f(z)$ is analytic in a domain if it is **analytic** at every point of the domain.

The point at which the function is not differentiable is called a **singular point** of the function.

An analytic function is also known as "holomorphic", "regular", "monogenic".

**Entire Function.** A function which is analytic everywhere (for all $z$ in the complex plane) is known as an entire function.

**For Example 1.** Polynomials rational functions are entire.

**2.** $|\bar{z}|^2$ is differentiable only at $z = 0$. So it is no where analytic.

**Note:** (*i*) An entire is always analytic, differentiable and continuous function. But converse is not true.

(*ii*) Analytic function is always differentiable and continuous. But converse is not true.

(*iii*) A differentiable function is always continuous. But converse is not true

## 7.10 THE NECESSARY CONDITION FOR F (Z) TO BE ANALYTIC

**Theorem.** *The necessary conditions for a function f (z) = u + iv to be analytic at all the points in a region R are*

(*i*) $\dfrac{\partial u}{\partial x} = \dfrac{\partial v}{\partial y}$ (*ii*) $\dfrac{\partial u}{\partial y} = -\dfrac{\partial v}{\partial x}$ provided $\dfrac{\partial u}{\partial x}, \dfrac{\partial u}{\partial y}, \dfrac{\partial v}{\partial x}, \dfrac{\partial v}{\partial y}$ exist.

**Proof:** Let $f(z)$ be an analytic function in a region $R$,

$$f(z) = u + iv,$$

where $u$ and $v$ are the functions of $x$ and $y$.

Let $\delta u$ and $\delta v$ be the increments of $u$ and $v$ respectively corresponding to increments $\delta x$ and $\delta y$ of $x$ and $y$.

$$\therefore \quad f(z + \delta z) = (u + \delta u) + i(v + \delta v)$$

Now $$\frac{f(z+\delta z) - f(z)}{\delta z} = \frac{(u+\delta u) + i(v+\delta v) - (u+iv)}{\delta z} = \frac{\delta u + i\delta v}{\delta z} = \frac{\delta u}{\delta z} + i\frac{\delta v}{\delta z}$$

$$\lim_{\delta z \to 0} \frac{f(z+\delta z) - f(z)}{\delta z} = \lim_{\delta z \to 0}\left(\frac{\delta u}{\delta z} + i\frac{\delta v}{\delta z}\right) \quad \text{or} \quad f'(z) = \lim_{\delta z \to 0}\left(\frac{\delta u}{\delta z} + i\frac{\delta v}{\delta z}\right) \quad \ldots (1)$$

since $\delta z$ can approach zero along any path.

(*a*) **Along real axis (*x*-axis)**

$$z = x + iy \qquad \text{but on } x\text{-axis, } y = 0$$

$$\therefore \quad z = x, \qquad \delta z = \delta x,\ \delta y = 0$$

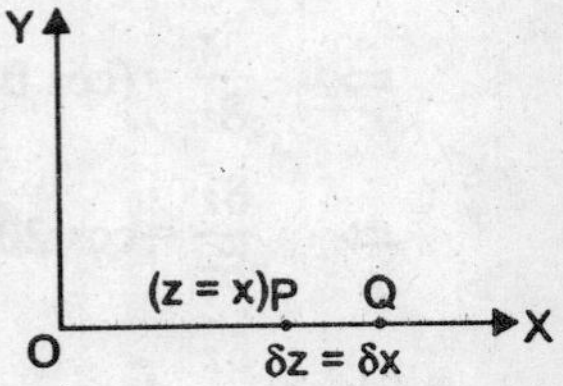


Putting these values in (1), we have

$$f'(z) = \lim_{\delta x \to 0}\left(\frac{\delta u}{\delta x} + i\frac{\delta v}{\delta x}\right) = \frac{\partial u}{\partial x} + i\frac{\partial v}{\partial x} \quad \ldots (2)$$

(*b*) **Along imaginary axis (*y*-axis)**

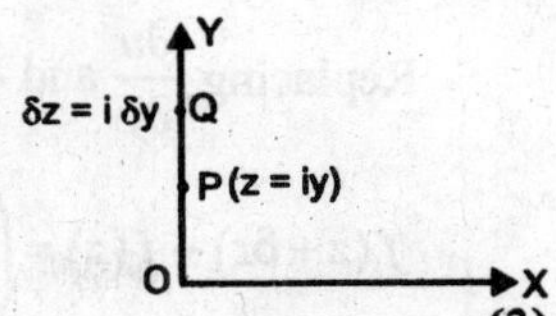


$$z = x + iy \qquad \text{but on } y\text{-axis, } x = 0$$

$$z = 0 + iy \qquad \delta x = 0,\ \delta z = i\delta y.$$

**Putting these values in (1), we get**

$$f'(z) = \lim_{\delta y \to 0}\left(\frac{\delta u}{i\delta y} + \frac{i\delta v}{i\delta y}\right) = \lim_{\delta y \to 0}\left(-i\frac{\delta u}{\delta y} + \frac{\delta v}{\delta y}\right) = -i\frac{\partial u}{\partial y} + \frac{\partial v}{\partial y} \qquad \ldots (3)$$

**If $f(z)$ is differentiable, then two values of $f'(z)$ must be the same.**
**Equating (2) and (3), we get**

$$\frac{\partial u}{\partial x} + i\frac{\partial v}{\partial x} = -i\frac{\partial u}{\partial y} + \frac{\partial v}{\partial y}$$

**Equating real and imaginary parts, we have**

$$\frac{\partial u}{\partial x} = \frac{\partial v}{\partial y}, \qquad \frac{\partial v}{\partial x} = -\frac{\partial u}{\partial y}$$

$$\boxed{\frac{\partial u}{\partial x} = \frac{\partial v}{\partial y}}$$

$$\boxed{\frac{\partial u}{\partial y} = -\frac{\partial v}{\partial x}}$$ **are known as Cauchy Riemann equations.**

## 7.11 SUFFICIENT CONDITION FOR F (Z) TO BE ANALYTIC

**Theorem.** ***The sufficient condition for a function $f(z) = u + iv$ to be analytic at all the points in a region R are***

(*i*) $\frac{\partial u}{\partial x} = \frac{\partial v}{\partial y}, \qquad \frac{\partial u}{\partial y} = -\frac{\partial v}{\partial x}$

(*ii*) $\frac{\partial u}{\partial x}, \frac{\partial u}{\partial y}, \frac{\partial v}{\partial x}, \frac{\partial v}{\partial y}$ *are continuous functions of x and y in region R.*

**Proof.** Let $f(z)$ be a single-valued function having

$$\frac{\partial u}{\partial x}, \frac{\partial u}{\partial y}, \frac{\partial v}{\partial x}, \frac{\partial v}{\partial y}$$

at each point in the region $R$. Then the $C - R$ equations are satisfied.
By Taylor's Theorem:

$$f(z+\delta z) = u(x+\delta x,\ y+\delta y) + iv(x+\delta x,\ y+\delta y)$$

$$= u(x,y) + \left(\frac{\partial u}{\partial x}\delta x + \frac{\partial u}{\partial y}\delta y\right) + \ldots + i\left[v(x,y) + \left(\frac{\partial v}{\partial x}\delta x + \frac{\partial v}{\partial y}\delta y\right) + \ldots\right]$$

$$= [u(x,y) + iv(x,y)] + \left[\frac{\partial u}{\partial x}\cdot\delta x + i\frac{\partial v}{\partial x}\cdot\delta x\right] + \left[\frac{\partial u}{\partial y}\delta y + i\frac{\partial v}{\partial y}\cdot\delta y\right] + \ldots$$

$$= f(z) + \left(\frac{\partial u}{\partial x} + i\frac{\partial v}{\partial x}\right)\delta x + \left(\frac{\partial u}{\partial y} + i\frac{\partial v}{\partial y}\right)\delta y + \ldots$$

**(Ignoring the terms of second power and higher powers)**

$$\Rightarrow \quad f(z+\delta z) - f(z) = \left(\frac{\partial u}{\partial x} + i\frac{\partial v}{\partial x}\right)\delta x + \left(\frac{\partial u}{\partial y} + i\frac{\partial v}{\partial y}\right)\delta y \qquad \ldots (1)$$

We know $C - R$ equations *i.e.*,

$$\frac{\partial u}{\partial x} = \frac{\partial v}{\partial y} \text{ and } \frac{\partial u}{\partial y} = -\frac{\partial v}{\partial x}$$

Replacing $\frac{\partial u}{\partial y}$ and $\frac{\partial v}{\partial y}$ by $-\frac{\partial v}{\partial x}$ and $\frac{\partial u}{\partial x}$ respectively in (1), we get

$$f(z+\delta z)-f(z)=\left(\frac{\partial u}{\partial x}+i\frac{\partial v}{\partial x}\right).\delta x+\left(-\frac{\partial v}{\partial x}+i\frac{\partial u}{\partial x}\right).\delta y \qquad \text{(taking } i \text{ common)}$$

$$=\left(\frac{\partial u}{\partial x}+i\frac{\partial v}{\partial x}\right).\delta x+\left(i\frac{\partial v}{\partial x}+\frac{\partial u}{\partial x}\right).i\delta y=\left(\frac{\partial u}{\partial x}+i\frac{\partial v}{\partial x}\right).(\delta x+i\delta y)=\left(\frac{\partial u}{\partial x}+i\frac{\partial v}{\partial x}\right).\delta z$$

$$\Rightarrow \quad \frac{f(z+\delta z)-f(z)}{\delta z}=\frac{\partial u}{\partial x}+i\frac{\partial v}{\partial x}$$

$$\Rightarrow \quad \lim_{\delta z\to 0}\frac{f(z+\delta z)-f(z)}{\delta z}=\frac{\partial u}{\partial x}+i\frac{\partial v}{\partial x}$$

$$\Rightarrow \quad \boxed{f'(z)=\frac{\partial u}{\partial x}+i\frac{\partial v}{\partial x}}$$

$$\Rightarrow \quad \boxed{f'(z)=\frac{\partial v}{\partial y}-i\frac{\partial u}{\partial y}}$$

**Proved.**

**Remember: 1. If a function is analytic in a domain $D$, then $u$, $v$ satisfy $C-R$ conditions at all points in $D$.**
**2. $C-R$ conditions are necessary but not sufficient for analytic function.**
**3. $C-R$ conditions are sufficient if the partial derivative are continuous.**

**Example 11.** *Determine whether $\frac{1}{z}$ is analytic or not?* ***(R.G.P.V. Bhopal, III Sem., June 2003)***

**Solution.** Let $w=f(z)=u+iv=\frac{1}{z} \quad\Rightarrow\quad u+iv=\frac{1}{x+iy}=\frac{x-iy}{x^2+y^2}$

Equating real and imaginary parts, we get

$$u=\frac{x}{x^2+y^2}, \qquad v=\frac{-y}{x^2+y^2}$$

$$\frac{\partial u}{\partial x}=\frac{(x^2+y^2).1-x.2x}{(x^2+y^2)^2}=\frac{y^2-x^2}{(x^2+y^2)^2}, \qquad \frac{\partial u}{\partial y}=\frac{-2xy}{(x^2+y^2)^2}.$$

$$\frac{\partial v}{\partial x}=\frac{2xy}{(x^2+y^2)^2}, \qquad \frac{\partial v}{\partial y}=\frac{y^2-x^2}{(x^2+y^2)^2}$$

Thus, $$\frac{\partial u}{\partial x}=\frac{\partial v}{\partial y} \text{ and } \frac{\partial u}{\partial y}=-\frac{\partial v}{\partial x}.$$

Thus C – R equations are satisfied. Also partial derivatives are continuous except at (0, 0).
Therefore $\frac{1}{z}$ is analytic everywhere except at $z=0$.

Also $$\frac{dw}{dz}=-\frac{1}{z^2}$$

This again shows that $\frac{dw}{dz}$ exists everywhere except at $z=0$. Hence $\frac{1}{z}$ is analytic everywhere except at $z=0$. **Ans.**

**Example 12.** *Show that the function $e^x$ (cos y + i sin y) is an analytic function, find its derivative.* *(R.G.P.V., Bhopal, III Semester, June 2008)*

**Solution.** Let $e^x(\cos y + i\sin y) = u + iv$

So, $e^x \cos y = u$ and $e^x \sin y = v$ then $\dfrac{\partial u}{\partial x} = e^x \cos y,\quad \dfrac{\partial v}{\partial y} = e^x \cos y$

$$\frac{\partial u}{\partial y} = -e^x \sin y, \qquad \frac{\partial v}{\partial x} = e^x \sin y$$

Here we see that $\dfrac{\partial u}{\partial x} = \dfrac{\partial v}{\partial y}, \qquad \dfrac{\partial u}{\partial y} = -\dfrac{\partial v}{\partial x}$

These are $C - R$ equations and are satisfied and the partial derivatives are continuous. Hence, $e^x(\cos y + i\sin y)$ is analytic.

$$f(z) = u + iv = e^x(\cos y + y\sin y) \text{ and } \frac{\partial u}{\partial x} = e^x\cos y, \qquad \frac{\partial v}{\partial x} = e^x \sin y$$

$$f'(z) = \frac{\partial u}{\partial x} + i\frac{\partial v}{\partial x} = e^x\cos y + ie^x\sin y = e^x(\cos y + i\sin y) = e^x.e^{iy} = e^{x+iy} = e^z.$$

Which is the required derivative. **Ans.**

**Example 13.** *Test the analyticity of the function $w = \sin z$ and hence derive that:*

$$\frac{d}{dz}(\sin z) = \cos z$$

**Solution.** $w = \sin z = \sin(x + iy)$

$= \sin x \cos iy + \cos x \sin iy$

$= \sin x \cosh y + i\cos x \sinh y$

$u = \sin x\cosh y,\ v = \cos x \sinh y$ $\quad \begin{bmatrix} \cos iy = \cosh y \\ \sin iy = i\sinh y \end{bmatrix}$

$$\frac{\partial u}{\partial x} = \cos x\cosh y, \frac{\partial u}{\partial y} = \sin x \sinh y$$

$$\frac{\partial v}{\partial x} = -\sin x\sinh y, \frac{\partial v}{\partial y} = \cos x\cosh y$$

Thus $\dfrac{\partial u}{\partial x} = \dfrac{\partial v}{\partial y}$ and $\dfrac{\partial u}{\partial y} = -\dfrac{\partial v}{\partial x}$

So $C - R$ equations are satisfied and partial derivatives are continuous.

Hence, $\sin z$ is an analytic function.

$$\frac{d}{dz}(\sin z) = \frac{d}{dz}[\sin x\cosh y + i\cos x\sinh y]$$

$$= \frac{\partial}{\partial x}(\sin x\cosh y + i\cos x \sinh y)$$

$$= \cos x\cosh y - i\sin x\sinh y = \cos x\cos iy - \sin x \sin iy$$

$$= \cos(x + iy) = \cos z$$

**Ans.**

$$\cosh x = \frac{e^x + e^{-x}}{2} \qquad \text{... (1)}$$

$$\sinh x = \frac{e^x - e^{-x}}{2} \qquad \text{... (2)}$$

$$\cos x = \frac{e^{ix} + e^{-ix}}{2} \qquad \text{... (3)}$$

$$\sin x = \frac{e^{ix} - e^{-ix}}{2i} \qquad \text{... (4)}$$

From (1) $\cosh ix = \dfrac{e^{ix} + e^{-ix}}{2} = \cos x$

From (3) $\cos ix = \dfrac{e^{i(ix)} + e^{-i(ix)}}{2}$

$$= \frac{e^x + e^{-x}}{2} = \cosh x$$

From (4) $\sin ix = \dfrac{e^{i(ix)} - e^{-i(ix)}}{2i}$

$$= i\,\frac{e^x - e^{-x}}{2} = i\sinh x$$

From (2) $\sinh ix = \dfrac{e^{ix} - e^{-ix}}{2} = i\sin x$

**Example 14.** *Show that the real and imaginary parts of the function $w = \log z$ satisfy the Cauchy-Riemann equations when $z$ is not zero. Find its derivative.*

**Solution.** To separate the real and imaginary parts of log $z$, we put $x = r\cos\theta;\ y = r\sin\theta$

$$w = \log z = \log(x + iy)$$

$$\Rightarrow\quad u + iv = \log(r\cos\theta + ir\sin\theta) = \log r(\cos\theta + i\sin\theta) = \log_e r.e^{i\theta}$$

$$= \log_e r + \log_e e^{i\theta} = \log r + i\theta = \log\sqrt{x^2+y^2} + i\tan^{-1}\frac{y}{x} \qquad \left[\begin{array}{l} r = \sqrt{x^2+y^2} \\ \theta = \tan^{-1}\dfrac{y}{x}\end{array}\right]$$

So $\quad u = \log\sqrt{x^2+y^2} = \dfrac{1}{2}\log(x^2+y^2),\ v = \tan^{-1}\dfrac{y}{x}$

On differentiating $u$, $v$, we get

$$\frac{\partial u}{\partial x} = \frac{1}{2}\frac{1}{x^2+y^2}.(2x) = \frac{x}{x^2+y^2} \qquad \text{... (1)}$$

$$\frac{\partial v}{\partial y} = \frac{1}{1+\frac{y^2}{x^2}}\left(\frac{1}{x}\right) = \frac{x}{x^2+y^2} \qquad \text{... (2)}$$

From (1) and (2), $\dfrac{\partial u}{\partial x} = \dfrac{\partial v}{\partial y}$ ... (A)

Again differentiating $u$, $v$, we have

$$\frac{\partial u}{\partial y} = \frac{1}{2}\frac{1}{x^2+y^2}(2y) = \frac{y}{x^2+y^2} \qquad \text{... (3)}$$

$$\frac{\partial v}{\partial x} = \frac{1}{1+\frac{y^2}{x^2}}\left(-\frac{y}{x^2}\right) = -\frac{y}{x^2+y^2} \qquad \text{... (4)}$$

From (3) and (4), we have

$$\frac{\partial u}{\partial y} = -\frac{\partial v}{\partial x} \qquad \text{... (B)}$$

Equations (A) and (B) are $C - R$ equations and partial derivatives are continuous.
Hence, $w = \log z$ is an analytic function except
when $\quad x^2 + y^2 = 0 \Rightarrow x = y = 0 \Rightarrow x + iy = 0 \Rightarrow z = 0$
Now $\quad w = u + iv$

$$\frac{dw}{dz} = \frac{\partial u}{\partial x} + i\frac{\partial v}{\partial x} = \frac{x}{x^2+y^2} - i\frac{y}{x^2+y^2} = \frac{x-iy}{x^2+y^2}$$

$$= \frac{x-iy}{(x+iy)(x-iy)} = \frac{1}{x+iy} = \frac{1}{z}$$

Which is the required derivative. **Ans.**

**Example 15.** *Discuss the analyticity of the function* $f(z) = z\bar{z}$.

**Solution.** $\quad f(z) = z\bar{z} = (x+iy)(x-iy) = x^2 - i^2y^2 = x^2 + y^2$

$$f(z) = x^2 + y^2 = u + iv.$$

$$u = x^2 + y^2,\ v = 0$$

At origin, $\quad \dfrac{\partial u}{\partial x} = \lim\limits_{h\to 0}\dfrac{u(0+h,\ 0) - u(0,\ 0)}{h} = \lim\limits_{h\to 0}\dfrac{h^2}{h} = 0$

$$\frac{\partial u}{\partial y} = \lim_{k\to 0} \frac{u(0,0+k)-u(0,0)}{k} = \lim_{k\to 0} \frac{k^2}{k} = 0$$

Also, $$\frac{\partial v}{\partial x} = \lim_{h\to 0} \frac{v(0+h,0)-v(0,0)}{h} = 0$$

$$\frac{\partial v}{\partial y} = \lim_{k\to 0} \frac{v(0,0+k)-v(0,0)}{k} = 0$$

Thus, $$\frac{\partial u}{\partial x} = \frac{\partial v}{\partial y} \text{ and } \frac{\partial u}{\partial y} = -\frac{\partial v}{\partial x}.$$

Hence, C – R equations are satisfied at the origin.

$$f'(0) = \lim_{z\to 0} \frac{f(z)-f(0)}{z} = \lim_{z\to 0} \frac{(x^2+y^2)-0}{x+iy}$$

Let $z \to 0$ along the line $y = mx$, then

$$f'(0) = \lim_{x\to 0} \frac{(x^2+m^2x^2)}{(x+imx)} = \lim_{x\to 0} \frac{(1+m^2)x}{1+im} = 0$$

Therefore, $f'(0)$ is unique. Hence the function $f(z)$ is analytic at $z = 0$. **Ans.**

**Example 16.** *Show that the function $f(z) = u + iv$, where*

$$f(z) = \begin{cases} \dfrac{x^3(1+i)-y^3(1-i)}{x^2+y^2}, & z \neq 0 \\ = 0, & z = 0 \end{cases}$$

*satisfies the Cauchy-Riemann equations at $z = 0$. Is the function analytic at $z = 0$? Justify your answer.* *(MDU Dec 2009)*

**Solution.** $$f(z) = \frac{x^3(1+i)-y^3(1-i)}{x^2+y^2} = u+iv$$

$$u = \frac{x^3-y^3}{x^2+y^2}, \qquad v = \frac{x^3+y^3}{x^2+y^2}$$

[By differentiation the value of $\frac{\partial u}{\partial x}, \frac{\partial u}{\partial y}, \frac{\partial v}{\partial x}, \frac{\partial v}{\partial y}$ at (0, 0) we get $\frac{0}{0}$, so we apply first principle method]

At the origin

$$\frac{\partial u}{\partial x} = \lim_{h\to 0} \frac{u(0+h,\ 0)-u(0,\ 0)}{h} = \lim_{h\to 0} \frac{\frac{h^3}{h^2}}{h} = 1 \qquad \text{(Along } x\text{- axis)}$$

$$\frac{\partial u}{\partial y} = \lim_{k\to 0} \frac{u(0,\ 0+k)-u(0,\ 0)}{k} = \lim_{k\to 0} \frac{\frac{-k^3}{k^2}}{k} = -1 \qquad \text{(Along } y\text{- axis)}$$

$$\frac{\partial v}{\partial x} = \lim_{h\to 0} \frac{v(0+h,\ 0)-v(0,\ 0)}{h} = \lim_{h\to 0} \frac{\frac{h^3}{h^2}}{h} = 1 \qquad \text{(Along } x\text{- axis)}$$

$$\frac{\partial v}{\partial y} = \lim_{k\to 0} \frac{v(0,\ 0+k)-v(0,\ 0)}{k} = \lim_{k\to 0} \frac{\frac{k^3}{k^2}}{k} = 1 \qquad \text{(Along } y\text{-axis)}$$

Thus we see that

$$\frac{\partial u}{\partial x} = \frac{\partial v}{\partial y} \qquad \text{and} \qquad \frac{\partial u}{\partial y} = -\frac{\partial v}{\partial x}$$

Hence, Cauchy-Riemann equations are satisfied at $z = 0$.

Again $$f'(0) = \lim_{z\to 0} \frac{f(0+z)-f(0)}{z} = \lim_{z\to 0}\left[\frac{\frac{x^3-y^3+i\,(x^3+y^3)}{x^2+y^2}-(0)}{x+i\,y}\right]$$

$$= \lim_{z\to 0}\left[\frac{x^3-y^3+i(x^3+y^3)}{x^2+y^2}\cdot\frac{1}{x+iy}\right]$$

Now let $z \to 0$ along $y = x$, then

$$f'(0) = \lim_{x\to 0}\frac{x^3-x^3+i(x^3+x^3)}{x^2+x^2}\left(\frac{1}{x+ix}\right)$$

$$= \frac{2i}{2(1+i)} = \frac{i}{1+i} = \frac{i(1-i)}{(1+i)(1-i)} = \frac{i+1}{1+1} = \frac{1}{2}(1+i) \quad \text{... (1)}$$

Again let $z \to 0$ along $y = 0$, then

$$f'(0) = \lim_{x\to 0}\frac{x^3+ix^3}{x^2}\cdot\frac{1}{x} = (1+i) \qquad [\text{Increment} = z] \quad \text{... (2)}$$

From (1) and (2), we see that $f'(0)$ is not unique. Hence the function $f(z)$ is not analytic at $z = 0$. **Ans.**

**Example 17.** *Show that the function*

$$f(z) = e^{-z^{-4}}, \quad (z \neq 0) \quad \text{and}$$
$$f(0) = 0$$

*is not analytic at $z = 0$,*
*although, Cauchy-Riemann equations are satisfied at the point. How would you explain this.*

**Solution.** $$f(z) = u + iv = e^{-z^{-4}} = e^{-(x+iy)^{-4}} = e^{-\frac{1}{(x+iy)^4}}$$

$$\Rightarrow \quad u + iv = e^{-\frac{(x-iy)^4}{(x^2+y^2)^4}} = e^{-\frac{1}{(x^2+y^2)^4}[(x^4+y^4-6x^2y^2)-i4xy(x^2-y^2)]}$$

$$\Rightarrow \quad u+iv = e^{-\frac{x^4+y^4-6x^2y^2}{(x^2+y^2)^4}}.e^{\frac{-i4xy(x^2-y^2)}{(x^2+y^2)^4}}$$

$$\Rightarrow \quad u+iv = e^{-\frac{x^4+y^4-6x^2y^2}{(x^2+y^2)^4}}\left[\cos\frac{4xy(x^2-y^2)}{(x^2+y^2)^4} - i\sin\frac{4xy(x^2-y^2)}{(x^2+y^2)^4}\right]$$

Equating real and imaginary parts, we get

$$u = e^{-\frac{x^4+y^4-6x^2y^2}{(x^2+y^2)^4}}\cos\frac{4xy(x^2-y^2)}{(x^2+y^2)^4}, \quad v = e^{-\frac{x^4+y^4-6x^2y^2}{(x^2+y^2)^4}}\sin\frac{4xy(x^2-y^2)}{(x^2+y^2)^4}$$

At $z = 0$ $$\frac{\partial u}{\partial x} = \lim_{h\to 0}\frac{u(0+h,\,0)-u(0,\,0)}{h} = \lim_{h\to 0}\frac{e^{-h^{-4}}}{h} = \lim_{h\to 0}\frac{1}{h\,e^{\frac{1}{h^4}}}$$

$$= \lim_{h\to 0}\left[\frac{1}{h\left[1+\frac{1}{h^4}+\frac{1}{2!h^8}+\frac{1}{3!h^{12}}+\ldots\right]}\right], \quad \left(e^x = 1+x+\frac{x^2}{2!}+\ldots\right)$$

$$= \lim_{h\to 0}\left[\frac{1}{\left[h+\frac{1}{h^3}+\frac{1}{2\,h^7}+\frac{1}{6h^{11}}\cdots\right]}\right] = \frac{1}{0+\infty} = \frac{1}{\infty} = 0$$

$$\frac{\partial u}{\partial y} = \lim_{k\to 0}\frac{u(0,0+k)-u(0,0)}{k} = \lim_{k\to 0}\frac{e^{-k^{-4}}}{k} = \lim_{k\to 0}\frac{1}{k\,\frac{1}{e^{k^4}}} = 0$$

$$\frac{\partial v}{\partial x} = \lim_{h\to 0}\frac{v(0+h,\,0)-v(0,0)}{h} = \lim_{h\to 0}\frac{0}{h} = 0$$

$$\frac{\partial v}{\partial y} = \lim_{k\to 0}\frac{v(0,0+k)-v(0,0)}{k} = \lim_{k\to 0}\frac{0}{k} = 0$$

Hence $\frac{\partial u}{\partial x} = \frac{\partial v}{\partial y}$ and $\frac{\partial u}{\partial y} = -\frac{\partial v}{\partial x}$ **($C-R$ equations are satisfied at $z = 0$)**

But $f'(0) = \lim_{z\to 0}\frac{f(z)-f(0)}{z} = \lim_{z\to 0}\frac{e^{-z^{-4}}}{z}$

Along $z = re^{i\frac{\pi}{4}}$ $f'(0) = \lim_{r\to 0}\frac{e^{-r^{-4}}\cdot e^{-\left(e^{i\frac{\pi}{4}}\right)^{-4}}}{re^{i\frac{\pi}{4}}} = \lim_{r\to 0}\frac{e^{-r^{-4}}\cdot e^{-\left(\cos\frac{\pi}{4}+i\sin\frac{\pi}{4}\right)^{-4}}}{re^{i\frac{\pi}{4}}}$

$$= \lim_{r\to 0}\frac{e^{-r^{-4}}\cdot e^{-\cos\pi}}{re^{i\frac{\pi}{4}}} = \lim_{r\to 0}\frac{e^{-r^{-4}}\cdot e}{re^{i\frac{\pi}{4}}} = \infty$$

**Showing that $f'(z)$ does not exist at $z = 0$. Hence $f(z)$ is not analytic at $z = 0$. Proved.**

**Example 18.** *Examine the nature of the function*

$$f(z) = \frac{x^2y^5(x+iy)}{x^4+y^{10}}; z \neq 0$$

$$f(0) = 0$$

*in the region including the origin.*

**Solution.** Here $f(z) = u + iv = \frac{x^2y^5(x+iy)}{x^4+y^{10}}; z \neq 0$

**Equating real and imaginary parts, we get**

$$u = \frac{x^3y^5}{x^4+y^{10}},\quad v = \frac{x^2y^6}{x^4+y^{10}}$$

$$\frac{\partial u}{\partial x} = \lim_{h\to 0}\frac{u(0+h,0)-u(0,0)}{h} = \lim_{h\to 0}\frac{\frac{0}{h^4}}{h} = \lim_{h\to 0}\frac{0}{h} = 0$$

$$\frac{\partial u}{\partial y} = \lim_{k\to 0}\frac{u(0,0+k)-u(0,0)}{k} = \lim_{k\to 0}\frac{\frac{0}{k^{10}}}{k} = \lim_{k\to 0}\frac{0}{k} = 0$$

$$\frac{\partial v}{\partial x} = \lim_{h\to 0} \frac{v(0+h,0)-v(0,0)}{h} = \lim_{h\to 0} \frac{\frac{0}{h^4}}{h} = \lim_{h\to 0} \frac{0}{h} = 0$$

$$\frac{\partial v}{\partial y} = \lim_{k\to 0} \frac{v(0,0+k)-v(0,0)}{k} = \lim_{k\to 0} \frac{\frac{0}{k^{10}}}{k} = \lim_{k\to 0} \frac{0}{k} = 0$$

From the above results, it is clear that

$$\frac{\partial u}{\partial x} = \frac{\partial v}{\partial y} \quad \text{and} \quad \frac{\partial u}{\partial y} = -\frac{\partial v}{\partial x}$$

Hence, C-R equations are satisfied at the origin.

But $$f'(0) = \lim_{z\to 0} \frac{f(0+z)-f(0)}{z} = \lim_{\substack{x\to 0\\ y\to 0}} \left[\frac{x^2y^5(x+iy)}{x^4+y^{10}} - 0\right]\cdot\frac{1}{x+iy} \quad \text{(Increment} = z)$$

$$= \lim_{\substack{x\to 0\\ y\to 0}} \frac{x^2y^5}{x^4+y^{10}}$$

Let $z \to 0$ along the radius vector $y = mx$, then

$$f'(0) = \lim_{x\to 0} \frac{m^5x^7}{x^4+m^{10}x^{10}} = \lim_{x\to 0} \frac{m^5x^3}{1+m^{10}x^6} = \frac{0}{1} = 0 \quad \text{... (1)}$$

Again let $z \to 0$ along the curve $y^5 = x^2$

$$f'(0) = \lim_{x\to 0} \frac{x^4}{x^4+x^4} = \frac{1}{2} \quad \text{... (2)}$$

(1) and (2) shows that $f'(0)$ does not exist. Hence, $f(z)$ is not analytic at origin although Cauchy-Riemann equations are satisfied there. **Ans.**

## 7.12 C–R EQUATIONS IN POLAR FORM

$$\frac{\partial u}{\partial r} = \frac{1}{r}\frac{\partial v}{\partial \theta}$$

$$\frac{\partial u}{\partial \theta} = -r\frac{\partial v}{\partial r}$$

*(MDU, Dec. 2010, RGPV., K.U. 2009, Bhopal, III Sem. Dec. 2007)*

**Proof.** We know that $x = r\cos\theta$, and $u$ is a function of $x$ and $y$.

$$z = x + iy = r(\cos\theta + i\sin\theta) = re^{i\theta}$$

$$u + iv = f(z) = f(re^{i\theta}) \quad \text{... (1)}$$

Differentiating (1) partially w.r.t., "$r$", we get

$$\frac{\partial u}{\partial r} + i\frac{\partial v}{\partial r} = f'(re^{i\theta}).e^{i\theta} \quad \text{... (2)}$$

Differentiating (1) w.r.t. "$\theta$", we get

$$\frac{\partial u}{\partial \theta} + i\frac{\partial v}{\partial \theta} = f'(re^{i\theta})re^{i\theta}i \quad \text{... (3)}$$

Substituting the value of $f'(re^{i\theta})e^{i\theta}$ from (2) in (3), we obtain

$$\frac{\partial u}{\partial \theta} + i\frac{\partial v}{\partial \theta} = r\left(\frac{\partial u}{\partial r} + i\frac{\partial v}{\partial r}\right)i \quad \text{or} \quad \frac{\partial u}{\partial \theta} + i\frac{\partial v}{\partial \theta} = ir\frac{\partial u}{\partial r} - r\frac{\partial v}{\partial r}$$

Equating real and imaginary parts, we get

$$\boxed{\frac{\partial u}{\partial \theta} = -r\frac{\partial v}{\partial r}} \Rightarrow \frac{\partial v}{\partial r} = \frac{-1}{r}\frac{\partial u}{\partial \theta}$$

And $$\boxed{\frac{\partial u}{\partial r} = \frac{1}{r}\frac{\partial v}{\partial \theta}}$$ **Proved.**

## 7.13 DERIVATIVE OF W OR F (Z) IN POLAR FORM

**We know that** $w = u + iv$, $\frac{\partial w}{\partial x} = \frac{\partial u}{\partial x} + i\frac{\partial v}{\partial x}$

**But**
$$\frac{dw}{dz} = \frac{\partial w}{\partial r}\frac{\partial r}{\partial x} + \frac{\partial w}{\partial \theta}\frac{\partial \theta}{\partial x} = \frac{\partial w}{\partial r}\cos\theta - \left(\frac{\partial u}{\partial \theta} + i\frac{\partial v}{\partial \theta}\right)\frac{\sin\theta}{r}$$

$$= \frac{\partial w}{\partial r}\cos\theta - \left(-r\frac{\partial v}{\partial r} + i\cdot r\frac{\partial u}{\partial r}\right)\frac{\sin\theta}{r} \qquad \frac{\partial u}{\partial \theta} = -r\frac{\partial v}{\partial r}$$

$$= \frac{\partial w}{\partial r}\cos\theta - i\left(\frac{\partial u}{\partial r} + i\frac{\partial v}{\partial r}\right)\sin\theta \qquad \frac{\partial v}{\partial \theta} = r\frac{\partial u}{\partial r}$$

$$= \frac{\partial w}{\partial r}\cos\theta - i\frac{\partial}{\partial r}(u+iv)\sin\theta = \frac{\partial w}{\partial r}\cos\theta - i\frac{\partial w}{\partial r}\sin\theta \qquad [\because w = u + iv]$$

$$= (\cos\theta - i\sin\theta)\frac{\partial w}{\partial r} \qquad \text{... (1)}$$

**Second form of** $\frac{dw}{dz}$

$$\frac{dw}{dz} = \frac{\partial w}{\partial r}\frac{\partial r}{\partial x} + \frac{\partial w}{\partial \theta}\frac{\partial \theta}{\partial x} = \frac{\partial(u+iv)}{\partial r}\cos\theta - \frac{\partial w}{\partial \theta}\frac{\sin\theta}{r} \qquad [w = u + iv]$$

$$= \left(\frac{\partial u}{\partial r} + i\frac{\partial v}{\partial r}\right)\cos\theta - \frac{\partial w}{\partial \theta}\frac{\sin\theta}{r}$$

$$= \left(\frac{1}{r}\frac{\partial v}{\partial \theta} - i\cdot\frac{1}{r}\cdot\frac{\partial u}{\partial \theta}\right)\cos\theta - \frac{\partial w}{\partial \theta}\cdot\frac{\sin\theta}{r}$$

$$= -\frac{i}{r}\left(\frac{\partial u}{\partial \theta} + i\frac{\partial v}{\partial \theta}\right)\cos\theta - \frac{\partial w}{\partial \theta}\left(\frac{\sin\theta}{r}\right)$$

$$= -\frac{i}{r}\frac{\partial}{\partial \theta}(u+iv)\cos\theta - \frac{\partial w}{\partial \theta}\left(\frac{\sin\theta}{r}\right) = -\frac{i}{r}\frac{\partial w}{\partial \theta}\cos\theta - \frac{\partial w}{\partial \theta}\frac{\sin\theta}{r} \qquad [w = u + iv]$$

$$= -\frac{i}{r}(\cos\theta - i\sin\theta)\frac{\partial w}{\partial \theta} \qquad \text{... (2)}$$

$$\boxed{\frac{dw}{dz} = (\cos\theta - i\sin\theta)\frac{\partial w}{\partial r}} \qquad \left[-\frac{i}{r}\frac{\partial w}{\partial \theta} = \frac{\partial w}{\partial r}\right]$$

$$\boxed{\frac{dw}{dz} = -\frac{i}{r}(\cos\theta - i\sin\theta)\frac{\partial w}{\partial \theta}}$$

**These are the two forms for** $\frac{dw}{dz}$.

## EXERCISE 7.3

**Determine which of the following functions are analytic:**

1. $x^2 + iy^2$ **Ans.** Analytic at all points $y = x$
2. $2xy + i(x^2 - y^2)$ **Ans.** Not analytic except $z = 0$
3. $\dfrac{x-iy}{x^2+y^2}$ **Ans.** Not analytic
4. $\dfrac{1}{(z-1)(z+1)}$ **Ans.** Analytic at all points, except $z = \pm 1$
5. $\dfrac{x-iy}{x-iy+a}$ **Ans.** Not analytic
6. $\sin x \cosh y + i \cos x \sinh y$ **Ans.** Yes, analytic
7. $xy + iy^2$ **Ans.** Yes, analytic at origin
8. Discuss the analyticity of the function $f(z) = z\bar{z} + \bar{z}^2$ in the complex plane, where $\bar{z}$ is the complex conjugate of $z$. Also find the points where it is differentiable but not analytic. **Ans.** Differentiable only at $z = 0$, No where analytic.
9. Show the function of $\bar{z}$ is not analytic any where.
10. If $f(z) = \begin{cases} \dfrac{x^2 y(y-ix)}{x^4+y^2}, & \text{when } z \neq 0 \\ 0, & \text{when } z = 0 \end{cases}$

    prove that $\dfrac{f(z)-f(0)}{z} \to 0$, as $z \to 0$, along any radius vector but not as $z \to 0$ in any manner. *(AMIETE, Dec. 2010)*
11. If $f(z)$ is an analytic function with constant modulus, show that $f(z)$ is constant. *(AMIETE, Dec. 2009)*

**Choose the correct answer :**

12. The Cauchy-Riemann equations for $f(z) = u(x, y) + iv(x, y)$ to be analytic are :

    (*a*) $\dfrac{\partial^2 u}{\partial x^2} + \dfrac{\partial^2 u}{\partial y^2} = 0,\ \dfrac{\partial^2 v}{\partial x^2} + \dfrac{\partial^2 v}{\partial y^2} = 0$ (*b*) $\dfrac{\partial u}{\partial x} = -\dfrac{\partial v}{\partial y},\ \dfrac{\partial u}{\partial y} = -\dfrac{\partial v}{\partial x}$

    (*c*) $\dfrac{\partial u}{\partial x} = \dfrac{\partial v}{\partial y},\ \dfrac{\partial u}{\partial y} = -\dfrac{\partial v}{\partial x}$ (*d*) $\dfrac{\partial u}{\partial x} = -\dfrac{\partial v}{\partial y},\ \dfrac{\partial u}{\partial y} = \dfrac{\partial v}{\partial x}$ **Ans.** (*c*)

    *(R.G.P.V., Bhopal, III Semester, Dec. 2006)*
13. Polar form of *C-R* equations are :

    (*a*) $\dfrac{\partial u}{\partial \theta} = \dfrac{1}{r}\dfrac{\partial v}{\partial r},\ \dfrac{\partial u}{\partial r} = r\dfrac{\partial v}{\partial \theta}$ (*b*) $\dfrac{\partial u}{\partial \theta} = r\dfrac{\partial v}{\partial r},\ \dfrac{\partial u}{\partial r} = \dfrac{1}{r}\dfrac{\partial v}{\partial \theta}$

    (*c*) $\dfrac{\partial u}{\partial r} = \dfrac{1}{r}\dfrac{\partial v}{\partial \theta},\ \dfrac{\partial u}{\partial \theta} = -r\dfrac{\partial v}{\partial r}$ (*d*) $\dfrac{\partial u}{\partial r} = r\dfrac{\partial v}{\partial \theta},\ \dfrac{\partial u}{\partial \theta} = -r\dfrac{\partial v}{\partial r}$ **Ans.** (*c*)

    *(R.G.P.V., Bhopal, III Semester, June, 2007)*
14. The curve $u(x, y) = C$ and $v(x, y) = C'$ are orthogonal if

    (*a*) $u$ and $v$ are complex functions (*b*) $u + iv$ is an analytic function.

    (*c*) $u - v$ is an analytic function. (*d*) $u + v$ is an analytic function

    **Ans.** (*b*) $\left[\dfrac{\partial u}{\partial \theta} = -r\dfrac{\partial v}{\partial r}\quad \dfrac{\partial v}{\partial \theta} = r\dfrac{\partial u}{\partial r}\right]$
15. If $f(z) = \dfrac{1}{2}\log_e(x^2+y^2) + i\tan^{-1}\left(\dfrac{\alpha x}{y}\right)$ be an analytic function then $\alpha$ is equal

    (*a*) $+1$ (*b*) $-1$ (*c*) $+2$ (*d*) $-2$ *(AMIETE, Dec. 2009)* **Ans.** (b)

## 7.14 ORTHOGONAL CURVES (U.P. III SEMESTER, JUNE 2009)

Two curves are said to be orthogonal to each other, when they intersect at right angle at each of their points of intersection.

The analytic function $f(z) = u + iv$ consists of two families of curves $u(x, y) = c_1$ and $v(x, y) = c_2$ which form an orthogonal system.

$$u(x, y) = c_1 \quad \text{...(1)}$$
$$v(x, y) = c_2 \quad \text{... (2)}$$

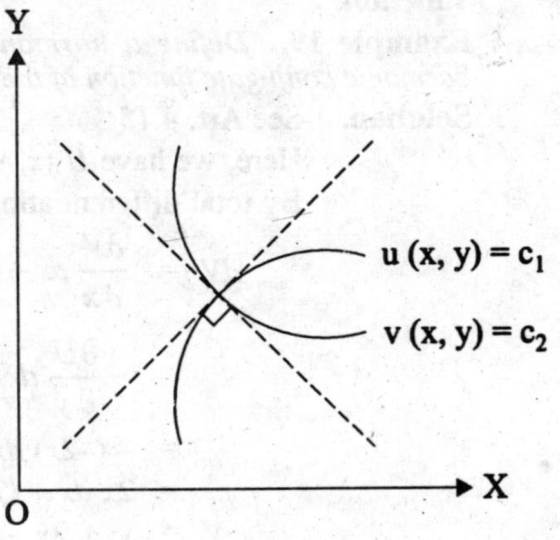


Differentiating (1), $\frac{\partial u}{\partial x}dx + \frac{\partial u}{\partial y}dy = 0$

$$\Rightarrow \quad \frac{dy}{dx} = -\frac{\frac{\partial u}{\partial x}}{\frac{\partial u}{\partial y}} = m_1 \text{ (say)}$$

Similarly from (2), $\frac{dy}{dx} = -\frac{\frac{\partial v}{\partial x}}{\frac{\partial v}{\partial y}} = m_2$ *(say)*

The product of two slopes

$$m_1 m_2 = \left(-\frac{\frac{\partial u}{\partial x}}{\frac{\partial u}{\partial y}}\right)\left(-\frac{\frac{\partial v}{\partial x}}{\frac{\partial v}{\partial y}}\right) = \left(-\frac{\frac{\partial u}{\partial x}}{\frac{\partial u}{\partial y}}\right)\left(-\frac{-\frac{\partial u}{\partial y}}{\frac{\partial u}{\partial x}}\right) \quad (C-R \text{ equations})$$
$$= -1$$

Since $m_1 m_2 = -1$, two curves $u = c_1$ and $v = c_2$ are orthogonal, and $c_1$, $c_2$ are parameters, $u = c_1$ and $v = c_2$ form an orthogonal system.

## 7.15 HARMONIC FUNCTION *(U.P., III Semester 2009-2010)*

Any function which satisfies the Laplace's equation is known as a harmonic function.

**Theorem.** If $f(z) = u + iv$ is an analytic function, then $u$ and $v$ are both harmonic functions.

**Proof.** Let $f(z) = u + iv$, be an analytic function, then we have

$$\left.\begin{aligned} \frac{\partial u}{\partial x} &= \frac{\partial v}{\partial y} \quad \text{...(1)} \\ \frac{\partial u}{\partial y} &= -\frac{\partial v}{\partial x} \quad \text{...(2)} \end{aligned}\right] \; C-R \text{ equations.}$$

Differentiating (1) with respect to $x$, we get $\frac{\partial^2 u}{\partial x^2} = \frac{\partial^2 v}{\partial x \partial y}$ ... (3)

Differentiating (2) w.r.t. '$y$' we have $\frac{\partial^2 u}{\partial y^2} = -\frac{\partial^2 v}{\partial y \partial x}$ ... (4)

Adding (3) and (4) we have $\frac{\partial^2 u}{\partial x^2} + \frac{\partial^2 u}{\partial y^2} = \frac{\partial^2 v}{\partial x \partial y} - \frac{\partial^2 v}{\partial y \partial x}$

$$\frac{\partial^2 u}{\partial x^2} + \frac{\partial^2 u}{\partial y^2} = 0 \qquad \left(\frac{\partial^2 v}{\partial x \partial y} = \frac{\partial^2 v}{\partial y \partial x}\right)$$

Similarly $$\frac{\partial^2 v}{\partial x^2} + \frac{\partial^2 v}{\partial y^2} = 0$$

Therefore both $u$ and $v$ are harmonic functions.

Such functions $u$, $v$ are called **Conjugate harmonic functions if $u + iv$ is also analytic function.**

**Example 19.** *Define a harmonic function and conjugate harmonic function. Find the harmonic conjugate function of the function $U(x, y) = 2x(1 - y)$. (U.P., III Semester Dec. 2009)*

**Solution.** See Art. 4.15

Here, we have $U(x, y) = 2x(1 - y)$. Let V be the harmonic conjugate of U.

By total differentiation

$$\left[\begin{array}{l} U = 2x - 2xy \\ \dfrac{\partial U}{\partial x} = 2 - 2y \\ \dfrac{\partial U}{\partial y} = -2x \end{array}\right]$$

$$\begin{aligned} dV &= \frac{\partial V}{\partial x}dx + \frac{\partial V}{\partial y}dy \\ &= -\frac{\partial U}{\partial y}dx + \frac{\partial U}{\partial x}dy \\ &= -(-2x)\,dx + (2 - 2y)\,dy + C \\ &= 2x\,dx + (2\,dy - 2y\,dy) + C \\ V &= x^2 + 2y - y^2 + C \end{aligned}$$

Hence, the harmonic conjugate of $U$ is $x^2 + 2y - y^2 + C$ **Ans.**

**Example 20.** *Prove that $u = x^2 - y^2$ and $v = \dfrac{y}{x^2 + y^2}$ are harmonic functions of $(x, y)$, but are not harmonic conjugates.*

**Solution.** We have, $u = x^2 - y^2$

$$\frac{\partial u}{\partial x} = 2x, \quad \frac{\partial^2 u}{\partial x^2} = 2, \quad \frac{\partial u}{\partial y} = -2y, \quad \frac{\partial^2 u}{\partial y^2} = -2$$

$$\frac{\partial^2 u}{\partial x^2} + \frac{\partial^2 u}{\partial y^2} = 2 - 2 = 0$$

$u(x, y)$ satisfies Laplace equation, hence $u(x, y)$ is harmonic

$$v = \frac{y}{x^2 + y^2}, \quad \frac{\partial v}{\partial x} = -\frac{2xy}{(x^2 + y^2)^2}$$

$$\frac{\partial^2 v}{\partial x^2} = \frac{(x^2 + y^2)^2(-2y) - (-2xy)\,2(x^2 + y^2)\,2x}{(x^2 + y^2)^4}$$

$$= \frac{(x^2 + y^2)(-2y) - (-2xy)4x}{(x^2 + y^2)^3} = \frac{6x^2y - 2y^3}{(x^2 + y^2)^3}$$

$$\frac{\partial v}{\partial y} = \frac{(x^2 + y^2)\cdot 1 - y\,(2y)}{(x^2 + y^2)^2} = \frac{x^2 - y^2}{(x^2 + y^2)^2} \qquad \ldots (1)$$

$$\frac{\partial^2 v}{\partial y^2} = \frac{(x^2 + y^2)^2(-2y) - (x^2 - y^2)\,2(x^2 + y^2)(2y)}{(x^2 + y^2)^4} = \frac{(x^2 + y^2)(-2y) - (x^2 - y^2)(4y)}{(x^2 + y^2)^3}$$

$$= \frac{-2x^2y - 2y^3 - 4x^2y + 4y^3}{(x^2 + y^2)^3} = \frac{-6x^2y + 2y^3}{(x^2 + y^2)^3} \qquad \ldots (2)$$

On adding (1) and (2), we get $\dfrac{\partial^2 v}{\partial x^2} + \dfrac{\partial^2 v}{\partial y^2} = 0$

$v(x, y)$ also satisfies Laplace equations, hence $v(x, y)$ is also harmonic function.

But $\quad \dfrac{\partial u}{\partial x} \neq \dfrac{\partial v}{\partial y} \quad$ and $\quad \dfrac{\partial u}{\partial y} \neq -\dfrac{\partial v}{\partial x}$

Therefore $u$ and $v$ are not harmonic conjugates. **Proved.**

**Example 21.** *If u(x, y) and v(x, y) are harmonic functions in a region R, prove that the function*

$$\left[\left(\frac{\partial u}{\partial y}-\frac{\partial v}{\partial x}\right)+i\left(\frac{\partial u}{\partial x}+\frac{\partial v}{\partial y}\right)\right]$$

*is an analytic function of $z = x + iy$.* **(*R.G.P.V., Bhopal, III Semester, Dec. 2004*)**

**Solution.** Since $u(x, y)$ and $v(x, y)$ are harmonic functions in a region $R$, therefore

$$\frac{\partial^2 u}{\partial x^2}+\frac{\partial^2 u}{\partial y^2} = 0 \quad \ldots (1) \qquad \text{and} \qquad \frac{\partial^2 v}{\partial x^2}+\frac{\partial^2 v}{\partial y^2} = 0 \quad \ldots (2)$$

Let $$F(z) = R+iS = \left(\frac{\partial u}{\partial y}-\frac{\partial v}{\partial x}\right)+i\left(\frac{\partial u}{\partial x}+\frac{\partial v}{\partial y}\right)$$

Equating real and imaginary parts, we get

$$R = \frac{\partial u}{\partial y}-\frac{\partial v}{\partial x},$$

$$\frac{\partial R}{\partial x} = \frac{\partial^2 u}{\partial x\,\partial y}-\frac{\partial^2 v}{\partial x^2} \quad \ldots (3) \qquad \frac{\partial R}{\partial y} = \frac{\partial^2 u}{\partial y^2}-\frac{\partial^2 v}{\partial x\,\partial y} \quad \ldots (4)$$

$$S = \frac{\partial u}{\partial x}+\frac{\partial v}{\partial y}$$

$$\frac{\partial S}{\partial x} = \frac{\partial^2 u}{\partial x^2}+\frac{\partial^2 v}{\partial x\,\partial y} \quad \ldots (5) \qquad \frac{\partial S}{\partial y} = \frac{\partial^2 u}{\partial x\,\partial y}+\frac{\partial^2 v}{\partial y^2} \quad \ldots (6)$$

Putting the value of $\frac{\partial^2 u}{\partial x^2}$ from (1) in (5), we get

$$\frac{\partial S}{\partial x} = -\frac{\partial^2 u}{\partial y^2}+\frac{\partial^2 v}{\partial x\,\partial y} \quad \ldots (7)$$

Putting the value of $\frac{\partial^2 v}{\partial y^2}$ from (2) in (6), we get

$$\frac{\partial S}{\partial y} = \frac{\partial^2 u}{\partial x\,\partial y}-\frac{\partial^2 v}{\partial x^2} \quad \ldots (8)$$

From (3) and (8), $$\frac{\partial R}{\partial x} = \frac{\partial S}{\partial y}$$

From (4) and (7), $$\frac{\partial R}{\partial y} = -\frac{\partial S}{\partial x}$$

Therefore, C-R equations are satisfied and hence the given function is analytic. **Proved.**

## 7.16 APPLICATION TO FLOW PROBLEMS

Consider two dimensional irrotational motion in a plane parallel to $xy$-plane.

The velocity $v$ of fluid can be expressed as

$$v = v_x\hat{i}+v_y\hat{j} \quad \ldots (1)$$

Since the motion is irrotational, a scalar function $\phi((x, y)$ gives the velocity components.

$$V = \nabla\phi(x, y) = \left(\hat{i}\frac{\partial}{\partial x}+\hat{j}\frac{\partial}{\partial y}\right)\phi = \hat{i}\frac{\partial\phi}{\partial x}+\hat{j}\frac{\partial\phi}{\partial y} \quad \ldots (2)$$

On comparing (1) and (2), we get

$$v_x = \frac{\partial \phi}{\partial x} \quad \text{and} \quad v_y = \frac{\partial \phi}{\partial y} \qquad \text{... (3)}$$

## 7.17 VELOCITY POTENTIAL FUNCTION

The scalar function $\phi(x, y)$ which gives the velocity component is called the velocity potential function.

As the fluid is incompressible

$$div\ v = 0$$

$$\nabla v = 0 \Rightarrow \left(\hat{i}\frac{\partial}{\partial x} + \hat{j}\frac{\partial}{\partial y}\right)\cdot(\hat{i}\, v_x + \hat{j}\, v_y) = 0$$

$$\frac{\partial v_x}{\partial x} + \frac{\partial v_y}{\partial y} = 0 \qquad \text{... (4)}$$

Putting the values of $v_x$ and $v_y$ from (3) in (4), we get

$$\frac{\partial}{\partial x}\left(\frac{\partial \phi}{\partial x}\right) + \frac{\partial}{\partial y}\left(\frac{\partial \phi}{\partial y}\right) = 0 \quad \Rightarrow \quad \frac{\partial^2 \phi}{\partial x^2} + \frac{\partial^2 \phi}{\partial y^2} = 0$$

This is Laplace equation. The function $\phi$ is harmonic and is a real part of analytic function

$$f(z) = \phi(x, y) + i\,\psi(x, y)$$

We know that

$$\frac{dy}{dx} = -\frac{\frac{\partial \psi}{\partial x}}{\frac{\partial \psi}{\partial y}} = \frac{\frac{\partial \phi}{\partial y}}{\frac{\partial \phi}{\partial x}} \qquad f(z) = \phi + i\psi \quad [CR\text{-equations}]$$

$$= \frac{v_y}{v_x} \qquad \text{[Using (3)]}$$

Here the resultant velocity $\sqrt{v_x^2 + v_y^2}$ of the fluid is along the tangent to the curve

$$\psi(x, y) = C'$$

Such curves are known as *stream lines* and $\psi(x, y)$ is known as *stream function.*

The curves represented by $\phi(x, y) = c$ are called *equipotential lines.*

As $\phi(x, y)$ and $\psi(x, y)$ are conjugates of analytic function $f(z)$. The equipotential lines $\phi(x, y) = C$ and the stream potential line $\psi(x, y) = C'$ intersect each other orthogonally.

$$f'(z) = \frac{\partial \phi}{\partial x} + i\frac{\partial \psi}{\partial x} = \frac{\partial \phi}{\partial x} - i\frac{\partial \psi}{\partial y} = v_x - iv_y$$

The magnitude of the resultant velocity $= \left|\frac{df}{dz}\right| = \sqrt{v_x^2 + v_y^2}$

The function $f(z)$ which represents the flow pattern is called the *complex potential.*

## 7.18 METHOD TO FIND THE CONJUGATE FUNCTION

**Case I. Given.** If $f(z) = u + iv$, and $u$ is known.

**To find.** $v$, conjugate function.

**Method.** We know that $\quad dv = \frac{\partial v}{\partial x}\cdot dx + \frac{\partial v}{\partial y}\cdot dy \qquad \text{... (1)}$

Replacing $\frac{\partial v}{\partial x}$ by $-\frac{\partial u}{\partial y}$ and $\frac{\partial v}{\partial y}$ by $\frac{\partial u}{\partial x}$ in (1), we get [C-R equations]

$$dv = -\frac{\partial u}{\partial y}\cdot dx + \frac{\partial u}{\partial x}\cdot dy$$

$$v = -\int \frac{\partial u}{\partial y} dx + \int \frac{\partial u}{\partial x}\cdot dy$$

$\Rightarrow$ $$v = \int M\,dx + \int N\,dy \qquad \text{... (2)}$$

where $M = -\frac{\partial u}{\partial y}$ and $N = \frac{\partial u}{\partial x}$

so that $$\frac{\partial M}{\partial y} = \frac{\partial}{\partial y}\left(-\frac{\partial u}{\partial y}\right) = -\frac{\partial^2 u}{\partial y^2} \text{ and } \frac{\partial N}{\partial x} = \frac{\partial}{\partial x}\left(\frac{\partial u}{\partial x}\right) = \frac{\partial^2 u}{\partial x^2}$$

since $u$ is a conjugate function, so $\frac{\partial^2 u}{\partial x^2} + \frac{\partial^2 u}{\partial y^2} = 0$

$\Rightarrow$ $$-\frac{\partial^2 u}{\partial y^2} = \frac{\partial^2 u}{\partial x^2} \quad \Rightarrow \quad \frac{\partial M}{\partial y} = \frac{\partial N}{\partial x} \qquad \text{... (3)}$$

Equation (3) satisfies the condition of an exact differential equation.
So equation (2) can be integrated and thus $v$ is determined.

**Case II.** Similarly, if $v = v(x, y)$ is given

To find out $u$.

We know that $$du = \frac{\partial u}{\partial x} dx + i\frac{\partial u}{\partial y} dy \qquad \text{... (4)}$$

On substituting the values of $\frac{\partial u}{\partial x}$ and $\frac{\partial u}{\partial y}$ in (4), we get

$$du = \frac{\partial v}{\partial y} dx - \frac{\partial v}{\partial x} dy$$

On integrating, we get

$$u = \int \frac{\partial v}{\partial y} dx - \int \frac{\partial v}{\partial x} dy \qquad \text{...(5)}$$

(since v is already known so $\frac{\partial v}{\partial y}, \frac{\partial v}{\partial x}$ on R.H.S. are also known)

Equation (5) is an exact differential equation. On solving (5), $u$ can be determined.
Consequently $f(z) = u + iv$ can also be determined.

**Example 22.** *Prove that $u = x^2 - y^2 - 2xy - 2x + 3y$ is harmonic. Find a function $v$ such that $f(z) = u + iv$ is analytic. Also express $f(z)$ in terms of $z$.*

*(R.G.P.V., Bhopal, III Semester, June 2005)*

**Solution.** We have,

$$u = x^2 - y^2 - 2xy - 2x + 3y$$

$$\frac{\partial u}{\partial x} = 2x - 2y - 2 \quad \Rightarrow \quad \frac{\partial^2 u}{\partial x^2} = 2$$

$$\frac{\partial u}{\partial y} = -2y - 2x + 3 \quad \Rightarrow \quad \frac{\partial^2 u}{\partial y^2} = -2$$

$$\frac{\partial^2 u}{\partial x^2} + \frac{\partial^2 u}{\partial y^2} = 2 - 2 = 0$$

Since Laplace equation is satisfied, therefore $u$ is harmonic. **Proved.**

We know that $dv = \frac{\partial v}{\partial x}dx + \frac{\partial v}{\partial y}dy$

$\Rightarrow \quad dv = -\frac{\partial u}{\partial y}dx + \frac{\partial u}{\partial x}dy \quad ...(1) \quad \left[\because \frac{\partial v}{\partial x} = -\frac{\partial u}{\partial y} \text{ and } \frac{\partial v}{\partial y} = \frac{\partial u}{\partial x}\right]$

Putting the values of $\frac{\partial u}{\partial y}$ and $\frac{\partial u}{\partial x}$ in (1), we get

$$dv = -(-2y - 2x + 3)\,dx + (2x - 2y - 2)dy$$

$\Rightarrow \quad v = \int(2y + 2x - 3)dx + \int(-2y - 2)dy + C$ (Ignoring $2x$)

Hence, $v = 2xy + x^2 - 3x - y^2 - 2y + C$ **Ans.**

Now,
$$\begin{aligned} f(z) &= u + iv \\ &= (x^2 - y^2 - 2xy - 2x + 3y) + i\,(2xy + x^2 - 3x - y^2 - 2y) + iC \\ &= (x^2 - y^2 + 2ixy) + (ix^2 - iy^2 - 2xy) - (2 + 3i)\,x - i\,(2 + 3i)\,y + iC \\ &= (x^2 - y^2 + 2ixy) + i\,(x^2 - y^2 + 2ixy) - (2 + 3i)\,x - i\,(2 + 3i)y + iC \\ &= (x + iy)^2 + i\,(x + iy)^2 - (2 + 3i)\,(x + iy) + iC \\ &= z^2 + iz^2 - (2 + 3i)\,z + iC \\ &= (1 + i)\,z^2 - (2 + 3i)\,z + iC \end{aligned}$$

Which is the required expression of $f(z)$ in terms of $z$. **Ans.**

**Example 23.** *If $w = \phi + i\psi$ represents the complex potential for an electric field and*

$$\psi = x^2 - y^2 + \frac{x}{x^2 + y^2},$$

*determine the function $\phi$.*

**Solution.** $w = \phi + i\psi$ and $\psi = x^2 - y^2 + \frac{x}{x^2 + y^2}$

$$\frac{\partial \psi}{\partial x} = 2x + \frac{(x^2 + y^2)\cdot 1 - x\cdot 2x}{(x^2 + y^2)^2} = 2x + \frac{y^2 - x^2}{(x^2 + y^2)^2}$$

$$\frac{\partial \psi}{\partial y} = -2y - \frac{x(2y)}{(x^2 + y^2)^2} = -2y - \frac{2xy}{(x^2 + y^2)^2}$$

We know that, $d\phi = \frac{\partial \phi}{\partial x}dx + \frac{\partial \phi}{\partial y}dy = \frac{\partial \psi}{\partial y}dx - \frac{\partial \psi}{\partial x}dy$

$$= \left(-2y - \frac{2xy}{(x^2 + y^2)^2}\right)dx - \left(2x + \frac{y^2 - x^2}{(x^2 + y^2)^2}\right)dy$$

$$\phi = \int\left[-2y - \frac{2xy}{(x^2 + y^2)^2}\right]dx + c$$

This is an exact differential equation.

$$\phi = -2xy + \frac{y}{x^2 + y^2} + C$$ **Ans.**

Which is the required function.

**Example 24.** *An electrostatic field in the xy-plane is given by the potential function $\phi = 3x^2y - y^3$, find the stream function.* (*R.G.P.V., Bhopal, III Semester, Dec. 2001*)

**Solution.** Let $\psi(x, y)$ be a stream function

We know that $d\psi = \frac{\partial \psi}{\partial x}dx + \frac{\partial \psi}{\partial y}dy = \left(-\frac{\partial \phi}{\partial y}\right)dx + \left(\frac{\partial \phi}{\partial x}\right)dy$ [C-R equations]

$$= \{-(3x^2 - 3y^2)\}\, dx + 6xy\, dy$$
$$= -3x^2\, dx + (3y^2\, dx + 6xy\, dy)$$
$$= -d(x^3) + 3d(xy^2)$$
$$\psi = \int -d(x^3) + 3d(xy^2) + c$$
$$\psi = -x^3 + 3xy^2 + c$$

$\psi$ is the required stream function. **Ans.**

**Example 25.** *Find the imaginary part of the analytic function whose real part is* $x^3 - 3xy^2 + 3x^2 - 3y^2$. *(R.G.P.V., Bhopal, III Semester, Dec. 2008, 2005)*

**Solution.** Let $u(x, y) = x^3 - 3xy^2 + 3x^2 - 3y^2$

$$\frac{\partial u}{\partial x} = 3x^2 - 3y^2 + 6x$$

$$\frac{\partial u}{\partial y} = -6xy - 6y$$

We know that

$$dv = \frac{\partial v}{\partial x} dx + \frac{\partial v}{\partial y} dy \quad \Rightarrow \quad dv = -\frac{\partial u}{\partial y} dx + \frac{\partial u}{\partial x} dy$$

$$\Rightarrow \quad dv = (6xy + 6y)\, dx + (3x^2 - 3y^2 + 6x)\, dy$$

This is an exact differential equation.

$$v = \int (6xy + 6y)\, dx + \int -3y^2 dy + C$$
$$= 3x^2 y + 6xy - y^3 + C$$

Which is the required imaginary part. **Ans.**

**Example 26.** *If* $u - v = (x - y)(x^2 + 4xy + y^2)$ *and* $f(z) = u + iv$ *is an analytic function of* $z = x + iy$, *find* $f(z)$ *in terms of* $z$.

**Solution.** $u + iv = f(z) \quad \Rightarrow \quad iu - v = if(z)$

Adding these, $(u - v) + i(u + v) = (1 + i) f(z)$

Let $U + iV = (1 + i) f(z)$ where $U = u - v$ and $V = u + v$

$$F(z) = (1 + i) f(z)$$
$$U = u - v = (x - y)(x^2 + 4xy + y^2)$$
$$= x^3 + 3x^2y - 3xy^2 - y^3$$
$$\frac{\partial U}{\partial x} = 3x^2 + 6xy - 3y^2$$
$$\frac{\partial U}{\partial y} = 3x^2 - 6xy - 3y^2$$

We know that $dV = \dfrac{\partial V}{\partial x} \cdot dx + \dfrac{\partial V}{\partial y} dy = -\dfrac{\partial U}{\partial y} \cdot dx + \dfrac{\partial U}{\partial x} \cdot dy$ [C-R equations]

On putting the values of $\dfrac{\partial U}{\partial x}$ and $\dfrac{\partial U}{\partial y}$, we get

$$= (-3x^2 + 6xy + 3y^2)\, dx + (3x^2 + 6xy - 3y^2) \cdot dy$$

Integrating, we get

$$V = \underset{(y \text{ as constant})}{\int (-3x^2 + 6xy + 3y^2) dx} + \underset{(\text{Ignoring terms of } x)}{\int (-3y^2) dy}$$

$$= -x^3 + 3x^2y + 3xy^2 - y^3 + c$$

$$F(z) = U + iV$$
$$= (x^3 + 3x^2y - 3xy^2 - y^3) + i(-x^3 + 3x^2y + 3xy^2 - y^3) + ic$$
$$= (1 - i)x^3 + (1 + i)3x^2y - (1 - i)3xy^2 - (1 + i)y^3 + ic$$
$$= (1 - i)x^3 + i(1 - i)3x^2y - (1 - i)3xy^2 - i(1 - i)y^3 + ic$$

$$= (1 - i)\,[x^3 + 3\,ix^2y - 3\,xy^2 - iy^3] + ic$$
$$= (1 - i)\,(x + iy)^3 + iC = (1 - i)\,z^3 + ic$$
$$(1 + i)\,f(z) = (1 - i)\,z^3 + ic, \qquad [F(z) = (1 + i)\,f(z)]$$
$$f(z) = \frac{1-i}{1+i}z^3 + \frac{ic}{1+i} = -\frac{i(1+i)}{(1+i)}z^3 + \frac{i(1-i)}{(1+i)(1-i)}c = -iz^3 + \frac{1+i}{2}c \qquad \textbf{Ans.}$$

**Example 27.** *If $f(z) = u + iv$ is an analytic function of $z = x + iy$ and*
*$u - v = e^{-x}[(x - y)\sin y - (x + y)\cos y]$*
*find $f(z)$.* (*U.P. III Semester, 2009-2010*)

**Solution.** We know that,

$$f(z) = u + iv \qquad \text{... (1)}$$
$$i\,f(z) = i\,u - v \qquad \text{... (2)}$$
$$F(z) = U + iV$$
$$U = u - v = e^{-x}[(x - y)\sin y - (x + y)\cos y]$$
$$\frac{\partial U}{\partial x} = -e^{-x}[(x - y)\sin y - (x + y)\cos y] + e^{-x}[\sin y - \cos y]$$
$$\frac{\partial U}{\partial y} = e^{-x}[(x - y)\cos y - \sin y - (x + y)(-\sin y) - \cos y]$$

We know that,

$$dV = \frac{\partial V}{\partial x}dx + \frac{\partial V}{\partial y}dy = -\frac{\partial U}{\partial y}dx + \frac{\partial U}{\partial x}dy \qquad \text{[C – R equations]}$$
$$= -e^{-x}[(x - y)\cos y - \sin y + (x + y)\sin y - \cos y]\,dx$$
$$-e^{-x}[(x - y)\sin y - (x + y)\cos y - \sin y + \cos y]\,dy$$
$$= -e^{-x}x\,\{(\cos y + \sin y)\,dx - e^{-x}(-y\cos y - \sin y + y\sin y - \cos y)\,dx$$
$$-e^{-x}[(x - y)\sin y - (x + y)\cos y - \sin y + \cos y]\,dy$$
$$V = (\cos y + \sin y)(x\,e^{-x} + e^{-x}) + e^{-x}(-y\cos y - \sin y + y\sin y - \cos y) + C$$
$$F(z) = U + iV$$
$$F(z) = e^{-x}[(x - y)\sin y - (x + y)\cos y] + i\,e^{-x}[x\cos y + \cos y + x\sin y + \sin y$$
$$-y\cos y - \sin y + y\sin y - \cos y] + iC$$
$$= e^{-x}[\{x\sin y - y\sin y - x\cos y - y\cos y\} + i\{x\cos y + x\sin y - y\cos y + y\sin y\}] + iC$$
$$= e^{-x}[(x + iy)\sin y - (x + iy)\cos y + (-y + ix)\sin y + (-y + ix)\cos y] + iC$$
$$= e^{-x}[(x + iy)\sin y - (x + iy)\cos y + i(x + iy)\sin y + i(x + iy)\cos y] + iC$$
$$= e^{-x}(x + iy)[\sin y - \cos y + i\sin y + i\cos y] + iC$$
$$= e^{-x}(x + iy)[(1 + i)\sin y + i(1 + i)\cos y] + iC$$
$$(1 + i)\,f(z) = e^{-x}(x + iy)(1 + i)(\sin y + i\cos y) + iC$$
$$f(z) = e^{-x}(x + iy)(\sin y + i\cos y) + \frac{iC}{1+i}$$
$$= i\,z\,e^{-x}(\cos y - i\sin y) + \frac{iC}{1+i}$$
$$= i\,z\,e^{-x}e^{-iy} = i\,z\,e^{-(x+iy)} = i\,z\,e^{-z} + \frac{iC}{1+i} \qquad \textbf{Ans.}$$

Let $\phi_1(x, y) = -e^{-x}[(x - y)\sin y - (x + y)\cos y] + e^{-x}[\sin y - \cos y]$
$$\phi_1(z, 0) = -e^{-z}[z\sin 0 - z\cos 0] + e^{-z}[\sin 0 + \cos 0]$$
$$= -e^{-z}[z - 1]$$
Let $\phi_2(x, y) = e^{-x}[(x - y)\cos y - \sin y + (x + y)\sin y - \cos y]$
$$\phi_2(z, 0) = e^{-z}[(z)\cos 0 - \sin 0 + z\sin 0 - \cos 0]$$

$$= e^{-z}[z-1]$$

$$F(z) = U + iV$$

$$F'(z) = \frac{\partial U}{\partial x} + i\frac{\partial V}{\partial x} = \frac{\partial U}{\partial x} - i\frac{\partial U}{\partial y} = f_1(z, 0) - i f_2(z, 0)$$

$$= e^{-z}(z-1) - i e^{-z}(z-1) = (1-i)e^{-z}(z-1) = (1-i)e^{-z}(z-1)$$

$$F(z) = (1-i)\left[z\frac{e^{-z}}{-1} - \int \frac{e^{-z}}{-1}dz\right] + C = (1-i)[-z e^{-z} - e^{-z}] + C$$

$$(1+i)f(z) = (-1+i)(z+1)e^{-z} + C$$

$$f(z) = \frac{(-1+i)}{1+i}(z+1)e^{-z} + C = \frac{(-1+i)(1-i)}{(1+i)(1-i)}(z+1)e^{-z} + C$$

$$= i(z+1)e^{-z} + C$$ **Ans.**

**Example 28.** *Let $f(z) = u(r, \theta) + iv(r, \theta)$ be an analytic function and $u = -r^3 \sin 3\theta$, then construct the corresponding analytic function $f(z)$ in terms of $z$.*

**Solution.** $u = -r^3 \sin 3\theta$

$$\frac{\partial u}{\partial r} = -3r^2 \sin 3\theta, \qquad \frac{\partial u}{\partial \theta} = -3r^3 \cos 3\theta$$

We know that $$dv = \frac{\partial v}{\partial r}dr + \frac{\partial v}{\partial \theta}d\theta$$

$$= \left(-\frac{1}{r}\frac{\partial u}{\partial \theta}\right)dr + \left(r\frac{\partial u}{\partial r}\right)d\theta \qquad \left(\begin{array}{c} C-R \text{ equations} \\ \frac{\partial u}{\partial r} = \frac{1}{r}\frac{\partial v}{\partial \theta} \\ \frac{\partial u}{\partial \theta} = -r\frac{\partial v}{\partial r} \end{array}\right)$$

$$= -\frac{1}{r}(-3r^3 \cos 3\theta)\,dr + r(-3r^2 \sin 3\theta)\,d\theta$$

$$= 3r^2 \cos 3\theta \cdot dr - 3r^3 \sin 3\theta\, d\theta$$

$$v = \int (3r^2 \cos 3\theta)dr - c = r^3 \cos 3\theta + c$$

$$f(z) = u + iv = -r^3 \sin 3\theta + ir^3 \cos 3\theta + ic = ir^3(\cos 3\theta + i\sin 3\theta) + ic$$

$$= ir^3 e^{i3\theta} + ic = i(re^{i\theta})^3 + ic = iz^3 + ic$$ **Ans.**

This is the required analytic function.

**Example 29.** *Find analytic function $f(z) = u(r, \theta) + iv(r, \theta)$ such that* $v(r, \theta) = r^2 \cos 2\theta - r\cos\theta + 2.$ *(A.M.I.E.T.E., June 2007)*

**Solution.** We have, $$v = r^2 \cos 2\theta - r\cos\theta + 2 \quad \text{... (1)}$$

Differentiating (1), we get

$$\frac{\partial v}{\partial \theta} = -2r^2 \sin 2\theta + r\sin\theta \quad \text{... (2)}$$

$$\frac{\partial v}{\partial r} = 2r\cos 2\theta - \cos\theta \quad \text{... (3)}$$

Using $C-R$ equations in polar coordinates, we get

$$r\frac{\partial u}{\partial r} = \frac{\partial v}{\partial \theta} = -2r^2 \sin 2\theta + r\sin\theta \quad \text{[From (2)]}$$

$$\Rightarrow \quad \frac{\partial u}{\partial r} = -2r\sin 2\theta + \sin\theta \quad \text{... (4)}$$

$$-\frac{1}{r}\frac{\partial u}{\partial \theta}=\frac{\partial v}{\partial r}=2r\cos 2\theta-\cos\theta \qquad \text{[From (3)]}$$

$$\Rightarrow \qquad \frac{\partial u}{\partial \theta}=-2r^2\cos 2\theta+r\cos\theta \qquad \text{... (5)}$$

By total differentiation formula

$$du=\frac{\partial u}{\partial r}dr+\frac{\partial u}{\partial \theta}d\theta=(-2r\sin 2\theta+\sin\theta)dr+(-2r^2\cos 2\theta+r\cos\theta)d\theta$$

$$=-[(2r\,dr)\sin 2\theta+r^2(2\cos 2\theta\,d\theta)]+[\sin\theta\cdot dr+r(\cos\theta\,d\theta)]$$

$$=-[(2r\,dr)\sin 2\theta-\sin\theta dr]+[-r^2 2\cos 2\theta\,d\theta+r\cos\theta\,d\theta]$$

$$=-d(r^2\sin 2\theta)+d(r\sin\theta) \qquad \text{(Exact differential equation)}$$

Integrating, we get

$$u=-r^2\sin 2\theta+r\sin\theta+c$$

Hence, $\quad f(z)=u+iv$

$$=(-r^2\sin 2\theta+r\sin\theta+c)+i(r^2\cos 2\theta-r\cos\theta+2)$$

$$=ir^2(\cos 2\theta+i\sin 2\theta)-ir(\cos\theta+i\sin\theta)+2i+c$$

$$=ir^2e^{2i\theta}-ir\,e^{i\theta}+2i+c=i(r^2e^{2i\theta}-r\,e^{i\theta})+2i+c. \qquad \textbf{Ans.}$$

This is the required analytic function.

**Example 30.** *Deduce the following with the polar form of Cauchy-Riemann equations :*

(*a*) $\frac{\partial^2 u}{\partial r^2}+\frac{1}{r}\frac{\partial u}{\partial r}+\frac{1}{r^2}\frac{\partial^2 u}{\partial \theta^2}=0$ (*MDU, Dec. 2010, K.U. 2009*) (*b*) $f'(z)=\frac{r}{z}\left[\frac{\partial u}{\partial r}+i\frac{\partial v}{\partial r}\right]$

**Solution.** We know that polar form of C-R equations are

$$\frac{\partial u}{\partial r}=\frac{1}{r}\frac{\partial v}{\partial \theta} \qquad \text{... (1)}$$

$$\frac{\partial u}{\partial \theta}=-r\frac{\partial v}{\partial r} \qquad \text{... (2)}$$

(*a*) Differentiating (1) partially w.r.t. r., we get

$$\frac{\partial^2 u}{\partial r^2}=-\frac{1}{r^2}\frac{\partial v}{\partial \theta}+\frac{1}{r}\frac{\partial^2 v}{\partial r\partial \theta} \qquad \text{... (3)}$$

Differentiating (2) partially w.r.t. θ, we have

$$\frac{\partial^2 u}{\partial \theta^2}=-r\frac{\partial^2 v}{\partial \theta\partial r} \qquad \text{...(4)}$$

Thus using (1), (3) and (4), we get

$$\frac{\partial^2 u}{\partial r^2}+\frac{1}{r}\frac{\partial u}{\partial r}+\frac{1}{r^2}\frac{\partial^2 u}{\partial \theta^2}=-\frac{1}{r^2}\frac{\partial v}{\partial \theta}+\frac{1}{r}\frac{\partial^2 v}{\partial r\partial \theta}+\frac{1}{r}\left(\frac{1}{r}\frac{\partial v}{\partial \theta}\right)+\frac{1}{r^2}\left(-r\frac{\partial^2 v}{\partial \theta\partial r}\right)=0 \qquad \left[\frac{\partial^2 v}{\partial r\partial \theta}=\frac{\partial^2 v}{\partial \theta\partial r}\right]$$

**Proved.**

(*b*) Now, $\quad r\left(\frac{\partial u}{\partial r}+i\frac{\partial v}{\partial r}\right)=r\left[\left(\frac{\partial u}{\partial x}\frac{\partial x}{\partial r}+\frac{\partial u}{\partial y}\frac{\partial y}{\partial r}\right)+i\left(\frac{\partial v}{\partial x}\frac{\partial x}{\partial r}+\frac{\partial v}{\partial y}\frac{\partial y}{\partial r}\right)\right]$

$$=r\left[\left(\frac{\partial u}{\partial x}\cos\theta+\frac{\partial u}{\partial y}\sin\theta\right)+i\left(\frac{\partial v}{\partial x}\cos\theta+\frac{\partial v}{\partial y}\sin\theta\right)\right]$$

$$= r\cos\theta\left(\frac{\partial u}{\partial x}+i\frac{\partial v}{\partial x}\right)+r\sin\theta\left(\frac{\partial u}{\partial y}+i\frac{\partial v}{\partial y}\right)$$

$$= x\left(\frac{\partial u}{\partial x}+i\frac{\partial v}{\partial x}\right)+iy\left(\frac{\partial v}{\partial y}-i\frac{\partial u}{\partial y}\right) \quad \text{(By C-R equations)}$$

$$= x f'(z) + iy f'(z) = (x+iy) f'(z) = z f'(z).$$

$$\therefore \quad f'(z)=\frac{r}{z}\left(\frac{\partial u}{\partial r}+i\frac{\partial v}{\partial r}\right) \quad \textbf{Proved.}$$

## 7.19 MILNE THOMSON METHOD (TO CONSTRUCT AN ANALYTIC FUNCTION)

By this method $f(z)$ is directly constructed without finding $v$ and the method is given below:

Since $z = x + iy$ and $\bar{z} = x - iy$

$$\therefore \quad x=\frac{z+\bar{z}}{2},\ y=\frac{z-\bar{z}}{2i}$$

$$f(z)\equiv u(x,y)+iv(x,y) \quad \text{... (1)}$$

$$f(z)\equiv u\left(\frac{z+\bar{z}}{2},\frac{z-\bar{z}}{2i}\right)+iv\left(\frac{z+\bar{z}}{2},\frac{z-\bar{z}}{2i}\right)$$

This relation can be regarded as a formal identity in two independent variables $z$ and $\bar{z}$. Replacing $\bar{z}$ by $z$, we get

$$f(z)\equiv u(z,0)+iv(z,0)$$

Which can be obtained by replacing $x$ by $z$ and $y$ by 0 in (1)

**Case I. If $u$ is given**

We have $f(z) = u + iv$

$$\therefore \quad f'(z)=\frac{\partial u}{\partial x}+i\frac{\partial v}{\partial x}, \qquad f'(z)=\frac{\partial u}{\partial x}-i\frac{\partial u}{\partial y} \quad (C-R \text{ equations})$$

If we write $\frac{\partial u}{\partial x}=\phi_1(x,y)$, $\frac{\partial u}{\partial y}=\phi_2(x,y)$

$$f'(z)=\phi_1(x,y)-i\phi_2(x,y) \text{ or } f'(z)=\phi_1(z,0)-i\phi_2(z,0)$$

On integrating $f(z)=\int\phi_1(z,0)\,dz-i\int\phi_2(z,0)dz+c$

**Case II. If $v$ is given**

$$f(z) = u + iv$$

$$f'(z) = \frac{\partial u}{\partial x}+i\frac{\partial v}{\partial x}=\frac{\partial v}{\partial y}+i\frac{\partial v}{\partial x} = \psi_1(x,y)+i\,\psi_2(x,y)$$

when $\psi_1(x,y)=\frac{\partial v}{\partial y}$, $\psi_2(x,y)=\frac{\partial v}{\partial x}$.

$$f(z)=\int\psi_1(z,0)\,dz+i\int\psi_2(z,0)\,dz+c$$

## 7.20 WORKING RULE: TO CONSTRUCT AN ANALYTIC FUNCTION BY MILNE THOMSON METHOD

**Case I.** When $u$ is given

**Step 1.** Find $\frac{\partial u}{\partial x}$ and equate it to $\phi_1(x,y)$.

**Step 2.** Find $\frac{\partial u}{\partial y}$ and equate it to $\phi_2(x,y)$.

**Step 3.** Replace $x$ by $z$ and $y$ by 0 in $\phi_1(x,y)$ to get $\phi_1(z,0)$.

**Step 3.** Replace $x$ by $z$ and $y$ by 0 in $\phi_1(x, y)$ to get $\phi_1(z, 0)$.
**Step 4.** Replace $x$ by $z$ and $y$ by 0 in $\phi_2(x, y)$ to get $\phi_2(z, 0)$.
**Step 5.** Find $f(z)$ by the formula $f(z) = \int\{\phi_1(z, 0) - i\phi_2(z, 0)\}\, dz + c$

**Case II.** When $v$ is given

**Step 1.** Find $\frac{\partial v}{\partial x}$ and equate it to $\psi_2(x, y)$.
**Step 2.** Find $\frac{\partial v}{\partial y}$ and equate it to $\psi_1(x, y)$.
**Step 3.** Replace $x$ by $z$ and $y$ by 0 in $\psi_1(x, y)$ to get $\psi_1(z, 0)$.
**Step 4.** Replace $x$ by $z$ and $y$ by 0 in $\psi_2(x, y)$ to get $\psi_2(z, 0)$.
**Step 5.** Find $f(z)$ by the formula

$$f(z) = \int\{\psi_1(z, 0) + i\psi_2(z, 0)\}\, dz + c$$

**Case III.** When $u - v$ is given.

We know that
$$f(z) = u + iv \qquad ...(1)$$
$$if(z) = iu - v \qquad ...(2) \text{ [Multiplying by } i]$$

Adding (1) and (2), we get
$$(1 + i) f(z) = (u - v) + i(u + v)$$
$$\Rightarrow \quad F(z) = U + iV$$
where
$$F(z) = (1 + i) f(z) \qquad ...(3) \quad \begin{bmatrix} U = u - v \\ V = u + v \end{bmatrix}$$
Here,
$$U = (u - v) \text{ is given}$$
Find out F($z$) by the method described in case I, then substitute the value of F($z$) in (3), we get
$$f(z) = \frac{F(z)}{1+i}$$

**Case IV.** When $u + v$ is given.

We know that
$$f(z) = u + iv \qquad ...(1)$$
$$if(z) = iu - v \qquad \text{[Multiplying by } i]...(2)$$

Adding (1) and (2), we get
$$(1 + i) f(z) = (u - v) + i(u + v)$$
$$\Rightarrow \quad F(z) = U + iV$$
where
$$F(z) = (1 + i) f(z) \qquad ...(3) \quad \begin{bmatrix} U = u - v \\ V = u + v \end{bmatrix}$$
Here,
$$V = (u + v) \text{ is given}$$
Find out F($z$) by the method described in case II, then substitute the value of F($z$) in (3), we get
$$f(z) = \frac{F(z)}{1+i}$$

**Example 31.** *If $u = x^2 - y^2$, find a corresponding analytic function.*

**Solution.**
$$\frac{\partial u}{\partial x} = 2x = \phi_1(x, y), \qquad \frac{\partial u}{\partial y} = -2y = \phi_2(x, y)$$
On replacing $x$ by $z$ and $y$ by 0, we have
$$f(z) = \int [\phi_1(z,0) - i\phi_2(z,0)]\, dz + C$$
$$= \int [2z - i(0)]dz + c = \int 2z\, dz + c = z^2 + C \qquad \textbf{Ans.}$$
This is the required analytic function.

**Example 32.** *Show that the function $u = e^{-2xy} \sin(x^2 - y^2)$ is harmonic. Find the conjugate function $v$ and express $u + iv$ as an analytic function of $z$.*

*(R.G.P.V. Bhopal, III Semester, June, 2007, Dec. 2006)*

**Solution.** We have,
$$u = e^{-2xy} \sin(x^2 - y^2) \qquad ...(1)$$
Differentiating (1), w.r.t. $x$, we get
$$\frac{\partial u}{\partial x} = 2x\, e^{-2xy} \cos(x^2 - y^2) - 2y e^{-2xy} \sin(x^2 - y^2)$$

$$\Rightarrow \quad \frac{\partial u}{\partial x} = e^{-2xy}\,[2x\cos(x^2-y^2) - 2y\sin(x^2-y^2)] = \phi_1(x, y) \quad \text{... (2)}$$

$\phi_1(z, 0) = 2z \cos z^2$

Differentiating (1), w.r.t. $y$, we get

$$\frac{\partial u}{\partial y} = -\,2ye^{-2xy}\cos(x^2-y^2) - 2xe^{-2xy}\sin(x^2-y^2)$$

$$\Rightarrow \quad \frac{\partial u}{\partial y} = e^{-2xy}\,[-2y\cos(x^2-y^2) - 2x\sin(x^2-y^2)] = \phi_2(x, y) \quad \text{... (3)}$$

$\phi_2(z, 0) = -\,2z \sin z^2$

Differentiating (2), w.r.t. '$x$', we get

$$\frac{\partial^2 u}{\partial x^2} = -2ye^{-2xy}\,[2x\cos(x^2-y^2) - 2y\sin(x^2-y^2)]$$
$$+\,e^{-2xy}\,[2\cos(x^2-y^2) + 2x\,(2x)\,\{-\sin(x^2-y^2)\} - 2y(2x)\cos(x^2-y^2)]$$

$$\Rightarrow \quad \frac{\partial^2 u}{\partial x^2} = e^{-2xy}\,[-4xy\cos(x^2-y^2) + 4y^2\sin(x^2-y^2) + 2\cos(x^2-y^2)$$
$$-4x^2\sin(x^2-y^2) - 4xy\cos(x^2-y^2)\,]$$

$$= e^{-2xy}\,[-8xy\cos(x^2-y^2) + 4y^2\sin(x^2-y^2) + 2\cos(x^2-y^2) - 4x^2\sin(x^2-y^2)] \quad \text{...(4)}$$

Differentiating (3), w.r.t. '$y$', we get

$$\frac{\partial^2 u}{\partial y^2} = -2xe^{-2xy}\,[-2y\cos(x^2-y^2) - 2x\sin(x^2-y^2)]$$
$$+\,e^{-2xy}\left[-2\cos(x^2-y^2) + 2y\,(-2y)\sin(x^2-y^2) - 2x(-2y)\cos(x^2-y^2)\right]$$

$$\Rightarrow \quad \frac{\partial^2 u}{\partial y^2} = e^{-2xy}\,[4xy\cos(x^2-y^2) + 4x^2\sin(x^2-y^2) - 2\cos(x^2-y^2)$$
$$-4y^2\sin(x^2-y^2) + 4xy\cos(x^2-y^2)]$$

$$\frac{\partial^2 u}{\partial y^2} = e^{-2xy}\,[8xy\cos(x^2-y^2) + 4x^2\sin(x^2-y^2) - 2\cos(x^2-y^2) - 4y^2\sin(x^2-y^2)] \quad \text{... (5)}$$

Adding (4) and (5), we get $\dfrac{\partial^2 u}{\partial x^2} + \dfrac{\partial^2 u}{\partial y^2} = 0$

Which proves that $u$ is harmonic.

Now we have to express $u + iv$ as a function of $z$

$$f(z) = \int [\phi_1(z, 0) - i\,\phi_2(z, 0)]\,dz = \int [2z\cos z^2 - i\,(-\,2z\sin z^2)]\,dz$$
$$= \sin z^2 - i\cos z^2 + C = -\,i\,(\cos z^2 + i\sin z^2) + C = -\,ie^{iz^2} + C$$ **Ans.**

**Example 33.** *If* $u = \dfrac{\sin 2x}{\cosh 2y + \cos 2x}$, *find* $f(z)$. *(R.G.P.V., Bhopal, III Semester, Dec. 2003)*

**Solution.** $\dfrac{\partial u}{\partial x} = \dfrac{(\cosh 2y + \cos 2x)2\cos 2x - \sin 2x(-2\sin 2x)}{(\cosh 2y + \cos 2x)^2}$

$$= \frac{2\cosh 2y\cos 2x + 2(\cos^2 2x + \sin^2 2x)}{(\cosh 2y + \cos 2x)^2} = \frac{2\cosh 2y\cos 2x + 2}{(\cosh 2y + \cos 2x)^2} = \phi_1(x, y) \quad \text{... (1)}$$

Now putting $x = z,\ y = 0$ in (1), we get

$$\phi_1(z, 0) = \frac{2\cos 2z + 2}{(1+\cos 2z)^2}$$

$$\frac{\partial u}{\partial y} = \frac{-\sin 2x(2\sinh 2y)}{(\cosh 2y+\cos 2x)^2} = \frac{-2\sin 2x \sinh 2y}{(\cosh 2y+\cos 2x)^2} = \phi_2(x, y) \quad \ldots (2)$$

Now putting $x = z$ and $y = 0$ in (2), we get

$$\phi_2(z, 0) = 0$$

$$f(z) = \int [\phi_1(z,0) - i\phi_2(z,0)]\, dz = \int \frac{(2\cos 2z+2)}{(1+\cos 2z)^2} dz = 2\int \frac{1}{1+\cos 2z} dz$$

$$= 2\int \frac{1}{2\cos^2 z} dz = \int \sec^2 z\, dz = \tan z + C \qquad \textbf{Ans.}$$

which is the required function.

**Example 34.** *Show that $e^x$ (x cos y − y sin y) is a harmonic function. Find the analytic function for which $e^x$ (x cos y − y sin y) is imaginary part.*

*(U.P., III Semester, June 2009, R.G.P.V., Bhopal, III Semester, June 2004)*

**Solution.** Here $v = e^x (x\cos y - y\sin y)$

Differentiating partially w.r.t. $x$ and $y$, we have

$$\frac{\partial v}{\partial x} = e^x (x\cos y - y\sin y) + e^x \cos y = \psi_2(x, y), \quad \text{(say)} \quad \ldots (1)$$

$$\frac{\partial v}{\partial y} = e^x (-x\sin y - y\cos y - \sin y) = \psi_1(x, y) \quad \text{(say)} \quad \ldots (2)$$

$$\frac{\partial^2 v}{\partial x^2} = e^x (x\cos y - y\sin y) + e^x \cos y + e^x \cos y$$

$$= e^x (x\cos y - y\sin y + 2\cos y) \quad \ldots (3)$$

and

$$\frac{\partial^2 v}{\partial y^2} = e^x (-x\cos y + y\sin y - 2\cos y) \quad \ldots (4)$$

Adding equations (3) and (4), we have

$$\frac{\partial^2 v}{\partial x^2} + \frac{\partial^2 v}{\partial y^2} = 0 \quad \Rightarrow v \text{ is a harmonic function.}$$

Now putting $x = z,\ y = 0$ in (1) and (2), we get

$$\psi_2(z, 0) = ze^z + e^z \qquad \psi_1(z, 0) = 0$$

Hence by Milne-Thomson method, we have

$$f(z) = \int [\psi_1(z, 0) + i\psi_2(z, 0)]\, dz + C$$

$$= \int [0 + i(ze^z + e^z)]\, dz + C = i(ze^z - e^z + e^z) + C = ize^z + C.$$

This is the required analytic function. **Ans.**

**Example 35.** *If $u - v = (x - y)(x^2 + 4xy + y^2)$ and $f(z) = u + iv$ is an analytic function of $z = x + iy$, find $f(z)$ in terms of z by Milne Thomson method.*

**Solution.** We know that

$$f(z) = u + iv \quad \ldots (1)$$

$$if(z) = iu - v \quad \ldots (2)$$

Adding (1) and (2), we get

$$(1 + i) f(z) = (u - v) + i(u + v)$$

$$F(z) = U + iV$$
$$U = u - v = (x-y)(x^2+4xy+y^2)$$
$$\frac{\partial U}{\partial x} = (x^2+4xy+y^2) + (x-y)(2x+4y)$$
$$= x^2+4xy+y^2+2x^2+4xy-2xy-4y^2 = 3x^2+6xy-3y^2$$
$$\phi_1(x, y) = 3x^2+6xy-3y^2$$
$$\phi_1(z, 0) = 3z^2$$
$$\frac{\partial U}{\partial y} = -(x^2+4xy+y^2)+(x-y)(4x+2y)$$
$$= -x^2-4xy-y^2+4x^2+2xy-4xy-2y^2 = 3x^2-6xy-3y^2$$
$$\phi_2(x, y) = 3x^2-6xy-3y^2$$
$$\phi_2(z, 0) = 3z^2$$
$$F(z) = U + iV$$
$$F'(z) = \frac{\partial U}{\partial x} + i\frac{\partial V}{\partial x} = \frac{\partial U}{\partial x} - i\frac{\partial U}{\partial y} = \phi_1(z, 0) - i\,\phi_2(z, 0) = 3z^2 - i\,3z^2$$
$$= 3(1-i)z^2$$
$$F(z) = (1-i)z^3 + C$$
$$(1+i)f(z) = (1-i)z^3 + C$$
$$f(z) = \frac{1-i}{1+i}z^3 + \frac{C}{1+i} = \frac{(1-i)(1-i)}{(1+i)(1-i)}z^3 + C_1$$
$$= \frac{1-2i+(-i)^2}{1+1}z^3 + C_1 = \frac{1-2i-1}{2}z^3 + C_1 = -i\,z^3 + C_1 \qquad \textbf{Ans.}$$

**Note:** This example has already been solved on page 529 as Example 26.

**Example 36.** *If $f(z) = u + iv$ is an analytic function of $z$ and $u - v = \dfrac{\cos x + \sin x - e^{-y}}{2\cos x - 2\cosh y}$, prove that*

$$f(z) = \frac{1}{2}\left[1 - \cot\frac{z}{2}\right] \text{ when } f\left(\frac{\pi}{2}\right) = 0. \quad \textit{(R.G.P.V. Bhopal, III Semester, Dec. 2007)}$$

**Solution.** We know that $f(z) = u + iv$

$\therefore$ $i\,f(z) = iu - v$ [Multiplying by $i$]

On adding, we get $(1+i)f(z) = (u-v) + i(u+v)$

Let $F(z) = U + iV$

We have, $U = u - v = \dfrac{\cos x + \sin x - e^{-y}}{2\cos x - 2\cosh y}$

$$\Rightarrow \quad U = \frac{\cos x + \sin x - \cosh y + \sinh y}{2\cos x - 2\cosh y} \qquad [\because e^{-y} = \cosh y - \sinh y]$$

$$= \frac{\cos x - \cosh y}{2(\cos x - \cosh y)} + \frac{\sin x + \sinh y}{2(\cos x - \cosh y)} = \frac{1}{2} + \frac{\sin x + \sinh y}{2(\cos x - \cosh y)} \qquad ...(1)$$

Differentiating (1) w.r.t. $x$ partially, we get

$$\frac{\partial U}{\partial x} = \frac{1}{2}\left[\frac{(\cos x - \cosh y)\cos x - (\sin x + \sinh y)(-\sin x)}{(\cos x - \cosh y)^2}\right]$$
$$= \frac{1}{2}\left[\frac{(\cos^2 x + \sin^2 x - \cosh y\cos x + \sinh y\sin x)}{(\cos x - \cosh y)^2}\right]$$
$$\phi_1(x, y) = \frac{1}{2}\left[\frac{1 - \cosh y\cos x + \sinh y\sin x}{(\cos x - \cosh y)^2}\right] = \frac{1}{2}\,\frac{1 - \cos iy\cos x - i\,\sin iy\sin x}{(\cos x - \cosh y)^2} \qquad ...(2)$$

Replacing $x$ by $z$ and $y$ by 0 in (2), we get

$$\phi_1(z, 0) = \frac{1}{2}\left[\frac{1-\cos z}{(\cos z-1)^2}\right] = \frac{-(\cos z-1)}{2(\cos z-1)^2} = \frac{-1}{2(\cos z-1)} = \frac{1}{2(1-\cos z)}$$

Differentiating (1) partially w.r.t. $y$, we get

$$\frac{\partial U}{\partial y} = \frac{1}{2}\left[\frac{(\cos x-\cosh y)\,.\cosh y-(\sin x+\sinh y)(-\sinh y)}{(\cos x-\cosh y)^2}\right]$$

$$= \frac{1}{2}\left[\frac{(\cos x\cosh y)+\sin x\sinh y-(\cosh^2 y-\sinh^2 y)}{(\cos x-\cosh y)^2}\right]$$

$$\phi_2(x, y) = \frac{1}{2}\left[\frac{\cos x\cosh y+\sin x\sinh y-1}{(\cos x-\cosh y)^2}\right] \quad ...(3)$$

Replacing $x$ by $z$ and $y$ by 0 in (3), we have

$$\phi_2(z, 0) = \frac{1}{2}\left[\frac{\cos z-1}{(\cos z-1)^2}\right] = \frac{1}{2}\cdot\frac{1}{\cos z-1} = \frac{1}{2}\cdot\left(\frac{-1}{1-\cos z}\right)$$

$$F'(z) = \frac{\partial U}{\partial x}+i\frac{\partial V}{\partial x} = \frac{\partial U}{\partial x}-i\frac{\partial U}{\partial y} \quad \text{[C–R equations]}$$

$$= \phi_1(z, 0) - i\,\phi_2(z, 0)$$

By Milne Thomson Method,

$$F(z) = \int[\phi_1(z, 0)-i\phi_2(z, 0)]\,dz$$

$$= \int\left[\frac{1}{2}\cdot\frac{1}{(1-\cos z)}+\frac{i}{2}\cdot\frac{1}{1-\cos z}\right]dz$$

$$= \frac{1+i}{2}\int\frac{1}{2\sin^2 z/2}\,dz = \frac{1+i}{4}\int\operatorname{cosec}^2(z/2)\,dz$$

$$= \left(\frac{1+i}{4}\right)\cdot\frac{(-\cot z/2)}{\left(\frac{1}{2}\right)}+C = -\left(\frac{1+i}{2}\right)\cot\frac{z}{2}+C$$

$$\Rightarrow \quad (1+i)\,f(z) = -\left(\frac{1+i}{2}\right)\cot\frac{z}{2}+C \quad \Rightarrow f(z) = -\frac{1}{2}\cot\frac{z}{2}+\frac{C}{1+i} \quad ...(4)$$

On putting $z = \frac{\pi}{2}$ in (4), we get

$$f\left(\frac{\pi}{2}\right) = -\frac{1}{2}\cot\frac{\pi}{4}+\frac{C}{1+i}$$

$$0 = -\frac{1}{2}+\frac{C}{1+i} \quad \Rightarrow \quad \frac{C}{1+i} = \frac{1}{2} \quad \left[f\left(\frac{\pi}{2}\right) = 0, \text{ given}\right]$$

On putting the value of $\frac{C}{1+i}$ in (4), we get

$$f(z) = -\frac{1}{2}\cot\frac{z}{2}+\frac{1}{2}$$

Hence, $$f(z) = \frac{1}{2}\left(1-\cot\frac{z}{2}\right)$$ **Proved.**

## EXERCISE 7.4

**Show that the following functions are harmonic and determine the conjugate functions.**

1. $u = 2x(1 - y)$ **Ans.** $v = x^2 - y^2 + 2y + C$

2. $u = 2x - x^3 + 3xy + 3xy^2$ **Ans.** $v = -\frac{3}{2}x^2 - 3x^2y + 2y + \frac{3}{2}y^2 + y^3 + C$

**Determine the analytic function, whose real part is**

3. $\log\sqrt{x^2+y^2}$ (*K.U., 2009*) **Ans.** $\log z + C$ 4. $\cos x \cosh y$ **Ans.** $\cos z + c$

5. $e^{-x}(\cos y + \sin y)$ (*AMIETE, June 2010*) **Ans.** $(1 + i)e^{-z} + C$

6. $e^{2x}(x\cos 2y - y\sin 2y)$ **Ans.** $ze^{2z} + iC$ 7. $e^{-x}(x\cos y + y\sin y)$ and $f(0) = i$. **Ans.** $ze^{-z} + i$

**Determine the analytic function, whose imaginary part is**

8. $v = \log(x^2+y^2) + x - 2y$ (*G.B.T.U., 2012*) **Ans.** $2i\log z - (2-i)z + C$

9. $v = \sinh x \cos y$ **Ans.** $\sin iz + C$ 10. $v = \left(r - \frac{1}{r}\right)\sin\theta$ **Ans.** $z + \frac{1}{z} + C$

11. Find the analytic function whose real part is $\frac{\sin 2x}{(\cosh 2y - \cos 2x)}$ (*MDU Dec. 2010*)

[**Hint:** See sloved Example 33 on page 535] **Ans.** $f(z) = \cot z + c$

12. If $f(z) = u + iv$ is an analytic function of $z = x + iy$ and $u - v = \frac{e^y - \cos x + \sin x}{\cosh y - \cos x}$, find $f(z)$ subject to the condition that $f\left(\frac{\pi}{2}\right) = \frac{3-i}{2}$. **Ans.** $f(z) = \cot\frac{z}{2} + \frac{1-i}{2}$

13. Find an analytic function $f(z) = u(r, \theta) + iv(r, \theta)$ such that $v(r, \theta) = r^2\cos 2\theta - r\cos\theta + 2$. **Ans.** $i[z^2 - z + 2]$

14. Show that the function $u = x^2 - y^2 - 2xy - 2x - y - 1$ is harmonic. Find the conjugate harmonic function $v$ and express $u + iv$ as a function of $z$ where $z = x + iy$. **Ans.** $(1 + i)z^2 + (-2 + i)z - 1$

15. Construct an analytic function of the form $f(z) = u + iv$, where $v$ is $\tan^{-1}(y/x)$, $x \neq 0, y \neq 0$ **Ans.** $\log cz$

16. Show that the function $u = e^{-2xy}\sin(x^2 - y^2)$ is harmonic. Find the conjugate function $v$ and express $u + iv$ as an analytic function of $z$. **Ans.** $v = e^{-2xy}\cos(x^2 - y^2) + C$
$f(z) = -ie^{iz^2} + C_1$

17. Show that the function $v(x, y) = e^x \sin y$ is harmonic. Find its conjugate harmonic function $u(x, y)$ and the corresponding analytic function $f(z)$. (*AMIETE, June 2009*) **Ans.** $u = e^x\cos y + C, f(z) = e^z + C$

**Choose the correct answer:**

18. The harmonic conjugate of $u = x^3 - 3xy^2$ is
(*a*) $y^3 - 3xy^2$ (*b*) $3x^2y - y^3$ (*c*) $3xy^2 - y^3$ (*d*) $3xy^2 - x^3$ (*AMIETE, June 2010*) **Ans.** (*b*)

## 7.21 PARTIAL DIFFERENTIATION OF FUNCTION OF COMPLEX VARIABLE

**Example 37.** *Prove that* $4\frac{\partial^2}{\partial z\,\partial\bar{z}} = \left[\frac{\partial^2}{\partial x^2} + \frac{\partial^2}{\partial y^2}\right]$

**Solution.** We know that

$$x + iy = z \quad \ldots (1), \qquad x - iy = \bar{z} \quad \ldots (2)$$

From (1) and (2), we get

$$x = \frac{1}{2}(z + \bar{z}), \qquad y = \frac{-i}{2}(z - \bar{z})$$

$$\Rightarrow \qquad \frac{\partial x}{\partial z} = \frac{1}{2}, \qquad \frac{\partial y}{\partial z} = -\frac{i}{2}$$

and $$\frac{\partial x}{\partial \bar{z}} = \frac{1}{2}, \qquad \frac{\partial y}{\partial \bar{z}} = \frac{i}{2}$$

We know that,

$$\frac{\partial}{\partial z} = \frac{\partial}{\partial x}\left(\frac{\partial x}{\partial z}\right) + \frac{\partial}{\partial y}\left(\frac{\partial y}{\partial z}\right) = \frac{\partial}{\partial x}\left(\frac{1}{2}\right) + \frac{\partial}{\partial y}\left(\frac{-i}{2}\right) = \frac{1}{2}\left(\frac{\partial}{\partial x} - i\frac{\partial}{\partial y}\right) \qquad \text{...(3)}$$

$$\frac{\partial}{\partial \bar{z}} = \frac{\partial}{\partial x}\left(\frac{\partial x}{\partial \bar{z}}\right) + \frac{\partial}{\partial y}\left(\frac{\partial y}{\partial \bar{z}}\right) = \frac{\partial}{\partial x}\left(\frac{1}{2}\right) + \frac{\partial}{\partial y}\left(\frac{i}{2}\right) = \frac{1}{2}\left(\frac{\partial}{\partial x} + i\frac{\partial}{\partial y}\right) \qquad \text{... (4)}$$

From (3) and (4), we get

$$\frac{\partial^2}{\partial z \partial \bar{z}} = \frac{\partial}{\partial z}\left(\frac{\partial}{\partial \bar{z}}\right) = \frac{1}{4}\left(\frac{\partial}{\partial x} - i\frac{\partial}{\partial y}\right)\left(\frac{\partial}{\partial x} + i\frac{\partial}{\partial y}\right)$$

$$= \frac{1}{4}\left(\frac{\partial^2}{\partial x^2} + \frac{\partial^2}{\partial y^2} + i\frac{\partial^2}{\partial x \partial y} - i\frac{\partial^2}{\partial x \partial y}\right) = \frac{1}{4}\left(\frac{\partial^2}{\partial x^2} + \frac{\partial^2}{\partial y^2}\right)$$

$$\Rightarrow \qquad 4\frac{\partial^2}{\partial z\, \partial \bar{z}} = \left[\frac{\partial^2}{\partial x^2} + \frac{\partial^2}{\partial y^2}\right]$$ **Proved.**

**Example 38.** *If f (z) is a harmonic function of z, show that*

$$\left\{\frac{\partial}{\partial x}|f(z)|\right\}^2 + \left\{\frac{\partial}{\partial y}|f(z)|\right\}^2 = |f'(z)|^2$$ *(K.U., 2009, U.P. III Semester, June 2009)*

**Solution.** Since $f(z) = u(x, y) + i\,v(x, y)$

so $$|f(z)| = \sqrt{u^2 + v^2} \qquad \text{... (1)}$$

Differentiating (1) partially w.r.t. '$x$', we get

$$\frac{\partial}{\partial x}|f(z)| = \frac{\partial}{\partial x}(\sqrt{u^2 + v^2})$$

$$= \frac{1}{2}(u^2 + v^2)^{\frac{-1}{2}}\left(2u\frac{\partial u}{\partial x} + 2v\frac{\partial v}{\partial x}\right) = \frac{u\frac{\partial u}{\partial x} + v\frac{\partial v}{\partial x}}{\sqrt{u^2 + v^2}} = \frac{u\frac{\partial u}{\partial x} + v\frac{\partial v}{\partial x}}{|f(z)|} \qquad \text{... (2)}$$

Similarly $$\frac{\partial}{\partial y}|f(z)| = \frac{u\frac{\partial u}{\partial y} + v\frac{\partial v}{\partial y}}{|f(z)|} \qquad \text{... (3)}$$

Squaring (2) and (3) adding, we get

$$\left[\frac{\partial}{\partial x}|f(z)|\right]^2 + \left[\frac{\partial}{\partial y}|f(z)|\right]^2 = \frac{\left(u\frac{\partial u}{\partial x} + v\frac{\partial v}{\partial x}\right)^2 + \left(u\frac{\partial u}{\partial y} + v\frac{\partial v}{\partial y}\right)^2}{|f(z)|^2}$$

$$= \frac{\left(u\frac{\partial u}{\partial x}\right)^2 + 2uv\frac{\partial u}{\partial x}\frac{\partial v}{\partial x} + \left(v\frac{\partial v}{\partial x}\right)^2 + \left(u\frac{\partial u}{\partial y}\right)^2 + 2u\frac{\partial u}{\partial y}.v\frac{\partial v}{\partial y} + \left(v\frac{\partial v}{\partial y}\right)^2}{|f(z)|^2}$$

By C-R equation $$\frac{\partial u}{\partial x} = \frac{\partial v}{\partial y} \quad \text{and} \quad \frac{\partial u}{\partial y} = -\frac{\partial v}{\partial x}$$

Now $\qquad 2uv\dfrac{\partial u}{\partial x}\dfrac{\partial v}{\partial x}=2uv\left(\dfrac{\partial v}{\partial y}\right)\left(-\dfrac{\partial u}{\partial y}\right)$

Putting the value of $2uv\dfrac{\partial u}{\partial x}.\dfrac{\partial v}{\partial x}=-2uv\dfrac{\partial v}{\partial y}.\dfrac{\partial u}{\partial y}$ in (4), we get

$$\left[\frac{\partial}{\partial x}|f(z)|\right]^2+\left[\frac{\partial}{\partial y}|f(z)|\right]^2=\frac{\left(u\frac{\partial u}{\partial x}\right)^2-2uv\frac{\partial v}{\partial y}.\frac{\partial u}{\partial y}+\left(v\frac{\partial v}{\partial x}\right)^2+\left(u\frac{\partial u}{\partial y}\right)^2+2uv\frac{\partial u}{\partial y}.\frac{\partial v}{\partial y}+\left(v\frac{\partial v}{\partial y}\right)^2}{|f(z)|^2}$$

$$=\frac{u^2\left(\frac{\partial u}{\partial x}\right)^2+u^2\left(\frac{\partial u}{\partial y}\right)^2+v^2\left(\frac{\partial v}{\partial x}\right)^2+v^2\left(\frac{\partial v}{\partial y}\right)^2}{|f(z)|^2}=\frac{u^2\left[\left(\frac{\partial u}{\partial x}\right)^2+\left(\frac{\partial u}{\partial y}\right)^2\right]+v^2\left[\left(\frac{\partial v}{\partial x}\right)^2+\left(\frac{\partial v}{\partial y}\right)^2\right]}{|f(z)|^2}$$

$$=\frac{u^2\left[\left(\frac{\partial u}{\partial x}\right)^2+\left(\frac{-\partial v}{\partial x}\right)^2\right]+v^2\left[\left(\frac{\partial v}{\partial x}\right)^2+\left(\frac{\partial u}{\partial x}\right)^2\right]}{|f(z)|^2}$$ [C – R equations]

$$=\frac{(u^2+v^2)\left[\left(\frac{\partial u}{\partial x}\right)^2+\left(\frac{\partial v}{\partial x}\right)^2\right]}{|f(z)|^2}=\frac{|f(z)|^2\left[\left(\frac{\partial u}{\partial x}\right)^2+\left(\frac{\partial v}{\partial x}\right)^2\right]}{|f(z)|^2}=\left(\frac{\partial u}{\partial x}\right)^2+\left(\frac{\partial v}{\partial x}\right)^2$$

[ $|f(x)|^2=u^2+v^2$ ]

$=|f'(z)|^2$ **Proved.**

**Example 39.** *Prove that* $\left(\dfrac{\partial^2}{\partial x^2}+\dfrac{\partial^2}{\partial y^2}\right)|u|^P=P(P-1)|u|^{P-2}|f'(z)|^2$

**Solution.** We know that $\dfrac{\partial^2}{\partial x^2}+\dfrac{\partial^2}{\partial y^2}=4\dfrac{\partial^2}{\partial z\,\partial \bar{z}}$ [Example 46, page 173]

$$\left(\frac{\partial^2}{\partial x^2}+\frac{\partial^2}{\partial y^2}\right)|u|^P=\frac{1}{2^P}\frac{4\partial^2}{\partial z\partial \bar{z}}[f(z)+f(\bar{z})]^P \qquad \left[\because\ u=\frac{1}{2}[f(z)+f(\bar{z})]\right]$$

$$=\frac{4}{2^P}\frac{\partial}{\partial \bar{z}}P[f(z)+f(\bar{z})]^{P-1}f'(z)=\frac{1}{2^{P-2}}P(P-1)[f(z)+f(\bar{z})]^{P-2}f'(z)f'(\bar{z})$$

$$=P(P-1)\left[\frac{1}{2}\{f(z)+f(\bar{z})\}\right]^{P-2}[f'(z)f'(\bar{z})]\ =P(P-1)|u|^{P-2}|f'(z)|^2$$ **Proved.**

**Example 40.** *Prove that*

$$\left(\frac{\partial^2}{\partial x^2}+\frac{\partial^2}{\partial y^2}\right)\log|f'(z)|=0$$

**Solution.** We have, $\left(\dfrac{\partial^2}{\partial x^2}+\dfrac{\partial^2}{\partial y^2}\right)=4\dfrac{\partial^2}{\partial z\partial \bar{z}}$ *(Example 37 on page 539)*

Hence $\left(\dfrac{\partial^2}{\partial x^2}+\dfrac{\partial^2}{\partial y^2}\right)\{\log|f'(z)|\}=4\dfrac{\partial^2}{\partial z\,\partial \bar{z}}\{\log|f'(z)|\}=4\dfrac{\partial^2}{\partial z\partial \bar{z}}\dfrac{1}{2}\log|f'(z)|^2$

$$=2\frac{\partial^2}{\partial z\partial \bar{z}}\log\{f'(z)f'(\bar{z})\}=2\frac{\partial^2}{\partial z\partial \bar{z}}[\log f'(z)+\log f'(\bar{z})]=2\frac{\partial}{\partial z}\left(0+\frac{1}{f'(\bar{z})}f''(\bar{z})\right)$$

$$= 2\frac{\partial}{\partial z}\frac{f''(\bar{z})}{f'(\bar{z})} \qquad \left[\begin{array}{l}\bar{z} \text{ is constant in regards to} \\ \text{differentiation } w.r.t.\ z\end{array}\right]$$

$= 2 \times 0$

$= 0$

**Proved.**

**Example 41.** *Prove that* $\left(\frac{\partial^2}{\partial x^2}+\frac{\partial^2}{\partial y^2}\right)|Rf(z)|^2 = 2|f'(z)|^2$

*(G.B.T.U., April 2012)*

**Solution.** $f(z) = u + iv$ or $R f(z) = u \Rightarrow$ Real part of $f(z) = u$

$$\frac{\partial}{\partial x}u^2 = 2u\frac{\partial u}{\partial x}$$

$$\frac{\partial^2}{\partial x^2}u^2 = 2\left(\frac{\partial u}{\partial x}\right)^2 + 2u\frac{\partial^2 u}{\partial x^2} \qquad \ldots (1)$$

Similarly, $$\frac{\partial^2}{\partial y^2}u^2 = 2\left(\frac{\partial u}{\partial y}\right)^2 + 2u\frac{\partial^2 u}{\partial y^2} \qquad \ldots (2)$$

Adding (1) and (2), we get

$$\left(\frac{\partial^2}{\partial x^2}+\frac{\partial^2}{\partial y^2}\right)u^2 = 2\left[\left(\frac{\partial u}{\partial x}\right)^2+\left(\frac{\partial u}{\partial y}\right)^2\right]+2u\left(\frac{\partial^2 u}{\partial x^2}+\frac{\partial^2 u}{\partial y^2}\right)$$

$$= 2\left[\left(\frac{\partial u}{\partial x}\right)^2+\left(\frac{\partial u}{\partial y}\right)^2\right]+0 = 2\left[\left(\frac{\partial u}{\partial x}\right)^2+\left(-\frac{\partial v}{\partial x}\right)^2\right] = 2|f'(z)|^2 \qquad \left(\frac{\partial u}{\partial y} = -\frac{\partial v}{\partial x}\right)$$

$$\Rightarrow \quad \left(\frac{\partial^2}{\partial x^2}+\frac{\partial^2}{\partial y^2}\right)|R f(z)|^2 = 2|f'(z)|^2$$

**Proved.**

**Example 42.** *If f (z) is regular function of z, show that*

$$\left(\frac{\partial^2}{\partial x^2}+\frac{\partial^2}{\partial y^2}\right)|f(z)|^2 = 4|f'(z)|^2$$ *(R.G.P.V., Bhopal, III Semester, June 2004)*

**Solution.** $$f(z) = u + iv$$

$$|f(z)|^2 = u^2 + v^2 \qquad \ldots (1)$$

Let $$\phi = u^2 + v^2$$

Differentiating (1) *w.r.t.* $x$, we get

$$\frac{\partial \phi}{\partial x} = 2u\frac{\partial u}{\partial x} + 2v\frac{\partial v}{\partial x}$$

$$\frac{\partial^2 \phi}{\partial x^2} = 2\left[u\frac{\partial^2 u}{\partial x^2}+\left(\frac{\partial u}{\partial x}\right)^2+v\frac{\partial^2 v}{\partial x^2}+\left(\frac{\partial v}{\partial x}\right)^2\right] \qquad \ldots(2)$$

Similarly, $$\frac{\partial^2 \phi}{\partial y^2} = 2\left[u\frac{\partial^2 u}{\partial y^2}+\left(\frac{\partial u}{\partial y}\right)^2+v\frac{\partial^2 v}{\partial y^2}+\left(\frac{\partial v}{\partial y}\right)^2\right] \qquad \ldots (3)$$

Adding (2) and (3), we get

$$\frac{\partial^2 \phi}{\partial x^2}+\frac{\partial^2 \phi}{\partial y^2} = 2\left[u\left(\frac{\partial^2 u}{\partial x^2}+\frac{\partial^2 u}{\partial y^2}\right)+v\left(\frac{\partial^2 v}{\partial x^2}+\frac{\partial^2 v}{\partial y^2}\right)+\left\{\left(\frac{\partial u}{\partial x}\right)^2+\left(\frac{\partial u}{\partial y}\right)^2+\left(\frac{\partial v}{\partial x}\right)^2+\left(\frac{\partial v}{\partial y}\right)^2\right\}\right] \qquad \ldots (4)$$

By C – R equations $$\left(\frac{\partial u}{\partial x}\right)^2 = \left(\frac{\partial v}{\partial y}\right)^2$$

$$\left(\frac{\partial u}{\partial y}\right)^2 = \left(-\frac{\partial v}{\partial x}\right)^2 = \left(\frac{\partial v}{\partial x}\right)^2 \qquad \ldots (5)$$

By Laplace equations $\frac{\partial^2 u}{\partial x^2}+\frac{\partial^2 u}{\partial y^2}=0 \quad \Rightarrow \quad \frac{\partial^2 v}{\partial x^2}+\frac{\partial^2 v}{\partial y^2}=0$

On putting the values of $\left(\frac{\partial^2 u}{\partial x^2}+\frac{\partial^2 u}{\partial y^2}\right),\left(\frac{\partial^2 v}{\partial x^2}+\frac{\partial^2 v}{\partial y^2}\right),\frac{\partial v}{\partial x},\frac{\partial v}{\partial y}$ from (5) in (4), we get

$$\frac{\partial^2 \phi}{\partial x^2}+\frac{\partial^2 \phi}{\partial y^2}=4\left[\left(\frac{\partial u}{\partial x}\right)^2+\left(\frac{\partial v}{\partial y}\right)^2\right],\ \left(\frac{\partial^2}{\partial x^2}+\frac{\partial^2}{\partial y^2}\right)\phi=4\left|\frac{\partial u}{\partial x}+i\frac{\partial v}{\partial x}\right|^2$$

$$\left(\frac{\partial^2}{\partial x^2}+\frac{\partial^2}{\partial y^2}\right)|f(z)|^2=4\,|f'(z)|^2$$ **Proved**

**Example 43.** *If |f(z)| is constant, prove that f(z) is also constant.*

**Solution.**

$$f(z)=u+iv$$
$$|f(z)|^2=u^2+v^2$$
$$|f(z)|=\text{constant}=c \text{ (given)}$$
$$u^2+v^2=c^2 \qquad \text{... (1)}$$

Differentiating (1) w.r.t. $x$, we get $2u\frac{\partial u}{\partial x}+2v\frac{\partial v}{\partial x}=0 \quad \Rightarrow \quad u\frac{\partial u}{\partial x}+v\frac{\partial v}{\partial x}=0$ ... (2)

Differentiating (1) w.r.t. '$y$', we get $2u\frac{\partial u}{\partial y}+2v\frac{\partial v}{\partial y}=0$

$$-u\frac{\partial v}{\partial x}+v\frac{\partial u}{\partial x}=0 \qquad \text{... (3)}$$

Squaring (2) and (3) and then adding, we get

$$u^2\left(\frac{\partial u}{\partial x}\right)^2+v^2\left(\frac{\partial v}{\partial x}\right)^2+u^2\left(\frac{\partial v}{\partial x}\right)^2+v^2\left(\frac{\partial u}{\partial x}\right)^2=0$$

$$(u^2+v^2)\left[\left(\frac{\partial u}{\partial x}\right)^2+\left(\frac{\partial v}{\partial x}\right)^2\right]=0$$

$$\left(\frac{\partial u}{\partial x}\right)^2+\left(\frac{\partial v}{\partial x}\right)^2=0$$

As $$f(z)=u+iv \Rightarrow f'(z)=\frac{\partial u}{\partial x}+i\frac{\partial v}{\partial x}$$

$$\overline{f'(z)}=\frac{\partial u}{\partial x}-i\frac{\partial v}{\partial x}$$

$$|f'(z)|^2=\left(\frac{\partial u}{\partial x}\right)^2+\left(\frac{\partial v}{\partial x}\right)^2=0$$

$$|f'(z)|^2=0 \quad \Rightarrow \quad f(z) \text{ is constant.}$$ **Proved.**

## EXERCISE 7.5

1. If $f(z)=u+iv$ is an analytic function of $z=x+iy$, and $\psi$ is any function of $x$ and $y$ with differential coefficients of the first two orders, then show that

$$\left(\frac{\partial \psi}{\partial x}\right)^2+\left(\frac{\partial \psi}{\partial y}\right)^2=\left[\left(\frac{\partial \psi}{\partial u}\right)^2+\left(\frac{\partial \psi}{\partial v}\right)^2\right]|f'(z)|^2$$

and $$\frac{\partial^2 \psi}{\partial x^2}+\frac{\partial^2 \psi}{\partial y^2}=\left(\frac{\partial^2 \psi}{\partial u^2}+\frac{\partial^2 \psi}{\partial v^2}\right)|f'(z)|^2.$$

2. If $|f'(z)|$ is the product of a function of $x$ and a function of $y$, then show that

$$f'(z) = \exp(\alpha z^2 + \beta z + \gamma)$$

where $\alpha$ is a real and $\beta$ and $\gamma$ are complex constants.

**Choose the correct alternative:**

3. If $|f(z)|$ is constant then $f(z)$ is

(*a*) Variable (*b*) Partially variable and constant (*c*) Constant (*d*) None of these **Ans.** (*c*)

4. If $f(z) = u + iv$ then $|f(z)|$ is equal to

(*a*) $\sqrt{u^2+v^2}$ (*b*) $u^2 + v^2$ (*c*) $u + v$ (*d*) $\sqrt{u^2-v^2}$ **Ans.** (*a*)

5. If $z = r(\cos\theta + i\sin\theta)$ then $|z|^3$ is equal to

(*a*) $(\cos\theta + i\sin\theta)^3$ (*b*) $r^3(\cos\theta + i\sin\theta)^3$ (*c*) $r^3/2$ (*d*) $r^3$ **Ans.** (*d*)

## 7.22 INTRODUCTION (LINE INTEGRAL)

In case of real variable, the path of integration of $\int_a^b f(x)\,dx$ is always along the $x$-axis from $x = a$ to $x = b$. But in case of a complex function $f(z)$ the path of the definite integral $\int_a^b f(z)\,dz$ can be along any curve from $z = a$ to $z = b$.

$$z = x + iy \quad \Rightarrow \quad dz = dx + i\,dy \;\ldots (1) \quad dz = dx \text{ if } y = 0 \;\ldots (2) \quad dz = i\,dy \text{ if } x = 0 \;\ldots (3)$$

In (1), (2), (3) the direction of $dz$ are different. Its value depends upon the path (curve) of integration. But the value of integral from $a$ to $b$ remains the same along any regular curve from $a$ to $b$.

In case the initial point and final point coincide so that $c$ is a closed curve, then this integral is called *contour integral* and is denoted by $\oint_C f(z)\,dz$.

If $f(z) = u(x, y) + iv(x, y)$, then since $dz = dx + i\,dy$, we have

$$\oint_C f(z)\,dz = \int_C (u + iv)(dx + i\,dy) = \int_C (u\,dx - v\,dy) + i\int_C (v\,dx + u\,dy)$$

which shows that the evaluation of the line integral of a complex function can be reduced to the evaluation of two line integrals of real functions.

Let us consider a few examples:

**Complex Integral (Line Integral)**

**Example 44.** *Evaluate* $\int_0^{2+i} (\bar{z})^2 dz$ *along the real axis from* $z = 0$ *to* $z = 2$ *and then along a line parallel to y-axis from* $z = 2$ *to* $z = 2 + i$.

*(R.G.P.V., Bhopal, III Semester, June 2005)*

**Solution.** $\int_0^{2+i} (\bar{z})^2 dz = \int_0^{2+i} (x - iy)^2 (dx + i\,dy)$

$$= \int_{OA} (x^2)\,dx + \int_{AB} (2 - iy)^2\, i\,dy$$

[Along $OA$, $y = 0$, $dy = 0$, $x$ varies 0 to 2.
Along $AB$, $x = 2$, $dx = 0$ and $y$ varies 0 to 1]

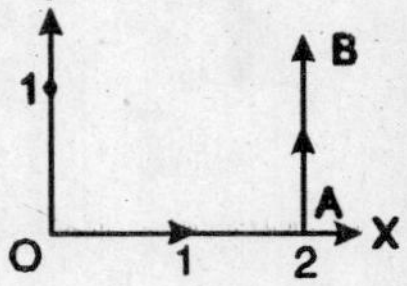


$$= \int_0^2 x^2 dx + \int_0^1 (2 - iy)^2\, i\,dy$$

$$= \left[\frac{x^3}{3}\right]_0^2 + i\int_0^1 (4 - 4iy - y^2)\,dy = \frac{8}{3} + i\left[4y - 2iy^2 - \frac{y^3}{3}\right]_0^1$$

$$= \frac{8}{3} + i\left[4 - 2i - \frac{1}{3}\right] = \frac{8}{3} + \frac{i}{3}(11 - 6i) = \frac{1}{3}(8 + 11i + 6) = \frac{1}{3}(14 + 11i)$$

Which is the required value of the given integral. **Ans.**

**Example 45.** *Evaluate* $\int_0^{1+i} (x^2 - iy)\, dz$, *along the path*

(*a*) $y = x$ *(R.G.P.V., Bhopal, III Semester, Dec. 2007)* (*b*) $y = x^2$.

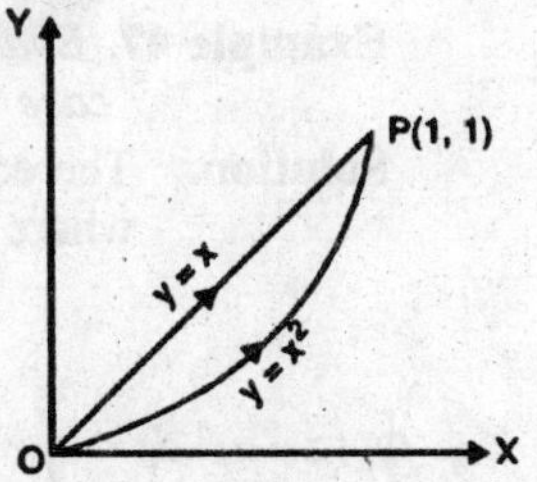


**Solution.** (*a*) Along the line $y = x$,

$dy = dx$ so that $dz = dx + idy$

$\Rightarrow$ $dz = dx + idx = (1 + i)\, dx$

$\therefore$ $\int_0^{1+i} (x^2 - iy)\, dz$

[On putting $y = x$ and $dz = (1 + i)dx$]

$$= \int_0^1 (x^2 - ix)(1 + i)\, dx$$

$$= (1+i)\left[\frac{x^3}{3} - i\frac{x^2}{2}\right]_0^1 = (1+i)\left(\frac{1}{3} - \frac{1}{2}i\right) = \frac{(1+i)(2-3i)}{6} = \frac{5}{6} - \frac{1}{6}i.$$

Which is the required value of the given integral. **Ans.**

(*b*) Along the parabola $y = x^2$, $dy = 2x\, dx$ so that

$dz = dx + idy \Rightarrow dz = dx + 2ix\, dx = (1 + 2ix)\, dx$

and $x$ varies from 0 to 1.

$$\therefore \int_0^{1+i} (x^2 - iy)\, dx = \int_0^1 (x^2 - ix^2)(1 + 2ix)dx = \int_0^1 x^2 (1 - i)(1 + 2ix)\, dx$$

$$= (1 - i)\int_0^1 x^2 (1 + 2ix)\, dx = (1-i)\left[\frac{x^3}{3} + i\frac{x^4}{2}\right]_0^1$$

$$= (1-i)\left[\frac{1}{3} + \frac{1}{2}i\right] = \frac{(1-i)(2+3i)}{6} = \frac{1}{6}(2 + 3i - 2i + 3) = \frac{5}{6} + \frac{1}{6}i$$

Which is the required value of the given integral. **Ans.**

**Example 46.** *Evaluate* $\int_C (12z^2 - 4iz)\, dz$ *along the curve C joining the points (1, 1) and (2, 3)* *(U.P., III Semester, Dec. 2009)*

**Solution.** Here, we hae

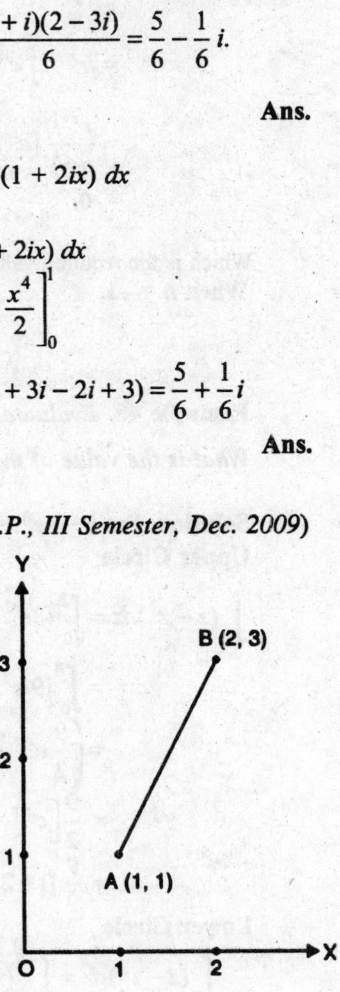


$$\int_C (12z^2 - 4iz)\, dz = \int_C [12\,(x + iy)^2 - 4i\,(x + iy)]\,(dx + i\, dy)$$

$$= \int_C [12\,(x^2 - y^2 + 2i\, xy) - 4ix + 4y]\,(dx + i\, dy)$$

$$= \int_C (12\, x^2 - 12y^2 + 24\, ixy - 4ix + 4y)\,(dx + i\, dy) \quad \ldots (1)$$

Equation of the line AB passing through (1, 1) and (2, 3) is

$$y - 1 = \frac{3-1}{2-1}(x - 1)$$

$$y - 1 = 2\,(x - 1) \Rightarrow y = 2x - 1 \Rightarrow dy = 2\, dx$$

Putting the values of $y$ and $dy$ in (1), we get

$$= \int_1^2 (12x^2 - 12\,(2x - 1)^2 + 24\, ix\,(2x - 1) - 4ix + 4\,(2x - 1)]\,(dx + 2i\, dx)$$

$$= \int_1^2 [12x^2 - 48x^2 + 48x - 12 + 48\, ix^2 - 24\, ix - 4ix + 8x - 4]\,(1 + 2i)\, dx$$

$$= (1 + 2i)\int_1^2 [-36 + 48i)\, x^2 + (56 - 28i)\, x - 16]\, dx$$

$$= (1 + 2i)\left[(-36 + 48i)\frac{x^3}{3} + (56 - 28i)\frac{x^2}{2} - 16x\right]_1^2$$

$$= (1+2i)\left[(-36+48i)\frac{8}{3}+(56-28i)\,2-16\times 2-(36+48i)\frac{1}{3}-(56-28i)\frac{1}{2}+16\right]$$

$$= (1+2i)\,(-96+128i+112-56i-32+12-16i-28+14\,i+16$$

$$= (1+2i)\,(-16+70i) = -16+70i-32i-140 = -156+38i$$ **Ans.**

**Example 47.** *Evaluate* $\int (z-a)^n\, dz$ *where c is the circle with centre a and r. Discuss the case when n = –1.*

**Solution.** The equation of circle $C$ is $|z-a| = r$ or $z-a = re^{i\theta}$ where $\theta$ varies from 0 to $2\pi$

$$dz = ire^{i\theta}d\theta$$

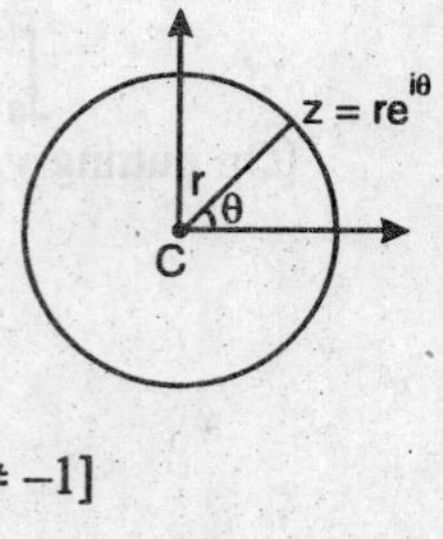


$$\oint_C (z-a)^n\, dz = \int_0^{2\pi} r^n e^{in\theta}.ire^{i\theta}\, d\theta$$

$$= ir^{n+1}\int_0^{2\pi} e^{i(n+1)\theta}\, d\theta = ir^{n+1}\left[\frac{e^{i(n+1)\theta}}{i(n+1)}\right]_0^{2\pi} \qquad [\because n \neq -1]$$

$$= \frac{r^{n+1}}{n+1}[e^{i2(n+1)\pi}-1] = \frac{r^{n+1}}{n+1}[\cos 2(n+1)\,\pi + i\sin 2(n+1)\pi - 1] = \frac{r^{n+1}}{n+1}[1+0i-1]$$

$$= 0. \qquad [\text{When } n \neq -1]$$

Which is the required value of the given integral.

When $n = -1$,

$$\oint_C \frac{dz}{z-a} = \int_0^{2\pi}\frac{ire^{i\theta}d\theta}{re^{i\theta}} = i\int_0^{2\pi} d\theta = 2\pi i.$$ **Ans.**

**Example 48.** *Evaluate* $\int_C (z-z^2)dz$, *where C is the upper half of the circle |z – 2| = 3. What is the value of the integral if C is the lower half of the above given circle?*

*(MDU, Dec 2009)*

**Solution.** Put $z = re^{i\theta} = 3e^{i\theta} \Rightarrow dz = 3i.e^{i\theta}\, d\theta$

**Upper Circle**

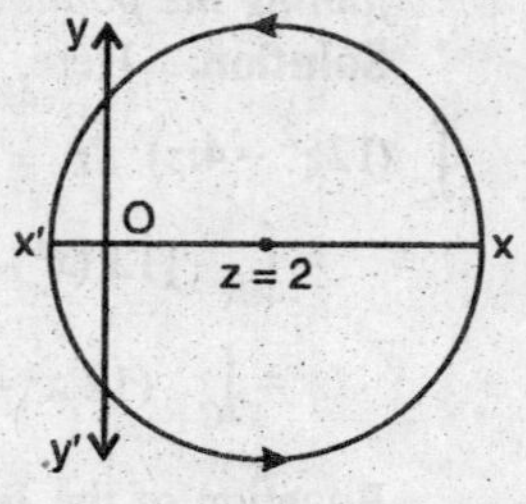


$$\int_C (z-z^2)dz = \int_0^{\pi}(3e^{i\theta}-9e^{2i\theta})3ie^{i\theta}d\theta$$

$$= \int_0^{\pi}[9ie^{2i\theta}-27ie^{3i\theta}]d\theta = \left[\frac{9}{2}e^{2i\theta}-\frac{27}{3}e^{3i\theta}\right]_0^{\pi}$$

$$= \left(\frac{9}{2}e^{2i\pi}-9e^{3i\pi}\right)-\left(\frac{9}{2}-9\right)$$

$$= \frac{9}{2}\left[e^{2i\pi}-2e^{3i\pi}+1\right] \qquad \left(\begin{aligned} e^{i\theta} &= \cos\theta + i\sin\theta \\ \therefore\ e^{i2\pi} &= 1,\ \ e^{i3\pi} = -1\end{aligned}\right)$$

$$= \frac{9}{2}[1+2+1] = 18$$ **Ans.**

**Lower Circle**

$$\int_C (z-z^2)dz = \int_{\pi}^{2\pi}(9ie^{2i\theta}-27ie^{3i\theta})\,d\theta$$

$$= \left[\frac{9}{2}e^{2i\theta}-9e^{3i\theta}\right]_{\pi}^{2\pi} = \left(\frac{9}{2}e^{4\pi i}-9e^{6\pi i}\right)-\left(\frac{9}{2}e^{2\pi i}-9e^{3\pi i}\right)$$

$$= \left(\frac{9}{2}-9\right)-\left(\frac{9}{2}+9\right) = -18$$ **Ans.**

## EXERCISE 7.6

**1.** Integrate $f(z) = x^2 + ixy$ from $A$ (1, 1) to $B$ (2, 8) along

(*i*) the straight line $AB$; (*ii*) the curve $C, x = t, y = t^3$. **Ans.** (*i*) $-\frac{1}{3}(147-71)i$ (*ii*) $-\left(\frac{1094}{21}-\frac{124i}{5}\right)$

**2.** Evaluate $\int_{1-i}^{2+i} (2x+iy+1)\,dz$ along

(*i*) $x = t+1,\ y = 2t^2 - 1$; (*ii*) the straight line joining $1 - i$ and $2 + i$. **Ans.** (*i*) $4+\frac{25}{3}i$ (*ii*) $4 + 8i$

(*R.G.P.V., Bhopal, Dec. 2008*)

**3.** Evalute the line integral $\int_C z^2 dz$ where $C$ is the boundary of a triangle with vertices $0,\ 1 + i.\ -1 + i$ clockwise. **Ans. 0**

**4.** Evaluate $\int_C (z+1)^2 dz$ where $C$ is the boundary of the rectangle with vertices at the points $a + ib,\ -a + ib,\ -a - ib,\ a - ib$. **Ans. 0**

**5.** Evaluate the integral $\int_c |z|\,dz$, where $C$ is the straight line from $z = -i$ to $z = i$. **Ans.** $i$

**6.** Evaluate the integral $\int_c |z|\,dz$, where $C$ is the left half of the unit circle $|z| = 1$ from $z = -i$ to $z = i$. **Ans.** $2i$

**7.** Evaluate the integral $\int_c \log z\,dz$, where $C$ is the unit circle $|z| = 1$. **Ans.** $2\pi i$

**8.** Integrate $xz$ along the straight line from $A$ (1, 1) to $B$ (2, 4) in the complex plane. Is the value the same if the path of integration from $A$ to $B$ is along the curve $x = t, y = t^2$ ?

**Ans.** $-\frac{151}{15}+\frac{45i}{4}$

**9.** Evaluate $\int_0^{2+i} (\bar{z})^2 dz$, along

(*i*) the real axis to 2 and then vertically to $2 + i$, (*ii*) the line $y = x/2$.

(*U.P., III Semester, June 2009*) **Ans.** (*i*) $\frac{1}{3}(14+i)$, (*ii*) $\frac{5}{3}(2-i)$

**Choose the correct answer:**

**10.** The value of $\int_C \frac{4z^2+z+5}{z-4}\,dz$, where C : $9x^2 + 4y^2 = 36$

(*i*) $-1$ (*ii*) 1 (*iii*) 2 (*iv*) 0 (*AMIETE, June 2009*) **Ans.** (*iv*)

## 7.23 IMPORTANT DEFINITIONS

(*i*) **Simply connected Region.** A connected region is said to be a simply connected if all the interior points of a closed curve C drawn in the region D are the points of the region D.

(*ii*) **Multi-Connected Region.** Multi-connected region is bounded by more than one curve. We can convert a multi-connected region into a simply connected one, by giving it one or more cuts.

**Note.** A function $f(z)$ is said to be **meromorphic** in a region $R$ if it is analytic in the region $R$ except at a finite number of poles.

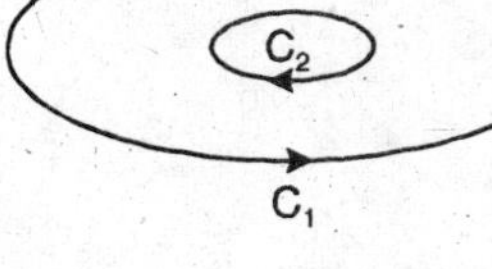


Multi-Connected Region

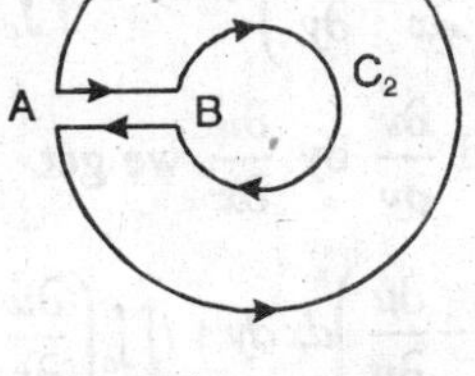


Simply Connected Region

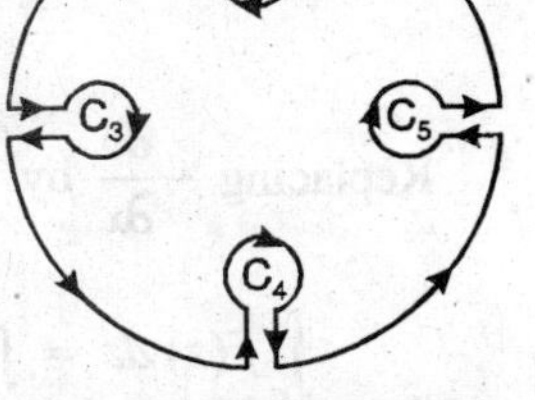


Simply Connected Region

(*iii*) **Single-valued and Multi-valued function**

If a function has only one value for a given value of $z$, then it is a single valued function.

*For example* $\quad f(z) = z^2$

If a function has more than one value, it is known as multi-valued function,

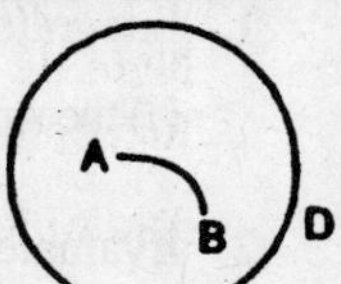


*For example* $\quad f(z) = z^{\frac{1}{2}}$

(*iv*) **Limit of a function**

A function $f(z)$ is said to have a limit $l$ at a point $z = z_0$, if for a given an arbitrary chosen positive number $\varepsilon$, there exists a positive number $\delta$, such that

$$|f(z)-l| < \varepsilon \text{ for } |z-z_0| < \delta$$

It may be written as $\lim\limits_{z\to z_0} f(z) = l$

(*v*) **Continuity**

A function $f(z)$ is said to be continuous at a point $z = z_0$ if for a given an arbitrary positive number $\varepsilon$, there exists a positive number $\delta$, such that

$$|f(z)-l| < \varepsilon \text{ for } |z-z_0| < \delta$$

In other words, a function $f(z)$ is continuous at a point $z = z_0$ if

(*a*) $f(z_0)$ exists $\qquad$ (*b*) $\lim\limits_{z\to z_0} f(z) = f(z)_{z=0}$

(*vi*) **Multiple point.** If an equation is satisfied by more than one value of the variable in the given range, then the point is called a multiple point of the arc.

(*vii*) **Jordan arc.** A continuous arc without multiple points is called a Jordan arc.

**Regular arc.** If the derivatives of the given function are also continuous in the given range, then the arc is called a regular arc.

(*viii*) **Contour.** A contour is a Jordan curve consisting of continuous chain of a finite number of regular arcs.

The contour is said to be closed if the starting point A of the arc coincides with the end point B of the last arc.

(*ix*) **Zeros of an Analytic function.**

The value of $z$ for which the analytic function $f(z)$ becomes zero is said to be the zero of $f(z)$. **For example,** (1) Zeros of $z^2 - 3z + 2$ are $z = 1$ and $z = 2$.

(2) Zeros of $\cos z$ is $\pm\ (2n-1)\dfrac{\pi}{2}$, where $n=1, 2, 3$.........

## 7.24 CAUCHY'S INTEGRAL THEOREM

(*AMIETE, Dec. 2009, U.P. III Semester, 2009-2010, R.G.P.V., Bhopal, III Semester, Dec. 2002*)

If a function $f(z)$ is analytic and its derivative $f'(z)$ continuous at all points inside and on a simple closed curve $c$, then $\int_c f(z)\,dz = 0$.

**Proof.** Let the region enclosed by the curve $c$ be $R$ and let

$$f(z) = u + iv, \quad z = x + iy, \; dz = dx + idy$$

$$\int_c f(z)\,dz = \int_c (u+iv)(dx+idy) = \int_c (u\,dx - v\,dy) + i\int_c (v\,dx + u\,dy)$$

$$= \iint_R \left(-\frac{\partial v}{\partial x} - \frac{\partial u}{\partial y}\right) dx\,dy + i\iint_c \left(\frac{\partial u}{\partial x} - \frac{\partial v}{\partial y}\right) dx\,dy \qquad \text{(By Green's theorem)}$$

Replacing $-\dfrac{\partial v}{\partial x}$ by $\dfrac{\partial u}{\partial y}$ and $\dfrac{\partial v}{\partial y}$ by $\dfrac{\partial u}{\partial x}$ we get

$$\int_c f(z)\,dz = \iint_R \left(\frac{\partial u}{\partial y} - \frac{\partial u}{\partial y}\right) dx\,dy + i\iint_c \left(\frac{\partial u}{\partial x} - \frac{\partial u}{\partial x}\right) dx\,dy = 0 + i0 = 0$$

$\Rightarrow \quad \int_C f(z)\, dz = 0$ **Proved.**

**Note.** If there is no pole inside and on the contour then the value of the integral of the function is zero.

**Example 49.** *Find the integral* $\int_c \dfrac{3z^2+7z+1}{z+1}\, dz$, where $C$ is the circle $|z| = \dfrac{1}{2}$.

**Solution.** Poles of the integrand are given by putting the denominator equal to zero.

$z + 1 = 0 \quad \Rightarrow \quad z = -1$

The given circle $|z| = \dfrac{1}{2}$ with centre at $z = 0$ *and radius* $\dfrac{1}{2}$

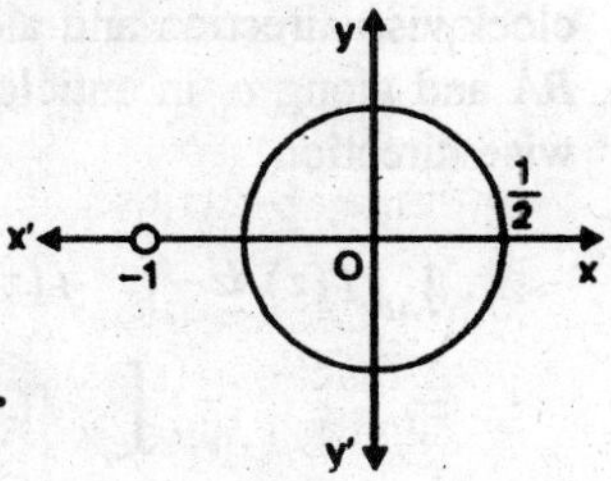


does not enclose any singularity of the given function.

$\int_C \dfrac{3z^2+7z+1}{z+1}\, dz = 0$ *(By Cauchy Integral theorem)* **Ans.**

**Example 50.** *Evaluate* $\oint_C \dfrac{dz}{z^2+9}$, *where C is*

(*i*) $|z + 3i| = 2$ (*ii*) $|z| = 5$ *(M.D.U. May 2009)*

**Solution.** Here $f(z) = \dfrac{1}{z^2+9}$

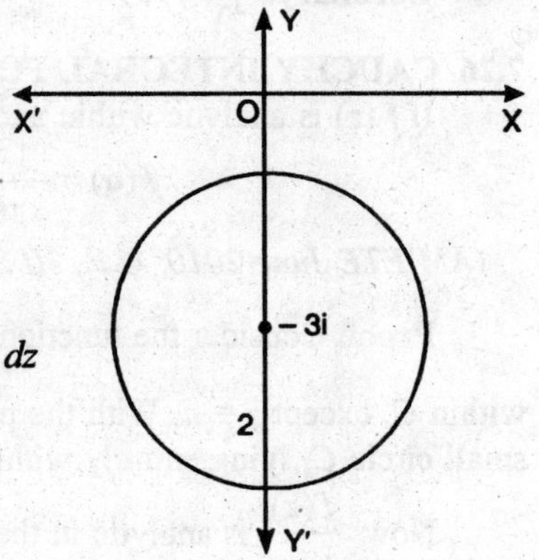


The poles of $f(z)$ can be determined by equating the denominator equal to zero.

(*i*) $\therefore \quad z^2 + 9 = 0 \quad \Rightarrow \quad z = \pm 3i$

Pole at $z = -3i$ lies in the given circle $C$.

$$\int_C f(z)\, dz = \int_C \frac{1}{z^2+9}\, dz = \int_C \frac{1}{(z+3i)(z-3i)}\, dz$$

$$= \int_C \frac{\frac{1}{z-3i}}{z+3i}\, dz$$

$$= 2\pi i \left[\frac{1}{z-3i}\right]_{z=-3i}$$

$$= 2\pi i \left[\frac{1}{-3i-3i}\right] = \frac{-2\pi i}{6i} = -\frac{\pi}{3} \quad \textbf{Ans.}$$

(*ii*) Both the poles $z = 3i$ and $z = -3i$ lie inside the given contour

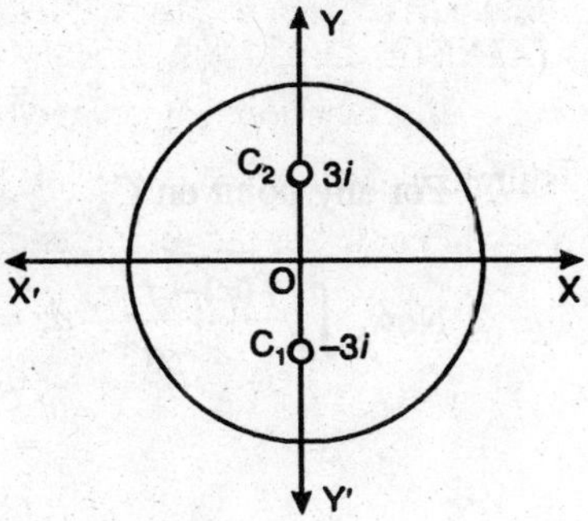


$$\int_C f(z)\, dz = \int_C \frac{1}{z^2+9}\, dz = \int_C \frac{1}{(z+3i)(z-3i)}\, dz$$

$$= \int_{C_1} \frac{\frac{1}{z-3i}}{z+3i}\, dz + \int_{C_2} \frac{\frac{1}{z+3i}}{z-3i}\, dz$$

$$= 2\pi i \left[\frac{1}{z-3i}\right]_{z=-3i} + 2\pi i \left[\frac{1}{z+3i}\right]_{z=3i}$$

$$= 2\pi i \left[\frac{1}{-3i-3i}\right] + 2\pi i \left[\frac{1}{3i+3i}\right] = -\frac{\pi}{3} + \frac{\pi}{3} = 0 \quad \textbf{Ans.}$$

## 7.25 EXTENSION OF CAUCHY'S THEOREM TO MULTIPLE CONNECTED REGION

If $f(z)$ is analytic in the region $R$ between two simple closed curves $c_1$ and $c_2$ then

$$\int_{c_1} f(z)dz = \int_{c2} f(z)dz$$

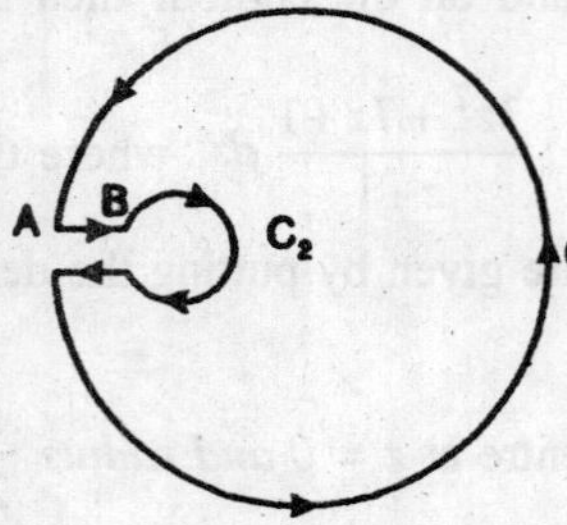


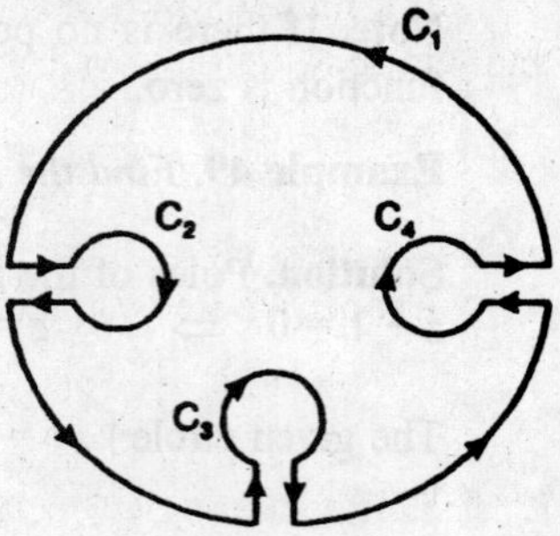


**Proof.** $\int f(z)\,dz = 0$
where the path of integration is along $AB$, and curves $c_2$ in clockwise direction and along $BA$ and along $c_1$ in anticlockwise direction.

$$\int_{AB} f(z)\,dz - \int_{c_2} f(z)\,dz + \int_{BA} f(z)\,dz + \int_{c_1} f(z)dz = 0$$

$$\Rightarrow \quad -\int_{c_2} f(z)\,dz + \int_{c_1} f(z)\,dz = 0 \qquad \text{as} \int_{AB} f(z)\,dz = -\int_{BA} f(z)\,dz$$

$$\int_{c_1} f(z)\,dz = \int_{c_2} f(z)\,dz$$

**Proved.**

**Corollary.** $\int_{c_1} f(z)\,dz = \int_{c_2} f(z)\,dz + \int_{c_3} f(z)\,dz + \int_{c_4} f(z)\,dz$

## 7.26 CAUCHY INTEGRAL FORMULA

If $f(z)$ is analytic within and on a closed curve $C$, and if $a$ is any point within $C$, then

$$f(a) = \frac{1}{2\pi i}\int_z \frac{f(z)}{z-a}\,dz$$

*(AMIETE June 2010, U.P., III Semester Dec. 2009 R.G.P.V., Bhopal, III Semester, June 2008)*

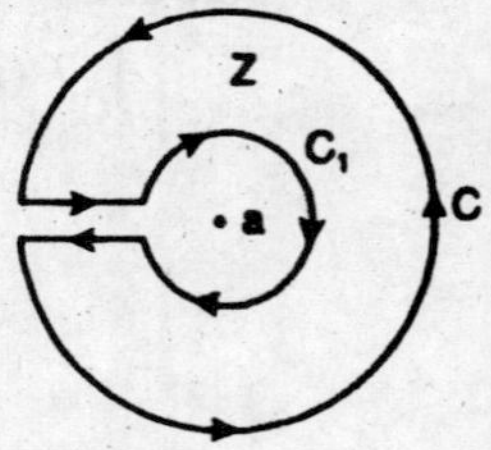


**Proof.** Consider the function $\dfrac{f(z)}{z-a}$, which is analytic at all points within $C$, except $z = a$. With the point $a$ as centre and radius $r$, draw a small circle $C_1$ lying entirely within $C$.

Now $\dfrac{f(z)}{z-a}$ is analytic in the region between $C$ and $C_1$; hence by Cauchy's Integral Theorem for multiple connected region, we have

$$\int_c \frac{f(z).dz}{z-a} = \int_{c_1} \frac{f(z)}{z-a}\,dz = \int_{c_1} \frac{f(z) - f(a) + f(a)}{z-a}\,.\,dz$$

$$= \int_{c_1} \frac{f(z) - f(a)}{z-a}\,dz + f(a)\int_{c_1} \frac{dz}{z-a} \qquad \text{... (1)}$$

For any point on $C_1$

$$\text{Now,} \int_{c_1} \frac{f(z)-f(a)}{z-a}\,dz = \int_0^{2\pi} \frac{f(a+re^{i\theta}) - f(a)}{re^{i\theta}}\, ire^{i\theta}\,d\theta \qquad [\,z-a = re^{i\theta} \text{ and } dz = ire^{i\theta}\,d\theta\,]$$

$$= \int_0^{2\pi} [f(a+re^{i\theta}) - f(a)]\; id\theta = 0 \qquad \text{(where } r \text{ tends to zero).}$$

$$\int_{c_1} \frac{dz}{z-a} = \int_0^{2\pi} \frac{ir\,e^{i\theta}\,d\theta}{r\,e^{i\theta}} = \int_0^{2\pi} id\theta = i[\theta]_0^{2\pi} = 2\pi i$$

Putting the values of the integrals in R.H.S. of (1), we have

$$\int_c \frac{f(z)\,dz}{z-a} = 0 + f(a)\,(2\pi i) \Rightarrow f(a) = \frac{1}{2\pi i}\int_c \frac{f(z)}{z-a}\,dz$$

**Proved.**

## 7.27 CAUCHY INTEGRAL FORMULA FOR THE DERIVATIVE OF AN ANALYTIC FUNCTION

*(R.G.P.V., Bhopal, III Semester, Dec. 2007)*

**If a function $f(z)$ is analytic in a region $R$, then its derivative at any point $z = a$ of $R$ is also analytic in $R$, and is given by**

$$f'(a)=\frac{1}{2\pi i}\int_c \frac{f(z)}{(z-a)^2}dz$$

**where $c$ is any closed curve in $R$ surrounding the point $z = a$.**

**Proof.** We know Cauchy's Integral formula

$$f(a)=\frac{1}{2\pi i}\int_c \frac{f(z)}{(z-a)}dz \quad \text{... (1)}$$

Differentiating (1) w.r.t. '$a$', we get

$$f'(a)=\frac{1}{2\pi i}\int_c f(z)\frac{\partial}{\partial a}\left(\frac{1}{z-a}\right)dz$$

$$\boxed{f'(a)=\frac{1}{2\pi i}\int_c \frac{f(z)}{(z-a)^2}dz}$$

Similarly, $$\boxed{f''(a)=\frac{2!}{2\pi i}\int_c \frac{f(z)dz}{(z-a)^3}} \Rightarrow \boxed{f^n(a)=\frac{n!}{2\pi i}\int_c \frac{f(z)dz}{(z-a)^{n+1}}}$$

**Example 51.** *Evaluate* $\int_C \frac{e^{3z}}{(z-\log 2)^4}dz$, *where C is the square with vertices at ± 1, ± i*

*(M.D.U. Dec. 2009)*

**Solution.** Here we have $\int_C \frac{e^{3z}}{(z-\log 2)^4}dz$

The pole is found by putting the denominator equal to zero

$(z - \log 2)^4 = 0 \Rightarrow z = \log 2$

The integral has a pole of fourth order.

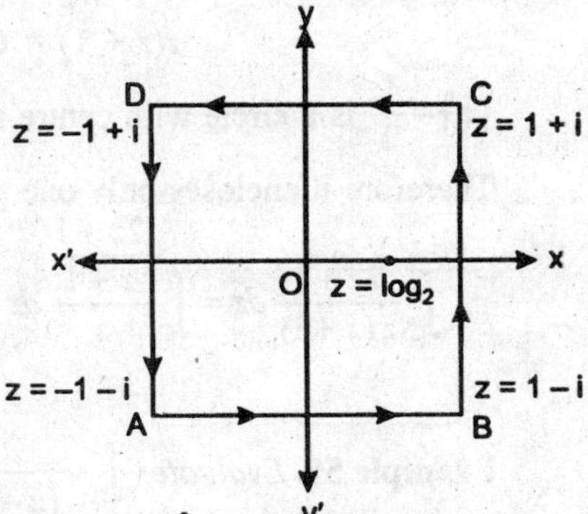


$$\int_C \frac{e^{3z}}{(z-\log_2)^4}dz=\frac{2\pi i}{3!}\frac{d^3}{dz^3}(e^{3z})_{z=\log 2}$$

[By Cauchy formula]

$$=\frac{2\pi i}{3!}\,3.3.3(e^{3z})_{z=\log 2}=9\pi i\,e^{3\log 2}=9\pi i\,e^{\log 2^3}=9\pi i(2)^3=72\pi i \quad \textbf{Ans.}$$

**Example 52.** *Prove that* $\int_C \frac{dz}{z-a}=2\pi i$, *where C is the circle* $|z-a| = r$

*(R.G.P.V., Bhopal, III Semester, Dec. 2006)*

**Solution.** We have,

$$\int_C \frac{dz}{z-a}, \text{ where } C \text{ is the circle with centre } (a, 0) \text{ and radius } r.$$

By Cauchy Integral Formula

$$\left[\int_C \frac{f(z)}{z-a}dz=2\pi i\ f(a)\right]$$

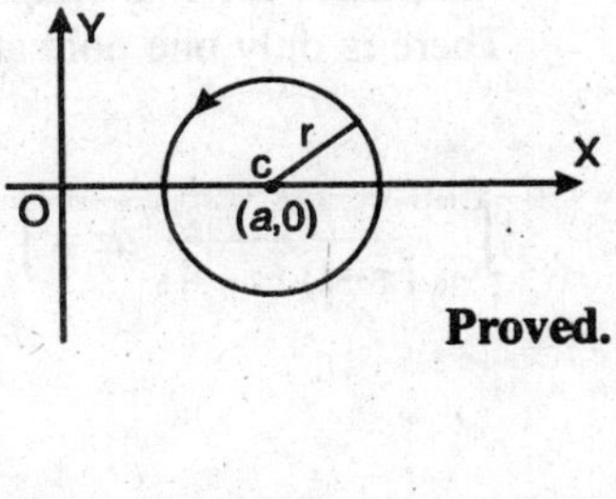


$$\int_C \frac{dz}{z-a}=2\pi i\ (1)$$

$$\Rightarrow \int_C \frac{dz}{z-a}=2\pi i$$

**Proved.**

**Example 53.** *Use Cauchy's integral formula to evaluate* $\int_c \frac{z}{(z^2-3z+2)}\,dz$ *where c is the circle* $|z-2| = \frac{1}{2}$ *(U.P. III Semester, June 2009)*

**Solution.** Here, we have

$$\int_c \frac{z}{(z^2-3z+2)}\,dz$$

The poles are determined by putting the denominator equal to zero

i.e.; $z^2 - 3z + 2 = 0 \Rightarrow (z-1)\ (z-2) = 0$

$\Rightarrow \qquad z = 1, 2$

So, there are two poles $z = 1$ and $z = 2$.

There is only one pole at $z = 2$ inside the given circle.

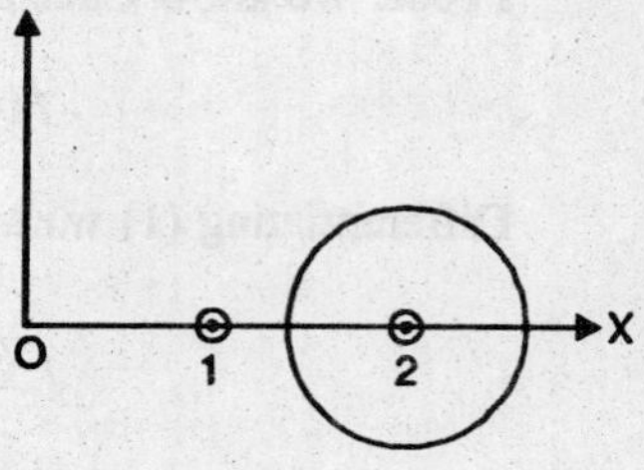


$$\int_c \frac{z}{(z^2-3z+2)}\,dz = \int_c \frac{z}{(z-1)\,(z-2)}\,dz$$

$$= \int_c \frac{\frac{z}{z-1}}{z-2}\,dz \qquad \left[\int_c \frac{f(z)}{z-a}\,dz = 2\pi i\, f(a)\right]$$

$$= 2\pi i\left[\frac{z}{z-1}\right]_{z=2} = 2\pi i\left(\frac{2}{2-1}\right) = 4\,\pi\, i$$ **Ans.**

**Example 54.** *Use Cauchy's integral formula to calculate*

$\int_C \frac{2z+1}{z^2+z}\,dz$ *where C is* $|z| = \frac{1}{2}$. *(AMIETE, Dec. 2009)*

**Solution.** Poles are given by

$$z^2 + z = 0$$

$\Rightarrow \qquad z(z+1) = 0 \qquad \Rightarrow \qquad z = 0, -1$

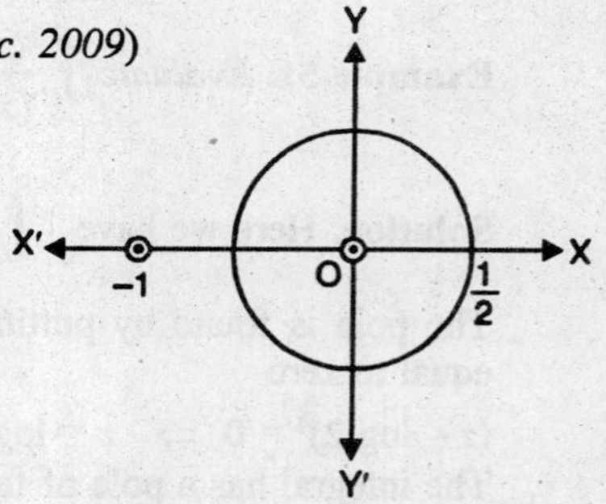


$|z| = \frac{1}{2}$ is a circle with centre at origin and radius $\frac{1}{2}$.

Therefore it encloses only one pole $z = 0$.

$$\therefore \int_C \frac{2z+1}{z(z+1)}\,dz = \int_C \frac{\frac{2z+1}{z+1}}{z}\,dz = 2\pi i\left[\frac{2z+1}{z+1}\right]_{z=0} = 2\pi i$$ **Ans.**

**Example 55.** *Evaluate:* $\int_C \frac{e^z}{(z-1)\,(z-4)}\,dz$ *where C is the circle* $|z| = 2$ *by using Cauchy's Integral Formula.* *(R.G.P.V., Bhopal, III Semester, June 2006)*

**Solution.** We have,

$\int_C \frac{e^z}{(z-1)\,(z-4)}\,dz$ where C is the circle with centre at origin and radius 2.

Poles are given by putting the denominator equal to zero.

$$(z-1)(z-4) = 0$$

$\Rightarrow \qquad z = 1, 4$

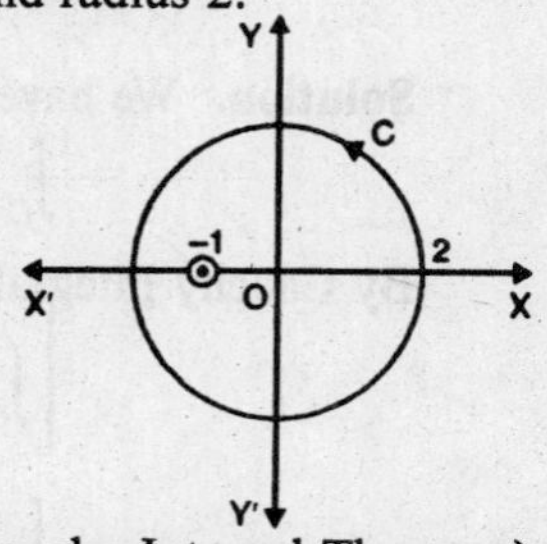


Here there are two simple poles at $z = 1$ and $z = 4$.

There is only one pole at $z = 1$ inside the contour. Therefore

$$\int_C \frac{e^z}{(z-1)\,(z-4)}\,dz = \int \frac{\frac{e^z}{(z-4)}}{(z-1)}\,dz = 2\pi i\left[\frac{e^z}{z-4}\right]_{z=1}$$ (By Cauchy Integral Theorem)

$$= 2\pi i\left(\frac{e}{1-4}\right) = -\frac{2\pi i e}{3}$$

Which is the required value of the given integral. **Ans.**

**Example 56.** *State the Cauchy's integral formula. Show that* $\int_C \frac{e^{zt}}{z^2+1}\,dz = 2\pi i \sin t$

*if t > 0 and C is the circle | z | = 3* *(U.P., III Semester, Dec. 2009)*

**Solution.** Here, we have $\int_C \frac{e^{zt}}{z^2+1}\,dz$

The poles are determined by putting the denominator equal to zero.

i.e., $z^2 + 1 = 0$

$\Rightarrow \quad z^2 = -1$

$\Rightarrow \quad z = \pm\sqrt{-1} = \pm i$

$\Rightarrow \quad z = i, -i$

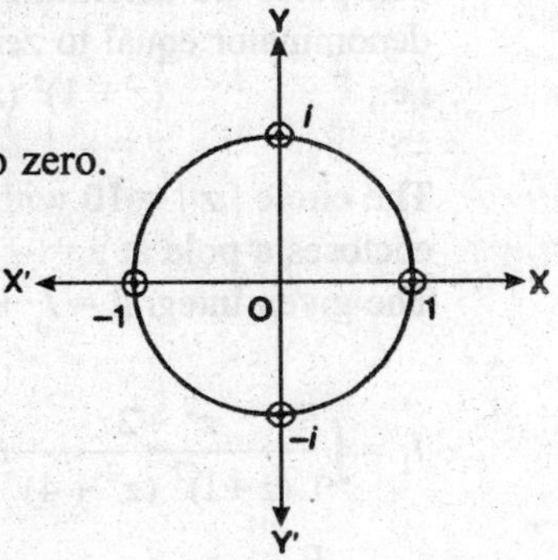


The integrand has two simple poles at $z = i$ and at $z = -i$. Both poles are inside the given circle with centre at origin and radius 3.

Now, $\int_C \frac{e^{zt}}{z^2+1}\,dz = \frac{1}{2i}\int_C\left(\frac{e^{zt}}{z-i} - \frac{e^{zt}}{z+i}\right)dz$ [By partial fraction]

$$= \frac{1}{2i}\left[\int_{C_1}\frac{e^{zt}}{z-i}\,dz - \int_{C_2}\frac{e^{zt}}{z+i}\,dz\right] = \frac{1}{2i}\left[2\pi i\,(e^{zt})_{z=i} - 2\pi i\,(e^{zt})_{z=-i}\right]$$

$$= \frac{2\pi i}{2i}\left[e^{ti} - e^{-ti}\right] = 2\pi i \sin t$$

**Example 57.** *Evaluate the following integral using Cauchy integral formula*

$$\int_c \frac{4-3z}{z(z-1)(z-2)}\,dz \text{ where } c \text{ is the circle } |z| = \frac{3}{2}.$$

*(AMIETE, Dec. 2009, R.G.P.V., Bhopal, III Semester, June 2008)*

**Solution.** Poles of the integrand are given by putting the denominator equal to zero.

$z(z-1)(z-2)$ or $z = 0, 1, 2$

The integrand has three simple poles at $z = 0, 1, 2$.

The given circle $|z| = \frac{3}{2}$ with centre at $z = 0$ and radius $= \frac{3}{2}$ encloses two poles $z = 0$, and $z = 1$.

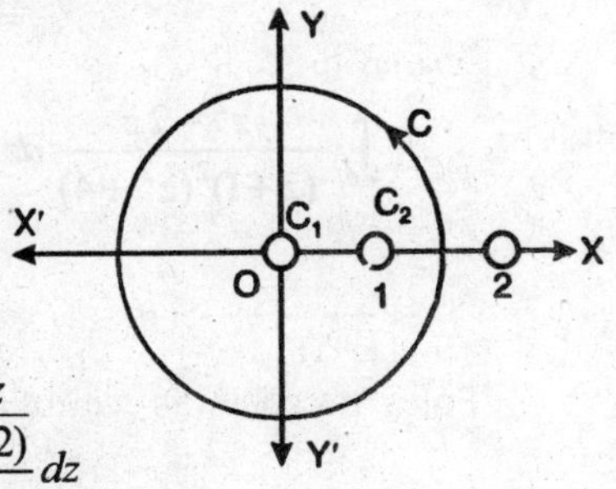


$$\int_C \frac{4-3z}{z(z-1)(z-2)}\,dz = \int_{C_1}\frac{\frac{4-3z}{(z-1)(z-2)}}{z}\,dz + \int_{C_2}\frac{\frac{4-3z}{z(z-2)}}{z-1}\,dz$$

$$= 2\pi i\left[\frac{4-3z}{(z-1)(z-2)}\right]_{z=0} + 2\pi i\left[\frac{4-3z}{z(z-2)}\right]_{z=1}$$

$$= 2\pi i.\frac{4}{(-1)(-2)} + 2\pi i\frac{4-3}{1(1-2)} = 2\pi i(2-1) = 2\pi i$$

Which is the required value of the given integral. **Ans.**

**Example 58.** *Evaluate* $\int_c \frac{z^2-2z}{(z+1)^2\,(z^2+4)}\,dz$

*where c is the circle* | z | = *10.* *(U.P. III Semester, June 2009)*

**Solution.** Here, we have

$$\int_c \frac{z^2-2z}{(z+1)^2\,(z^2+4)}\,dz$$

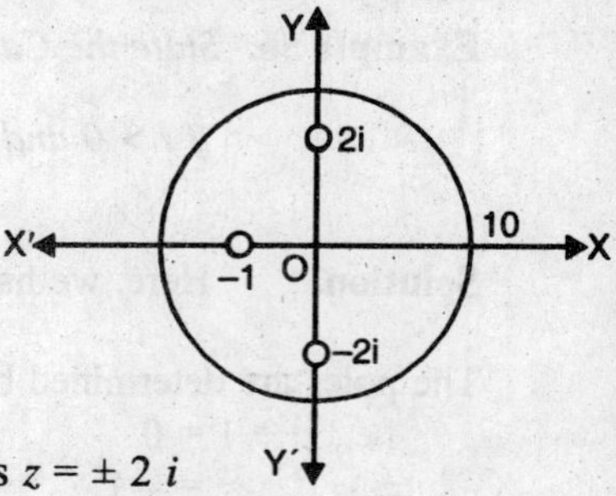


The poles are determined by putting the denominator equal to zero.

i.e.; $(z+1)^2\,(z^2+4)=0$

$\Rightarrow$ $z=-1,-1$ and $z=\pm 2\,i$

The circle | z | = 10 with centre at origin and radius = 10.

encloses a pole at $z=-1$ of second order and simple poles $z=\pm 2\,i$

The given integral $= I_1+I_2+I_3$

$$I_1=\int_{c_1}\frac{z^2-2z}{(z+1)^2\,(z^2+4)}\,dz=\int_{c_1}\frac{\frac{z^2-2z}{z^2+4}}{(z+1)^2}\,dz=2\pi i\left[\frac{d}{dz}\,\frac{z^2-2z}{z^2+4}\right]_{z=-1}$$

$$=2\pi i\left[\frac{(z^2+4)\,(2z-2)-(z^2-2z)\,2z}{(z^2+4)^2}\right]_{z=-1}=2\pi i\left[\frac{(1+4)\,(-2-2)-(1+2)\,2(-1)}{(1+4)^2}\right]$$

$$=2\pi i\left(-\frac{14}{25}\right)=\frac{-28\pi i}{25}$$

$$I_2=\int_{c_2}\frac{\frac{z^2-2z}{(z+1)^2\,(z+2i)}}{(z-2i)}\,dz=2\pi i\left[\frac{z^2-2z}{(z+1)^2\,(z+2i)}\right]_{z=2i}=2\pi i\left[\frac{-4-4i}{(2i+1)^2\,(2i+2i)}\right]=2\pi i\,\frac{(1+i)}{4+3i}$$

$$I_3=\int_{c_3}\frac{\frac{z^2-2z}{(z+1)^2\,(z-2i)}}{(z+2i)}\,dz=2\pi i\left[\frac{z^2-2z}{(z+1)^2\,(z-2i)}\right]_{z=-2i}$$

$$=2\pi i\left[\frac{-4+4i}{(-2i+1)^2\,(-2i-2i)}\right]=2\pi i\,\frac{(i-1)}{(3i-4)}$$

$$\int_c\frac{z^2-2z}{(z+1)^2(z^2+4)}\,dz=I_1+I_2+I_3$$

$$=\frac{-28\pi i}{25}+2\pi i\left(\frac{1+i}{4+3i}\right)+2\pi i\left(\frac{i-1}{3i-4}\right)$$

$$=2\pi i\left[\frac{-14}{25}+\frac{1+i}{(4+3i)}+\frac{(i-1)}{(3i-4)}\right]$$

$$=2\pi i\left[\frac{-14}{25}+\frac{(1+i)\,(3i-4)+(i-1)\,(4+3i)}{(-9-16)}\right]$$

$$=\frac{2\pi i}{-25}\,[14+(3i-4-3-4i)+(4i-3-4-3i)]$$

$$=0$$ **Ans.**

**Example 59.** *Find the value of* $\int_C \frac{3z^2+z}{z^2-1}\,dz$.

*If c is circle* $|z-1|=1$ *(R.G.P.V., Bhopal, III Semester, June 2007)*

**Solution.** Poles of the integrand are given by putting the denominator equal to zero.

$$z^2-1=0,\ z^2=1,\ z=\pm 1$$

The circle with centre $z=1$ and radius unity encloses a simple pole at $z=1$.

By Cauchy Integral formula

$$\int_C \frac{3z^2+z}{z^2-1}\,dz=\int_C \frac{\frac{3z^2+z}{z+1}}{z-1}\,dz = 2\pi i\left[\frac{3z^2+z}{z+1}\right]_{z=1} = 2\pi i\left(\frac{3+1}{1+1}\right)=4\pi i$$

Which is the required value of the given integral. **Ans.**

**Example 60.** *Use Cauchy integral formula to evaluate.*

$$\int_C \frac{\sin \pi z^2+\cos \pi z^2}{(z-1)(z-2)}\,dz$$

*where c is the circle* $|z|=3$.

*(AMIETE, Dec. 2010, R.G.P.V., Bhopal, III Semester, June 2003)*

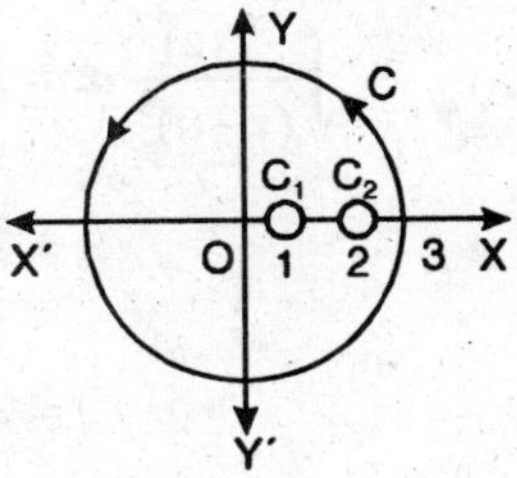


**Solution.** $\oint \frac{\sin \pi z^2+\cos \pi z^2}{(z-1)(z-2)}\,dz$

Poles of the integrand are given by putting the denominator equal to zero.

$$(z-1)(z-2)=0 \ \Rightarrow\ z=1,2$$

The integrand has two poles at $z=1, 2$.

The given circle $|z|=3$ with centre at $z=0$ and radius 3 encloses both the poles $z=1$, and $z=2$.

$$\oint_C \frac{\sin \pi z^2+\cos \pi z^2}{(z-1)(z-2)}\,dz=\int_{C_1} \frac{\frac{\sin \pi z^2+\cos \pi z^2}{(z-2)}}{(z-1)}\,dz+\int_{C_2} \frac{\frac{\sin \pi z^2+\cos \pi z^2}{(z-1)}}{(z-2)}\,dz$$

$$= 2\pi i\left[\frac{\sin \pi z^2+\cos \pi z^2}{z-2}\right]_{z=1}+2\pi i\left[\frac{\sin \pi z^2+\cos \pi z^2}{z-1}\right]_{z=2}$$

$$= 2\pi i\left(\frac{\sin \pi+\cos \pi}{1-2}\right)+2\pi i\left(\frac{\sin 4\pi+\cos 4\pi}{2-1}\right) = 2\pi i\left(\frac{-1}{-1}\right)+2\pi i\left(\frac{1}{1}\right)=4\pi i$$

Which is the required value of the given integral. **Ans.**

**Example 61.** *Derive Cauchy Integral Formula.*

*Evaluate* $\int_C \frac{e^{3iz}}{(z+\pi)^3}\,dz$

*where C is the circle* $|z-\pi|=3.2$

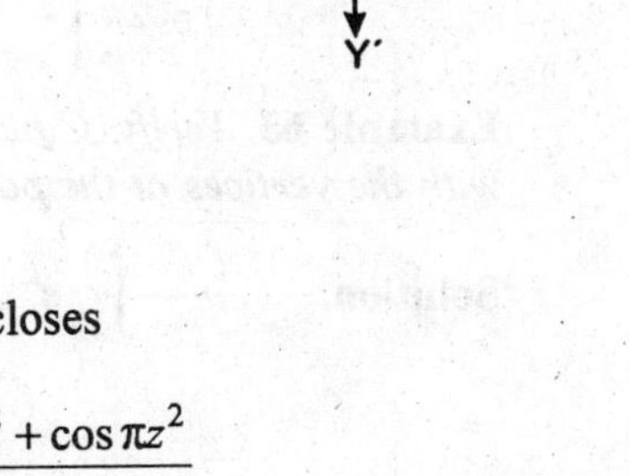


**Solution.** Here, $I=\int_C \frac{e^{3iz}}{(z+\pi)^3}\,dz$

Where $C$ is a circle $\{|z-\pi|=3.2\}$ with centre $(\pi, 0)$ and radius 3.2.

Poles are determined by putting the denominator equal to zero.

$$(z+\pi)^3 = 0 \quad \Rightarrow \quad z = -\pi, -\pi, -\pi$$

There is a pole at $z - \pi$ of order 3. But there is no pole within $C$.

By Cauchy Integral Formula $\int_C \frac{e^{3iz}}{(z+\pi)^3} dz = 0$ **Ans.**

**Example 62.** *Evaluate, using Cauchy's integral formula,*

$$\int_C \frac{\log z}{(z-1)^3} dz \quad where\ C\ is\ |z-1| = \frac{1}{2}. \qquad (MDU.\ Dec.\ 2010)$$

**Solution.** Using Cauchy's Integral formula,

$$\int_C \frac{\log z}{(z-1)^3} dz \qquad C: |z-1| = \frac{1}{2}$$

Poles are determined by putting denominater equal to zero.

$$(z-1)^3 = 0 \quad \Rightarrow \quad z = 1, 1, 1$$

There is one pole of order three at $z = 1$ which is inside the circle $C$.

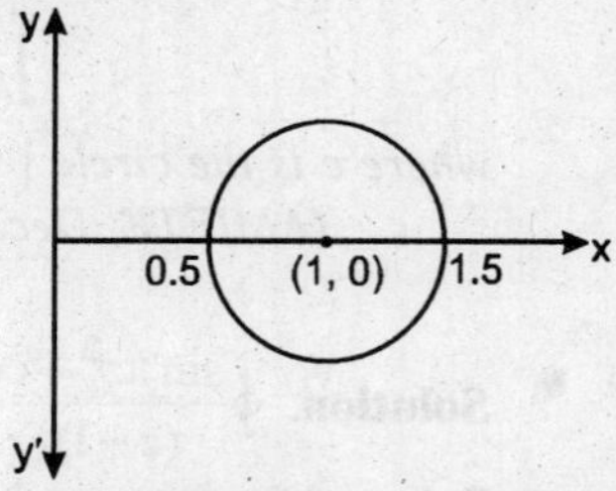


$$\int \frac{f(z)}{(z-a)^3} dz = 2\pi i f^2(a)$$

$$= 2\pi i \left[\frac{d^2}{dz^2} \log z\right]_{z=1} = 2\pi i \left[\frac{d}{dz}\left(\frac{1}{z}\right)\right]_{z=1}$$

$$= 2\pi i \left(-\frac{1}{z^2}\right)_{z=1} = -2\pi i$$

**Example 63.** *Verify, Cauchy theorem by integrating $e^{iz}$ along the boundary of the triangle with the vertices at the points $1 + i$, $-1 + i$ and $-1 - i$.*

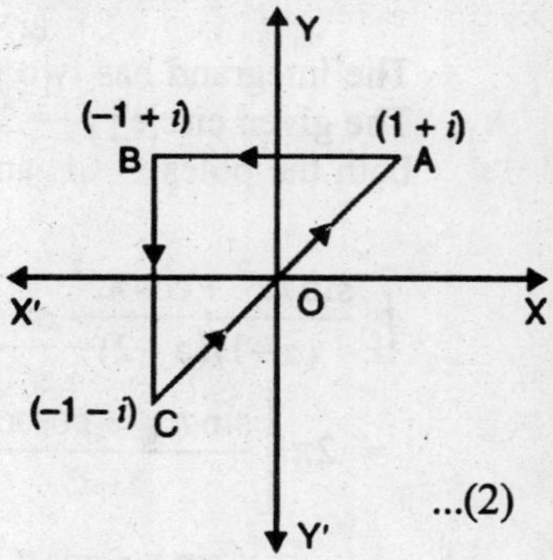


**Solution.**

$$\int_{AB} e^{iz} dz = \left(\frac{e^{iz}}{i}\right)_{1+i}^{-1+i} = \frac{1}{i}\left[e^{i(-1+i)} - e^{i(1+i)}\right]$$

$$= \frac{1}{i}\left[e^{-i-1} - e^{i-1}\right] \quad ...(1)$$

$$\int_{BC} e^{iz} dz = \left[\frac{e^{iz}}{i}\right]_{-1+i}^{-1-i} = \frac{1}{i}\left[e^{i(-1-i)} - e^{i(-1+i)}\right]$$

$$= \frac{1}{i}\left[e^{-i+1} - e^{-i-1}\right] \quad ...(2)$$

$$\int_{CA} e^{iz} dz = \left[\frac{e^{iz}}{i}\right]_{-1-i}^{1+i} = \frac{1}{i}\left[e^{i(1+i)} - e^{i(-1-i)}\right] = \frac{1}{i}\left[e^{i-1} - e^{-i+1}\right] \quad ...(3)$$

On adding (1), (2) and (3), we get

$$\int_{AB} e^{iz} dz + \int_{BC} e^{iz} dz + \int_{CA} e^{iz} dz = \frac{1}{i}\left[\left(e^{-i-1} - e^{i-1}\right) + \left(e^{-i+1} - e^{-i-1}\right) + \left(e^{i-1} - e^{-i+1}\right)\right]$$

$$\Rightarrow \quad \int_{\Delta ABC} e^{iz} dz = 0 \quad ...(4)$$

The given function has no pole. So there is no pole in Δ ABC.

The given function $e^{iz}$ is analytic inside and on the triangle ABC.

By Cauchy Theorem, we have $\int_C e^{iz} dz = 0$ ...(5)

From (4) and (5) theorem is verified.

## EXERCISE 7.7

**Evaluate the following:**

**1.** $\int_C \frac{1}{z-a}\,dz$, where $c$ is a simple closed curve and the point $z = a$ is
(*i*) outside $c$; (*ii*) inside $c$. **Ans.** (*i*) 0 (*ii*) $2\pi i$

**2.** $\int_c \frac{e^z}{z-1}\,dz$, where $c$ is the circle $|z| = 2$. **Ans.** $2\pi ie$

**3.** $\int_C \frac{\cos \pi z}{z-1}\,dz$, where $c$ is the circle $|z| = 3$. **Ans.** $-2\pi i$

**4.** $\int_C \frac{\cos \pi z^2}{(z-1)(z-2)}\,dz$, where $c$ is the circle $|z| = 3$. **Ans.** $4\pi i$

**5.** $\int_C \frac{e^{-z}}{(z+2)^5}\,dz$, where $c$ is the circle $|z| = 3$. **Ans.** $\frac{\neq ie^2}{12}$

**6.** $\int_C \frac{e^{2z}}{(z+1)^4}\,dz$, where $c$ is the circle $|z| = 2$. **Ans.** $\frac{8\pi}{3} ie^{-2}$

**7.** $\int_C \frac{2z^2+z}{z^2-1}\,dz$ where $c$ is the circle $|z-1| = 1$ **Ans.** $3\pi i$

**8.** $\int_C \frac{e^z}{z^2(z+1)^3}\,dz$, $C : |z| = 2$. (*AMIETE, June 2009*) **Ans.** $\pi i\left[-4+\frac{11}{e}\right]$

**9.** Evaluate $\oint_C \frac{z^3+z+1}{z^2-3z+2}\,dz$, where $C$ is the ellipse $4x^2 + 9y^2 = 1$.
(*M.D.U. Dec. 2005, May 2008*) **Ans.** 0

**10.** Evaluate $\int_C \frac{\sin^2 z}{\left(z-\frac{\pi}{6}\right)^3}\,dz$, where $C$ is $|z| = 1$. (*M.D.U. May 2006, Dec. 2006*) **Ans.** $\pi i$

**11.** Evaluate $\oint \frac{\sin 6z}{\left(z-\frac{\pi}{6}\right)^3}\,dz$, where $C$ is the circle $|z| = 1$. (*M.D.U. May 2005*) **Ans.** $\frac{21}{16}\neq i$

**12.** If $f(\xi) = \oint_C \frac{4z^2+z+5}{z-\xi}\,dz$, where $C$ is the ellipse $\frac{x^2}{4}+\frac{y^2}{9} = 1$, find $f(1), f(i), f'(-1)$ and $f''(-i)$.
(*J.N.T.U. 2005; M.D.U., Dec. 2007*) **Ans.** $20\pi i$, $2\pi(i-1)$, $-14\pi i$, $16\pi i$

**13.** $\int_C \frac{z}{z^2-3z+2}\,dz$, where $C$ is $|z-2| = \frac{1}{2}$. (*U.P.T.U. 2009; M.D.U. 2007*) **Ans.** $4\pi i$

**14.** $\int_C \frac{e^z\,dz}{(z+1)^2}$, where $C$ is $|z-1| = 3$. (*M.D.U. Day 2006*) **Ans.** $\frac{2\neq i}{e}$

**Choose the correct alternative:**

**15.** The value of the integral $\int_C \frac{z^2+1}{(z+1)(z+2)}\,dz$, where $C$ is $|z| = \frac{3}{2}$ is

(*i*) $\pi i$ (*ii*) 0 (*iii*) $2\pi i$ (*iv*) $4\pi i$ **Ans.** (*iv*)

(*AMIETE, June 2010*)

**16.** Cauchy's Integral formula states that if $f(z)$ is analytic within $a$ and on $a$ closed curve $C$ and if $a$ is any point within $C$ then $f(a) =$ : (*R.G.P.V., Bhopal, III Semester, June 2007*)

(*i*) $\frac{1}{2\pi i}\oint \frac{f(z)\,dz}{z-a}$ (*ii*) $\frac{1}{2\pi i}\oint f(z)\,dz$ (*iii*) $\frac{1}{2\pi i}\oint \frac{dz}{z-a}$ (*iv*) none of these. **Ans.** (*i*)

**17.** The value of $\int_C \frac{z^2-z+1}{z-1} dz$, $C$ being $|z| = \frac{1}{2}$ is :

(*i*) $2\pi i$ (*ii*) $\frac{1}{2\pi i}$ (*iii*) 0 (*iv*) $\pi i$ (*R.G.P.V., Bhopal, III Sem., Dec. 2006*) **Ans.** (*iii*)

**18.** If $f(z) = \frac{z^2}{(z-1)^2(z+2)}$, then Res. $f(-2)$ is :

(*i*) $\frac{5}{9}$ (*ii*) $\frac{4}{9}$ (*iii*) $\frac{1}{9}$ (*iv*) $\frac{3}{9}$ (*RGPV, Bhopal, III Sems, Dec. 2006*) **Ans.** (*ii*)

**19.** Let $f(z) = \frac{1}{(z-2)^4(z+3)^6}$, then $z = 2$ and $z = -3$ are the poles of order :

(*i*) 6 and 4 (*ii*) 2 and 3 (*iii*) 3 and 4 (*iv*) 4 and 6 (*RGPV, Bhopal, III Sem., June 2006*) **Ans.** (*iv*)

**20.** The value of the integral $\int_C \frac{z+1}{z^3-2z^2} dz$, where $C$ is the circle $|z| = 1$ is equal to.

(*i*) $2\pi i$ (*ii*) $-\frac{2}{3}\pi i$ (*iii*) zero (*iv*) $-\frac{3}{2}\pi i$ (*AMIETE, Dec. 2010*) **Ans.** (*iv*)

## 7.28 GEOMETRICAL REPRESENTATION

To draw a curve of complex variable $(x, y)$ on z-plane we take two axes *i.e.*, one real axis and the other imaginary axis. A number of points $(x, y)$ are plotted on z-plane, by taking different value of $z$ (different values of $x$ and $y$). The curve $C$ is drawn by joining the plotted points. The diagram obtained is called *Argand diagram* in z-plane.

But a complex function $w = f(z)$ *i.e.*, $(u + iv) = f(x + iy)$ involves four variables $x$, $y$ and $u$, $v$.

A figure of only three dimensions $(x, y, z)$ is possible in a plane. A figure of four dimensional region for $x, y, u, v$ is not possible.

So, we choose two complex planes z-plane and w-plane. In the w-plane we plot the corresponding points $w = u + iv$. By joining these points we have a corresponding curve $C'$ in w-plane.

## 7.29 TRANSFORMATION

For every point $(x, y)$ in the z-plane, the relation $w = f(z)$ defines a corresponding point $(u, v)$ in the w-plane. We call this "transformation or mapping of z-plane into w-plane". If a point $z_0$ maps into the point $w_0$, $w_0$ is also known as the image of $z_0$.

If the point $P(x, y)$ moves along a curve $C$ in z-plane, the point $P'(u, v)$ will move along a corresponding curve $C'$ in w-plane, then we say that a curve $C$ in the z-plane is mapped into the corresponding curve $C'$ in the w-plane by the relation $w = f(z)$.

**Example 64.** *Transform the rectangular region ABCD in z-plane bounded by $x = 1$, $x = 3$; $y = 0$ and $y = 3$. Under the transformation $w = z + (2 + i)$.*

**Solution.** Here, $w = z + (2 + i)$

$$\Rightarrow \quad u + iv = x + iy + (2 + i)$$
$$= (x + 2) + i(y + 1)$$

By equating real and imaginary quantities, we have $u = x + 2$ and $v = y + 1$.

| z-plane | w-plane | z-plane | w-plane |
|---|---|---|---|
| $x$ | $u = x + 2$ | $y$ | $v = y + 1$ |
| 1 | $= 1 + 2 = 3$ | 0 | $= 0 + 1 = 1$ |
| 3 | $= 3 + 2 = 5$ | 3 | $= 3 + 1 = 4$ |

Here the lines $x = 1, x = 3; y = 0$ and $y = 1$ in the $z$-plane are transformed onto the line $u = 3, u = 5; v = 1$ and $v = 4$ in the $w$-plane. The region $ABCD$ in $z$-plane is transformed into the region $EFGH$ in $w$-plane. **Ans.**

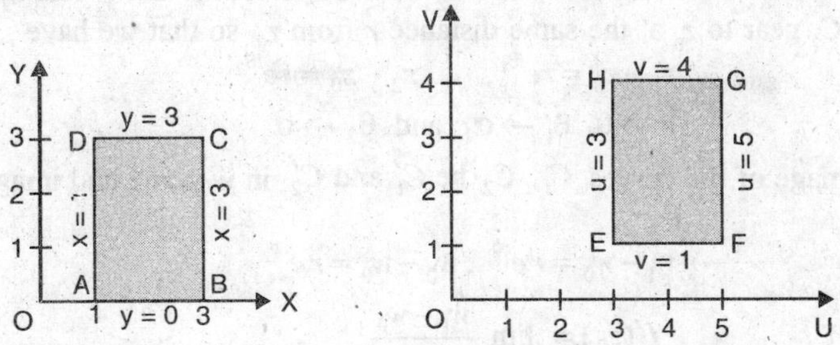


**Example 65.** *Transform the curve $x^2 - y^2 = 4$ under the mapping $w = z^2$.*

**Solution.** $w = z^2 = (x + y)^2, \quad u + iv = x^2 - y^2 + 2ixy$

This gives $u = x^2 - y^2$ and $v = 2xy$

**Table of (x, y) and (u, v)**

| | | | | | | | |
|---|---|---|---|---|---|---|---|
| $x$ | 2 | 2.5 | 3 | 3.5 | 4 | 4.5 | 5 |
| $y = \pm\sqrt{x^2 - 4}$ | 0 | ± 1.5 | ± 2.2 | ± 2.9 | ± 3.5 | ± 4.1 | ± 4.6 |
| $u = x^2 - y^2$ | 4 | 4 | 4 | 4 | 4 | 4 | 4 |
| $v = 2xy$ | 0 | ± 7.5 | ± 13.2 | ± 20.3 | ± 28 | ± 36.9 | ± 46 |

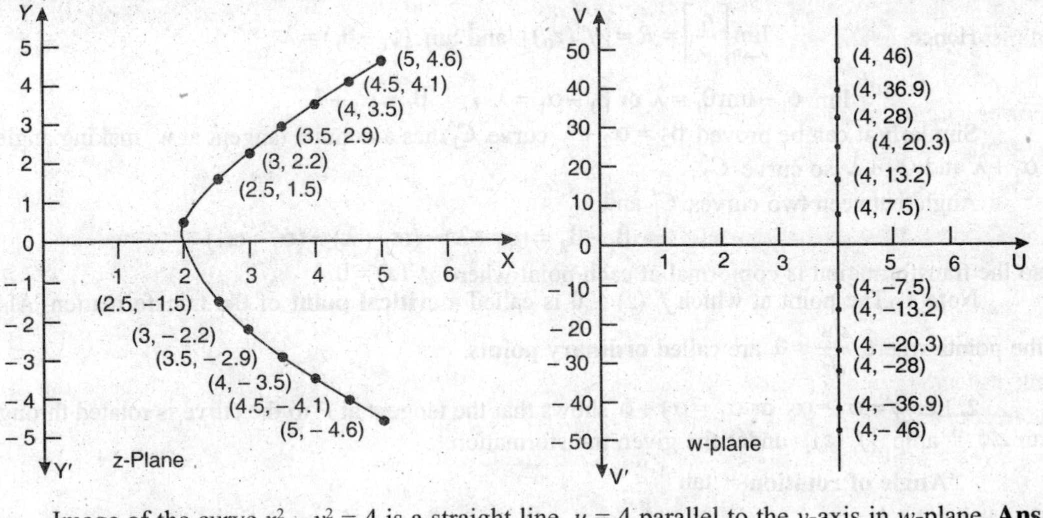


Image of the curve $x^2 - y^2 = 4$ is a straight line, $u = 4$ parallel to the $v$-axis in $w$-plane. **Ans.**

## 7.30 CONFORMAL TRANSFORMATION *(U.P. III Semester Dec., 2006, 2005)*

Let two curves $C, C_1$ in the $z$-plane intersect at the point $P$ and the corresponding curve $C', C'_1$ in the $w$-plane intersect at $P'$. **If the angle of intersection of the curves at $P$ in $z$-plane is the same as the angle of intersection of the curves of $w$-plane at $P'$ in magnitude and sense, then the transformation is called conformal:**

**conditions:** (*i*) $f(z)$ is analytic. (*ii*) $f'(z) \neq 0$ Or

If the sense of the rotation as well as the magnitude of the angle is preserved, the transformation is said to be **conformal.**

If only the magnitude of the angle is preserved, transformation is **Isogonal.**

**7.31 THEOREM. If $f(z)$ is analytic, mapping is conformal** (*U.P. III Semester Dec. 2005*)

**Proof.** Let $C_1$ and $C_2$ be the two curves in the $z$-plane intersecting at the point $z_0$ and let the tangents at this point make angles $\alpha_1$ and $\alpha_2$ with the real axis. Let $z_1$ and $z_2$ be the points on the curves $C_1$ and $C_2$ near to $z_0$ at the same distance $r$ from $z_0$, so that we have

$$z_1 - z_0 = re^{i\theta_1}, \qquad z_2 - z_0 = re^{i\theta_2}$$

As $\qquad r \to 0,\ \theta_1 \to \alpha_1$ and $\theta_2 \to \alpha_2$

Let the image of the curves $C_1$, $C_2$ be $C_1'$ and $C_2'$ in $w$-plane and images of $z_0$, $z_1$ and $z_2$ be $w_0$, $w_1$ and $w_2$.

Let $\qquad w_1 - w_0 = r.e^{i\phi_1},\ w_2 - w_0 = r.e^{i\phi_2}$

$$f'(z_0) = \lim_{z_1 \to z_0} \frac{w_1 - w_0}{z_1 - z_0}$$

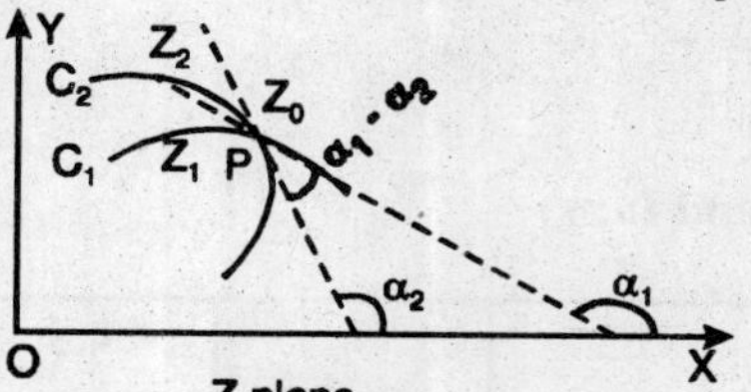

Z-plane

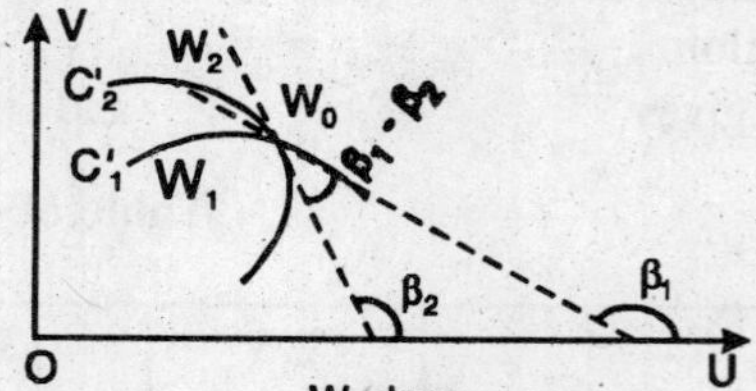

W-plane

$$Re^{i\lambda} = \lim_{r \to 0} \frac{r_1 e^{i\phi_1}}{re^{i\theta_1}} \qquad \text{(since } f'(z_0) = Re^{i\lambda}\text{)}$$

$$Re^{i\lambda} = \frac{r_1}{r} e^{i\phi_1 - i\theta_1} = \frac{r_1}{r} e^{i(\phi_1 - \theta_1)}$$

Hence $\qquad \lim_{r \to 0} \left[\frac{r_1}{r}\right] = R = |f'(z_0)|$ and $\lim\ (\phi_1 - \theta_1) = \lambda$

$\Rightarrow \qquad \lim \phi_1 - \lim \theta_1 = \lambda$ or $\beta_1 - \alpha_1 = \lambda$ *i.e.*, $\beta_1 = \alpha_1 + \lambda$

Similarly it can be proved $\beta_2 = \alpha_2 + \lambda$ curve $C_1'$ has a definite tangent at $w_0$ making angles $\alpha_1 + \lambda$ and $\alpha_2 + \lambda$ so curve $C_2'$.

Angle between two curves $C_1'$ and $C_2'$

$$= \beta_1 - \beta_2 = (\alpha_1 + \lambda) - (\alpha_2 + \lambda) = (\alpha_1 - \alpha_2)$$

so the transformation is conformal at each point where $f'(z) = 0$

**Note 1.** The point at which $f'(z) = 0$ is called a **critical point** of the transformation. Also the points where $\frac{dw}{dz} \neq 0$ are called **ordinary points.**

**2.** Let $\phi = \alpha_1 - \alpha_2$ or $\alpha_1 = \alpha_2 + \phi$ shows that the tangent at $P$ to the curve is rotated through an $\angle\phi$ = amp $\{f'(z)\}$ under the given transformation.

**Angle of rotation** $= \tan^{-1} \frac{v}{u}$.

**3.** In formal transformation, element of arc passing through $P$ is magnified by the factor $|f'(z)|$. The area element is also magnified by the factor $|f'(z)|$ or $J = \frac{\partial(u, v)}{\partial(x, y)}$ in a conformal transformation.

$$J = \frac{\partial(u, v)}{\partial(x, y)} = \begin{vmatrix} \frac{\partial u}{\partial x} & \frac{\partial u}{\partial y} \\ \frac{\partial v}{\partial x} & \frac{\partial v}{\partial y} \end{vmatrix} = \begin{vmatrix} \frac{\partial u}{\partial x} & -\frac{\partial v}{\partial x} \\ \frac{\partial v}{\partial x} & \frac{\partial u}{\partial x} \end{vmatrix} = \left(\frac{\partial u}{\partial x}\right)^2 + \left(\frac{\partial v}{\partial x}\right)^2$$

$$= \left|\frac{\partial u}{\partial x} + i\frac{\partial v}{\partial x}\right|^2 = |f'(z)|^2 = |f'(x+iy)|^2$$

$|f'(z)|$ is called the **coefficient of magnification.**

**4.** Conjugate functions remain conjugate functions after conformal transformation. A function which is the solution of Laplace's equation, its transformed function again remains the solution of Laplace's equation after conformal transformation.

### 7.32 THEOREM

*Prove that an analytic function f (z) ceases to be conformal at the points where $f'(z) = 0$.* *(U.P. III Semester, Dec. 2006)*

**Proof.** Let $f'(z) = 0$ and $f'(z_0) = 0$ at $z = z_0$

Suppose that $f'(z_0)$ has a zero of order $(n - 1)$ at the point $z_0$, then first $(n - 1)$ derivatives of $f(z)$ vanish at $z_0$ but $f^n(z_0) \neq 0$, hence at any point $z$ in the neighbourhood of $z_0$, we have by Taylor's Theorem.

$$f(z) = f(z_0) + a_n (z - z_0)^n + \dots\dots$$

where $\quad a_n = \dfrac{f^n(z_0)}{n!}$, so that $a_n \neq 0$.

Hence, $\quad f(z_1) - f(z_0) = a_n (z_1 - z_0)^n + \dots\dots$

i.e. $\quad w_1 - w_0 = a_n (z_1 - z_0)^n + \dots.$

or $\quad \rho_1 e^{i\phi_1} = |a_n| . r^n e^{i(n\theta_1 + \lambda)} + \dots\dots \quad$ where $\lambda = \text{amp } a_n$

Hence, $\quad \text{Lim } \phi_1 = \text{Lim } (n\theta_1 + \lambda) = n\alpha_1 + \lambda$

Similary, $\quad \text{Lim } \phi_2 = n\alpha_2 + \lambda.$

Thus the curves $\gamma_1$ and $\gamma_2$ still have definite tangents at $w_0$.

But the angle between the tangents

$$= \text{Lim } \phi_2 - \text{Lim } \phi_1 = n(d_2 - d_1)$$

So magnitude of the angle is not preserved.

Also the linear magnification $R = \text{Lim } (\rho_1 / r) = 0.$

Hence, the conformal property does not hold good at a point where $f'(z) = 0$.

**Example 66.** *If $u = 2x^2 + y^2$ and $v = \dfrac{y^2}{x}$, show that the curves u = constant and v= constant cut orthogonally at all intersections but that the transformation w = u + iv is not conformal.* *(Q. Bank U.P. III Semester 2002)*

**Solution.** For the curve, $2x^2 + y^2 = u$

$$2x^2 + y^2 = \text{constant} = k_1 \text{ (say)} \qquad \dots(1)$$

Differentiating (1), we get

$$4x + 2y\frac{dy}{dx} = 0 \quad \Rightarrow \quad \frac{dy}{dx} = -\frac{2x}{y} = m_1 \text{ (say)} \qquad \dots(2)$$

$$\frac{y^2}{x} = v$$

For the curve, $\dfrac{y^2}{x} = \text{constant} = k_2$ (say),

$$\Rightarrow \quad y^2 = k_2 x. \qquad \dots(3)$$

Differentiating (3), we get

$$2y\frac{dy}{dx} = k_2 \quad \Rightarrow \quad \frac{dy}{dx} = \frac{k_2}{2y} = \frac{y^2}{x} \times \frac{1}{2y} = \frac{y}{2x} = m_2 \text{ (say)} \qquad \dots(4)$$

From (2) and (4), we see that

$$m_1 m_2 = \left(\frac{-2x}{y}\right)\left(\frac{y}{2x}\right) = -1$$

Hence, two curves cut orthogonally.

However, since $\quad \frac{\partial u}{\partial x} = 4x, \quad \frac{\partial u}{\partial y} = 2y$

$$\frac{\partial v}{\partial x} = -\frac{y^2}{x^2}, \quad \frac{\partial v}{\partial y} = \frac{2y}{x}$$

The Cauchy-Riemann equations are not satisfied by $u$ and $v$.

Hence, the function $u + iv$ is not analytic. So, the transformation is not conformal. **Proved**

**Example 67.** *For the conformal transformation $w = z^2$, show that*

*(a) The coefficient of magnification at $z = 2 + i$ is $2\sqrt{5}$*

*(b) The angle of rotation at $z = 2 + i$ is $\tan^{-1} 0.5$.*

*(c) The co-efficient of magnification at $z = 1 + i$ is $2\sqrt{2}$.*

*(d) The angle of rotation at $z = 1 + i$ is $\frac{\pi}{4}$.*

*(Q. Bank U.P. III Semester 2002)*

**Solution.** (*i*) Let $\qquad w = f(z) = z^2$

$\therefore \qquad f'(z) = 2z$

$$f'(2 + i) = 2(2 + i) = 4 + 2i.$$

(*a*) Coefficient of magnification at $z = 2 + i$ is $= |f'(2+i)| = |4 + 2i| = \sqrt{16+4} = 2\sqrt{5}$.

(*b*) Angle of rotation at $z = 2 + i$ is $=$ amp. $f'(2 + i) = (4 + 2i) = \tan^{-1}\left(\frac{2}{4}\right) = \tan^{-1}(0.5)$.

and $\qquad f'(1 + i) = 2(1 + i) = 2 + 2i$

$\therefore$ (*c*) The co-efficient of magnification at $z = 1 + i$ is $|f'(1+i)| = |2 + 2i| = \sqrt{4+4} = 2\sqrt{2}$

(*d*) The angle of rotation at $z = 1 + i$ is amp. $|f'(1+i)| = 2 + 2i = \tan^{-1}\frac{2}{2} = \frac{\pi}{4}$ **Ans.**

**Standard transformations**

## 7.33 TRANSLATION $\qquad w = z + C,$

where $\qquad C = a + ib$

$$u + iv = a + iy + a + ib$$

$$u = x + a \text{ and } v = y + b$$

On substituting the values of $x$ and $y$ in the equation of the curve to be transformed, we get the equation of the image in the $w$-plane.

The point $P(x, y)$ in the $z$-plane is mapped onto the point $P'(x + a, y + b)$ in the $w$-plane. Similarly other points of $z$-plane are mapped onto $w$-plane. Thus if $w$-plane is superimposed on the $z$-plane, the figure of $w$-plane is shifted through a vector $C$.

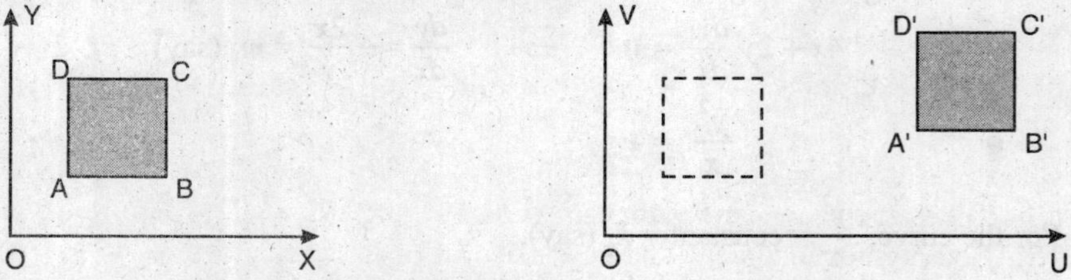


In other words the transformation is mere translation of the axes.

## 7.34 ROTATION $w = ze^{i\theta}$

The figure in *z*-plane rotates through an angle θ in anticlockwise in *w*-plane.

**Example 68.** *Consider the transformation $w = ze^{i\pi/4}$ and determine the region R′ in w-plane corresponding to the triangular region R bounded by the lines x = 0, y = 0 and x + y = 1 in z-plane.*

**Solution.**
$$w = ze^{i\pi/4}$$
$$w = (x+iy)\left(\cos\frac{\pi}{4} + i\sin\frac{\pi}{4}\right)$$
$$\Rightarrow \quad u + iv = (x+iy)\left(\frac{1+i}{\sqrt{2}}\right) = \frac{1}{\sqrt{2}}[x - y + i(x+y)]$$

Equating real and imaginay parts, we get
$$\Rightarrow \quad u = \frac{1}{\sqrt{2}}(x-y), \qquad v = \frac{1}{\sqrt{2}}(x+y) \qquad \ldots (1)$$

(*i*) Put $x = 0$, $\quad u = -\frac{1}{\sqrt{2}}y, \quad v = \frac{1}{\sqrt{2}}y$ or $v = -u$

(*ii*) Put $y = 0$, $\quad u = \frac{1}{\sqrt{2}}x, \quad v = \frac{1}{\sqrt{2}}x$ or $v = u$

(*iii*) Putting $x + y = 1$ in (1), we get $v = \frac{1}{\sqrt{2}}$

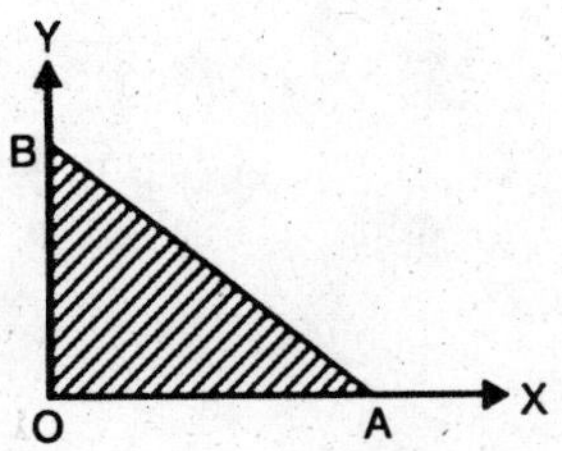


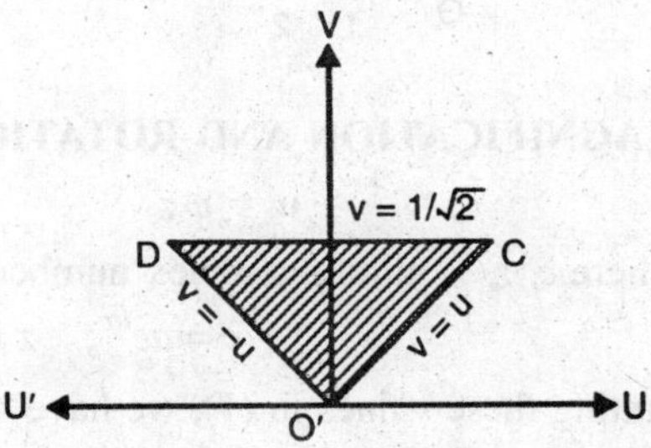


Hence the triangular region *OAB* in z-plane is mapped on a triangular region O′CD of *w*-plane bounded by the lines $v = u$, $v = -u$, $v = \frac{1}{\sqrt{2}}$.

$$f'(z) = \frac{1}{\sqrt{2}}(1+i)$$
$$f(z) = \frac{1}{\sqrt{2}}[(x-y) + i(x+y)]$$

Amp. $f'(z) = \tan^{-1}(1) = \frac{\pi}{4}$

The mapping $w = ze^{i\pi/4}$ performs a rotation of *R* through an angle π/4. **Ans.**

## 7.35 MAGNIFICATION

$$\boxed{w = cz}$$

where *c* is a real quantity.

(*i*) The figure in *w*-plane is magnified *c*-times the size of the figure in *z*-plane.

(*ii*) Both figures in *z*-plane and *w*-plane are singular.

**Example 69.** *Determine the region in w-plane on the transformation of rectangular region enclosed by x = 1, y = 1, x = 2 and y = 2 in the z-plane. The transformation is w = 3z.*

**Solution.** We have, $w = 3z$

$$u + iv = 3(x + iy)$$

Equating the real and imaginary parts, we get

$$u = 3x \quad \text{and} \quad v = 3y$$

| *z*-plane | | *w*-plane | |
|---|---|---|---|
| $x$ | $y$ | $u = 3x$ | $v = 3y$ |
| 1 | 1 | 3 | 3 |
| 2 | 2 | 6 | 6 |

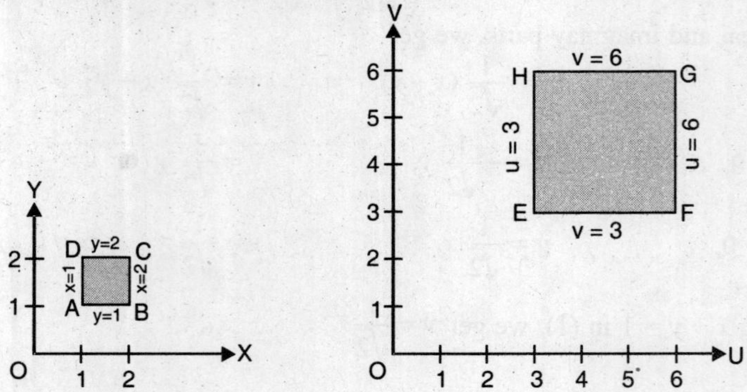


## 7.36 MAGNIFICATION AND ROTATION

$$\mathbf{w = c\,z} \qquad \dots (1)$$

where *c, z, w* all are complex numbers.

$$c = ae^{i\alpha}, \quad z = re^{i\theta}, \quad w = Re^{i\phi}$$

Putting these values in (1), we have

$$Re^{i\phi} = (ae^{i\alpha})(re^{i\theta}) = are^{i(\theta+\alpha)}$$

*i.e.* $R = ar$ and $\phi = \theta + \alpha$

Thus we see that the transform $w = c\,z$ corresponding to a rotation, together with magnification.

Algebraically $\quad w = c\,z \quad$ or $\quad u + iv = (a + ib)(x + iy)$

$\Rightarrow \quad u + iv = ax - by + i(ay + bx)$

$$u = ax - by \text{ and } v = ay + bx$$

On solving these equations, we can get the values of $x$ and $y$.

$$x = \frac{au + bv}{a^2 + b^2}, \; y = \frac{-bu + av}{a^2 + b^2}$$

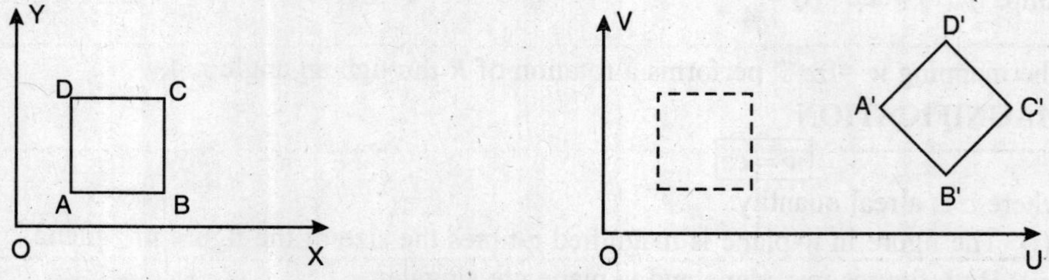


On putting values of $x$ and $y$ in the equation of the curve to be transformed we get the equation of the image.

**Example 70.** *Find the image of the triangle with vertices at i, 1 + i, 1 − i in the z-plane, under the transformation* (*i*) $w = 3z + 4 - 2i$, (*ii*) $w = e^{\frac{5\pi i}{3}} . z - 2 + 4i$

**Solution.** (*i*) $w = 3z + 4 - 2i$

$\Rightarrow \quad u + iv = 3(x + iy) + 4 - 2i \quad \Rightarrow \quad u = 3x + 4,\ v = 3y - 2$

| S. No. | $x$ | $y$ | $u = 3x + 4$ | $v = 3y - 2$ |
|---|---|---|---|---|
| 1. | 0 | 1 | 4 | 1 |
| 2. | 1 | 1 | 7 | 1 |
| 3. | 1 | –1 | 7 | – 5 |

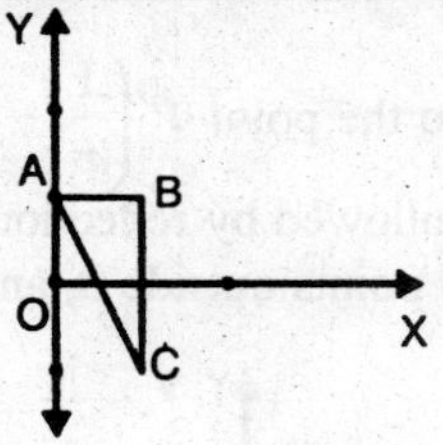


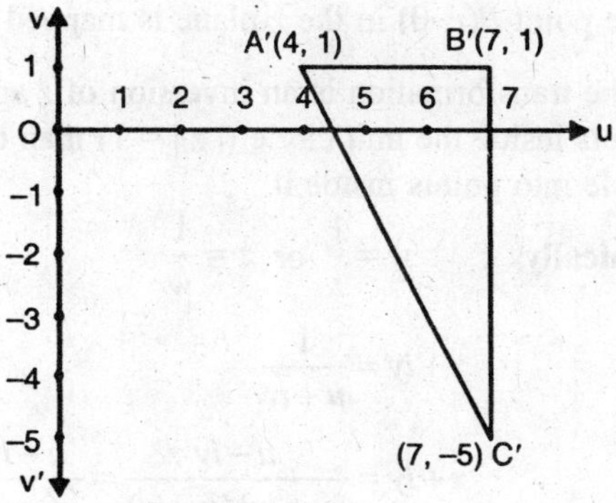


(*ii*) $w = e^{\frac{5\pi i}{3}} . z - 2 + 4i$

$$\Rightarrow \quad u + iv = \left(\cos\frac{5\pi}{3} + i\sin\frac{5\pi}{3}\right)(x + iy) - 2 + 4i$$

$$= \left(\frac{1}{2} - \frac{\sqrt{3}}{2} i\right)(x + iy) - 2 + 4i$$

$$= \frac{x}{2} - 2 + \frac{\sqrt{3}}{2} y + i\left(-\frac{\sqrt{3}}{2} x + \frac{y}{2} + 4\right)$$

$$\Rightarrow \quad u = \frac{x}{2} - 2 + \frac{\sqrt{3}}{2} y \quad \text{and} \quad v = -\frac{\sqrt{3}}{2} x + \frac{y}{2} + 4$$

| *S.No.* | *z-Plane* | | *w-plane* | |
|---|---|---|---|---|
| | $x$ | $y$ | $u = \frac{x}{2} - 2 + \frac{\sqrt{3}}{2} y$ | $v = -\frac{\sqrt{3}}{2} x + \frac{y}{2} + 4$ |
| 1. | 0 | 1 | $-2 + \frac{\sqrt{3}}{2}$ | $\frac{9}{2}$ |
| 2. | 1 | 1 | $-\frac{3}{2} + \frac{\sqrt{3}}{2}$ | $-\frac{\sqrt{3}}{2} + \frac{9}{2}$ |
| 3. | 1 | –1 | $-\frac{3}{2} - \frac{\sqrt{3}}{2}$ | $-\frac{\sqrt{3}}{2} + \frac{7}{2}$ |

## 7.37 INVERSION AND REFLECTION

$$w=\frac{1}{z} \quad \text{... (1)}$$

If $z=re^{i\theta}$ and $w=Re^{i\phi}$

Putting these values in (1), we get

$$Re^{i\phi}=\frac{1}{re^{i\theta}}=\frac{1}{r}e^{-i\theta}$$

On equating, $R=\frac{1}{r}$ and $\phi=-\theta$

Thus the point $P(r, \theta)$ in the $z$-plane is mapped onto the point $P'\left(\frac{1}{r}, -\theta\right)$ in the $w$-plane.

Hence the transformation is an inversion of $z$ and followed by reflection into the real axis. The points inside the unit circle ($|z|=1$) map onto points outside it, and points outside the unit circle into points inside it.

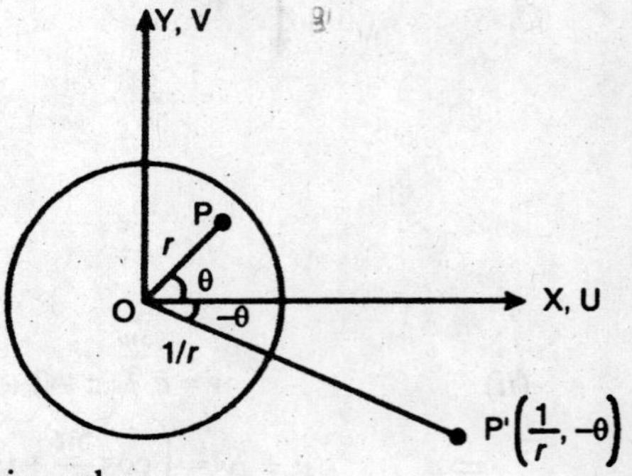


**Algebraically** $w=\frac{1}{z}$ or $z=\frac{1}{w}$

$$x+iy=\frac{1}{u+iv}$$

$$\Rightarrow \quad x+iy=\frac{u-iv}{(u+iv)(u-iv)}=\frac{u-iv}{u^2+v^2}$$

$$x=\frac{u}{u^2+v^2}, \quad y=-\frac{v}{u^2+v^2}$$

Let the circle $x^2+y^2+2gx+2fy+c=0$ ... (1) be in $z$-plane.

On substituting the values of $x$ and $y$ in (1), we get

$$\frac{u^2}{(u^2+v^2)^2}+\frac{v^2}{(u^2+v^2)^2}+2g\frac{u}{u^2+v^2}+2f\frac{(-v)}{u^2+v^2}+c=0$$

This is the equation of circle in $w$-plane. This shows that a circle in $z$-plane transforms to another circle in $w$-plane.

But a circle through origin transforms into a straight line.

**Example 71.** *Under the transformation* $w=\frac{1}{z}$, *find the image of* $y-x+1=0$ *(PTU May 2007)*

**Solution.** Here the equation of straight line may be given

$$y-x+1=0$$

$$w=\frac{1}{z} \Rightarrow z=\frac{1}{w}$$

$$\Rightarrow \quad x+iy=\frac{1}{u+iv}=\frac{u+iv}{u^2+v^2}$$

so that $x=\frac{u}{u^2+v^2}$ and $y=-\frac{v}{u^2+v^2}$

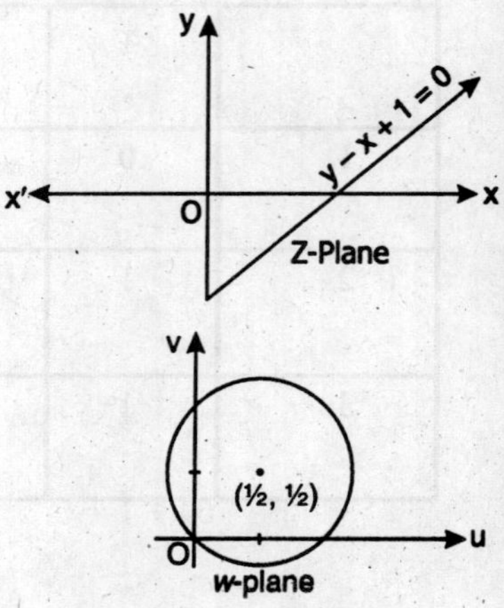


Putting the values of $x$, $y$ in terms of $u$, $v$, we get

$$-\frac{v}{u^2+v^2}-\frac{u}{u^2+v^2}+1=0$$

$$\Rightarrow \quad -u-v+u^2+v^2=0$$

$$\Rightarrow \quad u^2+v^2-u-v=0$$

This is the equation of a circle with centre at $\left(\frac{1}{2}, \frac{1}{2}\right)$ and radius $= \frac{1}{\sqrt{2}}$ **Ans.**

**Example 72.** *Find the image of* $|z - 3i| = 3$ *under the mapping* $w = \frac{1}{z}$.

*(Uttarakhand, III Semester 2008)*

**Solution.** $w = \frac{1}{z} \quad \Rightarrow \quad z = \frac{1}{w}$

$$\Rightarrow \quad x + iy = \frac{1}{u+iv} = \frac{u-iv}{(u+iv)(u-iv)} = \frac{u-iv}{u^2+v^2}$$

$$\Rightarrow \quad x = \frac{u}{u^2+v^2} \quad \text{and} \quad y = -\frac{v}{u^2+v^2} \qquad \ldots (1)$$

The given curve is $|z - 3i| = 3$

$\Rightarrow \quad |x + iy - 3i| = 3 \quad \Rightarrow \quad x^2 + (y-3)^2 = 9 \qquad \ldots (2)$

On substituting the values of $x$ and $y$ from (1) into (2), we get

$$\frac{u^2}{(u^2+v^2)^2} + \left(\frac{-v}{u^2+v^2} - 3\right)^2 = 9$$

$$\frac{u^2}{(u^2+v^2)^2} + \frac{(-v-3u^2-3v^2)^2}{(u^2+v^2)^2} = 9$$

$$\Rightarrow \quad u^2 + (-v-3u^2-3v^2)^2 = 9(u^2+v^2)^2$$

$$\Rightarrow \quad u^2 + v^2 + 9u^4 + 9v^4 + 6u^2v + 6v^3 + 18u^2v^2 = 9u^4 + 18u^2v^2 + 9v^4$$

$$\Rightarrow \quad u^2 + v^2 + 6u^2v + 6v^3 = 0$$

$$\Rightarrow \quad u^2 + v^2 + 6v(u^2+v^2) = 0$$

$$\Rightarrow \quad (u^2+v^2)(6v+1) = 0$$

$\Rightarrow \quad 6v + 1 = 0$ is the equation of the image. **Ans.**

**Second Method.** $|z - 3i| = 3, \quad z = \frac{1}{w}$

$$\left|\frac{1}{w} - 3i\right| = 3 \quad \Rightarrow \quad |1 - 3iw| = 3|w|$$

$$\Rightarrow \quad |1 - 3i(u+iv)| = 3|u+iv| \quad \Rightarrow \quad |1 + 3v - 3iu| = 3|u+iv|$$

$$\Rightarrow \quad (1+3v)^2 + 9u^2 + 9(u^2+v^2) \quad \Rightarrow \quad 1 + 6v + 9v^2 + 9u^2 = 9(u^2+v^2)$$

$\Rightarrow \quad 1 + 6v = 0$ **Ans.**

**Third Method.** $|z - 3i| = 3 \quad \Rightarrow \quad z - 3i = 3e^{i\theta} \quad \Rightarrow \quad z = 3i + 3e^{i\theta}$

$$w = \frac{1}{z} = \frac{1}{3i + 3e^{i\theta}} \quad \Rightarrow \quad 3w = \frac{1}{i + e^{i\theta}}$$

$$3(u+iv) = \frac{1}{i + \cos\theta + i\sin\theta}$$

$$\Rightarrow \quad (3u + 3iv) = \frac{\cos\theta - i(1+\sin\theta)}{\cos^2\theta + (1+\sin\theta)^2} \quad \Rightarrow \quad 3v = -\frac{1+\sin\theta}{2+2\sin\theta} = -\frac{1}{2}$$

$6v + 1 = 0$ **Ans.**

**Example 73.** *Image of* $|z+1| = 1$ *under the mapping* $w = \frac{1}{z}$ is

(*a*) $2v + 1 = 1$ (*b*) $2v - 1 = 1$ (*c*) $2u + 1 = 0$ (*d*) $2u - 1 = 0$ (*AMIETE, June 2009*)

**Solution.** $w = \frac{1}{z} \Rightarrow u + iv = \frac{1}{x + iy} = \frac{x - iy}{x^2 + y^2}$

$$\Rightarrow \quad u = \frac{x}{x^2 + y^2}, \quad v = -\frac{y}{x^2 + y^2}.$$

Given $|z + 1| = 1 \Rightarrow |x + iy + 1| = 1 \Rightarrow (x + 1)^2 + y^2 = 1$

$$\Rightarrow \quad x^2 + y^2 + 2x = 0 \Rightarrow x^2 + y^2 = -2x \Rightarrow \frac{1}{2} = -\frac{x}{x^2 + y^2} = -u$$

$$\Rightarrow \quad \frac{1}{2} = -u \Rightarrow 2u + 1 = 0$$

Hence (*c*) is correct answer. **Ans.**

**Example 74.** *Show that under the transformation* $w = \frac{1}{z}$, *the image of the hyperbola* $x^2 - y^2 = 1$ *is the lemniscate* $R^2 = \cos 2\phi$.

**Solution.** $x^2 - y^2 = 1$

Putting $x = r\cos\theta$ and $y = r\sin\theta$

$$\Rightarrow \quad r^2\cos^2\theta - r^2\sin^2\theta = 1 \Rightarrow r^2(\cos^2\theta - \sin^2\theta) = 1$$

$$\Rightarrow \quad r^2\cos 2\theta = 1 \qquad \ldots (1)$$

And $w = \frac{1}{z} \Rightarrow z = \frac{1}{w} \Rightarrow re^{i\theta} = \frac{1}{Re^{i\phi}} \Rightarrow re^{i\theta} = \frac{1}{R}e^{-i\phi}$

Equating real and imaginary parts, we get

$$\therefore \quad r = \frac{1}{R} \quad \text{and} \quad \theta = -\phi$$

Putting the values of $r$ and $\theta$ in (1), we get

$$\frac{1}{R^2}\cos 2(-\phi) = 1 \quad \Rightarrow \quad R^2 = \cos 2\phi$$ **Proved.**

## EXERCISE 7.8

1. Find the image of the semi infinite, strip $x > 0$, $0 < y < 2$ under the transformation $w = iz + 1$.

   **Ans.** Strip $-1 < u < 1$, $v > 0$

2. Determine the region in the $w$-plane in which the rectangle bounded by the lines $x = 0$, $y = 0$, $x = 2$ and $y = 1$ is mapped under the transformation $w = \sqrt{2}\, e^{i\pi/4}\, z$.

   (*Q. Bank U.P. III Semester 2002*)

   **Ans.** Region bounded by the lines $v = -u$, $v = u$, $u + v = 4$ and $v - u = 2$.

3. Show that the condition for transformation $w = a^2 + blcz + d$ to make the circle $|w| = 1$ correspond to a straight line in the $z$-plane is $(a) = (c)$.

4. What is the region of the $w$-plane in two ways the rectangular region in the $z$-plane bounded by the lines $x = 0$, $y = 0$, $x = 1$ and $y = 2$ is mapped under the transformation $w = z + (2 - i)$?

   **Ans.** Region bounded by $u = 2$, $v = -1$, $u = 3$ and $v = 1$.

5. Find the image of $|z - 2i|$ under the mapping $w = \frac{1}{z}$ **Ans.** $v + \frac{1}{4} = 0$

**6.** For the mapping $w(z) = 1/z$, find the image of the family of circles $x^2 + y^2 = ax$, where $a$ is real.

**Ans.** $u = \frac{1}{a}$ is a straight line || to $v$-axis.

**7.** Show that the function $w = \frac{4}{z}$ transforms the straight line $x = c$ in the $z$-plane into a circle in the $w$-plane.

**8.** If $(w+1)^2 = \frac{4}{z}$, then prove that the unit circle in the $w$-plane corresponds to a parabola in the $z$-plane, and the inside of the circle to the outside of the parabola.

**9.** Find the image of $|z - 2i| = 2$ under the mapping $w = \frac{1}{z}$

(*Q. Bank U.P. 2002*) **Ans.** $4v + 1 = 0$

**10.** The image of the circle $|z - 1| = 1$ in the complex plane, under the mapping $w = u + iv = \frac{1}{z}$ is

(*i*) $|w - 1| = 1$ (*ii*) $u^2 + v^2 = 1$ (*iii*) $u = \frac{1}{2}$ (*iv*) $v = \frac{1}{2}$ **Ans.** (*iii*)

**11.** Inverse transformation $w = \frac{1}{z}$ transforms the straight line $ay + bx = 0$ into

(*i*) Circle (*ii*) straight line through the origin
(*iii*) straight line (*iv*) none of these

**12.** The analytic function $f(z)$, which maps the angular region $0 \le \theta \le \pi/4$ onto the region $\pi/4 \le \phi \le \pi/2$ is

(*i*) $ze^{i\pi/4}$ (*ii*) $z + \pi/4$ (*iii*) $iz$ (*iv*) $e^{z+i\pi/4}$ **Ans.**

## 7.38 BILINEAR TRANSFORMATION (Mobius Transformation)

$$\boxed{w = \frac{az+b}{cz+d}} \qquad ad - bc \neq 0 \qquad \text{... (1)}$$

(1) is known as bilinear transformation.

If $ad - bc \neq 0$ then $\frac{dw}{dz} \neq 0$ *i.e.* transformation is conformal.

From (1), $$z = \frac{-dw+b}{cw-a} \qquad \text{...(2)}$$

This is also bilinear except $w = \frac{a}{c}$.

**Note.** From (1), every point of $z$-plane is mapped into unique point in $w$-plane except $z = -\frac{d}{c}$

From (2), every point of $w$-plane is mapped into unique point in $z$-plane except $w = \frac{a}{c}$.

## 7.39 INVARIANT POINTS OF BILINEAR TRANSFORMATION

We know that $$w = \frac{az+b}{cz+d} \qquad \text{... (1)}$$

If $z$ maps into itself, then $w = z$

(1) becomes $$z = \frac{az+b}{cz+d} \qquad \text{... (2)}$$

Roots of (2) are the invariants or fixed points of the bilinear transformation.
If the roots are equal, the bilinear transformation is said to be parabolic.

## 7.40 CROSS-RATIO

If there are four points $z_1, z_2, z_3, z_4$ taken in order, then the ratio $\dfrac{(z_1-z_2)(z_3-z_4)}{(z_2-z_3)(z_4-z_1)}$ is called the cross-ratio of $z_1; z_2, z_3, z_4$.

## 7.41 THEOREM

A bilinear transformation preserves cross-ratio of four points

**Proof.** We know that $w = \dfrac{az+b}{cz+d}$.

As $w_1, w_2, w_3, w_4$ are images of $z_1, z_2, z_3, z_4$ respectively, so

$$w_1 = \frac{az_1+b}{cz_1+d}, \quad w_2 = \frac{az_2+b}{cz_2+d}$$

$$\therefore \qquad w_1 - w_2 = \frac{(ad-bc)}{(cz_1+d)(cz_2+d)}(z_1-z_2) \qquad ...(1)$$

Similarly

$$w_2 - w_3 = \frac{ad-bc}{(cz_2+d)(cz_3+d)}(z_2-z_3) \qquad ...(2)$$

$$w_3 - w_4 = \frac{ad-bc}{(cz_3+d)(cz_4+d)}(z_3-z_4) \qquad ...(3)$$

$$w_4 - w_1 = \frac{ad-bc}{(cz_4+d)(cz_1+d)}(z_4-z_1) \qquad ...(4)$$

From (1), (2), (3) and (4), we have

$$\boxed{\frac{(w-w_1)(w_2-w_3)}{(w-w_3)(w_2-w_1)} = \frac{(z-z_1)(z_2-z_3)}{(z-z_3)(z_2-z_1)}}$$

$$\Rightarrow \qquad (w_1, w_2, w_3, w_4) = (z_1, z_2, z_3, z_4).$$

## 7.42 PROPERTIES OF BILINEAR TRANSFORMATION

**1.** A bilinear transformation maps circles into circles.

**2.** A bilinear transformation preserves cross ratio of four points.

If four points $z_1, z_2, z_3, z_4$ of the $z$-plane map onto the points $w_1, w_2, w_3, w_4$ of the $w$-plane respectively.

$$\Rightarrow \qquad \frac{(w_1-w_2)(w_3-w_4)}{(w_1-w_4)(w_3-w_2)} = \frac{(z_1-z_2)(z_3-z_4)}{(z_1-z_4)(z_3-z_2)}$$

Hence, under the **bilinear** transform of four points cross-ratio is preserved.

## 7.43 METHODS TO FIND BILINEAR TRANSFORMATION

**1.** By finding $a, b, c, d$ for $\dfrac{az+b}{cz+d}$ with the given conditions.

**2.** With the help of cross-ratio

$$\boxed{\frac{(w_1-w_2)(w_3-w_4)}{(w_1-w_4)(w_3-w_2)} = \frac{(z-z_1)(z_2-z_3)}{(z-z_3)(z_2-z_1)}}$$

**Example 75.** *Find the bilinear transformation which maps the points* $z = 1, i, -1$ *into the poin's* $w = i, 0, -i$.

*Hence find the image of* $|z| < 1$. *(U.P., III Semester, 2008, Summer 2002)*
*(U.P. (Agri. Engg.) 2002)*

**Solution.** Let the required transformation be $w = \dfrac{az+b}{cz+d}$

or $w = \dfrac{\frac{a}{d}z + \frac{b}{d}}{\frac{c}{d}z + 1} = \dfrac{pz+q}{rz+1}$ ... (1) $\left[p = \frac{a}{d}, q = \frac{b}{d}, r = \frac{c}{d}\right]$

| z | w |
|---|---|
| 1 | i |
| i | 0 |
| −1 | −i |

On substituting the values of $z$ and corresponding values of $w$ in (1), we get

$$i = \frac{p+q}{r+1} \Rightarrow p + q = ir + i \quad \text{... (2)}$$

$$0 = \frac{pi+q}{ri+1} \Rightarrow pi + q = 0 \quad \text{... (3)}$$

$$-i = \frac{-p+q}{-r+1} \Rightarrow -p + q = ir - i \quad \text{... (4)}$$

On subtracting (4) from (2), we get $2p = 2i$ or $p = i$

On putting the value of $p$ in (3), we have $i\,(i) + q = 0$ or $q = 1$

On substituting the values of $p$ and $q$ in (2), we obtain

$$i + 1 = i\,r + i \quad \text{or} \quad 1 = i\,r \quad \text{or} \quad r = -i$$

Putting the values of $p, q, r$ in (1), we have

$$w = \frac{iz+1}{-iz+1}$$

$$u + iv = \frac{i(x+iy)+1}{-i(x+iy)+1} = \frac{(ix-y+1)(ix+y+1)}{(-ix+y+1)(ix+y+1)} = \frac{-x^2-y^2+1+2ix}{x^2+(y+1)^2}$$

Equating real parts, we get

$$u = \frac{-x^2-y^2+1}{x^2+(y+1)^2} \quad \text{... (5)}$$

But $|z| < 1 \Rightarrow x^2 + y^2 < 1 \Rightarrow 1 - x^2 - y^2 > 0$

From (5) $u > 0$ As denominator is positive. **Ans.**

**Example 76.** *Find the bilinear transformation which maps the points $z = 0, -1, i$ onto $w = i, 0, \infty$. Also find the image of the unit circle $|z| = 1$.*

[*Uttarakhand, III Semester 2008, U.P. III semester (C.O.) 2003*]

**Solution.** On putting $z = 0, -1, i$ into $w = i, 0, \infty$ respectively in

$$\frac{(w-w_1)(w_2-w_3)}{(w-w_3)(w_2-w_1)} = \frac{(z-z_1)(z_2-z_3)}{(z-z_3)(z_2-z_1)} \quad \text{...(1)}$$

| z | w |
|---|---|
| 0 | i |
| −1 | 0 |
| i | ∞ |

$$\Rightarrow \frac{(w-w_1)\left(\frac{w_2}{w_3}-1\right)}{\left(\frac{w}{w_3}-1\right)(w_2-w_1)} = \frac{(z-z_1)(z_2-z_3)}{(z-z_3)(z_2-z_1)}$$

$$\frac{(w-i)(-1)}{(-1)(0-i)} = \frac{(z-0)(-1-i)}{(z-i)(-1-0)} \Rightarrow \left(\frac{w-i}{-i}\right) = \frac{z(1+i)}{z-i}$$

$$\Rightarrow w - i = \frac{(-i+1)z}{z-i} \Rightarrow w = \frac{(1-i)z}{z-i} + i = \frac{(1-i)z+iz+1}{z-i}$$

$$\Rightarrow w = \frac{z+1}{z-i} \quad \text{...(2)}$$ **Ans.**

From (2) $z = \dfrac{iw+1}{w-1}$ ...(3) $\left[\text{Inverse transformation is } z = \dfrac{-dw+b}{cw-a}\right]$

And $|z| = 1$

$\Rightarrow \quad \left|\dfrac{iw+1}{w-1}\right| = 1 \qquad \Rightarrow \quad |1 + iw| = |w - 1|$

$\Rightarrow \quad |1 + i(u + iv)| = |u + iv - 1| \quad \Rightarrow \quad |1 - v + iu| = |u - 1 + iv|$

$\Rightarrow \quad (1 - v)^2 + u^2 = (u - 1)^2 + v^2 \quad \Rightarrow \quad 1 + v^2 - 2v + u^2 = u^2 + 1 - 2u + v^2$

$\Rightarrow \quad u - v = 0 \quad \Rightarrow \quad v = u$ **Ans.**

**Example 77.** *Find the fixed points and the normal form of the following bilinear transformations.*

(*a*) $w = \dfrac{3z-4}{z-1}$ and (*b*) $w = \dfrac{z-1}{z+1}$

*Discuss the nature of these transformations.*

**Solution.** (*a*) The fixed points are obtained by

$$z = \frac{3z-4}{z-1} \quad \text{or} \quad z^2 - 4z + 4 = 0 \quad \text{or} \quad (z-2)^2 = 0 \Rightarrow z = 2$$

$z = 2$ is the only fixed point. This transformation is parabolic.

**Normal Form**

$$w = \frac{3z-4}{z-1} \Rightarrow \frac{1}{w-2} = \frac{1}{\frac{3z-4}{z-1} - 2} = \frac{z-1}{3z-4-2z+2} = \frac{z-1}{z-2}$$

and
$$\frac{1}{w-2} = \frac{1}{z-2} + 1$$

(*b*) The fixed points are obtained by

$$z = \frac{z-1}{z+1} \quad \Rightarrow \quad z^2 + z = z - 1 \Rightarrow z^2 = -1 \Rightarrow z = \pm i$$

Hence $\pm\, i$ are the two fixed points.

**Normal Form**

$$w = \frac{z-1}{z+1}$$

$$w - i = \frac{z-1}{z+1} - i = \frac{z-1-i(z+1)}{z+1} \qquad \text{... (1)}$$

and
$$w + i = \frac{z-1}{z+1} + i = \frac{z-1+i\,(z+1)}{z+1} \qquad \text{...(2)}$$

On dividing (1) by (2), we get

$$\frac{w-i}{w+i} = \frac{z-1-iz-i}{z-1+iz+i} = \frac{(1-i)(z-i)}{(1+i)(z+i)} = \frac{(-i^2-i)(z-i)}{(1+i)(z+i)}$$

$$\frac{w-1}{w+1} = -i\left(\frac{z-i}{z+i}\right) = k\left(\frac{z-i}{z+i}\right) \quad \text{where } k = -\,i$$

The transformation is elliptic. **Ans.**

**Example 78.** *The fixed points of the transformation* $w = \dfrac{2z-5}{z+4}$ *are given by*:

(*a*) $\left(\dfrac{5}{2}, 0\right)$ (*b*) $(-4, 0)$ (*c*) $(-1 + 2i, \; -1 - 2i)$ (*d*) $(-1+\sqrt{6}, -1-\sqrt{6})$

(*AMIETE, Dec. 2010*)

**Solution.** Here $f(z) = \dfrac{2z-5}{z+4}$

In the case of fixed point $z = \dfrac{2z-5}{z+4}$

$\Rightarrow z^2 + 4z = 2z - 5 \quad \Rightarrow \quad z^2 + 2z + 5 = 0$

$\Rightarrow z = \dfrac{-2 \pm \sqrt{4-20}}{2} = \dfrac{-2 \pm 4i}{2} = -1 \pm 2i$

Thus, $z = -1 \pm 2i$ are the only fixed points.

Hence (c) is correct answer. **Ans.**

**Example 79.** *Show that the transformation* $w = i\dfrac{1-z}{1+z}$ *transforms the circle* $|z| = 1$ *onto the real axis of the w-plane and the interior of the circle into the upper half of the w-plane.*

*(U.P., III Semester, Dec. 2003)*

**Solution.** $w = i\left(\dfrac{1-z}{1+z}\right)$

$$(u+iv) = i\left(\frac{1-(x+iy)}{1+(x+iy)}\right) = \frac{(i-ix+y)}{[1+(x+iy)]}\frac{[(1+x)-iy]}{[(1+x)-iy]} \quad \text{(Rationalizing)}$$

$$= \frac{i+ix+y-ix-ix^2-xy+y+xy-iy^2}{(1+x)^2+y^2} = \frac{y-xy+y+xy+i+ix-ix-ix^2-iy^2}{(1+x)^2+y^2}$$

$$= \frac{2y+i(1-x^2-y^2)}{(1+x)^2+y^2}$$

Equating the real and imaginary parts, we get

$$u = \frac{2y}{(1+x)^2+y^2} \quad \text{... (1)}$$

and $$v = \frac{1-(x^2+y^2)}{(1+x)^2+y^2} \quad \text{... (2)}$$

when $x^2 + y^2 = 1$, then $v = \dfrac{1-1}{(1+x)^2+y^2} = 0$

$v = 0$ is the equation of the real axis in the $w$ – plane. **Proved.**

(*b*) Now the equation of the interior of the circle is $x^2 + y^2 < 1$.

Dividing (1) by (2), we get

$$\frac{u}{v} = \frac{2y}{1-(x^2+y^2)}, \qquad u - u(x^2+y^2) = 2vy, \qquad u(x^2+y^2) = u - 2vy$$

$$x^2+y^2 = 1 - \frac{2vy}{u}, \quad 1 - \frac{2vy}{u} < 1 \qquad [\text{as } x^2+y^2 < 1]$$

$$-\frac{2vy}{u} < 0, \qquad 2vy > 0$$

$$v > 0$$

$v > 0$ is the equation of the upper half of w-plane. **Proved.**

**Example 80.** *Show that* $\omega = \dfrac{i-z}{i+z}$ *maps the real axis of the z-plane into the circle* $|\,w\,| = 1$ *and (ii) the half-plane* $y > 0$ *into the interior of the unit circle* $|\omega| < 1$ *in the w-plane.*

*(U.P., III Semester, Dec. 2005, 2002)*

**Solution.** We have $\omega = \dfrac{i-z}{i+z}$

$$|\omega| = \left|\frac{i-z}{i+z}\right| = \frac{|i-z|}{|i+z|} = \frac{|i-x-iy|}{|i+x+iy|}$$

$$|\omega| = \left|\frac{-x+i(1-y)}{x+i(1+y)}\right|, \quad |\omega| = \frac{\sqrt{x^2+(1-y)^2}}{\sqrt{x^2+(1+y)^2}}$$

Now the real axis in $z$-plane *i.e.i* $y = 0$, transform into

$$|\omega| = \frac{\sqrt{x^2+1}}{\sqrt{x^2+1}} = 1, \qquad |\omega| = 1 \qquad |z| = 1$$

Hence the real axis in the z-plane is mapped into the circle $|\omega| = 1$.

(*ii*) The interior of the circle *i.e.* $|w| < 1$ gives.

$$\frac{\sqrt{x^2+(1-y)^2}}{\sqrt{x^2+(1+y)^2}} < 1 \;\Rightarrow\; \frac{x^2+(1-y)^2}{x^2+(1+y)^2} < 1 \qquad \Rightarrow\; x^2+(1-y)^2 < x^2+(1+y)^2$$

$$\Rightarrow \qquad 1+y^2-2y < 1+y^2+2y \;\Rightarrow\; -4y < 0 \;\Rightarrow\; y > 0.$$

Thus the upper half of the $z$-plane corresponds to the interior of the circle $|w| = 0$. **Proved..**

**Example 81.** *Show that the transformation* $w = \dfrac{3-z}{z-2}$ *transforms the circle with centre* $\left(\dfrac{5}{2}, 0\right)$ *and radius* $\dfrac{1}{2}$ *in the z-plane into the imaginary axis in the w-plane and the interior of the circle into the right half of the plane.* (*A.M.I.E.T.E., Summer 2007, 2000*)

**Solution.** $w = \dfrac{3-z}{z-2} \;\Rightarrow\; u+iv = \dfrac{3-x-iy}{x+iy-2} \;\Rightarrow\; (u+iv)(x+iy-2) = 3-x-iy$

$$\Rightarrow \qquad ux + iuy - 2u + ivx - vy - 2iv = 3 - x - iy$$

$$\Rightarrow \qquad ux - 2u - vy + i(uy + vx - 2v) = 3 - x - iy$$

Equating real and imaginary quantities, we have

$$ux - vy - 2u = 3 - x \quad \text{and} \quad vx - 2v + uy = -y$$

$$\Rightarrow \qquad (u+1)x - vy = 2u + 3 \text{ and } vx + (u+1)y = 2v$$

On solving the equations for $x$ and $y$, we have

$$x = \frac{2u^2+2v^2+5u+3}{u^2+v^2+2u+1}, \quad y = \frac{-v}{u^2+v^2+2u+1}$$

Here, the equation of the given circle is $\left(x-\dfrac{5}{2}\right)^2 + y^2 = \dfrac{1}{4}$ ... (1)

Putting the values of $x$ and $y$ in (1), we have

$$\left(\frac{2u^2+2v^2+5u+3}{u^2+v^2+2u+1} - \frac{5}{2}\right)^2 + \left(\frac{-v}{u^2+v^2+2u+1}\right)^2 = \frac{1}{4}$$

$$\Rightarrow \qquad \left(\frac{-u^2-v^2+1}{2(u^2+v^2+2u+1)}\right)^2 + \left(\frac{-v}{u^2+v^2+2u+1}\right)^2 = \frac{1}{4}$$

$$\Rightarrow \qquad (-u^2-v^2+1)^2 + 4v^2 = (u^2+v^2+2u+1)^2$$

$$\Rightarrow \qquad (u^2+v^2-1)^2 + 4v^2 = [(u^2+v^2-1)+(2u+2)]^2$$

$$\Rightarrow \qquad (u^2+v^2-1)^2 + 4v^2 = (u^2+v^2-1)^2 + (2u+2)^2 + 2(u^2+v^2-1)(2u+2)$$

$$\Rightarrow \qquad v^2 = (u+1)^2 + (u^2+v^2-1)(u+1)$$

$$\Rightarrow \qquad v^2 = u^2+2u+1+u^3+uv^2-u+u^2+v^2-1$$

$$\Rightarrow \qquad 0 = u^3+2u^2+u+uv^2$$

$$\Rightarrow \qquad u(u^2+2u+1+v^2)=0 \;\Rightarrow\; u=0 \text{ i.e., equation of imaginary axis.}$$

Equation of the interior of the circle is $\left(x-\frac{5}{2}\right)^2+y^2<\frac{1}{4}$.
Then corresponding equation in $u$, $v$ is

$$u(u^2+2u+1+v^2)>0 \quad \text{or} \quad u[(u+1)^2+v^2]>0$$

As $(u+1)^2+v^2>0$ so $u=0$ *i.e.*, equation of the right half plane. **Ans.**

## 7.44 INVERSE POINT WITH RESPECT TO A CIRCLE

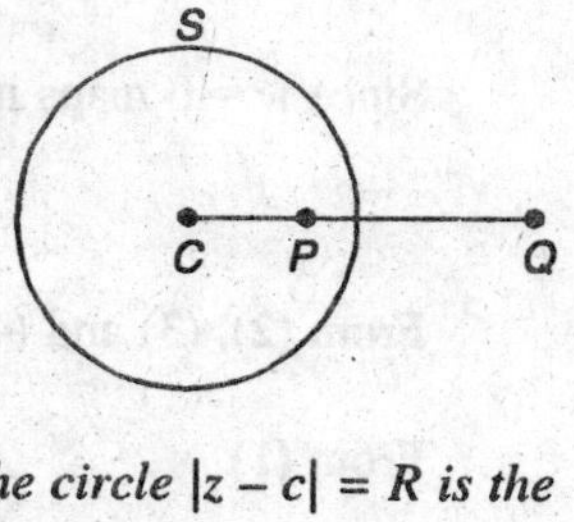


Two points $P$ and $Q$ are said to be the inverse points with respect to a circle $S$ if they are collinear with the centre $C$ on the same side of it, and if the product of their distances from the centre is equal to $r^2$ where $r$ is the radius of the circle.

Thus when $P$ and $Q$ are the inverse points of the circle, then the three points $C$, $P$, $Q$ are collinear, and also $CP.CQ = r^2$

**Example 82.** *Show that the inverse of a point a, with respect to the circle $|z-c|=R$ is the point* $c+\dfrac{R^2}{\overline{a}-\overline{c}}$

**Solution.** Let $b$ be the inverse point of the point $a'$ with respect to the circle $|z-c|=R$.

**Condition I.** The points $a$, $b$, $c$ are collinear. Hence

$$\arg(b-c)=\arg(a-c)=-\arg\overline{(a-c)} \qquad (\text{since } \arg z=-\arg\overline{z})$$

$$\Rightarrow \qquad \arg(\overline{b}-\overline{c})+\arg(\overline{a}-\overline{c})=0 \quad \text{or} \quad \arg(\overline{b}-\overline{c})(\overline{a}-\overline{c})=0$$

$\therefore \quad (\overline{b}-\overline{c})(\overline{a}-\overline{c})$ is real, so that

$$(\overline{b}-\overline{c})(\overline{a}-\overline{c})=|(\overline{b}-\overline{c})(\overline{a}-\overline{c})|$$

**Condition II.** $|\overline{b}-\overline{c}||\overline{a}-\overline{c}|=R^2 \;\Rightarrow\; |\overline{b}-\overline{c}||\overline{a}-\overline{c}|=R^2 \qquad \{|z|=|\overline{z}|\}$

$$|(\overline{b}-\overline{c})(\overline{a}-\overline{c})|=R^2 \;\Rightarrow\; (\overline{b}-\overline{c})(\overline{a}-\overline{c})=R^2 \;\Rightarrow\; \overline{b}-\overline{c}=\frac{R^2}{\overline{a}-\overline{c}}$$

$$\Rightarrow \qquad \overline{b}=\overline{c}+\frac{R^2}{\overline{a}-\overline{c}}$$ **Proved.**

**Example 83.** *Find a Mobius transformation which maps the circle $|w|\leq 1$ into the circle $|z-1|<1$ and maps $w=0$, $w=1$ respectively into $z=\frac{1}{2}$, $z=0$.*

**Solution.** Let the transformation be,

$$w=\frac{az+b}{cz+d} \qquad \dots (1)$$

| z | w |
|---|---|
| ½ | 0 |
| 0 | 1 |

Since, $w=0$ maps into $z=\frac{1}{2}$,
From (1), we get

$$0=\frac{\frac{a}{2}+b}{\frac{c}{2}+d} \;\Rightarrow\; \frac{a}{2}+b=0 \;\Rightarrow\; b=\frac{-a}{2} \qquad \dots (2)$$

Since $w=1$ maps into $z=0$, from (1), we get

$$1=\frac{0+b}{0+d} \;\Rightarrow\; b=d \qquad \dots (3)$$

Here $\qquad |w|=1$ corresponding to $|z-1|=1$

Therefore points $w$, $\frac{1}{w}$ inverse with respect to the circle $|w|=1$ correspond to the points

$z, 1+\dfrac{1}{z-1}$ inverse with respect to the circle $|z-1| = 1$

[z and $a+\dfrac{R^2}{z-a}$ are inverse points on the circle $|z-a| = R$]

*Particular* $w = 0$ and $\infty$ correspond to

$$z = \frac{1}{2}, 1+\frac{1}{\frac{1}{2}-1} \Rightarrow z = \frac{1}{2}, -1$$

Since $w = 0$ maps into $z = -1$, from (1), we get

$$\infty = \frac{-a+b}{-c+d} \Rightarrow -c+d = 0 \Rightarrow c = d \qquad \ldots (4)$$

From (2), (3) and (4), $\quad b = -\dfrac{a}{2}, \qquad b = c = d$

From (1) $\quad w = \dfrac{az+b}{cz+d} = \dfrac{-2bz+b}{bz+b} = \dfrac{-2z+1}{z+1}$ **Ans.**

**Example 84.** *Show that bilinear transformation of a circle of z-plane into a circle of w-plane and inverse points are transformed into inverse points.*

*In particular case in which the circle in the z-plane transform into a straight line in the w-plane, the inverse points transform into points symmetrical about this line.*

**Solution.** The equation of a circle is

$$\frac{|z-p|}{|z-q|} = k \qquad \ldots (1) \text{ with inverse points } p, q, k \neq 1.$$

Let the bilinear transformation is $w = \dfrac{az+b}{cz+d}$ ... (2)

Under this transformation points $p, q$ in the $z$-plane map into $\dfrac{aq+b}{cq+d}$ and $\dfrac{aq+b}{cq+d}$ in the $w$-plane.

From (2), we get $\quad z = \dfrac{dw-b}{-cw+a}$ ... (3)

Putting the value of $z$ from (3) into (1), we get

$$\frac{\left|\dfrac{dw-b}{-cw+a} - p\right|}{\left|\dfrac{dw-b}{-cw+a} - q\right|} = k \Rightarrow \frac{\left|w - \dfrac{ap+b}{cp+d}\right|}{\left|w - \dfrac{aq+b}{cq+d}\right|} = k\frac{|cq+d|}{|cp+d|} \qquad \ldots (4)$$

This is the equation of circle in $w$-plane. Its inverse points are

$$\frac{cp+b}{cp+d} \text{ and } \frac{aq+b}{cq+d}.$$

**Particular case.** If $\quad k\dfrac{|cp+d|}{|cq+d|} = 1$

then equation (4), becdomes

$$\frac{\left|w - \dfrac{ap+b}{cp+d}\right|}{\left|w - \dfrac{aq+b}{cp+d}\right|} = 1 \qquad \ldots (5)$$

(5) is the equation of a line bisecting at right angles to the join of the points $\frac{ap+b}{cp+d}$ and $\frac{aq+b}{cq+d}$.

**Example 85.** *Find two bilinear transformations whose fixed points are 1 and 2.*

*(Q. Bank U.P.T.U. 2002)*

**Solution.** We have, $$w=\frac{az+b}{cz+d} \quad \text{... (1)}$$

Fixed points are given by

$$z=\frac{az+b}{cz+d}$$

$$\Rightarrow \quad cz^2-(a-d)\,z-b=0 \quad \Rightarrow \quad z^2-\frac{(a-d)}{c}z-\frac{b}{c}=0 \quad \text{... (2)}$$

Fixed points are 1 and 2, so

$$(z-1)(z-2)=0$$

$$\Rightarrow \quad z^2-3z+2=0 \quad \text{... (3)}$$

Equating the coefficeints of $z$ and constants in (2) and (3), we get

$$\therefore \quad \frac{a-d}{c}=3 \quad \text{and} \quad -\frac{b}{c}=2$$

$$\Rightarrow \quad b=-2c \quad \text{and} \quad d=a-3c$$

Putting the values of $b$ and $d$ in (1), we get

$$w=\frac{az-2c}{cz+a-3c} \text{ has its fixed points at } z=1 \text{ and } z=2.$$

Taking $a=1$, $c=-1$ and $a=2$, $c=-1$, we have

$$w=\frac{z+2}{4-z} \quad \text{and} \quad w=\frac{2(z+1)}{5-z}$$ **Ans.**

**Example 86.** *Show that the transformation* $w=\frac{2z+3}{z-4}$ *maps the circle* $x^2+y^2-4x=0$ *onto the straight line* $4u+3=0$.

**Solution.** We have, $$w=\frac{2z+3}{z-4}$$

The inverse transformation is $$z=\frac{4w+3}{w-2} \quad \text{... (1)}$$

Now the circle $x^2+y^2-4x=0$ can be written as $z\bar{z}-2(z+\bar{z})=0$ $\quad \left[\begin{array}{l} z=x+iy \\ \bar{z}=x-iy \end{array}\right]$

Substituting for $z$ and $\bar{z}$ from (1), we get

$$\frac{4w+3}{w-2}\cdot\frac{4\bar{w}+3}{\bar{w}-2}-2\left(\frac{4w+3}{w-2}+\frac{4\bar{w}+3}{\bar{w}-2}\right)=0$$

$$\Rightarrow \quad 16w\bar{w}+12w+12\bar{w}+9-2(4w\bar{w}+3\bar{w}-8w-6+4w\bar{w}+3w-8\bar{w}-6)=0$$

$$\Rightarrow \quad 22(w+\bar{w})+33=0 \quad \Rightarrow \quad 22(2u)+33=0 \Rightarrow 4u+3=0 \quad \left[\begin{array}{l} w=u+iv \\ \bar{w}=u-iv \end{array}\right]$$

Thus, circle is transformed into a straight line. **Ans.**

**Example 87.** *If a is any real positive number, show that the transformation* $w=\frac{z-a}{z+a}$ *transforms conformally the plane* $x>0$ *to the unit circle* $|w|<1$. *What are the transforms of* $|w|=$ *constant and* $\arg w=$ *constant in z-plane?* *(Q. Bank U.P. III Semester 2002)*

**Solution.** We have, $$w=\frac{z-a}{z+a}$$

(*i*) $$|w|<1$$

$$\Rightarrow \quad \left|\frac{z-a}{z+a}\right| < 1$$

$$\Rightarrow \quad |z-a| < |z+a|$$

$$\Rightarrow \quad |x-a+iy| < |x+a+iy|$$

$$\Rightarrow \quad (x-a)^2 + y^2 < (x+a)^2 + y^2$$

$$\Rightarrow \quad -2ax < 2ax$$

$$\Rightarrow \quad 4ax > 0$$

$$\Rightarrow \quad x > 0$$

(*ii*) Hence the transformation (1) transforms conformally the plane $x > 0$ to the unit circle $|w| < 1$.

The circle $|w| = k$ transform into

$$\left|\frac{z-a}{z+a}\right| = k \quad \Rightarrow \quad \left|\frac{x-a+iy}{x+a+iy}\right| = k \quad \Rightarrow \quad \frac{(x-a)^2+y^2}{(x+a)^2+y^2} = k^2$$

$$(x-a)^2 + y^2 = k^2[(x+a)^2 + y^2]$$

$$\Rightarrow \quad x^2 + y^2 + a^2 - 2ax\left(\frac{1+k^2}{1-k^2}\right) = 0 \qquad \text{... (2)}$$

There is a series of coaxal circles is $z$-plane.

(*iii*) From (1), we have

$$\text{Re}^{i\phi} = \frac{z-a}{z+a}$$

$$\Rightarrow \quad \text{Re}^{i\phi} = \frac{(x-a)+iy}{(x+a)+iy} \qquad \text{... (3)}$$

Take logarithm of both sides of (3), we get

$$\log R + i\phi = \log\{(x-a)+iy\} - \log\{(x+a)+iy\}$$

$$\therefore \quad \phi = \tan^{-1}\left(\frac{y}{x-a}\right) - \tan^{-1}\left(\frac{y}{x+a}\right)$$

$$\Rightarrow \quad \tan\phi = \frac{\frac{y}{x-a} - \frac{y}{x+a}}{1 + \frac{y^2}{x^2-a^2}} = \frac{y(x+a) - y(x-a)}{x^2 - a^a + y^2}$$

So, the lines $\phi = \alpha$ transform into

$$\tan\alpha = \frac{2ay}{x^2+y^2-a^2}$$

$\Rightarrow$ $x^2 + y^2 - a^2 - 2ay \cot\alpha = 0$ which are coaxal circles orthgonal to (2). **Ans.**

## EXERCISE 7.9

1. Find the bilinear transformation that maps the points $z_1 = 2$, $z_2 = i$, $z_3 = -2$ into the points $w_1 = 1$, $w_2 = i$ and $w_3 = -1$ respectively. **Ans.** $w = \dfrac{3z+2i}{iz+6}$
2. Determine the bilinear transformation which maps $z_1 = 0$, $z_2 = 1$, $z_3 = \infty$ onto $w_1 = i$, $w_2 = -1$, $w_3 = -i$ respectively. **Ans.** $w = \dfrac{z-i}{iz-1}$
3. Verify that the equation $w = \dfrac{1+iz}{1+z}$ maps the exterior of the circle $|z| = 1$ into the upper half plane $v > 0$.
4. Find the bilinear transformation which maps $1, i, -1$ to $2, i, -2$ respectively. Find the fixed

and critical points of the transformation. **Ans.** $i, 2i$

**5.** Show that the transformation $w = \dfrac{i(1-z)}{1+z}$ maps the circle $|z| = 1$ into the real axis of the $w$-plane and the interior of the circle $|z| < 1$ into the upper half of the $w$-plane.

**6.** Show that the transformation $w = \dfrac{iz+2}{4z+i}$ transforms the real axis in the $z$-plane into circle in the $w$-plane. Find the centre and the radius of this circle. *(A.M.I.E.T.E., Winter 2000)*

**Ans.** $\left(0, \frac{7}{8}\right), \frac{9}{8}$

**7.** Show that the transformation $w = \dfrac{2z+3}{z-4}$ maps the circle $x^2 + y^2 - 4x = 0$ onto the straight line $4u + 3 = 0$

**8.** If $z_0$ is the upper half of the $z$-plane show that the bilinear transformation

$$w = e^{i\alpha}\left(\frac{z - z_0}{z - \bar{z}_0}\right)$$

maps the upper half of the $z$-plane into the interior of the unit circle at the origin in the $w$-plane.

**9.** Find the condition that the transformation $w = \dfrac{az+b}{cz+d}$ transforms the unit circle in the $w$-plane into straight line in the $z$-plane. **Ans.** If $\left|\dfrac{c}{a}\right| = 1$ or $|a| = |c|$

**10.** Prove that $w = \dfrac{z}{1-z}$ maps the upper half of the $z$-plane onto the upper half of the w-plane. What is the image of the circle $|z| = 1$ under this transformation ?

**Ans.** Straight line $2u + 1 = 0$

**11.** Show that the map of the real axis of the z-plane on the w-plane by the transformation is a circle and find its centre and radius. **Ans.** Centre $\left(0, -\frac{1}{2}\right)$, Radius $= \frac{1}{2}$

**12.** Find the invariant points of the transformation $w = -\left(\dfrac{2z+4i}{iz+1}\right)$. Prove also that these two points together with any point $z$ and its image $w$, form a set of four points having a constant cross ratio. **Ans.** $4i$ and $-i$

**13.** Show that under the transformation $w = \dfrac{z-i}{z+i}$, the real axis in z-plane is mapped into the circle $|w| = 1$. What portion of the $z$-plane corresponds to the interior of the circle ?
**Ans.** The half $z$-plane above the real axis corresponds to the interior of the circle $|w| = 1$.

**14.** Discuss the application of the transformation $w = \dfrac{iz+1}{z+i}$ to the areas in the $z$-plane which are respectively inside and outside the unit circle with its centre at the origin.

**15.** What is the form of a bilinear transformation which has one fixed point $\alpha$ and the other fixed point $\infty$?

**16.** Prove that, in general, in the bilinear transformation $w = \dfrac{az+b}{cz+d}$, there are two values of $z$ (invariant points) for which $w = z$ but there is only one value if $(a-d)^2 + 4bc = 0$.

**Choose the correct alternative:**

**17.** The fixed points of the mapping w = $(5z + 4)/(z + 5)$ are
(*i*) $-4/5, -5$ (*ii*) $2, 2$ (*iii*) $-2, -2$ (*iv*) $2, -2$ **Ans.** (*iv*)

**18.** The fixed points of the mapping $f(z) = \dfrac{3iz+13}{z-3i}$ are
(*i*) $3i \pm 2$ (*ii*) $3 \pm 2i$ (*iii*) $2 \pm 3i$ (*iv*) $-2 \pm 3i$ **Ans.** (*i*)

**7.45 TRANSFORMATION:** $w = z^2$

**Solution.** $w = z^2$

$$u + iv = (x + iy)^2 = x^2 - y^2 + 2ixy$$

Equating real and imaginary parts, we get $u = x^2 - y^2$, $v = 2xy$

(*i*) (*a*) Any line parallel to *x*-axis, *i.e.*, $y = c$, maps into

$$u = x^2 - c^2, \quad v = 2cx$$

Eliminating *x*, we get $v^2 = 4c^2 (u + c^2)$ ... (1) which is a parabola.

(*b*) Any line parallel to *y*-axis, *i.e.*, $x = b$, maps into a curve

$$u = b^2 - y^2, \; v = 2by$$

Eliminating *y*, we get $v^2 = -4b^2(u - b^2)$, ... (2) which is a parabola.

(*c*) The rectangular region bounded by the lines $x = 1$, $x = 2$, and $y = 1$, $y = 2$ maps into the region bounded by the parabolas.

By putting $x = 1 = b$ in (2) we get $v^2 = -4(u - 1)$,,

By putting $x = 2 = b$ in (2) we get $v^2 = -16(u - 4)$

By putting $y = 1 = c$ in (1) we get $v^2 = 4(u + 1)$,

By putting $y = 2 = c$ in (1) we get $v^2 = 16(u + 4)$

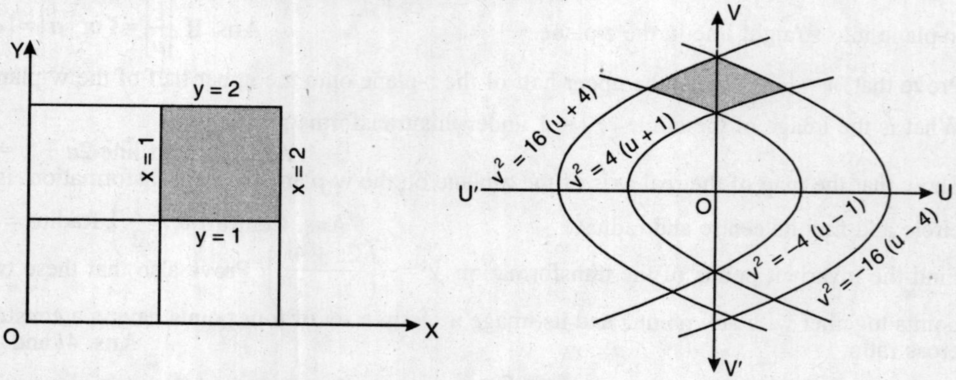


(*ii*) (*a*) In polar co-ordinates: $z = re^{i\theta}$, $w = Re^{i\phi}$

$$w = z^2$$

$$Re^{i\phi} = r^2 e^{2i\theta}$$

Then $\quad R = r^2, \; \phi = 2\theta$

In *z*-plane, a circle $r = a$ maps into $R = a^2$ in *w*-plane.

Thus, circles with centre at the origin map into circles with centre at the origin.

(*b*) If $\theta = 0$, $\phi = 0$ *i.e.*, real axis in *z*-plane maps into real axis in *w*-plane

If , $\theta = \frac{\pi}{2}$, $\phi = \pi$ *i.e.*, the positive imaginary axis in *z*-plane maps into negative real axis in *w*-plane.

Thus, the first quadrant in *z*-plane $0 \le \theta \le \frac{\pi}{2}$, maps into upper half of *w*-plane $0 \le \phi \le \pi$.

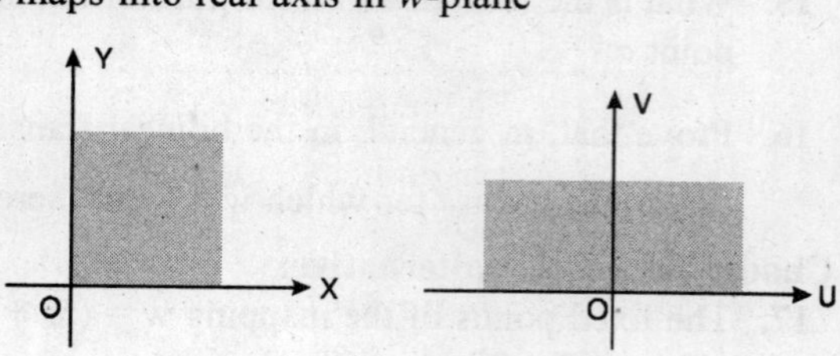


The angles in *z*-plane at origin maps into double angle in *w*-plane at origin.

Hence, the mapping $w = z^2$ is not conformal at the origin.

It is conformal in the entire *z*-plane except origin. Since $\frac{dw}{dz} = 2z = 0$ for $z = 0$, therefore, it is critical point of mapping.

**Example 88.** *For the conformal transformation $w = z^2$, show that*
*(a) the coefficient of magnification at $z = 2 + i$ is* $2\sqrt{5}$
*(b) the angle of rotation at $z = 2 + i$ is $\tan^{-1}$ (0.5).*

**Solution.**
$$z = 2 + i$$
$$f(z) = w = z^2$$
$$= (2 + i)^2 = 4 - 1 + 4i = 3 + 4i$$
$$f'(z) = 2z = 2(2 + i) = 4 + 2i$$

(*a*) Coefficient of magnification $= |f'(z)| = \sqrt{4^2 + 2^2} = \sqrt{20} = 2\sqrt{5}$ **Proved.**

(*b*) The angle of rotation $= \tan^{-1}\dfrac{v}{u} = \tan^{-1}\dfrac{2}{4} = \tan^{-1}$ (0.5) **Proved.**

**Example 89.** *For the conformal transformation $w = z^2$, show that the circle $|z - 1| = 1$ transforms into the cardioid $R = 2(2 + \cos\phi)$ where $Re^{i\phi}$ in the w-plane.*

**Solution.** $|z - 1| = 1$ ... (1)

Equation (1) represents a circle with centre at (1, 0) and radius 1.
Shifting the pole to the point (1, 0), any point on (1) is $1 + e^{i\theta}$
Transformation is under $w = z^2$.

$$Re^{i\phi} = (1 + e^{i\theta})^2$$
$$= e^{i\theta}\left(e^{\frac{i\theta}{2}} + e^{-\frac{i\theta}{2}}\right)^2$$
$$= e^{i\theta}\left(2\cos\frac{\theta}{2}\right)^2 = 4e^{i\theta}\cos^2\frac{\theta}{2}$$

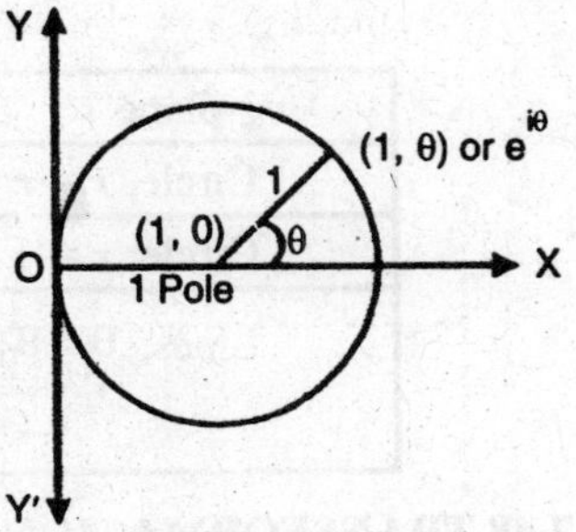


This gives $$R = 4\cos^2\frac{\theta}{2},$$

$\Rightarrow$ $$R = 2\left(2\cos^2\frac{\phi}{2}\right)$$ $[\phi = \theta]$

$\Rightarrow$ $$R = 2(\cos\phi + 1)$$ **Proved.**

## 7.46 TRANSFORMATION: $w = z^n$ $(n \in N)$

$$Re^{i\phi} = (re^{i\theta})^n = r^n e^{in\theta}$$

Hence, $$R = r^n, \ \phi = n\theta$$

**Mapping of simple figures**

| *z-plane* | *w-plane* |
|---|---|
| Circle, $r = a$ | Circle, $R = a^n$ |
| The initial line, $\theta = 0$ | The initial line, $\phi = 0$ |
| The straight line, $\theta = \theta_0$ | The straight line, $\phi = n\,\theta_0$ |

## 7.47 TRANSFORMATION: $w = z + \dfrac{1}{z}$

$$\frac{dw}{dz} = 1 - \frac{1}{z^2}$$

At $z = \pm 1$, $\dfrac{dw}{dz} = 0$, so transformation is not conformal at $z = \pm 1$.

$$w = z + \frac{1}{z} = r(\cos\theta + i\sin\theta) + \frac{1}{r(\cos\theta + i\sin\theta)}$$

$$= r(\cos\theta + i\sin\theta) + \frac{1}{r}(\cos\theta - i\sin\theta)$$

$$u + iv = \left(r + \frac{1}{r}\right)\cos\theta + i\left(r - \frac{1}{r}\right)\sin\theta$$

$$u = \left(r + \frac{1}{r}\right)\cos\theta \quad \text{and} \quad v = \left(r - \frac{1}{r}\right)\sin\theta$$

$$\frac{u}{r + \frac{1}{r}} = \cos\theta \quad \text{and} \quad \frac{v}{r - \frac{1}{r}} = \sin\theta$$

$$\sin^2\theta + \cos^2\theta = \frac{u^2}{\left(r + \frac{1}{r}\right)^2} + \frac{v^2}{\left(r - \frac{1}{r}\right)^2} \Rightarrow 1 = \frac{u^2}{\left(r + \frac{1}{r}\right)^2} + \frac{v^2}{\left(r - \frac{1}{r}\right)^2}$$

| *z*-plane | *w*-plane |
|---|---|
| Circle, $r = r$ | Ellipses |
| Circle, $r = 1$ | Lines $u = 2$ |
| Lines, $\theta = \theta_0$ | Hyperbola : $\frac{u^2}{4\cos^2\theta} - \frac{v^2}{4\sin^2\theta} = 1$ |

## 7.48 TRANSFORMATION: $w = e^z$

$$u + iv = e^{x+iy} = e^x.e^{iy} = e^x(\cos y + i\sin y)$$

Equating real and imaginary parts, we have

$$u = e^x \cos y, \qquad v = e^x \sin y$$

Again $\quad w = e^z$

$$Re^{i\phi} = e^{x+iy} = e^x.e^{iy}$$

Hence $\quad R = e^x$ or $x = \log_e R$ and $y = \phi$

**Mapping of simple figures**

| *z-plane* | *w-plane* |
|---|---|
| The straight line $x = c$ | Circle $R = e^c$ |
| *y*-axis ($x = 0$) | Unit Circle $R = e^0 = 1$ |
| Region between $y = 0, y = \pi$ | Upper half plane |
| Region between $y = 0, y = -\pi$ | Lower half plane |
| Region between the lines $y = c$ and $y = c + 2\pi$ | Whole plane |

**Example 90.** *Find the image and draw a rough sketch of the mapping of the region $1 \le x \le 2$ and $2 \le y \le 3$ under the mapping $w = e^z$.*

**Solution.** $\quad z = x + iy$

Let $\quad w = Re^{i\phi}$ ... (1)

But $\quad w = e^z + e^{x+iy}$ ... (2)

From (1) and (2); $\quad Re^{i\phi} = e^{x+iy} = e^x.e^{iy}$

Equating real and imaginary parts, we get $R = e^x$ ... (3) and $\phi = y$

(*i*) Here $1 \le x$, then $R = e^1$ is circle of radius $e^1 = 2.7$

$x = 2$, then $R = e^2$ represents a circle of radius $e^2 = 7.4$

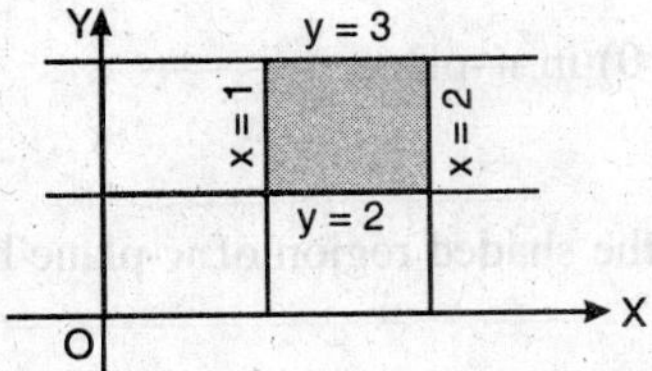


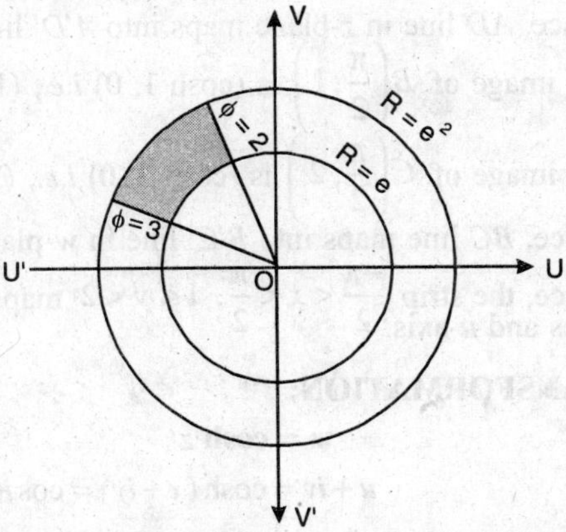


(*ii*) $y = 2$ and $\phi = 2$ represents radial line making an angle of 2 radians with the $x$-axis.
$y = 3$, then $\phi = 3$ represents radial line making an angle 3 radians with $x$-axis.
Hence, the mapping of the region $1 \le x \le 2$ and $2 \le y \le 3$ maps the shaded sectors in the figure. **Ans.**

**Example 91.** *Find the image of the strip* $-\frac{\pi}{2} < x < \frac{\pi}{2}, 1 < y < 2$ *under the mapping* $w(z) = \sin z$.

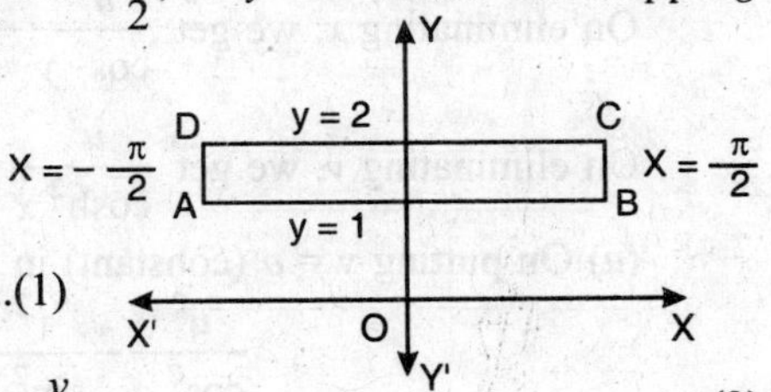


**Solution.** $w(z) = \sin z = \sin(x + iy)$

$$= \sin x \cos iy + \cos x \sin iy$$

$$u + iv = \sin x \cosh y + i \cos x \sinh y$$

$$u = \sin x \cosh y \quad \Rightarrow \sin x = \frac{u}{\cosh y} \quad \text{...(1)}$$

$$v = \cos x \sinh y \quad \Rightarrow \cos x = \frac{v}{\sinh y} \quad \text{...(2)}$$

Eliminating $x$ from (1) and (2), we get

$$\sin^2 x + \cos^2 x = \frac{u^2}{\cosh^2 y} + \frac{v^2}{\sinh^2 y} \Rightarrow 1 = \frac{u^2}{\cosh^2 y} + \frac{v^2}{\sinh^2 y}$$

Hence $\quad y = 2$, maps into the ellipse

$$\frac{u^2}{\cosh^2 2} + \frac{v^2}{\sinh^2 2} = 1 \Rightarrow \frac{u^2}{14.15} + \frac{v^2}{13.15} = 1$$

Also $\quad y = 1$, maps into the ellipse.

$$\frac{u}{\cosh^2 1} + \frac{v^2}{\sinh^2 1} = 1 \Rightarrow \frac{u^2}{2.38} + \frac{v^2}{1.38} = 1$$

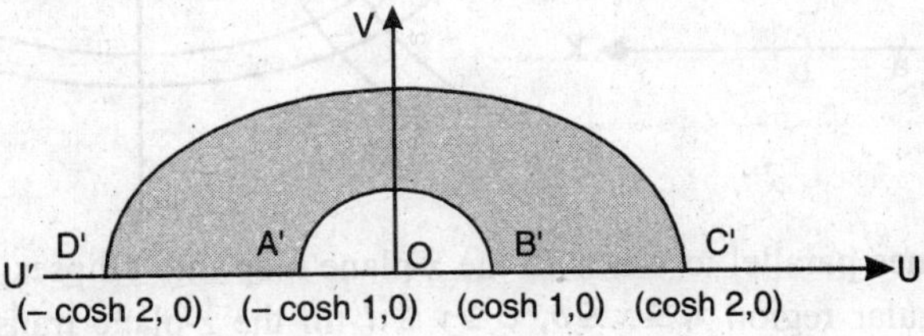


The image of $A\left(\frac{-\pi}{2}, 1\right)$ in $z$-plane is $(-\cosh 1, 0)$ *i.e.* $(-1.543, 0)$ in $w$-plane

The image of the point $D\left(-\frac{\pi}{2},\ 2\right)$ in $z$-plane is (– cosh 2, 0) *i.e.*, (– 3.762, 0).

Hence, *AD* line in $z$-plane maps into *A'D'* line in $w$-plane.

The image of $B\left(\frac{\pi}{2}, 1\right)$ is (cosh 1, 0) *i.e.*, (1.543, 0) in $w$-plane.

The image of $C\left(\frac{\pi}{2}, 2\right)$ is (cosh 2, 0) *i.e.*, (3.762, 0) in $w$-plane.

Hence, *BC* line maps into *B'C'* line in $w$-plane.

Hence, the strip $\frac{-\pi}{2} < x < \frac{\pi}{2},\ 1 < y < 2$ maps into the shaded region of $w$-plane bounded by the ellipses and $u$-axis. **Ans.**

**7.49 TRANSFORMATION:**

$$w = \cosh z$$

$$u + iv = \cosh(x + iy) = \cos i(x + iy) = \cos(ix - y)$$

$$= \cos ix \cos y + \sin ix \sin y = \cosh x \cos y + i \sinh x \sin y$$

So $\quad u = \cosh x \cos y, \quad v = \sinh x \sin y$

$\Rightarrow \quad \cosh x = \frac{u}{\cos y} \quad$ and $\quad \sinh x = \frac{v}{\sin y}$

On eliminating $x$, we get $\frac{u^2}{\cos^2 y} - \frac{v^2}{\sin^2 y} = 1$ ... (1) $(\cosh^2 x - \sinh^2 x = 1)$

On eliminating $y$, we get $\frac{u^2}{\cosh^2 x} + \frac{v^2}{\sinh^2 x} = 1$ ... (2) $(\cos^2 y + \sin^2 y = 1)$

(*a*) On putting $y = a$ (constant) in (1), we get

$$\frac{u^2}{\cos^2 a} - \frac{v^2}{\sin^2 a} = 1 \quad \text{i.e., Hyperbola.}$$

It shows that the lines parallel to $x$-axis in the $z$-plane map into hyperbola in the $w$-plane.

(*b*) On substituting $x = b$ (constant) in (2), we obtain

$$\frac{u^2}{\cosh^2 b} + \frac{v^2}{\sinh^2 b} = 1$$

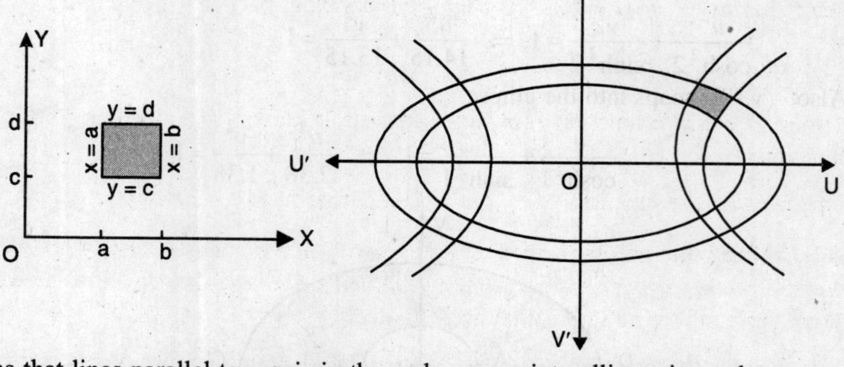


It means that lines parallel to $y$-axis in the $z$-plane map into ellipses in $w$-plane.

(*c*) The rectangular region $a \le x \le b,\ c \le y \le d$ in the $z$-plane transforms into the shaded portion in the $w$-plane.

## EXERCISE 7.10

1. Determine the region of the w-plane into which the region bounded by $x = 1$, $y = 1$, $x + y = 1$ is mapped by the transformation $w = z^2$. (*U.P. III Semester Dec. 2004*)
**Ans.** $4u + v^2 = 4,\ 4u - v^2 = -4,\ u^2 - 2v^2 = 1$

2. By the transformation $w = z^2$, show that the circle $|z - a| = c$ in the $z$-plane correspond to the limacon in the $w$-plane. **Ans.** $R = 2c(a + c\cos\phi)$

## 7.50 ZERO OF ANALYTIC FUNCTION

A zero of analytic function $f(x)$ is the value of $z$ for which $f(z) = 0$.

**Example 92.** *Find out the zeros and discuss the nature of the singularities of*

$$f(z) = \frac{(z-2)}{z^2}\sin\left(\frac{1}{z-1}\right)$$ (*R.G.P.V. Bhopal, III Semester, Dec. 2004*)

**Solution.** Poles of $f(z)$ are given by equating to zero the denominator of $f(z)$ i.e. $z = 0$ is a pole of order two.

zeros of $f(z)$ are given by equating to zero the numerator of $f(z)$ *i.e.*, $(z-2)\sin\left(\frac{1}{z-1}\right) = 0$

$\Rightarrow$ Either $z - 2 = 0$ or $\sin\left(\frac{1}{z-1}\right) = 0$

$\Rightarrow$ $z = 2$ and $\frac{1}{z-1} = n\pi$

$\Rightarrow$ $z = 2,$ $z = \frac{1}{n\pi} + 1,\ n = \pm 1, \pm 2, \ldots\ldots\ldots$

Thus, $z = 2$ is a simple zero. The limit point of the zeros are given by

$$z = \frac{1}{n\pi} + 1 \quad (n = \pm 1, \pm 2, \ldots\ldots\ldots\ldots) \text{ is } z = 1.$$

Hence $z = 1$ is an isolated essential singularity. **Ans.**

## 7.51 PRINCIPAL PART

$$\text{If } f(z) = \sum_{n=0}^{\infty} a_n (z - z_0)^n + \sum_{n=1}^{\infty} b_n (z - z_0)^{-n}$$

then the term $\sum_{n=1}^{\infty} b_n (z - z_0)^{-n}$ is called the principal part of the function $f(z)$ at $z = z_0$

## 7.52 SINGULAR POINT

A point at which a function $f(z)$ is not analytic is known as a singular point or **singularity** of the function.

*For example*, the function $\frac{1}{z-2}$ has a singular point at $z - 2 = 0$ or $z = 2$.

**Isolated singular point.** If $z = a$ is a singularity of $f(z)$ and if there is no other singularity within a small circle surrounding the point $z = a$, then $z = a$ is said to be an isolated singularity of the function $f(z)$; otherwise it is called non-isolated.

*For example*, the function $\frac{1}{(z-1)(z-3)}$ has two isolated singlar points, namely $z = 1$ and $z = 3$. [ Put $(z-1)(z-3) = 0 \Rightarrow z = 1, 3$].

**Example of non-isolated singularity.** Function $\frac{1}{\sin\frac{\pi}{z}}$ is not analytic at the points where

$\sin\frac{\pi}{z}=0$, *i.e.*, at the points $\frac{\pi}{z}=n\pi$ *i.e.*, the points $z=\frac{1}{n}$ $(n=1,2,3,...)$. Thus $z=1,\frac{1}{2},\frac{1}{3},...,z=0$ are the points of singularity. $z = 0$ is the **non-isolated singularity** of the function $\frac{1}{\sin\frac{\pi}{z}}$ because in the neighbourhood of $z = 0$, there are infinite number of other singularities $z=\frac{1}{n}$, where $n$ is very large.

**Pole of order *m*.** Let a function $f(z)$ have an isolated singular point $z = a$, $f(z)$ can be expanded in a Laurent's series around $z = a$, giving

$$f(z)=a_0+a_1(z-a)+a_2(z-a)^2+...$$
$$+\frac{b_1}{z-a}+\frac{b_2}{(z-a)^2}+...+\frac{b_m}{(z-a)^m}+\frac{b_{m+1}}{(z-a)^{m+1}}+\frac{b_{m+2}}{(z-a)^{m+2}}+... \quad ...(1)$$

In some cases it may happen that the coefficients $b_{m+1}=b_{m+2}=b_{m+3}=0$, then (1) reduces to

$$f(z)=a_0+a_1(z-a)+a_2(z-a)^2+...+\frac{b_1}{(z-a)}+\frac{b_2}{(z-a)^2}+...+\frac{b_m}{(z-a)^m}$$

$$f(z)=a_0+a_1(z-a)+a_2(z-a)^2+.....+\frac{1}{(z-a)^m}\{b_1(z-a)^{m-1}+b_2(z-a)^{m-2}$$
$$+b_3(z-a)^{m-3}+....+b_m\}$$

then $z = a$ is said to be a **pole of order *m*** of the function $f(z)$, when $m = 1$, the pole is said to be **simple pole**. In this case

$$f(z)=a_0+a_1(z-a)+a_2(z-a)^2+...+\frac{b_1}{z-a}$$

If the number of the terms of negative powers in expansion (1) is infinite, then $z = a$ is called an essential singular point of $f(z)$.

## 7.53 REMOVABLE SINGULARITY

If $f(z)=\sum_{n=0}^{\infty}a_n(z-a)^n$

$$\Rightarrow f(z)=a_0+a_1(z-a)+a_2(z-a)^2+.....+a_n(z-a)^n+.....$$

**Here the coefficients of negative powers are zero *i.e.* Laurent series does not contain negative power of $(z-a)$ then $z = a$ is called a removable signularity *i.e.*, $f(z)$ can be made analytic by redefining $f(a)$ suitably *i.e.* if $\lim_{x\to 0} f(z)$ exists.**

**Remark.** This type of singularity can be made to disappear by defining the function suitably

*e.g.*, $f(z)=\frac{\sin(z-a)}{z-a}$ has removable singularity at $z = a$ because

$$\frac{\sin(z-a)}{z-a}=\frac{1}{z-a}\left\{(z-a)-\frac{(z-a)^3}{3!}+\frac{(z-a)^5}{5!}......\infty\right\}=1-\frac{(z-a)^2}{3!}+\frac{(z-a)^4}{5!}.....\infty$$

**has no term containing negative powers of $(z-a)$. However this singularity can be removed and the function can be made analytic by defining $f(z)=\frac{\sin(z-a)}{z-a}=1$ at $z = a$**

## 7.54 WORKING RULE TO FIND SINGULARITY

**Step 1.** If $\underset{z\to a}{\text{Lt}}\, f(z)$ exists and is finite then $z = a$ is a **removable singular point.**

**Step 2.** If $\underset{z \to a}{\text{Lt}}\ f(z)$ does not exist then $z = a$ is an **essential singular point.**

**Step 3.** If $\underset{z \to a}{\text{Lt}}\ f(z)$ is infinite then $f(z)$ has a **pole at $z = a$.** The order of the pole is same as the number of negative power terms in the series expansion of $f(z)$.

**Example 93.** *Define the singularity of a function. Find the singularity (ties) of the functions*

(*i*) $f(z) = \sin \frac{1}{z}$ (*ii*) $g(z) = \frac{e^{\frac{1}{z}}}{z^2}$ (*U.P. III Semester, 2009-2010*)

**Solution.** See Art. 7.52 on page 585 for definition.

(*i*) We know that

$$\sin \frac{1}{z} = \frac{1}{z} - \frac{1}{3!\, z^3} + \frac{1}{5!\, z^5} + \ldots\ldots\ldots + (-1)^n \frac{1}{(2n+1)!\, z^{2n+1}}$$

Obviously, there is a number of singularity.

$$\sin \frac{1}{z} \text{ is not analytic at } z = 0. \qquad \left(\frac{1}{z} = \infty \text{ at } z = 0\right)$$

Hence, $\sin \frac{1}{z}$ has a singularity at $z = 0$.

(ii) Here, we have $g(z) = \frac{e^{\frac{1}{z}}}{z^2}$

We know that, $\left(\frac{1}{z^2}\right)\left(e^{\frac{1}{z}}\right) = \frac{1}{z^2}\left(1 + \frac{1}{z} + \frac{1}{2!\, z^2} + \frac{1}{3!\, z^3} + \ldots + \frac{1}{n!\, z^n} + \ldots\right)$

$$= \frac{1}{z^2} + \frac{1}{z^3} + \frac{1}{2!\, z^4} + \frac{1}{3!\, z^5} + \ldots\ldots \frac{1}{n!\, z^{n+2}} + \ldots\ldots$$

Here, $f(z)$ has infinite number of terms in negative powers of $z$.

Hence, $f(z)$ has essential singularity at $z = 0$. **Ans.**

**Example 94.** *Find the pole of the function* $\frac{e^{z-a}}{(z-a)^2}$

**Solution.** $\frac{e^{z-a}}{(z-a)^2} = \frac{1}{(z-a)^2}\left[1 + (z-a) + \frac{(z-a)^2}{2!} + \ldots\right]$

The given function has negative power 2 of $(z - a)$.

So, the given function has a pole at $z = a$ of order 2. **Ans.**

**Example 95.** *Find the poles of* $f(z) = \sin\left(\frac{1}{z-a}\right)$

**Solution.** $\sin\left(\frac{1}{z-a}\right) = \frac{1}{z-a} - \frac{1}{3!}\frac{1}{(z-a)^3} + \frac{1}{5!}\frac{1}{(z-a)^5} - \ldots..$

The given function $f(z)$ has infinite number of terms in the negative powers of $z - a$.

So, $f(z)$ has essential singularity at $z = a$. **Ans.**

**Example 96.** *Discuss singularity of* $\frac{1}{1-e^z}$ *at* $z = 2\pi i$.

**Solution.** We have, $f(z) = \frac{1}{1-e^z}$

The poles are determined by putting the denominator equal to zero.

*i.e.,* $1 - e^z = 0$

$\Rightarrow$ $e^z = 1 = (\cos 2n\pi + i \sin 2n\pi) = e^{2n\pi i}$

$\Rightarrow \qquad z = 2n\pi i \qquad (n = 0, \pm 1, \pm 2, \ldots\ldots)$

Clearly $z = 2\pi i$ is a simple pole. **Ans.**

**Example 97.** *Discuss singularity of* $\dfrac{\cot \pi z}{(z-a)^2}$ *at* $z = a$ *and* $z = \infty$.
*(R.G.P.V., Bhopal, III Semester, Dec. 2002)*

**Solution.** Let $\qquad f(z) = \dfrac{\cot \pi z}{(z-a)^2} = \dfrac{\cos \pi z}{\sin \pi z\,(z-a)^2}$

The poles are given by putting the denominator equal to zero.

*i.e.*, $\sin \pi z\,(z-a)^2 = 0 \quad \Rightarrow \quad (z-a)^2 = 0 \quad$ or $\quad \sin \pi z = 0 = \sin n\pi$

$\Rightarrow \qquad z = a, \ \pi z = n\pi, \qquad (n \in \mathrm{I})$

$\Rightarrow \qquad z = a, n$

$f(z)$ has essential singularity at $z = \infty$.

Also, $z = a$ being repeated twice gives the double pole. **Ans.**

**Example 98.** *Determine the poles of the function*

$$f(z) = \frac{1}{z^4+1}$$

*(R.G.P.V., Bhopal, III Semester, June 2003)*

**Solution.** $\qquad f(z) = \dfrac{1}{z^4+1}$

The poles of $f(z)$ are determined by putting the denominator equal to zero.

*i.e.*, $\qquad z^4 + 1 = 0 \quad \Rightarrow \quad z^4 = -1$

$$z = (-1)^{\frac{1}{4}} = (\cos \pi + i \sin \pi)^{\frac{1}{4}}$$

$$= [\cos(2n+1)\pi + i \sin(2n+1)\pi]^{\frac{1}{4}} \quad \text{[By De Moiver's theorem]}$$

$$= \left[\cos\frac{(2n+1)\pi}{4} + i \sin\frac{(2n+1)\pi}{4}\right]$$

If $n = 0$, Pole at $\qquad z = \left[\cos\dfrac{\pi}{4} + i\sin\dfrac{\pi}{4}\right] = \left(\dfrac{1}{\sqrt{2}} + i\dfrac{1}{\sqrt{2}}\right)$

If $n = 1$, Pole at $\qquad z = \left[\cos\dfrac{3\pi}{4} + i\sin\dfrac{3\pi}{4}\right] = \left(-\dfrac{1}{\sqrt{2}} + i\dfrac{1}{\sqrt{2}}\right)$

If $n = 2$, Pole at $\qquad z = \left[\cos\dfrac{5\pi}{4} + i\sin\dfrac{5\pi}{4}\right] = \left(-\dfrac{1}{\sqrt{2}} - i\dfrac{1}{\sqrt{2}}\right)$

If $n = 3$, Pole at $\qquad z = \left[\cos\dfrac{7\pi}{4} + i\sin\dfrac{7\pi}{4}\right] = \left(\dfrac{1}{\sqrt{2}} - i\dfrac{1}{\sqrt{2}}\right)$ **Ans.**

**Example 99.** *Show that the function* $e^z$ *has an isolated essential singularity at* $z = \infty$.
*(R.G.P.V., Bhopal, III Semester, Dec. 2003)*

**Solution.** Let $f(z) = e^z$

Putting $\quad z = \dfrac{1}{t}$, we get $\ f\left(\dfrac{1}{t}\right) = e^{\frac{1}{t}} = 1 + \dfrac{1}{t} + \dfrac{1}{2!\,t^2} + \dfrac{1}{3!\,t^3} + \ldots\ldots$

Here, the principal part of $f\left(\dfrac{1}{t}\right)$;

$$\frac{1}{t} + \frac{1}{2!\,t^2} + \frac{1}{3!\,t^3} + \ldots$$

Contains infinite number of terms.

Hence $t = 0$ is an isolated essential singularity of $e^{\frac{1}{t}}$ and $z = \infty$ is an isolated essential singularity of $e^z$. **Ans.**

## EXERCISE 7.11

**Find the poles or singularity of the following functions:**

1. $\dfrac{1}{(z-2)(z-3)}$ **Ans.** 2 simple poles at $z = 2$ and $z = 3$.

2. $\dfrac{e^z}{(z-2)^3}$ **Ans.** Pole at $z = 2$ of order 3.

3. $\dfrac{1}{\sin z - \cos z}$ **Ans.** Simple pole at $z = \dfrac{\pi}{4}$

4. $\cot \dfrac{1}{z}$ **Ans.** Essential singularity at $z = 0$

5. $z \operatorname{cosec} z$ **Ans.** Non-isolated essential singularity

6. $\sin \dfrac{1}{z}$ **Ans.** Essential singularity

**Choose the correct alternative:**

7. Let $f(z) = \dfrac{1}{(z-2)^4 (z+3)^6}$, then $z = 2$ and $z = -3$ are the poles of order :

(*a*) 6 and 4 (*b*) 2 and 3 (*c*) 3 and 4 (*d*) 4 and 6 **Ans.** (*d*)

(*R.G.P.V., Bhopal III Semester, June 2007*)

## 7.55 THEOREM

If $f(z)$ has a pole at $z = a$, then $|f(z)| \to \infty$ as $z \to a$.

**Proof.** Let $z = a$ be a pole of order $m$ of $f(z)$. Then by Laurent's theorem

$$f(z) = \sum_0^\infty a_n (z-a)^n + \sum_1^m b_n (z-a)^{-n}$$

$$= \sum_0^\infty a_n (z-a)^n + \frac{b_1}{z-a} + \frac{b_2}{(z-a)^2} + \ldots\ldots + \frac{b_m}{(z-a)^m}$$

$$= \sum_0^\infty a_n (z-a)^n + \frac{1}{(z-a)^m}[b_1(z-a)^{m-1} + b_2(z-a)^{m-2} + \ldots\ldots + b_{m-1}(z-a) + b_m]$$

$$= \sum_0^\infty a_n (z-a)^n + \frac{\varphi(z)}{(z-a)^m}$$

Now $\varphi(z) \to b_m$ as $z \to a$.

Hence $|f(z)| \to \infty$ as $z \to a$. **Proved.**

**Example 100.** *If an analytic function $f(z)$ has a pole of order m at $z = a$, then $\dfrac{1}{f(z)}$ has a zero of order m at $z = a$.*

**Solution.** If $f(z)$ has a pole of order $m$ at $z = a$, then

$$f(z) = \frac{\varphi(z)}{(z-a)^m} \quad \text{where } \varphi(z) \text{ is analytic and non-zero at } z = a.$$

$$\therefore \quad \frac{1}{f(z)} = \frac{(z-a)^m}{\varphi(z)}$$

Clearly, $\dfrac{1}{f(z)}$ has a zero of order $m$ at $z = a$, since $\phi(a) \neq 0$. **Proved.**

## 7.56 DEFINITION OF THE RESIDUE AT A POLE

Let $z = a$ be a pole of order $m$ of a function $f(z)$ and $C_1$ circle of radius $r$ with centre at $z = a$ which does not contain any other singularities except at $z = a$ then $f(z)$ is analytic within the annulus $r < |z - a| < R$ can be expanded within the annulus. Laurent's series:

$$f(z) = \sum_{n=0}^{\infty} a_n (z-a)^n + \sum_{n=1}^{\infty} b_n (z-a)^{-n} \quad \ldots(1)$$

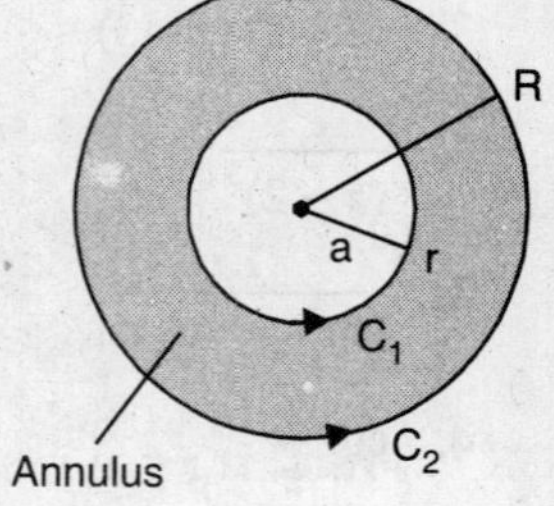


where $$a_n = \frac{1}{2\pi i} \int_C \frac{f(z)}{(z-a)^{n+1}}\, dz \quad \ldots(2)$$

and $$b_n = \frac{1}{2\pi i} \int_{C_1} \frac{f(z)}{(z-a)^{-n+1}}\, dz \quad \ldots(3)$$

$|z - a| = r$ being the circle $C_1$.

Particularly, $$b_1 = \frac{1}{2\pi i} \int_{C_1} f(z)\, dz$$

The coefficient $b_1$ is called residue of $f(z)$ at the pole $z = a$. It is denoted by symbol Res. $(z = a) = b_1$.

## 7.57 RESIDUE AT INFINITY

Residue of $f(z)$ at $z = \infty$ is defined as $-\frac{1}{2\pi i} \int_C f(z)\, dz$ where the integration is taken round $C$ in anti-clockwise direction.

where $C$ is a large circle containing all finite singularities of $f(z)$.

## 7.58 METHOD OF FINDING RESIDUES

**(a) Residue at simple pole**

(i) If $f(z)$ has a simple pole at $z = a$, then

$$\text{Res } f(a) = \lim_{z \to a} (z-a) f(z)$$

**Proof.** $$f(z) = a_0 + a_1(z-a) + a_2(z-a)^2 + \ldots + \frac{b_1}{z-a}$$

$$\Rightarrow \quad (z-a) f(z) = a_0(z-a) + a_1(z-a)^2 + a_2(z-a)^3 + \ldots + b_1$$

$$\Rightarrow \quad b_1 = (z-a) f(z) - [a_0(z-a) + a_1(z-a)^2 + a_2(z-a)^3 + \ldots]$$

Taking limit as $z \to a$, we have $b_1 = \lim_{z \to a} (z-a) f(z)$

$$\textbf{Res } (at\ z = a) = \lim_{z \to a} (z-a) f(z)$$ **Proved.**

(ii) If $f(z)$ is of the form $f(z) = \frac{\phi(z)}{\psi(z)}$ where $\psi(a) = 0$, but $\phi(a) \neq 0$

$$\textbf{Res } (at\ z = a) = \frac{\phi(a)}{\psi'(a)}$$

**Proof.** $$f(z) = \frac{\phi(z)}{\psi(z)}$$

$$\text{Res (at } z = a) = \lim_{z \to a} (z-a) f(z) = \lim_{z \to a} (z-a) \frac{\phi(z)}{\psi(z)}$$

$$= \lim_{z \to a} \frac{(z-a)[\phi(a) + (z-a)\phi'(a) + \ldots]}{\psi(a) + (z-a)\psi'(a) + \frac{(z-a)^2}{2!}\psi''(a) + \ldots} \quad \text{(By Taylor's Theorem)}$$

$$= \lim_{z \to a} \frac{(z-a)[\phi(a) + (z-a)\phi'(a) + \ldots]}{(z-a)\psi'(a) + \frac{(z-a)^2}{2!}\psi''(a) + \ldots} \quad [\text{since } \Psi(a) = 0]$$

$$= \lim_{z\to a} \frac{\phi(a)+(z-a)\,\phi'(a)+\ldots}{\psi'(a)+\frac{z-a}{2!}\psi''(a)+\ldots}$$

$$\textbf{Res (at } z = a) = \frac{\phi(a)}{\psi'(a)}$$ **Proved.**

(*b*) **Residue at a pole of order *n*.** If $f(z)$ has a pole of order $n$ at $z = a$, then

$$\textbf{Res (at } z = a) = \frac{1}{(n-1)!}\left\{\frac{d^{n-1}}{dz^{n-1}}[(z-a)^n f(z)]\right\}_{z=a}$$

**Proof.** If $z = a$ is a pole of order $n$ of function $f(z)$, then by Laurent's theorem

$$f(z) = a_0 + a_1(z-a) + a_2(z-a)^2 + \ldots + \frac{b_1}{z-a} + \frac{b_2}{(z-a)^2} + \ldots + \frac{b_n}{(z-a)^n}$$

Multiplying by $(z-a)^n$, we get

$$(z-a)^n f(z) = a_0(z-a)^n + a_1(z-a)^{n+1} + a_2(z-a)^{n+2} + \ldots$$
$$+\; b_1(z-a)^{n-1} + b_2(z-a)^{n-2} + b_3(z-a)^{n-3} + \ldots + b_n$$

Differentiating both sides w.r.t. '$z$' $n-1$ times and putting $z = a$, we get

$$\left\{\frac{d^{n-1}}{dz^{n-1}}[(z-a)^n f(z)]\right\}_{z=a} = (n-1)!\,b_1$$

$$\Rightarrow \qquad b_1 = \frac{1}{(n-1)!}\left\{\frac{d^{n-1}}{dz^{n-1}}[(z-a)^n f(z)]\right\}_{z=a}$$

$$\textbf{Residue } f(a) = \frac{1}{(n-1)!}\left\{\frac{d^{n-1}}{dz^{n-1}}[(z-a)^n f(z)]\right\}_{z=a}$$ **Proved.**

(*c*) **Residue at a pole $z = a$ of any order (simple or of order *m*)**

$$\textbf{Res } f(a) = \textbf{coefficient of } \frac{1}{t}$$

**Proof.** If $f(z)$ has a pole of order $m$, then by Laurent's theorem

$$f(z) = a_0 + a_1(z-a) + a_2(z-a)^2 \ldots + \frac{b_1}{z-a} + \frac{b_2}{(z-a)^2} + \ldots + \frac{b_m}{(z-a)^m}$$

If we put $\quad z - a = t \quad$ or $\quad z = a + t$, then

$$f(a+t) = a_0 + a_1 t + a_2 t^2 + \ldots + \frac{b_1}{t} + \frac{b_2}{t^2} + \ldots + \frac{b_m}{t^m}$$

Res $f(a) = b_1$, **Res $f(a)$ = coefficient of $\frac{1}{t}$** **Proved.**

**Rule.** Put $z = a + t$ in the function $f(z)$, expand it in powers of $t$. Coefficient of $\frac{1}{t}$ is the residue of $f(z)$ at $z = a$.

(*d*) **Residue of $f(z)$ (at $z = \infty$)** $= \lim_{z\to\infty} \{-z\,f(z)\}$

or $\quad$ The residue of $f(z)$ at infinity $= -\dfrac{1}{2\pi i}\displaystyle\int_c f(z)\,dz$

## 7.59 RESIDUE BY DEFINITION

**Example 101.** *Find the residue at $z = 0$ of $z \cos\frac{1}{z}$.*

**Solution.** Expanding the function in powers of $\frac{1}{z}$, we have

$$z\cos\frac{1}{z} = z\left[1-\frac{1}{2!\,z^2}+\frac{1}{4!\,z^4}-.....\right] = z-\frac{1}{2z}+\frac{1}{24z^3}-......$$

This is the Laurent's expansion about $z = 0$.

The coefficient of $\frac{1}{z}$ in it is $-\frac{1}{2}$. So the residue of $z\cos\frac{1}{z}$ at $z = 0$ is $-\frac{1}{2}$. **Ans.**

**Example 102.** *Find the residue of* $f(z)=\frac{z^3}{z^2-1}$ *at* $z = \infty$.

**Solution.** We have, $f(z) = \frac{z^3}{z^2-1}$

$$f(z) = \frac{z^3}{z^2\left(1-\frac{1}{z^2}\right)} = z\left(1-\frac{1}{z^2}\right)^{-1} = z\left(1+\frac{1}{z^2}+\frac{1}{z^4}+.....\right) = z+\frac{1}{z}+\frac{1}{z^3}+.....$$

$$\text{Residue at infinity} = -\left(\text{coeff. of } \frac{1}{z}\right) = -1.$$ **Ans.**

## 7.60 FORMULA: RESIDUE $= \lim_{z\to a} (z-a)\, f(z)$

**Example 103.** *Determine the pole and residue at the pole of the function* $f(z)=\frac{z}{z-1}$

**Solution.** The poles of $f(z)$ are given by putting the denominator equal to zero.

$\therefore \quad z - 1 = 0 \;\Rightarrow\; z = 1$

The function $f(z)$ has a simple pole at $z = 1$.

Residue is calculated by the formula

$$\text{Residue} = \lim_{z\to a} (z-a)\, f(z)$$

$$\text{Residue of } f(z) \text{ (at } z = 1) = \lim_{z\to 1} (z-1)\left(\frac{z}{z-1}\right) = \lim_{z\to 1} (z) = 1$$

Hence, $f(z)$ has a simple pole at $z = 1$ and residue at the pole is 1. **Ans.**

**Example 104.** *Evaluate the residues of* $\frac{z^2}{(z-1)(z-2)(z-3)}$ *at* $z = 1, 2, 3$ *and infinity and show that their sum is zero.* *(R.G.P.V., Bhopal, III Semester Dec. 2002)*

**Solution.** Let $f(z) = \frac{z^2}{(z-1)(z-2)(z-3)}$

The poles of $f(z)$ are determined by putting the denominator equal to zero.

$$\therefore (z-1)(z-2)(z-3) = 0 \Rightarrow z = 1,\ 2,\ 3$$

$$\text{Residue of } f(z) \text{ at } (z = 1) = \lim_{z\to 1} (z-1)\, f(z) = \lim_{z\to 1} (z-1)\,.\,\frac{z^2}{(z-1)(z-2)(z-3)}$$

$$= \lim_{z\to 1} \frac{z^2}{(z-2)(z-3)} = \frac{1}{2}$$

$$\text{Residue of } f(z) \text{ at } (z = 2) = \lim_{z\to 2} (z-2)\, f(z) = \lim_{z\to 2} (z-2)\, \frac{z^2}{(z-1)(z-2)(z-3)}$$

$$= \lim_{z\to 2} \frac{z^2}{(z-1)(z-3)} = \frac{4}{(1)(-1)} = -4$$

Residue of $f(z)$ at $(z = 3)$ $= \lim_{z\to 3} (z-3)\, f(z)$

$$= \lim_{z\to 3} (z-3) \frac{z^2}{(z-1)(z-2)(z-3)} = \lim_{z\to 3} \frac{z^2}{(z-1)(z-2)} = \frac{9}{2}$$

Residue of $f(z)$ at $(z = \infty)$ $= \lim_{z\to\infty} -z\, f(z) = \frac{-z(z^2)}{(z-1)(z-2)(z-3)}$

$$= \lim_{z\to\infty} \frac{-1}{\left(1-\frac{1}{z}\right)\left(1-\frac{2}{z}\right)\left(1-\frac{3}{z}\right)} = -1$$

Sum of the residues at all the poles of $f(z) = \frac{1}{2} - 4 + \frac{9}{2} - 1 = 0$

Hence, the sum of the residues is zero. **Proved.**

## 7.61 FORMULA: RESIDUE OF $f(a) = \frac{1}{(n-1)!}\left\{\frac{d^{n-1}}{dz^{n-1}}[(z-a)^n f(z)]\right\}_{z=a}$

**Example 105.** *Find the residue of a function*

$$f(z) = \frac{z^2}{(z+1)^2 (z-2)} \quad \textit{at its double pole.}$$

**Solution.** We have, $f(z) = \frac{z^2}{(z+1)^2 (z-2)}$

Poles are determined by putting denominator equal to zero.

*i.e.*; $(z+1)^2 (z-2) = 0$

$\Rightarrow$ $z = -1, -1$ and $z = 2$

The function has a double pole at $z = -1$

$$\text{Residue at } (z=-1) = \lim_{z\to -1} \frac{1}{(2-1)!}\left[\frac{d}{dz}\left\{(z+1)^2 \frac{z^2}{(z+1)^2 (z-2)}\right\}\right]$$

$$= \left[\frac{d}{dz}\left(\frac{z^2}{z-2}\right)\right]_{z=-1} = \left(\frac{(z-2)\,2z - z^2\cdot 1}{(z-2)^2}\right)_{z=-1} = \left[\frac{z^2-4z}{(z-2)^2}\right]_{z=-1} = \frac{(-1)^2-4(-1)}{(-1-2)^2}$$

$$\text{Residue at } (z=-1) = \frac{1+4}{9} = \frac{5}{9}$$ **Ans.**

**Example 106.** *Find the residue of* $\frac{1}{(z^2+1)^3}$ *at* $z = i$.

**Solution.** Let $f(z) = \frac{1}{(z^2+1)^3}$

The poles of $f(z)$ are determined by putting denominator equal to zero.

*i.e.*; $(z^2+1)^3 = 0$

$\Rightarrow$ $(z+i)^3 (z-i)^3 = 0$

$\Rightarrow$ $z = \pm i$

Here, $z = i$ is a pole of order 3 of $f(z)$.

Residue at $z = i$:

$$= \lim_{z\to i} \frac{1}{(3-1)!}\left\{\frac{d^{3-1}}{dz^{3-1}}\left[(z-i)^3 \frac{1}{(z^2+1)^3}\right]\right\} = \lim_{z\to i} \frac{1}{2!}\left\{\frac{d^2}{dz^2}\left(\frac{1}{(z+i)^3}\right)\right\}$$

$$= \lim_{z \to i} \frac{1}{2}\left(\frac{3 \times 4}{(z+i)^5}\right) = \frac{1}{2} \times \frac{12}{(i+i)^5} = \frac{6}{32i} = \frac{3}{16i} = -\frac{3i}{16}$$

Hence, the residue of the given function at $z = i$ is $-\frac{3i}{16}$. **Ans.**

## 7.62 FORMULA: RES. (AT $z = a$) = $\frac{\phi(a)}{\psi'(a)}$

**Example 107.** *Determine the poles and residue at each pole of the function $f(z) = \cot z$.*

**Solution.** $f(z) = \cot z = \frac{\cos z}{\sin z}$

The poles of the function $f(z)$ are given by

$\sin z = 0, \; z = n\pi, \;$ where $n = 0, \pm 1, \pm 2, \pm 3 \ldots$

Residue of $f(z)$ at $z = n\pi$ is $= \frac{\cos z}{\frac{d}{dz}(\sin z)} = \frac{\cos z}{\cos z} = 1$ $\left[\text{Res. at } (z = a) = \frac{\phi(a)}{\psi'(a)}\right]$ **Ans.**

**Example 108.** *Determine the poles of the function and residue at the poles.*

$$f(z) = \frac{z}{\sin z}$$

**Solution.** $f(z) = \frac{z}{\sin z}$

Poles are determined by putting $\sin z = 0 = \sin n\pi \quad \Rightarrow z = n\pi$

$$\text{Residue} = \left(\frac{z}{\cos z}\right)_{z=n\pi} \qquad \left[\text{Residue} = \frac{\phi(a)}{\psi'(a)}\right]$$

$$= \frac{n\pi}{\cos n\pi} = \frac{n\pi}{(-1)^n}$$

Hence, the residue of the given function at pole $z = n\pi$ is $\frac{n\pi}{(-1)^n}$. **Ans.**

## 7.63 FORMULA: RESIDUE = COEFFICIENT OF $\frac{1}{t}$

**where $z = \frac{1}{t}$**

**Example 109.** *Find the residue of $\frac{z^3}{(z-1)^4(z-2)(z-3)}$ at a pole of order 4.*

**Solution.** The poles of $f(z)$ are determined by putting the denominator equal to zero.

$\therefore \quad (z-1)^4 (z-2)(z-3) = 0 \quad \Rightarrow \quad z = 1, 2, 3$

Here $\quad z = 1$ is a pole of order 4.

$$f(z) = \frac{z^3}{(z-1)^4(z-2)(z-3)} \qquad \ldots(1)$$

Putting $z - 1 = t$ or $z = 1 + t$ in (1), we get

$$f(1+t) = \frac{(1+t)^3}{t^4(t-1)(t-2)} = \frac{1}{t^4}(t^3 + 3t^2 + 3t + 1)(1-t)^{-1}\frac{1}{2}\left(1 - \frac{t}{2}\right)^{-1}$$

$$= \frac{1}{2}\left(\frac{1}{t} + \frac{3}{t^2} + \frac{3}{t^3} + \frac{1}{t^4}\right)(1 + t + t^2 + t^3 + \ldots) \times \left(1 + \frac{t}{2} + \frac{t^2}{4} + \frac{t^3}{8} \ldots\right)$$

$$=\frac{1}{2}\left(\frac{1}{t}+\frac{3}{t^2}+\frac{3}{t^3}+\frac{1}{t^4}\right)\left(1+\frac{3}{2}t+\frac{7}{4}t^2+\frac{15}{8}t^3+\ldots\right)=\frac{1}{2}\left(\frac{1}{t}+\frac{9}{2t}+\frac{21}{4}\frac{1}{t}+\frac{15}{8}\frac{1}{t}\right)+\ldots$$

$$=\frac{1}{2}\left(1+\frac{9}{2}+\frac{21}{4}+\frac{15}{8}\right)\frac{1}{t} \qquad \left[\text{Res } f(a)=\text{coeffi. of } \frac{1}{t}\right]$$

Coefficient of $\frac{1}{t}=\frac{1}{2}\left(1+\frac{9}{2}+\frac{21}{4}+\frac{15}{8}\right)=\frac{101}{16}$,

Hence, the residue of the given function at a pole of order 4 is $\frac{101}{16}$. **Ans.**

**Example 110.** *Find the residue of* $f(z)=\frac{ze^z}{(z-a)^3}$ *at its pole.*

**Solution.** The pole of $f(z)$ is given by $(z-a)^3=0$ *i.e.*, $z=a$

Here $z=a$ is a pole of order 3.

Putting $z-a=t$ where $t$ is small.

$$f(z)=\frac{ze^z}{(z-a)^3} \Rightarrow f(z)=\frac{(a+t)e^{a+t}}{t^3}=\left(\frac{a}{t^3}+\frac{1}{t^2}\right)e^{a+t}=e^a\left(\frac{a}{t^3}+\frac{1}{t^2}\right)e^t \qquad (z=a+t)$$

$$=e^a\left(\frac{a}{t^3}+\frac{1}{t^2}\right)\left(1+\frac{t}{1!}+\frac{t^2}{2!}+\ldots\right)=e^a\left[\frac{a}{t^3}+\frac{a}{t^2}+\frac{a}{2t}+\frac{1}{t^2}+\frac{1}{t}+\frac{1}{2}+\ldots\right]$$

$$=e^a\left[\frac{1}{2}+\left(\frac{a}{2}+1\right)\frac{1}{t}+(a+1)\frac{1}{t^2}+(a)\frac{1}{t^3}+\ldots\right]$$

Coefficient of $\frac{1}{t}=e^a\left(\frac{a}{2}+1\right)$

Hence the residue at $z=a$ is $e^a\left(\frac{a}{2}+1\right)$. **Ans.**

## EXERCISE 7.12

1. Determine the poles of the following functions. Find the order of each pole.

(*i*) $\frac{z^2}{(z-a)(z-b)(z-c)}$ **Ans.** Simple poles at $z=a$, $z=b$, $z=c$

(*ii*) $\frac{z-3}{(z-2)^2(z+1)}$ **Ans.** Pole at $z=2$ of second order and $z=-1$ of first order.

(*iii*) $\frac{ze^{iz}}{z^2+a^2}$ **Ans.** Poles at $z=\pm ia$, order 1.

(*iv*) $\frac{1}{(z-1)(z-2)}$ **Ans.** $z=2, z=1$

**Find the residue of**

2. $\frac{z^3}{(z-2)(z-3)}$ at its poles. **Ans.** 27, – 8

3. $\frac{z^2}{z^2+a^2}$ at $z=ia$. **Ans.** $\frac{1}{2}ia$

4. $\frac{1}{(z^2+a^2)^2}$ at $z=ia$ **Ans.** $-\frac{i}{4a^3}$

**5.** tan $z$ at its pole. **Ans.** $f\left(n+\frac{\pi}{2}\right) = -1$ at its pole

**6.** $z^2 e^{1/z}$ at the point $z = 0$. **Ans.** $\frac{1}{6}$ **7.** $z^2 \sin\left(\frac{1}{z}\right)$ at $z = 0$ **Ans.** $-\frac{1}{6}$

**8.** $\frac{1}{z^2(z-i)}$ at $z = i$ **Ans.** $-1$ **9.** $\frac{e^{2z}}{1-e^z}$ at its pole **Ans.** $-1$

**10.** $\frac{1+e^z}{\sin z + z\cos z}$ at $z = 0$ **Ans.** 1 **11.** $\frac{1}{z(e^z-1)}$ at its poles. **Ans.** $-\frac{1}{2}$

## 7.64 CAUCHY'S RESIDUE THEOREM

*(MDU, DEC. 2008)*

If $f(z)$ is analytic in a closed curve $C$, except at a finite number of poles within $C$, then $\int_c f(z)\,dz = 2\pi i$ **(sum of residues at the poles within** $C$**).**

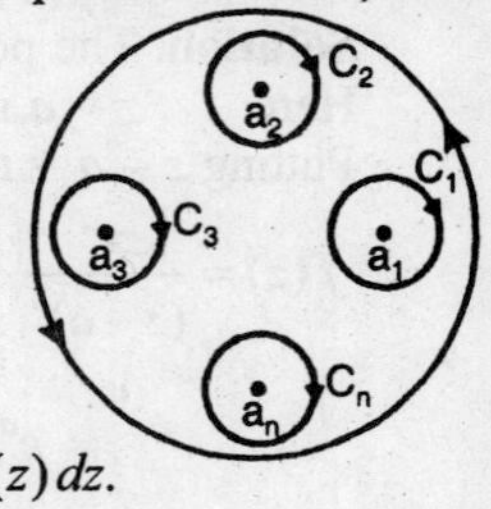


**Proof.** Let $C_1, C_2, C_3, \ldots, C_n$ be the non-intersecting circles with centres at $a_1, a_2, a_3, \ldots, a_n$ respectively, and radii so small that they lie entirely within the closed curve $C$. Then $f(z)$ is analytic in the multiple connected region lying between the curves $C$ and $C_1, C_2, \ldots, C_n$.

Applying Cauchy's theorem

$$\int_c f(z)\,dz = \int_{c_1} f(z)\,dz + \int_{c_2} f(z)\,dz + \int_{c_3} f(z)\,dz + \ldots + \int_{c_n} f(z)\,dz.$$

$$= 2\pi i\,[\text{Res } f(a_1) + \text{Res } f(a_2) + \text{Res } f(a_3) + \ldots + \text{Res } f(a_n)]$$ **Proved.**

**Example 111.** *Evaluate the following integral using residue theorem*

$$\int_c \frac{1+z}{z(2-z)}\,dz$$

*where c is the circle* $|z| = 1$.

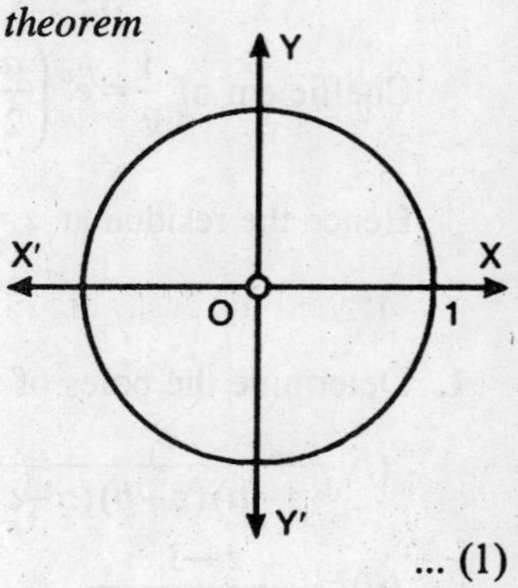


**Solution.** The poles of the integrand are given by putting the denominator equal to zero.

$$z(2-z) = 0 \quad \text{or} \quad z = 0, 2$$

The integrand is analytic on $|z| = 1$ and all points inside except $z = 0$, as a pole at $z = 0$ is inside the circle $|z| = 1$.

Hence by residue theorem

$$\int_c \frac{1+z}{z(2-z)}\,dz = 2\pi i \quad [\text{Res } f(0)] \qquad \ldots (1)$$

$$\text{Residue } f(0) = \lim_{z\to 0} z.\frac{1+z}{z(2-z)} = \lim_{z\to 0}\frac{1+z}{2-z} = \frac{1}{2}$$

Putting the value of Residue $f(0)$ in (1), we get

$$\int_c \frac{1+z}{z(2-z)}\,dz = 2\pi i\left(\frac{1}{2}\right) = \pi i$$ **Ans.**

**Example 112.** *Determine the poles of the following function and residue at each pole:*

$$f(z) = \frac{z^2}{(z-1)^2(z+2)}$$ *and hence evaluate*

$$\int_c \frac{z^2\,dz}{(z-1)^2(z+2)}$$ *where c:* $|z| = 3$. *(R.G.P.V. Bhopal, III Sem. Dec. 2007)*

Poles of $f(z)$ are given by $(z-1)^2(z+2)=0$ *i.e.* $z = 1, 1, -2$
The pole at $z = 1$ is of second order and the pole at $z = -2$ is simple.

$$\text{Residue of } f(z) \text{ (at } z = 1) = \lim_{z\to 1} \frac{1}{(2-1)!}\frac{d}{dz}\frac{(z-1)^2 z^2}{(z-1)^2(z+2)}$$

$$= \lim_{z\to 1} \frac{d}{dz}\frac{z^2}{z+2} = \lim_{z\to 1}\frac{(z+2)\,2z - 1.z^2}{(z+2)^2}$$

$$= \lim_{z\to 1}\frac{z^2+4z}{(z+2)^2} = \frac{1+4}{(1+2)^2} = \frac{5}{9}$$

$$\text{Residue of } f(z) \text{ (at } z = -2) = \lim_{z\to -2}\frac{(z+2)z^2}{(z-1)^2(z+2)} = \lim_{z\to -2}\frac{z^2}{(z-1)^2} = \frac{4}{(-2-1)^2} = \frac{4}{9}$$ **Ans.**

$$\int_C \frac{z^2\,dz}{(z-1)^2(z+2)} = 2\pi i\left(\frac{5}{9}+\frac{4}{9}\right) = 2\pi i$$ **Ans.**

**Example 113.** *Using Residue theorem, evaluate* $\frac{1}{2\pi i}\int_C \frac{e^{zt}\,dz}{z^2(z^2+2z+2)}$
*where C is the cirlce* $|z| = 3$. (*U.P., III Semester, Dec. 2009*)

**Solution.** Here, we have

$$\frac{1}{2\pi i}\int_C \frac{e^{zt}\,dz}{z^2(z^2+2z+2)}$$

Poles are given by

$z = 0$ (double pole)

$z = -1 \pm i$ (simple poles)

All the four poles are inside the given circle.

$|z| = 3$

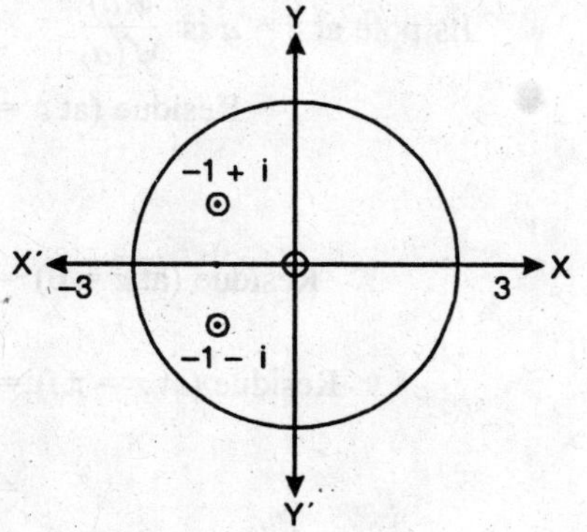


$$\text{Residue at } z = 0 \text{ is } \lim_{z\to 0}\frac{d}{dz} z^2 \frac{e^{zt}}{z^2(z^2+2z+2)}$$

$$= \lim_{z\to 0}\frac{d}{dz}\frac{e^{zt}}{z^2+2z+2}$$

$$= \lim_{z\to 0}\frac{(z^2+2z+2)\,t\,e^{zt} - (2z+2)e^{zt}}{(z^2+2z+2)^2}$$

$$= \frac{2te^0 - 2e^0}{4} = \frac{(t-1)}{2}$$

$z^2+2z+2 = 0$
$\Rightarrow z^2+2z+1 = -1$
$\Rightarrow (z+1)^2 = -1$
$\Rightarrow z+1 = \pm i$
$\Rightarrow z = -1 \pm i$

Residue at $z = -1 + i$

$$= \lim_{z\to -1+i}\frac{(z+1-i)\,e^{zt}}{z^2(z+1-i)(z+1+i)} = \lim_{z\to -1+i}\frac{e^{zt}}{z^2(z+1+i)}$$

$$= \frac{e^{(-1+i)t}}{(-1+i)^2(-1+i+1+i)} = \frac{e^{(-1+i)t}}{(1-2i-1)(2i)} = \frac{e^{(-1+i)t}}{4}$$

$$\int \frac{e^{zt}}{z^2\,(z^2+2z+2)}\,dz = 2\pi i \quad \text{(Sum of the Residues)}$$

$$\Rightarrow \quad \frac{1}{2\pi i}\int \frac{e^{zt}}{z^2\,(z^2+2z+2)}\,dz = \frac{t-1}{2} + \frac{e^{(-1+i)t}}{4} + \frac{e^{(-1-i)t}}{4}$$

$$= \frac{t-1}{2} + \frac{e^{-t}}{4}\,(e^{it} + e^{-it}) = \frac{t-1}{2} + \frac{e^{-t}}{4}\,(2\cos t)$$

$$= \frac{t-1}{2} + \frac{e^{-t}}{2}\cos t$$ **Ans.**

**Example 114.** *Evaluate* $\oint_C \frac{1}{\sinh z}\,dz$, *where C is the circle* $|z| = 4$.

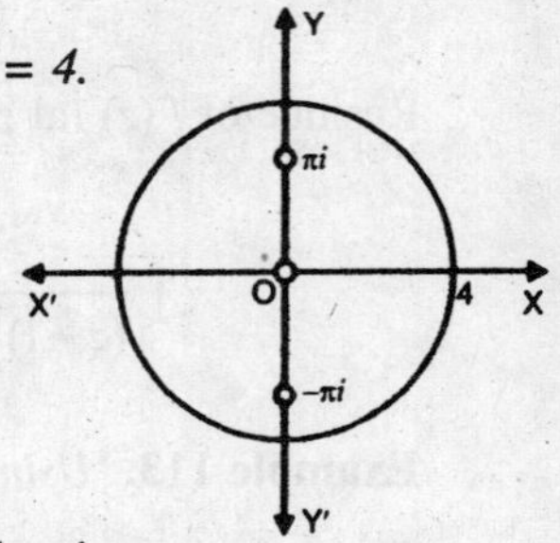


**Solution.** Here, $\quad f(z) = \frac{1}{\sinh z}$.

Poles are given by

$$\sinh z = 0$$

$$\Rightarrow \quad \sin iz = 0$$

$$\Rightarrow \quad z = n\pi i \text{ where } n \text{ is an integer.}$$

Out of these, the poles $z = -\pi i$, 0 and $\pi i$ lie inside the circle $|z| = 4$.

The given function $\frac{1}{\sinh z}$ is of the form $\frac{\phi(z)}{\psi(z)}$

Its pole at $z = a$ is $\frac{\phi(a)}{\psi'(a)}$.

Residue (at $z = -\pi i$)

$$= \frac{1}{\cosh(-\pi i)} = \frac{1}{\cos i(-\pi i)} = \frac{1}{\cos \pi} = \frac{1}{-1} = -1$$

$$\text{Residue (at } z = 0) = \frac{1}{\cosh 0} = \frac{1}{1} = 1$$

$$\text{Residue (at } z = \pi i) = \frac{1}{\cosh(\pi i)} = \frac{1}{\cos i\,(\pi i)} = \frac{1}{\cos(-\pi)}$$

$$= \frac{1}{\cos \pi} = \frac{1}{-1} = -1$$

Residue at $-\pi i$, 0, $\pi i$ are respectively $-1$, 1 and $-1$.
Hence, the required integral $= 2\pi i\,(-1 + 1 - 1) = -2\pi i$. **Ans.**

**Example 115.** *Evaluate* $\int_c \frac{dz}{z\sin z}$ : *c is the unit circle about origin.*

**Solution.**
$$\frac{1}{z\sin z} = \frac{1}{z\left[z - \frac{z^3}{3!} + \frac{z^5}{5!} - \ldots\right]} = \frac{1}{z^2\left[1 - \frac{z^2}{3!} + \frac{z^4}{5!} - \ldots\right]}$$

$$= \frac{1}{z^2}\left[1 - \left(\frac{z^2}{6} - \frac{z^4}{120}\ldots\right)\right]^{-1} = \frac{1}{z^2}\left[1 + \left(\frac{z^2}{6} - \frac{z^4}{120}\ldots\right) + \left(\frac{z^2}{6} - \frac{z^4}{120}\ldots\right)^2 \ldots\right]$$

$$= \frac{1}{z^2}\left[1 + \frac{z^2}{6} - \frac{z^4}{120} + \frac{z^4}{36} + \ldots\right] = \frac{1}{z^2} + \frac{1}{6} - \frac{z^2}{120} + \frac{z^2}{36}\ldots = \frac{1}{z^2} + \frac{1}{6} + \frac{7}{360}z^2\ldots$$

This shows that $z = 0$ is a pole of order 2 for the function $\frac{1}{z \sin z}$ and the residue at the pole is zero, (coefficient of $\frac{1}{z}$).

Now the pole at $z = 0$ lies within $C$.

$$\therefore \quad \int \frac{1}{z \sin z} dz = 2\pi i \text{ (Sum of Residues)} = 0 \qquad \textbf{Ans.}$$

**Example 116.** *Evaluate* $\int_C \frac{e^z}{\cos \pi z} dz$, *where C is the unit circle* $|z| = 1$.

(*M.D.U. 2005, 2007,2008*)

**Solution.** Here $f(z) = \frac{e^z}{\cos \pi z}$

$$= \frac{e^z}{\left(1 - \frac{(\pi z)^2}{2!} + \frac{(\pi z)^4}{4!} - \ldots\right)}$$

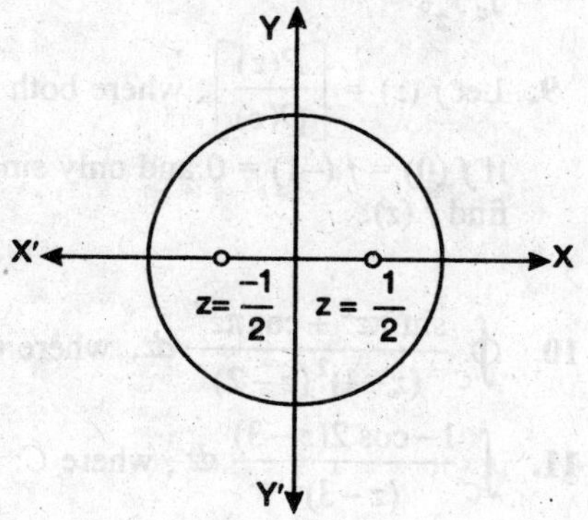


It has simple poles at $z = \pm\frac{1}{2}, \pm\frac{3}{2}, \pm\frac{5}{2}, \ldots\ldots$, of which only $z = \pm\frac{1}{2}$ lie inside the circle $|z| = 1$.

Residue of $f(z)$ at $z = \frac{1}{2}$ is

$$\lim_{z \to \frac{1}{2}} \left(z - \frac{1}{2}\right) f(z) = \lim_{z \to \frac{1}{2}} \frac{\left(z - \frac{1}{2}\right) e^z}{\cos \pi z} \qquad \left[\text{Form } \frac{0}{0}\right]$$

$$= \lim_{z \to \frac{1}{2}} \frac{\left(z - \frac{1}{2}\right) e^z + e^z}{-\pi \sin \pi z} \qquad \text{[By L' Hopital's Rule]}$$

$$= \frac{e^{1/2}}{-\pi}.$$

Similarly, residue of $f(z)$ at $z = -\frac{1}{2}$ is $\frac{e^{-1/2}}{\pi}$.

$$\therefore \text{ By residue theorem } \oint_C \frac{e^z}{\cos \pi z} dz = 2\pi i \text{ (sum of residues)}$$

$$= 2\pi i \left(-\frac{e^{1/2}}{\pi} + \frac{e^{-1/2}}{\pi}\right) = -4i \left(\frac{e^{1/2} - e^{-1/2}}{2}\right) = -4i \sinh \frac{1}{2}. \qquad \textbf{Ans.}$$

## EXERCISE 7.13

**Evaluate the following complex integrals:**

1. $\int_c \frac{1 - 2z}{z(z-1)(z-2)} dz$, where $C$ is the circle $|z| = 1.5$ (*MDU Dec. 2006*) **Ans.** $3\pi i$
2. $\int_c \frac{z^2 e^{zt}}{z^2 + 1} dz$, where $C$ is the circle $|z| = 2$ **Ans.** $-2\pi i \sin t$

3. $\int_C \frac{z-1}{(z+1)^2(z-2)}dz$, where $C$ is the circle $|z-i|=2$. **Ans.** $-\frac{2\pi i}{9}$

4. $\int_C \frac{2z^2+z}{z^2-1}dz$, where $C$ is the circle $|z-1|=1$. **Ans.** $3\pi i$

5. $\int_C \frac{e^{2z}+z^2}{(z-1)^5}dz$, where $C$ is the circle $|z|=2$ **Ans.** $\frac{4\pi e^2 i}{3}$

6. $\int_C \frac{dz}{(z^2+1)(z^2-4)}$, where $C$ is the circle $|z|=1.5$ **Ans.** 0

7. $\int_C \frac{4z^2-4z+1}{(z-2)(z^2+4)}dz$, where $C$ is the circle $|z|=1$ **Ans.** 0

8. $\int_C \frac{\sin z}{z^6}dz$, where $C$ is the circle $|z|=2$ **Ans.** $\frac{\pi i}{60}$

9. Let $f(z)=\left[\frac{P(z)}{Q(z)}\right]$, where both $P(z)$ and $Q(z)$ are complex polynomial of degree two. If $f(0)=f(-1)=0$ and only singularity of $f(z)$ is of order 2 at $z=1$ with residue $-1$, then find $f(z)$.

**Ans.** $f(z)=-\frac{1}{3}\frac{z(z+1)}{(z-1)^2}$

10. $\oint_C \frac{\sin \pi z^2+\cos \pi z^2}{(z-1)^2(z-2)}dz$, where $C$ is the circle $|z|=3$ (*MDU. Dec. 2008*) **Ans.** $4\pi i(\pi+1)$

11. $\int_C \frac{1-\cos 2(z-3)}{(z-3)^3}dz$, where $C$: $|z-3|=1$. (*MDU. Dec. 2004*) **Ans.** $4\pi i$

## 7.65 EVALUATION OF REAL DEFINITE INTEGRALS BY CONTOUR INTEGRATION

A large number of real definite integrals, whose evaluation by usual methods become sometimes very tedious, can be easily evaluated by using Cauchy's theorem of residues. For finding the integrals we take a closed curve $C$, find the poles of the function $f(z)$ and calculate residues at those poles only which lie within the curve $C$.

$$\int_C f(z)\,dz = 2\pi i \text{ (sum of the residues of } f(z) \text{ at the poles within } C)$$

We call the curve, a contour and the process of integration along a contour is called contour integration.

## 7.66 INTEGRATION ROUND UNIT CIRCLE OF THE TYPE

$$\int_0^{2\pi} f(\cos\theta, \sin\theta)\,d\theta$$

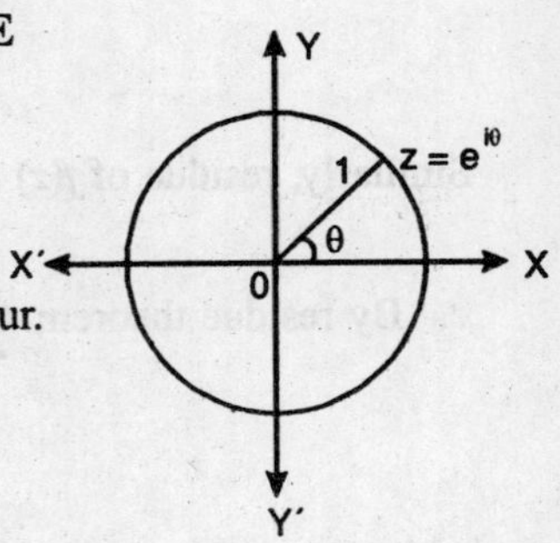


where $f(\cos\theta, \sin\theta)$ is a rational function of $\cos\theta$ and $\sin\theta$.

Convert $\sin\theta$, $\cos\theta$ into $z$.

Consider a circle of unit radius with centre at origin, as contour.

$$\sin\theta = \frac{e^{i\theta}-e^{-i\theta}}{2i} = \frac{1}{2i}\left[z-\frac{1}{z}\right], \qquad z = re^{i\theta} = 1.\,e^{i\theta} = e^{i\theta}$$

$$\cos\theta = \frac{e^{i\theta}+e^{-i\theta}}{2} = \frac{1}{2}\left[z+\frac{1}{z}\right]$$

As we know

$$z = e^{i\theta} \Rightarrow dz = e^{i\theta}\,i\,d\theta = z\,i\,d\theta \text{ or } d\theta = \frac{dz}{iz}$$

The integrand is converted into a function of $z$.

Then apply Cauchy's residue theorem to evaluate the integral.

Some examples of these are illustrated below.

**Example 117.** *Evaluate the integral:*

$$\int_0^{2\pi} \frac{d\theta}{5-3\cos\theta}$$ *(R.G.P.V., Bhopal, III Semester, June 2007)*

**Solution.** $$\int_0^{2\pi} \frac{d\theta}{5-3\cos\theta} = \int_0^{2\pi} \frac{d\theta}{5-3\left(\frac{e^{i\theta}+e^{-i\theta}}{2}\right)}$$

$$= \int_0^{2\pi} \frac{2\,d\theta}{10-3e^{i\theta}-3e^{-i\theta}} \qquad \left[\begin{array}{l} e^{i\theta}=z \Rightarrow i.e.^{i\theta}\, d\theta = dz \\ d\theta = \frac{dz}{iz}\end{array}\right]$$

$$= \int_C \frac{2}{10-3z-\frac{3}{z}} \frac{dz}{iz} = \frac{1}{i}\int_C \frac{2\,dz}{10z-3z^2-3}$$

[$C$ is the unit circle $|z| = 1$]

$$= -\frac{2}{i}\int_C \frac{dz}{3z^2-10z+3}$$

$$= -\frac{2}{i}\int_C \frac{dz}{(3z-1)(z-3)} = 2i\int_C \frac{dz}{(3z-1)(z-3)}$$

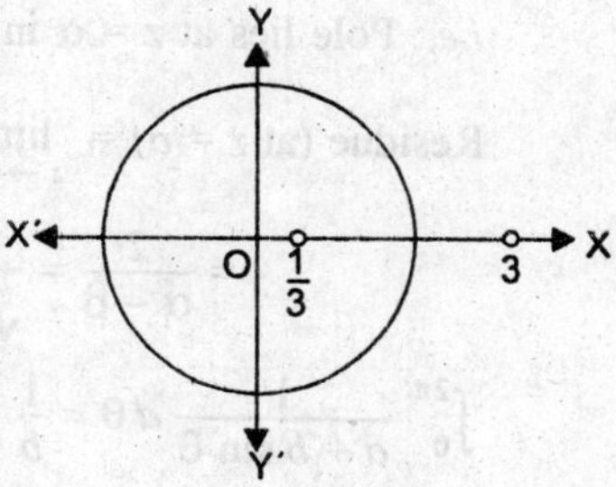


Let $$I = 2i\int_C \frac{dz}{(3z-1)(z-3)}$$

Poles of the integrand are given by

$$(3z-1)(z-3) = 0 \quad\Rightarrow\quad z = \frac{1}{3}, 3$$

There is only one pole at $z = \frac{1}{3}$ inside the unit circle $C$.

Residue at $\left(z = \frac{1}{3}\right) = \lim\limits_{z\to\frac{1}{3}} \left(z-\frac{1}{3}\right) f(z) = \lim\limits_{z\to\frac{1}{3}} 2i \frac{\left(z-\frac{1}{3}\right)}{(3z-1)(z-3)} = \lim\limits_{z\to\frac{1}{3}} 2i \frac{1}{3(z-3)}$

$$= \frac{2i}{3\left(\frac{1}{3}-3\right)} = -\frac{2i}{8} = -\frac{i}{4}$$

Hence, by Cauchy's Residue Theorem

$$I = 2\pi i \text{ (Sum of the residues within Contour)} = 2\pi i\left(-\frac{i}{4}\right) = \frac{\pi}{2}$$

$$\int_0^{2\pi} \frac{d\theta}{5-3\cos\theta} = \frac{\pi}{2}$$ **Ans.**

**Example 118.** *Evaluate* $\int_0^{2\pi} \frac{d\theta}{a+b\sin\theta}$ if $a > |b|$

*(G.B.T.U., III 2012, U.P. III Semester 2009-2010)*

**Solution.** Let $$I = \int_0^{2\pi} \frac{d\theta}{a+b\sin\theta}$$

$$= \int_0^{2\pi} \frac{1}{a+b\frac{e^{i\theta}-e^{-i\theta}}{2i}}\, d\theta \qquad \left[\text{Writing } e^{i\theta} = z,\ d\theta = \frac{dz}{iz}\right]$$

$$= \int_C \frac{1}{a + \frac{b}{2i}\left(z - \frac{1}{z}\right)} \frac{dz}{iz}$$ (where $C$ is the unit circle $|z| = 1$)

$$= \int_c \frac{1}{2iaz + bz^2 - b} dz = \frac{1}{b}\int \frac{2}{z^2 + \frac{2aiz}{b} - 1} dz$$

$$= \frac{1}{b}\int_c \frac{2}{(z-\alpha)(z-\beta)} dz \qquad \left[bz^2 + 2aiz - b = b\left\{z^2 + \frac{2aiz}{b} - 1\right\}\right]$$

Where $\alpha + \beta = -\dfrac{2ai}{b}$

$|\alpha\beta| = -1$

$|\alpha| < 1$ then $|\beta| > 1$

*i.e.*; Pole lies at $z = \alpha$ in the unit circle.

$$\left[\begin{aligned}(\alpha-\beta)^2 &= (\alpha+\beta)^2 - 4\alpha\beta\\ &= -\frac{4a^2}{b^2} + 4\\ \alpha - \beta &= 2\frac{\sqrt{b^2 - a^2}}{b}\end{aligned}\right]$$

Residue (at $z = a$) $= \lim_{z \to \alpha} (z-\alpha) \dfrac{2}{(z-\alpha)(z-\beta)}$

$$= \frac{2}{\alpha - \beta} = \frac{b}{\sqrt{b^2 - a^2}} = \frac{b}{i\sqrt{a^2 - b^2}}$$

$$\int_0^{2\pi} \frac{1}{a + b\sin\theta} d\theta = \frac{1}{b}\int_c \frac{2}{z^2 + 2\frac{aiz}{b} - 1} dz = 2\pi i \frac{b}{bi\sqrt{a^2 - b^2}} = \frac{2\pi}{\sqrt{a^2 - b^2}}$$ **Ans.**

**Example 119.** *Use the complex variable technique to find the value of the integral :*

$$\int_0^{2\pi} \frac{d\theta}{2 + \cos\theta}.$$ (*R.G.P.V., Bhopal, III Semester, Dec. 2003*)

**Solution.** Let $I = \displaystyle\int_0^{2\pi} \frac{d\theta}{2+\cos\theta} = \int_0^{2\pi} \frac{d\theta}{2 + \frac{e^{i\theta} + e^{-i\theta}}{2}} = \int_0^{2\pi} \frac{2d\theta}{4 + e^{i\theta} + e^{-i\theta}}$

Put $e^{i\theta} = z$ so that $e^{i\theta}(i\,d\theta) = dz \Rightarrow iz\,d\theta = dz \Rightarrow d\theta = \dfrac{dz}{iz}$

$$I = \int_C \frac{2\frac{dz}{iz}}{4 + z + \frac{1}{z}}$$ where $C$ denotes the unit circle $|z| = 1$.

$$= \frac{1}{i}\int_c \frac{2\,dz}{z^2 + 4z + 1}$$

The poles are given by putting the denominator equal to zero.

$$z^2 + 4z + 1 = 0 \text{ or } z = \frac{-4 \pm \sqrt{16-4}}{2} = \frac{-4 \pm 2\sqrt{3}}{2} = -2 \pm \sqrt{3}$$

The pole within the unit circle $C$ is a simple pole at $z = -2 + \sqrt{3}$. Now we calculate the residue at this pole.

$$\text{Residue at } (z = -2+\sqrt{3}) = \lim_{z \to (-2+\sqrt{3})} \frac{1}{i} \frac{(z + 2 - \sqrt{3})2}{(z + 2 - \sqrt{3})(z + 2 + \sqrt{3})}$$

$$= \lim_{z \to (-2+\sqrt{3})} \frac{2}{i(z + 2 + \sqrt{3})} = \frac{2}{i(-2 + \sqrt{3} + 2 + \sqrt{3})} = \frac{1}{\sqrt{3}i}$$

Hence by Cauchy's Residue Theorem, we have

$$\int_0^{2\pi} \frac{d\theta}{2+\cos\theta} = 2\pi i \text{ (sum of the residues within the contour)}$$

$$= 2\pi i \frac{1}{i\sqrt{3}} = \frac{2\pi}{\sqrt{3}}$$ **Ans.**

**Example 120.** *Using complex variable techniques evaluate the real integral*

$$\int_0^{2\pi} \frac{\sin^2\theta}{5-4\cos\theta}\, d\theta$$

**Solution.** If we write $z = e^{i\theta}$

$$\cos\theta = \frac{1}{2}\left(z+\frac{1}{z}\right), \qquad d\theta = \frac{dz}{iz}$$

and so $$I = \int_0^{2\pi} \frac{\sin^2\theta}{5-4\cos\theta}\, d\theta = \frac{1}{2}\int_0^{2\pi} \frac{1-\cos 2\theta}{5-4\cos\theta}\, d\theta$$

$$I = \text{Real part of } \frac{1}{2}\int_0^{2\pi} \frac{1-\cos 2\theta - i\sin 2\theta}{5-4\cos\theta}\, d\theta$$ [where $C$ is a circle of unit radius with centre $z = 0$]

$$= \text{Real part of } \frac{1}{2}\int_0^{2\pi} \frac{1-e^{2i\theta}}{5-4\cos\theta}\, d\theta$$

$$= \text{Real part of } \frac{1}{2}\int_c \frac{1-z^2}{5-2(z+\frac{1}{z})}\left(\frac{dz}{iz}\right) = \text{Real part of } \frac{1}{2i}\int_c \frac{1-z^2}{5z-2z^2-2}\, dz$$

$$= \text{Real part of } \frac{1}{2i}\int_c \frac{z^2-1}{2z^2-5z+2}\, dz$$

Poles are determined by $2z^2-5z+2=0$ or $(2z-1)(z-2) = 0$ or $z = \frac{1}{2}, 2$

So inside the contour $C$ there is a simple pole at $z = \frac{1}{2}$

Residue at the simple pole $\left(z=\frac{1}{2}\right) = \lim_{z\to\frac{1}{2}}\left(z-\frac{1}{2}\right)\frac{z^2-1}{(2z-1)(z-2)}$

$$= \lim_{z\to\frac{1}{2}} \frac{z^2-1}{2(z-2)} = \frac{\frac{1}{4}-1}{2\left(\frac{1}{2}-2\right)} = \frac{1}{4}$$

$$I = \text{Real part of } \frac{1}{2i}\int_c \frac{(z^2-1)}{2z^2-5z+2}\, dz = \frac{1}{2i} 2\pi i \text{ (sum of the residues)}$$

$$\Rightarrow \quad \int_0^{2\pi} \frac{\sin^2\theta}{5-4\cos\theta}\, d\theta = \pi\left(\frac{1}{4}\right) = \frac{\pi}{4}$$ **Ans.**

**Example 121.** *Using contour integration, evaluate the real integral*

$$\int_0^{\pi} \frac{1+2\cos\theta}{5+4\cos\theta}\, d\theta$$ *(R.G.P.V., Bhopal, III Semester, Dec. 2004)*

**Solution.** Let $I = \int_0^{\pi} \frac{1+2\cos\theta}{5+4\cos\theta}\, d\theta = \frac{1}{2}\int_0^{2\pi} \frac{1+2\cos\theta}{5+4\cos\theta}\, d\theta$ (Even Function)

$$= \text{Real part of } \frac{1}{2}\int_0^{2\pi} \frac{1+2e^{i\theta}}{5+4\cos\theta}\, d\theta$$

$$= \text{Real part of } \frac{1}{2}\int_0^{2\pi} \frac{1+2e^{i\theta}}{5+2(e^{i\theta}+e^{-i\theta})}\, d\theta$$

writing $e^{i\theta} = z,\ d\theta = \frac{dz}{iz}$ where $C$ is the unit circle $|z| = 1$.

$$= \text{Real part of } \frac{1}{2}\int_C \frac{1+2z}{5+2\left(z+\frac{1}{z}\right)} \frac{dz}{iz} = \text{Real part of } \frac{1}{2}\int_C \frac{-i(1+2z)}{2z^2+5z+2}\, dz$$

$$= \text{Real part of } \frac{1}{2}\int_C \frac{-i(2z+1)}{(2z+1)(z+2)}\, dz = \text{Real part of } -\frac{i}{2}\int_C \frac{1}{z+2}\, dz$$

Pole is given by $z + 2 = 0$ *i.e.* $z = -2$.
Thus there is no pole of $f(z)$ inside the unit circle $C$. Hence $f(z)$ is analytic in C.

By Cauchy's Theorem $\int_C f(z)\, dz = 0$ if $f(z)$ is analytic in $C$.

$\therefore \qquad I =$ Real part of zero $= 0$

Hence, the given integral = 0 **Ans.**

**Example 122.** *Using complex variables, evaluate the real integral*

$$\int_0^{2\pi} \frac{d\theta}{1-2p\sin\theta+p^2},\quad \text{where } p^2 < 1.$$

*(Kerala 2005; MDU Dec. 2008)*

**Solution.** $\int_0^{2\pi} \frac{d\theta}{1-2p\sin\theta+p^2} = \int_0^{2\pi} \frac{d\theta}{1-2p\frac{(e^{i\theta}-e^{-i\theta})}{2i}+p^2}$

Let $\qquad I = \int_0^{2\pi} \frac{d\theta}{1+ip(e^{i\theta}-e^{-i\theta})+p^2}$

Writing $\qquad z = e^{i\theta},\ dz = ie^{i\theta} d\theta = i z\, d\theta,\ d\theta = \frac{dz}{zi}$

$$I = \int_C \frac{1}{1+ip\left(z-\frac{1}{z}\right)+p^2} \frac{dz}{zi}$$ where $c$ is the unit circle $|z| = 1$.

$$= \int_C \frac{dz}{zi-pz^2+p+p^2zi} = \int_C \frac{dz}{-pz^2+ip^2z+zi+p} = \int_C \frac{dz}{(iz+p)(izp+1)}$$

Poles are given by $(iz+p)(ipz+1) = 0$

$\Rightarrow \qquad z = -\frac{p}{i} = ip$ and $z = -\frac{1}{pi} = \frac{i}{p}$ $\qquad |ip| < 1$ and $\left|\frac{i}{p}\right| > 1$ as $p^2 < 1$

$pi$ is the only pole inside the unit circle.

Residue $(z = pi) = \lim_{z\to pi} \frac{(z-pi)}{(iz+p)(izp+1)} = \lim_{z\to pi}\left[\frac{1}{i(izp+1)}\right] = \frac{1}{i}\frac{1}{(-p^2+1)}$

Hence by Cauchy's residue theorem

$$\int_0^{2\pi} \frac{d\theta}{1-2p\sin\theta+p^2} = 2\pi i\left(\frac{1}{i}\frac{1}{1-p^2}\right) = \frac{2\pi}{1-p^2}$$ **Ans.**

**Example 123.** *Apply calculus of residue to prove that:*

$$\int_0^{2\pi} \frac{\cos 2\theta\, d\theta}{1-2a\cos\theta+a^2} = \frac{2\pi a^2}{1-a^2}, \qquad (a^2 < 1)$$

*(MDU. May 2007, 2003, R.G.P.V., Bhopal, III Semester, June 2003)*

**Solution.** Let $I = \int_0^{2\pi} \frac{\cos 2\theta\, d\theta}{1-2a\cos\theta+a^2} = \int_0^{2\pi} \frac{\cos 2\theta\, d\theta}{1-a\,(e^{i\theta}+e^{-i\theta})+a^2}$

$$= \text{Real part of } \int_0^{2\pi} \frac{e^{2i\theta}}{(1-ae^{i\theta})\,(1-ae^{-i\theta})}\, d\theta$$

$$= \text{Real part of } \oint_C \frac{z^2}{(1-az)\left(1-\frac{a}{z}\right)}\frac{dz}{iz} \qquad \left[\text{Put } e^{i\theta} = z \text{ so that } d\theta = \frac{dz}{iz}\right]$$

$$= \text{Real part of } \oint_C \frac{-iz^2}{(1-az)\,(z-a)}\, dz \qquad [C \text{ is the unit circle } |z| = 1]$$

Poles of $\frac{-iz^2}{(1-az)\,(z-a)}$ are given by

$$(1-az)\,(z-a) = 0$$

Thus, $z = \frac{1}{a}$ and $z = a$ are the simple poles. Only $z = a$ lies within the unit circle $C$ as $a < 1$.

The residue of $f(z)$ at $(z = a)$ $= \lim_{z\to a} (z-a)\frac{-iz^2}{(1-az)\,(z-a)} = \lim_{z\to a}\frac{-iz^2}{(1-az)} = -\frac{ia^2}{1-a^2}$

Hence, by Cauchy's Residue Theorem, we have

$$\oint_C f(z)\, dz = 2\pi i \text{ [Sum of residues within the contour]}$$

$$= 2\pi i\left(-\frac{ia^2}{1-a^2}\right) = \frac{2\pi a^2}{1-a^2} \text{ which is purely real.}$$

Thus, $I = \text{Real part of } \oint_C f(z)\,dz = \frac{2\pi a^2}{1-a^2}$

Hence, $\int_0^{2\pi} \frac{\cos 2\theta}{1-2a\cos\theta+a^2} = \frac{2\pi a^2}{1-a^2}$. **Proved.**

**Example 124.** *Evaluate:* $\int_0^{2\pi} \frac{\cos 2\theta}{5+4\cos\theta}\, d\theta$ *by using contour integration.*

*(R.G.P.V., Bhopal, III Semester, June 2007)*

**Solution.**

Let $I = \int_0^{2\pi} \frac{\cos 2\theta}{5+4\cos\theta}\, d\theta$

$$= \text{Real part of } \int_0^{2\pi} \frac{\cos 2\theta + i\sin 2\theta}{5+4\cos\theta}\, d\theta$$

$$= \text{Real part of } \int_0^{2\pi} \frac{e^{2i\theta}}{5+2\,(e^{i\theta}+e^{-i\theta})}\, d\theta \qquad \left[\begin{array}{r} e^{i\theta} = z \\ \Rightarrow \quad i.e.^{i\theta}\, d\theta = dz \\ \Rightarrow d\theta = \dfrac{dz}{ie^{i\theta}} = \dfrac{dz}{iz} \end{array}\right]$$

$$= \text{Real part of } \oint_C \frac{z^2}{5+2\left(z+\frac{1}{z}\right)} \frac{dz}{iz}$$

[$C$ is the unit circle $|z| = 1$]

$$= \text{Real part of } \oint_C \frac{z^2}{5z+2z^2+2} \frac{dz}{i}$$

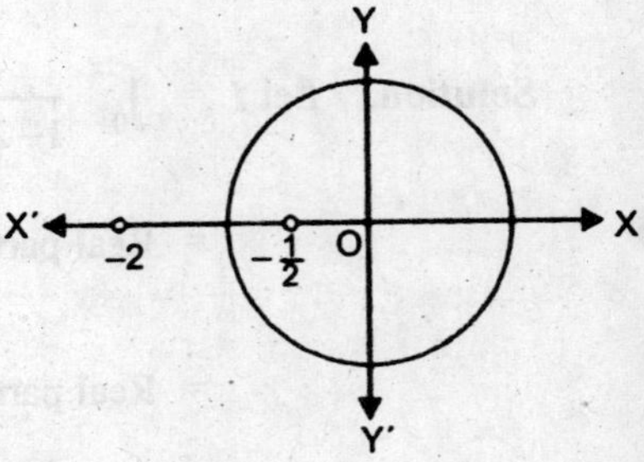


$$= \text{Real part of } \oint_C \frac{-iz^2}{2z^2+5z+2}\, dz$$

$$= \text{Real part of } \oint_C \frac{-iz^2}{(2z+1)(z+2)}\, dz$$

Poles are determined by putting denominator equal to zero.

$$(2z+1)(z+2) = 0 \quad \Rightarrow \quad z = -\frac{1}{2}, -2$$

The only simple pole at $z = -\frac{1}{2}$ is inside the contour.

$$\text{Residue at } \left(z = -\frac{1}{2}\right) = \lim_{z \to -\frac{1}{2}} \left(z+\frac{1}{2}\right) f(z) = \lim_{z \to -\frac{1}{2}} \left(z+\frac{1}{2}\right) \frac{-iz^2}{(2z+1)(z+2)}$$

$$= \lim_{z \to -\frac{1}{2}} \frac{-iz^2}{2(z+2)} = \frac{-i\left(-\frac{1}{2}\right)^2}{2\left(-\frac{1}{2}+2\right)} = \frac{-i}{12}$$

By Cauchy's Integral Theorem

$$\int_C f(z)\, dz = 2\pi i \text{ (Sum of the residues within } C)$$

$$= 2\pi i \left(\frac{-i}{12}\right) = \frac{\pi}{6}, \text{ which is real}$$

$$\int_0^{2\pi} \frac{\cos 2\theta}{5+4\cos\theta}\, d\theta = \frac{\pi}{6}$$ **Ans.**

**Example 125.** *Evaluate contour integration of the real integral*

$$\int_0^{2\pi} \frac{\cos 3\theta}{5-4\cos\theta}\, d\theta.$$ *(U.P., III Sem., 2009, R.G.P.V., Bhopal, III Semester, Dec. 2007)* *(MDU, Dec. 2010)*

**Solution.** $\displaystyle\int_0^{2\pi} \frac{\cos 3\theta}{5-4\cos\theta}\, d\theta = \text{Real part of } \int_0^{2\pi} \frac{e^{3i\theta}}{5-4\cos\theta}\, d\theta$

$$= \text{Real part of } \int_0^{2\pi} \frac{e^{3i\theta}}{5-2(e^{i\theta}+e^{-i\theta})}\, d\theta \qquad \text{On writing } z = e^{i\theta} \text{ and } d\theta = \frac{dz}{iz}$$

$$= \text{Real part of} \int_c \frac{z^3}{5 - 2\left(z + \frac{1}{z}\right)} \frac{dz}{iz} \qquad C \text{ is the unit circle.}$$

$$= \text{Real part of } \frac{1}{i}\int_c \frac{z^3}{5z - 2z^2 - 2} dz \;= \text{Real part of } -\frac{1}{i}\int \frac{z^3}{2z^2 - 5z + 2} dz$$

$$= \text{Real part of } i\int \frac{z^3}{(2z-1)(z-2)} dz$$

Poles are given by $(2z-1)(z-2) = 0$ *i.e.* $z = \frac{1}{2}, z = 2$

$z = \frac{1}{2}$ is the only pole inside the unit circle.

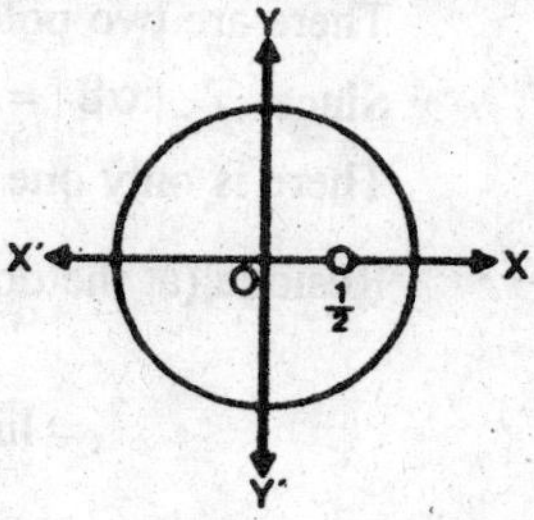

$$\text{Residue}\left(\text{at } z = \frac{1}{2}\right) = \lim_{z \to \frac{1}{2}} \frac{i\left(z - \frac{1}{2}\right)z^3}{(2z-1)(z-2)}$$

$$= \lim_{z \to \frac{1}{2}} \frac{iz^3}{2\,(z-2)} = \frac{i\frac{1}{8}}{2\left(\frac{1}{2} - 2\right)} = -\frac{i}{24}$$

$$\int_0^{2\pi} \frac{\cos 3\theta}{5 - 4\cos\theta} d\theta = \text{Real part of } 2\pi i\left(\frac{-i}{24}\right) = \frac{\pi}{12}$$ **Ans.**

**Question.** *Evaluate* : $\int_0^{\infty} \frac{\cos 3\theta}{5 + 4\cos\theta} d\theta$ *(U.P. III Semester, Dec. 2008, 2006)*

**Example 126.** *Use the residue theorem to show that*

$$\int_0^{2\pi} \frac{d\theta}{(a + b\cos\theta)^2} = \frac{2\pi a}{(a^2 - b^2)^{3/2}} \qquad \textit{where } a > 0, b > 0, a > b.$$

*(R.G.P.V., Bhopal, III Semester, June 2004)*

**Solution.** $$\int_0^{2\pi} \frac{d\theta}{(a + b\cos\theta)^2} = \int_0^{2\pi} \frac{d\theta}{\left(a + b\cdot\frac{e^{i\theta} + e^{-i\theta}}{2}\right)^2}$$

Put $e^{i\theta} = z$, so that $e^{i\theta}(i d\theta) = dz \;\Rightarrow\; iz d\theta = dz \Rightarrow d\theta = \frac{dz}{iz}$

$$= \int_c \frac{1}{\left\{a + \frac{b}{2}\left(z + \frac{1}{z}\right)\right\}^2} \frac{dz}{iz} \qquad \text{where } C \text{ is the unit circle } |z| = 1$$

$$\int_c \frac{1}{\left(a + \frac{bz}{2} + \frac{b}{2z}\right)^2} \frac{dz}{iz} = \int_c \frac{-4iz}{\left(a + \frac{bz}{2} + \frac{b}{2z}\right)^2} \frac{dz}{(2z)^2}$$

$$= \int_c \frac{-4iz dz}{(bz^2 + 2az + b)^2} = \frac{-4i}{b^2}\int_c \frac{z\,dz}{\left(z^2 + \frac{2az}{b} + 1\right)^2}$$

The poles are given by putting the denominator equal to zero.

*i.e.,* $\left(z^2+\frac{2a}{b}z+1\right)^2=0$

$\Rightarrow$ $(z-\alpha)^2(z-\beta)^2=0$

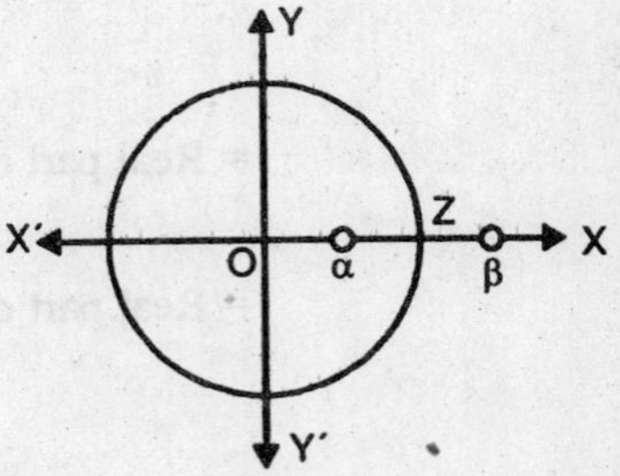


where $$\alpha=\frac{-\frac{2a}{b}+\sqrt{\frac{4a^2}{b^2}-4}}{2}=\frac{-a+\sqrt{a^2-b^2}}{b}$$

$$\beta=\frac{-\frac{2a}{b}-\sqrt{\frac{4a^2}{b^2}-4}}{2}=\frac{-a-\sqrt{a^2-b^2}}{b}$$

There are two poles, at $z=\alpha$ and at $z=\beta$, each of order 2.

Since $|\alpha\beta|=1$ or $|\alpha|\,|\beta|=1$ if $|\alpha|<1$ then $|\beta|>1$.

There is only one pole [$|\alpha|<1$] of order 2 within the unit circle $C$.

Residue (at the double pole $z=\alpha$) $=\lim\limits_{z\to\alpha}\frac{d}{dz}(z-\alpha)^2\frac{(-4iz)}{b^2(z-\alpha)^2(z-\beta)^2}$

$$=\lim_{z\to\alpha}\frac{d}{dz}\frac{-4iz}{b^2(z-\beta)^2}$$

$$=-\frac{4i}{b^2}\lim_{z\to\alpha}\frac{(z-\beta)^2.1-2(z-\beta)z}{(z-\beta)^4}=\frac{-4i}{b^2}\lim_{z\to\alpha}\frac{z-\beta-2z}{(z-\beta)^3}=\frac{-4i}{b^2}\lim_{z\to\alpha}\frac{-(z+\beta)}{(z-\beta)^3}$$

$$=\frac{4i}{b^2}\frac{(\alpha+\beta)}{(\alpha-\beta)^3}=\frac{4i}{b^2}\frac{\alpha+\beta}{[(\alpha+\beta)^2-4\alpha\beta]^{\frac{3}{2}}}=\frac{4i}{b^2}\frac{-\frac{2a}{b}}{\left[\left(-\frac{2a}{b}\right)^2-4(1)\right]^{\frac{3}{2}}}$$

$$=\frac{-8ai}{(4a^2-4b^2)^{\frac{3}{2}}}=-\frac{ai}{(a^2-b^2)^{\frac{3}{2}}}$$

Hence, $\int_0^{2\pi}\frac{d\theta}{(a+b\cos\theta)^2}=2\pi i\times\frac{-ai}{(a^2-b^2)^{3/2}}=\frac{2\pi a}{(a^2-b^2)^{3/2}}$ **Proved.**

**Example 127.** *Evaluate by Contour integration:*

$$\int_0^{2\pi}e^{\cos\theta}\cos(\sin\theta-n\theta)\,d\theta.$$

**Solution.** Let $I=\int_0^{2\pi}e^{\cos\theta}[\cos(\sin\theta-n\theta)+i\sin(\sin\theta-n\theta)]\,d\theta$

$$=\int_0^{2\pi}e^{\cos\theta}e^{i(\sin\theta-n\theta)}d\theta=\int_0^{2\pi}e^{\cos\theta+i\sin\theta}\cdot e^{-ni\theta}d\theta$$

$$=\int_0^{2\pi}e^{e^{i\theta}}\cdot e^{-in\theta}d\theta \qquad ...(1)$$

Put $e^{i\theta}=z$ so that $d\theta=\frac{dz}{iz}$ then,

$$I=\int_C e^z\cdot\frac{1}{z^n}\cdot\frac{dz}{iz}=-i\int_C\frac{e^z}{z^{n+1}}dz$$

Pole is at $z = 0$ of order $(n + 1)$.
It lies inside the unit circle.
Residue of $f(z)$ at $z = 0$ is

$$= \frac{1}{(n+1-1)!}\left[\frac{d^n}{dz^n}\left\{z^{n+1}\cdot\frac{-ie^z}{z^{n+1}}\right\}\right]_{z=0} = \frac{-i}{n!}\left[\frac{d^n}{dz^n}(e^z)\right]_{z=0} = \frac{-i}{n!}\left(e^z\right)_{z=0} = \frac{-i}{n!}$$

$\therefore$ By Cauchy's Residue theorem,

$$I = 2\pi i\left(\frac{-i}{n!}\right) = \frac{2\pi}{n!}$$

Comparing real part of $\int_0^{2\pi} e^{\cos\theta}\,[\cos(\sin\theta - n\theta) + i\sin(\sin\theta - n\theta)]\,d\theta = \frac{2\pi}{n!}$,

we have

$$\int_0^{2\pi} e^{\cos\theta}\cos(\sin\theta - n\theta)\,d\theta = \frac{2\pi}{n!}$$ **Ans.**

## EXERCISE 7.14

**Evaluate the following integrals:**

1. $\int_0^{2\pi} \frac{\sin^2\theta}{a + b\cos\theta}\,d\theta$ *(R.G.P.V., Bhopal, III Semester, June 2008)* **Ans.** $\frac{2\pi}{b^2}\{a - \sqrt{a^2 - b^2}\},\ a > b > 0$

2. $\int_0^{2\pi} \frac{(1 + 2\cos\theta)^n\cos n\theta}{3 + 2\cos\theta}\,d\theta$ **Ans.** $\frac{2\pi}{\sqrt{5}}(3 - \sqrt{5})^n,\ n > 0$

3. $\int_0^{2\pi} \frac{d\theta}{2 + \cos\theta}$ **Ans.** $\frac{2\pi}{\sqrt{3}}$

4. $\int_0^{2\pi} \frac{4}{5 + 4\sin\theta}\,d\theta$ **Ans.** $\frac{8\pi}{3}$

5. $\int_0^{\pi} \frac{d\theta}{17 - 8\cos\theta}$ **Ans.** $\frac{\pi}{15}$

6. $\int_0^{\pi} \frac{d\theta}{a + b\cos\theta}$, *where* $a > |b|$. *Hence or otherwise evaluate* $\int_0^{2\pi} \frac{d\theta}{\sqrt{2} - \cos\theta}$. **Ans.** $\frac{\pi}{\sqrt{a^2 - b^2}};\ \pi$

## 7.67 EVALUATION OF $\int_{-\infty}^{\infty} \frac{f_1(x)}{f_2(x)}\,dx$ where $f_1(x)$ and $f_2(x)$ are polynomials in $x$.

Such integrals can be reduced to contour integrals, if
(*i*) $f_2(x)$ has no real roots.
(*ii*) the degree of $f_2(x)$ is greater than that of $f_1(x)$ by at least two.

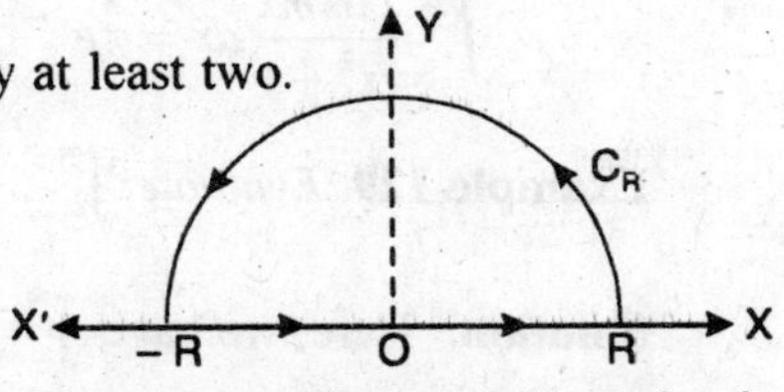


**Procedure:** Let $f(x) = \frac{f_1(x)}{f_2(x)}$

Consider $\int_C f(z)\,dz$
where $C$ is a curve, consisting of the upper half $C_R$ of the circle $|z| = R$, and part of the real axis from $-R$ to $R$.

If there are no poles of $f(z)$ on the real axis, the circle $|z| = R$ which is arbitrary can be taken such that there is no singularity on its circumference $C_R$ in the upper half of the plane, but possibly some poles inside the contour $C$ specified above.

Using Cauchy's theorem of residues, we have

$$\int_C f(z)\,dz = 2\pi i \times \text{(sum of the residues of } f(z) \text{ at the poles within } C)$$

*i.e.* $\int_{-R}^{R} f(x)\,dx + \int_{CR} f(z)\,dz = 2\pi i$ (sum of residues within $C$)

$\Rightarrow$ $\int_{-R}^{R} f(x)\,dx = -\int_{CR} f(z)\,dz + 2\pi i$ (sum of residues within $C$)

$$\therefore \quad \lim_{R\to\infty}\int_{-R}^{R} f(x)\,dx = -\lim_{R\to\infty}\int_{CR} f(z)\,dz + 2\pi i \text{ (sum of residues within } C) \quad \ldots (1)$$

Now, $\lim_{R\to\infty}\int_{CR} f(z)\,dz = \int_0^{\pi} f(Re^{i\theta})\,Ri\,e^{i\theta}\,d\theta$

$= 0$ when $R \to \infty$

(1) reduces $\int_{-\infty}^{\infty} f(x)\,dx = 2\pi i$ (sum of residues within $C$)

**Example 128.** *Evaluate* $\int_0^{\infty} \frac{\cos mx}{(x^2+1)}\,dx$. *(R.G.P.V., Bhopal, III Semester, Dec. 2006)*

**Solution.** $\int_0^{\infty} \frac{\cos mx}{x^2+1}\,dx$

Consider the integral $\int_C f(z)\,dz$, where

$f(z) = \frac{e^{imz}}{z^2+1}$, taken round the closed contour $c$ consisting of the upper half of a large circle $|z| = R$ and the real axis from $-R$ to $R$.

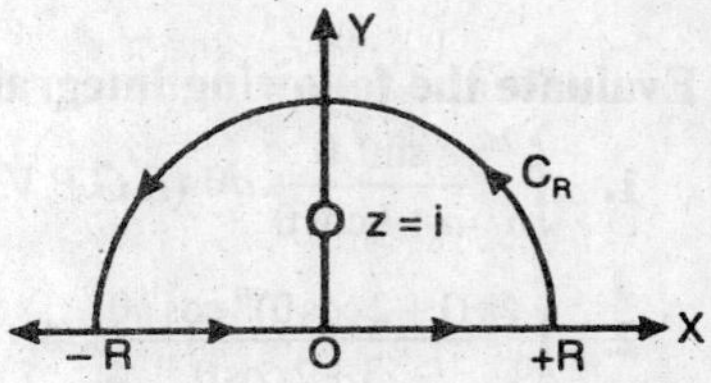


Poles of $f(z)$ are given by

$z^2 + 1 = 0$ *i.e.* $z^2 = -1$ *i.e.* $z = \pm i$

The only pole which lies within the contour is at $z = i$.

The residue of $f(z)$ at $z = i$

$$= \lim_{z\to i}\frac{(z-i)e^{imz}}{(z^2+1)} = \lim_{z\to i}\frac{e^{imz}}{z+i} = \frac{e^{-m}}{2i}$$

Hence by Cauchy's residue theorem, we have

$$\int_C f(z)\,dz = 2\pi i \times \text{ sum of the residues}$$

$$\Rightarrow \quad \int_C \frac{e^{imz}}{z^2+1}\,dz = 2\pi i \times \frac{e^{-m}}{2i} \quad \Rightarrow \quad \int_{-R}^{R}\frac{e^{imx}}{x^2+1}\,dx = \pi e^{-m}$$

Equating real parts, we have

$$\int_{-\infty}^{\infty}\frac{\cos mx}{x^2+1}\,dx = \pi e^{-m} \quad \Rightarrow \quad \int_0^{\infty}\frac{\cos mx}{x^2+1}\,dx = \frac{\pi e^{-m}}{2}$$ **Ans.**

**Example 129.** *Evaluate* $\int_{-\infty}^{\infty}\frac{x\sin\pi x}{x^2+2x+5}\,dx$ *(U.P. III Semester 2009-2010)*

**Solution.** Here, we have $\int_{-\infty}^{\infty}\frac{x\sin\pi x}{x^2+2x+5}\,dx$

Let us consider $\int_c \frac{z\sin\pi z}{z^2+2z+5}\,dz$

The pole can be determined by putting the denominator equal to zero.

$$z^2+2z+5=0 \quad \Rightarrow \quad z = \frac{-2\pm\sqrt{4-20}}{2} \quad \Rightarrow \quad z = -1 \pm 2i$$

Out of two poles, only $z = -1 + 2i$ is inside the contour.

Residue at $z = -1 + 2i$

$$= \lim_{z\to -1+2i}(z+1-2i)\frac{z\sin\pi z}{z^2+2z+5} = \lim_{z\to -1+2i}(z+1-2i)\frac{z\sin\pi z}{(z+1-2i)(z+1+2i)}$$

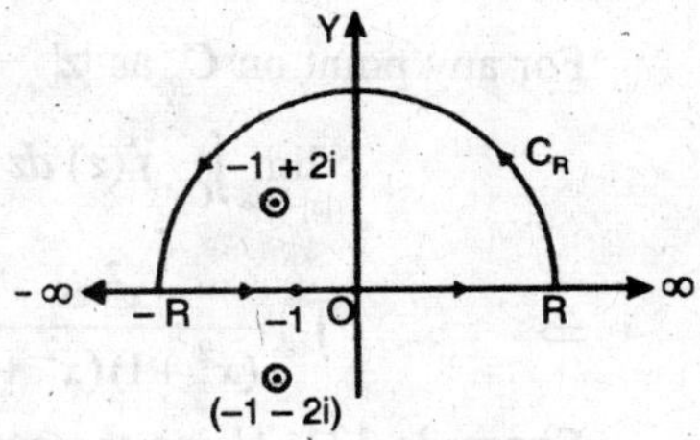


$$= \lim_{z \to -1+2i} \frac{z \sin \pi z}{(z+1+2i)} = \frac{(-1+2i) \sin \pi(-1+2i)}{(-1+2i+1+2i)}$$

$$= \frac{(-1+2i) \sin \pi(-1+2i)}{4i}$$

$$\int_{-R}^{R} \frac{z \sin \pi z}{z^2+2z+5} dz = 2\pi i \text{ (Residue)}$$

$$= 2\pi i \frac{(-1+2i) \sin \pi(-1+2i)}{4i} = \frac{\pi}{2} \quad (2i-1) \sin(-\pi + 2\pi i)$$

$$= \frac{\pi}{2}(2i-1)(-\sin 2\pi i) \qquad \left[\begin{aligned} \sin(-\pi+\theta) &= -\sin(\pi-\theta) \\ &= -\sin\theta \end{aligned}\right.$$

$$= \frac{\pi}{2}(1-2i) \sin 2\pi i = \frac{\pi}{2}(1-2i)\, i \sinh 2\pi$$

$$= \frac{\pi}{2}(i+2) \sinh 2\pi \qquad \text{(Taking real parts)}$$

Hence $\int_{-\infty}^{\infty} \frac{x \sin \pi x}{x^2+2x+5} = \pi \sinh 2\pi$ **Ans.**

**Example 130.** *Evaluate* $\int_{-\infty}^{\infty} \frac{x^2 dx}{(x^2+1)(x^2+4)}$. *(MDU, Dec. 2006)*

**Solution.** We consider $\int_C \frac{z^2 dz}{(z^2+1)(z^2+4)} = \int_C f(z)\, dz$

where $C$ is the contour consisting of the semi-circle $C_R$ of radius $R$ together with the part of the real axis from $-R$ to $+R$.

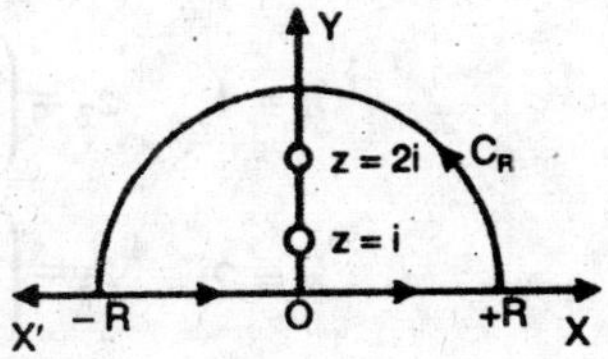


The integral has simple poles at

$$z = \pm i, \; z = \pm 2i$$

of which $z = i$, $2i$ only lie inside $C$.

The residue (at $z = i$) $= \lim_{z \to i}(z-i) \frac{z^2}{(z+i)(z-i)(z^2+4)}$

$$= \lim_{z \to i} \frac{z^2}{(z+i)(z^2+4)} = \frac{-1}{2i(-1+4)} = \frac{-1}{6i}$$

The residue (at $z = 2i$) $= \lim_{z \to 2i}(z-2i) \cdot \frac{z^2}{(z^2+1)(z+2i)(z-2i)}$

$$= \lim_{z \to 2i} \frac{z^2}{(z^2+1)(z+2i)} = \frac{(2i)^2}{(-4+1)(2i+2i)} = \frac{1}{3i}$$

By theorem of residue;

$$\int_C f(z)\, dz = 2\pi i[\text{Res } f(i) + \text{Res } f(2i)] = 2\pi i\left(-\frac{1}{6i} + \frac{1}{3i}\right) = \frac{\pi}{3}$$

*i.e.* $\int_{-R}^{R} f(x) dx + \int_{C_R} f(z)\, dz = \frac{\pi}{3}$ ... (1)

Hence by making $R \to \infty$, relation (1) becomes

$$\int_{-\infty}^{\infty} f(x) dx + \lim_{z \to \infty} \int_{C_R} f(z)\, dz = \frac{\pi}{3}$$

Now $R \to \infty$, $\int_{C_R} f(z) dz$ vanishes.

For any point on $C_R$ as $|z| \to \infty$, $f(z) = 0$

$$\lim_{|z|\to\infty} \int_{C_R} f(z)\, dz = 0, \quad \int_{-\infty}^{\infty} f(x)\, dx = \frac{\pi}{3}$$

$$\Rightarrow \qquad \int_{-\infty}^{\infty} \frac{x^2\, dx}{(x^2+1)(x^2+4)} = \frac{\pi}{3}$$ **Ans.**

**Example 131.** *Using the complex variable techniques, evaluate the integral*

$$\int_{-\infty}^{\infty} \frac{1}{x^4+1}\, dx$$ *(AMIETE, June 2010, U.P. III Semester, Dec. 2006)*

**Solution.** $\int_{-\infty}^{\infty} \frac{1}{x^4+1}\, dx$

Consider $\int_C f(z)\, dz$, where $f(z) = \frac{1}{z^4+1}$

taken around the closed contour consisting of real axis and upper half $C_R$, *i.e.* $|z| = R$.
Poles of $f(z)$ are given by

$$z^4 + 1 = 0 \text{ i.e. } z^4 = -1 = (\cos\pi + i\sin\pi)$$

$$\Rightarrow \qquad z^4 = [\cos(2n+1)\pi + i\sin(2n+1)\pi]$$

$$z = [\cos(2n+1)\pi + i\sin(2n+1)\pi]^{\frac{1}{4}} = \left[\cos(2n+1)\frac{\pi}{4} + i\sin(2n+1)\frac{\pi}{4}\right]$$

If $n = 0$, $\quad z_1 = \left(\cos\frac{\pi}{4} + i\sin\frac{\pi}{4}\right) = \left(\frac{1}{\sqrt{2}} + i\frac{1}{\sqrt{2}}\right) = e^{i\frac{\pi}{4}}$

$n = 1$, $\quad z_2 = \left(\cos\frac{3\pi}{4} + i\sin\frac{3\pi}{4}\right) = \left(-\frac{1}{\sqrt{2}} + i\frac{1}{\sqrt{2}}\right) = e^{i\frac{3\pi}{4}}$

$n = 2$, $\quad z_3 = \left(\cos\frac{5\pi}{4} + i\sin\frac{5\pi}{4}\right) = \left(-\frac{1}{\sqrt{2}} - i\frac{1}{\sqrt{2}}\right)$

$n = 3$, $\quad z_4 = \left(\cos\frac{7\pi}{4} + i\sin\frac{7\pi}{4}\right) = \left(\frac{1}{\sqrt{2}} - i\frac{1}{\sqrt{2}}\right)$

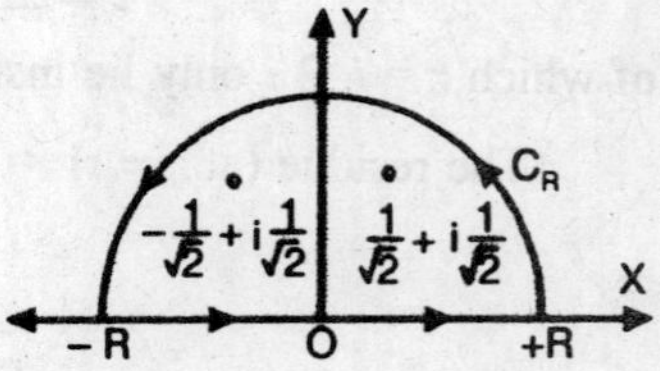


There are four poles, but only two poles at $z_1$ and $z_2$ lie within the contour.

$$\text{Residue}\left(\text{at } z = e^{\frac{i\pi}{4}}\right) = \left[\frac{1}{\frac{d}{dz}(z^4+1)}\right]_{z=e^{\frac{i\pi}{4}}} = \left[\frac{1}{4z^3}\right]_{z=e^{\frac{i\pi}{4}}} = \frac{1}{4\left(e^{i\frac{\pi}{4}}\right)^3} = \frac{1}{4e^{i\frac{3\pi}{4}}}$$

$$= \frac{1}{4}e^{-i\frac{3\pi}{4}} = \frac{1}{4}\left[\cos\frac{3\pi}{4} - i\sin\frac{3\pi}{4}\right] = \frac{1}{4}\left[-\frac{1}{\sqrt{2}} - i\frac{1}{\sqrt{2}}\right]$$

$$\text{Residue}\left(\text{at } z = e^{\frac{3i\pi}{4}}\right) = \left[\frac{1}{\frac{d}{dz}(z^4+1)}\right]_{z=e^{\frac{3i\pi}{4}}} = \frac{1}{[4z^3]_{z=\frac{3i\pi}{4}}} = \frac{1}{4\left(e^{i\frac{3\pi}{4}}\right)^3} = \frac{1}{4e^{i\frac{9\pi}{4}}}$$

$$= \frac{1}{4} e^{-i\frac{9\pi}{4}} = \frac{1}{4}\left(\cos\frac{9\pi}{4} - i\sin\frac{9\pi}{4}\right) = \frac{1}{4}\left(\frac{1}{\sqrt{2}} - i\frac{1}{\sqrt{2}}\right)$$

$$\int_C f(z)\,dz = 2\pi i \text{ (sum of residues at poles within c)}$$

$$\int_{-R}^{R} f(z)\,dz + \int_{C_R} f(z)\,dz = 2\pi i \text{ (sum of the residues)}$$

$$\int_{-R}^{R} \frac{1}{x^4+1}\,dx + \int_{C_R} \frac{1}{z^4+1}\,dz = 2\pi i \text{ (sum of the residues)}$$

Now, $$\left|\int_{C_R} \frac{1}{z^4+1}\,dz\right| \le \int_{C_R} \frac{1}{|z^4+1|}\,|dz|$$

$$\le \int_{C_R} \frac{1}{(|z^4|-1)}\,|dz| \quad [\text{Since } z = Re^{i\theta}, |dz| = |Re^{i\theta}\, i\, d\theta| = R\,d\theta]$$

$$\le \int_0^{\pi} \frac{1}{R^4-1}\,R\,d\theta \le \frac{R}{R^4-1}\int_0^{\pi} d\theta$$

$$\le \frac{R\pi}{R^4-1} = \frac{\pi/R^3}{1-1/R^4} \quad \text{which} \to 0$$

as $R \to \infty$.

Hence, $$\int_{-R}^{R} \frac{1}{x^4+1}\,dx = 2\pi i \text{ (Sum of the residues within contour)}$$

As $R \to \infty$

Hence, $$\int_{-\infty}^{\infty} \frac{1}{x^4+1}\,dx = 2\pi i \text{ (Sum of the residues within contour)} \quad \ldots (1)$$

$$\int_{-\infty}^{\infty} \frac{1}{x^4+1}\,dx = 2\pi i\left[\frac{1}{4}\left(-\frac{1}{\sqrt{2}} - i\frac{1}{\sqrt{2}}\right) + \frac{1}{4}\left(\frac{1}{\sqrt{2}} - i\frac{1}{\sqrt{2}}\right)\right]$$

$$= \frac{\pi}{2} i\left(-\frac{1}{\sqrt{2}} - i\frac{1}{\sqrt{2}} + \frac{1}{\sqrt{2}} - i\frac{1}{\sqrt{2}}\right) = \frac{\pi i}{2}\left(-i\frac{2}{\sqrt{2}}\right) = \frac{\pi}{\sqrt{2}}$$

Hence, the given integral $= \dfrac{\pi}{\sqrt{2}}$ **Ans.**

**Example 132.** *Using complex variable techniques, evaluate the real integral*

$$\int_0^{\infty} \frac{dx}{1+x^6}$$ *(MDU May, 2006)*

**Solution.** Let $f(z) = \dfrac{1}{1+z^6}$

We consider $\displaystyle\int_C \frac{1}{1+z^6}\,dz$

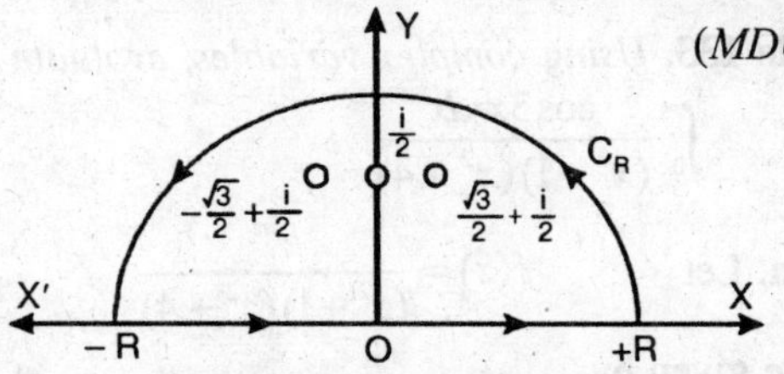


where $C$ is the contour consisting of the semi-circle $C_R$ of radius $R$ together with the part of real axis from $-R$ to $R$.

Poles are given by $1 + z^6 = 0$

$$z^6 = -1 = \cos\pi + i\sin\pi = \cos(2n\pi+\pi) + i\sin(2n\pi+\pi)$$

$$= e^{(2n+1)\pi i}$$

$$z = e^{\frac{2n+1}{6}\pi i} = \left[\cos\frac{2n\pi+\pi}{6} + i\sin\frac{2n\pi+\pi}{6}\right] \quad \text{where } n = 0, 1, 2, 3, 4, 5$$

If $\quad n = 0, \quad z = e^{\frac{\pi i}{6}} = \cos\frac{\pi}{6} + i\sin\frac{\pi}{6} = \frac{\sqrt{3}}{2} + \frac{i}{2}$

If $\quad n = 1, \quad z = e^{\frac{i\pi}{2}} = \cos\frac{\pi}{2} + i\sin\frac{\pi}{2} = i$

If $\quad n = 2, \quad z = e^{i\frac{5\pi}{6}} = \cos\frac{5\pi}{6} + i\sin\frac{5\pi}{6} = \frac{-\sqrt{3}}{2} + \frac{i}{2}$

If $\quad n = 3, \quad z = e^{i\frac{7\pi}{6}} = \cos\frac{7\pi}{6} + i\sin\frac{7\pi}{6} = \frac{-\sqrt{3}}{2} - \frac{i}{2}$

If $\quad n = 4, \quad z = e^{i\frac{3\pi}{2}} = \cos\frac{3\pi}{2} + i\sin\frac{3\pi}{2} = -i$

If $\quad n = 5, \quad z = e^{i\frac{11\pi}{6}} = \cos\frac{11\pi}{6} + i\sin\frac{11\pi}{6} = \frac{\sqrt{3}}{2} - \frac{i}{2}$

Only first three poles *i.e.*, $e^{i\frac{\pi}{6}}, e^{i\frac{\pi}{2}}, e^{i\frac{5\pi}{6}}$ are inside the contour.

$$\text{Residue}\left(\text{at } z = e^{\frac{i\pi}{6}}\right) = \lim_{z \to e^{\frac{i\pi}{6}}} \frac{1}{\frac{d}{dz}(1+z^6)} = \lim_{z \to e^{\frac{i\pi}{6}}} \frac{1}{6z^5} = \frac{1}{6} e^{\frac{-i5\pi}{6}}$$

$$\text{Residue}\left(\text{at } z = e^{\frac{i\pi}{2}}\right) = \lim_{z \to e^{\frac{i\pi}{2}}} \frac{1}{\frac{d}{dz}(1+z^6)} = \lim_{z \to e^{\frac{i\pi}{2}}} \frac{1}{6z^5} = \frac{1}{6} e^{\frac{-i5\pi}{2}}$$

$$\text{Residue}\left(\text{at } z = e^{\frac{i5\pi}{6}}\right) = \lim_{z \to e^{\frac{i5\pi}{6}}} \frac{1}{\frac{d}{dz}(1+z^6)} = \lim_{z \to e^{\frac{i5\pi}{6}}} \frac{1}{6z^5} = \frac{1}{6} e^{\frac{-i25\pi}{6}}$$

$$\text{Sum of the residues} = \frac{1}{6}\left[e^{\frac{-5i\pi}{6}} + e^{-\frac{i5\pi}{2}} + e^{-\frac{i25\pi}{6}}\right] = \frac{1}{6}\left(-\frac{\sqrt{3}}{2} - \frac{i}{2} + 0 - i + \frac{\sqrt{3}}{2} - \frac{i}{2}\right) = \frac{1}{6}(-2i) = -\frac{i}{3}$$

$$\Rightarrow \quad \int_C \frac{dz}{1+z^6} = 2\pi i \text{ (Residue)} = 2\pi i\left(-\frac{i}{3}\right) = \frac{2\pi}{3}$$

$$\Rightarrow \quad \int_{-\infty}^{\infty} \frac{dx}{1+x^6} = \frac{2\pi}{3} \quad \Rightarrow \quad \int_0^{\infty} \frac{dx}{1+x^6} = \frac{\pi}{3} \quad \textbf{Ans.}$$

**Example 133.** *Using complex variables, evaluate the real integral*

$$\int_0^{\infty} \frac{\cos 3x \, dx}{(x^2+1)(x^2+4)}$$

**Solution.** Let $\quad f(z) = \dfrac{e^{3iz}}{(z^2+1)(z^2+4)}$

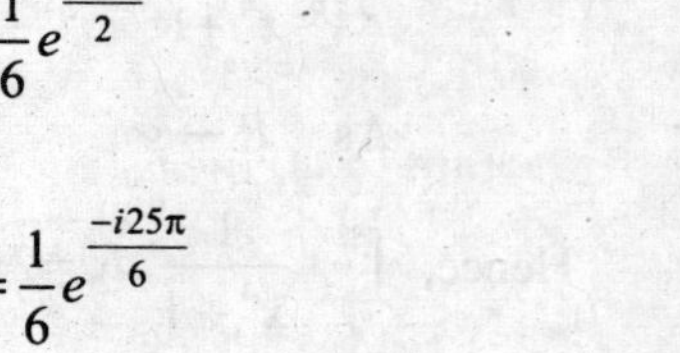


Poles are given by

$$(z^2+1)(z^2+4) = 0$$

*i.e.,* $\quad z^2 + 1 = 0$ or $z = \pm i$

$$z^2 + 4 = 0 \text{ or } z = \pm 2i$$

Let $C$ be a closed contour consisting of the upper half $C_R$ of a large circle $|z| = R$ and the real axis from $-R$ to $+R$. The poles at $z = i$ and $z = 2i$ lie within the contour.

$$\text{Residue (at } z = i) = \lim_{z\to i}\frac{(z-i)e^{3iz}}{(z^2+1)(z^2+4)} = \lim_{z\to i}\frac{e^{3iz}}{(z+i)(z^2+4)} = \frac{e^{-3}}{6i}$$

$$\text{Residue (at } z = 2i) = \lim_{z\to 2i}\frac{(z-2i)e^{3iz}}{(z^2+1)(z^2+4)} = \lim_{z\to 2i}\frac{e^{3iz}}{(z^2+1)(z+2i)} = \frac{e^{-6}}{-12i}$$

By theorem of Residue $\int_C f(z)dz = 2\pi i$ [Sum of Residues]

$$\int_{-R}^{R}\frac{e^{3iz}dz}{(z^2+1)(z^2+4)} + \int_{C_R}\frac{e^{3iz}dz}{(z^2+1)(z^2+4)} = 2\pi i\left[\frac{e^{-3}}{6i} + \frac{e^{-6}}{-12i}\right]$$

$$\left[\int_{C_R}\frac{e^{3iz}\,dz}{(z^2+1)(z^2+4)} = 0 \text{ as } z = Re^{i\theta} \text{ and } R\to\infty\right]$$

$$\int_{-R}^{R}\frac{e^{3ix}}{(x^2+1)(x^2+4)}dx = \pi\left[\frac{e^{-3}}{3} - \frac{e^{-6}}{6}\right]$$

$$\int_0^{\infty}\frac{\cos 3x\,dx}{(x^2+1)(x^2+4)} = \text{Real part of } \frac{1}{2}\int_{-\infty}^{\infty}\frac{e^{3ix}dx}{(x^2+1)(x^2+4)}$$

$$= \text{Real part of } \frac{\pi}{2}\left(\frac{e^{-3}}{3} - \frac{e^{-6}}{6}\right)$$

Hence, given integral $= \dfrac{\pi}{2}\left[\dfrac{e^{-3}}{3} - \dfrac{e^{-6}}{6}\right]$ **Ans.**

**Example 134.** *Evaluate:* $\displaystyle\int_0^{\infty}\frac{dx}{(a^2+x^2)^2}$ *(MDU. Dec. 2009)*

**Sol.** Consider the integral $\displaystyle\int_C f(z)\,dz$ where $f(z) = \dfrac{1}{(a^2+z^2)^2}$

Poles of $f(z)$ are given by putting denominator equal to zero.

$(a^2 + z^2)^2 = 0 \Rightarrow a^2 + z^2 = 0 \Rightarrow z = \pm ai$ each repeated twice

Sine there is no pole on the real axis, therefore we may take the contour $C$ consisting of the semicirle $C_R$ which is the upper half of a large circle $|z| = R$, and the real axis from $-R$ to $R$.

Here by Cauchy's residues theorem we have

$$\oint_C f(z)dz = \int_{-R}^{R} f(x)\,dx + \int_{C_R} f(z)\,dz = 2\pi i \text{ (sum of residues)}$$

or
$$\int_{-R}^{R}\frac{1}{(a^2+x^2)^2}dx + \int_{C_R}\frac{dz}{(a^2+z^2)^2} = 2\pi i \text{ (sum of residues)} \quad \ldots (1)$$

The only pole within the contour $C$ is $z = ai$, and is of order 2.

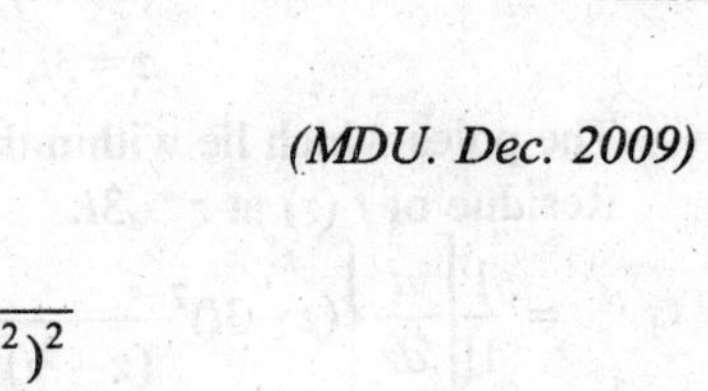


Here $f(z) = \dfrac{1}{(z-ai)^2(z+ai)^2} = \dfrac{\phi(z)}{(z-ai)^2}$

where $\phi(z) = \dfrac{1}{(z+ai)^2} \Rightarrow \phi'(z) = -\dfrac{2}{(z+ai)^3}$

$\therefore$ Residue at the double pole $(z = ai) = \dfrac{\phi'(ai)}{1!} = -\dfrac{2}{(2ai)^3} = -\dfrac{i}{4a^3}$

and
$$\left|\int_{C_R}\frac{1}{(a^2+z^2)^2}dz\right| \le \int_{C_R}\frac{|dz|}{|a^2+z^2|^2} \le \int_{C_R}\frac{|dz|}{(|z|^2-a^2)^2} = \int_0^{\pi}\frac{R\,d\theta}{(R^2-a^2)^2}$$

$$= \frac{\pi R}{(R^2 - a^2)^2} \to 0 \text{ and } R \to \infty \text{ since } z = Re^{i\theta}$$

Hance when $R \to \infty$, relation (1) becomes

$$\int_{-\infty}^{\infty} \frac{1}{(a^2 + x^2)^2} dx = 2\pi i \text{ (residue)} = 2\pi i \left( \frac{-i}{4a^3} \right) = \frac{\pi}{2a^3}$$

$$\Rightarrow \quad \int_0^{\infty} \frac{1}{(a^2 + x^2)^2} dx = \frac{\pi}{4a^3}$$ **Ans.**

**Example 135.** *Using complex variable techniques, evaluate the real integral*

$$\int_0^{\infty} \frac{\cos 2x}{(x^2 + 9)^2 (x^2 + 16)} dx$$

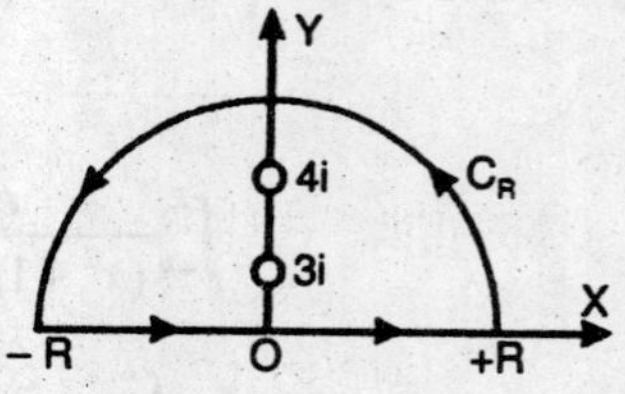


**Solution.** Consider the integral $\int_C f(z)\, dz$,

where $$f(z) = \frac{e^{2iz}}{(z^2 + 9)^2 (z^2 + 16)},$$

taken around the closed contour $C$ consisting of the upper half of a large circle $|z| = R$ and the real axis from $-R$ to $R$.

Poles of $f(z)$ are given by

$$(z^2 + 9)^2 (z^2 + 16) = 0$$

*i.e.* $$(z + 3i)^2 (z - 3i)^2 (z + 4i)(z - 4i) = 0$$

*i.e.* $$z = 3i, -3i, 4i, -4i$$

The poles which lie within the contour are $z = 3i$ of the second order and $z = 4i$ simple pole.

Residue of $f(z)$ at $z = 3i$

$$= \frac{1}{1!} \left[ \frac{d}{dz} \left\{ (z - 3i)^2 \frac{e^{2iz}}{(z - 3i)^2 (z + 3i)^2 (z^2 + 16)} \right\} \right]_{z=3i} = \left[ \frac{d}{dz} \left\{ \frac{e^{2iz}}{(z + 3i)^2 (z^2 + 16)} \right\} \right]_{z=3i}$$

$$= \left[ \frac{(z + 3i)^2 (z^2 + 16) 2ie^{2iz} - e^{2iz} [2(z + 3i)(z^2 + 16) + 2z(z + 3i)^2]}{(z + 3i)^4 (z^2 + 16)^2} \right]_{z=3i}$$

$$= \left[ \frac{(z + 3i)(z^2 + 16) 2i e^{2iz} - e^{2iz} [2(z^2 + 16) + 2z(z + 3i)]}{(z + 3i)^3 (z^2 + 16)^2} \right]_{z=3i}$$

$$= \frac{6i \times 7 \times 2i e^{-6} - e^{-6} (2 \times 7 + 6i \times 6i)}{(6i)^3 (7)^2} = \frac{e^{-6} [-84 + 22] i}{216 \times 49} = \frac{e^{-6} (-62) i}{216 \times 49} = -\frac{i31e^{-6}}{108 \times 49}$$

$$\text{Residue of } f(z) \text{ at } (z = 4i) = \lim_{z \to 4i} (z - 4i) \frac{e^{2iz}}{(z^2 + 9)^2 (z - 4i)(z + 4i)}$$

$$= \frac{e^{-8}}{(-16 + 9)^2 (4i + 4i)} = \frac{e^{-8}}{49 \times 8i} = \frac{-ie^{-8}}{392}$$

$$\text{Sum of the residues} = -\frac{i31e^{-6}}{108 \times 49} - \frac{ie^{-8}}{392}$$

Hence by Cauchy's Residue Theorem, we have

$$\int_C f(z)\, dz = 2\pi i \times \text{Sum of the residues within } C$$

*i.e.* $$\int_{-R}^{R} f(x) dx + \int_{C_R} f(z)\, dz = 2\pi i \times \text{sum of residues}$$

or $$\int_{-R}^{R} \frac{e^{2ix}}{(x^2+9)^2(x^2+16)}dx + \int_{C_R} \frac{e^{2iz}}{(z^2+9)^2(z^2+16)}dz = 2\pi i \times \text{ Sum of residues} \quad \text{... (1)}$$

Now let $R \to \infty$, so as to show that the second integral in above relation vanishes. For any point on $C_R$, *as* $|z| \to \infty$

Let $$F(z) = \frac{1}{z^6} \frac{e^{2iz}}{\left(1+\frac{9}{z^2}\right)^2\left(1+\frac{16}{z^2}\right)}$$

$$\lim_{|z|\to\infty} F(z) = 0 \quad \text{or} \quad \int_{C_R} \frac{e^{2iz}}{(z^2+9)^2(z^2+16)}dz = 0 \text{ as } z \to \infty$$

Hence by making $R \to \infty$, relation (1) becomes

$$\therefore \quad \int_{-\infty}^{\infty} \frac{e^{2ix}}{(x^2+9)^2(x^2+16)}dx = 2\pi i\left[\frac{-i31e^{-6}}{108\times 49} - i\frac{e^{-8}}{392}\right] = \frac{2\pi}{196}\left[\frac{31e^{-6}}{27}+\frac{e^{-8}}{2}\right]$$

Equating real parts, we have

$$\int_{-\infty}^{\infty} \frac{\cos 2x dx}{(x^2+9)^2(x^2+16)} = \frac{\pi}{98}\left(\frac{31e^{-6}}{27}+\frac{e^{-8}}{2}\right)$$

$$\int_0^{\infty} \frac{\cos 2x}{(x^2+9)^2(x^2+16)}dx = \frac{\pi}{196}\left(\frac{31e^{-6}}{27}+\frac{e^{-8}}{2}\right) \quad \left[\because \int_{-\infty}^{\infty} f(x)dx = 2\int_0^{\infty} f(x)dx \text{ If } f(x) \text{ is even function.}\right] \quad \textbf{Ans.}$$

## EXERCISE 7.15

**Evaluate the following :**

**1.** $\int_0^{\infty} \frac{1}{1+x^2}dx$ **Ans.** $\frac{\pi}{2}$ **2.** $\int_{-\infty}^{\infty} \frac{1}{(x^2+1)^2}dx$ **Ans.** $\frac{\pi}{2}$

**3.** $\int_0^{\infty} \frac{x^3 \sin x}{(x^2+a^2)(x^2+b^2)}dx$ **Ans.** $\frac{\pi}{2(a^2-b^2)}[a^2e^{-a}-b^2e^{-b}]$

**4.** $\int_{-\infty}^{\infty} \frac{\cos x}{(x^2+a^2)(x^2+b^2)}dx, \quad a > b > 0$ **Ans.** $\frac{\pi}{a^2-b^2}\left(\frac{e^{-b}}{b}-\frac{e^{-a}}{a}\right)$

**5.** Show that $\int_0^{\infty} \frac{\cos x}{x^2+a^2}dx = \frac{\pi e^{-a}}{2a}$

**6.** Show that $\int_0^{\infty} \frac{x^3 \sin x}{(x^2+a^2)}dx = -\frac{\pi}{4}(a-2)\, a^{-a},\ a > 0$

**Evaluate the following :**

**7.** $\int_{-\infty}^{\infty} \frac{\sin mx}{x(x^2+a^2)}dx, \quad m > 0, a > 0$ **Ans.** $\frac{\pi}{a^2}(2-e^{-ma})$

**8.** $\int_0^{\infty} \frac{x^2}{x^6+1}dx$ (*MDU, 2008, 2005*) **Ans.** $\frac{\pi}{2}$ **9.** $\int_0^{\infty} \frac{x \sin ax}{x^4+a^4}dx$ **Ans.** $\frac{\pi}{2a^2}e^{-\frac{a^2}{\sqrt{2}}}\sin\frac{a^2}{\sqrt{2}}$

**10.** $\int_0^{\infty} \frac{x^6}{(a^4+x^4)^2}dx$ **Ans.** $\frac{3\pi\sqrt{2}}{16a}, a > 0$ **11.** $\int_0^{\infty} \frac{\cos x^2+\sin x^2-1}{x^2}dx$

**12.** $\int_0^{\infty} \frac{\cos mx}{x^4+x^2+1}dx$ **Ans.** $\frac{\pi}{\sqrt{3}}\sin\frac{1}{2}\left(m+\frac{\pi}{3}\right)e^{-\frac{1}{2}m\sqrt{3}}$ **13.** $\int_0^{\infty} \frac{\log(1+x^2)}{1+x^2}dx$ **Ans.** $\pi \log 2$

# 8

# Special Functions

## 8.1. SPECIAL FUNCTIONS

Algebraic, trigonometric, exponential and logarithmic functions are the elementary functions. Bessel and Legendre functions are the special functions of mathematics.

## 8.2. POWER SERIES SOLUTIONS OF DIFFERENTIAL EQUATIONS

We know that the solution of the differential equation

$$\frac{d^2y}{dx^2}-y=0$$

are $\quad y=e^x=1+\frac{x}{1!}+\frac{x^2}{2!}+\frac{x^3}{3!}+...$ and $y=e^{-x}=1-\frac{x}{1!}+\frac{x^2}{2!}-\frac{x^3}{3!}+...$

These are power series solution of the given differential equation.

Another example of the differential equation

$$\frac{d^2y}{dx^2}+y=0$$

is satisfied by the power series

$$y=\sin x=x-\frac{x^3}{3!}+\frac{x^5}{5!}-\frac{x^7}{7!}+...$$

This idea leads to the methods of obtaining the solution of a linear differential equation of second order in series form.

If the solution of a differential equation involves log $x$, $e^x$ etc. which cannot be expanded in the series of ascending power of $x$, but they can be expanded in descending powers of $x$.

Then the solution of the differential equation will be a series of descending powers of $x$, the infinite series solution obtained will have its own region of convergence or validity.

## 8.3 ORDINARY POINT

Consider the equation

$$P_0\frac{d^2y}{dx^2}+P_1\frac{dy}{dx}+P_2y=0$$

where $P_0$, $P_1$, $P_2$ are polynomials in $x$.

$x = a$ is an ordinary point of the above equation if $P_0$ does not vanish for $x = a$

**Singular point.** If $P_0$ vanishes for $x = a$, then $x = a$ is a singular point of the above equation.

The general solution of a linear differential equation of second order will consist of two series say $y_1$ and $y_2$. Then the general solution will be $y = ay_1 + by_2$, where $a$ and $b$ are arbitrary constants. The two infinite series are said to be linearly *dependent* if one is multiple of the other, otherwise they are said to be linearly *independent.*

4.3 *Solution of the differential equation when $x = 0$ is an ordinary point i.e. when $P_0$ does not vanish for $x = 0$.*

(*i*) Let $y = a_0 + a_1x + a_2x^2 + a_3x^3 + ....... + a_kx^k + .... = \sum_{k=0}^{\infty} a_k x^k$ be the solution of the given differenftial equation.

(*ii*) Find $\frac{dy}{dx}, \frac{d^2y}{dx^2}$ etc.

$$\frac{dy}{dx} = a_1 + 2a_2x + 3a_3x^2 + .... + ka_kx^{k-1} + ... + ... = \sum_{k=1}^{\infty} ka_kx^k$$

$$\frac{d^2y}{dx^2} = 2a_2 + 2.3\, a_3x + ..... + a_k\; k(k-1)\, x^{K-2} + .... = \sum_{k=2}^{\infty} a_k \,.\, k\,(k-1)\,.\, x^{k-2}$$

(*iii*) Substitute the expressions of $y, \frac{dy}{dx}, \frac{d^2y}{dx^2}$ etc. in the given differential equation.

(*iv*) Calculate $a_0, a_1, a_2$ .... coefficients of various powers of $x$ by equating the coefficients to zero.

(*v*) Substitute the values of $a_0, a_1, a_2$ ... in the differential equation to get the required series solution.

**Example 1.** *Solve in series the equation* $\frac{d^2y}{dx^2} + x^2y = 0$

**Solution.** $\frac{d^2y}{dx^2} + x^2y = 0$ ... (1)

Let $y = a_0 + a_1x + a_2x^2 + a_3x^3 + a_4x^2 + a_5x^5 + a_6x^6 + a_7x^7 + {}_7{}_8x^8 + ... + a_nx^n + ... ...$ (2)

Since $x = 0$ is the ordinary point of the equation (1).

Then $\frac{dy}{dx} = a_1 + 2a_2x + 3\,a_3\,x^2 + 4\,a_4\,x^3 + 5\,a_5\,x^4 + 6\,a_6\,x^5 + 7\,a_7\,x^6 + 8\,a_8\,x^7 + ...$

$$\frac{d^2y}{dx^2} = 2a_2 + 2\,.\,3\,a_3\,x + 3\,.\,4\,a_4\,x^2 + 4\,.\,5\,a_5\,x^3 + 5\,.\,6\,a_6\,x^4 + 6\,.\,7\,a_7\,x^5 + 7\,.\,8\,a_8\,x^6 + ...$$

Substituting in (1), we get

$$2\,a_2 + 2.3\,a_3\,x + 3.4\,a_4\,x^2 + 4.5\,x_5\,x^3 + 5.6\,a_6\,x^4 + 6.7\,a_7\,x^5 + 7.8\,a_8\,x^6 + ...$$
$$+ x^2\,(a_0 + a_1\,x + a_2\,x^2 + a_3\,x^3 + a_4\,x^4 + a_5\,x^5 + a_6\,x^6 + a_7\,x^7 + ...) = 0$$

$$2\,a_2 + 6\,a_3\,x + (a_0 + 12\,a_4)\,x^2 + (a_1 + 20a_5)\,x^3 + (a_2 + 30\,a_6)\,x^4 + ...$$
$$+ [a_{n-2} + (n+2)\,(n+1)\,a_{n+2}]\,x^n + ... = 0$$

Equating to zero the coefficients of the various powers of $x$, we obtain

$$a_2 = 0,\; a_3 = 0$$

$$a_0 + 12a_4 = 0 \;\; i.e.\;\; a_4 = -\frac{1}{12}a_0$$

$$a_1 + 20\,a_5 = 0 \;\; i.e.\;\; a_5 = -\frac{1}{20}a_1$$

$$a_2 + 30\,a_6 = 0 \;\; i.e.\;\; a_6 = -\frac{1}{30}\,a_2 = 0 \qquad (a_2 = 0)$$

and so on. In general $a_{n-2} + (n+2)\,(n+1)\,a_{n+2} = 0$ or $a_{n+2} = -\frac{a_{n-2}}{(n+1)(n+2)}$

Putting $n = 5$, $a_7 = -\frac{a_3}{6\times 7} = 0 \qquad (a_3 = 0)$

Putting $n = 6$, $$a_8 = -\frac{a_4}{7\times8} = \frac{a_0}{12\times7\times8}$$

Putting $n = 7$, $$a_9 = -\frac{a_5}{8\times9} = \frac{a_1}{20\times8\times9}$$

Putting $n = 8$, $$a_{10} = -\frac{a_6}{9\times10} = 0 \qquad (a_6 = 0)$$

Putting $n = 9$, $$a_{11} = -\frac{a_7}{11\times10} = 0 \qquad (a_7 = 0)$$

Putting $n = 10$, $$a_{12} = -\frac{a_8}{12\times11} = -\frac{a_0}{12\times8\times7\times11\times12}$$

Substituting these values in (2), we get

$$y = a_0 + a_1 x - \frac{1}{12}a_0 x^4 - \frac{a_1}{20}x^5 + \frac{a_0}{12\times7\times8}x^8 + \frac{a_1}{20\times8\times9}x^9 - \frac{a_0}{12\times8\times7\times11\times12}x^{12} + \ldots$$

$$y = a_0\left(1 - \frac{1}{12}x^4 + \frac{x^8}{12\times7\times8} - \frac{x^{12}}{12\times8\times7\times11\times12} + \ldots\right) + a_1\left(x - \frac{x^5}{20} + \frac{x^9}{20\times8\times9}\ldots\right) \quad \textbf{Ans.}$$

**Example 2.** *Find the power series solution of $(1 - x^2)\, y'' - 2\,x\,y' + 2\,y = 0$ about $x = 0$.*

*(A.M.I.E.T.E., Winter 2000)*

**Solution.** Let $y = a_0 + a_1 x + a_2 x^2 + a_3 x^3 + \ldots + a_n x^n + \ldots$ be the required solution. Since $x = 0$ is an ordinary point of the given equation, this can be written as

$$y = \sum_{k=0}^{\infty} a_k x^k$$

Then $$\frac{dy}{dx} = \sum_{k=1}^{\infty} a_k \cdot k\, x^{k-1}, \qquad \frac{d^2y}{dx^2} = \sum_{k=2}^{\infty} a_k \cdot k\,(k-1)\, x^{k-2}$$

Substituting the values of $y \cdot \frac{dy}{dx}$, and $\frac{d^2y}{dx^2}$ in the given equation we get

$$(1 - x^2)\,\Sigma a_k\, k \cdot (k-1)\, x^{k-2} - 2x\,\Sigma a_k \cdot k\, x^{k-1} + 2\Sigma a_k\, x^k = 0$$

$$\Rightarrow \quad \Sigma a_k \cdot k \cdot (k-1) x^{k-2} - \Sigma a_k k(k-1) x^k - 2\,\Sigma a_k . k x^k + 2\Sigma a_k x^k = 0$$

$$\Rightarrow \quad \Sigma a_k \cdot k \cdot (k-1) x^{k-2} - \Sigma[k(k-1) + 2k - 2] a_k x^k = 0$$

$$\Rightarrow \quad \Sigma a_k \cdot k \cdot (k-1) x^{k-2} - \Sigma(k^2 + k - 2) a_k x^k = 0$$

where the first summation extends over all values of $k$ from 2 to $\infty$ and the second from $k = 0$ to $\infty$.

Now equating the coefficient of $x^k$ equal to zero, we have

$$\Rightarrow \quad (k+2)\,(k+1) a_{k+2} - (k^2 + k - 2) a_k = 0$$

$$\Rightarrow \quad a_{k+2} = \frac{k^2 + k - 2}{(k+2)(k+1)} a_k = \frac{(k+2)(k-1)}{(k+2)(k+1)} a_k$$

$$\Rightarrow \quad a_{k+2} = \frac{k-1}{k+1} a_k$$

For $k = 0$ $$a_2 = -a_0,\; a_3 = 0,\; a_4 = \frac{a_2}{3} = -\frac{a_0}{3},\; a_5 = \frac{2}{4}a_3 = 0$$

For $k = 4$ $$a_6 = \frac{3}{5}a_4 = \frac{3}{5}\left(-\frac{a_0}{3}\right) = -\frac{a_0}{5},\ a_7 = \frac{4}{5}a_5 = 0, \text{ etc.}$$

$$y = a_0 + a_1x + a_2x^2 + a_3x^3 + a_4x^4 + a_5x^5 + a_6x^6 + a_7x^7 + ...$$

$$\Rightarrow \quad y = a_0 + a_0x - a_0x^2 + 0 - \frac{a_0}{3}x^4 + 0 - \frac{a_0}{5}x^6 + 0 + ...$$

$$\Rightarrow \quad y = a_0\left[1 - x^2 - \frac{x^4}{3} - \frac{x^6}{5} + ...\right] + a_1x$$ **Ans.**

**Example 3.** *Solve*

$$(1+x^2)\frac{d^2y}{dx^2} + x\frac{dy}{dx} - y = 0 \quad ...(1)$$

**Solution.** Let the solution of the given differential equation he

$$y = a_0 + a_1x + a_2x^2 + a_3x^3 + ...$$

Since $x = 0$ is the ordinary point of the given equation.

$$\frac{dy}{dx} = a_1 + 2a_2x + 3a_3x^2 + 4a_4x^3 + ....$$

$$\frac{d^2y}{dx^2} = 2a_2 + 6a_3x + 12a_4x^2 + 20a_5x^3 + ...$$

Substituting for $y$, $\frac{dy}{dx}$, and $\frac{d^2y}{dx^2}$ in the given differential equation, we have

$$(1 + x^2)(2a_2 + 6a_3x + 12a_4x^2 + 20a_5x^3 + ...) + x(a_1 + 2a_2x + 3a_3x^2 + 4a_4x^3 + ...)$$
$$- (a_0 + a_1x + a_2x^2 + a_3x^3 + ...) = 0$$

$$(2a_2 - a_0) + (6a_3 + a_1 - a_1)x + (2a_2 + 12a_4 + 2a_2 - a_2)x^2$$
$$+ (20a_5 + 6a_3 + 3a_3 - a_3)x^3 + ... = 0$$

Equating the coefficients of various powers of $x$ to zero, we obtain

$2a_2 - a_0 = 0$ or $a_2 = \frac{1}{2}a_0$ (Constant term)

$6a_3 = 0$ or $a_3 = 0$ (Coefficient of $x$)

$12a_4 + 3a_2 = 0$ or $a_4 = \frac{-1}{4}a_2 = -\frac{1}{8}a_0$ (Cofficient of $x^2$)

$20a_5 + 8a_3 = 0$ or $a_5 = -\frac{2}{5}a_3 = 0$ (Coefficient of $x^3$)

So solution is $$y = a_0\left[1 + \frac{x^2}{2} - \frac{x^4}{8} ....\right] + a_1x$$ **Ans.**

**Note.** In the above examples we get found the solutions about $x = 0$, which are valid in the finite region around $x = 0$.

We can also find the solution about a point other than $x = 0$, say about $x = c$. In this case we have to find out the series solution of powers of $(x - c)$, and the series in valid (convergent) around the point $x = c$.

In this method first we shift the origin to the point $x = c$, by putting $x = t + c$. The differential equation so obtained is solved by the method already discussed.

## EXERCISE 8.1

Solve the following differential equation by power series method :

1. $\dfrac{d^2y}{dx^2}+xy=0$ **Ans.** $y=a_0\left(1-\dfrac{x^3}{3!}+\dfrac{4x^6}{6!}-\dfrac{7.4x^9}{9!}+...\right)+a_1\left(x-\dfrac{2x^4}{4!}+\dfrac{5.2x^7}{7!}+...\right)$

2. $y''+xy'+x^2y=0$ **Ans.** $y=a_0\left(1-\dfrac{1}{12}x^4-...\right)+a_1\left(x-\dfrac{1}{6}x^3-\dfrac{1}{40}x^5...\right)$

3. $(x^2+1)y''+xy'-xy=0$ **Ans.** $y=a_0\left(1+\dfrac{x^3}{6}-\dfrac{3}{40}x^5+...\right)+a_1\left(x-\dfrac{x^3}{6}+\dfrac{x^4}{12}+\dfrac{3}{40}x^5....\right)$

4. $y''-2x^2y'+4xy=x^2+2x+4$

**Ans.** $y=a_0\left(1-\dfrac{2}{3}x^2-\dfrac{2}{45}x^6-\dfrac{2}{405}x^9....\right)+a_1\left(x-\dfrac{1}{6}x^4-\dfrac{1}{63}x^7-\dfrac{1}{567}x^{10}...\right)$
$+2x^2+\dfrac{1}{3}x^3+\dfrac{1}{12}x^4+\dfrac{1}{45}x^6+\dfrac{1}{126}x^7+\dfrac{1}{405}x^9+\dfrac{1}{1134}x^{10}+....$

5. $(1-x^2)\dfrac{d^2y}{dx^2}-x\dfrac{dy}{dx}+4y=0$ ( ) **Ans.** $y=90\,(1-2x^2)+91x(1-\dfrac{x^4}{8}-\dfrac{x^6}{16}+......$

6. $(x^2+2)y''+xy'-(1+x)y=0$ **Ans.** $y=a_0\left(1+\dfrac{1}{4}x^2+\dfrac{1}{12}x^3-\dfrac{1}{32}x^4...\right)+a_1\left(x+\dfrac{1}{24}x^4+....\right)$

## 8.4 SOLUTION ABOUT SINGULAR POINTS

There are two types of singular points. (1) Regular singular point, (2) Irregular singular points.

**Definition.** Consider the equation

(1) $$y''+P_1(x)\,y'+P_2(x)y=0 \qquad ...(i)$$

and assume that at least one of the functions $P_1$ and $P_2$ is not analytic ($P_1=\infty$ or $P_2=\infty$) at $x=a$, so that $x=a$ is a *singular point of* (*i*)

Consider

$$Q_1(x)=(x-a)P_1(x), \qquad Q_2(x)=(x-a)^2P_2(x)$$

If $Q_1$ and $Q_2$ are analytic (not $\infty$) at $x=a$, then $x=a$ is called a ***regular singular point,*** other irregular.

**Example 4.** *Find regular singular points of the defferential equation.*

$$2x^2\frac{d^2y}{dx^2}+3x\frac{dy}{dx}+(x^2-4)\,y=0 \qquad ...(1)$$

**Solution.** $$\frac{d^2y}{dx^2}+\frac{3}{2x}\frac{dy}{dx}+\frac{x^2-4}{2x^2}y=0$$

$$P_1=\frac{3}{2x} \text{ and } P_2=\frac{x^2-4}{2x^2}$$

$$Q_1=x\cdot P_1=x\left(\frac{3}{2x}\right)=\frac{3}{2}, \qquad Q_2=x^2P_2=x^2\cdot\frac{x^2-4}{2x^2}=\frac{1}{2}(x^2-4)$$

Since both $P_1$ and $P_2$ are not analytic ($P_1=\infty$, $P_2=\infty$) at $x=0$ therefore $x=0$ is a singular point of (*1*). Moreover both $Q_1$ and $Q_2$ are analytic ($Q_1\neq\infty$, $Q_2\neq\infty$) at $x=0$. Hence $x=0$ is a regular singular point of (*1*).

**Example 5.** Find regular singular points of the differential equation.

$$x^2(x-2)^2y''+2(x-2)y'+(x+3)y=0 \qquad ...(1)$$

**Solution.** $$P_1=\frac{2(x-2)}{x^2(x-2)^2}=\frac{2}{x^2(x-2)} \text{ and } P_2=\frac{x+3}{x^2(x-2)^2}$$

$P_1$ and $P_2$ are not analytic ($P_1 = \infty$, $P_2 = \infty$) at $x = 0$. and $x = 2$. Hence both these points are singular points of (1).

(*i*) At $x = 0$ $\quad Q_1 = x \cdot P_1 = \dfrac{2}{x(x-2)}$

$$Q_2 = x^2 \,.\, P_2 = x^2 . \frac{(x+3)}{x^2(x-2)^2} = \frac{x+3}{(x-2)^2}$$

Since $Q_1$ is not analytic ($Q_1 = \infty$) at $x = 0$, so $x = 0$ is an irregular singular point.

(*ii*) At $x = 2$

$$Q_1 = (x-2)P_1 = (x-2) \cdot \frac{2(x-2)}{x^2(x-2)^2} = \frac{2}{x^2}$$

$$Q_2 = (x-2)^2 P_2 = (x-2)^2 \frac{(x+3)}{x^2(x-2)^2} = \frac{x+3}{x^2}$$

Since both $Q_1$ and $Q_2$ are analytic ($Q_1 \neq \infty, Q_2 \neq \infty$) at $x = 2$, so $x = 2$ is a regular singular point.

The solution of a differential equation about a regular singular point can be obtained. The cases of irregular singular points are beyond the scope of this book.

**8.5 FROBENIUS METHOD :** If $x = 0$ is a regular singularity of the equation.

$$\frac{d^2 y}{dx^2} + P_1(x)\frac{dy}{dx} + P_2(x)y = 0 \qquad \text{... (1)} \qquad [P(0) = 0]$$

Then the series solution is $y = x^m(a_0 + a_1 x + a_2 x^2 + a_3 x^3 + ...) = \sum_{k=0}^{\infty} a_k x^{m+k}$

The value of $m$ will be determined by substituting the expressions for $y, \dfrac{dy}{dx}, \dfrac{d^2y}{dx^2}$ in (1), we get the identity.

On equating the coefficient of lowest power of $x$ in the identity to zero, a quadratic equation in $m$ (**indicial equation**) is obtained.

Thus, we will get two values of $m$. The series solution of (1) will depend on the nature of the roots of the indicial equation.

(*i*) **Case 1 : When roots $m_1$, $m_2$ are distinct and not differing by an integer)** $m_1 - m_2 \neq 0$ or a positive integer. *e.g.*, $m_1 = \dfrac{1}{2}, m_2 = 2$ .

The complete solution is $\quad y = c_1(y)_{m_1} + c_2(y)_{m_2}$

(*ii*) **Case 2 : When roots $m_1$, $m_2$ *are equal i.e.* $m_1 = m_2$**

$$y = c_1(y)_{m_1} + c_2\left(\frac{\partial y}{\partial m}\right)_{m_1}$$

(*iii*) **Case 3 : When roots $m_1$, $m_2$ are distinct and differ by an integer ($m_1 < m_2$)**

*e.g.*, $m_1 = \dfrac{3}{2}, m_2 = \dfrac{5}{2}$ or $m_1 = 2, m_2 = 4$.

If some of the coefficients of $y$ series become infinite when $m = m_1$, to overcome this difficulty, replace $a_0$ by $b_0$ $(m - m_1)$. We get a solution which is only a constant multiple of the first solution.

Complete solution is $y = c_1(y)_{m_1} + c_2\left(\dfrac{\partial y}{\partial m}\right)_{m_2}$

(*iv*) **Case 4 : Roots are distinct and differing by an integer, making some coefficient indeterminate**

Complete solution is $y = c_1(y)_{m_1} + c_2(y)_{m_2}$

if the coefficients do not become infinite when $m = m_2$.

**Case I : When the roots are distinct and not differing by an integer.**

**Example 5.** *Find solution in generalized series form about x = 0 of the differential equation*

$$3x\frac{d^2y}{dx^2}+2\frac{dy}{dx}+y=0$$

**Solution.**
$$3x\frac{d^2y}{dx^2}+2\frac{dy}{dx}+y=0 \quad \text{... (1)}$$

Since $x = 0$ is a regular singular point, we assume the solution in the form

$$y=\sum_{k=0}^{\infty} a_k x^{m+k}$$

Such that $\frac{dy}{dx}=\sum_{k=0}^{\infty} a_k(m+k)x^{m+k-1}$ $\quad \frac{d^2y}{dx^2}=\sum_{k=0}^{\infty} a_k(m+k)(m+k-1)x^{m+k-2}$

Substituting for $y$, $\frac{dy}{dx}$ and $\frac{d^2y}{dx^2}$ in the given equation (1) we get

$$3\Sigma a_k(m+k)(m+k-1)x^{m+k-1}+2\Sigma a_k(m+k)x^{m+k-1}+\Sigma a_k x^{m+k}=0$$

or
$$\Sigma a_k[3(m+k)(m+k-1)+2(m+k)]x^{m+k-1}+\Sigma a_k x^{m+k}=0 \quad \text{... (2)}$$

The coefficient of the lowest degree term $x^{m-1}$ in the identity (2) is obtained by putting $k = 0$ in first summation only and equating it to zero. Then the **indicial equation** is

$$a_0[3m(m-1)+2m]=0 \quad \text{or} \quad a_0[3m^2-m]=0 \quad \text{or} \quad a_0 m(3m-1)=0$$

Since $\quad a_0 \neq 0,\ m = 0, \text{ or } 1/3$

The coefficient of next lowest degree term $x^m$ in the identity (2) is obtained by putting $k = 1$ in first summation and $k = 0$ in the second summation and equating it to zero.

$$a_1[3(m+1)m+2(m+1)]+a_0=0$$

or
$$a_1[3m^2+5m+2]+a_0=0 \quad \text{or} \quad a_1(3m+2)(m+1)+a_0=0$$

$$a_1=-\frac{1}{(3m+2)(m+1)}a_0$$

Equating to zero the coefficient of $x^{m+k}$, the recurrence relation is given by

$$a_{k+1}[3(m+k+1)(m+k)+2(m+k+1)]+a_k=0.$$

or
$$a_{k+1}(m+k+1)(3m+3k+2)+a_k=0 \quad \text{or} \quad a_{k+1}=\frac{-1}{(m+k+1)(3m+3k+2)}a_k$$

This gives

For $k = 0$, $\quad a_1=\frac{-1}{(m+1)(3m+2)}a_0$

For $k = 1$, $\quad a_2=\frac{-1}{(m+2)(3m+5)}a_1=\frac{1}{(m+1)(m+2)(3m+2)(3m+5)}a_0$

For $k = 2$, $\quad a_3=\frac{-1}{(m+3)(3m+8)}a_2$

$$=\frac{-1}{(m+1)(m+2)(m+3)(3m+2)(3m+5)(3m+8)}a_0$$

**For $m = 0$**

$$a_1 = -\frac{1}{2}a_0, \qquad a_2 = \frac{1}{20}a_0, \qquad a_3 = -\frac{1}{480}a_0$$

Hence for $m = 0$, $y_1 = a_0\left(1 - \frac{1}{2}x + \frac{1}{20}x^2 - \frac{1}{480}x^3 + \ldots\right)$

**For** $m = \frac{1}{3}$

$$a_1 = -\frac{1}{4}a_0, \qquad a_2 = \frac{1}{56}a_0, \qquad a_3 = \frac{-1}{1680}a_0$$

Hence for $m = \frac{1}{3}$, the second solution is

$$y_2 = a_0\left(x^{\frac{1}{3}} - \frac{1}{4}x^{\frac{4}{3}} + \frac{1}{56}x^{\frac{7}{3}} - \frac{1}{1680}x^{\frac{10}{3}} + \ldots\right)$$

Thus the complete solution is

$$y = Ay_1 + By_2$$

$$y = a_0\left(1 - \frac{x}{2} + \frac{x^2}{20} - \frac{x^3}{480} + \ldots\right) + b_0 x^{1/3}\left(1 - \frac{x}{4} + \frac{x^2}{56} - \frac{x^3}{1680} + \ldots\right) \quad \textbf{Ans.}$$

**Example 6.** *Solve* $x(x-1)y'' + (3x-1)y' + y = 0$ *(A.M.I.E.T.E., Summer 2004)*

**Solution.** $x(x-1)y'' + (3x-1)y' + y = 0$ ...(1)

Since, $x = 0$ is a regular singular point, we assume the solution in the form

$$y = \sum_{k=0}^{\infty} a_k x^{m+k}$$

such that $\frac{dy}{dx} = \sum_{k=0}^{\infty} a_k (m+k) x^{m+k-1}$, $\quad \frac{d^2y}{dx^2} = \sum_{k=0}^{\infty} a_k (m+k)(m+k-1) x^{m+k-2}$

Substituting the expressions for $y, \frac{dy}{dx}, \frac{d^2y}{dx^2}$ in (1) we have

$$\Sigma x(x-1) a_k (m+k)(m+k-1) x^{m+k-2} + (3x-1)\Sigma a_k (m+k) x^{m+k-1} + \Sigma a_k x^{m+k} = 0$$

or $\Sigma a_k (m+k)(m+k-1) x^{m+k} - \Sigma a_k (m+k)(m+k-1) x^{m+k-1} + 3\Sigma a_k (m+k) x^{m+k} - \Sigma a_k (m+k) x^{m+k-1} + \Sigma a_k x^{m+k} = 0$

or $\Sigma a_k [(m+k)(m+k-1) + 3(m+k) + 1] x^{m+k} - \Sigma a_k [(m+k)(m+k-1) + (m+k)] x^{m+k-1} = 0$

or $\Sigma a_k [(m+k)(m+k+2) + 1] x^{m+k} - \Sigma a_k (m+k)^2 x^{m+k-1} = 0$ ...(2)

The coefficient of lowest degree term $x^{m-1}$ in (2) is obtained by putting $k = 0$ in the second ımmation only of (2) and equating it to zero. Then the **indicial equation** is

$$a_0 (m+0)^2 = 0 \quad \Rightarrow \quad m = 0,\ 0 \text{ as } a_0 \neq 0$$

The coefficient of the next lowest degree term $x^m$ in (2) is obtained by putting $k = 0$ in the rst summation and $k = 1$ in the second summation only of (2) and equating it to zero, we ge‹

$$a_1 [(m+0)(m+2) + 1] - a_0 (m+0)^2 = 0$$

$$a_1 - a_0 = 0 \quad \Rightarrow \quad a_1 = a_0 \quad (\text{as } m = 0)$$

Equating the coefficient of $x^{m+k}$ to zero, the recurrence relation is given by

$$a_k [(m+k)(m+k+2)+1] - a_{k+1} (m+k+1)^2 = 0$$

$$a_k (m+k+1)^2 - a_{k+1} (m+k+1)^2 = 0$$

Hence $a_{k+1} = a_k$

$$y = x^m [a_0 + a_1 x + a_2 x^2 + .........]$$

$$y = a_0 x^m [1 + x + x^2 + x^3 + ........]$$

$$(m = 0)$$

when $m = 0, 0$, this gives only one solution instead of two.

Second solution is given by

$$\left(\frac{\partial y}{\partial m}\right)_{m=0} \quad \text{and} \quad y_1 = a_0 (1 + x + x^2 + x^3)$$

$$\frac{\partial y}{\partial m} = a_0 x^m \log x [1 + x + x^2 + x^3 + .....]$$

$$y_2 = a_0 \log x [1 + x + x^2 + x^3 + .......] \qquad m = 0$$

$$y_1 = a_0 [1 + x + x^2 + x^3 + .......] \qquad m = 0$$

$$y = A y_1 + B y_2$$

$$y = A [1 + x + x^2 + x^3 + ......] + B \log x (1 + x + x^2 + x^3 + .......)$$ **Ans.**

**Example 7.** *Using extended power series method find one solution of the differential equation $xy'' + y' + x^2 y = 0$. Indicate the form of a second solution which is linearly independent of the first obtained above.*

**Solution.** $x\frac{d^2y}{dx^2} + \frac{dy}{dx} + x^2 y = 0$ ... (1)

Let $y = \Sigma a_k x^{m+k}$

$$\frac{dy}{dx} = \Sigma a_k (m+k) x^{m+k-1}, \qquad \frac{d^2y}{dx^2} = \Sigma a_k (m+k)(m+k-1) x^{m+k-2}$$

Substituting the values of $y$, $\frac{dy}{dx}$ and $\frac{d^2y}{dx^2}$ in (1), we get

$$x \Sigma a_k (m+k)(m+k-1) x^{m+k-2} + \Sigma a_k (m+k) x^{m+k-1} + x^2 \Sigma a_k x^{m+k} = 0$$

or $\Sigma a_k (m+k)(m+k-1) x^{m+k-1} + \Sigma a_k (m+k) x^{m+k-1} + \Sigma a_k x^{m+k+2} = 0$

or $\Sigma a_k [(m+k)(m+k-1) + (m+k)] x^{m+k-1} + \Sigma a_k x^{m+k+2} = 0$

or $\Sigma a_k (m+k)^2 x^{m+k-1} + \Sigma a_k x^{m+k+2} = 0$

The coefficient of lowest degree term $x^{m-1}$ in (2) is obtained by putting $k = 0$ in first summation of (2) only and equating it to zero. Then the **indicial equation** is

$$a_0 m^2 = 0 \Rightarrow m^2 = 0 \text{ or } m = 0, 0$$

The coefficient of the next lowest degree term $x^m$ in (2) is obtained by putting $k = 1$ in first summation only and equating it to zero.

$$a_1 (m+1)^2 = 0 \quad \text{or} \quad a_1 \Rightarrow 0$$

Equating the coefficient of $x^{m+1}$ for $k = 2$ we get

$$a_2 (m+2)^2 = 0 \Rightarrow a_2 = 0$$

Equating the coefficient of $x^{m+k+2}$ to zero, we have

$$a_{k+3}(m+k+3)^2 + a_k = 0$$

$$a_{k+3} = -\frac{a_k}{(m+k+3)^2}$$

$k = 0,\quad a_3 = -\frac{1}{(m+3)^2} a_0$

$k = 1,\quad a_4 = -\frac{1}{(m+4)^2} a_1 = 0,\ a_7 = 0,\ a_{10} = 0$

$k = 2,\quad a_5 = -\frac{1}{(m+5)^2} a_2 = 0,\ a_8 = 0,\ a_{11} = 0$

$k = 3,\quad a_6 = -\frac{1}{(m+6)^2} a_3 = \frac{1}{(m+3)^2(m+6)^2} a_0$

$$a_9 = -\frac{1}{(m+9)^2} a_6 = -\frac{1}{(m+3)^2(m+6)^2(m+9)^2} a_0$$

$$y = x^m a_0 \left[1 - \frac{x^3}{(m+3)^2} + \frac{x^6}{(m+3)^2(m+6)^2} - \frac{x^9}{(m+3)^2(m+6)^2(m+9)^2} + \ldots\right] \quad \ldots (3)$$

To get the first solution, let $m = 0$ in (3), then

$$y_1 = a_0 \left[1 - \frac{x^3}{3^2} + \frac{x^6}{3^2 \times 6^2} - \frac{x^9}{3^2 \times 6^2 \times 9^2} + \ldots\right] \quad \ldots (4)$$

To get the second independent solution, differentiate (3) *w.r.t. m*. Then

$$\frac{\partial y}{\partial m} = (x^m \log x)\, a_0 \left[1 - \frac{x^3}{(m+3)^2} + \frac{x^6}{(m+3)^2(m+6)^2} - \frac{x^9}{(m+3)^2(m+6)^2(m+9)^2} + \ldots\right]$$

$$+ x^m a_0 \left[\frac{2x^3}{(m+3)^3} - \frac{2x^6}{(m+3)^3(m+6)^2} - \frac{2x^6}{(m+3)^2(m+6)^3} + \frac{2x^9}{(m+3)^3(m+6)^2(m+9)^2} + \frac{2x^9}{(m+3)^2(m+6)^3(m+9)^2} + \frac{2x^9}{(m+3)^2(m+6)^2(m+9)^3} + \ldots\right] \quad \ldots (5)$$

Putting $m = 0$ in (5), we get

$$y_2 = (\log x) a_0 \left[1 - \frac{x^3}{3^2} + \frac{x^6}{3^2 \times 6^2} - \frac{x^9}{3^2 \times 6^2 \times 9^2} + \ldots\right]$$

$$+ a_0 \left[\frac{2x^3}{3^3} - \frac{2x^6}{3^3 \times 6^2} - \frac{2x^6}{3^2 \times 6^3} + \frac{2x^9}{3^3 \times 6^2 \times 9^2} + \frac{2x^9}{3^2 \times 6^3 \times 9^2} + \frac{2x^9}{3^2 \times 6^2 \times 9^3} + \ldots\right] \quad \ldots (6)$$

hence the general solution is given by (4) and (6)

$$y = c_1 y_1 + c_2 y_2$$

$$y = c_1 a_0 \left[1 - \frac{x^3}{3^2} + \frac{x^6}{3^2 \times 6^2} - \frac{x^9}{3^2 \times 6^2 \times 9^2} + \ldots\right]$$

$$+ c_2 (\log x) a_0 \left[1 - \frac{x^3}{3^2} + \frac{x^6}{3^2 \times 6^2} - \frac{x^9}{3^2 \times 6^2 \times 9^2} + \ldots\right]$$

$$+c_2a_0\left[\frac{2x^3}{3^3}-\frac{2x^6}{3^2\times 6^6}\left(\frac{1}{3}+\frac{1}{6}\right)+\frac{2\,x^9}{3^2\times 6^2\times 9^2}\left(\frac{1}{3}+\frac{1}{6}+\frac{1}{9}\right)+...\right]$$

$$=(c_1+c_2\log x)\,a_0\left[1-\frac{x^3}{3^2}+\frac{x^6}{3^2\times 6^2}-\frac{x^9}{3^2\times 6^2\times 9^2}+...\right]$$

$$+2c_2a_0\left[\frac{x^3}{3^3}-\frac{x^6}{3^5\times 2^2}\left(1+\frac{1}{2}\right)+\frac{2x^9}{3^9\times 2^2}\left(1+\frac{1}{2}+\frac{1}{3}\right)+...\right] \quad \textbf{Ans.}$$

**Case III : When $m_1$ and $m_2$ are distinct and differing by an integer, then**

$$y=c_1(y)_{m_1}+c_2\left(\frac{\partial y}{\partial m}\right)_{m_2} \qquad \left[\begin{array}{l}\text{If coefficient } = \infty\\ \text{when } m=m_2\end{array}\right]$$

**Example 8.** *Solve*

$$x^2\frac{d^2y}{dx^2}+x\frac{dy}{dx}+(x^2-4)y=0 \quad ...(1)$$

**Solution.** Let $y=\Sigma a_k\, x^{m+k}$

$$\frac{dy}{dx}=\Sigma a_k(m+k)x^{m+k-1}, \qquad \frac{d^2y}{dx^2}=\Sigma a_k(m+k)(m+k-1)\,x^{m+k-2}$$

Substituting the values of $\frac{d^2y}{dx^2}$, $\frac{dy}{dx}$ and $y$ in (1), we get

$$x^2\,\Sigma a_k(m+k)(m+k-1)\,x^{m+k-2}+x\Sigma a_k(m+k)\,x^{m+k-1}+(x^2-4)\,\Sigma a_k x^{m+k}=0$$

$$\Sigma a_k[(m+k)(m+k-1)+(m+k)-4]\,x^{m+k}+\Sigma a_k x^{m+k+2}=0$$

$$\Sigma a_k(m+k+2)(m+k-2)\,x^{m+k}+\Sigma a_k\,x^{m+k+2}=0 \quad ...(2)$$

The coefficient of lowest degree term $x^m$ in (2) is obtained by putting $k=0$ in first summation only and equating it to zero. Then the indicial equation is

$$a_0(m+2)(m-2)=0 \Rightarrow m=2,-2$$

The coefficient of next lowest term $x^{m+1}$ in (2) is obtained by putting $k=1$ in first summation only and equating it to zero.

$$a_1(m+3)(m-1)=0 \Rightarrow a_1=0$$

Equating the coefficient of $x^{m+k+2}$

$$a_{k+2}\,(m+k+4)\,(m+k)+a_k=0 \quad \text{or} \quad a_{k+2}=-\frac{a_k}{(m+k+4)(m+k)}$$

$$a_1=a_3=a_5=...=0$$

$$a_2=-\frac{a_0}{m(m+4)}$$

$$a_4=-\frac{a_2}{(m+2)(m+6)}=\frac{a_0}{m(m+2)(m+4)(m+6)}$$

$$a_6=-\frac{a_4}{(m+4)(m+8)}=-\frac{a_0}{m(m+2)(m+4)^2(m+6)(m+8)}$$

Hence $$y = a_0 x^m \left[1 - \frac{x^2}{m(m+4)} + \frac{x^4}{m(m+2)(m+4)(m+6)} - \frac{x^6}{m(m+2)(m+4)^2(m+6)(m+8)} + \ldots\right] \quad \ldots (3)$$

Putting $m = 2$ in (3), we get

$$y_1 = a_0 x^2 \left[1 - \frac{x^2}{2\times 6} + \frac{x^4}{2\times 4\times 6\times 8} - \frac{x^6}{2\times 4\times 6^2\times 8\times 10} + \ldots\right] \quad \ldots (4)$$

Coefficient of $x^4$, $x^6$ etc. in (3) becomes infinite on putting $m = -2$. To overcome this difficulty we put

$$a_0 = b_0(m+2) \text{ in (3) and we get}$$

$$y = b_0 x^m \left[(m+2) - \frac{(m+2)x^2}{m(m+4)} + \frac{x^4}{m(m+4)(m+6)} - \frac{x^6}{m(m+4)^2(m+6)(m+8)} + \ldots\right] \quad \ldots (5)$$

On differentiating (5) w.r.t. '$m$', we get

$$\frac{\partial y}{\partial m} = b_0(x^m \cdot \log x)\left[(m+2) - \frac{(m+2)x^2}{m(m+4)} + \frac{x^4}{m(m+4)(m+6)} - \frac{x^6}{m(m+4)^2(m+6)(m+8)} + \ldots\right]$$
$$+ b_0 x^m \left[1 - \frac{(m+2)x^2}{m(m+4)}\left(\frac{1}{m+2} - \frac{1}{m} - \frac{1}{m+4}\right) + \frac{x^4}{m(m+4)(m+6)}\left(-\frac{1}{m} - \frac{1}{m+4} - \frac{1}{m+6}\right) + \ldots\right]$$

On replacing $m$ by $-2$, we get

$$\left(\frac{\partial y}{\partial m}\right)_{m=-2} = (b_0 x^{-2}\log x)\left[0 - 0 + \frac{x^4}{(-2)(2)(4)} - \frac{x^6}{(-2)(2)^2(4)(6)} + \ldots\right]$$
$$+ b_0 x^{-2}\left[1 + \frac{x^2}{2^2} + \frac{x^4}{2^2\times 4}\left(\frac{1}{4}\right) + \ldots\right]$$

or

$$y_2 = b_0 x^2 \log x\left(-\frac{1}{2^2\times 4} + \frac{x^2}{2^3\times 4\times 6} - \frac{x^4}{2^3\times 4^2\times 6\times 8} + \ldots\right) + b_0 x^{-2}\left(1 + \frac{x^2}{2^2} + \frac{x^4}{2^2\times 4^2} + \ldots\right)$$

General solution is $y = c_1 y_1 + c_2 y_2$

$$y = c_1 x^2\left(1 - \frac{x^2}{2\times 6} + \frac{x^4}{2\times 4\times 6\times 8} - \frac{x^6}{2\times 4\times 6^2\times 8\times 10} + \ldots\right)$$
$$+ c_2\left[x^2 \log x\left(-\frac{1}{2^2\times 4} + \frac{x^2}{2^3\times 4\times 6} - \frac{x^4}{2^3\times 4^2\times 6\times 8} + \ldots\right) + x^{-2}\left(1 + \frac{x^2}{2^2} + \frac{x^4}{2^2\times 4^2} + \ldots\right)\right]$$

**Ans.**

**Case IV. If the roots differ by an integer such that one or more coefficients are indeterminate.**

**Example 9.** *Find the extended power series solution of the differential equation*

$$x^2 y'' + 4xy' + (x^2 + 2)\, y = 0 \quad \ldots (1)$$

**Solution.** Let $y = \Sigma\, a_k x^{m+k}$ be the required solution of the given equation.

The $$\frac{dy}{dx} = \Sigma a_k (m+k) x^{m+k-1}$$

$$\frac{d^2y}{dx^2} = \Sigma a_k (m+k)(m+k-1) x^{m+k-2}$$

Substituting the value of $y$, $\frac{dy}{dx}$ and $\frac{d^2y}{dx^2}$ in the given equation.

$$x^2 \Sigma a_k (m+k)(m+k-1) x^{m+k-2} + 4x \Sigma a_k (m+k) x^{m+k-1} + (x^2+2) \Sigma a_k x^{m+k} = 0$$

$$\Sigma a_k (m+k)(m+k-1) x^{m+k} + 4 \Sigma a_k (m+k) x^{m+k} + \Sigma a_k x^{m+k+2} + \Sigma 2 a_k x^{m+k} = 0$$

$$\Sigma a_k [(m+k)(m+k-1) + 4(m+k) + 2] x^{m+k} + \Sigma a_k x^{m+k+2} = 0$$

$$\Sigma a_k [(m+k)^2 + 3(m+k) + 2] x^{m+k} + \Sigma a_k x^{m+k+2} = 0 \qquad \text{... (2)}$$

The coefficient of lowest degree term $x^m$ in (2) is obtained by putting k = 0 in first summation only and equating it to zero. Then the indicial equation is

$$a_0 (m^2 + 3m + 2) = 0$$

$$a_0 \neq 0,\ m^2 + 3m + 2 = 0 \quad \text{or} \quad (m+1)(m+2) = 0,\ m = -1, -2$$

The coefficient of next lowest degree term $x^{m+1}$ in (2) is obtained by putting $k = 1$ in first summation only and equating it to zero.

$$a_1 [m^2 + 5m + 6] = 0 \text{ or } a_1 (m+2)(m+3) = 0 \Rightarrow a_1 = \frac{0}{(m+2)(m+3)}$$

when $m = -2$, $a_1$ becomes indeterminate $\left(\frac{0}{0}\right)$. But in this case we get the identity $a_1$ (0) = 0 which is satisfied by every value of $a_1$. Therefore in this case we can take $a_1$ as arbitrary constant

Equating the coefficient of $x^{m+k+2}$

$$a_{k+2} [m^2 + (2k+4+3)m + (k+2)^2 + 3(k+2) + 2] + a_k = 0$$

$$a_{k+2} [m^2 + (2k+7)m + k^2 + 7k + 12] + a_k = 0$$

$$a_{k+2} = -\frac{1}{m^2 + (2k+7)m + k^2 + 7k + 12} a_k$$

For $k = 0$, $$a_2 = -\frac{1}{m^2 + 7m + 12} a_0 = -\frac{1}{(m+3)(m+4)} a_0$$

$k = 1$ $$a_3 = -\frac{1}{m^2 + 9m + 20} a_1 = -\frac{1}{(m+4)(m+5)} a_1$$

$$a_4 = -\frac{1}{m^2 + 11m + 30} a_2 = \frac{1}{(m+3)(m+4)(m+5)(m+6)} a_0$$

$$a_5 = -\frac{1}{m^2 + 13m + 42} a_3 = \left\{\frac{1}{(m+4)(m+5)(m+6)(m+7)} a_1\right\}$$

For $m = -1$

$$a_2 = -\frac{1}{6}a_0, \quad a_3 = \frac{1}{12}a_1, \qquad a_4 = \frac{1}{120}a_0, \qquad a_5 = \frac{1}{360}a_1$$

Hence for $\quad m = -1,$

$$y_1 = x^{-1}\left[1 - \frac{1}{6}x^2 + \frac{1}{120}x^4 + \ldots\right]a_0 + \left[1 - \frac{1}{12}x^2 + \frac{x^4}{360} + \ldots\right]a_1$$

For $\quad m = -2$

$$a_2 = -\frac{1}{2}a_0, \quad a_3 = -\frac{1}{6}a_1, \qquad a_4 = \frac{1}{24}a_0, \qquad a_5 = \frac{1}{120}a_1$$

Hence for $m = -2$, second solution is

$$y_2 = x^{-2}\left[1 - \frac{x^2}{2} + \frac{x^4}{24} + \ldots\right]a_0 + \left[\frac{1}{x} - \frac{x}{6} + \frac{x^3}{120} + \ldots\right]a_1$$

$$y_2 = x^{-2}\left[\left\{1 - \frac{x^2}{2} + \frac{x^4}{4} + \ldots\right\}a_0 + \left\{x - \frac{x^3}{3} + \frac{x^5}{5} + \ldots\right\}a_1\right]$$

$$y_2 = x^{-2}\left[a_0 \cos x + a_1 \sin x\right]$$

**Ans.**

Thus the complete solution is $y = Ay_1 + By_2$

## EXERCISE 8.2

Solve the following differential equations by power series method :

**1.** $x^2y'' + 6xy' + (6 - x^2)y = 0.$ *(A.M.I.E.T.E., Dec. 2005)*

**Ans.** $y = Ay_1 + By_2$ when $y_1 = a_0x^{-2}\left[1 + \frac{x^2}{3} + \frac{x^4}{5} + \ldots\ldots\right] + a_1x^{-2}\left[x + \frac{x^3}{3.4} + \frac{x^5}{3.4.5.6} + \ldots\ldots\right]$

$$y_2 = a_0x^{-3}\left[1 + \frac{x^2}{2} + \frac{x^4}{4} + \ldots\right] + a_1x^{-3}\left[x + \frac{x^3}{3} + \frac{x^5}{5} + \ldots\right]$$

**2.** $2x^2\frac{d^2y}{dx^2} - x\frac{dy}{dx} + (x^2 + 1)y = 0$ **Ans.** $y = a_0x\left(1 - \frac{x^2}{10} + \frac{x^4}{360} + \ldots\right) + a_0x^{1/2}\left(1 + \frac{x^2}{6} + \frac{x^4}{168} + \ldots\right)$

**3.** $2x(1 - x)\frac{d^2y}{dx^2} + (5 - 7x)\frac{dy}{dx} - 3y = 0$ **Ans.** $a_0\left(1 + \frac{3}{5}x + \frac{3}{7}x^2 + \frac{3}{9}x^3 + \ldots\right) + b_0x^{-3/2}$

**4.** $2x(1 - x)\frac{d^2y}{dx^2} + (1 - x)\frac{dy}{dx} + 3y = 0$ **Ans.** $y = a_0\sqrt{x}\,(1 - x) + a_1\left(1 - 3x + \frac{3x^2}{1.3} + \frac{3x^3}{3.5} + \frac{3x^4}{5.7}\ldots\right)$

**5.** $x\frac{d^2y}{dx^2} + 3\frac{dy}{dx} + 4x^3y = 0$ **Ans.** $a_0x^{-2}\left(1 - \frac{x^4}{2} + \frac{x^8}{4} + \ldots\right) a_2x^{-2}\left(x^2 - \frac{x^{o}}{3} + \frac{x^{10}}{5} + \ldots\right)$

**6.** $xy'' + (x - 1)y' - y = 0$ **Ans.** $A\left(1 - x + \frac{x^2}{2} - \frac{x^3}{3} + \ldots\right) + B\left(x^2 - 2\frac{x^3}{3} + \frac{2x^4}{4} - 2\frac{x^5}{5}\ldots\right)$

7. $xy'' + y' + xy = 0$ about $x = 0$

**Ans.** $y = A\left[1 - \frac{1}{2^2}x^2 + \frac{x^4}{2^2 \cdot 4^2} - \frac{x^6}{2^2 \cdot 4^2 \cdot 6^2} + \ldots\ldots\right]$

$$+ B\left[y_1 \log x + a_0\left\{\frac{x^2}{2^2} - \frac{1}{2^2 \cdot 4^2}\left(1 + \frac{1}{2}\right)x^4 + \frac{1}{2^2 \cdot 4^2 \cdot 6^2}\left(1 + \frac{1}{2} + \frac{1}{3}\right)x^6 + \ldots\ldots\right\}\right]$$

(*A.M.I.E.T.E., Summer 2001, 2000*)

8. $x(1-x)y'' + 4y' + 2y = 0$

**Ans.** $y_1 = a_0\left[1 - \frac{x}{2} + \frac{x^2}{10}\right]$, $y_2 = a_0 x^{-3}\,[1 - 5x + 10x^2 - 10x^3 + 5x^4 - x^5]$

(*A.M.I.E.T.E., Summer 2001, 2000, Winter 2003*)

9. $x^2\frac{d^2y}{dx^2} + 5x\frac{dy}{dx} + x^2y = 0$

[**Hint** $y = \Sigma a_k\,(m + k0\;(m + k + 4)\;x^{m+k} + \Sigma a_k x^{m+k+2}]$

Indicial eq. $a_0\;m\,(m + 4) = 0$

**Ans.** $= a_0\left[1 - \frac{x^2}{12} + \frac{x^4}{364} - \ldots\ldots\right] + b_0 x^{-4}\log x\left[1 - \frac{x^4}{16} - \frac{x^6}{16}\ldots\ldots\right] + b_0 x^{-}2\left[\frac{1}{4} + \frac{x^2}{64}\;\ldots\right]$

## 8.6 BESSEL'S EQUATION

The differential equation

$$x^2\frac{d^2y}{dx^2} + x\frac{dy}{dx} + (x^2 - n^2)\,y = 0$$

is called the *Bessel's differential equation,* and particular solutions of this equation are called Bessel's functions of order *n*.

## 8.7 SOLUTION OF BESSEL'S EQUATION

$$x^2\frac{d^2y}{dx^2} + x\frac{dy}{dx} + (x^2 - n^2)\,y = 0. \qquad \ldots(1)$$

Let $\quad y = \Sigma a_r x^{m+r}$ or $y = a_0 x^m + a_1 x^{m+1} + a_2 x^{m+2} + \ldots \qquad \ldots(2)$

so that $\frac{dy}{dx} = \Sigma a_r\,(m + r)\,x^{m+r-1}$ and $\frac{d^2y}{dx^2} = \Sigma a_r\,(m + r)\,(m + r - 1)\,x^{m+r-2}$

Substituting these values in the equation, we have

$$x^2\,\Sigma a_r\,(m + r)\,(m + r - 1)\,x^{m+r-2} + x\,\Sigma a_r\,(m + r)\,x^{m+r-1} + (x^2 - n^2)\,\Sigma a_r x^{m+r} = 0$$

or $\quad \Sigma a_r\,(m + r)\,(m + r - 1)\,x^{m+r} + \Sigma a_r\,(m + r)\,x^{m+r} + \Sigma a_r x^{m+r+2} - n^2\,\Sigma a_r x^{m+r} = 0$

or $\quad \Sigma a_r\,[(m + r)\,(m + r - 1) + (m + r) - n^2]\,x^{m+r} + \Sigma a_r x^{m+r+2} = 0$

or $\quad \Sigma a_r\,[(m + r)^2 - n^2]\,x^{m+r} + \Sigma a_r x^{m+r+2} = 0.$

Equating the coefficient of $x^m$ to zero, we get

$$a_0\,[(m + 0)^2 - n^2] = 0. \qquad (r = 0)$$

or $\quad m^2 = n^2$ *i.e.* $m = n \qquad a_0 \neq 0$

Equating the coefficient of $x^{m+1}$ $\qquad r = 1$

$$a_1\,[(m + 1)^2 - n^2] = 0 \text{ i.e. } a_1 = 0. \qquad \text{since } (m + 1)^2 - n^2 \neq 0$$

Equating the coefficient of $x^{m+r+2}$ to zero, to find relation in successive coefficients, we get

$$a_{r+2}\,[(m + r + 2)^2 - n^2] + a_r = 0 \qquad \text{or} \qquad a_{r+2} = -\frac{1}{(m + r + 2)^2 - n^2}\cdot a_r$$

Therefore $a_3 = a_5 = a_7 = \ldots = 0$, since $a_1 = 0$page 571

If $r = 0$, $a_2 = -\frac{1}{(m+2)^2 - n^2} a_0$

If $r = 2$, $a_4 = -\frac{1}{(m+4)^2 - n^2} a_2 = \frac{1}{[(m+2)^2 - n^2][(m+4)^2 - n^2)]} a_0$ and so on

On substituting the values of the coefficients in (2) we have

$$y = a_0 x^m - \frac{a_0}{(m+2)^2 - n^2} x^{m+2} + \frac{a_0}{[(m+2)^2 - n^2][(m+4)^2 - n^2]} x^{m+4} + \ldots$$

$$y = a_0 x^m \left[ 1 - \frac{1}{(m+2)^2 - n^2} x^2 + \frac{1}{[(m+2)^2 - n^2][(m+4)^2 - n^2]} x^4 - \ldots \right]$$

For $m = n$

$$y = a_0 x^n \left[ 1 - \frac{1}{4(n+1)} x^2 + \frac{1}{4^2 \cdot 2!(n+1)(n+2)} x^4 - \ldots \right]$$

where $a_0$ is an arbitrary constant.

For $m = -n$

$$y = a_0 x^{-n} \left[ 1 - \frac{1}{4(-n+1)} x^2 + \frac{1}{4^2 \cdot 2!(-n+1)(-n+2)} x^4 - \ldots \right]$$

## 8.8 BESSEL FUNCTIONS, $J_n(x)$

The Bessel's equation is $x^2 \frac{d^2y}{dx^2} + x \frac{dy}{dx} + (x^2 - n^2) y = 0$ ...(1)

Solution of (1) is

$$y = a_0 x^n \left[ 1 - \frac{x^2}{2 \cdot 2(n+1)} + \frac{x^4}{2.4.2^2(n+1)(n+2)} - \ldots + (-1)^r \frac{x^{2r}}{(2^r r!) \cdot 2^r (n+1)(n+2) \ldots (n+r)} + \ldots \right]$$

$$= a_0 x^n \sum_{r=0}^{\infty} (-1)^r \frac{x^{2r}}{2^{2r} \cdot r!(n+1)(n+2) \ldots (n+r)}$$

where $a_0$ is an arbitrary constant.

If $a_0 = \frac{1}{2^n \Gamma(n+1)}$

The above solution is called Bessel's function denoted by $J_n(x)$.

Thus $J_n(x) = \frac{1}{2^n \Gamma(n+1)} \Sigma (-1)^r \frac{x^{n+2r}}{2^{2r} \cdot r!(n+1)(n+2) \ldots (n+r)}$

$(\Gamma n+1 = n!)$

$$J_n(x) = \left(\frac{x}{2}\right)^n \left\{ \frac{1}{\Gamma(n+1)} - \frac{1}{\lfloor 1 \Gamma(n+2)} \left(\frac{x}{2}\right)^2 + \frac{1}{\lfloor 2 \Gamma(n+3)} \left(\frac{x}{2}\right)^4 - \frac{1}{\lfloor 3 \Gamma(n+4)} \left(\frac{x}{2}\right)^6 + \ldots \right\}$$

or

$$J_n(x) = \sum_{r=0}^{\infty} \frac{(-1)^r}{r!(n+r)!} \left(\frac{x}{2}\right)^{n+2r} \quad \ldots(2)$$

If $n = 0, J_0(x) = \Sigma \frac{(-1)^r}{(r!)^2}\left(\frac{x}{2}\right)^{2r}$ or $J_0(x) = 1 - \frac{x^2}{2^2} + \frac{x^4}{2^2 \cdot 4^2} - \frac{x^6}{2^2 \cdot 4^2 \cdot 6^2} + \ldots$

If $n = 1$, $J_1(x) = \frac{x}{2} - \frac{x^3}{2^2 \cdot 4} + \frac{x^5}{2^2 \cdot 4^2 \cdot 6} - \ldots$

We draw the graphs of these two functions. Both the functions are oscillatory with a varying period and a decreasing amplitude.

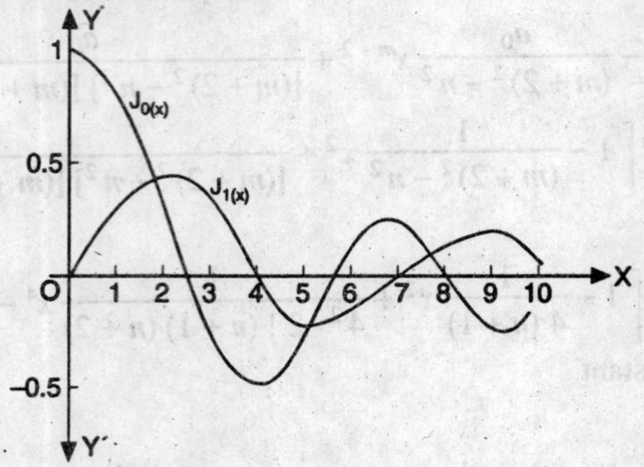


Replacing $n$ by $-n$ in (2), we get

$$J_{-n}(x) = \sum_{r=0}^{\infty} \frac{(-1)^r}{r!\,\Gamma(-n+r+1)}\left(\frac{x}{2}\right)^{-n+2r}$$

**General solution of Bessel's Equation is**

$$y = AJ_n(x) + BJ_{-n}(x)$$

**Example 10.** *Prove that*

$$J_{-n}(x) = (-1)^n J_n(x)$$

*where n is a positive integer.* ***(A.M.I.E.T.E., Winter 2001)***

**Solution.**
$$J_{-n}(x) = \sum_{r=0}^{\infty} (-1)^r \frac{1}{r!\,\Gamma(r-n+1)}\left(\frac{x}{2}\right)^{-n+2r}$$

$$= \sum_{r=0}^{n-1} \frac{(-1)^r\left(\frac{x}{2}\right)^{-n+2r}}{r!\,\Gamma(-n+r+1)} + \sum_{r=n}^{\infty} \frac{(-1)^r\left(\frac{x}{2}\right)^{-n+2r}}{r!\,\Gamma(-n+r+1)}$$

$$= 0 + \sum_{r=n}^{\infty} \frac{(-1)^r\left(\frac{x}{2}\right)^{-n+2r}}{r!\,\Gamma(-n+r+1)} \qquad \text{since } \Gamma \text{ – ve integer} = \infty$$

On putting $r = n + k$

$$J_{-n}(x) = \sum_{k=0}^{\infty} \frac{(-1)^{n+k}\left(\frac{x}{2}\right)^{n+2k}}{(n+k)!\,\Gamma(k+1)}$$

$$= (-1)^n \sum_{k=0}^{\infty} \frac{(-1)^k\left(\frac{x}{2}\right)^{n+2k}}{(n+k)!\,k!}$$

$$= (-1)^n J_n(x)$$ **Proved.**

**Example 11.** *Prove that*

(*a*) $J_{1/2}(x) = \sqrt{\left(\frac{2}{\pi x}\right)} \sin x$ (*b*) $J_{-1/2}(x) = \sqrt{\left(\frac{2}{\pi x}\right)} \cos x$

**Solution.** *We know that*

$$J_n(x) = \frac{x^n}{2^n \,\overline{|n+1}}\left[1 - \frac{x^2}{2.2\,(n+1)} + \frac{x^4}{2.4.2^2\,(n+1)\,(n+2)} \ldots\right] \quad \ldots(1)$$

(*a*) Substituting $n = \frac{1}{2}$ in (1) we obtain

$$J_{1/2}(x) = \frac{x^{1/2}}{2^{1/2}\,\overline{\left|\frac{1}{2}+1\right.}}\left[1 - \frac{x^2}{2.2.\left(\frac{1}{2}+1\right)} + \frac{x^4}{2.4.2^2\left(\frac{1}{2}+1\right)\left(\frac{1}{2}+2\right)} \ldots\right]$$

$$= \frac{\sqrt{x}}{\sqrt{2}\,\overline{|3/2}}\left[1 - \frac{x^2}{2.3} + \frac{x^4}{2.3.4.5} \ldots\right] = \frac{\sqrt{x}}{\sqrt{2}\cdot\frac{1}{2}\,\overline{\left|\frac{1}{2}\right.}}\,\frac{1}{x}\left[x - \frac{x^3}{3\,!} + \frac{x^5}{5\,!} - \ldots\right]$$

$$= \frac{1}{\sqrt{2x}\cdot\frac{1}{2}\sqrt{\pi}} \sin x = \sqrt{\left(\frac{2}{\pi x}\right)} \sin x \qquad \left(\text{since } \Gamma\frac{1}{2} = \sqrt{\pi}\right)$$ **Proved.**

(*b*) Again substituting $n = -\frac{1}{2}$ in (1), we have

$$J_{-1/2}(x) = \frac{x^{-1/2}}{2^{-1/2}\,\overline{\left|-\frac{1}{2}+1\right.}}\left[1 - \frac{x^2}{2.2.\left(-\frac{1}{2}+1\right)} + \frac{x^4}{2.4.2^2\left(-\frac{1}{2}+1\right)\left(-\frac{1}{2}+2\right)} - \ldots\right]$$

$$= \frac{\sqrt{2}}{\sqrt{x}\,\overline{\left|\frac{1}{2}\right.}}\left[1 - \frac{x^2}{2\,!} + \frac{x^4}{4\,!} - \ldots\right] = \sqrt{\left(\frac{2}{\pi x}\right)} \cos x \qquad \left(\text{since } \Gamma\frac{1}{2} = \sqrt{\pi}\right)$$ **Proved.**

## 8.9 RECURRENCE FORMULAE

**Formula I.** $x J_n' = n J_n - x J_{n+1}$

**Proof.** We know that

$$J_n = \sum_{r=0}^{\infty} \frac{(-1)^r}{r\,!\,\Gamma(n+r+1)}\left(\frac{x}{2}\right)^{n+2r}$$

Differentiating with respect to *x*, we get

$$J_n' = \sum \frac{(-1)^r\,(n+2r)}{r\,!\,\Gamma(n+r+1)}\left(\frac{x}{2}\right)^{n+2r-1} \frac{1}{2}.$$

$$x J_n' = n \sum \frac{(-1)^r}{r\,!\,\Gamma(n+r+1)}\left(\frac{x}{2}\right)^{n+2r} + x \sum \frac{(-1)^r \cdot 2r}{2 \cdot r\,!\,\Gamma(n+r+1)}\left(\frac{x}{2}\right)^{n+2r-1}$$

$$= n J_n + x \sum_{r=1}^{\infty} \frac{(-1)^r}{(r-1)\,!\,\Gamma(n+r+1)}\left(\frac{x}{2}\right)^{n+2r-1}$$

$$= n J_n + x \sum_{s=0}^{\infty} \frac{(-1)^{s+1}}{s!\,\Gamma(n+s+2)} \left(\frac{x}{2}\right)^{n+2s+1} \qquad \text{Putting } r-1=s$$

$$= n J_n + x \sum_{s=0}^{\infty} \frac{(-1)^{s}}{s!\,\Gamma[(n+1)+s+1]} \left(\frac{x}{2}\right)^{(n+1)+2s}$$

$$= n J_n - x J_{n+1}$$ **Proved.**

**Formula II.** $x J_n' = -n J_n + x J_{n-1}$

**Proof.** We have $J_n = \sum_{r=0}^{\infty} \frac{(-1)^r}{r!\,\Gamma(n+r+1)} \left(\frac{x}{2}\right)^{n+2r}$

Differentiating w.r.t. '$x$' we get $J_n' = \sum_{r=0}^{\infty} \frac{(-1)^r (n+2r)}{r!\,\Gamma(n+r+1)} \left(\frac{x}{2}\right)^{n+2r-1} \frac{1}{2}$

$$x J_n' = \sum_{r=0}^{\infty} \frac{(-1)^r (n+2r)}{r!\,\Gamma(n+r+1)} \left(\frac{x}{2}\right)^{n+2r} = \sum_{r=0}^{\infty} \frac{(-1)^r [(2n+2r)-n]}{r!\,\Gamma(n+r+1)} \left(\frac{x}{2}\right)^{n+2r}$$

$$= \sum_{r=0}^{\infty} \frac{(-1)^r (2n+2r)}{r!\,\Gamma(n+r+1)} \left(\frac{x}{2}\right)^{n+2r} - n \sum_{r=0}^{\infty} \frac{(-1)^r}{r!\,\Gamma(n+r+1)} \left(\frac{x}{2}\right)^{n+2r}$$

$$= \sum_{r=0}^{\infty} \frac{(-1)^r 2}{r!\,\Gamma(n+r)} \left(\frac{x}{2}\right)^{n+2r} - n J_n = x \sum_{r=0}^{\infty} \frac{(-1)^r}{r!\,\Gamma[(n-1)+r+1]} \left(\frac{x}{2}\right)^{(n-1+2r)} - n J_n$$

$$= x J_{n-1} - n J_n$$ **Proved.**

**Formula III.** $2 J_n' = J_{n-1} - J_{n+1}$

**Proof.** We know that

$$x J_n' = n J_n - x J_{n+1} \qquad \ldots(1) \qquad \text{Recurrence formula } I$$

$$x J_n' = -n J_n + x J_{n-1} \qquad \ldots(2) \qquad \text{Recurrence formula } II$$

Adding (1) and (2), we get

$$2 x J_n' = -x J_{n+1} + x J_{n-1} \quad \text{or} \quad 2 J_n' = J_{n-1} - J_{n+1}$$ **Proved.**

**Formula IV.** $2 n J_n = x (J_{n-1} + J_{n+1})$

**Proof.** We know that

$$x J_n' = n J_n - x J_{n-1} \qquad \text{Recurrence formula I}$$

$$x J_n' = -n J_n + x J_{n+1} \qquad \text{Recurrence formula II}$$

Subtracting (2) from (1), we get

$$0 = 2 n J_n - x J_{n+1} - x J_{n-1} \quad \text{or} \quad 2 n J_n = x (J_{n-1} + J_{n+1})$$ **Proved.**

**Formula V.** $\frac{d}{dx}(x^{-n} \cdot J_n) = -x^{-n} J_{n+1}$

**Proof.** We know that

$$x\,J_n' = n\,J_n - x\,J_{n+1} \qquad \text{Recurrence formula } I$$

Multiplying by $x^{-n-1}$, we obtain

$$x^{-n}\,J_n' = n\,x^{-n-1}\,J_n - x^{-n}\,J_{n+1}$$

*i.e.*, $$x^{-n}\,J_n' - n\,x^{-n-1}\,J_n = -x^{-n}\,J_{n+1}$$

or $$\frac{d}{dx}(x^{-n}\,J_n) = -x^{-n}\,J_{n+1}$$ **Proved.**

**Formula VI.** $\frac{d}{dx}(x^n\,J_n) = x^n\,J_{n-1}$

**Proof.** We know that

$$x\,J_n' = -n\,J_n + x\,J_{n-1} \qquad \text{Recurrence formula } II$$

Multiplying by $x^{n-1}$, we have

$$x^n\,J_n' = -n\,x^{n-1}\,J_n + x^n\,J_{n-1}$$

*i.e.*, $$x^n\,J_n' + n\,x^{n-1}\,J_n = x^n\,J_{n-1}$$

or $$\frac{d}{dx}(x^n\,J_n) = x^n\,J_{n-1}$$ **Proved.**

**Example 12.** *Prove that* $J_2'(x) = \left(1 - \frac{4}{x^2}\right) J_1(x) + \frac{2}{x} J_0(x)$ *where* $J_n(x)$ *is the Bessel function of first kind.* *(U.P. III Semester, Summer 2001)*

**Solution.** By recurrence formula *II*

$$x\,J_n' = -n\,J_n + x\,J_{n-1} \qquad ...(1)$$

On putting $n = 2$, in (1) we have $x\,J_2' = -2\,J_2 + x\,J_1$

or $$J_2' = -\frac{2}{x}J_2 + J_1 \qquad ...(2)$$

By recurrence formula *I*

$$x\,J_n' = n\,J_n - x\,J_{n+1} \qquad ...(3)$$

From (1) and (3) we have $-n\,J_n + x\,J_{n-1} = n\,J_n - x\,J_{n+1}$

On putting $n = 1$, $-J_1 + x\,J_0 = J_1 - x\,J_2$

or $$-\frac{1}{x}J_1 + J_0 = \frac{1}{x}J_1 - J_2 \quad \text{or} \quad J_2 = \frac{2}{x}J_1 - J_0 \qquad ...(4)$$

Putting the value of $J_2$ from (4) in (2) we get

$$J_2' = -\frac{2}{x}\left(\frac{2}{x}J_1 - J_0\right) + J_1 = -\frac{4}{x^2}J_1 + \frac{2}{x}J_0 + J_1$$

$$= \left(1 - \frac{4}{x^2}\right)J_1 + \frac{2}{x}J_0$$ **Proved.**

**Example 13.** *Using the recurrence relations, show that*

$$4\,J_n''(x) = J_{n-2}(x) - 2\,J_n(x) + J_{n+2}(x).$$

**Solution.** We know that the recurrence formula

$$2\,J_n' = J_{n-1} - J_{n+1} \qquad ...(1)$$

On differentiating again, we have

$$2J_n'' = J'_{n-1} - J'_{n+1} \qquad ...(2)$$

Replacing $n$ by $n-1$ and $n$ by $n+1$ in (1) we have

$$2J'_{n-1} = J_{n-2} - J_n \quad \text{or} \quad J'_{n-1} = \frac{1}{2}J_{n-2} - \frac{1}{2}J_n \qquad ...(3)$$

$$2J'_{n+1} = J_n - J_{n+2} \quad \text{or} \quad J'_{n+1} = \frac{1}{2}J_n - \frac{1}{2}J_{n+2} \qquad ...(4)$$

Putting the values of $J'_{n-1}$ and $J'_{n+1}$ from (3) and (4) in (2) we get

$$2J_n'' = \frac{1}{2}\left[J_{n-2} - J_n\right] - \frac{1}{2}\left[J_n - J_{n+2}\right]$$

or $$4J_n'' = J_{n-2} - J_n - J_n + J_{n+2}$$

or $$4J_n'' = J_{n-2} - 2J_n + J_{n+2}$$ **Proved.**

**Example 14.** *Prove that* $\frac{d}{dx}[x^n J_n(x)] = x^n J_{n-1}(x)$ *(A.M.I.E.T.E., Summer 2002)*

**Solution.** $$x^n J_n(x) = x^n \sum_{r=0}^{\infty} \frac{(-1)^r}{r!\,\Gamma(n+r+1)}\left(\frac{x}{2}\right)^{n+2r} = \sum \frac{(-1)^r x^{2n+2r}}{r!\,\Gamma(n+r+1)\cdot 2^{n+2r}}$$

$$\frac{d}{dx}[x^n J_n(x)] = \sum \frac{(-1)^r (2n+2r)\, x^{2n+2r-1}}{r!\,\Gamma(n+r+1)\cdot 2^{n+2r}}$$

$$= x^n \sum \frac{(-1)^r (n+r)}{r!\,\Gamma(n+r+1)}\left(\frac{x}{2}\right)^{n+2r-1} = x \sum \frac{(-1)^r}{r!\,\Gamma(n+r)}\left(\frac{x}{2}\right)^{n+2r-1}$$

$$= x^n \sum \frac{(-1)^r}{r!\,\Gamma(n-1+r+1)}\left(\frac{x}{2}\right)^{\overline{n-1+2r}}$$

$$= x^n J_{n-1}(x)$$ **Proved.**

Similarly we can prove that

$$\frac{d}{dx}[x^{-n} J_n(x)] = -x^{-n} J_{n+1}(x)$$

**Example 15.** *Prove that* $\frac{d}{dx}(x J_n J_{n+1}) = x(J_n^2 - J_{n+1}^2)$

**Solution.** $$\frac{d}{dx}(x J_n J_{n+1}) = J_n J_{n+1} + x\frac{d}{dx}(J_n J_{n+1})$$

$$= J_n J_{n+1} + x(J_n J'_{n+1} + J_n' J_{n+1})$$

$$= J_n J_{n+1} + (x J_n') J_{n+1} + J_n (x J'_{n+1}) \qquad ...(1)$$

Recurrence formula *I*, $\quad x J_n' = n J_n - x J_{n+1} \qquad ...(2)$

Recurrence formula *II*, $\quad x J_n' = -n J_n + x J_{n-1}$

Putting $n+1$ for $n$, $\quad x J'_{n+1} = -(n+1) J_{n+1} + x J_n \qquad ...(3)$

Putting the values of $x J_n'$ and $x J'_{n+1}$ from (2) and (3) in (1) we obtain

$$\frac{d}{dx}(x J_n J_{n+1}) = J_n J_{n+1} + (n J_n - x J_{n+1}) J_{n+1} + J_n[-(n+1) J_{n+1} + x J_n]$$

$$= (1 + n - n - 1) J_n J_{n+1} + x(J_n^2 - J_{n+1}^2)$$

$$= x\,(J_n^2 - J_{n+1}^2)$$ **Proved.**

**Example 16.** *Prove that*

$$J_3(x) + 3\,J_0'(x) + 4\,J_0'''(x) = 0$$ *(A.M.I.E.T.E., Summer 2000)*

**Solution.** We know that

$$J_0'(x) = -J_1(x) \quad \dots (1)$$

Differentiating (1) w.r.t., '$x$', we get

$$J_0''(x) = -J_1'(x) \quad \dots (2)$$

$$2\,J_n'(x) = J_{n-1}(x) - J_{n+1}(x) \quad \dots (3)$$

Taking $n = 1$ in (3), we have

$$2\,J_1'(x) = J_0(x) - J_2(x) \quad \dots (4)$$

Putting the value of $J_1'(x)$ from (2) in (4), we get

$$2\,J_0''(x) = J_2(x) - J_0(x) \quad \dots (5)$$

Differentiating (5), we have

$$2\,J_0'''(x) = J_2'(x) - J_0'(x) \quad \dots (6)$$

Now taking $n = 2$ in (3), we have

$$2\,J_2'(x) = J_1(x) - J_3(x) \quad \dots (7)$$

Putting the value of $J_2'(x)$ from (7) in (6), we obtain

$$4\,J_0'''(x) = [J_1(x) - J_3(x)] - 2\,J_0'(x)$$

$$= [-J_0'(x) - J_3(x)] - 2\,J_0'(x)$$

or $\quad 4\,J_0'''(x) + 3\,J_0'(x) + J_3(x) = 0$ **Proved.**

**Example 17.** *Prove that*

$$\int J_3(x)\,dx + J_2(x) + \frac{2}{x}\,J_1(x) = 0$$ *(A.M.I.E.T.E., Summer 2000)*

**Solution.** We know that

$$\frac{d}{dx}[x^{-n}\,J_n(x)] = -x^{-n}\,J_{n+1}(x)$$ (Recurrence Relation V)

Integrating above relation, we get

$$x^{-n}\,J_n(x) = -\int x^{-n}\,J_{n+1}(x)\,dx \quad \dots (1)$$

On taking $n = 2$ in (1), we have

$$\int x^{-2}\,J_3(x)\,dx = -x^{-2}\,J_2(x) \quad \dots (2)$$

Again $\quad \int J_3(x)\,dx = \int x^2\,(x^{-2})\,J_3(x)\,dx$

$$= x^2 \int (x^{-2})\,J_3(x)\,dx - \int 2x \int (x^{-2}\,J_3\,x)\,dx \quad (3)$$

Putting the value of $\int x^{-2}\,J_3(x)\,dx$ from (2) in (3), we get

$$\int J_3(x)\,dx = x^2\,(-x^{-2}\,J_2) - \int 2x\,(-x^{-2}\,J_2)\,dx$$

$$= -J_2 + 2\int x^{-1}\,J_2\,dx = -J_2 + 2\,(-x^{-1}\,J_1)$$

On using (1), again, when $n = 1$

Hence $\int J_3(x)\,dx + J_2 + \frac{2}{x} J_1 = 0$ **Proved.**

## 8.10 EQUATIONS REDUCIBLE TO BESSEL'S EQUATION

There are some differential equations which can be reduced to Bessel's equation and, therefore, can be solved.

(*a*) We shall reduce the following differential equation to Bessel's equation.

$$x^2 \frac{d^2y}{dx^2} + x\frac{dy}{dx} + (k^2x^2 - n^2)\,y = 0 \quad ...(1)$$

Put $t = kx,\ \frac{dt}{dx} = k,\ \frac{dy}{dx} = \frac{dy}{dt}\frac{dt}{dx} = k\frac{dy}{dt}$

$$\frac{d^2y}{dx^2} = \frac{d}{dx}\left(k\frac{dy}{dt}\right) = \frac{d}{dt}\left(k\frac{dy}{dt}\right)\frac{dt}{dx} = k^2\frac{d^2y}{dt^2}$$

Thus (1) becomes

$$\left(\frac{t^2}{k^2}\right)\left(k^2\frac{d^2y}{dt^2}\right) + \left(\frac{t}{k}\right)\left(k\frac{dy}{dt}\right) + (t^2 - n^2)\,y = 0 \quad \text{or} \quad t^2\frac{d^2y}{dt^2} + t\frac{dy}{dt} + (t^2 - n^2)\,y = 0$$

∴ Its solution is $y = c_1 J_n(t) + c_2 J_{-n}(t)$,

Hence solution of (1) is $y = c_1 J_n(kx) + c_2 J_{-n}(kx)$.

(*b*) Let us reduce the following differential equation to Bessel's equation.

$$x\frac{d^2y}{dx^2} + a\frac{dy}{dx} + k^2xy = 0 \quad ...(2)$$

Put $y = x^n z,\ \frac{dy}{dx} = x^n\frac{dz}{dx} + n\,x^{n-1}\cdot z$

$$\frac{d^2y}{dx^2} = x^n\frac{d^2y}{dx^2} + n\,x^{n-1}\frac{dz}{dx} + n\,x^{n-1}\frac{dz}{dx} + n(n-1)\cdot x^{n-2}\cdot z$$

$$= x^n\frac{d^2y}{dx^2} + 2\,n\,x^{n-1}\frac{dz}{dx} + n(n-1)\,x^{n-2}\cdot z.$$

Then (2) becomes

$$x\left[x^n\frac{d^2z}{dx^2} + 2\,n\,x^{n-1}\frac{dz}{dx} + n(n-1)\,x^{n-2}z\right] + a\left[x^n\frac{dz}{dx} + n\,x^{n-1}z\right] + k^2\cdot x\,.\,x^n z = 0$$

or $$x^{n+1}\frac{d^2z}{dx^2} + (2n+a)\,x^n\frac{dz}{dx} + [k^2x^2 + n^2 + (a-1)\,n]\,x^{n-1}z = 0 \quad ...(3)$$

Dividing (3) by $x^{n-1}$, we get

$$x^2\frac{d^2z}{dx^2} + (2n+a)\,x\frac{dz}{dx} + [k^2x^2 + n^2 + (a-1)\,n]\,z = 0$$

Let us put $2n + a = 1$, then $x^2\frac{d^2z}{dx^2} + x\frac{dz}{dx} + (k^2x^2 - n^2)\,z = 0$

Its solution is $z = c_1 J_n(kx) + c_2 J_{-n}(kx)$

Hence the solution of (2) is $y = x^n[c_1 J_n(kx) + c_2 J_{-n}(kx)]$ $\quad n \notin 1$

where $n = \frac{1-a}{2}$

(*c*) To reduce the following differential equation to Bessel's equation.

$$x\frac{d^2y}{dx^2}+c\frac{dy}{dx}+k^2x^ry=0 \qquad ...(4)$$

Put $\qquad x=t^m,\ t=x^{1/m}$

So that $\qquad \frac{dy}{dx}=\frac{dy}{dt}\frac{dt}{dx}=\frac{dy}{dt}\left(\frac{1}{m}t^{1-m}\right)$

$$\frac{d^2y}{dx^2}=\frac{d}{dx}\left(\frac{1}{m}t^{1-m}\frac{dy}{dt}\right)=\frac{d}{dt}\left(\frac{1}{m}t^{1-m}\frac{dy}{dt}\right)\frac{dt}{dx}$$

$$=\left(\frac{1}{m}t^{1-m}\frac{d^2y}{dt^2}+\frac{1}{m}(1-m)\,t^{-m}\frac{dy}{dt}\right)\frac{1}{m}t^{1-m}=\frac{1}{m^2}t^{2-2m}\frac{d^2y}{dt^2}+\frac{1-m}{m^2}t^{1-2m}\frac{dy}{dt}$$

Now (4) becomes.

$$t^m\left[\frac{1}{m}t^{2-2m}\frac{d^2y}{dt^2}+\frac{1-m}{m^2}t^{1-2m}\frac{dy}{dt}\right]+c\frac{1}{m}t^{1-m}\frac{dy}{dt}+k^2t^{mr}y=0$$

or $$\frac{1}{m^2}t^{2-m}\frac{d^2y}{dt^2}+\frac{1-m+cm}{m^2}t^{1-m}\frac{dy}{dt}+k^2t^{mr}y=0$$

On multiplying by $\frac{m^2}{t^{1-m}}$

$$t\frac{d^2y}{dt^2}+(1-m+cm)\frac{dy}{dt}+(km)^2t^{mr+m-1}y=0 \qquad ...(5)$$

Let us put $\quad a=1-m+cm$ and $m=\frac{1}{r+1}\Rightarrow\ mr+m-1=\frac{r}{r+1}+\frac{1}{r+1}-1=0$

Thus (5) becomes $\quad t\frac{d^2y}{dt^2}+a\frac{dy}{dt}+(km)^2y=0$

Its solution is $\qquad y=t^n[c_1J_n(kmt)+c_2J_{-n}(kmt)]$

Solution of (4) is $\qquad y=x^{n/m}[c_1J_n(kmx^{1/m})+c_2J_{-n}(kmx^{1/m})]$ **Ans.**

## 8.11 ORTHOGONALITY OF BESSEL FUNCTIONS

$$\int_0^1 xJ_n(\alpha x)\cdot J_n(\beta x)\,dx=0$$

where $\alpha$ and $\beta$ are the roots of $\quad J_n(x)=0.$

**Proof.** We know that

$$x^2\frac{d^2y}{dx^2}+x\frac{dy}{dx}+(\alpha^2x^2-n^2)y=0 \qquad ...(1)$$

$$x^2\frac{d^2z}{dx^2}+x\frac{dz}{dx}+(\beta^2x^2-n^2)z=0 \qquad ...(2)$$

Solutions of (1) and (2) are $y=J_n(\alpha x),\quad z=J_n(\beta x)$ respectively.

Multiplying (1) by $\frac{z}{x}$ and (2) by $-\frac{y}{x}$ and add, we get

$$x\left(z\frac{d^2y}{dx^2}-y\frac{d^2z}{dx^2}\right)+\left(z\frac{dy}{dx}-y\frac{dz}{dx}\right)+(\alpha^2-\beta^2)xyz=0.$$

$$\frac{d}{dx}\left[x\left(z\frac{dy}{dx}-y\frac{dz}{dx}\right)\right]+(\alpha^2-\beta^2)xyz=0 \qquad ...(3)$$

Integrating (3) w.r.t. '$x$' between the limits 0 and 1, we get

$$\left[x\left(z\frac{dy}{dx}-y\frac{dz}{dx}\right)\right]_0^1+(\alpha^2-\beta^2)\int_0^1 xyz\,dx=0$$

$$(\beta^2-\alpha^2)\int_0^1 xyz\,dx=\left[x\left(z\frac{dy}{dx}-y\frac{dz}{dx}\right)\right]_0^1$$

$$=\left[z\frac{dy}{dx}-y\frac{dz}{dx}\right]_{x=1} \quad ...(4)$$

Putting the values of $y = J_n(\alpha x)$, $\frac{dy}{dx} = \alpha J_n'(\alpha x)$, $z = J_n(\beta x)$, $\frac{dz}{dx} = \beta, J_n'(\beta x)$ in (4) we get

$$(\beta^2-\alpha^2)\int_0^1 x J_n(\alpha x)\cdot J_n(\beta x)\,dx=\left[\alpha J_n'(\alpha x) J_n(\beta x)-\beta J_n'(\beta x) J_n(\alpha x)\right]_{x=1}$$

$$=\alpha J_n'(\alpha) J_n(\beta)-\beta J_n'(\beta) J_n(\alpha) \quad ...(5)$$

Since $\alpha, \beta$ are the roots of $J_n(x) = 0$, so $J_n(\alpha) = J_n(\beta) = 0$.

Putting the values of $J_n(\alpha) = J_n(\beta) = 0$ in (5), we get

$$(\beta^2-\alpha^2)\int_0^1 x J_n(\alpha x)\cdot J_n(\beta x)\,dx=0$$

or

$$\int_0^1 x J_n(\alpha x)\cdot J_n(\beta x)\,dx=0$$

**Proved.**

**Example 18.** *Prove that*

$$\int_0^1 x\,[J_n(\alpha x)]^2\,dx=\frac{1}{2}[J_{n+1}(\alpha)]^2.$$

**Solution.** (From (5) of articles 8.11) we know that

$$(\beta^2-\alpha^2)\int_0^1 x J_n(\alpha x)\cdot J_n(\beta x)\,dx=\alpha J_n'(\alpha)\cdot J_n(\beta)-BJ_n'(\beta)\cdot J_n(\alpha)$$

when $\beta = \alpha$

We also know that $J_n(\alpha) = 0$. Let $\beta$ be a neighbouring value of $\alpha$, which tends to $\alpha$. Then

$$\lim_{\beta\to\alpha}\int_0^1 x J_n(\alpha x) J_n(\beta x)\,dx=\lim_{\beta\to\alpha}\frac{0+\alpha J_n'(\alpha)\cdot J_n(\beta)}{\beta^2-\alpha^2}$$

As the limit is of the form $\frac{0}{0}$, we apply $L'$ Hopital's rule

$$\int_0^1 x J_n^2(\alpha x)\,dx=\lim_{\beta\to\alpha}\frac{0+\alpha J_n'(\alpha)\cdot J_n'(\beta)}{2\beta}=\frac{1}{2}[J_n'(\alpha)]^2$$

**Proved**

## 8.12 A GENERATING FUNCTION FOR $J_n(x)$

Prove that $J_n(x)$ is the coefficient of $z^n$ in the expansion of $e^{\frac{x}{2}\left(z-\frac{1}{z}\right)}$.

**Proof.** We know that $e^t = 1 + t + \frac{t^2}{2!} + \frac{t^3}{3!} + \ldots$

$$e^{\frac{xz}{2}} = 1 + \left(\frac{xz}{2}\right) + \frac{1}{2!}\left(\frac{xz}{2}\right)^2 + \frac{1}{3!}\left(\frac{xz}{2}\right)^3 + \ldots \qquad \ldots(1)$$

$$e^{-\frac{x}{2z}} = 1 - \left(\frac{x}{2z}\right) + \frac{1}{2!}\left(\frac{x}{2z}\right)^2 - \frac{1}{3!}\left(\frac{x}{2z}\right)^3 + \ldots \qquad \ldots(2)$$

On multiplying (1) and (2), we get

$$e^{\frac{x}{2}\left(z-\frac{1}{z}\right)} = \left[1 + \left(\frac{xz}{2}\right) + \frac{1}{2!}\left(\frac{xz}{2}\right)^2 + \frac{1}{3!}\left(\frac{xz}{2}\right)^3 + \ldots\right] \times$$

$$\left[1 - \frac{x}{2z} + \frac{1}{2!}\left(\frac{x}{2z}\right)^2 - \frac{1}{3!}\left(\frac{x}{2z}\right)^3 + \ldots\right] \qquad \ldots(3)$$

The coefficient of $z^n$ in the product of (3)

$$= \frac{1}{n!}\left(\frac{x}{2}\right)^n - \frac{1}{(n+1)!}\left(\frac{x}{2}\right)^{n+2} + \frac{1}{2!\,(n+2)!}\left(\frac{x}{2}\right)^{n+4} \ldots$$

$$= J_n(x)$$

Similarly coefficient of $z^{-n}$ in the product of equation (3) $= J_{-n}(x)$

$$\therefore \qquad e^{\frac{x}{2}\left(z-\frac{1}{z}\right)} = J_0 + zJ_1 + z^2 J_2 + z^3 J_3 + \ldots + z^{-1} J_{-1} + z^{-2} J_{-2} + z^{-3} J_{-3} + \ldots$$

$$= \sum_{n=-\infty}^{\infty} z^n J_n(x)$$

For this reason $e^{\frac{x}{2}\left(z-\frac{1}{z}\right)}$ is known as the generating function of Bessel functions.

**Cor.** In the expansion of (3), coefficient of $z^0$

$$= 1 - \frac{x^2}{2^2} + \frac{x^4}{2^2 \cdot 4^2} - \ldots \quad \text{or} \quad J_0 = 1 - \frac{x^2}{2^2} + \frac{x^4}{2^2 \cdot 4^2} - \ldots$$

## 8.13 TRIGONOMETRIC EXPANSION INVOLVING BESSEL FUNCTIONS

We know that

$$e^{\frac{x}{2}\left(z-\frac{1}{z}\right)} = J_0 + zJ_1 + z^2 J_2 + z^3 J_3 + \ldots + z^{-1} J_{-1} + z^{-2} J_{-2} + z^{-3} J_{-3} + \ldots \qquad \ldots(1)$$

Putting $z = e^{i\theta}$ in (1), we get

$$e^{\frac{x}{2}(e^{i\theta} - e^{-i\theta})} = J_0 + J_1 e^{i\theta} + J_2 e^{2i\theta} + J_3 e^{3i\theta} + \ldots$$

$$+ J_{-1} e^{-i\theta} + J_{-2} e^{-2i\theta} + J_{-3} e^{-3i\theta} + \ldots$$

$$\left(\text{since } \frac{e^{i\theta} - e^{-i\theta}}{2i} = \sin\theta\right)$$

$$e^{ix\sin\theta} = J_0 + J_1 e^{i\theta} + J_2 e^{2i\theta} + J_3 e^{3i\theta} + \ldots - J_1 e^{-i\theta} + J_2 e^{-2i\theta} - J_3 e^{-3i\theta} - \ldots$$

$$\left(\text{since } J_{-n} = (-1)^n J_n\right)$$

or $$\cos(x\sin\theta) + i\sin(x\sin\theta) = J_0 + J_1(e^{i\theta} - e^{-i\theta}) + J_2(e^{2i\theta} + e^{-2i\theta})$$

$$+ J_3(e^{3i\theta} - e^{-3i\theta}) + \ldots$$

or $$\cos(x\sin\theta) + i\sin(x\sin\theta) = J_0 + J_1(2i\sin\theta) + J_2(2\cos 2\theta) + J_3(2i\sin 3\theta) + \ldots$$

Now equating real and imaginary parts, we get

$$\cos(x\sin\theta) = J_0 + 2J_2\cos 2\theta + 2J_4\cos 4\theta + \ldots \quad \ldots(2)$$

$$\sin(x\sin\theta) = 2J_1\sin\theta + 2J_3\sin 3\theta + 2J_5\sin 5\theta + \ldots \quad \ldots(3)$$

On putting $\theta = \frac{\pi}{2} - \alpha$ in (2) and (3) we get

$$\cos(x\cos\alpha) = J_0 - 2J_2\cos 2\alpha + 2J_4\cos 4\alpha - \ldots$$

$$\sin(x\cos\alpha) = 2J_1\cos\alpha - 2J_3\cos 3\alpha + 2J_5\cos 5\alpha - \ldots$$

**Example 19.** *Prove that*

$$\cos x = J_0 - 2J_2 + 2J_4 - \ldots$$

$$\sin x = 2J_1 - 2J_3 + 2J_5 + \ldots$$

*(A.M.I.E.T.E., Dec. 2004)*

**Solution.** We know that

$$\cos(x\sin\theta) = J_0 + 2J_2\cos 2\theta + 2J_4\cos 4\theta + \ldots \quad \ldots(1)$$

$$\sin(x\sin\theta) = 2J_1\sin\theta + 2J_3\sin 3\theta + 2J_5\sin 5\theta + \ldots \quad \ldots(2)$$

Putting $\theta = \frac{\pi}{2}$ in (1) and (2), we get $\cos x = J_0 - 2J_2 + 2J_4 - \ldots$

and $\sin x = 2J_1 - 2J_3 + 2J_5 - \ldots$ **Proved.**

**Example 20.** *Prove that*

$$x\sin x = 2[2^2J_2 - 4^2J_4 + 6^2J_6 - \ldots]$$

$$x\cos x = 2[1^2J_1 - 3^2J_3 + 5^2J_5 + \ldots]$$

**Solution.** We know that

$$\cos(x\sin\theta) = J_0 + 2J_2\cos 2\theta + 2J_4\cos 4\theta + \ldots \quad \ldots(1)$$

and

$$\sin(x\sin\theta) = 2J_1\sin\theta + 2J_3\sin 3\theta + 2J_5\sin 5\theta + \ldots \quad \ldots(2)$$

Differentiating (1) w.r.t. "θ" we get

$$[-\sin(x\sin\theta)]\,x\cos\theta = 0 - 4J_2\sin 2\theta - 8J_4\sin 4\theta + \ldots \quad \ldots(3)$$

Again differentiating (3) w.r.t., "θ" we get

$$[-\sin(x\sin\theta)](-x\sin\theta) + [-\cos(x\sin\theta)(x\cos\theta)]\,x\cos\theta$$

$$= -8J_2\cos 2\theta - 32J_4\cos 4\theta + \ldots \quad \ldots(4)$$

Now putting $\theta = \frac{\pi}{2}$ in (4) we get

$$x\sin x = 8J_2 - 32J_4 + \ldots = 2[2^2J_2 - 4^2J_4 + 6^2J_6 - \ldots]$$

Similarly differentiating (2) twice and putting $\theta = \frac{\pi}{2}$, we have,

$$x\cos x = 2[1^2J_1 - 3^2J_3 + 5^2J_5 - \ldots]$$

**Proved.**

**Example 21.** *Prove that* $J_0^2 + 2J_1^2 + 2J_2^2 + \ldots = 1$ *(A.M.I.E.T.E., Dec. 2006)*

**Solution.** $J_0 + 2J_2\cos 2\theta + 2J_4\cos 4\theta + \ldots = \cos(x\sin\theta)$ (From At 8.13) ...(1)

$$2J_1\sin\theta + 2J_3\sin 3\theta + 2J_5\sin 5\theta + \ldots = \sin(x\sin\theta) \quad \ldots(2)$$

Now squaring (1) and integrating w.r.t. 'θ' between the limits 0 and π, we get

$$J_0^2\,\pi + 2\,J_2^2\,\pi + 2\,J_4^2\,\pi + \ldots = \int_0^\pi \cos^2 (x \sin \theta)\, d\theta \qquad \ldots(3)$$

$$\int_0^\pi 2 \sin^2 n\theta\, d\theta = \pi$$

$$\int_0^\pi 2 \cos^2 n\theta\, d\theta = \pi$$

$$\int_0^\pi 2 \sin n\theta \sin m\theta\, d\theta = 0$$

$$\int_0^\pi 2 \cos n\theta \cos m\theta\, d\theta = 0$$

Also squaring (2) and integrating w.r.t. "θ" between the limits 0 and π, we get

$$2\,J_1^2\,\pi + 2\,J_3^2\,\pi + 2\,J_5^2\,\pi + \ldots = \int_0^\pi \sin^2 (x \sin \theta)\, d\theta. \qquad \ldots(4)$$

Adding (3) and (4) we get

$$\pi\,[J_0^2 + 2\,J_1^2 + 2\,J_2^2 + 2\,J_3^2 + \ldots] = \int_0^\pi \cos^2 (x \sin \theta)\, d\theta + \int_0^\pi \sin^2 (x \sin \theta)\, d\theta = \int_0^\pi d\theta = \pi$$

or $$J_0^2 + 2\,J_1^2 + 2\,J_2^2 + 2\,J_3^2 + \ldots = 1$$ **Proved.**

## 8.14 BESSEL'S INTEGRAL

To prove that

$$(a)\ J_0(x) = \frac{1}{\pi}\int_0^\pi \cos (x \sin \theta)\, d\theta \quad (b)\ J_n(x) = \frac{1}{\pi}\int_0^\pi \cos (n\theta - x \sin \theta)\, d\theta$$

**Proof.** We know that

$$\cos (x \sin \theta) = J_0 + 2\,J_2 \cos 2\theta + 2\,J_4 \cos 4\theta + \ldots \qquad \ldots(1)$$

$$\sin (x \sin \theta) = 2\,J_1 \sin \theta + 2\,J_3 \sin 3\theta + 2\,J_5 \sin 5\theta + \ldots \qquad \ldots(2)$$

(*a*) Integrating (1) between the limits 0 and π, we have

$$\int_0^\pi \cos (x \sin \theta)\, d\theta = \int_0^\pi (J_0 + 2\,J_2 \cos 2\theta + 2\,J_4 \cos 4\theta + \ldots)\, d\theta$$

$$= J_0 \int_0^\pi d\theta + 2\,J_2 \int_0^\pi \cos 2\theta\, d\theta + 2\,J_4 \int_0^\pi \cos 4\theta\, d\theta + \ldots$$

$$= J_0\,\pi + 0 + 0$$

*i.e.* $$J_0 = \frac{1}{\pi}\int_0^\pi \cos (x \sin \theta)\, d\theta$$ **Proved.**

(*b*) Multiplying (1) by $\cos n\theta$ and integrating between the limits 0 and π, we have

$$\int_0^\pi \cos (x \sin \theta) \cos n\theta\, d\theta = \int_0^\pi [J_0 \cos n\theta + 2\,J_2 \cos 2\theta \cos n\theta + 2\,J_4 \cos 4\theta \cos n\theta + \ldots]\, d\theta$$

$$= 2\,J_0 \int_0^\pi \cos n\theta\, d\theta + 2\,J_2 \int_0^\pi \cos 2\theta \cos n\theta\, d\theta + \ldots$$

$$= 0 \quad \text{if } n \text{ is odd} \qquad \ldots(3)$$

$$= \pi\,J_n \quad \text{if } n \text{ is even} \qquad \ldots(4)$$

Again multiplying (2) by $\sin n\theta$ and integrating between the limits 0 and π, we have

$$\int_0^\pi \sin(x\sin\theta)\sin n\theta\, d\theta = \int_0^\pi (2J_1 \sin\theta \sin n\theta + 2J_3 \sin 3\theta \sin n\theta + \ldots)\, d\theta$$

$$= 2J_1 \int_0^\pi \sin\theta\sin n\theta\, d\theta + 2J_3 \int_0^\pi \sin 3\theta \sin n\theta\, d\theta + \ldots = 0 \quad \text{if } n \text{ is even} \quad \ldots(5)$$

$$= \pi J_n \quad \text{if } n \text{ is odd} \quad \ldots(6)$$

Adding (3) and (6) or (4) and (5), we get

$$\int_0^\pi [\cos(x\sin\theta)\cos n\theta + \sin(x\sin\theta)\sin n\theta]\, d\theta = \pi J_n$$

or $\displaystyle\int_0^\pi \cos(n\theta - x\sin\theta)\, d\theta = \pi J_n.$ or $\displaystyle J_n = \frac{1}{\pi}\int_0^\pi \cos(n\theta - x\sin\theta)\, d\theta$ **Proved.**

## Exercise 8.3

Prove that

1. $\sqrt{\left(\frac{\pi x}{2}\right)}\, J_{3/2}(x) = \frac{\sin x}{x} - \cos x$ *(A.M.I.E.T.E., June 2006)*

2. $\sqrt{\left(\frac{\pi x}{2}\right)}\, J_{-3/2}(x) = -\sin x - \frac{\cos x}{x}.$

3. $J_4(x) = \left(\frac{48}{x^3} - \frac{8}{x}\right) J_1(x) + \left(1 - \frac{24}{x^2}\right) J_0'(x).$

4. (*a*) $J_1(x) = \frac{x}{2} - \frac{x^3}{2^3 \cdot 1!\,2!} + \frac{x^5}{2^5 \cdot 2!\,3!} - \frac{x^7}{2^7 \cdot 3!\,4!} + \ldots$

   (*b*) $J_3(x) + 3J_0'(x) + 4J_0'''(x) = 0$ *(A.M.I.E.T.E., Summer 2000)*

5. (*a*) $J_0(2) = 0.224,$ (*b*) $J_1 = 0.44.$

6. (*a*) $J_0'(x) = -J_1(x)$ (*b*) $J_0''(x) = \frac{1}{2}[J_2(x) - J_0(x)].$ (*c*) $J_{5/2}(x) = \sqrt{\frac{2}{\pi x}}\left[\frac{3 - x^2}{x^2}\sin x - \frac{3\cos x}{x}\right]$

7. $\frac{d}{dx}[x J_1(x)] = x J_0(x).$

8. $\frac{d}{dx}(J_n^2(x) + J_{n+1}^2(x)) = 2\left(\frac{n}{x} J_n^2(x) - \frac{n+1}{x} J_{n+1}^2(x)\right)$

9. (*a*) $\int_0^\pi x^{n+1} J_n(x)\, dx = x^{n+1} J_{n+1}(x), \quad n > -1$

   (*b*) $\int_0^x x^{-n} j_{n+1}(x)\, dx = \frac{1}{2^n \sqrt{n+1}} - x^{-n} j_n(x), n > 1.$ *(A.M.I.E.T.E., Winter 2002)*

   (*c*) $x^2 J_n''(x) = (n^2 - n - x^2) J_n(x) + x J_{n+1}(x)$ *(A.M.I.E.T.E., Summer 2001)*

10. $\int x^2 J_0 J_1\, dx = \frac{1}{2} x^2 J_0' + c$

11. $J_{3/2}(x)\sin x - J_{-3/2}(x)\cos(x) = \frac{\sqrt{2}\,\pi}{x^3}$

12. $J_1'' = \left(\frac{2}{x^2} - 1\right) J_1(x) - \frac{1}{x} J_0(x)$

13. $4J_0'''(x) + 3J_0(x) + J_3(x) = 0$

14. If $J_0(2) = a, J_1(2) = b$ find $J_2(2), J_1'(2), J_2'(2)$ in terms of a and b where $J_n(x)$ is the Bessel function of first kind. *(A.M.I.E.T.E., Summer 1996)* **Ans.** $J_2(2) = b - a,\ J_1'(2) = a - \frac{b}{2},\ J_2'(2) = a$

15. (*a*) Integrate $\int x^3 J_0(x)\, dx$, where $J_n(x)$ is the Bessel function of first kind, in terms *of* $J_0(x)$, $J_1(x)$ and $J_2(x)$.

    (*b*) Express $J_4(x)$ in terms of $J_0(x)$ and $J_1(x)$. *(A.M.I.E.T.E., Summer 2003, 2002)*

16. Prove that (*a*) $\int J_3(x)\, dx = c - J_2(x) - \frac{2}{x} J_1(x)$

    (*b*) $\int x J_0^2(x)\, dx = \frac{1}{2} x^2 [J_0^2(x) + J_1^2(x)]$ *(A.M.I.E.T.E., Winter 2003)*

17. Show that $\int_0^\infty e^{-bx} J_0(ax)\, dx = \frac{1}{\sqrt{a^2 + b^2}}$ and hence deduce that $\int_0^\infty J_0(ax)\, dx = \frac{1}{a}.$

    *(A.M.I.E.T.E., June 2003)*

**18.** Evalute $\int x^{-1} J_4(x)\,dx$ (*A.M.I.E.T.E. Dec. 2005*) **Ans.** ( )

**19.** Choose the correct or the best of the answers/statements given in the following parts:

(*a*) If $J_0$ and $J_1$ are Bessel functions then $J_1'(x)$ is given by

(*i*) $J_0(x) - \frac{1}{x} J_1(x)$; (*ii*) $-J_0$; (*iii*) $J_0(x) + \frac{1}{x} J_1(x)$; (iv) $J_0(x) - \frac{1}{x^2} J_1(x)$.

(*b*) $J_{\frac{1}{2}}(x)$ is given by

(*i*) $\sqrt{(2\pi/x)}\sin x$. (*ii*) $\sqrt{(2\pi/x)}\cos x$. (*iii*) $\sqrt{(\pi/2x)}\cos x$. (*iv*) $\sqrt{(2/\pi x)}\sin x$.

(*c*) The value of integral $\int_0^{\pi} [J_{-2}(x) - J_2(x)]\,dx$. where $J_n(x)$ is the Bessel function of the first kind of order *n*, is equal to

(*i*) 0 (*ii*) – 2 (*iii*) 2 (*iv*) 1

(*d*) $\int \frac{1}{x} J_2(x)\,dx$ is

(*i*) $x J_1(x) + c$ (*ii*) $\frac{1}{x} J_1(x) + c$ (*iii*) $-x J_1(x) + c$ (*iv*) $-\frac{1}{x} J_1(x) + c$ (*A.M.I.E.T.E., June 2006*)

(*e*) The integral $\int x J_0(x)\,dx$ is equal to

(*i*) $x J_1(x) - J_0(x)$; (*ii*) $x J_1(x)$; (*iii*) $J_1(x)$; (*iv*) $x^2 J_n(x)$.

(*f*) If $J_n(x)$ is the Bessel function of the first kind, then $\int x^{-2} J_3(x)\,dx$, is

(*i*) $x^{-2} J_2(x) + c$; (*ii*) $x^2 J_2(x) + c$; (*iii*) $-x^{-2} J_3(x) + c$; (*iv*) $-x^{-1} J_3(x) + c_2$

(*g*) If $J_{n+1}(x) = \frac{2}{x} J_n(x) - J_0(x)$ where $J_n$ is the Besset function of first kind order *n*. Then *n* is

(*i*) 0, (*ii*) 2, (*iii*) –1, (*iv*) None of these (*A.M.I.E.T.E., Winter 2001*)

(*h*) The series $x - \frac{x^3}{2^2(1!)^2} + \frac{x^5}{2^4(2!)^2} - \frac{x^7}{2^6(3!)^2}$ ............∞ equals

(*i*) $J_{1/2}(x)$ (*ii*) $j_0(x)$ (*iii*) $x J_0(x)$ (iv) $x J_{1/2}(x)$ (*A.M.I.E.T.E., Winter 2003*)

(*i*) The value of Bessel function $J_2(x)$ in term of $J_1(x)$ and $J_0(x)$ is

(*i*) $2 J_1(x) - x J_0(x)$ (*ii*) $\frac{4}{x} J_1(x) - J_0(x)$

(*iii*) $\frac{2}{x} J_1(x) - \frac{2}{x} J_0(x)$ (*iv*) $\frac{2}{x} J_1(x) - J_0(x)$ (*A.M.I.E.T.E., Dec. 2005*)

**Ans.** (*a*) (i), (*b*) (*iv*), (*c*) (*i*), (*d*) (*iv*), (*e*) (*ii*), (*f*) (*i*), (*g*) (*iv*), (*h*) (*iii*), (*i*) (*iii*).

## 8.15. FOURIER–BESSEL EXPANSION

If a function $f(x)$ is continuous and has a finite number of oscillations in the interval $0 \le x \le a$, then $f(x)$ can be expanded in a series.

$$f(x) = C_1 J_n(\alpha_1 x) + C_2 J_n(\alpha_2 x) + C_3 J_n(\alpha_3 x) + \ldots + C_n J_n(\alpha_n x) + \ldots \text{or } f(x) = \sum_{i=1}^{\infty} C_i J_n(\alpha_i x).$$

where $\alpha_1, \alpha_2, \alpha_3$ are the roots of the equation $J_n(x) = 0$.

**Solution.** [The orthogonal property of Bessel functions enables us to expand a function in terms of Bessel function].

Let $$f(x) = \sum_{i=1}^{\infty} C_i J_n(\alpha_i x) \qquad \ldots (1)$$

Multiplying both sides of (1) by $x J_n(\alpha_j x)$, we get

$$x f(x) J_n(\alpha_j x) = \sum_{i=1}^{\infty} C_i x J_n(\alpha_j x) \cdot J_n(\alpha_i x) \quad \text{... (2)}$$

Integrating both sides of (2) from $x = 0$ to $x = a$, we have

$$\int_0^a x f(x) \cdot J_n(\alpha_j x)\, dx = \sum_{i=1}^{\infty} C_i \int_0^a x J_n(\alpha_j x) \cdot J_n(\alpha_i x)\, dx \quad \text{... (3)}$$

By orthogonal property of Bessel's function, we know that

$$\int_0^a x J_n(\alpha_i x) \cdot J_n(\alpha_j x)\, dx = \begin{cases} 0 & \text{if } i \neq j \\ \dfrac{a^2}{2} J^2_{(n+1)}(\alpha_i a) & \text{if } i = j \end{cases}$$

On applying this property on the right-hand side of (3), reduces to

$$\int_0^a x f(x) J_n(\alpha_i x)\, dx = C_i \cdot \frac{a^2}{2} J^2_{n+1}(\alpha_i a)$$

or
$$C_i = \frac{2 \int_0^a x f(x) J_n(\alpha_i x)\, dx}{a^2 \cdot J^2_{n+1}(\alpha_i a)}$$

By putting the values of the coefficients $C$'s in (1), we get the Fourier-Bessel Expansions. **Ans.**

**Example 22.** *Show that* $\sum_{i=1}^{\infty} \frac{2 J_0(a_i x)}{a_i J_1(a_1 i)} = 1$, *where* $a_1, a_2, a_3$ ..... *are the roots of* $J_0(x)$.

**Solution.** Let
$$f(x) = \sum_{i=1}^{\infty} C_i J_n(a_i x), \quad \text{... (1)}$$

then
$$C_n = \frac{2}{J^2_{n+1}(a_i)} \int_0^1 x J_n(a_i x) f(x)\, dx \quad \text{... (2)}$$

Putting $f(x) = 1$ and $n = 0$ in (1), we get

$$1 = \sum_{i=1}^{\infty} C_i J_0(a_i x)$$

$$C_i = \frac{2}{J_1^2(a_i)} \int_0^1 x J_0(a_i x)\, dx = \frac{2}{J_1^2(a_i)} \left[ \frac{J_1(a_i)}{a_i} \right] = \frac{2}{a_i J_1(a_i)}$$

substituting the value of $C_i f(x)$ and $n$ in (2), we obtain

$$1 = \sum_{i=1}^{\infty} \frac{2}{a_i J_1(a_i)} J_0(a_i x)$$

or
$$\sum \frac{2 J_0(a_i x)}{a_i J_1(a_i)} = 1$$
**Proved.**

**Example 23.** *Expand* $f(x) = x^2$ *in the interval* $0 < x < 2$ *in terms of* $J_2(\alpha_n x)$ *where* $\alpha_n$ *are the roots of* $J_2(2\alpha_n) = 0$.

**Solution.**
$$f(x) = x^2$$

$$x^2 = \sum_{i=1}^{\infty} C_i J_2(\alpha_i x) \quad \text{... (1)}$$

Multiplying both sides of (1) by $n J_2(\alpha_j x)$, we get

$$x^3 J_2(\alpha_j \cdot x) = \sum_{i=1}^{\infty} C_i\, x\, J_2(\alpha_i x) \cdot J_2(\alpha_j \cdot x) \qquad \text{... (2)}$$

Integrating (2) w.r.t. $x$ from $x = 0$ to $x = 2$, we get

$$\int_0^2 x^3 \cdot J_2(\alpha j\, x)\, dx = \sum_{i=1}^{\infty} C_i \int_0^2 x\, J_2(\alpha_i x) \cdot J_2(\alpha_j \cdot x)\, dx$$

$$\left[\frac{x^3 \cdot J_3(\alpha_i x)}{\alpha_i}\right]_0^2 = C_i \int_0^2 x\, J_2^2(\alpha_i x)\, dx \qquad (i = j) \text{ (other integrals are zero)}$$

$$\frac{8\, J_3(2\,\alpha_i)}{\alpha_i} = C_i \frac{2^2}{2} J_3^2(2\,\alpha_i)$$

$$C_i = \frac{8\, J_3(2\,\alpha_i)}{\alpha_i} \cdot \frac{2}{4\, J_3^2(2\,\alpha_i)} = \frac{4}{\alpha_i\, J_3(2\,\alpha_i)}$$

On putting the values of coefficients $C_i$ in (1), we get

Hence
$$x^2 = \sum_{i=1}^{\infty} \frac{4\, J_2(\alpha_i x)}{\alpha_i\, J_3(2\,\alpha_i)}$$
**Ans.**

### Exercise 8.4

1. Expand $f(x) = x^3$ in the interval $0 < x < 3$ in terms of Bessel functions $J_1(\alpha_n x)$ where $\alpha_n$ are the roots of $J_1(3\,\alpha) = 0$.

   **Ans.** $x^3 = \sum_{i=1}^{\infty} \frac{6}{\alpha_i^2} \frac{1}{J_2^2(3\,\alpha_i)} \{3\,\alpha_i J_2(3\,\alpha_i) - 2\, J_3(3\,\alpha_i)\}$

2. Show that the Fourier-Bessel series in $J_2(\alpha_i x)$ for $f(x) = x^2$, $(0 < x < a)$ where $\alpha_i$ are positive roots of $J_2(x) = 0$ is

$$x^2 = 2\,a^2 \sum_{i=1}^{\infty} \frac{J_2(\alpha_i x)}{a\,\alpha_i\, J_3(\alpha_i a)}$$

## 8.16. BER AND BEI FUNCTIONS

The following differential equation is useful in certain problems in electrical engineering.

$$x \frac{d^2y}{dx^2} + \frac{dy}{dx} - ixy = 0 \qquad \text{... (1)}$$

This equation (1) is a particular case of the differential equation (1) of article 8.8 on page 611.

On putting $n = 0$, $k^2 = -x$ in equation (1) of article 8.8.

$$x \frac{d^2y}{dx^2} + \frac{dy}{dx} - ixy = 0$$

Its solutions $y = J_0(k\,x) = J_0[(-1)^{1/2} x] = J_0(i^{3/2} x)$

$$y = J_0(i^{3/2} x) = 1 - \frac{i^3 x^2}{2^2} + \frac{i^6 x^4}{(2!)^2\, 2^4} - \frac{i^9 x^6}{(3!)^2\, 2^6} + \frac{i^{12} x^8}{(4!)^2\, 2^8} \ldots\ldots$$

$$= \left[1 - \frac{x^4}{2^2 \cdot 4^2} + \frac{x^8}{2^2 \cdot 4^2 \cdot 6^2 \cdot 8^2} \ldots\ldots\right] + i\left[\frac{x^2}{2^2} - \frac{x^6}{2^2 \cdot 4^2 \cdot 6^2} + \frac{x^{10}}{2^2 \cdot 4^2 \cdot 6^2 \cdot 8^2 \cdot 10^2} \ldots\ldots\right]$$

$$= 1 + \sum_{k=1}^{\infty} (-1)^k \frac{x^{4k}}{2^2 \cdot 4^2 \cdot 6^2 \ldots\ldots (4k)^2} + i\left[-\sum_{k=1}^{\infty} \frac{(-1)^k\, x^{4k-2}}{2^2 \cdot 4^2 \cdot 6^2 \cdot 8^2 \ldots\ldots (4k-2)^2}\right]$$

$$J_0(i^{3/2}x) = \text{Ber (Bessel real)} + \text{Bei (Bessel imaginary)}$$

$$\text{Ber}\, x = 1 + \sum_{k=1}^{\infty} (-1)^k \frac{x^{4k}}{2^2 \cdot 4^2 \cdot 6^2 \ldots\ldots (4k)^2}$$

$$\text{Bei}\, x = -\sum_{k=1}^{\infty} \frac{(-1)^k x^{4k-2}}{2^2 \cdot 4^2 \cdot 6^2 \ldots\ldots (4k-2)^2}$$

**Example 24.** *Show that* $\frac{d}{dx}(x \text{ Ber}' x) = -x \text{ Bei}\, x$.

**Solution.** We know that

$$\text{Ber}\, x = 1 - \frac{x^4}{2^2 \cdot 4^2} + \frac{x^8}{2^2 \cdot 4^2 \cdot 6^2 \cdot 8^2} - \ldots\ldots \infty$$

On differentiating, $\text{Ber}' x = -\frac{4x^3}{2^2 \cdot 4^2} + \frac{8x^7}{2^2 \cdot 4^2 \cdot 6^2 \cdot 8^2} - \ldots \infty$

$$(x \text{ Ber}' x) = -\frac{x^4}{2^2 \cdot 4} + \frac{x^8}{2^2 \cdot 4^2 \cdot 6^2 \cdot 8^2} - \ldots \infty$$

$$\frac{d}{dx}(x \text{ Ber}' x) = -\frac{x^3}{2^2} + \frac{x^7}{2^2 \cdot 4^2 \cdot 6^2} - \ldots\ldots \infty$$

$$= -x\left[\frac{x^2}{2^2} - \frac{x^6}{2^2 \cdot 4^2 \cdot 6^2} + \ldots\ldots \infty\right] = -x \text{ Bei}\, x$$

**Example 25.** *Show that* $\frac{d}{dx}(x \text{ Bei}' x) = x \text{ Ber}\, x$.

**Solution.** We know that $\text{Bei}\, x = \frac{x^2}{2^2} - \frac{x^6}{2^2 \cdot 4^2 \cdot 6^2} + \frac{x^{10}}{2^2 \cdot 4^2 \cdot 6^2 \cdot 8^2 \cdot 10^2} - \ldots \infty$ ... (1)

On differentiating (1) w.r.t. '$x$', we get

$$\text{Bei}' x = \frac{x}{2} - \frac{x^5}{2^2 \cdot 4^2 \cdot 6} + \frac{x^9}{2^2 \cdot 4^2 \cdot 6^2 \cdot 8^2 \cdot 10} - \ldots\ldots \infty$$

$$(x \text{ Bei}' x) = \frac{x^2}{2} - \frac{x^6}{2^2 \cdot 4^2 \cdot 6} + \frac{x^{10}}{2^2 \cdot 4^2 \cdot 6^2 \cdot 8^2 \cdot 10} - \ldots\ldots \infty$$

$$\frac{d}{dx}(x \text{ Bei}\, x) = x - \frac{x^5}{2^2 \cdot 4^2} + \frac{x^9}{2^2 \cdot 4^2 \cdot 6^2 \cdot 8^2} - \ldots\ldots \infty$$

$$= x\left[1 - \frac{x^4}{2^2 \cdot 4^2} + \frac{x^8}{2^2 \cdot 4^2 \cdot 6^2 \cdot 8^2} - \ldots\ldots \infty\right] = x \text{ Ber}\, x$$

**Example 26.** *Show that*

$$\int_0^a x\,[Ber^2 x + Bei^2 x]\, dx = a\,[Ber\, a\, Bei'\, a - Bei\, a\, Ber'\, a]$$

**Solution.** $\frac{d}{dx}(x \text{ Bei}' x) = x \text{ Ber}\, x$ ...(1) (See example 25)

and $\frac{d}{dx}(x \text{ Ber}' x) = -x \text{ Bei}\, x$ ... (2) (See example 24)

Multiplying (1) by Ber $x$ and (2) by Bei $x$ and subtracting, we get

$$\text{Ber}\, x \frac{d}{dx}(x \text{ Bei}' x) - \text{Bei}\, x \frac{d}{dx}(x \text{ Ber}' x) = x \text{ Ber}^2 x + x \text{ Bei}^2 x \quad \ldots (3)$$

Integrating both sides of (3) from 0 to $a$, we get

$$\int_0^a \left[\operatorname{Ber} x \frac{d}{dx}(x \operatorname{Bei}' x) - \operatorname{Bei} x \frac{d}{dx}(x \operatorname{Ber}' x)\right] dx = \int_0^a (x \operatorname{Ber}^2 x + x \operatorname{Bei}^2 x)\, dx$$

or

$$\int x(\operatorname{Ber}^2 x + \operatorname{Bei}^2 x)\, dx = \int_0^a \left[\operatorname{Ber} x \frac{d}{dx}(x \operatorname{Bei}' x) - \operatorname{Bei} x \frac{d}{dx}(x \operatorname{Ber}' x)\right] dx$$

On adding and subtracting $\operatorname{Bei}' x \frac{d}{dx}(x \operatorname{Ber} x)$ on R.H.S.

$$= \int_0^a \left[\operatorname{Ber} x \frac{d}{dx}(x \operatorname{Ber}' x) + x \operatorname{Bei}' x \frac{d}{dx}(\operatorname{Ber} x) - \operatorname{Bei} x \frac{d}{dx}(x \operatorname{Ber}' x) - x \operatorname{Bei}' x \frac{d}{dx}(\operatorname{Ber} x)\right] dx$$

$$= \int_0^a \frac{d}{dx}[x(\operatorname{Ber} x \operatorname{Bei}' x - \operatorname{Bei} x \operatorname{Ber}' x)]\, dx = [x(\operatorname{Ber} x \operatorname{Bei}' x - \operatorname{Bei} x \operatorname{Ber}' x)]_0^a$$

$= a\,[\operatorname{Ber} a \operatorname{Bei}' a - \operatorname{Bei} a \operatorname{Ber}' a]$ **Proved.**

## 8.17 LEGENDRE'S EQUATION

The differential equation $$(1-x^2)\frac{d^2y}{dx^2} - 2x\frac{dy}{dx} + n(n+1)y = 0 \qquad \ldots(1)$$

is known as Legendre's equation. The above equation can also be written as

$$\frac{d}{dx}\left\{(1-x^2)\frac{dy}{dx}\right\} + n(n+1)y = 0 \qquad n \in I$$

This equation can be integrated in series of ascending or descending powers of $x$. *i.e.*, series in ascending or descending powers of $x$ can be found which satisfy the equation (1).

Let the series in descending powers of $x$ be

$$y = x^m(a_0 + a_1 x^{-1} + a_2 x^{-2} + \ldots) \qquad \ldots(2)$$

or

$$y = \sum_{r=0}^{\infty} a_r x^{m-r}$$

so that

$$\frac{dy}{dx} = \sum_{r=0}^{\infty} a_r (m-r) x^{m-r-1}$$

and

$$\frac{d^2y}{dx^2} = \sum_{r=0}^{\infty} a_r (m-r)(m-r-1) x^{m-r-2}$$

Substituting these in (1), we have

$$(1-x^2)\sum_{r=0}^{\infty} a_r (m-r)(m-r-1)x^{m-r-2} - 2x\sum_{r=0}^{\infty} a_r(m-r)x^{m-r-1} + n(n+1)\sum_{0}^{\infty} a_r x^{m-r} = 0$$

or

$$\sum_{r=0}^{\infty} a_r (m-r)(m-r-1)x^{m-r-2} + \{n(n+1) - 2(m-r) - (m-r)(m-r-1)\}\, x^{m-r} a_r = 0$$

or

$$\sum_{r=0}^{\infty} [(m-r)(m-r-1)x^{m-r-2} + \{n(n+1) - (m-r)(m-r+1)\}\, x^{m-r}]\, a_r = 0 \qquad \ldots(3)$$

The equation (3) is an identity and therefore coefficients of various powers of $x$ must vanish. Now equating to zero the coefficients of $x^m$ from the above we have $(r = 0)$

$$a_0\{n(n+1) - m(m+1)\} = 0$$

But $a_0 \neq 0$, as it is the coefficient of the very first term in the series.

Hence $\quad n(n+1) - m(m+1) = 0$ ....(4)

*i.e.,* $\quad n^2 + n - m^2 - m = 0 \quad$ or $\quad (n^2 - m^2) + (n - m) = 0$

or $\quad (n - m)(n + m + 1) = 0$

which gives $\quad m = n \quad$ or $\quad m = -n - 1$ ...(5)

This is important as it determines the index.

Next, equating to zero the coefficient of $x^{m-1}$ by putting $r = 1$,

$$a_1 [n(n+1) - (m-1)m] = 0$$

or $\quad a_1 [(m+n)(m-n-1) = 0$

which gives $\quad a_1 = 0$ ...(6)

Since $\quad (m+n)(m-n-1) \neq 0. \quad$ by (5)

Again to find a relation in successive coefficients $a_r$, etc., equating the coefficient of $x^{m-r-2}$ to zero, we get

$$(m-r)(m-r-1)a_r + [n(n+1) - (m-r-2)(m-r-1)]a_{r+2} = 0$$

Now $n(n+1) - (m-r-2)(m-r-1) = n^2 + n - (m-r-1-1)(m-r-1)$

$$= -[(m-r-1)^2 - (m-r-1) - n^2 - n]$$

$$= -[(m-r-1+n)(m-r-1-n) - (m-r-1+n)]$$

$$= -[(m-r-1+n)(m-r-1-n-1)]$$

$$= (m-r+n-1)(m-r+n-2)$$

or $\quad (m-r)(m-r-1)a_r - (m-r+n-1)(m-r-n-2)a_{r+2} = 0$

or
$$a_{r+2} = \frac{(m-r)(m-r-1)}{(m-r+n-1)(m-r-n-2)} a_r \qquad \text{...(7)}$$

Now since $\quad a_1 = a_3 = a_5 = a_7 = \ldots = 0$

For the two values given by (5) there arises following two cases.

**Case I: When $m = n$**

$$a_{r+2} = -\frac{(n-r)(n-r-1)}{(2n-r-1)(r+2)} a_r \qquad \text{from (7)}$$

so that,
$$a_2 = -\frac{n(n-1)}{(2n-1)2} a_0,$$

$$a_4 = -\frac{(n-2)(n-3)}{(2n-3)\times 4} a_2 = \frac{n(n-1)(n-2)(n-3)}{(2n-1)(2n-3)\,2.4} a_0$$

and so on and $\quad a_1 = a_3 = a_5 = \ldots = 0$

Hence the series (2) becomes

$$y = a_0 \left[ x^n - \frac{n(n-1)}{(2n-1)\cdot 2} x^{n-2} + \frac{n(n-1)(n-2)(n-3)}{(2n-1)(2n-3)\,2.4} \cdot x^{n-4} - \ldots \right] \qquad \text{...(8)}$$

which is a solution of (1)

**Case II: When $m = -(n+1)$, we have**

$$a_{r+2} = \frac{(n+r+1)(n+r+2)}{(r+2)(2n+r+3)} a_r \qquad \text{from (7)}$$

so that $$a_2 = \frac{(n+1)(n+2)}{2(2n+3)} a_0;$$

$$a_4 = \frac{(n+1)(n+2)(n+3)(n+4)}{2.4(2n+3)(2n+5)} a_0 \qquad \text{and so on.}$$

Hence the series (2) in this case becomes

$$y = a_0 \left[ x^{-n-1} + \frac{(n+1)(n+2)}{2 \cdot (2n+3)} x^{-n-3} + \frac{(n+1)(n+2)(n+3)(n+4)}{2.4(2n+3)(2n+5)} \cdot x^{-n-5} + \dots \right] \quad ...(9)$$

This gives another solution of (1) in a series of descending powers of $x$.

**Note.** If we want to integrate the Legendre's equation in a series of ascending powers of $x$, we may proceed by taking

$$y = a_0 x^k + a_1 x^{k+1} + a_2 x^{k+2} + \dots = \sum_{r=0}^{\infty} a_r k^{k+r}$$

But integration in descending powers of $x$ is more important than that in ascending powers of $x$.

## 8.18 LEGENDRE'S POLYNOMIAL $P_n(x)$.

**Definition:**

The Legendre's Equation is

$$(1-x^2)\frac{d^2y}{dx^2} - 2x\frac{dy}{dx} + n(n+1)\, y = 0 \quad ...(1)$$

The solution of the above equation in the series of descending powers of $x$ is

$$y = a_0 \left[ x^n - \frac{n(n-1)}{(2n-1)2} x^{n-2} + \frac{n(n-1)(n-2)(n-3)}{(2n-1)(2n-3)2.4} x^{n-4} \dots \right]$$

where $a_0$ is an arbitrary constant.

Now if $n$ is a positive integer and $a_0 = \frac{1.3.5 \dots (2n-1)}{n!}$ the above solution is $P_n(x)$, so that

$$P_n(x) = \frac{1.3.5 \dots (2n-1)}{n!} \left[ x^n - \frac{n(n-1)}{(2n-1) \cdot 2} x^{n-2} + \dots \right]$$

**Note 1.** This is a terminating series.

When $n$ is even, it contains $\frac{1}{2}n + 1$ terms, the last term being

$$(-1)^{\frac{1}{2}n} \frac{n(n-1)(n-2) \dots 1}{(2n-1)(2n-3) \dots (n+1)\, 2.4.6 \dots n}$$

And when $n$ is odd it contains $\frac{1}{2}(n+1)$ terms and the last term in this case is

$$(-1)^{\frac{1}{2}(n-1)} \frac{n(n-1) \dots 3.2}{(2n-1)(2n-3) \dots (n+2)\, 2.4 \dots (n-1)} x$$

$P_n(x)$ is called the *Legendre's functions of the first kind.*

**Note.** $P_n(x)$ is that solution of Legendre's equation (1) which is equal to unity when $x = 1$.

## 8.19 LEGENDRE'S FUNCTION OF THE SECOND KIND *i.e.* $Q_n(x)$.

Another solution of Legendre's equation

$$(1-x^2)\frac{d^2y}{dx^2}-2x\frac{dy}{dx}+n(n+1)y=0$$

when $n$ is a positive integer

$$y=a_0\left[x^{-n-1}+\frac{(n+1)(n+2)}{2\cdot(2n+3)}x^{-n-3}+\dots\right]$$

If we take $\quad a_0=\dfrac{n!}{1.3.5\dots(2n+1)}$

the above solution is called $Q_n(x)$, so that

$$Q_n(x)=\frac{n!}{1.3.5.\dots(2n+1)}\left[x^{-n-1}+\frac{(n+1)(n+2)}{2\cdot(2n+3)}x^{-n-3}+\dots\right]$$

The series for $Q_n(x)$ is a non-terminating series.

## 8.20 GENERAL SOLUTION OF LEGENDRE'S EQUATION

Since $P_n(x)$ and $Q_n(x)$ are two independent solutions of Legendre's equation, therefore the most general solution of Legendre's equation is

$$y=A\,P_n(x)+B\,Q_n(x)$$ *(A.M.I.ET.E., Winter 2001*

where $A$ and $B$ are two arbitrary constants.

## 8.21 RODRIGUE'S FORMULA

$$P_n(x)=\frac{1}{2^n\cdot n!}\frac{d^n}{dx^n}(x^2-1)^n$$ *(A.M.I.ET.E., Winter 2001)*

**Proof.** Let $\quad v=(x^2-1)^n \quad$ ...(1)

Then $\quad \dfrac{dv}{dx}=n(x^2-1)^{n-1}(2x)$

Multiplying both sides by $(x^2-1)$, we get

$$(x^2-1)\frac{dv}{dx}=2n(x^2-1)^n x.$$

or $$(x^2-1)\frac{dv}{dx}=2nvx \quad \text{...(2)}$$

Now differentiating (2), $(n+1)$ times by Leibnitz's theorem, we have

$$(x^2-1)\frac{d^{n+2}v}{dx^{n+2}}+{}^{(n+1)}C_1(2x)\frac{d^{n+1}v}{dx^{n+1}}+{}^{(n+1)}C_2(2)\frac{d^nv}{dx^n}=2n\left[x\frac{d^{n+1}v}{dx^{n+1}}+{}^{(n+1)}C_1(1)\frac{d^nv}{dx^n}\right]$$

or $$(x^2-1)\frac{d^{n+2}v}{dx^{n+2}}+2x[{}^{n+1}C_1-n]\frac{d^{n+1}v}{dx^{n+1}}+2[{}^{n+1}C_2-n.{}^{(n+1)}C_1]\frac{d^nv}{dx^n}=0$$

or $$(x^2-1)\frac{d^{n+2}v}{dx^{n+2}}+2x\frac{d^{n+1}v}{dx^{n+1}}-n(n+1)\frac{d^nv}{dx^n}=0 \quad \text{...(3)}$$

If we put $\dfrac{d^nv}{dx^n}=y$, (3) becomes

$$(x^2-1)\frac{d^2y}{dx^2}+2x\frac{dy}{dx}-n(n+1)y=0$$

or $$(1-x^2)\frac{d^2y}{dx^2}-2x\frac{dy}{dx}+n(n+1)y=0$$

This shows that $y = \frac{d^n v}{dx^n}$ is a solution of Legendre's equation.

$$\therefore \qquad C\frac{d^n v}{dx^n} = P_n(x) \qquad \ldots(4)$$

where $C$ is a constant.

But $$v = (x^2-1)^n = (x+1)^n (x-1)^n$$

so that $$\frac{d^n v}{dx^n} = (x+1)^n \frac{d^n}{dx^n}(x-1)^n + {}^nC_1 \cdot n(x+1)^{n-1} \cdot \frac{d^{n-1}}{dx^{n-1}}(x-1)^n + \ldots + (x-1)^n \frac{d^n}{dx^n}(x+1)^n = 0$$

when $x = 1$, $$\frac{d^n v}{dx^n} = 2^n \cdot n!$$

All the other terms disappear as $(x-1)$ is a factor in every term except first.

Therefore when $x = 1$, (4) gives

$$C \cdot 2^n \cdot n! = P_n(1) = 1 \qquad P_n(1) = 1$$

$$C = \frac{1}{2^n \cdot n!}. \qquad \ldots (5)$$

Substituting the value of $C$ from (1) in (5) we have

$$P_n(x) = \frac{1}{2^n \cdot n!}\frac{d^n v}{dx^n}$$

$$P_n(x) = \frac{1}{2^n\, n!}\frac{d^n}{dx^n}(x^2-1)^n$$

**Example 27.** *Let $P_n(x)$ be the Legendre polynomial of degree n. Show that for any function, $f(x)$, for which the nth derivative is continuous,*

$$\int_{-1}^{1} f(x)\, P_n(x)\, dx = \frac{(-1)^n}{2^n\, n!}\int_{-1}^{1} (x^2-1)^n f^n(x)\, dx.$$

**Solution.** $$\int_{-1}^{1} f(x)\, P_n(x)\, dx = \int_{-1}^{+1} f(x)\cdot \frac{1}{2^n\, n!}\frac{d^n}{dx^n}(x^2-1)^n\, dx$$

$$\left[P_n(x) = \frac{1}{2^n\, n!}\frac{d^n}{dx^n}(x^2-1)^n\right]$$

$$= \frac{1}{2^n\, n!}\int_{-1}^{+1} f(x)\cdot \frac{d^n}{dx^n}(x^2-1)^n\, dx$$

Integrating by parts, we get

$$= \frac{1}{2^n\, n!}\left[f(x)\cdot \frac{d^{n-1}}{dx^{n-1}}(x^2-1)^n - \int f'(x)\cdot \frac{d^{n-1}}{dx^{n-1}}(x^2-1)^n\, dx\right]_{-1}^{+1}$$

$$= \frac{1}{2^n\, n!}\left[0 - \int_{-1}^{+1} f'(x)\cdot \frac{d^{n-1}}{dx^{n-1}}(x^2-1)^n\, dx\right]$$

$$= \frac{(-1)}{2^n\, n!}\int_{-1}^{+1} f'(x)\cdot \frac{d^{n-1}}{dx^{n-1}}(x^2-1)^n\, dx$$

Again integrating by parts, we have

$$= \frac{(-1)}{2^n\, n\,!}\left[f'(x)\,\frac{d^{\,n-2}}{dx^{n-2}}(x^2-1)^n - \int f''(x)\cdot\frac{d^{\,n-2}}{dx^{n-2}}(x^2-1)^n\,dx\right]_{-1}^{+1}$$

$$= \frac{(-1)^2}{2^n\, n\,!}\int_{-1}^{+1} f''(x)\,\frac{d^{\,n-2}}{dx^{n-2}}\cdot(x^2-1)^n\,dx$$

**Integrating $(n-2)$ times, by parts, we get**

$$= \frac{(-1)^n}{2^n\, n\,!}\int_{-1}^{+1} f^n(x)\,(x^2-1)^n\,dx$$

**Proved.**

## 8.22 LEGENDRE POLYNOMIALS

$$P_n(x) = \frac{1}{2^n\cdot n\,!}\cdot\frac{d^n}{dx^n}(x^2-1)^n \qquad \textbf{(Rodrigue's formula)}$$

**If $n = 0$,** $\quad P_0(x) = \dfrac{1}{2^0\cdot 0\,!} = 1$

**If $n = 1$,** $\quad P_1(x) = \dfrac{1}{2^1\cdot 1\,!}\dfrac{d}{dx}(x^2-1) = \dfrac{1}{2}(2\,x) = x$

**If $n = 2$,** $\quad P_2(x) = \dfrac{1}{2^2\cdot 2\,!}\dfrac{d^2}{dx^2}(x^2-1)^2 = \dfrac{1}{8}\dfrac{d}{dx}[2\,(x^2-1)\,(2\,x)]$

$$= \frac{1}{2}[(x^2-1)\cdot 1 + 2\,x\cdot x] = \frac{1}{2}(3\,x^2-1)$$

**Similarly** $\quad P_3(x) = \dfrac{1}{2}(5\,x^3-3\,x)$

$$P_4(x) = \frac{1}{8}(35\,x^4-30\,x^2+3)$$

$$P_5(x) = \frac{1}{8}(63\,x^5-70\,x^3+15\,x)$$

$$P_6(x) = \frac{1}{16}(231\,x^6-315\,x^4+105\,x^2-5)$$

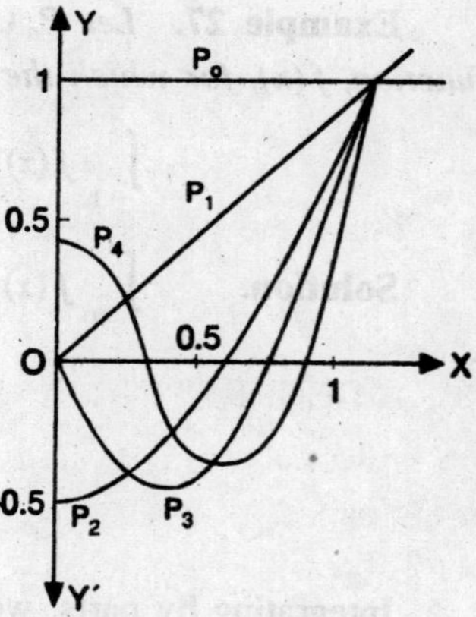


..............................................

$$P_n(x) = \sum_{r=0}^{N}\frac{(-1)^r\,(2\,n-2\,r)\,!}{2^n\cdot r\,!\,(n-r)\,!\,(n-2\,r)\,!}\,x^{n-2\,r}$$

where $\quad N = \dfrac{n}{2}$ if $n$ is even.

$$N = \frac{1}{2}(n-1) \text{ if } n \text{ is odd.}$$

**Note.** We can evaluate $P_n(x)$ by expanding $(x^2-1)^n$ by Binomial theorem.

$$(x^2-1)^n = \sum_{r=0}^{r=n} {}^nC_r\,(x^2)^{n-r}\,(-1)^r = \sum_{r=0}^{r=n}(-1)^r\,\frac{n\,!}{r\,!\,(n-r)\,!}\,x^{2n-2r}$$

$$P_n(x) = \frac{1}{2^n\cdot n!}\,\frac{d^n}{dx^n}(x^2-1)^n = \frac{1}{2^n\cdot n!}\sum_{r=0}^{r=n}(-1)^r\,\frac{n!}{r!\,(n-r)!}\,\frac{d^n}{dx^n}(x^{2n-2r})$$

$$= \sum_{r=0}^{N} \frac{(-1)^r (2n-2r)!}{2^n \cdot r!(n-r)!(n-2r)!} x^{n-2r}$$

Either $x^0$ or $x^1$ is in the last term.

$$\therefore \qquad n - 2r = 0 \quad \text{or} \quad r = \frac{n}{2} \qquad (n \text{ is even})$$

$$\text{or} \qquad n - 2r = 1 \quad \text{or} \quad r = \frac{1}{2}(n-1) \qquad (n \text{ is odd})$$

**Example 28.** *Express $f(x) = 4x^3 + 6x^2 + 7x + 2$ in terms of Legendre Polynomials*

**Solution.** Let

$$4x^3 + 6x^2 + 7x + 2 \equiv a P_3(x) + b P_2(x) + c P_1(x) + d P_0(x) \qquad ...(1)$$

$$\equiv a\left(\frac{5x^3}{2} - \frac{3x}{2}\right) + b\left(\frac{3x^2}{2} - \frac{1}{2}\right) + c(x) + d(1)$$

$$\equiv \frac{5ax^3}{2} - \frac{3ax}{2} + \frac{3bx^2}{2} - \frac{b}{2} + cx + d$$

$$\equiv \frac{5ax^3}{2} + \frac{3bx^2}{2} + \left(\frac{-3a}{2} + c\right)x - \frac{b}{2} + d.$$

Equating the coefficients of like powers of $x$, we have

$$4 = \frac{5a}{2}, \quad \text{or} \quad a = \frac{8}{5}$$

$$6 = \frac{3b}{2} \quad \text{or} \quad b = 4$$

$$7 = \frac{-3a}{2} + c \quad \text{or} \quad 7 = \frac{-3}{2}\left(\frac{8}{5}\right) + c \quad \text{or} \quad c = \frac{47}{5}$$

$$2 = \frac{-b}{2} + d \quad \text{or} \quad 2 = \frac{-4}{2} + d \quad \text{or} \quad d = 4$$

Putting the values of $a, b, c, d$ in (1), we get

$$4x^3 + 6x^2 + 7x + 2 = \frac{8}{5} P_3(x) + 4 P_2(x) + \frac{47}{5} P_1(x) + 4 P_0(x) \qquad \textbf{Ans.}$$

## 8.23 A GENERATING FUNCTION OF LEGENDRE'S POLYNOMIAL

Prove that $P_n(x)$ is the coefficient of $z^n$ in the expansion of $(1 - 2xz + z^2)^{-1/2}$ in ascendig powers of $z$.

**Proof.** $(1 - 2xz + z^2)^{-1/2} = [1 - z(2x - z)]^{-1/2}$

$$= 1 + \frac{1}{2} z(2x - z) + \frac{-\frac{1}{2}\left(-\frac{3}{2}\right)}{2!} z^2 (2x - z)^2 + \ldots$$

$$+ \frac{-\frac{1}{2}\left(-\frac{3}{2}\right)\left(-\frac{5}{2}\right)\ldots\left(-\frac{1}{2} - n + 1\right)}{n!} (-z)^n (2x - z)^n + \ldots \qquad ...(1)$$

Now coefficient of $z^n$ in

$$\frac{\left(-\frac{1}{2}\right)\left(-\frac{3}{2}\right)\left(-\frac{5}{2}\right)\ldots\left(-\frac{1}{2} - n + 1\right)}{n!} (-z)^n (2x - z)^n$$

$$= \frac{\left(-\frac{1}{2}\right)\left(-\frac{3}{2}\right)\left(-\frac{5}{2}\right)\cdots\left(-\frac{1}{2}-n+1\right)}{n\,!}(-1)^n(2x)^n$$

$$= \frac{1.3.5\ldots(2n-1)}{2^n \cdot n\,!}(2)^n \cdot x^n = \frac{1.3.5\ldots(2n-1)}{n\,!}x^n$$

Coefficient of $z^n$ in

$$\frac{\left(-\frac{1}{2}\right)\left(-\frac{3}{2}\right)\left(-\frac{5}{2}\right)\cdots\left(-\frac{1}{2}-n+2\right)}{(n-1)\,!}(-z)^{n-1}(2x-z)^{n-1}$$

$$= \frac{\left(-\frac{1}{2}\right)\left(-\frac{3}{2}\right)\left(-\frac{5}{2}\right)\cdots\left(-\frac{1}{2}-n+2\right)}{(n-1)\,!}(-1)^{n-1}[-(n-1)(2x)^{n-2}]$$

$$= \frac{1.3.5\ldots(2n-3)}{2^{n-1}\cdot(n-1)\,!}(2)^{n-2}(n-1)x^{n-1} = \frac{1.3.5\ldots(2n-3)}{2\cdot(n-1)\,!}(n-1)x^{n-2}$$

$$= \frac{1.3.5\ldots(2n-3)}{2\cdot(n-1)\,!}\times\frac{(2n-1)}{(2n-1)}(n-1)x^{n-2} = \frac{1.3.5\ldots(2n-3)(2n-1)}{n\,!}\times\frac{n(n-1)}{2(2n-1)}x^{n-2}$$

Coefficient of $z^n$ in

$$\frac{\left(-\frac{1}{2}\right)\left(-\frac{3}{2}\right)\left(-\frac{5}{2}\right)\cdots\left(-\frac{1}{2}-n+3\right)}{(n-2)\,!}z^{n-2}(2x-z)^{(n-2)}$$

$$= \frac{\left(-\frac{1}{2}\right)\left(-\frac{3}{2}\right)\left(-\frac{5}{2}\right)\cdots\left(-\frac{1}{2}-n+3\right)}{(n-2)\,!}\times(-1)^{n-2}\times\frac{(n-2)(n-3)}{2\,!}(2x)^{n-4}$$

$$= \frac{1.3.5.\ldots(2n-5)}{2^{n-2}(n-2)\,!}\times\frac{(n-2)(n-3)}{2\,!}(2x)^{n-4}$$

$$= \frac{1.3.5\ldots(2n-5)(2n-3)(2n-1)}{4(n-2)\,!}\times\frac{(n-2)(n-3)}{2(2n-3)(2n-1)}x^{n-4}$$

$$= \frac{1.3.5\ldots(2n-1)}{4n(n-1)(n-2)\,!}\times\frac{n(n-1)(n-2)(n-3)}{2(2n-3)(2n-1)}x^{n-4}$$

$$= \frac{1.3.5\ldots(2n-1)}{n\,!}\times\frac{n(n-1)(n-2)(n-3)}{2.4(2n-1)(2n-3)}x^{n-4}$$

and so on.

Thus coefficient of $z^n$ in the expansion of (1)

$$= \frac{1.3.5\ldots(2n-1)}{n\,!}\left[x^n - \frac{n(n-1)}{2(2n-1)}\cdot x^{n-2} + \frac{n(n-1)(n-2)(n-3)}{2.4.(2n-1)(2n-3)}x^{n-4} - \ldots\right]$$

$$= P_n(x)$$

Thus coefficients of $z, z^2, z^3$ ... etc. in (1) are $P_1(x), P_2(x), P_3(x)\ldots$

Hence

$$(1-2xz+z^2)^{-1/2} = P_0(x) + zP_1(x) + z^2P_2(x) + z^3P_3(x) + \ldots + z^nP_n(x) + \ldots$$

*i.e.,*
$$(1-2xz+z^2)^{-1/2} = \sum_{n=0}^{n=\infty} P_n(x)\cdot z^n.$$ **Proved.**

**Example 29.** *Prove that $P_n(1) = 1$.*

**Solution.** We know that

$$(1-2xz+z^2)^{-1/2} = 1 + zP_1(x) + z^2P_2(x) + z^3P_3(x) + \ldots + z^nP_n(x) + \ldots$$

Substituting 1 for $x$ in the above equation, we get

$$(1-2z+z^2)^{-1/2} = 1 + zP_1(1) + z^2P_2(1) + z^3P_3(1) + \ldots + z^nP_n(1) + \ldots$$

$$[(1-z)^2]^{-1/2} = \sum_{n=0}^{\infty} z^n P_n(1) \quad \text{or} \quad (1-z)^{-1} = \Sigma z^n P_n(1)$$

or
$$\Sigma z^n P_n(1) = (1-z)^{-1} = 1 + z + z^2 + z^3 + \ldots + z^n + \ldots$$

Equating the coefficients of $z^n$ on both sides we get

$$P_n(1) = 1$$ **Proved.**

**Example 30.** *Show that*

*(i)* $P_{2n}(0) = (-1)^n \dfrac{1.3.5\ldots(2n-1)}{2.4.6\ldots 2n}$ *(ii)* $P_{2n+1}(0) = 0.$

**Solution.** We know that

$$\Sigma z^{2n} P_{2n}(x) = (1-2xz+z^2)^{-1/2}$$

$$\Sigma z^{2n} P_{2n}(0) = (1+z^2)^{-1/2}$$

$$= 1 + \left(-\frac{1}{2}\right)z^2 + \frac{\left(-\frac{1}{2}\right)\left(-\frac{3}{2}\right)}{2!}(z^2)^2 + \frac{\left(-\frac{1}{2}\right)\left(-\frac{3}{2}\right)\left(-\frac{5}{2}\right)}{3!}(z^2)^3$$

$$+ \ldots + \frac{\left(-\frac{1}{2}\right)\left(-\frac{3}{2}\right)\left(-\frac{5}{2}\right)\ldots\left(-\frac{1}{2}-n+1\right)}{n!}(z^2)^n + \ldots$$

Equating the coefficient of $z^{2n}$ both sides we get

$$P_{2n}(0) = \frac{\left(-\frac{1}{2}\right)\left(-\frac{3}{2}\right)\left(-\frac{5}{2}\right)\ldots\left(-\frac{1}{2}-n+1\right)}{n!}$$

$$= (-1)^n \frac{1.3.5\ldots(2n-1)}{2^n\cdot n!}$$

$$= (-1)^n \frac{1.3.5\ldots(2n-1)}{2.4.6.8\ldots 2n}$$ **Proved.**

Coefficient of $z^{2n+1} = P_{2n+1}(0) = 0$ **Proved.**

## 8.24 ORTHOGONALITY OF LEGENDRE POLYNOMIALS

$$\int_{-1}^{+1} P_m(x)\cdot P_n(x)\,dx = 0 \qquad n \neq m$$

**Proof.** $P_n(x)$ is a solution of

$$(1-x^2)\frac{d^2y}{dx^2}-2x\frac{dy}{dx}+n(n+1)y=0 \qquad ...(1)$$

$P_m(x)$ is the solution of

$$(1-x^2)\frac{d^2z}{dx^2}-2x\frac{dz}{dx}+m(m+1)z=0 \qquad ...(2)$$

Multiplying (1) by z and (2) by y and subtracting, we get

$$(1-x^2)\left[z\frac{d^2y}{dx^2}-y\frac{d^2z}{dx^2}\right]-2x\left[z\frac{dy}{dx}-y\frac{dz}{dx}\right]+[n(n+1)-m(m+1)]yz=0$$

$$(1-x^2)\left[\left\{z\frac{d^2y}{dx^2}+\frac{dz}{dx}\times\frac{dy}{dz}\right\}-\left\{\frac{dy}{dx}\frac{dz}{dx}+y\frac{d^2z}{dx^2}\right\}-2x\left(z\frac{dy}{dx}-y\frac{dz}{dx}\right)+(n-m)(n+m+i)yz=0\right.$$

or $$\frac{d}{dx}\left[(1-x^2)\left(z\frac{dy}{dx}-y\frac{dz}{dx}\right)\right]+(n-m)(n+m+1)yz=0$$

Now integrating from − 1 to 1, we get

$$\left[(1-x^2)\left(z\frac{dy}{dx}-y\frac{dz}{dx}\right)\right]_{-1}^{+1}+(n-m)(n+m+1)\int_{-1}^{+1}y.z\,dx=0.$$

or $$0+(n-m)(n+m+1)\int_{-1}^{+1}y\cdot z\,dx=0$$

or $$\int_{-1}^{+1}P_n(x)\cdot P_m(x)\,dx=0 \qquad \text{if } n\neq m$$ **Proved**

**Example 31.** *Prove that*

$$\int_{-1}^{+1}[P_n(x)]^2\,dx=\frac{2}{2n+1}$$ *(U.P. III Semester, Summer, 2004, 2002)*

**Solution.** We know that $(1-2xz+z^2)^{-1/2}=\Sigma z^n P_n(x)$

Squaring both sides we get

$$(1-2xz+z^2)^{-1}=\Sigma z^{2n}P_n^2(x)+2\Sigma z^{m+n}P_m(x)\cdot P_n(x)$$

Integrating both sides between − 1 and + 1, we have

$$\int_{-1}^{+1}\Sigma z^{2n}\cdot P_n^2(x)\,dx+\int_{-1}^{+1}2\Sigma z^{m+n}\cdot P_m(x)\cdot P_n(x)\,dx=\int_{-1}^{+1}(1-2xz+z^2)^{-1}\,dx$$

$$\int_{-1}^{+1}\Sigma z^{2n}P_n^2(x)\,dx+0=\int_{-1}^{+1}\frac{1}{1-2xz+z^2}\,dx$$

or $$\Sigma z^{2n}\int_{-1}^{+1}P_n^2(x)\,dx=-\frac{1}{2z}\left[\log(1-2xz+z^2)\right]_{-1}^{+1}$$

$$=-\frac{1}{2z}\log\frac{1-2z+z^2}{1+2z+z^2}=-\frac{1}{2z}\log\left(\frac{1-z}{1+z}\right)^2$$

$$=\frac{1}{z}\log\frac{1+z}{1-z}=\frac{1}{z}[\log(1+z)-\log(1-z)]$$

$$= \frac{1}{z}\left[\left(z - \frac{z^2}{2} + \frac{z^3}{3} - \frac{z^4}{4} + \ldots + \frac{z^{2n+1}}{2n+1} + \ldots\right) - \left(-z - \frac{z^2}{2} - \frac{z^3}{3} - \frac{z^4}{4} - \ldots - \frac{z^{2n+1}}{2n+1} - \ldots\right)\right]$$

$$= \frac{2}{z}\left[z + \frac{z^3}{3} + \frac{z^5}{5} + \ldots + \frac{z^{2n+1}}{2n+1} + \ldots\right] = 2\left[1 + \frac{z^2}{3} + \frac{z^4}{5} + \ldots + \frac{z^{2n}}{2n+1} + \ldots\right]$$

Equating the coefficient of $z^{2n}$ on both sides, we have

$$\int_{-1}^{+1} [P_n(x)]^2\, dx = \frac{2}{2n+1}.$$ **Proved.**

Hence $$\int_{-1}^{+1} P_3^2(x)\, dx = \frac{2}{2\times 3+1} = \frac{2}{7}.$$

**Example 32.** *Assuming that a polynomial $f(x)$ of degree n can be written as*

$$f(x) = \sum_{0}^{\infty} C_m P_m(x),$$

*show that* $$C_m = \frac{2m+1}{2}\int_{-1}^{1} f(x)\, P_m(x)\, dx$$

**Solution.** $$f(x) = \sum_{0}^{\infty} C_m P_m(x)$$

$$= C_0P_0(x) + C_1P_1(x) + C_2P_2(x) + C_3P_3(x) + C_4P_4(x) + \ldots + C_mP_m(x) + \ldots$$

Multiplying both sides by $P_m(x)$, we get

$$P_m(x) f(x) = C_0P_0(x)\, P_m(x) + C_1P_1(x)\, P_m(x) + C_2P_2(x)\, P_m(x) + \ldots + C_mP_m^2(x) + \ldots$$

$$\int_{-1}^{+1} f(x)\, P_m(x)\, dx = \int_{-1}^{+1} [C_0P_0(x)\, P_m(x) + C_1P_1(x)\, P_m(x) + C_2P_2(x)\, P_m(x) + \ldots + C_mP_m^2(x) + \ldots]\, dx$$

$$= \left[0 + 0 + \ldots + C_m \frac{2}{2m+1} + ..\right] = \frac{2\, C_m}{2m+1}$$

$$C_m = \frac{2m+1}{2}\int_{-1}^{+1} f(x)\, P_m(x)\, dx$$ **Proved.**

**Example 33.** *Using the Rodrigue's formula for Legendre function, prove that*

$$\int_{-1}^{+1} x^m P_n(x)\, dx = 0, \text{ where } m, n \text{ are positive integers and } m < n.$$

**Solution.** $$\int_{-1}^{+1} x^m P_n(x)\, dx = \int_{-1}^{+1} x^m \frac{1}{2^n\, n!}\frac{d^n}{dx^n}(x^2-1)^n\, dx$$

$$= \frac{1}{2^n\, n!}\int_{-1}^{+1} x^m \frac{d^n}{dx^n}(x^2-1)^n\, dx$$

On integrating by parts we get

$$= \frac{1}{2^n n!}\left[\left\{x^m \frac{d^{n-1}}{dx^{n-1}}(x^2-1)^n\right\}_{-1}^{+1} - \int_{-1}^{+1} m\, x^{m-1} \frac{d^{n-1}}{dx^{n-1}}(x^2-1)^n\, dx\right]$$

$$= 0 - \frac{m}{2^n n!}\int_{-1}^{+1} x^{m-1} \frac{d^{n-1}}{dx^{n-1}}(x^2-1)^n\, dx$$

$$\int_{-1}^{+1} x^m P_n(x)\, dx = -\frac{(-1)^2 m(m-1)}{2^n n!}\int_{-1}^{+1} x^{m-2}\frac{d^{n-2}}{dx^{n-2}}(x^2-1)^n\, dx$$

**Integrating $m-2$ times, we get**

$$= (-1)^m \frac{m(m-1)\ldots 1}{2^n n!}\int_{-1}^{+1}\frac{d^{n-m}}{dx^{n-m}}(x^2-1)^n\, dx$$

$$= \frac{(-1)^m m!}{2^n n!}\int_{-1}^{+1}\frac{d^{n-m}}{dx^{n-m}}(x^2-1)^n\, dx$$

$$= \frac{(-1)^m m!}{2^n n!}\left[\frac{d^{n-m-1}}{dx^{n-m-1}}(x^2-1)^n\right]_{-1}^{+1} = 0$$ **Ans.**

## 8.25 RECURRENCE FORMULAE FOR $P_n(x)$

**Formula 1.** $nP_n = (2n-1)\, x P_{n-1} - (n-1) P_{n-2}$.

**Solution.** We know that $(1 - 2xz + z^2)^{-1/2} = \Sigma z^n P_n(x)$

Differentiating w.r.t. '$z$', we get

$$-\frac{1}{2}(1-2xz+z^2)^{-3/2}(-2x+2z) = \Sigma n z^{n-1} P_n(x)$$

Multiplying both sides by $(1-2xz+z^2)$, we get

$$(1-2xz+z^2)^{-1/2}(x-z) = (1-2xz+z^2)\,\Sigma n z^{n-1} P_n(x)$$

$$(x-z)\,\Sigma z^n P_n(x) = (1-2xz+z^2)\,\Sigma n z^{n-1} P_n(x) \quad \ldots(1)$$

Equating the coefficients of $z^{n-1}$ from both sides, we get

$$x P_{n-1} - P_{n-2} = nP_n - 2x(n-1) P_{n-1} + (n-2) P_{n-2}$$

or $$nP_n = (2n-1)\, x P_{n-1} - (n-1) P_{n-2}.$$ **Proved.**

**Formula II.** $x P_n' - P'_{n-1} = nP_n$.

**Solution.** We know that $(1-2xz+z^2)^{-1/2} = \Sigma z^n P_n(x)$ ..(1)

Differentiating (1) with respect to $z$, we get

$$-\frac{1}{2}(1-2xz+z^2)^{-3/2}(-2x+2z) = \Sigma n z^{n-1} P_n(x)$$

or $$(x-z)(1-2xz+z^2)^{-3/2} = \Sigma n z^{n-1} P_n(x) \quad \ldots(2)$$

Differentiating (1) with respect to $x$, we get

$$-\frac{1}{2}(1-2xz+z^2)^{-3/2}(-2z) = \Sigma z^n P'_n(x)$$

or $$z(1-2xz+z^2)^{-3/2} = \Sigma z^n P_n'(x) \quad \ldots(3)$$

Dividing (2) by (3), we get

$$\frac{x-z}{z} = \frac{\Sigma\, n z^{n-1} P_n(x)}{\Sigma\, z^n P_n'(x)}$$

or $$(x-z)\,\Sigma\, z^n P_n'(x) = \Sigma\, n z^n P_n(x)$$

Equating coefficients of $z^n$ from both sides, we get

$$x P_n'(x) - P'_{n-1}(x) = nP_n(x)$$ **Proved.**

**Formula III.** $P_n' - x P'_{n-1} = nP_{n-1}$

**Solution.** $nP_n = (2n-1)\, x P_{n-1} - (n-1)\, P_{n-2}$ Recurrence formula *I*

Differentiating the above formula w.r.t. '$x$', we get

$$nP_n' = (2n-1)\, P_{n-1} + (2n-1)\, x P'_{n-1} - (n-1)\, P'_{n-2}$$

or $$n\,[P_n' - x P'_{n-1}] - (n-1)\,[x P'_{n-1} - P'_{n-2}] = (2n-1)\, P_{n-1}$$

or $$n\,[P_n' - x P'_{n-1}] - (n-1)\,[(n-1)\, P_{n-1}] = (2n-1)\, P_{n-1}$$

(From formula *II*)

or $$n\,[P_n' - x P'_{n-1}] = [(n-1)^2 + (2n-1)]\, P_{n-1} = n^2 P_{n-1}$$

or $$P_n' - x P'_{n-1} = nP_{n-1}.$$ **Proved.**

**Formula IV.** $P'_{n+1} - P'_{n-1} = (2n+1)\, P_n$

**Solution.** $nP_n = (2n-1)\, x P_{n-1} - (n-1)\, P_{n-2}$ (Formula *I*)

Replacing $n$ by $(n+1)$,

$$(n+1)\, P_{n+1} = (2n+2-1)\, x P_n - nP_{n-1}$$

or $$(n+1)\, P_{n+1} = (2n+1)\, x P_n - nP_{n-1} \quad ...(1)$$

Differentiating (1) w.r.t. '$x$', we get

$$(n+1)\, P'_{n+1} = (2n+1)\, P_n + (2n+1)\, x P_n' - nP'_{n-1} \quad ...(2)$$

$$x P_n' - P'_{n-1} = nP_n$$ (Recurrence formula *II*) ...(3)

Substituting the value of $x P_n'$ from (3) into (2) we get

$$(n+1)\, P'_{n+1} = (2n+1)\, P_n + (2n+1)\,[nP_n + P'_{n+1}] - nP'_{n-1}$$

or $$(n+1)\, P'_{n+1} - (n+1)\, P'_{n-1} = (2n+1)\,(1+n)\, P_n$$

or $$P'_{n+1} - P'_{n-1} = (2n+1)\, P_n$$ **Proved.**

**Formula V.** $(x^2-1)\, P_n' = n\,[x P_n - P_{n-1}]$

**Solution.** $P_n' - x P'_{n-1} = nP_{n-1}$ ...(1) [Recurrence Formula *III*]

$x P_n' - P'_{n-1} = nP_n$ ...(2) (Recurrence Formula *II*)

Multiplying (2) by $x$ and subtracting from (1), we get

$$(1-x^2)\, P_n' = n\,(P_{n-1} - x P_n).$$ **Proved.**

**Formula VI.** $(x^2-1)\, P_n' = (n+1)\,(P_{n+1} - x P_n)$

**Solution.** $nP_n = (2n-1)\, x P_{n-1} - (n-1)\, P_{n-2}$ (Recurrence formula *I*)

Replacing $n$ by $(n+1)$, we get

$$(n+1)\, P_{n+1} = (2n+2-1)\, x P_n - nP_{n-1}$$

$$(n+1)\, P_{n+1} = (2n+1)\, x P_n - nP_{n-1}$$

which can be written as

$$(n+1)(P_{n+1} - x P_n) = n(x P_n - P_{n-1}) \quad \text{...(1)}$$

But $$(x^2 - 1) P_n' = n(x P_n - P_{n-1}). \quad \text{...(2)} \quad \text{(Recurrence formula V)}$$

From (1) and (2), we get

or $$(x^2 - 1) P_n' = (n+1)(P_{n+1} - x P_n). \qquad \textbf{Proved.}$$

**Example 34.** *Prove that*

$$\int_{-1}^{+1} x^2 P_{n+1}(x) \cdot P_{n-1}(x)\, dx = \frac{2n(n+1)}{(2n-1)(2n+1)(2n+3)} \qquad \textit{(Bhopal 2000)}$$

**Solution.** The recurrence formula *I* is

$$(2n+1) x P_n = (n+1) P_{n+1} + n P_{n-1}$$

Replacing $n$ by $(n+1)$ and $(n-1)$, we have

$$(2n+3) x P_{n+1} = (n+2) P_{n+2} + (n+1) P_n \quad \text{...(1)}$$

$$(2n-1) x P_{n-1} = n P_n + (n-1) P_{n-2} \quad \text{...(2)}$$

Multiplying (1) and (2) and integrating in the limits $-1$ to $+1$, we have

$$(2n+3)(2n-1)\int_{-1}^{+1} x^2 P_{n+1} \cdot P_{n-1}\, dx = n(n+1)\int_{-1}^{1} P_n^2\, dx + n(n+2)\int_{-1}^{+1} P_n \cdot P_{n+2}\, dx$$

$$+ (n^2 - 1)\int_{-1}^{+1} P_n P_{n-2}\, dx + (n-1)(n+2)\int_{-1}^{+1} P_{n+2} \cdot P_{n-2}\, dx$$

$$= n(n+1)\int_{-1}^{1} P_n^2\, dx + 0 + 0 + 0 \quad \text{(Orthogonality Property)}$$

$$= n(n+1) \cdot \frac{2}{(2n+1)}$$

or $$\int_{-1}^{+1} x^2 \cdot P_{n+1} \cdot P_{n-1}\, dx = \frac{2n(n+1)}{(2n-1)(2n+1)(2n+3)} \qquad \textbf{Proved.}$$

## Exercise 8.5

1. Express in terms of Legendre Polynomials.

(*a*) $1 + x - x^2$ (*b*) $x^4 + x^3 + x^2 + x + 1$ (*c*) $1 + 2x - 3x^2 + 4x^3$ (*d*) $x^3 + 1$ (*A.M.I.E.T.E., Dec. 2005*)

**Ans.** (*a*) $-\frac{2}{3} P_2(x) + P_1(x) + \frac{2}{3} P_0(x)$ (*b*) $\frac{8}{35} P_4(x) + \frac{2}{5} P_3(x) + \frac{26}{21} P_2(x) + \frac{8}{5} P_1(x) + \frac{23}{15} P_0(x)$

(*c*) $\frac{8}{5} P_3(x) - 2 P_2(x) + \frac{22}{5} P_1(x)$ (*d*) $\frac{2}{5} P_3(x) + \frac{3}{5} P_1(x) + P_0(x)$

Show that

2. (*a*) $x^5 = \frac{8}{63}\left[P_5(x) + \frac{7}{2} P_3(x) + \frac{27}{8} P_1(x)\right]$ (*b*) $x^3 = \frac{2}{5} P_3(x) + \frac{3}{5} P_1(x)$

3. $P_n(-x) = (-1)^n P_n(x)$.

4. $\dfrac{1+z}{z(1-2xz+z^2)^{\frac{1}{2}}} - \dfrac{1}{z} = \sum_{n=0}^{\infty} \left[P_n(x) + P_{n+1}(x)\right] z^n$ (*A.M.I.E.T.E., Summer 2001*)

5. $\dfrac{1-z^2}{(1-2xz+z^2)^{3/2}} = \sum_{n=0}^{\infty} (2n+1) P_n(x) z^n$ (*A.M.I.E.T.E., Winter 2002*)

**6.** (a) $\int_{-1}^{1} P_n(x)\,dx = 0 \quad (n \neq 0)$. (b) $\int_{-1}^{+1} x^3 . P_3(x)\,dx = \frac{4}{35}$ (*A.M.I.E.T.E., Summer 2001*)

**7.** $\int_{-1}^{+1} \frac{P_n(x)\,dx}{\sqrt{1-2xh+h^2}} = \frac{2h^n}{2n+1}$. (*A.M.I.E.T.E., Winter 2000*)

**8.** $\int_{-1}^{+1} x P_n(x) P_{n-1}(x)\,dx = \frac{2n}{4n^2-1}$.

**9.** $\int_{-1}^{+1} (1-x^2)\cdot P_m'(x)\cdot P_n'(x)\,dx \begin{cases} = 0 & \text{when } m \neq n \\ = \frac{2n(n+1)}{2n+1} & \text{when } m = n \end{cases}$ (*A.M.I.E.T.E., Dec. 2006*)

where $P_m(x)$ is the Legendre polynomial. You may assume the result.

$$\int_{-1}^{+1} P_n^2(x)\,dx = \frac{2}{2n+1}$$

**10.** Choose the correct or the best of the answers given in the following parts :

(*a*) The polynomial $2x^2 + x + 3$ in terms of Legendre polynomials is

(*i*) $\frac{1}{3}(4P_2 - 3P_1 + 11P_0)$ (*ii*) $\frac{1}{3}(4P_2 + 3P_1 - 11P_0)$

(*iii*) $\frac{1}{3}(4P_2 + 3P_1 + 11P_0)$ (*iv*) $\frac{1}{3}(4P_2 - 3P_1 - 11P_0)$

(*b*) Let $P_n(x)$ be the Legendre Polynomial. Then $P_n'(-x)$ is equal to

(*i*) $(-1)^{n+1} P_n'(x)$ (*ii*) $(-1)^n P_n'(x)$ (*iii*) $(-1)^n P_n(x)$ (*iv*) $P_n''(x)$

(*c*) If $P_n(x)$ is the Legendre polynomial of order *n*, then $3x^2 + 3x + 1$ can be expressed as

(*i*) $2P_2 + 3P_1$; (*ii*) $4P_2 + 2P_1 + P_0$; (*iii*) $3P_2 + 3P_1 + P_0$; (*iv*) $2P_2 + 3P_1 + 2P_0$

(*d*) If $\int_{-1}^{+1} P_n(x)\,dx = 2$, then *n* is

(*i*) 1 (*ii*) 0 (*iii*) $-1$ (*iv*) None of these (*A.M.I.E.T.E., Winter 2000*)

(*e*) Legendre polynomial $P_5(x) = k\left(x^5 - \frac{70}{63}x^3 + \frac{15}{63}x\right)$ where *k* is equal to

(*i*) 63/2 (*ii*) 63/5 (*iii*) 63/10 (*iv*) 63/8

(*f*) Let $P_n(x)$ be Legendre polynomial of degree $n > 1$. Then $\int_{-1}^{+1} (1+x) P_n(x)\,dx$ is equal to

(*i*) 0. (*ii*) $1/(2n+1)$. (*iii*) $2/(2n+1)$. (*iv*) $n/(2n+1)$.

(*g*) The value of $\int_{-1}^{1} (2x+1) P_3(x)\,dx$ where $P_3(x)$ is the third degree Legendre polynomial is

(*i*) 1 (*ii*) $-1$ (*iii*) 2 (*iv*) 0

(*h*) The value of integral $\int_{-1}^{1} x^3 P_3(x)\,dx$, where $P_3(x)$ is a Legendre polynomial of degree 3, equals

(*i*) $\frac{11}{35}$ (*ii*) 0, (*iii*) $\frac{2}{35}$ (*iv*) $\frac{4}{35}$ (*A.M.I.E.T.E., Winter 2003*)

(*j*) *The Legendre polynomial $P_n(x)$ has*

(*i*) *n* real zeros between 0 and 1, (*ii*) *n* zeros of which only one is between $-1$ and 1,

(*iii*) $2n - 1$ real zeros between $-1$ and 1, (*iv*) none of these

(*k*) *The incorrect equation among the following is*

(*i*) $P_0(x) = 1$ (*ii*) $P_1(x) = x$

(*iii*) $P_n(-x) = (-1)^{n+1} P_n(x)$ (*iv*) $(1-x^2)P_n''(x) - 2x P_n'(x) + n(n+1) P_n(x) = 0$

(*l*) The Rodrigue formula for $P_n(x)$, the Legendre Polynomial of degree *n* is $P_n(x) = k\dfrac{d^n\,|(x^2-1)^n|}{dx^n}$

(*i*) $k = n!/2^n$ (*ii*) $k = 2^n/n!$ (*iii*) $k = 1/2^n\,n!$ (*iv*) $1/2^n\,(n!)^2$.

(*m*) Using the recurrence relation, for Legendre's polynomial

$(n+1)\;P_{n+1}(x) = (2n+1)\,x\,P_n(x) - n\,P_{n-1}(x)$, the value of $P_2$ (1.5) equals to

(*i*) 1.5 (*ii*) 2.8 (*iii*) 2.875 (*iv*) 2.5 (*A.M.I.E.T.E. Dec. 2005*)

(*n*) Which of the following statement is correct:

(*a*) $P_2(x) = 3x\,P_1(x) + \frac{1}{2}P_0(x)$ (*b*) $P_2(x) = \frac{3}{2}x\,P_1(x) - \frac{1}{2}P_0(x)$

(*c*) $P_2(x) = \frac{3}{2}x\,P_1(x) + P_0(x)$ (*d*) $P_2(x) = \frac{1}{2}x\,P_1(x) + \frac{3}{2}P_0(x)$ (*AM.I.E.T.E., June 2006*)

**Ans.** (*a*) (*iii*), (*b*) (*i*), (*c*) (*iv*), (*d*) (*iii*), (*e*) (*iv*), (*f*) (*i*), (*g*) (*iv*), (*h*) (*iv*), (*j*) (*i*), (*k*) (*iii*), (*m*) (*iii*)

## 8.26 FOURIER-LEGENDRE EXPANSION

Let $f(x)$ be a function defined from $x = -1$ to $x = 1$.

The Fourier-Legendre expansion of $f(x)$

$$f(x) = C_1 P_1(x) + C_2 P_2(x) + C_3 P_3(x) + \ldots\ldots \qquad \ldots (1)$$

$$\Rightarrow \qquad f(x) = \sum_{n=0}^{\infty} C_n P_n(x)$$

Multiplying both sides of (1) by $P_n(x)$, we havc

$$f(x)\cdot P_n(x) = C_1 P_1(x)\cdot P_n(x) + C_2 P_2(x)\,P_n(x) + \ldots + C_n P_n^2(x) + \ldots \qquad \ldots (2)$$

Integrating both sides of (2), we get

$$\int_{-1}^{+1} f(x)\cdot P_n(x)\,dx = C_1\int_{-1}^{+1} P_1(x)\cdot P_n(x)\,dx + C_2\int_{-1}^{+1} P_2(x)\cdot P_n(x)\,dx + \ldots + C_n\int_{-1}^{+1} P_n^2(x)\,dx +$$

$$\int_{-1}^{+1} f(x)\cdot P_n(x)\,dx = C_n\int_{-1}^{+1} P_n^2(x)\,dx \qquad \text{(Other integrals are equal to zero)}$$

$$= C_n\,\frac{2}{2n+1}$$

$$\Rightarrow \qquad C_n = \frac{2n+1}{2}\int_{-1}^{+1} f(x)\cdot P_n(x)\,dx \qquad \Rightarrow \qquad C_n = \left(n+\frac{1}{2}\right)\int_{-1}^{+1} f(x)\cdot P_n(x)\,dx\,.$$

**Example 35.** *Expand the function*

$$f(x) = \begin{cases} 0 & -1 < x < 0 \\ 1 & 0 < x < 1 \end{cases}$$

*in terms of Legendre polynomials.*

**Solution.** Let $f(x) = \sum_{n=0}^{\infty} C_n P_n(x)\,dx$, then

$$C_n = \left(n+\frac{1}{2}\right)\int_{-1}^{+1} f(x)\,P_n(x)\,dx$$

$$= \left(n+\frac{1}{2}\right)\left[\int_{-1}^{0} 0\,P_n(x)\,dx + \int_{0}^{1} 1\,P_n(x)\,dx\right] = \left(n+\frac{1}{2}\right)\int_{0}^{1} P_n(x)\,dx.$$

$$= \left(n+\frac{1}{2}\right)\left[\int_0^1 P_0(x)\,dx + \int_0^1 P_1(x)\,dx + \int_0^1 P_2(x)\,dx + \int_0^1 P_3(x)\,dx + \ldots\right]$$

$$C_0 = \frac{1}{2}\int_0^1 P_0(x)\,dx = \frac{1}{2}\int_0^1 1\cdot dx = \frac{1}{2}(x)_0^1 = \frac{1}{2}$$

$$C_1 = \frac{3}{2}\int_0^1 P_1(x)\,dx = \frac{3}{2}\int_0^1 x\,dx = \frac{3}{2}\left[\frac{x^2}{2}\right]_0^1 = \frac{3}{4}$$

$$C_2 = \frac{5}{2}\int_0^1 P_2(x)\,dx = \frac{5}{2}\int_0^1 \frac{1}{2}(3x^2-1)\,dx = \frac{5}{4}\left(x^3-x\right)_0^1 = 0$$

$$C_3 = \frac{7}{2}\int_0^1 P_3(x)\,dx = \frac{7}{2}\int_0^1 \frac{1}{2}(5x^3-3x)\,dx = \frac{7}{4}\left[\frac{5x^4}{4}-\frac{3x^2}{2}\right]_0^1 = \frac{-7}{16}$$

$$C_4 = \frac{9}{2}\int_0^1 P_4(x)\,dx = \frac{9}{2}\int_0^1 \frac{1}{8}(35x^4-30x^2+3)\,dx = \frac{9}{16}\left[7x^5-10x^3+3x\right]_0^1 = 0$$

$$C_5 = \frac{11}{2}\int_0^1 P_5(x)\,dx = \frac{11}{2}\int_0^1 \frac{1}{8}(63x^5-70x^3+15x)\,dx = \frac{11}{16}\left[\frac{21}{2}x^6-\frac{35}{2}x^4+\frac{15}{2}x^2\right]_0^1 = \frac{11}{32}$$

Hence, $f(x) = \frac{1}{2}P_0(x) + \frac{3}{4}P_1(x) - \frac{7}{16}P_3(x) + \frac{11}{32}P_5(x) \ldots\ldots$ **Ans.**

**Example 36.** *Express the function*

$$f(x) = \begin{cases} 0 & -1 < x \le 0 \\ x & 0 < x < 1 \end{cases}$$

*in Fourier-Legendre expansion.*

**Solution.** Let $f(x) = \sum_{n=0}^{\infty} C_n P_n(x)$, then

$$C_n = \left(n+\frac{1}{2}\right)\int_{-1}^{+1} f(x)\,P_n(x)\,dx$$

$$C_n = \left(n+\frac{1}{2}\right)\int_{-1}^{0} 0\cdot P_n(x)\,dx + \left(n+\frac{1}{2}\right)\int_0^1 x\cdot P_n(x)\,dx = \left(n+\frac{1}{2}\right)\int_0^1 x\cdot P_n(x)\,dx$$

$$C_0 = \frac{1}{2}\int_0^1 x\cdot P_0(x)\,dx = \frac{1}{2}\int_0^1 x\cdot 1\,dx = \frac{1}{2}\int_0^1 x\,dx = \frac{1}{2}\left[\frac{x^2}{2}\right]_0^1 = \frac{1}{4}$$

$$C_1 = \frac{3}{2}\int_0^1 x\cdot P_1(x)\,dx = \frac{3}{2}\int_0^1 x\cdot x\,dx = \frac{3}{2}\int_0^1 x^2\,dx = \frac{3}{2}\left[\frac{x^3}{3}\right]_0^1 = \frac{1}{2}$$

$$C_2 = \frac{5}{2}\int_0^1 x\cdot P_2(x)\,dx = \frac{5}{2}\int_0^1 x\cdot\frac{3x^2-1}{2}\,dx = \frac{5}{4}\int_0^1 (3x^3-x)\,dx = \frac{5}{4}\left(\frac{3x^4}{4}-\frac{x^2}{2}\right) = \frac{5}{16}$$

$$C_3 = \frac{7}{2}\int_0^1 x\cdot P_3(x)\,dx = \frac{7}{2}\int_0^1 x\cdot\frac{5x^3-3x}{2}\,dx = \frac{7}{4}\int_0^1 (5x^4-3x^2)\,dx = \frac{7}{4}\left(x^5-x^3\right)_0^1 = 0$$

$$C_4 = \frac{9}{2}\int_0^1 x\cdot P_4(x)\,dx = \frac{9}{2}\int_0^1 x\cdot\frac{35x^4-30x^2+3}{8}\,dx$$

$$= \frac{9}{16}\int_0^1 (35x^5-30x^3+3x)\,dx = \frac{9}{16}\left[\frac{35x^6}{6}-\frac{15}{2}x^4+\frac{3x^2}{2}\right]_0^1 = \frac{-3}{12}$$

$$C_5 = \frac{11}{2}\int_0^1 x \cdot P_5(x)\,dx = \frac{11}{2}\int_0^1 x \cdot \frac{63x^5 - 70x^3 + 15x}{8}\,dx$$

$$= \frac{11}{16}\int_0^1 (63x^6 - 70x^4 + 15x^2)\,dx = \frac{11}{16}\left[9x^7 - 14x^5 + 5x^3\right]_0^1 = 0$$

Hence

$$f(x) = \frac{1}{4}P_0(x) + \frac{1}{2}P_1(x) + \frac{5}{16}P_2(x) - \frac{3}{32}P_4(x) + \ldots\ldots$$ **Ans.**

**Similar Question**

**Question.** Expand the function

$$f(x) = \begin{cases} 0 & , \quad -1 < x < 0 \\ x^2 & , \quad 0 < x < 1 \end{cases}$$

in terms of Legendre polynomials.

**Ans.** $f(x) = \frac{1}{6}P_0(x) + \frac{3}{8}P_1(x) + \frac{1}{3}P_2(x) + \frac{7}{48}P_3(x) - \frac{11}{384}P_5(x) + \ldots\ldots$

## 8.27 LAGUERRES DIFFERENTIAL EQUATION

$$x\frac{d^2y}{dx^2} + (1 - x)\frac{dy}{dx} + ny = 0$$

Its solution is given by Laguerres polynomial

$$L_n(x) = e^x \frac{d^n}{dx^n}(x^n \cdot e^{-x})$$

$$L_0(x) = e^x(x^0 e^{-x}) = 1$$

$$L_1(x) = e^x \frac{d}{dx}(x\,e^{-x}) = e^x[x(-e^{-x}) + (1)\,e^{-x}] = 1 - x$$

$$L_2(x) = e^x \frac{d^2}{dx^2}(x^2 e^{-x}) = e^x[x^2 e^{-x} - 4x\,e^{-x} + 2\,e^{-x}] = x^2 - 4x + 2$$

$$L_3(x) = e^x \frac{d^3}{dx^3}(x^3 e^{-x}) = e^x[-x^3 e^{-x} + 9x^2 e^{-x} - 18x\,e^{-x} + 6\,e^{-x}]$$

$$= -x^3 + 9x^2 - 18x + 6$$

(*b*) Generating function of Laguerres polynomials $L_n$ is

$$\frac{-\dfrac{-xt}{e^{1-t}}}{1-t} = \sum_{n=0}^{\infty} \frac{L_n(x)}{n!}t^n$$

(*c*) Orthogonal properties for Laguerres polynomials

$$\int_0^\infty e^{-x} L_n(x) \cdot L_n(x)\,dx = \begin{cases} 0 & , \quad m \neq n \\ (n\,!)^2 & , \quad m = n \end{cases}$$

## 8.28 STRUM-LIOUVILLE EQUATION

$$\frac{d}{dx}\left[p(x) \cdot \frac{dy}{dx}\right] + [\lambda\,q(x) + r(x)]\,y = 0$$

**Solution.** We know that Bessel's equation is

$$X^2\frac{d^2y}{dX^2} + X\frac{dy}{dX} + (X^2 - n^2)\,y = 0 \qquad \ldots (1)$$

Substituting $X = kx$ in (1), we get

$$\frac{dy}{dX} = \frac{dy}{dx}\frac{dx}{dX} = \frac{dy}{dx}\frac{1}{k} \quad \text{and} \quad \frac{d^2y}{dX^2} = \frac{d^2y}{dx^2}\frac{1}{k^2}$$

$$k^2x^2\left(\frac{d^2y}{dx^2}\frac{1}{k^2}\right) + (kx)\left(\frac{dy}{dx}\frac{1}{k}\right) + (k^2x^2 - n^2)\,y = 0$$

or

$$x^2\frac{d^2y}{dx^2} + x\frac{dy}{dx} + (k^2x^2 - n^2)\,y = 0$$

$$x\frac{d^2y}{dx^2} + \frac{dy}{dx} + \left(k^2x - \frac{n^2}{x}\right) = 0$$

$$\left(x\frac{d^2y}{dx^2} + \frac{dy}{dx}\right) + \left(\lambda x - \frac{n^2}{x}\right)y = 0 \qquad (\text{Put } k^2 = \lambda)$$

$$\frac{d}{dx}\left(x\frac{dy}{dx}\right) + \left(\lambda x - \frac{n^2}{x}\right)y = 0 \qquad \text{... (2)}$$

Equations (1) and (2) are of the form.

$$\frac{d}{dx}\left[p(x)\cdot\frac{dy}{dx}\right] + [\lambda\, q(x) + r(x)]\,y = 0 \qquad \text{... (3)}$$

Equation (3) is known as the Strum-Liouville equation.

Equation (3) with the following conditions is known as Strum-Liouville problem.

$$\alpha_1\, y(a) + \alpha_2\, y'(a) = 0$$

$$\beta_1\, y(b) + \beta_2\, y'(b) = 0$$

Solution of Strum-Liouville problem is called an eigen function where $\lambda$ is an eigen value.

Particular Case. Putting $p = 1, q = 1, r = 0$ in (3), we have

$$\frac{d^2y}{dx^2} + \lambda y = 0$$

Now taking conditions $\alpha_1 = \beta_1 = 1$ and $\alpha_2 = \beta_2 = 0$

$$y(a) = 0 \quad \text{and} \quad y(b) = 0$$

Hence $\left.\begin{array}{r} y'' + \lambda y = 0 \\ y(a) = 0, \quad y(b) = 0 \end{array}\right]$ simplest form of Strum-Liouville problem.

## 8.29 ORTHOGONALITY

$$\int_a^b P(x)\, y_m(x)\cdot y_n(x)\, dx = 0\ , \qquad m \neq n$$

$$\int_a^b P(x)\left[y_m(x)\right]^2 dx = \|y_m\|\ , \qquad m = n$$

where $\|y_m\|$ is the norm of $y_m$ ,

If the function is orthogonal and have norm equal to 1, then the function is known as orthogonal.

## 8.30 ORTHOGONALITY OF EIGEN FUNCTIONS

If $p$, $q$, $r$ and $r'$ are the functions in Strum-Liouville equation and $\lambda_m(x), \lambda_n(x)$ be the eigen functions of Strum-Liouville problem, then

$$\left(\lambda_m - \lambda_n\right) \int_a^b q\, y_m\, y_n = y_m\,(p\,y'\,n) - y_n\,(p\,y'\,m)$$

$$= \frac{d}{dx}\left[(p\,y'_n)\,y_m - (p\,y'_m)\,y_n\right]_a^b$$

$$= p\,(b)\left[y_n'\,(b)\,y_m\,(b) - y_m'\,(b)\,y_n\,(b)\right] - p\,(a)\left[y_n'\,(a)\,y_m\,(a) - y_n'\,(a)\,y_n\,(a)\right] \quad \ldots (A)$$

$$= 0 \text{ if}$$

(i) $y\,(a) = y\,(b)$ (ii) $y'\,(a) = y'\,(b)$

(iii) $\alpha_1\, y\,(a) + \alpha_2\, y'\,(a) = 0$ (iv) $\beta_1\, y\,(b) + \beta_2\, y'\,(b) = 0$

Equation (A) becomes

$$\int_a^b q\, y_m\, y_n = 0 \qquad (m \neq n)$$

It means that eigen functions $y_m, y_n$ are orthogonal with the weight $q\,(x)$.

## Exercise 8.6

1. The integral $\int_0^{\pi} P_n\,(\cos\theta)\,\sin 2\theta\, d\theta,\ n > 1$, where $P_n\,(x)$ is the Legendre's polynomial of degree $n$ equals.

(A) 1. (B) $\frac{1}{2}$ (C) 0 (D) 2.

(*A.M.I.E.T.E., Dec. 2004*) **Ans.** (*c*)

2. Show that under change of dependent variable $y$ defined by the substitution $y = u/\sqrt{t}$, the Bessel's equation of order $\nu$ becomes $\frac{d^2u}{dt^2} + \left(1 + \frac{1 - 4\nu^2}{4t^2}\right) u = 0$. Hence show that for large values of $t$, the solutions of Bessel's equation are described approximately by the expression of the form $C_1\,\frac{\sin t}{\sqrt{t}} + C_2\,\frac{\cos t}{\sqrt{t}}$.

(*A.M.I.E.T.E., June 2005*)

3. Using Rodrigues formula for Legendre polynomials $P_n\,(x)$, show that

$$\int_{-1}^{1} f(x)\,P_n(x)\,dx = \frac{(-1)^n}{2^n\,n!}\int_{-1}^{1} f^{(n)}\,(x)\,(x^2-1)^n\,dx$$

where $f$ is any function integrable on interval $[-1, 1]$. Hence show that

$$\int_{-1}^{1} P_m(m)\,P_n\,(x)\,dx = 0,\ m \neq n.$$

(*A.M.I.E.T.E., June 2005*)

# 9

# Partial Differential Equations

**9.1. PARTIAL DIFFERENTIAL EQUATIONS** are those equations which contain partial differential coefficients, independent variables and dependent variables.

The independent variables will be denoted by $x$ and $y$ and the dependent variable by $z$. The partial differential coefficients are denoted as follows:

$$\frac{\partial z}{\partial x} = p, \qquad \frac{\partial z}{\partial y} = q.$$

$$\frac{\partial^2 z}{\partial x^2} = r, \qquad \frac{\partial^2 z}{\partial x \partial y} = s, \qquad \frac{\partial^2 z}{\partial y^2} = t$$

**9.2. ORDER** of a partial differential equation is the same as that of the order of the highest differential coefficient in it.

**9.3 CLASSIFICATION**

Consider the equation. $$A\frac{\partial^2 u}{\partial x^2} + B\frac{\partial^2 u}{\partial x\, \partial y} + C\frac{\partial^2 u}{\partial y^2} + F(x, y, u, p, q) = 0 \qquad \text{... (1)}$$

Where $A$, $B$, $C$ may be constants or functions of $x$ and $y$. Now the equation (1) is

1. Parabolic; if $B^2 - 4AC = 0$
2. Elliptic; if $B^2 - 4AC < 0$
3. Hyperbolic; if $B^2 - 4AC > 0$

**9.4 METHOD OF FORMING PARTIAL DIFFERENTIAL EQUATIONS**

A partial differential equation is formed by two methods.

(i) By eliminating arbitrary constants.

(ii) By eliminating arbitrary functions.

**(*i*) Method of elimination of arbitrary constants**

**Example 1.** *Form a partial differential equation from*

$$x^2 + y^2 + (z - c)^2 = a^2.$$

**Solution.** $$x^2 + y^2 + (z - c)^2 = a^2 \qquad \text{...(1)}$$

(1) contains two arbitrary constants $a$ and $c$.

Differentiating (1) partially w.r.t. $x$ we get

$$2x + 2(z - c)\frac{\partial z}{\partial x} = 0$$

$$\Rightarrow \qquad x + (z - c)\,p = 0 \qquad \text{...(2)}$$

Differentiating (1) partially w.r.t. $y$ we get

$$2y + 2(z - c)\frac{\partial z}{\partial y} = 0$$

$$y + (z - c)\, q = 0 \qquad ...(3)$$

Let us eliminate $c$ from (2) and (3)

From (2) $\qquad (z - c) = -\dfrac{x}{p}$

Putting this value of $z - c$ in (3), we get $y - \dfrac{x}{p} q = 0$

or $\qquad yp - xq = 0$ **Ans.**

*(ii)* **Method of elimination of arbitrary functions**

**Example 2.** *Form the partial differential equation from*

$$z = f(x^2 - y^2)$$

**Solution.** $\qquad z = f(x^2 - y^2) \qquad ...(1)$

Differentiating (1) w.r.t $x$ and $y$

$$P = \frac{\partial z}{\partial x} = f'(x^2 - y^2)\, 2x \qquad ...(2)$$

$$q = \frac{\partial z}{\partial y} = f(x^2 - y^2)\,(-2y) \qquad ...(3)$$

Dividing (2) by (3) we get $\dfrac{p}{q} = \dfrac{-x}{y}$ or $py = -qx$

or $\qquad yp + xq = 0$ **Ans.**

## EXERCISE 9.1

Form the partial differential equation

1. $z = (x + a)(y + b)$ **Ans.** $pq = z$
2. $(x - h)^2 + (y - k)^2 + z^2 = a^2$ **Ans.** $z^2 (p^2 + q^2 + 1) = a^2$
3. $2z = (ax + y)^2 + b$ **Ans.** $px + qy = q^2$
4. $ax^2 + by^2 + z^2 = 1$ **Ans.** $z(px + qy) = z^2 - 1$
5. $x^2 + y^2 = (z - c)^2 \tan^2 \alpha$ **Ans.** $yp - xq = 0$
6. $z = f(x^2 + y^2)$ **Ans.** $yp - xq = 0$
7. $2z = \dfrac{x^2}{a^2} + \dfrac{y^2}{b^2}$ *(A.M.I.E., Winter 2001)* **Ans.** $2z = xp + yq$
8. $f(x + y + z,\ x^2 + y^2 + z^2) = 0$ **Ans.** $(y - z)p + (z - x)q = x - y$

## 9.5 SOLUTION OF EQUATION BY DIRECT INTEGRATION

**Example 3.** *Solve* $\qquad \dfrac{\partial^3 z}{\partial x^2 \partial y} = \cos(2x + 3y)$

**Solution.** $\qquad \dfrac{\partial^3 z}{\partial x^2 \partial y} = \cos(2x + 3y)$

Integrating w.r.t. '$x$', we get $\dfrac{\partial^2 z}{\partial x \partial y} = \dfrac{1}{2} \sin(2x + 3y) + f(y)$

Integrating w.r.t. $x$, we get $\dfrac{\partial z}{\partial y} = -\dfrac{1}{4} \cos(2x + 3y) + x \int f(y)\, dy + g(y)$

$$= -\frac{1}{4}\cos(2x+3y) + x\phi(y) + g(y)$$

Integrating w.r.t. 'y' we get

$$z = \frac{1}{12}\sin(2x+3y) + x\int\phi(y)\,dy + \int g(y)\,dy$$

$$z = -\frac{1}{12}\sin(2x+3y) + x\phi_1(y) + \phi_2(y)$$ **Ans.**

**Example 4.** *Solve* $\dfrac{\partial^2 z}{\partial x \partial y} = x^2 y$

*subject to the condition* $z(x, 0) = x^2$ *and* $z(1, y) = \cos y$.

**Solution.** $$\frac{\partial}{\partial x}\left(\frac{\partial z}{\partial y}\right) = x^2 y$$

On integrating w.r.t. $x$, we obtain $\dfrac{\partial z}{\partial y} = \dfrac{x^3}{3}y + f(y)$

Integrating w.r.t. $y$, we obtain $z = \dfrac{x^3}{3}\cdot\dfrac{y^2}{2} + \int f(y)\,dy + g(x)$

$$\left[F(y) = \int f(y)\,dy\right]$$

or $$z = \frac{x^3 y^2}{6} + F(y) + g(x) \quad \text{... (1)}$$

**Condition 1:** Putting $z = x^2$ and $y = 0$ in (1), we get

$$x^2 = 0 + F(0) + g(x)$$

Putting the value of $g(x)$ in (1), we get $z = \dfrac{x^3 y^2}{6} + F(y) + x^2 - F(0)$ ...(2)

**Condition 2:** $z(1, y) = \cos y$

Putting $x = 1$ and $z = \cos y$ in (2), we get

$$\cos y = \frac{y^2}{6} + F(y) + 1 - F(0)$$

Putting the value of $F(y)$ in (2), we obtain

$$z = \frac{1}{6}x^3y^2 + \cos y - \frac{1}{6}y^2 - 1 + F(0) + x^2 - F(0)$$

or $$z = \frac{1}{6}x^3y^2 + \cos y - \frac{1}{6}y^2 - 1 + x^2$$ **Ans.**

**Example 5.** *Solve* $\dfrac{\partial^2 z}{\partial y^2} = z$, *if* $y = 0$, $z = e^x$ *and* $\dfrac{\partial z}{\partial y} = e^{-x}$

**Solution.** If $z$ is a function of $y$ alone, then

$$z = \sinh y \,.\, f(x) + \cosh y \,.\, \phi(x) \quad \text{... (1)}$$

$$\frac{\partial^2 z}{\partial y^2} = z \Rightarrow (D^2 - 1)z = 0 \Rightarrow m = \pm 1$$

$$\Rightarrow \quad z = A e^{y} + B e^{-y} = A \sinh y + B \cosh y$$

$$= f(x)\sinh y + \phi(x)\,.\,\cosh y$$

On putting $y = 0$ and $z = e^x$ in (1), we obtain

$$e^x = \phi(x)$$

(1) becomes $\quad z = \sinh y \,.\, f(x) + \cosh y \,.\, e^x \quad$ ...(2)

On differentiating (2) w.r.t. $y$, we get

$$\frac{\partial z}{\partial y} = \cosh y \cdot f(x) + \sinh y \cdot e^x \quad \text{...(3)}$$

On putting $y = 0$ and $\frac{\partial z}{\partial y} = e^{-x}$ in (3), we obtain

$$e^{-x} = f(x)$$

(2) becomes, $\quad z = e^{-x} \sinh y + e^x \cosh y \quad$ **Ans.**

## EXERCISE 9.2

Solve the following:

1. $\frac{\partial^2 z}{\partial x \partial y} = xy^2$ **Ans.** $z = \frac{x^2 y^3}{6} + f(y) + \phi(x)$

2. $\frac{\partial^2 z}{\partial x \partial y} = e^y \cos x$ **Ans.** $z = e^y \sin x + f(y) + \phi(x)$

3. $\frac{\partial^2 z}{\partial x \partial y} = \frac{y}{x} + 2$ **Ans.** $z = \frac{y^2}{2} \log x + 2xy + f(v) + \phi(x)$

4. $\frac{\partial^2 z}{\partial x^2} = a^2 z$, when $x = 0$, $\frac{\partial z}{\partial x} = a \sin y$ and $\frac{\partial z}{\partial y} = 0$ **Ans.** $z = \sin x + e^y \cos x$

5. $\frac{\partial^2 z}{\partial x \partial y} = \sin x \sin y$ if $\frac{\partial z}{\partial y} = -2 \sin y$ when $x = 0$, and $z = 0$ when $y$ is an odd multiple of $\frac{\pi}{2}$.

**Ans.** $z = \cos x \cos y + \cos y$

6. The partial differential equation $y \frac{\partial^2 u}{\partial x^2} + 2x \frac{\partial^2 u}{\partial x \partial y} + y \frac{\partial^2 u}{\partial y^2} = 0$ is elliptic if

(*a*) $x^2 = y^2$ (*b*) $x^2 < y^2$ (*c*) $x^2 + y^2 > 1$ (*d*) $x^2 + y^2 = 1$

(*A.M.I.E.T.E., Dec. 2004*) **Ans.** (*b*)

## 9.6 LAGRANGE'S LINEAR EQUATION IS AN EQUATION OF THE TYPE

$$Pp + Qq = R$$

where $P, Q, R$ are the functions of $x, y, z$ and $p = \frac{\partial z}{\partial x}, q = \frac{\partial z}{\partial y}$

**Solution.** $\quad Pp + Qq = R \quad$ ...(1)

This form of the equation is obtained by eliminating an arbitrary function $f$ from

$$f(u, v) = 0 \quad \text{...(2)}$$

where $u, v$ are functions of $x, y, z$.

Differentiating (2) partially w.r.t. to $x$ and $y$.

$$\frac{\partial f}{\partial u}\left(\frac{\partial u}{\partial x} + \frac{\partial u}{\partial z}\frac{\partial z}{\partial x}\right) + \frac{\partial f}{\partial v}\left(\frac{\partial v}{\partial x} + \frac{\partial v}{\partial z}\frac{\partial z}{\partial x}\right) = 0 \quad \text{...(3)} \quad \text{and} \quad \frac{\partial f}{\partial u}\left(\frac{\partial u}{\partial y} + \frac{\partial u}{\partial z}\frac{\partial z}{\partial y}\right) + \frac{\partial f}{\partial v}\left(\frac{\partial v}{\partial y} + \frac{\partial v}{\partial z}\frac{\partial z}{\partial y}\right) = 0 \quad \text{...(4)}$$

Let us eliminate $\frac{\partial f}{\partial u}$ and $\frac{\partial f}{\partial v}$ from (3) and (4).

From (3), $\frac{\partial f}{\partial u}\left[\frac{\partial u}{\partial x}+\frac{\partial u}{\partial z}p\right]=-\frac{\partial f}{\partial v}\left[\frac{\partial v}{\partial x}+\frac{\partial v}{\partial z}p\right]$ ...(5)

From (4), $\frac{\partial f}{\partial u}\left[\frac{\partial u}{\partial y}+\frac{\partial u}{\partial z}q\right]=-\frac{\partial f}{\partial v}\left[\frac{\partial v}{\partial y}+\frac{\partial v}{\partial z}q\right]$ ...(6)

Dividing (5) by (6), we get $\dfrac{\frac{\partial u}{\partial x}+\frac{\partial u}{\partial z}.p}{\frac{\partial u}{\partial y}+\frac{\partial u}{\partial z}.q}=\dfrac{\frac{\partial v}{\partial x}+\frac{\partial v}{\partial z}.p}{\frac{\partial v}{\partial y}+\frac{\partial v}{\partial z}.q}$

or $\left[\frac{\partial u}{\partial x}+\frac{\partial u}{\partial z}.p\right]\left[\frac{\partial v}{\partial y}+\frac{\partial v}{\partial z}q\right]=\left[\frac{\partial u}{\partial y}+\frac{\partial u}{\partial z}.q\right]\left[\frac{\partial v}{\partial x}+\frac{\partial v}{\partial z}.p\right]$

$$\frac{\partial u}{\partial x}\times\frac{\partial v}{\partial y}+\frac{\partial u}{\partial x}\times\frac{\partial v}{\partial z}.q+\frac{\partial u}{\partial z}\times p\times\frac{\partial v}{\partial y}+\frac{\partial u}{\partial z}\times\frac{\partial v}{\partial z}.pq$$

or $$=\frac{\partial u}{\partial y}\times\frac{\partial v}{\partial x}+\frac{\partial u}{\partial y}\times\frac{\partial v}{\partial z}.p+\frac{\partial u}{\partial z}.q\times\frac{\partial v}{\partial x}+\frac{\partial u}{\partial z}\times\frac{\partial v}{\partial z}.pq$$

$$\left[\frac{\partial u}{\partial y}\times\frac{\partial v}{\partial z}-\frac{\partial u}{\partial z}\times\frac{\partial v}{\partial y}\right]p+\left[\frac{\partial u}{\partial z}\times\frac{\partial v}{\partial x}-\frac{\partial u}{\partial x}\times\frac{\partial v}{\partial z}\right]q=\frac{\partial u}{\partial x}\times\frac{\partial v}{\partial y}-\frac{\partial u}{\partial y}\times\frac{\partial v}{\partial x} \quad ...(7)$$

*If (1) and (7) are the same, then the coefficients of p, q are equal.*

$$P=\frac{\partial u}{\partial y}\times\frac{\partial v}{\partial z}-\frac{\partial u}{\partial z}\times\frac{\partial v}{\partial y}$$

$$Q=\frac{\partial u}{\partial z}\times\frac{\partial v}{\partial x}-\frac{\partial u}{\partial x}\times\frac{\partial v}{\partial z} \quad ...(8)$$

$$R=\frac{\partial u}{\partial x}\times\frac{\partial v}{\partial y}-\frac{\partial u}{\partial y}\times\frac{\partial v}{\partial x}$$

*Now suppose $u=c_1$ and $v=c_2$ are two solutions, where a, b are constants.*

*Differentiating $u=c_1$ and $v=c_2$*

$$\frac{\partial u}{\partial x}dx+\frac{\partial u}{\partial y}dy+\frac{\partial u}{\partial z}dz=0 \quad ...(9)$$

and $$\frac{\partial v}{\partial x}dx+\frac{\partial v}{\partial y}dy+\frac{\partial v}{\partial z}dz=0 \quad ...(10)$$

Solving (9) and (10), we get

$$\frac{dx}{\frac{\partial u}{\partial y}\times\frac{\partial v}{\partial z}-\frac{\partial u}{\partial z}\times\frac{\partial v}{\partial y}}=\frac{dy}{\frac{\partial u}{\partial z}\times\frac{\partial v}{\partial x}-\frac{\partial u}{\partial x}\times\frac{\partial v}{\partial z}}=\frac{dz}{\frac{\partial u}{\partial x}\times\frac{\partial v}{\partial y}-\frac{\partial u}{\partial y}\times\frac{\partial v}{\partial x}} \quad ...(11)$$

From (8) and (11) $\frac{dx}{P}=\frac{dy}{Q}=\frac{dz}{R}$

Solutions of these equations are $u=c_1$ and $v=C_2$

$\therefore$ *f(u, v) = 0 is the required solution of* (1).

## 9.7 WORKING RULE

**First step.** Write down the auxiliary equations

$$\frac{dx}{P}=\frac{dy}{Q}=\frac{dz}{R}$$

**Second step.** Solve the above auxiliary equations.

Let the two solutions be $u = c_1$ and $v = c_2$.

**Third step.** Then $f(u, v) = 0$ or $u = \phi(v)$ is the required solution of

$$Pp + Qq = R.$$

**Example 6.** *Solve the following partial differential equation*

$$yq - xp = z. \quad where\ P = \frac{\partial z}{\partial x}, q = \frac{\partial z}{\partial y}.$$

**Solution.** $yq - xp = z$

Here the auxiliary equations are

$$\Rightarrow \quad \frac{dx}{-x} = \frac{dy}{y} = \frac{dz}{z}$$

$$\Rightarrow \quad -\log x = \log y - \log a \qquad \text{(From first two equations)}$$

$$\Rightarrow \quad xy = a \qquad ...(1)$$

$$\Rightarrow \quad \log y = \log z + \log b \qquad \text{(From last two equations)}$$

$$\frac{y}{z} = b \qquad ...(2)$$

From (1) and (2)

Hence the solution is $f\left(xy, \frac{y}{z}\right) = 0$ **Ans.**

**Example 7.** *Solve* $y^2p - xyq = x(z - 2y)$ *(A.M.I.E., Summer 2001)*

**Solution.** $y^2p - xyq = x(z - 2y)$

The auxiliary equations are

$$\frac{dx}{y^2} = \frac{dy}{-xy} = \frac{dz}{x(z-2y)} \qquad ...(1)$$

Considering first two members of the equations

$$\frac{dx}{y} = \frac{dy}{-x} \qquad \Rightarrow x\,dx = -y\,dy$$

Integrating $\frac{x^2}{2} = -\frac{y^2}{2} + \frac{C_1}{2} \qquad \Rightarrow x^2 + y^2 = C_1 \qquad ...(2)$

From last two equations of (1)

$$-\frac{dy}{y} = \frac{dz}{z-2y}$$

$$\Rightarrow \quad -z\,dy + 2y\,dy = y\,dz \qquad \Rightarrow 2y\,dy = y\,dz + z\,dy$$

On integration, we get

$$y^2 = yz + C_2$$

$$y^2 - yz = C_2 \qquad ...(3)$$

From (2) and (3)

$$x^2 + y^2 = f(y^2 - yz)$$ **Ans.**

**Example 8.** *Solve* $(x^2 - yz)\, p + (y^2 - zx)\, q = z^2 - xy$ *(A.M.I.E., Summer 2001)*

**Solution.** $(x^2 - yz)p + (y^2 - zx)q = z^2 - xy$ ...(1)

The auxiliary equations are

$$\frac{dx}{x^2 - yz} = \frac{dy}{y^2 - zx} = \frac{dz}{z^2 - xy}$$

or
$$\frac{dx - dy}{x^2 - yz - y^2 + zx} = \frac{dy - dz}{y^2 - zx - z^2 + xy} = \frac{dz - dx}{z^2 - xy - x^2 + yz}$$

$$\frac{dx - dy}{(x-y)(x+y+z)} = \frac{dy - dz}{(x+y+z)(y-z)} = \frac{dz - dx}{(x+y+z)(z-x)}$$

$$\frac{dx - dy}{(x-y)} = \frac{dy - dz}{(y-z)} = \frac{dz - dx}{(z-x)} \qquad ...(2)$$

Intergrating first members of (2), we have

$$\log (x - y) = \log (y - z) + \log c_1$$

$$\log \frac{x-y}{y-z} = \log c_1 \quad \text{or} \quad \frac{x-y}{y-z} = c_1$$

Similarly from last two members of (2) ,we have

$$\frac{y-z}{z-x} = c_2$$

The required solution is

$$f\left[\frac{x-y}{y-z}, \frac{y-z}{z-x}\right] = 0$$ **Ans.**

## 9.8 METHOD OF MULTIPLIERS

Let the auxiliary equations be

$$\frac{dx}{p} = \frac{dy}{Q} = \frac{dz}{R}$$

*l, m, n* may be constants or functions of *x, y, z* then we have

$$\frac{dx}{p} = \frac{dy}{Q} = \frac{dz}{R} = \frac{l dx + m dy + n dz}{lp + mQ + nR}$$

*l, m, n* are chosen in such a way that

$$lP + mQ + nR = 0$$

Thus $\qquad l dx + m dy + n dz = 0$

Solve this differential equation, if the solution is $u = c_1$.

Similarly, choose another set of multipliers $(l_1, m_1, n_1)$ and if the second solution is $v = C_2$.

$\therefore$ Required solution is $f(u, v) = 0$.

**Example 9.** *Solve*

$$(mz - ny)\frac{\partial z}{\partial x} + (nx - lz)\frac{\partial z}{\partial y} = ly - mx$$ *(A.M.I.E. Winter 2001)*

**Solution.** $(mz - ny)\dfrac{\partial z}{\partial x} + (nx - lz)\dfrac{\partial z}{\partial y} = ly - mx$

Here, the auxiliary equations are

$$\frac{dx}{mz-ny}=\frac{dy}{nx-lz}=\frac{dz}{ly-mx}$$

Using multipliers *x, y, z* we get

$$\text{Each fraction} = \frac{x\,dx+y\,dy+z\,dz}{x(mz-ny)+y(nx-lz)+z(ly-mx)}=\frac{x\,dx+y\,dy+z\,dz}{0}$$

$$\therefore \quad \Rightarrow \qquad x\,dx+y\,dy+z\,dz=0$$

which on integration gives $x^2+y^2+z^2=c_1$ ...(1)

Again using multipliers, *l, m, n,* we get

$$\text{each fraction} = \frac{l\,dx+m\,dy+n\,dz}{l(mz-ny)+m(nx-lz)+n(ly-mx)}=\frac{l\,dx+m\,dy+n\,dz}{0}$$

$$\therefore \quad \Rightarrow \qquad l dx+m\,dy+n\,dz=0$$

which, on integration gives.

$$lx+m\,y+n\,z=c_2 \qquad ...(2)$$

Hence from (1) and (2), the required solution is $x^2+y^2+z^2=f(lx+my+nz)$ **Ans.**

**Example 10.** *Find the general solution of*

$$x(z^2-y^2)\frac{\partial z}{\partial x}+y(x^2-z^2)\frac{\partial z}{\partial y}=z(y^2-x^2)$$

**Solution.** $x(z^2-y^2)\dfrac{\partial z}{\partial x}+y(x^2-z^2)\dfrac{\partial z}{\partial y}=z(y^2-x^2)$

The auxiliary simultaneous equations are

$$\frac{dx}{x(z^2-y^2)}=\frac{dy}{y(x^2-z^2)}=\frac{dz}{z(y^2-x^2)} \qquad ...(1)$$

Using multipliers *x, y, z* we get

Each term of (1) is equal to

$$\frac{x\,dx+y\,dy+z\,dz}{x^2(z^2-y^2)+y^2(x^2-z^2)+z^2(y^2-x^2)}=\frac{xdx+ydy+zdz}{0}$$

$$\Rightarrow \qquad x\,dx+y\,dy+z\,dz=0$$

On integration $\qquad x^2+y^2+z^2=C_1$ ...(2)

Again (1) can be written as

$$\frac{\frac{dx}{x}}{z^2-y^2}=\frac{\frac{dy}{y}}{x^2-z^2}=\frac{\frac{dz}{z}}{y^2-x^2}=\frac{\frac{dx}{x}+\frac{dy}{y}+\frac{dz}{z}}{(z^2-y^2)+(x^2-z^2)+(y^2-x^2)}=\frac{\frac{dx}{x}+\frac{dy}{y}+\frac{dz}{z}}{0}$$

$$\Rightarrow \quad \frac{dx}{x}+\frac{dy}{y}+\frac{dz}{z}=0$$

$\Rightarrow$ $\log x+\log y+\log z=\log C_2$

$\Rightarrow$ $\log xyz=\log C_2 \qquad \Rightarrow \qquad xyz=C_2$ ...(3)

From (2) and (3), the general solution is $xyz=f(x^2+y^2+z^2)$ **Ans.**

**Example 11.** *Solve the partial differential equation*

$$\frac{y-z}{yz}p=\frac{z-x}{zx}q=\frac{x-y}{xy}$$ (*A.M.I.E., Winter 2001*)

**Solution.** $$\frac{y-z}{yz}p=\frac{z-x}{zx}q=\frac{x-y}{xy}$$

Multiplying by $xyz$, we get

$x(y-z)p+y(z-x)q=z(x-y)$

$$\frac{dx}{x(y-z)}=\frac{dy}{y(z-x)}=\frac{dz}{z(x-y)}=\frac{dz+dy+dz}{x(y-z)+y(z-x)+z(x-y)} \quad \dots (1)$$

$$=\frac{dx+dy+dz}{0}$$

$\therefore$ $$dx+dy+dz=0$$

Which on integration gives

$$x+y+z=a \quad \dots (2)$$

Again (1) can be written

$$\frac{\frac{dx}{x}}{y-z}=\frac{\frac{dy}{y}}{z-x}=\frac{\frac{dz}{z}}{x-y}=\frac{\frac{dx}{x}+\frac{dy}{y}+\frac{dz}{z}}{(y-z)+(z-x)+(x-y)}=\frac{\frac{dx}{x}+\frac{dy}{y}+\frac{dz}{z}}{0}$$

or $$\frac{dx}{x}+\frac{dy}{y}+\frac{dz}{z}=0$$

On integration we get

$$\log x+\log y+\log z=\log b \Rightarrow \log xyz=\log b \Rightarrow xyz=b \quad \dots (3)$$

From (2) and (3) the general solution is

$$xyz=f(x+y+z)$$ **Ans.**

**Example 12.** *Solve* $(x^2-y^2-z^2)p+2xy\,q=2xz$. (*A.M.I.E., Summer, 2004, 2000*)

**Solution.** $(x^2-y^2-z^2)p+2xyq=2xz$

Here the auxiliary equations are

$$\frac{dx}{x^2-y^2-z^2}=\frac{dy}{2xy}=\frac{dz}{2xz} \quad \dots (1)$$

From the last two members of (1) we have $dz$

$$\frac{dy}{y}=\frac{dz}{z}$$

which on integration gives

$$\log y=\log z+\log a \quad \text{or} \quad \log\frac{y}{z}=\log a$$

or $$\frac{y}{z}=a \quad \dots (2)$$

Using multipliers $x, y, z$ we have

$$\frac{dx}{x^2-y^2-z^2}=\frac{dy}{2xy}=\frac{dz}{2xz}=\frac{x\,dx+y\,dy+z\,dz}{x(x^2+y^2+z^2)}$$

$$\frac{2x\,dx+2y\,dy+2z\,dz}{(x^2+y^2+z^2)}=\frac{dz}{z}$$

which on integration gives

$$\log(x^2+y^2+z^2)=\log z+\log b$$

$$\frac{x^2+y^2+z^2}{z}=b \qquad ...(3)$$

Hence from (2) and (3), the required solution is

$$x^2+y^2+z^2=z\,f\left(\frac{y}{z}\right)$$ **Ans.**

**Example 13.** *Solve the differential equation*

$$x^2\frac{\partial z}{\partial x}+y^2\frac{\partial z}{\partial y}=(x+y)z.$$

**Solution.** $$x^2\frac{\partial z}{\partial x}+y^2\frac{\partial z}{\partial y}=(x+y)z. \qquad ...(1)$$

The auxiliary equations of (1) are

$$\frac{dx}{x^2}=\frac{dy}{y^2}=\frac{dz}{(x+y)z} \qquad ...(2)$$

Take first two members of (2) and integrate them

$$-\frac{1}{x}=-\frac{1}{y}+c$$

$$\frac{1}{x}-\frac{1}{y}=c_1 \qquad ...(3)$$

(2) can be written as $$\frac{\frac{dx}{x}}{x}=\frac{\frac{dy}{y}}{y}+\frac{\frac{dz}{z}}{x+y}=\frac{\frac{dx}{x}+\frac{dy}{y}-\frac{dz}{z}}{(x+y)-(x+y)}$$

or $$\frac{dx}{x}+\frac{dy}{y}-\frac{dz}{z}=0$$

On integration we get

or $$\log x+\log y-\log z=\log c_2$$

or $$\log\frac{xy}{z}=\log c_2 \qquad \text{or} \qquad \frac{xy}{z}=c_2 \qquad ...(4)$$

From (3) and (4) we have

$$f\left[\frac{1}{x}-\frac{1}{y},\frac{xy}{z}\right]=0$$ **Ans.**

**Example 14.** *Find the general solution of*

$$x\frac{\partial z}{\partial x}+y\frac{\partial z}{\partial y}+t\frac{\partial z}{\partial t}=xyt$$

**Solution.** The auxiliary equations are $$\frac{dx}{x}=\frac{dy}{y}=\frac{dt}{t}=\frac{dz}{xyt} \qquad ...(1)$$

Taking the first two members and integrating, we get

$$\log x=\log y+\log a$$

$\Rightarrow \quad \log x = \log\ ay \quad \Rightarrow \quad x = ay \quad \Rightarrow \quad y/x = a$ ...(2)

Similarly, from the 2nd and 3rd members

$$\frac{t}{y} = b \qquad ...(3)$$

Multiplying the equations (1) by *xyt*, we get

$$dz = \frac{tydx}{1} = \frac{txdy}{1} = \frac{xydt}{1} = \frac{tydx + txdy + xydt}{3}$$

Integrating,

$$z = \frac{1}{3}xyt + c \quad or \quad z - \frac{1}{3}xyt = c \qquad ...(4)$$

From (2), (3) and (4) the solution is

$$z - \frac{1}{3}xyt = f\left(\frac{y}{x}\right) + \phi\left(\frac{t}{y}\right)$$ **Ans.**

**Example 15.** *Solve* $(y + z)p - (x + z)\, q = x - y$

**Solution.** $(y+z)p - (x+z)q = x - y$ ... (1)

$\therefore$ The auxiliary equations are

$$\frac{dx}{y+z} = \frac{dy}{-(x+z)} = \frac{dz}{x-y} \qquad ...(2)$$

$$\Rightarrow \quad \frac{dx}{y+z} = \frac{dy}{-(x+z)} = \frac{dz}{x-y} = \frac{dx+dy+dz}{y+z-(x+z)+x-y}$$

$$\Rightarrow \quad \frac{dz}{x-y} = \frac{dx+dy+dz}{0}$$

Thus, we have $dx + dy + dz = 0$

which on integration gives $x + y + z = c_1$, ...(3)

Let us use multipliers $(x, y, -z)$ for (2)

$$\frac{dx}{y+z} = \frac{dy}{-(x+z)} = \frac{dz}{x-y} = \frac{x\,dx + y\,dy + zdz}{x(y+z) - y(x+z) - z(x-y)}$$

or

$$\frac{dx}{y+z} = \frac{dy}{-(x+z)} = \frac{dz}{x-y} = \frac{x\,dx + y\,dy - z\,dz}{0}$$

Integrating $x\,dx + y\,dy - z\,dz = 0$, we get

$$\frac{x^2}{2} + \frac{y^2}{2} - \frac{z^2}{2} = c_2$$

or

$$x^2 + y^2 - z^2 = 2c_2 \qquad ...(4)$$

From (3) and (4), we get the required solution

$$f(x + y + z,\ x^2 + y^2 - z^2) = 0$$ **Ans.**

**Example 16.** *Solve* $zp + yq = x$

**Solution.** $zp+yq = x$ ...(1)

The auxiliary equations are $\dfrac{dx}{z} = \dfrac{dy}{y} = \dfrac{dz}{x}$

(*i*) (*ii*) (*iii*)

From (*i*) and (*ii*) $\frac{dx}{z}=\frac{dz}{x}$ or $x.dx = z.\,dz$

$$\Rightarrow \quad \frac{x^2}{2}=\frac{z^2}{2}-\frac{c_1}{2} \quad \text{or} \quad x^2 = z^2 - c_1 \qquad ...(2)$$

$$\Rightarrow \quad z=\sqrt{x^2+c_1}$$

Putting the value of z in (1)

$$\frac{dx}{\sqrt{x^2+c_1}}=\frac{dy}{y}$$

$$\sinh^{-1}\frac{x}{\sqrt{c_1}}=\log y+c_2 \quad \text{or} \quad \sinh^{-1}\frac{x}{\sqrt{c_1}}-\log y=c_2 \qquad ...(3)$$

From (2) and (3), the required solution is

$$f\left(z^2-x^2\right)=\sinh^{-1}\frac{x}{\sqrt{c_1}}-\log y$$

**Ans.**

**Example 17.** *Solve* $px\,(z-2y^2)=(z-qy)(z-y^2-2x^3)$. (*A.M.I.E., Summer 2000*)

**Solution.** $px\,(z-2y^2)=(z-qy)\,(z-y^2-2x^3)$ ...(1)

$\Rightarrow$ $px\,(z-2y^2)+qy\,(z-y^2-2x^3)=z\,(z-y^2-2x^3)$

Here the auxiliary equations are

$$\frac{dx}{x\left(z-2y^2\right)}=\frac{dy}{y\left(z-y^2-2x^3\right)}=\frac{dz}{z\left(z-y^2-2x^3\right)} \qquad ...(2)$$

From the last two members of (2) we have

$$\frac{dy}{y}=\frac{dz}{z}$$

which gives on integration

$\log y=\log z+\log a$ or $y=a\,z$ ...(3)

From the first and third members of (2) we have

$$\frac{dx}{x(z-2y^2)}=\frac{dz}{z(z-y^2-2x^3)} \qquad \text{Put } y=az$$

$$\Rightarrow \quad \frac{dx}{x\left(z-2a^2z^2\right)}=\frac{dz}{z\left(z-a^2z^2-2x^3\right)}$$

$$\frac{dx}{x\left(1-2a^2z\right)}=\frac{dz}{z-a^2z^2-2x^3}$$

$$\Rightarrow \quad z\,dx-a^2z^2dx-2x^3dx=xdz-2a^2xz\,dz$$

$$\Rightarrow \quad (xdz-zdx)-a^2\,(2xz\,dz-z^2\,dx)+2x^3dx=0$$

On integrating, we have

$$\frac{z}{x}-a^2\frac{z^2}{x}+x^2=b \qquad ...(4)$$

From (3) and (4), we have

$$\frac{y}{z}=f\left(\frac{z}{x}-\frac{a^2z^2}{x}+x^2\right)$$

**Ans.**

## EXERCISE 9.3

Solve the following partial differential equations :

1. $p \tan x + q \tan y = \tan z$ **Ans.** $f\left(\dfrac{\sin x}{\sin y}, \dfrac{\sin y}{\sin z}\right) = 0$

2. $y\dfrac{\partial z}{\partial x} + y\dfrac{\partial z}{\partial y} = z^2 + 1$ *(AMIE. Winter 2002)* **Ans.** $f(x - y) = \log y - \tan^{-1} z$

3. $(y - z)p + (x - y)q = z - x$ **Ans.** $f(x + y + z, x^2 + 2yz) = 0$

4. $(y + zx)p - (x + yz)q = x^2 - y^2$ **Ans.** $f(x^2 + y^2 - z^2) = (x - y)^2 - (z + 1)^2$

5. $zx\dfrac{\partial z}{\partial x} - zy\dfrac{\partial z}{\partial y} = y^2 - x^2$ **Ans.** $f(x^2 + y^2 + z^2, xy) = 0$

6. $pz - qz = z^2 + (x + y)^2$ **Ans.** $[z^2 + (x + y)^2]\, e^{-2x} = f(x + y)$

7. $p + q + 2xz = 0$ **Ans.** $f(x - y) = x^2 + \log z$

8. $x^2p + y^2q + z^2 = 0$ **Ans.** $f\left(\dfrac{1}{y} - \dfrac{1}{x}, \dfrac{1}{y} + \dfrac{1}{z}\right) = 0$

9. $(x^2 + y^2)p + 2xyq = (x + y)z$ **Ans.** $f\left(\dfrac{x+y}{z}, \dfrac{2y}{x^2 - y^2}\right) = 0$

10. $\dfrac{\partial z}{\partial x} - 2\dfrac{\partial z}{\partial y} = 2x - e^y + 1$ **Ans.** $f(2x + y) = z - \dfrac{(2x+1)^2}{4} - \dfrac{e^y}{2}$

11. $p + 3q = 5z + \tan(y - 3x)$ **Ans.** $f(y - 3x) = \dfrac{e^{5x}}{5z + \tan(y - 3x)}$

12. $xp - yq + x^2 - y^2 = 0$ **Ans.** $f(xy) = \dfrac{x^2}{2} + \dfrac{y^2}{2} + z$

13. $(x+y)\left(\dfrac{\partial z}{\partial x} + \dfrac{\partial z}{\partial y}\right) = z - 1$ **Ans.** $f(x - y) = \dfrac{x+y}{(z-1)^2}$

14. $(x^3 + 3xy^2)\dfrac{\partial z}{\partial x} + (y^3 + 3x^2y)\dfrac{\partial z}{\partial y} = 2(x^2 + y^2)z$ **Ans.** $f\left(\dfrac{xy}{z^2}, (x - y)^{-2} - (x + y)^{-2}\right) = 0$

15. $(z^2 - 2yz - y^2)P + (xy + zx)q = xy - zx$ **Ans.** $(x^2 + y^2 + z^2) = f(y^2 - 2yz - z^2)$

16. Find the solution of the equation $\dfrac{x\partial z}{\partial y} - \dfrac{y\partial z}{\partial x} = 0$, which passes through the curve $z = 1$, $x^2 + y^2 = 4$ **Ans.** $f(x^2 + y^2 - 4, z - 1) = 0$

17. $2x(y + z^2)p + y(2y + z^2)q = z^3$ *(AMIE Winter 2003)*

18. $3\dfrac{\partial u}{\partial x} + 2\dfrac{\partial u}{\partial y} = 0$, $u(x, 0) = 4e^{-x}$ **Ans.** $u = ue^{-x+\frac{3y}{2}}$

19. $4\dfrac{\partial u}{\partial t} + \dfrac{\partial u}{\partial x} = 3u$, when $t = 0$, $u = 3e^{-x} - e^{-5x}$ **Ans.** $u = 3e^{-x+t} - 3e^{-5x+2t}$

## 9.9 PARTIAL DIFFERENTIAL EQUATIONS NON-LINEAR IN p AND q.

We give below the methods of solving non-linear partial differential equations in certain standard form only.

**Type I.** *Equation of the Type* $f(p, q) = 0$ *i.e.*, equations containing $p$ and $q$ only.

**Method.** Let the required solution be

$$z = ax + by + c \quad ...(1)$$

$$\therefore \quad \frac{\partial z}{\partial x} = a, \quad \frac{\partial z}{\partial y} = b.$$

On putting these values in $f(p, q) = 0$
we get $f(a, b) = 0,$
From this, find the value of $b$ in terms of $a$ and substitute the value of $b$ in (1), that will be the required solution.

**Example 18.** *Solve* $p^2 + q^2 = 1$ ...(1)
**Solution.** Let $z = ax + by + c$ ...(2)

$$\therefore \quad p = \frac{\partial z}{\partial x} = a, \qquad q = \frac{\partial z}{\partial y} = b$$

On substituting the values of $p$ and $q$ in (1), we have

$$\therefore \quad a^2 + b^2 = 1 \text{ or } \quad b = \sqrt{1-a^2}$$

Putting the value of $b$ in (2), we get $z = ax + \sqrt{1-a^2}\, y + c$

This is the required solution. **Ans.**

**Example 19.** *Solve* $x^2p^2 + y^2 q^2 = z^2$. *(RGPV, Bhopal, Feb. 2008)*
**Solution.** This equation can be transformed in the above type.

$$\frac{x^2}{z^2} p^2 + \frac{y^2}{z^2} q^2 = 1$$

$$\Rightarrow \quad \left(\frac{x}{z}\frac{\partial z}{\partial x}\right)^2 + \left(\frac{y}{Z}\frac{\partial z}{\partial y}\right)^2 = 1 \Rightarrow \left(\frac{\frac{\partial z}{z}}{\frac{\partial x}{x}}\right)^2 + \left(\frac{\frac{\partial z}{z}}{\frac{\partial y}{y}}\right)^2 = 1 \quad \text{...(1)}$$

Let $\quad \frac{\partial z}{z} = \partial Z, \quad \frac{\partial x}{x} = \partial X, \quad \frac{\partial y}{y} = \partial Y,$

$\therefore \quad \log z = Z, \quad \log x = X, \quad \log y = Y$
$\therefore$ (1) can be written as

$$\left(\frac{\partial Z}{\partial X}\right)^2 + \left(\frac{\partial Z}{\partial Y}\right)^2 = 1 \quad \text{...(2)}$$

$$\Rightarrow \quad P^2 + Q^2 = 1$$

Let the required solution be

$$Z = aX + bY + c$$

$$P = \frac{\partial Z}{\partial X} = a, \; Q = \frac{\partial Z}{\partial Y} = b$$

From (2) we have

$$a^2 + b^2 = 1 \text{ or } b = \sqrt{1-a^2}$$

$$Z = aX + \sqrt{1-a^2}\; Y + c$$

$$\log z = a \log x + \sqrt{1-a^2}\; \log y + c$$ **Ans.**

## EXERCISE 9.4

Solve the following partial differential equations

1. $pq = 1$ **Ans.** $z = ax + \frac{1}{a} y + c$
2. $\sqrt{p} + \sqrt{q} = 1$ **Ans.** $z = ax + (1 - \sqrt{a})^2 y + c$
3. $p^2 - q^2 = 1$ **Ans.** $z = ax - \sqrt{(a^2 - 1)}\; y + c$
4. $pq + p + q = 0$ **Ans.** $z = ax - \frac{a}{1+a} y + c$

**Type II.** *Equation of the type*

$$z = px + qy + f(p,q)$$

Its solution is $z = ax + by + f(a, b)$

**Example 20.** *Solve* $z = px + qy + p^2 + q^2$

**Solution.** $z = px + qy + p^2 + q^2$ $\qquad p = a, q = b$

Its solution is $z = ax+by+a^2 + b^2$ **Ans.**

**Example 21.** *Solve* $z = px + qy + 2\sqrt{pq}$

**Solution.** $z = px+ qy + 2\sqrt{pq}$

Its solution is $z = ax+by+2\sqrt{ab}$ **Ans.**

**Type III. Equation of the type f(z, p, q) = 0** equations not containing $x$ and $y$.

Let $z$ be a function of $u$ where

$$u = x + ay.$$

$$\frac{\partial u}{\partial x} = 1 \quad \text{and} \quad \frac{\partial u}{\partial y} = a$$

Then
$$p = \frac{\partial z}{\partial x} = \frac{dz}{du}.\frac{\partial u}{\partial x} = \frac{dz}{du}$$

$$q = \frac{\partial z}{\partial y} = \frac{dz}{du}.\frac{\partial u}{\partial y} = \frac{dz}{du}(a)$$

**On putting the values of $p$ and $q$ in the given equation $f(z, p, q) = 0$, it becomes**

$f\left(z, \frac{dz}{dy}, a\frac{dz}{du}\right) = 0$ **which is an ordinary differential equation of the first order.**

**Rule.** Assume $u = x + ay$; **replace $p$ and $q$ by $\frac{dz}{du}$ and $a\frac{dz}{du}$ in the given equation and then solve the ordinary differential equation obtained.**

**Example 22.** *Solve*

$$p(1 + q) = qz$$

**Solution.** $p(1 + q) = qz$ ... (1)

Let $u = x + ay \Rightarrow \frac{\partial u}{\partial x} = 1, \quad \frac{\partial u}{\partial y} = a$

$$p = \frac{\partial z}{\partial x} = \frac{dz}{du}\frac{\partial u}{\partial x} = \frac{dz}{du} \quad \text{and} \quad q = \frac{\partial z}{\partial y} = \frac{dz}{du}\frac{\partial u}{\partial y} = a\frac{dz}{dv}$$

(1) becomes

$$\frac{dz}{du}\left(1 + a\frac{dz}{du}\right) = a\frac{dz}{du}z \quad \text{or} \quad 1 + a\frac{\partial z}{\partial u} = az$$

$$\Rightarrow \quad a\frac{dz}{du} = az - 1 \Rightarrow du = \frac{a\,dz}{az - 1}$$

Integrating, we get

$$u = \log(az - 1) + \log c$$

$$x + ay = \log c\,(az - 1)$$ **Ans.**

**Example 23.** *Solve* $p(1+q^2) = q(z-a)$.

**Solution.** Let $u = x + by$

So that $$p = \frac{dz}{du} \quad \text{and} \quad q = b\frac{dz}{du}$$

Substituting these values of $p$ and $q$ in the given equation, we have

$$\frac{dz}{du}\left[1+b^2\left(\frac{dz}{du}\right)^2\right] = b\frac{dz}{du}(z-a)$$

$$1+b^2\left(\frac{dz}{du}\right)^2 = b(z-a) \quad \text{or} \quad b^2\left(\frac{dz}{du}\right)^2 = bz-ab-1$$

$$\frac{dz}{du} = \frac{1}{b}\sqrt{bz-ab-1}$$

$$\int\frac{b\,dz}{\sqrt{bz-ab-1}} = \int du + c$$

$$2\sqrt{bz-ab-1} = u+c$$

$$4(bz-ab-1) = (u+c)^2$$

$$4(bz-ab-1) = (x+by+c)^2$$ **Ans.**

**Example 24.** *Solve* $z^2(p^2x^2+q^2) = 1$ ...(1)

**Solution.** $z^2(p^2x^2+q^2) = 1$

$$\Rightarrow \quad z^2\left[\left(x\frac{\partial z}{\partial x}\right)^2+\left(\frac{\partial z}{\partial y}\right)^2\right] = 1 \quad \Rightarrow \quad z^2\left[\left(\frac{\frac{\partial z}{\partial x}}{x}\right)^2+\left(\frac{\partial z}{\partial y}\right)^2\right] = 1$$

$$\Rightarrow \quad z^2\left[\left(\frac{\partial z}{\partial X}\right)^2+\left(\frac{\partial z}{\partial y}\right)^2\right] = 1 \qquad ...(2)$$

where $\frac{\partial x}{x} = \partial X$ or $\log x = X$

Let $$u = X + ay$$

$$\frac{\partial z}{\partial X} = \frac{dz}{du} \quad \text{and} \quad \frac{\partial z}{\partial y} = a\frac{dz}{du}$$

Then (2) becomes

$$z^2\left[\left(\frac{dz}{du}\right)^2+\left(a\frac{dz}{du}\right)^2\right] = 1 \Rightarrow \left(\frac{dz}{du}\right)^2+a^2\left(\frac{dz}{du}\right)^2 = \frac{1}{z^2}$$

$$\Rightarrow \quad \left(\frac{dz}{du}\right)^2 = \frac{1}{z^2(1+a^2)} \Rightarrow \frac{dz}{du} = \frac{1}{z\sqrt{1+a^2}} \Rightarrow z\,dz = \frac{du}{\sqrt{1+a^2}}$$

$$\Rightarrow \quad \int z\,dz = \int\frac{du}{\sqrt{1+a^2}}+c \quad \text{or} \quad \frac{z^2}{2} = \frac{u}{\sqrt{1+a^2}}+c$$

$$\sqrt{1+a^2}\,\frac{z^2}{2} = u + c\sqrt{1+a^2}$$

$$= X + ay + c\sqrt{1+a^2}$$

$$= \log x + ay + c\sqrt{1+a^2}$$ **Ans.**

## EXERCISE 9.5

Solve

1. $z^2(p^2+q^2+1)=1$ **Ans.** $(1-z^2)^{\frac{1}{2}}=-\dfrac{x+ay}{\sqrt{1+a^2}}+c$

2. $1+q^2=q(z-a)$ **Ans.** $\dfrac{x+by}{b}+\dfrac{1}{4}(z-a)^2=\dfrac{1}{4}(z-a)\sqrt{(z-a)^2-2^2}+4\cosh^{-1}\left(\dfrac{z-a}{2}\right)$

3. $x^2p^2+y^2q^2=z$ **Ans.** $2\sqrt{z}=\dfrac{\log x+a\log y}{\sqrt{1+a^2}}+c$

**Type IV.** *Equation of the type* $f_1(x,p)=f_2(y,q)$

**In these equations, $z$ is absent and the terms containing $x$ and $p$ can be written on one side and the terms containing $y$ and $q$ can be written on the other side.**

**Method.** Let $f_1(x,p)=f_2(y,q)=a$

$f_1(x,p)=a$, solve it for $p$. Let $p=F_1(x)$

$f_2(y,q)=a$, solve it for $q$. Let $q=F_2(y)$

Since $dz=\dfrac{\partial z}{\partial x}dx+\dfrac{\partial z}{\partial y}dy \Rightarrow dz=p\,dx+q\,dy$

$\Rightarrow dz=F_1(x)\,dx+F_2(y)\,dy \Rightarrow z=\int F_1(x)\,dx+\int F_2(y)\,dy+c$

**Example 25.** *Solve* $p-x^2=q+y^2$.

**Solution.** $p-x^2=q+y^2=c$ (say)

*i.e.* $p=x^2+c$ and $q=c-y^2$

Putting these values of $p$ and $q$ in

$$dz=pdx+qdy=(x^2+c)\,dx+(c-y^2)\,dy$$

$$z=\left(\frac{x^3}{x}+cx\right)+\left(cy-\frac{y^3}{3}\right)+c_1$$ **Ans.**

**Example 26.** *Solve* $p^2+q^2=z^2(x+y)$.

**Solution.** $p^2+q^2=z^2(x+y) \Rightarrow \left(\dfrac{p}{z}\right)^2+\left(\dfrac{q}{z}\right)^2=(x+y)$

$\Rightarrow \left(\dfrac{1}{z}\dfrac{\partial z}{\partial x}\right)^2+\left(\dfrac{1}{z}\dfrac{\partial z}{\partial y}\right)^2=x+y \Rightarrow \left(\dfrac{\frac{\partial z}{z}}{\partial x}\right)^2+\left(\dfrac{\frac{\partial z}{z}}{\partial y}\right)^2=x+y$

$\Rightarrow \left(\dfrac{\partial z}{\partial x}\right)^2+\left(\dfrac{\partial z}{\partial y}\right)^2=x+y$ where $\dfrac{\partial z}{z}=\partial Z$ or $\log z=Z$

$\Rightarrow p^2+Q^2=x+y \Rightarrow p^2-x=y-Q^2=a$

$P^2-x=a \Rightarrow P=\sqrt{a+x}$

$y-Q^2=a \Rightarrow Q=\sqrt{y-a}$

Therefore, the equation $dZ=\dfrac{\partial Z}{\partial x}dx+\dfrac{\partial Z}{\partial y}dy$

$dZ=Pdx+Qdy$ gives

$$dZ=\sqrt{a+x}\,dx+\sqrt{y-a}\,dy$$

$$z = \int \sqrt{a+x}\, dx + \int \sqrt{y-a}\, dy + c$$

$$\Rightarrow \quad \log z = \frac{2}{3}(a+x)^{3/2} + \frac{2}{3}(y-a)^{3/2} + c \qquad \textbf{Ans.}$$

## EXERCISE 9.6

Solve

1. $q - p + x - y = 0$ **Ans.** $2z = (x+a)^2 + (y+a)^2 + b$

2. $\sqrt{p} + \sqrt{q} = 2x$ **Ans.** $z = \frac{1}{6}(2x - a)^3 + a^2 y + b$

3. $q = xp + p^2$ **Ans.** $z = -\frac{x^2}{4} + \left\{\frac{x\sqrt{x^2+4a}}{4} + a\log\left(x + \sqrt{x^2+4a}\right)\right\} + ay + b$

4. $z^2(p^2 + q^2) = x^2 + y^2$

**Ans.** $z^2 = x\sqrt{x^2+a} + a\log(x + \sqrt{x^2+a}) + y\sqrt{y^2-a} - a\log(y + \sqrt{y^2-a}) + 2b$

5. $z(p^2 + q^2) = x - y$ **Ans.** $z^{3/2} = (x+a)^{3/2} + (y+a)^{3/2} + b$

6. $p^2 - q^2 = x - y$ **Ans.** $z = \frac{2}{3}(x+c)^{3/2} + \frac{2}{3}(y+c)^{3/2} + c_1$

7. $(p^2 + q^2)y = qz$ **Ans.** $z^2 = (cx + a)^2 + c^2y^2$

8. Tick √ the correct answer.

(*a*) The partial differential equation from $z = (a + x)^2 + y$ is

(*i*) $z = \frac{1}{4}\left(\frac{\partial z}{\partial x}\right)^2 + y$ (*ii*) $z = \frac{1}{4}\left(\frac{\partial z}{\partial y}\right)^2 + y$ (*iii*) $z = \left(\frac{\partial z}{\partial x}\right)^2 + y$ (*iv*) $z = \left(\frac{\partial z}{\partial y}\right)^2 + y$

(*b*) The solution of $xp + yq = z$ is

(*i*) $f(x, y) = 0$ (*ii*) $f\left(\frac{x}{y}, \frac{y}{z}\right) = 0$ (*iii*) $f(xy, yz) = 0$ (*iv*) $f(x^2, y^2) = 0$

(*c*) The solution of $p + q = z$ is

(*i*) $f(x + y, y + \log z) = 0$ (*ii*) $f(xy, y\log z) = 0$

(*iii*) $f(x - y, y - \log z) = 0$ (*iv*) None of these

(*d*) The solution of $(y - z)p + (z - x)q = x - y$ is

(*i*) $f(x + y + z) = xyz$ (*ii*) $f(x^2 + y^2 + z^2) = xyz$

(*iii*) $f(x^2 + y^2 + z^2, x^2y^2z^2) = 0$ (*iv*) $f(x+y+z) = x^2 + y^2 + z^2$

**Ans.** (*a*) (*i*), (*b*) (*ii*), (*c*) (*iii*), (*d*), (*iv*)

## 9.10 CHARPIT'S METHOD

General method for solving partial differential equation with two independent variables.

**Solution.** Let the general partial differential equation be

$$f(x, y, z, p, q) = 0 \qquad \text{... (1)}$$

Since $z$ depends on $x, y$, we have

$$dz = \frac{\partial z}{\partial x}dx + \frac{\partial z}{\partial y}dy$$

$$dz = p\,dx + q\,dy \qquad \text{... (2)}$$

The main aim in Charpits method is to find another relation between the variables $x, y, z$ and $p, q$. Let the relation be

$$\phi(x, y, z, p, q) = 0 \qquad \text{...(3)}$$

On solving (1) and (3), we get the values of $p$ and $q$.

These values of $p$ and $q$ when substituted in (2), it becomes integrable.

To determine $\phi$, (1) and (3) are differentiated w.r.t. $x$ and $y$ giving

$$\left.\begin{aligned}&\frac{\partial f}{\partial x}+\frac{\partial f}{\partial z}p+\frac{\partial f}{\partial p}\frac{\partial p}{\partial x}+\frac{\partial f}{\partial q}\frac{\partial q}{\partial x}=0\\&\frac{\partial \phi}{\partial x}+\frac{\partial \phi}{\partial z}p+\frac{\partial \phi}{\partial p}\frac{\partial p}{\partial x}+\frac{\partial \phi}{\partial q}\frac{\partial q}{\partial x}=0\end{aligned}\right\}\text{ w.r.t. x, (First pair)}$$

$$\left.\begin{aligned}&\frac{\partial f}{\partial y}+\frac{\partial f}{\partial z}q+\frac{\partial f}{\partial p}\frac{\partial p}{\partial y}+\frac{\partial f}{\partial q}\frac{\partial q}{\partial y}=0\\&\frac{\partial \phi}{\partial y}+\frac{\partial \phi}{\partial z}q+\frac{\partial \phi}{\partial p}\frac{\partial p}{\partial y}+\frac{\partial \phi}{\partial q}\frac{\partial q}{\partial y}=0\end{aligned}\right\}\text{ w.r.t. } y\text{, (Second pair)}$$

Eliminating $\frac{\partial p}{\partial x}$ between the equation of first pair, we have

$$-\frac{\partial p}{\partial x}=\frac{\frac{\partial f}{\partial x}+\frac{\partial f}{\partial z}p+\frac{\partial f}{\partial q}\frac{\partial q}{\partial x}}{\frac{\partial f}{\partial p}}=\frac{\frac{\partial \phi}{\partial x}+\frac{\partial \phi}{\partial z}p+\frac{\partial \phi}{\partial q}\frac{\partial q}{\partial x}}{\frac{\partial \phi}{\partial p}}$$

or $$\left(\frac{\partial f}{\partial x}\frac{\partial \phi}{\partial p}-\frac{\partial \phi}{\partial x}\frac{\partial f}{\partial p}\right)+p\left(\frac{\partial f}{\partial z}\frac{\partial \phi}{\partial p}-\frac{\partial \phi}{\partial z}\frac{\partial f}{\partial p}\right)+\frac{\partial q}{\partial y}\left(\frac{\partial f}{\partial q}\frac{\partial \phi}{\partial p}-\frac{\partial \phi}{\partial q}\frac{\partial f}{\partial p}\right)=0 \quad ...(4)$$

On eliminating $\frac{\partial q}{\partial y}$ between the equations of second pair, we have

$$\left(\frac{\partial f}{\partial y}\frac{\partial \phi}{\partial q}-\frac{\partial \phi}{\partial y}\frac{\partial f}{\partial q}\right)+q\left(\frac{\partial f}{\partial z}\frac{\partial \phi}{\partial q}-\frac{\partial \phi}{\partial z}\frac{\partial f}{\partial q}\right)+\frac{\partial q}{\partial y}\left(\frac{\partial f}{\partial p}\frac{\partial \phi}{\partial q}-\frac{\partial \phi}{\partial p}\frac{\partial f}{\partial q}\right)=0 \quad ...(5)$$

Adding (4) and (5) and keeping in view the relation on, the terms of the last brackets of (4) and(5) cancel. On rearranging, we get

$$\frac{\partial \phi}{\partial f}\left(\frac{\partial f}{\partial x}+p\frac{\partial f}{\partial z}\right)+\frac{\partial \phi}{\partial q}\left(\frac{\partial f}{\partial y}+q\frac{\partial f}{\partial z}\right)+\frac{\partial \phi}{\partial z}\left(-p\frac{\partial f}{\partial p}-q\frac{\partial f}{\partial q}\right)+\left(-\frac{\partial f}{\partial p}\right)\frac{\partial \phi}{\partial x}+\left(-\frac{\partial f}{\partial q}\right)\frac{\partial \phi}{\partial y}=0$$

or $$\left(-\frac{\partial f}{\partial p}\right)\left(\frac{\partial \phi}{\partial x}\right)+\left(-\frac{\partial f}{\partial q}\right)\frac{\partial \phi}{\partial y}+\left(-p\frac{\partial f}{\partial p}-q\frac{\partial f}{\partial q}\right)\frac{\partial \phi}{\partial z}+\left(\frac{\partial f}{\partial x}+p\frac{\partial f}{\partial z}\right)+\left(\frac{\partial f}{\partial y}+q\frac{\partial f}{\partial z}\right)\frac{\partial \phi}{\partial q}=0 \quad ...(6)$$

Equation (6) is a Lagrange's linear equation of the first order with $x, y, z, p, q$ as independent variables and 4) as dependent variable. Its subsidiary equations are

$$\frac{dx}{-\frac{\partial f}{\partial p}}=\frac{dy}{-\frac{\partial f}{\partial q}}=\frac{dz}{-p\frac{\partial f}{\partial p}--q\frac{\partial f}{\partial q}}=\frac{dp}{\frac{\partial f}{\partial x}+p\frac{\partial f}{\partial z}}=\frac{dq}{\frac{\partial f}{\partial y}+q\frac{\partial f}{\partial z}}=\frac{\partial \phi}{0} \quad ...(7)$$

(Commit to memory)

Any of the integrals of (7) satisfies (6). Such an integral involving $p$ or $q$ or both may be taken as assumed relation (3). However, we should choose the simplest integral involving $p$ and $q$ derived from (7). This relation and equation (1) gives the values of $p$ and $q$. The values of $p$ and $q$ are substituted in (2). On integration new eq. (2) gives the solution of (1).

**Example 27.** *Solve* $px + qy = pq$

**Solution.** $f(x, y, z, p, q) = 0$ is $px + qy - pq = 0$ ...(1)

$$\frac{\partial f}{\partial x}=p,\quad \frac{\partial f}{\partial y}=q,\quad \frac{\partial f}{\partial z}=0,\quad \frac{\partial f}{\partial p}=x-q,\quad \frac{\partial f}{\partial q}=y-p$$

Charpits' equations are

$$\frac{dx}{-\frac{\partial f}{\partial p}}=-\frac{dy}{-\frac{\partial f}{\partial q}}=\frac{dz}{-p\frac{\partial f}{\partial p}--q\frac{\partial f}{\partial q}}=\frac{dp}{\frac{\partial f}{\partial x}+p\frac{\partial f}{\partial z}}=\frac{dq}{\frac{\partial f}{\partial y}+q\frac{\partial f}{\partial z}}=\frac{\partial\phi}{0}$$

$$\frac{dx}{-(x-q)}=\frac{dy}{-(y-p)}=\frac{dz}{-p(x-q)-q(y-p)}=\frac{dp}{p}=\frac{dq}{q}=\frac{d\phi}{0}$$

We have to choose the simplest integral involving $p$ and $q$

$\Rightarrow$ $$\frac{dp}{p}=\frac{dq}{q} \text{ or } \log p=\log q+\log a \Rightarrow p=aq$$

Putting for $p$ in the given equation (1), we get

$$q\,(ax+y)=aq^2 \qquad \therefore\quad q=\frac{y+ax}{a}$$

$\therefore$ $$p=aq=y+ax$$

Now $$dz=pdx+qdy \qquad ...(2)$$

Putting for $p$ and $q$ in (2), we get

$$dz=(y+ax)\,dx+\frac{y+ax}{a}dy$$

$$adz=(y+ax)+(y+ax)\,dy$$

$$adz=(y+ax)\,(adx+dy)$$

Integrating $$az=\frac{(y+ax)^2}{2}+b$$ **Ans.**

**Example 28**. *Solve* $(p^2+q^2)y=qz.$ ... (1)

**Solution.** $f(x,y,z,p,q)=0$ is $(p^2+q^2)y-qz=0$

$$\frac{\partial f}{\partial x}=0,\quad \frac{\partial f}{\partial x}=p^2+q^2,\quad \frac{\partial f}{\partial z}=-q,\quad \frac{\partial f}{\partial p}=2py,\quad \frac{\partial f}{\partial q}=2qy-z$$

Now Charpits equations are

$$\frac{dx}{-\frac{\partial f}{\partial p}}=\frac{dy}{-\frac{\partial f}{\partial q}}=\frac{dz}{-p\frac{\partial f}{\partial p}-q\frac{\partial f}{\partial q}}=\frac{dp}{\frac{\partial f}{\partial x}+p\frac{\partial f}{\partial z}}=\frac{dq}{\frac{\partial f}{\partial y}+q\frac{\partial f}{\partial z}}$$

$$\Rightarrow\quad \frac{dx}{-2py}=\frac{dy}{-2q+z}=\frac{dz}{-2p^2y-2q^2y+qz}=\frac{dp}{-pq}=\frac{dq}{p^2+q^2-q^2}=\frac{d\phi}{0}$$

We have to choose the simplest integral involving $p$ and $q$.

$$\frac{dp}{-pq}=\frac{dq}{p^2} \Rightarrow \frac{dp}{q}=\frac{dq}{p} \Rightarrow pdp+qdp=0$$

Integrating $$p^2+q^2=a^2\text{(say)}$$

Putting for $p^2+q^2$ in the equation (1), we get

$$a^2y=qz \Rightarrow q=\frac{a^2y}{z} \quad\text{so}\quad p=\sqrt{a^2-q^2}=\sqrt{a^2-\frac{a^4y^2}{z^2}}$$

$$p=\frac{a}{z}\sqrt{z^2-a^2y^2}$$

Now $$dz = p\,dx + q\,dy \qquad ...(2)$$

Putting for $p$ and $q$ in (2), we get,

$$dz = \frac{a}{z}\sqrt{z^2 - a^2y^2}\,dx + \frac{a^2y}{z}dy \quad \Rightarrow \quad zdz = a\sqrt{z^2 - a^2y^2} + a^2y\,dy$$

$$\frac{zdz - a^2y\,dy}{\sqrt{z^2 - a^2y^2}} = a.dx$$

Integrating, we get, $\frac{1}{2}\frac{2}{1}\sqrt{z^2 - a^2y^2} = ax + b$

On squaring, $z^2 - a^2y^2 = (ax+b)^2$ **Ans.**

## EXERCISE 9.7

**Solve the following:**

1. $z = p \cdot q$ **Ans.** $2\sqrt{az} = ax + y + \sqrt{ab}$
2. $(p+q)(px+qy) - 1 = 0$ **Ans.** $z\sqrt{(1+a)} = 2\sqrt{(ax+y)} + b$
3. $z = px + gy + p^2 + q^2$ **Ans.** $z = ax + by + a^2 + b^2$
4. $z = p^2x + q^2y$ **Ans.** $(1+a)z = \left[\sqrt{ax} + \sqrt{(b+y)}\right]^2$
5. $z^2 = pq\,xy$ **Ans.** $z = ax^b y^{1/b}$
6. $px + pq + qy = yz$ **Ans.** $\log(z - ax) = y - a\log(a+y) + b$
7. $q + xp = p^2$ **Ans.** $z = ax\,e^{-y} - \frac{1}{2}a^2e^{-2y} + b$

## 9.11 LINEAR HOMOGENEOUS PARTIAL DIFFERENTIAL EQUATIONS OF nTH ORDER WITH CONSTANT COEFFICIENTS

An equation of the type

$$a_0\frac{\partial^n z}{\partial x^n} + a_1\frac{\partial^n z}{\partial x^{n-1}\partial y} + ..... + a_n\frac{\partial^n z}{\partial y^n} = F(x,y) \qquad ..(1)$$

is called a homogeneous linear partial differential equation of nth order with constant coefficients. It is called homogeneous because all the terms contain derivatives of the same order.

Putting $\frac{\partial}{\partial x} = D$ and $\frac{\partial}{\partial y} = D'$, (1) becomes

$$(a_0 + D^n + a_1D^{n-1}D' + ...... + a_n D'^n)z = F(x,y)$$

or
$$f(D, D')z = F(x,y)$$

## 9.12 RULES FOR FINDING THE COMPLEMENTARY FUNCTION

Consider the equation

$$a_0\frac{\partial^2 z}{\partial x^2} + a_1\frac{\partial^2 z}{\partial x\partial y} + a_2\frac{\partial^2 z}{\partial y^2} = 0 \qquad \text{or} \qquad (a_0D^2 + a_1DD' + a_2D'^2)z = 0$$

**1st step :** Put $D = m$ and $D' = 1$

$$a_0m^2 + a_1m + a_2 = 0$$

This is the auxiliary equation.

**2nd step :** Solve the auxiliary equation.

**Case 1.** If the roots of the auxiliary equation are real and different; say $m_1, m_2$

Then C.F. $= f_1(y + m_1x) + f_2(y + m_2x)$.

**Case 2.** If the roots are equal; say $m$

Then C.F. $= f_1(y + mx) + xf_2(y + mx)$

**Example 29.** *Solve* $(D^3 - 4D^2 D' + 3DD'^2)z = 0$.

**Solution.** $(D^3 - 4D^2 D' + 3DD'^2)z = 0$ $\qquad [D = m,\ D' = 1]$

Its auxiliary equation is

$$m^3 - 4m^2 + 3m = 0 \quad \Rightarrow \quad m(m^2 - 4m + 3) = 0$$

$$m(m-1)(m-3) = 0 \quad \Rightarrow \quad m = 0, 1, 3$$

The required solution is $z = f_1(y) + f_2(y + x) + f_3(y + 3x)$ **Ans.**

**Example 30.** *Solve* $\dfrac{\partial^2 z}{\partial x^2} - 4\dfrac{\partial^2 z}{\partial x \partial y} + 4\dfrac{\partial^2 z}{\partial y^2} = 0$

**Solution.** $(D^2 - 4DD' + 4D'^2)z = 0$

Its auxiliary equation is $[D = m,\ D' = 1]$

$$m^2 - 4m + 4 = 0 \quad \Rightarrow \quad (m-2)^2 = 0 \Rightarrow m = 2, 2$$

The required solution is $z = f_1(y + 2x) + xf_2(y + 2x)$ **Ans.**

## EXERCISE 9.8

Solve the following equations :

1. $\dfrac{\partial^2 z}{\partial x^2} + \dfrac{4\partial^2 z}{\partial x \partial y} - 5\dfrac{\partial^2 z}{\partial y^2} = 0$ **Ans.** $z = f_1(y + x) + f_2(y - 5x)$

2. $2\dfrac{\partial^2 z}{\partial x^2} + 5\dfrac{\partial^2 z}{\partial x \partial y} + 2\dfrac{\partial^2 z}{\partial y^2} = 0$ **Ans.** $z = f_1(2y - x) + f_2(y - 2x)$

3. $(D^3 - 6D^2 D' + 11DD'^2 - 6D'^3)z =$ **Ans.** $z = f_1(y + x) + f_2(y + 2x) + f_3(y + 3x)$

4. $\dfrac{\partial^2 z}{\partial x^2} - 2\dfrac{\partial^2 z}{\partial x \partial y} + \dfrac{\partial^2 z}{\partial y^2} = 0$ **Ans.** $z = fi(y + x) + xf_2(y + x)$

5. $(D^3 - 6.D^2 D' + 12DD'^2 - 8D'^3)z = 0$ **Ans.** $z = f_1(y + 2x) + xf_2(y + 2x) + x^2 f_3(y + 2x)$

6. $\dfrac{\partial^4 z}{\partial x^4} - \dfrac{\partial^4 z}{\partial y^4} = 0$ **Ans.** $z = f_1(y + x) + f_2(y - x) + f_3(y + ix) + f_4(y - ix)$

7. $\dfrac{\partial^2 u}{\partial x^2} + \dfrac{\partial^2 u}{\partial y^2} = 0$, when $u = \sin y$, $x = 0$ for all $y$ and $u \to 0$ when $x \to \infty$.

   **Ans.** $u = f_1(y + ix) + f_2(y - ix)$

## 9.13. RULES FOR FINDING THE PARTICULAR INTEGRAL

Given partial differential equation is

$$f(D, D')z = F(x, y)$$

$$P.I. = \frac{1}{f(D, D')} F(x, y)$$

(*i*) **When** $F(x,y) = e^{ax+by}$

$$P.I. = \frac{1}{f(D, D')} e^{ax+by} = \frac{e^{ax+by}}{f(a, b)} \qquad [\text{Put } D = a,\ D' = b]$$

(*ii*) When F(*x*,*y*) = sin(*ax* + *by*) or cos (*ax* + *by*)

$$P.I. = \frac{1}{f(D^2, DD', D'^2)} \sin(ax+by) \text{ or } \cos(ax+by)$$

$$= \frac{\sin(ax+by) \, or \cos(ax+by)}{f(-a^2, -ab, -b^2)} \qquad \left[\begin{array}{l} \text{Put } D^2 = -a^2 \\ DD' = -ab, \; D'^2 = -b^2 \end{array}\right]$$

(*iii*) When F $(x,y) = x^m y^n$

$$P.I. = \frac{1}{f(D,D')} x^m y^n = [f(D,D')]^{-1} x^m y^n$$

Expand $[f(D, D')]^{-1}$ in ascending power of $D$ or $D'$ and operate on $x^m y^n$ term by term.

(*iv*) When = Any function $F(x, y)$

$$P.I. = \frac{1}{f(D,D')} F(x,y)$$

Resolve $\frac{1}{f(D,D')}$ into partial fractions

Considering $f(D, D')$ as a function of $D$ alone

$$P.I. = \frac{1}{D - mD'} F(x,y) = \int F(x, c - mx)\, dx$$

where c is replaced by $y + mx$ after integration.

**Case 1.** *When R.H. S.* $= e^{ax+by}$

**Example 31.** *Solve* : $\frac{\partial^3 z}{\partial x^3} - 3\frac{\partial^3 z}{\partial x^2 \partial y} + 4\frac{\partial^3 z}{\partial y^3} = e^{x+2y}$

**Solution.** $$\frac{\partial^3 z}{\partial x^3} - 3\frac{\partial^3 z}{\partial x^2 \partial y} + 4\frac{\partial^3 z}{\partial y^3} = e^{x+2y}$$

Given equation in symbolic form is

$$(D^3 - 3D^2 D' + 4D'^3)z = e^{x+2y}$$

Its A.E. is $m^3 - 3m^2 + 4 = 0$ whence, $m = -1, 2, 2.$

$$\text{C.F.} = f_1(y - x) + f_2(y + 2x) + x f_3(y + 2x)$$

$$P.I. = \frac{1}{D^3 - 3D^2 D' + 4D'^3} e^{x+2y}$$

$$= \frac{1}{1 - 6 + 32} e^{x+2y} = \frac{e^{x+2y}}{27} \qquad [\text{Put } D = 1, D' = 2]$$

Hence complete solution is

$$z = f_1(y - x) + f_2(y + 2x) + x f_3(y + 2x) + \frac{e^{x+2y}}{27}$$ **Ans.**

## EXERCISE 9.9

**Solve the following equations:**

**1.** $\frac{\partial^2 z}{\partial x^2} - \frac{\partial^2 z}{\partial y^2} = e^{x+2y}$ **Ans.** $z = f_1(y + x) + f_2(y - x) - \frac{e^{x+2y}}{3}$

**2.** $\frac{\partial^2 z}{\partial x^2} - 5\frac{\partial^2 z}{\partial x \partial y} + 6\frac{\partial^2 z}{\partial y^2} = e^{x+y}$ **Ans.** $z = f_1(y + 2x) + f_2(y + 3x) + \frac{1}{2} e^{x+y}$

3. $\dfrac{\partial^2 z}{\partial x^2}-4\dfrac{\partial^2 z}{\partial x\partial y}+4\dfrac{\partial^2 z}{\partial y^2}=e^{2x+y}$ **Ans.** $z=f_1(y+2x)+xf_2(y+2x)+\dfrac{x^2}{2}e^{2x+y}$

4. $\dfrac{\partial^2 z}{\partial x^2}-7\dfrac{\partial^2 z}{\partial x\partial y}+12\dfrac{\partial^2 z}{\partial y^2}=e^{x-y}$ **Ans.** $z=f_1(y+3x)+f_2(y+4x)+\dfrac{1}{20}e^{x-y}$

5. $\dfrac{\partial^3 z}{\partial x^3}-2\dfrac{\partial^3 z}{\partial x^2\partial y}=2e^{2x-y}$ **Ans.** $z=f_1(y)+x f_2(y)+f_3(y+2x)+\dfrac{1}{8}e^{2x-y}$

6. $(D^2-2DD'+D'^2)z=e^{x+2y}$ **Ans.** $z=f_1(y+x)+xf_2(y+x)+e^{x+2y}$

7. $\dfrac{\partial^2 z}{\partial x^2}-\dfrac{\partial^2 z}{\partial y^2}-2\dfrac{\partial z}{\partial x}+2\dfrac{\partial z}{\partial y}=e^{2x+3y}$ **Ans.** $z=f_1(y+x)+e^{2x}f_2(y-x)-\dfrac{1}{3}e^{2x+3y}$

8. $\dfrac{\partial^2 z}{\partial x^2}-5\dfrac{\partial^2 z}{\partial x\partial y}+6\dfrac{\partial^2 z}{\partial y^2}=\exp(3x-2y)$ **Ans.** $z=f_1(y+2x)+f_2(y+3x)+\dfrac{1}{63}e^{3x-2y}$

**Case II.** When R.H.S. = sin $(ax+by)$ or cos $(ax+by)$

**Example 32.** *Solve* $\dfrac{\partial^3 z}{\partial x^3}-4\dfrac{\partial^3 z}{\partial x^2\partial y}+4\dfrac{\partial^3 z}{\partial x\partial y^2}=2\sin(3x+2y)$

**Solution.** $\dfrac{\partial^3 z}{\partial x^3}-4\dfrac{\partial^3 z}{\partial x^2\partial y}+4\dfrac{\partial^3 z}{\partial x\partial y^2}=2\sin(3x+2y)$

Putting $\dfrac{\partial}{\partial x}=D,\quad \dfrac{\partial}{\partial y}=D'$

$$D^3z-4D^2D'z+4DD'^2z=2\sin(3x+2y)$$

A.E. is $D^3-4D^2D'+4DD'^2=0 \Rightarrow D(D^2-4DD'+4D'^2)=0$

Put $D=m,\ D'=1$

$$m(m^2-4m+4)=0 \Rightarrow m(m-2)^2=0 \Rightarrow m=0,2,2$$

C.F. is $f_1(y)+f_2(y+2x)+xf_3(y+2x)$

$$P.I.=\frac{1}{D^3-4D^2D'+4DD'^2}2\sin(3x+2y)=2.\frac{1}{D(D^2-4DD'+4D'^2)}\sin(3x+2y)$$

$$=2.\frac{1}{D[-9-4(-6)+4(-4)]}\sin(3x+2y)=-\frac{2}{D}\sin(3x+2y)$$

$$=-\frac{2}{3}[-\cos(3x+2y)]=\frac{2}{3}\cos(3x+2y)$$

General solution is

$$z=f_1(y)+f_2(y+2x)+xf_3(y+2x)+\frac{2}{3}\cos(3x+2y)$$ **Ans.**

**Example 33.** *Solve* $\dfrac{\partial^2 z}{\partial x^2}-\dfrac{\partial^2 z}{\partial x\partial y}=\sin x\cos 2y$

**Solution.** $\dfrac{\partial^2 z}{\partial x^2}-\dfrac{\partial^2 z}{\partial x\partial y}=\sin x\cos 2y$

The given equation can be written in the form

$$(D^2-DD')z=\sin x\cos 2y \qquad \text{where } D=\frac{\partial}{\partial x},\ D'=\frac{\partial}{\partial y}$$

Writing $D = m$ and $D' = 1$, the auxiliary equation is

$$m^2 - m = 0 \Rightarrow m(m-1) = 0 \Rightarrow m = 0, 1$$

$$C.F. = f_1(y) + f_2(y+x)$$

$$P.I. = \frac{1}{D^2 - DD'} \sin x \cos 2y = \frac{1}{D^2 - DD'} \frac{1}{2} [\sin(x+2y) + \sin(x-2y)]$$

$$= \frac{1}{2} \frac{1}{D^2 - DD'} \sin(x+2y) + \frac{1}{2} \frac{1}{D^2 - DD'} \sin(x-2y)$$

Put $D^2 = -1$, $DD' = -2$ in the first integral and $D^2 = -1$, $DD' = 2$ in the second integral.

$$\text{P.I.} = \frac{1}{2} \frac{\sin(x+2y)}{-1-(-2)} + \frac{1}{2} \frac{\sin(x-2y)}{-1-(2)} = \frac{1}{2} \sin(x+2y) - \frac{1}{6} \sin(x-2y)$$

Hence the complete solution is $z$ = C.F. + P.I.

i.e. $$z = f_1(y) + f_2(y+x) + \frac{1}{2} \sin(x+2y) - \frac{1}{6} \sin(x-2y)$$ **Ans.**

**Example 34.** *Solve* $(D^2 + DD' - 6D'^2) z = \cos(2x+y)$

**Solution.** $(D^2 + DD' - 6D'^2) z = \cos(2x+y)$

A.E. is $m^2 + m - 6 = 0 \Rightarrow m = 2, -3$

$$\text{C.F.} = f_1(y+2x) + f_2(y-3x)$$

$$\text{P.I.} = \frac{1}{D^2 + DD' - 6D'^2} \cos(2x+y)$$

$$D^2 + DD' - 6D'^2 = -4 - 2 - 6(-1) = 0$$

$\therefore$ It is a case of failure.

Now $$P.I. = \frac{1}{D^2 + DD' - 6D'^2} \cos(2x+y)$$ (Case IV)

$$= x \frac{1}{2D + D'} \cos(2x+y) = x \frac{D}{2D^2 + DD'} \cos(2x+y)$$

$$= x \frac{D}{2(-4) - 2} \cos(2x+y) = -\frac{x}{10} D \cos(2x+y)$$

$$= 2 \frac{x}{10} \sin(2x+y) = \frac{x}{5} \sin(2x+y)$$

$$z = f_1(y+2x) + f_2(y-3x) + \frac{x}{5} \sin(2x+y)$$ **Ans.**

**Example 35.** *Solve the equation*

$$(D^3 - 7DD'^2 - 6D'^3) z = \sin(x+2y) + e^{2x+y}.$$

**Solution** $(D^3 - 7DD'^2 - 6D'^3) z = \sin(x+2y) + e^{2x+y}$ ...(1)

Its auxiliary equation is

$$m^3 - 7m - 6 = 0 \Rightarrow (m+1)(m+2)(m-3) = 0 \Rightarrow m = -1, -2, 3$$

$$\text{C.F.} = f_1(y-x) + f_2(y-2x) + f_3(y+3x)$$

$$\text{P.I.} = \frac{1}{D^3 - 7DD'^2 - 6D'^3} \left[\sin(x+2y) + e^{2x+y}\right]$$

$$=\frac{1}{D^3-7DD'^2-6D'^3}\sin(x+2y)+\frac{1}{D^3-7DD'^2-6D'^3}e^{2x+y}$$

$$=\frac{1}{D^2.D-7DD'^2-6D'^2D'}\sin(x+2y)+\frac{e^{2x+y}}{(2)^3-7(2)(1)^2-6(1)^3}$$

Put $D^2=-1,\ D'^2=-2^2$

$$=\frac{1}{-D-7D(-4)-6(-4)D'}\sin(x+2y)+\frac{e^{2x+y}}{8-14-6}$$

$$=\frac{1}{27D+24D'}\sin(x+2y)-\frac{1}{12}e^{2x+y}=\frac{1}{3}\frac{1}{9D+8D'}\sin(x+2y)-\frac{1}{12}e^{2x+y}$$

$$=\frac{1}{3}\frac{D}{9D^2+8DD'}\sin(x+2y)-\frac{1}{12}e^{2x+y}=\frac{1}{3}\frac{D}{9(-1)+8(-2)}\sin(x+2y)-\frac{1}{12}e^{2x+y}$$

$$=-\frac{1}{75}D\sin(x+2y)-\frac{1}{12}e^{2x+y}=-\frac{1}{75}\cos(x+2y)-\frac{1}{12}e^{2x+y}$$

Hence the complete solution is

$$z=f_1(y-x)+f_2(y-2x)+f_3(y+3x)-\frac{1}{75}\cos(x+2y)-\frac{1}{12}e^{2x+y}$$ **Ans.**

## EXERCISE 9.10

**Solve the following equations :**

1. $\frac{\partial^2 z}{\partial x^2}-2\frac{\partial^2 z}{\partial x\partial y}+\frac{\partial^2 z}{\partial y^2}=\sin x$ **Ans.** $z=f_1(y+x)+xf_2(y+x)-\sin x$

2. $[2D^2-5DD'+2D'^2]\,z=5\sin(2x+y)$. **Ans.** $z=f_1(y+2x)+f_2(2y+x)-\frac{5}{3}x\cos(2x+y)$

3. $\frac{\partial^2 z}{\partial x^2}-\frac{\partial^2 z}{\partial x\partial y}=\cos(x+2y)$ **Ans.** $z=f_1(y)+f_2(y+x)+\cos(x+2y)$

4. $(D^2-DD')\,z=\cos x\cos 2y$ **Ans.** $z=f_1(y)+f_2(y+x)+\frac{1}{2}\cos(x+2y)-\frac{1}{6}\cos(x-2y)$

5. $(D^2+2D'D+D'^2)\,z=\sin(x+2y)$ **Ans.** $z=f_1(y-x)+xf_2(y-x)-\frac{1}{9}\sin(x+2y)$

6. $\frac{\partial^2 z}{\partial x^2}-3\frac{\partial^2 z}{\partial x\partial y}+2\frac{\partial^2 z}{\partial y^2}=e^{2x+3y}+\sin(x+2y)$

   **Ans.** $z=c_1f(y+x)+f_2(y+2x)+\frac{1}{4}e^{2x+3y}-\frac{1}{15}\sin(x-2y)$

**Case III.** When R.H.S. $=x^m y^n$

**Example 36.** *Find the general integral of the equation*

$$\frac{\partial^2 z}{\partial x^2}+3\frac{\partial^2 z}{\partial x\partial y}+2\frac{\partial^2 z}{\partial y^2}=x+y$$

**Solution.** $$\frac{\partial^2 z}{\partial x^2}+3\frac{\partial^2 z}{\partial x\partial y}+2\frac{\partial^2 z}{\partial y^2}=x+y$$

with $D=\frac{\partial}{\partial x},D'=\frac{\partial}{\partial y}$, the given equation can be written in the form

$$(D^2+3DD'+2D'^2)\,z=x+y$$

Writing $D=m$ and $D'=1$, the auxiliary equation is

$$m^2+3m+2=0\Rightarrow(m+1)(m+2)=0\ \Rightarrow\ m=-1,-2$$

$$\therefore \quad \text{C.F.} = f_1(y-x) + f_2(y-2x)$$

$$\text{P.I.} = \frac{1}{D^2+3DD'+2D'^2}(x+y)$$

$$= \frac{1}{D^2}\left(1+\frac{3D'}{D}+\frac{2D'^2}{D^2}\right)^{-1}(x+y) = \frac{1}{D^2}\left(1-\frac{3D'}{D}\ldots\right)(x+y)$$

$$= \frac{1}{D^2}\left[x+y-3\frac{1}{D}(1)\right] = \frac{1}{D^2}[x+y-3x]$$

$$= \frac{1}{D^2}[y-2x] = \frac{x^2}{2}y - \frac{x^3}{3}$$

Hence the complete solution is $\quad z = f_1(y-x) + f_2(y-2x) + \frac{x^2y}{2} - \frac{x^3}{3}$ **Ans.**

**Example 37.** *Solve* $\frac{\partial^2 z}{\partial x^2} + \frac{\partial^2 z}{\partial x \partial y} - 6\frac{\partial^2 z}{\partial y^2} = x + y$

**Solution.** With $D = \frac{\partial}{\partial x}, D' = \frac{\partial}{\partial y}$, the given equation can be written in the form

$$\left(D^2 + DD' - 6D'^2\right)z = x + y$$

Writing $D = m$ and $D' = 1$, the auxiliary equation is $m^2 + m - 6 = 0$

$$\Rightarrow \quad (m+3)(m-2) = 0 \quad \Rightarrow \quad m = -3, 2$$

$$\therefore \quad \text{C.F.} = f_1(y-3x) + f_2(y+2x)$$

$$\therefore \quad \text{P.I.} = \frac{1}{D^2+DD'-6D'^2}(x+y)$$

$$= \frac{1}{D^2}\left(1+\frac{D'}{D}-\frac{6D'^2}{D^2}\right)^{-1}(x+y) = \frac{1}{D^2}\left[1-\frac{D'}{D}+\ldots\right](x+y)$$

$$= \frac{1}{D^2}\left(x+y-\frac{1}{D}(1)\right) = \frac{1}{D^2}(x+y-x) = \frac{1}{D^2}y = \frac{yx^2}{2}$$

The complete solution is

$$z = f_1(y-3x) + f_2(y+2x) + \frac{yx^2}{2}$$ **Ans.**

**Example 38.** *Solve* $\frac{\partial^3 z}{\partial x^3} - 2\frac{\partial^3 z}{\partial x^2 \partial y} = 2e^{2x} + 3x^2y$ *(A.M.I.E., Summer 2004, 2001)*

**Solution.** $\frac{\partial^3 z}{\partial x^3} - 2\frac{\partial^3 z}{\partial x^2 \partial y} = 2e^{2x} + 3x^2y$

$$\Rightarrow \quad (D^3 - 2D^2D')z = 2e^{2x} + 3x^2y$$

Its auxiliary equation is

$$m^3 - 2m^2 = 0$$

$$\Rightarrow \quad m^2(m-2) = 0$$

$$\Rightarrow \quad m = 0, 0, 2.$$

$$\text{C.F.} = f_1(y) + xf_2(y) + f_3(y+2x)$$

$$\text{P.I.} = \frac{1}{D^3-2D^2D'}\left(2e^{2x}+3x^2y\right)$$

$$=\frac{1}{D^3-2D^2D'}2e^{2x}+\frac{1}{D^3-2D^2D'}3x^2y$$

$$=2\frac{e^{2x}}{(2)^3-2(2)^2(0)}+3.\frac{1}{D^3\left(1-\frac{2D'}{D}\right)}x^2y=\frac{2e^{2x}}{8}+\frac{3}{D^3}\left(1-\frac{2D'}{D}\right)^{-1}x^2y$$

$$=\frac{e^{2x}}{4}+\frac{3}{D^3}\left(1+\frac{2D'}{D}\cdots\right)x^2y=\frac{e^{2x}}{4}+\frac{3}{D^3}\left[x^2y+\frac{2}{D}x^2\right]=\frac{e^{2x}}{4}+\frac{3}{D^3}\left(x^2y+\frac{2x^3}{3}\right)$$

$$=\frac{e^{2x}}{4}+3y\frac{1}{D^3}x^2+\frac{2}{D^3}x^3=\frac{e^{2x}}{4}+3y\frac{x^5}{3.4.5}+2\frac{x^6}{4.5.6}=\frac{e^{2x}}{4}+\frac{x^5y}{20}+\frac{x^6}{60}$$

$$=\frac{1}{60}\left(15e^{2x}+3x^5y+x^6\right)$$

Hence the complete solution is

$$z=f_1(y)+xf_2(y)+f_3(y+2x)+\frac{1}{60}(15e^{2x}+3x^5y+x^6)$$ **Ans.**

## EXERCISE 9.11

**Solve the following equations :**

**1.** $\frac{\partial^2 z}{\partial x^2}-\frac{\partial^2 z}{\partial y^2}=x-y$ **Ans.** $z=f_1(y-x)+f_2(y+x)+\frac{x^3}{6}-\frac{x^2y}{2}$

**2.** $\frac{\partial^2 z}{\partial x^2}+\frac{3\partial^2 z}{\partial x\partial y}+\frac{2\partial^2 z}{\partial y^2}=12xy$ *(A.M.I.E., Winter 2001)*

**Ans.** $z=f_1(y-x)+f_2(y-2x)+2x^3y-\frac{3x^4}{2}$

**3.** $\frac{\partial^2 z}{\partial x^2}-\frac{\partial^2 z}{\partial x\partial y}-6\frac{\partial^2 z}{\partial y^2}=xy$ **Ans.** $z=f_1(y-2x)+f_2(y+3x)+\frac{x^3y}{6}+\frac{x^4}{24}$

**4.** $r+2s+t=2(y-x)+\sin(x-y)$ **Ans.** $z=f_1(y-x)+xf_2(y-x)+x^2y-x^3+\frac{x^2}{2}\sin(x-y)$

**5.** $\frac{\partial^2 z}{\partial x^2}-a^2\frac{\partial^2 z}{\partial y^2}=x^2$ **Ans.** $z=f_1(y+ax)+f_2(y-ax)+\frac{x^4}{12}$

**6.** $\frac{\partial^2 z}{\partial x^2}-2\frac{\partial^2 z}{\partial x\partial y}+\frac{\partial^2 z}{\partial y^2}=x^2+y$ **Ans.** $z=f_1(y+x)+xf_2(y+x)+\frac{x^4}{12}+\frac{x^2y}{2}+\frac{x^3}{3}$

**7.** $\frac{\partial^2 z}{\partial x^2}+3\frac{\partial^2 z}{\partial x\partial y}-4\frac{\partial^2 z}{\partial y^2}=x+\sin y$ **Ans.** $z=f_1(y+x)+f_2(y-4x)+\frac{x^3}{6}+\frac{1}{4}\sin y$

**8.** $(D^3-3D^2D')z=x^2y$ **Ans.** $z=f_1(y)+xf_2(y)+f_3(y+3x)+\frac{x^5y}{60}+\frac{x^6}{120}$

**Case IV. When R.H.S. = Any function**

**Example 39.** *Solve* $(D^2-DD'-2D'^2)z=(y-1)e^x$

**Solution.** $(D^2-DD'-2D'^2)z=(y-1)e^x$

A.E. is $D^2-DD'-2D'^2=0 \Rightarrow m^2-m-2=0$

$\Rightarrow (m-2)(m+1)=0 \Rightarrow m=2,-1$

$$\text{C.F.}=f_1(y+2x)+f_2(y-x)$$

$$\text{P.I.}=\frac{1}{D^2-DD'-2D'^2}(y-1)e^x$$

$$= \frac{1}{(D+D')(D-2D')}(y-1)e^x = \frac{1}{D+D'}\int\left[(c-2x-1)e^x dx\right] \qquad \text{[Put } y = c-2x]$$

$$= \frac{1}{D+D'}\left[(c-2x-1)e^x + 2e^x\right]$$

$$= \frac{1}{D+D'}\left[ce^x - 2xe^x + e^x\right] \qquad \text{[Put } c = y+2x]$$

$$= \frac{1}{D+D'}\left[(y+2x)e^x - 2xe^x + e^x\right] = \frac{1}{D+D'}\left[ye^x + e^x\right]$$

$$= \int\left[(c+x)e^x + e^x\right]dx \qquad \text{[Put } y = c+x]$$

$$= (c+x)e^x - e^x + e^x$$

$$= ce^x + xe^x = (y-x)e^x + xe^x \qquad \text{[Put } c = y-x]$$

$$= y\,e^x$$

Hence complete solution is $z = f_1(y+2x) + f_2(y-x) + ye^x$ **Ans.**

**Example 40.** *Solve* $\frac{\partial^2 z}{\partial x^2} + \frac{\partial^2 z}{\partial x \partial y} - 6\frac{\partial^2 z}{\partial y^2} = y\cos x$

**Solution.** $\frac{\partial^2 z}{\partial x^2} + \frac{\partial^2 z}{\partial x \partial y} - 6\frac{\partial^2 z}{\partial y^2} = y\cos x$

$$(D^2 + DD' - 6D'^2) = y\cos x$$

Its auxiliary equation is $m^2 + m - 6 = 0$

$$(m+3)(m-2) = 0$$

$$m = 2, -3$$

$$\text{C.F.} = f_1(y+2x) + f_2(y-3x)$$

$$\text{P.I.} = \frac{1}{D^2 + DD' - 6D'^2}y\cos x = \frac{1}{(D-2D')(D+3D')}y\cos x$$

$$= \frac{1}{D-2D'}\int(c+3x)\cos x\,dx \qquad \text{Put } y = c+3x$$

$$= \frac{1}{D-2D'}[(c+3x)\sin x + 3\cos x] = \frac{1}{D-2D'}[y\sin x + 3\cos x] \qquad \text{Put } c+3x = y$$

$$= \int[(c-2x)\sin x + 3\cos x]\,dx \qquad \text{Put } y = c-2x$$

$$= (c-2x)(-\cos x) - 2\sin x + 3\sin x = -y\cos x + \sin x \qquad \text{Put } c-2x = y$$

Hence the complete solution is

$$z = f_1(y+2x) + f_2(y-3x) + \sin x - y\cos x$$ **Ans.**

## EXERCISE 9.12

**Solve the following equations:**

1. $(D-D')(D+2D')z = (y+1)e^x$ **Ans.** $z = f_1(y+x) + f_2(y-2x) + ye^x$

2. $\frac{\partial^2 z}{\partial x^2} - \frac{\partial^2 z}{\partial y^2} = \tan^3 x\tan y - \tan x\tan^3 y$ **Ans.** $z = f_1(y+x) + f_2(x-y) + \frac{1}{2}\tan x\tan y$

3. $(D^2 - DD' - 2D'^2)z = (2x^2 + xy - y^2)\sin xy - \cos xy$ **Ans.** $z = f_1(y+2x) + f_2(y-x) + \sin xy$

**4. Tick √ the correct answer :**

(*a*) The solution of $\frac{\partial^3 z}{\partial x^3} = 0$ is

(*i*) $z = f_1(y) + xf_2(y) + x^2 f_3(y)$ (*ii*) $z = (1 + x + x^2) f(y)$

(*iii*) $z = f_1(x) + yf_2(x) + y^2 f_3(x)$ (*iv*) $z = (1 + y + y^2) f(x)$

(*b*) The solution of $\frac{\partial^2 z}{\partial x^2} - \frac{\partial^2 z}{\partial y^2} = 0$ is

(*i*) $z = f_1(y + x) + f_1(y - x)$ (*ii*) $z = f_1(y + x) + f_2(y - x)$

(*iii*) $z = f_2(y + x) + f_2(y - x)$ (*iv*) $z = f(x^2 - y^2)$

(*c*) Particular integral of $(2D^2 - 3DD' + D'^2) z = e^{x+2y}$ is

(*i*) $xe^{x+2y}$ (*ii*) $\frac{1}{2} e^{x+2y}$ (*iii*) $-\frac{x}{2} e^{x+2y}$ (*iv*) $\frac{x^2}{2} e^{x+2y}$

(*d*) Particular integral of $(D^2 - D'^2) z = \cos(x + y)$ is

(*i*) $\frac{x}{2} \cos(x + y)$ (*ii*) $x \sin(x + y)$ (*iii*) $x \cos(x + y)$ (*iv*) $\frac{x}{2} \sin(x + y)$

**Ans.** *(a) (i), (b) (ii), (c) (iii), (d) (iv).*

## 9.14 NON-HOMOGENEOUS LINEAR EQUATIONS

The linear differential equations which are not homogeneous are called Non-homogeneous Linear Equations.

For example,

$$3\frac{\partial^2 z}{\partial x^2} + 2\frac{\partial^2 z}{\partial x \partial y} + 4\frac{\partial^2 z}{\partial y^2} + 5\frac{\partial z}{\partial x} + \frac{\partial z}{\partial y} + z = 0$$

$$f(D, D') = f_1(x, y)$$

Its solution, $z = \text{C.F.} + \text{P.I.}$

Complementary Function: Let the non-homogeneous equation be

$$(D - mD' - a) z = 0 \Rightarrow \frac{\partial z}{\partial x} - m\frac{\partial z}{\partial y} - az = 0$$

$$p - mq = az$$

The Lagrange's subsidiary equations are

$$\frac{dx}{1} = \frac{dy}{-m} = \frac{dz}{az}$$

From first two relations we have, $-mdx = dy$

$$dy + mdx = 0 \Rightarrow y + mx = c_1 \quad \text{... (1)}$$

and from first and third relation, $dx = \frac{dz}{az} \Rightarrow x = \frac{1}{a} \log z + c_2 \Rightarrow z = c_3 e^{ax}$ ... (2)

From (1) and (2), we have $z = e^{ax} \phi(y + mx)$

Similarly the solution of $(D - mD' - a)^2 Z = 0$ is

$$z = e^{ax} \phi_1(y + mx) + xe^{ax} \phi_2(y + mx)$$

**Example 41.** *Solve* $(D + D' - 2)(D + 4D' - 3) z = 0$

**Solution.** The equation can be rewritten as $\{D - (-D)' - 2\}\{D - (-4D') - 3\} z = 0$

Hence the solution is

$$z = e^{2x}\,\phi_1(y - mx) + e^{3x}\,\phi_2\,(y - 4mx)$$ **Ans.**

**Example 42.** *Solve* $(D + 3\,D' + 4)^2\, z = 0$

**Solution.** The equation is rewritten as

$$[D - (-3\,D') - (-4)]^2\, z = 0$$

Hence the solution is

$$z = e^{-4x}\phi_1\,(y - 3x) + x\,e^{-4x}\,\phi_2\,(y - 3x)$$ **Ans.**

**Example 43.** *Solve* $r + 2s + t + 2p + 2q + z = 0$

**Solution.** The equation is rewritten as

$$(D^2 + 2\,D\,D' + D^2 + 2D + 2\,D' + 1)\,z = 0$$

$$\Rightarrow \quad [(D + D')^2 + 2\,(D + D') + 1]\,z = 0$$

$$\Rightarrow \quad (D + D' + 1)^2\, z = 0$$

$$\Rightarrow \quad [D - (-D') - (-1)]^2\, z = 0$$

Hence the solution is

$$z = e^{-x}\phi_1 y\,(y - x) + xe^{-x}\phi_2(y - x)$$

**Example 44.** *Solve* $r - t + p - q = 0$

**Solution.** The equation is rewritten as

$$(D^2 - D'^2 + D - D')\,z = 0$$

$$\Rightarrow \quad [(D - D')\,(D + D') + 1\,(D - D')]\,z = 0$$

$$\Rightarrow \quad (D - D')\,(D + D' + 1)z = 0$$

Hence the solution is

$$z = \phi_1(y + x) + e^{-x}\,\phi_2\,(y - x).$$ **Ans.**

**Particular Integral**

**Case 1.** $$\frac{1}{F(D, D')}e^{ax+by} = \frac{1}{F(a, b)}e^{ax+by}$$

**Example 45.** Solve $(D - D' - 2)\,(D - D' - 3)\,z = e^{3x - 2y}$

The complementary function is

$$e^{2x}\,\phi_1\,(y + x) + e^{3x}\,\phi_2\,(y + x)$$

$$\text{P.I.} = \frac{1}{(D - D' - 2)(D - D' - 3)}e^{3x-2y} = \frac{1}{[3-(-2)-2][3-(-2)-3]}e^{3x-2y} = \frac{1}{6}e^{3x-2y}$$

Hence the complete solution is

$$z = e^{2x}\phi_1(y + x) + e^{3x}\phi_2\,(y + x) + \frac{1}{6}e^{3x-2y}$$ **Ans.**

**Case 2.** $$\frac{1}{F(D^2, DD', D'^2)}\sin(ax + by) = \frac{1}{F(-a^2, -ab, -b^2)}\sin(ax + by)$$

**Example 46.** *Solve* $(D + 1)\,(D + D' - 1)\,z = \sin\,(x + 2y)$

**Solution.** $\quad \text{C.F.} = e^{-x}\phi\,(y) + e^{-x}\phi_2(y - x)$

$$\text{P.I.} = \frac{1}{(D+1)(D+D'-1)}\sin(x+2y) = \frac{1}{D^2 + DD' + D' - 1}\sin(x+2y)$$

$$= \frac{1}{-1+(-2)+D'-1}\sin(x+2y) = \frac{1}{D'-4}\sin(x+2y)$$

$$= \frac{D'+4}{(D'^2-16)}\sin(x+2y) = \frac{D'+4}{(-4-16)}\sin(x+2y)$$

$$= -\frac{1}{20}(D'+4)\sin(x+2y) = -\frac{1}{20}\left[D'\sin(x+2y)+4\sin(x+2y)\right]$$

$$= -\frac{1}{20}\left[2\cos(x+2y)+4\sin(x+2y)\right]$$

Hence, the solution is $z = e^{-x}\phi\ (y) + e^{-x}\phi_2\ (y-x) - \frac{1}{10}[\cos\ (x+2y) + 2\ \sin(x+2y)]$ **Ans.**

**Case 3.** $$\frac{1}{F(D,D')}x^m y^n = \left[F(D,D')\right]^{-1} x^m y^n$$

**Example 47.** *Solve* $[D^2 - D'^2 + D + 3\ D' - 2]\ z = x^2\ y$

**Solution.** $(D - D' + 2)\ (D + D' - 1)\ z = 0$

C.F. $= e^{-2x}\ \phi_1(y+x) + e^x\ \phi_2\ (y-x)$

P.I. $= \dfrac{1}{(D-D'+2)(D+D'-1)}x^2 y$

$$= \frac{1}{D^2-D'^2+D+3D'-2}x^2y = -\frac{1}{2}\frac{1}{1-\frac{3D'}{2}-\frac{D}{2}+\frac{D'^2}{2}-\frac{D^2}{2}}x^2y$$

$$= -\frac{1}{2}\left[1-\frac{1}{2}\left(3D'+D-D'^2\right)+D^2\right]^{-1} x^2 y$$

$$= -\frac{1}{2}\left[1+\frac{1}{2}\left(3D'+D-D'^2+D^2\right)+\frac{1}{4}\left(3D'+D-D'^2+D^2\right)^2 + \frac{1}{8}\left(3D'+D-D'^2+D^2\right)^3 + \right]x^2y$$

$$= -\frac{1}{2}\left[1+\frac{1}{2}\left(3D'+D-D'^2+D^2\right)+\frac{1}{4}\left(9D'^2+D^2+6DD'+6D^2D'\right) + \frac{1}{8}\left(9D^2D'\right)+\ldots\right]x^2y$$

$$= -\frac{1}{2}\left[x^2y+\frac{1}{2}\left(3x^2+2xy-0+2y\right)+\frac{1}{4}\left(0+2y+12x+12\right)+\frac{1}{8}(18)\right]$$

$$= -\frac{1}{2}\left[x^2y+\frac{3x^2}{2}+xy+y+\frac{y}{2}+3x+3+\frac{9}{4}\right] = -\frac{1}{2}\left(x^2y+\frac{3x^2}{2}+xy+\frac{3y}{2}+3x+\frac{21}{4}\right)$$

Hence, the complete solution is

$$z = e^{-2x}\phi_1(y+x)+e^x\phi_2(y-x)-\frac{1}{2}\left(x^2y+\frac{3x^2}{2}+xy+\frac{3y}{2}+3x+\frac{21}{4}\right)$$ **Ans.**

**Case 4.** $$\frac{1}{F(D,D')}\left[e^{ax+by}\phi(x,y)\right] = e^{ax+by}\frac{1}{F(D+a,D'+b)}\phi(x,y)$$

**Example 48.** *Solve* $(D-3D'-2)^2 z = 2e^{2x}\sin(y+3x)$

**Solution.** A.E. is $(D-3D'-2)^2 = 0$

$$\text{C.F.} = e^{2x}\phi_1(y+3x) + x\,e^{2x}\phi_2(y+3x)$$

$$\text{P.I.} = \frac{1}{(D-3D'-2)^2}2e^{2x}\cdot\sin(y+3x)$$

$$= 2e^{2x}\frac{1}{(D+2-3D'-2)^2}\sin(y+3x) = 2e^{2x}\frac{1}{(D-3D')^2}\sin(y+3x)$$

$$= 2e^{2x}\cdot x\frac{1}{2(D-3D')}\sin(y+3x) \text{ (As denominator becomes zero)}$$

$$= 2x^2e^{2x}\frac{1}{2}\sin(y+3x) \qquad \text{(Again differentiate)}$$

$$= x^2e^{2x}\sin(y+3x)$$

Hence the complete solution is

$$z = e^{2x}\phi(y+3x) + x\,e^{2x}\phi_2(y+3x) + x^2e^{2x}\sin(y+3x)$$ **Ans.**

**Example 49.** *Solve* $(D^2 + DD' - 6D'^2)z = x^2 \sin(x+y)$

**Solution.** $(D^2 + DD' - 6D'^2)z = x^2\sin(x+y)$

For complementary function

$$(D^2 + DD' - 6D'^2) = 0 \Rightarrow (D-2D')(D+3D') = 0$$

$$\text{C.F.} = \phi_1(y+2x) + \phi_2(y-3x)$$

$$\text{P.I.} = \frac{1}{D-DD'-6D'^2}x^2\sin(x+y)$$

$$= \text{Imaginary part of } \frac{1}{D^2-DD'-6D'^2}x^2[\cos(x+y) + i\sin(x+y)]$$

$$= \text{Imaginary part of } \frac{1}{D^2-DD'-6D'^2}x^2e^{i(x+y)} = \text{Imaginary part of } e^{iy}\frac{1}{D^2-Di-6(i)^2}x^2e^{ix}$$

$$= \text{Imaginary part of } e^{i(x+y)}\frac{1}{(D+i)^2+(D+i)i+6}x^2$$

$$= \text{Imaginary part of } e^{i(x+y)}\frac{1}{D^2+3iD+4}x^2 = \text{Imaginary part of } \frac{e^{i(x+y)}}{4}\frac{1}{1+\frac{3iD}{4}+\frac{D^2}{4}}x^2$$

$$= \text{Imaginary part of } \frac{e^{i(x+y)}}{4}\left[1+\frac{3iD}{4}+\frac{D^2}{4}\right]^{-1}x^2$$

$$= \text{Imaginary part of } \frac{e^{i(x+y)}}{4}\left[1-\frac{3iD}{4}-\frac{D^2}{4}-\frac{9D^2}{16}\cdots\right]x^2$$

$$= \text{Imaginary part of } \frac{e^{i(x+y)}}{4}\left[x^2-\frac{3ix}{2}-\frac{2}{4}-\frac{9}{16}(2)\right]$$

$$= \text{Imaginary part of } \frac{1}{4}[\cos(x+y)+i\sin(x+y)]\left[x^2-\frac{3ix}{2}-\frac{13}{8}\right]$$

$$= \frac{1}{4}\left[\sin(x+y)\left(x^2 - \frac{13}{8}\right) - \frac{3}{2}x\cos(x+y)\right] = \frac{1}{4}\sin(x+y)\left(x^2 - \frac{13}{8}\right) - \frac{3x}{8}\cos(x+y)$$

Hence, the complete solution is

$$z = \phi_1(y+2x) + \phi_2(y-3x) + \frac{1}{4}\sin(x+y)\left(x^2 - \frac{13}{8}\right) - \frac{3x}{8}\cos(x+y) \qquad \textbf{Ans.}$$

**EXERCISE 9.13**

Solve the following equations:

1. $(D^2 + 2DD' + D'^2 - 2D - 2D')z = 0.$ **Ans.** $z = f_1(x-y) + e^{2x} f_2(x-y)$

2. $(D^2 - D'^2 - 3D + 3D')z = e^{x-2y}$ **Ans.** $z = \phi_1(y+x) + e^{3x}\phi_2(y-x) - \frac{1}{12}e^{x-2y}$

3. $(D - D' - 1)(D + D' - 2)z = e^{2x-y}$ **Ans.** $z = e^x\phi_1(x+y) + e^{2x}\phi_2(y-x) - \frac{1}{2}e^{2x-y}$

4. $(D^2 - D'^2 - 3D + 3D')z = e^{x+2y}$ **Ans.** $z = \phi_1(y+x) + e^{3x}\phi_1(x-y) - xe^{x+2y}$

5. $(D + D')(D + D' - 2)z = \sin(x+2y)$

**Ans.** $z = \phi_1(y-x) + e^{2x}(y-x) + \frac{1}{117}[6\cos(x+2y) - 9\sin(x+2y)]$

6. $(D^2 - DD' - 2D)z = \cos(3x+4y)$

**Ans.** $z = \phi_1(y) + e^{2x}\phi_2(y+x) + \frac{1}{15}[\cos(3x+4y) - 2\sin(3x+4y)]$

7. $(DD' + D - D' - 1)z = xy$ **Ans.** $z = e^{-y}\phi_1(x) + e^x\phi_2(y) - (xy + y - x - 1)$

8. $(D + D' - 1)(D + 2D' - 3)z = 4 + 3x + 6y$ **Ans.** $z = e^x\phi_1(x-y) + e^{3x}\phi_2(2x-y) + 6 + x + 2y$

9. $\frac{\partial^2 z}{\partial x^2} - \frac{\partial^2 z}{\partial y^2} - 3\frac{\partial z}{\partial x} + 3\frac{\partial z}{\partial y} = xy + e^{x+2y}$ *(UP. HI Semester, Summer 2002)*

**Ans.** $z = f_1(y+x) + e^{3x}f_2(y-x) - \frac{1}{3}\left(\frac{x^2y}{3} + \frac{x^3}{6} + \frac{x^2}{3} + \frac{xy}{3} + \frac{2x}{9}\right) - xe^{2x-y}$

10. $(D - D' - 1)(D - D' - 2)z - e^{2x-y}$ **Ans.** $z = e^x f_1(y+x) + e^{2x}f_2(y+x) + \frac{1}{2}e^{2x-y}$

11. $D(D + D' - 1)(D + 3D' - 2)z = x^2 - 4xy + 2y^2$

**Ans.** $z = \phi_1(y) + e^x\phi_2(x-y) + e^{2x}\phi_3(3x-y) + \frac{1}{2}\left[\frac{x^3}{3} - 2x^2y + 2xy^2 - \frac{7}{2}x^2 + 4xy + \frac{x}{2}\right]$

12. $(D - D' + 2)(D + D' - 1)z = e^{x-y} - x^2y$

**Ans.** $z = e^{2y}\phi_1(x+y)e^x\phi_2(x-y) - \frac{e^{x-y}}{4} + \frac{1}{2}\left[x^2y + xy + \frac{3x^2}{2} + \frac{3}{2}y + 3x + \frac{21}{4}\right]$

13. $(D^2 - DD' - 2D'^2 + 2D' + 2D)z = e^{2x+3y} + \sin(2x + y) + xy$

**Ans.** $z = \phi_1(x-y) + e^y\phi_2(2x+y) - \frac{1}{10}e^{2x+3y} - \frac{1}{6}\cos(2x+y) + \frac{x}{24}(6xy - 6y + 9x - 2x^2 - 12)$

## 9.15 MONGE'S METHOD (Non linear equation of the second order)

Let the equation be $Rr + Ss + Tt = V$ ... (1)

where $R, S, T, V$ are functions of $x, y, z, p$ and $q$. $r = \frac{\partial^2 f}{\partial x^2}$, $s = \frac{\partial^2 f}{\partial x \partial y}$, $t = \frac{\partial^2 f}{\partial y^2}$

We have $$dp = \frac{\partial p}{\partial x}dx + \frac{\partial p}{\partial y}dy = r\,dx + s\,dy \qquad ...(2)$$

and $$dq = \frac{\partial q}{\partial x}dx + \frac{\partial q}{\partial y}dy = s\,dx + t\,dy \quad ...(3)$$

From (2) and (3), we have $r = \dfrac{dp - s\,dy}{dx}$ and $t = \dfrac{dq - s\,dx}{dy}$

Putting the value of $r$ and $t$ in (1), we get

$$R\left(\frac{dp - s\,dy}{dx}\right) + Ss + T\left(\frac{dq - s\,dx}{dy}\right) = V$$

$$\Rightarrow \quad R\,dp\,dy + T\,dq\,dx - V\,dx\,dy - s(R\,dy^2 - S\,dx\,dy + T\,dx^2) = 0 \quad ... (4)$$

Equation (4) is satisfied if

$$R\,dp\,dy + T\,dq\,dx - V\,dx\,dy = 0 \quad ...(5)$$

$$R\,dy^2 - S\,dx\,dy + T\,dx^2 = 0 \quad ... (6)$$

Equations (5) and (6) are called **Monge's equations.**

Since (6) can be factorised into two equations.

$$dy - m_1\,dx = 0 \text{ and } dy - m_2\,dx = 0$$

Now combine $dy - m_1\,dx = 0$ and equation (5). If need be, we may also use the relation $dz = p \cdot dx + q \cdot dy$ while solving (5) and (6). The solution leads to two integrals

$$u\,(x, y, z, p, q) = a \text{ and } V\,(x, y, z, p, q) = b$$

Then we get a relation between $u$ and $v$. $V = f_1(u)$ ...(7)

Equation (7) is further integrated by methods of first order equations.

**Note.** If the intermediate solution is of the form $Pr + Qq = R$, then we use lagrange's equation.

**Example 50.** *Solve* $r = a^2 t$.

**Solution.** We have $dp = r\,dx + s\,dy$ and $dq = s\,dx + t\,dy$ which gives

$$r = \frac{dp - s\,dy}{dx} \text{ and } t = \frac{dq - s\,dx}{dy}$$

Putting these values of $r$ and $t$ in $r = a^2 t$, we get $\dfrac{dp - s\,dy}{dx} = a^2\dfrac{dq - s\,dx}{dy}$

$$\Rightarrow \quad dp\,dy - a^2 dx\,dq - s\,(dy^2 - a^2 dx^2) = 0$$

Thus, the Monges' equations are

$$dp\,dy - a^2\,dx\,dq = 0 \quad ...(1)$$

$$dy^2 - a^2\,dx^2 = 0 \quad ... (2)$$

(2) can be resolved into factors

$$dy - a\,dx = 0 \quad ...(3)$$

and $$dy + a\,dx = 0 \quad ...(4)$$

Combining (3) with (1), we get

$$dp\,(a\,dx) - a^2\,dx\,dq = 0 \text{ or } dp - a\,dq = 0 \quad ...(5)$$

(3) and (5) on integration give respectively

$$\left.\begin{aligned} y - ax &= A \\ \text{and } p - aq &= B \end{aligned}\right] \Rightarrow p - aq = f_1(y - ax) \quad ... (6)$$

Similarly combining (4) and (1)

$$p + aq = f_2\,(y + ax) \quad ...(7)$$

Adding and subtracting (6) and (7), we get

$$p=\frac{1}{2}\left[f_1(y-ax)+f_2(y+ax)\right], q=\frac{1}{2a}\left[f_2(y+ax)-f_1(y-ax)\right]$$

Substituting these values in $dz = p\,dx + q\,dy$

$$dz=\frac{1}{2}\left[f_1(y-ax)+f_2(y+ax)\right]dx+\frac{1}{2a}\left[f_2(y+ax)-f_1(y-ax)\right]dy$$

$$dz=\frac{1}{2a}(dy+a\,dx)\,f_2(y+ax)-\frac{1}{2a}(dy-a\,dx)\,f_1(y-ax)$$

Integrating, $$z=\frac{1}{2a}\phi_1(y+ax)-\frac{1}{2a}\phi_2(y-ax)$$

$\Rightarrow$ $$z=F_1(y+ax)+F_2(y-ax)$$ **Ans.**

**Example 51.** *Solve* $r - t\cos^2 x + p\tan x = 0$

**Solution.** $r=\dfrac{dp-s\,dy}{dx}$ *and* $t=\dfrac{dq-s\,dx}{dy}$

Putting for $r$ and $t$ in the given equation, we get

$$\frac{dp-s\,dy}{dx}-\frac{dq-s\,dx}{dy}\cos^2 x+p\tan x=0$$

$\Rightarrow$ $$dp\,dy - s\,dy^2 - dx\,dq\cos^2 x + s\,dx^2\cos^2 x + p\,dx\,dy\tan x = 0$$

$\Rightarrow$ $$dp\,dy - dx\,dq\cos^2 x + p\,dx\,dy\tan x - s\,(dy^2 - dx^2\cos^2 x) = 0$$

Monge's equations are

$$dp\,dy - dx\,dq\cos^2 x + p\,dx\,dy\tan x = 0 \quad (1)$$

$$dy^2 - dx^2\cos^2 x = 0 \quad \text{...(2)}$$

Eq. (2) is factorised $(dy + dx\cos x)(dy - dx\cos x) = 0$

$$dy - dx\cos x = 0 \quad \text{...(3)}$$

$$dy + dx\cos x = 0 \quad \text{...(4)}$$

Integrating (3) and (4), we get

$$y - \sin x = A \quad \text{...(5)}$$

$$y + \sin x = B \quad \text{...(6)}$$

Combining (3) and (1), we get

$$dp - dq\cdot\cos x + p\tan x\,dx = 0$$

$\Rightarrow$ $$(dp\sec x + p\sec x\tan x\,dx) - dq = 0$$

Integrating $$p\sec x - q = B \quad \text{...(7)}$$

Combining (5) and (7), we have

$$p\sec x - q = f_1(y - \sin x) \quad \text{...(8)}$$

On combining (6) and (7), we get

$$p\sec x + q = f_2(y + \sin x) \quad \text{...(9)}$$

From (5) and (9)

$$p=\frac{1}{2}\cos x[f_1(y-\sin x)+f_2(y+\sin x)] \quad \text{and} \quad q=\frac{1}{2}[f_2(y+\sin x)-f_1(y-\sin x)]$$

Putting for $p$ and $q$ in $dz = p\,dx + q\,dy$, we get

$$dz=\frac{1}{2}\cos x[f_1(y-\sin x)+f_2(y+\sin x)]\,dx+\frac{1}{2}[f_2(y+\sin x)-f_1(y-\sin x)]\,dy$$

$\Rightarrow$ $$dz=\frac{1}{2}f_2(y+\sin x)[dy+\cos x\,dx]-\frac{1}{2}f_1(y-\sin x)[dy-\cos x\,dx]$$

Integrating we get $$z=\frac{1}{2}F_2(y+\sin x)+F_1(y-\sin x)$$ **Ans.**

## EXERCISE 9.14

**Solve**

**1.** $r+(a+b)s+abt=xy$ **Ans.** $z=\frac{1}{6}x^3y-(a+b)\frac{x^4}{24}+F_1(y-ax)+F_2(y-bx)$

**2.** $y^2r-2ys+t=p+6y$ **Ans.** $z=y^3-yF_1(y^2+2x)+F_2(y^2+2x)$

**3.** $xy(t-r)+(x^2-y^2)(s-2)=py-qx$ **Ans.** $z=xy+F_1(x^2+y^2)+F_2\left(\frac{y}{x}\right)$

**4.** $(1+q)^2r-2(1+p+q+pq)s+(1+p)^2t=0$ **Ans.** $z=F_1(x+y+z)+xF_2(x+y+z)$

**5.** $t-r\sec^4y=2q\tan y$ **Ans.** $z=F_1(x-\tan y)+F_2(x+\tan y)$

**6.** $(q+1)s=(p+1)t$ **Ans.** $z=f_1(x)+f_2(x+y+z)$

**7.** $(r-s)y+(s-t)x+q-p=0$ **Ans.** $z=f_1(x+y)+f_2(x^2-y^2)$

## Partial Differential Equations in Practical Problems

## 9.16 INTRODUCTION

In practical problems, the following types of equations are generally used

(*i*) *Wave equation* : $$\frac{\partial^2 u}{\partial t^2}=c^2\frac{\partial^2 u}{\partial x^2}$$

(*ii*) *One-dimensional heat flow* : $$\frac{\partial u}{\partial t}=c^2\frac{\partial^2 u}{\partial x^2}$$

(*iii*) *Two-dimensional heat flow* : $$\frac{\partial^2 u}{\partial x^2}+\frac{\partial^2 u}{\partial y^2}=0$$

(*iv*) *Radio equations* : $$-\frac{\partial V}{\partial x}=L\frac{\partial I}{\partial t},\ -\frac{\partial I}{\partial x}=C\frac{\partial V}{\partial t}$$

## 9.17 METHOD OF SEPARATION OF VARIABLES

In this method, we assume that the dependent variable is the product of two functions, each of which involves only one of the independent variables. So two ordinary differential equations are formed.

**Example 1.** *Using the method of separation of variables, solve*

$$\frac{\partial u}{\partial x}=2\frac{\partial u}{\partial t}+u$$

where $u(x,0)=6e^{-3x}$ (*A.M.I.E.T.E., Summer 2002*)

**Solution.** $$\frac{\partial u}{\partial x}=2\frac{\partial u}{\partial t}+u \quad \text{... (1)}$$

Let $$u=X(x).T(t) \quad \text{... (2)}$$

where $X$ is a function of $x$ only and $T$ is a function of $t$ only.

Putting the value of $u$ in (1), we get

$$\frac{\partial(X.T)}{\partial x}=2\frac{\partial}{\partial t}(X.T)+X.T$$

$$T\frac{dX}{dx}=2X\frac{dT}{dt}+X.T \Rightarrow TX'=2X.T'+X.T \Rightarrow T.\frac{X'}{X}=2\frac{T'}{T}+1=c\text{ (say)}$$

(a) $\frac{X'}{X} = c \Rightarrow \frac{1}{X}\frac{dX}{dx} = c \Rightarrow \frac{dX}{X} = c\,dx$

On integration $\log X = cx + \log a. \Rightarrow \log\frac{X}{a} = cx \Rightarrow \frac{X}{a} = e^{cx} \Rightarrow X = ae^{cx}$

(b) $\frac{2T'}{T} + 1 = c \Rightarrow \frac{T'}{T} = \frac{1}{2}(c-1) \Rightarrow \frac{1}{T}\frac{dT}{dt} = \frac{1}{2}(c-1) \Rightarrow \frac{dT}{T} = \frac{1}{2}(c-1)dt$

On integration $\log T = \frac{1}{2}(c-1)t + \log b \Rightarrow \log\frac{T}{b} = \frac{1}{2}(c-1)t$

$\Rightarrow \frac{T}{b} = e^{\frac{1}{2}(c-1)t} \Rightarrow T = be^{\frac{1}{2}(c-1)t}$

Putting the value of $X$ and $T$ in (2), we have

$$u = ae^{cx}.be^{\frac{1}{2}(c-1)t}$$

$$\Rightarrow \quad u = ab\,e^{cx+\frac{1}{2}(c-1)t} \quad ...(3)$$

$$\Rightarrow \quad u(x,0) = ab\,e^{cx}$$

But $u(x,0) = 6e^{-3x}$

*i.e.* $ab\,e^{cx} = 6e^{-3x} \Rightarrow ab = 6$ and $c = -3$

Putting the value of $ab$ and $c$ in (3), we have

$$u = 6e^{-3x+\frac{1}{2}(-3-1)t}$$

$$u = 6e^{-3x-2t}$$

which is the required solution. **Ans.**

**Example 2.** *Use the method of separation of variables to solve the equation :*

$$\frac{\partial^2 v}{\partial x^2} = \frac{\partial v}{\partial t}$$

*given that* $v = 0$ *when* $t \to \infty$, *as well as* $v = 0$ *at* $x = 0$ *and* $x = l$.

**Solution.** $\frac{\partial^2 v}{\partial x^2} = \frac{\partial v}{\partial t}$ ...(1)

Let us assume that $v = XT$ where $X$ is a function of $x$ only and $T$ that of $t$ only.

$$\frac{\partial v}{\partial t} = X\frac{dT}{dt} \quad \text{and} \quad \frac{\partial^2 v}{\partial x^2} = T\frac{d^2X}{dx^2}$$

Substituting these values in (1), we get

$$X\frac{dT}{dt} = T\frac{d^2X}{dx^2}$$

Let each side of (2) be equal to a constant $(-p^2)$

$$\Rightarrow \quad \frac{1}{T}\frac{dT}{dt} = \frac{1}{X}\frac{d^2X}{dx^2} = -p^2 \quad ...(2)$$

$$\frac{1}{T}\frac{dT}{dt} = -p^2 \Rightarrow \frac{dT}{dt} + p^2T = 0 \quad ...(3)$$

and $\frac{1}{X}\frac{d^2T}{dx^2} = -p^2 \Rightarrow \frac{d^2X}{dx^2} + p^2X = 0$ ...(4)

Solving (3) and (4), we have

$$T = C_1e^{-p^2t}$$

$$X = C_2 \cos px + C_3 \sin px \qquad ...(5)$$

$\therefore \quad v = C_1 e^{-p^2 t}(C_2 \cos px + C_3 \sin px)$

Putting $x = 0$, $v = 0$ in (5), we get

$0 = C_1 e^{-p^2 t} C_2 \qquad \therefore C_2 = 0$, since $C_1 \neq 0$

On putting the value of $C_2$ in (5), we get

$$v = C_1 e^{-p^2 t} C_3 \sin px \qquad ..(6)$$

Again putting $x = l$, $v = 0$ in (6), we get

$$0 = C_1 e^{-p^2 t} . C_3 \sin pl$$

Since $C_3$ cannot be zero.

$\therefore \sin pl = 0 = \sin n\pi \quad \therefore \qquad p = \frac{n\pi}{l}$, $n$ is any integer.

On putting the value of $p$ in (6) it becomes

$$v = C_1 C_3 e^{-\frac{n^2\pi^2 t}{l^2}} \sin \frac{n\pi x}{l}$$

Hence $\qquad v = b_n e^{\frac{-n^2\pi^2 t}{l^2}} \sin \frac{n\pi x}{l} \qquad$ where $b_n = C_1 C_3$

This equation satisfies the given condition for all integral values of $n$. Hence taking $n = 1, 2, 3, \ldots\ldots$, *the most general solution is*

$$v = \sum_{n=1}^{\infty} b_n e^{-\frac{n^2\pi^2 t}{l^2}} \frac{\sin n\pi x}{l} \qquad \textbf{Ans.}$$

**Exercise 9.15**

Using the method of separation of variables, find the solution of the following equations

1. $2x\frac{\partial z}{\partial x} - 3y\frac{\partial z}{\partial y} = 0$ **Ans.** $z = c x^{\frac{k}{2}} y^{\frac{k}{3}}$

2. $\frac{\partial u}{\partial x} + u = \frac{\partial u}{\partial t}$ if $u = 4e^{-3x}$, when $t = 0$ **Ans.** $u = 4e^{-3x-2t}$

3. $4\frac{\partial u}{\partial x} + \frac{\partial u}{\partial y} = 3u$, and $u = e^{-5y}$, when $x = 0$. **Ans.** $u = e^{2x-5y}$

4. $4\frac{\partial u}{\partial t} + \frac{\partial u}{\partial x} = 3u, u = 3e^{-x} - e^{-5x}$ at $t = 0$ (*A.M.I.E.T.E, Winter 2002, 2000*) **Ans.** $u = 3e^{t-x} - e^{2t-5x}$

5. $3\frac{\partial u}{\partial x} + 2\frac{\partial u}{\partial y} = 0;\ u(x,0) = 4e^{-x}$ (*A.M.I.E.T.E, Summer, 2000*) **Ans.** $u = 4e^{-x+3/2y}$

6. $\frac{\partial u}{\partial x} + \frac{\partial u}{\partial y} = 2(x+y)u$ **Ans.** $u = ce^{x^2+y^2+k(x-y)}$

7. $\frac{\partial u}{\partial t} = 4\frac{\partial^2 u}{\partial x^2}$ If $u(x,0) = 4x - \frac{1}{2}x^2$ **Ans.** $u = \left(4x - \frac{x^2}{2}\right)e^{-p^2 t}$

8. $\frac{\partial^2 u}{\partial x^2} = 2\frac{\partial u}{\partial t}$ if $u(x,0) = x(4-x)$ **Ans.** $u = x(4-x)e^{-\frac{p^2 t}{2}}$

9. $\frac{\partial u}{\partial t} = \frac{\partial^2 u}{\partial x^2}$ if $u(x,0) = 2x$ when $0 \le x \le \frac{l}{2}$

$\qquad = 2(l-x)$ when $\frac{l}{2} \le x \le l$

**Ans.** $u = 2xe^{-k^2 t}$ for $0 \le x \le \frac{l}{2}, u = 2(l-x)e^{-k^2 t}$ for $\frac{l}{2} \le x \le l$

10. $\dfrac{\partial u}{\partial t} = \dfrac{\partial^2 u}{\partial x^2}$ if $u(x,0) = \sin \pi x$ **Ans.** $u = \sin \pi x . e^{-p^2 t}$

11. $\dfrac{\partial^2 u}{\partial x^2} = \dfrac{\partial u}{\partial t}$ if $u(x,0) = x^2(25 - x^2)$ **Ans.** $u = x^2(25 - x^2)e^{-p^2 t}$

12. $x^2 u_{xx} + 3y^2 u = 0$

13. $\dfrac{\partial^2 z}{\partial x^2} - 2\dfrac{\partial z}{\partial x} + \dfrac{\partial z}{\partial y} = 0$ **Ans.** $z = c_1 e^{[1+\sqrt{(1-p)^2}]x + p^2 y} + c_2 e^{[1-\sqrt{1-p^2}]x + p^2 y}$

14. $\dfrac{\partial^2 u}{\partial t^2} = \dfrac{\partial^2 u}{\partial x^2}$ If $u(x,0) = \dfrac{1}{2}x(1-x)$ **Ans.** $u = \dfrac{x}{2}(1-x)\cos pt + c_2 \sin pt(c_3 \cos px + c_4 \sin px)$

15. $16\dfrac{\partial^2 u}{\partial x^2} = \dfrac{\partial^2 u}{\partial t^2}$ if $u(x,0) = x^2(5-x)$ **Ans.** $u = x^2(5-x)\cos pt + c_4 \sin pt\left(c_1 \cos\dfrac{px}{4} + c_2 \sin\dfrac{px}{4}\right)$

16. $\dfrac{\partial^2 u}{\partial x^2} = \dfrac{\partial u}{\partial y} + 2u$ if $u = 0$, **Ans.** $u = (c_1 \cos px + c_2 \sin px)\, c_3\, e^{-(p^2+2)y}$

17. $x\dfrac{\partial u}{\partial x} + y\dfrac{\partial u}{\partial y} = 0$ **Ans.** $u = A\, e^{1/2(x^2 - y^2)k}$

18. $\dfrac{\partial u}{\partial x} = \dfrac{\partial u}{\partial y}$, **Ans.** $u = Ae^{k(x+y)}$

19. $\dfrac{\partial u}{\partial x} = 2\dfrac{\partial u}{\partial y} + u$, $u(x,0) = 4e^{-3x}$ *(A.M.I.E.T.E., Summer 2001)* **Ans.** $u = 4e^{-(3x+2y)}$

20. $2\dfrac{\partial u}{\partial x} + 3\dfrac{\partial u}{\partial y} + 5u = 0$, $u(0,y) = 2e^{-y}$ **Ans.** $u = 2e^{-x-y}$

21. $\dfrac{\partial u}{\partial x} + \dfrac{\partial u}{\partial y} = 3u$ and $u = e^{-5y}$ when $x = 0$ **Ans.** $u = e^{8x-5y}$

22. $\dfrac{\partial u}{\partial x} - 2\dfrac{\partial u}{\partial y} = u$, given that $u(x,0) = 3e^{-5x} + 2e^{-3x}$ **Ans.** $u = 3e^{-5x-3y} + 2e^{-3x-2y}$

## 9.18 EQUATION OF VIBRATING STRING

Consider an elastic string tightly stretched between two points O and A. Let O be the origin and OA as $x$-axis. On giving a small displacement to the string, perpendicular to its length (parallel to the $y$-axis). Let $y$ be the displacement at the point $P(x, y)$ at any time. The wave equation.

$$\frac{\partial^2 y}{\partial t^2} = c^2 \frac{\partial^2 y}{\partial x^2}$$

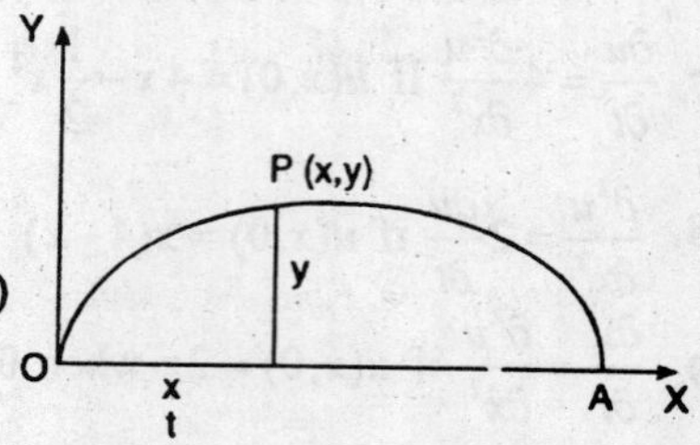


**Example 3.** *Obtain the solution of the wave equation*

$$\frac{\partial^2 y}{\partial t^2} = c^2 \frac{\partial^2 y}{\partial x^2}$$

*(A.M.I.E.T.E., Summer 2002)*

*using the method of separation of variables.*

**Solution.** $$\frac{\partial^2 y}{\partial t^2} = c^2 \frac{\partial^2 y}{\partial x^2}$$

Let $y = XT$ where $X$ is a function of $x$ only and $T$ is a function of $t$ only.

$$\frac{\partial y}{\partial t} = X\frac{dT}{dt} \quad \text{and} \quad \frac{\partial y}{\partial x} = T\frac{dX}{dx}$$

Since $T$ and $X$ are functions of a single variable only.

$$\frac{\partial^2 y}{\partial t^2} = X\frac{d^2T}{dt^2} \qquad \text{and} \qquad \frac{\partial^2 y}{\partial x^2} = T\frac{d^2X}{dx^2}$$

Substituting these values in the given equation, we get

$$X\frac{d^2T}{dt^2} = c^2 T\frac{d^2X}{dx^2}$$

By separating the variables, we get

$$\frac{\frac{d^2T}{dt^2}}{c^2T} = \frac{\frac{d^2X}{dx^2}}{X} = k \qquad \text{(say).}$$

(Each side is constant, since the variables $x$ and $y$ are independent).

$$\therefore \qquad \frac{d^2T}{dt^2} - k\,c^2T = 0 \qquad \text{and} \qquad \frac{d^2X}{dx^2} - kX = 0$$

Auxiliary equations are

$$m^2 - kc^2 = 0 \Rightarrow m = \pm c\sqrt{k} \quad \textit{and} \quad m^2 - k = 0 \Rightarrow m = \pm\sqrt{k}$$

*Case 1. If* $k > 0$.

$$T = C_1 e^{c\sqrt{k}t} + C_2 e^{-c\sqrt{k}t}$$
$$X = C_3 e^{c\sqrt{k}x} + C_4 e^{-\sqrt{k}x}$$

*Case 2. If* $k < 0$.

$$T = C_5 \cos c\sqrt{k}t + C_6 \sin c\sqrt{k}t$$
$$X = C_7 \cos\sqrt{k}x + C_8 \sin\sqrt{k}x$$

*Case 3. If* $k = 0$.

$$T = C_9 t + C_{10}$$
$$X = C_{11}x + C_{12}$$

These are the three cases depending upon the particular problems. Here we are dealing with wave motion ($k < 0$).

$$y = TX$$

$$y = \left(C_5 \cos c\sqrt{k}t + C_6 \sin c\sqrt{k}t\right)\times\left(C_7 \cos\sqrt{k}x + C_8 \sin\sqrt{k}x\right) \qquad \textbf{Ans.}$$

**Example 4.** *Find the solution of the wave equation*

$$\frac{\partial^2 y}{\partial t^2} = c^2 \frac{\partial^2 y}{\partial x^2}$$

***such that*** $y = P_o \cos pt$, ($P_o$ *is a constant*) *when* $x = l$ *and* $y = 0$ *when* $x = 0$.

**Solution.** $$\frac{\partial^2 y}{\partial t^2} = c^2 \frac{\partial^2 y}{\partial x^2} \qquad \text{...(1)}$$

Its solution is as given in Example 3 on page 710.

$$y=\left(c_1\cos c\sqrt{kt}+c_2\sin c\sqrt{kt}\right)\left(c_3\cos\sqrt{kx}+c_4\sin\sqrt{kx}\right) \quad ...(2)$$

Put $y=0$, when $x=0$

$$0=\left(c_1\cos c\sqrt{kt}+c_2\sin c\sqrt{kt}\right)c_3 \quad \Rightarrow c_3=0$$

(2) is reduced to

$$y=\left(c_1\cos c\sqrt{kt}+c_2\sin c\sqrt{kt}\right)c_4\sin\sqrt{kx}$$

$$y=c_1c_4\cos c\sqrt{kt}\sin\sqrt{kx}+c_2c_4\sin c\sqrt{kt}\sin\sqrt{kx} \quad ...(3)$$

put $y=P_0\cos pt$ *when* $x=l$

$$P_0\cos pt=c_1c_4\cos c\sqrt{kt}\sin\sqrt{kl}+c_2c_4\sin c\sqrt{kt}\sin\sqrt{kl}$$

Equating the coefficient of sin and cos on both sides

$$P_0=c_1c_4\sin\sqrt{kl} \quad \Rightarrow c_1c_4=\frac{P_0}{\sin\sqrt{k}\,l}$$

$$0=c_2c_4\sin\sqrt{kl} \quad \Rightarrow c_2=0$$

And $p=c\sqrt{k}$ $\quad \Rightarrow \frac{p}{c}=\sqrt{k}$

(3) becomes $\quad y=\frac{P_0}{\sin\sqrt{kl}}\cos pt\sin\frac{p}{c}x$

$$y=\frac{P_0}{\sin\frac{p}{c}l}\cos pt\sin\frac{p}{c}x$$

**Ans.**

**Example 5.** *A string is stretched and fastened to two points l apart Motion is started by displacing the string in the form* $y=a\sin\frac{\pi x}{l}$ *from which it is released at a time* $t=0$. *Show that the displacement of any point at a distance x from one end at time t is given by*

$$y(x,t)=a\sin\left(\frac{\pi x}{l}\right)\cos\left(\frac{\pi ct}{l}\right)$$

*(A.M.I.E.T.E., Winter 2003, A.M.I.E., Winter 2001)*

**Solution.** The vibration of the string is given by:

$$\frac{\partial^2 y}{\partial t^2}=C^2\frac{\partial^2 y}{\partial x^2} \quad ...(1)$$

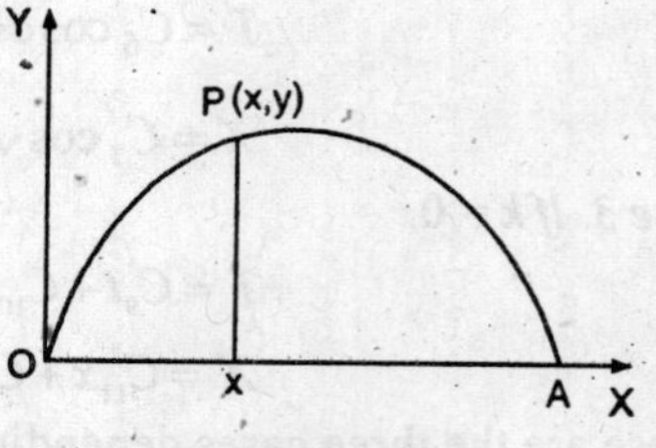


As the end points of the string are fixed, for all time,

$$y(0,t)=0 \quad ...(2)$$

and $\quad y(l,t)=0 \quad ...(3)$

Since the initial transverse velocity of any point of the string is zero, therefore,

$$\left(\frac{\partial y}{\partial t}\right)_{t=0}=0 \quad ...(4)$$

Also $\quad y(x,0)=a\sin\frac{\pi x}{l} \quad ...(5)$

Now we have to solve (1), subject to the above boundary conditions. Since the vibration of the string is periodic, therefore, the solution of (1) is of the form

$$y(x,t)=(C_1\cos px+C_2\sin px)(C_3\cos Cpt+C_4\sin Cpt) \quad ...(6)$$

By (2) $\quad y(0,t)=C_1(C_3\cos Cpt+C_4\sin Cpt)=0$

For this to be true for all time, $C_1 = 0$.

Hence $\qquad y(x,t) = C_2 \sin px(C_3 \cos Cpt + C_4 \sin Cpt)$ ...(7)

and $\qquad \dfrac{\partial y}{\partial t} = C_2 \sin px\left[C_3(-Cp \sin Cpt) + C_4(Cp \cos Cpt)\right]$

$\therefore$ By (4) $\qquad \left(\dfrac{\partial y}{\partial t}\right)_{t=0} = C_2 \sin px(C_4 Cp) = 0$

When $\qquad C_2 C_4 Cp = 0$

If $C_2 = 0$, (7) will lead to the trivial solution $y(x, t) = 0$.

$\therefore$ the only possibility is that $C_4 = 0$

Thus (7) becomes

$$y(x, t) = C_2 C_3 \sin px \cos Cpt \qquad ...(8)$$

If $x = l$ then $y = 0, 0 = C_2 C_3 \sin p\, l \cos Cpt,$ for all t.

Since $C_2$ and $C_3 \neq 0$, we have $\sin p\, l = 0 \;\therefore\; p\, l = n\pi$

*i.e.* $\qquad p = \dfrac{n\pi}{l}$, where $n$ is an integer.

Hence (8) reduces to

$$y(x,t) = C_2 C_3 \sin \frac{n\pi x}{l} \cos \frac{n\pi Ct}{l} \qquad ...(9)$$

Finally imposing the last condition (5), we have

$$y(x,0) = C_2 C_3 \sin \frac{n\pi x}{l} = a \sin \frac{\pi x}{l}$$

which will be satisfied by taking $C_2 C_3 = a$ and $n = 1$

Hence the required solution is

$$y(x,t) = a \sin \frac{\pi x}{l} \cos \frac{\pi C t}{l}$$ **Proved.**

**Example 6.** *The vibrations of an elastic string is governed by the partial differential equation*

$$\frac{\partial^2 u}{\partial t^2} = \frac{\partial^2 u}{\partial x^2}$$

*The length of the string is $\pi$ and the ends are fixed. The initial velocity is zero and the initial deflection is u (x, 0) = 2(sin x + sin 3x). Find the deflection u (x, t) of the vibrating string for* $t \geq 0$.

**Solution.** $\qquad \dfrac{\partial^2 u}{\partial t^2} = \dfrac{\partial^2 u}{\partial x^2}$

$\Rightarrow \qquad u = (c_1 \cos pt + c_2 \sin pt)(c_3 \cos px + c_4 \sin px)$ ...(1)

On putting $x = 0, u = 0$ in (1), we get

$$0 = (c_1 \cos pt + c_2 \sin pt)c_3 \Rightarrow c_3 = 0$$

On putting $c_3 = 0$ in (1), it reduces

$$u = (c_1 \cos pt + c_2 \sin pt)c_4 \sin px \qquad ...(2)$$

On putting $x=\pi$ and $u=0$ in (2), we have

$$0=(c_1\cos pt+c_2\sin pt)c_4\sin p\pi$$

$$\sin p\pi=0=\sin n\pi \qquad n=1,2,3,4..$$

$$\therefore \quad p\pi=n\pi \quad \text{or} \quad p=n$$

On substituting the value of $p$ in (2), we get

$$u=(c_1\cos nt+c_2\sin nt)c_4\sin nx \qquad ...(3)$$

On differentiating (3) w.r.t. "$t$", we get

$$\frac{du}{dt}=(-c_1 n\sin nt+c_2 n\cos nt)c_4\sin nx \qquad ...(4)$$

On putting $\frac{du}{dt}=0$, $t=0$ in (4) we have

$$0=(c_2 n)(c_4\sin nx)\Rightarrow c_2=0$$

On putting $c_2=0$, (3) becomes

$$u=(c_1\cos nt)(c_4\sin nx)$$

$$u=c_1c_4\cos nt\sin nx \qquad ...(5)$$

given $u(x,0)=2(\sin x+\sin 3x)$

On putting $t=0$ in (5), we have

$$u=(x,0)=c_1c_4\sin nx$$

$$2(\sin x+\sin 3x)=c_1c_4\sin nx$$

$$4\sin 2x\cos x=c_1c_4\sin nx$$

$$c_1c_4=4\cos x \qquad 2=n$$

On substituting the value of $c_1c_4$ and $n=2$,(5) becomes

$$u(x,t)=4\cos x\cos 2t\sin 2x$$ **Ans**

**Example 7.** *A tightly stretched string with fixed end points $x=0$ and $x=l$ is initially in a position given by $y=y_0\sin^3\left(\frac{\pi x}{l}\right)$. If it is released from rest from this position, find the displacement $y(x,t)$.*

**Solution.** Let the equation to the vibrating string be

$$\frac{\partial^2 y}{\partial t^2}=c^2\frac{\partial^2 y}{\partial x^2} \qquad ...(1)$$

Here the initial conditions are

$$y(0,t)=0,\; y(l,t)=0$$

$$\frac{\partial y}{\partial t}=0 \text{ at } t=0,\; y(x,0)=y_0\sin^3\frac{\pi x}{l}$$

The solution of (1) is of the form

$$y=(c_1\cos px+c_2\sin px)(c_3\cos pct+c_4\sin pct) \qquad ...(2)$$

Now $y(0,t)=0$ gives

$$0=c_1(c_3\cos pct+c_4\sin pct)\Rightarrow c_1=0$$

Hence (2) becomes

$$y=c_2\sin px(c_3\cos pct+c_4\sin pct) \qquad ...(3)$$

$y(l,t)=0$ gives

$$0 = c_2 \sin pl\left(c_3 \cos pct + c_4 \sin pct\right)$$

$\therefore$ $\sin pl = 0 = \sin n\pi$ or $pl = n\pi$, or $p = \dfrac{n\pi}{l}$ where $n = 0, 1, 2, 3......$

On putting the value of $p$ in (3), we get

$$y = c_2 \sin\frac{n\pi x}{l}\left(c_3 \cos\frac{n\pi ct}{l} + c_4 \sin\frac{n\pi ct}{l}\right) \quad ...(4)$$

Now $\dfrac{\partial y}{\partial t} = c_2 \sin\dfrac{n\pi x}{l}\left(-\dfrac{n\pi c}{l} c_3 \sin\dfrac{n\pi c}{l} t + c_4 \dfrac{n\pi c}{l} \cos\dfrac{n\pi ct}{l}\right)$

Since $\dfrac{\partial y}{\partial t} = 0$ when $t = 0$, we have

$$0 = c_2 \sin\frac{n\pi x}{l} c_4 \frac{n\pi c}{l} \qquad \Rightarrow c_4 = 0$$

Now (4) reduces to

$$y = c_2 c_3 \sin\frac{n\pi x}{l} \cos\frac{n\pi ct}{l}$$

$$\Rightarrow \quad y = b_n \sin\frac{n\pi x}{l} \cos\frac{n\pi c}{l} t \qquad (b_n = c_2 c_3)$$

$$\Rightarrow \quad y = \sum_{n=1}^{\infty} b_n \sin\frac{n\pi x}{l} \cos\frac{n\pi c}{l} t \quad ...(5)$$

But $\quad y(x, 0) = y_0 \sin^3\dfrac{\pi x}{l} = \dfrac{y_0}{4}\left(3 \sin\dfrac{\pi x}{l} - \sin\dfrac{3\pi x}{l}\right)$ (given) ...(6)

On putting $t = 0$ in (5), we get, $y(x, 0) = \displaystyle\sum_{n=1}^{\infty} b_n \sin\frac{n\pi x}{l}$ ...(7)

From (6) and (7), we have

$$y = \sum_{n=1}^{\infty} b_n \sin\frac{n\pi x}{l} = \frac{y_0}{4}\left(3 \sin\frac{\pi x}{l} - \sin\frac{3\pi x}{l}\right)$$

$$y = b_1 \sin\frac{\pi x}{l} + b_2 \sin\frac{2\pi x}{l} + b_3 \sin\frac{3\pi x}{l} = \frac{3y_0}{4} \sin\frac{\pi x}{l} - \frac{y_0}{4} \sin\frac{3\pi x}{l}$$

Comparing the coefficients, we have

$$b_1 = \frac{3y_0}{4}, \qquad b_3 = -\frac{y_0}{4}$$

and all others b's are zero.
Hence (5) becomes

$$y = \frac{y_0}{4}\left(3 \sin\frac{\pi x}{1} \cos\frac{c\pi t}{l} - \sin\frac{3\pi x}{l} \cos\frac{3c\pi t}{l}\right) \quad \textbf{Ans.}$$

**Example 8.** *Solve the wave equation*

$$\frac{\partial^2 u}{\partial t^2} = a^2 \frac{\partial^2 u}{\partial x^2}$$

*under the condition:*

$$u = 0 \text{ when } x = 0 \text{ and } x = \pi$$

$$\frac{\partial u}{\partial t} = 0 \text{ when } t = 0 \text{ and } u(x, 0) = x, 0 < x < \pi.$$

**Solution.** The solution is of the form

$$u(x, t) = (c_1 \cos px + c_2 \sin px)(c_3 \cos a pt + c_4 \sin a pt) \quad ...(1)$$

Since $u(0, t) = 0$.

$$0 = c_1\left(c_3 \cos a pt + c_4 \sin a pt\right) \Rightarrow \quad c_1 = 0$$

Then (1) becomes

$$u(x, t) = c_2 \sin px\,(c_3 \cos a\,pt + c_4 \sin a\,pt) \quad ...(2)$$

$$u(\pi, t) = 0$$

$$0 = c_2 \sin p\pi\,(c_3 \cos a\,pt + c_4 \sin a\,pt) \Rightarrow \sin p\,\pi = 0 = \sin n\,\pi \text{ or } p = n$$

Thus $u(x, t) = c_2 \sin nx\,(c_3 \cos a\,nt + c_4 \sin a\,n\,t)$

$$u(x, t) = \sin n\,x\,(b_1 \cos a\,nt + b_2 \sin a\,n\,t) \quad ...(3)$$

Now $\dfrac{\partial u}{\partial t} = \sin nx\ (-ab_1\,n \sin ant + ab_2 n \cos ant)$

As $\dfrac{\partial u}{\partial t} = 0$ when $t = 0$ we have

$$0 = \sin nx(ab_2 n) \qquad \Rightarrow \qquad b_2 = 0$$

$$u(x,t) = \sin nx(b_1 \cos ant) \text{ or } u(x, t) = b_n \sin nx \cos a\,nt \quad ...(4)$$

$$u(x,t) = \sum_{n=1}^{\infty} b_n \sin nx \cos ant \quad ...(5)$$

On putting $u(x, 0) = x$, we have

$$x = \sum_{n=1}^{\infty} b_n \sin nx, \quad \text{where } b_n = \frac{2}{\pi}\int_0^{\pi} x \sin nx\, dx$$

$$= \frac{2}{\pi}\left[x\left(-\frac{\cos nx}{n}\right) - (1)\left(-\frac{\sin nx}{n^2}\right)\right]_0^{\pi} \quad ..(6)$$

$$= \frac{2}{\pi}\left[-\frac{\pi}{n}\cos n\pi\right] = -\frac{2}{n}(-1)^n$$

Hence, the required solution is

$$u(x,t) = -2\sum_{n=1}^{\infty} \frac{(-1)^n}{n} \sin nx \cos nat$$ **Ans.**

**Example 9.** *A string is stretched and fastened to two points l apart. Motion is started by displacing the string into the form $y = k(lx - x^2)$ from which it is released at time $t = 0$. Find the displacement of any point on the string at a distance of x from one end at time t.*

*(A.M.I.E.T.E., Dec. 2006, Summer 2000)*

**Solution.** The vibration of the string is given by

$$\frac{\partial^2 y}{\partial t^2} = c^2 \frac{\partial^2 y}{\partial x^2} \quad ...(1)$$

As the end points of the string are fixed for all time,

$$y(0, t) = 0 \quad ...(2)$$

and $$y(l, t) = 0 \quad ...(3)$$

Since the initial transverse velocity of any point of the string is zero, therefore,

$$\left(\frac{\partial y}{\partial t}\right)_{t=0} = 0 \quad ...(4)$$

and $$y(x, 0) = k(lx - x^2) \quad ...(5)$$

Solution of (1) is $$y = (c_1 \cos px + c_2 \sin px)(c_3 \cos c\,pt + c_4 \sin cpt) \quad ...(6)$$

By (2), $y(0, t) = 0$

$$0 = c_1 (c_3 \cos c\,pt + c_4 \sin c\,pt) \quad \therefore \quad c_1 = 0$$

Hence (6) becomes $\quad y = c_2 \sin px(c_3 \cos cpt + c_4 \sin cpt)$ ...(7)

$$\frac{\partial y}{\partial t} = c_2 \sin px(-c_3 cp \sin cpt + c_4 cp \cos cpt)$$

By (4) $\left(\frac{\partial y}{\partial t}\right)_{t=0} = 0$

$$0 = c_2 \sin px(c_4 cp) \quad \Rightarrow \quad c_4 = 0 \quad \text{(since } c_2 \neq 0\text{)}$$

Hence (7) is reduced to

$$y = c_2 \sin px(c_3 \cos cpt)$$

$$y = c_2 c_3 \sin px \cos cpt \qquad ...(8)$$

$$y(l, t) = 0$$

On putting $x = l$ in equation (8), we get

$$0 = c_2 c_3 \sin pl \cos cpt \quad \Rightarrow \quad 0 = \sin pl$$

$\Rightarrow \quad \sin n\pi = \sin pl \quad$ or $\quad pl = n\pi, \quad p = \frac{n\pi}{l}$ where $n = 1, 2, 3....$

On putting $p = \frac{n\pi}{1}$, equation (8) becomes $\quad (c_2 c_3 = b_n)$

$$\Rightarrow \quad y = b_n \sin \frac{n\pi x}{l} \cos \frac{n\pi c}{l} t$$

We can have any number of solutions by taking different integral values of $n$ and the complete solution will be the sum of these solutions. Thus,

$$y(x,t) = \sum_{n=1}^{\infty} b_n \sin \frac{n\pi x}{l} \cos \frac{n\pi c}{l} t \qquad ... (9)$$

$$y(x,0) = \sum_{n=1}^{\infty} b_n \sin \frac{n\pi x}{l}$$

$$lx - x^2 = \sum_{n=1}^{\infty} b_n \sin \frac{n\pi x}{l} \qquad \text{[Using (5)]} \qquad ...(10)$$

Now it is clear that (10) represents the expansion of $f(x)$ in the form of a Fourier sine series and consequently

$$b_n = \frac{2}{l} \int_0^l f(x) \sin \frac{n\pi x}{l} dx \qquad ...(11)$$

$$b_n = \frac{2}{l} \int_0^l (lx - x^2) \sin \frac{n\pi x}{l} dx$$

$$= \frac{2}{l} \left[ (lx - x^2)\left(-\cos \frac{n\pi x}{l}\right) \frac{l}{n\pi} - (l - 2x)\left(-\sin \frac{n\pi x}{l}\right) \frac{l^2}{n^2\pi^2} + (-2)\left(\cos \frac{n\pi x}{l}\right) \frac{l^3}{n^3\pi^3} \right]_0^l$$

$$= \frac{2}{l} \left[ (-1)^{n+1} \frac{2l^3}{n^3\pi^3} + \frac{2l^3}{n^3\pi^3} \right] = \frac{8l^2}{n^3\pi^3}, \text{ when } n \text{ is odd}$$

$$= 0, \quad \text{when } n \text{ is even}$$

Putting the value of $b_n$ in (9), we get

$$y = \sum \frac{8l^2}{n^3\pi^3}\sin\frac{n\pi x}{l}\cos\frac{n\pi c}{l}t \text{ when } n \text{ is odd.}$$ **Ans.**

## 9.19. SOLUTION OF WAVE EQUATION BY D'ALMBERT'S METHOD

$$\frac{\partial^2 y}{\partial t^2} = c^2\frac{\partial^2 y}{\partial x^2} \quad \ldots(1)$$

Let us introduce the two new independent variables $u = x + ct,$ $v = x - ct$

So that $y$ becomes a function of $u$ and $v$

$$\frac{\partial y}{\partial x} = \frac{\partial y}{\partial u}\frac{\partial u}{\partial x} + \frac{\partial y}{\partial v}\frac{\partial v}{\partial x} = \frac{\partial y}{\partial u}(1) + \frac{\partial y}{\partial v}(1) = \frac{\partial y}{\partial u} + \frac{\partial y}{\partial v}$$

$$\frac{\partial}{\partial x} = \frac{\partial}{\partial u} + \frac{\partial}{\partial v}$$

$$\frac{\partial^2 y}{\partial x^2} = \frac{\partial}{\partial x}\left(\frac{\partial y}{\partial x}\right) = \left(\frac{\partial}{\partial u} + \frac{\partial}{\partial v}\right)\left(\frac{\partial y}{\partial u} + \frac{\partial y}{\partial v}\right)$$

$$= \frac{\partial}{\partial u}\left(\frac{\partial y}{\partial u} + \frac{\partial y}{\partial v}\right) + \frac{\partial}{\partial v}\left(\frac{\partial y}{\partial u} + \frac{\partial y}{\partial v}\right) = \frac{\partial^2 y}{\partial u^2} + 2\frac{\partial^2 y}{\partial u\partial v} + \frac{\partial^2 y}{\partial v^2} \quad \ldots(2)$$

$$\frac{\partial y}{\partial t} = \frac{\partial y}{\partial u}\frac{\partial u}{\partial t} + \frac{\partial y}{\partial v}\frac{\partial v}{\partial t} = \frac{\partial y}{\partial u}c + \frac{\partial y}{\partial v}(-c) = c\left(\frac{\partial y}{\partial u} - \frac{\partial y}{\partial v}\right) \quad \left[\because \frac{\partial u}{\partial t} = c, \frac{\partial v}{\partial t} = -c\right]$$

$$\frac{\partial}{\partial t} = c\left(\frac{\partial}{\partial u} - \frac{\partial}{\partial v}\right)$$

$$\frac{\partial^2 y}{\partial t^2} = \frac{\partial}{\partial t}\left(\frac{\partial y}{\partial t}\right) = c\left(\frac{\partial}{\partial u} - \frac{\partial}{\partial v}\right)c\left(\frac{\partial y}{\partial u} - \frac{\partial y}{\partial v}\right)$$

$$= c^2\left(\frac{\partial^2 y}{\partial u^2} - 2\frac{\partial^2 y}{\partial u\partial v} + \frac{\partial^2 y}{\partial v^2}\right) \quad \ldots(3)$$

Substituting the values of $\frac{\partial^2 y}{\partial x^2}$ and $\frac{\partial^2 y}{\partial t^2}$ from (2) and (3) in (1), we get

$$c^2\left(\frac{\partial^2 y}{\partial u^2} - 2\frac{\partial^2 y}{\partial u\partial v} + \frac{\partial^2 y}{\partial v^2}\right) = c^2\left(\frac{\partial^2 y}{\partial u^2} + 2\frac{\partial^2 y}{\partial u\partial v} + \frac{\partial^2 y}{\partial v^2}\right) \text{ or } \frac{\partial^2 y}{\partial u\partial v} = 0 \quad \ldots(4)$$

Integrating (4) *w.r.t* $v$, we get $\frac{\partial y}{\partial u} = f(u)$ ... (5)

where $f(u)$ is constant in respect of $v$. Again integrating (5) *w. r.t* '$u$,' we get

$$y = \int f(u)du + \psi(v)$$

where $\psi(v)$ is constant in respect of $u$

$$y = \phi(u) + \psi(v) \text{ where } \phi(u) = \int f(u)du$$

$$\Rightarrow \quad y(x, t) = \phi(x + ct) + \psi(x - ct) \quad \ldots(6)$$

This is D'Alemberts solution of wave equations (1)

To determine $\phi, \psi$ let us apply initial conditions, $y(x, 0) = f(x)$ and $\frac{\partial y}{\partial t} = 0$ when $t = 0$.

Differentiating (6) w. r.t. "$t$", we get

$$\frac{\partial y}{\partial t} = c\phi'(x+ct) - c\psi'(x-ct) \quad ...(7)$$

Putting $\frac{\partial y}{\partial t} = 0$, and $t = 0$ in (7) we get $\quad 0 = c\phi'(x) - c\psi'(x)$

$$\Rightarrow \quad \phi'(x) = \psi'(x) \Rightarrow \phi(x) = \psi(x) + b$$

Again substituting $y = f(x)$ and $t = 0$ in (6) we get

$$f(x) = \phi(x) + \psi(x) \Rightarrow f(x) = [\psi(x) + b] + \psi(x)$$

$$\Rightarrow \quad f(x) = 2\psi(x) + b$$

so that $\psi(x) = \frac{1}{2}[f(x) - b]$ and $\phi(x) = \frac{1}{2}[f(x) + b]$

On putting the values of $\phi(x+ct)$ and $\psi(x-ct)$ in (6), we get

$$y(x,t) = \frac{1}{2}[f(x+ct) + b] + \frac{1}{2}[f(x-ct) - b]$$

$$\Rightarrow \quad y(x,t) = \frac{1}{2}[f(x+ct) + f(x-ct)]$$ **Ans.**

## EXERCISE 9.16

1. Solve the wave equation $\frac{\partial^2 u}{\partial t^2} = c^2 \frac{\partial^2 u}{\partial x^2}$

under the conditions $u = 0$, when $x = 0$ and $x = \pi$

$\frac{\partial u}{\partial t} = 0$ when $t = 0$ and $u(x, 0) = x, 0 < x < \pi$. **Ans.** $u = \frac{2\sum(-1)^{n+1}}{n} \sin nx \cos nct$

2. Using the transformations $v = x + ct$ and $z = x - ct$, solve the following :

$\frac{\partial^2 u}{\partial t^2} = c^2 \frac{\partial^2 u}{\partial x^2}$; $\quad \frac{\partial u}{\partial t}(x,0) = 0$; $\quad u(x,0) = f(x)$. **Ans.** $u(x,t) = \frac{1}{2}[f(x+ct) + f(x-ct)]$

3. A string of length $l$ is initially at rest in equilibrium position and each of its points given velocity,

$$\left(\frac{\partial y}{\partial t}\right)_{t=0} = b \sin^3 \frac{\pi x}{l}$$

Find the displacement $y(x, t)$. (*A.M.I.E.T.E., Summer 2001*) **Ans.** $y(x,t) = \sum b_n \sin \frac{n\pi x}{l} \sin \frac{nc\pi}{l} t$

4. Find the solution of the equation of a vibrating string of length $l$ satisfying the initial conditions :

$$y = f(x) \text{ when } t = 0, \text{ and } \frac{\partial y}{\partial t} = g(x) \text{ when } t = 0$$

It is assumed that the equation of a vibrating string is $\frac{\partial^2 y}{\partial t^2} = a^2 \frac{\partial^2 y}{\partial x^2}$

**Ans.** $y(x,t) = \sum_{n=1}^{\infty} (b_n \cos \frac{n\pi at}{l} + c_n \sin \frac{n\pi at}{l}) \sin \frac{n\pi}{l} x$

where $b_n = \frac{2}{l} \int_0^l f(x) \sin \frac{n\pi x}{l} dx$, $C_n = \frac{2}{cn\pi} \int_0^l g(x) \sin \frac{n\pi x}{l} dx$

5. A tightly stretched string with fixed end points $x = 0$ and $x = l$ is initially at rest in its equilibrium position. If it is set vibrating by giving to each of its points a velocity $\lambda x(l - x)$, find the displacement of the string at any distance $x$ from one end at any time $t$.

**Ans.** $y = \frac{8\lambda l^3}{c\pi^4} \sum_{(n=1)}^{n} \frac{1}{(2n-1)^4} \sin \frac{(2n-1)\pi x}{l} \sin \frac{(2n-1)\pi ct}{l}$

6. A tightly stretched string of length $l$ fastened at both ends, is disturbed from the position of equilibrium by imparting to each of its points an initial velocity of magnitude $f(x)$. Show that the solution of the problem is

$$u(x,t)=\frac{2}{l}\sum_{n=1}^{\infty}\left[\int_0^l f(x)\sin\frac{n\pi}{l}x\,dx\right]\sin\frac{n\pi x}{l}\cos\frac{n\pi at}{l}$$

7. A tightly stretched string with fixed end points $x=0$ *and* $x=\pi$ is initially at rest in its equilibrium position. If it is set vibrating by giving each point a velocity $\left(\frac{\partial y}{\partial t}\right)_{t=0}=0.03\sin x-0.04\sin 3x$ then find the displacement $y(x,\ t)$ at any point of the string at any time $t$.

8. Find the solution of the equation $\frac{\partial^2 u}{\partial t^2}=a^2\frac{\partial^2 u}{\partial x^2}$ which satisfies the conditions.

$u(0,t)=0,\ u(l,\ t)=0,\ u(x,\ 0)=\phi(x),\ u_t(x,\ 0)=0$ **Ans.** $u(x,\ t)=\sum_{n=1}^{\infty}b_n\cos\frac{n\pi at}{l}\sin\frac{n\pi x}{l}$

9. Find the solution of the equation $\frac{\partial^2 y}{\partial x^2}=\frac{1}{c^2}\frac{\partial^2 y}{\partial t^2}$ subject to the boundary conditions.

$$y(0,t)=0,\qquad y(l,t)=0,\qquad y(x,0)=\phi(x),\qquad \frac{\partial y}{\partial t}(x,0)=\psi(x)$$

**Ans.** $y=\phi(x).\cos\frac{n\pi ct}{l}+\frac{l\psi(x)}{n\pi c}.\frac{\sin\frac{n\pi ct}{l}}{\sin\frac{n\pi x}{l}}$

10. The vibrations of an elastic string of length $l$ are governed by the one-dimensional wave equation $\frac{\partial^2 u}{\partial t^2}=c^2\frac{\partial^2 u}{\partial x^2}$. The string is fixed at the ends.

$u(0,t)=0=u(l,t)$ for all $t$. The initial deflection is

$$u(x,\ 0)=x;\ 0<x<l/2,\qquad u(x,\ 0)=l-x;\ \frac{l}{2}\le x\le l$$

and the initial velocity is zero. Find the deflection of the string at any instant of time.

*(A.M.I.E.T.E., Summer 2001, ,A.M.I.E. Summer 2001)* **Ans.** $\frac{4l}{\pi^2}\sum_{n=1}^{\infty}\frac{\sin\frac{n\pi}{2}}{n^2}\sin\frac{n\pi x}{l}\cos\frac{n\pi ct}{l}$

## 9.20 ONE DIMENSIONAL HEAT FLOW

Let heat flow along a bar of uniform cross-section, in the direction perpendicular to the cross-section. Take one end of the bar as origin and the direction of heat flow is along x-axis.

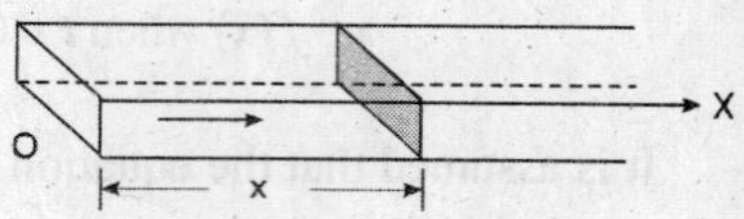


Let the temperature of the bar at any time $t$ at a point $x$ distance from the origin be $u(x,t)$. Then the equation of one dimensional heat flow is $\frac{\partial u}{\partial t}=c^2\frac{\partial^2 u}{\partial x^2}$

**Example 10.** *A rod of length l with insulated sides is initially at a uniform temperature u. Its ends are suddenly cooled to O°C and are kept at that temperature. Prove that the temperature function u (x, t) is given by*

$$u(x,t)=\sum_{n=1}^{\infty}b_n\sin\frac{n\pi x}{l}.e^{-\frac{c^2\pi^2n^2t}{l^2}}$$

*where $b_n$ is determined from the equation.*

$$U_0 = \sum_{n=1}^{\infty} b_n \sin\frac{n\pi x}{l}$$

**Solution.** Let the equation for the conduction of heat be

$$\frac{\partial u}{\partial t} = c^2 \frac{\partial^2 u}{\partial x^2} \qquad \text{...(1)}$$

Let us assume that $u = XT$, where $X$ is a function of $x$ alone and $T$ that of $t$ alone.

$$\therefore \quad \frac{\partial u}{\partial t} = X\frac{dT}{dt} \quad \text{and} \quad \frac{\partial^2 u}{\partial x^2} = T\frac{d^2X}{dx^2}$$

Substituting these values in (1), we get $X\frac{dT}{dt} = c^2 T \frac{d^2X}{dx^2}$

i.e. $$\frac{1}{c^2 T}\frac{dT}{dt} = \frac{1}{X}\frac{d^2X}{dx^2} \qquad \text{...(2)}$$

Let each side be equal to a constant $(-p^2)$.

$$\frac{1}{c^2 T}\frac{dT}{dt} = -p^2 \quad \Rightarrow \quad \frac{dT}{dt} + p^2c^2T = 0 \qquad \text{...(3)}$$

and $$\frac{1}{X}\frac{d^2X}{dx^2} = -p^2 \quad \Rightarrow \quad \frac{d^2X}{dx^2} + p^2X = 0 \qquad \text{...(4)}$$

Solving (3) and (4) we have

$$T = c_1 e^{-p^2c^2t} \quad \text{and} \quad X = c_2 \cos px + c_3 \sin px$$

$$\therefore \quad u = c_1 e^{-p^2c^2t}(c_2 \cos px + c_3 \sin px) \qquad \text{...(5)}$$

Putting $x = 0$, $u = 0$ in (5), we get

$$0 = c_1 e^{-p^2c^2t}(c_2) \quad \Rightarrow \quad c_2 = 0 \text{ since } c_1 \neq 0$$

(5) becomes $$u = c_1 e^{-p^2c^2t} c_3 \sin px \qquad \text{...(6)}$$

Again putting $x = l$, $u = 0$ in (6), we get

$$0 = c_1 e^{-p^2c^2t}.c_3 \sin pl \Rightarrow \quad \sin pl = 0 = \sin n\pi$$

$$\Rightarrow \quad pl = n\pi \quad \Rightarrow \quad p = \frac{n\pi}{l}, n \text{ is any integer}$$

Hence (6) becomes $u = c_1 c_3 e^{\frac{-n^2c^2\pi^2}{l^2}t} \sin\frac{n\pi x}{l} = b_n e^{\frac{n^2c^2\pi^2t}{l^2}} \sin\frac{n\pi}{l}x, \qquad b_n = c_1c_3$

This equation satisfies the given conditions for all integral values of $n$. Hence taking $n = 1, 2, 3, \ldots\ldots$, *the most general solution is*

$$u = \sum_{n=1}^{\infty} b_n e^{\frac{-n^2c^2\pi^2t}{l^2}} \sin\frac{n\pi}{l}x$$

By initial conditions $u = U_0$ when $t = 0$

$$U_0 = \sum_{n=1}^{\infty} b_n \sin\frac{n\pi x}{l}$$

**Proved.**

**Example 11.** *Find the solution of*

$$\frac{\partial^2 u}{\partial x^2} = h^2 \frac{\partial u}{\partial t}$$

*for which* $u(0, t) = u(l, t) = 0$, $u(x, 0) = \sin \frac{\pi x}{l}$ *by method of variable separable.*

**Solution.** $$\frac{\partial^2 u}{\partial x^2} = h^2 \frac{\partial u}{\partial t} \quad \text{... (1)}$$

In example 10 the given equation was

$$\frac{\partial^2 u}{\partial x^2} = \frac{1}{c^2} \frac{\partial u}{\partial t} \quad \text{... (2)}$$

On comparing (1) and (2) we get $h^2 = \frac{1}{c^2}$

Thus solution of (1) is

$$u = (c_2 \cos px + c_3 \sin px)\, c_1 e^{-\frac{p^2 t}{h^2}} \quad \text{[Using (5) of example (10)]} \quad \text{...(3)}$$

On putting $x = 0$, $u = 0$ in (3), we get

$$0 = c_1 c_2 e^{-\frac{p^2 t}{h^2}} \qquad c_1 \neq 0 \Rightarrow c_2 = 0$$

(3) is reduced to

$$u = c_3 \sin px c_1 e^{-\frac{p^2 t}{h^2}} \quad \text{...(4)}$$

On putting $x = l$ and $u = 0$ in (4), we get

$$0 = c_3 \sin pl c_1 e^{-\frac{p^2 t}{h^2}}$$

$$c_3 \neq 0, \quad c_1 \neq 0 \qquad [\therefore \sin pl = 0 = \sin n\pi \quad \Rightarrow \quad p = \frac{n\pi}{l}]$$

Now (4) is reduced to

$$u = c_1 c_3 \sin \frac{n\pi x}{l} e^{-\frac{n^2 \pi^2 t}{h^2 l^2}} \quad \text{...(5)}$$

On putting $t = 0$, $u = \sin \frac{\pi x}{l}$ in (5) we get

$$\sin \frac{\pi x}{l} = c_4 \sin \frac{n\pi x}{l} \qquad [\text{put } c_1 c_3 = c_4]$$

This equation will be satisfied if

$$n = 1 \text{ and } c_4 = 1$$

On putting the values of $c_4$ and $n$ in (5), we have

$$u = \sin \frac{\pi x}{l} e^{-\frac{\pi^2 t}{h^2 l^2}}$$ **Ans.**

**Example 12.** *The ends A and B of a rod 20 cm long have the temperatures at 30° C and at 80°C until steady state prevails. The temperature of the ends are changed to 40°C and 60°C respectively. Find the temperature distribution in the rod at time t.*

**Solution.** The initial temperature distribution in the rod is

$$u = 30 + \frac{50}{20} x \qquad i.e., u = 30 + \frac{5}{2} x$$

and the final distribution (*i.e.* in steady state)is

$$u = 40 + \frac{20}{20}x = 40 + x$$

To get $u$ in the intermediate period, reckoning time from the instant when the end temperature were changed, we assumed

$$u = u_1(x, t) + u_2(x)$$

where $u_2(x)$ is the steady state temperature distribution in the rod (*i.e.* temperature after a sufficiently long time) and $u_1(x, t)$ is the transient temperature distribution which tends to zero as $t$ increases.

Thus $\qquad u_2(x) = 40 + x$

Now $u_1(x, t)$ satisfies the one-dimensional heat-flow equation

$$c^2 \frac{\partial^2 u}{\partial x^2} = \frac{\partial u}{\partial t}$$

Hence $u$ is of the form

$$u = 40 + x + \sum (a_k \cos kx + b_k \sin kx) e^{-c^2 k^2 t}$$

Since $\qquad u = 40°$, when $x = 0$ and $u = 60°$, when $x = 20$, we get

$$a_k = 0, k = \frac{n\pi}{20}$$

Hence $\qquad u = 40 + x + \sum_{n=1}^{\infty} b_n \sin \frac{n\pi x}{20} e^{-c^2 \left(\frac{n\pi}{20}\right)^2 t} \qquad \text{...(1)}$

Using the initial condition *i.e.*,

$$u = 30 + \frac{5}{2}x \text{ when } t = 0\text{, we get}$$

$$30 + \frac{5}{2}x = 40 + x + \sum b_n \sin \frac{n\pi}{20}x \quad \Rightarrow \quad \frac{3}{2}x - 10 = \sum b_n \sin \frac{n\pi x}{20}$$

Hence $\qquad b_n = \frac{2}{20}\int_0^{20} \left(\frac{3}{2}x - 10\right) \sin \frac{n\pi x}{20}\, dx$

$$= \frac{1}{10}\left[\left(\frac{3x}{2} - 10\right)\left(-\frac{20}{n\pi}\cos \frac{n\pi x}{20}\right) - \frac{3}{2}\left(\frac{-400}{n^2\pi^2}\sin \frac{n\pi x}{20}\right)\right]_0^{20}$$

$$= \frac{1}{10}\left[-20\left(\frac{20}{n\pi}\right)(-1)^n - (-10)\left(\frac{20}{n\pi}\right)\right] = -\frac{20}{n\pi}\left[2(-1)^n + 1\right]$$

Putting this value of $b_n$ in (1), we get

$$\therefore \qquad u = 40 + x - \frac{20}{\pi}\sum \left[\left(\frac{2(-1)^n}{n}\right) \sin \frac{n\pi x}{20} . e^{-\left(\frac{cn\pi}{20}\right)^2 t}\right] \qquad \textbf{Ans.}$$

## EXERCISE 9.17

1. Solve the following boundary value problem which arises in the heat conduction in a rod :

$$\frac{\partial u}{\partial t} = c^2 \frac{\partial^2 u}{\partial x^2}, \quad u(0, t) = u(l, t) = 0 \qquad \textit{(A.M.I.E.T.E., Summer 2002)}$$

$$u(x, 0) = 100\frac{x}{l} \qquad \textbf{Ans.} \quad u(x, t) = \frac{200}{\pi}\sum_{n=1}^{\infty} \frac{(-1)^{n+1}}{n} \sin \frac{n\pi x}{l} e^{-\frac{c^2 n^2 \pi^2}{l^2} t}$$

2. Determine the solution of one-dimensional heat equation $\frac{\partial u}{\partial t} = c^2 \frac{\partial^2 u}{\partial x^2}$ subject to the boundary conditions $u(0, t) = 0$, $u(l, t) = 0$ $(t > 0)$ and the initial condition $u(x, 0) = x$, $l$ being the length of the bar.

**Ans.** $y = \sum_{n=1}^{\infty} (-1)^{n+1} \frac{2l}{n\pi} \sin \frac{n\pi x}{l} e^{-\frac{n^2\pi^2c^2t}{l^2}}$

3. Solve the non-homogeneous heat conduction equation $u = a^2 u_{xx} + \sin 3\pi x$ subject to the following conditions :

$$u(x, 0) = \sin 2\pi x, u(0, t) = u(l, t) = 0$$

4. Solve $\frac{\partial u}{\partial t} = a^2 \frac{\partial^2 u}{\partial x^2}$, given that (*i*) $u = 0$ when $x = 0$ and $x = l$ for all t

(*ii*) $u = 3\sin \frac{\pi x}{l}$, when $t = 0$ for all $x$, $0 < x < l$. **Ans.** $u = 3\sin \frac{\pi x}{l} e^{\frac{-a^2\pi^2 t}{l^2}}$

5. (*a*) Find by the method of separation of variables the solution of $U(x, t)$ of the boundary value problem

$$\frac{\partial U}{\partial t} = 3\frac{\partial^2 U}{\partial x^2}, \quad t > 0, 0 < \alpha < 2$$

$U(0, t) = 0$, $U(2, t) = 0$

$U(x, 0) = x$, $0 < x < 2$

**Ans.** $U = \sum_{n=1}^{\infty} \frac{\sin \frac{n\pi x}{2}}{\sin \frac{n\pi}{2}} e^{-\frac{3n^2\pi^2 t}{4}}$

(*b*) The ends A and B of a rod 30 cm long have their temperatures kept at 20°C and 80°C respectively until steady-state conditions prevail. The temperature of the end B is suddenly reduced to 60°C and kept so while at the end A temperature is raised to 40°C. Find temperature distribution in the rod at time *t*. (*A.M.I.E.T.E., Winter 2002*)

6. The equation $\frac{\partial u}{\partial t} = \mu \frac{\partial^2 u}{\partial x^2}$ refers to the conduction of heat along a bar, given that $u = u_0 \sin nt$ when $x = 0$, for all values of $t$ and $u = 0$ when $x$ is very large.

Without radiation, show that if $u = Ae^{-gx} \sin (nt - gx)$, where $A$, $g$ and $n$ are positive constants, then $g = \sqrt{\frac{n}{2\mu}}$

7. An insulated rod of length $l$ has its ends $A$ and $B$ maintained at 0°C and 100°C respectively until steadystate condition prevails. If $B$ is suddenly reduced to 0°C and maintained at 0°C, find the temperature at a distance $x$ from $A$ at time $t$, solve the equation of heat

$$\frac{\partial u}{\partial t} = C^2 \frac{\partial^2 u}{\partial x^2}$$

by the method of separation of variables and obtain the solution. (*A.M.I.E., Summer 2004*)

**Ans.** $u(x,t) = \frac{200}{\pi} \sum_{n=1}^{\infty} \frac{(-1)^{n+1}}{n} \sin \frac{n\pi x}{l} . e^{-\frac{c^2 n^2 \pi^2}{l^2} . t}$

8. Solve $\frac{\partial u}{\partial t} = a^2 \frac{\partial^2 u}{\partial x^2}$ under the conditions

$u'(0, t) = 0 \quad t > 0$

$u'(\pi, t) = 0$

$u(x, 0) = x^2, \; 0 < x < \pi$

**Ans.** $u(x,t) = \frac{\pi^2}{3} + 4\sum_{n=1}^{\infty} \frac{(-1)^n}{n^2} \cos nx e^{-a^2 n^2 t}$

9. A rod of length $l$ has its lateral surface insulated and is so thin that heat flow in the rod can be regarded as one dimensional. Initially the rod is at the temperature 100 throughout. At $t = 0$ the temperaure at the left end of the rod is suddenly reduced to 50 and maintained thereafter at this value, while the right end is maintained at 100. Let $u(x, t)$ be the temperature at point $x$ in the rod at any subsequent time $t$.

(*i*) Write down the appropriate partial differential equation for $u(x, t)$ with initial and boundary conditions.

(*ii*) Solve the differential equation in (*i*) above using method of separation of variables and show that

$$u(x,t) = 50 + \frac{50x}{l} + \frac{100}{\pi}\sum_{n=1}^{\infty}\frac{1}{n}\sin\frac{n\pi x}{l}\exp\frac{-n^2\pi^2 t}{a^2 l^2}$$

Where $a^2$ is the constant involved in the partial differential equation. (*A.M.I.E.T.E., Dec. 2004*)

**10.** A uniform rod of length $a$ whose surface is thermally insulated is initially at temperature $\theta = \theta_0$. At time $t = 0$, one end is suddenly cooled to $\theta = 0$ and subsequently maintained at this temperature, the other end remains thermally insulated. Find the temperature $\theta(x, t)$.

$$\textbf{Ans. } \theta(x,t) = \frac{4\theta}{\pi}\sum_{n=1}^{\infty}\frac{\sin\frac{(2n+1)\pi x}{2a}}{2n+1}e^{-\frac{(2n+1)^2\pi^2c^2t}{4a^2}}$$

**11.** Solve $\frac{\partial U}{\partial t} = a^2\frac{\partial^2 U}{\partial x^2}$ under the conditions

(*i*) $U \neq \infty$ if $t \to \infty$; (*ii*) $U(0,t) = U(\pi,t) = 0$; (*iii*) $U(x,0) = \pi x - x^2$

$$\textbf{Ans. } u = \frac{8}{\pi}\sum_{n=1}^{\infty}\frac{\sin(2n-1)x}{(2n-1)^3}e^{-a^2}(2n-1)^2 t$$

**12.** The temperature distribution in a bar of length $\pi$, which is prefectly insulated at the ends $x = 0$ and $x = \pi$ is governed by the partial differential equation $\frac{\partial u}{\partial t} = \frac{\partial^2 u}{\partial x^2}$. Assuming the initial temperature as $u(x,0) = f(x) = \cos 2x$, find the temprature distribution at any instant of time.

**Ans.** $u = e^{4t}\cos 2x$

**13.** The heat flow in a bar of length 10 cm of homogeneous material is governed by the partial differential equation $u_t = c^2 u_{xx}$. The ends of the bar are kept at temperature 0°C, and the initial temperature is $f(x) = x(10 - x)$. Find the temperature in the bar at any instant of time.

**14.** Find the temperature $u(x, t)$ in a bar of length $\pi$ which is perfectly insulated everywhere including the ends $x = 0$ and $x = \pi$. This leads to the conditions $\frac{\partial u}{\partial x}(0,t) = 0, \frac{\partial u}{\partial x}(\pi,t) = 0$. Further the initial conditions are as given below:

$$u(x,0) = \begin{cases} x, 0 < x < \pi/2 \\ \pi x, \pi/2 \le x < \pi \end{cases}$$

Find the solution by the separation of variables.

## 9.21 TWO DIMENSIONAL HEAT FLOW

Consider the heat flow in a metal plate of uniform thickness, in the directions parallel to length and breadth of the plate. There is no heat flow along the normal to the plane of the rectangle.

Let $u(x, y)$ be the temperature at any point $(x, y)$ of the plate at time $t$ is given by

$$\frac{\partial u}{\partial t} = c^2\left(\frac{\partial^2 u}{\partial x^2} + \frac{\partial^2 u}{\partial y^2}\right) \quad ...(1)$$

In the steady state, $u$ does not change with $t$.

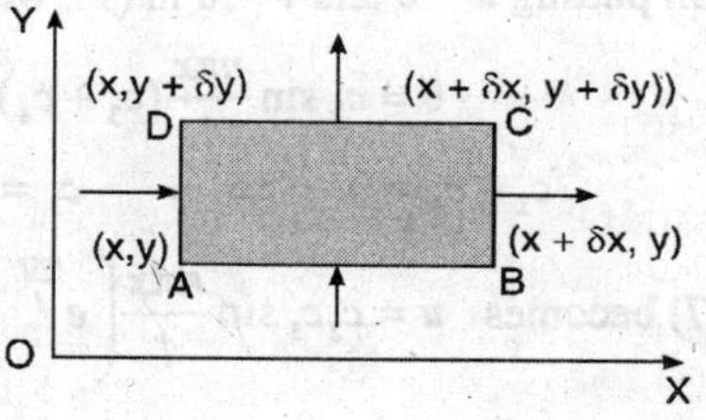


$$\therefore \quad \frac{\partial u}{\partial t} = 0$$

(1) becomes $\frac{\partial^2 u}{\partial x^2} + \frac{\partial^2 u}{\partial y^2} = 0$

This is called Laplace equation in two dimensions.

**Example 13.** *Solve* $\frac{\partial^2 u}{\partial x^2}+\frac{\partial^2 u}{\partial y^2}=0$ *which satisfies the conditions*

$$u(0,y)=u(l,y)=u(x,0)=0$$

*and* $u\,(x,\,a)=\sin\frac{n\pi x}{l}$ *(A.M.I.E.T.E., Winter 2000)*

**Solution.** $$\frac{\partial^2 u}{\partial x^2}+\frac{\partial^2 u}{\partial y^2}=0 \quad \text{... (1)}$$

Let $$u = X(x)\,.\,Y(y) \quad \text{...(2)}$$

Putting the values of $\frac{\partial^2 u}{\partial x^2}$ and $\frac{\partial^2 u}{\partial y^2}$ in (1), we have

$$X''Y+XY''=0$$

$$\Rightarrow \quad \frac{X''}{X}=-\frac{Y''}{Y}=-p^2 \quad \text{(say)}$$

$$\therefore \quad X''=-p^2X \quad \Rightarrow \quad X''+p^2X=0 \quad \text{...(3)}$$

and $$Y''=p^2Y \quad \Rightarrow \quad Y''-p^2Y=0 \quad \text{...(4)}$$

A.E. of (3) is $m^2+p^2=0 \quad \Rightarrow \quad m=\pm ip$

$$\therefore \quad X=c_1\cos px+c_2\sin px$$

A.E. of (4) is $m^2-p^2=0 \quad \Rightarrow \quad m=\pm p$

$$\therefore \quad Y=c_3e^{py}+c_4e^{-py}$$

Putting the values of $X$ and $Y$ in (2), we have

$$u=(c_1\cos px+c_2\sin px)(c_3e^{py}+c_4e^{-py}) \quad \text{...(5)}$$

Putting $x=0, u=0$ in (5), we have

$$0=c_1(c_3e^{py}+c_4e^{-py})$$

$$\therefore \quad c_1=0$$

(5) is reduaced to $u=c_2\sin px(c_3e^{py}+c_4e^{-py})$ ... (6)

On putting $x=l$, $u=0$, we have

$$0=c_2\sin pl(c_3e^{py}+c_4e^{-py})$$

$$c_2\neq 0 \quad \therefore \sin p\,l=0=\sin n\,\pi$$

$$\Rightarrow \quad pl=n\pi \quad \Rightarrow \quad p=\frac{n\pi}{l}$$

Now (6) becomes

$$u=c_2\sin\frac{n\pi x}{l}\left(c_3e^{\frac{n\pi y}{l}}+c_4e^{\frac{-n\pi y}{l}}\right) \quad \text{... (7)}$$

On putting $u=0$ and $y=0$ in (7), we have

$$0=c_2\sin\frac{n\pi x}{l}(c_3+c_4)$$

$$\therefore \quad c_3+c_4=0 \quad \Rightarrow \quad c_3=-c_4$$

(7) becomes $$u=c_2c_3\sin\frac{n\pi x}{l}\left(e^{\frac{n\pi y}{l}}-e^{\frac{-n\pi y}{l}}\right) \quad \text{...(8)}$$

On putting $y = a$ and $u = \sin\dfrac{n\pi x}{l}$ in (8), we get

$$\sin\frac{n\pi x}{l} = c_2 c_3 \sin\frac{n\pi x}{l}\left(e^{\frac{n\pi a}{l}} - e^{\frac{-n\pi a}{l}}\right) \quad \text{i.e.} \quad c_2 c_3 = \frac{1}{e^{\frac{n\pi a}{l}} - e^{\frac{-n\pi a}{l}}}$$

Putting this value in (8), we have

$$u = \sin\frac{n\pi x}{l}\,\frac{e^{\frac{n\pi y}{l}} - e^{\frac{-n\pi y}{l}}}{e^{\frac{n\pi a}{l}} - e^{\frac{-n\pi a}{l}}} \Rightarrow u = \sin\frac{n\pi x}{l}\,\frac{\sinh\frac{n\pi y}{l}}{\sinh\frac{n\pi a}{l}} \qquad \textbf{Ans.}$$

**Example 14.** *A rectangular plate with insulated surfaces is 10 cm wide and so long compared to its width that it may be considered infinite in length without introducing an appreciable error. If the temperature along the short edge y = 0 is given by*

$$u(x, 0) = 20x, \quad 0 < x \le 5$$
$$= 20(10 - x), \; 5 < x < 10$$

*while the two long edges x = 0 and x = 10 as well as the other short edges are kept at 0°C. Find the steady state temperature at any point (x, y) of the plate.*

**Solution.** In the steady state, the temperature $u(x, y)$ at any point $p(x, y)$ satisfy the equation

$$\frac{\partial^2 u}{\partial x^2} + \frac{\partial^2 u}{\partial y^2} = 0 \qquad \text{...(1)}$$

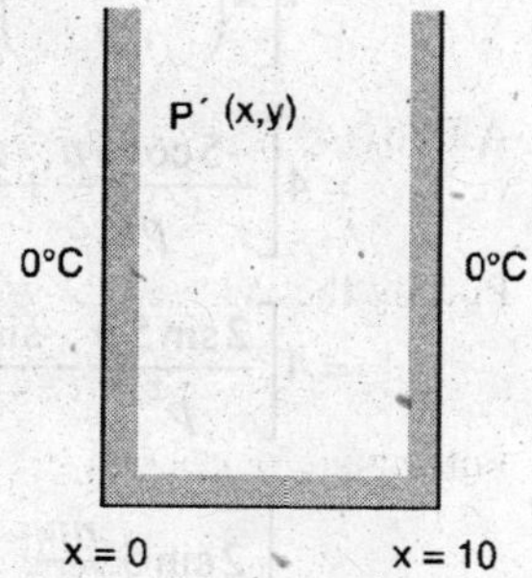


The boundary conditions are

$u(0, y) = 0$ for all values of $y$ ...(2)

$u(10, y) = 0$ for all values of $y$ ...(3)

$u(x, \infty) = 0$ for all values of $x$ ...(4)

$u(x, 0) = 20x \qquad 0 < x \le 5$

$\qquad = 20(10 - x) \qquad 5 < x < 10$ ...(5)

Now three possible solutions of (1) are

$$u = (C_1 e^{px} + C_2 e^{-px})(C_3 \cos py + C_4 \sin py) \qquad \text{...(6)}$$
$$u = (C_5 \cos px + C_6 \sin px)(C_7 e^{px} + C_8 e^{-py}) \qquad \text{...(7)}$$
$$u = (C_9 x + C_{10})(C_{11} y + C_{12}) \qquad \text{...(8)}$$

Of these, we have to choose that solution which is consistent with the physical nature of the problem. The solution (6) and (8) cannot satisfy the condition (2), (3)and (4). Thus, only possible solution is (7) *i.e.*, of the form.

$$u(x, y) = (C_1 \cos px + C_2 \sin px)(C_3 e^{py} + C_4 e^{-py}) \qquad \text{...(9)}$$

By (2) $\quad u(0, y) = C_1(C_3 e^{py} + C_4 e^{-py}) = 0$ for all values of $y$

$\therefore \qquad C_1 = 0$

$\therefore$ (9) reduces to $u(x, y) = C_2 \sin px\,(C_3 e^{py} + C_4 e^{-py})$ ...(10)

By (3) $u(10, y) = C_2 \sin 10p\,(C_3 e^{py} + C_4 e^{-py}) = 0 \qquad C_2 \ne 0$

$\therefore \quad \sin 10p = 0 \quad \Rightarrow \quad 10p = n\pi \quad \Rightarrow \quad p = \dfrac{n\pi}{10}$

Also to satisfy the condition (4) *i.e.*, $u = 0$ as $y \to \infty$

$$C_3 = 0$$

Hence (10) takes the form $u(x, y) = C_2\, C_4 \sin px . e^{-py}$

$\Rightarrow \quad u(x, y) = b_n \sin px . e^{-py}$ where $b_n = C_2\, C_4$

$\therefore$ The most general solution that satisfies (2), (3) & (4) is of the form

$$u(x, y) = \sum_{n=1}^{\infty} b_n \sin px e^{-py} \qquad ...(5)$$

Putting $y = 0$, $\quad u(x, 0) = \sum_{n=1}^{\infty} b_n \sin px \quad$ where $p = \dfrac{n\pi}{10}$

This requires the expansion of $u$ in Fourier series in the interval $x = 0$ and $x = 5$ and from $x = 5$ to $x = 10$.

$$b_n = \frac{2}{10}\int_0^5 20x \sin px dx + \frac{2}{10}\int_5^{10} 20\,(10 - x) \sin px dx$$

$$b_n = 4\int_0^5 x \sin px dx + 4\int_5^{10} (10 - x) \sin px dx$$

$$= 4\left[x\left(\frac{-\cos px}{p}\right) - (1)\left(\frac{-\sin px}{p^2}\right)\right]_0^5 + 4\left[(10 - x)\left(\frac{-\cos px}{p}\right) - (-1)\left(\frac{-\sin px}{p^2}\right)\right]_5^{10}$$

$$= 4\left[\frac{-5\cos 5p}{p} + \frac{\sin 5p}{p^2}\right] + 4\left[0 - \frac{\sin 10p}{p^2} + \frac{5\cos 5p}{p} + \frac{\sin 5p}{p^2}\right]$$

$$= 4\left[\frac{2\sin 5p}{p^2} - \frac{\sin 10p}{p^2}\right] \qquad \left(p = \frac{n\pi}{10}\right)$$

$$= 4\left[\frac{2\sin 5.\frac{n\pi}{10}}{\frac{n^2\pi^2}{100}} - \frac{\sin 10.\frac{n\pi}{10}}{\frac{n^2\pi^2}{100}}\right] = \frac{800}{n^2\pi^2}\sin\frac{n\pi}{2} - \frac{400}{n^2\pi^2}\sin n\pi$$

$$= \frac{800}{n^2\pi^2}\sin\frac{n\pi}{2} = 0 \text{ if n is even.} = \pm\frac{800}{n^2\pi^2} \text{ if n is odd.} \quad \text{or} \quad b_n = \frac{(-1)^{n+1}\,800}{(2n-1)^2\pi^2}$$

On putting the value of $b_n$ in (5) the temperature at any point $(x, y)$ is given by

$$u(x, y) = \frac{800}{\pi^2}\sum_{n=1}^{\infty}\frac{(-1)^{n+1}}{(2n-1)^2}\sin\frac{(2n-1)\pi x}{10}\, e^{-\frac{(2n-1)\pi y}{10}}$$ **Ans.**

## Exercise 9.18

1. Find by the method of separation of variables, a particular solution of the equation

$$\frac{\partial^2 u}{\partial x^2} + \frac{\partial^2 u}{\partial y^2} = 0$$

that tends to zero as $x$ tends to infinity and is equal to 2 cos $y$ when $x = 0$ **Ans.** $u = 2e^{-x}\cos y$

2. Solve the equation : $\quad u_{xx} + u_{yy} = 0$

$u = (0, y) = u(\pi, y) = 0$ for all $y$,

$u(x, 0) = k, \qquad 0 < x < \pi$

$\lim_{y\to\infty} u(x,y) = 0 \qquad 0 < x < \pi$ (*A.M.I.E.T.E., Summer 2003*)

**Ans.** $u(x,y) = \sum_{n=1}^{\infty} b_n \sin nx e^{-ny},\quad k = \sum_{n=1}^{\infty} b_n \sin nx$

3. Find the solution of Laplace's equation $\nabla^2 \psi = 0$ in cartesian coordinates in the region $0 \le x \le a, 0 \le y \le b$, to satisfying the conditions $y = 0$ on $x = 0$, $x = a$, $y = 0$, $y = b$ and $\psi = x(a-x), 0 < x < a$. (*A.M.I.E.T.E., Winter 2001*)

$$\textbf{Ans. } \psi = \frac{8a^2}{\pi^3}\sum_{n=0}^{\infty}\frac{1}{(2n+1)^3}\frac{\sin\frac{\pi x}{a}\sinh\frac{(2n+1)\pi y}{a}}{\sinh\frac{(2n+1)\pi b}{a}}$$

4. An infinitely long uniform plate is bounded by two parallel edges and an end at right angles to them. The breadth is $\pi$. This end is maintained at a temperature $u_0$ at all points and other edges are at zero temperature. Determine the temperature at any point of the plate in the steady state.

(*A.M.I.E.T.E., Dec. 2005*) **Ans.** $u(x,y) = \frac{4u_0}{\pi}\left[e^{-y}\sin x + \frac{1}{3}e^{-3y}\sin 3x + \frac{1}{5}e^{-5y}\sin 5x + \ldots\right]$

5. Solve $\frac{\partial^2 V}{\partial x^2} + \frac{\partial^2 V}{\partial y^2} = 0$, given that

(*i*) $V = 0$ when $x = 0$ and $x = c$ (*ii*) $V \to 0$ as $y \to \infty$; (*iii*) $V = V_0$ when $y = 0$.

**Ans.** $V(x,y) = \sum_{n=1}^{\infty} b_n \sin\frac{n\pi x}{c} e^{-\frac{n\pi y}{c}},\ V_0 = \sum_{n=1}^{\infty} b_n \sin\frac{n\pi x}{c}$

6. The steady state temperature distribution in a thin plate bounded by the lines $x = 0$, $x = a$, $y = 0$ and $y = \infty$, is governed by the partial differential equation

$$\frac{\partial^2 u}{\partial x^2} + \frac{\partial^2 u}{\partial y^2} = 0.$$

Obtain the steady state temperature distribution under the conditions

$$u(0,y) = 0, \qquad u(a,y) = 0, \qquad u(x,\infty) = 0$$

$$u(x,0) = x, \qquad 0 \le x \le a/2$$

$$= a - x \qquad a/2 \le x \le a$$

7. The points of trisection of a tightly stretched string are pulled aside through the same distance on opposite sides of the position of equilibrium and the string is released from rest. Derive an expression for the displacement of the string at subsequent time and show that the mid-point of the string always remains at rest.

## 9.22. LAPLACE EQUATION IN POLAR CO-ORDINATES

**Example 15.** *Solve* $\frac{\partial^2 u}{\partial r^2} + \frac{1}{r}\frac{\partial u}{\partial r} + \frac{1}{r^2}\frac{\partial^2 u}{\partial \theta^2} = 0$ *by the method of separation of variables.*

**Solution.** $\frac{\partial^2 u}{\partial r^2} + \frac{1}{r}\frac{\partial u}{\partial r} + \frac{1}{r^2}\frac{\partial^2 u}{\partial \theta^2} = 0 \quad\Rightarrow\quad r^2\frac{\partial^2 u}{\partial r^2} + r\frac{\partial u}{dr} + \frac{\partial^2 u}{\partial \theta^2} = 0$ ...(1)

Let $u = R(r).\,T(\theta)$

$$\frac{\partial u}{\partial r} = \frac{dR}{dr}.T(\theta) \quad \text{and} \quad \frac{\partial^2 u}{\partial r^2} = \frac{d^2 R}{dr^2}.T(\theta)$$

$$\frac{\partial u}{\partial \theta} = R(r).\frac{dT}{d\theta} \quad \text{and} \quad \frac{\partial^2 u}{\partial \theta^2} = R(r).\frac{d^2 T}{d\theta^2}$$

Putting the values of $\frac{\partial^2 u}{\partial r^2}$, $\frac{\partial u}{\partial r}$ and $\frac{\partial^2 u}{\partial \theta^2}$ in (1), we get

$$r^2.\frac{d^2 R}{dr^2}.T(\theta) + r\frac{dR}{dr}.T(\theta) + R(r)\frac{d^2 T}{d\theta^2} = 0$$

$$\left(r^2 \frac{d^2 R}{dr^2} + r.\frac{dR}{dr}\right).T + R\frac{d^2 T}{d\theta^2} = 0$$

$$\frac{r^2 \frac{d^2 R}{dr^2} + r.\frac{dR}{dr}}{R} = -\frac{1}{T}\frac{d^2 T}{d\theta^2} = h \quad \text{(say)}$$

| | |
|---|---|
| $r^2 \frac{d^2 R}{dr^2} + r.\frac{dR}{dr} - hR = 0$ | $\frac{d^2 T}{d\theta^2} + hT = 0$ |
| Put $r = e^z$ | $(D^2 + h)T = 0$ |
| $(D(D-1) + D - h)R = 0$ | $D^2 + h = 0$ or $D = \pm i\sqrt{h}$ |
| $D^2 - h = 0 \rightarrow D = \pm\sqrt{h}$ | $T = c_3 \cos(\sqrt{h}\theta) + c_4 \sin(\sqrt{h}\theta)$ |
| $R = c_1 e^{\sqrt{h}z} + c_2 e^{-\sqrt{h}z}$ | |
| $R = c_1 r^{\sqrt{h}} + c_2 r^{-\sqrt{h}}$ | |

$$u = (c_1 r^{\sqrt{h}} + c_2 r^{-\sqrt{h}})[c_3 \cos(\sqrt{h}\theta) + c_4 \sin(\sqrt{h}\theta)] \quad ...(2)$$

**Case 1.** If $h = k^2$

(2) becomes $u = (c_1 r^k + c_2 r^{-k})[c_3 \cos(k\theta) + c_4 \sin(k\theta)]$

**Case 2.** If $h = 0$

(2) becomes $u = (c_5 + z\, c_6)\,(c_7 + \theta c_8)$

$= [C_5 + (\log r)\, c_6]\,[c_7 + \theta c_8]$

**Case 3.** If $h = -p^2$

(2) becomes $u = (c_9 \cos pz + c_{10} \sin pz)(c_{11} e^{p\theta} + c_{12} e^{-p\theta})$

Then there are three possible solutions

$u = (c_1 r^k + c_2 r^{-k})[c_3 \cos(k\theta) + c_4 \sin(k\theta)]$

$u = (c_5 + c_6 \log r)(c_7 + c_8 \theta)$

$u = [c_9 \cos(p \log r) + c_{10} \sin(p \log r)][c_{11} e^{p\theta} + c_{12} e^{-p\theta}]$ **Ans.**

**Example 16.** *The diameter of a semicircular plate of radius a is kept at 0°C and the temperature at the semicircular boundary is T° C. Find the steady state temperature in the plate.*

**Solution**. Let the centre $O$ of the semicircular plate be the pole and the bounding diameter be as the initial line. Let $u(r, \theta)$ be the steady state temperature at any point $p(r, \theta)$ and $u$ satisfies the equation

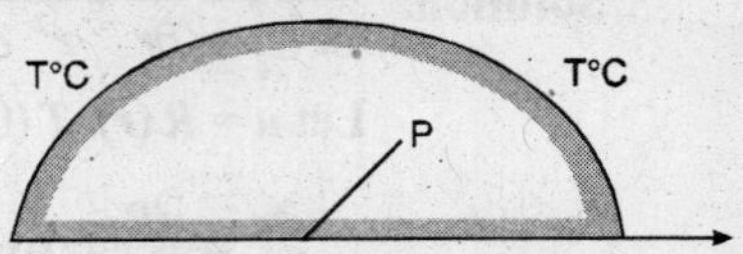


$$r^2 \frac{\partial^2 u}{\partial r^2} + r\frac{\partial u}{\partial r} + \frac{\partial^2 u}{\partial \theta^2} = 0 \quad ...(1)$$

The boundary conditions are

(i) $u(r,0)=0$ $\quad 0 \le r \le a$

(ii) $u(r,\pi)=0$ $\quad 0 \le r \le a$

(iii) $u(a,\theta)=T.$

From conditions (ii) and (iii), we have $u \to 0$ as $r \to 0$. Hence the appropriate solution of (i) is as solved in example 15.

$$u=(c_1 r^p + c_2 r^{-p})(c_3 \cos p\theta + c_4 \sin p\theta) \quad ...(2)$$

Putting $u(r,0)=0$ in (2), we get

$$0=(c_1 r^p + c_2 r^{-p})c_3 \to c_3 = 0$$

(2) becomes

$$u=(c_1 r^p + c_2 r^{-p})c_4 \sin p\theta \quad ...(3)$$

Putting $u(r,\pi)=0$ in (3), we get

$$0=(c_1 r^p + c_2 r^{-p})c_4 \sin p\pi \Rightarrow \sin p\pi = 0 = \sin n\pi$$

$$\Rightarrow \quad p\pi = n\pi \quad \Rightarrow \quad p = n$$

(3) becomes, on putting $p=n$

$$u=(c_1 r^n + c_2 r^{-n})c_4 \sin n\theta \quad ...(4)$$

Since, $u=0$ when $r=0$

$$0 = c_2$$

(4) becomes, $u = c_1 c_4 r^n \sin n\theta$

The most general solution of (1) is

$$u(r,\theta)=\sum_{n=1}^{\infty} b_n r^n \sin n\theta \quad ...(5)$$

Putting $r=a$ and $u=T$ in (5), we have

$$T=\sum_{n=1}^{\infty} b_n a^n \sin n\theta$$

By Fourier half range series, we get

$$b_n a^n = \frac{2}{\pi}\int_0^{\pi} T \sin n\theta \, d\theta = \frac{2}{\pi} T\left(\frac{-\cos n\theta}{n}\right)_0^{\pi} = \frac{2T}{n\pi}[-(-1)^n+1]$$

$b_n a^n = 0,$ **When $n$ is even.**

$b_n a^n = \dfrac{4T}{n\pi},$ **When $n$ is odd.**

$$\Rightarrow \quad b_n = \frac{4T}{n\pi a^n}$$

Hence, (5) becomes

$$u(r,\theta)=\frac{4T}{\pi}\left[\frac{r/a}{1}\sin\theta + \frac{(r/a)^3}{3}\sin 3\theta + \frac{(r/a)^5}{5}\sin 5\theta + ...\right]$$ **Ans.**

## Exercise 9.19

1. Solve the steady-state temperature equation

$$\frac{\partial^2 T}{\partial r^2}+\frac{1}{r}\frac{\partial T}{\partial r}+\frac{1}{r^2}\frac{\partial^2 T}{\partial \theta^2}=0;\quad 10\le r\le 20, 0\le\theta\le 2\pi$$

subject to the following conditions:

$$T(10,\theta)=15\cos\theta \text{ and } T(20,\theta)=30\sin\theta$$

**Ans.** $T(r,\theta)=\frac{4T}{\pi}\left[\frac{r}{a}\sin\theta+\frac{1}{3}\left(\frac{r}{a}\right)^3\sin 3\theta+.....\right]$

2. A semi-circular plate of radius $a$ has its circumference kept at temperature $u(a,\theta)=k\theta(\pi-\theta)$ while the boundary diameter is kept at zero temperature. Find the steady state temperature distribution $u(r,\theta)$ of the plate assuming the lateral surfaces of the plate to be insulated.

**Ans.** $u(r,\theta)=\frac{8k}{\pi}\sum_1^{\infty}\left(\frac{r}{a}\right)^{2n-1}\frac{\sin(2n-1)\theta}{(2n-1)^3}$

3. Find the steady state temperature in a circular plate of radius $a$ which has one half of its circumference at 0°C and the other half at 60°C.

**Ans.** $u(r,\theta)=50-\frac{200}{n}\sum_{n=1}^{\infty}\frac{1}{2n-1}\left(\frac{r}{a}\right)^{2n-1}\sin(2n-1)\theta.$

## 9.23 TRANSMISSION LINE EQUATIONS

$$\frac{\partial^2 V}{\partial x^2}=RC\frac{\partial V}{\partial t}$$

$$\frac{\partial^2 i}{\partial x^2}=RC\frac{\partial i}{\partial t}$$

are called telegraph equations,

where $V$ = potential, $i$ = current, C = capacitance, $L$ = inductance

$$\frac{\partial^2 V}{\partial x^2}=LC\frac{\partial^2 V}{\partial t^2}$$

$$\frac{\partial^2 i}{\partial x^2}=LC\frac{\partial^2 i}{\partial t^2}$$

are called radio equations.

**Example 17.** *Find the current i and voltage v in a transmission line of length l,t seconds after the ends are suddenly grounded given that* $i(x,0)=i_0, v(x,0)=v_0\sin\left(\frac{\pi x}{l}\right)$ *and that R and G are negligible.*

**Solution.** $\frac{\partial^2 v}{\partial x^2}=LC\frac{\partial^2 v}{\partial t^2}$

Let $v=XT$ where $X$ and $T$ are the functions of $x$ and $t$ respectively.

$$\frac{\partial^2 v}{\partial x^2}=T\frac{d^2X}{dx^2}\quad\text{and}\quad\frac{\partial^2 v}{\partial t^2}=X\frac{d^2T}{dt^2}$$

$$T\frac{d^2X}{dx^2}=LCX\frac{d^2T}{dt^2}$$

$$\frac{\frac{d^2X}{dx^2}}{X} = LC\,\frac{\frac{d^2T}{dt^2}}{T} = -p^2 \quad \text{say}$$

Since the initial conditions suggest the values of $v$ and $i$ are periodic functions,

$$\therefore \qquad X = c_1 \cos px + c_2 \sin px$$

$$T = c_3 \cos\frac{pt}{\sqrt{LC}} + c_4 \sin\frac{pt}{\sqrt{LC}}$$

$$v = XT$$

$$\Rightarrow \qquad v = (c_1 \cos px + c_2 \sin px)\left(c_3 \cos\frac{pt}{\sqrt{LC}} + c_4 \sin\frac{pt}{\sqrt{LC}}\right) \qquad ...(1)$$

When $t = 0,\ v = v_0 \sin\frac{\pi x}{l}$

$$v_0 \sin\frac{\pi x}{l} = (c_1 \cos px + c_2 \sin px)c_3 \qquad ...(2)$$

On equating the coefficients, we get

$$c_1 c_3 = 0 \quad \Rightarrow \quad c_1 = 0 \qquad \text{and} \qquad c_2 c_3 = v_0, \quad p = \frac{\pi}{l}$$

(1) becomes

$$v = \sin\frac{\pi x}{l}\left[v_0 \cos\frac{pt}{\sqrt{LC}} + c_2 c_4 \sin\frac{pt}{\sqrt{LC}}\right] \qquad ...(3)$$

Now when $t = 0, i = i_0$ (constant)

Hence $$\frac{\partial i}{\partial x} = 0$$

$$\frac{\partial i}{\partial x} = \frac{-c\,\partial v}{\partial t} \qquad \therefore \frac{\partial v}{\partial t} = 0 \ \text{ when } t = 0$$

Now $$\frac{\partial v}{\partial t} = \sin\frac{\pi x}{l}\left(\frac{p}{\sqrt{LC}}\right)\left[-v_0 \sin\frac{pt}{\sqrt{LC}} + c_2 c_4 \cos\frac{pt}{\sqrt{LC}}\right] \qquad ...(4)$$

On putting $\frac{\partial v}{\partial t} = 0$ and $t = 0$ in (4), we get $c_2 c_4 = 0 \Rightarrow c_4 = 0$

(3) is reduced to $v = v_0 \sin\frac{\pi x}{l} \cos\frac{\pi t}{l\sqrt{LC}}$

$$\Rightarrow \qquad \frac{\partial v}{\partial x} = \frac{x}{l} v_0 \cos\frac{\pi x}{l} \cos\frac{\pi t}{l\sqrt{LC}} = -L\frac{\partial i}{\partial t} \qquad ...(5)$$

and $$\frac{\partial v}{\partial t} = -\frac{v_0 \pi}{l\sqrt{LC}} \sin\frac{\pi x}{l} \sin\frac{\pi t}{t\sqrt{LC}} = \frac{-1}{C}\frac{\partial \iota}{\partial x} \qquad ...(6)$$

Integrating (5) and (6), we get

$$i = -v_0\sqrt{\frac{C}{L}} \cos\frac{\pi x}{l} \sin\frac{\pi t}{l\sqrt{LC}} + f(x)$$

$$i = -v_0\sqrt{\frac{C}{L}} \cos\frac{\pi x}{l} \sin\frac{\pi t}{l\sqrt{LC}} + F(t)$$

$\therefore$ $f(x)$ and $F(t)$ must be constant only, since $i = i_0$ when $t = 0$

$\therefore$ Constant $= i_0 = f(x)$

Hence $$i = i_0 - v_0\sqrt{\frac{C}{L}} \cos\frac{\pi x}{l} \sin\frac{\pi t}{\sqrt{LC}}.$$ **Ans.**

## Exercise 9.20

1. A transmission line 1,000 miles long is initially under steady state condition with potential 1,300 volts at the sending end ($x = 0$) and 1,200 volts at the receiving end ($x = 1000$). The terminal end of the line is suddenly grounded but the potential at the source is kept at 1,300 volts. Assuming the inductance and leakage to neglible, find the potential $v(x, t)$, if it atisfies the equation

$$v_t = \left(\frac{1}{RC}\right) v_{xx}.$$

**Ans.** $v(x,y) = \sum_1^{\infty} b_n \sin nx\, e^{-ny}$ *and* $k = \sum_1^{\infty} b_n \sin nx$

2. Obtain a solution of the telegraph equation

$$\frac{\partial^2 e}{\partial x^2} = RC\frac{\partial e}{\partial t}$$

suitable for the case when e decays with the time and when there is steady fall of potential from $e_0$ to 0 along the line of length $l$ initially and the sending end is suddenly earthed.

**Ans.** $e(x,t) = \dfrac{2e_0}{\pi}\sum_{n=1}^{\infty}\dfrac{1}{n}\sin\dfrac{n\pi x}{l}e^{\frac{-n^2\pi^2 l}{CRt^2}}$

3. Fill in the blanks :

(*a*) The general solution of the equation $\dfrac{\partial^2 z}{\partial x^2} - \dfrac{\partial^2 z}{\partial t^2} = 0$ is ..........................

(*b*) The general solution of the equation $\dfrac{\partial^2 z}{\partial x \partial y} = 0$ is ..........................

(*c*) The solution of $z(x,y)$ of the equation $\dfrac{\partial z}{\partial x} - y\dfrac{\partial z}{\partial y} = 0$ is ..........................

(*d*) The solution of $3x\dfrac{\partial z}{\partial x} - 5y\dfrac{\partial z}{\partial y} = 0$ is ..........................

(*e*) The solution of $\dfrac{\partial^2 z}{\partial x^2} = \sin(x\,y)$ is ..........................

(*f*) The solution of $\dfrac{\partial u}{\partial t} = 2\dfrac{\partial^2 u}{\partial x^2}$ if $u(0, t) = u(3, t) = 0$ and $u(x, 0) = 5\sin 4\pi x - 3\sin 8\pi x$ is......................

(*g*) If the unknown function in a differential equation depends on more than one independent variables, then the differential equation is said to be (*A.M.I.E., Winter 2001*)

**Ans.** (*a*) $(C_1 \cos px + C_2 \sin px)(C_3 \cos pt + C_4 \sin pt)$

(*b*) $f_1(x) + f_2(y)$

(*c*) $f(x + \log y, z) = 0$

(*d*) $f(x^5 y^3, z) = 0$

(*e*) $-\dfrac{1}{y^2}\sin(xy) + x\, f_1(y) + f_2(y)$

(*f*) $(5\sin 4\pi x e^{-32x^2 t} - 3\sin 8\pi x\, e^{-128\pi n^2 t})$

(*g*) Partial Differential Equation

# 10

# Statistics

**10.1 STATISTICS** is a branch of science dealing with the collection of data, organising, summarising, presenting and analysing data and drawing valid conclusions and thereafter making reasonable decisions on the basis of such analysis.

**10.2 FREQUENCY DISTRIBUTION** is the arranged data, summarised by distributing it into classes or categories with their frequencies.

**Wages of 100 workers**

| Wages in ₹ | 0-10 | 10-20 | 20-30 | 30-40 | 40-50 |
|---|---|---|---|---|---|
| Numbers of workers | 12 | 23 | 35 | 20 | 10 |

**10.3 GRAPHICAL REPRESENTATION.** It is often useful to represent frequency distribution by means of a diagram. The different types of diagrams are

1. Histogram
2. Frequency polygon
3. Frequency curve
4. Cumulative frequency curve or Ogive
5. Bar chart
6. Circles or Pie diagrams.

1. **Histogram** consists of a set of rectangles having their heights proportional to the class-frequencies, for equal class-intervals. For unequal class-interval, the areas of rectangles are proportional to the frequencies.

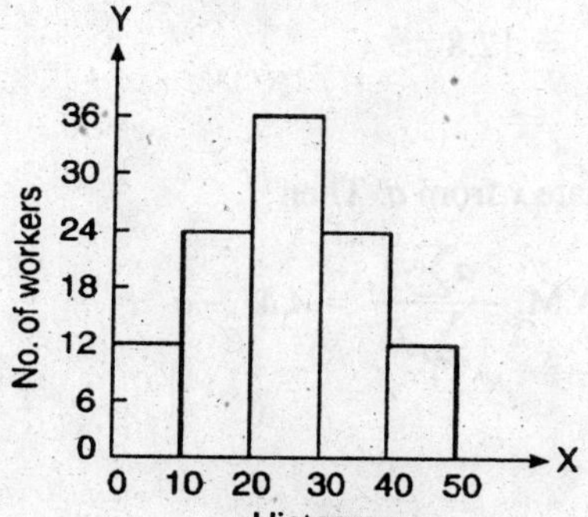

Histogram

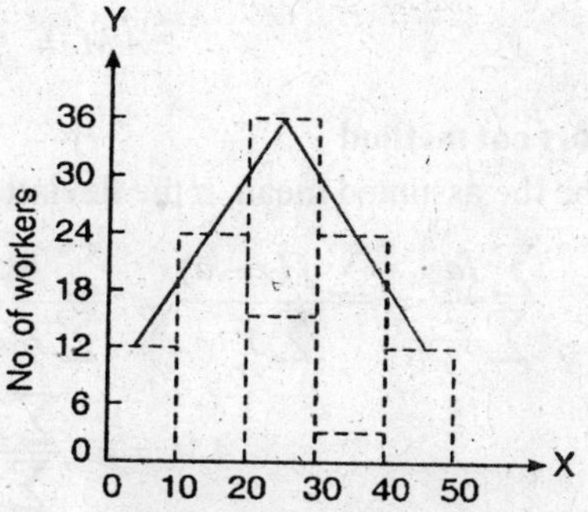

Frequency polygon

2. **Frequency Polygon** is a line graph of class-frequency plotted against class-mark. It can be obtained by connecting mid-points on the tops of the rectangles in the histogram.

3. **Cumulative Frequency curve or the Ogive.** If the various points are plotted according to the upper limit of the class as $x$ co-ordinate and the cumulative frequency as $y$ co-ordinate and these points are joined by a free hand smooth curve, the curve obtained is known as cumulative frequency curve or the Ogive.

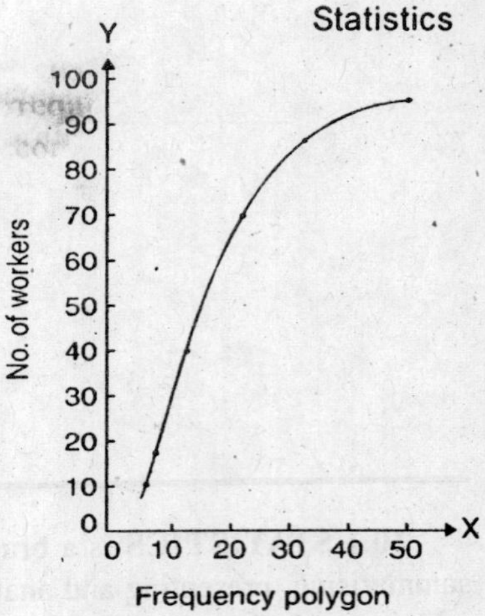

Frequency polygon

## 10.4 AVERAGE OR MEASURES OF CENTRAL TENDENCY

An average is a value which is representative of a set of data. Average value may also be termed as measures of central tendency. There are five types of averages in common.

(*i*) Arithmetic average or mean (*ii*) Median (*iii*) Mode (*iv*) Geometric Mean (*v*) Harmonic Mean

## 10.5 ARITHMETIC MEAN

If $x_1, x_2, x_3, \ldots x_n$ are $n$ numbers, then their arithmetic mean (A.M.) is defined by

$$AM = \frac{x_1 + x_2 + x_3 + \ldots\ldots + x_n}{n} = \frac{\sum x}{n}$$

If the number $x_1$ occurs $f_1$ times, $x_2$ occurs $f_2$ times and so on, then

$$AM = \frac{f_1 x_1 + f_2 x_2 + \ldots\ldots + f_n x_n}{f_1 + f_2 + \ldots\ldots + f_n} = \frac{\sum fx}{\sum f}$$

This is known as direct method.

**Example 1.** *Find the mean of 20, 22, 25, 28, 30.*

**Solution.** $$A.M. = \frac{20+22+25+28+30}{5} = \frac{125}{5} = 25$$ **Ans.**

**Example 2.** *Find the mean of the following :*

| *Numbers* | *8* | *10* | *15* | *20* |
|---|---|---|---|---|
| *Frequency* | *5* | *8* | *8* | *4* |

**Solution.** $\sum fx = 8 \times 5 + 10 \times 8 + 15 \times 8 + 20 \times 4 = 40 + 80 + 120 + 80 = 320$

$$\sum f = 5 + 8 + 8 + 4 = 25$$

$$A.M. = \frac{\sum fx}{\sum f} = \frac{320}{25} = 12.8$$ **Ans.**

(*b*) **Short cut method**

Let $a$ be the assumed mean, $d$ the deviation of the variate $x$ from $a$. Then

$$\frac{\sum fd}{\sum f} = \frac{\sum f(x-a)}{\sum f} = \frac{\sum fx}{\sum f} - \frac{\sum fa}{\sum f} = \text{A.M.} - \frac{a\sum f}{\sum f} = A.M. - a$$

$$\therefore \quad A.M. = a + \frac{\sum fd}{\sum f}$$

**Example 3.** *Find the arithmetic mean for the following distribution:*

| *Class* | *0 – 10* | *10 – 20* | *20 – 30* | *30 – 40* | *40 – 50* |
|---|---|---|---|---|---|
| *Frequency* | *7* | *8* | *20* | *10* | *5* |

**Solution.** Let assumed mean $(a) = 25$.

| *Class* | *Mid-value* $x$ | *Frequency* $f$ | $x-25=d$ | *fd* |
|---|---|---|---|---|
| 0 – 10 | 5 | 7 | –20 | –140 |
| 10 – 20 | 15 | 8 | –10 | –80 |
| 20 – 30 | 25 | 20 | 0 | 0 |
| 30 – 40 | 35 | 10 | + 10 | + 100 |
| 40 – 50 | 45 | 5 | + 20 | + 100 |
| Total | | 50 | | – 20 |

$$A.M. = a + \frac{\sum fd}{\sum f} = 25 + \frac{-20}{50} = 24.6$$

**Ans.**

(*c*) **Step deviation method**

Let *a* be the assumed mean, *i* the width of the class interval and

$$D = \frac{x-a}{i}, A.M. = a + \frac{\sum fD}{\sum f} i$$

**Example 4.** *Find the arithmetic mean of the data given in example 3 by step deviation method*

**Solution.** Let $a = 25$

| *Class* | *Mid-value* $x$ | *frequency* $f$ | $D = \frac{x-a}{i}$ | $f.D$ |
|---|---|---|---|---|
| 0 – 10 | 5 | 7 | –2 | –14 |
| 10 – 20 | 15 | 8 | –1 | –8 |
| 20 – 30 | 25 | 20 | 0 | 0 |
| 30 – 40 | 35 | 10 | + 1 | + 10 |
| 40 – 50 | 45 | 5 | + 2 | + 10 |
| Total | | 50 | | – 2 |

$$A.M = a + \frac{\sum fD}{\sum f} i = 25 + \frac{-2}{50} \times 10 = 24.6$$

**Ans.**

## 10.6 MEDIAN

**Median is defined as the measure of the central item when they are arranged in ascending or descending order of magnitude.**

**When the total number of the items is odd and equal to say *n*, then the value of $\frac{1}{2}(n+1)$ th item gives the median.**

**When the total number of the frequencies is even, say *n*, then there are two middle items, and so the mean of the values of $\frac{1}{2}$ *n*th and $\left(\frac{1}{2}n+1\right)$ th items is the median.**

**Example 5.** *Find the median of 6, 8, 9, 10, 11, 12, 13.*

**Solution.** Total number of items = 7

$$\text{The middle item} = \frac{1}{2}(7+1)^{th} = 4th$$

$$\text{Median} = \text{Value of the 4th item} = 10$$

**Ans.**

For grouped data, Median $= l + \dfrac{\frac{1}{2}N - F}{f} . i$

where $l$ is the lower limit of the median class, $f$ is the frequency of the class, $i$ is the width of the class-interval, $F$ is the total of all the preceeding frequencies of the median-class and $N$ is total frequency of the data.

**Example 6.** *Find the value of Median from the following data:*

| No. of days for which absent (less than) | 5 | 10 | 15 | 20 | 25 | 30 | 35 | 40 | 45 |
|---|---|---|---|---|---|---|---|---|---|
| No. of students | 29 | 224 | 465 | 582 | 634 | 644 | 650 | 653 | 655 |

**Solution.** The given cumulative frequency distribution will first be converted into ordinary frequency as under

| *Class- Interval* | *Cumulative frequency* | *Ordinary frequency* |
|---|---|---|
| 0 − 5 | 29 | 29 = 29 |
| 5 − 10 | 224 | 224 − 29 = 195 |
| 10 − 15 | 465 | 465 − 224 = 241 |
| 15 − 20 | 582 | 582 − 465 = 117 |
| 20 − 25 | 634 | 634 − 582 = 52 |
| 25 − 30 | 644 | 644 − 634 = 10 |
| 30 − 35 | 650 | 650 − 644 = 6 |
| 35 − 40 | 653 | 653 − 650 = 3 |
| 40 − 45 | 655 | 655 − 653 = 2 |

Median = size of $\dfrac{655}{2}$ or 327.5th item

327.5th item lies in 10-15 which is the median class.

$$M = l + \frac{\frac{N}{2} - C}{f} i$$

where $l$ *stands* for lower limit of median class,

$N$ stands for the total frequency,

$C$ stands for the cumulative frequency just preceeding the median class,

$i$ stands for class interval

$f$ stands for frequency for the median class.

$$\text{Median} = 10 + \frac{\frac{655}{2} - 224}{241} \times 5$$

$$= 10 + \frac{103.5 \times 5}{241} = 10 + 2.15 = 12.15 \qquad \textbf{Ans.}$$

## 10.7 MODE

Mode is defined to be the size of the variable which occurs most frequently.

**Example 7.** *Find the mode of the following items :*

*0, 1, 6, 7, 2, 3, 7, 6, 6, 2, 6, 0, 5, 6, 0.*

**Solution.** 6 occurs 5 times and no other item occurs 5 or more than 5 times, hence the mode is 6. **Ans.**

For grouped data, $Mode = l + \dfrac{f - f_{-1}}{2f - f_{-1} - f_1}.i$

where $l$ is the lower limit of the modal class, $f$ is the frequency of the modal class, $i$ is the width of the class, $f_{-1}$ is the frequency before the modal class and $f_1$ is the frequency after the modal class.

**Emperical formula**

**Mean – Mode = 3 [Mean – Median]**

**Example 8.** *Find the mode from the following data:*

| Age | 0 – 6 | 6 – 12 | 12 – 18 | 18 – 24 | 24 – 30 | 30 – 36 | 36 – 42 |
|---|---|---|---|---|---|---|---|
| Frequency | 6 | 11 | 25 | 35 | 18 | 12 | 6 |

**Solution.**

| *Age* | *Frequency* | *Cumulative frequency* |
|---|---|---|
| 0 – 6 | 6 | 6 |
| 6 – 12 | 11 | 17 |
| 12 – 18 | $25 = f_{-1}$ | 42 |
| 18 – 24 | $35 = f$ | 77 |
| 24 – 30 | $18 = f_1$ | 95 |
| 30 – 36 | 12 | 107 |
| 36 – 42 | 6 | 113 |

$$\text{Mode} = l + \frac{f - f_{-1}}{2f - f_{-1} - f_1} \times i$$

$$= 18 + \frac{35 - 25}{70 - 25 - 18} \times 6$$

$$= 18 + \frac{60}{27} = 18 + 2.22 = 20.22$$ **Ans.**

## 10.8 GEOMETRIC MEAN

If $x_1, x_2, x_3, \ldots\ldots, x_n$ be $n$ values of variates $x$, then the geometric mean

$$G = (x_1 \times x_2 \times x_3 \times x_4 \times \ldots\ldots \times x_n)^{\frac{1}{n}}$$

**Example 9.** *Find the geometric mean of 4, 8, 16.*

**Solution.** $G.M. = (4 \times 8 \times 16)^{1/3} = 8.$ **Ans.**

## 10.9 HARMONIC MEAN

Harmonic mean of a series of values is defined as the reciprocal of the arithmetic mean of their reciprocals. Thus if $H$ be the harmonic mean, then

$$\frac{1}{H} = \frac{1}{n}\left[\frac{1}{x_1} + \frac{1}{x_2} + \ldots\ldots + \frac{1}{x_n}\right]$$

**Example 10.** *Calculate the harmonic mean of 4, 8, 16.*

**Solution.** $\frac{1}{H} = \frac{1}{3}\left[\frac{1}{4} + \frac{1}{8} + \frac{1}{16}\right] = \frac{7}{48}$

$$H = \frac{48}{7} = 6.853$$ **Ans.**

## 10.10 AVERAGE DEVIATION OR MEAN DEVIATION

It is the mean of the absolute values of the deviations of a given set of numbers from their arithmetic mean.

If $x_1, x_2, x_3, ....., x_n$ be a set of numbers with frequencies $f_1, f_2, .....f_n$ respectively. Let $x$ be the arithmetic mean of the numbers $x_1, x_2, ...., x_n$, then

$$\text{Mean deviation} = \frac{\sum f_i |x_i - \bar{x}|}{\sum f_i}$$

**Example 11.** *Find the mean deviation of the following frequency distribution.*

| Class | 0 – 6 | 6–12 | 12 –18 | 18 – 24 | 24 –30 |
|---|---|---|---|---|---|
| Frequency | 8 | 10 | 12 | 9 | 5 |

**Solution.** Let $a = 15$

| *Class* | *Mid- value* $x$ | *Frequency* $f$ | $d = x - a$ | *fd* | $\lvert x - 14\rvert$ | $f\lvert x - 14\rvert$ |
|---|---|---|---|---|---|---|
| 0–6 | 3 | 8 | –12 | –96 | 11 | 88 |
| 6–12 | 9 | 10 | –6 | –60 | 5 | 50 |
| 12–18 | 15 | 12 | 0 | 0 | 1 | 12 |
| 18–24 | 21 | 9 | +6 | 54 | 7 | 63 |
| 24–30 | 27 | 5 | + 12 | 60 | 13 | 65 |
| Total | | 44 | | – 42 | | 278 |

$$\text{Mean} = a + \frac{\sum fd}{\sum f} = 15 - \frac{42}{44} = 14 \text{ nearly}$$

$$\text{Average deviation} = \frac{\sum f|x - \bar{x}|}{\sum f} = \frac{278}{44} = 6.3$$ **Ans.**

## 10.11 STANDARD DEVIATION

Standard deviation is defined as the square root of the mean of the square of the deviation from the arithmetic mean.

$$S.D. = \sigma = \sqrt{\frac{\sum f(x - \bar{x})^2}{\sum f}}$$

**Note.** 1. The square of the standard deviation $\sigma^2$ is called variance.

2. $\sigma^2$ is called the second moment about the mean and is denoted by $\mu_2$.

## 10.12 SHORTEST METHOD FOR CALCULATING STANDARD DEVIATION

We know that $\sigma^2 = \frac{1}{N}\sum f(x - \bar{x})^2 = \frac{1}{N}\sum f(x - a - \overline{x - a})^2$

$$= \frac{1}{N}\sum f(d - \overline{x - a})^2 \qquad \text{Where } x - a = d$$

$$= \frac{1}{N}\sum fd^2 - 2(\overline{x}-a)\frac{1}{N}\sum fd + (\overline{x}-a)^2 \frac{1}{N}\sum f \quad \sum f = N$$

$$= \frac{1}{N}\sum fd^2 - 2(\overline{x}-a)\frac{1}{N}\sum fd + (\overline{x}-a)^2$$

$$\overline{x} = a + \frac{\sum fd}{N} \qquad \text{or} \qquad \overline{x} - a = \frac{\sum fd}{N}$$

$$\sigma^2 = \frac{1}{N}\sum fd^2 - 2\left(\frac{\sum fd}{N}\right)\left(\frac{1}{N}\sum fd\right) + \left(\frac{\sum fd}{N}\right)^2$$

$$= \frac{1}{N}\sum fd^2 - \left(\frac{\sum fd}{N}\right)^2$$

$$S.D. = \sigma = \sqrt{\frac{\sum fd^2}{N} - \left(\frac{\sum fd}{N}\right)^2}$$

**Note.** Coefficient of variation $= \frac{\sigma}{\overline{x}} \times 100$

**Example 12.** *Calculate the mean and standard deviation for the following data :*

| Size of item | 6 | 7 | 8 | 9 | 10 | 11 | 12 |
|---|---|---|---|---|---|---|---|
| Frequency | 3 | 6 | 9 | 13 | 8 | 5 | 4 |

**Solution.** Assumed mean = 9 *(A.M.I.E., Winter 2001)*

| $x$ | $f$ | $d = x - a$ | $f.d.$ | $f.\ d^2$ |
|---|---|---|---|---|
| 6 | 3 | –3 | – 9 | 27 |
| 7 | 6 | –2 | –12 | 24 |
| 8 | 9 | –1 | – 9 | 9 |
| 9 | 13 | 0 | 0 | 0 |
| 10 | 8 | +1 | 8 | 8 |
| 11 | 5 | +2 | 10 | 20 |
| 12 | 4 | +3 | 12 | 36 |
| | $\sum f = 48$ | | $\sum fd = 0$ | $\sum fd^2 = 124$ |

$$\text{Mean} = a + \frac{\sum fd}{\sum f} = 9 + 0 = 9$$

$$S.D. = \sqrt{\frac{\sum f(x-\overline{x})^2}{\sum f}}$$

$$= \sqrt{\frac{\sum fd^2}{\sum f} - \left(\frac{\sum fd}{\sum f}\right)^2} = \sqrt{\frac{124}{48}} = 1.6$$ **Ans.**

**Example 13.** *From the following frequency distribution, compute the standard deviation of 100 students :*

| Mass in kg | 60 – 62 | 63 – 65 | 66 – 68 | 69 – 71 | 72 – 74 |
|---|---|---|---|---|---|
| Number of students | 5 | 18 | 42 | 27 | 8 |

**Solution.** Assumed mean = 67

| *Mass in kg* | *Number of students f* | *x* | *d = x—67* | *f. d* | *f. d²* |
|---|---|---|---|---|---|
| 60 – 62 | 5 | 61 | – 6 | – 30 | 180 |
| 63 – 65 | 18 | 64 | – 3 | – 54 | 162 |
| 66 – 68 | 42 | 67 | 0 | 0 | 0 |
| 69 – 71 | 27 | 70 | 3 | 81 | 243 |
| 72 –74 | 8 | 73 | 6 | 48 | 288 |
| | $\sum f = 100$ | | | $\sum fd = 45$ | $\sum f\,d^2 = 873$ |

$$S.D. = \sqrt{\frac{\sum fd^2}{\sum f} - \left(\frac{\sum fd}{\sum f}\right)^2} = \sqrt{\frac{873}{100} - \left(\frac{45}{100}\right)^2}$$

$$= \sqrt{8.73 - 0.2025} = \sqrt{8.5275} = 2.9202$$ **Ans.**

**Example 14.** *Compute the standard deviation for the following frequency distribution:*

| Class interval | 0 – 4 | 4 – 8 | 8 – 12 | 12 – 16 |
|---|---|---|---|---|
| Frequency | 4 | 8 | 2 | 1 |

**Solution.** Assumed mean = 6

| *Class interval* | *f* | *x* | *d = x—6* | *fd* | *fd²* |
|---|---|---|---|---|---|
| 0 – 4 | 4 | 2 | – 4 | – 16 | 64 |
| 4 – 8 | 8 | 6 | 0 | 0 | 0 |
| 8 – 12 | 2 | 10 | + 4 | 8 | 32 |
| 12 – 16 | 1 | 14 | + 8 | 8 | 64 |
| | $\sum f = 15$ | | | $\sum fd = 0$ | $\sum f\,d^2 = 160$ |

$$S.D. = \sqrt{\frac{\sum fd^2}{\sum f} - \left(\frac{\sum fd}{\sum f}\right)^2} = \sqrt{\frac{160}{15}} - 0 = 3.266$$ **Ans.**

## 10.13 MOMENTS

The *r*th moment of a variable *x* about the mean $\bar{x}$ is usually denoted by $\mu_r$ is given by

$$\mu_r = \frac{1}{N}\sum f_i(x_i - \bar{x})^r, \qquad \sum f_i = N$$

The *r*th moment of a variable *x* about any point *a* is defined by

$$\mu'_r = \frac{1}{N} f_i(x_i - a)^r$$

In particular
$$\mu_0 = \frac{1}{N}\sum f_i(x - \bar{x})^0 = \frac{1}{N}\sum f_i = \frac{N}{N} = 1$$

$$\mu'_0 = \frac{1}{N}\sum f_i(x - a)^0 = \frac{1}{N}\sum f_i = \frac{N}{N} = 1$$

$$\mu_1 = \frac{1}{N}\sum f_i(x-\bar{x}) = 0, \quad \mu'_1 = \frac{1}{N}\sum f_i(x-a) = \bar{x}-a$$

$$\mu_2 = \frac{1}{N}\sum f_i(x-\bar{x})^2 = \sigma^2.$$

**Relation between moments about mean and moment about any point.**

$$\mu_r = \frac{1}{N}\sum f_i(x-\bar{x})^r = \frac{1}{N}\sum f_i\left[(x-a)-(\bar{x}-a)\right]^r$$

$$= \frac{1}{N}\sum f_i(X_i-d)^r \quad \text{where } X_i = x-a \text{ and } d = \bar{x}-a$$

$$= \frac{1}{N}\left[\sum f_i X_i^r - {}^rC_1\left(\sum f_i X_i^{r-1}\right)d + {}^rC_2\left(\sum f_i x_i^{r-2}\right)d^2 - {}^rC_3\left(\sum f_i x_i^{r-3}\right)d^3 + \ldots\ldots\right]$$

$$= \mu'_r - {}^rC_1 d\mu'_{r-1} + {}^rC_2 d^2\mu'_{r-2} - {}^rC_3 d^3\mu'_{r-3} + \ldots\ldots$$

In particular

$$\mu_2 = \mu'_2 - \mu'^2_1$$

$$\mu_3 = \mu'_3 - 3\mu'_2\mu'_1 + 2\mu'^3_1$$

$$\mu_4 = \mu'_4 - 4\mu'_3\mu'_1 + 6\mu'_2\mu'^2_1 - 3\mu'^4_1$$

**Note.** 1. The sum of the coefficients of the various terms on the right-hand side is zero.

2. The dimension of each term on right-hand side is the same as that of terms on the left.

## 10.14 MOMENT GENERATING FUNCTION

The moment generating function of the variate $x$ about $x = a$ is defined as the expected value of $e^{t(x-a)}$ and is denoted by $M_a(t)$.

$$M_a(t) = \sum P_i e^{t(x_i-a)}$$

$$= \sum P_i + t\sum P_i(x_i-a) + \frac{t^2}{2!}\sum P_i(x_i-a)^2 + \ldots \frac{t^r}{r!}\sum P_i(x_i-a)^r + \ldots$$

$$= 1 + t\mu'_1 + \frac{t^2}{2!}\mu'_2 + \ldots + \frac{t^r}{r!}\mu'_r + \ldots$$

where $\mu_r'$ is the moment of order $r$ about $a$

Hence $\qquad \mu_r' = \text{coefficient of } \dfrac{t^r}{r!} \quad \text{or} \quad \mu_r' = \left[\dfrac{d^r}{dt^r}M_a(t)\right]_{t=0}$

again $\qquad M_a(t) = \sum P_i e^{t(x_i-a)} = e^{-at}\sum P_i e^{tx_i} = e^{-at}M_0(t)$

Thus the moment generating function about the point $a = e^{-at}$ moment generating function about the origin.

## 10.15 (1) SKEWNESS

Skewness denotes the opposite of symmetry. It is lack of symmetry. In a symmetrical series, the mode, the median, and the arithmetic average are identical.

$$\text{Coefficient of skewness} = \frac{\text{Mean} - \text{Mode}}{\text{standard deviation}}$$

**(2) KURTOSIS.** It measures the degree of peakedness of a distribution and is given by Measure of kurtosis

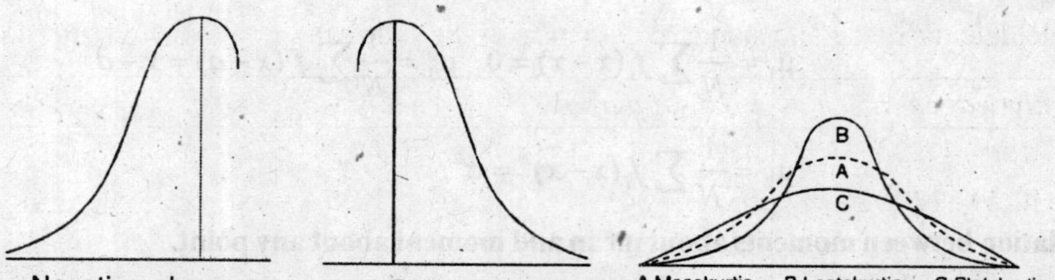


$$\beta_2 = \frac{\mu_4}{\mu_2^2},\ \mu_2 = \frac{\sum(x-\bar{x})^2}{N},\ \mu_4 = \frac{\sum(x-\bar{x})^4}{N},$$

If $\beta_2 = 3$, the curve is normal or mesokurtic.
If $\beta_2 > 3$, the curve is peaked or leptokurtic.
If $\beta_2 < 3$, the curve is flat topped or platykurtic.

## Exercise 10.1

**1.** Marks obtained by 9 students in statistics are given below

52, 57, 40, 70, 43, 40, 65, 35, 48

Calculate the arithmetic mean. **Ans.** 50.

**2.** Calculate the mean of the following:

| Height in cm | 65 | 66 | 67 | 68 | 69 | 70 | 71 | 72 | 73 |
|---|---|---|---|---|---|---|---|---|---|
| Number of plants | 1 | 4 | 5 | 7 | 11 | 10 | 6 | 4 | 2 |

**Ans.** 69.18.

**3.** Find the mean for the following distribution :

| *Marks* | *No. of students* | *Marks* | *No. of Students* |
|---|---|---|---|
| 0-10 | 3 | 50 – 60 | 15 |
| 10-20 | 5 | 60 – 70 | 12 |
| 20-30 | 7 | 70 – 80 | 6 |
| 30-40 | 10 | 80 – 90 | 2 |
| 40-50 | 12 | 90 – 100 | 8 |

**Ans.** 51.75

**4.** Determine the mode from the following figures:

25, 15, 23, 40, 27, 25, 23, 25, 20. **Ans.** 25.

**5.** Find the median of the following :

20, 18, 22, 27, 25, 12, 15. **Ans.** 20.

**6.** The Mean of 200 items was 50. Later on it was discovered that two items were misread as 92 and 8 instead of 192 and 88. Find the corrected mean **Ans.** 53.6

7. Calculate the mean and standard deviation of the following frequency distribution.

| *Weekly wages in ₹* | *No. of persons* | *Weekly wages in ₹* | *No of persons* |
|---|---|---|---|
| 4.5 – 12.5 | 4 | 44.5 – 52.5 | 3 |
| 12.5 – 20.5 | 24 | 52.5 – 60.5 | 5 |
| 20.5 – 28.5 | 21 | 60.5 – 68.5 | 8 |
| 28.5 – 36.5 | 18 | 68.5 – 76.5 | 2 |
| 36.5 – 44.5 | 5 | | |

**Ans.** Mean = 31.34, S.D. = 16.67

8. Compute the standard deviation from the following distribution of marks obtained by 90 students:

| Marks | 20 – 29 | 30 – 39 | 40 – 49 | 50 – 59 | 60 – 69 | 70 – 79 | 80 – 89 | 90 – 99 |
|---|---|---|---|---|---|---|---|---|
| No. of students | 5 | 12 | 15 | 20 | 18 | 10 | 6 | 4 |

**Ans.** 17.65.

9. The following table shows the Marks obtained by 100 candidates in an examination. Calculate the mean, median and standard deviation.

| Marks | 1 – 10 | 11 – 20 | 21 – 30 | 31 – 40 | 41 – 50 | 51 – 60 |
|---|---|---|---|---|---|---|
| No. of candidates | 3 | 16 | 26 | 31 | 16 | 8 |

**Ans.** Mean = 32, S.D. = 12.36, Median = 32.11

10. Fill in the blanks :

(*a*) Average value may be termed as measure of..... (*b*) $\beta_2$ = ..............

(*c*) $\mu_2$ =............. (*d*) The curve is normal if $\beta_2$ = ........

(*e*) The value of $\sum f(x-\bar{x})$ =...............

(*f*) The measure of central item is called as...........
(*g*) The size of the variable which occurs most frequently is known as...........
(*h*) Coefficient of skewness = ..........
(*i*) The ratio of the standard deviation to the mean is known as..........
(*j*) The square of standard deviation is known as the..........

**Ans.** (*a*) Central tendency, (*b*) $\frac{\mu_4}{\mu_2^2}$, (*c*) $\frac{\sum (x-\bar{x})^2}{N}$, (*d*) 3, (*e*) 0, (*f*) median,

(*g*) mode, (*h*) $\frac{\text{Mean} - \text{Mode}}{\text{Standard deviation}}$, (*i*) Coefficient of standard deviation, (*j*) Variance

11. The expected value of a random variable $X$ is 2 and its variance is 1, then variance of $3X + 4$ is
(*a*) 9 (*b*) 7 (*c*) 3 (*d*) 13 (*A.M.I.E.T.E., Dec. 2004*) **Ans.** (*a*)

12. The expected value of a random variable $X$ is 3 and its variance is 2. Then the variance of $2X + 5$ is
(*a*) 8 (*b*) 9 (*c*) 10 (*d*) 11 (*A.M.I.E.T.E., June 2006*) **Ans.** (*a*)

## 10.16 CORRELATION

Whenever two variables x and y are so related that an increase in the one is accompanied by an increase or decrease in the other, then the variables are said to be correlated.

For example, the yield of crop varies with the amount of rainfall.

If an increase in one variable corresponds to an increase in the other, the correlation is said to be positive. If increase in one corresponds to the decrease in the other the correlation is said to be negative. If there is no relationship between the two variables, they are said to be independent.

**Perfect Correlation:** If two variables vary in such a way that their ratio is always constant, then the correlation is said to be perfect.

## 10.17 SCATTER OR DOT-DIAGRAM

When we plot the corresponding values of two variables, taking one on $x$-axis and the other along $y$-axis, it shows a collection of dots.

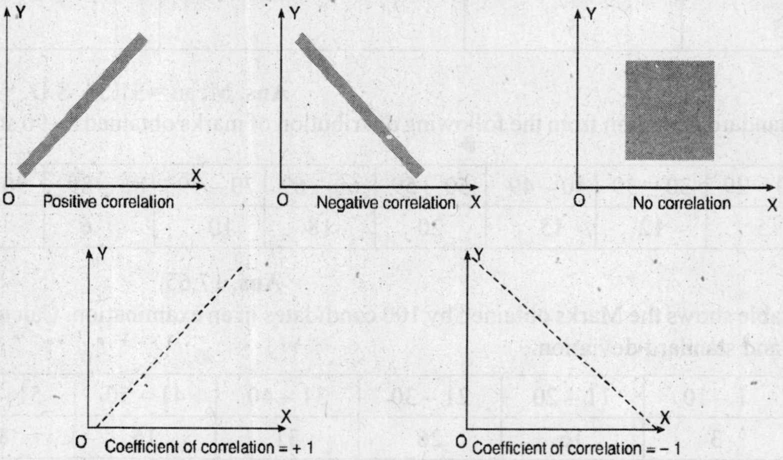


This collection of dots is called a dot diagram or a scatter diagram

**Methods of Determining Simple Correlation**

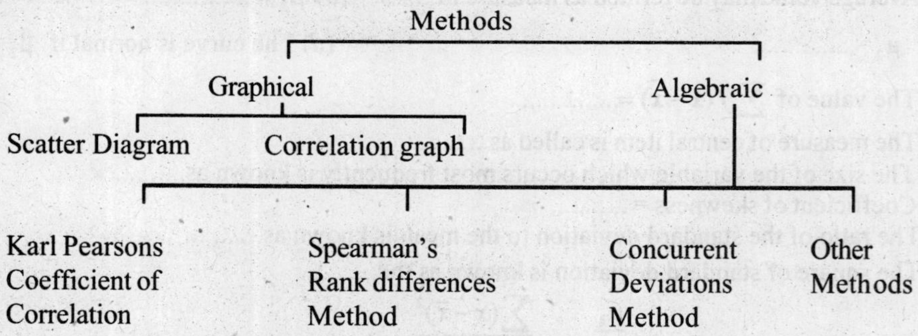


## 10.18 KARL PEARSON'S COEFFICIENT OF CORRELATION

$r$ between two variables $x$ and y is defined by the relation

$$r = \frac{\sum XY}{\sqrt{\left(\sum X^2\right)\left(\sum Y^2\right)}} = \frac{P}{\sigma_x \sigma_y} = \frac{Covariance(x, y)}{\sqrt{variance\,x}\sqrt{variance\,y}},$$

where $X = x - \bar{x}$, $Y = y - \bar{y}$

*i.e.* $X$, Y are the deviations measured from their respective means,

$$p = \left(\frac{\sum XY}{n}\right) = \text{co - variance}$$

and $\sigma_x, \sigma_y$ being the standard deviations of these series.

**Example 15.** *Ten students got the following percentage of marks in Economics and Statistics.*

| Roll No. | 1 | 2 | 3 | 4 | 5 | 6 | 7 | 8 | 9 | 10 |
|---|---|---|---|---|---|---|---|---|---|---|
| Marks in Economics | 78 | 36 | 98 | 25 | 75 | 82 | 90 | 62 | 65 | 39 |
| Marks in Statistics | 84 | 51 | 91 | 60 | 68 | 62 | 86 | 58 | 53 | 47 |

*Calculate the coefficient of correlation.*

**Solution.** Let the marks of two subjects be denoted by $x$ and $y$ respectively.

Then the mean for $x$ marks $= \frac{650}{10} = 65$ and the mean of $y$ marks $= \frac{660}{10} = 66$

If $X$ and $Y$ are deviations of $x$'s and $y$'s from their respective means, then the data may be arranged in the following form :

| $x$ | $y$ | $X = x - 65$ | $Y = y - 66$ | $X^2$ | $Y^2$ | $XY$ |
|---|---|---|---|---|---|---|
| 78 | 84 | 13 | 18 | 169 | 324 | 234 |
| 36 | 51 | −29 | −15 | 841 | 225 | 435 |
| 98 | 91 | 33 | 25 | 1089 | 625 | 825 |
| 25 | 60 | −40 | −6 | 1600 | 36 | 240 |
| 75 | 68 | 10 | 2 | 100 | 4 | 20 |
| 82 | 62 | 17 | −4 | 289 | 16 | −68 |
| 90 | 86 | 25 | 20 | 625 | 400 | 500 |
| 62 | 58 | −3 | −8 | 9 | 64 | 24 |
| 65 | 53 | 0 | −13 | 0 | 169 | 0 |
| 39 | 47 | −26 | −19 | 676 | 361 | 494 |
| 650 | 660 | 0 | 0 | 5398 | 2224 | 2704 |

Here $\sum X^2 = 5398, \sum Y^2 = 2224, \sum XY = 2704$

$$\therefore \quad r = \frac{\sum XY}{\sqrt{\left(\sum X^2\right)\left(\sum Y^2\right)}} = \frac{2704}{\sqrt{5398 \times 2224}}$$

$$= \frac{2704}{73.4 \times 47.1} = \frac{2704}{3457} = 0.78$$ **Ans.**

**Example 16.** *Find the coefficient of correlation between the age and the sum assured from the following table.*

| | *Sum assured (in ₹)* | | | | |
|---|---|---|---|---|---|
| *Age-group* | *10,000* | *20,000* | *30,000* | *40,000* | *50,000* |
| 20 – 30 | 4 | 6 | 3 | 7 | 1 |
| 30 – 40 | 2 | 8 | 15 | 7 | 1 |
| 40 – 50 | 3 | 9 | 12 | 6 | 2 |
| 50 – 60 | 8 | 4 | 2 | — | — |

**Solution.** Let the sum assured denote by $x$ and the age group by $y$.

$$x' = \frac{x - 30,000}{10,000}, \quad y' = \frac{y - 45}{10}$$

| y \ x | | | 10,000 | | 20,000 | | 30,000 | | 40,000 | | 50,000 | | | | | |
|---|---|---|---|---|---|---|---|---|---|---|---|---|---|---|---|---|
| | | x' / y' | −2 | | −1 | | 0 | | 1 | | 2 | | Σf (Rows) | f.y' | f.y'^2 | f.x'y' |
| | | | f | fx'y' | f | fx'y' | f | fx'y' | f | fx'y' | f | fx'y' | | | | |
| 20–30 | 25 | −2 | 4 | 16 | 6 | 12 | 3 | 0 | 7 | −14 | 1 | −4 | 21 | −42 | 84 | +10 |
| 30–40 | 35 | −1 | 2 | 4 | 8 | 8 | 15 | 0 | 7 | −7 | 1 | −2 | 33 | −33 | 33 | +3 |
| 40–50 | 45 | 0 | 3 | 0 | 9 | 0 | 12 | 0 | 6 | 0 | 2 | 0 | 32 | 0 | 0 | 0 |
| 50–60 | 55 | 1 | 8 | −16 | 4 | −4 | 2 | 0 | – | 0 | – | 0 | 14 | 14 | 14 | −20 |
| | | Σf (column) | 17 | | 27 | | 32 | | 20 | | 4 | | N = 100 | Σfy' = −61 | Σfy'^2 =131 | Σfx'y' = −7 |
| | | fx' | −34 | | −27 | | 0 | | 20 | | 8 | | Σfx' = −33 | | | |
| | | fx'^2 | 68 | | 27 | | 0 | | 20 | | 16 | | Σfx'^2 = 131 | | | |
| | | fx'y' | | 4 | | 16 | | 0 | | −21 | | −6 | Σfx'y' = −7 | | | |

$$r=\frac{N\sum fx'y'-\sum f.x'\sum f.y'}{\sqrt{N\sum f.x'^2-(\sum f.x')^2}\sqrt{N\sum f.x''^2-(\sum f.y')^2}}$$

$$=\frac{100(-7)-(-33)(-61)}{\sqrt{100(131)-(-33)^2}\sqrt{100(131)-(-61)^2}}=\frac{-700-2013}{\sqrt{13100-1089}\sqrt{13100-3721}}$$

$$=\frac{-2713}{\sqrt{12011}\sqrt{9379}}=\frac{-2713}{109.59\times 96.85}=\frac{-2713}{10613.7915}=-0.2556$$

**Hence, the age and sum assured are negatively correlated, *i.e.*, as age goes up the sum assured comes down.** **Ans.**

## 10.19 SHORT-CUT METHOD

$$r=\frac{\frac{\sum X'Y'}{N}-\left(\frac{\sum X'}{N}\right)\left(\frac{\sum Y'}{N}\right)}{\sqrt{\left\{\frac{\sum X'^2}{N}-\left(\frac{\sum X'}{N}\right)^2\right\}}\sqrt{\left\{\frac{\sum Y'^2}{N}-\left(\frac{\sum Y'}{N}\right)^2\right\}}}$$

**where $r$ is the coefficient of correlation.**

$X'$ = deviation from *assumed* mean of $x = x-a$

$Y'$ = deviation from *assumed* mean of $y = y-b$

$N$ = Total number of items.

**Example 17.** *Calculate the coefficient of correlation for the following table :*

| *x–age* / *marks* | *0 – 4* | *4 – 8* | *8 – 12* | *12 – 16* |
|---|---|---|---|---|
| 0 – 5 | 7 | — | — | — |
| 5 – 10 | 6 | 8 | — | — |
| 10 – 15 | — | 5 | 3 | — |
| 15 – 20 | — | 7 | 2 | — |
| 20 – 25 | — | — | — | 9 |

**Solution.** Replace the class-interval for x and *y* by their mid-points and then let

$$X' = \frac{x-10}{4} \text{ and } Y' = \frac{y-12.5}{5}$$

| *y* | | *x* / *X'* / *Y'* | 2 | | 6 | | 10 | | 14 | | Σ*f* (Row) | *fY'* | *fY'*² | *fX'Y'* |
|---|---|---|---|---|---|---|---|---|---|---|---|---|---|---|
| | | | –2 | | –1 | | 0 | | 1 | | | | | |
| | | | *f* | *fX'Y'* | *f* | *fX'Y'* | *f* | *fX'Y'* | *f* | *fX'Y'* | | | | |
| 0–5 | 2.5 | –2 | 7 | 28 | | | | | | | 7 | –14 | –28 | 28 |
| 5–10 | 7.5 | –1 | 6 | 12 | 8 | 8 | | | | | 14 | –14 | –14 | 20 |
| 10–15 | 12.5 | 0 | | | 5 | 0 | 3 | 0 | | | 8 | 0 | 0 | 0 |
| 15–20 | 17.5 | 1 | | | 7 –7 | | 2 | 0 | | | 9 | 9 | 9 | –7 |
| 20–25 | 22.5 | 2 | | | | | | | 9 | 18 | 9 | 18 | 36 | 18 |
| Σ*f* | | | 13 | | 20 | | 5 | | 9 | | 47 | Σ*fY'* = –1 | Σ*fY'*² =87 | Σ*fX'Y'* = 59 |
| *fX'* | | | –26 | | –20 | | 0 | | 9 | | Σ*fX'* = –37 | | | |
| *fX'*² | | | 52 | | 20 | | 0 | | 9 | | Σ*fX'*² = 81 | | | |
| *fX'Y'* | | | | 40 | | 1 | | 0 | | 18 | Σ*fX'Y'* = 59 | | | |

Here, $\sum fX' = -37, \sum fX'^2 = 81, \sum fY' = -1, \sum fY'^2 = 87, \Sigma fX'Y' = 59$

$$r = \frac{\dfrac{\sum fX'Y'}{N} - \left(\dfrac{\sum fX'}{N}\right)\left(\dfrac{\sum fY'}{N}\right)}{\sqrt{\dfrac{\sum fX'^2}{N} - \left(\dfrac{\sum fX'}{N}\right)^2}\sqrt{\dfrac{\sum fY'^2}{N} - \left(\dfrac{\sum fY'}{N}\right)^2}}$$

$$= \frac{\dfrac{59}{47} - \left(\dfrac{-37}{47}\right)\left(\dfrac{-1}{47}\right)}{\sqrt{\left\{\dfrac{81}{47} - \left(\dfrac{-37}{47}\right)^2\right\}}\sqrt{\left\{\dfrac{87}{47} - \left(\dfrac{-1}{47}\right)^2\right\}}} = \frac{1.255 - 0.017}{\sqrt{1.732 - 0.620}\sqrt{1.851 - 0.0005}}$$

$$= \frac{1.238}{\sqrt{1.103}\sqrt{1.8505}} = \frac{1.238}{1.05 \times 1.36} = \frac{1.238}{1.428} = 0.87$$

**Ans.**

## 10. 20 SPEARMAN'S RANK CORRELATION

$$r = 1 - \frac{6\sum d^2}{n(n^2-1)}$$

**Solution.** Let $(x_1, y_1), (x_2, y_2)...(x_n, y_n)$ be the ranks of $n$ individuals corresponding to two characteristics.

Assuming nor two individuals are equal in either classification, each individual takes the values 1, 2, 3, ... $n$ and hence their arithmetic means are, each

$$= \frac{\sum n}{n} = \frac{1}{n}\frac{n(n+1)}{2} = \frac{n+1}{2}$$

Let $x_1, x_2, x_3,... x_n$ be the values of variable $X$ and $y_1, y_2, y_3,... y_n$ those of $Y$.

Then $$d = X - Y = \left(x - \frac{n+1}{2}\right) - \left(y - \frac{n+1}{2}\right) = x - y$$

where $X$ and $Y$ are deviations from the mean.

$$\sum X^2 = \sum\left(x - \frac{n+1}{2}\right)^2 = \sum x^2 - (n+1)\sum x + \sum\left(\frac{n+1}{2}\right)^2$$

$$= \frac{n(n+1)(2n+1)}{6} - \frac{(n+1)n(n+1)}{2} + n\left(\frac{n+1}{2}\right)^2$$

$$= \frac{n(n^2-1)}{12}$$

Clearly, $$\sum X = \sum Y \quad \text{and} \quad \sum X^2 = \sum Y^2$$

$\therefore$ $$\sum Y^2 = \frac{n(n^2-1)}{12}$$

Hence $$\sum d^2 = \sum (x-y)^2 = \sum x^2 + \sum y^2 - 2\sum xy$$

$\therefore$ $$\sum XY = \frac{1}{2}\left[\frac{n(n^2-1)}{6} - \sum d^2\right]$$

$$= \frac{1}{12}n(n^2-1) - \frac{1}{2}\sum d^2$$

Putting these values in $$r = \frac{\sum XY}{\sqrt{\sum X^2}\sqrt{\sum Y^2}}$$

$$= \frac{\frac{1}{12}n(n^2-1) - \frac{1}{2}\sum d^2}{\frac{n(n^2-1)}{12}}$$

$$= 1 - \frac{6\sum d^2}{n(n^2-1)}$$ **Ans.**

## 10.21 SPEARMAN'S RANK CORRELATION COEFFICIENT

$$r = 1 - \frac{6\sum d^2}{n(n^2-1)}$$

where $r$ denotes rank coefficient of correlation and $d$ refers to the difference of ranks between paired items in two series.

**Example 18.** *Compute Spearman's rank correlation coefficient r for the following data:*

| *Person* | *A* | *B* | *C* | *D* | *E* | *F* | *G* | *H* | *I* | *J* |
|---|---|---|---|---|---|---|---|---|---|---|
| *Rank in statistics* | *9* | *10* | *6* | *5* | *7* | *2* | *4* | *8* | *1* | *3* |
| *Rank in income* | *1* | *2* | *3* | *4* | *5* | *6* | *7* | *8* | *9* | *10* |

**Solution.**

| *Person* | *Rank in statistics* | *Rank in income* | $d = R_1 - R_2$ | $d^2$ |
|---|---|---|---|---|
| A | 9 | 1 | 8 | 64 |
| B | 10 | 2 | 8 | 64 |
| C | 6 | 3 | 3 | 9 |
| D | 5 | 4 | 1 | 1 |
| E | 7 | 5 | 2 | 4 |
| F | 2 | 6 | –4 | 16 |
| G | 4 | 7 | –3 | 9 |
| H | 8 | 8 | 0 | 0 |
| I | 1 | 9 | –8 | 64 |
| J | 3 | 10 | –7 | 49 |
| | | | | $\Sigma d^2 = 280$ |

$$r = 1 - \frac{6\sum d^2}{n(n^2-1)}$$

$$r = 1 - \frac{6 \times 280}{10(100-1)} = 1 - 1.697 = -0.697$$ **Ans.**

**Example 19.** *Establish the formula* $\sigma^2_{x-y} = \sigma^2_x + \sigma^2_y - 2r\sigma_x\sigma_y$
*where r is the correlation coefficient between x and y.*

**Solution.** We know that $\sigma^2_x = \frac{\sum(x-\bar{x})^2}{n}$

$$\therefore \quad \sigma^2_{x-y} = \frac{\sum[(x-y) - \overline{(x-y)}]^2}{n}$$

$\overline{x-y} =$ mean of $(x-y)$ series = mean of $x$ – mean of $y = \bar{x} - \bar{y}$

$$\sigma^2_{x-y} = \frac{\sum[(x-y)-(\bar{x}-\bar{y})]^2}{n} = \frac{\sum[(x-\bar{x})-(y-\bar{y})]^2}{n}$$

$$= \frac{\sum[(x-\bar{x})^2 + (y-\bar{y})^2 - 2(x-\bar{x})(y-\bar{y})]}{n}$$

$$= \frac{\sum(x-\bar{x})^2}{n} + \frac{\sum(y-\bar{y})^2}{n} - \frac{2\sum(x-\bar{x})(y-\bar{y})}{n}$$

$$= \sigma^2_x + \sigma^2_y - \frac{2\sum(x-\bar{x})(y-\bar{y})}{n} \qquad ...(1)$$

We know that $r = \frac{\sum(x-\bar{x})(y-\bar{y})}{n\sigma_x\sigma_y}$ or $\frac{\sum(x-\bar{x})(y-\bar{y})}{n} = r\sigma_x\sigma_y$

Putting this value in (1), we get $\sigma^2_{x-y} = \sigma^2_x + \sigma^2_y - 2r\sigma_x\sigma_y$ **Proved.**

**Example 20.** *If X and Y are uncorrelated random variables, find the coefficient of correlation between X + Y and X – Y.*

**Solution.**

Let $u = X + Y$ and $v = X - Y$

Then $$r = \frac{\sum (u-\bar{u})(v-\bar{v})}{n\sigma_u \sigma_v}$$

Now $u = X + Y, \bar{u} = \bar{X} + \bar{Y}$

Similarly $\bar{v} = \bar{X} - \bar{Y}$

Now
$$\sum (u-\bar{u})(v-\bar{v})$$
$$= \sum (X - \bar{X} + Y - \bar{Y})[(X-\bar{X})-(Y-\bar{Y})]$$
$$= \sum (x+y)(x-y)$$
$$= \sum x^2 - \sum y^2$$
$$= n\sigma_x^2 - n\sigma_y^2$$

Also $$\sigma_u^2 = \frac{\sum (u-\bar{u})^2}{n} = \frac{1}{n}\sum [(X-\bar{X})+(Y-\bar{Y})]^2$$
$$= \frac{1}{n}\sum (x+y)^2$$
$$= \frac{1}{n}\left(\sum x^2 + \sum y^2 + 2\sum xy\right)$$
$$= \sigma_x^2 + \sigma_y^2$$ (As $X$ and Y are not correlated, we have $\sum xy = 0$)

Similarly $\sigma_v^2 = \sigma_x^2 + \sigma_y^2$

$\therefore$ $$r = \frac{\sum (u-\bar{u})(v-\bar{v})}{n\sigma_u \sigma_v}$$
$$= \frac{n(\sigma_x^2 - \sigma_y^2)}{\sqrt{n(\sigma_x^2+\sigma_y^2)}\sqrt{n(\sigma_x^2+\sigma_y^2)}}$$
$$= \frac{\sigma_x^2 - \sigma_y^2}{\sigma_x^2 + \sigma_y^2}$$ **Ans.**

## 10.22 REGRESSION

If the scatter diagram indicates some relationship between two variables $x$ and $y$, then the dots of the scatter diagram will be concentrated round a curve. This curve is called the *curve of regression.*

Regression analysis is the method used for estimating the unknown values of one variable corresponding to the known value of another variable.

## 10.23 LINE OF REGRESSION

When the curve is a straight line, it is called a line of regression. A line of regression is the straight line which gives the best fit in the least square sense to the given frequency.

## 10.24 EQUATIONS TO THE LINES OF REGRESSION

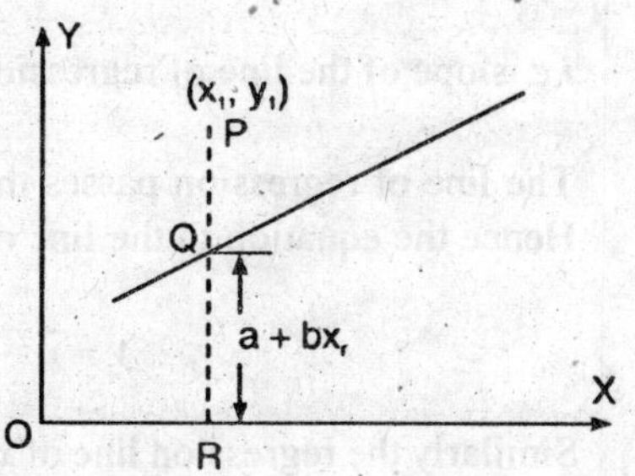


Let $y = a + bx$ ... (1)

be the equation of the line of regression of $y$ on $x$.

Let $(x_r, y_r)$ be any point of dot.

From the figure

$$PR = y_r$$

$$QR = a + bx_r$$

$$PQ = PR - QR = y_r - a - bx_r$$

Let $S$ be the sum of the squares of such distances, then

$$S = \sum (y - a - bx)^2$$

According to the principle of least squares, we have to choose $a$ and $b$ so that $S$ is minimum. The method of least square gives the condition for minimum value of $S$.

$$\frac{\partial S}{\partial a} = -2\sum (y - a - bx), \quad \frac{\partial S}{\partial b} = -2\sum (y - a - bx)x$$

$$\frac{\partial S}{\partial a} = 0, \frac{\partial S}{\partial b} = 0, \text{ for } S \text{ minimum}$$

*i.e.* $\sum (y - a - bx) = 0 \Rightarrow \sum y - na - b\sum x = 0$

$\Rightarrow$ $\sum y = na + b\sum x$ ... (2)

and $\sum (xy - ax - bx^2) = 0 \Rightarrow \sum xy - a\sum x - b\sum x^2 = 0$

$\sum xy = a\sum x + b\sum x^2$ ...(3)

Dividing (2) by $n$, we get

$$\frac{\sum y}{n} = a + b\frac{\sum x}{n} \qquad \left(\bar{y} = \frac{\sum y}{n}, \ \bar{x} = \frac{\sum x}{n}\right)$$

$$\bar{y} = a + b\bar{x}$$

where $\bar{x}$ and $\bar{y}$ are the means of $x$ series and $y$ series.

This shows that $(\bar{x}, \bar{y})$ lie on the line of regression (1), shifting the origin to $(\bar{x}, \bar{y})$, the equation (3) becomes

$$\sum (x - \bar{x})(y - \bar{y}) = a\sum (x - \bar{x}) + b\sum (x - \bar{x})^2$$

But $\sum (x - \bar{x}) = 0$ *i.e.* $\sum (x - \bar{x})(y - \bar{y}) = b\sum (x - \bar{x})^2$

$\Rightarrow$ $$b = \frac{\sum (x - \bar{x})(y - \bar{y})}{\sum (x - \bar{x})^2} = \frac{\sum XY}{\sum X^2}$$ ...(4)

We know $$r = \frac{\sum XY}{\sqrt{\sum X^2}\sqrt{\sum Y^2}} = \frac{\sum XY}{n\sqrt{\frac{\sum X^2}{n}}\sqrt{\frac{\sum Y^2}{n}}} = \frac{\sum XY}{n\sigma_x\sigma_y}$$

$\Rightarrow$ $$\sum XY = nr\sigma_x\sigma_y$$

Putting the value of $\sum XY$ in (4), we get

$$b = \frac{nr\sigma_x\sigma_y}{\sum X^2} = \frac{r\sigma_x\sigma_y}{\frac{\sum X^2}{n}} = \frac{r\sigma_x\sigma_y}{\sigma_x^2} = \frac{r\sigma_y}{\sigma_x}$$

*i.e.* slope of the line of regression $= b = r\dfrac{\sigma_y}{\sigma_x}$

The line of regression passes through $(\bar{x}, \bar{y})$.
Hence the equation to the line of regression is

$$y - \bar{y} = r\frac{\sigma_y}{\sigma_x}(x - \bar{x})$$

Similarly the regression line of $x$ on $y$ is

$$x - \bar{x} = r\frac{\sigma_x}{\sigma_y}(y - \bar{y}).$$

**Note.** $b_{yx} = r\dfrac{\sigma_y}{\sigma_x}$ and $b_{xy} = r\dfrac{\sigma_x}{\sigma_y}$ are known as the coefficients of regression.

$$b_{yx} . b_{xy} = \left(r\frac{\sigma_y}{\sigma_x}\right)\left(r\frac{\sigma_x}{\sigma_y}\right) = r^2$$

**Example 21.** *If $\theta$ be the acute angle between the two regression lines in the case of two variables $x$ and $y$, show that*

$$\tan\theta = \frac{1-r^2}{r}\frac{\sigma_x\sigma_y}{\sigma_x^2+\sigma_y^2}$$

*where $r, \sigma_x, \sigma_y$ have their usual meanings. Explain the significance where $r = 0$ and $r = \pm 1$.*

*(A.M.I.E., Winter 2001)*

**Solution.** Lines of regression are

$$y - \bar{y} = r\frac{\sigma_y}{\sigma_x}(x - \bar{x}) \quad \text{...(1)} \qquad \therefore m_1 = r\frac{\sigma_y}{\sigma_x}$$

and

$$x - \bar{x} = r\frac{\sigma_x}{\sigma_y}(y - \bar{y}) \quad \text{...(2)} \qquad \therefore m_2 = \frac{1}{r}\frac{\sigma_y}{\sigma_x}$$

$$\tan\theta = \frac{m_2 - m_1}{1 + m_1 m_2}$$

$$= \frac{\dfrac{1}{r}\dfrac{\sigma_y}{\sigma_x} - r\dfrac{\sigma_y}{\sigma_x}}{1 + r\dfrac{\sigma_y}{\sigma_x}\times\dfrac{1}{r}\dfrac{\sigma_y}{\sigma_x}} = \frac{\left(\dfrac{1}{r} - r\right)\dfrac{\sigma_y}{\sigma_x}}{1 + \dfrac{\sigma_y^2}{\sigma_x^2}}$$

$$= \frac{1-r^2}{r} . \frac{\left(\dfrac{\sigma_y}{\sigma_x}\right)\sigma_x^2}{\sigma_x^2+\sigma_y^2} = \frac{1-r^2}{r}\frac{\sigma_x\sigma_y}{\sigma_x^2+\sigma_y^2} \quad \text{...(3)}$$

**Proved.**

(*a*) If $r = 0$, then there is no relationship between the two variables and they are independent.

On putting the value of $r = 0$ in (3), we get $\tan\theta = \infty, \theta = \dfrac{\pi}{2}$. So the lines (1) and (2) are perpendicular.

*(A.M.I.E., Summer 1998)*

(*b*) If $r = 1$ or $-1$

On putting these values of $r$ in (3) we get, $\tan\theta = 0$ or $\theta = 0$

*i.e.* lines (1) and (2) coincide.

The correlation between the variables is perfect.

**Ans.**

**Example 22.** *Find the correlation coefficient between x and y, when the lines of regression are:*

$$2x-9y+6=0 \text{ and } x-2y+1=0$$

**Solution.** Let the line of regression of $x$ on $y$ be $2x-9y+6=0$

Then, the line of regression of $y$ on $x$ is $x-2y+1=0$

$\therefore$ $$2x-9y+6=0 \Rightarrow x=\frac{9}{2}y-3 \Rightarrow b_{xy}=\frac{9}{2}$$

and $$x-2y+1=0 \Rightarrow y=\frac{1}{2}x+\frac{1}{2} \Rightarrow b_{yx}=\frac{1}{2}$$

$$r=\sqrt{b_{xy}\cdot b_{yx}}=\sqrt{\frac{9}{2}\times\frac{1}{2}}=\frac{3}{2}>1 \text{ which is not possible.}$$

So our choice of regression line is incorrect.

$\therefore$ The regression line of $x$ on $y$ is $x-2y+1=0$

And, the regression line of $y$ on $x$ is $2x-9y+6=0$

$\therefore$ $$x-2y+1=0 \Rightarrow x=2y-1 \Rightarrow b_{xy}=2$$

And $$2x-9y+6=0 \Rightarrow y=\frac{2}{9}x+\frac{2}{3} \Rightarrow b_{yx}=\frac{2}{9}$$

$$r=\sqrt{b_{xy}\cdot b_{yx}}=\sqrt{2\times\frac{2}{9}}=\frac{2}{3}$$

Hence the correlation coefficient between $x$ and $y$ is $\frac{2}{3}$.

**Example 23.** *The following regression equations were obtained from a correlation table:*

$$y=0.516x+33.73, \quad x=0.512y+32.52$$

*Find the value of (a) the correlation coefficient, (b) the mean of x's and (c) the mean of y's*

**Solution.** $$y=0.516x+33.73 \quad \text{...(1)}$$

$$x=0.512y+32.52 \quad \text{...(2)}$$

(*a*) From (1), $$r\frac{\sigma_y}{\sigma_x}=0.516 \quad \text{...(3)}$$

From (2), $$r\frac{\sigma_x}{\sigma_y}=0.512 \quad \text{...(4)}$$

From (3) and (4)

$$\left(r\frac{\sigma_y}{\sigma_x}\right)\left(r\frac{\sigma_x}{\sigma_y}\right)=(0.516)(0.512)$$

$$r^2=0.516\times 0.512 \quad \Rightarrow \quad r=0.514$$

Coefficient of correlation $=0.514$. **Ans.**

(*b*) (1) and (2) pass through the point $(\bar{x},\bar{y})$.

$\therefore$ $$\bar{y}=0.516\bar{x}+33.73 \quad \text{...(5)}$$

$$\bar{x}=0.512\bar{y}+32.52 \quad \text{...(6)}$$

On solving (5) and (6), we get

$$\bar{x}=67.6, \quad \bar{y}=68.61$$ **Ans.**

**Example 24.** *The two regression equations of the variables x and y are*

$$x = 19.13 - 0.87y \text{ and } y = 11.64 - 0.50x.$$

*Find (i) Mean of x's; (ii) Mean of y's; (iii) The correlation coefficient between x and y.*

**Solution.**

$$x = 19.13 - 0.87y \quad ...(1)$$

$$y = 11.64 - 0.50x \quad ...(2)$$

As (1) and (2) pass through $(\bar{x}, \bar{y})$:

$$\bar{x} = 19.13 - 0.87\bar{y} \quad ...(3)$$

$$\bar{y} = 11.64 - 0.50\bar{x} \quad ...(4)$$

On solving (3) and (4) we get

$$\bar{x} = 15.935, \ \bar{y} = 3.67$$

From (1) $$r\frac{\sigma_x}{\sigma_y} = -0.87 \quad ...(5)$$

From (2) $$r\frac{\sigma_y}{\sigma_x} = -0.50 \quad ...(6)$$

As $\sigma_x$ and $\sigma_y$ are always positive, so $r$ is negative.

Multiplying (5) and (6) we get

$$r\frac{\sigma_x}{\sigma_y} \cdot r\frac{\sigma_y}{\sigma_x} = -0.87 \times (-0.50)$$

$\Rightarrow$ $$r^2 = 0.435 \Rightarrow r = -0.66$$ **Ans.**

**Example 25.** *The regression equations calculated from a given set of observations for two random variables are*

$$x = -0.4y + 6.4 \text{ and } y = -0.6x + 4.6$$

*Calculate* $\bar{x}, \bar{y}$ *and r.*

**Solution.** The regression equations are

$$x = -0.4y + 6.4 \quad ...(1)$$

$$y = -0.6x + 4.6 \quad ...(2)$$

From (1) coefficient of regression of $x$ on $y = r\frac{\sigma_x}{\sigma_y} = -0.4$ ...(3)

From (2) coefficient of regression of $y$ on $x = r\frac{\sigma_y}{\sigma_x} = -0.6$ ...(4)

From (3) and (4)

$$\left(r\frac{\sigma_x}{\sigma_y}\right)\left(r\frac{\sigma_y}{\sigma_x}\right) = (-0.4)(-0.6)$$

$\Rightarrow$ $$r^2 = 0.24$$

$\Rightarrow$ $$r = \pm 0.49$$

In (3) and (4), $\sigma_x$ and $\sigma_y$ are (always) positive so $r$ is negative

$$r = -0.49$$

To find $\bar{x}$ and $\bar{y}$ we solve the equations (1) and (2) simultaneously. Their point of intersection is $(\bar{x}, \bar{y})$

$$\bar{x} = 6, \qquad \bar{y} = 1 \qquad \textbf{Ans.}$$

**Example 26.** *Show that the geometric mean of the coefficients of regression is the coefficient of correlation.*

**Solution.** The coefficients of regressions are $r\dfrac{\sigma_y}{\sigma_x}$ and $r\dfrac{\sigma_x}{\sigma_y}$

*i.e.*
$$G.M. = \sqrt{r\frac{\sigma_y}{\sigma_x} . r\frac{\sigma_x}{\sigma_y}} = r$$

= coefficient of correlation. **Proved.**

**Example 27.** *Prove that arithmetic mean of the coefficients of regression is greater than the coefficient of correlation.* (*A.M.I.E., Summer 2000*)

**Solution.** Cefficients of regression are $r\dfrac{\sigma_y}{\sigma_x}$ and $r\dfrac{\sigma_x}{\sigma_y}$

We have prove that $A.M. > r$

$$\Rightarrow \quad \frac{1}{2}\left[r\frac{\sigma_y}{\sigma_x} + r\frac{\sigma_x}{\sigma_y}\right] > r \Rightarrow \frac{1}{2}\left[\frac{\sigma_y}{\sigma_x} + \frac{\sigma_x}{\sigma_y}\right] > 1$$

$$\Rightarrow \quad \frac{\sigma_y}{\sigma_x} + \frac{\sigma_x}{\sigma_y} - 2 > 0 \Rightarrow \frac{1}{\sigma_x\sigma_y}\left[\sigma_x^2 + \sigma_y^2 - 2\sigma_x\sigma_y\right] > 0$$

$$\Rightarrow \quad \frac{1}{\sigma_x\sigma_y}\left[\sigma_x - \sigma_y\right]^2 > 0 \quad \text{which is true.} \qquad \textbf{Proved.}$$

**Example 28.** *Find the regression line of y on x for the following data*

| x | 1 | 3 | 4 | 6 | 8 | 9 | 11 | 14 |
|---|---|---|---|---|---|---|---|---|
| y | 1 | 2 | 4 | 4 | 5 | 7 | 8 | 9 |

*Estimate the value of y, when x = 10.*

**Solution.**

| S. No. | x | y | xy | $x^2$ |
|---|---|---|---|---|
| 1 | 1 | 1 | 1 | 1 |
| 2 | 3 | 2 | 6 | 9 |
| 3 | 4 | 4 | 16 | 16 |
| 4 | 6 | 4 | 24 | 36 |
| 5 | 8 | 5 | 40 | 64 |
| 6 | 9 | 7 | 63 | 81 |
| 7 | 11 | 8 | 88 | 121 |
| 8 | 14 | 9 | 126 | 196 |
| Total | 56 | 40 | 364 | 524 |

Let $y = a + bx$ be the line of regression of $y$ on $x$, where $a$ and $b$ are given by the following equations :

$$\sum y = na + b\sum x \quad \text{or} \quad 40 = 8a + 56b \quad ...(1)$$

$$\sum xy = a\sum x + b\sum x^2 \quad \text{or} \quad 364 = 56a + 524b \quad ...(2)$$

On solving (1) and (2) we get,

$$a = \frac{6}{11} \text{ and } b = \frac{7}{11}$$

The equation of the required line is

$$y = \frac{6}{11} + \frac{7}{11}x \quad \text{or} \quad 7x - 11y + 6 = 0$$ **Ans.**

If $x = 10, y = \frac{6}{11} + \frac{7}{11}(10) = \frac{76}{11} = 6\frac{10}{11}$ **Ans.**

**Example 29.** *In a study between the amount of rainfall and the quantity of air pollution removed the following data were collected.*

| *Daily Rainfall in 0.01 cm* | *4.3* | *4.5* | *5.9* | *56* | *61* | *5.2* | *3.8* | *2.1* |
|---|---|---|---|---|---|---|---|---|
| *Pollution Removed (mg/m³)* | *12.6* | *121* | *11.6* | *11.8* | *11.4* | *11.8* | *13.2* | *14.1* |

*Find the regression line of y on x.* *(A.M.I.E., Summer 2000)*

**Solution.**

| *S.No.* | *x (metre)* | *y* | *xy* | $x^2$ |
|---|---|---|---|---|
| 1 | 4.3 | 12.6 | 54.18 | 18.49 |
| 2 | 4.5 | 12.1 | 54.45 | 20.25 |
| 3 | 5.9 | 11.6 | 68.44 | 34.81 |
| 4 | 5.6 | 11.8 | 66.08 | 31.36 |
| 5 | 6.1 | 11.4 | 69.54 | 37.21 |
| 6 | 5.2 | 11.8 | 61.36 | 27.04 |
| 7 | 3.8 | 13.2 | 50.16 | 14.44 |
| 8 | 2.1 | 14.1 | 29.61 | 4.41 |
| | 37.5 | 98.6 | 453.82 | 188.01 |

Let $y = a + bx$ be the equation of the line of regression of $y$ on $x$, where $a$ and $b$ are given by the following equations.

$$\sum y = na + b\sum x \quad \Rightarrow \quad 98.6 = 8a + 37.5b \quad ...(1)$$

$$\sum xy = a\sum x + b\sum x^2 \quad \Rightarrow \quad 453.82 = 37.5a + 188.01b \quad ...(2)$$

On solving (1) and (2), we get $a = 15.49$ and $b = -0.675$.

The equation of the line of regression is $y = 15.49 - 0.675x$ **Ans.**

**Example 30.** *The following data regarding the heights (y) and the weights (x) of 100 college students are given :*

$$\sum x = 15000, \quad \sum x^2 = 2272500$$

$$\sum y = 6800, \quad \sum y^2 = 46.3025$$

$$\sum xy = 1022250$$

Find the correlation coefficient between height and weight and state the equation of regression of height on weight

**Solution.** $\bar{x} = \frac{\sum x}{n} = \frac{15000}{100} = 150,\ \bar{y} = \frac{\sum y}{n} = \frac{6800}{100} = 68$

$$\sigma_x = \sqrt{\frac{\sum x^2}{n} - \left(\frac{\sum x}{n}\right)^2} = \sqrt{\frac{2272500}{100} - \left(\frac{15000}{100}\right)^2}$$

$$\sigma_x = \sqrt{22725 - 22500} = \sqrt{225} = 15$$

$$\sigma_y = \sqrt{\frac{\sum y^2}{n} - \left(\frac{\sum y}{n}\right)^2} = \sqrt{\frac{463025}{100} - \left(\frac{6800}{100}\right)^2}$$

$$= \sqrt{4630.25 - 4624} = \sqrt{6.25} = 2.5$$

$$r = \frac{\frac{\sum xy}{n} - (\bar{x})(\bar{y})}{(\sigma_x)(\sigma_y)} = \frac{\frac{1022250}{100} - (150)(68)}{15 \times 2.5}$$

$$= \frac{10222.5 - 10200}{15 \times 2.5} = \frac{22.5}{15 \times 2.5} = \frac{1.5}{2.5} = 0.6$$

Regression equation of $y$ on $x$ we have

$$y - \bar{y} = r\frac{\sigma_y}{\sigma_x}(x - \bar{x}),\ y - 68 = 0.6\left(\frac{2.5}{15}\right)(x - 150)$$

$$y - 68 = \frac{1}{10}(x - 150) \quad \text{or} \quad 10y - 680 = x - 150$$

$$10y = x + 530$$

**Ans.**

## 10.25 ERROR OF PREDICTION

The deviation of the predicted value from the observed value is known as the standard error of prediction. It is given by

$$E_{yx} = \sqrt{\frac{\sum (y - y_r)^2}{n}}$$

where $y$ is the actual value and $y_r$ the predicted value.

**Example 31.** *Prove that*

(i) $E_{yx} = \sigma_y.\sqrt{1 - r^2}$ (ii) $E_{xy} = \sigma_x\sqrt{1 - r^2}$

**Solution.** The equation of the line of regression of $y$ on $x$ is .

$$y - \bar{y} = r\frac{\sigma_y}{\sigma_x}(x - \bar{x})$$

$$y_r = \bar{y} + r\frac{\sigma_y}{\sigma_x}(x - \bar{x})$$

So, $$E_{yx} = \sqrt{\frac{\sum (y - y_r)^2}{n}} = \left[\frac{1}{n}\sum\left\{y - \bar{y} - r\frac{\sigma_y}{\sigma_x}(x - \bar{x})\right\}^2\right]^{1/2}$$

$$= \left[\frac{1}{n}\cdot\Sigma\left\{(y - \bar{y})^2 + \frac{r^2\sigma_y^2}{\sigma_x^2}(x - \bar{x})^2 - \frac{2r\sigma_y}{\sigma_x}(x - \bar{x})(y - \bar{y})\right\}\right]^{1/2}$$

$$= \left[\sum \frac{(y-\bar{y})^2}{n} + r^2 \frac{\sigma_y^2}{\sigma_x^2} \sum \frac{(x-\bar{x})^2}{n} - 2r\frac{\sigma_y}{\sigma_x} \sum \frac{(x-\bar{x})(y-\bar{y})}{n}\right]^{1/2}$$

$$= \left[\sigma_y^2 + r^2 \frac{\sigma_y^2}{\sigma_x^2} . \sigma_x^2 - 2r\frac{\sigma_y}{\sigma_x} r . \sigma_x . \sigma_y\right]^{1/2}$$

$$= \left[\sigma_y^2 + r^2\sigma_y^2 - 2r^2\sigma_y^2\right]^{1/2} = \left[\sigma_y^2 - r^2\sigma_y^2\right]^{1/2}$$

$$= \sigma_y\sqrt{1-r^2}$$ **Proved.**

(*ii*) Similarly (*ii*) may be proved.

**Example 32.** *Find the standard error of estimate of y on x for the data given below:*

| | | | | | | | | |
|---|---|---|---|---|---|---|---|---|
| x | 1 | 3 | 4 | 6 | 8 | 9 | 11 | 14 |
| y | 1 | 2 | 4 | 4 | 5 | 7 | 8 | 9 |

**Solution.** The equation of the line of regression of *y* on *x* is

$$y = \frac{7}{11}x + \frac{6}{11}. \quad \text{So} \quad y_r = \frac{7x}{11} + \frac{6}{11}$$ (*See Example 28 on Page 757*)

| S.No. | x | y | $y_r$ | $(y-y_r)$ | $(y-y_r)^2$ |
|---|---|---|---|---|---|
| 1 | 1 | 1 | $\frac{13}{11}$ | $-\frac{2}{11}$ | $\frac{4}{121}$ |
| 2 | 3 | 2 | $\frac{27}{11}$ | $-\frac{5}{11}$ | $\frac{25}{121}$ |
| 3 | 4 | 4 | $\frac{34}{11}$ | $\frac{10}{11}$ | $\frac{100}{121}$ |
| 4 | 6 | 4 | $\frac{48}{11}$ | $-\frac{4}{11}$ | $\frac{16}{121}$ |
| 5 | 8 | 5 | $\frac{62}{11}$ | $-\frac{7}{11}$ | $\frac{49}{121}$ |
| 6 | 9 | 7 | $\frac{69}{11}$ | $\frac{8}{11}$ | $\frac{64}{121}$ |
| 7 | 11 | 8 | $\frac{83}{11}$ | $\frac{5}{11}$ | $\frac{25}{121}$ |
| 8 | 14 | 9 | $\frac{104}{11}$ | $-\frac{5}{11}$ | $\frac{25}{121}$ |
| | | | | | $\Sigma(y-y_r)^2 = \frac{308}{121}$ |

$$E_{yx} = \sqrt{\frac{\sum(y-y_r)^2}{n}} = \sqrt{\frac{308}{121 \times 8}} = \sqrt{\frac{7}{22}} = 0.564$$ **Ans.**

## Exercise 10.2

**1. Find the coefficient of correlation between *x* and *y* from the table of their values :**

| | | | | | | | | |
|---|---|---|---|---|---|---|---|---|
| x | 1 | 3 | 4 | 6 | 8 | 9 | 11 | 14 |
| y | 1 | 2 | 4 | 4 | 5 | 7 | 8 | 9 |

**Ans.0.977.**

**2.** Find the coefficient of correlation of the following data taking new origin of $x$ at 70 and for $y$ at 67.

| | | | | | | | | |
|---|---|---|---|---|---|---|---|---|
| $x$ | 67 | 68 | 64 | 68 | 72 | 70 | 69 | 70 |
| $y$ | 65 | 66 | 67 | 67 | 68 | 69 | 71 | 73 |

(*AMIE winter 2002*) **Ans.** 0.472

**3.** $x$ and $y$ are two random variables with the same standard deviation and correlation coefficient $r$. Show that the coefficient of correlation between $x$ and $x + y$ is $\sqrt{\frac{1+r}{2}}$

**4.** Find the regression line of $y$ on $x$ for the data :

| | | | | | |
|---|---|---|---|---|---|
| $x$ | 1 | 4 | 2 | 3 | 5 |
| $y$ | 3 | 1 | 2 | 5 | 4 |

**Ans.** $y = 2.7 + 0.1x$

**5.** Find the correlation coefficient and the equations of regression lines from the following data :

| | | | | | |
|---|---|---|---|---|---|
| $x$ | 1 | 2 | 3 | 4 | 5 |
| $y$ | 2 | 5 | 3 | 8 | 7 |

**Ans.** $r = 0.81, x = 0.5y + 0.5,\ y = 1.3x + 1.1$

**6.** Find the regression line of $y$ on $x$ if

| | | | | | | | | | | |
|---|---|---|---|---|---|---|---|---|---|---|
| $x$ | 40 | 70 | 50 | 60 | 80 | 50 | 90 | 40 | 60 | 60 |
| $y$ | 2.5 | 6.0 | 4.5 | 5.0 | 4.5 | 2.0 | 5.5 | 3.0 | 4.5 | 3.0 |

**Ans.** $y = 0.55 + 0.0583\,x$

**7.** The following marks have been obtained by a class of students in statistics.

| | | | | | | | | | | | |
|---|---|---|---|---|---|---|---|---|---|---|---|
| Paper I | 80 | 45 | 55 | 56 | 58 | 60 | 65 | 68 | 70 | 75 | 85 |
| Paper II | 81 | 56 | 50 | 48 | 60 | 62 | 64 | 65 | 70 | 74 | 90 |

Compute the coefficient of correlation for the above data. Find the lines of regression.

**Ans.** $r = 0.918,\ y - 65.45 = 0.981\,(x - 65.18)$
$x - 65.18 = 0.859\,(y - 65.45)$

**8.** Find the equations to the lines of regression and the coefficient of correlation for the following data:

| | | | | | | |
|---|---|---|---|---|---|---|
| $x$ | 2 | 4 | 5 | 6 | 8 | 11 |
| $y$ | 18 | 12 | 10 | 8 | 7 | 5 |

**Ans.** $y - 10 = -1.34\,(x - 6), x - 6 = -0.632\,(y - 10), r = -0.92$

**9.** The following results were obtained from lineups in Applied Mechanics and Engineering Mathematics in an examination :

| | *Applied Mechanics* ($x$) | *Engg. Maths.* ($y$) |
|---|---|---|
| Mean | 47.5 | 10.5 |
| Standard deviation | 16.8 | 10.8 |

$r = 0.95$

Find both the regresssion equations. Also estimate the value of $y$ for $x = 30$.

**Ans.** $y = 0.611x + 10.5, x = 1.478y - 1.143, y = 28.83$

**10.** The following results were obtained from records of age ($x$) and systolic blood pressure ($y$) of a group of 10 men:

| | $x$ | $y$ |
|---|---|---|
| Mean | 53 | 142 |
| Variance | 130 | 165 |

and $\sum(x-\bar{x})(y-\bar{y}) = 1220$

Find the appropriate regression equation and use it to estimate the blood pressure of a man whose age is 45. **Ans.** $y = 0.94\,x + 92.26$, Blood pressure $= 134.56$

**11.** The regression equation are : $7x - 16y + 9 = 0$, $5y - 4x - 3 = 0$ find $\bar{x}, \bar{y}$ and $r$

(*AMIE, Winter 2003*) **Ans.** $\bar{x} = -\dfrac{3}{29}, \bar{y} = \dfrac{15}{19}, r = \dfrac{3}{4}$

**12.** If two regression coefficients are 0.8 and 0.2, what would be the value of coefficient of correlation? **Ans.** $r = 0.4$

**13.** In a partially destroyed laboratory record of an analysis of corrélation data, the following results only are legible :

Variance of $x = 9$

Regression equations: $8x - 10y + 66 = 0$, $40x - 18y - 214 = 0$.

What were (*a*) the mean values of $x$ and $y$, (*b*) the standard deviation of $y$, and (*c*) the coefficient of correlation between $x$ and $y$. (*A.M.I.E, .Summer 2001 2002*)

**Ans.** $\bar{x} = 13, \bar{y} = 17, y = 0.8x + 6.6, x = 0.45y - 5.35, r = 0.6, \sigma_y = 4.$

**14.** The following regression equations and variances are obtained from a correlation table :

$20x - 9y - 107 = 0, 4x - 5y + 33 = 0,$ variance of $x = 9$.

Find (*i*) the mean values of $x$ and $y$, (*ii*) the standard deviation of y. **Ans.** $\bar{x} = 13, \bar{y} = 17, \sigma_y = 4.$

(*A.M.I.E., Winter 2000*)

**15.** Two random variables have the least square regression lines with equations $3x + 2y = 26$ and $6x + y = 31$. Find mean values and correlation coefficient between $x$ and $y$.

**Ans.** $\bar{x} = 4, \bar{y} = 7, r = -0.5$

**16.** Find the Standard error of estimate of $y$ on $x$ for the data given below **Ans.** 1.349

**17.** Fill in the blanks :

(*a*) The correlation coefficient is the .............mean between the regression coefficients.

(*b*) The lines of regression always pass through a point ..............

(*c*) Arithmetic mean of the coefficients of regressions is ............than the coefficient of correltion.

| $x$ | 1 | 2 | 3 | 4 | 5 |
|---|---|---|---|---|---|
| $y$ | 2 | 5 | 3 | 8 | 7 |

(*A.M.I.E., Summer 2000*)

(*d*) The value of coefficient of correlation lies between .........and...........

(*e*) If the two regression lines are perpendicular to each other, then the coefficient of correlation is equal to..................

(*f*) If two regression coefficients are, – 0.1 and – 0.9, the value of r is ............

(*g*) The normal equations for fitting a curve of the form $y = a + bx + cx^2$ are ................,and.............

(*h*) If $r_1$ and $r_2$ are two regression coefficients, then signs of $r_1$, $r_2$ depend on ............

(*i*) If coefficient of correlation $r = 0$, the two lines of regression are..................

(*j*) If two regression lines concide then the coefficient of correlation is .........

(*A.M.I.E., Winter 2000*)

**Ans.** (*a*) geometric, (*b*) $(\bar{x}, \bar{y},)$ (*c*) greater, (*d*) –1 and 1, (*e*) 0, (*f*) – 0.3,

(*g*) $\Sigma y = na + b\Sigma x + c\,\Sigma x^2, \Sigma x y = a\Sigma x + b\Sigma x^2 + c\Sigma x^3$ and $\Sigma x^2 y = a\Sigma x^2 + b\Sigma x^3 + c\,\Sigma x^4$.

(*h*) Coefficient of regression, (*i*) perpendicular, (*j*) ±1

# 11

# Probability

## 11.1 PROBABILITY

Probability is a concept which numerically measure the degree of uncertainty and therefore, of certainity of the occurrence of events.

If an event $A$ can happen in $m$ ways, and fail in $n$ ways, all these ways being equally likely to occur, then the probability of the happening of $A$ is

$$= \frac{\text{Number of favourable cases}}{\text{Total number of mutually exclusive and equally likely cases}} = \frac{m}{m+n}$$

and that of its failing is defined as $\dfrac{n}{m+n}$

If the probability of the happening $= p$

and the probability of not happening $= q$

then $\quad p+q = \dfrac{m}{m+n} + \dfrac{n}{m+n} = \dfrac{m+n}{m+n} = 1$ or $p + q = 1$.

For instance, on tossing a coin, the probability of getting a head is $\dfrac{1}{2}$.

## 11.2 DEFINITIONS

1. **Die :** It is a small cube. Dots are . .. ... :: :·: ::: marked on its faces. Plural of the die is dice. On throwing a die, the outcome is the number of dots on its upper face.
2. **Cards :** A pack of cards consists of four suits *i.e.* Spades, Hearts, Diamonds and Clubs. Each suit consists of 13 cards, nine cards numbered 2, 3, 4, ..., 10, an Ace, a King, a Queen and a Jack or Knave. Colour of Spades and Clubs is black and that of Hearts and Diamonds is red. Kings, Queens and Jacks are known as *face* cards.
3. **Exhaustive Events or Sample Space :** The set of all possible outcomes of a single performance of an experiment is exhaustive events or sample space. Each outcome is called a sample point. In case of tossing a coin once, $S = (H, T)$ is the *sample space.* Two outcomes - Head and Tail - constitute an exhaustive event because no other outcome is possible.
4. **Random Experiment :** There are experiments, in which results may be altogether different, even though they are performed under identical conditions. They are known as random experiments. Tossing a coin or throwing a die is random experiment.
5. **Trial and Event :** Performing a random experiment is called a *trial* and outcome is termed as *event.* Tossing of a coin is a trial and the turning up of head or tail is an event.
6. **Equally likely events :** Two events are said to be *'equally likely'*, if one of them cannot be expected in preference to the other. For instance, if we draw a card from well-shuffled pack, we may get any card, then the 52 different cases are equally likely.
7. **Independent events :** Two events may be *independent,* when the actual happening of one does not influence in any way the probability of the happening of the other.

**Example.** The event of getting head on first coin and the event of getting tail on the second coin in a simultaneous throw of two coins are independent.

8. **Mutually Exclusive events :** Two events are known as *mutually exclusive,* when the occurrence of one of them excludes the occurrence of the other. For example, on tossing of a coin, either we get head or tail, but not both.
9. **Compound Event :** When two or more events occur in composition with each other, the simultaneous occurrence is called a compound event. When a die is thrown, getting a 5 or 6 is a compound event.
10. **Favourable Events :** The events, which ensure the required happening, are said to be favourable events. For example, in throwing a die, to have the even numbers, 2, 4 and 6 are favourable cases.
11. **Conditional Probability :** The probability of happening an event *A*, such that event *B* has already happened, is called the conditional probability of happening of *A* on the condition that *B* has already happened. It is usually denoted by $P(A/B)$.
12. **Odds in favour of an event and odds against an event**

    If number of favourable ways = $m$, number of not favourable events = $n$

    (*i*) Odds in favour of the event = $\frac{m}{n}$, (*ii*) Odds against the event = $\frac{n}{m}$.
13. **Classical Definition of Probability.** If there are $N$ equally likely, mutually, exclusive and exhaustive of events of an experiment and m of these are favourable, then the probability of the happening of the event is defined as $\frac{m}{N}$.
14. **Expected value.** If $p_1, p_2, p_3 \ldots p_n$ of the probabilities of the events $x_1, x_2, x_3 \ldots x_n$ respectively then expected value

$$E(x) = p_1 x_1 + p_2 x_2 + p_3 x_3 + \ldots + p_n x_n = \sum_{r=1}^{n} p_r x_r$$

**Example 1.** *Find the probability of throwing (a) 5, (b) an even number with an ordinary six faced die.*

**Solution.** (*a*) There are 6 possible ways in which the die can fall and there is only one way of throwing 5.

$$\text{Probability} = \frac{\text{Number of favourable ways}}{\text{Total number of equally likely ways}} = \frac{1}{6} \qquad \textbf{Ans.}$$

(*b*) Total number of ways of throwing a die = 6

Number of ways falling 2, 4, 6 = 3

$$\text{The required probability} = \frac{3}{6} = \frac{1}{2} \qquad \textbf{Ans.}$$

**Example 2.** *Find the probability of throwing 9 with two dice.*

**Solution.** Total number of possible ways of throwing two dice

$= 6 \times 6 = 36.$

Number of ways getting 9 *i.e.*, $(3+6), (4+5), (5+4), (6+3) = 4$.

$$\therefore \quad \text{The required probability} = \frac{4}{36} = \frac{1}{9} \qquad \textbf{Ans.}$$

**Example 3.** *From a pack of 52 cards, one is drawn at random. Find the probability of getting a king.*

**Solution.** A king can be chosen in 4 ways. But a card can be drawn in 52 ways.

$$\therefore \quad \text{The required probability} = \frac{4}{52} = \frac{1}{13} \qquad \textbf{Ans.}$$

## Exercise 11.1

1. In a class of 12 students, 5 are boys and the rest are girls. Find the probability that a student selected will be a girl. **Ans.** $\frac{7}{12}$

2. A bag contains 7 red and 8 black balls. Find the probability of drawing a red ball. **Ans.** $\frac{7}{15}$

3. Three of the six vertices of a regular hexagon are choosen at random. Find the probability that the triangle with three vertices is equilateral. **Ans.** $\frac{1}{10}$

4. What is the probability that a leap year, selected at random, will contain 53 Sundays.
*(A.M.I.E.T.E., Summer 2002, 2001)* **Ans.** $\frac{2}{7}$

5. **Choose the correct answer :**
(a) In solving any problem, odds against A are 4 to 3 and odds in favour of B in solving the same problem are 7 to 5. The probability that the problem will be solved is
(i) $\frac{5}{21}$ (*ii*) $\frac{16}{21}$ (*iii*) $\frac{15}{84}$ (*iv*) $\frac{69}{84}$ *(A.M.I.E.T.E., Winter 2003)* **Ans.** (*ii*)
(b) In a given race, the odds in favour of horses A, *B, C, D* are 1 : 3, 1 : 4, 1 : 5, 1 : 6 respectively. The probability that horse *C* wins the race is
(*i*) $\frac{1}{4}$ (*ii*) $\frac{1}{5}$ (*iii*) $\frac{1}{6}$ (*iv*) $\frac{1}{7}$ **Ans.** (*iii*)
(c) In tossing a fair die, the probability of getting an odd number or a number less than 4 is
(*i*) 2 (*ii*) 1/2 (*iii*) 2/3 (*iv*) 3/4 **Ans.** (*iii*)
(d) An unbiased coin is tossed 3 times. The probability of obtaining two heads is
(*i*) $\frac{1}{2}$ (*ii*) $\frac{3}{8}$ (*iii*) 1 (*iv*) $\frac{1}{8}$ *(A.M.I.E.T.E., Winter 2002)* **Ans.** (*ii*)

6. Fill in the blanks with appropriate correct answer
(*a*) Chance of throwing 6 at least once in four throws with single die is .........
*(A.M.I.E., Summer 2000)* **Ans.** $\frac{671}{1296}$
(*b*) A pair of fair dice is thrown and one die shows a four. The probability that the other die shows 5 is ..
*(A.M.I.E., Summer 2000)* **Ans.** $\frac{1}{36}$

## 11.3 ADDITION LAW OF PROBABILITY

If $p_1, p_2, ......, p_n$ be separate probabilities of mutually exclusive events, then the probability *P*, that any of these events will happen is given by $P = p_1 + p_2 + p_3 + ...... p_n$

**Proof.** Let *A, B, C,* ...... be the events, where probabilities are respectively $p_1, p_2, ...... p_n$. Let *n* be the total number of favourable cases to either *A* or *B* or C or......

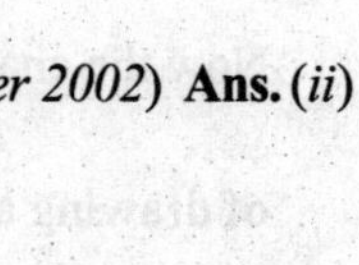


$$= m_1 + m_2 + m_3 + ....... + m_n$$

Hence
$$P(A+B+C...) = \frac{m_1 + m_2 + m_3 + ...... + m_n}{n}$$
$$= \frac{m_1}{n} + \frac{m_2}{n} + \frac{m_3}{n} + ...... + \frac{m_n}{n}$$
$$= P(A) + P(B) + P(C) + ....$$
$$P = p_1 + p_2 + p_3 + ...... + p_n$$
**Proved.**

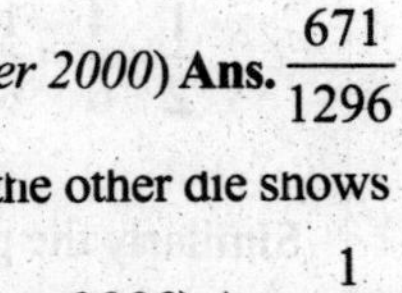

**NOT MUTUALLY EXCLUSIVE EVENTS**

Consider the case where two events $A$ and $B$ are not mutually exclusive. The probability of the event that either $A$ or $B$ or both occur is given as

$P(A \cup B) = P(A) + P(B) - P(A \cap B)$

**Example 4.** *An urn contains 10 black and 10 white balls. Find the probability of drawing two balls of the same colour.*

**Solution.** Probability of drawing two black balls $= \dfrac{{}^{10}C_2}{{}^{20}C_2}$

$\therefore$ Probability of drawing two red balls $= \dfrac{{}^{10}C_2}{{}^{20}C_2}$

$\therefore$ Probability of drawing two balls of the same colour

$$= \frac{{}^{10}C_2}{{}^{20}C_2} + \frac{{}^{10}C_2}{{}^{20}C_2} = 2.\frac{{}^{10}C_2}{{}^{20}C_2} = 2.\frac{\frac{10\times 9}{2\times 1}}{\frac{20\times 19}{2\times 1}}$$

$$= \frac{9}{19}$$ **Ans.**

**Example 5.** *A bag contains four white and two black balls and a second bag contains three of each colour. A bag is selected at random, and a ball is then drawn at random from the bag chosen. What is the probability that the ball drawn is white?*

**Solution.** There are two mutually exclusive cases,

(*i*) when the first bag is chosen, (*ii*) when the second bag is chosen.

Now the chance of choosing the first bag is $\frac{1}{2}$ and if this bag is chosen, the probability of drawing a white ball is 4/6. Hence the probability of drawing a white ball from first bag is

$$\frac{1}{2}\times\frac{4}{6} = \frac{1}{3}$$

Similarly the probability of drawing a white ball from second bag is $\frac{1}{2}\times\frac{3}{6} = \frac{1}{4}$

Since the events are mutually exclusive the required probability

$$= \frac{1}{3} + \frac{1}{4} = \frac{7}{12}$$ **Ans.**

**Example 6.** *Three machines I, II and III manufacture respectively 0.4, 0.5 and 0.1 of the total production. The percentage of defective items produced by I, II and III is 2, 4 and 1 per cent respectively. For an item chosen at random, what is the probability it is defective?*

**Solution.** The defective item produced by machine I $= \dfrac{0.4\times 2}{100} = \dfrac{0.8}{100}$

The defective item produced by machine II $= \dfrac{0.5\times 4}{100} = \dfrac{2}{100}$

The defective item produced by machine III $= \dfrac{0.1\times 1}{100} = \dfrac{0.1}{100}$

$$= \frac{0.8}{100} + \frac{2}{100} + \frac{0.1}{100} = \frac{2.9}{100} = 0.029$$

The required probability $= \frac{0.029}{1} = 0.029$ **Ans.**

## 11.4 MULTIPLICATION LAW OF PROBABILITY

**If there are two independent events the respective probabilities of which are known, then the probability that both will happen is the product of the probabilities of their happening respectively.**

$$P(AB) = P(A) \times P(B)$$

**Proof.** Suppose $A$ and $B$ are two independent events. Let $A$ happen in $m_1$ ways and fail in $n_1$ ways.

$$\therefore \quad P(A) = \frac{m_1}{m_1 + n_1}$$

Also let $B$ happen in $m_2$ ways and fail in $n_2$ ways.

$$P(B) = \frac{m_2}{m_2 + n_2}$$

Now there are four possibilities

$A$ and $B$ both may happen, then the number of ways $= m_1 \cdot m_2$.
$A$ may happen and $B$ may fail, then the number of ways $= m_1 . n_2$.
$A$ may fail and $B$ may happen, then the number of ways $= n_1 . m_2$.
$A$ and $B$ both may fail, then the number of ways $= n_1 . n_2$.

Thus, the total number of ways $= m_1 m_2 + m_1 n_2 + n_1 m_2 + n_1 n_2$

$$= (m_1 + n_1)(m_2 + n_2)$$

Hence the probabilities of the happening of both $A$ and $B$

$$P(AB) = \frac{m_1 m_2}{(m_1 + n_1)(m_2 + n_2)} = \frac{m_1}{m_1 + n_1} \cdot \frac{m_2}{m_2 + n_2}$$

$$= P(A) . P(B)$$ **Proved.**

**Example 7.** *An article manufactured by a company consists of two parts A and B. In the process of manufacture of part A, 9 out of 100 are likely to be defective. Similarly, 5 out of 100 are likely to be defective in the manufacture of part B. Calculate the probability that the assembled article will not be defective (assuming that the events of finding the part A non-defective and that of B are independent).*

**Solution.** Probability that part $A$ will be defective $= \frac{9}{100}$

Probability that part $A$ will not be defective $= \left(1 - \frac{9}{100}\right) = \frac{91}{100}$

Probability that part $B$ will be defective $= \frac{5}{100}$

Probability that part $B$ will not be defective $= \left(1 - \frac{5}{100}\right) = \frac{95}{100}$

Probability that the assembled article will not be defective = (Probability that part A will not be defective) × (Probability that part $B$ will not be defective)

$$= \left(\frac{91}{100}\right) \times \left(\frac{95}{100}\right) = 0.8645$$ **Ans.**

**Example 8.** *The probability that machine A will be performing an usual function in 5 years' time is $\frac{1}{4}$, while the probability that machine B will still be operating usefully at the end of the same period is $\frac{1}{3}$.*

*Find the probability in the following cases that in 5 years time:*

(*i*) *Both machines will be performing an usual function.*

(*ii*) *Neither will be operating.*

(*iii*) *Only machine B will be operating.*

(*iv*) *At least one of the machines will be operating.*

**Solution.** $P$ (*A* operating usefully) $= \frac{1}{4}$, $\quad q(A) = 1 - \frac{1}{4} = \frac{3}{4}$

$P$ (*B* operating usefully) $= \frac{1}{3}$ $\quad$ so $q(B) = 1 - \frac{1}{3} = \frac{2}{3}$

(*i*) $P$ (Both *A* and *B* will operate usefully) $= P(A) . P(B) = \left(\frac{1}{4}\right) \times \left(\frac{1}{3}\right) = \frac{1}{12}$

(*ii*) $P$ (Neither will be operating) $= q(A)\ q(B) = \left(\frac{3}{4}\right)\left(\frac{2}{3}\right) = \frac{1}{2}$

(*iii*) $P$ (Only *B* will be operating) $= p(B) \times q(A) = \left(\frac{1}{3}\right) \times \left(\frac{3}{4}\right) = \frac{1}{4}$

(*iv*) $P$ (At least one of the machines will be operating)

$= 1 - P$ (none of them operates)

$= 1 - \frac{1}{2} = \frac{1}{2}$ **Ans.**

**Example 9.** *There are two groups of subjects one of which consists of 5 science and 3 engineering subjects and the other consists of 3 science and 5 engineering subjects. An unbiased die is cast. If number 3 or number 5 turns up, a subject is selected at random from the first group, otherwise the subject is selected at random from the second group. Find the probability that an engineering subject is selected ultimately.* (*A.M.I.E.T.E., Summer 2000*)

**Solution.** Probability of turning up 3 or 5 $= \frac{2}{6} = \frac{1}{3}$

Probability of selecting engineering subject from first group $= \frac{3}{8}$

Now the probability of selecting engineering subject from first group on turning up 3 or 5

$$= \left(\frac{1}{3}\right) \times \left(\frac{3}{8}\right) = \frac{1}{8} \qquad \text{... (1)}$$

Probability of not turning up 3 or 5 $= 1 - \frac{1}{3} = \frac{2}{3}$

Probability of selecting engineering subject from second group $= \frac{5}{8}$

Now Probability of the selection of engineering subject from second group on not turning up 3 or 5

$$= \frac{2}{3} \times \frac{5}{8} = \frac{5}{12} \qquad \text{...(2)}$$

Probability of the selection of engineering subject $= \frac{1}{8} + \frac{5}{12}$ [From (1) and (2)]

$$= \frac{13}{24}$$ **Ans**

**Example 10.** *An urn contains nine balls, two of which are red three blue and four black. Three balls are drawn from the urn at random. What is the probability that*

(*i*) *the three balls are of different colours?*

(*ii*) *the three balls are of the same colour?* (*A.M.I.E., Summer 2000*)

**Solution.**

Urn contains 2 Red balls, 3 Blue balls and 4 Black balls.

(*i*) Three balls will be of different colours if one ball is red, one blue and one black ball are drawn.

$$\text{Required probability} = \frac{{}^2C_1 \times {}^3C_1 \times {}^4C_1}{{}^9C_3} = \frac{2 \times 3 \times 4}{84} = \frac{2}{7}$$ **Ans.**

(*ii*) Three balls will be of the same colour if either 3 blue balls or 3 black balls are drawn.

$P$ (3 Blue balls or 3 Black balls) = $P$ (3 Blue balls) + $P$ (3 Black balls)

$$= \frac{{}^3C_3}{{}^9C_3} + \frac{{}^4C_3}{{}^9C_3} = \frac{1+4}{84} = \frac{5}{84}$$ **Ans.**

**Example 11.** *An urn A contains 2 white and 4 black balls. Another urn B contains 5 white and 7 black balls. A ball is transferred from the urn A to the urn B, then a ball is drawn from urn B. Find the probability that it is white.*

**Solution.** Urn $A$ contains 2 white and 4 black balls.

Urn $B$ contains 5 white and 7 black balls.

Now there are two cases of transferring a ball from $A$ to $B$.

**Case I.** When a white ball is transferred from $A$ to $B$

$$P\text{ (Transfer of a white ball)} = \frac{2}{2+4} = \frac{1}{3}$$

After transfer of a white ball, urn $B$ contains 6 white balls and 7 black balls.

$P$ (Drawing a white ball from urn $B$ after transfer)

$$= P\text{ (Transfer of a white ball)} \times P\text{ (Drawing of a white ball)}$$

$$= \left(\frac{1}{3}\right)\left(\frac{6}{6+7}\right) = \frac{1}{3} \times \frac{6}{13} = \frac{2}{13}$$

**Case II.** When a black ball is transferred from $A$ to $B$.

$$P\text{ (Transfer of a black ball)} = \frac{4}{2+4} = \frac{2}{3}$$

After transfer of a black ball, urn $B$ contains 5 white and 8 black balls.

$P$ (Drawing a white ball from urn $B$ after transfer)

$$= P\text{ (Transfer of a black ball)} \times P\text{ (Drawing of a white ball)}$$

$$\text{Required probability} = \frac{2}{13} + \frac{10}{39} = \frac{16}{39}$$ **Ans.**

**Example 12.** *A bag contains 10 white and 15 black balls. Two balls are drawn in succession. What is the probability that first is white and second is black ?*

**Solution.** Probability of drawing one white ball = $\frac{10}{25}$

Probability of drawing one black ball without replacement = 15/24

Required probability of drawing first white ball and second black ball = $\frac{10}{25} \times \frac{15}{24} = \frac{1}{4}$ **Ans.**

**Example 13.** *A committee is to be formed by choosing two boys and four girls out of a group of five boys and six girls. What is the probability that a particular boy named A and a particular girl named B are selected in the committee?*

**Solution.** Two boys are to be selected out of 5 boys. A particular boy A is to be included in the committee. It means that only 1 boy is to be selected out of 4 boys.

Number of ways of selection = ${}^4C_1$

Similarly a girl B is to be included in the committee.

Then only 3 girls are to be selected out of 5 girls.

Number of ways of selection = ${}^5C_3$

$$\text{Required probability} = \frac{{}^4C_1 \times {}^5C_3}{{}^5C_2 + {}^6C_4} + = \frac{4 \times 10}{10 \times 15} = \frac{4}{15}$$ **Ans**

**Example 14.** *Three groups of children contain respectively 3 girls and 1 boy; 2 girls and 2 boys; 1 girl and 3 boys. One child is selected at random from each group. Find the chance of selecting 1 girl and 2 boys.*

**Solution.** There are three ways of selecting 1 girl and two boys.

**I way :** Girl is selected from first group, boy from second group and second boy from third group.

$$\text{Probability of the selection of (Girl + Boy + Boy)} = \frac{3}{4} \times \frac{2}{4} \times \frac{3}{4} = \frac{18}{64}$$

**II way :** Boy is selected from first group, girl from second group and second boy from third group.

$$\text{Probability of the selection of (Boy + Girl + Boy)} = \frac{1}{4} \times \frac{2}{4} \times \frac{3}{4} = \frac{6}{64}$$

**III way :** Boy is selected from first group, second boy from second group and the girl from the third group.

$$\text{Probability of selection of (Boy + Boy + Girl)} = \frac{1}{4} \times \frac{2}{4} \times \frac{1}{4} = \frac{2}{64}$$

$$\text{Total probability} = \frac{18}{64} + \frac{6}{64} + \frac{2}{64} = \frac{26}{64} = \frac{13}{32}$$ **Ans.**

**Example 15.** *The number of children in a family in a region are either 0, 1 or 2 with probability 0.2, 0.3 and 0.5 respectively. The probability of each child being a boy or girl 0.5. Find the probability that a family has no boy.*

**Solution.** Here there are three types of families

(*i*) Probability of zero child (boys) = 0.2

(*ii*)

| *Boy* | *Girl* |
|---|---|
| 0 | 1 |
| 1 | 0 |

Probability of zero boy in case II

$$= 0.3 \times 0.5 = 0.15$$

(*iii*)

| *Boy* | *Girl* |
|---|---|
| 0 | 2 |
| 1 | 1 |
| 2 | 0 |

In this case probability of zero boy $= 0.5 \times \frac{1}{3} = 0.167$

Considering all the three cases, the probability of zero boy

$$= 0.2 + 0.15 + 0.167 = 0.517$$ **Ans.**

**Example 16.** *A husband and wife appear in an interview for two vacancies in the same post. The probability of husband's selection is* $\frac{1}{7}$ *and that of wife 's selection is* $\frac{1}{5}$. *What is the probability that*

*(i) both of them will be selected. (ii) only one of them will be selected and*
*(iii) none of them will be selected ?*

**Solution.** $P$ (husband's selection) $= \frac{1}{7}$, $P$ (wife's selection) $= \frac{1}{5}$

(*i*) $P$ (both selected) $= \frac{1}{7} \times \frac{1}{5} = \frac{1}{35}$

(*ii*) $P$ (only one selected) = $P$ (only husband's selection) + $P$ (only wife's selection)

$$= \frac{1}{7} \times \frac{4}{5} + \frac{1}{5} \times \frac{6}{7} = \frac{10}{35} = \frac{2}{7}$$

(*iii*) $P$ (none of them will be selected) $= \frac{6}{7} \times \frac{4}{5} = \frac{24}{35}$ **Ans.**

**Example 17.** *A problem of statistics is given to three students A, B and C whose chances of solving it are* $\frac{1}{2}, \frac{3}{4}$ *and* $\frac{1}{4}$ *respectively. What is the probability that the problem will be solved?*

**(A.M.I.E., Winter 2001)**

**Solution.** The probability that $A$ can solve the problem $= \frac{1}{2}$

The probability that $A$ cannot solve the problem $= 1 - \frac{1}{2}$.

Similarly the probability that $B$ and C cannot solve the problem are $\left(1 - \frac{3}{4}\right)$ and $\left(1 - \frac{1}{4}\right)$

∴ The probability that $A, B, C$ cannot solve the problem

$$= \left(1 - \frac{1}{2}\right) \times \left(1 - \frac{3}{4}\right) \times \left(1 - \frac{1}{4}\right) = \frac{1}{2} \times \frac{1}{4} \times \frac{3}{4} = \frac{3}{32}.$$

Hence the probability that the problem can be solved

$$=1-\frac{3}{32}=\frac{29}{32}$$ **Ans.**

**Example 18.** *A student takes his examination in four subjects $\alpha, \beta, \gamma, \delta$. He estimates his chances of passing in $\alpha$ as $\frac{4}{5}$, in $\beta$ as $\frac{3}{4}$, in $\gamma$ as $\frac{5}{6}$ and in $\delta$ as $\frac{2}{3}$. To qualify, he must pass in $\alpha$ and at least two other subjects. What is the probability that he qualifies ?*

**Solution.** $P(\alpha)=\frac{4}{5}, P(\beta)=\frac{3}{4}, P(\gamma)=\frac{5}{6}, P(\delta)=\frac{2}{3}$.

There are four possibilities of passing at least two subjects

(*i*) Probability of passing $\beta, \gamma$ and failing $\delta$

$$=\frac{3}{4}\times\frac{5}{6}\times\left(1-\frac{2}{3}\right)=\frac{3}{4}\times\frac{5}{6}\times\frac{1}{3}=\frac{5}{24}$$

(*ii*) Probability of passing $\gamma$, $\delta$ and failing $\beta$

$$=\frac{5}{6}\times\frac{2}{3}\times\left(1-\frac{3}{4}\right)=\frac{5}{6}\times\frac{2}{3}\times\frac{1}{4}=\frac{5}{36}$$

(*iii*) Probability of passing $\delta$, $\beta$ and failing $\gamma$

$$=\frac{2}{3}\times\frac{3}{4}\times\left(1-\frac{5}{6}\right)=\frac{2}{3}\times\frac{3}{4}\times\frac{1}{6}=\frac{1}{12}$$

(*iv*) Probability of passing $\beta, \gamma, \delta$ $=\frac{3}{4}\times\frac{5}{6}\times\frac{2}{3}=\frac{5}{12}$.

Probability of passing at least two subjects

$$=\frac{5}{24}+\frac{5}{36}+\frac{1}{12}+\frac{5}{12}=\frac{61}{72}$$

Probability of passing $\alpha$ and at least two subjects

$$=\frac{4}{5}\times\frac{61}{72}=\frac{61}{90}$$ **Ans.**

**Example 19.** *There are 6 positive and 8 negative numbers. Four numbers are chosen at random, without replacement, and multiplied. What is the probability that the product is a positive number ?*

**Solution.** To get from the product of four numbers, a positive number, the possible combinations are as follows :

| *S. No.* | *Out of 6 Positive Numbers* | *Out of 8 Negative Numbers* | *Positive Numbers* |
|---|---|---|---|
| 1. | 4 | 0 | ${}^6C_4\times{}^8C_0=\frac{6\times5}{1\times2}\times1=15$ |
| 2. | 2 | 2 | ${}^6C_2\times{}^8C_2=\frac{6\times5}{1\times2}\times\frac{8\times7}{1\times2}=420$ |
| 3. | 0 | 4 | ${}^6C_0\times{}^8C_4=1\times\frac{8\times7\times6\times5}{1\times2\times3\times4}=70$ |
| | | | Total = 505 |

$$\text{Probability} = \frac{{}^6C_4 \times {}^8C_0 + {}^6C_2 \times {}^8C_2 + {}^6C_0 \times {}^8C_4}{{}^{14}C_4}$$

$$= \frac{15+420+70}{\frac{14\times13\times12\times11}{1\times2\times3\times4}} = \frac{505\times4\times3\times2\times1}{14\times13\times12\times11} = \frac{505}{1001}$$ **Ans**

**Example 20.** *A six-faced die is so biased that, when thrown, it is twice as likely to show an even number than an odd number. If it is thrown twice, what is the probability that the sum of two numbers thrown is odd ?*

**Solution.** A biased die, when thrown, shows even number twice than an odd number.

$$\text{Probability of showing even number} = \frac{2}{2+1} = \frac{2}{3}$$

$$\text{Probability of showing odd number} = \frac{1}{2+1} = \frac{1}{3}$$

Sum of two numbers is odd if the first is even and the second is odd or vice versa.

Probability of sum to be odd = Probability of an even number × Probability of an odd number + Probability of an odd number × Probability of an even number.

$$= \frac{2}{3}\times\frac{1}{3}+\frac{1}{3}\times\frac{2}{3} = \frac{2}{9}+\frac{2}{9} = \frac{4}{9}$$ **Ans.**

**Example 21.** *A can hit a target 3 times in 5 shots, B 2 times in 5 shots and C three times in 4 shots. All of them fire one shot each simultaneously at the target. What is the probability that (i) 2 shots hit (ii) At least two shots hit?* *(A.M.I.E.T.E., Summer 2003)*

**Solution.** Probability of A hitting the target $= \frac{3}{5}$

Probability of B hitting the target $= \frac{2}{5}$

Probability of C hitting the target $= \frac{3}{4}$

Probability that 2 shots hit the target

$$= P(A)\,P(B)\,q(C) + P(A)\,P(C)\,q(B) + P(B)\,P(C)\,q(A)$$

$$= \frac{3}{5}\times\frac{2}{5}\times\left(1-\frac{3}{4}\right)+\frac{3}{5}\times\frac{3}{4}\times\left(1-\frac{2}{5}\right)+\frac{2}{5}\times\frac{3}{4}\times\left(1-\frac{3}{5}\right)$$

$$= \frac{6}{25}\times\frac{1}{4}+\frac{9}{20}\times\frac{3}{5}+\frac{6}{20}\times\frac{2}{5} = \frac{6+27+12}{100} = \frac{45}{100} = \frac{9}{20}$$ **Ans.**

(*ii*) Probability of at least two shots hitting the target

= Probability of 2 shots + probability of 3 shots hitting the target

$$= \frac{9}{20} + P(A)\,P(B)\;P(C) = \frac{9}{20}+\frac{3}{5}\times\frac{2}{5}\times\frac{3}{4} = \frac{63}{100}$$ **Ans.**

**Example 22.** *A and B take turns in throwing two dice, the first to throw 10 being awarded the prize. Show that if A has the first throw, their chances of winning are in the ratio* 12:11.

**Solution.** The combinations of throwing 10 from two dice can be (6+ 4), (4 + 6), (5 + 5).

The number of combinations is 3.

Total combinations from two dice = 6 × 6 = 36.

$\therefore$ The probability of throwing $10 = p = \dfrac{3}{36} = \dfrac{1}{12}$

The probability of not getting $10 = q = 1 - \left(\dfrac{1}{12}\right) = \dfrac{11}{12}$

If $A$ is to win, he should throw 10 in either the first, the third, the fifth, ... throws.

Their respective probabilities are $= p, q^2 p, q^4 p, \ldots\ldots = \dfrac{1}{12}, \left(\dfrac{11}{12}\right)^2 \dfrac{1}{12}, \left(\dfrac{11}{12}\right)^4 \dfrac{1}{12} \ldots$

A's total probability of winning $= \dfrac{1}{12} + \left(\dfrac{11}{12}\right)^2 \cdot \dfrac{1}{12} + \left(\dfrac{11}{12}\right)^4 \dfrac{1}{12} + \ldots$

$$= \frac{\frac{1}{12}}{1-\left(\frac{11}{12}\right)^2} = \frac{12}{23} \qquad \left[\text{This is infinite G.P. Its sum} = \frac{a}{1-r}\right]$$

$B$ can win in either 2nd, 4th, 6th ... throws.
So B's total chance of winning $= qp + q^3 p + q^5 p + \ldots\ldots$

$$= \left(\frac{11}{12}\right)\left(\frac{1}{12}\right) + \left(\frac{11}{12}\right)^3\left(\frac{1}{12}\right) + \left(\frac{11}{12}\right)^5\left(\frac{1}{12}\right) + \ldots = \frac{\left(\frac{11}{12}\right)\left(\frac{1}{12}\right)}{1-\left(\frac{11}{12}\right)^2} = \frac{11}{23}$$

Hence A's chance to $B$'s chance is $\dfrac{12}{23} : \dfrac{11}{23} = 12 : 11$. **Proved.**

**Example 24.** *A and B throw alternatively a pair of dice. A wins if he throws 6 before B throws 7 and B wins if he throws 7 before A throws 6. Find their respective chances of winning, if A begins.* *(A.M.I.E.T.E., Summer 2002)*

**Solution.** Number of ways of throwing 6

*i.e.* $(1+5), (2+4), (3+3), (4+2), (5+1) = 5.$

Probability of throwing $6 = \dfrac{5}{36} = p_1, \quad q_1 = \dfrac{31}{36}$

Number of ways of throwing 7.

*i.e.* $(1+6), (2+5), (3+4), (4+3), (5+2), (6+1) = 6$

Probability of throwing $6 = \dfrac{6}{36} = \dfrac{1}{6} = P_2, \quad q_2 = \dfrac{5}{6}$

$$P(A) = p_1 + q_1 q_2 p_1 + q_1^2 q_2^2 p_1 + \ldots\ldots$$

$$P(B) = q_1 p_2 + q_1^2 q_2 p_2 + q_1^3 q_2^2 p_2 + \ldots\ldots$$

Probability of A's winning $= p_1 + q_1 q_2 p_1 + q_1^2 q_2^2 p_1 + \ldots\ldots$

$$= \frac{p_1}{1 - q_1 q_2} = \frac{\frac{5}{36}}{1 - \frac{31}{36} \times \frac{5}{6}} = \frac{5}{36} \times \frac{36 \times 6}{61} = \frac{30}{61}$$

Probability of B's winning $= q_1 p_2 + q_1^2 q_2 p_2 + q_1^3 q_2^2 p_2 + \ldots$

$$= \frac{q_1 p_2}{1-q_1 q_2} = \frac{\frac{31}{36}\times\frac{1}{6}}{1-\left(\frac{31}{36}\right)\left(\frac{5}{6}\right)} = \frac{31}{36\times 6}\times\frac{36\times 6}{61} = \frac{31}{61}$$ **Ans.**

## Exercise 11.2.

1. The probability that Nirmal will solve a problem is $\frac{2}{3}$ and the probability that Satyajit will solve it is $\frac{3}{4}$.

   What is the probability that

   (*a*) the problem will be solved

   (*b*) neither can solve it. **Ans.** (*a*) $\frac{11}{12}$, (*b*) $\frac{1}{12}$

2. Two persons A and B toss an unbiased coin alternately on the understanding that the first who gets the head wins. If A starts the game, then what are their respective chances of winning ? (*A.M.I.E.T.E. Summer 2004*) **Ans.** 4 : 1

3. Four persons are chosen at random from a group containing 3 men, 2 women, and 4 children. Show that the probability that exactly two of them will be children is $\frac{10}{21}$.

4. A five digit number is formed by using the digits 0, 1, 2, 3, 4 and 5 without repetition. Find the probability that the number is divisible by 6. **Ans.** $\frac{4}{25}$

5. The chances that doctor *A* will diagnose a disease *X* correctly is 60%. The chances that a patient will die by his treatment after correct diagnosis is 40% and the chances of death by wrong diagnosis is 70%. A patient of doctor *A*, who had disease *X*, died, what is the chance that his disease was diagnosed correctly ? **Ans.** $\frac{6}{13}$

6. An anti-aircraft gun can take a maximum of four shots on enemy's plane moving from it. The probabilities of hitting the plane at first, second, third and fourth shots are 0.4, 0.3, 0.2 and 0.1 respectively. Find the probability that the gun hits the plane. **Ans.** 0.6976.

7. An electronic component consists of three parts. Each part has probability 0.99 of performing satisfactorily. The component fails if two or more parts do not perform satisfactorily. Assuming that the parts perform independently, determine the probability that the component does not perform satisfactorily. **Ans.** 0.000298

8. The face cards are removed from a full pack. Out of the remaining 40 cards, 4 are drawn at random. What is the probability that they belong to different suits ? **Ans.** $\frac{1000}{9139}$

9. Of the cigarette smoking population, 70% are men and 30% women, 10% of these men and 20% of these women smoke 'WILLS.' What is the probability that a person seen smoking a 'WILLS' will be a man. **Ans.** $\frac{7}{13}$

10. A machine contains a component C that is vital to its operation. The reliability of component C is 80%. To improve the reliability of a machine, a similar component is used in parallel to form a system *S*. The machine will work provided that one of these components functions correctly. Calculate the reliability of the system *S*. **Ans.** 96%

11. In a bolt factory, machines *A*, *B* and *C* manufacture 25%, 35% and 40% of the total output respectively. Of their outputs, 5%, 4% and 2% are defective bolts. A bolt is chosen at random and found to be defective. What is the probability that the bolt came from machine *A* ? *B* ? *C* ? **Ans.** $\frac{25}{69}, \frac{28}{69}, \frac{16}{69}$

**12.** One bag contains four white and two black beads and another contains three of each colour. A bead is drawn from each bag. What is the probability that one is white and one is black ? **Ans.** $\frac{1}{2}$.

**13.** The odds that a book will be favourably reviewed by three independent critics are 5 to 2, 4 to 3, 3 to 4 respectively. What is the probability that of the three reviews, a majority will be favourable ?

(*A.M.I.E., Summer 2004*) **Ans.** $\frac{209}{343}$

**14.** Let E and F be independent events. The probability that both E and F happen is $\frac{1}{12}$ and the probability that neither *E* nor *F* happen is $\frac{1}{2}$. Then find *P* (*E*) and P (*F*). **Ans.** $P(E) = \frac{1}{3}$, $P(F) = \frac{1}{4}$

**15.** Given a random variable whose range set is ( 1, 2) and whose probability is $f(1) = \frac{1}{4}$ and $f(2) = \frac{3}{4}$. Find the mean and variance of the distribution. **Ans.** Mean $= \frac{7}{4}$, Var $= \frac{3}{16}$

**16.** A man takes a step forward with probability 0.4 and backward with probability 0.6. Find the probability that at the end of 11 steps, he is just one step away from the starting point. **Ans.** 0.210677186

**17.** What would be the expectation of the number of failures preceding the first success in an infinite series of independent trials with the constant probability of success *p* ?

**Solution.** The probabilities of success in 1st, 2nd, 3rd trials respectively are $p, qp, q^2p, q^3p, ......$

The expected number of failures preceding the first success

$$E(x) = (0.p) + (1.qp) + (2.q^2 p) + ...... \infty = qp\,[1 + 2q + 3q^2 + ...... \infty\,] \text{ where } q < 1.$$

$$= \frac{qp}{(1-q)^2} = \frac{qp}{p^2} = \frac{q}{p}$$ **Ans.**

**18.** The probability of an airplane engine failure (independent of other engines) when the aircraft is in flight is (1-P). For a successful flight at least 50% of the airplane engines should remain operational. For which values of P would you prefer a four engine airplane to a two engine one

(*A.M.I.E.T.E., Dec. 2004*)

**19.** A person plays *m* independent games. The probability of his winning any game is $\frac{a}{a+b}$ (*a*, *b* are positive numbers). Show that probability that the person wins an odd number of games is $\frac{1}{2}\,[(b+a)^m - (b-a)^m]/(b+a)^m$.

**20.** Fill in the blanks :

(*a*) If the probabilities of *n* independent events are $p_1, p_2, p_3, ..., p_n$, then the probability that at least one of the event will happen is .............

(*b*) For a biased die, the probabilities for the different faces to turn up are given below :

| Face | 1 | 2 | 3 | 4 | 5 | 6 |
|---|---|---|---|---|---|---|
| Prob. | 0.1 | 0.32 | 0.21 | 0.15 | 0.05 | 0.17 |

The die is tossed and you are told that either face 1 or face 2 has turned up. Then the probability that it is face 1, is.....

(*c*) The probability of getting a ticket of number of multiple of 5 in a random draw from a bag containing tickets of even numbers from 1 to 100, is ............

(*d*) A town has two doctors X and *Y* operating independently. If the prob. the doctor X is available, is 0.9 and that for *Y* is 0.8, then the prob. that at least one doctor is available, when needed is .......

(*e*) From a pack of well shuffled cards, one card is drawn randomly. A gambler bets it as a diamond or a king. The odds in favour of his winning the bet are ...........

(*f*) From a pack of cards, 2 cards are drawn, the first being replaced before the second is drawn. The probability that the first is a diamond and the second is a king will be ............

(g) From an urn containing 12 white and 8 black balls two balls are drawn at random. The probability that both the balls will turn to be black is............

(*h*) A ball is taken out of a pot containing 6 white and 12 red balls. The probability that the ball is white is ............

(*i*) A speaks truth in 75% and B in 80% of the cases. The percentage of cases in which they likely to contradict each other narrating the same incident is ............

**Ans.** (*a*) $1-(1-p_1)(1-p_2)....(1-p_n)$, (b) $\frac{5}{21}$ (*c*) $\frac{1}{5}$ (*d*) 0.98, (*e*) 4 : 9,

(*f*) $\frac{1}{52}$, (g) $\frac{14}{95}$, (h) $\frac{1}{3}$, (*i*) 35 %

**21. Tick √ the correct answer :**

(*i*) The probability that at least one of the events A and B occurs is 0.8 and the probability that both the events occur simultaneously is 0.25. The probability $P(A) + P(B)$ is:
(*i*) 0.65 (*ii*) 0.75 (*iii*) 0.85 (*iv*) 0.95

(*ii*) A, B, C are independent events such that $P(A) = P(B)$ and probability that at least one of them happens is 1/2. The probability that A or B happens given that at least one of A, B, or C happens is $\frac{2}{9}$. Find $P(A)$ and P (C). **Ans.** $P(A) = 1-\frac{\sqrt{7}}{3}$ –, $P(c) = \frac{5}{14}$

(*iii*) An unbiased coin is tossed five times. Given that heads were obtained in two of the tosses, the probability that these were obtained in the first two tosses is
(*a*) 1/10 (*b*) 1/4 (*c*) 1/32 (*d*) None of these.

(*iv*) Groups are formed of 4 persons out of 12 persons. The probability that one particular person is never included is
(*a*) 2/3 (*b*) 1/3 (*c*) 1/4 (*d*) none of these

(*v*) 50 tickets are serially numbered 1 to 50. One ticket is drawn from these at random. The probability of its being a multiple of 3 or 4 is
(*a*) 12/25 (*b*) 14/25 (*c*) 2/5 (*d*) none of these

(*vi*) The probabilities of occurring of two events *E*, *F* are 0.25 and 0.5 respectively and of occurring both simultaneously is 0.14. Then the probability of the occurrence of the neither event is
(*a*) 0.61 (*b*) 0.39 (*c*) 0.89 (*d*) none of these

(*vii*) A bag contains 5 black and 4 white balls. Two balls are drawn at random. The probability that they match, is
(*a*) 7/12 (*b*) 5/8 (*c*) 5/9 (*d*) 4/9

(*viii*) A, B, C in order toss a coin. The first to throw a head wins. Assuming if A begins and the game continues indefinitely their respective chances of winning the games are:
(*a*) $\frac{4}{7}, \frac{2}{7}, \frac{1}{7}$ (*b*) $\frac{1}{7}, \frac{4}{7}, \frac{2}{7}$, (*c*) $\frac{2}{7}, \frac{4}{7}, \frac{1}{7}$ (*d*) None of these

(*ix*) A purse contains 4 copper coins, 3 silver coins, the second purse contains 6 copper coins and 2 silver coins. A coin is taken out of any purse, the probability that it is a copper coin is:
(*a*) 4/7 (*b*) 3/4 (*c*) 3/7 (*d*) 37/56

(*x*) In rolling two fair dice, the probability of getting equal numbers or numbers with an even product is
(*a*) 6/36 (*b*) 30/36 (*c*) 27/36 (*d*) 3/36

(*xi*) One of the two events must occur. If the chance of one is 2/3 of the other, then odds in favour of the other are
(*a*) 1 : 3 (*b*) 2 : 3 (*c*) 3 : 1 (*d*) none of these

(*xii*) The probability that a certain beginner at golf gets a good shot if he uses the correct club is 1/3, and the probability of a good shot with an incorrect club is 1/4. In his bag are 5 different clubs, only one of which is correct for the shot in question. If he chooses a club at random and takes a stroke, the probability that he gets a good shot is

(*a*) $\frac{1}{3}$ (*b*) $\frac{1}{12}$ (*c*) $\frac{4}{15}$ (*d*) $\frac{7}{12}$

(*xiii*) India plays two matches each with West Indies and Australia. In any match, the probabilities of India getting points 0, 1 and 2, are 0.45, 0.05 and 0.50 respectively. Assuming that the outcomes are independent, the probability of India getting at least 7 points is

(*a*) 0.8750 (*b*) 0.0875 (*c*) 0.625 (*d*) 0.0250 (*A.M.I.E.T.E., Summer 2001*)

(*xiv*) A bag contains 10 bolts, 3 of which are defective. Two bolts are drawn without replacement. The probability that both the bolts drawn are not defective is

(*a*) $\frac{49}{100}$ (*b*) $\frac{7}{15}$ (*c*) $\frac{4}{9}$ (*d*) $\frac{3}{10}$

(*xv*) The probability that a family has *k* children is $(0.5)^{k+1}$, $k = 0, 1, 2, \ldots$ If four families are chosen at random, the probability that each family has at least one child is

(*a*) 1/16 (*b*) 1/256 (*c*) 3/16 (*d*) 3/256

(*xvi*) Two distinguishable dice are tossed simultaneously. The probability that multiple of 2 does not occur on the first die or multiple of 3 does not occur on the second die is

(*a*) $\frac{5}{36}$ (*b*) $\frac{10}{36}$ (*c*) $\frac{20}{36}$ (*d*) $\frac{30}{36}$

(*xvii*) An unbiased die with faces marked 1, 2, 3, 4, 5, 6 is rolled 4 times, out of four face values obtained, the probability that the minimum face value is not less than 2 and the maximum face value is not greater than 5 is then

(*a*) $\frac{16}{81}$ (*b*) $\frac{2}{9}$ (*c*) $\frac{80}{81}$ (*d*) $\frac{8}{9}$ (*A.M.I.E.T.E., Summer 2000*)

(*xviii*) There are q persons sitting in a row. Two of them are selected at random, the probability that the two selected persons are not together is

(*a*) $\frac{2}{q}$ (*b*) $1-\frac{2}{q}$ (*c*) $\frac{q(q-1)}{(q+1)(q+2)}$ (*d*) None of these

**Ans.** (*i*) (*b*), (*ii*) (*b*), (*iii*) (*a*), (*iv*) (*a*), (*v*) (*a*), (*vi*) (*b*), (*vii*) (*d*), (*viii*) (*a*), (*ix*) (*d*), (*x*) (*b*), (*xi*) (*d*), (*xii*) (*c*), (*xiii*) (*b*), (*xiv*) (*b*), (*xv*) (*a*), (*xvi*) (*d*), (*xvii*) (*a*), (*xviii*) (*b*)

## 11.5 CONDITIONAL PROBABILITY

Let *A* and *B* be two events of a sample space *S* and let $P(B) \neq 0$. Then conditional probability of the event *A*, given *B*, denoted by $P(A/B)$, is defined by

$$P(A/B) = \frac{P(A \cap B)}{P(B)} \qquad \text{...(1)}$$

**Theorem.** If the events *A* and *B* defined on a sample space S of a random experiment are independent, then

$$P(A/B) = P(A) \text{ and } P(B/A) = P(B)$$

**Proof.** *A* and *B* are given to be independent events,

$$P(A \text{ and } B) = P(A) \cdot P(B)$$

$\Rightarrow \qquad P(A/B) = \dfrac{P(A \cap B)}{P(B)} = \dfrac{P(A).P(B)}{P(B)} = P(A)$

$\Rightarrow \qquad P(B/A) = \dfrac{P(B \cap A)}{P(A)} = \dfrac{P(B).P(A)}{P(A)} = P(B)$

## 11.6 BAYES THEOREM

If $B_1, B_2, B_3, B_n$ are mutually exclusive events with $P(B_i) \neq 0$, $(i = 1, 2, \ldots n)$ of a random experiment then for any arbitrary event $A$ of the sample space of the above experiment with $P(A) > 0$, we have

$$P(B_i / A) = \frac{P(B_i)P(A/B_i)}{\sum_{i=1}^{n} P(B_i)P(A/B_i)} \qquad \text{(for } n = 3)$$

$$P(B_2 / A) = \frac{P(B_2)P(A/B_2)}{P(B_1)P(A/B_1) + P(B_2)P(A/B_2) + P(B_3)P(A/B_3)}$$

**Proof.** Let $S$ be the sample space of the random experiment.

The events $B_1, B_2, \ldots, B_n$ being exhaustive

$$S = B_1 \cup B_2 \cup \ldots \cup B_n \qquad [\because A \subset S]$$

$$\therefore \qquad A = A \cap S$$
$$= A \cap (B_1 \cup B_2 \cup \ldots \cup B_n)$$
$$= (A \cap B_1) \cup (A \cap B_2) \cup \ldots \cup (A \cap B_n) \qquad \text{[Distributive Law]}$$

$$\Rightarrow \qquad P(A) = P(A \cap B_1) + P(A \cap B_2) + \ldots + P(A \cap B_n)$$
$$= P(B_1)P(A/B_1) + P(B_2)\,P(A/B_2) + \ldots + P(B_n)P(A/B_n)$$
$$= \sum_{i=1}^{n} P(B_i)P(A/B_i) \qquad \ldots (1)$$

Now $\qquad P(A \cap B_i) = P(A)P(B_i / A)$

$$\Rightarrow \qquad P(B_i / A) = \frac{P(A \cap B_i)}{P(A)} = \frac{P(B_i)P(A/B_i)}{\sum_{i=1}^{n} P(B_i)P(A/B_i)} \qquad \text{[Using (1)]}$$

**Note.** $P(B)$ is the probability of occurrence $B$. If we are told that the event $A$ has already occurred. On knowing about the event $A$, $P(B)$ is changed to $P(B/A)$. With the help of Baye's theorem we can calculate $P(B/A)$.

**Example (A)** *An urn I contains 3 white and 4 red balls and an urn II contains 5 white and 6 red balls. One ball is drawn at random from one of the urns and is found to be white. Find the probability that it was drawn from urn I.*

**Solution.** Let $\quad U_1$: the ball is drawn from urn I

$U_2$ : the ball is drawn from urn II

$W$ : the ball is white.

We have to find $P(U_1/W)$

By Baye's Theorem

$$P(U_1 / W) = \frac{P(U_1)P(W/U_1)}{P(U_1)P(W/U_1) + P(U_2)P(W/U_2)} \qquad \ldots (1)$$

Since two urns are equally likely to be selected, $P(U_1) = P(U_2) = \frac{1}{2}$

$P(W/U_1) = P$ (a white ball is drawn from urn I) $= \frac{3}{7}$

$P(W/U_2) = P$ (a white ball is drawn from urn II) $= \frac{5}{11}$

$\therefore$ From (1), $$P(U_1/W) = \frac{\frac{1}{2}\times\frac{3}{7}}{\frac{1}{2}\times\frac{3}{7}+\frac{1}{2}\times\frac{5}{11}} = \frac{33}{68}$$ **Ans.**

**Example (B)** *Three urns contains 6 red, 4 black; 4 red, 6 black; 5 red, 5 black balls respectively. One of the urns is selected at random and a ball is drawn from it. If the ball drawn is red find the probability that it is drawn from the first urn.*

**Solution.** Let $U_1$: the ball is drawn from *urn I*.

$U_2$: the ball is drawn from *urn II*.

$U_3$: the ball is drawn from *urn III*.

$R$: the ball is red.

We have to find $P(U_1/R)$.

By Baye's Theorem,

$$P(U_1/R) = \frac{P(U_1)P(R/U_1)}{P(U_1)P(R/U_1)+P(U_2)P(R/U_2)+P(U_3)P(R/U_3)} \quad \dots (1)$$

Since the three urns are equally likely to be selected $P(U_1) = P(U_2) = P(U_3) = \frac{1}{3}$

Also $P(R/U_1) = P$ (a red ball is drawn from urn I) $= \frac{6}{10}$

$P(R/U_2) = P$ (a red ball is drawn from urn II) $= \frac{4}{10}$

$P(R/U_3) = P$ (a red ball is drawn from urn III) $= \frac{5}{10}$

$\therefore$ From (1), we have $= P(U_1/R) = \dfrac{\frac{1}{3}\times\frac{6}{10}}{\frac{1}{3}\times\frac{6}{10}+\frac{1}{3}\times\frac{4}{10}+\frac{1}{3}\times\frac{5}{10}} = \dfrac{2}{5}$ **Ans.**

**Example (C)** *In a bolt factory, machines A, B and C manufacture respectively 25%, 35% and 40% of the total. If their output 5, 4 and 2 per cent are defective bolts. A bolt is drawn at random from the product and is found to be defective. What is the probability that it was manufactured by machine B?*

**Solution.** $A$ : bolt is manufactured by machine $A$.

$B$ : bolt is manufactured by machine $B$.

$C$ : bolt is manufactured by machine $C$.

$P(A) = 0.25, P(B) = 0.35, P(C) = 0.40$

The probability of drawing a defective bolt manufactured by machine $A$ is $P(D/A) = 0.05$

Similarly, $P(D/B) = 0.04$ and $P(D/C) = 0.02$

By Baye's theorem

$$P(B/D) = \frac{P(B)P(D/B)}{P(A)P(D/A) + P(B)P(D/B) + P(C)P(D/C)}$$

$$= \frac{0.35 \times 0.04}{0.25 \times 0.05 + 0.35 \times 0.04 + 0.40 \times 0.02} = 0.41 \quad \textbf{Ans.}$$

**BINOMIAL DISTRIBUTION**

## 11.7 DISCRETE PROBABILITY DISTRIBUTION

$P(X = x_i) = Pi$ or $p(x_i)$ for $i = 1, 2, \ldots$ where (i) $p(x_i) \geq 0$ for all values of $i$,

(ii) $\sum p(x_i) = 1$.

The set of values $x_i$ with their probabilities $p_i$ constitute a discreate probability distribution of the distrete variate $X$.

## 11.8 BINOMIAL DISTRIBUTION $P(r) = {}^nC_r p^r . q^{n-r}$

To find the probability of the happening of an event once, twice, thrice, ...... $r$ times ..... exactly in $n$ trials.

Let the probability of the happening of an event $A$ in one trial be $p$ and its probability of not happening be $1 - p = q$.

We assume that there are $n$ trials and the happening of the event $A$ is $r$ times and its not happening is $n - r$ times.

This may be shown as follows

$$\underbrace{AA.....A}_{r \text{ times}} \qquad \underbrace{\overline{A}\,\overline{A}.......\overline{A}}_{n-r \text{ times}} \qquad \ldots(1)$$

$A$ indicates its happening, $\overline{A}$ its failure and $P(A) = p$ and $P(\overline{A}) = q$.

We see that (1) has the probability

$$\underbrace{pp.....p}_{r \text{ times}}\ \underbrace{q.q\ .....\ q}_{n-r \text{ times}} = p^r . q^{n-r} \qquad \ldots(2)$$

Clearly (1) is merely one order of arranging $r$ $A$'s.

The probability of (1) $= p^r q^{n-r} \times$ Number of different arrangements of $r$ $A$'s *and* $(n - r)$ $\overline{A}$'s.

The number of different arrangements of $r$ $A$'s and $(n - r)$ $\overline{A}$'s $= {}^nC_r$.

$\therefore$ Probability of the happening of an event $r$ times $= {}^nC_r p^r q^{n-r}$.

$= (r + 1)$th term of $(q + p)^n$ $(r = 0, 1, 2, \ldots\ldots, n)$.

If $r = 0$, probability of happening of an event 0 times $= {}^nC_0\, q^n p^0 = q^n$

If $r = 1$, probability of happening of an event 1 time $= {}^nC_1 q^{n-1} p$

If $r = 2$, probability of happening of an event 2 times $= {}^nC_2 q^{n-2} p^2$

If $r = 3$, probability of happening of an event 3 times $= {}^nC_3 q^{n-3} p^3$ and so on.

These terms are clearly the successive terms in the expansion of $(q + p)^n$.

Hence it is called Binomial Distribution.

**Example 24.** *Find the probability of getting 4 heads in 6 tosses of a fair coin.*

**Solution.** $p = \frac{1}{2},\ q = \frac{1}{2},\ n = 6,\ r = 4.$

We know that $P(r) = {}^nC_r q^{n-r} p^r \Rightarrow P(4) = {}^6C_4 q^{6-4} p^4$

$$= \frac{6 \times 5}{1 \times 2}\left(\frac{1}{2}\right)^2\left(\frac{1}{2}\right)^4 = 15 \times \left(\frac{1}{2}\right)^6 = \frac{15}{64} \quad \textbf{Ans.}$$

**Example 25.** *If on an average one ship in every ten is wrecked, find the probability that out of 5 ships expected to arrive, 4 at least will arrive safely.*

**Solution.** Out of 10 ships, one ship is wrecked.

*i.e.*, Nine ships out of ten ships are safe. $P$ (safety) $= \frac{9}{10}$

$P$ (At least 4 ships out of 5 are safe) $= P(4 \text{ or } 5) = P(4) + P(5)$

$$= {}^5C_4 p^4 q^{5-4} + {}^5C_5 p^5 q^0 = 5\left(\frac{9}{10}\right)^4\left(\frac{1}{10}\right)+\left(\frac{9}{10}\right)^5 = \left(\frac{9}{10}\right)^4\left(\frac{5}{10}+\frac{9}{10}\right) = \frac{7}{5}\left(\frac{9}{10}\right)^4 \quad \textbf{Ans.}$$

**Example 26.** *The overall percentage of failures in a certain examination is 20. If six candidates appear in the examination, what is the probability that at least five pass the examination?*

**Solution.** Probability of failures $= 20\% = \frac{20}{100} = \frac{1}{5}$

Probability of $(P) = 1 - \frac{1}{5} = \frac{4}{5}$

Probability of at least five pass $= P$ (5 or 6)

$$= P(5) + P(6) = {}^6C_5 p^5 q + {}^6C_6 p^6 q^0$$

$$= 6\left(\frac{4}{5}\right)^5\left(\frac{1}{5}\right)+\left(\frac{4}{5}\right)^6 = \left(\frac{4}{5}\right)^5\left[\frac{6}{5}+\frac{4}{5}\right] = 2\left(\frac{4}{5}\right)^5 = \frac{2048}{3125} = 0.65536 \quad \textbf{Ans.}$$

**Example 27.** *Ten percent of screws produced in a certain factory turn out to be defective. Find the probability that in a sample of 10 screws chosen at random, exactly two will be defective.*

**Solution.** $p = \frac{1}{10},\ q = \frac{9}{10},\ n = 10,\ r = 2 P(r) = {}^nC_r p^r q^{n-r}$

$$P(2) = {}^{10}C_2\left(\frac{1}{10}\right)^2\left(\frac{9}{10}\right)^{10-2} = \frac{10\times 9}{1\times 2}\left(\frac{1}{10}\right)^2\left(\frac{9}{10}\right)^8 = \frac{1}{2}\cdot\left(\frac{9}{10}\right)^9 = 0.1937 \quad \textbf{Ans.}$$

**Example 28.** *The probability that a man aged 60 will live to be 70 is 0.65. What is the probability that out of 10 men, now 60, at least 7 will live to be 70 ?*

**Solution.** The probability that a man aged 60 will live to be 70

$$= p = 0.65$$

$$q = 1 - p = 1 - 0.65 = 0.35$$

Number of men $= n = 10$

Probability that at least 7 men will live to 70 = ( 7 or 8 or 9 or 10 )

$$= P(7) + P(8) + P(9) + P(10) = {}^{10}C_7 q^3 p^7 + {}^{10}C_8 q^2 p^8 + {}^{10}C_9 q p^9 + p^{10}$$

$$= \frac{10\times 9\times 8}{1\times 2\times 3}(0.35)^3(0.65)^7 + \frac{10\times 9}{1\times 2}(0.35)^2(0.65)^8 + 10(0.35)(0.65)^9 + (0.65)^{10}$$

$$= (.65)^7\,[120\,(0.35)^3 + 45\,(0.35)^2(0.65) + 10\,(0.35)(0.65)^2 + (0.65)^3]$$

$$= (0.65)^7 \times 125\,[120\times(0.07)^3 + 45\times(0.07)^2(0.13) + 10\,(0.07)(0.13)^2 + (0.13)^3]$$

$$= 0.04901 \times 125\,[0.04116 + 0.028665 + 0.011830 + 0.002197]$$

$$= 6.12625 \times 0.083852 = 0.5137 \quad \textbf{Ans.}$$

**Example 29.** *If 10% of bolts produced by a machine are defective. Determine the probability that out of 10 bolts, chosen at random (i) 1 (ii) none (iii) at most 2 bolts will be defective.*

**Solution.** Probability of defective bolts $= p = 10\% = 0.1$

Probability of not defective bolts $= q = 1 - p = 1 - 0.1 = 0.9$

Total number of bolts $= n = 10$

(*i*) Probability of 1 defective bolt $= {}^{10}C_1 (0.1)^1 (0.9)^9 = 0.3874$

(*ii*) Probability that none is defective = Probability of 0 defective bolt

$$= P(0) = {}^{10}C_0 (0.1)^0 (0.9)^{10} = 0.3487$$

(*iii*) Probability of 2 defective $= {}^{10}C_2 (0.1)^2 (0.9)^8 = 0.1937$

Probability of at most 2 defective $= P$ (0 or 1 or 2)

$$= P(0) + P(1) + P(2) = 0.3487 + 0.3874 + 0.1937$$

$$= 0.9298$$ **Ans.**

**Example 30.** *A die is thrown 8 times and it is required to find the probability that 3 will show (i) Exactly 2 times (ii) At least seven times (iii) At least once.*

**Solution.** The probability of throwing 3 in a single trial $= p = \frac{1}{6}$

The probability of not throwing 3 in a single trial $= q = \frac{5}{6}$

(*i*) $P$ (getting 3, exactly 2 times) $= {}^8C_2 q^6 p^2 = 28\left(\frac{5}{6}\right)^6\left(\frac{1}{6}\right)^2 = \frac{28 \times 5^6}{6^8}$

(*ii*) $P$ (getting 3, at least seven times) = $P$ (getting 3, at 7 or 8 times)

$$= P(7) + P(8) = {}^8C_7 q^1 p^7 + {}^8C_8 q^0 p^8 = 8\left(\frac{5}{6}\right)\left(\frac{1}{6}\right)^7 + \left(\frac{1}{6}\right)^8 = \frac{41}{6^8}$$

(*iii*) $P$ (getting 3 at least once)

$= P$ (getting 3, at 1 or 2 or 3 or 4 or 5 or 6 or 7 or 8 times)

$= P(1) + P(2) + P(3) + P(4) + P(5) + P(6) + P(7) + P(8)$

$= 1 - P(0) = 1 - {}^8C_0 q^8 p^0$

$= 1 - \left(\frac{5}{6}\right)^8$ **Ans.**

**Example 31.** *An underground mine has 5 pumps installed for pumping out storm water; the probability of any one of the pumps failing during the storm is $\frac{1}{8}$. What is the probability that (i) at least 2 pumps will be working; (ii) all the pumps will be working during a particular storm?*

**Solution.** (*i*) Probability of pump failing $= \frac{1}{8}$

Probability of pump working $= 1 - \frac{1}{8} = \frac{7}{8}, \quad p = \frac{7}{8}, \quad q = \frac{1}{8}, \quad n = 5$

(*i*) $P$ (At least 2 pumps working) = $P$ (2 or 3 or 4 or 5 pumps working)

$$= P(2) + P(3) + P(4) + P(5) = {}^5C_2\, p^2 q^3 + {}^5C_3\, p^3 q^2 + {}^5C_4\, p^4 q + {}^5C_5\, p^5 q^0$$

$$= 10\left(\frac{7}{8}\right)^2\left(\frac{1}{8}\right)^3 + 10\left(\frac{7}{8}\right)^3\left(\frac{1}{8}\right)^2 + 5\left(\frac{7}{8}\right)^4\left(\frac{1}{8}\right) + \left(\frac{7}{8}\right)^5$$

$$= \frac{1}{8^5}[10\times 49 + 10\times 343 + 5\times 2401 + 16807]$$

$$= \frac{1}{8^5}[490 + 3430 + 12005 + 16807] = \frac{32732}{8^5} = \frac{8183}{8192}$$

(*ii*) $P$ (All the 5 pumps working) $= P(5) = {}^5C_5\, p^5 q^0 = \left[\frac{7}{8}\right]^5 = \frac{16807}{32768}$ **Ans.** (*i*) $\frac{8183}{8192}$ (*ii*) $\frac{16807}{32768}$

**Example 32.** *Assuming that 20% of the population of a city are literate, so that the chance of an individual being literate is* $\frac{1}{5}$ *and assuming that 100 investigators each take 10 individuals to see whether they are literate, how many investigators would you expect to report 3 or less were literate.* (*A.M.I.E.T.E., Summer 2000*)

**Solution.** $p = \frac{1}{5}, \qquad n = 10$

$$P(3 \text{ or less}) = P(0 \text{ or } 1 \text{ or } 2 \text{ or } 3) = P(0) + P(1) + P(2) + P(3)$$

$$= {}^{10}C_0\left(\frac{1}{5}\right)^0\left(\frac{4}{5}\right)^{10} + {}^{10}C_1\left(\frac{1}{5}\right)^1\left(\frac{4}{5}\right)^9 + {}^{10}C_2\left(\frac{1}{5}\right)^2\left(\frac{4}{5}\right)^8 + {}^{10}C_3\left(\frac{1}{5}\right)^3\left(\frac{4}{5}\right)^7$$

$$= \left(\frac{4}{5}\right)^{10} + \frac{10}{5}\left(\frac{4}{5}\right)^9 + \frac{45}{25}\left(\frac{4}{5}\right)^8 + \frac{120}{125}\left(\frac{4}{5}\right)^7 = \left(\frac{4}{5}\right)^7 [(0.8)^3 + 2(0.8)^2 + 1.8(0.8) + 0.96]$$

$$= 0.2097152\,[0.512 + 1.28 + 1.44 + 0.96] = 0.2097152 \times 4.192 = 0.879126118$$

Required number of investigators $= 0.879126118 \times 100 = 87.9126118$

$= 88$ approximate **Ans.**

**Example 33.** *Assuming half the population of a town consumes chocolates and that 100 investigators each take 10 individuals to see whether they are consumers, how many investigators would you expect to report that three people or less were consumers?*

**Solution.** The chance for an individual to be consumer is $p = \frac{1}{2}$

The chance of not being a consumer $= q = 1 - \frac{1}{2} = \frac{1}{2}$

Here we have to find the probabilities of 0, 1, 2 and 3 successes.

$$p(r \le 3) = P(0) + P(1) + P(2) + P(3) = q^0 + {}^{10}C_1 q^9 p^1 + {}^{10}C_2 q^8 p^2 + {}^{10}C_3 q^7 p^3$$

$$= \left(\frac{1}{2}\right)^{10} + 10\left(\frac{1}{2}\right)^9\left(\frac{1}{2}\right) + 45\left(\frac{1}{2}\right)^8\left(\frac{1}{2}\right)^2 + 120\left(\frac{1}{2}\right)^7\left(\frac{1}{2}\right)^3 = \left(\frac{1}{2}\right)^{10}[1 + 10 + 45 + 120] = \frac{176}{1024}$$

The number of investigators to report that three or less people were consumers of chocolates is given by

$$\frac{176}{1024} \times 100 = 17.2$$

Hence, 17 investigators would report that 3 or less people are consumers. **Ans.**

**Example 34.** *For special security in a certain protected area, it was decided to put three lighting bulbs on each pole. If each bulb has a probability p of burning out in the first 100 hours of service,*

*calculate the probability that at least one of them is still good after 100 hours.*
*If P = 0.3, how many bulbs would be needed on each pole to ensure 99% safety so that at least one is good after 100 hours?*

**Solution.** Probability of burning out in the first 100 hours of service $= p = 0.3$
Probability of working good in the first 100 hours

$$q = 1 - p = 1 - 0.3 = 0.7$$

(*i*) Probability that at least one of them is still good after 100 hours

$$= {}^3C_1 q\, p^2 + {}^3C_2 q^2 p^1 + {}^3C_3\, q^3 p^0$$

$$= \left[{}^3C_0 q^0 p^3 + {}^3C_1 q p^2 + {}^3C_2 q^2 p + {}^3C_3 q^3 p^0\right] - {}^3c_0 q^0 p^3$$

$$= 1 - p^3 = 1 - (0.3)^3 = 1 - 0.027 = 0.973$$ **Ans.**

(*ii*) Let the number of bulbs required be *n*.
$P$ (At least one bulb is good) $= 1 - p^n$

$$\Rightarrow \quad 0.99 = 1 - (0.3)^n \quad \Rightarrow \quad (0.3)^n = 1 - 0.99$$

$$\Rightarrow \quad (0.3)^n = 0.01 \quad \Rightarrow \log (0.3)^n = \log 0.01$$

$$\Rightarrow \quad n \log 0.3 = \log 0.01 \Rightarrow n = \frac{\log 0.01}{\log 0.3}$$

$$\Rightarrow \quad n = \frac{2.000}{1.477} = \frac{-2.000}{-0.523} = 3.8 \cong 4 \text{ Bulbs}$$ **Ans.**

## Exercise 11.3

**1.** If 20% of the bolts produced by a machine are defective, determine the probability that out of 4 bolts chosen at random
(*a*) 1 (*b*) 0 (*c*) At most 2
bolts will be defective. **Ans.** (*a*) 0.4096, (*b*) 0.4096, (*c*) 0.9728.

**2.** Six dice are thrown 729 times. How many times do you expect at least three dice to show a five or a six ? **Ans. 233**

**3.** If the chance that any one of the 10 telephone lines is busy at any instant is 0.2, what is the chance that 5 of the lines are busy ? What is the probability that all the lines are busy?
**Ans.** $^{10}C_5 (0.2)^5 (0.8)^5, (0.2)^{10}$

**4.** An insurance salesman sells policies to 5 men, all of identical age in good health. According to the actuarial tables the probability that a man of this particular age will be alive 30 years hence is $\frac{2}{3}$. Find the probability that in 30 years.
(*a*) All 5 men (*b*) At least 3 men (*c*) Only 2 men (*d*) At least 1 man
will be alive. **Ans.** (*a*) $\frac{32}{243}$ (*b*) $\frac{192}{243}$ (*c*) $\frac{40}{243}$ (*d*) $\frac{242}{243}$

**5.** Assuming a Binomial distribution, find the probability of obtaining at least two "six" in rolling a fair die 4 times. **Ans.** $\frac{171}{1296}$

**6.** If successive trials are independent and the probability of success on any trial is *p*, show that the probability that the first success occurs on the *n*th trial is

$$p(1-p)^{n-1}, \qquad n = 1, 2, 3 \ldots$$

**7.** Consider an urn in which 4 balls have been placed by the following scheme : A fair coin is tossed; if the coin falls head, a white ball is placed in the urn, and if the coin falls tail, a red ball is placed in urn. (*i*) What is the probability that the urn will contain exactly 3 white balls ? (*ii*) What is the probability that the urn will contain exactly 3 red balls, given that the first ball placed was red? **Ans.** (*i*) $\frac{1}{8}$ (*ii*) $\frac{3}{8}$

8. A box contains 10 screws, 3 of which are defective. Two screws are drawn at random without replacement. Find the probability that none of the two screws is defective. **Ans.** $\frac{7}{15}$

9. Out of 800 families with four children each, how many families would be expected to have : (*i*) 2 boys and 2 girls; (*ii*) at least one boy; (*iii*) no girl; (*iv*) at most two girls? Assume equal probabilities for boys and girls. **Ans.** (*i*) 300, (*ii*) 750, (*iii*) 50, (*iv*) 550.

10. In a hurdle race, a player has to cross 10 hurdles. The probability that he will clear each hurdle is 5/6. What is the probability that he will knock down less than 2 hurdles ? **Ans.** $\frac{8}{3}\left(\frac{5}{6}\right)^9$

11. An electronic component consists of three parts. Each part has probability 0.99 of performing satisfactorily. The component fails if 2 or more parts do not perform satisfactorily. Assuming that the parts perform independently, determine the probability that the component does not perform satisfactorily. **Ans.** 0.000298

12. Find the binomial distribution whose mean is 5 and variance is 10/3. **Ans.** $^{15}C_r\left(\frac{1}{3}\right)^r\left(\frac{2}{3}\right)^{15-r}$

13. The probability that on, joining Engineering College, a student will successfully complete the course of studies is $\frac{3}{5}$. Determine the probability that out of 5 students joining the College (*i*) none and (*ii*) at least two will successfully complete the course. **Ans.** (*i*) $\frac{32}{3125}$ (*ii*) $\frac{2853}{3125}$

14. A carton contains 20 fuses, 5 of which are defective. Three fuses are chosen at random and inspected. What is the probability that at most one defective fuse is found? **Ans.** $\frac{27}{32}$

15. A bag contains three coins, one of which is coined with two heads, while the other two coins are normal and not biased. A coin is thrown at random from the bag and tossed three times in succession. If heads turn up each time, what is the probability that this is the two-headed coin? **Ans.** $\frac{4}{5}$

16. In sampling a large number of parts manufactured by a machine, the mean number of defectives in a sample of 20 is 2. Out of 1,000 such samples, how many would be expected to contain at least 3 defective parts? **Ans.** 324

17. The incidence of occupational disease in an industry is such that the workers have 20% chance of suffering from it. What is the probability that out of 6 workers 4 or more will catch the disease? **Ans.** $\frac{53}{3125}$

18. If the probability of hitting a target is 10% and 10 shots are fired independently, what is the probability that the target will be hit at least once ? **Ans.** $1-(0.9)^{10}=0.65$ nearly

19. Among 10,000 random digits, find the probability $p$ that the digit 3 appears at most 950 times. (*A.M.I.E., Summer 2003*) **Ans.** $^{10,000}C_r\left(\frac{1}{10}\right)^r\left(\frac{9}{10}\right)^{10,000-r}$

20. A fair coin is tossed 400 times. Using normal approximation to the binomial, find the probability that a head will occur (*a*) more than 180 times and (*b*) less than 195 times. (*A.M.I.E. Winter 2004*) **Ans.** (*a*) $1-\left(\frac{1}{2}\right)^{221}$ (b) $1-\left(\frac{1}{2}\right)^{195}$

**21.** Four coins were tossed 200 times. The number of tosses showing 0, 1, 2, 3 and 4 heads were found to be as under. Fit a binomial distribution to these observed results. Find the expected frequencies.

| No. of heads: | 0 | 1 | 2 | 3 | 4 |
|---|---|---|---|---|---|
| No. of tosses: | 15 | 35 | 90 | 40 | 20 |

**22.** A firm plans to bid ₹ 300 per tonne for a contract to supply 1000 tonnes of a metal. It has two competitors A and B and it assumes that the probability that A will bid less than 300/- per tonne is 0.3 and that B will bid less than ₹ 300 per tonne is 0.7. If the lowest bidder gets all the business and the firms bid independently, what is the expected value of business in rupees to the firm. *(A.M.I.E.T.E., Dec. 2006)*

## 11.9 MEAN OF BINOMIAL DISTRIBUTION *(GBTU, 2012, A.M.I.E.T.E., Dec. 2006)*

$$(q+p)^n = q^n + {}^nC_1 q^{n-1} p^1 + {}^nC_2 q^{n-2} p^2 + {}^nC_3 q^{n-3} p^3 + \ldots + {}^nC_r q^{n-r} p^r + \ldots + p^n$$

| *Successes r* | *Frequency f* | *rf* |
|---|---|---|
| 0 | $q^n$ | 0 |
| 1 | $nq^{n-1}p$ | $nq^{n-1}p$ |
| 2 | $\frac{n(n-1)}{2} q^{n-2} p^2$ | $n(n-1)q^{n-2}p^2$ |
| 3 | $\frac{n(n-1)(n-2)}{6} q^{n-3} p^3$ | $\frac{n(n-1)(n-2)}{2} q^{n-3} p^3$ |
| ...... | ....... | .... |
| $n$ | $p^n$ | $np^n$ |

$$\Sigma fr = nq^{n-1}p + n(n-1)q^{n-2}p^2 + \frac{n(n-1)(n-2)}{2} q^{n-3}p^3 + \ldots + np^n$$

$$= np\left[q^{n-1} + \frac{(n-1)}{1!} q^{n-2} p + \frac{(n-1)(n-2)}{2} q^{n-3} p^2 + \ldots + p^{n-1}\right]$$

$$= np\,(q+p)^{n-1} = np \qquad \text{(since } q+p=1\text{)}$$

$$\Sigma f = q^n + nq^{n-1}p + \frac{n(n-1)}{2} q^{n-2} p^2 + \ldots + p^n$$

$$= (q+p)^n = 1 \qquad \text{(since } q+p=1\text{)}$$

Hence Mean $= \frac{\Sigma fr}{\Sigma f} = np$ **Ans.**

## 11.10 STANDARD DEVIATION OF BINOMIAL DISTRIBUTION *(A.M.I.E.T.E., Dec. 2006)*

| *Successess r* | *Frequency f* | $r^2 f$ |
|---|---|---|
| 0 | $q^n$ | 0 |
| 1 | $nq^{n-1}p$ | $nq^{n-1}p$ |
| 2 | $\frac{n(n-1)(n-2)}{2} q^{n-2} p^2$ | $2n(n-1)q^{n-2}p^2$ |
| 3 | $\frac{n(n-1)(n-2)}{6} q^{n-3} p^3$ | $\frac{3n(n-1)(n-2)}{2} q^{n-3} p^3$ |
| ...... | ....... | .... |
| $n$ | $p^n$ | $n^2p^n$ |

We know that $\sigma^2 = \frac{\Sigma f r^2}{\Sigma f} - \left(\frac{\Sigma f r}{\Sigma f}\right)^2$ ...(1)

*r* is the deviation of items (successes) from 0.

$$\Sigma f = 1, \quad \Sigma fr = np$$

$$\Sigma fr^2 = 0 + nq^{n-1}p + 2n(n-1)q^{n-2}p^2 + \frac{3n(n-1)(n-2)}{2}q^{n-3}p^3 + .... + n^2p^n$$

$$= np[q^{n-1} + \frac{2(n-1)}{1!}q^{n-2}p + \frac{3(n-1)(n-2)}{2!}q^{n-3}p^2 + .... + np^{n-1}]$$

$$= np[q^{n-1} + \frac{(n-1)q^{n-2}p}{1!} + \frac{(n-1)(n-2)}{2!}q^{n-3}p^2 + .... + p^{n-1}$$

$$\left. + \frac{(n-1)q^{n-2}p}{1!} + \frac{2(n-1)(n-2)}{2!}q^{n-3}p^2 + .... + (n-1)p^{n-1}\right]$$

$$= np\left[\left\{q^{n-1} + (n-1)q^{n-2}p + \frac{(n-1)(n-2)}{2!}q^{n-3}p^2 + .... + p^{n-1}\right\}\right.$$

$$\left. + (n-1)p\left\{q^{n-2} + (n-2)q^{n-3}p + \frac{(n-2)(n-3)}{2!}q^{n-4}p^2 + .... + p^{n-2}\right\}\right]$$

$$= np[\{(q+p)^{n-1}\} + (n-1)p(q+p)^{n-2}] = np[1 + (n-1)p]$$

$$= np[np + (1-p)] = np[np + q] = n^2p^2 + npq$$

Putting these values in (1), we have

$$\text{Variance} = \sigma^2 = \frac{n^2p^2 + npq}{1} - \left(\frac{np}{1}\right)^2 = npq,$$

$$S.D. = \sigma = \sqrt{npq}$$

Hence for the binomial distribution, Mean = $np$, $\mu_2 = \sigma^2 = npq$

**Example 35.** *Find the first four moments of the binomial distribution.* (*AMIETE, Summer 2000*)

**Solution. First moment** about the origin

$$\mu_1' = \sum_{r=0}^{n} {}^nC_r p^r q^{n-r} . r = \sum_{r=0}^{n} r . \frac{n(n-1)(n-2).......(n-r+1)}{r!} p^r q^{n-r}$$

$$= n\sum_{r=1}^{n} \frac{(n-1)(n-2).......(n-r+1)}{(n-r)!} p^r q^{n-r} = np\sum_{r=1}^{n} {}^{n-1}C_{r-1} p^{r-1} q^{n-r}$$

$$= np(q+p)^{n-1} = np$$

Thus, the mean of the Binomial distribution is np.

**Second moment** about the origin

$$\mu_2' = \sum_{r=0}^{n} {}^nC_r p^r q^{n-r} . r^2 \qquad [r^2 = r(r-1) + r]$$

$$= \sum_{r=0}^{n} \{r(r-1) + r\} {}^nC_r p^r q^{n-r} = \sum_{r=0}^{n} r(r-1) {}^nC_r p^r q^{n-r} + \sum_{r=0}^{n} r \cdot {}^nC_r p^r q^{n-r}$$

$$= \sum_{r=0}^{n} \frac{r(r-1)n(n-1)(n-2).......(n-r+1)}{r!} p^r q^{n-r}$$

$$= \sum_{r=0}^{n} \frac{r\,n(n-1)(n-2).......(n-r+1)}{r!} p^r q^{n-r}$$

$$= n(n-1)p^2 \sum_{r=2}^{n} \frac{(n-2)(n-3).......(n-r+1)}{(r-2)!} p^{r-2} q^{n-r}$$

$$+ np\Sigma \frac{(n-1)(n-2).......(n-r+1)}{(r-1)!} p^{r-1} q^{n-r}$$

$$= n(n-1)p^2(q+p)^{n-2} + np\,(q+p)^{n-1} = n(n-1)p^2 + np$$

**Third moment** about the origin

$$\mu_3' = \sum_{r=0}^{n} {}^nC_r\, p^r q^{n-r} \cdot r^3$$

[Let $r^3 = Ar(r-1)(r-2) + Br(r-1) + Cr$

By putting $r = 1, 2, 3$, we get A = 1, B = 3, C = 1]

$$\mu_3' = \sum_{r=0}^{n} \{r(r-1)(r-2) + 3r(r-1) + r\}\, {}^nC_r\, p^r q^{n-r}$$

$$= \sum_{r=0}^{n} r(r-1)(r-2)\, {}^nC_r\, p^r q^{n-r} + 3\sum_{r=0}^{n} r(r-1) \cdot {}^nC_r\, p^r q^{n-r} + \sum_{r=0}^{n} r \cdot {}^nC_r\, p^r q^{n-r}$$

$$= \sum_{r=0}^{n} \frac{r(r-1)(r-2) \cdot n(n-1) \ldots\ldots (n-r+1)}{r!} p^r q^{n-r}$$

$$+ 3\sum_{r=0}^{n} \frac{r(r-1) \cdot n(n-1) \ldots\ldots (n-r+1)}{r!} p^r q^{n-r} + \sum_{r=0}^{n} r\frac{n(n-1) \ldots\ldots (n-r+1)}{r!} p^r q^{n-r}$$

$$= \sum_{r=3}^{n} \frac{n(n-1)(n-2)(n-3) \ldots\ldots (n-r+1)}{(r-3)!} p^r q^{n-r}$$

$$+ 3\sum_{r=2}^{n} \frac{n(n-1)(n-2)(n-3) \ldots\ldots (n-r+1)}{(r-2)!} p^r q^{n-r}$$

$$+ \sum_{r=1}^{n} \frac{n(n-1)(n-2) \ldots\ldots (n-r+1)}{(r-1)!} p^r q^{n-r}$$

$$= n(n-1)(n-2)p^3 \sum_{r=3}^{n} {}^{n-3}C_{r-3}\, p^{r-3} q^{n-3} + 3n(n-1)p^2 \sum_{r=2}^{n} {}^{n-2}C_{r-2}\, p^{r-2} q^{n-2}$$

$$+ np \sum_{r=1}^{n} {}^{(n-1)}C_{r-1}\, p^{r-1} q^{n-1}$$

$$= n(n-1)(n-2)\,p^3(q+p)^{n-3} + 3\,n(n-1)\,p^2(q+p)^{n-2} + np\,(q+p)^{n-1}$$

$$= n(n-1)(n-2)\,p^3 + 3\,n(n-1)\,p^2 + np$$

**Fourth Moment**

$$\mu_4' = \sum_{r=0}^{n} {}^nC_r\, p^r q^{n-r} \cdot r^4$$

[Let $r^4 = Ar(r-1)(r-2)(r-3) + Br(r-1)(r-2) + Cr(r-1) + Dr$

By putting r = 1, 2, 3, 4, we get A = 1, B = 6, C = 7, D = 1]

$$\mu_4' = \sum_{r=0}^{n} r(r-1)(r-2)(r-3) \cdot {}^nC_r\, p^r q^{n-r} + \sum_{r=0}^{n} 6r(r-1)(r-2) \cdot {}^nC_r\, p^r q^{n-r}$$

$$+ \sum_{r=0}^{n} 7r(r-1) \cdot {}^nC_r\, p^r q^{n-r} + \sum_{r=0}^{n} r \cdot {}^nC_r\, p^r q^{n-r}$$

$$= \sum_{r=0}^{n} \frac{r(r-1)(r-2)(r-3) \cdot n(n-1) \ldots\ldots (n-r+1)}{r!} p^r q^{n-r}$$

$$+ 6\sum_{r=0}^{n} \frac{r(r-1)(r-2) \cdot n(n-1) \ldots\ldots (n-r+1)}{r!} p^r q^{n-r}$$

$$+ 7\sum_{r=0}^{n} \frac{r(r-1) \cdot n(n-1) \ldots\ldots (n-r+1)}{r!} p^r q^{n-r} + \sum_{r=0}^{n} \frac{r \cdot n(n-1) \ldots\ldots (n-r+1)}{r!} p^r q^{n-r}$$

$$= \sum_{r=4}^{n} \frac{n(n-1)(n-2)(n-3)(n-4)...(n-r+1)}{(r-4)!} p^r q^{n-r}$$

$$+6\sum_{r=3}^{n} \frac{n(n-1)(n-2)(n-3)...(n-r+1)}{(r-3)!} p^r q^{n-r}$$

$$+7\sum_{r=2}^{n} \frac{n(n-1)(n-2)...(n-r+1)}{(r-2)!} p^r q^{n-r} + \Sigma \frac{n(n-1)...(n-r+1)}{(r-1)!} p^r q^{n-r}$$

$$= n(n-1)(n-2)(n-3)\sum_{r=4}^{n} {}^{n-4}C_{r-4}\, p^r q^{n-r} + 6n(n-1)(n-2)\sum_{r=3}^{n} {}^{n-3}C_{r-3} \cdot p^r q^{n-r}$$

$$+7n(n-1)\sum_{r=2}^{n} {}^{n-2}C_{r-2} \cdot p^r q^{n-r} + n\sum_{r=1}^{n} {}^{n-1}C_{r-1} \cdot p^r q^{n-r}$$

$$= n(n-1)(n-2)(n-3)p^4(q+p)^{n-4} + 6n(n-1)(n-2)p^3(q+p)^{n-3}$$
$$+ 7n(n-1)p^2(q+p)^{n-2} + np(q+p)^{n-1}$$

$$= n(n-1)(n-2)(n-3)p^4 + 6n(n-1)(n-2)p^3 + 7n(n-1)p^2 + np$$

## 11.11 CENTRAL MOMENTS : (Moments about the mean)

Now, the first four central moments are obtained as follows:

Second Central Moment

$$\mu_2 = \mu_2' - \mu_1' = [n(n-1)p^2 + np] - n^2p^2 = np[(n-1)p + 1 - np] = np(1-p) = npq$$

Variance of Binomial distribution is $npq$

**Third Central Moment**

$$\mu_3 = \mu_3' - 3\mu_2'\mu_1' + 2\mu_1'^3$$
$$= \{n(n-1)(n-2)p^3 + 3n(n-1)p^2 + np\} - 3\{n^2p^2 + npq)np\} + 2n^3p^3$$
$$= np[-3np^2 + 3np + 2p^2 - 3p + 1 - 3npq]$$
$$= np[3np(1-p) + 2p^2 - 3p + 1 - 3npq]$$
$$= np[3npq + 2p^2 - 3p + 1 - 3npq] = np[2p^2 - 3p + 1] = np[2p^2 - 2p + q]$$
$$= np[-2p(1-p) + q] = np(-2pq + q) = npq(1-2p) = npq(q-p)$$

**Fourth Central Moment**

$$\mu_4 = \mu_4' - 4\mu_3'\mu_1' + 6\mu_2'\mu_1'^2 - 3\mu_1'^4$$
$$= n(n-1)(n-2)(n-3)p^4 + 6n(n-1)(n-2)p^3 + 7n(n-1)p^2$$
$$+ np - 4[n(n-1)(n-2)p^3 + 3n(n-1)p^2 + np]np$$
$$+ 6[n(n-1)p^2 + np]n^2p^2 - 3n^4p^4$$
$$= np[(n-1)(n-2)(n-3)p^3 + 6(n-1)(n-2)p^2 + 7(n-1)p$$
$$+ 1 - 4\{n(n-1)(n-2)p^3 + 3n(n-1)p^2 + np\}$$
$$+ 6\{n(n-1)p^2 + np\}np - 3n^3p^3]$$
$$= np[(n^3 - 6n^2 + 11n - 6)p^3 + (6n^2 - 18n + 12)p^2 + 7np - 7p + 1$$
$$+ \{(-4n^3 + 12n^2 - 8n)p^3 - 4(3n^2 - 3n)p^2 - 4np\}$$
$$+ \{(6n^3 - 6n^2)p^3 + 6n^2p^2\} - 3n^3p^3]$$
$$= np[(n^3 - 6n^2 + 11n - 6 - 4n^3 + 12n^2 - 8n + 6n^3 - 6n^2 - 3n^3)p^3$$
$$+ (6n^2 - 18n + 12 - 12n^2 + 12n + 6n^2)p^2 + (7n - 7 - 4n)p + 1]$$
$$= np[(3n-6)p^3 + (-6n+12)p^2 + (3n-7)p + 1]$$
$$= np[3np^3 - 6p^3 - 6np^2 + 12p^2 + 3np - 7p + 1]$$

$= np\,[\,3n\,p^3 - 3\,n\,p^2 - 6\,p^3 + 6\,p^2 - 3n\,p^2 + 3np + 6\,p^2 - 6p - p + 1]$

$= np\,[-3n\,p^2\,(1-p) + 6\,p^2\,(1-p) + 3np\,(1-p) - 6p\,(1-p) + (1-p)]$

$= np\,[-3n\,p^2\,q + 6\,p^2\,q + 3n\,pq - 6pq + q] = npq\,[-3n\,p^2 + 6p^2 + 3np - 6p + 1]$

$= npq\,[\,3np\,(1-p) - 6p\,(1-p) + 1] = npq\,[\,3npq - 6pq + 1\,]$

$= npq\,[\,1 + 3\,(n-2)\,pq\,]$ **Ans.**

## 11.12 MOMENT GENERATING FUNCTIONS OF BINOMIAL DISTRIBUTION ABOUT ORIGIN

$$M_0\,(t) = E\,(e^{tx}) = \Sigma\,{}^nC_x p^x q^{n-x} \cdot e^{tx}$$

$$= \Sigma\,{}^nC_x\,(p\,e^t)^x\,q^{n-x} = (q + pe^t)^n$$

Differentiating w.r.t. 't' we get $M'_0\,(t) = n\,(q + p\,e^t)^{n-1}\,p \cdot e^t$

On putting $t = 0$, we get $\mu_1' = n\,(q + p)^{n-1}\,p$

$$\mu_1' = np$$

Since $\qquad M_a\,(t) = e^{-at}\,M_0\,(t)$

Moment generating function of the Binomial distribution about its mean $(m) = np$ is given by

$$M_m\,(t) = e^{-npt}\,M_0\,(t)$$

$$M_m\,(t) = e^{-npt}(q + p\,e^t)^n = (qe^{-pt} + pe^{-pt+t})^n = (qe^{-pt} + pe^{(1-p)t})^n$$

$$= \left[q(1 - pt + \frac{p^2t^2}{2!} - \frac{p^3t^2}{2!} + \frac{p^4t^4}{4!} + \ldots) + p(1 + qt + \frac{q^2t^2}{2!} + \frac{q^3t^3}{3!} + \frac{p^4t^4}{4!} + \ldots.)\right]^n$$

$$= \left[1 + pq\frac{t^2}{2!} + pq(q^2 - p^2)\frac{t^3}{3!} + pq(q^3 + p^3)\frac{t^4}{4!} + \ldots.\right]^n$$

$$1 + \mu_1 t + \mu_2\frac{t^2}{2!} + \mu_3\frac{t^3}{3!} + \mu_4\frac{t^4}{4!} + \ldots..$$

$$= 1 + npq\frac{t^2}{2!} + npq(q-p)\frac{t^3}{3!} + npq[1 + 3(n-2)pq]\frac{t^4}{4!} + \ldots..$$

Equating the coefficients of like powers of t on both sides, we get

$$\mu_2 = npq, \qquad \mu_3 = npq\,(q-p),\ \mu_4 = npq\,[\,1 + 3\,(n-2)\,pq]$$

Hence the moment coefficient of skewness is

$$\beta_1 = \frac{\mu_3^2}{\mu_2^3} = \frac{[npq\,(q-p)]^2}{(npq)^3} = \frac{(q-p)^2}{npq}\ ; \quad \gamma_1 = \sqrt{\beta_1} = \frac{q-p}{\sqrt{npq}}$$

Coefficient of Kurtosis is given by

$$\beta_2 = \frac{\mu_4}{\mu_2^2} = \frac{npq\,[1 + 3pq\,(n-2)]}{(npq)^2} = 3 + \frac{1 - 6pq}{npq}\ ; \quad \gamma_2 = \beta_2 - 3 = \frac{1 - 6pq}{npq}$$

**Example 36.** *If the probability of a defective bolt is 0.1, find*
*(a) the mean (b) the standard deviation for the distribution bolts in a total of 400.*

**Solution.** $n = 400,\ p = 0.1,$ Mean $= np = 400 \times 0.1 = 40$

$$\text{Standard deviation} = \sqrt{npq} = \sqrt{400 \times 0.1(1 - 0.1)}$$

$$= \sqrt{400 \times 0.1 \times 0.9} = 20 \times 0.3 = 6$$ **Ans.**

**Example 37.** *A die is tossed thrice. A success is getting 1 or 6 on a toss. Find the mean and variance of the number of successes.* **(AMIETE, Dec. 2010)**

**Solution.** $\qquad n = 3,\ p = \frac{1}{3},\ q = \frac{2}{3}$

$$\text{Mean} = np = 3\times\frac{1}{3} = 1$$

$$\text{Variance} = npq = 3\times\frac{1}{3}\times\frac{2}{3} = \frac{2}{3}$$ **Ans..**

## 11.13 RECURRENCE RELATION FOR THE BINOMIAL DISTRIBUTION

By Binomial distribution, $P(r) = {}^nC_r\, p^r\, q^{n-r}$ ...(1) *(A.M.I.E., Summer 2002)*

$P(r+1) = {}^nC_{r+1}\, p^{r+1}\, q^{n-r-1}$ ...(2)

On dividing (2) by (1), we get

$$\frac{P(r+1)}{P(r)} = \frac{{}^nC_{r+1}}{{}^nC_r}\,\frac{p^{r+1}q^{n-r-1}}{p^r q^{n-r}}$$

$$= \frac{n(n-1)(n-2).....(n-r)}{(r+1)!}\,\frac{r!}{n(n-1)(n-2).....(n-r+1)}\,\frac{p}{q}$$

$$\frac{P(r+1)}{P(r)} = \frac{n-r}{r+1}\,\frac{p}{q} \Rightarrow P(r+1) = \frac{n-r}{r+1}\,\frac{p}{q}\,P(r)$$ **Ans.**

### Exercise 11.4

**1.** Fit a binomial distribution to the following frequency data:

| $x$ | 0 | 1 | 3 | 4 |
|---|---|---|---|---|
| $f$ | 28 | 62 | 10 | 4 |

*(U. P III Sem. Dec. 2004)*

**Ans.** $P(r) = {}^{104}C_r (0.00999)^r (0.99111)^{104-r}$

**2.** Fill in the blanks :

(*a*) A coin is biased so that a head is twice as likely to occur as a tail. If the coin is tossed 3 times, the prob. of getting exactly 2 tails, is ..........

(b) The probability of getting number 5 exactly two times in five throws of an unbiased die is ......

(*c*) A die is thrown 6 times. The probability to get greater than 4 appears at least once is...

(*d*) For what, one should be?

(i) Obtaining 6 at least once in 4 throws of a die.

or (*ii*) obtaining a double-six at least once in 24 throws with two dice.

(*e*) The probability of producing a defective bolt is 0.1. The probability that out of 5 bolts one will be defective is .......

(*f*) If the probability of hitting a target is 5% and 5 shots are fired independently, the probability that the target will be hit at least once is .......

(*g*) If $n$ and $p$ are the parameters of a binomial distribution the standard deviation is ....

(*h*) The mean, standard deviation and skewness of Binomial distribution are ..... and ..

*(A.M.I.E., Summer 2001)*

(*i*) If three persons selected at random are stopped on a street, then the probability that all of them were born on Sunday is ..... *(A.M.I.E., Winter 2001)*

**Ans.** (*a*) $\frac{2}{9}$, (*b*) $10.\frac{5^3}{6^5}$, (*c*) $\frac{665}{729}$, (*d*) (*i*) $\frac{671}{1296}$, (*ii*) $1-\left(\frac{35}{36}\right)^{24}$, (*e*) $\frac{1}{2}\left(\frac{9}{10}\right)^4$, (*f*) $1-(0.95)^5$,

(*g*) $\sqrt{npq}$ (*h*) $np, \sqrt{npq}, \frac{q-p}{\sqrt{npq}}$ (*i*) $\frac{1}{343}$

**3. Tick √ the correct answer :**

(*a*) If a coin is tossed 6 times in succession, the probability of getting at least one head is

(*i*) 1/64 (*ii*) 3/32 (*iii*) 63/64 (*iv*) 1/2

(*b*) A coin is tossed until a tail appears or at the most five times. Given that the tail does not appear on the first two tosses, the probability that the coin will be tossed 5 times, is

(*i*) 1/2 (*ii*) 3/5 (iii) 1/3 (*iv*) 1/4

(*c*) In a certain manufacturing process it is known that on an average, 1 in every 100 items is defective. What is the probability that 5 items are inspected before a defective item is found?

(*i*) 0.0096 (*ii*) 0.96 (*iii*) 0.096 (*iv*) none of these

(*d*) The probability that a marksman will hit a target is given as $\frac{1}{5}$. Then his probability of at least one hit in 10 shots is

(*i*) $1-\left(\frac{4}{5}\right)^{10}$ (*ii*) $\frac{1}{5^{10}}$ (*iii*) $1-\frac{1}{5^{10}}$ (*iv*) None of these

(*e*) The probability of having at least one tail in 4 throws with a coin is

(*i*) $\frac{15}{16}$ (*ii*) $\frac{1}{16}$ (*iii*) $\frac{1}{4}$ (*iv*) 1.

(*f*) A coin is tossed 3 times. The probability of obtaining two heads will be

(*i*) $\frac{3}{8}$ (*ii*) $\frac{1}{2}$. (*iii*) 1 (*iv*) 2.

(*g*) 8 coins are tossed simultaneously. The probability of getting at least 6 heads is

(*i*) $\frac{57}{64}$ (*ii*) $\frac{229}{256}$ (*iii*) $\frac{7}{64}$ (*iv*) $\frac{37}{256}$

(*h*) Three unbiased coins are tossed simultaneously. This is repeated four times. The probability of getting at least one head each time is

(*i*) $\left(\frac{3}{4}\right)^4$ (*ii*) $\left(\frac{7}{8}\right)^4$ (*iii*) $\left(\frac{1}{8}\right)^4$ (*iv*) $\left(\frac{1}{4}\right)^4$

(*i*) In rolling two fair dice, the probability of getting equal numbers or numbers with an even product is

(*i*) $\frac{6}{36}$ (*ii*) $\frac{30}{36}$ (*iii*) $\frac{27}{36}$ (*iv*) $\frac{3}{36}$

(*j*) In a binomial distribution the sum and the product of the mean and variance are $\frac{25}{3}$ and $\frac{50}{3}$ respectively. The distribution is

(*i*) $\left(\frac{4}{5}+\frac{1}{5}\right)^{15}$ (*ii*) $\left(\frac{2}{3}+\frac{1}{3}\right)^{15}$ (*iii*) $\left(\frac{3}{4}+\frac{1}{4}\right)^{15}$ (*iv*) None of these.

(*k*) A room has three lamp sockets. From a collection of 10 light bulbs of which only 6 are good. A person selects 3 at random and puts them in a socket. What is the probability that room will have light.

(*i*) 29/120 (*ii*) 39/60 (*iii*) 19/30 (*iv*) 29/30 (*A.M.I.E.T.E. Dec 2005*)

(*l*) The inequality between mean and variance of Binomial distribution which is true is

(*a*) Mean < Variance (*b*) Mean = Variance

(*c*) Mean > Variance (*d*) Mean × Variance = 1 (*A.M.I.E.T.E Dec. 2006*)

**Ans.** (*a*) (*iii*), (*b*) (*iv*), (*c*) (*i*), (*d*) (*i*), (*e*) (*i*), (*f*) (*i*), (*g*) (*iv*), (*h*) (*ii*), (*i*) (*ii*), (*j*) (*ii*), (*l*) (*c*),

4. Find the Binomial distribution whose mean is 5 and variance is $\frac{10}{3}$. **Ans.** ${}^{15}C_r\left(\frac{1}{3}\right)^r\left(\frac{2}{3}\right)^{15-r}$

5. (*a*) The mean, standard deviation and skewness of Binomial distribution are ......, .... and ....
(*AMIE Summer 2001*) **Ans.** $np, \sqrt{npq}, 0$

(*b*) If three persons selected at random are stopped on a street, then the probability that all of them were born on Sunday is ......... (*A.M.I.E, Winter 2001*) **Ans.** $\frac{1}{343}$

## POISSON DISTRIBUTION

### 11.14 POISSON DISTRIBUTION

Poisson distribution is a particular limiting form of the Binomial distribution when $p$ (or $q$) is very small and $n$ is large enough.

Poisson distribution is

$$\boxed{P(r) = \frac{m^r e^{-m}}{r!}}$$

where $m$ is the mean of the distribution.

**Proof.** In Binomial distribution.

$$P(r) = {}^nC_r q^{n-r} p^r = {}^nC_r (1-p)^{n-r} p^r$$

$$\left(\text{since mean} = m = np \Rightarrow p = \frac{m}{n}\right)$$

$$= {}^nC_r\left(1-\frac{m}{n}\right)^{n-r}\left(\frac{m}{n}\right)^r \qquad (m \text{ is constant})$$

$$= \frac{n(n-1)(n-2)\ldots\overline{(n-r-1)}}{r!}\left(\frac{m}{n}\right)^r\left(1-\frac{m}{n}\right)^{n-r}$$

$$= \frac{\frac{n}{n}\left(\frac{n}{n}-\frac{1}{n}\right)\left(\frac{n}{n}-\frac{2}{n}\right)\ldots\left(\frac{n}{n}-\frac{r-1}{n}\right)m^r\left(1-\frac{m}{n}\right)^n}{r!\left(1-\frac{m}{n}\right)^r}$$

$$= \frac{1\left(1-\frac{1}{n}\right)\left(1-\frac{2}{n}\right)\ldots\left(1-\frac{r-1}{n}\right)m^r\left(1-\frac{m}{n}\right)^n}{r!\left(1-\frac{m}{n}\right)^r}$$

$$\left[\begin{array}{l}\text{Taking limits, when n tends to infinity} \\ \lim\limits_{n\to\infty}\left(1-\frac{m}{n}\right)^n = \lim\limits_{n\to\infty}\left[\left(1-\frac{m}{n}\right)^{-\frac{n}{m}}\right]^{-m} = e^{-m} \quad \text{as} \quad \lim\limits_{n\to\infty}\left(1-\frac{m}{n}\right)^{-\frac{n}{m}} = e\end{array}\right]$$

$$P(r) = \frac{m^r}{r!}e^{-m}$$

$$\boxed{P(r) = \frac{e^{-m}.m^r}{r!}}$$

## 11.15 MEAN OF POISSON DISTRIBUTION

$$P(r) = \frac{e^{-m} . m^r}{r!}$$ (*A.M.I.E.T.E., Summer 2004, 2002*)

| *Successes r* | *Frequency f* | *f.r* |
|---|---|---|
| 0 | $\frac{e^{-m} m^0}{0!}$ | 0 |
| 1 | $\frac{e^{-m} m^1}{1!}$ | $e^{-m}. m$ |
| 2 | $\frac{e^{-m} m^2}{2!}$ | $e^{-m}. m^2$ |
| 3 | $\frac{e^{-m} m^3}{3!}$ | $\frac{e^{-m} . m^3}{2!}$ |
| ... | ... | ... |
| r | $\frac{e^{-m} m^r}{r!}$ | $\frac{e^{-m} .m^r}{(r-1)!}$ |
| ... | | ... |

$$\Sigma f r = 0 + e^{-m} .m + e^{-m} .m^2 + e^{-m} .\frac{m^3}{2!} + ... + e^{-m} \frac{m^r}{(r-1)!} + ... = e^{-m} . m \left[1 + \frac{m}{1!} + \frac{m^2}{2!} + ... + \frac{m^{r-1}}{(r-1)!} + ...\right]$$

$$= m . e^{-m}. [e^m] = m$$

$$\text{Mean} = \frac{\Sigma fr}{\Sigma f} = \frac{m}{1},$$

**Mean = m.** **Ans.**

## 11.16 STANDARD DEVIATION OF POISSON DISTRIBUTION

$$P(r) = \frac{e^{-m} m^r}{r!}$$ (*A.M.I.E.T.E., Summer 2002*)

| *Successes r* | *Frequency f* | *Product rf* | *Product $r^2f$* |
|---|---|---|---|
| 0 | $\frac{e^{-m} m^0}{0!}$ | 0 | 0 |
| 1 | $\frac{e^{-m} m^1}{1!}$ | $e^{-m}. m$ | $e^{-m}.m$ |
| 2 | $\frac{e^{-m} m^2}{2!}$ | $e^{-m}. m^2$ | $2e^{-m}. m^2$ |
| 3 | $\frac{e^{-m} m^3}{3!}$ | $\frac{e^{-m}.m^3}{2!}$ | $3e^{-m}.\frac{m^3}{2!}$ |
| ........ | ........ | ........ | ........ |
| r | $\frac{e^{-m} m^r}{r!}$ | $\frac{e^{-m}.m^r}{(r-1)!}$ | $\frac{re^{-m}.m^r}{(r-1)!}$ |
| ........ | ........ | ........ | ........ |

$\Sigma f = 1,$ $\Sigma fr = m$

$$\Sigma f r^2 = 0+e^{-m}.m+2e^{-m}.m^2+3.e^{-m}.\frac{m^3}{2}+...+\frac{r\,e^{-m}.m^r}{(r-1)!}+..$$

$$= m.e^{-m}\left[1+2m+\frac{3m^2}{2!}+\frac{4m^3}{3!}+...\frac{r.m^{r-1}}{(r-1)!}+..\right]$$

$$= m.e^{-m}\left[1+m+\frac{m^2}{2!}+\frac{m^3}{3!}+...\frac{m^{r-1}}{(r-1)!}+....+m+2\frac{m^2}{2!}+\frac{3m^3}{3!}+...+\frac{(r-1)m^{r-1}}{(r-1)!}+...\right]$$

$$= m.e^{-m}\left[\left\{1+m+\frac{m^2}{2!}+\frac{m^3}{3!}+...+\frac{m^{r-1}}{(r-1)!}+...\right\}+m\left\{1+\frac{m}{1!}+\frac{m^2}{2!}+...+\frac{m^{r-2}}{(r-2)!}+...\right\}\right]$$

$$= m.e^{-m}\,[e^m + m\,e^m] = m + m^2$$

$$\sigma^2 = \frac{\Sigma fr^2}{\Sigma f}-\left(\frac{\Sigma fr}{\Sigma f}\right)^2 = \frac{m+m^2}{1}-(m)^2 = m \quad \text{or} \quad \sigma = \sqrt{m}$$

**S. D. = $\sqrt{m}$**

Hence mean and variance of a Poisson distribution are each equal to $m$. Similarly we can obtain,

$$\mu_3 = m, \quad \mu_4 = 3m^2 + m$$

$$\beta_1 = \frac{1}{m}, \quad \beta_2 = 3+\frac{1}{m}$$

$$\gamma_1 = \frac{1}{\sqrt{m}}, \quad \gamma_2 = \frac{1}{m}$$

## 11.17 MEAN DEVIATION

Show that in a Poisson distribution with unit mean, and the mean deviation about the mean is $\left(\frac{2}{e}\right)$ times the standard deviation. *(A.M.I.E.T.E., Dec. 2005)*

**Solution.** $P(r)=\frac{m^r}{r!}e^{-m}$ But mean = 1 *i.e.* $m = 1$ and S.D. = $\sqrt{m} = 1$

Hence, $P(r)=\frac{e^{-m}}{r!}m^r = \frac{e^{-1}}{r!} = \frac{1}{e}.\frac{1}{r!}$

| $r$ | $P(r)$ | $\lvert r-1\rvert$ | $P(r)\,\lvert r-1\rvert$ |
|---|---|---|---|
| 0 | $\frac{1}{e}$ | 1 | $\frac{1}{e}$ |
| 1 | $\frac{1}{e}$ | 0 | 0 |
| 2 | $\frac{1}{e}\frac{1}{2!}$ | 1 | $\frac{1}{e}\frac{1}{2!}$ |
| 3 | $\frac{1}{e}\frac{1}{3!}$ | 2 | $\frac{1}{e}\frac{2}{3!}$ |
| 4 | $\frac{1}{e}\frac{1}{4!}$ | 3 | $\frac{1}{e}\frac{3}{4!}$ |
| ......... | ......... | ......... | ......... |
| $r$ | $\frac{1}{e}\frac{1}{r!}$ | $r-1$ | $\frac{1}{e}\frac{r-1}{r!}$ |

Mean Deviation $= \Sigma P(r)|r-1| = \frac{1}{e} + 0 + \frac{1}{e}\frac{1}{2!} + \frac{1}{e}\frac{2}{3!} + \frac{1}{e}\frac{3}{4!} + \ldots + \frac{1}{e}\frac{r-1}{r!} + \ldots$

$$= \frac{1}{e}\left[1 + 0 + \frac{1}{2!} + \frac{2}{3!} + \frac{3}{4!} + \ldots + \frac{r-1}{r!} + \ldots\right]$$

$$= \frac{1}{e}\left[1 + \left(\frac{1}{1!} - \frac{1}{1!}\right) + \left(\frac{2}{2!} - \frac{1}{2!}\right) + \left(\frac{3}{3!} - \frac{1}{3!}\right) + \left(\frac{4}{4!} - \frac{1}{4!}\right) + \ldots + \left(\frac{r}{r!} - \frac{1}{r!}\right) + \ldots\right]$$

$$= \frac{1}{e}\left[1 + \frac{1}{1!} + \frac{2}{2!} + \frac{3}{3!} + \frac{4}{4!} \times \ldots + \frac{r}{r!} + \ldots - \frac{1}{1!} - \frac{1}{2!} - \frac{1}{3!} - \frac{1}{4!} \ldots - \frac{1}{r!} - \ldots\right]$$

$$= \frac{1}{e}\left[1 + \left\{1 + \frac{1}{1!} + \frac{1}{2!} + \frac{1}{3!} + \ldots + \frac{1}{(r-1)!} + \ldots\right\} - \left\{1 + \frac{1}{1!} + \frac{1}{2!} + \frac{1}{3!} + \frac{1}{4!} + \ldots + \frac{1}{r!} \ldots\right\} + 1\right]$$

$$= \frac{1}{e}[1 + e - e + 1] = \frac{2}{e}$$

$$= \frac{2}{e}(1) = \frac{2}{e} \text{ S.D.}$$

**Proved.**

## 11.18 MOMENT GENERATING FUNCTION OF POISSON DISTRIBUTION

(*A.M.I.E., Summer 2000*)

**Solution.** $P(r) = \frac{e^{-m} m^r}{r!}$

Let $M_x(t)$ be the moment generating function, then

$$M_x(t) = \sum_{r=0}^{\infty} e^{tr} \frac{e^{-m}.m^r}{r!} = \sum_{r=0}^{\infty} e^{-m}.\frac{(me^t)^r}{r!} = e^{-m}\left[1 + me^t + \frac{(me^t)^2}{2!} + \frac{(me^t)^3}{3!} + \ldots\right] = e^{-m}.e^{me^t} = e^{m(e^t-1)}$$

## 11.19 CUMULANTS

The cumulant generating function $K_x(t)$ is given by

$$K_x(t) = \log_e M_x(t) = \log_e e^{m(e^t-1)} = m(e^t - 1)\log_e e$$

$$= m(e^t - 1) = m\left[1 + t + \frac{t^2}{2!} + \frac{t^3}{3!} + \ldots + \frac{t^r}{r!} + \ldots - 1\right]$$

$$= m\left[t + \frac{t^2}{2!} + \frac{t^3}{3!} + \ldots + \frac{t^r}{r!} + \ldots\right]$$

Now $K_r = r$ th cumulant = coefficient of $\frac{t^r}{r!}$ in $K(t) = m$

*i.e.*, $k_r = m$, where $r = 1, 2, 3, \ldots$

Hence, all the cumulants of the Poisson distribution are equal. In particular, we have

Mean $= K_1 = m$, $\mu_2 = K_2 = m$, $\mu_3 = K_3 = m$

$$\mu_4 = K_4 + 3K_2^2 = m + 3m^2$$

$$\beta_1 = \frac{\mu_3^2}{\mu_2^3} = \frac{m^2}{m^3} = \frac{1}{m}, \quad \beta_2 = \frac{\mu_4}{\mu_2^2} = \frac{m + 3m^2}{m^2} = \frac{1}{m} + 3$$

## 11.20 RECURRENCE FORMULA FOR POISSON DISTRIBUTION

**Solution.** By Poisson distribution

$$P(r) = \frac{e^{-m}.m^r}{r!} \quad \text{... (1)}$$

$$P(r+1) = \frac{e^{-m}m^{r+1}}{(r+1)!} \quad \text{... (2)}$$

On dividing (2) by (1), we get

$$\frac{P(r+1)}{P(r)} = \frac{e^{-m}m^{r+1}}{(r+1)!}\,\frac{r!}{e^{-m}.m^r} = \frac{m}{r+1}$$

$$P(r+1) = \frac{m}{r+1}P(r) \qquad \textbf{Ans.}$$

**Example 38.** *If the variance of the Poisson distribution is 2, find the probabilities for r = 1, 2, 3, 4 from the recurrence relation of the Poisson distribution. Also find $P(r \geq 4)$.*

**Solution.** Variance = $m$ = 2;
Mean = 2

$$P(r+1) = \frac{m}{r+1}P(r) \qquad \text{[Recurrence relation]}$$

Now $$P(r+1) = \frac{2}{r+1}P(r) \qquad (m = 2)$$

If $r = 0,\ P(1) = \frac{2}{0+1}P(0) = \frac{2}{0+1}(0.1353) = 0.2706 \qquad [P(0) = e^{-m} = e^{-2} = 0.1353]$

If $r = 1,\ P(2) = \frac{2}{1+1}P(1) = \frac{2}{2}(0.2706) = 0.2706$

If $r = 2,\ P(3) = \frac{2}{2+1}P(2) = \frac{2}{3}(0.2706) = 0.1804$

If $r = 3,\ P(4) = \frac{2}{3+1}P(3) = \frac{1}{2}(0.1804) = 0.0902$

$$P(r \geq 4) = P(4) + P(5) + P(6) + \ldots$$
$$= 1 - [P(0) + P(1) + P(2) + P(3)]$$
$$= 1 - [0.1353 + 0.2706 + 0.2706 + 0.1804]$$
$$= 1 - 0.8569 = 0.1431 \qquad \textbf{Ans.}$$

**Example 39.** *Assume that the probability of an individual coal miner being killed in a mine accident during a year is $\frac{1}{2400}$. Use appropriate statistical distribution to calculate the probability that in a mine employing 200 miners, there will be at least one fatal accident in a year.* *(A.M.I.E.T.E., Summer 2001)*

**Solution.** $P = \frac{1}{2400},\quad n = 200$

$$m = np = \frac{200}{2400} = \frac{1}{12}$$

P (At least one) $= P(1 \text{ or } 2 \text{ or } 3 \text{ or } \ldots\ldots \text{ or } 200)$
$= P(1) + P(2) + P(3) + \ldots + P(200)$

$$= 1 - P(0) = 1 - \frac{e^{-m}.m^0}{0!}$$

$$= 1 - e^{-\frac{1}{12}} = 1 - 0.92 = 0.08$$ **Ans.**

**Example 40.** *Suppose 3% of bolts made by a machine are defective, the defects occurring at random during production. If bolts are packaged 50 per box, find*
*(a) exact probability and*
*(b) Poisson approximation to it, that a given box will contain 5 defectives.*

**Solution.** $$p = \frac{3}{100} = 0.03$$

$$q = 1 - p = 1 - 0.03 = 0.97$$

(*a*) Hence the probability for 5 defective bolts in a lot of 50

$$= {}^{50}C_5\,(0.03)^5\,(0.97)^{45} = 0.013074 \text{ (Binomial Distribution)}$$

(*b*) To get Poisson approximation $m = n\,p = 50 \times \frac{3}{100} = \frac{3}{2} = 1.5$

Required Poisson approximation $= \frac{m^r e^{-m}}{r!} = \frac{(1.5)^5 e^{-1.5}}{5!} = 0.01412$ **Ans.**

**Example 41.** *The number of arrivals of customers during any day follows Poisson distribution with a mean of 5. What is the probability that the total number of customers on two days selected at random is less than 2?*

**Solution.** $m = 5$

$$P(r) = \frac{e^{-m} m^r}{r!}, \quad P(r) = \frac{e^{-5}(5)^r}{r!}$$

If the number of customers on two days < 2 = 1 or 0

| *First day* | *Second Day* | *Total* |
|---|---|---|
| 0 | 0 | 0 |
| 0 | 1 | 1 |
| 1 | 0 | 1 |

Required probability $= P(0)\,P(0) + P(0)\,P(1) + P(1)\,P(0)$

$$= \frac{e^{-5}(5)^0}{0!}.\frac{e^{-5}(5)^0}{0!} + \frac{e^{-5}(5)^0}{0!}.\frac{e^{-5}(5)^1}{1!} + \frac{e^{-5}(5)^1}{1!}\frac{e^{-5}(5)^0}{0!}$$

$$= e^{-5}\cdot e^{-5} + e^{-5}\cdot e^{-5}\cdot 5 + e^{-5}\cdot 5\cdot e^{-5}$$

$$= e^{-10}\,[1 + 5 + 5] = 11e^{-10} = 11 \times 4.54 \times 10^{-5}$$

$$= 4.994 \times 10^{-4}$$ **Ans.**

**Example 42.** *Using Poisson distribution, find the probability that the ace of spades will be drawn from a pack of well-shuffled cards at least once in 104 consecutive trials.*

**Solution.** Probability of the ace of spades $= P = \frac{1}{52}$, $n = 104$

$$m = np = 104 \times \frac{1}{52} = 2$$

$$P(r) = e^{-m}.\frac{m^r}{r!} = e^{-2}.\frac{2^r}{r!} = \frac{1}{e^2}\frac{2^r}{r!}$$

$P$ (At least once) = $P(1) + P(2) + P(3) + \ldots + P(104) = 1 - P(0)$

$$= 1 - \frac{1}{e^2} \times \frac{2^0}{0!} = 1 - \frac{1}{e^2} = 1 - 0.135 = 0.865$$ **Ans.**

**Example 43.** *In a certain factory producing cycle tyres, there is a small chance of 1 in 500 tyres to be defective. The tyres are supplied in lots of 10. Using Poisson distribution, calculate the approximate number of lots containing no defective, one defective and two defective tyres, respectively, in a consignment of 10,000 lots.*

**Solution.** $p = \frac{1}{500}, \quad n = 10$

$$m = np = 10 \cdot \frac{1}{500} = \frac{1}{50} = 0.02, \quad P(r) = \frac{e^{-m} . m^r}{r!}$$

| S.No. | Proability of defective | Number of lots containing defective |
|---|---|---|
| 1 | $P(0) = \frac{e^{-0.02}(0.02)^0}{0!} = e^{-0.02} = 0.9802$ | $10{,}000 \times 0.9802 = 9802$ lots |
| 2 | $P(1) = \frac{e^{-0.02}(0.02)^1}{1!}$ $= 0.9802 \times 0.02 = 0.019604$ | $10{,}000 \times 0.019604 = 196$ lots |
| 3. | $P(2) = \frac{e^{-0.02}(0.02)^2}{2!}$ $= 0.9802 \times 0.0002 = 0.00019604$ | $10{,}000 \times 0.000196 = 2$ lots |

**Ans.**

**Example 44.** *A car hire firm has two cars which it hires out day by day. The number of demands for a car on each day is distributed as a Poisson distribution with mean 1.5. Calculate the number of days in a year on which*

*(i) neither car is on demand* $(e^{-1.5} = 0.2231)$

*(ii) a car demand is refused.*

*(MDU, Dec. 2010, A.M.I.E., Summer 2004 Winter 2001, June 2009)*

**Solution.** $m = 1.5$

**(*i*) If the car is not used, then demand ($r$) = 0**

$$P(r) = \frac{e^{-m} . m^r}{r!}, \quad P(0) = \frac{e^{-1.5}(1.5)^0}{0!} = e^{-1.5} = 0.2231$$

**Number of days in a year when the demand is zero = 365 × 0.2231 = 81.4315** **Ans. 81 days**

**(*ii*) Some demand is refused if the number of demands is more than two *i.e.* $r > 2$.**

$$P(r > 2) = P(3) + P(4) + \ldots = 1 - [P(0) + P(1) + P(2)]$$

$$= 1 - \left[\frac{e^{-1.5}(1.5)^0}{0!} + \frac{e^{-1.5}(1.5)^1}{1!} + \frac{e^{-1.5}(1.5)^2}{2!}\right]$$

$$= 1 - [e^{-1.5} + e^{-1.5} \times 1.5 + e^{-1.5} \times 1.125]$$

$$= 1 - e^{-1.5}[1 + 1.5 + 1.125] = 1 - e^{-1.5} \times 3.625$$

$$= 1 - 0.2231 \times 3.625 = 1 - 0.8087375$$

$$= 0.1912625$$ **Ans.**

**Number of days in a year when some demand of car is refused**

**= 365 × 0.1912625 = 69.81 = 70 days** **Ans.**

**Example 45.** *If the probability that an individual suffers a bad reaction from a certain injection is 0.001, determine the probability that out of 2000 individuals*

*(a) exactly 3 (b) more than 2 individuals (c) None (d) More than one individual will suffer a bad reaction.* (A.M.I.E.T.E., June 2007, Winter, 2002, 2000)

**Solution.** $p = 0.001, \quad n = 2000$

$$m = np = 2000 \times 0.001 = 2$$

$$\therefore \quad P(r) = \frac{e^{-m} m^r}{r!} = e^{-2}\frac{2^r}{r!} = \frac{1}{e^2} \times \frac{2^r}{r!}$$

(a) $P\text{ (Exactly 3)} = P(3) = \frac{1}{e^2} \cdot \frac{2^3}{3!} = \frac{1}{(2.718)^2} \times \frac{8}{6} = (0.135) \times \frac{4}{3} = 0.18$

(b) $P\text{ (more than 2)} = P(3) + P(4) + P(5) + \dots + P(2000)$

$$= 1 - [P(0) + P(1) + P(2)]$$

$$= 1 - \left[\frac{e^{-2}(2)^0}{0!} + \frac{e^{-2}(2)^1}{1!} + \frac{e^{-2}(2)^2}{2!}\right]$$

$$= 1 - e^{-2}[1 + 2 + 2] = 1 - \frac{5}{e^2}$$

$$= 1 - 5 \times 0.135 = 1 - 0.675 = 0.325 \qquad \textbf{Ans.}$$

(c) $P\text{ (none)} = P(0) = \frac{e^{-2}(2)^0}{0!} = 0.135$

(d) $P\text{ (more than 1)} = P(2) + P(3) + P(4) + \dots + P(2000) = 1 - [P(0) + P(1)]$

$$= 1 - \left[\frac{e^{-2}(2)^0}{0!} + \frac{e^{-2}(2)^1}{1!}\right] = 1 - 3e^{-2} = 1 - 3 \times 0.135 = 1 - 0.405 = 0.595 \textbf{ Ans.}$$

**Example 46.** *A manufacturer knows that the razor blades he makes contain on an average 0.5% of defectives. He packs them in packets of 5. What is the probability that a packet picked at random will contain 3 or more faulty blades ?*

**Solution.** $p = 0.5\% = 0.005, n = 5$

$$m = np = 5 \times 0.005 = 0.025$$

$$p(r) = \frac{e^{-m} \cdot m^r}{r!} = \frac{e^{-0.025}(0.025)^r}{r!}$$

$$P\text{ (3 or more)} = P(3) + P(4) + P(5) = \frac{e^{-0.025}(0.025)^3}{3!} + \frac{e^{-0.025}(0.025)^4}{4!} + \frac{e^{-0.025}(0.025)^5}{5!}$$

$$= \frac{e^{-0.025}(0.025)^3}{5!}[20 + 5(0.025) + (0.025)^2]$$

$$= \frac{0.975 \times 0.000015625 \times 20.125625}{120}$$

$$= 0.000002555. \qquad \textbf{Ans.}$$

**Example 47.** *Suppose that a book of 600 pages contains 40 printing mistakes. Assume that these errors are randomly distributed throughout the book and x, the number of errors per page has a Poisson distribution. What is the probability that 10 pages selected at random will be free of errors ?*

**Solution.** $p = \frac{40}{600} = \frac{1}{15}$, $n = 10$, $m = np = 10 \times \frac{1}{15} = \frac{2}{3}$

$$P(r) = \frac{e^{-m} \cdot m^r}{r!} = \frac{e^{-2/3}\left(\frac{2}{3}\right)^r}{r!}$$

$$P(0) = \frac{e^{-2/3}\left(\frac{2}{3}\right)^0}{0!} = e^{-2/3} = 0.51 \quad \textbf{Ans.}$$

**Example 48.** *A manufacturer knows that the condensers he makes contain on an average 1% of defectives. He packs them in boxes of 100. What is the probability that a box picked out at random will contain 4 or more faulty condensers?*

**Solution.** $p = 1\% = 0.01,\ n = 100,\quad m = np = 100 \times 0.01 = 1$

$$P(r) = \frac{e^{-m}.(m)^r}{r!} = \frac{e^{-1}(1)^r}{r!} = \frac{e^{-1}}{r!}$$

$$P(4 \text{ or more faulty condensers}) = P(4) + P(5) + \ldots + P(100) = 1 - [P(0) + P(1) + P(2) + P(3)]$$

$$= 1 - \left[\frac{e^{-1}}{0!} + \frac{e^{-1}}{1!} + \frac{e^{-1}}{2!} + \frac{e^{-1}}{3!}\right] = 1 - e^{-1}[1 + 1 + \frac{1}{2} + \frac{1}{6}] = 1 - \frac{8}{3e} = 1 - 0.981 = 0.019 \quad \textbf{Ans.}$$

**Example 49.** *An insurance company found that only 0.01% of the population is involved in a certain type of accident each year. If its 1000 policy holders were randomly selected from the population, what is the probability that not more than two of its clients are involved in such an accident next year? (given that $e^{-0.1} = 0.9048$)*

**Solution.** $p = 0.01\% = \frac{1}{100} \times \frac{1}{100} = \frac{1}{10000}$, $n = 1000$

$$m = np = (1000) \times \frac{1}{10000} = \frac{1}{10} = 0.1$$

$$P(r) = \frac{e^{-m} m^r}{r!}$$

$$P(\text{not more than } 2) = P(0, 1 \text{ and } 2) = P(0) + P(1) + P(2)$$

$$= \frac{e^{-0.1}(0.1)^0}{0!} + \frac{e^{-0.1}(0.1)^1}{1!} + \frac{e^{-0.1}(0.1)^2}{2!}$$

$$= e^{-0.1}\left(1 + 0.1 + \frac{0.01}{2}\right) = 0.9048 \times 1.105 = 0.9998 \quad \textbf{Ans.}$$

**Example 50.** *A skilled typist, on routine work kept a record of mistakes made per day during 300 working days.*

| *Mistakes per day* | 0 | 1 | 2 | 3 | 4 | 5 | 6 |
|---|---|---|---|---|---|---|---|
| No. of days | 143 | 90 | 42 | 12 | 9 | 3 | 1 |

*Fit a Poisson distribution to the above data and hence calculate the theoretical frequencies.*

**Solution.** The mean number of mistakes

$$= \frac{1}{300}(143 \times 0 + 90 \times 1 + 42 \times 2 + 12 \times 3 + 9 \times 4 + 3 \times 5 + 1 \times 6)$$

$$= \frac{1}{300}(90 + 84 + 36 + 36 + 15 + 6) = \frac{267}{300} = 0.89$$

| Number of mistakes | Probability $P(r) = \frac{e^{-0.89} \times (0.89)^r}{r!}$ | Theoretical frequency |
|---|---|---|
| 0 | $\frac{e^{-0.89} \times (0.89)^0}{0!} = 0.411$ | $0.411 \times 300 = 123.3 = 123$ (say) |
| 1 | $\frac{e^{-0.89} \times (0.89)^1}{1!} = 0.365$ | $0.365 \times 300 = 109.5 = 110$ (say) |
| 2 | $\frac{e^{-0.89} \times (0.89)^2}{2!} = 0.163$ | $0.163 \times 300 = 48.9 = 49$ (say) |
| 3 | $\frac{e^{-0.89} \times (0.89)^3}{3!} = 0.048$ | $0.048 \times 300 = 14.4 = 14$ (say) |
| 4 | $\frac{e^{-0.89} \times (0.89)^4}{4!} = 0.011$ | $0.011 \times 300 = 3.3 = 3$ (say) |
| 5 | $\frac{e^{-0.89} \times (0.89)^5}{5!} = 0.002$ | $0.002 \times 300 = 0.6 = 1$ (say) |
| 6 | $\frac{e^{-0.89} \times (0.89)^6}{6!} = 0.0003$ | $0.0003 \times 300 = 0.09 = 0$ (say) |

**Example 51.** *Fit a Poisson distribution to the following data which gives the number of yeast cells per square for 400 squares.*

| *No. of cells per square (x)* | *0* | *1* | *2* | *3* | *4* | *5* | *6* | *7* | *8* | *9* | *10* | *Total* |
|---|---|---|---|---|---|---|---|---|---|---|---|---|
| No. of squares | 103 | 143 | 98 | 42 | 8 | 4 | 2 | 0 | 0 | 0 | 0 | 400 |

*It is given that* $e^{-1.32} = 0.2674$ *(A.M.I.E., Summer 2000)*

**Solution.**

| *x* | *0* | *1* | *2* | *3* | *4* | *5* | *6* | *7* | *8* | *9* | *10* | *Total* |
|---|---|---|---|---|---|---|---|---|---|---|---|---|
| *f* | *103* | *143* | *98* | *42* | *8* | *4* | *2* | *0* | *0* | *0* | *0* | *400* |
| *f.x* | *0.* | *143* | *196* | *126* | *32* | *20* | *12* | *0* | *0* | *0* | *0* | *529* |

$$m = \text{Mean} = \frac{\Sigma f.x}{\Sigma f} = \frac{529}{400} = 1.32$$

**But Poisson distribution is** $P(x) = \frac{e^{-m} \cdot m^x}{r!} = \frac{e^{-1.32}(1.32)^x}{x!} \Rightarrow P(r) = \frac{0.2674(1.32)^x}{x!}$

| *Number of cell* | Probability $P(x) = \frac{0.2674(1.32)^x}{x!}$ | Theoretical frequency | |
|---|---|---|---|
| 0 | $\frac{0.2674(1.32)^0}{0!} = 0.2674$ | $0.2674 \times 400 = 107$ | |
| 1 | $\frac{0.2674(1.32)^1}{1!} = 0.353$ | $0.353 \times 400 = 141$ | |
| 2 | $\frac{0.2674(1.32)^2}{2!} = 0.233$ | $0.233 \times 400 = 93.2$ | |
| 3 | $\frac{0.2674(1.32)^3}{3!} = 0.1025$ | $0.1025 \times 400 = 41$ | |
| 4 | $\frac{0.2674(1.32)^4}{4!} = 0.0338$ | $0.0338 \times 400 = 13.52$ | i.e., 14 |
| 5 | $\frac{0.2674(1.32)^5}{5!} = 0.00893$ | $0.00893 \times 400 = 3.57$ | i.e., 4 |
| 6 | $\frac{0.2674(1.32)^6}{6!} = 0.00196$ | $0.00196 \times 400 = 0.784$ | i.e., 1 |
| 7 | $\frac{0.2674(1.32)^7}{7!} = 0.00037$ | $0.00037 \times 400 = 0.148$ | i.e., 0 |
| 8 | $\frac{0.2674(1.32)^8}{8!} = 0.00006$ | $0.00006 \times 400 = 0.24$ | i.e., 0 |
| 9 | $\frac{0.2674(1.32)^9}{9!} = 0.00000897$ | $0.00000897 \times 400 = 0.003588$ | i.e., 0 |
| 10 | $\frac{0.2674(1.32)^{10}}{10!} = 0.00000118$ | $0.00000118 \times 400 = 0.000472$ | i.e.,0 |

## Exercise 11.5

**1.** Find the probability that at most 5 defective fuses will be found in a box of 200 fuses if experience shows that 2 per cent of such fuses are defective. **Ans. 0.785**

**2.** The number of accidents during a year in a factory has the Poisson distribution with mean 1.5. The accidents during different years are assumed independent. Find the probability that only 2 accidents take place during 2 years time. **Ans. 0.224**

**3.** A manufacturer of cotter pins knows that 5% of his product is defective. If he sells cotter pins in boxes of 100 and guarantee that not more than 10 pins will be defective, what is the approximate probability that a box will fail to meet the guaranteed quality. $[e^{-5} = 0.006738]$ **Ans. 0.0136875**

**4.** Suppose the number of telephone calls on an operator received from 9.00 to 9.05 follow a Poisson distribution with mean 3. Find the probability that

(*i*) the operator will receive no calls in that time interval tomorrow,

(*ii*) in the next three days the operator will receive a total of 1 call in that time interval.

$[e^{-3} = 0.04978]$ **Ans.** (*i*) $e^{-3}$ (*ii*) $3 \times (e^{-3})^2 (e^{-3}.3)$

**5.** On the basis of past record it has been found that there is a 70% chance of power-cut in a city on any particular day. What is the probability that from the first to the 10th day of the month, there are 5 or more days without power cut. (*A.M.I.E.T.E., Summer 2001*)

**Ans.** $\left(\frac{3^5}{5!}+\frac{3^6}{6!}+\frac{3^7}{7!}+\frac{3^8}{8!}+\frac{3^9}{9!}+\frac{3^{10}}{10!}\right)e^{-3}$

6. The distribution of typing mistakes committed by a typist is given below. Assuming a Poisson model, find out the expected frequencies:

| Mistakes per pages | 0 | 1 | 2 | 3 | 4 | 5 |
|---|---|---|---|---|---|---|
| No. of pages | 142 | 156 | 69 | 27 | 5 | 1 |

**Ans.** 147, 147, 74, 25, 6, 1 pages.

7. Let $x$ be the number of cars per minute passing a certain crossing of roads between 5.00 P.M. and 7.00 P.M. on a holiday. Assume $x$ has a Poisson distribution with mean 4. Find the probability of observing atmost 3 cars during any given minute between 5.00 P.M. and 7 P.M. (given $e^{-4} = 0.0183$)

**Ans.** 0.4331

8. Let $x$ be the number of cars, passing a certain point, per minute at a particular time. Assuming that $x$ has a poisson distribution with mean 0.5, find the probability of observing 3 or fewer cars during any given minute. **Ans.** 0.998

9. Number of customers arriving at a service counter during a day has a Poisson distribution with mean 100. Find the probability that at least one customer will arrive on each day during a period of five days. Also find the probability that exactly 3 customers will arrive during two days.

**Ans.** $(1 - e^{-100})^5,\ e^{-200} \times \dfrac{4(100)^3}{3}$

10. The random variable $X$ has a Poisson distribution. If

$P(X = 1) = 0.01487, P(X = 2) = 0.04461$. Then find $P(X = 3)$. **Ans.** 0.08922

11. A source of water is known to contain bacteria with mean number of bacteria per cc equal to 2. Five 1 cc test tubes were filled with water. Assuming that Poisson distribution is applicable, calculate the probability that exactly 2 test tubes contain at least 1 bacterium each.

**Ans.** $\dfrac{2}{5}(1 - e^{-2}) = 0.3459$

12. In a normal summer, a truck driver gets on an average one puncture in 1000 km. Applying Poisson distribution, find the probability that he will have

(*i*) no puncture, (*ii*) two punctures in a journey of 3000 kms. **Ans.** (*i*) $e^{-3}$ (*ii*) $4.5\ e^{-3}$

13. Wireless sets are manufactured with 25 soldered joints each. On the average, 1 joint in 500 is defective. How many sets can be expected to be free from defective joints in a consignment of 10000 sets ?

**Ans.** 9512

14. In a certain factory turning out razor blades, there is small chance $\dfrac{1}{500}$ for any blade to be defective.

The blades are supplied in packets of 10. Using Poisson's distribution, calculate the approximate number of packets containing (i) no defective (ii) one defective and (iii) two defective blades respectively in a consignment of 10,000 packets. ($e^{-0.02} = 0.9802$). **Ans.** (*i*) 9802 (*ii*) 196 (*iii*) 2

15. If m and $\mu_r$ denote by the mean and central rth moment of a Poisson distribution, then prove that

$$\mu_{r+1} = r m \mu_{r-1} + m\frac{d\mu_r}{dm}. \qquad \left[\textbf{Hint. } \mu_r = \sum_{n=0}^{\infty} (x - m)^r \frac{e^{-m} m^x}{x!}, \text{ find } \frac{d\mu_r}{dm}\right]$$

**16.** The random variable $x$ has a Poisson distribution. If $P(x=3)=\frac{1}{6}$, $P(x=2)=\frac{1}{3}$, then $P(x=0)$ is

(*i*) exp (–3/2) (*ii*) exp (3/2) (*iii*) exp (–3) (*iv*) exp (–1/2) **Ans.** (*i*)

**17.** Suppose that on an average 1 house in 1000 houses gets fire in a year in a district. If there are 2000 houses in that district find the probability that exactly 5 houses will have fire during the year. Also find approximate probability using Poisson distribution. (*A.M.I.E.T.E., Dec. 2006*)

**18.** Assuming that the probability of a fatal accident in a factory during the year is $\frac{1}{1200}$. Calculate the probability that in a factory employing 300 workers, there will be at least two fatal accidents in a year ($e^{-0.25} = -0.770$)

**19.** An insurance company found that only 0.01% of the population is involved in a certain type of accident each year. If 1000 policy holders were randomly selected from the population, what is the probability that not more than 2 of its clients are involved in such an accident next year. (*A.M.I.E. Summer 2001*) **Ans. 0.9998**)

**20.** Fill in the blanks :

(*a*) If a random variable $x$ follows Poisson distribution such that $P(x=1) = P(x=2)$, then the mean of the distribution is ......

(*b*) Mean and variance of a Poisson distribution are ......

(*c*) If the probability of a defective fuse is 0.05, the variance for the distribution of defective fuses in a total of 40 is .....

(*d*) The probability of the king of hearts drawn from a pack of cards once in 52 trials is ...

(*e*) If the standard deviation of the Poisson distribution is $\sqrt{2}$, the probability for $r = 2$ is .....

(*f*) If $x$ has a modified Poisson distribution.

$$P_k = P_r(x=k) = \frac{(e^m-1)^{-1}m^k}{k!}, (k=1,2,3....),$$ then the expectation of $x$ is ....

(*g*) If $x$ has a poisson distribution such that $P(x=k) = P(x=k+1)$ for some positive integer $k$ then mean of $x$ is ... (*A.M.I.E., Summer 2000*)

**Ans.** (*a*) 2, (*b*) equal, (*c*) 2, (*d*) $\frac{1}{e}$, (*e*) $\frac{2}{e^2}$, (*f*) $\frac{me^m}{e^m-1}$ (g) $k+1$.

**21.** Choose the correct answer:

(*a*) Let $X$ be a Poisson random variable, such that $2P(X=0) = P(X=2)$. Then standard deviation of $x$ is

(*i*) 4. (*ii*) 2. (*iii*) $-\sqrt{2}$ (*iv*) $\sqrt{2}$ **Ans.** (*iv*)

(*b*) A card is drawn from a well shuffled pack of cards. A sequence of 156 consecutive trials are made. Using Poisson distribution, the probability that the Queen of clubs will be drawn at least once is obtained as

(*i*) $e^{-3}$ (*ii*) $1-e^{-3}$ (*iii*) $e^{-\frac{1}{3}}$ (*iv*) $1-e^{-\frac{1}{3}}$ **Ans.** (*ii*)

## NORMAL DISTRIBUTION

### 11.21 CONTINUOUS DISTRIBUTION

So far we have dealt with discrete distributions where the variate takes only the integral values. But the variates like temperature, heights and weights can take all values in a given interval. Such variables are called continuous variables.

**Distribution function.**

If $F(x) = P(X \le x) = \int_{-\infty}^{x} f(x)dx$, then $f(x)$ is defined as the Distribution Function.

Let $f(x)$ be a continuous function, then Mean $= \int_{-\infty}^{+\infty} x f(x) dx$

$$\text{Variance} = \int_{-\infty}^{+\infty} (x-\bar{x})^2 . f(x) dx. \qquad (\bar{x} = \text{mean})$$

**Note.** $f(x)$ is called probability density function if

(1) $f(x) \geq 0$ for every value of $x$. (2) $\int_{-\infty}^{\infty} f(x) dx = 1$ (3) $\int_a^b f(x) dx = P, (a < x < b)$

**Example 52.** *A function* $f(x)$ *is defined as follows*

$$f(x) = \begin{cases} 0, & x < 2 \\ \frac{1}{18}(2x+3), & 2 \leq x \leq 4 \\ 0, & x > 4 \end{cases}$$

*Show that it is a probability density function.*

**Solution.**

$$f(x) = \begin{cases} 0, & x < 2 \\ \frac{1}{18}(2x+3), & 2 \leq x \leq 4 \\ 0, & x > 4 \end{cases}$$

*If $f(x)$ is a probability density function, then*

(*i*) $\int_{-\infty}^{\infty} f(x) dx = 1$

Here $\int_2^4 \frac{1}{18}(2x+3) dx = \frac{1}{18}[x^2+3x]_2^4 = \frac{1}{18}(16+12-4-6) = 1$

(*ii*) $f(x) > 0$ for $2 \leq x \leq 4$

Hence, the given function is a probability density function. **Proved.**

**Example 53.** *The diameter of an electric cable is assumed to be continuous random variate with probability density function:*

$f(x) = 6x(1-x), \; 0 \leq x \leq 1$

(*i*) *verify that above is a p.d.f.* (*ii*) *find the mean and variance.*

**Solution.** (*i*) $\int_{-\infty}^{\infty} f(x) dx = \int_0^1 6x(1-x) dx = \int_0^1 (6x - 6x^2) dx$

$$= (3x^2 - 2x^3)_0^1 = 3 - 2 = 1$$

Secondly $f(x) > 0$ for $0 \leq x \leq 1$.

Hence the given function is a probability density function.

(*ii*) Mean $= \int_{-\infty}^{\infty} x . f(x) dx = \int_0^1 x . 6x(1-x) dx$

$$= \int_0^1 (6x^2 - 6x^3) dx = \left(2x^3 - \frac{3}{2}x^4\right)_0^1 = 2 - \frac{3}{2} = \frac{1}{2} \qquad \textbf{Ans.}$$

$$\text{Variance} = \int_{-\infty}^{\infty} (x-\bar{x})^2 . f(x) dx = \int_0^1 \left(x - \frac{1}{2}\right)^2 .6x(1-x) dx$$

$$= \int_0^1 \left(x^2 - x + \frac{1}{4}\right)(6x - 6x^2) dx = \int_0^1 \left(12x^3 - 6x^4 - \frac{15}{2}x^2 + \frac{3}{2}x\right) dx$$

$$= \left(3x^4 - \frac{6}{5}x^5 - \frac{5}{2}x^3 + \frac{3x^2}{4}\right)_0^1 = \left(3 - \frac{6}{5} - \frac{5}{2} + \frac{3}{4}\right) = \frac{1}{20}$$ **Ans.**

**Example 54.** *If the probability density function of a random variable x is*

$$f(x) = \begin{cases} kx^{\alpha-1}(1-x)^{\beta-1}, & \textit{for } 0 < x < 1, \alpha > 0, \beta > 0 \\ 0, & \textit{otherwise} \end{cases}$$

*Find k and mean of x.*

**Solution.** If $f(x)$ is a probability density function,

Then $\int_{-\infty}^{\infty} f(x)dx = 1$

Here $\int_0^1 kx^{\alpha-1}(1-x)^{\beta-1}dx = 1$ [$f(x)$ is a beta function.]

$$\Rightarrow \quad k\frac{\overline{|\alpha}\,\overline{|\beta}}{\overline{|\alpha+\beta}} = 1 \qquad \Rightarrow \qquad k = \frac{\overline{|\alpha+\beta}}{\overline{|\alpha}\,\overline{|\beta}}$$ **Ans.**

$$\text{Mean} = \int_{-\infty}^{\infty} x.f(x)dx = \int_0^1 x.kx^{\alpha-1}(1-x)^{\beta-1}dx = k\int_0^1 x^{\alpha+1-1}(1-x)^{\beta-1}dx$$

$$= \frac{\overline{|\alpha+\beta}}{\overline{|\alpha}\,\overline{|\beta}}.\frac{\overline{|\alpha+1}\,\overline{|\beta}}{\overline{|\alpha+\beta+1}} = \frac{\overline{|\alpha+\beta}}{\overline{|\alpha}\,\overline{|\beta}}\frac{\alpha\overline{|\alpha}\,\overline{|\beta}}{(\alpha+\beta)\overline{|\alpha+\beta}} = \frac{\alpha}{\alpha+\beta}$$ **Ans.**

### Exercise 11.6

1. The two equal sides of an isosceles triangle are of length $a$ each and the angle $\theta$ between them has a probability density function proportional to $\theta(\pi - 0)$ in the range $\left(0, \frac{\pi}{2}\right)$ and zero otherwise. Find the mean value and variance of area of triangle. *(AMIETE Dec. 2005)*
2. Suppose that certain bolts have length L = 400 + X mm, where X is a random variable with probability distribution function.

$$f(x) = \frac{3}{4}(1-x^2), \ -1 \geq x \leq 1 \text{ and } 0, \text{ otherwise}$$

(i) Determine C so that with probability $\frac{11}{16}$, a bolt will have length between 400 – C and 400 + C

(ii) Find the mean and variance of bolt length L. Also find mean and variance of $(2L + 5)$.

3. Let $f(x)$ be a function defined as $f(x) = e^{-x}$, for $x \geq 0$ and $f(x) = 0$ for $x < 0$, then the value of probability distribution function $x = 2$.

(a) $1 + e^{-2}$ (b) $1 - e^{-2}$ (c) $1 + e^2$ (d) $1 + e^{-2.5}$ *(A.M.I.E.T.E., Dec. 2006)* **Ans. (b)**

## 11.22 MOMENT GENERATING FUNCTION OF THE CONTINUOUS PROBABILITY DISTRIBUTION ABOUT $x = a$ is given by

$$M_a(t) = \int_{-\infty}^{\infty} e^{t(x-a)} f(x)dx \qquad \text{where } f(x) \text{ is p.d.f.}$$

**Example 55.** *Find the moment generating function of the exponential distribution*

$$f(x) = \frac{1}{c}e^{-x/c} \qquad 0 \leq x \leq \infty, \ c > 0$$

*Hence find its mean and S.D.*

**Solution.** The moment generating function about origin is

$$M_0(t) = \int_0^{\infty} e^{tx}\frac{1}{c}e^{-x/c}dx = \frac{1}{c}\int_0^{\infty} e^{(t-1/c)}x dx$$

$$= \frac{1}{c}\left[\frac{e^{(t-1/c)x}}{t-\frac{1}{c}}\right]_0^\infty = \frac{1}{c}\left[-\frac{1}{t-\frac{1}{c}}\right] = \frac{1}{1-ct} = (1-ct)^{-1}$$

$$= 1 + ct + c^2t^2 + c^3t^3 + c^4t^4 + \ldots$$

$$\mu_1' = \frac{d}{dt}[M_0(t)]_{t=0} = [c + 2c^2t + 3c^3t^2 + 4c^4t^3 + \ldots]_{t=0} = c$$

$$\mu_2' = \frac{d^2}{dt^2}[M_0(t)]_{t=0} = [2c^2 + 6c^3t + 12c^4t^2 + \ldots]_{t=0} = 2c^2$$

$$\mu_2 = \mu_2' - (\mu_1')^2 = 2c^2 - c^2 = c^2$$ **Ans.**

S.D. $= c$

Mean $= c$, S.D. $= c$

## 11.23 NORMAL DISTRIBUTION

**Normal distribution is a continuous distribution. It is derived as the limiting form of the Binomial distribution for large values of *n* and *p* and *q* are not very small.**

**The normal distribution is given by the equation**

$$f(x) = \frac{1}{\sigma\sqrt{(2\pi)}} e^{-\frac{(x-\mu)^2}{2\sigma^2}} \quad \ldots (1)$$

**where $\mu$ = mean, $\sigma$ = standard deviation, $\pi = 3.14159 \ldots$, $e = 2.71828 \ldots$**

$$P(x_1 < x < x_2) = \int_{x_1}^{x_2} \frac{1}{\sigma\sqrt{2\pi}} e^{-\frac{(x-\mu)^2}{2\sigma^2}} dx$$

**On substitution $z = \frac{x-\mu}{\sigma}$ in (1), we get $f(z) = \frac{1}{\sqrt{2\pi}} e^{-\frac{1}{2}z^2}$** ... (2)

**Here mean = 0, standard deviation = 1.**

**(2) is known as standard form of normal distribution.**

## 11.24 NORMAL CURVE

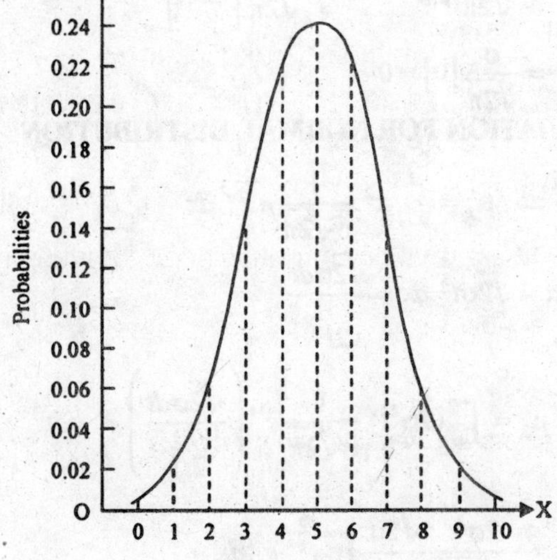

Let us show binomial distribution graphically. The probabilities of heads in 1 tosses are
$^{10}C_0q^{10}p^0,\ ^{10}C_1q^9p^1,\ ^{10}C_2q^8p^2,\ ^{10}C_3q^7p^3,$
$^{10}C_4q^6p^4,\ ^{10}C_5q^5p^5,\ ^{10}C_6q^4p^6,\ ^{10}C_7q^3p^7,\ ^{10}C_8q^2p^8,\ ^{10}C_9q^1p^9,\ ^{10}C_{10}q^0p^{10}.$

$p=\frac{1}{2}, q=\frac{1}{2}$. It is shown in the given figure.

If the variates (heads here) are treated as if they were continuous, the required probability curve will be a *normal curve* as shown in the above figure by dotted lines.

**Properties of the normal curve.** $y = y_0 e^{-\frac{x^2}{2\sigma^2}}$

1. The curve is symmetrical about the $y$-axis. The mean, median and mode coincide at the origin.
2. The curve is drawn, if mean (origin of $x$) and standard deviation are given. The value of $y_0$ can be calculated from the fact that the area of the curve must be equal to the total number of observations.
3. $y$ decreases rapidly as $x$ increases numerically. The curve extends to infinity on either side of the origin.
4. (*a*) $P(\mu-\sigma < x < \mu+\sigma) = 68\%$

   (*b*) $P(\mu-2\sigma < x < \mu+2\sigma) = 95.5\%$

   (*c*) $P(\mu-3\sigma < x < \mu+3\sigma) = 99.7\%$

Hence (*a*) About $\frac{2}{3}$ of the values will lie between $(\mu-\sigma)$ and $(\mu+\sigma)$.

(b) About 95% of the values will lie between $(\mu-2\sigma)$ and $(\mu+2\sigma)$.

(c) About 99.7 % of the values will be between $(\mu-3\sigma)$ and $(\mu+3\sigma)$.

### 11.25 MEAN FOR NORMAL DISTRIBUTION

$$\text{Mean} = \int_{-\infty}^{+\infty} \frac{1}{\sigma\sqrt{2\pi}} e^{-\frac{x^2}{2\sigma^2}} \cdot x\,dx \qquad \left[\text{Putting } \frac{x}{\sigma} = t\right]$$

$$= \frac{1}{\sigma\sqrt{2\pi}} \int_{-\infty}^{+\infty} e^{-\frac{t^2}{2}} (t\sigma)(\sigma dt)$$

$$= \frac{\sigma}{\sqrt{2\pi}} \int_{-\infty}^{+\infty} t e^{-\frac{t^2}{2}} dt = \frac{\sigma}{\sqrt{2\pi}} \left[ e^{-\frac{t^2}{2}} \right]_{-\infty}^{+\infty}$$

$$= \frac{\sigma}{\sqrt{2\pi}} [0] = 0$$

### 11.26 STANDARD DEVIATION FOR NORMAL DISTRIBUTION

$$\mu_2' = \int x^2 . f(x) dx \Rightarrow \mu_2' = \int_{-\infty}^{+\infty} x^2 . \frac{1}{\sigma\sqrt{2\pi}} e^{-\frac{x^2}{2\sigma^2}} dx$$

$$\text{Put } \frac{x^2}{2\sigma^2} = t \Rightarrow x = \sqrt{2}\sigma t^{\frac{1}{2}} . dx = \frac{\sqrt{2}\sigma dt}{2t^{\frac{1}{2}}}$$

$$\Rightarrow \qquad \mu_2' = \int_{-\infty}^{+\infty} (2\sigma^2 t) . \frac{1}{\sigma\sqrt{2\pi}} e^{-t} \left( \frac{\sqrt{2}\sigma dt}{2t^{\frac{1}{2}}} \right)$$

$$= \frac{2\sigma^2}{\sigma\sqrt{(2\pi)}} \frac{\sqrt{2}\sigma}{2} \int_{-\infty}^{+\infty} t^{\frac{3}{2}-1} e^{-t} dt, \qquad \left[ \int_0^{\infty} x^{n-1} e^{-x} dx = \overline{|n} \right]$$

$$= \frac{\sigma^2}{\sqrt{\pi}}.2\overline{\left|\frac{3}{2}\right.} = 2\frac{\sigma^2}{\sqrt{\pi}}.\frac{1}{2}\overline{\left|\frac{1}{2}\right.} = \frac{\sigma^2}{\sqrt{\pi}}\sqrt{\pi} = \sigma^2$$

$$\mu_2 = \mu_2' - (\mu_1)^2 = \sigma^2 - (0)^2 = \sigma^2$$

$$S.D. = \sigma$$ **Ans.**

## 11.27 MEDIAN OF THE NORMAL DISTRIBUTION

If $a$ is the median, then it divides the total area into two equal halves so that

$$\int_{-\infty}^{a} f(x)dx = \frac{1}{2} = \int_{a}^{\infty} f(x)dx$$

where $$f(x) = \frac{1}{\sigma\sqrt{2\pi}} e^{-\frac{(x-\mu)^2}{2\sigma}}$$

Suppose $a >$ mean, $\mu$ then

$$\int_{-\infty}^{\mu} f(x)dx + \int_{a}^{\infty} f(x)dx = \frac{1}{2} \qquad \left[\text{But} \int_{-\infty}^{\mu} f(x)dx = \frac{1}{2}\right]$$

$$\frac{1}{2} + \int_{\mu}^{a} f(x)dx = \frac{1}{2} \qquad (\mu = \text{mean})$$

$$\int_{\mu}^{a} f(x)dx = 0$$

Thus $a = \mu$

Similarly, when $a <$ mean, we have $a = \mu$

Thus, median = mean = $\mu$.

**Q.** Let $X$ be a random variable having a normal distribution.
If $P(x < 0) = P(X > 2) = 0.4$, then mean value of $X$ is.

(*a*) 0 (*b*) 1 (*c*) 1.5 (*d*) 2 (*A.M.I.E.T.E., Dec. 2004*) **Ans. (*b*)**

## 11.28 MEAN DEVIATION ABOUT THE MEAN $\mu$

Mean deviation $= E|x - \mu|$

$$= \int_{-\infty}^{\infty} |x-\mu| \frac{1}{\sigma\sqrt{2\pi}} e^{-\frac{(x-\mu)^2}{2\sigma^2}} dx$$

$$= \sigma \cdot \frac{1}{\sqrt{2\pi}} \int_{-\infty}^{\infty} |z| e^{-\frac{z^2}{2}} dz \qquad \text{where } z = \frac{x-\mu}{\sigma}$$

$$= \sigma \frac{1}{\sqrt{2\pi}} \left[\int_{-\infty}^{0} -ze^{-\frac{z^2}{2}} dz + \int_{0}^{\infty} ze^{-\frac{z^2}{2}} dz\right]$$

$$= \frac{2\sigma}{\sqrt{2\pi}} \int_{0}^{\infty} ze^{-\frac{z^2}{2}} dz \qquad \text{(as the function is given)}$$

$$= \sigma\sqrt{\frac{2}{\pi}} = \frac{4}{5}\sigma \text{ approximately.}$$

## 11.29 MODE OF THE NORMAL DISRIBUTION

We know that mode is the value of the variate $x$ for which $f(x)$ is maximum. Thus, by differential calculus $f(x)$ is maximum if $f'(x) = 0$ and $f''(x) < 0$

where $$f(x) = \frac{1}{\sigma\sqrt{2\pi}} e^{-\frac{(x-\mu)^2}{2\sigma^2}}$$

Clearly $f(x)$ will be maximum when the exponent will be maximum which will be the case when $x = \mu$.

Thus mode is $\mu$ and modal ordinate $= \frac{1}{\sigma\sqrt{2\pi}}$

## 11.30 MOMENT OF NORMAL DISTRIBUTION

$$\mu_{2n+1} = \int_{-\infty}^{\infty} (x-\mu)^{2n+1} f(x) dx$$ (*A.M.I.E., Winter 2001*)

$$= \frac{1}{\sigma\sqrt{2\pi}} \int_{-\infty}^{\infty} (x-\mu)^{2n+1} e^{\frac{-(x-\mu)^2}{2\sigma^2}} dx$$

$$= \frac{1}{\sqrt{2\pi}} \int_{-\infty}^{\infty} (\sigma z)^{2n+1} = e^{-\frac{z^2}{2}} dz \qquad \left[z = \frac{x-\mu}{\sigma}\right]$$

$$= \frac{\sigma^{2n+1}}{\sqrt{2\pi}} \int_{-\infty}^{\infty} z^{2n+1} e^{-\frac{z^2}{2}} dz$$

$$= 0$$ (Since $z^{2n+1} e^{\frac{-z^2}{2}}$ is an odd function)

$$\mu_{2n} = \int_{-\infty}^{\infty} (x-\mu)^{2n} f(x) dx$$

$$= \frac{1}{\sqrt{2\pi}} \int_{-\infty}^{\infty} (\sigma z)^{2n} = e^{-\frac{z^2}{2}} dz = \frac{\sigma^{2n}}{\sqrt{2\pi}} \int_{-\infty}^{\infty} z^{2n} e^{-\frac{z^2}{2}} dz$$

$$= \frac{2\sigma^{2n}}{\sqrt{2\pi}} \int_{0}^{\infty} z^{2n} e^{-\frac{z^2}{2}} dz \qquad \left[z^{2n}.e^{-\frac{z^2}{2}} \text{ is an even function}\right]$$

$$= \frac{2^n \sigma^{2n}}{\sqrt{\pi}} \int_{0}^{\infty} e^{-t} t^{\left(n-\frac{1}{2}\right)} dt \qquad \left[\frac{z^2}{2} = t\right]$$

$$= \frac{2^n \sigma^{2n}}{\sqrt{\pi}} \Gamma\left(n+\frac{1}{2}\right)$$

Changing $n$ to $(n-1)$, we get

$$\mu_{2n-2} = \frac{2^{n-1}\sigma^{2n-2}}{\sqrt{\pi}} \Gamma\left(n-\frac{1}{2}\right)$$

On dividing, we get

$$\frac{\mu_{2n}}{\mu_{2n-2}} = 2\sigma^2 \frac{\Gamma\left(n+\frac{1}{2}\right)}{\Gamma\left(n-\frac{1}{2}\right)} = \frac{2\sigma^2\left(n-\frac{1}{2}\right)\Gamma\left(n-\frac{1}{2}\right)}{\Gamma\left(n-\frac{1}{2}\right)} = 2\sigma^2\left(n-\frac{1}{2}\right)$$

$$\mu_{2n} = \sigma^2(2n-1)\mu_{2n-2}$$

which gives the recurrence relation for the moments of normal distribution.

$$\mu_{2n} = [(2n-1)\sigma^2][(2n-3)\sigma^2]\mu_{2n-4}$$
$$= [(2n-1)\sigma^2][(2n-3)\sigma^2][(2n-5)\sigma^2]\mu_{2n-6}$$
$$= [(2n-1)\sigma^2][(2n-3)\sigma^2][(2n-5)\sigma^2]-(3\sigma^2)(1\cdot\sigma^2)\mu_0$$
$$= (2n-1)(2n-3)(2n-5)\ldots.1\cdot\sigma^{2n}$$
$$= 1.3.5.7\ldots.(2n-5)(2n-3)(2n-1)\sigma^{2n}$$

**Example 56.** *Fit a normal curve to the following data :*

| *Length of line (in cm)* | 8.60 | 8.59 | 8.58 | 8.57 | 8.56 | 8.55 | 8.54 | 8.53 | 8.52 |
|---|---|---|---|---|---|---|---|---|---|
| *Frequency* | 2 | 3 | 4 | 9 | 10 | 8 | 4 | 1 | 1 |

**Solution.** Let $a = 8.56$

| $x$ | $f$ | $d = (x-a)$ | $fd$ | $fd^2$ |
|---|---|---|---|---|
| 8.60 | 2 | 0.04 | 0.08 | 0.0032 |
| 8.59 | 3 | 0.03 | 0.09 | 0.0027 |
| 8.58 | 4 | 0.02 | 0.08 | 0.0016 |
| 8.57 | 9 | 0.01 | 0.09 | 0.0009 |
| 8.56 | 10 | 0 | 0 | 0 |
| 8.55 | 8 | –0.01 | –0.08 | 0.0008 |
| 8.54 | 4 | –0.02 | –0.08 | 0.0016 |
| 8.53 | 1 | –0.03 | –0.03 | 0.0009 |
| 8.52 | 1 | –0.04 | –0.04 | 0.0016 |
| | $\Sigma f = 42$ | | $\Sigma fd = 0.11$ | $\Sigma fd^2 = 0.0133$ |

$$\text{Mean } (\mu) = a + \frac{\Sigma fd}{\Sigma f} = 8.56 + \frac{0.11}{42} = 8.56262$$

$$S.D.(\sigma) = \sqrt{\frac{\Sigma fd^2}{\Sigma f} - \left(\frac{\Sigma fd}{\Sigma f}\right)^2}$$

$$\sigma = \sqrt{\frac{0.0133}{42} - \left(\frac{0.11}{42}\right)^2} = \sqrt{0.00031666 - 0.00000686}$$

$$= 0.0176$$

**Hence the equation of the normal curve fitted to the given data is**

$$P(x) = \frac{1}{\sigma\sqrt{2\pi}} e^{-(x-\mu)^2/2\sigma^2}, -\infty \le x \le \infty$$

**where $\mu = 8.56262$, $\sigma = 0.0176$** **Ans.**

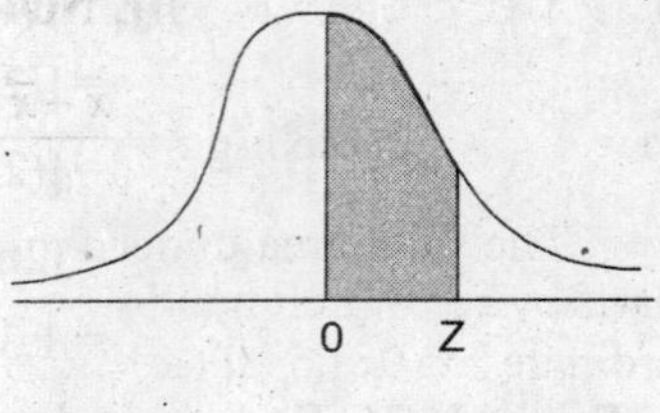


Table

Area under standard normal curve from 0 to $\frac{x-\mu}{\sigma}$

| $\frac{x-\mu}{\sigma}$ | 0.00 | 0.01 | 0.02 | 0.03 | 0.04 | 0.05 | 0.06 | 0.07 | 0.08 | 0.09 |
|---|---|---|---|---|---|---|---|---|---|---|
| 0.0 | .0000 | .0040 | .0080 | .0120 | .0159 | .0199 | .0239 | .0279 | .0319 | .0359 |
| 0.1 | .0398 | .0438 | .0478 | .0517 | .0557 | .0596 | .0636 | .0675 | .0714 | .0753 |
| 0.2 | .0793 | .0832 | .0871 | .0910 | .0948 | .0987 | .1026 | .1084 | .1103 | .1141 |
| 0.3 | .1179 | .1217 | .1255 | .1293 | .1331 | .1368 | .1406 | .1443 | .1480 | .1517 |
| 0.4 | .1554 | .1591 | .1628 | .1664 | .1700 | .1736 | .1772 | .1808 | .1844 | .1879 |
| 0.5 | .1915 | .1950 | .1985 | .2019 | .2054 | .2088 | .2123 | .2157 | .2190 | .2224 |
| 0.6 | .2257 | .2291 | .2324 | .2357 | .2389 | .2422 | .2454 | .2486 | .2518 | .2549 |
| 0.7 | .2580 | .2611 | .2642 | .2671 | .2704 | .2734 | .2764 | .2794 | .2823 | .2852 |
| 0.8 | .2881 | .2910 | .2939 | .2967 | .2995 | .3023 | .3051 | .3078 | .3106 | .3133 |
| 0.9 | .3159 | .3186 | .3212 | 3238 | .3264 | .3289 | .3315 | .3340 | .3365 | .3389 |
| 1.0 | .3413 | .3438 | .3461 | .3485 | .3508 | .3531 | .3554 | .3577 | .3599 | .3621 |
| 1.1 | .3643 | .3665 | .3686 | .3708 | .3729 | .3749 | .3770 | .3790 | .3810 | .3830 |
| 1.2 | .3849 | .3869 | .3888 | .3907 | .3925 | .3944 | .3962 | .3980 | .3997 | .4015 |
| 1.3 | .4032 | .4049 | .4066 | .4082 | .4099 | .4115 | .4131 | .4147 | .4162 | .4177 |
| 1.4 | .4192 | .4207 | .4222 | .4236 | .4251 | .4265 | .4279 | .4292 | .4306 | .4319 |
| 1.5 | .4232 | .4345 | .4357 | .4370 | .4382 | .4394 | .4406 | .4418 | .4430 | .4441 |
| 1.6 | .4452 | .4463 | .4474 | .4485 | .4495 | .4505 | .4515 | .4525 | .4535 | .4545 |
| 1.7 | .4554 | .4564 | .4573 | .4582 | .4591 | .4599 | .4608 | .4616 | .4625 | .4633 |
| 1.8 | .4641 | .4649 | .4656 | .4664 | .4671 | .4678 | .4686 | .4693 | .4699 | .4706 |
| 1.9 | .4713 | .4719 | .4726 | .4732 | .4738 | .4744 | .4750 | .4756 | .4762 | .4767 |
| 2.0 | .4772 | .4778 | .4783 | .4788 | .4793 | .4798 | .4803 | .4808 | .4812 | .4817 |
| 2.1 | .4821 | .4826 | .4830 | .4834 | .4838 | .4842 | .4846 | .4850 | .4854 | .4857 |
| 2.2 | .4861 | .4865 | .4868 | .4871 | .4875 | .4878 | .4881 | .4884 | .4887 | .4990 |
| 2.3 | .4893 | .4896 | .4898 | .4901 | .4904 | .4906 | .4909 | .4911 | .4913 | .4916 |
| 2.4 | .4918 | .4920 | .4922 | .4925 | .4927 | .4929 | .4931 | .4932 | .4934 | .4936 |
| 2.5 | .4938 | .4940 | .4941 | .4943 | .4945 | .4946 | .4948 | .4949 | .4951 | .4952 |
| 2.6 | .4953 | .4955 | .4956 | .4957 | .4959 | .4960 | .4961 | .4962 | .4963 | .4964 |
| 2.7 | .4965 | .4966 | .4967 | .4968 | .4969 | .4970 | .4971 | .4972 | .4973 | .4974 |
| 2.8 | .4974 | .4975 | .4976 | .4977 | .4977 | .4978 | .4979 | .4980 | .4980 | .4981 |
| 2.9 | .4981 | .4982 | .4983 | .4983 | .4984 | .4984 | .4985 | .4985 | .4986 | .4986 |
| 3.0 | .49865 | .4987 | .4987 | .4988 | .4988 | .4989 | .4989 | .4989 | .4990 | .4990 |
| 3.1 | .49903 | .4991 | .4991 | .4991 | .4992 | .4992 | .4992 | .4992 | .4993 | .4993 |

## 11.31 AREA UNDER THE NORMAL CURVE

By taking $z = \frac{x - \overline{x}}{\sigma}$, the standard normal curve is formed.

The total area under this curve is 1. The area under the curve is divided into two equal parts by $z = 0$. Left hand side area and right hand side area to $z = 0$ is 0.5. The area between the ordinate $z = 0$ and any other ordinate can be noted from the table given on the next page.

**Example 57.** *On a final examination in mathematics, the mean was 72, and the standard deviation was 15. Determine the standard scores of students receiving graders.*

(*a*) 60 (*b*) 93 (*c*) 72

**Solution.**

(*a*) $z = \frac{x - \overline{x}}{\sigma} = \frac{60 - 72}{15} = -0.8$ (*b*) $z = \frac{93 - 72}{15} = +1.4$ (*c*) $z = \frac{72 - 72}{15} = 0$ **Ans.**

**Example 58.** *Find the area under the normal curve in each of the cases*

(*a*) *z = 0 and z = 1.2;* (*b*) *z = – 0.68 and z = 0;*

(*c*) *z = – 0.46 and z = 2.21;* (*d*) *z = 0.81 and z = 1.94;*

(*e*) *To the left of z = – 0.6;* (*f*) *Right of z = –1.28.*

**Solution.**

(*a*) Area between $z = 0$ and $z = 1.2$ $= 0.3849$

(*b*) Area between $z = 0$ and $z = -0.68$ $= 0.2518$

(*a*)

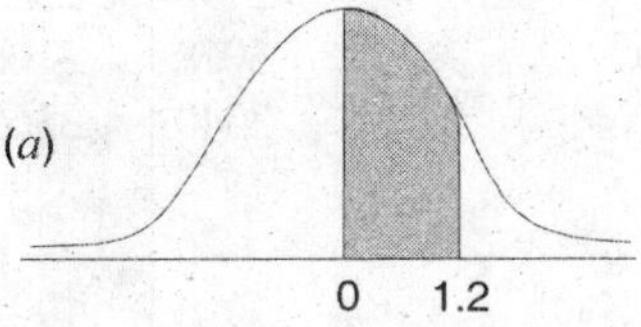


(*b*)

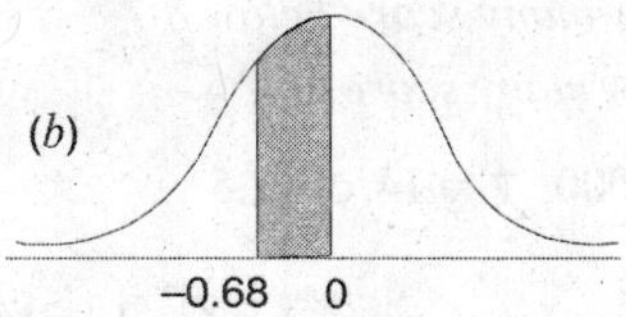


(*c*) Required area = (Area between $z = 0$ and $z = 2.21$)

+ (Area between $z = 0$ and $z = -0.46$)

= (Area between $z = 0$ and $z = 2.21$)

+ (Area between $z = 0$ and $z = 0.46$)

$= 0.4865 + 0.1772 = 0.6637.$

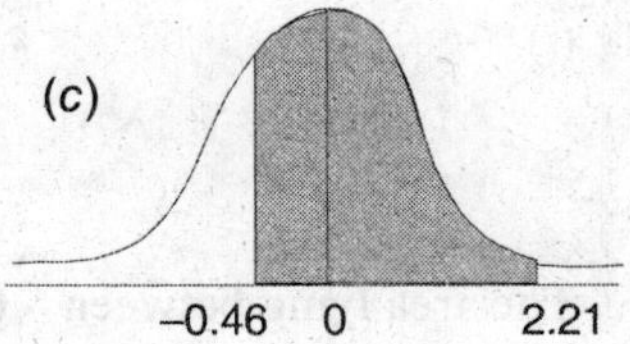


(*d*) Required area = (Area between $z = 0$ and $z = 1.94$) – (Area between $z = 0$ and $z = 0.81$)

$= 0.4738 - 0.2910 = 0.1828$

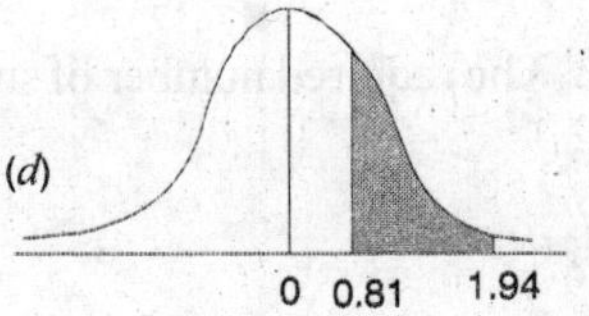


(*e*) Required area = 0.5 – (Area between $z = 0$ and $z = 0.6$)

$= 0.5 - 0.2257 = 0.2743$

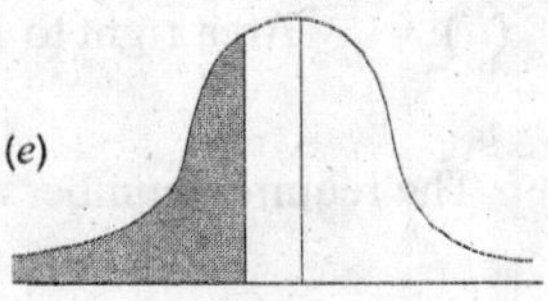


(*f*) Required area = (Area between $z = 0$ and $z = -1.28$) + 0.5

$= 0.3997 + 0.5$

$= 0.8997.$

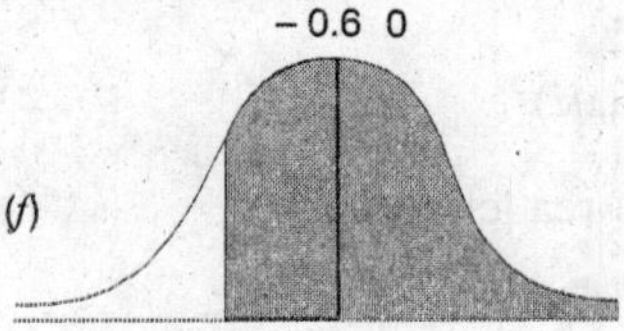

**Example 59.** *Find the value of z in each of the cases*

(*a*) *Area between 0 and z is 0.3770*

(*b*) *Area to the left of z is 0.8621*

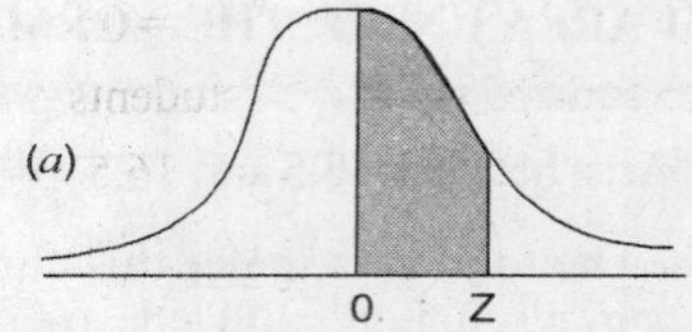


**Solution.**

(*a*) $z = \pm 1.16$

(*b*) Since the area is greater than 0.5.

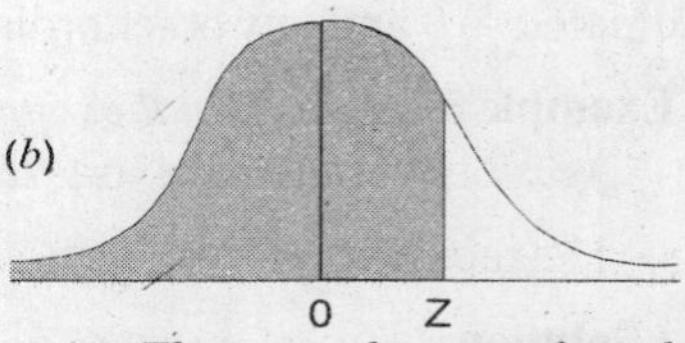


Area between 0 and *z*.

$= 0.8621 - 0.5 = 0.3621$

from which $z = 1.09$ **Ans.**

**Example 60.** *Students of a class were given an aptitude test. Their marks were found to be normally distributed with mean 60 and standard deviation 5. What percentage of students scored more than 60 marks?*

**Solution.** $x = 60, \quad \bar{x} = 60, \quad \sigma = 5$

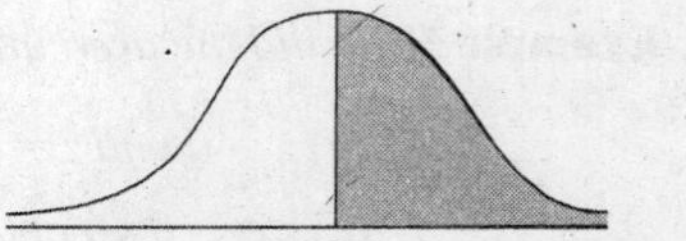


$$z = \frac{x - \bar{x}}{\sigma} = \frac{60 - 60}{5} = 0$$

if $x > 60$ then $z > 0$

Area lying to the right of $z = 0$ is 0.5.

The percentage of students getting more than 60 marks = 50 % **Ans.**

**Example 61.** *In a sample of 1000 cases, the mean of a certain test is 14 and standard deviation is 2.5. Assuming the distribution to be normal, find*

(*i*) *how many students score between 12 and 15 ?*

(*ii*) *how many score above 18 ?*

(*iii*) *how many score below 8 ?*

(*iv*) *how many score 16 ?*

**Solution.** $n = 1000, \ \bar{x} = 14, \sigma = 2.5$

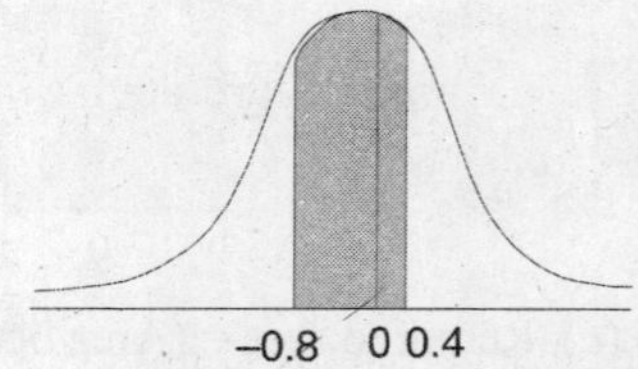


(*i*) $$z_1 = \frac{x - \bar{x}}{\sigma} = \frac{12 - 14}{2.5} = -0.8$$

$$z_2 = \frac{15 - 14}{2.5} = \frac{1}{2.5} = 0.4$$

The area lying between – 0.8 to 0.4 = Area from 0 to – 0.8 + area from 0 to 0.4

$= 0.2881 + 0.1554 = 0.4435$

The required number of students $= 1000 \times 0.4435 = 443.5 = 444$ (say)

$$z_1 = \frac{18 - 14}{2.5} = \frac{4}{2.5} = 1.6$$

(*ii*) Area right to 1.6 $= 0.5 -$ Area between 0 and 1.6

$= 0.5 - 0.4452 = 0.0548$

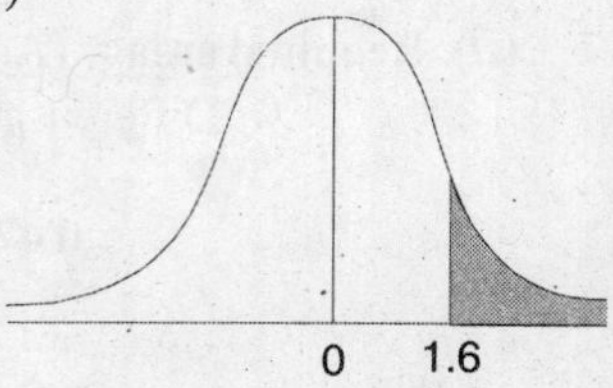


The required number of students

$= 1000 \times 0.0548 = 54.8 = 55$ (say)

(*iii*) $$z = \frac{8 - 14}{2.5} = -\frac{6}{2.5} = -2.4$$

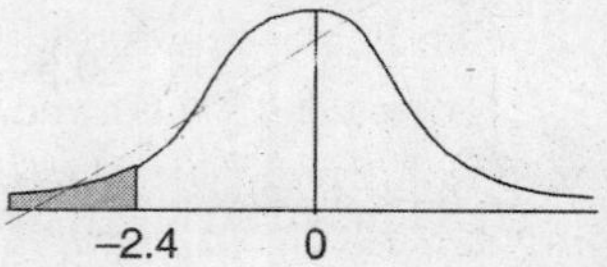


Area left to – 2.4 $= 0.5 -$ area between 0 and – 2.4

$= 0.5 - 0.4918 = 0.0082$

The required number of students $= 1000 \times 0.0082 = 8.2 = 8$ (say)

(*iv*) Area between 15.5 and 16.5

$$z_1 = \frac{15.5 - 14}{2.5} = 0.6$$

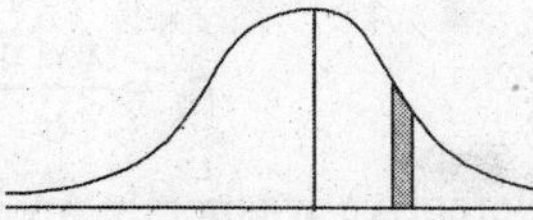


and $$z_2 = \frac{16.5 - 14}{2.5} = 1$$

Area between 0.6 and $1 = 0.3413 - 0.2257 = 0.1156$

The required number of students $= 0.1156 \times 1000$

$= 115.6 = 116$ say **Ans.**

**Example 62.** *Five thousand candidates appeared in a certain examination carrying a maximum of 100 marks. It was found that the marks were normally distributed with mean 39.5 and with standard deviation 12.5. Determine approximately the number of candidates who secured a first class for which a minimum of 60 marks is necessary. You may see the table given below (x denotes the deviation from the mean).*

*The proportion A of the whole area of the normal curve lying to the left of the ordinate at the deviation* $\frac{x}{\sigma}$ *is :*

| $\frac{x}{\sigma}$ | *1.5* | *1.6* | *1.7* | *1.8* |
|---|---|---|---|---|
| *A.* | *0.93319* | *0.94520* | *0.95543* | *0.96407* |

**Solution.** Mean $= \bar{x} = 39.5$

Standard deviation $= \sigma = 12.5$

$$\frac{x}{\sigma} = \frac{60 - 39.5}{12.5} = \frac{20.5}{12.5} = \frac{41}{25} = 1.64$$

We have to find out area for $\frac{x}{\sigma} = 1.64$.

Area for $\frac{x}{\sigma} = 1.6$ is 0.94520.

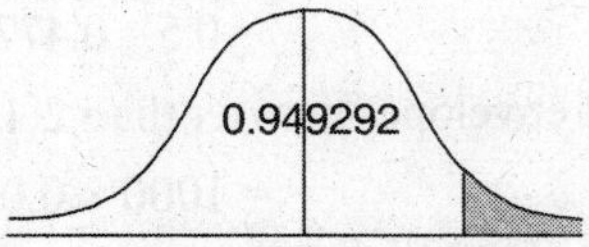


Area for $\frac{x}{\sigma} = 1.7$ is 0.95543.

Difference for 0.1 = 0.01023

Difference for 0.04 = 0.004092

$\therefore$ Area for $\frac{x}{\sigma} = 1.64$ is $0.94520 + 0.004092 = 0.949292$.

Area right to ordinate $1.64 = 1 - 0.949292 = 0.050708$

Number of candidates who secured marks 60 or more $= 5000 \times 0.050708 = 253.54$

Candidates securing first division $= 254$ **Ans.**

**Example 63.** *The mean inside diameter of a sample of 200 washers produced by a machine is 0.0502 cm and the standard deviation is 0.005 cm. The purpose for which these washers are intended allows a maximum tolerance in the diameter of 0.496 to 0.508 cm, otherwise the washers are considered defective. Determine the percentage of defective washers produced by the machine, assuming the diameters are normally distributed* (*A.M.I.E., Summer 2001*)

**Solution.** $z_1 = \dfrac{x-\bar{x}}{\sigma} = \dfrac{0.496-0.502}{0.005} = -1.2$

$z_2 = \dfrac{x-\bar{x}}{\sigma} = \dfrac{0.508-0.502}{0.005} = +1.2$

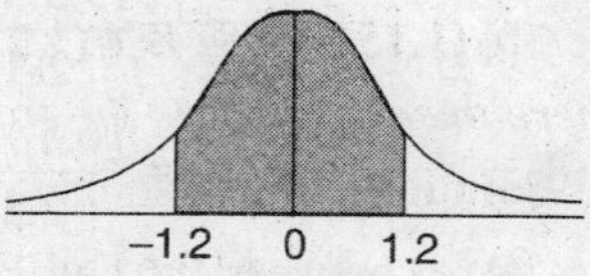


Area for non-defective washers = Area between $z = -1.2$
and $z = +1.2$
= 2 Area between $z = 0$
and $z = 1.2$.
$= 2 \times (0.3849) = 0.7698 = 76.98\%$

Percentage of defective washers = 100 – 76.98
= 23.02% **Ans.**

**Example 64.** *A manufacturer of envelopes knows that the weight of the envelopes is normally distributed with mean 1.9 gm and variance 0.01 gm. Find how many envelopes weighing (i) 2 gm or more, (ii) 2.1 gm or more, can be expected in a given packet of 1000 envelopes. [Given : if t is the normal variable, then* $\phi(o \le t \le 1) = 0.3413$ *and* $\phi(0 \le t \le 2) = 0.4772$].

**Solution.** $\mu = 1.9$ gm, Variance = 0.01 gm

(*i*) $x = 2$ gms or more

$z = \dfrac{x-\mu}{\sigma} = \dfrac{2-1.9}{0.1} = \dfrac{0.1}{0.1} = 1$

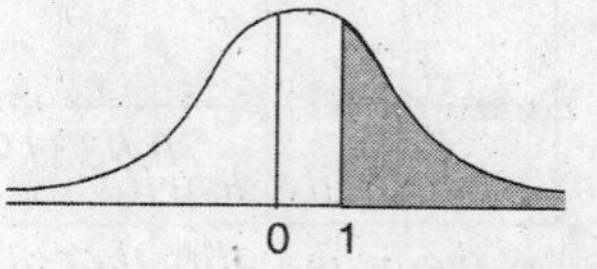


$P(z > 2)$ = Area right to $z = 1$
$= 0.5 - 0.3413 = 0.1587$

Number of envelopes heavier than 2 gm in a lot of 1000
$= 1000 \times 0.1587 = 158.7 = 159$ (app)

(*ii*) $z = \dfrac{2.1-1.9}{0.1} = \dfrac{0.2}{0.1} = 2$

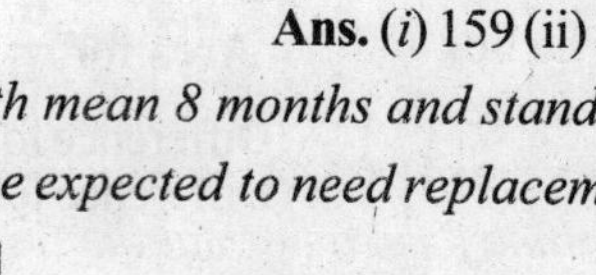


$P(z > 2)$ = Area right to $z = 2$
$= 0.5 - 0.4772 = 0.0228$

Number of envelopes heavier than 2.1 gm in a lot of 1000
$= 1000 \times 0.0228 = 22.8 = 23$ (app) **Ans.** (*i*) 159 (ii) 23

**Example 65.** *The life of army shoes is 'normally' distributed with mean 8 months and standard deviation 2 months. If 5000 pairs are issued how many pairs would be expected to need replacement after 12 months? [Given that* $P(z \ge 2) = 0.0228$ *and* $z = (x - \mu)/\sigma$ ]

**Solution.** Mean $(\mu) = 8$,
Standard deviation $(\sigma) = 2$
Number of pairs of shoes = 5000
Total months $(x) = 12$

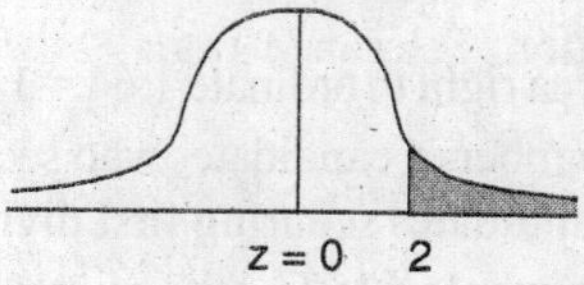


When $z = \dfrac{x-\mu}{\sigma} = \dfrac{12-8}{2} = 2$

Area $(z \ge 2) = 0.0228$

Number of pairs whose life is more than 12 months $(z > 2)$

$$= 5000 \times 0.0228 = 114$$

Replacement after 12 months = 5000 – 114 = 4886 pairs of shoes **Ans.**

**Example 66.** *In a male population of 1000, the mean height is 68.16 inches and standard deviation is 3.2 inches. How many men may be more than 6 feet (72 inches)?*

[$\phi(1.15) = 0.8749, \phi(1.2) = 0.8849, \phi(1.25) = 0.8944$]

*where the argument is the standard normal variable.*

**Solution.** Male population = 1000

Mean height = 68.16 inches

Standard deviation = 3.2 inches

Men more than 7.2 inches = ?

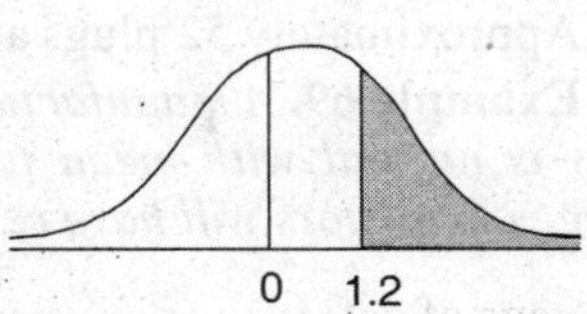


$$\phi(1.15) = 0.8749, \phi(1.2) = 0.8849$$

$$\phi(1.25) = 0.8944$$

$$z = \frac{x - \bar{x}}{\sigma} = \frac{72 - 68.16}{3.2} = 1.2$$

$$\phi(1.2) = 0.8849$$

$\phi$ for more than 1.2 = 1 – 0.8849 = 0.1151

Men more than 72 inches = 1000 × 0.1151 = 115.1

= 115 (say) **Ans.**

**Example 67.** *Pipes for tobacco are being packed in fancy plastic boxes. The length of the pipes is normally distributed with* $\mu = 5''$ *and* $\sigma = 0.1''$. *The internal length of the boxes is* $5.2''$. *What is the probability that the box would be small for the pipe?*

[*given that* $\phi(1.8) = 0.9641$, $\phi(2) = 0.9772$, $\phi(2.5) = 0.9938$]

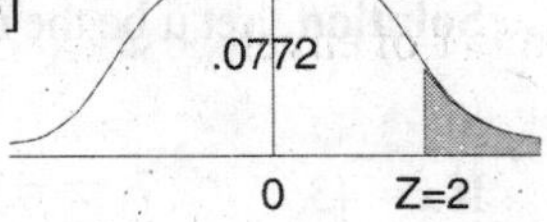


**Solution.** $\mu = 5''$, $\sigma = 0.1''$, $x = 5.2''$

$$\phi(1.8) = 0.9641, \phi(2) = 0.9722, \phi(2.5) = 0.9938$$

$$z = \frac{x - \mu}{\sigma} = \frac{5.2 - 5}{0.1} = 2$$

$$\phi(2) = 0.9772$$

$$\phi(z > 2) = 1 - 0.9772 = 0.0228$$

The box will be small if the length of the pipe is more than 5.2" ($z = 2$).

Hence the probability is 0.0228 **Ans.**

**Example 68.** *Assuming that the diameters of 1,000 brass plugs taken consecutively from a machine form a normal distribution with mean 0. 7515 cm and standard deviation 0. 0020 cm, how many of the plugs are likely to be rejected if the approved diameter is* $0.752 \pm 0.004$ *cm ?*

**Solution.** Tolerance limits of the diameter of non-defective plugs are

0.752 – 0.004 = 0.748 cm and

0.752 + 0.004 = 0.756 cm

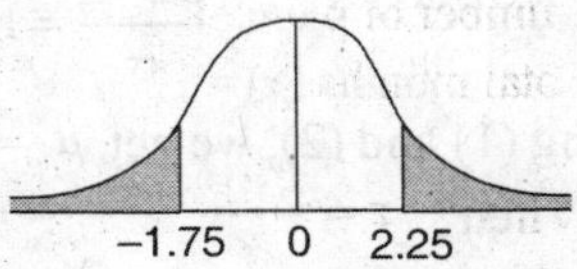


$$z = \frac{x - \mu}{\sigma}$$

If $x_1 = 0.748$, $$z_1 = \frac{0.748 - 0.7515}{0.002} = -1.75$$

If $x_2 = 0.756$, $\quad z_2 = \dfrac{0.756 - 0.7515}{0.002} = 2.25$

Area under $\quad z_1 = -1.75$ to $z_2 = 2.25$

$= (\text{Area from } z=0 \text{ to } z_1 = -1.75) + (\text{Area from } z = 0 \text{ to } z_2 = 2.25) = 0.4599 + 0.4878 = 0.9477$

Number of plugs likely to be rejected $= 1000(1 - 0.9477) = 1000 \times 0.0523 = 52.3$

Approximately 52 plugs are likely to be rejected. **Ans.**

**Example 69.** *A manufacturer knows from experience that the resistance of resistors he produces is normal with mean $\mu = 100$ ohms and standard deviation $\sigma = 2$ ohms. What percentage of resistors will have resistance between 98 ohms and 102 ohms ?*

*(A.M.I.E.T.E., Winter 2003)*

**Solution.** $\mu = 100$ ohms, $\sigma = 2$ ohms

$x_1 = 98, x_2 = 102$

$z = \dfrac{x - \mu}{\sigma}, z_1 = \dfrac{98 - 100}{2} = -1$

$z_2 = \dfrac{102 - 100}{2} = +1$

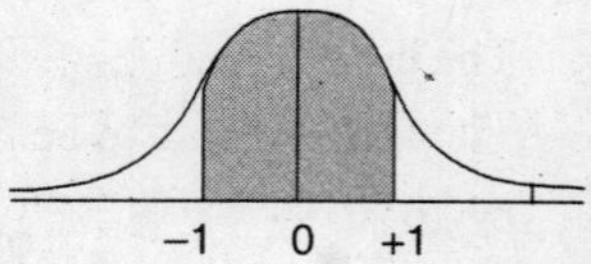


Area between $\quad z_1 = -1$ and $z_2 = +1$

$= (\text{Area betweem } z = 0 \text{ and } z = -1)$

$+ (\text{Area between } z = 0 \text{ and } z = +1)$

$= 2\,(\text{Area between } z = 0 \text{ and } z = +1) = 2 \times 0.3413 = 0.6826$

Percentage of resistors having resistance between 98 ohms and 102 ohms = 68.26 **Ans.**

**Example 70.** *In a normal distribution, 31% of the items are under 45 and 8% are over 64. Find the mean and standard deviation of the distribution. (A.M.I.E.T.E., Winter 2003)*

**Solution.** Let $\mu$ be the mean and $\sigma$ the S.D.

If $x = 45$, $\quad z = \dfrac{45 - \mu}{\sigma}$

If $x = 64$, $\quad z = \dfrac{64 - \mu}{\sigma}$

Area between 0 and $\dfrac{45 - \mu}{\sigma} = 0.50 - 0.31 = 0.19$

[From the table, for the area 0.19, $z = 0.496$]

$$\frac{45 - \mu}{\sigma} = -0.496 \qquad \text{...(1)}$$

Area between $z = 0$ and $z = \dfrac{64 - \mu}{\sigma} = 0.5 - 0.08 = 0.42$.

(From the table, for area 0.42, $z = 1.405$)

$$\frac{64 - \mu}{\sigma} = 1.405 \qquad \text{...(2)}$$

Solving (1) and (2), we get $\mu = 50$, $\sigma = 10$. **Ans.**

## Exercise 11.7

1. In a regiment of 1000, the mean height of the soldiers is 68.12 units and the standard deviation is 3.374 units. Assuming a normal distribution, how many soldiers could be expected to be more than 72 units? It is given that

$P(z=1.00)=0.3413, P(z=1.15)=0.3749$ and

$P(z=1.25)=0.3944$, where $z$ is the standard normal variable. **Ans.** 125

2. If the height of 300 students are normally distributed with mean 64.5 inches and standard deviation 3.3 inches, find the height below which 99% of the students lie. **Ans.** 68.7295 inches
3. The lifetime of radio tubes manufactured in a factory is known to have an average value of 10 years. Find the probability that the lifetime of a tube taken randomly (*i*) exceeds 15 years, (*ii*) is less than 5 years, assuming that the exponential probability law is followed. **Ans.** (*i*) 0.2231, (*ii*) 0.3935.
4. Analysis of past data shows that hub thickness of a particular type of gear is normally distributed about a mean thickness of 2.00 cm with a standard deviation of 0.04 cm.
   (*i*) What is the probability that a gear chosen at random will have a thickness greater than 2.06 cm?
   (*ii*) How many gears will have a thickness between 1.89 and 1.95 cm?
   Given $\phi(1.5)=0.4332, \phi(2.75)=6.4970, \phi(1.25)=0.3944$.
   (*A.M.I.E., Summer 2001*) **Ans.** (*i*) 0.068 (*ii*) 62
5. The breaking strength $X$ of a cotton fabric is normally distributed with $E(x)=16$ and $\sigma(x)=1$. The fabric is said to be good if $X \geq 14$. What is the probability that a fabric chosen at random is good. Given that $\phi(2)=0.9772$ **Ans.** 0.9772
6. A manufacturer knows from experience that the resistance of resistors he produces is normal with mean $\mu=140\ \Omega$ and standard deviation $\sigma=5\ \Omega$. Find the percentage of the resistors that will have resistance between $138\ \Omega$ and $142\ \Omega$. (given $\phi(0.4)=0.6554$, where $z$ is the standard normal variate). **Ans.** 31.08%
7. A manufacturing company packs pencils in fancy plastic boxes. The length of the pencils is normally distributed with $\mu=6"$ and $\sigma=0.2"$. The internal length of the boxes is 6.4". What is the probability that the box would be too small for the pencils (Given that a value of the standardized normal distribution function is $\phi(2)=0.9772$). **Ans.** 0.0228.
8. A manufacturer produces airmail envelopes, whose weight is normal with mean $\mu=1.95$ gm and standard deviation $\sigma=0.05$ gm. The envelopes are sold in lots of 1000. How many envelopes in a lot will be heavier than 2 gm? Use the fact that
   $$\frac{1}{\sqrt{2\pi}}\int_0^1 \exp(-x^2/2)dx = 0.3413$$ **Ans.** 159
9. A sample of 100 dry battery cells tested to find the length of life produced the following results. $\bar{x}=12$ hours, $\sigma=3$ hours. Assuming the data to be normally distributed, what percentage of battery cells expected to have life?
   (*i*) more than 15 hours. (*ii*) less than 6 hours.
   [Given $P(0<z<1)=0.3413$ and $P(0<z<2)=0.4772$.] **Ans.** (*i*) 15.87% (*ii*) 2.28%
10. Find the mean and variance of the density function $f(x)=\lambda e^{-\lambda x}$ **Ans.** $\frac{1}{\lambda}, \frac{1}{\lambda^2}$.
11. Fill in the blanks :
   (*a*) The mean of the marks obtained by the students is 50 and the variance is 25. If a student gets 60 marks, his standard score is.........
   (*b*) If $f(x)=\frac{1}{\sqrt{2\pi}}e^{-\frac{x^2}{2}}$, then its mean is and standard deviation is.........
   (*c*) In the standard normal curve the area between $z=-1$ and $z=1$ is nearly.........
   (*d*) If $\sigma=2$, $\bar{x}=5$, the equation of normal distribution is..........
   (*e*) The marks obtained were found normally distributed with mean 75 and variance 100. The percentage of students who scored more than 75 marks is........
   (*f*) The mean, median and mode of a normal distribution are...... (*A.M.I.E., Summer 2000*)

(*g*) Exponential distribution $f(x)$ is defined by $f(x) = a\,e^{-2r}, 0 < x < \infty$, then a =
(*h*) The probability density function of beta distribution with $\alpha = 1, \beta = 4$ is $f(x)$ =......
(*i*) For a standard normal variate $z$ $P(-0.72 \le z \le 0)$ = ......... (*A.M.I.E., Winter 1997*) **Ans.** 0.2642

**Ans.** (*a*) 2, (*b*) 0.1 , (*c*) 68%, (*d*) $f(x) = \frac{1}{\sqrt{2\pi}} e^{-\frac{(x-5)^2}{8}}$, (*e*) 50%, (*f*) zero, (*g*) 2, (*h*) $4(1-x)^3$

**12.** The probability density function $f(x)$ of a continuous random variable $x$ is defined by

$$f(x) = \frac{A}{x^3}, \quad 5 \le x \le 10$$
$$= 0, \quad \text{elsewhere}$$

The value of $A$ is

(*i*) 50, (*ii*) 1, (*iii*) –200, (*iv*) $\frac{200}{3}$ **Ans.** (*iv*)

**13.** The cumulative distribution function $F$ of a continuous variate x is such that $F(a) = 0.5$ and $F(b) = 0.7$. then value of $P(a \le X \le b)$ is given as

(*a*) 0 (*b*) 0.5 (*c*) 0.2 (*d*) 0.7 ... (*A.M.I.E.T.E. Dec, 2005*)

**14.** A discrete random variate X has probability mass function f which is positive at x = – 1, 0, 1 and is zero elsewhere. If $f(0) = \frac{1}{2}$, then the value of $E(x^2)$ is

(*a*) 1 (*b*) 0 (*c*) $\frac{1}{2}$ (*d*) $= -\frac{1}{2}$ (*A.MIE.T.E. Dec. 2005*)

**15.** If $x$ is normally distributed with mean 1 and variance 4,
(*i*) *Find* $Pr(-3 \le x \le 3)$; (*ii*) Obtain $k$ if $Pr(x \le k) = 0.90$ **Ans.** (*i*) .8185, (ii) 3.56.

**16.** A normal variable × has mean 1 and variance 4. Find the probability that $x \ge 3$. (Given: z is the standard normal variable and $\phi(0) = 0.5, \phi(0.5) = 0.6915, \phi(1) = 0.8413, \phi(1.5) = 0.9332$) **Ans.** 0.1587

**17.** (*a*) If $x$ is normally distributed with mean 4 and variance 9; find
(*i*) $Pr(2.5 \le x \le 5.5)$. (*ii*) Obtain $k$ if $Pr(x \le k) = 0.9$.
Use $Pr(z \le .5) = 0.691$ and $Pr(z \le 1.3) = 0.90$. **Ans.** (*i*) 0.382 (*ii*) 7.9.

(*b*) If $\log_e x$ is normally distributed with mean 1 and variance 4, find $P\left(\frac{1}{2} < x < 2\right)$, given that $\log_e 2 = 0.693$. **Ans.** 0.24.

(*c*) For a standard normal variate $z$ $P(-0.72 \le z \le 5\,0)$ =.... **Ans.** 0.2642

**18.** The random variable $x$ is normally distributed with $E(x) = 2$ and variance $V(x) = 4$. Find a number $p$ (approximately), such that $P(x > p) = 2P(x \le p)$. [The values of the standard normal distribution are $\phi(-0.43) = 0.3336$, and $\phi(-0.44) = 0.3300$].

(*A.M.I.E.T.E., Summer 1995*) **Ans.** 1.13834

**19.** The continuous random variable $x$ is normally distributed with $E(x) = \mu$ and $V(x) = \sigma^2$. If $Y = cx + d$, then find $V(Y)$. **Ans.** $c^2\sigma^2$

**20.** The pdf of $X$ is given by $f(X) = \lambda e^{-\lambda X}$, $x \ge 0 \lambda > 0$.
Calculate $Pr[X > E(X)]$. If $X \sim N(75, 25)$, find $Pr[X > 80/X > 77]$.

If $X \sim N(10, 4)$ find $Pr[|X| \ge 5]$. **Ans.** $\frac{1}{e}, \frac{1}{5\sqrt{2\pi}} e^{-\frac{(x-75)^2}{2(0.5)}}$, 0.062

**21.** If the resistance X of certain wires in an electrical networks have a normal distribution with mean of 0.01 ohm and a standard deviation of 0.001 ohm, and specification requires that the wires should have resistance between 0.009 ohm and 0.011 ohms, then find the expected number of wires in a sample of 1000 that are within the specification. Also, find the expected number among 1000 wires that cross the upper specification. (You may use normal table values $\Phi(0.5) = 0.6915, \Phi(1) = 0.8413, \Phi(1.5) = 0.9332, \Phi(2) = 0.9772$ .......................................................... (*A.M.I.E.T.E., Dec. 2004*)

22. A random variable $x$ has a standard normal distribution $\phi$. Prove $Pr(|x|>k)=2[1-\phi(k)]$

23. The random variable $x$ has the probability density function $f(x) = kx$ if $0 \le x \le 2$. Find $k$. Find $x$ such that

(i) $Pr(X \le x) = 0.1$ (ii) $Pr(X \le x) = 0.95$ **Ans.** $k = \frac{1}{2}$, (i) $x = 0.632$ (ii) $x = 1.949$

24. For a normal curve, show that $\mu_{2n+1} = 0$ and $\mu_{2n} = (2n-1)\,\sigma^2\,\mu_{2n-2}$.

25. In a normal distribution, 7% of the items are under 35 and 89% are under 63. Determine mean and variance of distribution. [Area of z for 0.43 = 1.48. Area of z for 0.39 = 1.23] (*A.M.I.E.T.E., Winter 2001*) **Ans.** $\mu = 50.29, \sigma^2 = 106.73$

26. The length of an item manufactured on an automatic machine tool is a normally distributed randam variable with parameters $m(x) = 10$, and $\sigma^2 = \frac{1}{200}$. Find the probability of defective production of the tolerance is $10 \pm 0.05$. (*A.M.I.E.T.E., Winter 2001*) **Ans.** 0.04798

27. In a mathematics examination, the average grade was 82 and the standard deviation was 5. All the students with grades from 88 to 94 received a grade B. If the grades are normally distributed and 8 students received a B grade, find how many students took the examination.
Given:

| x/6 | 1.20 | 2.00 | 2.40 | 2.45 |
|---|---|---|---|---|
| A | 0.3849 | 0.4772 | 0.4918 | 0.4929 |

(*A.M.I.E., Winter 2001*) **Ans.** 75 students

28. The income of a group of 10,000 persons was found to be normally distributed with mean ₹ 750 p.m. and standard deviation of ₹ 50. Show that, of this group, about 95% had income exceeding ₹ 668 and only 5% had income exceeding ₹ 832. Also find the lowest income among the richest 100. (*U.P. III Semester, Dec. 2004*) **Ans.** ₹ 866

29. A continuous type random variable X has probability density $f(x)$ which is proportional to $x^2$ and X takes values in the interval [0, 2]. Find the distribution function of the random variable use this to find $P(X>1.2)$ and conditional probability $P(X>1.2/X>1)$. (*A.M.I.E.T.E., Dec. 2006*)

## 11.32 OTHER DISTRIBUTIONS

**(1) Uniform (or Rectangular) Distribution**

Here $P(x) = \frac{1}{n}$ $\Sigma p(x) = n\left(\frac{1}{n}\right) = 1$

The value of probability for all variates $x_1, x_2, ...., x_n$ is the same $\frac{1}{n}$.

**(2) Geometric Distribution**

Let $r$ be the number of failures preceding the first success

$p(r) = q^r p$ where $r = 0, 1, 2, 3 ......, q = 1 - p$

$$\Sigma p(r) = \Sigma q^r p = p(1 + q + q^2 + ..... + q^r ...) \quad \text{(Geometric series)}$$

$$= p\frac{1}{1-q} = \frac{p}{p} = 1 \qquad \text{Mean} = \frac{q}{p}, \qquad \text{Variance} = \frac{q}{p^2}$$

**(3) Negative Binomial Distribution**

The probability of the event that occurs for the kth time on the rth trial

$$p(k, r) = {}^{r-1}C_{k-1}\, p^k q^{r-k}$$

For $k = 1$, the negative Binomial distribution becomes geometric distribution.

**(4) Hypergeometric Distribution**

Let the number $y$ white balls be $m$ and $n$ black balls in a bag. If $r$ balls are drawn at a time with replacement

$$p\,(k \text{ white}) = {}^mC_k \frac{{}^nC_{r-k}}{{}^{m+n}C_r} \text{ where } k = 0,1,2,\ldots,r, \quad r \le m,\ r \le n$$

$$\sum_{k=0}^{r} p(k) = 1 \text{ since } \sum {}^mC_k\, {}^nC_{r-k} = (m+n)C_r$$

**(5) Exponential Distribution**

Let $f(x)$ be a continuous distribution

$$f(x) = e^{-cx} \qquad \text{from } x > 0$$

Hence mean $= \frac{1}{C}$ = standard deviation $= \frac{1}{C}$

**(6) Weibull Distribution is given by** $f(x) = \frac{\alpha}{C} x^{\alpha-1} e^{-\frac{x^\alpha}{c}}, \ x > 0,\ c > 0$

**where $C$ is a scale parameter and $\alpha$ is a shape parameter.**

**This distribution is used for**

**(1) variation in the fatigue resistance of steel and its elastic limits.**

**(2) variation of length of service of radio service equipment.**

**Exercise 11.8**

**Find the mean and variance for the following distributions**

1. Rectangular distribution **Ans.** $\frac{1}{2}, \frac{1}{12}$
2. Uniform distribution $f(x) = \frac{1}{n}$ $x = 1, 2, \ldots n$ **Ans.** $\frac{1}{2}(n+1), \frac{1}{12}(n^2-1)$
3. Geometric distribution $p\,(r) = 2^r$, $r = 1,2,3 \ldots$
4. Exponential distribution $p(x) = \lambda e^{-\lambda x}$ **Ans.** $\frac{1}{\lambda}, \frac{1}{\lambda}$

## SAMPLING OF VARIABLES

### 11.33 POPULATION (Universe)

Before giving the notion of sampling, we will first define *population*. The group of individuals under study is called *population* or *universe*. It may be finite or infinite.

### 11.34 SAMPLING

A part selected from the population is called *a sample*. The process of selection of a sample is called sampling. A *Random sample* is one in which each member of population has an equal chance of being included in it. There are ${}^NC_n$ different samples of size $n$ that can be picked up from a population of size $N$.

### 11.35 PARAMETERS AND STATISTICS

The statistical constants of the population such as mean ($\mu$), standard deviation ($\sigma$) are called parameters. Parameters are denoted by Greek letters.

The mean ($\bar{x}$), standard deviation $|S|$ of a sample are known as statistics. Statistics are denoted by Roman letters.

**Symbols for Population and Samples**

| Characteristic | Population<br>Parameter | Sample<br>Statistic |
|---|---|---|
| Symbols | population size = $N$<br>population mean = $\mu$<br>population standard deviation = $\sigma$<br>population proportion = $p$ | sample size = $n$<br>sample mean = $\bar{x}$<br>sample standard deviation = $s$<br>sample proportion = $\tilde{p}$ |

## 11.36 AIMS OF A SAMPLE

The population parameters are not known generally. Then the sample characteristics are utilised to approximately determine or estimate of the population. Thus, static is an estimate of the parameter. To what extent can we depend on the sample estimates?

The estimate of mean and standard deviation of the population is a primary purpose of all scientific experimentation. The logic of the sampling theory is the logic of *induction.* In induction, we pass from a particular (sample) to general (population). This type of generalization here is known as *statistical inference.* The conclusion in the sampling studies are based not on certainties but on probabilities.

## 11.37 TYPES OF SAMPLING

Following types of sampling are common:

(1) Purposive sampling (2) Random sampling

(3) Stratified sampling (4) Systematic sampling

## 11.38 SAMPLING DISTRIBUTION

From a population a number of samples are drawn of equal size *n.* Find out the mean of each sample. The means of samples are not equal. The means with their respective frequencies are grouped. The frequency distribution so formed is known as *sampling distribution of the mean.* Similarly, sampling distribution of standard deviation we can have.

**11.39 STANDARD ERROR (S.E.)** is the standard deviation of the sampling distribution. For assessing the difference between the expected value and observed value, standard error is used. Reciprocal of standard error is known as *precision.*

## 11.40 SAMPLING DISTRIBUTION OF MEANS FROM INFINITE POPULATION

Let the population be infinitely large and having a population mean of $\mu$ and a population variance of $\sigma^2$. If $x$ is a random variable denoting the measurement of the characteristic, then

Expected value of $x$, $E(x) = \mu$

Variance of $x$, $\text{Var}(x) = \sigma^2$

The sample mean $\bar{x}$ is the sum of $n$ random variables, *viz.*, $x_1, x_2, \ldots, x_n$, each being divided by $n$. Here, $x_1, x_2, \ldots, x_n$ are independent random variables from the infinitely large population.

$\therefore$ $E(x_1) = \mu$ and $Var(x_1) = \sigma^2$

$E(x_2) = \mu$ and $Var(x_2) = \sigma^2$ and so on

Finally
$$E(\bar{x}) = E\left[\frac{x_1 + x_2 + \ldots + x_n}{n}\right]$$
$$= \frac{1}{n}Ex_1 + \frac{1}{n}E(x_2) + \ldots + \frac{1}{n}E(x_n)$$
$$= \frac{1}{n}\mu + \frac{1}{n}\mu + \ldots + \frac{1}{n}\mu$$
$$= \mu$$

and
$$Var(\bar{x}) = Var\left[\frac{x_1 + x_2 + \ldots + x_n}{n}\right]$$
$$= Var\left(\frac{x_1}{n}\right) + Var\left(\frac{x_2}{n}\right) + \ldots + Var\left(\frac{x_n}{n}\right)$$
$$= \frac{1}{n^2}Var(x_1) + \frac{1}{n^2}Var(x_2) + \ldots + \frac{1}{n^2}Var(x_n)$$

$$=\frac{1}{n^2}.\sigma^2+\frac{1}{n^2}.\sigma^2+...+\frac{1}{n^2}.\sigma^2$$

$$=\frac{n\sigma^2}{n^2}=\frac{\sigma^2}{n}$$

The expected value of the sample mean is the same as population mean. The variance of the sample mean is the variance of the population divided by the sample size.
The average value of the sample tends to true population mean. If sample size *(n)* is increased then variance of $\bar{x}$, $\left(\frac{\sigma^2}{n}\right)$ gets reduced, by taking large value of *n*, the variance $\left(\frac{\sigma^2}{n}\right)$ of $\bar{x}$ can be made as small as desired. The standard deviation $\left(\frac{\sigma}{\sqrt{n}}\right)$ of $\bar{x}$ is also called **standard error of the mean.**

It is denoted by $\sigma_{\bar{x}}$.

**Sampling with Replacement**

When the sampling is done with replacement, so that the population is back to the same form before the next sample member is picked up. We have

$$E(\bar{x})=\mu$$

$$Var\ (\bar{x})=\frac{\sigma^2}{n}\ \text{or}\ \sigma_{\bar{x}}=\frac{\sigma}{\sqrt{n}}$$

**Sampling without replacement from Finite population**

When a sample is picked up without replacement from a finite population, the probability distribution of second random variable depends on the outcome of the first pick up. *n* sample members do not remain independent. Now we have

$$E\ (\bar{x})=\mu$$

and $$Var(\bar{x})=\sigma_{\bar{x}}^{\ 2}=\frac{\sigma^2}{n}\cdot\frac{N-n}{N-1}\quad\text{or}\quad\sigma_{\bar{x}}=\frac{\sigma}{\sqrt{n}}\cdot\frac{\sqrt{N-n}}{N-1}$$

$$=\frac{\sigma}{\sqrt{n}}\ \text{app}\qquad(\text{if}\ \frac{n}{N}\ \text{isverysmall})$$

**Sampling from Normal Population**

If $x\sim N\left(\mu,\sigma^2\right)$ then it follows that $\bar{x}\sim N(\mu,\frac{\sigma^2}{n})$

**Example 71.** *The diameter of a component produced on a semi-automatic machine is known to be distributed normally with a mean of 10 mm and a standard deviation of* 0.1 *mm. If we pick up a random sample of size 5, what is the probability that the same mean will be between 9.95 and 10.05 mm?*

**Solution.** Let *x* be a random variable representing the diameter of one component picked up at random.

Here $x$ ~N(10, 0.01), Therefore, $\bar{x}\sim N\left(10,\frac{0.01}{5}\right)$ $\qquad\left[\bar{x}=N\left(\bar{x},\frac{\sigma^2}{n}\right)\right]$

$$Pr\{9.95\le\bar{x}\le10.05\}=2\times Pr\{10\le\bar{x}\le10.05\}\qquad\left\{z=\frac{\bar{x}-\mu}{\frac{\sigma}{\sqrt{n}}}\right\}$$

$$= 2 \times Pr\left\{\frac{10-\mu}{\frac{\sigma}{\sqrt{n}}} \le \frac{\bar{x}-\mu}{\frac{\sigma}{\sqrt{n}}} \le \frac{10.05-\mu}{\frac{\sigma}{\sqrt{n}}}\right\}$$

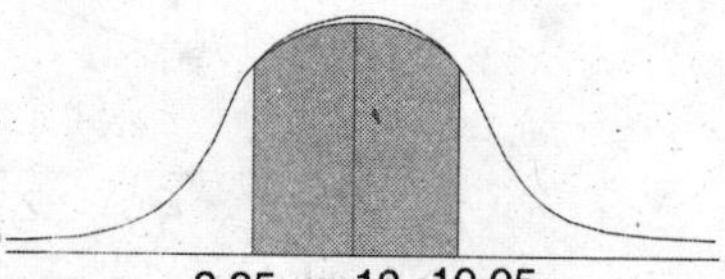


$$= 2 \times Pr\left\{0 \le z \le \frac{10.05-10}{\frac{0.1}{\sqrt{5}}}\right\}$$

$$= 2 \times Pr\{0 \le z \le 1.12\}$$
$$= 2 \times 0.3686$$
$$= 0.7372$$

**Ans.**

**Similar Question**

A sample of size 25 is picked up at random from a population which is normally distributed with a mean 100 and a variance of 36. Calculate (*a*) $Pr\{\bar{x} \le 99\}$, (*b*) $Pr\{98 \le \bar{x} \le 100\}$

**Ans.** (*a*) 0.2023 (*b*) 0.4522

## 11.41 SAMPLING DISTRIBUTION OF THE VARIANCE

We use a sample statistic called the sample variance to estimate the population variance. The sample variance is usually denoted by $s^2$

$$s^2 = \frac{\sum_{i=1}^{n}(x_i - \bar{x})^2}{n-1}$$

## 11.42 TESTING A HYPOTHESIS

On the basis of sample information, we make certain decisions about the population. In taking such decisions we make certain assumptions. These assumptions are known as *statistical hypothesis.* These hypothesis are tested. Assuming the hypothesis correct we calculate the probability of getting the observed sample. If this probability is less than a certain assigned value, the hypothesis is to be rejected.

## 11.43 NULL HYPOTHESIS ($H_0$)

Null hypothesis is based for analysing the problem. Null hypothesis is the *hypothesis of no difference.* Thus, we shall persume that there is no significant difference between the observed value and expected value. Then, we shall test whether this hypothesis is satisfied by the data or not. If the hypothesis is not approved the difference is considered to be significant. If hypothesis is approved then the difference would be described as due to sampling fluctuation. Null hypothesis is denoted by $H_0$.

## 11.44 ERRORS

In sampling theory to draw valid inferences about the population parameter on the basis of the sample results.

We decide to accept or to reject the lot after examining a sample from it. As such, we are liable to commit the following two types of errors.

**Type 1 Error.** If $H_0$ is rejected while it should have been accepted.

**Type II Error.** If $H_0$ is accepted while it should have been rejected.

## 11.45 LEVEL OF SIGNIFICANCE

There are two critical regions which cover 5% and 1% areas of the normal curve. The shaded portions are the critical regions.

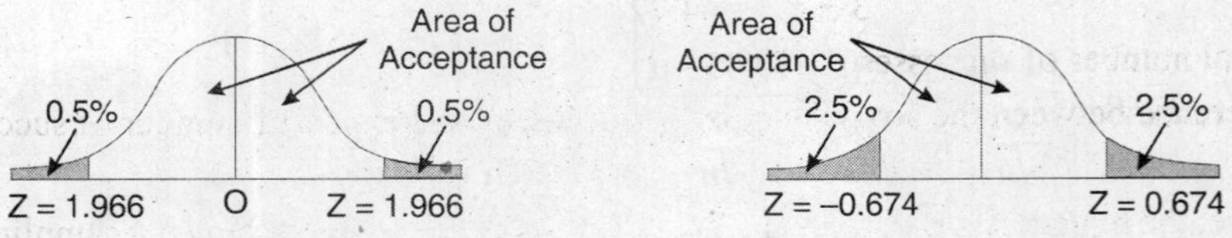


Thus, the probability of the value of the variate falling in the critical region is the level of significance. If the variate falls in the critical area, the hypothesis is to be rejected.

## 11.46 TEST OF SIGNIFICANCE

The tests which enables us to decide whether to accept or to reject the null hypothesis is called the tests of significance. If the difference between the sample values and the population values are so large (lies in critical area), it is to be rejected

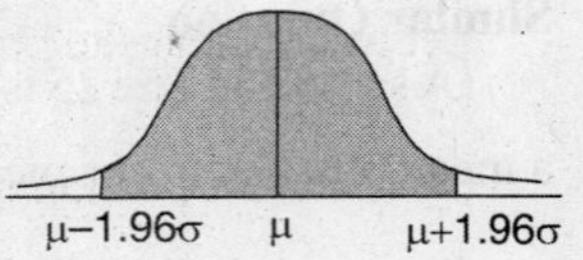


## 11.47 CONFIDENCE LIMITS

$\mu - 1.96\ \sigma, \mu + 1.96\sigma$ are 95% confidence limits as the area between $\mu - 1.96\ \sigma$ and $\mu + 1.96\ \sigma$ is 95%. If a sample statistics lies in the interval $\mu - 1.96\ \sigma$, $\mu + 1.96\ \sigma$, we call 95% confidence interval.

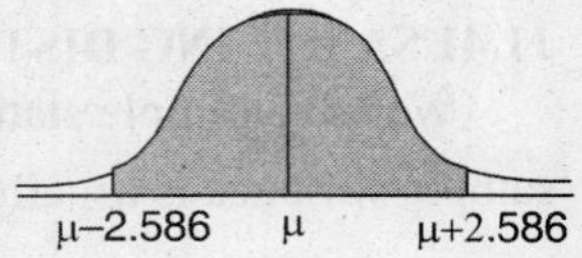


Similarly, $\mu - 2.58\ \sigma$, $\mu + 2.58\ \sigma$ is 99% confidence limits as the area between $\mu - 2.58\ \sigma$ and $\mu + 2.58\ \sigma$ is 99%. The numbers 1.96, 2.58 are called confidence coefficients.

## 11.48 TEST OF SIGNIFICANCE OF LARGE SAMPLES ($N > 30$)

Normal distribution is the limiting case of Binomial distribution when $n$ is large enough. For normal distribution 5% of the items lie outside $\mu \pm 1.96\ \sigma$ while only 1% of the items lie outside $\mu \pm 2.586\ \sigma$.

$$z = \frac{x - \mu}{\sigma}$$

where $z$ is the standard normal variate and $x$ is the observed number of successes.

First we find the value of $z$. Test of significance depends upon the value of $z$.

(*i*) (*a*) If $|z| < 1.96$, difference between the observed and expected number of successes is not significant at the 5% level of significance.

(*b*) If $|z| > 1.96$, difference is significant at 5% level of significance.

(*ii*) (*a*) If $|z| < 2.58$, difference between the observed and expected number of successes is not significant at 1% level of significance.

(*b*) If $|z| > 2.58$, difference is significant at 1% level of significance.

**Example 72.** *A cubical die was thrown 9,000 times and 1 or 6 was obtained 3120 times. Can the deviation from expected value lie due to fluctuations of sampling?*

**Solution.** Let us consider the hypothesis that the die is an unbiased one and hence the probability of obtaining 1 or $6 = \frac{2}{6} = \frac{1}{3}$ *i.e.*, $p = \frac{1}{3}$, $q = \frac{2}{3}$

The expected value of the number of successes $= np = 9000 \times \frac{1}{3} = 3000$

Also $\sigma = \text{S.D.} = \sqrt{npq} = \sqrt{9000 \times \frac{1}{3} \times \frac{2}{3}} = \sqrt{2000} = 44.72$

$$3\sigma = 3 \times 44.72 = 134.16$$

Actual number of successes = 3120

Difference between the actual number of successes and expected number of successes

$$= 3120 - 3000 = 120 \text{ which is } <3\sigma$$

Hence, the hypothesis is correct and the deviation is due to fluctuations of sampling due randon causes. **Ans.**

## 11.49 SAMPLING DISTRIBUTION OF THE PROPORTION

A simple sample of $n$ items is drawn from the population. It is same as a series of $n$ independent trials with the probability $P$ of success. The probabilities of 0, 1, 2, ..., $n$ success are the terms in the binomial expansion of $(q + p)^n$.

Here mean = $np$ and standard deviation = $\sqrt{npq}$.

Let us consider the proportion of successes, then

(*a*) Mean proportion of successes $= \dfrac{np}{n} = p$

(*b*) Standard deviation (standard error) of proportion of successes $= \dfrac{\sqrt{npq}}{n} = \sqrt{\dfrac{pq}{n}}$

(*c*) Precision of the proportion of success $= \dfrac{1}{\text{S.E.}} = \sqrt{\dfrac{n}{pq}}$.

**Example 73.** *A group of scientific mens reported 1705 sons and 1527 daughters. Do these figures conform to the hypothesis that the sex ratio is* $\dfrac{1}{2}$.

**Solution.** The total number of observations = 1705 + 1527 = 3232

The number of sons = 1705

Therefore, the observed male ratio $= \dfrac{1705}{3232} = 0.5175$

On the given hypothesis the male ratio = 0.5000

Thus, the difference between the observed ratio and theoretical ratio

$$= 0.5275 - 0.5000$$
$$= 0.0275$$

The standard deviation of the proportion $= s = \sqrt{\dfrac{pq}{n}} = \sqrt{\dfrac{\frac{1}{2} \times \frac{1}{2}}{3232}} = 0.0088$

The difference is more than 3 times of standard deviation.

Hence, it can be definitely said that the figures given do not conform to the given hypothesis.

## 11.50 ESTIMATION OF THE PARAMETERS OF THE POPULATION

The mean, standard deviation etc. of the population are known as parameters. They are denoted by $\mu$ and $\sigma$. Their estimates are based on the sample values. The mean and standard deviation of a sample are denoted by $\bar{x}$ and $s$ respectively. Thus, a static is an estimate of the parameter. There are two types of estimates.

(*i*) ***Point estimation:*** An estimate of a population parameter given by a single number is called a point estimation of the parameter. For example,

$$(\text{S.D.})^2 = \frac{\sum (x - \bar{x})^2}{n-1}$$

(*ii*) *Interval estimation:* An interval in which population parameter may be expected to lie with a given degree of confidence. The intervals are

(*i*) $\bar{x} - \sigma_s$ to $\bar{x} + \sigma_s$ (68.27% confidence level)

(*ii*) $\bar{x} - 2\sigma_s$ to $\bar{x} + 2\sigma_s$ (95.45% confidence level)

(*iii*) $\bar{x} - 3\sigma_s$ to $\bar{x} + 3\sigma_s$ (99.73% confidence level)

$\bar{x}$ and $\sigma_s$ are the mean and S.D. of the sample.

Similarly, $\bar{x} \pm 1.96\sigma_s$, $\bar{x} \pm 2.58\sigma_s$ are 95% and 99% confidence of limits for $\mu$.

$\bar{x} \pm 1.96\dfrac{\sigma}{\sqrt{n}}$ and $\bar{x} \pm 2.58\dfrac{\sigma}{\sqrt{n}}$ are also the intervals as $\sigma_s = \dfrac{\sigma}{\sqrt{n}}$

## 11.51 COMPARISON OF LARGE SAMPLES

Let two large samples of size $n_1$, $n_2$ be drawn from two populations of proportions of attributes A's as $P_1$, $P_2$ respectively.

(*i*) *Hypothesis:* As regards the attribute A, the two populations are similar. On combining the two samples

$$P = \frac{n_1 P_1 + n_2 P_2}{n_1 + n_2}$$

where $p$ is the common proportion of attributes.

Let $e_1$, $e_2$ be the standard errors in the two samples, then

$$e_1^2 = \frac{pq}{n_1} \text{ and } e_2^2 = \frac{pq}{n_2}$$

If $e$ be the standard error of the combined samples, then

$$e = e_1^2 + e_2^2 = \frac{pq}{n_1} + \frac{pq}{n_2} = pq\left[\frac{1}{n_1} + \frac{1}{n_2}\right]$$

$$z = \frac{P_1 - P_2}{e}$$

1. If $z > 3$, the difference between $P_1$ and $P_2$ is significant.
2. If $z < 2$, the difference may be due to fluctuations of sampling.
3. If $2 < z < 3$, the difference is significant at 5% level of significance.

(*ii*) *Hypothesis.* In the two populations, the proportions of attribute $A$ are not the same, then standard error $e$ of the difference $p_1 - p_2$ is

$$e^2 = p_1 + p_2$$

$$= \frac{P_1\ q_1}{n_1} + \frac{P_2 - q_2}{n_2},\ z = \frac{P_1 - P_2}{e} < 3,$$

Difference is due to fluctuations of samples.

**Example 74.** *In a sample of 600 men from a certain city, 450 are found smokers. In another sample of 900 men from another city, 450 are smokers. Do the data indicate that the cities are significantly different with respect to the habit of smoking among men.*

**Solution.** $n_1 = 600$ men, Number of smokers= 450, $P_1 = \dfrac{450}{600} = 0.75$

$n_2 = 900$ men, Number of smokers= 450, $P_2 = \dfrac{450}{900} = 0.5$

$$P = \frac{n_1 P_1 + n_2 P_2}{n_1 + n_2} = \frac{600 \times 0.75 + 900 \times 0.5}{600 + 900} = \frac{900}{1500} = 0.60$$

$$q = 1 - p = 1 - 0.6 = 0.4$$

$$e^2 = P_1^2 + P_2^2 = pq\left(\frac{1}{n_1} + \frac{1}{n_2}\right)$$

$$e^2 = 0.6 \times 0.4\left(\frac{1}{600} + \frac{1}{900}\right) = 0.000667$$

$$e = 0.02582$$

$$z = \frac{P_1 - P_2}{e} = \frac{0.75 - 0.50}{0.02582} = 9.682$$

$z > 3$ so that the difference is significant. **Ans.**

**Example 75.** *One type of aircraft is found to develop engine trouble in 5 flights out of a total of 100 and another type in 7 flights out of a total of 200 flights. Is there a significant difference in the two types of aircrafts so far as engine defects are concerned.*

**Solution.** $n_1 = 100$ flights, Number of troubled flights = 5, $p_1 = \frac{5}{100} = \frac{1}{20}$

$n_2 = 200$ flights, Number of troubled flights = 7, $p_2 = \frac{7}{200}$

$$e^2 = \frac{p_1 q_1}{n_1} + \frac{p_2 q_2}{n_2} = \frac{0.05 \times 0.95}{100} + \frac{0.035 \times 0.965}{200}$$

$$= 0.000475 + 0.0001689 = 0.0006439$$

$$e = 0.0254$$

$$z = \frac{0.05 - 0.035}{0.0254} = 0.59$$

$z < 1$, Difference is not significant. **Ans.**

## 11.52 THE t-DISTRIBUTION (For small sample)

The students distribution is used to test the significance of

(*i*) The mean of a small sample.

(*ii*) The difference between the means of two small samples or to compare two small samples.

(*iii*) The correlation coefficient.

Let $x_1, x_2, x_3, ..., x_n$, be the members of random sample drawn from a normal population with mean $\mu$. If $\bar{x}$ be the mean of the sample then

$$t = \frac{\bar{x} - \mu}{\frac{s}{\sqrt{n}}} \quad \text{where } s^2 = \frac{\sum (x - \bar{x})^2}{n - 1}$$

**Example 76.** *A machine which produces mica insulaung washers for use in electric device to turn out washers having a thickness of 10 mm. A sample of 10 washers has an average thickness 9.52 mm with a standard deviation of 0.6 mm. Find out t.*

**Solution.** $\bar{x} = 9.52$, $M = 10$, $S' = 0.6$, $n = 10$

$$t = \frac{\bar{x} - M}{\frac{s}{\sqrt{n}}} = \frac{9.52 - 10}{\frac{0.6}{\sqrt{10}}} = -\frac{0.48\sqrt{10}}{0.6} = -\frac{4}{5}\sqrt{10}$$

$$= -0.8 \times 3.16 = -2.528$$ **Ans.**

**Example 77.** *Compute the students t for the following values in a sample of eight: −4, −2, −2, 0, 2, 2, 3, 3 taking the mean of universe to be zero.*

**Solution.** $\mu = 0$

| S.No. | $x$ | $x-\bar{x} = \left(x - \frac{1}{4}\right)$ | $(x-\bar{x})^2 = \left(x - \frac{1}{4}\right)^2$ |
|---|---|---|---|
| 1 | −4 | $-\frac{17}{4}$ | $\frac{289}{16}$ |
| 2 | −2 | $-\frac{9}{4}$ | $\frac{81}{16}$ |
| 3 | −2 | $-\frac{9}{4}$ | $\frac{81}{16}$ |
| 4 | 0 | $-\frac{1}{4}$ | $\frac{1}{16}$ |
| 5 | 2 | $\frac{7}{4}$ | $\frac{49}{16}$ |
| 6 | 2 | $\frac{7}{4}$ | $\frac{49}{16}$ |
| 7 | 3 | $\frac{11}{4}$ | $\frac{121}{16}$ |
| 8 | 3 | $\frac{11}{4}$ | $\frac{121}{16}$ |
| $n = 8$ | $\sum x = 2$ | | $\sum (x-\bar{x})^2 = \frac{792}{16}$ |

$$\bar{x} = \frac{2}{8} = \frac{1}{4}$$

$$\text{S.D.} = s = \sqrt{\frac{(x-\bar{x})^2}{n-1}} = \sqrt{\frac{792}{16 \times 7}} = \sqrt{7.07} = 2.66$$

$$t = \frac{\bar{x} - \mu}{\frac{S''}{\sqrt{n}}} = \frac{\frac{1}{4} - 0}{\frac{2.66}{\sqrt{8}}} = \frac{\sqrt{8}}{4(2.66)} = \frac{2.83}{10.64} = 0.266$$ **Ans.**

## 11.53 WORKING RULE

**To calculate significance of sample mean at 5% level.**

**Calculate $t = \frac{\bar{x} - \mu}{s}\sqrt{n}$ and compare it to the value of $t$ with $(n-1)$ degrees of freedom at 5% level, obtained from the table. Let this tabulated value of $t$ be $t_1$.**

**If $t < t_1$, then we accept the hypothesis *i.e.*, we say that the sample is drawn from the population.**

If $t > t_1$, we compare it with the tabulated value of $t$ at 1% level of significance for $(n-1)$ degrees of freedom. Denote it by $t_2$. If $t_1 < t < t_2$ then we say that the value of $t$ is significant.

If $t > t_1$, we reject the hypothesis and the sample is not drawn from the population.

**Example 78.** *A manufacturer intends that his electric bulbs have a life of 1000 hours. He tests a sample of 20 bulbs, drawn at random from a batch and discovers that the mean life of the sample bulbs is 990 hours with a s.d of 22 hours. Does this signify that the batch is not up to the standard?*

**[Given:** *The table value of t at 1% level is significance with 19 degrees of freedom is 2.539***]**

**Solution.** $\bar{x} = 990, \sigma = 22, x = 1000$

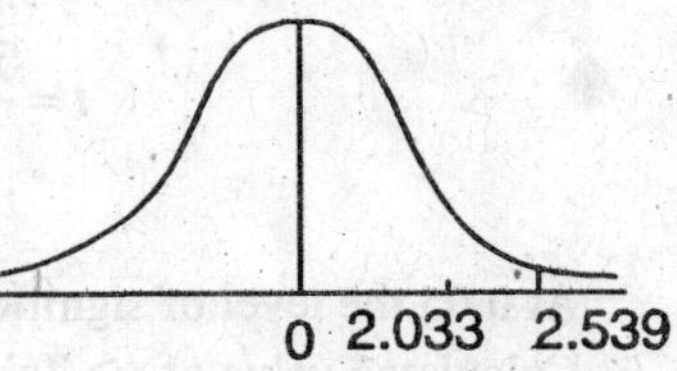


$$t = \frac{\bar{x}-\mu}{\frac{s}{\sqrt{n}}} = \frac{1000-990}{\frac{22}{\sqrt{20}}}$$

$$= \frac{10\sqrt{20}}{22} = \frac{22.36}{11} = 2.033$$

Since the calculated value of $t$ (2.032) is less than the value of $t$ (2.539) from the table. Hence, it is not correct to say that this batch is not upto this standard. **Ans.**

**Example 79.** *Ten individuals are chosen at random from a population and their heights are found to be in inches 63, 63, 64, 65, 66, 69, 69, 70, 70, 71. Discuss the suggestion that the Mean height of universe is 65.*

*For 9 degree of freedom t at 5% level of significance = 2.262.*

**Solution.**

| x | $x-67$ | $(x-67)^2$ |
|---|---|---|
| 63 | −4 | 16 |
| 63 | −4 | 16 |
| 64 | −3 | 9 |
| 65 | −2 | 4 |
| 66 | −1 | 1 |
| 69 | +2 | 4 |
| 69 | +2 | 4 |
| 70 | +3 | 9 |
| 70 | +3 | 9 |
| 71 | +4 | 16 |
| $\sum x = 670$ | | $\sum (x-\bar{x})^2 = 88$ |

$$\bar{x} = \frac{\sum x}{n} = \frac{670}{10} = 67, \quad s = \sqrt{\frac{\sum (x-\bar{x})^2}{n-1}} = \sqrt{\frac{88}{9}} = 3.13$$

$$t = \frac{\bar{x}-\mu}{\frac{s}{\sqrt{n}}} = \frac{67-65}{\frac{3.13}{\sqrt{10}}}$$

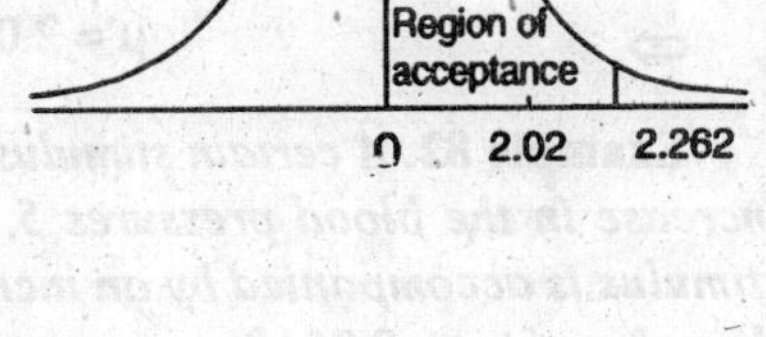


$$= \frac{2\sqrt{10}}{3.13} = 2.02$$

$$2.02 < 2.262$$

Calculated value of $t$ (2.02) is less than the table value of $t$ (2.262). The hypothesis is accepted the mean height of universe is 65 inches. **Ans.**

**Example 80.** *The mean life time of sample of 100 fluorescent light bulbs produced by a company is computed to be 1570 hours with a standard deviation of 120 hours. The company claims that the average life of the bulbs produced by it is 1600 hours. Using the level of significance of 0.05, is the claim acceptable?*

**Solution.** $\bar{x} = 1570,\ S = 120,\ n = 100,\ \mu = 1600$

$$t = \frac{\bar{x}-\mu}{\frac{s}{\sqrt{n}}} = \frac{1570-1600}{\frac{120}{\sqrt{100}}} = \frac{1570-1600}{12} = 2.5$$

At 0.05 the level of significance, $t = 1.96$

Calculated value of $t >$ Table value of $t$.

$$2.5 > 1.96$$

Hence the claim is to be rejected. **Ans.**

**Example 81.** *A sample of 6 persons in an office revealed an average daily smoking of 10, 12, 8, 9, 16, 5 cigarettes. The average level of smoking in the whole office has to be estimated at 90% level of confidence.*

*$t = 2.015$ for 5 degree of freedom*

**Solution.**

| $x$ | $x-10$ | $(x-10)^2$ |
|---|---|---|
| 10 | 0 | 0 |
| 12 | 2 | 4 |
| 8 | −2 | 4 |
| 9 | −1 | 1 |
| 16 | +6 | 36 |
| 5 | −5 | 25 |
| Total | 0 | $\sum (x-10)^2 = 70$ |

$$\text{Mean} = a + \frac{\sum fd}{\sum f} = 10 + \frac{0}{6} = 10$$

$$s = \sqrt{\frac{\sum (x-\bar{x})^2}{n-1}} = \sqrt{\frac{70}{5}} = 3.74$$

At 90% level of confidence, $t = \pm 2.015$.

$$t = \frac{\bar{x}-\mu}{\frac{s}{\sqrt{n}}} \Rightarrow \pm 2.015 = \frac{10-\mu}{\frac{3.74}{\sqrt{6}}}$$

$$\Rightarrow \quad \mu = 2.015 \times \frac{3.74}{\sqrt{6}} + 10 = 6.92, 13.08$$ **Ans.**

**Example 82.** *A certain stimulus administered to each of 12 patients resulted in the following increase in the blood pressures 5, 2, 8, −1, 3, 0, 6, −2, 1, 5, 0, 4. Can it be calculated that stimulus is accompanied by an increase in blood pressure given that for 11 degrees of freedom the value of $t_{0.5}$ is 2.201?*

**Solution.**

$$\bar{x}=\frac{5+2+8-1+3+0+6-2+1+5+0+4}{12}$$

$$=\frac{31}{12}=2.583=2.6 \text{ approx.}$$

| $x$ | $x-2.6$ | $(x-2.6)^2$ |
|---|---|---|
| 5 | 2.4 | 5.76 |
| 2 | −0.6 | 0.36 |
| 8 | 5.4 | 29.16 |
| −1 | −3.6 | 12.96 |
| 3 | 0.4 | 0.16 |
| 0 | −2.6 | 6.76 |
| 6 | 3.4 | 11.56 |
| −2 | −4.6 | 21.16 |
| 1 | −1.6 | 2.56 |
| 5 | 2.4 | 5.76 |
| 0 | −2.6 | 6.76 |
| 4 | 1.4 | 1.96 |
| $\sum x=12$ | $\sum(x-2.6)$ | $\sum(x-2.6)^2=104.92$ |

$$s^2=\frac{\sum(x-\bar{x})^2}{n-1}=\frac{104.92}{12-1}=9.54$$

$$s=3.08$$

**Assuming that the stimulus will not be accompanied by increase in blood pressure, i.e., the mean of increase in blood pressure for the population is zero, we have**

$$t=\frac{\bar{x}-\mu}{s}\sqrt{n}=\frac{2.6-0}{3.08}\sqrt{12}=\frac{2.6}{3.08}\times 3.464=2.924$$

**As the computed value of *t, i.e.,* 2.924 is greater than $t_{0.05}$, *i.e.,* 2.201 we find that our assumption is wrong and we conclude that as a result of the stimulus blood pressure will increase. Ans.**

**Example 83.** ***A fertiliser mixing machine is set to give 12 kg of nitrate for quintal bag of fertiliser: Ten 100 kg bags are examined. The percentages of nitrate per bag are as follows:***

***11, 14, 13, 12, 13, 12, 13, 14, 11, 12***

***Is there any reason to believe that the machine is defective? Value of t for 9 degrees of freedom is 2.262.***

**Solution.**

**The calculation of $\bar{x}$ and *s* is given in the following table:**

| $x$ | $d=x-12$ | $d^2$ |
|---|---|---|
| 11 | −1 | 1 |
| 14 | 2 | 4 |
| 13 | 1 | 1 |
| 12 | 0 | 0 |
| 13 | 1 | 1 |

| | | |
|---|---|---|
| 12 | 0 | 0 |
| 13 | 1 | 1 |
| 14 | 2 | 4 |
| 11 | −1 | 1 |
| 12 | 0 | 0 |
| $\sum x = 125$ | $\sum d = 5$ | $\sum d^2 = 13$ |

$$\mu = 12 \text{ kg},\ n = 10,\ \bar{x} = \frac{\sum x}{n} = \frac{125}{10} = 12.5$$

$$s^2 = \frac{\sum d^2}{n} - \left(\frac{\sum d^2}{n}\right) = \frac{13}{10} - \left(\frac{5}{10}\right)^2 = \frac{13}{10} - \frac{1}{4} = \frac{21}{20} = \frac{105}{100}$$

$$s = 1.024$$

Value of $t$ for 9 degrees of freedom = 2.262

Also
$$t = \frac{\bar{x} - \mu}{s}\sqrt{n}$$

$$= \frac{12.5 - 12}{1.024}\sqrt{10} = 1.54$$

Since the value of t is less than 2.262, there in no reason to believe that machine is defective.

**Ans.**

**Example 84.** *A random sample of size 16 values from a normal population showed a mean of 53 and a sum of squares of deviation from the mean equals to 150. Can this sample be regarded as taken from the population having 56 as mean? Obtain 95% and 99% confidence limits of the mean of the population.*

$$\gamma = 15,\ \alpha = 0.05,\ t = 2.131$$
$$\alpha = 0.01,\ t = 2.947$$

**Solution.**
$$\mu = 56,\ n = 16,\ \bar{x} = 53,\ \sum (x - \bar{x})^2 = 150$$

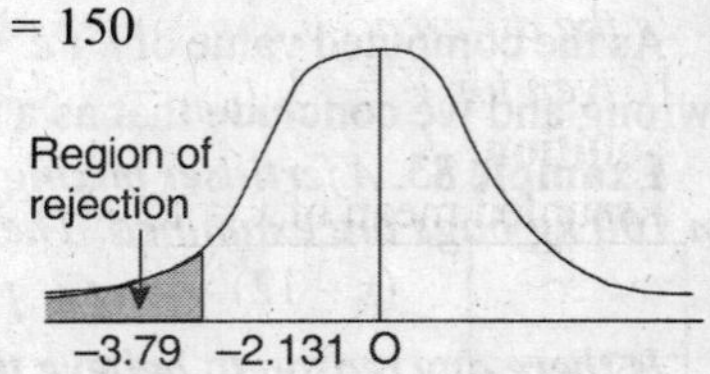


$$s^2 = \frac{\sum (x - \bar{x})^2}{n - 1} = \frac{150}{15} = 10$$

$$s = \sqrt{10}$$

$$t = \frac{\bar{x} - \mu}{\frac{s}{\sqrt{n}}} = \frac{53 - 56}{\frac{\sqrt{10}}{\sqrt{16}}} = \frac{-3 \times 4}{\sqrt{10}} = -3.79$$

$$t = 3.79$$

3.79 > 2.131 and also 3.79 > 72.947.

Thus, the sample cannot be regarded as taken from the population. **Ans.**

## 11.54 TESTING FOR DIFFERENCE BETWEEN MEANS OF TWO SMALL SAMPLES

Let the mean and variance of the first population be $\mu_1$ and $\sigma_1^2$ and $\mu_2$ . $\sigma_2^2$ be the mean and variance of the second population.

Let $\bar{x}_1$ be the mean of small sample of size $n_1$ from first population and $\bar{x}_2$ the mean of a sample of size $n_2$ from second population.

We know that

$$E(\bar{x}_1) = \mu_1 \text{ and } Var\,(\bar{x}_1) = \frac{\sigma_1^2}{n_1}$$

$$E(\bar{x}_2) = \mu_2 \text{ and } Var\,(\bar{x}_2) = \frac{\sigma_2^2}{n_2}$$

If the samples are independent, then $(\bar{x}_1)$ and $(\bar{x}_2)$ are also independent.

$$E(\bar{x}_1 - \bar{x}_2) = \mu_1 - \mu_2 \text{ and } Var\,(\bar{x}_1 - \bar{x}_2) = \frac{\sigma_1^2}{n_1} + \frac{\sigma_2^2}{n_2}$$

$$\bar{x}_1 \sim N\left(\mu_1, \frac{\sigma_1^2}{n_1}\right) \text{ and } \bar{x}_2 \sim N\left(\mu_2, \frac{\sigma_2^2}{n_2}\right)$$

then $$(\bar{x}_1 - \bar{x}_2) \sim N\left(\mu_1 - \mu_2, \frac{\sigma_1^2}{n_1} + \frac{\sigma_2^2}{n_2}\right)$$

$$t = \frac{(\bar{x}_1 - \bar{x}_2) - (\mu_1 - \mu_2)}{\sqrt{\dfrac{\sigma_1^2}{n_1} + \dfrac{\sigma_2^2}{n_2}}}$$

If the population is the same then

$$t = \frac{\bar{x}_1 - \bar{x}_2}{\sigma\sqrt{\dfrac{1}{n_1} + \dfrac{1}{n_2}}} \qquad (\mu_1 - \mu_2 = \mu_1 - \mu_1 = 0)$$

**Example 85.** *Two independent samples of 8 and 7 items respectively had the following values of the variable (weight in ounces):*

*Sample 1: 9 11 13 11 15 9 12 14*

*Sample 2: 10 12 10 14 9 8 10*

*Is the difference between the means of the sample significant?*

*Is the difference between the means of the sample significant?*

*[Given for $V = 13$, $t_{0.05} = 2.16$]*

**Solution.**

Assumed mean of $x = 12$, Assumed mean of $y = 10$

| $x$ | $(x-12)$ | $(x-12)^2$ | $y$ | $(y-10)$ | $(y-10)^2$ |
|---|---|---|---|---|---|
| 9 | −3 | 9 | 10 | 0 | 0 |
| 11 | −1 | 1 | 12 | 2 | 4 |
| 13 | 1 | 1 | 10 | 0 | 0 |
| 11 | −1 | 1 | 14 | 4 | 16 |
| 15 | 3 | 9 | 9 | −1 | 1 |
| 9 | −3 | 9 | 8 | −2 | 4 |
| 12 | 0 | 0 | 10 | 0 | 0 |
| 14 | 2 | 4 | — | — | — |
| $\Sigma x = 94$ | $\Sigma(x-12) = -2$ | $\Sigma(x-12)^2 = 34$ | $\Sigma y = 73$ | $\Sigma(y-10) = 3$ | $\Sigma(y-10)^2 = 25$ |

$$\bar{x} = \frac{\sum x}{n} = \frac{94}{8} = 11.75$$

$$\sigma_x^2 = \frac{\sum(x-12)^2}{n} - \left(\frac{\sum(x-12)}{n}\right)^2 = \frac{34}{8} - \left(\frac{-2}{8}\right)^2 = 4.1875$$

$$\bar{y} = \frac{\sum y}{n} = \frac{73}{7} = 10.43$$

$$\sigma_y^2 = \frac{\sum (y-10)^2}{n} - \left[\sum \frac{(y-10)}{n}\right]^2 = \frac{25}{7} - \left(\frac{3}{7}\right)^2 = 3.438$$

$$s = \sqrt{\frac{\sum (x-\bar{x})^2 + \sum (y-\bar{y})^2}{n_1+n_2-2}} = \sqrt{\frac{34+25}{8+7-2}} = \sqrt{\frac{59}{13}} = \sqrt{4.54} = 2.13$$

$$t = \frac{\bar{x}-\bar{y}}{S\sqrt{\frac{1}{n_1}+\frac{1}{n_2}}} = \frac{11.75-10.43}{2.13\sqrt{\frac{1}{8}+\frac{1}{7}}} = \frac{1.32}{2.13\sqrt{0.268}} = \frac{1.32}{2.13 \times 0.518}$$

$$= \frac{1.32}{1.103} = 1.12$$

The 5% value of $t$ for 13 degree of freedom is given to be 2.16. Since calculated value of $t$ is 1.12 is less than 2.16, the difference between the means of samples is not significant. **Ans.**

**Exercise 11.9**

**1.** A random sample of six steel beams has mean compressive strength of 58.392 psi (pounds per square inch) with a standard deviation of $s$ = 648 psi. Test the null hypothesis $H_0 = \mu$ = 58,000 psi against the alternative hypothesis $H_1$: $\mu >$ 58,000 psi at 5% level of significance (value for $t$ at 5 degree of freedom and 5% significance level is 2.0157). Here $\mu$ denotes the population mean. *(A.M.I.E., Summer 2000)*

**2.** A certain cubical die was thrown 96 times and shows 2 upwards 184 times. Is the die biased?

**Ans.** die is biased.

**3.** In a sample of 100 residents of a colony 60 are found to be wheat eaters and 40 rice eaters. Can we assume that both food articles are equally popular?

**4.** Out of 400 children, 150 are found to be under weight. Assuming the conditions of simple sampling, estimate the percentage of children who are underweight in, and assign limits within which the percentage probably lies. **Ans.** 37.5% approx. Limits = 37.5 ± 3 (2.4)

**5.** 500 eggs are taken at random from a large consignment, and 50 are found to be bad. Estimate the percentage of bad eggs in the consignment and assign limits within which the percentage probably lies. **Ans.** 10%, 10 ± 3.9

**6.** A machine puts out 16 imprefect articles in a sample of 500. After the machine is repaired, puts out 3 imprefect articles in a batch of 100. Has the machine been improved?

**Ans.** The machine has not been improved.

**7.** In a city $A$, 20% of a random sample of 900 school boys had a certain slight physical defect. In another city $B$, 18.5% of a random sample of 1600 school boys had the same defect. Is the difference between the proportions significant?

**Ans.** $z$ = 0.37, Difference between proportions is significant.

**8.** In two large populations there are 30% and 25% respectively of fair haired people. Is this difference likely to be hidden in samples of 1200 and 900 respectively from the two populations? **Ans.** $z$ = 2.5, not hidden at 5% level of significance.

9. One thousand articles from a factory are examined and found to be three percent defective. Fifteen hundred similar articles from a second factory are found to be only 2 percent defective. Can it reasonably be concluded that the product of the first factory is inferior to the second?

**Ans.** It cannot be reasonable concluded that the product of the first factory is inferior to that of the second.

10. A manufacturing company claims 90% assurance that the capacitors manufactured by them will show a tolerance of better than 5%. The capacitors are packaged and sold in lots of 10. Show that about 26% of his customers ought to complain that capacitors do not reach the specified standard.

11. An experiment was conducted on nine individuals. The experiment showed that due to smoking, the pulse rate increased in the following order:

$$5, 3, 4, -1, 2, -3, 4, 3, 1.$$

Can you maintain that smoking leads to an increase in the pulse rate?

($t$ for 8 d.f. at 5% level of significance = 2.31). **Ans.** Yes.

12. Nine patients to whom a certain drink was administered registered the following in blood pressure: 7, 3, –1, 4, –3, 5, 6, –4, 1. Show that the data do not indicate that the drink was responsible for these increments.

13. A machine has produced washers having a thickness of 0.50 mm. To determine whether the machine is in proper working order, a sample of 10 washers is chosen for which the mean thickness is 0.53 mm. and the standard deviation is 0.03 mm. Test the hypothesis that the machine is in proper working order using a level of significance *(a)* 0.05 *(b)* 0.01.

**Ans.** (*a*) The machine is not in proper working order at 0.05 level of significance.
(*b*) The machine is in proper working order at 0.01 level of significance.

14. Eleven school boys were given a test in drawing. They were given a months further tuition and a second test of equal difficulty was held at the end of it. Do the marks give evidence that the students have benefitted by extra coaching.

| Boys | 1 | 2 | 3 | 4 | 5 | 6 | 7 | 8 | 9 | 10 | 11 |
|---|---|---|---|---|---|---|---|---|---|---|---|
| Marks I Test | 23 | 20 | 19 | 21 | 18 | 20 | 18 | 17 | 23 | 16 | 19 |
| Marks II Test | 24 | 19 | 22 | 18 | 20 | 22 | 20 | 20 | 23 | 20 | 17 |

**Ans.** $t$ = 1.48, The value of $t$ is not significant at 5% level of significance. (i.e., the test, i.e., the students) no evidence that the students have benefitted by extra coaching.

15. Two horses $A$ and $B$ were tested according to the time (in seconds) to run a particular race with the following results:

| | | | | | | | |
|---|---|---|---|---|---|---|---|
| Horse A | 28 | 30 | 32 | 33 | 33 | 29 | 39 |
| Horse B | 9 | 30 | 30 | 24 | 27 | 29 | |

Test whether you can discriminate between two horses? **Ans.** Yes with 75% confidence.

## 11.55 THE CHI-SQUARE DISTRIBUTION

Chi-square is a measure of actual divergence of the observed and expected frequencies. If $f_0$ is the observed frequency and $f_e$ the expected frequency of a class interval, then $x^2$ is defined as

$$x^2 = \sum \frac{(f_o - f_e)^2}{f_e}$$

## 11.56 DEGREE OF FREEDOM (*df*)

The degree of freedom refers to the number of "independent constrants" in a set of data. We shall illustrate this concept with example. There is a 2 x 2 association table and the actual frequencies as under:

Let the two attributes *A* and *B* be independent.

$$\text{Expected frequency of } (AB) = \frac{30 \times 60}{100} = 18$$

After findingthefrequency of (AB), the expected frequencies of the remaining three classes are automatically fixed.

Expected frequency of $(\alpha B) = 60 - 18 = 42$

Expected frequency of $(A\beta) = 30 - 18 = 12$

Expected frequency of $(\alpha\beta) = 70 - 42 = 28$

It means that only one choice is fixing of frequency of *AB* is independent choice. Frequencies of the remaining three classes depend on the frequency of (AB). It means, we have only one degree of freedom.

Degree of freedom = $(r-1)(c-1)$

where *r* is the number of rows and *c* is the number of columns.

## 11.57 $\chi^2$ – CURVE

Let $x_1, x_2, ...., x_n$ be *n* standard variates with mean zero and S.D. unity. Then $x^2$-distribution has the $x^2_1, x^2_2, ...., x^2_n$ random variates.

Equation of the $\chi^2$-curve is

$$y = y_0 e^{\frac{x^2}{2}} (x^2)^{\frac{r-1}{2}}, \quad r = n - 1$$

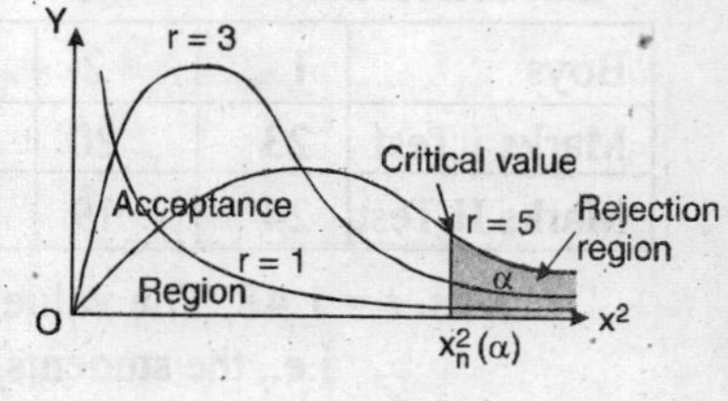


where *r* is the degree of freedom.

Since this equation does not have any parameter. So it can be used for every problem of chi-square. $\chi^2_n(\alpha)$ denote the value of chi-square for *n* degree of freedom such that the area to the right of this point is a.

## 11.58 GOODNESS OF FIT

The value of $\chi^2$ is used to find the divergence of the observed frequency from the expected frequency.

If the value of *P* is high the fit is said to be good. It means that there no significant divergence between observed and expected data.

If the curve of the expected frequency is super imposed on the curve of observed frequencies there would not be much divergence between the two. The fit would be good. If the value of *P* is small, the fit is said to be poor.

## 11.59 STEPS FOR TESTING

(*i*) First calculate the value of $\chi^2$.

(*ii*) From the table read the value of $\chi^2$ for a given degree of freedom.

(*iii*) Find out the probability $P$ corresponding to the calculated values of $\chi^2$.

(*iv*) If $P > 0.05$, the value is not significant and it is a good fit.

(*v*) If $P < 0.05$, the deviations are significant.

**Example 86.** *The following table is given*

| | | *Eye colour infathers* | | |
|---|---|---|---|---|
| | | *not light* | *light* | |
| *Eye colour in* | *Not light* | *230* | *148* | *378* |
| *fathers* | *light* | *251* | *471* | *622* |
| | | *381* | *619* | *1000* |

*Test whether the colour of the son's eyes is associated with that of the fathers.* Given: *value of $\chi^2$ is 3.84 for 1 degree of freedom.*

**Solution.**

**Hypothesis:** Let the eye colour of sons and the eye colour of fathers independent.

| | | *Eye colour in sons* | |
|---|---|---|---|
| | | *not light* | *light* |
| Eye colour | Not light | $\frac{378 \times 381}{1000} = 144$ | $\frac{378 \times 619}{1000} = 234$ |
| in fathers | light | $\frac{622 \times 381}{1000} = 237$ | $\frac{622 \times 619}{1000} = 385$ |

$$\chi^2 = \Sigma \frac{(f_0 - f_e)^2}{f_e}$$

$$\chi^2 = \frac{(230-144)^2}{144} + \frac{(148-234)^2}{234} + \frac{(151-237)^2}{237} + \frac{(471-385)^2}{385}$$

$$= (86)^2 \left[\frac{1}{144} + \frac{1}{234} + \frac{1}{237} + \frac{1}{385}\right] = 133.37$$

The degree of freedom $= (c-1)(r-1) = (2-1)(2-1) = 1$

The value of $\chi^2$ at 5% level of significance for 1 degree of freedom is 3.841 and the calculated value is 133.37

$$133.37 > 3.841$$

This leads to the conclusion that the hypothesis is wrong and the colour of son's eyes is associated with that of the fathers to a great extent. **Ans.**

**Example 87.** *From the following table, showing the number of plants having certain characters, test the hypothesis that the flower colour is independent of flatness of leaf*

| | *Flat leaves* | *Curled leaves* | *Total* |
|---|---|---|---|
| White Flowers | 99 | 36 | 135 |
| Red Flowers | 20 | 5 | 25 |
| Total | 119 | 41 | 160 |

**Solution. Null Hypothesis:** The flower colour is dependent of flatness of leaf. The following table shows the theoretical frequencies.

| | *Flat leaves* | *Curled leaves* | *Total* |
|---|---|---|---|
| White flowers | $\frac{135\times 119}{160}=100$ | $\frac{135\times 41}{160}=35$ | 135 |
| Red flowers | $\frac{25\times 119}{160}=19$ | $\frac{25\times 41}{160}=6$ | 25 |
| Total | 119 | 41 | 160 |

$$\chi^2 = \Sigma\frac{(f_0-f_e)^2}{f_e}$$

$$\chi^2 = \frac{(90-100)^2}{100}+\frac{(36-35)^2}{35}+\frac{(20-19)^2}{19}+\frac{(5-6)^2}{6}$$

$$\chi^2 = \frac{1}{100}+\frac{1}{35}+\frac{1}{19}+\frac{1}{6} = 0.2579$$

Degree of freedom = $(r-1)(c-1) = (2-1)(2-1) = 1$

We have $\chi^2 = 0.0158$ at 0.1 level of significance.

$0.2579 > 0.0158$

This leads to the conclusion that the hypothesis is wrong and the flower colour is independent of flatness of leaf at the 0.1 level of significance. **Ans.**

**Example 88**. *The following table gives the number of air craft accidents that occurs during various days of the week. Find whether the accidents are uniformly distributed over the week.*

| *Days* | *Sun.* | *Mon.* | *Tues.* | *Wed.* | *Thurs.* | *Fri.* | *Sat.* |
|---|---|---|---|---|---|---|---|
| *No. of accidents* | *14* | *16* | *8* | *12* | *11* | *9* | *14* |

**Given:** *The values of chi-square significant at 5, 6, 7, are respectively 11.07, 12.59, 14.07 at the 5% level of significance.*

**Solution. Null Hypothesis:** The accidents are uniformly distributed over the week.

Expected frequencies of the accidents are given below:

| Days | Sun. | Mon. | Tues. | Wed. | Thurs. | Fri. | Sat. | Total |
|---|---|---|---|---|---|---|---|---|
| No. of accidents | 12 | 12 | 12 | 12 | 12 | 12 | 12 | 84 |

$$\chi^2 = \frac{(14-12)^2}{12}+\frac{(16-12)^2}{12}+\frac{(8-12)^2}{12}+\frac{(12-12)^2}{12}+\frac{(11-12)^2}{12}+\frac{(9-12)^2}{12}+\frac{(14-12)^2}{12}$$

$$= \frac{1}{12}[4+16+16+0+1+9+4] = \frac{50}{12} = 4.17$$

The number of degrees of freedom = Number of observations – Number of independent constants = 7 – 1 = 6.

The tabulated $\chi^2_{0.05}$ for 6 d.f. = 12.59

Since the calculated $\chi^2$ is much less than the tabulated value, we accept the null hypothesis. Hence, the accidents are uniformly distributed over the week. **Ans.**

**Example 89.** *A set of five similar coins is tossed 320 times and the result is*

| *No. of heads* | *0* | *1* | *2* | *3* | *4* | *5* |
|---|---|---|---|---|---|---|
| *Frequency* | *6* | *27* | *72* | *112* | *71* | *32* |

*Test the hypothesis that the data follow a binomial distribution.*

**Solution.** $P$ (Head) $= \frac{1}{2}$, $q = 1 - \frac{1}{2} = \frac{1}{2}$

Theoretical frequencies are

$$P(0\,H) = q^5 = \left(\frac{1}{2}\right)^5 = \left(\frac{1}{32}\right), \text{ Frequency of 0 head} = \frac{320}{32} = 10$$

$$P(1\,H) = {}^5C_1\, pq^4 = {}^5C_1\left(\frac{1}{2}\right)\left(\frac{1}{2}\right) = \frac{5}{32} \text{ Frequency of 1 head} = \frac{5}{32}\times 320 = 50$$

$$P(2\,H) = {}^5C_2\, p^2q^3 = 10\left(\frac{1}{2}\right)^2\left(\frac{1}{2}\right)^3 = \frac{10}{32}, \text{ Frequency of 2 heads} = \frac{10}{32}\times 320 = 100$$

$$P(3\,H) = {}^5C_3\, p^3q^2 = 10\left(\frac{1}{2}\right)^3\left(\frac{1}{2}\right)^2 = \frac{10}{32}, \text{ Frequency of 3 heads} = \frac{10}{32}\times 320 = 100$$

$$P(4\,H) = {}^5C_4\, p^4q = 5\left(\frac{1}{2}\right)^4\left(\frac{1}{2}\right) = \frac{5}{32}, \text{ Frequency of 4 heads} = \frac{5}{32}\times 320 = 50$$

$$P(5\,H) = {}^5C_5\, p^5q^0 = \left(\frac{1}{2}\right)^5 = \frac{1}{32}, \text{ Frequency of 5 heads} = \frac{1}{32}\times 320 = 10$$

$$\chi^2 = \frac{(6-10)^2}{10} + \frac{(27-50)^2}{50} + \frac{(72-100)^2}{100} + \frac{(112-100)^2}{100} + \frac{(71-50)^2}{50} + \frac{(32-10)^2}{10}$$

$$= \frac{1}{100}[160 + 1058 + 784 + 144 + 882 + 4840] = \frac{7868}{100} = 78.68$$

Degree of freedom = 6 – 1 = 5

For 5 degree of freedom, $\chi^2 = 11.07$

Since the calculated value of $\chi^2$ is much greater than that of $\chi^2$ at 5% level of significance, the hypothesis that the data follow the binomial law is rejected. **Ans.**

**Example 90.** *Fit a Poisson distribution to the following data and test the goodness of fit.*

| *x* | 0 | 1 | 2 | 3 | 4 | 5 | 6 |
|---|---|---|---|---|---|---|---|
| *f* | 275 | 72 | 30 | 7 | 5 | 2 | 1 |

**Solution.**

Mean, $$m = \frac{\Sigma fx}{\Sigma f} = \frac{0+72+60+21+20+10+6}{392} = \frac{189}{392} = 0.482$$

Poisson distribution

$$P(r)=\frac{e^{-m}m^r}{r!} \quad \text{or} \quad P(r)=\frac{e^{-0.482}(0.482)^r}{r!}$$

$$P(0)=e^{-0.482}=0.6175, \quad f(0)=392\times0.6175=242.1$$

$$P(1)=\frac{e^{-0.482}(0.482)^1}{1!}=0.2976, \quad f(1)=392\times0.2976=116.7$$

$$P(2)=\frac{e^{-0.482}(0.482)^2}{2!}=0.0717, \quad f(2)=392\times0.0717=28.1$$

$$\left.\begin{aligned} p(3)&=\frac{e^{-0.482}(0.482)^3}{3!}=0.0115, \quad f(3)=392\times0.0115=4.5\\ p(4)&=\frac{e^{-0.482}(0.482)^4}{4!}=0.00139, \quad f(4)=392\times0.0139=0.5\\ P(5)&=\frac{e^{-0.482}(0.482)^5}{5!}=0.0001, \quad f(5)=392\times0.0001=0.1\\ P(6)&=\frac{e^{-0.482}(0.482)^6}{6!}=0.00001, \quad f(6)=392\times0.00001=0 \end{aligned}\right\} \quad (5.1)$$

$$\chi^2=\frac{(275-242.1)^2}{242.1}+\frac{(72-116.7)^2}{116.7}+\frac{(30-28.1)^2}{28.1}+\frac{[(7+5+2+1)-(4.5+0.5+0.1)]^2}{4.5+0.5+0.1}$$

$$\Rightarrow \quad \chi^2=4.471+17.122+0.128+19.217=40.938$$

Degree of freedom = 7 – 1 – 1 – 3 = 2

[One d.f. being lost because $\Sigma 0=\Sigma E$; 1 d.f. is lost because the parameter $m$ has been estimated; 3 d.f. are lost because of pooling the last four expected cell frequencies which are less than 5]

Tabulated value of $\chi^2$ for 2 d.f. at 5% level of significance = 5.99.

Since, the calculated value of $\chi^2$ (40.938) is much greater than 5.99, it is highly significant. Hence, Poisson distribution is not good fit. **Ans.**

**Exercise 11.10**

1. The following information is obtained concerning an investigation of 50 ordinary shops of small size.

| | Shops | | Total |
|---|---|---|---|
| | In Town | In Villages, | |
| Run by men | 17 | 18 | 35 |
| Run by women | 3 | 12 | 15 |
| Total | 20 | 30 | 50 |

Can it be inferred that shops run by women are relatively more in villages than in towns? Use $\chi^2$ test.

**Ans.** $\chi^2=3.57$, Hypothesis is wrong.

2. Of a group of patients who complained they did not sleep well, some were given sleeping pills while others were given sugar pills (although they all thought they were getting sleeping pills). They were later asked whether the pills helped them or not. The result of their responses are shown in the table given below. Assuming that all patients told the truth, test the hypothesis that there is no difference between sleeping pills and sugar pills at a significance level of 0.05.

| | *Slept well* | *Did not sleep well* |
|---|---|---|
| Took sleeping pills | 44 | 10 |
| Took sugar pills | 81 | 35 |

**Ans.** The hypothesis cannot be rejected at the 0.05 level.

3. In an experiment on immunization of cattle from tuberculosis the following results were obtained

| | *Died* | *Unaffected* |
|---|---|---|
| Inoculated | 12 | 26 |
| Not inoculated | 16 | 6 |

Examine the effect of vaccine in controlling susceptibility to tuberculosis.

**Ans.** $\chi^2 = 9.367$, vaccine is effective.

4. Genetic theory states that children having one parent of blood type *M* and other blood type *N* will always be one of three types *M, MN, N* and that the proportions of these types will on average be 1 : 2 : 1. A report states that out of 300 children having one *M* parent and one *N* parent, 30% were found to be of type *M*, 45% of type *MN* and remainder of type *N*. Test the hypothesis by $\chi^2$ test. **Ans.** Hypothesis is correct.

5. In an experiment on pea-breeding, Mendal obtained the following frequencies of seeds; 315 round and yellow, 101 wrinkled and yellow; 108 round and green, 32 wrinkled and green. Total 556. Theory predicts that the frequencies should be in the proportion 9 : 3 : 3 : 1 respectively. Set up proper hypothesis and test it at 10% level of significance.

**Ans.** $\chi^2 = 0.51$. There seems to be good correspondence between theory and experiment.

6. On a particular proposal of national importance, political party *A* and party *B* cast votes as given in the table. At a level of significance of *(a)* 0.01 and *(b)* 0.05, test the hypothesis that there is no difference between the two parties in so far as this proposal is concerned.

| | In Favour | Opposed | Undecided |
|---|---|---|---|
| Party A | 85 | 78 | 37 |
| Party B | 118 | 61 | 25 |

**Ans.** The hypothesis can be rejected at both levels.

7. The table shows the relation between the performance in mathematics and electronics, using a (*a*) 0.05 (*b*) 0.01 significance level.

| *Mathematics* | *Electronics* | | |
|---|---|---|---|
| | *High marks* | *Medium marks* | *Low marks* |
| High marks | 56 | 71 | 12 |
| Medium marks | 47 | 163 | 38 |
| Low marks | 14 | 42 | 85 |

**Ans.** The hypothesis can be rejected at both levels.

8. The results of a survey made to determine whether the age of a driver 21 years of age and older has any effect on the number of automobile accidents in which he is involved (including minor accidents) are given in the table below. At a level of significance of (*a*) 0.05 and (*b*) 0.01, test the hypothesis that number of accidents is independent of the age of the driver.

| | | *Age of the driver* | | | | |
|---|---|---|---|---|---|---|
| | | *21-30* | *31-40* | *41-50* | *51-60* | *61-70* |
| Number of accidents | 0 | 748 | 821 | 786 | 720 | 672 |
| | 1 | 74 | 60 | 51 | 66 | 50 |
| | 2 | 31 | 25 | 22 | 16 | 15 |
| | more than 2 | 9 | 10 | 6 | 5 | 7 |

**Ans.** The hypothesis cannot be rejected at either level.

9. A die is thrown 60 times with the following results.

| Face | 1 | 2 | 3 | 4 | 5 | 6 |
|---|---|---|---|---|---|---|
| Frequency | 8 | 7 | 12 | 8 | 14 | 11 |

Test at 5% level of significance if the die is honest, assuming that $P(\chi^2 > 11.1) = 0.05$ with 5 d.f

10. Fit a Binomial Distribution to the data

| $x$ | 0 | 1 | 2 | 3 | 4 | 5 |
|---|---|---|---|---|---|---|
| $f$ | 38 | 144 | 342 | 287 | 164 | 25 |

and test for goodness of fit at the level of significance 0.05.

**Ans.** $\chi^2 = 7.97$, Binomial distribution gives a good fit at 5% level.

11. Fit a Poisson distribution to the following data and test for its goodness of fit at level of significance 0.05.

| $x$ | 0 | 1 | 2 | 3 | 4 |
|---|---|---|---|---|---|
| $f$ | 419 | 352 | 154 | 56 | 19 |

**Ans.** Poisson distribution can be fitted to the data.

12. A bird watching sitting in a park has spotted a number of birds belonging to 6 categories. The exact classification is given below:

| Category | 1 | 2 | 3 | 4 | 5 | 6 |
|---|---|---|---|---|---|---|
| Frequency | 6 | 7 | 13 | 17 | 6 | 5 |

Test at 5% level of significance whether or not the data is compatible with the assumption that this particular park is visited by birds belonging to these six categories in the proportion $= 1 : 1 : 2 : 3 : 1 : 1$.

Given $P(\chi^2 = 1.07) = 0.05$ for 5 degree of freedom.

13. Two hundred digits were chosen at random from a set of tables. The frequencies of the digits were as follows:

| Digits | 0 | 1 | 2 | 3 | 4 | 5 | 6 | 7 | 8 | 9 |
|---|---|---|---|---|---|---|---|---|---|---|
| Frequency | 18 | 19 | 23 | 21 | 16 | 25 | 22 | 20 | 21 | 15 |

Use $\chi^2$ test to assess the corrections of hypothesis that the digits were distributed in equal numbers in the table from which they were chosen.

Given that the values of $\chi^2$ are respectively 16.9, 18.3., 19.7 for 9, 10 and 11 degrees of freedom at 5% level of significance. **Ans.** $\chi^2 = 4.3$, the hypothesis seems reasonable correct.

14. A survey of 320 families with 5 children each revealed the following distribution:

| No. of boys | 5 | 4 | 3 | 2 | 1 | 0 |
|---|---|---|---|---|---|---|
| No. of girls | 0 | 1 | 2 | 3 | 4 | 5 |
| No. of families | 14 | 56 | 110 | 88 | 40 | 40 |

Is this result consistent with the hypothesis that male and female births are equally probable?

**Ans.** $\chi^2 = 7.16$, Equal probability for male and female births may be accepted.

## 11.60 F-DISTRIBUTION HAS THE FOLLOWING APPLICATION

### F-Test for Equality of Population Variances

Suppose we want to test

(*i*) Whether two independent samples $x_1, x_2, \ldots x_n$ and $y_1, y_2, \ldots, y_n$ have been drawn from the normal population with the same variance $\sigma^2$.

(*ii*) Whether the two independent estimates of the population variance are homogeneous or not.

Under the null hypothesis $(H_0)$

(*i*) $\sigma_x^2 = \sigma_y^2 = \sigma^2$, population variances are equal.

(*ii*) Two independent estimates of population variance are homogeneous, *F* is given by

$$F = \frac{S_x^2}{S_y^2}$$

where $S_x^2 = \dfrac{\sum_{n=1}^{n_2}(x-\bar{x})^2}{n_1 - 1}$, $S_y^2 = \dfrac{\sum_{n=1}^{n_2}(y-\bar{y})^2}{n-1}$

and $\bar{x}, \bar{y}$ are sample means, $S_1^2, S_2^2$ are unbiased estimates of two samples from two normal population with $(\sigma_1 = \sigma_2)$.

The distributions of variance ratio *F* with $r_1$ and $r_2$ is

$$y = y_0 . F^{\frac{r_1 - r_2}{2}} \left[1 + \frac{r_1}{r_2} F\right]^{-\frac{(r_1 + r_2)}{2}}$$

where $r_1 = n_1 - 1$, $r_2 = n_2 = 1$, $r_1$ and $r_2$ are d.f. of two samples.

**Note. (*i*)** Greater of the two variances $S_x^2, S_y^2$ is to be taken in the numerator and $n_1$ corresponds to the greater variance. Calculated value of *F* is compared with tabulated value of *F* at certain level of significance, $H_0$ is either rejected or accepted.

**(*ii*)** *Significative Test*

If the calculated value of *F* is higher than this table value of *F* for the given degree of freedom at 5% level of significance as such the difference is significant. It means that the variance between the samples is significantly greater than variance within the samples. In other words, the samples are not picked up from the same population or the mean value of various samples are significantly different from each other.

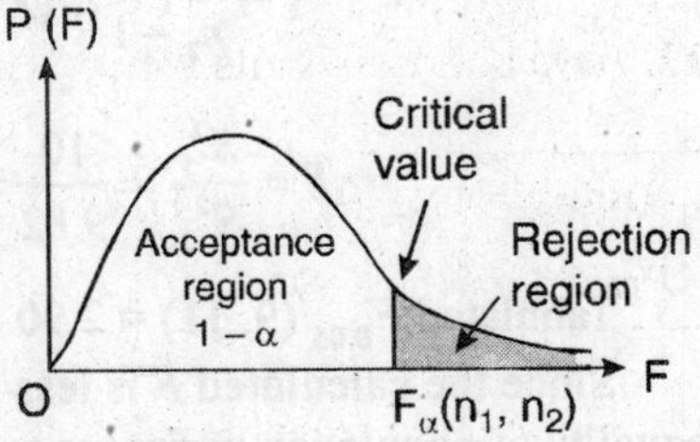


## 11.61 FISHER'S Z-DISTRIBUTION

On substitution $z = \frac{1}{2}\log_e F$ or $F = e^{2z}$ in the *F*-distribution, we have the Fisher's *z*-distribution.

It is of the form

$$y = y_0 e^{r_1 z}(r_1 e^{2z} + r_2)$$

The curve is more symmetrical than *F*-distribution curve.

**Example 91.** *The I.Q.'s of 25 students from one college showed a variance of 16 and those of an equal number from the other college had a variance of 8. Discuss whether there is any significant difference in variability of intelligence.*

**Given:** *F* (5%) = 1.98, *F* (1%) = 2.62

**Solution.** $\sigma_1^2 = 16$, $\sigma_2^2 = 8$

$$F = \frac{\sigma_1^2}{\sigma_2^2} = \frac{16}{8} = 2$$

Tabulated value of *F* at 5% level of significance = 1.98

Calculated value of *F* (2) Tabulated value of *F* (1.98)

Hence, variability of intelligence is just significant at 5% level of significant.

Tabulated value of $F$ at 1% level of significance = 2.62

Calculated value $F$ (2) < Tabulated value of $F$ (2.62)

Hence, variability of intelligence is not significant at 1% level of significance. **Ans.**

**Example 92.** *Two random samples are:*

| *Sample* | *Size* | *Sum of squares of deviations from the mean* |
|---|---|---|
| *1* | *10* | *90* |
| *2* | *12* | *108* |

*Test whether the samples come from the same normal population at 5% level of significance.*

**[Given:** $F_{0.05}(9, 11) = 2.90$, $F_{0.05}(11, 9) = 3.10$]

**Solution. Null Hypothesis:** The two samples have been drawn from the same normal population.

$$n_1 = 10, \ \Sigma(x-\bar{x})^2 = 90$$

$$n_2 = 12, \ \Sigma(y-\bar{y})^2 = 108$$

$$S_1^2 = \frac{1}{n_1 - 1}\Sigma(x-\bar{x})^2 = \frac{90}{9} = 10$$

$$S_2^2 = \frac{1}{n_2 - 1}\Sigma(y-\bar{y})^2 = \frac{108}{11} = 9.82$$

$$F = \frac{S^2_1}{S^2_2} = \frac{10}{9.82} = 1.018$$

Tabulated $F_{0.05}$ (9, 11) = 2.90

Since the calculated $F$ is less than tabulated $F$, it is not significant. Hence, null hypothesis of equality of population variances may be accepted.

**Example 93.** *Two random samples from two normal populations are given below:*

| | | | | | | |
|---|---|---|---|---|---|---|
| Sample *I* | 16 | 26 | 27 | 23 | 24 | 22 |
| Sample *II* | 33 | 42 | 35 | 32 | 28 | 31 |

*Do the estimates of population variances differ significantly?*

| | | | |
|---|---|---|---|
| *Degree of Freedom* | *(5, 5)* | *(5, 6)* | *(6, 5)* |
| *5% value of F* | *5.05* | *4.39* | *4.95* |

**Solution.**

| *Sample I* $x$ | $x-\bar{x}$ $x-23$ | $(x-\bar{x})^2$ | *Sample II* | $y-\bar{y}$ $y-33.5$ | $(y-\bar{y})^2$ |
|---|---|---|---|---|---|
| 16 | −7 | 49 | 31 | −0.5 | 0.25 |
| 26 | 3 | 9 | 42 | 8.5 | 72.25 |
| 27 | 4 | 16 | 35 | 1.5 | 2.25 |
| 23 | 0 | 0 | 32 | −1.5 | 2.25 |
| 24 | 1 | 1 | 28 | −5.5 | 30.25 |
| 22 | −1 | 1 | 31 | −2.5 | 6.25 |
| 138 | 7 | 6 | 201 | $\sigma_{76}$ | 113.50 |

$$\bar{x} = \frac{\Sigma x}{n} = \frac{138}{6} = 23, \ \bar{y} = \frac{\Sigma y}{n} = \frac{201}{6} = 33.5$$

$$S_1^2 = \frac{\Sigma(x-\bar{x})^2}{n-1} = \frac{76}{5} = 15.2$$

$$S_2^2 = \frac{\Sigma(y-\bar{y})^2}{n-1} = \frac{113.50}{5} = 22.7$$

$$F = \frac{S_2^2}{S_1^2} = \frac{22.7}{15.2} = 1.4934$$

**Tabulated value of $F_{0.05} = 5.05$**

**Since the calculated value (1.4934) of *F is* less than the tabulated value of *F (5.05)*. Hence, the difference is not significant.** **Ans.**

## EXERCISE 11.11

**1.** **The diameters of two random samples, each of size 10, of bulbs produced by two machines have standard deviations $S_1 = 0.01$ and $S_2 = 0.015$. Assuming that the diameters have independent distributions, test the hypothesis that, the two machines are equally good by testing.**

**Ans.** $F$**=1.5,yes hypothesis is correct.**

**2.** **The mean diameter of rivets produced by two firms *A* and *B* are practically the same but their standard deviations are different. For 16 rivets manufactured by firm *A*, the S.D. is 3.8 mm while for 22 rivets manufactured by firm *B* is 2.9 mm. Do you think products from firm A are better quality than those of firm *B*** **Ans. Yes**

**3.** **Mango-trees were grown under two experimental conditions. Two random samples of 11 and 9 mango-trees show the samples standard deviations of their weights as 0.8 and 0.5 respectively. Assuming that the weight distributions are normal, test the hypothesis that the true variances are equal, against the alternative that they are not, at the 10% level.**

**[Assume that $P(F_{10,8} > 3.35) = 0.05$ and $P(F_{8,10} > 3.07) = 0.05$]**

**Ans.** $F = 2.5$**, Not significant, hence null hypothesis of equality of population variances may be accepted at level of significance $\alpha = 1.0$.**

**4.** **Two random samples drawn from two normal populations are:**

| | | | | | | | | | | | | |
|---|---|---|---|---|---|---|---|---|---|---|---|---|
| Sample *I* | 20 | 16 | 26 | 27 | 23 | 22 | 18 | 29 | 25 | 19 | | |
| Sample *II* | 27 | 33 | 42 | 35 | 32 | 34 | 38 | 28 | 41 | 23 | 30 | 37 |

**Obtain the estimates of the variances of the populations and test whether the population have the same variance.**

**(Given: $F_{0.05} = 3.11$ for 11 and 9 d.f.)**

**Ans.** $F = 2.368$**, The hypothesis seems to be correct at the 0.05 level of significance.**

# 12

# Fourier Series

## 12.1 PERIODIC FUNCTIONS

If the value of each ordinate $f(t)$ repeats itself at equal intervals in the abscissa, then $f(t)$ is said to be a periodic function.

If $f(t) = f(t + T) = f(t + 2T) = \ldots$ .... then $T$ is called the period of the function $f(t)$.

For example :

$\sin x = \sin (x + 2\pi) = \sin (x + 4\pi) = \ldots$ so $\sin x$ is a periodic function with the period $2\pi$. This is also called sinusoidal periodic function.

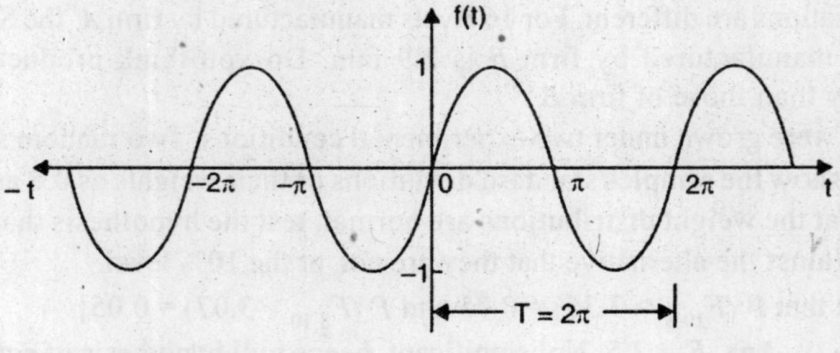


## 12.2 FOURIER SERIES

Here we will express a non-sinusoidal periodic function into a fundamental and its harmonics. A series of sines and cosines of an angle and its multiples of the form.

$$\frac{a_0}{2} + a_1 \cos x + a_2 \cos 2x + a_3 \cos 3x + \ldots + a_n \cos nx + \ldots$$

$$+ b_1 \sin x + b_2 \sin 2x + b_3 \sin 3x + \ldots + b_n \sin nx + \ldots$$

$$= \frac{a_0}{2} + \sum_{n=1}^{\infty} a_n \cos nx + \sum_{n=1}^{\infty} b_n \sin nx.$$

is called the *Fourier series, where* $a_1, a_2, \ldots a_n \ldots b_1, b_2, b_3 \ldots b_n \ldots$ are constants.

A periodic function $f(x)$ can be expanded in a Fourier Series. The series consists of the following:

(*i*) A constant term $a_0$ (called d.c. component in electrical work).

(*ii*) A component at the fundamental frequency determined by the values of $a_1$, $b_1$.

(*iii*) Components of the harmonics (multiples of the fundamental frequency) determined by $a_2, a_3 \ldots b_2, b_3 \ldots$. And $a_0, a_1, a_2 \ldots, b_1, b_2 \ldots$ are known as *Fourier coefficients* or Fourier constants.

## 12.3. DIRICHLET'S CONDITIONS FOR A FOURIER SERIES

If the function $f(x)$ for the interval $(-\pi, \pi)$

(1) is single-valued (2) is bounded

(3) has at most a finite number of maxima and minima.

(4) has only a finite number of discontinuous

(5) is $f(x+2\pi) = f(x)$ for values of x outside $[-\pi, \pi]$, then

$$S_p(x) = \frac{a_0}{2} + \sum_{n=1}^{P} a_n \cos nx + \sum_{n=1}^{P} b_n \sin nx$$

converges to $f(x)$ as $P \to \infty$ at values of $x$ for which $f(x)$ is continuous and to $\frac{1}{2}[f(x+0)+f(x-0)]$ at points of discontinuity.

## 12.4. ADVANTAGES OF FOURIER SERIES

1. Discontinuous function can be represented by Fourier series. Although derivatives of the discontinuous functions do not exist. (This is not true for Taylor's series).
2. The Fourier series is useful in expanding the periodic functions since outside the closed interval, there exists a periodic extension of the function.
3. Expansion of an oscillating function by Fourier series gives all modes of oscillation (fundamental and all overtones) which is extremely useful in physics.
4. Fourier series of a discontinuous function is not uniformly convergent at all points.
5. Term by term integration of a convergent Fourier series is always valid, and it may be valid if the series is not convergent. However, term by term, differentiation of a Fourier series is not valid in most cases.

## 12.5 USEFUL INTEGRALS

The following integrals are useful in Fourier Series.

(*i*) $\int_0^{2\pi} \sin nx\, dx = 0$ (*ii*) $\int_0^{2\pi} \cos nx\, dx = 0$

(*iii*) $\int_0^{2\pi} \sin^2 nx\, dx = \pi$ (*iv*) $\int_0^{2\pi} \cos^2 nx\, dx = \pi$

(*v*) $\int_0^{2\pi} \sin nx \cdot \sin mx\, dx = 0$ (*vi*) $\int_0^{2\pi} \cos nx \cos mx\, dx = 0$

(*vii*) $\int_0^{2\pi} \sin nx \cdot \cos mx\, dx = 0$ (*viii*) $\int_0^{2\pi} \sin nx \cdot \cos nx\, dx = 0$

(*ix*) $\int uv\, dx = uv_1 - u'v_2 + u''v_3 - u'''v_4 + \ldots$

where $v_1 = \int v\, dx,\ v_2 = \int v_1\, dx$ and so on $u' = \frac{du}{dx},\ u'' = \frac{d^2u}{dx^2}$ and so on and

(*x*) $\sin n\pi = 0,\ \cos n\pi = (-1)^n$ where $n \in I$

## 12.6 DETERMINATION OF FOURIER COEFFICIENTS (EULER'S FORMULAE)

$$f(x) = \frac{a_0}{2} + a_1 \cos x + a_2 \cos 2x + \ldots + a_n \cos nx + \ldots$$

$$+ b_1 \sin x + b_2 \sin 2x + \ldots + b_n \sin nx + \ldots \quad \ldots (1)$$

(*i*) **To find $a_0$:** Integrate both sides of (1) from $x = 0$ to $x = 2\pi$.

$$\int_0^{2\pi} f(x)dx = \frac{a_0}{2}\int_0^{2\pi} dx + a_1\int_0^{2\pi} \cos x\, dx + a_2\int_0^{2\pi} \cos 2x\, dx + .... + a_n\int_0^{2\pi} \cos nx\, dx + ...$$

$$+b_1\int_0^{2\pi} \sin x\, dx + b_2\int_0^{2\pi} \sin 2dx + ... + b_n\int_0^{2\pi} \sin nx\, dx + ...$$

$$= \frac{a_0}{2}\int_0^{2\pi} dx, \quad \text{(other integrals = 0 by formulae (i) and (ii) of Art. 12.5)}$$

$$\int_0^{2\pi} f(x)\, dx = \frac{a_0}{2} 2\pi, \quad \Rightarrow \quad a_0 = \frac{1}{\pi}\int_0^{2\pi} f(x)\, dx \quad ... (2)$$

(*ii*) **To find $a_n$:** Multiply each side of (1) by cos $nx$ and integrate from $x = 0$ to $x = 2\pi$.

$$\int_0^{2\pi} f(x)\cos nx\, dx = \frac{a_0}{2}\int_0^{2\pi} \cos nx\, dx + a_1\int_0^{2\pi} \cos x \cos nx\, dx + ... + a_n\int_0^{2\pi} \cos^2 nx\, dx ...$$

$$+b_1\int_0^{2\pi} \sin x \cos nx\, dx + b_2\int_0^{2\pi} \sin 2x \cos nx\, dx + ...$$

$$= a_n\int_0^{2\pi} \cos^2 nx\, dx = a_n \pi \quad \text{(Other integrals = 0, by formulae on Page 851)}$$

$$\therefore \quad a_n = \frac{1}{\pi}\int_0^{2\pi} f(x)\cos nx\, dx$$

By taking $n = 1, 2$ ... we can find the values of $a_1, a_2$....

(*iii*) **To find $b_n$:** Multiply each side of (1) by sin $nx$ and integrate from $x = 0$ to $x = 2\pi$.

$$\int_0^{2\pi} f(x)\sin nx\, dx = \frac{a_0}{2}\int_0^{2\pi} \sin nx\, dx + a_1\int_0^{2\pi} \cos x \sin nx\, dx + ... + a_n\int_0^{2\pi} \cos nx \sin nx\, dx + ...$$

$$+b_1\int_0^{2\pi} \sin x \sin nx\, dx + ... b_n\int_0^{2\pi} \sin^2 nx\, dx + ...$$

$$= b_n\int_0^{2\pi} \sin^2 nx\, dx \quad \text{(All other integrals = 0, Article No. 12.5)}$$

$$= b_n \pi$$

$$\Rightarrow \quad b_n = \frac{1}{\pi}\int_0^{2\pi} f(x)\sin nx\, dx \quad ... (4)$$

**Note :** To get similar formula of $a_0$, $\frac{1}{2}$ has been written with $a_0$ in Fourier series.

**Example 1.** *Find the Fourier series representing*

$$f(x) = x, \quad 0 < x < 2\pi$$

*and sketch its graph from $x = -4\pi$ to $x = 4\pi$.*

**Solution.** Let $f(x) = \frac{a_0}{2} + a_1 \cos x + a_2 \cos 2x + ..... + b_1 \sin x + b_2 \sin 2x + ...$ ... (1)

Hence $a_0 = \frac{1}{\pi}\int_0^{2\pi} f(x)dx = \frac{1}{\pi}\int_0^{2\pi} x\,dx = \frac{1}{\pi}\left[\frac{x^2}{2}\right]_0^{2\pi} = 2\pi$

$$a_n = \frac{1}{\pi}\int_0^{2\pi} f(x)\cos nx\,dx = \frac{1}{\pi}\int_0^{2\pi} x\cos nx\,dx$$

$$= \frac{1}{\pi}\left[x\frac{\sin nx}{n} - 1.\left(-\frac{\cos nx}{n^2}\right)\right]_0^{2\pi} = \frac{1}{\pi}\left[\frac{\cos 2n\pi}{n^2} - \frac{1}{n^2}\right] = \frac{1}{n^2\pi}(1-1) = 0$$

$$b_n = \frac{1}{\pi}\int_0^{2\pi} f(x)\sin nx\,dx = \frac{1}{\pi}\int_0^{2\pi} x\sin nx\,dx$$

$$= \frac{1}{\pi}\left[x\left(-\frac{\cos nx}{n}\right) - 1.\left(\frac{-\sin nx}{n^2}\right)\right]_0^{2\pi} = \frac{1}{\pi}\left[\frac{-2\pi\cos 2n\pi}{n}\right] = -\frac{2}{n}$$

**Substituting the values of $a_0$, $a_n$, $b_n$ in (1), we get**

$$x = \pi - 2\left[\sin x + \frac{1}{2}\sin 2x + \frac{1}{3}\sin 3x + \ldots\right] \quad \textbf{Ans.}$$

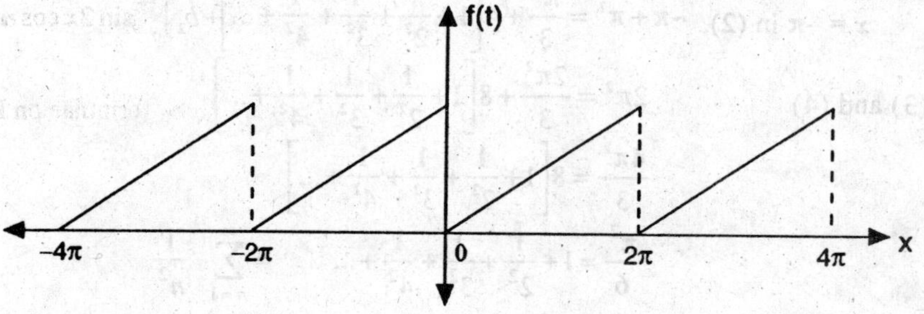


**Example 2.** *Given that $f(x) = x + x^2$ for $-\pi < x < \pi$, find the Fourier expression of $f(x)$.*

*Deduce that* $\frac{\pi^2}{6} = 1 + \frac{1}{2^2} + \frac{1}{3^2} + \frac{1}{4^2} + \ldots$ *(UP. II Semester; Summer 2003)*

**Solution.** Let $x + x^2 = \frac{a_0}{2} + a_1\cos x + a_2\cos 2x + \ldots + b_1\sin x + b_2\sin 2x + \ldots$ ...(1)

$$a_0 = \frac{1}{\pi}\int_{-\pi}^{\pi} f(x)dx = \frac{1}{\pi}\int_{-\pi}^{\pi}(x+x^2)dx \quad (f(x) = x \text{ odd function})$$

$$= \frac{2}{\pi}\left[\frac{x^3}{3}\right]_0^{\pi} = \frac{2}{\pi}\left[\frac{\pi^3}{3}\right] = \frac{2\pi^2}{3}$$

$$a_n = \frac{1}{\pi}\int_{-\pi}^{\pi} f(x)\cos nx\,dx = \frac{1}{\pi}\int_{-\pi}^{\pi}(x+x^2)\cos nx\,dx = \frac{2}{\pi}\int_0^{\pi} x^2\cos nx\,dx$$

(*x* cos *nx* is odd function)

$$= \frac{2}{\pi}\left[x^2\frac{(\sin nx)}{n} - (2x)\frac{(-\cos nx)}{n^2} + (2)\left(-\frac{\sin nx}{n^3}\right)\right]_0^{\pi}$$

$$= \frac{2}{\pi}\left[\pi^2\frac{\sin n\pi}{n} - 2\pi\left(\frac{-\cos n\pi}{n^2}\right) + 2\left(-\frac{\sin n\pi}{n^3}\right)\right] = \frac{4(-1)^n}{n^2}$$

$$b_n = \frac{1}{\pi}\int_{-\pi}^{\pi} f(x)\sin nx\,dx = \frac{1}{\pi}\int_{-\pi}^{\pi}(x+x^2)\sin nx\,dx$$

$$= \frac{2}{\pi}\int_0^{\pi} x \sin nx\, dx \qquad (x^2 \sin nx \text{ is an odd function})$$

$$= \frac{2}{\pi}\left[(x)\left(-\frac{\cos nx}{n}\right)-(1)\left(\frac{-\sin nx}{n^2}\right)\right]_0^{\pi} = \frac{2}{\pi}\left[-(\pi)\frac{\cos n\pi}{n}+2\frac{\sin n\pi}{n^3}\right]$$

$$= \frac{2}{\pi}\left[-\frac{\pi}{n}\cos n\pi\right] = -\frac{2}{n}(-1)^n = \frac{2}{n}(-1)^{n+1}$$

Substituting the values of $a_0$, $a_n$, $b_n$ in (1) we get

$$x+x^2 = \frac{\pi^2}{3}+4\left[-\cos x+\frac{1}{2^2}\cos 2x-\frac{1}{3^2}\cos 3x+\ldots\right]+2\left[\sin x-\frac{1}{2}\sin 2x+\frac{1}{3}\sin 3x-\ldots\right] \quad \ldots (2)$$

Put $x = \pi$ in (2),

$$\pi+\pi^2 = \frac{\pi^2}{3}+4\left[1+\frac{1}{2^2}+\frac{1}{3^2}+\frac{1}{4^2}+\ldots\right] \quad \ldots (3)$$

Put $x = -\pi$ in (2), $-\pi+\pi^2 = \dfrac{\pi^2}{3}+4\left[1+\dfrac{1}{2^2}+\dfrac{1}{3^2}+\dfrac{1}{4^2}+\ldots\right]$ ... (4)

Adding (3) and (4) $\qquad 2\pi^2 = \dfrac{2\pi^2}{3}+8\left[1+\dfrac{1}{2^2}+\dfrac{1}{3^2}+\dfrac{1}{4^2}+\ldots\right]$

$$\frac{4\pi^2}{3} = 8\left[1+\frac{1}{2^2}+\frac{1}{3^2}+\frac{1}{4^2}+\ldots\right]$$

$$\frac{\pi^2}{6} = 1+\frac{1}{2^2}+\frac{1}{3^2}+\frac{1}{4^2}+\ldots = \sum_{n=1}^{\infty}\frac{1}{n^2}$$ **Ans.**

## Exercise 12.1

1. Find a Fourier series to represent, $f(x) = \pi - x$ for $0 < x < 2\pi$.

**Ans.** $2\left[\sin x+\frac{1}{2}\sin 2x+\frac{1}{3}\sin 3x+\ldots+\frac{1}{n}\sin nx+\ldots\right]$

2. Find a Fourier series to represent $x - x^2$ from $x = -\pi$ to $\pi$ and show that

$$\frac{\pi^2}{12} = \frac{1}{1^2}-\frac{1}{2^2}+\frac{1}{3^2}-\frac{1}{4^2}+\ldots$$

**Ans.** $-\dfrac{\pi^2}{3}+4\left[\dfrac{\cos x}{1^2}-\dfrac{\cos 2x}{2^2}+\dfrac{\cos 3x}{3^2}-\dfrac{\cos 4x}{4^2}+\ldots\right]+2\left[\dfrac{\sin x}{1}-\dfrac{\sin 2x}{2}+\dfrac{\sin 3x}{3}-\dfrac{\sin 4x}{4}+\ldots\right]$

3. Find a Fourier series to represent: $f(x) = x \sin x$, for $0 < x < 2\pi$.

**Ans.** $-1+\pi\sin x-\dfrac{1}{2}\cos x+2\left[\dfrac{\cos 2x}{2^2-1}+\dfrac{\cos 3x}{3^2-1}+\dfrac{\cos 4x}{4^2-1}+\ldots\right]$

4. Find a Fourier series to represent the function $f(x) = e^x$, for $-\pi < x < \pi$ and hence derive a series for $\dfrac{\pi}{\sinh \pi}$.

**Ans.** $\dfrac{2\sinh\pi}{\pi}\left[\left(\dfrac{1}{2}-\dfrac{1}{1^2+1}\cos x+\dfrac{1}{2^2+1}\cos 2x-\dfrac{1}{3^2+1}\cos 3x+\ldots\right)\right]$

$+\left[\dfrac{1}{1^2+1}\sin x-\dfrac{2}{2^2+1}\sin 2x+\dfrac{3}{3^2+1}\sin 3x\ldots\right]$ and

$$\frac{\pi}{\sinh\pi} = 1+2\left[-\frac{1}{2}+\frac{1}{5}-\frac{1}{10}+\ldots\right]$$

5. Obtain the Fourier series for $f(x) = e^{-x}$ in the interval $0 \le x < 2\pi$.

**Ans.** $\dfrac{1-e^{-2\pi}}{\pi}\left[\dfrac{1}{2}+\dfrac{1}{2}\cos x+\dfrac{1}{5}\cos 2x+\dfrac{1}{10}\cos 3x+\dfrac{1}{2}\sin x+\dfrac{2}{5}\sin 2x+\dfrac{3}{10}\sin 3x+...\right]$

6. If $f(x)=\left(\dfrac{\pi-x}{2}\right)^2$, $0<x<2\pi$, show that $f(x)=\dfrac{\pi^2}{12}+\sum_{n=1}^{\infty}\dfrac{\cos nx}{n^2}$

7. Prove that $x^2=\dfrac{\pi^2}{3}+4\sum_{n=1}^{\infty}(-1)\dfrac{\cos nx}{n^2}$, $-\pi<x<\pi$

Hence show that (*i*) $\sum\dfrac{1}{n^2}=\dfrac{\pi}{6}$ (*ii*) $\sum\dfrac{1}{(2n-1)^2}=\dfrac{\pi^2}{8}$ (*iii*) $\sum\dfrac{1}{n^4}=\dfrac{\pi^4}{90}$

8. If $f(x)$ is a periodic function defined over a period $(0, 2\pi)$ $f(x)=\dfrac{(3x^2-6x\pi+2\pi^2)}{12}$

Prove that $f(x)=\sum_{n=1}^{\infty}\dfrac{\cos nx}{n^2}$ and hence show that $\dfrac{\pi^2}{6}=1+\dfrac{1}{2^2}+\dfrac{1}{3^2}+...$

## 12.7FUNCTION DEFINED IN TWO OR MORE SUB-RANGES

**Example 3.** *Find the Fourier series of the function*

$$f(x)=\begin{cases}-1 & \text{for } -\pi<x<-\dfrac{\pi}{2}\\ 0 & \text{for } -\dfrac{\pi}{2}<x<\dfrac{\pi}{2}\\ +1 & \text{for } \dfrac{\pi}{2}<x<\pi\end{cases}$$

**Solution.** Let $f(x)=\dfrac{a_0}{2}+a_1\cos x+a_2\cos 2x+...+b_1\sin x+b_2\sin 2x+...$ ...(1)

$$a_0=\frac{1}{\pi}\int_{-\pi}^{\pi}f(x)dx=\frac{1}{\pi}\int_{-\pi}^{-\pi/2}(-1)dx+\frac{1}{\pi}\int_{-\pi/2}^{\pi/2}0\,dx+\frac{1}{\pi}\int_{\pi/2}^{\pi}1\,dx$$

$$=\frac{1}{\pi}[-x]_{-\pi}^{-\pi/2}+\frac{1}{\pi}[x]_{\pi/2}^{\pi}=\frac{1}{\pi}\left[\frac{\pi}{2}-\pi-\frac{\pi}{2}\right]=0$$

$$a_n=\frac{1}{\pi}\int_{-\pi}^{+\pi}f(x)\cos nx\,dx$$

$$=\frac{1}{\pi}\int_{-\pi}^{-\pi/2}(-1)\cos nx\,dx+\frac{1}{\pi}\int_{-\pi/2}^{\pi/2}(0)\cos nx\,dx+\frac{1}{\pi}\int_{\pi/2}^{\pi}(1)\cos nx\,dx$$

$$=-\frac{1}{\pi}\left[\frac{\sin nx}{n}\right]_{-\pi}^{-\pi/2}+\frac{1}{\pi}\left[\frac{\sin nx}{n}\right]_{\pi/2}^{\pi}=-\frac{1}{\pi}\left[-\frac{\sin\frac{n\pi}{2}}{n}+\frac{\sin n\pi}{n}\right]+\frac{1}{\pi}\left[\frac{\sin n\pi}{n}-\frac{\sin\frac{n\pi}{2}}{n}\right]=0$$

$$b_n=\frac{1}{\pi}\int_{-\pi}^{\pi}f(x)\sin nx\,dx$$

$$=\frac{1}{\pi}\int_{-\pi}^{-\pi/2}(-1)\sin nx\,dx+\frac{1}{\pi}\int_{-\pi/2}^{\pi/2}(0)\sin nx\,dx$$

$$+\frac{1}{n}\int_{\pi/2}^{\pi}(1)\sin nx\,dx$$

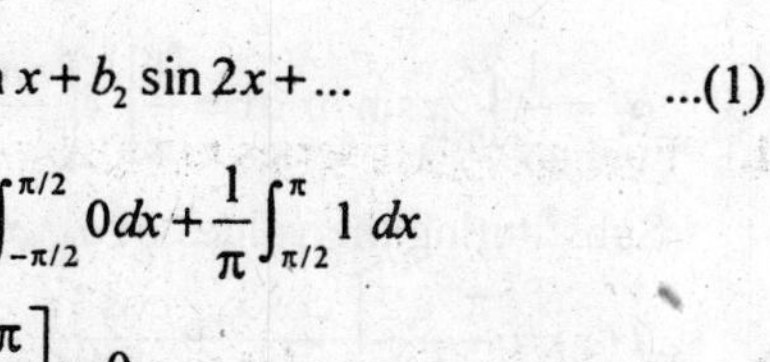

$$= \pi\left[\frac{\cos nx}{n}\right]_{-\pi}^{-\pi/2} - \frac{1}{\pi}\left[\frac{\cos nx}{n}\right]_{\pi/2}^{\pi}$$

$$= \frac{1}{n\pi}\left[\cos\frac{n\pi}{2} - \cos n\pi\right] - \frac{1}{n\pi}\left(\cos n\pi - \cos\frac{n\pi}{2}\right) = \frac{2}{n\pi}\left[\cos\frac{n\pi}{2} - \cos n\pi\right]$$

$$b_1 = \frac{2}{\pi}, \qquad b_2 = -\frac{2}{\pi}, \qquad b_3 = \frac{2}{3\pi}$$

Putting the values of $a_0$, $a_n$, $b_n$ in (1) we get $f(x) = \frac{1}{\pi}\left[2\sin x - 2\sin 2x + \frac{2}{3}\sin 3x + \ldots\right]$ **Ans.**

**Example 4.** *Find the Fourier series for the periodic function*

$$f(x) = \begin{cases} 0, & -\pi < x < 0 \\ x, & 0 < x < \pi \end{cases}$$

$$f(x + 2\pi) = f(x)$$

**Solution.** Let $f(x) = \frac{a_0}{2} + a_1 \cos x + a_0 \cos 2x + \ldots + v_1 \sin x + b_2 \sin 2x + \ldots$ ... (1)

$$a_0 = \frac{1}{\pi}\int_{-\pi}^{0} 0.dx + \frac{1}{\pi}\int_0^{\pi} x\,dx = \frac{1}{\pi}\left[\frac{x^2}{2}\right]_0^{\pi} = \frac{1}{\pi}\left(\frac{\pi^2}{2}\right) = \frac{\pi}{2}$$

$$a_n = \frac{1}{\pi}\int_0^{\pi} x\cos nx\,dx = \frac{1}{\pi}\left[x.\frac{\sin nx}{n} - (1)\left(-\frac{\cos nx}{n^2}\right)\right]_0^{\pi} = \frac{1}{\pi}\left(\frac{\cos n\pi}{n^2}\right)_0^{\pi}$$

$$= \frac{1}{\pi}\left[\frac{(-1)^n}{n^2} - \frac{1}{n^2}\right] = -\frac{2}{n^2\pi} \text{ when } n \text{ is odd}$$

$$= 0, \text{ when } n \text{ is even.}$$

$$b_n = \frac{1}{\pi}\int_0^{\pi} x\sin nx\,dx = \frac{1}{\pi}\left[x\left(-\frac{\cos nx}{n}\right) - (1)\left(-\frac{\sin nx}{n^2}\right)\right]_0^{\pi} = \frac{1}{\pi}\left[-\pi\frac{(-1)^n}{n}\right] = \frac{(-1)^{n+1}}{n}$$

Substituting the values of $a_0, a_1, a_2 \ldots b_1, b_2, \ldots$ in (1), we get

$$f(x) = \frac{\pi}{4} - \frac{2}{\pi}\left[\frac{\cos x}{1^2} + \frac{\cos 3x}{3^2} + \frac{\cos 5x}{5^2} \ldots\right] + \left[\frac{\sin x}{1} - \frac{\sin 2x}{2} + \frac{\sin 3x}{3} + \ldots\right]$$ **Ans.**

## DISCONTINUOUS FUNCTIONS

At a point of discontinuity, Fourier series gives the value of $f(x)$ as the arithmetic mean of left and right limits.

At the point of discontinuity, $x = c$

$$\text{At } x = c,\ f(x) = \frac{1}{2}[f(c-0) + f(c+0)]$$

**Example 5.** *Find the Fourier series for f(x), if* $f(x) = \begin{cases} -\pi, & -\pi < x < 0 \\ x, & 0 < x < \pi \end{cases}$

Deduce that $\frac{1}{1^2} + \frac{1}{3^2} + \frac{1}{5^2} + \ldots = \frac{\pi^2}{8}$

**Solution.** Let $f(x) = \frac{a_0}{2} + a_1 \cos x + a_2 \cos 2x + \ldots + a_n \cos nx + \ldots$

$$+ b_1 \sin x + b_2 \sin 2x + \ldots + b_n \sin nx + \ldots \qquad \ldots (1)$$

$$a_0 = \frac{1}{\pi}\int_{-\pi}^{\pi} f(x)\,dx$$

Then $a_0 = \frac{1}{\pi}\left[\int_{-\pi}^{0}(-\pi)\,dx + \int_0^{\pi} x\,dx\right] = \frac{1}{\pi}\left[-\pi(x)_{-\pi}^{0} + (x^2/2)_0^{\pi}\right] = \frac{1}{\pi}(-\pi^2 + \pi^2/2) = -\frac{\pi}{2};$

$$a_n = \frac{1}{\pi}\int_{-\pi}^{\pi} f(x)\cos nx\,dx$$

$$a_n = \frac{1}{\pi}\left[\int_{-\pi}^{0}(-\pi)\cos nx\,dx + \int_{0}^{\pi} x\cos nx\,dx\right] = \frac{1}{\pi}\left[-\pi\left(\frac{\sin nx}{n}\right)_{-\pi}^{0} + \left(\frac{x\sin nx}{n} + \frac{\cos nx}{n^2}\right)_{0}^{\pi}\right]$$

$$= \frac{1}{\pi}\left[0 + \frac{1}{n^2}\cos n\pi - \frac{1}{n^2}\right] = \frac{1}{\pi n^2}(\cos n\pi - 1) = \frac{1}{n^2\pi}[(-1)^n - 1] = \frac{-2}{n^2\pi} \text{ when n is odd}$$

$$b_n = \frac{1}{\pi}\int_{-\pi}^{\pi} f(x)\sin nx\,dx$$

$$b_n = \frac{1}{\pi}\left[\int_{-\pi}^{0}(-\pi)\sin nx\,dx + \int_{0}^{\pi} x\sin nx\,dx\right] = \frac{1}{\pi}\left[\left(\frac{\pi\cos nx}{n}\right)_{-\pi}^{0} + \left(-x\frac{\cos nx}{n} + \frac{\sin nx}{n^2}\right)_{0}^{\pi}\right]$$

$$= \frac{1}{\pi}\left[\frac{\pi}{n}(1 - \cos n\pi) - \frac{\pi}{n}\cos n\pi\right] = \frac{1}{n}(1 - 2\cos n\pi) = \frac{1}{n}(1 - 2(-1)^n$$

$b_n = \dfrac{3}{n}$ when $n$ is odd

$= \dfrac{-1}{n}$ when $n$ is even

$$f(x) = -\frac{\pi}{4} - \frac{2}{\pi}\left(\cos x + \frac{\cos 3x}{3^2} + \frac{\cos 5x}{5^2} + \dots\right) + 3\sin x - \frac{\sin 2x}{2} + \frac{3\sin 3x}{3} - \frac{\sin 4x}{4} + \dots \quad \dots (2)$$

Putting $x = 0$ in (2), we get $f(0) = -\dfrac{\pi}{4} - \dfrac{2}{\pi}\left(1 + \dfrac{1}{3^2} + \dfrac{1}{5^2} + \dots\infty\right)$ ... (3

Now $f(x)$ is discontinuous at $x = 0$.

But $f(0-0) = -\pi$ and $f(0+0) = 0$

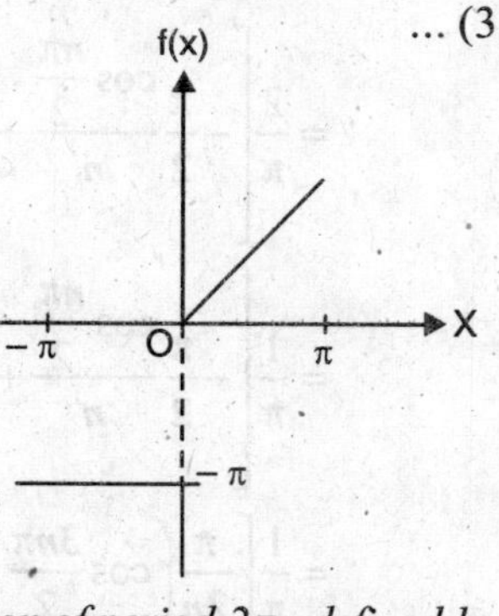


$$\therefore f(0) = \frac{1}{2}[f(0-0) + f(0+0)] = -\pi/2$$

From (3), $-\dfrac{\pi}{2} = -\dfrac{\pi}{4} - \dfrac{2}{\pi}\left[\dfrac{1}{1^2} + \dfrac{1}{3^2} + \dfrac{1}{5^2} + \dots\right]$

or $\dfrac{\pi^2}{8} = \dfrac{1}{1^2} + \dfrac{1}{3^2} + \dfrac{1}{5^2} + \dots$ **Proved.**

**Example 6.** *Find the Fourier series expansion of the periodic function of period 2π-, defined by*

$$f(x) = \begin{cases} x & \text{if } -\dfrac{\pi}{2} < x < \dfrac{\pi}{2} \\ \pi - 1 & \text{if } \dfrac{\pi}{2} < x < \dfrac{3\pi}{2} \end{cases}$$

**Solution.** Let $f(x) = \dfrac{a_0}{2} + a_1\cos x + a_2\cos 2x + \dots + b_1\sin x + b_2\sin 2x + \dots$

Now $$a_0 = \frac{1}{\pi}\int_{-\pi/2}^{\pi/2} x\,dx + \frac{1}{\pi}\int_{\pi/2}^{\frac{3\pi}{2}}(\pi - x)dx = \frac{1}{\pi}\left(\frac{x^2}{2}\right)_{-\pi/2}^{\pi/2} + \frac{1}{\pi}\left(\pi x - \frac{x^2}{2}\right)_{\pi/2}^{\frac{3\pi}{2}}$$

$$= \frac{1}{\pi}\left(\frac{\pi^2}{8} - \frac{\pi^2}{8}\right) + \frac{1}{\pi}\left(\frac{3\pi^2}{2} - \frac{9\pi^2}{8} - \frac{\pi^2}{2} + \frac{\pi^2}{8}\right) = \pi\left(\frac{3}{2} - \frac{9}{8} - \frac{1}{2} + \frac{1}{8}\right) = 0$$

$$a_n = \frac{1}{\pi}\int_{-\pi/2}^{\pi/2} x\cos nx\,dx + \frac{1}{\pi}\int_{\pi/2}^{\frac{3\pi}{2}} (\pi - x)\cos nx\,dx$$

$$= \frac{1}{\pi}\left[x\frac{\sin nx}{n} - (1)\left(-\frac{\cos nx}{n^2}\right)\right]_{-\pi/2}^{\pi/2} + \frac{1}{\pi}\left[(\pi - x)\frac{\sin nx}{n} - (-1)\left(-\frac{\cos nx}{n^2}\right)\right]_{\pi/2}^{\frac{3\pi}{2}}$$

$$= \frac{1}{\pi}\left[\frac{\pi}{2}\frac{\sin\frac{n\pi}{2}}{n} + \frac{\cos\frac{n\pi}{2}}{n^2} - \frac{\pi}{2}\frac{\sin\frac{n\pi}{2}}{n} - \frac{\cos\frac{n\pi}{2}}{n^2}\right]$$

$$+ \frac{1}{\pi}\left[-\frac{\pi}{2}\frac{\sin\frac{3n\pi}{2}}{n} + \frac{\cos\frac{3n\pi}{2}}{n^2} - \frac{\pi}{2}\frac{\sin\frac{n\pi}{2}}{n} + \frac{\cos\frac{n\pi}{2}}{n^2}\right]$$

$$= \frac{1}{\pi}\left[-\frac{\pi}{2n}\left(\sin\frac{3n\pi}{2} + \sin\frac{n\pi}{2}\right) - \frac{1}{n^2}\left(\cos\frac{3n\pi}{2} - \cos\frac{n\pi}{2}\right)\right]$$

$$= \frac{1}{\pi}\left[-\frac{\pi}{n}\sin n\pi\cos\frac{n\pi}{2} + \frac{2}{n^2}\sin\frac{n\pi}{2}\sin n\pi\right] = 0$$

$$b_n = \frac{1}{\pi}\int_{-\pi/2}^{\pi/2} x\sin nx\,dx + \frac{1}{\pi}\int_{\pi/2}^{3\pi/2} (\pi - x)\sin nx\,dx = \frac{2}{\pi}\int_0^{\pi/2} x\sin nx\,dx + \frac{1}{\pi}\int_{\pi/2}^{\frac{3\pi}{2}} (\pi - x)\sin nx\,dx$$

$$= \frac{2}{\pi}\left[x\left(-\frac{\cos nx}{n}\right) - (1)\left(-\frac{\sin nx}{n^2}\right)\right]_0^{\pi/2} + \frac{1}{\pi}\left[(\pi - x)\left(-\frac{\cos nx}{n}\right) - (-1)\left(-\frac{\sin nx}{n^2}\right)\right]_{\pi/2}^{\frac{3\pi}{2}}$$

$$= \frac{2}{\pi}\left[-\frac{\pi}{2}\frac{\cos\frac{n\pi}{2}}{n} + \frac{\sin\frac{n\pi}{2}}{n^2}\right] + \frac{1}{\pi}\left[\frac{\pi}{2}\frac{\cos\frac{3n\pi}{2}}{n} - \frac{\sin\frac{3n\pi}{2}}{n^2} + \frac{\pi}{2}\frac{\cos\frac{n\pi}{2}}{n} + \frac{\sin\frac{n\pi}{2}}{n^2}\right]$$

$$= \frac{1}{\pi}\left[-\frac{\pi}{2}\frac{\cos\frac{n\pi}{2}}{n} + \frac{3\sin\frac{n\pi}{2}}{n^2} + \frac{\pi}{2}\frac{\cos 3\frac{n\pi}{2}}{n} - \frac{\sin\frac{3n\pi}{2}}{n^2}\right]$$

$$= \frac{1}{\pi}\left[\frac{\pi}{2n}\left(\cos\frac{3n\pi}{2} - \cos\frac{n\pi}{2}\right) + \frac{3}{n^2}\sin\frac{n\pi}{2} - \frac{1}{n^2}\sin\frac{3n\pi}{2}\right]$$

$$= \frac{1}{\pi}\left[-\frac{\pi}{n}\sin\frac{n\pi}{2}\sin n\pi + \frac{3}{n^2}\sin\frac{n\pi}{2} - \frac{1}{n^2}\sin\frac{3n\pi}{2}\right] = \frac{1}{n^2\pi}\left[3\sin\frac{n\pi}{2} - \sin\frac{3n\pi}{2}\right]$$

Substituting the values of $a_0, a_1, a_2 \ldots b_1, b_2 \ldots$ we get $f(x)\frac{4}{\pi}\left[\frac{\sin x}{1^2} - \frac{\sin 3x}{3^2} + \frac{\sin 5x}{5^2} - \ldots\right]$ **Ans.**

**Example 7.** *Find the Fourier series of the function defined as*

$$f(x) = \begin{cases} x + \pi & \text{for } 0 < x < \pi \\ -x - \pi & -\pi < x < 0 \end{cases} \quad \textit{and} \quad f(x + 2\pi) = f(x).$$

**Solution.** $a_0 = \frac{1}{\pi}\int_{-\pi}^{\pi} f(x)dx = \frac{1}{\pi}\int_{-\pi}^{0} f(x)dx - \frac{1}{\pi}\int_0^{\pi} f(x)dx$

$$= \frac{1}{\pi}\int_{-\pi}^{0} (-x - \pi)dx + \frac{1}{\pi}\int_0^{\pi} (x + \pi)dx = \frac{1}{\pi}\left(-\frac{x^2}{2} - \pi x\right)_{-\pi}^{0} + \frac{1}{\pi}\left(\frac{x^2}{2} + \pi x\right)_0^{\pi} -$$

$$=\frac{1}{\pi}\left(\frac{\pi^2}{2}-\pi^2\right)+\frac{1}{\pi}\left(\frac{\pi^2}{2}+\pi^2\right)=\pi\left(\frac{1}{2}-1\right)+\pi\left(\frac{1}{2}+1\right)=\pi$$

$$a_n=\frac{1}{\pi}\int_{-\pi}^{\pi}f(x)\cos nx\,dx=\frac{1}{\pi}\int_{-\pi}^{0}f(x)\cos nx\,dx+\frac{1}{\pi}\int_{0}^{\pi}f(x)\cos nx\,dx$$

$$=\frac{1}{\pi}\int_{-\pi}^{0}(-x-\pi)\cos nx\,dx+\frac{1}{\pi}\int_{0}^{\pi}(x+\pi)\cos nx\,dx$$

$$=\frac{1}{\pi}\left[(-x-\pi)\frac{\sin nx}{n}-(-1)\left\{-\frac{\cos nx}{n^2}\right\}\right]_{-\pi}^{0}+\frac{1}{\pi}\left[(x+\pi)\frac{\sin nx}{n}-(1)\left\{-\frac{\cos nx}{n^2}\right\}\right]_{0}^{\pi}$$

$$=\frac{1}{\pi}\left[-\frac{1}{n^2}+\frac{(-1)^n}{n^2}\right]+\frac{1}{\pi}\left[-\frac{(-1)^n}{n^2}-\frac{1}{n^2}\right]=\frac{2}{n^2\pi}[(-1)^n-1]$$

$a_n=\dfrac{-4}{n^2\pi}$, If $n$ is odd.

and $a_n=0$ if $n$ is even.

$$b_n=\frac{1}{\pi}\int_{-\pi}^{\pi}f(x)\sin nx\,dx=\frac{1}{\pi}\int_{-\pi}^{0}f(x)\sin nx\,dx+\frac{1}{\pi}\int_{0}^{\pi}f(x)\sin nx\,dx$$

$$=\frac{1}{\pi}\int_{-\pi}^{0}(-x-\pi)\sin nx\,dx+\frac{1}{\pi}\int_{0}^{\pi}(x+\pi)\sin nx\,dx$$

$$=\frac{1}{\pi}\left[(-x-\pi)\left(-\frac{\cos nx}{n}\right)-(-1)\left(-\frac{\sin nx}{n^2}\right)\right]_{-\pi}^{0}+\frac{1}{\pi}\left[(x+\pi)\left(-\frac{\cos nx}{n}\right)-(1)\left(-\frac{\sin nx}{n^2}\right)\right]_{0}^{\pi}$$

$$=\frac{1}{\pi}\left[\frac{\pi}{n}\right]+\frac{1}{\pi}\left[-\frac{2\pi}{n}(-1)^n+\frac{\pi}{n}\right]=\frac{1}{n}[(1)-2(-1)^n+(1)]=\frac{2}{n}[1-(-1)^n]$$

$=\dfrac{4}{n}$, if $n$ is odd.

$=0$, if $n$ is even.

Fourier series is $\quad f(x)=\dfrac{a_0}{2}+a_1\cos x+a_2\cos 2x+\ldots+b_1\sin x+b_2\sin 2x+\ldots$

$$f(x)=\frac{\pi}{2}-\frac{4}{\pi}\left(\frac{\cos x}{1^2}+\frac{\cos 3x}{3^2}+\ldots\right)+4\left(\frac{\sin x}{1}+\frac{\sin 3x}{3}+\ldots\right)$$ **Ans.**

## Exercise 12.2

**1.** Find the Fourier series of the function

$$f(x)=\begin{cases}-1 & \text{for } -\pi<x<0\\ 1 & \text{for } 0<x<\pi\end{cases}$$

where $f(x+2\pi)=f(x)$. **Ans.** $\dfrac{4}{\pi}\left[\dfrac{1}{1}\sin x+\dfrac{1}{3}\sin 3x+\dfrac{1}{5}\sin 5x+\dfrac{1}{7}\sin 7x+\ldots\right]$

**2.** Find the Fourier series for the function

$$f(x)=\begin{cases}-\dfrac{\pi}{4} & \text{for } -\pi<x<0\\ \dfrac{\pi}{4} & \text{for } 0<x<\pi\end{cases}$$

and $f(-\pi)=f(0)=f(\pi)=0$, $\quad f(x)=f(x+2\pi)$ for all $x$.

Deduce that $\dfrac{\pi}{4}=1-\dfrac{1}{3}+\dfrac{1}{5}-\dfrac{1}{7}+\ldots$ **Ans.** $\dfrac{\sin x}{1}+\dfrac{\sin 3x}{3}+\dfrac{\sin 5x}{5}+\dfrac{\sin 7x}{7}+\ldots..$

3. Find the Fourier series of the function

$$f(x)=\begin{cases}0 & \text{for } -\pi\le x\le 0\\ 1 & \text{for } 0<x<\dfrac{\pi}{2}\\ 0 & \text{for } \dfrac{\pi}{2}\le x\le \pi\end{cases}$$

**4. Obtain a Fourier series to represent the following periodic function**

$$f(x) = 0 \text{ when } 0<x<\pi$$
$$f(x) = 1 \text{ when } \pi<x<2\pi$$

**Ans.** $\frac{1}{2}-\frac{2}{\pi}\left(\sin x+\frac{1}{3}\sin 3x+\frac{1}{5}\sin 5x+...\right)$

**5. Find the Fourier expansion of the function defined in a single period by the relations.**

$$f(x)=\begin{cases}1 & \text{for } 0<x<\pi\\ 2 & \text{for } \pi<x<2\pi\end{cases}$$

and from it deduce that $\frac{\pi}{4}=1-\frac{1}{3}+\frac{1}{5}-\frac{1}{7}+...$

**Ans.** $\frac{3}{2}-\frac{2}{\pi}\left(\sin x+\frac{1}{3}\sin 3x+\frac{1}{5}\sin 5x+...\right)$

**6. Find a Fourier series to represent the function**

$$f(x)=\begin{cases}0 & \text{for } -\pi<x\le 0\\ \frac{1}{4}\pi x & \text{for } 0<x<\pi\end{cases}$$

and hence deduce that $\frac{\pi^2}{8}=\frac{1}{1^2}+\frac{1}{3^2}+\frac{1}{5^2}+...$

**Ans.** $\frac{\pi^2}{16}+\sum_{n=1}^{\infty}\left(\frac{[(-1)^n-1]}{4n^2}\cos nx-\frac{(-1)^n\pi}{4n}\sin nx+...\right)$

7. Find the Fourier series for $f(x)$, if

$$f(x) = -\pi \quad \text{for } -\pi<x\le 0$$
$$= x \quad \text{for } 0<x<\pi$$
$$= \frac{-\pi}{2} \quad \text{for } x=0$$

Deduce that $\frac{1}{1^2}+\frac{1}{3^2}+\frac{1}{5^2}+...=\frac{\pi^2}{8}$

**Ans.** $-\frac{\pi}{4}-\frac{2}{\pi}\left(\cos x+\frac{\cos 3x}{3^2}+\frac{\cos 5x}{5^2}+...\right)+3\sin x-\frac{1}{2}\sin 2x+\frac{3}{3}\sin 3x-\frac{1}{4}\sin 4x+...$

8. Obtain a Fourier series to represent the function

$$f(x) = |x| \text{ for } -\pi<x<\pi$$

and hence deduce $\frac{\pi^2}{8}=\frac{1}{1^2}+\frac{1}{3^2}+\frac{1}{5^2}+...$

**Ans.** $\frac{\pi}{2}-\frac{4}{\pi}\left[\cos x+\frac{1}{3^2}\cos 3x+\frac{1}{5^2}\cos 5x+...\right]$

**9.** Expand as a Fourier series, the function $f(x)$ defined as

$$f(x) = \pi + x \quad \text{for} \quad -\pi < x < -\frac{\pi}{2}$$
$$= \frac{\pi}{2} \quad \text{for} \quad -\frac{\pi}{2} < x < \frac{\pi}{2}$$
$$= \pi - x \quad \text{for} \quad \frac{\pi}{2} < x < \pi$$

**Ans.** $\frac{3\pi}{8} + \frac{2}{\pi}\left[\frac{1}{1^2}\cos x - \frac{2}{2^2}\cos 2x + \frac{1}{3^2}\cos 3x + \ldots\right]$

**10.** Obtain a Fourier series to represent the function

$$f(x) = |\sin x| \text{ for } -\pi < x < \pi \qquad \left\{\begin{array}{lll}\textbf{Hint} & f(x) = -\sin x & \text{for } -\pi < x < 0 \\ & \quad\;\;\, = \sin x & \text{for } \quad 0 < x < \pi\end{array}\right\}$$

**Ans.** $\frac{2}{\pi} - \frac{4}{\pi}\left[\frac{1}{3}\cos 2x + \frac{1}{15}\cos 4x + \frac{1}{35}\cos 6x + \ldots\right]$

**11.** An alternating current after passing through a rectifier has the form

$$i = I\sin\theta \quad \text{for } 0 < \theta < \pi$$
$$= 0 \quad \text{for } \pi < \theta < 2\pi$$

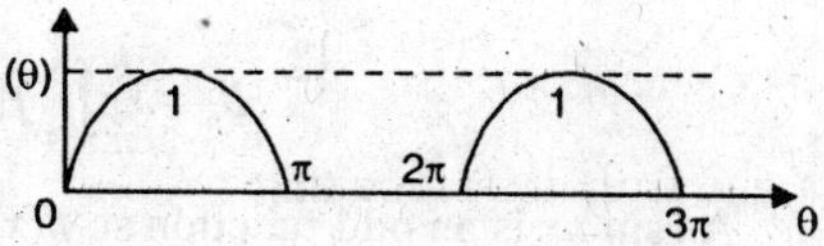


Find the Fourier series of the function.

**Ans.** $\frac{I}{\pi} - \frac{2I}{\pi}\left(\frac{\cos 2\theta}{3} + \frac{\cos 4\theta}{15} + \ldots\right) + \frac{I}{2}\sin\theta$

**12.** If $f(x) = 0 \quad$ for $-\pi < x < 0$

$= \sin x \quad$ for $0 < x < \pi$

Prove that $f(x) = \frac{1}{\pi} + \frac{\sin x}{2} - \frac{2}{\pi}\sum_{m=1}^{\infty}\frac{\cos 2mx}{4m^2 - 1}$.

Hence show that $\frac{1}{1.3} - \frac{1}{3.5} + \frac{1}{5.7} \ldots \infty = \frac{1}{4}(\pi - 2)$

## 12.8(a) EVEN FUNCTION

A function $f(x)$ is said to be even (or symmetric) function if, $f(-x) = f(x)$

**The graph of such a function is symmetric with respect to *y*-axis [*f*(*x*) axis]. Here *y*-axis is a mirror for the reflection of the curve.**

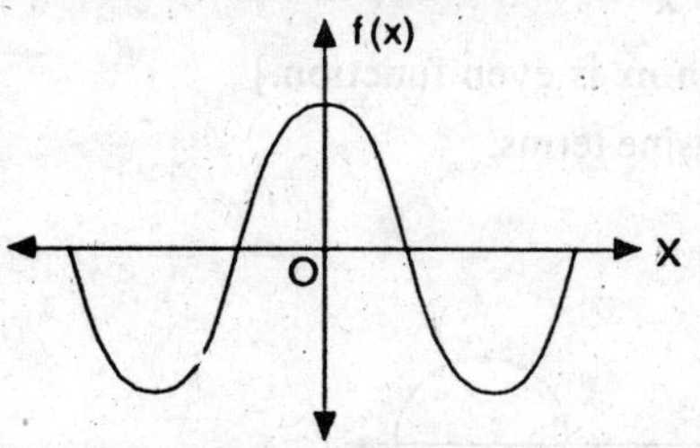


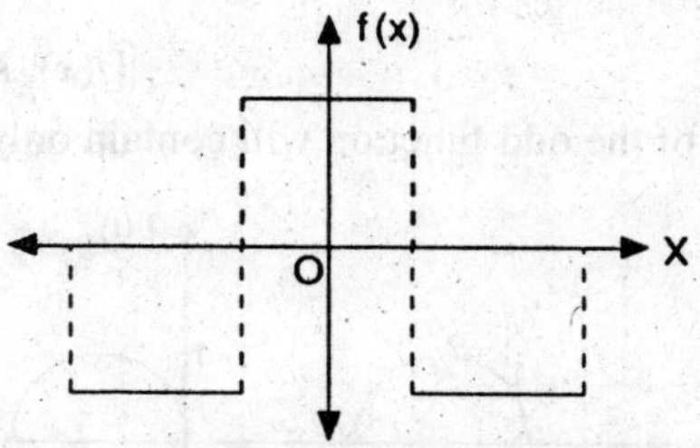


**The area under such a curve from $-\pi$ to $\pi$ is double the area from 0 to $\pi$.**

$$\therefore \quad \int_{-\pi}^{\pi} f(x)\,dx = 2\int_0^{\pi} f(x)\,dx$$

**(b) Odd Function**

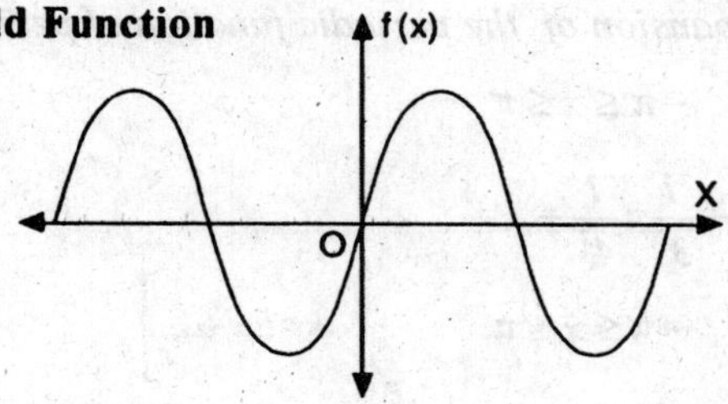


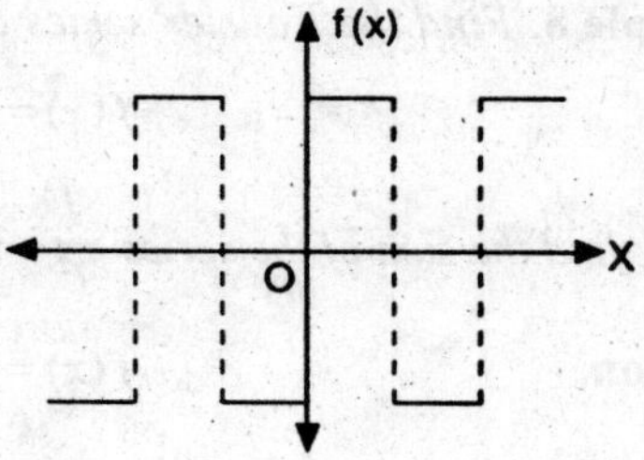

A function $f(x)$ is called odd (or skew symmetric) function if

$$f(-x) = -f(x)$$

Here the area under the curve from $-\pi$ to $\pi$ is zero.

$$\int_{-\pi}^{\pi} f(x)dx = 0$$

**Expansion of an even function:**

$$a_0 = \frac{1}{\pi}\int_{-\pi}^{\pi} f(x)dx = \frac{2}{\pi}\int_0^{\pi} f(x)dx$$

$$a_n = \frac{1}{\pi}\int_{-\pi}^{\pi} f(x)\cos nx\, dx = \frac{2}{\pi}\int_0^{\pi} f(x)\cos nx\, dx$$

As $f(x)$ and $\cos nx$ are both even functions.

$\therefore$ The product of $f(x)$. $\cos nx$ is also an even function. page 846

$$b_n = \frac{1}{\pi}\int_{-\pi}^{\pi} f(x)\sin nx\, dx = 0$$

As $\sin nx$ is an odd function so $f(x)$. $\sin nx$ is also an odd function. We need not to calculate $b_n$. It saves our labour a lot.

The series of the even function will contain only cosine terms.

**Expansion of an odd function :**

$$a_0 = \frac{1}{\pi}\int_{-\pi}^{\pi} f(x)dx = 0$$

$$a_n = \frac{1}{\pi}\int_{-\pi}^{\pi} f(x)\cos nx\, dx = 0 \qquad [f(x).\cos nx \text{ is odd function.}]$$

$$b_n = \frac{1}{\pi}\int_{-\pi}^{\pi} f(x)\sin nx\, dx = \frac{2}{\pi}\int_0^{\pi} f(x)\sin nx\, dx$$

[$f(x)$. $\sin nx$ is even function.]

The series of the odd function will contain only sine terms.

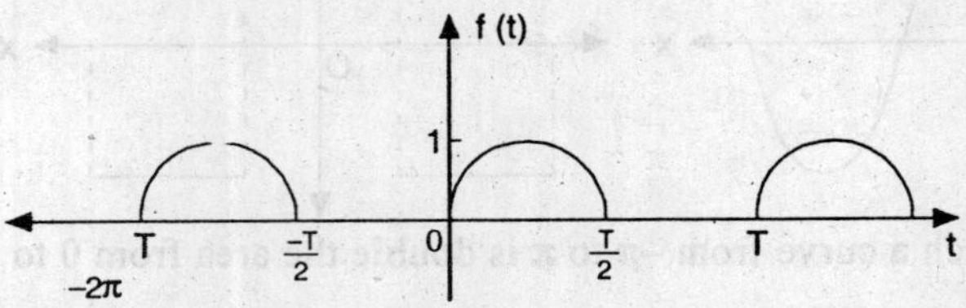


The function shown below is neither odd nor even so it contains both sine and cosine terms

**Example 8.** *Find the Fourier series expansion of the periodic function of period $2\pi$*

$$f(x) = x^2, \ -\pi \le x \le \pi$$

*Hence, find the sum of the series* $\frac{1}{1^2} - \frac{1}{2^2} + \frac{1}{3^2} - \frac{1}{4^2} + \ldots$

**Solution.** $\qquad f(x) = x^2, \ -\pi \le x \le \pi$

This is an even function. $\therefore \quad b_n = 0,$

$$a_0 = \frac{2}{\pi}\int_0^{\pi} f(x)dx = \frac{2}{\pi}\int_0^{\pi} x^2 dx = \frac{2}{\pi}\left[\frac{x^3}{3}\right]_0^{\pi} = \frac{2\pi^2}{3}$$

$$a_n = \frac{2}{\pi}\int_0^{\pi} f(x)\cos nx\,dx = \frac{2}{\pi}\int_0^{\pi} x^2 \cos nx\,dx$$

$$= \frac{2}{\pi}\left[x^2\left(\frac{\sin nx}{n}\right) - (2x)\left(-\frac{\cos nx}{n^2}\right) + (2)\left(-\frac{\sin nx}{n^3}\right)\right]_0^{\pi}$$

$$= \frac{2}{\pi}\left[\frac{\pi^2 \sin n\pi}{n} + \frac{2\pi\cos n\pi}{n^2} - \frac{2\sin n\pi}{n^3}\right] = \frac{4(-1)^n}{n^2}$$

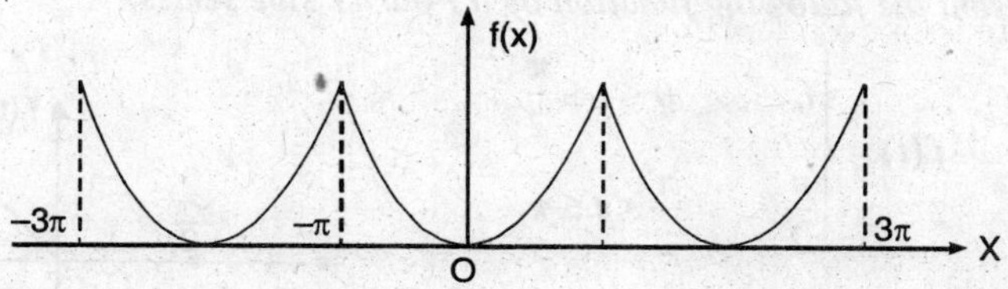


Fourier series is $\quad f(x) = \frac{a_0}{2} + a_1 \cos x + a_2 \cos 2x + a_3 \cos 3x + ... + a_n \cos nx + ...$

$$x^2 = \frac{\pi^2}{3} - 4\left[\frac{\cos x}{1^2} - \frac{\cos 2x}{2^2} + \frac{\cos 3x}{3^3} - \frac{\cos 4x}{4^2} + ...\right]$$

On putting $x = 0$, we have

$$0 = \frac{\pi^2}{3} - 4\left[\frac{1}{1^2} - \frac{1}{2^2} + \frac{1}{3^2} - \frac{1}{4^2} ...\right]$$

$$\frac{1}{1^2} - \frac{1}{2^2} + \frac{1}{3^2} - \frac{1}{4^2} ... = \frac{\pi^2}{12}$$ **Ans.**

**Example 9.** *Obtain a Fourier expression for*

$$f(x) = x^3 \quad for\ -\pi < x < \pi.$$

**Solution.** $f(x) = x^3$ is an odd function.

$\therefore \quad a_0 = 0$ and $a_n = 0$

$$b_n = \frac{2}{\pi}\int_0^{\pi} f(x)\sin nx\,dx = \frac{2}{\pi}\int_0^{\pi} x^3 \sin nx\,dx$$

$$\left[\int uv = uv_1 - u'v_2 + u''v_3 - u'''v_4 + ...\right]$$

$$= \frac{2}{\pi}\left[x^3\left(\frac{\cos nx}{n}\right) - 3x^2\left(-\frac{\sin nx}{n^2}\right) + 6x\left(\frac{\cos nx}{n^3}\right) - 6\left(\frac{\sin nx}{n^4}\right)\right]_0^{\pi}$$

$$= \frac{2}{\pi}\left[-\frac{\pi^3 \cos n\pi}{n} + \frac{6\pi\cos n\pi}{n^3}\right] = 2.(-1)^n\left[-\frac{\pi^2}{n} + \frac{6}{n^3}\right]$$

$$\therefore \quad x^3 = 2\left[-\left(\frac{\pi^2}{1} + \frac{6}{1^3}\right)\sin x + \left(-\frac{\pi^2}{2} + \frac{6}{2^3}\right)\sin 2x - \left(-\frac{\pi^2}{3} + \frac{6}{3^3}\right)\sin 3x ...\right]$$ **Ans.**

## 12.9 HALF-RANGE SERIES, PERIOD 0 TO π

The given function is defined in the interval (0, π) and it is immaterial whatever the function may be outside the interval (0, π). To get the series of cosines only we assume that $f(x)$ is an even function in the interval ( –π , π).

$$a_n = \frac{2}{\pi}\int_0^{\pi} f(x)\cos nx\, dx \quad \text{and} \quad b_n = 0$$

To expand $f(x)$ as a sine series we extend the function in the interval ( –π , π) as an odd function.

$$b_n = \frac{2}{\pi}\int_0^{\pi} f(x)\sin nx\, dx \quad \text{and} \quad a_n = 0$$

**Example 10.** *Represent the following function by a Fourier sine series:*

$$f(t) = \begin{cases} t, & 0 < t \le \frac{\pi}{2} \\ \frac{\pi}{2}, & \frac{\pi}{2} < t \le \pi \end{cases}$$

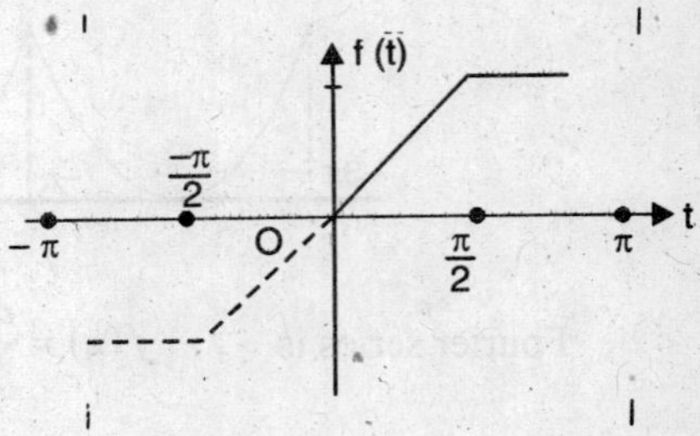


**Solution.** $b_n = \frac{2}{\pi}\int_0^{\pi} f(t)\sin nt\, dt$

$$= \frac{2}{\pi}\int_0^{\pi/2} t\sin nt\, dt + \frac{2}{\pi}\int_{\pi/2}^{\pi} \frac{\pi}{2}\sin nt\, dt$$

$$= \frac{2}{\pi}\left[t\left(-\frac{\cos nt}{n}\right) - (1)\left(-\frac{\sin nt}{n^2}\right)\right]_0^{\pi/2} + \frac{2}{\pi}\frac{\pi}{2}\left[-\frac{\cos nt}{n}\right]_{\pi/2}^{\pi}$$

$$= \frac{2}{\pi}\left[-\frac{\pi}{2}\frac{\cos\frac{n\pi}{2}}{n} + \frac{\sin\frac{n\pi}{2}}{n^2}\right] + \left[-\frac{\cos n\pi}{n} + \frac{\cos\frac{n\pi}{2}}{n}\right]$$

$$b_1 = \frac{2}{\pi}\left[-\frac{\pi}{2}\cos\frac{\pi}{2} + \sin\frac{\pi}{2}\right] + \left[-\cos\pi + \cos\frac{\pi}{2}\right] = \frac{2}{\pi}[0+1] + [1] = \frac{2}{\pi} + 1$$

$$b_2 = \frac{2}{\pi}\left[-\frac{\pi}{2}\frac{\cos\pi}{2} + \frac{\sin\pi}{2}\right] + \left[-\frac{\cos 2\pi}{2} + \frac{\cos\pi}{2}\right] = \frac{2}{\pi}\left[-\frac{\pi}{2}\frac{(-1)}{2} + 0\right] + \left[-\frac{1}{2} - \frac{1}{2}\right]$$

$$= \frac{2}{\pi}\left[\frac{\pi}{4}\right] - 1 = \frac{1}{2} - 1 = -\frac{1}{2}$$

$$b_3 = \frac{2}{\pi}\left[-\frac{\pi}{2}\frac{\cos\frac{3\pi}{2}}{3} + \frac{\sin\frac{3\pi}{2}}{3^2}\right] + \left[-\frac{\cos 3\pi}{3} + \frac{\cos\frac{3\pi}{2}}{3}\right]$$

$$= \frac{2}{\pi}\left[-\frac{\pi}{2}(0) - \frac{1}{9}\right] + \left[\frac{1}{3} + 0\right] = -\frac{2}{9\pi} + \frac{1}{3}$$

$$f(t) = \left(\frac{2}{\pi} + 1\right)\sin t - \frac{1}{2}\sin 2t + \left(-\frac{2}{9\pi} + \frac{1}{3}\right)\sin 3t + \ldots$$ **Ans.**

**Example 11.** *Find the Fourier sine series for the function*

$$f(x) = e^{ax} \text{ for } 0 < x < \pi$$

*where a is constant*

**Solution.** $b_n = \dfrac{2}{\pi}\int_0^{\pi} e^{ax} \sin nx \, dx$

$$= \frac{2}{\pi}\left[\frac{e^{ax}}{a^2+n^2}(a\sin n\pi - n\cos nx)\right]_0^{\pi}$$

$$\left(\int e^{ax} \sin bx \, dx = \frac{e^{ax}}{a^2+b^2}[a\sin bx - b\cos bx]\right)$$

$$= \frac{2}{\pi}\left[\frac{e^{a\pi}}{a^2+n^2}(a\sin n\pi - n\cos n\pi) + \frac{n}{a^2+n^2}\right]$$

$$= \frac{2}{\pi}\frac{n}{a^2+n^2}\left[-(-1)e^{a\pi}+1\right] = \frac{2n}{(a^2+n^2)\pi}\left[1-(-1)^n e^{a\pi}\right]$$

$$b_1 = \frac{2(1+e^{a\pi})}{(a^2+1^2)\pi}, \qquad b_2 = \frac{2.2.1(1-e^{a\pi})}{(a^2+2^2)\pi}$$

$$e^{ax} = \frac{2}{\pi}\left[\frac{1+e^{a\pi}}{a^2+1^2}\sin x + \frac{2(1-e^{a\pi})}{a^2+2^2}\sin 2x + \ldots\right]$$ **Ans.**

**Exercise 12.3**

**1.** Find the Fourier cosine series for the function

$$f(x) = \begin{cases} 1 & \text{for } 0 < x < \dfrac{\pi}{2} \\ 0 & \text{for } \dfrac{\pi}{2} < x < \pi. \end{cases}$$

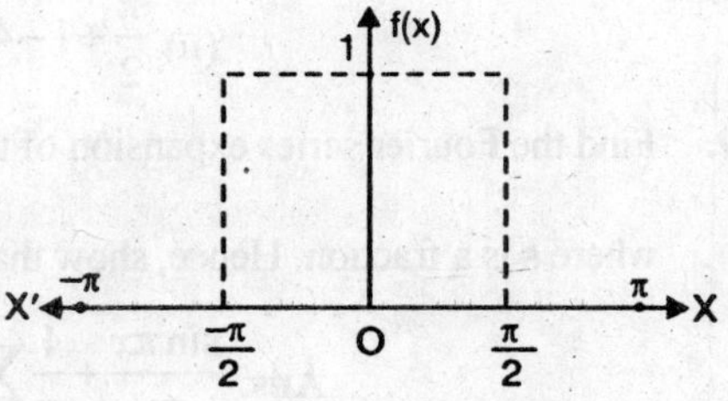


**Ans.** $\dfrac{1}{2} + \dfrac{2}{\pi}\left[\cos x - \dfrac{1}{3}\cos 3x + \dfrac{1}{5}\cos 5x - \ldots\right]$

**2.** Find a series of cosine of multiples of $x$ which will represent $f(x)$ in $(0, \pi)$ where

$$f(x) = \begin{cases} 0 & \text{for } 0 < x < \dfrac{\pi}{2} \\ \dfrac{\pi}{2} & \text{for } \dfrac{\pi}{2} < x < \pi \end{cases}$$

Deduce that $1 - \dfrac{1}{3} + \dfrac{1}{5} - \dfrac{1}{7} + \ldots \infty = \dfrac{\pi}{4}$ **Ans.** $\dfrac{\pi}{4} - \cos x + \dfrac{1}{3}\cos 3x - \dfrac{1}{5}\cos 5x + \ldots$

**3.** Express $f(x) = x$ as a sine series in $0 < x < \pi$. **Ans.** $2\left[\sin x - \dfrac{1}{2}\sin 2x + \dfrac{1}{3}\sin 3x - \ldots\right]$

**4.** Find the cosine series for $f(x) = \pi - x$ in the interval $0 \le x \le \pi$.

**Ans.** $\dfrac{\pi}{2} + \dfrac{4}{\pi}\left(\cos x + \dfrac{\cos 3x}{3^2} + \dfrac{\cos 5x}{5^2} + \ldots\right)$

**5.** If $f(x) = \begin{cases} 0 & \text{for } 0 < x < \dfrac{\pi}{2} \\ \pi - x, & \text{for } \dfrac{\pi}{2} < x < \pi \end{cases}$

Show that: (*i*) $f(x)=\frac{4}{\pi}\left(\sin x-\frac{1}{3^2}\sin 3x+\frac{1}{5^2}\sin 5x-...\right)$

(*ii*) $f(x)=\frac{\pi}{4}-\frac{2}{\pi}\left(\frac{1}{1^2}\cos 2x+\frac{1}{3^2}\cos 6x+\frac{1}{5^2}\cos 10x+...\right)$

(*i*) $f(x)=\frac{4}{\pi}\left(\sin x-\frac{1}{3^2}\sin 3x+\frac{1}{5^2}\sin 5x-...\right)$

(*ii*) $f(x)=\frac{\pi}{4}-\frac{2}{\pi}\left(\frac{1}{1^2}\cos 2x+\frac{1}{3^2}\cos 6x+\frac{1}{5^2}\cos 10x+...\right)$

6. Obtain the half-range cosine series for $f(x)=x^2$ in $0<x<\pi$.

**Ans.** $\frac{\pi^2}{3}-\frac{4}{\pi}\left(\cos x-\frac{1}{2^2}\cos 2x+\frac{1}{3^2}\cos 3x-...\right)$

7. Find (*i*) sine series and (*ii*) cosine series for the function

$f(x)=e^x$ for $0<x<\pi$.

**Ans.** (*i*) $\frac{2}{\pi}\sum_1^\infty n\left[\frac{1-(-1)^n e^\pi}{n^2+1}\right]\sin nx$ (*ii*) $\frac{e^\pi-1}{\pi}-\frac{2}{\pi}\sum_1^\infty\frac{1-(-1)^n e^\pi}{n^2+1}\cos nx$

8. If $f(x)=x+1$, for $0<x<\pi$, find its Fourier (*i*) sine series (*ii*) cosine series. Hence deduce that

(*i*) $1-\frac{1}{3}+\frac{1}{5}-\frac{1}{7}+...=\frac{\pi}{4}$ (*ii*) $1-\frac{1}{3^2}+\frac{1}{5^2}-\frac{1}{7^2}+...=\frac{\pi^2}{8}$

**Ans.** (*i*) $\frac{2}{\pi}\left[(\pi+2)\sin x-\frac{\pi}{2}\sin 2x+\frac{1}{3}(\pi+2)\sin 3x-\frac{\pi}{4}\sin 4x+...\right]$

(*ii*) $\frac{\pi}{2}+1-4\left(\cos x+\frac{\cos 3x}{3^2}+\frac{\cos 5x}{5^2}+...\right)$

9. Find the Fourier series expansion of the function $f(x)=\cos(sx)$, $-\pi\le x\le\pi$

where s is a fraction. Hence, show that $\cos\theta=\frac{1}{\theta}+\frac{2\theta}{\theta^2-\pi^2}+\frac{2\theta}{\theta^2-4\pi^2}+...$

**Ans.** $\frac{\sin\pi x}{\pi s}+\frac{1}{\pi}\sum\left(\frac{\sin(s\pi+n\pi)}{s+n}+\frac{\sin(s\pi-n\pi)}{s-n}\right)\cos nx$

## 12.10 CHANGE OF INTERVAL AND FUNCTIONS HAVING ARBITRARY PERIOD

In electrical engineering problems, the period of the function is not always $2\pi$ but $T$ or $2c$. This period must be converted to the length $2\pi$. The independent variable $x$ is also to be changed proportionally.

Let the function $f(x)$ be defined in the interval $(-c, c)$. Now we want to change the function to the period of $2\pi$ so that we can use the formulae of $a_n$, $b_n$ as discussed in article 12.6.

$\because$ $2c$ is the interval for the variable $x$.

$\therefore$ 1 is the interval for the variable $=\frac{x}{2c}$

$\therefore$ $2\pi$ is the interval for the variable $=\frac{x2\pi}{2c}=\frac{\pi x}{c}$

so put $z=\frac{\pi x}{c}$ or $x=\frac{zc}{\pi}$

Thus the function $f(x)$ of period $2c$ is transformed to the function

$f\left(\frac{cz}{\pi}\right)$ or the period of $F(z)$ is $2\pi$

$F(z)$ can be expanded in the Fourier series.

$$F(z)=f\left(\frac{cz}{\pi}\right)=\frac{a_0}{2}+a_1\cos z+a_2\cos 2z+\ldots\ldots b_1\sin z+b_2\sin 2z+\ldots$$

where $a_0=\frac{1}{\pi}\int_0^{2\pi}F(z)dz=\frac{1}{\pi}\int_0^{2\pi}f\left(\frac{cz}{\pi}\right)dz$

$$=\frac{1}{\pi}\int_0^{2c}f(x)d\left(\frac{\pi x}{c}\right)=\frac{1}{c}\int_0^{2c}f(x)dx \qquad \left[\text{Put } z=\frac{\pi x}{c}\right]$$

$$a_n=\frac{1}{\pi}\int_0^{2\pi}F(z)\cos nz\,dz=\frac{1}{\pi}\int_0^{2\pi}f\left(\frac{cz}{\pi}\right)\cos nz\,dz$$

$$=\frac{1}{\pi}\int_0^{2c}f(x)\cos\frac{n\pi x}{c}\,d\left(\frac{\pi x}{c}\right)=\frac{1}{c}\int_0^{2c}f(x)\cos\frac{n\pi x}{c}dx \qquad \left[\text{Put } z=\frac{\pi x}{c}\right]$$

Similarly, $b_n=\frac{1}{c}\int_0^{2c}f(x)\sin\frac{n\pi x}{c}dx.$

**Cor.** Half range series [Interval $(0, c)$]

**Cosine series:**

$$f(x)=\frac{a_0}{2}+a_1\cos\frac{\pi x}{c}+a_2\cos\frac{2\pi x}{c}+\ldots+a_n\cos\frac{n\pi x}{c}+\ldots$$

where $a_0=\frac{2}{c}\int_0^c f(x)dx,\ a_n=\frac{2}{c}\int_0^c f(x)\cos\frac{n\pi x}{c}dx$

**Sine series:** $f(x)=b_1\sin\frac{\pi x}{c}+b_2\sin\frac{2\pi x}{c}+\ldots+b_n\sin\frac{n\pi x}{c}+\ldots$

where $b_n=\frac{2}{c}\int_c^2 f(x)\sin\frac{n\pi x}{c}dx.$

**Example 12.** *A periodic function of period 4 is defined as*

$$f(x)=|x|,\quad -2<x<2.$$

*Find its Fourier series expansion.*

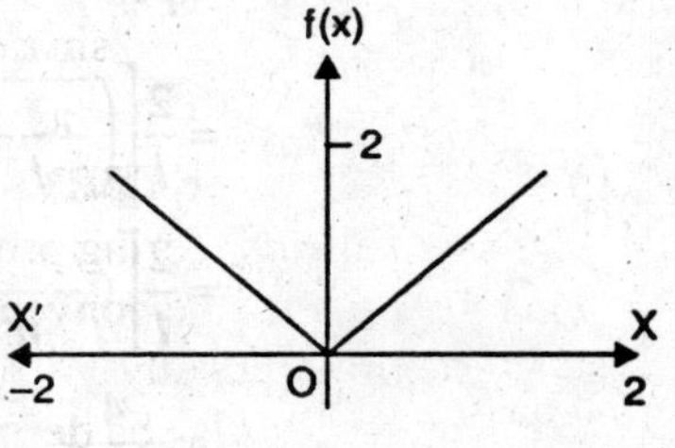


**Solution.** $f(x)=|x| \quad -2<x<2.$

$f(x)=x \quad 0<x<2$

$=-x \quad -2<x<0$

$$a_0=\frac{1}{c}\int_{-c}^{c}f(x)dx=\frac{1}{2}\int_0^2 x\,dx+\frac{1}{2}\int_{-2}^{0}(-x)dx$$

$$=\frac{1}{2}\left[\frac{x^2}{2}\right]_0^2+\frac{1}{2}\left[\frac{-x^2}{2}\right]_{-2}^{0}=\frac{1}{4}(4-0)+\frac{1}{4}(0+4)=2$$

$$a_n=\frac{1}{c}\int_{-c}^{c}f(x)\cos\frac{n\pi x}{c}dx=\frac{1}{2}\int_0^2 x\cos\frac{n\pi x}{2}dx+\frac{1}{2}\int_{-2}^{0}(-x)\cos\frac{n\pi x}{2}dx$$

$$=\frac{1}{2}\left[x\left(\frac{2}{n\pi}\sin\frac{n\pi x}{2}\right)-(1)\left(-\frac{4}{n^2\pi^2}\cos\frac{n\pi x}{2}\right)\right]_0^2$$

$$+\frac{1}{2}\left[(-x)\left(\frac{2}{n\pi}\sin\frac{n\pi x}{2}\right)-(-1)\left(-\frac{4}{n^2\pi^2}\right)\cos\frac{n\pi x}{2}\right]_{-2}^{0}$$

$$=\frac{1}{2}\left[0+\frac{4}{n^2\pi^2}(-1)^n-\frac{4}{n^2\pi^2}\right]+\frac{1}{2}\left[0-\frac{4}{n^2\pi^2}+\frac{4}{n^2\pi^2}(-1)^n\right]$$

$$=\frac{1}{2}\frac{4}{n^2\pi^2}[(-1)^n-1-1+(-1)^n]=\frac{4}{n^2\pi^2}[(-1)^n-1]$$

$$=-\frac{8}{n^2\pi^2} \qquad \text{if } n \text{ is odd.}$$

$$=0 \qquad \text{if } n \text{ is even}$$

$b_n = 0$ as $f(x)$ is even function.

Fourier series is

$$f(x)=\frac{a_0}{2}+a_1\cos\frac{\pi x}{c}+c_2\cos\frac{2\pi x}{c}+\ldots+b_1\sin\frac{\pi x}{c}+b_2\sin\frac{2\pi x}{c}+\ldots$$

$$f(x)=1-\frac{8}{\pi^2}\left[\frac{\cos\frac{\pi x}{2}}{1^2}+\frac{\cos\frac{3\pi x}{2}}{3^2}+\frac{\cos\frac{5\pi x}{2}}{5^2}+\ldots\right] \qquad \textbf{Ans.}$$

**Example 13.** *Find Fourier half-range even expansion of the function,*

$$f(x)=\left(\frac{-x}{l}+1\right), \qquad 0\le x\le l$$

**Solution.** $a_0=\frac{2}{l}\int_0^l f(x)dx=\frac{2}{l}\int_0^l\left(-\frac{x}{l}+1\right)dx$

$$=\frac{2}{l}\left[-\frac{x^2}{2l}+x\right]_0^l=\frac{2}{l}\left[-\frac{l^2}{2l}+1\right]=\frac{2l}{l}\left[-\frac{1}{2}+1\right]=1$$

$$a_n=\frac{2}{l}\int_0^l f(x)\cos\frac{n\pi x}{l}dx=\frac{2}{l}\int_0^l\left(-\frac{x}{l}+1\right)\cos\frac{n\pi x}{l}dx$$

$$=\frac{2}{l}\left[\left(-\frac{x}{l}+1\right)\left(\frac{l}{n\pi}\sin\frac{n\pi x}{l}\right)-\left(-\frac{1}{l}\right)\left(-\frac{l^2}{n^2\pi^2}\cos\frac{n\pi x}{l}\right)\right]_0^l$$

$$=\frac{2}{l}\left[0-\frac{l}{n^2\pi^2}\cos n\pi+\frac{l}{n^2\pi^2}\right]=\frac{2}{l}\frac{l}{n^2\pi^2}[-(-1)^n+1]=\frac{2}{n^2\pi^2}[1-(-1)^n]$$

$$=\frac{4}{n^2\pi^2} \qquad \text{when } n \text{ is odd.}$$

$$=0 \qquad \text{when } n \text{ is even.}$$

$$f(x)=\frac{1}{2}+\frac{4}{\pi^2}\left[\frac{1}{1^2}\cos\frac{\pi x}{l}+\frac{1}{3^2}\cos\frac{3\pi x}{l}+\frac{1}{5^2}\cos\frac{5\pi x}{l}\ldots\right] \qquad \textbf{Ans.}$$

**Example 14.** *Find the Fourier half-range cosine series of the function*

$$f(t)=2t, \qquad 0<t<1$$
$$=2(2-t), \quad 1<t<2$$

**Solution.** $f(t)=2t, \qquad 0<t<1$

$$=2(2-t), \quad 1<t<2$$

Let $$f(t)=\frac{a_0}{2}+a_1\cos\frac{\pi t}{c}+a_2\cos\frac{2\pi t}{c}+a_3\cos\frac{3\pi t}{c}+\ldots$$

$$+b_1\sin\frac{\pi t}{c}+b_2\sin\frac{2\pi t}{c}+b_3\sin\frac{3\pi t}{c}+\ldots \quad \ldots(1)$$

**Hence $c = 2$, because it is half range series.**

Here $$a_0=\frac{2}{c}\int_0^c f(t)\,dt=\frac{2}{2}\int_0^1 2t\,dt+\frac{2}{2}\int_1^2 2(2-t)\,dt$$

$$=\left[t^2\right]_0^1+\left[2\left(2t-\frac{t^2}{2}\right)\right]_1^2=1+[(4t-t^2)]_1^2=1+(8-4-4+1)=2$$

$$a_n=\frac{2}{c}\int_0^c f(t)\cos\frac{n\pi t}{c}\,dt=\frac{2}{2}\int_0^1 2t\cos\frac{n\pi t}{2}dt+\frac{2}{2}\int_1^2 2(2-t)\cos\frac{n\pi t}{2}\,dt$$

$$=\left[2t\left(\frac{2}{n\pi}\sin\frac{n\pi t}{2}\right)-(2)\left(-\frac{4}{n^2\pi^2}\cos\frac{n\pi t}{2}\right)\right]_0^1$$

$$+\left[(4-2t)\left(\frac{2}{n\pi}\sin\frac{n\pi t}{2}\right)-(-2)\left(-\frac{4}{n^2\pi^2}\cos\frac{n\pi t}{2}\right)\right]_1^2$$

$$=\left[\frac{4}{n\pi}\sin\frac{n\pi}{2}+\frac{8}{n^2\pi^2}\cos\frac{n\pi}{2}-\frac{8}{n^2\pi^2}\right]+\left[0-\frac{8}{n^2\pi^2}\cos n\pi-\frac{4}{n\pi}\sin\frac{n\pi}{2}+\frac{8}{n^2\pi^2}\cos\frac{n\pi}{2}\right]$$

$$=\frac{16}{n^2\pi^2}\cos\frac{n\pi}{2}-\frac{8}{n^2\pi^2}-\frac{8}{n^2\pi^2}\cos n\pi=\frac{8}{n^2\pi^2}\left[2\cos\frac{n\pi}{2}-1-\cos n\pi\right]$$

$$f(t)=1+\frac{8}{\pi^2}\sum_{n=1}^{\infty}\frac{1}{n^2}\left[2\cos\frac{n\pi}{2}-1-\cos n\pi\right]\cos\frac{n\pi t}{2}$$ **Ans.**

**Example 15.** *Obtain the Fourier cosine series expansion of the periodic function defined by*

$$f(t)=\sin\left(\frac{\pi t}{l}\right),\quad 0<t<l$$

**Solution.** $$f(t)=\sin\left(\frac{\pi t}{l}\right),\quad 0<t<l$$

$$a_0=\frac{2}{l}\int_0^l\sin\left(\frac{\pi t}{l}\right)dt=\frac{2}{l}\left(-\frac{l}{\pi}\cos\frac{\pi t}{l}\right)_0^l=-\frac{2}{\pi}(\cos\pi-\cos 0)=-\frac{2}{\pi}(-1-1)=\frac{4}{\pi}$$

$$a_n=\frac{2}{l}\int_0^l\sin\left(\frac{\pi t}{l}\right)\cos\frac{n\pi t}{l}\,dt=\frac{1}{l}\int_0^1\left[\sin\left(\frac{\pi t}{l}+\frac{n\pi t}{l}\right)-\sin\left(\frac{n\pi t}{l}-\frac{\pi t}{l}\right)\right]dt$$

$$= \frac{1}{l}\int_0^l \sin(n+1)\frac{\pi t}{l}\,dt - \frac{1}{l}\int_0^l \sin(n-1)\frac{\pi t}{l}\,dt$$

$$= \frac{1}{l}\left[-\frac{l}{(n+1)\pi}\cos\frac{(n+1)\pi t}{l}\right]_0^l - \frac{1}{l}\left[\frac{l}{(n-1)\pi}\cos\frac{(n-1)\pi t}{l}\right]_0^l$$

$$= \frac{-1}{(n+1)\pi}[\cos(n+1)\pi - \cos 0] + \frac{1}{(n-1)\pi}[\cos(n-1)\pi - \cos 0]$$

$$= \frac{1}{(n+1)\pi}[(-1)^{n-1} - 1] + \frac{1}{(n-1)\pi}[(-1)^{n+1} - 1]$$

$$= (-1)^{n+1}\left[-\frac{1}{(n+1)\pi} + \frac{1}{(n-1)\pi}\right] + \frac{1}{(n+1)\pi} - \frac{1}{(n-1)\pi}$$

$$= (-1)^{n+1}\frac{2}{(n^2-1)\pi} - \frac{2}{(n^2-1)\pi} = \frac{2}{(n^2-1)\pi}\left[(-1)^{n+1} - 1\right]$$

$$= \frac{-4}{(n^2-1)\pi} \qquad \text{when } n \text{ is even}$$

$$= 0 \qquad \text{when } n \text{ is odd.}$$

The above formula for finding the value of $a_1$ is not applicable.

$$a_1 = \frac{2}{l}\int_0^l \sin\frac{\pi t}{l}\cos\frac{\pi t}{l}\,dt = \frac{1}{l}\int_0^l \sin\frac{2\pi t}{l}\,dt$$

$$= \frac{1}{l}\left(-\frac{l}{2\pi}\cos\frac{2\pi t}{l}\right)_0^l = -\frac{l}{2\pi l}(\cos 2\pi - \cos 0) = \frac{1}{2\pi}(1-1) = 0$$

$$f(t) = \frac{a_0}{2} + a_1\cos\frac{\pi t}{l} + a_2\cos\frac{2\pi t}{l} + a_3\cos\frac{3\pi t}{l} + a_4\cos\frac{4\pi t}{l} + \ldots$$

$$= \frac{2}{\pi} - \frac{4}{\pi}\left[\frac{1}{3}\cos\frac{2\pi t}{l} + \frac{1}{15}\cos\frac{4\pi t}{l} + \frac{1}{35}\cos\frac{6\pi t}{l} + \ldots\right] \qquad \textbf{Ans.}$$

**Example 16.** *Find the Fourier series expansion of the periodic function of period 1*

$$f(x) = \frac{1}{2} + x, \qquad -\frac{1}{2} < x \le 0$$

$$= \frac{1}{2} - x, \qquad 0 < x < \frac{1}{2}$$

**Solution.** Let $f(x) = \frac{a_0}{2} + a_1\cos\frac{\pi x}{c} + a_2\cos\frac{2\pi x}{c} + \ldots$

$$+ b_1\sin\frac{\pi x}{c} + b_2\sin 2\frac{\pi x}{c} + b_3\sin\frac{3\pi x}{c} + \ldots \qquad \ldots (1)$$

Here $2c = 1$ or $c = \frac{1}{2}$

$$a_0 = \frac{1}{c}\int_{-c}^{c} f(x)\,dx = \frac{1}{1/2}\int_{-1/2}^{0}\left(\frac{1}{2} + x\right)dx + \frac{1}{1/2}\int_0^{1/2}\left(\frac{1}{2} - x\right)dx$$

$$= 2\left[\frac{x}{2} + \frac{x^2}{2}\right]_{-1/2}^{0} + \left[\frac{x}{2} - \frac{x^2}{2}\right]_0^{l/2} = 2\left[\frac{1}{4} - \frac{1}{8}\right] + \left[\frac{1}{4} - \frac{1}{8}\right] = \frac{1}{2}$$

$$a_n = \frac{1}{c}\int_{-c}^{c} f(x)\cos\frac{n\pi x}{c}\,dx$$

$$= \frac{1}{1/2}\int_{-1/2}^{0}\left(\frac{1}{2}+x\right)\cos\frac{n\pi x}{1/2}\,dx + \frac{1}{1/2}\int_{0}^{1/2}\left(\frac{1}{2}-x\right)\cos\frac{n\pi x}{1/2}\,dx$$

$$= 2\int_{-1/2}^{0}\left(\frac{1}{2}+x\right)\cos 2n\pi x\,dx + 2\int_{0}^{1/2}\left(\frac{1}{2}-x\right)\cos 2n\pi x\,dx$$

$$= 2\left[\left(\frac{1}{2}+x\right)\frac{\sin 2n\pi x}{2n\pi} - (1)\left(-\frac{\cos 2n\pi x}{4n^2\pi^2}\right)\right]_{-1/2}^{0}$$

$$+2\left[\left(\frac{1}{2}-x\right)\frac{\sin 2n\pi x}{2n\pi} - (-1)\left(\frac{-\cos 2n\pi x}{4n^2\pi^2}\right)\right]_{0}^{1/2}$$

$$= 2\left[0+\frac{1}{4n^2\pi^2}-\frac{(-1)^n}{4n^2\pi^2}\right]+2\left[0-\frac{(-1)^n}{4n^2\pi^2}+\frac{1}{4n^2\pi^2}\right] = \frac{1}{\pi^2}\left[\frac{1}{n^2}-\frac{(-1)^n}{n^2}\right]$$

$$= \frac{2}{n^2\pi^2} \qquad \text{if } n \text{ is odd}$$

$$= 0 \qquad \text{if } n \text{ is even}$$

$$b_n = \frac{1}{c}\int_{-c}^{c} f(x)\sin\frac{n\pi x}{c}\,dx$$

$$= \frac{1}{1/2}\int_{-1/2}^{0}\left(\frac{1}{2}+x\right)\sin\frac{n\pi x}{1/2}\,dx + \frac{1}{1/2}\int_{0}^{1/2}\left(\frac{1}{2}-x\right)\sin\frac{n\pi x}{1/2}\,dx$$

$$= 2\int_{-1/2}^{0}\left(\frac{1}{2}+x\right)\sin 2n\pi x\,dx + 2\int_{0}^{1/2}\left(\frac{1}{2}-x\right)\sin 2n\pi x\,dx$$

$$= 2\left[\left(\frac{1}{2}+x\right)\left(\frac{-\cos 2n\pi x}{2n\pi}\right) - (1)\left(-\frac{\sin 2n\pi x}{4n^2\pi^2}\right)\right]_{-1/2}^{0}$$

$$+2\left[\left(\frac{1}{2}-x\right)\left(-\frac{\cos 2n\pi x}{2n\pi}\right) - (-1)\left(-\frac{\sin 2n\pi x}{4n^2\pi^2}\right)\right]_{0}^{1/2}$$

$$= 2\left[-\frac{1}{4n\pi}\right]+\left[\frac{1}{4n\pi}\right]=0$$

**Substituting the values of $a_0, a_1, a_2, a_3, \ldots b_1, b_2, b_3 \ldots$ in (1) we have**

$$f(x) = \frac{1}{4}+\frac{2}{\pi^2}\left[\frac{\cos 2\pi x}{1^2}+\frac{\cos 6\pi x}{3^2}+\frac{\cos 10\pi x}{5^2}+\ldots\right] \qquad \textbf{Ans.}$$

**Example 17.** *Prove that* $\frac{l}{2}-x = \frac{l}{\pi}\sum_{1}^{\infty}\frac{1}{n}\sin\frac{2n\pi x}{l}, \quad 0<x<l$

**Solution.** $f(x) = \frac{1}{2}-x$

$$a_0 = \frac{1}{1/2}\int_0^l f(x)\,dx = \frac{2}{l}\int_0^l\left(\frac{1}{2}-x\right)dx = \frac{2}{l}\left[\frac{lx}{2}-\frac{x^2}{2}\right]_0^l = 0$$

$$a_n=\frac{1}{1/2}\int_0^l f(x)\cos\frac{n\pi x}{1/2}\,dx=\frac{2}{l}\int_0^1\left(\frac{1}{2}-x\right)\cos\frac{2n\pi x}{1}dx$$

$$=\frac{2}{1}\left[\left(\frac{1}{2}-x\right)\frac{1}{2n\pi}\sin\frac{2n\pi x}{1}-(-1)-\frac{1^2}{4n^2\pi^2}\cos\frac{2n\pi x}{1}\right]_0^1$$

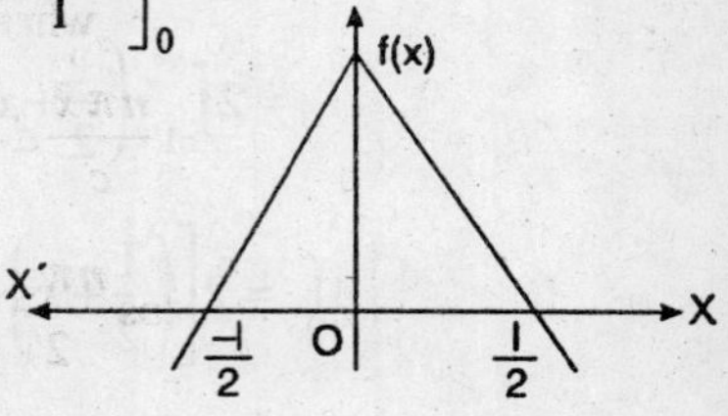


$$=\frac{2}{1}\left[0-\frac{1^2}{4n^2\pi^2}\cos 2n\pi+\frac{1^2}{4n^2\pi^2}\right]$$

$$=\frac{2}{1}\frac{1^2}{4n^2\pi^2}(-\cos 2n\pi+1)=\frac{1}{2n^2\pi^2}(-1+1)=0$$

$$b_n=\frac{1}{1/2}\int_0^1 f(x)\sin\frac{n\pi x}{1/2}\,dx=\frac{2}{1}\int_0^1\left(\frac{1}{2}-x\right)\sin\frac{2n\pi x}{1}dx$$

$$=\frac{2}{1}\left[\left(\frac{1}{2}-x\right)\left(-\frac{1}{2n\pi}\cos\frac{2n\pi x}{1}\right)-(-1)\left(-\frac{1^2}{4n^2\pi^2}\sin\frac{2n\pi x}{1}\right)\right]_0^1$$

$$=\frac{2}{1}\left[\frac{1}{2}\frac{1}{2n\pi}\cos 2n\pi-\frac{1}{2}.\frac{1}{2n\pi}(1)\right]=\frac{2}{1}\left[\frac{1^2}{2n\pi}\right]=\frac{1}{n\pi}$$

Fourier series is

$$f(x)=\frac{a_0}{2}+a_1\cos\frac{n\pi x}{1/2}+a_2\cos\frac{2n\pi x}{1/2}+a_3\cos\frac{3n\pi x}{1/2}+\ldots$$

$$+b_1\sin\frac{n\pi x}{1/2}+b_2\sin\frac{2n\pi x}{1/2}+b_3\sin\frac{3n\pi x}{1/2}+\ldots$$

$$\frac{1}{2}-x=\frac{1}{\pi}\sin\frac{2\pi x}{1}+\frac{1}{2\pi}\sin\frac{4\pi x}{1}+\frac{1}{3\pi}\sin\frac{6\pi x}{1}+\ldots$$

$$=\frac{1}{\pi}\sum_1^{\infty}\frac{1}{n}\sin\frac{2n\pi x}{1}$$

**Prove**

**Example 18.** *Find the Fourier series corresponding to the function f (x) defined in (−2, 2) as follows*

$$f(x)=\begin{cases}2 & \text{in } -2\le x\le 0\\ x & \text{in } \quad 0<x<2\end{cases}$$

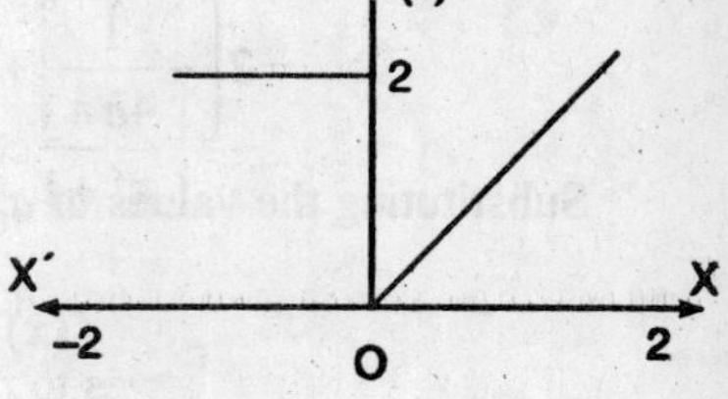


**Solution.** Here the interval is ( −2, 2) and $c = 2$

$$a_0=\frac{1}{c}\int_{-c}^{c}f(x)dx=\frac{1}{2}\left[\int_{-2}^{0}2\,dx+\int_{-2}^{0}x\,dx\right]$$

$$=\frac{1}{2}\left[[2x]_{-2}^{0}+\left(\frac{x^2}{2}\right)_0^2\right]=\frac{1}{2}[4+2]=3$$

$$a_n=\frac{1}{c}\int_{-c}^{c}f(x)\cos\left(\frac{n\pi x}{c}\right)dx=\frac{1}{2}\left[\int_{-2}^{0}2.\cos\frac{n\pi x}{2}dx+\int_0^2 x\cos\frac{n\pi x}{2}dx\right]$$

$$=\frac{1}{2}\left[\frac{4}{n\pi}\left(\sin\frac{n\pi x}{2}\right)_{-2}^{0}+\left(x\frac{2}{n\pi}\sin\frac{n\pi x}{2}+\frac{4}{n^2\pi^2}\cos\frac{n\pi x}{2}\right)_0^2\right]$$

$$= \frac{1}{2}\left[\frac{4}{n^2\pi^2}\cos n\pi - \frac{4}{n^2\pi^2}\right] = \frac{2}{n^2\pi^2}[(-1)^n - 1]$$

$$= \frac{4}{n^2\pi^2} \quad \text{when } n \text{ is odd}$$

$$= 0 \quad \text{when } n \text{ is even.}$$

$$b_n = \frac{1}{c}\int_{-c}^{c} f(x)\sin\frac{n\pi x}{c}\,dx = \frac{1}{2}\int_{-2}^{0} 2\sin\frac{n\pi x}{2}dx + \frac{1}{2}\int_0^2 x\sin\frac{n\pi x}{2}dx$$

$$= \frac{1}{2}\left[2\left(-\frac{2}{n\pi}\cos\frac{n\pi x}{2}\right)\right]_{-2}^{0} + \frac{1}{2}\left[x\left(-\frac{2}{n\pi}\cos\frac{n\pi x}{2}\right) + (1)\frac{4}{n^2\pi^2}\sin\frac{n\pi x}{2}\right]_0^2$$

$$= \frac{1}{2}\left[-\frac{4}{n\pi} + \frac{4}{n\pi}\cos n\pi\right] + \frac{1}{2}\left[-\frac{4}{n\pi}\cos n\pi + \frac{4}{n^2\pi^2}\sin n\pi\right] = \frac{1}{2}\left[-\frac{4}{n\pi}\right] = -\frac{2}{n\pi}$$

$$f(x) = \frac{a_0}{2} + a_1\cos\frac{\pi x}{c} + a_2\cos\frac{2\pi x}{c} + a_3\cos\frac{3\pi x}{c} + \ldots$$

$$+ b_1\sin\frac{\pi x}{c} + b_2\sin\frac{2\pi x}{c} + b_3\sin\frac{3\pi x}{c} + \ldots$$

$$= \frac{3}{2} - \frac{4}{\pi^2}\left\{\frac{1}{1^2}\cos\frac{\pi x}{2} + \frac{1}{3^2}\cos\frac{3\pi x}{2} + \ldots\right\}$$

$$- \frac{2}{\pi}\left\{\frac{1}{1}\sin\frac{\pi x}{2} + \frac{1}{2}\sin\frac{2\pi x}{2} + \frac{1}{2}\sin\frac{3\pi x}{2} + \ldots\right\} \quad \textbf{Ans.}$$

**Example 19.** *Expand $f(x) = e^x$ in a cosine series over (0, 1).*

**Solution.** $f(x) = e^x$ and $c = 1$

$$a_0 = \frac{2}{c}\int_0^c f(x)dx = \frac{2}{1}\int_0^1 e^x dx = 2(e-1)$$

$$a_n = \frac{2}{c}\int_0^c f(x)\cos\frac{n\pi x}{c}dx = \frac{2}{1}\int_0^1 e^x\cos\frac{n\pi x}{1}dx$$

$$= 2\left[\frac{e^x}{n^2\pi^2+1}(\cos n\pi x + n\pi\sin n\pi x)\right]_0^1$$

$$= \frac{2}{n^2\pi^2+1}[(-1)^n e - 1]$$

$$f(x) = \frac{a_0}{2} + a_1\cos\pi x + a_2\cos 2\pi x + a_3\cos 3\pi x + \ldots.$$

$$e^x = e - 1 + 2\left[\frac{-e-1}{\pi^2+1}\cos\pi x + \frac{e-1}{4\pi^2+1}\cos 2\pi x + \frac{-e-1}{9\pi^2+1}\cos 3\pi x + \ldots\right] \quad \textbf{Ans.}$$

## Exercise 12.4

1. Find the Fourier series to represent $f(x)$, where

$$f(x) = -a \qquad -c < x < 0$$
$$= a \qquad 0 < x < c$$

**Ans.** $\frac{4a}{\pi}\left[\sin\frac{\pi x}{c} + \frac{1}{3}\sin\frac{3\pi x}{c} + \frac{1}{5}\sin\frac{5\pi x}{c} + \ldots\right]$

2. Find the half-range sine series for the function

$$f(x) = 2x - 1 \qquad 0 < x < 1.$$

**Ans.** $-\frac{2}{\pi}\left[\sin\pi x + \frac{1}{2}\sin 4\pi x + \frac{1}{3}\sin 6\pi x + \ldots\right]$

3. Express $f(x) = x$ as a cosine, half range series in $0 < x < 2$.

$$\textbf{Ans. } 1 - \frac{8}{\pi^2}\left[\frac{1}{1^2}\cos\frac{\pi x}{2} + \frac{1}{3^2}\cos\frac{3\pi x}{2} + \frac{1}{5^2}\cos\frac{5\pi x}{2} + \ldots\right]$$

4. Find the Fourier series of the function

$$f(x) = \begin{bmatrix} -2 & \text{for} & -4 < x < -2 \\ x & \text{for} & -2 < x < 2 \\ 2 & \text{for} & 2 < x < 4 \end{bmatrix}$$

$$\textbf{Ans. } \frac{4}{\pi} + \frac{8}{\pi^2}\sin\frac{\pi x}{4} - \frac{2}{\pi}\sin\frac{2\pi x}{4} + \left(\frac{4}{3\pi} - \frac{8}{3^2\pi}\right)\sin\frac{3\pi x}{4} - \frac{2}{2\pi}\sin\frac{4\pi x}{4} + \ldots$$

5. Find the Fourier series to represent

$$f(x) = x^2 - 2 \qquad \text{from} \qquad -2 < x < 2.$$

$$\textbf{Ans. } -\frac{2}{3} - \frac{16}{\pi^2}\left[\cos\frac{\pi x}{2^2} - \frac{1}{4}\cos\pi x + \frac{1}{9}\cos\frac{3\pi x}{2} \ldots\right]$$

6. If $f(x) = e^{-x} - c < x < c$, show that

$$f(x) = (e^c - e^{-c})\left\{\frac{1}{2c} - c\left(\frac{1}{c^2+\pi^2}\cos\frac{\pi x}{c} - \frac{1}{c^2+4\pi^2}\cos\frac{2\pi x}{c} + \ldots\right)\right.$$
$$\left. -\pi\left(\frac{1}{c^2+\pi^2}\sin\frac{\pi x}{c} - \frac{1}{c^2+4\pi^2}\sin\frac{2\pi x}{c} + \ldots\right)\right\}$$

7. A sinusodial voltage E sin $\omega t$ is passed through a half wave rectifier which clips the negative portion of the wave. Develop the resulting portion of the function

$$u(t) = 0 \qquad \text{when} \qquad -\frac{T}{2} < t < 0$$
$$= E\sin\omega t \qquad \text{when} \qquad 0 < t < \frac{T}{2} \qquad \left(T = \frac{2\pi}{\omega}\right)$$

$$\textbf{Ans. } \frac{E}{\pi} + \frac{E}{2}\sin\omega t - \frac{2E}{\pi}\left[\frac{1}{1.3}\cos 2\omega t + \frac{1}{3.5}\cos 4\omega t + \frac{1}{5.7}\cos 6\omega t + \ldots\right]$$

8. A periodic square wave has a period 4. The function generating the square is

$$f(t) = 0 \quad \text{for} \quad -2 < t < -1$$
$$= k \quad \text{for} \quad -1 < t < 1$$
$$= 0 \quad \text{for} \quad 1 < t < 2$$

Find the Fourier series of the function. $\textbf{Ans. } f(t) = \frac{k}{2} + \frac{2k}{\pi}\left[\cos\frac{\pi t}{2} - \frac{1}{3}\cos\frac{3\pi t}{2} + \ldots\right]$

9. Find a Fourier series to represent $x^2$ in the interval $(-l, l)$.

$$\textbf{Ans. } \frac{l^2}{3} - \frac{4l^2}{\pi^2}\left[\cos\pi x - \frac{\cos\pi x}{2^2} + \frac{\cos 3\pi x}{3^2} \ldots\right]$$

## 12.11. PARSEVAL'S FORMULA

$$\boxed{\int_{-c}^{c}[f(x)]^2\,dx = c\left\{\frac{1}{2}a_0^2 + \sum_{n=1}^{\infty}\left(a_n^2 + b_n^2\right)\right\}}$$

**Solution.** We know that $f(x) = \frac{a_0}{2} + \sum_{n=1}^{\infty}\left(a_n\cos\frac{n\pi x}{c} + b_n\sin\frac{n\pi x}{c}\right)$ .... (1)

Multiplying (1) by $f(x)$, we get

$$[f(x)]^2 = \frac{a_0}{2} f(x) + \sum_{n=1}^{\infty} a_n f(x) \cos\frac{n\pi x}{c} + \sum_{n=1}^{\infty} b_n f(x) \sin\frac{n\pi x}{c} \qquad \text{.... (2)}$$

Integrating term by term from $-c$ to $c$, we have

$$\int_{-c}^{c} [f(x)]^2\, dx = \frac{a_0}{2}\int_{-c}^{c} f(x) dx + \sum_{n=1}^{\infty} a_n \int_{-c}^{c} f(x) \cos\frac{n\pi x}{c}$$

$$+ \sum_{n=1}^{\infty} b_n \int_{-c}^{c} f(x) \sin\frac{n\pi x}{c}\, dx \qquad \text{.... (3)}$$

We have the following results

$$a_0 = \frac{1}{c}\int_{-c}^{c} f(x)\, dx \quad \Rightarrow \quad \int_{-c}^{c} f(x) dx = c\, a_0$$

$$a_n = \frac{1}{c}\int_{-c}^{c} f(x) = \cos\frac{n\pi x}{c}\, dx \quad \Rightarrow \quad \int_{-c}^{c} f(x) \cos\frac{n\pi x}{c}\, dx = c\, a_n$$

$$b_n = \frac{1}{c}\int_{-c}^{c} f(x) \sin\frac{n\pi x}{c}\, dx \quad \Rightarrow \quad \int_{-c}^{c} f(x) \sin\frac{n\pi x}{c}\, dx = c\, b_n$$

On putting these integrals in (3), we get

$$\int_{-c}^{c} [f(x)]^2 dx = c\frac{a_0^2}{2} + \sum_{n=1}^{\infty} c\, a_n^2 + \Sigma\, c b_n^2 = c\left[\frac{a_0^2}{2} + \sum_{n=1}^{\infty}\left(a_n^2 + b_n^2\right)\right]$$

This is the Parseval's formula

**Note.** 1. If $0 < x < 2c$, then $\int_0^{2c} [f(x)]^2 dx = c\left[\frac{a_0^2}{2} + \sum_{n=1}^{\infty}\left(a_n^2 + b_n^2\right)\right]$

2. If $0 < x < c$ (Half range cosine series), $\int_0^{c} [f(x)]^2 = \frac{c}{2}\left[\frac{a_0^2}{2} + \sum_{n=1}^{\infty} a_n^2\right]$

3. If $0 < x < c$ (Half range sine series), $\int_0^{c} [f(x)]^2 dx = \frac{c}{2}\left[\frac{a_0^2}{2} + \sum_{n=1}^{\infty} b_n^2\right]$

4. R.M.S. $= \left\{\frac{\int_a^b [f(x)]^2 dx}{b-a}\right\}^{\frac{1}{2}}$

**Example 20.** *By using the sine series for $f(x) = 1$ in $0 < x < \pi$ show that*

$$\frac{\pi^2}{8} = 1 + \frac{1}{3^2} + \frac{1}{5^2} + \frac{1}{7^2} + \ldots.$$

**Solution.** sine series is $f(x) = \Sigma\, b_n \sin nx$

$$b_n = \frac{2}{\pi}\int_0^{\pi} f(x) \sin nx\, dx$$

$$= \frac{2}{\pi}\int_0^{\pi} (1) \sin nx\, dx = \frac{2}{\pi}\left(\frac{-\cos nx}{n}\right)_0^{\pi} = \frac{-2}{n\pi}[\cos n\pi - 1] = \frac{-2}{n\pi}[(-1)^n - 1]$$

$= \frac{2}{n\pi}$ if $n$ is odd.

$= 0$ if n is even

Then, the sine series is

$$1=\frac{4}{\pi}\sin x+\frac{4}{3\pi}\sin 3x+\frac{4}{5\pi}\sin 5x+\frac{4}{7\pi}\sin 7x+.....$$

$$\int_0^c [f(x)]^2\,dx=\frac{c}{2}\left[b_1^2+b_2^2+b_3^2+b_4^2+b_5^2+.....\right]$$

$$\int_0^\pi (1)^2\,dx=\frac{\pi}{2}\left[\left(\frac{4}{\pi}\right)^2+\left(\frac{4}{3\pi}\right)^2+\left(\frac{4}{5\pi}\right)^2+\left(\frac{4}{7\pi}\right)^2+.....\right]$$

$$[x]_0^\pi=\left(\frac{\pi}{2}\right)\left(\frac{16}{\pi^2}\right)\left[1+\frac{1}{3^2}+\frac{1}{5^2}+\frac{1}{7^2}+.....\right]$$

$$\pi=\frac{\pi}{2}\left(\frac{16}{\pi^2}\right)\left[1+\frac{1}{3^2}+\frac{1}{5^2}+\frac{1}{7^2}+.....\right]$$

$$\frac{\pi^2}{8}=1+\frac{1}{3^2}+\frac{1}{5^2}+\frac{1}{7^2}+.....$$ **Proved.**

**Example 21.** *If* $f(x)=\begin{cases}\pi x, & 0<x<1\\ \pi(2-x), & 1<x<2\end{cases}$

*using half range cosine series, show that* $\frac{1}{1^4}+\frac{1}{3^4}+\frac{1}{5^4}+.....=\frac{\pi^4}{96}$

**Solution.** Half range cosine series is

$$f(x)=\frac{a_0}{2}+\Sigma a_n\cos\frac{n\pi x}{c}$$

where $a_0=\frac{2}{c}\int_0^c f(x)dx=\frac{2}{2}\left[\int_0^1 \pi x\,dx+\int_1^2 \pi(2-x)\,dx\right]$

$$=\pi\left(\frac{x^2}{2}\right)_0^1+\pi\left(2x-\frac{x^2}{2}\right)_1^2=\frac{\pi}{2}+\pi\left[(4-2)-\left(2-\frac{1}{2}\right)\right]$$

$$=\pi$$

$$a_n=\frac{2}{c}\int_0^c f(x)\cos\frac{n\pi x}{c}\,dx$$

$$=\frac{2}{2}\left[\int_0^1 \pi x\cos\frac{n\pi x}{2}\,dx+\int_1^2 \pi\,(2-x)\cos\frac{n\pi x}{2}\,dx\right]$$

$$=\pi\left[x\frac{\sin\frac{n\pi x}{2}}{\frac{n\pi}{2}}-\left(\frac{-\cos\frac{n\pi x}{2}}{\frac{n^2\pi^2}{4}}\right)\right]_0^1+\pi\left[(2-c)\frac{\sin\frac{n\pi x}{2}}{\frac{n\pi}{2}}-(-1)\left(\frac{-\cos\frac{n\pi x}{2}}{\frac{n^2\pi^2}{4}}\right)\right]_1^2$$

$$=\pi\left[\frac{2}{n\pi}\sin\frac{n\pi}{2}+\frac{4}{n^2\pi^2}\cos\frac{n\pi}{2}-\frac{4}{n^2\pi^2}\right]+\pi\left[0-\frac{4}{n^2\pi^2}\cos n\pi-\frac{2}{n\pi}\sin\frac{n\pi}{2}+\frac{4}{n^2\pi^2}\cos\frac{n\pi}{2}\right]$$

$$=\pi\left[\frac{8}{n^2\pi^2}\cos\frac{n\pi}{2}-\frac{4}{n^2\pi^2}-\frac{4}{n^2\pi^2}\cos n\pi\right]=\frac{4}{n^2\pi}\left[2\cos\frac{n\pi}{2}-1-\cos n\pi\right]$$

$$a_1 = 0,\ a_2 = \frac{-4}{\pi},\ a_3 = 0,\ a_4 = 0,\ a_5 = 0,\ a_6 = \frac{-4}{9\pi}$$

$$\int_0^c [f(x)^2\,dx = \frac{c}{2}\left[\frac{a_0^2}{2} + a_1^2 + a_2^2 + a_3^2 + ....\right]$$

$$\int_0^1 (\pi x)^2\,dx + \int_1^2 \pi^2(2-x)^2\,dx = \frac{2}{2}\left[\frac{\pi^2}{2} + \frac{16}{\pi^2} + \frac{16}{81\pi^2} + ....\right]$$

$$\pi^2\left[\frac{x^3}{3}\right]_0^1 - \pi^2\left[\frac{(2-x)^3}{3}\right]_1^2 = \frac{\pi^2}{2} + \frac{16}{\pi^2} + \frac{16}{81\pi^2} + ....$$

$$\frac{\pi^2}{3} - \pi^2\left(0 - \frac{1}{3}\right) = \frac{\pi^2}{2} + \frac{16}{\pi^2}\left[1 + \frac{1}{81} + .....\right]$$

$$\frac{2\pi^2}{3} - \frac{\pi^2}{2} = \frac{16}{\pi^2}\left[1 + \frac{1}{3^4} + \frac{1}{5^4} + .....\right]$$

$$\frac{\pi^2}{6} = \frac{16}{\pi^2}\left[1 + \frac{1}{3^4} + \frac{1}{5^4} + .....\right]$$

$$\frac{\pi^4}{96} = 1 + \frac{1}{3^4} + \frac{1}{5^4} + .....$$ **Ans.**

**Example 22.** *Prove that for $0 < x < \pi$*

(*a*) $$x(\pi - x) = \frac{\pi^2}{6} - \left[\frac{\cos x}{1^2} + \frac{\cos 4x}{2^2} + \frac{\cos 6x}{3^2} + ....\right]$$

(*b*) $$x(\pi - x) = \frac{8}{\pi}\left[\frac{\sin x}{1^2} + \frac{\sin 3x}{3^2} + \frac{\sin 5x}{5^2} + ....\right]$$

*Deduce from* (*a*) *and* (*b*) *respectively that*

(*c*) $\displaystyle\sum_{n=1}^{\infty}\frac{1}{n^4} = \frac{\pi^4}{90}$ (*d*) $\displaystyle\sum_{n=1}^{\infty}\frac{1}{n^6} = \frac{\pi}{945}$

**Solution.** Half range cosine series

$$a_0 = \frac{2}{\pi}\int_0^{\pi} x(\pi - x) = \frac{2}{\pi}\left[\frac{\pi x^2}{2} - \frac{x^3}{3}\right]_0^{\pi} = \frac{2}{\pi}\left[\frac{\pi^3}{2} - \frac{\pi^3}{3}\right] = \frac{\pi^2}{3}$$

$$a_n = \frac{2}{\pi}\int_0^{\pi} x(\pi - x)\cos nx\,dx$$

$$= \frac{2}{\pi}\left[(\pi x - x^2)\frac{\sin nx}{n} - (\pi - 2x)\left(\frac{-\cos nx}{n^2}\right) + (-2)\left(\frac{-\sin nx}{n^3}\right)\right]_0^{\pi}$$

$$= \frac{2}{\pi}\left[0 - \frac{\pi(-1)^n}{n^2} + 0 - \frac{\pi}{n^2}\right] = \frac{2}{\pi}\left(\frac{\pi}{n^2}\right)[-(-1)^n - 1]$$

$$= -\frac{4}{n^2} \quad \text{when } n \text{ is even}$$

$$= 0 \quad \text{when } n \text{ is odd}$$

Hence, $x(\pi - x) = \dfrac{\pi^2}{6} - \left[\dfrac{\cos 2x}{1^2} + \dfrac{\cos 4x}{2^2} + ...\right] \Rightarrow x(\pi - x) = \dfrac{\pi^2}{6} - 4\left[\dfrac{\cos 2x}{2^2} + \dfrac{\cos 4x}{4^2} + ....\right]$

By Parseval's formula

$$\frac{2}{\pi}\int_0^{\pi} x^2(\pi-x)^2\,dx = \frac{a_0^2}{2}+\sum a_n^2$$

$$\frac{2}{\pi}\int_0^{\pi}(\pi^2x^2-2\pi x^3+x^4)\,dx = \frac{1}{2}\left(\frac{\pi^4}{9}\right)+16\left[\frac{1}{2^4}+\frac{1}{4^4}+\frac{1}{6^4}+....\right]$$

$$\frac{2}{\pi}\left[\frac{\pi^2x^3}{3}-\frac{2\pi x^4}{4}+\frac{x^5}{5}\right]_0^{\pi} = \frac{\pi^4}{18}+\left[\frac{1}{1^4}+\frac{1}{2^4}+\frac{1}{3^4}+.....\right]$$

$$-\frac{2}{\pi}\left[\frac{\pi^5}{3}-\frac{2\pi^5}{4}+\frac{x^5}{5}\right] = \frac{\pi^4}{18}+\left[\frac{1}{1^4}+\frac{1}{2^4}+\frac{1}{3^4}+.....\right]$$

$$\frac{\pi^4}{15} = \frac{\pi^4}{18}+\sum_{n=1}^{\infty}\frac{1}{n^4} \quad\Rightarrow\quad \sum_{n=1}^{\infty}\frac{1}{n^4} = \frac{\pi^4}{90}$$

(*b*) Half range sine series

$$b_n = \frac{2}{\pi}\int_0^{\pi} x(\pi-x)\sin nx\,dx$$

$$= \frac{2}{\pi}\left[(\pi x-x^2)\left(\frac{-\cos nx}{n}\right)-(\pi-2x)\left(\frac{-\sin nx}{n^2}\right)+(-2)\frac{\cos nx}{n^3}\right]_0^{\pi}$$

$$= \frac{2}{\pi}\left[-2\frac{(-1)^n}{n^3}+\frac{2}{n^3}\right] = \frac{4}{\pi n^3}\left[-(-1)^n+1\right]$$

$$= \frac{8}{n^3\pi} \qquad \text{when } n \text{ is odd}$$

$$= 0 \qquad \text{when } n \text{ is even.}$$

$$\therefore \quad x(\pi-x) = \frac{8}{\pi}\left[\frac{\sin x}{1^3}+\frac{\sin 3x}{3^3}+\frac{\sin 5x}{5^3}+....\right]$$

By Parseval's formula

$$\frac{2}{\pi}\int_0^{\pi} x^2(\pi-x)^2\,dx = \sum b_n^2$$

$$\frac{\pi^2}{15} = \frac{64}{\pi^2}\left[\frac{1}{1^6}+\frac{1}{3^6}+\frac{1}{5^6}+....\right]$$

$$\frac{\pi^4}{960} = \frac{1}{1^6}+\frac{1}{3^6}+\frac{1}{5^6}$$

Let $S = \frac{1}{1^6}+\frac{1}{2^6}+\frac{1}{3^6}+\frac{1}{4^6}+... = \left(\frac{1}{1^6}+\frac{1}{3^6}+\frac{1}{5^6}+.....\right)+\left(\frac{1}{2^6}+\frac{1}{4^6}+\frac{1}{6^6}+....\right)$

$$S = \frac{\pi^4}{960}+\left(\frac{1}{2^6}+\frac{1}{4^6}+\frac{1}{6^6}+...\right) = \frac{\pi^4}{960}+\frac{1}{2^6}\left[\frac{1}{1^6}+\frac{1}{2^6}+\frac{1}{3^6}+.....\right]$$

$$S = \frac{\pi^4}{960}+\frac{S}{64}$$

$$S-\frac{S}{64} = \frac{\pi^4}{960} \quad\text{or}\quad \frac{63S}{64} = \frac{\pi^4}{960}$$

$$S = \frac{\pi^4}{960} \times \frac{64}{63} = \frac{\pi^4}{945}$$

$$\sum_{n=1}^{\infty} \frac{1}{n^6} = \frac{\pi^4}{945}$$

**Proved.**

**Exercise 12.5**

1. Prove that $0 < x < c$,

$$x = \frac{c}{2} - \frac{4c}{\pi^2}\left(\cos\frac{\pi x}{c} + \frac{1}{3^2}\cos\frac{3\pi x}{c} + \frac{1}{5^2}\cos\frac{5\pi x}{c} + \ldots\right)$$

and deduce that

(i) $\frac{1}{1^4} + \frac{1}{3^4} + \frac{1}{5^4} + \ldots = \frac{\pi^4}{96}$ (ii) $\frac{1}{1^4} + \frac{1}{2^4} + \frac{1}{3^4} + \frac{1}{4^4} + \ldots = \frac{\pi^4}{90}$

## 12.12. FOURIER SERIES IN COMPLEX FORM

Fourier series of a function $f(x)$ of period $2l$ is

$$f(x) = \frac{a_0}{2} + a_1 \cos\frac{\pi x}{l} + a_2 \cos\frac{2\pi x}{l} + \ldots + a_n \cos\frac{n\pi x}{l} + \ldots$$

$$+ b_1 \sin\frac{\pi x}{l} + b_2 \sin\frac{2\pi x}{l} + \ldots + b_n \sin\frac{n\pi x}{l} + \ldots \quad \ldots (1)$$

We know that $\cos x = \frac{e^{ix} + e^{-ix}}{2}$ and $\sin x = \frac{e^{ix} - e^{-ix}}{2i}$

On putting the values of $\cos x$ and $\sin x$ in (1), we get

$$f(x) = \frac{a_0}{2} + a_1 \frac{e^{\frac{i\pi x}{l}} + e^{\frac{-i\pi x}{l}}}{2} + a_2 \frac{e^{\frac{2i\pi x}{l}} + e^{\frac{-2i\pi x}{l}}}{2} + \ldots + b_1 \frac{e^{\frac{i\pi x}{l}} - e^{\frac{i\pi x}{l}}}{2i} + b_2 \frac{e^{\frac{2i\pi x}{l}} - e^{\frac{-2i\pi x}{l}}}{2i} + \ldots$$

$$= \frac{a_0}{2} + (a_1 - ib_1)e^{\frac{i\pi x}{l}} + (a_2 - ib_2)e^{\frac{2i\pi x}{l}} + \ldots + (a_1 + ib_1)e^{\frac{-i\pi x}{l}} + (a_2 + ib_2)e^{\frac{-2i\pi x}{l}} + \ldots$$

$$= c_0 + c_1 e^{\frac{i\pi x}{l}} + c_2 e^{\frac{2i\pi x}{l}} + \ldots + c_{-1} e^{\frac{-i\pi x}{l}} + c_{-2} e^{\frac{2i\pi x}{l}} + \ldots$$

$$= c_0 + \sum_{n=1}^{\infty} c_n e^{\frac{in\pi x}{l}} + \sum_{n=1}^{\infty} c_{-n} e^{\frac{-in\pi x}{l}}$$

$$c_n = \frac{1}{2}(a_n - ib_n), \quad c_{-n} = \frac{1}{2}(a_n + ib_n)$$

where $c_0 = \frac{a_0}{2} = \frac{1}{2}\frac{1}{l}\int_0^{2l} f(x)\,dx$

$$c_n = \frac{1}{2}\left[\frac{1}{l}\int_0^{2l} f(x)\cos\frac{n\pi x}{l}\,dx - \frac{i}{l}\int_0^{2l} f(x)\sin\frac{n\pi x}{l}\,dx\right] \Rightarrow c_n = \frac{1}{2}\frac{1}{l}\int_0^{2l} f(x)\left\{\cos\frac{n\pi x}{l} - i\sin\frac{n\pi x}{l}\right\}dx$$

$$c_n = \frac{1}{2l}\int_0^{2l} f(x)e^{\frac{-in\pi x}{l}}\,dx$$

$$c_{-n} = \frac{1}{2l}\int_0^{2l} f(x)e^{\frac{in\pi x}{l}}\,dx$$

**Example 23.** *Obtain the complex form of the Fourier series of the function*

$$f(x)=\begin{cases}0 & -\pi\le x\le 0\\ 1 & 0\le x\le \pi\end{cases}$$

**Solution.** $$c_0=\frac{1}{2\pi}\int_0^{\pi}dx=\frac{1}{2}$$

$$c_n=\frac{1}{2\pi}\int_{-\pi}^{\pi}f(x)\,e^{-inx}dx$$

$$=\frac{1}{2\pi}\left[\int_{-\pi}^{0}0.e^{-inx}\,dx+\int_0^{\pi}1.e^{-inx}\,dx\right]=\frac{1}{2\pi}\int_0^{\pi}e^{-inx}\,dx=\frac{1}{2\pi}\left[\frac{e^{-inx}}{-in}\right]_0^{\pi}$$

$$=-\frac{1}{2n\pi i}\left[e^{-in\pi}-1\right]=\frac{1}{2n\pi i}[\cos n\pi-i\sin n\pi-1]=-\frac{1}{2n\pi i}\left[(-1)^n-1\right]$$

$$=\begin{cases}\dfrac{1}{in\pi}, & n \text{ is odd}\\ 0, & n \text{ is even}\end{cases}$$

$$f(x)=\frac{1}{2}+\frac{1}{i\pi}\left[\frac{e^{ix}}{1}+\frac{e^{3ix}}{3}+\frac{e^{5ix}}{5}+\ldots\right]+\frac{1}{i\pi}\left[\frac{e^{-ix}}{-1}+\frac{e^{-3ix}}{-3}+\frac{e^{-5ix}}{-5}+\ldots\right]$$

$$=\frac{1}{2}-\frac{1}{i\pi}\left[\left(e^{ix}-e^{-ix}\right)+\frac{1}{3}\left(e^{3ix}-e^{-3ix}\right)+\frac{1}{5}\left(e^{5ix}-e^{-5ix}\right)+\ldots\right]$$ **Ans.**

### Exercise 12.6

Find the complex form of the Fourier series

1. $f(x)=e^{-x}, -1\le x\le 1$ **Ans.** $\sum_{n=-\infty}^{\infty}\frac{(-1)^n(1-in\pi)}{1+n^2\pi^2}\sinh 1.e^{in\pi x}$

2. $f(x)=e^{ax}, -1<x<1$ **Ans.** $\frac{2}{\pi}-\frac{2}{\pi}\left[\frac{e^{2it}+e^{-2it}}{1.3}+\frac{e^{4it}+e^{-4it}}{3.5}+\frac{e^{6it}+e^{-6it}}{5.7}+\ldots\right]$

3. $f(x)=\cos ax, -\pi<x<\pi$ **Ans.** $\frac{a}{\pi}\sin a\pi\sum_{-\infty}^{\infty}\frac{(-1)^n e^{inx}}{a^2-n^2}$

## 12.13 PRACTICAL HARMONIC ANALYSIS

Sometimes the function is not given by a formula, but by a graph or by a table of corresponding values. The process of finding the Fourier series for a function given by such values of the function and independent variable is known as **Harmonic Analysis**. The Fourier constants are evaluated by the following formulae :

(1) $$a_0=\frac{1}{\pi}\int_0^{2\pi}f(x)dx$$

$$=2\frac{1}{2\pi-0}\int_0^{2\pi}f(x)dx \qquad \left[\text{Mean}=\frac{1}{b-a}\int_a^b f(x)dx\right]$$

or $$a_0=2\,[\text{mean value of } f(x) \text{ in } (0, 2, \pi)]$$

(2) $$a_n=\frac{1}{\pi}\int_0^{2\pi}f(x)\cos nx\,dx=2\frac{1}{2\pi-0}\int_0^{2\pi}f(x)\cos nx\,dx$$

$$b_n=2\,[\text{mean value of } f(x)\cos nx \text{ in } (0, 2\pi)]$$

(3) $$b_n = \frac{1}{\pi}\int_0^{2\pi} f(x)\sin nx\,dx = 2\frac{1}{2\pi - 0}\int_0^{2\pi} f(x)\sin nx\,dx$$

$$b_n = 2\ [\text{mean value of } f(x) \sin nx \text{ in } (0, 2\pi)]$$

**Fundamental of first harmonic.** The term ($a_1 \cos x + b_1 \sin x$) in Fourier series is called the fundamental or first harmonic.

**Second harmonic.** The term ($a_2 \cos 2x + b_2 \sin 2x$) in Fourier series is called the second harmonic and so on.

**Example 24.** *Find the Fourier series as far as the second harmonic to represent the function given by table below :*

| $x$ | 0° | 30° | 60° | 90° | 120° | 150° | 180° | 210° | 240° | 270° | 300° | 330° |
|---|---|---|---|---|---|---|---|---|---|---|---|---|
| $f(x)$ | 2.34 | 3.01 | 3.69 | 4.15 | 3.69 | 2.20 | 0.83 | 0.51 | 0.88 | 1.09 | 1.19 | 1.64 |

**Solution**

| $x°$ | $\sin x$ | $\sin 2x$ | $\cos x$ | $\cos 2x$ | $f(x)$ | $f(x) \sin x$ | $f(x) \sin 2x$ | $f(x) \cos x$ | $f(x) \cos 2x$ |
|---|---|---|---|---|---|---|---|---|---|
| 0° | 0 | 0 | 1 | 1 | 2.34 | 0 | 0 | 2.340 | 2.340 |
| 30° | 0.50 | 0.87 | 0.87 | 0.50 | 3.01 | 1.505 | 2.619 | 2.619 | 1.505 |
| 60° | 0.87 | 0.87 | 0.50 | −0.50 | 3.69 | 3.210 | 3.210 | 1.845 | −1.845 |
| 90° | 1.00 | 0 | 0 | −1.00 | 4.15 | 4.150 | 0 | 0 | −4.150 |
| 120° | 0.87 | −0.87 | −0.50 | −0.50 | 3.69 | 3.210 | −3.210 | −1.845 | −1.845 |
| 150° | 0.50 | −0.87 | −0.87 | 0.50 | 2.20 | 1.100 | −1.914 | −1.914 | 1.100 |
| 180° | 0 | 0 | −1 | 1.00 | 0.83 | 0 | 0 | −0.830 | 0.830 |
| 210° | −0.50 | 0.87 | −0.87 | 0.50 | 0.51 | −0.255 | 0.444 | −0.444 | 0.255 |
| 240° | −0.87 | 0.87 | −0.50 | −0.50 | 0.88 | −0.766 | 0.766 | −0.440 | −0.440 |
| 270° | −1.00 | 0 | 0 | −1.00 | 1.09 | −1.090 | 0 | 0 | −1.090 |
| 300° | −0.87 | −0.87 | 0.50 | −0.50 | 1.19 | −1.035 | −1.035 | 0.595 | −0.595 |
| 330° | −0.50 | −0.87 | 0.87 | 0.50 | 1.64 | −0.820 | −1.427 | 1.427 | 0.820 |
| | | | | | 25.22 | 9.209 | −0.547 | 3.353 | −3.115 |

$$a_0 = 2 \times \text{Mean of } f(x) = 2 \times \frac{25.22}{12} = 4.203$$

$$a_1 = 2 \times \text{Mean of } f(x) \cos x = 2 \times \frac{3.353}{12} = 0.559$$

$$a_2 = 2 \times \text{Mean of } f(x) \cos 2x = 2 \times \frac{-3.115}{12} = -0.519$$

$$b_1 = 2 \times \text{Mean of } f(x) \sin x = 2 \times \frac{9.209}{12} = 1.535$$

$$b_2 = 2 \times \text{Mean of } f(x) \sin 2x = 2 \times \frac{-0.547}{12} = -0.091$$

Fourier series is

$$f(x) = \frac{a_0}{2} + a_1 \cos x + a_2 \cos 2x + \ldots + b_1 \sin x + b_2 \sin 2x + \ldots$$

$$= 2.1015 + 0.559 \cos x - 0.519 \cos 2x + \ldots + 1.535 \sin x - 0.091 \sin 2x + \ldots$$ **Ans.**

**Example 31.** *A machine completes its cycle of operations every time as certain pulley completes a revolution. The displacement $f(x)$ of a point on a certain portion of the machine is given in the table given below for twelve positions of the pulley, x being the angle in degree turned through by the pulley. Find a Fourier series to represent $f(x)$ for all values of x.*

| $x$ | 30° | 60° | 90° | 120° | 150° | 180° | 210° | 240° | 270° | 300° | 330° | 360° |
|---|---|---|---|---|---|---|---|---|---|---|---|---|
| $f(x)$ | 7.976 | 8.026 | 7.204 | 5.676 | 3.674 | 1.764 | 0.552 | 0.262 | 0.904 | 2.492 | 4.736 | 6.824 |

**Solution.**

| $x$ | $\sin x$ | $\sin 2x$ | $\sin 3x$ | $\cos x$ | $\cos 2x$ | $\cos 3x$ | $f(x)$ | $f(x) \times \sin x$ | $f(x) \times \sin 2x$ | $f(x) \times \sin 3x$ | $f(x) \times \cos x$ | $f(x) \times \cos 2x$ | $f(x) \times \cos 3x$ |
|---|---|---|---|---|---|---|---|---|---|---|---|---|---|
| 30° | 0.50 | 0.87 | 1 | 0.87 | 0.50 | 0 | 7.976 | 3.988 | 6.939 | 7.976 | 6.939 | 3.988 | 0 |
| 60° | 0.87 | 0.87 | 0 | 0.50 | − 0.50 | − 1 | 8.026 | 6.983 | 6.983 | 0 | 4.013 | 4.013 | − 8.026 |
| 90° | 1.00 | 0 | − 1 | 0 | − 1 | 0 | 7.204 | 7.204 | 0 | − 7.204 | 0 | − 7.204 | 0 |
| 120° | 0.87 | − 0.87 | 0 | − 0.50 | − 0.50 | 1 | 5.676 | 4.938 | − 4.939 | 0 | − 2.838 | − 2.838 | 5.676 |
| 150° | 0.50 | − 0.87 | 1 | − 0.87 | 0.50 | 0 | 3.674 | 1.837 | − 3.196 | − 3.196 | − 3.196 | 1.837 | 0 |
| 180° | 0 | 0 | 0 | − 1 | 1 | − 1 | 1.764 | 0 | 0 | − 1.764 | − 1.764 | 1.764 | − 1.764 |
| 210° | − 0.50 | 0.87 | − 1 | − 0.87 | 0.50 | 0 | 0.552 | − 0.276 | 0.480 | 0.480 | −0.480 | 0.276 | 0 |
| 240° | − 0.87 | 0.87 | 0 | − 0.50 | − 0.50 | 1 | 0.262 | − 0.228 | 0.228 | − 0.131 | − 0.131 | 0.131 | 0.262 |
| 270° | − 1.00 | 0 | 1 | 0 | − 1.00 | 0 | 0.904 | − 0.904 | 0 | 0 | 0 | − 0.904 | 0 |
| 300° | − 0.87 | − 0.87 | 0 | 0.50 | − 0.50 | − 1 | 2.492 | − 2.168 | − 2.168 | 1.246 | 1.246 | −1.296 | − 2.492 |
| 330° | − 0.50 | − 0.87 | − 1 | 0.87 | 0.50 | 0 | 4.736 | −2.368 | − 4.120 | 4.120 | 4.120 | 2.368 | 0 |
| 360° | 0 | 0 | 0 | 1 | 1 | 1 | 6.824 | 0 | 0 | 0 | 6.824 | 6.824 | 6.824 |
| | | | | | | Σ | 50.09 | 19.206 | 0.207 | 0.062 | 14.733 | 0.721 | 0.460 |

$$a_0 = 2 \times \text{Mean value of } f(x) = 2 \times \frac{50.09}{12} = 8.34$$

$$a_1 = 2 \times \text{Mean value of } f(x) \cos x = 2 \times \frac{14.733}{12} = 2.45$$

$$a_2 = 2 \times \text{Mean value of } f(x) \cos 2x = 2 \times \frac{0.721}{12} = 0.12$$

$$a_3 = 2 \times \text{Mean value of } f(x) \cos 3x = 2 \times \frac{0.460}{12} = 0.08$$

$$b_1 = 2 \times \text{Mean value of } f(x) \sin x = 2 \times \frac{19.206}{12} = 3.16$$

$$b_2 = 2 \times \text{Mean value of } f(x) \sin 2x = 2 \times \frac{0.207}{12} = 0.03$$

$$b_3 = 2 \times \text{Mean value of } f(x) \sin 3x = 2 \times \frac{0.062}{12} = 0.01$$

Fourier series is

$$f(x) = \frac{a_0}{2} + a_1 \cos x + a_2 \cos 2x + a_3 \cos 3x + \ldots + b_1 \sin x + b_2 \sin 2x + b_3 \sin 3x + \ldots$$

$$= 4.17 + 2.45\cos x + 0.12 \cos 2x + 0.08 \cos 3x + \ldots$$

$$+ 3.16 \sin x + 0.03 \sin 2x + 0.01 \sin 3x + \ldots$$ **Ans.**

**Example 32.** *Obtain the constant terms and the coefficients of the first sine and cosine terms in the Fourier series of $f(x)$ as given in the following table.*

| $x$ | 0 | 1 | 2 | 3 | 4 | 5 |
|---|---|---|---|---|---|---|
| $f(x)$ | 9 | 18 | 24 | 28 | 26 | 20 |

**Solution.**

| $x$ | $\frac{x\pi}{3}$ | $\sin\frac{\pi x}{3}$ | $\cos\frac{\pi x}{3}$ | $f(x)$ | $f(x)\sin\frac{\pi x}{3}$ | $f(x)\cos\frac{\pi x}{3}$ |
|---|---|---|---|---|---|---|
| 0 | 0 | 0 | 1.0 | 9 | 0 | 9 |
| 1 | $\frac{\pi}{3}$ | 0.87 | 0.5 | 18 | 15.66 | 9 |
| 2 | $\frac{2\pi}{3}$ | 0.87 | – 0.5 | 24 | 20.88 | – 12 |
| 3 | $\frac{3\pi}{3}$ | 0 | – 1.0 | 28 | 0 | – 28 |
| 4 | $\frac{4\pi}{3}$ | – 0.87 | – 0.5 | 26 | – 22.62 | – 13 |
| 5 | $\frac{5\pi}{3}$ | – 0.87 | 0.5 | 20 | – 17.4 | 10 |
| | | | | $\Sigma = 125$ | $\Sigma = -3.468$ | $\Sigma = 25$ |

$$a_0 = 2 \text{ Mean value of } f(x) = 2 \times \frac{125}{6} = 41.67$$

$$a_1 = 2 \text{ Mean value of } f(x) \cos \frac{\pi x}{3} = 2 \times \frac{-25}{6} = -8.33$$

$$b_1 = 2 \text{ Mean value of } f(x) \sin \frac{\pi x}{3} = 2 \times \frac{-3.48}{6} = -1.16$$

Fourier series is $$f(x) = \frac{a_0}{2} + a_1 \cos \frac{\pi x}{3} + \ldots + b_1 \sin \frac{\pi x}{3} + \ldots$$

$$= 20.84 - 8.33 \cos \frac{\pi x}{3} + \ldots - 1.16 \sin \frac{\pi x}{3} + \ldots$$ **Ans.**

## Exercise 12.7

**1.** In a machine the displacement $f(x)$ of a given point is given for a certain angle $x°$ as follows:

| $x°$ | 0° | 30° | 60° | 90° | 120° | 150° | 180° | 210° | 240° | 270° | 300° | 330° |
|---|---|---|---|---|---|---|---|---|---|---|---|---|
| $f(x)$ | 7.9 | 8.0 | 7.2 | 5.6 | 3.6 | 1.7 | 0.5 | 0.2 | 0.9 | 2.5 | 4.7 | 6.8 |

Find the coefficient of sin $2x$ in the Fourier series representing the above variations.

**Ans.** –0.072

**2.** The displacement $f(x)$ of a part of a machine is tabulated with corresponding angular moment '$x$' of the crank. Express $f(x)$ as a Fourier series upto third harmonic.

| $x°$ | 0° | 30° | 60° | 90° | 120° | 150° | 180° | 210° | 240° | 270° | 300° | 330° |
|---|---|---|---|---|---|---|---|---|---|---|---|---|
| $f(x)$ | 1.80 | 1.10 | 0.30 | 0.16 | 0.50 | 1.30 | 2.16 | 1.25 | 1.30 | 1.52 | 1.76 | 2.00 |

**Ans.** $f(x) = 1.26 + 0.04 \cos x + 0.53 \cos 2x - 0.01 \cos 3x + ....$
$-0.63 \sin x - 0.23 \sin 2x + 0.085 \sin 3x + ...$

**3.** The following values of $y$ give the displacement in cms of a certain machine part of the rotation $x$ of the flywheel. Expand $f(x)$ in the form of a Fourier series.

| $x$ | 0 | $\frac{\pi}{6}$ | $\frac{2\pi}{6}$ | $\frac{3\pi}{6}$ | $\frac{4\pi}{6}$ | $\frac{5\pi}{6}$ |
|---|---|---|---|---|---|---|
| $f(x)$ | 0 | 9.2 | 14.4 | 17.8 | 17.3 | 11.7 |

**Ans.** $f(x) = 11.733 - 7.733 \cos 2x - 2.833 \cos 4x + .....$
$-1.566 \sin 2x - 0.116 \sin 4x + ...$

**4.** Analyse harmonically the data given below and express $y$ in Fourier series upto the second harmonic.

| $x$ | 0 | $\frac{\pi}{3}$ | $\frac{2\pi}{3}$ | $\pi$ | $\frac{4\pi}{3}$ | $\frac{5\pi}{3}$ | $2\pi$ |
|---|---|---|---|---|---|---|---|
| $y$ | 1.0 | 1.4 | 1.9 | 1.7 | 1.5 | 1.2 | 1.0 |

# 13

# Laplace Transformation

## 13.1 INTRODUCTION

Laplace transforms help in solving the differential equations with boundary values without finding the general solution and the values of the arbitrary constants.

## 13.2 LAPLACE TRANSFORM

**Definition.** Let $f(t)$ be function defined for all positive values of $t$, then

$$F(s) = \int_0^\infty e^{-st} f(t)\,dt$$

provided the integral exists, is called the **Laplace Transform** of $f(t)$. It is denoted as

$$L[f(t)] = F(s) = \int_0^\infty e^{-st} f(t)\,dt$$

## 13.3 IMPORTANT FORMULAE

1. $L(1) = \dfrac{1}{s}$

2. $L(t^n) = \dfrac{n!}{s^{n+1}}$, when $n = 0, 1, 2, 3 \ldots$

3. $L(e^{at}) = \dfrac{1}{s-a}$ $\quad (s > a)$

4. $L(\cosh at) = \dfrac{s}{s^2 - a^2}$ $\quad (s^2 > a^2)$

5. $L(\sinh at) = \dfrac{a}{s^2 - a^2}$ $\quad (s^2 > a^2)$

6. $L(\sin at) = \dfrac{a}{s^2 + a^2}$ $\quad (s > 0)$

7. $L(\cos at) = \dfrac{s}{s^2 + a^2}$ $\quad (s > 0)$

**1.** $\boxed{L(1) = \dfrac{1}{s}}$

**Proof.** $L(1) = \int_0^\infty 1 \cdot e^{-st}\,dt = \left[\dfrac{e^{-st}}{-s}\right]_0^\infty = -\dfrac{1}{s}\left[\dfrac{1}{e^{st}}\right]_0^\infty = -\dfrac{1}{s}[0-1] = \dfrac{1}{s}$

Hence $\quad L(1) = \dfrac{1}{s}$ **Proved.**

**2.** $\boxed{L(t^n) = \dfrac{n!}{s^{n+1}}}$ where $n$ and $s$ are positive.

**Proof.** $\quad L(t^n) = \int_0^\infty e^{-st} t^n\,dt$

Putting $\quad st = x \Rightarrow t = \dfrac{x}{s} \Rightarrow dt = \dfrac{dx}{s}$

Thus, we have $L(t^n) = \int_0^\infty e^{-x} \left(\frac{x}{s}\right)^n \frac{dx}{s} \Rightarrow L(t^n) = \frac{1}{s^{n+1}} \int_0^\infty e^{-x} . x^n \, dx$

$$\Rightarrow L(t^n) = \frac{\sqrt{n+1}}{s^{n+1}} \Rightarrow L\left(t^n\right) = \frac{n!}{s^{n+1}} \quad \left[\begin{array}{l} \sqrt{n+1} = \int_0^\infty e^{-x} . x^n \, dx \\ \text{and} \quad \sqrt{n+1} = n! \end{array}\right]$$ **Proved.**

3. $\boxed{L\left(e^{at}\right) = \frac{1}{s-a}}$, where $s > a$

**Proof.** $L(e^{at}) = \int_0^\infty e^{-st} . e^{at} dt = \int_0^\infty e^{-st+at} . dt$

$$= \int_0^\infty e^{(-s+a)t} . dt = \int_0^\infty e^{-(s-a)t} . dt = \left[\frac{e^{-(s-a)t}}{-(s-a)}\right]_0^\infty = -\frac{1}{s-a}\left[\frac{1}{e^{(s-a)t}}\right]_0^\infty$$

$$= \frac{-1}{(s-a)}(0-1) = \frac{1}{s-a}$$ **Proved.**

4. $\boxed{L(\cosh at) = \frac{s}{s^2 - a^2}}$

**Proof.** $L(\cosh at) = L\left[\frac{e^{at} + e^{-at}}{2}\right]$ $\left(\therefore \cosh at = \frac{e^{at} + e^{-at}}{2}\right)$

$$= \frac{1}{2} L\left(e^{at}\right) + \frac{1}{2} L\left(e^{-at}\right) = \frac{1}{2}\left[\frac{1}{s-a} + \frac{1}{s+a}\right] \quad \left[L\left(e^{at}\right) = \frac{1}{s-a}\right]$$

$$= \frac{1}{2}\left[\frac{s+a+s-a}{s^2-a^2}\right] = \frac{s}{s^2-a^2}$$ **Proved.**

5. $\boxed{L(\sinh at) = \frac{a}{s^2 - a^2}}$

**Proof.** $L(\sinh at) = L\left[\frac{1}{2}\left(e^{at} - e^{-at}\right)\right]$

$$= \frac{1}{2}[L(e^{at}) - L(e^{-at})] = \frac{1}{2}\left[\frac{1}{s-a} - \frac{1}{s+a}\right] = \frac{1}{2}\left[\frac{s+a-s+a}{s^2-a^2}\right]$$

$$= \frac{a}{s^2-a^2}$$ **Proved.**

6. $\boxed{L(\sin at) = \frac{a}{s^2 + a^2}}$

**Proof.** $L(\sin at) = L\left[\frac{e^{iat} - e^{-iat}}{2i}\right]$ $\left[\because \sin at = \frac{e^{iat} - e^{-iat}}{2i}\right]$

$$= \frac{1}{2i}\left[L\left(e^{iat} - e^{-iat}\right)\right] = \frac{1}{2i}\left[L\left(e^{iat}\right) - L\left(e^{-iat}\right)\right]$$

$$= \frac{1}{2i}\left[\frac{1}{s-ia} - \frac{1}{s+ia}\right] = \frac{1}{2i} \frac{s+ia-s+ia}{s^2+a^2} = \frac{1}{2i} \frac{2ia}{s^2+a^2} = \frac{a}{s^2+a^2}$$ **Proved.**

7. $\boxed{L(\cos at) = \frac{s}{s^2 + a^2}}$

**Proof.** $L(\cos at) = L\left(\frac{e^{iat} + e^{-iat}}{2}\right)$ $\left[\because \cos at = \frac{e^{iat} + e^{-iat}}{2}\right]$

$$= \frac{1}{2}[L(e^{iat} + e^{-iat})] = \frac{1}{2}[L(e^{iat}) + L(e^{-iat})] = \frac{1}{2}\left[\frac{1}{s-ia} + \frac{1}{s+ia}\right] = \frac{1}{2}\frac{s+ia+s-ia}{s^2+a^2}$$

$$= \frac{s}{s^2+a^2}$$ **Proved.**

**Example 1.** *Find the Laplace transform of f(t) defined as*

$$f(t) = \begin{cases} \frac{t}{k}, & when \ 0 < t < k \\ 1, & when \ t > k \end{cases}$$

**Solution.** $L[f(t)] = \int_0^k \frac{t}{k} e^{-st} dt + \int_k^\infty 1.e^{-st} dt = \frac{1}{k}\left[\left(t\frac{e^{-st}}{-s}\right)_0^k - \int_0^k \frac{e^{-st}}{-s} dt\right] + \left[\frac{e^{-st}}{-s}\right]_k^\infty$

$$= \frac{1}{k}\left[\frac{ke^{-ks}}{-s} - \left(\frac{e^{-st}}{s^2}\right)_0^k\right] + \frac{e^{-ks}}{s} = \frac{1}{k}\left[\frac{ke^{-ks}}{-s} - \frac{e^{-sk}}{s^2} + \frac{1}{s^2}\right] + \frac{e^{-ks}}{s}$$

$$= -\frac{e^{-sk}}{s} - \frac{1}{k}\frac{e^{-ks}}{s^2} + \frac{1}{k}\frac{1}{s^2} + \frac{e^{-ks}}{s} = \frac{1}{ks^2}[-e^{-ks} + 1]$$ **Ans.**

**Example 2.** *From the first principle, find the Laplace transform of (1 + cos 2 t).*

**Solution.** Laplace transform of (1 + cos 2 *t*)

$$= \int_0^\infty e^{-st}(1+\cos 2t)dt = \int_0^\infty e^{-st}\left(1 + \frac{e^{2it} + e^{-2it}}{2}\right)dt$$

$$= \frac{1}{2}\int_0^\infty \left[2e^{-st} + e^{(-s+2i)t} + e^{(-s-2i)t}\right]dt = \frac{1}{2}\left[\frac{2e^{-st}}{-s} + \frac{e^{(-s+2i)t}}{-s+2i} + \frac{e^{(-s-2i)t}}{-s-2i}\right]_0^\infty$$

$$= \frac{1}{2}\left[\left(0 + \frac{2}{s}\right) + \frac{1}{-s+2i}(0-1) + \frac{1}{-s-2i}(0-1)\right]$$

$$= \frac{1}{2}\left[\frac{2}{s} + \frac{1}{s-2i} + \frac{1}{s+2i}\right] = \frac{1}{2}\left[\frac{2}{s} + \frac{2s}{s^2+4}\right]$$

$$= \frac{1}{s} + \frac{s}{s^2+4} = \frac{2s^2+4}{s(s^2+4)}$$ **Ans.**

## 13.4 PROPERTIES OF LAPLACE TRANSFORMS

**(1)** $L[af_1(t) + bf_2(t)] = a\,L[f_1(t)] + b\,L[f_2(t)]$

**Proof.** $L[af_1(t) + bf_2(t)] = \int_0^\infty e^{-st}[af_1(t) + bf_2(t)]\,dt$

$$= a\int_0^\infty e^{-st} f_1(t)dt + b\int_0^\infty e^{-st} f_2(t)dt$$

$$= a\,L[f_1(t)] + b\,L[f_2(t)]$$ **Proved.**

**(2) First Shifting Theorem.** If $L f(t) = F(s)$, then

$$L[e^{at} f(t)] = F(s-a)$$

**Proof.** $L[e^{at} f(t)] = \int_0^\infty e^{-st} \cdot e^{at} f(t)\,dt = \int_0^\infty e^{-(s-a)t} f(t)dt$

$$= \int_0^\infty e^{-rt} f(t)dt \qquad \text{where } r = s - a$$

$$= F(r) = F(s-a)$$ **Proved.**

With the help of this property, we can have the following important results :

(1) $L\left(e^{at}t^n\right)=\dfrac{n!}{(s-a)^{n+1}}$ $\left[L\left(t^n\right)=\dfrac{n!}{s^{n+1}}\right]$

(2) $L\left(e^{at}\cosh bt\right)=\dfrac{s-a}{(s-a)^2-b^2}$ (3) $L\left(e^{at}\sinh bt\right)=\dfrac{b}{(s-a)^2-b^2}$

(4) $L\left(e^{at}\sin bt\right)=\dfrac{b}{(s-a)^2+b^2}$ (5) $L\left(e^{at}\cos bt\right)=\dfrac{s-a}{(s-a)^2+b^2}$

**Example 3.** *Find the Laplace transform of cas² t.*

**Solution.** $\cos 2t = 2\cos^2 t - 1$

$\therefore$ $\cos^2 t = \dfrac{1}{2}[\cos 2t + 1]$

$$L(\cos^2 t) = L\left[\frac{1}{2}(\cos 2t+1)\right]=\frac{1}{2}\left[L(\cos 2t)+L(1)\right]$$

$$=\frac{1}{2}\left[\frac{s}{s^2+(2)^2}+\frac{1}{s}\right]=\frac{1}{2}\left[\frac{s}{s^2+4}+\frac{1}{s}\right]$$ **Ans.**

**Example 4.** *Find the Laplace Transform of* $t^{-\frac{1}{2}}$.

**Solution.** We know that $L\left(t^n\right)=\dfrac{\overline{|n+1}}{s^{n+1}}$

Put $n=-\dfrac{1}{2}$, $L\left(t^{-1/2}\right)=\dfrac{\overline{\left|-\frac{1}{2}+1\right.}}{s^{-1/2+1}}=\dfrac{\overline{\left|\frac{1}{2}\right.}}{\sqrt{s}}=\dfrac{\sqrt{\pi}}{\sqrt{s}}$, where $\overline{\left|\frac{1}{2}\right.}=\sqrt{\pi}$ **Ans.**

**Example 5.** *Find the Laplace Transform of t sin at.*

**Solution.** $$L(t\sin at)=L\left(t\,\frac{e^{iat}-e^{-iat}}{2i}\right)=\frac{1}{2i}\left[L\left(t.e^{iat}\right)-L\left(t.e^{-iat}\right)\right]$$

$$=\frac{1}{2i}\left[\frac{1}{(s-ia)^2}-\frac{1}{(s+ia)^2}\right]=\frac{1}{2i}\left[\frac{(s+ia)^2-(s-ia)^2}{(s-ia)^2(s+ia)^2}\right]$$

$$=\frac{1}{2i}\frac{\left(s^2+2ias-a^2\right)-\left(s^2-2ias-a^2\right)}{\left(s^2+a^2\right)^2}$$

$$=\frac{1}{2i}\frac{4ias}{\left(s^2+a^2\right)^2}=\frac{2as}{\left(s^2+a^2\right)^2}$$ **Ans.**

**Example 6.** *Find the Laplace Transform of t² cos at.*

**Solution.** $$L\left(t^2\cos at\right)=L\left(t^2\,\frac{e^{iat}+e^{-iat}}{2}\right)=\frac{1}{2}\left[L\left(t^2.e^{iat}\right)+L\left(t^2e^{-iat}\right)\right]$$

$$=\frac{1}{2}\left[\frac{2!}{(s-ia)^3}+\frac{2!}{(s+ia)^3}\right]=\frac{(s+ia)^3+(s-ia)^3}{(s-ia)^3(s+ia)^3}$$

$$= \frac{\left(s^3+3ias^2-3a^2s-ia^3\right)+\left(s^3-3ias^2-3a^2s+ia^3\right)}{\left(s^2+a^2\right)^3}$$

$$= \frac{2s^3-6a^2s}{\left(s^2+a^2\right)^3} = \frac{2s\left(s^2-3a^2\right)}{\left(s^2+a^2\right)^3}$$ **Ans.**

## Exercise 13.1

Find the Laplace transforms of the following:

**1.** $t + t^2 + t^3$ **Ans.** $\frac{1}{s^2}+\frac{2}{s^3}+\frac{6}{s^4}$  **2.** $\sin t \cos t$ **Ans.** $\frac{1}{s^2+4}$

**3.** $t^{7/2} e^{5t}$ (*M.D.U. Dec. 2009*) **Ans.** $\frac{105\sqrt{\pi}}{16\,(s-5)^{9/2}}$

**4.** $\sin^3 2t$ **Ans.** $\frac{48}{\left(s^2+4\right)\left(s^2+36\right)}$

**5.** $e^{-t}\cos^2 t$ **Ans.** $\frac{1}{2s+2}+\frac{s+1}{2s^2+4s+10}$  **6.** $\sin 2t \cos 3t$ **Ans.** $\frac{2(s^2-5)}{\left(s^2+1\right)\left(s^2+25\right)}$

**7.** $\sin 2t \sin 3t$ **Ans.** $\frac{12s}{\left(s^2+1\right)(s^2+25)}$

**8.** $\cos at \sinh at$ **Ans.** $\frac{1}{2}\left[\frac{s-a}{(s-a)^2+a^2}-\frac{s+a}{(s+a)^2+a^2}\right]$

**9.** $\sinh^3 t$ **Ans.** $\frac{6}{\left(s^2-1\right)\left(s^2-9\right)}$  **10.** $\cos t \cos 2t$ **Ans.** $\frac{s\left(s^2+5\right)}{(s^2+1)(s^2+9)}$

**11.** $\cosh at \sin at$ **Ans.** $\frac{a\left(s^2+2a^2\right)}{s^4+4a^4}$

**12.** $f(t)=\begin{cases}\cos\left(t-\frac{2\pi}{3}\right), & t>\frac{2\pi}{3}\\ 0, & t<\frac{2\pi}{3}\end{cases}$ **Ans.** $e^{\frac{-2\pi s}{3}}\frac{s}{s^2+1}$

## 13.5 LAPLACE TRANSFORM OF THE DERIVATIVE OF $f(t)$

$$L\,[f'(t)] = s\,L\,[f(t)] - f(0) \qquad \text{where } L\,[f(t)] = F(s).$$

**Proof.** $L[f'(t)] = \int_0^\infty e^{-st} f'(t)\,dt$

**Integrating by parts, we get**

$$L[f'(t)] = \left[e^{-st}\cdot f(t)\right]_0^\infty - \int_0^\infty \left(-se^{-st}\right) f(t)\,dt$$

$$= -f(0) + s\int_0^\infty e^{-st} f(t)\,dt \qquad \left(e^{-st} f(t) = 0, \text{ when } t=\infty\right)$$

$$= -f(0) + sLf(t)$$

$\Rightarrow$ $L\,[f'(t)] = s\,L[f(t)] - f(0)$ **Proved.**

**Note.** Roughly, Laplace transform of derivative of $f(t)$ corresponds to multiplication of the Laplace transform of $f(t)$ by $s$.

## 13.6 LAPLACE TRANSFORM OF DERIVATIVE OF ORDER $n$.

$$L[f^n(t)] = s^n L[f(t)] - s^{n-1}f(0) - s^{n-2}f'(0) - s^{n-3}f''(0) - \ldots - f^{n-1}(0)$$

**Proof.** We have already proved in Article 13.5 that

$$L[f'(t)] = s\,L[f(t)] - f(0) \qquad ...(1)$$

Replacing $f(t)$ by $f'(t)$ and $f'(t)$ by $f''(t)$ in (1), we get

$$L[f''(t)] = s\,L[f'(t)] - f'(0) \qquad ...(2)$$

Putting the value of $L[f'(t)]$ from (1) in (2), we have

$$L[f''(t)] = s\,[s\,L[f(t)] - f(0)] - f'(0)$$

$$\Rightarrow \quad L[f''(t)] = s^2 L[f(t)] - sf(0) - f'(0)$$

Similarly, $L[f'''(t)] = s^3 L[f(t)] - s^2 f(0) - s\,f'(0) - f''(0)$

$$L[f^{iv}(t)] = s^4 L[f(t)] - s^3 f(0) - s^2 f'(0) - s\,f''(0) - f'''(0)$$

$$\boxed{L[f^n(t)] = s^n L[f(t)] - s^{n-1} f(0) - s^{n-2} f'(0) - s^{n-3} f''(0) + \ldots\ldots - f^{n-1}(0)}$$

## 13.7 LAPLACE TRANSFORM OF INTEGRAL OF $f(t)$

$$L\left[\int_0^t f(t)\,dt\right] = \frac{1}{s}F(s), \qquad \text{where } L\,[f(t)] = F(s)$$

**Proof.** Let $\phi(t) = \int_0^t f(t)\,dt$ and $\phi(0) = 0$ then $\phi'(t) = f(t)$

We know the formula of Laplace transforms of $\phi'(t)$ *i.e.*

$$L[\phi'(t)] = s\,L\,[\phi(t)] - \phi(0)$$

$$\Rightarrow \quad L\,[\phi'(t)] = s\,L\,[\phi(t)] \qquad [\,\phi(0) = 0]$$

$$\Rightarrow \quad L\,[\phi(t)] = \frac{1}{s}L\,[\phi'(t)]$$

Putting the values of $\phi(t)$ and $\phi'(t)$, we get

$$L\left[\int_0^t f(t)\,dt\right] = \frac{1}{s}L[f(t)] \quad \text{or} \quad L\left[\int_0^t f(t)\,dt\right] = \frac{1}{s}F(s)$$

**Proved.**

**Note:** (1) Laplace Transform of Integral of $f(t)$ corresponds to the division of the Laplace transform of $f(t)$ by $s$.

(2) $\int_0^t f(t)\,dt = L^{-1}\left[\frac{1}{s}F(s)\right]$

## 13.8 LAPLACE TRANSFORM OF $t.\,f(t)$ (Multipication by $t$)

If $L[f(t)] = F(s)$, then

$$L\,[t^n f(t)] = (-1)^n \frac{d^n}{ds^n}[F(s)].$$

**Proof.** $L\,[f(t)] = F(s) = \int_0^\infty e^{-st} f(t)\,dt \qquad ...(1)$

Differentiating (1) w.r.t. "$s$" we get

$$\therefore \quad \frac{d}{ds}[F(s)] = \frac{d}{ds}\left[\int_0^\infty e^{-st} f(t)\,dt\right] = \int_0^\infty \frac{\partial}{\partial s}\left(e^{-st}\right) f(t)\,dt$$

$$= \int_0^\infty \left(-te^{-st}\right) f(t)\,dt = \int_0^\infty e^{-st}\left[-t \cdot f(t)\right] dt$$

$$= L\left[-t\,f(t)\right] \quad \text{or} \quad L\left[tf(t)\right] = (-1)^1 \frac{d}{ds}\left[F(s)\right]$$

Similarly $$L[t^2 f(t)]=(-1)^2 \frac{d^2}{ds^2}[F(s)]$$

$$L[t^3 f(t)]=(-1)^3 \frac{d^3}{ds^3}[F(s)]$$

... ... .... ... ...

$$L[t^n f(t)]=(-1)^n \frac{d^n}{ds^n}[F(s)]$$ **Proved.**

**Example 7.** *Find the Laplace transform of t sinh at.*

**Solution.** $L(\sinh at) = \frac{a}{s^2-a^2}$

$\therefore$ $L[t \sinh at] = -\frac{d}{ds}\left(-\frac{a}{s^2-a^2}\right)$

$\Rightarrow$ $L[t \sinh at] = \frac{2as}{(s^2-a^2)}$ **Ans.**

**Example 8.** *Find the Laplace transform of $t^2$ cos at*

**Solution.** $L(\cos at) = \frac{a}{s^2+a^2}$

$$L(t^2 \cos at) = (-1)^2 \frac{d^2}{ds^2}\left[\frac{s}{s^2+a^2}\right] = \frac{d}{ds}\frac{(s^2+a^2).1-s(2s)}{(s^2+a^2)^2} = \frac{d}{ds}\frac{a^2-s^2}{(s^2+a^2)^2}$$

$$= \frac{(s^2+a^2)^2(-2s)-(a^2-s^2).2(s^2+a^2)(2s)}{(s^2+a^2)^4} = \frac{-2s^3-2a^2s-4a^2s+4s^3}{(s^2+a^2)^3}$$

$$= \frac{2s(s^2-3a^2)}{(s^2+a^2)^3}$$ **Ans.**

**Example 9.** *Obtain the Laplace transform of*

*$t^2 e^t$. sin 4t*

**Solution.** $L(\sin 4t) = \frac{4}{s^2+16}$, $L(e^t \sin 4t) = \frac{4}{(s-1)^2+16}$

$$L(t e^t \sin 4t) = -\frac{d}{ds}\frac{4}{s^2-2s+17} = \frac{4(2s-2)}{(s^2-2s+17)^2}$$

$$L(t^2 e^t \sin 4t) = -4\frac{d}{ds}\frac{2s-2}{(s^2-2s+17)^2}$$

$$= -4\frac{(s^2-2s+17)^2 2-(2s-2)2(s^2-2s+17)(2s-2)}{(s^2-2s+17)^4}$$

$$= \frac{-4(2s^2-4s+34-8s^2+16s-8)}{(s^2-2s+17)^3}$$

$$= \frac{-4(-6s^2+12s+26)}{(s^2-2s+17)^3} = \frac{8(3s^2-6s-13)}{(s^2-2s+17)^3}$$ **Ans.**

## Exercise 13.2

Find the Laplace transforms of the following :

1. $t \sin 2t$ (*Madras 2006*) **Ans.** $\dfrac{4s}{(s^2+4)^2}$ 2. $t \sin at$ **Ans.** $\dfrac{2as}{(s^2+a^2)^2}$

3. $t \cosh at$ **Ans.** $\dfrac{s^2+a^2}{(s^2-a^2)^2}$ 4. $t \cos t$ **Ans.** $\dfrac{s^2-1}{(s^2+1)^2}$

5. $t \cosh t$ **Ans.** $\dfrac{s^2+1}{(s^2-1)^2}$ 6. $t^2 \sin t$ **Ans.** $\dfrac{2(3s^2-1)}{(s^2+1)^3}$

7. $t^3 e^{-3t}$ **Ans.** $\dfrac{6}{(s+3)^4}$ 8. $t \sin^2 3t$ **Ans.** $\dfrac{1}{2}\left[\dfrac{1}{s^2}-\dfrac{s^2-36}{(s^2+36)^2}\right]$

9. $t e^{at} \sin at$ **Ans.** $\dfrac{2a(s-a)}{(s^2-2as+2a^2)^2}$

10. $\int_0^t e^{-2t} t \sin^3 t\,dt$ **Ans.** $\dfrac{3(s+2)}{2s}\left[\dfrac{1}{[(s+2)^2+9]^2}-\dfrac{1}{[(s+2)^2+1]^2}\right]$

11. $t e^{-t} \cosh t$ **Ans.** $\dfrac{s^2+2s+2}{(s^2+2s)^2}$

12. $t^2 e^{-2t} \cos t$ **Ans.** $\dfrac{2(s^3+6s^2+9s+2)}{(s^2+4s+5)^3}$

13. (*a*) Laplace transform of $t^n e^{-at}$ is

(*i*) $\dfrac{\lceil n}{(s+a)^n}$ (*ii*) $\dfrac{(n+1)!}{(s+a)^{n+1}}$ (*iii*) $\dfrac{n!}{(s+a)^n}$ (*iv*) $\dfrac{\lceil n+1}{(s+a)^{n+1}}$ **Ans.** (*iv*)

(*b*) Laplace transform of $f(t) = t\,e^{at} \cdot \sin(at)$, $t > 0$

(*i*) $\dfrac{2a(s-a)}{[(s-a)^2+a^2]^2}$ (*ii*) $\dfrac{a(s-a)}{(s-a)^2+a^2}$ (*iii*) $\dfrac{s-a}{(s-a)^2+a^2}$ (*iv*) $\dfrac{(s-a)^2}{(s-a)^2+a^2}$ **Ans.** (*i*)

(*c*) If $f(x) = x^4 P(x)$, where $P(x)$ has derivatives of all orders, then $L\left[\dfrac{d^4 f(x)}{dx^4}\right]$ is given by

(*i*) $s^3 L[f(x)]$ (*ii*) $s^4 L f(x)$

(*iii*) $s^4 L[f^3(x)]$ (*iv*) none of these. **Ans.** (*ii*)

(*d*) The Laplace transform of $te^{-t} \cosh 2t$ is

(*i*) $\dfrac{s^2+2s+5}{(s^2+2s-3)^2}$ (*ii*) $\dfrac{s^2-2s+5}{(s^2+2s-3)^2}$

(*iii*) $\dfrac{4s+4}{(s^2+2s-3)^2}$ (*iv*) $\dfrac{4s-4}{(s^2+2s-3)^2}$ **Ans.** (*i*)

## 13.9 LAPLACE TRANSFORM OF $\frac{1}{t}f(t)$ (Division by t)

If $\quad L[f(t)] = F(s)$, then $L\left[\frac{1}{t}f(t)\right] = \int_s^\infty F(s)\,ds$

**Proof.** $\quad L[f(t)] = F(s) \quad \Rightarrow \quad F(s) = \int_0^\infty e^{-t} f(t)\,dt \qquad ...(1)$

**Integrating (1) w.r.t. '$s$', we have**

$$\int_s^\infty F(s)\,ds = \int_s^\infty\left[\int_0^\infty e^{-st} f(t)\,dt\right]ds$$

$$= \int_0^\infty\left[\int_s^\infty e^{-st} f(t)\,ds\right]dt = \int_0^\infty\left[\frac{e^{-st}f(t)}{-t}\right]_s^\infty dt$$

$$= \int_0^\infty \frac{-f(t)}{t}\left[e^{-st}\right]_s^\infty dt = \int_0^\infty \frac{-f(t)}{t}\left[0 - e^{-st}\right]dt$$

$$= \int_0^\infty e^{-st}\left\{\frac{1}{t}f(t)\right\}dt = L\left[\frac{1}{t}f(t)\right]$$

$$\Rightarrow \qquad L\left[\frac{1}{t}f(t)\right] = \int_s^\infty F(s)\,ds.$$ **Proved.**

**Cor.** $\quad L^{-1}\int_s^\infty F(s)\,ds = \frac{1}{t}f(t)$

**Example 10.** *Find the Laplace transform of* $\frac{\sin 2t}{t}$.

**Solution.** $\quad L(\sin 2t) = \frac{2}{s^2+4}$

$$L\left(\frac{\sin 2t}{t}\right) = \int_s^\infty \frac{2}{s^2+4}\,ds = 2\frac{1}{2}\left[\tan^{-1}\frac{s}{2}\right]_s^\infty$$

$$= \left[\tan^{-1}\infty - \tan^{-1}\frac{s}{2}\right] = \frac{\pi}{2} - \tan^{-1}\frac{s}{2}$$

$$= \cot^{-1}\frac{s}{2}$$ **Ans.**

**Example 11.** *Find the Laplace transform of* $f(t) = \int_0^t \frac{\sin t}{t}\,dt$.

**Solution.** $\quad L \sin t = \frac{1}{s^2+1}$

$$L\frac{\sin t}{t} = \int_s^\infty \frac{1}{s^2+1}\,ds = \left[\tan^{-1}s\right]_s^\infty = \frac{\pi}{2} - \tan^{-1}s = \cot^{-1}s$$

$$L\int_0^t \frac{\sin t}{t}\,dt = \frac{1}{s}\cot^{-1}s$$ **Ans.**

**Example 12.** *Find the Laplace transform of* $\frac{1-\cos t}{t^2}$.

**Solution.** $L(1-\cos t) = L(1) - L(\cos t) = \frac{1}{s} - \frac{s}{s^2+1}$

$$L\frac{(1-\cos t)}{t} = \int_s^\infty\left(\frac{1}{s}-\frac{s}{s^2+1}\right)ds = \left[\log s - \frac{1}{2}\log\left(s^2+1\right)\right]_s^\infty$$

$$= \frac{1}{2}\left[\log s^2 - \log\left(s^2+1\right)\right]_s^\infty = \frac{1}{2}\left[\log\frac{s^2}{s^2+1}\right]_s^\infty$$

$$= \frac{1}{2}\left[\log\frac{s^2}{s^2\left(1+\frac{1}{s^2}\right)}\right]_s^\infty = \frac{1}{2}\left[0-\log\frac{s^2}{s^2+1}\right] = -\frac{1}{2}\log\frac{s^2}{s^2+1}$$

Again, $L\left[\frac{1-\cos t}{t^2}\right] = -\frac{1}{2}\int_s^\infty \log\frac{s^2}{s^2+1}ds = -\frac{1}{2}\int_s^\infty\left(\log\frac{s^2}{s^2+1}\cdot 1\right)ds$

Integrating by parts, we have

$$= -\frac{1}{2}\left[\log\frac{s^2}{s^2+1}\cdot s - \int\frac{s^2+1}{s^2}\frac{\left(s^2+1\right)2s - s^2(2s)}{\left(s^2+1\right)^2}\cdot s\,ds\right]_s^\infty$$

$$= -\frac{1}{2}\left[s\log\frac{s^2}{s^2+1} - 2\int\frac{1}{s^2+1}ds\right]_s^\infty = -\frac{1}{2}\left[s\log\frac{s^2}{s^2+1} - 2\tan^{-1}s\right]_s^\infty$$

$$= -\frac{1}{2}\left[0-2\left(\frac{\pi}{2}\right) - s\log\frac{s^2}{s^2+1} + 2\tan^{-1}s\right] = -\frac{1}{2}\left[-\pi - s\log\frac{s^2}{s^2+1} + 2\tan^{-1}s\right]$$

$$= \frac{\pi}{2} + \frac{s}{2}\log\frac{s^2}{s^2+1} - \tan^{-1}s$$

$$= \left(\frac{\pi}{2} - \tan^{-1}s\right) + \frac{s}{2}\log\frac{s^2}{s^2+1} = \cot^{-1}s + \frac{s}{2}\log\frac{s^2}{s^2+1}.$$ **Ans.**

**Example 13.** *Evaluate* $L\left[e^{-4t}\frac{\sin 3t}{t}\right]$.

**Solution.** $L\sin 3t = \frac{3}{s^2+3^2} \Rightarrow L\frac{\sin 3t}{t} = \int_s^\infty \frac{3}{s^2+9}ds = \left[\frac{3}{3}\tan^{-1}\frac{s}{3}\right]_s^\infty$

$$= \frac{\pi}{2} - \tan^{-1}\frac{s}{3} = \cot^{-1}\frac{s}{3}$$

$$L\left[e^{-4t}\frac{\sin 3t}{t}\right] = \cot^{-1}\frac{s+4}{3} = \tan^{-1}\frac{3}{s+4}$$ **Ans.**

## Exercise 13.3

Find Laplace transform of the following:

1. $\frac{1}{t}(1-e^t)$ **Ans.** $\log\frac{s-1}{s}$

2. $\frac{1}{t}\left(e^{-at}-e^{-bt}\right)$ **Ans.** $\log\frac{s+b}{s+a}$

3. $\frac{1}{t}(1-\cos at)$ **Ans.** $-\frac{1}{2}\log\frac{s^2}{s^2+a^2}$

4. $\frac{1}{t}\sin^2 t$ **Ans.** $\frac{1}{4}\log\frac{s^2+4}{s^2}$

5. $\frac{1}{t}\sinh t$ **Ans.** $-\frac{1}{2}\log\frac{s-1}{s+1}$

6. $\frac{1}{t}\left(e^{-t}\sin t\right)$ **Ans.** $\cot^{-1}(s+1)$ 7. $\frac{1}{t}(1-\cos t)$ **Ans.** $\frac{1}{2}[\log(s^2+1)-\log s^2]$

8. $\int_0^\infty \frac{1}{t}e^{-2t}\sin t\,dt$ **Ans.** $\frac{1}{s}\cot^{-1}(s+2)$ 9. $\int_0^\infty \frac{e^{-t}-e^{-3t}}{t}dt$ **Ans.** $\log 3$

10. $\frac{1}{t}$ $(\cos at-\cos bt)$ **Ans.** $-\frac{1}{2}\log\frac{s^2+a^2}{s^2+b^2}$

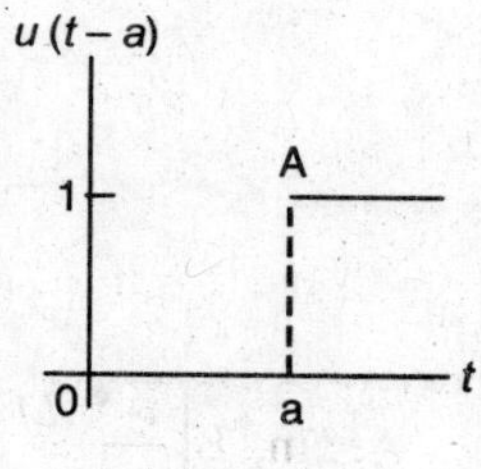


## 13.10 UNIT STEP FUNCTION

With the help of unit step functions, we can find the inverse transform of functions, which cannot be determined with previous methods.

The unit step functions $u(t-a)$ is defined as follows:

$$u(t-a)=\begin{cases}0 & \text{when } t<a\\ 1 & \text{when } t\ge a\end{cases}\qquad \text{where } a\ge 0.$$

**Example 14.** *Express the following function in terms of units step functions and find its Laplace transform:*

$$f(t)=\begin{cases}8, & t<2\\ 6, & t>2\end{cases}$$

**Solution.** $$f(t)=\begin{cases}8+0, & t<2\\ 8-2, & t>2\end{cases}$$

$$=8+\begin{cases}0, & t<2\\ -2, & t>2\end{cases}=8+(-2)\begin{cases}0, & t<2\\ 1, & t>2\end{cases}$$

$$=8-2u(t-2)$$

$$Lf(t)=8\,L(1)-2\,L\,u(t-2)=\frac{8}{s}-2\frac{e^{-2s}}{s}$$ **Ans.**

**Example 15.** *Draw the graph of* $u(t-a)-u(t-b)$

**Solution.** As in Art 13.10 the graph of $u(t-a)$ is a straight line from A to $\infty$. Similarly, the graph of $u(t-b)$ a straight line from B to $\infty$.

Hence, the graph of $u[t-a]-u[t-b]$ is AB.

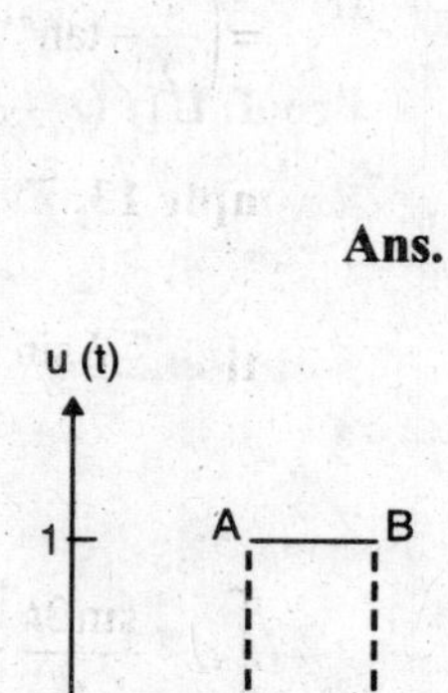


**Example 16.** *Express the following function in terms of unit step function and find its Laplace transform :*

$$f(t)=\begin{cases}E, & a<t<b\\ 0, & t>b\end{cases}$$

**Solution.** $$f(t)=E\begin{cases}1, & a<t<b\\ 0, & t>b\end{cases}.=E\,[u(t-a)-u(t-b)]$$

$$Lf(t)=E\left[\frac{e^{-as}}{s}-\frac{e^{-bs}}{s}\right]$$ **Ans.**

**Example 17.** *Express the following function in terms of unit step function :*

$$f(t) = \begin{cases} t-1, & 1<t<2 \\ 3-t, & 2<t<3 \end{cases}$$

*and find its Laplace transform.*

**Solution.** $$f(t) = \begin{cases} t-1, & 1<t<2 \\ 3-t, & 2<t<3 \end{cases}$$

$$= (t-1)[u(t-1)-u(t-2)] + (3-t)[u(t-2)-u(t-3)]$$

$$= (t-1)u(t-1)-(t-1)u(t-2) + (3-t)\,u(t-2) + (t-3)u(t-3)$$

$$= (t-1)u(t-1)-2(t-2)\,u(t-2) + (t-3)\,u(t-3)$$

$$Lf(t) = \frac{e^{-s}}{s^2} - 2\frac{e^{-2s}}{s^2} + \frac{e^{-3s}}{s^2}$$ **Ans.**

**Laplace Transform of unit function**

$$L[u(t-a)] = \frac{e^{-as}}{s}.$$

**Proof.**

$$L[u(t-a)] = \int_0^\infty e^{-st} u(t-a)dt$$

$$= \int_0^a e^{-st}\, 0.dt + \int_a^\infty e^{-st}.1\, dt = 0 + \left[\frac{e^{-st}}{-s}\right]_a^\infty$$

$\therefore$ $$L[u(t-a)] = \frac{e^{-as}}{s}$$ **Proved.**

## 13.11 SECOND SHIFTING THEOREM

If $L[f(t)] = F(s)$, then $\mathbf{L[f(t-a).u(t-a)] = e^{-as} F(s)}$.

**Proof.** $$L[f(t-a).u(t-a)] = \int_0^\infty e^{-st}\,[f(t-a).u(t-a)]\,dt$$

$$= \int_0^a e^{-st} f(t-a).0\,dt + \int_0^\infty e^{-st} f(t-a)(1)\,dt$$

$$= \int_0^\infty e^{-st} f(t-a)\,dt$$

$$= \int_0^\infty e^{-s(u+a)} f(u)\,du \quad \text{where } u = t-a$$

$$= e^{-sa}\int_0^\infty e^{-su}\cdot f(u)\,du = e^{-sa}F(s)$$ **Proved.**

## 13.12 THEOREM

$$L f(t)\, u(t-a) = e^{-as}\, L[f(t+a)]$$

**Proof.** $$Lf(t).u(t-a) = \int_0^\infty e^{-st}\,[f(t).u(t-a)]\,dt$$

$$= \int_0^a e^{-st}[f(t).u(t-a)]\,dt + \int_a^\infty e^{-st}.[f(t).u(t-a)]\,dt$$

$$= 0 + \int_a^\infty e^{-st}.f(t)(1)\,dt$$

$$= \int_0^\infty e^{-s(y+a)}\cdot f(y+a)\,dy = e^{-as}\int_0^\infty e^{-sy}\cdot f(y+a)\,dy \qquad (t-a=y)$$

$$= e^{-as}\int_0^\infty e^{-st}\cdot f(t+a)\,dt = e^{-as}Lf(t+a)$$ **Proved.**

**Example 18.** *Find the Laplace Transform of $t^2\,u(t-3)$.*

**Solution.** $t^2.u(t-3) = [(t-3)^2 + 6(t-3) + 9]u(t-3)$

$$= (t-3)^2.\,u(t-3) + 6(t-3).\,u(t-3) + 9u(t-3)$$

$$L\,t^2.u(t-3) = L\,(t-3)^2.\,u(t-3) + 6L(t-3).\,u(t-3) + 9L\,u(t-3)$$

$$= e^{-3s}\left[\frac{2}{s^3}+\frac{6}{s^2}+\frac{9}{s}\right]$$ **Ans.**

**Aliter** $L\,t^2\,u(t-3) = e^{-3s}L\,(t+3)^2 = e^{-3s}L\,[t^2+6t+9]$

$$= e^{3s}\left[\frac{2}{s^3}+\frac{6}{s^2}+\frac{9}{s}\right]$$ **Ans.**

**Example 19.** *Find the Laplace transform of $e^{-2t}u_\pi(t)$.*

where $u_\pi(t) = \begin{cases} 0: & t<\pi \\ 1: & t>\pi \end{cases}$

**Solution.** $u_\pi(t) = \begin{cases} 0: & t<\pi \\ 1: & t>\pi \end{cases}$

$$= u(t-\pi)$$

$$Le^{-2t}u_\pi(t) = Le^{-2t}u(t-\pi) \qquad f(t) = e^{-2t}$$

$$= e^{-\pi s}Lf(t+\pi) \qquad f(t+\pi) = e^{-2(t+\pi)}$$

$$= e^{-\pi s}Le^{-2(t+\pi)} = e^{-\pi s}e^{-2\pi}Le^{-2t}$$

$$= e^{-(\pi s+2\pi)}\frac{1}{s+2}$$

$$= \frac{e^{-\pi(s+2)}}{s+2}$$ **Ans.**

**Example 20.** *Represent $f(t) = \sin 2t$, $2\pi < t < 4\pi$ and $f(t) = 0$ otherwise, in terms of unit step function and then find its Laplace transform.*

**Solution.** $f(t) = \begin{cases} \sin 2t, & 2\pi<t<4\pi \\ 0, & \text{otherwise} \end{cases}$

$$f(t) = \sin 2t\,[u(t-2\pi) - u(t-4\pi)]$$

$$Lf(t) = L\,[\sin 2t\,.\,u(t-2\pi)] - L\,[\sin 2t\,.\,u(t-4\pi)]$$

$$= e^{-2\pi s}L\,[\sin 2(t+2\pi)] - e^{-4\pi s}L\,[\sin 2(t+4\pi)]$$

$$= e^{-2\pi s}L\,[\sin 2t] - e^{-4\pi s}L\,[\sin(2t)]$$

$$= e^{-2\pi s}\frac{2}{s^2+4} - e^{-4\pi s}\frac{2}{s^2+4}$$

$$= (e^{-2\pi s} - e^{-4\pi s})\frac{2}{s^2+4}$$ **Ans.**

## Exercise 13.4

**Find the Laplace transform of the following:**

1. $f(t)=\begin{cases} t-1, & 1<t<2 \\ 0, & \text{otherwise} \end{cases}$ **Ans.** $\dfrac{e^{-s}-e^{-2s}}{s^2}-\dfrac{e^{-2s}}{s}$

2. $e^t u(t-1)$ **Ans.** $\dfrac{e^{-(s-1)}}{s-1}$

3. $\dfrac{1-e^{2t}}{t}+tu(t)+\cosh t.\cos t$ **Ans.** $\log\dfrac{s-2}{s}+\dfrac{1}{s^2}+\dfrac{s^3}{s^4+4}$

4. $t^2u(t-2)$ **Ans.** $\dfrac{e^{-2s}}{s^3}\left(4s^2+4s+2\right)$

5. $\sin t\, u(t-4)$ **Ans.** $\dfrac{e^{-4s}}{s^2+1}[\cos 4+s\sin 4]$

6. $f(t)=K(t-2)[u(t-2)-u(t-3)]$ **Ans.** $\dfrac{K}{s^2}\left[e^{-2s}-(s+1)e^{-3s}\right]$

7. $f(t)=K\dfrac{\sin \pi t}{T}[u(t-2T)-u(t-3T)]$ **Ans.** $\dfrac{K\pi T}{s^2T^2+\pi^2}\left(e^{-2sT}-e^{-3sT}\right)$

**Express the following in terms of unit step functions and obtain Laplace transforms.**

8. $f(t)=\begin{cases} t, & 0<t<2 \\ 0, & 2<t \end{cases}$ **Ans.** $u(t)-u(t-2), \dfrac{1-(2s+1)e^{-2s}}{s^2}$

9. $f(t)=\begin{cases} \sin t, & 0<t<\pi \\ t, & t>\pi \end{cases}$ **Ans.** $\dfrac{1+e^{-\pi s}}{s^2+1}+\dfrac{e^{-\pi s}(\pi s+1)}{s^2}$

10. $f(t)=\begin{cases} 4, & 0<t<1 \\ -2, & 0<t<3 \\ 5, & t>3 \end{cases}$ **Ans.** $\dfrac{4-6e^{-s}+7e^{-3s}}{s}$

11. The Laplace transform of $t\,u_2(t)$ is

(*i*) $\left(\dfrac{1}{s^2}+\dfrac{2}{s}\right)e^{-2s}$ (*ii*) $\dfrac{1}{s^2}e^{-2s}$ (*iii*) $\left(\dfrac{1}{s^2}-\dfrac{2}{s}\right)e^{-2s}$ (*iv*) $\dfrac{e^{-2s}}{s^2}$ **Ans.** (*i*)

## 13.13 (1) IMPULSE FUNCTION

When a large force acts for a short time, then the product of the force and the time is called impulse in applied mechanics. The unit impulse function is the limiting function.

$$\delta(t-1)=\frac{1}{\varepsilon}, a<t<a+\varepsilon$$

$$=0, \quad \text{otherwise}$$

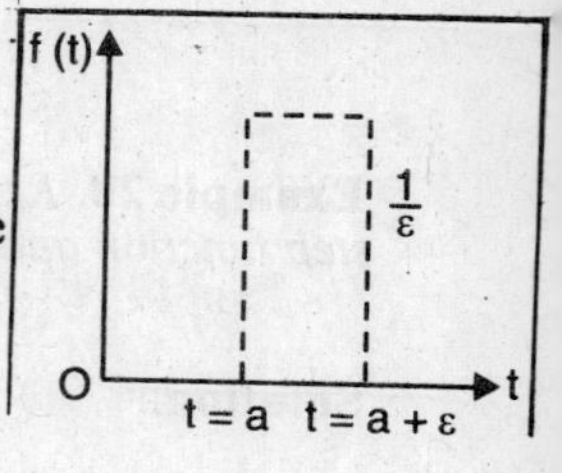


The value of the function (height of the strip in the figure) becomes infinite as a $\varepsilon \to 0$ and the area of the rectangle is unity.

**(2) The Unit Impulse function** is defined as follows:

$$\delta(t-a)=\begin{cases} \infty & \text{for } t=a \\ 0 & \text{for } t\neq a. \end{cases}$$

and $\int_0^\infty \delta(t-a).dt=1$ [Area of strip = 1]

**(3) Laplace Transform of unit Impulse function**

$$\int_0^\infty f(t)\,\delta(t-a)dt=\int_a^{a+\varepsilon} f(t).\frac{1}{\varepsilon}dt$$

$\left\{\begin{array}{l}\text{Mean value Throrem}\\ \int_a^b f(t)dt=(b-a)f(\eta)\end{array}\right.$

$$= (a+\varepsilon-a)\, f(\eta) \cdot \frac{1}{\varepsilon} \quad \text{where } a < \eta < a+\varepsilon$$

$$= f(\eta)$$

**Property I:** $\int_0^{\infty} f(t)\,\delta(t-a)\,dt = f(a)$ as $\varepsilon \to 0$

**Note.** If $f(t) = e^{-st}$ and $L[\delta(t-a)] = e^{-as}$

**Example 21.** *Evaluate* $\int_{-\infty}^{\infty} e^{-5t}\,\delta(t-2)$.

**Solution.** $\int_{-\infty}^{\infty} e^{-5t}\,\delta(t-2)\, e^{-5\times 2} = e^{-10}$ **Ans.**

**Property II:** $\int_{-\infty}^{\infty} f(t)\,\delta'(t-a)\,dt = -f'(a)$

**Proof.** $\int_{-\infty}^{\infty} f(t)\,\delta'(t-a)\,dt = \left[f(t)\cdot\delta(t-a)\right]_{-\infty}^{\infty} - \int_{-\infty}^{\infty} f'(t)\,\delta(t-a)\,dt$

$$= 0 - 0 - f'(a) = -f'(a)$$

**Example 22.** *Find the Laplace transform of* $t^3\,\delta(t-4)$.

**Solution.** $L\, t^3\,\delta(t-4) = \int_0^{\infty} e^{-st}\, t^3\,\delta(t-4)\,dt$

$$= 4^3\, e^{-4s}$$ **Ans.**

**Exercise 13.5**

Evaluate the following :

1. $\int_0^{\infty} e^{-3t}\,\delta(t-4)\,dt$ **Ans.** $e^{-12}$

2. $\int_{-\infty}^{\infty} \sin 2t\,\delta\left(t-\frac{\pi}{4}\right)$ **Ans.** 1.

3. $\int_{-\infty}^{\infty} e^{-3t}\,\delta'(t-2)$ **Ans.** $3e^{-6}$

4. $\frac{\delta(t-4)}{t}$ **Ans.** $\frac{e^{-4s}}{4}$

5. Laplace transforms of $\cos t \log t\, \delta(t-\pi)$ **Ans.** $-e^{-\pi s}\log\pi$

6. $e^{-4t}\,\delta(t-3)$ **Ans.** $e^{-3(s+4)}$

## 13.14 PERIODIC FUNCTIONS

Let $f(t)$ be a periodic function with Period $T$, then

$$L[f(t)] = \frac{\int_0^T e^{-st} f(t)\,dt}{1-e^{-sT}}$$

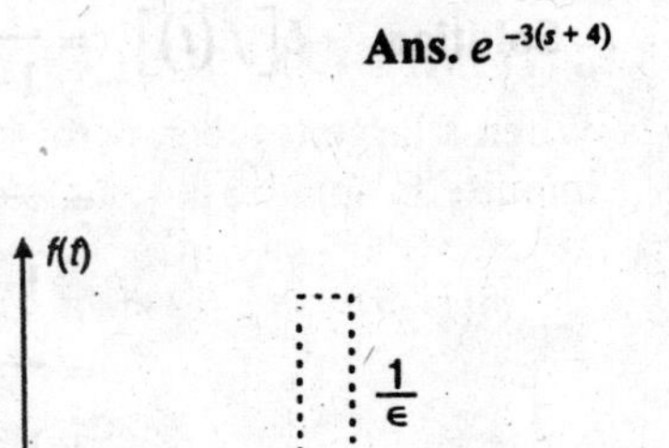


**Proof.** $L[f(t)] = \int_0^{\infty} e^{-st} f(t)\,dt$

$$= \int_0^T e^{-st} f(t)\,dt + \int_T^{2T} e^{-st} f(t)\,dt + \int_{2T}^{3T} e^{-st} f(t)\,dt + ..$$

Substituting $t = u+T$ in second integral and $t = u + 2T$ in third integral, and so on.

$$L[f(t)] = \int_0^T e^{-st} f(t)\,dt + \int_0^T e^{-s(u+T)} f(u+T)\,du + \int_0^T e^{-s(u+2T)} f(u+2T)\,du + ....$$

$$= \int_0^T e^{-st} f(t)\,dt + e^{-sT}\int_0^T e^{-su} f(u)\,du + e^{-2sT}\int_0^T e^{-su} f(u)\,du + ...$$

$$\left[f(u)=f(u+T)=f(u+2T)=f(u+3T)=....\right]$$

$$=\int_0^T e^{-st}f(t)\,dt+e^{-sT}\int_0^T e^{-st}f(t)\,dt+e^{-2sT}\int_0^T e^{-st}f(t)\,dt+...$$

$$=\left[1+e^{-sT}+e^{-2sT}+e^{-3sT}+...\right]\int_0^T e^{-st}f(t)\,dt \qquad \left[1+a+a^2+a^3+...=\frac{1}{1-a}\right]$$

$$=\frac{1}{1-e^{-sT}}\int_0^T e^{-st}f(t)\,dt.$$

**Proved.**

**Example 23.** *Find the Laplace transform of the waveform*

$$f(t)=\left(\frac{2t}{3}\right),\ 0\le t\le 3.$$

**Solution.**

$$L\left[f(t)\right]=\frac{1}{1-e^{-sT}}\int_0^T e^{-st}f(t)\,dt$$

$$L\left[\frac{2t}{3}\right]=\frac{1}{1-e^{-3s}}\int_0^3 e^{-st}\left(\frac{2}{3}t\right)dt=\frac{1}{1-e^{-3s}}\frac{2}{3}\left[\frac{te^{-st}}{-s}-(1)\frac{e^{-st}}{s^2}\right]_0^3$$

$$=\frac{2}{3}\frac{1}{1-e^{-3s}}\left[\frac{3e^{-3s}}{-s}-\frac{e^{-3s}}{s^2}+\frac{1}{s^2}\right]=\frac{2}{3}.\frac{1}{1-e^{-3s}}\left[\frac{3e^{-3s}}{-s}+\frac{1-e^{-3s}}{s^2}\right]$$

$$=\frac{2e^{-3s}}{-s\left(1-e^{-3s}\right)}+\frac{2}{3s^2}$$

**Ans.**

**Example 24.** *Find the Laplace transform of the function (Half wave rectifier)*

$$f(t)=\begin{cases}\sin\omega t & \text{for } 0<t<\dfrac{\pi}{\omega}\\ 0 & \text{for } \dfrac{\pi}{\omega}<t<\dfrac{2\pi}{\omega}.\end{cases}$$

*(U.P. II Semester, 2010, Summer 2002)*

**Solution.**

$$L\left[f(t)\right]=\frac{1}{1-e^{-sT}}\int_0^T e^{-st}f(t)\,dt$$

$$=\frac{1}{1-e^{-\frac{2\pi s}{\omega}}}\int_0^{2\pi/\omega} e^{-st}f(t)\,dt \qquad \left[\begin{array}{l}f(t)\text{ is a periodic function}\\ T=\dfrac{2\pi}{\omega}\end{array}\right]$$

$$=\frac{1}{1-e^{-\frac{2\pi s}{\omega}}}\left[\int_0^{\pi/\omega} e^{-st}\sin\omega t\,dt+\int_{\pi/\omega}^{2\pi/\omega} e^{-st}\times 0\times dt\right]$$

$$=\frac{1}{1-e^{-\frac{2\pi s}{\omega}}}\int_0^{\pi/\omega} e^{-st}\sin\omega t\,dt$$

$$\left[\int e^{ax}\sin bx\,dx=e^{ax}\frac{(a\sin bx-b\cos bx)}{a^2+b^2}\right]$$

$$L[f(t)]=\frac{1}{1-e^{-\frac{2\pi s}{\omega}}}\left[\frac{e^{-st}(-s\sin\omega t-\omega\cos\omega t)}{s^2+\omega^2}\right]_0^{\pi/\omega}$$

$$=\frac{1}{1-e^{-\frac{2\pi s}{\omega}}}\left[\frac{\omega e^{-\frac{\pi}{\omega}s}+\omega}{s^2+\omega^2}\right]=\frac{\omega\left[1+e^{-\frac{\pi s}{\omega}}\right]}{(s^2+\omega^2)\left[1-e^{-\frac{2\pi s}{\omega}}\right]}$$

$$= \frac{\omega\left[1+e^{-\frac{\pi}{\omega}s}\right]}{(s^2+\omega^2)\left(1-e^{-\frac{\pi}{\omega}s}\right)\left(1+e^{-\frac{\pi}{\omega}s}\right)}$$

$$= \frac{\omega}{(s^2+\omega^2)\left[1-e^{-\frac{\pi s}{\omega}}\right]}$$ **Ans.**

**Example 25.** *Find the Laplace Transform of the Periodic function (saw tooth wave)*

$$f(t) = \frac{kt}{T} \text{ for } 0 < t < T, \qquad f(t+T) = f(t)$$

**Solution.** $$L[f(t)] = \frac{1}{1-e^{-sT}}\int_0^T e^{-st} f(t)\,dt = \frac{1}{1-e^{-sT}}\int_0^T e^{-st}\frac{kt}{T}\,dt$$

$$= \frac{1}{1-e^{-sT}}\frac{k}{T}\int_0^T e^{-st}.t\,dt = \frac{k}{T\left(1-e^{-sT}\right)}\left[t\frac{e^{-st}}{-s} - \int 1.\frac{e^{-st}}{-s}\,dt\right]_0^T$$

Integrating by parts

$$= \frac{k}{T\left(1-e^{-sT}\right)}\left[\frac{te^{-st}}{-s} - \frac{e^{-st}}{s^2}\right]_0^T = \frac{k}{T\left(1-e^{-sT}\right)}\left[\frac{Te^{-sT}}{-s} - \frac{e^{-sT}}{s^2} + \frac{1}{s^2}\right]$$

$$= \frac{k}{T\left(1-e^{-sT}\right)}\left[\frac{Te^{-sT}}{-s} + \frac{1}{s^2}\left(1-e^{-sT}\right)\right] = -\frac{ke^{-sT}}{s\left(1-e^{-sT}\right)} + \frac{k}{Ts^2}$$ **Ans.**

**Example 26.** *Obtain Laplace transform of rectangular wave given by*

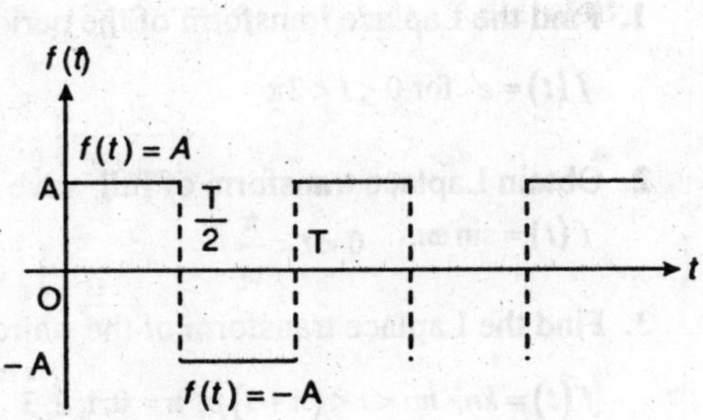


**Solution.** $$Lf(t) = \frac{\int_0^T e^{-st} f(t)\,dt}{1-e^{-sT}}$$

$$= \frac{\int_0^{\frac{T}{2}} e^{-st} A\,dt + \int_{\frac{T}{2}}^{T} e^{-st}(-A)\,dt}{1-e^{-sT}}$$

$$= A\frac{\left[\frac{e^{-st}}{-s}\right]_0^{\frac{T}{2}} - \left[\frac{e^{-st}}{-s}\right]_{\frac{T}{2}}^{T}}{1-e^{-sT}}$$

$$= \frac{A}{1-e^{-sT}}\left[-\frac{e^{-\frac{sT}{2}}}{s} + \frac{1}{s} + \frac{e^{-sT}}{s} - \frac{e^{-\frac{sT}{2}}}{s}\right]$$

$$= \frac{A}{s(1-e^{-sT})}\left[1-2e^{-\frac{sT}{2}} + e^{-sT}\right] = \frac{A}{s(1-e^{-sT})}\left[1-e^{-\frac{sT}{2}}\right]^2$$

$$= \frac{A\left[1-e^{-\frac{sT}{2}}\right]^2}{s(1+e^{-\frac{sT}{2}})(1-e^{-\frac{sT}{2}})} = \frac{A}{s}\frac{\left(1-e^{-\frac{sT}{2}}\right)}{\left(1+e^{-\frac{sT}{2}}\right)}$$

$$= \frac{A}{s} \frac{\left(e^{\frac{sT}{4}} - e^{-\frac{sT}{4}}\right)}{\left(e^{\frac{sT}{4}} + e^{-\frac{sT}{4}}\right)} = \frac{A}{s} \tanh \frac{sT}{4}$$ **Ans.**

**Example 27.** *A periodic square wave function f(t), in terms of unit step functions, is written as*

$$f(t) = k[u_0(t) - 2u_a(t) + 2u_{2a}(t) - 2u_{3a}(t) + \ldots]$$

*Show that the Laplace transform of f(t) is given by*

$$L[f(t)] = \frac{k}{s} \tanh\left(\frac{as}{2}\right)$$

**Solution.**

$$f(t) = k[u_0(t) - 2u_a(t) + 2u_{2a}(t) - 2u_{3a}(t) + \ldots]$$

$$Lf(t) = k[Lu_0(t) - 2Lu_a(t) + 2Lu_{2a}(t) - 2Lu_{3a}(t) + \ldots]$$

$$= k\left[\frac{1}{s} - 2\frac{e^{-as}}{s} + 2\frac{e^{-2as}}{s} - 2\frac{e^{-3as}}{s} + \ldots\right]$$

$$= \frac{k}{s}\left[1 - 2e^{-as} + 2e^{-2as} - 2e^{-3as} + \ldots\right]$$

$$= \frac{k}{s}\left[1 - 2(e^{-as} - e^{-2as} + e^{-3as} - \ldots)\right]$$

$$= \frac{k}{s}\left[1 - 2\frac{e^{-as}}{1 + e^{-as}}\right] = \frac{k}{s}\left[\frac{1 + e^{-as} - 2e^{-as}}{1 + e^{-as}}\right]$$

$$= \frac{k}{s}\left[\frac{1 - e^{-as}}{1 + e^{-as}}\right] = \frac{k}{s}\left[\frac{e^{\frac{as}{2}} - e^{-\frac{as}{2}}}{e^{\frac{as}{2}} + e^{-\frac{as}{2}}}\right] = \frac{k}{s} \tanh \frac{as}{2}$$ **Ans.**

## Exercise 13.6

**1.** Find the Laplace transform of the periodic function

$f(t) = e^t$ for $0 < t < 2\pi$ **Ans.** $\dfrac{e^{2(1-s)\pi} - 1}{(1-s)\left(1 - e^{-2\pi s}\right)}$

**2.** Obtain Laplace transform of full wave rectified sine wave given by

$f(t) = \sin \omega t, \quad 0 < t < \dfrac{\pi}{\omega}$ **Ans.** $\dfrac{\omega}{\left(s^2 + \omega^2\right)} \coth \dfrac{\pi s}{2\omega}$

**3.** Find the Laplace transform of the staircase function

$f(t) = kn, \; np < t < (n+1)p, \; n = 0, 1, 2, 3$ **Ans.** $\dfrac{ke^{ps}}{s\left(1 - e^{-ps}\right)}$

**Find Laplace transform of the following:**

**4.** $f(t) = t^2, \; 0 < t < 2, \; f(t+2) = f(t)$ **Ans.** $\dfrac{2 - e^{-2s} - 4se^{-2s} - 4s^2e^{-2s}}{s^3\left(1 - e^{-2s}\right)}$

**5.** $f(t) = \begin{cases} 1, & 0 \le t \le \dfrac{a}{2} \\ -1, & \dfrac{a}{2} \le t < a \end{cases}$ *(U.P. II Semester, 2004)* **Ans.** $\dfrac{1}{s} \tanh \dfrac{as}{4}$

**6.** $f(t) = \begin{cases} \cos \omega t, & 0 < t < \dfrac{\pi}{\omega} \\ 0, & \dfrac{\pi}{\omega} < t < \dfrac{2\pi}{\omega} \end{cases}$ **Ans.** $\dfrac{s}{\left(s^2 + \omega^2\right)\left(1 - e^{-\frac{\pi s}{\omega}}\right)}$

7. $f(t)=\begin{cases} t, & 0<t<1 \\ 0, & 1<t<2 \end{cases}$ $\quad f(t+2)=f(t)$ **Ans.** $\dfrac{1-e^{-s}(s+1)}{s^2\left(1-e^{-2s}\right)}$

8. $f(t)=\begin{cases} \dfrac{2t}{T}, & 0\le t\le \dfrac{T}{2} \\ \dfrac{2}{T}(T-t), & \dfrac{T}{2}\le t\le T \end{cases}$ $\quad f(t+T)=f(t)$ **Ans.** $\dfrac{2}{Ts^2}\tanh\dfrac{sT}{4}-\dfrac{1}{s\left(e^{\frac{sT}{2}}+1\right)}$

## 13.15 CONVOLUTION THEOREM

If $L[f_1(t)]=F_1(s)$ and $L[f_2(t)]=F_2(s)$

then $L\left\{\int_0^t f_1(x)f_2(t-x)dx\right\}=F_1(s).F_2(s)$

or

$L^{-1}F_1(s).F_2(s)=\int_0^t f_1(x)f_2(t-x)\,dx$

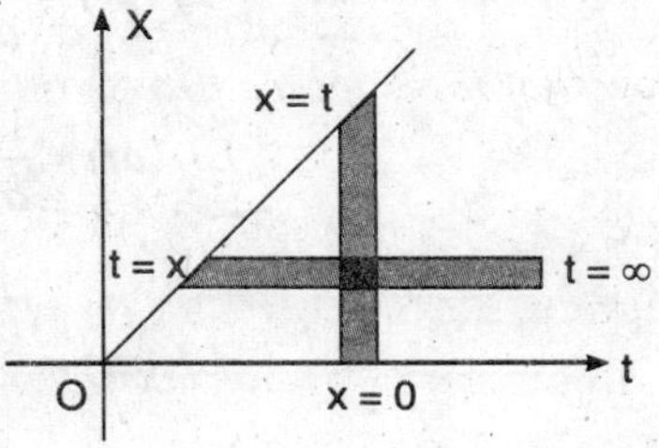


**Proof.** We have $L\left\{\int_0^{\infty} f_1(x)f_2(t-x)\,dx\right\}=\int_0^{\infty}e^{-st}\int_0^t f_1(x)f_2(t-x)\,dx\,dt$

$$=\int_0^{\infty}\int_0^t e^{-st}f_1(x)f_2(t-x)dx\,dt$$

where the double integral is taken over the infinite region in the first quadrant lying between the lines $x=0$ and $x=t$.

On changing the order of integration, the above integral becomes

$$\int_0^{\infty}\int_0^{\infty}e^{-st}f_1(x)f_2(t-x)dt\,dx$$

$$=\int_0^{\infty}e^{-sx}f_1(x)\,dx\int_x^{\infty}e^{-s(t-x)}f_2(t-x)\,dt$$

$$=\int_0^{\infty}e^{-sx}f_1(x)\,dx\int_0^{\infty}e^{-sz}f_2(z)\,dz\text{, on putting } t-x=z$$

$$=\int_0^{\infty}e^{-sx}f_1(x)F_2(s)\,dx=\left[\int_0^{\infty}e^{-sx}f_1(x)\,dx\right]F_2(s)$$

$=F_1(s)\,F_2(s)$ **Proved.**

## 13.16 LAPLACE TRANSFORM OF BESSEL FUNCTIONS $J_0(x)$ AND $J_1(x)$

**Solution.** We know that

$$J_0(t)=\left[1-\frac{t^2}{2^2}+\frac{t^4}{2^2.4^2}-\frac{t^6}{2^2.4^2.6^2}+\dots\right]$$

Taking Laplace transforms of both sides, we have

$$LJ_0(t)=\frac{1}{s}-\frac{1}{2^2}.\frac{2!}{s^3}+\frac{1}{2^2.4^2}\frac{4!}{s^5}-\frac{1}{2^2.4^2.6^2}\frac{6!}{s^7}+\dots$$

$$=\frac{1}{s}\left[1-\frac{1}{2}\left(\frac{1}{s^2}\right)+\frac{1.3}{2.4}\left(\frac{1}{s^4}\right)-\frac{1.3.5}{2.4.6}\left(\frac{1}{s^6}\right)+\dots\right]$$

$$=\frac{1}{s}\left[1+\left(-\frac{1}{2}\right)\left(\frac{1}{s^2}\right)+\frac{\left(-\frac{1}{2}\right)\left(-\frac{3}{2}\right)}{2!}\left(\frac{1}{s^2}\right)^2+\frac{\left(-\frac{1}{2}\right)\left(-\frac{3}{2}\right)\left(-\frac{5}{2}\right)}{3!}\left(\frac{1}{s^2}\right)^3+\dots\right]$$

$$= \frac{1}{s}\left[1+\frac{1}{s^2}\right]^{-\frac{1}{2}}$$ (By Binomial theorem)

$$= \frac{1}{\sqrt{s^2+1}}$$ **Ans.**

We know that $Lf(at) = \frac{1}{a}F\left(\frac{s}{a}\right)$

$$\therefore \quad LJ_0(at) = \frac{1}{a}\frac{1}{\sqrt{\frac{s^2}{a^2}+1}} = \frac{1}{\sqrt{s^2+a^2}}$$

$$LJ_1(at) = -LJ_0'(x) = -[sLJ_0(x) - J_0(0)]$$

$$= -\left[s.\frac{1}{\sqrt{s^2+1}} - 1\right] = 1 - \frac{s}{\sqrt{s^2+1}}$$ **Ans.**

## 13.17 EVALUATION OF INTEGRALS

We can evaluate number of integrals having lower limit 0 and upper limit ∞ by the help of Laplace transform.

**Example 28.** *Evaluate* $\int_0^\infty te^{-3t}\sin t\,dt.$

**Solution.** $\int_0^\infty te^{-3t}\sin t\,dt. = \int_0^\infty te^{-st}\sin t\,dt$ (s = 3)

$$= L(t\sin t) = -\frac{d}{ds}\left(\frac{1}{s^2+1}\right) = \frac{2s}{(s^2+1)^2}$$

$$= \frac{2\times 3}{(3^2+1)^2} = \frac{6}{100} = \frac{3}{50}$$ **Ans.**

**Example 29.** *Evaluate* $\int_0^\infty \frac{e^{-t}\sin t}{t}dt$ *and* $\int_0^\infty \frac{\sin t}{t}dt.$

**Solution.** $\int_0^\infty \frac{e^{-t}\sin t}{t}dt = \int_0^\infty e^{-st}\frac{\sin t}{t}dt.$ (s = 1)

$$= L\left[\frac{\sin t}{t}\right] = \int_s^\infty \frac{1}{s^2+1}ds = \left[\tan^{-1}s\right]_s^\infty$$

$$= \frac{\pi}{2} - \tan^{-1}s \quad ...(1) \quad = \frac{\pi}{2} - \tan^{-1}(1)$$ (s = 1)

$$= \frac{\pi}{2} - \frac{\pi}{4} = \frac{\pi}{4}$$ **Ans.**

On putting $s = 0$ in (1), we get

$$\int_0^\infty \frac{\sin t}{t}dt = \frac{\pi}{2} - \tan^{-1}(0) = \frac{\pi}{2}$$ **Ans.**

## EXERCISE 13.7

**Evaluate the following by using Laplace Transform:**

1. $\int_0^\infty t e^{-4t} \sin t \, dt$ **Ans.** $\frac{8}{289}$
2. $\int_0^\infty \frac{e^{-2t} \sinh t \sin t}{t} dt$ **Ans.** $\frac{1}{2} \tan^{-1} \frac{1}{2}$
3. $\int_0^\infty \frac{\sin^2 t}{t^2} dt$ **Ans.** $i\frac{5}{2}$
4. $\int_0^\infty \frac{e^{-t} - e^{-4t}}{t} dt$ **Ans.** log 4

## 13.18 FORMULATION OF LAPLACE TRANSFORM

| S.No. | $f(t)$ | $F(s)$ |
|---|---|---|
| 1. | $e^{at}$ | $\frac{1}{s-a}$ |
| 2. | $t^n$ | $\frac{\lfloor n+1}{s^{n+1}}$ or $\frac{n!}{s^{n+1}}$ |
| 3. | $\sin at$ | $\frac{a}{s^2+a^2}$ |
| 4. | $\cos at$ | $\frac{s}{s^2+a^2}$ |
| 5. | $\sinh at$ | $\frac{a}{s^2-a^2}$ |
| 6. | $\cosh at$ | $\frac{s}{s^2-a^2}$ |
| 7. | $u(t-a)$ | $\frac{e^{-as}}{s}$ |
| 8. | $\delta(t-a)$ | $e^{-as}$ |
| 9. | $e^{bt} \sin at$ | $\frac{a}{(s-b)^2+a^2}$ |
| 10. | $e^{bt} \cos at$ | $\frac{s-b}{(s-b)^2+a^2}$ |
| 11. | $\frac{t}{2a} \sin at$ | $\frac{s}{(s^2+a^2)^2}$ |
| 12. | $t \cos at$ | $\frac{s^2-a^2}{(s^2+a^2)^2}$ |
| 13. | $\frac{1}{2a^3}(\sin at - at \cos at)$ | $\frac{1}{(s^2+a^2)^2}$ |
| 14. | $\frac{1}{2a}(\sin at + at \cos at)$ | $\frac{s^2}{(s^2+a^2)^2}$ |

## 13.19 PROPERTIES OF LAPLACE TRANSFORM

| *S.No.* | *Property* | *f (t)* | *F (s)* |
|---|---|---|---|
| 1. | Scaling | $f(at)$ | $\frac{1}{a}F\left(\frac{s}{a}\right), \quad a>0$ |
| 2. | Derivative | $\frac{df(t)}{dt}$ <br> $\frac{d^2 f(t)}{dt^2}$ <br> $\frac{d^3 f(t)}{dt^3}$ | $sF(s)-f(0), \quad s>0$ <br> $s^2F(s)-sf(0)-f'(0), \quad s>0$ <br> $s^3F(s)-s^2f(0)-sf'(0)-f''(0), s>0$ |
| 3. | Integral | $\int_0^t f(t)\,dt$ | $\frac{1}{s}F(s), \quad s>0$ |
| 4. | Initial Value | $\lim_{t\to 0} f(t)$ | $\lim_{s\to\infty} sF(s)$ |
| 5. | Final Value | $\lim_{t\to\infty} f(t)$ | $\lim_{s\to\infty} sF(s)$ |
| 6. | First shifting | $e^{-at}f(t)$ | $F(s+a)$ |
| 7. | Second shifting | $f(t)\,u(t-a)$ | $e^{-a}\,L\,f(t+a)$ |
| 8. | Multiplication by $t$ | $t\,f(t)$ | $-\frac{d}{ds}F(s)$ |
| 9. | Multiplication by $t^n$ | $t^n f(t)$ | $(-1)^n \frac{d^n}{ds^n}F(s)$ |
| 10. | Division by $t$ | $\frac{1}{t}f(t)$ | $\int_s^\infty F(s)\,ds$ |
| 11. | Periodic function | $f(t)$ | $\frac{\int_0^T e^{-st}f(t)}{1-e^{-st}} \quad f(t+T)=f(t)$ |
| 12. | Convolution | $f(t) * g(t)$ | $F(s)\,G(s)$ |

## 13.20 INVERSE LAPLACE TRANSFORMS

Now we obtain $f(t)$ when $F(s)$ is given,then we say that inverse Laplace transform of $F(s)$ is $f(t)$.

If $L[f(t)] = F(s)$ then $L^{-1}[F(s)] = f(t)$.
where $L^{-1}$ is called the inverse Laplace transform operator.

**From the application point of view, the inverse Laplace transform is very useful.**

## 13.21 IMPORTANT FORMULAE

1. $L^{-1}\left(\frac{1}{s}\right) = 1$

2. $L^{-1}\frac{1}{s^n} = \frac{t^{n-1}}{(n-1)!}$

3. $L^{-1}\frac{1}{s-a} = e^{at}$

4. $L^{-1}\frac{s}{s^2-a^2} = \cosh at$

5. $L^{-1}\frac{1}{s^2-a^2} = \frac{1}{a}\sinh at$

6. $L^{-1}\frac{1}{s^2+a^2} = \frac{1}{a}\sin at$

7. $L^{-1}\frac{s}{s^2+a^2} = \cos at$

8. $L^{-1}F(s-a) = e^{at}f(t)$

9. $L^{-1}\frac{1}{(s-a)^2+b^2} = \frac{1}{b}e^{at}\sin bt$

10. $L^{-1}\frac{s-a}{(s-a)^2+b^2} = e^{at}\cos bt$

11. $L^{-1}\frac{1}{(s-a)^2-b^2} = \frac{1}{b}e^{at}\sinh bt$

12. $L^{-1}\frac{s-a}{(s-a)^2-b^2} = e^{at}\cosh bt$

13. $L^{-1}\frac{1}{(s^2+a^2)^2} = \frac{1}{2a^3}(\sin at - at\cos at)$

14. $L^{-1}\frac{s}{(s^2+a^2)^2} = \frac{1}{2a}t\sin at$

15. $L^{-1}\frac{s^2-a^2}{(s^2+a^2)^2} = t\cos at$

16. $L^{-1}(1) = s(t)$

17. $L^{-1}\frac{s^2}{(s^2+a^2)^2} = \frac{1}{2a}[\sin at + at\cos at]$

18. $L^{-1}\left\{\frac{1}{s}F(s)\right\} = \int_0^t f(t)\,dt$

**Example 30.** *Find the inverse Laplace Transform of the following:*

(*i*) $\frac{1}{s-2}$ (*ii*) $\frac{1}{s^2-9}$ (*iii*) $\frac{s}{s^2-16}$ (*iv*) $\frac{1}{s^2+25}$ (*v*) $\frac{s}{s^2+9}$

(*vi*) $\frac{1}{(s-2)^2+1}$ (*vii*) $\frac{s-1}{(s-1)^2+4}$ (*viii*) $\frac{1}{(s+3)^2-4}$ (*ix*) $\frac{s+2}{(s+2)^2-25}$ (*x*) $\frac{1}{2s-7}$

**Solution.** (*i*) $L^{-1}\frac{1}{s-2} = e^{2t}$ (*ii*) $L^{-1}\frac{1}{s^2-9} = L^{-1}\frac{1}{3}\cdot\frac{3}{s^2-(3)^2} = \frac{1}{3}\sinh 3t$

(*iii*) $L^{-1}\frac{s}{s^2-16} = L^{-1}\frac{s}{s^2-(4)^2} = \cosh 4t$ (*iv*) $L^{-1}\frac{1}{s^2+25} = \frac{1}{5}\frac{5}{s^2+(5)^2} = \frac{1}{5}\sin 5t$

(*v*) $L^{-1}\frac{s}{s^2+9} = \frac{s}{s^2+(3)^2} = \cos 3t$ (*vi*) $L^{-1}\frac{1}{(s-2)^2+1} = e^{2t}\sin t$

(*vii*) $L^{-1}\frac{s-1}{(s-1)^2+4} = e^{t}\cos 2t$ (*viii*) $L^{-1}\frac{1}{(s+3)^2-4} = \frac{1}{2}\frac{2}{(s+3)^2-(2)^2} = \frac{1}{2}e^{-3t}\sinh 2t$

(*ix*) $L^{-1}\frac{s+2}{(s+2)^2-25} = L^{-1}\frac{(s+2)}{(s+2)^2-(5)^2} = e^{-2t}\cosh 5t$

(*x*) $L^{-1}\frac{1}{2s-7} = \frac{1}{2}e^{\frac{7}{2}t}$ $\left[L^{-1}F(as) = \frac{1}{a}f\left(\frac{t}{a}\right)\right]$ **Ans.**

**Example 31.** *Find the inverse Laplace transform of*

(*i*) $\frac{s^2+s+2}{s^{3/2}}$ (*ii*) $\frac{2s-5}{9s^2-25}$ (*iii*) $\frac{s-2}{6s^2+20}$

**Solution.** (*i*) $L^{-1}\frac{s^2+s+2}{s^{3/2}} = L^{-1}s^{1/2} + L^{-1}S^{-1/2} + L^{-1}\frac{2}{s^{3/2}}$

$$= L^{-1}\frac{1}{s^{-1/2}} + L^{-1}\frac{1}{s^{1/2}} + L^{-1}\frac{2}{s^{3/2}} = \frac{t^{-1/2-1}}{\left|\overline{-\frac{1}{2}}\right.} + \frac{t^{1/2-1}}{\left|\overline{-\frac{1}{2}}\right.} + \frac{2t^{3/2-1}}{\left|\overline{-\frac{3}{2}}\right.}$$

$$= \frac{1}{\left|\overline{-\frac{1}{2}}\right. t^{3/2}} + \frac{1}{\sqrt{\pi t}} + \frac{4\sqrt{t}}{\sqrt{\pi}}$$ **Ans.**

(*ii*) $$L^{-1}\frac{2s-5}{9s^2-25} = L^{-1}\left[\frac{2s}{9s^2-25} - \frac{5}{9s^2-25}\right] = L^{-1}\left[\frac{2s}{9\left[s^2-\left(\frac{5}{3}\right)^2\right]} - \frac{5}{9\left[s^2-\left(\frac{5}{3}\right)^2\right]}\right]$$

$$= \frac{2}{9}\cosh\frac{5}{3}t - \frac{1}{3}L^{-1}\left(\frac{\frac{5}{3}}{s^2-\left(\frac{5}{3}\right)^2}\right) = \frac{2}{9}\cosh\frac{5t}{3} - \frac{1}{3}\sin\frac{5t}{3}$$ **Ans.**

(*iii*) $$L^{-1}\frac{s-2}{6s^2+20} = L^{-1}\frac{s}{6s^2+20} - L^{-1}\frac{2}{6s^2+20} = \frac{1}{6}L^{-1}\frac{s}{s^2+\frac{10}{3}} - \frac{1}{3}L^{-1}\frac{1}{s^2+\frac{10}{3}}$$

$$= \frac{1}{6}\cos\sqrt{\frac{10}{3}}t - \frac{1}{3}\times\sqrt{\frac{3}{10}}L^{-1}\frac{\sqrt{\frac{10}{3}}}{s^2+\frac{10}{3}} = \frac{1}{6}\cos\sqrt{\frac{10}{3}}t - \frac{1}{\sqrt{30}}\sin\sqrt{\frac{10}{3}}t$$ **Ans.**

### Exercise 13.8

Find the inverse Laplace transform of the following:

1. $\dfrac{3s-8}{4s^2+25}$ **Ans.** $\dfrac{3}{4}\cos\dfrac{5t}{2} - \dfrac{4}{5}\sin\dfrac{5t}{2}$

2. $\dfrac{3(s^2-2)^2}{2s^5}$ **Ans.** $\dfrac{3}{2} - 3t^2 + \dfrac{1}{2}t^4$

3. $\dfrac{2s-5}{4s^2+25} + \dfrac{4s-18}{9-s^2}$ **Ans.** $\dfrac{1}{2}\left(\cos\dfrac{5t}{2} - \sin\dfrac{5t}{2}\right) - 4\cosh 3t + 6\sinh 3t$

4. $\dfrac{5s-10}{9s^2-16}$ **Ans.** $\dfrac{5}{9}\cosh\dfrac{4}{3}t - \dfrac{5}{6}\sinh\dfrac{4}{3}t$

5. $\dfrac{1}{4s} + \dfrac{16}{1-s^2}$ **Ans.** $\dfrac{1}{4} - 16\sinh t$

## 13.22 MULTIPLICATION by *s*

$$L^{-1}[sF(s)] = \frac{d}{dt}f(t) + f(0)\delta(t)$$

**Example 32.** *Find the inverse Laplace transform of*

(*i*) $\dfrac{s}{s^2+1}$ (*ii*) $\dfrac{s}{4s^2-25}$ (*iii*) $\dfrac{3s}{2s+9}$

**Solution.** (*i*) $$L^{-1}\frac{1}{s^2+1} = \sin t$$

$$L^{-1}\frac{s}{s^2+1} = \frac{d}{dt}(\sin t) + \sin(0)\delta(t)$$

$$= \cos t$$ **Ans.**

(*ii*) $$L^{-1}\frac{1}{4s^2-25} = \frac{1}{4}L^{-1}\frac{1}{s^2-\frac{25}{4}} = \frac{1}{4}\cdot\frac{2}{5}L^{-1}\frac{\frac{5}{2}}{s^2-\left(\frac{5}{2}\right)^2} = \frac{1}{10}\sinh\frac{5}{2}t$$

$$L^{-1}\frac{s}{4s^2-25}=\frac{1}{10}\frac{d}{dt}\sinh\frac{5}{2}t+\frac{1}{10}\sinh\frac{5}{2}(0)$$

$$=\frac{1}{10}\left(\frac{5}{2}\right)\cosh\frac{5}{2}t=\frac{1}{4}\cosh\frac{5}{2}t$$ **Ans.**

(*iii*) $$L^{-1}\frac{3}{2s+9}=\frac{3}{2}L^{-1}\frac{1}{s+\frac{9}{2}}=\frac{3}{2}e^{-9/2t}$$

$$L^{-1}\frac{3s}{2s+9}=\frac{3}{2}\frac{d}{dt}\left(e^{-9/2t}\right)+\frac{3}{2}e^{-9/2(0)}=\frac{3}{2}\left(-\frac{9}{2}\right)e^{-\frac{11}{2}t}+\frac{3}{2}$$

$$=-\frac{27}{4}e^{-11/2t}+\frac{3}{2}$$ **Ans.**

## Exercise 13.9

*Find the inverse Laplace transform of the following:*

1. $\dfrac{s}{s+5}$ **Ans.** $-5e^{-5t}$
2. $\dfrac{2s}{3s+6}$ **Ans.** $-\dfrac{4}{3}e^{-2t}$
3. $\dfrac{s}{2s^2-1}$ **Ans.** $\dfrac{1}{2}\cosh\dfrac{t}{2}$
4. $\dfrac{s^2}{s^2+a^2}$ **Ans.** $-a\sin at+1$
5. $\dfrac{s^2+4}{s^2+9}$ **Ans.** $-\dfrac{5}{3}\sin 3t+1$
6. $\dfrac{1}{(s-3)^2}$ (*Madras, 2006*) **Ans.** $e^{3t}.t$
7. $L^{-1}\dfrac{s^2}{(s^2+4)^2}$ is

(*i*) $\sin 2t+\dfrac{t}{2}\cos 2t$ (*ii*) $\dfrac{1}{4}\sin 2t+\dfrac{t}{2}\cos 2t$ (*iii*) $\dfrac{1}{4}\sin 2t+t\cos 2t$ (*iv*) $\dfrac{1}{4}\sin 2t+\dfrac{t}{4}\cos 2t$

**Ans.** (*ii*)

## 13.23 DIVISION BY *s* (multiplication by $\frac{1}{s}$)

$$L^{-1}\left[\frac{F(s)}{s}\right]=\int_0^t L^{-1}[F(s)]dt=\int_0^t f(t)dt$$

**Example 33.** *Find the inverse Laplace transform of*

(*i*) $\dfrac{1}{s(s+a)}$ (*ii*) $\dfrac{1}{s(s^2+1)}$ (*iii*) $\dfrac{s^2+3}{s(s^2+9)}$

**Solution.** (*i*) $L^{-1}\left(\dfrac{1}{s+a}\right)=e^{-at}$

$$L^{-1}\left[\frac{1}{s(s+a)}\right]=\int_0^t L^{-1}\left(\frac{1}{s+a}\right)dt=\int_0^t e^{-at}dt=\left[\frac{e^{-at}}{-a}\right]_0^t$$

$$=\frac{e^{-at}}{-a}+\frac{1}{a}=\frac{1}{a}\left[1-e^{-at}\right]$$ **Ans.**

(*ii*) $$L^{-1}\frac{1}{s^2+1}=\sin t$$

$$L^{-1}\frac{1}{s}\left(\frac{1}{s^2+1}\right)=\int_0^t L^{-1}\left(\frac{1}{s^2+1}\right)dt=\int_0^t \sin t\,dt=\left[-\cos t\right]_0^t=-\cos t+1$$

**Ans.**

(*iii*) $$L^{-1}\frac{s^2+3}{s(s^2+9)}=L^{-1}\left[\frac{s^2+9-6}{s(s^2+9)}\right]=L^{-1}\left[\frac{1}{s}-\frac{6}{s(s^2+9)}\right]$$

$$=1-2\int_0^t \sin 3t\,dt=1-\int_0^t L^{-1}\left(\frac{6}{s^2+9}\right)ds=1+2\times\frac{1}{3}[\cos 3t]_0^t=1+\frac{2}{3}\cos 3t-\frac{2}{3}$$

$$=\frac{2}{3}\cos 3t+\frac{1}{3}=\frac{1}{3}[2\cos 3t+1]$$ **Ans.**

**Exercise 13.10**

Find the inverse Laplace transform of the following:

1. $\frac{1}{2s(s-3)}$ **Ans.** $\frac{1}{2}\left[\frac{e^{3t}}{3}-1\right]$

2. $\frac{1}{s(s+2)}$ **Ans.** $\frac{1-e^{-2t}}{2}$

3. $\frac{1}{s(s^2-16)}$ **Ans.** $\frac{1}{16}[\cosh 4t-1]$

4. $\frac{1}{s(s^2+a^2)}$ **Ans.** $\frac{1-\cos at}{a^2}$

5. $\frac{s^2+2}{s(s^2+4)}$ **Ans.** $\cos^2 t$

6. $\frac{1}{s^2(s+1)}$ **Ans.** $t-1+e^{-t}$

7. $L^{-1}\frac{s^2}{s(s^2+1)}$ **Ans.** $\frac{t^2}{2}+\cos t-1$

8. $L^{-1}\frac{s^2}{s(s^2+1)}$ is

(*i*) $1-\cos t$ (*ii*) $1+\cos t$ (*iii*) $1-\sin t$ (*iv*) $1+\sin t$ **Ans.** (*i*)

## 13.24 FIRST SHIFTING PROPERTY

If $L^{-1}F(s)=f(t)$ then $L^{-1}F(s+a)=e^{-at}L^{-1}[F(s)]$

**Example 34.** *Find the inverse Laplace transform of*

(*i*) $\frac{1}{(s+2)^5}$ (*ii*) $\frac{s}{s^2+4s+13}$ (*iii*) $\frac{1}{9s^2+6s+1}$ (*iv*) $\frac{s-1}{s^2-6s+25}$ (*iv*) $\frac{s-1}{s^2-6s+25}$

**Solution.** (*i*) $$L^{-1}\frac{1}{s^5}=\frac{t^4}{4!}$$

then $$L^{-1}\frac{1}{(s+2)^5}=e^{-2t}\frac{t^4}{4!}$$ **Ans.**

(*ii*) $$L^{-1}\left(\frac{s}{s^2+4s+13}\right)=L^{-1}\frac{s+2-2}{(s+2)^2+(3)^2}=L^{-1}\frac{s+2}{(s+2)^2+(3)^2}-L^{-1}\frac{2}{(s+2)^2+3^2}$$

$$=e^{-2t}L^{-1}\frac{s}{s^2+3^2}-e^{-2t}L^{-1}\frac{2}{3}\left(\frac{3}{s^2+3^2}\right)=e^{-2t}\cos 3t-\frac{2}{3}e^{-2t}\sin 3t$$ **Ans.**

(*iii*) $$L^{-1}\frac{1}{9s^2+6s+1}=L^{-1}\frac{1}{(3s+1)^2}=\frac{1}{9}L^{-1}\frac{1}{\left(s+\frac{1}{3}\right)^2}=\frac{1}{9}e^{-t/3}L^{-1}\frac{1}{s^2}=\frac{1}{9}e^{-t/3}t=\frac{te^{-t/3}}{9}$$ **Ans.**

(*iv*) $$L^{-1}\left(\frac{s-1}{s^2-6s+25}\right)=L^{-1}\left[\frac{s-1}{(s-3)^2+(4)^2}\right]=L^{-1}\left[\frac{s-3+2}{(s-3)^2+(4)^2}\right]$$

$$=L^{-1}\left[\frac{s-3}{(s-3)^2+(4)^2}\right]+\frac{1}{2}L^{-1}\left[\frac{4}{(s-3)^2+(4)^2}\right]$$

$$=e^{3t}\cos 4t+\frac{1}{2}e^{3t}\sin 4t$$ **Ans.**

### Exercise 13.11

Obtain the inverse Laplace transform of the following:

1. $\dfrac{s+8}{s^2+4s+5}$ Ans. $e^{-2t}(\cos t+6\sin t)$

2. $\dfrac{s}{(s+3)^2+4}$ Ans. $e^{-3t}(\cos 2t-1.5\sin 2t)$

3. $\dfrac{s}{(s+7)^4}$ Ans. $e^{-7t}\dfrac{t^2}{6}(3-7t)$

4. $\dfrac{s+2}{s^2-2s-8}$ Ans. $e^{t}(\cosh 3t+\sinh 3t)$

5. $\dfrac{s}{s^2+6s+25}$ Ans. $e^{-3t}\left[\cos 4t-\dfrac{3}{4}\sin 4t\right]$

6. $\dfrac{1}{2(s-1)^2+32}$ Ans. $\dfrac{e^{t}}{8}\sin 4t$

7. $\dfrac{s-4}{4(s-3)^2+16}$ Ans. $\dfrac{1}{4}e^{3t}\cos 2t-\dfrac{1}{8}e^{3t}\sin 2t$

## 13.25 SECOND SHIFTING PROPERTY

$$L^{-1}\left[e^{-as}F(s)\right]=f(t-a)U(t-a)$$

**Example 35.** *Obtain inverse Laplace transform of*

(*i*) $\dfrac{e^{-\pi s}}{(s+3)}$ (*ii*) $\dfrac{e^{-s}}{(s+1)^3}$

**Solution.** (*i*) $L^{-1}\dfrac{1}{s+3}=e^{-3t}$

$$\Rightarrow \quad L^{-1}\frac{e^{-\pi s}}{s+3}=e^{-3(t-\pi)}U(t-\pi) \qquad \textbf{Ans.}$$

(*ii*) $L^{-1}\dfrac{1}{s^3}=\dfrac{t^2}{2!}$

$$\Rightarrow \quad L^{-1}\frac{1}{(s+1)^3}=e^{-t}\frac{t^2}{2!}$$

$$\Rightarrow \quad L^{-1}\frac{e^{-s}}{(s+1)^3}=e^{-(t-1)}.\frac{(t-1)^2}{2!}.U(t-1) \qquad \textbf{Ans.}$$

**Example 36.** *Find the inverse Laplace transform of* $\dfrac{se^{-s/2}+\pi e^{-s}}{s^2+\pi^2}$ *in terms of unit step functions.*

**Solution.** $L^{-1}\dfrac{\pi}{s^2+\pi^2}=\sin \pi t$

$$L^{-1}\left[e^{-s}\frac{\pi}{s^2+\pi^2}\right]=\sin \pi(t-1).u(t-1) =-\sin(\pi t).u(t-1) \qquad \dots(1)$$

and $L^{-1}\dfrac{s}{s^2+\pi^2}=\cos \pi t$

$$L^{-1}\left[e^{-s/2}\frac{s}{s^2+\pi^2}\right]=\cos \pi\left(t-\frac{1}{2}\right).u\left(t-\frac{1}{2}\right) =\sin \pi t.u\left(t-\frac{1}{2}\right) \qquad \dots(2)$$

On adding (1) and (2), we get

$$L^{-1}\left[\frac{e^{-s/2}s+e^{-s}.\pi}{s^2+\pi^2}\right]=\sin(\pi t).u\left(t-\frac{1}{2}\right)-\sin(\pi t).u(t-1)$$

$$=\sin \pi t\left[u\left(t-\frac{1}{2}\right)-u(t-1)\right] \qquad \textbf{Ans.}$$

**Example 37.** *Find the value of* $L^{-1}\left\{\dfrac{1}{(s^2+a^2)^2}\right\}$.

**Solution.** $$\frac{1}{(s^2+a^2)^2}=\frac{1}{s}\cdot\frac{s}{(s^2+a^2)^2}=-\frac{1}{2s}\frac{d}{ds}\left(\frac{1}{s^2+a^2}\right)$$

$$\Rightarrow\quad L^{-1}\left\{\frac{1}{(s^2+a^2)^2}\right\}=L^{-1}\left\{-\frac{1}{2s}\frac{d}{ds}\left(\frac{1}{s^2+a^2}\right)\right\}$$

$$=-\frac{1}{2s}\left\{-t\frac{1}{a}\sin at\right\}=\frac{1}{2a}\cdot\frac{1}{s}\,[t\sin at]$$

$$=\frac{1}{2a}\cdot\int_0^t t\sin at\,dt=\frac{1}{2a}\left[t\left(-\frac{\cos at}{a}\right)-\int\left(-\frac{\cos at}{a}\right)dt\right]_0^t$$

$$=\frac{1}{2a}\left[-\frac{t}{a}\cos at+\frac{\sin at}{a^2}\right]_0^t$$

$$=\frac{1}{2a^3}\left[-at\cos at+\sin at\right]$$ **Ans.**

## Exercise 13.12

Obtain inverse Laplace transform of the following:

1. $\dfrac{e^{-s}}{(s+2)^3}$ **Ans.** $e^{-(t-2)}\dfrac{(t-2)^2}{2}U(t-2)$

2. $\dfrac{e^{-2s}}{(s+1)(s^2+2s+2)}$ **Ans.** $e^{-(t-2)}\{1-\cos(t-2)\}u(t-2)$

3. $\dfrac{e^{-s}}{\sqrt{s+1}}$ **Ans.** $\dfrac{e^{-(t-1)}}{\sqrt{\pi(t-1)}}.U(t-1)$

4. $\dfrac{e^{-\frac{\pi}{2}s}+e^{-\frac{3\pi}{2}s}}{s^2+1}$ **Ans.** $\cot t\left[U\left(t-\dfrac{3\pi}{2}\right)-U\left(t-\dfrac{\pi}{2}\right)\right]$

5. $\dfrac{e^{-4s}(s+2)}{s^2+4s+5}$ **Ans.** $e^{-2(t-u)}\cos(t-u)U(t-4)$

6. $\dfrac{e^{-as}}{s^2}$ **Ans.** $f(t)=t-a$ when $t>a$
$\quad = 0$ when $t<a$

7. $\dfrac{e^{-\pi s}}{s^2+1}$ **Ans.** $-\sin t.u(t-\pi)$

**Tick (√) the correct answers:**

8. (*a*) The inverse Laplace transform of $(e^{-3s})/s^3$, is

(*i*) $(t-3)u_3(t)$ (*ii*) $(t-3)^2u_3(t)$ (*iii*) $(t-3)^2u_3(t)$ (*iv*) $(t+3)^2u_3(t)$ **Ans.** (*iv*)

(*b*) If Laplace transform of a function $f(t)$ equals $(e^{-2s}-e^{-s})/s$, then

(*i*) $f(t)=1,\ t>1$;

(*ii*) $f(t)=1$, when $1<t<2$, and 0 otherwise ;

(*iii*) $f(t)=-1$, when $1<t<3$, and 0 otherwise ;

(*iv*) $f(t)=-1$, when $1<t<2$, and 0 otherwise. **Ans.** (*iv*)

(c) The Laplace inverse $L^{-1}\left[\frac{2}{s}(e^{-2s}-e^{-4s})\right]$ equals

(i) 2, if $0<t<4$ ; 0 otherwise, (ii) 2, if $t>0$

(iii) 2, if $0<t<2$ ; 0 otherwise, (iv) 2, if $2<t<4$; 0 otherwise **Ans.** (iv)

(d) The Laplace transform of $tu_2(t)$ is

(i) $\left(\frac{1}{s^2}+\frac{2}{2}\right)e^{-2s}$ (ii) $\frac{1}{s^2}e^{-2s}$ (iii) $\left(\frac{1}{s^2}-\frac{2}{s}\right)e^{-2s}$ (iv) $\frac{1}{s^2}e^{-2s}$ **Ans.** (i)

(e) The inverse Laplace transform of $\frac{Ke^{-as}}{S^2+k^2}$ is

(i) sin $kt$ (ii) cos $kt$ (iii) $u(t-a)\sin kt$ (iv) none of these. **Ans.** (iv)

(f) Inverse Laplace's transform of 1 is:

(i) 1 (ii) $\delta(t)$ (iii) $\delta(t-1)$ (iv) $u(t)$ **Ans.** (ii)

## 13.26 INVERSE LAPLACE TRANSFORMS OF DERIVATIVES

$$L^{-1}\left[\frac{d}{ds}F(s)\right]=-tL^{-1}[F(s)]=-tf(t) \quad \text{or} \quad L^{-1}[F(s)]=-\frac{1}{t}L^{-1}\left[\frac{d}{ds}F(s)\right]$$

**Example 38.** *Find inverse Laplace transform of* $\tan^{-1}\frac{1}{s}$

**Solution.**
$$L^{-1}\left(\tan^{-1}\frac{1}{s}\right)=\frac{1}{t}L^{-1}\left[\frac{d}{ds}\tan^{-1}\frac{1}{s}\right]$$
$$=-\frac{1}{t}L^{-1}\left[\frac{1}{1+\frac{1}{s^2}}\left(-\frac{1}{s^2}\right)\right]=\frac{1}{t}L^{-1}\left[\frac{1}{1+s^2}\right]$$
$$=\frac{\sin t}{t}$$
**Ans.**

**Example 39.** *Obtain the inverse Laplace transform of* $\log\frac{s^2-1}{s^2}$.

**Solution.**
$$L^{-1}\left[\log\frac{s^2-1}{s^2}\right]=-\frac{1}{t}L^{-1}\left[\frac{d}{ds}\log\frac{s^2-1}{s^2}\right]$$
$$=-\frac{1}{t}L^{-1}\left[\frac{d}{ds}\{\log(s^2-1)-2\log s\}\right]=-\frac{1}{t}L^{-1}\left[\frac{2s}{s^2-1}-\frac{2}{s}\right]=-\frac{1}{t}[2\cosh t-2]$$
$$=\frac{2}{t}[1-\cosh t]$$

**Example 40.** *Find* $L^{-1}\left[\cot^{-1}(1+s)\right]$.

**Solution.**
$$L^{-1}\left[\cot^{-1}(1+s)\right]=-\frac{1}{t}L^{-1}\left[\frac{d}{ds}\cot^{-1}(1+s)\right]$$
$$=-\frac{1}{t}L^{-1}\left[\frac{-1}{1+(s+1)^2}\right]=\frac{1}{t}L^{-1}\left[\frac{1}{(s+1)^2+1}\right]$$
$$=\frac{1}{t}e^{-t}\sin t$$
**Ans.**

## Exercise 13.13

Obtain inverse Laplace transform of the following:

1. $\log\left(1+\frac{\omega^2}{s^2}\right)$ **Ans.** $\frac{2}{t}\cos\omega t+2$
2. $\frac{s}{1+s^2+s^4}$ **Ans.** $\frac{2}{\sqrt{3}}\sin\frac{\sqrt{3}}{2}t\sinh\frac{t}{2}$
3. $\frac{s}{(s^2+a^2)^2}$ **Ans.** $\frac{t\sin at}{2a}$
4. $s\log\frac{s}{\sqrt{s^2+1}}+\cot^{-1}s$ **Ans.** $\frac{1-\cos t}{t^2}$
5. $\frac{1}{2}\log\left\{\frac{s^2+b^2}{(s-a)^2}\right\}$ **Ans.** $\frac{e^{-at}-\cos bt}{t}$
6. $\tan^{-1}(s+1)$ **Ans.** $-\frac{1}{t}e^{-t}\sin t$

## 13.27 INVERSE LAPLACE TRANSFORM OF INTEGRALS

$$L^{-1}\left[\int_s^\infty f(s)ds\right]=\frac{f(t)}{t}=\frac{1}{t}L^{-1}[F(s)] \quad \text{or} \quad L^{-1}[F(s)]=tL^{-1}\left[\int_s^\infty F(s)ds\right]$$

**Example 41.** *Obtain* $L^{-1}\frac{2s}{(s^2+1)^2}$.

**Solution.** $L^{-1}\frac{2s}{(s^2+1)^2}-tL^{-1}\int_s^\infty\frac{2sds}{(s^2+1)^2}=tL^{-1}\left[-\frac{1}{s^2+1}\right]_s^\infty=tL^{-1}\left[-0+\frac{1}{s^2+1}\right]$

$= t\sin t$ **Ans.**

## 13.28 PARTIAL FRACTIONS METHOD

**Example 42.** *Find the inverse transforms of*

$$\frac{1}{s^2-5s+6}.$$

**Solution.** Let us convert the given function into partial fractions.

$$L^{-1}\left[\frac{1}{s^2-5s+6}\right]=L^{-1}\left[\frac{1}{s-3}-\frac{1}{s-2}\right]$$

$$=L^{-1}\left(\frac{1}{s-3}\right)-L^{-1}\left(\frac{1}{s-2}\right)=e^{3t}-e^{2t}$$ **Ans.**

**Example 43.** *Find the inverse Laplace transforms of* $\frac{s+4}{s(s-1)(s^2+4)}$

**Solution.** Let us first resolve $\frac{s+4}{s(s-1)(s^2+4)}$ into partial fractions.

$$\frac{s+4}{s(s-1)(s^2+4)}\equiv\frac{A}{s}+\frac{B}{s-1}+\frac{Cs+D}{s^2+4} \quad \text{...(1)}$$

$$s+4\equiv A(s-1)(s^2+4)+Bs(s^2+4)+(Cs+D)s(s-1)$$

Putting $s = 0$, we get $4 = -4A$ or $A = -1$

Putting $s = 1$, we get $5 = B . 1 . (1 + 4) \Rightarrow B = 1$

Equating the coefficients of $s^3$ on both sides of (1), we have

$$0 = A + B + C \quad \Rightarrow \quad 0 = -1 + 1 + C \quad \Rightarrow \quad C = 0.$$

Equating the coefficients of $s$ on both sides of (1), we get

$$1 = 4A + 4B - D \quad \Rightarrow \quad 1 = -4 + 4 - D \quad \Rightarrow \quad D = -1$$

On putting the values of $A, B, C, D$ in (1), we get

$$\frac{s+4}{s(s-1)(s^2+4)}=-\frac{1}{s}+\frac{1}{s-1}-\frac{1}{s^2+4}$$

$$\therefore \quad L^{-1}\left[\frac{s+4}{s(s-1)(s^2+4)}\right] = L^{-1}\left[-\frac{1}{s}+\frac{1}{s-1}-\frac{1}{s^2+4}\right]$$

$$= -L^{-1}\left(\frac{1}{s}\right)+L^{-1}\left(\frac{1}{s-1}\right)-\frac{1}{2}L^{-1}\left(\frac{2}{s^2+2^2}\right)$$

$$= -1+e^t-\frac{1}{2}\sin 2t \qquad \textbf{Ans.}$$

**Example 44.** *Find the Laplace inverse of*

$$\frac{s^2}{(s^2+a^2)(s^2+b^2)}$$

**Solution.** Let us convert the given function into partial fractions.

$$L^{-1}\left[\frac{s^2}{(s^2+a^2)(s^2+b^2)}\right] = L^{-1}\left[\frac{a^2}{a^2-b^2}\cdot\frac{1}{s^2+a^2}-\frac{b^2}{a^2-b^2}\cdot\frac{1}{s^2+b^2}\right]$$

$$= \frac{1}{a^2-b^2}L^{-1}\left[\frac{a^2}{s^2+a^2}-\frac{b^2}{s^2+b^2}\right] = \frac{1}{a^2-b^2}\left[a^2\left(\frac{1}{a}\sin at\right)-b^2\left(\frac{1}{b}\sin bt\right)\right]$$

$$= \frac{1}{a^2-b^2}[a\sin at - b\sin bt] \qquad \textbf{Ans.}$$

### Exercise 13.14

Find the inverse transform of:

1. $\dfrac{s^2+2s+6}{s^3}$ **Ans.** $1+2t+3t^2$
2. $\dfrac{1}{s^2-7s+12}$ **Ans.** $e^{4t}-e^{3t}$
3. $\dfrac{s+2}{s^2-4s+13}$ **Ans.** $e^{2t}\cos 3t+\frac{4}{3}e^{2t}\sin 3t$
4. $\dfrac{3s+1}{(s-1)(s^2+1)}$ **Ans.** $e^t-2\cos t+\sin t$
5. $\dfrac{11s^2-2s+5}{2s^3-3s^2-3s+2}$ **Ans.** $2e^{-t}+5e^{2t}-\frac{3}{2}e^{t/2}$
6. $\dfrac{2s^2-6s+5}{(s-1)(s-2)(s-3)}$ **Ans.** $\frac{1}{2}e^t-e^{2t}+\frac{5}{2}e^{3t}$
7. $\dfrac{s-4}{(s-4)^2+9}$ **Ans.** $e^{4t}\cos 3t$
8. $\dfrac{16}{(s^2+2s+5)^2}$ **Ans.** $e^{-t}(\sin 2t-2t\cos 2t)$
9. $\dfrac{1}{(s+1)(s^2+2s+2)}$ **Ans.** $e^{-t}(1-\cos t)$
10. $\dfrac{1}{(s-2)(s^2+1)}$ **Ans.** $\frac{1}{5}e^{2t}-\frac{1}{5}\cos t-\frac{2}{5}\sin t$
11. $\dfrac{s^2-6s+7}{(s^2-4s+5)^2}$ **Ans.** $e^{2t}[t\cos t-\sin t]$

## 13.29 INVERSE LAPLACE TRANSFORM BY CONVOLUTION

$$L\left\{\int_0^t f_1(x)*f_2(t-x)dx\right\}=F_1(s).F_2(s) \Rightarrow \int_0^t f_1(x).f_2(t-x)dx = L^{-1}F_1(s).F_2(s)$$

**Example 45.** *Using the convolution theorem, find*

$$L^{-1}\left\{\frac{s^2}{(s^2+a^2)(s^2+b^2)}\right\},\ a \neq b$$

**Solution.** We have

$$L(\cos at)=\frac{s}{s^2+a^2} \text{ and } L(\cos bt)=\frac{s}{s^2+b^2}$$

Hence by the convolution theorem

$$L\left\{\int_0^t \cos ax \cos b(t-x)dx\right\}=\frac{s^2}{(s^2+a^2)(s^2+b^2)}$$

Therefore,

$$L^{-1}\left\{\frac{s^2}{(s^2+a^2)(s^2+b^2)}\right\}=\int_0^t \cos ax \cos b(t-x)dx$$

$$=\frac{1}{2}\int_0^t \{\cos(ax+bt-bx)+\cos(ax-bt+bx)\}dx$$

$$=\frac{1}{2}\int_0^t \cos[(a-b)x+bt]dx+\frac{1}{2}\int_0^t \cos[(a+b)x-bt]dx$$

$$=\left[\frac{\sin[(a-b)x+bt]}{2(a-b)}\right]_0^t+\left[\frac{\sin[(a+b)x-bt]}{2(a+b)}\right]_0^t=\frac{\sin at-\sin bt}{2(a-b)}+\frac{\sin at+\sin bt}{2(a+b)}$$

$$=\frac{a\sin at-b\sin bt}{a^2-b^2}$$ **Ans.**

**Example 46.** *Obtain* $L^{-1}\dfrac{1}{s(s^2+a^2)}$

**Solution.** $L^{-1}\dfrac{1}{s}=1$ and $L^{-1}\dfrac{1}{(s^2+a^2)}=\dfrac{\sin at}{a}$

Hence by the convolution theorem

$$L\int_0^t\left\{1.\frac{\sin a(t-x)}{a}dx\right\}=\left(\frac{1}{s}\right)\left(\frac{1}{s^2+a^2}\right)$$

$$L^{-1}\left\{\frac{1}{s(s^2+a^2)}\right\}=\int_0^t 1.\frac{\sin a(t-x)}{a}dx=\left[\frac{-\cos(at-ax)}{-a^2}\right]_0^t=\frac{1}{a^2}[1-\cos at]$$ **Ans.**

### Exercise 13.15

Obtain the inverse Laplace transform by convolution.

1. $\dfrac{s^2}{(s^2+a^2)^2}$ **Ans.** $\dfrac{1}{2}t\cos at+\dfrac{1}{2a}\sin at$
2. $\dfrac{1}{(s^2+1)^3}$ **Ans.** $\dfrac{1}{8}\{(3-t^2)\sin t-3t\cos t\}$
3. $\dfrac{s}{(s^2+a^2)^2}$ **Ans.** $\dfrac{t\sin at}{2a}$
4. $\dfrac{1}{s^2(s^2-a^2)}$ **Ans.** $\dfrac{1}{a^3}[-at+\sinh at]$
5. $\dfrac{1}{(s+1)(s^2+1)}$ **Ans.** $\dfrac{1}{2}(\cos t-\sin t-e^{-t})$

## 13.30. SOLUTION OF DIFFERENTIAL EQUATIONS BY LAPLACE TRANSFORMS

Ordinary linear differential equations with constant coefficients can be easily solved by the Laplace Transform method, without finding the general solution and the arbitrary constants.

The method will be clear from the following examples:

**Example 47.** *Using Laplace transforms, find the solution of the initial value problem*

$$y'' - 4y' + 4y = 64 \sin 2t$$
$$y(0) = 0, \; y'(0) = 1.$$

**Solution.** Here, we have $y'' - 4y' + 4y = 64 \sin 2t$ ... (1)

$$y(0) = 0, \; y'(0) = 1.$$

Taking Laplace transform of both sides of (1), we have

$$[s^2 \bar{y} - sy(0) - y'(0)] - 4[s\bar{y} - y(0)] + 4\bar{y} = \frac{64 \times 2}{s^2 + 4} \quad \text{... (2)}$$

On putting the values of $y(0)$ and $y'(0)$ in (2), we get

$$s^2 \bar{y} - 1 - 4s\bar{y} + 4\bar{y} = \frac{128}{s^2 + 4}$$

$$\Rightarrow \quad (s^2 - 4s + 4)\bar{y} = 1 + \frac{128}{s^2 + 4}, \quad \Rightarrow \quad (s-2)^2 \bar{y} = 1 + \frac{128}{s^2 + 4}$$

$$\Rightarrow \quad \bar{y} = \frac{1}{(s-2)^2} + \frac{128}{(s-2)^2 (s^2 + 4)} = \frac{1}{(s-2)^2} - \frac{8}{s-2} + \frac{16}{(s-2)^2} + \frac{8s}{s^2 + 4}$$

$$\Rightarrow \quad y = L^{-1}\left[-\frac{8}{s-2} + \frac{17}{(s-2)^2} + \frac{8s}{s^2 + 4}\right]$$

$$\Rightarrow \quad y = -8e^{2t} + 17t\,e^{2t} + 8 \cos 2t$$ **Ans.**

**Example 48.** *Using the Laplace transforms, find the solution of the initial value problem*

$$y'' + 25y = 10 \cos 5t$$
$$y(0) = 2, \; y'(0) = 0$$

**Solution.** Taking Laplace transform of the given differential equation, we get

$$[s^2 \bar{y} - sy(0) - y'(0)] + 25\bar{y} = 10 \frac{s}{s^2 + 25}$$

$$\Rightarrow \quad s^2 \bar{y} - 2s + 25\bar{y} = \frac{10s}{s^2 + 25}$$

$$\Rightarrow \quad (s^2 + 25)\bar{y} = 2s + \frac{10s}{s^2 + 25}$$

$$\Rightarrow \quad \bar{y} = \frac{2s}{s^2 + 25} + \frac{10s}{s^2 + 25}$$

$$\Rightarrow \quad y = L^{-1}\left[\frac{2s}{s^2 + 25} + \frac{10s}{(s^2 + 25)^2}\right] = 2 \cos 5t + L^{-1}\left[\frac{10s}{(s^2 + 25)^2}\right]$$

$$= 2 \cos 5t - L^{-1} \frac{d}{ds}\left[\frac{-5}{(s^2 + 25)}\right]$$

$$= 2 \cos 5t + t \sin 5t$$ **Ans.**

**Example 49.** *Applying convolution, solve the following initial value problem*

$$y'' + y = \sin 3t$$
$$y(0) = 0, \qquad y'(0) = 0.$$

**Solution.** $y'' + y = \sin 3t$

Taking Laplace transform of both the sides, we have

$$[s^2\bar{y} - sy(0) - y'(0)] + \bar{y} = \frac{3}{s^2+9} \qquad ...(1)$$

On putting the values of $y(0), y'(0)$ in (1), we get

$$s^2\bar{y} + \bar{y} = \frac{3}{s^2+9} \quad \Rightarrow \quad (s^2+1)\bar{y} = \frac{3}{s^2+9}$$

$$\Rightarrow \qquad \bar{y} = \frac{3}{(s^2+1)(s^2+9)} = \frac{3}{8}\left[\frac{1}{s^2+1} - \frac{1}{s^2+9}\right]$$

Taking the inversion transform, we get

$$y = \frac{3}{8}L^{-1}\frac{1}{s^2+1} - \frac{3}{8}L^{-1}\frac{1}{s^2+9}$$

$$y = \frac{3}{8}\sin t - \frac{3}{8}\times\frac{1}{3}\sin 3t = \frac{3}{8}\sin t - \frac{1}{8}\sin 3t \qquad \textbf{Ans.}$$

**Example 50.** *Solve* $[tD^2 + (1-2t)D - 2]y = 0$, *where* $y(0) = 1, y'(0) = 2$.

(*R.G.P.V. June, 2002*)

**Solution.** Here, $tD^2y + (1-2t)Dy - 2y = 0 \Rightarrow \quad ty'' + y' - 2ty' - 2y = 0$

Taking Laplace transform of given differential equation, we get

$$L(ty'') + L(y') - 2L(ty') - 2L(y) = 0 \Rightarrow -\frac{d}{ds}L\{y''\} + L\{y'\} + 2\frac{d}{ds}L(y') - 2L(y) = 0$$

$$-\frac{d}{ds}[s^2\bar{y} - sy(0) - y'(0)] + [s\bar{y} - y(0)] + 2\frac{d}{ds}[s\bar{y} - y(0)] - 2\bar{y} = 0$$

Putting the values of $y(0)$ and $y'(0)$, we get

$$-\frac{d}{ds}(s^2\bar{y} - s - 2) + (s\bar{y} - 1) + 2\frac{d}{ds}(s\bar{y} - 1) - 2\bar{y} = 0 \qquad [\because y(0) = 1, y'(0) = 2]$$

$$\Rightarrow -\frac{s^2 d\bar{y}}{ds} - 2s\bar{y} + 1 + s\bar{y} - 1 + 2\left(s\frac{d\bar{y}}{ds} + \bar{y}\right) - 2\bar{y} = 0 \quad \Rightarrow -(s^2 - 2s)\frac{d\bar{y}}{ds} - s\bar{y} = 0$$

$$\Rightarrow -\frac{d\bar{y}}{\bar{y}} + \frac{1}{s-2}ds = 0 \qquad \text{(Separating the variables)}$$

$$\Rightarrow \int\frac{d\bar{y}}{\bar{y}} + \int\frac{ds}{s-2} = 0 \quad \Rightarrow \quad \log\bar{y} + \log(s-2) = \log C$$

$$\Rightarrow \bar{y}(s-2) = C \Rightarrow \quad \bar{y} = \frac{C}{s-2} \quad \Rightarrow y = CL^{-1}\left\{\frac{1}{s-2}\right\} \Rightarrow y = Ce^{2t} \qquad ...(1)$$

Putting $y(0) = 1$ in (1), we get

$$1 = Ce^0 \quad \Rightarrow \quad C = 1$$

Putting $C = 1$ in (1), we get $y = e^{2t}$

This is the required solution. **Ans.**

**Example 51.** *Using Laplace transform technique solve the following initial value problem*

$$\frac{d^2y}{dt^2} + 2\frac{dy}{dt} + 2y = 5\sin t, \textit{ where } y(0) = y'(0) = 0$$

**Solution.** We have, $\quad y'' + 2y' + 2y = 5\sin t$

$$y(0) = y'(0) = 0$$

Taking the Laplace Transform of both sides, we have

$$[s^2\bar{y} - sy(0) - y'(0)] + 2[s\bar{y} - y(0)] + 2\bar{y} = 5\times\frac{1}{s^2+1} \qquad ...(1)$$

On substituting the values of $y(0)$, and $y'(0)$ in (1), we get

$$s^2\bar{y} + 2s\bar{y} + 2\bar{y} = \frac{5}{s^2+1} \Rightarrow [s^2+2s+2]\bar{y} = \frac{5}{s^2+1}$$

$$\Rightarrow \quad \bar{y} = \frac{5}{(s^2+2s+2)(s^2+1)}$$

Resolving into partial fractions, $y = \frac{2s+3}{s^2+2s+2} + \frac{-2s+1}{s^2+1}$

Taking the inverse transform, we get

$$y = L^{-1}\left[\frac{2s+3}{s^2+2s+2}\right] + L^{-1}\left(\frac{-2s+1}{s^2+1}\right) = L^{-1}\left[\frac{2(s+1)+1}{(s+1)^2+1}\right] + L^{-1}\left(\frac{-2s}{s^2+1}\right) + L^{-1}\left(\frac{1}{s^2+1}\right)$$

$$= L^{-1}\left[\frac{2(s+1)}{(s+1)^2+1}\right] + L^{-1}\left(\frac{1}{(s+1)^2+1}\right) - 2\cos t + \sin t$$

$$= 2e^{-t}\cos t + e^{-t}\sin t - 2\cos t + \sin t \qquad \textbf{Ans.}$$

**Example 52.** *Solve the initial value problem*

$$2y'' + 5y' + 2y = e^{-2t}, \quad y(0) = 1, \; y'(0) = 1,$$

*using the Laplace transforms.*

**Solution.** $2y'' + 5y' + 2y = e^{-2t}, y(0) = 1, y'(0) = 1$

Taking the Laplace Transform of both sides, we get

$$2[s^2\bar{y} - sy(0) - y'(0)] + 5[s\bar{y} - y(0)] + 2\bar{y} = \frac{1}{s+2} \qquad \text{...(1)}$$

On substituting the values of $y(0)$ and $y'(0)$ in (1), we get

$$2[s^2\bar{y} - s - 1] + 5[s\bar{y} - 1] + 2\bar{y} = \frac{1}{s+2}$$

$$[2s^2+5s+2]\bar{y} - 2s - 2 - 5 = \frac{1}{s+2}$$

$$\bar{y} = \frac{1}{(s+2)(2s^2+5s+2)} + \frac{2s+7}{2s^2+5s+2} = \frac{1+2s^2+7s+4s+14}{(2s^2+5s+2)(s+2)} = \frac{2s^2+11s+15}{(2s+1)(s+2)^2}$$

$$= \frac{4/9}{2s+1} - \frac{11/9}{s+2} - \frac{1/3}{(s+2)^2} = \frac{4}{9}\frac{1}{2}\frac{1}{s+\frac{1}{2}} - \frac{11}{9}\frac{1}{s+2} - \frac{1}{3}\frac{1}{(s+2)^2}$$

$$y = \frac{2}{9}e^{-\frac{1}{2}t} - \frac{11}{9}e^{-2t} - \frac{1}{3}te^{-2t} \qquad \textbf{Ans.}$$

**Example 53.** *Solve* $\frac{d^2y}{dx^2} + 2\frac{dy}{dx} + 5y = e^{-x}\sin x$ *where* $y(0) = 0, \; y'(0) = 1$

**Solution.** $\frac{d^2y}{dx^2} + 2\frac{dy}{dx} + 5y = e^{-x}\sin x$

Taking the Laplace Transform of both the sides, we get

$$[s^2\bar{y} - sy(0) - y'(0)] + 2[s\bar{y} - y(0)] + 5\bar{y} = L(e^{-x}\sin x)$$

$$[s^2\bar{y} - sy(0) - y'(0)] + 2[s\bar{y} - y(0)] + 5\bar{y} = \frac{1}{(s+1)^2+1} \qquad \text{... (1)}$$

On substituting the values of $y(0)$ and $y'(0)$ in (1), we get

$$(s^2\bar{y}-1)+2(s\bar{y})+5\bar{y}=\frac{1}{s^2+2s+2}$$

$$\left(s^2+2s+5\right)\bar{y}=1+\frac{1}{s^2+2s+2}=\frac{s^2+2s+3}{s^2+2s+2}$$

$$\bar{y}=\frac{s^2+2s+3}{\left(s^2+2s+5\right)\left(s^2+2s+2\right)}$$

On resolving the R.H. S. into partial fractions, we get

$$\bar{y}=\frac{2}{3}\frac{1}{s^2+2s+5}+\frac{1}{3}\frac{1}{s^2+2s+2}$$

On inversion, we obtain

$$y=\frac{2}{3}L^{-1}\frac{1}{s^2+2s+5}+\frac{1}{3}L^{-1}\frac{1}{s^2+2s+2}$$

$$\Rightarrow \quad y=\frac{1}{3}L^{-1}\frac{2}{(s+1)^2+(2)^2}+\frac{1}{3}L^{-1}\frac{1}{(s+1)^2+(1)^2}$$

$$\Rightarrow \quad y=\frac{1}{3}e^{-x}\sin 2x+\frac{1}{3}e^{-x}\sin x$$

$$\Rightarrow \quad y=\frac{1}{3}e^{-x}(\sin x+\sin 2x)$$ **Ans.**

**Example 54.** *Using Laplace transforms, find the solution of the initial value problem*

$$y''+9y=9u(t-3),\ y(0)=y'(0)=0$$

*where $u(t-3)$ is the unit step function.*

**Solution.** $y''+9y=9u(t-3)$ ...(1)

Taking Laplace transform of (1), we have

$$s^2\bar{y}-sy(0)-y'(0)+9\bar{y}=9\frac{e^{-3s}}{s} \quad ...(2)$$

Putting the values of $y(0)=0$ and $y'(0)=0$ in (2), we get

$$s^2\bar{y}+9\bar{y}=\frac{9e^{-3s}}{s}$$

$$\Rightarrow \quad \left(s^2+9\right)\bar{y}=9\frac{e^{-3s}}{s}$$

$$\Rightarrow \quad \bar{y}=\frac{9e^{-3s}}{s\left(s^2+9\right)} \Rightarrow y=L^{-1}\frac{9e^{-3s}}{s\left(s^2+9\right)}$$

$$\Rightarrow \quad L^{-1}\frac{3}{s^2+9}=\sin 3t$$

$$\Rightarrow \quad 3L^{-3}\frac{3}{s\left(s^2+9\right)}=3\int_0^t \sin 3t\,dt=-\left[\cos 3t\right]_0^t=1-\cos 3t$$

$$\Rightarrow \quad y=L^{-1}\frac{9e^{-3s}}{s\left(s^2+9\right)}$$

$$\Rightarrow \quad y=\left[1-\cos 3(t-3)\right]u(t-3)$$ **Ans.**

**Example 55.** *A resistance R in series with inductance L is connected with e.m.f. E. (t). The current i is given by*

$$L\frac{di}{dt}+Ri=E(t)$$

*If the switch is connected at t = 0 and disconnected at t = a, find the current i in terms of t.*

*(UP., II Semester; Summer 2001)*

**Solution.** Conditions under which current $i$ flows are $i = 0$ at $t = 0$,

$$E(t)=\begin{cases}E, & 0<t<a\\ 0, & t>a\end{cases}$$

Given equation is $\quad L\frac{di}{dt}+Ri=E(t)$ ...(1)

Taking Laplace transform of (1), we get

$$L\,[s\bar{i}-i(0)]+R\bar{i}=\int_0^\infty e^{-st}E(t)\,dt$$

$$Ls\bar{i}+R\bar{i}=\int_0^\infty e^{-st}E(t)\,dt \qquad [i(0)=0]$$

$$(Ls+R)\bar{i}=\int_0^\infty e^{-st}.E\,dt=\int_0^a e^{-st}E\,dt+\int_a^\infty e^{-st}E\,dt$$

$$=E\left[\frac{e^{-st}}{-s}\right]_0^a+0=\frac{E}{s}\left[1-e^{-as}\right]=\frac{E}{s}-\frac{E}{s}e^{-as}$$

$$\bar{i}=\frac{E}{s(Ls+R)}-\frac{Ee^{-as}}{s(Ls+R)}$$

On inversion, we obtain $\quad i=L^{-1}\left[\frac{E}{s(Ls+R)}\right]-L^{-1}\left[\frac{Ee^{-as}}{s(Ls+R)}\right]$ ...(2)

Now we have to find the value of $L^{-1}\left[\frac{E}{s(Ls+R)}\right]$

$$L^{-1}\left[\frac{E}{s(Ls+R)}\right]=\frac{E}{L}L^{-1}\left[\frac{E}{s\left(s+\frac{R}{L}\right)}\right] \qquad \text{(Resolving into partial fractions)}$$

$$=\frac{E}{L}\frac{L}{R}L^{-1}\left[\frac{1}{s}-\frac{1}{s+\frac{R}{L}}\right]=\frac{E}{R}\left[1-e^{-\frac{R}{L}t}\right]$$

and $\quad L^{-1}\left[\frac{Ee^{-as}}{s(Ls+R)}\right]=\frac{E}{R}\left[1-e^{-\frac{R}{L}(t-a)}\right]u(t-a)$

[By the second shifting theorem]

On substituting the values of the inverse transforms in (2) we get

$$i=\frac{E}{R}\left[1-e^{-\frac{R}{L}t}\right]-\frac{E}{R}\left[1-e^{-\frac{R}{L}(t-a)}\right]u(t-a)$$

Hence $\quad i=\frac{E}{R}\left[1-e^{-\frac{R}{L}t}\right]$ for $0<t<a$, $\qquad [u(t-a)=0]$

$$i=\frac{E}{R}\left[1-e^{-\frac{R}{L}t}\right]-\frac{E}{R}\left\{1-e^{-\frac{R}{L}(t-a)}\right\} \qquad \text{for } t>a \quad [u(t-a)=1]$$

$$=\frac{E}{R}\left[e^{-\frac{R}{L}(t-a)}-e^{-\frac{R}{L}t}\right]=\frac{E}{R}e^{-\frac{R}{L}t}\left\{e^{\frac{Ra}{L}}-1\right\}$$ **Ans.**

**Example 56.** *Using the Laplace transform, find the current i (t) in the LC - circuit. Assuming L = 1 henry, C = 1 farad, zero initial current and charge on the capacitor, and*

*v (t) = t, when 0 < t < 1*

*= 0 otherwise.*

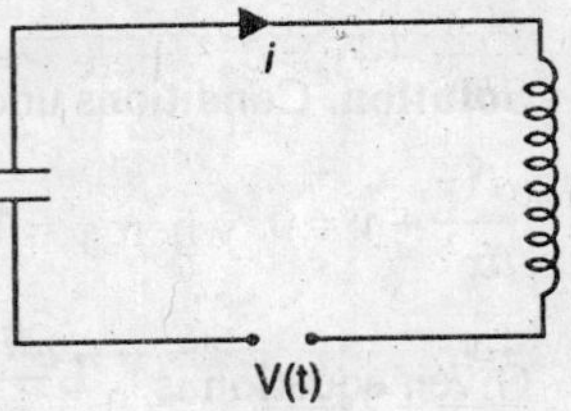


**Solution.** The differential equation for L and C circuit is given by

$$L\frac{d^2q}{dt^2}+\frac{q}{C}=E \qquad \text{... (1)}$$

Putting $L = 1$, $C = 1$, $E = v(t)$ in (1), we get $\frac{d^2q}{dt^2}+q=v(t)$ ... (2)

Taking Laplace Transform of (2), we have

$$s^2\bar{q}-sq(0)-q'(0)+\bar{q}=\int_0^\infty v(t)\,e^{-st}\,dt$$

Substituting $q(0) = 0$, and $q'(0) = 0$, we get

$$s^2\bar{q}+\bar{q}=\int_0^1 te^{-st}\,dt+\int_1^\infty 0\,e^{-st}\,dt$$

$$\Rightarrow \quad (s^2+1)\,\bar{q}=\left[t\frac{e^{-st}}{-s}\right]_0^1-\int_0^1\frac{e^{-st}}{-s}\,dt=\frac{e^{-s}}{-s}-\left[\frac{e^{-st}}{s^2}\right]_0^1=-\frac{e^{-s}}{s}-\frac{e^{-s}}{s^2}+\frac{1}{s^2}$$

$$\Rightarrow \quad \bar{q}=\frac{1}{s^2+1}\left[-\frac{e^{-s}}{s}-\frac{e^{-s}}{s^2}+\frac{1}{s^2}\right]$$

$$\Rightarrow \quad \bar{q}=\frac{-e^{-s}}{s\left(s^2+1\right)}-\frac{e^{-s}}{s^2\left(s^2+1\right)}+\frac{1}{s^2\left(s^2+1\right)}$$

Taking Inverse Laplace Transform, we get

$$q=L^{-1}\frac{-e^{-s}}{s\,(s^2+1)}-L^{-1}\frac{e^{-s}}{s^2\,(s^2+1)}+L^{-1}\frac{1}{s^2\,(s^2+1)} \qquad \text{... (3)}$$

We know that

$$L^{-1}\,[e^{-as}F(s)]=f(t-a)\,u(t-a)$$

$$L^{-1}\left[\frac{1}{s(s^2+1)}\right]=\int_0^t\sin t\,dt=[-\cos t]_0^t=1-\cos t \qquad \text{... (4)}$$

$$L^{-1}\left[\frac{1}{s^2\,(s^2+1)}\right]=\int_0^t(1-\cos t)\,dt=t-\sin t \qquad \text{... (5)}$$

In view of this, we have

$$L^{-1}\left[\frac{-e^{-s}}{s\,(s^2+1)}\right]=-[1-\cos(t-1)]\,u(t-1) \qquad \text{[From (4)]}$$

$$L^{-1}\;L^{-1}\left[\frac{e^{-s}}{s^2\,(s^2+1)}\right]=[(t-1)-\sin(t-1)]\,u(t-1) \qquad \text{[From (5)]}$$

Putting the above values in (3), we get

$$q=-[1-\cos(t-1)]\,u(t-1)-[(t-1)-\sin(t-1)]\,u(t-1)+t-\sin t$$ **Ans.**

## EXERCISE 13.16

Solve the following differential equations:

1. $\frac{d^2y}{dx^2}+y=0$, where $y = 1$ and $\frac{dy}{dx}=-1$ at $x = 0$. **Ans.** $y = \cos x - \sin x$

2. $\frac{d^2y}{dx^2}-4y=0$, where $y = 0$ and $\frac{dy}{dx}=-6$ at $x = 0$. **Ans.** $y=-\frac{3}{2}e^{2x}+\frac{3}{2}e^{-2x}$

3. $\frac{d^2y}{dx^2}+y=0$, where $y = 1$, $\frac{dy}{dx}=1$ at $x = 0$. **Ans.** $y = \sin x + \cos x$

4. $\frac{d^2y}{dx^2}+2\frac{dy}{dx}+5y=0$, where $y = 2$, $\frac{dy}{dx}=-4$ at $x = 0$. **Ans.** $y = e^{-x}(2\cos 2x - \sin 2x)$

5. $\frac{d^3y}{dx^3}+2\frac{d^2y}{dx^2}-\frac{dy}{dx}-2y=0$, given $y=\frac{dy}{dx}=0$, $\frac{d^2y}{dx^2}=6$ at $x = 0$. **Ans.** $y = e^x - 3e^{-x} + 2e^{-2x}$

6. $\frac{d^2y}{dx^2}+y=3\cos 2x$, where $y=\frac{dy}{dx}=0$ at $x = 0$. **Ans.** $y = \cos x - \cos 2x$.

7. $\frac{d^2y}{dx^2}+\frac{dy}{dx}-2y=1-2x$, given $y = 0$. $\frac{dy}{dx}=4$ at $x = 0$. **Ans.** $y = e^x - e^{-2x} + x$

8. $\frac{d^2y}{dx^2}-3\frac{dy}{dx}+2y=4e^{2x}$, *given* $y = -3$, and $\frac{dy}{dx}=5$ at $x = 0$. **Ans.** $y = -7e^x + 4e^{2x} + 4x\,e^{2x}$

9. $\frac{d^2y}{dx^2}-3\frac{dy}{dx}+2y=4x+e^{2x}$, where $y = 1$, $\frac{dy}{dx}=-1$ at $x = 0$ **Ans.** $y=3+2x+\frac{1}{2}e^{2x}-2e^{2x}-\frac{1}{2}e^x$.

10. $\frac{d^3y}{dx^3}+2\frac{d^2y}{dx^2}-\frac{dy}{dx}-2y=0$, where $y = 1$, $\frac{dy}{dx}=2$, $\frac{d^2y}{dx^2}=2$ at $x = 0$ **Ans.** $\frac{5}{3}e^x-e^{-x}+\frac{1}{3}e^{-2x}$

11. $(D^2 - D - 2)\,x = 20\sin 2t$, $x_0 = -1$, $x_1 = 2$ **Ans.** $x = 2e^{2t} - 4e^{-t} + \cos 2t - 3\sin 2t$

12. $(D^3 + D^2)\,x = 6t^2 + 4$, $x(0) = 0$, $x'(0) = 2$, $x''(0) = 0$ **Ans.** $x=\frac{1}{2}t^4-2t^3+8t^2-16t+16-16e^{-t}$

13. $\frac{d^2x}{dt^2}-2\frac{dx}{dt}+x=e^t$, where $x(0) = 2$, $\frac{dx}{dt}=-1$ at $t = 0$ **Ans.** $x=2e^t-3te^t+\frac{1}{2}t^2e^t$

14. $(D^2 + n^2)\,x = a\sin(nt + \alpha)$ where $x = Dx = 0$ at $t = 0$.

**Ans.** $x = an\cos\alpha\,(\sin nt - nt\cos nt)+\frac{a\sin 2\alpha}{2n}(t\sin nt)$

15. $y'' + 2y' + y = t\,e^{-t}$ if $y(0) = 1$, $y'(0) = -2$. **Ans.** $y=\left(1-t+\frac{t^3}{6}\right)e^{-t}$

16. $\frac{d^2y}{dx^2}+y=x\cos 2x$, where $y=\frac{dy}{dx}=0$ at $x = 0$. **Ans.** $y=\frac{4}{9}\sin 2x-\frac{5}{9}\sin x-\frac{x}{3}\cos 2x$

17. $\frac{d^3y}{dx^3}-3\frac{d^2y}{dx^2}+3\frac{dy}{dx}-y=x^2e^{2x}$, where $y = 1$, $\frac{dy}{dx}=0$, $\frac{d^2y}{dx^2}=-2$ at $x = 0$.

**Ans.** $y = e^{2x}(x^2 - 6x + 12) - e^x(15x^2 + 7x + 11)$

18. $y'' + 4y' + 3y = t$, $t > 0$; given that $y(0) = 0$ and $y'(0) = 1$. **Ans.** $y=-\frac{4}{9}+\frac{t}{6}+e^{-t}-\frac{5}{9}e^{-3t}$

**19.** $y'' + 2y = r(t)$, $y(0) = 0$, $y'(0) = 0$ where $r(t) = \begin{cases} 0, & t \geq 1 \\ 1, & 0 \leq t < 1 \end{cases}$ **Ans.** $y = \frac{1}{2} - \frac{1}{2}\cos\sqrt{2}\,t$.

**20.** $\frac{d^2y}{dt^2} + 4y = u(t-2)$, where $u$ is unit step function

$y(0) = 0$ and $y'(0) = 1$ **Ans.** $y = \frac{1}{2}\sin 2t$ for $t < 2$

**21.** $\frac{d^2y}{dx^2} + y = u(t-\pi) - u(t-2\pi)$, $y(0) = y'(0) = 0$

**Ans.** $y = (1+\cos t)\, u(t-\pi) - (1-\cos t)u(t-2\pi)$

**22.** A condenser of capacity C is charged to potential E and discharged at $t = 0$ through an inductance L and resistance R. The charge $q$ at lime $t$ is governed by the differential equation

$$L\frac{d^2q}{dt^2} + R\frac{dq}{dt} + \frac{q}{C} = E$$

Using Laplace transforms, show that the charge $q$ is given by

$$q = \frac{CE}{n} e^{-\mu t}[\mu \sin nt + n\cos nt] \text{ where } \mu = \frac{R}{2L} \text{ and } \eta^2 = \frac{1}{LC} - \frac{R^2}{4L^2}$$

## 13.31 SOLUTION OF SIMULTANEOUS DIFFERENTIAL EQUATIONS BY LAPLACE TRANSFORMS

Simultaneous differential equations can also be solved by Laplace Transform method.

**Example 57.** *Solve* $\frac{dx}{dt} + y = 0$ *and* $\frac{dy}{dt} - x = 0$ *under the condition* $x(0) = 1$, $y(0) = 0$

**Solution.** $x' + y = 0$ ...(1)

$y' - x = 0$ ...(2)

Taking the Laplace transform of (1) and (2) we get

$[s\bar{x} - x(0)] + \bar{y} = 0$ ... (3)

$[s\bar{y} - y(0)] - \bar{x} = 0$ ...(4)

On substituting the values of $x(0)$ and $y(0)$ in (3) and (4), we get

$s\bar{x} - 1 + \bar{y} = 0$ ...(5)

$s\bar{y} - \bar{x} = 0$ ...(6)

Solving (5) and (6) for $\bar{x}$ and $\bar{y}$ we get

$$\bar{x} = \frac{s}{s^2+1}, \quad \bar{y} = \frac{1}{s^2+1}$$

On inversion, we obtain $x = L^{-1}\left(\frac{s}{s^2+1}\right)$, $y = L^{-1}\left(\frac{1}{s^2+1}\right)$

$x = \cos t$, $y = \sin t$ **Ans.**

**Example 58.** *Using Laplace transforms, solve the differential equations*

$(D+1)y_1 + (D-1)y_2 = e^{-t}$

$(D+2)y_1 + (D+1)y_2 = e^{t}$

*where* $D = d/dt$ *and* $y_1(0) = 1$, $y_2(0) = 0$

**Solution.** $(D+1)y_1 + (D-1)y_2 = e^{-t}$ ...(1)

$(D+2)y_1 + (D+1)y_2 = e^{t}$ ...(2)

Multiply (1) by $(D+1)$ and (2) by $(D-1)$ we get

$(D+1)^2 y_1 + (D^2-1)y_2 = (D+1)e^{-t}$ ...(3)

$$(D-1)(D+2)y_1+(D^2-1)y_2=(D-1)e^t \quad ...(4)$$

Subtracting (4) from (3) we get

$$(D^2+2D+1-D^2-D+2)y_1=(-e^{-t}+e^{-t})-(e^t-e^t)$$

$$\Rightarrow \quad (D+3)y_1=0 \quad \Rightarrow \quad Dy_1+3y_1=0$$

Taking Laplace transform we have $s\bar{y}_1-y_1(0)+3\bar{y}_1=0$

$$(s+3)\bar{y}_1=1 \quad \Rightarrow \quad \bar{y}_1=\frac{1}{s+3} \quad \Rightarrow y_1=e^{-3t}$$

Putting the value of $y_1$ in (1) we get

$$(D+1)e^{-3t}+(D-1)y_2=e^{-t}$$

$$-3e^{-3t}+e^{-3t}+(D-1)y_2=e^{-t}$$

$$(D-1)y_2=e^{-t}+2e^{-3t} \quad \Rightarrow \quad Dy_2-y_2=e^{-t}+2e^{-3t}$$

Taking Laplace transform, we get

$$s\bar{y}_2-y_2(0)-\bar{y}_2=\frac{1}{s+1}+\frac{2}{s+3}$$

$$(s-1)\bar{y}_2=\frac{1}{s+1}+\frac{2}{s+3}$$

$$\bar{y}_2=\frac{1}{s^2-1}+\frac{2}{s^2+2s-3}$$

$$y_2=L^{-1}\left[\frac{1}{s^2-1}+\frac{2}{(s+1)^2-(2)^2}\right]$$

$$y_2=\sinh t+e^{-t}\sinh 2t$$

$$y_1=e^{-3t} \text{ and } y_2=\sinh t+e^{-t}\sinh 2t \quad \textbf{Ans.}$$

**Example 59.** *Solve* $\frac{dx}{dt}-y=e^t,\ \frac{dy}{dt}+x=\sin t$

*given* $x(0)=1,\ y(0)=0$

**Solution.** $x'-y=e^t \quad ...(1)$

$y'+x=\sin t \quad ...(2)$

Taking the Laplace Transform of (1) and (2), we get

$$[s\bar{x}-x(0)]-\bar{y}=\frac{1}{s-1} \quad ...(3)$$

$$[s\bar{y}-y(0)]+\bar{x}=\frac{1}{s^2+1} \quad ...(4)$$

On substituting the values of $x(0)$ and $y(0)$ in (3) and (4), we get

$$s\bar{x}-1-\bar{y}=\frac{1}{s-1} \quad ....(5)$$

$$s\bar{y}+x=\frac{1}{s^2+1} \quad ...(6)$$

On solving (5) and (6), we get

$$\bar{x}=\frac{s^4+s^2+s-1}{(s-1)(s^2+1)^2}=\frac{1}{2}\frac{1}{s-1}+\frac{1}{2}\frac{s+1}{s^2+1}+\frac{1}{(s^2+1)^2} \quad ...(7)$$

$$\bar{y}=\frac{-s^3+s^2-2s}{(s-1)(s^2+1)^2}=\frac{1}{2}\frac{1}{s-1}+\frac{1}{2}\frac{s-1}{(s^2+1)}+\frac{s}{(s^2+1)^2} \quad ..(8)$$

On inversion of (7), we obtain

$$x=\frac{1}{2}L^{-1}\frac{1}{s-1}+\frac{1}{2}L^{-1}\frac{s}{s^2+1}+\frac{1}{2}L^{-1}\frac{1}{s^2+1}+L^{-1}\frac{1}{(s^2+1)^2}$$

$$= \frac{1}{2}e^t + \frac{1}{2}\cos t + \frac{1}{2}\sin t + \frac{1}{2}(\sin t - t\cos t) = \frac{1}{2}[e^t + \cos t + 2\sin t - t\cos t]$$

On inversion of (8), we get

$$y = -\frac{1}{2}L^{-1}\frac{1}{s-1} + \frac{1}{2}L^{-1}\frac{s}{s^2+1} - \frac{1}{2}L^{-1}\frac{1}{s^2+1} + L^{-1}\frac{s}{(s^2+1)^2}$$

$$\Rightarrow \quad y = -\frac{1}{2}e^t + \frac{1}{2}\cos t - \frac{1}{2}\sin t + \frac{1}{2}t\sin t$$

$$\Rightarrow \quad y = \frac{1}{2}[-e^t - \sin t + \cos t + t\sin t]$$ **Ans.**

**Example 60.** *Using the Laplace transforms, solve the initial value problem*

$$y_1'' = y_1 + 3y_2$$

$$y_2'' = 4y_1 - 4e^t$$

$$y_1(0) = 2,\ y_1'(0) = 3,\ y_2(0) = 1,\ y_2'(0) = 2$$

**Solution.** $$y_1'' = y_1 + 3y_2 \quad \ldots (1)$$

$$y_2'' = 4y_1 - 4e^t \quad \ldots (2)$$

Taking the Laplace transform of (1) and (2), we get

$$s^2\bar{y}_1 - sy_1(0) - y_1'(0) = \bar{y}_1 + 3\bar{y}_2 \quad \ldots (3)$$

$$s^2\bar{y}_2 - sy_2(0) - y_2'(0) = 4\bar{y}_1 - \frac{4}{s-1} \quad \ldots (4)$$

Putting the values of $y_1(0), y'(0), y_2(0), y_2'(0)$ in (3) and (4), we get

$$s^2\bar{y}_1 - 2s - 3 = \bar{y}_1 + 3\bar{y}_2 \quad \Rightarrow \quad (s^2-1)\,\bar{y}_1 - 3\bar{y}_2 = 2s+3 \quad \ldots (5)$$

$$\Rightarrow \quad s^2\bar{y}_2 - s - 2 = 4\bar{y}_1 - \frac{4}{s-1} \quad \Rightarrow \quad 4\bar{y}_1 - s\bar{y}_2 = \frac{4}{s-1} - s - 2 \quad \ldots (6)$$

On solving (5) and (6), we get

$$\bar{y}_1 = \frac{(2s-3)(s^2+3)(s+2)}{(s-1)(s^2+3)(s^2-4)} = \frac{2s-3}{(s-1)(s-2)} = \frac{1}{s-1} + \frac{1}{s-2}$$

$$y_1 = e^t + e^{2t}$$

$$\bar{y}_2 = \frac{(s+2)(s^2+3)}{(s^2+3)(s^2-4)} = \frac{1}{s-2} \Rightarrow y_2 = e^{2t}$$ **Ans.**

## Exercise 13.17

**Solve the following :**

1. $\frac{dx}{dt} + 4y = 0,\ \frac{dy}{dt} - 9x = 0$ Given $x = 2$ and $y = 1$ at $t = 0$.

**Ans.** $x = -\frac{2}{3}\sin 6t + 2\cos 6t,\ y = \cos 6t + 3\sin 6t$

2. $4\frac{dy}{dt} + \frac{dx}{dt} + 3y = 0,\ \frac{3dx}{dt} + 2x + \frac{dy}{dt} = 1$

under the condition $x = y = 0$ at $t = 0$. **Ans.** $x = \frac{1}{2} - \frac{1}{5}e^{-t} - \frac{3}{10}e^{-\frac{6}{11}t},\ y = \frac{1}{5}e^{-t} - \frac{1}{5}e^{-\frac{6}{11}t}$

3. $\frac{dx}{dt} + 5x - 2y = t,\ \frac{dy}{dt} + 2x + y = 0$ being given $x = y = 0$ when $t = 0$.

**Ans.** $x = -\frac{1}{27}(1+6t)e^{-3t} + \frac{1}{27}(1+3t),\ y = -\frac{2}{27}(2+3t)e^{-3t} - \frac{2t}{9} + \frac{4}{27}$

4. $\dfrac{dx}{dt}+y=\sin t,\ \dfrac{dy}{dt}+x=\cos t$

given that $x = 2$, and $y = 0$ when $t = 0$. **Ans.** $x = e^t + e^{-t},\ y= e^{-t} - e^t + \sin t$

5. $(D - 1)x - 2y = t,\ -2x+(D - 1)y = t,\ t > 0$

where $D = d/dt$ and $x(0) = 2,\ y(0) = 4$

6. The small oscillations of a certain system with two degrees of freedom are given by the equations

$$D^2x + 3x - 2y = 0,\quad D^2x+D^2y - 3x+5y= 0$$

If $x = 0, y = 0,\ Dx = 3,\ Dy = 2$ when $t = 0$.

**Ans.** $x=-\dfrac{11}{4}\sin t+\dfrac{1}{12}\sin 3t,\ y=\dfrac{11}{4}\sin t-\dfrac{1}{4}\sin 3t$

7. $3\dfrac{dx}{dt}+3\dfrac{dy}{dt}+5x=25\cos t,\ 2\dfrac{dx}{dt}-3\dfrac{dy}{dt}=5\sin t$ with $x(0) = 2,\ y(0) = 3$.

**Ans.** $x = 2\cos t + 3\sin t,\ y = 3\cos t + 2\sin 2t$

## METHODS TO FIND OUT RESIDUES ON PAGE 590 (Art. 7.58)

## 13.32 INVERSION FORMULA FOR THE LAPLACE TRANSFORM

$f(x)$ = sum of the residues of $e^{sx}F(s)$ at the poles of $F(s)$.

**Proof.** The Laplace Transform of $f(x)$ is defined by

$$F(s)=\int_0^\infty e^{-st}\cdot f(t)\,dt$$

Multiplying by $e^{sx}$

$$e^{sx}F(s)=e^{sx}\int_0^\infty e^{-st}\cdot f(t)\,dt$$

Integrating w.r.t. '$s$' between the limits $a + ir$ and $a - ir$, we have

$$\int_{a-ir}^{a+ir} e^{sx}F(s)\,ds=\int_{a-ir}^{a+ir} e^{sx}ds\int_0^\infty e^{-st}\cdot f(t)\,dt$$

Putting $s = a - ip,\ ds = -\,idp = -i\int_r^{-r} e^{x(a-ip)}\int_0^\infty f(t)e^{-(a-ip)t}\,dt\,dp$

$$= ie^{ax}\int_{-r}^{r} e^{-ipx}dp\int_0^\infty f(t)e^{-at}\cdot e^{ipt}\,dt. \quad \text{...(1)}$$

Let us now define $\phi(x)$ as $\phi(x)=\begin{cases} e^{-ax}f(x) & \text{when } x\ge 0\\ 0 & \text{when } x<0\end{cases}$

The Fourier complex integral of $\phi(x)$ is

$$\phi(x)=\frac{1}{2\pi}\int_{-\infty}^{\infty} e^{ipx}\int_{-\infty}^{\infty}\phi(t)e^{ipt}\,dt\,dp$$

$$\Rightarrow\quad e^{-ax}f(x)=\frac{1}{2\pi}\int_{-\infty}^{\infty} e^{-ipx}\int_0^\infty\left[e^{-at}f(t)\right]e^{ipt}\,dt\,dp \quad \text{... (2)}$$

In the limiting case when $r\to\infty$, (1) becomes

$$\int_{a-i\infty}^{a+i\infty} e^{sx}F(s)\,ds=ie^{ax}\int_{-\infty}^{\infty} e^{-ipx}dp\int_{-\infty}^{\infty} f(t)e^{-at}\cdot e^{ipt}\,dt \quad \text{...(3)}$$

Substituting the value of the integral from (2) in (3), we get

$$\int_{a-i\infty}^{a+i\infty} e^{sx}F(s)\,ds=ie^{ax}\left[2\pi e^{-ax}f(x)\right]=2\pi i f(x)$$

$$\Rightarrow \qquad f(x)=\frac{1}{2\pi i}\int_{a-i\infty}^{a+i\infty} e^{sx}F(s)\,ds \qquad ...(4)$$

Equation (4) is called the inversion formula for the Laplace transform.

To obtain $f(x)$, the integration is performed along a line $AB$ parallel to imaginary axis in the complex plane such that all the singularities of $F(s)$ lie to its left. The contour $c$ includes the line $AB$ and the semicircle $c'$ (*i.e.* $BDA$).

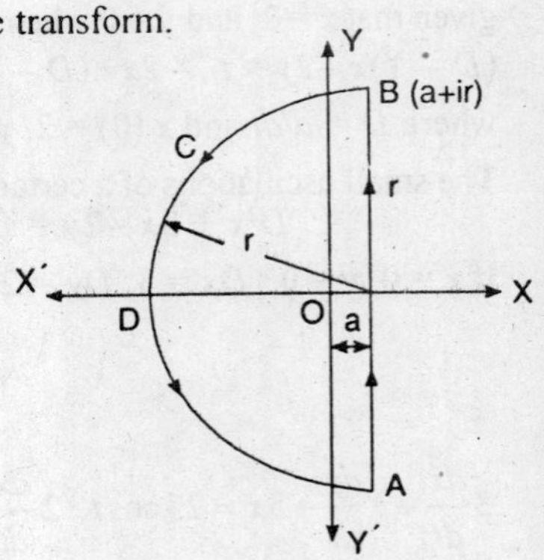


From (4)

$$f(x)=\frac{1}{2\pi i}\int_{AB} e^{sx}F(x)\,ds$$

$$=\frac{1}{2\pi i}\int_{c} e^{sx}F(s)\,ds$$

$$-\frac{1}{2\pi i}\int_{c'} e^{sx}F(s)\,ds$$

The integral over $c'$ tends to zero as $r\to\infty$. Therefore,

$$f(x)=\lim_{r\to\infty}\frac{1}{2\pi i}\int_{c} e^{sx}F(s)\,ds$$

**$f(x)$ = sum of the residue of $e^{sx}F(s)$ at the poles of $F(s)$.**

**Note.** Methods for finding the residue: See article 7.58 on page 590.

**Example 61.** *Obtain the inverse Laplace transform of* $\dfrac{s+1}{s^2+2s}$

**Solution.** Let $\qquad F(s)=\dfrac{s+1}{s^2+2s} \qquad ...(1)$

$$L^{-1}\left[\frac{s+1}{s^2+2s}\right]=\text{Sum of the residues of } e^{st}.\frac{s+1}{s^2+2s} \text{ at the poles.} \qquad ...(2)$$

The poles of (1) are determined by equating the denominator to zero, *i.e.*

$$s^2+2s=0 \quad \text{or} \quad s(s+2)=0 \qquad i.e.\ s=0,-2$$

There are two simple poles at $s=0$ and $s=-2$.

$$\text{Residue of } e^{st}.F(s) \text{ (at } s=0) = \lim_{s\to 0}\left[(s-0)\frac{e^{st}(s+1)}{s^2+2s}\right]=\lim_{s\to 0}\left[\frac{e^{st}(s+1)}{(s+2)}\right]=\frac{1}{2}$$

$$\text{Residue of } e^{st}.F(s) \text{ (at } s=-2) = \lim_{s\to -2}\left[\frac{(s+2)e^{st}(s+1)}{s(s+2)}\right]$$

$$=\lim_{s\to -2}\left[\frac{e^{st}(s+1)}{s}\right]=\frac{e^{-2t}(-2+1)}{-2}=\frac{e^{-2t}}{2}$$

$$\text{Sum of the residue [at } s=0 \text{ and } s=-2]=\frac{1}{2}+\frac{e^{-2t}}{2}$$

Putting the value of residues in (2), we get

$$L^{-1}\left[\frac{s+1}{s^2+2s}\right]=\frac{1}{2}+\frac{e^{-2t}}{2}$$ **Ans.**

**Example 62.** *Find the inverse Laplace transform of* $\dfrac{1}{(s+1)(s^2+1)}$.

**Solution.** Let $$F(s)=\frac{1}{(s+1)(s^2+1)} \qquad ...(1)$$

$$L^{-1}\left[\frac{1}{(s+1)(s^2+1)}\right] = \text{sum of residues of } e^{st} F(s) \text{ at the poles.} \qquad ...(2)$$

The poles of (1) are obtained by equating the denominator equal to zero, *i.e.*,

$$(s+1)(s^2+1)=0 \quad \Rightarrow \quad s=-1,\ +i, -i$$

There are three poles of $F(s)$ at $s=-1$, $s=+i$ and $s=-i$.

Residue of $e^{st}.F(s)$ (at $s=-1$) $= \lim\limits_{s\to -1}(s+1)\dfrac{e^{st}}{(s+1)(s^2+1)} = \lim\limits_{s\to -1}\dfrac{e^{-t}}{s^2+1} = \dfrac{e^{-t}}{2}$

Residue of $e^{st}.F(s)$ (at $s=i$) $= \lim\limits_{s\to i}(s-i)\dfrac{e^{st}}{(s+1)(s^2+1)}$

$$= \lim_{s\to i}\frac{e^{st}}{(s+1)(s+i)} = \frac{e^{it}}{(i+1)(2i)} = -i\frac{e^{it}}{2}\frac{1-i}{(i+1)(1-i)} = -\frac{e^{it}}{4}(1+i)$$

Residue of $e^{st}.F(s)$ (at $s=-i$) $= \lim\limits_{s\to -i}(s+i)\dfrac{e^{st}}{(s+1)(s^2+1)}$

$$= \lim_{s\to -i}\frac{e^{st}}{(s+1)(s-i)} = \frac{e^{-it}}{(-i+1)(-2i)} = \frac{e^{-it}(i-1)}{4}$$

Substituting the values of the residues in (2), we get

$$L^{-1}\left[\frac{1}{(s+1)(s^2+1)}\right] = \frac{e^{-t}}{2} - \frac{e^{it}(1+i)}{4} + \frac{e^{-it}(i-1)}{4}$$

$$= \frac{e^{-t}}{2} + \frac{-e^{it} - ie^{it} + ie^{-it} - e^{-it}}{4} = \frac{e^{-t}}{2} - \frac{e^{it}+e^{-it}}{4} - \frac{i}{2}\frac{e^{it}-e^{-it}}{2}$$

$$= \frac{e^{-t}}{2} - \frac{1}{2}\cos t + \frac{1}{2}\sin t \qquad \textbf{Ans.}$$

**Example 63.** *Find the inverse Laplace transform of* $\dfrac{s^2-1}{(s^2+1)^2}$

**Solution.** Let $$F(s) = \frac{s^2-1}{(s^2+1)^2} \qquad ...(1)$$

$$L^{-1}\left[\frac{s^2-1}{(s^2+1)^2}\right] = \text{sum of residues of } e^{st}.F(s) \text{ at the poles} \qquad ...(2)$$

The poles of (1) are obtained by equating denominator to zero.

$$(s^2+1)^2 = 0 \quad i.e.\ s = i, -i$$

There are two poles of second order at $s=i$ and $s=-i$

Residue of $e^{st}.F(s)$ (at $s=i$) $= \dfrac{d}{ds}\left[(s-i)^2\dfrac{e^{st}(s^2-1)}{(s^2+1)^2}\right]_{s=i} = \dfrac{d}{ds}\left[\dfrac{e^{st}(s^2-1)}{(s+i)^2}\right]_{s=i}$

$$= \left[\frac{(s+i)^2\left[e^{st}t(s^2-1)+e^{st}2s\right]-2(s+i)e^{st}(s^2-1)}{(s+i)^4}\right]_{s=i}$$

$$=\left[\frac{(s+i)\left[e^{st}.t(s^2-1)+e^{st}.2s\right]+e^{st}(s^2-1)}{(s+i)^3}\right]_{s=i}$$

$$=\frac{2i\left[e^{it}.t(-2)+e^{it}2i\right]-2e^{it}(-2)}{(2i)^3}=\frac{-4ite^{it}}{-8i}=\frac{te^{it}}{2}$$

Residue of $e^{st}.F(s)$ (at $s=-i$) $=\frac{d}{ds}\left[(s+i)^2\frac{e^{st}(s^2-1)}{(s^2+1)^2}\right]_{s=-i}=\frac{d}{ds}\left[\frac{e^{st}.(s^2-1)}{(s-i)^2}\right]_{s=-i}$

$$=\left[\frac{(s-i)^2\left[e^{st}.t(s^2-1)+2s\,e^{st}\right]-e^{st}.(s^2-1)2(s-i)}{(s-i)^4}\right]_{s=-i}$$

$$=\left[\frac{(s-i)\left[e^{st}.(s^2-1)+2s\,e^{st}\right]-e^{st}(s^2-1)2}{(s-i)^3}\right]_{s=-i}$$

$$=\frac{-2i\left[e^{-it}.t(-2)-2ie^{-it}\right]-e^{-it}(-2)2}{(-2i)^3}=\frac{4it.e^{-it}}{(-2i)^3}=\frac{t.e^{-it}}{2}$$

Sum of the residues at ($s=i$ and $s=-i$)

$$=\frac{t.e^{it}}{2}+\frac{t.e^{-it}}{2}=t\frac{e^{it}+e^{-it}}{2}=t\cos t. \qquad ...(3)$$

Putting the value of sum of residues from (3) in (2), we get

$$L^{-1}\left[\frac{s^2-1}{(s^2+1)^2}\right]=t\cos t$$ **Ans.**

**Example 64.** *Obtain the inverse Laplace Transform of* $\frac{e^{-b\sqrt{s}}}{s}$

**Solution.** Let $$F(s)=\frac{e^{-b\sqrt{s}}}{s} \qquad ...(1)$$

$$L^{-1}\left(\frac{e^{-b\sqrt{s}}}{s}\right)=\frac{1}{2\pi i}\int_{a-i\infty}^{a+i\infty}e^{st}.\frac{e^{-b\sqrt{s}}}{s}ds \qquad ...(2)$$

The simple pole of $F(s)$ is at $s=0$. Let us have a contour $ABCDEF$ excluding the pole at $x=0$. The contour encloses no singularity, therefore, by Cauchy theorem.

$$\int_{ABCDEFA}e^{st}.F(s)=0$$

$$\Rightarrow \int_{AB}e^{st}.F(s)ds+\int_{Bc}e^{st}.F(s)ds+\int_{CD}e^{st}.F(s)ds+\int_{DE}e^{st}.F(s)ds+$$

$$\int_{EF}e^{st}.F(s)ds+\int_{FA}e^{st}.F(s)ds=0 \qquad ...(3)$$

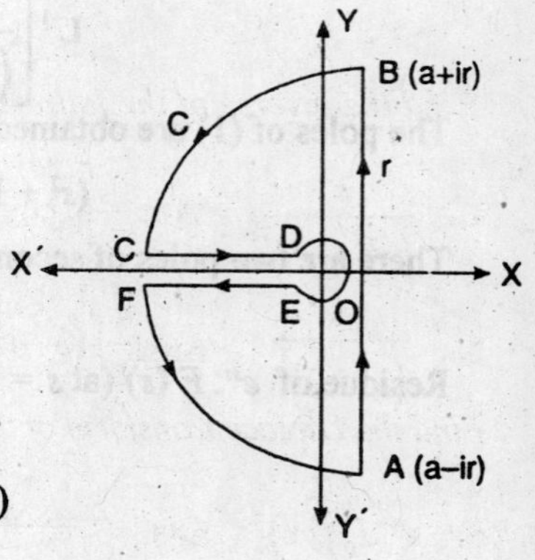


Let $OC=\rho$, $OD=\varepsilon$, then along $CD$, $s=Re^{i\pi}$

$$\int_{CD}e^{sx}.F(s)ds=\int_{p}^{\varepsilon}e^{-xR}\frac{e^{-ib\sqrt{r}}}{R}dR$$

$$\int_{EF}e^{sx}.F(s)ds=\int_{\varepsilon}^{p}e^{-xR}\frac{e^{-ib\sqrt{R}}}{R}dR \qquad (S=Re^{-i\pi} \text{ along } EF)$$

$$\int_{DE} e^{sx}.F(s)\,ds = \int_{\pi}^{-\pi} \frac{1}{\varepsilon e^{i\theta}}\left(\varepsilon e^{i\theta} i d\theta\right) \qquad \left\{\begin{array}{l} S=\varepsilon e^{i\theta} \text{ along } DE \\ e^{\pi S}=1 \\ e^{-b\sqrt{s}}=1 \end{array}\right.$$

$$= -2\pi i$$

$$\int_{BC} e^{sx}.F(s)\,ds = 0,\ \int_{FA} e^{sx}.F(s)\,ds = 0$$

On putting the values of the integrals in (3), we have

$$\int_{a-ir}^{a+ir} \frac{e^{xs-b\sqrt{s}}}{s}\,ds + \int_{\varepsilon}^{p} e^{-xR}\,\frac{e^{ib\sqrt{R}}-e^{-ib\sqrt{R}}}{R}\,dR - 2\pi i = 0$$

$$\Rightarrow \quad \int_{a-i\infty}^{a+i\infty} \frac{e^{xs-b\sqrt{s}}}{s}\,ds = 2\pi i - 2i\int_0^{\infty} e^{-xR}\,\frac{\sin b\sqrt{R}}{R}\,dR \qquad \left(\begin{array}{l}\varepsilon \to 0\\ p \to \infty\end{array}\right)$$

$$\Rightarrow \quad \frac{1}{2\pi i}\int_{a-i\infty}^{a+i\infty} \frac{e^{xs-b\sqrt{s}}}{s}\,ds = 1 - \frac{2}{\pi}\int_0^{\infty} e^{-u^2}\,\frac{\sin\left(\frac{bu}{\sqrt{x}}\right)}{u}\,du \qquad \left(R=\frac{u^2}{x}\right) \qquad \ldots(4)$$

We know that $\int_0^{\infty} e^{-u^2}\cos 2bu\,du = \frac{1}{2}\sqrt{\pi}\ e^{-b^2}$

Integrating both sides w.r.t., "b"

$$\int_0^{\infty} e^{-u^2}\left[\frac{\sin 2bu}{2u}\right] du = \frac{1}{2}\sqrt{\pi}\ \int e^{-b^2}\,db$$

Taking limits 0 to $\frac{b}{2\sqrt{x}}$, we have

$$\int_0^{\infty} e^{-u^2}\left(\frac{\sin 2bu}{2u}\right)_0^{\frac{b}{2\sqrt{x}}} du = \frac{\sqrt{\pi}}{2}\int_0^{\frac{b}{2\sqrt{x}}} e^{-b^2}\,db$$

$$\int_0^{\infty} e^{-u^2}\sin\frac{\frac{bu}{\sqrt{x}}}{u}\,du = \sqrt{\pi}\cdot\frac{\sqrt{x}}{2}\,e.r.f.\left(\frac{b}{2\sqrt{x}}\right) \qquad \left[e.r.f.x = \frac{2}{\sqrt{\pi}}\int_0^x e^{-u^2}\,du\right]$$

$$= \frac{\pi}{2}\,e.r.f.\left(\frac{b}{2\sqrt{x}}\right)$$

Putting the value of the above integral in (4), we have

$$\frac{1}{2\pi i}\int_{a-i\infty}^{a+i\infty} e^{xs}\,\frac{e^{-b\sqrt{s}}}{s}\,ds = 1 - \frac{2}{\pi}\frac{\pi}{2}\,e.r.f.\left(\frac{b}{2\sqrt{x}}\right)$$

$$= 1 - e.r.f.\left(\frac{b}{2\sqrt{x}}\right)$$

**Ans.**

## Exercise 13.18

*Find the inverse of the following by convolution theorem*

1. $\frac{s^2}{\left(s^2+a^2\right)^2}$ **Ans.** $\frac{1}{2}\left[t\cos at + \frac{1}{2a}\sin at\right]$

2. $\frac{1}{s\left(s^2+a^2\right)}$ **Ans.** $\frac{1-\cos at}{a^2}$

3. $\frac{1}{\left(s^2+1\right)^3}$ **Ans.** $\frac{1}{8}\left[\left(3-t^2\right)\sin t - 3t\cos t\right]$

4. $\frac{s}{\left(s^2+a^2\right)^2}$ **Ans.** $\frac{1}{2a}t\sin at$

*Find the Laplace transform of the following.*

5. $e^{ax}J_0(bx)$ **Ans.** $\frac{1}{\sqrt{s^2+2as+a^2+b^2}}$

6. $xJ_0(ax)$ **Ans.** $\frac{s}{\left(s^2+a^2\right)^{3/2}}$

7. $xJ_1(x)$ **Ans.** $\dfrac{1}{(s^2+1)^{3/2}}$

*Find the inverse Laplace transform of the following by residue method:*

8. $\dfrac{1}{(s+1)(s+2)}$ **Ans.** $e^{-t}-e^{-2t}$

9. $\dfrac{1}{(s-1)(s^2+1)}$ **Ans.** $\dfrac{1}{2}(e^t-\sin t-\cos t)$

10. $\dfrac{4s+5}{(s+2)(s-1)^2}$ **Ans.** $3te^t+\dfrac{1}{3}e^t-\dfrac{1}{3}e^{-2t}$

11. $\dfrac{2}{(s-1)^2(s^2+1)}$ **Ans.** $e^t(t-1)+\cos t$

12. $\dfrac{\cosh x\sqrt{s}}{s\cosh\sqrt{s}}, 0<x<1$ **Ans.** $\dfrac{4}{\pi}\sum_{n=1}^{\infty}\dfrac{(-1)^n}{2n-1}.e^{\frac{-(2n-1)^2\pi^2 t}{4}}\cos\left(n-\dfrac{1}{2}\right)\pi x+1$

13. $\dfrac{\sinh x\sqrt{s}}{x\sinh\sqrt{s}}$, **Ans.** $x+\dfrac{2}{\pi}\sum_{n=1}^{\infty}\dfrac{(-1)^n}{n}.e^{-n^2\pi^2 t}.\sin n\pi x$

14. Prove that $L^{-1}\left[\dfrac{e^{-a\sqrt{s}}}{\sqrt{s}}\right]=\dfrac{1}{\sqrt{\pi t}}.e^{-\frac{a^2}{4t}}$

15. $e^{\sqrt{-s}}$ **Ans.** $\dfrac{1}{2\sqrt{\pi}}.t^{-\frac{3}{2}}e^{-\frac{1}{4t}}$

## 13.33 HEAVISIDE's Inverse Formula of $\dfrac{F(s)}{G(s)}$

If $F(s)$ and $G(s)$ be two polynomials in $s$. The degree of $F(s)$ is less than that of $G(s)$. Let $\alpha_1, \alpha_2, \alpha_3, \ldots\ldots\ldots \alpha_n$ be $n$ roots of the equation $G(s)=0$

Inverse Laplace formula of $\dfrac{F(s)}{G(s)}$ is given by $L^{-1}\left\{\dfrac{F(s)}{G(s)}\right\}=\sum_{i=1}^{n}\dfrac{F(\alpha_i)}{G'(\alpha_i)}e^{\alpha_i t}$

**Example 65.** *Find* $L^{-1}\left\{\dfrac{2s^2+5s-4}{s^3+s^2-2s}\right\}$

**Solution.** Let $F(s)=2s^2+5s-4$

and
$$G(s)=s^3+s^2-2s=s(s^2+s-2)=s(s+2)(s-1)$$
$$G'(s)=3s^2+2s-2$$
$G(s)=0$ has three roots, 0, 1, –2.

or
$$\alpha_1=0,\ \alpha_2=1,\ \alpha_3=-2$$

By Heaviside's Inverse formula $L^{-1}\left\{\dfrac{F(s)}{G(s)}\right\}=\sum_{i=1}^{n}\dfrac{F(\alpha_i)}{G'(\alpha_i)}e^{\alpha_i t}$

$$=\frac{F(\alpha_1)}{G'(\alpha_1)}e^{t\alpha_1}+\frac{F(\alpha_2)}{G'(\alpha_2)}e^{t\alpha_2}+\frac{F(\alpha_3)}{G'(\alpha_3)}e^{t\alpha_3}=\frac{F(0)}{G'(0)}e^0+\frac{F(1)}{G'(1)}e^t+\frac{F(-2)}{G'(-2)}e^{-2t}$$

$$=\frac{-4}{-2}e^0+\frac{3}{3}e^t+\frac{(-6)}{6}e^{-2t}=2+e^t-e^{-2t}$$ **Ans.**

### Exercise 13.19

Using Heaviside's expansion formula, find the inverse Laplace transform of the following:

1. $\dfrac{s-1}{s^2+3s+2}$ **Ans.** $-2e^{-t}+3e^{-2t}$

2. $\dfrac{s}{(s-1)(s-2)(s-3)}$ **Ans.** $\dfrac{1}{2}e^t-2e^{2t}+\dfrac{3}{2}e^{3t}$

3. $\dfrac{2s+3}{(s-2)(s-3)(s-4)}$ **Ans** $\dfrac{7}{2}e^{2t}-9e^{3t}+\dfrac{11}{2}e^{4t}$

4. $\dfrac{11s^2-2s+5}{2s^3-3s^2-3s+2}$ **Ans.** $2e^{-2t}+5.e^{2t}-\dfrac{3}{2}e^{\frac{t}{2}}$

# 14

# Integral Transforms

## 14.1 INTRODUCTION

Integral transforms are used in the solution of partial differential equations. The choice of a particular transform to be used for the solution of a differential equations depends upon the nature of the boundary conditions of the equation and the facility with which the transform $F(s)$ can be converted to give $f(x)$.

## 14.2 INTEGRAL TRANSFORMS

The integral transform $F(s)$ of a function $f(x)$ with the Kernel $k(s, k)$ is defined as

$$I[f(x)] = F(s) = \int_a^b f(x)\,k(s,x)\,dx.$$

For example

1. Laplace transform with the kernel $k(s, x) = e^{-sx}$

$$L[f(x)] = F(s) = \int_0^\infty f(x).e^{-sx}\,dx$$

2. Fourier Complex transform with the kernel $k(s, k) = e^{-isx}$

$$F[f(x)] = F(s) = \frac{1}{\sqrt{2\pi}}\int_{-\infty}^{\infty} f(x)e^{isx}\,dx.$$

$$f(x) = \frac{1}{\sqrt{2\pi}}\int_{-\infty}^{\infty} F(s)e^{-isx}\,dx \qquad \text{(Inversion formula)}$$

3. Fourier Sine transform with the kernel $k(s, x) = \sin sx$

$$F_s[f(x)] = F(s) = \sqrt{\frac{2}{\pi}}\int_0^\infty f(x)\,\sin sx.dx$$

$$f(x) = \sqrt{\frac{2}{\pi}}\int_0^\infty F_s(s)\,\sin sx\,ds \qquad \text{(Inversion formula)}$$

4. Fourier Cosine transform with the kernel $k(s, x) = \cos sx$

$$F_c[f(x)] = F(s) = \sqrt{\frac{2}{\pi}}\int_0^\infty f(x)\,\cos sx.dx$$

$$f(x) = \sqrt{\frac{2}{\pi}}\int_0^\infty F_c(s)\,\cos sx\,ds \qquad \text{(Inversion formula)}$$

5. Hankel Transform with the kernel $(k, s) = x\,J_n(sx)$

$$H[f(x)] = F(s) = \int_0^\infty f(x).x J_n(sx)\,dx$$

$$f(x) = \int_0^\infty s\,F(s)\,J_n(sx)\,dx$$ **(Inversion formula)**

6. Hilbert Transform with the kernel $k(s, x) = \dfrac{1}{s-x}$

$$F(s) = \frac{1}{\pi}\int_{-\infty}^{\infty} \frac{f(x)}{s-x}\,dx$$

$$f(x) = \frac{-1}{\pi}\int_{-\infty}^{\infty} \frac{F(s)}{s-x}\,ds$$ **(Inversion formula)**

7. Mellin transform with the kernel $k(s, x) = x^{s-1}$

$$M[f(x)] = F(s) = \int_0^\infty f(x).x^{s-1}\,dx.$$

The students have already done "Laplace transform" and also learnt to solve the ordinary differential equations by using Laplace transforms.

Integral transforms are used in solving the partial differential equation with boundary conditions.

**List of Formulae of Fourier Integrals**

1. Fourier Integral for $f(x)$ is $\quad f(x) = \dfrac{1}{\pi}\displaystyle\int_0^\infty \int_{-\infty}^{\infty} f(t)\,\cos u(t-x)\,du\,dt$

2. Fourier Sine Integral for $f(x)$ is $\quad f(x) = \dfrac{2}{\pi}\displaystyle\int_0^\infty \int_0^\infty f(t)\,\sin ut\,\sin ux\,du\,dt$

3. Fourier Cosine Integral for $f(x)$ is $\quad f(x) = \dfrac{2}{\pi}\displaystyle\int_0^\infty \int_0^\infty f(t)\,\cos ut\,\cos ux\,du\,dt$

## 14.3 FOURIER INTEGRAL THEOREM

It states that $f(x) = \dfrac{1}{\pi}\displaystyle\int_0^\infty \int_{-\infty}^{\infty} f(t)\cos u(t-x)\,dt\,du$

**Proof.** We know that Fourier series of a function $f(x)$ in $(-c, c)$ is given by

$$f(x) = \frac{a_0}{2} + \sum_{n=1}^{\infty} a_n \cos\frac{n\pi x}{c} + \sum_{n=1}^{\infty} b_n \sin\frac{n\pi x}{c} \quad \text{... (1)}$$

where $a_0$, $a_n$ and $b_n$ are given by

$$a_0 = \frac{1}{c}\int_{-c}^{c} f(t)dt, \quad a_n = \frac{1}{c}\int_{-c}^{c} f(t)\cos\frac{n\pi t}{c}\,dt$$

$$b_n = \frac{1}{c}\int_{-c}^{c} f(t)\sin\frac{n\pi t}{c}\,dt$$

Substituting the values of $a_0$, $a_n$ and $b_n$ in (1) we get

$$f(x) = \frac{1}{2c}\int_{-c}^{c} f(t)\,dt + \sum_{n=1}^{\infty}\frac{1}{c}\int_{-c}^{c} f(t)\cos\frac{n\pi t}{c}\cos\frac{n\pi x}{c}\,dt + \sum_{n=1}^{\infty}\frac{1}{c}\int_{-c}^{c} f(t)\sin\frac{n\pi t}{c}\sin\frac{n\pi x}{c}\,dt$$

$$= \frac{1}{2c}\int_{-c}^{c} f(t)\,dt + \sum_{n=1}^{\infty}\frac{1}{c}\int_{-c}^{c} f(t)\left[\cos\frac{n\pi t}{c}\cos\frac{n\pi x}{c} + \sin\frac{n\pi t}{c}\sin\frac{n\pi x}{c}\right]dt$$

$$= \frac{1}{2c}\int_{-c}^{c} f(t)\,dt + \sum_{n=1}^{\infty}\frac{1}{c}\, f(t)\cos\frac{n\pi}{c}(t-x)\,dt$$

$$= \frac{1}{2c}\int_{-c}^{c} f(t)\left\{1 + 2\sum_{n=1}^{\infty}\cos\frac{n\pi}{c}(t-x)\right\}dt \qquad \text{... (2)}$$

Since cosine functions are even functions *i.e.*, $\cos(-\theta) = \cos\theta$ the expression

$$1 + 2\sum_{n=1}^{\infty}\cos\frac{n\pi}{c}(t-x) = \sum_{n=-\infty}^{\infty}\cos\frac{n\pi}{c}(t-x)$$

Therefore, (2) becomes

$$f(x) = \frac{1}{2c}\int_{-c}^{c} f(t)\left\{\sum_{-\infty}^{\infty}\cos\frac{n\pi}{c}(t-x)\right\}dt$$

$$= \frac{1}{2\pi}\int_{-c}^{c} f(t)\left\{\frac{\pi}{c}\sum_{-\infty}^{\infty}\cos\frac{n\pi}{c}(t-x)\right\}dt \qquad \text{... (3)}$$

Let us now assume that $c$ increases indefinitely, so that we may write $\frac{n\pi}{c} = u$ and $\frac{\pi}{c} = du$.

This assumption gives

$$\lim_{c\to\infty}\left\{\frac{\pi}{c}\sum_{-\infty}^{\infty}\cos\frac{n\pi}{c}(t-x)\right\} = \int_{-\infty}^{\infty}\cos u(t-x)\,du$$

$$= 2\int_{0}^{\infty}\cos u(t-x)\,du \qquad \text{... (4)}$$

Substituting in (3) from (4), we obtain

$$f(x) = \frac{1}{2\pi}\int_{-\infty}^{\infty} f(t)\left\{2\int_{0}^{\infty}\cos u(t-x)\,du\right\}dt \qquad \text{...(5)}$$

Thus $$f(x) = \frac{1}{\pi}\int_{0}^{\infty}\int_{-\infty}^{\infty} f(t)\cos u(t-x)\,du\,dt$$ **Proved.**

**Note.** We have assumed the following conditions on $f(x)$.

(*i*) $f(x)$ is defined as single-valued except at finite points in $(-c, c)$.

(*ii*) $f(x)$ is periodic outside $(-c, c)$ with period $2c$.

(*iii*) $f(x)$ and $f'(x)$ are sectionally continuous in $(-c, c)$.

(*iv*) $\int_{-\infty}^{\infty}|f(x)|\,dx$ converges, *i.e.*, $f(x)$ is absolutely integrable in $(-\infty, \infty)$.

## 14.4 FOURIER SINE AND COSINE INTEGRALS

$$f(x) = \frac{2}{\pi}\int_{0}^{\infty}\sin ux\,du\int_{0}^{\infty} f(t)\sin ut\,dt$$ **(Fourier Sine Integral)**

$$f(x) = \frac{2}{\pi}\int_{0}^{\infty}\cos ux\,du\int_{0}^{\infty} f(t)\cos ut\,dt$$ **(Fourier Cosine Integral)**

**Proof.** We know that

$$\cos u(t-x) = \cos(ut - ux)$$

or

$$\cos u(t-x) = \cos ut \cos ux + \sin ut \sin ux$$

Then equation (5) of article 14.3, can be written as

$$f(x) = \frac{1}{\pi}\int_0^\infty \int_{-\infty}^\infty f(t)\,(\cos ut \cos ux + \sin ut \sin ux)\,du\,dt$$

$$f(x) = \frac{1}{\pi}\int_0^\infty \int_{-\infty}^\infty f(t) \cos ut \cos ux \,du\,dt + \frac{1}{\pi}\int_0^\infty \int_{-\infty}^\infty f(t) \sin ut \sin ux\,du\,dt \quad \text{... (6)}$$

**Case 1.** When $f(t)$ is odd.

$\therefore$ $f(t)\cos u\,t$ is odd hence $\int_0^\infty \int_{-\infty}^\infty f(t) \cos ut \cos ux\, d\,u\,dt = 0$

$$\left[\begin{array}{ll} \text{For odd function} & \int_{-a}^{a} f(x)dx = 0 \\ \text{For even function} & \int_{-a}^{a} f(x)\,dx = 2\int_0^a f(x)\,dx \end{array}\right]$$

From (6) we have

$$\therefore \quad f(x) = \frac{2}{\pi}\int_0^\infty \sin ux\,du \int_0^\infty f(t) \sin ut\,dt \quad \text{... (7)}$$

The relation (7) is called **Fourier sine integral.**

**Case 2.** When $f(t)$ is even.

$\therefore$ $f(t)\sin u\,t$ is odd. $\int_0^\infty \int_{-\infty}^\infty f(t) \sin ut \sin ux\,d\,u\,dt = 0$

$\therefore$ $f(t)\cos ut$ is even.

From (6) we have

$$f(x) = \frac{2}{\pi}\int_0^\infty \cos ux\,du \int_0^\infty f(t)\cos ut\,dt \quad \text{... (8)}$$

The relation (8) is known as Fourier cosine integral.

## 14.5 FOURIER'S COMPLEX INTEGRAL

$$f(x) = \frac{1}{2\pi}\int_{-\infty}^\infty e^{-iux}\,du \int_{-\infty}^\infty f(t)e^{-iut}\,dt$$

**Proof.** We know that $\int_{-a}^{a} f(x)\,dx = 0$ if $f(x)$ is odd function.

$$\therefore \quad \int_{-\infty}^\infty \sin u(t-x)\,du = 0 \qquad \text{[since } \sin u\,(t-x) \text{ is odd.]}$$

Obviously we have

$$\frac{1}{2\pi}\int_{-\infty}^\infty f(t)\,dt \int_{-\infty}^\infty \sin u(t-x)\,du = 0$$

or

$$\frac{i}{2\pi}\int_{-\infty}^\infty f(t)\,dt \int_{-\infty}^\infty \sin u(t-x)\,du = 0 \quad \text{... (9)}$$

(Multiplying by $i$)

On adding (5) and (9) we have

$$f(x) = \frac{1}{2\pi}\int_{-\infty}^\infty \int_{-\infty}^\infty f(t) \cos u(t-x)\,du\,dt + \frac{i}{2\pi}\int_{-\infty}^\infty f(t)\,dt \int_{-\infty}^\infty \sin u(t-x)\,du$$

$$= \frac{1}{2\pi}\int_{-\infty}^{\infty} f(t)\,dt \int_{-\infty}^{\infty} [\cos u(t-x) + i \sin u(t-x)]\,du$$

$$= \frac{1}{2\pi}\int_{-\infty}^{\infty} f(t)\,dt \int_{-\infty}^{\infty} e^{iu(t-x)}\,du$$

or $$f(x) = \frac{1}{2\pi}\int_{-\infty}^{\infty} e^{-iux}\,du \int_{-\infty}^{\infty} f(t)\,e^{iut}\,dt \qquad \text{... (10)}$$

Relation (10) is called Fourier's Complex Integral.

**Example 1.** *Express the function*

$$f(x) = \begin{cases} 1 & \text{when } |x| \le 1 \\ 0 & \text{when } |x| > 1 \end{cases}$$

*as a Fourier integral. Hence evaluate*

$$\int_0^{\infty} \frac{\sin\lambda \cos\lambda x}{\lambda}\,d\lambda$$

**Solution.** The Fourier Integral for $f(x)$ is

$$f(x) = \frac{1}{\pi}\int_0^{\infty}\int_{-\infty}^{\infty} f(t)\cos\lambda(t-x)\,dt\,d\lambda$$

$$= \frac{1}{\pi}\int_0^{\infty}\int_{-1}^{1} \cos\lambda(t-x)\,dt\,d\lambda \qquad \text{(since } f(t) = 1\text{)}$$

$$= \frac{1}{\pi}\int_0^{\infty}\left[\frac{\sin\lambda(t-x)}{\lambda}\right]_{-1}^{1} d\lambda$$

$$= \frac{1}{\pi}\int_0^{\infty}\frac{\sin\lambda(1-x) + \sin\lambda(1+x)}{\lambda}\,d\lambda \qquad \text{By } \sin C + \sin D \text{ formula}$$

Thus $$f(x) = \frac{2}{\pi}\int_0^{\infty}\frac{\sin\lambda\cos\lambda x}{\lambda}\,d\lambda \qquad \textbf{Ans.}$$

or $$\int_0^{\infty}\frac{\sin\lambda\cos\lambda x}{\lambda}\,d\lambda = \frac{\pi}{2} f(x)$$

or $$\int_0^{\infty}\frac{\sin\lambda\cos\lambda x}{\lambda}\,d\lambda = \begin{cases} \frac{\pi}{2} & \text{for } |x| < 1 \\ 0 & \text{for } |x| > 1 \end{cases}$$

For $|x| = 1$, which is a point of discontinuity of $f(x)$, value of integral $= \frac{\pi/2 + 0}{2} = \frac{\pi}{4}$ **Ans.**

**Example 2.** *Find the Fourier sine integral for*

$$f(x) = e^{-\beta x} \qquad (\beta > 0)$$

*hence show that* $\frac{\pi}{2}e^{-\beta x} = \int_0^{\infty}\frac{\lambda\sin\lambda x}{\beta^2 + \lambda^2}\,d\lambda$

**Solution.** The Fourier sine transform of $f(x)$ is

$$f(x) = \frac{2}{\pi}\int_0^{\infty}\sin\lambda x\,d\lambda \int_0^{\infty} f(t)\sin\lambda t\,dt \qquad \text{... (1)}$$

Putting the value of $f(x)$ in (1) we get

$$e^{-\beta x}=\frac{2}{\pi}\int_0^{\infty}\sin\lambda x\,d\lambda\int_0^{\infty}e^{-\beta t}\sin\lambda t\,dt$$

$$=\frac{2}{\pi}\int_0^{\infty}\sin\lambda x\,d\lambda\left[\frac{e^{-\beta t}}{(\beta^2+\lambda^2)}(-\beta\sin\lambda t-\lambda\cos\lambda t)_0^{\infty}\right]$$

$$=\frac{2}{\pi}\int_0^{\infty}\sin\lambda x\,d\lambda\left[0+\frac{\lambda}{\beta^2+\lambda^2}\right]$$

$$e^{-\beta x}=\frac{2}{\pi}\int_0^{\infty}\frac{\lambda\sin\lambda x}{\beta^2+\lambda^2}d\lambda \quad \text{or} \quad \frac{\pi}{2}e^{-\beta x}=\int_0^{\infty}\frac{\lambda\sin\lambda x}{\beta^2+\lambda^2}d\lambda.$$ **Proved.**

**Example 3.** *Using Fourier cosine integral representation of an appropriate function, show that*

$$\int_0^{\infty}\frac{\cos wx}{k^2+w^2}dw=\frac{\pi e^{-kx}}{2k}, \qquad x>0, k>0.$$

**Solution.** We know that Fourier integral is

$$f(x)=\frac{2}{\pi}\int_0^{\infty}\cos ux\,du\int_0^{\infty}f(t)\cos ut\,dt$$

Putting the value of $f(t)$ and replacing $u$ by $w$ we get

$$e^{-kx}=\frac{2}{\pi}\int_0^{\infty}\cos wx\,dw\int_0^{\infty}e^{-kt}\cos wt\,dt$$

$$=\frac{2}{\pi}\int_0^{\infty}\cos wx\,dw\left[\frac{e^{-kt}}{k^2+w^2}\{-k\cos wt+w\sin wt\}_0^{\infty}\right]$$

$$=\frac{2}{\pi}\int_0^{\infty}\cos wx\,dw\left[0+\frac{k}{k^2+w^2}\right]=\frac{2k}{\pi}\int_0^{\infty}\frac{\cos wx\,dw}{k^2+w^2}$$

or $$\int_0^{\infty}\frac{\cos wx}{k^2+w^2}=\frac{\pi e^{-kx}}{2k}$$ **Proved.**

## 14.6 FOURIER TRANSFORMS

We have done in Article 14.5 that

$$f(x)=\frac{1}{2\pi}\int_{-\infty}^{\infty}e^{-iux}du\int_{-\infty}^{\infty}f(t)e^{iut}dt$$

$$=\frac{1}{2\pi}\int_{-\infty}^{\infty}e^{-isx}ds\int_{-\infty}^{\infty}f(t)e^{ist}dt \qquad (u=s) \quad \ldots(1)$$

Putting $\int_{-\infty}^{\infty}f(t)e^{ist}dt=F(s)$ in (1) we get ... (2)

or $$f(x)=\frac{1}{2\pi}\int_{-\infty}^{\infty}e^{-isx}.F(s)\,ds \qquad \ldots(3)$$

In (2) $F(s)$ is called the **Fourier transform** of $f(x)$.
In (3) $f(x)$ is called the **inverse Fourier transform** of $F(s)$.

For reasons of symmetry, we multiply both $F(x)$ and $F(s)$ by $\sqrt{\frac{1}{2\pi}}$ instead of having the factor $\frac{1}{2\pi}$ in only one function. Thus, we obtain the definition of Fourier transforms as

$$F(s)=\frac{1}{\sqrt{2\pi}}\int_{-\infty}^{\infty} f(t)e^{ist}dt$$

$$f(x)=\frac{1}{\sqrt{2\pi}}\int_{-\infty}^{\infty} F(s)e^{-isx}ds$$

## 14.7 FOURIER SINE AND COSINE TRANSFORMS

From equation (7) of Article 14.4 we know that

$$f(x)=\frac{2}{\pi}\int_0^{\infty}\sin sx\,ds\int_0^{\infty} f(t)\sin st\,dt \qquad (s=u)$$

$$f(x)=\sqrt{\frac{2}{\pi}}\int_0^{\infty}\sin sx\,ds\,F(s) \qquad \text{... (1)}$$

$$F(s)=\sqrt{\frac{2}{\pi}}\int_0^{\infty} f(t)\sin st\,dt \qquad \text{... (2)}$$

In equation (2), $F(s)$ is called **Fourier sine transform** of $f(x)$.
In equation (1), $f(x)$ is called the **Inverse Fourier sine transform** of $F(s)$
From equation (8) of Article 14.4, we have

$$f(x)=\sqrt{\frac{2}{\pi}}\int_0^{\infty}\cos sx\,F(s)\,dx \qquad \text{... (3)}$$

$$F(s)=\sqrt{\frac{2}{\pi}}\int_0^{\infty} f(t)\cos st\,dt \qquad \text{... (4)}$$

In equation (4), $F(s)$ is called **Fourier cosine transform** of $F(x)$.
In equation (3), $f(x)$ is called the **inverse Fourier cosine transform of** $F(s)$.

**Example 4.** *Find the Fourier transform of*

$$f(x)=\begin{cases}1 \text{ for } |x|<a\\ 0 \text{ for } |x|>a\end{cases}$$

**Solution.** The Fourier transform of a function $f(x)$ is given by

$$F(s)=\frac{1}{\sqrt{2\pi}}\int_{-\infty}^{\infty} f(x).e^{isx}dx$$

Substituting the value of $f(x)$, we get

$$F(s)=\frac{1}{\sqrt{2\pi}}\int_{-a}^{a} 1.e^{isx}dx=\left[\frac{e^{isx}}{is}\right]_{-a}^{a}=\frac{1}{\sqrt{2\pi}}\frac{1}{(is)}\left[e^{ias}-e^{-ias}\right]$$

$$=\frac{1}{\sqrt{2\pi}}\frac{2}{s}.\frac{e^{ias}-e^{-ias}}{2i}=\frac{1}{\sqrt{2\pi}}\frac{2\sin sa}{s}=\sqrt{\frac{2}{\pi}}\frac{\sin sa}{s}$$ **Ans.**

**Example 5.** *Find the Fourier transform of*

$$f(x)=\begin{cases}1-x^2 \text{ if } |x|\le 1.\\ 0 \text{ if } |x|>1.\end{cases}$$

**Solution.** $$f(x)=\begin{cases}1-x^2 & -1<x<1\\ 0 & |x|>1\end{cases}$$

The Fourier transform of a function $f(x)$ is given by

$$F(s)=\frac{1}{\sqrt{2\pi}}\int_{-\infty}^{\infty} f(x).e^{isx}.dx \qquad \ldots (1)$$

Substituting the values of $f(x)$ in (1), we get

$$F(s)=\frac{1}{\sqrt{2\pi}}\int_{-1}^{1}(1-x^2)\,e^{isx}\,dx$$

Integrating by parts, we get $\left[[uv]_1 = uv_1 - u'v_2 + u''v_3 \ldots\right]$

$$F(s)=\frac{1}{\sqrt{2\pi}}\left[(1-x^2)\frac{e^{isx}}{is}-(-2x)\frac{e^{isx}}{(is)^2}+(-2)\frac{e^{isx}}{(is)^3}\right]_{-1}^{1}$$

$$=\frac{1}{\sqrt{2\pi}}\left[-2\frac{e^{is}}{s^2}+2\frac{e^{is}}{is^3}-2\frac{e^{is}}{s^2}-\frac{e^{-is}}{is^3}\right]$$

$$=\frac{1}{\sqrt{2\pi}}\left[-\frac{2}{s^2}(e^{is}+e^{-is})+\frac{2}{is^3}(e^{is}-e^{-is})\right]$$

$$=\frac{1}{\sqrt{2\pi}}\left[-\frac{2}{s^2}(2\cos s)+\frac{2}{is^3}(2i\sin s)\right]$$

$$=\frac{1}{\sqrt{2\pi}}\frac{4}{s^3}[-s\cos s+\sin s]$$ **Ans.**

**Example 6.** *Find the Fourier sine and cosine transforms of* $f(x)=e^{-ax}$.

**Solution.** The Fourier sine transform of $f(x)$ is given by

$$F(s)=\sqrt{\frac{2}{\pi}}\int_0^{\infty} f(x)\sin sx\,dx$$

Putting the value of $f(x)$ we get

$$F(s)=\sqrt{\frac{2}{\pi}}\int_0^{\infty} e^{-ax}\sin sx\,dx$$

$$=\sqrt{\frac{2}{\pi}}\frac{e^{-ax}}{a^2+s^2}[-a\sin sx-s\cos sx]_0^{\infty}$$

$$=-\sqrt{\frac{2}{\pi}}\frac{e^{-ax}}{a^2+s^2}[a\sin sx+s\cos sx]_0^{\infty}$$

$$=\sqrt{\frac{2}{\pi}}\left[-0+\frac{1}{a^2+s^2}\times s\right]=\sqrt{\frac{2}{\pi}}\frac{s}{a^2+s^2}$$ **Ans.**

The Fourier cosine transform is

$$F(s)=\sqrt{\frac{2}{\pi}}\int_0^{\infty} e^{-ax}\cos sx\,dx$$

$$=\sqrt{\frac{2}{\pi}}\left[\frac{e^{-ax}}{a^2+s^2}\{-a\cos sx+s\sin sx\}\right]_0^{\infty}=\sqrt{\frac{2}{\pi}}\left[0+\frac{a}{a^2+s^2}\right]$$

$$= \sqrt{\frac{2}{\pi}} \frac{a}{a^2 + s^2}$$ **Ans.**

**Example 7.** *Find Fourier sine transform of* $\frac{1}{x}$.

**Solution.** $F_s\left(\frac{1}{x}\right) = \sqrt{\frac{2}{\pi}} \int_0^\infty \frac{\sin sx}{x} dx$

$$= \sqrt{\frac{2}{\pi}} \int_0^\infty \frac{\sin \theta}{\frac{\theta}{s}} \frac{d\theta}{s}$$ Putting $s\,x = \theta$ so that $s\,dx = d\theta$

$$= \sqrt{\frac{2}{\pi}} \int_0^\infty \frac{\sin \theta}{\theta} d\theta = \sqrt{\frac{2}{\pi}} \left(\frac{\pi}{2}\right) = \sqrt{\frac{\pi}{2}}$$ **Ans.**

**Example 8.** *Find the Fourier cosine transform of*
$f(x) = e^{-2x} + 4\,e^{-3x}$

**Solution.** The Fourier cosine transform of $f(x)$ is given by

$$F(s) = \sqrt{\frac{2}{\pi}} \int_0^\infty f(x) \cos sx\, dx$$

Putting the value of $f(x)$ we get

$$F(s) = \sqrt{\frac{2}{\pi}} \int_0^\infty (e^{-2x} + 4e^{-3x}) \cos sx\, dx$$

$$= \sqrt{\frac{2}{\pi}} \int_0^\infty e^{-2x} \cos sx\, dx + \sqrt{\frac{2}{\pi}} \int_0^\infty 4e^{-3x} \cos sx\, dx$$

$$\left( \because \int e^{-ax} \cdot \cos bx\, dx = \frac{e^{-ax}}{a^2 + b^2} [b \sin bx - a \cos bx] \right)$$

$$= \sqrt{\frac{2}{\pi}} \left[ \frac{2}{s^2 + 4} + 4 \cdot \frac{3}{s^2 + 9} \right] = 2\sqrt{\frac{2}{\pi}} \left[ \frac{1}{s^2 + 4} + \frac{6}{s^2 + 9} \right]$$ **Ans.**

**Example 9.** *Find the Fourier sine transform of*

$$f(x) = \frac{e^{-ax}}{x}.$$

**Solution.** The sine transform of the function $f(x)$ is given by

$$F(s) = \sqrt{\frac{2}{\pi}} \int_0^\infty f(x) \sin sx\, dx$$

Substituting the value of $f(x)$, we get

$$F(s) = \sqrt{\frac{2}{\pi}} \int_0^\infty \frac{e^{-ax}}{x} \sin sx\, dx$$

Differentiating both sides w.r.t. '*s*' we get

$$\frac{d}{dx}[F(s)] = \sqrt{\frac{2}{\pi}} \int_0^\infty \frac{e^{-ax}}{x} (x \cos sx)\, dx$$

$$= \sqrt{\frac{2}{\pi}} \int_0^\infty e^{-ax} \cos ax\, dx = \sqrt{\frac{2}{\pi}} \frac{a}{s^2 + a^2}$$

Integrating w.r.t. '$s$' we get

$$F(s) = \sqrt{\frac{2}{\pi}} \int \frac{a}{s^2 + a^2} ds = \sqrt{\frac{2}{\pi}} \tan^{-1} \frac{s}{a} + c$$

For $s = 0$, $F(s) = 0$

Putting these values in the above equation we get

$$0 = 0 + c \quad \text{or} \quad c = 0 \qquad \therefore \quad F(s) = \sqrt{\frac{2}{\pi}} \tan^{-1} \frac{s}{a}$$

**Ans.**

**Example 10.** *Find the Fourier sine and cosine transforms of*

$$f(x) = \begin{cases} 1, & 0 < x < a \\ 0, & x > a \end{cases}$$

**Solution.** Fourier sine Transform

$$F(s) = \sqrt{\frac{2}{\pi}} \int_0^\infty f(x) \sin sx\, dx = \sqrt{\frac{2}{\pi}} \int_0^a 1 \sin sx\, dx$$

$$F(s) = \sqrt{\frac{2}{\pi}} \left[ -\frac{\cos sx}{s} \right]_0^a = \sqrt{\frac{2}{\pi}} \left[ -\frac{\cos as}{s} + \frac{1}{s} \right]$$

Fourier cosine transform

$$F(s) = \sqrt{\frac{2}{\pi}} \int_0^\infty f(x) \cos sx\, dx = \sqrt{\frac{2}{\pi}} \int_0^a 1 \cos sx\, dx = \sqrt{\frac{2}{\pi}} \left[ \frac{\sin sx}{s} \right]_0^a$$

$$= \sqrt{\frac{2}{\pi}} \frac{\sin as}{s}$$

**Ans.**

**Example 11.** *Find the Fourier cosine transform of*

$$f(x) = \begin{cases} x & \text{for} \quad 0 < x < \frac{1}{2} \\ 1 - x & \text{for} \quad \frac{1}{2} < x < 1 \\ 0 & \text{for} \quad x > 1. \end{cases}$$

**Solution.** Fourier cosine transform

$$F(s) = \sqrt{\frac{2}{\pi}} \int_0^\infty f(x) \cos sx\, dx$$

$$= \sqrt{\frac{2}{\pi}} \int_0^{1/2} x \cos sx\, dx + \sqrt{\frac{2}{\pi}} \int_{1/2}^1 (1 - x) \cos sx\, dx$$

$$= \sqrt{\frac{2}{\pi}} \left[ x \frac{\sin sx}{s} - \left( \frac{-\cos sx}{s^2} \right) \right]_0^{1/2} + \sqrt{\frac{2}{\pi}} \left[ (1 - x) \frac{\sin sx}{s} - (-1) \frac{(-\cos sx)}{s^2} \right]_{1/2}^1$$

$$= \sqrt{\frac{2}{\pi}} \left[ \frac{1}{2} \frac{\sin s/2}{s} + \frac{\cos s/2}{s^2} - \frac{1}{s^2} \right] + \sqrt{\frac{2}{\pi}} \left[ -\frac{\cos s}{s^2} - \frac{1}{2} \frac{\sin s/2}{s} + \frac{\cos s/2}{s^2} \right]$$

$$= \sqrt{\frac{2}{\pi}} \left[ -\frac{\cos s}{s^2} + \frac{2 \cos s/2}{s^2} - \frac{1}{s^2} \right]$$

**Ans.**

**Example 12.** *Find the Fourier transform of the function*

$$f(x) = \begin{cases} 1+\dfrac{x}{a}, & -a < x < 0 \\ 1-\dfrac{x}{a} & 0 < x < a \\ 0 & \text{otherwise} \end{cases}$$

*(U.P., III Semester, Summer 2002)*

**Solution.** Fourier tranform of $f(x)$ is given by

$$F(s) = \frac{1}{\sqrt{2\pi}} \int_{-\infty}^{\infty} f(x) \cdot e^{isx} dx$$

$$= \frac{1}{\sqrt{2\pi}} \int_{-a}^{0} \left(1+\frac{x}{a}\right) e^{isx} dx + \frac{1}{\sqrt{2\pi}} \int_{0}^{a} \left(1-\frac{x}{a}\right) e^{isx} dx$$

$$= \frac{1}{\sqrt{2\pi}} \left[\left(1+\frac{x}{a}\right) \times \frac{e^{isx}}{is} - \left(\frac{1}{a}\right)\frac{e^{isx}}{-s^2}\right]_{-a}^{0} + \frac{1}{\sqrt{2\pi}} \left[\left(1-\frac{x}{a}\right)\frac{e^{isx}}{is} - \left(-\frac{1}{a}\right)\frac{e^{isx}}{-s^2}\right]_{0}^{a}$$

$$= \frac{1}{\sqrt{2\pi}} \left[\frac{1}{is} + \frac{1}{a}\frac{1}{s^2} + \frac{1}{a}\frac{e^{-isa}}{-s^2}\right] + \frac{1}{\sqrt{2\pi}} \left[\frac{1}{a} \cdot \frac{e^{isa}}{-s^2} - \frac{1}{is} + \frac{1}{a}\frac{1}{s^2}\right]$$

$$= \frac{1}{\sqrt{2\pi}} \left[\frac{2}{as^2} + \frac{1}{-as^2}\left(e^{isa} + e^{-isa}\right)\right]$$

$$= \frac{1}{\sqrt{2\pi}} \left[\frac{2}{as^2} - \frac{2}{as^2}\cos sa\right] = \frac{1}{\sqrt{2\pi}} \frac{2}{as^2} [1-\cos as]$$

$$= \frac{2}{\sqrt{2\pi}\, as^2} 2\sin^2\frac{as}{2} = \frac{2\sqrt{2}\sin^2\dfrac{as}{2}}{\sqrt{\pi}\, as^2}$$ **Ans.**

**Example 13.** *Find Fourier sine and cosine transform of (a)* $x^{n-1}$. *(b)* $\dfrac{1}{\sqrt{x}}$.

**Solution.** (*a*) $F_s(x^{n-1}) = \sqrt{\dfrac{2}{\pi}} \displaystyle\int_0^{\infty} \sin sx \cdot x^{n-1}\, dx$

$$F_c(x^{n-1}) = \sqrt{\frac{2}{\pi}} \int_0^{\infty} \cos sx \cdot x^{n-1} dx$$

$$F_c(x^{n-1}) + F_s(x^{n-1}) = \frac{2}{\sqrt{\pi}} \int_0^{\infty} (\cos sx + i \sin sx)\, x^{n-1} dx$$

$$= \sqrt{\frac{2}{\pi}} \int_0^{\infty} e^{isx} x^{n-1} dx$$

$$= \sqrt{\frac{2}{\pi}} \int_0^{\infty} e^{-t} \left(-\frac{t}{is}\right)^{n-1} \left(-\frac{dt}{is}\right)$$

$\left[\begin{array}{l} \textbf{Putting } isx = -t, \\ x = -\dfrac{t}{is}, \\ dx = -\dfrac{dt}{is} \end{array}\right.$

$$= \sqrt{\frac{2}{\pi}} \frac{1}{(is)^n}(-1)^n \int_0^{\infty} e^{-t} t^{n-1} dt$$

$$= \sqrt{\frac{2}{\pi}} \frac{(i)^{2n}}{(i)^n s^n} \lceil n = \sqrt{\frac{2}{\pi}} \frac{(i)^n}{s^n} \lceil n$$

$$= \sqrt{\frac{2}{\pi}} \frac{\left(\cos\frac{\pi}{2} + i\sin\frac{\pi}{2}\right)^n \lceil n}{s^n} = \sqrt{\frac{2}{\pi}} \frac{\left(\cos\frac{n\pi}{2} + i\sin\frac{n\pi}{2}\right)\lceil n}{s^n}$$

Equating real and imaginary parts we get

$$F_c(x^{n-1}) = \sqrt{\frac{2}{\pi}} \frac{\lceil n}{s^n} \cos\frac{n\pi}{2}$$

$$F_s(x^{n-1}) = \sqrt{\frac{2}{\pi}} \frac{\lceil n}{s^n} \sin\frac{n\pi}{2}$$ **Ans.**

(*b*) $$n = \frac{1}{2}$$

$$F_c\left(\frac{1}{\sqrt{x}}\right) = \sqrt{\frac{2}{\pi}} \frac{\left\lceil \frac{1}{2} \right.}{s^{\frac{1}{2}}} \cos\frac{\pi}{4} = \sqrt{\frac{2}{\pi}} \frac{\sqrt{\pi}}{\sqrt{s}} \frac{1}{\sqrt{2}} = \frac{1}{\sqrt{s}}$$

$$F_s\left(\frac{1}{\sqrt{x}}\right) = \sqrt{\frac{2}{\pi}} \frac{\left\lceil \frac{1}{2} \right.}{s^{\frac{1}{2}}} \sin\frac{\pi}{4} = \sqrt{\frac{2}{\pi}} \frac{\sqrt{\pi}}{\sqrt{s}} \frac{1}{\sqrt{2}} = \frac{1}{\sqrt{s}}$$ **Ans.**

**Example 14.** *Find the complex Fourier transform of dirac delta function* $\delta(t-a)$.

**Solution.** $$F\{\delta(t-a)\} = \frac{1}{\sqrt{2\pi}} \int_{-\infty}^{\infty} e^{ist} \delta(t-a)\, dt$$

$$= \frac{1}{\sqrt{2\pi}} \underset{h \to 0}{\text{Lt}} \int_a^{a+h} \frac{1}{h} e^{ist} dt$$

$$= \frac{1}{\sqrt{2\pi}} \underset{h \to 0}{\text{Lt}} \frac{1}{h} \left(\frac{e^{ist}}{is}\right)_a^{a+h}$$

$$= \frac{1}{\sqrt{2\pi}} \underset{h \to 0}{\text{Lt}}\, e^{isa} \left(\frac{e^{ish} - 1}{ish}\right)$$

$$= \frac{e^{isa}}{\sqrt{2\pi}} \qquad \text{since } \underset{\theta \to 0}{\text{Lt}} \frac{e^{\theta} - 1}{\theta} = 1$$ **Ans.**

**Note.** Dirac delta function $\delta(t-a)$ is defined as

$$\delta(t-a) = \underset{h \to 0}{\text{Lt}}\, I(h\, t-a) \text{ where}$$

$$I(h, t-a) = \frac{1}{h} \qquad \text{for } a < t < a+h$$

$$= 0 \qquad \text{for } t < a \text{ and } t > a+h$$ **Ans.**

**Example 15.** *Show that*

(*a*) $F_s[x f(x)] = -\frac{d}{ds} F_c(s)$ (*b*) $F_c[x f(x)] = \frac{d}{ds} F_s(s)$

and hence find Fourier cosine and sine transform of $xe^{-ax}$.

**Solution.** (*a*) $F_c(s) = \sqrt{\frac{2}{\pi}} \int_0^\infty f(x) \cos sx\, dx$

$$\frac{d}{ds} F_c(s) = -\sqrt{\frac{2}{\pi}} \int_0^\infty x f(x) \sin sx\, dx$$

$$= -\sqrt{\frac{2}{\pi}}\, F_s\{x\, f(x)\}$$

(*b*) $F_s(s) = \sqrt{\frac{2}{\pi}} \int_0^\infty f(x) \sin sx\, dx$

$$\frac{d}{ds} F_s(s) = \sqrt{\frac{2}{\pi}} \int_0^\infty x f(x) \cos sx\, dx$$

$$= \sqrt{\frac{2}{\pi}}\, F_c\{x\, f(x)\}$$

(*c*) $F_c(xe^{-ax}) = \frac{d}{ds} F_s(e^{-ax})$

$$= \frac{d}{ds}\left[\sqrt{\frac{2}{\pi}} \frac{s}{a^2+s^2}\right] \quad \text{(Using example 6)}$$

$$= \sqrt{\frac{2}{\pi}} \frac{(a^2+s^2) - s(2s)}{(a^2+s^2)^2} = \sqrt{\frac{2}{\pi}} \frac{a^2 - s^2}{(a^2+s^2)^2}$$

(*d*) $F_s(xe^{-ax}) = -\frac{d}{dx} F_c(e^{-ax}) = -\frac{d}{ds}\left[\sqrt{\frac{2}{\pi}} \frac{a}{a^2+s^2}\right]$ **(Using example 6)**

$$= \sqrt{\frac{2}{\pi}} \frac{2as}{(a^2+s^2)^2}$$ **Ans.**

**Example 16.** *Find Fourier cosine transform of* $e^{-a^2x^2}$ *and hence evaluate Fourier sine ransform of* $xe^{-a^2x^2}$

**Solution.** $F_c(e^{-a^2x^2}) = \sqrt{\frac{2}{\pi}} \int_0^\infty e^{-a^2x^2} \cdot \cos sx\, ds$

$$= \text{Real part of } \sqrt{\frac{2}{\pi}} \int_0^\infty e^{-a^2x^2} \cdot e^{isx}\, ds$$

$$= \text{Real part of } \sqrt{\frac{2}{\pi}} \int_0^\infty e^{-a^2x^2 + isx}\, ds$$

$$= \frac{1}{a\sqrt{2}} e^{-\frac{s^2}{4a^2}} \quad \text{(Refer example)}$$

We know that

$$F_s(x f(x)) = -\frac{d}{ds} F_c f(x)$$

$$F_s(xe^{-a^2x^2}) = -\frac{d}{ds} F_c(e^{-a^2x^2}) \qquad f(x) = e^{-a^2x^2}$$

$$= -\frac{d}{ds} \frac{1}{a\sqrt{2}} e^{-\frac{s^2}{4a^2}} = \frac{1}{a\sqrt{2}} e^{-\frac{s^2}{4a^2}} \frac{s}{2a^2}$$

$$= \frac{1}{2a^3\sqrt{2}} e^{-\frac{s^2}{4a^2}}$$ **Ans.**

## EXERCISE 14.1

1. Express $f(x) = \begin{cases} 1 \text{ for } 0 \le x \le \pi \\ 0 \text{ for } x > \pi \end{cases}$ as a Fourier sine integral and hence evaluate

$$\int_0^\infty \frac{1-\cos \pi\lambda}{\lambda} \sin \lambda x \, d\lambda$$ **Ans.** $\frac{\pi}{4}$

2. Find the Fourier's cosine integral of the function $e^{-ax}$. Hence show that

$$\int_0^\infty \frac{\cos \lambda x}{\lambda^2 + 1^2} d\lambda = \frac{\pi}{2} e^{-x}, \quad x \ge 0$$ **Ans.** $\frac{2a}{\pi} \int_0^\infty \frac{\cos \lambda x}{\lambda^2 + a^2} d\lambda$

3. Show that the Fourier transform of

$$f(x) = 0 \text{ for } x < \alpha$$
$$= 1 \text{ for } \alpha < x < \beta$$
$$= 0 \text{ for } x > \beta$$

is $\frac{1}{\sqrt{2\pi}} \left( \frac{e^{i\beta s} - e^{i\alpha s}}{is} \right)$

4. Find the Fourier transform of $f(x)$ if

$$f(x) = \begin{cases} x, & |x| \le a \\ 0, & |x| > a \end{cases}$$ **Ans.** $\frac{1}{\sqrt{2\pi}} \frac{2i}{s^2} (as \cos as - \sin as)$

5. Show that the Fourier transform of

$$f(x) = \begin{cases} a - |x| & \text{for } |x| < a \\ 0 & \text{for } |x| > a > 0 \end{cases}$$

is $\sqrt{\frac{2}{\pi}} \frac{1-\cos as}{s^2}$. Hence show that $\int_0^\infty \left( \frac{\sin t}{t} \right)^2 dt = \pi/2$

6. Show that the Fourier transform of

$$f(x)=\begin{cases}\dfrac{\sqrt{2\pi}}{2a} & \text{for } |x|\le a\\ 0 & \text{for } |x|>a\end{cases}$$

is $\dfrac{\sin sa}{sa}$

7. Show that the transform of $e^{-\frac{x^2}{2}}$ is $e^{-\frac{s^2}{2}}$ by finding the Fourier transform of $e^{-a^2x^2}$, $a>0$.

8. Show that the Fourier transform of $e^{\frac{-x^2}{2}}$ is self-reciprocal.

9. Find Fourier transform of $e^{-a|x|}$ is $a>0$. **Ans.** $\dfrac{2a}{a^2+s^2}$

10. Find Fourier transform of $\dfrac{1}{\sqrt{|x|}}$.

11. Find the Fourier tranform of $f(x)=e^{ikx}$, $a<x<b$
$= 0,\quad x<a$ and $x>b$

**Ans.** $\dfrac{i}{\sqrt{2\pi}\,(k+s)}\left[e^{i(k+s)a}-e^{i(k+s)b}\right]$

12. Find the Fourier sine transform of $e^{-|x|}$. Hence evaluate

$$\int_0^{\infty}\frac{x\sin mx}{1+x^2}\,dx.$$

**Ans.** $\dfrac{s}{1+s^2}, \dfrac{\pi}{2}e^{-m}$

13. Show that the Fourier sine transform of

$$f(x)=\begin{cases}x & \text{for } 0<x<1\\ 2-x & \text{for } 1<x<2\\ 0 & \text{for } x>2\end{cases}$$

is $2\sin s\,(1-\cos s)/s^2$.

14. Show that the Fourier sine transform of $\dfrac{x}{1+x^2}$ is $\sqrt{\dfrac{\pi}{2}}\,as\,e^{-as}$

15. Show that the Fourier cosine transform of $\dfrac{1}{1+x^2}$ is $\sqrt{\dfrac{\pi}{2}}\,e^{-s}$

16. Find the sine and cosine transforms of $e^{-ax}$ $(a>0)$ **Ans.** $\dfrac{s}{a^2+s^2}, \dfrac{a}{a^2+s^2}$

17. Find the Fourier sine and cosine transform of $\cosh x-\sinh x$

18. Find the Fourier sine and cosine transform of $ae^{-\alpha x}+be^{-\beta x}$, $\alpha,\beta>0$.

**Ans.** $\dfrac{as}{s^2+\alpha^2}+\dfrac{bs}{s^2+\beta^2}, \dfrac{a\alpha}{s^2+\alpha^2}+\dfrac{b\beta}{s^2+\beta^2}$

## 14.8 PROPERTIES OF FOURIER TRANSFORMS

(1) **LINEAR PROPERTY.** If $F_1(s)$ and $F_2(s)$ are Fourier transforms of $f_1(x)$ and $f_2(x)$ respectively, then

$$F[af_1(x)+b\,f_2(x)]=a\,F_1(s)+b\,F_2(s)$$

where $a$ and $b$ are constants.

We know that $$F_1(s) = \frac{1}{\sqrt{2\pi}} \int_{-\infty}^{\infty} e^{isx} \cdot f_1(x)\,dx$$

and $$F_2(s) = \frac{1}{\sqrt{2\pi}} \int_{-\infty}^{\infty} e^{isx} f_2(x)\,dx$$

$$F[af_1(x) + b\,f_2(x)] = \frac{1}{\sqrt{2\pi}} \int_{-\infty}^{\infty} e^{isx}\left[af_1(x) + b\,f_2(x)\right]dx$$

$$= a\frac{1}{\sqrt{2\pi}} \int_{-\infty}^{\infty} e^{isx} f_1(x)\,dx + b\int_{-\infty}^{\infty} e^{isx} f_2(x)\,dx$$

$$= aF_1(s) + bF_2(s)$$ **Proved.**

**(2) CHANGE OF SCALE PROPERTY**

**If $F(s)$ is the complex Fourier transform of $f(x)$, then**

$$F\{f(ax)\} = \frac{1}{a}F\left(\frac{s}{a}\right)$$

**Proof.** We know that $$F(s) = \frac{1}{\sqrt{2\pi}} \int_{-\infty}^{\infty} e^{isx} f(x)\,dx$$

$$F\{f(ax)\} = \frac{1}{\sqrt{2\pi}} \int_{-\infty}^{\infty} e^{isx} f(ax)\,dx$$ Put $ax = t \Rightarrow dx = \dfrac{dt}{a}$

$$= \frac{1}{\sqrt{2\pi}} \int_{-\infty}^{\infty} e^{is\frac{t}{a}} f(t)\frac{dt}{a}$$

$$= \frac{1}{a}\frac{1}{\sqrt{2\pi}} \int_{-\infty}^{\infty} e^{i\left(\frac{s}{a}\right)t} f(t)\,dt$$

$$= \frac{1}{a}F\left(\frac{s}{a}\right)$$ **Proved.**

**(3) SHIFTING PROPERTY**

**If $F(s)$ is the complex Fourier transform of $f(x)$, then**

$$F\{f(x-a)\} = e^{isa}\,F(s)$$

**Proof.** $$F(s) = \int_{-\infty}^{\infty} e^{isx} f(x)\,dx$$

$$F\{f(x-a)\} = \frac{1}{\sqrt{2\pi}} \int_{-\infty}^{\infty} e^{isx} f(x-a)\,dx$$ [Put $x - a = t$, so that $dx = dt$]

$$= \frac{1}{\sqrt{2\pi}} \int_{-\infty}^{\infty} e^{is(t+a)} f(t)\,dt = \frac{e^{isa}}{\sqrt{2\pi}} \int_{-\infty}^{\infty} e^{ist} f(t)\,dt$$

$$= e^{isa}\,F(s)$$ **Proved.**

(4) $\boldsymbol{F\left\{e^{iax} f(x)\right\} = F(s+a)}$

**Proof.** $$F\{e^{iax} f(x)\} = \frac{1}{\sqrt{2\pi}} \int_{-\infty}^{\infty} e^{iax} f(x) e^{isx} dx = \frac{1}{\sqrt{2\pi}} \int_{-\infty}^{\infty} e^{i(s+a)x} f(x)\, dx$$
$$= F(s+a)$$

**Proved.**

**(5) MODULATION THEOREM**

If $F(s)$ is the complex Fourier transform of $f(x)$, then

$$F\{f(x)\cos ax\} = \frac{1}{2}[F(s+a) + F(s-a)]$$

**Proof.** We know that $F(s) = \frac{1}{\sqrt{2\pi}} \int_{-\infty}^{\infty} e^{isx} f(x)\, dx$

$$F\{f(x)\cos ax\} = \frac{1}{\sqrt{2\pi}} \int_{-\infty}^{\infty} e^{isx} f(x) \cos ax\, dx$$

$$= \frac{1}{\sqrt{2\pi}} \int_{-\infty}^{\infty} e^{isx} f(x) \frac{e^{iax} + e^{-iax}}{2} dx$$

$$= \frac{1}{2} \frac{1}{\sqrt{2\pi}} \int_{-\infty}^{\infty} e^{isx} f(x) e^{iax} dx + \frac{1}{2} \frac{1}{\sqrt{2\pi}} \int_{-\infty}^{\infty} e^{isx} e^{-iax} dx$$

$$= \frac{1}{2} \frac{1}{\sqrt{2\pi}} \int_{-\infty}^{\infty} e^{i(s+a)x} f(x)\, dx + \frac{1}{2} \frac{1}{\sqrt{2\pi}} \int_{-\infty}^{\infty} e^{i(s-a)x} f(x) dx$$

$$= \frac{1}{2} F(s+a) + \frac{1}{2} F(s-a)$$

$$= \frac{1}{2}[F(s+a) + F(s-a)]$$

**Proved.**

(6) **If $F\{f(x)\} = F(s)$, then $F\{x^n f(x)\} = (-i)^n \frac{d^n}{ds^n} F(s)$.**

**Proof.** We know that

$$F(s) = \frac{1}{\sqrt{2\pi}} \int_{-\infty}^{\infty} e^{isx} f(x)\, dx \qquad \ldots (1)$$

Differentiating (1) w.r.t. $s$ both sides, $n$ times, we get

$$\frac{d^n F(s)}{ds^n} = \frac{1}{\sqrt{2\pi}} \int_{-\infty}^{\infty} (ix)^n e^{isx} f(x)\, dx$$

$$= (i)^n \frac{1}{\sqrt{2\pi}} \int_{-\infty}^{\infty} x^n e^{isx} \cdot f(x) \cdot dx$$

$$= (i)^n F(x^n f(x))$$

$$F(x^n f(x)) = (-i)^n \frac{d^n \{F(s)\}}{ds^n}$$

(7) $\boldsymbol{F\{f'(x)\} = is F(s)}$ **if $f(x) \to 0$ as $x \to \pm\infty$**

**Proof.** $$F\{f'(x)\} = \frac{1}{\sqrt{2\pi}} \int_{-\infty}^{\infty} e^{isx} f'(x)\, dx$$

$$= \frac{1}{\sqrt{2\pi}} \int_{-\infty}^{\infty} e^{isx}\, d\{f(x)\}\, dx$$

$$= \frac{1}{\sqrt{2\pi}} \left[ \left\{ e^{isx} f(x) \right\}_{-\infty}^{\infty} e^{isx} f(x)\, dx \right]$$

$$= \frac{1}{\sqrt{2\pi}} \left[ 0 - is \int_{-\infty}^{\infty} e^{is} f(x)\, dx \right]$$

$$= -is\, F(s).$$ **Proved.**

(8) $F\left\{ \int_a^x f(x)\, dx \right\} = \dfrac{F(s)}{(-is)}$

**Proof.** Let $f_1(x) = \int_a^x f(x)\, dx \quad \Rightarrow \quad f_1'(x) = f(x)$

$$F\{f'(x)\} = (-is) F_1(s) = (-is) F\{f_1(x)\}$$

$$= -is\, F\left\{ \int_a^x f(x)\, dx \right\}$$

$$F\left\{ \int_a^x f(x)\, dx \right\} = \frac{1}{(-is)} F\{f_1'(x)\}$$

$$= \frac{1}{(-is)} F\{f(x)\} = \frac{F(s)}{(-is)}$$ **Proved.**

**Note.** $F_s(s)$ and $F_c(s)$ are Fourier sine and cosine transforms of $f(x)$ respectively.

**Properties.**

1. $F_s\{af(x) + bg(x)\} = aF_s\{f(x)\} + b\, F_s\{g(x)\}$
2. $F_c\{af(x) + bg(x)\} = aF_c\{f\{x\}\} + b\, F_c\{g(x)\}$
3. $F_s\{f(ax)\} = \dfrac{1}{a} F_s\left(\dfrac{s}{a}\right)$
4. $F_c\{f(ax)\} = \dfrac{1}{a} F_c\left(\dfrac{s}{a}\right)$
5. $F_s[f(x)\sin ax] = \dfrac{1}{2}[F_c(s-a) - F_c(s+a)]$
6. $F_c\{f(x)\sin ax\} = \dfrac{1}{2}[F_s(s+a) - F_s(s-a)]$
7. $F_s\{f(x)\cos ax\} = \dfrac{1}{2}[F_s(s+a) + F_s(s-a)]$

**Proof of (5) :** $F_s\{f(x)\sin ax\} = \int_0^\infty f(x)\sin ax \, . \, \sin sx\, dx$

$$= \frac{1}{2} \int_0^\infty f(x)\{\cos(s-a)x - \cos(s+a)x\}\, dx$$

$$= \frac{1}{2} \left[ \int_0^\infty f(x)\cos(s-a)x\, dx - \int_{0.}^\infty f(x)\cos(s+a)x\, dx \right]$$

$$= \frac{1}{2}[F_c(s-a) - F_c(s+a)]$$ **Proved**

## 14.9 CONVOLUTION

The Convolution of two functions $f(x)$ and $g(x)$ is defined as

$$f(x) * g(x) = \int_{-\infty}^{\infty} f(x)\, g(x-u)\, du$$

**Convolution Theorem on Fourier Transform**

The Fourier transform of the convolution of $f(x)$ and $g(x)$ is the product of their Fourier transforms, *i.e.*,

$$F[f(x) * g(x)] = F[f(x)] \cdot F[g(x)]$$

**Proof.** We know that

$$f(x) * g(x) = \frac{1}{\sqrt{2\pi}} \int_{-\infty}^{\infty} f(u) \cdot g(x-u)\, du \qquad \text{... (1)}$$

Taking Fourier transform of both sides of (1), we have

$$F[f(x) * g(x)] = F\left[\int_{-\infty}^{\infty} f(u) \cdot g(x-u)\, du\right]$$

$$= \frac{1}{\sqrt{2\pi}} \int_{-\infty}^{\infty} \left[\frac{1}{\sqrt{2\pi}} \int_{-\infty}^{\infty} f(u) \cdot g(x-u)\, du\right] e^{isx} dx$$

$$= \frac{1}{\sqrt{2\pi}} \int_{-\infty}^{\infty} f(u)\, du\, \frac{1}{\sqrt{2\pi}} \left\{\int_{-\infty}^{\infty} g(x-u)\, e^{isx} dx\right\}$$

$$= \frac{1}{\sqrt{2\pi}} \int_{-\infty}^{\infty} \{f(u) \cdot du \cdot Fg(x-u)\}$$

$$= \frac{1}{\sqrt{2\pi}} \int_{-\infty}^{\infty} f(u) du \cdot e^{ius} G(s) \qquad \text{(using shifting property)}$$

$$= G(s) \cdot \frac{1}{\sqrt{2\pi}} \int_{-\infty}^{\infty} f(u)\, e^{ius} du$$

$$= G(s) \cdot F(s)$$

$$= F(s) \cdot G(s)$$ **Proved.**

By inversion

$$F^{-1}\{F(s) \cdot G(s)\} = f * g = F^{-1}\{F(s)\} * F^{-1}\{G(s)\}.$$

## 14.10 PARSEVAL'S IDENTITY FOR FOURIER TRANSFORMS

If the fourier transform of $f(x)$ and $g(x)$ be $F(s)$ and $G(s)$ respectively, then

(*i*) $\int_{-\infty}^{\infty} F(s)\overline{G}(s)\, ds = \int_{-\infty}^{\infty} f(x)\, \overline{g}(x)\, dx$

where $\overline{G}(s)$ is the complex conjugate of $G(s)$ and $\overline{g}(x)$ is the complex conjugate of $g(x)$

(*ii*) $\int_{-\infty}^{\infty} [F(s)]^2 ds = \int_{-\infty}^{\infty} |f(x)|^2 dx$

**Proof.** $\int_{-\infty}^{\infty} [f(x)\overline{g}(x)]\, dx = \int_{-\infty}^{\infty} f(x)\left[\frac{1}{\sqrt{2\pi}} \int_{-\infty}^{\infty} \overline{G}(s)\, d^{isx} ds\right] ds$

Since $$\bar{g}(x) = \frac{1}{\sqrt{2\pi}} \int_{-\infty}^{\infty} \bar{G}(s) e^{isx} ds$$

$$\int_{-\infty}^{\infty} f(x)\bar{g}(x)\, dx = \frac{1}{\sqrt{2\pi}} \int_{-\infty}^{\infty} \bar{G}(s) ds \cdot \int_{-\infty}^{\infty} f(x) e^{isx} dx$$

$$\left[\text{since } \frac{1}{\sqrt{2\pi}} \int_{-\infty}^{\infty} f(x) e^{isx} dx = F(s)\right] \text{ Fourier Transform}$$

$$= \int_{-\infty}^{\infty} \bar{G}(s)\, F(s)\, ds \quad \text{... (1)}$$

Putting $g(x) = f(x)$ in (1) we get

$$\int_{-\infty}^{\infty} F(s) \cdot \bar{F}(s)\, ds = \int_{-\infty}^{\infty} f(x) \cdot \bar{f}(x)\, dx$$

or $$\int_{-\infty}^{\infty} [F(s)]^2 ds = \int_{-\infty}^{\infty} [f(x)]^2 dx$$ **Proved.**

## 14.11 PARSEVAL'S IDENTITY FOR COSINE TRANSFORM

(*i*) $\frac{2}{\pi} \int_0^{\infty} F_C(s) \cdot G_C(s) ds = \int_0^{\infty} f(x) \cdot g(x) dx$ (*ii*) $\frac{2}{\pi} \int_0^{\infty} |F_c(s)|^2 dx = \int_0^{\infty} |f(x)|^2 dx$

## 14.12 PARSEVAL'S IDENTITY FOR SINE TRANSFORM

(*i*) $\frac{2}{\pi} \int_0^{\infty} F_s(s) \cdot G_s(s) ds = \int_0^{\infty} f(x) \cdot g(x) dx$ (*ii*) $\frac{2}{\pi} \int_0^{\infty} |F_s(s)|^2 ds = \int_0^{\infty} |f(x)|^2 dx$

**Example. 17.** *Using Parseval's identity, show that*

$$\int_0^{\infty} \frac{dx}{(x^2+1)^2} = \frac{\pi}{4}$$

**Solution.** Let $f(x) = e^{-x}$ so that $F_c(s) = \frac{1}{1+s^2}$

By Parseval's identity for cosine transformation

$$\frac{2}{\pi} \int_0^{\infty} [F_c(s)]^2 ds = \int_0^{\infty} |f(x)|^2 dx$$

$$\frac{2}{\pi} \int_0^{\infty} \left| \frac{1}{(1+s^2)^2} \right|^2 ds = \int_0^{\infty} |e^{-x}|^2 dx = \int_0^{\infty} |e^{-2x}|\, dx = \left[ \frac{e^{-2x}}{-2} \right]_0^{\infty}$$

$$\int_0^{\infty} \left| \frac{1}{(1+s^2)} \right|^2 ds = \frac{1}{2} \frac{\pi}{2} = \frac{\pi}{4}$$ **Ans.**

**Example 18.** *Using Parseval's identity, show that*

$$\int_0^{\infty} \frac{x^2 dx}{(x^2+1)^2} = \frac{\pi}{4}$$

**Solution.** Let $f(x) = \frac{x}{x^2+1}$ so that $F_s(s) = \frac{\pi}{2} e^{(-s)}$

By Parseval's identity for sine transformation

$$\frac{2}{\pi} \int_0^{\infty} |F_s(s)|^2 ds = \int_0^{\infty} |f(x)|^2 dx$$

$$\int_0^\infty \left|\frac{x}{x^2+1}\right|^2 dx = \frac{2}{\pi}\int_0^\infty \left|\frac{\pi}{2}e^{-x}\right|^2 ds$$

$$= \left(\frac{2}{\pi}\right)\left(\frac{\pi^2}{4}\right)\int_0^\infty \left|e^{-2s}\right| ds = \frac{\pi}{2}\left[\frac{e^{-2s}}{-2}\right]_0^\infty$$

$$= \frac{\pi}{2}\left[0+\frac{1}{2}\right] = \frac{\pi}{4}$$ **Proved**

**Example 19.** *Using Parseval's identity, prove that*

$$\int_0^\infty \frac{dt}{(a^2+t^2)(b^2+t^2)} = \frac{\pi}{2ab(a+b)}$$

**Solution.** Let $f(x) = e^{-ax}$ $g(x) = e^{-bx}$

Then $F_c(s) = \frac{a}{a^2+s^2}$, $G_c(s) = \frac{b}{b^2+s^2}$

By Parseval's identity for Fourier cosine transformation

$$\frac{2}{\pi}\int_0^\infty F_c(s)G_c(s)\,ds = \int_0^\infty f(x)\cdot g(x)\,dx \qquad \text{...(1)}$$

On substitution in (1), we get

$$\frac{2}{\pi}\int_0^\infty \left(\frac{a}{a^2+s^2}\right)\left(\frac{b}{b^2+s^2}\right) ds = \int_0^\infty e^{-ax}\cdot e^{-bx}\,dx$$

$$\frac{2}{\pi}\int_0^\infty \frac{ab}{(a^2+s^2)}\frac{ds}{(b^2+s^2)} = \int_0^\infty e^{-(a+b)x}\,dx$$

$$= \left[\frac{e^{-(a+b)x}}{-(a+b)}\right]_0^\infty = \left[0+\frac{1}{a+b}\right]$$

$$\int_0^\infty \frac{ds}{(a^2+s^2)(b^2+s^2)} = \frac{\pi}{2ab}\frac{1}{a+b}$$

$$\int_0^\infty \frac{dt}{(a^2+t^2)(b^2+t^2)} = \frac{\pi}{2ab(a+b)}$$ **Proved**

**Example 20.** *Using Parseval's identity, prove* $\int_0^\infty \left(\frac{\sin t}{t}\right)^2 dt = \frac{\pi}{2}$.

**Solution.** By example we know that

if $$f(x) = \begin{cases} 1 & \text{for } |x|<0 \\ 0 & \text{for } |x|>a>0 \end{cases}$$

then $$F(s) = \sqrt{\frac{2}{\pi}}\frac{\sin as}{s}$$

Using Parseval's identity

$$\int_{-\infty}^\infty |f(t)|^2\,dt = \int_{-\infty}^\infty |F(s)|^2\,ds$$

$$\therefore \quad \int_{-a}^{a} (1)^2\,dt = \int_{-\infty}^{\infty} \frac{2}{\pi}\left(\frac{\sin as}{s}\right)^2 ds$$

$$2a = \frac{2}{\pi}\int_{-\infty}^{\infty}\left(\frac{\sin as}{s}\right)^2 ds$$

Putting $as = t$, we get $\quad 2a = \frac{2}{\pi}\int_{-\infty}^{\infty}\left(\frac{\sin t}{\frac{t}{a}}\right)^2 \frac{dt}{a} \qquad \left(a\,ds = dt,\ ds = \frac{dt}{a}\right)$

$$a\pi = a\int_{-\infty}^{\infty}\left(\frac{\sin t}{t}\right)^2 dt$$

$$\int_0^{\infty}\left(\frac{\sin t}{t}\right)^2 dt = \pi$$ **Proved.**

**Example 21.** *Find the Fourier transform of*

$$f(x) = \begin{cases} 1-|x| & \text{for } |x|<1 \\ 0 & \text{for } |x|>1 \end{cases}$$

*and hence find the value of* $\int_0^{\infty}\left(\frac{\sin t}{t}\right)^4 dt$.

**Solution.** $\quad F\{f(x)\} = \frac{1}{\sqrt{2\pi}}\int_{-\infty}^{\infty} f(x)\,e^{isx}\,dx$

$$= \frac{1}{\sqrt{2\pi}}\int_{-1}^{1}(1-|x|)(\cos sx + i\sin sx)\,dx$$

$$= \frac{1}{\sqrt{2\pi}}\int_{-1}^{1}(1-|x|)\cos sx\,dx + \frac{i}{\sqrt{2\pi}}\int_{-1}^{1}(1-|x|)\sin sx\,dx$$

(Even function) (odd function)

$$= \frac{2}{\sqrt{2\pi}}\int_0^1 (1-x)\cos sx\,dx + 0$$

$$= \sqrt{\frac{2}{\pi}}\left[\left\{(1-x)\frac{\sin sx}{s}\right\}_0^1 + \int_0^1 \frac{\sin sx}{s}\,dx\right]$$

$$= \sqrt{\frac{2}{\pi}}\left[0 + \left\{\frac{-\cos sx}{s^2}\right\}_0^1\right] = \sqrt{\frac{2}{\pi}}\left(\frac{1-\cos s}{s^2}\right)$$

Using Parseval's identity, we get

$$\int_{-\infty}^{\infty} |F(s)|^2\,ds = \int_{-\infty}^{\infty} |f(t)|^2\,dt$$

$$\frac{2}{\pi}\int_{-\infty}^{\infty}\frac{(1-\cos s)^2}{s^4}\,ds = \int_{-1}^{+1}(1-|x|)^2\,dx$$

$$\frac{4}{\pi}\int_0^{\infty}\frac{\left(1-1+2\sin^2\frac{s}{2}\right)^2}{s^4}ds=\int_{-1}^{+1}(1+x^2-2x)\,dx \quad \text{(Odd function)}$$

$$\frac{16}{\pi}\int_0^{\infty}\frac{\sin^4\frac{s}{2}}{s^4}ds=2\int_0^1(1+x^2)\,dx=2\left(x+\frac{x^3}{3}\right)_0^1=\frac{2}{3}$$

Putting $\frac{s}{2}=x$, we get

$$\int_0^{\infty}\frac{\sin^4 x}{x^4}dx=\frac{\pi}{3}$$ **Ans.**

**Example 22.** *Solve for f(x) from the integral equation*

$$\int_0^{\infty} f(x)\cos sx\,dx=e^{-s}$$

**Solution.** $$\int_0^{\infty} f(x)\cos x\,dx=e^{-s} \quad \text{... (1)}$$

Multiplying (1) by $\sqrt{\frac{2}{\pi}}$, we get

$$\sqrt{\frac{2}{\pi}}\int_0^{\infty} f(x)\cos sx\,dx=\sqrt{\frac{2}{\pi}}\,e^{-s}$$

$$F_c\{f(x)\}=\sqrt{\frac{2}{\pi}}\,e^{-s}$$

$$f(x)=F_c^{-1}\left[\sqrt{\frac{2}{\pi}}e^{-s}\right]$$

$$=\sqrt{\frac{2}{\pi}}\left[\sqrt{\frac{2}{\pi}}\int_0^{\infty}e^{-s}\cos sx\,ds\right]$$

$$=\frac{2}{\pi}\int_0^{\infty}e^{-s}\cos sx\,ds$$

$$=\frac{2}{\pi}\left[\frac{e^{-s}}{1+x^2}\{\cos sx+s\sin sx\}\right]_0^{\infty}$$

$$=\frac{2}{\pi}\frac{1}{1+x^2}$$ **Ans.**

**Example. 23.** *Solve for f (x) from the integral equation*

$$\int_0^{\infty} f(x)\sin sx\,dx=\begin{cases}1 \text{ for } 0\le s<1\\ 2 \text{ for } t\le s<2\\ 0 \text{ for } \quad s\ge 2\end{cases}$$

**Solution.** Multiplying by $\sqrt{\frac{2}{\pi}}$ both sides of the given equation, we get

$$\sqrt{\frac{2}{\pi}}\int_0^\infty f(x)\sin sx\,dx = \begin{cases} \sqrt{\dfrac{2}{\pi}} & \text{for } 0 \le s < 1 \\ 2\sqrt{\dfrac{2}{\pi}} & \text{for } 1 \le s < 2 \\ 0 & \text{for } s \ge 2 \end{cases}$$

$$F_s f(x) = \begin{cases} \sqrt{\dfrac{2}{\pi}} & \text{for } 0 \le s < 1 \\ 2\sqrt{\dfrac{2}{\pi}} & \text{for } 1 \le s < 2 \\ 0 & \text{for } s \ge 2 \end{cases}$$

$$f(x) = F_s^{-1} \text{ (R.H.S.)}$$

$$= \frac{2}{\pi}\int_0^1 \sin sx\,ds + \frac{4}{\pi}\int_1^2 \sin sx ds$$

$$= \frac{2}{\pi}\left(\frac{1-\cos x}{x}\right) + \frac{4}{\pi}\left(\frac{\cos x - \cos 2x}{x}\right)$$

$$= \frac{2}{\pi x}[1 - \cos x + 2\cos x - 2\cos 2x]$$

$$f(x) = \frac{2}{\pi x}(1 + \cos x - 2\cos 2x)$$ **Ans.**

**Example 24.** *Find the function if its sine transform is* $\dfrac{e^{-s}}{s}$.

**Solution.** Let $F_s(f(x)) = \dfrac{e^{-as}}{s}$

Then, $$f(x) = \sqrt{\frac{2}{\pi}}\int_0^\infty \frac{e^{-as}}{s}\sin sx\,ds \quad \text{... (1)}$$

$$\frac{df}{dx} = \sqrt{\frac{2}{\pi}}\int_0^\infty e^{-as}\cos sx\,ds$$

$$= \sqrt{\frac{2}{\pi}}\cdot\frac{a}{a^2+x^2}$$

$$f(x) = \sqrt{\frac{2}{\pi}}\cdot a\int\frac{dx}{a^2+x^2}$$

$$= \sqrt{\frac{2}{\pi}}\tan^{-1}\frac{x}{a} + c \quad \text{(2)}$$

At $x = 0$, $f(0) = 0$ using (1)

Using this in (2), $c = 0$

Hence, $$f(x) = \sqrt{\frac{2}{\pi}}\tan^{-1}\frac{x}{a}.$$

Putting $a = 0$, we get

$$F_s^{-1}\left(\frac{1}{s}\right) = \sqrt{\frac{2}{\pi}} \cdot \frac{\pi}{2} = \sqrt{\frac{\pi}{2}} \textbf{ Ans.} \qquad \dots (3)$$

**Example 25.** *Prove (i)* $F\{x^n f(x)\} = (-i)^n \dfrac{d^n F(s)}{ds^n}$ *and (ii)* $F\{f^n(x)\} = (-is)^n F(s)$

*(iii) Hence solve for f (x) if* $\int_{-\infty}^{\infty} f(t) e^{|x-r|} dt = \phi(x)$ *is known.*

**Proof.** (*i*)

$$F(s) = \frac{1}{\sqrt{2\pi}} \int_{-\infty}^{\infty} e^{isx} f(x)\, dx$$

$$\frac{d^n}{ds^n} F(s) = \frac{1}{\sqrt{2\pi}} \int_{-\infty}^{\infty} (ix)^n e^{isx} f(x)\, dx$$

$$(-i)^n \frac{d^n}{ds^n} F(s) = \frac{1}{\sqrt{2\pi}} \int_{-\infty}^{\infty} x^n e^{isx} f(x)\, dx$$

$$= F\{x^n f(x)\}$$

(*ii*) Similarly,

$$F\{f^n(x)\} = \frac{1}{\sqrt{2\pi}} \int_{-\infty}^{\infty} e^{isx} \frac{d^n}{dx^n} f(x) \cdot dx$$

$$= (-is)^n F(s).$$

Using integration by parts successively and making assumptions that $f, f', \dots, f^{(n-1)} \to 0$ as $f(x) \to \pm\infty$.

(*iii*)

$$\frac{1}{\sqrt{2\pi}} \phi(x) = \frac{1}{\sqrt{2\pi}} \int_{-\infty}^{\infty} f(t) e^{|x-t|} dt, \text{ from the given equation}$$

$$= f(x) * e^{-|x|}$$

By convolution theorem,

$$\frac{1}{\sqrt{2\pi}} \bar{\phi}(s) = F(s) \cdot \sqrt{\frac{2}{\pi}} \frac{1}{1+s^2}$$

$$F(s) = \frac{1}{2}(1+s^2)\bar{\phi}(s)$$

$$= \frac{1}{2}[\bar{\phi}(s) - (-is)^2 \bar{\phi}(s)]$$

$$\therefore \qquad f(x) = \frac{1}{2}\phi(x) - \frac{1}{2}\phi''(x) \text{ using the result derived in } (ii)$$

## EXERCISE 14.2

Using Parseval's identity

**1.** Prove that $\displaystyle\int_0^{\infty} \frac{\sin at}{t(a^2+t^2)} dt = \frac{\pi}{2} \frac{1-e^{-a^2}}{a^2}$

**2.** Evaluate $\displaystyle\int_0^{\infty} \left(\frac{1-\cos x}{x}\right)^2 dx$ **Ans.** $\dfrac{\pi}{2}$

## 14.13. FOURIER TRANSFORM OF DERIVATIVES

We have already seen that,

$$F\{f^n(x)\} = (-is)^n \; F(s)$$

(i) $\therefore \; F\left(\dfrac{\partial^2 u}{\partial x^2}\right) = (-is)^2 \, F\{u(x)\} = -s^2 \bar{u}$ where $\bar{u}$ is Fourier transform of $u$ w.r.t. $x$.

(ii) $$F_c\{f'(x)\} = -\sqrt{\frac{2}{\pi}} f(o) + s\, F_s(s)$$

$$\text{L.H.S.} = \sqrt{\frac{2}{\pi}} \int_0^\infty f'(x) \cdot \cos sx \, dx = \sqrt{\frac{2}{\pi}} \int_0^\infty \cos sx \, d\{f(x)\}$$

$$= \sqrt{\frac{2}{\pi}} \left[ \{f(x) \cos sx\}_0^\infty + s \int_0^\infty f(x) \sin sx \, dx \right]$$

$$= s\, F_s(s) - \sqrt{\frac{2}{\pi}} f(0) \quad \text{assuming } f(x) \to 0 \text{ as } x \to \infty$$

(iii) $$F_s\{f'(x)\} = \sqrt{\frac{2}{\pi}} \int_0^\infty \sin sx \, d\,[f(x)]$$

$$= \sqrt{\frac{2}{\pi}} \left[ \{f(x) \sin sx\}_0^\infty - s \int_0^\infty f(x) \cos sx \, dx \right]$$

$$= -s\, f_c(s)$$

(iv) $$F_c\{f''(x)\} = \sqrt{\frac{2}{\pi}} \int_0^\infty \cos sx \, d\,[f'(x)]$$

$$= \sqrt{\frac{2}{\pi}} \left[ \{f'(x) \cos sx\}_0^\infty + s \int_0^\infty f'(x) \sin sx \, dx \right]$$

$$= -\sqrt{\frac{2}{\pi}} f'(0) + s\, F_s\{f'(x)\}$$

$$= -s^2 F_c(s) - \sqrt{\frac{2}{\pi}} f'(0) \quad \text{assuming } f(x),\ f'(x) \to 0 \text{ as } x \to \infty$$

(v) $$F_s\{f''\}(x) = \sqrt{\frac{2}{\pi}} \left[ \int_0^\infty \sin sx \, d[f'(x)] \right]$$

$$= \sqrt{\frac{2}{\pi}} \left[ (f'(x) \sin sx)_0^\infty - s \int_0^\infty f'(x) \cos sx \, dx \right]$$

$$= -s\, F_c\{f'(x)\} = -s\left[s\, F_s(s) - \sqrt{\frac{2}{\pi}} f(0)\right]$$

$$= -s^2 F_s(s) + \sqrt{\frac{2}{\pi}}\, s\, f(0) \text{ assuming } f(x), f'(x) \to 0 \text{ as } x \to \infty.$$

## 14.14. RELATIONSHIP BETWEEN FOURIER AND LAPLACE TRANSFORMS

Consider

$$f(t) = \begin{cases} e^{-st} g(t) & \text{for } t > 0 \\ 0 & \text{for } t < 0 \end{cases} \quad \text{... (1)}$$

Then the Fourier transform of $f(t)$ is given by

$$F\{f(t)\} = \frac{1}{\sqrt{2\pi}} \int_{-\infty}^{\infty} e^{ist} f(t)\, dt$$

$$= \frac{1}{\sqrt{2\pi}} \int_{0}^{\infty} e^{(is-x)t} g(t)\, dt = \frac{1}{\sqrt{2\pi}} \int_{0}^{\infty} e^{-pt} g(t)\, dt \quad \text{where } p = x - is$$

$$= \frac{1}{\sqrt{2\pi}} L\{g(t)\}$$

$\therefore$ Fourier transform of $f(t) = \dfrac{1}{\sqrt{2\pi}} \times$ Laplace transform of $g(t)$ defined by (1).

## 14.15. SOLUTION OF BOUNDARY VALUE PROBLEMS BY USING INTEGRAL TRANSFORM

**Solution of heat conduction problems by Laplace transform.**

**Example 26.** *A semi-infinite solids $x > 0$ is initially at temperature zero. At time $t = 0$, a constant temperature $u_0$ is applied and maintained at the face $x = 0$. Find the temperature at any point of the solid and at any time $t > 0$.*

**Solution.** Let $u(x, t)$ be the temperature at any point $x$ and at any time $t$. The equation governing the flow of heat in the solid is given by

$$\frac{\partial u}{\partial t} = c^2 \frac{\partial^2 u}{\partial x^2} \quad x > 0, t > 0 \quad \text{... (1)}$$

The initial and boundary conditions are

$$u = 0 \text{ when } t = 0 \text{ for all } x (x \geq 0) \quad \text{... (2)}$$

$$u = u_0 \text{ when } x = 0 \text{ for all } t, \quad \text{... (3)}$$

$$u \text{ is finite for all } x \text{ and for all } t, \quad \text{... (4)}$$

Multiplying equation (1) by $e^{-st}$ and integrate w.r.t. '$t$' from 0 to $\infty$,

$$\int_0^{\infty} \frac{\partial u}{\partial t} e^{-st}\, dt = c^2 \int_0^{\infty} \frac{\partial^2 u}{\partial x^2} e^{-st}\, dt \text{ or } s\bar{u} = c^2 \frac{d^2 \bar{u}}{dx^2} \quad \text{... (5)}$$

($\bar{u}$ = Laplace transform of $u$)

Similarly Laplace transformation of equation (3) gives

$$s\bar{u} = u_0 \quad \text{when } x = 0 \text{ or } \bar{u} = \frac{u_0}{s} \quad \text{... (6)}$$

Equation (5) is an ordinary differential equation and its solution is given by

$$\bar{u} = Ae^{\frac{\sqrt{s}}{c}x} + Be^{-\frac{\sqrt{s}}{c}x} \quad \text{... (7)}$$

According to condition (4), $u$ is finite at $x \to \infty$.

$\therefore \quad A = 0$

So (7) becomes $\quad \bar{u} = Be^{-\frac{\sqrt{s}}{c}x}$

Using (6) equation $\quad \bar{u} = \frac{u_0}{s}$ when $x = 0$, $u_0 / s = B$

Thus (8) becomes $\quad \bar{u} = \frac{u_0}{s} e^{-\frac{\sqrt{s}}{c}x}$

To get $u$ from $\bar{u}$, we invert the transformation.

$$u = u_0\left(1 - erf\frac{x}{2c\sqrt{t}}\right) \quad \textbf{Ans.}$$

**Solution of wave equation by Laplace transform**

**Example 27.** *An infinitely long string having one end at x = 0 is initialy at rest along x-axis. The end x = 0 is given a transverse displacement f(t), when t > 0. Find the displacement of any point of the string at any time.*

**Solution.** Let $y(x, t)$ be the displacement, then wave equation is

$$\frac{\partial^2 y}{\partial x^2} = c^2 \frac{\partial^2 y}{\partial t^2} \quad \text{... (1)}$$

subject to the conditions

$y(x, 0) = 0 \quad \text{... (2)} \qquad \frac{\partial y}{\partial t}(x,0) = 0 \quad \text{... (3)}$

$y(0, t) = f(t) \quad \text{... (4)} \qquad y(x, t)$ is bounded ... (5)

On taking Laplace transform of (1), we have

$$L\left(\frac{\partial^2 y}{\partial x^2}\right) = c^2 L\frac{\partial^2 y}{\partial t^2}$$

$$s^2\bar{y} - sy(x,0) - \frac{\partial y}{\partial t}(x,0) = c^2\frac{d^2\bar{y}}{dx^2} \quad \text{... (6)}$$

On putting $y(x, 0) = 0$, $\frac{\partial y}{\partial t}(x, 0) = 0$ in (6) we get

$$s^2\,\bar{y} = c^2\frac{d^2\bar{y}}{dx^2} \quad \text{or} \quad \frac{d^2\bar{y}}{dx^2} = \left(\frac{s}{c}\right)^2 \bar{y} \quad \text{... (7)}$$

Laplace transform of (4), $\bar{y}(0, s) = \bar{f}(s)$ at $x = 0$ ... (8)

On solving (7), we get $\bar{y} = Ae^{\frac{sx}{c}} + Be^{-\frac{sx}{c}}$

According to condition (5), $y$ is finite at $x \to \infty$, this gives $A = 0$ so (9) becomes

$$\bar{y} = Be^{-\frac{sx}{c}} \quad \text{... (10)}$$

Putting the value of $\bar{y}(0, s) = \bar{f}(s)$ at $x = 0$ in (10), we get $\bar{f}(s) = B$

Thus (10) becomes $$\bar{y} = \bar{f}(s) \cdot e^{-\frac{sx}{c}}$$

To get $y$ from $\bar{y}$, we use complex inversion formula

$$y = \frac{1}{2\pi i} \int_{a-i\infty}^{a+i\infty} e^{\left(t-\frac{x}{c}\right)s} - f(s)\, ds$$

Hence $$y = f\left(t - \frac{x}{c}\right) \quad \textbf{Ans.}$$

**Example 28.** *A uniform rod of length $l$ is at rest in its equilibrium position with the end $x = 0$ fixed. At $t = 0$, a constant force $F_0$ per unit area is applied at the free end. Determine the motion of the rod for $t > 0$.*

**Solution.** Let $y(x, t)$ be the displacement in the rod. Equation of motion is given by

$$\frac{\partial^2 y}{\partial t^2} = c^2 \frac{\partial^2 y}{\partial x^2} \quad l > x > 0, \quad t > 0 \qquad \ldots (1)$$

subject to the conditions

$$y(x, 0) = 0 \quad \ldots (2) \qquad \frac{\partial y}{\partial t}(x, 0) = 0 \qquad \ldots (3)$$

$$y(0, t) = 0 \quad \ldots (4) \qquad \frac{\partial y}{\partial x}(l, t) = \frac{F_0}{E} \qquad \ldots (5)$$

where $E$ = Young's modulus.

Applying Laplace transform on (1) $$c^2 \frac{d^2 \bar{y}}{dx^2} = s^2 \bar{y} \qquad \ldots (6)$$

Eq. (6) is an ordinary differential equation and its solution is

$$\bar{y} = A e^{\frac{sx}{c}} + B e^{-\frac{sx}{c}} \qquad \ldots (7)$$

Putting $y = 0, x = 0$ from (2) in (7), we get

$0 = A + B$ or $B = -A$, then (7) becomes

$$\bar{y} = A\left(e^{\frac{sx}{c}} - e^{-\frac{sx}{c}}\right) \qquad \ldots (8)$$

Laplace transform of (3) $$\frac{d\bar{y}}{dx} = \frac{F_0}{Es} \text{ at } x = l \qquad \ldots (9)$$

Differentiating (8) w.r.t. '$x$' we get

$$\frac{d\bar{y}}{dx} = A\left(\frac{s}{c} e^{\frac{sx}{c}} + \frac{s}{c} e^{-\frac{sx}{c}}\right) \qquad \ldots (10)$$

Putting the value of $\frac{d\bar{y}}{dx}$ from (9) in (10), we have

$$\frac{F_0}{ES} = A \frac{s}{c}\left(e^{\frac{s}{c}l} + e^{-\frac{s}{c}l}\right) \quad \text{or} \quad A = \frac{F_0}{Es} \frac{c}{s} \frac{1}{e^{\frac{s}{c}l} + e^{-\frac{s}{c}l}}$$

Putting the value of $A$ in (8) we obtain

$$\bar{y}=\frac{cF_0}{Es^2}\,\frac{e^{\frac{s}{c}x}-e^{-\frac{s}{c}x}}{e^{\frac{sl}{c}}+e^{-\frac{sl}{c}}}=\frac{cF_0}{E.s^2}\,\frac{1-e^{-\frac{2s}{c}x}}{1+e^{-\frac{2sl}{c}}}\,\frac{e^{\frac{s}{c}x}}{e^{\frac{s}{c}l}}$$

$$=\frac{cF_0}{Es^2}\left[\left(1-e^{\frac{-2s}{c}x}\right)\left(1+e^{-\frac{2sl}{c}}\right)^{-1}\right]e^{\frac{s}{c}(x-l)}$$

$$=\frac{c\,F_0}{Es^2}\left[\left(1-e^{-\frac{2s}{c}x}\right)\left(1-e^{-\frac{2sl}{c}}+\ldots\right)\right]\times e^{\frac{s}{c}(x-l)}$$

$$\bar{y}=\frac{c\,F_0}{Es^2}\left[1-e^{-\frac{2s}{c}x}-e^{-\frac{2sl}{c}}+e^{-\frac{2s}{c}(x+l)}+\ldots\right]e^{\frac{s}{c}(x-l)} \qquad \ldots (11)$$

Putting $x = l$ in (11) we get

$$x=l\ \bar{y}=\frac{cF_0}{Es^2}\left[1-e^{-\frac{2s}{c}l}-e^{-\frac{2sl}{c}}+e^{-\frac{2s(l+l)}{c}}\ \ldots.\right] \qquad \ldots (12)$$

Applying inversion transformation on (12) we get

$$y=\frac{F_0c}{E}t, \qquad 0<t<\frac{2l}{c} \qquad \ldots (13)$$

$$y=\frac{F_0c}{E}t-\frac{2F_0c}{E}\left(t-\frac{2l}{c}\right); \qquad \frac{2l}{c}<t<\frac{4l}{c}$$

Putting $\frac{2l}{c}=\lambda$ in (13), we have

$$y=\begin{cases} At & 0<t<\lambda \\ At-2A\,(t-\lambda), \text{ where } A=\dfrac{F_0C}{E}, & \lambda<t<2\lambda \end{cases} \quad \textbf{Ans.}$$

**Solution of Transmission Lines equations by Laplace Transformations.**

**Example 29.** *A semi-infinite transmission line, of negligible inductance and leakage per unit length has its voltage and current equal to zero. A constant voltage $v_0$ is applied at the sending end ($x = 0$) at $t = 0$. Find the voltage and current at any point ($x > 0$) at any instant.*

**Solution.** Let $v$ and $i$ be the voltage and current at any point $x$ and at any time $t$.

$$\left.\begin{aligned} -\frac{\partial v}{\partial x}&=Ri+L\frac{\partial i}{\partial t}\\ -\frac{\partial i}{\partial x}&=c\frac{\partial v}{\partial t}+GV \end{aligned}\right]$$

On putting $L = 0$, $G = 0$ in above equations we get

$$\frac{\partial v}{\partial x}=-Ri \qquad \ldots (1)$$

$$\frac{\partial i}{\partial x} = -c\frac{\partial v}{\partial t} \quad \text{... (2)}$$

Conditions are $v(x, 0) = 0$ ... (3)

$i(x, 0) = 0$ ... (4)

$v(0, t) = v_0$ ... (5)

$v(x, t)$ finite for all $x$ and $t$. ... (6)

Applying Laplace transform of (1) and (2), we get

$$\frac{d\bar{v}}{dx} = -R\bar{i} \quad \text{... (7)}$$

$$\frac{di}{dx} = -c(s\bar{v} - v) \text{ and } v(x, 0) = 0 \text{ or } \frac{di}{dx} = -cs\bar{v} \quad \text{... (8)}$$

Differentiating (7) w.r.t. '$x$' we get $\dfrac{d^2\bar{v}}{dx^2} = -R\dfrac{di}{dx}$

or $$\frac{d^2\bar{v}}{dx^2} = -R(-cs\bar{v}) \qquad \left(\frac{d\bar{v}}{dx} = -cs\bar{v}\right)$$

$$\frac{d^2\bar{v}}{dx^2} = Rcs\bar{v} \text{ or } \frac{d^2\bar{v}}{dx^2} - Rcs\bar{v} = 0 \quad \text{... (9)}$$

Laplace transform of (5) is $\bar{v}(0,s) = \dfrac{v_0}{s}$ ... (10)

And Laplace transform of (6) is $\bar{v}(x,s)$ remains finite as $x \to \infty$. ... (11)

Equation (9) is an ordinary differential equation and its solution is

$$\bar{v} = Ae^{\sqrt{Rcs}\,x} + Be^{-\sqrt{Rcs}\,x} \quad \text{.. (12)}$$

As $x \to \infty$, $\bar{v}$ remains finite only when $A = 0$.

So (12) becomes $$\bar{v} = Be^{-\sqrt{Rcs}\,x} \quad \text{... (13)}$$

Putting $\bar{v} = \dfrac{v_0}{s}$ and $x = 0$ in (13) we get $\dfrac{v_0}{s} = B$

Substituting the value of $B$ in (13) we have

$$\bar{v} = \frac{v_0}{s}e^{\sqrt{Rcs}\,x}$$

On applying inversion transform we get

$$v = v_0\, L^{-1}\left[\frac{e^{-\sqrt{Rcs}\,x}}{s}\right] = v_0\, erf\, c\left[\frac{x\sqrt{Rc}}{2\sqrt{t}}\right]$$

$$v = v_0 \frac{x\sqrt{Rc}}{2\sqrt{\pi}} \int_0^t u^{-\frac{3}{2}} e^{-\left(\frac{Rcx^2}{4u}\right)} \cdot du \quad \text{... (14)}$$

From (1) $$i = \frac{-1}{r}\frac{\partial v}{\partial x} \quad \text{... (15)}$$

On differentiating (14) we get

$$\frac{\partial v}{\partial x} = \frac{v_0\, x\sqrt{Rc}}{2\sqrt{\pi}} t^{-\frac{3}{2}} e^{-\frac{Rcx^2}{4t}}$$

Putting the value of $\frac{\partial v}{\partial x}$ in (15), we obtain

$$i = \frac{v_0 x}{2\sqrt{\pi}}\sqrt{\frac{c}{R}} \cdot t^{-\frac{3}{2}} e^{-\frac{Rcx^2}{4t}}$$ **Ans.**

**Solution of partial differential Equations by Fourier Transform**

**Example 30.** *Solve* $\frac{\partial^2 u}{\partial t^2} = \alpha^2 \frac{\partial^2 u}{\partial x^2}, -\infty < x < \infty, t \geq 0$ *with conditions* $u(x, 0) = f(x)$, $\frac{\partial u}{\partial t}(x, 0) = g(x)$ *and assuming* $u, \frac{\partial u}{\partial x} \to 0$ *as* $x \to \pm\infty$.

**Solution.** Taking Fourier transform on both sides of the differential equation.

$$\frac{d^2\bar{u}}{dt^2} = \alpha^2(-s^2\bar{u})$$ where $\bar{u}$ is Fourier transform of $u$ with respect to $x$,.

$$\frac{d^2\bar{u}}{dt^2} + \alpha^2 s^2 \bar{u} = 0$$

Auxiliary equation is $m^2 + \alpha^2 s^2 = 0 \quad \Rightarrow \quad m = \pm i\alpha s$

$$\therefore \quad \bar{u}(s,t) = Ae^{i\alpha st} + Be^{-i\alpha st} \quad \text{... (1)}$$

Since $u(x, 0) = f(x)$ and $\frac{\partial u}{\partial t}(x,0) = g(x)$,

$\bar{u}(s,0) = F(s)$ and $\frac{d\bar{u}}{dt}(s,o) = G(s)$ on taking transform.

Using these condition in (1),

$$\bar{u}(s,0) = A + B = F(s) \quad \text{... (2)}$$

$$\frac{d\bar{u}}{dt}(s, 0) = i\alpha s(A - B) = G(s) \quad \text{...(3)}$$

Solving $$A = \frac{1}{2}\left[F(s) + \frac{G(s)}{i\alpha s}\right]$$

$$B = \frac{1}{2}\left[F(s) - \frac{G(s)}{i\alpha s}\right]$$

Using these values in (1),

$$\bar{u}(s,t) = \frac{1}{2}\left[F(s) + \frac{G(s)}{i\alpha s}\right]e^{i\alpha st} + \frac{1}{2}\left[f(s) - \frac{G(s)}{i\alpha s}\right]e^{-i\alpha st} \quad \text{... (4)}$$

By inversion theorem, (4) reduces to,

$$u(x, t) = \frac{1}{2}\left[f(x - \alpha t) - \frac{1}{\alpha}\int_{\alpha}^{x-\alpha t} g(\theta)\, d\theta\right] + \frac{1}{2}\left[f(x + \alpha t) + \frac{1}{\alpha}\int_{\alpha}^{x+\alpha t} g(\theta)\, d\theta\right]$$

Using the result

$$F\left(\int_{\alpha}^{x} f(t)\, dt\right) = \frac{F(s)}{(-is)}$$ **Ans.**

## 14.16 FOURIER TRANSFORMS OF PARTIAL DERIVATIVE OF A FUNCTION

$$F_f\left[\frac{\partial^2 u}{\partial x^2}\right] = -s^2\, F(u) \quad \text{where } F(u) \text{ is Fourier transform of } u \text{ w.r.t. } x.$$

$$F_s\left[\frac{\partial^2 u}{\partial x^2}\right] = s\,(u)_{x=0} - s^2\, F_s(u) \qquad \text{(sine transform)}$$

$$F_c\left[\frac{\partial^2 u}{\partial x^2}\right] = -\left[\frac{\partial u}{\partial x}\right]_{x=0} - s^2\, F_c(u) \qquad \text{(cosine transform)}$$

**Proof.** Let $F[u(x, t)]$ be the Fourier transform of the function $u(x, t)$, *i.e.*

$$F[u(x,t)] = \int_{-\infty}^{\infty} e^{isx}\, u(x,t)\, dx$$

The Fourier transform of $\frac{\partial^2 u}{\partial x^2}$ is given by

$$F\left[\frac{\partial^2 u}{\partial x^2}\right] = \int_{-\infty}^{\infty} e^{isx}\, \frac{\partial^2 u}{\partial x^2}\, dx.$$

Integrating by parts, we have

$$F\left[\frac{\partial^2 u}{\partial x^2}\right] = \left[e^{isx}\,\frac{\partial u}{\partial x} - \int ise^{isx}\,\frac{\partial u}{\partial x}\, dx\right]_{-\infty}^{\infty}$$

$$= \left[e^{isx}\,\frac{\partial u}{\partial x} - is\, e^{isx} u + \int (is)^2\, e^{isx} u\, dx\right]_{-\infty}^{\infty} \qquad \text{Again integrating}$$

$$= \left[0 + 0 - s^2 \int_{-\infty}^{\infty} e^{isx}.u\,dx\right] \qquad \left[u = 0,\ \frac{\partial u}{\partial x} = 0 \text{ when } x \to \infty\right]$$

Thus

$$F\left[\frac{\partial^2 u}{\partial x^2}\right] = -s^2\, F[u(x,t)]$$

Similarly the Fourier sine transform of $\frac{\partial^2 u}{\partial x^2}$ is given by

$$F_s\left[\frac{\partial^2 u}{\partial x^2}\right] = \int_0^{\infty} \frac{\partial^2 u}{\partial x^2} \sin sx\, dx$$

or

$$F_s\left[\frac{\partial^2 u}{\partial x^2}\right] = s\,[u(x,t)]_{x=0} - s^2\, F_s\,[u(s,t)] \qquad \text{(sine transform)}$$

and

$$F_c\left[\frac{\partial^2 u}{\partial x^2}\right] = -\left[\frac{\partial u}{\partial x}\right]_{x=0} - s^2\, F_c\,[u(s,t)] \qquad \text{(cosine transform)}$$

**Solution of heat conduction problems by Fourier sine Transforms**

**Example 31.** *Solve the equation*

$$\frac{\partial u}{\partial t} = \frac{\partial^2 u}{\partial x^2}, \qquad x > 0, t > 0$$

*subject to the conditions*

*(i)* $u = 0$ *when* $x = 0, t > 0$

*(ii)* $u = \begin{cases} 1 & 0 < x < 1 \\ 0 & x \geq 1 \end{cases}$ *when* $t = 0$

*(iii)* $u(x, t)$ *is bounded.*

(**Note.** If $u$ at $x = 0$ is given, take Fourier sine transform and if $\frac{\partial u}{\partial x}$ at $x = 0$ is given, use Fourier cosine transform.)

**Solution.** In view of the initial conditions, we apply Fourier sine transform

$$\int_0^\infty \frac{\partial u}{\partial t} \sin sx \, dx = \int_0^\infty \frac{\partial^2 u}{\partial x^2} \sin sx \, dx$$

$$\frac{\partial}{\partial t} \int_0^\infty u \sin sx \, dx = -s^2 \bar{u}(s) + s\, u(0) \qquad u = 0 \text{ when } x = 0$$

$$\frac{\partial \bar{u}}{\partial t} = -s^2 \bar{u} \quad \text{or} \quad \frac{\partial \bar{u}}{\partial t} + s^2 \bar{u} = 0$$

$$\therefore \qquad \bar{u} = A e^{-s^2 t} \qquad \text{... (1)}$$

$$\bar{u} = \bar{u}(s, t) = \int_0^\infty u(x, t) \sin sx \, dx$$

$$\bar{u} = \bar{u}(s, 0) = \int_0^\infty u(x, 0) \sin sx \, dx$$

$$\bar{u}(s, 0) = \int_0^\infty 1 \cdot \sin sx \, dx = \left[\frac{-\cos sx}{s}\right]_0^1 = \frac{1 - \cos s}{s} \qquad \text{... (2)}$$

From (2) putting the value of $\bar{u}$ $(s, 0)$ in (1) we get $\frac{1 - \cos s}{s} = A$

$$\therefore \qquad \bar{u} = \frac{1 - \cos s}{s} e^{-s^2 t} \qquad \text{or} \qquad u = \frac{2}{\pi} \int_0^\infty \left(\frac{1 - \cos s}{s}\right) e^{-s^2 t} \sin xs \, ds$$

**Example 32.** *Solve* $\frac{\partial u}{dt} = k \frac{\partial^2 u}{\partial x^2}$ *for* $x \geq 0, t \geq 0$ *under the given conditions* $u = u_0$ *at* $x = 0$, $t > 0$ *with initial condition* $u(x, 0) = 0$, $x \geq 0$

**Solution.** Taking Fourier sine transforms

$$F_s\left(\frac{\partial u}{\partial t}\right) = F_s\left(k \frac{\partial^2 u}{\partial x^2}\right)$$

$$\frac{d}{dt} \bar{u} = k\left[-s^2 \bar{u} + \sqrt{\frac{2}{\pi}}\, s\, u(0, t)\right]$$

$$= -ks^2\bar{u} + \sqrt{\frac{2}{\pi}}\, k s u_0 \text{ where } \bar{u} \text{ is the Fourier sine transform of } u.$$

$$\frac{d\bar{u}}{dt} + sk^2\bar{u} = \sqrt{\frac{2}{\pi}}\, s k u_0$$

This is linear in $\bar{u}$ .

$$\therefore \quad \bar{u}\, e^{ks^2 t} = \sqrt{\frac{2}{\pi}} k u_0 \int s\, e^{ks^2 t} dt = \sqrt{\frac{2}{\pi}} \frac{u_0}{s} e^{ks^2 t} + c \qquad \text{... (1)}$$

Since, $u\ (x, 0)= 0$, $\bar{u}\ (s, 0) = 0$. Using this in (1)

$$0 = \sqrt{\frac{2}{\pi}} \frac{u_0}{s} + c \quad \Rightarrow \quad c = -\sqrt{\frac{2}{\pi}} \frac{u_0}{s}$$

$$e^{ks^2 t} = \sqrt{\frac{2}{\pi}} \frac{u_0}{s}\left(e^{ks^2 t} - 1\right) \quad \Rightarrow \quad \bar{u} = \sqrt{\frac{2}{\pi}} \frac{u_0}{s}\left(1 - e^{ks^2 t}\right)$$

By inversion theorem,

$$u(x,t) = \frac{2u_0}{\pi} \int_0^\infty \left(\frac{1 - e^{ks^2 t}}{s}\right) \sin sx\, ds. \qquad \textbf{Ans.}$$

**Example 33.** *Solve* $\dfrac{\partial u}{\partial t} = k\dfrac{\partial^2 u}{\partial x^2}$ *for* $0 \le x < \infty, t > 0$ *given the conditions*

*(i)* $u\ (x, 0) = 0$ *for* $x \ge 0$ *(ii)* $\dfrac{\partial u}{\partial x}(0,t) = -a$ *(constant)*

*(iii)* $u(x, t)$ *is bounded.*

**Solution.** In this problem, $\dfrac{\partial u}{\partial x}$ at $x = 0$ is given. Hence, take Fourier cosine transform on both sides of the given equation.

$$F_c\left(\frac{\partial u}{\partial t}\right) = F_c\left(k\frac{\partial^2 u}{\partial x^2}\right)$$

$$\frac{d\bar{u}}{dt} = k\left(-s^2\bar{u} - \sqrt{\frac{2}{\pi}} \cdot \frac{\partial u}{\partial x}(0, t)\right)$$

$$= -ks^2\bar{u} + \sqrt{\frac{2}{\pi}}\, ka \qquad \text{using condition } (ii)$$

$$\frac{d\bar{u}}{dt} + ks^2\bar{u} = \sqrt{\frac{2}{\pi}}\, ka$$

This is linear in $\bar{u}$. Therefore, solving

$$\bar{u} e^{ks^2 t} = \int \sqrt{\frac{2}{\pi}}\, ka\, e^{ks^2 t} dt = \sqrt{\frac{2}{\pi}} ka \frac{e^{ks^2 t}}{ks^2} + c$$

$$\bar{u}(s,t) = \sqrt{\frac{2}{\pi}} \frac{a}{s^2} + c\, e^{-ks^2 t} \qquad \text{... (1)}$$

Since $u(x, 0) = 0$ for $x \geq 0.$

$\bar{u}(s, 0) = 0.$

Using this in (1), we get

$$\bar{u}(s,0) = c + \sqrt{\frac{2}{\pi}}\frac{a}{s^2} = 0$$

$$\therefore \quad c = -\sqrt{\frac{2}{\pi}}\frac{a}{s^2}$$

Substituting this in (1)

$$\bar{u}(s, t) = \sqrt{\frac{2}{\pi}}\frac{a}{s^2}\left(1 - e^{-ks^2 t}\right)$$

By inversion theorem,

$$u(x,t) = \frac{2}{\pi}\cdot a\int_0^\infty \frac{1 - e^{-ks^2 t}}{s^2}\cos sx\, ds.$$ **Ans.**

## EXERCISE 14.3

**1.** Use Fourier sine transform to solve the equation

$$\frac{\partial u}{\partial t} = 2\frac{\partial^2 u}{\partial x^2}$$

Under the conditions $u(0, t) = 0$, $u(x, 0) = e^{-x}$, $u(x, t)$ is bounded.

**Ans.** $u(x,t) = \frac{2}{\pi}\int_0^\infty \frac{s}{s^2+1}e^{-s^2 t}\sin sx\, ds$

**2.** A tightly stretched string with fixed end points $x = b$ and $x = c$ is initially in a position given by $y = b\sin\left(\frac{\pi x}{c}\right)$. It is released from rest in this position. Show by the method of Laplace transform that the displacement $y$ at any distance $x$ from one end and at any time $t$ is given by

$$y = b\sin\frac{\pi x}{c}\cos\frac{\pi q}{c}t.$$

and $y$ satisfies the equation $\frac{\partial^2 y}{\partial x^2} = a^2\frac{\partial^2 y}{\partial t^2}$

**3.** A string is stretched tightly between $x = 0$ and $x = l$ and both ends are given displacement $y = a\sin pt$ perpendicular to the string. If the string satisfies the differential equation

$$\frac{\partial^2 y}{\partial x^2} = \frac{1}{c^2}\frac{\partial^2 y}{\partial t^2}$$

Show that the oscillations of the string are given by

$$y = a\sec\frac{Pl}{2c}\cos\left(\frac{Px}{c} - \frac{Pl}{2c}\right)\sin pt.$$

**4.** An infinite cable with resistance $R$ ohms/km, capacitance $C$ Farads/km, and negligible inductance and leakage is subjected to constant E.M.F. $E_0$ at the home end at time $t = 0$. Using the operational method show that the entering current at any subsequent time $t$ is

$$I(t) = E_0 \left( \frac{C}{\pi R t} \right)^{1/2}$$

**5.** Solve the equation for high voltage semi-infinite line with the following initial and boundary conditions

$v(x, t) = 0$ and $i(x, 0) = 0$, $v(0, t) = v_0\, u(t)$, $v(x, t)$ is finite as $x \to \infty$.

**Ans.** $v = v_0\, u[t - x\sqrt{LC}]$, for $x \le \dfrac{t}{\sqrt{LC}}$ and

$v = 0$ for $x > \dfrac{t}{\sqrt{LC}}$

**6.** Solve $\dfrac{\partial u}{\partial t} = \dfrac{\partial^2 u}{\partial x^2}$ if

(*i*) $\dfrac{\partial u}{\partial x}(0,t) = 0$ for $t > 0$. (*ii*) $u(x, 0) = \begin{cases} x & 0 \le x \le 1 \\ 0 & x > 1 \end{cases}$

(*iii*) and $u(x, t)$ is bounded for $x > 0$, $t > 0$

**Ans.** $\left[ u(x,t) = \dfrac{2}{\pi} \int_0^\infty \left( \dfrac{\sin s}{s} + \dfrac{\cos s - 1}{s^2} \right) e^{-s^2 t} \cos sx\, ds \right]$

## 14.17. FINITE FOURIER TRANSFORMS

Let $f(x)$ denote a function which is sectionally continuous over the range $(0, l)$. Then the **finite Fourier sine transform** of $f(x)$ on this interval is defined as

$$F_s(p) = \bar{f}_s(P) = \int_0^l f(x) \sin \frac{p\pi x}{l} dx$$

where $p$ is an integer (Instead of $s$, we take $p$ as a parameter)

**Inversion formula for sine transform**

If $f_s(p) = F_s(p)$ is the finite Fourier sine transform of $f(x)$ in $(0, l)$ then the inversion formula for sine transform is

$$f(x) = \frac{2}{l} \sum_{p=l}^{\infty} \bar{f}_s(p) \sin \frac{p\pi x}{l}$$

**Proof.** For the given function $f(x)$ in $(0, l)$, if we find the half range Fourier sine series, we get.

$$f(x) = \sum_{n=1}^{\infty} b_n \sin \frac{n\pi x}{l} \qquad \text{... (1)}$$

where $$b_n = \frac{2}{l} \int_0^l f(x) \sin \frac{n\pi x}{l} dx$$

$$\therefore \quad b_p = \frac{2}{l} \int_0^l f(x) \sin \frac{p\pi x}{l} dx = \frac{2}{l} \bar{f}_s(p) \text{ by definition}$$

Substituting in (1), we get

Substituting in (1), we get

$$\therefore \quad f(x) = \frac{2}{l}\sum_{p=1}^{\infty} f_s(p)\sin\frac{p\pi x}{l} \qquad \textbf{Ans.}$$

**Finite Fourier Cosine Transform**

Let $f(x)$ denote a sectionally continuous function in $(0, l)$.

Then the Finite Fourier cosine transform of $f(x)$ over $(0, l)$ is defined as

$$F_c(p) = \bar{f}_c(p) = \int_0^l f(x)\cos\frac{p\pi x}{l}\,dx \quad \text{where } p \text{ is an integer.}$$

**Inversion formula for cosine transform**

If $\bar{f}_c(P)$ is the finite Fourier cosine transform of $f(x)$ in $(0, l)$. then the inversion formula for cosine transform is

$$f(x) = \frac{1}{l}\bar{f}_c(0) + \frac{2}{l}\sum_{p=l}^{\infty} \bar{f}_c(p)\cos\frac{p\pi x}{l}$$

where $$\bar{f}_c(0) = \int_0^l f(x)\,dx.$$

**Proof.** If we find half range Fourier cosine series for $f(x)$ in $(0, l)$, we obtain.

$$f(x) = \frac{a_0}{2} + \sum_{n=1}^{\infty} a_n\cos\frac{n\pi x}{l} \qquad \text{... (2)}$$

where $$a_n = \frac{2}{l}\int_0^l f(x)\cos\frac{n\pi x}{l}\,dx$$

$$\therefore \quad a_p = \frac{2}{l}\bar{f}_c(p)$$

$$a_0 = \frac{2}{l}\int_0^l f(x)\,dx$$

$$= \frac{2}{l}\bar{f}_c(0).$$

Substituting in (2), we get,

$$f(x) = \frac{1}{l}\bar{f}_c(0) + \frac{2}{l}\sum_{p=1}^{\infty} \bar{f}_c(p)\cos\frac{p\pi x}{l}$$

**Example. 34.** *Find the finite Fourier sine and cosine transforms of*

| | |
|---|---|
| (i) $f(x) = 1$ | in $(0, \pi)$ |
| (ii) $f(x) = x$ | in $(0, l)$ |
| (iii) $f(x) = x^2$ | in $(0, l)$ |
| (iv) $f(x) = 1$ | in $0 < x < \pi/2$ |
| $= -1$ | in $\pi/2 < x < \pi$ |
| (v) $f(x) = x^3$ | in $(0, l)$ |
| (vi) $f(x) = e^{ax}$ | in $(0, l)$ |

(i) $$\bar{f}_s(p) = F_s(1) = \int_0^{\pi} 1.\sin\frac{p\pi x}{\pi}dx = \left(-\frac{\cos px}{p}\right)_0^{\pi} = \frac{1-\cos p\pi}{p} \quad \text{if } p \neq 0$$

$$\bar{f}_c(p) = \int_0^{\pi} 1\cdot\cos px\,dx = \left(\frac{\sin px}{p}\right)_0^{\pi} = \frac{1}{p}(0-0) = 0$$

(ii) $$\bar{f}_s(p) = F_s(p) = \int_0^{l} x\sin\frac{p\pi x}{l}dx$$

$$= \left[(x)\left(\frac{-\cos\frac{p\pi x}{l}}{\frac{p\pi}{l}}\right) - (1)\left(-\frac{\sin\frac{p\pi x}{l}}{\frac{p^2\pi^2}{l^2}}\right)\right]_0^l = \frac{-l}{p\pi}(l\cos p\pi)$$

$$= \frac{-l^2}{p\pi}(-1)^P \quad \text{if } p \neq 0$$

$$\bar{f}_c(p) = F_c(x) = \int_0^{l} x\cos\frac{p\pi x}{l}dx$$

$$= \left[(x)\left(\frac{\sin\frac{p\pi x}{l}}{\frac{p\pi}{l}}\right) - (1)\left(-\frac{\cos\frac{p\pi x}{l}}{\frac{p^2\pi^2}{l^2}}\right)\right]_0^1 = \frac{l^2}{p^2\pi^2}[(-1)^p - 1] \text{ if } p \neq 0$$

(iii) $$\bar{f}_s(p) = F_s(x^2) = \int_0^{l} x^2\sin\frac{p\pi x}{l}dx$$

$$= \left[(x^2)\left(-\frac{\cos\frac{p\pi x}{l}}{\frac{p\pi}{l}}\right) - (2x)\left(-\frac{\sin\frac{p\pi x}{l}}{\frac{p^2\pi^2}{l^2}}\right) + (2)\left(\frac{\cos\frac{p\pi x}{l}}{\frac{p^3\pi^3}{l^3}}\right)\right]_0^l$$

$$= \frac{-l^3}{p\pi}(-1)^p + \frac{2l^3}{p^3\pi^3}\left[(-1)^p - 1\right] \quad \text{if } p \neq 0$$

$$\bar{f}_c(p) = \int_0^{l} (x^2)\cos\frac{p\pi x}{l}dx$$

$$= \left[(x^2)\left(\frac{\sin\frac{p\pi x}{l}}{\frac{p\pi}{l}}\right) - (2x)\left(-\frac{\cos\frac{p\pi x}{l}}{\frac{p^2\pi^2}{l^2}}\right) + (2)\left(-\frac{\sin\frac{p\pi x}{l}}{\frac{p^3\pi^3}{l^3}}\right)\right]_0^1$$

$$= \frac{2l^3}{p^2\pi^2}(-1)^p \qquad \text{if } p \neq 0$$

(iv) $$F_s\{f(x)\} = \int_0^{\pi/2} \sin px\, dx + \int_{\pi/2}^{\pi} (-1)\sin px\, dx$$

$$= \left(-\frac{\cos px}{p}\right)_0^{\pi/2} + \left(\frac{\cos px}{p}\right)_{\pi/2}^{\pi}$$

$$= -\frac{1}{p}\left(\cos\frac{p\pi}{2} - 1\right) + \frac{1}{p}\left(\cos p\pi - \cos\frac{p\pi}{2}\right)$$

$$= -\frac{1}{p}\left(\cos p\pi - 2\cos\frac{p\pi}{2} + 1\right) \qquad \text{if } p \neq 0$$

$$F_c(f(x)) = \int_0^{\pi/2} \cos px\, dx - \int_{\pi/2}^{\pi} \cos px\, dx$$

$$= \left(\frac{\sin px}{p}\right)_0^{\pi/2} - \left(\frac{\sin px}{p}\right)_{\pi/2}^{\pi} = \frac{2}{p}\sin\frac{p\pi}{2} \qquad \text{If } p \neq 0$$

(v) $$F_s(x^3) = \int_0^l x^3 \sin\frac{p\pi x}{l}\, dx$$

$$= \left[(x^3)\left(-\frac{\cos\frac{p\pi x}{l}}{\frac{p\pi}{l}}\right) - (3x^2)\left(-\frac{\sin\frac{p\pi x}{l}}{\frac{p^2\pi^2}{l^2}}\right) + (6x)\left(\frac{\cos\frac{p\pi x}{l}}{\frac{p^3\pi^3}{l^3}}\right) - (6)\left(-\frac{\sin\frac{p\pi x}{l}}{\frac{p^4\pi^4}{l^4}}\right)\right]_0^l$$

$$= -\frac{l^p}{p\pi}(-1)^p + \frac{6\,l^4}{p^3\pi^3}(-1)^P \qquad \text{if } p \neq 0$$

$$F_c(x^3) = \int_0^l x^3 \cos\frac{p\pi x}{l}\, dx$$

$$= \left[(x^3)\left(\frac{\sin\frac{p\pi x}{l}}{\frac{p\pi}{l}}\right) - (3x^2)\left(-\frac{\cos\frac{p\pi x}{l}}{\frac{p^2\pi^2}{l^2}}\right) + (6x)\left(\frac{-\sin\frac{p\pi x}{l}}{\frac{p^3\pi^3}{l^3}}\right) - (6)\left(\frac{\cos\frac{p\pi x}{l}}{\frac{p^4\pi^4}{l^4}}\right)\right]_0^l$$

$$= \frac{3l^4}{\pi^2 p^2}(-1)^p - \frac{6l^4}{p^4\pi^4}[(-1)^p - 1] \qquad \text{if } p \neq 0$$

(vi) $$F_s(e^{ax}) = \int_0^l e^{ax} \sin\frac{p\pi x}{l}\, dx$$

$$= \left\{ \frac{e^{ax}}{a^2 + \frac{p^2\pi^2}{l^2}} \left[ a \sin\frac{p\pi x}{l} - \frac{p\pi}{l}\cos\frac{p\pi x}{l} \right] \right\}_0^l$$

$$= \frac{e^{ax}}{a^2 + \frac{p^2 a^2}{l^2}} \cdot \left( -\frac{p\pi}{l}(-1)^p \right) + \frac{1}{a^2 + \frac{p^2\pi^2}{l^2}} \left( \frac{p\pi}{l} \right)$$

$$F_c(e^{ax}) = \left\{ \frac{e^{ax}}{a^2 + \frac{p^2\pi^2}{l^2}} \left[ a \cos\frac{p\pi x}{l} + \frac{p\pi}{l}\sin\frac{p\pi x}{l} \right] \right\}_0^l = \frac{e^{al}}{a^2 + \frac{p^2\pi^2}{l^2}} a(-1)^p - \frac{1}{a^2 + \frac{p^2\pi^2}{l^2}}(a)$$

**Example 35.** *Find f (x) if its finite Fourier sine transform is* $\frac{2\pi}{p^3}(-1)^{p-1}$ *for p = 1, 2, ..., $0 < x < \pi$.*

**Solution.** By inversion Theorem,

$$f(x) = \frac{2}{\pi}\sum_{p=1}^{\infty} \frac{2\pi}{p^3}(-1)^{p-1} \sin px \quad = 4\sum_{p=1}^{\infty} \frac{(-1)^{p-1}}{p^3} \sin px$$

**Example 36.** *Find f (x) if its finite Fourier sine transform is given by*

(i) $F_s(p) = \frac{1 - \cos p\pi}{p^2 \pi^2}$ *for p = 1, 2, 3, ... and $0 < x < \pi$*

(ii) $F_s(p) = \frac{16(-1)^{p-1}}{p^3}$ *for p = 1, 2, 3, .... and $0 < x < 8$*

(iii) $F_s(p) = \frac{\cos\frac{2\pi p}{3}}{(2p+1)^2}$ *for p = 1, 2, 3, ... and $0 < x < 1$.*

**Solution.** By inversion theorem

(i) $f(x) = \frac{2}{\pi}\sum_{p=1}^{\infty}\left( \frac{1 - \cos p\pi}{p^2\pi^2} \right).\sin px \quad = \frac{2}{\pi^3}\sum_{p=1}^{\infty}\left( \frac{1 - \cos p\pi}{p^2} \right).\sin px$

(ii) $f(x) = \frac{2}{l}\sum_{p=1}^{\infty} F_s(p)\sin\left( \frac{p\pi x}{l} \right)$

$$= \frac{2}{8}\sum_{p=1}^{\infty} \frac{16(-1)^{p-1}}{p^3}\sin\left( \frac{p\pi x}{8} \right) \quad = 4\sum_{p=1}^{\infty} \frac{(-1)^{p-1}}{p^3}\sin\left( \frac{p\pi x}{8} \right) \quad \text{(since } l = 8\text{)}$$

(iii) $f(x) = \frac{2}{l}\sum_{p=1}^{\infty} F_s(p)\sin\left( \frac{p\pi x}{l} \right) = 2\sum_{p=1}^{\infty} \frac{\cos\left( \frac{2\pi p}{3} \right)}{(2p+1)^2} \sin(p\pi x)$ (since $l = 1$) **Ans.**

**Example 37.** *Find f (x) if its finite Fourier cosine transform is*

(*i*) $F_c(p) = \frac{1}{2p}\left(\frac{p\pi}{2}\right)$ *for p = 1, 2, 3, ....*

$= \frac{\pi}{4}$ *for p = 0 given* $0 < x < 2\pi$

(*ii*) $F_c(p) = \frac{6\sin\frac{p\pi}{2} - \cos p\pi}{(2p+1)\pi}$ *for p = 1, 2, 3, ....*

$= \frac{2}{\pi}$ *for p = 0 given* $0 < x < 4$

(*iii*) $F_c(p) = \frac{\cos\left(\frac{2p\pi}{3}\right)}{(2p+1)^2}$ *for p = 1, 2, 3, ...*

$= 1$ *for p = 0 given* $0 < x < 1$

**Solution:** By inversion theorem,

$$f(x) = \frac{1}{l}F_c(0) + \frac{2}{l}\sum_{p=1}^{\infty} F_c(p).\cos\frac{p\pi x}{l}.$$

(*i*) Here $F_c(0) = \pi/4$ and $l = 2\pi$

$$\therefore \quad f(x) = \frac{1}{2\pi}\left(\frac{\pi}{4}\right) + \frac{2}{2\pi}\sum_{p=1}^{\infty}\frac{1}{2p}\sin\left(\frac{p\pi}{2}\right)\cos\left(\frac{p\pi x}{2\pi}\right)$$

$$= \frac{1}{8} + \frac{1}{2\pi}\sum_{p=1}^{\infty}\frac{1}{p}\sin\left(\frac{p\pi}{2}\right)\cos\left(\frac{px}{2}\right)$$

(*ii*) Here $F_c(0) = \frac{2}{\pi}$ and $l = 4$

$$f(x) = \frac{1}{4}\left(\frac{2}{\pi}\right) + \frac{2}{4}\sum_{p=1}^{\infty}\frac{\left(6\sin\frac{p\pi}{2} - \cos p\pi\right)}{(2p+1)\pi}\cos\left(\frac{p\pi x}{4}\right)$$

$$= \frac{1}{2\pi} + \frac{1}{2\pi}\sum_{p=1}^{\infty}\frac{\left(6\sin\frac{p\pi}{2} - \cos p\pi\right)}{(2p+1)}.\cos\left(\frac{p\pi x}{4}\right)$$

(*iii*) Here $F_c(0) = 1, l = 1$

$$\therefore \quad f(x) = \frac{1}{1} + \frac{2}{1}\sum_{p=1}^{\infty}\frac{1}{(2p+1)^2}\cos\left(\frac{2p\pi}{3}\right).\cos(p\pi x)$$

$$= 1 + 2\sum_{p=1}^{\infty}\frac{\cos\left(\frac{2p\pi}{3}\right)}{(2p+1)^2}\cos(p\pi x)$$ **Ans.**

**Example 38.** *Find the finite Fourier sine transform of $f(x) = 1$ in $(0, \pi)$. Use the inversion theorem and find Fourier series for $f(x) = 1$ in $(0, \pi)$. Hence prove*

(i) $1-\frac{1}{3}+\frac{1}{5}-\frac{1}{7}+....=\pi/4$  (ii) $\frac{1}{1^2}+\frac{1}{3^2}+\frac{1}{5^2}+....=\pi^2/8$

**Solution.** $$F_s(1) = \int_0^{\pi} 1.\sin\left(\frac{p\pi x}{\pi}\right)dx$$

$$\bar{f}_s(p) = \frac{1-\cos p\pi}{p} \quad \text{if } p \neq 0$$

By inversion theorem,

$$f(x) = \frac{2}{l}\sum_{p=1}^{\infty} F_s\,(p)\sin\frac{p\pi x}{l}$$

$$1 = \frac{2}{\pi}\sum_{p=1}^{\infty}\frac{1-(-1)^p}{p}.\sin px \qquad \text{since } l = \pi$$

$$1 = \frac{4}{\pi}\left[\sin x + \frac{1}{3}\sin 3x + \frac{1}{5}\sin 5x + ...\right] \qquad .... (1)$$

This is the half range Fourier sine series for $f(x) = 1$ in $(0, \pi)$ getting $x = \pi/2$.

On putting $x = \frac{\pi}{2}$ in (1) we get

$$\frac{4}{\pi}\left[1-\frac{1}{3}+\frac{1}{5}-\frac{1}{7}+....\right] = 1$$

$$\therefore \quad 1-\frac{1}{3}+\frac{1}{5}-\frac{1}{7}+....=\pi/4$$

In the half Fourier sine series $l_n = \frac{4}{\pi}\frac{1}{n}$ for n odd

By using Parseval's Theorem

$$\text{(Range)}\left[\frac{1}{2}\sum b_n^2\right] = \int_0^{\pi}(1)^2\,dx$$

$$\pi\left[\frac{1}{2}.\frac{16}{\pi^2}\sum_{n=1.3.5....}^{\pi}\frac{1}{n^2}\right] = \pi$$

*i.e.,* $$\sum_{n=1.3.5}^{\infty}\frac{1}{n^2} = \frac{\pi^2}{8}$$ **Ans.**

## EXERCISE 14.4

Find the finite Fourier sine and cosine transforms of

1. $f(x) = 2x$ in (0, 4) **Ans.** $F_s(s) = \begin{cases}\frac{32}{s\pi}(1-\cos s\pi), & s \neq 0\\ 0, & s = 0\end{cases}$, $F_c(s) = \begin{cases}\frac{32}{s^2\pi^2}(\cos s\pi - 1), & s \neq 0\\ 16. & s = 0\end{cases}$

2. $f(x) = x$ in $(0, \pi)$ **Ans.** $F_s(s) = \begin{cases} \frac{\pi}{s}(-1)^{s+1}, & s \neq 0 \\ 0, & s = 0 \end{cases}$, $F_c(s) = \begin{cases} \frac{(-1)^s - 1}{s^2}, & s \neq 0 \\ \frac{\pi^2}{2}, & s = 0 \end{cases}$

3. $f(x) = \cos ax$ in $(0, \pi)$ **Ans.** $F_s(s) = \frac{s}{s^2 - a^2}[1 - (-1)^s \cos a\pi]$, $F_c(s) = 0$

4. $f(x) = 1 - \frac{x}{\pi}$ in $(0, \pi)$ **Ans.** $F_s(s) = \frac{1}{s}$, $F_c(s) = \frac{-1}{s^2 \pi}[1 - \cos s\pi]$

5. $f(x) = \begin{cases} x \text{ in } (0, \pi/2) \\ \pi - x \text{ in } (\pi/2, \pi) \end{cases}$ **Ans.** $F_s(s) = \frac{2}{s^2} \sin \frac{s\pi}{2}$, $F_c(s) = \frac{1}{s^2}[1 + \cos s\pi]$

6. Find finite Fourier cosine transform of $\left(1 - \frac{x}{\pi}\right)^2 . 0 < x < \pi$. **Ans.** $F_c(s) = \begin{cases} \frac{2}{\pi s^2}, & s \neq 0 \\ \frac{\pi}{3}, & s = 0 \end{cases}$

7. Find $f(x)$ if $\bar{f}_c(p) = \dfrac{\sin\left(\frac{p\pi}{2}\right)}{2p}$, $p = 1, 2, 3 \ldots$ and

$= \frac{\pi}{4}$ if $p = 0$ given $0 < x < 2\pi$. **Ans.** $\frac{1}{8} + \frac{1}{\pi} \sum_{p=1}^{\infty} \frac{\sin \frac{p\pi}{2}}{2p} \cos \frac{px}{2}$

8. $f(x) = \frac{\pi}{3} - x + \frac{x^2}{2\pi}$ in $[0, \pi]$ **Ans.** $F_s(s) = \frac{1}{6s^3\pi}[\pi^2 \cos p\pi + 6 \cos p\pi, + 2p^2\pi - 6]$, $F_c(s) = \frac{1}{s^2}$

## 14.18 FINITE FOURIER SINE AND COSINE TRANSFORMS OF DERIVATIVES

Using the definition and the integration by parts, we can easily prove the following results.

For $0 \le x \le l$,

(i) $F_s(f'(x)) = -\frac{p\pi}{l} \bar{f}_c(p)$

(ii) $F_c\{f'(x)\} = f(l)(-1)^p - f(0) + \frac{p\pi}{l} \bar{f}_s(p)$

(iii) $F_s\{f''(x)\} = -\frac{p^2\pi^2}{l^2} \bar{f}_s(p) + \frac{p\pi}{l}\left[f(0) - (-1)^p f(l)\right]$

(iv) $F_c\{f''(x)\} = -\frac{p^2\pi^2}{l^2} \bar{f}_c(p) + f'(l)(-1)^p - f'(0)$

**Proof:** *(i)* $F_s(f'(x)) = \int_0^l f'(x) \sin \frac{p\pi x}{l} dx = \int_0^l \sin \frac{p\pi x}{l} \cdot d\{f(x)\}$

$$= \left(f(x) \sin \frac{p\pi x}{l}\right)_0^l - \int_0^l f(x) . \cos \frac{p\pi x}{l} \frac{p\pi}{l} dx$$

$$= -\frac{p\pi}{l}\bar{f}_c(p) \qquad \dots (1)$$

(ii) $$F_c\{f'(x)\} = \int_0^l f'(x)\cos\frac{p\pi x}{l}dx = \left(f(x)\cos\frac{p\pi x}{l}\right)_0^l - \int_0^l f(x).\frac{p\pi}{l}\sin\frac{p\pi x}{l}dx$$

$$= (-1)^p \bar{f}(l) - f(0) + \frac{p\pi}{l}\bar{f}_s(p) \qquad \dots (2)$$

(iii) $$F_s[f''(x)] = \int_0^l \sin\frac{p\pi x}{l}\, d[f'(x)]$$

$$= \left(f'(x)\sin\frac{p\pi x}{l}\right)_0^l - \frac{p\pi}{l}\int_0^l f'(x)\cos\frac{p\pi x}{l}ds$$

$$= -\frac{p\pi}{l}\left[(-1)^p f'(l) - f(0) + \frac{p\pi}{l}\bar{f}_s(p)\right] \qquad \text{[Using (2)]}$$

$$= -\frac{p^2\pi^2}{l^2}\bar{f}_s(p) + \frac{p\pi}{l}\left[f(0) - (-1)^p f(l)\right] \qquad \dots (3)$$

(iv) $$F_c\{f''(x)\} = \int_0^l \cos\frac{p\pi x}{l}\, d[f'(x)]$$

$$= \left[f'(x)\cos\frac{p\pi x}{l}\right]_0^l + \frac{p\pi}{l}\int_0^l f'(x)\sin\frac{p\pi x}{l}dx$$

$$= (-1)^p f'(l) - f'(0) + \frac{p\pi}{l}\left[-\frac{p\pi}{l}\bar{f}_c(p)\right] \qquad \text{[Using (1)]}$$

$$= -\frac{p^2\pi^2}{l^2}\bar{f}_c(p) + f'(l)(-1)^p - f'(0) \qquad \dots (4)$$

**Note.** If $u = u(x, t)$, then

$$F_s\left[\frac{\partial u}{\partial x}\right] = \frac{-p\pi}{l}F_c(u)$$

$$F_c\left[\frac{\partial u}{\partial x}\right] = \frac{p\pi}{l}F_s(u) - u(0, t) + (-1)^p u(l, t)$$

$$F_s\left[\frac{\partial^2 u}{\partial x^2}\right] = \frac{p^2\pi^2}{l^2}F_s(u) + \frac{p\pi}{l}[u(0, t) - (-1)^p u(l, t)]$$

$$F_c\left[\frac{\partial^2 u}{\partial x^2}\right] = -\frac{p^2\pi^2}{l^2}F_c(u) + \frac{\partial u}{\partial x}(l, t)\cos p\pi - \frac{\partial u}{\partial x}(0, t)$$

**Example 39.** *Using finite Fourier transform, solve*

$$\frac{\partial u}{\partial t} = \frac{\partial^2 u}{\partial x^2} \text{ given } u(0, t) = 0 \text{ and } u(4, t) = 0$$

*and* $u(x, 0) = 2x$ *where* $0 < x < 4,\ t > 0$

**Solution.** Since $u(0, t)$ given, take finite Fourier sine transform.

$$\int_0^4 \frac{\partial u}{\partial t} \sin\frac{p\pi x}{4}\,dx = \int_0^4 \frac{\partial^2 u}{\partial x^2} \sin\frac{p\pi x}{4}\,dx$$

$$\frac{d}{dt}\bar{u}_s = F_s\left(\frac{\partial^2 u}{\partial x^2}\right) = -\frac{p^2\pi^2}{16}\bar{u}_s + \frac{p\pi}{4}[u(0, t) - (-1)^p u(4, t)]$$

$$= -\frac{p^2\pi^2}{16}\bar{u}_s \text{ [using } u(0, t) = 0,\ u(4, t) = 0]$$

$$\frac{d\bar{u}_s}{\bar{u}_s} = -\frac{p^2\pi^2}{16}dt$$

Integrating $\qquad \log \bar{u}_s = -\dfrac{p^2\pi^2}{16}t + c$

$$\bar{u}_s = Ae^{-\frac{\pi^2 p^2}{16}t} \qquad \text{... (1)}$$

Since $u(x, 0) = 2x$

$$\bar{u}_s(p, 0) = \int_0^4 (2x)\sin\left(\frac{p\pi x}{4}\right)dx = -\frac{32}{p\pi}\cos p\pi \qquad \text{... (2)}$$

Using (2) in (1),

$$\bar{u}_s(p, 0) = A = -\frac{32}{p\pi}\cos p\pi\,.$$

Substituting in (1),

$$\therefore \qquad \bar{u}_s = -\frac{32}{p\pi}(-1)^p\, e^{-\frac{p^2\pi^2}{16}t}$$

By inversion Theorem,

$$u(x, t) = \frac{2}{4}\sum_{p=1}^{\infty}\frac{32}{p\pi}(-1)^{p+1}e^{-\frac{p^2\pi^2}{16}}\sin\left(\frac{p\pi x}{4}\right) \qquad \textbf{Ans.}$$

**Example 40.** *Solve* $\dfrac{\partial u}{\partial t} = \dfrac{\partial^2 u}{\partial x^2}$, $0 < x < 6,\ t > 0$

*given* $\qquad \dfrac{\partial u}{\partial x}(0, t) = 0,\ \dfrac{\partial u}{\partial x}(6, t) = 0$ *and* $u(x, 0) = 2x$.

**Solution.** Since $\dfrac{\partial u}{\partial x}(0, t)$ is given, use finite Fourier cosine transform

$$\int_0^6 \frac{\partial u}{\partial t}\cos\frac{p\pi x}{6}\,dx = \int_0^6 \frac{\partial^2 u}{\partial x^2}\cos\frac{p\pi x}{6}\,dx$$

$$\frac{d}{dt}\bar{u}_c = -\frac{p^2\pi^2}{36}\bar{u}_c + \frac{\partial u}{\partial x}(6, t)\cos p\pi - \frac{\partial u}{\partial x}(0, t) = -\frac{p^2\pi^2}{36}\bar{u}_c$$

$$\Rightarrow \qquad \frac{d\bar{u}_c}{\bar{u}_c} = -\frac{p^2\pi^2}{36}dt$$

$$\Rightarrow \qquad \log \bar{u}_c = -\frac{p^2\pi^2}{36}t + c$$

$$\bar{u}_c = A\, e^{-\frac{p^2\pi^2}{36}t} \qquad \text{... (1)}$$

$$u(x, 0) = 2x.$$

$\therefore$ At $t = 0$

$$\bar{u}_c(p, 0) = \int_0^6 (2x)\cos\frac{p\pi x}{6}dx = \frac{72}{p^2\pi^2}(\cos p\pi - 1) \qquad \text{... (2)}$$

Using this in (1), we get $\bar{u}_c(p, 0) = A = \frac{72}{p^2\pi^2}(\cos p\pi - 1)$

Substituting in (1), we get $\bar{u}_c(p, t) = \frac{72}{p^2\pi^2}(\cos p\pi)$

By inversion theorem,

$$u(x, t) = \frac{1}{l}\bar{f}_c(0) + \frac{2}{1}\sum_{p=1}^{\infty}\bar{f}_c(p)\cos\left(\frac{p\pi x}{l}\right)$$

$$= \frac{1}{6}\int_0^6 (2x)\,dx + \frac{2}{6}\sum_{p=1}^{\infty}\frac{72}{p^2\pi^2}(\cos p\pi - 1)\,e^{-\frac{p^2\pi^2}{36}t}.\cos\left(\frac{p\pi x}{6}\right)$$

$$= 6 + \frac{24}{\pi^2}\sum_{p=1}^{\infty}\frac{(\cos p\pi - 1)}{p^2}e^{-\frac{p^2\pi^2}{36}t}\cos\left(\frac{p\pi x}{6}\right). \qquad \textbf{Ans.}$$

**Example 41.** *Solve* $\frac{\partial u}{\partial t} = 2\frac{\partial^2 u}{\partial x^2}, 0 < x < 4, t > 0$

*given* $u(0, t) = 0$; $u(4, t) = 0$; $u(x, 0) = 3\sin \pi x - 2\sin 5\pi x$.

**Solution.** sin $u(0, t)$ is given, take finite Fourier sine transform. The equation becomes (as in example 39 on page 977)

$$\frac{d}{dt}\bar{u}_c = 2\left[-\frac{p^2\pi^2}{16}\bar{u}_s + \frac{p\pi}{4}\{u(0, t) - (-1)^p u(4, t)\}\right] = -\frac{p^2\pi^2}{8}\bar{u}_s$$

Solving we get, $\bar{u}_s = A e^{-\frac{p^2\pi^2}{8}t}$ ... (1)

$$u(x, 0) = 3\sin\pi x - 2\sin 5\pi x$$

Taking sine Transform,

$$\bar{u}_c(p, 0) = \int_0^4 (3\sin\pi x - 2\sin 5\pi x)\sin\frac{p\pi x}{4}dx = 0 \text{ if } p \neq 4, p \neq 20.$$

If $p = 4$, $\bar{u}_s(4, 0) = 6$

If $p = 20$, $\bar{u}_s(20, 0) = -4$

$$u(x, t) = \frac{2}{4}\sum_{p=1}^{\infty}\bar{u}_s(p, t)\sin\left(\frac{p\pi x}{4}\right)$$

$$= \frac{1}{2}[6\, e^{-\frac{p^2\pi^2}{8}t} \sin \pi x - 4\, e^{-\frac{p^2\pi^4}{8}t} \sin 5\pi x]$$

where $p$ in the first term is 4 and $p$ in the second term is 20

$$= 3e^{-2\pi^2 t} \sin \pi x - 2e^{-50\pi^2 t} \sin 5\pi x .$$ **Ans.**

## EXERCISE 14.5

**1.** Solve $\frac{\partial u}{\partial t} = \frac{\partial^2 u}{\partial x^2}, 0 < x < 6, t > 0$ given that $u(0, t) = 0 = u(6, t)$ and $u(x, 0) = \begin{cases} 1 \text{ for } 0 < x < 3 \\ 0 \text{ for } 3 < x < 6 \end{cases}$

**Ans.** $u(x,t) = \frac{2}{\pi} \sum_{p=1}^{\pi} \left( \frac{1 - \cos \frac{p\pi}{2}}{p} \right) e^{-\frac{p^2\pi^2 t}{36}} \sin \left( \frac{p\pi x}{6} \right)$

**2.** Solve $\frac{\partial v}{\partial t} = \frac{\partial^2 v}{\partial x^2}$ subject to conditions $v(0, t) = 1$, $v(\pi, t) = 3$

$v(x, 0) = 1$ for $0 < x < \pi$, $t > 0$

**Ans.** $v(x, t) = \frac{4}{\pi} \sum_{p=1}^{\infty} \frac{\cos p\pi}{p} e^{-p^2 t} \sin px + 1 + \frac{2x}{\pi}$

**3.** Solve $\frac{\partial \theta}{\partial t} = \frac{\partial^2 \theta}{\partial x^2}$ **Ans.** $\theta(x, t) = \pi \sum_{p=1}^{\infty} (-1)^{p+1} e^{-p^2 t} \sin px$

given $\theta(0, t) = 0$, $\theta(\pi, t) = 0$, $\theta(x, 0) = 2x$ for $0 < x < \pi$, $t > 0$

## TABLE

| *Function f (x)* | *Fourier Sine Transform $F_s(s)$* |
|---|---|
| $\begin{cases} 1 & 0 < x < b \\ 0 & x > b \end{cases}$ | $\frac{1 - \cos bs}{s}$ |
| $x^{-1}$ | $\frac{x}{2}$ |
| $\frac{x}{x^2 + b^2}$ | $\frac{\pi}{2} e^{-bs}$ |
| $e^{-bx}$ | $\frac{s}{s^2 + b^2}$ |
| $x^{n-1} e^{-bx}$ | $\frac{\Gamma(n) \sin(n \tan^{-1} s/b)}{(s^2 + b^2)^{n/2}}$ |
| $x e^{-bx^2}$ | $\frac{\sqrt{\pi}}{4b^{3/2}} s e^{-s^2/4b}$ |
| $x^{-1/2}$ | $\sqrt{\frac{\pi}{2s}}$ |
| $x^{-n}$ | $\frac{\pi s^{n-1} \sec(n\pi/2)}{2\Gamma(n)}, \quad 0 < n < 2$ |

| | |
|---|---|
| $\frac{\sin bx}{x}$ | $\frac{1}{2}\ln\left(\frac{s+b}{s-b}\right)$ |
| $\frac{\cos bx}{x}$ | $\begin{cases} 0 & s<b \\ \pi/4 & s=b \\ \pi/2 & s>b \end{cases}$ |
| $\tan^{-1}(x/b)$ | $\frac{\pi}{2s}e^{-bs}$ |
| $\csc bx$ | $\frac{\pi}{2b}\tanh\frac{\pi s}{2b}$ |
| $\frac{1}{e^{2x}-1}$ | $\frac{\pi}{4}\cot h\left(\frac{\pi s}{2}\right)-\frac{1}{2s}$ |
| $\begin{cases} 1 & 0<x<b \\ 0 & x>b \end{cases}$ | $\frac{\sin bs}{s}$ |
| $\frac{1}{x^2+b^2}$ | $\frac{\pi e^{-bs}}{2b}$ |
| $x^{-n}$ | $\frac{\pi s^{n-1}\sec(n\pi/2)}{2\Gamma(n)}$, $0<n<1$ |
| $\ln\left(\frac{x^2+b^2}{x^2+c^2}\right)$ | $\frac{e^{-cs}-e^{-bs}}{\pi s}$ |
| $\frac{\sin bx}{x^2}$ | $\begin{cases} \pi/2 & s<b \\ \pi/4 & s=b \\ 0 & s>b \end{cases}$ |
| $\sin bx^2$ | $\frac{\sqrt{\pi}}{8b}\left(\cos\frac{s^2}{4b}-\sin\frac{s^2}{4b}\right)$ |
| $\cos bx^2$ | $\sqrt{\frac{\pi}{8b}}\left(\cos\frac{s^2}{4b}+\sin\frac{s^2}{4b}\right)$ |
| $\operatorname{sech} bx$ | $\frac{\pi}{2b}\operatorname{sech}\frac{s\pi}{2b}$ |
| $\frac{\cosh(\sqrt{\pi}x/2)}{\cosh(\sqrt{\pi}\,x)}$ | $\frac{\sqrt{\pi}}{2}\frac{\cosh(\sqrt{\pi}s/2)}{\cosh(\sqrt{\pi}\,s)}$ |
| $\frac{e^{-b\sqrt{x}}}{\sqrt{x}}$ | $\sqrt{\frac{\pi}{2s}}\{\cos(2b\sqrt{s})-\sin(2b\sqrt{s})\}$ |

# 15

# Numerical Techniques

## 15.1 INTRODUCTION

In this chapter, we shall deal with the methods for solving the equations. Sometimes, a rough approximation of a root can be found by graph and more accurate results by the following methods:

(*i*) Newton Raphson method or successive substitution method.

(*ii*) Rule of false position method (Regula falsi method).

(*iii*) Iteration method.

## 15.2 SOLUTION OF THE EQUATIONS GRAPHICALLY

**Step1.** Find a small interval $(a, b)$ between wich the root of the equation lies.

Let $$f(x) = 0 \quad \ldots(1)$$

and $$f(a) = -\text{ve and } f(b) = +\text{ ve}$$

then the root of the equation (1) lies between a and b.

For example $$f(x) = 2x^2 + x - 15 = 0$$

$$f(2) = 8 + 2 - 15 = -5 = -\text{ve}$$

$$f(3) = 18 + 3 - 15 = +6 = +\text{ ve}$$

$\therefore$ The root of the equation lies between 2 and 3.

**Step 2.** Write the equation $f(x) = 0$ as $\phi(x) = \psi(x)$

For example $2x^2 + x - 15 = 0$ or $2x^2 = 15 - x$

**Step 3.** Prepare two tables for $y = \phi(x)$ and $y = \psi(x)$ taking values of $x$ between $a$ and $b$.

**Step 4.** Plot these points and join them to get smooth curves

**Step 5.** Note down the abscissa of the point of intersection of the curves $y = \phi(x)$ and $y = \psi(x)$. This is the required root of the equation $f(x) = 0$

**Note.** Sometimes we do not write $f(x) = 0$ as $\phi(x) = \psi(x)$. We adopt the following method:

(*i*) Find a small interval $(a, b)$between which the root lies $f(a)$ and $f(b)$ are of opposite signs.

(*ii*) Prepare a table of the different values of $x$ between a and $b$, for $y = f(x)$.

(*iii*) Plot these points and join them to get smooth curve.

(*iv*) The real root of the equation $f(x) = 0$ is the abscissa where the curve cuts the x– axis Note it.

**Example 1.** *Find graphically the positive root of the equation $x^3 - 6x - 13 = 0$.*

**Solution.**

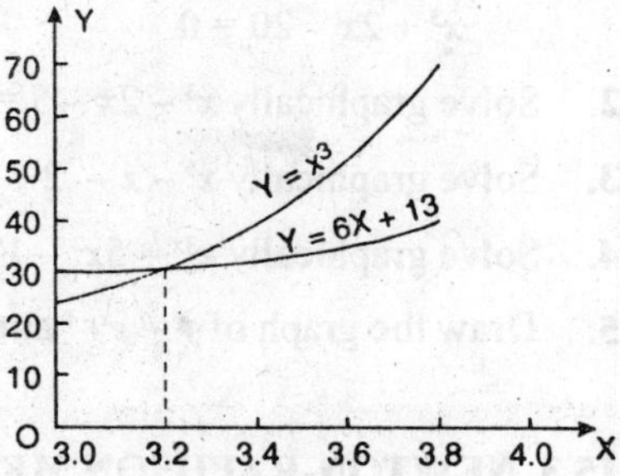


$f(x) = x^3 - 6x - 13 = 0.$ ...(1)

$f(3) = 27 - 18 - 13 = -4 = -ve$

$f(4) = 64 - 24 - 13 = 27 = +ve$

The root of (1) lies between 3 and 4 as $f(3)$ and $f(4)$ are opposite in sign.

(1) is written as $x^3 = 6x + 13$, $y = x^3$

and $y = 6x + 13$

Let us draw two curves for $y = x^3$ and $y = 6x + 13$.

$y = x^3$

| $x$ | 3 | 3.2 | 3.4 | 3.6 | 3.8 | 4.0 |
|---|---|---|---|---|---|---|
| $y$ | 27 | 32.8 | 39.3 | 46.7 | 54.9 | 64 |

$y = 6x + 13$

| $x$ | 3 | 3.2 | 3.4 | 3.6 | 3.8 | 4 |
|---|---|---|---|---|---|---|
| $y$ | 31 | 32.2 | 33.4 | 34.6 | 35.8 | 37 |

Let the origin be (3,0).

The graphs of $y = x^3$ and $y = 6x + 13$ are sketched in the figure. The abscissa of the point of intersection of two curves is 3.2.

$\therefore$ The root of the given equation is 3.2 **Ans.**

**Example 2.** *Solve graphically the equation $x - 1 = \sin x$.*

**Solution.** $x - 1 = \sin x$

We take two equations $y = x - 1$ and $y = \sin x$. Let us find out the abscissa of the point of intersection of the line $y = x - 1$ and the curve $y = \sin x$ and give a rough estimate of the root.

For the straight line $y = x - 1$, we have the table:

| $x$ | 0 | 1 | $3\pi/4$ |
|---|---|---|---|
| $y = x - 1$ | $-1$ | 0 | 1.4 |

For the sine curve, we have the following table:

| $x$ | 0 | $\pi/4$ | $\pi/2$ | $3\pi/4$ | $\pi$ |
|---|---|---|---|---|---|
| $y = \sin x$ | 0 | 0.71 | 1.00 | 0.71 | 0 |

On the same axes, and with same scale construct the graphs of $y = x - 1$ and $y = \sin x$.

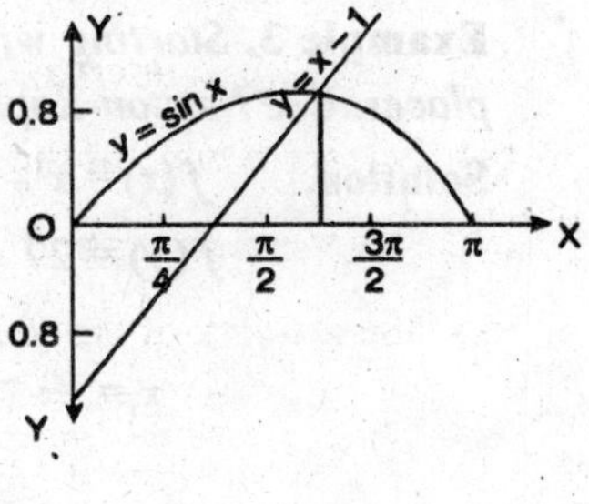


From the graph, we get $x = 1.95$ radians approximately.

**Ans.**

## EXERCISE 15.1

1. Draw the graph of $y = x^3$ and $y = -2x + 20$ and find the apporoximate solution of the equation $x^3 + 2x - 20 = 0$. **Ans.** 2.47
2. Solve graphically $x^3 - 2x - 5 = 0$. **Ans.** 2.099
3. Solve graphically $x^5 - x - .2 = 0$. **Ans.** – 0.2
4. Solve graphically $e^{3x} - 5x^2 - 17 = 0$. **Ans.** 1.04
5. Draw the graph of $y = e^{x-1}$ and find graphically the values of the root equation $3 - x = e^{x-1}$

   **Ans.** 1.44

## 15.3 NEWTON-RAPHSON METHOD OR SUCCESSIVE SUBSTITUTION METHOD

By this method, we get closer approximation of the root of an equation if we already know its approximate root.

Let the equation be $f(x) = 0$. ...(1)

Let its approximate root be $a$ and better approximate root be $a + h$.

Now we proceed to find $h$.

$$f(a+h) = 0 \text{ approximately } \quad [\text{as } a + h \text{, is the root of } f(x) = 0] \qquad ...(2)$$

By Taylor's theorem

$$f(a+h) = f(a) + hf'(a) + \frac{h^2}{2} f''(a) + ...$$

or $$f(a+h) = f(a) + h\, f'(a) \qquad ...(3)$$

Since $h$ is small, we neglect the $h^2$ and higher power of $h$.

From (2) and (3), we have

$$0 = f(a) + hf'(a) \Rightarrow h = -\frac{f(a)}{f'(a)}$$

or $$a + h = a - \frac{f(a)}{f'(a)} = a_1 \qquad [\text{First approximate root} = a]$$

Second approximate root $a_2 = a_1 - \dfrac{f(a_1)}{f'(a_1)}$

Similarly third approximate root, $a_3 = a_2 - \dfrac{f(a_2)}{f'(a_2)}$

By repeating this operation, we get closer approximation of the root.

**Note.** (1) In the beginning, we guess two numbers $b$ and $c$ such that $f(b)$ and $f(c)$ are of opposite sign. Then the first approximate root $a$ lies between $b$ and $c$.

(2) If $f'(x)$ is zero or nearly zero, this method fails.

**Example 3.** *Starting with $x_0 = 3$, find a root of $x^3 - 3x - 5 = 0$, correct to three decimal places. Use Newton-Raphson method*

**Solution.** $f(x) = x^3 - 3x - 5 = 0, \qquad f'(x) = 3x^2 - 3$

$f(3) = 27 - 9 - 5 = 13, \qquad f'(3) = 27 - 3 = 24$

$$x_1 = x_0 - \frac{f(x_0)}{f'(x_0)} = 3 - \frac{f(3)}{f'(3)} = 3 - \frac{13}{24} = 3 - 0.5417 = 2.4583$$

$$x_2 = 2.4583 - \frac{f(2.4583)}{f'(2.4583)} = 2.4583 - \frac{2.4812}{15.1297} = 2.4583 - 0.1640 = 2.2943$$

$$x_3 = 2.2943 - \frac{f(2.2943)}{f'(2.2943)} = 2.2943 - \frac{0.1939}{12.7914} = 2.2791$$

$$f(2.2791) = 0.0010$$

Hence the required root = 2.2791 **Ans.**

**Example 4.** *Find the real root of the following equation, correct to three decimal places, using Newton-Raphson method*

$$x^3 - 2x - 5 = 0$$

**Solution.** $x^3 - 2x - 5 = 0$ ...(1)

Let $f(x) = x^3 - 2x - 5$

$$f(2) = 8 - 4 - 5 = -1$$

$$f(2.5) = (2.5)^3 - 2(2.5) - 5 = +5.625$$

Since $f(2)$ and $f(2.5)$ are, of opposite sign, the root of (1) lies between 2 and 2.5 ; $f(2)$ is near to zero than $f(2.5)$, so 2 is better appropriate root than 2.5.

$$f'(x) = 3x^2 - 2 \qquad f'(2) = 12 - 2 = 10$$

Let 2 be an approximate root of (1). By Newton-Raphson method

$$a_1 = a - \frac{f(a)}{f'(a)} = 2 - \frac{f(2)}{f'(2)} = 2 - \frac{-1}{10} = 2.1$$

$$f(2.1) = (2.1)^3 - 2(2.1) - 5 = 9.261 - 4.2 - 5 = 0.061$$

$$f'(2.1) = 3(2.1)^2 - 2 = 11.23$$

$$a_2 = 2.1 \frac{f(2.1)}{f'(2.1)} = 2.1 - \frac{0.061}{11.23} = 2.1 - 0.00543 = 2.09457$$

$$f(2.09457) = (2.09457)^3 - 2(2.09457) - 5$$

$$= 9.1893 - 4.18914 - 5 = -0.00016$$

$$f'(2.09457) = 3(2.09457)^2 - 2 = 13.16167 - 2 = 11.16167$$

$$a_3 = 2.09457 - \frac{f(2.09457)}{f'(2.09457)} = 2.09457 - \frac{-0.00016}{11.16167} = 2.09457 + 0.000014 = 2.09456$$

Hence, $a_3 = a_2$ correct upto four places of decimal, so the root of (1) is 2.0945 **Ans.**

**Example 5.** *Find an interval of length 1, in which the root of $f(x) = 3x^3 - 4x^2 - 4x - 7 = 0$ lies. Take the middle point of this interval as the starting approximation and iterate two times, using the Newton- Raphson method.*

**Solution.** $f(x) = 3x^3 - 4x^2 - 4x - 7 = 0$ ...(1)

$$f(2) = 24 - 16 - 8 - 7 = -7$$

$$f(3) = 81 - 36 - 12 - 7 = +26$$

The root of (1) lies between 2 and 3 as $f(2)$ and $f(3)$ are of opposite signs.

The middle point of this interval is 2.5.

$$f(2.5) = 46.875 - 25 - 10 - 7 = 4.875 \text{ and } f'(x) = 9x^2 - 8x - 4$$

$$f'(2.5) = 56.25 - 20 - 4 = 32.25$$

By Newton-Raphson method

$$a_1 = a - \frac{f(a)}{f'(a)}$$

$$a_1 = 2.5 - \frac{f(2.5)}{f'(2.5)} = 2.5 - \frac{4.875}{32.25} = 2.5 - 0.15 = 2.35$$

$$f(2.35) = 38.93 - 22.09 - 9.4 - 7 = 0.44$$

$$f'(2.35) = 49.7 - 18.8 - 4 = 26.9$$

$$a_2 = 2.35 - \frac{f(2.35)}{f'(2.35)} = 2.35 - \frac{0.44}{26.9} = 2.35 - 0.016 = 2.334$$

$$f(2.334) = 38.14 - 21.79 - 9.34 - 7 = 0.01 \text{ which is nearly zero.}$$

Hence the required root is 2.334 **Ans.**

**Example 6.** *By using Newton-Raphson's method find the root of $x^4 - x - 10 = 0$, which is near to $x = 2$ correct to three places of decimal.*

**Solution.** $f(x) = x^4 - x - 10 = 0$, and $f'(x) = 4x^3 - 1$

$f(2) = 16 - 2 - 10 = 4$ and $f'(2) = 32 - 1 = 31$

By Newton-Raphson's method

$$a_1 = a - \frac{f(a)}{f'(a)} = 2 - \frac{f(2)}{f'(2)} = 2 - \frac{4}{31} = 2 - .129 = 1.871$$

$$f(1.871) = (1.871)^4 - 1.871 - 10 = 12.25 - 1.871 - 10 = 0.379$$

$$f'(1.871) = 4(1.871)^3 - 1 = 4 \times 6.5497 - 1 = 25.1988.$$

$$a_2 = 1.871 - \frac{f(1.871)}{f'(1.871)} = 1.871 - \frac{0.379}{25.1988} = 1.871 - 0.0150 = 1.856$$

$$f(1.856) = (1.856)^4 - (1.856) - 10 = 11.8662 - 11.856 = 0.0102$$

$$f'(1.856) = 4(1.856)^3 - 1 = 4 \times 6.3934 - 1 = 24.5736$$

$$a_3 = 1.856 - \frac{f(1.856)}{f'(1.856)} = 1.856 - \frac{0.0102}{24.5736} = 1.856 - 0.00042 = 1.8556$$

$$f(1.8556) = (1.8556)^4 - 1.8556 - 10 = 0.00038$$

$$f'(1.8556) = 4(1.8556)^3 - 1 = 24.5572$$

$$a_4 = 1.8556 - \frac{f(1.8556)}{f'(1.8556)} = 1.8556 - \frac{0.00038}{24.5572} = 1.8556 - 0.00002 = 1.85558$$

Required root = 1.85558 **Ans.**

**Example 7.** *Determine the root of $x^4 + x^3 - 7x^2 - x + 5 = 0$ which lies between 2 and 3 correct to three decimal places.*

**Solution.** $f(x) = x^4 + x^3 - 7x^2 - x + 5 = 0$

$\Rightarrow$ $f(2) = 16 + 8 - 28 - 2 + 5 = -1$

$\Rightarrow$ $f(3) = 81 + 27 - 63 - 3 + 5 = +47.$

Root lies between 2 and 3.

Taking $x_1 = 2$ as first approximate root. $\left[\begin{array}{l} f'(x) = 4x^3 + 3x^2 - 14x - 1 \\ f'(2) = 32 + 12 - 28 - 1 = 15 \end{array}\right]$

$$x_2 = x_1 - \frac{f(x_1)}{f'(x_1)} = 2 - \frac{f(2)}{f'(2)}$$

$$x_2 = 2 - \frac{-1}{15} = 2\frac{1}{15} = 2.067$$

$$x_3 = x_2 - \frac{f(x_2)}{f'(x_2)} = 2.067 - \frac{f(2.067)}{f'(2.067)}$$

$$= 2.067 - \frac{-0.0028}{18.422} = 2.067 + 0.0001519 = 2.0671519$$ **Ans.**

**Example 8.** *Using Newton-Raphson method evaluate to two decimal figures, the root of the equation $e^x = 3x$ lying between 0 and 1.*

**Solution.** $f(x) = e^x - 3x = 0$

$f(0) = 1$

$f(1) = e^1 - 3 = -0.2817$

The middle point of the interval (0, 1) is 0.5.

$$f(0.5) = e^{0.5} - 3(0.5) = 1.649 - 1.5 = 0.149$$

$$f'(x) = e^x - 3,\ f'(0.5) = e^{0.5} - 3 = 1.649 - 3 = -1.351$$

By Newton - Raphson method $a_1 = a - \dfrac{f(a)}{f'(a)}$

so $$a_1 = 0.5 - \frac{f(0.5)}{f'(0.5)} = 0.5 - \frac{0.149}{-1.351} = 0.5 + 0.11 = 0.61$$

$$f(0.61) = e^{0.61} - 3(0.61) = 1.84 - 1.83 = 0.01$$

$$f'(0.61) = e^{0.61} - 3 = 1.84 - 3 = -1.16$$

$$a_2 = 0.61 - \frac{f(0.61)}{f'(0.61)} = 0.61 - \frac{0.01}{-1.16} = 0.61 + 0.0086 = 0.6186$$

$$f(0.6186) = e^{0.6186} - 3(0.6186) = 1.8563 - 1.8558 = 0.0005$$

$$x = 0.6186$$ **Ans.**

**Example 9.** *Compute the real root of $x \log_{10} x - 1.2 = 0$*

**Solution.** $x \log_{10} x - 1.2 = 0$

Let $f(x) = x \log_{10} x - 1.2$ or $f(3) = 3 \log_{10} 3 - 1.2$

$\Rightarrow$ $f(3) = 3 \times 0.4771 - 1.2 = 1.4313 - 1.2 = +0.2313$

and $f(2) = 2 \times 0.3010 - 1.2 = 0.6020 - 1.2 = -0.5980$

$f(3)$ is $+ve$ and $f(2)$ is $-ve$, so the root of the given equation lies between 3 and 2.

$$f(x) = x \log_{10} x - 1.2 = 0.4343\, x \log_e x - 1.2$$

$$f'(x) = 0.4343 \log_e x + 0.4343 = \log_{10} x + 0.4343$$

Taking $x_1 = 3$ as first approximation, we have

$$x_2 = x_1 - \frac{f(x_1)}{f'(x_1)} \qquad \text{(By Newton's method)}$$

$$x_2 = 3 - \frac{3\log_{10}3 - 1.2}{\log_{10}3 + 0.4343} = 3 - \frac{0.2313}{0.9114} = 3 - 0.2538 = 2.7462$$

$$x_3 = 2.7462 - \frac{2.7462\log_{10}2.7462 - 1.2}{\log_{10}2.7462 + 0.4343} = 2.7462 - \frac{0.0048}{0.8730} = 2.7407$$ **Ans.**

**Example 10.** *Write the Newton-Raphson procedure for finding $\sqrt[3]{N}$, where N is a real number. Use it to find $\sqrt[3]{18}$ correct to 2 decimals, assuming 2.5 as the initial approximation.*

**Solution.** Let $x = \sqrt[3]{N} \Rightarrow x^3 = N \Rightarrow x^3 - N = 0$

Let $f(x) = x^3 - N = 0 \Rightarrow f'(x) = 3x^2$

By Newton – Raphson Method, $x_{n+1} = x_n - \dfrac{f(x_n)}{f'(x_n)}$

$$x_{n+1} = x_n - \frac{x_n^3 - N}{3x_n^2} = \frac{2x_n^3 + N}{3x_n^2}, \quad n = 0, 1, 2, \ldots$$

Let $N = 18$, $x$ = app. cube root of $18 = 2.5$

$$x_1 = \frac{2(2.5)^3 + 18}{3 \times (2.5)^2} = 2.62667$$

Repeat this method.

### Exercise 15.2

Solve the following equations by Newton's methods

1. $x^3 - 2x - 5 = 0$ **Ans. 2.0946**
2. $x^3 - 2x + 0.5 = 0$ **Ans. 0.2578**
3. $3x^3 + 8x^2 + 8x + 5 = 0$ **Ans. –1.67**
4. $x^3 - 5x + 3 = 0$ **Ans. 0.6565**
5. $x - 2\sin x = 0$ **Ans. 1.8955**
6. $xe^x - 2 = 0$ **Ans. 0.853**
7. $x^2 - 4\sin x = 0$ **Ans. 1.9337**
8. Apply Newton-Raphson method to find an approximate solution of the equation $e^x - 3^x = 0$ correct upto three significant figures (assume $x = 0.4$ as an approximate root of the equation). **Ans. 0.619**
9. Determine approximately the root of the equation $x + \log_{10} x = 3.375$ correct to two significant figures. **Ans. 2.911**
10. Determine approximately the smallest positive root of the equation $x^2 + 2x - 2 = 0$, correct to two significant figures using Newton-Raphson method. **Ans. 0.7482**
11. Design a Newton-Raphson iteration to compute the cube-root of a positive number, $N$. Perform two iterations of this method to compute $(2)^{1/3}$ starting from $x_0 = 1$. **Ans. 1.264**
12. A root of the equation $e^x = 1 + x + \dfrac{x^2}{2} + \dfrac{x^2 e^{0.3x}}{6}$ is close to 2.5. Find this root to three decimal places, using Newton-Raphson method. **Ans. 2.364**

## 15.4 RULE OF FALSE POSITION (REGULA FALSI)

Let $f(x) = 0$ ...(1)

Let $y = f(x)$ be represented by the curve $AB$.

The curve $AB$ cuts the $x$-axis at $P$.

The real root of (1) is $OP$.

The false position of the curve $AB$ is taken as the chord $AB$. The chord $AB$ cuts the $x$-axis at $Q$. The approximate root of $f(x) = 0$ is $OQ$.

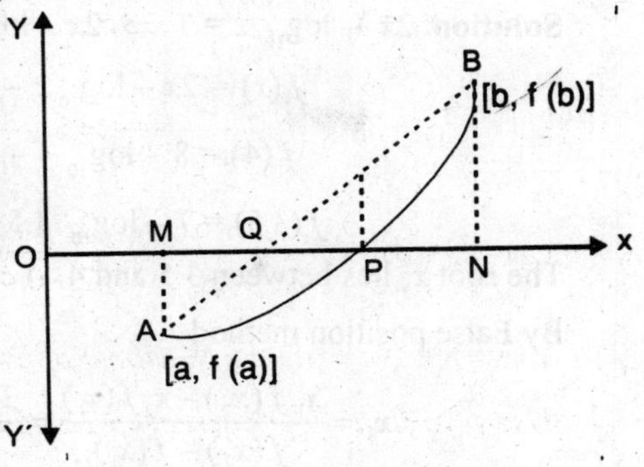


By this method, we find $OQ$.

Let $A\ [a, f(a)]$, $B\ [b, f(b)]$ be the extremities of the chord $AB$.

The equation of the chord $AB$ is

$$y - f(a) = \frac{f(b) - f(a)}{b - a}(x - a) \qquad \text{(Two points form)}$$

To find $OQ$, put $y = 0$, $-f(a) = \dfrac{f(b) - f(a)}{b - a}(x - a)$

$$(x - a) = \frac{-(b - a) f(a)}{f(b) - f(a)} \Rightarrow x = a + \frac{(a - b) f(a)}{f(b) - f(a)}$$

$$x = \frac{af(b) - bf(a)}{f(b) - f(a)}$$

Repeat the above rule.

**Example 11.** *Find an approximate value of the root of the equation $x^3 + x - 1 = 0$ near $x = 1$, using the method of false position (regula falsi) two times.*

**Solution.** $f(x) = x^3 + x - 1 = 0$

$f(1) = 1 + 1 - 1 = +1$

$f(0.5) = (0.5)^3 + (0.5) - 1 = -0.375, \qquad f(1) . f(0.5) < 0$

The root lies between 0.5 and 1.

Let $a = 0.5$ and $b = 1$

$$x_1 = \frac{a f(b) - b f(a)}{f(b) - f(a)} \Rightarrow x_1 = \frac{0.5 f(1) - 1 f(0.5)}{f(1) - f(0.5)}$$

$$= \frac{0.5 (1) - 1 (-0.375)}{1 + 0.375} = 0.6363$$

Now $f(0.6363) = (0.6364)^3 + 0.6364 - 1 = -0.1059$

and $f(1) = 1$

$\therefore$ Root lies between 0.6363 and 1. $\qquad f(0.6363) . f(1) < 0$

$a = 0.6363, \qquad b = 1$

$$x_2 = \frac{0.6363 f(1) - 1 f(0.6363)}{f(1) - f(0.6363)} = \frac{0.6363 - 1(-0.1059)}{1 + 0.1059}$$

$$= 0.6712$$

Now, $f(0.6712) = -0.0264$ and $f(1) = 1$

$a = 0.6712$ and $b = 1 \qquad [f(0.6712) . f(1) < 0]$

$$x_3 = \frac{0.6712 f(1) - 1 f(0.6712)}{f(1) - f(0.6712)} = \frac{0.6712 - (-0.0264)}{1 - (-0.0264)} = 0.6797 \quad \textbf{Ans.}$$

**Example 12.** *Find the root of the equation $2x - \log_{10} x = 7$ which lies between 3.5 and 4, correct to five places of decimal, using method of false position.*

**Solution.** $2x - \log_{10} x = 7 \Rightarrow 2x - \log_{10} x - 7 = 0$

$$f(x) = 2x - \log_{10} x - 7$$

$$f(4) = 8 - \log_{10} 4 - 7 = 1 - 0.60206 = 0.39794$$

$$f(3.5) = 7 - \log_{10} 3.5 - 7 = -0.54407$$

The root $x_3$ lies between 3.5 and 4.

By False position method

$$x_3 = \frac{x_1 f(x_2) - x_2 f(x_1)}{f(x_2) - f(x_1)} = \frac{3.5 f(4) - 4 f(3.5)}{f(4) - f(3.5)}$$

$$= \frac{3.5(3.39794) - 4(-0.54407)}{0.39794 - (-0.54407)} = \frac{1.39279 + 2.17628}{0.94201} = \frac{3.56907}{0.94201} = 3.78878$$

$$f(3.78878) = 7.57756 - 0.57850 - 7 = -0.00094$$

Again applying False position method

$$x_4 = \frac{3.78878 f(4) - 4 f(3.78878)}{f(4) - f(3.78878)} = \frac{3.78878 \times 0.39794 - 4 \times (-0.00094)}{0.39794 - (-0.00094)}$$

$$= \frac{1.50771 + 0.00376}{0.39888} = \frac{1.51147}{0.39888} = 3.78928$$ **Ans.**

**Example 13.** *Find by the method of Regula Falsi a root of the equation*
*$x^3 + x^2 - 3x - 3 = 0$ lying between 1 and 2.*

**Solution.** $f(x) = x^3 + x^2 - 3x - 3 = 0$

$$f(1) = 1 + 1 - 3 - 3 = -4 = -ve$$

$$f(2) = 8 + 4 - 6 - 3 = +3 = +ve$$

The root lies between 1 and 2 as $f(1)$ is –ve and $f(2)$ is + ve.

By Regula Falsi method:

$$x_1 = \frac{1 f(2) - 2 f(1)}{f(2) - f(1)} = \frac{1 \times 3 - 2 \times -4}{3 - (-4)} = \frac{11}{7} = 1.571$$

$$f(1.571) = (1.571)^3 + (1.571)^2 - 3(1.571) - 3$$

$$= 3.877 + 2.468 - 4.713 - 3 = -1.368 = -ve$$

The root lies between 1.571 and 2 as $f(1.571)$ is –ve and $f(2)$ is + ve.

$$x_2 = \frac{1.571 f(2) - 2 f(1.571)}{f(2) - f(1.571)}$$

$$= \frac{1.571 \times 3 - 2 \times (-1.368)}{3 - (-1.368)} = \frac{4.713 + 2.736}{4.368} = 1.705$$

$$f(1.705) = (1.705)^3 + (1.705)^2 - 3(1.705) - 3 = 4.960 + 2.908 - 5.115 - 3$$

$$= -0.252 = -ve.$$

The root lies between 1.705 and 2 as $f(1.705)$ is – ve and $f(2)$ is + ve.

$$x_3 = \frac{1.705 f(2) - 2 f(1.705)}{f(2) - f(1.705)} = \frac{1.705 \times 3 - 2 \times (-0.252)}{3 - (-0.252)} = 1.728$$ **Ans.**

**Example 14.** *Find the approximate value, correct to three places of decimals, of the real root which lies between – 2 and –3 of the equation $x^3 - 3x + 4 = 0$, using the method of false position three times in succession.*

**Solution.** $f(x) = x^3 - 3x + 4 = 0$

$$x_1 = -2,\ x_2 = -3$$

$$f(x_1) = f(-2) = (-2)^3 - 3(-2) + 4 = -8 + 6 + 4 = 2.$$

$$f(x_2) = f(-3) = (-3)^3 - 3(-3) + 4 = -27 + 9 + 4 = -14$$

$$x = \frac{x_1 f(x_2) - x_2 f(x_1)}{f(x_2) - f(x_1)} = \frac{-2f(-3) - (-3)f(-2)}{f(-3) - f(-2)} = \frac{-2(-14) - (-3)(2)}{(-14) - (2)} = \frac{28 + 6}{-16} = \frac{34}{-16}$$

$$= -2.125$$

$$f(-2.125) = (-2.125)^3 - 3(-2.125) + 4 = -9.596 + 6.375 + 4 = +0.779$$

$$f(-3) = -14 \text{ and } f(-2.125) = +0.779$$

$\therefore$ Root lies between – 2.125 and – 3.

$$x = \frac{(-2.125)f(-3) - (-3)f(-2.125)}{f(-3) - f(-2.125)} = \frac{(-2.125)(-14) - (-3)(0.779)}{(-14) - (0.779)}$$

$$= \frac{29.750 + 2.337}{-14.779} = \frac{32.087}{-14.779} = -2.171$$

$$f(-2.171) = (-2.171)^3 - 3(-2.171) + 4 = -10.22 + 6.513 + 4 = +0.293.$$

$$f(-3) = -14 \text{ and } f(-2.171) = +0.293$$

$\therefore$ Root lies between – 3 and – 2.171.

$$x = \frac{(-2.171)f(-3) - (-3)f(-2.171)}{f(-3) - f(-2.171)} = \frac{(-2.171)(-14) - (-3)(0.293)}{-14 - 0.293}$$

$$= \frac{30.494 + 0.879}{-14.293} = \frac{31.273}{-14.293} = -2.188$$ **Ans.**

**Example 15.** *The negative root of the equation $3x^3 + 8x^2 + 8x + 5 = 0$ is to be determined. Find the root by Regula Falsi method. Stop iteration when $f(x_2) < 0.02$* *(A.M.I.E., Summer 2001)*

**Solution.** $f(x) = 3x^3 + 8x^2 + 8x + 5 = 0$

$$f(-1) = -3 + 8 - 8 + 5 = +2$$
$$f(-1.5) = 3(-1.5)^3 + 8(-1.5)^2 + 8(-1.5) + 5 = -10.125 + 18 - 12 + 5 = +0.875$$
$$f(-1.6) = 3(-1.6)^3 + 8(-1.6)^2 + 8(-1.6) + 5 = -12.288 + 20.48 - 12.8 + 5 = +0.392$$
$$f(-1.7) = 3(-1.7)^3 + 8(-1.7)^2 + 8(-1.7) + 5 = -14.739 + 23.12 - 13.6 + 5 = -0.219$$

**Since $f(-1.6)$ and $f(-1.7)$ are of opposite signs so the root lies between –1.6 and – 1.7**
**By Regula Falsi method:**

$$a_1 = \frac{-1.6f(-1.7) - (-1.7)f(-1.6)}{f(-1.7) - f(-1.6)} = \frac{-1.6(-0.219) - (-1.7)(0.392)}{-0.219 - (0.392)}$$

$$= \frac{0.3504 + 0.6664}{-0.611} = -\frac{1.0168}{0.611} = -1.664$$

$$f(-1.664) = 3(-1.664)^3 + 8(-1.664)^2 + 8(-1.664) + 5$$

$$= -13.822 + 22.151 - 13.312 + 5 = 0.017$$

$$f(-1.664) = 0.017 < 0.02$$

Hence the negative root of the given equation is – 1.664 **Ans.**

**Example 16.** *Determine the root of*

$$x^4 + x^3 - 7x^2 - x + 5 = 0$$

*which lies between 2 and 3 correct to three decimal places.*

**Solution.**

$$f(x) = x^4 + x^3 - 7x^2 - x + 5 = 0$$

$$f(2) = 16 + 8 - 28 - 2 + 5 = -1$$

$$f(3) = 81 + 27 - 63 - 3 + 5 = +47$$

$f(2) = -1$ is nearer to zero than $+47$.

∴ Root is near to 2.

Let us try on 2.1.

$$f(2.1) = (2.1)^4 + (2.1)^3 - 7(2.1)^2 - (2.1) + 5 = 19.4481 + 9.261 - 30.87 - 2.1 + 5 = +0.7391.$$

Now the root lies between 2 and 2.1.

By the method of False position :

$$x_1 = \frac{af(b) - bf(a)}{f(b) - f(a)} = \frac{2f(2.1) - 2.1f(2)}{f(2.1) - f(2)} = \frac{2(0.739) - 2.1(-1)}{(0.739) - (-1)}$$

$$= \frac{1.4782 + 2.1}{1.739} = \frac{3.5782}{1.739} = 2.0576$$

$$f(2.0576) = (2.0576)^4 + (2.0576)^3 - 7(2.0576)^2 - (2.0576) + 5$$

$$= 17.9244 + 8.7113 - 29.6360 - 2.0576 + 5 = -0.0579$$

$$x_2 = \frac{2.0576 f(2.1) - 2.1 f(2.0576)}{f(2.1) - f(2.0576)} = \frac{2.0576\,(0.7391) - 2.1(-0.0579)}{0.7391 - (-0.0579)}$$

$$= \frac{1.5208 + 0.1216}{0.7970} = \frac{1.6424}{0.7970} = 2.0607$$

$$f(2.0607) = (2.0607)^4 + (2.0607)^3 - 7(2.0607)^2 - (2.0607) + 5$$

$$= 18.0326 + 8.7507 - 29.7254 - 2.0607 + 5 = -0.0028.$$

$$x_3 = \frac{2.0607 f(2.1) - 2.1 f(2.0607)}{f(2.1) - f(2.0607)} = \frac{2.0607\,(0.7391) - 2.1(-0.0028)}{0.7391 - (-0.0028)}$$

$$= \frac{1.5231 + 0.0059}{0.7419} = \frac{1.5290}{0.7419} = 2.0609$$

The root of the given equation is 2.0609. **Ans.**

## Exercise 15.3

Solve the following equations by Regula Falsi method :

**1.** $x^3 - 2x - 5 = 0$ **Ans.** 2.0946 **2.** $x^3 - 10x^2 + 40x - 35 = 0$ **Ans.** 1.1975.

**3.** $x^3 + x^2 + 3x + 4 = 0$ **Ans.** – 1.22248 **4.** $x^6 - x^4 - x^3 - 1 = 0$ **Ans.** 1.4036

**5.** $x^3 - 9x + 1 = 0$ (Root between 2 & 3) **Ans.** 2.9416 **6.** $x^3 - 5x - 7 = 0$ **Ans.** 2.746

**7.** $x^3 - x - 1 = 0$ **Ans.** 1.315 **8.** $3x^3 - 5x^2 + 3x - 5 = 0$ **Ans.** 1.6629

9. The smallest positive root of the equation $x = e^{-x}$ is to be determined. Show that the root lies in (0,1). Using the Regula Falsi method, find the root correct to three decimals. **Ans.** 0.6065

10. Obtain a root of the equation $x^3 - 4x - 9 = 0$, correct to three decimal places using the method of false position. **Ans.** 2.7064

11. Use the method of false position to find the root of the equation $x^3 - 18 = 0$, given that it lies between 2 and 3. Write down three steps of the procedure. **Ans.** 2.621

12. Find the root of the equation $\tan x + \tanh x = 0$ which lies in the interval (1.6, 3.0) correct to four significant digits using any one of the numerical methods. **Ans.** 2.365 app.

## 15.5 ITERATION METHOD

Let $f(x) = 0$ ...(1)

(1) can be written as $x = \phi(x)$ ...(2)

where $|\phi'(x)| < 1$

$\therefore$ Let first approximate root be $x_1 = a$

**Second Approximation $x_2$**

Putting $x = x_1$ in R.H.S. of (2), we have $x_2 = \phi(x_1)$

Similarly $x_3 = \phi(x_2)$

By repeating this method, we get the better approximation of the root.

**Example 17.** *Use the method of iteration to solve the equation $x = \exp(-x)$, starting with $x = 1.00$. Perform four iterations, taking the readings upto four decimal places.*

**Solution.** $x = e^{-x}$ ...(1)

$$\phi(x) = e^{-x}, \quad \phi'(x) = -e^{-x} \qquad |\phi'(x)| = e^{-x}$$

$$|\phi'(1)| = e^{-1} = \frac{1}{e} = 0.3679 < 1$$

Putting $x = 1$ in (1) we get $x_1 = e^{-1} = 0.3679$

Putting $x = 0.3679$ in (1) we have $x_2 = e^{-0.3679} = 0.692$

Putting $x = 0.692$ in (1), we obtain $x_3 = e^{-0.692} = 0.5$

Putting $x = 0.5$ in (1), we get $x_4 = e^{-0.5} = 0.6065$ **Ans.**

**Example 18.** *Find a real root of the equation $x^3 + x^2 - 1 = 0$ by the method of iteration.*

**Solution.** $f(x) = x^3 + x^2 - 1$ ...(1)

$$f(0.7) = 0.343 + 0.49 - 1 = -0.167 = -ve$$

$$f(0.8) = 0.512 + 0.64 - 1 = +0.152 = +ve$$

As $f(0.7)$ and $f(0.8)$ are of opposite sign, so that root lies between 0.7 and 0.8. Let the first approximate root be 0.7.

$$x^3 + x^2 - 1 = 0 \text{ or } x^3 = 1 - x^2$$

$$x = (1 - x^2)^{1/3} x = \phi(x) \qquad \text{where } \phi(x) = (1 - x^2)^{1/3}$$

$$|\phi'(0.7)| = 0.73 < 1$$

$$x_1 = [1 - (0.7)^2]^{1/3} = (1 - 0.49)^{1/3} = (0.51)^{1/3} = 0.799$$

$$x_2 = [1 - (0.799)^2]^{1/3} = (1 - 0.63840)^{1/3} = (0.316)^{1/3} = 0.712$$

$$x_3 = [1-(0.712)^2]^{1/3} = [1-0.507]^{1/3} = (0.493)^{1/3} = 0.79$$

$$x_4 = [1-(0.79)^2]^{1/3} = [1-0.6341]^{1/3} = (0.3759)^{1/3} = 0.722$$

$$x_5 = [1-(0.722)^2]^{1/3} = 1-0.5212]^{1/3} = (0.4788)^{1/3} = 0.782$$

$$f(0.782) = (0.782)^3 + (0.782)^2 - 1 = 0.478 + 0.611 - 1 = 0.089$$

Root = 0.782 **Ans.**

**Example 19.** *Find a solution of $x^3 + x - 1 = 0$ by iteration.*

**Solution.** $f(x) = x^3 + x - 1 = 0$

The approximate root of the given equation is 1 as shown by rough sketch. We can write the equation in the form

$$x = \frac{1}{1+x^2} \text{ thus } x_{n+1} = \frac{1}{1+x_n^2}$$

Let $$\phi(x) = \frac{1}{1+x^2},\ |\phi'(x)| = \left|\frac{2x}{(1+x^2)^2}\right|,\quad |\phi'(1)| = \frac{1}{2} < 1$$

$$x_1 = \frac{1}{1+1^2} = 0.5$$

$$x_2 = \frac{1}{1+1(0.5)^2} = 0.800$$

$$x_3 = \frac{1}{1+(0.8)^2} = 0.610$$

$$x_4 = \frac{1}{1+(0.610)^2} = 0.729$$

Similarly $x_5 = 0.653,\ x_6 = 0.701.$

The exact root is 0.682328. **Ans.**

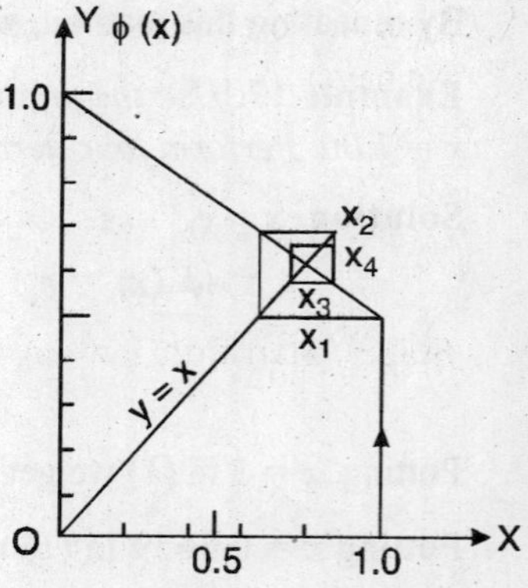


**Exercise 15.4**

Solve by iteration method

1. $1 + \log x = \dfrac{x}{2}$ **Ans.** 5.36

2. $\sin x = \dfrac{x+1}{x-1}$ [**Hint.** Approximate root = – 5.5] **Ans.** – 5.5174

3. Use the method of iteration to find a root, near 2, of the equation $x^3 = x^2 + x + 1$. Carry out five iterations. **Ans.** 1.8408

## 15.6 SOLUTION OF LINEAR SYSTEMS

Here we shall discuss two methods for solving the linear systems i.e., Gauss-Seidel and Crout's methods.

### Gauss method

By this method elimination of unknown is done more systematically and we have a check to detect the errors. The method is explained in the following example :

**Example 20.** *Solve the following simultaneous equations*

$$2x + 3y + z = 13$$
$$x - y - 2z = -1$$
$$3x + y + 4z = 15$$

**Step 1.** We write the equation, first, which has unity as coefficient of $x$, otherwise divide the equation by the coefficient of $x$ to make it unity. Thus

$$x - y - 2z = -1$$

$$2x + 3y + z = 13$$

$$3x + y + 4z = 15$$

**Step 2.** To eliminate x, subtract suitable multiples of first equation from the remaining equations, and we get

$$x - y - 2z = -1$$

$$5y + 5z = 15 \qquad (2) - 2(1)$$

$$4y + 10z = 18 \qquad (3) - 3(1)$$

**Step 3.** The coefficient of $y$ is made unity in none of the resulting equations and we have

$$x - y - 2z = -1$$

$$y + z = 3 \qquad 1/5\,(2)$$

$$4y + 10z = 18$$

**Step 4.** To eliminate $y$, subtract suitable multiple of second equation from the third. Thus we have

$$x - y - 2z = -1$$

$$y + z = 3$$

$$6z = 6$$

**Step 5.** Start from bottom and substitute.

$$6z = 6 \Rightarrow z = 1$$

$$y + z = 3 \Rightarrow y + 1 = 3 \Rightarrow y = 2$$

$$x - y - 2z = -1 \Rightarrow x - 2 - 2 = -1 \Rightarrow x = 3$$

Now the solution is presented in the table given below in a compact form. It contains only the coefficients of the unknowns and constant term from step 1. One additional column contains the sum of all the numbers appearing in each row.

| Step | *No. of row* | *x* | *y* | *z* | *Constant* | *Check sum* | *Explanation* |
|---|---|---|---|---|---|---|---|
| 1 | 1 | 1 | –1 | –2 | – 1 | – 3 | |
| | 2 | 2 | 3 | 1 | 13 | 19 | |
| | 3 | 3 | 1 | 4 | 15 | 23 | |
| 2 | 4 | 1 | –1 | – 2 | – 1 | – 3 | |
| | 5 | 0 | 5 | 5 | 15 | 25 | (2) – 2 (1) |
| | 6 | 0 | 4 | 10 | 18 | 32 | (3) – 3 (1) |
| | 7 | | 1 | 1 | 3 | 5 | 1/5 (5) |
| | 8 | | 0 | 6 | 6 | 12 | (6) – 4 (7) |
| | 9 | | | 1 | 1 | 2 | 1/6 (8) |
| | Check sum | 7 | 12 | 24 | 69 | 112 | |

**From (9),** $z = 1$

**From (7),** $y + z = 3 \Rightarrow y + 1 = 3 \Rightarrow y = 2$

**From (4)** $x - y - 2z = -1 \Rightarrow x - 2 - 2 = -1 \Rightarrow x = 3$

**Note:** (1) Sum of the numbers in the check column should be equal to the sum of the numbers in last check row. If the two do not tally some mistake has been made and must be rectified.

(2) In each step the coefficient of the first unknown is called the pivotal coefficient.

## 15.7 CROUT'S METHOD

In Gauss elimination method, the number of steps increase rapidly with the number of unknowns, that method is laborious and time-consuming. To save labour, there is one method known as *Crout's method* or *Cholesky's method.* It will greatly facilitate solution if we could transform the equations.

$$\left.\begin{aligned} a_{11}\,x + a_{12}\,y + a_{13}\,z &= b_1 \\ a_{21}\,x + a_{22}\,y + a_{23}\,z &= b_2 \\ a_{31}\,x + a_{32}\,y + a_{33}\,z &= b_3 \end{aligned}\right\} \qquad \text{...(1)}$$

into the equations of the triangular form

$$\left.\begin{aligned} x + u_{12}\,y + u_{13}\,z &= c_1 \\ y + u_{23}\,z &= c_2 \\ z &= c_3 \end{aligned}\right\} \qquad \text{...(2)}$$

Equation (2) on backward substitution gives the values of $x, y, z$.

Let us write down equation (1) and (2) in the matrix form as upper triangular matrices.

$$AX = B \qquad \text{...(3)}$$

and $$UX = C \qquad \text{...(4)}$$

where $$A = \begin{bmatrix} a_{11} & a_{12} & a_{13} \\ a_{21} & a_{22} & a_{23} \\ a_{31} & a_{32} & a_{33} \end{bmatrix} \text{ and } U = \begin{bmatrix} 1 & u_{12} & u_{13} \\ 0 & 1 & u_{23} \\ 0 & 0 & 1 \end{bmatrix}$$

are upper triangular matrices.

$$X = \begin{bmatrix} x \\ y \\ z \end{bmatrix},\ B = \begin{bmatrix} b_1 \\ b_2 \\ b_3 \end{bmatrix},\ C = \begin{bmatrix} c_1 \\ c_2 \\ c_3 \end{bmatrix}$$

Now if $A = LU$

where $L$ is a lower triangular matrix.

On putting the value of $A = LU$ in (3), we get

$$LUX = B \quad \text{or} \quad LC = B \qquad [\text{Since } UX = C, \text{ from (4)}]$$

Since $LU = A$ and $LC = B$,

By combining these two relations, we get $L\,(U/C) = (A/B)$ ...(5)

where $(U/C)$ are augmented matrix of (3) and (4).

From (5), we have

$$\begin{bmatrix} l_{11} & 0 & 0 \\ l_{21} & l_{22} & 0 \\ l_{31} & l_{32} & l_{33} \end{bmatrix} \begin{bmatrix} 1 & u_{12} & u_{13} & c_1 \\ 0 & 1 & u_{23} & c_2 \\ 0 & 0 & 1 & c_3 \end{bmatrix} = \begin{bmatrix} a_{11} & a_{12} & a_{13} & b_1 \\ a_{21} & a_{22} & a_{23} & b_2 \\ a_{31} & a_{32} & a_{33} & b_3 \end{bmatrix}$$

On multiplication, we have

$$\begin{bmatrix} l_{11} & l_{11}\,u_{12} & l_{11}\,u_{13} & l_{11}\,c_1 \\ l_{21} & l_{21}\,u_{12} + l_{22} & l_{21}\,u_{13} + l_{22}\,u_{23} & l_{21}\,c_1 + l_{22}\,c_2 \\ l_{31} & l_{31}\,u_{12} + l_{32} & l_{31}\,u_{13} + l_{32}\,u_{23} + l_{33} & l_{31}\,c_1 + l_{32}\,c_2 + l_{33}\,c_3 \end{bmatrix} = \begin{bmatrix} a_{11} & a_{12} & a_{13} & b_1 \\ a_{21} & a_{22} & a_{23} & b_2 \\ a_{31} & a_{32} & a_{33} & b_3 \end{bmatrix}$$

$$l_{11} = a_{11},\ l_{21} = a_{21},\ l_{31} = a_{31}$$

$$l_{11}u_{12} = a_{12},\ l_{11}u_{13} = a_{12},\ l_{11}c_1 = b_1$$

$$\Rightarrow \quad u_{12} = \frac{a_{12}}{l_{11}}, \text{ or } u_{13} = \frac{a_{12}}{l_{11}} \text{ or } c_1 = \frac{b_1}{l_{11}};$$

$$\Rightarrow \quad l_{21}u_{12} + l_{22} = a_{23},\quad l_{31}u_{12} + l_{32} = a_{32},$$

$$\Rightarrow \quad l_{22} = a_{23} - l_{21}u_{12}, \quad \text{or} \quad l_{32} = a_{32} - l_{31}u_{12};$$

$$\Rightarrow \quad l_{21}u_{13} + l_{22}u_{23} = a_{23},\quad l_{21}c_1 + l_{22}c_2 = b_2$$

$$u_{23} = (a_{23} - l_{21}u_{13})/l_{22} \quad \text{or} \quad c_2 = (b_2 - l_{21}c_1)/l_{22};$$

$$\Rightarrow \quad l_{31}u_{13} + l_{32}u_{23} + l_{33} = a_{33},\quad l_{31}c_1 + l_{32}c_2 + l_{33}c_3 = b_3;$$

$$\Rightarrow \quad l_{33} = a_{33} - l_{31}u_{13} - l_{32}u_{23} \quad \text{or} \quad c_3 = (b_3 - l_{31}c_1 - l_{32}c_2)/l_{33}$$

Thus values of u's and c's are known. On substitution in (2) we get the values of $x, y, z$.

A table is prepared. The upper half of the table contains the coefficients of the original equations. The lower half of the table contains the element of $L$ and $U$. An additional column in the table is for check sums.

After the upper half is completed, the entries in the lower half are made in the flollowing order :

(1) First column (2) First row (3) Second column (4) Second row and so on.

General formulae for calculating $l$'s, $u$'s and $c$'s are

$$l_{rs} = a_{rs} - \sum_{i=1}^{s-1} l_{ri}u_{is}\ (r \geq s),$$

$$u_{rs} = \left[a_{rs} - \sum_{i=1}^{r-1} l_{ri}u_{is}\right]/l_{rr} \quad (r < s),$$

$$c_r = \left[b_r - \sum_{i=1}^{r-1} l_{ri}c_i\right]/l_{rr}$$

| | | | | |
|---|---|---|---|---|
| $a_{11}$ | $a_{12}$ | $a_{13}$ | $b_1$ | $B_1$ |
| $a_{21}$ | $a_{22}$ | $a_{23}$ | $b_2$ | $B_2$ |
| $a_{31}$ | $a_{32}$ | $a_{33}$ | $b_3$ | $B_3$ |
| $l_{11}$ | $u_{12}$ | $u_{13}$ | $c_1$ | $C_1$ |
| $l_{21}$ | $l_{22}$ | $u_{23}$ | $c_2$ | $C_2$ |
| $l_{31}$ | $l_{32}$ | $l_{33}$ | $c_3$ | $C_3$ |

B's are the check sums and C's are calculated in the following two ways :

(*i*) $C_1 = \frac{B_1}{l_{11}},\ C_2 = (B_2 - l_{21}C_1)/l_{22}$

(*ii*) $C_1 = 1 + u_{12} + u_{13} + c_1,\quad C_2 = 1 + u_{23} + c_2$

If the values of $C$ obtained by two different ways differ, some mistake has been made in computation, and must be rectified before we proceed further.

**Example 21.** *Solve the following equations by Crout's method:*

$$4x + y - z = 13,\quad 3x + 5y + 2z = 21,\ 2x + y + 6z = 14.$$

**Solution.** The above equations are written in the form of

$$\left.\begin{aligned} x + u_{12}\,y + u_{13}\,z &= c_1 \\ y + u_{23}\,z &= c_2 \\ z &= c_3 \end{aligned}\right] \qquad \ldots(1)$$

$$l_{11} = a_{11} = 4,\ l_{21} = a_{21} = 3,\quad l_{31} = a_{31} = 2$$

$$u_{12} = \frac{a_{12}}{l_{11}} = \frac{1}{4}, u_{13} = \frac{a_{13}}{l_{11}} = \frac{-1}{4}, \; c_1 = \frac{b_1}{l_{11}} = \frac{13}{4}$$

$$l_{22} = a_{22} - l_{21}u_{12} = 5 - 3 \times \frac{1}{4} = \frac{17}{4},$$

$$l_{32} = a_{32} - l_{31}u_{12} = 1 - 2 \times \frac{1}{4} = \frac{1}{2}$$

$$u_{23} = (a_{23} - l_{21}u_{13}) / l_{22} = \left[2 - 3\left(-\frac{1}{4}\right)\right] / \left(\frac{17}{4}\right) = \frac{11}{17}$$

$$c_2 = (b_2 - l_{21}c_1) / l_{22} = \left[21 - 3\left(\frac{13}{4}\right)\right] / \frac{17}{4} = \frac{45}{17}$$

$$l_{33} = a_{33} - l_{31}u_{13} - l_{32}u_{23} = 6 - 2\left(-\frac{1}{4}\right) - \frac{1}{2}\left(\frac{11}{17}\right) = \frac{105}{17}$$

$$c_3 = (b_{33} - l_{31}c_1 - l_{32}c_2) / l_{33} = \left(14 - 2 \times \frac{13}{4} - \frac{1}{2} \times \frac{45}{17}\right) / \frac{105}{17} = 1$$

**On putting the values of u's and c's in (1), we get**

$$x + \frac{1}{4}y - \frac{1}{4}z = \frac{13}{4}, y + \frac{11}{17}z = \frac{45}{17}, \; z = 1$$

**By backward substitution, we get**

$$z = 1$$

$$y + \frac{11}{17}z = \frac{45}{17} \Rightarrow y + \frac{11}{17} = \frac{45}{17} \Rightarrow y = 2$$

$$x + \frac{1}{4}y - \frac{1}{4}z = \frac{13}{4} \Rightarrow x + \frac{1}{2} - \frac{1}{4} = \frac{13}{4} \Rightarrow x = 3$$

$$x = 3, y = 2, z = 1$$

**Ans.**

**The table is filled as follows:**

**(*i*) The upper half of the table is the augmented matrix of the given system of equations.**

**(*ii*) The lower half is completed in the following order:**

| | | | | |
|---|---|---|---|---|
| 4 | 1 | −1 | 13 | 17 |
| 3 | 5 | 2 | 21 | 31 |
| 2 | 1 | 6 | 14 | 23 |
| 4 | $\frac{1}{4}$ | $-\frac{1}{4}$ | $\frac{13}{4}$ | $\frac{17}{4}$ |
| 3 | $\frac{17}{4}$ | $\frac{11}{17}$ | $\frac{45}{17}$ | $\frac{73}{17}$ |
| 2 | $\frac{1}{2}$ | $\frac{105}{17}$ | 1 | 2 |

**First column** **is the same as first column of upper half.**

$$l_{11} = 4; \; l_{12} = 3, \; l_{13} = 2.$$

**First row.** **First element $l_{11}$ is 4. The other elements are obtained by dividing the corresponding elements of the first row of upper half by $l_{11}$.**

**Second column.** **First element $u_{21} = \frac{1}{4}$**

**Second element** $\quad l_{22} = a_{22} - l_{21}u_{12}$

**= corresponding element in the upper half**

**– (Product of the elements of the left to $l_{22}$ and $u_{12}$).**

**Third element** $\quad l_{33} = a_{33} - l_{31}u_{12}$

**= corresponding element in the upper half**

**– (Product of the elements of the left to $l_{32}$ and $u_{12}$).**

**and so on.**

**Example 22.** *Solve the following equations by Crout's method:*

$$4x + 3y + z - w = 14$$
$$2x + 5y + 2z + w = 17$$
$$x + 4y + 4z + 6w = 20$$
$$3x + y - z + 5w = 12$$

**Solution.** The given equations can be written as

$$x + u_{12}y + u_{13}z + u_{14}w = c_1$$
$$y + u_{23}z + u_{24}w = c_2$$
$$z + u_{34}w = c_3$$
$$w = c_4$$

where $l$'s and therefore $u$'s and $c$'s are obtained by the following formulae

$$l_{rs} = a_{rs} - \sum_{i=1}^{s-1} l_{ri}u_{is},\quad u_{rs} = \left(a_{rs} - \sum_{i=1}^{r-1} l_{ri}u_{is}\right)/l_{rr},\quad c_r = \left(b_r - \sum_{i=1}^{r-1} l_{ri}c\right)/l_{rr}$$

The entries in the table given below are made with the above formulae.

| | | | | | |
|---|---|---|---|---|---|
| 4 | 3 | 1 | −1 | 14 | 21 |
| 2 | 5 | 2 | 1 | 17 | 27 |
| 1 | 4 | 4 | 6 | 20 | 35 |
| 3 | 1 | −1 | 5 | 12 | 20 |
| 4 | $\frac{3}{4}$ | $\frac{1}{4}$ | $-\frac{1}{4}$ | $\frac{7}{2}$ | $\frac{21}{4}$ |
| 2 | $\frac{7}{2}$ | $\frac{3}{7}$ | $\frac{3}{7}$ | $\frac{20}{7}$ | $\frac{33}{7}$ |
| 1 | $\frac{13}{4}$ | $\frac{33}{14}$ | $\frac{68}{33}$ | $\frac{101}{33}$ | $\frac{202}{33}$ |
| 3 | $-\frac{5}{4}$ | $-\frac{17}{14}$ | $\frac{2030}{231}$ | 1 | 2 |

Therefore $\quad x + \frac{3}{4}y + \frac{z}{4} - \frac{\omega}{4} = \frac{7}{2}$

$$y + \frac{3}{7}z + \frac{3}{7}w = \frac{20}{7}$$

$$z + \frac{68}{33}w = \frac{101}{33}$$

$$w = 1$$

Now $w = 1$, $\quad z + \frac{68}{33} = \frac{101}{33} \Rightarrow z = 1$

$$y + \frac{3}{7} + \frac{3}{7} = \frac{20}{7} \Rightarrow y = 2$$

$$x + \frac{3}{2} + \frac{1}{4} - \frac{1}{4} = \frac{7}{2} \Rightarrow x = 2$$

$x = 2,\ y = 2,\ z = 1,\ w = 1$ **Ans.**

**Example 23.** *Solve the following system by Crout's method*

$x + 2y + z = 4,\ 2x - 3y - z = -3,$
$3x + y + 2z = 3$

**Solution.** By Crout's method

$$\begin{bmatrix} l_{11} & 0 & 0 \\ l_{21} & l_{22} & 0 \\ l_{31} & l_{32} & l_{33} \end{bmatrix} \begin{bmatrix} 1 & u_{12} & u_{13} & c_1 \\ 0 & 1 & u_{23} & c_2 \\ 0 & 0 & 1 & c_3 \end{bmatrix} = \begin{bmatrix} 1 & 2 & 1 & 4 \\ 2 & -3 & -1 & -3 \\ 3 & 1 & 2 & 3 \end{bmatrix}$$

$$\begin{bmatrix} l_{11} & l_{11}u_{12} & l_{11}u_{13} & l_{11}c_1 \\ l_{21} & l_{21}u_{12} + l_{22} & l_{21}u_{13} + l_{22}u_{23} & l_{21}c_1 + l_{22}c_2 \\ l_{31} & l_{31}u_{12} + l_{32} & l_{31}u_{13} + l_{32}u_{23}\ l_{33} & l_{31}c_1 + l_{32}c_2 + l_{33}c_3 \end{bmatrix} = \begin{bmatrix} 1 & 2 & 1 & 4 \\ 2 & -3 & -1 & -3 \\ 3 & 1 & 2 & 3 \end{bmatrix}$$

$$l_{11} = 1,\ l_{11}u_{12} = 2,\ l_{11}u_{13} = 1,\ l_{11}c_1 = 4$$
$$l_{21} = 2,\ l_{21}u_{12} + l_{22} = -3,\ l_{21}u_{13} + l_{22}u_{23} = -1,\ l_{21}c_1 + l_{22}c_2 = -3$$
$$l_{31} = 3,\ l_{31}u_{12} + l_{32} = 1,\ l_{31}u_{13} + l_{32}u_{23} + l_{33} = 2,\ l_{31}c_1 + l_{32}c_2 + l_{33}c_3 = -3$$

The following enteries are made in the following order :

(1) First column (2) First row (3) Second Column (4) Second row and so on.

| | | | |
|---|---|---|---|
| $l_{11} = 1$ | $u_{12} = 2$ | $u_{13} = 1$ | $c_1 = 4$ |
| $l_{21} = 1$ | $l_{22} = -7$ | $u_{23} = \frac{3}{7}$ | $c_2 = \frac{11}{7}$ |
| $l_{31} = 3$ | $l_{32} = -5$ | $l_{33} = \frac{8}{7}$ | $c_3 = -1$ |

$$\begin{bmatrix} 1 & 2 & 1 \\ 0 & 1 & \frac{3}{7} \\ 0 & 0 & 1 \end{bmatrix} \begin{bmatrix} x \\ y \\ z \end{bmatrix} = \begin{bmatrix} 4 \\ \frac{11}{7} \\ -1 \end{bmatrix} \quad \text{or} \quad \begin{aligned} x + 2y + z &= 4 \\ x + 2y + z &= 4 \\ y + \frac{3}{7} z &= \frac{11}{7}, \\ z &= -1 \end{aligned}$$

$$y - \frac{3}{7} = \frac{11}{7} \Rightarrow y = 2$$

$$x + 4 - 1 = 4 \Rightarrow x = 1$$

**Ans.** $x = 1, y = 2, z = -1$

## 15.8 ITERATIVE METHODS OR INDIRECT METHODS

**We start with an approximation to the true solution and by applying the method repeatedly we get better and better approximation till accurated solution is achieved.**

**There are two iterative methods for solving simultaneous equations.**

**(1) Jacobi's method (Method of simultaneous correction).**

**(2) Gauss-Seidel method (Method of successive correction).**

## 15.9 JACOBI'S METHOD

**The method is illustrated by taking an example.**

**Let**

$$\left.\begin{aligned} a_{11} x + a_{12} y + a_{13} z &= b_1 \\ a_{21} x + a_{22} y + a_{23} z &= b_2 \\ a_{31} x + a_{32} y + a_{33} z &= b_3 \end{aligned}\right\} \quad \text{...(1)}$$

**After division by suitable constants and transposition, the equations can be written as**

$$\left.\begin{aligned} x &= c_1 - k_{12} y - k_{13} z \\ y &= c_2 - k_{21} x - k_{23} z \\ z &= c_3 - k_{31} x - k_{32} y \end{aligned}\right\} \quad \text{...(2)}$$

**Let us assume $x = 0, y = 0$ and $z = 0$ as first approximation, substituting the values of $x, y, z$ on the right hand side of (2), we get $x = c_1, y = c_2, z = c_3$. This is the second approximation to the solution of the equations.**

**Again substituting these values of $x, y, z$ in (2) we get a third approximation.**

**The process is repeated till two successive approximations are equal or nearly equal.**

**Note: Condition for using the iterative methods is that the coefficients in the leading diagonal are large compared to the other. If are not so, then on interchanging the equation we can make the leading diagonal dominant diagonal.**

**Example 24.** *Solve by Jacobi's method*

$$\begin{aligned} 4x + y + 3z &= 17 \\ x + 5y + z &= 14 \\ 2x - y + 8z &= 12 \end{aligned}$$

**Solution.** The above equations can be written as

$$\left.\begin{aligned} x &= \frac{17}{4} - \frac{y}{4} - \frac{3z}{4} \\ y &= \frac{14}{5} - \frac{x}{5} - \frac{z}{5} \\ z &= \frac{3}{2} - \frac{x}{4} + \frac{y}{8} \end{aligned}\right] \quad \ldots(1)$$

On substituting x = y = z = 0 on the right hand side of (1), we get $x = \frac{17}{4}$, $y = \frac{14}{5}$, $z = \frac{3}{2}$

Again substituting these values of *x*, *y*, *z* on R.H.S. of (1), we obtain

$$x = \frac{17}{4} - \frac{7}{10} - \frac{9}{8} = \frac{97}{40}$$

$$y = \frac{14}{5} - \frac{17}{20} - \frac{3}{10} = \frac{33}{20}$$

$$z = \frac{3}{2} - \frac{17}{16} + \frac{7}{20} = \frac{63}{80}$$

Again putting these values on R.H.S. of (1) we get next approximations.

$$x = \frac{17}{4} - \frac{33}{80} - \frac{189}{320} = \frac{1039}{320} = 3.25$$

$$y = \frac{14}{5} - \frac{97}{200} - \frac{63}{400} = \frac{863}{400} = 2.16$$

$$z = \frac{3}{2} - \frac{97}{160} + \frac{33}{160} = \frac{176}{160} = 1.1$$

Substituting, again, the values of *x*, *y*, *z* on R.H.S. of (1) we get

$$x = \frac{17}{4} - \frac{2.16}{4} - \frac{3(1.1)}{4} = 2.885$$

$$y = \frac{14}{5} - \frac{3.25}{5} - \frac{1.1}{5} = 1.93$$

$$z = \frac{3}{2} - \frac{3.25}{4} + \frac{2.16}{8} = 0.96$$

Repeating the process for *x* = 2.885, *y* = 1.93, *z* = 0.96 we have

$$x = \frac{17}{4} - \frac{1.93}{4} - \frac{3}{4}(0.96) = 4.25 - 0.48 - 0.72 = 3.05$$

$$y = \frac{14}{5} - \frac{2.885}{5} - \frac{0.96}{5} = 2.8 - 0.577 - 0.192 = 2.03$$

$$z = \frac{3}{2} - \frac{2.885}{4} + \frac{1.93}{8} = 1.5 - 0.721 + 0.241 = 1.02$$

This can be written in a table

| *Iterations* | *1* | *2* | *3* | *4* | *5* | *6* |
|---|---|---|---|---|---|---|
| $x = \frac{17}{4} - \frac{y}{4} - \frac{3z}{4}$ | 0 | $\frac{17}{4} = 4.25$ | $\frac{97}{40} = 2.425$ | $\frac{1039}{320} = 3.25$ | 2.885 | 3.05 |
| $y = \frac{14}{5} - \frac{x}{5} - \frac{z}{5}$ | 0 | $\frac{14}{5} = 2.8$ | $\frac{33}{20} = 1.65$ | $\frac{863}{400} = 2.16$ | 1.93 | 2.03 |
| $z = \frac{3}{2} - \frac{x}{4} + \frac{y}{8}$ | 0 | $\frac{3}{2} = 1.5$ | $\frac{63}{80} = 0.7875$ | $\frac{176}{160} = 1.1$ | 0.96 | 1.02 |

After 6th iteration

$$x = 3.05, \quad y = 2.03, z = 1.02$$

The actual values are

$$x = 3, \ y = 2, \ z = 1.$$ **Ans.**

## 15.10 GAUSS-SEIDEL METHOD

Gauss-Seidel method is a modification of Jacobi's method. In place of substituting the same set of values in all the three equations (2) of Article 15.9, we use in each step the value obtained in the earlier step.

**Step 1.** First we put $y = z = 0$ in first of the equation (2) of Article 15.9 and $x = c_1$. Then in second equation we put this value of $x$ *i.e.*, $c_i$ and $z = 0$ and obtain $y$. In the third equation we use the values of $x$ and $y$ obtained earlier get $z$.

**Step 2.** We repeat the above procedure. In the first equation we put the values of $y$ and $z$ obtained in step 1 and redetermine $x$. By using the new value of $x$ and value of $z$ obtained in step 1 we redetermine $y$ and so on.

In other words, the latest values of the unknowns are used in each step.

**Example 25.** *Solve by Gauss-Seidel method*

$$6x + y + z = 105$$

$$4x + 8y + 3z = 155$$

$$5x + 4y - 10z = 65.$$

**Solution.** The numbers 6, 8, – 10 in the leading diagonal are the largest, so we can apply Gauss-Seidel method to solve the given equations.

The above equations can be written as

$$x = \frac{35}{2} - \frac{1}{6}y - \frac{1}{6}z \quad \text{...(1)}$$

$$y = \frac{155}{8} - \frac{1}{2}x - \frac{3}{8}z \quad \text{...(2)}$$

$$z = -\frac{13}{2} + \frac{1}{2}x + \frac{2}{5}y \quad \text{...(3)}$$

Putting $y = z = 0$ in equation (1), we get

$$x = \frac{35}{2}$$

Putting $x = \frac{35}{2}$ and $z = 0$ in equation (2), we have

$$y = \frac{155}{8} - \frac{35}{4} - 0 = \frac{85}{8}$$

Substituting $x = \frac{35}{2}, y = \frac{85}{8}$ in equation (3), we obtain

$$z = -\frac{13}{2} + \frac{35}{4} + \frac{17}{4} = \frac{13}{2}$$

Again starting from equaion (1) and putting $y = \frac{85}{8}, z = \frac{13}{2}$, we get

$$x = \frac{35}{2} - \frac{85}{48} - \frac{13}{12} = \frac{703}{48} = 14.64$$

Similarly the process is carried on and the roots so obtained are given in the following table :

| *Iterations* | *1* | 2 | 3 | 4 |
|---|---|---|---|---|
| $x=\frac{35}{2}-\frac{y}{6}-\frac{z}{6}$ | $\frac{35}{2}=17.5$ | 14.64 | 15.12 | 14.98 |
| $y=\frac{155}{8}-\frac{x}{2}-\frac{3}{8}z$ | $\frac{85}{8}=10.6$ | 9.62 | 10.06 | 9.98 |
| $z=-\frac{13}{2}+\frac{x}{2}+\frac{2}{5}y$ | $\frac{13}{2}=6.5$ | 4.67 | 5.084 | 4.98 |

At the end of fourth iteration the roots are $x = 14.98, y = 9.98, z = 4.98$.

But the actual roots are $x = 15,\ \ y = 10,\ \ z = 5$ **Ans.**

**Example 26.** *With the following system of equations*

$$3x + 2y = 4.5$$
$$2x + 3y - z = 5$$
$$-y + 2z = -0.5,$$

*set up the Gauss-Seidel iteration scheme for solution. Iterate two times, using the initial approximation as $x_0 = 0.4,\ y_0 = 1.6,\ z_0 = 0.4$*

**Solution.** The given equations can be written as

$$x = 1.5 - \frac{2}{3}y \qquad \text{...(1)}$$

$$y = \frac{5}{3} - \frac{2}{3}x + \frac{z}{3}$$

$$z = -0.25 + 0.5y \qquad \text{... (2)}$$

$$x_0 = 0.4,\ \ y_0 = 1.6,\ \ z_0 = 0.4 \qquad \text{... (3)}$$

Putting $y = 1.6$ in (1) we get $x = 0.433$

Putting $x = 0.433$ and $z = 0.4$ in (2), we have $y = 1.511$

Putting $y = 1.511$ in (3), we get $z = 0.506$

Again starting from equation (1) and putting $y = 1.511$ in (1), we get $x = 0.493$

Similarly the process is carried on and the roots so obtained are given in the following table :

| *Iterations* | *1* | 2 | 3 | 4 |
|---|---|---|---|---|
| $x = 1.5-\frac{2}{3}y$ | 0.433 | 0.493 | 0.495 | 0.4973 |
| $y=\frac{5}{3}-\frac{2}{3}x+\frac{z}{3}$ | 1.511 | 1.507 | 1.505 | 1.504 |
| $z=-0.25+0.5y$ | 0.506 | 0.504 | 0.503 | 0.502 |

At the end of fourth iteration the roots are

$$x = 0.497,\ y = 1.504,\ z = 0.502$$

But the actual roots are

$$x = 0.5,\ \ y = 1.5,\ \ z = 0.5$$ **Ans.**

**Example 27.** *The following system of equations is given:*

$$2x_1 - x_2 + 2x_3 = 3$$
$$x_1 + 3x_2 + 3x_3 = -1$$
$$x_1 + 2x_2 + 5x_3 = 1$$

*Iterate two times using the Gauss-Seidel method, starting with the initial approximations $x_1 = 0.3$, $x_2 = -0.8$ and $x_3 = 0.3$.*

**Solution.** $x_1 = 0.3, x_2 = -0.8, \quad x_3 = 0.3$

$$2x_1 - x_2 + 2x_3 = 3$$
$$x_1 + 3x_2 + 3x_3 = -1$$
$$x_1 + 2x_2 + 5x_3 = 1$$

The above equations can be written as

$$x_1 = \frac{3}{2} + \frac{x_2}{2} - x_3 \qquad \dots (1)$$

$$x_2 = -\frac{1}{3} - \frac{x_1}{3} - x_3 \qquad \dots (2)$$

$$x_3 = \frac{1}{5} - \frac{x_1}{5} - \frac{2}{5}x_2 \qquad \dots (3)$$

Putting $x_2 = -0.8$ and $x_3 = 0.3$ in eq. (1) we get

$$x_1 = 1.5 - 0.4 - 0.3 = 0.8$$

Putting $x_1 = 0.8$ and $x_3 = 0.3$ in eq. (2) we get

$$x_2 = -\frac{1}{3} - \frac{0.8}{3} - 0.3 = -0.9$$

Putting $x_1 = 0.8$ and $x_2 = -0.9$ in eq. (3), we get

$$x_3 = 0.2 - \frac{0.8}{5} - \frac{2}{5}(-0.9) = 0.4$$

Again starting from eq (1) and putting $x_2 = -0.9$, $x_3 = 0.4$, we get

$$x_1 = 1.5 - \frac{0.9}{2} - 0.4 = 0.65$$

Similarly the process is carried on and the roots so obtained are given in the following table :

| *Iterations* | *1* | *2* | *3* | *4* | *5* | *6* |
|---|---|---|---|---|---|---|
| $x_1 = \frac{3}{2} + \frac{x_2}{2} - x_3$ | 0.8 | 0.65 | 0.575 | 0.538 | 0.518 | 0.509 |
| $x_2 = -\frac{1}{3} - \frac{x_1}{3} - x_3$ | – 0.9 | – 0.95 | – 0.975 | – 0.988 | – 0.994 | – 0.997 |
| $x_3 = \frac{1}{5} - \frac{x_1}{5} - \frac{2}{5}x_2$ | 0.388 | 0.45 | 0.475 | 0.488 | 0.494 | 0.497 |

Exact roots are $x_1 = 0.5$, $x_2 = -1$, $x_3 = 0.5$ **Ans.**

**Example 28.** *Apply Gauss-Seidel method to solve*

$$5x+2y+z=12$$
$$x+4y+2z=15$$
$$x+2y+5z=20$$

*correct upto decimal places, taking* $x_0=y_0=z_0=0$

**Solution.** The given equations can be written as

$$x=\frac{12}{5}-\frac{2}{5}y-\frac{z}{5} \qquad ...(1)$$

$$y=\frac{15}{4}-\frac{x}{4}-\frac{z}{2} \qquad ...(2)$$

$$z=4-\frac{x}{5}-\frac{2y}{5} \qquad ...(3)$$

Putting $y=z=0$ in eq. (1) we have $x=2.4$

Putting $x=2.4$, $z=0$ in eq. (2), we have

$$y=\frac{15}{4}-\frac{2.4}{4}-0=3.15$$

Putting $x=2.4$, $y=3.15$ in eq. (3), we have

$$z=4-\frac{2.4}{5}-\frac{2\times 3.15}{5}=2.26$$

Again starting from eq. (1) and putting $y=3.15$, $z=2.26$, we get

$$x=\frac{12}{5}-\frac{2}{5}\times 3.15-\frac{2.26}{5}=0.688$$

Similarly the process is carried on and the roots so obtained are given in the following table:

| *Iterations* | *1* | 2 | 3 | 4 | 5 |
|---|---|---|---|---|---|
| $x=\frac{12}{5}-\frac{2}{5}y-\frac{z}{5}$ | 2.4 | 0.688 | 0.84416 | 0.962612 | 0.99426864 |
| $y=\frac{15}{4}-\frac{x}{4}-\frac{z}{2}$ | 3.15 | 2.448 | 2.09736 | 2.013237 | 2.00034144 |
| $x=4-\frac{x}{5}-\frac{2y}{5}$ | 2.26 | 2.8832 | 2.99222 | 3.0021828 | 3.001009696 |

**Ans. Exact roots** $x=1, y=2, z=3$

## EXERCISE 15.5

**1.** Solve by Gauss elimination method

$$6x-y-z=19$$
$$3x+4y+z=26$$
$$x+2y+6z=22$$

**Ans.** $x=4$, $y=3$, $z=2$

**2.** Use Gauss-Seidel method to solve the system of equations

$$3x+y+z=1$$
$$x+3y-z=11$$
$$x-2y+4y=21$$

**Ans.** $x=7, y=10, z=12$

**3.** Solve by Crout's method

$$\begin{bmatrix}3 & 1 & 1\\ 1 & 2 & 2\\ 2 & 1 & 3\end{bmatrix}\begin{bmatrix}x_1\\ x_2\\ x_3\end{bmatrix}=\begin{bmatrix}4\\ 3\\ 4\end{bmatrix}$$

**Ans.** $x_1=1, x_2=\frac{1}{2}, x_3=\frac{1}{2}$

4. Use Crout's method to solve

$$2x + 3y - 4z + 2w = -4$$
$$x + 2y + 3z - 4w = 7$$
$$4x - y + 2z - 2w = 7$$
$$3x + 5y - z + 6w = 5.$$

**Ans.** $x = 1, y = \frac{1}{2}, z = 2, w = \frac{1}{4}$

5. Use Jacobi's method to solve

$$10x - 2y - 3z = 205$$
$$2x - 10y + 2z = -154$$
$$2x + y - 10z = -120$$

upto the end of sixth iteration. **Ans.** $x = 32, y = 26, z = 21$

6. Use Jacobi's method to solve

$$5x + 2y + z = 12$$
$$x + 4y + 2z = 15$$
$$x + 2y + 5z = 20$$

upto the end of eighth iteration. **Ans.** $x = 1.08, y = 1.95, z = 3.16$

7. Solve Question (6) by Gauss-Seidel method upto fifth iteration. **Ans.** $x = 1, y = 2, z = 3$

8. Solve by Gauss-Seidel method

(*A*)
$$13x + 5y - 3z + w = 18$$
$$2x + 12y + z - 4w = 13$$
$$3x - 4y + 10z + w = 29$$
$$2x + y - 3z + 9w = 31$$

**Ans.** Exact roots $x = 1, y = 2, z = 3, w = 4.$

(*B*)
$$6x + y + z = 6$$
$$x + 8y + 2z = 4$$
$$3x + 2y + 10z = -1$$

Initial values $x = 0.8,\ y = 0.4, z = -0.45$

**Ans.** $x = 1,\ y = \frac{1}{2},\ z = -\frac{1}{2}$

9. Determine how many iterations of Gauss-Seidel method are needed in order to find solution of the system of equations :

$$9.9x_1 - 1.5x_2 + 2.6x_3 = 0$$
$$0.4x_1 + 13.6x_2 - 4.2x_3 = 8.2$$
$$0.7x_1 + 0.4x_2 + 7.1x_3 = -1.3$$

with an accuracy of $10^{-4}$

## 15.11 SOLUTION OF ORDINARY DIFFERENTIAL EQUATIONS

A number of differential equations cannot be solved by analytical methods. It is, therefore, imperative to solve them by numerical methods. We will discuss the following methods :

(1) Taylor's series method

(2) Picard's method

(3) Runge-Kutta method.

## 15.12 TAYLOR'S SERIES METHOD

Let us consider the first order differential equation

$$\frac{dy}{dx} = f(x, y) \qquad \text{... (1)}$$

under the condition $y = 0$ for $x = x_0$.

**Method.**

On differentiating (1) again and again, we get $\frac{d^2y}{dx^2}, \frac{d^3y}{dx^3}, \frac{d^4y}{dx^4}$ *etc*.

On putting $x = x_0$ and $y = 0$ in the above equations we get the values of

$$\frac{dy}{dx}, \frac{d^2y}{dx^2}, \frac{d^3y}{dx^3}, \frac{d^4y}{dx^4} \ldots\ldots$$

Substituting the values of $y', y'', y''', y''''$ ... in Taylor's series, we get

$$y = y_0 + (x - x_0)[y'(x_0)] + \frac{(x-x_0)^2}{2!}[y''(x_0)] + \frac{(x-x_0)^3}{3!}[y'''(x_0)] + \ldots$$

Thus we can obtain a power series for $y(x)$ in powers of $(x - x_0)$.

The method is illustrated by the example.

**Example 29.** *Using Taylor's series method, obtain the solution of* $\frac{dy}{dx} = 3x + y^2$ *and* $y = 1$, *when* $x = 0$

*Find the value of y for x = 0.1, correct to four places of decimals.*

**Solution.**
$$\frac{dy}{dx} = 3x + y^2 \qquad \ldots(1)$$

$$y(0) = 1 \qquad \ldots(2)$$

Differentiating (1) w.r.t '$x$', we get

$$\frac{d^2y}{dx^2} = 3 + 2y\frac{dy}{dx} \qquad \ldots(3)$$

$$\frac{d^3y}{dx^3} = 2y\frac{d^2y}{dx^2} + 2\left(\frac{dy}{dx}\right)^2 \qquad \ldots(4)$$

$$\frac{d^4y}{dx^4} = 2y\frac{d^3y}{dx^3} + 2\left(\frac{dy}{dx}\right)\left(\frac{d^2y}{dx^2}\right) + 4\left(\frac{dy}{dx}\right)\frac{d^2y}{dx^2} \qquad \ldots(5)$$

and so on

From (1), $\frac{dy}{dx} = 0 + (1)^2 = 1$

From (3), $\frac{d^2y}{dx^2} = 3 + 2(1)(1) = 5$

From (4), $\frac{d^3y}{dx^3} = 2(1)(5) + 2(1)^2 = 12$

From (5), $\frac{d^4y}{dx^4} = 2(1)(12) + 2(1)(5) + 4(1)(5) = 54$

We know by Taylor's series expansion

$$y = y_0 + (x - x_0)(y')_0 + \frac{(x-x_0)^2}{2!}(y'')_0 + \frac{(x-x_0)^3}{3!}(y''')_0 + \frac{(x-x_0)^4}{4!}(y^{iv})_0 + \ldots \qquad \ldots(6)$$

On substituting the values of $y(0)$, $y'(0)$, $y''(0)$, $y'''(0)$, $y^{iv}(0)$ etc. in (6), we get

$$y = 1 + x + \frac{x^2}{2!}(5) + \frac{x^3}{3!}(12) + \frac{x^4}{4!}(54) + \ldots$$

$$\Rightarrow \qquad y(x) = 1 + x + \frac{5}{2}x^2 + 2x^3 + \frac{9}{4}x^4 + \ldots$$

$$y(0.1) = 1 + 0.1 + \frac{5}{2}(0.01) + 2(0.001) + \frac{9}{4}(0.0001) + \ldots$$

$$= 1 + 0.1 + 0.025 + 0.002 + 0.000225 = 1.127225$$ **Ans.**

**Example 30.** *Use Taylor's series method to solve the equation*

$$\frac{dy}{dx} = -xy, \quad y(0) = 1$$

**Solution.** $y' = -xy$ ... (1)

$y'(0) = 0.$

Differentiating (1) repeatedly, we find

$$y'' = -xy' - y, \qquad y''(0) = -1$$

$$y''' = -xy'' - 2y', \qquad y'''(0) = 0$$

$$y^{iv} = -xy''' - 3y'', \qquad y^{iv}(0) = 3$$

$$y^{v} = -xy^{iv} - 4y''', \qquad y^{v}(0) = 0$$

$$y^{vi} = -xy^{v} - 5y^{iv}, \qquad y^{vi}(0) = -15$$

By Taylor's series expansion

$$y(x) = y(0) + \frac{x^2}{1!}y'(0) + \frac{x^2}{2!}y''(0) + \frac{x^3}{3!}y'''(0) + \frac{x^4}{4!}y^{iv}(0) + \ldots$$

$$= 1 + 0 + \frac{x^2}{2!}(-1) + 0 + \frac{x^4}{4!}(3) + 0 + \frac{x^6}{6!}(-15) + \ldots$$

$$= 1 - \frac{x^2}{2} + \frac{x^4}{8} - \frac{x^6}{48} + \ldots$$ **Ans.**

**Example 31.** *Apply Taylor series method of second order to integrate $y' = 2t + 3y$. $y(0) = 1$, $t \in [0, 0.1]$ with $h = 0.1$.*

**Solution.** $\dfrac{dy}{dt} = 2t + 3y, \qquad \dfrac{dy}{dt} = 0 + 3(1) = 3, \qquad y(0) = 1$

$$\frac{d^2y}{dt^2} = 2 + 3\frac{dy}{dt}, \qquad \frac{d^2y}{dt^2} = 2 + 3 \times 3 = 11$$

$$\frac{d^3y}{dt^3} = 3\frac{d^2y}{dt^2}, \qquad \frac{d^3y}{dt^3} = 3 \times 11 = 33$$

$$\frac{d^4y}{dt^4} = 3\frac{d^3y}{dt^3}, \qquad \frac{d^4y}{dt^4} = 3 \times 33 = 99$$

$$\frac{d^5y}{dt^5} = 3\frac{d^4y}{dt^4}, \qquad \frac{d^5y}{dt^5} = 3 \times 99 = 297$$

and so on.

We know by Taylor's series expansion

$$y = y_0 + (t-t_0)\left(\frac{dy}{dt}\right)_{t=0} + \frac{(t-t_0)^2}{2!}\left(\frac{d^2y}{dt^2}\right)_{t=0} + \frac{(t-t_0)^3}{3!}\left(\frac{d^3y}{dt^3}\right)_{t=0}$$
$$+ \frac{(t-t_0)^4}{4!}\left(\frac{d^4y}{dt^4}\right)_{t=0} + \frac{(t-t_0)^5}{5!}\left(\frac{d^5y}{dt^5}\right)_{t=0} + \ldots$$

$$y = 1 + 3t + \frac{11t^2}{2} + \frac{33t^3}{3!} + \frac{99t^4}{4!} + \frac{297t^5}{5!} + \ldots$$

$$y = 1 + 3(0.1) + \frac{11}{2}(0.1)^2 + \frac{33}{6}(0.1)^3 + \frac{99}{24}(0.1)^4 + \frac{297}{120}(0.1)^5$$

$$y(0.1) = 1 + 3(0.1) + \frac{11}{2}(0.1)^2 + \frac{11}{2}(0.1)^3 + \frac{33}{8}(0.1)^4 + \frac{99}{40}(0.1)^5 + \ldots$$

$$y(0.1) = 1 + 0.3 + 0.055 + 0.0055 + 0.0004125 + 0.00002475$$

$$y(0.1) = 1.36093725$$ **Ans.**

**Example 32.** *Find the solutions u(0.1) and u (0.2), of the initial value problem*

$$u' = x(1-2u^2)\ ;\ u(0) = 1$$

*using the first three non zero terms of the Taylor Series method and* $h = 0.1$.

**Solution.** $\frac{du}{dx} = x(1-2u^2)$

$$\Rightarrow \quad \frac{d^2u}{dx^2} = (1-2u^2) + x\left(-4u\frac{du}{dx}\right)$$

$$\Rightarrow \quad \frac{d^3u}{dx^3} = -4u\frac{du}{dx} - 4u\frac{du}{dx} + x\left[-4\left(\frac{du}{dx}\right)^2 - 4u\frac{d^2u}{dx^2}\right] = -8u\frac{du}{dx} + x\left[-4\left(\frac{du}{dx}\right)^2 - 4u\frac{d^2u}{dx^2}\right]$$

$$\Rightarrow \quad \frac{d^4u}{dx^4} = -8\left(\frac{du}{dx}\right)^2 - 8u\frac{d^2u}{dx^2} + \left[-4\left(\frac{du}{dx}\right)^2 - 4u\frac{d^2u}{dx^2}\right] + x\left[-8\left(\frac{du}{dx}\right)\frac{d^2u}{dx^2} - 4\frac{du}{dx}\frac{d^2u}{dx^2} - 4u\frac{d^3u}{dx^3}\right]$$

Putting $x = 0$, $u = 1$ in $\frac{du}{dx}, \frac{d^2u}{dx^2}, \frac{d^3u}{dx^3}, \frac{d^4u}{dx^4}$, we get

$$\frac{du}{dx} = 0, \quad \frac{d^2u}{dx^2} = [1-2(1)^2] + 0 = -1$$

$$\frac{d^3u}{dx^3} = 0 + 0 = 0,$$

$$\frac{d^4u}{dx^4} = 0 - 8(+1)(-1) + [0 - 4 \times 1(-1)] + 0 = 12$$

**By Taylor's series**

$$u(x) = u(x_0) + \frac{(x-x_0)}{1!}u'(x_0) + \frac{(x-x_0)^2}{2!}u''(x_0) + \frac{(x-x_0)^3}{3!}u'''(x_0) + \frac{(x-x_0)^4}{4!}u^{iv}(x_0) + \ldots$$

$$u(0.1) = u(0) + (0.1-0)u'(0) + \frac{(0.1-0)^2}{2!}u''(0) + \frac{(0.1-0)^3}{3!}u'''(0) + \frac{(0.1-0)^4}{4!}u^{iv}(0) + \ldots$$

$$= 1 + 0 + \frac{0.01}{2}(-1) + 0 + \frac{0.0001}{24}(12) = 1 - 0.005 + 0.00005 = 0.99505$$

Putting $x = 0.1$ and $u = 0.99505$ in $\frac{du}{dx}$ and $\frac{d^2u}{dx^2}$, we get

$$\frac{du}{dx} = x(1-2u^2)$$

$$\frac{du}{dx} = 0.1\ [1 - 2\ (0.99505)^2] = 0.1\ (-0.98025) = -0.098025$$

$$\frac{d^2u}{dx^2} = (1-2u^2) + x\left(-4u\frac{du}{dx}\right)$$

$$\frac{d^2u}{dx^2} = [1 - 2\ (0.99505)^2] + 0.1\ (-4 \times 0.99505 \times -0.098024)$$

$$= -0.98025 + 0.03902 = -0.94123$$

Putting $x = 0.2$, $x_0 = 0.1$, $u'\ (0.1)$ and $u''\ (0.1)$ in Taylor's series (1), we get

$$u\ (0.2) = u\ (0.1) + (0.2 - 0.1)\ u'\ (0.1) + \frac{(0.2-0.1)^2}{2!}u''\ (0.1) + \ldots$$

$$= 0.99505 + 0.1\ (-0.098025) + \frac{0.01}{2}\ (-0.94123)$$

$$= 0.99505 - 0.0098025 - 0.005 \times 0.94123$$

$$= 0.98054135$$

**Exercise 15.6**

Using Taylor's method, solve the following differential equations :

1. $\frac{dy}{dx} = x + y^2$, given $y(0) = 0$. **Ans.** $y = \frac{1}{2}x^2 + \frac{1}{20}x^5 + \ldots$

2. $\frac{d^2y}{dx^2} + xy = 0$, subject to $x = 0$, $y = c$ and $\frac{dy}{dx} = 0$. **Ans.** $y = c\left(1 - \frac{x^3}{3!} + \frac{1\times4}{6!}x^6 + \frac{1\times4\times7}{9!}x^9 + \ldots\right)$

3. $\frac{dy}{dx} = x^2y - 1$, given $y\ (0) = 1$, and find $y\ (0.03)$. **Ans.** $y = 1 - x + \frac{x^2}{3} - \frac{x^4}{4} + \ldots, 0.97001$

4. $\frac{dy}{dx} - y^2 - x = 0$, for $y\ (0) = 0$, find $y$ when $x = 0.2$ **Ans.** $y = 0.020016$

## 15.13 PICARD'S METHOD OF SUCCESSIVE APPROXIMATIONS

Let us consider the first order differential equation

$$\frac{dy}{dx} = f(x, y) \qquad \ldots(1)$$

and $y = y_0$ for $x = x_0$

**Method.** Integrating (1) between the limits $x_0$ and $x$, we get

$$\int_{y_0}^{y} dy = \int_{x_0}^{x} f(x,y)dx \quad \text{or} \quad y - y_0 = \int_{x_0}^{x} f(x,y)\,dx$$

$$y = y_0 + \int_{x_0}^{x} f(x,y)dx \qquad \ldots(2)$$

Equation (2) is the solution of (1). But (2) contains the unknown $y$ under the integral sign on right hand side.

On putting $y_0$ for $y$ on R.H.S. of (2), we get a first approximation $y_1$.

$$y_1 = y_0 + \int_{x_0}^{x} f(x, y_0)dx \qquad \ldots(3)$$

From(3) we get the value of $y_1$ and we put $y_1$ for $y$ on R.H.S.of (2) to get second approximation $y_2$.

Thus $$y_2 = y_0 + \int_{x_0}^{x} f(x, y_1)dx$$

Similarly third approximation is $y_3 = y_0 + \int_{x_0}^{x} f(x, y_2)\, dx$ and so on.

In this way we get a better approximation each time than the preceding one.

**Note.** This method is used to solve the differential equation if the succession integration can be performed easily.

The method is now illustrated by an example.

**Example 33.** *Using Picard's method, find a solution of* $\frac{dy}{dx} = 1 + xy$ *upto the third approximation, when* $x_0 = 0, y_0 = 0$

**Solution.** $$\frac{dy}{dx} = 1 + xy \qquad ...(1)$$

Integrating (1) w.r.t. '$x$' between the limits 0 and $x$, we get

$$\int_0^y dy = \int_0^x (1 + xy)\, dx \Rightarrow \quad y = \int_0^x (1 + xy)dx \qquad ...(2)$$

On putting $y$ (0) = 0 for $y$ on R.H.S. of (2), we have

$$y_1 = \int_0^x (1 + 0)\, dx \quad \Rightarrow \quad y_1 = x$$

On substituting $y_1 = x$ for $y$ on R.H.S. of (2), we obtain

$$y_2 = \int_0^x (1 + x^2)\, dx = x + \frac{x^3}{3}$$

$$y_3 = \int_0^x \left[1 + x\left(x + \frac{x^3}{3}\right)\right] dx = \int_0^x \left[1 + x^2 + \frac{x^4}{3}\right] dx$$

$$y_3 = x + \frac{x^3}{3} + \frac{x^5}{15}.$$ **Ans.**

**Example 34.** *Given the differential equation*

$$\frac{dy}{dx} = \frac{x^2}{1 + y^2}$$

*with the initial condition* $y = 0$ *when* $x = 0$, *use Picard's method to obtain* $y$ *for* $x = 0.25, 0.5$ *and 1.0 correct to three places of decimals.*

**Solution.** $$\frac{dy}{dx} = \frac{x^2}{1 + y^2}$$

$$y(x) = y_0 + \int_0^x \frac{x^2}{1 + y^2} dx = 0 + \int_0^x x^2 dx = \frac{x^3}{3}$$

Now using this value of $y$, we have

$$y = 0 + \int_0^x \frac{x^2}{1+\left(\frac{x^3}{3}\right)^2} dx = \tan^{-1}\frac{x^3}{3}$$

If $x = 0.25$, then $y = \tan^{-1}\frac{(0.25)^3}{3} = \tan^{-1}\frac{0.015625}{3} = \tan^{-1} 0.005208 = 0.0052$

If $x = 0.5$, then $y = \tan^{-1}\frac{(0.5)^3}{3} = \tan^{-1}\frac{0.125}{3} = \tan^{-1} 0.042 = 0.0420$

If $x = 1.0$, then $y = \tan^{-1}\frac{1}{3} = \tan^{-1} 0.3333 = 0.3218$ **Ans.**

**Example 35.** *Perform two iterations of Picard's method to find an approximate solution of the initial value problem*

$$y' = x + y^2 ;\ y(0) = 1.$$

**Solution.** $\frac{dy}{dx} = x + y^2$ ...(1) $y(0) = 1$

Integrating (1) w.r.t. '$x$' between 0 and $x$, we get

$$\int_0^y dy = \int_0^x (x + y^2) dx$$

$$y = y_0 + \int_0^x (x + y^2) dx \qquad ...(2)$$

On putting $y(0) = 1$ for $y$ on R.H.S. of (2), we have

$$y_1 = 1 + \int_0^x (x+1) dx = 1 + \frac{x^2}{2} + x$$

On substituting the value of $y_1$ for $y$ on R.H.S of (2), we get

$$y_2 = 1 + \int_0^x \left[x + \left(1 + x + \frac{x^2}{2}\right)^2\right] dx$$

$$y_2 = 1 + \int_0^x \left(x + 1 + x^2 + \frac{x^4}{4} + 2x + x^2 + x^3\right) dx$$

$$y_2 = 1 + \int_0^x \left(1 + 3x + 2x^2 + x^3 + \frac{x^4}{4}\right) dx$$

$$y_2 = 1 + x + \frac{3x^2}{2} + \frac{2x^3}{3} + \frac{x^4}{4} + \frac{x^5}{20}$$ **Ans.**

**Example 36.** *Using three successive approximations of Picard's method. Obtain approximate solution of the differential equation $y' = x^2 + y^2$ satisfying the initial condition $y(0) = 0$*

**Solution.** $\frac{dy}{dx} = x^2 + y^2$ ...(1) $y(0) = 0$

Integrating (1) w.r.t, '$x$' between the limit 0 and $x$, we get

$$\int_0^y dy = \int_0^x (x^2 + y^2) dx \qquad ... (2)$$

On putting $y(0) = 0$ for $y$ on R.H.S. of (2), we have

$$y_1 = \int_0^x (x^2 + 0) dx = \frac{x^3}{3}$$

On putting $\frac{x^3}{3}$ for $y$ on R.H.S. of (2), we obtain

$$y_2 = \int_0^x \left(x^2 + \frac{x^6}{9}\right) dx = \frac{x^3}{3} + \frac{x^7}{63}$$

$$y_3 = \int_0^x \left[x^2 + \left(\frac{x^3}{3} + \frac{x^7}{63}\right)^2\right] dx = \int_0^x \left(x^2 + \frac{x^6}{9} + \frac{2x^{10}}{189} + \frac{x^{14}}{3969}\right) dx$$

$$= \frac{x^3}{3} + \frac{x^7}{63} + \frac{2x^{11}}{2079} + \frac{x^{15}}{59535} \qquad \textbf{Ans.}$$

**Example 37.** *Use Picard's method to solve* $\frac{dy}{dx} = -xy$, $y(0) = 1$

**Solution.** $y(x) = y_0 - \int_0^x xy\, dx$

$$= 1 - \int_0^x x(1) dx = 1 - \int_0^x x dx = 1 - \frac{x^2}{2}$$

Now using this value of $y$, we have

$$y = 1 - \int_0^x x\left(1 - \frac{x^2}{2}\right) dx = 1 - \int_0^x \left(x - \frac{x^3}{2}\right) dx = 1 - \frac{x^2}{2} + \frac{x^4}{8}$$

$$= 1 - \int_0^x x\left(1 - \frac{x^2}{2} + \frac{x^4}{8}\right) dx = 1 - \int_0^x \left(x - \frac{x^3}{2} + \frac{x^5}{8}\right) dx = 1 - \frac{x^2}{2} + \frac{x^4}{8} - \frac{x^6}{48}$$

Repeating once again we shall obtain

$$y = 1 - \frac{x^2}{2} + \frac{x^4}{8} - \frac{x^6}{48} + \frac{x^8}{384} \qquad \textbf{Ans.}$$

**Example 38.** *Use Picard's method to solve the equations*

$$\frac{dx}{dt} = -y, \quad \frac{dy}{dt} = x$$

*given that* $x = 1$, $y = 0$ *when* $t = 0$.

**Solution.** $$\frac{dx}{dt} = -y \qquad \text{...(1)}$$

$$\frac{dy}{dt} = x \qquad \text{...(2)}$$

Integrating (1) w.r.t '$t$' from $t = 0$ to $t$, we get

$$[x]_1^x = -\int_0^t y\, dt \Rightarrow x - 1 = -\int_0^t y\, dt$$

$$x = 1 - \int_0^t y\, dt \qquad \text{...(3)}$$

Integrating (2), w.r.t., '$t$' from $t = 0$ to $t$, we get

$$[y]_0^y = \int_0^x x\, dt \Rightarrow y - 0 = \int_0^t x\, dt$$

$$y = \int_0^t x\, dt \qquad \text{...(4)}$$

Replacing $y$ by 0 in (3) and $x$ by 1 in (4), we have

$$x = 1 - \int_0^t 0\, dt = 1 \quad \text{and} \quad y = \int_0^t 1 .\, dt = t$$

$$x = 1 - \int_0^t t\,dt = 1 - \frac{t^2}{2}, \quad y = \int_0^t \left(1 - \frac{t^2}{2}\right) dt = t - \frac{t^3}{6}$$

$$x = 1 - \int_0^t \left(t - \frac{t^3}{6}\right) dt = 1 - \frac{t^2}{2} + \frac{t^4}{24}$$

$$y = \int_0^t \left(1 - \frac{t^2}{2} + \frac{t^4}{24}\right) dt = t - \frac{t^3}{6} + \frac{t^5}{120}, \qquad \textbf{Ans.}$$

and so on

**Exercise 15.7**

Using Picard's method, solve the following

1. $\frac{dy}{dx} = x + y^2$, given $y(0) = 0$. Upto third approximation.

**Ans.** $y = \frac{1}{2}x^2 + \frac{1}{20}x^5 + \frac{1}{160}x^8 + \frac{1}{4400}x^{11}$

2. Apply Picard's iteration method to find approximate solutions to the initial value problem

$$y' = 1 + y^2, \; y(0) = 0$$

3. $\frac{dy}{dx} = x - y$, given $y(0) = 1$ and find $y(0.2)$ to five places of decimals.

**Ans.** $y = 1 - x + x^2 - \frac{x^3}{3} + \frac{x^4}{12} - \frac{x^5}{60} + \frac{x^6}{720}$, 0.83746

4. $\frac{dy}{dx} = y + x$, given $y(0) = 1$, find $y(1)$, **Ans.** $y = 1 + x + x^2 + \frac{x^3}{3} + \frac{x^4}{12} + \frac{x^5}{60} + \frac{x^6}{120}$, 3.434.

5. $\frac{dy}{dx} = x^2 + y^2$, for $y(0) = 0$, find $y(0.4)$. **Ans.** 0.0214.

6. $\frac{dy}{dx} = 2y + z$, $\frac{dz}{dx} = y + 2z$ given $y(0) = 0$, $z(0) = 1$

**Ans.** $y = x + 2x^2 + \frac{13}{6}x^3 + \frac{5}{3}x^4 + \ldots, z = 1 + 2x + \frac{5}{2}x^2 + \frac{7}{3}x^3 + \frac{41}{40}x^4 + \ldots$

## 15.14 EULER'S METHOD

This is purely numerical method for solving the first order differential equations. This is an elementary method and which will demonstrate the procedure underlying these methods. This method should not be used for practical solution.

Consider the differential equation

$$\frac{dy}{dx} = f(x, y) \qquad \ldots(1)$$

Let $y = \phi(x)$ be the solution of (1). ...(2)

Let $(x_0, y_0), (x_1, y_1) \ldots (x_n, y_n), (x_{n+1}, y_{n+1})$ be the points on the curve of (2).

$x_0, x_1, \ldots x_n, x_{n+1} \ldots$ are equispaced at equal interval $h$.

$$y_{n+1} = \phi(x_{n+1}) \qquad [(x_{n+1}, y_{n+1}) \text{ lies on (2).}]$$

$$= \phi(x_n + h) \qquad (x_{n+1} = x_n + h)$$

$$= \phi(x_n) + h\,\phi'(x_n) + \frac{1}{2}h^2\phi''(x_n) + \ldots \qquad \ldots(3)$$

$$= \phi(x_n) + h\phi'(x_n) \qquad (h \text{ is very small})$$

$$= \phi(x_n) + hf(x_n, y_n) \qquad \left[\text{since } \frac{dy}{dx} = f(x,y)\right]$$

$$y_{n+1} = y_n + hf(x_n, y_n) \qquad [\text{since } y_n = \phi(x_n) \text{ from (2)}] \quad ...(4)$$

This formula (4) can be used to find $y_{n+1}$, where $y_n$ is known.

On substituting the value of $y_0$, $(n = 0)$ in (4) we get $y_1$,

Similarly putting the value of $y_1$, $(n = 1)$ in (4), we obtain $y_2$ and so on.

**Note.** Since we have neglected $1/2\ h^2 \phi''(x_n)$ and higher powers of $h$ from formula (4) there will be a larger error in $y_{n+1}$. Therefore it is not used in practical problems.

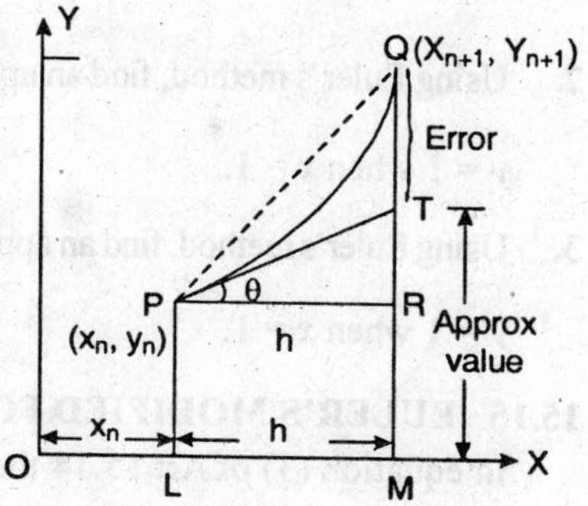


**Geometrically**

Let $y = \phi(x)$ be a solution curve $PQ$. The ordinate of $P$ *i.e.* $y_n$ is known.

Now we have to find the ordinate $y_{n+1}$ of any point $Q$.

$$y_{n+1} = MQ = MR + RQ = PL + RT + TQ \qquad (TQ = \text{Error})$$

$$= y_n + h \tan\theta = y_n + h\left(\frac{dy}{dx}\right) = y_n + hf(x_n, y_n)$$

**Example 39.** *Using Euler's method, find an approximate value of y corresponding to $x = 2$, given that $\frac{dy}{dx} = x + 2y$ and $y = 1$ when $x = 1$.*

**Solution.** $f(x, y) = x + 2y$

$$y_{n+1} = y_n + hf(x_n, y_n) = y_n + 0.1\,(x + 2y)$$

**Method:** In column 3 we record the value of $x + 2y$ and in column 4 we enter the sum of the value of $y$ and the product of 0.1 with the value of $x + 2y$. This value entered in 4th column is transfered to second column for the next calculation.

| $x$ | $y$ | $x + 2y = \frac{dy}{dx}$ | old $y + 0.1\left(\frac{dy}{dx}\right)$ = new $y$ |
|---|---|---|---|
| 1.0 | 1.00 | 3.00 | 1.0 + 0.1 (3) = 1.30 |
| 1.1 | 1.3 | 3.70 | 1.3 + 0.1 (3.7)= 1.67 |
| 1.2 | 1.67 | 4.54 | 1.67 + 0.1 (4.54) = 2.12 |
| 1.3 | 2.12 | 5.54 | 2.12 + 0.1 (5.54) = 2.67 |
| 1.4 | 2.67 | 6.74 | 2.67 + 0.1 (6.74) = 3.34 |
| 1.5 | 3.34 | 8.18 | 3.34 + 0.1 (8.18) = 4.16 |
| 1.6 | 4.16 | 9.92 | 4.16 + 0.1 (9.92) =5.15 |
| 1.7 | 5.15 | 12.00 | 5.15 + 0.1 (12.0) =6.35 |
| 1.8 | 6.35 | 14.50 | 6.35 + 0.1 (14.50) =7.80 |
| 1.9 | 7.80 | 17.50 | 7.80 + 0.1 (17.50) = 9.55 |
| 2.0 | 9.55 | | |

**Thus the required approximate value of $y = 9.55$** **Ans.**

## Exercise 15.8

1. Using Euler's method, find an approximate value of corresponding to $x = 1$, given that $\frac{dy}{dx} = x + y$ and $y = 1$ when $x = 0$. **Ans.** 3.18

2. Using Euler's method, find an approximate value of $y$ corresponding to $x = 1.4$, given $\frac{dy}{dx} = xy^{\frac{1}{2}}$ and $y = 1$ when $x = 1$. **Ans.** 1.49857.

3. Using Euler's method, find an approximate value of $y$ corresponding to $x = 1.6$, given $\frac{dy}{dx} = y^2 - \frac{y}{x}$ and $y = 1$ when $x = 1$. **Ans.** 1.1351

## 15.15 EULER'S MODIFIED FORMULA

In equation (3) of Art 15.14 the expansion of $y_{n+1}$ is

$$y_{n+1} = y_n + hf(x_n, y_n) + \frac{1}{2}h^2\phi''(x_n, y_n) + \frac{1}{3!}h^3\phi'''(x_n, y_n) + \ldots \quad \ldots(1)$$

In Euler's formula we omit $\frac{1}{2}h^2\phi''(x_n, y_n)$ and higher powers of $h$.

The error due to this omission is called **Truncation error.**

Now the formula is derived with small error.

Differentiating (1) w.r.t. $x$ we get

$$\left(\frac{dy}{dx}\right)_{n+1} = \left(\frac{dy}{dx}\right)_n + hf'(x_n, y_n) + \frac{1}{2}h^2\phi'''(x_n, y_n) + \ldots$$

$$\therefore \quad f(x_{n+1}, y_{n+1}) = f(x_n, y_n) + hf'(x_n, y_n) + \frac{1}{2}h^2\phi'''(x_n, y_n) + \ldots \quad \ldots (2)$$

$$= f(x_n, y_n) + h\phi''(x_n, y_n) + \frac{1}{2}h^2\phi'''(x_n, y_n) + \ldots$$

Multiplying (2) by $\frac{h}{2}$ and subtracting from (1) we get :

$$y_{n+1} - \frac{1}{2}hf(x_{n+1}, y_{n+1}) = y_n + \frac{h}{2}f(x_n, y_n) - \frac{h^3}{12}\phi'''(x_n, y_n)$$

Neglecting terms containing $h^3$ and higher powers of $h$, we obtain

$$y_{n+1} = y_n + h\left[\frac{f(x_n, y_n) + f(x_{n+1}, y_{n+1})}{2}\right] \quad \ldots (3)$$

Equation (3) is the Euler's modified formula.

But $f(x_{n+1}, y_{n+1})$ which occurs on the right hand side of equation (3), cannot be calculated since $y_{n+1}$ is unknown. So first we calculate $y_{n+1}$ from Euler's first formula.

$$y_{n+1} = y_n + hf(x_n, y_n)$$

Thus for each stage we use the following two formulae.

$$y_{n+1} = y_n + hf(x_n, y_n)$$

$$y_{n+1} = y_n + \frac{h}{2}[f(x_n, y_n) + f(x_{n+1}, y_{n+1})]$$

**Example 40.** *Apply Euler's modified method to solve* $\frac{dy}{dx} = x + 3y$ *subject to* $y(0) = 1$ *and hence find an approximate value of* $y$ *when* $x = 1$.

**Solution.** $f(x,y) = x + 3y$

$$y_{n+1} = y_n + hf(x_n, y_n)$$

$$y_{n+1} = y_n + \frac{h}{2}[f(x_n, y_n) + f(x_{n+1}, y_{n+1})]$$

This gives $y_{n+1} = y_n + 0.1(x_n + 3y_n)$

$$y_{n+1} = y_n + 0.05\,[(x_n + 3y_n) + (x_{n+1} + 3y_{n+1})].$$

The following table shows the computation work.

| $n$ | $x_n$ | $y_n$ | $x_n + 3y_n$ | *Euler's formula* $y_{n+1}$ | $x_{n+1}$ | $x_{n+1} + 3y_{n+1}$ | *Euler's* modified $y_{n+1}$ |
|---|---|---|---|---|---|---|---|
| 0 | 0.0 | 1 | 3 | 1.3 | 0.1 | 4 | 1.35 |
| 1 | 0.1 | 1.35 | 4.15 | 1.765 | 0.2 | 5.495 | 1.832 |
| 2 | 0.2 | 1.832 | 5.695 | 2.402 | 0.3 | 7.506 | 2.492 |
| 3 | 0.3 | 2.492 | 7.776 | 3.270 | 0.4 | 10.21 | 3.391 |
| 4 | 0.4 | 3.391 | 10.573 | 4.448 | 0.5 | 13.844 | 4.612 |
| 5 | 0.5 | 4.612 | 14.336 | 6.046 | 0.6 | 18.738 | 6.266 |
| 6 | 0.6 | 6.266 | 19.398 | 8.206 | 0.7 | 25.318 | 8.502 |
| 7 | 0.7 | 8.502 | 26.206 | 11.123 | 0.8 | 34.169 | 11.521 |
| 8 | 0.8 | 11.521 | 35.363 | 15.057 | 0.9 | 46.071 | 15.593 |
| 9 | 0.9 | 15.593 | 47.679 | 20.361 | 1.0 | 62.083 | 21.081 |
| 10 | 1.0 | 21.081 | | | | | |

Hence the required value of $y$ at $x = 1$ is 21.081.

The exact solution gives $y = 21.873$ for $x = 1$. The error is 0.792 *i.e.*, 3.6%. **Ans.**

**Procedure.** We calculated $y_{n+1}$ by Euler's formula *i.e.*, $y_{n+1} = y_n + 0.1\,(x_n + 3\,y_n)$ and entered in 5th column. In 7th column we record the sum *i.e.* $x_{n+1} + 3y_{n+1}$. Then we computed the value of $y_{n+1}$ by Euler's modified formula *i.e.*,

$$y_{n+1} = y_n + \frac{0.1}{2}\,[(x_n + 3y_n) + (x_{n+1} + 3y_{n+1})].$$ and entered in 8th column.

## EXERCISE 15.9

1. Using Euler's modified formula, find an approximate value of $y$ when $x = 0.3$, given that $\frac{dy}{dx} = x + y$ and $y = 1$ when $x = 0$. **Ans.** 1.3997

2. Using Euler's modified formula, find an approximate value of $y$ when $x = 0.06$, given that $\frac{dy}{dx} = x^2 + y$ and $y\,(0) = 1$, taking the interval 0.02. **Ans.** 1.0619

3. Using Euler's modified formula, solve $\frac{dy}{dx} = 1 - 2\,xy$ given $y = 0$ at $x = 0$ from $x = 0$ to 0.6 taking the interval $h = 0.2$. **Ans.** 0.4748

## 15.16 RUNGE'S FORMULA

Euler's modified formula is $y_{n+1} = y_n + \frac{h}{2}[f(x_n, y_n) + f(x_{n+1}, y_{n+1})]$

$$\Rightarrow \quad y_{n+1} = y_n + \frac{h}{2}[f(x_n, y_n) + f(x_n + h, y_n + hf_n)] \qquad \dots (1)$$

Let $k_1 = hf(x_n, y_n)$

and $k_2 = hf[x_n + h, y_n + hf(x_n, y_n)] \Rightarrow k_2 = hf(x_n + h, y_n + k_1)$

Putting the values of $k_1$ and $k_2$ in (1), we get

$$y_{n+1} = y_n + \frac{1}{2}(k_1 + k_2) \qquad ...(2)$$

This is known as Runge's formula of order 2.

**Example 41.** *Apply Runge 's formula of order 2 approximate value of y when x = 1.1, given* $\frac{dy}{dx} = 3x + y^2$ *and y = 1.2 when x = 1.*

**Solution.** Here we have $x_0 = 1, y_0 = 1.2, h = 0.1$

$$f(x,y) = 3x + y^2, f(x_0, y_0) = 3(1) + (1.2)^2 = 4.44$$

$$k_1 = hf(x_0, y_0) = 0.1 \times 4.44 = 0.444$$

$$k_2 = hf(x_0 + h, y_0 + k_1) = 0.1f(1.1, 1.2 + 0.444) = 0.1f(1.1, 1.644)$$

$$= 0.1[3 \times 1.1 + (1.644)^2] = 0.600$$

$$y_{n+1} = y_n + \frac{1}{2}(k_1 + k_2)$$

$$y_1 = 1.2 + \frac{1}{2}(0.444 + 0.600) = 1.722$$ **Ans.**

## EXERCISE 15.10

1. Apply Runge's formula of second order to find approximate value of $y$ when $x = 1.1$, given that $\frac{dy}{dx} = x - y$ and $y = 1$ when $x = 1$. **Ans.** 1.005

2. Apply Runge's formula of second order to find approximate value of $y$ when $x = 0.02$, given that $\frac{dy}{dx} = x^2 + y$ and $y(0) = 1$. **Ans.** 1.0202.

## 15.17 RUNGE'S FORMULA (THIRD ORDER)

$$y_1 = y_0 + \frac{1}{6}(k_1 + 4k_4 + k_3)$$

where $k_1 = hf(x_0, y_0),\ k_2 = hf\left(x_0 + \frac{h}{2}, y_0 + \frac{k_1}{2}\right)$

$$k_3 = hf(x_0 + h, y_0 + 2k_2 - k_1),\ y_1 = y_0 + \frac{1}{6}(k_1 + 4k_2 + k_3)$$

This is the Runge's Formula (third order) with an error of the order $h^4$.

**Example 42.** *Using Runge 's Formula (third order), solve the differential equation* $\frac{dy}{dx} = x - y$ *subject to y = 1 when x = 1.*

**Solution.** $f(x, y) = x - y$

Here $h = 0.1,\ x_0 = 1, y_0 = 1$

$$k_1 = hf(x_0, y_0) = 0.1(x - y) = 0.1(1 - 1) = 0$$

$$k_2 = hf\left(x_0 + \frac{h}{2}, y_0 + \frac{k_1}{2}\right) = 0.1f(1.05, 1 + 0) = 0.1(1.05 - 1) = 0.005$$

$$k_3 = hf(x_0 + h, y_0 + 2k_2 - k_1) = 0.1f(1.1, 1.01) = 0.1(1.1 - 1.01) = 0.009$$

$$y_1 = y + \frac{1}{6}(k_1 + 4k_4 + k_3)$$

$$y(0.1) = 1 + \frac{1}{6}(0 + 0.02 + 0.009) = 1 + 0.004833 = 1.004833$$ **Ans.**

## 15.18 RUNGE-KUTTA FORMULA (FOURTH ORDER)

A fourth order Runge-Kutta Formula for solving the differential equation is

$$y = y_0 + \frac{1}{6}(k_1 + 2k_2 + 2k_3 + k_4)$$

where $k_1 = hf(x_0, y_0),\quad k_2 = hf\left(x_0 + \frac{h}{2}, y_0 + \frac{k_1}{2}\right)$

$$k_3 = hf\left(x_0 + \frac{h}{2}, y_0 + \frac{k_2}{2}\right),\quad k_4 = hf(x_0 + h, y_0 + k_3)$$

$$y = y_0 + \frac{1}{6}[k_1 + 2k_2 + 2k_3 + k_4]$$

This is known as Runge-Kutta Formula. The error in this formula is of the order $h^5$. This method have greater accuracy. No deviatives are required to be tabulated.
It requires only functional values at some selected points on the sub interval.

**Example 43.** *Apply Runge-Kutta method to find an approximate value of y when x = 0.2, given that*

$$\frac{dy}{dx} = x + y, \qquad y = 1 \text{ when } x = 0$$

**Solution.** Let $h = 0.1$

Here $x_0 = 0,\ y_0 = 1,\qquad f(x, y) = x + y$

Now $k_1 = hf(x_0, y_0) = 0.1\,(0 + 1) = 0.1$

$$k_2 = hf\left(x_0 + \frac{h}{2}, y_0 + \frac{k_1}{2}\right) = 0.1\,f(0 + 0.05, 1 + 0.05) = 0.1[0.05 + 1.05] = 0.11$$

$$k_3 = hf\left(x_0 + \frac{h}{2}, y_0 + \frac{k_2}{2}\right) = 01\,f(0 + 0.05, 1 + 0.055) = 0.1(0.05 + 1.055] = 0.1105$$

$$k_4 = hf(x_0 + h, y_0 + k_3) = 0.1f(0 + 0.1,\ 1 + 0.1105)$$

$$= 0.1f(0.1, 1.1105) = 0.1(0.1 + 1.1105) = 0.12105$$

According to Runge-Kutta (Fourth order) formula

$$y = y_0 + \frac{1}{6}[k_1 + 2k_2 + 2k_3 + k_4]$$

$$y_{0.1} = 1 + \frac{1}{6}(0.1 + 0.22 + 0.221 + 0.12105) = 1 + \frac{1}{6}(0.66205) = 1.11034$$

**For the second step**

$$x_0 = 0.1,\quad y_0 = 1.11034,\qquad h = 0.1$$

$$k_1 = hf(x_0, y_0) = 0.1(0.1 + 1.11034) = 0.121034$$

$$k_2 = hf\left(x_0 + \frac{h}{2}, y_0 + \frac{k_1}{2}\right) = 0.1\,f(0.1 + 0.05, 1.11034 + 0.060517)$$

$$= 0.1\,(0.15 + 1.170857) = 0.1320857$$

$$k_3 = hf\left(x_0 + \frac{h}{2}, y_0 + \frac{k_2}{2}\right) = 0.1 f(0.1 + 0.05, 1.11034 + 0.0660428)$$

$$= 0.1\ (0.15 + 1.1763828) = 0.13263828$$

$$k_4 = hf(x_0 + h, y_0 + k_3) = 0.1\ (0.1 + 0.1, 1.11034 + 0.13263828)$$

$$= 0.1\ (0.2 + 1.24297828) = 0.144297828$$

$$y_1 = y_0 + \frac{1}{6}[k_1 + 2k_2 + 2k_3 + k_4]$$

$$= 1.11034 + \frac{1}{6}\ [0.121034 + 2 \times 0.1320857 + 2 \times 0.13263828 + 0.144297828]$$

$$= 1.11034 + \frac{1}{6}\ [0.121034 + 0.2641714 + 0.26527656 + 0.144297828]$$

$$= 1.11034 + \frac{1}{6} \times 0.794779788 = 1.11034 + 0.132463298$$

$$= 1.242803298$$ **Ans.**

**Example 44.** *Apply Runge-Kutta method (fourth order), to find an approximate value of y when x = 0.2, given that* $\frac{dy}{dx} = x + y^2$ *and y = 1 when x = 0.*

**Solution.** Let $h = 0.1$,

Here $x_0 = 0, \quad y_0 = 1, \quad f(x, y) = x + y^2$

Now $k_1 = hf(x_0, y_0) = 0.1\ (0 + 1) = 0.1$

$$k_2 = hf\left(x_0 + \frac{h}{2}, y_0 + \frac{k_1}{2}\right) = 0.1 f(0 + 0.05, 1 + 0.05) = 0.1[0.05 + (1.05)^2] = 0.11525$$

$$k_3 = hf\left(x_0 + \frac{h}{2}, y_0 + \frac{k_2}{2}\right) = 0.1 f(0 + 0.05, 1 + 0.057625)$$

$$= 0.1[0.05 + (1.057625)^2] = 0.11685$$

$$k_4 = hf(x_0 + h, y_0 + k_3) = 0.1 f(0 + 0.1, 1 + 0.11685) = 0.1\ [0.1 + (0.11685)^2] = 0.13474$$

According to Runge-Kutta (fourth order) formula

$$y_1 = y_0 + \frac{1}{6}(k_1 + 2k_2 + 2k_3 + k_4)$$

$$y_{0.1} = 1 + \frac{1}{6}[0.1 + 2(0.11525) + 2(0.11685) + 0.13474]$$

$$y_{0.1} = 1 + 0.1165 = 1.1165$$

For the second step $x_0 = 0.1, y_0 = 1.1165$

$$k_1 = 0.1\ (0.1 + 1.2466) = 0.1347$$

$$k_2 = 0.1\ (0.15 + 1.4014) = 0.1551$$

$$k_3 = 0.1\ (0.15 + 1.4259) = 0.1576$$

$$k_4 = 0.1\ (0.2 + 1.6233) = 0.1823$$

$$y_{0.2} = y_{0.1} + \frac{1}{6}(k_1 + 2k_2 + 2k_3 + k_4)$$

$$= 1.1165 + \frac{1}{6}[0.1347 + 2(0.1551) + 2(0.1576) + 0.1823]$$

$$= 1.1165 + 0.1571 = 1.2736$$ **Ans.**

**Example 45.** *Use the fourth order Runge-Kutta method to find u (0, 2), of the initial value problem u' = –2 t u², u (0) = 1, using h = 0.2.*

**Solution.** $h = 0.2$

Here $t = 0,\ u = 1,\ f(t, u) = -2\ tu^2$

$$k_1 = hf(t_0, u_0) = 0.2\ (-2\ tu^2) = 0.2\ (0) = 0$$

$$k_2 = hf\left(t_0 + \frac{h}{2},\ u_0 + \frac{k_1}{2}\right)$$

$$= 0.2f(0.1,\ 1 + 0) = 0.2\ f(0.1,\ 1) = 0.2\ (-2 \times 0.1 \times 1^2) = -0.04$$

$$k_3 = hf\left(t_0 + \frac{h}{2},\ u_0 + \frac{k_2}{2}\right) = 0.2f(0 + 0.1,\ 1 - 0.02) = 0.2f(0.1,\ 0.98)$$

$$= 0.2f\,[-2 \times 0.1 \times (0.98)^2] = -0.2[0.2 \times 0.9604] = -\,0.038416$$

$$k_4 = hf(t_0 + h,\ u_0 + k_3) \quad = 0.2\ f(0.2,\ 1 - 0.038416) = 0.2(-2) \times (0.2) \times (0.961584)^2$$

$$= -\,0.08 \times 0.924644 = -\,0.073972$$

$$u = u_0 + \frac{1}{6}[k_1 + 2k_2 + 2k_3 + k_4] = 1 + \frac{1}{6}[0 + 2(-0.04) + 2(-0.038416) + (-0.073972)]$$

$$= 1 - \frac{1}{6}[0.08 + 0.076832 + 0.073972]$$

$$= 1 - \frac{1}{6}(0.230804) = 1 - 0.038467 = 0.961533$$

**Ans.**

**Example 46.** *Find the solution y (0.1) of the initial value problem y' = –2 t y², y (0) = 1, with h = 0.1, using*

*(i) Taylor series method of order four, and*

*(ii) Runge-Kutta method of order four*

**Solution.** $\dfrac{dy}{dt} = -2ty^2,\ y(0) = 1,\ h = 0.1$, If $t = 0$ then $\dfrac{dy}{dt} = -2(0)(1)^2 = 0$

$$\frac{d^2y}{dt^2} = -2y^2 - 4ty\frac{dy}{dt},$$

If $t = 0$,
$$\frac{d^2y}{dt^2} = -2(1)^2 - 0 = -2$$

$$\frac{d^3y}{dt^3} = -4y\frac{dy}{dt} - 4y\frac{dy}{dt} - 4t\left(\frac{dy}{dt}\right)^2 - 4ty\frac{d^2y}{dt^2}$$

$$\frac{d^3y}{dt^3} = -8y\frac{dy}{dt} - 4t\left(\frac{dy}{dt}\right)^2 - 4ty\frac{d^2y}{dt^2}$$

If $t = 0$,
$$\frac{d^3y}{dt^3} = -8(1)(0) - 4(0)(0)^2 - 4(0)(1)(-2) = 0$$

$$\frac{d^4y}{dt^4} = -8\left(\frac{dy}{dt}\right)^2 - 8y\frac{d^2y}{dt} - 4\left(\frac{dy}{dt}\right)^2 - 4t\left(\frac{dy}{dt}\right)^3 - 4y\frac{d^2y}{dt^2} - 4t\frac{dy}{dt}\frac{d^2y}{dt^2} - 4ty\frac{d^3y}{dt^3}$$

If $t = 0$,
$$\frac{d^4y}{dt^4} = -8(0)^2 - 8(1) - (-2) - 4(0)^2 - 4(0)(0)^3$$

$$-4\,(1)\,(-2) - 4(0)\,(0)\,(-2) - 4\,(0)\,(1)\,(0) = 24$$

(*i*) By Taylor series method

$$y = y_0 + \frac{(t-0)}{1!}\frac{dy}{dt} + \frac{(t-0)^2}{2!}\frac{d^2y}{dt^2} + \frac{(t-0)^3}{3!}\frac{d^3y}{dt^3} + \frac{(t-0)^4}{4!}\frac{d^4y}{dt^4} + \ldots$$

$$y = 1 + t(0) + \frac{t^2}{2}(-2) + \frac{t^3}{6}(0) + \frac{t^4}{24}(24) = 1 - t^2 + t^4$$

$$y(0.1) = 1 - (0.1)^2 + (0.1)^4 = 1 - 0.01 + 0.0001 = 0.9901$$ **Ans.**

(*ii*) By Runge-Kutta method of order four

$$y = y_0 + \frac{1}{6}(k_1 + 2k_2 + 2k_3 + k_4)$$

where $k_1 = hf(t_0, y_0) = h(-2\,ty^2) = 0.1[-2 \times 0 \times (1)^2] = 0$

$$k_2 = hf\left(t_0 + \frac{h}{2},\ y_0 + \frac{k_1}{2}\right) = 0.1 f\left(0 + 0.05,\ 1 + \frac{0}{2}\right)$$

$$= 0.1 f(0.05,\ 1) = 0.1[-2\,(0.05)\,(1)^2] = -0.01$$

$$k_3 = hf\left(t_0 + \frac{h}{2},\ y_0 + \frac{k_2}{2}\right) = 0.1 f\left(0 + 0.05,\ 1 + \frac{-0.01}{2}\right)$$

$$= 0.1\, f\,(0.05,\ 0.995) = 0.1\,[-2\,(0.05)\,(0.995)^2] = -0.00990025$$

$$k_4 = hf(t_0 + h,\ y_0 + k_3) = 0.1\, f(0 + 0.1,\ 1 - 0.00990025)$$

$$= 0.1 f(0.1,\ 0.99009975) = 0.1\,[-2\,(0.1)\,(0.99009975)^2] = -0.01960595$$

$$y = y_0 + \frac{1}{6}(k_1 + 2k_2 + 2k_3 + k_4)$$

$$= 1 + \frac{1}{6}[0 + 2(-0.01) + 2(-0.00990025) - 0.01960595]$$

$$= 1 + \frac{1}{6}[-0.02 - 0.01980050 - 0.01960595]$$

$$= 1 + \frac{1}{6}[-0.05940645] = 1 - 0.009901075 = 0.99009892$$ **Ans.**

**Example 47.** *Use the Runge-Kutta fourth order method to find* $y(0.2)$ with $h = 0.1$ *for the initial value problem.*

$$dy/dx = \sqrt{x+y},\ y(0) = 1$$

**Solution.** $h = 0.1,\ x_0 = 0,\ y_0 = 1,\ f(x, y) = \sqrt{x+y}$

Now $k_1 = hf(x_0, y_0) = 0.1\ \sqrt{0+1} = 0.1$

$$k_2 = hf\left(x_0 + \frac{h}{2},\ y_0 + \frac{k_1}{2}\right) = 0.1 f(0 + 0.05,\ 1 + 0.05) = 0.1 f(0.05,\ 1.05)$$

$$= 0.1\sqrt{0.05 + 1.05} = 0.1\sqrt{1.1} = 0.10488$$

$$k_3 = hf\left(x_0 + \frac{h}{2},\ y_0 + \frac{k_2}{2}\right) = 0.1 f(0 + 0.05,\ 1 + 0.05244)$$

$$= 0.1 f(0.05,\ 1.05244) = 0.1\sqrt{0.05 + 1.05244} = 0.1\sqrt{1.10244} = 0.104997$$

$$k_4 = hf(x_0 + h,\ y_0 + k_3) = 0.1 f(0 + 0.1,\ 1 + 0.104997)$$

$$= 0.1 f(0.1,\ 1.104997) = 0.1\sqrt{0.1 + 1.104997} = 0.1\sqrt{1.204997} = 0.10977$$

According to Runge-Kutta (fourth order) formula

$$y_1 = y_0 + \frac{1}{6}(k_1 + 2k_2 + 2k_3 + k_4)$$

$$y_{0.01} = 1 + \frac{1}{6}(0.1 + 0.20976 + 0.209994 + 0.10977)$$

$$= 1 + \frac{1}{6}(0.629524) = 1 + 0.10492 = 1.10492$$

For the second step $x_0 = 0.1, y_0 = 1.10492,\ h = 0.1$

$$k_1 = hf(x_0, y_0) = 0.1\sqrt{0.1 + 1.10492} = 0.1\sqrt{1.20492} = 0.109769$$

$$k_2 = hf\left(x_0 + \frac{h}{2},\ y_0 + \frac{k_1}{2}\right) = 0.1f(0.1 + 0.05,\ 1.10492 + 0.0548845)$$

$$= 0.1f(0.15,\ 1.1598045) = 0.1\sqrt{0.15 + 1.1598045} = 0.1\sqrt{1.3098045} = 0.114447$$

$$k_3 = hf\left(x_0 + \frac{h}{2},\ y_0 + \frac{k_2}{2}\right) = 0.1f(0.1 + 0.05,\ 1.10492 + 0.0572235)$$

$$= 0.1f(0.15,\ 1.1621435) = 0.1\sqrt{0.15 + 1.1621435} = 0.1\sqrt{1.3121435} = 0.1145488$$

$$k_4 = hf(x_0 + h,\ y_0 + k_3) = 0.1f(0.1 + 0.1,\ 1.10492 + 0.1145488)$$

$$= 0.1f(0.2,\ 1.2194688) = 0.1\sqrt{0.2 + 1.2194688} = 0.1\sqrt{1.4194688} = 0.1191415$$

$$y_{0.2} = y_0 + \frac{1}{6}(k_1 + 2k_2 + 2k_3 + k_4)$$

$$= 1.10492 + \frac{1}{6}(0.109769 + 0.228894 + 0.2290976 + 0.1191415)$$

$$= 1.10492 + \frac{1}{6} \times 0.6869021 = 1.10492 + 0.1144837$$

$$= 1.2194037$$

**Ans.**

## EXERCISE 15.11

1. The initial value problem $y' = x(y+x) - 2, y(1) = 2$ is given. Find the value of $y(1.2)$ with $h = 0.2$ using the Runge-Kutta method of fourth order. **Ans.** $y(1.2) = 2.3138$
2. Use the Runge-Kutta method of fourth order to find $y(0.8)$ with $h = 0.2$ for the initial value problem.

$$\frac{dy}{dx} = \sqrt{x+y},\ y(0.4) = 0.41$$

**Ans.** 0.8489912

3. Find $y(0.2)$ for the equation

$$\frac{dy}{dx} = -xy,\quad y(0) = 1, \text{ using Runge-Kutta method.}$$

4. Apply the Runge-Kutta method to obtain $y(1.1)$ from the differential equation

$$\frac{dy}{dx} = xy^{1/3},\quad y(1) = 1, \text{ taking } h = 0.1.$$

5. Apply Runge-Kutta (fourth order) formula to find an approximate value of $y$ when $x = 1.1$, given that

$$\frac{dy}{dx} = x - y \text{ and } y = 1 \text{ at } x = 1.$$

**Ans.** 1.004837

## 15.19 HIGHER ORDER DIFFERENTIAL EQUATIONS

Let $\frac{dy}{dx} = f(x, y, z),\ \frac{dz}{dx} = g(x, y, z),\ y(x_0) = y_0,\ z(x_0) = z_0$

Formulae for the application of Runge-Kutta method are as follows :

$$k_1 = hf(x_n, y_n, z_n),\ m_1 = hg(x_n, y_n, z_n)$$

$$k_2 = \left(x_n + \frac{h}{2}, y_n + \frac{k_1}{2}, z_n + \frac{m_1}{2}\right)$$

$$m_2 = hg\left(x_n + \frac{h}{2}, y_n + \frac{k_1}{2}, z_n + \frac{m_1}{2}\right)$$

$$k_3 = hf\left(x_n + \frac{h}{2}, y_n + \frac{k_2}{2}, z_n + \frac{m_2}{2}\right)$$

$$m_3 = hg\left(x_n + \frac{h}{2}, y_n + \frac{k_2}{2}, z_n + \frac{m_2}{2}\right)$$

$$k_4 = hf(x_n + h, y_n + k_3, z_n + m_3)$$

$$m_4 = hg(x_n + h, y_n + k_3, z_n + m_3)$$

$$x_{n+1} = x_n + h$$

$$y_{n+1} = y_n + \frac{1}{6}(k_1 + 2k_2 + 2k_3 + k_4)$$

$$z_{n+1} = z_n + \frac{1}{6}(m_1 + 2m_2 + 2m_3 + m_4)$$

**Higher order differential equations are best treated by transforming the given equation into a system of first order simultaneous equations which can be solved by one of the aforesaid methods.**

**Consider, for example the second order differential equation :**

$$\frac{d^2y}{dx^2} = f\left(x, y, \frac{dy}{dx}\right);\ y(x_0) = y_0, \left(\frac{dy}{dx}\right)_{x=x_0} = y_0'$$

**Substituting** $\frac{dy}{dx} = z$, **we get**

$$\frac{dz}{dx} = f(x, y, z)$$

$$y(x_0) = y_0,\ z(x_0) = y_0'$$

**These constitute the equivalent system of simultaneous equations.**

**Example 48.** *Use Runge-Kutta method to find y (0.2) for the equation*

$$\frac{d^2y}{dx^2} = x\frac{dy}{dx} - y$$

*given that y = 1,* $\frac{dy}{dx} = 0$ *when x = 0.*

**Solution.** Substituting $\frac{dy}{dx} = z = f(x, y, z)$

**The given equation reduces to** $\frac{d^2y}{dx^2} = xz - y = g(x, y, z)$

**The initial conditions are given by** $x = 0, y = 1, z = 0$.

Also $\quad h = 0.2$

$$k_1 = hf(x, y, z) = hz = 0.2 \times 0 = 0$$

$$m_1 = hg(x, y, z) = h(xz - y) = 0.2(0 \times 0 - 1) = -0.2$$

$$k_2 = hf\left(x+\frac{h}{2},\ y+\frac{k_1}{2},\ z+\frac{m_1}{2}\right) = h\left(z+\frac{m_1}{2}\right) = 0.2\left(0-\frac{0.2}{2}\right) = -0.02$$

$$m_2 = hg\left(x+\frac{h}{2},\ y+\frac{k_1}{2},\ z+\frac{m_1}{2}\right) = h\left[\left(x+\frac{h}{2}\right)\left(z+\frac{m_1}{2}\right)-\left(y+\frac{k_1}{2}\right)\right]$$

$$= 0.2\left[\left(0+\frac{0.2}{2}\right)\left(0-\frac{0.02}{2}\right)-\left(1+\frac{0}{2}\right)\right] = 0.2[-0.01-1] = -0.202$$

$$k_3 = hf\left(x+\frac{h}{2},\ y+\frac{k_2}{2},\ z+\frac{m_2}{2}\right) = h\left(z+\frac{m_2}{2}\right) = 0.2\left(0-\frac{0.202}{2}\right) = -0.0202$$

$$m_3 = hg\left[x+\frac{h}{2},\ y+\frac{k_2}{2},\ z+\frac{m_2}{2}\right] = h\left[\left(x+\frac{h}{2}\right)\left(z+\frac{m_2}{2}\right)-\left(y+\frac{k_2}{2}\right)\right]$$

$$= 0.2\left[\left(0+\frac{0.2}{2}\right)\left(0-\frac{0.202}{2}\right)-\left(1-\frac{0.02}{2}\right)\right] = 0.2[-0.0101-0.99] = -0.20002$$

$$k_4 = hf(x+h,\ y+k_3,\ z+m_3) = h(z+m_3) = 0.2\,(0-0.20002) = -0.040004$$

$$m_4 = hg(x+h,\ y+k_3,\ z+m_3) = h[(x+h)(z+m_3)-(y+k_3)]$$

$$= 0.2\,[(0.2)(-0.20002)-(1-0.0202)] = 0.2\,[-0.040004-0.9798] = -0.2039608$$

This gives, at $x = 0.2$

$$y(0.2) = y(0) + \frac{1}{6}[k_1+2k_2+2k_3+k_4] = 1 + \frac{1}{6}[0+2(-0.02)+2(-0.0202)+(-0.040004)]$$

$$= 1 + \frac{1}{6}(-0.04-0.0404-0.040004) = 1 + \frac{1}{6}(-0.120404) = 0.97993266$$

$$z(0.2) = z(0) + \frac{1}{6}(m_1+2m_2+2m_3+m_4)$$

$$= 0 + \frac{1}{6}[-0.2+2(-0.202)+2(-0.20002)-0.2039608]$$

$$= \frac{1}{6}[-0.2-0.404-0.40004-0.2039608]$$

$$= \frac{1}{6}(-1.2080008) = -0.201333466$$

**Ans.**

## EXERCISE 15.12

1. Find $y(0.4)$ for the equation $\frac{dy}{dx} = x^2 + y^2$, $y(0) = 0$ by Picard's method. **Ans.** 0.0214.

2. Use Picard's method to solve $\frac{dy}{dx} = 2y+z$, $\frac{dz}{dx} = y+2z$; given that $y(0) = 0$, $z(0) = 1$.

   **Ans.** $y = x+2x^2+\frac{13}{6}x^3+\frac{5}{3}x^4+\ldots,\ z = 1+2x+\frac{5}{2}x^2+\frac{7}{3}x^3+\frac{41}{24}x^4+\ldots$

3. Employ Runge-Kutta method to find $y$ for $x = 0.2$ from $\frac{d^2y}{dx^2} = x\left(\frac{dy}{dx}\right)^2 - y^2$

   given that $y = 1, \frac{dy}{dx} = 0$ for $x = 0$. **Ans.** $y(0.2) = 0.9801$; $y'(0.2) = -0.1970$

4. Describe Runge-Kutta method (4th order) for obtaining solution of initial value problem :

$$y'' = f(x, y, y'),\ y(x_0) = y_0,\ y'(x_0) = y_0'$$

5. State clearly the conditions under which the method is applicable.

# 16

# Numerical Methods for Solution of Partial Differential Equations

## 16.1 GENERAL LINEAR PARTIAL DIFFERENTIAL EQUATIONS

General partial differential equation is of the form

$$A(x,y)\frac{\partial^2 u}{\partial x^2}+B(x,y)\frac{\partial^2 u}{\partial x \partial y}+C(x,y)\frac{\partial^2 u}{\partial y^2}+D(x,y)\frac{\partial u}{\partial x}+E(x,y)\frac{\partial u}{\partial y}+F(x,y)u+G(x,y)=0$$

This equation is called

(*i*) *Elliptic*, if $B^2 - 4AC < 0$

*e.g.* $$\frac{\partial^2 u}{\partial x^2}+\frac{\partial^2 u}{\partial y^2}=0 \qquad \text{Laplace Equation}$$

$$\frac{\partial^2 u}{\partial x^2}+\frac{\partial^2 u}{\partial y^2}=f(x,y) \qquad \text{Poisson's Equation}$$

(*ii*) *Parabolic*, if $B^2 - 4AC = 0$ *e.g.* $\frac{\partial u}{\partial t}=C^2\frac{\partial^2 u}{\partial x^2}$

One dimensional heat conduction equation.

(*iii*) *Hyperbolic*, if $B^2 - 4AC > 0$ *e.g.* $\frac{\partial^2 u}{\partial t^2}=C^2\frac{\partial^2 u}{\partial x^2}$

**Example 1.** *Determine the type of* $x^2\frac{\partial^2 u}{\partial y^2}+2xy\frac{\partial^2 u}{\partial x \partial y}+y^2\frac{\partial^2 u}{\partial x^2}=0$ (*A.M.I.E.T.E., Dec. 2006*)

**Solution.** Here, $A = x^2$, $B=2xy$, $C= y^2$.

$B^2 - 4AC = 4x^2y^2 - 4x^2y^2 = 0$. Hence, it is a parabolic equation. **Ans.**

## 16.2 FINITE-DIFFERENCE APPROXIMATION TO DERIVATIVES

By Taylor formula

$$u(x+h,y)=u(x,y)+\frac{h\partial u}{\partial x}+\frac{1}{2!}h^2\frac{\partial^2 u}{\partial x^2}+\frac{1}{3!}h^3\frac{\partial^3 u}{\partial x^3}+\ldots \qquad \ldots(1)$$

$$u(x-h,y)=u(x,y)-h\frac{\partial u}{\partial x}+\frac{1}{2!}h^2\frac{\partial^2 u}{\partial x^2}-\frac{1}{3!}h^3\frac{\partial^3 u}{\partial x^3}+\ldots \qquad \ldots(2)$$

From (1), neglecting $h^2$ and higher powers of $h$, we get

$$\frac{\partial u}{\partial x}\approx\frac{u(x+h,y)-u(x,y)}{h} \qquad \text{(Forward difference formula)} \quad \ldots(3)$$

Form (2), neglecting $h^2$ and higher powers of $h$, we have

$$\frac{\partial u}{\partial x}\approx\frac{u(x,y)-u(x-h,y)}{h} \qquad \text{(Backward difference formula)} \quad \ldots(4)$$

Subtracting (2) from (1) and neglecting $h^3$ and higher powers of $h$, we get

$$u(x+h,y)-u(x-h,y)\approx 2h\frac{\partial u}{\partial x}$$

$$\Rightarrow \quad \frac{\partial u}{\partial x} \approx \frac{1}{2h}[u(x+h, y) - u(x-h, y)] \quad \text{(Central difference formula)} \qquad ...(5)$$

Similarly,

$$\frac{\partial u}{\partial y} = \frac{u(x, y+k) - u(x, y)}{k} = \frac{u(x, y) - u(x, y-k)}{k} = \frac{u(x, y+k) - u(x, y-k)}{2k} \qquad ...(6)$$

Adding (1) and (2), neglecting $h^5$ and higher powers of $h$, we get

$$u(x+h, y) + u(x-h, y) = 2u(x, y) + h^2 \frac{\partial^2 u}{\partial x^2}$$

$$\frac{\partial^2 u}{\partial x^2} \approx \frac{1}{h^2}[u(x+h, y) - 2u(x, y) + u(x-h, y)] \qquad ...(7)$$

Similarly
$$\frac{\partial^2 u}{\partial y^2} \approx \frac{1}{k^2}[u(x, y+k) - 2u(x, y) + u(x, y-k)] \qquad ...(8)$$

and
$$\frac{\partial^2 u}{\partial x \partial y} \approx \frac{1}{4hk}[u(x+h, y+k) - u(x-h, y+k) - u(x+h, y-k) + u(x-h, y-k)] \qquad ...(9)$$

The given region (rectangle *ABCD*) is divided into smaller rectangles of sides $\delta x = h$ and $\delta y = k$. The origin is taken at the centre of the rectangle and the coordinates axes are drawn. The rectangle is divided into 36 small rectangles. Here there are 49 mesh-points or lattices or nodal points or grid points. The values of the function $u$ are $u_{ij}$, $u_{i+1, j}$, $u_{i+2, j}$... $u_{i, j+1}$, $u_{i, j+2}$... at the mesh-points.

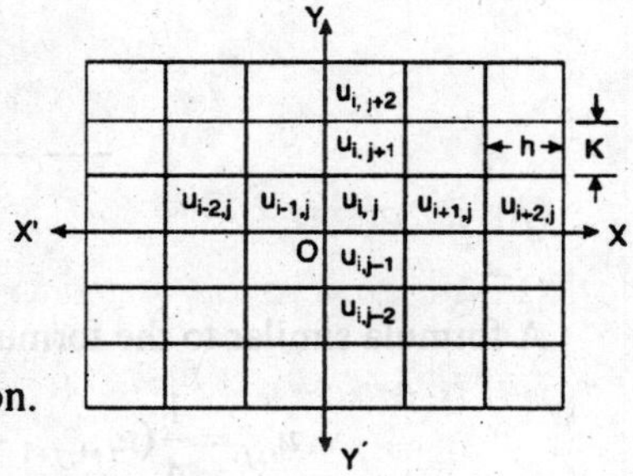


Let these values satisfy the given partial differential equation.
At the centre of the rectangle:
Equations (5), (6), (7), (8) and (9) are rewritten on the nodal points as below:

$$\frac{\partial u}{\partial x} = \frac{1}{2h}(u_{i+1,j} - u_{i-1,j}), \qquad \frac{\partial u}{\partial y} = \frac{1}{2k}(u_{i,j+1} - u_{i,j-1})$$

$$\frac{\partial^2 u}{\partial x^2} = \frac{1}{h^2}(u_{i+1,j} - 2u_{i,j} + u_{i-1,j})$$

$$\frac{\partial^2 u}{\partial y^2} = \frac{1}{k^2}(u_{i,j+1} - 2u_{i,j} + u_{i,j-1})$$

$$\frac{\partial^2 u}{\partial x \partial y} = \frac{1}{4hk}(u_{i+1,j+1} - u_{i+1,j-1} - u_{i-1,j+1} + u_{i-1,j-1})$$

## 16.3 SOLUTION OF PARTIAL DIFFERENTIAL EQUATION (LAPLACE EQUATION)

$$\frac{\partial^2 u}{\partial x^2} + \frac{\partial^2 u}{\partial y^2} = 0 \qquad ...(1)$$

On substituing the values of $\frac{\partial^2 u}{\partial x^2}$ and $\frac{\partial^2 u}{\partial y^2}$, we get

$$\frac{1}{h^2}[u(x+h, y) - 2u(x, y) + u(x-h, y)] + \frac{1}{k^2}[u(x, y+k) - 2u(x, y) + u(x, y-k)] = 0$$

For values of $h = k$ *i.e.* for square grid of the mesh size $h$, the above equation can be written as

$$u(x+h, y) - 2u(x, y) + u(x-h, y) + u(x, y+h) - 2u(x, y) + u(x, y-h) = 0$$

$$u(x, y) = \frac{1}{4}[u(x+h, y) + u(x-h, y) + u(x, y+h) + u(x, y-h)]$$

Denoting any mesh point $(x, y) = (ih, jh)$ as simply $i, j$, the above difference equation can be written as

$$u_i = \frac{1}{4}(u_{i+1,j} + u_{i-1,j} + u_{i,j+1} + u_{i,j-1}) \quad \text{...(2)}$$

Equation (2) shows that the value of $u(x, y)$ is the average of its four neighbours to the East, West, North and South. The formula (2) is called the Standard five points formula and is written as

$$u_{i+1,j} + u_{i-1,j} + u_{i,j+1} + u_{i,j-1} - 4u_{i,j} = 0$$

This formula is also known as Liebman's averaging procedure.

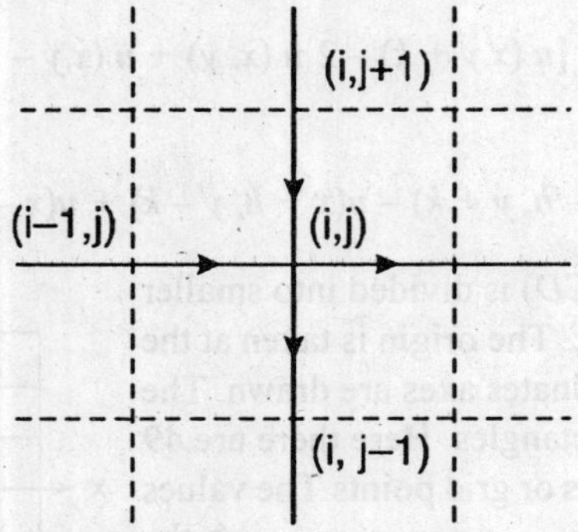


A formula similar to the formula (2) is sometimes used with convenience. It is given as

$$u_{i,j} = \frac{1}{4}(u_{i+1,j+1} + u_{i+1,j-1} + u_{i-1,j+1} + u_{i-1,j-1}) \quad \text{...(3)}$$

This is known as *diagonal five-point formula* as these points lie on the diagonals. Although formula (3) is less accurate than formula (2), still it is a good approximation for obtaining as starting values in the iteration procedure.

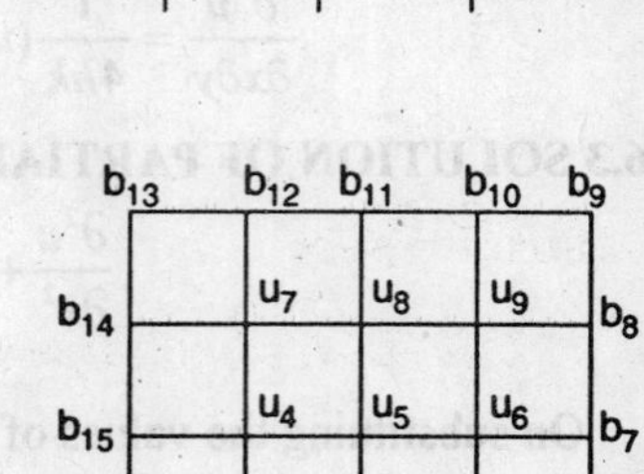


Whenever possible, Standard five-point formula is preferred in all commutations.

**Procedure**. We use the following diagonal five point formula to get the initial value of $u$ at the centre.

$$u_5 = \frac{1}{4}[b_1 + b_5 + b_9 + b_{13}]$$

Then the approximate values of diagonal interior points $u_1$, $u_3$, $u_7$, $u_9$ are calculated by the diagonal five-point formula

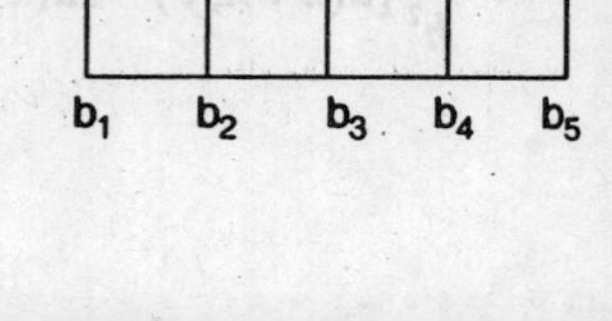


$$u_1 = \frac{1}{4}[b_1 + b_3 + u_5 + b_{15}], \qquad u_3 = \frac{1}{4}[b_3 + b_5 + b_7 + u_5]$$

$$u_7 = \frac{1}{4}[b_{15} + u_5 + b_{11} + b_{13}] \qquad u_9 = \frac{1}{4}[u_5 + b_7 + b_9 + b_{11}]$$

The values of the remaining interior points *i.e.* $u_2$, $u_4$, $u_6$ and $u_8$ are obtained by the standard five point formula.

$$u_2 = \frac{1}{4}[b_3 + u_3 + u_5 + u_1], \qquad u_4 = \frac{1}{4}[u_1 + u_5 + u_7 + b_{15}]$$

$$u_6 = \frac{1}{4}[u_3 + b_7 + u_9 + u_5] \qquad u_8 = \frac{1}{4}[u_5 + u_9 + b_{11} + u_7]$$

Having obtained all values $u_1, u_2, \ldots u_9$ once, their accuracy can be improved by the repeated application of either Jacobi's iteration formula or Gauss-Seidel iteration formula.

## 16.4 JACOBI'S ITERATION FORMULA

Let $u^n_{i,j}$ be the $n$th iterative value of $u_{i,j}$. Then Jacobi's iterative procedure is given below.

$$u^{(n+1)}_{i,j} = \frac{1}{4}[u^{(n)}_{i-1,j} + u^{(n)}_{i+1,j} + u^{(n)}_{i,j-1} + u^{(n)}_{i,j+1}]$$

## 16.5 GAUSS-SEIDEL METHOD

This method utilises the latest iterative value available and scans the mesh points symmetrically from left to right along successive rows. The formula is given below.

$$u^{(n+1)}_{i,j} = \frac{1}{4}[u^{(n+1)}_{i-1,j} + u^{(n)}_{i+1,j} + u^{(n+1)}_{i,j-1} + u^{(n)}_{i,j+1}]$$

## 16.6 SUCCESSIVE OVER-RELAXATION OR S.O.R. METHOD

Gauss-Seidel formula can be written as

$$u^{(n+1)}_{i,j} = u^{(n)}_{i,j} + \frac{1}{4}[u^{(n+1)}_{i-1,j} + u^{(n)}_{i+1,j} + u^{(n+1)}_{i,j-1} + u^{(n)}_{i,j+1} - 4u^{(n)}_{i,j}] = u^{(n)}_{i,j} + \frac{1}{4}R_{i,j}.$$

It gives the change $\frac{1}{4}R_{i,j}$ in the value of $u_{i,j}$ for one Gauss-Seidel iteration. In the *S.O.R.* method, larger change than this is given to $u^{(n)}_{i,j}$ and the iteration formula is given below:

$$u^{(n+1)}_{i,j} = u^{(n)}_{i,j} + \frac{1}{4}wR_{i,j} = \frac{1}{4}w[u^{(n+1)}_{i-1,j} + u^{(n)}_{i+1,j} + u^{(n+1)}_{i,j-1} + u^{(n)}_{i,j+1}] + (1-w)u^{(n)}_{i,j}$$

Here $w$ is called the accelerating factor and lies between 1 and 2.

**Example 1.** *Solve* $\frac{\partial^2 u}{\partial x^2} + \frac{\partial^2 u}{\partial y^2} = 0$ *in the domain of the figure given below by Gauss-Seidel method.*

**Solution.** Initially $u_1 = u_2 = u_3 = u_4 = 0$

$$u_1^{(n+1)} = \frac{1}{4}\left(1+1+u_2^{(n)}+u_1^{(n)}\right)$$

$$u_2^{(n+1)} = \frac{1}{4}\left(2+2+u_1^{(n+1)}+u_3^{(n)}\right)$$

$$u_3^{(n+1)} = \frac{1}{4}\left(2+2+u_1^{(n+1)}+u_3^{(n+1)}\right)$$

$$u_4^{(n+1)} = \frac{1}{4}[2+2+u_1^{(n+1)}+u_3^{(n+1)}]$$

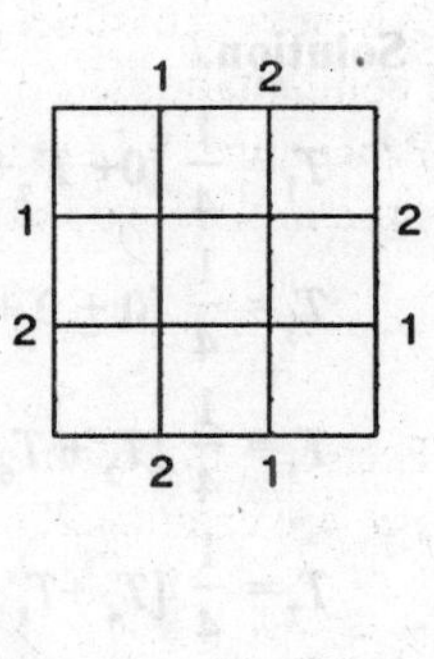


**First iteration**

$$u_1^{(1)} = \frac{1}{4}(1+0+1+0) = 0.5 \qquad u_2^{(1)} = \frac{1}{4}(2+0+2+0.5) = 1.125$$

$$u_3^{(1)} = \frac{1}{4}(1+1.125+1+0) = 0.781 \qquad u_4^{(1)} = \frac{1}{4}(2+0.781+2+0.5) = 1.320$$

**Second iteration**

$$u_1^{(2)} = \frac{1}{4}[1+1+1.125+1.320] = 1.111 \qquad u_2^{(2)} = \frac{1}{4}[2+2+1.111+0.781] = 1.473$$

$$u_3^{(2)} = \frac{1}{4}[1+1+1.473+1.320] = 1.198 \qquad u_4^{(2)} = \frac{1}{4}[2+2+1.111+1.198] = 1.577$$

**Third iteration**

$$u_1^{(3)} = \frac{1}{4}[1+1+1.473+1.577] = 1.263 \qquad u_2^{(3)} = \frac{1}{4}[2+2+1.263+1.198] = 1.615$$

$$u_3^{(3)} = \frac{1}{4}[1+1+1.615+1.577] = 1.298 \qquad u_4^{(3)} = \frac{1}{4}[2+2+1.263+1.298] = 1.640$$

**Fourth iteration**

$$u_1^{(4)} = \frac{1}{4}[1+1+1.615+1.640] = 1.314 \qquad u_2^{(4)} = \frac{1}{4}[2+2+1.314+1.298] = 1.653$$

$$u_3^{(4)} = \frac{1}{4}[1+1+1.653+1.640] = 1.323 \qquad u_4^{(4)} = \frac{1}{4}[2+2+1.314+1.323] = 1.659$$

**Fifth iteration**

$$u_1^{(5)} = \frac{1}{4}[1+1+1.653+1.659] = 1.328 \qquad u_2^{(5)} = \frac{1}{4}[2+2+1.328+1.323] = 1.663$$

$$u_3^{(5)} = \frac{1}{4}[1+1+1.663+1.659] = 1.331 \qquad u_4^{(5)} = \frac{1}{4}[2+2+1.328+1.331] = 1.665$$

**Sixth iteration**

$$u_1^{(6)} = \frac{1}{4}[1+1+1.663+1.665] = 1.333 \qquad u_2^{(6)} = \frac{1}{4}[2+2+1.332+1.331] = 1.666$$

$$u_3^{(6)} = \frac{1}{4}[1+1+1.666+1.665] = 1.333 \qquad u_4^{(6)} = \frac{1}{4}[2+2+1.332+1.333] = 1.666$$

$$u_1 = 1.333,\ u_2 = 1.667, \qquad u_3 = 1.333,\ u_4 = 1.667$$ **Ans.**

**Example 2.** *Solve* $\frac{\partial^2 T}{\partial x^2} + \frac{\partial^2 T}{\partial y^2} = 0$ *in the domain of the figure given below by Gauss-Seidel method.*

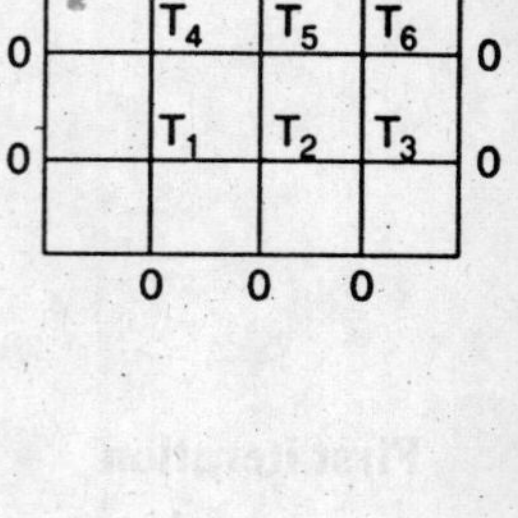


**Solution.**

$$T_1 = \frac{1}{4}[0 + T_2 + T_4 + 0] \qquad T_2 = \frac{1}{4}[0 + T_3 + T_5 + T_1]$$

$$T_3 = \frac{1}{4}[0 + 0 + T_2 + T_6] \qquad T_4 = \frac{1}{4}[T_1 + T_5 + T_7 + 0]$$

$$T_5 = \frac{1}{4}[T_2 + T_6 + T_8 + T_4] \qquad T_6 = \frac{1}{4}[T_3 + 0 + T_9 + T_5],$$

$$T_7 = \frac{1}{4}[T_4 + T_8 + 1 + 0] \qquad T_8 = \frac{1}{4}[T_5 + T_9 + 1 + T_7]$$

$$T_9 = \frac{1}{4}[T_6 + 0 + 1 + T_8]$$

**Gauss-Seidel Method**

**Initial approximations are**

$$T_1 = T_2 = T_3 = T_4 = T_5 = T_6 = T_7 = T_8 = T_9 = 0$$

Ten successive iterates are given below:

**First Iteration**

$$T_1^{(n+1)} = \frac{1}{4}[0+T_4^{(n)}+T_2^{(n)}+0], \qquad T_1^{(1)} = \frac{1}{4}[0+0+0+0] = 0$$

$$T_2^{(n+1)} = \frac{1}{4}[T_1^{(n+1)}+T_5^{(n)}+T_3^{(n)}+0], \qquad T_2^{(1)} = \frac{1}{4}[0+0+0+0] = 0$$

$$T_3^{(n+1)} = \frac{1}{4}[T_2^{(n+1)}+T_6^{(n)}+0+0], \qquad T_3^{(1)} = \frac{1}{4}[0+0+0+0] = 0$$

$$T_4^{(n+1)} = \frac{1}{4}[0+T_7^{(n)}+T_5^{(n)}+T_1^{(n+1)}], \qquad T_4^{(1)} = \frac{1}{4}[0+0+0+0] = 0$$

$$T_5^{(n+1)} = \frac{1}{4}[T_4^{(n+1)}+T_8^{(n)}+T_6^{(n)}+T_2^{(n+1)}], \qquad T_5^{(1)} = \frac{1}{4}[0+0+0+0] = 0$$

$$T_6^{(n+1)} = \frac{1}{4}[T_5^{(n+1)}+T_9^{(n)}+0+T_3^{(n+1)}], \qquad T_6^{(1)} = \frac{1}{4}[0+0+0+0] = 0$$

$$T_7^{(n+1)} = \frac{1}{4}[0+1+T_8^{(n)}+T_4^{(n+1)}], \qquad T_7^{(1)} = \frac{1}{4}[0+1+0+0] = 0.25$$

$$T_8^{(n+1)} = \frac{1}{4}[T_7^{(n+1)}+1+T_9^{(n)}+T_5^{(n+1)}], \qquad T_8^{(1)} = \frac{1}{4}[0.25+1+0+0] = 0.312$$

$$T_9^{(n+1)} = \frac{1}{4}[T_8^{(n+1)}+1+0+T_6^{(n+1)}] \qquad T_9^{(1)} = \frac{1}{4}[0.312+1+0+0] = 0.328$$

**Second Iteration**

$$T_1^{(2)} = \frac{1}{4}[0+0+0+0] = 0, \qquad T_2^{(2)} = \frac{1}{4}[0+0+0+0] = 0$$

$$T_3^{(2)} = \frac{1}{4}[0+0+0+0] = 0, \qquad T_4^{(2)} = \frac{1}{4}[0+0.25+0+0] = 0.062$$

$$T_5^{(2)} = \frac{1}{4}[0+0.312+0+0] = 0.078, \qquad T_6^{(2)} = \frac{1}{4}[0+0.328+0+0] = 0.082$$

$$T_7^{(2)} = \frac{1}{4}[0+1+0.312+0] = 0.328, \qquad T_8^{(2)} = \frac{1}{4}[0.25+1+0.328+0] = 0.394$$

$$T_9^{(2)} = \frac{1}{4}[0.312+1+0+0] = 0.328$$

**Third Iteration**

$$T_1^{(3)} = \frac{1}{4}[0+0.062+0+0] = 0.016, \qquad T_2^{(3)} = \frac{1}{4}[0.016+0.078+0+0] = 0.024$$

$$T_3^{(3)} = \frac{1}{4}[0.024+0.082+0+0] = 0.027, \qquad T_4^{(3)} = \frac{1}{4}[0+0.328+0.078+0.016] = 0.106$$

$$T_5^{(3)} = \frac{1}{4}[0.106+0.394+0.082+0.024] = 0.152,$$

$$T_6^{(3)} = \frac{1}{4}[0.152+0.328+0+0.027] = 0.127$$

$$T_7^{(3)} = \frac{1}{4}[0+1+0.394+0.106] = 0.375 \qquad T_8^{(3)} = \frac{1}{4}[0.375+1+0.328+0.152] = 0.464$$

$$T_9^{(3)} = \frac{1}{4}[0.464+1+0+0.127] = 0.398$$

and so on. **Ans.**

| *Iteration* | $T_1$ | $T_2$ | $T_3$ | $T_4$ | $T_5$ | $T_6$ | $T_7$ | $T_8$ | $T_9$ |
|---|---|---|---|---|---|---|---|---|---|
| 4 | 0.032 | 0.053 | 0.045 | 0.140 | 0.196 | 0.160 | 0.401 | 0.499 | 0.415 |
| 5 | 0.048 | 0.072 | 0.058 | 0.161 | 0.223 | 0.174 | 0.415 | 0.513 | 0.422 |
| 6 | 0.058 | 0.085 | 0.065 | 0.174 | 0.236 | 0.181 | 0.422 | 0.520 | 0.425 |
| 7 | 0.065 | 0.092 | 0.068 | 0.181 | 0.244 | 0.184 | 0.425 | 0.524 | 0.427 |
| 8 | 0.068 | 0.095 | 0.070 | 0.184 | 0.247 | 0.186 | 0.427 | 0.525 | 0.428 |
| 9 | 0.070 | 0.097 | 0.071 | 0.186 | 0.249 | 0.187 | 0.428 | 0.526 | 0.428 |
| 10 | 0.071 | 0.098 | 0.071 | 0.187 | 0.250 | 0.187 | 0.428 | 0.526 | 0.428 |

**Example 3.** *Solve* $\dfrac{\partial^2 u}{\partial x^2} + \dfrac{\partial^2 u}{\partial y^2} = 0$ *by Leibman's iteration process for the domain of the figure given below:*

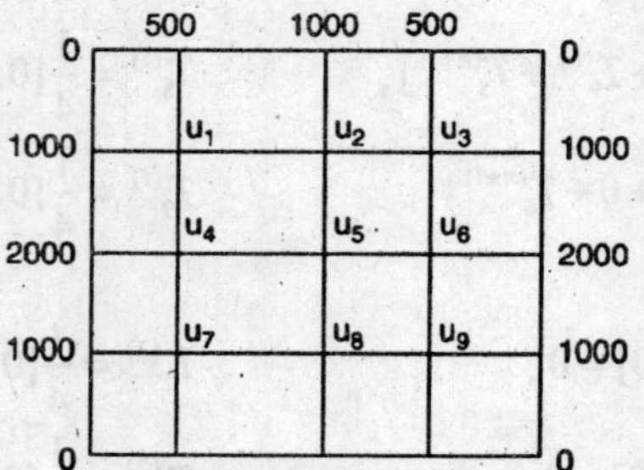


**Solution.** Values given on the figure are symmetrical about middle line.

$$\therefore \quad u_1 = u_3 = u_9 = u_7$$

$$u_2 = u_8,\ u_4 = u_6$$

$$u_5 = \frac{1}{4}(2000 + 2000 + 1000 + 1000) = 1500 \qquad \textbf{(Standard formula)}$$

$$u_1 = \frac{1}{4}[0 + 1000 + 1500 + 2000] = 1125 \qquad \textbf{(Diag. formula)}$$

Similarly $\quad u_1 = u_3 = u_9 = u_7 = 1125$

$$u_2 = \frac{1}{4}(1000 + 1125 + 1500 + 1125) \approx 1188 \qquad \textbf{(Standard formula)}$$

Similarly $\quad u_8 = u_2 = 1188$

$$u_4 = \frac{1}{4}[1125 + 2000 + 1125 + 1500] \approx 1438 \qquad \textbf{(Standard formula)}$$

Similarly $\quad u_4 = u_6 = 1438$

So $u_1 = 1125,\ u_2 = 1188,\ u_3 = 1125,\ u_4 = 1438,\ u_5 = 1500,\ u_6 = 1438,\ u_7 = 1125,\ u_8 = 1188,\ u_9 = 1125$

**Gauss-Seidel Method:**

$$u^{(n+1)}_{i,j} = \frac{1}{4}[u^{(n+1)}_{i-1,j} + u^{(n)}_{i+1,j} + u^{(n+1)}_{i,j-1} + u^{(n)}_{i,j+1}]$$

$$u_1^{(n+1)} = \frac{1}{4}[1000 + u_2^{(n)} + 500 + u_4^{(n)}] = u_3^{(n+1)} = u_5^{(n+1)} = u_7^{(n+1)}$$

$$u_2^{(n+1)} = \frac{1}{4}[u_1^{(n+1)} + u_3^{(n+1)} + 1000 + u_5^{(n+1)}] = u_8^{(n+1)}$$

$$u_4^{(n+1)} = \frac{1}{4}[2000 + u_5^{(n)} + u_1^{(n+1)} + u_7^{(n+1)}] = u_6^{(n+1)}$$

$$u_5^{(n+1)} = \frac{1}{4}[u_4^{(n+1)} + u_6^{(n+1)} + u_2^{(n+1)} + u_8^{(n+1)}]$$

**First Iteration**

$$u_1^{(1)} = \frac{1}{4}[1000 + 1188 + 500 + 1438] \approx 1032 = u_3^{(1)} = u_9^{(1)} = u_7^{(1)}$$

$$u_2^{(1)} = \frac{1}{4}[1032 + 1032 + 1000 + 1500] = 1141 = u_8^{(1)}$$

$$u_4^{(n+1)} = \frac{1}{4}[2000 + 1500 + 1032 + 1032] = 1391 = u_6^{(1)}$$

$$u_5^{(n+1)} = \frac{1}{4}[1091 + 1391 + 1141 + 1141] = 1266$$

**Second Iteration**

$$u_1^{(2)} = \frac{1}{4}[1000 + 1141 + 500 + 1391] = 1008 = u_3^{(2)} = u_9^{(2)} = u_7^{(2)}$$

$$u_2^{(2)} = \frac{1}{4}[1008 + 1008 + 1000 + 1266] = 1069 = u_8^{(2)}$$

$$u_4^{(2)} = \frac{1}{4}[2000 + 1266 + 1008 + 1008] = 1321 = u_6^{(2)}$$

$$u_5^{(2)} = \frac{1}{4}[1321 + 1321 + 1069 + 1069] = 1195$$

Similarly

| *Iteration* | $u_1 = u_3 = u_9 = u_7$ | $u_2 = u_8$ | $u_4 = u_6$ | $u_5$ |
|---|---|---|---|---|
| Third | 973 | 1035 | 1288 | 1162 |
| Fourth | 956 | 1019 | 1269 | 1144 |
| Fifth | 947 | 1010 | 1260 | 1135 |
| Sixth | 942 | 1005 | 1255 | 1130 |
| Seventh | 940 | 1003 | 1253 | 1128 |
| Eighth | 939 | 1002 | 1252 | 1127 |
| Ninth | 939 | 1001 | 1251 | 1126 |

Very small difference is in the eighth and ninth iteration

Thus, $u_1 = u_3 = u_7 = u_9 = 939$

$u_2 = u_8 = 1001,$

$u_4 = u_6 = 1251,$

$u_5 = 1126$ **Ans.**

**Example 4.** *Solve* $\frac{\partial^2 u}{\partial x^2}+\frac{\partial^2 u}{\partial y^2}=0$ *in the domain of the figure given below by*

*(a) Jacobi's method, (b) Gauss-Seidel method and*

*(c) Successive Over-Relaxation method*

**Solution.** **(a) Jacobi's Method**

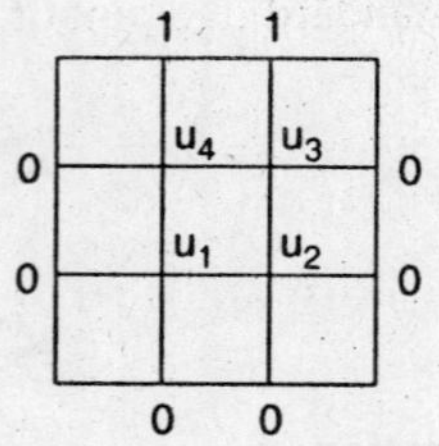


$$u_1^{(1)}=\frac{1}{4}[0+0+0+1]=0.25$$

$$u_2^{(2)}=\frac{1}{4}[0+0+0+1]=0.25$$

$$u_3^{(1)}=\frac{1}{4}[1+1+0+0]=0.5$$

$$u_4^{(1)}=\frac{1}{4}[1+1+0+0]=0.5$$

Seven successive iterates are given below:

| $u_1$ | $u_2$ | $u_3$ | $u_4$ |
|---|---|---|---|
| 0.1875 | 0.1875 | 0.4375 | 0.4375 |
| 0.15625 | 0.15625 | 0.40625 | 0.40625 |
| 0.14062 | 0.14062 | 0.39062 | 0.39062 |
| 0.13281 | 0.13281 | 0.38281 | 0.38281 |
| 0.12891 | 0.12891 | 0.37891 | 0.37891 |
| 0.12695 | 0.12695 | 0.37695 | 0.37695 |
| 0.12598 | 0.12598 | 0.37598 | 0.37518 |

**(b) Gauss-Seidel Method**

Five successive iterates are given below:

| $u_1$ | $u_2$ | $u_3$ | $u_4$ |
|---|---|---|---|
| 0.25 | 0.3125 | 0.5625 | 0.46875 |
| 0.21875 | 0.17187 | 0.42187 | 0.39844 |
| 0.14844 | 0.13672 | 0.38672 | 0.38086 |
| 0.13086 | 0.12793 | 0.37793 | 0.37646 |
| 0.12646 | 0.12573 | 0.37573 | 0.37537 |

**(c) Successive Over-Relaxation method.** Three successive iterates are given below:

| $u_1$ | $u_2$ | $u_3$ | $u_4$ |
|---|---|---|---|
| 0.275 | 0.35062 | 0.35062 | 0.35062 |
| 0.16534 | 0.10683 | 0.38183 | 0.37432 |
| 0.11785 | 0.12181 | 0.37216 | 0.37341 |

## 16.7 POISSON EQUATION

$$\nabla^2 u=\frac{\partial^2 u}{\partial x^2}+\frac{\partial^2 u}{\partial y^2}=f(x,y)$$

In this case the standard five-point formula is of the form

$$u_{i-1,j}+u_{i+1,j}+u_{i,j-1}+u_{i,j+1}-4u_{i,j}=h^2 f(ih,jh)$$

On applying the above formula we get equations. These equations can be solved by Gauss-Seidel iteration method.

**Example 5.** *Solve the Poisson's equation* $\frac{\partial^2 u}{\partial x^2}+\frac{\partial^2 u}{\partial y^2}=8x^2y^2$ *for the square mesh of the figure given below with u (x, y) = 0 on the boundary and mesh length = 1.*

**Solution.** Here $h = 1$

The standard five-point formula for the given equation is

$$u_{i-1,j}+u_{i+1,j}+u_{i,j+1}+u_{i,j-1}-4u_{i,j}=8i^2j^2 \qquad ...(1)$$

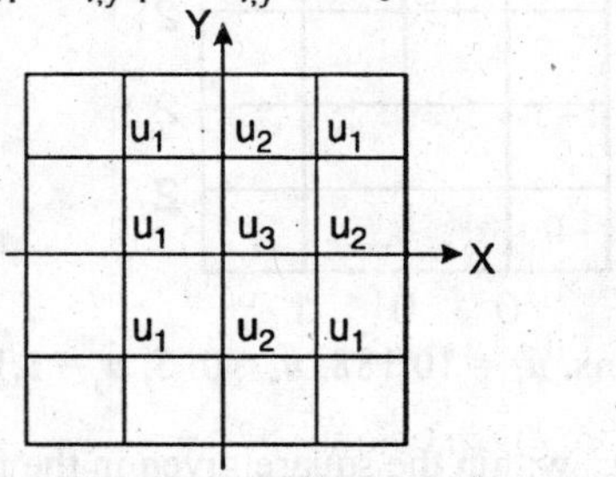


For $u_1$ ($i = -1, j = +1$), equation (1) becomes

$$0+u_2+0+u_2-4u_1=8(-1)^2(1)^2 \Rightarrow 4u_1=2u_2-8 \qquad ...(2)$$

For $u_2$ ($i = 0, j = 1$), equation (1) becomes

$$u_1+u_1+0+u_3-4u_2=0 \Rightarrow 4u_2=2u_1+u_3 \qquad ...(3)$$

For $u_3$ ($i = 0, j = 0$), equation (1) becomes

$$u_2+u_2+u_2+u_2-4u_3=0 \Rightarrow 4u_3=4u_2 \Rightarrow u_3=u_2 \qquad ...(4)$$

Putting $u_2$ for $u_3$ in (3), we get

$$4u_2=2u_1+u_2 \text{ or } 3u_2=2u_1$$

Putting $\frac{2u_1}{3}$ for $u_2$ in (2), we get

$$4u_1=\frac{4u_1}{3}-8 \Rightarrow 12u_1=4u_1-24$$

$$8u_1=-24 \Rightarrow u_1=-3$$

$$u_2=\frac{2u_1}{3}=\left(\frac{2}{3}\times -3\right)=-2, \qquad u_3=u_2=-2$$

$$u_1=-3, \qquad u_2=-2, \; u_3=-2$$ **Ans.**

## Exercise 16.1

1. Given the values of $u(x, y)$ on the boundary of the square in the figure given below, evaluate the function $u(x, y)$ satisfying the Laplace equation

$$\frac{\partial^2 u}{\partial x^2}+\frac{\partial^2 u}{\partial y^2}=0$$

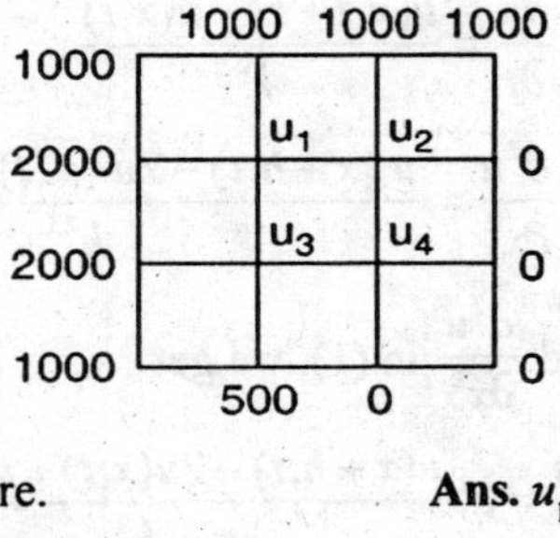


at pivotal points of this figure. **Ans.** $u_1=1208, u_2=792, u_3=1042, u_4=458$

2. Solve $\frac{\partial^2 u}{\partial x^2}+\frac{\partial^2 u}{\partial y^2}=0$

for the square mesh with boundary values as shown in the figure given below. Iterate until the maximum difference between successive values at any point is less than 0.005.

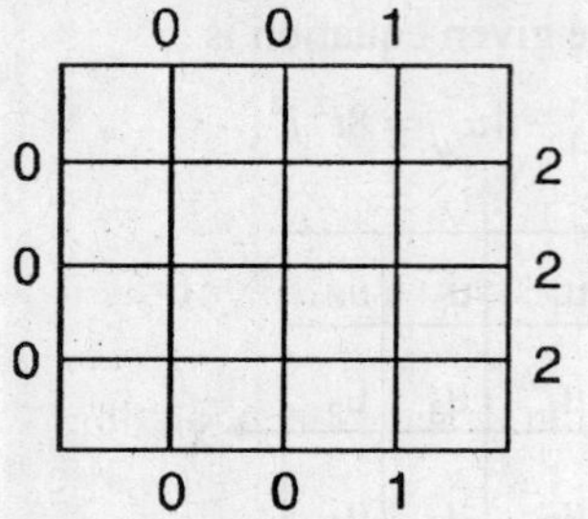


**Ans.** $u_1 = 10.188, u_2 = 0.5, u_3 = 1.188, u_4 = 0.25, u_5 = 0.625, u_6 = 1.25.$

3. Solve $\frac{\partial^2 u}{\partial x^2}+\frac{\partial^2 u}{\partial t^2}=0$ within the square given in the figure below.

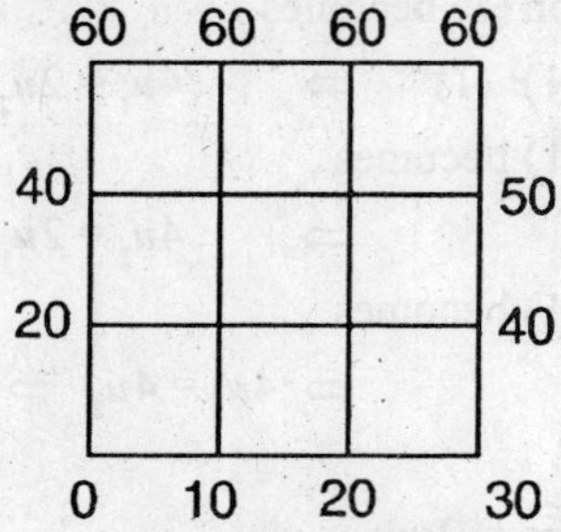


**Ans.** $u_1 = 26.66, u_2 = 33.33, u_3 = 43.33, u_4 = 46.66.$

4. Solve $\frac{\partial^2 u}{\partial x^2}+\frac{\partial^2 u}{\partial y^2}=-10(x^2+y^2+10)$

over the square with $x = 0 = y$, $x = 3 = y$ with $u = 0$ on the boundary and mess length = 1.

**Ans.** $u_1 = 75, u_2 = 82.5, u_3 = 67.05, u_4 = 75.$

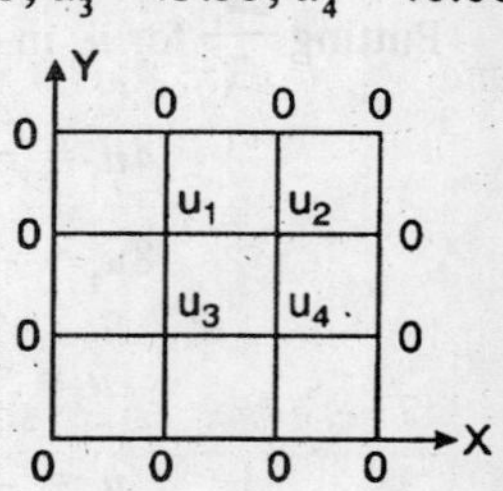


5. Solve $\frac{\partial^2 u}{\partial x^2}+\frac{\partial^2 u}{\partial y^2}=0$ with boundary values as shown in the figure given below.

**Ans.** $u_1 = 10.188, u_2 = 0.5, u_3 = 1.188$

$u_4 = 0.25, u_5 = 0.625, u_6 = 1.25$

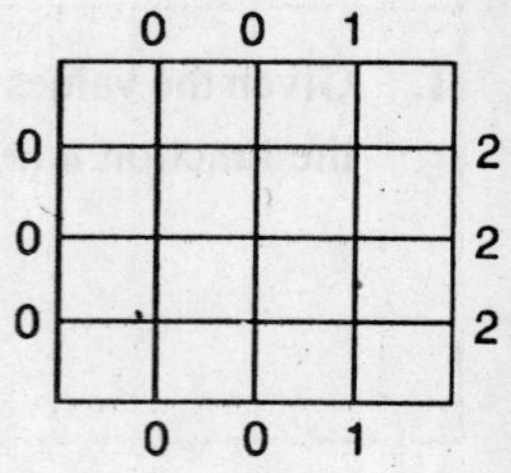


## 16.8 HEAT EQUATION (PARABOLIC EQUATION)

$$\frac{\partial u}{\partial t}=c^2\frac{\partial^2 u}{\partial x^2}$$

We know that
$$\frac{\partial u}{\partial t}=\frac{u(x,t+k)-u(x,t)}{k}$$

$$\frac{\partial^2 u}{\partial x^2}=\frac{u(x+h,t)-2u(x,t)+u(x-h,t)}{h^2}$$

On putting the values of $\frac{\partial u}{\partial t}$ and $\frac{\partial^2 u}{\partial x^2}$ in (1), we get

$$\frac{u(x,t+k)-u(x,t)}{k}=c^2\frac{u(x+h,t)-2u(x,t)+u(x-h,t)}{h^2}$$

$$u(x,\ t+k) = \frac{c^2k}{h^2}u(x+h,t) - \frac{2c^2k}{h^2}u(x,t) + u(x,t) + \frac{c^2k}{h^2}u(x-h,t)$$

$$\Rightarrow\ u(x,\ t+k) = au(x+h,\ t) + (1-2a)u(x,\ t) + au(x-h,t)$$

If $a = \frac{1}{2}$

$$u(x,\ t+k) = \frac{1}{2}u(x+h,\ t) + \frac{1}{2}u(x-h,t) \Rightarrow u_{i,\,j+1} = \frac{1}{2}[u_{i+1,j} + u_{i-1,\,j}]$$

It means that the value of $u$ at $x_i$, at time $t$ is the mean of the values of $u$ at $x_{i-1}$ and $x_{i+1}$ at the previous time $t_j$.

This relation is known as Bendre-Schimidt recurrence relation.

**Example 6.** *Find the values of u (x, t) satisfying the parabolic equation* $\frac{\partial u}{\partial t} = 4\frac{\partial^2 u}{\partial x^2}$ *with boundary conditions* $u(0,\ t) = 0 = u(8,\ t)$ *and* $u(x,\ 0) = 4x - \frac{1}{2}x^2$ *at the points.*

$$x = i : i = 0, 1, 2, 3, ..., 7 \quad \text{and} \quad t = \frac{1}{8}j : j = 0, 1, 2, 3, ..., 5.$$

**Solution.** $c^2 = 4,\ h = 1,\ k = \frac{1}{8}, \qquad a = \frac{c^2k}{h^2} = \frac{4\times 1/8}{(1)} = \frac{1}{2}$

Then the equation is

$$u_{i,j+1} = \frac{1}{2}(u_{i-1,j} + u_{i+1,j}) \qquad \text{... (1)}$$

Given $u(0,\ t) = 0 = u(8,\ t)$

$\Rightarrow \qquad u(0, j) = 0 = u(8, j)$ for all values of $j = 1, 2, 3, 4, 5$

and $\qquad u(x,\ 0) = 4x - \frac{1}{2}x^2, \quad u_{i,\,0} = 4i - \frac{1}{2}i^2$

$u_{0,0} = 0, \qquad u_{1,0} = 4(1) - \frac{1}{2}(1)^2 = 3.5, \qquad u_{2,0} = 4(2) - \frac{1}{2}(2)^2 = 6$

$u_{3,\,0} = 7.5, \qquad u_{4,\,0} = 8, \quad u_{5,\,0} = 7.5, \qquad u_{6,\,0} = 6,\ u_{7,\,0} = 3.5.$

These entries are shown in the following table :

| j \ i | 0 | 1 | 2 | 3 | 4 | 5 | 6 | 7 | 8 |
|---|---|---|---|---|---|---|---|---|---|
| 0 | 0 | 3.5 | 6 | 7.5 | 8 | 7.5 | 6 | 3.5 | 0 |
| 1 | 0 | 3 | 5.5 | 7 | 7.5 | 7 | 5.5 | 3 | 0 |
| 2 | 0 | 2.75 | 5 | 6.5 | 7 | 6.5 | 5 | 2.75 | 0 |
| 3 | 0 | 2.5 | 4.625 | 6 | 6.5 | 6 | 4.625 | 2.5 | 0 |
| 4 | 0 | 2.3125 | 4.25 | 5.5625 | 6 | 5.5625 | 4.25 | 2.3125 | 0 |
| 5 | 0 | 2.125 | 3.9375 | 5.125 | 5.5625 | 5.125 | 3.9375 | 2.125 | 0 |

Putting $j = 0$ in (1) we have

$$u_{i,\,1} = \frac{1}{2}(u_{i-1,\,0} + u_{i+1,\,0})$$

$$u_{1,\,1} = \frac{1}{2}(u_{0,\,0} + u_{2,\,0}) = \frac{1}{2}(0+6) = 3$$

$$u_{2,\,1} = \frac{1}{2}(u_{1,\,0} + u_{3,\,0}) = \frac{1}{2}(3.5+7.5) = 5.5$$

$$u_{3,1} = \frac{1}{2}(u_{2,0} + u_{4,0}) = \frac{1}{2}(6 + 8) = 7$$

$$u_{4,1} = \frac{1}{2}(u_{3,0} + u_{5,0}) = \frac{1}{2}(7.5 + 7.5) = 7.5$$

$$u_{5,1} = \frac{1}{2}(u_{4,0} + u_{6,0}) = \frac{1}{2}(8 + 6) = 7$$

$$u_{6,1} = \frac{1}{2}(u_{5,0} + u_{7,0}) = \frac{1}{2}(7.5 + 3.5) = 5.5$$

$$u_{7,1} = \frac{1}{2}(u_{6,0} + u_{8,0}) = \frac{1}{2}(6 + 0) = 3$$

Putting $j = 1$ in (1), we get

$$u_{i,2} = \frac{1}{2}(u_{i-1,1} + u_{i+1,1})$$

$$u_{1,2} = \frac{1}{2}(u_{0,1} + u_{2,0}) = \frac{1}{2}(0 + 5.5) = 2.75$$

$$u_{2,2} = \frac{1}{2}(u_{1,1} + u_{3,1}) = \frac{1}{2}(3 + 7) = 5$$

$$u_{3,2} = 6.5,\ u_{4,2} = 7,\ u_{5,2} = 6.5,\ u_{6,2} = 5,\ u_{7,2} = 2.75$$

Putting $j = 2$ in (1), we get

$$u_{i,3} = \frac{1}{2}(u_{i-1,2} + u_{i+1,2})$$

$$u_{1,3} = \frac{1}{2}(u_{0,2} + u_{2,2}) = \frac{1}{2}(0 + 5) = 2.5$$

$$u_{2,3} = \frac{1}{2}(u_{1,2} + u_{3,2}) = \frac{1}{2}(2.75 + 6.5) = 4.625$$

$$u_{3,3} = 6,\ u_{4,3} = 6.5,\ u_{5,3} = 6,\ u_{6,3} = 4.625,\ u_{7,3} = 2.5$$

Putting $j = 3$ in (1), we get

$$u_{i,4} = \frac{1}{2}(u_{i-1,3} + u_{i+1,3})$$

$$u_{1,4} = \frac{1}{2}(u_{0,3} + u_{2,3}) = \frac{1}{2}(0 + 4.625) = 2.3125$$

$$u_{2,4} = \frac{1}{2}(u_{1,3} + u_{3,2}) = \frac{1}{2}(2.5 + 6) = 4.25$$

$$u_{3,4} = 0.5625,\ u_{4,4} = 6,\ u_{5,4} = 5.5625,\ u_{6,4} = 4.25,\ u_{7,4} = 2.3125$$

Putting $j = 4$ in (1), we get

$$u_{i,5} = \frac{1}{2}(u_{i-1,4} + u_{i+1,4})$$

$$u_{1,5} = \frac{1}{2}(u_{0,4} + u_{2,4}) = \frac{1}{2}(0 + 4.25) = 2.125$$

$$u_{2,5} = \frac{1}{2}(u_{1,4} + u_{3,4}) = \frac{1}{2}(2.125 + 5.5625) = 3.9375$$

$$u_{3,5} = 5.125,\ u_{4,5} = 5.625,\ u_{5,5} = 5.125,\ u_{6,5} = 3.9375,\ u_{7,5} = 2.125$$

## 16.9 WAVE EQUATION (HYPERBOLIC EQUATION)

$$\frac{\partial^2 u}{\partial t^2} = c^2 \frac{\partial^2 u}{\partial x^2} \quad ...(1)$$

We know that

$$\frac{\partial^2 u}{\partial t^2} = \frac{u(x,t+k) - 2u(x,t) + u(x,t-k)}{k^2} \text{ and } \frac{\partial^2 u}{\partial x^2} = \frac{u(x+h,t) - 2u(x,t) + u(x-h,t)}{h^2}$$

Putting the values of $\frac{\partial^2 u}{\partial t^2}$ and $\frac{\partial^2 u}{\partial x^2}$ in (1) we have

$$\frac{u(x,t+k) - 2u(x,t) + u(x,t-k)}{k^2} = c^2 \frac{u(x+h,t) - 2u(x,t) + u(x-h,t)}{h^2}$$

$$\Rightarrow \quad u(x, t+k) - 2u(x, t) + u(x, t-k) = \frac{c^2 k^2}{h^2}[u(x+h, t) - 2u(x, t) + u(x-h, t)]$$

$$\Rightarrow \quad u_{i,j+1} - 2u_{i,j} + u_{i,j-1} = a^2c^2[u_{i+1,j} - 2u_{i,j} + u_{i-1,j}] \quad \left(a = \frac{k}{h}\right)$$

$$\Rightarrow \quad u_{i,j+1} = 2(1 - a^2c^2)u_{i,j} + a^2c^2(u_{i-1,j} + u_{i+1,j}) - u_{i,j-1} \quad ...(2)$$

If $a^2c^2 = 1$, Equation (2) reduces to

$$u_{i,j+1} = u_{i-1,j} + u_{i+1,j} - u_{i,j-1} \quad ...(3)$$

**Example 7.** *Solve* $\frac{\partial^2 u}{\partial t^2} = \frac{\partial^2 u}{\partial x^2}$ *with conditions* $u(0, t) = u(1, t) = 0$, $u(x, 0) = 1/2\, x(1-x)$ *and* $u(x, 0) = 0$, *taking* $h = k = 0.1$ *for* $0 \le t \le 0.4$. *Compare your solution with the exact solution* $x = 0.5$ *and* $t = 0.3$.

**Solution.** $c^2 = 1$. The difference equation for the given equation is

$$u_{i,j+1} = 2(1-\alpha^2)u_{i,j} + \alpha^2(u_{i-1,j} + u_{i+1,j}) - u_{i,j-1} \quad ...(1)$$

where $\alpha = \frac{k}{h}$. But $\alpha = \frac{0.1}{0.1} = 1$

Equation (1) reduces to

$$u_{i,j+1} = u_{i-1,j} + u_{i+1,j} - u_{i,j-1} \quad ...(2)$$

$$u(0,t) = u(1,t) = 0, u_{0j} = 0 \text{ and } u_{10,j} = 0$$

*i.e.*, the entries in the first column are zero.

since $\quad u(x, 0) = \frac{1}{2}x(1-x) \quad\quad u(i, 0) = \frac{1}{2}\, i(1-i)$

$= 0.045, 0.08, 0.105, 0.120, 0.125, 0.120, 0.105$ for $i = 0.1, 0.2, 0.3, 0.4, 0.5, 0.6, 0.7$ at $t = 0$

These are the entries of the fist row

Finally since $u_x(x, 0) = 0$

$$\therefore \quad \frac{u_{i,j+1} - u_{i,j}}{k} = 0 \text{ for } j = 0\ (t = 0), \quad u_{i,1} = u_{i,9}$$

Putting $j = 0$ in equation (2), we get

$$u_{i,1} = u_{i-1,0} + u_{i+1,0} - u_{i,-1}$$

$$= u_{i-1,0} + u_{i+1,0} - u_{i,1} \qquad (u_{i,1} = u_{i,-1})$$

$$2u_{i,1} = u_{i-1,0} + u_{i+1,0}, \qquad u_{i,1} = \frac{1}{2}[u_{i-1,0} + u_{i+1,0}]$$

For $i = 1$ $\quad u_{1,1} = \frac{1}{2}[u_{0,0} + u_{2,0}] = \frac{1}{2}[0 + 0.080] = 0.040$

For $i = 2$ $\quad u_{2,1} = \frac{1}{2}[u_{1,0} + u_{3,0}] = \frac{1}{2}[0.045 + 0.105] = 0.075$

For $i = 3$ $\quad u_{3,1} = [u_{2,0} + u_{4,0}] = \frac{1}{2}[0.08 + 0.120] = 0.100$

For $i = 4$ $\quad u_{4,1} = \frac{1}{2}[u_{3,0} + u_{5,0}] = \frac{1}{2}[0.105 + 0.125] = 0.115$

For $i = 5$ $\quad u_{5,1} = \frac{1}{2}[u_{4,0} + u_{6,0}] = \frac{1}{2}[0.120 + 0.120] = 0.120$

For $i = 6$ $\quad u_{6,1} = \frac{1}{2}[u_{5,0} + u_{7,0}] = \frac{1}{2}[0.125 + 0.105] = 0.115$

Putting $j = 1$ in equation (2), we get

$$u_{i,2} = u_{i-1,1} + u_{i+1,1} - u_{i,0}$$

For $i = 1$ $\quad u_{1,2} = u_{0,1} + u_{2,1} - u_{1,0} = 0 + 0.075 - 0.045 = 0.03$

For $i = 2$ $\quad u_{2,2} = u_{1,1} + u_{3,1} - u_{2,0} = 0.040 + 0.100 - 0.08 = 0.060$

For $i = 3$ $\quad u_{3,2} = u_{2,1} + u_{4,1} - u_{3,0} = 0.075 + 0.115 - 0.105 = 0.085$

For $i = 4$ $\quad u_{4,2} = u_{3,1} + u_{5,1} - u_{4,0} = 0.100 + 0.120 - 0.120 = 0.100$

For $i = 5$ $\quad u_{5,2} = u_{4,1} + u_{6,1} - u_{5,0} = 0.115 + 0.115 - 0.125 = 0.105$

Similarly for $\quad j = 2$

$$u_{i,3} = u_{i-1,2} + u_{i+1,2} - u_{i,1}$$

$$u_{1,3} = 0.020,\ u_{2,3} = 0.040,\ u_{3,3} = 0.060,\ u_{4,3} = 0.075,\ u_{5,3} = 0.80$$

For $j = 3$ $\quad u_{i,4} = u_{i-1,3} + u_{i+1,3} - u_{i,2}$

$$u_{1,4} = 0.010, \quad u_{2,4} = 0.02,$$

$$u_{3,4} = 0.030, \quad u_{4,4} = 0.040, \quad u_{5,4} = 0.048$$

| | | *0* | *0.1* | *0.2* | *0.3* | *0.4* | *0.5* | *0.6* |
|---|---|---|---|---|---|---|---|---|
| | *i* / *j* | *0* | *1* | 2 | 3 | 4 | 5 | 6 |
| *0* | *0* | 0 | 0.045 | 0.080 | 0.105 | 0.120 | 0.125 | 0.120 |
| 0.1 | 1 | 0 | 0.040 | 0.075 | 0.100 | 0.115 | 0.120 | 0.115 |
| 0.2 | 2 | 0 | 0.030 | 0.060 | 0.085 | 0.100 | 0.105 | |
| 0.3 | 3 | 0 | 0.020 | 0.040 | 0.060 | 0.075 | 0.080 | |
| 0.4 | 4 | 0 | 0.010 | 0.020 | 0.030 | 0.040 | 0.048 | |

The analytical (exact) solution of the given equation is

$$u = \frac{2}{\pi^3} \sum_{n=1}^{\infty} \frac{1}{n^3}(1 - \cos n\pi) \sin n\pi x \cos n\pi t$$

Comparison of two solutions is given below:

| $t$ = 0.3 | $x$= | 0.1 | 0.2 | 0.3 | 0.4 | 0.5 |
|---|---|---|---|---|---|---|
| Numerical solution | $u$ = | 0.02 | 0.04 | 0.06 | 0.075 | 0.08 |
| Exact solution | $u$= | 0.02 | 0.04 | 0.06 | 0.075 | 0.08 |

**Ans.**

## EXERCISE 16.2

**1.** Solve $\dfrac{\partial^2 u}{\partial x^2} = 2\dfrac{\partial u}{\partial t}$

under conditions $u(0, t) = u(4,t)=0$ and $u(x,0)= x(4-x)$, taking $h=1$, find the values upto $t = 5$.

**Ans.** $u_{1,0} = 3, u_{2,0} = 4, u_{3,0} = 3;\ u_{1,1} = 2, u_{2,1} = 3, u_{3,1} = 2$
$u_{1,2} = 1.5, u_{2,2} = 2, u_{3,2} = 1.5;\ u_{1,3} = 1, u_{2,3} = 1.5, u_{3,3} = 1$
$u_{1,4} = 0.75, u_{2,4} = 1, u_{3,4} = 0.75;\ u_{1,5} = 0.5, u_{2,5} = 0.75, u_{3,5} = 0.50$

**2.** Solve $\dfrac{\partial u}{\partial t} = \dfrac{\partial^2 u}{\partial x^2};\ 0 \le x \le 1, t \ge 0$ under the conditions that $u(0, t) = u(1, t) = 0$ and $u(x, 0) = 2x$ for $0 \le x \le \dfrac{1}{2} = (1-x)$ for $\dfrac{1}{2} \le x \le 1$.

**Ans.** $u_1 = 0.1989, u_2 = 0.3956, u_3 = 0.5834, u_4 = 0.7381, u_6 = 0.7591$

**3.** Solve $\dfrac{\partial u}{\partial t} = \dfrac{\partial^2 u}{\partial x^2};\ 0 \le x \le 1,\ t \ge 0$ under the conditions that

$$\left.\begin{array}{l} u = 0, \text{ at } x = 0 \\ u = 0, \text{ at } x = 1 \end{array}\right] t \ge 0$$

$u = \sin \pi x$ at $t = 0,\ 0 \le x \le 1$

find $u$ for $x = 0.8$ at $t = 1$. **Ans.** 0.4853

**4.** Solve $\dfrac{\partial^2 u}{\partial x^2} = \dfrac{\partial u}{\partial t}$ under the conditions that $u(0, t) = u(5, t) = 0,\ u(x, 0) = x^2(25 - x^2)$ taking $h = 1$ and $k = \dfrac{1}{2}$.

**Ans.** $u_{1,0} = 24,\quad u_{2,0} = 84, u_{3,0} = 144, u_{4,0} = 144$
$u_{1,1} = 42, u_{2,1} = 78, u_{3,1} = 78, u_{4,1} = 57$
$u_{1,2} = 39, u_{2,2} = 60, u_{3,2} = 67.5, u_{4,2} = 39$
$u_{1,3} = 30, u_{2,3} = 53.25, u_{3,3} = 49.5, u_{4,3} = 33.75$
$u_{1,4} = 26.625, u_{2,4} = 39.75, u_{3,4} = 43.5, u_{4,4} = 24.75$
$u_{1,5} = 19.875, u_{2,5} = 35.06, u_{3,5} = 32.25, u_{4,5} = 21.75$

**5.** Solve $16\dfrac{\partial^2 u}{\partial x^2} = \dfrac{\partial^2 u}{\partial t^2}$ taking $h = 1$, upto $t = 1.25$, under the conditions $u(0, t) = u(5, t) = 0,\ u_t(x, 0) = 0$ and $u(x, 0) = x^2(5 - x)$.

**Ans.** $u_{1,0} = 4,\ u_{2,0} = 12,\ u_{3,0} = 18,\ u_{4,0} = 16$
$u_{1,1} = 4,\ u_{2,1} = 12,\ u_{3,1} = 18,\ u_{4,1} = 16$
$u_{1,2} = 8,\ u_{2,2} = 10,\ u_{3,2} = 10,\ u_{4,2} = 2$
$u_{1,3} = 6,\ u_{2,3} = 6,\ u_{3,3} = -6,\ u_{4,3} = -6$
$u_{1,4} = -2,\ u_{2,4} = -10,\ u_{3,4} = -10,\ u_{4,4} = -8$
$u_{1,5} = -16,\ u_{2,5} = -18,\ u_{3,5} = -12,\ u_{4,5} = -4$

# 17

# Calculus of Variation

## 17.1 INTRODUCTION

The calculus of variations primarily deals with finding maximum or minimum value of a definite integral involving a certain function.

## 17.2 FUNCTIONALS

A simple example of functional is the shortest length of a curve through two points $A(x_1, y_1)$ and $B(x_2, y_2)$. In other words, the determination of the curve $y = y(x)$ for which $y = (x_1) = y_1,\ y(x_2) = y_2$ such that

$$\int_{x_1}^{x_2} \sqrt{1+\left(\frac{dy}{dx}\right)} dx \qquad ...(1)$$

is a minimum.

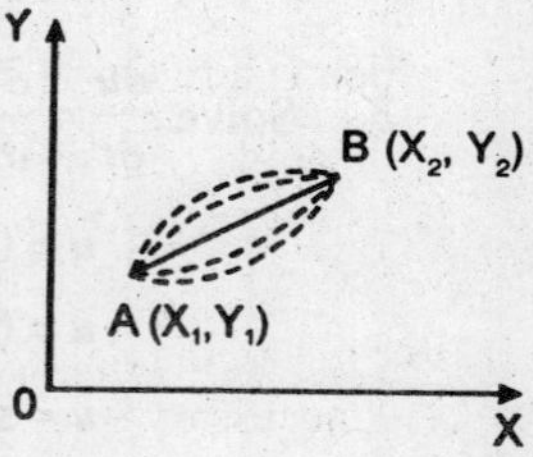


An integral such as (1) is called a *Functional.*

In general, it is required to find the curve $y = y(x)$ where $y(x_1) = y_1$ and $y(x_2) = y_2$ such that for a given function $f\left(x, y, \frac{dy}{dx}\right)$,

$$\int_{x_1}^{x_2} f\left(x, y, \frac{dy}{dx}\right) dx \qquad ...(2)$$

is maximum or minimum.

Integral (2) is known as the functional.

In differential calculus, we find the maximum or minimum value of functions. But the calculus of variations deals with the problems of maxima or minima of functionals.

A functional $I[y(x)]$ is said to be linear if it satisfies.

(*i*) $I[cy(x)] = c\,I[y(x)]$, where $c$ is an arbitrary constant.

(*ii*) $I[y_1(x) + y_2(x)] = I[y_1(x)] + I[y_2(x)]$, where $y_1(x) \in M$ and $y_2(x) \in M$.

## 17.3 DEFINITION

A functional $I[y(x)]$ is maximum on a curve $y = y(x)$, if the values of $I[y(x)]$ on any curve close to $y = y_1(x)$ do not exceed $I[y_1(x)]$. It means $\Delta I = I[y(x)] - I[y_1(x)] \le 0$ and $\Delta I = 0$ *on* $y = y_1(x)$.

In case of minimum of $I[y(x)]$, $\Delta I = 0$.

**Extremal:** A function $y = y(x)$ which extremizes a functional is called extremal or extremizing function.

## 17.4 EULER'S EQUATION IS

$$\frac{\partial f}{\partial y} - \frac{d}{dx}\left(\frac{\partial f}{\partial y}\right) = 0$$

This is the necessary condition for $I=\int_{x_1}^{x_2} f(x,y,y')\,dx$ to be maximum or minimum.

**Proof:** Let $y = y(x)$ be the curve AB which makes the given function I an extremum. Consider a family of neighbouring curves

$$Y = y(x)+\alpha\eta(x) \qquad \ldots(1)$$

where a is a parameter, $-r\alpha r$ and $\eta(x)$ is an arbitrary differentiable function.

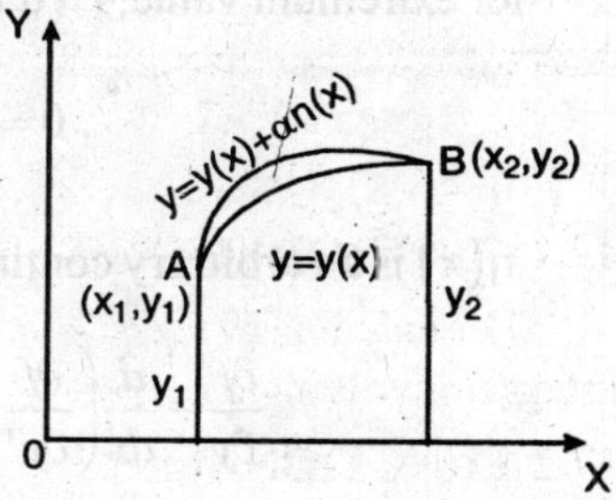


At the end points A and B,

$$\eta(x_1)=\eta(x_2)=0$$

when $\alpha = 0$, neighbouring curves become $y = y(x)$, which is extremal.

The family of neighbouring curves is called the family of *comparison functions.*

If in the functional $\int_{x_1}^{x_2} f(x,y,y')\,dx$ We replace $y$ by Y, we get

$$\int_{x_1}^{x^2} f(x,Y,Y')\,dx=\int_{x_1}^{x_2} f\left[x,y(x)+\alpha\eta(x),y'(x)+\alpha\eta'(x)\right]dx.$$

which is a function of $\alpha$, say $I(\alpha)$.

$$\therefore \qquad I(\alpha)=\int_{x_1}^{x_2} f(x,Y,Y')\,dx$$

For $\alpha = 0$, the neighbouring curves become the extremal, an extremum for $\alpha = 0$.
The necessary condition for this is $I'(\alpha) = 0$ ...(2)
Differentiating $I$ under the integral sign by Leibnitz's rule, we have

$$I'(\alpha)=\int_{x_1}^{x_2}\left(\frac{\partial f}{\partial x}\frac{\partial x}{\partial\alpha}+\frac{\partial f}{\partial Y}\frac{\partial Y}{\partial\alpha}+\frac{\partial f}{\partial Y'}\frac{\partial Y'}{\partial\alpha}\right)dx$$

$$I'(\alpha)=\int_{x_1}^{x_2}\left(\frac{\partial f}{\partial Y}\frac{\partial Y}{\partial\alpha}+\frac{\partial f}{\partial Y'}\frac{\partial Y'}{\partial\alpha}\right)dx \qquad \left(\frac{\partial x}{\partial\alpha}=0 \text{ as } \alpha \text{ is independent of } x\right) \qquad \ldots(3)$$

On differentiating (1), *w. r. t.* '$x$', we get, $Y'=y'(x)+\alpha\eta'(x)$

Again differentiating *w.r. t.* '$\alpha$', we get $\dfrac{\partial Y'}{\partial\alpha}=\eta'(x)$

Differentiating (1), *w. r. t.* $\alpha$, we get $\dfrac{\partial Y}{\partial\alpha}=\eta(x)$

Now (3) becomes $I'(\alpha)=\int_{x_1}^{x_2}\left[\frac{\partial f}{\partial Y}\eta(x)+\frac{\partial f}{\partial Y'}\eta'(x)\right]dx$

Integrating the second term on the right by parts, we get

$$=\int_{x_1}^{x_2}\frac{\partial f}{\partial Y}\eta(x)\,dx+\left[\left\{\frac{\partial f}{\partial Y'}\eta(x)\right\}_{x_1}^{x_2}-\int_{x_1}^{x_2}\frac{d}{dx}\left(\frac{\partial f}{\partial Y'}\right)\eta(x)\,dx\right]$$

$$= \int_{x_1}^{x_2} \frac{\partial f}{\partial Y} \eta(x)\,dx + \left[\frac{\partial f}{\partial Y'}\eta(x_2) - \frac{\partial f}{\partial Y'}\eta(x_1)\right] - \int_{x_1}^{x_2} \frac{d}{dx}\left[\frac{\partial f}{\partial Y'}\right]\eta(x)\,dx$$

$$= \int_{x_1}^{x_2} \frac{\partial f}{\partial Y}\eta(x)\,dx + 0 - \int_{x_1}^{x_2} \frac{d}{dx}\left(\frac{\partial f}{\partial Y'}\right)\eta(x)\,dx = \int_{x_1}^{x_2}\left[\frac{\partial f}{\partial Y} - \frac{d}{dx}\left(\frac{\partial f}{\partial Y'}\right)\right]\eta(x)\,dx \left[\eta(x_1) = \eta(x_2) = 0\right]$$

for extremum value, $I'(\alpha) = 0$

$$0 = \int_{x_1}^{x_2}\left[\frac{\partial f}{\partial Y} - \frac{d}{dx}\left(\frac{\partial f}{\partial Y'}\right)\right]\eta(x)\,dx$$

$\eta(x)$ is an arbitrary continuous function.

$\therefore \quad \dfrac{\partial f}{\partial y} - \dfrac{d}{dx}\left(\dfrac{\partial f}{\partial y'}\right) = 0$ which is a required Euler's equation.

**Note:** Other Forms of Euler's equation

**1.** $\dfrac{d}{dx} f(x, y, y') = \dfrac{\partial f}{\partial x}\dfrac{dx}{dx} + \dfrac{\partial f}{\partial y}\dfrac{dy}{dx} + \dfrac{\partial f}{\partial y'}\dfrac{dy'}{dx}$

or
$$\frac{df}{dx} = \frac{\partial f}{\partial x} + \frac{\partial f}{\partial y} y' + \frac{\partial f}{\partial y'} y'' \qquad \text{...(4)}$$

But
$$\frac{d}{dx}\left(y'\frac{\partial f}{\partial y'}\right) = y'\frac{d}{dx}\left(\frac{\partial f}{\partial y'}\right) + \frac{\partial f}{\partial y'} y'' \qquad \text{...(5)}$$

On subtracting ( 5 ) from ( 4 ), we have

$$\frac{df}{dx} - \frac{d}{dx}\left(y'\frac{\partial f}{\partial y'}\right) = \frac{\partial f}{\partial x} + \frac{\partial f}{\partial y} y' - y'\frac{d}{dx}\left(\frac{\partial f}{\partial y'}\right)$$

$$\frac{d}{dx}\left[f - y'\frac{\partial f}{\partial y'}\right] - \frac{\partial f}{\partial x} = y'\left[\frac{\partial f}{\partial y} - \frac{d}{dx}\frac{\partial f}{\partial y'}\right] = (y')(0) = 0 \qquad [\textit{Euler's equation}]$$

Hence
$$\frac{d}{dx}\left[f - y'\frac{\partial f}{\partial y'}\right] - \frac{\partial f}{\partial x} = 0 \qquad \text{...(6)}$$

Which is an another form of Euler's equation.

**2.** We know that $\dfrac{\partial f}{\partial y'}$ is also a function of $x, y, y'$ say $\phi(x, y, y')$.

$$\frac{d}{dx}\left(\frac{\partial f}{\partial y'}\right) = \frac{\partial \phi}{\partial x}\frac{dx}{dx} + \frac{\partial \phi}{\partial y}\frac{dy}{dx} + \frac{\partial \phi}{\partial y'}\frac{dy'}{dx} = \frac{\partial \phi}{\partial x} + \frac{\partial \phi}{\partial y} y' + \frac{\partial \phi}{\partial y'} y''$$

$$\frac{\partial}{\partial x}\left(\frac{\partial f}{\partial y'}\right) + \frac{\partial}{\partial y}\left(\frac{\partial f}{\partial y'}\right) y' + \frac{\partial}{\partial y'}\left(\frac{\partial f}{\partial y'}\right) y'' = \frac{\partial^2 f}{\partial x \partial y'} + y'\frac{\partial^2 f}{\partial y \partial y'} + y''\frac{\partial^2 f}{\partial y'^2}$$

Putting the value of $\dfrac{d}{dx}\left(\dfrac{\partial f}{\partial y'}\right)$ in Euler's equation, we get

$$\frac{\partial f}{\partial y} - \frac{\partial^2 f}{\partial x \partial y'} - y'\frac{\partial^2 f}{\partial y \partial y'} - y''\frac{\partial^2 f}{\partial y'^2} = 0 \qquad \text{...(7)}$$

This is the third form of Euler's equation.

## 17.5 EXTREMAL

Any function which satisfies Euler's equation is known as Extremal. Extremal is obtained by solving the Euler's equation.

**Case 1.** If $f$ is independent of $x$, *i.e.*, $\frac{\partial f}{\partial x} = 0$.

On substituting the value of $\frac{\partial f}{\partial x}$ in (6), we have $\frac{d}{dx}\left[f - y'\frac{\partial f}{\partial y'}\right] = 0$

Integrating, we get $f - y'\frac{\partial f}{\partial y'}$ = constant

**Case 2.** When $f$ is independent of $y$, *i.e.*, $\frac{\partial f}{\partial y} = 0$.

Putting the value of $\frac{\partial f}{\partial y}$ in Euler's equation, we get

$\frac{d}{dx}\left(\frac{\partial f}{\partial y'}\right) = 0$, Integrating we get $\frac{\partial f}{\partial y'}$ = **constant**

**Case 3.** If $f$ is an independent of $y'$, *i.e.*, $\frac{\partial f}{\partial y'} = 0$. On substituting the value of $\frac{\partial f}{\partial y'}$ in the Euler's equation, we get $\frac{\partial f}{\partial y} = 0$

This is the desired solution.

**Case 4.** If $f$ is independent of $x$ and $y$,

we have $\frac{\partial f}{\partial x} = 0$ and $\frac{\partial f}{\partial y} = 0$ or $\frac{\partial^2 f}{\partial x \partial y'} = 0$ and $\frac{\partial^2 f}{\partial y \partial y'} = 0$

Putting these value in Euler's equation (7), we have $y''\frac{\partial^2 f}{\partial y'^2} = 0$

If $\frac{\partial^2 f}{\partial y'^2} \neq 0$ then $y'' = 0$ whose solution is $y = ax + b$.

**Example 1.** *Test for an extremum the functional*

$$I[y(x)] = \int_0^1 (xy + y^2 - 2y^2 y')\,dx,\ y(0) = 1, y(1) = 2$$

**Solution.** Euler's equation

$$\frac{\partial f}{\partial y} - \frac{d}{dx}\left(\frac{\partial f}{\partial y'}\right) = 0 \qquad \text{...(1)}$$

Here $f = xy + y^2 - 2y^2 y'$

$$\frac{\partial f}{\partial y} = x + 2y - 4yy' \text{ and } \frac{\partial f}{\partial y'} = -2y^2$$

$$\frac{d}{dx}\left(\frac{\partial f}{\partial y'}\right) = \frac{d}{dx}\left(-2y^2\right) = -4yy'$$

Putting these values in (1), we get $x + 2y - 4yy' - (-4yy') = 0$

or

$x + 2y = 0$ *or* $y = -\frac{x}{2}$ At $x = 0, y = 0$; At $x = 1, y = -\frac{1}{2}$.

This extremal does not satisfy the boundary conditions $y(0) = 1$, $y(1) = 2$.
Hence there is no extremal. **Ans.**

**Example 2.** *Prove that the shortest distance between two points is along a straight line.*

**Solution.** Let $A\left(x_1, y_1\right)$ and $B\,(x_2, y_2)$ be the two given points and $s$ the length of the arc joining these points.

Then
$$s=\int_{x_1}^{x_2} ds=\int_{x_1}^{x_2}\sqrt{1+\left(\frac{dy}{dx}\right)^2}\,dx=\int_{x_1}^{x_2}\sqrt{\left(1+y'^2\right)}\,dx \quad \text{... (1)}$$
$$y(x_1)=y_1, \quad y(x_2)=y_2$$

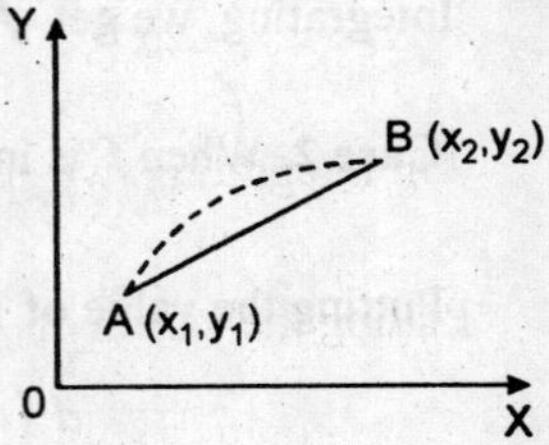


If $s$ satisfies the Euler's equation, then it will be minimum
$$\frac{\partial f}{\partial y}-\frac{d}{dx}\left(\frac{\partial f}{\partial y'}\right)=0 \quad \text{(Euler's equation)}$$

Here in (1), $f=\sqrt{\left(1+y'^2\right)}$

$f$ is independent of $y$, *i.e.*, $\dfrac{\partial f}{\partial y}=0$
$$\frac{d}{dx}\left(\frac{\partial f}{\partial y'}\right)=\frac{d}{dx}\left(\frac{\partial}{\partial y'}\sqrt{\left(1+y'^2\right)}\right)=\frac{d}{dx}\left[\frac{1}{2}\left(1+y'^2\right)^{-\frac{1}{2}}2y'\right]=\frac{d}{dx}\frac{y'}{\sqrt{\left(1+y'^2\right)}}$$

Putting these values in Euler's Equation, we have
$$0-\frac{d}{dx}\frac{y'}{\sqrt{(1+y')}}=0 \text{ or } \frac{d}{dx}\frac{y'}{\sqrt{\left(1+y'^2\right)}}=0$$

On integrating $\dfrac{y'}{\sqrt{\left(1+y'^2\right)}}$ constant (c), *i.e.*, $(y')^2=c^2\left(1+y'^2\right)$

or
$$y'^2\left(1-c^2\right)=c^2 \text{ or } y'^2=\frac{c^2}{1-c^2}=m^2 \text{ or } y'=m \text{ or } \frac{dy}{dx}=m$$

Integrating $\quad y = mx + c \quad$ ... (2)

which is a straight line. **Ans.**

Now
$$y(x_1)=y_1 \quad \text{and} \quad y(x_2)=y_2$$
$$mx_1+c=y_1 \text{ and } mx_2+c=y_2 \quad \text{...(3)}$$

on subtracting, we get

or
$$y_2-y_1=m(x_2-x_1) \text{ or } m=\frac{y_2-y_1}{x_2-x_1}$$

Subtracting (3) from (2), we get
$$y-y_1=m(x-x_1)$$
$$y-y_1=\frac{y_2-y_1}{x_2-x_1}(x-x_1)$$
**Proved.**

**Example 3.** *Find the curve connecting the points $(x_1, x_2)$ and $(x_2, y_2)$ which when rotated about the x-axis gives a minimum surface.*

*Find the external of the functional.*
$$\int_{x_1}^{x_2} 2\pi y\, ds \text{ or } 2\pi\int_{x_1}^{x_2} y\sqrt{\left(1+y'^2\right)}\,dx$$

*Subject to* $y(x_1)=y_1, y(x_2)=y_2$

**Solution.** $2\pi$ is constant so we have to find the extremal of

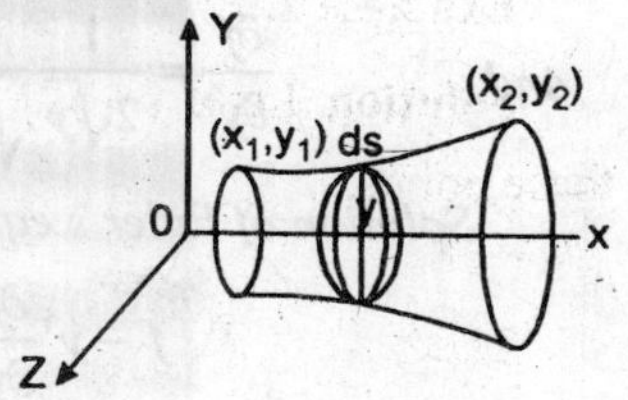


$$\int_{x_1}^{x_2} y\sqrt{(1+y'^2)}\,dx$$

Here $f = y\sqrt{(1+y'^2)}$ which is independent of $x$.

One form of Euler's equation is

$$\frac{d}{dx}\left[f - y'\frac{\partial f}{\partial y'}\right] - \frac{\partial f}{\partial x} = 0 \qquad \frac{d}{dx}\left[f - y'\frac{\partial f}{\partial y'}\right] = 0 \qquad \left(\frac{\partial f}{\partial x} = 0\right)$$

On integrating, we get, $f - y'\frac{\partial f}{\partial y'} = \text{constant } (c)$ ...(1)

Putting the values of $f$ and $\frac{\partial f}{\partial y'}$ in (1), we have

$$y = \sqrt{(1+y'^2)} - y'\frac{2y'}{2\sqrt{(1+y'^2)}}y = c$$

or $\quad y\sqrt{(1+y'^2)} - \frac{yy'^2}{\sqrt{(1+y'^2)}} = c$ or $y(1+y'^2) - yy'^2 = c\sqrt{(1+y'^2)}$.

$$y = c\sqrt{(1+y'^2)} \quad or \quad y^2 = c^2(1+y'^2)$$

or

$$y'^2 = \frac{y^2 - c^2}{c^2} \quad or \quad y' = \frac{\sqrt{y^2-c^2}}{c} \quad or \quad \frac{dy}{dx} = \frac{\sqrt{y^2-c^2}}{c}$$

$$\frac{dy}{\sqrt{y^2-c^2}} = \frac{dx}{c} \Rightarrow \int\frac{dy}{\sqrt{y^2-c^2}} = \int\frac{dx}{c} \Rightarrow \cosh^{-1}\frac{y}{c} = \frac{x}{c} + b$$

$y = c\cosh\left(\frac{x}{c} + b\right)$ which is the equation of catenary. This is the required extremal. **Ans.**

**Example 4.** *Find the curve connecting two points (not on a vertical line), such that a particle sliding down this curve under gravity (in absence of resistance) from one point to another reaches in the shortest time. (Brachistochrone problem).*

**Solution.** Let the particle slide on the curve OA from O with zero velocity. Let OP = $s$ and time taken from 0 to P = $t$. By the law of conservation of energy, we have

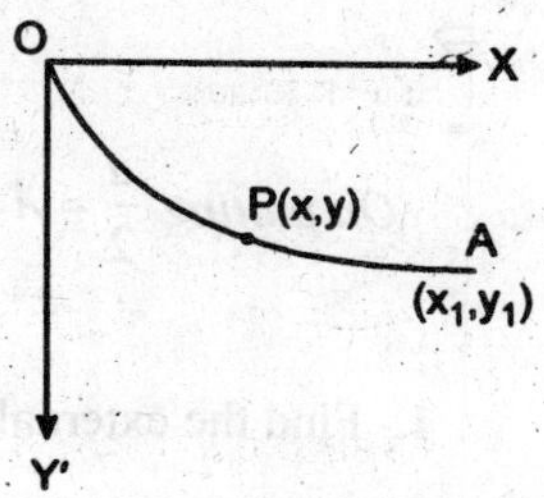


K.E. at $P$ – K.E. at O = potential energy at $P$.

$$\frac{1}{2}mv^2 - 0 = mgh$$

or $\quad \frac{1}{2}m\left(\frac{ds}{dt}\right)^2 = mgh \quad or \quad \frac{ds}{dt} = \sqrt{(2gy)}$

Time taken by the particle to move from O to A

$$T = \int_0^T dt = \int_0^{x_1}\frac{ds}{\sqrt{(2gy)}} = \frac{1}{\sqrt{(2g)}}\int_0^{x_1}\frac{ds}{\sqrt{y}} = \frac{1}{\sqrt{(2g)}}\int_0^{x_1}\frac{\sqrt{(1+y'^2)}}{\sqrt{y}}dx$$

Here, $f = \frac{\sqrt{(1+y'^2)}}{\sqrt{y}}$ *which is independent of x, i.e.,* $\frac{\partial f}{\partial x} = 0$.

*and*
$$\frac{\partial f}{\partial y'}=\frac{1}{2\sqrt{y}}\frac{2y'}{\sqrt{(1+y'^2)}}=\frac{y'}{\sqrt{y}\sqrt{(1+y'^2)}}$$

*Solution of Euler's equation is*

$$f-y'\frac{\partial f}{\partial y'}=\text{constant } c$$

*On substituting the values of f and* $\frac{\partial f}{\partial y'}$, *we get*

$$\frac{\sqrt{(1+y'^2)}}{\sqrt{y}}-y'\frac{y'}{\sqrt{y}\sqrt{(1+y'^2)}}=c$$

$$\Rightarrow \quad \sqrt{1+y^2}-\frac{y^2}{\sqrt{(1+y^2)}}=c\sqrt{y} \quad \text{or} \quad 1+y'^2-y'^2=c\sqrt{(1+y'^2)}\sqrt{y}$$

$$\Rightarrow \quad 1=c\sqrt{y(1+y'^2)} \quad \text{or} \quad 1+\left(\frac{dy}{dx}\right)^2=\frac{1}{yc^2} \quad \text{or} \quad \frac{dy}{dx}=\frac{\sqrt{1-yc^2}}{yc^2}$$

$$\Rightarrow \quad \frac{dy}{dx}=\frac{\sqrt{1/c^2-y}}{y}=\frac{\sqrt{a-y}}{y} \qquad \left(\frac{1}{c^2}=a\right)$$

$$dx=\sqrt{\frac{y}{a-y}}\,dy$$

$$\int_0^x x\,dx=\int_0^y\sqrt{\left(\frac{y}{a-y}\right)}dy \qquad \text{Put } y=a\sin^2\theta$$
$$dy=2a\sin\theta\cos\theta\,d\theta$$

$$x=\int_0^\theta\sqrt{\left(\frac{a\sin^2\theta}{a-a\sin^2\theta}\right)}2a\sin\theta\cos\theta\,d\theta=\int_0^\theta\left(\frac{\sin\theta}{\cos\theta}\right)2a\sin\theta\cos\theta d\theta=\int_0^\theta 2a\sin^2\theta d\theta$$

$$=a\int_0^\theta(1-\cos 2\theta)\,d\theta=a\left(\theta-\frac{\sin 2\theta}{2}\right)_0^\theta$$

$$\Rightarrow \quad x=\frac{a}{2}(2\theta-\sin 2\theta) \text{ and } y=a\sin^2\theta=\frac{a}{2}(1-\cos 2\theta)$$

*On putting* $\frac{a}{2}=A$ *and* $2\theta=\Theta$
$$\left.\begin{aligned}x&=A(\Theta-\sin\Theta)\\ y&=A(1-\cos\Theta)\end{aligned}\right] \text{ which is a cycloid.}$$

## EXERCISE 17.1

1. Find the external of the functional

$$I[y(x)]=\int_{x_0}^{x_1}\frac{1+y^2}{y'^2}dy$$ **Ans.** $y=\sinh(c_1x+c^2)$

2. Solve the Euler's equations for $\int_{x_0}^{x_1}(x+y')y'dx$. **Ans.** $y=-\frac{x^2}{4}+c_1x+c_2$

3. Solve the Euler's equation for $\int_{x_0}^{x_1}(1+x^2y')y'\,dx$ **Ans.** $y = cx^{-1} + c_2$

Find the extremals of the functional and extremism value of the following:

4. $I[y(x)] = \int_{x_0}^{x_1}\frac{1+y^2}{y'^2}\,dx$ **Ans.** $y = \sinh(c_1x + c_2)$

5. $I[y(x)] = \int_{\frac{1}{2}}^{1} x^2y^2\,dx$ subject to $y\left(\frac{1}{2}\right) = 1, y(2) = 4.$ **Ans.** $y = -\frac{c}{x} + d$, value $= 1$

6. $I[y(x)] = \int_0^2 (x-y')^2\,dx$ subject to $y(0) = 0, y(2) = 4.$ **Ans.** $y = \frac{x^2}{2} + cx + d$, value $= 2$

7. $\int_0^{\frac{\pi}{2}}(y'^2 - y^2)\,dx$ subject to $y(0), y\left(\frac{\pi}{2}\right) = 1$ **Ans.** $y = \sin x$, value $= 0$

8. $\int_0^1 (y'^2 + 12xy)\,dx$ subject to $y(0) = 0, y(1) = 1$ **Ans.** $y = x^3$, value $= \frac{21}{5}$

9. $\int_1^2 \frac{\sqrt{(1+y'^2)}}{x}\,dx$ subject to $y(1) = 0, y(2) = 1.$ **Ans.** $y = x^3$

## 17.6 ISOPERIMETRIC PROBLEMS

The determination of the shape of a closed curve of the given perimeter enclosing maximum area is the example of isoperimetric problem. In certain problems it is necessary to make a given integral.

$$I = \int_{x_1}^{x_2} f(x, y, y')\,dx \quad \text{....(1)}$$

maximum or minimum while keeping another integral

$$I = \int_{x_1}^{x_2} g(x, y, y')\,dx = K\,(\text{Constant}) \quad \text{... (2)}$$

Problems of this type are solved by Lagrange's multipliers method. We multiply (2) by $\lambda$ and add to (1) to extremize (1)

$$I^* = \int_{x_1}^{x_2} f(x, y, y')\,dx + \lambda\int_{x_1}^{x_2} g(x, y, y')\,dx = \int_{x_1}^{x_2} F\,dx \text{ (say)}$$

Then by Euler's equation $\frac{\partial F}{\partial y} - \frac{d}{dx}\left(\frac{\partial F}{\partial y'}\right) l = 0$

**Note.** Isoperimetric problem. To find out possible curves having the same perimeter, the one which encloses the maximum area.

**Example 5.** *Find the shape of the curve of the given perimeter enclosing maximum area.*
**Solution.** Let $P$ be the perimeter of the closed curve,

Then $$P = \int_{x_1}^{x_2}\sqrt{1+y'^2}\,dx \quad \text{...(1)}$$

The area enclosed by the curve, $x$-axis and two perpendicular lines is

$$A = \int_{x_1}^{x_2} y\,dx \qquad \text{....(2)}$$

We have to find the maximum value of (2) under the condition (1).

By Lagrange's multiplier method.

$$F = y + \lambda\sqrt{1+y'^2}$$

For maximum or minimum value of $A$, $F$ must satisfy Euler's equation

$$\frac{\partial F}{\partial y} - \frac{d}{dx}\left(\frac{\partial F}{\partial y'}\right) = 0$$

$$1 - \lambda\frac{d}{dx}\left[\frac{1}{2}(1+y'^2)^{-\frac{1}{2}}(2y')\right] = 0 \text{ or } 1 - \lambda\frac{d}{dx}\left(\frac{y'}{\sqrt{1+y'^2}}\right) = 0$$

Integrating w.r.t. '$x$', we get $x - \dfrac{\lambda y'}{(1+y'^2)} = c'$

or $$\frac{\lambda y'}{\sqrt{(1+y'^2)}} = x - a \text{ or } \lambda^2 y'^2 = (1+y'^2)(x-a)^2$$

$$[\lambda^2 - (x-a)^2]\,y'^2 = (x-a)^2$$

or $$y' = \frac{x-a}{\sqrt{[\lambda^2 - (x-a)^2]}} \qquad \text{or} \qquad \frac{dy}{dx} = \frac{x-a}{\sqrt{[\lambda^2 - (x-a)^2]}}$$

Integrating w.r.t. ($x$), we obtain

$$y = -\sqrt{[\lambda^2 - (x-a)^2]} + b$$

or $$y - b = -\sqrt{[\lambda^2 - (x-a)^2]} \quad (y-b)^2 = \lambda^2 - (x-a)^2 \quad \text{or} \quad (x-a)^2 + (y-b)^2 = l^2$$

This is the equation of a circle whose centre is $(a, b)$ and radius $\lambda$. **Ans.**

**Example 6.** *Find the extremal of the functional* $A = \int_{t_1}^{t_2} \frac{1}{2}(x\dot{y} - y\dot{x})\,dt$ *subject to the integral constraint* $\int_{t_1}^{t_2} \frac{1}{2}\sqrt{(\dot{x}^2 - \dot{y}^2)}\,dt = l.$

**Solution.** Here $f = \dfrac{1}{2}(x\dot{y} - \dot{x}y)$, $g = \sqrt{\dot{x}^2 - \dot{y}^2}$

$$F = f + \lambda g$$

$$F = \frac{1}{2}(x\dot{y} - y\dot{x}) + \lambda\sqrt{\dot{x}^2 + \dot{y}^2}$$

For $A$ to have extremal $F$ must satisfy the Euler's equation

$$\frac{\partial F}{\partial x} - \frac{d}{dx}\left[\frac{\partial F}{\partial \dot{x}}\right] = 0 \qquad \text{...(1)}$$

$$\frac{\partial F}{\partial x} - \frac{d}{dx}\left[\frac{\partial F}{\partial \dot{x}}\right] = 0 \qquad ...(1)$$

$$\frac{\partial F}{\partial x} - \frac{d}{dx}\left[\frac{\partial F}{\partial y}\right] = 0 \qquad ...(2)$$

From (1)

$$\frac{1}{2}\dot{y} - \frac{d}{dt}\left(-\frac{y}{2} + \frac{\lambda 2\dot{x}}{2\sqrt{\dot{x}^2+\dot{y}^2}}\right) = 0$$

$$\frac{d}{dt}\left(y - \frac{\lambda \dot{x}}{2\sqrt{\dot{x}^2+\dot{y}^2}}\right) = 0 \qquad ...(3)$$

From (2)

$$-\frac{1}{2}\dot{x} - \frac{d}{dt}\left[\frac{x}{2} + \frac{\lambda \dot{y}}{\sqrt{\dot{x}^2+\dot{y}^2}}\right] = 0 \qquad ...(4)$$

$$\frac{d}{dt}\left[x - \frac{\lambda \dot{y}}{\sqrt{\dot{x}^2+\dot{y}^2}}\right] = 0$$

Integrating (3) and (4), we have

$$y - \frac{\lambda \dot{x}}{\sqrt{\dot{x}^2+\dot{y}^2}} = c_1 \quad \text{or} \quad y - c_1 = \frac{\lambda \dot{x}}{\sqrt{\dot{x}^2+\dot{y}^2}} \qquad ...(5)$$

$$x - \frac{\lambda \dot{y}}{\sqrt{\dot{x}^2+\dot{y}^2}} = c_1 \quad \text{or} \quad x - c_2 = \frac{\lambda \dot{y}}{\sqrt{\dot{x}^2+\dot{y}^2}} \qquad ...(6)$$

Squaring (5), (6) and adding, we get

$$(x-c_2)^2 + (y-c_1)^2 = \lambda^2\left(\frac{\dot{x}^2+\dot{y}^2}{\dot{x}^2+\dot{y}^2}\right)$$

$$(x-c_2)^2 + (y-c_1)^2 = \lambda^2$$

This is the equation of circle. **Ans.**

**Example 7.** *Find the solid of maximum volume formed by the revolution of a given surface area.*

**Solution.** Let the curve PA pass through origin and it is rotated about the x - axis.

$$S = \int_0^a 2\pi y\, ds$$

$$S = \int_0^a 2\pi y\sqrt{(1+y'^2)}\,dx \qquad ...(1)$$

$$V = \int_0^a \pi y^2 dx \qquad ...(2)$$

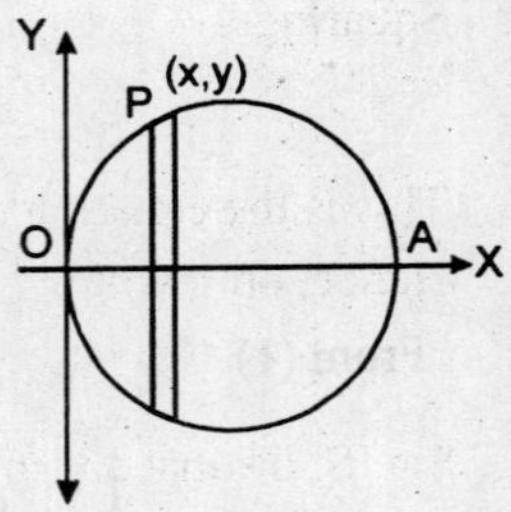


Here we have to extremize $V$ with the given $S$.

Here $\qquad f = \pi y^2,\ g = 2\pi y\sqrt{(1-y'^2)}$

$$F = f + \lambda g$$

$$F = \pi y^2 + \lambda 2\pi y\sqrt{(1+y'^2)}$$

For maximum $V$, $F$ must satisfy Euler's equation. But $F$ does not contain $x$.

$$\therefore \qquad F - y'\frac{\partial F}{\partial y'} = C$$

or $$\pi y^2 + \lambda 2\pi y\sqrt{(1+y'^2)} - y'\frac{1}{2}\frac{2\pi y\ \lambda 2y'}{\sqrt{(1+y'^2)}} = C$$

or $$\pi y^2 + 2\pi y\lambda\sqrt{(1+y'^2)} - \frac{2\pi\lambda y y'^2}{\sqrt{(1+y'^2)}} = C$$

or $$\pi y^2 + \frac{2\pi y\lambda}{\sqrt{(1+y'^2)}} = C$$

As the curve passes through origin (0, 0), so $C = 0$.

$$\pi y^2 + \frac{2\pi y\lambda}{\sqrt{(1+y'^2)}} = 0$$

or $$y + \frac{2\lambda}{\sqrt{(1+y'^2)}} = 0 \quad \text{or} \quad y\sqrt{(1+y'^2)} = -2\lambda$$

or $$1 + y'^2 = \frac{4\lambda^2}{y^2} \quad \text{or} \quad y'^2 = \frac{4\lambda^2}{y^2} - 1 = \frac{4\lambda^2 - y^2}{y^2}$$

or $$\frac{dy}{dx} = \frac{\sqrt{(4\lambda^2 - y^2)}}{y}$$

$$\int \frac{y\,dy}{\sqrt{(4\lambda^2 - y^2)}} = \int dx + C$$

$$-\sqrt{4\lambda^2 - y^2} = x + C \qquad ...(1)$$

or $$\sqrt{4\lambda^2 - y^2} = -x - C$$

The curve passes through (0, 0). On putting $x = 0,\ y = 0$ in (1) we get

$$-C = 2\lambda$$

(1) becomes $$\sqrt{4\lambda^2 - y^2} = -x + 2\lambda$$

Squaring $$4\lambda^2 - y^2 = (x-2\lambda)^2$$

or $$(x-2\lambda)^2 + y^2 = 4\lambda^2$$

This is the equation of a circle.

Hence, on revolving the circle about $x$- axis, the solid formed is a sphere. **Ans.**

## EXERCISE 17.2

1. Show that an isosceles triangle has the smallest perimeter for a given area and a given base.
2. Find the external in the isoperimetric problem of the extremum of

$$\int_0^1 (y'^2 + z'^2 - 4xz' - 4z)\,dx$$

subject to $\int_0^1 (y'^2 + xy' - z'^2)\,dx = 2, y(0) = 0, z(0) = 0, y(1) = 1, z\ (1) = 1.$

**Ans.** $y = \dfrac{-5x^2}{2} + \dfrac{7x}{2}, z = x.$

3. Find the surface with the smallest area which encloses a given volume. **Ans.** Sphere
4. Find the external of the functional $\int_{t_1}^{t_2} \sqrt{x^2 + y^2 + z^2}\,dt$ subject to $x^2 + y^2 + z^2 = a^2$.

**Ans.** Arc of a great circle of a sphere.

5. Find the extremals of the isoperimetric problem $\int_{x_0}^{x_1} y'^2\,dx$ subject to $\int_{x_0}^{x_1} y\,dx = c.$ **Ans.** $y = x^2 + ax + b$

## 17.7 FUNCTIONALS OF SECOND ORDER DERIVATIVES

Let us consider the extremum of a functional.

$$\int_{x_1}^{x_2} [f(x, y, y', y'')]\ dx \qquad \text{... (1)}$$

The necessary condition for the above mentioned functional to be extremum is

$$\frac{\partial f}{\partial y} - \frac{d}{dx}\left(\frac{\partial f}{\partial y'}\right) + \frac{d^2}{dx^2}\left(\frac{\partial f}{\partial y''}\right) = 0$$

**Proof.** Let the boundary conditions be

$$y(x_1) = y_1,\ y(x_2) = y_2,\ y'(x_1) = y'_1,\ y'(x_2) = y'_2$$

Let $\alpha$ be a parameter and $\eta(x)$ is a differentiable function.

At the end points $\eta(x_1) = \eta(x_2) = 0$ and $\eta'(x_1) = \eta'(x_2) = 0$

Putting $y + \alpha\,\eta(x)$ for $y$ in (1), we have

$$\int_{x_1}^{x_2} f[x, y+\alpha\,\eta(x), y'+\alpha\,\eta'(x), y''+\alpha\,\eta''(x)]\,dx$$

Writing $$\int_{x_1}^{x_2} f[x, y+\alpha\,\eta(x), y'+\alpha\,\eta'(x), y''+\alpha\,\eta''(x)]\,dx = \int_{x_1}^{x_2} F dx = 1$$

For extremum value of (1)

$$\frac{dI}{d\alpha}=0$$

$$\frac{dI}{d\alpha}=\int_{x_1}^{x_2}\frac{\partial F}{\partial\alpha}dx$$

Differentiating under the sign of integral, we get

$$=\int_{x_1}^{x_2}\left(\frac{\partial F}{\partial y}\frac{\partial y}{\partial\alpha}+\frac{\partial F}{\partial y'}\frac{\partial y'}{\partial\alpha}+\frac{\partial F}{\partial y''}\frac{\partial y''}{\partial\alpha}\right)dx=\int_{x_1}^{x_2}\left(\frac{\partial F}{\partial y}\frac{\partial(\alpha n)}{\partial\alpha}+\frac{\partial F}{\partial y'}\frac{\partial(\alpha n')}{\partial\alpha}+\frac{\partial F}{\partial y''}\frac{\partial(\alpha n'')}{\partial\alpha}\right)dx$$

But $\frac{dI}{d\alpha}=0$ when $\alpha = 0$

$$0=\int_{x_1}^{x_2}\left[\frac{\partial f}{\partial y}\eta+\frac{\partial f}{\partial y}\eta'+\frac{\partial f}{\partial y}\eta''\right]dx \quad \text{or} \quad \int_{x_1}^{x_2}\frac{\partial f}{\partial y}\eta\,dx+\int_{x_1}^{x_2}\frac{\partial f}{\partial y'}\eta' dx+\int_{x_1}^{x_2}\frac{\partial f}{\partial y''}\eta'' dx=0$$

Integrating by parts, w.r.t. '$x$', we have

$$\int_{x_1}^{x_2}\frac{\partial f}{\partial y}\eta\,dx+\left[\frac{\partial f}{\partial y}\eta-\int_{x_1}^{x_2}\frac{\partial}{\partial x}\left(\frac{\partial f}{\partial y'}\right).\eta\,dx\right]_{x_1}^{x_2}+\left(\frac{\partial f}{\partial y''}\eta'-\frac{d}{dx}\left(\frac{\partial f}{\partial y''}\right).\eta+\frac{d^2}{dx^2}\left(\frac{\partial f}{\partial y''}\right)\int_{x_1}^{x_2}\eta dx\right)_{x_1}^{x_2}=0$$

But $n(x_1)=n(x_2) = 0$ and $\eta'(x_1)=\eta'(x_2)=0$

So $\int_{x_1}^{x_2}\left[\frac{\partial f}{\partial y}-\frac{d}{dx}\left(\frac{\partial f}{\partial y'}\right)+\frac{d^2}{dx^2}\left(\frac{\partial f}{\partial y''}\right)\right]\eta(x)dx=0 \Rightarrow \frac{\partial f}{\partial y}-\frac{d}{dx}\left(\frac{\partial f}{\partial y'}\right)+\frac{d^2}{dx^2}\left(\frac{\partial f}{\partial y''}\right)=0$ **Proved.**

## EXERCISE 17.3

1. Find the extremal of $\int_{x_0}^{x_1}(16y^2-y''^2+x^2)\,dx$. **Ans.** $y=c_1e^{2x}+c_2e^{-2x}+c_3\cos 2x+c_4\sin 2x$

2. Find the extremal of $\int_{-c}^{c}(ay+\frac{1}{2}by''^2)\,dx$ subject to $y(-c)=0, y'(-c)=0,$ $y(c)=0, y'(c)=0.$ **Ans.** $y=-\frac{a}{24b}(x^2-c^2)^2$

3. Find the extremal of $\int_0^x y''^2\,dx$ subject to $\int_0^x y^2\,dx=1, y(0)=y(p)=0, y''(p)=0.$ **Ans.** $y=a_1\sin x+a_2\sin 2x+\ldots\ldots$

4. Find the extremal of $\int_{x_0}^{x_1}(2xy+y'''^2)\,dx$. **Ans.** $y=\frac{x^7}{7!}+c_1x^5+c_2x^4+c_3x^3+c_4x^2+c_5x+c_6$

# 18

# Tensor Analysis

## 18.1 INTRODUCTION

*Scalars* are specified by magnitude only, *vectors* have magnitude as well as direction. *Tensors* are associated with magnitude and two or more directions. For example, the stress of an elastic solid at a point depends upon two directions. One of the directions is given by the normal to the area, while the other is that of the force on it.

Tensors are similar to vectors. A vector can be specified by its components (Magnitude and direction). A tensor can be specified only by its components which depend upon the system of reference. The components of the same tensor will be different for two different sets of axes with different orientations.

Tensors analysis is suitable for mathematical formulation of natural laws in forms which are invariant with respect to different frames of reference. That is why Einstein used tensors for the formulation of his Theory of Relativity.

## 18.2 CO-ORDINATE TRANSFORMATION

If we have two systems of rectangular co-ordinate axes *OX, OY, OZ; OX'*, or, *OZ'*; having same origin such that the direction cosines of the lines

*OX', OY', OZ'* relative to the system *OX YZ* are

$$l_1, m_1, n_1;\ l_2, m_2, n_2;\ l_3, m_3, n_3$$

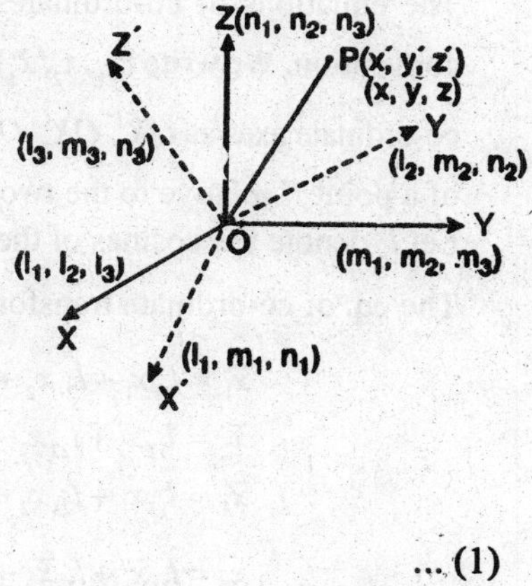


Two equivalent systems of transformation equations express $x', y', z'$ in terms of *x, y, z* and vice versa.

$$\left.\begin{aligned} x' &= l_1 x + m_1 y + n_1 z \\ y' &= l_2 x + m_2 y + n_2 z \\ z' &= l_3 x + m_3 y + n_3 z \end{aligned}\right\} \qquad \ldots (1)$$

$$\left.\begin{aligned} x &= l_1 x' + l_2 y' + l_3 z' \\ y &= m_1 x' + m_2 y' + m_3 z' \\ z &= n_1 x' + n_2 y' + n_3 z' \end{aligned}\right\} \qquad \ldots (2)$$

where $(x', y', z')$ and $(x, y, z)$ are co-ordinates of point *P* relative to two systems of co-ordinate axes. System of transformation eq. shown above in (1) and (2) can be written as

| | $x$ | $y$ | $z$ |
|---|---|---|---|
| $x'$ | $l_1$ | $m_1$ | $n_1$ |
| $y'$ | $l_2$ | $m_2$ | $n_2$ |
| $z'$ | $l_3$ | $m_3$ | $n_3$ |

## 18.3 SUMMATION CONVENTION

The sum of the following $a_1 x_1 + a_2 x_2 + a_3 a_3 + .... + a_n x_n$ ...(1)

can be written in brief as $\sum_{i=1}^{i=n} a_i x_i$ ... (2)

More simplified and compact notation for (2) used by Einstein is $a_i x^i$. ...(3)

In (3) we have omitted $\Sigma$-sign.

$$a_i x^i = a_1 x^1 + a_2 x^2 + a_3 x_3^3 + ....a_n x^n$$

We write $x_1, x_2, x_3, ... x_n$ as $x^1, x^2, x^3, ... x^n$ in tenser analysis. These superscripts donot stand for powers of $x$ but indicate different symbols. The power of $x^i$ is written as $(x^i)^2, (x^i)^3 ....$

**Example 1.** *Write out* $a_{rs} x^s = b_r (r, s = 1, 2, 3....n)$ *in full:*

**Solution.** $a_{rs} x^s = b_r$

$a_{is} x^s + a_{2s} x^s + a_{3s} x^s + ... + a_{ns} x^n = b_1 + b_2 + b_3 + .... + b_n$ ($r$ occurs 1 to $n$)

$(a_{11} x^1 + a_{12} x^2 + a_{13} x^3 + .... a_{1n} x^n) + (a_{21} x^1 + a_{22} x^2 + a_{23} x^3 + ... + a_{2n} x^n)$
$+ (a_{31} x^1 + a_{32} x^2 + a_{33} x^3 + ... +....a_{3n} x^n) + ... = b_1 + b_2 + b_3 +....+ b_n$

**Example 2.** If $f = f(x^1, x^2, x^3, ... x^n)$ then show that $df = \frac{\partial f}{dx^i} dx^i$

**Solution.** $df = \frac{\partial f}{dx^1} dx^1 + \frac{\partial f}{dx^2} dx^2 + ... + \frac{\partial f}{dx^n} dx^n = \frac{\partial f}{\partial x^i} dx^i$ **Proved.**

## 18.4 SUMMATION OF CO-ORDINATES

The equations of co-ordinates can be written in very compact form in terms of summation convention. We write $(x_1, x_2, x_3)$ and $(\bar{x}_1, \bar{x}_2, \bar{x}_3)$ instead of $(x, y, z)$ and $(x', y', z')$ and denote the co-ordinate axes as $OX_1, OX_2, OX_3$ and $O\bar{X}_1, O\bar{X}_2, O\bar{X}_3$. Also we denote $x_i, \bar{x}_j$ as the co-ordinates of a point $P$ relative to the two systems of axes; where $i = 1, 2, 3, j = 1, 2, 3$.

Let $l_{ij}$ denote the cosines of the angles between $OX_i, O\bar{X}_j$. In general $l_{ij} \neq l_{ji}$

The eq. of co-ordinate transformation can be written as

$$\left.\begin{aligned} \bar{x}_1 &= l_{11} x_1 + l_{21} x_2 + l_{31} x_3 \\ \bar{x}_2 &= l_{12} x_1 + l_{22} x_2 + l_{32} x_3 \\ \bar{x}_3 &= l_{13} x_1 + l_{23} x_2 + l_{33} x_3 \end{aligned}\right\} \quad ... (1a)$$

$$\left.\begin{aligned} x_1 &= l_{11} \bar{x}_1 + l_{12} \bar{x}_2 + l_{13} \bar{x}_3 \\ x_2 &= l_{21} \bar{x}_1 + l_{22} \bar{x}_2 + l_{23} \bar{x}_3 \\ x_3 &= l_{31} \bar{x}_1 + l_{32} \bar{x}_2 + l_{33} \bar{x}_3 \end{aligned}\right\} \quad ... (1b)$$

These equations of co-ordinate transformation can be represented by means of a table form such that

| | $x_1$ | $x_2$ | $x_3$ |
|---|---|---|---|
| $\bar{x}_1$ | $l_{11}$ | $l_{21}$ | $l_{31}$ |
| $\bar{x}_2$ | $l_{12}$ | $l_{22}$ | $l_{32}$ |
| $\bar{x}_3$ | $l_{13}$ | $l_{23}$ | $l_{33}$ |

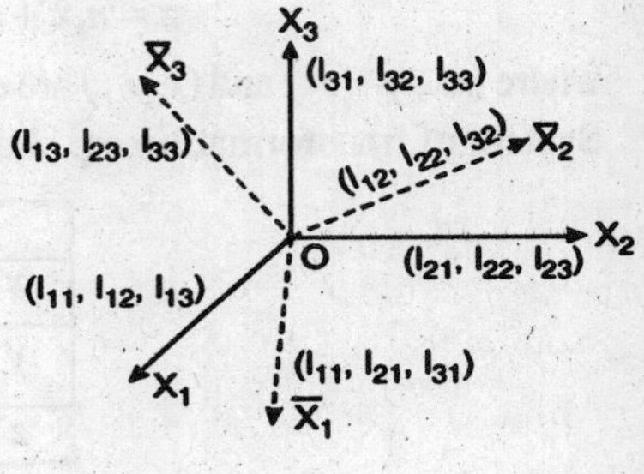


Adopting summation on convention *i.e.*,

$a_{11} + a_{22} + a_{33} = a_{ii}$

$a_{usip} b_{iq} = a_{1p} b_{1q} + a_{2p} b_{2q} + a_{3p} b_3 q$ we re-write above equations as

$\bar{x}_1 = l_{i1} x_i \qquad x_1 = l_{1j} \bar{x}_j$

$\bar{x}_2 = l_{i2} x_i \qquad x_2 = l_{2j} \bar{x}_j$

$\bar{x}_3 = l_{i3} x_i \qquad x_3 = l_{3j} \bar{x}_j$

We can re-write these equations in single equation in the form.

$\bar{x}_j = l_{ij} x_i, \; x_i = l_{ij} \bar{x}_j$

which are complete equivalents of the equations of co-ordinate transformation from either system to another.

## 18.5 RELATION BETWEEN THE DIRECTION COSINES OF THREE MUTUALLY PERPENDICULAR STRAIGHT LINES

The direction cosines of any three mutually perpendicular straight lines $O\bar{X}_1, O\bar{X}_2, O\bar{X}_3$ relative to the system $OX_1, OX_2, OX_3$ are $l_{11}, l_{21}, l_{31}, l_{12}, l_{22}, l_{32}, l_{23}, l_{33}$.

The relation between these direction cosines are

$l_{11}l_{11} + l_{21}l_{21} + l_{31}l_{31} = l_{j1}\, l_{j1} = 1 \qquad l_{12}l_{12} + l_{22}l_{22} + l_{32}l_{32} = l_{j2} l_{j2} = 1$

$l_{13}l_{13} + l_{23}l_{23} + l_{33}l_{33} = l_{j3}\, l_{j3} = 1.$

Similarly,

$l_{11}l_{12} + l_{21}l_{22} + l_{31}l_{32} = l_{j1}\, l_{j2} = 0 \qquad l_{12}l_{13} + l_{22}l_{23} + l_{32}l_{33} = l_{j2} l_{j3} = 0$

$l_{13}l_{11} + l_{23}l_{21} + l_{33}l_{31} = l_{j3}\, l_{j1} = 0$

Finally, we can write these equations by means of a single equation as

$$l_{ij} l_{kj} = \begin{cases} 1, \text{ when } i = k \\ 0, \text{ when } i \neq k \end{cases} \quad \text{or} \quad \delta_{ik} = \begin{cases} 1, \text{ when } i = k \\ 0, \text{ when } i \neq k \end{cases}$$

where $\delta_{ik}$ is the kronecker delta.

or $\qquad \delta_{ik} = l_{ij} l_{kj}$

Now, we know that $\bar{x}_j = l_{ij} x_i$

Multiplying both sides by $l_{jk}$ then

or $\qquad l_{jk} \bar{x}_j = l_{ij} l_{jk} x_i \Rightarrow l_{jk} \bar{x}_j = \delta_{ik} x_i$

putting $i = k$ *i.e.*, $\delta_{ik} = 1$ when $i = k$

$$\delta_{kk} x_k = l_{jk} \bar{x}_j \Rightarrow x_k = l_{jk} \bar{x}_j$$

## 18.6 TRANSFORMATION OF VELOCITY COMPONENTS ON CHANGE FROM ONE SYSTEM OF RECTANGULAR AXES TO ANOTHER

We know that with the help of parallelogram law of velocities, that any given velocity can be represented by means of its three components along three mutually perpendicular lines and the three components characterise velocity completely. The components change as we pass from one system of mutually perpendicular lines to another.

Let $OX_1, OX_2, OX_3$ and $O\bar{X}_1, O\bar{X}_2, O\bar{X}_3$ are two systems of rectangular axes and suppose that $l_i, \bar{l}_j$ are the direction cosines of the line of action of the velocity and $v$, denote the magnitude of the velocity. Then

$$v_i = l v,\ \bar{v}_j = v\bar{l}_j \quad \text{... (1)}$$

where $v_i$ and $\bar{v}_j$, denotes the components of velocity relative to the two systems of axes.

By the equation of co-ordinate transformation, we have

$$\bar{l}_j = l_{ij} l_i,\ l_i = l_{ij}\bar{l}_j \quad \text{... (2)}$$

From (1) and (2).

$$\frac{\bar{v}_j}{v} = l_{ij}\frac{v_i}{v},\ \frac{v_i}{v} = l_{ij}\frac{\bar{v}_j}{v}. \qquad i.e.,\ \bar{v}_j = l_{ij} v_i$$

Thus we see equation of transformation of velocity components are same as for the transformation of co-ordinate of points.

## 18.7 RANK OF A TENSOR

The rank of a tensor is the number (without counting an index which appears once as a subscript) of indices in the symbol representing a tensor. For example

| *Tensor* | *Symbol* | *Rank* |
|---|---|---|
| Scalar | $A$ | zero |
| Contravariant Tensor | $B^i$ | 1 |
| Covariant Tensor | $C_k$ | 1 |
| Covariant Tensor | $D_y$ | 2 |
| Mixed Tensor | $E^{il}_{jkl}$ | 4 |

In an $n$-dimensional space, the number of components of a tensor of rank $r$ is $n^r$.

## 18.8 FIRST ORDER TENSORS

**Definition.** Any entity representable by a set of three numbers (called components) relatively to a system of rectangular axes is called first order tensors, if its components $a_i$, $a_j$ relatively to any two systems of rectangular axes $OX_1, OX_2, OX_3, O\bar{X}_1, O\bar{X}_2, O\bar{X}_3$ are connected by the relation, $\bar{a}_j = l_{ij} a_i$ ...(3)

$\Rightarrow \quad a_i = l_{ij}\,\bar{a}_j$

$l_{ij}$ being cosines of angle between $OX_i$ and $O\bar{X}_j$. A tensor of first order is also called a *vector.*

**Note.** Consider any two tensors of first order and let $a_i, b_j, \bar{a}_p, \bar{b}_q$; be the components of the same relatively to two different systems of axes, we have

$$\bar{a}_p = l_{ip} a_i,\ \bar{b}_q = l_{jq} b_j$$

where $l_{ip}$ and $l_{jq}$ have their usual meanings. This gives

$$\bar{a}_p \bar{b}_q = l_{ip} a_i l_{jq} b_j = l_{ip} l_{jq} a_i b_j \quad \text{... (1)}$$

The R.H.S. of (1) denotes the sum of 9 terms obtained by giving all possible pair of values to the dummy suffixes $i,j$ so that each components of $\bar{a}_p\bar{b}_q$ is expressed as a linear combination of nine components of the set $a_i$, $b_j$; the coefficient being dependent only upon the positions of the two systems of axes relative to each other and not on the components of the sets $\bar{a}_p\bar{b}_q$, $a_i b_j$

## 18.9 SECOND ORDER TENSORS

**Definition.** Any entity representable by a two suffixes set relatively to a system of rectangular axes is called a second order tensor, if the sets $a_{ij}, \bar{a}_{pq}$ representing the entity relative to any two systems of rectangular axes $OX_1, OX_2, OX_3, O\overline{X}_1, O\overline{X}_2, O\overline{X}_3$ are connected by relation.

$$\overline{a}_{pq} = l_{ip}\, l_{jp}\, a_{ij}$$

## 18.10 TENSORS OF ANY ORDER

**Definition.** Any entity representable by a set with *m,* suffixes relatively to a system of rectangular co-ordinate axes is called a tensor of order *m,* if the set $a_{ijkl}$ ...... , $\overline{a}_{pqrs}$ .... representing the entity relatively to any two systems of rectangular axes $OX_1, OX_2, OX_3, O\overline{X}_1, O\overline{X}_2, O\overline{X}_3$ are connected by the relation

$$\overline{a}_{pqrs} \ldots = l_{ip} l_{jq} l_{kr} l_{ls} \ldots a_{ijkl} \ldots$$

We say that $a_{ijkl}$..... are the components of tensor relatively to the rectangular system of axes $OX_1, OX_2, OX_3$.

## 18.11 TENSOR OF ZERO ORDER

**Definition.** Any entity representable by a single number such that the same number represents the entity irrespective of any underlying system of axes is called a tensor of order zero. A tensor of order zero is also called a *scalar.*

## 18.12 ALGEBRAIC OPERATIONS ON TENSORS

**Theorem.** If $a_{ijkl}$ ...., $b_{ijkl}$ .... are two tensors of the same order then

$$c_{ijkl} \ldots = a_{ijkl} \ldots + b_{ijkl} \ldots$$

is a tensor of the same order.

**Proof.** Let $a_{ijkl}$ ...., $b_{ijkl}$ ..... and $\overline{a}_{pqrs}$ .....,$\overline{b}_{pqrs}$ ... be the components of the given tensors relatively to two systems $OX_1, OX_2, OX_3, O\overline{X}_1, O\overline{X}_2, O\overline{X}_3$.

We write

$$c_{ijkl} \ldots = a_{ijkl} \ldots + b_{ijkl} \ldots,$$

$$\overline{c}_{pqrs} \ldots = \overline{a}_{pqrs} \ldots + \overline{b}_{pqrs} \ldots$$

Let $l_{ij}$ denote the cosine of the angle between $OX_i$ and $OX_j$.

$$\overline{c}_{pqrs} \ldots = l_{ip} l_{iq} l_{kr} \ldots c_{ijkl} \ldots \quad \ldots (1)$$

As $a_{ijkl}$..... and $b_{ijkl}$ ...... are tensors, we have

$$\overline{a}_{pqrs} \ldots = l_{ip} l_{jq} l_{kr} l_{ls} \ldots a_{ijkl} \ldots \quad \ldots (2)$$

$$\overline{b}_{pqrs} \ldots = l_{ip} l_{jq} l_{kr} l_{ls} \ldots b_{ijkl} \ldots \quad \ldots (3)$$

Adding (2) and (3), we obtain (1).

Hence the theorem

Similarly, we can show for difference

$$\overline{a}_{pqrs} \ldots - \overline{b}_{pqrs} \ldots = l_{ip} l_{jq} l_{kr} l_{ls} \ldots \left[ a_{ijkl} \ldots - b_{ijkl} \ldots \right]$$

$$\overline{d}_{pqrs} \ldots = l_{ip} l_{jq} l_{kr} l_{ls} \ldots d_{ijkl} \ldots$$

## 18.13 PRODUCT OF TWO TENSORS

**Theorem.** If $a_{ijkl}$ ...., $b_{pqrs}$ ....be two tensors of order α and β respectively, then $c_{ijkl \ldots pqrs} \ldots = a_{ijkl} \ldots b_{pqrs} \ldots$ is a tensor of order α + β.

**Proof.** Let $a_{ijkl}$....., $b_{pqrs}$ ...... and $\overline{a}_{i_1 j_1 k_1 l_1}$ .....,$\overline{b}_{p_1 j_1 k_1 l_1}$....., $\overline{b}_{p_1 q_1 r_1 s_1}$ ..... be the components of given tensor relatively to two systems $OX_1$, $OX_2$, $OX_3$, $O\overline{X}_1, O\overline{X}_2, O\overline{X}_3$ we write

$$c_{ijkl\ldots pqrs} = a_{ijkl}\ldots\, b_{pqrs}$$
$$\overline{c}_{i_1 j_1 k_1 l_1 \ldots\ldots p_1 q_1 r_1 s_1}\ldots.. = \overline{a}_{i_1 j_1 k_1 l_1}\ldots..\overline{b}_{p_1 q_1 r_1 s_1}\ldots..$$

Let $l_{ij}$ be the direction cosines of the angle between $OX_i$ and $O\overline{X}_j$, then

$$\overline{c}_{i_1 j_1 k_1 l_1 \ldots\ldots p_1 q_1}\ldots.. = l_{ii_1} l_{jj_1}\ldots.. l_{pp_1} l_{qq_1}\ldots. c_{ijk\ldots pqr\ldots\ldots} \quad \ldots (1)$$

As $a_{ijkl}$ ..... and $b_{pqrs}$ ...... are tensors we have

$$\overline{a}_{i_1 j_1 k_1 l_1}\ldots.. = l_{ii_1} l_{jj_1} l_{kk_1}\ldots.. a_{ijkl}\ldots.. \quad \ldots (2)$$

$$\overline{b}_{p_1 q_1 r_1 s_1}\ldots.. = l_{pp_1} l_{qq_1} l_{rr_1}\ldots.. b_{pqrs}\ldots. \quad \ldots (3)$$

Multiplying (2) and (3) we get (1). The new tensor obtained is called product of the tensors.

## 18.14 QUOTIENT LAW OF TENSORS

**Thorem.** If there be an entity representable by a multisuffix set $a_{ij}$ relatively to any given system of rectangular axes and if $a_{ij} b_i$ is a vector, where $b_i$ is any arbitrary vector whatsoever then $a_{ij}$ is a tensor of order two.

**Proof.** $a_{ij} b_i = c_j$ so that $c_j$ is a vector. Let $a_{ij} b_i$, $c_j$ and $\overline{a}_{pq}$, $\overline{b}_p$, $\overline{c}_q$ be the components of the given entity and two vectors relatively to two systems of axes $OX_1$, $OX_2$, $OX_3$, $O\overline{X}_1, O\overline{X}_2, O\overline{X}_3$, then we have

$$a_{ij} b_i = c_j \quad \ldots (1)$$

$$\overline{a}_{pq} \overline{b}_p = \overline{c}_q \quad \ldots (2)$$

Also, $b_i$, $c_j$ being vectors, we have

$$\overline{c}_q = l_{jq} c_j \quad \ldots (3)$$

$$b_i = l_{ip} \overline{b}_p \quad \ldots (4)$$

From these, we have

$$\overline{a}_{pq} \overline{b}_p = \overline{c}_q = l_{jq} c_j - l_{jq} a_{ij} b_i = l_{jq} a_{ij} l_{ip} \overline{b}_p = l_{ip} l_{jq} a_{ij} \overline{b}_p$$

*i.e.,* $\left(\overline{a}_{pq} - l_{ip} l_{iq} a_{ij}\right) \overline{b}_p = 0$

As the vector $\overline{b}_p$ is arbitrary, we consider three vectors whose components relatively to $O\overline{X}_1, O\overline{X}_2, O\overline{X}_3$ are 1, 0, 0 ; 0, 1, 0 ; 0, 0, 1.

For these vectors, we have from (5)

$\overline{a}_{1q} - l_{il} l_{jq} a_{ij} = 0, \qquad \overline{a}_{2q} - l_{i2} l_{jq} a_{ij} = 0, \qquad \overline{a}_{3q} - l_{i3} l_{jq} a_{ij} = 0$

These are equivalent to

$$\overline{a}_{pq} - l_{ip} l_{jq} a_{ij} = 0$$

*i.e.,* $\overline{a}_{pq} = l_{ip} l_{jq} a_{ij}$, [This shows that $a_{ij}$ is of second order]

so that the components of the given entity obey the tensorial transformation laws. Hence the result.

## 18.15 CONTRACTION THEOREM

**Theorem.** If $a_{ijkl}$ ...... is a tensor of order *m*, then the set obtained on identifying any two suffixes is a tensor of order (*m* –2).

**Proof.** Let $a_{ijkl}$ ..... , $\overline{a}_{pqrs}$ ....., be the components of the given tensor relatively to two coordinate systems of axes $OX_1$, $OX_2$, $OX_3$, $O\overline{X}_1, O\overline{X}_2, O\overline{X}_3$, so that we have,

$$\bar{a}_{pqrs}..... = l_{ip} l_{jq} l_{kr} ....a_{ijkl}..... \quad ...(1)$$

Let us identify $q$ and $s$ then,

$$\bar{a}_{pqrs}..... = l_{ip} l_{js} l_{kr} l_{ls} ....a_{ijkl}.....$$

$$\bar{a}_{pr}..... = l_{ip} l_{kr} \delta_{jl} ....a_{ijkl}..... \quad \left[\because \begin{array}{l} \delta_{jl} = 0, \ j \neq l \\ \delta_{jl} = 1, \ j = l \end{array}\right]$$

$$= l_{ip} l_{kr} ....a_{ilkl}.... = l_{ip} l_{kr} ....a_{ik}....$$

This shows that the order of the tensor reduces by two.

Hence the theorem.

## 18.16 SYMMETRIC AND ANTISYMMETRIC TENSORS

If $\quad A_{rs}^{k} = A_{sr}^{k} \quad$ or $\quad \left(A_{k}^{r,s} = A_{k}^{s,r}\right)$

then $A_{rs}^{k}$ (or $A_{sr}^{k}$) are said to be symmetric tensors.

If $\quad B_{rs}^{k} = -B_{sr}^{k} \quad$ or $\quad \left(B_{k}^{rs} = -B_{k}^{sr}\right)$

then $B_{rs}^{k}$ (or $B_{k}^{sr}$) are known as antisymmetric tensors.

The symmetric (or antisymmetric) property is conserved under a transform of co-ordinates.

## 18.17 SYMMETRIC AND SKEW SYMMETRIC TENSORS

Invariance of the symmetric and skew-symmetric character of the sets of components of tensors

**Theorem.** Show that if $a_{ijkl}.....$ is symmetric (skew-symmetric) in any two suffixes, then so is also $\bar{a}_{pqrs}.....$ in the same suffixes.

**Proof.** Let $a_{ijkl}.....$, $\bar{a}_{pqrs}.....$ be the components of a tensor respectively to two systems of axes $OX_1$, $OX_2$, $OX_3$, $O\bar{X}_1, O\bar{X}_2, O\bar{X}_3$. Then we have

$$\bar{a}_{pqrs}..... = l_{ip} l_{jq} l_{kr} l_{ls} ....a_{ijkl}..... \quad ...(1)$$

Now, suppose that $a_{ijkl}......$ is symmetric in the second and fourth suffixes. Interchanging $q$ and $s$ on the two sides of (1) we obtain

$$\bar{a}_{psrq}..... = l_{ip} l_{js} l_{kr} l_{lq} ....a_{ijkl}..... \quad ...(2)$$

As $j$ and $l$ are dummy, we can interchange them. Then interchanging $j$ and $l$ on the R.H.S. of (2) we get

$$\bar{a}_{psrq}..... = l_{ip} l_{ls} l_{kr} l_{jq} ....a_{ijkl}..... \quad ...(3)$$

$$\bar{a}_{psrq}..... = l_{ip} l_{jq} l_{kr} l_{ls} ....a_{ijkl}.....$$

The set $a_{ijkl}.........$ is symmetric in the second and fourth suffixes.

Now from (1) and (3) we have

$$\bar{a}_{pqrs}..... = \bar{a}_{psrq}......$$

Hence the result.

**Definition.** A tensor is said to be symmetric (skew-symmetric) in any two suffixes if its components relatively to every co-ordinate system are symmetric (skew-symmetric) in the two suffixes, in question.

A tensor is said to be symmetric (skew-symmetric) if it is so in every pair of suffixes, *e.g.*,

If $u_i$, $v_j$ be any two vectors then the two second order tensors $u_i v_j + u_j v_i$, $u_i v_j - u_j v_i$ are respectively symmetric and skew-symmetric.

### 18.18 THEOREM

Every second order tensor can be expressed as the sum of a symmetric and a skew-symmetric tensor.

**Proof.** Let $a_{ij}$ be any tensor of order 2. Now,

$$\overline{a}_{pq} = l_{ip} l_{jq} a_{ij} = l_{jp} l_{iq} a_{ji} \qquad ...(1)$$

where we have interchanged the two dummy suffixes $i$ and $j$. Then (1) shows that $a_{ij}$ is also a tensor of order two.

Now,
$$a_{ij} = \frac{1}{2}\left[a_{ij} + a_{ji}\right] + \frac{1}{2}\left[a_{ij} - a_{ji}\right]$$

$$= \text{symmetric} + \text{skew-symmetric}$$

Thus $a_{ij}$ is the sum of symmetric and skew symmetric tensors.

### 18.19 A FUNDAMENTAL PROPERTY OF TENSORS

**Theorem.** If the components of a tensor relatively to any one system of co-ordinate axes are all zero, then the components relatively to every system of co-ordinate axes are also zero.

**Proof.** Consider a tensor whose components relatively to the systems of axes $OX_1$, $OX_2$, $OX_3$, $O\overline{X}_1, O\overline{X}_2, O\overline{X}_3$ are $a_{ijkl}$ ...., $\overline{a}_{pqrs}$ ......and let $a_{ijkl}$ ....= 0 for every system of values of $i, j, k, l$...... we have

$$\overline{a}_{pqrs}..... = l_{ip} l_{jq}\, l_{kr}\, l_{ls} ....a_{ijkl}..... = 0$$

for every system of values of $p, q, r, s$......

### 18.20 ZERO TENSOR

**Def.** A tensor whose components relatively to one co-ordinate system and, also relatively to every co-ordinate system are all zero is known as zero tensor.

A zero tensor of every order is denoted by 0.

## EXERCISE 18.1

1. Write the following using summation convention:

   (a) $(x^1)^1 + (x^1)^2 + (x^1)^3$ ..... $(x^1)^n$ **Ans.** $(x^1)^i$

   (b) $(x^1)^2 + (x^2)^2 + (x^3)^2 + ... (x^n)^2$ **Ans.** $(x^i)^2$

   (c) $\dfrac{df}{dt} = \dfrac{\partial f}{\partial x^1}\dfrac{dx^1}{dt} + \dfrac{\partial f}{\partial x^2}\dfrac{dx^2}{dt} + ..... + \dfrac{\partial f}{\partial x^n}\dfrac{dx^n}{dt}$ **Ans.** $\dfrac{df}{dt} = \dfrac{\partial f}{\partial x^i}\dfrac{dx^i}{dt}$

2. Write out all the tensor in $S = a_{ij} x^i x^j$ taking $n = 3$.

   **Ans.** $S = (a_{11}x^1x^1 + a_{12}x^1x^2 + a_{13}x^1x^3) + (a_{21}x^2x^1 + a_{22}x^2x^2 + a_{23}x^2x^3) + (a_{31}x^3x^1 + a_{32}x^3x^2 + a_{33}x^3x^3)$

3. Write the tensor contained in $x^{pq}$, $x_{qr}$ if $n = 2$

   **Ans.** $(x^{11} + x^{21})x_{11} + (x^{11} + x^{21})x_{12} + (x^{12} + x^{22})x_{21} + (x^{12} + x^{22})x_{22}$

4. How many equations in a four dimensional space are represented by $R^{\alpha}_{\beta rp} = 0$ **Ans.** 8

5. Show that every tensor can be expressed as the sum of two tensors, one of which is symmetric and the other skew-symmetric in a pair of covariant or contravariant indices.

6. Show that the symmetric (or antisymmetric) property of a tensor is conserved under a transformation of co-ordinates.

**7.** If $A^i$ and $B_j$ are components of a contravariant and covariant tensor of rank one, then show that $C^i_j = A^i B_j$ are the components of a mixed tensor of rank 2.

**8.** Write down the laws of transformation for the tensors $A^{ij}_k$ and $B^{ij}_{klm}$

**Ans.** $\bar{A}^{ij}_k = \frac{\partial \bar{x}^i}{\partial x^p}\frac{\partial \bar{x}^j}{\partial x^q}\frac{\partial x^r}{\partial \bar{x}^k} A^{pq}_r$, $\bar{B}^{ij}_{klm} = \frac{\partial \bar{x}^i}{\partial x^p}\frac{\partial \bar{x}^j}{\partial x^q}\frac{\partial x^r}{\partial \bar{x}^k}\frac{\partial x^s}{\partial \bar{x}^l}\frac{\partial x^t}{\partial \bar{x}^m} B^{rq}_{rst}$

**9.** Evaluate (*a*) $\delta^i_j \delta^j_k$ (*b*) $\delta^i_j \delta^j_k \delta^k_l$ **Ans.** (*a*) $\delta^i_k$ (*b*) $\delta^i_j$

**10.** Show that the velocity of a fluid at any point is a contravariant tensor of rank one.

## 18.21 TWO SPECIAL TENSORS

### 1. Alternate tensor

Consider an abstract entity of order 3 and dimension 3 such that its components relatively to every system of co-ordinate axes are the same and given by $\in_{ijk}$ where

$$\in_{ijk} = \begin{cases} 0 & \text{if any two of } ijk \text{ are equal} \\ 1 & \text{if } ijk \text{ is a cyclic permutation } 1, 2, 3 \\ -1 & \text{if } ijk \text{ is an anti cyclic permutation } 1, 2, 3 \end{cases}$$

Thus for unequal values of the suffixes, we have

$$\in_{123} = \in_{312} = \in_{231} = 1, \ \in_{132} = \in_{213} = \in_{321} = -1$$

Let $OX_1, OX_2, OX_3, O\bar{X}_1, O\bar{X}_2, O\bar{X}_3$ be two systems of rectangular axes. We want to show that $\in_{ijk}$ is a tensor of order three. Consider, now expression

$$l_{ip} l_{jq} l_{kr} \in_{ijk} \qquad \ldots (1)$$

For any given system of values *p, q, r,* the expression (1) consists of a sum of $3^3 = 27$ terms of which 6 only are non-zero, for the other 21 terms corresponds to a case when atleast two of *i, j, k* are equal. The expression (1) can be written as in the form of determinant

$$\begin{vmatrix} l_{ip} & l_{2p} & l_{3p} \\ l_{1q} & l_{2q} & l_{3q} \\ l_{1r} & l_{2r} & l_{3r} \end{vmatrix}$$

From properties of determinants,

$$\text{Above determinant} = \begin{cases} 0 \text{ if any two of } p, q, r \text{ have equal value.} \\ 1 \text{ if } p, q, r \text{ is a cyclic permutation of } 1, 2, 3 \\ -1 \text{ if } p, q, r \text{ is a non cyclic permutation of } 1, 2, 3 \end{cases}$$

Thus we see that the components of the given entity in any two systems of rectangular axes satisfy the tensorial transformation equations so that the entity is a tensor. This tensor is known as *Alternate tensor.* Thus, we see alternate tensor is same as skew-symmetric tensor. $\in_{ijk}$, always denote the alternate tensor.

## 18.22 KRONECKER TENSOR

The symbol $\delta^k_i$ kronecker delta is defined as

$$\delta^k_i = 0 \text{ when } k \neq i$$

$$\delta^k_i = 1 \text{ when } k = i$$

It mean $\delta^1_1 = \delta^2_2 = \ldots = \delta^n_n = 1$ and $\delta^2_1 = \delta^3_1 = \delta^1_2 = \delta^3_2 = \ldots = 0$

In general $A_{ij}\delta^i_k = A_{i1}\delta^1_k = A_{i2}\delta^2_k + \ldots + A_{ik}\delta^k_k + \ldots + A_{1n}\delta^n_k$

$$= 0 + 0 + \ldots + A_{ik}(1) + \ldots + 0 = A_{ik}$$

**Example 3.** *If $A^{ij}$ are the cofactors of $a^{ij}$ in a deteminant $\Delta$ of order three, then show that*

$$a_{ij}A^{kj} = \Delta\delta_i^k$$

**Solution.** We know that

$$a_{11}A^{11} + a_{12}A^{12} + a_{13}A^{13} = \Delta \quad \text{... (1)}$$

$$a_{11}A^{21} + a_{12}A^{22} + a_{13}A^{23} = 0 \quad \text{... (2)}$$

$$a_{11}A^{31} + a_{12}A^{32} + a_{13}A^{33} = 0 \quad \text{... (3)}$$

These three equations can be written in brief as

$$a_{ij}A^{1j} = \Delta \quad \text{... (4)} \qquad a_{ij}A^{2j} = 0 \quad \text{... (5)} \qquad a_{ij}A^{3j} = 0 \quad \text{... (6)}$$

Using kronecker delta, equations (4), (5), (6) can be combined into a single equation:

$$a_{ij}A^{kj} = \Delta\delta_1^k \quad \text{... (7)}$$

Similarly six more equations are given by

$$a_{2j}A^{kj} = \Delta\delta_2^k \quad \text{.... (8)} \quad \text{and} \quad a_{3j}A^{kj} = \Delta\delta_3^k \quad \text{... (9)}$$

Equations (7), (8), (9) can be written as a single equation.

$$a_{ij}A^{kj} = \Delta\delta_i^k \quad \text{... (10)}$$

All the nine equations of the determinant are included in one equation (10).

## 18.23 ISOTROPIC TENSOR

A tensor which has the same set of components relatively to every system of co-ordinate axes is called an *Isotropic tensor.*

## 18.24 RELATION BETWEEN ALTERNATE AND KRONECKER TENSOR

Prove that $\in_{ijm}\in_{klm} = \delta_{ik}\delta_{jl} - \delta_{il}\delta_{jk}$. Here each side is a tensor of order 4 so that tensor equality is equivalent to set of 81 scalar equality. We have to prove that

$$\in_{ij1}\in_{kl1} + \in_{ij2}\in_{kl2} + \in_{ij3}\in_{kl3} = \delta_{ik}\delta_{jl} - \delta_{il}\delta_{jk}$$

**Proof: Case I.** When $i = j$ or $k = l$. There will be 45 such equations and for all these equations L.H.S. = 0 = R.H.S.

**Case II.** If the pair $(i, j)$ such that $i \neq j$ is different from the pair $(l, k)$, $l \neq k$ we see that there will be 24 such scalar equations for which L.H.S. = 0 = R.H.S.

**Ex.** $(i, j) = (1, 2), (j, k) = (1, 3), (3, 1), (2, 3), (3, 2)$

$(i, j) = (2, 1), (j, k) = (1, 3), (3, 1), (2, 3), (3, 2).$

**Case III.** Thus we are left to consider the possibility when $i, j$ and $k, l$ take the pairs of values,

$$(1, 2); (1, 3); (2, 3); (2, 1); (3, 1); (3, 2).$$

Consider the first case we have

$i = 1, j = 2, k = 1, l = 2$; $i = 1, j = 2, k = 2, l = 1$; $i = 2, j = 1, k = 1, l = 2$; $i = 2, j = 1, k = 2, l = 1$.

Each pair of $(i, j)$ i.e (1, 2) gives two scalar equations. Thus 6 pairs of $(i, j)$ give 12 such scalar equations. In these cases we have

L.H.S. = 1 = R.H.S., L.H.S. = –1 = R.H.S.

L.H.S. = –1 = R.H.S., L.H.S. = 1 = R.H.S.

This result is also true for other five cases. Hence we have the result.

**Example 4.** *Prove that* $\in_{ilm}\in_{jlm} = 2\delta_{ij}$

**Proof.** We know $\in_{ilm}\in_{jkm} = \delta_{ij}\delta_{lk} - \delta_{ik}\delta_{lj}$

Taking $k = l$ we get

$$\in_{ilm} \in_{jlm} = \delta_{ij}\delta_{ll} - \delta_{il}\delta_{lj}$$

Now $\quad \delta_{ll} = \delta_{11} + \delta_{22} + \delta_{33} = 1+1+1 = 3$

$$\delta_{il}\delta_{lj} = \delta_{ij} \qquad \because \delta_{ij} a_{im} = a_{jm}$$

$\therefore \quad \in_{ilm} \in_{jlm} = 3\delta_{ij} - \delta_{ij} = 2\delta_{ij}$ **Proved.**

**Example 5.** *Prove that* $\in_{ijk} \in_{ijk} = 6$

**Proof.** $\quad \in_{ilm} \in_{jkm} = \delta_{ij}\delta_{lk} - \delta_{ik}\delta_{lj}$

Taking $k = l$, we get

$$\in_{ilm} \in_{jlm} = 2\delta_{ij}$$

Taking $i = j$ (Contraction)

$$\in_{ilm} \in_{ilm} = 2\delta_{ii} = 2 \times 3 = 6$$ **Proved.**

## 18.25 MATRICES AND TENSORS OF FIRST AND SECOND ORDER

Consider any vector. Its components $a_i$ relatively to any system of axes may be written in the form of a row or a column matrix as

$$[a_1 a_2 a_3] \text{ or } \begin{bmatrix} a_1 \\ a_2 \\ a_3 \end{bmatrix}$$

We shall be writing $a_i = [a_1 a_2 a_3]$ or $a_i = \begin{bmatrix} a_1 \\ a_2 \\ a_3 \end{bmatrix}$

Now consider second order tensor. Its components $a_{ij}$ relatively to any system of rectangular axes can be written as the form of matrix such that $a_{ij}$ occurs at the intersection of the $i^{th}$ row and $j^{th}$ column. Thus we shall write

$$\left[a_{ij}\right] = \begin{bmatrix} a_{11} & a_{12} & a_{13} \\ a_{21} & a_{22} & a_{23} \\ a_{31} & a_{32} & a_{33} \end{bmatrix}$$

A matrix obtained by interchanging rows and columns of a given matrix is called the transpose of the same. Transpose of $[a_{ij}]$ will be denoted by $[a_{ij}]'$.

Sum of two matrices of the same type is the matrix whose elements are the sums of the corresponding elements of two matrices.

## 18.26 SCALAR AND VECTOR PRODUCTS OF TWO VECTORS

**Def. 1.** *Scalar product.* The scalar $u_i v_i$ is called the scalar product of the two vectors $u_i, v_j$.
Thus the scalar product $= u_1v_1 + u_2 v_2 + u_3v_3$.

**Def. 2.** *Vector product* The vector $E_{ijk}u_iv_j$ is called vector product of two vectors $u_i, v_j$ taken in this order. Components of these vectors are $u_2 v_3 = u_3v_2, u_3v_1 = u_1v_3, u_1v_2 = u_2 v_1$.

## 18.27 THE THREE SCALAR INVARIANTS OF A SECOND ORDER TENSOR

**I.** $\quad a_{ii}$ or $a_{11} + a_{22} + a_{33}$

**II.** $\quad \frac{1}{2}(a_{ii}a_{jj} - a_{ij}a_{ji})$ or $\quad a_{11}a_{22} + a_{22}a_{33} + a_{33}a_{11} - a_{12}a_{21} - a_{23}a_{32} - a_{31}a_{13}$

**III.** $|a_{ij}|$ or $\begin{vmatrix} a_{11} & a_{12} & a_{13} \\ a_{21} & a_{22} & a_{23} \\ a_{31} & a_{32} & a_{33} \end{vmatrix}$

**Proof.** I. Identifying *i, j* we see that $a_{ii}$ is scalar. Thus

$$a_{ii} = a_{11} + a_{22} + a_{23} \quad \text{...(1)}$$

is invariant.

II. Consider now the tensor of 4th order, $a_{ij}a_{pq}$. Identifying *i* with *q* and *j* with *p* we see that $a_{ij}a_{ji}$ is scalar.

Thus $$a_{ij}a_{ji} = (a_{11})^2 + (a_{22})^2 + (a_{33})^2 + 2a_{12}a_{21} + 2a_{23}a_{32} + 2a_{31}a_{13} \quad \text{... (2)}$$

is invariant.Subtracting(2) from square of (1) and dividing by 2. We establish invariance of II.

(III) If $a_{ij}$, $\bar{a}_{pq}$ denote the components of tensor relatively to any two co-ordinate systems of axes, then is the usual notation, we have

$$\bar{a}_{pq} = l_{ip}l_{jq}a_{ij}$$

$$|\bar{a}_{pq}| = |l_{ip}||l_{jq}||a_{ij}|$$

since $$|l_{ip}| = |l_{jq}|$$

$\therefore$ $$|\bar{a}_{pq}| = |l_{ip}|^2 |a_{ij}| \quad \text{but} \quad |l_{ip}|^2 = 1$$

$\therefore$ $$|\bar{a}_{pq}| = |a_{ij}|$$

Hence it is an invariance.

## 18.28 SINGULAR AND NON-SINGULAR TENSORS OF SECOND ORDER

A tensor of second order is said to be singular or non-singular according as its determinant is zero or non zero.

## 18.29 RECIPROCAL OF A NON-SINGULAR TENSOR

Suppose $a_{ij}$ be a second order tensor such that $|a_{ij}| \neq 0$

**Lemma 1.** We form another matrix

$$A_{ij} = \frac{\text{Cofactor of } a_{ij} \text{ in the determinant } a_{ij}}{|a_{ij}|}$$

Now, by theory of determinants, we know

$$A_{ki}\, a_{ij} = \delta_{kj} \quad \text{...(1)}$$

We shall now show that $A_{ij}$ is a second order tensor, we can not do so, by using Quotient law, from equation (1) since $a_{ij}$ is not an arbitrary tensor. Let $c_j$ be an arbitrary vector, then

$$c_j a_{ij} = d_i \quad \text{... (2)}$$

So that $d_i$ is also a vector. We shall prove that this is an arbitrary vector. Now (2) is equivalent to a set of 3 linear equations in the components of $c_j$ and as the determinant of $a_{ij} \neq 0$, we may assign any arbitrary values to $d_i$ and the resulting equations can be uniquely solved for the components of $c_j$. Thus $d_i$ is an arbitrary vector. We now have

$$d_i A_{ki} = a_{ij}c_j A_{ki} = A_{ki}a_{ij}c_j = \delta_{kj}c_j = c_k$$

$$c_k = d_i a_{ki} \quad \text{..(3)}$$

Therefore by quotient law $A_{ki}$ is a second order tensor.

**Lemma 2.** $e_{ij} = \dfrac{\text{Cofactor of } A_{ij} \text{ in the detterminant } A_{ij}}{|A_{ij}|}$

We know from the theory of determinants.

$$|A_{ij}||a_{ij}| = 1 \qquad \text{But} \qquad |a_{ij}| \neq 0$$

Hence determinant $|A_{ij}| \neq 0$

We shall now show that $e_{ij} = a_{ij}$

$$e_{ki} A_{ij} = \delta_{kj}$$

Take inner product with $a_{ji}$

$$e_{ki} A_{ij} a_{jl} = \delta_{ki} a_{jl}$$

$$e_{ki} \delta_{il} = a_{kl}$$

$$e_{kl} = a_{kl}$$

**Def.** Two second order non-singular tensors $a_{ij}$ and $A_{ij}$ are said to be conjugate (or reciprocal) tensors if they satisfy the equation

$$A_{ki} a_{ij'} = \delta_{kj}$$

## 18.30 EIGEN VALUES AND EIGEN VECTORS OF A TENSOR OF SECOND ORDER

**Def.** A scalar, $\lambda$, is called an eigen value of second order tensor $a_{ij'}$ if there exists a non-zero vector $x$, such that $a_{ij} x_j = \lambda x_i$. This equation is equivalent to

$$a_{ij} x_j = \lambda \delta_{ij} x_j$$

or $$\left(a_{ij} - \lambda \delta_{ij}\right) x_j = 0 \qquad \text{...(1)}$$

since $x_j \neq 0$, Hence $|a_{ij} - \lambda \delta_{ij}| = 0$ ...(2)

This is a necessary condition for $\lambda$, to be eigen value. Eq. (2) is cubic eq. in $\lambda$ and therefore in general will give us three eigen values may not all be distinct corresponding to the tensor $a_{ij}$.

Consider now any system of co-ordinate axes $OX_1$, $OX_2$, $OX_3$, and let $a_{ij}$ be the component of the given tensor in this system. Consider now a vector $x_i$, whose components relatively to $OX_1$, $OX_2$, $OX_3$ are given on solving (1). As the components of $x_i$ are not zero relatively to one system $OX_1$, $OX_2$, $OX_3$, this vector can be zero vector *i.e.* its components relatively to any system of axes can not all be zero.

The tensor eq. (1) being true for one system $OX_1$, $OX_2$, $OX_3$ will be true for every system of axes.

Thus we see that every second order tensor possesses three eigen values, not necessarily all distinct. These eigen values are the roots of the cubic $|a_{ij} - \lambda \delta_{ij}| = 0$ in $\lambda$. Also to each eigen value corresponds an eigen vector. The vector $x_i$ corresponding to eigen value $\lambda$ is called an eigen vector.

## 18.31 THEOREM

Orthogonality of eigen vectors corresponding to distinct eigen values of a symmetric second order tensor.

**Proof.** Let $a_{ij}$ be a symmetric second order tensor, and let $x_i$ and $y_i$ be the eigen vectors corresponding to the distinct eigen values $\lambda_1$ and $\lambda_2$ $(\lambda_1 \neq \lambda_2)$ we have

$$a_{ij}x_j = \lambda_1 x_i \qquad ...(1)$$

$$a_{ij}y_j = \lambda_2 y_i \qquad ...(2)$$

Now, $\lambda_1 x_i y_i = a_{ij} x_j y_i$

$= a_{ji} x_j y_i$ $\quad [\because a_{ij} = a_{ji}]$

$= a_{ij} y_j x_i$ (Interchanging dummy indices)

$\lambda_1 x_i y_i = a_{ij} y_j x_i$

$\therefore \quad \lambda_1 x_i y_i = \lambda_2 y_i x_i \quad$ or $\quad (\lambda_1 - \lambda_2)(x_i y_i) = 0$

Since $\lambda_1 - \lambda_2 \neq 0 \;\; \therefore \; x_i y_i = 0$

Thus $x$ and $y_i$ are orthogonal *i.e*. the eigen vectors are orthogonal.

## 18.32 REALITY OF THE EIGEN VALUES

**Theorem.** The eigen values of symmetric second order tensor are real

**Proof.** Let $\lambda$ be any eigen value so that we have a relation

$$a_{ij} x_j = \lambda x_i \qquad ... (1)$$

Here the components of $x_j$ cannot be assumed to be all real. Taking complex conjugate (denoted by bar) in (1), we get

$$\bar{a}_{ij} \bar{x}_j = \bar{\lambda} \bar{x}_i$$

$$a_{ij} \bar{x}_j = \bar{\lambda} \bar{x}_i$$

$\left[\because a_{ij}\right.$ is symmetric $\therefore a_{ij} = a_{ji}$

$\bar{a}_{ij} = a_{ij}$ (all elements are real)

Take inner product by $x_i$

$$a_{ij} \bar{x}_j x_i = \bar{\lambda} \bar{x}_i x_i$$

$\bar{\lambda} \{\bar{x}_i x_i\} = a_{ji} (\bar{x}_j x_i)$ [By symmetry]

$= a_{ij} \bar{x}_i x_j$ [Interchanging dummy indices]

$= a_{ij} x_i \bar{x}_j$ $[\bar{x}_i x_i = \text{real}]$

$= \bar{a}_{ij} x_i \bar{x}_j = \text{real}$ $[\because (a - ib)(a + ib) = a^2 + b^2$ which is real$]$

This shows that the right hand side is real. Hence $\bar{\lambda}$ is real. Thus $\lambda$ is real *i.e.* eigen values are real.

**Proved.**

## 18.33 ASSOCIATION OF A SKEW SYMMETRIC TENSORS OF ORDER TWO AND VECTORS

We associate the skew symmetric tensor of order two.

$$a_{ij} = \in_{ijk} a_k \qquad ... (1)$$

The tensor $a_{ij}$ is skew symmetric for

$$a_{ji} = \in_{jik} a_k = - \in_{ijk} a_k = -a_{ij}$$

The relation (1) is equivalent to statements

$$a_{23} = a_1,\ a_{32} = -a_1;\ a_{31} = a_2,\ a_{13} = -a_2;\ a_{12} = a_3,\ a_{21} = -a_3;\ a_{11} = 0, a_{22} = 0; a_{33} = 0.$$

On the inner multiplication with $\in_{ijm}$ we obtain from (1)

$\in_{ijm} a_{ij} = \in_{ijm} \in_{ijk} a_k \qquad \in_{ijk} \in_{pqk} = \delta_{ip}\delta_{jq} - \delta_{iq}\delta_{jp}$

$= 2\delta_{mk} a_k \qquad \in_{ijk} \in_{pjk} = \delta_{ip}\delta_{jj} - \delta_{jp}\delta_{ij}$

$= 2a_m$ when $k = m \qquad = 3\delta_{ip} - \delta_{ip} = 2\delta_{ip}$

Hence $a_m = \dfrac{1}{2} \in_{ijm} a_{ij}$

This shows that association is one-one.

## 18.34 TENSOR FIELDS

A tensor field or a tensor point function is said to be defined when there is given a law which associates to each point of given region of space a tensor of the same order. Thus a tensor field $a_{ij}$...... of any order is defined if the components $a_{ij}$...... are functions of $x_1, x_2, x_3$.

## 18.35 GRADIENT OF TENSOR FIELDS:GRADIENT OF A SCALAR FUNCTION.

Let $u$ be a scalar point function so that there is a value of $u$ associated with each point of a given region of space. Thus if $OX_1, OX_2, OX_3$ and $O\overline{X}_1, O\overline{X}_2, O\overline{X}_3$ be any two systems, then $u$ is a function of $x_i$ and $\overline{x}_p$ which are co-ordinates of any point $P$ relatively to the two systems of axes. For any point $P$, $x_i, \overline{x}_p$ are different but the values of $u$ are same. Consider now two sets of first order

$$\frac{\partial u}{\partial x_i}, \frac{\partial u}{\partial \overline{x}_p} \qquad \text{we have } \frac{\partial u}{\partial x_i} = \frac{\partial u}{\partial \overline{x}_p}\frac{\partial \overline{x}_p}{\partial x_i}$$

We know that $\overline{x}_p = l_{ip}x_i$

$$\therefore \quad \frac{\partial \overline{x}_p}{\partial x_i} = l_{ip} \qquad \therefore \quad \frac{\partial u}{\partial x_i} = l_{ip}\frac{\partial u}{\partial \overline{x}_p}$$

Thus we see that $\frac{\partial u}{\partial x_i}$ is a tensor of order one *i.e.* a vector. This is usually denoted by $u, i$,

$$\text{grad } u = u, i$$

If components $\frac{\partial u}{\partial x_i}$ and $\frac{\partial u}{\partial \overline{x}_p}$ relatively to two systems of axes $OX_1, OX_2, OX_3$ and $O\overline{X}_1, O\overline{X}_2, O\overline{X}_3$ obey the tensorial transformation law. This vector is called the *gradient of scalar u.*

## 18.36 GRADIENT OF VECTOR

Consider now any tensor field $u_i$ of order one. If $u_i$, $\overline{u}_p$ be the components relatively to two systems of axes $OX_1, OX_2, OX_3, O\overline{X}_1, O\overline{X}_2, O\overline{X}_3$, we have $\overline{u}_p = l_{ip}u_i$

$$\therefore \quad \frac{\partial \overline{u}_p}{\partial \overline{x}_j} = l_{ip}\frac{\partial u_i}{\partial \overline{x}_j} = l_{ip}\frac{\partial u_i}{dx_k}\cdot\frac{\partial x_k}{\partial \overline{x}_j} = l_{ip}l_{kj}\frac{\partial u_i}{\partial x_k} \qquad \left[\begin{array}{l} x_k = l_{kj}\overline{x}_j \\ \dfrac{\partial x_k}{\partial \overline{x}_j} = l_{kj} \end{array}\right.$$

We see $\frac{\partial u_i}{\partial x_k}$ is a tensor of second order. It is denoted by symbol $u_{i,j}$ and is called the *gradient of* $u_{i,j}$.

## 18.37 DIVERGENCE OF VECTOR POINT FUNCTION

The scalar of the gradient of a vector point function is called the divergence of the point function. Thus if $u_i$ is a vector point function so that

$$u_{i,j} = \frac{\partial u_i}{\partial x_j} \text{ is its gradient,then } u_{i,i} = \frac{\partial u_i}{\partial x_i} = \frac{\partial u_1}{\partial x_1} + \frac{\partial u_2}{\partial x_2} + \frac{\partial u_3}{\partial x_3} \text{ is called div } u_i$$

$$\text{div } u_i = u_{i,i}$$

## 18.38 CURL OF A VECTOR POINT FUNCTION

The vector of the gradient of a vector point function is called the curl of the point function.

Thus if $u_i$ is a vector point function so that $u_{i,j} = \dfrac{\partial u_i}{\partial x_j}$ is its gradient, then the vector of a tensor *i.e.* the vector $\in_{jik} u_{i,j}$ is called the curl of $u_i$ denoted by the symbol curl id, $u_i = \in_{jik} u_{i,j}$.

**Example 6.** *Prove the following results*

(*i*) grad $(\phi\psi) = \phi$ grad $\psi + \psi$ grad $\phi$ (*ii*) grad $(\vec{f}.\vec{g}) = \vec{f} \times \text{curl } \vec{g} + \vec{g} \times \text{curl } \vec{f} + \vec{f}.\nabla\vec{g} + \vec{g}.\nabla\vec{f}$

(*iii*) div $(\phi\vec{f}) = \phi$ div $\vec{f} + \vec{f}$ grad $\phi$ (*iv*) div $(\vec{f} \times \vec{g}) = \vec{g}.\text{ curl } \vec{f} - \vec{f}.\text{curl } \vec{g}$

(*v*) curl $(\phi\vec{f}) = \text{grad}\,\phi \times \vec{f} + \phi \text{ curl } \vec{f}$ (*vi*) curl $(\vec{f} \times \vec{g}) = \vec{f} \text{ div } \vec{g} - \vec{g} \text{ div } \vec{f} + \vec{g}.\nabla\vec{f} - \vec{f}.\nabla\vec{g}$

**Proof.** (*i*) grad $(\phi\psi) = (\phi\psi),\ i = \phi\psi,\ i + \psi\phi,\ i = \phi$ grad $\psi + \psi$ grad $\phi$.

(*ii*) grad $(\vec{f}.\vec{g}) = (f_i g_i)_{,j}$

$$= f g_{i,j} + g f_{i,j} \qquad \text{... (1)}$$

Now $\vec{f} \times \text{curl } \vec{g} = \in_{pkm} f_p \in_{jik} g_{i,j} = -\in_{pmk} \in_{jik} f_p g_{i,j}$

$$= -\left[\delta_{pj}\delta_{mi} - \delta_{pi}\delta_{mj}\right] f_p g_{i,j} = -\delta_{pj}\delta_{mi} f_p g_{i,j} + \delta_{pi}\delta_{mj} f_p g_{i,j}$$

Identifying *p, j* and *m, i* in first part and *p, i* and *m, j* in second part, we get

$$\vec{f} \times \text{curl } \vec{g} = -\delta_{pp}\delta_{mm} f_p g_{m,p} + \delta_{pp}\delta_{mm} f_p g_{p,m} = -f_p g_{m,p} + f_p g_{p,m}$$

$$= -\vec{f}.\nabla\vec{g} + f_p g_{p,m} \qquad \text{... (2)}$$

Similarly, $\vec{g} \times \text{curl } \vec{f} = -g.\nabla\vec{f} + g_p f_{p,m}$ ... (3)

Substituting the values of $f_p g_{p,m}$, $g_p f_{p,m}$ from (2) and (3) into (1), we get

$$\text{grad}(\vec{f}.\vec{g}) = \vec{g} \times \text{curl } \vec{f} + \vec{g}.\nabla\vec{f} + \vec{f} \times \text{curl } \vec{g} + \vec{f}.\nabla\vec{g}$$

or $\text{grad}(\vec{f}.\vec{g}) = \vec{g} \times \text{curl } \vec{f} + \vec{f} \times \text{curl } \vec{g} + \vec{g}.\nabla\vec{f} + \vec{f}.\nabla\vec{g}$

(*iii*) $\text{div}(\phi\vec{f}) = (\phi f_i)_{,i} = \phi f_{i,i} + \phi_{,i} f_i = \phi \,\text{div}\,\vec{f} + f_i \phi_{,i} = \phi \,\text{div}\,\vec{f} + \vec{f}.\text{grad}\,\phi$

(*iv*) $\text{div}(\vec{f} \times \vec{g}) = (\in_{ijk} f_i g_j)_{,k} = \in_{ijk} \{f_i g_{j,k} + g_j f_{i,k}\} = \in_{ijk} f_i g_{j,k} + \in_{ijk} g_j f_{i,k}$

$$= -\in_{kji} f_i g_{j,k} + \{\in_{kij} g_j f_{i,k}\} = -f_i \in_{kji} g_{j,k} + g_j \in_{kij} f_{i,k}$$

$$= -\vec{f}.(\text{curl } \vec{g}) + \vec{g}.(\text{curl } \vec{f}) = \vec{g}.(\text{curl } \vec{f}) - \vec{f}.(\text{curl } \vec{g})$$

(*v*) $\text{curl}(\phi\vec{f}) = \in_{jik} (\phi f_i)_{,j} = \in_{jik} \phi f_{i,j} + \in_{jik} \phi_{,j} f_i$

$$= \phi \in_{jik} f_{i,j} + \in_{jik} f_i \phi_{,j} = \phi(\text{curl } \vec{f}) + (\text{grad } \phi) \times \vec{f}$$

(*vi*) curl $(\vec{f} \times \vec{g}) = \in_{mkn} (\in_{ijk} f_i g_j)_{,m} = -\in_{mnk} \in_{ijk} [f_i g_{j,m} + g_j f_{i,m}]$

$$= -[\delta_{mi}\delta_{nj} - \delta_{mj}\delta_{nj}][f_i g_{j,m} + g_j f_{i,m}] = -[\delta_{mi}\delta_{nj} f_i g_{j,m} - \delta_{mj}\delta_{ni} f_i g_{j,m} + \delta_{mi}\delta_{nj} g_j f_{i,m} - \delta_{mj}\delta_{ni} g_j f_{i,m}]$$

$$= -\delta_{mm}\delta_{nn} f_m g_{n,m} + \delta_{mm}\delta_{nn} f_n g_{m,m} - \delta_{mm}\delta_{nn} g_n f_{m,m} + \delta_{mm}\delta_{nn} g_m f_{n,m}$$

$$= -f_m g_{n,m} + f_n g_{m,m} - g_n f_{m,m} + g_m f_{n,m} = -\vec{f}.\nabla\vec{g} + (\text{div } \vec{g})\vec{f} - (\text{div } \vec{f})\vec{g} + \nabla\vec{f}.\vec{g}$$

$$= \vec{f} \text{ div} \vec{g} - \vec{g} \text{ div } \vec{f} + \vec{g}.\nabla\vec{f} - \vec{f}.\nabla\vec{g}$$

## 18.39 SECOND ORDER DIFFERENTIAL OPERATORS

(*i*) div (grad $\phi$) = $\nabla^2\phi$  (*ii*) curl (grad $\phi$) = 0

(*iii*) div (curl $\vec{f}$) = 0  (*iv*) grad (div $\vec{f}$) = curl(curl $\vec{f}$) + $\nabla^2\vec{f}$

**Proof.** (*i*) div (grad $\phi$) = $(\phi_{,i})_{,i} = \phi_{,ii} = \nabla^2\phi$

(*ii*) $$\text{curl (grad } \phi) = \in_{jik} (\phi_{,i})_{,i}$$

$$= \in_{jik} \phi_{,ij} = I \text{ ,say} \quad \text{... (1)}$$

$$\phi_{,ij} = \phi_{,ji} \qquad \left[\frac{\partial^2\phi}{\partial x\partial y} = \frac{\partial^2\phi}{\partial y\partial x}\right]$$

Now, $$I = \in_{jik} \phi_{,ji} = \in_{ijk} \phi_{,ij} \quad \text{...(2)}$$

From (1) and (2) $$2I = (\in_{jik} + \in_{ijk})\phi_{,ij}$$

$$2I = (-\in_{ijk} + \in_{ijk})\phi_{,ij} = 0$$

Hence, $I = 0$ or curl (grad $\phi$) = 0

(*iii*) $$\text{div (curl } \vec{f}) = (\in_{jik} f_{i,j})_{,k} = \in_{jik} f_{i,jk} = I \text{ (say)} \quad \text{... (1)}$$

Because $$f_{i,jk} = f_{i,kj}$$

Then $$I = \in_{jik} f_{i,kj} = \in_{kij} f_{i,jk} \quad \text{...(2)}$$

From (1) and (2) $$2I = (\in_{jik} + \in_{kij}) f_{i,jk} = (\in_{jik} - \in_{jik}) f_{i,jk} = 0$$

Hence, $I = 0$ or div (curl $\vec{f}$) = 0

(*iv*) $$\text{grad (div } \vec{f}) = (f_{i,i})_{,j} = f_{i,ij} \quad \text{... (1)}$$

$$curl(curl\vec{f}) = \in_{mkn} (\in_{jik} f_{i,j})_m = \in_{nmk} \in_{jik} f_{i,jm} = (\delta_{nj}\delta_{mi} - \delta_{ni}\delta_{mj}) f_{i,jm}$$

$$= (\delta_{nn}\delta_{mm} f_{m,nm} - \delta_{nn}\delta_{mm} f_{n,mm}) = f_{m,nm} - f_{n,mm}$$

$$= f_{m,mn} - f_{n,mm} = \text{grad (div } \vec{f}) = \nabla^2\vec{f} \quad \text{[From (1)]}$$

Thus, $$\text{grad (div } \vec{f}) = curl(curl\vec{f}) + \nabla^2\vec{f}$$

## 18.40 TENSORIAL FORM OF GAUSS'S AND STOKE'S THEOREM

**Gauss's divergence theorem.**

If $\vec{F}$, is a continuously differentiable vector point function and *S* is a closed surface enclosing a region *V*, then

$$\oint_s \vec{F}.\hat{n}ds = \int_V \text{div } \vec{F}\, dv \quad \text{...(1)}$$

where $\hat{\eta}$ is a unit vector $\oint_s \overrightarrow{F}_i\, n_i ds = \int_V F_{i,i} dV = \int \frac{\partial F_i}{\partial x_i} dV$

## 18.41 STOKE'S THEOREM

If $\vec{F}$ is any continuously differentiable vector point function and *S* is a surface bounded by a curve *c*, then

$$\oint_c \vec{F}.dr = \int_S \text{curl}\,\vec{F}.\hat{\eta}ds \qquad ...(2)$$

where $\hat{\eta}$ is a unit vector $\qquad \oint_c \vec{F}_i\, dr_i = \int_S (\in_{jik} F_{i,j})n_k ds$

**Example 7.** *By means of divergence theorem of Gauss's, show that*

$$\oint_S \in_{qpi}\, n_p\, \in_{ijk}\, a_j x_k ds = 2a_q V$$

*where V is the volume enclosed by the surface S, having the outward drawn normal n. The position vector to any point in V is* $x_i$ and $a_i$, *is an arbitrary constant vector.*

**Proof.** $\qquad$ L.H.S. $= \oint_S \in_{ipq} \in_{ijk}\, n_p a_j x_k ds$

$$= \oint_S (\delta_{qj}\delta_{pk} - \delta_{qk}\delta_{pj})n_p a_j x_k ds = \oint_S n_k a_q x_k ds - \oint_S n_j a_j x_q ds$$

$$= a_q \oint_S n_k x_k ds - a_j \oint_S n_j x_q ds = a_q \oint_V \frac{\partial x_k}{\partial x_k} dv - a_j \oint_V \frac{\partial x_q}{\partial x_j} dv \; = a_q \delta_{kk} v - a_j \delta_{qj} v$$

$$= 3a_q v - a_q v = 2a_q v$$

**Proved.**

**Example 8.** *If* $\vec{q} = \vec{w}\times\vec{r}$, *show that* $2\vec{w} = \nabla\times\vec{q}$ *using the index notation. The vector* $\vec{w}$ *is a constant.*

**Solution.** $\qquad q_k = \in_{ijk} w_i x_j \qquad$ (given)

$$\left[\nabla\times\vec{q}\right]_m = \in_{lkm} q_{k,l} = \in_{lkm} \in_{ijk} (w_i x_j)_{,l}$$

$$\left[\nabla\times\vec{q}\right]_m = \in_{mlk} \in_{ijk} (w_i x_{j,l} + x_j w_{i,l}) = \in_{mlk} \in_{ijk} w_i x_{j,l}$$

Since $\vec{w}$ is a constant vector

$$\therefore \quad w_{i,l} = 0 = (\delta_{mi}\delta_{lj} - \delta_{mj}\delta_{li})w_i x_{j,l} = w_m x_{l,l} - w_l x_{m,l} = w_m \delta_{ll} - w_l \delta_{ml} = 3w_m - w_m$$

$$\left[\nabla\times\vec{q}\right]_m w = 2w_m$$

Hence $\qquad \nabla\times\vec{q} = 2\vec{w}.$ **Proved.**

## 18.42 RELATION BETWEEN ALTERNATE AND KRONECKER TENSOR

$$\in_{ijk}\in_{lmn} = \begin{vmatrix} l_{1i} & l_{2i} & l_{3i} \\ l_{1j} & l_{2j} & l_{3j} \\ l_{1k} & l_{2k} & l_{3k} \end{vmatrix} \times \begin{vmatrix} l_{1l} & l_{2l} & l_{3l} \\ l_{1m} & l_{2m} & l_{3m} \\ l_{1n} & l_{2n} & l_{3n} \end{vmatrix} = \begin{vmatrix} \delta_{il} & \delta_{im} & \delta_{in} \\ \delta_{jl} & \delta_{jm} & \delta_{jn} \\ \delta_{kl} & \delta_{km} & \delta_{kn} \end{vmatrix}$$

Identifying *k* and *1*, we get

$$\in_{ijk}\in_{kmn} = \begin{vmatrix} \delta_{ik} & \delta_{im} & \delta_{in} \\ \delta_{jk} & \delta_{jm} & \delta_{jn} \\ \delta_{kk} & \delta_{km} & \delta_{kn} \end{vmatrix} = \begin{vmatrix} \delta_{ik} & \delta_{im} & \delta_{in} \\ \delta_{jk} & \delta_{jm} & \delta_{jn} \\ 3 & \delta_{km} & \delta_{kn} \end{vmatrix}$$

$$\because \quad \delta_{kk} = \delta_{11} + \delta_{22} + \delta_{33} = 1 + 1 + 1 = 3.$$

Expanding the determinant, we have

$$\in_{ijk}\in_{kmn} = \delta_{ik}(\delta_{jm}\delta_{kn} - \delta_{jn}\delta_{km}) + \delta_{im}\,(3\,\delta_{jn} - \delta_{jk}\,\delta_{kn}) + \delta_{in}\,(\delta_{jk}\,\delta_{km} - 3\delta_{jm})$$

$$= \delta_{jm}\,\delta_{in} - \delta_{jn}\,\delta_{im} + 3\delta_{im}\,\delta_{jn} - \delta_{im}\,\delta_{jn} + \delta_{in}\,\delta_{jm} - 3\delta_{in}\,\delta_{jm}$$

$$\in_{ijk} \in_{kmn} = \delta_{im}\, \delta_{jn} - \delta_{in}\, \delta_{jm}$$

## 18.43 THE THREE SCALAR INVARIANTS OF A SECOND ORDER TENSOR

Let $a_{ij}$ be a second order tensor

*(i)* $a_{ii}$

**Proof.** $\bar{a}_{pq} = l_{ip}\, l_{jq}\, a_{ij}$

Identifying $p$ and $q$, we have $\bar{a}_{pp} = l_{ip}\, l_{jp}\, a_{ij}$

$\bar{a}_{pp} = \delta_{ij}\, a_{ij} = a_{ii}$ Hence $a_{ii}$ is an invariant.

*(ii)* $\frac{1}{2}\,(a_{ii}\, a_{jj} - a_{ij}\, a_{ij})$

We know that $a_{ii}$ and $a_{jj}$ are invariants. Now we have to show that $a_{ij}\, a_{ji}$ is also an invariant. Then $(a_{ii}\, a_{jj} - a_{ij}\, a_{ij})$ will also be invariant.

Let $\bar{a}_{pq} = l_{ip}\, l_{jp}\, a_{ij}$ and $\bar{a}_{rs} = l_{mr}\, l_{ns}\, a_{mn}$

Now consider the tensor of 4[th] order

$$\bar{a}_{pq}\, \bar{a}_{rs} = l_{ip}\, l_{jq}\, l_{mr}\, l_{ns}\, a_{ij}\, a_{mn}$$

First identifying $r$ and $q$ and then identifying $p$ and $s$ we have

$$\bar{a}_{pq}\, \bar{a}_{qp} = l_{is}\, l_{jr}\, l_{mr}\, l_{ns}\, a_{ij}\, a_{mn} = \delta_{in}\, \delta_{jm}\, a_{ij}\, a_{mn} = a_{ij}\, a_{ji}$$

Hence $a_{ij}\, a_{ji}$ is an invariant. Therefore $\frac{1}{2}\,(a_{ii}\, a_{jj} - a_{ij}\, a_{ji})$ is invariant

*(iii)* $|\, a_{ij} |$ **Proof.** $\bar{a}_{pq} = l_{ip}\, l_{jq}\, a_{ij} \Rightarrow |\bar{a}_{pq}| = |l_{ip}|\, |l_{jq}|\, |a_{ij}|$

We know by the property of determinants $|l_{ip}|\, |l_{jq}| = 1 \Rightarrow |\bar{a}_{pq}| = |a_{ij}|$

Hence $|\, a_{ij}|$ is an invariant.

## 18.44 TENSOR ANALYSIS

**Example 9.** *What is a mixed tensor of second rank? Prove that* $\delta_q^{\,p}$ *is a mixed tensor of the second rank*

**Solution.** The $N^2$ quantities $A_s^{\,q}$ are called components of a mixed tensor of the second rank if $\bar{A}_r^{\,p} = \dfrac{\partial \bar{x}^p}{\partial x^q} \dfrac{\partial x^s}{\partial \bar{x}^r} A_s^{\,q}$

Now, if $\delta_s^{\,q}$ defined by $\delta_s^{\,q} = \begin{cases} 0 & \text{if } p \neq q \\ 1 & \text{if } p = q \end{cases}$

is a mixed tensor of second rank, it must transform according to the rule $\bar{\delta}_k^i = \dfrac{\partial \bar{x}^i}{\partial x^p} \dfrac{\partial x^q}{\partial \bar{x}^k} \delta_q^p$

The right side equals $\dfrac{\partial \bar{x}^j}{\partial x^p} \dfrac{\partial x^p}{\partial \bar{x}^k} = \delta_k^j$

since $\bar{\delta}_k^j = \delta_k^j = 1$ if $j = k$, and 0 if $j \neq k$, if follows that $\delta_r^p$ is a mixed tensor of rank two.

**Example 10.** *Evaluate (i)* $\delta_q^p\; A_s^{qr}$ *(ii)* $\delta_q^p\; \delta_r^p$

**Solution.** (*i*) $\delta^p_q \; A^{qr}_s = \delta^p_q \; A^{pr}_s = A^{pr}_s$

(*ii*) $\delta^p_q \; \delta^p_r = \delta^p_p \; \delta^p_r = \delta^p_r$ $\qquad \because \delta^p_q = 1$

**Example 11.** *Show that every tensor can be expressed as the sum of two tensors one of which is symmetric and the other skew-symmetric in a pair of covariant or contravariant indices.*

**Solution.** Consider the tensor $B^{pq}$, we have

$$B^{pq} = \frac{1}{2}(B^{pq} + B^{qp}) + \frac{1}{2}(B^{pq} - B^{qp})$$

But $\qquad R^{pq} = \frac{1}{2}(B^{pq} + B^{qp}) = R^{qp}$ $\qquad$ is symmetric and

$$S^{pq} = \frac{1}{2}(B^{pq} - B^{qp}) = R^{qp}$$ is skew-symm.

Thus $B^{pq}$ = symm tensor + skew-symm tensor.

By similar reasoning the result is seen to be true for any tensor.

**Example 12.** *What is contraction as applied to tensors? Prove that the contraction of the tensor $A^p_q$ is a scalar or invariant.*

**Solution.** *Contraction.* If one contravariant and one covariant index of a tensor are set equal, the result indicates that a summation over the equal indices is to be taken according to the summation convention. This resulting sum is a tensor of rank two less than that of the original tensor. The process is called contraction. For example, in the tensor of rank 3, $B^{mp}_q$, set $p = q$ we get $B_q^{\;mp} = C^m$, a tensor of rank 1.

To prove that contraction of $A^p_q$ is a scalar or invariant.

We have $\qquad \bar{A}^{\,j}_{\;k} = \dfrac{\partial \bar{x}^j}{\partial x^p}\dfrac{\partial x^q}{\partial \bar{x}^k} A^p_q$

Putting $j = k$, $\qquad \bar{A}^j_j = \dfrac{\partial \bar{x}^j}{\partial x^p}\dfrac{\partial x^q}{\partial \bar{x}^j} A^p_q = \delta^q_{\;p} A^p_p = A^p_p$

Then $\bar{A}^j_j = A^p_q$ and it follows that $A_p^{\;p}$ must be an invariant. Since $A^p_q$ is a tensor of rank two and contraction with respect to a single index lowers the rank by two. Therefore, an invariant is a tensor of rank zero. **Proved.**

**Example 13.** *A covariant tensor has components xy, $2y - z^2$, xz in rectangular co-ordinates. Find its covariant components in spherical co-ordinates.*

**Solution.** Let $A_j$ denote the covariant component in rectangular co-ordinates

$$x^1 = x,\ x^2 = y,\ x^3 = z.$$

Then $\qquad A_1 = xy = x^1 x^2$

$$A_2 = 2y - z^2 = 2x^2 - (x^3)^2$$

$$A_3 = xz = x^1 x^3$$

Let $\bar{A}_k$ denote the covariant component in spherical co-ordinates $\bar{x}^1 = r, \bar{x}^2 = \theta, \bar{x}^3 = \phi$

Then $\qquad \bar{A}_k = \dfrac{\partial \bar{x}^j}{\partial \bar{x}^k} A_j \qquad \ldots (1)$

In spherical co-ordinates

$$x = r \sin\theta \cos\phi$$

or $$x^1 = \bar{x}^1 \sin \bar{x}^2 \cos \bar{x}^3 \qquad \text{...(2)}$$

$$y = r \sin\theta \sin\phi$$

or $$x^2 = \bar{x}^1 \sin \bar{x}^2 \sin \bar{x}^3 \qquad \text{...(3)}$$

$$z = r \cos\theta$$

or $$x^3 = \bar{x}^1 \cos \bar{x}^2 \qquad \text{...(4)}$$

Therefore equation (1) yields the covariant component.

$$\bar{A}_1 = \frac{\partial x^1}{\partial \bar{x}^1} A_1 + \frac{\partial x^2}{\partial \bar{x}^1} A_2 + \frac{\partial x^3}{\partial \bar{x}^1} A_3$$

$$= (\sin \bar{x}^2 \cos \bar{x}^3)(x^1 x^2) + (\sin \bar{x}^2 \sin \bar{x}^3) \times [(2x^2 - (x^3)^2)] + (\cos \bar{x}^2)(x^1 x^3)$$

$$= (\sin\theta \cos\phi))(r^2 \sin^2\theta \sin\phi \cos\phi) + (\sin\theta \sin\phi)(2r \sin\theta \sin\phi - r^2 \cos^2\theta)$$
$$+ (\cos\theta)(r^2 \sin\theta \cos\theta \cos\phi)$$

$$\bar{A}_2 = \frac{\partial x^1}{\partial \bar{x}^2} A_1 + \frac{\partial x^2}{\partial \bar{x}^2} A_2 + \frac{\partial x^3}{\partial \bar{x}^2} A_3$$

$$= (\bar{x}^1 \cos \bar{x}^2 \cos \bar{x}^3)(x^1 x^2) + (\bar{x}^1 \cos \bar{x}^2 \sin \bar{x}^3)[(2x^2 - (x^3)^2)] + \bar{x}^1(-\sin \bar{x}^2)(x^1 x^2)$$

or $$\bar{A}_2 = (r \cos\theta \cos\phi)(r^2 \sin^2\theta \sin\phi \cos\phi) + (r \cos\theta \sin\phi)(2r \sin\theta \sin\phi - r^2 \cos^2\theta)$$

$$+ (-r \sin\theta)(r^2 \sin\theta \cos\theta \cos\phi) \quad \bar{A}_3 = \frac{\partial x^1}{\partial \bar{x}^3} A_1 + \frac{\partial x^2}{\partial \bar{x}^3} A_2 + \frac{\partial x^3}{\partial \bar{x}^3} A_3$$

$$= (-r \sin\theta \sin\phi)(r^2 \sin^2\theta \sin\phi \cos\phi) + (r \sin\theta \cos\phi)(2r \sin\theta \sin\phi - r^2 \cos^2\theta).$$ **Ans.**

**Example 14.** *Define symmetric and skew-symmetric tensors. Prove that a symmetric tensor of rank two has at most* $\frac{N(N+1)}{2}$ *different components in N-dimensional space* $V_N$

**Solution.** Symmetric Tensor. A tensor is called symmetric with respect to two contravariant or two covariant indices if its components remain unaltered upon interchange of the indices.

Thus if $A_{qs}^{mpr} = A_{qs}^{pmr}$, the tensor is symmetric in $m$ and $p$. If a tensor is symmetric with respect to any two contravariant and any two covariant indices, it is called symmetric.

**Skew-symmetric.** A tensor is called skew-symmetric with respect to two contravariant or two covariant indices if its component change sign upon interchange of the indices. Thus, if $A_{qs}^{mpr} = -A_{qs}^{pmr}$ the tensor is skew symmetric in $m$ and $p$. If a tensor is skew-symmetric with respect to any two contravariant and any two covariant indices it is called skew-symmetric.

Let $A_{pq}$ be a tensor of rank 2. The number of its all components in $V_N$ is $N^2$.

The components of $A_{pq}$ are

$$A_{11} \quad A_{12} \quad A_{13} \ \ldots\ldots\ldots\ldots \ A_{1N}$$

$$A_{21} \quad A_{22} \quad A_{23} \ldots\ldots\ldots\ldots A_{2N}$$

$$\ldots\ldots\ldots\ldots\ldots\ldots\ldots\ldots\ldots\ldots\ldots\ldots\ldots\ldots$$

$$A_{N1} \quad A_{N2} \quad A_{N3} \ldots\ldots\ldots\ldots A_{NN}$$

There are $N$ independent components of the form

$$A_{11}, A_{22}, A_{33} \ldots\ldots\ldots\ldots A_{NN}$$

Hence number of components of the form $A_{12}, A_{23}, A_{34}, \ldots\ldots\ldots\ldots$ in which there are distinct subscripts will be $N^2 - N$. But these component are symmetric. *i.e.*, $A_{12} = A_{21}$ etc.

Hence number of different component of this form are $\frac{1}{2}(N^2 - N)$

$\therefore$ Total number of different components are

$$= \frac{1}{2}(N^2 - N) + N = \frac{N^2}{2} + \frac{N}{2} = \frac{N(N+1)}{2}$$

**Example 15.** *Define a metric or fundamental tensor. Determine the components of the fundamental tensor in cylindrical co-ordinates.*

**Solution. Metric or Fundamental Tensor.**

In rectangular coordinates $(x, y, z)$ the differential of arc length $ds$ is obtained from $ds^2 = dx^2 + dy^2 + dz^2$. By transforming to general curvilinear co-ordinates this becomes

$$ds^2 = \sum_{p=1}^{3} \sum_{q=1}^{3} g_{pq}\, du_p\, du_q$$

Such spaces are called three-dimensional Euclidean spaces. We define the line element $ds$ in this space to be given by the quadratic form, called the metric form or metric,

$$ds^2 = \sum_{p=1}^{N} \sum_{q=1}^{N} g_{pq}\, dx^p\, dx^q \qquad \ldots(1)$$

$$\Rightarrow \qquad ds^2 = g_{pq}\, dx^p\, dx^q$$

The quantities $g_{pq}$ are the components of a covariant tensor of rank 2 called the *metric* tensor or *fundamental* tensor.

We know that $ds^2 = dx^2 + dy^2 + dz^2$

In cylindrical co-ordinates,

$$x = r\cos\theta,\ y = r\sin\theta,\ z = z$$

$$\therefore \qquad dx = -r\sin\theta\, d\theta + \cos\theta\, dr$$

$$dy = r\cos\theta\, d\theta + \sin\theta\, dr$$

$$dz = dz$$

$$ds^2 = dx^2 + dy^2 + dz^2$$

$$= (-r\sin\theta\, d\theta + \cos\theta\, dr)^2 + (r\cos\theta\, d\theta + \sin\theta\, dr)^2 + (dz)^2$$

or $$ds^2 = (dr)^2 + r^2 (d\theta)^2 + (dz)^2$$

Also metric is given by

$$ds^2 = g_{pq}\, dx^p\, dx^q \qquad \ldots(2)$$

If $x^1 = r, x^2 = \theta, x^3 = z$

then comparing (1) & (2), we have

$$g_{11} = 1,\ g_{22} = r^2,\quad g_{33} = 1,\ g_{12} = g_{21} = 0,\ g_{13} = g_{31} = 0,\ g_{23} = g_{32} = 0.$$

Metric tensor is given by $\begin{pmatrix} g_{11} & g_{12} & g_{13} \\ g_{21} & g_{22} & g_{23} \\ g_{31} & g_{32} & g_{33} \end{pmatrix}$

Metric tensor in cylindrical co-ordinates $= \begin{pmatrix} 1 & 0 & 0 \\ 0 & r^2 & 0 \\ 0 & 0 & 1 \end{pmatrix}$ **Ans.**

**Example 16.** *Define what is meant by invariant? Show that the contraction of the outer product of the tensors* $A^p$ and $B_q$ *is an invariant.*

**Solution. Scalar or Invariant.** Suppose $\phi$ is a function of the co-ordinates $x^k$, and let $\phi$ denote the functional value under a transformation to a new set of co-ordinates $\bar{x}^k$. Then $\phi$ is called a *scalar* or *invariant* with respect to the co-ordinate transformation if $\phi = \bar{\phi}$.

Since $A^p$ and $B_q$ are tensors.

$$\bar{A}^j = \frac{\partial \bar{x}^j}{\partial x^p} A^p,\ \bar{B}_k = \frac{\partial x^q}{\partial \bar{x}^k} B_q$$

Then $$\bar{A}^j\, \bar{B}_k = \frac{\partial \bar{x}^j}{\partial x^p}\frac{\partial x^q}{\partial \bar{x}^k} A^p B_q$$

By contraction (putting $j = k$)

$$\bar{A}^j\, \bar{B}_j = \frac{\partial \bar{x}^j}{\partial x^p}\frac{\partial x^q}{\partial \bar{x}^j} A^p B_q = \delta^q_p A^p B_q = A^p B_p$$ **Proved.**

and so $A^p\, B_p$ is an invariant.

**Example 17.** *What do you understand by associated tensors ?*

**Solution. Associated Tensors.** Given a tensor we can derive other tensors by raising or lowering indices. For example, given the tensor $A_{pq}$ we obtain by raising the index $p$, the tensor $A^p_{\cdot q}$, the dot indicating the original position of the moved index. By raising the index $q$ also we obtain $A^{p\,q}_{\cdot\,\cdot}$. We shall often write $A^{pq}$. These derived tensors can be obtained by forming inner products of the given tensor with the metric tensor $g_{pq}$ or its conjugate $g^{pq}$. Thus, for example

$$A^p_q = g^{rq} A_{rq},\ A^{pq} = g^{rp} g^{sq} A_{rs}$$

All tensors obtained from a given tensor by forming inner products with the metric tensor and its conjugate are called *associated tensors* of the given tensor. For example; $A^m$ and $A_m$ are associated tensors, the first are contravariant and the second covariant components. The relation between them is given by

$$A_p = g^{pq} A^q \quad \Rightarrow \quad A^p = g^{pq} A_q$$

For rectangular co-ordinates $g_{pq} = 1$ if $p = q$, and 0 if $p \neq q$, so that $A_p = A^p$, which explains why no distinction was made between contravariant and covariant components of a vector for rectangular co-ordinates.

## 18.45 CONJUGATE OR RECIPROCAL TENSORS

Let $g = |g_{pq}|$ denote the determinant with elements $g_{pq}$ and suppose $g \neq 0$. Define $g^{pq}$ by

$$g^{pq} = \frac{\text{cofactor of } g_{pq}}{g}$$

Then $g^{pq}$ is a symmetric contravariant tensor of rank two called the conjugate or Reciprocal tensor of $g_{pq}$.

Also $$g^{pq}\, g_{rq} = \delta^p_r$$

## 18.46 CHRISTOFFEL SYMBOLS

The symbols $$[pq,r] = \frac{1}{2}\left(\frac{\partial g_{pr}}{\partial x^q} + \frac{\partial g_{qr}}{\partial x^p} - \frac{\partial g_{pq}}{\partial x^r}\right); \left\{\begin{matrix} s \\ pq \end{matrix}\right\} = g^{sr}\,[pq,r]$$

are called the Christoffel symbols of the first and second kind respectively.

**Example 18.** *Prove that* $[pq, r] = g_{rs}\left\{\begin{matrix} s \\ pq \end{matrix}\right\}$

**Solution.** $$g_{ks}\left\{\begin{matrix} s \\ pq \end{matrix}\right\} = g_{ks}\, g^{sr}[pq,r] = \delta^r_k\,[pq, r] = [pq, k]$$

$$\Rightarrow \qquad [pq,\ k] = g_{ks}\left\{\begin{matrix} s \\ pq \end{matrix}\right\}$$

*ie.* $$[pq,\ r] = g_{rs}\left\{\begin{matrix} s \\ pq \end{matrix}\right\}$$ **Proved.**

**Example 19.** *Prove that (i) $[pq, r] = [qp, r]$ (ii)* $\left\{\begin{matrix} s \\ pq \end{matrix}\right\} = \left\{\begin{matrix} s \\ qp \end{matrix}\right\}$.

**Solution.** (*i*) $$[pq,r] = \frac{1}{2}\left(\frac{\partial g_{pr}}{\partial x^q} + \frac{\partial g_{qr}}{\partial x^p} - \frac{\partial g_{pq}}{\partial x^r}\right)$$

$$= \frac{1}{2}\left(\frac{\partial g_{qr}}{\partial x^p} + \frac{\partial g_{pr}}{\partial x^q} - \frac{\partial g_{qp}}{\partial x^r}\right)$$

$$[pq, r] = [qp, r]$$

(*ii*) $$\left\{\begin{matrix} s \\ pq \end{matrix}\right\} = g^{sr}\,[pq,r] = g^{sr}[qp,r] = \left\{\begin{matrix} s \\ qp \end{matrix}\right\}$$ **Proved.**

**Example 20.** *Prove that* $\dfrac{\partial g_{pq}}{\partial x^m} = [pm, q] + [qm, p]$

**Solution.** $[pm, q] + [qm, p]$

$$= \frac{1}{2}\left(\frac{\partial g_{pq}}{\partial x^m} + \frac{\partial g_{mq}}{\partial x^p} - \frac{\partial g_{pm}}{\partial x^q}\right) + \frac{1}{2}\left(\frac{\partial g_{qp}}{\partial x^m} + \frac{\partial g_{mp}}{\partial x^q} - \frac{\partial g_{qm}}{\partial x^p}\right) = \frac{1}{2}\frac{\partial g_{pq}}{\partial x^m} + \frac{1}{2}\frac{\partial g_{qp}}{\partial x^m} = \frac{\partial g_{pq}}{\partial x^m}$$

**Example 21.** *Prove that* $\dfrac{\partial g^{pq}}{\partial x^m} = -g^{pn}\left\{\begin{matrix} q \\ mn \end{matrix}\right\} - g^{qn}\left\{\begin{matrix} p \\ mn \end{matrix}\right\}$

**Solution.** $$\frac{\partial}{\partial x^m}(g^{jk}g_{ij}) = \frac{\partial}{\partial x^m}(\delta_i^k) = 0$$

Then $$g^{jk}\frac{\partial g_{ij}}{\partial x^m} + \frac{\partial g^{jk}}{\partial x^m}g_{ij} = 0 \Rightarrow g^{ij}\frac{\partial g^{jk}}{\partial x^m} = -g^{jk}\frac{\partial g_{ij}}{\partial x^m}$$

Multiplying by $g^{ir}$

*i.e.* $$\delta_j^r\frac{\partial g^{jk}}{\partial x^m} = -g^{ir}g^{jk}[im, j] + [jm, i]$$

$$\Rightarrow \quad \frac{\partial g^{rk}}{\partial x^m} = -g^{ir}\begin{Bmatrix} k \\ im \end{Bmatrix} - g^{jk}\begin{Bmatrix} r \\ jm \end{Bmatrix}$$

Replace $r, k, i, j$ by $p, q, n, n$, we get $\frac{\partial g^{pq}}{\partial x^m} = -g^{pn}\begin{Bmatrix} q \\ mn \end{Bmatrix} - g^{qn}\begin{Bmatrix} p \\ mn \end{Bmatrix}$ **Proved.**

**Example 22.** *Derive transformation laws for the christoffel symbols of the first and the second kind.*

**Solution.** Since $\bar{g}_{jk} = \frac{\partial x^p}{\partial \bar{x}^j}\frac{\partial x^q}{\partial \bar{x}^k}g_{pq}$

$$\therefore \quad \frac{\partial \bar{g}_{jk}}{\partial \bar{x}^m} = \frac{\partial x^p}{\partial \bar{x}^j}\frac{\partial x^q}{\partial \bar{x}^k}\frac{\partial g_{pq}}{\partial x^r}\frac{\partial x^r}{\partial \bar{x}^m} + \frac{\partial x^p}{\partial \bar{x}^j}\frac{\partial^2 x^q}{\partial \bar{x}^m \partial \bar{x}^k}g_{pq} + \frac{\partial^2 x^p}{\partial \bar{x}^m \partial \bar{x}^j}\frac{\partial x^q}{\partial \bar{x}^k}g_{pq} \quad \text{... (1)}$$

By cyclic permutation of indices $n$, $k$, $m$ and $p$, $q$, $r$

$$\frac{\partial \bar{g}_{km}}{\partial \bar{x}^j} = \frac{\partial x^q}{\partial \bar{x}^k}\frac{\partial x^r}{\partial \bar{x}^m}\frac{\partial g_{qr}}{\partial x^p}\frac{\partial x^p}{\partial \bar{x}^j} + \frac{\partial x^q}{\partial \bar{x}^k}\frac{\partial^2 x^r}{\partial \bar{x}^j \partial \bar{x}^m}g_{qr} + \frac{\partial^2 x^q}{\partial \bar{x}^j \partial \bar{x}^k}\frac{\partial x^r}{\partial \bar{x}^m}g_{pq} \quad \text{...(2)}$$

$$\frac{\partial \bar{g}_{mj}}{\partial \bar{x}^k} = \frac{\partial x^r}{\partial \bar{x}^m}\frac{\partial x^p}{\partial \bar{x}^j}\frac{\partial g_{rp}}{\partial x^q}\frac{\partial x^q}{\partial \bar{x}^k} + \frac{\partial x^r}{\partial \bar{x}^m}\frac{\partial^2 x^p}{\partial \bar{x}^k \partial \bar{x}^j}g_{rp} + \frac{\partial^2 x^r}{\partial \bar{x}^k \partial \bar{x}^m}\frac{\partial x^p}{\partial \bar{x}^j}g_{rp} \quad \text{...(3)}$$

Subtracting (1) from the sum of (2) and (3) and multiplying by $\frac{1}{2}$, we obtain on using the definition of the Christoffel symbols of the first kind,

$$\overline{[jk, m]} = \frac{\partial x^p}{\partial \bar{x}^j}\frac{\partial x^q}{\partial \bar{x}^k}\frac{\partial x^r}{\partial \bar{x}^m}[pq, r] + \frac{\partial^2 x^p}{\partial \bar{x}^j \partial \bar{x}^k}\frac{\partial x^q}{\partial \bar{x}^m}g_{pq} \quad \text{... (4)}$$

## 18.47 TRANSFORMATION LAW FOR SECOND KIND

Multiplying (4) by $\bar{g}^{nm}$

$$\bar{g}^{nm} = \frac{\partial \bar{x}^n}{\partial x^s}\frac{\partial \bar{x}^m}{\partial x^t}g^{st} \qquad \text{we get}$$

$$\bar{g}^{nm}\,\overline{[jk, m]} = \frac{\partial x^p}{\partial \bar{x}^j}\frac{\partial x^q}{\partial \bar{x}^k}\frac{\partial x^r}{\partial \bar{x}^m}\frac{\partial \bar{x}^n}{\partial x^s}\frac{\partial \bar{x}^m}{\partial x^t}g^{st}[pq, r] + \frac{\partial^2 x^p}{\partial \bar{x}^j \partial \bar{x}^k}\frac{\partial x^q}{\partial \bar{x}^m}\frac{\partial \bar{x}^n}{\partial x^s}\frac{\partial \bar{x}^m}{\partial x^t}g^{st}g_{pq}$$

Then $$\overline{\begin{Bmatrix} n \\ jk \end{Bmatrix}} = \frac{\partial x^p}{\partial \bar{x}^j}\frac{\partial x^q}{\partial \bar{x}^k}\frac{\partial \bar{x}^n}{\partial x^s}\delta_t^r g^{st}[pq, r] + \frac{\partial^2 x^p}{\partial \bar{x}^j \partial \bar{x}^k}\frac{\partial \bar{x}^n}{\partial x^s}\delta_t^q g^{st}g_{pq}$$

$$\overline{\begin{Bmatrix} n \\ jk \end{Bmatrix}} = \frac{\partial x^p}{\partial \bar{x}^j}\frac{\partial x^q}{\partial \bar{x}^k}\frac{\partial \bar{x}^n}{\partial x^s}\begin{Bmatrix} s \\ pq \end{Bmatrix} + \frac{\partial^2 x^p}{\partial \bar{x}^j \partial \bar{x}^k}\frac{\partial \bar{x}^n}{\partial x^p} \quad \ldots (5)$$

(4) and (5) are required transformation laws.

**Example 23.** *If* $(ds)^2 = r^2 (d\theta)^2 + r^2 \sin^2 \theta \, (d\phi)^2$, *find the value of*

(*a*) [22, 1] (*b*) [12, 2] (*c*) [1, 22] (*d*) [2, 12].

**Solution.** $(ds)^2 = r^2 (d\theta)^2 + r^2 \sin^2 \theta \, (d\phi)^2$

$$g_{11} = r^2, \; g_{22} = r^2 \sin^2 \theta, \; g_{12} = 0 = g_{21}$$

$$g = \begin{vmatrix} g_{11} & g_{12} \\ g_{21} & g_{22} \end{vmatrix} = \begin{vmatrix} r^2 & 0 \\ 0 & r^2 \sin^2 \theta \end{vmatrix} = r^4 \sin^2 \theta$$

$$g^{11} = \frac{\text{cofactor of } g_{11}}{g} = \frac{r^2 \sin^2 \theta}{r^4 \sin^2 \theta} = \frac{1}{r^2}$$

$$g^{22} = \frac{\text{cofactor of } g_{22}}{g} = \frac{r^2}{r^4 \sin^2 \theta} = \frac{1}{r^2 \sin^2 \theta}$$

The christoffel symbols of first kind are

$$[ij, k] = \frac{1}{2}\left[\frac{\partial g_{jk}}{\partial x^i} + \frac{\partial g_{ik}}{\partial x^j} - \frac{\partial g_{ij}}{\partial x^k}\right]$$

(*a*)
$$[22,1] = \frac{1}{2}\left[\frac{\partial g_{21}}{\partial x^2} + \frac{\partial g_{21}}{\partial x^2} - \frac{\partial g_{22}}{\partial x^1}\right] = \frac{1}{2}\left[\frac{\partial(0)}{\partial \phi} + \frac{\partial(0)}{\partial \phi} - \frac{\partial(r^2 \sin^2 \theta)}{\partial \theta}\right]$$

$$= r^2 \sin\theta \cos\theta$$

(*b*)
$$[12,2] = \frac{1}{2}\left[\frac{\partial g_{22}}{\partial x^1} + \frac{\partial g_{12}}{\partial x^2} - \frac{\partial g_{12}}{\partial x^2}\right] = \frac{1}{2}\left[\frac{\partial(r^2 \sin^2 \theta)}{\partial \theta} + \frac{\partial(0)}{\partial \phi} - \frac{\partial(0)}{\partial \phi}\right]$$

$$= r^2 \sin\theta \cos\theta$$

(*c*) The christoffel symbols of second kind are

$$\begin{Bmatrix} k \\ ij \end{Bmatrix} = g^{kl}\,[ij, l]$$

$$\begin{Bmatrix} 1 \\ 22 \end{Bmatrix} = g^{ll}[22, l] = g^{11}\,[22, 1]\, g^{12}\,[22, 2]$$

$$= \frac{1}{r^2}\,[-r^2 \sin\theta \cos\theta] + 0 \quad (g^{12} = 0) = -\sin\theta\cos\theta$$

(*d*)
$$\begin{Bmatrix} 2 \\ 12 \end{Bmatrix} = g^{2l}\,[22, l] = g^{21}\,[12, 1] + g^{22}\,[12, 2]$$

$$= 0 + \frac{1}{r^2 \sin^2 \theta}[r^2 \sin\theta\cos\theta] = \frac{\cos\theta}{\sin\theta} = \cot\theta = r^4 \sin\theta\cos\theta$$ **Ans.**

**Example 24.** *If* $(ds)^2 = (dr)^2 + r^2 (d\theta)^2 + r^2 \sin^2\theta\, (d\phi)^2$, *find the value of*

(*a*) [22, 1], (*b*) [33, 1], (*c*) [13, 3], (d) [23, 3],

(*e*) $\begin{Bmatrix} 1 \\ 22 \end{Bmatrix}$, (*f*) $\begin{Bmatrix} 1 \\ 33 \end{Bmatrix}$, (g) $\begin{Bmatrix} 3 \\ 13 \end{Bmatrix}$, (*h*) $\begin{Bmatrix} 3 \\ 23 \end{Bmatrix}$.

**Solution.** $(ds)^2 = (dr)^2 + r^2 (d\theta)^2 + r^2 \sin^2\theta\, (d\phi)^2$

$x_1 = r,\ x_2 = \theta,\ x_3 = \phi$

$g_{11} = 1,\ g_{22} = r^2,\ g_{33} = r^2 \sin\theta$

$g_{12} = 0 = g_{13} = .....$

$$g = \begin{vmatrix} g_{11} & g_{12} & g_{13} \\ g_{21} & g_{22} & g_{23} \\ g_{31} & g_{32} & g_{33} \end{vmatrix} = \begin{vmatrix} 1 & 0 & 0 \\ 0 & r^2 & 0 \\ 0 & 0 & r^2 \sin^2\theta \end{vmatrix} = r^4 \sin^2\theta$$

$$g^{11} = \frac{\text{cofactor of } g_{11}}{g} = \frac{\begin{vmatrix} r^2 & 0 \\ 0 & r^2 \sin^2\theta \end{vmatrix}}{r^4 \sin^2\theta} = \frac{r^4 \sin^2\theta}{r^4 \sin^2\theta} = 1$$

$$g^{22} = \frac{\text{cofactor of } g_{22}}{g} = \frac{\begin{vmatrix} 1 & 0 \\ 0 & r^2 \sin^2\theta \end{vmatrix}}{r^4 \sin^2\theta} = \frac{r^2 \sin^2\theta}{r^4 \sin^2\theta} = \frac{1}{r^2}$$

$$g^{33} = \frac{\text{cofactor of } g_{33}}{g} = \frac{\begin{vmatrix} 1 & 0 \\ 0 & r^2 \end{vmatrix}}{r^4 \sin^2\theta} = \frac{r^2}{r^4 \sin^2\theta} = \frac{1}{r^2 \sin^2\theta}$$

The christoffel symbols of the first kind are

$$[ij, k] = \frac{1}{2}\left[\frac{\partial g_{jk}}{\partial x^i} + \frac{\partial g_{ik}}{\partial x^i} - \frac{\partial g_{ij}}{\partial x^k}\right]$$

(*a*) $$[22, 1] = \frac{1}{2}\left[\frac{\partial g_{21}}{\partial x^2} + \frac{\partial g_{21}}{\partial x^2} - \frac{\partial g_{22}}{\partial x^1}\right] = \frac{1}{2}\left[\frac{\partial(0)}{\partial\theta} + \frac{\partial(0)}{\partial\theta} - \frac{\partial(r^2)}{\partial r}\right] = \frac{-1}{2}(-2r) = -r$$

(*b*) $$[33, 1] = \frac{1}{2}\left[\frac{\partial g_{31}}{\partial x^3} + \frac{\partial g_{31}}{\partial x^3} - \frac{\partial g_{33}}{\partial x^1}\right] = \frac{1}{2}\left[\frac{\partial(0)}{\partial\phi} + \frac{\partial(0)}{\partial\phi} - \frac{\partial(r^2 \sin^2\theta)}{\partial r}\right] = \frac{1}{2}(-2r\sin^2\theta) = -r\sin^2\theta$$

(*c*) $$[13, 3] = \frac{1}{2}\left[\frac{\partial g_{33}}{\partial x^1} + \frac{\partial g_{13}}{\partial x^3} - \frac{\partial g_{13}}{\partial x^3}\right] = \frac{1}{2}\frac{\partial g_{33}}{\partial x^1} = \frac{1}{2}\left[\frac{\partial(r^2 \sin^2\theta)}{\partial r}\right] = r\sin^2\theta$$

(*d*) $$[23, 3] = \frac{1}{2}\left[\frac{\partial g_{33}}{\partial x^2} + \frac{\partial g_{23}}{\partial x^3} - \frac{\partial g_{23}}{\partial x^3}\right] = \frac{1}{2}\frac{\partial g_{33}}{\partial x^2} = \frac{1}{2}\frac{\partial}{\partial\theta}(r^2 \sin^2\theta) = r^2 \sin\theta\cos\theta$$

The christoffel symbols of the second kind are

$$\begin{Bmatrix} k \\ ij \end{Bmatrix} = g^{kl}[ij, l]$$

(e) $\begin{Bmatrix} 1 \\ 22 \end{Bmatrix} = g^{1l}[22,l] = g^{11}[22,1] + g^{12}[22,2] + g^{13}[22,3] = (1)(-r) + 0 + 0 = -r$

(f) $\begin{Bmatrix} 1 \\ 33 \end{Bmatrix} = g^{1l}[33,l] = g^{11}[33,1] + g^{12}[33,2] + g^{13}[33,3] = (1)(-r\sin^2\theta) + 0 + 0 = -r\sin^2\theta$

(g) $\begin{Bmatrix} 3 \\ 13 \end{Bmatrix} = g^{3l}[13,l] = g^{31}[13,1] + g^{32}[13,2] + g^{33}[13,3]$

$$= 0\,(r\sin^2\theta) + 0\,[0] + \frac{1}{r^2\sin^2\theta}[r\sin^2\theta] = \frac{1}{r}$$

(h) $\begin{Bmatrix} 3 \\ 23 \end{Bmatrix} = g^{3l}[23,l] = g^{31}[23,1] + g^{32}[23,2] + g^{33}[23,3]$

$$= 0\,[0] + 0\,[0] + \frac{1}{r^2\sin^2\theta}(r\sin\theta\cos\theta) = \cot\theta$$ **Ans.**

**Example 25.** *Prove that* $\begin{Bmatrix} p \\ pq \end{Bmatrix} = \frac{\partial}{\partial x^q}\log\sqrt{g}$

**Solution.** $g = g_{jk}\,G(j,k)$ (Sum over $k$ only)

where $G(j,k)$ is the cofactor of $g_{jk}$ in the determinant $g = |g_{jk}| \neq 0$ since $G(j,k)$ does not contain $g_{jk}$ explicitly,

$$\frac{\partial g}{\partial g_{jr}} = G(j,r)$$

Then, summing over $j$ and $r$

$$\frac{\partial g}{\partial x^m} = \frac{\partial g}{\partial g_{jr}}\frac{\partial g_{jr}}{\partial x^m} = G(j,r)\frac{\partial g_{jr}}{\partial x^m} = g\,g^{jr}\frac{\partial g_{jr}}{\partial x^m} = g\,g^{jr}([jm,r] + [rm,j])$$

$$\frac{\partial g}{\partial x^m} = g\left(\begin{Bmatrix} j \\ jm \end{Bmatrix} + \begin{Bmatrix} r \\ rm \end{Bmatrix}\right) = 2g\begin{Bmatrix} j \\ jm \end{Bmatrix}$$

Thus $$\frac{1}{2g}\frac{\partial g}{\partial x^m} = \begin{Bmatrix} j \\ jm \end{Bmatrix} \quad \text{or} \quad \begin{Bmatrix} j \\ jm \end{Bmatrix} = \frac{\partial}{\partial x^m}\log\sqrt{g}$$

Replacing $j$ by $p$ and $m$ by $q$, $\begin{Bmatrix} p \\ pq \end{Bmatrix} = \frac{\partial}{\partial x^q}\log\sqrt{g}$ **Proved.**

## 18.48 CONTRAVARIANT, COVARIANT AND MIXED TENSOR

If $A_i$ be a set of $n$ functions of the co-ordinates $x^1, x^2, x^3 \ldots x^n\,(x^i)$. They are transformed in another system of co-ordinates $\bar{x}^1, \bar{x}^2, \bar{x}^3 \ldots \bar{x}^n$ according to $\bar{A}_i = \frac{\partial x^j}{\partial \bar{x}^i}A_j$

$A_i$ are called the components of a covariant tensor.

If $\phi\,(x^1, x^2 \ldots x^n)$ be a scalar functions, then $\frac{\partial\phi}{\partial \bar{x}^i} = \frac{\partial\phi}{\partial x^1}\frac{\partial x^1}{\partial \bar{x}^i} + \frac{\partial\phi}{\partial x^2}\frac{\partial x^2}{\partial \bar{x}^i} + \ldots + \frac{\partial\phi}{\partial x^n}\frac{\partial x^n}{\partial \bar{x}^i}$

then $\frac{\partial \phi}{\partial x^1}, \frac{\partial \phi}{\partial x^2}, \ldots \frac{\partial \phi}{\partial x^n}$ are the components of a covariant vector.

Since $x$ is a function of $x^i$ (*i.e.*, $x^1, x^2, \ldots x^n$)

so $$d\bar{x}^i = \frac{\partial \bar{x}^i}{\partial x^1} dx^1 + \frac{\partial \bar{x}^i}{\partial x^2} dx^2 + \ldots + \frac{\partial \bar{x}^i}{\partial x^n} dx^n \quad \ldots (2) \quad = \frac{\partial \bar{x}^i}{\partial x^1} dx^1$$

On comparing (1) and (2) we can say that $dx^1, dx^2, \ldots dx^n$ is an example of a contravariant tensor.

If $q$ and $s$ varry from 1 to $n$, then $A^{qs}$ will be $n^2$ functions.

If $N^2$ quantities $A^{qs}$ in a co-ordinate system $(x^1, x^2, \ldots x^N)$ are related to $N^2$ other quantities $\bar{A}^{pr}$ in another system $(\bar{x}^1, \bar{x}^2, \ldots \bar{x}^N)$ by the transformation equations

$$\bar{A}^{pr} = \sum_{s=1}^{N} \sum_{q=1}^{N} \frac{\partial \bar{x}^p}{\partial x^q} \frac{\partial \bar{x}^r}{\partial x^s} A^{qs} \quad \Rightarrow \quad \bar{A}^{pr} = \frac{\partial \bar{x}^p}{\partial x^q} \frac{\partial \bar{x}^r}{\partial x^s} A^{qs}$$

they are called *contravariant* components of a tensor of the second rank.

If the transformation law is $\bar{A}^{pr} = \frac{\partial x^q}{\partial \bar{x}^p} \frac{\partial x^s}{\partial \bar{x}^r} A_{qs},$

then quantities $A_{qs}$ are called components of *covariant* tensor of second rank.

The $N^2$ quantities $A^q_s$ are called components of a *mixed* tensor of second rank if

$$\bar{A}^p_r = \frac{\partial \bar{x}^p}{\partial x^q} \frac{\partial x^s}{\partial \bar{x}^r} A^q_s$$

## EXERCISE 18.2

1. If $A^i$ are the components of an absolute contravariant tensor of rank one, show that $\frac{\partial A_i}{\partial x_j}$ are the components of a mixed tensor.

2. If $A^{ji}$ and $A_{ij}$ are reciprocal symmetric tensors and $x_i$ are the components of a covariant tensor of rank one, show that $A_{ij}\, x^i x^j = A^{ij}\, x_i x_j$, where $x^i = A^{i\alpha} x_\alpha$ .

3. If the components of a tensor are zero in one co-ordinate system, then prove that the components are zero in all co-ordinate systems.

4. Show that the expression $A\,(i, j, k)$ is a tensor if its inner product with an arbitrary tensor $B^{jl}_k$ is a tensor.

5. $A^{ij}$ is a contravariant tensor and $B_i$ a covariant tensor. Show that $A^{ij} B_k$ is a tensor of rank three, but $A^{ij} B_j$ is a tensor of rank one.

6. If $g_{ij}$ denotes the components of a covariant tensor of rank two, show that the product $g_{ij} dx^i dx^j$ is an invariant scalar.

7. Find $g$ and $g^{ij}$ corresponding to the metric

$$ds^2 = 5\,(dx^1)^2 + 3\,(dx^2)^2 + 4\,(dx^3)^2 - 6\, dx^1\, dx^2 + 4 dx^2\, dx^3.$$

**Ans.** $g = 4, g^{11} = 2, g^{22} = 5, g^{33} = 1.5, g^{12} = 3, g^{23} = -2.5, g^{13} = -1.5$

**8.** Find the values of $g$ and $g^{ij}$, if

$$ds^2 = \frac{dr^2}{1-\frac{r^2}{R^2}} + r^2(d\theta^2 + \sin^2\theta d\phi^2), \text{ where R is constant}$$

**Ans.** $g = \frac{r^4 \sin^2\theta}{1-\frac{r^2}{R^2}}; g^{11} = 1-\frac{r^2}{R^2}, g^{22} = \frac{1}{r^2}, g^{33} = \frac{1}{r^2\sin^2\theta}, g^{ij} = 0\ (i \neq j)$

**9.** Prove that the angle $\theta_{12}, \theta_{23}, \theta_{31}$ between the co-ordinate curves in a three dimensional co-ordinate system are given by

$$\cos\theta_{12} = \frac{g_{12}}{\sqrt{g_{11}g_{22}}}, \cos\theta_{23} = \frac{g_{23}}{\sqrt{g_{22}\,g_{33}}}, \cos\theta_{31} = \frac{g_{31}}{\sqrt{g_{33}\,g_{11}}}$$

**10.** Prove that for an orthogonal co-ordinate system

(a) $g_{12} = g_{23} = g_{31} = 0$ (b) $g^{11} = \frac{1}{g_{11}}, g^{22} = \frac{1}{g_{22}}, g^{33} = \frac{1}{g_{33}}$

**11.** Surface of a sphere is a two dimensional Riemannian space. Find its fundamental metric tensor. If $a$ be the fixed radius of the sphere.

**Ans.** $g_{11} = a^2$, $g_{22} = a^2\sin^2\theta$, $g = a^4\sin^2\theta$

$g^{11} = \frac{1}{a^2}, g^{22} = \frac{1}{a^2\sin^2\theta}, g^{12} = 0 = g^{21}$.

# 19

# Z - Transform

## 19.1 INTRODUCTION

Z-transform plays an important role in discrete analysis. Its role in discrete analysis is the same as that of Laplace and Fourier transforms in continuous system. Communication is one of the field whose development is based on discrete analysis. Difference equations are also based on discrete system and their solutions and analysis are carried out by Z- transform.

## 19.2 SEQUENCE

Sequence $\{f(k)\}$ is an ordered list of real or complex numbers.

## 19.3 REPRESENTATION OF A SEQUENCE

**First Method.**

The elementary way is to list all the members of the sequence such as

$$\{f(k)\} = \{15, 10, 7, \underset{\uparrow}{4}, 1, -1, 0, 3, 6\}$$

The symbol ↑ is used to denote the term in zero position *i.e.*, $k = 0$, $k$ is an index of position of a term in the sequence.

$$\{g(k)\} = \{15, 10, 7, 4, 1, \underset{\uparrow}{-1}, 0, 3, 6\}$$

Two sequences $\{f(k)\}$ and $\{g(k)\}$ have the same terms but these sequences are not identically the same as the zeroth term of those sequences are different.

In case the symbol ↑ is not given then left hand end term is considered as the term corresponding to $k = 0$.

In sequence $\{8, 6, 3, -1, 0, 1, 4, 5\}$

the zero$^{th}$ term is 8, the left hand end term.

**Second Method.**

The second way of specifying the sequence is to define the general term of the sequence $\{f(k)\}$ as function of $k$.

*For example,* $$f(k) = \frac{1}{3^k}$$

This sequence represents $\left\{\ldots\frac{1}{3^{-3}}, \frac{1}{3^{-2}}, \underset{\substack{\uparrow \\ k=0}}{\frac{1}{3^{-1}}}, 1, \frac{1}{3}, \frac{1}{3^2}, \frac{1}{3^3}\ldots\right\}$

If $\quad f(k)=\frac{1}{3^k},\ -4\le k\le 3 \qquad \left\{\frac{1}{3^{-4}},\frac{1}{3^{-3}},\frac{1}{3^{-2}},\frac{1}{3^{-1}},1,\frac{1}{3^1},\frac{1}{3^2},\frac{1}{3^3}\right\}$

## 19.4 BASIC OPERATIONS ON SEQUENCES

Let $\{f(k)\}$ and $\{g(k)\}$ be two sequences having same number of terms.

**Addition.** $\{f(k)\}+\{g(k)\}=\{f(k)+g(k)\}$

**Multiplication.** Let $a$ be a scalar, then $a\{f(k)\}=\{af(k)\}$

**Linearity.** $a\{f(k)\}+b\{g(k)\}=\{af(k)+bg(k)\}$

### EXERCISE 19.1

**1.** Write down the term corresponding to $k=2$

$$(6, 7, 5, 1, 0, \underset{\uparrow}{4}, 6, 8, 10)$$

**Ans.** 8

**2.** Write down the term corresponding to $k=-3$.

$$(20, 16, 14, 13, 12, \underset{\uparrow}{10}, 5, 1, 0)$$

**Ans.** 14

**3.** Write down the sequence $\{f(k)\}$ where $f(k)=\frac{1}{2^k}$ **Ans.** $\left\{\ldots\frac{1}{2^{-3}},\frac{1}{2^{-2}},\frac{1}{2^{-1}},1,\frac{1}{2^1},\frac{1}{2^2},\frac{1}{2^3}\ldots\right\}$

**4.** Write down the sequence $\{f(k)\}$ where $f(k)=\frac{1}{4^k},\quad -3\le k\le 4$

**Ans.** $\left\{\frac{1}{4^{-3}},\frac{1}{4^{-2}},\frac{1}{4^{-1}},1,\frac{1}{4^1},\frac{1}{4^2},\frac{1}{4^3},\frac{1}{4^4}\right\}$

**5.** Write down the sequence $\frac{1}{2}\{f(k)\}$, where $f(k)=\frac{1}{3^k}$

**Ans.** $\left\{\frac{27}{2},\frac{9}{2},\frac{3}{2},\frac{1}{2},\frac{1}{6},\frac{1}{18},\frac{1}{54}\ldots\right\}$

**6.** What sequence is generated when $f(k)=\begin{cases}0 & , \quad k<0\\ \cos\frac{k}{2}, & k\ge 0\end{cases}$

**Ans.** $\left\{\ldots 0,0,1,\cos\frac{1}{2},\cos 1,\cos\frac{3}{2}\ldots\right\}$

**7.** Write the sequence $\frac{1}{3}\{f(k)\}+\frac{1}{4}\{g(k)\}=\{F(k)\}$. Where $f(k)=\frac{1}{3^k}$ and

$$g(k)=\begin{cases}0, & k<0\\ 4, & k\ge 0\end{cases}$$

**Ans.** $\{F(k)\}=\begin{cases}\frac{1}{3^{k+1}}, & k<0\\ \frac{1}{3^{k+1}}+1, & k\ge 0\end{cases}$

## 19.5 Z-TRANSFORM

**Definition.** The $Z$- transform of a sequence $\{f(k)\}$ is denoted as $Z[\{f(k)\}]$.

It is defined as $Z[\{f(k)\}]=F(z)=\sum_{k=-\infty}^{\infty} f(k)z^{-k}=\sum_{k=-\infty}^{\infty}\frac{f(k)}{z^k}$

where **1.** $Z$ is a complex number.

**2.** $Z$ is an operator of Z-transform.

**3.** $F(z)$ is the Z-transform of $\{f(k)\}$.

**Example 1.** *If* $f(k) = \{15, 10, 7, \underset{\uparrow}{4}, 1, -1, 0, 3, 6\}$, *then*

$$Z[\{f(k)\}] = F(z) = 15z^3 + 10z^2 + 7z + 4 + \frac{1}{z} - \frac{1}{z^2} + 0 + \frac{3}{z^4} + \frac{6}{z^5}$$

**Example 2.** *If* $g(k) = \{15, 10, 7, 4, 1, -1, 0, \underset{\uparrow}{3}, 6\}$, then

$$Z[\{g(k)\}] = F(z) = 15z^7 + 10z^6 + 7z^5 + 4z^4 + z^3 - z^2 + 0 + 3 + \frac{6}{z}$$

**Example 3.** *The Z-transform of the sequence* $\{8, 6, 3\ -1, 0, 14, 5\}$ *is*

$$8 + \frac{6}{z} + \frac{3}{z^2} - \frac{1}{z^3} + 0 + \frac{14}{z^5} + \frac{5}{z^6}$$

**Example 4.** *If* $f(k) = \frac{1}{3^k}$ *then* $Z[\{f(k)\}] = \ldots + 27z^3 + 9z^2 + 3z + 1 + \frac{1}{3z} + \frac{1}{9z^2} + \frac{1}{27z^3} + \ldots$

**Example 5.** *If* $f(k) = \frac{1}{3^k},\ -4 \le k \le 3$, *then*

$$Z[\{f(k)\}] = 81z^4 + 27z^3 + 9z^2 + 3z + 1 + \frac{1}{3z} + \frac{1}{9z^2} + \frac{1}{27z^3}$$

**Example 6.** *Find Z-transform of the sequence* $\left\{\frac{1}{2^k}\right\}, -4 \le k \le 4$

**Solution.** $F(z) = \sum_{k=-4}^{4} \frac{1}{2^k} z^{-k} = 16z^4 + 8z^3 + 4z^2 + 2z + 1 + \frac{1}{2z} + \frac{1}{4z^2} + \frac{1}{8z^3} + \frac{1}{16z^4}$ **Ans.**

**Example 7.** *Find Z-transform of the sequence* $\{a^k\},\ k \ge 0$.

**Solution.** $F(z) = \sum_{k=0}^{\infty} a^k z^{-k} = 1 + \frac{a}{z} + \frac{a^2}{z^2} + \frac{a^3}{z^3} + \ldots$

This is a Geometrical series whose sum $= \frac{a}{1-r} = \frac{1}{1-\frac{a}{z}} = \frac{z}{z-a}$ **Ans.**

## 19.6 PROPERTIES OF Z-TRANSFORMS

**Linearity.**

**Theorem 1:** If $\{f(k)\}$ and $\{g(k)\}$ are such that they can be added and $a$ and $b$ are constants, then

$$Z\{af(k) + bg(k)\} = aZ[\{f(k)\}] + bZ[\{g(k)\}]$$

**Proof.** $Z[\{af(k) + bg(k)\}] = \sum_{k=-\infty}^{\infty} [af(k) + bg(k)]z^{-k}$ [By definition]

$$= \sum_{k=-\infty}^{\infty} [af(k)z^{-k} + bg(k)z^{-k}] = a\sum_{k=-\infty}^{\infty} f(k)z^{-k} + b\sum_{k=-\infty}^{\infty} g(k)z^{-k}$$

$$= aZ[\{f(k)\}] + bZ[\{g(k)\}]$$ **Proved.**

**Example 8.** *Find the Z transform of* $\{f(k)\}$, *where*

$$f(k) = \begin{cases} 5^k, & k < 0 \\ 3^k, & k \ge 0 \end{cases}$$

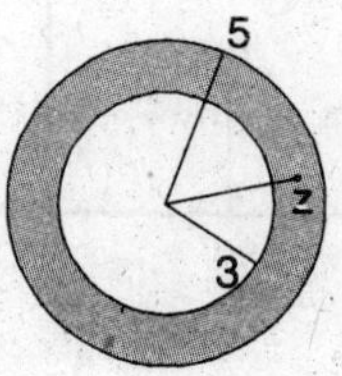

**Solution.** $$Z[\{f(k)\}] = \sum_{k=-\infty}^{-1} 5^k z^{-k} + \sum_{k=0}^{\infty} 3^k z^{-k}$$

$$= [....+ 5^{-3} z^3 + 5^{-2} z^2 + 5^{-1} z^1] + \left[1 + \frac{3}{z^{-1}} + \frac{9}{z^{-2}} + \frac{27}{z^{-3}} + ....\right] \quad \text{[G. P.]}$$

$$= \frac{5^{-1} z}{1 - 5^{-1} z} + \frac{1}{1 - \frac{3}{z^{-1}}} = \frac{z}{5 - z} + \frac{z}{z - 3} \qquad \left[S = \frac{a}{1 - r}\right]$$

$$= \frac{z^2 - 3z + 5z - z^2}{(5 - z)(z - 3)} = \frac{-2z}{z^2 - 8z + 15}, \qquad \left|\frac{z}{5}\right| < 1, \; \left|\frac{3}{z}\right| < 1$$

Two series are convergent in annulus. Here $3 < |z|$ and $|z| < 5$. **Ans.**

**Example 9.** *Find the Z-transform of* $\{a^{|k|}\}$

**Solution.** $$Z\left[\{a^{|k|}\}\right] = \sum_{k=-\infty}^{\infty} a^{|k|} z^{-k} = \sum_{k=-\infty}^{-1} a^{-k} z^{-k} + \sum_{k=0}^{\infty} a^k z^{-k}$$

$$= [\,...+ a^3 z^3 + a^2 z^2 + az] + [1 + az^{-1} + a^2 z^{-2} + a^3 z^{-3} + ....\,]$$

These are two G.P. and sum of G.P. $= \dfrac{a}{1-r}$

$$= \frac{az}{1 - az} + \frac{1}{1 - az^{-1}}, \qquad |az| < 1 \text{ and } |az^{-1}| < 1$$

$$= \frac{az}{1 - az} + \frac{z}{z - a} = \frac{az(z - a) + z(1 - az)}{(1 - az)(z - a)} = \frac{az^2 - a^2 z + z - az^2}{(1 - az)(z - a)} = \frac{z - a^2 z}{(1 - az)(z - a)} \quad \textbf{Ans.}$$

**Example 10.** *Find the Z-transform of* $\left\{\left(\frac{1}{2}\right)^{|k|}\right\}$.

**Solution.** $$Z\left[\left\{\left(\frac{1}{2}\right)^{|k|}\right\}\right] = \sum_{-\infty}^{\infty} \left(\frac{1}{2}\right)^{|k|} z^{-k} = \sum_{k=-\infty}^{-1} \left(\frac{1}{2}\right)^{-k} z^{-k} + \sum_{k=0}^{\infty} \left(\frac{1}{2}\right)^{-k} z^{-k}$$

$$= \left(....+ \frac{z^4}{16} + \frac{z^3}{8} + \frac{z^2}{4} + \frac{z}{2}\right) + \left(1 + \frac{1}{2z} + \frac{1}{4z^2} + \frac{1}{8z^3} + ...\right)$$

These infinite series are G.P., and sum of a G.P. $= \dfrac{a}{1-r}$

$$= \frac{\frac{z}{2}}{1 - \frac{z}{2}} + \frac{1}{1 - \frac{1}{2z}} = \frac{z}{2 - z} + \frac{2z}{2z - 1} = \frac{2z^2 - z + 4z - 2z^2}{(2 - z)(2z - 1)} = \frac{3z}{(2 - z)(2z - 1)} \quad \textbf{Ans.}$$

**Example 11.** *Express the signals shown below in terms of unit impulse functions and hence find the Z-transform.*

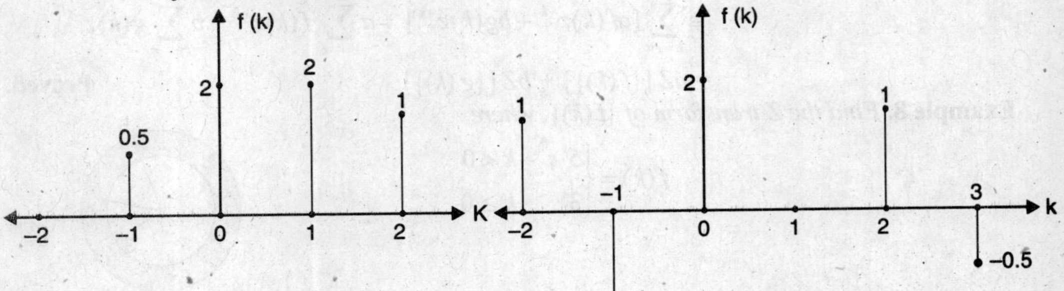

**Solution.** (*i*) $f(k) = 0.5\,\delta(k+1) + 2\,\delta(k) + 2\,\delta(k-1) + \delta(k-2)$

(*ii*) $f(k) = \delta(k+2) - \delta(k+1) + 2\,\delta(k) + \delta(k-2) - 0.5\,\delta(k-3)$

**Example 12.** *Find the Z-transform of unit impulse*

$$\delta(k) = \begin{cases} 1, & k = 0 \\ 0, & k \neq 0 \end{cases}$$

**Solution.** $Z[\{f(k)\}] = \sum_{k=-\infty}^{\infty} \delta(k) z^{-k} = [.....0 + 0 + 1 + 0 + 0....] = 1$ **Ans.**

**Example 13.** *Find the Z-transform of discrete unit step*

$$U(k) = \begin{cases} 0, & k < 0 \\ 1, & k \geq 0 \end{cases}$$

**Solution.** $Z[\{U(k)\}] = \sum_{k=0}^{\infty} U(k) z^{-k} = [1 + z^{-1} + z^{-2} + z^{-3} + ...]$

$$= \frac{1}{1-z^{-1}} = \frac{1}{1-\frac{1}{z}} = \frac{z}{z-1}$$ [This is G.P., its sum is $\frac{a}{1-r}$] **Ans.**

**Example 14.** *Find the Z-transform of* $\sin \alpha k$, $k \geq 0$

**Solution.** $Z[\{\sin \alpha k\}] = \sum_{k=0}^{\infty} \sin \alpha k\, z^{-k} = \sum_{k=0}^{\infty} \frac{e^{i\alpha k} - e^{-i\alpha k}}{2i} z^{-k}$

$$= \frac{1}{2i} \sum_{k=0}^{\infty} e^{i\alpha k} z^{-k} - \frac{1}{2i} \sum_{k=0}^{\infty} e^{-i\alpha k} z^{-k} = \frac{1}{2i} \sum_{k=0}^{\infty} (e^{i\alpha} z^{-1})^k - \frac{1}{2i} \sum_{k=0}^{\infty} (e^{-i\alpha} z^{-1})^k$$

$$= \frac{1}{2i}\left[1 + (e^{i\alpha} z^{-1}) + (e^{i\alpha} z^{-1})^2 + ...\right] - \frac{1}{2i}\left[1 + (e^{-i\alpha} z^{-1}) + (e^{-i\alpha} z^{-1})^2 + ...\right]$$

These infinite series are G.P. and sum of a G.P. $= \frac{a}{1-r}$

$$= \frac{1}{2i} \frac{1}{1-e^{i\alpha} z^{-1}} - \frac{1}{2i} \frac{1}{1-e^{-i\alpha} z^{-1}} = \frac{1}{2i} \frac{z}{z-e^{i\alpha}} - \frac{1}{2i} \frac{z}{z-e^{-i\alpha}}$$

$$= \frac{1}{2i}\left[\frac{z}{z-e^{i\alpha}} - \frac{z}{z-e^{-i\alpha}}\right] = \frac{1}{2i} \frac{z(z-e^{-i\alpha}) - z(z-e^{i\alpha})}{(z-e^{i\alpha})(z-e^{-i\alpha})}$$

$$= \frac{1}{2i} \frac{z(e^{i\alpha} - e^{-i\alpha})}{z^2 - z(e^{i\alpha} + e^{-i\alpha}) + 1} = \frac{z \sin \alpha}{z^2 - 2z \cos \alpha + 1}$$ **Ans.**

**Example 15.** *Find the Z-transform of* $c^k \cosh \alpha k$. $(k \geq 0)$.

**Solution.** $Z[\{c^k \cosh \alpha k\}] = \sum_{k=0}^{\infty} (c^k \cosh \alpha k) z^{-k} = \sum_{k=0}^{\infty} \frac{c^k}{2} (e^{\alpha k} + e^{-\alpha k}) z^{-k}$

$$= \frac{1}{2} \sum_{k=0}^{\infty} (ce^{\alpha} z^{-1})^k + \frac{1}{2} \sum_{k=0}^{\infty} \left(ce^{-\alpha} z^{-1}\right)^k$$

$$= \frac{1}{2}\left[1 + (ce^{\alpha} z^{-1}) + (ce^{\alpha} z^{-1})^2 + ....\right] + \frac{1}{2}\left[1 + (ce^{-\alpha} z^{-1}) + (ce^{-\alpha} z^{-1})^2 + ....\right] \quad \left[|z| > |ce^{\alpha}|, |z| > |ce^{-\alpha}|\right]$$

$$=\frac{1}{2}\frac{1}{1-ce^{\alpha}z^{-1}}+\frac{1}{2}\frac{1}{1-ce^{-\alpha}z^{-1}}=\frac{1}{2}\frac{1-ce^{-\alpha}z^{-1}+1-ce^{\alpha}z^{-1}}{(1-ce^{\alpha}z^{-1})(1-ce^{-\alpha}z^{-1})}$$

$$=\frac{1-c\left(\frac{e^{\alpha}+e^{-\alpha}}{2}\right)z^{-1}}{1+c^2z^{-2}-ce^{\alpha}z^{-1}-ce^{-\alpha}z^{-1}}=\frac{1-(c\cosh\alpha)z^{-1}}{1+c^2z^{-2}-2cz^{-1}\cosh\alpha} \quad \text{(Multiply by } z^2\text{)}$$

$$=\frac{z(z-c\cosh\alpha)}{z^2+c^2-2cz\cosh\alpha}$$ **Ans.**

**Corollary.** If $c = 1$, then

$$Z[\{\cosh\alpha k\}]=\frac{z(z-\cosh\alpha)}{z^2-2z\cosh\alpha+1}$$

**Example 16.** *Find the Z-transform of* $\cosh\left(\frac{k\pi}{2}+\alpha\right)$ *(U.P., III Semester, 2008)*

**Solution.** $$F(z)=\sum_{k=0}^{\infty}\cosh\left(\frac{k\pi}{2}+\alpha\right)z^{-k}=\sum_{k=0}^{\infty}\frac{e^{\frac{k\pi}{2}+\alpha}+e^{-\left(\frac{k\pi}{2}+\alpha\right)}}{2}z^{-k}$$

$$=\frac{1}{2}\sum_{k=0}^{\infty}e^{\frac{k\pi}{2}+\alpha}z^{-k}+\frac{1}{2}\sum_{k=0}^{\infty}e^{-\frac{k\pi}{2}-\alpha}z^{-k}=\frac{1}{2}e^{\alpha}\sum_{k=0}^{\infty}\left(e^{\frac{\pi}{2}}z^{-1}\right)^k+\frac{1}{2}e^{-\alpha}\sum_{k=0}^{\infty}\left(e^{\frac{-\pi}{2}}z^{-1}\right)^k$$

$$=\frac{1}{2}e^{\alpha}\left[1+\left(e^{\frac{\pi}{2}}z^{-1}\right)+\left(e^{\frac{\pi}{2}}z^{-1}\right)^2+\ldots\right]+\frac{1}{2}e^{-\alpha}\left[1+\left(e^{-\frac{\pi}{2}}z^{-1}\right)+\left(e^{-\frac{\pi}{2}}z^{-1}\right)^2+\ldots\right]$$

(Sum of geometrical series $=\frac{a}{1-r}$)

$$=\frac{1}{2}e^{\alpha}\frac{1}{1-e^{\frac{\pi}{2}}z^{-1}}+\frac{1}{2}e^{-\alpha}\frac{1}{1-e^{-\frac{\pi}{2}}z^{-1}}=\frac{1}{2}\frac{e^{\alpha}\left(1-e^{-\frac{\pi}{2}}z^{-1}\right)+e^{-\alpha}\left(1-e^{\frac{\pi}{2}}z^{-1}\right)}{\left(1-e^{\frac{\pi}{2}}z^{-1}\right)\left(1-e^{-\frac{\pi}{2}}z^{-1}\right)}$$

$$=\frac{\frac{e^{\alpha}+e^{-\alpha}}{2}-\frac{e^{\alpha-\frac{\pi}{2}}+e^{-\alpha+\frac{\pi}{2}}}{2}.z^{-1}}{1-e^{\frac{\pi}{2}}z^{-1}-e^{-\frac{\pi}{2}}z^{-1}+z^{-2}}=\frac{\cosh\alpha-\cosh\left(\alpha-\frac{\pi}{2}\right).z^{-1}}{1-\left(2\cosh\frac{\pi}{2}\right)z^{-1}+z^{-2}}=\frac{z^2\cosh\alpha-z\cosh\left(\frac{\pi}{2}-\alpha\right)}{z^2-2z\cosh\frac{\pi}{2}+1}$$ **Ans.**

**Example 17.** *Find Z-transform of sin* $(3k+5)$.

**Solution.** $$F(z)=\sum_{k=0}^{\infty}\sin(3k+5)z^{-k}=\sum_{k=0}^{\infty}\frac{e^{i(3k+5)}-e^{-i(3k+5)}}{2i}z^{-k}$$

$$=\frac{1}{2i}\sum_{k=0}^{\infty}e^{i(3k+5)}z^{-k}-\frac{1}{2i}\sum_{k=0}^{\infty}e^{-i(3k+5)}z^{-k}=\frac{1}{2i}e^{i5}\sum_{k=0}^{\infty}(e^{3i}z^{-1})^k-\frac{1}{2i}e^{-5i}\sum_{k=0}^{\infty}(e^{-3i}z^{-1})^k$$

$$=\frac{1}{2i}e^{5i}[1+(e^{3i}z^{-1})+(e^{3i}z^{-1})^2+\ldots]-\frac{1}{2i}e^{-5i}[1+(e^{-3i}z^{-1})+(e^{-3i}z^{-1})^2+\ldots]$$

$$= \frac{e^{i5}}{2i}\frac{1}{1-e^{3i}z^{-1}} - \frac{1}{2i}e^{-5i}\frac{1}{1-e^{-3i}z^{-1}} \qquad \left[S = \frac{a}{1-r}\right]$$

$$= \frac{1}{2i}\frac{e^{i5}(1-e^{-3i}z^{-1}) - e^{-5i}(1-e^{3i}z^{-1})}{(1-e^{3i}z^{-1})(1-e^{-3i}z^{-1})} = \frac{1}{2i}\frac{(e^{i5}-e^{-5i}) - e^{2i}z^{-1} + e^{-2i}z^{-1}}{1-e^{3i}z^{-1} - e^{-3i}z^{-1} + z^{-2}}$$

$$= \frac{\dfrac{e^{i5}-e^{-5i}}{2i} - z^{-1}\dfrac{e^{2i}-e^{-2i}}{2i}}{1-(e^{3i}+e^{-3i})z^{-1} + z^{-2}} = \frac{\sin 5 - z^{-1}\sin 2}{1-(2\cos 3)z^{-1} + z^{-2}}$$

$$= \frac{z^2 \sin 5 - z \sin 2}{z^2 - 2z\cos 3 + 1} \qquad |z| > 1$$ **Ans.**

**Example 18.** *Find the Z-transform of $c^k \cos \alpha k,\ k \geq 0$.* *(U.P., III Semester, Dec. 2004)*

**Solution.** $F(z) = \sum_{k=0}^{\infty} [c^k \cos \alpha k] z^{-k}$

$$= \sum_{k=0}^{\infty} c^k \left[\frac{e^{i\alpha k} + e^{-i\alpha k}}{2}\right] z^{-k} = \sum_{k=0}^{\infty}\left(\frac{1}{2}c^k e^{i\alpha k}\right) z^{-k} + \sum_{k=0}^{\infty}\left(\frac{1}{2}c^k e^{-i\alpha k}\right) z^{-k}$$

$$= \frac{1}{2}\sum_{k=0}^{\infty}(ce^{i\alpha}z^{-1})^k + \frac{1}{2}\sum_{k=0}^{\infty}(ce^{-i\alpha}z^{-1})^k \qquad ({}^nC_r = {}^nC_{n-r})$$

$$= \frac{1}{2}\left[1 + (ce^{i\alpha}z^{-1}) + (ce^{i\alpha}z^{-1})^2 + \ldots\right] + \frac{1}{2}\left[1 + (ce^{-i\alpha}z^{-1}) + (ce^{-i\alpha}z^{-1})^2 + \ldots\right]$$

This is a Geometric series whose sum is $\dfrac{a}{1-r}$

$$= \frac{1}{2}\left[\frac{1}{1-ce^{i\alpha}z^{-1}}\right] + \frac{1}{2}\left[\frac{1}{1-ce^{-i\alpha}z^{-1}}\right], \qquad [|z| > |c|]$$

$$= \frac{1}{2}\frac{1-ce^{-i\alpha}z^{-1} + 1 - ce^{i\alpha}z^{-1}}{(1-ce^{i\alpha}z^{-1})(1-ce^{-i\alpha}z^{-1})} = \frac{1}{2}\frac{2 - c(e^{i\alpha}+e^{-i\alpha})z^{-1}}{\left[1-ce^{-i\alpha}z^{-1} - ce^{i\alpha}z^{-1} + c^2z^{-2}\right]}$$

$$= \frac{1 - c\cos\alpha . z^{-1}}{1 - 2cz^{-1}\cos\alpha + c^2 z^{-2}} \qquad [|z| > |c|]$$

$$= \frac{z^2 - cz\cos\alpha}{z^2 - 2cz\cos\alpha + c^2}$$ **Ans.**

**Corollary.** If $c = 1$, then

$$Z[\{\cos\alpha k\}] = \frac{z^2 - z\cos\alpha}{z^2 - 2z\cos\alpha + 1}$$

**Example 19.** *Find the Z-transform of* $\left\{\cos\left(\dfrac{k\pi}{8} + \alpha\right)\right\}$.

**Solution.** $Z\left[\left\{\cos\left(\dfrac{k\pi}{8}+\alpha\right)\right\}\right] = \sum \cos\left(\dfrac{k\pi}{8}+\alpha\right) z^{-k} = \sum\left\{\cos\dfrac{k\pi}{8}\cos\alpha - \sin\dfrac{k\pi}{8}\sin\alpha\right\} z^{-k}$

$$= \sum \cos\frac{k\pi}{8}\cos\alpha z^{-k} - \sum \sin\frac{k\pi}{8}\sin\alpha z^{-k} = \cos\alpha \sum \cos\frac{k\pi}{8} z^{-k} - \sin\alpha \sum \sin\frac{k\pi}{8} z^{-k}$$

$$= \cos\alpha \frac{z^2 - z\cos\frac{\pi}{8}}{z^2 - 2z\cos\frac{\pi}{8} + 1} - \sin\alpha \frac{z\sin\frac{\pi}{8}}{z^2 - 2z\cos\frac{\pi}{8} + 1}$$ [See Example 37, 41]

$$= \frac{(z^2 - z\cos\frac{\pi}{8})\cos\alpha - z\sin\frac{\pi}{8}\sin\alpha}{z^2 - 2z\cos\frac{\pi}{8} + 1} = \frac{z^2\cos\alpha - z[\cos\frac{\pi}{8}\cos\alpha + \sin\frac{\pi}{8}\sin\alpha]}{z^2 - 2z\cos\frac{\pi}{8} + 1}$$

$$= \frac{z^2\cos\alpha - z\cos(\frac{\pi}{8} - \alpha)}{z^2 - 2z\cos\frac{\pi}{8} + 1}$$ **Ans.**

**Example 20.** *Find the Z-transform of* $\{{}^nC_k\}$ $(0 \le k \le n)$.

**Solution.** $Z\left[\{{}^nC_k\}\right] = \sum_{k=0}^{n} {}^nC_k z^{-k} = 1 + {}^nC_1 z^{-1} + {}^nC_2 z^{-2} + {}^nC_3 z^{-3} + \ldots + {}^nC_n z^{-n}$

$= (1 + z^{-1})^n$ (This is the expansion of Binomial theorem.) **Ans.**

**Example 21.** *Find Z-transform of* $\{{}^{k+n}C_n\}$

**Solution.** $Z\left[\{{}^{k+n}C_n\}\right] = \sum_{k=0}^{\infty} {}^{k+n}C_n z^{-k}$ $\left(\begin{matrix} k+n > n \\ k > 0 \end{matrix}\right)$

$$= \sum_{k=0}^{\infty} {}^{k+n}C_k z^{-k} \qquad ({}^nC_r = {}^nC_{n-r})$$

$$= 1 + {}^{n+1}C_1 z^{-1} + {}^{n+2}C_2 z^{-2} + {}^{n+3}C_3 z^{-3} + \ldots.$$

$$= 1 + (n+1)z^{-1} + \frac{(n+2)(n+1)}{2!} z^{-2} + \frac{(n+3)(n+2)(n+1)}{3!}(z^{-3}) + \ldots$$

$$= 1 + (-n-1)(-z^{-1}) + \frac{(-n-1)(-n-2)}{2!}(-z^{-1})^2 + \frac{(-n-1)(-n-2)(-n-3)}{3!}(-z^{-1})^3 + \ldots$$

This is the expansion of Binomial theorem.

$= (1 - z^{-1})^{-n-1} = (1 - z^{-1})^{-(n+1)}$ **Ans.**

**Example 22.** *Find the Z-transform of* $\{{}^{k+n}C_n a^k\}$.

**Solution.** $Z\left[\{{}^{k+n}C_n a^k\}\right] = \sum_{k=0}^{\infty} {}^{k+n}C_n a^k z^{-k} = \sum_{k=0}^{\infty} {}^{k+n}C_k a^k z^{-k}$ $({}^nC_r = {}^nC_{n-r})$

$= \sum_{k=0}^{\infty} {}^{k+n}C_k (az^{-1})^k$ , $|z| > |a|$

$= (1 - a z^{-1})^{-(n+1)}$ (See Example 21)

**Corollary.** If $n = 1$

$$Z\left[\{{}^{k+1}C_1 a^k\}\right] = (1 - az^{-1})^{-2} = \frac{z^2}{(z-a)^2}$$

If $n = 2$

$$Z\left[\{{}^{k+2}C_2 a^k\}\right] = (1 - az^{-1})^{-3} = \frac{z^3}{(z-a)^3}$$

If $n \to n-1$

$$Z\left[\left\{{}^{k+n-1}C_{n-1}a^k\right\}\right] = (1-az^{-1})^{-(n-1+1)} = (1-az^{-1})^{-n} = \frac{z^n}{(z-a)^n}$$ **Ans.**

**Example 23.** *Find the Z-transform of* $\left\{\frac{a^k}{k!}\right\}$. $(k \geq 0)$

**Solution.** $$Z\left[\left\{\frac{a^k}{k!}\right\}\right] = \sum_{k=0}^{\infty} \frac{a^k}{k!} z^{-k} = \sum_{k=0}^{\infty} \frac{(az^{-1})^k}{k!} = 1 + \frac{az^{-1}}{1!} + \frac{(az^{-2})^2}{2!} + \frac{(az^{-1})^3}{3!} + \ldots$$

This is exponential series.

$$= e^{az^{-1}} = e^{\frac{a}{z}}$$ **Ans.**

**Example 24.** *Find the Z-transform of* $f(k) = \frac{1}{(k+1)!}, k \geq 0.$ *(Q. Bank U.P., III Sem. 2002)*

**Solution.** $$Z\left[\frac{1}{(k+1)!}\right] = \sum_{k=0}^{\infty} \frac{1}{(k+1)!} z^{-k} = 1 + \frac{1}{2!}z^{-1} + \frac{1}{3!}z^{-2} + \ldots$$

$$= z\left[z^{-1} + \frac{1}{2!}z^{-2} + \frac{1}{3!}z^{-3} + \ldots\right] = z\left[1 + z^{-1} + \frac{1}{2!}z^{-2} + \frac{1}{3!} + z^{-3} \ldots\ldots -1\right]$$

$$= z\,(e^{z^{-1}} - 1) = z\,(e^{\frac{1}{z}} - 1)$$ **Ans.**

**Example 25.** *Find the Z-transform of*

(*i*) $u(k-1)$ (*ii*) $4k\,\delta(k-1)\,;\, k \geq 0$ (*iii*) $\delta(k-n)\,;\, k \geq 0$

**Solution.**

(*i*) $$Z\{f(k)\} = Zu(k-1) = \sum_{k=1}^{\infty} 1 . z^{-k} = \frac{1}{z} + \frac{1}{z^2} + \frac{1}{z^3} + \ldots = \frac{1}{z}\left(1 + \frac{1}{z} + \frac{1}{z^2} + \ldots\right) [u(k-1) = 1]$$

This infinite series is a G.P., whose sum is $\frac{a}{1-r}$

$$= \frac{1}{z} \frac{1}{\left(1 - \frac{1}{z}\right)} \qquad \left[\text{if } \left|\frac{1}{z}\right| < 1\right]$$

$$= \frac{1}{z-1} \text{ if } |z| > 1$$

(*ii*) $$Z\{f(k)\} = Z\{4^k\,\delta(k-1)\} = \sum_{k=0}^{\infty} 4^k . \delta(k-1)\, z^{-k} = \frac{4}{z}. \qquad [\delta(k-1) \Rightarrow k = 1]$$

(*iii*) $$Z\{f(k)\} = Z[\delta(k-n)] = \sum_{k=0}^{\infty} \delta(k-n)\, z^{-k} = \frac{1}{z^n},$$ $n$ is (+)ve integer. $[\delta(k-n) \Rightarrow k = n]$

**Example 26.** *Determine the Z-transform of*

$f(k) = \delta(k+1)\} + 3\delta(k) + 6\delta(k-3) - \delta(k-4)$

**Solution.** By linearity property, we have

$F(z) = Z\{f(k)\} = Z\{\delta(k+1)\} + 3\,Z\{\delta(k)\} + 6\,Z\{\delta(k-3)] - Z\{\delta(k-4)\}$

$[\delta(k+1) \Rightarrow k = -1, \delta(k) \Rightarrow k = 0, \delta(k-3) \Rightarrow k = 3, \delta(k-4) \Rightarrow k = 4]$

$= z + 3 + 6z^{-3} - z^{-4}$ **Ans.**

## EXERCISE 19.2

Find the $Z$-transform of the following for $(k \geq 0)$ :

1. $2^k$ **Ans.** $\dfrac{z}{z-2}, \quad |z|>2$

2. $\sin 2k$ **Ans.** $\dfrac{z \sin 2}{z^2 - 2z\cos 2 + 1}$

3. $\sin \alpha k$, **Ans.** $\dfrac{z \sin \alpha}{z^2 - 2z\cos\alpha + 1}$

4. $\sinh \dfrac{k\pi}{2}$ **Ans.** $\dfrac{z \sinh \frac{\pi}{2}}{z^2 - 2z\cosh\frac{\pi}{2} + 1}$

5. $\sin(3k+5)$ **Ans.** $\dfrac{z^2 \sin 5 - z\sin 2}{z^2 - 2z\cos 3 + 1}, \quad |z|>1$

6. $\sin\left(\dfrac{k\pi}{2}+\alpha\right)$ **Ans.** $\dfrac{z^2\sin\alpha + z\cos\alpha}{z^2+1}, \quad |z|>1$

7. $c^k \sinh \alpha k$ **Ans.** $\dfrac{cz\sinh\alpha}{z^2 - 2cz\cosh\alpha + 1}$

8. $\cos \alpha k$ **Ans.** $\dfrac{z(z-\cos\alpha)}{z^2 - 2z\cos\alpha + 1}$

9. $\cos\left(\dfrac{k\pi}{2}+\dfrac{\pi}{4}\right)$ **Ans.** $\dfrac{z^2 - z}{\sqrt{2}(z^2+1)}$

10. $\cos\left(\dfrac{k\pi}{8}+\alpha\right)$ **Ans.** $\dfrac{z^2\cos\alpha - z\cos\left(\frac{\pi}{8}-\alpha\right)}{z^2 - 2z\cos\frac{\pi}{8} + 1}$

11. $\cosh \alpha k$ **Ans.** $\dfrac{z(z-\cosh\alpha)}{z^2 - 2z\cosh\alpha + 1}$

12. $a\cos\alpha k + b\sin\alpha k$ **Ans.** $\dfrac{a - z^{-1}(a\cos\alpha - b\sin\alpha)}{1 - 2z^{-1}\cos\alpha + z^{-2}}, \quad |z|>1$

13. $\dfrac{a^k}{k!}$ **Ans.** $e^{\frac{a}{z}}$

14. $\dfrac{1}{k}, k>0$ **Ans.** $-\log(1-z^{-1})$

15. $a^k, \quad k<0$ **Ans.** $\dfrac{-1}{(1-az^{-1})}$

16. $2^{|k|}$ **Ans.** $\dfrac{-3z}{(1-2z)(z-2)}$

17. $f(k) = \begin{cases} 1, & \text{for } k \geq 0 \\ 0, & \text{for } k < 0 \end{cases}$ **Ans.** $\dfrac{z}{z-1}, \quad |z|>1$

## 19.7 CHANGE OF SCALE

**Theorem.** If $Z[\{f(k)\}] = F(z)$ then $Z[\{a^k f(k)\}] = F\left(\dfrac{z}{a}\right)$

**Proof.**

$$F(z) = Z[\{f(k)\}] = \sum_{k=-\infty}^{\infty} f(k) z^{-k}$$

Substituting $\dfrac{z}{a}$ for $z$, we get $F\left(\dfrac{z}{a}\right) = \displaystyle\sum_{k=-\infty}^{\infty} f(k)\left(\dfrac{z}{a}\right)^{-k}$ ...(1)

But

$$Z\left[\left\{a^k f(k)\right\}\right] = \sum_{k=-\infty}^{\infty} a^k f(k) z^{-k} = \sum_{k=-\infty}^{\infty} f(k)\left(\frac{z}{a}\right)^{-k} \quad ...(2)$$

From (1) and (2), we get $Z\left[\left\{a^k f(k)\right\}\right] = F\left(\dfrac{z}{a}\right)$ **Proved.**

**Example 27.** *Find the Z-transform of $a^k$, $k \geq 0$.*

**Solution.** We know that

$$Z[\{1\}] = \frac{z}{z-1}$$ (See example 13)

For the given sequence, by the scale change formula the *Z*-transform

$$Z\left[\{a^k.1\}\right] = \frac{\frac{z}{a}}{\frac{z}{a}-1} = \frac{z}{z-a}$$ **Ans.**

**Example 28.** *Find the Z-transform of $c^k \sin \alpha k$, $k \geq 0$.*

**Solution.** We know that

$$Z\left[\{\sin \alpha k\}\right] = \frac{z \sin \alpha}{z^2 - 2z\cos\alpha + 1}$$ (See example 14)

By applying the formula of change of scale, we get

$$Z\left[\{c^k \sin \alpha k\}\right] = \frac{\frac{z}{c}\sin\alpha}{\left(\frac{z}{c}\right)^2 - 2\left(\frac{z}{c}\right)\cos\alpha + 1} = \frac{cz\sin\alpha}{z^2 - 2cz\cos\alpha + c^2}$$ **Ans.**

## 19.8 SHIFTING PROPERTY

**Theorem.** If $Z[\{f(k)\}] = F(z)$,

then $Z[\{f(k \pm n)\}] = z^{\pm n} F(z)$

**Proof.** $$Z[\{f(k \pm n)\}] = \sum_{k=-\infty}^{\infty} f(k \pm n) z^{-k} = z^{\pm n} \sum_{k=-\infty}^{\infty} f(k \pm n) z^{-(k \pm n)} \qquad (r = k \pm n)$$

$$= z^{\pm n} \sum_{k=-\infty}^{\infty} f(r) z^{-r} = z^{\pm n} F(z)$$ **Proved.**

**Case I.** $Z[\{f(k+n)\}] = f(k+n)z^{-k},\ k \geq 0 = z^n \sum_{k=0}^{\infty} f(k+n) z^{-(k+n)}$

Put $r = k + n$

$$= z^n \sum_{r=n}^{\infty} f(r) z^{-r} = z^n \sum_{r=0}^{\infty} f(r) z^{-r} - z^n \sum_{r=0}^{n-1} f(r) z^{-r} = z^n F(z) - z^n \sum_{r=0}^{n-1} f(r) z^{-r}$$

$$= z^n F(z) - \sum_{r=0}^{n-1} f(r) z^{n-r} = z^n F(z) - \sum_{r=0}^{n-1} f(r) z^{n-r}$$ **Ans.**

**Case II.** $$Z[\{f(k-n)\}] = \sum_{k=0}^{\infty} f(k-n) z^{-k} = z^{-n} \sum_{k=0}^{\infty} f(k-n) z^{-(k-n)}, \qquad k \geq 0$$

Put $r = k - n$

$$= z^{-n} \sum_{r=-n}^{\infty} f(r) z^{-r}$$

$$= z^{-n} \sum_{r=0}^{\infty} f(r) z^{-r} + z^{n} \sum_{r=-n}^{-1} f(r) z^{-r} = 2^{-n} F(z) + \sum_{r=-n}^{-1} f(r) z^{-n-r}$$ Put $r = -m$

$$= z^{-n} F(z) + \sum_{m=1}^{n} f(-m) z^{-n+m}$$ **Ans.**

**Corollary 1.** If $\{f(k)\}$ is casual sequence, then

$$Z[\{f(k-n)\}] = z^{-n} F(z)$$

Since $f(-1)=f(-2)=f(-3)=\ldots=f(-n)=0$

**Corollary 2.** For casual sequence

$$Z[\{f(k-1)\}]=z^{-1}F(z) \text{ as } f(-1)=0$$
$$Z[\{f(k+1)\}]=z\,F(z)-z\,f(0)$$
$$Z[\{f(k+2)\}]=z^2F(z)-z^2f(0)-z\,f(1)$$

## 19.9 INVERSE Z-TRANSFORM

Finding the sequence $\{f(k)\}$ from $F(z)$ is defined as inverse Z-transform. It is denoted as

$$Z^{-1}F(z)=\{f(k)\} \qquad Z^{-1} \text{ is the inverse Z-transform.}$$

## 19.10 THEOREM

If $\{f(k)\}=F(z)$, $\{g(k)\}=G(z)$, $a$ and $b$ are constants,

then $Z^{-1}[a\,F(z)+b\,G(z)]=aZ^{-1}[F(z)]+bZ^{-1}[G(z)]$

**Proof.** We know that

$$Z[\{a\,f(k)+bg(k)\}]=a\,Z[\{f(k)\}]+b\,Z[\{g(k)\}]=aF(z)+b\,G(z)$$
$$\therefore\quad Z^{-1}[a\,F(z)+bG(z)]=\{af(k)+bg(k)\}=a\,\{f(k)\}+b\{g(k)\}$$
$$=a\,Z^{-1}[F(z)]+bZ^{-1}[G(z)]$$

**Proved.**

**Example 29.** *Find the inverse Z-transform of* $\frac{1}{z-2}$

**Solution.** $F(z)=\frac{1}{z-2}$

**Case I.** If $\left|\frac{2}{z}\right|<1$, $F(z)=\frac{1}{z}\frac{1}{1-2z^{-1}}=z^{-1}(1-2z^{-1})^{-1}=z^{-1}[1+2z^{-1}+2^2z^{-2}+\ldots]$

$$=z^{-1}+2\,z^{-2}+2^2z^{-3}+\ldots$$

$$\{f(k)\}=\{2^{k-1}\},\qquad k\geq 1$$

**Case II.** If $\left|\frac{z}{2}\right|<1$

$$F(z)=\frac{1}{z-2}=-\frac{1}{2}\frac{1}{1-\frac{z}{2}}=\frac{-1}{2}\left(1-\frac{z}{2}\right)^{-1}=-\frac{1}{2}\left[1+\frac{z}{2}+\frac{z^2}{2^2}+\ldots\right]$$

$$\{f(k)\}=\{-2^{k-1}\},\qquad k\leq 0$$

**Note.** The inverse Z-transform can only be settled when region of convergence (ROC) is given.

## 19.11 SOLUTION OF DIFFERENCE EQUATIONS

**Example 30.** *Solve the difference equation*

$$y_{k+1}-2y_{k-1}=0,\quad k\geq 1,\quad y_{(0)}=1$$

**Solution.** $y_{k+1}-2y_{k-1}=0$ ...(1)

Taking the $Z$-transform of both sides of (1), we get

$$Z[y_{k+1}-2y_{k-1}]=0$$
$$Z[y_{k+1}]-2Z[y_{k-1}]=0$$
$$zY(z)-y_0z-2Y(z)=0 \qquad (y_0=1)$$
$$(z-2)\,Y(z)-z=0$$

$$Y(z) = \frac{z}{z-2}$$

$$\{y_{(k)}\} = Z^{-1}\left[\frac{z}{z-2}\right] = Z^{-1}\left[\frac{1}{1-2z^{-1}}\right]$$

$$= Z^{-1}[1-2z^{-1}]^{-1} = 1 + 2z^{-1} + (2z^{-1})^2 + \ldots = \{2^k\}, \quad k \geq 0 \quad \textbf{Ans.}$$

## 19.12 MULTIPLICATION BY *K*

**Theorem.** If $Z[\{f(k)\}] = F(z)$, then $Z[\{k f(k)\}] = -z \frac{d}{dz} F(z)$

**Proof.** $$Z[\{kf(k)\}] = \sum_{k=-\infty}^{\infty} kf(k)z^{-k} = -z\sum_{k=-\infty}^{\infty} -kf(k)z^{-k-1} = -z\sum_{k=-\infty}^{\infty} f(k)(-kz^{-k-1})$$

$$= -z\sum_{k=-\infty}^{\infty} f(k)\frac{d}{dz}(z^{-k}) = -z\frac{d}{dz}\sum_{k=-\infty}^{\infty} f(k)z^{-k} = -z\frac{d}{dz}F(z) \quad \textbf{Proved.}$$

In general $$Z[\{k^n f(k)\}] = \left(-z\frac{d}{dz}\right)^n F(z)$$

**Corollary 1.** If $f(k) = 1$ then $Z[\{1\}] = (1 - z^{-1})^{-1}$, $k \geq 0$

Putting these values in the above theorem, we get

$$Z[\{k\}] = -z\frac{d}{dz}(1-z^{-1})^{-1} = z(1-z^{-1})^{-2}\left(\frac{1}{z^2}\right) = z^{-1}(1-z^{-1})^{-2} \quad \textbf{Ans.}$$

**Corollary 2.** If $f(k) = 1$, then $Z[\{1\}] = (1-z^{-1})^{-1} = \dfrac{z}{z-1}$ and $n = 2$

$$= Z[\{k^2\}] = \left(-z\frac{d}{dz}\right)^2 \frac{z}{z-1} = \left(-z\frac{d}{dz}\right)\left(-z\frac{d}{dz}\right)\frac{z}{z-1}$$

$$= -z\frac{d}{dz}\left[-z\frac{(z-1).1-z.1}{(z-1)^2}\right] = -z\frac{d}{dz}\left[\frac{z}{(z-1)^2}\right] = -z\frac{(z-1)^2.1-z.2(z-1)}{(z-1)^4} = -z\frac{z-1-2z}{(z-1)^3}$$

$$= -z\frac{-z-1}{(z-1)^3} = \frac{z^2+z}{(z-1)^3} = \frac{z(z+1)}{(z-1)^3} \quad \textbf{Ans.}$$

## 19.13 DIVISION BY *K*

**Theorem.** If $Z[\{f(k)\}] = F(z)$, then $Z\left[\left\{\frac{f(k)}{k}\right\}\right] = -\int^z z^{-1}F(z)dz$

**Proof.** $$Z\left[\left\{\frac{f(k)}{k}\right\}\right] = \sum_{k=-\infty}^{\infty} \frac{f(k)}{k} z^{-k}$$

$$= \sum_{k=-\infty}^{\infty} f(k)\left(\frac{1}{k}z^{-k}\right) = -\sum_{k=-\infty}^{\infty} f(k)\int^z z^{-k-1}dz$$

$$= -\int^z \sum_{k=-\infty}^{\infty} f(k)z^{-k-1}dz = -\int^z z^{-1}\sum_{k=-\infty}^{\infty} f(k)z^{-k}dz = -\int^z z^{-1}F(z)dz$$

$$Z\left[\left\{\frac{f(k)}{k}\right\}\right] = -\int^z z^{-1}F(z)dz$$

## 19.14 INITIAL VALUE

**Theorem.** If $Z[\{f(k)\}] = F(z), \qquad k \geq 0$

then $$f(0) = \lim_{z\to\infty} F(z).$$

**Proof.** $$Z[\{f(k)\}] = \sum_{k=0}^{\infty} f(k) z^{-k} = F(z)$$

$$f(0) + f(1)\, z^{-1} + f(2)\, z^{-2} + \ldots = F(z)$$

Taking the limit, $z \to \infty$, we get

$$f(0) = \lim_{z\to\infty} F(z)$$ **Proved.**

## 19.15 FINAL VALUE

**Theorem.** $$\lim_{k\to\infty} f(k) = \lim_{z\to 1} (z-1) F(z)$$

**Proof.** $$Z[\{f(k+1) - f(k)\}] = \sum_{k=0}^{\infty} [f(k+1) - f(k)] z^{-k}$$

$$zF(z) - f(0) - F(z) = \lim_{n\to\infty} \sum_{k=0}^{n} [f(k+1) - f(k)] z^{-k}$$

$$\lim_{z\to 1} (z-1) F(z) = f(0) + \lim_{z\to 1} \lim_{n\to\infty} \sum [f(k+1) - f(k)] z^{-k}$$

By changing the order of limits, we get

$$\lim_{z\to 1} (z-1) F(z) = f(0) + \lim_{n\to\infty} \sum_{k=0}^{n} \lim_{z\to 1} [f(k+1) - f(k)] z^{-k}$$

$$= \lim_{n\to\infty} \left[ f(0) + \sum_{k=0}^{n} \{f(k+1) - f(k)\} \right]$$

$$= \lim [f(0) - f(0) + f(1) - f(1) + f(2) - f(2) + \ldots + f(n+1) - f(n)]$$

$$= \lim_{n\to\infty} f(n+1) = \lim_{n\to\infty} f(n) = \lim_{k\to\infty} f(k)$$

## 19.16 PARTIAL SUM

**Theorem.** If $Z[\{f(k)\}] = F(z)$, then $$Z\left[\left\{ \sum_{n\to -\infty}^{k} f(n) \right\}\right] = \frac{F(z)}{1 - z^{-1}}$$

**Proof.** Let $\{g(k)\}$ be a sequence such that $g(k) = \sum_{n=-\infty}^{k} f(n)$

We are required to find $Z[\{g(k)\}]$,

We know that $g(k) - g(k-1) = \sum_{n=-\infty}^{k} f(n) - \sum_{n=-\infty}^{k-1} f(n) = -f(k)$

$$Z[\{g(k)\} - \{g(k-1)\}] = Z[\{f(k)\}]$$

$$Z[\{g(k)\}] - Z[\{g(k-1)\}] = F(z)$$

$$G(z) - z^{-1} G(z) = F(z)$$

$$\sum_{n=-\infty}^{k} f(k) = G(z) = \frac{F(z)}{1 - z^{-1}}$$ **Proved.**

## 19.17 CONVOLUTION

Let two sequences be $\{f(k)\}$ and $\{g(k)\}$ and the convolution of $\{f(k)\}$ and $\{g(k)\}$ be $\{h(k)\}$ and denoted as

$$\{h(k)\} = \{f(k)\} * \{g(k)\}$$

where $$\{h(k)\} = \sum_{n=-\infty}^{\infty} f(n) g(k-n) \qquad \ldots(1)$$

$$= \sum_{n=-\infty}^{\infty} g(n) f(k-n) = \{g(k)\} * \{f(k)\}$$

**Proof.** Z-transform of (1) is

$$Z\{h(k)\} = \sum_{k=-\infty}^{\infty}\left(\sum_{n=-\infty}^{\infty} f(n)g(k-n)\right)z^{-k} = \sum_{n=-\infty}^{\infty}\sum_{k=-\infty}^{\infty} f(n)g(k-n)z^{-k}$$

$$= \sum_{n=-\infty}^{\infty} f(n)z^{-n} \sum_{k=-\infty}^{\infty} g(k-n)z^{-(k-n)} = \left[\sum_{n=-\infty}^{\infty} f(n)z^{-n}\right] G(z) = F(z)G(z)$$

**Note.** Region of convergence of $H(z)$ is the common region of convergence of $F(z)$ and $G(z)$.

## 19.18 CONVOLUTION PROPERTY OF CASUAL SEQUENCE

$$F(z) = \{f(0) + f(1)z^{-1} + f(2)z^{-2} + f(3)z^{-3} + \ldots\}$$

and $$G(z) = \{g(0) + g(1)z^{-1} + g(2)z^{-2} + g(3)z^{-3} + \ldots\}$$

Now

$$F(z)G(z) = \{f(0) + f(1)z^{-1} + f(2)z^{-2} + f(3)z^{-3} + \ldots \{g(0) + g(1)z^{-1} + g(2)z^{-2} + g(3)z^{-3} + \ldots\}$$

$$= f(0)g(0) + \{f(1)g(0) + f(0)g(1)\}z^{-1} + \{f(0)g(2) + f(1)g(1) + f(2)g(0)\}z^{-2} + \ldots$$

$$= h(0) + h(1)z^{-1} + h(2)z^{-2} + \ldots = Z\{h(k)\}$$

$$Z\{h(k)\} = Z\{f(k)\} * \{g(k)\}$$ **Proved.**

**Example 31.** *Evaluate the Z-transform of the sequence*

$$\{f(k)\} = \sum_{k=0}^{\infty} 2^k \sum_{k=0}^{\infty} 3^k$$

**Solution.** $Z\left[\{2^k\}\right] = 1 + 2z^{-1} + 2^2 z^{-2} + 2^3 z^{-3} + \ldots = \dfrac{1}{1-2z^{-1}}$

Similarly, $$Z\left[\{3^k\}\right] = \frac{1}{1-3z^{-1}}$$

$$Z\left[\{f(k)\}\right] = Z\{2^k\}\{3^k\} = \frac{1}{1-2z^{-1}} \cdot \frac{1}{1-3z^{-1}} = \frac{1}{(1-2z^{-1})(1-3z^{-1})}$$ **Ans.**

## EXERCISE 19.3

Find the Z-transform of the following for $(k > 0)$ :

1. $e^{\alpha k}$ **Ans.** $\dfrac{1}{1-z^{-1}e^{\alpha k}}$

2. $\sin 5k$ **Ans.** $\dfrac{z\sin 5}{z^2 - 2z\cos 5 + 1}$

3. $\cos 3k$ **Ans.** $\dfrac{z(z-\cos 3)}{z^2 - 2z\cos 3 + 1}$

4. $\sinh 7k$ **Ans.** $\dfrac{z\sinh 7}{z^2 - 2z\cosh 7 + 1}$

5. $\cosh 9k$ **Ans.** $\dfrac{z(z-\cosh 9)}{z^2 - 2z\cosh 9 + 1}$

6. $\sin(5k+3)$ **Ans.** $\dfrac{z^2\sin 3 + z\sin 2}{z^2 - 2z\cos 5 + 1}$

**7.** $\cos\left(\frac{k\pi}{3}+5\right)$ **Ans.** $\dfrac{z^2\cos 5 - z\cos(\frac{\pi}{3}-5)}{z^2-2z\cos\frac{\pi}{3}+1}$

**8.** $\cos\left(\frac{k\pi}{5}+6\right)$ **Ans.** $\dfrac{z^2\cosh 6 - z\cosh(\frac{\pi}{5}-6)}{z^2-2z\cos\frac{\pi}{5}+1}$ **9.** $3^k\cosh 5k$ **Ans.** $\dfrac{z(z-3\cosh 5)}{z^2-2z\cosh 5+9}$

## 19.19 TRANSFORM OF IMPORTANT SEQUENCES

| *S. No.* | *Sequence* | *Z-transform* | |
|---|---|---|---|
| 1 | $\{f(k)\}$ | $F(z)$ | |
| 2 | $\delta(k)$ | 1 | |
| 3 | $U(k)$ or 1 | $(1-z^{-1})^{-1}$ | |
| 4 | $k$ | $-z\frac{d}{dz}(1-z^{-1})^{-1}$ | $\lvert z\rvert > 1$ |
| 5 | $k^n$ | $\left(-z\frac{d}{dz}\right)^n(1-z^{-1})^{-1}$ | $\lvert z\rvert > 1$ |
| 6 | ${}^kC_n$ | $z^{-n}(1-z^{-1})^{-(n+1)}$ | $\lvert z\rvert > 1$ |
| 7 | ${}^{k+n}C_n.a^k$ | $(1-az^{-1})^{-(n+1)}$ | $\lvert z\rvert > \lvert a\rvert$ |
| 8 | ${}^nC_k$ | $(1-z^{-1})^n$ | $0\le k\le n, \lvert z\rvert > 0$ |
| 9 | $a^k$ | $(1-az^{-1})^{-1}$ | $\lvert z\rvert > \lvert a\rvert$ |
| 10 | $k^n a^k$ | $\left(-z\frac{d}{dz}\right)^n(1-az^{-1})^{-1}$ | $\lvert z\rvert > \lvert a\rvert$ |
| 11 | $a^k$ | $-(1-az^{-1})^{-1}$ | $\lvert z\rvert > \lvert a\rvert$ |
| 12 | $k^n a^k$ | $-\left(-z\frac{d}{dz}\right)^n(1-az^{-1})^{-1}$ | $\lvert z\rvert < \lvert a\rvert$ |
| 13 | $a^{\lvert k\rvert}$ | $\frac{(1-a^2)(1-az)}{(1-az^{-1})^{-1}}$ | $\lvert a\rvert < \lvert z\rvert < \frac{1}{\lvert a\rvert}$ |
| 14 | $\sin\alpha k$ | $\frac{z\sin\alpha}{z^2-2z\cos\alpha+1}$ | $k\ge 0$ |
| 15 | $c^k\sin\alpha k$ | $\frac{cz\sin\alpha}{z^2-2cz\cos\alpha+c^2}$ | $k\ge 0$ |
| 16 | $\cos\alpha k$ | $\frac{z^2-z\cos\alpha}{z^2-2z\cos\alpha+1}$ | $k\ge 0$ |
| 17 | $c^k\cos\alpha k$ | $\frac{z^2-cz\cos\alpha}{z^2-2cz\cos\alpha+c^2}$ | $k\ge 0$ |
| 18 | $\cosh\alpha k$ | $\frac{z^2-z\cosh\alpha}{z^2-2z\cosh\alpha+1}$ | $k\ge 0$ |

| | | | |
|---|---|---|---|
| 19 | $c^k \cosh \alpha k$ | $\dfrac{z^2 - cz\cosh\alpha}{z^2 - 2cz\cosh\alpha + c^2}$ | $k \geq 0$ |
| 20 | $\sinh \alpha k$ | $\dfrac{z\sinh\alpha}{z^2 - 2z\cosh\alpha + 1}$ | $k \geq 0$ |
| 21 | $c^k \sinh \alpha k$ | $\dfrac{cz\sinh\alpha}{z^2 - 2cz\cosh\alpha + c^2}$ | $k \geq 0$ |
| 22 | Change of scale | $a^k f(k)$ | $F\left(\dfrac{z}{a}\right)$ |
| 23 | Shifting | $f(k \pm n)$ | $z^{\pm n} F(z)$ |
| 24 | | $f(k-n)$ | $z^{-n} F(z) + \sum_{r=1}^{n} f(-r) z^{-n+r}$, $k < 0$ |
| 25 | Casual sequence | $f(k+n)$ | $z^n F(z) - \sum_{r=0}^{n-1} f(r) z^{n-r}$ |
| 26 | | $f(k-n)$ | $z^{-n} F(z)$ |
| 27 | Multiplication by $k$ | $k f(k)$ | $-z \dfrac{d}{dz} F(z)$ |
| 28 | Multiplication by $k$ | $k^n f(k)$ | $\left(-z \dfrac{d}{dz}\right)^n F(z)$ |
| 29 | Division by $k$ | $\dfrac{1}{k} f(k)$ | $-\int^{z} z^{-1} F(z) dz$ |
| 30 | Initial value theorem | $f(0)$ | $\lim_{z \to \infty} F(z)$ |
| 31 | Final value theorem | $\lim_{k \to \infty} f(k)$ | $\lim_{z \to 1} (z-1) F(z)$ |
| 32 | Partial sum | $\sum_{n \to -\infty}^{k} f(n)$ | $\dfrac{F(z)}{z-1}$ |
| 33 | | $\sum_{n \to -\infty}^{\infty} f(n)$ | $F(1)$ |
| 34 | Convolution | $f(k)^* g(k)$ | $F(z) \cdot G(z)$ |

**Inverse Z-transforms**

| S. No | | *Inverse Z-transform* | |
|---|---|---|---|
| | | $\|z\| > \|a\|$, $k > 0$ | $\|z\| < \|a\|$, $k < 0$ |
| 1 | $\dfrac{z}{z-a}$ | $a^k U(k)$ | $-a^k$ |
| 2 | $\dfrac{z^2}{(z-a)^2}$ | $(k+1) a^k$ | $-(k+1) a^k$ |

| | | | |
|---|---|---|---|
| 3 | $\dfrac{z^3}{(z-a)^3}$ | $\dfrac{1}{2!}(k+1)(k+2)a^k U(k)$ | $-\dfrac{1}{2!}(k+1)(k+2)a^k U(-k+2)$ |
| 4 | $\dfrac{z^n}{(z-a)^n}$ | $\dfrac{1}{(n-1)!}(k+1)...(k+n-1)$ $a^k U(k)$ | $-\dfrac{1}{(n-1)!}(k+1)...(k+n-1)a^k$ |
| 5 | $\dfrac{1}{z-a}$ | $a^{k-1}U(k-1)$ | $-a^{k-1}U(-k)$ |
| 6 | $\dfrac{1}{(z-a)^2}$ | $(k-1)a^{k-2}U(k-2)$ | $-(k-1)a^{k-2}U(-k+1)$û |
| 7 | $\dfrac{1}{(z-a)^3}$ | $\dfrac{1}{2}(k-2)(k-1)a^{k-3}U(k-3)$ | $-\dfrac{1}{2}(k-2)(k-)a^{k-3}U(-k+2)$ |

## 19.20 INVERSE OF Z-TRANSFORM BY DIVISION

From *Z*-transform *F* (*z*), we find the sequence $\{f(k)\}$ if $F(z)$ is a rational function of *z*. Region of convergence must be given.

**1. By Direct Division**

**Example 32.** *Find* $Z^{-1}\left[\dfrac{1}{z-2}\right]$

**Solution. Case I.** $|z|>2$

$$\begin{array}{r|l} & \dfrac{1}{z}+\dfrac{2}{z^2}+\dfrac{4}{z^3} \\ \hline z-2 & 1 \\ & 1-\dfrac{2}{z} \\ \hline & \dfrac{2}{z} \\ & \dfrac{2}{z}-\dfrac{4}{z^2} \\ \hline & \dfrac{4}{z^2} \\ & \dfrac{4}{z^2}-\dfrac{8}{z^3} \\ \hline & \dfrac{8}{z^3} \end{array}$$

$$\frac{1}{z-2}=\frac{1}{z}+\frac{2}{z^2}+\frac{4}{z^3}+\frac{8}{z^4}+\ldots$$

$$= z^{-1}+2z^{-2}+4z^{-3}+8z^{-4}+\ldots+2^{k-1}z^{-k}+\ldots$$

$$= \{2^{k-1}\}\, z^{-k}$$

$$Z^{-1}\left[\frac{1}{z-2}\right]=\{2^{k-1}\}$$ **Ans.**

**Case II.** $|z|<2$

$$\begin{array}{r|l} & -\dfrac{1}{2}-\dfrac{z}{4}-\dfrac{z^2}{8}-\dfrac{z^3}{16} \\ \hline -2+z & 1 \\ & 1-\dfrac{z}{2} \\ \hline & \dfrac{z}{2} \\ & \dfrac{z}{2}-\dfrac{z^2}{4} \\ \hline & \dfrac{z^2}{4} \\ & \dfrac{z^2}{4}-\dfrac{z^3}{8} \\ \hline & \dfrac{z^3}{8} \\ & \dfrac{z^3}{8}-\dfrac{z^4}{16} \\ \hline & \dfrac{z^4}{16} \end{array}$$

$$\frac{1}{z-2}=-\frac{1}{2}-\frac{z}{4}-\frac{z^2}{8}-\frac{z^3}{16}-\ldots$$

$$=-\frac{1}{2}\left[1+\frac{z}{2}+\frac{z^2}{2^2}+\frac{z^3}{2^3}+\ldots+\frac{z^k}{2^k}+\ldots\right]$$

$$=-\frac{1}{2}-\frac{z}{2^2}-\frac{z^2}{2^3}-\frac{z^3}{2^4}-\ldots-\frac{z^k}{2^{k+1}}-\ldots=\{2^{-k-1}\}z^k$$

$$Z^{-1}\left[\frac{1}{z-2}\right]=\{2^{-k-1}\}$$ **Ans.**

## 19.21 BINOMIAL EXPANSION AND PARTIAL FRACTION

**Example 33.** *Find the inverse Z-transform of* $\dfrac{4z}{z-a}$

(*i*) $|z|>|a|$ (*ii*) $|z|<|a|$

**Solution. Case I.** $|z|>|a|$

$$\frac{4z}{z-a}=\frac{4z}{z}\frac{1}{1-\frac{a}{z}}=4\left(1-\frac{a}{z}\right)^{-1}=4\left[1+\frac{a}{z}+\frac{a^2}{z^2}+\frac{a^3}{z^3}+\ldots\right]$$

$$=4+4a\,z^{-1}+4a^2z^{-2}+4a^3z^{-3}+\ldots+4a^kz^{-k}+\ldots=\{4a^k\}z^{-k}$$

$$Z^{-1}\left(\frac{4z}{z-a}\right)=\{4a^k\}$$ **Ans.**

**Case II.** $|z|<|a|$

$$\frac{4z}{z-a}=-\frac{1}{a}\frac{z}{\left(1-\frac{z}{a}\right)}=\frac{-z}{a}\left(1-\frac{z}{a}\right)^{-1}$$

$$=-\frac{z}{a}\left[1+\frac{z}{a}+\frac{z^2}{a^2}+\frac{z^3}{a^3}+\ldots\right]=-\frac{z}{a}-\frac{z^2}{a^2}-\frac{z^3}{a^3}-\frac{z^4}{a^4}\ldots$$

$$\{f(k)\}=\left\{\ldots-\frac{1}{a^4},-\frac{1}{a^3},-\frac{1}{a^2},-\frac{1}{a}\right\}$$ **Ans.**

### EXERCISE 19.4

**Find the inverse of the *Z*-transform of the following :**

1. $\dfrac{z^2}{\left(z-\frac{1}{4}\right)\left(z-\frac{1}{5}\right)}$, $|z|<\dfrac{1}{5}$ **Ans.** $4\left(\dfrac{1}{5}\right)^k-5\left(\dfrac{1}{4}\right)^k$

2. $\dfrac{z}{\left(z-\frac{1}{4}\right)\left(z-\frac{1}{5}\right)}$, $|z|>\dfrac{1}{4}$ **Ans.** $20\left[\left(\dfrac{1}{4}\right)^k-\left(\dfrac{1}{5}\right)^k\right]U(k)$

3. $\dfrac{z^3}{\left(z-\frac{1}{2}\right)^2(z-1)}$, $|z|<\dfrac{1}{2}$ **Ans.** $-4(1)^k+(k+3)\left(\dfrac{1}{2}\right)^k$, $(k<0)$

4. $\dfrac{z+1}{z^2-2z+1}$, $|z|>1$ **Ans.** $(2k+1)\,U(k)$

## 19.22 INVERSE OF Z-TRANSFORM BY PARTIAL FRACTIONS

Let $f(z)=\dfrac{R(z)}{D(z)}$ [If the degree of Numerator < the degree of Denominator.]

and $F(z)=Q(z)+\dfrac{R(z)}{D(z)}$ [If the degree of Numerator > the degree of Denominator.]

$\dfrac{R(z)}{D(z)}$ is then expressed into partial fractions.

We convert $\dfrac{F(z)}{z}$ into partial fractions and not that of $F(z)$.

Let $\dfrac{F(z)}{z}=\dfrac{A}{z-a}+\dfrac{B}{z-b}+\dfrac{C}{z-c}+\dfrac{D}{(z-c)^2}+\dfrac{E}{(z-c)^3}+\dfrac{Mz+N}{z^2+pz+q}$

then $F(z)=A\dfrac{z}{z-a}+B\dfrac{z}{z-b}+C\dfrac{z}{z-c}+D\dfrac{z}{(z-c)^2}+E\dfrac{z}{(z-c)^3}+\dfrac{z(Mz+N)}{z^2+pz+q}$

$$Z^{-1}F(z)=AZ^{-1}\frac{z}{z-a}+BZ^{-1}\frac{z}{z-b}+CZ^{-1}\frac{z}{z-c}+DZ^{-1}\frac{z}{(z-c)^2}+EZ^{-1}\frac{z}{(z-c)^3}+Z^{-1}\frac{z(Mz+N)}{z^2+pz+q}$$

**(*i*) Linear non-repeated factor**

Let the linear non-repeated factor be $\dfrac{z}{z-a}$.

$$Z^{-1}\left(\frac{z}{z-a}\right)=Z^{-1}\frac{1}{1-az^{-1}}=\{a^k\} \qquad \text{If } |z|>|a|$$

$$Z^{-1}\left(\frac{z}{z-a}\right)=-Z^{-1}\frac{\frac{z}{a}}{1-\frac{z}{a}}=\{-a^k\},\ k<0, \qquad \text{If } |z|<|a|$$

**(*ii*) Linear repeated factor**

Let the linear repeated factor be $\dfrac{z}{(z-b)^r}$. $\qquad r\ge 2, |z|>|b|$

$$Z^{-1}\frac{z}{(z-b)^r}=Z^{-1}\left[z^{-(r-1)}\frac{z^r}{(z-b)^r}\right]$$

$$=\frac{(k+2-r)(k+3-r)...kU(k-r+1)b^{(k-r+1)}}{(r-1)!}$$

$$Z^{-1}\frac{z}{(z-b)^r}=-\frac{(k+2-r)(k+3-r)......kb^{k-r+1}}{(r-1)!}U(-k-r-1), \text{ If } |z|<|b|$$

**(*iii*) Quadratic non-repeated factor**

Let the quadratic non-repeated factor be

$$\frac{Mz^2+Nz}{z^2+pz+q}... \qquad ...(1) \qquad \text{If } |z|>0$$

Compare (1) with $Z[\{c^k\cos\alpha k\}]=\dfrac{z^2-cz\cos\alpha}{z^2-2cz\cos\alpha+c^2}$

$$p=-2c\cos\alpha,\ q=c^2$$

or with $Z[\{c^k\cosh\alpha k\}]=\dfrac{z^2-cz\cosh\alpha}{z^2-2cz\cosh\alpha+c^2}$

or $\qquad p=2c\cosh\alpha, \qquad q=c^2$

$\dfrac{p}{-2c}=\cos\alpha \Rightarrow \left|\dfrac{p}{2c}\right|<1 \text{ or } >1$, $c$ is given by $(-2c\cos\alpha)$ and $\alpha$ by $(-2c\cosh\alpha)$

**Case I.** If $\left|\dfrac{p}{2c}\right|<1$

$$\frac{Mz^2+Nz}{z^2+pz+q}=\frac{Mz(z-c\cos\alpha)+\dfrac{Mc\cos\alpha+N}{c\sin\alpha}(cz\sin\alpha)}{z^2-2cz\cos\alpha+c^2}$$

$$=\frac{M(z^2-cz\cos\alpha)}{z^2-2cz\cos\alpha+c^2}+\frac{Mc\cos\alpha+N}{c\sin\alpha}\cdot\frac{cz\sin\alpha}{z^2-2cz\cos\alpha+c^2}$$

$$Z^{-1}\frac{Mz^2+Nz}{z^2+pz+q}=M\{c^k\cos\alpha k\}+\left(\frac{Mc\cos\alpha+N}{c\sin\alpha}\right)[c^k\sin\alpha k]$$

**Case II.** If $\left|\dfrac{p}{2c}\right|>1$

$$\frac{Mz^2+Nz}{z^2+pz+q}=\frac{Mz(z-c\cosh\alpha)+\dfrac{Mc\cosh\alpha+N}{c\sinh\alpha}(cz\sinh\alpha)}{z^2-2cz\cosh\alpha+c^2}$$

$$=\frac{Mz(z-c\cosh\alpha)}{z^2-2cz\cosh\alpha+c^2}+\frac{Mc\cosh\alpha+N}{c\sinh\alpha}\cdot\frac{cz\sinh\alpha}{z^2-2cz\cosh\alpha+c^2}$$

$$=\frac{M(z^2-cz\cosh\alpha)}{z^2-2cz\cosh\alpha+c^2}+\frac{Mc\cosh\alpha+N}{c\sinh\alpha}\frac{cz\sinh\alpha}{z^2-2cz\cosh\alpha+c^2}$$

Similarly, $Z^{-1}\left[\dfrac{Mz^2+Nz}{z^2+pz+q}\right]=M\{c^k\cosh\alpha k\}+\dfrac{Mc\cosh\alpha+N}{c\sinh\alpha}[c^k\sinh\alpha]$

**Example 34.** *Find the inverse Z-transform of* $\dfrac{1}{(z-3)(z-2)}$

(*i*) $|z|<2$ (*ii*) $2<|z|<3$ (*iii*) $|z|>3$

**Solution.** $F(z)=\dfrac{1}{(z-3)(z-2)}=\dfrac{1}{z-3}-\dfrac{1}{z-2}$

**Case (*i*).** $|z|<2$

$$F(z)=-\frac{1}{3}\frac{1}{1-\frac{z}{3}}+\frac{1}{2}\frac{1}{1-\frac{z}{2}}=-\frac{1}{3}\left(1-\frac{z}{3}\right)^{-1}+\frac{1}{2}\left(1-\frac{z}{2}\right)^{-1}$$

$$=-\frac{1}{3}\left(1+\frac{z}{3}+\frac{z^2}{3^2}+\frac{z^3}{3^3}+\ldots\right)+\frac{1}{2}\left[1+\frac{z}{2}+\frac{z^2}{2^2}+\frac{z^3}{2^3}+\ldots\right]$$

$$=-\frac{1}{3}-\frac{z}{3^2}-\frac{z^2}{3^3}-\frac{z^3}{3^4}-\ldots+\frac{1}{2}+\frac{z}{2^2}+\frac{z^2}{2^3}+\frac{z^3}{2^4}+\ldots\ldots\ldots$$ **Ans.**

**Case (*ii*).** $2<|z|<3$

$$F(z)=-\frac{1}{3}\frac{1}{1-\frac{z}{3}}-\frac{1}{z}\frac{1}{1-\frac{2}{z}}=-\frac{1}{3}\left[1-\frac{z}{3}\right]^{-1}-\frac{1}{z}\left[1-\frac{2}{z}\right]^{-1}$$

$$=-\frac{1}{3}\left(1+\frac{z}{3}+\frac{z^2}{3^2}+\frac{z^3}{3^3}+\ldots\right)-\frac{1}{z}\left[1+\frac{2}{z}+\frac{2^2}{z^2}+\frac{2^3}{z^3}+\ldots\right]$$

$$=-\frac{1}{3}-\frac{z}{3^2}-\frac{z^2}{3^3}-\frac{z^3}{3^4}-\ldots-\frac{1}{z}-\frac{2}{z^2}-\frac{2^2}{z^3}-\frac{2^3}{z^4}-\ldots$$

$$=\ldots\ldots-\frac{z^3}{3^4}-\frac{z^2}{3^3}-\frac{z}{3^2}-\frac{1}{3}-\frac{1}{z}-\frac{2}{z^2}-\frac{2^2}{z^3}-\frac{2^3}{z^4}-\ldots$$

$$=\ldots\ldots-\frac{z^3}{3^4}-\frac{z^2}{3^3}-\frac{z}{3^2}-\frac{1}{3}-z^{-1}-2z^{-2}-2^2z^{-3}-2^3z^{-4}-\ldots\ldots$$

$$\Rightarrow f(k) = \{-2^{k-1}\},\quad k > 0 \Rightarrow f(k) = \{-3^{k-1}\},\quad k \le 0$$ **Ans.**

**Case (*iii*).** $|z| > 3$

$$F(z) = \frac{1}{z}\frac{1}{1-\frac{3}{z}} - \frac{1}{z}\frac{1}{1-\frac{2}{z}} = \frac{1}{z}\left(1-\frac{3}{z}\right)^{-1} - \frac{1}{z}\left(1-\frac{2}{z}\right)^{-1}$$

$$= \frac{1}{z}\left[1+\frac{3}{z}+\frac{3^2}{z^2}+\frac{3^3}{z^3}+\ldots\right] - \frac{1}{z}\left(1+\frac{2}{z}+\frac{2^2}{z^2}+\frac{2^3}{z^3}+\ldots\right)$$

$$= \frac{1}{z}+\frac{3}{z^2}+\frac{3^2}{z^3}+\frac{3^3}{z^4}+\ldots - \frac{1}{z}-\frac{2}{z^2}-\frac{2^2}{z^3}-\frac{2^3}{z^4}+\ldots$$

$$= \{3^{k-1} - 2^{k-1}\}\, z^{-k}, \qquad k \ge 1$$

$$= 0 \qquad\qquad , \qquad k \le 1$$ **Ans.**

**Example 35.** *Find the inverse of Z-transform of* $\dfrac{1}{(z-5)^3}$, $|z| > 5$

**Solution.**
$$F(z) = \frac{1}{(z-5)^3} = \frac{1}{z^3}\frac{1}{\left(1-\frac{5}{z}\right)^3} = \frac{1}{z^3}(1-5z^{-1})^{-3}$$

$$= z^{-3}\left[1 + 15z^{-1} + 6\,(5z^{-1})^2 + 10(5\,z^{-1})^3 + \ldots\ldots + \frac{(n+1)(n+2)}{2}(5z^{-1})^n + \ldots..\right]$$

$$= z^{-3}\left\{\frac{(k+1)(k+2)}{2}5^k\right\}z^{-k} = \left\{\frac{(k+1)(k+2)}{2}5^k\right\}z^{-k-3}$$

Replacing $k$ by $k-3$ we get

$$= \left\{\frac{(k-3+1)(k-3+2)}{2}5^{k-3}\right\}z^{-k} = \left\{\frac{(k-2)(k-1)}{2}5^{k-3}\right\}z^{-k}$$

$$Z^{-1}F(z) = f(k) = \begin{cases} \dfrac{(k-2)(k-1)}{2}5^{k-3}, & k \ge 3 \\ 0 & , \quad k < 3 \end{cases}$$ **Ans.**

**Example 36.** *Obtain* $Z^{-1}\dfrac{1}{\left(z-\frac{1}{2}\right)\left(z-\frac{1}{3}\right)}$. *When (i)* $\frac{1}{3} < |z| < \frac{1}{2}$ *(ii)* $\frac{1}{2} < |z|$

**Solution.**
$$F(z) = \frac{1}{\left(z-\frac{1}{2}\right)\left(z-\frac{1}{3}\right)} = 6\left[\frac{1}{\left(z-\frac{1}{2}\right)} - \frac{1}{\left(z-\frac{1}{3}\right)}\right] = \frac{6}{z-\frac{1}{2}} - \frac{6}{z-\frac{1}{3}}$$

(*i*) $\frac{1}{3} < |z| < \frac{1}{2}$,
$$F(z) = \frac{6}{-\frac{1}{2}(1-2z)} - \frac{1}{z}\frac{6}{(1-\frac{1}{3}z^{-1})} = -12(1-2z)^{-1} - \frac{6}{z}(1-\frac{1}{3}z^{-1})^{-1}$$

$$= -12[1+(2z)+(2z)^2+(2z)^3+\ldots] - \frac{6}{z}\left(1+\frac{1}{3z}+\frac{1}{(3z)^2}+\frac{1}{(3z)^3}+\ldots\right)$$

$$= -12(2z)^k - \frac{6}{z}\left(\frac{1}{3z}\right)^k = -12(2z)^k - 6\left(\frac{1}{3^k z^{k+1}}\right)$$

$$f(k) = -\frac{6}{3^{k-1}}, \qquad \text{if } k > 0$$

$$f(k) = -12\,.\,2^{-k}, \qquad \text{if } k < 0$$

(*ii*) $\frac{1}{2} < |z|$

$$F(z) = \frac{6}{z-\frac{1}{2}} - \frac{6}{z-\frac{1}{3}} = \frac{1}{z}\frac{6}{(1-\frac{1}{2z})} - \frac{1}{z}\frac{6}{(1-\frac{1}{3z})} = 6z^{-1}\left(1-\frac{1}{2}z^{-1}\right)^{-1} - 6z^{-1}\left(1-\frac{1}{3}z^{-1}\right)^{-1}$$

$$= 6z^{-1}\left[1+\frac{1}{2}z^{-1}+\left(\frac{1}{2}\right)^2 z^{-2}+\left(\frac{1}{2}\right)^3 z^{-3}+\dots\right] - 6z^{-1}\left[1+\frac{1}{3}z^{-1}+\left(\frac{1}{3}\right)^2 z^{-2}+\left(\frac{1}{3}\right)^3 z^{-3}+\dots\right]$$

$$= 6\left[z^{-1}+\frac{1}{2}z^{-2}+\left(\frac{1}{2}\right)^2 z^{-3}+\left(\frac{1}{2}\right)^3 z^{-4}+\dots\right] - 6\left[z^{-1}+\frac{1}{3}z^{-2}+\left(\frac{1}{3}\right)^2 z^{-3}+\left(\frac{1}{3}\right)^3 z^{-4}+\dots\right]$$

$$= 6\left\{\frac{1}{2^{k+1}}\right\}z^{-k} - 6\left\{\frac{1}{3^{k+1}}\right\}z^{-k} \Rightarrow Z^{-1}F(z) = f(k) = 6\left[\frac{1}{2^{k-1}} - \frac{1}{3^{k-1}}\right], \quad k \geq 1$$

**Ans.**

**Example 37.** *Obtain* $Z^{-1}\dfrac{2z^2-10z+13}{(z-3)^2(z-2)}$, *when* $2 < |z| < 3$

**Solution.** Let $\dfrac{2z^2-10z+13}{(z-3)^2(z-2)} = \dfrac{A}{(z-3)^2} + \dfrac{B}{z-3} + \dfrac{C}{z-2}$

Converting into partial fractions, we get

$$= \frac{1}{(z-3)^2} + \frac{1}{z-3} + \frac{1}{z-2}$$

$$= \frac{1}{9}\frac{1}{\left(1-\frac{z}{3}\right)^2} - \frac{1}{3}\frac{1}{1-\frac{z}{3}} + \frac{1}{z}\frac{1}{1-\frac{2}{z}} = \frac{1}{9}\left(1-\frac{z}{3}\right)^{-2} - \frac{1}{3}\left(1-\frac{z}{3}\right)^{-1} + \frac{1}{z}\left(1-\frac{2}{z}\right)^{-1}$$

$$= \frac{1}{9}\left[1+\frac{2z}{3}+\frac{3z^2}{9}+\frac{4z^3}{27}+\dots\right] - \frac{1}{3}\left[1+\frac{z}{3}+\frac{z^2}{9}+\frac{z^3}{27}+\dots\right] + \frac{1}{z}\left[1+\frac{2}{z}+\frac{4}{z^2}+\frac{8}{z^3}+\dots\right]$$

$$= \frac{1}{3^2}+\frac{2z}{3^3}+\frac{3z^2}{3^4}+\frac{4z^3}{3^5}\dots - \frac{1}{3} - \frac{z}{3^2} - \frac{z^2}{3^3} - \frac{z^3}{3^4} - \dots + \frac{1}{z}+\frac{2}{z^2}+\frac{4}{z^3}+\frac{8}{z^4}+\dots$$

$$= \{2^{k-1}\}\,z^{-k}, \qquad k \geq 1 \qquad \dots(1)$$

$$= \left[\frac{k+1}{3^{k+2}} - \frac{1}{3^{k+1}}\right]z^k,\ k < 0 \quad \text{or} = \frac{-k-2}{3^{-k+2}},\ k \leq 0 \qquad \dots(2)$$

Hence, $Z^{-1}[F(z)] = 2^{k-1}$ if $k \geq 1$, and $Z^{-1}[F(z)] = -(k+2)\,3^{k-2},\ k \leq 0$ **Ans.**

**Example 38.** *Find* $Z^{-1}\dfrac{3z^2+4z}{z^2-z+1}$ $\quad |z| > 1.$

**Solution.** Let $c^2 = 1 \Rightarrow c = \pm 1$

If $c = 1$, $\left|\dfrac{p}{2c}\right| = \left|\dfrac{-1}{2\times 1}\right| = \dfrac{1}{2} < 1$

So $-1 = -2c\cos\alpha$ or $-1 = -2 \times 1 \times \cos\alpha \Rightarrow \cos\alpha = \dfrac{1}{2}$

$$\alpha = \frac{\pi}{3}, \quad \sin\alpha = \sin\frac{\pi}{3} = \frac{\sqrt{3}}{2}$$

$$\frac{3z^2+4z}{z^2-z+1} = \frac{3z(z-c\cos\alpha) + \frac{(3c\cos\alpha+4)}{c\sin\alpha}cz\sin\alpha}{z^2-2cz\cos\alpha+c^2}$$

$$\frac{3z^2+4z}{z^2-z+1}=\frac{3z(z-c\cos\alpha)}{z^2-2cz\cos\alpha+c^2}+\frac{\dfrac{3c\cos\alpha+4}{c\sin\alpha}cz\sin\alpha}{z^2-2cz\cos\alpha+c^2}$$

Putting the values of $c$ and $\alpha$, in the coefficient of $cz \sin \alpha$, we get

$$=\frac{3z(z-c\cos\alpha)}{z^2-2cz\cos\alpha+c^2}+\frac{\dfrac{3\times1\times\frac{1}{2}+4}{1\times\frac{\sqrt{3}}{2}}cz\sin\alpha}{z^2-2cz\cos\alpha+c^2}$$

$$Z^{-1}\left[\frac{3z^2+4z}{z^2-z+1}\right]=\left[\left\{3c^k\cos\alpha k\right\}+\left\{\frac{11}{\sqrt{3}}c^k\sin\alpha k\right\}\right]U_k$$

Putting the values of $c$ and $\alpha$, we get

$$Z^{-1}\left[\frac{3z^2+4z}{z^2-z+1}\right]=\left[3(1)^k\left\{\cos\frac{\pi}{3}k\right\}+\frac{11}{3}(1)^k\left\{\sin\frac{\pi}{3}k\right\}\right]U_k$$

$$=\left[3\left\{\cos\frac{\pi k}{3}\right\}+\frac{11}{\sqrt{3}}\left\{\sin\frac{\pi}{3}k\right\}\right]U_k$$ **Ans.**

**Example 39.** *Obtain* $Z^{-1}\dfrac{2z^2+3z}{z^2+z+\frac{1}{9}}$ $\left(|z|>\dfrac{1}{3}\right)$

**Solution.** $c^2=\dfrac{1}{9}\Rightarrow c=\pm\dfrac{1}{3},\ p=1$

If $\qquad c=-\dfrac{1}{3},\ \left|\dfrac{p}{2c}\right|=\left|\dfrac{1}{2\times-\frac{1}{3}}\right|=\dfrac{3}{2}>1$

Hence, $\qquad 1=-2c\cosh\alpha \quad$ or $\quad 1=-2\left(-\dfrac{1}{3}\right)\cosh\alpha$

or $\qquad \cosh\alpha=\dfrac{3}{2},\ \sinh\alpha=\dfrac{\sqrt{5}}{2} \qquad (\cosh^2\alpha-\sinh^2\alpha=1)$

$$\frac{2z^2+3z}{z^2+z+\frac{1}{9}}=\frac{2z(z-c\cosh\alpha)+\dfrac{(2c\cosh\alpha+3)}{c\sinh\alpha}cz\sinh\alpha}{z^2-2cz\cosh\alpha+c^2}$$

Putting the values of $c$ and $\alpha$ in the coefficient of $cz \sin \alpha$, we get

$$=2\frac{z(z-c\cosh\alpha)}{z^2-2cz\cosh\alpha+c^2}+\frac{\dfrac{\left[2\left(-\frac{1}{3}\right)\left(\frac{3}{2}\right)+3\right]}{-\frac{1}{3}\times\frac{\sqrt{5}}{2}}cz\sinh\alpha}{z^2-2cz\cosh\alpha+c^2}$$

$$=2\frac{z(z-c\cosh\alpha)}{z^2-2cz\cosh\alpha+c^2}+\frac{\left(\dfrac{-1+3}{\frac{-\sqrt{5}}{6}}\right)cz\sinh\alpha}{z^2-2cz\cosh\alpha+c^2}=2\frac{z(z-c\cosh\alpha)}{z^2-2cz\cosh\alpha+c^2}-\frac{12}{\sqrt{5}}\frac{cz\sinh\alpha}{z^2-2cz\cosh\alpha+c^2}$$

$$Z^{-1}\frac{2z^2+3z}{z^2+z+\frac{1}{9}} = 2\left(-\frac{1}{3}\right)^k \cosh \alpha k - \frac{12}{\sqrt{5}}\left(-\frac{1}{3}\right)^k \sinh \alpha,\ k \ge 0$$

where $\cosh \alpha = 3/2$. **Ans.**

## EXERCISE 19.5

**Evaluate inverse *Z*-transform of the following:**

1. $\dfrac{z}{3-z},\quad |z|<3$ **Ans.** $\{3^k\}, k<0$

2. $\dfrac{2z^2-5z}{(z-2)(z-3)},\quad |z|>3$ **Ans.** $\{2^k+3^k\}, k \ge 0$

3. $\dfrac{z}{z-e^{\alpha}}$ **Ans.** $\{e^{k\alpha}\}$

4. $\dfrac{z^3}{z^3-27},\quad |z|>3$ **Ans.** $\{f(k)\} = \{3^k\} = 0$, for $k = 0, 3, 6, 9$

5. $\dfrac{2z}{(z-2)^2},\quad |z|>2$ **Ans.** $\{k\,2^k\}, k \ge 0$

6. $\dfrac{z^2+z\cos\alpha}{z^2-2z\cos\alpha+1},\quad |z|>1$ **Ans.** $\{\cos \alpha k\}, k \ge 0$

7. $-\log(1-2z^{-1}),\ |z|>2$ **Ans.** $\left\{\dfrac{2^k}{k}\right\}, k \ge 1$

8. $\dfrac{4z}{4z+1}+\dfrac{5z}{5z+1},\quad \dfrac{1}{5}<|z|<\dfrac{1}{4}$ **Ans.** $f(k) = \begin{cases} -\left(-\dfrac{1}{4}\right)^k, & k<0 \\ \left(\dfrac{1}{5}\right)^k, & k \ge 0 \end{cases}$

9. $\dfrac{z(z^2+4z+1)}{(z-1)^4},\quad |z|>1$ **Ans.** $\{k^3\}$

10 $\dfrac{ze^{-a}}{(z-e^{-a})^2},\quad |z|>|e^{-a}|$ **Ans.** $\{k\,e^{-\alpha k}\}$

11. $\dfrac{ze^{-a}\sin b}{z^2-2e^{-a}z\cos b+e^{-2a}},\quad |z|>|e^{-a}|$ **Ans.** $\{e^{-ak}\sin bk\}$

12. $\dfrac{z(z-e^{-a}\cos b)}{z^2-2e^{-a}z\cos b+e^{-2a}}$ **Ans.** $\{e^{-ak}\cos bk\}$

13. $\dfrac{z^2\sin\beta+\frac{z}{\sqrt{2}}(\cos\beta-\sin\beta)}{z^2-\sqrt{2}z+1},\quad |z|>1$ **Ans.** $\left\{\sin\left(\dfrac{k\pi}{4}+\beta\right)\right\},\quad k \ge 0$

14. $\dfrac{z(z+a)}{(z-a)^3},\quad |z|>a$ **Ans.** $\{k^2 a^{k-1} U(k-1)\}, k \ge 0$

15. $\dfrac{15z}{(4-z)(4z-1)},\quad \dfrac{1}{4}<|z|<4$ **Ans.** $\left\{\left(\dfrac{1}{4}\right)^{|k|}\right\}$

## 19.23 INVERSION OF Z-TRANSFORM BY RESIDUE METHOD

Take the contour $c$ such that all the poles of the function $z$ lie within the contour. Then by Residue Method

$f(k) = \Sigma$ Residue of $z^{k-1} F(z)$ at its poles.

where Residue for simple pole $z = z_i$ is $= \left[(z - z_i) z^{k-1} F(z)\right]_{z = z_i}$

Residue of order $r$ at the pole $z = z_i$

$$\left[\frac{1}{(r-1)!} \frac{d^{r-1}}{dz^{r-1}} (z - z_i)^r z^{k-1} F(z)\right]_{z = z_i}$$

**Example 40.** *Obtain* $Z^{-1} \dfrac{z}{(z-2)(z-3)}$

**Solution.** The poles are determined by

$(z-2)(z-3) = 0 \quad \Rightarrow \quad z = 2, 3$

There are two poles. Let us consider the contour $|z| > 3$.

$$\text{Residue at } (z = 2) = \left[(z-2) z^{k-1} \frac{z}{(z-2)(z-3)}\right]_{z=2} = \left[\frac{z^k}{z-3}\right]_{z=2} = \frac{2^k}{-1}$$

$$\text{Residue at } (z = 3) = \left[(z-3) z^{k-1} \frac{z}{(z-2)(z-3)}\right]_{z=3} = \left[\frac{z^k}{z-2}\right]_{z=3} = \frac{3^k}{1}$$

Hence, $f(k) =$ Sum of the residues $= 3^k - 2^k$. **Ans.**

**Example 41.** *Obtain* $Z^{-1} \dfrac{3z^2 - 18z + 26}{(z-2)(z-3)(z-4)}$

**Solution.** The poles are determined by $(z-2)(z-3)(z-4) = 0 \Rightarrow z = 2, 3, 4$.

There are three poles. Let us consider the contour $|z| > 4$.

$$\text{Residue at } (z = 2) = \left[\frac{(z-2) z^{k-1} [3z^2 - 18z + 26]}{(z-2)(z-3)(z-4)}\right]_{z=2}$$

$$= \left[\frac{3z^{k+1} - 18z^k + 26z^{k-1}}{(z-3)(z-4)}\right]_{z=2} = \frac{3.2^{k+1} - 18.2^k + 26.2^{k-1}}{(-1)(-2)}$$

$$= 3.2^k - 9.2^k + 13.2^{k-1} = -6.2^k + 13.2^{k-1} = -12.2^{k-1} + 13.2^{k-1} = 2^{k-1}$$

$$\text{Residue at } (z = 3) = \left[\frac{(z-3) z^{k-1} [3z^2 - 18z + 26]}{(z-2)(z-3)(z-4)}\right]_{z=3} = \left[\frac{3z^{k+1} - 18z^k + 26z^{k-1}}{(z-2)(z-4)}\right]_{z=3}$$

$$= \frac{3.3^{k+1} - 18.3^k + 26.3^{k-1}}{1(-1)} = -3.3^{k+1} + 18.3^k - 26.3^{k-1}$$

$$= -27.3^{k-1} + 54.3^{k-1} - 26.3^{k-1} = 3^{k-1}$$

$$\text{Residue at } (z = 4) = \left[(z-4) \frac{z^{k-1} [3z^2 - 18z + 26]}{(z-2)(z-3)(z-4)}\right]_{z=4} = \left[\frac{3z^{k+1} - 18z^k + 26z^{k-1}}{(z-2)(z-3)}\right]_{z=4}$$

$$= \frac{3.4^{k+1} - 18.4^k + 26.4^{k-1}}{(2)(1)} = \frac{3}{2}.4^{k+1} - 9.4^k + 13.4^{k-1}$$

$$= 24.4^{k-1} - 36.4^{k-1} + 13.4^{k-1} = 4^{k-1}$$

Hence $f(k) =$ Sum of the residues $= 2^{k-1} + 3^{k-1} - 4^{k-1}, \; k > 0$ **Ans.**

**Example 42.** *Obtain* $Z^{-1}\dfrac{z(3z^2-6z+4)}{(z-1)^2(z-2)}$

**Solution.** The poles are determined by $(z-1)^2(z-2)=0 \quad\Rightarrow\quad z=1,1,2$

There are two poles, simple pole at $z=2$ and pole of order 2 at $z=1$. Let us consider the contour $|z|>2$.

Residue at $(z=2)$ is $\left[\dfrac{(z-2)z^{k-1}.z[3z^2-6z+4]}{(z-1)^2(z-2)}\right]_{z=2}=\left[\dfrac{3z^{k+2}-6z^{k+1}+4z^k}{(z-1)^2}\right]_{z=2}$

$$=3.2^{k+2}-6.2^{k+1}+4.2^k=12.2^k-12.2^k+4.2^k=4.2^k=2^{k+2}$$

Residue at $(z=1)=\dfrac{d}{dz}\left[(z-1)^2\dfrac{z^{k-1}.z(3z^2-6z+4)}{(z-1)^2(z-2)}\right]_{z=1}=\dfrac{d}{dz}\left[\dfrac{3z^{k+2}-6z^{k+1}+4z^k}{z-2}\right]_{z=1}$

$$=\left[\frac{(z-2)[3(k+2)z^{k+1}-6(k+1)z^k+4kz^{k-1}]-(3z^{k+2}-6z^{k+1}+4z^k).1}{(z-2)^2}\right]_{z=1}$$

$$=-\{(3k+6)-6k-6+4k\}-\{3-6+4\}$$

$$=-3k-6+6k+6-4k-3+6-4=-k-1$$

$f(k)=$ Sum of the residues $=[2^{k+2}-k-1]U(k)$ **Ans.**

**Example 43.** *Find* $Z^{-1}\dfrac{9z^3}{(3z-1)^2(z-2)}$

**Solution.** The poles are determined by putting the denominator equal to zero, *i.e.*,

$$(3z-1)^2(z-2)=0 \quad\Rightarrow z=\frac{1}{3},\frac{1}{3},2$$

There are two poles *i.e.*, one simple pole at $z=2$ and second pole of order 2 at $z=\dfrac{1}{3}$.

Let us consider the contour $|z|>2$.

Residue at $(z=2)=\left[(z-2).z^{k-1}.\dfrac{9z^3}{(3z-1)^2(z-2)}\right]_{z=2}=\left[\dfrac{9z^{k+2}}{(3z-1)^2}\right]_{z=2}=\dfrac{9.2^{k+2}}{25}$

Residue at $\left(z=\dfrac{1}{3}\right)=\dfrac{d}{dz}\left[\dfrac{\left(z-\frac{1}{3}\right)^2.z^{k-1}.9z^3}{(3z-1)^2.(z-2)}\right]_{z=\frac{1}{3}}=\dfrac{d}{dz}\left[\dfrac{9z^{k+2}}{9(z-2)}\right]_{z=\frac{1}{3}}=\dfrac{d}{dz}\left(\dfrac{z^{k+2}}{z-2}\right)_{z=\frac{1}{3}}$

$$=\left[\frac{(z-2)(k+2)z^{k+1}-z^{k+2}}{(z-2)^2}\right]_{z=\frac{1}{3}}=\left[\frac{(k+2)z^{k+2}-2(k+2)z^{k+1}-z^{k+2}}{(z-2)^2}\right]_{z=\frac{1}{3}}$$

$$=\left[\frac{z^{k+2}(k+2-1)-2(k+2)z^{k+1}}{(z-2)^2}\right]_{z=\frac{1}{3}}$$

$$=\left[\frac{(k+1)z^{k+2}-2(k+2)z^{k+1}}{(z-2)^2}\right]_{z=\frac{1}{3}}=\left[\frac{z^{k+1}\{(k+1)z-2(k+2)\}}{(z-2)^2}\right]_{z=\frac{1}{3}}$$

$$= \frac{\left(\frac{1}{3}\right)^{k+1}\left\{(k+1)\frac{1}{3}-2(k+2)\right\}}{\frac{25}{9}} = \frac{\left(\frac{1}{3}\right)^{k+1}.9.\frac{1}{3}[k+1-6(k+2)]}{25}$$

$$= \left(\frac{1}{3}\right)^k\left(\frac{k+1-6k-12}{25}\right) = \frac{\left(\frac{1}{3}\right)^k(-5k-11)}{25}$$

Hence, $f(k)$ = Sum of residues $= \frac{9}{25}.2^{k+2} - \left(\frac{1}{3}\right)^k\frac{(5k+11)}{25}$. **Ans.**

**Example 44.** *Find* $Z^{-1}\frac{z^2}{z^2+4}$, $\quad |z|>2$

**Solution.** The poles are determined by $z^2+4=0 \quad \Rightarrow \quad z=\pm 2i$

There are two poles at $z = 2i$ and $z = -2i$.

Let us consider a contour $|z|>2$

Residue at $(z=2i)$ $= \left[\frac{(z-2i)z^{k-1}.z^2}{z^2+4}\right]_{z=2i} = \left[\frac{z^{k+1}}{z+2i}\right]_{z=2i} = \frac{(2i)^{k+1}}{4i} = 2^{k-1}(i)^k$

$$= 2^{k-1}\left[\cos\frac{\pi}{2}+i\sin\frac{\pi}{2}\right]^k = 2^{k-1}\left[\cos\frac{k\pi}{2}+i\sin\frac{k\pi}{2}\right]$$

Residue at $(z=-2i)$ $= \left[(z+2i)z^{k-1}\frac{z^2}{z^2+4}\right]_{z=-2i} = \left[\frac{z^{k+1}}{z-2i}\right]_{z=-2i} = \frac{(-2i)^{k+1}}{-4i}$

$$= 2^{k-1}(-i)^k = 2^{k-1}\left[\cos\frac{\pi}{2}-i\sin\frac{\pi}{2}\right]^k = 2^{k-1}\left[\cos\frac{k\pi}{2}-i\sin\frac{k\pi}{2}\right]$$

$f(k)$ = Sum of the residues $= 2^{k-1}\left[\cos\frac{k\pi}{2}+i\sin\frac{k\pi}{2}\right]+2^{k-1}\left[\cos\frac{k\pi}{2}-i\sin\frac{k\pi}{2}\right]$

$$f(k) = 2^{k-1}\left[2\cos\frac{k\pi}{2}\right] = 2^k\cos\frac{k\pi}{2}$$ **Ans.**

## 19.24 CONVOLUTION METHOD

We know that $\quad Z\{f*g\} = F(z)\,G(z)$

$$Z^{-1}\{F(z).\,G(z)\} = f*g = \sum_{n=0}^{k} f(n)\,g(k-n).$$

## 19.25 INVERSE OF Z-TRANSFORM BY CONVOLUTION METHOD

**Example 45.** *Using Convolution theorem, evaluate*

$$Z^{-1}\left\{\frac{z^2}{(z-1)(z-3)}\right\}.$$

(*Q. Bank U.P., III Semester 2002*)

**Solution.** We know that

$$Z^{-1}\{F(z).\,G(z)\} = f*g$$

Let $\quad F(z) = \frac{z}{z-1} \quad \therefore \; f(k) = (1)^k$

$$G(z)=\frac{z}{z-3} \qquad \therefore\ g(k)=(3)^k$$

Now, $Z^{-1}\{F(z)\,.\,G(z)\}=(1)^k*(3)^k$

$$=\sum_{n=0}^{k}1^n\,3^{k-n}=3^k\sum_{n=0}^{k}\left(\frac{1}{3}\right)^n \qquad (a\text{ G.P.})$$

$$=3^k\left\{\frac{\left(\frac{1}{3}\right)^{k+1}-1}{\frac{1}{3}-1}\right\}=\frac{(1)^{k+1}-(3)^{k+1}}{1-3}=\left\{\frac{1}{2}(3^{k+1}-1)\right\} \qquad \textbf{Ans.}$$

## 19.26 INVERSE OF Z-TRANSFORM BY POWER SERIES METHOD

**Example 46.** *Find* $Z^{-1}\left[\log\frac{z}{z+1}\right]$ *by power series method.*

**Solution.** Let $F(z)=\log\left(\frac{z}{z+1}\right)=\log\left(\frac{1}{1+\frac{1}{z}}\right)=-\log\left(1+\frac{1}{z}\right)$

$$=-\frac{1}{z}+\frac{1}{2}\left(\frac{1}{z}\right)^2-\frac{1}{3}\left(\frac{1}{z}\right)^3+\frac{1}{4}\left(\frac{1}{z}\right)^4+\ldots..$$

$$=-\frac{1}{z}+\frac{1}{2z^2}-\frac{1}{3z^3}+\frac{1}{4z^4}+\ldots...+\frac{(-1)^k}{k\,z^k}+\ldots..$$

$$f(k)=Z^{-1}[F(z)]=\begin{cases}0\ , & \text{for } k=0\\ \frac{(-1)^k}{k}, & \text{othersie}\end{cases} \qquad \textbf{Ans.}$$

### EXERCISE 19.6

**Find the inverse Z-transform of the following functions by residue method:**

1. $\frac{z}{(z-1)(z-2)}$ **Ans.** $2^k-1,\ k\ge 0$
2. $\frac{3z^2+2z}{z^2-3z+2},\ 1<|z|<2$ **Ans.** $\{-5\}\ (k\ge 0),\ -8\,(2)^k,\ k<0$
3. $\frac{2z^2+3z}{z^2+z+1},\ |z|>1$ **Ans.** $\left[2\{\cos\frac{2\pi k}{3}\}+\frac{4}{\sqrt{3}}\{\sin\frac{2\pi k}{3}\}\right]U(k)$
4. $\frac{z(z+1)}{(z-1)(z^2+z+1)},\ |z|>1$ **Ans.** $\frac{2}{3}\left[1-\cos\frac{2\pi k}{3}\right]U(k)$
5. $\frac{16z^3}{(4z-1)^2\,(z-1)}$ **Ans.** $\left[\frac{16}{9}-\frac{1}{9}(3k+7)\left(\frac{1}{4}\right)^k\right]U(k)$
6. $\frac{z^2}{z^2+1}$ **Ans.** $\cos\frac{k\pi}{2}$

## 19.27 SOLUTION OF A DIFFERENCE EQUATION BY Z TRANSFORM

A solution of a difference equation is any function which satisfies the given equation. The general solution of a difference equation is defined as the solution which involves as many arbitrary constants as the order of the difference equation.

The particular solution is a solution obtained from the general solution by assigning particular values to periodic constants.

## 19.28 THEOREM

Prove that $Z(y_{k+n}) = z^n\left(\bar{y} - y_0 - \frac{y_1}{z} - \dots - \frac{y_{n-1}}{z^{n-1}}\right)$, where $Z(y_k) = \bar{y}$

**Proof.** L.H. S. $= Z(y_{k+n}) = \sum_{k=0}^{\infty} y_{k+n} z^{-k} = z^n \sum_{k=0}^{\infty} y_{k+n} z^{-(n+k)}$ ...(1)

On putting $m = n + k$ in (1), we get

$$Z(y_{k+n}) = z^n \sum_{m=n}^{\infty} y_m z^{-m} = z^n \left[\sum_{m=0}^{\infty} y_m z^{-m} - \sum_{m=0}^{n-1} y_m z^{-m}\right]$$

$$= z^n \left[\bar{y} - y_0 - \frac{y_1}{z} - \frac{y_2}{z^2} - \dots - \frac{y_{n-1}}{z^{n-1}}\right]$$

Remember

**For $n = 1$,** $Z(y_{k+1}) = (z\bar{y} - z y_0)$; **For $n = 2$,** $Z(y_{k+2}) = (z^2 \bar{y} - z^2 y_0 - z y_1)$

**For $n = 3$,** $Z(y_{k+3}) = (z^3 \bar{y} - z^3 y_0 - z^2 y_1 - z y_2)$

**Note.** If $Z(y_k) = \bar{y}$, then $Z(y_{k-n}) = Z^{-n}\bar{y}$

**Example 47.** *Solve the difference equation*

$6y_{k+2} - y_{k-1} - y_k = 0$, $y(0) = 0, y(1) = 1$ *by Z-transform.*

**Solution.** $6y_{k+2} - y_{k+1} - y_k = 0$ ...(1)

Taking the Z-transform of both sides of (1), we get

$$Z[6y_{k+2} - y_{k+1} - y_k] = 0$$

$$Z(6y_{k+2}) - Z(y_{k+1}) - Z(y_k) = 0$$

$$6[z^2\bar{y} - z^2 y(0) - z\,y(1)] - [z\bar{y} - zy(0)] - \bar{y} = 0 \quad \text{... (2)}$$

On putting the values of $y(0)$ and $y(1)$ in (2), we get

$$6z^2\bar{y} - 6z - z\bar{y} - \bar{y} = 0 \quad \Rightarrow \quad (6z^2 - z - 1)\bar{y} = 6z$$

$$\bar{y} = \frac{6z}{6z^2 - z - 1} = \frac{6z}{(3z+1)(2z-1)} = \frac{z^{-1}}{\left(1+\frac{z^{-1}}{3}\right)\left(1-\frac{z^{-1}}{2}\right)} = \frac{\frac{6}{5}}{1-\frac{z^{-1}}{2}} - \frac{\frac{6}{5}}{1+\frac{z^{-1}}{3}}$$

$$y_k = Z^{-1}\left[\frac{\frac{6}{5}}{1-\frac{z^{-1}}{2}}\right] - Z^{-1}\left[\frac{\frac{6}{5}}{1+\frac{z^{-1}}{3}}\right] = \frac{6}{5}\left(\frac{1}{2}\right)^k - \frac{6}{5}\left(-\frac{1}{3}\right)^k = \frac{6}{5}\left[\left(\frac{1}{2}\right)^k - \left(-\frac{1}{3}\right)^k\right]$$ **Ans.**

**Example 48.** *Solve the difference equation*

$$y_{k+3} - 3y_{k+2} + 3y_{k+1} - y_k = U(k)$$
$$y(0) = y(1) = y(2) = 0 \quad \textit{by Z-transforms.}$$

**Solution.** $y_{k+3} - 3y_{k+2} + 3y_{k+1} - y_k = U(k)$ ...(1)

Taking the $Z$-transform of both sides of (1), we get

$$Z[y_{k+3} - 3y_{k+2} + 3y_{k+1} - y_k] = Z\,U(k),$$

$$\Rightarrow Z[y_{k+3}] - 3Z[y_{k+2}] + 3Z[y_{k+1}] - Z[y_k] = Z\,U(k)$$

$$\Rightarrow [z^3\bar{y} - z^3 y(0) - z^2 y(1) - zy(2)] - 3[z^2\bar{y} - z^2 y(0) - z\,y(1)]$$
$$+ 3[z\bar{y} - z\,y(0)] - \bar{y} = Z\,U(k)$$

Putting the values of $y(0) = y(1) = y(2) = 0$ in the above equation, we get

$$z^3\bar{y} - 3z^2\bar{y} + 3z\bar{y} - \bar{y} = \frac{1}{1-z^{-1}}$$

$$[z^3 - 3z^2 + 3z - 1]\,\bar{y} = \frac{1}{1-z^{-1}} \Rightarrow (z-1)^3\,\bar{y} = \frac{1}{1-z^{-1}}$$

$$\bar{y} = \frac{1}{(z-1)^3(1-z^{-1})} = \frac{1}{z^3(1-z^{-1})^3(1-z^{-1})} = z^{-3}(1-z^{-1})^{-4}$$

$$y_k = \text{coeff. of } z^{-k} \text{ in } z^{-3}(1-z^{-1})^{-4} = \text{coeff. of } z^{-k-3} \text{ in } (1-z^{-1})^{-4}$$

$$= \frac{(k-2)(k-1)k}{6},\ k \ge 3$$ **Ans.**

**Example 49.** *Solve by Z-transform.* $y_{k+1} + \frac{1}{4}y_k = \left(\frac{1}{4}\right)^k$, $(k \ge 0), y(0) = 0$

**Solution.** $y_{k+1} + \frac{1}{4}y_k = \left(\frac{1}{4}\right)^k$ ... (1)

Taking Z-transform of both sides of (1), we get

$$Z[y_{k+1} + \frac{1}{4}y_k] = Z\left[\left(\frac{1}{4}\right)^k\right]$$

$$Z[y_{k+1}] + Z\left[\frac{1}{4}y_k\right] = Z\left[\left(\frac{1}{4}\right)^k\right]$$

$$z\bar{y} - zy(0) + \frac{1}{4}\bar{y} = \frac{1}{1-\frac{1}{4}z^{-1}}, \qquad |z| > \frac{1}{4}$$

$$z\bar{y} - 0 + \frac{1}{4}\bar{y} = \frac{1}{1-\frac{1}{4}z^{-1}} \qquad \Rightarrow \qquad (z+\frac{1}{4})\,\bar{y} = \frac{1}{1-\frac{1}{4}z^{-1}}$$

$$\bar{y} = \frac{1}{z+\frac{1}{4}} \times \frac{1}{1-\frac{1}{4}z^{-1}} = \frac{z^{-1}}{1+\frac{1}{4}z^{-1}} \times \frac{1}{1-\frac{1}{4}z^{-1}} = \frac{-2}{1+\frac{1}{4}z^{-1}} + \frac{2}{1-\frac{1}{4}z^{-1}}$$

$$y_{(k)} = Z^{-1}\left[\frac{-2}{1+\frac{1}{4}z^{-1}}\right] + Z^{-1}\left[\frac{2}{1-\frac{1}{4}z^{-1}}\right] = Z^{-1}\left[-2(1+\frac{1}{4}z^{-1})^{-1}\right] + Z^{-1}\left[2(1-\frac{1}{4}z^{-1})^{-1}\right]$$

$$= -2\left(-\frac{1}{4}\right)^k + 2\left(\frac{1}{4}\right)^k$$ **Ans.**

**Example 50.** *Solve* $y_k + \frac{1}{4}y_{k-1} = U_{(k)} + \frac{1}{3}U_{(k-1)}$

**Solution.** $y_k + \frac{1}{4}y_{k-1} = U_{(k)} + \frac{1}{3}U_{(k-1)}$ ...(1)

Taking the *Z*-transform of both sides of (1), we get

$$Z[\{y_k\}] + \frac{1}{4}Z[\{y_{k-1}\}] = Z[\{U_k\}] + \frac{1}{3}[\{U_{k-1}\}]$$

$$\bar{y} + \frac{1}{4}z^{-1}\,\bar{y} = \left[1+\frac{1}{3}z^{-1}\right]U(k), \qquad \left[1+\frac{1}{4}z^{-1}\right]\bar{y} = (1+\frac{1}{3}z^{-1})\,U(k)$$

$$\bar{y} = \frac{1+\frac{1}{3}z^{-1}}{1+\frac{1}{4}z^{-1}} = \frac{z+\frac{1}{3}}{z+\frac{1}{4}}$$

There is only one simple pole at $z = -\frac{1}{4}$

Let us consider the contour $|z| > \frac{1}{4}$.

$$\text{Residue at }\left(z = -\frac{1}{4}\right) = \left[(z+\frac{1}{4})z^{k-1}\frac{z+\frac{1}{3}}{z+\frac{1}{4}}\right]_{z=-\frac{1}{4}} = \left[z^k + \frac{z^{k-1}}{3}\right]_{z=-\frac{1}{4}}$$

$$= \left(-\frac{1}{4}\right)^k + \frac{1}{3}\left(-\frac{1}{4}\right)^{k-1} = -\frac{1}{4}\left(-\frac{1}{4}\right)^{k-1} + \frac{1}{3}\left(-\frac{1}{4}\right)^{k-1} = \frac{1}{12}\left(-\frac{1}{4}\right)^{k-1}$$

Hence, $y_k = \text{Residue} = \frac{1}{12}\left(-\frac{1}{4}\right)^{k-1}$

**Example 51.** *Solve* $y_k + \frac{1}{25}y_{k-2} = \left(\frac{1}{5}\right)^k \cos\frac{k\pi}{2}$ $(k \geq 0)$ ***by residue method.***

**Solution.** $y_k + \frac{1}{25}y_{k-2} = \left(\frac{1}{5}\right)^k \cos\frac{k\pi}{2}$ ... (1)

Taking Z-transform of both sides of (1), we obtain

$$Z\left[y_k + \frac{1}{25}y_{k-2}\right] = Z\left(\frac{1}{5}\right)^k \cos\frac{k\pi}{2} \quad \Rightarrow \quad \bar{y} + \frac{1}{25}z^{-2}\,\bar{y} = \frac{z^2}{z^2+\frac{1}{25}}$$

$$\Rightarrow \quad \left[1+\frac{1}{25}z^{-2}\right]\bar{y}=\frac{z^2}{z^2+\frac{1}{25}} \quad \Rightarrow \bar{y}=\frac{z^2}{\left(1+\frac{1}{25}z^{-2}\right)\left(z^2+\frac{1}{25}\right)}=\frac{z^4}{\left(z^2+\frac{1}{25}\right)^2}$$

There are two poles of second order at $z=\frac{i}{5}$ and $z=-\frac{i}{5}$

Let us consider a contour $|z|>\left|\frac{1}{5}\right|$

$$\text{Residue at } \left(z=\frac{i}{5}\right)=\left[\frac{d}{dz}(z-\frac{i}{5})^2\frac{z^{k-1}z^4}{(z^2+\frac{1}{25})^2}\right]_{z=\frac{i}{5}}=\left[\frac{d}{dz}\frac{z^{k+3}}{(z+\frac{i}{5})^2}\right]_{z=\frac{i}{5}}$$

$$=\left[\frac{(z+\frac{i}{5})^2(k+3)z^{k+2}-z^{k+3}2(z+\frac{i}{5})}{(z+\frac{i}{5})^4}\right]_{z=\frac{i}{5}}=\left[\frac{(z+\frac{i}{5})(k+3)z^{k+2}-2z^{k+3}}{(z+\frac{i}{5})^3}\right]_{z=\frac{i}{5}}$$

$$=\frac{\left(\frac{2i}{5}\right)(k+3)\left(\frac{i}{5}\right)^{k+2}-2\left(\frac{i}{5}\right)^{k+3}}{\left(\frac{2i}{5}\right)^3}=\left(\frac{5}{2i}\right)^3[(2k+6)\left(\frac{i}{5}\right)^{k+3}-2\left(\frac{i}{5}\right)^{k+3}]$$

$$=\left(\frac{5}{2i}\right)^3\left[(2k+4)\left(\frac{i}{5}\right)^{k+3}\right]=\left(\frac{1}{8}\right)(2k+4)\left(\frac{i}{5}\right)^k=\frac{1}{4}(k+2)\left(\frac{1}{5}\right)^k[\cos\frac{\pi}{2}+i\sin\frac{\pi}{2}]^k$$

$$=\frac{1}{4}(k+2)\left(\frac{1}{5}\right)^k\left[\cos\frac{k\pi}{2}+i\sin\frac{k\pi}{2}\right]$$

$$\text{Residue at } \left(z=-\frac{i}{5}\right)=\frac{1}{4}(k+2)\left(\frac{1}{5}\right)^k\left[\cos\frac{k\pi}{2}-i\sin\frac{k\pi}{2}\right]$$

$(i \to -i)$

$$y_k = \text{Sum of the residues}$$

$$\Rightarrow \quad y_k=\frac{1}{2}(k+2)\frac{1}{5^k}\cos\frac{k\pi}{2} \qquad \textbf{Ans.}$$

## EXERCISE 19.7

**Solve the following difference equations by Z-transform:**

**1.** $y_k-\frac{5}{6}y_{k-1}+\frac{1}{6}y_{k-2}=U_{(k)}$ **Ans.** $y_k=\left[3-3\left(\frac{1}{2}\right)^k+\left(\frac{1}{3}\right)^k\right]U(k)$

**2.** $6y_{k+2}+5y_{k+1}-y_k=6U(k)$ **Ans.** $y_k=\left[\frac{6}{7}(k+1)-\frac{78}{49}+\frac{36}{49}\left(-\frac{1}{6}\right)^k\right]U(k)$

**3.** $y_{k+1}-y_{k-1}=U(k),\ y(0)=0$ **Ans.** $\left[\frac{k}{2}+\frac{1}{4}\right]U(k)$

4. $y_{k+1} - 2y_k + y_{k-1} = a^k,\ a \neq 1$

**Ans.** $y_k = \dfrac{1}{a}(k+1)U(k) - \dfrac{a}{(a-1)^2}U(k) + \dfrac{a}{(a-1)^2}a^k U(k) + \dfrac{1}{1-a}kU(k-1)$

5. $y_k + \dfrac{1}{9}y_{k-2} = \left(\dfrac{1}{3}\right)^k \cos\dfrac{k\pi}{2},\ (k \geq 0)$ **Ans.** $y_k = \left(\dfrac{k+2}{2}\right)\left(\dfrac{1}{3}\right)^k \cos\dfrac{k\pi}{2}U(k)$

6. $y_{k+2} - 3y_{k+1} - 4y_k = 0;\ y_0 = 3,\ y_1 = -2$ **Ans.** $y_k = \dfrac{1}{5}(4)^k + \dfrac{14}{5}(-1)^k$

7. $y_{k+2} + y_{k+1} - 2y_k = 0;\ y_0 = 4,\ y_1 = 0$ **Ans.** $y_k = \dfrac{8}{3} + \dfrac{4}{3}(-2)^k$

8. $y_{k+2} - 2y_{k+1} + y_k = k;\ y_{(0)} = 0,\ y_{(1)} = 0$ **Ans.** $y_k = \dfrac{k-1}{4}\{1-(-1)^k\}$

9. $y_{k+2} - 4y_k = 0;\ y_{(0)} = 0,\ y_{(1)} = 2$ **Ans.** $y_k = 2^{k-1} + (-2)^{k-1}$

(*Uttarakhand III Semester, 2008*)

10. $y_{k+2} - 2y_{k+1} + y_k = 2^k;\ y_{(0)} = 2,\ y_{(1)} = 1$ **Ans.** $y_k = 1 - 2k + 2^k$

11. $8y_{k+2} - 6y_{k+1} + y_k = 5\sin\left(\dfrac{k\pi}{2}\right)$ (*U.P. III Semester, Dec. 2006*)

# 20

# Infinite Series

## 20.1 SEQUENCE

A *sequence* is a succession of numbers or terms formed according to some definite rule. The $n$th term in a sequence is denoted by $u_n$.

*For example,* if $u_n = 2n + 1$.

By giving different values of $n$ in $u_n$, we get different terms of the sequence.

Thus, $u_1 = 3,\ u_2 = 5,\ u_3 = 7, \ldots$

A sequence having unlimited number of terms is known as an *infinite sequence.*

## 20.2 LIMIT

If a sequence tends to a limit $l$, then we write $\lim\limits_{n \to \infty} (u_n) = l$

## 20.3 CONVERGENT SEQUENCE

If the limit of a sequence is finite, the sequence is *convergent.* If the limit of a sequence does not tend to a finite number, the sequence is said to be *divergent.*

*e.g.,* $1, \frac{1}{4}, \frac{1}{9}, \frac{1}{16}, \ldots, \frac{1}{n^2} + \ldots$ . is a convergent sequence.

$3, 5, 7, \ldots, (2n + 1), \ldots$ is a divergent sequence.

## 20.4 BOUNDED SEQUENCE

$u_1, u_2, u_3 \ldots, u_n \ldots$ is a bounded sequence if $u_n < k$ for every $n$.

## 20.5 MONOTONIC SEQUENCE

The sequence is either increasing or decreasing, such sequences are called *monotonic.*

*e.g.,* $1, 4, 7, 10, \ldots$ is a monotonic sequence.

$1, \frac{1}{2}, \frac{1}{3}, \frac{1}{4}, \ldots$ is also a monotonic sequenece.

$1, -1, 1, -1, 1, \ldots$ is not a monotonic sequence.

A sequence which is monotonic and bounded is a convergent sequence.

### EXERCISE 20.1

**Determine the general term of each of the following sequence. Prove that the following sequences are convergent.**

1. $\frac{1}{2}, \frac{1}{4}, \frac{1}{8}, \frac{1}{16}, \ldots$ **Ans.** $\frac{1}{2^n}$

2. $\frac{1}{2}, \frac{2}{3}, \frac{3}{4}, \frac{4}{5}, \ldots$ **Ans.** $\frac{n}{n+1}$

3. $1, -1, 1, -1, \ldots$ **Ans.** $(-1)^{n-1}$

4. $\frac{1^2}{1!}, \frac{2^2}{2!}, \frac{3^2}{3!}, \frac{4^2}{4!}, \frac{5^2}{5!}, \ldots$ **Ans.** $\frac{n^2}{n!}$

Which of the following sequences are convergent ?

5. $u_n = \dfrac{n+1}{n}$ **Ans.** Convergent
6. $u_n = 3n$ **Ans.** Divergent
7. $u_n = n^2$ **Ans.** Divergent
8. $u_n = \dfrac{1}{n}$ **Ans.** Convergent

## 20.6 REMEMBER THE FOLLOWING LIMITS

(i) $\lim_{n \to \infty} x^n = 0$ if $x < 1$ and $\lim_{n \to \infty} x^n = \infty$ if $x > 1$

(ii) $\lim_{n \to \infty} \dfrac{x^n}{n!} = 0$ for all values of $x$ (iii) $\lim_{n \to \infty} \dfrac{\log n}{n} = 0$

(iv) $\lim_{n \to \infty} \left(1 + \dfrac{1}{n}\right)^n = e$ (v) $\lim_{n \to \infty} (n)^{1/n} = 1$

(vi) $\lim_{n \to \infty} [n!]^{1/n} = \infty$ (vii) $\lim_{n \to \infty} \left[\dfrac{(n!)}{n}\right]^{1/n} = \dfrac{1}{e}$

(viii) $\lim_{n \to \infty} n\, x^n = 0$ if $x < 1$ (ix) $\lim_{n \to \infty} n^h = \infty$

(x) $\lim_{n \to \infty} \dfrac{1}{n^h} = 0$

(xi) $\lim_{x \to \infty} \left[\dfrac{a^x - 1}{x}\right] = \log a$ or $\lim_{n \to \infty} \dfrac{a^{1/n} - 1}{1/n} = \log a$

(xii) $\lim_{x \to 0} \dfrac{\sin x}{x} = 1$ (xiii) $\lim_{x \to 0} \dfrac{\tan x}{x} = 1$

## 20.7 SERIES

A *series* is the sum of a sequence.

Let $u_1, u_2, u_3, \ldots\ldots, u_n, \ldots\ldots$ be a given sequence. Then, the expression
$u_1 + u_2 + u_3 + \ldots\ldots + u_n + \ldots\ldots$ is called the series associated with the given sequence.
*For example,* $1 + 3 + 5 + 7 + \ldots$ is a series.

If the number of terms of a series is limited, the series is called *finite.* When the number of terms of a series are unlimited, it is called an *infinite series.*

$$u_1 + u_2 + u_3 + u_4 + \ldots + u_n + \ldots \infty$$

is called an infinite series and it is denoted by $\sum_{n=1}^{\infty} u_n$ or $\Sigma\, u_n$. The sum of the first $n$ terms of a series is denoted by $S_n$.

## 20.8 CONVERGENT, DIVERGENT AND OSCILLATORY SERIES

Consider the infinite series $\Sigma\, u_n = u_1 + u_2 + u_3 + \ldots + u_n + \ldots \infty$

$$S_n = u_1 + u_2 + u_3 + \ldots + u_n$$

Three cases arise:

(i) If $S_n$ tends to a finite number as $n \to \infty$, the series $\Sigma\, u_n$ is said to be *convergent.*

(ii) If $S_n$ tends to infinity as $n \to \infty$, the series $\Sigma\, u_n$ is said to be *divergent.*

(iii) If $S_n$ does not tend to a unique limit, finite or infinite, the series $\Sigma\, u_n$ is called *oscillatory.*

## 20.9 PROPERTIES OF INFINITE SERIES

1. The nature of an infinite series does not change:
   (*i*) by multiplication of all terms by a constant $k$.
   (*ii*) by addition or deletion of a finite number of terms.
2. If two series $\Sigma\, u_n$ and $\Sigma\, v_n$ are convergent, then $\Sigma\,(u_n + v_n)$ is also convergent.

**Example 1.** *Examine the nature of the series* $1 + 2 + 3 + 4 + \ldots + n + \ldots \infty$.

**Solution.** Let $$S_n = 1 + 2 + 3 + 4 + \ldots + n = \frac{n(n+1)}{2} \qquad \text{[Series in A.P.]}$$

Since $$\lim_{n \to \infty} S_n = \lim_{n \to \infty} \frac{n(n+1)}{2} \Rightarrow \infty$$

Hence, this series is divergent. **Ans.**

**Example 2.** *Test the convergence of the series* $1 + \frac{1}{2} + \frac{1}{4} + \frac{1}{8} + \ldots \infty$

**Solution.** Let $$S_n = 1 + \frac{1}{2} + \frac{1}{4} + \frac{1}{8} + \ldots \infty \qquad \text{[Series in G.P.]}$$

$$= \frac{1}{1 - \frac{1}{2}} = 2 \qquad \left(S_n = \frac{a}{1-r}\right)$$

$$\lim_{n \to \infty} S_n = 2$$

Hence, the series is convergent. **Ans.**

**Example 3.** *Prove that the following series:*

$\frac{2}{3!} + \frac{3}{4!} + \frac{4}{5!} + \ldots$ *is convergent and find its sum.* (*M.U. 2008*)

**Solution.** Here, $$u_n = \frac{n+1}{(n+2)!} = \frac{n+2-1}{(n+2)!} = \frac{n+2}{(n+2)!} - \frac{1}{(n+2)!}$$

$$= \frac{1}{(n+1)!} - \frac{1}{(n+2)!}$$

$$S_n = \left(\frac{1}{2!} - \frac{1}{3!}\right) + \left(\frac{1}{3!} - \frac{1}{4!}\right) + \left(\frac{1}{4!} - \frac{1}{5!}\right) + \ldots$$

$$+ \left(\frac{1}{(n+1)!} - \frac{1}{(n+2)!}\right) = \frac{1}{2!} - \frac{1}{(n+2)!}$$

$$\lim_{n \to \infty} S_n = \lim_{n \to \infty} \left[\frac{1}{2!} - \frac{1}{(n+2)!}\right] = \frac{1}{2}$$

$\therefore$ $\Sigma\, u_n$ converges and its limit is $\frac{1}{2}$. **Ans.**

**Example 4.** *Discuss the nature of the series* $2 - 2 + 2 - 2 + 2 - \ldots \infty$.

**Solution.** Let $$S_n = 2 - 2 + 2 - 2 + 2 - \ldots \infty$$
$$= 0 \text{ if } n \text{ is even}$$
$$= 2 \text{ if } n \text{ is odd.}$$

Hence, $S_n$ does not tend to a unique limit, and, therefore, the given series is oscillatory.

**Ans.**

## EXERCISE 20.2

**Discuss the nature of the following series:**

**1.** $1 + 4 + 7 + 10 + \dots \infty$ **Ans.** Divergent

**2.** $1 + \frac{5}{4} + \frac{6}{4} + \frac{7}{4} + \dots \infty$ **Ans.** Divergent

**3.** $6 - 5 - 1 + 6 - 5 - 1 + 6 - 5 - 1 + \dots \infty$ **Ans.** Oscillatory

**4.** $3 + \frac{3}{2} + \frac{3}{2^2} + \dots \infty$ **Ans.** Convergent

**5.** $1^2 + 2^2 + 3^2 + 4^2 + \dots \infty$ **Ans.** Divergent

**6.** $1 - \frac{1}{2} + \frac{1}{4} - \frac{1}{8} + \dots \infty$ **Ans.** Convergent

**7.** $\frac{1}{1.3} + \frac{1}{3.5} + \frac{1}{5.7} + \dots \infty$ **Ans.** Convergent

**8.** $\frac{1}{2!} + \frac{2}{3!} + \frac{3}{4!} + \dots \infty$ **Ans.** Convergent

**9.** $\log 3 + \log \frac{4}{3} + \log \frac{5}{4} + \dots \infty$ **Ans.** Divergent

**10.** $\sum \log \frac{n}{n+1}$ **Ans.** Divergent

**11.** $\Sigma (\sqrt{n+1} - \sqrt{n})$ **Ans.** Divergent

**12.** $\sum \frac{1}{n(n+2)}$ **Ans.** Convergent

**13.** $\sum \frac{1}{n(n+1)(n+2)(n+3)}$ **Ans.** Convergent

**14.** $\sum \frac{n}{(n+1)(n+2)(n+3)}$ **Ans.** Convergent

**15.** $\sum \frac{2n+1}{n^2(n+1)^2}$ **Ans.** Convergent

## 20.10 PROPERTIES OF GEOMETRIC SERIES

**The series $1 + r + r^2 + r^3 + \dots \infty$ is**
**(*i*) convergent if $|r| < 1$ (*ii*) divergent if $r \geq 1$ (*iii*) oscillatory if $r \leq -1$.**

**Proof.**
$$S_n = 1 + r + r^2 + \dots + r^{n-1} = \frac{1 - r^n}{1 - r}$$

(*i*) When $|r| < 1$,
$$\lim_{n \to \infty} r^n = 0$$
$$\lim_{n \to \infty} S_n = \lim_{n \to \infty} \frac{1 - r^n}{1 - r} = \frac{1 - 0}{1 - r} = \frac{1}{1 - r}$$

Hence, the series is convergent.

(*ii*) (*a*) When $r > 1$,
$$\lim_{n \to \infty} r^n = \infty \quad \therefore \quad \lim_{n \to \infty} S_n = \lim_{n \to \infty} \frac{r^n - 1}{r - 1} \Rightarrow \infty$$

Hence, the series is divergent.

(*b*) When $r = 1$, the series becomes $1 + 1 + 1 + 1 + \dots \infty$
$$S_n = 1 + 1 + 1 + 1 + \dots = n$$
$$\lim_{n \to \infty} S_n = \lim_{n \to \infty} n = \infty$$

Hence, the series is divergent.

(*iii*) (*a*) When $r = -1$, the series becomes $1 - 1 + 1 - 1 + 1 - \dots \infty$
$$S_n = 0 \text{ if } n \text{ is even}$$
$$= 1 \text{ if } n \text{ is odd}$$

Hence, the series is oscillatory.

(*b*) When $r < -1$, let $r = -k$ where $k > 1$.

$$r^n = (-k)^n = (-1)^n k^n$$

$$\lim_{n \to \infty} S_n = \lim_{n \to \infty} \frac{1 - r^n}{1 - r} = \lim_{n \to \infty} \frac{1 - (-1)^n k^n}{1 - (-k)}$$

$$= +\infty \text{ if } n \text{ is odd}$$

$$= -\infty \text{ if } n \text{ is even}$$

Hence, the series is oscillatory. **Proved.**

## EXERCISE 20.3

**Test the nature of the following series :**

**1.** $1 + \frac{1}{2} + \frac{1}{2^2} + \frac{1}{2^3} + \ldots \infty$ **Ans.** Convergent **2.** $1 + \frac{3}{4} + \frac{9}{16} + \frac{27}{64} + \ldots \infty$ **Ans. Convergent**

**3.** $1 - \frac{1}{3} + \frac{1}{9} - \frac{1}{27} + \ldots \infty$ **Ans.** Convergent **4.** $1 - 2 + 4 - 8 + \ldots \infty$ **Ans.** Oscillatory

**5.** $2 + 3 + \frac{9}{2} + \frac{27}{4} + \ldots \infty$ **Ans.** Divergent **6.** $1 + \frac{4}{3} + \left(\frac{4}{3}\right)^2 + \left(\frac{4}{3}\right)^3 + \ldots \infty$ **Ans. Divergent**

**7.** State, which one of the alternatives in the following is correct:
The series $1 - 1 + 1 - 1 + \ldots$ is

(*i*) Convergent with its sum equal to 0. (*ii*) Convergent with its sum equal to 1.
(*iii*) Divergent. (*iv*) Oscillatory. **Ans.** Oscillatory series

## 20.11 POSITIVE TERM SERIES

If all terms after few negative terms in an infinite series are positive, such a series is a positive term series.

*e.g.*, $-10 - 6 - 1 + 5 + 12 + 20 + \ldots$ is a positive term series.

By omitting the negative terms, the nature of a positive term series remains unchanged.

## 20.12 NECESSARY CONDITIONS FOR CONVERGENT SERIES

For every convergent series $\Sigma u_n$.

$$\lim_{n \to \infty} u_n = 0$$

**Solution.** Let $S_n = u_1 + u_2 + u_3 + \ldots + u_n$

$$\lim_{n \to \infty} S_n = k \qquad \text{(a finite quantity)}$$

Also $\lim_{n \to \infty} S_{n-1} = k$ (a finite quantity)

$$S_n = S_{n-1} + u_n$$

$$u_n = S_n - S_{n-1}$$

$$\lim_{n \to \infty} u_n = \lim_{n \to \infty} [S_n - S_{n-1}] = 0$$

$$\lim_{n \to \infty} u_n = 0$$

**Corollary.** Converse of the above theorem is not true.

*e.g.*, $1 + \frac{1}{\sqrt{2}} + \frac{1}{\sqrt{3}} + \frac{1}{\sqrt{4}} + \ldots + \frac{1}{\sqrt{n}} + \ldots \infty$ is divergent.

$$S_n = 1 + \frac{1}{\sqrt{2}} + \frac{1}{\sqrt{3}} + \frac{1}{\sqrt{4}} + \ldots + \frac{1}{\sqrt{n}}$$

$$> \frac{1}{\sqrt{n}} + \frac{1}{\sqrt{n}} + \frac{1}{\sqrt{n}} + \frac{1}{\sqrt{n}} + \ldots + \frac{1}{\sqrt{n}}$$

$$> \frac{n}{\sqrt{n}} > \sqrt{n}$$

$$\lim_{n\to\infty} S_n = \lim_{n\to\infty} \sqrt{n} = \infty$$

Thus, the series is divergent, although $\lim\limits_{n\to\infty} u_n = \lim\limits_{n\to\infty} \frac{1}{\sqrt{n}} = 0$

So $\lim\limits_{n\to\infty} u_n = 0$ is a necessary condition but not a sufficient condition for convergence.

**Note: 1. Test for divergence**

| |
|---|
| *If* $\lim\limits_{n\to\infty} u_n \neq 0$, *the series* $\Sigma\, u_n$ *must be divergent.* |

**2.** ***To determine the nature of a series we have to find*** $S_n$***. Since it is not possible to find*** $S_n$ ***for every series, we have to device tests for convergence without involving*** $S_n$***.***

## 20.13 CAUCHY'S FUNDAMENTAL TEST FOR DIVERGENCE

If $\lim\limits_{n\to\infty} u_n \neq 0$, the series is divergent.

**Example 5.** *Test for convergence of the series* $1 + \frac{2}{3} + \frac{3}{4} + \frac{4}{5} + \ldots + \frac{n}{n+1} + \ldots \infty$

**Solution.** Here, $\lim\limits_{n\to\infty} u_n = \lim\limits_{n\to\infty} \frac{n}{n+1} = \lim\limits_{n\to\infty} \frac{1}{1+\frac{1}{n}} = 1 \neq 0$

Hence, by **Cauchy's Fundamental Test** for divergence, the series is divergent. **Ans.**

**Example 6.** *Test for convergence the series* $1 + \frac{3}{5} + \frac{8}{10} + \frac{15}{17} + \ldots + \frac{2^n - 1}{2^n + 1} + \ldots \infty$

**Solution.** Here, $\lim\limits_{n\to\infty} u_n = \lim\limits_{n\to\infty} \frac{2^n - 1}{2^n + 1} = \lim\limits_{n\to\infty} \frac{1 - \frac{1}{2^n}}{1 + \frac{1}{2^n}} = 1 \neq 0$

Hence, by **Cauchy's Fundamental Test** for divergence the series is divergent. **Ans.**

**Example 7.** Test the convergence of the following series:

$$\sqrt{\frac{1}{4}} + \sqrt{\frac{2}{6}} + \sqrt{\frac{3}{8}} + \cdots + \sqrt{\frac{n}{2(n+1)}} + \cdots$$ *(M.D.U., 2000)*

**Solution.** Here, we have

$$\sqrt{\frac{1}{4}} + \sqrt{\frac{2}{6}} + \sqrt{\frac{3}{8}} + \cdots + \sqrt{\frac{n}{2(n+1)}} + \cdots$$

$$u_n = \sqrt{\frac{n}{2(n+1)}} = \sqrt{\frac{1}{2\left(1+\frac{1}{n}\right)}}$$

$$\lim_{n\to\infty} u_n = \lim_{n\to\infty} \frac{1}{\sqrt{2\left(1+\frac{1}{n}\right)}} = \frac{1}{\sqrt{2}} \neq 0$$

$\Rightarrow \Sigma\, u_n$ does not converge.

The given series is a series of + ve terms,

Hence by Cauchy fundamental test for divergence, the series is divergent. **Ans.**

## EXERCISE 20.4

**Examine for convergence:**

1. $\frac{1}{\sqrt{2}}+\frac{2}{\sqrt{5}}+\frac{4}{\sqrt{17}}+\ldots+\frac{2^n}{\sqrt{4^n+1}}+\ldots\infty$ **Ans.** Divergent

2. $\sum_{n=1}^{\infty}\frac{n}{n+1}$ **Ans.** Divergent

3. $\sum_{n=1}^{\infty}\sqrt{\frac{n}{n+1}}$ **Ans.** Divergent

4. $\sum \cos\frac{1}{n}$ **Ans.** Divergent

5. $1+\frac{1}{2}+2,+\frac{1}{3}+3+\frac{1}{4}+4+\ldots$ **Ans.** Divergent

6. $\Sigma\,(6-n^2)$ **Ans.** Divergent

7. $\Sigma\,(-2^n)$ **Ans.** Divergent

8. $\Sigma\, 3^{n+1}$ **Ans.** Divergent

## 20.14 *p*-SERIES

*The series* $\frac{1}{1^p}+\frac{1}{2^p}+\frac{1}{3^p}+\ldots\infty$ *is* (*i*) *convergent if* $p>1$ (*ii*) *Divergent if* $p\le 1$.

(*MDU, Dec. 2010*)

**Solution. Case 1: ($p > 1$)**

The given series can be grouped as

$$\frac{1}{1^p}+\left(\frac{1}{2^p}+\frac{1}{3^p}\right)+\left(\frac{1}{4^p}+\frac{1}{5^p}+\frac{1}{6^p}+\frac{1}{7^p}\right)+\left(\frac{1}{8^p}+\frac{1}{9^p}+\frac{1}{10^p}+\frac{1}{11^p}+\frac{1}{12^p}+\frac{1}{13^p}+\frac{1}{14^p}+\frac{1}{15^p}\right)+\ldots$$

Now
$$\frac{1}{1^p}=1 \qquad \ldots(1)$$

$$\frac{1}{2^p}+\frac{1}{3^p}<\frac{1}{2^p}+\frac{1}{2^p}=\frac{2}{2^p} \qquad \ldots(2)$$

$$\frac{1}{4^p}+\frac{1}{5^p}+\frac{1}{6^p}+\frac{1}{7^p}<\frac{1}{4^p}+\frac{1}{4^p}+\frac{1}{4^p}+\frac{1}{4^p}=\frac{4}{4^p} \qquad \ldots(3)$$

$$\frac{1}{8^p}+\frac{1}{9^p}+\ldots+\frac{1}{15^p}<\frac{1}{8^p}+\frac{1}{8^p}+\ldots+\frac{1}{8^p}=\frac{8}{8^p} \qquad \ldots(4)$$

On adding (1), (2), (3) and (4), we get:

$$\frac{1}{1^p}+\left(\frac{1}{2^p}+\frac{1}{3^p}\right)+\left(\frac{1}{4^p}+\frac{1}{5^p}+\frac{1}{6^p}+\frac{1}{7^p}\right)+\left(\frac{1}{8^p}+\frac{1}{9^p}+\ldots+\frac{1}{15^p}\right)+\ldots$$

$$<\frac{1}{1^p}+\frac{2}{2^p}+\frac{4}{4^p}+\frac{8}{8^p}+\ldots$$

$$<1+\left(\frac{1}{2}\right)^{p-1}+\left(\frac{1}{2}\right)^{2p-2}+\left(\frac{1}{2}\right)^{3p-3}+\ldots$$

$$<\frac{1}{1-\left(\frac{1}{2}\right)^{p-1}} \qquad \left[\text{G.P., } r=\left(\frac{1}{2}\right)^{p-1},\ S=\frac{1}{1-r}\right]$$

< Finite number if $p>1$

Hence, the given series is convergent when $p > 1$.

**Case 2: $p = 1$**

When $p = 1$, the given series becomes

$$1 + \frac{1}{2} + \left(\frac{1}{3} + \frac{1}{4}\right) + \left(\frac{1}{5} + \frac{1}{6} + \frac{1}{7} + \frac{1}{8}\right) + \left(\frac{1}{9} + \frac{1}{10} + \ldots + \frac{1}{16}\right) + \ldots$$

$$1 + \frac{1}{2} = 1 + \frac{1}{2} \qquad \ldots(1)$$

$$\frac{1}{3} + \frac{1}{4} > \frac{1}{4} + \frac{1}{4} = \frac{1}{2} \qquad \ldots(2)$$

$$\frac{1}{5} + \frac{1}{6} + \frac{1}{7} + \frac{1}{8} > \frac{1}{8} + \frac{1}{8} + \frac{1}{8} + \frac{1}{8} = \frac{4}{8} = \frac{1}{2} \qquad \ldots(3)$$

$$\frac{1}{9} + \frac{1}{10} + \ldots + \frac{1}{16} > \frac{1}{16} + \frac{1}{16} + \ldots + \frac{1}{16} = \frac{8}{16} = \frac{1}{2} \qquad \ldots(4)$$

On adding (1), (2), (3) and (4), we get

$$1 + \frac{1}{2} + \left(\frac{1}{3} + \frac{1}{4}\right) + \left(\frac{1}{5} + \frac{1}{6} + \frac{1}{7} + \frac{1}{8}\right) + \left(\frac{1}{9} + \frac{1}{10} + \ldots + \frac{1}{16}\right) + \ldots$$

$$> 1 + \frac{1}{2} + \frac{1}{2} + \frac{1}{2} + \frac{1}{2} + \ldots$$

$$> 1 + \frac{n}{2} \quad (n \to \infty)$$

$$> \infty$$

Hence, the given series is divergent when $p = 1$.

**Case 3: $p < 1$**

$$\frac{1}{2^p} > \frac{1}{2}, \qquad \frac{1}{3^p} > \frac{1}{3}, \qquad \frac{1}{4^p} > \frac{1}{4} \text{ and so on}$$

Therefore, $\frac{1}{1^p} + \frac{1}{2^p} + \frac{1}{3^p} + \frac{1}{4^p} + \ldots > 1 + \frac{1}{2} + \frac{1}{3} + \frac{1}{4} + \ldots$

> divergent series ($p = 1$) [From Case 2]

$\left[\text{As the series on R.H.S.} \left(1 + \frac{1}{2} + \frac{1}{3} + \frac{1}{4} + \ldots\right) \text{ is divergent}\right]$

Hence, the given series is divergent when $p < 1$.

## 20.15 COMPARISON TEST

If two positive terms $\Sigma\, u_n$ and $\Sigma\, v_n$ be such that

**$\lim\limits_{n \to \infty} \frac{u_n}{v_n} = k$ (finite number), then both series converge or diverge together.**

**Proof.** By definition of limit there exists a positive number $\varepsilon$, however small, such that

$$\left|\frac{u_n}{v_n} - k\right| < \varepsilon \text{ for } n > m \qquad i.e., \ -\varepsilon < \frac{u_n}{v_n} - k < +\varepsilon$$

$$k - \varepsilon < \frac{u_n}{v_n} < k + \varepsilon \text{ for } n > m$$

Ignoring the first $m$ terms of both series, we have

$$k - \varepsilon < \frac{u_n}{v_n} < k + \varepsilon \text{ for all } n. \quad \text{...(1)}$$

**Case 1.** $\Sigma\, v_n$ is convergent, then

$$\lim_{n \to \infty} (v_1 + v_2 + \dots v_n) = h \text{ (say)} \qquad \text{where } h \text{ is a finite number.}$$

From (1), $u_n < (k + \varepsilon)\, v_n$ for all $n$.

$$\lim_{n \to \infty} (u_1 + u_2 + \dots + u_n) < (k + \varepsilon) \lim_{n \to \infty} (v_1 + v_2 + \dots + v_n) = (k + \varepsilon)h$$

Hence, $\Sigma\, u_n$ is also convergent.

**Case 2.** $\Sigma\, v_n$ is divergent, then

$$\lim_{n \to \infty} (v_1 + v_2 + \dots + v_n) \to \infty \quad \text{...(2)}$$

Now from (1) $\qquad k - \varepsilon < \dfrac{u_n}{v_n}$

$$u_n > (k - \varepsilon) v_n \text{ for all } n$$

$$\lim_{n \to \infty} (u_1 + u_2 + \dots + u_n) > (k - \varepsilon) \lim_{n \to \infty} (v_1 + v_2 + \dots + v_n)$$

From (2), $\lim_{n \to \infty} (u_1 + u_2 + \dots + u_n) \to \infty$

Hence, $\Sigma\, u_n$ is also divergent.

**Note.** For testing the convergence of a series, this Comparison Test is very useful. We choose $\Sigma\, v_n$ ($p$-series) in such a way that

$$\lim_{n \to \infty} \frac{u_n}{v_n} = \text{finite number.}$$

Then the nature of both the series is the same. The nature of $\Sigma\, v_n$ ($p$-series) is already known, so the nature of $\Sigma\, u_n$ is also known.

**Example 8.** *Test the series* $\displaystyle\sum_{n=1}^{\infty} \frac{1}{n + 10}$ *for convergence or divergence.*

**Solution.** Here, $\qquad u_n = \dfrac{1}{n + 10}$

Let $\qquad v_n = \dfrac{1}{n}$

$$\lim_{n \to \infty} \frac{u_n}{v_n} = \lim_{n \to \infty} \frac{n}{n + 10} = \lim_{n \to \infty} \frac{1}{1 + \frac{10}{n}} = 1 = \text{finite number.}$$

According to Comparison Test both series converge or diverge together, but $\Sigma\, v_n$ is divergent as $p = 1$.

$\therefore \quad \Sigma\, u_n$ is also divergent. **Ans.**

**Example 9.** *Test the convergence of the following series:*

$$\frac{1}{\sqrt{1} + \sqrt{2}} + \frac{1}{\sqrt{2} + \sqrt{3}} + \frac{1}{\sqrt{3} + \sqrt{4}} + \cdots \qquad \textit{(M.D.U., 2000)}$$

**Solution.** Here, we have

$$\frac{1}{\sqrt{1} + \sqrt{2}} + \frac{1}{\sqrt{2} + \sqrt{3}} + \frac{1}{\sqrt{3} + \sqrt{4}} + \cdots$$

$$u_n = \frac{1}{\sqrt{n}+\sqrt{n+1}} = \frac{1}{\sqrt{n}\left[1+\sqrt{1+\frac{1}{n}}\right]}$$

Let us compare $\Sigma\, u_n$ with $\Sigma\, v_n$, where

$$v_n = \frac{1}{\sqrt{n}}$$

$$\frac{u_n}{v_n} = \frac{1}{\sqrt{n}\left[1+\sqrt{1+\frac{1}{n}}\right]} \cdot \frac{\sqrt{n}}{1} = \frac{1}{1+\sqrt{1+\frac{1}{n}}}$$

$$\lim_{n\to\infty} \frac{u_n}{v_n} = \lim_{n\to\infty} \frac{1}{1+\sqrt{1+\frac{1}{n}}} = \frac{1}{1+1} = \frac{1}{2}$$

Which is finite and non-zero.

$\therefore\ \Sigma\, u_n$ and $\Sigma\, v_n$, converge or diverge together since $\Sigma\, v_n, = \sum \frac{1}{n^{\frac{1}{2}}}$ is of the form $\sum \frac{1}{n^p}$.

$$P = \frac{1}{2} < 1$$

$\therefore\ \Sigma\, v_n$ is divergent $\Rightarrow \Sigma\, u_n$ is also divergent. **Ans.**

**Example 10.** *Examine the convergence of the series:*

$$\sum\left(\sqrt[3]{n^3+1}-n\right)$$ *(M.D.U. 2003)*

**Solution.** Here, we have $\sum\left(\sqrt[3]{n^3+1}-n\right)$

$$u_n = (n^3+1)^{\frac{1}{3}} - n = \left[n^3\left(1+\frac{1}{n^3}\right)\right]^{\frac{1}{3}} - n$$

$$= n\left[\left(1+\frac{1}{n^3}\right)^{\frac{1}{3}} - 1\right] = n\left[1+\frac{1}{3}\cdot\frac{1}{n^3}+\frac{\frac{1}{3}\left(\frac{1}{3}-1\right)}{2!}\left(\frac{1}{n^3}\right)^2+\ldots-1\right]$$

$$= \frac{n}{n^3}\left[\frac{1}{3}-\frac{1}{9}\cdot\frac{1}{n^3}\ldots\right] = \frac{1}{n^2}\left[\frac{1}{3}-\frac{1}{9}\cdot\frac{1}{n^3}\ldots\right]$$

Let $$v_n = \frac{1}{n^2}$$

$$\frac{u_n}{v_n} = \frac{1}{n^2}\left[\frac{1}{3}-\frac{1}{9}\cdot\frac{1}{n^3}\ldots\right]\cdot\frac{n^2}{1} = \left[\frac{1}{3}-\frac{1}{9}\cdot\frac{1}{n^3}\ldots\right]$$

$$\Rightarrow \quad \lim_{n\to\infty}\frac{u_n}{v_n} = \lim_{n\to\infty}\left(\frac{1}{3}-\frac{1}{9}\cdot\frac{1}{n^3}\ldots\right) = \frac{1}{3}$$

which is finite and non-zero.

$\therefore\ \sum u_n$ and $\sum v_n$ converge or diverge together.

Since $\sum v_n = \sum \frac{1}{n^2}$ is of the form $\sum \frac{1}{n^p}$ with $p = 2 > 1$

$\therefore \quad \sum v_n$ is convergent $\Rightarrow \sum u_n$ is convergent. **Ans.**

**Example 11.** *Test the convergence of the following series*

$$\frac{1}{1+2^{-1}}+\frac{2}{1+2^{-2}}+\frac{3}{1+2^{-3}}+\cdots$$ *(M.D.U., 2000)*

**Solution.** Here, we have

$$\frac{1}{1+2^{-1}}+\frac{2}{1+2^{-2}}+\frac{3}{1+2^{-3}}+\cdots$$

Here $$u_n = \frac{n}{1+2^{-n}} = \frac{n}{1+\frac{1}{2^n}}$$

Let $$v_n = n$$

Let us compare $\sum u_n$ with $\sum v_n$,

$$\frac{u_n}{v_n} = \frac{n}{1+\frac{1}{2^n}}\cdot\frac{1}{n} = \frac{1}{1+\frac{1}{2^n}}$$

$$\lim_{n\to\infty}\frac{u_n}{v_n} = \lim_{n\to\infty}\frac{1}{1+\frac{1}{2^n}} = \frac{1}{1+0} = 1$$

Which is finite and non-zero.

$\therefore \sum u_n$ and $\sum v_n$, converge or diverge together since $\sum v_n = \sum \frac{1}{n}$ is of the form $\sum \frac{1}{n^p}$ with $p = 1$.

$\therefore \sum v_n$ is divergent $\Rightarrow \sum u_n$ is also divergent. **Ans.**

**Example 12.** *Examine the convergence of the series* $\frac{\sqrt{2}-1}{3^3-1}+\frac{\sqrt{3}-1}{4^3-1}+\frac{\sqrt{4}-1}{5^3-1}+\cdots$

*(M.D.U., 2000)*

**Solution.** Here, we have

$$\frac{\sqrt{2}-1}{3^3-1}+\frac{\sqrt{3}-1}{4^3-1}+\frac{\sqrt{4}-1}{5^3-1}+\cdots$$

Here $$u_n = \frac{\sqrt{n+1}-1}{(n+2)^3-1} = \frac{\sqrt{n}\left(\sqrt{1+\frac{1}{n}}-\frac{1}{\sqrt{n}}\right)}{n^3\left[\left(1+\frac{2}{n}\right)^3-\frac{1}{n^3}\right]} = \frac{\sqrt{1+\frac{1}{n}}-\frac{1}{\sqrt{n}}}{n^{\frac{5}{2}}\left[\left(1+\frac{2}{n}\right)^3-\frac{1}{n^3}\right]}$$

Let $$v_n = \frac{1}{n^{\frac{5}{2}}}$$

Let us compare $\sum u_n$ with $\sum v_n$,

$$\frac{u_n}{v_n} = \frac{\sqrt{1+\frac{1}{n}}-\frac{1}{\sqrt{n}}}{n^{\frac{5}{2}}\left[\left(1+\frac{2}{n}\right)^3-\frac{1}{n^3}\right]}\times\frac{n^{\frac{5}{2}}}{1} = \frac{\sqrt{1+\frac{1}{n}}-\frac{1}{\sqrt{n}}}{\left[\left(1+\frac{2}{n}\right)^3-\frac{1}{n^3}\right]}$$

$$\lim_{n\to\infty}\frac{u_n}{v_n} = \lim_{n\to\infty}\frac{\sqrt{1+\frac{1}{n}}-\frac{1}{\sqrt{n}}}{\left[\left(1+\frac{2}{n}\right)^3-\frac{1}{n^3}\right]} = \frac{\sqrt{1+0}-0}{(1-0)^3-0} = 1$$

Which is finite and non-zero.

$\therefore \sum u_n$ and $\sum v_n$, converge or diverge together since $\sum v_n = \sum \frac{1}{n^{\frac{5}{2}}}$ is of the form

$\sum \frac{1}{n^p}$. where $p = \frac{5}{2} > 1$.

$\therefore \sum v_n$ is convergent $\Rightarrow \sum u_n$ is convergent. **Ans.**

**Example 13.** *Test the convergence and divergence of the following series.*

$$\sum_{n=1}^{\infty}\frac{2n^2+3n}{5+n^5}$$ *(Gujarat, I Semester, Jan. 2009)*

**Solution.** Here, $$u_n = \frac{2n^2+3n}{5+n^5} = \frac{n^2\left(2+\frac{3}{n}\right)}{n^5\left(\frac{5}{n^5}+1\right)} = \frac{1}{n^3}\,\frac{2+\frac{3}{n}}{\frac{5}{n^5}+1}$$

Let $$v_n = \frac{1}{n^3}$$

**By Comparison Test**

$$\lim_{n\to\infty}\frac{u_n}{v_n} = \lim_{n\to\infty}\frac{n^3\left(2+\frac{3}{n}\right)}{n^3\left(\frac{5}{n^5}+1\right)} = \lim_{n\to\infty}\frac{2+\frac{3}{n}}{\frac{5}{n^5}+1} = 2 = \text{Finite number.}$$

According to comparison test both series converge or diverge together but $\Sigma v_n$ is convergent as $p = 2$.

Hence, the given series is convergent. **Ans.**

**Example 14.** *Test the following series for convergence* $\frac{2}{1^p}+\frac{3}{2^p}+\frac{4}{3^p}+\frac{5}{4^p}+\ldots$

**Solution.** Given series is $\frac{2}{1^p}+\frac{3}{2^p}+\frac{4}{3^p}+\frac{5}{4^p}+\ldots$

Here $$u_n = \frac{n+1}{n^p} = \frac{1+\frac{1}{n}}{n^{p-1}}.$$

Let $$v_n = \frac{1}{n^{p-1}} \quad \therefore \quad \frac{u_n}{v_n} = \frac{1+\frac{1}{n}}{n^{p-1}}\times\frac{n^{p-1}}{1} = 1+\frac{1}{n}$$

$$\lim_{n\to\infty}\frac{u_n}{v_n} = 1$$

Therefore, both the series are either convergent or divergent.

But $\Sigma v_n$ is convergent if $p-1>1$, *i.e.*, if $p>2$ (P series)

and is divergent if $p-1\le 1$, *i.e.*, if $p\le 2$

$\therefore$ The given series is convergent if $p>2$ and divergent if $p\le 2$. **Ans.**

## EXERCISE 20.5

**Examine the convergence or divergence of the following series:**

1. $2 + \frac{3}{2}.\frac{1}{4} + \frac{4}{3}.\frac{1}{4^2} + \frac{5}{4}.\frac{1}{4^3} + \ldots \infty$ **Ans.** Convergent

2. $1 + \frac{1.2}{1.3} + \frac{1.2.3}{1.3.5} + \frac{1.2.3.4}{1.3.5.7} + \ldots \infty$ **Ans.** Convergent

3. $\frac{1}{1.2} + \frac{2}{3.4} + \frac{3}{5.6} + \ldots \infty$ **Ans.** Divergent

4. $\frac{1}{1.2.3} + \frac{1}{2.3.4} + \frac{1}{3.4.5} + \ldots\ldots\ldots \infty$ **Ans.** Convergent (*M.D. University, Dec. 2004*)

5. $1 + \frac{2^2}{2!} + \frac{3^2}{3!} + \frac{4^2}{4!} + \ldots \infty$ **Ans.** Convergent

6. $\frac{1}{1+2} + \frac{2}{1+2^2} + \frac{3}{1+2^3} + \ldots$ **Ans.** Convergent (*M.D. University, 2001*)

7. $\frac{1}{3} + \frac{2!}{3^2} + \frac{3!}{3^3} + \ldots \infty$ **Ans.** Convergent

8. $\sum_{n=1}^{\infty} \frac{1}{\sqrt{n} + \sqrt{n+1}}$ **Ans.** Divergent

9. $\sum_{n=1}^{\infty} \frac{2n^3 + 5}{4n^5 + 1}$ **Ans.** Convergent

10. $\sum_{n=1}^{\infty} \frac{a^n}{x^n + n^a}$ **Ans.** If $x > a$, convergent; if $x \le a$, Divergent

11. $\sum_{n=1}^{\infty} \frac{\sqrt{n}}{n^2 + 1}$ **Ans.** Convergent

12. $\sum_{n=1}^{\infty} \sqrt{(n^2 + 1)} - n$ **Ans.** Divergent

13. $\sum_{n=1}^{\infty} \left[\sqrt{(n^4 + 1)} - \sqrt{(n^4 - 1)}\right]$ **Ans.** Convergent

14. $\sum_{n=1}^{\infty} \frac{2^n + 1}{3^n + n}$ **Ans.** Convergent

15. $\sum_{n=1}^{\infty} \frac{n^n}{n!}$ **Ans.** Convergent

16. $\sum_{n=1}^{\infty} \frac{n^2}{e^n}$ **Ans.** Convergent

## 20.16 D'ALEMBERT'S RATIO TEST

**Statement.** ***If $\Sigma u_n$ is a positive term series such that $\lim_{n \to \infty} \frac{u_{n+1}}{u_n} = k$ then***

***(i) the series is convergent if $k < 1$ (ii) the series is divergent if $k > 1$***

**Solution.**

**Case I.** When $\lim_{n \to \infty} \frac{u_{n+1}}{u_n} = k < 1$

By definition of a limit, we can find a number $r$ $(< 1)$ such that

$$\frac{u_{n+1}}{u_n} < r \text{ for all } n \ge m \qquad \left[\frac{u_2}{u_1} < r, \frac{u_3}{u_2} < r, \frac{u_4}{u_3} < r \ldots\right]$$

Omitting the first $m$ terms, let the series be

$$u_1 + u_2 + u_3 + u_4 + \ldots \infty$$

$$= u_1\left(1 + \frac{u_2}{u_1} + \frac{u_3}{u_1} + \frac{u_4}{u_1} + \ldots\right) = u_1\left(1 + \frac{u_2}{u_1} + \frac{u_3}{u_2}.\frac{u_2}{u_1} + \frac{u_4}{u_3}.\frac{u_3}{u_2}.\frac{u_2}{u_1} + \ldots \infty\right)$$

$$< u_1 (1 + r + r^2 + r^3 + \ldots \infty) \qquad (r < 1)$$

$$= \frac{u_1}{1-r}, \text{ which is a finite quantity.}$$

Hence, $\Sigma\, u_n$ is convergent.

**Case 2.** When $\lim\limits_{n\to\infty} \frac{u_{n+1}}{u_n} = k > 1$

By definition of limit, we can find a number $m$ such that $\frac{u_{n+1}}{u_n} \geq 1$ for all $n \geq m$

$$\frac{u_2}{u_1} > 1, \qquad \frac{u_3}{u_2} > 1, \qquad \frac{u_4}{u_3} > 1$$

Ignoring the first $m$ terms, let the series be

$$u_1 + u_2 + u_3 + u_4 + \ldots \infty$$

$$= u_1\left(1 + \frac{u_2}{u_1} + \frac{u_3}{u_1} + \frac{u_4}{u_1} + \ldots\right) = u_1\left(1 + \frac{u_2}{u_1} + \frac{u_3}{u_2}\cdot\frac{u_2}{u_1} + \frac{u_4}{u_3}\cdot\frac{u_3}{u_2}\cdot\frac{u_2}{u_1} + \ldots \infty\right)$$

$$\geq u_1\,(1 + 1 + 1 + 1 \ldots \text{ to } n \text{ terms}) = nu_1$$

$$\left[\because \lim_{n\to\infty} (u_1 + u_2 + \ldots + u_n) = nu_1\right]$$

$$\lim_{n\to\infty} S_n \geq \lim_{n\to\infty} nu_1 = \infty$$

Hence, $\Sigma\, u_n$ is divergent.

**Note.** When $\frac{u_{n+1}}{u_n} = 1 \quad (k = 1)$

The ratio test fails.

***For Example.*** Consider the series whose $n^{\text{th}}$ term $= \frac{1}{n}$

$$\lim_{n\to\infty} \frac{u_{n+1}}{u_n} = \lim_{n\to\infty} \frac{\frac{1}{n+1}}{\frac{1}{n}} = \lim_{n\to\infty} \frac{n}{n+1} = \lim_{n\to\infty} \frac{1}{1+\frac{1}{n}} = 1 \qquad \ldots(1)$$

Consider the second series whose $n^{\text{th}}$ term is $\frac{1}{n^2}$.

$$\lim_{n\to\infty} \frac{u_{n+1}}{u_n} = \lim_{n\to\infty} \frac{\frac{1}{(n+1)^2}}{\frac{1}{n^2}} = \lim_{n\to\infty} \left(\frac{n}{n+1}\right)^2 = 1 \qquad \ldots(2)$$

Thus, from (1) and (2) in both cases $\lim\limits_{n\to\infty} \frac{u_{n+1}}{u_n} = 1$

But we know that the first series is divergent as $p = 1$.

The second series is convergent as $p = 2$.

Hence, when $\lim\limits_{n\to\infty} \frac{u_{n+1}}{u_n} = 1$, the series may be convergent or divergent.

Thus, ratio test fails when $k = 1$.

**Example 15.** *Test for convergence of the series whose $n^{th}$ term is* $\frac{n^2}{2^n}$.

**Solution.** Here, we have $u_n = \frac{n^2}{2^n}, \qquad u_{n+1} = \frac{(n+1)^2}{2^{n+1}}$

**By D'Alembert's Test**

$$\lim_{n\to\infty} \frac{u_{n+1}}{u_n} = \lim_{n\to\infty} \frac{(n+1)^2}{2^{n+1}}\cdot\frac{2^n}{n^2} = \lim_{n\to\infty} \frac{1}{2}\left(1 + \frac{1}{n}\right)^2 = \frac{1}{2} < 1$$

Hence, the series is convergent by D'Alembert's Ratio Test. **Ans.**

**Example 16.** *Test for convergence the series whose $n^{th}$ term is* $\dfrac{2^n}{n^3}$.

**Solution.** Here, we have $u_n = \dfrac{2^n}{n^3}$, $\quad u_{n+1} = \dfrac{2^{n+1}}{(n+1)^3}$

**By D'Alembert's Ratio Test**

$$\frac{u_{n+1}}{u_n} = \frac{2^{n+1}}{(n+1)^3}\cdot\frac{n^3}{2^n} = \frac{2}{\left(1+\dfrac{1}{n}\right)^3} \Rightarrow \lim_{n\to\infty}\frac{u_{n+1}}{u_n} = \lim_{n\to\infty}\frac{2}{\left(1+\dfrac{1}{n}\right)^3} = 2 > 1$$

Hence, the series is divergent. **Ans.**

**Example 17.** *Discuss the convergence of the series*:

$$\sum\frac{\sqrt{n}}{\sqrt{n^2+1}}x^n. \quad (x > 0)$$ (*M.D.University, Dec., 2001*)

**Solution.** Here, we have

$$u_n = \sqrt{\frac{n}{n^2+1}}\,x^n$$

$$\therefore \quad u_{n+1} = \sqrt{\frac{n+1}{(n+1)^2+1}}.x^{n+1}$$

$$\frac{u_n}{u_{n+1}} = \sqrt{\frac{n}{n+1}}.\sqrt{\frac{n^2+2n+2}{n^2+1}}.\frac{1}{x} = \sqrt{\frac{1}{1+\dfrac{1}{n}}.\frac{\left(1+\dfrac{2}{n}+\dfrac{2}{n^2}\right)}{\left(1+\dfrac{1}{n^2}\right)}}.\frac{1}{x}$$

$$\lim_{n\to\infty}\frac{u_n}{u_{n+1}} = \lim_{n\to\infty}\sqrt{\frac{1}{1+\dfrac{1}{n}}.\frac{\left(1+\dfrac{2}{n}+\dfrac{2}{n^2}\right)}{\left(1+\dfrac{1}{n^2}\right)}}.\frac{1}{x} = \frac{1}{x}$$

$\therefore$ By D' Alembert's Ratio Test, $\sum u_n$ converges if $\dfrac{1}{x} > 1$, *i.e.* $x < 1$ and diverges if $\dfrac{1}{x} < 1$ *i.e.*, $x > 1$.

When $x = 1$, the Ratio Test fails.

When $x = 1$, $u_n = \sqrt{\dfrac{n}{n^2+1}} = \sqrt{\dfrac{n}{n^2\left(1+\dfrac{1}{n^2}\right)}} = \dfrac{1}{\sqrt{n}}.\dfrac{1}{\sqrt{1+\dfrac{1}{n^2}}}$

$$v_n = \frac{1}{\sqrt{n}},$$

$$\frac{u_n}{v_n} = \frac{1}{\sqrt{n}}.\frac{1}{\sqrt{1+\dfrac{1}{n^2}}}.\frac{\sqrt{n}}{1} = \frac{1}{\sqrt{1+\dfrac{1}{n^2}}}$$

$$\lim_{n\to\infty}\frac{u_n}{v_n} = \lim_{n\to\infty}\frac{1}{\sqrt{1+\dfrac{1}{n^2}}} = 1$$

Which is finite and non-zero.

$\therefore$ By comparison test, $\sum u_n$ and $\sum v_n$ converge or diverge together.

Since $\sum v_n = \sum \frac{1}{\sqrt{n}}$ is of the form $\sum \frac{1}{n^p}$ with $p = \frac{1}{2} < 1$.

$\sum v_n$ diverges $\Rightarrow \sum u_n$ diverges.

Hence, the given series $\sum u_n$ converges if $x < 1$ and diverges if $x \geq 1$. **Ans.**

## EXERCISE 20.6

**Test the convergence for series:**

1. $\sum_{n=1}^{n} \frac{n^2}{3^n}$ **Ans.** Convergent
2. $\sum_{n=1}^{n} \frac{n!}{n^n}$ **Ans.** Convergent
3. $\left(\frac{1}{3}\right)^2 + \left(\frac{1.2}{3.5}\right)^2 + \left(\frac{1.2.3}{3.5.7}\right)^2 + \ldots \infty$ **Ans.** Convergent
4. $\frac{2}{1} + \frac{2.5.8}{1.5.9} + \frac{2.5.8.11}{1.5.9.13} + \ldots \infty$ **Ans.** Convergent
5. $\sum_{n=1}^{n} \frac{n!.2^n}{n^n}$ **Ans.** Convergent
6. $\sum_{n=1}^{\infty} \frac{x^{n-1}}{n.3^n}$ **Ans.** Convergent if $x > 3$, Divergent if $x < 3$
7. Prove that, if $u_{n+1} = \frac{k}{1 + u_n}$, where $k > 0$, $u_1 > 0$, then the series $\Sigma\, u_n$ converges to the positive root of the equation $x^2 + x = k$.

## 20.17 RAABE'S TEST (HIGHER RATIO TEST)

***If $\Sigma\, u_n$ is a positive term series such that $\lim_{n \to \infty} n\left(\frac{u_n}{u_{n+1}} - 1\right) = k$, then***

***(i) the series is convergent if $k > 1$ (ii) the series is divergent if $k < 1$.***

**Proof. Case I.** $k > 1$

Let $p$ be such that $k > p > 1$ and compare the given series $\Sigma\, u_n$ with $\sum \frac{1}{n^p}$ which is convergent as $p > 1$.

$$\frac{u_n}{u_{n+1}} > \frac{(n+1)^p}{n^p} \quad \text{or} \quad \left(\frac{u_n}{u_{n+1}}\right) > \left(1 + \frac{1}{n}\right)^p > 1 + \frac{p}{n} + \frac{p(p-1)}{2!}\frac{1}{n^2} + \ldots$$

(Binomial Theorem)

$$n\left(\frac{u_n}{u_{n+1}} - 1\right) > p + \frac{p\,(p-1)}{2!}\frac{1}{n} + \ldots.$$

If $$\lim_{n \to \infty} n\left(\frac{u_n}{u_{n+1}} - 1\right) > p$$

and $k > p$ which is true as $k > p > 1$; $\Sigma u_n$ is convergent when $k > 1$.

**Case II.** $k < 1$ Same steps as in Case I.

**Notes:**

1. Raabe's Test fails if $k = 1$
2. Raabe's Test is applied only when D'Alembert's Ratio Test fails.

**Example 18.** *Test the convergence for the series* $\frac{x}{1.2}+\frac{x^2}{3.4}+\frac{x^3}{5.6}+\frac{x^4}{7.8}+\ldots$ (*M.U. 2009*)

**Solution.** Here, $u_n = \frac{x^n}{(2n-1)\,2n}$ and $u_{n+1} = \frac{x^{n+1}}{(2n+1)\,(2n+2)}$

**By D'Alembert's Test**

$$\frac{u_{n+1}}{u_n} = \frac{x^{n+1}}{(2n+1)\,(2n+2)} \times \frac{(2n-1)2n}{x^n} = \frac{x\left(1-\frac{1}{2n}\right)}{\left(1+\frac{1}{2n}\right)\left(1+\frac{2}{2n}\right)}$$

$$\lim_{n\to\infty} \frac{u_{n+1}}{u_n} = x$$

(*i*) If $x < 1$, $\Sigma\, u_n$ is convergent (*ii*) If $x > 1$, $\Sigma\, u_n$ is divergent (*iii*) If $x = 1$, Test fails.
Let us apply **Raabe's Test** when $x = 1$

$$\lim_{n\to\infty} n\left(\frac{u_n}{u_{n+1}}-1\right) = \lim_{n\to\infty} n\left[\frac{(2n+1)(2n+2)}{2n(2n-1)}-1\right]$$

$$= \lim_{n\to\infty} n\left[\frac{(2n+1)(2n+2)-2n(2n-1)}{2n\,(2n-1)}\right]$$

$$= \lim_{n\to\infty} n\left[\frac{(8n+2)}{(2n)\,(2n-1)}\right] = \lim_{n\to\infty} \frac{\left(2+\frac{2}{4n}\right)}{1\left(1-\frac{1}{2n}\right)} = 2$$

So the series is convergent.
Hence we can say that the given series is convergent if $x \le 1$ and divergent, if $x > 1$. **Ans.**

**Example 19.** *Test the following series for convergence* $\sum \frac{1}{\sqrt{n+1}-1}$.

**Solution.** Here, $u_n = \frac{1}{\sqrt{n+1}-1}$, $u_{n+1} = \frac{1}{\sqrt{n+2}-1}$

$$\frac{u_{n+1}}{u_n} = \frac{\sqrt{n+1}-1}{\sqrt{n+2}-1}$$

$$\lim_{n\to\infty} \frac{u_{n+1}}{u_n} = \lim_{n\to\infty} \frac{\sqrt{1+\frac{1}{n}}-\frac{1}{\sqrt{n}}}{\sqrt{1+\frac{2}{n}}-\frac{1}{\sqrt{n}}} = 1$$

D'Alembert's test fails.
**By Raabe's test.**

$$\lim_{n\to\infty} n\left(\frac{u_n}{u_{n+1}}-1\right) = \lim_{n\to\infty} n\left(\frac{\sqrt{n+2}-1}{\sqrt{n+1}-1}-1\right)$$

$$= \lim_{n\to\infty} n\left[\frac{\sqrt{n+2}-1-\sqrt{n+1}+1}{\sqrt{n+1}-1}\right] = \lim_{n\to\infty} n\left[\frac{\sqrt{n+2}-\sqrt{n+1}}{\sqrt{n+1}-1}\right]$$

$$= \lim_{n \to \infty} n \left[ \frac{\sqrt{1 + \frac{2}{n}} - \sqrt{1 + \frac{1}{n}}}{\sqrt{1 + \frac{1}{n} - \frac{1}{n^2}}} \right] = 0 < 1$$

Hence, $\Sigma u_n$ is divergent. **Ans.**

**Example 20.** *Discuss the convergence of the series:*

$$\frac{x}{1} + \frac{1}{2} \cdot \frac{x^3}{3} + \frac{1 \cdot 3}{2 \cdot 4} \frac{x^5}{5} + \frac{1 \cdot 3 \cdot 5}{2 \cdot 4 \cdot 6} \cdot \frac{x^7}{7} + \ldots (x > 0)$$ *(M.D.U. Dec., 2001)*

**Solution.** Here, we have

$$\frac{x}{1} + \frac{1}{2} \cdot \frac{x^3}{3} + \frac{1 \cdot 3}{2 \cdot 4} \frac{x^5}{5} + \frac{1 \cdot 3 \cdot 5}{2 \cdot 4 \cdot 6} \cdot \frac{x^7}{7} + \ldots$$

Neglecting the first term, we have

$$u_n = \frac{1 \cdot 3 \cdot 5 \ldots (2n-1)}{2 \cdot 4 \cdot 5 \ldots (2n)} \cdot \frac{x^{2n+1}}{2n+1}$$

and $$u_{n+1} = \frac{1 \cdot 3 \cdot 5 \ldots (2n-1)(2n+1)}{2 \cdot 4 \cdot 5 \ldots (2n)(2n+2)} \cdot \frac{x^{2n+3}}{2n+3}$$

$$\frac{u_n}{u_{n+1}} = \frac{(2n-1)\, x^{2n+1}}{2n\,(2n+1)} \times \frac{2n\,(2n+2)(2n+3)}{(2n-1)(2n+1)\, x^{2n+3}} = \frac{(2n+2)(2n+3)}{(2n+1)(2n+1)} \frac{1}{x^2}$$

$$\therefore \quad = \frac{2n+2}{2n+1} \cdot \frac{2n+3}{2n+1} \cdot \frac{1}{x^2} = \frac{2n\left(1+\frac{1}{n}\right).2n\left(1+\frac{3}{2n}\right)}{2n\left(1+\frac{1}{2n}\right).2n\left(1+\frac{1}{2n}\right)} \cdot \frac{1}{x^2} = \frac{\left(1+\frac{1}{n}\right)\left(1+\frac{3}{2n}\right)}{\left(1+\frac{1}{2n}\right)^2} \cdot \frac{1}{x^2}$$

$$\lim_{n\to\infty} \frac{u_n}{u_{n+1}} = \lim_{n\to\infty} \frac{\left(1+\frac{1}{n}\right)\left(1+\frac{3}{2n}\right)}{\left(1+\frac{1}{2n}\right)^2} \cdot \frac{1}{x^2} = \frac{1}{x^2}$$

$\therefore$ Ratio Test, $\sum u_n$ is convergent if $\frac{1}{x^2} > 1$.

*i.e.;* $x^2 < 1$ and divergent if $\frac{1}{x^2} < 1$. *i.e.,* $x^2 > 1$.

If $x^2 = 1$, then Ratio Test fails.

Now Raabe's test

When $x^2 = 1$, we have $$\frac{u_n}{u_{n+1}} = \frac{(2n+2)(2n+3)}{(2n+1)^2} = \frac{4n^2 + 10n + 6}{4n^2 + 4n + 1}$$

$$\lim_{n\to\infty} n\left(\frac{u_n}{u_{n+1}} - 1\right) = \lim_{n\to\infty} n\left(\frac{4n^2 + 10n + 6}{4n^2 + 4n + 1} - 1\right)$$

$$= \lim_{n\to\infty} \frac{6n^2 + 5n}{4n^2 + 4n + 1} = \lim_{n\to\infty} \frac{6 + \frac{5}{n}}{4 + \frac{4}{n} + \frac{1}{n^2}} = \frac{6}{4} = \frac{3}{2} > 1$$

$\therefore$ By Raabe's Test, the series converges.

Hence, $\sum u_n$ is convergent if $x^2 \le 1$ and divergent if $x^2 > 1$. **Ans.**

**Example 21.** *Test the following series for convergence*

$$\frac{1}{2}x + x^2 + \frac{9}{8}x^3 + x^4 + \frac{25}{32}x^5 + \dots \infty$$

**Solution.** Here, $u_n = \frac{n^2.x^n}{2^n}, \quad u_{n+1} = \frac{(n+1)^2.x^{n+1}}{2^{n+1}}$

**By D'Alembert's Test**

$$\frac{u_{n+1}}{u_n} = \frac{(n+1)^2 x^{n+1}}{2^{n+1}}.\frac{2^n}{n^2 x^n} = \left(\frac{n+1}{n}\right)^2 \frac{x}{2}$$

$$\lim_{n\to\infty} \frac{u_{n+1}}{u_n} = \lim_{n\to\infty}\left(1+\frac{1}{n}\right)^2 \frac{x}{2} = \frac{x}{2}$$

(*i*) If $\frac{x}{2} < 1$ or $x < 2$, then $\Sigma\, u_n$ is convergent. (*ii*) If $\frac{x}{2} > 1$ or $x > 2$, then $\Sigma\, u_n$ is divergent.

(*iii*) If $\frac{x}{2} = 1$ or $x = 2$, then the test fails.

Let us apply **Raabe's test**

$$n\left(\frac{u_n}{u_{n+1}} - 1\right) = n\left[\frac{n^2}{(n+1)^2}\frac{2}{2} - 1\right] = n\left[\frac{n^2 - n^2 - 2n - 1}{(n+1)^2}\right] = \frac{-2n^2 - n}{(n+1)^2}$$

$$\lim_{n\to\infty} n\left(\frac{u_n}{u_{n+1}} - 1\right) = \lim_{n\to\infty} \frac{-2-\frac{1}{n}}{\left(1+\frac{1}{n}\right)^2} = -2 < 1$$

Hence, $\Sigma\, u_n$ is divergent if $x \geq 2$, and convergent if $x < 2$. **Ans.**

**Example 22.** *Show that the series* $\frac{1}{x} + \frac{2!}{x(x+1)} + \frac{3!}{x(x+1)(x+2)} + \dots$ *converges if* $x > 2$ *and diverges if* $x < 2$.

**Solution.** Here, $u_n = \frac{n!}{x(x+1)\,(x+2)\dots(x+n-1)}$

$$u_{n+1} = \frac{(n+1)!}{x(x+1)\,(x+2)\dots(x+n-1)\,(x+n)}$$

**By D'Alembert's test**

$$\frac{u_{n+1}}{u_n} = \frac{n+1}{(x+n)}, \quad \lim_{n\to\infty}\frac{u_{n+1}}{u_n} = \frac{1+\frac{1}{n}}{1+\frac{x}{n}} = 1$$

Test fails. Let us apply **Raabe's Test.**

$$\lim_{n\to\infty} n\left(\frac{u_n}{u_{n+1}} - 1\right) = \lim_{n\to\infty} n\left(\frac{x+n}{n+1} - 1\right) = \lim_{n\to\infty} n\left(\frac{x-1}{n+1}\right) = \lim_{n\to\infty}\frac{x-1}{1+\frac{1}{n}} = x - 1$$

If $x - 1 > 1$ or $x > 2$, then $\Sigma\, u_n$ is convergent.
If $x - 1 < 1$ or $x < 2$, then $\Sigma\, u_n$ is divergent. **Ans.**

**Example 23.** *Discuss the convergence of the series* $\frac{x^2}{2\log 2} + \frac{x^3}{3\log 3} + \frac{x^4}{4\log 4} + \dots$

**Solution.** Here, we have $\dfrac{x^2}{2\log 2}+\dfrac{x^3}{3\log 3}+\dfrac{x^4}{4\log 4}+\ldots$

$$u_n=\frac{x^{n+1}}{(n+1)\log(n+1)},\qquad u_{n+1}=\frac{x^{n+2}}{(n+2)\log(n+2)}$$

**By D'Alembert's Test**

$$\lim_{n\to\infty}\frac{u_{n+1}}{u_n}=\lim_{n\to\infty}\frac{x^{n+2}}{(n+2)\log(n+2)}\times\frac{(n+1)\log(n+1)}{x^{n+1}}$$

$$=\lim_{n\to\infty}x\left(\frac{n+1}{n+2}\right)\frac{\log(n+1)}{\log(n+2)}$$

$$=\lim_{n\to\infty}x\left(\frac{1+\frac{1}{n}}{1+\frac{2}{n}}\right)\frac{\log n+\log\left(1+\frac{1}{n}\right)}{\log n+\log\left(1+\frac{2}{n}\right)}$$

$$=\lim_{n\to\infty}x\left(\frac{1+\frac{1}{n}}{1+\frac{2}{n}}\right)\left[\frac{\log n+\frac{1}{n}-\frac{1}{2}\frac{1}{n^2}+\ldots}{\log n+\frac{2}{n}-\frac{1}{2}\cdot\frac{4}{n^2}+\ldots}\right]$$

$$=\lim_{n\to\infty}x\left(\frac{1+\frac{1}{n}}{1+\frac{2}{n}}\right)\left[\frac{1+\frac{1}{n\log n}+\ldots}{1+\frac{2}{n\log n}+\ldots}\right]=x$$

(*i*) When $x<1$, the series is convergent (*ii*) When $x>1$, the series is divergent.
(*iii*) When $x=1$, the test fails.

**Let us apply Raabe's Test**

$$\frac{u_n}{u_{n+1}}=\left(\frac{n+2}{n+1}\right)\frac{\log(n+2)}{\log(n+1)}=\left(\frac{n+2}{n+1}\right)\frac{\log n+\log\left(1+\frac{2}{n}\right)}{\log n+\log\left(1+\frac{1}{n}\right)}$$

**By D'Alembert's Test**

$$=\left(\frac{n+2}{n+1}\right)\frac{\log n+\frac{2}{n}-\frac{1}{2}\cdot\frac{4}{n^2}+\ldots}{\log n+\frac{1}{n}-\frac{1}{2}\cdot\frac{1}{n^2}+\ldots}=\left(\frac{n+2}{n+1}\right)\frac{1+\frac{2}{n\log n}+\ldots}{1+\frac{1}{n\log n}+\ldots}$$

$$=\frac{n+2}{n+1}\left(1+\frac{2}{n\log n}\right)\left(1+\frac{1}{n\log n}\right)^{-1}=\frac{n+2}{n+1}\left(1+\frac{2}{n\log n}\right)\left(1-\frac{1}{n\log n}\right)$$

$$=\frac{n+2}{n+1}\left(1+\frac{2}{n\log n}-\frac{1}{n\log n}+\ldots\right)=\left(\frac{n+2}{n+1}\right)\left[1+\frac{1}{n\log n}\right]$$

$$\lim_{n\to\infty}\frac{u_n}{u_{n+1}}=\lim_{n\to\infty}=\left[\frac{1+\frac{2}{n}}{1+\frac{1}{n}}\right]\left[1+\frac{1}{n\log n}\right]=1+\frac{1}{n\log n}$$

$$n\left[\frac{u_n}{u_{n+1}}-1\right]=n\left[1+\frac{1}{n\log n}-1\right]=\frac{1}{\log n}=0<1$$

Thus the series is divergent when $x=1$.

Hence, the series converges if $x < 1$ and diverges if $x \geq 1$. **Ans.**

**Example 24.** *Test the series for convergence*

$$1 + \frac{\alpha.\beta}{1.\gamma}x + \frac{\alpha(\alpha+1).\beta(\beta+1)}{1.2.\gamma(\gamma+1)}x^2 + \frac{\alpha(\alpha+1)(\alpha+2).\beta(\beta+1)(\beta+2)}{1.2.3.\gamma(\gamma+1)(\gamma+2)}x^3 + \ldots$$

**Solution.** $u_n = \dfrac{\alpha(\alpha+1)(\alpha+2)\ldots[\alpha+(n-1)].\beta(\beta+1)\ldots[\beta+(n-1)]}{n!\,\gamma(\gamma+1)\ldots[\gamma+(n-1)]}x^n$

$$u_{n+1} = \frac{\alpha(\alpha+1)(\alpha+2)\ldots[\alpha+(n-1)](\alpha+n).\beta(\beta+1)\ldots[\beta+(n-1)](\beta+n)}{(n+1)!\,\gamma(\gamma+1)\ldots[\gamma+(n-1)](\gamma+n)}x^{n+1}$$

**By D'Alembert's Test**

$$\lim_{n\to\infty}\frac{u_{n+1}}{u_n} = \lim_{n\to\infty}\frac{(\alpha+n)(\beta+n)}{(n+1)(\gamma+n)}x = \lim_{n\to\infty}\frac{\left(1+\frac{\alpha}{n}\right)\left(1+\frac{\beta}{n}\right)}{\left(1+\frac{1}{n}\right)\left(1+\frac{\gamma}{n}\right)}.x$$

$$\lim_{n\to\infty}\frac{u_{n+1}}{u_n} = x$$

(*i*) If $x < 1$, the series is convergent. (*ii*) If $x > 1$, the series is divergent.
(*iii*) If $x = 1$, the test fails.

**Let us apply Raabe's Test**

$$n\left(\frac{u_n}{u_{n+1}} - 1\right) = n\left[\frac{(n+1)(\gamma+n)}{(\alpha+n)(\beta+n)} - 1\right] = n\left[\frac{n\gamma+n^2+\gamma+n-\alpha\beta-n\alpha-n\beta-n^2}{(\alpha+n)(\beta+n)}\right]$$

$$\lim_{n\to\infty} n\left(\frac{u_n}{u_{n+1}} - 1\right) = \lim_{n\to\infty}\frac{\gamma+\frac{\gamma}{n}+1-\frac{\alpha\beta}{n}-\alpha-\beta}{\left(\frac{\alpha}{n}+1\right)\left(\frac{\beta}{n}+1\right)} = \gamma+1-\alpha-\beta$$

(*i*) If $\gamma + 1 - \alpha - \beta > 1$ or $\gamma > \alpha + \beta$, then $\Sigma\, u_n$ is convergent.
(*ii*) If $\gamma + 1 - \alpha - \beta < 1$ or $\gamma < \alpha + \beta$, then $\Sigma\, u_n$ is divergent. **Ans.**

## EXERCISE 20.7

Determine the nature of the following series:

1. $1 + \frac{2!}{2^2} + \frac{3!}{3^2} + \frac{4!}{4^2} + \ldots \infty$ **Ans.** Divergent
2. $\frac{1}{1} + \frac{1.3}{1.4} + \frac{1.3.5}{1.4.7} + \frac{1.3.5.7}{1.4.7.10} + \ldots \infty$ **Ans.** Convergent
3. $1 + \frac{1+\alpha}{1+\beta} + \frac{(1+\alpha)(2+\alpha)}{(1+\beta)(2+\beta)} + \ldots \infty$ **Ans.** If $\beta - \alpha > 1$, convergent. If $\beta - \alpha \leq 1$, Divergent.
4. $\sum_{n=1}^{\infty}\frac{n^3}{e^n}$ **Ans.** Convergent
5. $x + \frac{2x^2}{2!} + \frac{3x^3}{3!} + \frac{4x^4}{4!} + \ldots \infty$ **Ans.** Convergent
6. $1 + \frac{x}{2} + \frac{x^2}{3} + \frac{x^3}{4} + \ldots$ **Ans.** Divergent
7. $1 + \frac{1}{2}x + \frac{1}{5}x^2 + \frac{1}{10}x^3 + \ldots$ **Ans.** Convergent if $-1 \leq x < 1$ and divergent if $|x| > 1$
8. $1 + \frac{(1!)^2}{2!}x^2 + \frac{(2!)^2}{4!}x^4 + \frac{(3!)^2}{6!}x^6 + \ldots \infty \; (x > 0)$

   **Ans.** If $x^2 < 4$, convergent; and divergent if $x^2 \geq 4$

**Find the values of $x$ for which the following series converges:**

9. $x^2 (\log 2)^q + x^3 (\log 3)^q + x^4 (\log 4)^q + \ldots$

**Ans.** If $x < 1$, convergent; and divergent if $x \geq 1$

10. $\sum_{n=0}^{\infty} \frac{(-1)^n \, n! x^n}{10^n}$

12. $\sum \frac{x^n}{2n\,(2n+1)}$ **Ans.** If $x \leq 1$, convergent; and if $x > 1$, divergent

11. $\sum \frac{1.2 \ldots n}{4.7 \ldots (3n+1)} x^n$ **Ans.** If $0 < x < 3$, convergent and divergent if $x \geq 3$.

12. $1 + \frac{(1!)^2}{2!} x + \frac{(2!)^2}{4!} x^2 + \frac{(3!)^2}{6!} x^3 + \ldots\ldots$ *(M.D.U., Dec. 2010)*

**Ans.** convergent if $x < 4$; divergent if $x \geq 4$.

## 20.18. GAUSS'S TEST

**Statement.** ***If $\Sigma u_n$ is a positive term series such that***

$$\frac{u_n}{u_{n+1}} = \alpha + \frac{\beta}{n} + \frac{\gamma}{n^2} \qquad \textit{where } \alpha > 0$$

***(i) if $\alpha > 1$, convergent     if $\alpha < 1$, divergent, whatever $\beta$ may be***

***(ii) if $\alpha = 1$ and*** $\begin{cases} \beta > 1, \textit{ convergent} \\ \beta \leq 1, \textit{ divergent} \end{cases}$

**Example 25.** *Test for convergence the series* $\frac{2^2.4^2}{3^2.5^2} + \frac{2^2.4^2.6^2}{3^2.5^2.7^2} + \frac{2^2.4^2.6^2.8^2}{3^2.5^2.7^2.9^2} + \ldots$

**Solution.** The given series is $\frac{2^2.4^2}{3^2.5^2} + \frac{2^2.4^2.6^2}{3^2.5^2.7^2} + \frac{2^2.4^2.6^2.8^2}{3^2.5^2.7^2.9^2} + \ldots$

$$+ \frac{2^2.4^2.6^2.8^2 \ldots (2n+2)^2}{3^2.5^2.7^2.9^2 \ldots (2n+3)^2} + \ldots \infty$$

$$u_{n+1} = \frac{2^2.4^2.6^2.8^2 \ldots (2n+2)^2 \, (2n+4)^2}{3^2.5^2.7^2.9^2 \ldots (2n+3)^2 \, (2n+5)^2}$$

**By D'Alembert's Test**

$$\frac{u_{n+1}}{u_n} = \frac{(2n+4)^2}{(2n+5)^2} = \frac{4n^2 + 16n + 16}{4n^2 + 20n + 25} = \frac{4 + \frac{16}{n} + \frac{16}{n^2}}{4 + \frac{20}{n} + \frac{25}{n^2}}$$

$$\lim_{n \to \infty} \frac{u_{n+1}}{u_n} = \lim_{n \to \infty} \frac{4 + \frac{16}{n} + \frac{16}{n^2}}{4 + \frac{20}{n} + \frac{25}{n^2}} = 1$$

D'Alembert's Test fails. Let us apply **Raabe's Test.**

$$\lim_{n \to \infty} n \left( \frac{u_n}{u_{n+1}} - 1 \right) = \lim_{n \to \infty} n \left( \frac{4n^2 + 20n + 25}{4n^2 + 16n + 16} - 1 \right)$$

$$= \lim_{n \to \infty} \left( \frac{4n^2 + 9n}{4n^2 + 16n + 16} \right) = \lim_{n \to \infty} \left[ \frac{4 + \frac{9}{n}}{4 + \frac{16}{n} + \frac{16}{n^2}} \right] = 1, \text{ Raabe's Test fails}$$

**Let us apply Gauss's Test**

$$\frac{u_n}{u_{n+1}} = \frac{(2n+5)^2}{(2n+4)^2} = \frac{\left(1+\frac{5}{2n}\right)^2}{\left(1+\frac{2}{n}\right)^2} = \left(1+\frac{5}{n}+\frac{25}{4n^2}\right)\left(1+\frac{2}{n}\right)^{-2}$$

$$= \left(1+\frac{5}{n}+\frac{25}{4n^2}\right)\left(1-\frac{4}{n}+\frac{(-2)\times(-3)}{2!}\frac{4}{n^2}+\ldots\right) = \left(1+\frac{5}{n}+\frac{25}{4n^2}\right)\left(1-\frac{4}{n}+\frac{12}{n^2}+\ldots\right)$$

$$= 1-\frac{4}{n}+\frac{12}{n^2}+\frac{5}{n}-\frac{20}{n^2}+\frac{25}{4n^2}+\ldots = 1+\frac{1}{n}-\frac{7}{n^2} \qquad \left(\frac{u_n}{u_{n+1}} = \alpha+\frac{\beta}{n}+\frac{\gamma}{n^2}\right)$$

Hence, $\alpha = 1$, $\beta = 1$. Thus, the series is divergent. **Ans.**

## 20.19 CAUCHY'S INTEGRAL TEST

**Statement.** *A positive term series $f(1) + f(2) + f(3) + \ldots + f(n) + \ldots$ where $f(n)$ decreases as $n$ increases, converges or diverges according to the integral*

$$\int_1^\infty f(x)\,dx$$

*is finite or infinite.*

**Proof.** In the figure, the area under the curve from $x = 1$ to $x = n + 1$ lies between the sum of the areas of small rectangles (small height) and sum of the areas of large rectangles (large height).

[$f(1), f(2)$ ... represent the height of the rectangles]

$$\Rightarrow \quad f(1) + f(2) + \ldots + f(n) \geq \int_1^{n+1} f(x)\,dx \geq f(2) + f(3) + \ldots + f(n+1)$$

$$S_n \geq \int_1^{n+1} f(x)\,dx \geq S_{n+1} - f(1)$$

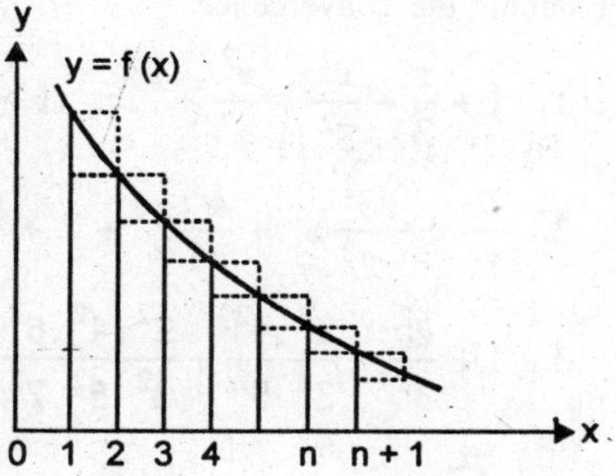


As $n \to \infty$, from the second inequality that if the integral has a finite value then $\lim_{n\to\infty} S_{n+1}$ is also finite, so $\Sigma f(n)$ is convergent.

Similarly, if the integral is infinite, then from the first inequality that $\lim_{n\to\infty} S_n \to \infty$, so the series is divergent.

**Example 26.** *Apply the integral test to determine the convergence of the p-series*

$$\frac{1}{1^p}+\frac{1}{2^p}+\frac{1}{3^p}+\ldots+\frac{1}{n^p}+\ldots\infty$$

**Solution.** (*i*) When $p > 1$, $\quad f(x) = \frac{1}{x^p}$

$$\int_1^\infty f(x)\,dx = \lim_{m\to\infty}\int_1^m \frac{1}{x^p}\,dx = \lim_{m\to\infty}\left[\frac{x^{1-p}}{1-p}\right]_1^m = \lim_{m\to\infty}\frac{1}{1-p}(m^{1-p}-1)$$

$$= \lim_{m\to\infty}\frac{1}{1-p}\left[\frac{1}{m^{p-1}}-1\right] = \frac{1}{p-1}, \text{ which is finite.}$$

By Cauchy's Integral Test, the series is convergent for $p > 1$.

(*ii*) When $p < 1$,

$$\int_1^{\infty} f(x)\,dx = \frac{1}{1-p}\left[\lim_{m\to\infty}(m^{1-p}-1)\right] \to \infty$$

Thus, the series is divergent, if $p < 1$.
(*iii*) When $p = 1$,

$$\int_1^{\infty} \frac{1}{x}\,dx = [\log x]_1^{\infty} \to \infty$$

Thus, the series is divergent.

Hence, $\Sigma \frac{1}{n^p}$ is convergent if $p > 1$ and divergent if $p \le 1$. **Ans.**

**Example 27.** *Examine the convergence of* $\sum_{n=2}^{\infty} \frac{1}{n \log n}$.

**Solution.** Here $f(x) = \frac{1}{x \log x}$

$$\int_2^{\infty} \frac{1}{x \log x}\,dx = \lim_{m\to\infty}\left[\log \log x\right]_2^m = \lim_{m\to\infty} [\log \log m - \log \log 2] \to \infty$$

By Cauchy's Integral Test the series is divergent. **Ans.**

**Example 28.** *Examine the convergence of* $\sum_{x=1}^{\infty} x\, e^{-x^2}$

**Solution.** Here $\quad f(x) = x\, e^{-x^2}$

Now, $\int_1^{\infty} x\, e^{-x^2}\,dx = \lim_{m\to\infty}\left[\frac{e^{-x^2}}{-2}\right]_1^m = \lim_{m\to\infty}\left[\frac{e^{-m^2}}{-2} + \frac{e^{-1}}{2}\right] = \frac{e^{-1}}{2} = \frac{1}{2e}$, which is finite.

Hence, the given series is convergent. **Ans.**

## EXERCISE 20.8

Examine the convergence:

1. $1 + \frac{x}{2} + \frac{x^2}{3^2} + \frac{x^2}{4^3} + \ldots \infty \quad (x > 0)$ **Ans.** Convergent
2. $\frac{2}{1^2}x + \frac{3^2}{2^3}x^2 + \frac{4^3}{3^4}x^3 + \ldots + \frac{(n+1)^n}{n^{n+1}}x^n + \ldots$ **Ans.** $x < 1$, convergent; $x \ge 1$, divergent
3. $1 + \frac{2^2}{3^2} + \frac{2^2.4^2}{3^2.5^2} + \frac{2^2.4^2.6^2}{3^2.5^2.7^2} + \ldots \infty$ **Ans.** Divergent
4. $\sum_{n=1}^{\infty} \frac{1}{\sqrt{n}}$ **Ans.** Divergent
5. $\sum_{n=1}^{\infty} \frac{1}{n^2+1}$ **Ans.** Convergent
6. $\sum_{n=1}^{\infty} \frac{1}{n^n}$ **Ans.** Convergent
7. $\sum_{n=1}^{\infty} n^2 e^{-n^3}$ **Ans.** Convergent
8. $\sum_{n=1}^{\infty} \frac{1}{n (\log n)^2}$ **Ans.** Convergent

## 20.20 CAUCHY'S ROOT TEST

**Statement.** *If* $\Sigma\, u_n$ *is positive term series such that* $\lim_{n\to\infty} (u_n)^{1/n} = k$, *then*

(*i*) *if* $k < 1$, *the series converges.* (*ii*) *if* $k > 1$, *the series diverges.*

**Proof.** By definition of limit

$$|(u_n)^{1/n} - k| < \varepsilon \text{ for } n > m$$

$$k - \varepsilon < (u_n)^{1/n} < k + \varepsilon \text{ for } n > m$$

(*i*) $$k < 1$$

$$k + \varepsilon < r < 1$$

$$(u_n)^{1/n} < k \quad \Rightarrow u_n < k^n$$

$$u_1 + u_2 + \ldots \infty < k + k^2 + \ldots + k^n + \ldots \infty$$

$$< \frac{1}{1-k} \text{ (a finite quantity)}$$

$\therefore$ The series is convergent.

(*ii*) $$k > 1$$

$$k - \varepsilon > 1$$

$$(u_n)^{1/n} > k - \varepsilon > 1$$

$$u_n > 1$$

$$S_n = u_1 + u_2 + \ldots u_n > n$$

$$\lim_{n \to \infty} S_n \to \infty$$

$\therefore$ The series is divergent.

(*iii*) $$k = 1$$

If $\lim_{n \to \infty} (u_n)^{1/n} = 1$, the test fails.

**For example,** $$\Sigma u_n = \sum \frac{1}{n^p}$$

$$\lim_{n \to \infty} (u_n)^{1/n} = \lim_{n \to \infty} \left(\frac{1}{n^p}\right)^{1/n} = \lim_{n \to \infty} \left(\frac{1}{n^{1/n}}\right)^{-p} = 1 \text{ for all } p, k = 1$$

But $\sum \frac{1}{n^p}$ is convergent for $p > 1$ and divergent for $p \le 1$.

Thus, we cannot say whether $\Sigma u_n$ is convergent or divergent for $k = 1$.

**Example 29.** *Examine the convergence of the series* $\sum \frac{1}{\left(1+\frac{1}{n}\right)^{n^2}}$ *(MDU, Dec. 2010)*

**Solution.** Here, $$u_n = \frac{1}{\left(1+\frac{1}{n}\right)^{n^2}} \Rightarrow (u_n)^{1/n} = \left[\frac{1}{\left(1+\frac{1}{n}\right)^{n^2}}\right]^{\frac{1}{n}} = \frac{1}{\left(1+\frac{1}{n}\right)^n}$$

$$\lim_{n \to \infty} (u_n)^{1/n} = \lim_{n \to \infty} \frac{1}{\left(1+\frac{1}{n}\right)^n} = \frac{1}{e} < 1$$

Hence, the given series is convergent. **Ans.**

**Example 30.** *Test the following series for convergence*

$$\sum_{n=1}^{\infty} \frac{(n+1)^n x^n}{n^{n+1}}$$

*(M.D.U. Dec., 2001)*

**Solution.** Here, we have

$$\sum_{n=1}^{\infty} \frac{(n+1)^n x^n}{n^{n+1}}$$

Here, $$u_n = \frac{(n+1)^n x^n}{n^{n+1}} = \left[\frac{(n+1)x}{n}\right]^n . \frac{1}{n}$$

$$\Rightarrow \quad (u_n)^{\frac{1}{n}} = \frac{(n+1)x}{n}.\frac{1}{\frac{1}{n^n}} = \left(1+\frac{1}{n}\right)x.\frac{1}{\frac{1}{n^n}}$$

$$\Rightarrow \quad \lim_{n\to\infty} (u_n)^{\frac{1}{n}} = \left[\lim_{n\to\infty}\left(1+\frac{1}{n}\right)x\right]\left[\lim_{n\to\infty}\frac{1}{\frac{1}{n^n}}\right]$$

$$= (1+0)\,x.\frac{1}{1} = x \qquad \left[\because \lim_{n\to\infty} n^{\frac{1}{n}} = 1\right]$$

$\therefore$ By Cauchy's root test, $\sum u_n$ is convergent if $x < 1$ and divergent if $x > 1$. The test fails when $x = 1$.

When $x = 1$, $\quad u_n = \dfrac{(n+1)^n}{n^{n+1}} = \dfrac{1}{n}.\dfrac{(n+1)^n}{n^n} = \dfrac{1}{n}\left(1+\dfrac{1}{n}\right)^n$

Let $\quad v_n = \dfrac{1}{n},$

$$\frac{u_n}{v_n} = \left(1+\frac{1}{n}\right)^n$$

$$\lim_{n\to\infty}\frac{u_n}{v_n} = \lim_{n\to\infty}\left(1+\frac{1}{n}\right)^n = e, \text{ which is finite and non-zero.}$$

$\therefore$ By camparison test, $\sum u_n$ and $\sum v_n$ converge or diverge together.

Since $\sum v_n = \sum \frac{1}{n}$ is of the form $\sum \frac{1}{n^p}$ with $p = 1$,

$\sum v_n$ divergnt $\Rightarrow \sum u_n$ also divergent

Hence, $\sum u_n$ is convergent is $x < 1$ and divergent if $x \geq 1$. **Ans.**

**Example 31.** *Discuss the convergence of the following series:*

$$\left(\frac{2^2}{1^2}-\frac{2}{1}\right)^{-1} + \left(\frac{3^3}{2^3}-\frac{3}{2}\right)^{-2} + \left(\frac{4^4}{3^4}-\frac{4}{3}\right)^{-3} + \dots \infty$$

**Solution.** Here, $u_n = \left[\dfrac{(n+1)^{n+1}}{n^{n+1}} - \dfrac{(n+1)}{n}\right]^{-n}$

$$[u_n]^{1/n} = \left[\left[\frac{(n+1)^{n+1}}{n^{n+1}} - \frac{n+1}{n}\right]^{-n}\right]^{\frac{1}{n}} = \left[\frac{(n+1)^{n+1}}{n^{n+1}} - \frac{n+1}{n}\right]^{-1}$$

$$\lim_{n\to\infty}(u_n)^{1/n} = \lim_{n\to\infty}\left[\left(1+\frac{1}{n}\right)^{n+1} - \left(1+\frac{1}{n}\right)\right]^{-1} = (e-1)^{-1} = \frac{1}{e-1} < 1$$

Hence, the given series is convergent. **Ans.**

## EXERCISE 20.9

**Discuss the convergence of the following series:**

1. $\displaystyle\sum_{n=1}^{\infty}\frac{1}{n^n}$ **Ans.** Convergent

2. $\displaystyle\sum_{n=1}^{\infty}\left(1+\frac{1}{n}\right)^{n^2}$ **Ans.** Divergent

3. $\sum_{n=1}^{\infty}\left(1+\frac{1}{\sqrt{n}}\right)^{-n^{3/2}}$ **Ans.** Convergent 4. $\sum_{n=1}^{\infty}\frac{1}{(\log n)^n}$ **Ans.** Convergent

5. $\Sigma\, n^{-k}$ **Ans.** If $k > 1$, convergent

6. $\sum_{n=2}^{\infty}\frac{1}{n(\log n)^k}$ **Ans.** If $k > 1$, convergent; and divergent if $k \le 1$.

7. $\Sigma\,(n \log n)^{-1}(\log \log n)^{-k}$ **Ans.** If $k > 1$, convergent; and divergent if $k \le 1$.

8. $\sum\left(1-\frac{1}{n}\right)^{n^2}$ **Ans.** Convergent 9. $\sum\frac{x^n}{n^n}$ **Ans.** Convergent

10. $(a+b)+(a^2+b^2)+(a^3+b^3)+\ldots$ **Ans.** Convergent if $a < 1, b < 1$; divergent if $a \ge 1, b \ge 1$

## 20.21 LOGARITHMIC TEST

**If $\Sigma\, u_n$ is a positive term series such that $\lim_{n\to\infty}\left(n \log \frac{u_n}{u_{n+1}}\right) = k$**

**(*i*) If $k > 1$, then the series is convergent. (*ii*) If $k < 1$, then the series is divergent.**

**Proof.** (*i*) If $k > 1$

Compare $\Sigma\, u_n$ with $\sum\frac{1}{n^p}$, if $k > p > 1$, then $\Sigma\, u_n$ converges.

$$\frac{u_n}{u_{n+1}} = \frac{(n+1)^p}{n^p} = \left(1+\frac{1}{n}\right)^p \quad \ldots(1)$$

Taking logarithm of both sides of (1), we have:

$$\log\frac{u_n}{u_{n+1}} > p\log\left(1+\frac{1}{n}\right) \quad \left[\log(1+x) = x - \frac{x^2}{2}+\frac{x^3}{3}-\frac{x^4}{4}+\ldots\right]$$

if $$\log\frac{u_n}{u_{n+1}} > p\left(\frac{1}{n}-\frac{1}{2n^2}+\frac{1}{3n^3}-\frac{1}{4n^4}+\ldots\right)$$

if $$n\log\frac{u_n}{u_{n+1}} > p\left(1-\frac{1}{2n}+\frac{1}{3n^2}-\frac{1}{4n^3}+\ldots\right)$$

$$\lim_{n\to\infty} n\log\frac{u_n}{u_{n+1}} > p$$

*i.e.,* $k > p$ which is true as $k > p > 1$. $\left[\lim_{n\to\infty} n\log\frac{u_n}{u_{n+1}} = k\right]$

Hence, $\Sigma\, u_n$ is convergent.

When $p < 1$

Similarly, when $p < 1$, $\Sigma\, u_n$ is divergent.

When $p = 1$, the test fails.

**Example 32.** *Test the convergence of the series* $x+\frac{2^2.x^2}{2!}+\frac{3^3.x^3}{3!}+\frac{4^4.x^4}{4!}+\ldots\infty$

(*MDU, Dec. 2010*)

**Solution.** Here the series is $x+\frac{2^2.x^2}{2!}+\frac{3^3.x^3}{3!}+\frac{4^4.x^4}{4!}+\ldots+\frac{n^n.x^n}{n!}+\ldots\infty$

$$u_n = \frac{n^n.x^n}{n!} \quad\text{and}\quad u_{n+1} = \frac{(n+1)^{n+1}.x^{n+1}}{(n+1)!}$$

$$\frac{u_n}{u_{n+1}} = \frac{n^n . x^n}{n!} \frac{(n+1)!}{(n+1)^{n+1} . x^{n+1}} = \frac{n^n}{(n+1)^n} \frac{1}{x} = \frac{1}{\left(1+\frac{1}{n}\right)^n} \frac{1}{x}$$

$$\lim_{n \to \infty} \frac{u_n}{u_{n+1}} = \lim_{n \to \infty} \frac{1}{\left(1+\frac{1}{n}\right)^n} \frac{1}{x} = \frac{1}{e} . \frac{1}{x}$$

If $\frac{1}{e x} > 1$ or $x < \frac{1}{e}$, the series is convergent.

If $\frac{1}{e x} < 1$ or $\frac{1}{e} < x$, the series is divergent. If $\frac{1}{e x} = 1$ or $x = \frac{1}{e}$, the test fails.

$$\log \frac{u_n}{u_{n+1}} = \log \frac{1}{\left(1+\frac{1}{n}\right)^n} . e = \log e - \log \left(1+\frac{1}{n}\right)^n$$

$$= 1 - n \log \left(1+\frac{1}{n}\right) = 1 - n \left[\frac{1}{n} - \frac{1}{2n^2} + \frac{1}{3n^3} + \ldots\right]$$

$$= 1 - 1 + \frac{1}{2n} - \frac{1}{3n^2} + \ldots = \frac{1}{2n} - \frac{1}{3n^2}$$

$$\lim_{n \to \infty} n \log \frac{u_n}{u_{n+1}} = \lim_{n \to \infty} \left[\frac{1}{2} - \frac{1}{3n}\right] = \frac{1}{2} < 1.$$

Thus, the series is divergent. **Ans.**

**Example 33.** *Discuss the convergence of the series:*

$$1 + \frac{x}{2} + \frac{2!}{3^2} x^2 + \frac{3!}{4^3} x^3 + \frac{4!}{5^4} x^4 + \ldots\ldots\ldots \infty \ (x > 0)$$ *(M.D.University, I Semester, 2009)*

**Solution.** Here, we have

$$1 + \frac{x}{2} + \frac{2!}{3^2} x^2 + \frac{3!}{4^3} x^3 + \frac{4!}{5^4} x^4 + \ldots\ldots\ldots \infty$$

Neglecting the first term, we get

$$u_n = \frac{n!}{(n+1)^n} x^n \quad \text{and} \quad u_{n+1} = \frac{(n+1)!}{(n+2)^{n+1}} x^{n+1}$$

$$\frac{u_n}{u_{n+1}} = \frac{n!}{(n+1)^n} . x^n . \frac{(n+2)^{n+1}}{(n+1)! . x^{n+1}} = \frac{(n+2)^{n+1}}{(n+1)^n . (n+1)} . \frac{1}{x}$$

$$\lim_{n\to\infty} \frac{u_n}{u_{n+1}} = \lim_{n\to\infty} \frac{(n+2)^{n+1}}{(n+1)^n . (n+1)} . \frac{1}{x}$$

$$= \lim_{n\to\infty} \frac{n^{n+1}\left(1+\frac{2}{n}\right)^{n+1}}{n^n \left(1+\frac{1}{n}\right)^n . n \left(1+\frac{1}{n}\right)} . \frac{1}{x} = \lim_{n\to\infty} \frac{\left(1+\frac{2}{n}\right)^n \left(1+\frac{2}{n}\right)}{\left(1+\frac{1}{n}\right)^n \left(1+\frac{1}{n}\right)} . \frac{1}{x} \quad \ldots(1)$$

$$= \frac{e^2}{e} . \frac{1}{x} = \frac{e}{x} . \left[\because \lim_{n\to\infty} \left(1+\frac{a}{n}\right)^n = \lim_{n\to\infty} \left\{\left(1+\frac{a}{n}\right)^{\frac{n}{a}}\right\}^a = e^a\right.$$

$$\left[\lim_{n\to\infty} \left(1+\frac{2}{n}\right)^n = \lim_{n\to\infty} \left[\left(1+\frac{2}{n}\right)^{\frac{n}{2}}\right]^2 = e^2\right]$$

$\therefore$ By D' Alembert's ratio test, the series converges if $1 < \frac{e}{x}$ or if $x < e$ and diverges if $\frac{e}{x} < 1$ or if $e < x$.

If $x = e$, the ratio test fails, $\because \lim_{n \to \infty} \frac{u_n}{u_{n+1}} = 1$

Now when $x = e$

Putting the value of $x$ in (1), we get

$$\frac{u_n}{u_{n+1}} = \frac{\left(1+\frac{2}{n}\right)^{n+1}}{\left(1+\frac{1}{n}\right)^{n+1}} \cdot \frac{1}{e}$$

Since the expression $\frac{u_n}{u_{n+1}}$ involves the number $e$, so we do not apply Raabe's test but apply logarithmic test.

$$\therefore \quad \log \frac{u_n}{u_{n+1}} = (n+1)\log\left(1+\frac{2}{n}\right) - (n+1)\log\left(1+\frac{1}{n}\right) - \log e$$

$$= (n+1)\left[\log\left(1+\frac{2}{n}\right) - \log\left(1+\frac{1}{n}\right)\right] - 1$$

$$= (n+1)\left[\left(\frac{2}{n} - \frac{1}{2}\cdot\frac{4}{n^2} + \frac{1}{3}\cdot\frac{8}{n^3} \ldots\right) - \left(\frac{1}{n} - \frac{1}{2n^2} + \frac{1}{3n^3} + \ldots\right)\right] - 1$$

$$= (n+1)\left[\frac{1}{n} - \frac{3}{2n^2} + \frac{7}{3n^3} - \ldots\ldots\right] - 1$$

$$= \left(1 - \frac{3}{2n}\right) + \left(\frac{1}{n} - \frac{3}{n^2} + \ldots\ldots\right) - 1 = \frac{1}{n} - \frac{3}{n^2} - \frac{3}{2n^2}$$

$$= 1 - \frac{1}{2n} - \frac{3}{2n^2} + \ldots\ldots - 1 = -\frac{1}{2n} - \frac{3}{2n^2}$$

$$\therefore \quad \lim_{n\to\infty} n \log \frac{u_n}{u_{n+1}} = \lim_{n\to\infty} n\left[-\frac{1}{2n} - \frac{3}{2n^2} + \ldots\ldots\right] = \lim_{n\to\infty}\left(-\frac{1}{2} - \frac{3}{2n} + \ldots\ldots\right) = -\frac{1}{2} < 1$$

$\therefore$ By log test, the series diverges.

Hence, the given series $\sum u_n$ converges if $x < e$ and diverges if $x \geq e$. **Ans.**

## EXERCISE 20.10

Examine the convergence for the following series :

1. $\frac{1^2}{4^2} + \frac{5^2}{8^2} + \frac{9^2}{12^2} + \frac{13^2}{16^2} + \ldots\infty$ **Ans.** Convergent

2. $1 + \frac{1!}{2}x + \frac{2!}{3^2}x^2 + \frac{3!}{4^3}x^3 + \ldots\infty$ **Ans** If $x < e$, convergent and divergent if $x \geq e$

3. $\frac{1^2}{2^2} + \frac{1^2 . 3^2}{2^2 . 4^2}x + \frac{1^2 . 3^2 . 5^2}{2^2 . 4^2 . 6^2}x^2 + \ldots\infty$ **Ans.** Convergent if $x < 1$, and divergent if $x \geq 1$

4. $\frac{a+x}{1!} + \frac{(a+2x)^2}{2!} + \frac{(a+3x)^3}{3!} + \ldots$ **Ans.** Convergent if $x < \frac{1}{e}$, divergent if $x \geq \frac{1}{e}$

## 20.22 DE MORGAN'S AND BERTRAND'S TEST

**If $\Sigma u_n$ is a positive term series such that**

$$\lim_{n \to \infty} \left[ \left\{ n\left( \frac{u_n}{u_{n+1}} - 1 \right) - 1 \right\} \log n \right] = k$$

**then the series is convergent if $k > 1$ and divergent if $k < 1$.**

**Example 34.** *Test for convergence the series* $1^p + \left(\frac{1}{2}\right)^p + \left(\frac{1 \times 3}{2 \times 4}\right)^p + \left(\frac{1 \times 3 \times 5}{2 \times 4 \times 6}\right)^p + \ldots$

**Solution.** The given series is :

$$1^p + \left(\frac{1}{2}\right)^p + \left(\frac{1\times3}{2\times4}\right)^p + \left(\frac{1\times3\times5}{2\times4\times6}\right)^p + \ldots$$

Here
$$u_n = \left[ \frac{1\times3\times5\times\ldots\times(2n-3)}{2\times4\times6\times\ldots\times(2n-2)} \right]^p$$

$$\therefore \quad u_{n+1} = \left[ \frac{1\times3\times5\times\ldots(2n-3)(2n-1)}{2\times4\times6\times\ldots\times(2n-2)(2n)} \right]^p$$

$$\therefore \quad \frac{u_{n+1}}{u_n} = \left( \frac{2n-1}{2n} \right)^p = \left( 1 - \frac{1}{2n} \right)^p$$

$$\lim_{n\to\infty} \frac{u_{n+1}}{u_n} = 1$$

$\therefore$ **D'Alembert's Test fails.**

Now let us apply Raabe's Test.

Here $n\left( \frac{u_n}{u_{n+1}} - 1 \right) = n\left[ \left( 1 - \frac{1}{2n} \right)^{-p} - 1 \right] = n\left[ 1 + \frac{p}{2n} + \frac{p(p+1)}{8n^2} + \ldots - 1 \right] = \frac{p}{2} + \frac{p(p+1)}{8n} + \ldots$

$$\therefore \quad \lim_{n\to\infty} n\left[ \frac{u_n}{u_{n+1}} - 1 \right] = \frac{p}{2}$$

If $\frac{p}{2} > 1, i.e., p > 2$, the series is convergent and divergent if $\frac{p}{2} < 1, i.e., p < 2.$

This test fails if $\frac{p}{2} = 1, i.e., p = 2.$

Now let us apply **De Morgan's Test.** When $p = 2$

$$n\left[ \frac{u_n}{u_{n+1}} - 1 \right] = 1 + \frac{3}{4n} + \ldots$$

Now, $\lim_{n\to\infty} \left[ n\left( \frac{u_n}{u_{n+1}} - 1 \right) - 1 \right] \log n = \lim_{n\to\infty} \left[ 1 + \frac{3}{4n} + \ldots - 1 \right] \log n$

$$= \lim_{n\to\infty} \frac{3}{4} \left[ \frac{\log n}{n} - \ldots \right] = 0 < 1 \qquad \left[ \lim_{n\to\infty} \frac{\log n}{n} = 0 \right]$$

$\therefore$ $\Sigma u_n$ is divergent when $p = 2$. **Ans.**

## 20.23 CAUCHY'S CONDENSATION TEST

**If $\phi(n)$ is positive for all positive integral values of $n$ and continually diminishes as $n$ increases and if $a$ be a positive integer greater than 1, then the two series $\Sigma\, \phi(n)$ and $\Sigma\, a^n \phi(a^n)$ are either both convergent or both divergent.**

**Example 35.** *Show that the series*

$$1+\frac{1}{2(\log 2)^p}+\frac{1}{3(\log 3)^p}+\ldots+\frac{1}{n(\log n)^p}+\ldots$$

*is convergent if $p > 1$ and divergent if $p = 1$ or $p < 1$.*

**Solution.** We apply Cauchy's Condensation Test.

Here $$\phi(n) = \frac{1}{n(\log n)^p}$$

$\therefore$ $n$th term of the second series $\Sigma\, a^n \phi(a^n)$ is :

$$a^n\left[\frac{1}{a^n(\log a^n)^p}\right] \text{ i.e., } \frac{1}{(\log a^n)^p} \text{ i.e., } \frac{1}{(n\log a)^p} \text{ i.e., } \frac{1}{(\log a)^p}\times\frac{1}{n^p}$$

$\therefore$ The given series will be convergent or divergent if $\sum\left[\frac{1}{(\log a)^p}\times\frac{1}{n^p}\right]$ is convergent or divergent, i.e., if $\sum\frac{1}{n^p}$ is convergent or divergent.

But we know that $\sum\frac{1}{n^p}$ is convergent when $p > 1$ and divergent if $p = 1$ or $< 1$.

Hence, the given series is convergent if $p > 1$ and divergent if $p = 1$ or $< 1$. **Proved.**

## 20.24 ALTERNATING SERIES

A series in which the terms are alternately negative is called the alternating series.

*e.g.,* $$u_1 - u_2 + u_3 - u_4 + \ldots\infty$$

## 20.25 LEIBNITZ'S RULE FOR CONVERGENCE OF AN ALTERNATING SERIES

**(*i*) Each term is numerically less than its preceeding term.**

**(*ii*) $\lim_{n\to\infty} u_n = 0$**

**Exmaple 36.** *Discuss the convergence of the series* $\sum_{n=1}^{\infty}(-1)^n\frac{n}{n^2+1}$.

**Solution.** The terms of the given series are alternately positive and negative;

(*i*) $$|u_n| = \frac{n}{n^2+1} \quad\text{and}\quad |u_{n+1}| = \frac{(n+1)}{(n+1)^2+1}$$

$$|u_n| - |u_{n+1}| = \frac{n}{n^2+1} - \frac{(n+1)}{(n+1)^2+1} = \frac{n(n+1)^2+n-(n+1)(n^2+1)}{(n^2+1)[(n+1)^2+1]}$$

$$= \frac{n^2+n-1}{(n^2+1)[(n+1)^2+1]} = +\text{ve}$$

and each term is numerically less than its preceeding term.

(*ii*) $$\lim_{n\to\infty} |u_n| = \lim_{n\to\infty} \frac{n}{n^2+1} = \lim_{n\to\infty} \frac{1}{n+\frac{1}{n}} = 0$$

Both conditions are satisfied.

Hence, by Leibnitz's rule, the given series is convergent. **Ans.**

**Example 37.** *Test the convergence of the series* $\frac{1}{6}-\frac{2}{11}+\frac{3}{16}-\frac{4}{21}+\frac{5}{26}-\ldots\infty$

**Solution.** The terms of the given series are alternately positive and negative.

$$u_n = (-1)^{n-1}\frac{n}{5n+1}$$

$$|u_n| = \frac{n}{5n+1} \quad \text{and} \quad |u_{n+1}| = \frac{n+1}{5(n+1)+1}$$

(*i*) $$|u_n| - |u_{n+1}| = \frac{n}{5n+1} - \frac{n+1}{5(n+1)+1} = \frac{5n^2+6n-5n^2-5n-n-1}{(5n+1)(5n+6)}$$

$$= \frac{-1}{(5n+1)(5n+6)}$$

$$\therefore \quad |u_n| > |u_{n+1}|$$

Thus each term is not numerically less than its preceeding terms.

(*ii*) $$\lim_{n\to\infty} u_n = \lim_{n\to\infty}\frac{n}{5n+1} = \lim_{n\to\infty}\frac{1}{5+\frac{1}{n}} = \frac{1}{5} \neq 0$$

$$\lim_{n\to\infty} u_n \neq 0$$

Both conditions for convergence are not satisfied.

Hence, the series is not convergent. It is oscillatory. **Ans.**

**Example 38.** *Test the following series for convergence and absolute convergence*:

$$1-\frac{1}{2\sqrt{2}}+\frac{1}{3\sqrt{3}}-\frac{1}{4\sqrt{4}}+\ldots$$ (*M.D.U. Dec., 2002*)

**Solution.** The given series is

$$\sum_{n=1}^{\infty} u_n = \sum_{n=1}^{\infty}(-1)^{n-1}.\frac{1}{n\sqrt{n}} = \sum_{n=1}^{\infty}(-1)^{n-1}\frac{1}{n^{\frac{3}{2}}} = \sum_{n=1}^{\infty}(-1)^{n-1}a_n$$

It is an alternating series.

Here, $$a_n = \frac{1}{n^{\frac{3}{2}}}$$

$$a_{n+1} = \frac{1}{(n+1)^{\frac{3}{2}}}$$

Since, $$\frac{1}{n^{\frac{3}{2}}} > \frac{1}{(n+1)^{\frac{3}{2}}}\ \forall\, n \qquad (\therefore a_n > a_{n+1}\ \forall\, n)$$

Also, $$\lim_{n\to\infty} a_n = \lim_{n\to\infty}\frac{1}{n^{\frac{3}{2}}} = 0$$

$\therefore$ By Leibnitz's test, the series $\sum u_n$ is convergent.

$$|u_n| = \frac{1}{n^{\frac{3}{2}}}$$

Now $\sum |u_n| = \sum \frac{1}{n^{\frac{3}{2}}}$ is convergent $\left(\because p = \frac{3}{2} > 1\right)$

Hence, the series $\sum u_n$ is absolutely convergent. **Ans.**

**Example 20.** *Test the convergence of the series:*

$$\sum_{n=1}^{\infty} \frac{(-1)^n (x+1)^n}{2^n n^2}$$

*(M.D. University, I Semester, 2009)*

**Solution.** Here, we have

$$\sum_{n=1}^{\infty} \frac{(-1)^n (x+1)^n}{2^n n^2} \qquad ...(1)$$

Here, $u_n = \frac{(-1)^n (x+1)^n}{2^n . n^2}$ and $u_{n+1} = \frac{(-1)^{n+1} (x+1)^{n+1}}{2^{n+1} .(n+1)^2}$

$$\frac{|u_n|}{|u_{n+1}|} = \frac{|x+1|^n}{2^n . n^2} . \frac{2^{n+1} .(n+1)^2}{|x+1|^{n+1}}$$

$$= 2\left(\frac{n+1}{n}\right)^2 . \frac{1}{|x+1|} = 2\left(1+\frac{1}{n}\right)^2 . \frac{1}{|x+1|}$$

$$\therefore \lim_{n\to\infty} \frac{|u_n|}{|u_{n+1}|} = \lim_{n\to\infty} 2\left(1+\frac{1}{n}\right)^2 . \frac{1}{|x+1|} = \frac{2}{|x+1|}$$

$\therefore$ By ratio test, the series $\sum |u_n|$ is convergent if

*i.e.,* $1 < \frac{2}{|x+1|}$ *i.e.,* if $|x + 1| < 2$

*i.e.,* if $- 2 < x + 1 < 2$ *i.e.,* if $- 3 < x < 1$

Also $\sum |u_n|$ is divergent if $\frac{2}{|x+1|} < 1$

*i.e.,* if $|x + 1| > 2$ *i.e.,* if $x + 1 > 2$ or $x + 1 < -2$ *i.e.,* if $x > 1$ or $x < -3$.

Ratio test fails when $x = 1$ or $- 3$.

When $x = 1$, $\sum u_n = \sum \frac{(-1)^n . 2^n}{2^n . n^2} = \sum \frac{(-1)^n}{n^2} = \sum (-1)^n . v_n$ From (1)

It is an alternating series.

Here, $v_n = \frac{1}{n^2}$, $v_{n+1} = \frac{1}{(n+1)^2}$

Clearly $v_n > v_{n+1}$ $\forall n$

Also $\lim_{n\to\infty} v_n = \lim_{n\to\infty} \frac{1}{n^2} = 0$

$\therefore$ By Leibnitz's test, $\sum u_n$ is convergent.

When $x = - 3$, $\sum u_n = \sum \frac{(-1)^n .(-2)^n}{2^n . n^2} = \sum \frac{(-1)^{2n} 2^n}{2^n . n^2} = \sum \frac{1}{n^2}$

Which is convergent.

Hence, the given series is convergent if $-3 \le x \le 1$ and divergent if $x > 1$ or $x < - 3$

## EXERCISE 20.11

Discuss the convergence of the following series :

1. $1-\frac{1}{\sqrt{2}}+\frac{1}{\sqrt{3}}-\frac{1}{\sqrt{4}}+\ldots$ **Ans.** Convergent
2. $1-2x+3x^2-4x^3+\ldots\infty\ (x<1)$ **Ans.** Convergent
3. $\frac{x}{1+x}-\frac{x^2}{1+x^2}+\frac{x^3}{1+x^3}-\frac{x^4}{1+x^4}+\ldots\infty\,(0<x<1)$ **Ans.** Convergent
4. $\frac{1}{1^p}-\frac{1}{2^p}+\frac{1}{3^p}-\frac{1}{4^p}+\ldots$ **Ans.** If $p>0$, convergent; oscillatory if $p<0$.
5. $\sum_{n=1}^{\infty}(-1)^{n-1}\frac{n}{2n-1}$ **Ans.** Oscillatory
6. Show that the series $\frac{1}{x+1}-\frac{1}{x+2}+\frac{1}{x+3}-\frac{1}{x+4}+\ldots$ is convergent for all real values of $x$ other than negative integers.
7. Prove that the series $x-\frac{x^2}{2}+\frac{x^3}{3}-\ldots$ converges if $-1\le x\le 1$.

## 20.26 ALTERNATING CONVERGENT SERIES

There are two types of alternating convergent series :

**(1) Absolutely convergent series. (2) Conditionally convergent series.**
**Absolutely convergent series. If $u_1+u_2+u_3+\ldots$ be such that $|u_1|+|u_2|+|u_3|+\ldots$ be convergent then $u_1+u_2+u_3+\ldots\infty$ is called absolutely convergent.**

**Example 40.** *Show that the series* $1-\frac{1}{2}+\frac{1}{3}-\frac{1}{4}+\frac{1}{5}-\frac{1}{6}+\ldots$ *is convergent but not absolutely convergent.*

**Solution.** $1-\frac{1}{2}+\frac{1}{3}-\frac{1}{4}+\ldots$

The terms of the series are alternately positive and negative.

(*i*) $|u_{n+1}|<|u_n|$ as $\frac{1}{n+1}<\frac{1}{n}$ (*ii*) $\lim_{n\to\infty}u_n=\lim_{n\to\infty}\frac{1}{n}=0$

Both conditions are satisfied. Hence, the given series is convergent.

But $1+\frac{1}{2}+\frac{1}{3}+\frac{1}{4}+\ldots\infty$ is divergent since in *p-series*, $p=1$.

Hence, the given series is conditionally convergent. **Ans.**

**Example 41.** *What can you say about the series* $1-\frac{1}{2^2}+\frac{1}{3^2}-\frac{1}{4^2}+\ldots\infty$ ?

**Solution.** $1-\frac{1}{2^2}+\frac{1}{3^2}-\frac{1}{4^2}+\ldots$

$$|u_n|=\frac{1}{n^2},\quad\text{and}\quad |u_{n+1}|=\frac{1}{(n+1)^2}$$

(*i*) $|u_{n+1}|<|u_n|$ (*ii*) $\lim_{n\to\infty}u_n=\lim_{n\to\infty}\frac{1}{n^2}=0$

Thus, the given series is convergent by Leibnitz's rule.

And $1+\frac{1}{2^2}+\frac{1}{3^2}+\frac{1}{4^2}+\ldots.$ is also convergent since in *p*-series, $p=2>1$.

Both the conditions are satisfied.
Hence, the given series is absolutely convergent. **Ans.**

**Example 42.** *Discuss the series for convergence* $1-\frac{1}{3}+\frac{1}{2}-\frac{1}{3^3}+\frac{1}{2^2}-\frac{1}{3^5}+\frac{1}{2^3}-\frac{1}{3^7}+\ldots$

**Solution.** The given series is rewitten as

$$1+\frac{1}{2}+\frac{1}{2^2}+\frac{1}{2^3}+\ldots-\left(\frac{1}{3}+\frac{1}{3^3}+\frac{1}{3^5}+\frac{1}{3^7}+\ldots\right)$$

$$\lim_{n\to\infty} S_n = \frac{1}{1-\frac{1}{2}}-\frac{\frac{1}{3}}{1-\frac{1}{3^2}}=2-\frac{3}{8}=1\frac{5}{8}$$

The given series is convergent.

Again $1+\frac{1}{2}+\frac{1}{2^2}+\frac{1}{2^3}+\ldots+\frac{1}{3}+\frac{1}{3^3}+\frac{1}{3^5}+\ldots$

$$\lim_{n\to\infty} S_n = \frac{1}{1-\frac{1}{2}}+\frac{\frac{1}{3}}{1-\frac{1}{3}} \qquad \left(\lim_{n\to\infty} S_n=\frac{a}{1-r}\right)$$

$$\lim_{n\to\infty} S_n = 2+\frac{3}{8}=\frac{19}{8}$$

Both the conditions are satisfied.
This series is also convergent.
Hence, the given series is absolutely convergent. **Ans.**

**Example 43.** *Test the convergence and divergence of the series*

$$5-\frac{10}{3}+\frac{20}{9}-\frac{40}{27}+\ldots\ldots\ldots$$ *(Gujarat, I Semester, Jan. 2009)*

**Solution.** The terms of the given series are alternately positive and negative and the given series is geometric infinite series.

(*i*) $$S=5-\frac{10}{3}+\frac{20}{9}-\frac{40}{27}+\frac{80}{81}+\ldots\ldots\ldots$$

Here $a=5$ and $r=-\frac{2}{3}$, $S=\frac{a}{1-r}$

$$S=\frac{5}{1-\left(-\frac{2}{3}\right)}=\frac{5}{1+\frac{2}{3}}=\frac{5}{\frac{5}{3}}=3$$

Sum of the series is finite.
Hence, the given series is convergent.

(*ii*) Again $5+\frac{10}{3}+\frac{20}{9}+\frac{40}{27}+\frac{80}{81}+\ldots$

This is also G.P.

Here, $a=5$ and $r=\frac{2}{3}$

$$S=\frac{a}{1-r}, \qquad S=\frac{5}{1-\frac{2}{3}}=\frac{5}{\frac{1}{3}}=15$$

Again sum of the positive terms is finite.
Thus the series is also convergent.
Both of the conditions are satisfied.
Hence, the given series is absolutely convergent. **Ans.**

**Example 44.** *Test the following series for convergence and divergence.*

$$\sum_{n=1}^{\infty} \tan^{-1}\left(\frac{1}{n^2+n+1}\right) \qquad \textit{(Gujarat, I Semester, Jan. 2009)}$$

**Solution.** Let

$$u_n = \tan^{-1}\frac{1}{1+n(n+1)}$$

$$u_n = \tan^{-1}\frac{(n+1)-n}{1+n(n+1)}$$

$$u_n = \tan^{-1}(n+1) - \tan^{-1}(n)$$

$$u_{n+1} = \tan^{-1}(n+2) - \tan^{-1}(n+1)$$

**By D' Alembert's test**

$$\lim_{n\to\infty}\frac{u_{n+1}}{u_n} = \lim_{n\to\infty}\frac{\tan^{-1}(n+2)-\tan^{-1}(n+1)}{\tan^{-1}(n+1)-\tan^{-1}n} = \lim_{n\to\infty}\frac{\frac{1}{1+(n+2)^2}-\frac{1}{1+(n+1)^2}}{\frac{1}{1+(n+1)^2}-\frac{1}{1+n^2}}$$

**[L'Hopital Rule]**

$$= \lim_{n\to\infty}\frac{\dfrac{1+(n+1)^2-1-(n+2)^2}{[1+(n+2)^2]\,[1+(n+1)^2]}}{\dfrac{1+n^2-1-(n+1)^2}{[1+(n+1)^2]\,[1+n^2]}} = \lim_{n\to\infty}\frac{(2n+3)(-1)}{(2n+1)(-1)}\times\frac{1+n^2}{1+(n+2)^2}$$

$$= \lim_{n\to\infty}\frac{(2n+3)(1+n^2)}{(2n+1)[n^2+4n+5]} = \lim_{n\to\infty}\frac{\left(2+\frac{3}{n}\right)\left(\frac{1}{n^2}+1\right)}{\left(2+\frac{1}{n}\right)\left(1+\frac{4}{n}+\frac{5}{n^2}\right)} = \frac{2}{2} = 1$$

Test fails.
Let us apply **Raabe's Test.**

$$\lim_{n\to\infty} n\left(\frac{u_n}{u_{n+1}}-1\right) = \lim_{n\to\infty} n\left[\frac{(2n+1)(n^2+4n+5)}{(2n+3)(n^2+1)}-1\right]$$

$$= \lim_{n\to\infty} n\left[\frac{(2n+1)(n^2+4n+5)-(2n+3)(n^2+1)}{(2n+3)(n^2+1)}\right]$$

$$= \lim_{n\to\infty} n\left[\frac{2n^3+8n^2+10n+n^2+4n+5-2n^3-2n-3n^2-3}{(2n+3)\,(n^2+1)}\right]$$

$$= \lim_{n\to\infty} n\left[\frac{6n^2+12n+2}{(2n+3)\,(n^2+1)}\right] = \lim_{n\to\infty}\frac{6+\frac{12}{n}+\frac{2}{n^2}}{\left(2+\frac{3}{n}\right)\left(1+\frac{1}{n^2}\right)} = \frac{6}{2} = 3 > 1$$

Hence, the given series is convergent. **Ans.**

## EXERCISE 20.12

Discuss the convergence of the following series :

1. $1-\frac{1}{2}+\frac{1}{4}-\frac{1}{8}+...\infty$ **Ans.** Absolutely convergent

2. $1-\frac{2}{5}+\frac{3}{10}-\frac{4}{17}+...+\frac{(-1)^{n-1}n}{n^2+1}+...\infty$ **Ans.** Conditionally convergent

3. $1-\frac{2}{3}+\frac{3}{3^2}-\frac{4}{3^3}+...$ **Ans.** Absolutely convergent

4. $1+\frac{1}{2^2}-\frac{1}{3^2}-\frac{1}{4^2}+\frac{1}{5^2}+\frac{1}{6^2}-\frac{1}{7^2}-\frac{1}{8^2}+...$ **Ans.** Absolutely convergent

5. $1-\frac{1}{\sqrt{2}}+\frac{1}{\sqrt{3}}-\frac{1}{\sqrt{4}}+...\infty$ **Ans.** Conditionally convergent

6. $\frac{\sin x}{1^3}-\frac{\sin 2x}{2^3}+\frac{\sin 3x}{3^3}+...$ **Ans.** Absolutely convergent

## 20.27 POWER SERIES IN $x$

$$a_0+a_1x+a_2x^2+...+a_nx^n+...$$

is power series in $x$, here $a$'s are independent of $x$.

**Proof.** $u_n = a_nx^n$ and $u_{n+1}=a_{n+1}x^{n+1}$

**D'Alembert's Ratio Test**

$$\lim_{n\to\infty}\frac{u_{n+1}}{u_n}=\lim_{n\to\infty}\frac{a_{n+1}x^{n+1}}{a_nx^n}=\lim_{n\to\infty}\frac{a_{n+1}}{a_n}x$$

If $\frac{a_{n+1}}{a_n}=k,$

If $|kx|<1 \Rightarrow |x|<\frac{1}{k}$, then the series is convergent.

Thus, the power series is convergent if $-\frac{1}{k}<x<\frac{1}{k}$.

Thus, the interval of the power series is $-\frac{1}{k}$ to $\frac{1}{k}$ for convergence. Outside this interval the series is divergent. **Ans.**

**Example 45.** *Find the values of x for which the series* $x-\frac{x^2}{2^2}+\frac{x^3}{3^2}-\frac{x^4}{4^2}+...\infty$ *converges.*

**Solution.** Here $u_n = (-1)^{n-1}\frac{x^n}{n^2}$, and $u_{n+1}=(-1)^n\frac{x^{n+1}}{(n+1)^2}$

$$\frac{u_{n+1}}{u_n}=-\frac{n^2}{(n+1)^2}x \Rightarrow \lim_{n\to\infty}\frac{u_{n+1}}{u_n}=-\lim_{n\to\infty}\frac{1}{\left(1+\frac{1}{n}\right)^2}x=-x$$

By **D'Alembert's Test** the given series is convergent for $|x|<1$ and divergent if $|x|>1$.

**At $x=1$.** The series becomes $1-\frac{1}{2^2}+\frac{1}{3^2}-\frac{1}{4^2}+...$

This is an alternately convergent series.

***At $x=-1$.*** The series becomes $-1-\frac{1}{2^2}-\frac{1}{3^2}-\frac{1}{4^2}+...$

This is also convergent series, $p = 2$

Hence, the interval of convergence is $-1 \le x \le 1$. **Ans.**

## 20.28 EXPONENTIAL SERIES

$$e^x = 1 + x + \frac{x^2}{2!} + \frac{x^3}{3!} + \ldots + \frac{x^n}{n!} + \ldots$$

**is convergent for all values of $x$.**

**Proof.** Here, we have $u_n = \dfrac{x^{n-1}}{(n-1)!}$, and $u_{n+1} = \dfrac{x^n}{n!}$

$$\lim_{n\to\infty} \frac{u_{n+1}}{u_n} = \lim_{n\to\infty} \frac{x^n}{n!} \cdot \frac{(n-1)!}{x^{n-1}} = \lim_{n\to\infty} \frac{x}{n} = 0 < 1$$

Hence, by D'Alembert's Test the exponential series is convergent for all values of $x$.

## 20.29 LOGARITHMIC SERIES

$$\log(1+x) = x - \frac{x^2}{2} + \frac{x^3}{3} - \frac{x^4}{4} + \ldots + (-1)^{n-1} \frac{x^n}{n} + \ldots$$

is convergent for $-1 < x \le 1$.

**Proof.** Here, $u_n = (-1)^{n-1} \dfrac{x^n}{n}$, and $u_{n+1} = (-1)^n \dfrac{x^{n+1}}{n+1}$

$$\lim_{n\to\infty} \frac{u_{n+1}}{u_n} = \lim_{n\to\infty} \frac{(-1)^n x^{n+1}}{n+1} \cdot \frac{n}{(-1)^{n-1} x^n} = -\lim_{n\to\infty} \left(\frac{n}{n+1}\right) x = -\lim_{n\to\infty} \left(\frac{1}{1+\frac{1}{n}}\right) x = -x$$

Thus, the series is convergent for $|x| < 1$ and divergent for $|x| > 1$.

*At $x = 1$.* The series becomes

$$1 - \frac{1}{2} + \frac{1}{3} - \frac{1}{4} + \ldots \infty \text{ which is convergent.}$$

*At $x = -1$.* The series becomes

$$-1 - \frac{1}{2} - \frac{1}{3} - \frac{1}{4} + \ldots \infty \text{ which is divergent.}$$

## 20.30 BINOMIAL SERIES

$$(1+x)^n = 1 + nx + \frac{n(n-1)}{2!} x^2 + \frac{n(n-1)(n-2)}{3!} x^3 + \ldots$$

is convergent for $|x| < 1$.

**Proof.** $u_r = \dfrac{n(n-1)\ldots(n-r+2)}{(r-1)!} x^{r-1}$

$$u_{r+1} = \frac{n(n-1)\ldots(n-r+1)}{r!} x^r$$

$$\lim_{r\to\infty} \frac{u_{r+1}}{u_r} = \lim_{r\to\infty} \frac{n-r+1}{r} x = \lim_{r\to\infty} \left(\frac{n+1}{r} - 1\right) x = -x \text{ for } r > n+1$$

If $|x| < 1$, the series is convergent by D'Alembert's Test.

## EXERCISE 20.13

Test the convergence of the following series

1. $1 + x + 2x^2 + 3x^3 + \ldots + nx^n + \ldots\infty$ **Ans.** Convergent if $-1 < x > 1$

2. $\dfrac{1}{2} + \dfrac{x}{3} + \dfrac{x^2}{4} + \dfrac{x^3}{5} + \ldots + \dfrac{x^n}{n+2} + \ldots\infty$ **Ans. Convergent for $-1 < x < 1$**

3. $1 + \dfrac{1}{1!} + \dfrac{1}{2!} + \dfrac{1}{3!} + \ldots\infty$ **Ans.** Convergent

4. $1 + \dfrac{4}{1!} + \dfrac{4^2}{2!} + \dfrac{4^3}{3!} + \ldots\infty$ **Ans. Convergent**

5. $3-\frac{3^2}{2}+\frac{3^3}{3}-\frac{3^4}{4}+\ldots\infty$ **Ans.** Convergent

6. $\sqrt{2}-\frac{2}{2}+\frac{2^{\frac{3}{2}}}{3}-\frac{2^2}{4}+\ldots\infty$ **Ans.** Convergent

7. $\frac{1}{1-x}+\frac{1}{2(1-x)^2}+\frac{1}{3(1-x)^3}+\ldots\infty$ **Ans.** Convergent, if $x \neq 1$

## 20.31 UNIFORM CONVERGENCE

**If for a given ε > 0, a number $N$ can be found *independent of* $x$, such that for every $x$ in the interval $(a, b)$, the series is said to be uniformly convergent in the interval $(a, b)$.**

**Example 46.** *Discuss the uniform convergence of the series*

$$1 + x + x^2 + \ldots\infty$$

**Solution.** $S_n = 1 + x + x^2 + \ldots + x^n = \frac{1-x^n}{1-x}$

$$S_\infty(x) = \lim_{n\to\infty} \frac{1-x^n}{1-x} = \frac{1}{1-x} \qquad \text{for } |x| < 1$$

$$|S_\infty(x) - S_n(x)| = \left|\frac{1}{1-x} - \frac{1-x^n}{1-x}\right| = \left|\frac{x^n}{1-x}\right| = \frac{|x|^n}{1-x} < \varepsilon$$

if $|x|^n < \varepsilon(1-x)$

Let $|x|^N = \varepsilon(1-x) \Rightarrow N = \frac{\log \varepsilon(1-x)}{\log|x|}$

In the interval $\left(-\frac{1}{2}, \frac{1}{2}\right)$, $N$ has maximum value. $N = \frac{\log\frac{\varepsilon}{2}}{\log\frac{1}{2}}$

Hence, the given series is uniformly convergent in the interval $\left(-\frac{1}{2}, \frac{1}{2}\right)$.

**Note.** The series is convergent in (– 1, 1) but not uniformly convergent.

## 20.32 ABEL'S TEST

If $v_n(x)$ be either monotonic decreasing in $n$ for each fixed $x$ in $(a, b)$ or monotonic increasing in $n$ for each fixed $x$ in $(a, b)$, $\Sigma\, a_n(x)\, v_n(x)$ is uniformly convergent in $(a, b)$ if

(*i*) $\Sigma\, a_n(x)$ is uniformly convergent in $(a, b)$.

(*ii*) There exists $k$ such that $|v_n(x)| < k$ for all $n$ when $a \le x < b$.

**Example 47.** *Prove that* $\frac{x^n}{n^3}$ *is uniformly convergent in (– 1, 1).*

**Solution.** (*i*) $\sum \frac{1}{n^3}$ is uniformly convergent. (*ii*) $|x^n| < k$ for all $n$ when $-1 < k < 1$

Hence, $\sum \frac{x^n}{n^3}$ is uniformly convergent by Abel's Test. **Ans.**

## EXERCISE 20.14

Prove that the following series are uniformly convergent in (– 1, 1).

1. $\sum \frac{x^n}{n^2}$

2. $\sum \frac{x^n}{n(n+1)}$

3. $\sum \frac{x^{2n}}{x^{2n}+n^2}$

4. Prove that the series $1+\frac{e^{-2x}}{2^2-1}-\frac{e^{-4x}}{4^2-1}+\frac{e^{-6x}}{6^2-1}+\ldots\infty$ is uniformly convergent with regard to $x$ if $x \ge 0$.

# 21

# Gama, Beta Function

## 21.1 GAMMA FUNCTION

*(U.P. I Semester Dec. 2007)*

$$\int_0^\infty e^{-x} x^{n-1} dx \qquad (n > 0)$$

is called gamma function of *n*. It is also written as $\Gamma n = \int_0^\infty e^{-x} x^{n-1} dx$

$$\boxed{\int_0^\infty e^{-x}\, x^{n-1}\, dx = \Gamma n}$$

**Example 1.** *Prove that* $\Gamma 1 = 1$

**Solution.** We know that, $\Gamma n = \int_0^\infty e^{-x} x^{n-1} dx$

Put $n = 1$, $\Gamma 1 = \int_0^\infty e^{-x}\, x^{1-1}\, dx = \int_0^\infty e^{-x} dx = \left[\frac{e^{-x}}{-1}\right]_0^\infty = 1$ **Proved.**

**Example 2.** *Prove that*

(*i*) $\Gamma(n+1) = n\Gamma n$ (*ii*) $\Gamma(n+1) = n!$ ***(Reduction formula)***

**Solution.**

(*i*) We know that, $\Gamma n = \int_0^\infty x^{n-1} e^{-x} dx$ ...(1)

Integrating by parts, we have

$$\Gamma n = \left[x^{n-1}\frac{e^{-x}}{-1}\right]_0^\infty - (n-1)\int_0^\infty x^{n-2}\frac{e^{-x}}{-1} dx$$

$$= \lim_{x \to 0}\left\{\left(1 - \frac{x}{1!} + \frac{x^2}{2!} - \dots + (-1)^n \frac{x^n}{n!} + \dots + \infty\right) x^{n-1}\right\} + (n-1)\int_0^\infty x^{n-2} e^{-x} dx = 0 + (n-1)\int_0^\infty x^{n-2} e^{-x} dx$$

$$\therefore \quad \Gamma n = (n-1)\Gamma(n-1) \qquad \text{...(2)}$$

$$\boxed{\Gamma(n+1) = n\Gamma n}$$ Replacing *n* by (*n* + 1) **Proved.**

(*ii*) Replacing *n* by $n - 1$ in (2), we get

$$\Gamma(n-1) = (n-2)\Gamma(n-2)$$

Putting the value $\Gamma(n-1)$ in (2), we get

$$\Gamma n = (n-1)(n-2)\Gamma(n-2)$$

Similarly, $\quad \overline{|n} = (n-1)(n-2) \ldots\ldots 3.2.1\, \overline{|1}$ ...(3)

Putting the value of $\overline{|1}$ in (3), we have

$$\overline{|n} = (n-1)(n-2) \ldots\ldots 3.2.1.1$$

$$\overline{|n} = (n-1)\,!$$

Replacing $n$ by $n+1$, we have

$$\overline{|n+1} = n\,!$$

**Proved.**

**Example 3.** *Evaluate* $\Gamma\left(\dfrac{-5}{2}\right)$ (*MTU 2014*)

**Solution.** $\overline{\left|-\frac{1}{2}+1\right.} = -\frac{1}{2}\overline{\left|-\frac{1}{2}\right.}$

$$\Rightarrow \quad \overline{\left|\frac{1}{2}\right.} = -\frac{1}{2}\overline{\left|-\frac{3}{2}+1\right.} \qquad \Rightarrow \quad \sqrt{\pi} = \left(-\frac{1}{2}\right)\left(-\frac{3}{2}\right)\overline{\left|-\frac{3}{2}\right.}$$

$$\Rightarrow \quad \sqrt{\pi} = \left(-\frac{1}{2}\right)\left(-\frac{3}{2}\right)\overline{\left|-\frac{5}{2}+1\right.} \qquad \Rightarrow \quad \sqrt{\pi} = \left(-\frac{1}{2}\right)\left(-\frac{3}{2}\right)\left(-\frac{5}{2}\right)\overline{\left|-\frac{5}{2}\right.}$$

$$\sqrt{\pi} = -\frac{15}{8}\overline{\left|-\frac{5}{2}\right.}$$

$$\overline{\left|-\frac{5}{2}\right.} = -\frac{8}{15}\sqrt{\pi}$$

**Ans.**

**Example 4.** *Evaluate* $\int_0^{\infty} \sqrt[4]{x}\, e^{-\sqrt{x}}\, dx$ .

**Solution.** Let $\quad I = \int_0^{\infty} x^{1/4}\, e^{-\sqrt{x}}\, dx$ ...(1)

Putting $\sqrt{x} = t \quad \Rightarrow x = t^2$ so that $dx = 2t\, dt$ in (1), we get

$$I = \int_0^{\infty} t^{1/2}\, e^{-t}\, 2t\, dt = 2\int_0^{\infty} t^{3/2} e^{-t} dt = 2\int_0^{\infty} t^{\frac{5}{2}-1}\, e^{-t}\, dt$$

$$= 2\,\overline{\left|\frac{5}{2}\right.}$$ [By definition]

$$= 2\cdot\frac{3}{2}\overline{\left|\frac{3}{2}\right.} = 2\cdot\frac{3}{2}\cdot\frac{1}{2}\overline{\left|\frac{1}{2}\right.} = \frac{3}{2}\sqrt{\pi}$$

**Ans.**

**Example 5.** *Evaluate* $\int_0^{\infty} \sqrt{x}\, e^{-\sqrt[3]{x}}\, dx.$

**Solution.** Let $I = \int_0^{\infty} \sqrt{x}\, e^{-\sqrt[3]{x}}\, dx$ ...(1)

Putting $\sqrt[3]{x} = t \quad \Rightarrow \quad x = t^3$ so that $dx = 3\, t^2\, dt$ in (1), we get

$$I = \int_0^{\infty} t^{3/2}\, e^{-t}\, 3t^2\, dt = 3\int_0^{\infty} t^{7/2} e^{-t} dt = 3\int_0^{\infty} t^{\frac{9}{2}-1}\, e^{-t}\, dt = 3\,\overline{\left|\frac{9}{2}\right.} = 3\cdot\frac{7}{2}\cdot\frac{5}{2}\cdot\frac{3}{2}\cdot\frac{1}{2}\overline{\left|\frac{1}{2}\right.} = \frac{315}{16}\sqrt{\pi}$$

**Ans.**

**Example 6.** *Evaluate* $\int_0^{\infty} x^{n-1}\, e^{-h^2x^2}\, dx.$

**Solution.** Let $\quad I = \int_0^{\infty} x^{n-1}\, e^{-h^2x^2}\, dx$ ...(1)

Putting $t = h^2 x^2 \Rightarrow x = \dfrac{\sqrt{t}}{h}$ so that $dx = \dfrac{dt}{2h\sqrt{t}}$, we get

$$I = \int_0^\infty \left(\frac{\sqrt{t}}{h}\right)^{n-1} e^{-t} \frac{dt}{2h\sqrt{t}} = \frac{1}{2h^n}\int_0^\infty t^{\frac{n-1}{2}} e^{-t} \frac{dt}{\sqrt{t}} = \frac{1}{2h^n}\int_0^\infty t^{\frac{n-2}{2}} e^{-t} dt$$

$$= \frac{1}{2h^n}\left\lceil \frac{n}{2} \right.$$ **Ans.**

**Example 7.** *Evaluate* $\int_0^\infty \frac{x^a}{a^x} dx.$ $(a > 1)$

**Solution.** Let $$I = \int_0^\infty \frac{x^a}{a^x} dx \qquad ...(1)$$

Putting $a^x = e^t \Rightarrow x \log a = t \Rightarrow x = \dfrac{t}{\log a}, \Rightarrow dx = \dfrac{dt}{\log a}$ in (1), we have

$$I = \int_0^\infty \left(\frac{t}{\log a}\right)^a e^{-t} \frac{dt}{\log a} = \frac{1}{(\log a)^{a+1}} \int_0^\infty e^{-t} t^a dt$$

$$= \frac{1}{(\log a)^{a+1}} \int_0^\infty t^{(a+1)-1} e^{-t} dt$$

$$= \frac{1}{(\log a)^{a+1}} \left\lceil a+1 \right.$$ **Ans.**

## 21.2 PROVE THAT

$$\int_0^1 x^m (\log x)^n dx = \frac{(-1)^n}{(m+1)^{n+1}} \left\lceil n+1 \right.$$

**Proof :** Put $\log x = -t$ so that $x = e^{-t} \Rightarrow dx = -e^{-t} dt$

$\therefore$ $$x^m = e^{-mt}$$

$$(\log x)^n = (-t)^n$$

Now, $$\int_0^1 x^m (\log x)^n dx = \int_\infty^0 e^{-mt} (-t)^n (-e^{-t})\, dt = \int_0^\infty (-1)^n e^{-mt-t} t^n dt$$

Putting $(m+1)\, t = u$ so that $(m+1)\, dt = du$, we get

$\therefore$ $$I = \int_0^\infty (-1)^n e^{-u} . \frac{u^n}{(m+1)^n} \frac{du}{(m+1)}$$

$$= \frac{(-1)^n}{(m+1)^{n+1}} \int_0^\infty e^{-u} u^n du = \frac{(-1)^n}{(m+1)^{n+1}} \int_0^\infty e^{-u} u^{(n+1)-1} du$$

$$= \frac{(-1)^n}{(m+1)^{n+1}} \left\lceil n+1 \right.$$ **Proved.**

**Example 8.** *Prove that* $\int_0^1 (x \log x)^4 dx = \dfrac{4!}{5^5}$ *(M.U. II Semester, 2009)*

**Solution.** We know that

$$\int_0^1 x^m (\log x)^n dx = \frac{(-1)^n}{(m+1)^{n+1}} \cdot \left\lceil n+1 \right. \qquad ...(1) \text{ [From Art 22.2]}$$

Now, $\int_0^1 (x\log x)^4 dx = \int_0^1 x^4(\log x)^4 dx$

Putting $m = n = 4$ in (1), we get

$$\int_0^1 x^4(\log x)^4 dx = \frac{(-1)^4}{(4+1)^{4+1}}\overline{|4+1}$$

$$= \frac{\overline{|5}}{5^5} = \frac{4!}{5^5}$$ **Proved.**

**Example 9.** *Evaluate* $\int_0^1 x^{n-1}\left[\log_e\left(\frac{1}{x}\right)\right]^{m-1} \cdot dx$

**Solution:** Put $\log_e \frac{1}{x} = t$ or $x = e^{-t}$ $\therefore dx = -e^{-t}\,dt$

$$\int_0^1 x^{n-1}\left[\log_e\left(\frac{1}{x}\right)\right]^{m-1} dx = \int_\infty^0 \left(e^{-t}\right)^{n-1}[t]^{m-1}\left(-e^{-t}dt\right) = \int_0^\infty e^{-nt}t^{m-1}\,dt$$

Putting $nt = u \Rightarrow t = \frac{u}{n}$ so that $dt = \frac{du}{n}$

$$= \int_0^\infty e^{-u}\left(\frac{u}{n}\right)^{m-1}\frac{du}{n} = \frac{1}{n^m}\int_0^\infty e^{-u}u^{m-1}du = \frac{1}{n^m}\overline{|m}$$ **Ans.**

## 21.3 TRANSFORMATION OF GAMMA FUNCTION

Prove that (*i*) $\int_0^\infty e^{-ky}y^{n-1}dy = \frac{\overline{|n}}{k^n}$ (*ii*) $\overline{\left|\frac{1}{2}\right.} = \sqrt{\pi}$

(*iii*) $\int_0^1\left(\log\frac{1}{y}\right)^{n-1} dy = \overline{|n}$ (*iv*) $\Gamma n = \frac{1}{n}\int_0^\infty e^{-x^{\frac{1}{n}}}dx$

**Solution:** We know that $\overline{|n} = \int_0^\infty x^{n-1}e^{-x}dx$ ...(1)

(*i*) Replace $x$ by $k\,y$, so that $dx = k\,dy$; then (1) becomes

$$\overline{|n} = \int_0^\infty (k\,y)^{n-1}e^{-ky}.k\,dy.$$

$$\overline{|n} = k^n\int_0^\infty e^{-ky}y^{n-1}dy$$

$\therefore$ $$\boxed{\int_0^\infty e^{-ky}y^{n-1}dy = \frac{\overline{|n}}{k^n}}$$ ...(2) **Proved.**

(*ii*) Replace $x^n$ by $y$, so that $n\,x^{n-1}\,dx = dy$ in (1), then

$$\overline{|n} = \int_0^\infty y^{\frac{n-1}{n}}e^{-y^{1/n}}\frac{dy}{nx^{n-1}}$$

$$= \int_0^\infty y^{\frac{n-1}{n}}e^{-y^{1/n}}\frac{dy}{ny^{\frac{n-1}{n}}} = \frac{1}{n}\int_0^\infty e^{-y^{1/n}}dy$$

When $n = \frac{1}{2}$, $$\overline{\left|\frac{1}{2}\right.} = \frac{1}{\frac{1}{2}}\int_0^\infty e^{-y^2}dy = 2\left[\frac{1}{2}\sqrt{\pi}\right]$$

$$\boxed{\overline{\left|\frac{1}{2}\right.} = \sqrt{\pi}}$$ **Proved.**

(*iii*) Putting $e^{-x} = y$, so that $-e^{-x}\,dx = dy$ and $-x = \log y$, $x = \log \dfrac{1}{y}$, (1) becomes

$$\overline{|n} = -\int_1^0 \left(\log\frac{1}{y}\right)^{n-1} y \cdot \frac{dy}{e^{-x}}$$

$$= \int_0^1 \left(\log\frac{1}{y}\right)^{n-1} y \cdot \frac{dy}{y} = \int_0^1 \left(\log\frac{1}{\mathbf{y}}\right)^{n-1} dy \qquad \textbf{Proved.}$$

(*iv*) We know that, $\Gamma(n) = \int_0^\infty e^{-x} . x^{n-1}\,dx$ ...(1)

Putting $x^n = y \Rightarrow x = y^{1/n}$ so that $dx = \dfrac{1}{n} y^{\frac{1}{n}-1}\,dy$ in (1), we get

$$\Gamma n = \int_0^\infty e^{-y^{1/n}}\, y^{\frac{n-1}{n}} . \frac{1}{n} y^{\frac{1}{n}-1}\,dy = \frac{1}{n}\int_0^\infty e^{-y^{\frac{1}{n}}}\,dy$$

$$\Gamma n = \frac{1}{n}\int_0^\infty e^{-x^{\frac{1}{n}}}\,dx. \qquad \textbf{Proved.}$$

## EXERCISE 21.1

**Evaluate:**

**1.** (*i*) $\overline{\left| -\frac{3}{2} \right.}$ (*ii*) $\overline{\left| -\frac{15}{2} \right.}$ (*iii*) $\overline{\left| \frac{7}{2} \right.}$ (*iv*) $\overline{|0}$

**Ans.** (*i*) $\frac{4}{3}\sqrt{\pi}$ (*ii*) $\dfrac{2^8\sqrt{\pi}}{15\times13\times11\times9\times7\times5\times3}$ (*iii*) $\dfrac{15\sqrt{\pi}}{8}$ (*iv*) $\infty$

**2.** $\int_0^\infty \sqrt{x}\, e^{-x} dx$ (*GBTU 2014*) **Ans.** $\overline{\left| \frac{3}{2} \right.}$

**3.** $\int_0^\infty x^4 e^{-x^2} dx$ **Ans.** $\dfrac{3\sqrt{\pi}}{8}$

**4.** $\int_0^\infty e^{-h^2x^2} dx$ (*UP. Jan. 2011*) **Ans.** $\dfrac{\sqrt{\pi}}{2h}$

**5.** $\int_0^\infty \int_0^\infty e^{-\left(ax^2+by^2\right)} x^{2m-1} y^{2n-1} dx\,dy, \quad a,b,m,n>0$ **Ans.** $\dfrac{\overline{|m}\,\overline{|n}}{4\,a^m b^n}$

**6.** $\int_0^1 \dfrac{dx}{\sqrt{-\log x}}$ **Ans.** $\sqrt{\pi}$

**7.** $\int_0^1 (x\log x)^3\,dx$ **Ans.** $-\dfrac{3}{128}$

**8.** $\int_0^1 \dfrac{dx}{\sqrt{x \log\frac{1}{x}}}$ **Ans.** $\sqrt{2\pi}$

**9.** Prove that 1.3.5.... $(2n - 1) = \dfrac{2^n \overline{\left| n+\frac{1}{2} \right.}}{\sqrt{\pi}}$

**10.** $\int_0^\infty e^{-y^{1/m}}\,dy = m\,\overline{|m}$

**11.** $\int_0^\infty \int_0^\infty e^{-(ax^2+by^2)}\, x^{2m-1}\, y^{2n-1}\,dx\,dy = \dfrac{\Gamma(m)\,\Gamma(n)}{4\,a^m b^n}$, where $a$, $b$, $m$, $n$ are positive.

**12.** $\int_0^{\pi/2} \dfrac{d\theta}{(a\cos^4\theta + b\sin^4\theta)} = \dfrac{(\Gamma\, 1/4)^2}{4(ab)^{1/4}\sqrt{\pi}}$ [**Hint.** Put $\tan\theta = t$ then $bt^4 = az$]

## 21.4 BETA FUNCTION

(*U.P. I Semester Dec. 2007*)

$$\int_0^1 x^{l-1}(1-x)^{m-1}dx \qquad (l > 0, m > 0)$$

is called the Beta function of *l, m*. It is also written as

$$\boxed{\beta(l, m) = \int_0^1 x^{l-1}(1-x)^{m-1}dx}$$

## 21.5 EVALUATION OF BETA FUNCTION

$$\beta(l, m) = \frac{\Gamma l\,\Gamma m}{\Gamma(l+m)}$$

**Solution.** We have, $\beta(l, m) = \int_0^1 x^{l-1}(1-x)^{m-1}dx = \int_0^1 (1-x)^{m-1}x^{l-1}dx$

Integrating by parts, we have

$$= \left[(1-x)^{m-1}\frac{x^l}{l}\right]_0^1 + (m-1)\int_0^1 (1-x)^{m-2}\left(\frac{x^l}{l}\right)dx$$

$$= \frac{(m-1)}{l}\int_0^1 (1-x)^{m-2}x^l dx$$

Again integrating by parts, we get

$$= \frac{(m-1)(m-2)}{l(l+1)}\int_0^1 (1-x)^{m-3}x^{l+1}dx$$

$$= \frac{(m-1)(m-2)...2.1}{l(l+1).....(l+m-2)}\int_0^1 x^{l+m-2}dx$$

$$= \frac{(m-1)(m-2)...2.1}{l(l+1).....(l+m-2)}\left[\frac{x^{l+m-1}}{l+m-1}\right]_0^1$$

$$= \frac{(m-1)(m-2)...2.1}{l(l+1).....(l+m-2)(l+m-1)}$$

$$= \frac{(m-1)!}{l(l+1).....(l+m-2)(l+m-1)} \times \frac{(l-1)(l-2).....1}{(l-1)(l-2).....1}$$

$$= \frac{(m-1)!\ (l-1)!}{1.2....(l-2)(l-1).l(l+1)...(l+m-2)(l+m-1)}$$

$$= \frac{(l-1)!(m-1)!}{(l+m-1)!}$$

$$= \frac{\Gamma l\,\Gamma m}{\Gamma(l+m)}$$

And if only *l* is positive integer and not *m* then

$$\boxed{\beta(l, m) = \frac{(l-1)!}{m(m+1)...(m+l-1)}}$$

**Ans.**

## 21.6 A PROPERTY OF BETA FUNCTION

$$\beta(l, m) = \beta(m, l)$$

**Solution.** We have

$$\beta(l, m) = \int_0^1 x^{l-1}(1-x)^{m-1}\,dx \qquad \left[\int_0^a f(x)\,dx = \int_0^a f(a-x)\,dx\right]$$

$$= \int_0^1 (1-x)^{l-1}\left[1-(1-x)\right]^{m-1} dx$$

$$= \int_0^1 (1-x)^{l-1} x^{m-1}\,dx = \int_0^1 x^{m-1}(1-x)^{l-1}\,dx$$

$l$ and $m$ are interchanged.

$$\boxed{\beta(l, m) = \beta(m, l)}$$ **Proved.**

**Example 10.** *Evaluate* $\int_0^1 x^4\left(1-\sqrt{x}\right)^5 dx$

**Solution.** Let $\sqrt{x} = t \Rightarrow x = t^2$ so that $dx = 2\,t\,dt$

$$\int_0^1 x^4\left(1-\sqrt{x}\right)^5 dx = \int_0^1 (t^2)^4(1-t)^5(2\,t\,dt)$$

$$= 2\int_0^1 t^9(1-t)^5\,dt = 2\,\beta(10, 6) = 2\frac{\boxed{10}\,\boxed{6}}{\boxed{16}} = 2\,.\frac{9!\,5!}{(15)!}$$

$$= 2.\frac{5!}{10\times11\times12\times13\times14\times15} = \frac{2\times1\times2\times3\times4\times5}{10\times11\times12\times13\times14\times15}$$

$$= \frac{1}{11\times13\times7\times15} = \frac{1}{15015}$$ **Ans.**

**Example 11.** *Evaluate* $\int_0^1 (1-x^3)^{-\frac{1}{2}}\,dx$

**Solution.** Let $x^3 = y \Rightarrow x = y^{1/3}$ so that $dx = \frac{1}{3}y^{-\frac{2}{3}}\,dy$

$$\int_0^1 (1-x^3)^{-\frac{1}{2}}\,dx = \int_0^1 (1-y)^{-\frac{1}{2}}\left(\frac{1}{3}y^{-\frac{2}{3}}\,dy\right)$$

$$= \frac{1}{3}\int_0^1 y^{-\frac{2}{3}}(1-y)^{-\frac{1}{2}}\,dy = \frac{1}{3}\beta\left(\frac{1}{3}, \frac{1}{2}\right) = \frac{1}{3}\frac{\boxed{\frac{1}{3}}\,\boxed{\frac{1}{2}}}{\boxed{\frac{5}{6}}}$$ **Ans.**

## 21.7 TRANSFORMATION OF BETA FUNCTION

We know that

$$\beta(l, m) = \int_0^1 x^{l-1}(1-x)^{m-1}\,dx,$$

Putting $x = \frac{1}{1+y}$ so that $dx = -\frac{1}{(1+y)^2}\,dy$ and $1-x = \frac{y}{1+y}$ in (1), we get

$$\beta(l, m) = \int_\infty^0 \left(\frac{1}{1+y}\right)^{l-1}\left(\frac{y}{1+y}\right)^{m-1}\left[-\frac{1}{(1+y)^2}\,dy\right]$$

$$= \int_0^\infty \frac{y^{m-1}}{(1+y)^{l+m}}\,dy$$

Since $l$, $m$ can be interchanged in $\beta\,(l, m)$,

$$\beta\,(l, m) = \int_0^\infty \frac{y^{l-1}}{(1+y)^{m+l}}\,dy \Rightarrow \beta\,(l, m) = \int_0^\infty \frac{x^{l-1}}{(1+x)^{m+l}}\,dx \quad ...(1)$$

**Example 12.** *Evaluate* $\int_0^1 \frac{x^{m-l} + x^{n-l}}{(1+x)^{m+n}}\,dx$ *(MTU, 2012)*

**Solution.** We know that

$$\beta\,(m, n) = \int_0^\infty \frac{x^{m-1}}{(1+x)^{m+n}}\,dx \Rightarrow \int_0^\infty \frac{x^{m-1}}{(1+x)^{m+n}}\,dx = \beta\,(m,n)$$

$$\Rightarrow \int_0^1 \frac{x^{m-1}}{(1+x)^{m+n}}\,dx + \int_1^\infty \frac{x^{m-1}}{(1+x)^{m+n}}\,dx = \beta\,(m, n) \quad ...(1)$$

Consider $\int_1^\infty \frac{x^{m-1}}{(1+x)^{m+n}}\,dx = \int_1^0 \frac{\left(\frac{1}{t}\right)^{m-1}}{\left(1+\frac{1}{t}\right)^{m+n}}\left(-\frac{1}{t^2}\,dt\right) = \int_0^1 \frac{\left(\frac{1}{t}\right)^{m-1}\frac{1}{t^2}}{\left(\frac{1}{t}\right)^{m+n}(t+1)^{m+n}}\,dt \quad \left(\text{Put } x = \frac{1}{t}\right)$

$$= \int_0^1 \frac{t^{n-1}}{(1+t)^{m+n}}\,dt = \int_0^1 \frac{x^{n-1}}{(1+x)^{m+n}}\,dx$$

Putting the value of $\int_1^\infty \frac{x^{m-1}}{(1+x)^{m+n}}\,dx$ in (1), we get

$$\int_0^1 \frac{x^{m-1}}{(1+x)^{m+n}}\,dx + \int_0^1 \frac{x^{n-1}}{(1+x)^{m+n}}\,dx = \beta\,(m, n)$$

$$\Rightarrow \int_0^1 \frac{x^{m-1} + x^{n-1}}{(1+x)^{m+n}}\,dx = \beta\,(m, n) \quad \textbf{Ans.}$$

## 21.8 RELATION BETWEEN BETA AND GAMMA FUNCTIONS

We know that, $\Gamma l = \int_0^\infty e^{-x}\,x^{l-1}dx$, [Put $zx = y$]

$$\frac{\Gamma l}{z^l} = \int_0^\infty e^{-zx}\,x^{l-1}dx$$

$$\Gamma l = \int_0^\infty z^l\,e^{-zx}x^{l-1}dx$$

Multiplying both sides by $e^{-z}\,z^{m-1}$, we have

$$\Gamma l\,.\,e^{-z}.\,z^{m-1} = \int_0^\infty e^{-z}.z^{m-1}.z^l.e^{-zx}.x^{l-1}dx$$

$$\Gamma l\,.\,e^{-z}.z^{m-1} = \int_0^\infty e^{-(1+x)z}.z^{l+m-1}.x^{l-1}dx$$

Integrating both sides w.r.t. '$z$', we get

$$\int_0^\infty \Gamma l\,e^{-z}z^{m-1}\,dz = \int_0^\infty\int_0^\infty e^{-(1+x)z}z^{l+m-1}x^{l-1}\,dx\,dz$$

$$\Gamma l\,\Gamma m = \int_0^\infty x^{l-1}\,dx \int_0^\infty e^{-(1+x)z}z^{l+m-1}dz$$

$$= \int_0^\infty x^{l-1} dx \cdot \frac{\Gamma(l+m)}{(1+x)^{l+m}} \qquad \text{[From (1), Art 21.7]}$$

$$\Gamma l \, \Gamma m = \Gamma(l+m) \int_0^\infty \frac{x^{l-1}}{(1+x)^{l+m}} dx = \Gamma(l+m) \cdot \beta(l, m)$$

$$\therefore \quad \boxed{\beta(l, m) = \frac{\Gamma l \, \Gamma m}{\Gamma(l+m)}}$$

This is the required relation.

## 21.9. SHOW THAT

$$\int_0^{\frac{\pi}{2}} \sin^p\theta \cos^q\theta d\theta = \frac{\Gamma\left(\frac{p+1}{2}\right) \Gamma\left(\frac{q+1}{2}\right)}{2\Gamma\left(\frac{p+q+2}{2}\right)}$$

**Solution.** We know that

$$\beta(m, n) = \int_0^1 x^{m-1} (1-x)^{n-1} dx \qquad \text{...(1)}$$

Putting $x = \sin^2\theta, \; dx = 2 \sin\theta \cos\theta \, d\theta$

and $1 - x = 1 - \sin^2\theta = \cos^2\theta$

Then (1) becomes

$$\beta(m,n) = \int_0^{\frac{\pi}{2}} \sin^{2m-2}\theta \cos^{2n-2}\theta \; 2\sin\theta\cos\theta \, d\theta$$

or

$$\frac{\Gamma m \, \Gamma n}{\Gamma(m+n)} = 2\int_0^{\frac{\pi}{2}} \sin^{2m-1}\theta \cos^{2n-1}\theta \, d\theta$$

Putting $2m - 1 = p, \quad i.e., \quad m = \frac{p+1}{2}$

and $2n - 1 = q, \quad i.e., \quad n = \frac{q+1}{2}$

$$\frac{\Gamma\frac{p+1}{2} \Gamma\frac{q+1}{2}}{\Gamma\frac{p+q+2}{2}} = 2\int_0^{\frac{\pi}{2}} \sin^p\theta \cdot \cos^q\theta \, d\theta$$

$$\boxed{\int_0^{\frac{\pi}{2}} \sin^p\theta \cos^q\theta \, d\theta = \frac{\Gamma\frac{p+1}{2} \Gamma\frac{q+1}{2}}{2\Gamma\frac{p+q+2}{2}}}$$

**Proved.**

**Example 13.** *Find the value of* $\Gamma\frac{1}{2}$.

**Solution.** We have already solved this problem in Art. 22.3(ii) Transformation of the Gamma Function.

Now, by **Second method:** We know that,

$$\int_0^{\frac{\pi}{2}} \sin^p\theta \cdot \cos^q\theta \, d\theta = \frac{\left|\frac{p+1}{2}\right. \left|\frac{q+1}{2}\right.}{2\left|\frac{p+q+2}{2}\right.}$$

Putting $p = q = 0$, we get $\int_0^{\frac{\pi}{2}} d\theta = \frac{\left|\frac{1}{2}\right. \left|\frac{1}{2}\right.}{2\left|1\right.}$

$$\Rightarrow \qquad [\theta]_0^{\pi/2} = \frac{1}{2}\left(\left|\frac{1}{2}\right.\right)^2 \Rightarrow \frac{\pi}{2} = \frac{1}{2}\left(\left|\frac{1}{2}\right.\right)^2$$

$$\Rightarrow \qquad \left(\left|\frac{1}{2}\right.\right)^2 = \pi \qquad \Rightarrow \left|\frac{1}{2}\right. = \sqrt{\pi}$$ **Ans.**

**Example 14.** *Show that*

$$\int_0^{\frac{\pi}{2}} \sqrt{\cot\theta} \, d\theta = \frac{1}{2} \left|\frac{1}{4}\right. \left|\frac{3}{4}\right.$$

**Solution.** We know that

$$\int_0^{\frac{\pi}{2}} \sin^p\theta \cdot \cos^q\theta \, d\theta = \frac{\left|\frac{p+1}{2}\right. \left|\frac{q+1}{2}\right.}{2\left|\frac{p+q+2}{2}\right.} \qquad \text{...(1)}$$

$$\int_0^{\frac{\pi}{2}} \sqrt{\cot\theta} \, d\theta = \int_0^{\frac{\pi}{2}} \frac{\cos^{1/2}\theta}{\sin^{1/2}\theta} d\theta = \int_0^{\frac{\pi}{2}} \sin^{-1/2}\theta \cos^{1/2}\theta \, d\theta$$

On applying formula (1), we have

$$\int_0^{\frac{\pi}{2}} \sqrt{\cot\theta} \, d\theta = \frac{\left|\frac{-\frac{1}{2}+1}{2}\right. \left|\frac{\frac{1}{2}+1}{2}\right.}{2\left|\frac{-\frac{1}{2}+\frac{1}{2}+2}{2}\right.} = \frac{\left|\frac{1}{4}\right. \left|\frac{3}{4}\right.}{2\left|1\right.} = \frac{1}{2}\left|\frac{1}{4}\right. \left|\frac{3}{4}\right.$$ **Proved.**

**Example 15.** *Using Beta and Gamma functions, evaluate*

$$\int_0^1 \left(\frac{x^3}{1-x^3}\right)^{\frac{1}{2}} dx$$

**Solution.**

$$\int_0^1 \left(\frac{x^3}{1-x^3}\right)^{\frac{1}{2}} dx \qquad \text{...(1)}$$

Putting $x^3 = \sin^2\theta$, so that $x = \sin^{\frac{2}{3}}\theta$, $dx = \frac{2}{3}\sin^{-\frac{1}{3}}\theta\cos\theta \, d\theta$ in (1), we get

$$\int_0^1 \left(\frac{x^3}{1-x^3}\right)^{\frac{1}{2}} dx = \int_0^{\pi/2} \left(\frac{\sin^2\theta}{1-\sin^2\theta}\right)^{\frac{1}{2}} \frac{2}{3}\sin^{-\frac{1}{3}}\theta\cos\theta \, d\theta$$

$$= \frac{2}{3}\int_0^{\pi/2} \left(\frac{\sin\theta}{\cos\theta}\right) \sin^{-\frac{1}{3}}\theta\cos\theta \, d\theta = \frac{2}{3}\int_0^{\pi/2} \sin^{\frac{2}{3}}\theta \, d\theta$$

$$= \frac{2}{3}\frac{\Gamma\left(\frac{\frac{2}{3}+1}{2}\right)\Gamma\left(\frac{0+1}{2}\right)}{\Gamma\left(\frac{\frac{2}{3}+1+1}{2}\right)} = \frac{2}{3}\frac{\Gamma\left(\frac{5}{6}\right)\Gamma\left(\frac{1}{2}\right)}{\Gamma\left(\frac{4}{3}\right)} = \frac{2}{3}\frac{\sqrt{\pi}\,\Gamma\left(\frac{5}{6}\right)}{\Gamma\left(\frac{4}{3}\right)}$$ **Ans.**

**Example 16.** *Evaluate* $\int_{-1}^{+1}(1+x)^{p-1}(1-x)^{q-1}\,dx$.

**Solution.** Put $x = \cos 2\theta$, then $dx = -2\sin 2\theta\, d\theta$

$$\begin{aligned}
\int_{-1}^{+1}(1+x)^{p-1}(1-x)^{q-1}\,dx &= \int_{\frac{\pi}{2}}^{0}(1+\cos 2\theta)^{p-1}(1-\cos 2\theta)^{q-1}(-2\sin 2\theta\, d\theta)\\
&= \int_{\frac{\pi}{2}}^{0}\left(1+2\cos^2\theta-1\right)^{p-1}\left(1-1+2\sin^2\theta\right)^{q-1}(-4\sin\theta\cos\theta\, d\theta)\\
&= 4\int_0^{\frac{\pi}{2}} 2^{p-1}\cos^{2p-2}\theta\,.\,2^{q-1}\sin^{2q-2}\theta\,.\sin\theta\cos\theta\, d\theta\\
&= 2^{p+q}\int_0^{\infty}\sin^{2q-1}\theta\cos^{2p-1}\theta\, d\theta\\
&= 2^{p+q}\frac{\Gamma\left(\frac{2q}{2}\right)\Gamma\left(\frac{2p}{2}\right)}{2\,\Gamma\left(\frac{2p+2q}{2}\right)} = 2^{p+q-1}\frac{\Gamma(p)\,\Gamma(q)}{\Gamma(p+q)}
\end{aligned}$$ **Ans.**

**Example 17.** *Show that* $\Gamma(n)\,\Gamma(1-n) = \frac{\pi}{\sin n\pi}$ $(0 < n < 1)$

**Solution.** We know that

$$\beta(m, n) = \int_0^{\infty}\frac{x^{n-1}}{(1+x)^{m+n}}\,dx \qquad \text{[From (1), Art 21.7]}$$

$$\frac{\Gamma(m)\,\Gamma(n)}{\Gamma(m+n)} = \int_0^{\infty}\frac{x^{n-1}}{(1+x)^{m+n}}\,dx$$

Putting $m + n = 1$ or $m = 1 - n$, we get

$$\frac{\Gamma(1-n)\,\Gamma(n)}{\Gamma(1)} = \int_0^{\infty}\frac{x^{n-1}}{(1+x)^1}\,dx$$

$$\Gamma(1-n)\,\Gamma(n) = \int_0^{\infty}\frac{x^{n-1}}{1+x}\,dx \qquad \left[\int_0^{\infty}\frac{x^{n-1}}{1+x}\,dx = \frac{\pi}{\sin n\pi}\right]$$

$$\Rightarrow \Gamma(n)\,\Gamma(1-n) = \frac{\pi}{\sin n\pi}$$ **Proved.**

**Example 18.** *Assuming* $\Gamma(n)\,\Gamma(1-n) = \pi \operatorname{cosec} n\pi$. $0 < n < 1$, *show that*

$$\int_0^{\infty}\frac{x^{p-1}}{1+x}\,dx = \left(\frac{\pi}{\sin p\pi}\right).\ 0 < p < 1$$ (*U.P., I Semester , Dec 2009*)

**Solution:** Here, we have $\pi \operatorname{cosec} n\pi = \Gamma(n)\,\Gamma(1-n)$

$$\Rightarrow \qquad \frac{\pi}{\sin n\pi} = \overline{|n}\,\overline{|1-n}$$

We know that $\displaystyle\int_0^\infty \frac{x^{n-1}}{1+x}dx = \frac{\overline{|1-n}\,\overline{|n}}{\overline{|1}}$ ...(1) (From example 10)

Setting $m + n = 1$ so that $m = 1 - n$ in (1), we get

$$\int_0^\infty \frac{x^{n-1}}{(1+x)^{m+n}}dx = \frac{\overline{|m}\,\overline{|n}}{\overline{|m+n}} = B(m,n)$$ **Proved.**

**Example 19.** *Using Beta and Gamma functions, evaluate* $\displaystyle\int_0^\infty \frac{dx}{1+x^4}$. *(GBTU, Dec. 2012)*

**Solution.** Let $\displaystyle I = \int_0^\infty \frac{dx}{1+x^4}$

Putting $x^4 = y$ so that $x = y^{\frac{1}{4}}$ and $dx = \frac{1}{4}y^{-\frac{3}{4}}\,dy$, we get

$$I = \int_0^\infty \frac{\frac{1}{4}y^{-\frac{3}{4}}}{1+y}\cdot dy = \frac{1}{4}\int_0^\infty \frac{y^{\frac{1}{4}-1}}{1+y}dy \qquad \left[\int_0^\infty \frac{x^{n-1}}{1+x}dx = \frac{\pi}{\sin n\pi}\right]$$

$$= \frac{1}{4}\frac{\pi}{\sin\frac{\pi}{4}} = \frac{\pi\sqrt{2}}{4}$$ **Ans.**

**Example 20.** *Prove that* $\overline{\left|\left(\frac{1}{4}\right)\right.}\,\overline{\left|\left(\frac{3}{4}\right)\right.} = \pi\sqrt{2}$

**Solution.** Putting $n = \frac{1}{4}$ in result of example 18, we obtain

$$\overline{\left|\left(\frac{1}{4}\right)\right.}\,\overline{\left|\left(1-\frac{1}{4}\right)\right.} = \frac{\pi}{\sin\frac{\pi}{4}}$$

$$\Rightarrow \qquad \overline{\left|\left(\frac{1}{4}\right)\right.}\,\overline{\left|\left(\frac{3}{4}\right)\right.} = \frac{\pi}{\left(\frac{1}{\sqrt{2}}\right)} \Rightarrow \overline{\left|\left(\frac{1}{4}\right)\right.}\,\overline{\left|\left(\frac{3}{4}\right)\right.} = \pi\sqrt{2}.$$ **Proved.**

**Example 21.** *Evaluate* $\displaystyle\int_0^1 \frac{dx}{(1-x^n)^{\frac{1}{n}}}$

**Solution.** Let $x^n = \sin^2\theta$ or $x = \sin^{2/n}\theta$

So that $\qquad dx = \frac{2}{n}\sin^{2/n-1}\theta\cos\theta\,d\theta$

$$\int_0^1 \frac{dx}{(1-x^n)^{\frac{1}{n}}} = \int_0^{\frac{\pi}{2}} \frac{\frac{2}{n}\sin^{2/n-1}\theta\cos\theta}{(1-\sin^2\theta)^{1/n}}d\theta = \frac{2}{n}\int_0^{\frac{\pi}{2}} \frac{\sin^{2/n-1}\theta\cos\theta}{(\cos^2\theta)^{1/n}}d\theta$$

$$= \frac{2}{n}\int_0^{\frac{\pi}{2}} \sin^{2/n-1}\theta\cos^{1-2/n}\theta\,d\theta$$

$$= \frac{2}{n}\frac{\left|\overline{\dfrac{\frac{2}{n}-1+1}{2}}\right.\left|\overline{\dfrac{1-\frac{2}{n}+1}{2}}\right.}{2\left|\overline{\dfrac{\frac{2}{n}-1+1+2-\frac{2}{n}}{2}}\right.} = \frac{1}{n}\frac{\left|\overline{\frac{1}{n}}\right.\left|\overline{\frac{n-1}{n}}\right.}{\left|\overline{1}\right.} \qquad \left(\left|\overline{\frac{1}{n}}\right.\left|\overline{1-\frac{1}{n}}\right. = \frac{\pi}{\sin\frac{\pi}{n}}\right)$$

$$= \frac{\pi}{n\sin\frac{\pi}{n}}$$ **Ans.**

**Example 22.** *Show that* $\int_0^a \frac{dx}{\sqrt[n]{(a^n - x^n)}} = \frac{\pi}{n}\, cosec\left(\frac{\pi}{n}\right)$, *where* $n > 1$.*(M.U. II Semester 2009)*

**Solution.** Let $x^n = a^n \sin^2\theta \quad \Rightarrow \quad x = a\sin^{\frac{2}{n}}\theta$

So that $\qquad dx = \frac{2a}{n}\sin^{\frac{2}{n}-1}\theta\cos\theta\, d\theta$

$$\therefore \int_0^a \frac{dx}{\sqrt[n]{a^n - x^n}} = \int_0^{\frac{\pi}{2}} \frac{a\times\frac{2}{n}\sin^{\frac{2}{n}-1}\theta\cos\theta}{(a^n - a^n\sin^2\theta)^{\frac{1}{n}}}\, d\theta = \frac{2}{n}\int_0^{\frac{\pi}{2}} \frac{\sin^{\frac{2}{n}-1}\theta\cos\theta}{\cos^{\frac{2}{n}}\theta}\, d\theta$$

$$= \frac{2}{n}\int_0^{\frac{\pi}{2}} \sin^{\frac{2}{n}-1}\theta\cos^{1-\frac{2}{n}}\theta\, d\theta$$

$$= \frac{2}{n}\frac{\left|\overline{\dfrac{\frac{2}{n}-1+1}{2}}\right.\left|\overline{\dfrac{1-\frac{2}{n}+1}{2}}\right.}{2\left|\overline{\dfrac{\frac{2}{n}-1+1+2-\frac{2}{n}}{2}}\right.} = \frac{1}{n}\frac{\left|\overline{\frac{1}{n}}\right.\left|\overline{\frac{n-1}{n}}\right.}{\left|\overline{1}\right.} \qquad \left[\left|\overline{\frac{1}{n}}\right.\left|\overline{1-\frac{1}{n}}\right. = \frac{\pi}{\sin\frac{\pi}{n}}\right]$$

$$= \frac{\pi}{n\sin\frac{\pi}{n}} = \frac{\pi}{n}\,\text{cosec}\left(\frac{\pi}{n}\right)$$ **Proved.**

**Example 23.** *Show that* $\int_0^{\frac{\pi}{2}} tan^P\theta\, d\theta = \frac{\pi}{2} sec\frac{P\pi}{2}$ *and indicate the restriction on the values of P.*

**Solution.** $$\int_0^{\frac{\pi}{2}} \tan^P\theta\, d\theta = \int_0^{\frac{\pi}{2}} \sin^P\theta\cos^{-P}\theta\, d\theta$$

$$= \frac{\left|\overline{\frac{P+1}{2}}\right.\left|\overline{\frac{-P+1}{2}}\right.}{2\left|\overline{\frac{P+1-P+1}{2}}\right.} \qquad \begin{bmatrix} 1-P>0 \\ 1>P \end{bmatrix}$$

$$= \frac{\left|\overline{\frac{P+1}{2}}\right.\left|\overline{\frac{-P+1}{2}}\right.}{2\left|\overline{1}\right.} \qquad \begin{bmatrix} 1+P>0 \\ P>-1 \end{bmatrix}$$

$$= \frac{1}{2}\Gamma\left(\frac{P+1}{2}\right)\Gamma\left(\frac{-P+1}{2}\right) \qquad \therefore 1 > P > -1$$

$$= \frac{1}{2}\frac{\pi}{\sin\frac{P+1}{2}\pi} = \frac{1}{2}\frac{\pi}{\cos\frac{P\pi}{2}} = \frac{\pi}{2}\sec\frac{P\pi}{2}$$ **Proved.**

## 21.10 DUPLICATION FORMULA

$$\Gamma(m)\,\Gamma\left(m+\frac{1}{2}\right) = \frac{\sqrt{\pi}}{2^{2m-1}}\Gamma(2m), \text{ where } m \text{ is positive.}$$

*Hence show that* $\beta(m, m) = 2^{1-2m}\,\beta\left(m, \frac{1}{2}\right)$ (*U.P., II Semester; Summer 2001*)

**Proof.** We know that

$$\frac{\Gamma\left(\frac{p+1}{2}\right)\Gamma\left(\frac{q+1}{2}\right)}{2\,\Gamma\left(\frac{p+q+2}{2}\right)} = \int_0^{\frac{\pi}{2}} \sin^p\theta\cdot\cos^q\theta\, d\theta$$

Putting $q = p$, we get

$$\frac{\Gamma\left(\frac{p+1}{2}\right)\Gamma\left(\frac{p+1}{2}\right)}{2\,\Gamma(p+1)} = \int_0^{\frac{\pi}{2}} \sin^p\theta\cdot\cos^p\theta\, d\theta = \int_0^{\frac{\pi}{2}} (\sin\theta\cos\theta)^p\, d\theta$$

$$= \int_0^{\frac{\pi}{2}} \frac{1}{2^p}(2\sin\theta\cos\theta)^p\, d\theta = \frac{1}{2^p}\int_0^{\frac{\pi}{2}} (\sin 2\theta)^p\, d\theta$$

$$\left[\text{Putting } 2\theta = t \Rightarrow d\theta = \frac{dt}{2}\right]$$

$$= \frac{1}{2^p}\int_0^{\pi} \sin^p t\,\frac{dt}{2}$$

$$= \frac{1}{2^p}\cdot\frac{1}{2}\cdot 2\int_0^{\frac{\pi}{2}} \sin^p t\, dt = \frac{1}{2^p}\int_0^{\frac{\pi}{2}} \sin^p t\cos^0 t\, dt$$

$$= \frac{1}{2^p}\frac{\Gamma\left(\frac{p+1}{2}\right)\Gamma\left(\frac{0+1}{2}\right)}{2\,\Gamma\left(\frac{p+2}{2}\right)}$$

$$\Rightarrow \quad \frac{\Gamma\left(\frac{p+1}{2}\right)\Gamma\left(\frac{p+1}{2}\right)}{2\,\Gamma(p+1)} = \frac{1}{2^p}\frac{\Gamma\left(\frac{p+1}{2}\right)\Gamma\left(\frac{1}{2}\right)}{2\,\Gamma\left(\frac{p+2}{2}\right)}$$

$$\Rightarrow \quad \frac{\Gamma\left(\frac{p+1}{2}\right)}{\Gamma(p+1)} = \frac{1}{2^p}\frac{\Gamma\left(\frac{1}{2}\right)}{\Gamma\left(\frac{p+2}{2}\right)}$$

$$\Rightarrow \quad \frac{\Gamma\left(\frac{p+1}{2}\right)}{\Gamma(p+1)} = \frac{1}{2^p}\frac{\sqrt{\pi}}{\Gamma\left(\frac{p+2}{2}\right)}$$

Take $\frac{p+1}{2} = m \quad \Rightarrow \quad p = 2m - 1$

$$\Rightarrow \qquad \frac{\Gamma m}{\Gamma 2m} = \frac{1}{2^{2m-1}}\frac{\sqrt{\pi}}{\Gamma\left(\frac{2m+1}{2}\right)} \qquad ...(1)$$

$$\Gamma m\,\Gamma\left(m+\frac{1}{2}\right) = \frac{\sqrt{\pi}}{2^{2m-1}}\,\Gamma 2m$$

**Proved.**

Multiplying both sides of (1) by $\Gamma m$, we have

$$\frac{\Gamma m\,\Gamma m}{\Gamma 2m} = 2^{1-2m}\frac{\Gamma\frac{1}{2}\cdot\Gamma m}{\Gamma\left(m+\frac{1}{2}\right)}$$

$$\boldsymbol{\beta(m, m) = 2^{1-2m}\,\beta\left(m, \frac{1}{2}\right)}$$

**Proved.**

**Example 24.** *For a β function, show that*

$$\beta(p, q) = \beta(p+1, q) + \beta(p, q+1)$$

(*U.P., Ist Semester, Dec 2008*)

**Solution.** $\beta(p+1, q) + \beta(p, q+1)$

$$= \int_0^1 x^p (1-x)^{q-1}\, dx + \int_0^1 x^{p-1}(1-x)^q\, dx$$

$$= \int_0^1 x^{p-1}(1-x)^{q-1}\,[x + 1 - x]\, dx \qquad \text{(Taking common)}$$

$$= \int_0^1 x^{p-1}(1-x)^{q-1}\, dx$$

$$= \beta(p, q)$$

**Proved.**

**Example 25.** *Prove that* $\beta(m, m) \times \beta\left(m+\frac{1}{2}, m+\frac{1}{2}\right) = \frac{\pi}{m}\,2^{1-4m}$ (*M.U. II, Semester, 2008*)

**Solution.**

$$\text{L.H.S.} = \beta(m, m) \times \beta\left(m+\frac{1}{2}, m+\frac{1}{2}\right)$$

$$= \frac{\Gamma m\,\Gamma m}{\Gamma 2m} \times \frac{\Gamma\left(m+\frac{1}{2}\right)\Gamma\left(m+\frac{1}{2}\right)}{\Gamma(2m+1)}$$

$$= \frac{(\Gamma m)^2}{\Gamma 2m} \times \frac{\left(\Gamma\left(m+\frac{1}{2}\right)\right)^2}{2m\,\Gamma 2m} \qquad [\because \Gamma(2m+1) = 2m\,\Gamma 2m]$$

$$= \left(\frac{\Gamma m\,\Gamma\left(m+\frac{1}{2}\right)}{\Gamma 2m}\right)^2 \cdot \frac{1}{2m} = \left(\frac{\sqrt{\pi}}{2^{2m-1}}\right)^2 \frac{1}{2m}$$

$$\left[\begin{array}{c}\text{By duplication formula} \\ 2^{2m-1}\cdot\Gamma m\cdot\Gamma\left(m+\frac{1}{2}\right) = \sqrt{\pi}\cdot\Gamma 2m\end{array}\right]$$

$$\frac{\Gamma m\,\Gamma\left(m+\frac{1}{2}\right)}{\Gamma 2m} = \frac{\sqrt{\pi}}{2^{2m-1}}$$

$$= \frac{\pi}{2^{4m-2}}\cdot\frac{1}{2m}$$

$$= \frac{\pi}{m}\cdot 2^{1-4m}$$

$$= \text{R.H.S.}$$

**Proved.**

## 21.11 TO SHOW THAT

$$\Gamma\left(\frac{1}{n}\right)\Gamma\left(\frac{2}{n}\right)\Gamma\left(\frac{3}{n}\right)\ldots\ldots\Gamma\left(\frac{n-1}{n}\right)=\frac{(2\pi)^{\left(\frac{n-1}{2}\right)}}{n^{1/2}}$$

where $n$ is a positive integer than one.

**Proof.** Let $$P=\Gamma\left(\frac{1}{n}\right)\Gamma\left(\frac{2}{n}\right)\Gamma\left(\frac{3}{n}\right)\ldots\ldots\Gamma\left(\frac{n-2}{n}\right)\Gamma\left(\frac{n-1}{n}\right)$$

$$=\Gamma\left(\frac{1}{n}\right)\Gamma\left(\frac{2}{n}\right)\Gamma\left(\frac{3}{n}\right)\ldots\ldots\Gamma\left(1-\frac{2}{n}\right)\Gamma\left(1-\frac{1}{n}\right) \qquad \ldots(1)$$

Writing the value of $P$ in the reverse order, we have

$$P=\Gamma\left(1-\frac{1}{n}\right)\Gamma\left(1-\frac{2}{n}\right)\ldots\ldots\Gamma\left(\frac{3}{n}\right)\Gamma\left(\frac{2}{n}\right)\Gamma\left(\frac{1}{n}\right) \qquad \ldots(2)$$

Multiplying (1) and (2), we get

$$P^2=\left(\Gamma\left(\frac{1}{n}\right)\Gamma\left(1-\frac{1}{n}\right)\right)\left(\Gamma\left(\frac{2}{n}\right)\Gamma\left(1-\frac{2}{n}\right)\right)\ldots\ldots$$
$$\left(\Gamma\left(1-\frac{2}{n}\right)\Gamma\left(\frac{2}{n}\right)\right)\left(\Gamma\left(1-\frac{1}{n}\right)\Gamma\left(\frac{1}{n}\right)\right)$$

$$P^2=\frac{\pi}{\sin\left(\frac{\pi}{n}\right)}\cdot\frac{\pi}{\sin\left(\frac{2\pi}{n}\right)}\cdot\frac{\pi}{\sin\left(\frac{3\pi}{n}\right)}\ldots\ldots\frac{\pi}{\sin\frac{(n-1)\pi}{n}} \qquad \left[\because \Gamma(n)\,\Gamma(1-n)=\frac{\pi}{\sin n\pi}\right]$$

$$\Rightarrow \qquad P^2=\frac{\pi^{n-1}}{\sin\left(\frac{\pi}{n}\right)\sin\left(\frac{2\pi}{n}\right)\sin\left(\frac{3\pi}{n}\right)\ldots\ldots\sin\left\{\frac{(n-1)\pi}{n}\right\}} \qquad \ldots(3)$$

But from Trigonometry, we know that

$$\frac{\sin n\theta}{\sin\theta}=2^{n-1}\sin\left(\theta+\frac{\pi}{n}\right)\sin\left(\theta+\frac{2\pi}{n}\right)\ldots\ldots\sin\left\{\theta+\frac{(n-1)\pi}{n}\right\} \qquad \ldots(4)$$

Take Limit as $\theta\to 0$,

$$\underset{\theta\to 0}{\text{Lt}}\ \frac{\sin n\theta}{\sin\theta}=\underset{\theta\to 0}{\text{Lt}}\left(n\cdot\frac{\sin n\theta}{n\theta}\cdot\frac{\theta}{\sin\theta}\right)=n$$

On putting this limit in (4), we get

$$n=2^{n-1}\sin\frac{\pi}{n}\sin\frac{2\pi}{n}\sin\frac{3\pi}{n}\ldots\sin\frac{(n-1)\pi}{n} \qquad \text{[From (4)]}$$

$$\Rightarrow \qquad \sin\frac{\pi}{n}\cdot\sin\frac{2\pi}{n}\cdot\sin\frac{3\pi}{n}\ldots\sin\frac{(n-1)\pi}{n}=\frac{n}{2^{n-1}}$$

Substituting this in equation (3), we obtain

$$P^2=\frac{\pi^{n-1}}{\left(\frac{n}{2^{n-1}}\right)}=\frac{(2\pi)^{n-1}}{n}$$

$$\therefore \qquad P = \frac{(2\pi)^{(n-1)/2}}{n^{1/2}}$$

$$\Rightarrow \qquad \boxed{\overline{\left|\left(\frac{1}{n}\right)\right.}\ \overline{\left|\left(\frac{2}{n}\right)\right.}\ \overline{\left|\left(\frac{3}{n}\right)\right.} \cdots\cdots \overline{\left|\left(\frac{n-1}{n}\right)\right.} = \frac{(2\pi)^{(n-1)/2}}{n^{1/2}}}$$

## 21.12 TO SHOW THAT

*(i)* $\displaystyle\int_0^\infty e^{-ax}\, x^{n-1} \cos bx\, dx = \frac{\Gamma(n) \cos n\,\theta}{(a^2+b^2)^{n/2}}$ [*U.P., I Semester, (C.O.) 2004*]

*(ii)* $\displaystyle\int_0^\infty e^{-ax}\, x^{n-1} \sin bx\, dx = \frac{\Gamma(n) \sin n\,\theta}{(a^2+b^2)^{n/2}}$ where $\theta = \tan^{-1}\left(\frac{b}{a}\right)$ [*U.P., I Semester, 2003*]

**Proof.** We know that $\displaystyle\int_0^\infty e^{-ax} . x^{n-1}\, dx = \frac{\Gamma(n)}{a^n}$, where $a$, $n$ are positive.

Put $ax = z$ so that $\qquad dx = \dfrac{dz}{a}$

$$\therefore \qquad \int_0^\infty e^{-ax}\, x^{n-1}\, dx = \int_0^\infty e^{-z} \left(\frac{z}{a}\right)^{n-1} . \frac{dz}{a} = \frac{1}{a^n} \int_0^\infty e^{-z}\, z^{n-1}\, dz = \frac{\Gamma(n)}{a^n}$$

Replacing $a$ by $(a + ib)$, we have

$$\int_0^\infty e^{-(a+ib)x}\, x^{n-1}\, dx = \frac{\Gamma\, n}{(a+ib)^n} \qquad \ldots(1)$$

Now $\qquad e^{-(a+ib)x} = e^{-ax} . e^{-ibx} = e^{-ax} (\cos bx - i \sin bx)$

Putting the value of $e^{-(a+ib)x}$ in (1), we get

$$\int_0^\infty e^{-ax} (\cos bx - i \sin bx)\, x^{n-1}\, dx = \frac{\overline{|n}}{(a+ib)^n} \qquad \ldots(2)$$

Putting $a = r \cos\theta$ and $b = r \sin\theta$ so that $r^2 = a^2 + b^2$ and $\theta = \tan^{-1} \dfrac{b}{a}$

$$(a+ib)^n = (r\cos\theta + ir\sin\theta)^n = r^n (\cos\theta + i\sin\theta)^n$$

$$(a+ib)^n = r^n (\cos n\theta + i \sin n\theta) \qquad \text{[De Moivre's Theorm]}$$

Putting the value of $(a + ib)^n$ in (2), we have

$$\int_0^\infty e^{-ax} (\cos bx - i \sin bx) x^{n-1}\, dx = \frac{\overline{|n}}{r^n (\cos n\theta + i \sin n\theta)}$$

$$= \frac{\overline{|n}}{r^n} (\cos n\,\theta + i \sin n\,\theta)^{-1}$$

$$= \frac{\overline{|n}}{r^n} (\cos n\theta - i \sin n\theta)$$

Now, equating real and imaginary parts on the two sides, we get

*(i)* $\qquad \displaystyle\int_0^\infty e^{-ax}\, x^{n-1} \cos bx\, dx = \frac{\overline{|n}}{r^n} \cos n\,\theta$ and

(*ii*) $$\int_0^\infty e^{-ax}\, x^{n-1} \sin bx\, dx = \frac{\boxed{n}}{r^n} \sin n\,\theta$$

where $r^2 = a^2 + b^2$ and $\theta = \tan^{-1} \dfrac{b}{a}$.

**Example 26.** *Evaluate:*

(*i*) $\int_0^\infty \cos x^2\, dx$ (*ii*) $\int_{-\infty}^\infty \cos \dfrac{\pi\, x^2}{2}\, dx.$

**Solution.** (*i*) We know that

$$\int_0^\infty e^{-ax} . x^{n-1} \cos bx\, dx = \frac{\Gamma\, n \cos n\theta}{(a^2+b^2)^{n/2}}, \text{ where } \theta = \tan^{-1}\left(\frac{b}{a}\right)$$

Put $a = 0$, $$\int_0^\infty x^{n-1} \cos bx\, dx = \frac{\Gamma\, n}{b^n} \cos \frac{n\pi}{2} \qquad \left[\begin{aligned}\theta &= \tan^{-1}\left(\frac{b}{0}\right)\\ &= \tan^{-1}(\infty) = \frac{\pi}{2}\end{aligned}\right]$$

Put $x^n = z$ so that $$x^{n-1}\, dx = \frac{dz}{n} \text{ and } x = z^{1/n}$$

then, $$\int_0^\infty \cos bz^{1/n}\, dz = \frac{n\,\Gamma\, n}{b^n} \cos \frac{n\pi}{2}$$

$$\Rightarrow \quad \int_0^\infty \cos (bx^{1/n})\, dx = \frac{\Gamma(n+1)}{b^n} \cos \frac{n\pi}{2} \qquad \text{...(1)}$$

Here $b = 1$, $n = \dfrac{1}{2}$

$$\therefore \quad \int_0^\infty \cos x^2\, dx = \Gamma\left(\frac{3}{2}\right) \cos \frac{\pi}{4} = \frac{\sqrt{\pi}}{2} \cdot \frac{1}{\sqrt{2}} = \frac{1}{2} \frac{\sqrt{\pi}}{\sqrt{2}} \qquad \textbf{Ans.}$$

$$\text{(ii) } I = \int_{-\infty}^\infty \cos \frac{\pi x^2}{2}\, dx = 2 \int_0^\infty \cos \frac{\pi x^2}{2}\, dx \qquad \text{...(2)}$$

Putting $b = \dfrac{\pi}{2}$ and $n = \dfrac{1}{2}$ in equation (1), we get

$$\int_0^\infty \cos\left(\frac{\pi}{2} x^2\right) dx = \frac{\Gamma\left(\frac{3}{2}\right)}{\left(\frac{\pi}{2}\right)^{1/2}} \cos \frac{\pi}{4}$$

$$\therefore \text{ From (2),} \quad I = 2 \frac{\Gamma\left(\frac{3}{2}\right)}{\left(\frac{\pi}{2}\right)^{1/2}} \cos \frac{\pi}{4} = 2 . \frac{1}{2} \sqrt{\pi} . \frac{\sqrt{2}}{\sqrt{\pi}} . \frac{1}{\sqrt{2}} = 1. \qquad \textbf{Ans.}$$

**Example 27.** *Evaluate:* $\int_0^1 \log \Gamma(x)\, dx$

**Solution.** Let

$$I = \int_0^1 \log \Gamma(x)\, dx$$

$$= \int_0^1 \log \Gamma(1-x)\, dx \qquad \left[\int_0^a f(x)\, dx = \int_0^a f(a-x)\, dx\right]$$

Adding (1) and (2), we get

$$2I = \int_0^1 (\log \Gamma(x) + \log \Gamma(1-x))\, dx$$

$$= \int_0^1 \log (\Gamma(x)\, \Gamma(1-x))\, dx = \int_0^1 \log\left(\frac{\pi}{\sin \pi x}\right) dx \qquad \left[\overline{|x}\,\overline{|1-x} = \frac{\pi}{\sin \pi x}\right]$$

$$= \int_0^1 (\log \pi - \log \sin \pi x)\, dx = \int_0^1 \log \pi\, dx - \int_0^1 \log \sin \pi x\, dx$$

$$= I_1 - I_2 \qquad \ldots(3)$$

where

$$I_1 = \int_0^1 \log \pi\, dx = \log \pi$$

$$I_2 = \int_0^1 \log \sin \pi x\, dx \qquad \left[\text{Put } \pi x = t \Rightarrow dx = \frac{1}{\pi} dt\right]$$

$$= \int_0^\pi \log \sin t \left(\frac{dt}{\pi}\right)$$

$$= \frac{1}{\pi} \cdot 2 \int_0^{\pi/2} \log \sin t\, dt = \frac{2}{\pi}\left(-\frac{\pi}{2} \log 2\right) = -\log 2$$

From (3),

$$2I = \log \pi + \log 2 = \log 2\pi$$

$$I = \frac{1}{2} \log 2\pi. \qquad \textbf{Ans.}$$

**Example 28.** *Prove that* $\int_0^\infty x^2 e^{-x^4}\, dx \times \int_0^\infty \frac{e^{-x^2}}{\sqrt{x}}\, dx = \frac{\pi}{4\sqrt{2}}$ *(M.U., II Semester, 2008)*

**Solution.** Here, we have

$$\int_0^\infty x^2 e^{-x^4}\, dx \qquad \ldots(1)$$

Putting $x^4 = t \Rightarrow x = t^{\frac{1}{4}}$, $\quad dx = \frac{1}{4} t^{-\frac{3}{4}}\, dt$ in (1), we get

$$\int_0^\infty t^{\frac{1}{2}} e^{-t} \left(\frac{1}{4} t^{\frac{-3}{4}}\, dt\right) = \frac{1}{4} \int_0^\infty e^{-t} t^{-\frac{1}{4}}\, dt = \frac{1}{4} \int_0^\infty e^{-t}\, t^{\frac{3}{4}-1}\, dt = \frac{1}{4} \overline{\left|\frac{3}{4}\right.} \qquad \ldots(2)$$

Now, we calculate the value of $\int_0^\infty \frac{e^{-x^2}}{\sqrt{x}}\, dx \qquad \ldots(3)$

Putting $x^2 = t$, $\quad x = t^{\frac{1}{2}} \quad \Rightarrow \quad dx = \frac{1}{2} t^{-\frac{1}{2}}\, dt$ in (2), we get

$$\int_0^\infty e^{-t} t^{-\frac{1}{4}} \left(\frac{1}{2} t^{\frac{-1}{2}}\, dt\right) = \frac{1}{2} \int_0^\infty e^{-t} t^{-\frac{3}{4}}\, dt$$

$$= \frac{1}{2}\int_0^{\infty} e^{-t}\, t^{\frac{1}{4}-1}\, dt = \frac{1}{2}\Gamma\left(\frac{1}{4}\right) \qquad ...(4)$$

From (3) and (4), we get

$$\int_0^{\infty} x^2 e^{-x^4} dx \,.\, \int_0^{\infty} \frac{e^{-x^2}}{\sqrt{x}}\, dx = \left(\frac{1}{4}\Gamma\left(\frac{3}{4}\right)\right)\left(\frac{1}{2}\Gamma\left(\frac{1}{4}\right)\right) = \frac{1}{8}\,\frac{\Gamma\left(\frac{3}{4}\right)\Gamma\left(\frac{1}{4}\right)}{\Gamma\left(\frac{3}{4}+\frac{1}{4}\right)}$$

[Duplication formula

$$2^{2m-1}\Gamma(m)\,\Gamma\left(m+\frac{1}{2}\right) = \sqrt{\pi}\,\Gamma(2m)$$

Putting $m = \frac{1}{4}$, we get

$$2^{-\frac{1}{2}}\Gamma\left(\frac{1}{4}\right)\Gamma\left(\frac{3}{4}\right) = \sqrt{\pi}\,\Gamma\left(\frac{1}{2}\right)$$

$$\Gamma\left(\frac{1}{4}\right)\Gamma\left(\frac{3}{4}\right) = \sqrt{\pi}\sqrt{\pi}\,(2)^{\frac{1}{2}}$$

$$= \pi\sqrt{2}\ ]$$

$$= \frac{1}{8}(\pi\sqrt{2}) = \frac{\pi}{4\sqrt{2}}$$

**Proved.**

**Example 29.** *Evaluate* $\iint_A \frac{dx\, dy}{\sqrt{xy}}$ *using the substitutions*

$$x = \frac{u}{1+v^2}, \qquad y = \frac{uv}{1+v^2}$$

*where A is bounded by* $x^2 + y^2 - x = 0,\ y = 0,\ y > 0.$

**Solution.** Here $\sqrt{xy} = \sqrt{\left(\frac{u}{1+v^2}\right)\left(\frac{uv}{1+v^2}\right)} = \frac{u\sqrt{v}}{1+v^2}$

$$dx\, dy = \left|\frac{\partial(x, y)}{\partial(u, v)}\right| du\, dv$$

$$= \begin{vmatrix} \frac{\partial x}{\partial u} & \frac{\partial x}{\partial v} \\ \frac{\partial y}{\partial u} & \frac{\partial y}{\partial v} \end{vmatrix} du\, dv = \begin{vmatrix} \frac{1}{1+v^2} & -\frac{2uv}{(1+v^2)^2} \\ \frac{v}{1+v^2} & \frac{u(1-v^2)}{(1+v^2)^2} \end{vmatrix} du\, dv$$

$$= \left[\frac{u(1-v^2)}{(1+v^2)^3} + \frac{2uv^2}{(1+v^2)^3}\right] du\, dv = \left[\frac{u - uv^2 + 2uv^2}{(1+v^2)^3}\right] du\, dv$$

$$= \frac{u(1+v^2)}{(1+v^2)^3}\, du\, dv = \frac{u}{(1+v^2)^2}\, du\, dv$$

Also the circle $x^2 + y^2 - x = 0$ is transformed into

$$\frac{u^2}{(1+v^2)^2} + \frac{u^2v^2}{(1+v^2)^2} - \frac{u}{1+v^2} = 0 \Rightarrow \frac{u^2(1+v^2)}{(1+v^2)^2} - \frac{u}{1+v^2} = 0$$

$$\Rightarrow \quad \frac{u^2}{1+v^2} - \frac{u}{1+v^2} = 0 \Rightarrow u^2 - u = 0 \Rightarrow u(u-1) = 0 \Rightarrow u = 0,\ u = 1$$

Further $\quad y = 0 \Rightarrow \frac{uv}{1+v^2} = 0 \Rightarrow u = 0,\ v = 0$

and $y > 0 \Rightarrow uv > 0$ either both $u$ and $v$ are positive or both negative.

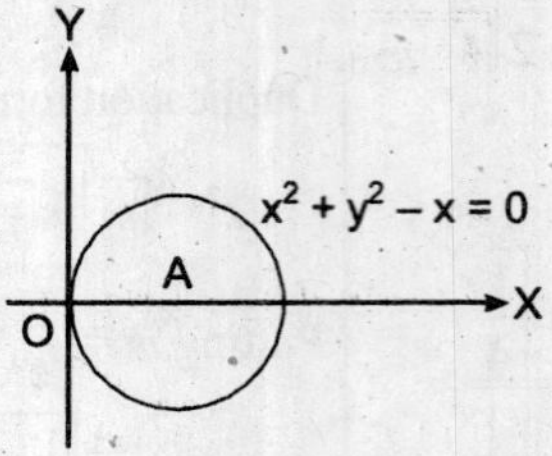


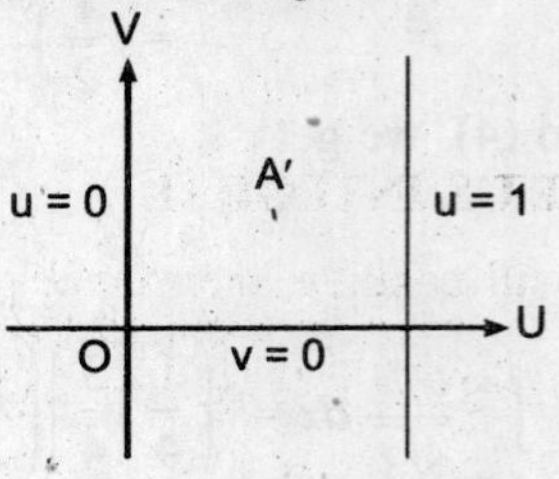


The area $A$, *i.e.*, $x^2 + y^2 - x = 0$ is transformed into $A'$ bounded by $u = 0$, $v = 0$ and $u = 1$ and $v = \infty$.

$$\iint \frac{dx\,dy}{\sqrt{x}} = \int_0^1 \int_0^\infty \frac{\dfrac{u}{(1+v^2)^2}}{\dfrac{u\sqrt{v}}{1+v^2}}\,du\,dv = \int_0^1 \int_0^\infty \frac{1}{\sqrt{v}(1+v^2)}\,dv\,du$$

On putting $v = \tan\theta$, $dv = \sec^2\theta\, d\theta$

$$= \int_0^1 \int_0^{\frac{\pi}{2}} \frac{\sec^2\theta}{\sqrt{\tan\theta}\left(1+\tan^2\theta\right)}\,d\theta\,du = \int_0^1 du \int_0^{\frac{\pi}{2}} \sqrt{\frac{\cos\theta}{\sin\theta}}\,d\theta = \int_0^1 du \int_0^{\frac{\pi}{2}} \sin^{-\frac{1}{2}}\theta\cos^{\frac{1}{2}}\theta\,d\theta$$

Duplication formula $\Gamma(m)\,\Gamma\left(m+\frac{1}{2}\right) = \frac{\sqrt{\pi}}{2^{2m-1}}\,\Gamma(2m)$

$$= \int_0^1 du\,\frac{\Gamma\left(\frac{-\frac{1}{2}+1}{2}\right)}{\Gamma(2)}\,\frac{\Gamma\left(\frac{\frac{1}{2}+1}{2}\right)}{\Gamma(2)} = \frac{1}{2}\int_0^1 du\,\Gamma\left(\frac{1}{4}\right)\Gamma\left(\frac{3}{4}\right) = \frac{1}{2}\int_0^1 du\left[\frac{\sqrt{\pi}}{2^{\frac{-1}{2}}}\,\Gamma\left(\frac{1}{2}\right)\right]$$

$$= \frac{1}{2}\int_0^1 du\,\sqrt{2}\sqrt{\pi}\cdot\sqrt{\pi} = \frac{\pi}{\sqrt{2}}\,[u]_0^1 = \frac{\pi}{\sqrt{2}}$$ **Ans.**

**Example 30.** *Prove that*

$$\iint_D x^{l-1}y^{m-1}\,dx\,dy = \frac{\Gamma(l)\,\Gamma(m)}{\Gamma(l+m+1)}\,h^{l+m}$$

*where D is the domain $x \ge 0$, $y \ge 0$ and $x + y \le h$.* (*U.P., I Semester, Dec. 2004*)

**Solution.** Putting $x = X h$ and $y = Yh$, $dx\,dy = h^2\,dX\,dY$

$$\iint_D x^{l-1}y^{m-1}\,dx\,dy = \iint_{D'} (X h)^{l-1}\,(Y h)^{m-1}\,h^2\,dX\,dY$$

where $D'$ is the domain

$X \ge 0, \qquad Y \ge 0, \qquad X + Y \le 1$

$$= h^{l+m}\int_0^1\int_0^{1-X} X^{l-1}Y^{m-1}\,dX\,dY = h^{l+m}\int_0^1 X^{l-1}\,dX\int_0^{1-X} Y^{m-1}\,dY$$

$$= h^{l+m}\int_0^1 X^{l-1}\,dX\left[\frac{Y^m}{m}\right]_0^{1-X} = \frac{h^{l+m}}{m}\int_0^1 X^{l-1}(1-X)^m\,dX$$

$$= \frac{h^{l+m}}{m}\,\beta(l, m+1) = \frac{h^{l+m}}{m}\,\frac{\Gamma(l)\,\Gamma(m+1)}{\Gamma(l+m+1)}$$

$$= \frac{h^{l+m}}{m} \frac{m\,\overline{|l}\,\overline{|m}}{\overline{|l+m+1}} = h^{l+m} \frac{\overline{|l}\,\overline{|m}}{\overline{|l+m+1}}$$ **Proved.**

## 21.13 DIRICHLET'S INTEGRAL

If $l$, $m$, $n$ are all positive, then the triple integral

$$\iiint_V x^{l-1} y^{m-1} z^{n-1} dx\, dy\, dz = \frac{\overline{|l}\,\overline{|m}\,\overline{|n}}{\overline{|l+m+n+1}}$$

*where V is the region $x \ge 0$, $y \ge 0$, $z \ge 0$ and $x + y + z \le 1$.*

**Proof.** Putting $y + z \le 1 - x = h$. Then $z \le h - y$

$$\iiint_V x^{l-1} y^{m-1} z^{n-1} dx\, dy\, dz = \int_0^1 x^{l-1} dx \int_0^{1-x} y^{m-1} dy \int_0^{1-x-y} z^{n-1}\, dz$$

$$= \int_0^1 x^{l-1} dx \left[\int_0^h \int_0^{h-y} y^{m-1} z^{n-1}\, dy\, dz\right]$$ [Put $x = h$]

$$= \int_0^1 x^{l-1}\, dx \left[\frac{\overline{|m}\,\overline{|n}}{\overline{|m+n+1}} h^{m+n}\right] = \int_0^1 x^{l-1} dx \left[\frac{\overline{|m}\,\overline{|n}}{\overline{|m+n+1}} h^{m+n}\right]$$

$$= \frac{\overline{|m}\,\overline{|n}}{\overline{|m+n+1}} \int_0^1 x^{l-1} (1-x)^{m+n} dx$$

$$= \frac{\overline{|m}\,\overline{|n}}{\overline{|m+n+1}} \beta(l, m+n+1)$$

$$= \frac{\overline{|m}\,\overline{|n}}{\overline{|m+n+1}} \frac{\overline{|l}\,\overline{|m+n+1}}{\overline{|l+m+n+1}}$$

$$\boxed{\iiint_V x^{l-1} y^{m-1} z^{n-1} dx\, dy\, dz = \frac{\overline{|l}\,\overline{|m}\,\overline{|n}}{\overline{|l+m+n+1}}}$$

**Note.** $$\iiint_V x^{l-1} y^{m-1} z^{n-1} dx\, dy\, dz = \frac{\overline{|l}\,\overline{|m}\,\overline{|n}}{\overline{|l+m+n+1}} h^{l+m+n}$$

where $V$ is the domain, $x \ge 0$, $y \ge 0$, $z \ge 0$ and $x + y + z \le h$.

**Corollary: Dirichlet's theorem for $n$ variables,** the theorem states that

$$\iiint \dots \int x_1^{l_1-1} x_2^{l_2-1} \dots x_n^{l_n-1}\, dx_1\, dx_2\, dx_3 \dots dx_n = \frac{\overline{|l_1}\,\overline{|l_2}\,\overline{|l_3} \dots \overline{|l_n}}{\overline{|1+l_1+l_2+\dots+l_n}} h^{l_1+l_2+\dots+l_n}$$

**Example 31.** *Show that* $\displaystyle\int_0^1 \frac{x^{m-1}(1-x)^{n-1}}{(a+x)^{m+n}} dx = \frac{\beta(m,n)}{a^n (1+a)^m}$

**Solution.** Put $\dfrac{x}{a+x} = \dfrac{t}{a+1}$

$$(a+1)\,x = t\,(a+x) \quad \Rightarrow \quad x = \frac{at}{a+1-t}$$

$$dx = \frac{(a+1-t)\,a\,dt - a\,t\,(-dt)}{(a+1-t)^2} = \frac{(a^2+a-at+at)}{(a+1-t)^2} dt = \frac{a(a+1)}{(a+1-t)^2} dt$$

$$\int_0^1 \frac{x^{m-1}(1-x)^{n-1}}{(a+x)^{m+n}}\,dx = \int_0^1 \frac{\left(\frac{at}{a+1-t}\right)^{m-1}\cdot\left(1-\frac{at}{a+1-t}\right)^{n-1}}{\left(a+\frac{at}{a+1-t}\right)^{m+n}}\frac{a(a+1)}{(a+1-t)^2}dt$$

$$= \int_0^1 \frac{(at)^{m-1}(a+1-t-at)^{n-1}}{(a^2+a-at+at)^{m+n}}a\,(a+1)\,dt$$

$$= \int_0^1 \frac{a^{m-1}\,t^{m-1}(a+1)^{n-1}(1-t)^{n-1}}{a^{m+n}(a+1)^{m+n}}a\,(a+1)\,dt$$

$$= \frac{1}{a^n(a+1)^m}\int_0^1 t^{m-1}(1-t)^{n-1}\,dt = \frac{1}{a^n(a+1)^m}\beta\,(m,n) \qquad \textbf{Proved.}$$

**Example 32.** *Prove that* $\int_0^\infty \frac{x^4(1+x^5)}{(1+x)^{15}}dx = \frac{1}{5005}$. (*M.U. II Semester, 2008*)

**Solution.** Let $I = \int_0^\infty \frac{x^4(1+x^5)}{(1+x)^{15}}dx$

$$\Rightarrow \quad I = \int_0^\infty \frac{x^4}{(1+x)^{15}}dx + \int_0^\infty \frac{x^9}{(1+x)^{15}}dx = I_1 + I_2 \qquad ...(1)$$

Now, put $x = \frac{t}{1-t}$, when $x = 0$, $t = 0$; when $x = \infty$, $t = 1$

$$1 + x = 1+\frac{t}{1-t} = \frac{1}{1-t} \quad \Rightarrow \quad dx = \frac{dt}{(1-t)^2}$$

$$\therefore \quad I_1 = \int_0^1 \left(\frac{t}{1-t}\right)^4 .(1-t)^{15}.\frac{1}{(1-t)^2}dt$$

$$= \int_0^1 t^4\,(1+t)^9\,dt = \beta\,(5,\,10) \qquad ...(2)$$

and

$$I_2 = \int_0^1 \left(\frac{t}{1-t}\right)^9 .(1-t)^{15}.\frac{dt}{(1-t)^2}$$

$$= \int_0^1 t^9(1-t)^4\,dt = \beta\,(10,\,5) \qquad ...(3)$$

$$\therefore \quad I = I_1 + I_2$$

$$= \beta\,(5,\,10) + \beta\,(10,\,5) \qquad \text{[Using (2) and (3)]}$$

$$= \beta\,(5,\,10) + \beta\,(5,\,10) \qquad [\because \beta(m,n) = \beta(n,m)]$$

$$= 2\,\beta\,(5,\,10)$$

$$= \frac{2\,\Gamma 5\,\Gamma 10}{\Gamma 15} = \frac{2.4!.9!}{14!}$$

$$= \frac{2\times4\times3\times2\times1\times9!}{14\times13\times12\times11\times10\times9!} = \frac{1}{7\times13\times11\times5} = \frac{1}{5005} \qquad \textbf{Proved.}$$

**Example 33.** *Find the value of*

$$\int_0^1 \frac{x^3-2x^4+x^5}{(1+x)^7}dx$$ (*M.U., II Semester, 2008*)

**Solution.** Let $$I = \int_0^1 \frac{x^3 - 2x^4 + x^5}{(1+x)^7}\, dx$$

$$\Rightarrow \quad I = \int_0^1 \frac{x^3}{(1+x)^7}\, dx - 2\int_0^1 \frac{x^4}{(1+x)^7}\, dx + \int_0^1 \frac{x^5}{(1+x)^7}\, dx$$

$$\Rightarrow \quad I = I_1 - 2\, I_2 + I_3 \qquad \text{...(1)}$$

Now put $x = \dfrac{t}{1-t}$ so that $1 + x = 1 + \dfrac{t}{1-t} \Rightarrow 1 + x = \dfrac{1}{1-t} \Rightarrow dx = \dfrac{dt}{(1-t)^2}$

$$\therefore \quad I_1 = \int_0^1 \left(\frac{t}{1-t}\right)^3 .(1-t)^7 . \frac{1}{(1-t)^2}\, dt$$

$$= \int_0^1 t^3 (1-t)^2\, dt = \int_0^1 t^{4-1} (1-t)^{3-1}\, dt = \beta\,(4, 3) \qquad \text{...(2)}$$

and $$I_2 = \int_0^1 \left(\frac{t}{1-t}\right)^4 (1-t)^7 . \frac{1}{(1-t)^2}\, dt$$

$$= \int_0^1 t^4 (1-t)\, dt = \int_0^1 t^{5-1} (1-t)^{2-1}\, dt = \beta\,(5, 2) \qquad \text{...(3)}$$

Also, $$I_3 = \int_0^1 \left(\frac{t}{1-t}\right)^5 (1-t)^7 \frac{dt}{(1-t)^2} = \int_0^1 t^5 (1-t)^0\, dt = \int_0^1 t^{6-1} (1-t)^{1-1}\, dt = \beta\,(6, 1) \qquad \text{...(4)}$$

Putting the values of $I_1$, $I_2$ and $I_3$ in (1), we get

$$I = \beta\,(4, 3) - 2\,\beta\,(5, 2) + \beta\,(6, 1)$$

$$= \frac{\overline{|4}\,\overline{|3}}{\overline{|4+3}} - 2\frac{\overline{|5}\,\overline{|2}}{\overline{|5+2}} + \frac{\overline{|6}\,\overline{|1}}{\overline{|6+1}} = \frac{\overline{|4}\,\overline{|3}}{\overline{|7}} - 2\frac{\overline{|5}\,\overline{|2}}{\overline{|7}} + \frac{\overline{|6}\,\overline{|1}}{\overline{|7}}$$

$$= \frac{\overline{|3}}{4\times5\times6} - 2\frac{1}{5\times6} + \frac{1}{6} = \frac{1}{60} - \frac{1}{15} + \frac{1}{6} = \frac{1-4+10}{60} = \frac{7}{60} \quad \textbf{Ans.}$$

## EXERCISE 21.2

**Prove that**

**1.** (a) $\displaystyle\int_0^{\frac{\pi}{2}} \sin^2\theta \cos^4\theta\, d\theta = \frac{\pi}{32}$ (b) $\displaystyle\int_0^{\frac{\pi}{2}} \sin^6\theta\, d\theta = \frac{5\pi}{32}$

**2.** (a) $\beta\,(m+1, n) = \dfrac{m}{m+n}\,\beta\,(m\,,\,n)$ (*U.P. I Sem. Jan 2011*) (b) $\beta\,(m, n+1) = \dfrac{n}{m+n}\,\beta\,(m\,,\,n)$

**3.** $\displaystyle\int_0^1 \sqrt{x}\,\sqrt[3]{1-x^2}\, dx = \frac{\overline{|\tfrac{3}{4}}\,\overline{|\tfrac{4}{3}}}{2\,\overline{|\tfrac{7}{12}}}$ **4.** $\displaystyle\int_0^1 (1-x^n)^{-\frac{1}{2}}\, dx = \frac{\overline{|\tfrac{1}{n}}\,\overline{|\tfrac{1}{2}}}{n\,\overline{|\tfrac{n+2}{2n}}}$

**5.** $\displaystyle\int_0^1 (1-x^{1/n})^m\, dx = \frac{\overline{|m}\,\overline{|n}}{\overline{|m+n}}$ **6.** $\displaystyle\int_1^\infty \frac{dx}{x^{P+1}(x-1)^q} = \beta(P+q, 1-q)$ if $-P < q < 1$

**7.** $\displaystyle\int_0^1 x^m (1-x^n)^P\, dx = \frac{1}{n}\frac{\overline{|\tfrac{m+1}{2}}\,\overline{|P+1}}{\overline{|\tfrac{m+1}{n}+P+1}}$ **8.** $\displaystyle\int_0^b (x-a)^m (b-x)^n\, dx = (b-a)^{m+n+1}\,.\,\beta\,(m+1, n+1)$

9. $\int_3^7 \sqrt[4]{(x-3)(7-x)}\, dx = \dfrac{2\left(\sqrt{\frac{1}{4}}\right)^2}{3\sqrt{\pi}}$ [**Hint.** Put $x = 4\,t + 3$]

10. $\int_0^{\frac{\pi}{2}} \dfrac{d\theta}{\sqrt{1-\frac{1}{2}\sin^2\theta}} = \dfrac{\left(\left|\frac{1}{4}\right.\right)^2}{4\sqrt{\pi}}$

11. If $\int_0^{\infty} e^{-x}\, x^{n-1}\, dx = I_n$ for $n > 0$ find $\dfrac{I_{n+1}}{I_n}$ (*A.M.I.E., Summer 2000*) **Ans.** $n$

**Show that :**

12. $\int_0^{\infty} \sqrt{x}e^{-x^2}\, dx \times \int_0^{\infty} \dfrac{e^{-x^2}}{\sqrt{x}}\, dx = \dfrac{\pi}{2\sqrt{2}}$

13. $\int_0^1 \dfrac{x^2 dx}{\sqrt{1-x^4}} \int_0^1 \dfrac{dx}{\sqrt{1+x^4}} = \dfrac{\pi}{4\sqrt{2}}$

14. $\int_0^{\infty} x^2 e^{-x^4}\, dx \times \int_0^{\infty} e^{-x^4}\, dx = \dfrac{\pi}{8\sqrt{2}}$

15. $\int_0^{\pi/2} \sin^3 x \cos^{5/2} x\, dx = \dfrac{8}{77}$

16. $\int_0^{\pi/2} \tan^n x\, dx = \dfrac{\pi}{2} \sec \dfrac{n\pi}{2}; -1 < n < 1.$

17. $\int_0^1 \dfrac{x^{m-1}(1-x)^{n-1}}{(a+bx)^{m+n}}\, dx = \dfrac{\beta(m, n)}{a^n (a+b)^m}$

18. $\dfrac{\beta(p, q+1)}{q} = \dfrac{\beta(p+1, q)}{p} = \dfrac{\beta(p, q)}{p+q}$

## 21.14 LIOUVILLE'S EXTENSION OF DIRICHLET THEOREM

**If the variables *x, y, z* are all positive such that $h_1 < x + y + z < h_2$, then**

$$\iiint f(x+y+z)\; x^{l-1} y^{m-1} z^{n-1}\, dx\, dy\, dz = \frac{\lceil l \lceil m \lceil n}{\lceil l+m+n} \int_{h_1}^{h_2} f(u)\, u^{l+m+n-1}\, du$$

**Proof.** By Dirichlet Theorem, we have

$$I = \iiint X^{l-1}\, Y^{m-1}\, Z^{n-1}\, dX\, dY\, dZ = \frac{\lceil l \lceil m \lceil n}{\lceil 1+l+m+n} \qquad ...(1)$$

Under the condition $x + y + z \le u \Rightarrow \dfrac{x}{u} + \dfrac{y}{u} + \dfrac{z}{u} \le 1$

Putting $X = \dfrac{x}{u}$, $Y = \dfrac{y}{u}$ and $Z = \dfrac{z}{u}$ so that

$dX = \dfrac{dx}{u}$, $dY = \dfrac{dy}{u}$, $dZ = \dfrac{dz}{u}$ in (1), we get

$$\iiint \left(\frac{x}{u}\right)^{l-1} \left(\frac{y}{u}\right)^{m-1} \left(\frac{z}{u}\right)^{n-1} \frac{dx}{u}\frac{dy}{u}\frac{dz}{u} = \frac{\lceil l \lceil m \lceil n}{\lceil 1+l+m+n}$$

$$\iiint x^{l-1}\, y^{m-1}\, z^{n-1}\, dx\, dy\, dz = u^{l+m+n} \frac{\lceil l \lceil m \lceil n}{\lceil 1+l+m+n}$$

Similarly, if $x + y + z \le u + \delta u$ then

$$\iiint x^{l-1}\, y^{m-1}\, z^{n-1}\, dx\, dy\, dz = (u+\delta u)^{l+m+n} \frac{\lceil l \lceil m \lceil n}{\lceil 1+l+m+n}$$

Hence, value of the integral I extended to all such values of the variables as make the sum of the variables lie between $u$ and $u + \delta u$ is given by

$$I = (u + \delta u)^{l+m+n} \frac{\overline{|l}\,\overline{|m}\,\overline{|n}}{\overline{|1+l+m+n}} - u^{l+m+n} \frac{\overline{|l}\,\overline{|m}\,\overline{|n}}{\overline{|1+l+m+n}}$$

$$\iiint x^{l-1} y^{m-1} z^{n-1}\, dx\, dy\, dz = \frac{\overline{|l}\,\overline{|m}\,\overline{|n}}{\overline{|1+l+m+n}} [(u+\delta u)^{l+m+n} - u^{l+m+n}]$$

$$= \frac{\overline{|l}\,\overline{|m}\,\overline{|n}}{\overline{|1+l+m+n}} u^{l+m+n} \left[\left(1+\frac{\delta u}{u}\right)^{l+m+n} - 1\right]$$

$$= \frac{\overline{|l}\,\overline{|m}\,\overline{|n}}{\overline{|1+l+m+n}} u^{l+m+n} \left[1 + (l+m+n)\frac{\delta u}{u} + \ldots\ldots - 1\right]$$

$$= \frac{\overline{|l}\,\overline{|m}\,\overline{|n}}{\overline{|1+l+m+n}} u^{l+m+n} (l+m+n) \frac{\delta u}{u} = \frac{\overline{|l}\,\overline{|m}\,\overline{|n}}{\overline{|l+m+n}} u^{l+m+n-1}\, \delta u$$

Let us consider $\iiint f(x+y+z)\, x^{l-1} y^{m-1} z^{n-1}\, dx\, dy\, dz$

Under the condition $h_1 \le x + y + z \le h_2$

When $x + y + z$ lies between $u$ and $u + \delta u$, the value of $f(x + y + z)$ can only differ from $f(u)$ by a small quantity of the same order as $\delta u$. Hence, neglecting square of $\delta u$, the part of the integral

$$\iiint f(x+y+z)\, x^{l-1} y^{m-1} z^{n-1}\, dx\, dy\, dz = \frac{\overline{|l}\,\overline{|m}\,\overline{|n}}{\overline{|l+m+n}} f(u) u^{l+m+n-1}\, \delta u$$

(supposing the sum of variables to be between $u$ and $u + \delta u$)

So $$\iiint f(x+y+z)\, x^{l-1} y^{m-1} z^{n-1}\, dx\, dy\, dz = \frac{\overline{|l}\,\overline{|m}\,\overline{|n}}{\overline{|(l+m+n)}} \int_{h_1}^{h_2} f(u)\, u^{l+m+n-1}\, du$$

**Example 34.** *Show that* $\iiint \frac{dx\, dy\, dz}{(x+y+z+1)^3} = \frac{1}{2}\log 2 - \frac{5}{16}$, *the integral being taken throughout the volume bounded by planes* $x = 0, y = 0, z = 0, x + y + z = 1$.

**Solution.** By Liouville's theorem when $0 < x + y + z < 1$

$$\iiint \frac{dx\, dy\, dz}{(x+y+z+1)^3} = \iiint \frac{x^{1-1} y^{1-1} z^{1-1}\, dx\, dy\, dz}{(x+y+z+1)^3} \qquad (0 \le x + y + z \le 1)$$

$$= \frac{\overline{|1}\,\overline{|1}\,\overline{|1}}{\overline{|1+1+1}} \int_0^1 \frac{1}{(u+1)^3} u^{3-1}\, du = \frac{1}{2} \int_0^1 \frac{u^2}{(u+1)^3}\, du$$

$$= \frac{1}{2} \int_0^1 \left[\frac{1}{u+1} - \frac{2}{(u+1)^2} + \frac{1}{(u+1)^3}\right] du \qquad \text{(Partial fractions)}$$

$$= \frac{1}{2} \left[\log(u+1) + \frac{2}{u+1} - \frac{1}{2(u+1)^2}\right]_0^1$$

$$= \frac{1}{2}\left[\log 2 + 2\left(\frac{1}{2} - 1\right) - \left(\frac{1}{8} - \frac{1}{2}\right)\right] = \frac{1}{2}\log 2 - \frac{5}{16}$$ **Proved.**

**Example 35.** *Find the value of* $\iiint \log(x+y+z)\,dx\,dy\,dz$ *the integral extending over all positive and zero values of x, y, z subject to the condition* $x+y+z<1$.

(*U.P., I Sem. 2001*)

**Solution.** By Liouville's theorem when $0 < x+y+z < 1$

$$\iiint \log(x+y+z)\,dx\,dy\,dz$$

$$= \iiint \log(x+y+z)\,x^{1-1}y^{1-1}z^{1-1}\,dx\,dy\,dz = \frac{\sqrt{1}\sqrt{1}\sqrt{1}}{\sqrt{1+1+1}}\int_0^1 (\log u)u^{1+1+1-1}\,du$$

$$= \frac{1}{\sqrt{3}}\int_0^1 u^2 \log u\,du = \frac{1}{2}\left[\log u\left(\frac{u^3}{3}\right) - \frac{1}{3}\frac{u^3}{3}\right]_0^1 = \frac{1}{2}\left(-\frac{1}{9}\right) = -\frac{1}{18} \qquad \textbf{Ans.}$$

**Example 36.** *Evaluate* $\iiint \frac{dx_1\,dx_2\ldots dx_n}{\sqrt{1-x_1^2-x_2^2\ldots-x_n^2}}$, *integral being extended to all positive values of the variables for which the expression is real.*

(*GBTU 2014, U.P., II Semester, Summer 2001*)

**Solution.** $\sqrt{1-x_1^2-x_2^2\ldots x_n^2}$ is real only when $x_1^2+x_2^2\ldots x_n^2<1$

Hence, the given integral is extended for all positive values of the variables $x_1, x_2 \ldots$ and $x_n$ such that $0 < x_1^2 + x_2^2 + \ldots + x_n^2 < 1$

Let us now put $x_1^2 = u_1$ *i.e.* $x_1 = u_1^{\frac{1}{2}}$ so that $dx_1 = \frac{1}{2}u_1^{-\frac{1}{2}}\,du_1$

$x_2^2 = u_2$ *i.e.* $x_2 = u_2^{\frac{1}{2}}$ so that $dx_2 = \frac{1}{2}u_2^{-\frac{1}{2}}\,du_2$

$x_n^2 = u_n$ *i.e.* $x_n = u_n^{\frac{1}{2}}$ so that $dx_n = \frac{1}{2}u_n^{-\frac{1}{2}}\,du_n$

Making these substitutions, the given condition becomes $0 < u_1 + u_2 + \ldots + u_n < 1$.

Hence, the required integral becomes

$$= \frac{1}{2^n}\iiint \frac{u_1^{-\frac{1}{2}}.u_2^{-\frac{1}{2}}\ldots u_n^{-\frac{1}{2}}.du_1.du_2\ldots du_n}{\sqrt{1-u_1-u_2\ldots-u_n}}$$

$$= \frac{1}{2^n}\iiint \frac{u_1^{\frac{1}{2}-1}.u_2^{\frac{1}{2}-1}\ldots u_n^{\frac{1}{2}-1}.du_1.du_2\ldots du_n}{\sqrt{1-u_1-u_2\ldots-u_n}}$$

$$= \frac{1}{2^n}\frac{\sqrt{\frac{1}{2}}.\sqrt{\frac{1}{2}}\ldots\sqrt{\frac{1}{2}}}{\sqrt{\frac{1}{2}+\frac{1}{2}+\ldots+\frac{1}{2}}}\int_0^1 \frac{1}{\sqrt{1-u}}.u^{\left(\frac{1}{2}+\frac{1}{2}+\ldots+\frac{1}{2}\right)-1}\,du$$

By Liouville's Extension of Dirichlet's Theorem

$$= \frac{1}{2^n} \frac{\left(\boxed{\frac{1}{2}}\right)^n}{\boxed{\frac{n}{2}}} \int_0^1 \frac{1}{\sqrt{1-u}} . u^{\frac{n}{2}-1} du$$

$$= \frac{1}{2^n} \frac{(\sqrt{\pi})^n}{\boxed{\frac{n}{2}}} \int_0^{\frac{\pi}{2}} \frac{1}{\sqrt{1-\sin^2 \theta}} . (\sin^2 \theta)^{\frac{n}{2}-1} \ 2 \sin \theta \cos \theta \, d\theta \qquad (u = \sin^2 \theta)$$

$$= \frac{1}{2^{n-1}} \frac{(\sqrt{\pi})^n}{\boxed{\frac{n}{2}}} \int_0^{\frac{\pi}{2}} \sin^{n-1} \theta \ \cos^{\circ}\theta d\theta = \frac{1}{2^{n-1}} \frac{\pi^{\frac{n}{2}}}{\boxed{\frac{n}{2}}} \frac{\boxed{\frac{n}{2}}\boxed{\frac{1}{2}}}{2\boxed{\frac{n+1}{2}}} = \frac{1}{2^n} \frac{\pi^{\frac{n+1}{2}}}{\boxed{\frac{n+1}{2}}} \qquad \textbf{Ans.}$$

**Example 37.** *Evaluate* $\iiint \sqrt{\frac{1-x^2-y^2-z^2}{1+x^2+y^2+z^2}}\, dx\, dy\, dz$, *integral being taken over all positive values of x, y, z such that* $x^2+y^2+z^2 \le 1$.

**Solution.** Putting $x^2 = u$, $y^2 = v$, $z^2 = w$ so that $u+v+w \le 1$

Also, $x = \sqrt{u} \quad \Rightarrow \quad dx = \frac{1}{2\sqrt{u}} du$

$y = \sqrt{v} \quad \Rightarrow \quad dy = \frac{1}{2\sqrt{v}} dv$

$z = \sqrt{w} \quad \Rightarrow \quad dz = \frac{1}{2\sqrt{w}} dw$

$\therefore$ The given integral

$$= \iiint \sqrt{\frac{1-(u+v+w)}{1+(u+v+w)}} \frac{du\, dv\, dw}{8\sqrt{uv\, w}}$$

$$= \frac{1}{8} \iiint u^{1/2-1}\, v^{1/2-1}\, w^{1/2-1} \sqrt{\frac{1-(u+v+w)}{1+(u+v+w)}}\, du\, dv\, dw$$

$$= \frac{1}{8} \frac{\boxed{\left(\frac{1}{2}\right)}\boxed{\left(\frac{1}{2}\right)}\boxed{\left(\frac{1}{2}\right)}}{\boxed{\left(\frac{1}{2}+\frac{1}{2}+\frac{1}{2}\right)}} \int_0^1 \sqrt{\frac{1-u}{1+u}}\, u^{\frac{1}{2}+\frac{1}{2}+\frac{1}{2}-1} du \quad \text{[Using Liouville's extension]}$$

$$= \frac{1}{8} \frac{\boxed{\left(\frac{1}{2}\right)}^3}{\frac{1}{2}\boxed{\left(\frac{1}{2}\right)}} \int_0^1 \frac{(1-u)}{\sqrt{1-u^2}}\, u^{1/2}\, du = \frac{\pi}{4} \int_0^1 \frac{(1-\sqrt{t})}{\sqrt{1-t}}\, t^{1/4} \frac{dt}{2\sqrt{t}} \qquad \text{where } u^2 = t$$

$$= \frac{\pi}{8} \int_0^1 \frac{(1-\sqrt{t})\, t^{-1/4}}{\sqrt{1-t}} dt = \frac{\pi}{8} \left[ \int_0^1 t^{\frac{3}{4}-1} (1-t)^{1/2-1}\, dt - \int_0^1 t^{5/4-1} (1-t)^{1/2-1}\, dt \right]$$

$$= \frac{\pi}{8} \left[ \beta\left(\frac{3}{4}, \frac{1}{2}\right) - \beta\left(\frac{5}{4}, \frac{1}{2}\right) \right] \qquad \textbf{Ans.}$$

**Example 38.** *Find the area and the mass contained in the first quadrant enclosed by the curve* $\left(\frac{x}{a}\right)^{\alpha}+\left(\frac{y}{b}\right)^{\beta}=1$ *where* $\alpha>0$, $\beta>0$ *given that density at any point p (xy) is* $k\sqrt{xy}$.

**Solution.** Here, we have $\left(\frac{x}{a}\right)^{\alpha}+\left(\frac{y}{b}\right)^{\beta}=1$

(*U.P. 1 Semester 2008*)

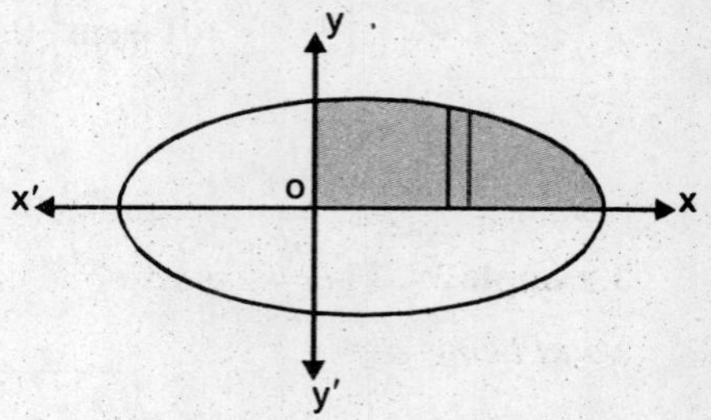


Put $\left(\frac{x}{a}\right)^{\alpha}=\cos^2 t$ and $\left(\frac{y}{b}\right)^{\beta}=\sin^2 t$

$\Rightarrow \quad x=a\cos^{\frac{2}{\alpha}} t$ $\qquad \Rightarrow y=b\sin^{\frac{2}{\beta}} t$

$$dx=\frac{2}{\alpha}a\cos^{\frac{2}{\alpha}-1} t\,(-\sin t)\,dt$$

$$\text{Area}=\int y\,dx$$

$$=\int_0^{\frac{\pi}{2}}\left(b\sin^{\frac{2}{\beta}} t\right)\left(-\frac{2a}{\alpha}\cos^{\frac{2}{\alpha}-1} t\sin t\right)dt=\frac{-2ab}{\alpha}\int_0^{\frac{\pi}{2}}\sin^{\frac{2}{\beta}+1} t\cos^{\frac{2}{\alpha}-1} t\,dt$$

$$=\frac{-2ab}{\alpha}\,\frac{\Gamma\left(\frac{\frac{2}{\beta}+1+1}{2}\right)\Gamma\left(\frac{\frac{2}{\alpha}-1+1}{2}\right)}{2\,\Gamma\left(\frac{\beta+1}{\beta}+\frac{1}{\alpha}\right)}=\frac{-ab}{\alpha}\,\frac{\Gamma\left(1+\frac{1}{\beta}\right)\Gamma\left(\frac{1}{\alpha}\right)}{\Gamma\left(\frac{\alpha+\beta+\alpha\beta}{\alpha\beta}\right)}$$

$$=-ab\,\frac{\frac{1}{\alpha}\Gamma\left(\frac{1}{\alpha}\right)\Gamma\left(1+\frac{1}{\beta}\right)}{\Gamma\left(\frac{\alpha+\beta+\alpha\beta}{\alpha\beta}\right)}=-ab\,\frac{\Gamma\left(\frac{1}{\alpha}+1\right)\Gamma\left(1+\frac{1}{\beta}\right)}{\Gamma\left(\frac{\alpha+\beta+\alpha\beta}{\alpha\beta}\right)}$$

$$\text{Required area}=\frac{\Gamma\left(\frac{\alpha+1}{\alpha}\right)\Gamma\left(\frac{\beta+1}{\beta}\right)}{\Gamma\left(\frac{\alpha+\beta+\alpha\beta}{\alpha\beta}\right)}$$ **Ans.**

$$\text{Density}=k\sqrt{xy}$$

$$\text{Mass}=\text{Area}\times\text{Density}$$

$$=\int\left[y\,dx\,k\sqrt{xy}\right]=4k\int_0^{\frac{\pi}{2}}x^{\frac{1}{2}}y^{\frac{3}{2}}\,dx$$

$$=4k\int_0^{\frac{\pi}{2}}\left(a\cos^{\frac{2}{\alpha}} t\right)^{\frac{1}{2}}\left(b\sin^{\frac{2}{\beta}} t\right)^{\frac{3}{2}}\frac{2}{\alpha}a\cos^{\frac{2}{\alpha}-1} t\,(-\sin t)\,dt$$

$$=4k\int_0^{\frac{\pi}{4}}a^{\frac{1}{2}}\cos^{\frac{1}{\alpha}} t\,b^{\frac{3}{2}}\sin^{\frac{3}{\beta}} t\,\frac{2}{\alpha}a\cos^{\frac{2}{\alpha}-1} t\,(-\sin t\,dt)$$

$$=\frac{8k}{\alpha}a^{\frac{3}{2}}b^{\frac{3}{2}}\int_0^{\frac{\pi}{2}}\sin^{\frac{3}{\beta}+1} t\cos^{\frac{2}{\alpha}-1} t\,dt$$ (–ve sign to be neglected)

$$= \frac{8k}{\alpha} a^{\frac{3}{2}} b^{\frac{3}{2}} \frac{\Gamma\left(\frac{\frac{3}{\beta}+1+1}{2}\right) \Gamma\left(\frac{\frac{2}{\alpha}-1+1}{2}\right)}{\Gamma\left(\frac{7}{4}\right)} = \frac{8k}{\alpha} a^{\frac{1}{2}} b^{\frac{3}{2}} \beta\left(\frac{5}{2}, \frac{2}{\alpha}-1\right)$$

$$= \frac{8k}{\alpha} a^{\frac{3}{2}} b^{\frac{3}{2}} \frac{\Gamma\left(\frac{\beta+1}{\beta}\right) \Gamma\left(\frac{1}{\alpha}\right)}{\Gamma\left(\frac{\beta+1}{\beta}+\frac{1}{\alpha}\right)}$$ **Ans.**

**Example 39.** *Find the mass of an octant of the ellipsoid* $\frac{x^2}{a^2}+\frac{y^2}{b^2}+\frac{z^2}{c^2}=1$, *the density at any point being* $\rho = k\,x\,y\,z$. *(U.P., I Semester, 2009)*

**Solution.** Mass $= \iiint \rho\, dv = \iiint (k\,x\,y\,z)\, dx\, dy\, dz$

$$= k \iiint (x\, dx)(y\, dy)\,(z\, dz) \quad \text{...(1)}$$

Putting $\frac{x^2}{a^2} = u,\ \frac{y^2}{b^2} = v,\ \frac{z^2}{c^2} = w$ and $u + v + w = 1$

so that $\frac{2x\, dx}{a^2} = du,\ \frac{2y\, dy}{b^2} = dv,\ \frac{2z\, dz}{c^2} = dw$

$$\text{Mass} = k \iiint \left(\frac{a^2 du}{2}\right)\left(\frac{b^2 dv}{2}\right)\left(\frac{c^2 dw}{2}\right)$$

$$= \frac{ka^2b^2c^2}{8} \iiint du\, dv\, dw, \qquad \text{where } u + v + w \le 1.$$

$$= \frac{ka^2b^2c^2}{8} \iiint u^{1-1}\, v^{1-1}\, w^{1-1}\, du\, dv\, dw$$

$$= \frac{ka^2b^2c^2}{8} \frac{\Gamma(1)\Gamma(1)\Gamma(1)}{\Gamma(3+1)} = \frac{ka^2b^2c^2}{8 \times 6} = \frac{ka^2b^2c^2}{48}$$ **Ans.**

**Example 40.** *Find the mass of a solid* $\left(\frac{x}{a}\right)^p + \left(\frac{y}{b}\right)^q + \left(\frac{z}{c}\right)^r = 1$, *the density at any point being* $\rho = kx^{l-1}y^{m-1}z^{n-1}$, *where x, y, z are all positive.* *(U.P., I Jan. 2011)*

**Solution.** Here, we have

$$\left(\frac{x}{a}\right)^p + \left(\frac{y}{b}\right)^q + \left(\frac{z}{c}\right)^r = 1$$

$$\text{Density} = k\, x^{l-1}\, y^{m-1}\, z^{n-1}$$

$$\text{Mass} = \text{Volume} \times \text{Density}$$

$$= \iiint dx\, dy\, dz.\, kx^{l-1} y^{m-1} z^{n-1} \quad \text{... (1)}$$

Put $\left(\frac{x}{a}\right)^p = u \quad \Rightarrow \quad x = au^{\frac{1}{p}} \quad \Rightarrow \quad dx = \frac{a}{p} u^{\frac{1}{p}-1}\, du$

$$\left(\frac{y}{b}\right)^q = v \quad \Rightarrow \quad y = bv^{\frac{1}{q}} \quad \Rightarrow \quad dy = \frac{b}{q}\, v^{\frac{1}{q}-1}\, dv$$

$$\left(\frac{z}{c}\right)^r = w \quad \Rightarrow \quad z = cw^{\frac{1}{r}} \quad \Rightarrow \quad dz = \frac{c}{r}\, w^{\frac{1}{r}-1}\, dw$$

*Then,* $u + v + w \le 1$. Also $u \ge 0, v \ge 0, w \ge 0$.

Sine $x, y, z$ are all positive.

Putting above values in (1), we get

$$\text{Mass} = k \iiint (au^{\frac{1}{p}})^{l-1} (bv^{\frac{1}{q}})^{m-1} (cw^{\frac{1}{r}})^{n-1} \left(\frac{a}{p} u^{\frac{1}{p}-1}\, du\right)\left(\frac{b}{q} v^{\frac{1}{q}-1}\, dv\right)\left(\frac{c}{r} w^{\frac{1}{r}-1}\, dw\right)$$

$$= k\,\frac{a^l b^m c^n}{pqr} \iiint u^{\frac{l}{p}-1}\, v^{\frac{m}{q}-1}\, w^{\frac{n}{r}-1}\, du\, dv\, dw = k\,\frac{a^l b^m c^n}{pqr}\,\frac{\left|\overline{\frac{l}{p}}\right.\left|\overline{\frac{m}{q}}\right.\left|\overline{\frac{n}{r}}\right.}{\left|\overline{\frac{l}{p}+\frac{m}{q}+\frac{n}{r}+1}\right.}$$ **Ans.**

## EXERCISE 21.3

**Evaluate:**

**1.** $\iiint e^{x+y+z}\, dx\, dy\, dz$ taken over the positive octant such that $x + y + z \le 1$. **Ans.** $\frac{e-2}{2}$

**2.** $\iiint \frac{dx\, dy\, dz}{(a^2 - x^2 - y^2 - z^2)}$ for all positive values of the variables for which the expression is real.

[**Hint.** $a^2 - x^2 - y^2 - z^2 > 0 \quad \Rightarrow \quad 0 < x^2 + y^2 + z^2 < a^2$] **Ans.** $\frac{\pi^2 a^2}{8}$

**3.** $\iiint_R (x + y + z + 1)^2\, dx\, dy\, dz$ where $R$ is defined by $x \ge 0, y \ge 0, z \ge 0, x + y + z \le 1$.

**Ans.** $\frac{31}{60}$

**4.** $\iiint x^{-\frac{1}{2}}\, y^{-\frac{1}{2}}\, z^{-\frac{1}{2}} (1 - x - y - z)^{\frac{1}{2}}\, dx\, dy\, dz,\ x + y + z \le 1, x > 0, y > 0, z > 0$ **Ans.** $\frac{\pi^2}{4}$

**5.** Show that $\iiint \frac{dx\, dy\, dz}{(x + y + z + 1)^2} = \frac{3}{4} - \log 2$, the integral being taken throughout the volume bounded by the planes $x = 0, y = 0, z = 0$ and $x + y + z = 1$.

# 22

# Chebyshev Polynomials

## 22.1 INTRODUCTION

In numerical analysis, main problem is approximating a function. For better approximation of a function the error should be minimum. To make the error minimum we use the least squares method. On the other hand, we may choose the approximation such that maximum component of error is minimised. This introduces Chebyshev polynomial. We apply Chebyshev polynomial in the economization of power series.

## 22.2 CHEBYSHEV POLYNOMIALS (Tchebcheff or Tschebyscheff polynomials)

The Chebyshev polynomials of first kind

$$T_n(x) = \cos(n\cos^{-1} x)$$

and the second kind

$$U_n(x) = \sin(n\cos^{-1} x)$$

where $n$ is non-negative integer

### 22.2.1. CHEBYSHEV EQUATION

$$(1-x^2)\frac{d^2y}{dx^2} - x\frac{dy}{dx} + n^2 y = 0 \qquad \text{... (1)}$$

Prove that $t_n^{(x)}$ and $U_n(x)$ are the independent solution of the Chebyshev equation

**Proof.**

Let $\qquad y = T_n(x) = \cos(n\cos^{-1} x) = \cos n\theta$

$$\frac{dy}{dx} = \frac{dy}{d\theta}\frac{d\theta}{dx} = -\frac{n\sin n\theta}{-\sin\theta}$$

$$\left\{\cos\theta = x \Rightarrow -\sin\theta\frac{d\theta}{dx} = 1 \text{ or } \frac{d\theta}{dx} = -\frac{1}{\sin\theta}\right\}$$

or

$$\frac{dy}{dx} = \frac{n\sin n\theta}{\sin\theta}$$

$$\frac{d^2y}{dx^2} = \frac{\sin\theta\,(n^2\cos n\theta) - n\sin n\theta\cdot\cos\theta}{\sin^2\theta}\frac{d\theta}{dx}$$

$$= \frac{n^2\sin\theta\cos n\theta - n\sin n\theta\cdot\cos\theta}{\sin^2\theta}\left(-\frac{1}{\sin\theta}\right)$$

$$\Rightarrow \qquad = \frac{-n^2 \cos n\theta + \left(\dfrac{n \sin n\theta}{\sin \theta}\right)(\cos \theta)}{\sin^2 \theta}$$

$$\Rightarrow \qquad \frac{d^2y}{dx^2} = \frac{-n^2 y + x\dfrac{dy}{dx}}{1-x^2}$$

$$\Rightarrow \qquad (1-x^2)\frac{d^2y}{dx^2} - x\frac{dy}{dx} + n^2 y = 0$$

Which is satisfied by $T_n(x)$.

Similarly we can prove that $U_n = \sin(n \cos^{-1} x)$ is a solution of Chebyshev equation.

Hence $T_n(x)$ and $U_n(x)$ both are the solutions of (1). But $U_n(x)$ can not be expressed as a constant multiple of $T_n(x)$ as shown below :

$$T_n(1) = \cos(n \cos^{-1} 1) = \cos n(0) = \cos 0 = 1$$

$$U_n(1) = \sin(n \cos^{-1} 1) = \sin n0 = 0$$

Thus $T_n(x)$ and $U_n(x)$ are independent solutions of Chebyhev's equation.

## 22.3 ORTHOGONAL PROPERTIES OF CHEBYSHEV POLYNOMIALS.

$$\int_{-1}^{1} \frac{T_m(x)T_n(x)}{\sqrt{1-x^2}}dx = \begin{cases} 0, & m \neq n \neq 0 \\ \dfrac{\pi}{2}, & m = n \neq 0 \\ \pi, & m = n = 0 \end{cases}$$

**Proof.** We know that

$$T_m(x) = \cos(m \cos^{-1} x) = \cos m\theta \qquad \begin{cases} \theta = \cos^{-1} x \\ x = \cos\theta \\ dx = -\sin\theta\, d\theta \end{cases}$$

$$T_n(x) = \cos(n \cos^{-1} x) = \cos n\theta$$

**(*a*) If $m \neq n \neq 0$**

$$\int_{-1}^{1} \frac{T_m(x).T_n(x)}{\sqrt{1-x^2}}dx = \int_{\pi}^{0} \frac{\cos m\theta \cos n\theta}{\sin\theta}(-\sin\theta\, d\theta)$$

$$= \int_0^{\pi} \cos m\theta \cos n\theta\, d\theta$$

$$= \frac{1}{2}\int_0^{\pi} [(\cos(m+n)\theta + \cos(m-n)\theta)]\, d\theta$$

$$= \frac{1}{2}\left[\frac{\sin(m+n)\theta}{m+n} + \frac{\sin(m-n)}{m-n}\right]_0^{\pi}$$

$$= 0 \qquad m \neq n$$

(*b*) If $m = n \neq 0$

$$\int_{-1}^{+1} \frac{T_n(x).T_n(x)dx}{\sqrt{1-x^2}} = \int_{\pi}^{0} \frac{\cos n\theta.\cos n\theta}{\sin\theta}(-\sin\theta\, d\theta)$$

$$= \int_0^{\pi} \cos^2 n\theta\, d\theta = \frac{1}{2}\int_0^{\pi} (\cos 2\theta + 1)\, d\theta$$

$$= \frac{1}{2}\left[\frac{\sin 2\theta}{2} + \theta\right]_0^{\pi} = \frac{\pi}{2}$$

(*c*) If $m = n = 0$

$$\int_{-1}^{+1} \frac{T_0(x)\, T_0(x)}{\sqrt{1-x^2}} dx = \int_{\pi}^{0} \frac{(1)(1)}{\sin\theta}(-\sin\theta\, d\theta) = \int_0^{\pi} d\theta = (\theta)_0^{\pi} = \pi$$

**Note:** (1) Similarly we can prove that $\int_{-1}^{+1} \frac{U_m(x) U_n(x)}{\sqrt{1-x^2}} dx = \begin{cases} 0, & m \neq n \\ \frac{\pi}{2}, & m = n \neq 0 \\ \pi, & m = n = 0 \end{cases}$

(2) The polynomials $T_n(x)$ are orthogonal with the function $\frac{1}{\sqrt{1-x^2}}$.

**Example 1.** *Prove that (a)* $T_{-n}(x) = T_n(x)$

(*b*) $T_0(x) = 1$ (*c*) $T_1(x) = x$

**Solution.** The Chebyshev polynomial of degree $n$ over the integral [–1, 1] is defined as

$$T_n(x) = \cos(n \cos^{-1} x) \quad \text{... (1)}$$

(*a*) On putting $-n$ for $n$ in (1), we get

$$T_{-n} = \cos(-n \cos^{-1} x)$$
$$= \cos(n \cos^{-1} x)$$
$$= T_n$$

$\Rightarrow$ $T_n = T_{-n}$

(*b*) Let $\cos^{-1} x = \theta$ so that $x = \cos\theta$

On putting $\cos^{-1} x = \theta$ in (1), it becomes

$$T_n(x) = \cos n\theta \quad \text{... (2)}$$

$$T_0(x) = \cos 0 = 1$$

On putting $n = 0$ in (1), we get $T(n) = \cos\theta = 1$

(*c*) If $n = 1$ On putting $n = 1$ in (1) we get, $T_1(x) = \cos\theta = x$

## 22.4 RECURRENCE RELATION OF CHEBYSHEV POLYNOMIALS

(I) **Formula I** $T_{n+1}(x) = 2x\, T_n(x) - T_{n-1}(x)$

$$\cos(n-1)\theta + \cos(n+1)\theta = 2\cos n\theta . \cos\theta \quad \text{... (1) (Trigonometric identity)}$$

On putting the values of $\cos(n-1)\theta$, $\cos(n+1)\theta$, $\cos n\theta$, $\cos\theta$ in (1) we get

$$T_{n-1}(x) + T_{n+1}(x) = 2x\, T_n(x)$$

$\Rightarrow$ $$T_{n+1}(x) = 2x\, T_n(x) - T_{n-1}(x) \quad \text{... (2)}$$

This is the required recurrence relation of Chebyshev polynomials

Similarly we can prove that $U_{n+1}(x) = 2x U_n(x) - U_{n-1}(x)$

On substituting $n = 1$, in (2) we have

$$T_2(x) = 2x\, T_1(x) - T_0(x)$$
$$= 2x(x) - 1 = 2x^2 - 1$$

If $n = 2$, $\quad T_3(x) = 2x\,T_2(x) - T_1(x) = 2x(2x^2 - 1) - x$

$$= 4x^3 - 3x$$

If $n = 3$, $\quad T_4(x) = 2x\,T_3(x) - T_2(x)$

$$= 2x(4x^3 - 3x) - (2x^2 - 1)$$

$$= 8x^4 - 8x^2 + 1$$

In this way

$$T_0(x) = 1$$

$$T_1(x) = x$$

$$T_2(x) = 2x^2 - 1$$

$$T_3(x) = 4x^3 - 3x$$

$$T_4(x) = 8x^4 - 8x^2 + 1$$

$$T_5(x) = 16x^5 - 20x^3 + 5x$$

$$T_6(x) = 32x^6 - 48x^4 + 18x^2 - 1$$

Here the coefficient of $x^n$ in $T_n(x)$ is always $2^{n-1}$.

**Graph of Chebyshev polynomials.**

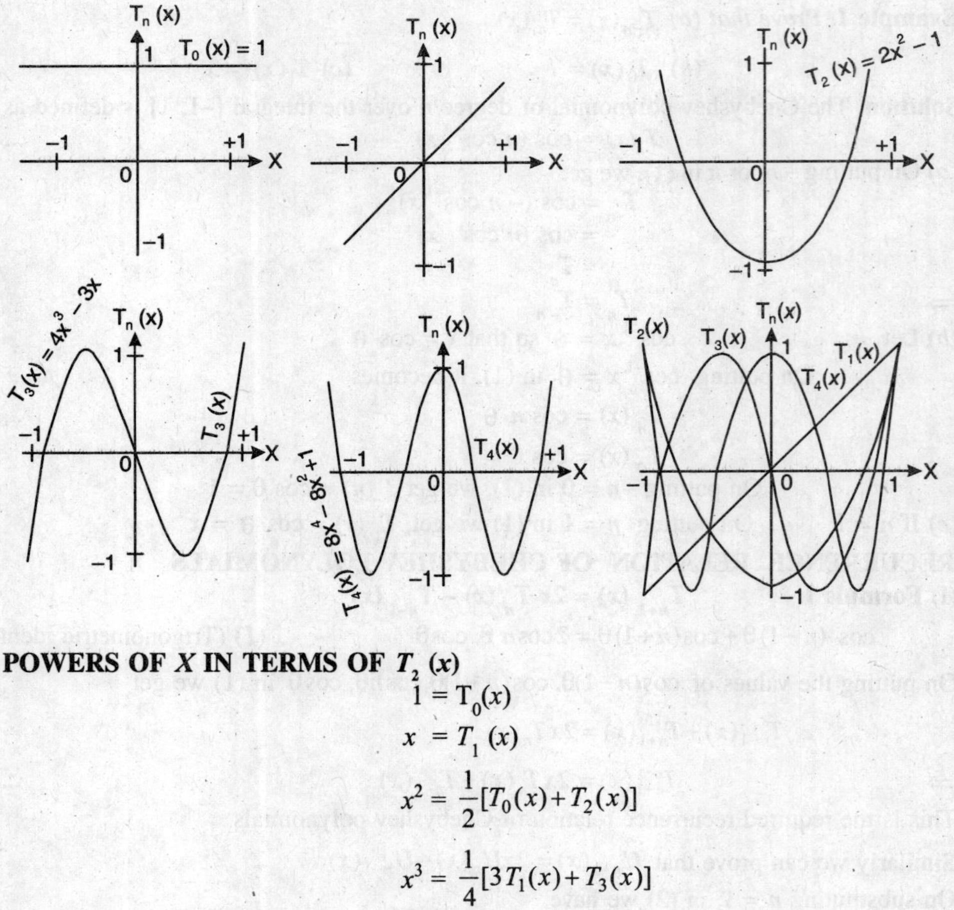


## 22.5 POWERS OF X IN TERMS OF $T_2(x)$

$$1 = T_0(x)$$

$$x = T_1(x)$$

$$x^2 = \frac{1}{2}[T_0(x) + T_2(x)]$$

$$x^3 = \frac{1}{4}[3T_1(x) + T_3(x)]$$

$$x^4 = \frac{1}{8}[3T_0(x) + 4T_2(x) + T_4(x)]$$

$$x^5 = \frac{1}{16}[10T_1(x)+5T_3(x)+T_5(x)]$$

$$x^6 = \frac{1}{32}[10T_0(x)+15T_2(x)+6T_4(x)+T_6(x)]$$

and so on.

**Recurrence Formula II**

$$(1-x^2)\,T_n'(x) = -nx\,T_n(x)+nT_{n-1}(x)$$

**Solution.** We know that

$$T_n(x) = \cos(n\cos^{-1}x) = \cos n\theta \qquad \theta = \cos^{-1}x \Rightarrow x = \cos\theta$$

$$T_n^1(x) = -\sin n\theta\left(n\frac{d\theta}{dx}\right) = \frac{-\sin n\theta}{1}\left(-\frac{n}{\sin\theta}\right) = \frac{n\sin n\theta}{\sin\theta} \qquad 1 = -\sin\theta\frac{d\theta}{dx}$$

$$\frac{d\theta}{dx} = -\frac{1}{\sin\theta}$$

Multiplying both sides by $(1-x^2)$ we get

$$(1-x^2)T_n'(x) = (1-x^2)\frac{n\sin n\theta}{\sin\theta} = \frac{(l-\cos^2\theta)n\sin n\theta}{\sin\theta} = \frac{(\sin^2\theta)(n\sin n\theta)}{\sin\theta} = n\sin n\theta\sin\theta \quad \text{... (1)}$$

$$-nxT_n(x)+nT_{n-1}(x) = -n\cos\theta\cos n\theta+n\cos(n-1)\theta$$

$$= n[-\cos\theta\cos n\theta+\cos(n\theta-\theta)]$$

$$= n[-\cos\theta\cos n\theta+\cos n\theta\cos\theta+\sin n\theta\sin\theta]$$

$$= n\sin n\theta.\sin\theta \quad \text{... (2)}$$

From (1) and (2)

$$(1-x^2)T_n'(x) = -nx\,T_n(x)+nT_{n-1}(x)$$ **Proved.**

**Example 2.** *Prove that*

$$[T_n(x)]^2 - T_{n+1}(x)T_{n-1}(x) = 1-x^2$$

**Solution.** Let $\cos^{-1}x = \theta \Rightarrow x = \cos\theta$

$$[T_n(x)]^2 - T_{n+1}(x)T_{n-1}(x) = [\cos(n\cos^{-1}x)]^2 - [\cos\{(n+1)\cos^{-1}x\}][\cos\{(n-1)\cos^{-1}x\}]$$

$$= \cos^2 n\theta - \cos(n+1)\theta\cos(n-1)\theta$$

$$= \cos^2 n\theta - \frac{1}{2}[\cos 2n\theta+\cos 2\theta]$$

$$= \cos^2 n\theta - \frac{1}{2}[2\cos^2 n\theta - 1 + 2\cos^2 - 1]$$

$$= \cos^2 n\theta - \cos^2 n\theta + \frac{1}{2} - \cos^2\theta + \frac{1}{2}$$

$$= 1-\cos^2\theta = 1-x^2$$ **Proved**

**Example 3.** *Prove that*

$$T_n(x) = \frac{1}{2}\left[\left\{x+i\sqrt{(1-x^2)}\right\}^n + \left\{x-i\sqrt{(1-x^2)}\right\}^n\right]$$

**Solution.** We know that

$$T_n(x) = \cos(n\cos^{-1} x) = \cos n\theta \quad \text{... (1)}$$

$$\theta = \cos^{-1} x \Rightarrow x = \cos\theta$$

$$\text{R.H.S.} = \frac{1}{2}[\{x+i\sqrt{(1-x^2)}\}^n + \{x-i\sqrt{(1-x^2)}\}^n]$$

$$= \frac{1}{2}[\{\cos\theta+i\sqrt{(1-\cos^2\theta)}\}^n + \{\cos\theta-i\sqrt{(1-\cos^2\theta)}\}^n]$$

$$= \frac{1}{2}[\{\cos\theta+i\sin\theta\}^n + \{\cos\theta-i\sin\theta\}^n]$$

$$= \frac{1}{2}[\cos n\theta+i\sin n\theta+\cos n\theta-i\sin n\theta] \quad \text{(De Moivre's theorem)}$$

$$= \cos n\theta \quad \text{... (2)}$$

From (1) and (2) we have

$$T_n(x) = \frac{1}{2}[\{x+i\sqrt{(1-x^2)}\}^n + \{x-i\sqrt{(1-x^2)}\}^n] \quad \textbf{Proved}$$

**Example 4.** *Prove that*

$$U_n(x) = -\frac{1}{2}i\left[\left\{x+i\sqrt{(1-x^2)}\right\}^n - \left\{x-i\sqrt{(1-x^2)}\right\}^n\right]$$

**Solution.** We know that

$$U_n(x) = \sin(n\cos^{-1} x) = \sin n\theta \quad \text{... (1)} \quad \cos^{-1} x = \theta \Rightarrow x = \cos\theta$$

$$\text{R.H.S.} = -\frac{1}{2}i[\{x+i\sqrt{(1-x^2)}\}^n - \{x-i\sqrt{(1-x^2)}\}^n]$$

$$= \frac{-i}{2}[\{\cos\theta+i\sqrt{(1-\cos^2\theta)}\}^n - \{\cos\theta-i\sqrt{(1-\cos^2\theta)}\}^n]$$

$$= -\frac{i}{2}[\{\cos\theta+i\sin\theta\}^n - \{\cos(-\theta)+i\sin(-\theta)\}^n]$$

$$= -\frac{i}{2}[\{\cos n\theta+i\sin n\theta\} - \{\cos(-n\theta)+i\sin(-n\theta)\}] \quad \text{(De Moivre's Theorem)}$$

$$= \frac{-i}{2}[\cos n\theta+i\sin n\theta-\cos n\theta+i\sin n\theta]$$

$$= \frac{-i}{2}(2i\sin n\theta) = \sin n\theta \quad \text{... (2)}$$

From (1) and (2) we have

$$U_n(x) = -\frac{1}{2}i[\{x+i\sqrt{(1-x^2)}\}^n - \{x-i\sqrt{(1-x^2)}\}^n] \quad \textbf{Proved.}$$

## 22.6 RECURRENCE FORMULAE FOR $u_n(x)$

**(I)** $u_{n+1}(x) - 2x\, u_n(x) + u_{n-1}(x) = 0$

**Proof** $u_n(x) = \sin(n\cos^{-1} x) = \sin n\theta$ $\quad (x = \cos\theta)$

Now $u_{n+1}(x) - 2x\, u_n(x) + u_{n-1}(x) = \sin(n+1)\theta - 2\cos\theta\sin n\theta + \sin(n-1)\theta$,

$$= \sin(n+1)\theta - [\sin(n+1)\theta + \sin(n-1)\theta] + \sin(n-1)\theta$$

$$= 0$$

**proved**

(ii) $(1-x^2)U_n'(x) = -n\,x\,U_n(x) + nU_{n-1}(x)$ (*U.P., III semestor 2002*)

**Proof.** We know that

$$U_n(x) = \sin\ (n\cos^{-1}x) = \sin n\theta$$

$$U_n'(x) = \cos n\theta\left(n\frac{d\theta}{dx}\right) = \cos n\theta\left(\frac{-n}{\sin\theta}\right) = -n\frac{\cos n\theta}{\sin\theta}$$

$$\theta = \cos^{-1}x,\quad x = \cos\theta,\quad 1 = -\sin\theta\frac{d\theta}{dx},\quad \frac{d\theta}{dx} = \frac{-1}{\sin\theta}$$

Multiplying both sides by $\sqrt{(1-x^2)}$ we get

$$\sqrt{(1-x^2)}\,U_n'(x) = -n\sqrt{(1-x^2)}\,\frac{\cos n\theta}{\sin\theta} = -n\sqrt{(1-\cos^2\theta)}\,\frac{\cos n\theta}{\sin\theta}$$

$$= \frac{n\sin^2\theta\cos n\theta}{\sin\theta} = -n\cos n\theta\sin\theta \qquad \text{... (1)}$$

$$\begin{aligned}
\text{R.H.S.} &= -nxU_n(x) + nU_{n-1}(x) \\
&= -n\cos\theta\sin n\theta + n\sin(n-1)\theta \\
&= -n[\cos\theta\sin n\theta - \sin(n\theta-\theta)] \\
&= -n[\cos\theta\sin n\theta - \sin n\theta\cos\theta + \cos n\theta\sin\theta] \\
&= -n(\cos n\theta\sin\theta) = -n\cos n\theta\sin\theta \qquad \text{... (2)}
\end{aligned}$$

From (1) and (2) we get

$$(1-x^2)U_n'(x) = -nxU_n(x) + nU_{n-1}(x)$$ **Proved.**

**Example 5.** *Show that* $\frac{1}{\sqrt{(1-x^2)}}U_n(x)$ *satisfies the differential equation*

$$(1-x^2)\frac{d^2y}{dx^2} - 3x\frac{dy}{dx} + (n^2-1)y = 0$$

**Solution.** Let $y = \frac{1}{\sqrt{(1-x^2)}}U_n(x)$

$$= \frac{\sin(n\cos^{-1}x)}{\sqrt{1-x^2}} = \frac{\sin n\theta}{\sqrt{1-x^2}}$$

$$\left[\cos^{-1}x = \theta,\quad \frac{-1}{\sqrt{1-x^2}} = \frac{d\theta}{dx}\right]$$

Differentiating both sides we get

$$\frac{dy}{dx} = \frac{\sqrt{1-x^2}\,.\cos n\theta\left(n\frac{d\theta}{dx}\right) - \frac{1}{2}(1-x^2)^{\frac{-1}{2}}(-2x).\sin n\theta}{(1-x^2)}$$

$$\Rightarrow \quad \frac{dy}{dx} = \frac{n\sqrt{1-x^2}\cos n\theta.\left(\frac{-1}{\sqrt{(1-x^2)}}\right) + \frac{x}{\sqrt{1-x^2}}\sin n\theta}{(1-x^2)}$$

$$\Rightarrow \quad (1-x^2)\frac{dy}{dx} = -n\cos n\theta + \frac{x\sin n\theta}{\sqrt{1-x^2}}$$

$$\Rightarrow \quad (1-x^2)\frac{dy}{dx} = -n\cos n\theta + xy$$

Again differentiating we get

$$(1-x^2)\frac{d^2y}{dx^2}-2x\frac{dy}{dx}=n^2\sin n\theta\frac{d\theta}{dx}+x\frac{dy}{dx}+1.y$$

$$\Rightarrow (1-x^2)\frac{d^2y}{dx^2}-3x\frac{dy}{dx}=-n^2\frac{\sin n\theta}{\sqrt{1-x^2}}+y$$

$$\Rightarrow \quad (1-x^2)\frac{d^2y}{dx^2}-3x\frac{dy}{dx} = -n^2y+y$$

$$\Rightarrow \quad (1-x^2)\frac{d^2y}{dx^2}-3\times\frac{dy}{dx}+(n^2-1)\,y=0$$ **Proved.**

**Example 6.** *Show that*

$$\sqrt{(1-x^2)}\,T_n(x)=U_{n+1}(x)-xU_n(x)$$

**Solution.** Let $\cos^{-1}x=\theta \Rightarrow x=\cos\theta$

$$\text{L.H.S.} = \sqrt{(1-x^2)}\;T_n(x)=\sqrt{(1-\cos^2\theta)}\cos(n\cos^{-1}x)=\sin\theta\cos n\theta$$

$$\begin{aligned}\text{R.H.S.} &= U_{n+1}(x)-xU_n(x)\\ &= \sin(n+1)\theta-\cos\theta\sin n\theta\\ &= \sin(n\theta+\theta)-\cos\theta\sin n\theta\\ &= \sin n\theta\cos\theta+\cos n\theta\sin\theta-\cos\theta\sin n\theta\\ &= \cos n\theta.\sin\theta\\ &= \text{L.H.S.}\end{aligned}$$ **Proved.**

**Example 7.** *Show that*

$$\sum_{r=0}^{n} T_{2r}(x)=\frac{1}{2}\left\{1+\frac{1}{\sqrt{(1-x^2)}}U_{2n+1}(x)\right\}$$

**Solution.** Let $\cos^{-1}x=\theta \quad \Rightarrow \quad x=\cos\theta$

$$\sum_{r=0}^{n} T_{2r}(x)=\sum_{r=0}^{n}\cos(2r\cos^{-1}x)=\sum_{r=0}^{n}\cos(2r\theta)$$

$$= \text{Real part of } \sum_{r=0}^{n}[(\cos(2r\theta)+i\sin(2r\theta))]$$

$$= \text{Real part of } \sum_{r=0}^{n} e^{i2r\theta}$$

$$= \text{Real part of } [1+e^{i2\theta}+e^{i4\theta}+e^{i6\theta}+....+e^{i2n\theta}]$$

(This is G.P. of $(n+1)$ terms with common ratio $e^{2i\theta}$)

$$= \text{Real part of } \frac{1-e^{(i2\theta)(n+1)}}{1-e^{i2\theta}} \qquad \left(\text{sum} = \frac{1-r^{n+1}}{1-r}\right)$$

$$= \text{Real part of } \frac{[1-e^{i(2n+2)\theta}][1-e^{-i2\theta}]}{(1-e^{i2\theta})(1-e^{-i2\theta})}$$

$$= \text{Real part of } \frac{1-e^{i(2n+2\theta)}-e^{-i2\theta}+e^{i2n\theta}}{1+1-e^{i2\theta}-e^{-i2\theta}}$$

$$= \text{Real part of } \frac{1-\{\cos(2n+2)\theta-i\sin(2n+2)\theta\}-\{\cos 2\theta-i\sin 2\theta\}+\{\cos 2n\theta-i\sin 2n\theta\}}{2-2\cos 2\theta}$$

$$= \frac{1-\cos(2n+2)\theta-\cos 2\theta+\cos 2n\theta}{2(1-\cos 2\theta)} \qquad =\frac{1}{2}\left[\frac{1-\cos 2\theta}{1-\cos\theta}-\frac{\cos(2n+2)}{1-\cos 2\theta}\cos 2n\theta\right]$$

$$= \frac{1}{2}\left[1-\frac{\cos(2n+2)\theta-\cos 2n\theta}{1-\cos 2\theta}\right]$$

$$= \frac{1}{2}\left[1+\frac{2\sin(2n+1)\theta\sin\theta}{2\sin^2\theta}\right]$$

$$= \frac{1}{2}\left[1+\frac{\sin(2n+1)\theta}{\sin\theta}\right]$$

$$= \frac{1}{2}\left[1+\frac{U_{(2n+1)}(x)}{\sqrt{1-x^2}}\right]$$

**Proved**

**Example 8.** *Prove that*

$$T_n(x)=\sum_{r=0}^{N}(-1)^r\frac{n!}{2r!\,(n-2r)!}(1-x^2)x^{n-2r} \qquad N=\frac{n}{2} \text{ if } n \text{ is even}$$

$$N=\frac{n-1}{2} \text{ if } n \text{ is odd}$$

**Solution.** We know that $T_n(x)= x^n-{}^nC_2x^{n-2}(1-x^2)+{}^nC_4x^{n-4}(1-x^2)^2+....$

$$T_n(x)=\frac{1}{2}\left[\left\{x+i\sqrt{(1-x^2)}\right\}^n+\left\{x-i\sqrt{(1-x^2)}\right\}^n\right] \qquad \text{... (1)}$$

On applying Binomial theorem on (1) we get

$$T_n=\frac{1}{2}\Big[x^n+{}^nC_1x^{n-1}(i\sqrt{1-x^2})+...+{}^nC_rx^{n-r}(-i\sqrt{1-x^2})^r+....$$

$$+\{x^n+{}^nC_1x^{n-1}(-i\sqrt{1-x^2})+...+{}^nC_rx^{n-r}(-i\sqrt{1-x^2})^r+....\}\Big]$$

$$=\frac{1}{2}\left[\left\{\sum_{r=0}^{n}{}^nC_rx^{n-r}(i\sqrt{1-x^2})^r\right\}+\left\{\sum_{r=0}^{n}{}^nC_rx^{n-r}(-i\sqrt{1-x^2})^r\right\}\right]$$

$$=\frac{1}{2}\sum_{r=0}^{n}{}^nC_r\,x^{n-r}(1-x^2)^{\frac{r}{2}}(i)^r\{1+(-1)^r\} \qquad \text{... (2)}$$

(*a*) If $r$ is odd or $r = 2s + 1$ $\qquad r\le n$

$$T_n=\sum_{s=0}^{\frac{n-1}{2}}{}^nC_{2s+1}\,x^{n-2s-1}(i)^{2s+1}[1+(-1)^{2s+1}] \qquad 2s+1\le n$$

$$T_n=\sum_{s=0}^{\frac{n-1}{2}}{}^nC_{2s+1}\,x^{n-2s-1}(i)^{2s+1}(1-1)=0 \qquad s\le\frac{n-1}{2}$$

(*b*) If $r$ is even or $r = 2s$

$$T_n = \frac{1}{2}\sum_{s=0}^{\frac{n}{2}} {}^nC_{2s}\, x^{n-2s}(1-x^2)^s (i)^{2s}\{1+(-1)^{2s}\} \qquad \left[\begin{array}{l} r \le n \\ 2s \le n \\ s \le \frac{n}{2} \\ \text{if } n \text{ is even}\end{array}\right.$$

$$= \frac{1}{2}\sum_{s=0}^{\frac{n}{2}} {}^nC_{2s}\, x^{n-2s}(1-x^2)^s (-1)^s\{1+1\}$$

$$= \sum_{s=0}^{\frac{n}{2}} (-1)^s\, {}^nC_{2s}\, x^{n-2s}(1-x^2)^s$$

$$= \sum_{s=0}^{\frac{n}{2}} (-1)^s \frac{n!}{(n-2s)!\, 2s!}(1-x^2)^s x^{n-2s}$$

(*ii*) If $n$ is odd or $n \equiv 2s+1$ $\qquad s = \frac{n-1}{2}$

$$T_n = \sum_{s=0}^{\frac{n-1}{2}} (-1)^s \frac{n!}{(n-2s)!\, 2s!}(1-x^2)^s x^{n-2s}$$

Hence $T_n = x^n - {}^nC_2\, x^{n-2}(1-x^2) + {}^nC_4\, x^{n-4}(1-x^2)^2 + \ldots$ **Proved.**

**Example 9.** *Find the value of the following:*

(*i*) $U_0(x)$ (*ii*) $U_1(x)$ (*iii*) $U_2(x)$ (*iv*) $U_3(x)$

**Solution.** We know that

$$U_n(x) = \sin(n\cos^{-1} x)$$

If $n = 0$, $\qquad U_0(x) = \sin(0) = 0$

If $n = 1$, $\qquad U_1(x) = \sin(\cos^{-1} x) = \sin\sin^{-1}\sqrt{(1-x^2)} = \sqrt{(1-x^2)}$

If $n = 2$, $U_2(x) = \sin(2\cos^{-1} x) = \sin 2\theta = 2\sin\theta\cos\theta = 2x\sqrt{1-x^2}$ $\qquad \left[\begin{array}{l}\theta = \cos^{-1} x \\ \cos\theta = x\end{array}\right.$

If $n = 3$, $U_3(x) = \sin(3\cos^{-1} x) = \sin 3\theta = 3\sin\theta - 4\sin^3\theta$

$$= 3\sqrt{1-x^2} - 4(\sqrt{1-x^2})^3 = \sqrt{1-x^2}\,[3-4(1-x^2)] = \sqrt{1-x^2}\,[4x^2-1]$$

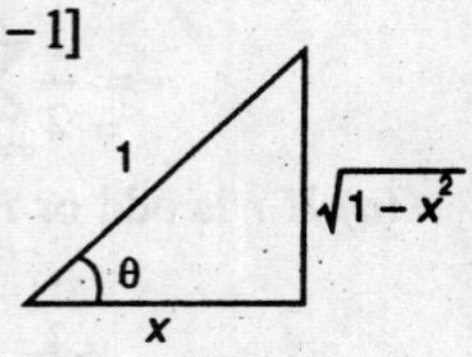


Here we see that $U_n(x)$ is not polynomial.

But if we define

$$U_n(x) = \frac{\sin\{(n+1)\cos^{-1} x\}}{\sin(\cos^{-1} x)} = \frac{U_{n+1}}{\sqrt{1-x^2}}$$

Then $U_n(x)$ is a polynomial in $x$ of degree $n$.

## 22.7 GENERATING FUNCTION FOR $T_n(x)$

$$\frac{1-t^2}{1-2tx+t^2}=T_0(x)+2\sum_{n=1}^{\infty} T_n(x).t^n$$

**Proof** L.H.S. $=\dfrac{1-t^2}{1-2tx+t^2}$ Put $x=\cos\theta=\dfrac{e^{i\theta}+e^{-i\theta}}{2}$

$$=\frac{1-t^2}{1-t(e^{i\theta}+e^{-i\theta})+t^2}$$

$$=\frac{1-t^2}{(1-te^{i\theta})-te^{-i\theta}+t^2}=\frac{1-t^2}{(1-te^{i\theta})-te^{-i\theta}(1-te^{i\theta})}$$

$$=\frac{1-t^2}{(1-te^{i\theta})(1-te^{-i\theta})}=(1-t^2)(1-te^{i\theta})^{-1}(1-te^{-i\theta})^{-1}$$

$$=(1-t^2)(1+te^{i\theta}+t^2e^{2i\theta}+...+t^re^{ri\theta}+...)[1+te^{-i\theta}+t^2e^{-2i\theta}+...+t^se^{-si\theta}+...]$$

$$=(1-t^2)\sum_{r=0}^{\infty}t^re^{ri\theta}\sum_{s=0}^{\infty}t^se^{-si\theta}=(1-t^2)\sum_{\substack{r=0\\s=0}}^{\infty}t^{r+s}e^{(r-s)i\theta}$$

$$=\sum_{\substack{r=0\\s=0}}^{\infty}i^{i(r-s)\theta}t^{r+s}-\sum_{\substack{r=0\\s=0}}^{\infty}i^{i(r-s)\theta}t^{r+s+2} \qquad ...(1)$$

On putting $r = s = 0$ in the first summation of (1) we get coefficient of $t_0=e^{i(0-0)0}=e^0=1=T_0(x)$.

On putting $r + s = n$ or $s = n - r$ in the first summation and $r + s + 2 = n$ or $s = n - r - 2$ in the second summation, we get

$$\text{Coefficient of } t^n=\sum_{r=0}^{n}e^{i[r-(n-r)]\theta}-\sum_{r=0}^{n-2}e^{i[r-(n-r-2)]\theta}$$

$$=\sum_{r=0}^{n}e^{i(-n+2r)\theta}-\sum_{r=0}^{n-2}e^{i[-n+2r+2]\theta}$$

$$=e^{-in\theta}\sum_{r=0}^{n}e^{i(2r)\theta}-e^{-i(n-2)\theta}\sum_{r=0}^{n-2}e^{i(2r)\theta}$$

$$=e^{-in\theta}[1+e^{2i\theta}+e^{4i\theta}+...\text{to}(n+1)\text{ terms}]-e^{-i(n-2)\theta}[1+e^{2i\theta}+e^{4i\theta}+...\text{to}(n-1)\text{ terms}]$$

$$=e^{-in\theta}\frac{1-(e^{2i\theta})^{n+1}}{1-e^{2i\theta}}-e^{-i(n-2)\theta}\frac{1-(e^{2i\theta})^{n-1}}{1-e^{2i\theta}} \qquad \text{[Sum of G.P.]}$$

$$=\frac{e^{-in\theta}-e^{i(n+2)\theta}}{1-e^{2i\theta}}-\frac{e^{-i(n-2)\theta}-e^{in\theta}}{1-e^{2i\theta}}$$

$$=\frac{e^{-in\theta}-e^{i(n+2)\theta}-e^{-i(n-2)\theta}+e^{in\theta}}{1-e^{2i\theta}}=\frac{e^{-in\theta}(1-e^{2i\theta})+e^{in\theta}(1-e^{2i\theta})}{1-e^{2i\theta}}$$

$$= \frac{(1-e^{2i\theta})(e^{in\theta}+e^{-in\theta})}{1-e^{2i\theta}} = (e^{in\theta}+e^{-in\theta}) = 2\cos n\theta = 2T_n(x)$$

or $$\frac{1-t^2}{1-2tx+t^2} = T_0(x) + 2\sum_{n=0}^{\infty} T_n(x)\, t^n$$ **Proved.**

**Example 10.** *Show that*

(i) $T_n(-1) = (-1)^n$ (ii) $T_{2n}(0) = (-1)^n$ (iii) $T_{2n+1}(0) = 0$

**Solution.** We know that

$$T_n(x) = \cos(n\cos^{-1} x) \qquad \dots (1)$$

(i) Replacing $x$ by $-1$ in (1) we get

$$T_n(-1) = \cos[n\cos^{-1}(-1)] = \cos[n\pi] = (-1)^n$$

(ii) Replacing $x$ by 0 and $n$ by $2n$ in (1) we get

$$T_{2n}(0) = \cos[2n\cos^{-1}(0)] = \cos\left(2n\frac{\pi}{2}\right) = \cos n\pi = (-1)^n$$

(iii) Replacing $x$ by 0 and $n$ by $2n+1$, we get

$$T_{2n+1}(0) = \cos[(2n+1)\cos^{-1}(0)] = \cos[(2n+1)\frac{\pi}{2}] = 0$$ **Proved.**

**Example 11.** *Show that*

(i) $U_n(1) = 0$ (ii) $U_n(-1) = 0$, (iii) $U_{2n}(0) = 0$, (iv) $U_{2n+1}(0) = (-1)^n$

**Solution.** We know that

$$U_n(x) = \sin(n\cos^{-1} x) \qquad \dots (1)$$

(i) Replacing $x$ by 1 in (1), we get

$$U_n(1) = \sin(n\cos^{-1} 1) = \sin(n \times 0) = 0$$

(ii) Replacing $x$ by $(-1)$ in (1) we get

$$U_n(-1) = \sin[n\cos^{-1}(-1)] = \sin(n\pi) = 0$$

(iii) Replacing $x$ by 0 and $n$ by $2n$ in (1) we get

$$U_{2n}(0) = \sin(2n\cos^{-1} 0) = \sin\left(2n\frac{\pi}{2}\right) = \sin(n\pi) = 0$$

(iv) Replacing $x$ by 0 and $n$ by $2n+1$, in (1) we get

$$U_{2n+1}(0) = \sin[(2n+1)\cos^{-1} 0] = \sin(2n+1)\frac{\pi}{2} = \sin\left(n\pi + \frac{\pi}{2}\right) = (-1)^n$$ **Proved.**

**Example 12.** *Show that*

$$2\{T_n(x)\}^2 = 1 + T_{2n}(x)$$

**Solution.** $$2[T_n(x)]^2 = 2[\cos(n\cos^{-1} x)]^2 \qquad \left[\begin{array}{l}\cos^{-1} x = 0 \\ \Rightarrow x = \cos\theta\end{array}\right]$$

$$= 2[\cos n\theta]^2$$

$$= \cos 2n\theta + 1$$

$$= \cos[2n\cos^{-1} x] + 1$$

$$= T_{2n}(x) + 1$$ **Proved.**

**Example 13.** *Show that*

$$T_{m+n}(x)+T_{m-n}(x)=2T_m(x)T_n(x)$$

**Solution.**
$$\begin{aligned} T_{m+n}(x)+T_{m-n}(x) &= \cos[(m+n)\cos^{-1}x]+\cos[(m-n)\cos^{-1}x] \\ &= \cos\,[m+n]\theta+\cos(m-n)\theta \qquad (\cos^{-1}x=\theta) \\ &= 2\cos m\theta\cos n\theta \\ &= 2\cos[m\cos^{-1}x]\cos[n\cos^{-1}x] \\ &= 2T_m(x).T_n(x) \end{aligned}$$

**Proved.**

**Example 14.** *Show that* $T_n'(x)=\dfrac{n}{\sqrt{1-x^2}}u_n(x)$

**Solution.** $T_n(x)=\cos(n\cos^{-1}x)=\cos n\theta \qquad \cos^{-1}x=\theta$

$$T_n'(x)=-\sin n\theta.n\frac{d\theta}{dx} \qquad \frac{-1}{\sqrt{1-x^2}}=\frac{d\theta}{dx}$$

$$= -\sin n\theta\left(-\frac{n}{\sqrt{1-x^2}}\right)$$

$$= \frac{n\sin n\theta}{\sqrt{1-x^2}}=\frac{n\sin[n\cos^{-1}x]}{\sqrt{1-x^2}}$$

$$= \frac{nu_n(x)}{\sqrt{1-x^2}}$$

**Ans.**

## EXERCISE 22.1

1. Express the following Chebyshev functions :
(*a*) $T_4(x)$ (*b*) $T_3(x)$ (c) $T_2(x)+2T_1(x)+2T_0(x)$
into ordinary polynomials.
**Ans.** (*a*) $8x^4-8x^2+1$ (*b*) $4x^3-3x$ (*c*) $2x^2+2x+1$

2. Express the polynomial
$12x^3+6x^2+4x+1$
in Chebyshev polynomical (*a*) first kind $T_n(x)$.
(*b*) second kind $U_n(x)=\dfrac{\sin\{(n+1)\cos^{-1}x\}}{\sin(\cos^{-1}x)}$

**Ans.** (*a*) $3T_3(x)+3T_2(x)+13T_1(x)+4T_0(x)$ (*b*) $\frac{3}{2}U_3(x)+\frac{3}{2}U_2(x)+5U_1(x)+\frac{5}{2}U_0(x)$

3. Express the polynomial
$16x^4+4x^3+2x^2+4x+5$ into the Chebyshev polynomial of first kind.
**Ans.** $2T_4(x)+T_3(x)+9T_2(x)+7T_1(x)+12T_0(x)$

4. If $U_n(x)=\dfrac{\sin\{(n+1)\cos^{-1}x\}}{\sin(\cos^{-1}x)}$, show that
(*a*) $U_n(-x)=(-1)^nU_n(x)$ (*b*) $U_n(1)=n+1$
(c) $U_n(-1)=(-1)^n(n+1)$

**5.** Prove that

$$U_{n+1}(x) - 2x\, U_n(x) + U_{n-1}(x) = 0$$

**6.** Prove that

(a) $\int_{-1}^{+1} \sqrt{1-x^2}\, U_m(x)\, U_n(x)\, dx = 0, \quad m \neq n$  (b) $\int_{-1}^{+1} \sqrt{1-x^2}\, (U_n(x))^2\, dx = \frac{\pi}{2}$

**8.** Prove that

$$U_n(x) = \sum_{r=0}^{\frac{1}{2}(n-1)} (-1)^r \frac{n!}{(2r+1)!\ (n-2r-1)!} (1-x^2)^{r+\frac{1}{2}}\, x^{n-2r-1}$$

$$= {}^{n+1}C_1 x^n - {}^{n+1}C_3\, x^{n-2}(1-x^2) + {}^{n+1}C_5 x^{n-4}(1-x^2)^2 \ldots\ldots$$

**7.** Prove that $T_n(x) = U_n(x) - xU_{n-1}(x)$

# 23

# Fuzzy Set

## 23.1. INTRODUCTION

If a doctor asks a patient "How are you" patient replies almost O.K."

The word "almost" is a vague term and not mathematical i.e. It means neither "yes" nor No" but between them.

**But the word "almost"** gives lots of information to the doctor & the doctor decides the **further future treatment of the** patient.

For this reason a **mathematical** modelling of vague knowledge is necessary. To convey such an information "Fuzzy set" is introduced. L.A. Zadeh in 1965 introduced this concept on the basis of membership function defined as:

$$\mu : X \to \{0,1\}$$

Here

$\mu(x) = 1$, means full membership

$\mu(x) = 0$, means non-membership and

$0 < \mu(x) < 1$, **means intermediate membership.**

Due to Zadeh's work, a theory of vagueness (fuzziness) is now fully developed. The concept of 'Fuzzy set" & membership degree were introduced to form a mathematical model of vagueness.

A set is collection of well-defined distinct objects. A set of intelligent students is not a set. Because the criteria to be intelligent is not well-defined. We cannot say whether a particular student belongs to or not. The belonginess is not clear but vague. In fuzzy set theory, we assume that all students are members, all belong to the set upto certain extent.

Let Among Anil, Rajiv and Suresh, Anil got 10 marks, Rajiv got 40 marks and Suresh get 90 marks out of 100. Hence in comparison to Anil Rajiv is intelligent but if compared to Suresh he is not intelligent. All the three can be said to be intelligent to some extent and hence all the three are the members of this set of intelligent students.

**For example:** We write

A = {0.5 Rita, 0.9 Kusum, 0.4 Suresh, 0.6 John, 0.2 Latif} for the set of rich people.

It indicates that Rita has 0.5 degree of membership in A, Kusum has 0.9 degree of membership in A. Suresh has a 0.4 degree of membership in A, John has a 0.6 degree of membership in A and Latif has a 0.2 degree of membership in A. Thus, Kusum is the richest and Latif is the poorest of these people.

## 23.2 FUZZY SET

**Definition.** Let $X$ be a non-zero set. A fuzzy set $A$ of this set $X$ is defined by the following set of pairs.

$$A = \{(x, \mu_A(x))\} : x \in X$$

Where, $\mu_A : X \to [0, 1]$

is a function called as the membership function of $A$ & $\mu_A(x)$ is the grade of membership or degree of belonginess or degree of membership of $x \in X$ in $A$.

Thus a fuzzy set is a set of pairs consisting of a particular element of the universe and its degree of membership.

A can also be written as

$$A = \{(x_1, \mu_A(x_1)), (x_2, \mu_A(x_2)), \ldots (x_n, \mu_A(x_n))\}$$

Symbolically we write

$$A = \left\{\frac{x_1}{\mu_A(x_1)} \cdot \frac{x_2}{\mu_A(x_2)} \cdots \frac{x_n}{\mu_A(x_n)}\right\}$$

## 23.3. EQUALITY OF TWO FUZZY SETS

**Example 1.** Let $X = \{2, 3, 4\}$

Consider the three fuzzy sets $A$, $B$, $C$ of $X$ as given below

$$A = \left\{\frac{2}{5}, \frac{3}{6}, \frac{4}{1}\right\}$$

$$B = \left\{\frac{2}{7}, \frac{4}{8}, \frac{3}{9}\right\}$$

$$C = \left\{\frac{4}{1}, \frac{2}{5}, \frac{3}{6}\right\}$$

Here in set 'A' and set "$C$" members and their degrees are same.

$\therefore$ $A = C$, But in set "$A$" & set "$B$" members are the same but their degrees of membership are not the same.

Hence $A \neq B$.

## 23.4. COMPLEMENT OF A 'FUZZY SET'

The component of a fuzzy set A is the set $\overline{A}$ with degree of the membership of an element in $\overline{A}$ is equal to one minus the degree of the membership of this element in A.

**Example.** The set "A" is written as

A = [0.9 Rama, 0.4 Manju, 0.8 Neera, 0.1 Jyoti] for the set of beautiful girls.

Thus $\overline{A}$ = {0.1 Rama, 0.6 Manju, 0.2 Neera, 0.9 Jyoti} for the set of girls who are not beautiful.

## 23.5. UNION OF TWO FUZZY SETS

The union of two fuzzy sets $A$ and $B$ is the fuzzy set $A \cup B$, where the degree of membership of an element in $A \cup B$ is the maximum of the degrees of membership of this element in $A$ and in $B$.

**Example.** We write

$A$ = {0.3 Radha, 0.9 Pawan, 0.6 Mahesh, 0.4 Kunal} for the set of rich people.

$B$ = {0.4 Radha, 0.8 Pawan, 0.2 Mahesh, 0.7 Kunal} for the set of famous people.

Here the degree in $A \cup B$ of each element is the maximum of degrees of membership of this element in $A$ & in $B$.

$A \cup B$ = {0.4 Radha, 0.9 Pawan, 0.6 Mahesh, 0.7 Kunal}

## 23.6. INTERSECTION OF TWO FUZZY SETS

The intersection of two fuzzy sets $A$ and $B$ is the fuzzy set $A \cap B$, where the degree of membership of an element in $A \cap B$ is the minimum of the degrees of membership of this element in $A$ and in $B$.

**Example:** Let $A$ = [0.5 Pushpa, 0.1 Suman, 0.8 Rani, 0.4 Kailash] for the set of fat people.

$B$ = [0.3 Pushpa, 0.6 Suman, 0.2 Rani, 0.7 Kailash } for the set of tall people.

Here the degree of membership in $A \cap B$ of each element is the minimum of degree of membership of this element in $A$ and in $B$.

$A \cap B$ = {0.3 Pushpa, 0.1 Suman, 0.2 Rani, 0.4 Kailash}

## 23.7. TRUTH VALUE

*(R.G.P.V., Bhopal I/II Sem. Summer 2004)*

**(i)** *Truth value of the negation of a proposition*

The truth value of the negation of a proposition in fuzzy logic is 1 minus the truth value of the proposition.

**Example.** If the truth value "Vimla is happy" is 0.9.

Then the truth value of the statement "Vimla is not happy" is 1–0.9 = 0.1

**Example.** If the truth value of the statement "Devendra is smart" is 0.8.

Then the truth value of the statement that Devendra is not smart" is 1–0.8 = 0.2

**(ii)** *Truth value of the conjuction of two prepositions.*

The truth value of the conjunction of two prepositions in the fuzzy logic is the minimum of the truth values of the two prepositions.

**Example.** If the truth value of the statement "Khan is brave" is 0.7 And the truth value of the statement "Kamal is brave". is 0.8.

Then the truth value of the statement "Khan and Kamal are brave is 0.7 (minimum of the two).

And the truth value of "neither Khan nor Kamal is brave" is 0.2 (As truth value of the negation of 1st statement is 1–0.7 = 0.3 and that of second statement is 1–0.8 = 0.2 & minimum of 0.3 and 0.2 is 0.2)

**(iii)** *The truth value of the disjunction of two prepositions.*

The truth value of the disjunction of two prepositions in fuzzy logic is the maximum of the truth values of the two prepositions.

**Example.** If the truth value of the statement "Neera is intelligent" is 0.9 and the truth value of the statement "Rekha is intelligent" is 0.6

Then the truth values of the statements

"Neera is intelligent, or "Rekha is intelligent" is 0.9 (Maximum of the two).

The truth values of the statements "Neera is not intelligent" or "Rekha is not intelligent" is 0.4 (Maximum of 1 – 0.9 = 0.1 and 1 –0.6 = 0.4).

## EXERCISE 23.1

**1.** Interpret the following:

(*i*) The set A = {0.7 Anu, 0.9 Rasika, 0.2 Sarita, 0.5 Kartik} for the set of honest people.

(*ii*) The set of B = [0.2 John, 0.4 Charu, 0.9 Medha, 0.8 Gagan] for the set of brave people.

**2.** What are the constituents of the pair in a fuzzy set.

**3.** Which of the two fuzzy sets are equal of the following:

$A$ = [0.3 Sonu, 0.8 Renu, 0.9 Paul, 0.5 Kunal]

$B$ = [0.6 Kunal, 0.9 Paul, 0.7 Renu, 0.3 Sonu]

$C$ = [0.8 Renu, 0.9 Paul, 0.3 Sonu, 0.5 Kunal]

**4.** Write down complement set of A, if

$A$ = [0.3 Krishna, 0.8 Kamal, 0.7 Rajnish, 0.6 Surendra]

**5.** Write down $A \cup B$ in fuzzy sets.

If fuzzy set = $A$ [0.5 $x_1$, 0.3 $x_2$, 0.7 $x_3$, 0.8 $x_4$]

Fuzzy set $B = [0.6\ x_1, \quad 0.4\ x_2, \quad 0.9\ x_3, \quad 0.1\ x_4]$

6. Write down $A \cap B$ in fuzzy sets.

If fuzzy set $A = [0.4$ P, 0.7 Q, 0.2 R, 0.5 S]

fuzzy set $B = [0.8$ P, 0.6 Q, 0.1 R, 0.4 S]

7. Find the truth value of the negation of the following prepositions.

(*i*) The truth value of "*A* is happy" is 0.8.

(*ii*) The truth value of "*B* is rich" is 0.7.

(*iii*) The truth value of "Kamla is beautiful" is 0.9.

8. Find the truth value of the conjunction of the two prepositions. If the truth value of the statements "Ranjeet is a good driver" is 0.7.

Latif is a good driver is 0.6.

9. Give the truth value of the disjunction of the two prepositions

If the truth value "Sarla has a good health" is 0.6.

And the truth value "Vijay possesses a good health" is 0.8.

10. Write short notes on "Fuzzy sets".

## ANSWERS

1. (*i*) Rasika is the most honest and Sarita is the least honest.

(*ii*) Medha is the bravest girl and John is the least.

2. Members and its degree of membership.

3. $A = C$

4. $\overline{A}$ = [0.7 Krishna, 0.2 Kamal, 0.3 Rajnish, 0.4 Surendra]

5. $A \cup B = [0.6\ x_1, \quad 0.4\ x_2, \quad 0.9\ x_3, \quad 0.8\ x_4]$

6. $A \cap B = [0.4$ P, 0.6 Q, 0.1 R, 0.4 S]

7. (*i*) The truth value of "*A* is not happy" is = 1 – 0.8 = 0.2.

(*ii*) The truth value of "*B* is not rich" is 1 – 0.7 = 0.3.

(*iii*) The truth value of "Kamla is not beautiful" is 1 – 0.9 = 0.1

8. The truth value of the conjunction, Ranjeet & Latif are good drivers is 0.6 (minimum of the two).

9. The truth value "Sarla has a good health" or "Vijay possesses a good health" is 0.8 (Maximum of the two).

## 23.8 APPLICATIONS

All engineering disciplines have already been affected to various degrees by new methodological possibilities opened by fuzzy sets, fuzzy measures.

(*i*) **Electrical Engineering**

By developing fuzzy controllers, electrical engineering was first engineering discipline within which the utility of fuzzy sets and fuzzy logic was recongised. Fuzzy image processing, electronic circuits for fuzzy logic or robbotics is also developed in electrical engineering.

(*ii*) **Civil Engineering**

In civil engineering, some initial ideas regarding the application of fuzzy sets emerged in 1970. There is the uncertainity in applying theoretical solution to civil engineering projects, designing at large. Designer deals with the uncertainity, in safety which is required in the construction of bridges; buildings, dams etc. Fuzzy set theory has already proven useful, consists of problems of assessing or evaluating existing constructions.

(*iii*) **Mechanical Engineering**

It was realised around mid-1980s that fuzzy set theory is eminently suited for mechanical engineering design.

A wide range of material might be used in mechanical engineering and the membership function is expressed in terms of corrosion, thermal expansion or some other measurable material property. A combination of several properties including the cost of different materials, may also be used.

(*iv*) **Industrial Engineering**

Two well-developed areas of fuzzy set theory that are directly relevant to industrial engineering are fuzzy control and fuzzy decision making.

Numerous their applications of fuzzy set theory in industrial engineering have also been explored to various degrees, Fuzzy set are convenient for estimating the service life of a given piece of equipment for various conditions under which it operates. In industrial environment, fuzzy sets are also applied in designing built-in tests for industrial systems.

(*v*) **Computer Engineering**

In mid 1980s, when the utility of fuzzy controllers became increasingly visible, the need for computer hardware to implement the various operations involved in fuzzy logic and approximate reasoning has been recognised. All inference rules of a complex fuzzy inferences engine are processed in parallel. This increases efficiency tremendously and extends the scope of applicability of fuzzy controllers, and potentially, other fuzzy expert systems. In digital mode, fuzzy sets are represented as vectors of numbers (0, 1). Analog fuzzy hardware is characterised by high speed and good compatibility with sensors, it is thus suitable for complex on-line fuzzy controllers.

(*vi*) **Reliability theory**

The classical theory of reliability is developed in world war II on the following assumptions.

(*a*) **Assumption of dichotomous states.** At any given time, the engineering products is either in functioning state or in failed state.

(*b*) **Probability assumption.** The behaviour of the engineering product with respect to the two critical states (functioning and failed) can adquately be characterised in terms of probability theory.

An alternative reliability theory, rooted in fuzzy sets and probability.

(*c*) **Assumption of fuzzy sets.** At any time the engineering products may be in functioning states to some degree and in failed state to another degree.

(*d*) **Possibility assumption.** The behaviour of the engineering product with respect to the two critical fuzzy states (fuzzy functioning state and fuzzy failed state) can adequately be characterised in terms of possibility theory, while second theory based on fuzzy sets is more meaningful.

(*vii*) **Robotics**

The fuzzy set theory that is relevant to robotics include approximate reasoning, fuzzy controllers and other kind of fuzzy systems, fuzzy pattern recognition and image processing, fuzzy data bases.

## EXERCISE 23.2

1. Write short note on the following:
Fuzzy logic affects many disciplines. (Rajiv Gandhi University, M.P. Summer 2001)
2. Define with example
Fuzzy graph, fuzzy relations. (Ravi Shanker Uni. I semester 2003)

# 24

# Hankel Transform

## 24.1 HANKEL TRANSFORM

If $J_n\,(sx)$ be the Bessel function of the first kind of order $n$, then the Hankel transform of a function $f\,(x)$, $(0 < x < \infty)$ denoted by $F\,(s)$ is defined as $H\,(s) = \int_0^\infty f(x).\; xJ_n(sx)\,dx$

Here $x\,J_n(sx)$ is the *Kernel of the transformation,*

## 24.2 THE FORMULAE USED IN FINDING THE HANKEL TRANSFORMS.

Recurrence relations for Bessel's functions

1. $x\,J_n' = nJ_n - x\,J_{n+1}$ 2. $xJ_n' = x\,J_{n-1} - nJ_n$

3. $2J_n' = J_{n-1} - J_{n+1}$ 4. $2n\,J_n = x[J_{n-1} + J_{n+1}]$

5. $\dfrac{d}{dx}(x^{-n}J_n) = -x^{-n}\,J_{n+1}$ 6. $\dfrac{d}{dx}(x^n\,J_n) = x^n\,J_{n-1}$ or $x^n\,J_n = \int x^n\,J_{n-1}\,dx$

From (6) Recurrence relation we can find the definite Integrals

7. $\int_0^\infty x^n J_{n-1}(x)\,dx = [x^n J_n(x)]_0^\infty$ ... (7)

In (7) we put $n = 1$ and substitute $J\,(x)$ by $J\,(sx)$, we get (8),

8. $\int_0^\infty x\,J_0\,(xs)\,dx = \left[\dfrac{x}{s}J_1(xs)\right]_0^\infty$ ... (8)

**Example 1.** *Evaluate* $\int_0^a x^2\,J_1(sx)\,dx$.

**Solution.** $\int_0^a x^2\,J_1(sx)\,dx$

$$= \left[\frac{x^2}{s}J_2(sx)_0^a\right] = \frac{a^2}{s}J_2(as)$$ **Ans.**

**Example 2.** *Evaluate* $\int x^3.J_0\,(sx)\,dx$

**Solution.** $\int_0^a x^3\,J_0\,(sx)\,dx = \int_0^a x^2.\{x\,J_0(sx)\}\,dx$

Integrating by parts we get

$$= \left[x^2.\left(\frac{x}{s}J_1(ax)\right)\right]_0^a - \int_0^a (2x)\left\{\frac{x}{s}J_1(sx)\right\}dx \qquad \left[\int x^n J_{n-1}(x)\,dx = x^n J_n(x)\right]$$

$$= \left[\frac{x^2}{s}J_1(sx)\right]_0^a - \left[2x.\left\{\frac{x}{s^2}J_2(sx)\right\}\right]_0^a$$

$$= \frac{a^3}{s}J_1(as) - \frac{2}{s}\int_0^a x^2 J_1(sx)dx$$

$$= \frac{a^3}{s}J_1(as) - \frac{2}{s^2}\left[x^2 J_2(sx)\right]_0^a$$

$$= \frac{a^3}{s}J_1(as) - \frac{2a^2}{s^2}J_2(as)$$ **Ans.**

**Example 3.** *Evaluate* $\int_0^a x(a^2 - x^2) J_0(sx)\,dx$

**Solution.** $\int_0^a x(a^2 - x^2)J_0(sx)\,dx = \int_0^a (a^2 - x^2).\{xJ_0(sx)\}\,dx$

Integrating by parts we get

$$= \left[(a^2 - x^2).\frac{x}{s}J_1(sx)\right]_0^a - \int_0^a (-2x)\frac{x}{s}J_1(sx)\,dx \qquad \left[\int x^n J_{n-1}(x)dx = x^n J_n(x)\right]$$

$$= 0 + 2\int_0^a x.\frac{x}{s}J_1(sx)\,dx$$

$$= \frac{2}{s}\int_0^a x^2 J_1(sx)dx$$

$$= \left[\frac{2}{s}\frac{x^2}{s}J_2(sx)\right]_0^a$$

$$= \left[\frac{2x^2}{s^2}J_2(sx)\right]_0^a = \frac{2a^2}{s^2}J_2(as)$$

$$= \frac{2a^2}{s^2}\left[\frac{2}{as}J_1(as) - J_0(as)\right] = \frac{4a}{s^3}J_1(as) - \frac{2a^2}{s^2}J_0(as)$$ **Ans.**

## 24.3 SOME MORE INTEGRALS INVOLVING EXPONENTIAL FUNCTIONS AND BESSEL'S FUNCTION

1. $\int_0^\infty e^{-ax}J_0(sx)\,dx = (a^2 + s^2)^{\frac{-1}{2}}$

2. $\int_0^\infty e^{-ax}J_1(sx)\,dx = \frac{1}{s} - \frac{a}{s\sqrt{a^2 + s^2}}$

3. $\int_0^\infty xe^{-ax}J_0(sx)\,dx = a(a^2 + s^2)^{-3/2}$

4. $\int_0^\infty xe^{-ax}J_1(sx)\,dx = s(a^2 + s^2)^{-3/2}$

5. $\int_0^\infty \frac{e^{-ax}}{x} J_1(sx)\,dx = \frac{(a^2+s^2)^{\frac{1}{2}} - a}{s}$

6. $J_n(x) = \frac{x^n}{2^n \lfloor (n+1)} \left[1 - \frac{x^2}{\lfloor 1.2^2 (n+1)} + \frac{x^4}{\lfloor 2.2^2.4^2 (n+1)(n+2)} \cdots\right]$

7. $J_{\frac{1}{2}}(x) = \sqrt{\frac{2}{\pi x}} \sin x$ 8. $J_{\frac{-1}{2}}(x) = \sqrt{\frac{2}{\pi x}} \cos x.$

## LINEARITY PROPERTY

**Theorem 1.** $H\{f(x) + g(x)\} = H\{f(x)\} + H\{g(x)\}$

**Proof.** $H\{f(x) + g(x)\} = \int_0^\infty x\{f(x) + g(x)\} J_n(sx)\,dx$

$$= \int_0^\infty x f(x).J_n(sx)\,dx + \int_0^\infty x g(x) J_n(sx)\,dx$$

$$= H\{f(x)\} + H\{g(x)\}$$ **Proved.**

**Theorem 2.** $H\{f(ax)\} = a^{-2} H\left(\frac{s}{a}\right)$ (Similarity Theorem)

**Proof.** We know that

$$H\{f(ax)\} = \int_0^\infty x f(ax) J_n(sx)\,dx$$

$$= \int_0^\infty (ax) f(ax) J_n\left(\frac{s}{a} ax\right) d(ax).\frac{1}{a^2}$$

$$= \frac{1}{a^2}\int_0^\infty t f(t) J_n\left(\frac{s}{a} t\right) dt$$ (Putting $t = ax$)

$$= a^{-2} H\left(\frac{s}{a}\right)$$ **Proved.**

**Example 4.** *Find the Hankel transform of the function*

$$f(x) = \begin{cases} 1 & 0 < x < a,\ n = 0 \\ 0 & x > a,\ n = 0 \end{cases}$$

**Solution.** Let $H(s)$ be the Hankel Transform of $f(x)$.

**Solution.** $H(s) = \int_0^\infty f(x)\, x J_0(sx)\,dx$

$$= \int_0^a 1.x J_0(sx)dx + \int_a^\infty a\, x J_0(sx)dx$$

$$= \int_0^a x J_0(sx)\,dx + 0$$

$$= \int_0^a x J_0(sx)\,dx = \left[\frac{x}{s} J_1(sx)\right]_0^a = \frac{a}{s} J_1(as)$$ **Ans.**

**Example 5.** *Find the Hankel Transform of the function*

$$f(x) = \begin{cases} x^n, 0 < x < a, & n > -1 \\ 0, \quad x > a, & n > -1 \end{cases}$$

**Solution.** Let $H(s)$ be the Hankel Transform of $f(x)$.

$$H(s) = \int_0^\infty f(x).x\, J_n(sx)\, dx$$

$$= \int_0^a x^n.x\, J_n(sx)\, dx + \int_a^\infty 0.x\, J_n(sx)\, dx$$

$$= \int_0^a x^{n+1} J_n(sx)\, dx$$

$$= \left[\frac{x^{n+1}}{s} J_{n+1}(sx)\right]_0^a$$

$$= \frac{a^{n+1}}{s} J_{n+1}(as)$$ **Ans.**

**Example 6.** *Find the Hankel Transform*

$$f(x) = \begin{cases} a^2 - x^2,\ 0 < x < a & n = 0 \\ 0, \quad x > a & n = 0 \end{cases}$$

**Solution.** Let $H(s)$ be the Hankel transform of $f(x)$.

$$H(s) = \int_0^\infty f(x).x\, J_n(sx)\, dx$$

$$= \int_0^a (a^2 - x^2).x\, J_0(sx)\, dx + \int_a^\infty 0.x\, J_0(sx)\, dx$$

$$= \int_0^a (a^2 - x^2)\, x\, J_0(sx)\, dx$$

$$= a^2 \int_0^a x\, J_0(sx)\, dx - \int_0^a x^3\, J_0(dx)\, dx \qquad \text{... (1)}$$

Let us find out the above two integrals

$$a^2 \int_0^a x\, J_0(sx) dx = a^2 \left[\frac{x}{s} J_1(sx)\right]_0^a = a^2 \frac{a}{s} J_1(as) = \frac{a^3}{s} J_1(as) \qquad \text{... (2)}$$

$$\int_0^a x^3\, J_0(sx) dx = \frac{a^3}{s} J_1(as) - \frac{2a^2}{s^2} J_2(as) \qquad \text{... (3)}$$

(See Example 2 on page 1208)

On putting these values from (2) and (3) in (1) we get

$$\text{H}(s) = \frac{a^3}{s} J_1(as) - \frac{a^3}{s} J_1(as) + \frac{2a^2}{s^2} J_2(as) = \frac{2a^2}{s^2} J_2(as)$$

$$= \frac{2a^2}{s^2}\left[\frac{2}{as} J_1(as) - J_0(as)\right] = \frac{4a^2}{s^3} J_1(as) - \frac{2a^2}{s^2} J_0(as)$$

**Example 7.** *Find the Hankel transform of* $\frac{e^{-ax}}{x^2}$, $n = 1$.

**Solution.** Let $H(s)$ be the Hankel Transform of $f(x)$

*i.e.*,
$$H(s) = \int_0^\infty f(x).x.J_n(sx)\,dx$$
$$= \int_0^\infty \frac{e^{-ax}}{x^2}.x.J_1(sx)\,dx$$
$$= \int_0^\infty \frac{e^{-ax}}{x}.J_1(sx)\,dx$$
$$= \frac{(a^2+s^2)^{\frac{1}{2}} - a}{s}$$ **Ans.**

**Example 8.** *Find the Hankel Transform of the function*
$$\frac{e^{-ax}}{x}, \quad n = 0$$

**Solution.** Let $H(s)$ be the Hankel Transform of this function $f(x)$.

*i.e.*
$$H(s) = \int_0^\infty f(x).xJ_n(sx)\,dx$$
$$= \int_0^\infty \frac{e^{-ax}}{x}.x\,J_0(sx)\,dx$$
$$= \int_0^\infty e^{-ax}.J_0(sx)\,dx$$
$$= (a^2+s^2)^{\frac{-1}{2}}$$ **Ans.**

**Example 9.** *Show that if n = 0, the Hankel transform*
$$H\left\{\frac{\sin ax}{x}\right\} = \begin{cases} 0 & \text{if } s > a \\ \dfrac{1}{\sqrt{a^2-s^2}} & \text{if } 0 < s < a \end{cases}$$

*(U.P. III Semester, Summer 2002)*

**Solution.**
$$H(s) = \int_0^\infty f(x).xJ_n(sx)dx$$
$$H\left\{\frac{\sin ax}{x}\right\} = \int_0^\infty \frac{\sin ax}{x}.x\,J_0(sx)\,dx = \int_0^\infty \sin ax\,J_0(sx)dx$$
$$= \text{Imaginary part of } \int_0^a -e^{-iax}J_0(sx)\,dx$$
$$= \text{Imaginary part of } \left\{-(i^2a^2+s^2)^{\frac{-1}{2}}\right\}$$
$$H\left(\frac{\sin ax}{x}\right) = \text{Imaginary part of } \frac{-1}{\sqrt{s^2-a^2}}$$

**Case 1.** $s > a$, $\quad H\left(\dfrac{\sin ax}{x}\right) = 0$

**Case 2** $0 < s < a$

$$H\left(\frac{\sin ax}{x}\right) = \text{Imaginary part of } \frac{-1}{i\sqrt{a^2 - s^2}} \qquad 0 < s < a$$

$$= \text{Imaginary part of } \frac{i}{\sqrt{a^2 - s^2}} = \frac{1}{\sqrt{a^2 - s^2}}$$

**Proved.**

**Example 10.** *Find the Hankel transform of the function* $\frac{e^{-ax}}{x}$, $n = 1$.

**Solution.** Let $H(s)$ be the Hankel Transform of the function $f(x)$.

*i.e.*

$$H(s) = \int_0^\infty f(x).x\, J_n(sx)\, dx$$

$$= \int_0^\infty \frac{e^{-ax}}{x}.x\;\; J_1(sx)(dx)$$

$$= \int_0^\infty e^{-ax}.J_1(sx)\, dx$$

$$= \frac{1}{s} - \frac{a}{s\left(s^2 + a^2\right)^{1/2}}$$

**Ans.**

**Example 11.** *Find the Hankel Transform of* $e^{-ax}$. $n = 0$.
**Solution.** Let $H(s)$ be the Hankel transform of $f(x)$,

$$H(s) = \int_0^\infty f(x)\, x\, J_n(x)\, dx.$$

$$H(s) = \int_0^\infty e^{-ax} x\, J_0(sx)\, dx$$

$$= \frac{a}{(a^2 + s^2)^{\frac{3}{2}}}$$

**Ans.**

**Example 12.** *Find the Hankel transform of* $e^{-ax}$, $n = 1$.
**Solution.** Let H (s) be the Hankel Transform of $f(x)$.

$$H(s) = \int_0^\infty f(x).x\, J_n(x)\, dx$$

*i.e.*

$$H(s) = \int_0^\infty e^{-ax}.x.J_1(sx)\, dx$$

$$= s\,(a^2 + s^2)^{\frac{-3}{2}}$$

**Ans.**

## 24.4 INVERSION FORMULA FOR HANKEL TRANSFORM

If $H(s)$ be the Hankel transform of the function $f(x)$ for $-\infty < x < \infty$

*i.e.*

$$H(s) = \int_{-\infty}^{\infty} f(x)\, x\, J_n(sx)\, dx$$

Then

$$f(x) = \int_{-\infty}^{\infty} H(s)\, s\, J_n(sx)\, ds$$

is said to be the inversion formula for the Hankel transform $H(s)$ and we may write

$$f(x) = H^{-1}[H(s)]$$

We know that in Fourier transform

$$F(s.t) = \int_{-\infty}^{\infty}\int_{-\infty}^{\infty} f(s,y)\, e^{i(sx+ty)}\, dx\, dy \qquad ...(1)$$

then

$$f(x, y) = \frac{1}{4\pi^2}\int_{-\infty}^{\infty}\int_{-\infty}^{\infty} F(s, t)\, e^{i(sx+ty)}\, ds\, dt$$

On putting $x = r\cos\theta,\ y = r\sin\theta,\ (s = p\cos\alpha),\ t = p\sin\alpha$ in (1), we get

$$F(p, \alpha) = \frac{1}{4\pi^2}\int_0^{\infty} r\, dr \int_0^{2\pi} f(r, \theta)\, e^{irp\cos(\theta-\alpha)}\, d\theta \qquad ... (2)$$

and

$$f(r, \theta) = \frac{1}{4\pi^2}\int_0^{\infty} p\, dp \int_0^{2\pi} F(\rho, \alpha)\, e^{-irp\cos(\theta-\alpha)}\, d\alpha \qquad ... (3)$$

On putting $f(r)e^{-in\theta}$ for $f(r, \theta)$ in (2), we get

$$F(p, \alpha) = \int_0^{\infty} f(r) r\, dr \int_0^{2\pi} e^{i\{-n\theta+pr\cos(\theta-\alpha)\}}\, d\theta \qquad ... (4)$$

In (4), we put $\phi = \alpha - \theta - \frac{\pi}{2}$, we get

$$F(p, \alpha) = \int_0^{\infty} f(r) r\, dr \int_0^{2\pi} e^{i\{n(\phi+\frac{\pi}{2}-\alpha)+pr\cos(\phi+\frac{\pi}{2})\}}\, d\phi$$

$$= \int_0^{\infty} f(r) r\, dr . e^{in(\frac{\pi}{2}-\alpha)} \int_0^{2\pi} e^{i(n\phi - pr\sin\phi)}\, d\phi$$

$$= \int_0^{\infty} f(r) r\, dr . 2\pi e^{in(\frac{\pi}{2}-\alpha)} J_n(pr)\, dr$$

$$\because \quad J_n(pr) = \frac{1}{2\pi}\int_0^{2\pi} e^{i(n\phi - pr\sin\phi)}\, d\phi$$

$$= 2\pi e^{in(\frac{\pi}{2}-\alpha)} \int_0^{\infty} f(r).r\, J_n(pr)\, dr$$

$$= 2\pi e^{in(\frac{\pi}{2}-\alpha)} F(p) \qquad ... (5)$$

Putting $f(r, \theta) = f(r)e^{-in\theta}$ and using (3) and (5), we have

$$f(r)e^{-in\theta} = \frac{1}{4\pi^2}\int_0^{\infty} p.dp \int_0^{2\pi} 2\pi e^{in(\frac{\pi}{2}-\alpha)} F(p)\, e^{-ipr\cos(\theta-\alpha)}\, d\alpha$$

$$= \frac{1}{2\pi}\int_0^{\infty} pF(p)\, dp \int_0^{2\pi} e^{i\{n(\frac{\pi}{2}-\alpha)-pr\cos(\theta-\alpha)\}}\, d\alpha$$

Substituting $\Psi = \theta - \alpha + \frac{\pi}{2}$,

$$f(r)e^{-in\theta} = \frac{1}{2\pi}\int_0^{\infty} pF(p)\, dp \int_0^{2\pi} e^{i\{n(\Psi-\theta-pr\cos\left(\frac{\pi}{2}-\Psi\right)}\, d\Psi$$

$$= \frac{1}{2\pi}\int_0^{\infty} p\, F(p)\, dp e^{-in\theta} \int_0^{2\pi} e^{i(n\Psi - pr\sin\Psi)} d\psi$$

$$f(r) = \frac{1}{2\pi}\int_0^{\infty} p F(p)\, dp . 2\pi\, J_n(pr)\, dp$$

$$f(r) = \int_0^{\infty} F(s) s J_n(sr)\, ds$$

or $$f(x) = \int_0^{\infty} F(s). s\, J_n(sx)\, ds$$

This is the required inversion formula.

**Example 13.** *Find $H^{-1}[e^{-as}]$, when $n = 0$.*

**Solution.** $$f(x) = \int_0^{\infty} s H(s) J_n(sx)\, ds$$

$$= \int_0^{\infty} s e^{-as} J_0(sx)\, ds$$

$$= \frac{a}{(a^2+x^2)^{3/2}}$$ **Ans.**

**Example 14.** *Find $H^{-1}[s^{-2} e^{-as}]$ when $n = 1$.*

**Solution.** $$H^{-1}[s^2 e^{-as}] = \int_0^{\infty} sH(s) J_n(sx) ds = \int_0^{\infty} s^{-2} e^{-as} s\, J_1(sx)\, ds$$

$$= \int_0^{\infty} \frac{1}{s} e^{-as} J_1(sx)\, ds$$

$$= \frac{(a^2+x^2)^{\frac{1}{2}} - a}{x}$$ **Ans.**

## 24.5 PARSIVAL'S THEOREM FOR HANKEL TRANSFORM

Let $F(s)$ and $G(s)$ be the Hankel Transforms of the functions $f(x)$ and $g(x)$. Then

$$\int_0^{\infty} x. f(x). g(x)\, dx = \int_0^{\infty} s F(s). G(s)\, ds$$

**Proof:** On putting the value of $G(s)$ in $\int_0^{\infty} s F(s). G.(s)\, ds$ we get

$$\int_0^{\infty} s. F(s). G(s) ds = \int_0^{\infty} s. F(s) ds \int_0^{\infty} g(x). x\, J_n(sx)\, dx$$

$$= \int_0^{\infty} x\, g(x)\, dx \int_0^{\infty} s F(s). J_n(sx)\, ds$$ (On changing the order of integration)

$$= \int_0^{\infty} x\, g(x)\, dx. f(x).$$ **Proved**

## 24.6 HANKEL TRANSFORMATION OF THE DERIVATIVE OF A FUNCTION.

$$H\left\{\frac{df}{dx}\right\}_n = -s\left[\frac{n+1}{2n} H\{f(x)\}_{n-1} - \frac{n-1}{2n} H_{n+1}\{f(x)\}\right]$$

**Proof.** If $H(s)$ be the Hankel transformation of order n of $f(x)$

*i.e.* $H(s) = \int_0^{\infty} x f(x) J_n(sx)\, dx$, then the Hankel transformation of $\frac{df}{dx}$ is

$$H\left\{\frac{df}{dx}\right\}_n = \int_0^\infty x\frac{df}{dx}J_n(sx)\,dx$$

On integrating by parts, we get

$$= [x\,f(x).J_n(sx)]_0^\infty - \int_0^\infty f(x)\frac{d}{dx}[x\,J_n(sx)]\,dx$$

$$= 0 - \int_0^\infty f(x)[1.J_n(sx) + xsJ'(sx)]\,dx \qquad ..(1)$$

Assuming that $x\,f(x) \to 0$ and $x \to 0, x \to \infty$

Putting $sx\,J_n'(sx) = sx\,J_{n-1}(sx) - n\,J_n(sx)$ in (1), we get

$$= -\int_0^\infty f(x)J_n(sx)\,dx - \int_0^\infty f(x)\{xs\,J_{n-1}(sx) - nJ_n(sx)\}dx$$

$$= (n-1)\int_0^\infty f(x)J_n(sx)\,dx - s\int_0^\infty x\,f(x)J_{n-1}(sx)\,dx$$

$$= (n-1)\int_0^\infty f(x)J_n(sx)\,dx - s\,H_{n-1}(s)$$

The recurrence relation (4) is

$$2n\;J_n(x) = x\,J_{n-1}(x) + x\,J_{n+1}(x)$$

On replacing $x$ by $sx$ we get

$$2n\,J_n(x) = sx\;J_{n-1}(x) + sx\,J_{n+1}(x)$$

$$2n\int_0^\infty f(x)\,J_n(sx)\,dx = s\left[\int_0^\infty x\,f\,(x)\,J_{n-1}(sx)dx + \int_0^\infty sx\,f(x)\,J_{n+1}(sx)\,dx\right]$$

$$= s\,H_{n-1}(s) + s\,H_{n+1}(s)$$

or
$$\int_0^\infty f(x)J_n(sx)\,dx = \frac{s}{2n}H_{n-1}(s) + \frac{s}{2n}H_{n+1}(s) \qquad ...(2)$$

On putting the value of $\int_0^\infty f(x)J_n(sx)\,dx$ from (2) in (1) we get

$$H_n\left\{\frac{df}{dx}\right\} = s\left[\frac{n-1}{2n}H_{n-1}(s) + \frac{n-1}{2n}H_{n+1}(s)\right] - sH_{n-1}(s)$$

$$H_n\left(\frac{df}{dx}\right) = -s\left[\frac{n+1}{2n}H_{n-1}(s) - \frac{n-1}{2n}H_{n+1}(s)\right] \qquad ...(3)$$

This is the required formula for the Hankel transform of $\frac{df}{dx}$. **Proved**

On replacing $n$ by $n-1$ in (3), we get

$$H_{n-1}\left(\frac{df}{dx}\right) = -s\left[\frac{n}{2(n-1)}H_{n-2}(s) - \frac{n-2}{2(n-1)}H_n(s)\right] \qquad ...(4)$$

Putting $(n+1)$ for n in (3) we get

$$H_{n+1}\left(\frac{df}{dx}\right) = -s\left[\frac{n+2}{2(n+1)}H_n(s) - \frac{n}{2(n+1)}H_{n+2}(s)\right] \qquad ...(5)$$

From (3), (4), (5), and replacing $\frac{df}{dx}$ by $\frac{d^2 f}{dx^2}$ we have

$$H_n\left(\frac{d^2 f}{dx^2}\right) = -s\left[\frac{n+1}{2n}H_{n-1}\left(\frac{df}{dx}\right) - \frac{n-1}{2n}H_{n+2}\left(\frac{df}{dx}\right)\right] \qquad \ldots (6)$$

$$= \frac{s^2}{4}\left[\frac{n+1}{n-4}H_{n-2}(s) - 2\frac{n^2-3}{n^2-1}H_n(s) + \frac{n-1}{n+1}H_{n+2}(s)\right] \qquad \ldots (7)$$

**Corollary.** Putting $n = 1, 2, 3,$ in (3) we get

$$H_1\left(\frac{df}{dx}\right) = -s\, H_0(s)$$

$$H_2\left(\frac{df}{dx}\right) = -s\left(\frac{3}{4}H_1(s) - \frac{1}{4}H_3(s)\right)$$

$$H_3\left(\frac{df}{dx}\right) = -s\left[\frac{2}{3}H_2(s) - \frac{1}{3}H_4(s)\right]$$

**Example 15.** *Find Hankel Transforms of the following*

(a) $\frac{d^2 f}{dx^2}$ (b) $\frac{d^2 f}{dx^2} + \frac{1}{x}\frac{df}{dx}$ (c) $\frac{d^2 f}{dx^2} + \frac{1}{x}\frac{df}{dx} - \frac{n^2}{x^2}f$

**Solution.**

(a) $H\left\{\frac{d^2 f}{dx^2}\right\} = \int_0^\infty \frac{d^2 f}{dx^2}.x\, J_n(sx)\, dx$ Integrating by parts we get

$$= \left[\frac{df}{dx}x.J_n(sx)\right]_0^\infty - \int_0^\infty \frac{df}{dx}.\frac{d}{dx}[x\, J_n(sx)]\, dx$$

Putting $x\, f(x) \to 0$ where $x \to 0$ or $x \to \infty$

$$H\left\{\frac{d^2 f}{dx^2}\right\} = 0 - \int_0^\infty \frac{df}{dx}[J_n(sx) + sx\, J'(sx)]\, dx$$

$$\int_0^\infty x\frac{d^2 f}{dx^2}.J_n(sx)\, dx = \int_0^\infty x\frac{d^2 f}{dx^2}.J_n(sx)\, dx$$

(b) $$\int_0^\infty x\left[\frac{d^2 f}{dx^2} + \frac{1}{x}\frac{df}{dx}\right]J_n(sx)\, dx = -s\int_0^\infty \frac{df}{dx}.x\, J_n'(sx)\, dx$$

$$= -s\left[f(x).xJ_n'(sx)\right]_0^\infty - \int_0^\infty f(x)\frac{d}{dx}\{xJ_n'(sx)\}dx$$

$$= s\int_0^\infty f(x)\frac{d}{dx}\{x\, J_n'(sx)\}dx$$

$\because\ x\, f(x) \to 0$ as $x \to 0$ or $x \to \infty$

But $J_n\,(sx)$ is the solution of Bessel's differential equation, so it satisfies Bessel's equation

$$\frac{d}{dx}\left(x\frac{dy}{dx}\right) + \left(1 - \frac{n^2}{x^2}\right)xy = 0$$

$$\frac{d}{dx}\left[x\frac{dy}{dx}J_n(x)\right]+\left(1-\frac{n^2}{x^2}\right)x.J_n(x)=0$$

On replacing $x$ by $sx$ we get

$$\frac{1}{s}\frac{d}{dx}[sx\,J_n'(sx)]=-\left(s^2-\frac{n^2}{x^2}\right)\frac{x}{s^2}$$

$$\frac{d}{dx}[x\,J_n'(sx)]=-\left(s^2-\frac{n^2}{x^2}\right)\frac{x}{s}J_n(sx)$$

On putting the value of $\frac{d}{dx}[x\,J_n'(sx)]$ in (1) we get

$$\int_0^\infty\left[\frac{d^2f}{dx^2}+\frac{1}{x}\frac{df}{dx}\right].x\,J_n(sx)\,dx=-s\int_0^\infty f(x)\left(s^2-\frac{n^2}{x^2}\right)\frac{x}{s}J_n(sx)\,dx$$

(c) $$\int_0^\infty\left[\frac{d^2f}{dx^2}+\frac{1}{x}\frac{df}{dx}-\frac{n^2}{x^2}f\right].x\,J_n(sx)\,dx=-s^2\int_0^\infty f(x).x\,J_n(sx)\,dx$$

$$=-s^2\,H(s) \qquad \text{... (2)}$$

**Deduction I.** On putting $n = 0$ in (2), we get

$$\int_0^\infty x\left(\frac{d^2f}{dx^2}+\frac{1}{x}\frac{df}{dx}\right)J_0(sx)\,dx=-s^2H_0(s)$$

**Deduction II.** On putting $n = 1$ in (2), we get

$$\int_0^\infty x\left(\frac{d^2f}{dx^2}+\frac{1}{x}\frac{df}{dx}-\frac{f}{x^2}\right)J_1(sx)\,dx=-s^2H_1(s)$$

or $$\int_0^\infty x\frac{df}{dx}J_1\,(sx)=-s\,H_0(s)\quad\text{where}\quad H_0(s)=\int_0^\infty x\,f(x)J_0(sx)\,dx$$

**Example 16.** Find $H\left\{\frac{\partial}{\partial x}\left(\frac{e^{-ax}}{x}\right)\right\}$ when $n = 1$

**Solution.** $$H\left\{\frac{\partial f}{\partial x}\right\}=\int_0^\infty x\frac{df}{dx}J_1(sx)dx=-sH_0(s)$$

$$=-s\int_0^\infty x\,f(x)\,J_0\,(sx)\,dx$$

$$=-s\int_0^\infty x\left(\frac{e^{-ax}}{x}\right)J_0(sx)\,dx$$

$$=-s\int_0^\infty e^{-ax}J_0(sx)\,dx$$

$$=\frac{-s}{(a^2+s^2)^{\frac{1}{2}}}$$ **Ans.**

**Example 17.** *Evaluate* $H\left\{\dfrac{\partial^2}{\partial t^2} f(x,t)\right\}$.

**Solution.**
$$\bar{H} + \left\{\frac{\partial^2}{\partial t^2} f(x,t)\right\} = \int_0^\infty x \frac{\partial^2 f}{\partial t^2} J_n(sx)\, dx$$
$$= \frac{\partial^2}{\partial t^2} \int_0^\infty x\, f(x,t)\, J_n(sx) dx$$
$$= \frac{\partial^2}{\partial t^2} H\{f(p,t)\} \text{ **Ans.**}$$

**Example 18.** *Find* $H\left\{\dfrac{d^2(e^{-ax})}{dx^2} + \dfrac{1}{x}\dfrac{d(e^{-ax})}{dx}\right\}$, *when* $n = 0$.

**Solution.**
$$H\left\{\frac{d^2(e^{-ax})}{dx^2} + \frac{1}{x}\frac{d}{dx}(e^{-ax})\right\} = \int_0^\infty \left\{\frac{d^2(e^{-ax})}{dx^2} + \frac{1}{x}\frac{d(e^{-ax})}{dx}\right\} x.\, J_0(sx)\, dx$$
$$= -s^2 H_0(s)$$
$$= -s^2 \int_0^\infty e^{-ax}.\, J_0(sx)\, dx = \frac{-s^2}{(a^2+s^2)^{\frac{1}{2}}} \quad \textbf{Ans.}$$

**Application if Hankel Transform to Boundary problems**

**Example 19.** *The magnetic potential V for a circular disc of radius a and strength w, magnetised parallel to its axis, satisfying Laplace's equation is equal to* $2\pi w$ *on the disc itself and vanishes at its exterior points in the plane of the disc. Show that at the points (r, z), Z > 0.*

$$V = 2\pi w \int_0^\infty e^{-sz} J_0(sr) J_1(sa) ds.$$

**Solution.** The magnetic potential $V$ satisfies the Laplace's equation.

$$\frac{\partial^2 V}{\partial r^2} + \frac{1}{r}\frac{\partial V}{\partial r} + \frac{\partial^2 V}{\partial z^2} = 0, \quad \text{..... (1)} \quad 0 < r < \infty$$

Boundary conditions are

$$V = 2\pi w,\ 0 \le r < a,\ z = 0 \text{ and } V = 0,\ r > a,\ z = 0.$$

Taking Hankel transform of (1), we have

$$\int_0^\infty \left(\frac{\partial^2 V}{\partial r^2} + \frac{1}{r}\frac{\partial V}{\partial r}\right) r\, J_0(sr)\, dr + \int_0^\infty \frac{\partial^2 V}{\partial z^2}.r\, J_0(sr)\, dr = 0$$

$$-s^2 H_0(V) + \frac{d^2}{dz^2} H_0(V) = 0 \text{ where } H_0(V) = \int_0^\infty V.r\, J_0(sr) dr$$

$$D^2 H_0(V) - s^2 H_0(V) = 0 \Rightarrow (D^2 - s^2) H_0(V) = 0$$

Its solution $\quad H_0(V) = Ae^{sz} + Be^{-sz}$ ... (2)

And $\quad H_0(V)_{z=0} = \int_0^a (V)_{z=0}.r\, J_0(sr) dr + \int_a^\infty (V)_{z=0}\, r\, J_0(sr) dr$

$$= \int_0^a 2\pi w r\, J_0(sr)dr + 0 = 2\pi w \int_0^a \frac{1}{s}\frac{d}{dr}(r\, J_1(sr))\, dr$$

$$= \frac{2\pi\, wa}{s} J_1(sa) \qquad \text{... (3)}$$

Putting the values of $H_0(V) = 0$ and $z = \infty$ in (2), we get

$$0 = A e^{s\infty} \Rightarrow A = 0$$

so (2) reduces to $\quad H_0(V) = B e^{-sz} \qquad$ ... (4)

On putting the value of $H(V)_z$ from (4) and $z = 0$ in (3), we get

$$\frac{2\pi wa}{f} J_1(sa) = B$$

On substituting the value of $B$ in (4), we have

$$H_0(V) = \frac{2\pi wa}{s} J_1(sa)\, e^{-sz}$$

By inversion formula, we get

$$V(r, z) = 2\pi wa \int_0^\infty e^{-sz} J_0(sr)\, J_1(as)\, ds \qquad \textbf{Ans.}$$

**Example 20.** *Find the potential $V(r, z)$ of a field due to a flat circular disc of unit radius with centre at origin and axis along z-axis, satisfying the differential equation*

$$\frac{\partial^2 V}{\partial r^2} + \frac{1}{r}\frac{\partial V}{\partial r} + \frac{\partial V}{\partial z^2} = 0,$$

$$0 \le r \le \infty, \quad z \ge 0 \text{ and}$$

(*i*) $V = V_0$ when $z = 0, \qquad 0 \le r < 1$

(*ii*) $\dfrac{\partial V}{\partial z} = 0$ when $z = 0, \qquad r > 1\cdot$

**Solution.** $\dfrac{\partial^2 V}{\partial r^2} + \dfrac{1}{r}\dfrac{\partial V}{\partial r} + \dfrac{\partial^2 V}{\partial z^2} = 0 \qquad$ ... (1)

Taking Hankel Transform of (1), we get

$$\int_0^\infty \left(\frac{\partial^2 V}{\partial r^2} + \frac{1}{r}\frac{\partial V}{\partial r}\right) r\, J_0(sr)dr + \int_0^\infty \frac{\partial^2 V}{\partial z^2} r\, J_0(sr)\, dr = 0$$

or $\quad -s^2 H_0(V) + \dfrac{\partial^2}{\partial z^2}\displaystyle\int_0^\infty Vr\, J_0(sr)\, dr = 0$

or $\quad -s^2 H_0(V) + \dfrac{\partial^2}{\partial z^2} H_0(V) = 0$

or $\quad \dfrac{d^2 H_0(V)}{dz^2} - s^2 H_0(V) = 0 \quad$ or $\quad (D^2 - s^2)\, H_0(v) = 0$

A.E. is $\quad m^2 - s^2 = 0 \quad$ or $\quad m = \pm s$

$$H_0(V) = A e^{sz} + B e^{-sz} \qquad \text{... (2)}$$

On putting $z=\infty$, $H_0(v)=0$ (as $V=0$) in (2), we get

$$0 = Ae^{\infty s}+0 \Rightarrow A=0$$

On putting $A = 0$ in (2), we have

$$H_0(v)=Be^{-sz} \qquad \text{... (3)}$$

Applying inversion formula to get $V$

$$V(r, z)=\int_0^{\infty} B(s)e^{-sz}s J_0(sr)\,ds \qquad \text{... (4)}$$

On putting $z = 0$ in (4), we have

$$V(r,0)=\int_0^{\infty} s\,B(s)J_0(sr)\,ds = V_0 \qquad 0 \le r \le 1$$

On differentiating (4), w.r.t. 'z', we obtain

$$\left(\frac{\partial V}{\partial z}\right)=\int_0^{\infty} B(s)\,(-s e^{-sz})s\,J_0(sr)\,ds \qquad \text{... (5)}$$

On putting z = 0 in (5), we get

$$\left(\frac{\partial V}{\partial z}\right)_{z=0}=\int_0^{\infty} -s^2 B(s)\,J_0(sr)\,ds=0,\ r>1 \qquad \text{... (6)}$$

On comparing (4) and (5), we get

$$\int_0^{\infty} J_0(sr)\frac{\sin s}{s}\,ds=\frac{\pi}{2}, \qquad 0 \le r \le 1 \text{ and}$$

$$\int_0^{\infty} J_0(sr)\sin s\,ds = 0 \qquad r>0$$

$$B(s)=\frac{2}{\pi}V_0\frac{\sin s}{s}$$

Hence, the required solution is

$$V(r, z)=\frac{2V_0}{\pi}\int_0^{\infty} e^{-sz}\frac{\sin s}{s}J_0(sr)\,ds$$

## 24.7 FINITE HANKEL TRANSMISSION FORMATION

If $f(x)$ be a function satisfying Dirichlet conditions in the interval $(0, a)$ then the

$$f(x)=\frac{2}{a^2}\sum_{i=0}^{\infty} H(si)\frac{J_n(s_i x)}{[J_n'(sia)^2]}, \text{ where } H_{(si)}=\int_0^{a} x\,f(x)\,J_n(s_i x)\,dx$$

Where $si$ is a root of the equation $J_n(a\,si)=0$

The upper limit $a$ is generally converted to 1 by suitable transformation. All the roots of $J_n(si)$ are real and distinct.

**Particular case**

If $n = 0$ and $a = 1$, then $J_0'(x)=-J_1(x)$

The inversion formula reduces to

$$f(x) = 2\sum H(si)\frac{J_0(s_i x)}{\{J_1(s_i)\}^2}$$

Where $si$ are the roots of $J_0(si) = 0$

**General Form**

$$f(x) = \sum_{i=0}^{\infty} C_i\ J_n(s_i x), \qquad 0 \le x \le a$$

where
$$C_i = \frac{2}{a^2 J_{n+1}^2(s_i a)}\int_0^a x\, f(x)\, J_n(s_i x)\, dx$$

$$= \frac{2H(si)}{a^2[J_{n+1}(sia)]^2} = \frac{2H(si)}{a^2[J_n'(sia)]^2}$$

so
$$f(x) = \frac{2}{a^2}\sum_{i=0}^{\infty} H(si)\frac{J_n(six)}{[J_n'(sia)]^2}$$

If $a = 1$, then

$$f(x) = 2\sum_{i=0}^{\infty} H(si)\frac{J_n(sir)}{J_n^2(si)}$$

and $\Sigma\ J_n(si) = 0$

## 24.8 ANOTHER FORM OF HANKEL TRANSFORM

If origin is not inluded in the interval and $f(x)$ satisfies the Dirichlets condition $0 < b \le x \le a$,

then $[H_n(x) = J_n(x) + iY_n(x)]$

$$H(si) = \int_a^b x\, f(x)[J_n(\text{six})Y_n(\text{sia}) - Y_n(\text{sia})\, J_n(\text{sia})]\, dx$$

Where $Y_n$ is the Bessel function of order $n$ of second kind and $si$ is the root of the equation

$$J_n(\text{sia})\, Y_n(\text{si}\,b) - J_n(\text{si}\,b)Y_n(\text{sia}) = 0$$

**Inversion formula**

$$f(x) = \sum \frac{2\text{si}^2 J_n^2(\text{sia})\, H(\text{si})}{J_n^2(\text{si}\,b) - J_n^2(\text{sia})}[J_n(\text{six})\, Y_n(\text{si}\,b) - J_n(sib)Y_n(sia)]$$

**Example 21.** *Find* $f(x)$ *if* $H\{f(x)\} = \frac{c}{s}J_1(sa)$, *s being the root of* $J_0(s\alpha)$.

**Solution.** We know the inversion formula.

$$f(x) = \frac{2}{a^2}\sum_{i=0}^{\infty} H_n(s)\frac{J_n(sx)}{[J_n'(sa)]^2}$$

$$= \frac{2}{a}\sum \frac{c}{s}\, J_1(sa)\frac{J_0(sx)}{[J_0'(sa)]^2}$$

$$= \frac{2}{a}\sum \frac{c}{s}\frac{J_1(sa)\,J_0(sx)}{[J_1(sa)]^2} \qquad [J_0'(x) = -J_1(x)]$$

$$= \frac{2}{a^2}\sum \frac{c}{s}\frac{J_n(sx)}{J_1(sa)} \qquad \textbf{Ans.}$$

**Example 22.** *Show that* $\int_0^a x\,J_0(sx)\,dx = J_1(as).\left(\frac{a}{s}\right)$

**Solution.** We know that

$$\frac{d}{dx}[x^n J_n(x)] = x^n J_{n-1}(x) \qquad \text{....(1) Recurrence Relation}$$

Replacing $x$ by $sx$ and putting $n = 1$ in (1), we get

$$\frac{1}{s}\frac{d}{dx}[sx\,J_1(sx)] = x\,J_0(sx)$$

$$\frac{1}{s}\frac{d}{dx}[x\,J_1(sx)] = x\,J_0(sx)$$

or $$\frac{1}{s}[x\,J_1(sx)]_0^a = \int_0^a x\,J_0(sx)\,dx$$

$\therefore$ $$\frac{a}{s}J_1(sa) = \int_0^a x\,J_0(sx)\,dx \qquad \textbf{Proved.}$$

**Example 23.** *Find* $H(x^{n-1})$, $n > 0$ *and* $x\,J_{n-1}(sx)$ *is the kernel of the transform.*

**Solution.** $$H(x^{n-1}) = \int_0^a x^{n-1}.x\,J_{n-1}(s_i x)\,dx$$

$$= \int_0^a x^n\,J_{n-1}(s_i x)\,dx$$

$$= \frac{1}{s_i}\int_0^a \frac{d}{dx}[x^n\,J_n(s_i x)]\,dx$$

$$= \frac{1}{s_i}[x^n J_n(s_i x)]_0^a$$

$$= \frac{1}{s_i}a^n\,J_n(s_i a) \qquad \textbf{Ans.}$$

**Example 24.** *Find* $H[x^n]$, $n > -1$ *and* $x\,J_n(si\,x)$ *is the kernel of the transform.*

**Solution.** $$H(x^n) = \int_0^a x^n\,x J_n(s_i\,x)\,dx$$

$$= \int_0^a x^{n+1}\,J_n(s_i\,x)\,dx$$

$$= \int_0^a \left[\frac{1}{\text{si}}\frac{d}{dx}x^{n+1}J_{n+1}(s_i\,x)\right]dx$$

$$= \left[\frac{1}{s_i}x^{n+1}\,J_{n+1}(s_i\,x)\right]_0^a$$

$$= \frac{a^{n+1}}{\text{si}} J_{n+1}(si\,a)$$ **Ans.**

**Example 25.** *Show that*

$$\int_0^a x^3 J_0(sx)\,dx = \frac{a^2}{s^2}[2\,J_0(sa) + \left(as - \frac{4}{as}\right) J_1(as)$$

**Solution.** We know that

$$\frac{d}{dx}[x^n J_n(x)] = x^n J_{n-1}(x) \quad \dots (1) \quad \text{(Recurrence Relation)}$$

Replacing $x$, by $sx$ and putting $n = 1$ in (1), we get

$$\frac{1}{s}\frac{d}{dx}[x\ J_1(sx)] = x\ J_0(sx)$$

Now $$\int_0^a x^3\ J_0(sx)\,dx = \int_0^a x^2.x\ J_0(sx)\,dx$$

$$= \int_0^a x^2\left[\frac{1}{s}\frac{d}{dx}[x\,J_1(sx)]\right] dx$$

Integrating by parts, we get

$$= \left[x^2.\frac{1}{s} x\,J_1(sx)\right]_0^a - \int_0^a \frac{2x}{s}.x\,J_1(sx)\,dx$$

$$= \frac{a^3}{s} J_1(sa) - \frac{2}{s}\int_0^a x^2\ J_1(sx)\,dx$$

$$= \frac{a^3}{s} J_1(sa) - \frac{2}{s}\int_0^a \frac{1}{s}\left[\frac{d}{dx}(x^2 J_2(sx)\right] dx$$

$$= \frac{a^3}{s} J_1\,(sa) - \frac{2}{s^2}\left[x^2 J_2(sx)\right]_0^a$$

$$= \frac{a^3}{s} J_1(sa) - \frac{2a^2}{s^2} J_2(as) \quad \dots (2)$$

We also know that $\frac{2n}{x} J_n(x) = J_{n-1}(x) + J_{n+1}(x)$ (Recurrence Relation) ... (3)

Replacing $x$ by $sa$ and putting $n = 1$ in (3) we get

$$\frac{2}{sa} J_1(sa) = J_0(sa) + J_2(sa)$$

or $$J_2(sa) = \frac{2}{sa} J_1(sa) - J_0(sa)$$

Substituting the value of $J_2(sa)$ in (2), we get

$$\int_0^a x^3 J_0(sx)\ dx = \frac{a^3}{s} J_1(sa) - \frac{2a^2}{s^2}\left[\frac{2}{sa} J_1(sa) - J_0(sa)\right]$$

$$= \frac{a^3}{s} J_1(sa) - \frac{4a}{s^3} J_1(sa) + \frac{2a^2}{s^2} J_0(sa)$$

$$= \left[\frac{a^3}{s} - \frac{4a}{s^3}\right] J_1(sa) + \frac{2a^2}{s^2} J_0(sa)$$

$$= \frac{a^2}{s^2}\left[2J_0(as) + \left(as - \frac{4}{as}\right) J_1(as)\right]$$ **Proved.**

**Example 26.** *Find* $H[1-x^2]$, $x\,J_0(sx)$ *being the kernel*

**Solution.**

$$H_0(1-x^2) = \int_0^a (1-x^2).x\,J_0(sx)\,dx$$

$$= \int_0^a x\,J_0(sx)\,dx - \int_0^a x^3 J_0(sx)\,dx$$

$$= \left[\frac{x}{s} J_1(sx)\right]_0^a - \frac{a^2}{s^2}\left[2J_0(sa) + \left(as - \frac{4}{as}\right) J_1(a)\right]$$

$$= \frac{a}{s} J_1(as) - \frac{a^2}{s^2}\left[2J_0(sa) + \left(as - \frac{4}{as}\right) J_1(a)\right]$$ **Ans.**

**Example 27.** *Prove that the finite Hankel transform of* $\frac{2^{1+n-m}}{\lceil m-n} x^n (1-x^2)^{m-n-1}$ *is* $s^{n-m}.J_m(s)$, *for* $0 \le x < 1$.

**Solution.** We know that $H[f(x)] = \int_0^a f(x).x\,J_n(sx)\,dx$.

$$H\left[\frac{2^{1+n-m}}{\lceil m-n} x^n (1-x^2)^{m-n-1}\right] = \int_0^1 \frac{2^{1+n-m}}{\lceil m-n} x^n (1-x^2)^{m-n-1}.x\,J_n(sx)\,dx$$

$$= \int_0^1 \frac{2^{1+n-m}}{\lceil m-n} x^n (1-x^2)^{m-n-1}.x \sum_{r=0}^{\infty} \frac{(-1)^r}{\lfloor r\,\lceil(n+r+1)} \left(\frac{sx}{2}\right)^{n+2r} dx$$

$$= \frac{1}{\lceil m-n} \sum_{r=0}^{\infty} \frac{(-1)^r s^{n+2r}}{\lfloor r\,\lceil n+r+1\, 2^{-1+m+2r}} \int_0^1 x^{2(n+r)}.(1-x^2)^{m-n-1}.x\,dx$$

[Put $x^2 = t$, $2x\,dx = dt$]

$$= \frac{1}{\lceil m-n} \sum_{r=0}^{\infty} \frac{(-1)^s s^{n+2r}}{\lfloor r\,\lceil n+r+1\, 2^{m+2r}} \int_0^1 t^{(n+r+1)-1}(1-t)^{(m-n)-1} dt$$

$$= \frac{1}{\lceil m-n} \sum_{r=0}^{\infty} \frac{(-1)^r s^{n+2r}}{\lfloor r\,\lceil n+r+1\; 2^{m+2r}} \frac{\lceil n+r+1\,\lceil m-n}{\lceil n+r+1+m-n} \qquad \int_0^a t^{m-1}(1-t)^{n-1} dt = \frac{\lceil m\,\lceil n}{\lceil m+n}$$

$$= s^{n-m}\sum_{r=0}^{\infty}\frac{(-1)^r}{\lceil m+r+1}\left(\frac{s}{2}\right)^{m+2r} = (s)^{n-m} J_m(s)$$ **Proved**

**Example 28.** *Find* $H_n\left[\dfrac{J_n(\alpha x)}{J_n(\alpha)}\right]$.

**Solution.** We know that $J_n(\alpha x)$ and $J_n(s\alpha)$ are the solutions of Bessel's equation.

$$\therefore \quad x^2\frac{d^2}{dx^2}J_n(\alpha x)+x\frac{d}{dx}J_n(\alpha x)+(\alpha^2x^2-n^2)J_n(\alpha x)=0 \quad \text{... (1)}$$

$$x^2\frac{d^2}{dx^2}J_n(sx)+x\frac{d}{dx}J_n(sx)+(s^2x^2-n^2)J_n(sx)=0 \quad \text{... (2)}$$

Multiplying (1) by $J_n(sx)$, (2) by $J_n(\alpha x)$ and subtracting, we get

$$(\alpha^2-s^2)\,x\,J_n(\alpha x)\,J_n(sx)=\frac{d}{dx}\left[x\left\{J_n(\alpha x)\frac{d}{dx}J_n(sx)-J_n(sx)\frac{d}{dx}J_n(\alpha x)\right\}\right]$$

Integrating with respect to $x$ from 0 to 1 and using $J_n(s)=0$, we get

$$(\alpha^2-s^2)\int_0^1 x\,J_n(\alpha x)\,J_n(sx)\,dx = s\left[x\left\{J_n(\alpha x)J_n'(sx)-J_n(sx)J_n'(\alpha x)\right\}\right]_0^1$$

$$= s[J_n(\alpha)J_n'(s)-J_n(s)J_n'(\alpha)]$$

$$= s\,J_n(\alpha)\,J_n'(s) \qquad J_n(s)=0$$

or
$$\int_0^1\frac{J_n(\alpha x)}{J_n(\alpha)}.x\,J_n(sx)\,dx=\frac{s}{\alpha^2-s^2}J_n'(s)$$

$$H_n\left\{\frac{J_n(\alpha x)}{J_n(\alpha)}\right\}=\frac{s}{\alpha^2-s^2}J_n'(s)$$ **Proved.**

If $n=0$,
$$H_n\left\{\frac{J_n(\alpha x)}{J_n(\alpha)}\right\}=\frac{s}{\alpha^2-s^2}J_0'(s) \qquad [J_0'(s)=-J_1(s)]$$

$$=\frac{-s}{\alpha^2-s^2}J_1(s).$$ **Ans.**

## EXERCISE 24.1

**1.** Prove that $H_0\left[\dfrac{a}{(a^2+x^2)^{3/2}}\right]=e^{-as}$.

**2.** Find the Hankel transform of $1 - x^2$, taking $x\,J_0\,(sx)$ as the kernel, where $0 \le x \le 1$. **Ans.** $\frac{4}{s^3} J_1(s)$

**3.** Find the Hankel transform of $\frac{\cos ax}{x}$ taking $x\,J_0(sx)$ as the kernel

**Ans.** (*i*) $(s^2 - a^2)^{-\frac{1}{2}}$ if $s > a$. (*ii*) 0 if $s < a$.

**4.** Prove that $H\left\{\frac{\sin ax}{a}\right\}_{n=1} = \frac{a}{s(s^2 - a^2)^{1/2}}$, $\quad s > a$

**5.** Prove that $H_n\left(\frac{df}{dx}\right) = \frac{s}{2n}[(n-1)H_{n+1}(f) - (n+1)H_{n-1}(f)]$

and if $n = 1$, then $H_1\left(\frac{df}{dx}\right) = -s\,H_0(f(x))$

**6.** Prove that $H_n\left\{\frac{d^2 f}{dx^2} + \frac{1}{x}\frac{df}{dx}\right\} = \frac{s}{2}\left[-H_{n-1}\left(\frac{df}{dx}\right) + H_{n+1}\left(\frac{df}{dx}\right)\right]$

When $s$ is a root of $J_n(s) = 0$

**7.** Prove that $H_0\left\{\frac{d^2 f}{dx^2} + \frac{1}{x}\frac{df}{dx}\right\} = -s^2 H_0\{f(x)\}$

When $s$ is a root of $s\,J_n'(s) + h\,J_n(s) = 0$

**8.** Prove that

$$H_n\left\{\frac{d^2 f}{dx^2} + \frac{1}{x}\frac{df}{dx}\left(\frac{n^2}{x^2} f(x)\right)\right\} = -s\,f(1)\,J_n'(s) - s^2 H_n\{f(x)\}.$$

**9.** Find the Hankel Transform of $\frac{d^2 f}{dx^2} + \frac{1}{x}\frac{df}{dx}$ if $s$ is the root of the equation

$$J_n(sa)Y_n(sb) - J_n(sb)Y_n(sa) = 0.$$

If $f(a) = 0 = f(b)$, then deduce that $H_0\left[\frac{d^2 f}{dx^2} + \frac{1}{x}\frac{df}{dx}\right] = -s^2 H_0\{f(x)\}$

**10.** Viscous fluid is contained between two infinitely long concentric circular cylinders of radii $a$ and $b$. The inner cylinder is kept at rest and outer cylinder suddenly starts rotating with uniform angular velocity $\omega$. Find the velocity $v$ of the fluid if the equation of motion is

$$\frac{\partial^2 v}{\partial r^2} + \frac{1}{r}\frac{\partial v}{\partial r} - \frac{v}{r^2} = \frac{1}{\nu}\frac{\partial v}{\partial t}, \qquad a < r < b,\ t > 0$$

$\nu$ being Kinematic viscosity.

**Hint:** Take $f^2(s) = \int_a^b f(r).r\,B_1(sr)\,dr$, $\quad b > a$

where $B_1(sr) = J_1(sr)Y_1(sa) - Y_1(sr)J_1(sa), Y_1(sr)$ being Bessel's function of second kind of order one, and $s$ is a positive root of

$J_1(sb)Y_1(sa) = Y_1(sb)J_1(sa)$.

Multiplying the given equation by $B_1(sr)$ and integrating w.r.t. '$r$' from $a$ to $b$ with boundary conditions

$v = b\omega$ when $r = b$

$v = 0$ when $r = a$, $v = 0$ when $t = 0$.

**Ans.** $$v = \pi b\omega \sum_p \frac{1 - e^{-vp^2 t}}{J_1^2(sa) - J_1^2(sb)} J_1(sa) J_1(sb) B_1(sr)$$

## HANKEL TRANSFORM

| *No.* | *Function f(x)* | *n* | *Hankel Transform F (s)* |
|---|---|---|---|
| 1. | $f(x)=\begin{cases}x^n, & 0<x<a\\ 0, & x>a\end{cases}$ | $n>-1$ | $\frac{a^{n+1}}{s}J_{n+1}(sx)$ |
| 2. | $f(x)=\begin{cases}1, & 0<x<a\\ 0, & x>a\end{cases}$ | $n=0$ | $\frac{a}{s}J_1(sx)$ |
| 3. | $f(x)=\begin{cases}a^2-x^2, & 0<x<a\\ 0, & x>a\end{cases}$ | $n=0$ | $\frac{4a}{s^3}J_1(sx)-\frac{2a^2}{s^2}J_0(sx)$ |
| 4. | $f(x)=x^{m-1}$ | $n>-1$ | $\frac{2m\sqrt{\left(\frac{1}{2}+\frac{1}{2}m+\frac{n}{2}\right)}}{s^{m+1}\sqrt{\left(\frac{1}{2}-\frac{m}{2}+\frac{n}{2}\right)}}$ |
| 5. | $x^2e^{-ax}$ | $n=1$ | $\frac{\sqrt{s^2+a^2}-a}{s}$ |
| 6. | $e^{-ax}$ | $n=0$ | $a(s^2+a^2)^{-3/2}$ |
| 7. | $e^{-ax}$ | $n=1$ | $s(s^2+a^2)^{-3/2}$ |
| 8. | $x^ne^{-qx^2}$ | $n>-1$ | $\frac{s}{(2q)^{n+1}}e^{\frac{-s^2}{4q}}$ |
| 9. | $\frac{e^{-ax}}{x}$ | $n=0$ | $(s^2+a^2)^{\frac{-1}{2}}$ |
| 10. | $\frac{e^{-ax}}{x}$ | $n=1$ | $\frac{1}{s}-\frac{a}{s(s^2+a^2)^{\frac{1}{2}}}$ |
| 11. | $\frac{\sin ax}{a}$ | $n=0$ | $\begin{cases}0, & s>a\\ (a^2-s^2)^{-\frac{1}{2}}, & 0<s<a\end{cases}$ |
| 12. | $\frac{\sin ax}{a}$ | $n=1$ | $\begin{cases}\frac{a}{(s^2-a^2)^{1/2}}, & s>a\\ 0, & s<a\end{cases}$ |
| 13. | $\frac{\sin x^2}{x^2}$ | $n=0$ | $\begin{cases}\sin^{-1}\frac{1}{s}, & s>1\\ \frac{\pi}{2}, & s<1\end{cases}$ |
| 14. | $\frac{a}{(a^2+x^2)^{3/2}}$ | $n=0$ | $e^{-as}$ |
| 15. | $x^{m-2}e^{-ax^2}$ | $n>-1$ | $\frac{\sqrt{s^n\,\frac{n}{2}+\frac{m}{2}}}{2^{n+1}s^{\frac{n}{2}+\frac{m}{2}}\,1+n}\times F\left\{\frac{n}{2}+\frac{m}{2},\ n+1,\ -\frac{s^2}{4a}\right\}$ |

# 25

# Hilbert Transform

## 25.1 INTRODUCTION

Method of separating signals is based on Phase selecting which use phase shifts between the pertinent signals to achieve the desired separation.

On shifting the phase angle of all components of a given signal through ± 90° degrees, the resultling function of time is called as the Hilbert transform of the signal.

The Hilbert transform of $f(t)$ is denoted by $Hi\ \{f(t)\}$ and is defined as

$$H_i\ \{f(t)\} = \frac{1}{\pi}\int_{-\infty}^{\infty} \frac{f(s)\,ds}{t-s}$$

**LINEARITY PROPERTY**

The Hilbert transform of $f(t)$ is a linear operation.

**Inverse of Hilbert Transform**

It is defined as

$$f(t) = -\frac{1}{\pi}\int_{-\infty}^{\infty} \frac{H_i\{f(t)\}}{t-s}\,ds$$

$f(t)$ and $Hi\ \{f(t)\}$ make a Hilbert-transform pair.

## 25.2. ELEMENTARY FUNCTIONS AND THEIR HILBERT TRANSFORM

| S.No. | Functions | Hilbert Transforms |
|---|---|---|
| 1. | $\cos t$ | $H_i\ \{\cos t\} = \frac{1}{\pi}\int_{-\infty}^{\infty} \frac{\sin s}{s-t}\,ds = -\sin t$ |
| 2. | $\sin t$ | $H_i\ \{\sin t\} \frac{1}{\pi}\int_{-\infty}^{\infty} \frac{\sin s}{s-t}\,ds = \cos t$ |
| 3. | $\frac{\sin t}{t}$ | $H_i \left\{\frac{\sin t}{t}\right\} = \frac{1}{\pi}\int_{-\infty}^{\infty} \frac{\frac{\sin s}{s}}{s-t}\,ds = \frac{\cos t - 1}{t}$ |
| 4. | $\frac{1}{1+t^2}$ | $H_i \left\{\frac{1}{1+t^2}\right\} = \frac{1}{\pi}\int_{-\infty}^{\infty} \frac{\frac{1}{1+t^2}}{s-t}\,ds = \frac{-t}{1+t^2}$ |
| 5. | $\delta\ (t)$ | $H_i\ \{\delta(t)\} = \frac{1}{\pi}\int_{-\infty}^{\infty} \frac{\delta}{s-t}\,ds = -\frac{1}{\pi t}$ |
| 6. | $\frac{1}{t}$ | $H_i \left\{\frac{1}{t}\right\} = \frac{1}{\pi}\int_{-\infty}^{\infty} \frac{\frac{1}{s}}{s-t}\,ds = -\pi\delta(t)$ |

## 25.3 PROPERTIES

This Hilbert transform differs from the Fourier transform in a way that it operates exclusively in the time domain. Some important properties are listed below :

1. The *amplitude spectrum* of a signal $f(t)$ and its Hilbert transform $H_i\{f(t)\}$ is the same.

$$H_i\{f(t)\} = \frac{1}{\pi t} * f(t) = \frac{1}{\pi}\int_{-\infty}^{\infty} \frac{f(s)}{s-t}\,ds$$

2. *Inverse Transform* :
   If $H_i\{f(t)\}$ is the Hilbert transform of $f(t)$, then the inverse Hilbert transform is given below.

$$\text{If } H_i\{f(t)\} = \frac{1}{\pi}\int_{-\infty}^{\infty} \frac{f(s)}{s-t}\,ds$$

$$f(t) = \frac{1}{\pi}\int_{-\infty}^{\infty} \frac{H_i\{f(s)\}}{s-t}\,ds$$

3. *Orthogonality*. A signal $f(t)$ and its Hilbert transform $H_i\{f(t)\}$ are orthogonal.

$$\int_{-\infty}^{\infty} f(t)\cdot H_i\{f(t)\}\,dt = 0$$

## 25.4 APPLICATIONS

1. **Phase selectively :** Hilbert transform is used to realise phase selectivity in the generation of a special kind of modulation (single side band modulation)
2. It gives a mathematical basis to represent band-past signals.

# 26

# Empirical Laws and Curve Fitting

## (Method of Least Squares)

### 26.1. EMPIRICAL LAW

A law which connects the two variables of a given data is known as *empirical law*. Several equations of different types can be obtained to express the given data approximately.

### 26.2. CURVE FITTING

To find the empirical law, a curve of "best fit" which can pass through most of the points of the given data or nearest to them is drawn. The process of finding such an equation of "best fit" is known as *curve fitting*. The equation of the curve (best fit) is used to predict the unknown values. To obtain an equation representing the data, we can apply Graphical method or Method of least-squares.

### 26.3. GRAPHICAL METHOD

If we are required to fit (a st. line) a linear law $y = a + bx$ to the given data with the help of graph, we proceed as follows:

(*i*) Plot the given points. (*ii*) Draw the straight line of best fit such that, it may pass through most of the points or nearest to them.

(*iii*) Substitute the co-ordinates of two suitable points in the equation $y = a + bx$. Two simultaneous equations are obtained. On solving these equations, we get, the values of $a$ and $b$, so the law $y = a + bx$ is determined.

If a straight line is not suitable to the points of the data, a smooth curve is drawn through them. From the shape of the curve, we can infer the approximate equation of the curve. The equation of the curve can be transformed to the form $y = a + bx$.

### 26.4. DETERMINATION OF OTHER EMPIRICAL LAWS REDUCIBLE TO LINEAR FORM

By simple transformation we can reduce the non-linear to linear form.

A list of transformation is given below:

| S.No. | *Equation of the curve* | *Substitution* | *Reduced linear equation* |
|---|---|---|---|
| 1. | $y = a + bx^2$ | $y = y,\ x^2 = X$ | $Y = a + bX$ |
| 2. | $xy = ax + by$ or $y = a + b\dfrac{y}{x}$ | $y = Y,\ \dfrac{y}{x} = X$ | $Y = a + bX$ |
| 3. | $y = a + x + bx^2$ or $y - x = a + bx^2$ | $y - x = Y,\ x^2 = X$ | $Y = a + bX$ |
| 4. | $y = Ax^n$<br>$\log y = \log A + n \log x$ | $\log y = Y,\ \log A = a$<br>$\log x = X$ | $Y = a + nX$ |

| S.No. | *Equation of the curve* | *Substitution* | *Reduced linear equation* |
|---|---|---|---|
| 5. | $yx^n = m$<br>$\log y + n \log x = \log m$ | $Y = \log y \quad X = \log x \quad a = \log m$ | $Y = a - nX$ |
| 6. | $y = A\,x^b$<br>$\log y = \log A + b \log x$ | $Y = \log y \quad a = \log A \quad X = \log x$ | $Y = a + bX$ |

**Example 1.** *In the following table some observed values of x and y are given:*

| | | | | | | |
|---|---|---|---|---|---|---|
| *x* | 2 | 3 | 4 | 5 | 6 | 7 |
| *y* | 4 | 5 | 5.71 | 6.25 | 6.67 | 7 |

The law connecting x and y is xy = ax + by. Find the best values of a and b.

**Solution.** $xy = ax + by$ or $y = a + b\dfrac{y}{x}$

Put $y = Y$ and $\dfrac{y}{x} = X$. The transformed equation is $Y = a + bX$.

The following tables gives the value of $X$ and $Y$.

| | | | | | | |
|---|---|---|---|---|---|---|
| $X = \dfrac{y}{x}$ | 2 | 1.66 | 1.43 | 1.25 | 1.11 | 1 |
| $Y = y$ | 4 | 5 | 5.71 | 6.25 | 6.67 | 7 |
| Points | $P_1$ | $P_2$ | $P_3$ | $P_4$ | $P_5$ | $P_6$ |

A best fit line $Y = a + bX$ is drawn. Two points (2, 4) and (1, 7) of this line satisfy the equation $Y = a + bX$.

$$4 = a + 2b, \quad 7 = a + b$$

On solving $a = 10,\ b = -3$

$$Y = 10 - 3X$$

$$y = 10 - 3\frac{y}{x} \text{ or } xy = 10x - 3y \quad \textbf{Ans.}$$

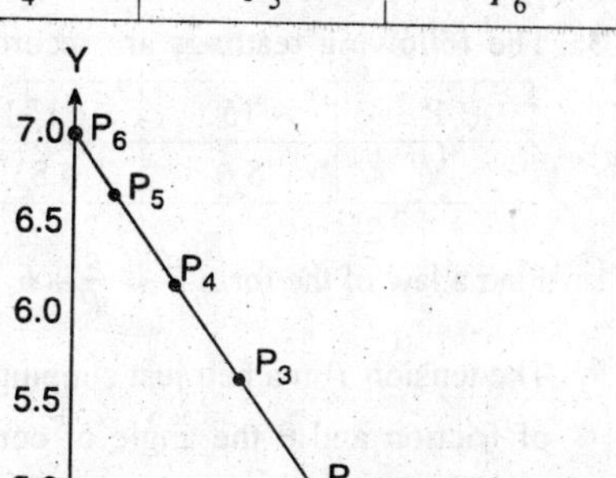


**Example 2.** *The following figures were obtained in a calibration test of the discharge of water through an orifice.*

| | | | | | | |
|---|---|---|---|---|---|---|
| Head H | 2.2 | 1.8 | 1.4 | 1.1 | 0.8 | 0.6 |
| Quantity Q | 8.9 | 8.03 | 7.23 | 6.4 | 5.5 | 4.85 |

The law connecting H and Q has the form $Q = AH^n$. *Find this law.*

**Solution.** $Q = AH^n$

$$\log_{10} Q = \log_{10} A + n \log_{10} H$$

Put $\log_{10} Q = y,\ \log_{10} A = a,\ \log_{10} H = x$

The equation becomes $y = a + nx$

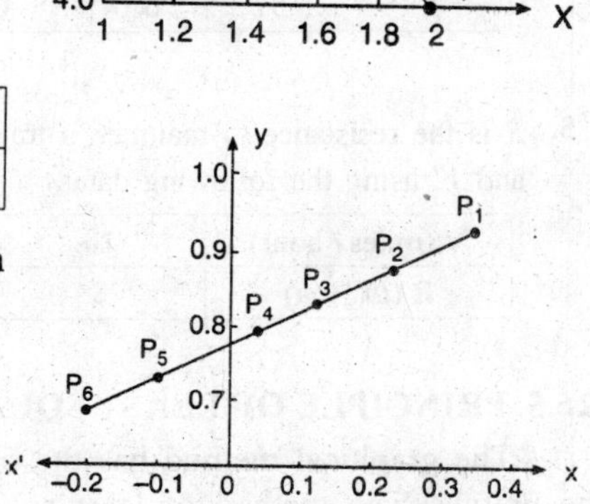

| $x (= \log H)$ | 0.34 | 0.26 | 0.15 | 0.04 | –0.10 | –0.22 |
|---|---|---|---|---|---|---|
| $y (= \log Q)$ | 0.95 | 0.91 | 0.86 | 0.81 | 0.74 | 0.69 |
| Points | $P_1$ | $P_2$ | $P_3$ | $P_4$ | $P_5$ | $P_6$ |

Two points (– 0.1, 0.74) and (0.34, 0.95) are taken on this line.

$$0.74 = a - 0.1\, n \qquad \text{... (1)}$$

$$0.95 = a + 0.34\, n \qquad \text{... (2)}$$

On solving (1) and (2), we get $a = 0.787$, $n = 0.47$

$$0.787 = \log a \quad \text{or} \quad a = 6.13$$

$$Q = aH^n \quad \text{or} \quad Q = 6.13\, H^{0.47}$$ **Ans.**

### Exercise 26.1

**1.** Fit a straight line to the following data $y$ as the dependent variable.

| x | 1 | 2 | 3 | 4 | 5 |
|---|---|---|---|---|---|
| y | 5 | 7 | 9 | 10 | 11 |

**Ans.** $y = 1.5,\ x - 0.3$

**2.** The following values of $x$ and $y$ obey the law $y = ae^{bx}$. Find the best values $a$ and $b$.

| x | 2.70 | 2.87 | 3.26 | 3.68 | 3.89 |
|---|---|---|---|---|---|
| y | 3.86 | 4.2 | 5.1 | 6.3 | 7 |

**Ans.** $a = 0.91,\ b = 0.53$

**3.** The following readings are recorded in an experiment designed to verify Boyle's law.

| P | 15 | 17 | 20 | 25 | 30 | 34 | 40 |
|---|---|---|---|---|---|---|---|
| V | 5.0 | 4.83 | 3.60 | 2.76 | 2.20 | 1.87 | 1.50 |

Find a law of the form $V = \dfrac{a}{P} + b$ **Ans.** $a = 84.5,\ b = -0.6$

**4.** The tension $T$ in a belt just slipping over a pulley satisfies the law $T = T_0\, e^{\mu\, \theta}$ where $\mu$ is the coefficient of friction and $\theta$ the angle of contact between the pulley and the belt in radian. If $T = 5$ kg for all values of $\theta$, find the best value of $\mu$ from the experimental results given below:

| $\theta$ | $\frac{\pi}{8}$ | $\frac{\pi}{4}$ | $\frac{3\pi}{8}$ | $\frac{\pi}{2}$ | $\frac{5\pi}{8}$ | $\frac{3\pi}{4}$ | $\frac{7\pi}{8}$ | $\pi$ |
|---|---|---|---|---|---|---|---|---|
| T | 5.57 | 6.14 | 6.84 | 7.61 | 8.47 | 9.34 | 10.32 | 11.45 |

**Ans.** $\mu = 0.27$

**5.** $R$ is the resistance to maintain a train at speed $V$, find a law of the type $R = a + bV^2$ connecting $R$ and $V$, using the following data:

| V (miles / hour) | 10 | 20 | 30 | 40 | 50 |
|---|---|---|---|---|---|
| R (*lb* / ton) | 8 | 10 | 15 | 21 | 30 |

**Ans.** $a = 7.35,\ b = 0.0085$

## 26.5. PRINCIPLE OF LEAST SQUARES

The graphical method has the obvious drawback in that the straight line drawn may not be unique. The method of least squares is probably the most systematic procedure to fit a *unique* curve through the given data points.

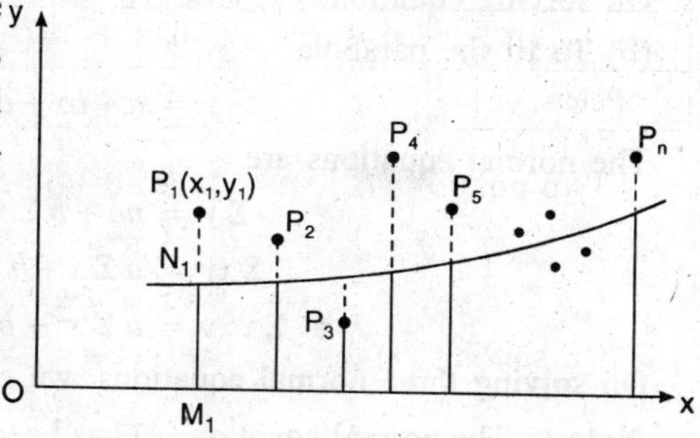


Let $y = f(x)$ be the equation of curve to be fitted to the given data points $(x_1, y_1), (x_2, y_2) \dots (x_n, y_n)$. At $x = x_i$, the observed (or experimental) value of the ordinate $PM$ is $y_i$ and the corresponding value on the fitting curve is $NM$, *i.e.*, $[f(x_1)]$. The difference of the observed and the expected (theoretical) value is $PN = P_1M_1 - N_1M_1 = e_1$. This difference is called the error.

$$e_1 = y_1 - f(x_1)$$

Similarly,

$$e_2 = y_2 - f(x_2)$$

$$e_3 = y_3 - f(x_3)$$

$$e_n = y_n - f(x_n)$$

Some of the errors $e_1, e_2, e_3, \dots e_n$ will be positive and others negative. To make all errors positive, we square each of the errors.

$$S = e_1^2 + e_2^2 + e_3^2 + \dots + e_n^2$$

The curve of the best fit is that for which the sum of the squares of errors (S) is minimum. This is called the principle of least squares.

## 26.6 METHOD OF LEAST SQUARES

Let

$$y = a + bx \qquad \dots (1)$$

be the straight line to be fitted to the given data points $(x_1, y_1), (x_2, y_2) \dots (x_n, y_n)$.

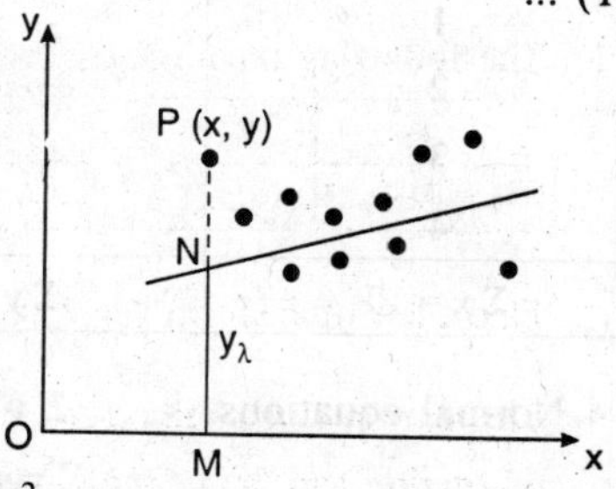


Let $y_{t_1}$ be the theoretical value for $x_1$.

Then

$$e_1 = y_1 - y_{t_1}$$

$$e_1 = y_1 - (a + bx_1)$$

$$e_1^2 = (y_1 - a - bx_1)^2$$

$$S = e_1^2 + e_2^2 + e_3^2 + \dots + e_n^2 = \Sigma e_i^2$$

$$S = \sum_{i=1}^{n} (y_i - a - bx_i)^2$$

For $S$ to be minimum

$$\frac{\partial S}{\partial a} = \sum_{i=1}^{n} 2\,(y_i - a - bx_i)(-1) = 0 \text{ or } \Sigma\,(y - a - bx) = 0 \qquad \dots (2)$$

[To generalise $y_i$, $y_i$ is written as $y$]

$$\frac{\partial S}{\partial b} = \sum_{i=1}^{n} 2\,(y_i - a - bx_i)(-x_i) \text{ or } \Sigma\,(xy - ax - bx^2) = 0 \qquad \dots (3)$$

On simplification equation (2) and (3) becomes

$$\Sigma y = na + b\,\Sigma x \qquad \dots (4)$$

$$\Sigma xy = a\,\Sigma x + b\,\Sigma x^2 \qquad \dots (5)$$

The equations (3) and (4) are known as Normal equations.

On solving equations (3) and (4), we get the values of $a$ and $b$.

(*b*) To fit the parabola:

$$y = a + bx + cx^2 \quad \text{... (6)}$$

The normal equations are

$$\Sigma y = na + b\,\Sigma x + c\,\Sigma x^2 \quad \text{... (7)}$$

$$\Sigma xy = a\,\Sigma x + b\,\Sigma x^2 + c\,\Sigma x^3 \quad \text{... (8)}$$

$$\Sigma x^2 y = a\,\Sigma x^2 + b\,\Sigma x^3 + c\,\Sigma x^4 \quad \text{... (9)}$$

On solving three normal equations, we get the values of $a$, $b$ and $c$.

**Note 1.** The normal equation (4) has been obtained by putting $\Sigma$ on both sides of equation (1). Equation (5) is obtained by multiplying $\Sigma x$ on both sides of (1).

**2.** The normal equation (7), (8), (9) are obtained by multiplying by $\Sigma$, $\Sigma x$ and $\Sigma x^2$ on both sides of equation (6).

**Example 3.** *Find the best values of a and b so that $y = a + bx$ fits the data given in the table.*

| $x$ | 0 | 1 | 2 | 3 | 4 |
|---|---|---|---|---|---|
| $y$ | 1.0 | 2.9 | 4.8 | 6.7 | 8.6 |

**Solution.** $y = a + bx$ ... (1)

| x | y | xy | $x^2$ |
|---|---|---|---|
| 0 | 1.0 | 0 | 0 |
| 1 | 2.9 | 2.9 | 1 |
| 2 | 4.8 | 9.6 | 4 |
| 3 | 6.7 | 20.1 | 9 |
| 4 | 8.6 | 13.4 | 16 |
| $\Sigma x = 10$ | $\Sigma y = 24.0$ | $\Sigma xy = 67.0$ | $\Sigma x^2 = 30$ |

Normal equations

$$\Sigma y = na + b\,\Sigma x \quad \text{... (2)}$$

$$\Sigma xy = a\,\Sigma x + b\,\Sigma x^2 \quad \text{... (3)}$$

On putting the values of $\Sigma x, \Sigma y, \Sigma xy, \Sigma x^2$ in (2) and (3), we have

$$24 = 5a + 10b \quad \text{... (4)}$$

$$67 = 10a + 30b \quad \text{... (5)}$$

On solving (4) and (5), we get

$$a = 1 \quad b = 1.9$$

On substituting the values of $a$ and $b$ in (1), we get

$$y = 1 + 1.9\,x$$ **Ans.**

**Example 4.** *By the method of least squares, find the straight line that best fits the following data:*

| $x$ | 1 | 2 | 3 | 4 | 5 |
|---|---|---|---|---|---|
| $y$ | 14 | 27 | 40 | 55 | 68 |

**Solution.** Let the equation of the straight line best fit be $y = a + bx$ ... (1)

| x | $y$ | $xy$ | $x^2$ |
|---|---|---|---|
| 1 | 14 | 14 | 1 |
| 2 | 27 | 54 | 4 |
| 3 | 40 | 120 | 9 |
| 4 | 55 | 220 | 16 |
| 5 | 68 | 340 | 25 |
| $\Sigma x = 15$ | $\Sigma y = 204$ | $\Sigma xy = 748$ | $\Sigma x^2 = 55$ |

Normal equations are $\Sigma y = na + b\,\Sigma x$ ... (2)

$$\Sigma xy = a\,\Sigma x + b\,\Sigma x^2 \quad \text{... (3)}$$

On putting the values of $\Sigma x, \Sigma y, \Sigma xy$ and $\Sigma x^2$ in (2) and (3), we have

$$204 = 5a + 15b \quad \text{... (4)}$$

$$748 = 15a + 55b \quad \text{... (5)}$$

On solving equations (4) and (5), we get

$$a = 0, \quad b = 13.6$$

On substituting the values of $a$ and $b$ in (1), we get

$$y = 13.6x$$

**Ans.**

**Example 5.** *Find least squares polynomial approximation of degree two to the data.*

| *x* | *0* | *1* | *2* | *3* | *4* |
|---|---|---|---|---|---|
| *y* | *– 4* | *–1* | *4* | *11* | *20* |

Also compute the least error.

Solution. Let the equation of the polynomial be

$$y = a + bx + cx^2 \quad \text{... (1)}$$

| x | $y$ | $xy$ | $x^2$ | $x^2 y$ | $x^3$ | $x^4$ |
|---|---|---|---|---|---|---|
| 0 | – 4 | 0 | 0 | 0 | 0 | 0 |
| 1 | – 1 | – 1 | 1 | – 1 | 1 | 1 |
| 2 | 4 | 8 | 4 | 16 | 8 | 16 |
| 3 | 11 | 33 | 9 | 99 | 27 | 81 |
| 4 | 20 | 80 | 16 | 320 | 64 | 256 |
| $\Sigma x = 10$ | $\Sigma y = 30$ | $\Sigma xy = 120$ | $\Sigma x^2 = 30$ | $\Sigma x^2 y = 434$ | $\Sigma x^3 = 100$ | $\Sigma x^4 = 354$ |

Normal equations are $\Sigma y = 5a + b\,\Sigma x + c\,\Sigma x^2$ ... (2)

$$\Sigma xy = a\,\Sigma x + b\,\Sigma x^2 + c\,\Sigma x^3 \quad \text{... (3)}$$

$$434 = 30a + 100b + 354c \quad \text{... (4)}$$

On putting the values of $\Sigma x, \Sigma y, \Sigma xy, \Sigma x^2, \Sigma x^2 y, \Sigma x^3, \Sigma x^4$ in equations (2), (3), (4), we obtain

$$30 = 5a + 10b + 30c \quad \text{... (5)}$$

$$120 = 10a + 30b + 100c \quad \text{... (6)}$$

$$434 = 30a + 100b + 354c \quad \text{... (4)}$$

On solving these equations, we get $a = -4$, $b = 2$, $c = 1$.

The required polynomial is

$$y = -4 + 2x + x^2, \quad \text{Error} = 0$$ **Ans.**

**Example 6.** *Find the least squares approximation of second degree for the discrete data:*

| $x$ | $-2$ | $-1$ | $0$ | $1$ | $2$ |
|---|---|---|---|---|---|
| $y$ | $15$ | $1$ | $1$ | $3$ | $19$ |

**Solution.** Let the equation of second degree polynomial be $y = a + bx + cx^2$... (1)

| $x$ | $y$ | $xy$ | $x^2$ | $x^2y$ | $x^3$ | $x^4$ |
|---|---|---|---|---|---|---|
| −2 | 15 | −30 | 4 | 60 | −8 | 16 |
| −1 | 1 | −1 | 1 | 1 | −1 | 1 |
| 0 | 1 | 0 | 0 | 0 | 0 | 0 |
| 1 | 3 | 3 | 1 | 3 | 1 | 1 |
| 2 | 19 | 38 | 4 | 76 | 8 | 16 |
| $\Sigma x = 0$ | $\Sigma y = 39$ | $\Sigma xy = 10$ | $\Sigma x^2 = 10$ | $\Sigma x^2 y = 140$ | $\Sigma x^3 = 0$ | $\Sigma x^4 = 34$ |

Normal equations are

$$\Sigma y = na + b\,\Sigma x + c\,\Sigma x^2 \quad \text{... (2)}$$

$$\Sigma xy = a\,\Sigma x + b\,\Sigma x^2 + c\,\Sigma x^3 \quad \text{... (3)}$$

$$\Sigma x^2 y = a\,\Sigma x^2 + b\,\Sigma x^3 + c\,\Sigma x^4 \quad \text{... (4)}$$

On putting the values of $\Sigma x, \Sigma y, \Sigma xy, \Sigma x^2, \Sigma x^2 y, \Sigma x^3, \Sigma x^4$ in equations (2), (3), (4), we have

$$39 = 5a + 0b + 10\,c \quad \text{... (5)}$$

$$10 = 0 + 10b + 0.c \quad \text{... (6)}$$

$$140 = 10a + 0b + 34\,c \quad \text{... (7)}$$

On solving (5), (6), (7), we get, $a = -\dfrac{37}{35}$, $b = 1$, $c = \dfrac{31}{7}$

The required polynomial of second degree is

$$y = -\frac{37}{35} + x + \frac{31}{7}x^2$$ **Ans.**

## 26.7 CHANGE OF SCALE

If the data is of equal interval in large numbers then we change the scale as $u = \dfrac{x - x_0}{h}$.

**Example 7.** Fit a second degree parabola to the following data by least squares method:

| $x$ | *1929* | *1930* | *1931* | *1932* | *1933* | *1934* | *1935* | *1936* | *1937* |
|---|---|---|---|---|---|---|---|---|---|
| $y$ | *352* | *356* | *357* | *358* | *360* | *361* | *361* | *360* | *359* |

**Solution.** Taking $x_0 = 1933$, $y_0 = 357$ *(U.P., II Semester, Summer 2001)*

Taking $u = x - x_0$, $v = y - y_0$

$u = x - 1933$, $v = y - 357$

The equation $y = a + bx + cx^2$ is transformed to $v = A + Bu + Cu^2$ ... (1)

| $x$ | $u = x - 1933$ | $y$ | $v = y - 357$ | $uv$ | $u^2$ | $u^2 v$ | $u^3$ | $u^4$ |
|---|---|---|---|---|---|---|---|---|
| 1929 | −4 | 352 | −5 | 20 | 16 | −80 | −64 | 256 |
| 1930 | −3 | 360 | −1 | 3 | 9 | −9 | −27 | 81 |
| 1931 | −2 | 357 | 0 | 0 | 4 | 0 | −8 | 16 |
| 1932 | −1 | 358 | 1 | −1 | 1 | 1 | −1 | 1 |
| 1933 | 0 | 360 | 3 | 0 | 0 | 0 | 0 | 0 |
| 1934 | 1 | 361 | 4 | 4 | 1 | 4 | 1 | 1 |
| 1935 | 2 | 361 | 4 | 8 | 4 | 16 | 8 | 16 |
| 1936 | 3 | 360 | 3 | 9 | 9 | 27 | 27 | 81 |
| 1937 | 4 | 359 | 2 | 8 | 16 | 32 | 64 | 256 |
| Total | $\Sigma u = 0$ | | $\Sigma v = 11$ | $\Sigma uv = 51$ | $\Sigma u^2 = 60$ | $\Sigma u^2 v = -9$ | $\Sigma u^3 = 0$ | $\Sigma u^4 = 708$ |

Normal equations are

$\Sigma v = 9A + B\Sigma u + C\Sigma u^2$ or $11 = 9A + 0B + 60C$ or $11 = 9A + 60C$

$\Sigma uv = A\Sigma u + B\Sigma u^2 + C\Sigma u^3$ or $51 = 0A + 60B + 0C$ or $51 = 60B$ or $B = \frac{17}{20}$

$\Sigma u^2 v = A\Sigma u^2 + B\Sigma u^3 + C\Sigma u^4$ or $-9 = 60A + 0B + 708C$ *or* $-9 = 60A + 708C$

On solving these equations, we get $A = \frac{694}{231}$, $B = \frac{17}{20}$, $C = -\frac{247}{924}$

$$V = \frac{694}{231} + \frac{17}{20}u - \frac{247}{924}u^2$$

$$y - 357 = \frac{694}{231} + \frac{17}{20}(x - 1933) - \frac{247}{924}(x - 1933)^2$$

$$= \frac{694}{231} + \frac{17x}{20} - \frac{32861}{20} - \frac{247x^2}{924} - \frac{247}{924}(-3866x) - \frac{247}{924} + (1933)^2$$

$$y = \frac{694}{231} - \frac{32861}{20} - \frac{247}{924}(1933)^2 + \frac{17}{20}x + \frac{247 \times 3866}{924}x - \frac{247}{924}x^2$$

$$y = 3 - 1643.05 - 998823.36 + 357 + .85x + 1033.44x - .267x^2$$

$$y = -1000106.41 + 1034.29x - 0.267x^2$$

Ans.

**Example 8.** *Find the least squares fit of the form $y = a + bx^2$ to the following data:*

| $x$ | −1 | 0 | 1 | 2 |
|---|---|---|---|---|
| y | 2 | 5 | 3 | 0 |

**Solution.** $y = a + bx^2$ ... (1)

Let $x^2 = z$, $y = a + bz$ ... (2)

| x | y | $z = x^2$ | $yz$ | $z^2$ |
|---|---|---|---|---|
| −1 | 2 | 1 | 2 | 1 |
| 0 | 5 | 0 | 0 | 0 |
| 1 | 3 | 1 | 3 | 1 |
| 2 | 0 | 4 | 0 | 16 |
| | $\Sigma y = 10$ | $\Sigma z = 6$ | $\Sigma yz = 5$ | $\Sigma z^2 = 18$ |

Normal equations are $\Sigma y = na + b\,\Sigma z$ ... (3)

$\Sigma yz = a\,\Sigma z + b\,\Sigma z^2$ ... (4)

On putting the values of $\Sigma y, \Sigma z, \Sigma yz, \Sigma z^2$ in (3) and (4), we have

$10 = 4a + 6b$ ... (5)

$5 = 6a + 18b$ ... (6)

On solving equations (5) and (6), we get, $a = \frac{25}{6}$, $b = \frac{10}{9}$

On substituting the values of $a$, $b$ in (2), we obtain

$$y = \frac{25}{6} + \frac{10}{9} z$$

or $$y = \frac{25}{6} + \frac{10}{9} x^2 \text{ or } 18y = 75 - 20x^2$$ **Ans.**

**Example 9.** *Use least-squares method to fit a curve of the form $y = ae^{bx}$ to the data.*

| x | 1 | 2 | 3 | 4 | 5 | 6 |
|---|---|---|---|---|---|---|
| y | 7.209 | 5.265 | 3.846 | 2.809 | 2.052 | 1.499 |

**Solution.** $y = ae^{bx}$ ... (1)

$\log_e y = \log_e a + bx$ ... (2)

On putting $\log_e y = Y$, $\log_e a = c$ in (2), it becomes

$Y = c + bx$ ... (3)

| x | $y$ | $Y = \log_e y$ | $xY$ | $x^2$ |
|---|---|---|---|---|
| 1 | 7.209 | 1.97533 | 1.97533 | 1 |
| 2 | 5.265 | 1.66108 | 3.32216 | 4 |
| 3 | 3.846 | 1.34703 | 4.04109 | 9 |
| 4 | 2.809 | 1.03283 | 4.13132 | 16 |
| 5 | 2.052 | 0.71881 | 3.59405 | 25 |
| 6 | 1.499 | 0.40480 | 2.4288 | 36 |
| $\Sigma x = 21$ | | $\Sigma y = 7.13988$ | $\Sigma xY = 19.49275$ | $\Sigma x^2 = 91$ |

Normal equations are $\Sigma Y = nc + b\,\Sigma x$ ... (4)

$\Sigma xY = c\,\Sigma x + b\,\Sigma x^2$ ... (5)

On putting the values of $\Sigma x, \Sigma xY$ and $\Sigma x^2$ in equations (4) and (5), we get

$7.13988 = 6c + 21b$ ... (6)

$19.49275 = 21c + 91b$ ... (7)

On solving (6) and (7), we obtain $b = -0.3141$, $c = 2.28933$

$c = \log_e a = 2.28933 \Rightarrow a = 9.86832$

– On putting the values of $a$ and $b$ in (1), we obtain

$$y = 9.86832\, e^{-0.3131x}$$ **Ans.**

## Exercise 26.2

1. Find the linear least square polynomials based on data

| | | | | |
|---|---|---|---|---|
| x | – 2 | – 1 | 0 | 1 |
| y | 6 | 3 | 2 | 2 |

is given. find the least square straight line approximation to the data. **Ans.** $y = \frac{13}{5} - \frac{13}{10}x$

2. The data

| | | | | |
|---|---|---|---|---|
| x | 1 | 2 | 3 | 4 |
| y | 3 | 7 | 13 | 21 |

is given. Find square straight line approximation to the data.

**Ans.** $y = -4 + 6x$

3. Use the least-square method to obtain a parabola that approximates the data

| | | | | | | |
|---|---|---|---|---|---|---|
| x | 1.0 | 1.2 | 1.4 | 1.6 | 1.8 | 2 |
| y | 2.345 | 2.419 | 2.592 | 2.863 | 3.233 | 3.702 |

Ans. $y = 3.124 - 2.477 + 1.458x^2$

4. Find the values of $a$, $b$, $c$ so that $y = a + bx + cx^2$ is the best fit to the data:

| | | | | | |
|---|---|---|---|---|---|
| x | 0 | 1 | 2 | 3 | 4 |
| y | 1 | 0 | 3 | 10 | 21 |

Ans. $a = 1$, $b = -3$, $c = 2$

5. Find the least squares approximation of the form $y = a + bx^2$, for the data

| | | | | | | |
|---|---|---|---|---|---|---|
| x | 0 | 0.1 | 0.2 | 0.3 | 0.4 | 0.5 |
| y | 1 | 1.01 | 0.99 | 0.85 | 0.81 | 0.75 |

**Ans.** $y = 1.0032 - 1.1081x^2$

6. If $V$ (km/hr) and $R$ (kg/ton) are related by a relation of type $R = a + bV^2$, find by the method of least squares, $a$ and $b$ with the help of the following table:

| | | | | | |
|---|---|---|---|---|---|
| V | 10 | 20 | 30 | 40 | 50 |
| R | 8 | 10 | 15 | 21 | 30 |

Ans. $a = 2$, $b = 3$

Ans. $a = 632$, $b = 0.0095$

7. Determine the constants a and b by the least-squares method such that $y = ae^{bx}$ fits the following data:

| | | | | |
|---|---|---|---|---|
| x | 1.0 | 1.2 | 1.4 | 1.6 |
| y | 40.170 | 73.196 | 133.372 | 243.02 |

8. The pressure and volume of a gas are related by the equation $PV^r = k$, $r$ and $k$ being constants. Fit this equation to the following set of observations:

| P (kg/cm²) | 0.5 | 1.0 | 1.5 | 2.0 | 2.5 | 3.0 |
|---|---|---|---|---|---|---|
| V (litres) | 1.62 | 1.00 | 0.75 | 0.62 | 0.52 | 0.46 |

**Ans.** $PV^{1.276} = 1.039$

**9.** Fit a least-square geometric curve $y = ax^b$ to the following data:

| x | 1 | 2 | 3 | 4 | 5 |
|---|---|---|---|---|---|
| y | 0.5 | 2 | 4.5 | 8 | 12.5 |

**Ans.** $a = 0.5012,\ b = 1.9977$

**10.** Using method of least-squares, fit a relation of the form $y = ab^x$ to the following data:

| x | 2 | 3 | 4 | 5 | 6 |
|---|---|---|---|---|---|
| y | 144 | 172.8 | 207.4 | 248.8 | 298.5 |

**Ans.** $a = 99.86,\ b = 1.2$

**11.** The following table gives the results of the measurements of train resistances; $V$ is the velocity in miles per hour, $R$ is the resistance in pounds per ton:

| V: | 20 | 40 | 60 | 80 | 100 | 120 |
|---|---|---|---|---|---|---|
| R: | 5.5 | 9.1 | 14.9 | 22.8 | 33.3 | 46.0 |

If $R$ is related to $V$ by the relation $R = a + bV + cV^2$, find $a$, $b$ and $c$ by using the method of least squares.

*(U.P., II Semester 2002)*

$$\left[\begin{aligned} \textbf{Hint. } 131.6 &= 6a + 420b + 36400\,c \\ 12042 &= 420\,a + 36400\,b + 3528000\,c \\ 1219720 &= 36400\,a + 3528000\,b + 364000000\,c \end{aligned}\right]$$

**Ans.** $R = 4.2247 + 0.0016\,V + .002877\,V^2$

**12.** Employ the method of least square to fit a parabola $y = a + bx + cx^2$ in the following data

$(x, y)$ : $(-1, 2)$, $(0, 0)$, $(0, 1)$, $(1, 2)$. *(U.P. III Semester Dec. 2004)*

**[Hint:** $5 = 4a + 0 + 2c$

$0 = 0 + 2b + 0$

$4 = 2a + 0 + 2c$

On solving, $a = \frac{1}{2},\ c = \frac{3}{2}$ **Ans.** $y = \frac{1}{2} + \frac{3}{2}x^3$

**13.** If $F$ is pull required to lift a load W by means of a pulley, fit a linear law $F = a - bw$ connecting F and W against the following data:

| *W* | 50 | 70 | 100 | 120 |
|---|---|---|---|---|
| *F* | 12 | 15 | 21 | 25 |

*(U.P. III Semester, Dec. 2004)*

**Ans.** $F = 2.2785 - 0.1879W$

# 27

# Linear Programming

## 27.1 INTRODUCTION

It will be of interest to know that linear programming had its origin during the second world war (1939-45). To fight the war man and material (resources) have to be maintained. There has to be efficient and safe land, sea and airtransport etc.

The government in England studied the problems during war particularly problems of armed forces, civil defence and navel strategy etc. The study for the solutions of the above problems resulted the linear programming.

*Linear programming* is the most popular mathematical technique which involve the limited resources in an optimal manner.

The term *programming* means planning to maximize profit or minimize cost or minimize loss or minimum use of resources or minimizing the time etc. Such problems are called *optimization Problem*. The term linear means that all equations or inequations involved are linear.

During the world war II Marshall K. Wood worked on the allocation of the resources for the United States. Methods were developed to allocate resources in such a way as to minimize or maximize the desired object of the problems. George B. Dantzig was a member of the air force group who devised the Simplex method in 1947.

Consider the following example:-

## TYPE I. MAXIMIZATION PROBLEM

**Example 1.** *A manufacturer produces two types of toys i.e., A and B. Each toy of type A requires 4 hours of moulding and two hours of polishing whereas each toy of type B requires 3 hours of moulding 5 hours of polishing. Moulder works for 80 hours in a week and polisher works for 180 hours in a week. Profit on a toy of type A is Rs. 3 and on a toy of type B is Rs. 4. In what way the manufacturer allocates his production capacity for the two types of toys so that he may make the maximum profit per week.*

**Solution.**

**Table.** The above information can be written in tabular form as follows:

| *Toy* \ *operation* | *Moulding (in hours)* | *Polishing (in hours)* | *Profit (in Rs.)* |
|---|---|---|---|
| *A* | 4 | 2 | 3 |
| *B* | 3 | 5 | 4 |
| Time available (in hours) | 80 | 180 | |

Let $x$ be the number of toys of type $A$ and $y$ be the number of toys of type $B$ produced per week.

Profit on one toy of type $A$ = Rs. 3

Profit on $x$ toys of type $A$ = Rs. $3x$

Profit on one toy of type $B$ = Rs. 4
Profit on $y$ toys of type $B$ = Rs. $4y$

Let $Z$ be the weekly profit.
Then the weekly profit in Rs. is

$$Z = 3x + 4y$$

Here, $Z$ is known as *objective function* which has to maximize/minimize.
One toy of type $A$ on moulding requires = 4 hours.
$x$ toys of type $A$ on moulding requires = $4x$ hours
One toy of type $B$ on moulding requires = 3 hours
$y$ toys of type $B$ on moulding requires = $3y$ hours
On moulding total time required = $4x + 3y$ hours
But moulder works for only 80 hours in a week.
So, $4x + 3y$ hours cannot exceed 80 hours.

$$\Rightarrow \quad 4x+3y \leq 80$$

This is known as *first constraint*.
One toy of type $A$ on polishing requires = 2 hours
$x$ toys of type $A$ on polishing requires = $2x$ hours
One toy of type $B$ on polishing requires = 5 hours
$y$ toys of type $B$ on polishing requires = $5y$ hours
On polishing total time required = $2x + 5y$ hours
But polisher works for only 180 hours in a week.
So, $2x + 5y$ hours cannot exceed 180 hours.

$$\Rightarrow \quad 2x+5y \leq 180$$

This is known as *second constraint.*
Since, the number of toys produced is non-negative.

$$\Rightarrow \quad x \geq 0 \text{ and } y \geq 0$$

This is known as *third constraint.*
Under these three constraints (conditions) we have to plan the system to get the maximum profit.
Now, we summarize the above informations in mathematical form as follows :

To maximize $\quad Z = 3x + 4y \quad$ ... (1)

Subject to the constraints :

$$4x+3y \leq 80 \quad \text{... (2)}$$

$$2x+5y \leq 180 \quad \text{... (3)}$$

$$x \geq 0, \quad y \geq 0 \quad \text{... (4)}$$

The above mathematical expression is known as *mathematical formulation.*
From the above inequations we find out the values of $x$ and $y$.
The values of $x$ and $y$ are substituted in the objective function $Z = 3x + 4y$.
The maximum/minimum value of the objective function is known as *optimal value.*

## 27.2. SOME DEFINITIONS

1. **Linear Programming Problem**
   Here, we have to optimize the linear function $Z$ subject to certain conditions. Such problems are called linear programming problems. As example 1 on page 1243.
2. **Objective functions**
   Objective function is a linear function of several variables, subject to the conditions that $Z = 3x + 4y$ in the previous example.

**3. Optimal Value**

Optimal value is a maximum or minimum value of a objective function to be calculated in a linear programming problem.

**Example 2.** *A manufacturer of leather belts makes three types of belts A, B and C which are processed on three machines $M_1$, $M_2$ and $M_3$. Belt A requires 2 hours on machine $M_1$ and 3 hours on machine $M_3$. Belt B requires 3 hours on machine $M_1$, 2 hours on machine $M_2$ and 2 hours on machine $M_3$ and belt C requires 5 hours on machine $M_2$ and 4 hours on machine $M_3$. There are 8 hours of time per day available on machine $M_1$, 10 hours of time per day available on machine $M_2$ and 15 hours of time per day available on machine $M_3$. The profit gained from belt A is Rs. 3.00 per unit, from belt B is Rs. 5.00 per unit and from belt C is Rs. 4.00 per unit. Formulate the L.P.P. to maximize the profit.*

**Solution.** The above information is given in the following Table:

**1. Table:**

| Belts \ Machines | $M_1$ (in hours) | $M_2$ (in hours) | $M_3$ (in hours) | Profit (in Rs.) |
|---|---|---|---|---|
| A | 2 | 0 | 3 | 3 |
| B | 3 | 2 | 2 | 5 |
| C | 0 | 5 | 4 | 4 |
| Available time (in hours) | 8 | 10 | 15 | |

**2. Decision Variables.** The decision variables are the number of belt *A*, belt *B* and belt *C*. Let the number of belt *A* be $x_1$, the number of belt *B* be $x_2$, and the number of belt *C* be $x_3$.

**3. Objective Function.** To maximise the profit.

Profit on 1 belt $A$ = Rs. 3 　　Profit on 1 belt $B$ = Rs. 5

Profit on $x_1$ belts = Rs. $3x_1$ 　　Profit on $x_2$ belts $B$ = Rs. $5x_2$

Profit on 1 belt $C$ = Rs. 4

Profit on $x_3$ belts $C$ = Rs. $4x_3$

Total profit = $3x_1 + 5x_2 + 4x_3$

$\Rightarrow \quad Z = 3x_1 + 5x_2 + 4x_3$

**4. Constraint (*i*)** The time available on machine $M_1$ = 8 hours per day.

Time required for 1 belt $A$ on machine $M_1$ = 2 hours per day

Time required for $x_1$ belts $A$ on machine $M_1$ = $2x_1$ hours per day

Time required for 1 belt $B$ on machine $M_1$ = 3 hours per day

Time required for $x_2$ belts $B$ on machine $M_1$ = $3x_2$ hours per day

$$2x_1 + 3x_2 \le 8$$

**Constraint (*ii*)** The time available on machine $M_2$ = 10 hours per day.

Time required for 1 belt B on machine $M_2$ = 2 hours per day

Time required for $x_2$ belts $B$ on machine $M_2$ = $2x_2$ hours day

Time required for 1 belt $C$ on machine $M_2$ = 5 hours per day

Time required for $x_3$ belts $C$ on machine $M_2$ = $5x_3$ hours per day

$$2x_2 + 5x_3 \le 10$$

**Constraint (*iii*)** The time available on machine $M_3$ = 15 hour per day.

Time required for 1 belt $A$ on machine $M_3$ = 3 hours per day

Time required for $x_1$ belts $A$ on machine $M_3$ = $3x_1$ hours per day

Time required for 1 belt $B$ on machine $M_3$ = 2 hours per day

Time required for $x_2$ belts $B$ on machine $M_3 = 2x_2$ hours per day
Time required for 1 belt $C$ on machine $M_3 = 4$ hours per day
Time required for $x_3$ belts $C$ on machine $M_3 = 4x_3$ hours per day

$$3x_1 + 2x_2 + 4x_3 \le 15$$

**Constraint (*iv*).** The number of belt $A$, belt $B$ and belt $C$ are non-negative.

$$x_1 \ge 0, \quad x_2 \ge 0 \quad \text{and} \quad x_3 \ge 0.$$

**5. Mathematical Formulation.**

The linear programming problem of the given problem is as follows

To maximise $Z = 3x_1 + 5x_2 + 4x_3$ ...(1)

Subject to the constraints $2x_1 + 3x_2 \le 8$ ...(2)

$2x_2 + 5x_3 \le 10$ ...(3)

$3x_1 + 2x_2 + 4x_3 \le 15$ ...(4)

$x_1 \ge 0, \quad x_2 \ge 0, \quad x_3 \ge 0$ ...(5)

**Ans.**

## TYPE II. DIET PROBLEMS

**Example 3.** *A dietician mixes together two kinds of food, say, X and Y in such a way that the mixture contains at least 6 units of vitamin A, 7 units of vitamin B, 12 units of vitamin C and 9 units of vitamin D. The vitamin contents of 1 kg of food X and 1 kg of food Y are given below:*

| *Cost* | *Vitamin A* | *Vitamin B* | *Vitamin C* | *Vitamin D* |
|---|---|---|---|---|
| *Food X* | *1* | *1* | *1* | *2* |
| *Food Y* | *2* | *1* | *3* | *1* |

*One kg of food X costs Rs. 5, whereas one kg of food Y costs Rs. 8. Formulate the linear programming problem.*

**Solution.**

**1. Decision Variables.** Decision Variables are units of food $X$ and food $Y$. Let food $X$ in the mixture be $x$ kg. and food $Y$ in the mixture be $y$ kg.

**2. Objective Function.** To minimise the cost

1 kg of food $X$ costs Rs. 5
$x$ kg of food $X$ costs Rs. $5x$
1 kg of food $Y$ costs Rs. 8
$y$ kg of food $Y$ costs Rs. $8y$

Total cost of food $X$ and $Y = 5x + 8y$

$$\Rightarrow \quad Z = 5x + 8y$$

**3. Constraint (*i*)** The mixture contains atleast 6 units of vitamin $A$.

1 kg of food $X$ contains 1 unit of vitamin $A$.
$x$ kg of food $X$ contains $x$ units of vitamin $A$.
1 kg of food $Y$ contains 1 unit of vitamin $A$.
$y$ kg of food $Y$ contains $2y$ units of vitamin $A$.

$$\Rightarrow \quad x + 2y \ge 6$$

**Constraint (*ii*)** The mixture contains atleast 7 units of vitamin *B*.

1 kg of food *X* contains 1 unit of vitamin *B*.

*x* kg of food *X* contains *x* units of vitamin *B*.

1 kg of food *Y* contains 1 unit of vitamin *B*.

*y* kg of food *Y* contains *y* units of vitamin *B*.

$\Rightarrow \quad x+y \geq 7$

**Constraint (*iii*)** The mixture contains at least 12 units of vitamin *C*.

1 kg of food *X* contains 1 unit of vitamin *C*.

*x* kg of food *X* contains *x* units of vitamin *C*.

1 kg of food *Y* contains 3 units of vitamin *C*.

*y* kg of food *Y* contains 3*y* units of vitamin *C*.

$\Rightarrow \quad x+3y \geq 12$

**Constraint (*iv*)** The mixture contains atleast 9 units of vitamin *D*.

1 kg of food *X* contains 2 units of vitamin *D*.

*x* kg of food *X* contains 2*x* units of vitamin *D*.

1 kg of food *Y* contains 1 unit of vitamin *D*.

*y* kg of food *Y* contains *y* units of vitamin *D*.

$\Rightarrow \quad 2x+y \geq 9$

**Constraint (*v*).** The number of kg of food *x* and *y* is non-negative.

$x \geq 0$

$y \geq 0$

**4. Mathematical Formulation.** The linear programming problem of the given problem is as follows

To minimise $Z = 5x + 8y$ ...(1)

Subject to the constraints $x+2y \geq 6$ ...(2)

$x+y \geq 7$ ...(3)

$x+3y \geq 12$ ...(4)

$2x+y \geq 9$ ...(5)

$x \geq 0, \quad y \geq 0$ ...(6)

**Ans.**

## TYPE III. TRANSPORTATION PROBLEM

**Example 4.** *There is a factory located at each of the two places P and Q. From these locations, a certain commodity is delivered to each of the three depots situated at A, B, and C. The weekly requirements of the depots are respectively 5, 5 and 4 units of the commodity while the production capacity of the factories at P and Q are 8 and 6 units respectively just sufficient for the requirement of depots. The cost of transportation per unit is given below:*

| To / From | Cost (in Rs.) A | B | C |
|---|---|---|---|
| P | 16 | 10 | 15 |
| Q | 10 | 12 | 10 |

*How many units should be transported from each factory to each depot in order that the transportation cost is minimum. Formulate the above as a linear programming problem.*

**Solution.** The given information is shown in the following figure

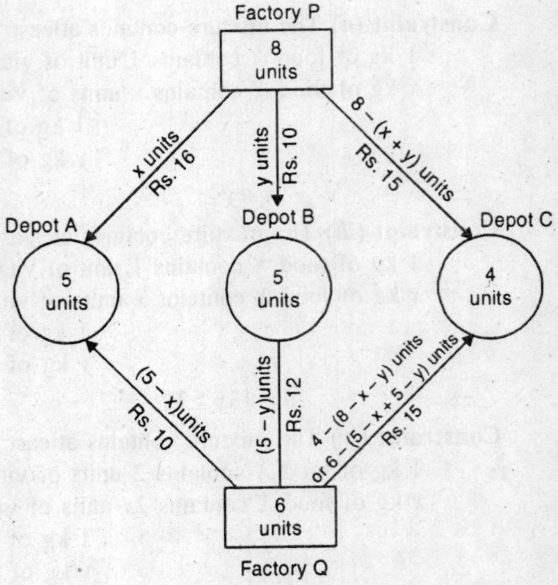


1. **Decision Variables.** Decision variables are units of commodity to be transported from the factories to the depots.
   Let the factory at $P$ transport $x$ units of commodity to depot at $A$ and $y$ units to depot at $B$ so the remaining units at $P$, $8 - (x + y)$ will be transported to depot at $C$.
2. **Constraint (*i*).** From Factory at $P$, $(8 - x - y)$ units will be transported to the depot at $C$.

$$8 - x - y \geq 0$$

$$\Rightarrow \quad x + y \leq 8 \quad \ldots (1)$$

**Constraint (*ii*).** $x$ and $y$ units are non-negative units.

$$x \geq 0 \quad \ldots (2)$$

$$y \geq 0 \quad \ldots (3)$$

**Constraint (*iii*).**
The remaining requirements $(5 - x)$ units are to be transported from the factory at $Q$ to the depot at $A$.

**Constraint (*iv*).** $5 - x \geq 0 \Rightarrow x \leq 5$ ... (4)

The remaining requirements $(5 - y)$ units are to be transported from factory $Q$ to the depot at $B$.

$$5 - y \geq 0 \Rightarrow y \leq 5 \quad \ldots (5)$$

**Constraint (*v*).**
The remaining requirements $4 - (5 - x + 5 - y)$ units of commodity will be transported from the factory at $Q$ to the depot at $C$.

$$4 - (5 - x + 5 - y) \geq 0$$

$$\Rightarrow \quad x + y - 4 \geq 0 \quad \Rightarrow \quad x + y \geq 4 \quad \ldots (6)$$

3. **Objective function.** The transportation cost from the factory at $P$ to the depots at $A$, $B$ and $C$ are respectively Rs. $16x$, $10y$ and $15(8 - x - y)$
   Total cost of transportation from factory at $P = 16x + 10y + 15(8 - x - y)$.
   Similarly, the transportation cost from the factory at $Q$ to the depots at $A$, $B$ and $C$ are Rs. $10(5 - x), 12(5 - y), 10(x + y - 4)$ respectively.

   Total transportation cost from factory $Q = 10(5 - x) + 12(5 - y) + 10(x + y - 4)$
   Grand total cost of transportation from both the factories at $P$ and $Q$ to all the depots

$$Z = 16x + 10y + 15(8 - x - y) + 10(5 - x) + 12(5 - y) + 10(x + y - 4)$$

$$\Rightarrow \quad Z = x - 7y + 190$$

4. **Mathematical Formulation.**
   To minimize $Z = x - 7y + 190$
   Subject to constraints $x + y \leq 8$

$$x + y \geq 4$$
$$x \leq 5$$
$$y \leq 5$$
$$x \geq 0, \quad y \geq 0.$$ **Ans.**

## EXERCISE 27.1

1. A furniture dealer deals in only two items *viz.*, tables and chairs. He has Rs. 11,000 to invest and a space to store at most 40 pieces. A table costs him Rs. 500 and a chair Rs. 200. He can sell a table at a profit of Rs. 50 and a chair at a profit of Rs. 15. Assume that he can sell all the items that he buys. Formulate this problem as an L.P.P, so that he can maximize the profit.

**Ans.** Maximize, $Z = 50x + 15y$

Subject to the constraints : $x + y \leq 40$, $500x + 200y \leq 11000$, $x \geq 0$, $y \geq 0$

2. A manufacturer produces nuts and bolts for industrial machinery. It takes 1 hour of work on machine $A$ and 3 hours on machine $B$ to produce a package of nuts; while it takes 3 hours on machine $A$ and 1 hour on machine $B$ to produce a package of bolts. He earns a profit of Rs. 2.50 per package on nuts and Re. 1 per package on bolts. Form a linear programming problem to maximize his profit, if he operates each machine for at the most 12 hours a day.

**Ans.** Maximize, $Z = 2.5x + y$

Subject to the constraints : $x + 3y \leq 12$, $3x + y \leq 12$, $x \geq 0$, $y \geq 0$

3. A factory produces two products $A$ and $B$. To manufacture one unit of product $A$, a machine has to work for $1\frac{1}{2}$ hours and a craftsman has to work for 2 hours. To manufacture one unit of product $B$, the machine has to work for 3 hours and the craftsman for one hour. In a week, the factory can avail of 80 hours of machine time and 70 hours of craftsman's time. The profits on each unit of $A$ and $B$ are Rs. 10 and Rs. 8 respectively. If the manufacturer can sell all the items produced, how many of each should be produced to get maximum profit per week? Formulate the above as a linear programming problem.

**Ans.** Maximize $Z = 10x + 8y$

Subject to the constraints : $1.5x + 3y \leq 80$, $2x + y \leq 70$, $x \geq 0$, $y \geq 0$

4. A manufacturer has 3 machines I, II and III. Machines I and II are capable of being operated for at the most 12 hours whereas machine III must be operated atleast for 5 hours a day. He produces two items each requiring the use of the three machines.
The number of hours required for producing I unit of each of the items $A$ and $B$ on the three machines are given below:

| *Items* | *Number of hours required on the machines* | | |
|---|---|---|---|
| | *I* | *II* | *III* |
| *A* | 1 | 2 | 1 |
| *B* | 2 | 1 | 5/4 |

He makes a profit of Rs. 60 on item $A$ and Rs. 40 on item $B$. Assuming that he can sell all that he produces, how many of each item should be produced so as to maximize the profit. Formulate the above as L.P.P.

**Ans.** Maximize $Z = 60x + 40y$

Subject to the constraints : $x + 2y \leq 12$, $2x + y \leq 12$, $x + \frac{5}{4}y \geq 5$, $x \geq 0$, $y \geq 0$

5. A factory manufactures two varieties of machines $A$ and $B$. Each type is made of certain alloy metal. The factory has only 500 kg of this metal available in a day. To manufacture machine A, 20 kg of metal is required and 50 kg is required for machine $B$. Each machine requires 20 minutes assembly time and the assembly department has only 280 minutes. Further the machines $A$ and $B$ require 10 minutes and 5 minutes respectively to be painted. The painting department is restricted not to use more than 200 minutes in a day. The factory earns profit of Rs. 1000 and Rs. 2000 on machines A and B respectively. Formulate the above as a linear programming problem to maximize the profit.

**Ans.** Maximize $Z = 1000x + 2000y$

Subject to the constraints : $2x+5y \le 50$, $x+y \le 14$, $2x+y \le 40$, $x \ge 0$, $y \ge 0$

6. A manufacturer produces two products $A$ and $B$ during a given period of time. Each of these products require four different operations, *viz.* Grinding, Turning, Assembly and Testing. The requirement in hours per unit of manufacturing of the product are given below:

| *Operation* | *A* | *B* |
|---|---|---|
| Grinding | 1 | 2 |
| Turning | 3 | 1 |
| Assembly | 4 | 3 |
| Testing | 5 | 4 |

The available capacities of these operations in hours for the given time period are:

| | | | |
|---|---|---|---|
| Grinding | 30 | Turning | 60 |
| Assembly | 200 | Testing | 200 |

Profit on each unit of A is Rs. 3 and Rs. 2 for each unit of B.
Formulate the problem as a linear programming model to maximize the profit assuming that the firm can sell all the items that it produces at the prevailing market price.

**Ans.** Maximize, $Z = 3x + 2y$

Subject to the constraints : $x+2y \le 30$, $3x+y \le 60$, $4x+3y \le 200$,

$5x+4y \le 200$, $x \ge 0$, $y \ge 0$

7. A toy company manufactures two types of dolls, $A$ and $B$. Each doll of type $B$ takes twice as long as to produce as one of type $A$. If the company produces only type $A$, it can make a maximum of 2000 dolls per day. The supply of plastic is sufficient to produce 1,500 dolls per day. Type $B$ requires a fancy dress which cannot be available for more than 600 per day. If the company makes profits of Rs. 3 and Rs. 5 per doll, respectively on dolls $A$ and $B$, how many of each should be produced per day in order to maximize the profit ?

**Ans.** Maximize, $Z = 3x + 5y$

Subject to the constraints : $x \le 2000$, $y \le 600$, $x+y \le 15000$

8. A person consumes two types of food, $A$ and $B$, everyday to obtain 8 units of protein, 12 units of carbohydrates and 9 units of fat which is his daily minimum requirements. 1 kg of food $A$ contains 2, 6, 1 units of protein, carbohydrates and fat, respectively. 1 kg of food $B$ contains 1, 1 and 3 units of protein, carbohydrates and fat, respectively. Food $A$ costs Rs. 8 per kg while food $B$ costs Rs. 5 per kg. Form an LPP to find how many kg of each food should he buy daily to minimize his cost of food and still meet minimal nutritional requirements.

**Ans.** Minimize $Z = 5x + 8y$

Subject to the constraints : $2x+y \ge 8$, $6x+y \ge 12$, $x+3y \ge 9$, $x \ge 0$, $y \ge 0$

**9.** A dietician wishes to mix two types of foods $F_1$ and $F_2$ in such a way that the vitamin contents of the mixture contains atleast 6 units of vitamin *A* and 8 units of vitamin *B*. Food $F_1$ contains 2 units/kg of vitamin A and 3 units/kg of vitamin B while food $F_2$ contains 3 units/kg of vitamin A and 4 units/kg of vitamin B. Food $F_1$ costs Rs. 50/kg and food $F_2$ costs Rs. 75/kg. Formulate the problem as a L.P.P. to minimize the cost of mixture.

**Ans.** Minimize $z = 50x + 75y$

Subject to the constraints : $2x + 3y \geq 6$, $3x + 4y \geq 8$, $x \geq 0$, $y \geq 0$

**10.** A producer has 50 and 85 units of labour and capital respectively which he can use to produce two types of goods *X* and *Y*. To produce one unit of *X*, 1 unit of labour and 2 units of capital are required. Similarly, 3 units of labour and 2 units of capital are required to produce one unit of *Y*. If *X* and *Y* are priced at Rs. 100 and Rs. 150 per unit respectively, how should the producer use his resources to maximise the total revenue? Formulate L.P.P.

**Ans.** Maximize $Z = 100x + 150y$

Subject to the constriants : $x + 3y \leq 50$, $2x + 2y \leq 85$, $x \geq 0$, $y \geq 0$

## 27.3. GRAPHICAL METHOD OF SOLVING LINEAR PROGRAMMING PROBLEMS

If a problem contains only two variables then we can solve the given problem by graphical method. There are two graphical method to solve a linear programming problem.

**1.** Corner point method

**2.** Iso-profit or iso-cost method

## 27.4. CORNER POINT METHOD

This method is based on the fundamental extreme point theorem.

In previous class we have learnt how to formulate a system of linear inequalities involving two variables *x* and *y* mathematically.

**Working Rule**

**Step 1.** Formulate the given L.P.P. in mathematical form.

**Step 2.** The inequations are converted into equations.

In the equation on putting $y = 0$ we get *x*-coordinate on *x*-axis. Similarly, putting $x = 0$ we get *y*-coordinate on *y*-axis. Join these two points to get the graph of the equation.

**Step 3.** The inequation of a line divides the plane into two half planes, to choose the plane of the inequation we put $x = 0$ and $y = 0$ in the inequation. If origin satisfies the inequation then the region containing the origin is the region represented by the given inequation. Otherwise the half plane not containing the origin is the region represented by the given inequation.

**Step 4.** The region satisfying all the inequations is the feasible region.

**Step 5.** The vertices (corner points) of the required region are known as extreme points of the set of all feasible solutions of the L.P.P.

**Step 6.** By putting the values of *x* and *y* of each corner point in the objective function we get the values of the objective function at each of the vertices of the feasible region. Out of all the values of the objective function, we get a point at which the objective function is optimum (maximum or minimum).

Consider the following example:-

**Example 1.** *Solve the following L.P.P. graphically :*

*Maximize,* $Z = 3x + 2y$

*Subject to the constraints*

$x + 2y \le 10, \quad 3x + y \le 15, \quad x \ge 0, \quad y \ge 0$

**Solution.** On changing the given inequations into equations, we have

$x + 2y = 10$ and $3x + y = 15$; $x = 0$ and $y = 0$

**Region Represented by $x + 2y \le 10$.**

The line $x + 2y = 10$ meets the *x*-axis at the point $A(10, 0)$ and meets *y*-axis at the point *B* (0, 5). Join *AB* to obtain the graph of $x + 2y = 10$.

Here, origin ($x = 0$, $y = 0$) satisfies the inequation $x + 2y \le 10$. So, the half plane containing the origin represents the solution set of the inequation $x + 2y \le 10$.

**Region Represented by $3x + y \le 15$.**

The line $3x + y = 15$ meets the *x*-axis at the point *C* (5, 0) and meets the *y*-axis at the point *D* (0, 15). Join *CD* to obtain the graph of the line $3x + y = 15$. Here, origin ($x = 0$, $y = 0$) satisfies the inequation $3x + y \le 15$. So, the half plane containing the origin represents the solution set of the inequation $3x + y \le 15$.

**Region represented by $x \ge 0$ and $y \ge 0$.**

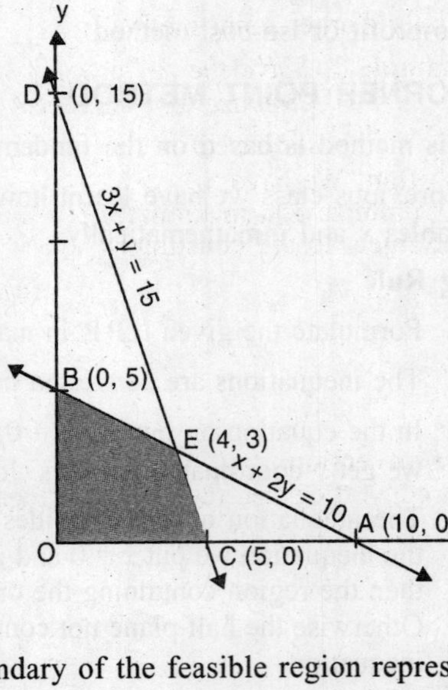


All the points in the first quadrant satisfies $x \ge 0$ and $y \ge 0$. So, the first quadrant is the region represented by $x \ge 0$ and $y \ge 0$.

The shaded region *O C E B O* represents the region satisfying all the inequations.

Each point of this region represents a feasible choice.

Every point of this region is called the feasible solution of the problem.

**Feasible region.** The shaded portion determined by all the constraints including non-negative constants of a linear programming problem is called the feasible region. In this example O BEC (shaded) is the feasible region for the problem.

**Feasible Solution.** Points within and on the boundary of the feasible region represent feasible solutions of the constraints.

Here, (5, 0), (4, 3) and (0, 5) are the feasible solutions. Any point outside the feasible region is called an infeasible solution. *For example,* (0, 15) and (10, 0) are infeasible solutions.

**Optimal Solution.** Any point in the feasible region that gives the optimal value (maximum or minimum) is called an optimal solution.

In the feasible region there are infinitely many points which satisfy all the constraints. It is not possible to check all the points for the maximum value of the objective function.

$$Z = 3x + 2y$$

For finding out the optimal solution we have to use the following theorems

**Theorem 1.** *The optimal value of objective function must occur at a corner point (vertex) of the feasible region.*

**Theorem 2.** *If the feasible region is bounded then the objective function Z has both maximum and minimum values on corner points of the bounded region.*

In the above example we have following table showing the value of the objective function at the corner points of the feasible region.

| *Vertex of the feasible region* | *Corresponding value of* $Z = 3x + 2y$ *(in Rs.)* |
|---|---|
| $O\ (0, 0)$ | 0 |
| $C\ (5, 0)$ | 15 |
| $E\ (4, 3)$ | 18 **Maximum** |
| $B\ (0, 5)$ | 10 |

We observe that the maximum value of the objective function is 18.

This method of solving linear programming problem is called *corner point method.*

**Procedure:** The solution of the given L.P.P. should be divided under the following heads:

1. Conversion of inequalities into equations.
2. Draw the graph of the lines and find regions represented by the inequations.
3. Apply Corner point method.

**Example 2.** *Solve the following linear programming problem graphically:*
*Minimize and maximize* $Z = x + 2y$
*subject to* $x+2y \geq 100,\ 2x-y \leq 0,\ 2x+y \leq 200,\ x \geq 0, y \geq 0.$

**Solution.** We have,

Minimize and maximize $Z = x + 2y$ ...(1)

Subject to the constraints

$$x+2y \geq 100 \quad ...(2)$$

$$2x-y \leq 0 \quad ...(3)$$

$$2x+y \leq 200 \quad ...(4)$$

$$x \geq 0,\ y \geq 0 \quad ...(5)$$

**1. Conversion.**

On converting the above inequations, we have the following equations :

$$x+2y=100$$

$$2x-y=0$$

$$2x+y=200$$

$$x=0, \quad y=0$$

**2. Drawing of graphs.**

**The region represented by the inequation** $x+2y \geq 100$.

$$x + 2y = 100$$

| | | |
|---|---|---|
| $x$ | 100 | 0 |
| $y$ | 0 | 50 |
| *Points* | *A* | *B* |

The line $x + 2y = 100$ meets the $x$-axis at $A$ (100, 0) and meets $y$-axis at $B$ (0, 50). On joining $A$, $B$ we get the graph of the line $x + 2y = 100$. Since (0, 0) does not satisfy the inequation $x+2y \geq 100$, so the half plane not containing origin represents the region of the solution set of the inequation $x+2y \geq 100$.

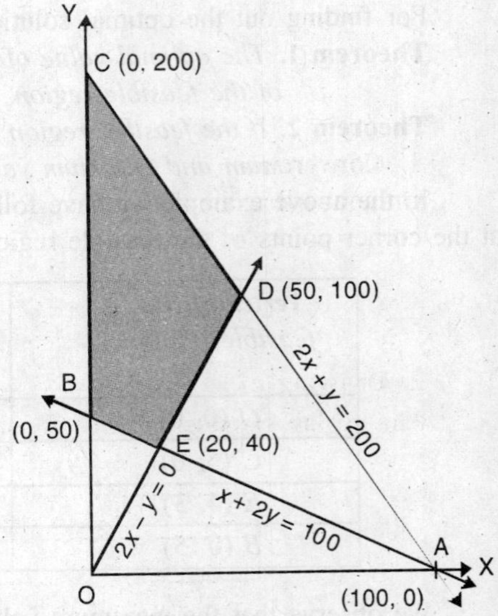


**The region represented by** $2x - y \leq 0$.

The line $2x - y = 0$ passes through origin.

Point $B$ (0, 50) satisfies the inequation $2x - y \leq 0$. So the half plane containing $B(0, 50)$ represents the region of the solution set of the inequation $2x - y \leq 0$.

**The region represented by** $2x + y \leq 200$.

$2x + y = 200$

| | | |
|---|---|---|
| $x$ | 100 | 0 |
| $y$ | 0 | 200 |
| *Point* | $A$ | $C$ |

The line $2x + y = 200$ meets the $x$-axis at $A$ (100, 0) and meets $y$-axis at the point $C$ (0, 200). On joining the points $A$ and $C$, we get the graph of the line $2x + y = 200$.

Since, (0, 0) satisfies the inequation $2x + y \leq 200$, so the half plane containing the origin represents the region of the solution set of the inequation $2x + y \leq 200$.

**The region represented by** $x \geq 0$ **and** $y \geq 0$.

$x \geq 0$ and $y \geq 0$ represent the first quadrant.

The shaded region $BCDEB$ bounded by the inequations (2) to (5) is the feasible region.

**3. Corner Point Method.** The coordinates of the corner points are $B$ (0, 50), $C$ (0, 200), $D$ (50, 100), $E$ (20, 40).

Now we evaluate $Z = x + 2y$ at the corner points.

| *Corner points of the feasible regions BCDEB* | *Corresponding value of* $Z = x + 2y$ |
|---|---|
| $B$ (0, 50) | 100 |
| $C$ (0, 200) | 400 |
| $D$ (50, 100) | 250 |
| $E$ (20, 40) | 100 |

The maximum value of the objective function at $C$ (0, 200) is 400.

The minimum value of the objective function at $E$ (20, 40) is 100.

**Example 3.** *Solve graphically the following linear programming problem to minimise the cost* $Z = 3x + 2y$ *subject to the following constraints:*

$5x + y \geq 10$; $\quad x + y \geq 6$; $\quad x + 4y \geq 12$; $\quad x \geq 0, y \geq 0$.

**Solution.** We have,

Minimise $\quad Z = 3x + 2y \quad$ ... (1)

Subject to the constraints

$$5x + y \geq 10 \quad \text{... (2)}$$

$$x+y\geq 6 \quad \text{... (3)}$$
$$x+4y\geq 12 \quad \text{... (4)}$$
$$x\geq 0,\ y\geq 0 \quad \text{... (5)}$$

**1. Conversion.**

On converting the above inequations, we have the following equations

$$5x + y = 10$$
$$x + y = 6$$
$$x + 4y = 12$$
$$x = 0,\ y = 0$$

**2. Drawing of graphs.**

**The region represented by $5x+y\geq 10$.**

$$5x + y = 10$$

| | | |
|---|---|---|
| $x$ | 2 | 0 |
| $y$ | 0 | 10 |
| *Point* | *A* | *B* |

On joining the points *A* and *B*, we get the graph of the line $5x + y = 10$. Put $x = 0$, $y = 0$ in $5x+y\geq 10$, then $0+0\geq 10$ which is false. So half plane not containing origin represents the region of the solution set of the inequation $5x+y\geq 10$.

**The region represented by $x+y\geq 6$**

$$x + y = 6$$

| | | |
|---|---|---|
| $x$ | 6 | 0 |
| $y$ | 0 | 6 |
| *Point* | *C* | *D* |

On joining the points *C* and *D*, we get the graph of the line $x + y = 6$.

Put $x = 0$, $y = 0$ in $x+y\geq 6$, then $0+0\geq 6$ which is false. So the half plane not containing origin represents the region of the solution set of $x+y\geq 6$.

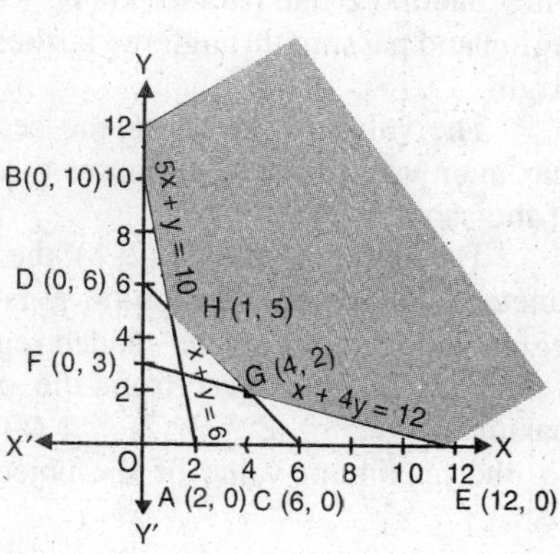


**The region represented by $x+4y\geq 12$**

$$x + 4y = 12$$

| | | |
|---|---|---|
| $x$ | 12 | 0 |
| $y$ | 0 | 3 |
| *Point* | *E* | *F* |

On joining the points *E* and *F* we get the graph of the line $x + 4y = 12$.

Put $x = 0$, $y = 0$ in $x+4y\geq 12$, then $0+0\geq 12$ which is false.

So, the half plane not containing origin represents the region of the solution set of the inequation $x+4y\geq 12$.

**The region represented by $x\geq 0$ and $y\geq 0$.**

The first quadrant is represented by $x\geq 0$ and $y\geq 0$.

The shaded region is represented by the inequations (2) to (5).

**3. Corner Point Method.** The coordinates of the vertices of the feasible region are $E$ (12, 0), $G$ (4, 2), $H$ (1, 5) and $B$ (0, 10) respectively.

Note that the coordinates of $G$ and $H$ are obtained by solving the equations $x + y = 6$, $x + 4y = 12$ and $x + y = 6$, $5x + y = 10$ respectively.

The value of the objective function at these points are given in the following table :

| *Corner Point (x, y) of the feasible region EGHB* | *Value of the objective function* $Z = 3x + 2y$ |
|---|---|
| $E$, (12, 0) | $3 \times 12 + 2 \times 0 = 36$ |
| $G$, (4, 2) | $3 \times 4 + 2 \times 2 = 12 + 4 = 16$ |
| $H$, (1, 5) | $3 \times 1 + 2 \times 5 = 3 + 10 = 13$ |
| $B$, (0, 10) | $3 \times 0 + 2 \times 10 = 0 + 20 = 20$ |

Clearly, $H$ is minimum when $x = 1$, $y = 5$.

So, $x = 1$, $y = 5$ is the optimal solution of the given L.P.P.

Hence, $Z$ is minimum when $x = 1$ and $y = 5$ and the minimum value of $Z$ is Rs. 13.

## 27.5 ISO-PROFIT OR ISO-COST METHOD (MAXIMUM Z)

Let $Z = 3x + 2y$ be the objective function.

On putting any value of $Z$ (say 12) in $Z = 3x + 2y$, we have

$$12 = 3x + 2y \qquad \ldots (1)$$

and draw the corresponding line of the objective function. This line is called Iso-profit or Iso-cost line, since every point on this line will give the same value of $Z$ (12) (same profit or same cost). Draw one more line parallel to (1), within the feasible region and passing through the farthest point from the origin.

The value of the $Z$ on the second line is the maximum value of the $Z$. It passes through one corner of the shaded region.

Through the point $E$ (2, 8), the line of objective function is passing. The point $E$ is on the shaded region and is vertex of the shaded region and the line $Z = 3x + 2y$ is farthest from the origin. Here the maximum value of $Z = 3(2) + 2(8) = 6 + 16 = 22$.

So, the maximum value of the objective function is 22.

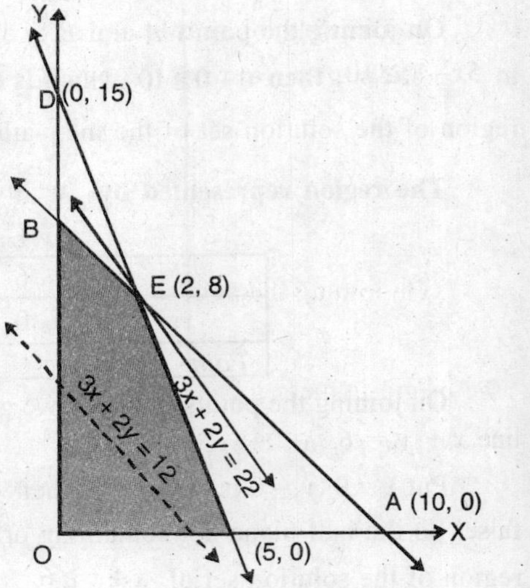


## 27.6. ISO-PROFIT OR ISO-COST METHOD (MINIMUM Z)

Let $Z = x + 2y$

Let us take three different values of $Z$; $Z_1$, $Z_2$ and $Z_3$, we get

$$Z_1 = x + 2y \qquad \ldots (1)$$

$$Z_2 = x + 2y \qquad \ldots (2)$$

$$Z_3 = x + 2y \qquad \ldots (3)$$

The above three lines are parallel to each other having the same slope

$$\left(-\frac{1}{2}\right)$$

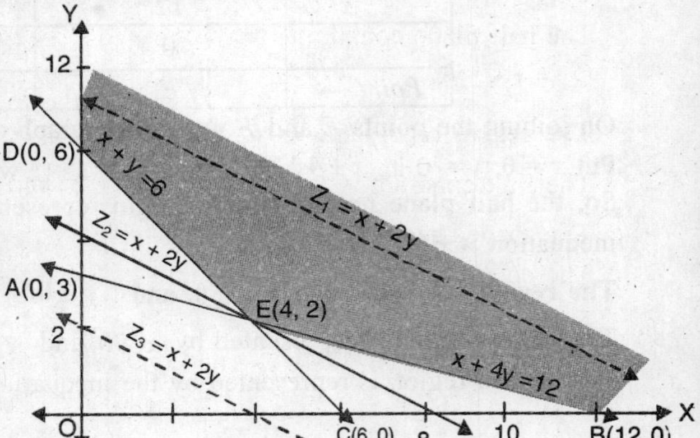

Out of these three lines, line (2) is the only line which passes through the point $E$ (4, 2) nearest to the origin and passing through the feasible region.

Thus, $Z_2$ is the minimum value of $Z$.

$x = 4$ and $y = 2$ gives the optimal solution.

$$Z_2 = 4 + 2(2) = 4 + 4 = 8$$

So, the minimum value of $Z$ is 8.

**Example 4.** *Solve graphically the following linear programming problem:*

*maximize* $Z = 5x + 3y$

*subject to the constraints* :

$$x + y \le 300, \quad 2x + y \le 360, \quad x \ge 0, \quad y \ge 0$$

**Solution.** We have,

Maximum $Z = 5x + 3y$ ... (1)

Subject to the constraints

$x + y \le 300$ ... (2)

$2x + y \le 360$ ... (3)

$x \ge 0$ and $y \ge 0$ ... (4)

**1. Conversion**

On changing the inequations into equations we get the following equations.

$x + y = 300$ and $2x + y = 360$

**2. Drawing of graphs**

**The region represented by $x + y \le 300$**

$x + y = 300$

| | | |
|---|---|---|
| *x* | 300 | 0 |
| *y* | 0 | 300 |
| *Point* | *A* | *B* |

On joining the points $A$ (300, 0) and $B$ (0, 300) we get the graph of the line $AB$, $x + y = 300$.

Put $x = 0$, $y = 0$ in $x + y \le 300$, then $0 + 0 \le 300$, which is true. The half plane containing the origin is the region of the solution set of the inequation $x + y \le 300$.

**The region represented by $2x + y \le 360$.**

$2x + y = 360$

| | | |
|---|---|---|
| *x* | 180 | 0 |
| *y* | 0 | 360 |
| *Point* | *C* | *D* |

On joining the points $C$ (180, 0) and $D$ (0, 360) we get the graph of the line $2x + y = 360$.

Put $x = 0$, $y = 0$ in $2x + y \le 360$, then $0 + 0 \le 360$, which is true.

The half plane containing the origin is the region of the solution set of the inequation $2x + y \le 360$.

**The region represented by $x \ge 0$ and $y \ge 0$.**

The first quadrant is represented by $x \ge 0$, and $y \ge 0$.

The shaded region $OABC$ is the feasible solution bounded by the inequations (2) to (4). The coordinates of the corner point of the feasible region are $A$ (180, 0), $B$ (60, 240) and $C$ (0, 300).

On solving the equations (2) and (3), we get the point of intersection B (60, 240).

**3. Iso-profit or Iso-cost Method.**

Now, we take a constant value (say 300) i.e.,

(20 times the *l.c.m.* of 5 and 3).

On putting $Z = 300$ in objective function (1), we get

$$300 = 5x + 3y \qquad \dots (5)$$

We draw the graph of the line

$$5x + 3y = 300.$$

Draw one more line parallel to (5) which is the farthest from the origin and has atleast one point of the feasible region.

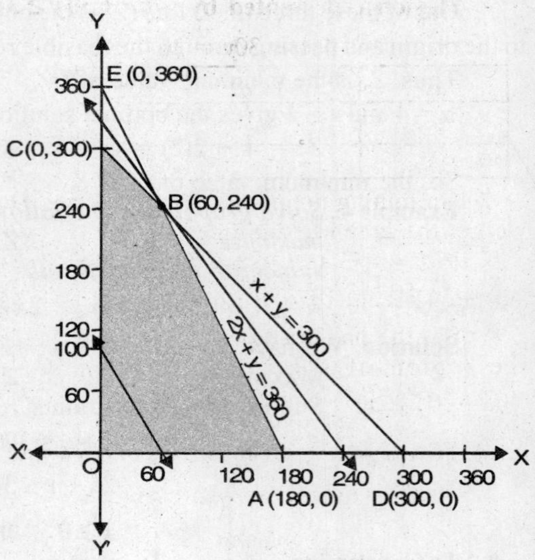


The parallel line to (5) passes through the farthest point B (60, 240) from the origin.

Here, $x = 60$ and $y = 240$ will give the maximum value of $Z$.

The maximum value of $Z$ is given by $Z = 5(60) + 3(240) = 300 + 720 = 1020$.

Hence, the maximum value of $Z$ is 1020. **Ans.**

**Example 5.** *Solve graphically the following L.P.P.*

*Minimize* $Z = 5x + 4y$

*subject to the constraints* :

$$80x + 100y \geq 88, \quad 40x + 30y \geq 36, \quad x \geq 0, \quad y \geq 0$$

**Solution.** We have,

Minimize $Z = 5x + 4y \qquad \dots (1)$

Subject to the constraints

$$80x + 100y \geq 88 \qquad \dots (2)$$

$$40x + 30y \geq 36 \qquad \dots (3)$$

$$x \geq 0, y \geq 0 \qquad \dots (4)$$

**1. Conversion**

On converting the inequations into equations, we get

$$80x + 100y = 88$$
$$40x + 30y = 36$$

**2. Drawing of graphs**

**Region represented by $80x + 100y \geq 88$.**

$$80x + 100y = 88$$

| *x* | 1.1 | 0.0 |
|---|---|---|
| *y* | 0 | 0.88 |
| *Point* | *A* | *B* |

On joining the points $A$ (1.1, 0) and $B$ (0.0, 0.88) we get the graph of the line $80x + 100y = 88$.

Put $x = 0, y = 0$ in the inequation $80x + 100y \geq 88$ then $0 + 0 \geq 88$, which is false.

So, the half plane not containing the origin is the region of solution set of the inequation $80x + 100y \geq 88$

**Region represented by $40x + 30y \geq 36$.**

$40x + 30y = 36$

| | | |
|---|---|---|
| $x$ | 0.90 | 0 |
| $y$ | 0 | 1.2 |
| *Point* | $C$ | $D$ |

On joining the points $C$ (0.90, 0) and $D$ (0, 1.2), we get the graph of the line $40x + 30y = 36$.

Put $x = 0$, $y = 0$ in the inequation $40x+30y \geq 36$, then $0+0 \geq 36$, which is false.

So, the half plane not containing the origin is the region of solution set of the inequation $40x+30y \geq 36$.

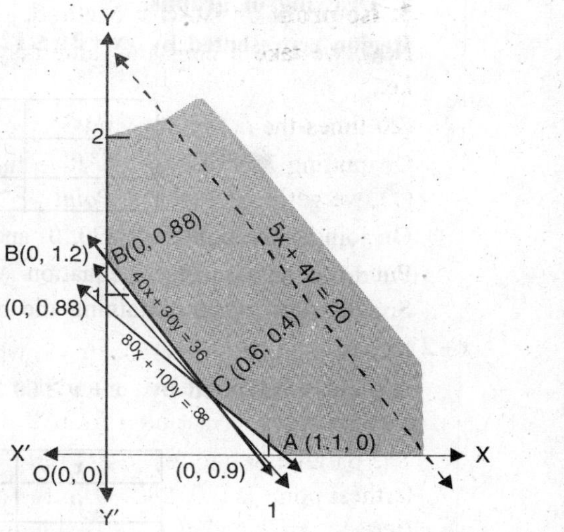


**The region represented by $x \geq 0$ and $y \geq 0$**

The first quadrant is represented by

$x \geq 0$ and $y \geq 0$.

The shaded region $ACB$ represents the feasible solution bounded by the inequations (2) to (4). On solving (2) and (3), we get the point of intersection of the lines (2) and (3), $C$ (0.6, 0.4).

**3. Iso-profit or Iso-cost Method**

Now, we take a constant = 20

On putting $Z = 20$ in (1), we get

$$5x + 4y = 20 \qquad \text{... (5)}$$

Now, we draw the graph of the line $5x + 4y = 20$.

Draw one more line parallel to (5), which is the nearest from the origin and has at least one point of the feasible region.

The parallel line to (5) passes through the nearest point C (0.6, 0.4) from the origin. Here, $x = 0.6$ and $y = 0.4$ will give the minimum value of $Z$. The minimum value of $Z$ is given by

$$Z = 5\ (0.6) + 4\ (0.4) = 3.0 + 1.6 = 4.6$$ **Ans.**

Hence, the minimum value of $Z$ is 4.6.

**Example 6.** *Solve graphically the following L.P.P.*

*Minimize and maximize* $Z = 5x + 10y$

*Subject to*

$x+2y \leq 120$, $x+y \geq 60$, $x-2y \geq 0$, $x \geq 0$, $y \geq 0$

**Solution.** We have,

Minimize and maximize $Z = 5x + 10y$ ... (1)

Subject to the constraints :

$$x+2y \leq 120 \qquad \text{... (2)}$$
$$x+y \geq 60 \qquad \text{... (3)}$$
$$x-2y \geq 0 \qquad \text{... (4)}$$
$$x \geq 0,\ y \geq 0 \qquad \text{... (5)}$$

**1. Conversion.**

On converting the inequations into equations, we get

$$x + 2y = 120 \qquad \text{... (6)}$$
$$x + y = 60 \qquad \text{... (7)}$$
$$x - 2y = 0 \qquad \text{... (8)}$$

**2. Drawing of graphs.**

**Region represented by** $x+2y \leq 120$.

$x + 2y = 120$

| | | |
|---|---|---|
| $x$ | 120 | 0 |
| $y$ | 0 | 60 |
| *Point* | *A* | *B* |

On joining the points *A* (120, 0) and *B* (0, 60), we get the graph of line $x + 2y = 120$.

Put $x = 0$, $y = 0$ in the inequation $x+2y \leq 120$, then $0+0 \leq 120$ which is true.

So, the half plane containing the origin is the region of solution set of the inequation $x+2y \leq 120$.

**Region represented by** $x+y \geq 60$.

$x + y = 60$

| | | |
|---|---|---|
| $x$ | 60 | 0 |
| $y$ | 0 | 60 |
| *Point* | *C* | *D* |

On joining the points *C* (60, 0) and *D* (0, 60), we get the graph of the line $x + y = 60$.

Put $x = 0$, $y = 0$ in the inequation $x+y \geq 60$, then $0+0 \geq 60$, which is false.

So, the half plane not containing the origin is the region of solution set of the inequation $x+y \geq 60$.

**Region represented by** $x-2y \geq 0$.

$x - 2y = 0$

| | | |
|---|---|---|
| $x$ | 0 | 6 |
| $y$ | 0 | 3 |
| *Point* | *O* | *F* |

On joining O (0, 0) and *F* (6, 3), we get the graph of the line OF *i.e.*, $x - 2y = 0$.

Put $x = 60$, $y = 0$ in the inequation $x-2y \geq 0$, then $60-2(0) \geq 0 \Rightarrow 60 \geq 0$, which is true.

So, the half plane containing the point *C* (60, 0) is the region of the solution set of the inequation $x-2y \geq 0$.

**Region represented by** $x \geq 0$ **and** $y \geq 0$.

The first quadrant is represented by $x \geq 0$ and $y \geq 0$.

The shaded region *ACDE* is the feasible solution bounded by the inequation (2) to (4).

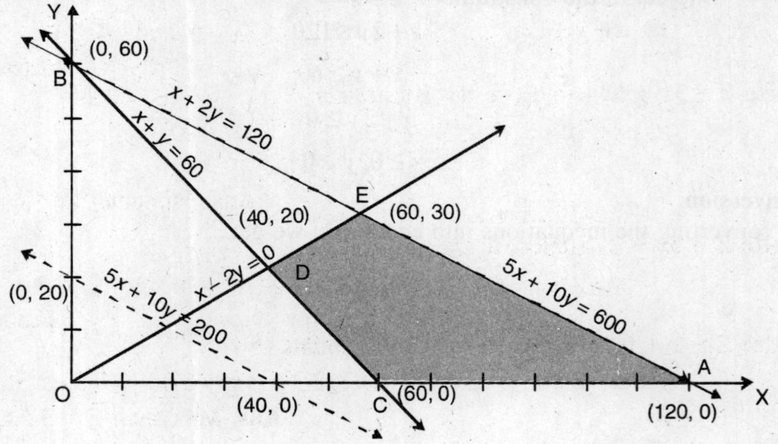

On solving (6) and (8), we get the point of intersection $E$ (60, 30). Also, solving (7) and (8), we get the point of intersection $D$ (40, 20).

The coordinates of the corner point of the feasible region are $A$ (120, 0), $C$ (60, 0), $D$ (40, 20) and $E$ (60, 30).

**3. Iso-profit or Iso-cost Method.**

Now, we take a constant value say 200 for $Z$.

Putting $Z = 200$ in (1), we get

$$5x + 10y = 200$$

We obtain the line $5x + 10y = 200$. This line meets $x$-axis at the point (40, 0) and meets $y$-axis at the point (0, 20). Join these points by dotted lines.

Now, drawn a line parallel to $5x + 10y = 200$, and passing through the shaded and also passing through a point to origin *i.e.*, $C$ (60, 0).

Hence, $x = 60$ and $y = 0$ give the minimum value of $Z$.

The minimum value of $Z$, is given by

$$Z = 5\ (60) + 10\ (0) = 300.$$ **Ans.**

**For maximum value.** Draw a line parallel to the objective function passing through the shaded region and a point $E$ (60, 30) farthest from the origin.

Hence, $x = 60$ and $y = 30$ gives the maximum value of $Z$.

The maximum value of $Z$ is given by

$$Z = 5\ (60) + 10\ (30)$$
$$= 300 + 300 = 600$$ **Ans.**

## EXERCISE 27.2

**Solve graphically each of the following linear programming problems :**

**1.** Maximize $Z = 10x + 6y$ subject to the constraints

$3x + y \leq 12, \quad 2x + 5y \leq 34, \quad x \geq 0, y \geq 0$

**Ans.** Maximum : $Z = 56$; $x = 2$, $y = 6$

**2.** Maximize $Z = 60x + 15y$ subject to the constraints

$x + y \leq 50, \quad 3x + y \leq 90, \quad x \geq 0, y \geq 0$

**Ans.** Maximum : $Z = 1650$, $x = 20$, $y = 30$

**3.** Minimize $Z = 18x + 10y$ subject to the constraints

$4x + y \geq 20, \quad 2x + 3y \geq 30, \quad x \geq 0, y \geq 0$

**Ans.** Minimum : $Z = 134$; $x = 3$, $y = 8$

**4.** Maximize $Z = 4x + 9y$ subject to the constraints

$x + 5y \leq 200, \quad 2x + 3y \leq 134, \quad x \geq 0, y \geq 0$

**Ans.** Maximum : $Z = 382$; $x = 10$, $y = 38$

**5.** Maximize $Z = 2x + 7y$ subject to the constraints

$x + y \leq 12, \quad 2x + y \geq 30, \quad x \geq 0$ and $y \geq 0$

**Ans.** No feasible region.

**6.** Minimize $Z = 3x + 5y$ subject to the constraints

$x + 3y \geq 3, \quad x + y \geq 2, \quad x, y \geq 0$

**Ans.** Minimum $Z = 7, x = \dfrac{3}{2}, y = \dfrac{1}{2}$

**7.** Maximize $Z = 5x + 2y$ subject to the constraints

$x + y \leq 2, \quad 3x + 3y \geq 12, \quad x, y \geq 0$

**Ans.** No. solution

**8.** Maximize $Z = x + 0.75\ y$ subject to the constraints

$x - y \geq 0, \quad -x + 2y \leq 2, \quad x, y \geq 0$

**Ans.** Maximum $Z = 3.5$, $x = 2$, $y = 2$

**9.** Minimize $Z = 20x + 10y$ subject to the constraints

$$x + 2y \le 40, \quad 3x + y \ge 30$$

**Ans.** Minimum $Z = 200$, $x = 10$, $y = 0$

**10.** Maximize $Z = 5x + 7y$ subject to the constraints

$$x + y \le 4, \quad 3x + 8y \le 24, \quad 10x + 7y \le 35, \quad x, y \ge 0$$

**Ans.** Maximum $Z = \frac{124}{5}, x = \frac{8}{5}, y = \frac{12}{5}$

**11.** Minimize $Z = 30x + 20y$ subject to the constraints

$$x + y \le 8, \quad x + 4y \ge 12, \quad x, y \ge 0$$

**Ans.** Minimum $Z = 60$, $x = 0, y = 3$

**12.** Maximize $Z = -x + 2y$ subject to the constraints

$$-x + 3y \le 10, \quad x + y \le 6, \quad x - y \le 2, \quad x, y \ge 0$$

**Ans.** Maximum $Z = \frac{20}{3}, x = 0, y = \frac{10}{3}$

**13.** Minimize $Z = x - 5y + 20$ subject to the constraints

$$x - y \ge 0, \quad -x + 2y \ge 2, \quad x \ge 3, \quad y \le 4, \quad x, y \ge 0$$

**Ans.** Minimum $Z = 4$, $x = 4$, $y = 4$

**14.** Maximize $Z = 8x + 6y$ subject to the constraints

$$2x + y \le 1000, \quad x + y \le 800, \quad x \le 400, \quad y \le 700, \quad x, y \ge 0$$

**Ans.** Maxmium $Z$ = Rs. 5,200, $x = 200$, $y = 600$

**15.** Maximize $Z = 2x + y$ subject to constraints

$$5x + 10y \le 50, \quad x + y \ge 1, \quad y \le 4, \quad x - y \le 0, \quad x \ge 0, \quad y \ge 0$$

**Ans.** Maximum $Z = 10; x = \frac{10}{3}, y = \frac{10}{3}$

**16.** Minimize and Maximize

$$Z = 3x + 9y$$

subject to constraints

$$x + 3y \le 60, \quad x + y \ge 10, \quad x \le y, \quad x \ge 0, y \ge 0$$

**Ans.** Minimum $Z = 60$, $x = 5$, $y = 5$
Maximum (Multiple optimal solutions)
$Z = 180$; $x = 15$, $y = 15$ and $x = 0$, $y = 20$.

**17.** Minimize $Z = -50x + 20y$ subject to the constraints

$$2x - y \ge -5, \quad 3x + y \ge 3, \quad 2x - 3y \le 12; x, y \ge 0$$

**Ans.** Minimum $Z = -300$; $x = 6$, $y = 0$.

**18.** Show the solution zone of the following inequalities on a graph paper:

$$5x + y \ge 10, \quad x + y \ge 6, \quad x + 4y \ge 12, \quad x \ge 0, y \ge 0$$

**Ans.** Minimum $Z = 13$, $x = 1$, $y = 5$

**Find $x$ and $y$ for which $3x + 2y$ is minimum subject to these inequalities. Use graphical method.**

**19.** Find the minimum value of $3x + 5y$ subject to the constraints:

$$-2x + y \le 4, \quad x + y \ge 3, \quad x - 2y \le 2, \quad x \ge 0, y \ge 0$$

**Ans.** Minimum $Z = -300$, $x = 6$, $y = 0$

**20.** Find the maximum value of $2x + y$ subject to the constraints:

$$x + 3y \ge 6, \quad x - 3y \le 3, \quad 3x + 4y \le 24$$

$$-3x + 2y \le 6, \quad 5x + y \ge 5, \quad x, y \ge 0$$

**Ans.** Maximum $Z = \frac{43}{3}, x = \frac{84}{13}, y = \frac{15}{3}$.

## 27.7 SOLUTION OF LINEAR PROGRAMMING PROBLEMS

Here we will solve the linear programming problems.

**Working Rule**

**Step 1.** Define the problem mathematically.

**Step 2.** Graph the constraint inequalities by converting them into equations. Find out their respective intercept on both the axes and connect them by straight lines.

**Step 3.** Find out the vertices of the feasible region.

**Step 4.** Find out the value of the objective function on the vertices.

**Step 5.** Find out the optimum value of the objective function.

**Procedure** . The solution of the given LPP should be divided under the following heads:

1. Prepare a *table* of the data given in the problem.
2. Write down the *decision variables*.
3. Form the *objective function*.
4. Write down the constraints.
5. Mathematical formulation.
6. Region represented by inequations.
7. Apply Corner point method/Iso-cost or iso-profit method.

**Type I. To maximize the objective Function (*Z*)**

**Example 1.** *If a young man rides his motor cycle at 25 km/hour he had to spend Rs. 2 per km on petrol. If he rides at a faster speed of 40 km/hour, the petrol cost increases at Rs. 5 per km. He has Rs. 100 to spend on petrol and wishes to find what is the maximum distance he can travel within one hour, express this as an L.P.P. and solve it graphically.*

**Solution.**

The above information are given in the following table :

**1. Table**

| *S.N.* | *Speeds (km per hour)* | *Consumption of petrol per km.* | *Total amount Spent on petrol* |
|---|---|---|---|
| 1. | 25 | Rs. 2 | Rs. 100 |
| 2. | 40 | Rs. 5 | |

**2. Decision Variables:** Let the number of km riding motorcycle at the speed of 25k/h = $x$ km

Let the number of km riding motor cycle at the speed of 40 km/hour = $y$ km

**3. Objective function**

To maximize the distance of the journey.

$$Z = x + y$$

**4. Constraint** (*i*). The young man has Rs. 100 to spend on petrol.

When the speed is 25 km/hour cost of petrol for 1 km = Rs. 2

When the speed is 25 km/hour cost of petrol for $x$ km = Rs. 2 $x$

When the speed is 40 km/hour cost of petrol for 1 km = Rs. 5

When the speed is 40 km/hour cost of petrol for $y$ km = Rs. 5 $y$.

$$\therefore \quad 2x + 5y \leq 100$$

**Constraint** (*ii*) $\qquad \text{Time} = \dfrac{\text{distance}}{\text{speed}}$

Time taken in the first journey $= \dfrac{x}{25}$

Time taken in the second journey $= \frac{y}{40}$

Total time given $= 1$ hour

$\therefore \quad \frac{x}{25}+\frac{y}{40} \le 1$

**Constraint (*iii*)** The distances in the journey are non-negative.

$\therefore \quad x \ge 0$ and $y \ge 0$

**5. Mathematical Formulation**

To maximize, $Z = x + y$ ... (1)

Subject to the constraints :

$2x + 5y \le 100$ ... (2)

$\frac{x}{25}+\frac{y}{40} \le 1 \Rightarrow 8x + 5y \le 200$ ... (3)

$x \ge 0, y \ge 0.$ ... (4)

**6. Region represented by the contraints**

$2x + 5y = 100$

| | | |
|---|---|---|
| $x$ | 50 | 0 |
| $y$ | 0 | 20 |
| *Point* | *A* | *B* |

$8x + 5y = 200$

| | | |
|---|---|---|
| $x$ | 25 | 0 |
| $y$ | 0 | 40 |
| *Point* | *C* | *D* |

We have drawn the graphs of the following lines :

$2x + 5y = 100$

$8x + 5y = 200$

$x = 0$

$y = 0$

Feasible region is represented by the shaded portion *OCEBO*.

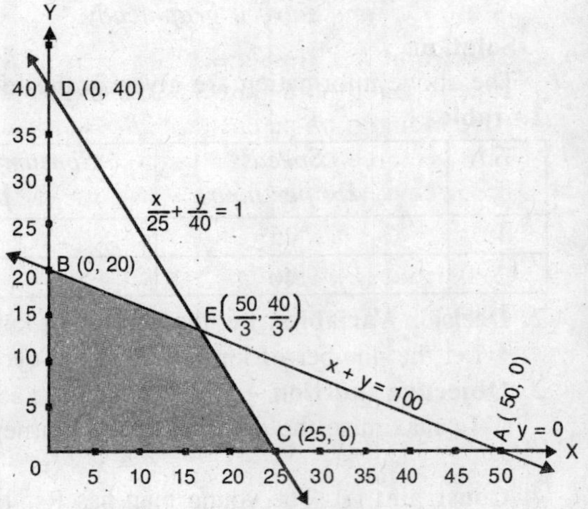


**7. Corner Point Method**

The coordinates of the vertices of feasible region *OCEBO* are *O* (0, 0), *C* (25, 0), $E\left(\frac{50}{3}, \frac{40}{3}\right)$ and *B* (0, 20).

The values of the objective function at these points are as follows :

| *Corner Point (x, y) of the feasible region OCEBO* | *Value of the objective function* $Z = x + y$ |
|---|---|
| $C$ (25, 0) | $25 + 0 = 25$ |
| $E\left(\frac{50}{3}, \frac{40}{3}\right)$ | $\frac{50}{3}+\frac{40}{3}=\frac{90}{3}=30$ **Maximum** |
| $B$ (0, 20) | $0 + 20 = 20$ |

Hence, $Z = 30$ is maximum when $x = \frac{50}{3}$ and $y = \frac{40}{3}$. **Ans.**

**Example 2.** *A factory manufactures two types of screws, A and B; each type requiring the use of two machines, an automatic and a hand operated. It takes 4 minutes on the automatic and 6 minutes on hand operated machines to manufacture a package of screws A, while it takes 6 minutes on automatic and 3 minutes on the hand operated machines to manufacture a package of screws B. Each machine is available for at the most 4 hours on any day. The manufacturer can sell a package of screws A at a profit of Rs. 7 and screws B at a profit of Rs 10. Assuming that he can sell all the screws he manufactures, how many packages of each type should the factory owner produce in a day in order to maximise his profit ? Determine the maximum profit.*

**Solution.** The given data can be put in the tabular form as :

**1. Table:**

| *Screws**Machine* | *Automatic Operated* | *Hand Operated* | *Profit (in Rs.)* |
|---|---|---|---|
| A | 4 min | 6 min | 7 |
| B | 6 min | 3 min | 10 |
| | 4 hours | 4 hours | |

**2. Decision Variables.** Let the manufacturer produce $x$ packages of screws $A$ and $y$ packages of screws $B$ per day respectively.

**3. Objective function.**

The profit on one package of screws $A$ type = Rs. 7

The profit on $x$ packages of screws $A$ type = Rs. $7x$

The profit on one package of screws $B$ type = Rs. 10

The profit on $y$ packages of screws $B$ type = Rs. $10y$

$\therefore$ Total profit = Rs. $(7x + 10y)$

Let $Z$ denotes the maximum profit, then

$$Z = 7x + 10y$$

**4. Constraint (*i*).** The time available on automatic machine is 4 hours i.e., 240 minutes.

Time required on automatic machine for one package of screws $A$ = 4 min.

Time required on automatic machine for $x$ packages of screws $A$ = $4x$ min.

Time required on automatic machine for one package of screws $B$ = 6 min.

Time required on automatic machine for $y$ packages of screws $B$ = $6y$ min.

$$4x + 6y \leq 240$$

**Constraint (*ii*).** Similarly, the time available on hand operated machine is 4 hours *i.e.*, 240 minutes.

$$\therefore \quad 6x + 3y \leq 240$$

**Constraint (*iii*).** Since the number of packages can not be negative, therefore

$x \geq 0$ and $y \geq 0$.

**5. Mathematical formulation.**

So, the mathematical form of the L.P.P. is as follows :

Maximize $\quad Z = 7x + 10y \quad$ ... (1)

subject to the constraints :

$$4x + 6y \leq 240 \quad \text{... (2)}$$

$$6x + 3y \leq 240 \quad \text{... (3)}$$

$$x \geq 0, y \geq 0 \quad \text{... (4)}$$

**6. Region represented by the inequations:**

$4x + 6y = 240$

| $x$ | 60 | 0 |
|---|---|---|
| $y$ | 0 | 40 |
| Points | $A$ | $B$ |

$6x + 3y = 240$

| $x$ | 40 | 0 |
|---|---|---|
| $y$ | 0 | 80 |
| Points | $C$ | $D$ |

We draw the lines $4x + 6y = 240$ and $6x + 3y = 240$ on suitable scales:

(*i*) Put $x = 0$, $y = 0$ in $4x + 6y \le 240$, then $0 + 0 \le 240$, which is true. We shade the half plane containing the origin.

(*ii*) Put $x = 0$, $y = 0$ is $6x + 3y \le 240$, then $0 + 0 \le 240$, which is true. We shade the half plane containing the origin.

The shaded region is represented by the inequations (2) to (4).

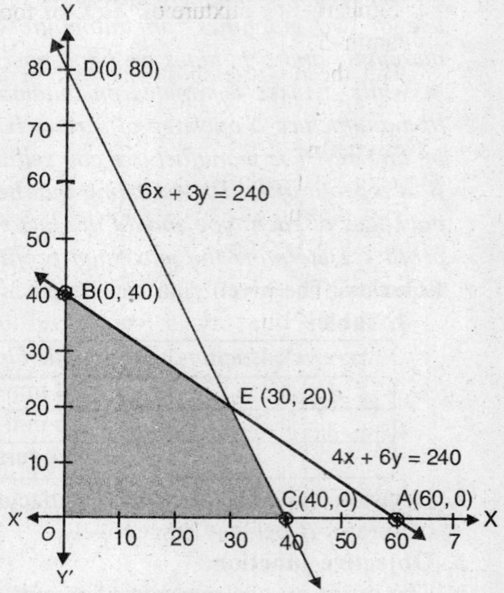


**7. Corner Point Method**

The coordinates of the vertices *O*, *C*, *E* and *B* of the feasible region *OCEBO* are *O* (0, 0), *C* (40, 0), *E* (30, 20) and *B* (0, 40). The coordinates of *E* are obtained by solving the equations $4x + 6y = 240$ and $6x + 3y = 240$.

The value of the objective function at these points are given in the following table.

| *Corner point (x, y) of the feasible region OCEBO.* | *Value of the objective function* $Z = 7x + 10y$ |
|---|---|
| *C* (40, 0) | $7(40) + 10(0) = 280$ |
| *E* (30, 20) | $7(30) + 10(20) = 410$ **Maximum** |
| *B* (0, 40) | $7(0) + 10(40) = 400$ |

Hence, *Z* is maximum when $x = 30$ and $y = 20$ and the maximum value of *Z* is Rs. 410.

**Ans.**

**Example 3.** *A dietician wishes to mix together two kinds of food X and Y in such a way that the mixture contains atleast 10 units of vitamin A, 12 units of vitamin B and 8 units of vitamin C. The vitamin contents of one kg. food is given below:*

| *Food* | *Vitamin A* | *Vitamin B* | *Vitamin C* | *Cost per kg.* |
|---|---|---|---|---|
| *X* | 1 | 2 | 3 | 16 |
| *Y* | 2 | 2 | 1 | 20 |
| *Mixture* | 10 | 12 | 8 | |

*One kg. of food X costs Rs. 16 and one kg. of food Y costs Rs. 20. Find the least cost of the mixture which will produce the required diet?*

**Solution.**

**1. Decision variables.**

Let *x* kg. of food *X* and *y* kg of food *Y* are mixed together to make the mixture.

**2. Objective function.**

It is given that one kg. of food *X* costs Rs. 16 and one kg. of food *Y* cost Rs. 20. So, *x* kg. of food *X* and *y* kg of food *Y* will cost Rs. $(16x + 20y)$.

$$\Rightarrow Z = 16x + 20y$$

**3. Constraint (*i*)** Since one kg. of food *X* contains one unit of vitamin *A*

$\therefore$ *x* kg. of food *X* contains *x* units of vitamin *A*

Since one kg. of food *Y* contains 2 units of vitamin *A*

$\therefore$ *y* kg. of food *Y* contains $2y$ units of vitamin *A*

Therefore, the mixture contains $x + 2y$ units of vitamin *A*. But the mixture should contain at least 10 units of vitamin *A*.

$$\therefore \qquad x + 2y \ge 10$$

**Constraint (*ii*).**

Similarly, the mixture of $x$ kg. of food $X$ and $y$ kg. of food $Y$ contains $(2x + 2y)$ units of vitamin $B$.

But the mixture should contain at least 12 units of vitamin $B$.

$\therefore \qquad 2x + 2y \geq 12$

**Constraint (*iii*).**

$x$ kg. of food $X$ and $y$ kg. of food $Y$ contains $3x + y$ units of vitamin $C$.

But the mixture should contain at least 8 units of vitamin $C$.

$\therefore \qquad 3x + y \geq 8.$

**Constraint (*iv*).**

Since the quantity of food $X$ and food $Y$ can not be negative:

$x \geq 0$ and $y \geq 0.$

**4. Mathematical Formulation.**

Thus the given L.P.P. is

Minimize $Z = 16x + 20y$ ...(1)

Subject to $x + 2y \geq 10$ ...(2)

$2x + 2y \geq 12$ ...(3)

$3x + y \geq 8$ ...(4)

and $x \geq 0,\quad y \geq 0$ ...(5)

**5. Region Represented by the inequations.**

To solve this L.P.P. we draw the lines

$x + 2y = 10$ $\qquad 2x + 2y = 12$ $\qquad$ and $3x + y = 8$

| $x$ | 10 | 0 |
|---|---|---|
| $y$ | 0 | 5 |
| *Points* | $A$ | $B$ |

| $x$ | 6 | 0 |
|---|---|---|
| $y$ | 0 | 6 |
| *Points* | $C$ | $D$ |

| $x$ | $\frac{8}{3}$ | 0 |
|---|---|---|
| $y$ | 0 | 8 |
| *Points* | $E$ | $F$ |

The feasible region of the L.P.P. is shaded in the adjoining figure.

**6. Corner Point Method**

The coordinates of the corner points are $A(10, 0)$, $Q(2, 4)$, $P(1, 5)$ and $F(0, 8)$.

Now, we evaluate $Z = 16x + 20y$ at the corner points.

| *Corner points of the feasible region* | *Corresponding values of* $Z = 16x + 20y$ |
|---|---|
| $A$ (10, 0) | $Z = 16(10) + 20(0) = 160$ |
| **$Q$ (2, 4)** | **$Z = 16(2) + 20(4) = 112$** |
| $P$ (1, 5) | $Z = 16(1) + 20(5) = 116$ |
| $F$ (0, 8) | $Z = 16(0) + 20(8) = 160$ |

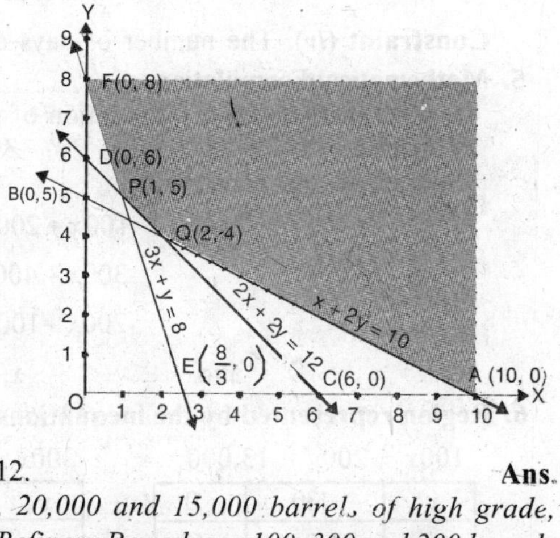


The minimum value of $Z = 112$ at the point $x = 2,\ y = 4$.

Hence, the least cost of the mixture is Rs. 112. **Ans.**

**Example 4.** *An oil company requires 13,000, 20,000 and 15,000 barrels of high grade, medium grade and low grade oil respectively. Refinery P produces 100, 300 and 200 barrels per day of high, medium and low grade oil respectively whereas the Refinery Q produces*

*produces 200, 400 and 100 barrels per day respectively. If P costs Rs. 400 per day and Q costs Rs. 300 per day to operate, how many days should each be run to minimise the cost of requirement?*

**Solution.** The given data can be put in the tabular form as:

**1. Table:**

| Refinery / Grade | P | Q | Minimum Requirement |
|---|---|---|---|
| High grade | 100 | 200 | 13,000 |
| Medium grade | 300 | 400 | 20,000 |
| Low grade | 200 | 100 | 15,000 |
| Cost per day | Rs. 400 per day | Rs. 300 per day | |

**2. Decision variables.**

Let the refineries $A$ and $B$ should run for $x$ and $y$ days respectively to minimize the total cost.

**3. Objective function.**

The cost of refinery $P$ running for one day = Rs. 400.

The cost of refinery $P$ running for $x$ days = Rs. $400x$.

The cost of refinery $Q$ running for one day = Rs. 300

The cost of refinery $Q$ running for $y$ days = Rs. $300y$

$\therefore$ Total cost = $400x + 300y$

Let $Z$ denote the minimum cost, then

$$Z = 400x + 300y$$

**4. Constraint (*i*).**

Since the minimum daily requirement of high grade oil is 13,000 barrels, then

$$100x + 200y \geq 13{,}000$$

**Constraint (*ii*).**

Since the minimum daily requirement of medium grade oil is 20,000 barrels, then

$$300x + 400y \geq 20{,}000$$

**Constraint (*iii*).**

Since, the minimum daily requirement of low grade oil is 15,000 barrels, then

$$200x + 100y \geq 15{,}000$$

**Constraint (*iv*).** The number of days can not be negative. *i.e.*, $x \geq 0,\ y \geq 0$

**5. Mathematical Formulation.**

So, the mathematical formulation of given L.P.P. is as follows:

Minimize $Z = 400x + 300y$ ...(1)

Subject to the constraints:

$$100x + 200y \geq 13{,}000 \quad \text{...(2)}$$

$$300x + 400y \geq 20{,}000 \quad \text{...(3)}$$

$$200x + 100y \geq 15{,}000 \quad \text{...(4)}$$

and $x, y \geq 0$ ...(5)

**6. Region represented by the inequations.**

$100x + 200y = 13{,}000$

| $x$ | 130 | 0 |
|---|---|---|
| $y$ | 0 | 65 |
| Point | $A$ | $B$ |

$300x + 400y = 20{,}000$

| $x$ | 200/3 | 0 |
|---|---|---|
| $y$ | 0 | 50 |
| Point | $C$ | $D$ |

$200x + 100y = 15{,}000$

| $x$ | 75 | 0 |
|---|---|---|
| $y$ | 0 | 150 |
| Point | $E$ | $F$ |

We draw the lines $100x + 200y = 13,000$; $300x + 400y = 20,000$ and $300x + 100y = 15,000$ on suitable scales.

Then the shaded region $AHF$ will be represented by the inequations (2) to (5).

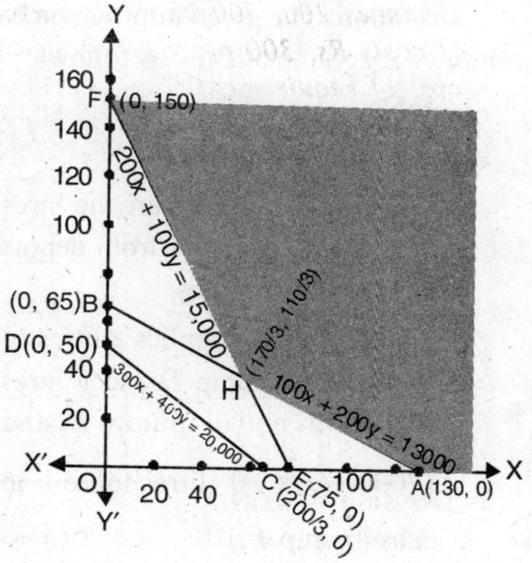


**7. Corner Point Method:** The coordinates of the vertices $A$, $H$ and $F$ of the feasible region $AHF$ are $A(130,0)$, $H\left(\frac{170}{3}, \frac{110}{3}\right)$ and $F(0, 150)$ respectively.

Note that the coordinates of $H$ are obtained by solving the equations

$100x + 200y = 13000$ and $200x + 100y = 15000$.

The value of the objective function at these points are given in the following table:

| *Corner Point (x, y) of the feasible region* | *Value of the objective function* $Z = 400x + 300y$ |
|---|---|
| $A(130, 0)$ | $400 \times 130 + 300(0) = 52000$ |
| $\boldsymbol{H\left(\frac{170}{3}, \frac{110}{3}\right)}$ | $400\times\frac{170}{3}+300\times\frac{110}{3}=\frac{68000}{3}+\frac{33000}{3}$ $= \boldsymbol{\frac{101000}{3} = 33666.67}$ |
| $F(0,150)$ | $400 \times 0 + 300 \times 150 = 45000$ |

Clearly, cost is minimum when $x = \frac{170}{3}$, $y = \frac{110}{3}$. Thus, $x = \frac{170}{3}$, $y = \frac{110}{3}$ is the optimal solution of the given L.P.P.

Hence, the total cost is minimum when the refinery $P$ will run for $x = \frac{170}{3}$ days and refinery $Q$ will run for $y = \frac{110}{3}$ days and the minimum cost = Rs. 33,666.67. **Ans.**

## TYPE III. TRANSPORTATION PROBLEMS

**Example 5.** *An oil company has two depots A and B with capacities of 7000 l and 4000 l respectively. The company is to supply oil to three petrol pumps, D, E and F, whose requirements are 4500 l, 3000 l and 3500 l respectively. The distances (in km) between the depots and the petrol pumps is given in the following table:*

*Distance in (km.)*

| *From/To* | *A* | *B* |
|---|---|---|
| *D* | 7 | 3 |
| *E* | 6 | 4 |
| *F* | 3 | 2 |

*Assuming that the transportation cost of 10 litres of oil is Re. 1 per km, how should the delivery be scheduled in order that the transportation cost is minimum ? What is the minimum cost?*

**Solution.** The given information can be exhibited diagrammatically as follows:

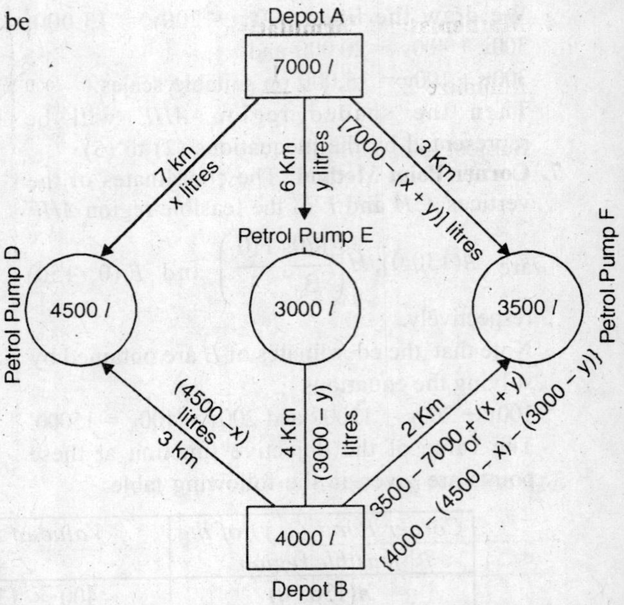


1. **Table.**
2. **Decision Variables.**

Decision variables are the litres of oil to be supplied from depots to the petrol pumps.

Let Depot $A$ supplies $x$ litres of oil to petrol pump $D$ and $y$ litres of oil to petrol pump $E$ and $[7000-(x+y)]$ litres of oil to petrol pump $F$.

**Constraint (*i*).** $7000-(x+y)\geq 0$

$$\Rightarrow \quad x+y\leq 7000$$

$$x\geq 0,\ y\geq 0$$

**Constraint (*ii*)** The remaining requirement $(4500-x)$ litres of oil is supplied from the depot $B$ to the petrol pump $D$, $(3000-y)$ litres of oil to petrol pump $E$ and $4000-[(4500-x)+(3000-y)]$ litres of oil to petrol pump $F$.

$$4500-x\geq 0 \quad\Rightarrow\quad x\leq 4500$$

$$3000-y\geq 0 \quad\Rightarrow\quad y\leq 3000$$

$$4000-4500+x-3000+y\geq 0 \quad\Rightarrow\quad x+y\geq 3500$$

3. **Objective function.**

The transportation cost from depot $A$ to the petrol pump $D$, $E$ and $F$ is Rs. $\frac{x}{10}$, Rs. $\frac{6y}{10}$ and $Rs.\ \frac{3}{10}(7000-x-y)$.

Similarly, the transportation cost from depot $B$ to petrol pump $D$, $E$ and $F$ is $Rs.\ \frac{3}{10}(4500-x)$, $Rs.\ \frac{4}{10}(3000-y)$ and Rs. $\frac{2}{10}[x+y-3500]$.

So, the total cost of transportation $= Z$

$$\Rightarrow \quad Z=\text{Rs.}\left(\frac{7x}{10}+\frac{6y}{10}+\frac{3}{10}(7000-x-y)+\frac{3}{10}(4500-x)+\frac{4}{10}(3000-y)+\frac{2}{10}(x+y-3500)\right)$$

$$\Rightarrow \quad Z=\left(\frac{7}{10}-\frac{3}{10}-\frac{3}{10}+\frac{2}{10}\right)x+\left(\frac{6}{10}-\frac{3}{10}-\frac{4}{10}+\frac{2}{10}\right)y+2100+1350+1200-700$$

$$=\text{Rs.}\left(\frac{3}{10}x+\frac{1}{10}y+3950\right)$$

**4. Mathematical formulation.**

Minimize $Z = \frac{3}{10}x + \frac{1}{10}y + 3950$ ...(1)

Subject to constraints

$x + y \leq 7000$ ...(2)

$x + y \geq 3500$ ...(3)

$x \leq 4500$ ...(4)

$y \leq 3000$ ...(5)

$x \geq 0,\ y \geq 0$ ...(6)

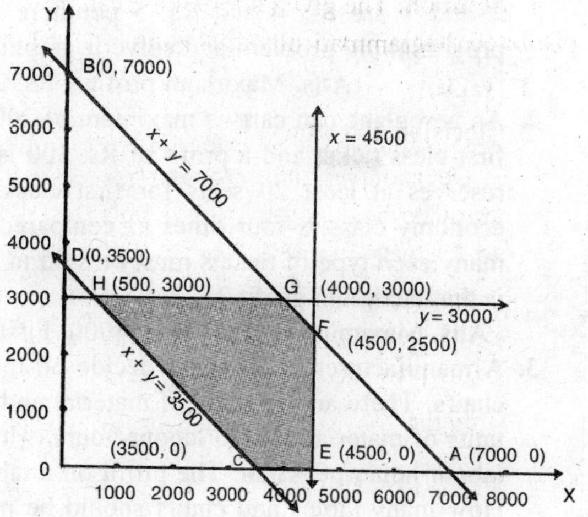


**5. Region represented by the inequations.**

$x + y = 7000$

| $x$ | 7000 | 0 |
|---|---|---|
| $y$ | 0 | 7000 |
| *Point* | *A* | *B* |

$x + y = 3500$

| $x$ | 3500 | 0 |
|---|---|---|
| $y$ | 0 | 3500 |
| *Point* | *C* | *D* |

The shaded region *C E F G H C* represented by the inequations (2) to (7).

**6. Corner Point method.**

The coordinates of the vertices *C, E, F, G, H* and *C* of the feasible region *CEFGHC* are *C*(3500, 0), *E*(4500, 0), *F*(4500, 2500), *G*(4000, 3000) and *H* (500, 3000) respectively. The coordinates of *F* are obtained by solving $x + y = 7000$ and $x = 4500$, coordinates of *G* are obtained by solving $x + y = 7000$ and $y = 3000$, the coordinates of *H* are obtained by solving $x + y = 3500$ and $y = 3000$. The value of the objective function at these points are given in the following table.

| *Corner points (x, y) of the feasible region C E F G H C* | *Value of the objective function* $Z = \frac{3}{10}x + \frac{1}{10}y + 3950$ |
|---|---|
| *C* (3500, 0) | $\frac{3}{10}(3500) + \frac{1}{10}(0) + 3950 = 5000$ |
| *E* (4500, 0) | $\frac{3}{10}(4500) + \frac{1}{10}(0) + 3950 = 5300$ |
| *F* (4500, 2500) | $\frac{3}{10}(4500) + \frac{1}{10}(2500) + 3950 = 5550$ |
| *G* (4000, 3000) | $\frac{3}{10}(4000) + \frac{1}{10}(3000) + 3950 = 5450$ |
| ***H* (500, 3000)** | $\mathbf{\frac{3}{10}(500) + \frac{1}{10}(3000) + 3950 = 4400}$ |

Hence, *Z* is minimum when $x = 500$ and $y = 3000$; and the minimum value of *Z* is 4400. **Ans.**

## EXERCISE 27.3

**1.** A manufacturer produces two items *X* and *Y*. *X* needs two hours on machine *A* and 2 hours on machine *B*. *Y* needs 3 hours on machine *A* and 1 hour on machine *B*. If machine *A* can run for a maximum of 12 hours per day and machine *B for* 8 hours per day and profits from

$X$ and $Y$ are Rs. 4 and Rs. 5 per item respectively. Formulate the problem as a linear programming problem and solve it graphically.

**Ans.** Maximum profit = Rs. 22, Number of item $x = 3$, Number of item $y = 2$.

2. An aeroplane can carry a maximum of 200 passengers. A profit of Rs. 400 is made on each first class ticket and a profit of Rs. 300 is made on each economy class ticket. The airline reserves at least 20 seats for first class. However, Passengers who prefer to travel by economy class is four times as compared to passengers to the first class. Determine how many each type of tickets must be sold in order to maximize the profit for the airline. What is the maximum profit ?

**Ans.** Maximum profit = Rs. 64000, First class tickets = 40. Economy class tickets = 160.

3. A manufacturer is trying to decide on the product quantities of two products, tables and chairs. There are 98 units of material and 80 labour-hours available. Each table requires 7 units of material and 10 labour-hours, while each chair requires 14 units of material and 8 labour-hours per chair. The profit on a table and a chair is Rs. 25 and Rs. 20, respectively. How many tables and chairs should be produced to have maximum profit? (**Hint**: Use Iso-profit method).

**Ans.** Maximum profit = Rs. 200, Tables = 8, chairs = 0 or Tables = 4, chairs = 5.

4. A firm manufactures two products, $X$ and $Y$, each requiring the use of three machines $M_1$, $M_2$ and $M_3$. The time required for each product in hours and total time available in hours on each machine are as follows:

| *Machine* | *Product X* | *Product Y* | *Available time (in hours)* |
|---|---|---|---|
| $M_1$ | 2 | 1 | 70 |
| $M_2$ | 1 | 1 | 40 |
| $M_3$ | 1 | 3 | 90 |

If the profit is Rs. 40 per unit for product $X$ and Rs. 60 per unit for product $Y$. How many units of each product should be manufactured to maximize profit ?

**Ans.** Maximum profit = Rs. 2100, Product $x = 15$, Product $y = 25$.

5. A company manufactures two types of novelty souvenirs made of plywood. Souvenirs of type A require 5 minutes each for cutting and 10 minutes each for assembling. Souvenirs of type B require 8 minutes each for cutting and 8 minutes each for assembling. There are 3 hours 20 minutes available for cutting and 4 hours for assembling. The profit is 50 paise each for type A and 60 paise each for type B souvenirs. How many souvenirs of each type should the company manufacture in order to maximize the profit ?

**Ans.** Maximum profit = Rs. 16, Type A = 8, Type B = 20.

6. A factory owner purchases two types of machines, A and B, for his factory. The requirements and limitations for the machines are as follows:

| *Machine* | *Area occupied by the machine* | *Labour force for each machine* | *Daily output in units* |
|---|---|---|---|
| *A* | 1000 sq. m | 12 men | 60 |
| *B* | 1200 sq. m | 8 men | 40 |

He has an area of 9000 sq. m available and 72 skilled men who can operate the machines. How many machines of each type should he buy to maximize the daily output ?

**Ans.** Maximum output = 360 units (*i*) Type A = 6, type B = 0.
(*ii*) Type A = 4, type B = 3.

7. A manufacturer makes two products. A and B. Product A sells at Rs. 200 each and takes $\frac{1}{2}$ hour to make. Product B sells at Rs. 300 each and takes 1 hour to make. There is a

permanent order for 14 units of product A and 16 units of product B. A working week consists of 40 hours and the weekly turnover must not be less than Rs. 10000. If the profit on each of product A is Rs. 20 and on product B is Rs. 30, then how many of each should be produced so that the profit is maximum ? Also find the maximum profit.

**Ans.** Maximum profit = Rs. 1440, Product A = 48, Product B = 16.

**8.** A toy company manufactures two types of dolls, A and B. Each doll of type B takes twice as long as to produce one of type A. If the Company produces only type A, it can make a maximum of 2000 dolls per day. The supply of plastic is sufficient to produce 1500 dolls per day. Type *B* requires a fancy dress which cannot be available for more than 600 dolls per day. If the company makes profits of Rs. 3 and Rs. 5 per doll respectively on dolls A and B, how many of each should be produced per day in order to maximize the profit ?

**Ans.** Maximum profit = Rs. 5500, Doll A = 1000, Doll B = 500.

**9.** A shopkeeper deals in two items, thermosflasks and air tight containers. A flask costs him Rs. 120 and an air tight container costs him Rs. 60. He has at the most Rs. 12,000 to invest and has space to store a maximum of 150 items. The profit on selling a flask in Rs. 20 and an air tight container is Rs. 15. Assuming that he will be able to sell all things he buys, how many of each item should he buy to maximise his profit? Solve the problem graphically.

**Ans.** Maximum profit = Rs. 2500, Flasks = 50, Containers = 100

**10.** Sudhanshu wants to invest atmost Rs. 12000 in Saving Certificate and Bonds. According to rules, he has to invest atleast Rs. 2000 in Certificates and atleast Rs. 4000 in Bonds. If the rates of interest in Certificates and Bonds are 8% and 10% p.a. respectively, how much money should he invest to earn maximum yearly income. Find also his maximum yearly income.

**Ans.** Certificates : Rs. 2000; Bonds: Rs. 10,000; Income: Rs. 1160.

**11.** A manufacturer has 3 machines installed in his factory. Machines I and II are capable of being operated for atmost 12 hours whereas machine III must operate atleast for 5 hours a day. He produces only two items, each requiring the use of three machines.

The number of hours required for producing 1 unit each of the items on the three machines is given in the following table:

| | *Number of hours required by the machines* | | |
|---|---|---|---|
| *Item* | *I* | *II* | *III* |
| *A* | 1 | 2 | 1 |
| *B* | 2 | 1 | 1.25 |

He makes a profit of Rs. 6 and Rs. 4 on items A and B respectively. Assuming that he can sell all that he produces, how many of each item should he produce so as to maximize his profit ? Determine his maximum profit.

**Ans.** 4 units of A; 4 units of B; Rs. 40

**12.** A company manufactures two types of toys–A and B. Toy A requires 4 minutes for cutting and 8 minutes for assembling and Toy B requires 8 minutes for cutting and 8 minutes for assembling. There are 3 hours and 20 minutes available in a day for cutting and 4 hours for assembling. The profit on a piece of toy A is Rs. 50 and that on toy B is Rs. 60. How many toys of each type should be made daily to have maximum profit? Solve the problem graphically.

**Ans.** Maximum profit = Rs. 1700, Toys A = 10, Toys B = 20

**13.** A firm manufactures two types of products A and B and sells them at a profit of Rs. 5 per unit of type A and Rs. 3 per unit of type B. Each product is processed on two machines $M_1$ and $M_2$. One unit of type A requires one minute of processing time on $M_1$ and two minutes of processing time on $M_2$; whereas one unit of type B requires one minute of processing time on $M_1$ and one minute on $M_2$. Machines $M_1$ and $M_2$ are respectively available for atmost 5 hours and 6 hours in a day. Find out how many units of each type of product should the firm produce a day in order to maximize the profit. Solve the problem graphically.

**Ans.** Maximum profit = Rs. 1020, 60 units of type A and 240 units of type B.

**14.** A company manufactures two articles A and B. There are two departments through which these articles are processed : (*i*) assembly and (*ii*) finishing departments. The maximum capacity of the first department is 60 hours a week and that of the other department is 48 hours a week. The production of each article A requires 4 hours in asembly and 2 hours in finishing and that of each unit of B requires 2 hours in assembly and 4 hours in finishing. If the profit is Rs. 6 for each unit of A and Rs. 8 for each unit of B, find the number of units of A and B to be produced per week in order to have maximum profit.

**Ans.** Maximum profit = Rs. 120, Articles A = 12, Articles B = 6

**15.** A factory owner wants to purchase two types of machines, A and B, for his factory. The machine A requires an area of 1000 $m^2$ and 12 skilled men for running it and its daily output is 50 units, whereas the machine B requires 1200 $m^2$ area and 8 skilled men, and its daily output is 40 units. If an area of 7600 $m^2$ and 72 skilled men be available to operate the machine, how many machines of each type should be bought to maximise the daily output?

**Ans.** Maximum output = 320, machine A = 4, machine B = 3

**16.** A manufacturer makes two types of cups, A and B. Three machines are required to manufacture the cups and the time in minutes required by each is as given below:

| *Type of Cup* | *Machines* | | |
|---|---|---|---|
| | *I* | *II* | *III* |
| *A* | 12 | 18 | 6 |
| *B* | 6 | 0 | 9 |

Each machine is available for a maximum period of 6 hours per day. If the profit on each cup A is 75 paise, and on B it is 50 paise, show that 15 cups of type A and 30 cups of type B should be manufactured per day to get the maximum profit.

**Ans.** Maximum profit = Rs. 26.25, Type A = 15, Type B = 30 cups

**17.** A company manufactures two types of toys A and B. Type A requires 5 minutes each for cutting and 10 minutes each for assembling. Type B requires 8 minutes each for cutting and 8 minutes each for assembling. There are 3 hours available for cutting and 4 hours available for assembling in a day. The profit is Rs. 50 each on type A and Rs. 60 each on type B. How many toys of each type should the company manufacture in a day to maximise the profit ?

**Ans.** Maximum profit = Rs. 1500, Type A = 12, Type B = 15.

**18.** A producer has 20 and 10 units of labour and capital respectively which he can use to produce two kinds of goods X and Y. To produce one unit of goods X, 2 units of capital and 1 unit of labour is required. To produce one unit of goods Y, 3 units of labour and 1 unit of capital is required. If X and Y are priced at Rs. 80 and Rs. 100 per unit respectively, how should the producer use his resources to maximize the total revenue? Solve the problem graphically.

**Ans.** Maximum Revenue: Rs. 760; X : 2 units; Y : 6 units.

**19.** A farm is engaged in breeding goats. The goats are fed on various products grown on the farm. They need certain nutrients, named as X, Y and Z. The goats are fed on two products A and B. One unit of product A contains 36 units of X, 3 units of Y and 20 units of Z, while one unit of product B contains 6 units of X, 12 units of Y and 10 units of Z. The minimum requirement of X, Y and Z is 108 units, 36 units and 100 units respectively. Product A costs

Rs. 20 per unit and product B costs Rs. 40 per unit. How many units of each product must be taken to minimize the cost?

**Ans.** Minimum cost = Rs. 160, Product A = 4 units, Product B = 2 units.

**20.** A company producing soft drinks has a contract which requires that a minimum of 80 units of chemical A and 60 units of chemical B are to go in each bottle of the drink. The chemicals are available in a prepared mix from two different suppliers. Supplier X has a mix of 4 units of A and 2 units of B that costs Rs. 10 and the supplier *Y* has a mix of 1 unit of A and 1 unit of *B* that costs Rs. 4. How many mixes from X and Y should the company purchase to honour contract requirement and yet minimize the cost?

**Ans.** Minimum cost = Rs. 260. Mix of type A = 10 units, Mix of type B = 40 units.

**21.** To maintain one's health, a person must fulfil minimum daily requirements for the following three nutrients–calcium, protein and calories. His diet consists of only food items I and II whose prices and nutrient contents are shown below:

| *Price* | *Food I* *Rs. 0.60 per unit* | *Food II* *Re. 1 per unit* | *Minimum requirements* |
|---|---|---|---|
| Calcium | 10 | 4 | 20 |
| Protein | 5 | 5 | 20 |
| Calories | 2 | 6 | 12 |

Find the combination of food items so that the cost may be minimum.

**Ans.** Minimum cost = Rs. 2.80, Food I = 3 units, Food II = 1 unit.

**22.** A diet for a sick person must contain at least 4000 units of vitamin, 50 units of minerals and 1400 units of calories. Two foods A and B are available at a cost of Rs. 4 and Rs. 3 per unit respectively. If one unit of A contains 200 units of vitamin, 1 unit of mineral and 40 units of calories and one unit of food B contains 100 units of vitamin, 2 units of minerals and 40 units of calories, find what combination of foods should be used to have the least cost?

**Ans.** Minimum cost = Rs. 110, Food A = 5 units, Food B = 30 units.

**23.** A dietician wishes to mix two types of food in such a way that the vitamin contents of the mixture contain at least 8 units of vitamin A and 10 units of vitamin C. Food 1 contains 2 units/ kg of vitamin A and 1 unit /kg of vitamin C while food II contains I unit/kg of vitamin A and 2 units/kg of vitamin C. It costs Rs. 5 per kg to purchase food I and Rs. 7 per kg to purchase food II. Determine the minium cost of such a mixture.

**Ans.** Minimum cost = Rs. 38, Food I = 2 kg, Food II = 4 kg.

**24.** A medical company has factories at two places, A and B. From these places, supply is made to each of its three agencies situated at P, Q and R. The monthly requirements of the agencies are, respectively, 40, 40 and 50 packets of the medicines, while the production capacity of factories, A and B are 60 and 70 packets, respectively. The transportation cost per packet from the factories to the agencies are given below:

**Transportation cost per packet (in Rs.)**

| *To* \ *From* | *A* | *B* |
|---|---|---|
| P | 5 | 4 |
| Q | 4 | 2 |
| R | 3 | 5 |

How many packets from each factory be transported to each agency so that the cost of transportation is minimum? Also find the minimum cost.

**Ans.** Minimum cost = Rs. 400, From A : 10 packets, 0 packets, 50 packets to P, Q, R, respectively, From B : 30 packets, 40 packets and 0 packets to P, Q, R, respectively.

**25.** Two godowns A and B have grain storage capacity of 100 quintals and 50 quintals respectively. They supply to three ration shopes D, E and F whose requirements are 60, 50 and 40 quintals respectively. The cost of transportation per quital from the godown to the shops are given in the following table:

| From / To | Godown A | Godown B |
|---|---|---|
| D | 6.00 | 4.00 |
| E | 3.00 | 2.00 |
| F | 2.50 | 3.00 |

How should the supplies we transported in order that the transportation cost is minimum.

**Ans.** From A : 10 quintals, 50 quintals and 40 quintals to D, E and F respectively
From B : 50 quintals, 0 quintal and 0 quintal to D, E and F respectively.

**26.** There is a factory located at each of the two places *P* and *Q*. From these locations, a certain commodity is delivered to each of these depots situated at *A*, *B* and *C*. The weekly requirements of the depots are respectively 5, 5 and 4 units of the commodity while the production capacity of the factories at *P* and *Q* are respectively 8 and 6 units. The cost of transportation per unit is given below:

| To / From | Cost (in Rs.) A | B | C |
|---|---|---|---|
| *P* | 16 | 10 | 15 |
| *Q* | 10 | 12 | 10 |

How many units should be transported from each factory to each depot in order that the transportation cost is minimum. Formulate the above L.P.P mathematically and then solve it.

**Ans.** Minimum cost = Rs. 155,
From *P* : 0, 5, 3 units to depots at A, B, C respectively
From *Q* : 5, 0 and 1 units to depots at A, B and C respectively.

**27.** A brick manufacturer has two depots, A and B with stock of 30,000 and 20,000 bricks respectively. He receives orders from three builders *P*, *Q* and *R* for 15,000, 20,000 and 15,000 bricks respectively. The cost in Rs. transporting 1000 bricks to the builders from the depots are given below:

| From / To | *P* | *Q* | *R* |
|---|---|---|---|
| *A* | 40 | 20 | 30 |
| *B* | 20 | 60 | 40 |

How should the manufacturer fulfil the orders so as to keep the cost of transportation minimum.

**Ans.** Minimum cost = Rs. 1200
From *A* : 0, 20 and 10 thousand bricks to builders P, Q and R.
From *B* : 15, 0 and 5 thousand bricks to builders P, Q and R.

**28.** A firm manufacturers headache pills in two sizes A and B. Size A contains 2 grains of aspirin 5 grains of bicarbonate and 1 grain of codeine; size *B* contains 1 grain of aspirin, 8 grains of bicarbonate and 66 grains of codeine. It has been found by users that it requires at least 12 grains of aspirin, 7.4 grains of bicarbonate and 24 grains of codeine for providing immediate relief. Determine graphically the least number of pills a patient should have to get immediate relief. Determine also the quantity of codeine consumed by patient.

**Ans.** 2 pills of size A, 8 pills of size B; Quantity of codeine = 50 grains

**29.** A manufacturer of patient medicines is preparing a production plan on medicines A and B. There are sufficient raw materials available to make 20,000 bottles of A and 40,000 bottles of B, but there are only 45000 bottles into which either of the medicines can be bottled. Further, it takes 3 hours to prepare enough material to fill 1000 bottles of A, it takes 1 hour to prepare enough material to fill 1000 bottles of B and there are 66 hours available for this operation. The profit is Rs. 8 per bottle for A and Rs. 7 per bottle for B. How should the manufacturer schedule his production in order to maximize his profit?

**Ans.** Maximum profit = Rs. 3,25,500, 10,500 bottles of A, 34,500 bottles of B.

**30.** A dietician has to develop a special diet using two foods P and Q. Each packet of food P contains 12 units of calcium, 4 units of iron, 6 units of cholesterol and 6 units of vitamin A, while each packet of the same quantity of food Q contains 3 units of calcium, 20 units of iron, 4 units of cholesterol, and 3 units of vitamin A. The diet requires atleast 240 units of calcium, atleast 460 units of iron and at most 300 units of cholesterol. How many packets of each food should be used to minimize the amount of vitamin A in the diet? What is the minimum amount of vitamin A?

**Ans.** Minimum amount of vitamin A = 150 units;
15 packets of food P
20 packets of food Q.

**31.** A manufacturing company makes two models A and B of a product. Each piece of Model A requires 9 labour hours for fabricating and 1 labour hour for finishing while each piece of Model B requires 12 labour hours for fabricating and 3 labour hours for finishing. For fabricating and finishing, the maximum labour hours available are 180 and 30 per week respectively. The company makes a profit of Rs. 8000 on each piece of model A and Rs. 12000 on each piece of Model B. How many pieces of Model A and Model B should be manufactured per week to realise maximum profit? What is the maximum profit for week?

**Ans.** Maximum profit = Rs. 1,68,000; 12 pieces of model A and 6 pieces of model B.

## 27.8 SIMPLEX METHOD

We have learnt to solve linear programming problems involving two variables $x$ and $y$ graphically. A linear programming problem involving more than two variables can be solved by algebraic method. This algebraic method is known as Simplex Method.

This method consists of a number of steps. In first step we get the value of $Z$ which is equal to the value at one vertex by graphical solution. In the next step the value of $Z$ will be better than previous one and is equal to at the next adjoining vertex and so on. Since, the number of vertices is finite, the Simplex Method also consists of finite number of steps to get the optimal solution. Consider the following problem:

Maximize $\quad Z = 4x + 5y \qquad$ ...(1)

Subject to the constraints:

$$2x+3y \le 12 \qquad ...(2)$$

$$2x+y \le 8 \qquad ...(3)$$

$$x \ge 0, \quad y \ge 0 \qquad ...(4)$$

Inequalities (1) and (2) are converted into equations by adding non negative quantities $s_1$ and $s_2$ is known as slack variables.

The original variables are decision variables.

$$Z = 4x + 5y + 0\ s_1 + 0\ s_2 = 0 \Rightarrow -4x - 5y - 0s_1 - 0s_2 + Z = 0 \quad ...(5)$$

$$2x + 3y + s_1 = 12 \quad ...(6)$$

$$2x + y + s_2 = 8 \quad ...(7)$$

where, $x \geq 0, \quad y \geq 0, \quad s_1 \geq 0 \quad \text{and} \quad s_2 \geq 0.$

Variables $s_1$ and $s_2$ each occur in exactly one equation and they have a coefficient + 1. We call these basic variables.

| Basic variables are those variables that have a coefficient of + 1 in only one equation and coefficient of zero in the remaining equations. |
|---|

The remaining variables are called non-basic variables (implicit variables).

**Note:** In inequalities sometimes some non-negative variables are subtracted to form equations. These variables are called Surplus Variables.

**Initial Basic Feasible solution:**

On putting decision variables $x$ and $y$ equal to zero in (5), (6) and (7), we get initial basic feasible solution.

$0 + 0 + s_1 = 12 \quad \Rightarrow \quad s_1 = 12$

$0 + 0 + s_2 = 8 \quad \Rightarrow \quad s_2 = 8$

$s_1 = 12$ and $s_2 = 8$

and the objective function $Z = 4x + 5y = 0 + 0 = 0 \quad \Rightarrow \quad Z = 0$

This information is given in the following table:

| | Coefficients of | | | | | |
|---|---|---|---|---|---|---|
| | $x$ | $y$ | $s_1$ | $s_2$ | $Z$ | Value |
| $Z$ | – 4 | – 5 | 0 | 0 | 1 | 0 |
| $s_1$ | 2 | 3 | 1 | 0 | 0 | 12 |
| $s_2$ | 2 | 1 | 0 | 1 | 0 | 8 |

... (8)

**Step 2.** If the $Z$ row of table (8) contains no negative entries of column $x$ and $y$ we get an optimal solution otherwise not.

To increase the value of $Z$ there are two possibilities

(*i*) Put $x = 0$ and $y$ greater than zero in (1) or first row of table (8).

(*ii*) Put $y = 0$ and $x$ greater than zero in (1) or first row of (8).

**To determine which of these alternatives (*i*) and (*ii*) is better**

(*i*) Put $x = 0$ and $y = 1$ in (1) or first row of (8), we get $z = 5$.

(*ii*) Put $y = 0$ and $x = 1$ in (1) or first row of (8), we get $z = 4$.

Thus, the better alternative is to keep $x$ fixed at $x = 0$ and increase $y$.

The variable ($y$) to be increased is called the entering variable. The entering variable is marked with an arrow at the top of the table.

The coefficient of entering variable in the first row of table (8) is the most negative.

**Step 3.** To increase $Z$ the value of entering variable ($y$) should be positive it means $y$ becomes basic variable (since basic variables are positive and non basic variables are zero). Now, a basic variables ($s_1$ or $s_2$) is to be converted as non-basic variable, and is called Departing Variable.

**Selection of departing variable.**

(*i*) In the second row, we find the ratio of the value in the last column and the coefficient of the entering variable *i.e.* $\frac{12}{3}=4$

(*ii*) Similarly, In the third row again we find the ratio of the last value and coefficient of the entering variable; i.e. $\frac{8}{1}=8$.

Here, the ratio are 4 and 8. Since, 4 is smaller non-negative of these ratios and is in the $s_1$-row, therefore the departing variable is $s_1$.

The departing variable is marked with an arrow on the left side of the table.

Entering Variable ↓ (y)

| | | $x$ | $y$ | $s_1$ | $s_2$ | $Z$ | Value | Ratio | |
|---|---|---|---|---|---|---|---|---|---|
| | $Z$ | $-4$ | $-5$ | 0 | 0 | 1 | 0 | | |
| Departing variable → | $s_1$ | 2 | [3] | 1 | 0 | 0 | 12 | $\frac{12}{3}=4$ | ... (9) |
| | $s_2$ | 2 | 1 | 0 | 1 | 0 | 8 | $\frac{8}{1}=8$ | |

The new basic variables are $Z$, $y$ and $s_2$ and non-basic variables are $x$ and $s_1$.

The intersection of column of the entering variable and the row of departing variable gives the pivot entry. Here the pivot entry is 3 which is placed in the box.

The pivot entry is to be converted into 1.

Here, we have to multiply the departing variable-row by $\frac{1}{3}$ to get 1 at the pivot entry.

| | $x$ | $y$ | $s_1$ | $s_2$ | $Z$ | *Value* | |
|---|---|---|---|---|---|---|---|
| $Z$ | $-4$ | $-5$ | 0 | 0 | 1 | 0 | |
| $y$ | $\frac{2}{3}$ | 1 | $\frac{1}{3}$ | 0 | 0 | 4 | ... (10) |
| $s_2$ | 2 | 1 | 0 | 1 | 0 | 8 | |

We multiply the second row by 5 and add to the first row to get 0 above the pivot entry. Again we subtract the second row from the third row to get 0 below the pivot entery.

| | $x$ | $y$ | $s_1$ | $s_2$ | $Z$ | *Value* | |
|---|---|---|---|---|---|---|---|
| $Z$ | $-\frac{2}{3}$ | 0 | $\frac{5}{3}$ | 0 | 1 | 20 | |
| $y$ | $\frac{2}{3}$ | 1 | $\frac{1}{3}$ | 0 | 0 | 4 | ... (11) |
| $s_2$ | $\frac{4}{3}$ | 0 | $-\frac{1}{3}$ | 1 | 0 | 4 | |

Since, in the first row ($Z$ row) there is a negative entry in the $x$-column. So, we have not obtained optimal solution. Since the coefficient of $x$ in the first row is $-\frac{2}{3}$ the most negative. Therefore, $x$ is the new Entering Variable.

The ratio of $y$ in the second row is $\frac{4}{\frac{2}{3}} = 6$.

In the third row the ratio is $\frac{4}{\frac{4}{3}} = 3$.

Here, 3 is the smallest non-negative ratio in the $s_2$-row. Thus, $s_2$ is the new departing variable.

Entering Variable ↓ ($x$)

| | $x$ | $y$ | $s_1$ | $s_2$ | $z$ | | Ratio |
|---|---|---|---|---|---|---|---|
| $Z$ | $-\frac{2}{3}$ | 0 | $\frac{5}{3}$ | 0 | 1 | 20 | |
| $y$ | $\frac{2}{3}$ | 1 | $\frac{1}{3}$ | 0 | 0 | 4 | $\frac{4}{\frac{2}{3}} = 6$ |
| Departing variable → $s_2$ | $\boxed{\frac{4}{3}}$ | 0 | $-\frac{1}{3}$ | 1 | 0 | 4 | $\frac{4}{\frac{4}{3}} = 3$ |

...(12)

Now third row is multiplied by $\frac{3}{4}$ to get 1 at the pivotal entry of $x$ – Column.

| | $x$ | $y$ | $s_1$ | $s_2$ | $z$ | Value |
|---|---|---|---|---|---|---|
| $Z$ | $-\frac{2}{3}$ | 0 | $\frac{5}{3}$ | 0 | 1 | 20 |
| $y$ | $\frac{2}{3}$ | 1 | $\frac{1}{3}$ | 0 | 0 | 4 |
| $s_2$ | 1 | 0 | $-\frac{1}{4}$ | $\frac{3}{4}$ | 0 | 3 |

... (13)

Multiplying third row by $\frac{2}{3}$ and adding to the first row to get zero in the $x$-column. Again we multiply the third row by $-\frac{2}{3}$ and add to second row to get zero in the $x$-column.

| | $x$ | $y$ | $s_1$ | $s_2$ | $Z$ | Value |
|---|---|---|---|---|---|---|
| $Z$ | 0 | 0 | $\frac{3}{2}$ | $\frac{1}{2}$ | 1 | 22 |
| $y$ | 0 | 1 | $\frac{1}{2}$ | $-\frac{1}{2}$ | 0 | 2 |
| $x$ | 1 | 0 | $-\frac{1}{4}$ | $\frac{3}{4}$ | 0 | 3 |

... (14)

Since, $Z$ row of table (14) has no negative entry in the columns of variables. Therefore, this is the case of optimal solution. From the last column of table (14), we have

$$x = 3, \quad y = 2, \quad s_1 = 0, \quad s_2 = 0$$

and the maximum value of $Z = 22$.

**Working Rule of Simplex Method**

**Step 1.** Construct the initial table by putting decision variables equal to zero.

**Step 2.** If we get optimal solution we stop, otherwise proceed to step 3.

**Step 3.** Find the Entering Variable whose coefficient in the *Z*-row is most negative.

**Step 4.** Find the Departing Variable i.e. the basic variable in the row where the quotient is as small as possible, yet non-negative.

**Step 5.** Find the pivotal entry at the intersection of the entering variable column and the departing variable-row.

**Step 6.** Use pivotal element for elemination (to get 0) and construct new table and return step 2.

**Example 1.** *Using Simplex method solve the L.P.P.*

*Maximize* $Z = 8x + 6y$ ...(1)

*subject to* $8x + 4y \le 18$ ...(2)

$x \le 2$ ...(3)

$y \le 1.25$ ...(4)

$x, y \ge 0.$

**Solution.** Inequalities (2), (3) and (4) are converted into equations by adding non-negative quantities $s_1$, $s_2$, $s_3$.

$$-8x - 6y + 0s_1 + 0s_2 + 0s_3 + Z = 0$$

$$8x + 4y + s_1 = 18 \quad ...(5)$$

$$x + s_2 = 2 \quad ...(6)$$

$$y + s_3 = 1.25 \quad ...(7)$$

where $x, y \ge 0$, $s_1 \ge 0$, $s_2 \ge 0$, $s_3 \ge 0$.

If $s_1$, $s_2$ and $s_3 = 0$ the optimal solution is

$$x = 2, y = 1.25$$

On putting decision variables *x* and *y* equal to zero in (5), (6) and (7), we get

$s_1 = 18$, $s_2 = 2$ and $s_3 = 1.25$ and $Z = 0$.

This is initial basic feasible solution.

**Step 1.** The above information is given in the following table.

| | Coefficients of | | | | | | |
|---|---|---|---|---|---|---|---|
| | $x$ | $y$ | $s_1$ | $s_2$ | $s_3$ | $Z$ | Value |
| $Z$ | – 8 | – 6 | 0 | 0 | 0 | 1 | 0 |
| $s_1$ | 8 | 4 | 1 | 0 | 0 | 0 | 18 |
| $s_2$ | 1 | 0 | 0 | 1 | 0 | 0 | 2 |
| $s_3$ | 0 | 1 | 0 | 0 | 1 | 0 | 1.25 |

... (8)

**Step 2.** The coefficient of *x* and *y* in the *Z*-row are – 8 and – 6. Since the coefficient (–8) of *x* is the most negative. Therefore, *x* is Entering variable.

**Step 3. Selection of Departing variable**

In the second row find the ratio of the value of the last row and the coefficient (–8) of the entering variable *i.e.* $\frac{18}{8} = \frac{9}{4}$.

In the third row $\frac{2}{1} = 2$.

Thus the smallest non-negative ratio is 2 in $s_2$ – row. Therefore, the departing element is $s_2$.

**Step 4.**

Entering variable ↓ (x)

| | $x$ | $y$ | $s_1$ | $s_2$ | $s_3$ | $Z$ | Value | Ratio |
|---|---|---|---|---|---|---|---|---|
| $Z$ | $-8$ | $-6$ | 0 | 0 | 0 | 1 | 0 | |
| $s_1$ | 8 | 4 | 1 | 0 | 0 | 0 | 18 | $\frac{18}{8}=\frac{9}{4}$ |
| Departing variable → $s_2$ | [1] | 0 | 0 | 1 | 0 | 0 | 2 | $\frac{2}{1}=2$ |
| $s_3$ | 0 | 1 | 0 | 0 | 1 | 0 | 1.25 | |

... (9)

**Step 5.**

On multiplying the third row by 8 and adding to the first row we get 0 in the $x$-column. On multiplying the third row by $-8$ and adding to second row we get 0, in the $x$-column.

| | $x$ | $y$ | $s_1$ | $s_2$ | $s_3$ | $Z$ | *Value* |
|---|---|---|---|---|---|---|---|
| $Z$ | 0 | $-6$ | 0 | 8 | 0 | 1 | 16 |
| $s_1$ | 0 | 4 | 1 | $-8$ | 0 | 0 | 2 |
| $x$ | 1 | 0 | 0 | 1 | 0 | 0 | 2 |
| $s_3$ | 0 | 1 | 0 | 0 | 1 | 0 | 1.25 |

... (10)

Since the coefficient of $y$ is $(-6)$ most negative. So the entering variable is $y$.

The ratio in the second row $= \frac{2}{4} = \frac{1}{2} = 0.5$

The ratio in the fourth row $= \frac{1.25}{1} = 1.25$

The smallest ratio is $\frac{1}{2}$ in the $s_1$-row. So, the departing element is $s_1$.

Entering variable ↓ (x)

| | $x$ | $y$ | $s_1$ | $s_2$ | $s_3$ | $Z$ | Value | Ratio |
|---|---|---|---|---|---|---|---|---|
| $Z$ | 0 | $-6$ | 0 | 8 | 0 | 1 | 16 | |
| Departing Variable → $s_1$ | 0 | 4 | 1 | $-8$ | 0 | 0 | 2 | $\frac{2}{4}=\frac{1}{2}$ |
| $x$ | 1 | 0 | 0 | 1 | 0 | 0 | 2 | |
| $s_3$ | 0 | 1 | 0 | 0 | 1 | 0 | 1.25 | $\frac{1.25}{1}=1.25$ |

...(11)

Dividing second row by 4, we get

| | $x$ | $y$ | $s_1$ | $s_2$ | $s_3$ | $Z$ | Value |
|---|---|---|---|---|---|---|---|
| $Z$ | 0 | $-6$ | 0 | 8 | 0 | 1 | 16 |
| $y$ | 0 | 1 | $\frac{1}{4}$ | $-2$ | 0 | 0 | $\frac{1}{2}$ |
| $x$ | 1 | 0 | 0 | 1 | 0 | 0 | 2 |
| $s_3$ | 0 | 1 | 0 | 0 | 1 | 0 | 1.25 |

... (12)

Multiplying second row by 6 and adding to the first row we get 0 in the $y$-column. Subtracting second row from the fourth row we get 0 in the $y$-column.

| | $x$ | $y$ | $s_1$ | $s_2$ | $s_3$ | $Z$ | Value |
|---|---|---|---|---|---|---|---|
| $Z$ | 0 | 0 | $\frac{3}{2}$ | $-4$ | 0 | 1 | 19 |
| $y$ | 0 | 1 | $\frac{1}{4}$ | $-2$ | 0 | 0 | $\frac{1}{2}$ |
| $x$ | 1 | 0 | 0 | 1 | 0 | 0 | 2 |
| $s_3$ | 0 | 0 | $-\frac{1}{4}$ | 2 | 1 | 0 | 0.75 |

... (13)

Here, the coefficient of $s_2$ is $-4$ most negative. So, $s_2$ is entering variable.

The ratio in the second row $= \frac{1}{2}\left(-\frac{1}{2}\right) = -\frac{1}{4}$

The ratio in the third row $= \frac{2}{1} = 2$

The ratio in the fourth row $= \frac{0.75}{2} = \frac{3}{8}$

Thus, $\frac{3}{8}$ is the smallest non-negative ratio. Therefore, the departing variable is $s_3$.

Entering variable ↓ ($s_2$)

| | | $x$ | $y$ | $s_1$ | $s_2$ | $s_3$ | $Z$ | Value | Ratio |
|---|---|---|---|---|---|---|---|---|---|
| | $Z$ | 0 | 0 | $\frac{3}{2}$ | $-4$ | 0 | 1 | 19 | |
| | $y$ | 0 | 1 | $\frac{1}{4}$ | $-2$ | 0 | 0 | $\frac{1}{2}$ | $\frac{\frac{1}{2}}{-2} = -\frac{1}{4}$ |
| | $x$ | 1 | 0 | 0 | 1 | 0 | 0 | 2 | $\frac{2}{1} = 2$ |
| Departing variable ← | $s_3$ | 0 | 0 | $-\frac{1}{4}$ | [2] | 1 | 0 | 0.75 | $\frac{0.75}{2} = \frac{3}{8}$ |

...(14)

On multiplying the fourth row by $\frac{1}{2}$ we get 1 in the pivot entry.

| | $x$ | $y$ | $s_1$ | $s_2$ | $s_3$ | $Z$ | Value |
|---|---|---|---|---|---|---|---|
| $Z$ | 0 | 0 | $\frac{3}{2}$ | $-4$ | 0 | 1 | 19 |
| $y$ | 0 | 1 | $\frac{1}{4}$ | $-2$ | 0 | 0 | $\frac{1}{2}$ |
| $x$ | 1 | 0 | 0 | 1 | 0 | 0 | 2 |
| $s_2$ | 0 | 0 | $-\frac{1}{8}$ | 1 | $\frac{1}{2}$ | 0 | .375 |

... (15)

On multiplying fourth row by 4 and adding to first row we get 0 in $s_1$ column.
On multiplying fourth row by 2 and adding to second row we get 0 in the column of $s_2$.
On subtracting fourth row from third row, we get 0 in $x$-row

| | $x$ | $y$ | $s_1$ | $s_2'$ | $s_3$ | $Z$ | Value |
|---|---|---|---|---|---|---|---|
| $Z$ | 0 | 0 | 1 | 0 | 2 | 1 | 20.5 |
| $y$ | 0 | 1 | 0 | 0 | 1 | 0 | $\frac{5}{4}$ |
| $x$ | 1 | 0 | $\frac{1}{8}$ | 0 | $-\frac{1}{2}$ | 0 | 1.625 |
| $s_2$ | 0 | 0 | $-\frac{1}{8}$ | 1 | $\frac{1}{2}$ | 0 | 0.375 |

... (16)

Since, $Z$ row of table (16) has non-negative entries in the column of variables.
Therefore, this is the case of optimal solution. From the last column of table (16), we have

$$x = 1.625, \qquad y = \frac{5}{4}$$

And the maximum value of $Z = 20.5$

**Example 2.** *Using Simplex method solve the following L.P.P.* **Ans.**

*Maximize* $Z = x_1 + x_2 + 3x_3$ ...(1)

*Subject to* $3x_1 + 2x_2 + x_3 \le 3$ ...(2)

$2x_1 + x_2 + 2x_3 \le 2$ ...(3)

$x_1, x_2, x_3 \ge 0.$ ...(4)

**Solution.** Inequalities (2), (3) and (4) are converted into equations by adding non-negative variables $s_1, s_2$.

$$Z = x_1 + x_2 + 3x_3 + 0s_1 + 0s_2 \Rightarrow -x_1 - x_2 - 3x_3 + 0s_1 + 0s_2 + Z = 0 \quad ...(5)$$

$$3x_1 + 2x_2 + x_3 + s_1 = 3 \quad ...(6)$$

$$2x_1 + x_2 + 2x_3 + s_2 = 0 \quad ...(7)$$

where, $x_1, \ge 0, \quad x_2 \ge 0, \quad s_1 \ge 0, \quad s_2 \ge 0.$

On putting decision variables $x_1, x_2, x_3$ equal to zero in (5), (6) and (7), we get

$$0-0-0-0-0+Z=0 \Rightarrow Z=0$$

$$0-0-0+s_1=3 \Rightarrow s_1=3$$

$$0+0+0+s_2=2 \Rightarrow s_2=2$$

Entering variable ↓ ($x_3$)

| | | $x_1$ | $x_2$ | $x_3$ | $s_1$ | $s_2$ | $Z$ | Value | Ratio |
|---|---|---|---|---|---|---|---|---|---|
| | $Z$ | $-1$ | $-1$ | **($-3$)** | 0 | 0 | 1 | 0 | |
| | $s_1$ | 3 | 2 | 1 | 1 | 0 | 0 | 3 | $\frac{3}{1}=3$ |
| Departing variable → | $s_2$ | 2 | 1 | [2] | 0 | 1 | 0 | 2 | $\frac{2}{2}=1$ |

... (8)

Since coefficient ($-3$) of $x_3$ is most negative, so $x_3$ is entering variable.
The smallest ratio is 1 in the $s_2$-row, therfore $s_2$ is departing variable. The pivot entry (2) is at the intersection of $x_3$-column and $s_2$-row.

To make pivot entry (1), we multiply third row by $\frac{1}{2}$, we get

| | $x_1$ | $x_2$ | $x_3$ | $s_1$ | $s_2$ | $Z$ | Value |
|---|---|---|---|---|---|---|---|
| $Z$ | $-1$ | $-1$ | $-3$ | 0 | 0 | 1 | 0 |
| $s_1$ | 3 | 2 | 1 | 1 | 0 | 0 | 3 |
| $x_3$ | 1 | $\frac{1}{2}$ | 1 | 0 | $\frac{1}{2}$ | 0 | 1 |

...(9)

Applying $R_1 \to R_1 + 3R_3$ and $R_2 \to R_2 - R_1$, we get

| | $x_1$ | $x_2$ | $x_3$ | $s_1$ | $s_2$ | $Z$ | Value |
|---|---|---|---|---|---|---|---|
| $Z$ | 2 | $\frac{1}{2}$ | 0 | 0 | $\frac{3}{2}$ | 1 | 3 |
| $s_1$ | 2 | $\frac{3}{2}$ | 0 | 1 | $-\frac{1}{2}$ | 0 | 2 |
| $x_3$ | 1 | $\frac{1}{2}$ | 1 | 0 | $\frac{1}{2}$ | 0 | 1 |

... (10)

Since, $Z$ row of the table (10) has non-negative entries in the column of variables, therefore, this is the case of optimal solution. From the last column of the table we have $x_1 = 0$, $x_2 = 0$ and $x_3 = 1$ and the maximum value of $Z = 3$. **Ans.**

**Example 3.** *Using Simplex Method solve the L.P.P.:*

*Maximize* $Z = 3x_1 + 2x_2$ ...(1)

*Subject to the constraints*

$$x_1 + x_2 \geq 1 \quad ...(2)$$

$$x_1 + x_2 \leq 7 \quad ...(3)$$

$$x_1 + 2x_2 \geq 10 \quad ...(4)$$

$$x_2 \leq 3 \quad ...(5)$$

$$x_1 \geq 0, \qquad x_2 \geq 0$$

**Solution.**

**1. Conversion of the inequality $\geq$ into $\leq$ inequality**

Convert (2) and (4) into ($\leq$) type, multiplying by $-1$.

$$-x_1 - x_2 \leq -1$$

$$-x_1 - 2x_2 \leq -10$$

**2. Express the problem in standard form:**

Introducing slack variables $s_1$, $s_2$, $s_3$, $s_4$, we get

Maximize $Z = 3x_1 + 2x_2 + 0s_1 + 0s_2 + 0s_3 + 0s_4$ ...(6)

$\Rightarrow \quad -3x_1 - 2x_2 - 0s_1 - 0s_2 - 0s_3 - 0s_4 + Z = 0$

Subject to the constraints:

$$-x_1 - x_2 + s_1 = -1 \quad ...(7)$$

$$x_1 + x_2 + s_2 = 7 \quad ...(8)$$

$$-x_1 - 2x_2 + s_3 = -10 \quad \text{...(9)}$$
$$x_2 + s_4 = 3 \quad \text{...(10)}$$
$$x_1, x_2, s_1, s_2, s_3, s_4 \geq 0.$$

**3. Initial Basic solution**

Putting the decision variables $x_1$ and $x_2$ equal to zero in (6), (7), (8), (9) and (10), we get

$$Z = -0 - 0 + 0 + 0 + 0 + 0 \quad \Rightarrow \quad Z = 0$$
$$-0 - 0 + s_1 = -1 \quad \Rightarrow \quad s_1 = -1$$
$$0 + 0 + s_2 = 7 \quad \Rightarrow \quad s_2 = 7$$
$$-0 - 0 + s_3 = -10 \quad \Rightarrow \quad s_3 = -10$$
$$0 + s_4 = 3 \quad \Rightarrow \quad s_4 = 3$$

**4.** Let us construct the *initial table*

Entering variable ↓ ($x_1$)

| | $x_1$ | $x_2$ | $s_1$ | $s_2$ | $s_3$ | $s_4$ | $Z$ | Value | Ratio |
|---|---|---|---|---|---|---|---|---|---|
| $Z$ | (–3) | – 2 | 0 | 0 | 0 | 0 | 1 | 0 | |
| Departing Variable → $s_1$ | [–1] | – 1 | 1 | 0 | 0 | 0 | 0 | –1 | $\frac{-1}{-1} = 1$ smallest |
| $s_2$ | 1 | 1 | 0 | 1 | 0 | 0 | 0 | 7 | $\frac{7}{1} = 7$ |
| $s_3$ | –1 | –2 | 0 | 0 | 1 | 0 | 0 | –10 | $\frac{-10}{-1} = 10$ |
| $s_4$ | 0 | 1 | 0 | 0 | 0 | 1 | 0 | 3 | |

In the above table coefficient of $x_1$ is most negative. So, it is entering variable. The ratio of $s_1$-row is the smallest non-negative. So $s_1$ in the departing variable.

Applying $R_2' \to -R_2$, $R_1 \to R_1 + 3R_2'$, $R_3 \to R_3 - R_2'$, $R_4 \to R_4 + R_2'$

Entering variable ↓ ($s_1$)

| | $x_1$ | $x_2$ | $s_1$ | $s_2$ | $s_3$ | $s_4$ | $Z$ | Value | Ratio |
|---|---|---|---|---|---|---|---|---|---|
| $Z$ | 0 | 1 | **– 3** | 0 | 0 | 0 | 1 | 3 | |
| $x_1$ | 1 | 1 | – 1 | 0 | 0 | 0 | 0 | 1 | $\frac{1}{-1} = -1$ |
| Departing Variable → $s_2$ | 0 | 0 | [1] | 1 | 0 | 0 | 0 | 6 | $\frac{6}{1} = 6$ smallest |
| $s_3$ | 0 | – 1 | – 1 | 0 | 1 | 0 | 0 | – 9 | $\frac{-9}{-1} = 9$ |
| $s_4$ | 0 | 1 | 0 | 0 | 0 | 1 | 0 | 3 | |

**Departing variable**

In the above table coefficient of $s_1$ is most negative. So $s_1$ is the entering variable. The ratio of $s_2$-row is the smallest non-negative. So $s_2$ is the departing variable.

Applying $R_1 \to R_1 + 3R_3$, $R_2 \to R_2 + R_3$, $R_4 \to (R_4 + R_3)$

| | $x_1$ | $x_2$ | $s_1$ | $s_2$ | $s_3$ | $s_4$ | $Z$ | Value |
|---|---|---|---|---|---|---|---|---|
| $Z$ | 0 | 1 | 0 | 3 | 0 | 0 | 1 | 21 |
| $x_1$ | 1 | 1 | 0 | 1 | 0 | 0 | 0 | 7 |
| $s_1$ | 0 | 0 | 1 | 1 | 0 | 0 | 0 | 6 |
| $s_3$ | 0 | –1 | 0 | 1 | 1 | 0 | 0 | – 3 |
| $s_4$ | 0 | 1 | 0 | 0 | 0 | 1 | 0 | 3 |

Since, the first row ($Z$-row) has no negative entry in the columns of variables. Therefore, this is the case of optimal solution. From the last column of above table, we have

$$x_1 = 7, \qquad x_2 = 0$$

Maximum value of $z = 21$ **Ans.**

## 27.9. DEGENERACY

In locating the pivot row we may face two difficulties.

(*i*) In an initial Simplex table one or more entries in the last column may be zero. If the variable to be replaced is already zero then it is difficult to construct next table.

(*ii*) If the ratio for two or more variables is indentical, then there is a problem of selecting the pivot row.

The above two conditions give rise to phenomenon called degeneracy.

Here, a degenerate linear programming problem can either be solved by an arbitrary selection of one of the tied variables in finite number of steps or problem will begin to cycle.

**Example 4.** *Maximize* $Z = 20x_1 + 6x_2 + 8x_3$ ... (1)

*Subject to the constraints:*

$$8x_1 + 2x_2 + 3x_3 \le 200 \qquad ...(2)$$

$$4x_1 + 3x_2 \le 150 \qquad ...(3)$$

$$2x_1 + x_3 \le 50 \qquad ...(4)$$

$$x_1, x_2, x_3 \ge 0$$

**Solution.** Inequalities (2), (3), (4) are converted into equations by adding non-negative variables $s_1$, $s_2$ and $s_3$.

$$Z = 20x_1 + 6x_2 + 8x_3 + 0s_1 + 0s_2 + 0s_3 \Rightarrow -20x_1 - 6x_2 - 8x_3 - 0s_1 - 0s_2 - 0s_3 + Z = 0 \qquad ...(5)$$

$$8x_1 + 2x_2 + 3x_3 + s_1 = 200 \qquad ...(6)$$

$$4x_1 + 3x_2 + s_2 = 150 \qquad ...(7)$$

$$2x_1 + x_3 + s_3 = 50 \qquad ...(8)$$

On putting decision variables $x_1$, $x_2$ and $x_3$ equal to zero in (5), (6), (7) and (8), we get

$$0 - 0 - 0 - 0 - 0 - 0 + Z = 0 \Rightarrow Z = 0$$

$$0 + 0 + 0 + s_1 = 200 \Rightarrow s_1 = 200$$

$$0 + 0 + s_2 = 150 \Rightarrow s_2 = 150$$

$$0 + 0 + s_3 = 50 \Rightarrow s_3 = 50$$

Let us construct initial feasible table.

| | Coefficient of | | | | | | | | |
|---|---|---|---|---|---|---|---|---|---|
| | $x_1$ | $x_2$ | $x_3$ | $s_1$ | $s_2$ | $s_3$ | $Z$ | Value | Ratio |
| $Z$ | **– 20** | – 6 | – 8 | 0 | 0 | 0 | 1 | 0 | |
| $s_1$ | 8 | 2 | 3 | 1 | 0 | 0 | 0 | 200 | $\frac{200}{8}=25$ |
| $s_2$ | 4 | 3 | 0 | 0 | 1 | 0 | 0 | 150 | $\frac{150}{4}=37.5$ |
| $s_3$ | 2 | 0 | 1 | 0 | 0 | 1 | 0 | 50 | $\frac{50}{2}=25$ |

...(9)

From the above table the ratios of the IInd and IVth rows are identical i.e. 25. So, there is a tie between the non-negative ratios of $s_1$-row and $s_3$-row.

**Case I.** Let us choose $s_3$ row as the pivot row.

Entering variable ↓ ($x_1$)

| | | $x_1$ | $x_2$ | $x_3$ | $s_1$ | $s_2$ | $s_3$ | $Z$ | *Value* | |
|---|---|---|---|---|---|---|---|---|---|---|
| | $Z$ | **– 20** | – 6 | – 8 | 0 | 0 | 0 | 1 | 0 | |
| | $s_1$ | 8 | 2 | 3 | 1 | 0 | 0 | 0 | 200 | |
| Departing | $s_2$ | 4 | 3 | 0 | 0 | 1 | 0 | 0 | 150 | |
| variable → | $s_3$ | [2] | 0 | 1 | 0 | 0 | 1 | 0 | 50 | Pivot row |

Pivot column ... (10)

Apply $R_4' \to \frac{1}{2} R_4,\ R_1' \to R_1 + 20R_4',\ R_2' \to R_2 - 8R_4',\ R_3' \to R_3 - 4R_4'$; this yields following table:

Entering variable ↓ ($x_2$)

| | | $x_1$ | $x_2$ | $x_3$ | $s_1$ | $s_2$ | $s_3$ | $Z$ | Value | Ratio |
|---|---|---|---|---|---|---|---|---|---|---|
| | $Z$ | 0 | **– 6** | 2 | 0 | 0 | 10 | 1 | 500 | |
| Departing variable → | $s_1$ | 0 | [2] | – 1 | 1 | 0 | – 4 | 0 | 0 | $\frac{0}{2}=0$ |
| | $s_2$ | 0 | 3 | – 2 | 0 | 1 | – 2 | 0 | 50 | $\frac{50}{3}$ |
| | $x_1$ | 1 | 0 | $\frac{1}{2}$ | 0 | 0 | $\frac{1}{2}$ | 0 | 25 | $\frac{25}{1}=25$ |

...(11)

Applying $R_2' \to \frac{1}{2} R_2,\quad R_1 \to R_1 + 6R_2',\quad R_3 \to R_3 - 3R_2'$; we get the following table:

Entering variable ↓ ($s_3$)

| | | $x_1$ | $x_2$ | $x_3$ | $s_1$ | $s_2$ | $s_3$ | $Z$ | *Value* | *Ratio* |
|---|---|---|---|---|---|---|---|---|---|---|
| | $Z$ | 0 | 0 | – 1 | 3 | 0 | **– 2** | 1 | 500 | |
| | $x_2$ | 0 | 1 | $-\frac{1}{2}$ | $\frac{1}{2}$ | 0 | – 2 | 0 | 0 | $\frac{0}{-2}=-ve$ |
| Departing variable → | $s_2$ | 0 | 0 | $-\frac{1}{2}$ | $-\frac{3}{2}$ | 1 | [4] | 0 | 50 | $\frac{50}{4}=12.5$ |
| | $x_1$ | 1 | 0 | $\frac{1}{2}$ | 0 | 0 | $\frac{1}{2}$ | 0 | 25 | $\frac{25}{½}=50$ |

...(12)

Applying $R_3' \to \frac{1}{4}R_3,\ R_1 \to R_1 + 2R_3',\ R_2 \to R_2 + 2R_3',\ R_4 \to R_4 - \frac{1}{2}R_3'$; this gives

Entering variable ↓ ($x_3$)

| | $x_1$ | $x_2$ | $x_3$ | $s_1$ | $s_2$ | $s_3$ | $Z$ | Value | Ratio |
|---|---|---|---|---|---|---|---|---|---|
| $Z$ | 0 | 0 | $-\frac{5}{4}$ | $\frac{9}{4}$ | $\frac{1}{2}$ | 0 | 1 | 525 | |
| $x_2$ | 0 | 1 | $-\frac{3}{4}$ | $-\frac{1}{4}$ | $\frac{1}{2}$ | 0 | 0 | 25 | $25 \div \left(-\frac{3}{4}\right) = -ve$ |
| $s_3$ | 0 | 0 | $-\frac{1}{8}$ | $-\frac{3}{8}$ | $\frac{1}{4}$ | 1 | 0 | $\frac{25}{2}$ | $\frac{25}{2} \div \left(-\frac{1}{8}\right) = -ve$ |
| Departing variable → $x_1$ | 1 | 0 | $\boxed{\frac{9}{16}}$ | $\frac{3}{16}$ | $-\frac{1}{8}$ | 0 | 0 | $\frac{75}{4}$ | $\frac{75}{4} \div \left(\frac{9}{16}\right) = \frac{100}{3}$ |

...(13)

Apply $R_4 \to \left(\frac{16}{9}\right)R_4',\ R_1 \to R_1 + \frac{5}{4}R_4,\ R_2 \to R_2 + \frac{3}{4}R_4,\ R_3 \to R_3 + \left(\frac{1}{8}\right)R_4$; this gives

| | $x_1$ | $x_2$ | $x_3$ | $s_1$ | $s_2$ | $s_3$ | $Z$ | Value |
|---|---|---|---|---|---|---|---|---|
| $Z$ | $\frac{20}{9}$ | 0 | 0 | $\frac{8}{3}$ | $\frac{2}{9}$ | 0 | 1 | $\frac{1700}{3}$ |
| $x_2$ | $\frac{4}{3}$ | 1 | 0 | 0 | $\frac{1}{3}$ | 0 | 0 | 50 |
| $s_3$ | $\frac{2}{9}$ | 0 | 0 | $-\frac{1}{3}$ | $\frac{2}{9}$ | 1 | 0 | $\frac{50}{3}$ |
| $x_3$ | $\frac{16}{9}$ | 0 | 1 | $\frac{1}{3}$ | $-\frac{2}{9}$ | 0 | 0 | $\frac{100}{3}$ |

... (14)

Since $Z$ row of table (14) has non-negative entries in the column of variables, therefore this is the case of optimal solution.

From the last column of the table, we have

$$x_1 = 0,\ x_2 = 50,\ x_3 = \frac{100}{3} \text{ and the maximum value of } Z = \frac{1700}{3}$$ **Ans.**

**Case II.** Now, we can choose $s_1$-row as pivot row, and we can solve L.P.P. by the method used in case I.

## MAXIMIZATION PRINCIPLE

If an objective function is to minimize then it can be converted into a maximization problem simply by multiplying the objective function by (–1).

**Example 5.**

Minimize $Z = -4x + 2y$

subject to constraints

$$6x + 2y \le 18$$

$$3x - 2y \le 6$$

$$x \ge 0,\ y \ge 0$$

**Solution.** Since the problem is that of minimizing the objective function, we convert it into that of maximizing by multiplying $Z$ by (–1).

$$-Z = \text{Maximizing } Z'$$

$$Z' = -Z = 4x - 2y$$

By adding slack variables, the inequalities are converted into equations.

$$Z' = 4x - 2y + 0s_1 + 0s_2 \quad \Rightarrow \quad -4x + 2y - 0s_1 - 0s_2 + Z' = 0 \qquad ...(1)$$

$$6x + 2y + s_1 + 0s_2 = 18 \qquad ...(2)$$

$$3x - 2y + 0s_1 + s_2 = 6 \qquad ...(3)$$

$$x \geq 0, y \geq 0, s_1 \geq 0, s_2 \geq 0$$

On putting decision variables $x = 0$, $y = 0$ in (1), (2) and (3), we get

$$0 - 0 - 0 - 0 + Z' = 0 \quad \Rightarrow \quad Z' = 0$$

$$0 + 0 + s_1 + 0 = 18 \quad \Rightarrow \quad s_1 = 18$$

$$0 - 0 + 0 + s_2 = 6 \quad \Rightarrow \quad s_2 = 6$$

The initial table is

Entering variable ↓ ($x$)

| | | $x$ | $y$ | $s_1$ | $s_2$ | $Z'$ | Value | Ratio |
|---|---|---|---|---|---|---|---|---|
| | $Z'$ | –4 | 2 | 0 | 0 | 1 | 1 | 0 |
| | $s_1$ | 6 | 2 | 1 | 0 | 0 | 18 | 18 ÷ 6 = 3 |
| Departing variable → | $s_2$ | [3] | –2 | 0 | 1 | 0 | 6 | 6 ÷ 3 = 2 |

... (4)

Since, the coefficient of $x$ is most negative, so $x$ is entering variable. Since the ratio (2) is the smallest in $s_2$-row, so $s_2$ is departing variable. Pivot entry (3) is at the intersection of $x$-row and $s_2$-column.

To make pivot entry (1) we multiply the third row by $\frac{1}{3}$ to get 1.

| | $x$ | $y$ | $s_1$ | $s_2$ | $Z$ | Value |
|---|---|---|---|---|---|---|
| $Z'$ | –4 | 2 | 0 | 0 | 1 | 0 |
| $s_1$ | 6 | 2 | 1 | 0 | 0 | 18 |
| $s_2$ | 1 | $-\frac{2}{3}$ | 0 | $\frac{1}{3}$ | 0 | 2 |

... (5)

Multiply the third row by 4 and add to the first row.
Multiply the third row by – 6 and add to the second row.

Entering variable ↓ ($y$)

| | | $x$ | $y$ | $s_1$ | $s_2$ | $Z$ | Value | Ratio |
|---|---|---|---|---|---|---|---|---|
| | $Z'$ | 0 | $-\frac{2}{3}$ | 0 | $\frac{4}{3}$ | 1 | 8 | |
| Departing variable → | $s_1$ | 0 | [6] | 1 | –2 | 0 | 6 | 6 ÷ 6 = 1 |
| | $x$ | 1 | $-\frac{2}{3}$ | 0 | $\frac{1}{3}$ | 0 | 2 | $2 \div \left(-\frac{2}{3}\right) = -3 = -ve$ |

... (6)

Since the coefficient $\left(-\frac{2}{3}\right)$ of $y$ is most negative, so $y$ is the entering variable.

Since the ratio (1) is the smallest in $s_1$-row, so Departing variable is $s_1$.
The pivot entry is at the intersection of $y$-column and $s_1$-row.

Multiply the second row by $\frac{1}{6}$ to get 1.

| | $x$ | $y$ | $s_1$ | $s_2$ | $Z'$ | Value |
|---|---|---|---|---|---|---|
| $Z'$ | 0 | $-\frac{2}{3}$ | 0 | $\frac{4}{3}$ | 1 | 8 |
| $s_1$ | 0 | [1] | $\frac{1}{6}$ | $-\frac{2}{3}$ | 0 | 1 |
| $x$ | 1 | $-\frac{2}{3}$ | 0 | $\frac{1}{3}$ | 0 | 2 |

...(7)

Multiplying the second row by $\frac{2}{3}$ and add to the first row

Multiplying the second row by $\frac{2}{3}$ and add to the third row

| | $x$ | $y$ | $s_1$ | $s_2$ | $Z'$ | value |
|---|---|---|---|---|---|---|
| $Z'$ | 0 | 0 | $\frac{1}{9}$ | $\frac{8}{9}$ | 1 | $\frac{26}{3}$ |
| $y$ | 0 | 1 | $\frac{1}{6}$ | $-\frac{2}{3}$ | 0 | 1 |
| $x$ | 0 | 0 | $\frac{1}{9}$ | $-\frac{1}{9}$ | 0 | $\frac{8}{3}$ |

...(8)

Since there is no negative entry in the $Z'$ row, we have arrived at an optimal solution.

Optimal solution is $x = \frac{8}{3}$, $y = 1$, $s_1 = 0$, $s_2 = 0$

The maximum value of $Z' = \frac{26}{3}$

or

the minimum value of $Z = -\frac{26}{3}$ **Ans.**

## EXERCISE 27.4

***Using Simplex method, solve the following L.P.P.***

**1.** Maximize $Z = x_1 + 3x_2$

Subject to $x_1 + 2x_2 \le 10,\ 0 \le x_1 \le 5,\ 0 \le x_2 \le 4$ **Ans.** $x_1 = 2$, $x_2 = 4$, Max. $Z = 14$

**2.** Maximize $Z = 4x_1 + 10x_2$

Subject to $2x_1 + x_2 \le 50, 2x_1 + 5x_2 \le 100,\ 2x_1 + 3x_2 \le 90, x_1, x_2 \ge 0$

**Ans.** $x_1 = 0$, $x_2 = 20$, Max. $Z = 200$

3. Maximize $Z = 4x_1 + 5x_2$
Subject to $x_1 - 2x_2 \le 2, \quad 2x_1 + x_2 \le 6, \quad x_1 + 2x_2 \le 5, \quad -x_1 + x_2 \le 2, \quad x_1, \quad x_2 \ge 0.$

**Ans.** $x_1 = \frac{7}{3}, \quad x_2 = \frac{4}{3}, \quad \text{Max.}\, Z = 16.$

4. Maximize $Z = 10x_1 + x_2 + 2x_3$
Subject to $x_1 + x_2 - 2x_2 \le 10, \quad 4x_1 + x_2 + x_3 \le 20, \quad x_1, x_2, x_3 \ge 0.$

**Ans.** $x_1 = 5, \quad x_2 = x_3 = 0;$ Max. $Z = 50$

5. Maximize $Z = 5x_1 + 3x_2$
Subject to $x_1 + x_2 \le 2, \quad 5x_1 + 2x_2 \le 10, \quad 3x_1 + 8x_2 \le 12, \quad x_1, \quad x_2 \ge 0.$

**Ans.** $x_1 = 2, \quad x_2 = 0;$ Max.$Z = 10.$

6. Maximize $Z = 10x_1 + 12x_2$
Subject to $x_1 + 2x_2 \le 150, \quad x_1 + x_2 \le 100, \quad x_1, x_2 \ge 0.$

**Ans.** Max.$Z = 1100, \quad x_1 = 50, \quad x_2 = 50$

7. Maximize $Z = 2x_1 + 5x_2$
Subject to the constraints $-4x_1 + x_2 \le 5, \quad -x_1 + x_2 \le 4, \quad x_1, \quad x_2 \ge 0.$

**Ans.** Unbounded solution

8. Maximize $Z = 3x_1 + 5x_2$
Subject to the constraints $2x_1 + 5x_2 \le 132, \quad 3x_1 + 2x_2 \le 100;$ $x_2 \ge 0$ and $x_1$ unrestricted in sign.

**Ans.** Max. $Z = \frac{1688}{11}, \quad x_1 = \frac{236}{11}, \quad x_2 = \frac{196}{11}$

9. Minimize $Z = 4x_1 + 3x_2$
Subject to constraints $200x_1 + 100x_2 \ge 4000, \quad x_1 + 2x_2 \ge 50, \quad 40x_1 + 40x_2 \ge 1400, \quad x_1, x_2 \ge 0$

**Ans.** Min. $Z = 110, x_1 = 5, x_2 = 30$

10. Minimize $Z = 5x + 4y$
Subject to the constraints $80x + 10y \ge 88, \quad 40x + 30y \ge 36, \quad x \ge 0, \quad y \ge 0$

**Ans.** Min. $Z = 4.6, x = 0.6, y = 0.4.$

## 27.10 DUALITY

With every L.P.P. there is always associated another L.P.P. called the dual problem. We call the given probelm as primal problem. If the primal problem requires maximization the dual problem is one of the minimizing problem and vice-versa.

## 27.11 DUAL OF L.P.P.

Suppose the primal problem is that of maximization of the total net revenue. The dual problem be that of minimization of cost of raw material in a factory.

**Example 1.** A goldsmith specilizes in the production of three products chain, ring and bangles. The three products require silver and labour where supplies are limited. The following table gives the details

| | *Units of silver required* | *Units of labour required* | *Price (in Rs.)* |
|---|---|---|---|
| Chain | 2 | 4 | 500 |
| Ring | 1 | 5 | 100 |
| Pair of Bangles | 3 | 6 | 600 |
| Available Resources | 1000 | 150 | |

**Solution.** Let $x_1$, $x_2$ and $x_3$ denote the number of chains, rings and pair of bangles respectively.

The mathmatical formulation of the above problem is:

Maximize $\qquad Z = 500x_1 + 100x_2 + 600x_3 \qquad \ldots (1)$

Subject to the constraints:

$$2x_1 + x_2 + 3x_3 \le 1000 \qquad \ldots (2)$$

$$4x_1 + 5x_2 + 6x_3 \le 150 \qquad \ldots (3)$$

$$x_1, x_2, x_3 \ge 0$$

This is a resource allocation problem.

Let $y_1$ and $y_2$ be the price of unit value of silver and labour.

The dual of the above problem would be :

Minimize $\qquad Z' = 1000y_1 + 150y_2 \qquad \ldots (4)$

Subject to constraints :

$$2y_1 + 4y_2 \ge 500 \qquad \ldots (5)$$

$$y_1 + 5y_2 \ge 100 \qquad \ldots (6)$$

$$3y_1 + 6y_2 \ge 600 \qquad \ldots (7)$$

$$y_1, y_2 \ge 0.$$

The 5th inequality, therefore, means that the total value of silver and labour required to produce a chain must be at least equal to the price of the chain. A similar interpretion can be given to 6th and 7th inequalities corresponding to the ring and pair of bangles.

## WORKING RULE FOR DUAL PROBLEM

**Step 1.** If the objective function Z of the primal problem is to be maximized, then the objective function $Z'$ of the dual problem is to be minimized and vice-versa.

**Step 2.** If in the primal problem the set of variables $x_1$, $x_2$, $x_3$ is used then the set of variables used in dual problem are $y_1$, $y_2$ and $y_3$.

**Step 3.** The inequalities $\le$ of the constraints must be $\ge$ in the dual problem and vice-versa.

**Step 4.** The constants on the right hand side of the constraints are written in a coloumn. These constants from top to bottom become the coefficients of $y_1$, $y_2$, $y_3$ in the objective function from left to right in a row.

**Step 5.** The coefficients in the constraints from left to right are placed from top to bottom *i.e.*, first row becomes the first column and second row becomes the second column and so on.

**Step 6.** Number of variables $x_1$, $x_2$, $x_3$ ... in primal problem = Number of constraints in the dual problem.

Number of constraints in the primal problem = Number of variables $y_1$, $y_2$, $y_3$ .... in the dual problem.

**Note :** The optimal value of $Z'$ in the dual is the optimal value of Z of the primal and vice-versa.

**Example 2.** *Write the dual of the following primal problems:*

*(i) Maximize* $Z = x_1 - x_2 + 3x_3$

*Subject to the constraints :*

$$x_1 + x_2 + x_3 \le 10$$

$$2x_1 - x_3 \le 2$$

$$2x_1 - 2x_2 + 3x_3 \le 6$$

$$x_1, x_2, x_3 \ge 0$$

(ii) *Minimize* $Z = 2x_1 + 2x_2$

*Subject to the constraints*

$$2x_1 + 4x_2 \ge 1$$

$$x_1 + 2x_2 \ge 1$$

$$2x_1 + x_2 \ge 1$$

$$x_1, x_2 \ge 0$$

**Solution**

**(i)** **Let $y_1, y_2$ and $y_3$ be the dual variables then the dual problem of the given primal problem is**

Minimize $\quad Z' = 10y_1 + 2y_2 + 6y_3$

Subject to the constraints:

$$y_1 + 2y_2 + 2y_3 \ge 1$$

$$y_1 - 2y_3 \ge -1$$

$$y_1 - y_2 + 3y_3 \ge 3$$

$$y_1, y_2, y_3 \ge 0$$

**(ii)** **Let $y_1, y_2$ and $y_3$ be the dual variables then the dual problem of the given primal probelm is**

Maximize $\quad Z' = y_1 + y_2 + y_3$

Subject to the constraints

$$2y_1 + y_2 + 2y_3 \le 2$$

$$4y_1 + 2y_2 + y_3 \le 2$$

$$y_1, y_2, y_3 \ge 0.$$

**Ans.**

**Example 3.** *Write the dual of the following primal problem and solve it.*

*Minimize* $\quad Z = 4x_1 + 2x_2$

*Subject to the constraints*

$$x_1 + x_2 \ge 3$$

$$x_1 - x_2 \ge 2$$

$$x_1, x_2 \ge 0$$

**Solution.** Let $y_1$ and $y_2$ be the dual variables of the given primal problem:

Maximize $\quad Z' = 3y_1 + 2y_2 \quad$ ... (1)

Subject to the constraints

$$y_1 + y_2 \le 4 \quad \text{... (2)}$$

$$y_1 - y_2 \le 2 \quad \text{... (3)}$$

$$y_1, y_2 \ge 0 \quad \text{... (4)}$$

Inequalities (2), (3) and (4) are converted into equations by adding slack variables $s_1$ and $s_2$.

Maximize $Z' = 3y_1 + 2y_2 + 0s_1 + 0s_2 \Rightarrow -3y_1 - 2y_2 - 0s_1 - 0s_2 + Z' = 0 \quad$ ... (5)

Subject to the constraints

$$y_1 + y_2 + s_1 = 4 \quad \text{... (6)}$$

$$y_1 - y_2 + s_2 = 2 \quad \text{... (7)}$$

$$y_1, y_2, s_1, s_2 \ge 0$$

On putting decision variables $y_1$, $y_2$ equal to zero in (5), (6) and (7), we get

$$Z' = 0$$
$$s_1 = 4$$
$$s_2 = 2$$

The initial table is constructed as below

Entering variable ↓ $y_1$

| | $y_1$ | $y_2$ | $s_1$ | $s_2$ | $Z'$ | Value | Ratio |
|---|---|---|---|---|---|---|---|
| $Z'$ | **–3** | –2 | 0 | 0 | 1 | 0 | |
| $s_1$ | 1 | 1 | 1 | 0 | 0 | 4 | $\frac{4}{1} = 4$ |
| Departing variable → $s_2$ | [1] | –1 | 0 | 1 | 0 | 2 | $\frac{2}{1} = 2$ |

Since the cofficient of $y_1$ is most negative, so $y_1$ is the entering variable. The ratio (2) is the smallest in $s_2$-row, so $s_2$ is departing variable.

(1) is the pivotal entry at the interection of $y$-column and $s_2$-row.

Apply $R_1 \to R_1 + 3R_3$, $\quad R_2 \to R_2 - R_3$

Entering variable ↓ $y_2$

| | $y_1$ | $y_2$ | $s_1$ | $s_2$ | $Z'$ | Value | Ratio |
|---|---|---|---|---|---|---|---|
| $Z'$ | 0 | **–5** | 0 | 3 | 1 | 6 | |
| Departing variable → $s_1$ | 0 | [2] | 1 | –1 | 0 | 2 | $\frac{2}{2} = 1$ |
| $y_1$ | 1 | –1 | 0 | 1 | 0 | 2 | $\frac{2}{-1} = -ve$ |

Since coefficient of $y_2$ is most negative, so $y_2$ is entering variable.
The ratio (1) is smallest in the $s_1$-row, so $s_1$ is departing variable.
The pivotal entry (2) is at the intersection of $y$-column and $s$-row.

Apply $R_2' \to \frac{1}{2} R_2$, $R_1 \to R_1 + 6R_2'$, $R_3 \to R_3 + R_2'$

| | $y_1$ | $y_2$ | $s_1$ | $s_2$ | $Z'$ | Value |
|---|---|---|---|---|---|---|
| $Z'$ | 0 | 0 | $\frac{5}{2}$ | $\frac{1}{2}$ | 1 | 11 |
| $y_2$ | 0 | 1 | $\frac{1}{2}$ | $-\frac{1}{2}$ | 0 | 1 |
| $y_1$ | 1 | 0 | $\frac{1}{2}$ | $\frac{1}{2}$ | 0 | 3 |

Since, Z row of above table has no negative entry in the column of variables.
Therefore this is the case of optimal solution from the last column.

The maximum value of $Z'$ is 11.
Therefore, the minimum value of $Z$ is also 11. **Ans.**

**Example 4.** *Using duality solve the following problem*

*Minimize* $Z = 0.7x_1 + 0.5x_2$

*Subject to the constraints*

$x_1 \geq 4, \quad x_2 \geq 6, \quad x_1 + 2x_2 \geq 20, \quad 2x_1 + x_2 \geq 18$

**Solution. 1.** The dual of the primal problem is

Maximize $Z' = 4y_1 + 6y_2 + 20y_3 + 18y_4$

Subject to the constraints

$y_1 + y_3 + 2y_4 \leq 0.7, \quad y_2 + 2y_3 + y_4 \leq 0.5, \quad y_1, y_2, y_3, y_4 \geq 0$

**2. Standard form** Introducing slack variables the dual probelm in the standard form becomes

Maximize $Z' = 4y_1 + 6y_2 + 20y_3 + 18y_4 + 0s_1 + 0s_2$ ... (1)

$\Rightarrow \quad -4y_1 - 6y_2 - 20y_3 - 18y_4 - 0s_1 - 0s_2 + Z' = 0$

Subject to the constraints

$y_1 + 0y_2 + y_3 + 2y_4 + s_1 + 0s_2 = 0.7$ ... (2)

$0y_1 + y_2 + 2y_3 + y_4 + 0s_1 + s_2 = 0.5$ ... (3)

$y_1, y_2, y_3, y_4 \geq 0$

**3. Initial Basic feasible solution**

Putting non-basic variables $y_1, y_2, y_3, y_4$ each equal to zero in equations (1), (2) and (3), we get

$y_1 = y_2 = y_3 = y_4 = 0$ (Non-basic variables)
$s_1 = 0.7, s_2 = 0.5$ (Basic variables)

**4. Initial table**

Entering variable ↓ ($y_3$)

| | | $y_1$ | $y_2$ | $y_3$ | $y_4$ | $s_1$ | $s_1$ | $Z'$ | Value | Ratio |
|---|---|---|---|---|---|---|---|---|---|---|
| | $Z'$ | –4 | –6 | **–20** | –18 | 0 | 0 | 1 | 0 | |
| | $s_1$ | 1 | 0 | 1 | 2 | 1 | 0 | 0 | 0.7 | $\frac{0.7}{1} = 0.7$ |
| Departing Variable → | $s_2$ | 0 | 1 | [2] | 1 | 0 | 1 | 0 | 0.5 | $\frac{0.5}{2} = 0.25$ |

... (4)

In the table-4 coefficient (–20) of $y_3$ is the most negative. So, $y_3$ is entering variable. The ratio of $s_2$-row is smallest non-negative. So, $s_2$ is the departing variable. The pivotal entry (2) is at the intersection of the $y_3$-column and $s_2$-row.

Applying $R_3' \to \frac{1}{2}R_3, R_1 \to R_1 + 20R_3, R_2 \to R_2 - R_3'$

Entering variable ↓ ($y_4$)

| | | $y_1$ | $y_2$ | $y_3$ | $y_4$ | $y_1$ | $y_2$ | $Z'$ | Value | Ratio |
|---|---|---|---|---|---|---|---|---|---|---|
| | $Z'$ | –4 | 4 | 0 | **–8** | 0 | 5 | 1 | 5 | |
| Departing variable → | $s_1$ | 1 | $-\frac{1}{2}$ | 0 | [$\frac{3}{2}$] | 1 | $-\frac{1}{2}$ | 0 | 0.45 | $\frac{0.45}{1.5} = 0.3$ smallest |
| | $y_3$ | 0 | $\frac{1}{2}$ | 1 | $\frac{1}{2}$ | 0 | $\frac{1}{2}$ | 0 | 0.25 | $\frac{0.25}{0.5} = 0.5$ |

... (5)

Since, the coefficient (–8) of $y_4$ is the most negative, so $y_4$ is the entering variable.

Since, the ratio (0.3) is the smallest ratio in $s_1$-row, so $s_1$ is the departing variable.

The pivotal entry $\left(\frac{3}{2}\right)$ is at the intersection of $y_4$-column and $s_1$-row.

Apply $R_2' \to \frac{2}{3}R_2, R_1 \to R_1 + 8R_2', R_3 \to R_3 - \frac{1}{2}R_2'$

| | $y_1$ | $y_2$ | $y_3$ | $y_4$ | $s_1$ | $s_2$ | $Z'$ | Value |
|---|---|---|---|---|---|---|---|---|
| $Z'$ | $\frac{4}{3}$ | $\frac{4}{3}$ | 0 | 0 | $\frac{16}{3}$ | $\frac{7}{3}$ | 1 | 7.4 |
| $y_4$ | $\frac{2}{3}$ | $-\frac{1}{3}$ | 0 | 1 | $\frac{2}{3}$ | $-\frac{1}{3}$ | 0 | 0.3 |
| $y_3$ | $-\frac{1}{3}$ | $\frac{1}{6}$ | 1 | 0 | $-\frac{1}{3}$ | $\frac{2}{3}$ | 0 | 0.1 |

... (6)

Since, the first row (Z row) has no negative entry in the column of the variables, therefore this is the case of optimal solution.

From the last column of the table (6), we have

$$y_1 = 0$$
$$y_2 = 0$$
$$y_3 = 0.1$$
$$y_4 = 0.3$$

Maximum value of $Z' = 7.4$

Hence, an optimal basic feasible solution to the given primal is

Minimum of $Z = 7.4$ **Ans.**

## TRANSPORTATION PROBLEMS

### 27.12. NORTH WEST CORNER METHOD

Let $a_i$ and $b_i$ be the capacities of the supplier $s_i$ and destinations $D_i$.

| | $D_1$ | $D_2$ | $D_3$ | $D_4$ | |
|---|---|---|---|---|---|
| $S_1$ | | | | | $a_1$ |
| $S_2$ | | | | | $a_2$ |
| $S_3$ | | | | | $a_3$ |
| $S_4$ | | | | | $a_4$ |
| | $b_1$ | $b_2$ | $b_3$ | $b_4$ | |

**Step 1.** Fill up the upper left hand (North-West) corner of the transportation table. The maximum feasible amount is allocated there.

$$(1, 1) \text{ Box} = \text{Min } (a_1, b_1)$$

If $a_1 > b_1$ then (1, 1) Box $= b_1$

In this way either the capacity of supplier $s_1$ is exhausted or the requirement of destination $D_1$.

| | $D_1$ | $D_2$ | $D_3$ | $D_4$ | |
|---|---|---|---|---|---|
| $S_1$ | $b_1$ | | | | $a_1$ |
| $S_2$ | | | | | $a_2$ |
| $S_3$ | | | | | $a_3$ |
| $S_4$ | | | | | $a_4$ |
| | | $b_2$ | $b_3$ | $b_4$ | |

**Step 2.** Fill up the box of first row and second column by $(a_1 - b_1)$.

| | $D_1$ | $D_2$ | $D_3$ | $D_4$ | |
|---|---|---|---|---|---|
| $S_1$ | $b_1$ | $a_1 - b_1$ | | | |
| $S_2$ | | | | | $a_2$ |
| $S_3$ | | | | | $a_3$ |
| $S_4$ | | | | | $a_4$ |
| | | $b_2$ | $b_3$ | $b_4$ | |

**Step 3.** If $b_2 > a_1 - b_1$, then $b_2 - (a_1 - b_1)$ is entry in the lower box i.e. of second row and second column.

| | $D_1$ | $D_2$ | $D_3$ | $D_4$ | |
|---|---|---|---|---|---|
| $S_1$ | $b_1$ | $a_1 - b_1$ | | | |
| $S_2$ | | $b_2 - (a_1 - b_1)$ | | | $a_2$ |
| $S_3$ | | | | | $a_3$ |
| $S_4$ | | | | | $a_4$ |
| | | | $b_3$ | $b_4$ | |

**Step 4.** If $a_2 > b_2 - a_1 + b_1$, then $a_2 - (b_2 - a_1 + b_1)$ is the entry in the box of second row and third column.

| | $D_1$ | $D_2$ | $D_3$ | $D_4$ | |
|---|---|---|---|---|---|
| $S_1$ | $b_1$ | $a_1 - b_1$ | | | |
| $S_2$ | | $b_2 - (a_1 - b_1)$ | $a_2 - (b_2 - a_1 + b_1)$ | $a_2$ | |
| $S_3$ | | | | | $a_3$ |
| $S_4$ | | | | $b_4$ | $a_4$ |

and so on.

**Example 5.** *Find the initial Basic feasible solution of the following transportation problem by North West corner method.*

| Warehouse / Factory | $W_1$ | $W_2$ | $W_3$ | $W_4$ | Productionof Factories |
|---|---|---|---|---|---|
| $F_1$ | 21 | 16 | 25 | 13 | 11 |
| $F_2$ | 17 | 18 | 14 | 23 | 13 |
| $F_3$ | 32 | 27 | 18 | 41 | 19 |
| Capacityof the warehouse | 6 | 10 | 12 | 15 | 43 |

**First Method. (North-West Corner Method)**

**Solution.** The first box occupies the upper left hand (North-West) corner of the transportation table. We allocate 6 units in the box of first row and first column. This is the requirement of ware house $W_1$. The remaining 5 units are to be allocated in the box of first row and second column.

The ware house $W_2$ requires 10 units of which 5 units have been supplied by factory $F_1$. Thus the remaining $(10 - 5)$ i.e. 5 units from factory $F_2$ to the box second row and second column

The remaining production of $F_2(13 - 5)$ i.e. 8 units are allocated to the box in the second row and third column.

The remaining capacity (12 – 8) i.e. 4 units of ware house $W_2$ is allocated to the box third row and third column.

The remaining production (19–4) *i.e.*; 15 units of factory $F_3$ is allocated to the box third row and fourth column.

| Factory \ Ware house | $W_1$ | $W_2$ | $W_3$ | $W_4$ | Production of Factories |
|---|---|---|---|---|---|
| $F_1$ | 21 (6) | 16 (5) | 25 | 13 | 11 |
| $F_2$ | 17 | 18 (5) | 14 (8) | 23 | 13 |
| $F_3$ | 32 | 27 | 18 (4) | 41 (15) | 19 |
| Capacity of the ware houses | 06 | 10 | 12 | 15 | 43 |

Clearly, the initial solution constraints of six boxes:

$x_{11} = 6, \quad x_{22} = 5, \quad x_{33} = 4$

$x_{12} = 5, \quad x_{23} = 8, \quad x_{34} = 15$

Total cost of transportation = Rs. $(21 \times 6) + (16 \times 5) + (18 \times 5) + (14 \times 8) + (18 \times 4) + (41 \times 15)$

= Rs. 1095 **Ans.**

## 27.13 VOGEL'S APPROXIMATION METHOD (VAM)

Let $S_i$ be the supplier, $D_i$ destination and $C_{ij}$ *is the cost of transportation from* $S_i$ to $D_j$. If the supply and demand are equal the problem is balanced.

**Step 1. Construct transportaton table.**

| | $D_1$ | $D_2$ | $D_3$ | $D_4$ | $D_5$ | |
|---|---|---|---|---|---|---|
| $S_1$ | $C_{11}$ | $C_{12}$ | $C_{13}$ | $C_{14}$ | $C_{15}$ | $a_1$ |
| $S_2$ | $C_{21}$ | $C_{22}$ | $C_{23}$ | $C_{24}$ | $C_{25}$ | $a_2$ |
| $S_3$ | $C_{31}$ | $C_{32}$ | $C_{33}$ | $C_{34}$ | $C_{35}$ | $a_3$ |
| $S_4$ | $C_{41}$ | $C_{42}$ | $C_{43}$ | $C_{44}$ | $C_{45}$ | $a_4$ |
| $S_5$ | $C_{51}$ | $C_{52}$ | $C_{53}$ | $C_{54}$ | $C_{55}$ | $a_5$ |
| | $b_1$ | $b_2$ | $b_3$ | $b_4$ | $b_5$ | |

**Step 2. Initial Basic Feasible Solution**

The initial allocation should satisfy the demand at each project site without violating the capacities of the suppliers and also meeting the restrictions.

The Vogel's Approximation Method (VAM) also takes care of the least cost of transportation.

(*i*) Write the difference between the least and the next to the least cost in each row, to the right of row in brackets. Similarly write the differences for each column below the column in brackets.

(*ii*) Identify the row or column with the largest-difference in row and column. If the largest difference corresponds to *i*th row and $C_{ij}$ is the lowest cost in the *i*th row, allocate min $(a_i, b_1)$ to $(i, j)$th box.

In case of a tie allocate to the box with next lower cost.

(*iii*) Repeat the above step 2 on the reduced table.

**Step 3.** Check for optimality.

**Note.** VAM is better than North-West Corner Method since in solving a transportation problem by VAM, lowest cost is also considered. But there is no consideration of cost in North West Corner Method.

**Example 6.** *Find the initial Basic feasible solution of the following transportation problem by VAM.*

| Ware-house / Factory | $W_1$ | $W_2$ | $W_3$ | $W_4$ | Production of Factories |
|---|---|---|---|---|---|
| $F_1$ | 21 | 16 | 25 | 13 | 11 |
| $F_2$ | 17 | 18 | 14 | 23 | 13 |
| $F_3$ | 32 | 27 | 18 | 41 | 19 |
| Capacity of the ware-house | 6 | 10 | 12 | 15 | 43 |

**Solution. Second Method (Vogel's approximation method) (VAM)**

**Step 1.** Here, the total production of the factories and total capacities of the ware-houses being the same i.e. 43, the problem is balanced.

**Step 2. Initial Basic Feasible solution.**

By VAM the difference between the smallest and next to the smallest costs in each row and each column are computed and written within brackets against the respective rows and column of table 1.

**Table -1**

| | | | | |
|---|---|---|---|---|
| 21 | 16 | 25 | [11] 13 | 11, (16 – 13 = 3) |
| 17 | 18 | 14 | 23 | 13, (17 – 14 = 3) |
| 32 | 27 | 18 | 41 | 19, (27 – 18 = 9) |
| 6 | 10 | 12 | 13 | |
| (21 – 17 = 4) | (18 – 16 = 2) | (18–14 = 4) | (23–13 = 10) | |

From table-1,

Largest difference (10) corresponds to 4th column. In this column, first row corresponds to lowest cost (13). So allocate to (1, 4) box with min (11, 23) ie 11.

The first row is exhausted. The reduced table 1 is written below

Table-2

| | | | | |
|---|---|---|---|---|
| 17 | 18 | 14 | [4] 23 | 13, (17 – 14 = 3) |
| 32 | 27 | 18 | 41 | 19, (27 – 18 = 9) |
| 6 | 10 | 12 | 15 – 11 = 4 | |
| (32 – 17 = 15) | (27 –18 = 9) | (18 – 14 = 4) | (41 – 23 = 18) | |

From table-2,

Largest difference 18 corresponds to 4th column. In this column first row corresponds to the lowest cost (13). So allocate to (1, 4) box min (4, 13) i.e. 4.

Fourth column is exhausted. The reduced table is written below

Table-3

| | | | |
|---|---|---|---|
| [6] 17 | 18 | 14 | 9 = 13 – 4, (17 – 14 = 3) |
| 32 | 27 | 18 | 19, (27 – 18 = 9) |
| 6 | 10 | 12 | |
| (32 – 17 = 15) | (27 – 18 = 9) | (18 – 14 = 4) | |

From table 3, largest difference (15) corresponds to first column. In this column first row corresponds to the lowest cost (17). So allocate to the (1, 1) box with the min (6, 9) ie 6.

First column of table - 3 is exhausted and the reduced table 4 is written below.

Table - 4

| | | |
|---|---|---|
| [3] 18 | 14 | 3 = 9 − 6, (18 − 14 = 4) |
| 27 | 18 | 19, (27 − 18 = 9) |
| 10 | 12 | |
| (27 − 18 = 9) | (18 − 14 = 4) | |

The largest difference (9) occurs at two places. So there is a tie here. Costs on both largest differences are identical choose any one. Allocate (1, 1) box with min (3, 10) i.e. 3.
First row of table-4 is exhausted. The reduced table 4 is written below:
From table 5, largest difference (27) corresponds to first column and is written in the box of first column and first row.

Table 5

| | | |
|---|---|---|
| [7] 27 | 18 | 19, (27 − 18 = 9) |
| 10 − 3 = 7 | 12 | |
| (27 − 0 = 27) | (18 − 0 = 18) | |

First column of table 5 is exhausted. The reduced table 6 has only one box.
From table 6, minimum of 12 and 12 is 12 and is written in this box.

Table 6

| | |
|---|---|
| [12] 18 | 19 − 7 = 12 |
| 12 | |

Finally the initial basic feasible solution is given in the table 7

Table 7

| | | | |
|---|---|---|---|
| 21 | 16 | 25 | [11] 13 |
| [6] 17 | [3] 18 | 14 | [4] 23 |
| 32 | [7] 27 | [12] 18 | 41 |

## Checking of optimality

We apply Modi-Method for checking of optimality as the number of allocation = $m + n -$ 1 (i.e. 6).

(*i*) Taking $r$ for row and $c$ for column, we have the following equations:

$r_2 + c_1 = 17$ Let $r_2 = 0,\ c_1 = 17$ $\quad r_2 + c_2 = 18 \Rightarrow \quad r_2 = 0,\ c_2 = 18$

$r_3 + c_2 = 27 \Rightarrow \quad c_2 = 18,\ r_3 = 9$ $\quad r_3 + c_3 = 18 \Rightarrow \quad r_3 = 9,\ c_3 = 9$

$r_2 + c_4 = 23 \Rightarrow \quad r_2 = 0,\ c_4 = 23$ $\quad r_1 + c_4 = 13 \Rightarrow \quad c_4 = 23,\ r_1 = -10$

(*ii*) Net Evaluation $E_{ij} = (r_i + c_j) - b_{ij}$ for all empty boxes

$E_{11} = r_1 + c_1 - b_{11} = -10 + 17 - 21 = -14$

$E_{12} = r_1 + c_2 - b_{12} = -10 + 18 - 16 = -8$

$E_{13} = r_1 + c_3 - b_{13} = -10 + 9 - 25 = -26$

$E_{23} = r_2 + c_3 - b_{23} = 0 + 9 - 14 = -5$

$E_{31} = r_3 + c_1 - b_{31} = \quad 9 + 17 - 32 = -6$

$E_{34} = r_3 + c_4 - b_{34} = \quad 9 + 23 - 41 = -9$

Since all the net evaluations are negative, so this solution is optimal.

**Table- 8**

| | | | |
|---|---|---|---|
| | | | 11 13 |
| 6 17 | 3 18 | | 4 23 |
| | 7 27 | 12 18 | |
| | | | |

Transportation charges = Transported units × transportation charges per unit

(*iii*) Optimal (minimum) transportation cost

= (11 × 13) + ( 6 × 17) + ( 3 × 18) + (4 × 23) + (7 × 27) + (12 × 18)

= 143 + 102 + 54 + 92 + 189 + 216

= Rs. 796 **Ans.**

## EXERCISE 27.5

***Obtain the dual of the following L.P.P.***

**1.** Maximize $Z = x_1 + 2x_2$ Subject to the constraints

$2x_1 - 3x_2 \le 3, \quad 4x_1 + x_2 \le -4, \quad x_1, x_2 \ge 0$

**Ans.** Minimize $Z' = 3y_1 - 4y_2$ Subject to the constraints

$2y_1 + 4y_2 \ge 1, \quad -3y_1 + y_2 \ge 2; \quad y_1, y_2 \ge 0.$

**2.** Maximize $Z = 5x_1 + 2x_2 + 6x_3 + 3x_4$ Subject to the constraints

$x_1 + x_2 + x_3 + x_4 \le 140, \quad 2x_1 + 5x_2 + 6x_3 + x_4 \le 260,$

$x_1 + 3x_2 + x_3 + 2x_4 \le 180, \quad x_1, x_2, x_3, x_4 \ge 0$

**Ans.** Minimize $Z' = 140y_1 + 260y_2 + 180y_3$ Subject to the constraints.

$y_1 + 2y_2 + y_3 \ge 5, \quad y_1 + 5y_2 + 3y_3 \ge 2, \quad y_1 + 6y_2 + y_3 \ge 6$

$y_1 + y_2 + 2y_3 \ge 3, \quad y_1, y_2, y_3 \ge 0$

**3.** Maximize $Z = 2x_1 + 2x_2 + 4x_3$ Subject to the constraints

$2x_1 + 3x_2 + 5x_3 \le 2, \quad 3x_1 + x_2 + 7x_3 \le 3, \quad x_1 + 4x_2 + 6x_3 \le 5, \quad x_1, x_2, x_3 \ge 0$

**Ans.** Minimize $Z' = 2y_1 + 3y_2 + 5y_3$ Subject to the constraints

$2y_1 + 3y_2 + y_3 \ge 2, \quad 3y_1 + y_2 + 4y_3 \ge 2, \quad 5y_1 + 7y_2 + 6y_3 \ge 4, \quad y_1 \ge 0, y_2 \ge 0, y_3 \ge 0$

**4.** Maximize $Z = 3x_1 - 2x_2 + 4x_3$ subject to the constraints

$3x_1 + 5x_2 + 4x_3 \le 7, \quad 6x_1 + x_2 + 3x_3 \le 4, \quad 7x_1 - 2x_2 - x_3 \le 10,$

$x_1 - 2x_2 + 5x_3 \le 3, \quad 4x_1 + 7x_2 - 2x_3 \le 2, \quad x_1, x_2, x_3 \ge 0$

**Ans.** Minimize $Z' = 7y_1 + 4y_2 + 10y_3 + 3y_4 + 2y_5$ Subject to the constraints

$3y_1 + 6y_2 + 7y_3 + y_4 + 4y_5 \ge 3, \quad 5y_1 + y_2 - 2y_3 - 2y_4 + 7y_5 \ge -2$

$4y_1 + 3y_2 - y_3 + 5y_4 - 2y_5 \ge 4, \quad y_1, y_2, y_3, y_4, y_5 \ge 0$

**5.** Maximize $Z = 3x_1 + 2x_2$
subject to the constraints

$x_1 + x_2 \le 1 \qquad x_1 + x_2 \le 7 \qquad x_1 + 2x_2 \le 10 \qquad x_1, x_2 \ge 0$

**Ans.** Minimize $Z' = y_1 + 7y_2 + 10y_3$ Subject to the constraints

$y_1 + y_2 + y_3 \ge 3, \quad y_1 + y_2 + 2y_3 \ge 2, \quad y_1, y_2, y_3 \ge 0$

**6.** Maximize $Z = 4x_1 + 3x_2$
subject to the constraints

$2x_1 + 9x_2 \le 180, \qquad 3x_1 + 6x_2 \le 120, \qquad x_1 + x_2 \le 180, \qquad x_1, x_2 \ge 0.$

**Ans.** Min $Z' = 180y_1 + 120y_2 + 180y_3$ Subject to the constraints

$2y_1 + 3y_2 + y_3 \ge 4, \quad 9y_1 + 6y_2 + y_3 \ge 3, \quad y_1 \ge 0, y_2 \ge 0$

**Using duality solve the following problems**

**7.** Maximize $Z = 2x_1 + x_2$ subject to the constraints

$x_1 + 2x_2 \le 10, x_1 + x_2 \le 6, x_1 - x_2 \le 2 \qquad x_1 - 2x_2 \le 1, x_1, x_2 \ge 0$

**Ans.** $x_1 = 4$, $x_2 = 2$, Maximum $Z = 10$

**8.** Maximize $Z = 3x_1 + 2x_2$ subject to the constraints

$x_1 + x_2 \ge 1, x_1 + x_2 \le 7, x_1 + 2x_2 \le 10, x_2 \le 3; \; x_1, x_2 \ge 0$

**Ans.** $x_1 = 7$, $x_2 = 0$, Max. $Z = 21$

**9.** Minimize $Z = 3x_1 + 4x_2$ subject to the constraints

$5x_1 + 10x_2 \ge 800, \qquad 15x_1 + 10x_2 \ge 1200, \qquad x_1, x_2 \ge 0$

**Ans.** Minimum $Z = 480$,

**10.** Maximize $Z = 5x_1 + 2x_2$ subject to the constratins

$10x_1 + 2x_2 \le 2100, \qquad x_1 + 0.5\,x_2 \le 600, \qquad x_2 \le 800, \qquad x_1, x_2 \ge 0$

**Ans.** Max. $Z = 1850$, $x_1 = 50$, $x_2 = 800$

**11.** *A company has factories $F_1$, $F_2$, $F_3$ which supply ware-houses at $W_1$, $W_2$ and $W_3$. Weekly factory capacities, weekly ware-house requirements and unit shipping costs (in rupees) are as follows:*

| Factories \ Ware-houses | $W_1$ | $W_2$ | $W_3$ | Supply |
|---|---|---|---|---|
| $F_1$ | 16 | 20 | 12 | 200 |
| $F_2$ | 14 | 8 | 18 | 160 |
| $F_3$ | 26 | 24 | 16 | 90 |
| Demand | 180 | 120 | 150 | 450 |

Determine the optimal distribution for this company to minimize shipping costs.

**Ans.** By North-West corner method, $x_{11} = 180$, $x_{12} = 20$, $x_{22} = 100$, $x_{23} = 60$, $x_{33} = 90$, Min. cost = Rs. 6600.

By V.A.M. $x_{11} = 140$, $x_{13} = 60$, $x_{21} = 40$, $x_{12} = 120$, $x_{33} = 90$, Min. cost = Rs. 5920.

**12.** *Solve the following transportation problem:*

| Consumers \ Suppliers | A | B | C | Available |
|---|---|---|---|---|
| I | 6 | 8 | 4 | 14 |
| II | 4 | 9 | 8 | 12 |
| III | 1 | 2 | 6 | 5 |
| Requirement | 6 | 10 | 15 | 31 |

**Ans.** By North-West Method : $x_{11} = 6$, $x_{12} = 8$, $x_{22} = 2$, $x_{23} = 10$, $x_{33} = 5$, Min. cost = Rs. 228.

By V.A.M: $x_{13} = 14$, $x_{21} = 6$, $x_{22} = 5$, $x_{22} = 5$, $x_{23} = 1$, $x_{32} = 5$, Min. cost = Rs. 143

# USEFUL FORMULAE

## TRIGONOMETRY

$\sin 2\theta = 2\sin\theta\cos\theta$, $\cos 2\theta = 2\cos^2\theta - 1$, $\cos 2\theta = 1 - 2\sin^2\theta$

| Angle | 0° | 30° | 45° | 60° | 90° | 180° |
|---|---|---|---|---|---|---|
| sin | 0 | $\frac{1}{2}$ | $\frac{1}{\sqrt{2}}$ | $\frac{\sqrt{3}}{2}$ | 1 | 0 |
| cos | 1 | $\frac{\sqrt{3}}{2}$ | $\frac{1}{\sqrt{2}}$ | $\frac{1}{2}$ | 0 | −1 |
| tan | 0 | $\frac{1}{\sqrt{3}}$ | 1 | $\sqrt{3}$ | ∞ | 0 |

$\sin(-\theta) = -\sin\theta$, $\cos(-\theta) = \cos\theta$

$\sin(90+\theta) = \cos\theta$ **(change)**

$\sin(\pi - \theta) = \sin\theta$ **(No change)**

**Hyperbolic Functions**

$$\sinh x = \frac{e^x - e^{-x}}{2}, \quad \cosh x = \frac{e^x + e^{-x}}{2}, \quad \cosh^2 x - \sinh^2 x = 1$$

$$\frac{d}{dx}(\sinh x) = \cosh x, \quad \frac{d}{dx}(\cosh x) = \sinh x,$$

$$\sin x = \frac{e^{ix} - e^{-ix}}{2i}, \quad \cos x = \frac{e^{ix} + e^{-ix}}{2}$$

$\sinh ix = i\sin x$, $i\sinh x = \sin ix$, cash $ix = \cos x$, $\cosh x = \cos ix$

**Binomial Theorem** $(1+x)^n = 1 + nx + \frac{n(n-1)}{2!}x^2 + \frac{n(n-1)(n-2)}{3!}x^3 + \ldots$

**Polar coordinates** $x = r\cos\theta$, $y = r\sin\theta$, $r^2 = x^2 + y^2$, $\theta = \tan^{-1}\frac{y}{x}$

$x = r\sin\theta\cos\phi$, $y = r\sin\theta\sin\phi$, $z = r\cos ix$

**Median** is the line joining the vertex to the mid point of the opposite side of a triangle.

**Centroid** or C.G. is the point of intersection of the medians of a triangle.

**Incentre** is the point of intersection of the bisectors of the angles of a triangle.

**Circumcentre** is the point of intersection of the perpendicular bisectors of the sides of a triangle.

**Orthocentre** is the point of intersection of the perpendiculars drawn from vertex to the opposite sides of a triangle.

**Asymptote** is the tangent to a curve at infinity.

## DIFFERENTIAL CALCULUS

$\frac{d}{dx}(x^n) = nx^{n-1}$ $\qquad$ $\frac{d}{dx}(\log_e x) = \frac{1}{x}$ $\qquad$ $\frac{d}{dx}(\sin x) = \cos x,$

$\frac{d}{dx}(\cos x) = -\sin x$ $\qquad$ $\frac{d}{dx}(\tan x) = \sec^2 x$ $\qquad$ $\frac{d}{dx}(\cot x) = -\text{cosec}^2 x$

$\frac{d}{dx}(\sec x) = \sec x \tan x$ $\qquad$ $\frac{d}{dx}(\text{cosec} x) = -\text{cosec}\, x \cot x$ $\qquad$ $\frac{d}{dx}(a^x) = a^x \log_e a$

$\frac{d}{dx}(e^x) = e^x$ $\qquad$ $\frac{d}{dx}(\sin^{-1} x) = \frac{1}{\sqrt{1-x^2}}$ $\qquad$ $\frac{d}{dx}(\cos^{-1} x) = \frac{-1}{\sqrt{1-x^2}}$

$\frac{d}{dx}(\tan^{-1} x) = \frac{1}{1+x^2}$ $\qquad$ $\frac{d}{dx}(\cot^{-1} x) = -\frac{1}{1+x^2}$ $\qquad$ $\frac{d}{dx}(\sec^{-1} x) = \frac{1}{x\sqrt{x^2-1}}$

$\frac{d}{dx}(\text{cosec}^{-1} x) = \frac{-1}{x\sqrt{x^2-1}}$ $\qquad$ $\frac{d}{dx}(\sinh x) = \cosh x$ $\qquad$ $\frac{d}{dx}(\cosh x) = \sinh x$

$\frac{d}{dx}(\tanh x) = \text{sech}^2 x$ $\qquad$ $\frac{d}{dx}(\coth x) = -\text{cosech}^2 x$

$\frac{d}{dx}(\text{sech}\,x) = -\text{sech}\,x \tanh x \qquad \frac{d}{dx}(\text{cosech}\,x) = -\text{cosech}\,x \coth x$

## INTEGRAL CALCULUS

$\int x^n dx = \frac{x^{n+1}}{n+1} \qquad \int \frac{1}{x} dx = \log_e x \qquad \int e^x dx = e^x$

$\int a^x dx = a^x \log_a e \qquad \int \sin x dx = -\cos x \qquad \int \cos x dx = \sin x$

$\int \tan x dx = \log \sec x \qquad \int \cot x dx = \log \sin x \qquad \int \frac{dx}{a^2 - x^2} = \frac{1}{2a} \log \frac{a+x}{a-x}$

$\int \sec x dx = \log \tan\left(\frac{x}{2} + \frac{\pi}{4}\right) = \log(\sec x + \tan x) \qquad \int \sinh x dx = \cosh x$

$\int \text{cosec}\, x dx = \log \tan \frac{x}{2} = \log(\text{cosec}\, x - \cot x) \qquad \int \text{cosech}^2 x dx = -\coth x$

$\int \sec x \tan x dx = \sec x \qquad \int \text{cosec} x \cot x dx = -\text{cosec} x$

$\int \frac{dx}{\sqrt{a^2 - x^2}} = \sin^{-1} \frac{x}{a} \qquad \int \frac{dx}{\sqrt{a^2 + x^2}} = \sinh^{-1} \frac{x}{a} \qquad \int \frac{dx}{\sqrt{x^2 - a^2}} = \cosh^{-1} \frac{x}{a}$

$\int \frac{dx}{a^2 + x^2} = \frac{1}{a} \tan^{-1} \frac{x}{a} \qquad \int \frac{-dx}{x^2 + a^2} = \frac{1}{a} \cot^{-1} \frac{x}{a} \qquad \int \frac{dx}{x\sqrt{x^2 - a^2}} = \frac{1}{a} \sec^{-1} \frac{x}{a}$

$\int \frac{-dx}{x\sqrt{x^2 - a^2}} = \frac{1}{a} \text{cosec}^{-1} \frac{x}{a} \qquad \int \cosh x dx = \sinh x \qquad \int \text{sech}^2 x dx = \tanh x$

$\int \text{sech}\, x \tanh x dx = -\text{sech}\, x \qquad \int \text{cosech}\, x \coth x dx = -\text{cosech}\, x$

$\int \sqrt{a^2 - x^2} dx = \frac{x}{2}\sqrt{a^2 - x^2} + \frac{a^2}{2} \sin^{-1} \frac{x}{a} \qquad \int \sqrt{a^2 + x^2} dx = \frac{x}{2}\sqrt{a^2 + x^2} + \frac{a^2}{2} \sinh^{-1} x$

$\int \sqrt{x^2 - a^2} dx = \frac{x}{2}\sqrt{x^2 - a^2} - \frac{a^2}{2} \cosh^{-1} \frac{x}{a}$

## PARTIAL DIFFERENTIATION

**Euler's Theorem:** If $z$ is a homogeneous function in $x$, $y$ of degree $n$,

then $x\frac{\partial z}{\partial x} + y\frac{\partial z}{\partial y} = nz; \qquad x^2 \frac{\partial^2 z}{\partial x^2} + 2xy\frac{\partial^2 z}{\partial x \partial y} + y^2 \frac{\partial^2 z}{\partial y^2} = n(n-1)z$

**Deduction I** If (*i*) $z = f(u)$ (*ii*) $z = x^n \phi\left(\frac{y}{x}\right)$, Then $x\frac{\partial u}{\partial x} + y\frac{\partial u}{\partial y} = n\frac{f(u)}{f'(u)}$

**Deduction II** $x^2 \frac{\partial^2 u}{\partial x^2} + 2xy\frac{\partial^2 u}{\partial x \partial y} + y^2 \frac{\partial^2 u}{\partial y^2} = g(u)[g'(u) - 1]$, Where $g(u) = n\frac{f(u)}{f'(u)}$

$$\frac{\partial z}{\partial t} = \frac{\partial z}{dx}\frac{dx}{dt} + \frac{dz}{dy}\frac{dy}{dt} \quad \text{if } z = f(x, y),\ x = \phi_1(t),\ y = \phi_2(t)$$

$$\frac{dy}{dx} = -\frac{\frac{\partial f}{\partial x}}{\frac{\partial f}{\partial y}} \quad \text{if } f(x, y) = c, \quad \frac{d^2 y}{dx^2} = -\frac{q^2 r - 2pqs + p^2 t}{q^3}$$

$$P = \frac{\partial f}{\partial x},\ q = \frac{\partial f}{\partial y},\ r = \frac{\partial^2 f}{\partial x^2},\ s = \frac{\partial^2 f}{\partial x \partial y},\ t = \frac{\partial^2 f}{\partial y^2}$$

**Taylor's Series:** $f(a+h, b+k) = f(a, b) + \left(h\frac{\partial}{\partial x} + k\frac{\partial}{\partial y}\right)f + \frac{1}{2!}\left(h\frac{\partial}{\partial x} + k\frac{\partial}{\partial y}\right)^2 f + \ldots\ldots$

$$f(x,y) = f(0,0) + \left(x\frac{\partial f}{\partial x} + y\frac{\partial f}{\partial y}\right)_{(0,0)} + \frac{1}{2!}\left(x^2\frac{\partial^2 f}{\partial x^2} + 2xy\frac{\partial^2 f}{\partial x \partial y} + y^2\frac{\partial^2 f}{\partial y^2}\right)_{(0,0)} + \ldots$$

**Maximum or minimum:** (*i*) $\frac{\partial f}{\partial x} = 0$, (*ii*) $\frac{\partial f}{\partial y} = 0$, (*iii*) $\frac{\partial^2 f}{\partial x^2}\cdot\frac{\partial^2 f}{\partial y^2} > \left(\frac{\partial^2 f}{\partial x \partial y}\right)^2$

**For Maximum:** $\frac{\partial^2 f}{\partial x^2} < 0$, for minimum: $\frac{\partial^2 f}{\partial x^2} > 0$

Lagranges Method for $f(x, y, z)$ to be maximum/minimum, if $\phi(x,y,z) = 0$

$$\frac{\partial f}{\partial x} + \lambda\frac{\partial \phi}{\partial x} = 0,\quad \frac{\partial f}{\partial y} + \lambda\frac{\partial \phi}{\partial y} = 0,\quad \frac{\partial f}{\partial z} + \lambda\frac{\partial \phi}{\partial z} = 0$$

**Jacobian**

$$J\left(\frac{u,v,w}{x,y,z}\right) = \frac{\partial(u,v,w)}{\partial(x,y,z)} = \begin{vmatrix} \frac{\partial u}{\partial x} & \frac{\partial u}{\partial y} & \frac{\partial u}{\partial z} \\ \frac{\partial v}{\partial x} & \frac{\partial v}{\partial y} & \frac{\partial v}{\partial z} \\ \frac{\partial w}{\partial x} & \frac{\partial w}{\partial y} & \frac{\partial w}{\partial z} \end{vmatrix} \qquad (i)\ \frac{\partial(u,v)}{\partial(x,y)} \times \frac{\partial(x,y)}{\partial(u,v)} = 1$$

(*ii*) $\frac{\partial(u,v)}{\partial(x,y)} = \frac{\partial(u,v)}{\partial(r,s)} \times \frac{\partial(r,s)}{\partial(x,y)}$, (*iii*) if $\frac{\partial(u,v,w)}{\partial(x,y,z)} = 0$ then $u$, $v$, $w$ are functionally related.

$dx\,dy = r\,d\theta\,dr \quad dxdydz = r^2 \sin\theta\, d\,r\, d\theta\, d\phi$

Area $= \int_{x_1}^{x_2}\int_{y_1}^{y_2} dx\,dy$, Volume $= \int_{x_1}^{x_2}\int_{y_1}^{y_2}\int_{z_1}^{z_2} dx\,dy\,dz$, $\quad dxdy = \frac{\partial(x,y)}{\partial(r,\theta)} dr\,d\theta$

Centre of gravity

$$\bar{x} = \frac{\iiint x\rho\,dx\,dy\,dz}{\iiint \rho\,dx\,dy\,dz},\quad \bar{y} = \frac{\iiint y\rho\,dx\,dy\,dz}{\iiint \rho\,dx\,dy\,dz},\quad \bar{z} = \frac{\iiint z\rho\,dx\,dy\,dz}{\iiint \rho\,dx\,dy\,dz}$$

**Moment of Inertia about $x$ – axis** $= \iiint \rho(y^2 + z^2)\,dx\,dy\,dz$

**Moment of Inertia about $y$ – axis** $= \iiint \rho(x^2 + z^2)\,dx\,dy\,dz$

**Moment of Inertia about $z$ – axis** $= \iiint \rho(x^2 + y^2)\,dx\,dy\,dz$

Centre of pressure $\quad \bar{x} = \frac{\iint x\rho\,d\,x\,d\,y}{\iint_A \rho\,dx\,dy},\quad \bar{y} = \frac{\iint_A y\rho\,dx\,dy}{\iint_A \rho\,dx\,dy}$

$$\Gamma(n+1) = \lfloor n,\qquad \Gamma(n+1) = n\,\Gamma n,\qquad \Gamma\left(\frac{1}{2}\right) = \sqrt{\pi}$$

**Gamma Function:** $\int_0^\infty e^{-x} x^{n-1}\,dx = \Gamma n$

**Beta Function:** $\beta(l,m) = \int_0^\infty x^{l-1}(1-x)^{m-1}\,dx = \frac{\Gamma l\,\Gamma m}{\Gamma(l+m)}$, $\quad \int_0^{\frac{\pi}{2}} \sin^m\theta \cos^n\theta\, d\theta = \frac{\Gamma\frac{m+1}{2}\,\Gamma\frac{n+1}{2}}{2\,\Gamma\frac{m+n+2}{2}}$

**Dirichlet's Integral** $= \iiint x^{l-1} y^{m-1} z^{n-1}\,dx\,dy\,dz = \frac{\Gamma l\,\Gamma m\,\Gamma n}{\Gamma(l+m+n+1)}$

**Liouville's Extension of Dirichlet theorem**

$$\iiint f(x+y+z)x^{l-1}.\,y^{m-1}.\,z^{n-1}dx\,dy\,dz = \frac{\Gamma l\,\Gamma m\,\Gamma n}{\Gamma l+m+n+1}\int_{h_1}^{h_2} f(u)u^{l+m+n-1}du$$

$$\textbf{Error function} = \frac{2}{\sqrt{\pi}}\int_0^x e^{-t^2}.dt$$

$$\frac{d}{d\alpha}\left\{\int_{\phi(\alpha)}^{\psi(\alpha)} f(x,y)dx\right\} = \frac{d}{dx}\left[\int_{\phi(\alpha)}^{\psi(\alpha)}\frac{\partial f(x,\alpha)}{\partial\alpha}dx - \frac{d\phi}{d\alpha}f[\phi(\alpha),\alpha]\right] + \frac{d\psi}{d\alpha}f[\psi(\alpha),\alpha]$$

**Differential equations**

(*i*) Variables separable: $f(y)dy = \phi(x)dx, \quad \int f(y)dy = \int \phi(x)dx + c$

(*ii*) Homogeneous Equation: $\frac{dy}{dx} = \frac{f(x)}{\phi(x)}$, where each term of $f(x)$ and $\phi(x)$ are of the same degree.

Put $y = vx$ so that $\frac{dy}{dx} = v + x.\frac{dv}{dx}$

(*iii*) Reducible to homogeneous: $\frac{dy}{dx} = \frac{ax+by+c}{Ax+By+C}$

Put $x = X+h,\ y = Y+k$ if $\frac{a}{A} \neq \frac{b}{B}$, Put $ax+by = z$ if $\frac{a}{A} = \frac{b}{B}$

Linear differential equation: $\frac{dy}{dx} + Py = Q$ where $P$ and $Q$ are not functions of $y$.

Integrating factor = $e^{\int pdx}$, then $y.e^{\int Pdx} = \int (Qe^{\int pdx})dx + c$

**Rules to find Complementary Function** $D \equiv \frac{d}{dz}$

1. when roots of A.E. = $m_1, m_2$; $\Rightarrow$ C.F. = $c_1e^{m_1x} + c_2e^{m_2x}$
2. when roots are equal; $\Rightarrow$ C.F. = $(c_1 + c_2x)e^{mx}$
3. when roots are complex $a \pm ib$; $\Rightarrow$ C.F. = $e^{ax}[c_1\cos bx + c_2\sin bx]$

**Rules to Find Particular Integral:**

(*i*) $\frac{1}{f(D)}e^{ax} = \frac{1}{f(a)}e^{ax}$, if $f(a) \neq 0$; $\frac{1}{f(x)}e^{ax} = x\frac{1}{f'(a)}e^{ax}$ if $f(a) = 0$

(*ii*) $\frac{1}{f(D)}x^n = [f(D)]^{-1}x^n$, Expand $[f(D)]^{-1}$ and then operate

(*iii*) $\frac{1}{f(D^2)}\sin ax = \frac{1}{f(-a^2)}\sin ax \qquad \frac{1}{f(D^2)}\cos ax = \frac{1}{f(-a^2)}\cos ax$

If $f(-a^2) = 0$ then $\frac{1}{f(D^2)}\sin ax = x\frac{1}{f'(-a^2)}\sin ax$ (*iv*) $\frac{1}{f(D)}e^{ax}.\phi(x) = e^{ax}.\frac{1}{f(D+a)}\phi(x)$

(*v*) $\frac{1}{f(D)}x\phi(x) = \left[x - \frac{1}{f(D)} \times f'(D)\right]\frac{1}{f(D)}\phi(x)$ (*vi*) $\frac{1}{D+a}\phi(x) = e^{-ax}\int e^{ax}\phi(x)dx$

Variation of Parameters C.F. = $Ay_1 + By_2$ then P.I. = $uv_1 + vy_2$

Where $u = -\int\frac{y_2Xdx}{y_1y_2' - y_1'y_2} \qquad v = \int\frac{y_1Xdx}{y_1y_2' - y_1'y_2}$

**Homogeneous Differential Equation** $x^2\frac{d^2y}{dx^2} - 2x\frac{dy}{dx} + 4y = x^4$

Put $x = e^z$, $x\frac{dy}{dx} = Dy$, $x^2\frac{d^2y}{dx^2} = D(D-1)y$, where $D \equiv \frac{d}{dz}$

**Matrices**

**Types of equations,** $Ax = B$; $C = \{A, B\}$

(1) ***Consistent equation*** If Rank of $C$ = Rank of $A$

(*a*) ***Unique solution*** If Rank of $C$ = Rank of A = Number of unknowns

(*b*) ***Infinite solution*** If Rank of $C$ = Rank of A < Number of unknowns

(2) **Inconsistent equation.** If Rank of $C \neq$ Rank of $A$.

***Eigen values*** are the roots of the characteristics equation $|A - \lambda I| = 0$

***Cayley Hamilton*** Theorem. Every square matrix satisfies its own characteristics equation.

***Diagonalisation*** $P^{-1}AP = D$, where $P$ is the modal matrix containing eigen vectors, $D$ is the diagonal matrix containing eigen values.

**VECTORS** If $\bar{r} = xi + yj + zk$ then $|\bar{r}| = \sqrt{x^2 + y^2 + z^2}$

$\overrightarrow{AB}$ = Position vector of $B$ – Position vector of $A$, Ratio formula $\bar{c} = \dfrac{m\bar{b} + n\bar{a}}{m+n}$

Scalar Product: $\bar{a}.\bar{b} = |\bar{a}|\,|\bar{b}|\cos\theta$, Work done = $\bar{F}.\bar{r}$, $\theta = \cos^{-1}\dfrac{\bar{a}.\bar{b}}{|\bar{a}||\bar{b}|}$

Vector Product $\bar{a} \times \bar{b} = |\bar{a}|\,|\bar{b}|\sin\theta.\hat{\eta}$.

Area of parallelogram = $\bar{a} \times \bar{b}$; Moment of a force = $\bar{r} \times \bar{F}$; $\vec{V} = \bar{w} \times \bar{r}$

$$\bar{a}.(\bar{b} \times \bar{c}) = \begin{bmatrix} a_1 & a_2 & a_3 \\ b_1 & b_2 & b_3 \\ c_1 & a_2 & c_3 \end{bmatrix},$$

$$\bar{a} \times (\bar{b} \times \bar{c}) = (\bar{a}.\bar{c})\bar{b} - (\bar{a}.\bar{b})\bar{c}$$

$$(\bar{a} \times \bar{b}).(\bar{c} \times \bar{d}) = \begin{vmatrix} \bar{a}.\bar{c} & \bar{a}.\bar{d} \\ \bar{b}.\bar{c} & \bar{b}.\bar{d} \end{vmatrix}$$

Volume of parallelopiped = $\bar{a}.(\bar{b} \times \bar{c})$, if $\bar{a}.(\bar{b} \times \bar{c}) = 0$ then $\bar{a}, \bar{b}, \bar{c}$ are coplanar.

$$(\bar{a} \times \bar{b}) \times (\bar{c} \times \bar{d}) - [\overline{abd}]\bar{c} - [\overline{abc}]\bar{d} = -[\overline{bcd}]\bar{a} + [\overline{acd}]\bar{b}$$

Velocity = $\dfrac{d\bar{r}}{dt}$, Acceleration = $\dfrac{d^2\bar{r}}{dt^2}$, Tangent vector = $\dfrac{d\bar{r}}{dt}$, Normal vector = $\bar{\nabla}\phi$

**Gradient** $\phi = \bar{\nabla}\phi$ Directional derivative = $\bar{\nabla}\phi$

**Divergence** $\bar{f} = \bar{\nabla}.\bar{f}$, If divergence $\bar{f} = 0$, then $f$ is called **solenoidal** vector.

**Curl** $\bar{f} = \bar{\nabla} \times \bar{f}$, If curl $\bar{f} = 0$, $\bar{f}$ is called **Irrotational** vector.

**Green's Theorem:** $\displaystyle\int_c \phi dx + \psi dy = \iint_s \left(\frac{\partial \psi}{\partial x} - \frac{\partial \phi}{\partial y}\right) dxdy$, **Stokes Theorem:** $\displaystyle\int_C \bar{F}.\bar{dr} = \iint_s \text{curl}\,\bar{F}.\hat{\eta}\,ds$

**Gauss theorem of Divergence:** $\displaystyle\iint_s \bar{F}.\eta ds = \iiint_v div\,\bar{f}\,dx\,dy\,dz$

**COMPLEX VARIABLE FUNCTION**

**Analytic function** *A* single valued function which is differentiable at $z = z_0$ is said to be analytic at the point $z = z_0$

***C – R* Equations:** $\dfrac{\partial u}{\partial x} = \dfrac{\partial v}{\partial y}$, $\dfrac{\partial u}{\partial y} = \dfrac{-\partial v}{\partial x}$. And $\dfrac{\partial u}{\partial r} = \dfrac{1}{r}\dfrac{\partial v}{\partial \theta}$; $\dfrac{\partial u}{\partial \theta} = -r\dfrac{\partial v}{\partial r}$

To find conjugate function, $dv = \dfrac{\partial v}{\partial x}dx + \dfrac{\partial v}{\partial y}dy = \dfrac{-\partial u}{\partial y}dx + \dfrac{\partial u}{\partial x}dy$ if $u$ is given

**Milne Thomson Method** $f(z) = \displaystyle\int \phi_1(z,0)\,dz + \int \phi_2(z,0)\,dz$, where $\phi_1(x,y) = \dfrac{\partial u}{\partial x}$, $\phi_2(x,y) = \dfrac{\partial u}{\partial y}$

$f(z) = \displaystyle\int \psi_1(z,0)dz + \int \psi_1(z,0)\,dz$, where $\psi_1(x,y) = \dfrac{\partial v}{\partial y}$, $\psi_2(x,y) = \dfrac{\partial v}{\partial x}$.

***Cauchy's Integral Theorem*** $\displaystyle\int_c f(z)\,dz = 0$ if $f(z)$ is analytic function within $C$.

***Cauchy's Integral formula*** $\displaystyle\int_c \frac{f(z)dz}{z-a} = 2\pi i f(a)$, if $f(z)$ is analytic in $c$, and a is a point within $C$.

**Residue** (*i*) Res $f(a) = \lim_{z \to a} (z-a) f(z)$, (*ii*) Res $(a) = \frac{\phi(a)}{\psi'(a)}$,

(*iii*) Res $(a) = \frac{1}{\lfloor n-1} \left\{ \frac{d^{n-1}}{dz^{n-1}} (z-a)^n f(z) \right\}$, (*iv*) Res $(a)$ = coefficient of $\frac{1}{t}$ where $t = z - a$

**Residue Theorem** $\int_c f(z)dz = 2\pi i$ (Sum of the residues at the poles written $C$)

$\int_0^{2\pi} f(\sin\theta, \cos\theta)\, d\theta$, put $\sin\theta = \frac{1}{2i}[z - \frac{1}{z}]$, $\cos\theta = \frac{1}{2}\left(z + \frac{1}{z}\right)$, $d\theta = \frac{dz}{iz}$

$C$ is the circle of radius one.

$\int_{-\infty}^{\infty} \frac{f_1(x)}{f_2(x)} dx$, consider $\int_c f(z)dz$ where $f(x) = \frac{f_1(x)}{f_2(x)}$ and $c$ is the semicircle with real axis.

**Bessel's Equation** $x^2 \frac{d^2y}{dx^2} + x\frac{dy}{dx} + (x^2 - n^2)y = 0$, $J_n(x) = \sum_{r=0}^{\infty} \frac{(-1)^r}{\lfloor r \lfloor n+r+1} \left(\frac{x}{2}\right)^{n+2r}$

**Recurrence Formula**

(*i*) $xJ_n' = nJ_n - xJ_{n+1}$ (*ii*) $xJ_n' = -nJ_n + xJ_{n-1}$ (*iii*) $2J_n' = J_{n-1} - J_{n+1}$

(*iv*) $2nJ_n = x(J_{n-1} + J_{n+1})$ (*v*) $\frac{d}{dx}(x^{-n}J_n) = -x^{-n}J_{n+1}$ (*vi*) $\frac{d}{dx}(x^n J_n) = x^n J_{n-1}$

**Legendre's Equation** $(1-x^2)\frac{d^2y}{dx^2} - 2x\frac{dy}{dx} + n(n-1)y = 0$

$$P_n(x) = \frac{1.3.5...(2n-1)}{\lfloor n}\left[x^n - \frac{n(n-1)}{(2n-1)2}x^{n-2} + \frac{n(n-1)(n-2)(r-3)}{(2n-1)(2n-3)2.4}x^{n-4} + ...\right]$$

***Rodrigue's formula*** $P_n(x) = \frac{1}{2^n \lfloor n} \frac{d^n}{dx^n}(x^2-1)^n$

***Generating Function*** $(1 - 2xz + z^2)^{-\frac{1}{2}} = \Sigma P_n(x) z^n$

***Orthogonality Property*** $\int_{-1}^{+1} P_n(x). P_m(x)dx = 0$. $m \neq n$ and $\int_{-1}^{+1} P_n^2(x)dx = \frac{2}{2n+1}$

***Recurrence Formulae*** (*i*) $(n+1)P_{n+1} = (2n+1)xP_n - nP_{n-1}$ (*ii*) $nP_n = xP_n' - P_{n-1}'$

(*iii*) $(2n+1)P_n = P_{n+1}' - P_{n-1}'$ (*iv*) $P_n' - xP_{n-1}' = nP_{n-1}$

(*v*) $(x^2-1)P_n' = n[xP_n - P_{n-1}]$ (*vi*) $(x^2-1)P_n' = (n+1)(P_{n+1} - xP_n)$

***Partial differential equation*** $Pp + Qq = R \Rightarrow \frac{dx}{P} = \frac{dy}{Q} = \frac{dz}{R}$ we can also use multipliers.

Homogeneous equations $a_0 \frac{\partial^2 z}{\partial x^2} + a_1 \frac{\partial^2 z}{\partial x \partial y} + a_2 \frac{\partial^2 z}{\partial y^2} = 0 \Rightarrow$ A.E. is $a_0 m^2 + a_1 m + a_2 = 0$

**Case I.** If $m = m_1$, $m = m_2$, C.F. $= f_1(y + m_1 x) + f_2(y + m_2 x)$

**Case II.** If $m = m_1 = m_2 \Rightarrow$ C.F. $= f_1(y + m_1 x) + x f_2(y + m_1 x)$

(*i*) Particular Integral $= \frac{1}{f(D, D')} e^{ax+by} = \frac{1}{f(a,b)} e^{ax+by}$

(*ii*) P.I. $= \frac{1}{f(D^2, DD', D^2)} \sin(ax+by) = \frac{\sin(ax+by)}{f(-a^2, -ab, -b^2)}$

(*iii*) P.I. $= \frac{1}{f(D, D')} f(x, y)$, use Binomial Theorem

(iv) P.I. $= \frac{1}{f(D+mD')} f(x,y) = \int f(x, c+mx)dx$

Non Homogeneous Equations Monges' Method $Rr + Ss + Tt = v$ (i)

$$r = \frac{dp - sdy}{dx}, \ t = \frac{dq - sdx}{dy}$$ substitute the value of $r$ and $t$ in (ii).

$$R\,dy^2 - S\,dxdy + T\,dx^2 = 0$$

**Applications** (i) $\frac{\partial^2 u}{\partial t^2} = c^2 \frac{\partial^2 u}{\partial x^2}$ (wave eq.) (ii) $\frac{\partial u}{\partial t} = c^2 \frac{\partial^2 u}{\partial x^2}$ (One dimension heat flow)

(iii) $\frac{\partial^2 u}{\partial x^2} + \frac{\partial^2 u}{\partial y^2} = 0$ (Two dimensions heat flow)

**STATISTICS** $A.M. = \frac{\Sigma fx}{\Sigma f}$, $A.M. = a + \frac{\Sigma fd}{\Sigma f}$, $A.M. = a + \frac{\Sigma f d'}{\Sigma f} i$

$$S.D. = \sqrt{\frac{\Sigma f(x-\bar{x})^2}{\Sigma f}}, \ S.D. = \sqrt{\frac{\Sigma fd^2}{\Sigma f} - \left(\frac{\Sigma fd}{\Sigma f}\right)^2}$$, Variance = $(S.D.)^2$

Coefficient of correlation $= \frac{\Sigma(x-\bar{x})(y-\bar{y})}{\sqrt{\Sigma(x-\bar{x})^2}\sqrt{\Sigma(y-\bar{y})^2}}$

Equation of line of regression of $y$ on $x$ is, $y - \bar{y} = r\frac{\sigma_y}{\sigma_x}(x - \bar{x})$

Equation of regression of $x$ on $y$ is $x - \bar{x} = r\frac{\sigma_x}{\sigma_y}(y - \bar{y})$

**Probability:** $P = \frac{\text{Number of favourable ways}}{\text{total number of equally likely ways}}$ $p + q = 1$,

$p$ (A or B) = $p$ (A) + $p$ (B), $p$ (A and B) = $p$ (A) . $p$ (B)

**Binomial Distribution:** $p(r) = nc_r p^r q^{n-r}$ Mean = $np$, Var = $npq$

**Poisson Distribution:** $p(r) = \frac{e^{-m}.m^r}{\lfloor r}$ Mean = $m$, var = $m$

**Normal Distribution:** $f(x) = \frac{1}{\sigma\sqrt{2\pi}} e^{-\frac{(x-\mu)^2}{2\sigma^2}}$ Standard Variate = $z = \frac{x-\mu}{\sigma}$

## FOURIER SERIES

$$f(x) = \frac{a_0}{2} + a_1 \cos x + a_2 \cos 2x + .... + a_n \cos nx + ... + b_1 \sin x + b_2 \sin 2x + ... + b_n \sin nx + ...$$

Where $a_0 = \frac{1}{\pi}\int_0^{2\pi} f(x)dx$, $a_n = \frac{1}{\pi}\int_0^{2\pi} f(x)\cos nx\,dx$, $b_n = \frac{1}{\pi}\int_0^{2\pi} f(x)\sin nx\,dx$

For even function: $a_0 = \frac{2}{\pi}\int_0^{\pi} f(x)dx$, $a_n = \frac{2}{\pi}\int_0^{\pi} f(x)\cos nx\,dx$, $b_n = 0$

For odd function: $a_0 = 0$, $a_n = 0$, $b_n = \frac{2}{\pi}\int_0^{\pi} f(x)\sin nx\,dx$

For arbitrary function:

$$a_0 = \frac{1}{c}\int_0^{2c} f(x)\,dx, \ a_n = \frac{1}{c}\int_0^{2c} f(x)\cos\frac{n\pi x}{c}dx \qquad b_n = \frac{1}{c}\int_0^{2c} f(x)\sin\frac{n\pi x}{c}dx$$

$\sin n\pi = 0$, $\cos n\pi = (-1)^n$

$\int uv\,dx = uv_1 - u'v_2 + u''v_3 - u'''v_4....$ (General formula for integration by parts)

**LAPLACE TRANSFORMATION**

1. $L(1) = \frac{1}{s}$ 2. $L(t^n) = \frac{n!}{s^{n+1}}$ 3. $L(e^{at}) = \frac{1}{s-a}$

4. $L(\cosh at) = \frac{s}{s^2 - a^2}$ 5. $L(\sinh at) = \frac{a}{s^2 - a^2}$ 6. $L(\sin at) = \frac{a}{s^2 + a^2}$

7. $L(\cos at) = \frac{s}{s^2 + a^2}$ 8. $Le^{at} f(t) = F(s-a)$ 9. $L^{-1} f'(t) = sLf(t) - f(0)$

10. $Lf''(t) = s^2 Lf(t) - sf(0) - f'(0)$ 11. $L\left[\int_0^1 f(t)dt\right] = \frac{1}{s}F(s)$ 12. $L[t^n f(t)] = (-1)^n \frac{d^n}{ds^n}[F(s)]$

13. $L\left[\frac{1}{t}f(t)\right] = \int_s^\infty F(s)ds$ 14. $u(t-a) = \begin{cases} 0 \text{ when } & t < a \\ 1 \text{ when } & t > a \end{cases}$ 15. $L[u(t-a)] = \frac{e^{-as}}{s}$

16 $L[f(t-a).u(t-a)] = e^{-as}F(s)$ 17. $L\delta(t-1) = \frac{1}{\in}$ 18. $L\delta(t-a) = e^{-as}$

19. $Lf(t) = \frac{\int_0^T e^{-st} f(t)dt}{1 - e^{-sT}}$ 20. $L\frac{t}{2a}\sin at = \frac{s}{(s^2 + a^2)^2}$ 21. $Lt\cos at = \frac{s^2 - a^2}{(s^2 + a^2)^2}$

22. $L\frac{1}{2a^3}(\sin at - at\cos at) = \frac{1}{(s^2 + a^2)^2}$ 23. $L\frac{1}{2a}(\sin at - at\cos at) = \frac{s^2}{(s^2 + a^2)^2}$

**CONVOLUTION THEOREM** $L\left[\int_0^t f_1(x) f_2(t-x)dx\right] = F_1(s) * F_2(s)$

**INVERSE LAPLACE TRANSFORM**

1. $L^{-1}\left(\frac{1}{s}\right) = 1$ 2. $L^{-1}\frac{1}{s^n} = \frac{t^{n-1}}{(n-1)!}$ 3. $L^{-1}\frac{1}{s-a} = e^{at}$

4. $L^{-1}\frac{s}{s^2 - a^2} = \cosh at$ 5. $L^{-1}\frac{1}{s^2 - a^2} = \frac{1}{a}\sinh at$ 6. $L^{-1}\frac{1}{s^2 + a^2} = \frac{1}{a}\sin at$

7. $L^{-1}\frac{s}{s^2 + a^2} = \cos at$ 8. $L^{-1}F(s-a) = e^{at} f(t)$

9. $L^{-1}\frac{1}{(s^2 + a^2)^2} = \frac{1}{2a^3}(\sin at - at\cos at)$

10. $L^{-1}\frac{s}{(s^2 + a^2)^2} = \frac{1}{2a}t\sin at$ 11. $L^{-1}\frac{s^2 - a^2}{(s^2 + a^2)^2} = t\cos at$

12. $L^{-1}\frac{s^2}{(s^2 + a^2)^2} = \frac{1}{2a}(\sin at + at\cos at)$

13. $L^{-1}[sF(s)] = \frac{d}{dt}f(t) + f(0)$ 14. $L^{-1}\left[\frac{1}{s}F(s)\right] = \int_0^t f(t)\,dt$

15. $L^{-1}F(s+a) = e^{-at} f(t)$ 16. $L^{-1}\left[e^{-as}F(s)\right] = f(t-a)u(t-a)$

17. $L^{-1}\left[\frac{d}{ds}F(s)\right] = -t\,f(t)$ 18. $L^{-1}\left[\int_s^\infty F(s)ds\right] = \frac{f(t)}{t}$

19. $L^{-1}\int_0^s f_1(x) f_2(t-x)dx = F_1(s).F_2(s)$ 20. $f(t)$ = sum of the residues of $e^{st} F(s)$ at the poles of $F(s)$

21. $L^{-1}\left[\frac{F(s)}{G(s)}\right] = \sum_{i=1}^{n} \frac{F(\alpha_i)}{G'(\alpha_i)} e^{\alpha_i t}$

## December 2014

# Engineering Mathematics-I

## AIMETE-ET/CS/IT (New Scheme)

*Time allowed : 3 hours ]* *[ Maximum Marks : 100*

**Note:** There are 9 Questions in all.

- **Question 1 is compulsory and carries 20 marks. Answer to Q.1 must be written in the space provided for it in the answer book supplied and nowhere else.**
- **The answer sheet for the Q. 1 will be collected by the invigilator after 45 minutes of the commencement of the examination.**
- **Out of the remaining Eight Questions answer any Five Questions. Each question carries 16 marks.**
- **Any required data not explicitly given, may be suitably assumed and stated.**

*Question numbers 1 to 6 carry 1 mark each.*

**1.** Choose the correct or the best alternative in the following: (2 × 10)

(*a*) If $x + y + z = \log z$, then the value of $\frac{\partial z}{\partial x}$ is

(*a*) $\frac{x}{1-x}$ (*b*) $\frac{z}{1-z}$ (*c*) $\frac{y}{1-y}$ (*d*) None of these

**Solution.** $x + y + z = \log z$

Differentiating w.r.t '*x*' we get

$$1 + 0 + \frac{\partial z}{\partial x} = \frac{1}{z}\frac{\partial I}{\partial x} \Rightarrow z + z\frac{\partial z}{\partial x} = \frac{\partial z}{\partial x} \Rightarrow z\frac{\partial z}{\partial x} - \frac{\partial z}{\partial x} = -z$$

$$\Rightarrow \quad (z-1)\frac{\partial z}{\partial x} = -z \Rightarrow \frac{\partial z}{\partial x} = \frac{-z}{z-1} = \frac{z}{1-z}$$ **Ans. (*b*)**

(*b*) The Jacobian $J\left(\frac{x, y}{r, \theta}\right)$ for the function $x = r\cos\theta$, $y = r\sin\theta$ is

(*a*) $r$ (*b*) 1 (*c*) $\frac{1}{r}$ (*d*) None of these

**Solution.** Refer to Example 62 on page 53 **Ans.** (*a*).

(*c*) Matrix has the value. This statement

(*a*) is always true (*b*) dependent upon the matrix

(*c*) is false (*d*) None of these

**Solution.** Matrix is an array of numbers and it has no value. It is false. **Ans** (*c*).

(*d*) The value of $\int_0^\infty e^{-x^2}\,dx$ is

(*a*) $\pi$ (*b*) $-\pi$ (*c*) $\frac{\sqrt{\pi}}{2}$ (*d*) None of these

**Solution.** We know that $\int_0^\infty e^{-x^2}\,dx = \frac{\sqrt{\pi}}{2}$ **Ans.** (*c*)

(*e*) If the Fourier series of $f(x)$ has **only sine terms**, then $f(x)$ must be

(*a*) Odd (*b*) Even (*c*) Both (*a*) and (*b*) (*d*) None of these

**Solution.** If the Fourier series of $f(x)$ **has only sin terms, then the function of** $f(x)$ **is odd.**

**Ans.** (*a*)

(*f*) The value of $J_{\frac{1}{2}}(x)$ is

(*a*) $\sqrt{\dfrac{2}{\pi x}}\cos x$ (*b*) $\sqrt{\dfrac{2}{\pi x}}\sin x$ (*c*) $\sqrt{\dfrac{2}{\pi}}\sin x$ (*d*) None of these

**Solution.** Refer Q.no. 1(*i*) of 2014 (**Old scheme**). **Ans.** (*a*)

(*g*) The value of $\int_{-1}^{+1} P_m(x)P_n(x)dx$ **for** $m \neq n$ **is**

(*a*) – 1 (*b*) 0 (*c*) 1 (*d*) None of these

**Solution.** $\int_{-1}^{+1} P_m(x)P_n(x)dx = 0$ **Ans.** (*b*)

(*h*) The order of convergent in Newton-Raphson method is

(*a*) 0 (*b*) 3 (*c*) 2 (*d*) None of these

**Solution.** The order of convergence in Newton's Raphson method is 2. **Ans.** (*c*)

(*i*) The solution of the differential equation $(D^2 + 6D + 9) = 5e^{3x}$ is

(*a*) $y = (c_1 + xc_2)e^{-3x} + \dfrac{5e^{3x}}{36}$ (*b*) $y = (c_1 + xc_2)e^{3x} - \dfrac{5}{36}$

(*c*) 0 (*d*) None of these

**Solution.** $(D^2 + 6D + 9)y = 5e^{3x}$

A.E. is $m^2 + 6m + 9 = 0 \Rightarrow (m + 3)^2 = 0,\ m = -3, -3$, C.F. $= (c_1 + c_2x)e^{-3x}$

$$\text{P.I.} = \frac{1}{D^2+6D+9}5e^{3x} = 5\cdot\frac{1}{9+18+9}e^{3x} = \frac{5}{36}e^{3x}$$

Hence $y = (c_1 + c_2x)e^{-3x} + \dfrac{5}{36}e^{3x}$ **Ans.** (*a*)

(*j*) The value of $\int_0^1\int_0^x (x^2 + y^2)dx\,dy$ is

(*a*) 5 (*b*) $\dfrac{1}{3}$ (*c*) $\dfrac{7}{2}$ (*d*) None of these

**Solution.** $\int_0^1\int_0^x (x^2 + y^2)dx\,dy = \int_0^1 \left(x^2y + \dfrac{y^3}{3}\right)_0^x dx$

$$= \int_0^1 \left(x^3 + \frac{x^3}{3}\right)dx = \frac{4}{3}\int_0^1 x^3\,dx = \frac{4}{3}\left(\frac{x^4}{4}\right)_0^1 = \frac{4}{3}\left(\frac{1}{4}\right) = \frac{1}{3}$$ **Ans.** (*b*)

**Answer any five questions out of eight questions.**
**Each question carries 16 marks**

**2.** (*a*) Find the minimum value of $x^2 + y^2 + z^2$ subject to the condition $xyz = a^3$. Find the minimum value of $x^2 + y^2 + z^2$. Subject to the condition $xyz = a^3$.

**Solution.** Let $f(x, y, z) = x^2 + y^2 + z^2$

$$\phi(x, y, z) = xyz - a^3 = 0 \quad \dots (i)$$

Let $\lambda$ be any non-zero arbitrary constant then by using method of Lagrange's multiplier.

$$\frac{\partial f}{\partial x} + \lambda \frac{\partial \phi}{\partial x} = 0$$

$$2x + \lambda.yz = 0 \quad \dots (2)$$

$$\frac{\partial f}{\partial y} + \lambda \frac{\partial \phi}{\partial y} = 0$$

$$2y + \lambda.xz = 0 \quad \dots (3)$$

$$\frac{\partial f}{\partial z} \times \lambda \frac{\partial \phi}{\partial z} = 0$$

$$\phi(x, y, z) \; \frac{2\partial}{\partial x} = yz$$

$$\frac{\partial \phi}{\partial y} = xz$$

$$\frac{\partial y}{\partial z} = xy$$

$$x^2 = y^2$$

$$2z + \lambda.xy = 0 \quad \dots (4),$$

$$3y^2 + \lambda xyz = 0 \dots \text{from (3)}$$

$$2z^2 + \lambda xyz = 0 \dots \text{from (4)}$$

$\Rightarrow$ $2y^2 - 2z^2 = 0 \Rightarrow y = z$ from (3) and (4)

Similarly $2x^2 - 2y^2 = 0 \rightarrow x = y$

From (1), $xyz = a^3$,

$\Rightarrow$ $(y)\, y(y) - a^3 \Rightarrow \pm y^3 = a^3, \Rightarrow y = \pm a \Rightarrow x = \pm a, \Rightarrow z = \pm a$

Minimum value of $f(x, y, z)$ is $f(\pm a, \pm a, \pm a) = (\pm a)^2 + (\pm a)^2 + (\pm a)^2 = a^2 + a^2 + a^2 = 3a^2$ **Ans.**

(*b*) Evaluate $\int_0^1 \frac{x^\alpha - 1}{\log x} dx, \alpha \geq 0,$ by using the method of differentiation under the sign of integration. ... (8)

**Solution.** Let $f(\alpha) = \int_0^1 \frac{x^\alpha - 1}{\log x} dx \quad \Rightarrow f(0) = 0$

$\because$ Variable of integral is $x$. So differentiating w.r to $\alpha$ doesn't effect integral.

$$\Rightarrow \quad f'(\alpha) = \int_0^1 \frac{x^\alpha \log x}{\log x} dx = \int x^\alpha dx = \left[\frac{x^{\alpha+1}}{\alpha+1}\right]_0^1 = \frac{1}{\alpha+1}$$

$$\because \quad f'(\alpha) = \frac{1}{\alpha+1}$$

Integrating both sides, we get

$$f(\alpha) = \log(\alpha + 1) + c$$

$$f(0) = \log 1 + c$$

$$\Rightarrow \quad 0 = 0 + c$$

$$f(\alpha) = \log(1 + \alpha)$$

$$\Rightarrow \quad \int_0^1 \frac{x^\alpha - 1}{\log x} dx = \log(1 + \alpha)$$ **Ans.**

**3.** (*a*) Expand $f(x) = x^3$ as a Fourier series in the interval $-\pi < x < \pi$. (8)

**Solution.** Here $f(x) = x^3$, $-\pi < x < \pi$ is odd function.

So, $a_0 = 0$ & $a_n = 0$, $b_n = \frac{2}{\pi}\int_0^\pi f(x) \sin nx\, dx = \frac{2}{\pi}\int_0^\pi x^3 \sin nx\, dx$

$$= \frac{2}{\pi}\left[x^3 \frac{(-\cos nx)}{n} - 3x^2 \frac{(-\sin nx)}{n^2} + 6x \cdot \frac{\cos nx}{n^3} - 6\frac{\sin nx}{n^4}\right]_0^\pi$$

$$= \frac{2}{\pi}\left[\left(-\frac{\pi^3 \cos n\pi}{n} + 0 + 6\pi\frac{\cos n\pi}{n^3} - 0\right) - (-0+0+0-0)\right]$$

$$= \frac{2}{\pi}\left[\frac{-\pi^3(-1)^n}{n} + \frac{6\pi(-1)^n}{n^3}\right] \qquad [\because \cos n\pi = (-1)^n]$$

$$= 2(-1)^n\left[-\frac{\pi^2}{n} + \frac{6}{n^3}\right]$$

Thus, $$f(x) = a_0 + \sum_{n=1}^{\infty} a_n \cos nx + \sum_{n=1}^{\infty} b_n \sin nx$$

Hence $$x^3 = 0 + 0 + \sum_{n=1}^{\infty} 2(-1)^n\left(-\frac{\pi^2}{n} + \frac{6}{n^3}\right)\sin nx$$

$$= 2\left[-\left(-\pi^2+6\right)\sin x + \left(\frac{-\pi^2}{2} + \frac{6}{2^3}\right)\sin 2x - \left(-\frac{\pi^2}{3} + \frac{6}{3^3}\right)\sin 3x + \ldots\right]$$ **Ans.**

(*b*) Obtain the half range cosine series for $\sin\left(\frac{\pi x}{l}\right)$ in the range $0 < x < l$. ...(8)

**Solution.** $$f(x) = \frac{a_0}{2} + \sum_{n=1}^{\infty} a_n \frac{\cos n\pi x}{l}$$

Where, $$a_0 = \frac{2}{l}\int_0^l f(x)dx = \frac{2}{l}\int_0^l \sin\frac{\pi x}{l}dx = \frac{2}{l}\left[\frac{-\cos\frac{\pi x}{l}}{\frac{\pi}{l}}\right]_0^l = \frac{2}{\pi}[-\cos\pi + 1] = \frac{4}{\pi}$$

$$a_n = \frac{2}{l}\int_0^l f(x)\cos\frac{n\pi}{l}dx = \frac{2}{l}\int_0^l \sin\frac{\pi x}{l}\cos\frac{n\pi}{l}dx$$

$$= \frac{2}{l}\int_0^l \frac{1}{2}\left(\sin(n+1)\frac{\pi x}{l} - \sin(n-1)\frac{\pi x}{l}\right)dx = \frac{1}{l}\left[\frac{-\cos(n+1)\frac{\pi x}{l}}{(n+1)\frac{\pi}{l}} + \frac{\cos(n-1)\frac{\pi x}{l}}{(n-1)\frac{\pi}{l}}\right]_0^l$$

$$= \frac{1}{l}\left[\left(\frac{-\cos(n+1)\pi}{(n+1)\frac{\pi}{l}} + \frac{\cos(n-1)\pi}{(n-1)\frac{\pi}{l}}\right) - \left(\frac{-l}{(n+1)\frac{\pi}{l}} + \frac{1}{(n-1)\frac{\pi}{l}}\right)\right]$$

$$= \frac{1}{l}\times\frac{l}{\pi}\left[\frac{-(-1)^{n+1}}{n+1} + \frac{(-1)^{n-1}}{n-1} + \frac{1}{n+1} - \frac{1}{n-1}\right]$$

$$= \frac{1}{\pi}\left[\frac{(-1)^{n+1}(-n+1+n+1)}{(n+1)(n-1)} + \frac{n-1-n-1}{(n+1)(n-1)}\right] = \frac{2}{(n+1)(n-1)\pi}[(-1)^{n+1} - 1]$$

$$\sin\frac{\pi x}{l} = \frac{2}{\pi} + \sum_{n=1}^{\infty}\frac{2}{(n^2-1)\pi}[(-1)^n - 1]\cos\frac{n\pi x}{l}$$ **Ans.**

**4.** (*a*) Find the Fourier sine transform of $\dfrac{1}{x(x^2+a^2)}$. (8)

**Solution.** $\because$ Formula for Fourier sine transform

$$f_s(f(x)) = \int_0^\infty f(x)\sin sx\,dx$$

$$\text{Let } I = \int_0^\infty \frac{\sin sx}{x(x^2+a^2)}dx \quad \text{... (1)}$$

$$\text{Differentiating eq. (1) w.r.t `}s\text{', we get } \frac{dI}{ds} = \int_0^\infty \frac{\cos sx}{(x^2+a^2)}dx \quad \text{... (2)}$$

$$\text{Again differentiating `}s\text{', we get } \frac{d^2I}{ds^2} = \int_0^\infty \frac{-x\sin sx}{(x^2+a^2)}dx = \int_0^\infty \frac{-x^2\sin sx}{x(x^2+a^2)}dx$$

$$= -\int_0^\infty \frac{(x^2+a^2-a^2)\sin sx}{x(x^2+a^2)}dx = -\int_0^\infty \frac{\sin sx}{x}dx + \int_0^\infty \frac{a^2}{x(x^2+a^2)}\sin sx\,dx$$

$$\frac{d^2I}{ds^2} = -\frac{\pi}{2} + a^2 I \Rightarrow \frac{d^2I}{ds^2} - a^2 I = -\frac{\pi}{2}$$

$$(D^2 - a^2)I = \frac{-\pi}{2}$$

A.E. is $m^2 - a^2 = 0 \quad \Rightarrow m = \pm a$

$$\text{C.F.} = c_1e^{as} + c_2e^{-as}$$

$$\text{P.I} = \frac{(-\pi/2)}{D^2 - a^2} = -\frac{\pi}{2}\frac{1}{D^2-a^2}e^{os} = \frac{-\pi/2}{-a^2} = \frac{\pi}{2a^2}$$

$$I = C.F + P.I$$

$$I = c_1e^{as} + c_2e^{-as} + \frac{\pi}{2a^2} \quad \text{... (3)}$$

$$\text{On putting } s = 0, \text{ we get } c_1 + c_2 + \frac{\pi}{2a^2} = 0 \quad \text{... (4)}$$

[From (1) If $s = 0$, then $I = 0$]

$$\frac{dI}{ds} = ac_1e^{as} - ac_2e^{-as} \quad \text{... (5)}$$

$$\text{On putting the value of } \frac{dI}{ds} \text{ from (2) in (5), we get } \int_0^\infty \frac{\cos sx}{x^2+a^2}dx = ac_1e^{as} - ac_2e^{-as}$$

$$\text{On putting } s = 0, \text{ we have } \int_0^\infty \frac{1}{x^2+a^2}dx = ac_1 - ac_2$$

$$\Rightarrow \quad \frac{1}{a}\left(\tan^{-1}\frac{x}{a}\right)_0^\infty = ac_1 - ac_2 \Rightarrow \frac{\pi}{2a} = ac_1 - ac_2$$

$$c_1 - c_2 = \frac{\pi}{2a^2} \quad \text{... (6)}$$

On adding (4) and (6), we get $2c_1 = 0 \;\; c_1 = 0$

From (3), $0 + c_2 + \dfrac{\pi}{2a^2} = 0 \Rightarrow c_2 = \dfrac{-\pi}{2a^2}$

On putting the values of $c_1$ and $c_2$ in (3), we get $I = -\dfrac{\pi}{2a^2}e^{-as} + \dfrac{\pi}{2a^2}$

$$I = \frac{\pi}{2a^2}(1 - e^{-as})$$ **Ans.**

(*b*) Find the Z-transform of sin (3*k* + 5)

**Solution.** Formula for Z-transform

$$f(z) = \sum_{k=0}^{\infty} f(k)z^{-k} = \sum_{k=0}^{\infty} \sin(3k+5)z^{-k} = \sum_{k=0}^{\infty}\left(\frac{e^{i(3k+5)} - e^{-i(3k+5)}}{2i}\right)z^{-k}$$

$$= \frac{1}{2i}\left[e^{5i}\sum_{k=0}^{\infty} e^{i3k}z^{-k}\right] - \frac{1}{2i}\left[e^{-5i}\sum_{k=0}^{\infty} e^{-3ik}z^{-k}\right] = \frac{1}{2i}\cdot e^{5i}\sum_{k=0}^{\infty}(e^{3i}z^{-1})^k - \frac{1}{2i}e^{-5i}\sum_{k=0}^{\infty}(e^{3i}z^{-1})^k$$

$$\left[\Sigma(e^{3i}z^{-1})^k \text{ in G.P and its sum} = \frac{a}{1-r}\right] \qquad \text{for } |2^{-1}| < 1$$

$$= \frac{1}{2i}\times e^{5i}\times\frac{1}{1-e^{3i}z^{-1}} - \frac{1}{2i}e^{-5i}\times\frac{1}{1-e^{-3i}z^{-1}} = \frac{1}{2i}\left\{\frac{e^{5i} - e^{5i}\cdot e^{-3i}z^{-1} - e^{-5i} + e^{-5i}\cdot e^{3i}z^{-1}}{(1-e^{3i}z^{-1})(1-e^{-3i}z^{-1})}\right\}$$

$$= \frac{1}{2i}\left\{\frac{e^{5i} - e^{-5i} - z^{-1}(e^{2i} - e^{-2i})}{1 - e^{-3i}z^{-1} - e^{3i}z^{-1} + e^{3i}z^{-1}\cdot e^{-3i}z^{-1}}\right\} = \frac{1}{2i}\left\{\frac{2i\sin 5 - 2iz^{-1}\sin 2}{1 - z - 1(e^{3i} + e^{-3i}) + z^{-2}}\right\}$$

$$= \frac{\sin 5 - z^{-1}\sin 2}{1 - 2\cos 3\, z^{-1} + z^{-2}} = \frac{z^2\sin 5 + z\sin 2}{z^2 - 2z\cos 3 + 1} \qquad \text{for } |z| > 1$$ **Ans.**

**5.** (*a*) Express $f(x) = 4x^3 - 2x^2 - 3x + 8$ in term of Legendre's polynomial.

**Solution.** Let $4x^3 - 2x^2 - 3x + 8 = aP_3(x) + bP_2(x) + cP_1(x) + dP_0(x)$

$$= a\left(\frac{5x^3}{2} - \frac{3}{2}x\right) + b\left(\frac{3}{2}x^2 - \frac{1}{2}\right) + c.x + d.1 = \frac{5a}{2}x^3 + \frac{3b}{2}x^2 + \left(\frac{-3a}{2} + c\right)x - \frac{b}{2} + d$$

Equating coeff. of $x^3$ on both sides, we get

$$4 = \frac{5a}{2} \Rightarrow a = \frac{8}{5}$$

Equating coeff. of $x^2$

$$-2 = \frac{3b}{2} \Rightarrow b = -\frac{4}{3}$$

Equating coeff. of $x$ on both sides, we get

$$-3 = -\frac{3a}{2} + c$$

$$= \frac{-3}{2}\times\frac{8}{5} + c$$

$$c = -3 + \frac{12}{5} = -\frac{3}{5}$$

Equating coefficient of $x^0$ in both sides

we get, $8 = \dfrac{-b}{2} + d \Rightarrow 8 = -\dfrac{1}{2}\times\dfrac{-4}{3} + d$

$$d = 8 - 2/3 \Rightarrow d = \frac{22}{3}$$

Putting the values of of $a$, $b$, $c$, $d$ in eq. (1), we get

$$4x^3 - 2x^2 - 3x + 8 = \frac{8}{5}P_3(x) - \frac{4}{3}P_2(x) - \frac{3}{5}P_1(x) + \frac{22}{3}$$ **Ans.**

(*b*) Prove that $J_{\frac{1}{2}}(x) = \sqrt{\frac{2}{\pi x}} \sin x$

**Solution.** Refer to Example 11(*a*) on page no. 635.

**6.** (*a*) Find the real root of $x^3 - 2x - 5 = 0$, correct to three decimal places using Newton-Raphson Method.(8)

**Solution.** Let $f(x) = x^3 - 2x - 5$, $f(0) = -5$, $f(1) = 1 - 2 - 5 = -6$

$f(2) = 8 - 4 - 5 = -1$, $f(3) = 27 - 6 - 5 = 16$

$\because$ $f(2)$ and $f(3)$ are of opposite sign.

So, one real root lie in the interval [2, 3]

$\because$ $f(2) = -1$ is nearer to zero.

$\Rightarrow$ Root $\alpha$ is near by $x = 2$

So, consider initial approximate $x_0 = 2.1$

$\because$ $f(x) = x^3 - 2x - 5, f'(x) = 3x^2 - 2$

Newton Raphson formula $x_{n+1} = x_n - \dfrac{f(x_n)}{f'(x_n)} = x_n - \dfrac{x_n^3 - 2x_n - 5}{3x_n^2 - 2} = \dfrac{3x_n^3 - 2x_n - x_n^3 + 2x_n + 5}{3x_n^2 - 2}$

$$x_{n+1} = \frac{2x_n^3 + 5}{3x_n^2 - 2} \quad \text{... (1)}$$

On putting the value of $x_0 = 2.1$ in (1) we get $x_1$ and so in

$x_0 = 2.1$

$x_1 = 2.09457$

$x_2 = 2.09455$ **Ans.**

Hence required root correct to four decimal places in $x = 2..0945$. **Ans.**

(*b*) Apply method of fourth order, to find an approximately value of $y$ when x = 0.1. Given that $y(0) = 1$. (8)

$$10\frac{dy}{dx} = x^2 + y^2,$$

**Solution.** $\dfrac{dy}{dx} = f(x, y) = \dfrac{1}{10}(x^2 + y^2)$, $x_0 = 0$, $y_0 = 1$, $h = 0.1$

$$k_1 = hf(x_0, y_0) = 0.1\, f(0, 1) = 0.1 \times \left(\frac{0^2 + 1^2}{10}\right) = 0.01$$

$$k_2 = hf\left(x_0 + \frac{h}{2}, y_0 + \frac{k_1}{2}\right) = 0.1 \times f\left(0 + \frac{0.1}{2}, 1 + \frac{0.01}{2}\right) = 0.1 \times f(0.05, 1.005)$$

$$= 0.1 \times \frac{1}{10}\,[(0.05)^2 + (1.005)^2\,] = 0.01013$$

$$k_3 = hf\left(x_0 + \frac{h}{2}, y_0 + \frac{k_2}{2}\right) = 0.1 \cdot f\left(0 + \frac{0.1}{2}, 1 + \frac{0.01013}{2}\right) = 0.1 \times f(0.05, 1.00507)$$

$$= 0.1 \times \frac{1}{10}[(0.05)^2 + (1.00507^2)] = 0.01013$$

$$k_4 = hf(x_0 + h, y_0 + k_3) = 0.1 \times f(0 + 0.1, 1 + 0.01013) = 0.1 \times f(0.1, 1.01013)$$

$$= 0.1 \times \frac{1}{10}\,[(0.1)^2 + (1.01013)^2] = 0.0103.$$

$$y_1 = y_0 + \frac{1}{6}(k_1 + 2k_2 + 2k_3 + k_4)$$

$$y(x_1) = 1 + \frac{1}{6}(0.01 + 2 \times 0.01013 + 2 \times 0.01013 + 0.0103),$$

$$y(x_0 + h) = 1.0101,\ y(0.1) = 1.0101.$$ **Ans.**

**7.** (*a*) Solve the differential equation $x^2\frac{d^2y}{dx^2} - 3x\frac{dy}{dx} + 5y = x \log x.$

**Solution.** Let $x = e^z \Rightarrow z = \log x \Rightarrow \frac{dy}{dx} = \frac{dy}{dz}\cdot\frac{dz}{dx} = \frac{1}{x}\frac{dy}{dz}$

Differentiating $x\frac{dy}{dx} = \frac{dy}{dz},\quad x\frac{d^2y}{dx^2} + \frac{dy}{dx} = \frac{d}{dz}\left(\frac{dy}{dz}\right)\cdot\frac{dz}{dx} = \frac{1}{x}\frac{d^2y}{dz^2}$

$$\Rightarrow \quad x^2\frac{d^2y}{dx^2} + x\frac{dy}{dx} = \frac{d^2y}{dz^2}$$

$$\Rightarrow \quad x^2\frac{d^2y}{dx^2} = \frac{d^2y}{dz^2} - x\frac{dy}{dx} = \frac{d^2y}{dz^2} - \frac{dy}{dz}$$

$$x^2\frac{d^2y}{dx^2} = D(D-1)\,y \qquad \left[D = \frac{d}{dz}\right]$$

The given differencial becomes

$D(D-1)y - 3\,Dy + 5y = ze^z \qquad (D^2 - 4D + 5)y = ze^z$

A.E. is $m^2 = 4m + 5 = 0,\ m = \frac{4 \pm \sqrt{16-20}}{2} = \frac{4 \pm 2i}{2} = 2 \pm i$

$C.F = e^{2z}(c_1 \cos z + c_2 \sin z)$

$$P.I = \frac{1}{D^2 - 4D + 5}ze^z = e^z\left[\frac{1}{(D+1)^2 - 4(D+1) + 5}z\right]$$

$$= e^z\left[\frac{1}{D^2 + 2D + 1 - 4D - 4 + 5}\right]z = e^z\left[\frac{1}{D^2 - 2D + 2}\right]z$$

$$= e^z\left(\frac{1}{2}\left(1 + \frac{D^2 - 2D}{2}\right)^{-1}z\right) = \frac{e^z}{2}\left(1 - \frac{D^2 - 2D}{2} + \ldots\right)z = \frac{e^z}{2}(z+1)$$

$$y = C.F + P.I = e^{2z}(c_1 \cos z + c_2 \sin z) + \frac{1}{2}e^{2z}(z+1)$$

$$= x^2[c_1 \cos(\log x) + c_2 \sin(\log x)] + \frac{1}{2}x(1 + \log x)$$ **Ans.**

(*b*) Solve $\frac{d^2y}{dx^2} - 2\frac{dy}{dx} + y = xe^x \sin x$ (8)

**Solution.** $(D^2 - 2D + 1)y = xe^x \sin x \qquad \left[\frac{d}{dx} = D,\right]$

A.E. $(m-1)^2 = 0,\ m = 1, 1\ C.F. = (c_1 + xc_2)e^x$

$$\text{P.I.} = \frac{1}{(D-1)^2}xe^x \sin x = e^x\frac{1}{(0-1+1)^2}x\sin x = \frac{1}{D^2}x\sin x$$

$$= e^x \cdot \int\left(\int x\sin x\,dx\right)dx = e^x\int[x(-\cos x) - 1\cdot(-\sin x)]\,dx$$

$$= e^x \int(-x\cos+\sin x)\,dx = e^x[-x\sin x + (-\cos x) - \cos x]$$

$$= e^x[-x\sin x+(-\cos x)-\cos x] = e^x[-x\sin x - 2\cos x]$$

$$y = \text{C.F.} + \text{P.I.} = (c_1 + xc_2)e^x + e^x(-x\sin x - 2\cos x)$$ **Ans.**

**8.** (*a*) Find the rank of the matrix $A$, where $A = \begin{bmatrix} 1 & 3 & 4 & 2 \\ 2 & -1 & 3 & 2 \\ 3 & -5 & 2 & 2 \\ 6 & -3 & 8 & 6 \end{bmatrix}$, by reducing it to a normal form.

**Solution.** $A = \begin{bmatrix} 1 & 3 & 4 & 2 \\ 2 & -1 & 3 & 2 \\ 3 & -5 & 2 & 2 \\ 6 & -3 & 8 & 6 \end{bmatrix}$, $A = \begin{bmatrix} 1 & 3 & 4 & 2 \\ 0 & -7 & -5 & -2 \\ 0 & -14 & -10 & -4 \\ 0 & -21 & -16 & -6 \end{bmatrix} \begin{matrix} R_2 \to R_2 - 2R_1 \\ R_3 \to R_3 - 3R_1 \\ R_4 \to R_4 - 6R_1 \end{matrix}$

$$A \sim \begin{bmatrix} 1 & 3 & 4 & 2 \\ 0 & -7 & -5 & -2 \\ 0 & 0 & 0 & 0 \\ 0 & 0 & -1 & 0 \end{bmatrix} \begin{matrix} R_3 \to R_3 - 2R_2 \\ R_4 \to R_4 - 3R2 \end{matrix}, \quad A \sim \begin{bmatrix} 1 & 3 & 4 & 2 \\ 0 & -7 & -5 & -2 \\ 0 & 0 & -1 & 0 \\ 0 & 0 & 0 & 0 \end{bmatrix} R_3 \leftrightarrow R_4$$

$$A \sim \begin{bmatrix} 1 & 0 & 0 & 0 \\ 0 & -7 & -5 & -2 \\ 0 & 0 & -1 & 0 \\ 0 & 0 & 0 & 0 \end{bmatrix} \begin{matrix} C_2 \to C_2 - 3C_1 \\ C_3 \to C_3 - 4C_1 \\ C_4 \to C_4 - 2C_1 \end{matrix}, \quad A \sim \begin{bmatrix} 1 & 0 & 0 & 0 \\ 0 & -7 & 0 & 0 \\ 0 & 0 & -1 & 0 \\ 0 & 0 & 0 & 0 \end{bmatrix} \begin{matrix} C_3 \to C_3 - \frac{5}{7}C_2 \\ C_4 \to C_4 - \frac{2}{7}C_2 \end{matrix}$$

$$A \sim \begin{bmatrix} 1 & 0 & 0 & 0 \\ 0 & 1 & 0 & 0 \\ 0 & 0 & 1 & 0 \\ 0 & 0 & 0 & 0 \end{bmatrix} \begin{matrix} R_2 \to \frac{-1}{7}R_2 \\ R_3 \to -R_3 \end{matrix} = \begin{bmatrix} I_3 & 0 \\ 0 & 0 \end{bmatrix}$$ **Ans.**

Hence Rank of $A = 3$

(*b*) Investigate the value of $\lambda$ and $\mu$ so that the equations
$2x + 3y + 5z = 9$; $7x + 3y - 2z = 8$; $2x + 3y + \lambda z = \mu$
(*i*) No solution, (*ii*) Unique solution, (*iii*) An infinite number of solutions. ... (8)

**Solution.** Given equation in matrix for $\begin{bmatrix} 2 & 3 & 5 \\ 7 & 3 & -2 \\ 2 & 3 & \lambda \end{bmatrix}\begin{bmatrix} x \\ y \\ z \end{bmatrix} = \begin{bmatrix} 9 \\ 8 \\ \mu \end{bmatrix} \Rightarrow AX = B$

Let Augmented matrix $C = (A : B) = \begin{bmatrix} 3 & 3 & 5 & : & 9 \\ 7 & 3 & -2 & : & 8 \\ 2 & 3 & \lambda & : & \mu \end{bmatrix}$

$$C = \begin{bmatrix} 2 & 3 & 5 & 9 \\ 0 & -15 & -39 & -47 \\ 0 & 0 & \lambda-5 & \mu-9 \end{bmatrix} \quad \begin{matrix} R_2 \to 2R_2 - 7R_1 \\ R_3 \to R_3 - R_1 \end{matrix}$$

(*i*) $\because$ for no. sol. $\Rightarrow$ Rank of $A \neq$ Rank of $C$

$\Rightarrow \lambda = 5$ & $\mu - 9 \neq 0 \Rightarrow \mu \neq 9$

(*ii*) For unique sol. $\Rightarrow$ Rank of $A$ = Rank of $C$ = 3, for this, $\lambda - 5 \neq 0 \quad \Rightarrow \lambda \neq 5$

(*iii*) For infinite no. of solutions. $\Rightarrow$ Rank of $A$ = Rank of $C$ = 2

For this $\lambda - 5 = 0$ and $\mu - 9 = 0 \Rightarrow \lambda = 5$ & $\mu = 9$ **Ans.**

**9.** (*a*) Evaluate $\int_0^\infty \int_0^\infty x e^{-\frac{x^2}{y}} \, dx dy$

**Solution.** $I = \int_0^\infty dy \int_0^\infty x e^{-\frac{x^2}{y}} dx$

Let $x^2 = t$

$$2x\,dx = dt = \int_0^\infty dy \int_0^\infty e^{-t/y} \frac{dt}{2} = \frac{1}{2}\int_0^\infty dy \left[\frac{e^{-t/y}}{-1/y}\right]_0^\infty = \frac{1}{2}\int_0^\infty dy[0 + y] = \frac{1}{2}\int_0^\infty y\,dy$$

$$= \frac{1}{2}\left[\frac{y^2}{2}\right]_0^\infty = \frac{1}{2} \times (\infty - 0) = \infty$$ **Ans.**

(*b*) Define Beta and Gamma functions. Prove that $\beta(m, n) = \frac{\Gamma(m)\,\Gamma(n)}{\Gamma(m+n)}$. (8)

**Solution.** Refer to article 21.7 on page no 1164.

## December 2014

## Engineering Mathematics-I

## AIMETE-ET/CS/IT (Old Scheme)

*Time allowed : 3 hours ]* *[ Maximum Marks : 100*

**1.** Choose the correct or the best alternative in the following:

(*a*) If $z = \sin^{-1}\left(\dfrac{x^2+y^2}{x+y}\right)$, then the value of $x\dfrac{\partial z}{\partial x}+y\dfrac{\partial z}{\partial y}$ is equal to

(*a*) $\sin z$ (*b*) $\cos z$ (*c*) $\tan z$ (*d*) $\cot z$

**Solution.** $z = \sin^{-1}\left(\dfrac{x^2+y^2}{x+y}\right) \Rightarrow \sin z = \dfrac{x^2+y^2}{x+y}$

Let $u = \sin z = \dfrac{x^2+y^2}{x+y} = xf\left(\dfrac{y}{x}\right)$, $u$ is degree of 1.

By Euler's theorem, we have

$$x\frac{\partial u}{\partial x}+y\frac{\partial u}{\partial y} = u \quad \Rightarrow \quad x\frac{\partial u}{\partial z}\frac{\partial z}{\partial x}+y\frac{\partial u}{\partial z}\frac{\partial z}{\partial y} = \sin z,$$

$$x\frac{\partial z}{\partial x}\cos z+y\frac{\partial z}{\partial y}\cos z = \sin z, \quad \Rightarrow \quad x\frac{\partial z}{\partial x}+y\frac{\partial z}{\partial y} = \tan z$$

**Ans. (*c*)**

(*b*) The value of integral $\iint xy\,dx\,dy$ over the positive quadrant of the circle $x^2 + y^2 = a^2$ is

(*a*) $a^4$ (*b*) $\dfrac{a^4}{2}$ (*c*) $\dfrac{a^4}{4}$ (*d*) $\dfrac{a^4}{8}$

**Solution.** $\displaystyle\int\int xy\,dy\,dx = \int_0^a x\,dx \int_0^{\sqrt{a^2-x^2}} y\,dy$

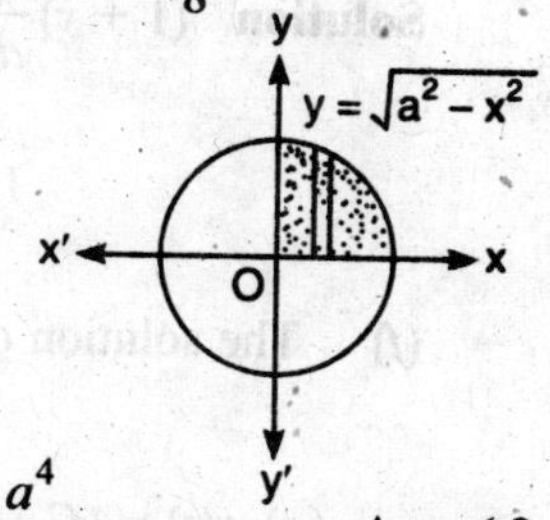


$$= \int_0^a x\,dx\left[\frac{y^2}{2}\right]_0^{\sqrt{a^2-x^2}} = \frac{1}{2}\int_0^a x\,d\,x(a^2-x^2)$$

$$= \frac{1}{2}\int(a^2x-x^3)\,dx = \frac{1}{2}\left[\frac{a^2x^2}{2}-\frac{x^4}{4}\right]_0^a = \frac{1}{2}\left[\frac{a^4}{2}-\frac{a^4}{4}\right] = \frac{1}{2}\frac{a^4}{4} = \frac{a^4}{8}$$

**Ans. (*d*)**

(*c*) If the product of two eigen values of the matrix $\begin{bmatrix} 6 & -2 & 2 \\ -2 & 3 & -1 \\ 2 & -1 & 3 \end{bmatrix}$ is 16, then the third eigen value is

(*a*) 2 (*b*) $-2$ (*c*) 3 (*d*) $-3$

**Solution.** Characteristic equation is $\begin{bmatrix} 6-\lambda & -2 & 2 \\ -2 & 3-\lambda & -1 \\ 2 & -1 & 3-\lambda \end{bmatrix} = 0$

$$(6-\lambda)\,[(3-\lambda)^2-1] + 2(-6+2\lambda+2) + 2(2-6+2\lambda)$$

$$\Rightarrow \quad (-\lambda + 6)(\lambda^2 - 6\lambda + 8) - 8 + 4\lambda + 4\lambda - 8 = 0$$

$$\Rightarrow \quad -\lambda^3 + 6\lambda^2 - 8\lambda + 6\lambda^2 - 36\lambda + 48 + 8\lambda - 16 = 0$$

$$\Rightarrow \quad -\lambda^3 + 12\lambda^2 - 36\lambda + 32 = 0$$

$$\lambda_1 \lambda_2 \lambda_3 = 32 \Rightarrow 16\lambda_3 = 32 \Rightarrow \lambda_3 = 2$$ **Ans. (a)**

(*d*) Taylor's series solution of $\frac{dy}{dx} = -xy$, $y(0) = 1$, upto $x^2$ is

(*a*) $y(x) = 1 + x + \frac{x^2}{2} + \ldots$ (*b*) $y(x) = 1 + x - \frac{x^2}{2} + \ldots$

(*c*) $y(x) = 1 + \frac{x^2}{2} + \ldots$ (*d*) $y(x) = 1 - \frac{x^2}{2} + \ldots$

**Solution.** Taylor's series is $f(x) = f(0) + xf'(0) + \frac{x^2}{2!}f''(0) + \ldots$ ... (5)

$$y = 1 \text{ when } x = 0, \; f(x) = \frac{dy}{dx} = -xy, \qquad f'(0) = 0 \text{ when } x = 0$$

$$\frac{d^2y}{dx^2} = -y - x\frac{dy}{dx} \quad \Rightarrow \quad f''(0) = -1 + 0 = -1$$

Putting three values in (1), we get $f(x) = 1 + x(0) + \frac{x^2}{2}(-1) = 1 - \frac{x^2}{2}$ **Ans. (d)**

(*e*) Integrating factor of the differential equation $(1 + x)\frac{dy}{dx} = 2y + (x + 1)^4$ is

(*a*) $(1 + x)^2$ (*b*) $\frac{1}{(1+x)^2}$ (*c*) $(1 + x)$ (*d*) $\frac{1}{1+x}$

**Solution.** $(1 + x)\frac{dy}{dx} = 2y + (x + 1)^4 \Rightarrow \frac{dy}{dx} - \frac{2y}{1+x} = (x + 1)^3$

$$\text{I.F.} = e^{\int \frac{-2}{1+x}dx} = e^{-2\log(1+x)} = e^{\log(1+x)^{-2}} = (1 + x)^{-2}$$ **Ans. (b)**

(*f*) The solution of $4\frac{d^3x}{dt^3} + 4\frac{d^2x}{dt^2} + \frac{dx}{dt} = 0$, is

(*a*) $x(t) = (C_1 + C_2t)^t + C_3$ (*b*) $x(t) = C_1 + C_2t + C_3e^{t/2}$

(*c*) $x(t) = (C_1 + C_2t)e^{-t/2} + C_3$ (*d*) $x(t) = (C_1 + C_2t + C_3e^{-t/2}$

**Solution.** $4\frac{d^3x}{dt^3} + 4\frac{d^2x}{dt^2} + \frac{dx}{dt} = 0$ or $(4D^3 + 4D^2 + D)x = 0$

A.E. is $4m^3 + 4m^2 + m = 0 \Rightarrow m(4m^2 + 4m + 1) = 0$

$$\Rightarrow \quad m(2m + 1)^2 = 0 \Rightarrow m = 0, m = -\frac{1}{2}, m = -\frac{1}{2}$$

C.F. is $c_1e^{ot} + (c_2 + tc_3)e^{-\frac{t}{2}}$, $x(t) = c_1 + (c_2 + tc_3)e^{-\frac{t}{2}}$ **Ans. (c)**

(*g*) Particular integral of the differential equation $(D - 2)^2 = e^{2x}$, is

(*a*) $\frac{x^2}{2}e^{2x}$ (*b*) $\frac{x^2}{4}e^{2x}$ (*c*) $xe^{2x}$ (*d*) $\frac{x}{2}e^{2x}$

**Solution.** $(D-2)^2 = e^{2x}$

$$\text{P.I.} = \frac{1}{(D-2)^2}e^{2x} = x\frac{1}{2(D-2)}e^{2x} = x^2\frac{1}{2}e^{2x} = \frac{x^2e^{2x}}{2}$$ **Ans. (a)**

(*h*) The value of $\beta\left(\frac{5}{2},\frac{5}{2}\right)$ is

(*a*) $\frac{3\pi}{192}$ (*b*) $\frac{3\pi}{128}$ (*c*) $\frac{3\pi}{64}$ (*d*) $\frac{3\pi}{50}$

**Solution.** $\beta\left(\frac{5}{2},\frac{5}{2}\right) = \frac{\left|\frac{5}{2}\right.\left|\frac{5}{2}\right.}{\left|5\right.} = \frac{\left(\frac{3}{2}\times\frac{1}{2}\sqrt{\pi}\right)^2}{4\times3\times2} = \frac{3\pi}{128}$ **Ans. (b)**

(*i*) $J_{1/2}(x)$ is equal to

(*a*) $J_{-\frac{1}{2}}(x)\cos x$ (*b*) $J_{-\frac{1}{2}}(x)\sin x$ (*c*) $J_{-\frac{1}{2}}(x)\cot x$ (*d*) $J_{-\frac{1}{2}}(x)\tan x$

**Solution.** We know that $J_n(x) = \frac{x^n}{2^n\left|n+1\right.}\left[1-\frac{x^2}{2.2(n+1)}+\frac{x^4}{2.4.2^2(n+1)(n+2)}\ldots\right]$

Putting $n = \frac{1}{2}$

$$J_{\frac{1}{2}}(x) = \frac{\sqrt{x}}{\sqrt{2}\left|\frac{3}{2}\right.}\left[1-\frac{x}{6}+\frac{x^4}{2.3.4.5}+\ldots\right] = J_{-\frac{1}{2}}\tan x$$ **Ans. (d)**

(*j*) The polynomial $3x^2 + 2x$ in terms of Legendre polynomials is

(*a*) $3P_2(x) + 2P_1(x)$ (*b*) $2P_2(x) + 2P_1(x)$

(*c*) $2P_2(x) + 2P_1(x) + P_0(x)$ (*d*) $2P_2(x) + P_1(x) + P_0(x)$

**Solution.** Let $3x^2 + 2x = aP_2(x) + bP_1(x) + cP_0(x)$

$$= a\left(\frac{3x^2}{2}-\frac{1}{2}\right)+b(x)+c(1) = \frac{3ax^2}{2} + bx + c - \frac{a}{2}$$

Equating like powers of $x$, we have

$$3 = \frac{3a}{2} \Rightarrow a = 2, \quad 2 = b \Rightarrow b = 2, \quad 0 = c - \frac{a}{2} \Rightarrow c = \frac{2}{2} = 1$$

Putting them values in (1), we get $3x^2 + 2x = 2P_2(x) + 2P_1(x) + P_0(x)$ **Ans. (c)**

**2.** (*a*) If $r^2 = x^2 + y^2$ and $V = r^m$, prove that $V_{xx} + V_{yy} + V_{zz} = m(m+1)r^{m-2}$

**Solution.** Refer to Example 14 on page 9.

(*b*) If $xyz = 8$, find the values of $x, y, z$ for which $u = \frac{5xyz}{x+2y+4z}$ is a maximum. (8)

**Solution.** $u = \frac{5xyz}{x+2y+4z}$

Let $\phi(x, y, z) = xyz - 8 = 0$ ... (1)

Consider $\lambda$ be any non-zero arbitrary constant then by using lagranges multiplier.

$$\frac{\partial u}{\partial x}+\lambda\frac{\partial\phi}{\partial x} = 0, \Rightarrow \frac{(x+2y+4z)\cdot5yz-5xyz\cdot1}{(x+2y+4z)^2}+\lambda yz = 0 \quad \ldots (2)$$

$$\frac{\partial u}{\partial y}+\lambda\frac{\partial\phi}{\partial y} = 0, \Rightarrow \frac{(x+2y+4z)\cdot5xz-5xyz\cdot2}{(x+2y+4z)^2}+\lambda xz = 0 \quad \ldots (3)$$

$$\frac{\partial u}{\partial z}+\lambda\frac{\partial \phi}{\partial z}=0,\quad \frac{(x+2y+4z)\cdot 5xy-5xyz\cdot 4}{(x+2y+4z)^2}+\lambda xy=0 \quad \text{... (4)}$$

Equation 2 × $x$ – equation (3) × $y$, we get $\dfrac{-5x^2yz+10xy^2z}{(x+2y+4z)^2}=0$

$\Rightarrow \quad -5xyz[x-2y]=0 \Rightarrow x=2y$

For $xyz \neq 0$ because we have to maximize $\dfrac{5xyz}{x+2y+4z}$

Eq. (3) × $y$ – Eq. (4) × $z$ we get $z=\dfrac{y}{2}$

But $xyz=8,\ (2y)y\left(\dfrac{y}{2}\right)=8 \Rightarrow y^3=8 \Rightarrow y=2,\ x=2y=2(2)=4$

$$z=\frac{y}{2}=\frac{2}{2}=1,\ x=4,\ y=2,\ z=1 \qquad \textbf{Ans.}$$

**3.** (*a*) Evaluate by changing order of integration of $\displaystyle\int_0^3\int_1^{\sqrt{4-y}}(x+y)\,dx\,dy$ ... (8)

**Solution.** $\displaystyle\int_0^3\int_1^{\sqrt{4-y}}(x+y)\,dx\,dy$

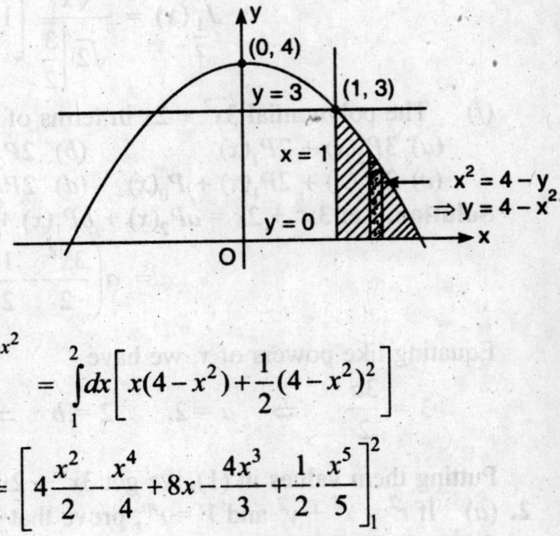


$y=0, \quad x=1$

$y=3, \quad x=\sqrt{4-y}$

$\Rightarrow \quad x^2=4-y$

Vertex = (0, 4)

Integral after changing order of integration

$$\int_1^2\int_0^{4-x^2}(x+y)\,dy\,dx=\int_1^2 dx\left[xy+\frac{y^2}{2}\right]_0^{4-x^2}=\int_1^2 dx\left[x(4-x^2)+\frac{1}{2}(4-x^2)^2\right]$$

$$=\int_1^2\left(4x-x^3+\frac{1}{2}(16-8x^2+x^4)\right)dx=\left[4\frac{x^2}{2}-\frac{x^4}{4}+8x-\frac{4x^3}{3}+\frac{1}{2}\cdot\frac{x^5}{5}\right]_1^2$$

$$=\left[2\times4-\frac{16}{4}+16-\frac{4}{3}\times8+\frac{1}{10}\times32\right]-\left[2-\frac{1}{4}+8-\frac{4}{3}+\frac{1}{10}\right]$$

$$=\left[8-4+16-\frac{28}{3}+\frac{16}{5}\right]-\left[10-\frac{1}{4}-\frac{4}{3}+\frac{1}{10}\right]=20-10-\frac{28}{3}+\frac{31}{10}+\frac{1}{4}=10-\frac{28}{3}+\frac{31}{10}+\frac{1}{4}$$

$$=\frac{600-560+186+15}{60}=\frac{40+201}{60}=\frac{241}{60} \qquad \textbf{Ans.}$$

(*b*) Find the volume of the tetrahedron bounded by the co-ordinate planes and the plane $\dfrac{x}{a}+\dfrac{y}{b}+\dfrac{z}{c}=1$.

**Solution.** Let *OABC* be the tetrahedron bounded by coordinate planes and plane

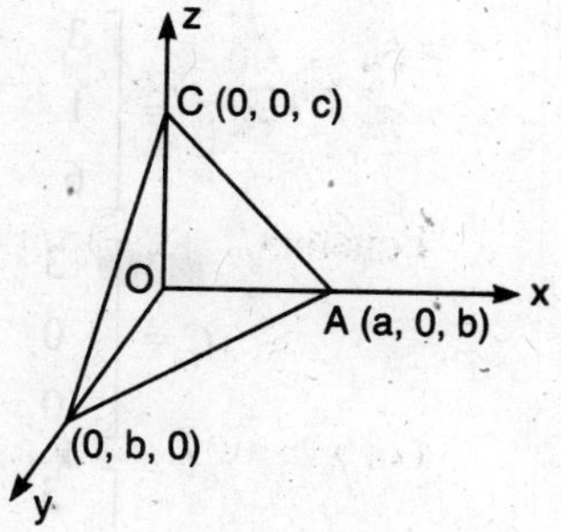


$$\frac{x}{a}+\frac{y}{b}+\frac{z}{c} = 1$$

Lower limits of the coordinate planes i.e., $z = 0, y = 0, x = 0$

Upper limit of $z$ from $\frac{x}{a}+\frac{y}{b}+\frac{z}{c} = 1$

$$\Rightarrow \quad \frac{z}{c} = 1 - \frac{x}{a}-\frac{y}{b} \Rightarrow z = c\left(1-\frac{x}{a}-\frac{y}{b}\right)$$

Lower limit of $y = 0$ (from coordinate plane) in xy-plane,

$$\frac{x}{a}+\frac{y}{b}+\frac{z}{c} = 1$$

Become $\frac{x}{a}+\frac{y}{b} = 1 \Rightarrow \frac{y}{b} = 1-\frac{x}{a} \Rightarrow y = b\left(1-\frac{x}{a}\right)$ is the upper limit

$$\text{Hence volume } = \int_0^a \int_0^{b\left(1-\frac{x}{a}\right)} \int_0^{c\left(1-\frac{x}{a}-\frac{y}{b}\right)} dx\,dy\,dz = \int_0^a dx \int_0^{b\left(1-\frac{x}{a}\right)} dy\,[z]_0^{c\left(1-\frac{x}{a}-\frac{y}{b}\right)}$$

$$= \int_0^a dx \int_0^{b\left(1-\frac{x}{a}\right)} c\left(1-\frac{x}{a}-\frac{y}{b}\right) dy = c\int_0^a dx\left[y-\frac{xy}{a}-\frac{1}{b}\frac{y^2}{2}\right]_0^{b\left(1-\frac{x}{a}\right)}$$

$$= c\int_0^a dx\left(\left(1-\frac{x}{a}\right)\cdot b\left(1-\frac{x}{a}\right)-\frac{1}{2b}b^2\left(1-\frac{x}{a}\right)^2\right) = c\int_0^a \frac{b}{2}\left(1-\frac{x}{a}\right)^2 dx = \frac{bc}{2}\left[\frac{\left(1-\frac{x}{a}\right)^2}{3\times-\frac{1}{a}}\right]_0^a$$

$$= \frac{bc}{2}\times\frac{-a}{3}[0-1] = \frac{abc}{6}$$ **Ans.**

(*b*) Find the volume of the tetrahedron bounded by the coordinate planes and the plane $\frac{x}{a}+\frac{y}{b}+\frac{z}{c}=1$. (8)

**Solution.** Refer to Example 41 on page 120.

**4.** (*a*) Show that if $\lambda \neq -5$, the system of equations, $3x - y + 4z = 3$, $x + 2y - 3z = -2$, $6x + 5y + \lambda z = -3$, have unique solution. If $\lambda = -5$, then show that the equations are consistent. Find the solution in each case.

**Solution.** The system of equation can be written in matrix form

$$\begin{bmatrix} 3 & -1 & 4 \\ 1 & 2 & -3 \\ 6 & 5 & \lambda \end{bmatrix}\begin{bmatrix} x \\ y \\ z \end{bmatrix} = \begin{bmatrix} 3 \\ -2 \\ -3 \end{bmatrix} \Rightarrow AX = B \text{ where } A = \begin{bmatrix} 3 & -1 & 4 \\ 1 & 2 & -3 \\ 6 & 5 & \lambda \end{bmatrix}$$

$$\text{Let} \quad C = [A : B] \quad X = \begin{bmatrix} x \\ y \\ z \end{bmatrix} \text{ and } B = \begin{bmatrix} 3 \\ -2 \\ -3 \end{bmatrix}$$

$$= \begin{bmatrix} 3 & -1 & 4 & : & 3 \\ 1 & 2 & -3 & : & -2 \\ 6 & 5 & \lambda & : & -3 \end{bmatrix}$$

$$C = \begin{bmatrix} 3 & -1 & 4 & : & 3 \\ 0 & 7 & -13 & : & -9 \\ 0 & 7 & \lambda-8 & : & -9 \end{bmatrix} \begin{matrix} \\ R_2 \to 3R_2 - R_1 \\ R_3 \to R_3 - 2R_1 \end{matrix}$$

$$C = \begin{bmatrix} 3 & -1 & 4 & : & 3 \\ 0 & 7 & -13 & : & -9 \\ 0 & 0 & \lambda+5 & : & 0 \end{bmatrix} R_3 \to R_3 \to R_2$$

Clearly, for $\lambda + 5 \neq 0 \Rightarrow \lambda \neq -5$

Rank of $C$ = Rank of $A$ = 3 (no. of 3 unkowns $(x, y, z)$)

So, system of equations have unique solution for solution $AX = B$.

$$\begin{bmatrix} 3 & -1 & 4 \\ 0 & 7 & -13 \\ 0 & 0 & \lambda+5 \end{bmatrix} \begin{bmatrix} x \\ y \\ z \end{bmatrix} = \begin{bmatrix} 3 \\ -9 \\ 0 \end{bmatrix}, \begin{bmatrix} 3 & -1 & 4 \\ 0 & 7 & -13 \\ 0 & 0 & 0 \end{bmatrix} \begin{bmatrix} x \\ y \\ z \end{bmatrix} = \begin{bmatrix} 3 \\ -9 \\ 0 \end{bmatrix} \qquad [\lambda + 5 = 0]$$

$\Rightarrow$ $3x - y + 4z = 3 \Rightarrow 7y - 13z = -9$ Let $z = k$

$$7y - 13k = -9 \Rightarrow y = -\frac{9}{7} + \frac{13k}{7},\; 3x + \frac{9}{7} - \frac{13k}{7} = 3$$

$$\Rightarrow \quad 3x = \frac{12}{7} + \frac{13k}{7} \Rightarrow x = \frac{4}{7} + \frac{13k}{21}$$ **Ans.**

(*b*) Use Gauss elimination and to solve the equations, $x - y + z = 6$

$2x + 4y + z = 3$

$3x + 2y - 2z = -2$ (8)

**Solution.** Give equation in matrix form $\begin{bmatrix} 1 & -1 & 1 \\ 2 & 4 & 1 \\ 3 & 2 & -2 \end{bmatrix} \begin{bmatrix} x \\ y \\ z \end{bmatrix} = \begin{bmatrix} 6 \\ 3 \\ -2 \end{bmatrix}$

$$\begin{bmatrix} 1 & -1 & 1 \\ 0 & 6 & -1 \\ 0 & 5 & -5 \end{bmatrix} \begin{bmatrix} x \\ y \\ z \end{bmatrix} = \begin{bmatrix} 6 \\ -9 \\ -20 \end{bmatrix} \begin{matrix} \\ R_2 \to R_2 - 2R_1 \\ R_3 \to R_3 - 3R_1 \end{matrix} \begin{bmatrix} 1 & -1 & 1 \\ 0 & 6 & -1 \\ 0 & 0 & -25 \end{bmatrix} \begin{bmatrix} x \\ y \\ z \end{bmatrix} = \begin{bmatrix} 6 \\ -9 \\ -75 \end{bmatrix} R_3 \to 6R_3 - 5R_2$$

Backward substitution $-25z = -75 \Rightarrow z = \dfrac{-75}{-25} = 3 \Rightarrow z = 3$

$\because$ $6y - z = -9 \Rightarrow 6y - 3 = -9 \Rightarrow 6y = -6 \Rightarrow y = -1$

$\because$ $x - y + z = 6,\; x + 1 + 3 = 6 \Rightarrow x = 2, x = 2, y = -1, z = 3$ **Ans.**

**5.** (*a*) Develop Newton-Raphson formula for finding $\sqrt{N}$, where $N$ is real number. Use it to find $\sqrt{41}$. Correct to 3 decimal places.

**Solution.** $x = \sqrt{N}, \Rightarrow x^2 = N \Rightarrow f(x) = x^2 - N = 0 \Rightarrow f'(x) = 2x$

Using Newton Raphson method.

Root of $f(x) = 0$, after $(n+1)$th iteration is $x_{n+1} = x_n - \dfrac{f(x_n)}{f'(x_n)} = x_n - \dfrac{x_n^2 - N}{2x_n} = \dfrac{2x_n^2 - x_n^2 + N}{2x_n}$

$$\Rightarrow x_{n+1} = \frac{x_n^2 + N}{2x_n} \Rightarrow x_{n+1} = \frac{1}{2}\left(x_n + \frac{N}{x_n}\right)$$ formula for find $\sqrt{N}$ by Newton-Raphson

method for find $\sqrt{41}$ put $N = 41$ i.e., $x_{n+1} = \frac{1}{2}\left(x_n + \frac{41}{x_n}\right)$

$\because$ Perfect square of 36 near 41 is $\sqrt{36} = 6$.

So, consider initial approximation $x_0 = 6$

$$x_1 = \frac{1}{2}\left(6 + \frac{41}{6}\right) = 6.41667,\ x_2 = \frac{1}{2}\left(6.41667 + \frac{41}{6.41667}\right) = 6.40314$$

$$x_3 = \frac{1}{2}\left(6.40314 + \frac{41}{6.40314}\right) = 6.40312$$

Hence value of $\sqrt{41}$ is 6.4031 (correct to four decimal places) **Ans.**

(*b*). Use Runge Kutta method of order four to find $y$ at $x = 0.2$ given that $\frac{dy}{dx} = \frac{y^2 - x^2}{y^2 + x^2}$ and $y(0) = 1$ taking $h = 0.2$. (8)

**Solution.** $f(x, y) = \frac{y^2 - x^2}{y^2 + x^2}$, $y(0) = 1$

$$\Rightarrow \quad x_0 = 0,\ y_0 = 1,\ h = 0.2,\ k_1 = hf(x_0, y_0) = 0.2f(0, 1) = 0.2\left(\frac{1^2 - 0^2}{1^2 + 0^2}\right) = 0.2$$

$$k_2 = hf\left(x_0 + \frac{h}{2}, y_0 + \frac{k_1}{2}\right) = 0.2f\left(0 + \frac{0.2}{2}, 1 + \frac{0.2}{2}\right) = 0.2 \times f(0.1, 1.1)$$

$$= 0.2 \times \left(\frac{1.1^2 - 0.1^2}{1.1^2 + 0.1^2}\right) = 0.1967$$

$$k_3 = h + \left(x_0 + \frac{h}{2}, y_0 + \frac{k_2}{2}\right) = 0.2 \times f\left(0 + 0.1, 1 + \frac{0.1967}{2}\right) = 0.2 \times f(0.1, 1.09835)$$

$$= 0.2 \times \left(\frac{1.09835^2 - 0.1^2}{1.09835^2 + 0.1^2}\right) = 0.1967$$

$$k_4 = hf(x_0 + h_1 y_0 + k_3) = 0.2 \times f(0 + 0.2, 1 + 0.1967) = 0.2 \times f(0.2, 1.1967) = 0.1891$$

$$\because \quad k = \frac{1}{6}(k_1 + 2k_2 + 2k_3 + k_4) = \frac{1}{6}(0.2 + 2 \times 0.1967 + 2 \times 0.1967 + 0.1891) = 0.19598$$

$$\because \quad y = y_0 + k,\ y(x_1) = 1 + 0.19598,\ y(x_0 + h) = 1.19598,\ y(0.2) = 1.19598$$ **Ans.**

**6.** (*a*) Solve the differential equation $(1 + x + y)^2 \frac{dy}{dx} = 1$

**Solution.** $(1 + x + y)^2 \frac{dy}{dx} = 1$ ... (1)

Put $1 + x + y = t$

Differentiating, we get $1 + \frac{dy}{dx} = \frac{dt}{dx} \Rightarrow \frac{dy}{dx} = \frac{dt}{dx} - 1$

On putting these values in (1), we get $t^2.\left(\frac{dt}{dx}-1\right)=1$

$$\frac{dt}{dx}-1=\frac{1}{t^2},\quad \frac{dt}{dx}=\frac{1}{t^2}+1=\frac{1+t^2}{t^2},\quad \frac{t^2}{1+t^2}dt=dx$$

$$\int\left(\frac{t^2+1-1}{1+t^2}\right)dt-\int dx+c,\quad \int\left(1-\frac{1}{1+t^2}\right)dt=x+c\Rightarrow t-\tan^{-1}t=x+c$$

$(1+x+y)-\tan^{-1}(1+x+y)=x+c,\ 1+y=\tan^{-1}(1+x+y)=c$ **Ans.**

(*b*) Show that the family of Paraboles $x^2=4a(y+a)$ is self orthogonal.

**Solution.** Try yourself.

**7.** (*a*) Solve the equation $(D^3+2D^2+D)y=x^2e^{2x}+\sin^2 x$

**Solution.** $(D^3+2D^2+D)y=x^2e^{2x}+\sin^2 x$

A.E. is $m^3+2m^2+m=0$

$\Rightarrow\quad m(m^2+2m+1)=0\Rightarrow m(m+1)^2=0\Rightarrow m=0,\ m=-1,-1$

C.F. is $c_1e^{0x}+(c_2+c_3x)e^{-x}$

$$\text{P.I.}=\frac{1}{(D^3+2D^2+D)}x^2e^{2x}+\frac{1}{D^3+2D^2+D}\sin^2 x$$

$$=e^{2x}\frac{1}{(D+2)^3+2(D+2)^2+(D+2)}x^2+\frac{1}{D^3+2D^2+D}\frac{1}{2}(1-\cos 2x)$$

$$=e^{2x}\frac{1}{D^3+6D^2+12D+8+2D^2+8D+8+D+2}x^2$$

$$+\frac{1}{2}\frac{1}{D^3+2D^2+D}1-\frac{1}{2}\frac{1}{D^3+2D^2+D}\cos 2x$$

$$=e^{2x}\frac{1}{D^3+8D^2+21D+18}x^2+\frac{x}{2}\frac{1}{3D^2+4D+1}e^{0x}-\frac{1}{2}\frac{1}{D^3+2D^2+D}\cos 2x$$

$$=e^{2x}\frac{1}{18\left(1+\frac{7}{6}D+\frac{4}{9}D^2+\frac{D^3}{18}\right)}x^2+\frac{7}{2}\frac{1}{D^3+2D^2+D}e^{0x}-\frac{1}{2}\frac{1}{-4D-8+D}\cos 2x$$

$$=e^{2x}\frac{1}{18}\left(1+\frac{7}{6}D+\frac{4}{9}D^2+\frac{D^3}{18}\right)^{-1}x^2+\frac{x}{2}\frac{1}{3D^2+4D+1}e^{ox}-\frac{1}{2}\frac{1}{-3D-8}\cos 2x$$

$$=\frac{e^{2x}}{18}\left[1-\frac{7}{6}D-\frac{4}{9}D^2-\frac{D^3}{18}+\frac{49}{36}D^2+\ldots\right]x^2+\frac{x}{2}+\frac{1}{2}\frac{3D-8}{9D^2-64}\cos 2x$$

$$=\frac{e^{2x}}{18}\left[x^2-\frac{7}{6}(2x)-\frac{4}{9}(2)+0+\frac{49}{36}(2)\right]+\frac{x}{2}+\frac{1}{2}\frac{3D-8}{-36-64}\cos 2x$$

$$=\frac{e^{2x}}{18}\left[x^2-\frac{7}{3}x-\frac{8}{9}+\frac{49}{18}\right]+\frac{x}{2}-\frac{1}{200}[-6\sin 2x-8\cos 2x]$$

$$=\frac{e^{2x}}{18}\left[x^2-\frac{7\pi}{3}+\frac{11}{6}\right]+\frac{x}{2}+\frac{1}{100}(3\sin 2x+4\cos 2x)$$ **Ans.**

$$y=c_1+c_2e^{-x}+c_3x\,e^{-x}+\frac{e^{2x}}{18}\left(x^2-\frac{7x}{3}+\frac{11}{6}\right)+\frac{x}{2}+\frac{1}{100}\ (3\sin 2x+4\cos 2x)$$ **Ans.**

(*b*) The method of variation of parameters to solve $\left(\frac{d^2y}{dx^2}-2\frac{dy}{dx}+y\right) = \frac{e^x}{x}$

**Solution.** $\left(\frac{d^2y}{dx^2}-2\frac{dy}{dx}+y\right) = \frac{e^x}{x}$

A.E. $m^2 - 2m + 1 = 0 \Rightarrow (m-1)^2 = 0 \Rightarrow m = 1, 1$

F. $= (c_1 + c_2x)e^x = c_1e^x + c_2xe^x$

P.I $= uy_1 + vy_2$ where $y_1 = e^x, y_2 = xe^x,\ u = \int \frac{-y_2}{y_1y_2' - y_1'y_2} X dx$

$$u = \int \frac{-xe^{x^2}}{e^x(e^x + xe^x) - e^x(xe^x)}\left(\frac{e^x}{x}\right) dx = \int \frac{-e^{2x}}{e^{2x}} dx = -\int dx = -x$$

$$v = \int \frac{e^x}{e^x(e^x + xe^x) - e^x(xe^x)} \frac{e^x}{x} dx = \int \frac{\frac{e^{2x}}{x}}{e^{2x}} dx = \int \frac{1}{x} dx = \log x$$

P.I. $= -xe^x + xe^x \log x = xe^x(\log x - 1)$,

$y =$ C.F. + P.I. $= (c_1 + c_2x)e^x + xe^x(\log x - 1)$ **Ans.**

**8.** (*a*) Find the series solution of the differential equation: $2x^2\frac{d^2y}{dx^2} - x\frac{dy}{dx} + (1-x^2)y = 0$

**Solution.** Here $x = 0$ is a singular point. So Frobenious method used

Let solution be $y = a_0x^m + a_1x^{m+1} + a_2x^{m+2} + \ldots$ $a_0 \neq 0$

$$= \sum_{k=0}^{\infty} a_kx^{m+k}, \qquad \frac{dy}{dx} = \sum_{k=0}^{\infty} a_k(m+k)x^{m+k-1}$$

$$\frac{d^2y}{dx^2} = \sum_{k=0}^{\infty} a_k\,(m+k)\,(m+k-1)\,x^{m+k-2}$$

Substituting these values in given different equation $2x^2\frac{d^2y}{dx^2} - x\frac{dy}{dx} + (1-x^2)y = 0$

$$2x^2\sum_{k=0}^{\infty} a_k(m+k)(m+k-1)x^{m+k-2} - x\sum_{k=0}^{\infty} a_k(m+k)x^{m+k-1} + (1-x^2)\sum_{k=0}^{\infty} a_kx^{m+k} = 0$$

$$2\sum_{k=0}^{\infty} a_k(m+k)(m+k-1)x^{m+k} - \sum_{k=0}^{\infty} a_k(m+k)x^{m+k} + \sum_{k=0}^{\infty} a_kx^{m+k} - \sum_{k=0}^{\infty} a_kx^{m+k+2} = 0$$

$$\Rightarrow \sum_{k=0}^{\infty} a_k[2(m+k)(m+k-1) - (m+k) + 1]x^{m+k} - \sum_{k=0}^{\infty} a_xx^{m+k+2} = 0$$

If $m + k = t$, then $2(m+x)(m+x-1) - (m+x) + 1$

$= 2t(t-1) - t + 1 = 2t(t-1) - 1(t-1) = (t-1)(2t-1)$

$= (m+k-1)[2(m+k) - 1]$

$$\Rightarrow \sum_{k=0}^{\infty} a_k(m+k-1)(2(m+k)-1)x^{m+k} - \sum_{k=0}^{\infty} a_kx^{m+k+2} = 0$$

Equating coefficient of $x^m$

$$a_0(m-1)(2m-1) = 0$$

$\because$ $a_0 \neq 0$ and $m = 1, \frac{1}{2}$

Equating coeff. of $x^{m+1}$, $a_1 m(2(m+1)-1 = 0 \Rightarrow a_1 m(2m+1) = 0$

$\because \quad m(2m+1) \neq 0 \Rightarrow a_1 = 0$

Equating coeff. of $x^{m+x+2}$

$$a_{k+2}(m+k+1)\,[2(m+k+2)-1] - a_k = 0$$

$$a_{k+2} = \frac{a_k}{(m+k+1)(2(m+k)+3)}$$

$$k = 0,\ a_2 = \frac{a_0}{(m+1)(2m+3)}$$

$$k = 1,\ a_3 = 0 \quad (\because a_1 = 0)$$

$$k = 2,\ a_4 = \frac{a_2}{(m+3)+[2(m+2)+3]} = \frac{a_0}{(m+3)(2m+7)(m+1)(2m+3)}$$

$$y = x^m(a_0 + a_1 x + a_2 x^2 + a_3 x^3 + \ldots)$$

$$= x^m\left(a_0 + 0 + \frac{a_0}{(m+1)(2m+3)}x^2 + 0 + \frac{a_0}{(m+1)(m+3)(2m+3)(2m+7)}x^4\right)$$

The solution is $\quad y = c_1(y)_{m=1} + c_2(y)_{m=\frac{1}{2}} + c_2$

$$= c_1 a_0 x\left(1 + \frac{1}{2\times 5}x^2 + \frac{1}{2\times 4\times 5\times 9}x^4 + \ldots\right) + c_2 a_0 x^{1/2}\left(1 + \frac{1}{\frac{3}{2}\times 4}x^2 + \frac{1}{\frac{3}{2}\times\frac{7}{2}\times 4\times 8}x^4 \ldots\right)$$

$$= Ax\left(1 + \frac{1}{10}x^2 + \frac{1}{360}x^4 + \ldots\right) + Bx^{1/2}\left(1 + \frac{1}{6}x^2 + \frac{1}{21\times 8}x^4 \ldots\right)$$ **Ans.**

(*b*) Show that $\int_a^b (x-a)^m (b-x)^n\,dx = (b-a)^{m+n+1}\ \beta(m+1, n+1)$

**Solution.** $\int_a^b (x-a)^m (b-x)^n\,dx$

Let $\quad x - a = t,\ dx = dt$

| $x$ | $a$ | $b$ |
|---|---|---|
| $t = x - a$ | 0 | $b - a$ |

$$\int_0^b (x-a)^m (b-x^n)\,dt = \int_0^{b=a} t^m[b-(a+t)]^n\,dt$$

Let $t = (b-a)p, \quad dt = (b-a)dp$

| $t$ | 0 | $b-a$ |
|---|---|---|
| $P = \frac{t}{b-a}$ | 0 | 1 |

$$= \int_0^1 (b-a)^m p^m (b-a-(b-a)p)^n\,dp(b-a) = (b-a)^{m+1}\int_0^1 \phi p^m.(b-a)^n(1-p)^n\,dp$$

$$= (b-a)^{m+n+1}\int_0^1 p^m(1-p)^n\,dp = (b-a)^{m+n+1}\int_0^1 p^{m+1-1}(1-p)^{n+1-1}\,dp$$

$$= (b-a)^{m+n+1}\ \beta(m+1, n+1)$$ **Ans.**

**9.** (*a*)

**Solution.** Refer to article 8.21 on page no. 654.

(*b*) Show that $\frac{d}{dx}\left[J_n^2(k)+J_{n+1}^2(x)\right]=2\left[\frac{n}{x}J_n^2(x)-\frac{n+1}{\pi}J_{n+1}^2(x)\right]$

**Solution.** Refer to Q.8 of Exercise 8.3 on page 646.

$$\text{L.H.S.} = \frac{d}{dx}\left[J_n^2(x)+J_{n+1}^2(x)\right]$$

$$= 2J_n(x)\, J'_n(x) + 2J_{n+1}(x)\, J'_{n+1}(x) \quad \begin{pmatrix} xJ'_n = nJ_n - xJ_{n+1} \\ xJ_n' = -nJ_n + xJ_{n-1} \end{pmatrix}$$

$$= -2J_n(x)\left[\frac{n}{x}J_n(x)-J_{n+1}(x)\right]+2J_{n+1}(x)\left[-\frac{(n+1)}{x}J_{n+1}(x)+J_n(x)\right]$$

$$= 2\left[\frac{n}{x}J_n^2(x)-\frac{(n+1)}{x}J_{n+1}^2(x)\right] = \text{R.H.S.}$$

**Proved.**

**December 2014**

# Engineering Mathematics-II

## AIMETE-ET/CS/IT (New Scheme)

*Time allowed : 3 hours ]* *[ Maximum Marks : 100*

**1.** Choose the correct or the best alternative in the following: (2 × 10)

(*a*) The value of $\int_c \log z\, dz$, where $c$ is the unit circle $|z| = 1$

(*a*) $2\pi i$ (*b*) $-2\pi i$ (*c*) 1 (*d*) $-1$

**Solution.** Refer to Q.7 on page 547. **Ans.** (*a*)

(*b*) The residue at simple pole of the function $f(z) = \dfrac{z^2}{(z-1)^2(z+2)}$ is

(*a*) $\dfrac{1}{2}$ (*b*) $\dfrac{-1}{2}$ (*c*) $\dfrac{4}{9}$ (*d*) 1

**Solution.** $f(z) = \dfrac{z^2}{(z-1)^2(z+2)}$

Poles are determined by $(z-1)^2(z+2) = 0$, $z = 1$ is pole of order 2, $z = -2$ simple pole.

Res. at $z = -2$ is $\lim\limits_{z \to -2} (z+2) \dfrac{Z^2}{(z+1)^2(z+2)} = \lim\limits_{z \to -2} \dfrac{z^2}{(z-1)^2} = \dfrac{(-2)^2}{(-2-1)^2} = \dfrac{4}{9}$ **Ans.** (*c*)

(*c*) If $\vec{F}$ is such that $\nabla \times \vec{F} = 0$ then $\vec{F}$ is called

(*a*) Rotational (*b*) Irrotational (*c*) Solenoidal (*d*) None of these

**Solution.** $\Delta \times \vec{F}$ is curl $\vec{F}$, $\nabla \times \vec{F} = 0$ is irrotational. **Ans.** (*b*)

(*d*) If $\Phi(x, y, z) = 3xyz^2 - y^3z^2$, then the value of grad $\Phi$ at point $(1, -2, -1)$ is

(*a*) 0 (*b*) 2 (*c*) $-2$ (*d*) $-6i - 9j - 4k$

**Solution.** $\phi(x, y, z) = 3xyz^2 - y^3z^2$ than grad $\phi = \nabla$ $\phi(x, y, z)$

$$= \left(i\frac{\partial}{\partial x} + j\frac{\partial}{\partial y} + k\frac{\partial}{\partial z}\right)(3xyz^2 - y^3z^2)$$

$$= i3yz^2 + j(3xz^2 - 3y^2z^2) + k(6xyz - 2y^3z)$$

Grad $\phi$ at $(1, -2, -1) = i3(-2)(-1)^2 + j[3(1)(-1)^2 - 3(-2)^2(-1)^2] + k[6(1)(-2)(-1) - 3(-2)^2(1)]$

$$= -6i + j(3 - 12) + k(12 - 16)$$

$$= -6i - 9j - 4k$$ **Ans.** (*d*)

(*e*) If $\vec{A} = xi - x^2j + (x-1)k$ and $\vec{B} = 2x^2i + 6xk$, then the value of $\int_0^2 \vec{A}\cdot\vec{B}\,dx$ is

(*a*) 5 (*b*) 12 (*c*) 7 (*d*) 5

**Solution.** $\vec{A} = xi - x^2 j + (x-1)\bar{k}$ , $\vec{B} = 2x^2 i + 6xk$

$$\vec{A}.\vec{B} = [xi - x^2 j + (x-1)k].[2x^2 i + 6x\bar{k}] = 2x^3 + 6x(x-1)$$

$$\int_0^2 \bar{A}.\bar{B}\,dx = \int_0^2 (2x^3 + 6x^2 - 6x)dx = \left(\frac{2x^4}{4} + \frac{6x^3}{3} - 6x\right)_0^2 = 8 + 16 - 12 = 12$$ **Ans.** (*b*)

(*f*) The value of $\int_0^6 \frac{dx}{1+x^2}$ by trapezoidal rule is

(*a*) 1 (*b*) – 1 (*c*) 1.4108 (*d*) 5

**Solution.**

| $x$ | 0 | 1 | 2 | 3 | 4 | 5 | 6 |
|---|---|---|---|---|---|---|---|
| $\frac{1}{1+x^2}$ | 1 | $\frac{1}{2}$ | $\frac{1}{5}$ | $\frac{1}{10}$ | $\frac{1}{17}$ | $\frac{1}{26}$ | $\frac{1}{37}$ |

Trapezoidal rule is $\int_0^6 (x)\ dx = \frac{h}{2}[(y_1 + y_6) + 2(y_2 + y_3 + y_4 + y_5)]$

$$\int_0^6 \frac{dx}{1+x^2} = \frac{1}{2}\left[\left(1+\frac{1}{37}\right) + 2\left(\frac{1}{2}+\frac{1}{5}+\frac{1}{10}+\frac{1}{17}+\frac{1}{26}\right)\right]$$

$$= \frac{1}{2}[(1 + .02703) + 2(0.5 + 0.2 + 0.1 + 0.05882 + 0.03846)]$$

$$= \frac{1}{2}[1.02703 + 2(0.89728)] = \frac{1}{2}[1.02703 + 1.79456] = \frac{1}{2}[2.82159] = 1.4108$$ **Ans.** (*c*)

(*g*) The solution of $xp + yq = z$ is

(*a*) $f\left(\frac{x}{y}, \frac{y}{z}\right) = 0$ (*b*) $f(x, y) = 0$ (*c*) $(xy, yz) = 0$ (*d*) $f(x^2 + y^2) = 0$

**Solution.** $xp + yq = z$, $\frac{dx}{x} = \frac{dy}{y} = \frac{dz}{z}$
(*i*) (*ii*) (*iii*)

$\log x = \log y + \log c_1$, from (*i*) and (*ii*)

$$\Rightarrow \quad \frac{x}{y} = c_1 \quad \text{... (1)}$$

$\log y = \log z + \log c_2$, from (*ii*) and (*iii*)

$$\Rightarrow \quad \frac{y}{z} = c_2 \quad \text{... (2)}$$

From (1) and (2), we have $f\left(\frac{x}{y}, \frac{y}{z}\right) = 0$ **Ans.** (*a*) (*h*) A five-digit number is formed by using the digit 0, 1, 2, 3 and 5 without repetition, the probability of the number that is divisible by 6 is

(*a*) 1 (*b*) – 1 (*c*) $\frac{25}{4}$ (*d*) none of these

**Solution.** **Ans (*d*)**

(*i*) If three persons selected at random are stopped on a street, then the probability that all of them were born on Sunday is

(*a*) $\frac{1}{35}$ (*b*) $\frac{1}{343}$ (*c*) $\frac{3}{352}$ (*d*) none of these

**Solution.** There are seven days in a week.

Probability of a person born on Sunday $= \frac{1}{7}$

Probability of three persons born on sunday $= \frac{1}{7} \times \frac{1}{7} \times \frac{1}{7} = \frac{1}{343}$ **Ans. (*b*)**

(*j*) In the Poisson distribution, if $2P(x = 1) = P(x = 2)$, then mean is

(*a*) 4 (*b*) 1 (*c*) – 1 (*d*) none of these

**Solution.** $2P(x = 1) = P(x = 2)$

$$\frac{2e^{-m} m}{\lfloor 1} = \frac{e^{-m} m^2}{\lfloor 2}, \Rightarrow 2m = \frac{m^2}{2} \Rightarrow 1 = \frac{m}{4} \Rightarrow m = 4$$

Mean $= m = 4$ **Ans. (*a*)**

**Answer any five questions out of eight questions. Each question carries 16 marks.**

**2.** (*a*) Show that the real and imaginary parts of the function $w = \log z$ satisfy the Cauchy – Riemann equations when z is not zero. Find its derivatives. (8)

**Solution.** Refer Example 14 on page 515.

(*b*) Evaluate $\int_0^{i+1} (x^2 - iy)\, dz$ along the path (*i*), $y = x$ and (*ii*) $y = x^2$ (8)

**Solution.** Refer Example 45 on page 545.

**3.** (*a*) Find the bilinear transformation, which maps $z_1 = 0,\ z_2 = 1,\ z_3 = \infty$ into $w_1 = i$, $w_2 = -1,\ w_3 = -i$.

**Solution.** Refer to Question 2 on page 578.

Using formula for finding bilinear transformation

$$\frac{(w - w_1)(w_2 - w_3)}{(w - w_3)(w_2 - w_1)} = \frac{(z - z_1)(z_2 - z_3)}{(z - z_3)(z_2 - z_1)} = \frac{(z - z_1)\left(\frac{z_2}{z_3} - 1\right)}{\left(\frac{z}{z_3} - 1\right)(z_2 - z_1)}$$

$$\Rightarrow \frac{(w - i)(-1 + i)}{(w + i)(-1 - i)} = \frac{(z - 0)(0 - 1)}{(0 - 1)(1 - 0)} \Rightarrow \frac{w - i}{w + i} \times \frac{1 - i}{1 + i} \times \frac{1 - i}{1 - i} = z \Rightarrow \frac{w - i}{w + i} \times \frac{1 + i^2 - 2i}{1^2 - i^2} = z$$

$$\Rightarrow \frac{w - i}{w + i} \times \frac{1 - 1 - 2i}{+2} = z \Rightarrow \frac{w - i}{w + i} = zi, \Rightarrow w - i = wzi - z, \Rightarrow w - wzi = i - z,$$

$$\Rightarrow \quad w = \frac{i-z}{1-iz}$$ **Ans.**

(*b*) Find the terms in the Laurent expansion of $f(z) = \dfrac{1}{(z+1)(z+3)}$, for the region.

(*i*) $1 < |z| < 3$ (*ii*) $|z| < 1$ (8)

**Solution.** $\because \quad \dfrac{1}{(z+1)(z+3)} = \dfrac{A}{z+1} + \dfrac{B}{z+3}$ (Apply partial fraction)

$$1 = A(z+3) + B(z+1)$$

$$1 = A(-1+3) \Rightarrow A = \frac{1}{2},$$

$$1 = B(-3+1) \Rightarrow B = -\frac{1}{2},$$

$$f(z) = \frac{1}{(z+1)(z+3)} = \frac{\frac{1}{2}}{z+1} - \frac{\frac{1}{2}}{z+3}$$

(*i*) Case I $1 < |z|$ & $|z| < 3 \Rightarrow \dfrac{1}{|z|} < 1$ & $\dfrac{|z|}{3} < 1$

So, $$f(z) = \frac{\frac{1}{2}}{z\left(1+\frac{1}{z}\right)} - \frac{\frac{1}{2}}{3\left(1+\frac{z}{3}\right)} = \frac{1}{2z}\left(1+\frac{1}{z}\right)^{-1} - \frac{1}{6}\left(1+\frac{z}{3}\right)^{-1}$$

$$= \frac{1}{2z}\left(1-\frac{1}{z}+\frac{1}{z^2}-\frac{1}{z^3}\ldots\right) - \frac{1}{6}\left(1-\frac{z}{3}+\frac{z^2}{9}-\frac{z^3}{27}\ldots\right)$$

$$= \frac{1}{2z} - \frac{1}{2z^2} + \frac{1}{2z^3} - \frac{1}{6} + \frac{z}{18} - \frac{z^2}{54} + \frac{z^3}{162} + \ldots$$ **Ans.**

(*ii*) When $|z| < 1 \Rightarrow \dfrac{|z|}{3} < 1$

$$f(z) = \frac{\frac{1}{2}}{1+z} - \frac{\frac{1}{2}}{3\left(1+\frac{z}{3}\right)} = \frac{1}{2}(1+z)^{-1} - \frac{1}{6}\left(1+\frac{z}{3}\right)^{-1}$$

$$= \frac{1}{2}(1-z+z^2-z^3\ldots) - \frac{1}{6}\left(1-\frac{z}{3}+\frac{z^2}{9}-\frac{z^3}{27}\ldots\right)$$

$$= \frac{1}{2} - \frac{1}{6} + z\left(-\frac{1}{2}+\frac{1}{18}\right) + z^2\left(\frac{1}{2}-\frac{1}{54}\right) + z^3\left(-\frac{1}{2}+\frac{1}{162}\right) = \frac{1}{3} - \frac{4}{9}z + \frac{13}{27}z^2 - \frac{40}{81}z^3$$ **Ans.**

**4.** (*a*) If $\vec{r} = xi + yj + zk$, $a = |\vec{r}|$ and $\vec{a}$ is constant vector, then find the value of

$$\text{div}\left[\frac{\vec{a}\times\vec{r}}{r^n}\right] \qquad (8)$$

**Solution.** Refer to Example 38 on page no. 400.

(*b*) Define CURL of a vector point function with physical interpretation. (8)

**Solution.** Refer to Article. 5.32 on page no. 403.

**5.** (*a*) Evaluate $\int \vec{F}\cdot d\vec{r}$, where $\vec{F} = \frac{iy - jx}{x^2+y^2}$ and $c$ is the circle $x^2 + y^2 = 1$ traversed counter clockwise. (8)

**Solution.** $I = \int \vec{F}.d\vec{r} = \int \frac{(y\hat{i} - x\hat{j})}{x^2+y^2}\cdot(dx\hat{i} + dy\hat{j} + dz\hat{k}) = \int \frac{ydx - xdy}{x^2+y^2}$

Let $x = \cos\theta \Rightarrow dx = -\sin\theta\, d\theta$ $(\because r = 1)$

$y = \sin\theta \Rightarrow dy = \cos\theta\, d\theta$

$$I = \int_0^{2\pi} \frac{\sin\theta(-\sin\theta)d\theta - \cos\theta\cdot\cos\theta\, d\theta}{\sin^2\theta + \cos^2\theta} = \int_0^{2\pi} -\frac{(\sin^2\theta + \cos^2\theta)}{1}d\theta$$

$$= -\int_0^{2\pi} d\theta = -[\theta]_0^{2\pi} = -2\pi \qquad \textbf{Ans.}$$

(*b*) Verify Stocke's Theorem for $\vec{F} = (x^2 + y - 4)i + 3xyj + (2xz + z^2)k$ over the surface of hemisphere $x^2 + y^2 + z^2 = 16$ above the $x$-$y$ plane. (8)

**Solution.** Refer to Example 97 on page no. 444.

**6.** (*a*) State and Prove Lagrange's interpolation formula. (8)

**Solution.** Interpolating polynomial $f(x)$ for data

| $x$ | $x_0$ | $x_1$ | $x_2$ | ... | $x_n$ |
|---|---|---|---|---|---|
| $f(x)$ | $f(x_0)$ | $f(x_1)$ | $f(x_2)$ | ... | $f(x_n)$ |

is given by

$$f(x) = \frac{(x-x_1)(x-x_2)...(x-x_n)}{(x_0-x_1)(x_0-x_2)...(x_0-x_n)}f(x_0) + \frac{(x-x_0)(x-x_2)...(x-x_n)}{(x_1-x_0)(x_1-x_2)...(x_1-x_n)}f(x_1)$$

$$+ ... + \frac{(x-x_0)(x-x_1)...(x-x_{n-1})}{(x_n-x_0)(x_n-x_1)...(x_n-x_{n-1})}f(x_n)$$

**Proof**

There are $(n + 1)$ points so $f(x)$ be a polynomial in $x$ of degree $n$.

i.e., $f(x) = A_0(x - x_1)(x - x_2) ... (x - x_n) + A_1(x - x_0)(x - x_2) ... (x - x_n)$
$+ ... + A_n(x - x_0)(x - x_1) ... (x - x_{n-1})$ ... (*i*)

Where $A_0, A_1, ... A_n$ are constant

Put $x = x_0$ in equation (1), we get

$$f(x_0) = A_0(x_0 - x_1)(x_0 - x_2) ... (x_0 - x_n) \Rightarrow A_0 = \frac{f(x_0)}{(x_0-x_1)(x_0-x_2)...(x_0-x_n)}$$

Put $x = x_1$ in equation (1), we get

$$f(x_1) = A_1(x_1 - x_0)(x_1 - x_2) \ldots (x_1 - x_n) \Rightarrow A_1 = \frac{f(x_1)}{(x_1 - x_0)(x_1 - x_2)\ldots(x_1 - x_n)}$$

Put $x = x_n$ in equation (1), we get

$$f(x_n) = A_n(x_n - x_0)(x_n - x_1) \ldots (x_n - x_{n-1}) \Rightarrow A_n = \frac{f(x_n)}{(x_n - x_0)(x_n - x_1)\ldots(x_n - x_{n-1})}$$

Substituting these values of $A_0$, $A_1$, $A_n$ in equation (1), we get

$$f(x) = \frac{(x - x_1)(x - x_2)\ldots(x - x_n)}{(x_0 - x_1)(x_0 - x_2)\ldots(x_0 - x_n)} f(x_0) + \frac{(x - x_0)(x - x_2)\ldots(x - x_n)}{(x_1 - x_0)(x_1 - x_2)\ldots(x_1 - x_n)} f(x_1)$$

$$+ \ldots + \frac{(x - x_0)(x - x_1)\ldots(x - x_{n-1})}{(x_n - x_0)(x_n - x_1)\ldots(x_n - x_{n-1})} f(x_n)$$ **Proved.**

(*b*) Evaluate $\int_{0.5}^{0.7} \sqrt{x e^{-x}}$ by Simpson's $\frac{1}{3}$ rule. (8)

**Solution.** Consider 4 sub interval

$$\text{Step size, } h = \frac{\text{Upper limit} - \text{Lower Limit}}{\text{No. of subinterval}} = \frac{0.7 - 0.5}{4} = 0.05$$

| $x$ | $x_0 = 0.5$ | $x_1 = 0.55$ | $x_2 = 0.6$ | $x_3 = 0.65$ | $x_4 = 0.7$ |
|---|---|---|---|---|---|
| $\sqrt{xe^{-x}}$ | 0.42888 | 0.42787 | 0.42511 | 0.42088 | 0.41547 |
| | $y_0$ | $y_1$ | $y_2$ | $y_3$ | $y_4$ |

Using Simpson's $\frac{1}{3}$rd rule, $I = \frac{h}{3}[(y_0 + y_4) + 4(y_1 + y_3) + 2y_2]$

$$= \frac{0.05}{3}[(0.42888 + 0.41547) + 4(0.42787 + 0.42088) + 2 \times 0.42511]$$

$$= \frac{0.05}{3} \times 5.08957 = 0.08483$$ **Ans.**

**7.** (*a*) Form the partial differential equation by eliminating the function *f* from the relation

$$z = y^2 + 2\int\left(\frac{1}{x} + \log y\right).$$ (8)

**Solution.** $\because\ z = y^2 + 2f\left(\frac{1}{x} + \log y\right)$, $p = \frac{\partial z}{\partial x} = 2f'\left(\frac{1}{x} + \log y\right) \times \left(-\frac{1}{x^2}\right)$

$$\Rightarrow \quad -px^2 = 2f'\left(\frac{1}{x} + \log y\right) \quad \ldots (2)$$

Differentiating equation (1), w.r.t. to *y*, we get

$$q = \frac{\partial z}{\partial y} = 2y + 2f'\left(\frac{1}{x} + \log y\right) \times \frac{1}{y} \qquad \Rightarrow q\,y = 2y^2 + 2f'\left(\frac{1}{x} + \log x\right)$$

$$\Rightarrow \quad qy - 2y^2 = 2f'\left(\frac{1}{x} + \log y\right) \quad \ldots (3)$$

Putting the value of $2f'\left(\frac{1}{x}+\log y\right)$ from (2) and (3), we get

$$\Rightarrow \quad qy - 2y^2 = -px^2 \Rightarrow px^2 + qy = 2y^2$$ **Ans.**

Is the required partial differential equation of 1st order.

(*b*) Solve $\frac{\partial^2 z}{\partial x^2}+\frac{\partial^2 z}{\partial x \partial y}-6\frac{\partial^2 z}{\partial y^2}=y\cos x$ ... (8)

**Solution.** Refer to Example 40 page no. 699.

**8.** (*a*) State and prove Baye's theorem.

**Solution.** Refer to Article 11.6 on page no. 779.

(*b*) A can hit a target 4 times in 5 shots, B 3 times in 4 shots, C twice in 3 shots. They fire a volley. What is the probability that at least two shots hit? (8)

**Solution.** Let $p(A)$ = probability of $A's$ hitting the target $= \frac{4}{5}$

$p(B)$ = probability of $B's$ hitting the target $= \frac{3}{4}$

$p(c)$ = probability of $C's$ hitting the target $= \frac{2}{3}$

Probability that at least two shots hit

$$= p(A)\,p(B)\,q(c) + p(A)\,q(b)\,p(c) + q(A)\,p(B)\,p(C) + p(A)\,p(B)\,p(C)$$

$$= \frac{4}{5}\times\frac{3}{4}\times\frac{1}{3}+\frac{4}{5}\times\frac{1}{4}\times\frac{2}{3}+\frac{1}{5}\times\frac{3}{4}\times\frac{2}{3}+\frac{4}{5}\times\frac{3}{4}\times\frac{2}{3}$$

$$= \frac{1}{5}+\frac{2}{15}+\frac{1}{10}+\frac{2}{5} = \frac{6+4+3+12}{30} = \frac{25}{30} = \frac{5}{6}$$ **Ans.**

**9.** (*a*) Out of 800 families with 4 children each, how many families would be expected to have
(*i*) 2 boys and 2 girls (*ii*) at least one boy
(*iii*) No girl (*iv*) at most two girls?
Assume equal probability for boys and girls

**Solution.** $p$ = probability of having boy $= \frac{1}{2}$, $q$ = probability of having girl $= \frac{1}{2}$

$n = 4,\ N = 800$

(*i*) Expected no. of families having 2 boys and 2 girls

$$= 800 \times {}^4C_2 \times \left(\frac{1}{2}\right)^2 \times \left(\frac{1}{2}\right)^2 = 800 \times 6 \times \frac{1}{4}\times\frac{1}{4} = 300$$

(*ii*) Expected no. of families have at least one boy

$$= 800\left(1 - {}^4C_0\left(\frac{1}{2}\right)^{\circ}\left(\frac{1}{2}\right)^4\right) = 800\left(1-\frac{1}{16}\right) = 800 \times \frac{15}{16} = 750$$

(*iii*) Expected no. of families having no. of girls

$$= 800 \times 4C_0\left(\frac{1}{2}\right)^{\circ}\left(\frac{1}{2}\right)^4 = 800 \times \frac{1}{16} = 50$$

(*iv*) Expected no. of families having at most two girls

$$= 800\left({}^4C_0\left(\frac{1}{2}\right)^{\circ}\left(\frac{1}{2}\right)^4 + {}^4C_1\left(\frac{1}{2}\right)^1\left(\frac{1}{2}\right)^3 + {}^4C_2\left(\frac{1}{2}\right)^2\left(\frac{1}{2}\right)^2\right)$$

$$= 800\left(\frac{1}{16} + 4\times\frac{1}{16} + 6\times\frac{1}{16}\right) = 50(1 + 4 + 6) = 550$$ **Ans.**

(*b*) **If there are 3 misprints in a book of 1000 pages find the probability that a given page will contain**

(*a*) no misprint (*b*) more than 2 misprints (8)

**Solution.** $p$ = probability of misprint $= \dfrac{3}{1000} = 0.003$

$$n = 1 \Rightarrow m = np = 0.003$$

(*i*) For no misprint $r = 0$ (using poisson distribution) $p(r) = \dfrac{e^{-m}\cdot m^r}{r!}$

$$\Rightarrow \quad p(0) = \frac{e^{-0.003}\cdot(0.003)^0}{0!} = 0.997$$

(*ii*) For more than 2 misprints ($r = 3$) $p(3) = \dfrac{e^{-0.003}\times(0.003)^3}{3!} = 4.5\times10^{-9}$ **Ans.**

# December 2014

## Engineering Mathematics-II

### AIMETE-ET/CS/IT

### (Old Scheme)

*Time allowed : 3 hours ]* *[ Maximum Marks : 100*

**1.** Choose the correct or the best alternative in the following: (2 × 10)

(*a*) If $\omega = \log z$, then $\dfrac{d\omega}{dz}$ is equal to

(*a*) $\dfrac{1}{z}$ (*b*) $-\dfrac{1}{z}$ (*c*) $z$ (*d*) $-z$

**Solution.** Refer to Example 14 on page 515. **Ans.** (*a*)

(*b*) The value of the integral $\int_C \dfrac{z^2 - z + 1}{z - 1}\, dz$, where $C$ is the circle and $|z| = 2$

(*a*) $\pi i$ (*b*) $2\pi i$ (*c*) $3\pi i$ (*d*) $4\pi i$

**Solution.** $\int_C \dfrac{z^2 - z + 1}{z - 1}\, dz$ $C: |z| = 2$

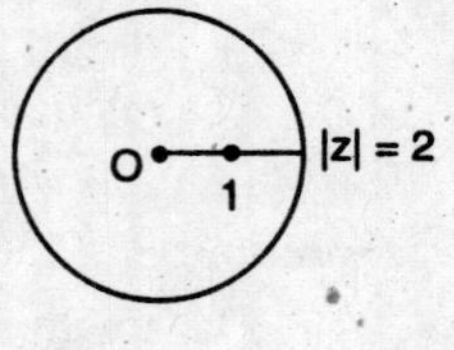


$z - 1 = 0 \Rightarrow z = 1$ is a simple pole

Residuce (at $z = 1$) $= \lim\limits_{z \to 1} \dfrac{(z-1)(z^2 - z + 1)}{(z-1)} = 1 - 1 + 1 = 1$

$$\int_C \frac{z^2 - z + 1}{z - 1}\, dz = 2\pi i(1) = 2\pi i$$ **Ans.** (*b*)

(*c*) The directional derivative of $\phi = 5x^2y - 5y^2z + 2.5z^2x$ at the point $P(1, 1, 1)$ in direction of the line $\dfrac{x-1}{2} = \dfrac{y-3}{-2} = z$, is

(*a*) $\dfrac{11}{3}$ (*b*) $\dfrac{11}{2}$ (*c*) $11\dfrac{2}{3}$ (*d*) $\dfrac{2}{3}$

**Solution.** $\phi = 5x^2y - 5y^2z + 2.5z^2x$

Directional derivative of $\phi = \left(\hat{i}\dfrac{\partial}{\partial x} + \hat{j}\dfrac{\partial}{\partial y} + \hat{k}\dfrac{\partial}{\partial z}\right)(5x^2y - 5y^2z + 2.5z^2x)$

$$= (10xy + 25\, z^2)\,\hat{i} + (5x^2 - 10yz)\,\hat{j} + (-5y^2 + 5zx)k$$

Directional derivative at $P(1, 1, 1) = 12.5i - 5j$

Directional derivative in the line $\dfrac{x-1}{-2} = \dfrac{y-3}{-2} = \dfrac{z}{1}$

$$= \frac{2}{3}i(12.5i) - \frac{2}{3}j(-5j) + \frac{1}{3}x0$$

$$= \frac{25}{3} + \frac{10}{3} = \frac{35}{3} = 11\frac{2}{3}$$ **Ans. (c)**

(*d*) The value of $\int_C \vec{F}\,\vec{dr}$ where $\vec{F} = 2x^2y\,\hat{i} + 3xy\,\hat{j}$ and $C$ is $y = 4x^2$ in the plane from (0, 0) to (1, 4) is

(*a*) $\frac{104}{9}$ (*b*) $\frac{104}{7}$ (*c*) $\frac{104}{3}$ (*d*) $\frac{104}{5}$

**Solution.** $\int_C \vec{F}\cdot\vec{dr} = \int(2x^2yi + 3xy\hat{j})\cdot(idx + jdy + kdz) = \int 2x^2y\,dx + 3xy\,dy$

$$= \int_0^1 2x^2(4x^2)\,dx + 3x(4x^2)(8x\,dx) \qquad y = 4x^2, \quad dy = 8x\,dx$$

$$= \int_0^1 \left(8x^4 + 96x^4\right)dx = 104\int_0^1 x^4\,dx = 104\left[\frac{x^5}{5}\right]_0^1 = \frac{104}{5}$$ **Ans. (d)**

(*e*) The relation between $E$ and $\Delta$ is

(*a*) $\Delta = 1 + E$ (*b*) $E = 1 - \Delta$ (*c*) $E = 1 + \Delta$ (*d*) $\Delta = -1 - E$

**Solution.** $E = 1 + \Delta$ **Ans. (c)**

(*f*) The solution of $(E^2 + 6E + 9)y_n = 0$ is

(*a*) $y_n = (c_1 + c_2n)(-3)^n$ (*b*) $y_n = (c_1 - c_2n)(3)^n$

(*c*) $y_n = (c_1 + c_2)(-3)^n$ (*d*) $y_n = (c_1 - c_2)(3)^n$

**Solution.** $(E_2 + 6E + 9)y_n = 0$

A.E. is $m^2 + 6m + 9 = 0$

$\Rightarrow$ $(m + 3)^2 = 0 \Rightarrow m = -3, -3$

$y = (c_1 + c_2n)(-3)^n$ **Ans. (a)**

(*g*) The solution of Partial Differential Equation of $p^2 + q^2 = 1$ is

(*a*) $z = ax - \sqrt{1-a^2}\cdot y + c$ (*b*) $z = ax + \sqrt{1-a^2}\cdot y + c$

(*c*) $z = ax + \sqrt{1+a^2}\cdot y + c$ (*d*) $z = ax - \sqrt{1+a^2}\cdot y + c$

**Solution.** Refer to Example 18 on page 684. **Ans. (b)**

(*h*) From a pack of 52 cards, one card is drawn at random, then the probability of getting a king is

(*a*) $\frac{1}{11}$ (*b*) $\frac{1}{12}$ (*c*) $\frac{1}{13}$ (*d*) $\frac{1}{14}$

**Solution.** There are 4 kings in a pack of 52 cards. $P(k) = \frac{4}{52} = \frac{1}{13}$ **Ans. (c)**

(*i*) The probability of getting 4 heads in 6 tosses of a fair coin is

(a) $\frac{15}{64}$ (b) $\frac{17}{64}$ (c) $\frac{19}{64}$ (d) $\frac{21}{64}$

**Solution.** $P(H) = \frac{1}{2}$ $q = \frac{1}{2}$ $n = 6, r = 4,$

$$P(r) = {}^nC_r P^r q^{n-r} = {}^6C_4 \left(\frac{1}{2}\right)^4 \left(\frac{1}{2}\right)^2 = 15\left(\frac{1}{2}\right)^6 = \frac{15}{64}$$ **Ans.** (a)

(j) The probability that the ace of spades will be drawn from a pack of well shuffled cards at least once in 104 consecutive trials is, (using Poisson distribution).

(a) 0.468 (b) 0.648 (c) 0.0864 (d) 0.864

**Solution.** $P(Ace) = \frac{1}{52}$, $n = 104,$ $m = np = 104\left[\frac{1}{52}\right] = 2,$

$$P(r) = \frac{e^{-m} \cdot m^2}{r!}$$

$$P(\text{At least once}) = 1 - \frac{e^{-2} \cdot (2)^{\circ}}{0!} = 1 - e^{-2} = 1 - 0.1353 = 0.8647 = 0.864$$ **Ans.** (d)

**Answer any five questions out of eight questions. Each question carries 16 marks.**

**2.** (a) Show that the function $f(z) = \sqrt{|xy|}$ is not analytic at the origin even though Cauchy-Riemann equations are satisfied thereof.

**Solution.** $\because u + iv = \sqrt{|xy|} \Rightarrow u = \sqrt{|xy|},\ v = 0$

$$\frac{\partial u}{\partial x} = \underset{x \to 0}{Lt} \frac{u(x+\Delta x.y) - u(x.y)}{\Delta x}, \qquad \left.\frac{\partial u}{\partial x}\right|_{\text{at }(0,0)} = \underset{x \to 0}{Lt} \frac{u(\Delta x, 0) - u(0,0)}{\Delta x}$$

At origin (0, 0) $$\frac{\partial u}{\partial x} = \underset{\Delta x \to 0}{Lt} \frac{0-0}{\Delta x} = 0 \quad \text{... }(i)$$

$$\frac{\partial u}{\partial y} = \underset{\Delta y \to 0}{Lt} \frac{u(0,\Delta y) - u(0,0)}{\Delta y} = \underset{\Delta y \to 0}{Lt} \frac{0-0}{\Delta y} = 0 \quad \text{... }(ii)$$

$$\frac{\partial u}{\partial x} = 0 \text{ ... }(iii) \quad (\because v = 0)$$

and $$\frac{\partial v}{\partial y} = 0 \quad \text{... }(iv)$$

from (i) & (iv)from (ii) & (iii) $\frac{\partial u}{\partial x} = \frac{\partial v}{\partial y}$ and $\frac{\partial u}{\partial y} = -\frac{\partial v}{\partial x}$

$\Rightarrow$ both CR equations are satisfied at origin

Also $$f'(z) = \underset{\Delta z \to 0}{Lt} \frac{f(z+\Delta z) - f(z)}{\Delta z}, f'(0) = \underset{\Delta z \to 0}{Lim} \frac{f(\Delta z) - f(0)}{\Delta z} = \underset{\Delta x \to \infty, \Delta y \to 0}{Lim} \frac{\sqrt{|\Delta x \cdot \Delta y|} - 0}{\Delta x + i\Delta y}$$

Consider a path $\Delta y = m\Delta x$

$$f'(0) = \lim_{\Delta x \to 0} \frac{\sqrt{|\Delta x . m\Delta x|}}{\Delta x + i m \Delta x} = \lim_{\Delta x \to 0} \frac{\sqrt{|m|}}{(1+i\sqrt{m})} = \frac{\sqrt{m}}{1+im} \text{ which is not unique}$$

$\Rightarrow$ $f'(0)$ doesn't exist $\Rightarrow f(z)$ is not analytic at $z = 0$. **Proved.**

(*b*) Find the bilinear transformation which maps the points $z_1 = 2$, $z_2 = i$ and $z_3 = -2$ into the points $w_1 = 1$, $w_2 = i$ and $w_3 = -1$. (8)

**Solution.** Refer to Q.1. on page 578.

Formula for bilinear transformation

$$\frac{(w-w_1)(w_2-w_3)}{(w-w_3)(w_2-w_1)} = \frac{(z-z_1)(z_2-z_3)}{(z-z_3)(z_2-z_1)} \Rightarrow \frac{(w-1)(i+1)}{(w+1)(i-1)} = \frac{(z-2)(i+2)}{(z+2)(i-2)}$$

$$\Rightarrow \frac{w-1}{w+1} = \frac{(z-2)}{z+2} \times \frac{(i+2)(i-1)}{(i-2)(i+1)} \Rightarrow \frac{z-2}{z+2} \times \frac{i^2-i+2i-2}{i^2+i-2i-2}$$

$$= \frac{z-2}{z+2} \times \frac{-3+i}{-3-i} \times \frac{-3+i}{-3+i} = \frac{z-2}{z+2} \times \frac{-3+i}{-3-i} \times \frac{-3+i}{-3+i}$$

$$= \frac{z-2}{z+2} \times \frac{9-1-6i}{10} = \frac{z-2}{z+2} \times \frac{4-3i}{5} = \frac{(z-2)(4-3i)}{5(z+2)}$$

Apply components and dividends

$$\frac{w-1+w+1}{w-1-w-1} = \frac{(z-2)(4-3i)+5(z+2)}{(z-2)(4-3i)-5(z+2)} \Rightarrow -w = \frac{z(4-3i+5)-8+6i+10}{z(4-3i-5)-8+6i-10}$$

$$= \frac{z(9-3i)+2+6i}{z(-1-3i)-18+6i},$$

$$w = -\frac{z(9-3i)+2+6i}{z(1+3i)+18-6i}$$ **Ans.**

**3.** (*a*) Find the Taylor's series expansion of a function of the complex variable.

$f(z) = \dfrac{1}{(z-1)(z-3)}$ about the point, $z = 4$. Find its region of convergence. (8)

**Solution.** $$f(z) = \frac{1}{(z-1)(z-3)} = \frac{\frac{1}{2}}{z-3} - \frac{\frac{1}{2}}{z-1}$$

Let $z - 4 = t \Rightarrow z = t + 4$

$$f(t+4) = \frac{\frac{1}{2}}{t+4-3} - \frac{\frac{1}{2}}{t+4-1} = \frac{\frac{1}{2}}{1+t} - \frac{\frac{1}{2}}{t+3}$$

$$= \frac{1}{2}(1+t)^{-1} - \frac{1}{6}\left(1+\frac{t}{3}\right)^{-1} \qquad |t| < 1, \left|\frac{t}{3}\right| < 1$$

$$= \frac{1}{2}(1 - t + t^2 - t^3 \ldots) - \frac{1}{6}\left(1 - \frac{t}{3} + \frac{t^2}{9} - \frac{t^3}{27} \ldots\right)$$

$$\text{ROC} = |t| < 1 \ \& \ \left|\frac{t}{3}\right| < 1$$

$$= \frac{1}{2} - \frac{1}{6} + t\left(-\frac{1}{2} + \frac{1}{18}\right) + t^2\left(\frac{1}{2} - \frac{1}{54}\right) \ldots,$$

$$|t| \ \& \ |t| < 3 = \frac{1}{3} - \frac{4}{9}t + \frac{13}{27}t^2 + \ldots \qquad \text{ROC: } |t| < 1$$

$$f(z) = \frac{1}{3} - \frac{4}{9}(z-4) + \frac{13}{27}(z-4)^2 + \ldots \qquad \text{ROC } |z-4| < 1$$

**Ans.**

(*b*) Determine the poles of the function $f(z) = \dfrac{z^2}{(z-1)^2(z+2)}$ and find residue of $f(z)$ at each pole.

**Solution.** Poles at $(z-1)^2(z+2) = 0$

$\Rightarrow$ Poles are at $z = 1$ of order 2

Poles is at $z = -2$ of order 1

$\because$ For poles of 1 order

Residue at $(z = a) = \lim_{z \to a} (z-a) f(z)$

$$\Rightarrow \text{Res(at } z = -2) \lim_{z \to -2} (z+2) \cdot \frac{z^2}{(z-1)^2(z+2)} = \frac{(-2)^2}{(-2-1)^2} = \frac{4}{9}$$

$\because$ For poles of order $n$

$$\text{Residue at } (z = a) = \frac{1}{(n-1)!} \frac{d^{n-1}}{dz^{n-1}} \left\{(z-a)^n f(z)\right\}_{\text{at } z=\alpha}$$

$$\text{Residue at } (z = 1) = \frac{1}{11} \frac{d}{dz}(z-1)^2 \left.\frac{z^2}{(z-1)^2(z+2)}\right\}_{\text{at } z=1} = \frac{d}{dz}\left(\frac{z^2}{z-2}\right)_{z=1} = \frac{(z+2)\cdot 2z - z^2 \cdot 1}{(z+2)^2}$$

$$= \frac{2z^2 + 4z - z^2}{(z+2)^2} = \left.\frac{z^2 + 4z}{(z+2)^2}\right|_{z=1} = \frac{1+4}{(1+2)^2} = \frac{5}{9}$$

**Ans.**

**4.** (*a*) Evaluate grad $e^{r^2}$ where $r^2 = x^2 + y^2 + z^2$.

**Solution.** $$\text{grad } e^{r^2} = \nabla(e^{r^2}) = \left(\frac{\partial}{\partial x}\hat{i} + \frac{\partial}{\partial y}\hat{j} + \frac{\partial}{\partial z}\hat{k}\right)(e^{r^2}) = \frac{\partial}{\partial x}\left(e^{r^2}\right)\hat{i} + \frac{\partial}{\partial y}\left(e^{r^2}\right)\hat{j} + \frac{\partial}{\partial z}\left(e^{r^2}\right)\hat{k}$$

$$= e^{r^2} \cdot 2r\frac{\partial r}{\partial x}i + e^{r^2} \cdot 2r\frac{\partial r}{\partial y}\hat{j} + e^{r^2} \cdot 2r\frac{\partial r}{\partial z}\hat{k} \qquad \ldots (1)$$

$\because$ $r^2 = x^2 + y^2 + z^2$

Differentiating both sides w.r.t to $x$.

$$2r\frac{\partial r}{\partial x} = 2x \Rightarrow \frac{\partial r}{\partial x} = \frac{x}{r}$$

Similarly, $\frac{\partial r}{\partial y} = \frac{y}{r}\cdot$ and $\frac{\partial r}{\partial z} = \frac{z}{r}$

From equation (1) grad $\left(e^{r^2}\right) = e^{r^2}\left(2r\cdot\frac{x}{r}\hat{i} + 2r\frac{y}{r}\hat{j} + 2r\cdot\frac{z}{r}\hat{k}\right) = 2e^{r^2}(x\hat{i} + y\hat{j} + z\hat{k})$

$= 2e^{r^2}\cdot\vec{\mathbf{r}}$ **Ans.**

(*b*) Prove that div $\left\{\frac{f(r)\cdot\vec{\mathbf{r}}}{r}\right\} = \frac{1}{r^2}\frac{d}{dr}(r^2 f)$ (8)

**Solution.** div. $\left(\frac{f(r)\cdot\vec{\mathbf{r}}}{r}\right) = \nabla\cdot\left(\frac{f(r)}{r}(x\hat{i} + y\hat{j} + z\hat{i})\right) = \left(\hat{i}\frac{\partial}{\partial x} + \hat{j}\frac{\partial}{\partial y} + k\frac{\partial}{\partial z}\right)\cdot\left(\frac{f(r)}{r}(xi + yj + zk)\right)$

$$= \frac{\partial}{\partial x}\left(\frac{f(r)}{r}x\right) + \frac{\partial}{\partial y}\left(\frac{f(r)}{r}y\right) + \frac{\partial}{\partial z}\left(\frac{f(r)}{r}z\right) \quad \text{... (1)}$$

$\because$ $\frac{\partial}{\partial x}\left(\frac{f(r)}{r}x\right) = \frac{f(r)}{r}\cdot 1 + xf(r)\cdot\frac{\partial}{\partial x}\left(\frac{1}{r}\right) + \frac{x}{r}\cdot\frac{\partial}{\partial x}(f(r))$

$$= \frac{f(r)}{r} + xf(r)\cdot\left(-\frac{1}{r^2}\right)\frac{\partial r}{\partial x} + \frac{x}{r}\cdot f'(r)\cdot\frac{\partial r}{\partial x}$$

$$= \frac{f(r)}{r} - \frac{x}{r^2}f(r)\cdot\frac{x}{r} + \frac{x}{r}f'(r)\cdot\frac{x}{r} = \frac{f(r)}{r} - \frac{x^2}{r^3}f(r) + \frac{x^2}{r^2}f'(r)$$

Gim, $\frac{\partial}{\partial y}\left(\frac{f(r)}{r}y\right) = \frac{f(r)}{r} - \frac{y^2}{r^3}f(r) + \frac{y^2}{r^2}f'(r)$

$$\frac{\partial}{\partial z}\left(\frac{f(r)}{r}z\right) = \frac{f(r)}{r} - \frac{z^2}{r^3}f(r) + \frac{z^2}{r^2}f'(r)$$

From equation (1)

$$\left(\frac{f(r)}{r}\cdot\vec{\mathbf{r}}\right) = \frac{3f(r)}{r} - \frac{(x^2+y^2+z^2)}{r^3}f(r) + \frac{x^2+y^2+z^2}{r^2}f'(r)$$

$$= \frac{3f(r)}{r} - \frac{r^2}{r^3}f(r) + \frac{r^2}{r^2}f'(r) = \frac{3}{r}f(r) - \frac{1}{r}f(r) + f'(r)$$

$$= \frac{2}{r}f(r) + f'(r) = \frac{1}{r^2}\{2r\,f(r) + r^2 f'(r)\} = \frac{1}{r^2}\frac{d}{dr}\left(r^2 f(r)\right) = \text{R.H.S.}$$

**5.** (*a*) Apply Green's theorem to evaluate $\oint_C 2y^2dx + 3xdy$ Where C is the boundary of closed region bounded bewtween $y = x$ and $y = x^2$. (8)

**Solution.** Green's theorem, $\oint_C fdx + gdy = \iint_R\left(\frac{\partial g}{\partial x} - \frac{\partial f}{\partial y}\right)dx\,dy$

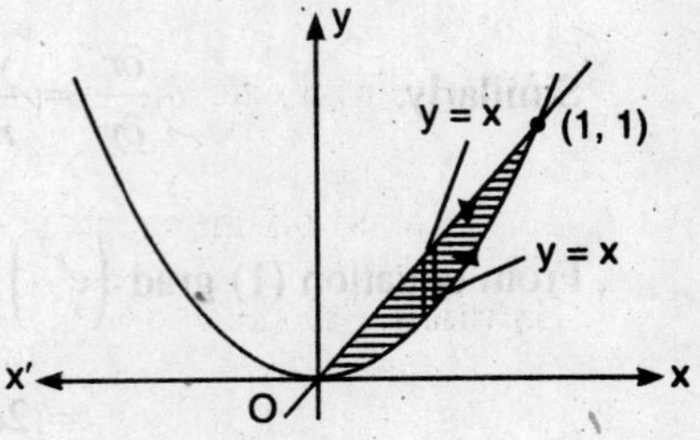


Here $f = 2y^2,\ \dfrac{\partial f}{\partial y} = 4y,$

$$g = 3x\ \frac{\partial g}{\partial x} = 3$$

$$\Rightarrow \oint_C 2y^2 dx + 3x dy = \iint_R \left(\frac{\partial g}{\partial x} - \frac{\partial f}{\partial y}\right) dx\,dy$$

$$= \iint_R (3-4y)\,dx\,dy = \int_0^1 \int_{y=x^2}^{x} (3-4y)\,dx\,dy = -\int_0^1 dx \left[3y - \frac{4y^2}{2}\right]_x^{x^2}$$

$$= -\int_0^1 dx\,[3x^2 - 2x^4 - 3x + 2x^2] = -\int_0^1 (5x^2 - 2x^4 - 3x)\,dx$$

$$= -\left[\frac{5x^3}{3} - \frac{2x^5}{5} - \frac{3x^2}{2}\right]_0^1 = -\left(\frac{5}{3} - \frac{2}{5} - \frac{3}{2}\right) = -\left(\frac{50-12-45}{30}\right) = \frac{7}{30} \quad \textbf{Ans.}$$

(*b*) Apply Stoke's theorem to calculate $\oint_C (4y\,dx + 2z\,dy + 6y\,dz)$ where $C$ is the curve of intersection of $x^2 + y^2 + z^2 = 6z$ and $z = x + 3$ **(8)**

**Solution.** Refer to Example 94 on page no. 442.

**6.** (*a*) Using Langrange's interpolation formula, find the value of $y$ corresponding to $x = 10$ from the following table: **(8)**

| | | | | |
|---|---|---|---|---|
| $x$ | 5 | 6 | 9 | 11 |
| $y$ | 12 | 13 | 14 | 16 |

**Solution.** Formula for Lagrange's interpolation

$$f(x) = \frac{(x-x_1)(x-x_2)(x-x_3)}{(x_0-x_1)(x_0-x_2)(x_0-x_3)} f(x_0) + \frac{(x-x_0)(x-x_2)(x-x_3)}{(x_1-x_0)(x_1-x_2)(x_1-x_3)} f(x_1)$$

$$+ \frac{(x-x_0)(x-x_1)(x-x_3)}{(x_2-x_0)(x_2-x_1)(x_2-x_3)} f(x_2) + \frac{(x-x_0)(x-x_1)(x-x_2)}{(x_3-x_0)(x_3-x_1)(x_3-x_2)} + f(x_3)$$

Hence $x_0 = s,\ x_1 = 6,\ x_2 = 9,\ x_3 = 11$

$$f(x_0) = 12,\ f(x_1) = 13,\ f(x_2) = 14,\ f(x_3) = 16,\ x = 10$$

$$f(x) = \frac{(10-6)(10-9)(10-11)}{(5-6)(5-9)(5-11)} \times 12 + \frac{(10-5)(10-9)(10-11)}{(6-5)(6-9)(6-11)} \times 13$$

$$+ \frac{(10-5)(10-6)(10-11)}{(9-5)(9-6)(9-11)} \times 14 + \frac{(10-5)(10-6)(10-9)}{(11-5)(11-6)(11-9)} \times 16$$

$$= \frac{4\times1\times-1}{-1\times-4\times-6} \times 12 + \frac{5\times1\times-1}{1\times-3\times-5} \times 13 + \frac{5\times4\times-1}{4\times3\times-2} \times 14 + \frac{5\times4\times1}{6\times5\times2} \times 16$$

$$= \frac{12}{6} - \frac{13}{3} + \frac{35}{3} + \frac{16}{3} = \frac{44}{3} \quad \textbf{Ans.}$$

(*b*) Approximate the integral, $I = \int_0^1 e^x dx$ using Simpson's 1/3[rd] Rule with $n = 8$ interval. Round off the result at 4 digits.

**Solution.** Step size $= \dfrac{\text{upper lim} - \text{lower limit}}{\text{no. of interval}} = \dfrac{1-0}{8} = 0.125$

| | | | | | | | | | |
|---|---|---|---|---|---|---|---|---|---|
| $x$ | 0 | 0.125 | 0.26 | 0.3675 | 0.5 | 0.625 | 0.75 | 0.875 | 1 |
| $f(x) = e^x$ | 1 | 1.133 | 1.284 | 1.455 | 1.649 | 1.868 | 2.117 | 2.399 | 2.718 |
| | $y_0$ | $y_1$ | $y_2$ | $y_3$ | $y_4$ | $y_5$ | $y_6$ | $y_7$ | $y_8$ |

Apply Simpson's $\frac{1}{3}^{\text{rd}}$ rule

$$I = \frac{h}{3}[(y_0 + y_8) + 4(y_1 + y_3 + y_5 + y_4) + 2(y_2 + y_4 + y_6)]$$

$$= \frac{0.125}{3}[1 + 2.718 + 4(1.333 + 1.455 + 1.868 + 2.399) + 2(1.284 + 1.649 + 2.117)]$$

$= 1.718$ **Ans.**

**7.** (*a*) Apply Charpit method to solve the equation $px + qy = pq$. (8)

**Solution.** Refer to Example 27 on page no. 689.

(*b*) Using the method of separation of variables, solve $\dfrac{\partial u}{\partial x} = 2\dfrac{\partial u}{\partial t} + u$, where $u(x, 0) = 6e^{-3x}$ (8)

**Solution.** Refer to Example 1 on page no. 707.

**8.** (*a*) An urn '*A*' contains 2 white and 4 black balls. Another urn '*B*' contain 5 white and 7 black balls. A ball is transferred from the urn '*A*' to the urn '*B*', then a ball is drawn from urn '*B*'. Find the probability that it is white. (8)

**Solution.** Refer to Example 11 on page no. 769.

(*b*) Assuming half the population of a town consumes chocolates and that 100 investigators each take 10 individuals to see whether they are consumers, how many investigators would you expect to report that three people or less were consumers?

**Solution.** Refer to Example 33 on page no. 784.

**9.** (*a*) If the variance of the Poisson distribution is 2, find the probabilities for r = 1, 2, 3, 4 from the recurrence relation of the poisson distribution. Also find $p(r \geq 4)$. (8)

**Solution.** Refer to Example 38 page no. 798.

(*b*) The life of army shoes is 'normally distributed with mean 8 months and standard deviation 2 months. If 5000 pairs are issued how many pairs would be expected to need replacement after 12 months? [Given that $P(z \geq 2) = 0.0228$ and $z = \dfrac{x - \mu}{\sigma}$] (8)

**Solution.** Refer to Example 65 on page no. 818.

# INDEX